Encyclopaedia of the History of Science, Technology, and Medicine in Non-Western Cultures

Encyclopaedia of the History of Science, Technology, and Medicine in Non-Western Cultures

Encyclopaedia of the History of Science, Technology, and Medicine in Non-Western Cultures

Editor

HELAINE SELIN
*Science Librarian, Hampshire College,
Amherst, Massachusetts, U.S.A.*

Kluwer Academic Publishers
DORDRECHT / BOSTON / LONDON

A C.I.P. Catalogue record for this book is available from the Library of Congress

ISBN 0-7923-4066-3 (HB)

Published by Kluwer Academic Publishers,
P.O. Box 17, 3300 AA Dordrecht, The Netherlands

Sold and distributed in the U.S.A. and Canada
by Kluwer Academic Publishers,
101 Philip Drive, Norwell, MA 02061, U.S.A.

In all other countries, sold and distributed
by Kluwer Academic Publishers,
P.O. Box 322, 3300 AH Dordrecht, The Netherlands

Printed on acid-free paper

Printed in the Netherlands

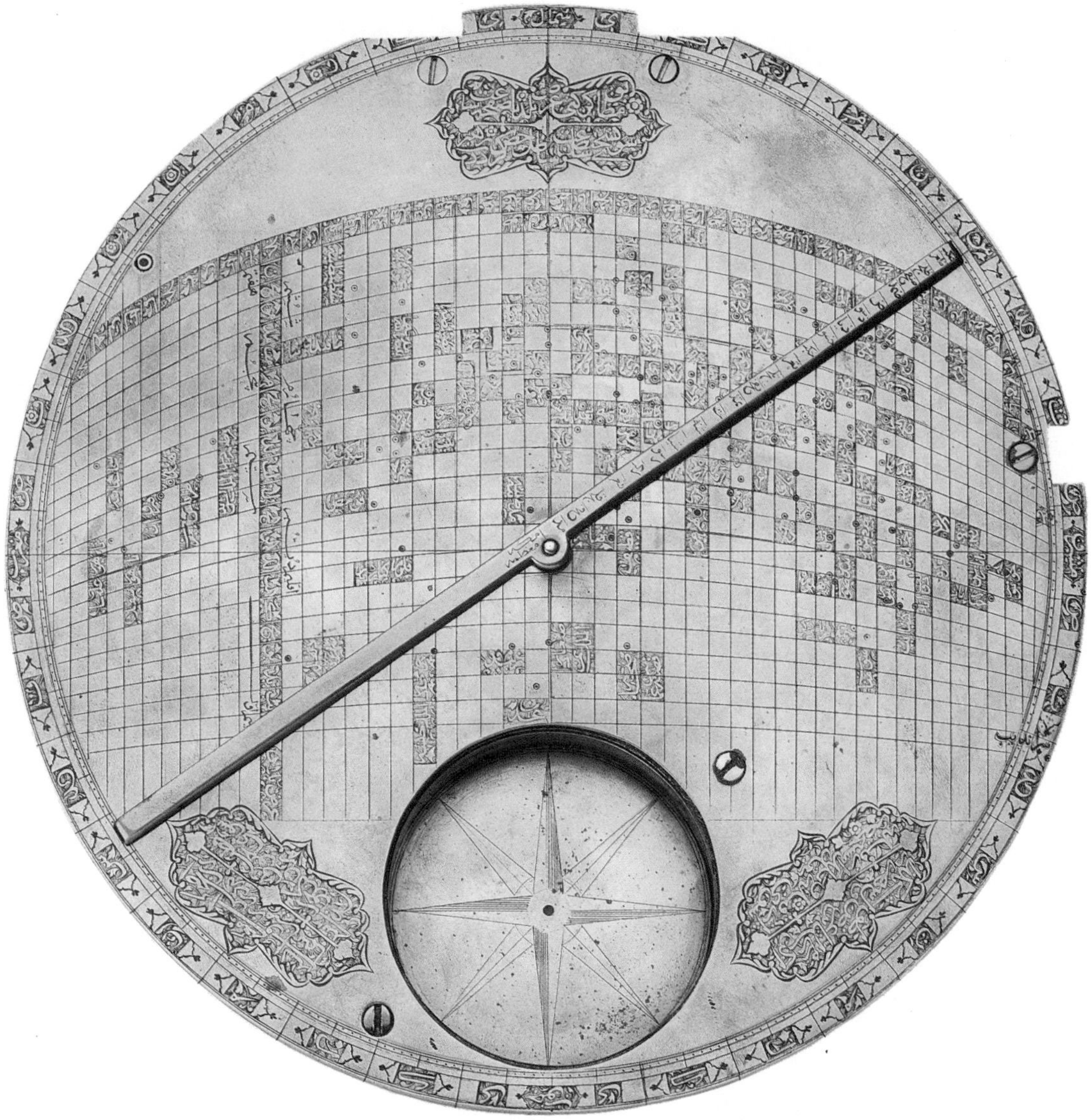

A remarkable world-map centred on Mecca with a highly sophisticated mathematical grid for finding the direction and distance to Mecca from any location in the Islamic Commonwealth. The map is engraved on brass and was made in Iran ca. 1700. However, the geographical data on the map was compiled some 250 years previously, and the mathematics underlying the cartographical grid were known to Muslim scientists such as Habash in the 9th century and al-Biruni in the 11th century. (See: "Maps and Mapmaking: Islamic World Maps Centered on Mecca". Private collection, courtesy of the owner and D.A. King, contributor. Photo by Christies of London.)

In Memoriam

Madilyn J. Engvall
1936–1994

The process
Is old light truth passing through new glass

– Tim Davis

TABLE OF CONTENTS

PREFACE

The *Encyclopaedia* fills a gap in both the history of science and in cultural studies. Reference works on other cultures tend either to omit science completely or pay little attention to it, and those on the history of science almost always start with the Greeks, with perhaps a mention of the Islamic world as a translator of Greek scientific works. The purpose of the *Encyclopaedia* is to bring together knowledge of many disparate fields in one place and to legitimize the study of other cultures' science. Our aim is not to claim the superiority of other cultures, but to engage in a mutual exchange of ideas. The Western academic divisions of science, technology, and medicine have been united in the *Encyclopaedia* because in ancient cultures these disciplines were connected. This work contributes to redressing the balance in the number of reference works devoted to the study of Western science, and encourages awareness of cultural diversity. The *Encyclopaedia* is the first compilation of this sort, and it is testimony both to the earlier Eurocentric view of academia as well as to the widened vision of today. There is nothing that crosses disciplinary and geographic boundaries, dealing with both scientific and philosophical issues, to the extent that this work does.

PERSONAL NOTE FROM THE EDITOR

Many years ago I taught African history at a secondary school in Central Africa. A few years before, some of the teachers in the country had designed a syllabus that included pre-European history, since the curriculum, left over from colonial days, did not include any mention of Africa before the Portuguese. After a year of teaching from this revised version, I asked my students what they thought was the most significant moment in African history, and virtually all of them said it was the arrival of David Livingstone.

It may well be that that was the most important moment for Africa, but it shocked me at the time that no one considered any African achievements worth mentioning. Over the years I have come to see, with the help of scholars like Michael Adas, that the dominance of the West means not only that Westerners disparage the rest of the world but also that the rest of the world sees itself as inferior to the West. This book is meant to take one step towards rectifying that, by describing the scientific achievements of those who have been overlooked or undervalued by scholars in both the West and the East.

The book is more than just a compilation of 600 disparate articles; it is a glimpse into how people describe and perceive and order the world. I hope the reader will do some exploring. In addition to reading about Maya astronomy, one can read about Mesoamerican mathematics and medicine, as well as a general article on Magic and Science, because all the fields are interrelated and entwined. It might be useful to read about astronomy in Africa and in Australia, to see how similar and different these cultures are. One can travel across disciplines, following the achievements of one culture, and across cultures, comparing the same discipline. And then it would be useful to read an essay on Transmission of Knowledge, or Rationality and Method, to put the articles and their contents in a broader philosophical and social context.

My hope, and that of the advisors and contributors to the project, is that the *Encyclopaedia* will expand the horizons of scholars, teachers, and students by illustrating how extensive the accomplishments of non-Western scientists are. May our future students never believe that science is limited to a fraction of the world.

—·—·—

A note about the authors' names, especially Asian ones: I made many embarrassing errors confusing peoples' surnames and given names, but I was reluctant to change authors' names to conform to the Western style, as it went against the spirit of the *Encyclopaedia*. Therefore, I have left the names as the authors wrote them.

ACKNOWLEDGEMENTS

My greatest thanks go to the scholars who participated in the project. They very generously gave me their work, their advice, their suggestions, and their time. Some members of the advisory board were more than just advisors; they helped to shape the work and give it clarity. Among the contributors and advisors, I must especially thank: Ho Peng Yoke, David Turnbull, Jan Hogendijk, H T. Huang, David King, Gloria Emeagwali, Ruben Mendoza, Cai Jingfeng, K.V. Sarma, Paul Kunitzsch, Boris Rosenfeld, Tzvi Langermann, Ruth Hendricks Willard, and Gregg de Young.

I must of course thank my editor, Annie Kuipers, and her wonderful assistant Evelien Bakker, for taking on such a big project with enthusiasm and affection. I have never encountered two people who work so carefully and so hard. Annie believed in the project, and in me, from the very beginning of our association, and it is certainly true that credit for the quality of the work goes largely to her. I would also like to thank Kennie Lyman, my first editor, who offered support, ideas and advice, and the feeling that we were doing something very important together.

At Hampshire College, I wish to thank my friends and colleagues Bonnie Vigeland, Serena Smith and Ann McNeil. They listened to my endless stories and complaints, and laughed and commiserated with me through all the ups and downs of this very complex undertaking. Tom Hart ably assisted with computer graphics, and Ken Hoffman helped with some of the mathematics. Amanda Seigel, a first year undergraduate student, was terrific both with the routine tasks and with fact and spelling checks. She became quite adept with Arabic and Sanskrit and did all the printing and reprinting with endless good cheer.

Joseph Needham, editor of *Science and Civilisation in China*, died in Cambridge in March 1995, just as I was finishing compiling the encyclopedia. All of us in the field of non-Western science owe him an enormous debt for bringing the intellectual worlds of the East and West together. In a sense all of our work follows from his.

Finally, I wish to thank my family for putting up with me. A project of this magnitude takes over your consciousness and your time. I'm sure I've been a horrible bore for six years but my lovely family has been very kind about it. I have kept Tim's poem ("I swallowed two pills that claim to heal unspecificity") on the wall in front of the computer, and Lisa's delicate flower drawings on the desk for inspiration. My deepest gratitude goes to my husband, Bob Rakoff; I would never have had the confidence to undertake this or the fortitude to complete it without him.

H. Selin

INTRODUCTION

In order to study the history of the science of non-Western cultures we must define both non-Western and science. The term non-Western is not a geographical designation; it is a cultural one. We use it to describe people outside of the Euro-American sphere, including the native cultures of the Americas. The fact that the majority of the world's population is defined by not being something (in this case non-Western) is testimony to the power of European and American colonialism and to the cultural domination of the Western world today.

In fact, for most of our recorded history the flow of knowledge, art, and power went the other way. Edwin Van Kley talks about this in his essay on East and West.

> By 500 BC the globe supported four major centers of civilization: the Chinese, the Indian, the Near Eastern, and the West, considering Greek culture as antecedent to what eventually became the West. Of the four the West was probably the least impressive in terms of territory, military power, wealth, and perhaps even traditional culture. Certainly this was the case after the fall of the Roman Empire in the fifth century AD. From that time until about AD 1500 the West probably should be regarded as a frontier region compared to the other centers of civilization. No visitor from Mars would likely have predicted that the West would eventually dominate the globe.

The gap between the rich and poor nations has widened over the years, as has the notion of Western science as the only science. In this view, Western science is science; everything else is anthropology. Although Eurocentrism has been challenged in many fields, especially in the arts and humanities, the challenge has not extended into science. If we wish to study science in non-Western cultures, we need to take several intellectual steps. First, we must accept that every culture has a science, a way of defining, controlling, and predicting events in the natural world. Then we must accept that every science is legitimate in terms of the culture from which it grew. We must extend this view to our own science, recognizing that it too is a reflection of its culture, and that culture plays a role in every step of doing science: in what we choose to study, how we collect the data, and how we interpret them. We say that Western science is superior because we consider it rational, objective, and value-free, and we look disparagingly at others' science and call it magic. The transformation of the word science as a distinct rationality valued above magic is uniquely European. It is not common to most non-Western societies, where magic and science and religion can easily coexist, even today. For example, the practice of *feng shui*, or geomancy, the art of finding the spiritually correct location for a building, is practiced in China and in Chinese communities all over the world. If we are to study this subject open-mindedly, we must see that even the concept of rationality is problematic, as it stems from Western ideas about what it is to be a knowing, moral, sane individual. David

Turnbull discusses this in his essay on Rationality, Objectivity, and Method. Certainly we accept that concepts of morality and sanity vary enormously in different cultures; we have to extend the same acceptance to the concept of knowing. Even naming this book the *Encyclopaedia of the History of Science, Technology, and Medicine in Non-Western Cultures* makes a value judgment about how knowledge is organized. There is no reason to assume that most cultures recognize the distinctions in those fields the way contemporary academics do.

We must even be careful about the words we use to describe cultures other than our own. Historians of science looked disparagingly upon Chinese science because in traditional China knowledge did not come under the same groupings as in the West. The Chinese had a more holistic world view, so that material on natural history might be included in a pharmacopoeia, or the works of a great poet might contain information on astronomy and alchemy. Works on military science might mention meteorology, firearms, magic, and divination at the same time. In fact, divination and mathematics, as well as astronomy and music, were the same word. The same is true with astrology and astronomy. In both Arabic and Chinese, there is one word for both sciences. What is noteworthy is that astronomy (naming the stars), the mere observing and recording of celestial phenomena, was considered inferior to astrology (studying the stars), the art of interpreting the data. Another example of this occurs among the Aztecs. The involved system of calendrics was largely based on astronomical observations, but the calendars themselves were applied both to ritual and practical ends. Medicine combined pragmatic remedies with shamanism and divination. Writing, architecture, and stone-, feather-, and metalworking all relied on sophisticated technologies; the resulting works served secular goals and/or displayed a complex religious symbolism. The empirical, scientific realm of understanding and inquiry was not readily separable from a more abstract, religious realm. Karen Jolly talks more about this in her essay on Magic and Science.

Even when we use the same conceptualizations, the meanings may not be identical, and our understanding may be limited by our ethnocentric conception. Jens Høyrup, in an essay on Near Eastern Geometry, discusses the mathematical concept π (*pi*). If we assume that the Babylonians or the Egyptians looked for the ratio between the circumference and the diameter of a circle, then we misunderstand a basic element of their own mathematics. "To the Babylonians, the fundamental ratio was the ratio between the area of the circle and the area of the square on the circumference, and the terminology they used demonstrates that they really thought of this as a constructed geometrical square. The Egyptians, for their part, were interested in the ratio between the sides of the squared circle and the circumscribed square. Both conceptualizations are fully legitimate, but they are certainly different from ours."

We know that in academia a process of questioning the literary and historical canon has begun. We must extend this to the study of science, especially in the cases where the science practiced was not a precursor to our own. This is a particularly important contribution that the *Encyclopaedia* makes; it brings together the mathematics of the Aztecs and the Australian aboriginal people in the same space as that of the Indians and Muslims. We have always paid tribute to Hindu–Arabic mathematics, as some of it formed the basis for Western mathematics. But we certainly have not recognized the equally important numeric systems of cultures with very different structures from ours. This is the result of seeing the world as a continuous progression to higher levels. Most other cultures created science and technology in response to their needs, so had no use for constant improvements. It is only in our time that this has worked the other way: that we create needs to meet the advancing technology. We have used this to disparage the lack of "achievement" of many of the world's peoples.

It is interesting that in some areas people have become more receptive to other ways of knowing. In medicine, patients, especially those in pain, have begun to question the exclusive superiority of Western medicines, and to incorporate other medical traditions into their treatment. Medical schools have recently added courses in complementary medicine to their curricula. We know that laboratory medicine has not been completely successful in curing many contemporary diseases, and we are aware of the contributions that other cultures' medicine might make. In agriculture, we are beginning to admit that techniques presumed to be inferior may have superior results, if not for yield, then at least for the land, and maybe, incidentally, for the health of the people who work the land. We are less impressed with big science, and no longer see it as a force only for good and progress. This provides an opening for the study of the science of other cultures.

In this project we are not trying to claim the superiority of other cultures; we wish to engage in a mutual exchange of ideas. In editing the *Encyclopaedia*, I tried to avoid ethnic cheerleading, and to edit out as many phrases as possible that fell into the "we were the first", "we were the best", "we were the only" categories. In the end, what does it matter who discovered gunpowder first? In fact, for many years, Western academics used the fact that the Chinese discovered gunpowder and did not "do" anything with it as proof of the Chinese people's essential lack of scientific acumen. For years scholars debated the question of why there was no scientific revolution in China. If we see that the Chinese used gunpowder, or paper or clocks or astronomical observatories, for the needs that they had at the time, we can marvel at their ability to respond to their needs without questioning why they didn't make new ones.

The *Encyclopaedia of the History of Science, Technology, and Medicine in Non-Western Cultures* includes a range of essays from short biographical and descriptive ones to long philosophical ones. These more general articles cover topics such as Colonialism and Science, Magic and Science, East and West, Technology and Culture, Science as a Western Phenomenon, Values and Science, and Rationality, Objectivity, and Method. Since the study of non-Western science is not just a study of facts, but a study of culture and philosophy, we included these articles in order to make the entries on Indian trigonometry or Pacific Island medicine more meaningful.

This project arose originally from a course on Comparative Scientific Traditions taught at Hampshire College in Amherst, Massachusetts, USA. The Ford Foundation financed the course, and provided some extra money for library acquisitions to support it. As I am the Science Librarian, I bought the books for the library, and then produced an annotated bibliography of about 800 books called *Science Across Cultures*, published by Garland Publishing of New York in 1992. The encyclopaedia project grew from that.

We had several goals for the encyclopaedia. The first was the simplest: to bring together knowledge of many disparate fields in one place. We united the Western academic divisions of science, technology, and medicine, because in ancient cultures these were connected; the study of the stars was interrelated to the study of the soil, navigation, mathematics, and healing. We also wanted to redress the balance in the number of reference works devoted to the study of Western science, and to encourage awareness of cultural diversity. We wanted to recognize the true value of the intellectual property of indigenous people. It is very important that the study of the science of non-Western people be added to the curriculum as a legitimate study in its own right and not just as a curiosity. Twenty years ago you could not study jazz in a major American conservatory, not to mention studying African or Brazilian music. Eventually, courses on ethnomusicology were introduced, as rather quaint but not quite serious additions. And today we know that there are complex patterns of drumming that Beethoven could not have conceived of, and you probably can not study music in an American conservatory without studying jazz. The same process happened with literature, from having a professor read a *haiku* poem in a class, to courses on Literature of the Other, to Nobel prizes for Chilean poets. We hope the same process will occur with science. Some people now incorporate bits and pieces of information about other cultures' science into their courses; we hope that in ten years minds and curricula will have expanded to include much more of this material.

The field of the history of non-Western science is not one without controversy. History is not fact objectively related; it is open to interpretation, and the interpretations change. When I began working on the book, I was quite innocent of these controversies, but I now know that many scholars disagree quite strongly with others in the field. In Islamic science, for instance, there seemed to be enough dissension in the scholarly community to include an article on the current debate in Islamic science.

We believe this is the first compilation of this sort, and it is testimony both to the Eurocentricity of academia before, and to the growing widening of its vision that we can produce it now. The history of the science of non-Western cultures is a relatively new field to Western academics, and it is a rich and fascinating one. The *Encyclopaedia* can be used to provide both factual information about the practices and practitioners of the sciences as well as insights into the world views and philosophies of the cultures that produced them. There are also many articles that in a sense provide the background to studying these sciences. Given the disparity in the number of articles on some cultures, such as the Chinese and Islamic, it might appear that they had more to contribute to world knowledge. The cultures that had writing and were less exploited by warfare and colonialism have left more behind for us to study. Surely there were mathematicians in the Pacific Island countries and in the Americas with skill equal to Zhu Shijie or Ibn al-Haytham. But no records survive. We hope that this apparent lack of balance will not be seen as a failing of the *Encyclopaedia*, but as an impetus for further research.

We hope that readers will achieve a deeper understanding of the relationship between science and culture and a new respect for the accomplishments of these ancient civilizations. If we continue to think of science as a purely Western phenomenon, we eliminate a world of possibilities and preserve a narrow view of life. As the Bantu proverb says, "He who never goes visiting thinks mother is the only cook".

HELAINE SELIN
Amherst, Massachusetts
Spring, 1997

LIST OF ENTRIES

A

ABACUS The abacus has taken several physical forms: a board covered with dust or wax, or simply a patch of sand on the ground which could be marked out; a tabulation or table marked out in columns within which counters (usually discs but in China bamboo rods) could be placed and moved about; a series of grooves in which counters could slide; and a frame with rods or strings on which beads could be strung so as to slide along.

These all have in common a principle of 'placing'. The columns or rods represent units, tens, hundreds, thousands, and so on. Or they may represent units of currency of increasing value, such as the former English pence, shillings, and pounds. A counter placed in the units column represents one, in the tens column ten, and in the hundreds column a hundred. Calculations then could be performed relatively mechanically by literally adding to or taking away counters. To multiply was to perform successive additions, to divide, successive subtractions. Many more sophisticated techniques evolved for performing the arithmetical manipulations, with accessory aids such as multiplication tables being found in the manuscripts. But the only fundamental rule which had to be grasped was that of the hierarchy of columnar values.

An abacus working on this system was commonplace in the Mediterranean world several centuries BC. The Greek historian Herodotus in the fifth century says that the Egyptians arranged their columns from right to left while the Greeks worked in the opposite direction. Roman authors mention the abacus too, describing the sand or dust board, or its variant the wax tablet, the table with lines drawn on it, and also the board with grooves. Examples survive.

In the Middle East, a rod (or wire or string) and bead abacus was the normal type. This is also found in Russia. In the Far East the Chinese at first (from at least 600 BC) seem to have used bamboo rods instead of round counters and to have drawn up the board differently. This method spread to Korea and Japan, where it persisted for centuries. The Chinese themselves adopted a bead abacus in about the twelfth century, which evolved and in its turn made its way to Japan.

Mathematically and conceptually speaking, some of the most advanced work was done after the importation into the West of Arabic works on the abacus from the tenth or eleventh century. This was also the period of transition from the use of Roman numerals to the actual adoption of Arabic numerals in northern Europe. The numerals had been known since the end of the tenth century, partly through the work of Gerbert of Aurillac (Pope Sylvester II). We find manuscripts describing the abacus and how to use it still employing counters with Roman numerals until several generations after his death, however.

The crucial addition in shifting thinking on to new ground was the counter for zero. This first appears in these texts with various names (such as *sipos, rota*) along with the Arabic numerals, in a purely decorative row at the tops of the columns. It then became apparent that if a counter was placed in a column to represent zero, and Arabic numerals used, the columns themselves ultimately could be dispensed with, for it became possible to write, say, 306, with a built-in 'placing' of the units, tens, and hundreds.

It is puzzling that these Western mediaeval abacus treatises normally give twenty-seven columns. It would thus be possible to calculate very large numbers with their aid, far bigger than any figure for which contemporaries can conceivably have had a use. There was some discussion of fractions in the same treatises, and a general indication that the interrelationships of numbers were beginning to be more thoroughly understood. Boethius' *Arithmetica* was still the theoretical treatise of choice, but these practical manuals sometimes import scraps of his work into their introductory discussions. The *rithmochachia*, a mediaeval arithmetical board-game which uses the Boethian patterns, is commonly found with treatises on the abacus of this location and period.

G.R. EVANS

REFERENCES

Barnard, F.P. *The Casting-Counter and the Counting-Board.* Oxford: Clarendon, 1916.

Evans, G.R. "*Difficillima et Ardua:* Theory and Practice in Treatises on the Abacus, 950–1150." *Journal of Mediaeval History* 3: 21–38, 1977.

Evans, G.R. "*Duc Oculum.* Aids to Understanding in Some Mediaeval Treatises on the Abacus." *Centaurus* 19: 252–63, 1976.

Evans, G.R. "From Abacus to Algorism: Theory and Practice in Mediaeval Arithmetic." *The British Journal for the History of Science* 10(35): 114–31, 1977.

Evans, G.R. "Schools and Scholars: the Study of the Abacus in English Schools, c. 980–c. 1150." *The English Historical Review* 94: 71–89, 1979.

Knott, C.G. "The Abacus in its Historic and Scientific Aspects." *Transactions of the Asiatic Society of Japan* xiv: 18, 1886.

Kojima, T. *The Japanese Abacus: its Use and Theory.* Tokyo: Charles Tuttle Co., 1954.

Smith, D.E. *History of Mathematics*. Boston: Ginn, 1925.

See also: Computation: Chinese Counting Rods

ABORTION Abortion, along with circumcision, is amongst the oldest operations known to human kind. While abortion meets an extraordinarily important need of the individual, society commonly treats it differently from other aspects of medical science. Usually, there is little honor to be gained by improving abortion technologies.

The explanations and beliefs different cultures develop concerning the origin and maturation of pregnancy help determine attitudes to abortion. The Holy *Quʾrān*, for example, describes pregnancy as a process of increasing complexity progressing from "A drop of seed, in a safe lodging, firmly affixed" to a "lump" with "bones, clothed with flesh," and some Islamic theologians permit abortion early in pregnancy. Most abortions are performed because the woman feels she cannot support the child if born, although unmarried women, especially in non-Western traditional societies, may abort for fear of punishment. In the ancient Ugandan royal household, abortion was carried out on princesses so as not to divide the kingdom.

Abortion is known in practically all cultures from preliterate societies to the most industrialized. Abortionists are often also traditional birth attendants or medicine men. A spectrum of techniques is used which differ in their complexity and in their consequences, and which are greatly influenced by the stage of pregnancy when the procedure is carried out. In non-Western societies most techniques fall into one of three categories: herbal remedies, abdominal massage, and the insertion of foreign bodies into the uterine cavity.

In many contemporary non-Western cultures a variety of brews and potions are concocted to bring on a late menstrual period. Such methods were also widely sold in Western cultures until the reform of the abortion law. The open sale of emmenogogues (medicines to bring on a late period) in Manila, Lima, or Dacca today finds a close parallel in Boston in the nineteenth century United States, or Birmingham, England in the middle of this century. The *Jamu* remedies sold every day in Indonesia include a number of emmenogoges. The use of a tea brewed with the spiny nettle (*Urtica magellanica*) by the Aymara Indians of Bolivia living near Lake Titicaca and the juice of hibiscus leaves (*Abelomschus diversifolius*) in the Pacific islands are two examples from among many herbal remedies in preliterate societies.

The use for abortifacient medications is usually limited to the first six or eight weeks after the last menstrual period. Many do not work, but rely on the fact that spontaneous abortion is common and will often be ascribed to a traditional remedy when one has been used. Others are physiologically active, either on uterine muscle or the embryo, although they often need to be used in doses that may be toxic to the woman. In the 1970s and 1980s the Human Reproduction Program of the World Health Organization tested a number of such abortifacients from around the world, and at least one, from Mexico, underwent preliminary screening by a pharmaceutical company. The time of collection, the method of preparation, and details of use may all be critically important in determining the outcome. An alternative technique for studying traditional abortifacients was developed by Moira Gallen in the Philippines, who worked with vendors of traditional abortifacients and then followed up the women who used them. The data suggest some herbal remedies do indeed bring on a late period.

An amalgamation of Western and non-Western cultures has taken place in Brazil where the Western drug *Cytotec*, an oral form of prostaglandin, is sold illicitly to women with an unintended pregnancy. It is estimated there are one to four million illegally induced abortions in Brazil each year, and each woman has one to three abortions in a lifetime. *Cytotec* produces bleeding from the uterus which, although it does not always lead to abortion, is usually sufficient to take the woman to the hospital, where the abortion is invariably completed in the operating theater. It is relatively safe but can be very painful. Until restrictions were placed on sales in 1991, 50,000 packets a month were being sold.

The second set of abortion technologies, with a history stretching back to preliterate societies, involves physical trauma to the woman's body. Many cultures associate falls and physical violence with abortion, as did the ancient Hebrews (*Bible*: Exodus 21:22). The oldest visual representation of an abortion anywhere in the world is on a bas relief in the great temples of Ankor Wat built by King Suryavarman II (AD 1130–1150). Massage abortion remains common from Burma, through Thailand and Malaysia to the Philippines and Indonesia. Traditional birth attendants use their hands, elbows, bare feet, or a wooden mallet (as portrayed on the Ankor reliefs) to pound the uterus and terminate an unwanted pregnancy. Operators begin by asking the woman to empty her bladder and then try to draw the uterus from beneath the pubic bone so they can apply pressure to the abdominal wall. Sometimes these procedures lead to vaginal bleeding and abortion with relative ease; on other occasions the pain may be so severe that the operator has to stop and return some other time. A study in Thailand estimated that 250,000 such massage abortions take place in the villages of Thailand each year. Gynecologists in Malaysia have described how women are sometimes admitted to hospital for what appear to be the symptoms of appendicitis, with fever and rigidity of their abdominal muscles; when the abdomen is open, the uterus is found to be so bruised and damaged that it may be necessary to do a hysterectomy.

This specialized technique, which has to be learnt from

generation to generation, is largely limited to Southeast Asia, although the American Indian Crows and Assiniboines used a board, on which two women jumped, placed across the abdomen of a recumbent pregnant woman, and Queensland native Australians used a thick twine wound around the abdomen combined with 'punching' the abdominal wall.

The third types of abortion techniques are the most common and are found in all continents. They involve passing a foreign body through the uterine cervix in order to dislodge the placenta and cause an abortion. In traditional societies a twig or root may be used and it may take a day or more for the procedure to work. The major risks are infection and hemorrhage. The Smith Sound Inuit used the thinned down rib of a seal with the point cased in a protective cover of tanned seal skin, which could be withdrawn by a thread when the instrument had been inserted into the uterus. The Fijians fashioned a similar instrument from *losilosi* wood, but without the protective cover for insertion, and the Hawaiians used a wooden dagger-shaped object up to 22 cm long which was perceived as an idol *Kapo*. In contemporary Latin America and much of urban Africa, the commonest method of inducing abortion is to pass soft urinary catheters, or *sonda*, such as those used by doctors when men have enlarged prostates. Such catheters are readily available, although traditional abortions do not always use adequate sterile techniques and, even under the best of conditions, leaving such a catheter in place can be associated with infection.

Epidemiological studies show beyond all doubt that the safest way of inducing first trimester abortion is through the use of vacuum aspiration, and most legal abortions in the Western world are done using this technique. A small tube, varying in diameter from something slightly larger than a drinking straw to about one centimeter, is passed through the cervix, and attached to a vacuum pump. In the first three months of pregnancy such a procedure generally takes about five minutes and is commonly done as an out patient procedure under local anesthesia.

Vacuum aspiration abortion was described in nineteenth century Scotland, but the technique used today was invented in China sometime in the 1950s by Wu and Wu. The method spread across certain parts of the Soviet Union and into Czechoslovakia and some other areas of Eastern Europe. In the 1960s a non-medically qualified practitioner from California called Harvey Karman invented a piece of hand-held vacuum aspiration equipment. Karman got the idea from descriptions of procedure performed in China, Russia, and Eastern Europe, and the flow of ideas has gone full circle with the syringe equipment now being widely used in many non-Western countries, such as Bangladesh, Vietnam, and Sri Lanka.

In the Ankor reliefs the women having abortions are surrounded by the flames of hell. Although abortion was disapproved of in the East, it was still considered a crime against the family, not against the state as it is commonly perceived in the West. Abortion before the felt fetal movements was legal in Britain and all states of the United States in 1800 and illegal in those same places by 1900. With the expansion of colonialism, Western abortion laws were imposed upon all colonized nations of the Third World. The nations of the then British Empire either adopted a form of the 1861 Offenses Against the Person Act of Queen Victoria's England or a version of the Indian penal code. French colonies enacted the code of Napoleon and even countries that were not colonized, such as Thailand and Japan, adopted some form of restrictive abortion legislation in the nineteenth century derived from Western statutes.

The second half of the twentieth century has seen a reversal of many restrictive laws. The majority of the world's population now lives in countries which have access to safe abortion on request. The technologies used all over the world owe a great deal to non-Western philosophies and inventiveness.

MALCOLM POTTS

REFERENCES

Arilha, Margaret and Barbosa, Regina Maria. "Cytotec in Brazil: 'At Least It Doesn't Kill'." *Reproductive Health Matters* 2: 41–52, 1993.

Devereux, George. *Abortion in Primitive Societies.* New York: Julian Press, 1955.

Gallen, Moira. "Abortion in the Philippines: a Study of Clients and Practitioners." *Studies in Family Planning* 13: 35–44, 1982.

Narkavonkit, Tongplaew. "Abortion in Rural Thailand: A Survey of Practitioners." *Studies in Family Planning* 10: 223–229, 1979.

Wu, Y.T. and Wu, H.C. "Suction Curettage for Artificial Abortion: Preliminary Report of 300 Cases." *Chinese Journal of Obstetrics* 6: 26, 1958.

See also: *Jamu* – Ethnobotany – Childbirth

ABŪ JAʿFAR AL-KHĀZIN Abū Jaʿfar Muḥammad ibn al-Ḥusayn al-Khāzin was a mathematician and astronomer who lived in the early tenth century AD in Khorasān. Until recently, it was believed that there were two different mathematicians in the same period, namely Abū Jaʿfar al-Khāzin and Abū Jaʿfar Muḥammad ibn al-Ḥusayn, but in 1978 Anbouba and Sezgin showed that they are the same person.

In mathematics, Abū Jaʿfar al-Khāzin is mainly known because he was the first to realize that a cubic equation could be solved geometrically by means of conic sections. Al-Māhānī (ca. AD 850) had shown that an auxiliary problem in Archimedes' *On the Sphere and Cylinder* II:4, which Archimedes had left unsolved, could be reduced to a cubic equation of the form $x^3 + c = ax^2$. Abū Jaʿfar knew the commentary to Archimedes' work by Eutocius of Ascalon (fifth century AD), in which Eutocius discusses a solution of the same auxiliary problem by means of conic sections. Abū Jaʿfar drew the conclusion that the equation $x^3 + c = ax^2$ could also be solved by means of conic sections. Abū Jaʿfar also studied a number of other mathematical problems. He stated that the equation $x^3 + y^3 = z^3$ did not have a solution in positive integers, but he was unable to give a correct proof. He also worked on the isoperimetric problem, and he wrote a commentary to Book X of Euclid's *Elements*.

In astronomy, Abū Jaʿfar's main work was the *Zīj al-ṣafāʾiḥ*, the Astronomical Handbook of Plates. This work is lost, but we have some references to it in the work of later authors. The work dealt with a strange variant of the astrolabe. One such instrument, made in the twelfth century, was still extant in the beginning of this century in Germany, but it has since disappeared. Photographs of this instrument have been published by David King. Abū Jaʿfar developed a homocentric solar model, in which the sun moves in a circle with the earth as its center, in such a way that its motion is uniform with respect to a point which does not coincide with the center of the earth.

JAN P. HOGENDIJK

REFERENCES

Anbouba, A. "L'Algèbre arabe; note annexe: identité d'Abū Jaʿfar al-Khāzin." *Journal for History of Arabic Science* 2: 98–100, 1978.

King, D.A. "New Light on the *Zīj al-ṣafāʾiḥ*." Centaurus 23: 105–117, 1980. Reprinted in D.A. King, *Islamic Astronomical Instruments*. London: Variorum, 1987.

Lorch, R. "Abū Jaʿfar al-Khāzin on Isoperimetry and the Archimedean Tradition." *Zeitschrift für Geschichte der arabisch-islamischen Wissenschaften* 3: 150–229, 1986.

J. Samsó. "A Homocentric Solar Model by Abū Jaʿfar al-Khāzin." *Journal for History of Arabic Science* 1: 268–275, 1977.

See also: al-Māhānī

ABŪ KĀMIL Abū Kāmil, Shujāʿ ibn Aslam (ca. 850–ca. 930), also known as "the Egyptian Reckoner" (*al-ḥāsib al-miṣrī*) was, according to the encyclopedist Ibn Khaldūn's report on algebra in his *Muqaddima*, chronologically the second greatest algebraist after al-Khwārizmī. He was certainly one of the most influential. The peak of his activity seems to have been at the end of the ninth century.

Although at the beginning of his *Kitāb fī 'l-jabr wa 'l-muqābala* (Algebra) he refers to al-Khwārizmī's similar work, Abū Kāmil's purpose is radically different, for he is addressing an audience of mathematicians presumed to have a thorough knowledge of Euclid's *Elements*. His *Algebra* consists of four main parts.

(a) Like his predecessor, Abū Kāmil begins by explaining how to solve the six standard equations and to deal with algebraic expressions involving an unknown and square roots. The next section (Book II) contains, as in his predecessor's work, six examples of problems and the resolutions of various questions, but the whole is notably more elaborate and geometrical illustrations or proofs are systematically appended. With Book III comes a difference, already hinted at in the introduction to the treatise: the problems now contain quadratic irrationalities, both as solutions and coefficients, and require notable proficiency in computing. Quadratic irrationalities may thus be said to enter definitely the domain of mathematics and no longer be confined to their Euclidean representation as line segments.

(b) These extensions found in Book III are put to immediate use in the resolutions of problems involving polygons in which the link between their sides and the radii of a circumscribed circle is reducible to a quadratic equation — since they are all constructible by ruler and compass.

(c) The subsequent set of quadratic indeterminate equations and systems is most interesting. The methods presented are similar to those of Diophantus's *Arithmetica*, but there are new cases and the problems are presented in a less particular form. Abū Kāmil surely relied on some Greek material unknown to us.

(d) A set of problems which are, broadly speaking, applications of algebra to daily life are directly appended to the former. Some of these, which correspond to highly unrealistic situations, belong more to recreational mathematics. The inclusion of such problems in an algebraic textbook was to become a medieval custom, with both mathematical and didactical motives. The *Algebra* ends with the classical problem of summing the successive powers of 2, which, from the ninth century on, became attached to the 64 cells of the chessboard.

Abū Kāmil's Arabic text is preserved by a single, but excellent, manuscript. The *Algebra* was commented upon several times, in particular in Spain, and the first large mathematical book of Christendom, Johannes Hispalensis's *Liber mahameleth*, is basically a development and improvement of parts *a* and *d*. Despite the *Algebra*'s importance in Spain, no Latin translation was undertaken until the fourteenth century, when Guillelmus (presumably: de Lunis) translated half of it (up to the beginning of part *c*). This translation is better than Mordekhai Finzi's fifteenth century Hebrew one, which, however, covers the whole work. Since these translations were late, Abū Kāmil had no direct influence in the Christian West. Similar material, however, may be found in writings of Leonardo (Fibonacci) of Pisa (fl. 1220).

Abū Kāmil also wrote *Kitāb al-ṭair* (Book of the Birds), a small treatise consisting of an introduction and six problems all dealing with the purchase of different kinds of birds, of which one knows the price per unit, the total number bought and the amount spent (both taken to be 100). Since there are more unknown kinds (3 to 5) than equations, these linear problems are all indeterminate. Abū Kāmil undertook to determine their number of (positive integral) solutions, which he found to be respectively 1, 6, 96 (correct: 98), 304, 0, 2676 (correct: 2678). Although such problems are frequently met in the medieval world, Abū Kāmil remained seemingly unparalleled in his search for all solutions in various cases.

Kitāb al-misāḥa wa'l-handasa, or *Kitāb misāḥat al-araḍīn* (On Measurement and Geometry) is an elementary treatise on calculating surfaces and volumes of common geometrical figures. Since it is meant for beginners, demonstrations and algebra are left out. Why Abū Kāmil found it necessary to write such an elementary book becomes clear when he describes some of the formulae then in use by official land surveyors in Egypt.

Finally, *Kitāb al-waṣāyābi'l-juḏūr* or *Kitāb al-waṣāyā bi'l-jabr wa'l-muqābala* (Estate Sharing Using Unknowns, or using Algebra) applies mathematics to inheritance problems. Abū Kāmil begins by explaining the requirements of the Muslim laws of inheritance and discussing the opinions of known jurists.

Bibliographic sources inform us that Abū Kāmil also wrote another treatise on algebra, and a further one on the rule of false position.

Note that two of the subjects included by al-Khwārizmī in his *Algebra* were treated by Abū Kāmil in separate works. From that time onward, algebra textbooks adopted the same form as his.

JACQUES SESIANO

REFERENCES

Abū Kāmil. *The Book of Algebra* [Reproduction of the Arabic manuscript]. Frankfurt: Institut für Geschichte der arabisch-islamischen Wissenschaften, 1986.

Levey, Martin. *The Algebra of Abū Kāmil* [(bd) Ed. of the Hebrew translation of part a]. Madison: University of Wisconsin, 1966.

Lorch, Richard and Jacques Sesiano. [Edition of the Latin translation]. In *Vestigia mathematica: Studies in Medieval and Early Modern Mathematics in Honor of H.L.L. Busard*. Ed. M. Folkerts and J. Hogendijk. Amsterdam: Rodopi, 1993, pp. 215–252 and 315–452.

Sesiano, Jacques. "Les Méthodes d'analyse indéterminée chez Abū Kāmil" [On part c of the *Algebra*]. *Centaurus* 21: 89–105, 1977.

Sesiano, Jacques. "Le *Kitāb al-Misāḥa* d'Abū Kāmil." *Centaurus* 38:1–21, 1996.

Suter, Heinrich. "Die Abhandlung des Abū Kāmil über das Fünfeck und Zehneck" [On part b of the *Algebra*]. *Bibliotheca Mathematica* 10: 15–42, 1909/10.

Suter, Heinrich. "Das Buch der Seltenheiten der Rechenkunst von Abū Kāmil el-Miṣrī" [On the *Book of the Birds*]. *Bibliotheca Mathematica* 11: 100–120, 1910/11.

See also: Algebra – Mathematics, Recreational – Surveying – Algebra, Surveyor's – Islamic number theory

ABŪ MAʿSHAR Arabic sources such as the *Kitāb al-Fihrist* give the date of Abū Maʿshar al-Balkī's death as 273 of the Hegira, which is AD 886, stating that he was over one hundred years old. Since he was of Persian (Afghan) origin these may well be solar years rather than the lunar years of the Muslim calendar. Therefore his age at his death could have been reckoned as "over" one hundred years if counted in lunar years in Muslim fashion, or else as one hundred solar years. David Pingree claimed to have found the exact date of his birth to be August 10, 787 in a natal horoscope in Abū Maʿshar's *Nativities*. The trouble with this calculation is that Abū Maʿshar himself, in *Mudhākarāt* (Recollections), lamented the fact that he did not know the exact date of his birth and had to rely therefore on a "universal" horoscope he had drawn up. The *Mudhākarāt* then supplies the basic elements of this universal horoscope. Matching these data with Pingree's, it is probably safe to consider the year 171 H/AD 787 as his birth date.

Abū Maʿshar, known in the West as Albumasar, was born in Balkhī in Khurāsān, actually Afghanistan, and seems to have lived there and acquired a reputation as an astrologer much before he came to settle in Baghdad during the reign of the Caliph al-Maʾmūn (813–833), shortly after 820. He lived in Iraq until the end of his life in 886. He must have traveled, at least up and down the Tigris, since in the *Mudhākarāt* he is shown refusing to embark on the stormy waters of the Tigris. He died in Wasit, a city situated in the Sawād, midway between Baghdad and Basra.

Abū Maʿshar's early years are clouded in confusion because of an error committed by the earliest and most important Arab bibliographer, al-Nadīm, in his *Kitāb al-Fihrist*. Writing in the late tenth century, nearly a century after Abū Maʿshar's death, al-Nadīm (d. ca. AD 987) apparently confused Abū Maʿshar the astrologer from Balkh with another

Abū Maʿshar, called an-Najīḥ, who lived in Medina but died in Baghdad AD 787, the year of Abū Maʿshar's birth.

Once settled in Baghdad, where he spent the remaining sixty years of his life, Abū Maʿshar seems to have been involved in its cultural activity, but also in its tumultuous civic life at a time when "nationalisms" in the form of the *Shuʿūbiya* (non-Arabs) were raising their aspirations to cultural parity with the dominant Arabs. Abū Maʿshar's reputation as an astrologer, his newly found friendship with al-Kindī, and the credit he gained through astrological predictions assessing the power of rulers must have opened for him the doors of the political and learned elite of Baghdad. Al-Nadīm relates an episode in which Abū Maʿshar was punished with lashes by the Caliph for a realistic prediction that the Caliph disliked. Both Abū Maʿshar and al-Kindī, using an intricate system of astral conjunctions inherited from the Sassanian tradition, attempted to anticipate the duration of the Arab rule. In his *Risāla* (Epistle) on the duration of the rule of Islam, al-Kindī tried to comfort the ruling Caliph by giving the Arab rule a minimum span of some 693 years, longer than Abū Maʿshar's prediction. In fact Abū Maʿshar gives a total of 693 years, just as al-Kindī did. Still, in combination with the parallel scheme of the two maleficent planets Saturn and Mars affecting the meaning of the conjunctions of Saturn and Jupiter, Abū Maʿshar tended to reduce the duration to some 310 or 330 years only, which would bring the end of Arab rule closer, thus encouraging the aspirations of the *shuʿūbiyya*.

The *Mudākarāt* of Abū Saʿid further tell us that, along with other astrologers, Abū Maʿshar accompanied the army of al-Muwaffaq in its campaign against the rebellious Zanj. Astrologers were used by both sides during these civic troubles. Abū Maʿshar's credit as an astrologer served him in these circumstances, for he may have been consulted by both sides in the Rebellion. At any rate he died in the city of Wasit, south of Baghdad, a city which had seen the farthest advance of the rebellious Zanj army and had been reconquered only shortly before by al-Muwaffaq.

The *Fihrist* credits Abū Maʿshar with 36 works, to which David Pingree adds six more. The list has remained fairly the same for all later bibliographers. This holds true for the original Arabic works as well as for the numerous translations into Latin, Greek, Hebrew, and medieval Romance languages. The uncertainty is due to a number of factors. Abū Maʿshar may have produced some works in several versions or editions. He has been imitated in a number of ways by later Arab authors, some of whom displayed his name prominently at the beginning of their own work, thus complicating the task of the bibliographer. Above all there is a lack of any systematic survey of Abū Maʿshar's production. In addition to the confusion still affecting the Arabic originals, a number of Abū Maʿshar's works were translated into so many media during the Middle Ages that the task of surveying the authentic remains is enormous. This illustrates the immense popularity enjoyed by Abū Maʿshar in the West.

RICHARD LEMAY

REFERENCES

Primary sources

Ibn al-Nadīm. *Kitāb al-Fihrist*. Trans. Bayard Dodge. New York: Columbia University Press, 1970. Vol. II, pp. 656–658.

Abū Maʿshar. *Albumasaris Flores astrologiae*. Augsburg: Erhard Ratdolt, 1488.

Abū Maʿshar. *Introductorium in astronomiam Albumasaris Abalachi octo continens libros partiales*. Augsburg: Erhard Ratdolt, 1489.

Abū Maʿshar. *Albumasar de magnis coniunctionibus et annorum revolutionibus ac earum profectionibus*. Augsburg: Erhard Ratdolt, 1489.

Abū Maʿshar. *Albumasaris. De revolutionibus nativitatum*. Ed. by D. Pingree. Berlin: Bibliotheca Scriptorum Graecorum et Romanorum Teubneriana, 1968.

Secondary sources

Dunlop, D.M. "The *Mudhākarāt fīʿIlm an-nujūm* (Dialogues on Astrology) attributed to Abū Maʿshar al-Balkhī (Albumasar)." In *Iran and Islam, in Memory of the Late Vladimir Minorsky*. Ed. C.E. Bosworth. Edinburgh: Edinburgh University Press, 1971, pp. 229–246.

Lemay, R. *Abu Maʿshar and Latin Aristotelianism in the Twelfth Century*. Beirut: American University of Beirut, 1962.

Millas Vallicrosa, J.M. "Abū Maʿshar." *Encyclopedia of Islam*. Leiden: E.J. Brill, 1913–36.

Pingree, D. "Abū Maʿshar." In *Dictionary of Scientific Biography*. Ed. Charles C. Gillispie. New York: Charles Scribner's Sons, 1975.

Pingree, D. *The Thousands of Abū Maʿshar*. London: The Warburg Institute, 1968.

Thorndike, L. "Albumasar in Sadan." *Isis* 45: 22–32, 1954.

See also: Astrology

ABŪ 'L-BARAKĀT Abū al-Barakāt al-Baghdādī (d. 1164 or 1165) was one of the most original thinkers of the medieval period. Born a Jew in about 1080, but converted late in life to Islam, Abū 'l-Barakāt was a prominent physician and natural philosopher who achieved considerable fame during his own lifetime, as his appellation *awḥad al-zamān* (unique of his age) attests. His numerous insights into physics and metaphysics have been elucidated by the late Shlomo Pines in a number of brilliant studies, on which this résumé depends in large measure.

Abū 'l-Barakāt's contributions are all contained in his *chef d'œuvre*, *al-Muʿtabar* (That Which has been Attained by Reflection). Although there may be some doctrinal discrepancies between various passages in the book, which may be due to the fact that the work is actually a collection of notes compiled over a considerable period of time, each section by itself displays a very clear and systematic exposition, surveying earlier opinions on the subject, objections to these, and possible answers to the objections (including the occasional concession that the objection is valid and necessitates a revision of the original idea), followed by Abū 'l-Barakāt's own opinion. Abū 'l-Barakāt exhibits a remarkable ability to disentangle issues that had become densely intertwined through centuries of debate, for example the three notions of time, space, and the infinite. Particularly significant are the occasions when the author gives great, occasionally decisive, weight to "common opinion", on the grounds that the issue at hand — the notions of time and space are the most important to fall into this group — involve *a priori* concepts which must be elucidated by examining how people actually perceive, rather than *a posteriori* academic analysis.

Some of the ideas which Abū 'l-Barakāt advances in the course of his discussions prefigure much later notions which proved to be correct: for example, the idea that a constant velocity applied to a moving body causes it to accelerate. Others, by contrast, showed themselves to be wrong: for instance, the idea that every type of body has a characteristic velocity which reaches its maximum when the resistance is zero. (In this way, Abū'l-Barakāt answers the objection that bodies moving in a vacuum would have an infinite velocity.) All in all, the work of Abū'l-Barakāt and its continuation by his student Fakhr al-Dīn al-Rāzī constituted a most serious challenge to the formulations of Ibn Sīnā, which then dominated physical and metaphysical thought in Near East.

All that we possess in the way of medical writings by Abū'l-Barakāt are a few prescriptions for remedies. These remain in manuscript and are as yet unstudied.

Y. TZVI LANGERMANN

REFERENCES

Abū 'l-Barakāt. *al-Muʾtabar* (That Which has been Attained by Reflection). Ed. in three volumes by Serefeddin Yaltkaya. Hyderabad, 1938–40.

Pines, Shlomo. *Studies in Abu'l-Barakāt al-Baghdādī, Physics and Metaphysics.* [The Collected Works of Shlomo Pines, volume 1]. Jerusalem: Magnes Press, and Leiden: E.J. Brill, 1979.

See also: al-Rāzī – Ibn Sīnā – Abū'l-Fidā'

ABŪ'L-FIDĀ' Abū'l-Fidā' Īsmāʿīl ibn ʿalī ibn Maḥmūd ibn Muḥammad ibn ʿUmar ibn Shahanshāh ibn Ayyūb, ʿImād al-Dīn al-Ayyūbī was a prince, historian, and geographer belonging to the Ayyūbid family. He was born in Damascus, Syria in AD 1273 and soon began his military career against the Crusaders and the Mongols. In AD 1299 he entered the service of the Sultan al-Malik al-Nāṣir and, after 12 years, he was installed as governor of Ḥamā. Two years later he received the title of al-Malik al-Ṣāliḥ. In AD 1320 he accompanied the Sultan Muḥammad on the pilgrimage to Mecca and was given the title of al-Malik al-Muʾayyad. He died at Ḥamā, Syria in AD 1331.

Abū'l-Fidā' is the author of some poetic productions, such as the version in verse of al-Māwardī's juridical work *al-Ḥāwī*. However, his celebrity is based on two works which can be considered basically compilations of earlier works which he elaborated and completed. One of them is the *Mukhtaṣar Taʾrīkh al-bashar* (A Summary on the History of Humanity) written in AD 1315 as a continuation of the *Kāmil fī-l-taʾrīkh* of Ibn al-Athīr. It was divided into two parts: the first was devoted to pre-Islamic Arabia and the second to the history of Islam until AD 1329. It was kept up to date until AD 1403 by other Arabic historians. It was translated into Western languages and became the basis for several historical syntheses by eighteenth-century Orientalists.

Abū'l-Fidā''s most important scientific work is *Taqwīm al-buldān* (A Sketch of the Countries) written between AD 1316 and 1321. It consists of a general geography in 28 chapters.

This book includes the problems and results of mathematical and physical geography without touching upon human geography or geographical lexicography. There is a table of the longitudes and latitudes of a number of cities, including the differing results found in the sources, setting up a comparative table for geographical coordinates. Among the sources of the book are geographers such as Ibn Ḥawqal and Ibn Saʿīd al-Maghribī.

This work was translated into German, Latin, and French between the sixteenth and the nineteenth centuries, making a significant contribution to the development of geography.

EMILIA CALVO

REFERENCES

Gibb, H.A.R. "Abū l-Fidā'." In *Encyclopédie de l'Islam*, 2nd ed., vol. I. Leiden: E.J. Brill; Paris: G.P. Maisonneuve, 1960.

Reinaud, J.-T. *Géographie d'Aboulféda traduite de l'arabe en francais et accompagnée de notes et d'éclaircissements*. 2 vols. Paris,

1848. Repr. Frankfurt: Institut für Geschichte der Arabisch-Islamischen Wissenschaften, 1985.

Sarton, George. *Introduction to the History of Science*, vol. III. Baltimore: Williams and Wilkins, 1947.

Sezgin, Fuat. *The Contribution of the Arabic-Islamic Geographers to the Formation of the World Map*. Frankfurt: Institut für Geschichte der Arabisch-Islamischen Wissenschaften, 1987.

Vernet, Juan. "Abū' l-Fidā᾿." In *Dictionary of Scientific Biography*, vol. I. New York: Charles Scribner's Sons, 1970.

ABŪ'L-ṢALT is known as Abū'l-Ṣalt al-Dānī. He was an Andalusian polymath born in Denia in 1067. In about 1096 he went to Egypt where he lived for sixteen years. An unsuccessful attempt to rescue a boat loaded with copper which had sunk in the harbor of Alexandria cost him three years in prison, after which he migrated to Mahdiyya (Tunis) where he died in 1134.

He wrote about pharmacology (a treatise on simple drugs, *Kitāb al-adwiya al-mufrada*, translated into Latin by Arnold of Vilanova), music, geometry, Aristotelian physics, and astronomy, and he seems to have been interested in a physical astronomy, different from the Ptolemaic mathematical astronomy which predominated in al-Andalus. His treatise on the use of the astrolabe, *Risāla fī-l-ʿamal bi-l-asṭurlāb*, written while he was in prison, probably introduced into Eastern Islam the characteristic Andalusian and Maghribi device, present in the back of Western instruments, which establishes the relation between the date of the Julian year and the solar longitude. He is also the author of a short treatise on the construction and use of the equatorium, *Ṣifat ʿamal ṣafīḥa-jāmiʿa tuqawwim bi-hā-jamiʿ al-kawākib al-sabʿa* (Description of the Way to Use a General Plate With Which to Calculate the Positions of the Seven Planets) which follows the techniques developed in al-Andalus by Ibn al-Samḥ (d. 1035) and Ibn al-Zarqāllu (d. 1100), although it also presents original details which show his ingenuity. He probably reintroduced this instrument in the Islamic East where it had appeared in the tenth century but was later forgotten until it was recovered by al-Kāshī (d. 1429).

JULIO SAMSÓ

REFERENCES

Comes, Mercè. *Ecuatorios andalusíes. Ibn al-Samḥ al-Zarqāllu y Abū'l Ṣalt*. Barcelona: Instituto de Cooperación con el Mundo Arabe and Universidad de Barcelona, 1991, pp. 139–157, 237–251.

Samsó, Julio. *Las Ciencias de los Antiguos en al-Andalus*. Madrid: Mapfre, 1992, pp. 313–317.

See also: Ibn al-Zarqāllu

ABŪ'L-WAFĀ᾿ Abū'l-Wafā᾿ al-Būzjānī, Muḥammad ibn Muḥammad ibn Yaḥyā ibn Ismāʿīl ibn al-ʿAbbās, was born in Būzjān [now in Iran] on 10 June 940. After he moved to Baghdad in 959, he wrote important works on arithmetic, trigonometry, and astronomy.

Abū'l-Wafā᾿ provided new solutions to many problems in spherical trigonometry and computed trigonometric tables with an accuracy that had not been achieved until his time. He made astronomical observations and assisted at observations in the garden of the palace of Sharaf al-Dawla. Finally, he wrote two astronomical handbooks, the *Wāḍiḥ Zīj* and *al-Majisṭī* (*Almagest*). More information about Abū'l-Wafā᾿'s tables must be obtained from *zīj*es that have incorporated material from his works, such as the *Baghdādī Zīj*, compiled shortly before the year 1285 by Jamāl al-Dīn al-Baghdādī. A solar equation table attributed to Abū'l-Wafā᾿ occurs in it.

Several later *zīj*es incorporate Abū'l-Wafā᾿'s mean motion parameters. Various sine and cotangent values which he gave in the extant part of *al-Majisṭī* are equal to the values found in al-Baghdādī's sine and cotangent tables. Furthermore, al-Baghdādī's table for the equation of daylight was computed by means of inverse linear interpolation in a sine table with accurate values to four sexagesimal places for every 15° of the argument, and Abū'l-Wafā᾿ is known to have computed an accurate sine table with just that format.

In *Risāla fī iqāmat al-burhān ʿalā 'l-dā'ir min al-falak min qaws al-nahār wa'rtifāʿ niṣf al-nahār wa'rtifāʿ al-waqt* (On Establishing the Proof of the [rule for finding the] Arc of Revolution from the Day Arc, the Noon Altitude, and the Altitude at the Time), Abū'l-Wafā᾿ deals with a fundamental problem of ancient astronomy, that of finding the time in terms of solar altitude. He mentions in the introduction that the formula stated by Ḥabash al-Ḥāsib (fl. 850) is only approximate. Abū'l-Wafā᾿ gives three proofs of the formula. The procedure of the first two proofs deals entirely with rectilinear configurations inside the sphere, in spite of the fact that the relation being investigated concerns arcs on the surface of the sphere. This technique was characteristic of Hindu spherical astronomy, as well as that of the Greeks prior to the application of Menelaus' Theorem. The method used in the third proof consists essentially of two applications of the Transversal Theorem. Contrary to Hindu trigonometry and most Islamic astronomers, Abū'l-Wafā᾿ was one of the few who defined the trigonometric functions with respect to the unit circle, as is the case nowadays. In *Fī ḥirāfat al-abʿād bain al-masākin* (On the Determination of the Distances between Localities) Abū'l-Wafā᾿ gives two rules for calculating the great circle distance between a pair of points on the earth's surface. He applies both to a worked example: given the terrestrial coordinates of Baghdad and Mecca he calculates the distance between them, a matter of some interest to Iraqi Muslims undertaking the pilgrimage. The first method employs standard medieval spherical trigonometry and can be regarded as a byproduct of a common procedure for calculating the *qibla*, the direction of Muslim prayer. It is called by al-Bīrūnī "the method of the *zīj*es". The second method is less ordinary and its validity is not obvious. In addition to the tangent function, it employs the versed sine, a term Abū'l-Wafā' does not use in the treatise studied above, but which appears frequently in the literature. The origin of this second rule might stem from the so-called analemma method, a common and useful ancient technique for solving spherical astronomical problems. The general idea was to project or rotate elements of the given solid configuration down into a single plane, where the desired magnitude appeared in its true size. The resulting plane configuration was then solved by constructions to scale or by plane trigonometry. Aside from the trigonometry, the text is of interest as an intact example of medieval computational mathematics. Numbers are represented in Arabic alphabetical sexagesimals throughout. The results of the multiplications suggest that all operations were carried out in sexagesimal arithmetic, with none of the very common intermediate transformations into decimal integers. Trigonometric functions and their inverses are carried out to four sexagesimal places. Al-Ḥubūbī challenged Abū'l-Wafā' to produce and prove a rule for calculating the area of a triangle in terms of its sides. In his *Jawāb Abī al-Wafā'Muḥ ibn Muḥ al-Būzjānī ʿammā sa'alahu al-Faqīh Abū ʿAlī al-Ḥasan ibn Ḥārith al-Ḥub ūbī fī misāḥat al-muthallathāt* (Answer of Abū'l-Wafā' to the Question Put to Him by the Jurist Abū 'Alī al-Ḥasan al-Ḥubūbī on Measuring the Triangle), Abū'l-Wafā' gives three such rules. None of these is identical with "Heron's Rule", but all are equivalent to it. The earliest work on finger reckoning that has survived is Abū'l-Wafā''s *Fīmāyaḥtāju ilaihi l-kuttāb wa-l-'ummāl min 'ilm al-ḥisāb* (On what Scribes and Officials Need from the Science of Arithmetic). As the name implies, it was written for state officials, and therefore gives an insight into tenth-century life in Islam from the administrative point of view. Three more works demonstrate Abū'l-Wafā''s interest in practical mathematics: *Fīmā yaḥtāju ilaihi aṣ-ṣāniʿmin aʿmāl al-handasīya* (On the Geometrical Constructions Necessary for the Craftsman), written after 990; *al-Mudkhal al-ḥifẓī ilā ṣināʿat al-arithmāṭiqī* (Introduction to Arithmetical Constructions); and *Risāla al-shamsīya fī l-fawā'id al-ḥisābīya* (On the Benefit of Arithmetic).

On What Scribes and Officials Need, written between 961 and 976, enjoyed widespread fame. The first three parts, "On Ratio", "On Multiplication and Division", and "Mensuration", are purely mathematical. The other four contain solutions of practical problems with regard to taxes, problems related to harvest, exchange of money units, conversion of payment in kind to cash, problems related to mail, weight units, and five problems concerning wells. In this compendium Abū'l-Wafā' systematically sets forth the methods of calculation used by merchants, by clerks in the departments of finance, and by land surveyors in their daily work; he also introduces refinements of commonly used methods and criticizes some for being incorrect. Considering the habits of the readers for whom the textbook was written, Abū'l-Wafā' completely avoids the use of numerals. Numbers are written in words, and their calculations are performed mentally. To remember the results of intermediary steps, calculators bent their finger joints in conventional ways which enabled them to indicate whole numbers from 1 to 9999. This same device was repeated to indicate numbers from ten thousand onward. All procedures, often quite complex, are only described by words. This treatise on practical arithmetic provides the model for all the treatises on the subject from the tenth to the sixteenth centuries.

He is cited as a source or an authority, but more often can only be discerned underneath. In the *Geometrical Constructions Necessary for the Craftsman* Abū'l-Wafā' discusses a host of interesting geometrical constructions and proofs. He constructs a regular pentagon, a regular octagon, and a regular decagon. The construction of the regular pentagon with a "rusty" compass is especially noteworthy. Such constructions are found in the writings of the ancient Hindus and Greeks, but Abū'l-Wafā' was the first to solve a large number of problems using this compass with fixed opening.

Renaissance Europe had a great interest in these constructions. The possible practical applications (such as making decorative patterns) may have been an additional motivation for studying things like a regular (or perhaps equilateral) pentagon inscribed in a square. However, the importance of such applications should not be overestimated. In proposing his original and elegant constructions, Abū'l-Wafā' simultaneously proved the inaccuracy of some practical methods used by the craftsmen.

To honor Abū'l-Wafā', a crater on the moon was named after him. He died in Baghdad in 997 or 998.

YVONNE DOLD-SAMPLONIUS

REFERENCES

Abū'l-Wafā' al-Būzjānī. *Knowledge of Geometry Necessary for the Craftsman*. Ed. S. A. al-ʿAlī. Baghdad: University of Baghdad Centre for the Revival of the Arabic Scientific Heritage, 1979 (in Arabic).

Ehrenkreutz, Andrew S. "The Kurr System in Medieval Iraq." *Journal of the Economic and Social History of the Orient* 5: 309–314, 1962.

Kennedy, Edward S. and Mustafa Mawaldi. "Abū'l-Wafā' and the Heron Theorems." *Journal of the History of Arabic Science* 3: 19–30, 1979.

Kennedy, Edward S. "Applied Mathematics in the Tenth Century: Abū'lWafā' Calculates the Distance Baghdad-Mecca." *Historia Mathematica* 11: 192–206, 1984.

Nadir, Nadir. "Abū'l-Wafā' on the Solar Altitude." *The Mathematics Teacher* 53: 460–463, 1960.

Saidan, A. S. *The Arithmetic of Abū al-Wafā' al-Būzajānī.* Amman: al-Lajnah al-Urdunnīyah lil-Ta'rīb wa-al-Naṣr wa-al-Tarjamah, 1973.

Saidan, Ahmad S. "The Arithmetic of Abū'l-Wafā'." *Isis* 65: 367–375, 1974.

Youschkevitch, A.P. "Abū'l Wafā'." *Dictionary of Scientific Biography*, vol. 1. Ed. Charles C. Gillispie. New York: Charles Scribner's Sons, 1975, pp. 39–43.

See also: *Qibla* – Ḥabash al-Ḥāsib – Algebra, Surveyor's – Mathematics, Recreational and Practical

ACOUSTICS IN CHINESE CULTURE A sound is perceived in terms of its pitch, loudness, and tone quality characteristics. The Chinese term for pitch is *yīn lü* or simply *lü*. Early mentions of pitch in connection to its function in ode singing, in musicology, and in the standardization of measures and weights are found in the *Yú Shū* (*The Book of Yú*). The Chinese used the terms *qīng* (clear) and *zhuó* (muddy) to describe, respectively, the high and low pitches. In remonstrating the decision of the High King Jǐng of Zhōu to have the Wǔ-Yì bell melted down and converted into the Dà-līn bell of lower pitch, minister Shàn Mū Gōng stated in 522 BC:

> The ear functions harmoniously within a certain range of high and low pitches.
> The determination of this pitch-range should not be left for individuals.
> For this reason bells constructed by our ancient kings never exceeded their corresponding size in *jūn* (unit) and weight in *dàn* (unit). This is where the specifications of measures and weights for pitches originated.

This statement, recorded in the *Guó Yǔ* (Discourses on States), reveals that the early Chinese were aware of the existence of an audible pitch-range.

A description on the development of the twelve pitches preserved in bells from antiquity is provided by music master Zhōu Jiū. He says that the reason that ranges and degrees can be established for pitch is due to Shén Gǔ of antiquity, who investigated and determined the *zhōng shēng* (middle tone) as the reference. The degrees of the pitches and tuned bells are the standards observed by all officials. From the *zhōng shēng*, one first establishes three pitches, then levels them out evenly into six pitches, and finally brings them to completion in twelve pitches. This is the *dāo* of nature.

This suggests that the twelve-pitch system was derived from 'trichords'. An early example of a trichord is provided by three jade stone-chimes unearthed from a pit of the Yin ruins (ca. thirteenth century BC) at Anyang; they not only are capable of producing tones but also have their pitch names engraved on the stones.

Early evidence for pitch standardization mentioned in the Chinese text is provided by the common notes found among the unearthed musical instruments. The most significant archaeological evidence is provided by the pitch-pipes unearthed in 1986 from a Chǔ tomb (M21) of the Warring States period located at the present Yǔtáishān in Jiānglíng, Hūběi province. These pitch-pipes, made of nodeless bamboo (open ends) with different lengths and diameters, are the earliest specimens currently available. Though broken, four of the pitchpipes still had readable inscriptions, providing not only the names of the pitches, but also explicit statements on assigning pitches through the usage of the character *dìng* (literally to fix or determine). From the measured frequencies of the unearthed Marquis Yǐ set-bells, one obtains for the Huáng-zhōng pitch a measured frequency of 410.1 vibs/sec in the fifth century BC.

In Chinese, the louder sound is called *dà* and the softer one *xì*. The modern Chinese term for loudness in acoustics is *yīn-liàng*. An early discussion on loudness is found in the *Guó Yǔ* on the arrangement of instrument in an orchestra, in which harmony and balance are considered essential.

An obvious question on loudness is its role in audibility. Lǎo Zǐ made some interesting observations which seem to have some bearing on the threshold of hearing. We have from the *Dào Dé Jīng* (Canon of the Virtue of Dào) the statement: That which is listened to but not heard is called *xī*. The term *xī*, as it is defined here, relates to the term *xī-shēng*, which Needham translates as 'tenuous note'.

At the time of Lǎo Zǐ, concepts such as frequency and intensity had not yet been developed. Lǎo Zǐ could not have distinguished the audibility of *xī-shēng* in relation to its frequency and intensity. But he probably was aware that most *xī-shēng* are low-pitch sounds and that their audibility depends sensitively on loudness. Thus, when he said that the loud sound contains *xī-shēng,* he was probably referring to the audibility of those *inaudible* low-pitch sounds at a louder level.

Tone quality is the perception of sound in relation to the dynamic structure of the sound. The Chinese term for tone quality is *yīn-zhì* and in classic usage simply *yīn*. The early Chinese acousticians identified tone quality with the sound-producing material and began to classify sounds in accordance with such materials. Eight distinct tone qualities known simply as *bā-yīn* (eight tones) are identified with eight such sound-producing materials.

Early mention of the 'eight tones' is found in the *Yú Shū* and *Zhŏ Zhuàn*. Other than being responsible for instituting music with the two sets of six pitches, the Grand Music Masters (*Dà-Shī*) were also responsible for making sure that all music was composed in pentatonic intonations: *gōng*, *shāng*, *jué*, *zhǐ*, and *yǔ*, and that all music was performed in eight tones: *jīn* (metal), *shí* (stone), *tǔ* (clay), *gé* (skin), *sī* (silk), *mù* (wood), *páo* (gourd), and *zhū* (bamboo).

Each of these sound-producing materials represents a basic tone quality. In 1936, Schaeffner commented that the *bā-yīn* was "probably the oldest extant classification of musical instruments in any civilization". Needham and Robinson compared the *bā-yīn* classification with the Greco–Roman classification of musical instruments, namely wind, stringed, and percussion instruments, and concluded that the Greco–Roman classification was more scientific. The point that needs to be emphasized here is that the *bā-yīn* classification was based on tone qualities and not on musical instruments, even though there is an intimate connection between the two. It is important to note that each complex tone has its own unique characteristic harmonic spectrum and wave form. There does not yet exist a satisfactory system for classifying tone quality. The fivefold classification of ideophones, membramophones, chordophones, electrophones and aerophones is again a classification based on musical instruments.

In addition to the tone quality of sound due to the sound-producing materials, the ancient Chinese were also interested in the variation in tone quality coming from the configuration of musical instruments and the different ways of playing musical instruments. Twelve sounds are specified in the *Zhōu Lǐ* to identify tone qualities with the configurations of the musical instruments. According to the *Lǚ-Shì Chūn-Qiū* of 239 BC, the techniques of exploiting timbre with overtones had reached a high level of art in *qín* (half-tube zither) playing. Indeed, in the playing the ancient lute (*gǔ qín*) with no frets and markings, each note could be played with a variety of subtleties in touch.

The Physical Nature of Sound

As described in the *Kǎo Gōng Jì* (The Artificers' Record), sound is produced by vibrations, and there is a relationship between the thickness of the vibrating walls and the pitch of the sound produced by the vibration.

In Chinese classic usage, the character *jí* means rapid (or fast) and *shū* means slow. Thus, sound produced by rapid vibration is called *yīn jí* (or *shēng jí*) and by slow vibration *yīn shū* (or *shēng shū*). Such technical terms are found in the description of acoustics of bells, stone-chimes, and drums in the *Kǎo Gōng Jì*. It is important to appreciate the relation between this terminology and the terminology for pitch. The terms *qīng* and *zhuó* for high and low pitches represent the perceptive description of sound in relation to the rate of vibration, while the terms *yīn jí* and *yīn shū* represent the physical description of sound in relation to the rate of vibration.

An explicit statement on the direct relation between the physical and perceptive descriptions of sound is found in the *Guăn Zĭ* (The Book of Master Guăn), in which it is stated that the sound of rapid vibration has a high pitch.

According to the *Kǎo Gōng Jì*, Chinese bell-makers examined the audibility of a bell's sound at a distance and discovered that the audible distance of a strike tone depends on the interplay of the diameter and the length of the bell. Such an experimental investigation on the dependence of the audible distance of a bell's sound on the dimensions of the bell was a significant step toward a scientific inquiry into the physical nature of sound propagation. The early Chinese probably looked to the propagation of a disturbance in water as a mental image for the propagation of sound. This is suggested by the hydraulic terms, *qīng* (clear) and *zhuó* (muddy). Thus, the analogy between the expanding pattern of ripples on water and the propagation of sound in the air probably also began early in Chinese civilization. However, no extant record with explicit mention of this mental connection between waves in water and in air is available earlier than the work of Wáng Chōng of the first century AD.

An important physical phenomenon of sound that was discovered early in Chinese civilization is resonance. In the *Zhuāng Zĭ* (The Book of Master Zhuāng), there is a passage attributed to Lŭ Jù of Western Zhōu, in which he says that the striking of the *gōng* note of one zither causes the *gōng* note of the other zither to vibrate, and that the same is true of the *juē* note, because the notes are of the same pitch.

The resonance phenomenon of Lŭ Jù achieved wide recognition in ancient times. The principle of resonance was later summarized in the statement "*shēng-bǐ zē-yìng*". This contains two key technical terms, *bǐ* and *yìng*. The term *bǐ*, which literally means 'comparison', is coined to represent 'matching in pitch', a condition for resonance first pointed out by Lŭ Jù, and the term *yìng*, which literally means 'respond', is coined to emphasize the sympathetic aspects of the vibrations in resonance. These technical terms became common in subsequent accounts of resonance in sound.

CHEN CHENG-YIH

REFERENCES

Chen Cheng-Yih. "The Generation of Chromatic Scales in the Chinese Bronze Set-Bells of the 5th Century". In *Science and Technology in Chinese Civilization*. Ed. Cheng-Yih Chen. Singapore: World Scientific, 1987, pp. 155–197.

Chen Cheng-Yih. "The Significance of the Marquis Yĭ Set-Bells on the History of Acoustics." In *Two-Tone Set Bells of Marquis Yĭ*. Ed. Cheng-Yih Chen. Singapore: World Scientific, 1994, pp. 145–244.

Lǐ Chún-Yī. *Zhōng-Guó Gǔ-Dài Yīn-Yuè Shǐ Gǎo* (Notes on the History of Music in Ancient China). Beijing: Yin-Yue Publisher, 1981.

Needham, Joseph. *Science and Civilisation in China*. Cambridge: Cambridge University Press, 1962, vol. 4, part I, sec. 26, pp. 126–228.

Schaeffner, A. *Origine des Instruments de Musique*. Paris: Payot, 1936.

Tán Wéi-Sì. "Jiānglíng Yǔtáishān 21-Hào Chǔ-Mǔ Lǜ-Guǎn Qiǎn-Lùn." (A Discussion of the Pitch-Pipes of the M21 Chǔ-Tomb of Yǔtáishān in Jiānglíng). *Cultural Relics* 4: 39–42, 1988.

ACUPUNCTURE Acupuncture is a branch of the Chinese medical system dating back as far as the early Zhou period in the first millennium BC. It has been constantly evolving, particularly during the last 300 years, and more specifically since 1950, when the art of acupuncture came to be widely developed, in both theory and practice. As indicated by the name, it is performed by implanting very thin and sharp needles into subcutaneous connective tissues and muscles at precisely specified points on the surface of the human body. In the Neolithic age, the earliest needles were made of flint, and subsequently, with the discovery of metal, bronze, silver, and gold. Nowadays acupuncture needles are made of stainless steel and are very thin, much thinner than the familiar hypodermic needles.

The earliest known text on acupuncture is the *Huangdi Neijing* (Inner Classic of the Yellow Emperor) divided into two parts, namely *Shuwen* (Candid Questions) and *Lingshu* (Vital Axis) written in the second century BC and in the first century BC, respectively. The text represents the accumulated knowledge and experiences of native practitioners over a long period of time. It deals with medical physiology and anatomy as well as clinical medicine. Throughout the centuries, this text has remained the main source of reference on acupuncture and is the foundation for all its subsequent development.

The next important text on acupuncture is the *Zhenjiu Jia Yi Jing* (Treatise on Acupuncture and Moxibustion) by Huangfu Mi of the third century. Suffering from severe rheumatism or neuralgia, and desperately seeking an efficacious treatment, Huangfu Mi had consulted well-known acupuncturists of his time and did extensive research on the existing medical texts himself. The result is the text on acupuncture and moxibustion dealing systematically with aspects of physiology, pathology, diagnosis, and therapy as well as prophylaxis. Huangfu Mi identified some 349 acu-points, grouping them separately, giving practical details on the implantation of needles, the duration required, and the desirable depths at specific acu-points. *Zhenjiu Jia Yi Jing* is a specialist manual on acupuncture, and had indeed exerted considerable influence on the development of acupuncture down to the tenth century. It also had far-reaching influence on the study and practice of acupuncture abroad, notably in Japan and Korea.

The eleventh century saw the first casting of life-size human acupuncture figures in bronze by Wang Weiyi. The bronze figures were pierced with small holes corresponding to the acupoints, then covered with wax, filled with water, and used for examination purposes. Accurate insertion of needles through the specific holes on the bronze figure would cause the water to flow out. Wang Weiyi also wrote an illustrated manual to accompany the bronze figures.

In the following centuries, several important texts on acupuncture appeared. They began with Wang Zhizong's *Zhenjiu Zi Sheng Jing* (Manual of Acupuncture and Moxibustion for Health) in the thirteenth century, followed by Hua Shou's *Shisi Jing Fahui* (An Elaboration of the Fourteen Acu-tracts) in the fourteenth century, Gao Wu's *Zhenjiu Ju Ying* (A Selection of Texts on Acupuncture and Moxibustion) in 1529, and Yang Jizhou's *Zhenjiu Da Cheng* (Principles of Acupuncture and Moxibustion) in 1601. After that, acupuncture became stagnant until its revival at the turn of the twentieth century.

In the *Huangdi Neijing* and all other medical texts, the human being is seen as a microcosm of the universe and subjected to the same tension and disruptions as nature itself. Just as the natural world adjusts to seasonal and climatic changes, the human body, too, is supposed to adjust to the inner forces within the body and to the outer forces of nature. Therefore, the concept of health is to seek moderation and harmony with the universal forces. One has to adapt oneself to nature's rhythmicity in accordance with the interplay of *Yin* and *Yang*, the Five Phases (*Wuxing*), and the circulation of *qi* through the various channels and tracts within the body.

In acupuncture, the circulation of *qi* through a network of channels and tracts round and round the body is termed the *Jinglo* system. The Chinese consider the human body a structural unity in which various tissues and organs have closely knit connections. The system can be likened to a kind of communication network with stems and branches. The inner parts of the *Jinglo* system originate at the inner organs, while their outer parts are on the surface areas of the skin. The system consists of twelve tracts running on either side of the body that represent the organs, and two "extra" tracts, one along the center at the front of the body and the other at the back. Besides the *Jinglo* system of acu-tracts, there is also the *Jingmo* system (tract-and-channel network system) which controls the circulation not only of the *qi* in the tracts but also of the blood in the blood vessels. The flow of blood is maintained by *qi*, and the motion of the *qi* depends on the blood. Hence, because of this connection running between the interior and exterior of the body, illness in the interior can get to the surface of the body, making certain skin areas particularly sensitive. It was the observation of these hypersensitive skin areas in the presence of diseases that led to the recognition of a series of points along the tracts. These tracts are, therefore, lines that connect the points associated with their respective organs. The number of points on each of the tracts varies from 9 (the heart acu-tract) to 67 (the bladder acu-tract). For the general purposes of treatment, only some 360 or 365 common acu-points are used. However, if the acu-points on the secondary channels and their small branches are taken into consideration, the number of acu-points can range from 500 to 800. As these acu-points are situated at specific places where *qi* is transmitted, they are also called "*qi*-points". All acu-points are designated by specific names indicative of their respective characteristics and therapeutic functions.

In a state of good health, the *yinyang* forces are in equilibrium, thus enabling the *qi* to promote and regulate the various functions of the inner organs. If the state of normality in the inner organs breaks down, sickness results. Different manifestations of a disease affecting any inner organ can be seen in various body areas interconnected by divisions of the various acu-tracts. Diseases of the inner organ therefore affect the circulation of the *qi* in the related acu-tracts. The principal function of acupuncture is to insert needles at specific acu-points in order to normalize the flow of *qi* in the inner organs.

Apart from the treatment of diseases, acupuncture can also be used for sedation and analgesia. This aspect of acupuncture, interestingly enough, is a recent innovation. In acupuncture analgesia, a single or several needles are inserted at certain points in the patient's body. Analgesia follows a period of stimulation caused by twirling the needles. Precisely how acupuncture works to kill pain is largely a matter of conjecture. The most popular explanation is the "gate control" theory. This theory holds that stimulation of the large so-called A-delta fibers in the sensory nerves (as produced, presumably, by acupuncture needles) closes a hypothetical gate in the spinal cord. This would block pain impulses, which travel along a different set of smaller nerve fibers, from traveling up the cord into the brain. However, this would only block pain traveling along peripheral nerves leading to and from the spinal cord. Following this, a secondary theory has been proposed, stating that there may be a second gate in the thalamus of the brain that closes to prevent pain sensations from reaching the cerebral cortex from below or above the spinal cord. A final word, however, has yet to be said.

ANG TIAN SE

REFERENCES

Chen, Danan. *Zhonguo Zhenjiu Xue* (Chinese Acupuncture and Moxibustion). Beijing: Renmin Weisheng, 1956.

Croizier, Ralph C. *Traditional Medicine in China: Science, Nationalism, and the Tensions of Cultural Change*. Cambridge, Massachusetts: Harvard East Asian Series, 34, 1968.

Guo, Aichun. *Huangdi Neijing Suwen Jiaozhu Yuyi* (A Critical Edition of the Candid Questions of the Inner Classic of Yellow Emperor with vernacular translation). Tianjing: Kexue Jishu, 1980.

Guo, Aichun. *Lingshu Jing Jiaoshi* (A Collated Version of the Vital Axis in the Inner Classic of Yellow Emperor). Beijing: Renmin Weisheng, 2 vols, 1982.

Needham, Joseph and Lu, Gwei-Djen. *Celestial Lancets; A History and Rationale of Acupuncture and Moxa*. Cambridge: Cambridge University Press, 1980.

Porkert, Manfred. *The Theoretical Foundations of Chinese Medicine Systems of Correspondence*. Cambridge, Massachusetts: M.I.T. Press, 1974.

Ren, Yingqiu. *Yun Qi Xueshuo* (Theory of the Circulation of the *Qi*). Shanghai: Kexue Jishu, 1983.

Sivin, Nathan. *Traditional Medicine in Contemporary China*. Ann Arbor: The University of Michigan Center for Chinese Studies, 1987.

Shandong Zhongyi Xueyuan (Shandong College of Chinese Medicine) Ed. *Zhenjiu Jiayi Jing Jiaoshi* (A Collated Version of the Treatise on Acupuncture and Moxibustion by Huangfu Mi). Beijing: Renmin Weisheng, 2 vols, 1979.

Tan, Leong T., Tan, Margaret Y.C. and I. Veith. *Acupuncture Therapy; Current Chinese Practice*. Philadelphia: Temple University Press, 1973.

Veith, Ilza. *The Yellow Emperor's Classic of Internal Medicine*. New Edition. Berkeley: University of California Press, 1966.

See also: *Huangdi Neijing* – *Zhenjiu Jia Yijing* – Medical Texts, Chinese – Moxibustion – Medicine in China – *Qi* – *Yinyang* – Five Phases (*Wuxing*) – Huangfu Mi

ACYUTA PIṢĀRAṬI Acyuta Piṣāraṭi (ca. 1550–1621), author of over ten texts, was an astronomer from Kerala in South India. It was he who enunciated, for the first time in Indian astronomy, the planetary correction called "reduction to the ecliptic" (the band of the zodiac through which the Sun apparently moves in its yearly course), in his work *Sphuṭanirṇaya* (Determination of True Planets), composed before 1593, which was later expanded into a full-fledged tract called *Rāśigolasphuṭānīti* (True Longitude Computation of the Sphere of the Zodiac).

Acyuta was popularly known as Tṛ-k-kaṇ ṭiyūr Acyuta Piṣāraṭi. The first term indicated his birthplace, Kuṇḍapura, and the last the sect of Kerala temple functionaries to which he belonged. He was the pupil of Jyeṣṭhadeva, author of *Yuktibhāṣā*, an elaborate treatise on mathematical and astronomical rationale. He was a protégé of Ravivarmaṇ, King of Veṭṭaṭṭunāṭu in Kerala and the teacher of the poet and grammarian Nārāyaṇa Bhaṭṭatiri of Melputtūr. Apart from one work, entitled *Praveśaka*, which was an instructive epitome of Sanskrit grammar, all his works were on astronomy. Acyuta is also referred to as a master of medicine and poetics.

Acyuta's work *Karaṇottama* is an astronomical manual in about a hundred verses, which, in five chapters, deals with the derivation of the mean and true planets, the eclipses, gnomonical shadow, and the phenomenon of the complementary situation of the Sun and the Moon. This work and its auto-commentary are significant in that they set out several computational methods for securing greater accuracy, which had been kept secret by earlier teachers and were intended to be transmitted only orally to disciples. Acyuta's *Uparāgakriyākrama* (Method of Computing Eclipses) is also significant for its containing methodologies unknown to other texts. The *Horāsāroccaya* of Acyuta, in seven chapters, is an instructive epitome of the astrological text *Jātakapaddhati* (A Course of Horoscopical Science) of Śrīpati (eleventh century).

It is noteworthy that Acyuta engendered an astronomical lineage, and the trail blazed by him was continued by several lines of scholars. One such line, of which there is documentary evidence, which continued for more than two hundred and fifty years, runs thus: Acyuta (1550–1621); pupil: Tṛppāṇikkara Potuvāl (sixteenth century); pupil: Nāvāyikkulattu āzhāti (seventeenth century); pupil: Pulimukhattu Potti (1686–1758); pupil: Rāmanāśān (eighteenth century); son-pupil: Kṛṣṇadāsa (Kṛṣṇanāśān (1786–1812). Later astronomers took up the innovative trends in Acyuta's works for rationalization and elaboration through short tracts, as evidenced by nine such tracts which have been identified and added as supplements to the edition of Acyuta's *Sphuṭanirṇaya*. This points to the impact made by Acyuta in the astronomical tradition of Kerala.

K.V. SARMA

REFERENCES

Sphuṭanirṇaya (Determination of True Planets). Ed. K.V. Sarma. Hoshiarpur: Vishveshvaranand Institute, 1974.

Rāśigolasphuṭānīti (True Longitude Computation of the Sphere of the Zodiac) Ed. K.V. Sarma. Hoshiarpur: Vishveshvaranand Institute, 1977.

Iyer, S.V. "Acyuta Piṣāraṭi and his Time and Works." *Journal of Oriental Research* (Madras) 22: 40–46, 1951–52.

Pingree, David. "Acyuta Piṣāraṭi." In *Census of Exact Sciences in Sanskrit*, Ser. I, Vol. I. Philadelphia: American Philosophical Society, 1970, pp. 36–38.

Sarma, K.V. *A Bibliography of Kerala and Kerala-based Astronomy and Astrology*. Hoshiarpur: Vishveshvaranand Institute, 1972.

See also: Astronomy in India – Śrīpati

AGRICULTURE IN AFRICA According to the philosopher Hegel, Africa was the inert mass around which the history of consciousness pivoted in its journey from East to West. Even today one still encounters the view that sub-Saharan Africa is exceptionally backward in science and technology, and has no indigenous intellectual history worth the name. Judged by the conventional standards of modernity, e.g. statistics on literacy and education, Africa does indeed seem a "backward" continent; but against this we must consider the crucial part played by the African savannas in the story of human evolution; human intellectual development was shaped by challenges set by the African environment. The legacy is still perhaps apparent in the continent's exceptional linguistic diversity and an enduring facility among its peoples for coping with severe environmental challenge. Until recently, however, African indigenous knowledge of environmental resources has failed to register in conventional histories of science and technology, due in large part to the distinctive resource endowments and consequent agrarian history of large parts of the continent.

Population density, historically, has been low over much of sub-Saharan Africa, and was depressed further by the slave trade and the wars and epidemics associated with colonial conquest. This has meant an emphasis (contrary to the trend of agrarian history in Asia) on land-management strategies that efficiently deploy scarce labor, but in settings ill-suited to plough agriculture or mechanization (contrary to experience in Europe and North America). In place of labor-intensive leveling, drainage and installation of irrigation, for example, African cultivators typically have sought to make use of diverse soil and land conditions as they find them. This requires emphasis on what might be termed "mix-and-match" approaches, e.g. maintaining different animal species in the pastoral herd, selecting a range of crop types adapted to different soil and land conditions, using different crop types in the same field (inter-cropping), or by ingenious dovetailing of a complex and varied portfolio of productive activities (hunting and gathering, shifting cultivation, tree-crop cultivation).

This emphasis on versatility above specialization has been of particular importance in those extensive tropical regions in sub-Saharan Africa blighted by insect-borne disease (notably malaria and sleeping sickness), climatic irregularities, poor soils, and lack of irrigation opportunities. Here, characteristically, human groups tend to invest heavily in the social "software" of agrarian relations rather than the "hardware" of technology and land improvement. Groups facing periodic drought may prefer to devote attention to the cooperative social relations that sustain an "optimal foraging strategy" rather than tie up large amounts of labor in costly land improvements such as irrigation systems. In an outright disaster social investments are portable, but land improvement is fixed and may have to be abandoned.

The knowledge and mental attitudes that support versatility among hazard-prone resource users do not lend themselves readily to the writing of conventional history of technology. Historians whose ideas have been formed against a background of more steadily evolving agrarian technological traditions in Europe and Asia have been at times tempted to conclude that African agriculture is deficient in technical expertise. Local knowledge tends to be regarded as makeshift and perhaps even irrational. Shifting cultivators and pastoralists, more anxious to meet the next challenge than celebrate past achievements, may have unwittingly reinforced these misperceptions. For Mende rice farmers in Sierra Leone several "traditional" techniques are colonial innovations — the local notion of tradition is simply "something that works". Sometimes, sensing potential damage to a good practical skill from wordy rationalization, African resource users may even seek to deny that they know anything useful about the topic under discussion. Outsiders end up perceiving a technological void where none exists.

Many schemes were set up during the colonial period to assist African farmers to climb across these imagined gaps in what was thought to be a fixed ladder of agro-technological progress. Colonial reformers concentrated at first on European innovations such as sickle and plough, only to be beaten back by African preferences for panicle selection in harvesting as a means to maintain the varietal distinctiveness of planting materials, or for the hand hoe as a superior means to maintain soil physical quality under difficult tropical climatic conditions. Later, it was supposed that Asia was the proper yardstick against which to measure the backwardness of African agrarian technology, and first steps were taken to transform African agriculture according to Asian experience. This culminated in efforts in the 1970s and 1980s to replicate the Asian Green Revolution in Africa, using fertilizer-responsive high-yield crop types under intensive management. Asia-to-Africa technology transfer repeatedly foundered on the issue of labor. Innovations for Asian farmers needed to be labor-absorbing, but for African farmers (not threatened, historically, to anything like the same extent by population pressure on land) labor-efficiency was often the criterion of greatest relevance.

During the colonial period in Africa a significant minority of long-serving agricultural officers came to appreciate that rural people did in fact have a considerable fund of valid practical knowledge of agriculture and the environment. This coincided with the rise of scientific ecology and related disciplines during the 1930s and 1940s; foresters, economic botanists, soil scientists, and veterinary officers were particularly active in recording African indigenous knowledge of agriculture and the environment, and in drawing parallels between these local concepts and emergent ideas in ecology and related disciplines. Perhaps the single most striking of these instances is the letter to the scientific journal *Nature* in 1936 by the Tanganyika-based soil scientist G. Milne, proposing the concept of the soil catena as a regularly recurring chain of soil types controlled by topography. African soils tended to be very old, and showed the influence of underlying rock types much less than in Europe. The concept of soil catena, by emphasizing topography and downplaying the role of geology, provided a much better guide to the way the soils had formed, and to how African farmers typically used their soils. Throughout the tropical zone, but especially where seasonal variations in rainfall distribution are most marked, farmers secure food supplies during the pre-harvest hungry season and spread their labor burdens by systematically planting up and down slopes, carefully matching different crops or crop types to the different soils within the catenary sequence. The catena rapidly became established as a basic organizing concept in tropical soil science throughout the world; Milne's letter to *Nature* makes clear its African roots by using the Sukuma terms employed by local farmers to categorize the different soils within the chain.

This is in stark contrast to earlier official thinking about land systems on the other side of the continent, in Sierra Leone. After famine in 1919, caused not by local agricultural incompetence but the loss of harvest labor resulting from the influenza that tracked colonial troops to their homes from the battlefields of Europe, the colonial governor Wilkinson sought (as he thought) to transform Sierra Leonean rice farmers from being shifting cultivators farming rain-fed uplands into permanent cultivators of irrigated valley-bottom lands along Asian lines. Moving local farmers from the tops to the bottoms of their valley slopes was for Wilkinson a shift of epochal proportions, through which the "ignorant" African would be able to catch up a thousand years of agrarian technological history in a matter of a few years. The official of the Madras Department of Agriculture in India employed by the Government of Sierra Leone to set things in motion requested to be sent home after nearly two years working with local farmers, on the grounds there was little if anything he could teach them that they did not know already, and that in any case they obtained better yields from similar resource endowments than farmers in Madras. However, the belief that there was something decisive about a shift from upland to valley-bottom farming survived into the modern period, and received a new boost from the example of the Green Revolution in the 1960s. This insistent categorical contrast on "uplands" and "wetlands" remains in stark contrast to local farmers' knowledge and practice in which seed types and labor resources are invested up and down catenary sequences, and across the upland-wetland divide, according to circumstances. Standing on the boundary between the rain-fed and water-logged soil types in their farms, Sierra Leonean peasants repudiate any gap between themselves and more technologically advanced farmers in Asia, and see instead only an opportunity for flexible adjustment to changing conditions. Recently, mathematicians have begun to provide formal tools with which to grasp the fuzzy logic that underpins this kind of cognitive flexibility, typical of African indigenous knowledge of environmental resources.

Rice farming on the western coast of West Africa is of considerable antiquity, and based on the domestication of the African species of rice (*Oryza glaberrima*). Due to contacts arising from the slave trade we have quite rich documentary sources concerning indigenous agricultural knowledge for this part of Africa at an early date, including a number of accounts from the seventeenth and eighteenth centuries specifying the way in which farmers matched planting materials to different soils, so spreading labor peaks and minimizing pre-harvest hunger. Today, rice farmers in the region continue to research this relationship whenever they encounter new material (e.g. accidental introductions, or new types that arise as spontaneous crosses). This knowledge of and interest in management of crop genetic resources is widespread in Africa, and is perhaps the single most important aspect of the legacy of indigenous agro-technological knowledge on the continent. Today it is threatened by social dislocation (including the effects of warfare) and agricultural modernization (including the spread of modern cultivars and labor-efficient harvesting technology).

African contributions to crop biodiversity management have been undervalued in the past, with some exceptions, because several of Africa's indigenous food crops are not widely known elsewhere (e.g. *Digitaria* millet in West Africa, *teff* in Ethiopia, and finger millet in eastern and central Africa). Vavilov recognized the Ethiopian Highlands as a major center of crop biodiversity, but a number of important crops originating in Africa are scattered more widely (e.g. African rice, white yam, sorghum, and oil palm). Greater recognition is still needed for the historical role played by Africa's farming populations in identifying, shaping, and conserving these genetic resources. In this context it is interesting to note the systematic efforts made by Thomas Jefferson in the 1790s to establish African rice in the United States. Jefferson was convinced that the hardy dryland cultivars selected and maintained by West African farmers would help reduce some of the health problems associated with rice farming in the coastal zone of South Carolina, and had a cask of upland rice imported from the coast of what is today the Republic of Guinea for distribution among inland planters in South Carolina and Georgia. Earlier, South Carolina rice planters had shown a preference for slaves from the coastal rice-growing regions of West Africa, and it is possible that the tidal-pumped wetland rice cultivation systems of the tidewater zone drew upon African technological expertise in this field. Historical examples such as these help correct the erroneous notion that the transfer of agricultural knowledge and technology between Africa and the rest of the world has been a one-way process.

During a period of aggressive modernization of African agriculture following the end of colonial rule the work of documenting and understanding African systems of resource knowledge and management, begun by ecologically-oriented technologists in the colonial period, was kept alive only by a handful of enthusiasts. Special mention should be made of African pioneers such as George Benneh in Ghana and Uzo Igbozurike in Nigeria. However, interest in local knowledge systems expanded enormously in the 1980s and 1990s, after the failure of many "high-tech" schemes to promote rapid change in African agriculture. A number of agricultural and other scientists now see indigenous knowledge as a resource for orthodox science (farmer experimentation attracts particular attention). Recent studies have highlighted the specialist knowledge of Africa's women farmers, pointed to the complex ways in which indigenous technical knowledge of the environment is bound up with social relationships of production and consumption, and drawn attention to local knowledge in biodiversity conservation. The ratification of an international convention on biodiversity, adopted by the United Nations' Rio de Janiero Conference on Environment and Development in 1992, gives indigenous agricultural knowledge a new visibility, and status in international law. Questions of ownership and preservation of Africa's abundant legacy of indigenous knowledge now attract attention, though some concern has been expressed that this might serve to ossify such knowledge, and reduce its practical utility.

PAUL RICHARDS

REFERENCES

Igbozurike, U. *Agriculture at the Crossroads*. Ile-Ife: University of Ife Press, 1977.

Milne, G. "Normal Erosion as a Factor in Soil Profile Development." *Nature* 138: 548–9, September 26, 1936.

Richards, P. *Indigenous Agricultural Revolution*. London: Hutchinson, 1985.

See also: Local Knowledge – Colonialism and Science – Environment and Nature – Ethnobotany – East and West – Food Technology in Africa

AGRICULTURE IN CHINA China is a vast country covering roughly the area of Europe, about ten per cent of which is suitable for farming. China has two distinct farming traditions that correspond to climatic zones. North China has sparse and irregular rain that falls mostly in the summer; the winters are long and cold. The uplands of the interior are formed of thick deposits of fertile loess; over the millennia the Yellow River has eaten away at the primary loess and deposited it as silt on the alluvial plains downstream. The main constraints on agricultural productivity in the northern region are the relatively short growing season (between five and eight months) and the lack of water. It is seldom possible to grow more than one crop a year. The typical crops are those which do well with little water: millets, wheat, sorghum, cotton, and beans.

From the Yangzi plains south, the climate is semi-tropical. Rainfall is much heavier and spread throughout the year. As a result, much of the natural soil fertility has been leached away, but irrigated rice fields counteract this effect. They build up their own fertile microecology that allows the same field to produce two or even three harvests a year, depending on the latitude. The growing season ranges from nine months to year round. Rice is the staple, grown everywhere in the southern zone; other important crops are winter wheat, maize, sweet potato, sugar, tea, and cotton along the Yangzi.

The distinction between a northern and a southern tradition has its roots in prehistoric times. The earliest Chinese farming sites date from the sixth millennium BC, later than those of the Fertile Crescent. Archaeologists used to consider that farming diffused throughout the Old World from that single center; the current evidence suggests however that within China itself there were at least two independent centers of plant domestication. At the 5000 BC neolithic site of Banpo, in the loesslands of Northwest China, the dryland crop of Setaria millet was grown. The village of Hemudu, built in the marshes near the mouth of the Yangzi at the same period, grew large quantities of wet rice.

Land was a scarce resource from very early times. Unlike the pattern of development in the West, where capital was invested in draft animals, equipment, and machines to substitute for labor, in China the historical trend was towards increasing the productivity of land through the application of skilled labor and cheap small-scale inputs. With one or two rare exceptions, economies of scale did not apply. The roots of this relationship between land, population, and labor developed very early in China's history.

From the sixth to the fourth centuries BC, several states battled for control of all China; the states with most men to fight and most grain to feed them emerged as victors. A fiscal policy based on peasant contributions of grain, textiles, and services became the norm. A strong state was one with a large population of skilled farmers. This remained the basic view of Chinese statesmen through the unification of the empire in 221 BC right up to the end of the Maoist era. For over two thousand years officials encouraged farming and took an active role in developing and diffusing knowledge and techniques. They fostered labor-intensive peasant farming and tried to control the accumulation of land in the hands of the rich. Successive medieval regimes confiscated land from the wealthy and redistributed it to peasants to ensure that it provided a livelihood for as many people as possible.

In part this intensity of land use was made possible by the particularities of Chinese farming systems. No arable land was wasted. There were few pastures in China proper. The main sources of animal protein were pigs, poultry and fish, and draft animals (oxen and mules in the north, water buffalo in the south) grazed on rough land. The practice of fallowing seems to have died out as early as two thousand years ago in parts of North China. Crop rotations alternated soil-enriching crops like beans with cereals; all human and animal waste was composted and returned to the fields. Medieval states in the north allocated approximately six acres of land to support a family. Land was even more intensively used in the south. By the seventeenth century, some fields in the south produced two crops of rice and another of tobacco or vegetables each year. In the early twentieth century families in the most densely populated regions lived off under a tenth of an acre of rice land.

The tax system meant that improving agriculture was a key concern in Chinese state policy from the very beginning, and many of the most important works on agriculture were written by members of the official elite in their capacity as civil servants. Imperial compilations and works by civil servants aimed at a readership of local officials who would pass on the information to the farmers under their jurisdiction. They tended to stress practical details of husbandry, including innovations that could be introduced, and they emphasized subsistence production. Other works, written by landowners who ran their own farms, were written for fellow landowners; these works usually included discussions of labor, prices and estate management.

The earliest extant agricultural treatise belongs to the second category. It is entitled *Qimin yaoshu* (Essential Techniques for the Peasantry); the author, Jia Sixie, completed its ten volumes in AD 535. It describes the agriculture of the dry regions of the Yellow River plains. Perhaps its most striking features are the detailed descriptions of how careful and repeated tillage techniques (different depths of plowing, sowing with a seed-drill, harrowing and hoeing) were used to conserve soil moisture in a dry climate, and crop rotations were used to increase fertility. We can reconstruct much of the equipment mentioned by Jia from somewhat earlier tomb paintings, as well as archaeological discoveries of huge state iron foundries from the first century AD, that mass-produced cast iron plowshares, moldboards, and the iron shoes of seed-drills. Together with Jia's numerous citations from earlier northern works, this evidence confirms that the *Qimin yaoshu* represented the culmination of a long northern tradition of productive estate farming that was dependent on large acreages and heavy capital investment in equipment and draft animals. It was a form of centralized estate farming that peasant farmers could not afford or compete with. The *Qimin yaoshu* was the last of the great works focused on the northern system.

Repeated wars and invasions ravaged the northern plains in medieval times. In the eleventh century the loss of the north to the Khitan finally established the Yangzi region as the political and economic center of Song dynasty China. The estates of the northern aristocrats disintegrated, and as under-equipped peasant farmers did their best to scratch a living from a few acres of dry soil, the north became a backward region compared to the fruitful south. Migrants flooded into the southern provinces from the north, and the state sought to encourage more productive agriculture by every possible means. Irrigation works were improved, seeds were handed out, information distributed, and cheap loans and tax breaks offered — the scope was similar to the Green Revolution of the 1970s, as was the impact on production. One particularly fruitful venture was the introduction of quick-ripening rices from Vietnam, that allowed rice farmers to double crop their fields, alternating winter wheat or barley with rice.

Several important agricultural works were published during the Song, but the landmark of the period dates from the Yuan (Mongol) dynasty. In 1313 Wang Zhen published a *Nongshu* (Treatise on Agriculture) several hundred pages long. His aim was to describe local innovations so that they could be adopted elsewhere. He included detailed woodblock illustrations of farming tools, machinery, irrigation equipment, and various types of terraced or dyked fields that permitted the extension of farming into mountainous or marshy lands. The treatise depicts a system of family farming in which poor peasants with little capital invested intensively in labor, skills, and low-cost inputs. This was the farming system that formed the basis for the commoditization and expansion of the rural economy in succeeding centuries.

Wang Zhen's *Treatise* was the paradigm for later works, including imperial compilations and the magisterial *Nongzheng quanshu* (Complete Treatise on Agricultural Administration) by the statesman Xu Guangqi, completed in 1639. Xu was a polymath and Christian convert who served for some years as Grand Minister of China. The *Nongzheng quanshu* advocates a balance between the production of essentials (cereals for food and fiber crops for textiles) necessary for the health of the central state, and cash crops and handicrafts that would ensure a prosperous rural economy. Xu was also preoccupied with population pressure and the need to expand the arable area and improve yields; he devotes long sections of his treatise to land reclamation and improved irrigation techniques, to his own experiments with manures and commercial fertilizers such as lime and bean cake, and to crops like sweet potatoes that can be grown on poor land.

A number of other works in the southern tradition were written by landowners. In striking contrast to the *Qimin yaoshu*, however, the landowners themselves farmed only a few acres and rented the rest out to tenants. The main criteria for selecting a suitable tenant were his skills and experience, and his assiduity at work; capital assets were not a consideration as they would have been for a contemporaneous English capitalist landowner. The main difference between landowner and tenant lay in the ownership of land; it did not extend to differences in the scale of the farm, or in the range and size of equipment. In this highly intensive and skilled farming system there were no economies of scale, and anyone who owned more than an acre of rice land would seek a tenant to farm the surplus. If levels of agricultural expertise were reckoned simply by the complexity of farm machinery or by levels of capital investment, then the farming methods of the eighteenth and nineteenth century might be reckoned a decline from those of eight hundred years earlier, for as average farm size diminished many peasants abandoned animal-drawn implements in favor of hand tools. However, if we look at the productivity of land, we see a different picture. Improvements in water control, in fertilizing, in the spacing of plants and the breeding of varieties enabled China's peasant farmers to increase the total output of crops at a rate that kept up with population growth until about 1800. This intensive small-scale farming also supported a diversified rural economy of small industries and handicrafts that fed into national and international trade networks.

The great inventions of Chinese agriculture are anonymous, the collective achievements of peasants recorded by servants of the state. Many important innovations occurred in the densely populated heartlands where pressure on land was most intense. There has been a tendency for historians to assume that political and technical superiority went together, and that the Chinese taught civilization to the barbarians. However, several features crucial to Chinese high farming came not from the center but from the periphery. The technique of transplanting rice, fundamental for the development of intensive wet-farming, was practiced by Thai-speaking populations in the Canton and Tonkin regions when they were conquered by China two thousand years ago. Terraced fields probably spread northwards into China from Vietnam and Yunnan, reaching the Yangzi region by the fourteenth century. Tea was introduced from Tibet and Western Sichuan some time before the eighth century. And the techniques of cotton cultivation and processing were introduced to the Lower Yangzi from Hainan island around the thirteenth century.

Although there was no indigenous development of experimental agricultural science or engineering, Western agronomists found much to admire in traditional Chinese farming in the early twentieth century, particularly its careful husbandry and sustainability. Also, traditional agronomic strengths, like crop breeding, have allied fruitfully with modern science. For example, Chinese geneticists used their vast range of local rice varieties to develop the first semi-dwarf high-yielding *indica* strains in the 1960s, several years before "miracle rices" were released by the International Rice Research Institute in the Philippines.

FRANCESCA BRAY

REFERENCES

Bray, Francesca. "Agriculture." *Science and Civilisation in China*, vol.VI pt.2, Ed. Joseph Needham. Cambridge: Cambridge University Press, 1984.

Buck, John Lossing. *Land Utilization in China*. Shanghai: Commercial Press, 1937.

Elvin, Mark. *The Pattern of the Chinese Past*. London: Methuen, 1973.

King, F.H. *Farmers of Forty Centuries, or Permanent Agriculture in China, Korea and Japan*. London: Jonathan Cape, 1926; repr. Emmaus, Pennsylvania: Rodale Press, 1972.

Huang, Philip C.C. *The Peasant Family and Rural Development in the Yangzi Delta, 1350–1988*. Stanford, California: Stanford University Press, 1990.

Perkins, Dwight H. *Agricultural Development in China 1368–1968*. Edinburgh: Edinburgh University Press, 1969.

See also: Food Technology in China

AGRICULTURE IN INDIA The Harappan culture related to the earliest agricultural settlements in the Indian subcontinent is dated between 2300 and 1700 BC. The crops of the Harappan period were chiefly of West Asian origin. They included wheat, barley, and peas. Of indigenous Indian origin were rice, tree cotton, and probably sesame. Rice first appeared in Gujarat and Bihar, not in the center of the Harappan culture in the Indus Valley. There is some rather doubtful evidence that African crops were also grown by the Harappans. There is a record of sorghum (*jowar*) from Sind and *Pennisetum* (*bajra*) from Gujarat. The earliest record of the African cereal, *Eleusine coracana* (*ragi*) is from Mysore, about 1899 BC. The Southeast Asian crops of importance to India are sugar-cane and banana, and they both appear in the early literary record. Crops of American origin include maize, grain amaranths, and potato. The dating of the introduction of maize is uncertain, the characteristics and distribution of some forms being such as to lend support to the view that they reached India in pre-Columbian times. Crops of the Indian subcontinent have influenced the agricultural development of ancient Egyptian, Assyrian, Sumerian, and Hittite civilizations through their early spread to these regions of the Old World. The Buddhists took several Indian crops and plants to Southeast Asian countries, and there was much early exchange of plant material with Africa. The Arabs distributed crops such as cotton, jute, and rice to the Mediterranean region in the eighth to tenth centuries AD. There was also a reciprocal exchange of several New World domesticates.

Agriculture Today

India is characterized by a wide variety of climates, soils, and topographies. It is rich in biodiversity and a seat of origin and diversification for several crop plants such as rice, millets, pigeon pea, okra, eggplant, loofah, gourds, pumpkin, ginger, turmeric, citrus, banana, tamarind, coconut, and black pepper. Because India is ethnically diverse, traditional agriculture is still practiced in many places. There are about 100 million operational holdings, and the country has over 20% of the world's farming population.

India has different ecosystems such as irrigated, rain fed, lowland, upland, semi-deep/deep water, and wasteland. Agriculture is primarily rain fed (rain dependent); it supports 40% of the human population, 60% of cattle, and contributes 44% to the total food production. Owing to differences in latitude, altitude, variation in rainfall, temperature and edaphic diversity, great variety exists in crops and cropping patterns.

There are two important growing seasons in India: the *Kharif* or the summer season, especially important for rice; and the *Rabi* or the winter season in which major crop grown is wheat. The *Kharif* crop is primarily rain dependent, and the *Rabi* is relatively more reliant on irrigation. The *Kharif*/rainy season cropping patterns include major crops such as rice, sorghum, pearl millet, maize, groundnut, and cotton. The *Rabi*/winter season cropping patterns include important crops like wheat, barley and to some extent oats, sorghum, and gram/chickpea. Mixed cropping is also practiced, especially during the kharif season. Pulses, grain legumes, and oilseeds are grown with maize, sorghum, and pearl millet. Brassica and safflower are grown mixed with gram or even with wheat. Under subsistence farming, on small holdings, mixed cropping provides food security and is consumption oriented.

India is the major producer of a number of agricultural commodities including rice, groundnut, sugar-cane, and tea. Food grains constitute roughly two-thirds of the total agricultural output. These consist of cereals, principally rice, wheat, maize, sorghum, and minor millets. India is the second largest producer of vegetables next to China. Mango accounts for almost half of the area and over a third of production; banana is the second largest and is followed by citrus, apple, guava, pineapple, grape, and papaya. Of non-food cash crops, the most important are oilseeds especially groundnut, short staple cotton, jute, sugar-cane, and tea.

Owing to improvements in recent years there has been widening of inter-regional disparities in agricultural production and productivity. Regions such as the north and north-west and the delta regions of peninsular India have prospered under assured irrigation, but dryland and semiarid regions have not done so well.

R.K. ARORA

REFERENCES

Arora, R.K. "Plant Diversity in the Indian Gene Centre." In *Plant Genetic Resources: Conservation and Management.* Ed. R.S. Paroda and R.K. Arora. New Delhi: International Board for Plant Genetic Resources, 1992, pp. 25–54.

Forty Years of Agricultural Research and Education in India. New Delhi: Indian Council of Agricultural Research, 1989.

Handbook of Agriculture. New Delhi: Indian Council of Agricultural Research, 1992.

"India." In *Regional Surveys of the World. The Far East and Australasia*. 24th ed. London: Europa Publications, 1993, pp. 275–335.

AGRICULTURE IN THE ISLAMIC WORLD The success of classical Islamic agriculture is due to the adaptation of agrarian techniques to local needs, and this adaptation itself is due to a spectacular cultural union of scientific knowledge from the past and the present, from the Near East, the Maghreb, and Andalusia. A culmination more subtle than a simple accumulation of techniques, it has been an enduring ecological success, proven by the course of human history.

In the definitions which open the *Kitāb al-filāḥah* (Book of Agriculture), this function is said to be blessed by God because it has as its end the production of the sustenance of life. Agriculture consists of restoring to the earth what has been furnished by harvesting from it, by fertilizing, watering, and making efforts to avoid the problems caused by excessive heat. This restoration to the soil implies a knowledge of the whole — the soils, the plants, the most suitable tools. Balance (*mizān*) is the aim, or reciprocity between what is taken from the earth and what must be given back in order to make this vital alliance with Nature endure.

The complex union of facts with the general conjunction of the Mediterranean world between the eleventh and the fifteenth centuries means that a de-positioning of history is indispensable for understanding a crisis as well as a success. No progress is linear, and it is always useful to draw inspiration from the aleatory nature of history, in order that this discipline, fundamentally cultural, may also have a practical impact.

The successional right of the four Islamic judicial schools permits "holdings" according to a customary right which is similar to the right to private property of the Romans. Royal power encouraged territorial expansion among the princes of the blood and high officials of the state.

From a historical point of view, the important thing is the fact of reciprocal information throughout the *Dār al-Islām* (the Islamic world). There emerges the impression of a coherent school and a general movement — of people, goods, and ideas — from the East to the Islamic West.

The ancient tradition, prolonged and recovered from the ancients (*al-Alwālī*), integrated the ideas of fourth century scholars like Aristotle, Dioscorides, Galen, and Anatolius Democritus with those of the botanists of the ninth century and contemporary scholars of geoponics, the art and science of cultivating the earth. If the mysterious *Filāḥat al-Nabaṭiyyah* (Nabataean Agriculture, by Ibn Waḥshiyyah, ca. 1000) traces the origins of agriculture back to Adam, those who lived in the classical age were equally inspired by knowledge obtained from anonymous farmers who retained the memory of ancient ways. Tradition and scientific curiosity have not always been at odds with each other.

On a religious level, the earth and water, as in the Hebraic tradition, belong to no one person; they belong to god. Historical accounts of Islamic expansion distinguish between Arab lands and lands situated in the conquered countries. Under the early caliphs, Arab lands were surveyed and registered. A basic tax was established, 10 *dirhams* for a *jerib* of grapes, 8 *dirhams* for a *jerib* of palm trees, 6 on a *jerib* of sugar cane, 2 on barley. The *jerib*, a unit of measure, equaled 360 cubits, according to al-Mawardī. The lands of people who freely converted to Islam were subject to a deduction of a tenth, *dīme*, varying according to province and century from the eighth to the twentieth centuries. In Andalusia, the tax was one-fifth. After the Reconquista, the farmers there continued to be called *quinteros*.

Lands which were forcibly conquered were redistributed. Al-Wansharīsī, a fifteenth century Maghreb legal expert, notes numerous examples for studying classical Islamic agriculture in Andalusia and the Maghreb.

Collective lands, *jemaʾa*, existed in certain parts of *Dār al-Islām*, used for the movement of flocks of large and small grazing stock. Cultivated land was also divided up legally and parceled out.

"Tributary" lands were conquered lands lying outside Arabia, beginning with Syria [Sham]. They were considered lands not belonging to anyone, the property of the state or the caliph. By contrast, they were left to their former owners, according to a right of use. A land tax called *kharaj* was paid to the Treasury (*bayt al-māl*) and the amount of tax was set according to the quality of the soil. From the 700s, the principle was applied in the form of a supplemental tax to Jews, Christians and Sabeans, being the *Ahl al-kitāb* (People of the Book), or *dhimmis*.

Individual land holdings were called *iqtā*; they were lands grafted to a private individual according to clauses which were more or less restrictive depending on rents. All lands which the caliph gave to his subjects so that they might transform and cultivate them were so designated. *Waqf* lands were lands granted by private individuals to mosques, hospitals, schools, and other charitable institutions. They are often translated as charitable properties. They were not subject to land speculation.

All these lands, except for the *waqf* properties, could be the object of commercial transaction — sale, rent, or purchase — which had nothing to do with feudal land statutes. In the twelfth century, the Sevillian Ibn Abdūn, in his *Treaty of Ḥisbā* encouraged personal appropriation of land as a means of stimulating economic growth.

The information which appears in the documentation needs examining. It is based on experimentation which resulted in the shattering of prior philosophical premises. Empiricism appeared to be the condition of renewing knowledge and techniques. Ibn al-ʿAwwām writes: "I affirm nothing which seems right to me without having proven it in numerous experiments." In agriculture, the results refer to the practical successes of sheaves of grain, fruits, or taxes.

The theory of climates, *al-iklim*, compares Andalusia with Iraq and makes pertinent constant reference to Nabatean agriculture.

"Indeed," writes Ibn al-ʿAwwām, "what suits our country is the result of what comes from the concordance of tradition with experimental results." The respect for ecological balance — *mizān* — between the soil, the micro-climate, and various cultivated plants guarantees the success of the harvest. The "weather" governs the results, and the seasons are stated in the "Calendar of Seville" according to their names in Syrian, Persian, Hebrew, and in indigenous romance languages. More subtle than a syncretism, it was a question of a whole society.

Islamic agriculture had, at first, been Arab agriculture, since Islam first appeared in the Arabian peninsula, among the Bedouins and camel drivers. Around the big cities, agriculture was the agriculture of the oasis, a natural miracle brought about by the presence of water in a desert of sand and stones. The history of agriculture in Islamic countries, established over a long duration as Fernand Braudel has described, is made up of a fundamental unity. The first great geo-climatic regions were sub-arid dominants around the Mediterranean basin. The Ummayyad empire, then the Abbasid empire, finally integrated the sub-tropical regions with temperate ones. However, the essential originality of Islamic agriculture is still linked to the Mediterranean regions and to the fluvial valleys. The first Islamic empires and the caliphate of Andalusia owed their agriculture to the great rivers carrying water and fertilizing silt (alluvium); the Tigris, Euphrates, Nile, Guadalquivir, and the Guadiana all gave both soil and waterless, sun baked lands.

Between the seventh and the thirteenth centuries, the displacement of peoples and technical skills gave rise to a migration of cultivated plants, from the East towards the West, from subtropical zones towards the Mediterranean basin, from the monsoon regions to semi-arid lands; from China towards Persia, passing through India; and from Afghanistan towards the Fertile Crescent and the Maghreb, creating in its passing the gardens of Sicily and of Andalusia. Just as the ancient Romans constructed aqueducts and waterworks to provide food on a scale for the cities and municipalities which were their centers of power, so the Islamic empire, founded on caravan cities, also wove a net across the countryside of hydraulic equipment for agricultural adaptation, for example *acequias*, *qanats*, and *norias*. In spite of the progressive climatic diversification which occurred as the area ruled by Islamic law increased, from the Sudan to the Caspian seaports, from the Straits of Gibraltar to the boundaries of the Ottoman Empire and to India, the determining character of their agricultural system remained the adaptation of irrigation to local and regional needs and the spread of plant species away from their original ecosystems.

The spread of agricultural land and the intensification of irrigation in sub-Iberian regions which tended to be hot and very humid sub-arid areas were spectacular. Legal aspects of land holding were closer to those of Rome and Byzantium than the medieval West. Individual property ownership was actively encouraged.

Technological and cultural methodologies were informed by the need for renovation while remaining empirical. Among the agrarian jobs, that of the autumn harvest is characteristic in that the human job prevails over the financial investment. The swing plough was preferred over the heavy Brabout plough of the French colonist; not exposing the deep beds of cultivated land to erosion and intense heat was the golden rule of ecology in Andalusia. In the golden Andalusian age, this protection of the Mediterranean soils was subject to laws of a scrupulously careful ecology. The *mishā*, a heavy, hand-held spade was the tool for restoring the soil. The object of such agriculture was closing the soil, not opening it.

Among agricultural systems, the biennial rotation of crops is essential for maintaining fallow fields. Biennially or by a more complex number of years, but always by an even number, the rotation of crops shows a deep understanding of the plant world. The refertilization of soil, the base of all agriculture, comes about through the joint knowledge of plants and soils, the mastery of botanical and edaphic science. In Andulusia, well before the era of the English physiocrats of the 1800s, this agriculture revolution was closely based on high level of knowledge of the life sciences and on a love of nature which was the common gift of both the Islamic and the Hebraic tradition.

Certain plants modified dietary habits, for example sorghum, a common basis of diet in Asia and Africa; rice in flooded areas; sugar-cane which is used for preserving and for therapeutics; the eggplant transported by the Jews in the second diaspora; citrus fruits from China; durum wheat from Africa, a nutritional mainstay in the form of bulgur and couscous; watermelons; and banana trees, acclimated in Egypt before arriving in the Maghreb and Andalusia.

The cultivation of other plants influenced styles and types of clothing throughout the Islamic regions. There was cotton, introduced to the Andalusia after the arrival of the Kurd Ziriyab in the 900s, dye plants which brought a passion for Persian and Indian colors to the puritanical Berbers, and perfume which supplied the base of a whole range of products, such as lotions, salves, and soaps, and which was manufactured from the almost limitless supplies of fragrant flowers from Turkey and Morocco.

With a deep love of nature and a relaxed way of life, classical Islamic society achieved ecological balance, a successful average economy of operation, based not on theory but on the acquired knowledge of many civilized traditions, a society which wanted to live without the specter of famine and hunger.

Colonialism seriously upset the traditional agricultural balance in order to increase profitability for the colonizers. This has been widely written about and is proven today by global economic realities. In the 1900s, colonial settlers and city authorities (wrongly) interpreted the indigenous practice of transhumance (moving flocks of animals from one area to another) as non-ownership of land. This distortion of a multisecular custom of complement between plant and animal husbandry (for example, in the Maghreb) caused grave damage to the native economy. Colonial agriculture thus found pseudo-legal advantages in a vast redistribution of land, which brought great economic benefit to the colonial settlers who had come out from the cities and towns.

We are witness today to the slow recovery of agrarian balance in former colonies like Algeria.

LUCIE BOLENS

REFERENCES

al-Hassan, A.Y. and D.R. Hill. *Islamic Technology*. Paris: UNESCO, 1986.

Bolens, L. *Agronomes andalous du Moyen Age*. Genève: Droz, 1981.

Bolens, L. "La Révolution agriciole andalouse du XIè siècle." *Studia Islamica* 47: 121–141, 1978.

Fahd, T. "L'Histoire de l'agriculture en Irak: al-Filāḥa an-Nabatiyya." *Handbuch der Orientalistik* 6, 1952–59, pp. 276–377.

Fahd, T. "Filāḥa." In *Encyclopedia of Islam*. Leiden: Brill, 1965, pp. 920–932.

Glick, T.F. *Islamic and Christian Spain in the Early Islamic World*. Princeton, New Jersey: Princeton University Press, 1979.

Watson, A. *Agricultural Innovation in the Early Islamic World*. Cambridge: Cambridge University Press, 1983.

See also: Technology in the Islamic World – Irrigation in the Islamic World – *Qanat*

AGRICULTURE IN JAPAN The later Neolithic culture in Japan is known as *Yayoi*. It can be best characterized by the introduction of rice cultivation. It is thought to have spread to Japan around the first century BC from rice growing areas of the Asiatic mainland. It probably reached western Japan via streams of immigrants from the south along the Ryukyu islands, or by way of Korea. It might possibly have been carried on the ocean stream directly from south China. Rice was probably at first grown on naturally marshy land. Settlers scattered their rice seeds there, and harvested ears at the neck of the paddy crop by stone knives, according to the custom in their native land. Being favored with adequate and well-distributed rainfall and moderate air temperatures, rice cultivation extended gradually eastward to Kinki (adjoining Osaka) in the first century AD and to Kanto (adjoining Tokyo) in the third century AD. In addition to rice, wheat, small red beans, and various kinds of millet were grown in small quantities for the cultivators' own use.

Overlapping with the Yayoi culture was the arrival of settlers from northern Asia who traveled to Japan across the Korean strait. They brought ironware which included weapons and implements with them in the fifth century A.D. The introduction of ironware was followed by the arrival of Korean smiths, who learned the advanced manufacturing methods from the Asiatic mainland. The adoption of farming implements such as hoes, spades, and sickles made a significant impact on the growth of agricultural productivity. Iron hoes were used to dig wells, ponds, reservoirs, and ditches to supply water to paddy fields. They also served to till the soil of already cultivated land deeply and minutely. Ear-harvesting by stone knives was replaced by a new method by which the rice crop was reaped at the lower stem by a sickle. On the whole, iron implements enabled the transformation of uncultivated wilderness into cultivated land. With a gradual but steady increase in population, the expansion of paddy fields was necessary, because, for the same acreage, rice cultivation could support much more population than other crops. Rice cultivation extended northeastward in the eighth century AD and reached the northern extremity of the mainland in the thirteenth. The details concerning the switch from broad seeding to transplanting, which began in the eighth century, remain obscure. It is said that multiple cropping and application of grass, leaves, and ashes gathered from waste and forest land dated from the eighth century.

The most epoch-making institutional innovation until the medieval ages was the Taika Reform and the land surveys at the turn of the sixteenth and seventeenth centuries. The latter especially affected heavily the framework of agrarian structure from that time on.

Tokugawa Period

During the three hundred years of the Tokugawa period (1603–1868), peasants were subject to the stringent constraints of a feudal system. They were tied to their land and in principle were not to leave their villages. Their title to land was so restrictive that they could not sell or mortgage it. Nor were they free to choose which crop to plant.

On the other hand, peasants and rural communities were drawn into the process of commercialization. As urban population grew and internal travel increased, cities provided a stimulus to the development of commercial agricultural production. In addition, with the development of inland transport and the spread of local markets during the seventeenth century, commercial fertilizers such as dried fish and oil cakes were applied in major commercial crops in the most advanced regions around the cities. The new fertilizers not only raised crop yield, but also permitted more intensive use of land. Double and triple cropping became common in western regions. In addition to an increased use of commercial fertilizers, famous varieties from the localities that had been developed spread to other areas through farmers' informal activities.

One of the most impressive changes during the Tokugawa period was the extension of irrigation. As the power of the feudal lord under the Tokugawa government became more firmly established at the beginning of the seventeenth century, it became institutionally possible to mobilize resources on a large scale for the construction of a more complex and wide-scale irrigation system. The creation of new paddy fields within the domains was for feudal lords a prime means of increasing revenue. In the meantime, engineering technology also advanced. Cultivation in alluvial plains of larger areas was made possible as the ability to control rivers improved, and was further extended downstream.

Agricultural technology in the Tokugawa period, especially since the later half of the eighteenth century, progressed by and large in the trend toward small farming units. In the earlier period, land surveys often showed that there were a few large holdings, somewhat more middle holdings and a large number of small holdings. Some large holdings were many times the size of the small ones in most villages. As these large and middle holdings were tended by members of extended families, they were gradually reduced by the dissolution of the families or by renting land to the ex-servant class. Thus, the small-scale family farmers became a dominant mode in rural Japan except in the very rural regions. At the same time the landlord–tenant system was developed through various kinds of transactions.

The earliest farming book in Japan was the *Seiryoki* (The Biography of Seiryo Doi) which treated the subjects of agricultural policy and farming practices from the viewpoint of a feudal lord. Among a considerable number of farming books published by Confucians, agronomists, local officers, and expert peasants in Tokugawa period, the *Nogyo Zensho* (The Handbook on Farming) by Antei Miyazaki has been recognized as the first scientific work in Japan. The topics he covered in this book represented the entire range of Japanese agriculture. Many general books on farming appeared following Antei's classic. Whereas his sources of knowledge stemmed from Chinese farming books, which included the *Nong Zheng Juan Shu* (Complete Treatise on Agriculture) by Xu Guang Qi, and from his own experience in farming practices and observations during travel, Antei's work merited attention with respect to intensive agriculture in small-scale farming.

Eijo Okura was one of the authors who published his works to meet a demand for fuller treatment of particular farming subjects. He paid attention to industrial crops such as tobacco and their cultivation method, and emphasized the application of commercial fertilizers. He was also known as a pioneer, having acquired the new knowledge from the Netherlands, the only country trading with Japan at that time.

Modernization Process

The so-called modernization of Japan began in 1868 with the Meiji Restoration that aimed at the creation of a modern state. The Land Tax Revision was indispensable to strengthening the financial basis for the newly formed government. Private ownership of land was recognized, and the government converted the feudal land tax in kind into money.

Prior to the Restoration, Japan switched from the closed-door policy which had isolated the country for more than two hundred years to the open-door policy. During that time the Japanese people were thirsty for foreign culture. The government worked hard in its attempt to transplant foreign crops, livestock, and large-scale farm machines, and to invite agronomy instructors from Western countries. After much trial and error, the Japanese learned that rationalization of indigenous technology through Western scientific minds and methods was more effective in the development of agriculture than anything else. The Itinerant Instructor System and increased applicability of indigenous techniques through various experimental tests provided good examples. The adoption of superior seed varieties resulted in an increased application of commercial fertilizers, such as herring meal, soybean cake, and guano. The progress in seed-fertilizer technology, combined with the improvement of irrigation and drainage, was a crucial element in the rapid growth of agricultural productivity until the end of an early phase in the modern economic development in Japan.

Indigenous techniques, however, which depended largely on personal skill and experience, could not last long. They must eventually give way to the results of Western-style scientific research. As the network system of the Agricultural Experiment Station became well organized, the National Agricultural Experiment Station began to direct its resources towards more basic research. A study to develop new rice varieties by artificial crossbreeding based on Mendelism was launched. A great number of new varieties of rice were developed under such studies, and propagated successfully throughout the country.

The research of agronomy not only in plant breeding but also in the field of genetics, plant pathology, soil science, and agricultural chemistry produced remarkable results during 1920–1940. Most of them were concerned with rice and rice cultivation. The Agricultural Land Reform, enforced under the direction of the Allied Forces after World War II, established a ceiling on land holdings and fixed the structure of small-scale owner farms. Under such a structure, selection of better seeds, increased use of chemical inputs, and intensive cultivation were accelerated to raise land productivity. However, the framework of small-scale farm units has faced many problems, including a rural–urban income gap, increased competitiveness, and structural improvement in the process of since the mid-1950s.

SEIJI SAKIURA

REFERENCES

Hayami, Yujiro, et al. *A Century of Agricultural Growth in Japan.* Tokyo: University of Tokyo Press, 1975.

Smith, Thomas C. *The Agrarian Origins of Modern Japan.* Stanford, California: Stanford University Press, 1959.

Xu Guang Qi. *Nong Zheng Juan Shu* (Complete Treatise on Agriculture). Taipei: Taiwan shang wu yin shu kuan, 1968.

AGRICULTURE OF THE MAYA The Classic lowland Maya long remained an enigma among early civilizations, having occupied a tropical environment with all the presumed constraints or limits of that environment on sociopolitical development. Some of the mystery has dissipated over the last twenty-five years by research directed to human–environment relationships among the Maya. New evidence and understanding have emerged of the means by which the Maya used the varied micro-environments of their homelands to sustain their growth and development over some

3000 years of occupation until their still unexplained collapse and depopulation between AD 800–1000.

The ancient lowland Maya occupied the greater Yucatán peninsular region (the modern states of Yucatán, Campeche, and Quintana Roo in México, Petén in Guatemala, and Belize), extending westward into Chiapas, Mexico, and southward into the Río Copán valley of Honduras. Agricultural villages of the Maya appeared in this region during the Early Preclassic Period (ca. 3000 BC or 4950 BP) in an environment that was apparently dominated by a mature, wet–dry tropical forest differentiated along a northwest to southeast rainfall gradient across the peninsula and by local soil and moisture variation. By the Late Preclassic Period (ca. 400 BC to AD 1), significant growth in population and city states was paralleled by major environmental changes, particularly the loss of forest and the manipulation of the wetlands that exist throughout the central and southern portions of the region. During the Late and Terminal Classic periods (ca. AD 510 to 900), major deforestation and other indicators of land stress were apparent throughout the peninsula, particularly in the central and southern lowlands (from central Quintana Roo and Campeche, México, into southern Petén) where populations reached substantial proportions and the civilization its material zenith. The end of this period brought the collapse of the Classic lowland Maya and the depopulation of the central and southern areas. In the north of the peninsula, however — arguably a far more difficult environment in which to sustain large populations, because of its extremely thin soils and paucity of surface water — the Maya remained entrenched and strongly active during the subsequent Postclassic Period and into the early colonial era. The Classic Maya collapse allowed the tropical forests to reassert themselves in the central and southern lowlands, albeit modified by past land uses. Much of the "jungle" of Petén through which the Spanish conquistadores struggled in the sixteenth century was probably not much older than 900 years and was not virgin.

Some early twentieth-century field workers in the Maya lowlands noted various features that they recognized as indicators of ancient land use, especially cultivation. These features included stone walls throughout the rolling, inland terrain and terracing in the western foothills of the Maya Mountains of Belize. Others speculated on the possible use of shallow lakes and wetlands by the Maya. These observations and insights, however, were peripheral to the dominant thesis that the ancient Maya lived primarily by swidden (slash-and-burn or shifting) cultivation. This thesis was supported by records of historic period cultivation in the Maya lowlands and assumptions about ancient population sizes that were consistent with this land-extensive form of cultivation.

Maya settlement studies, which increased significantly in the late 1960s, uncovered evidence indicating that large populations lived in the lowlands as early as the Late Preclassic Period. The number of people was sufficiently large in many areas to have put stress on any known swidden system of agriculture. Overall land pressures reached their peak during the Late Classic Period, a time when much of the Maya homelands was densely peopled.

Such evidence ran counter to the swidden thesis and was paralleled by emerging evidence of "non-swidden" agriculture. Beginning in the 1970s, repeated field studies began to document a wide variety and the geographic distribution of intensive Maya agricultures. Upland terracing and wetland cultivation drew much attention, in part because relic features of both have persisted in the landscape. Attention to agroforestry, orchard-gardens, and other kinds of land uses implied by botanical remains and species composition in contemporary forests followed.

The Maya used two basic forms of terraces to plant crops on rolling hills and uplands. Weir terraces are barriers placed across runoff channels. They capture sediments and moisture, typically providing a level planting surface behind each wall. Viewed from above, they form a ladder-pattern following the path of the drainage. Linear sloping dry field terraces are barriers that roughly contour the long shallow slopes of the uplands. A small level surface behind the wall grades into the slope. These terraces perhaps served to arrest the downslope loss of soil and soil nutrients. Because dry field terraces are usually associated with "other rock" walls positioned up and down slope, viewed from above they present a lattice or box pattern. The purpose of the up-down slope walls is uncertain, but they may have served as plot boundaries, wind breaks, and/or walkways.

Both terrace types display a similar wall structure in which large blocks of limestone front a rearward fill of cobble that extends under the blocks as well. The cobble fill served as a ballast, reducing pressures on the fronting wall and probably allowing excessive water in the wet season to filter through the cobble under the wall and onto the next terrace.

The linear sloping terrace is the more common of the two. Both types have been documented in the Puuc Hills in the north of the peninsula, along the central Belize–Petén border, and in southern Petén, although they are most extensive across the Campeche–Quintana Roo–northern Petén area (the central uplands) and on the Vaca Plateau of west-central Belize. Recent evidence suggests that dry field terracing is associated with calcareous soils on shallow slopes (<10°), although it has been reported on steeper slopes as well. The large-scale use of terracing in the central Maya lowlands appears to date to the Classic Period, although it probably began much earlier.

Maya use of wetlands has been the subject of far greater controversy than has their use of terracing, perhaps because of the social significance attached to hydraulic cultivation by some prehistorians. Aerial observations in the 1970s revealed ground patterns in swamps and bajos (seasonal wetlands) on the eastern and western coastal flanks of the central Maya uplands, with subsequent work suggesting possible use of wetlands within the uplands proper. Specific documentation of large systems exists for Pulltrouser Swamp and along the Río Hondo in northern Belize, the Bajo de Morocoy in southeastern Quintana Roo, and the Candelaria drainage in Campeche; smaller systems have been examined at Cobweb and Douglas swamps in Belize and near Río Azul in northeastern Petén. For the most part, confirmed ground patterns exist for wetlands located near sea level to about one hundred meters above main sea level. Many patterns indicative of wetland use observed from the air have yet to be studied.

Wetland cultivation was brought to an art form by the pre-Columbian peoples throughout the Americas. The most simple kind involved planting in the moist soils of the nearshore zone as waters receded during the dry season, facilitated by using drainage ditches where slopes were shallow. The most elaborate systems involved the construction of artificial island platforms and adjacent canals or ditches within the wetlands, the cropping platforms fertilized by muck from the canals. The Maya used various kinds.

The early work on ground patterns in Maya wetlands interpreted the features to be the more elaborate, island-platform kind. Subsequent work challenged this interpretation based on more thorough and deeper soil analysis, suggesting that surface ground patterns reflected natural deposition over a buried ditched (but not raised) system. It now appears that the Maya began to use wetlands during the Late Preclassic Period, if not earlier, as evidenced by the remains found in the buried, organic-rich soils below much of the ground patterning. Subsequent use is debated. One position postulates a rise in sea level or increased precipitation in the Late Preclassic that caused thick clays to be deposited over the ditched systems, burying them and their use but leaving the surface imprint. Another view holds that these clays were not deposited until the time of collapse or thereafter, allowing for wetland use through the Classic Period. Yet another view posits that as the clays were deposited they were incorporated into the wetland systems and that the surface expressions today represent the abandoned raised fields and canals.

Each of these interpretations draws largely on the same evidence but interprets it differently. A consensus has been reached that most of the rectilinear surface patterns in the wetlands in question are of Maya origin, that the systems were used as early as the Late Preclassic Period, and that clays were deposited over the earlier systems. All other claims remain contested, including ones dealing with the manner of wetland cultivation. One view holds that wetlands were used for dry season cultivation because of flooding problems during the wet season; another holds that fully developed systems may have alleviated flood problems and made cultivation feasible for most seasons. Regardless of the interpretations taken, it is recognized that from the Late Classic Period onwards the populations around wetlands with field and canal systems, especially in northern Belize, were astoundingly large.

The Spanish conquistadores reported extensive areas of "plantations" of shrub and tree crops among the Maya of northern Yucatán, and the discovery of relic boundary walls throughout the northern forests has led some to suggest the prehispanic use of managed forests and/or extensive orchard-gardens. Since the environments of the north are largely ill-suited for terraces and wetland fields, the orchard-garden is one of the few systems of cultivation observed by early chroniclers that was also appropriate for the central and southern lowlands. The suggestion that the Maya widely used such systems is old. Careful examination was stimulated in the 1970s by claims that the abundance of *Brosimum alicastrum* (ramón) among Maya ruins was the result of the past culling and planting of this species for their "Maya breadnut," stored in underground chambers (*chultunes*) also abundant among Maya ruins. Subsequent work, however, showed that the abundance of ramón is probably an edaphic response to the highly disturbed soils of ruins and that *chultunes* were used to store cured maize (corn).

The paleo-botanical evidence of the extensive use by the ancient Maya of fruits and fibers from trees common to contemporary Maya orchard-gardens continues to increase, however. Both orchard-garden and agroforestry systems are widespread in tropical agriculture and in the greater region. The pollen evidence of major deforestation in the central and southern lowlands, however, is more suggestive of orchard-gardens (small household spaces) than of agroforestry (large stands of managed forest).

Besides the array of Maya cropping systems requiring "landesque capital" (permanent/semipermanent land improvements, such as terraces), many requiring no such capital appear to have been used, though they left few remains easily detectable. Undoubtedly rain-fed cultivation of major staples (e.g. maize, beans, and squash) was employed without terraces where feasible. The intensity of this cultivation varied by levels of demand and the character of the land, but it ranged from land-rotational systems — known locally as *milpa* but more generically as long- to short-fallow swidden or slash-and-burn — to annual cultivation. The more intensive swiddens and annual modes of cultivation probably required considerable weeding, mulching, crop shading, and crop rotations.

Precisely where and when these systems were used are difficult to determine. Most of the upland calcareous soils prevalent throughout the region could have been cultivated, while the less abundant granitic based soils were most likely avoided because of their low fertility. Matching the calcareous soils with the time and places of settlement intensity provides strong clues about the general intensity of agriculture, with or without landesque capital features.

The lowland Maya Classic civilization no longer remains an enigma in terms of its agricultural (and population) base. Its tropical lowland realm contains fertile agricultural soils, and the Maya farmer developed sophisticated managerial and technological strategies for their use. These agricultural achievements help to explain the consistently large populations found in settlement studies of the Maya and the evidence that over time the Maya of the central and southern lowlands grew in density to in excess of 200 people/km^2 over some extended areas. They also help to explain the large number of city-states that emerged in the Classic Period and the ability of the elite to garner wealth and political power.

The material base of the Maya so understood takes some mystery out of the civilization, but not all. The collapse and depopulation of the central and southern lowlands remain a difficult puzzle to solve. Maya agriculture has been linked to the collapse in two opposed ways. In the first, the scale of agriculture is seen as having ultimately degraded the land, perhaps in concert with climate change, beyond a threshold of exhaustion. In the second, this scale of agriculture is seen as having been sustainable as long as the high levels of inputs on which it was based were maintained. In this view, the central challenge is to identify the social (or environmental) reasons that the inputs ceased. Whatever the view taken, including variants that merge the two polar views, the significant and sustained depopulation of the central and southern lowlands until the latter stages of this century is yet more puzzling. If the Maya degraded their lands so badly, how did the northern lowlands, with the poorest agricultural soils in the peninsula, manage to maintain a large population well into the conquest period? If the land was not degraded, why were the good soils of the central and southern areas abandoned for nearly a millennium?

B.L. TURNER II

REFERENCES

Culbert, T. Patrick and Don S. Rice, eds. *Precolumbian Population History in the Maya Lowlands*. Albuquerque: University of New Mexico Press, 1990.

Dunning, Nicholas P. *Lords of the Hills: Ancient Maya Settlement in the Puuc Region, Yucatán, Mexico*. Monographs in World Archaeology, No. 15. Madison, Wisconsin: Prehistory Press, 1992.

Harrison, Peter D. and B. L. Turner II, eds. *Pre-Hispanic Maya Agriculture*. Albuquerque: University of New Mexico Press, 1978.

Pohl, Mary D., ed. *Prehistoric Lowland Maya Environment and Subsistence Economy*. Cambridge, Massachusetts: Peabody Museum, Harvard University, 1985.

Pohl, Mary D., ed. *Ancient Maya Wetland Agriculture: Excavation on Albion Island, Northern Belize*. Boulder, Colorado: Westview Press, 1990.

Turner II, B. L. "The Rise and Fall of Maya Population and Agriculture, 1000 BC to Present: The Malthusian Perspective Reconsidered." In *Hunger and History: Food Shortages, Poverty and Deprivation*, Ed. L. Newman. Oxford: Basil Blackwell, 1990, pp. 178–211.

Turner II, B. L. *Once Beneath the Forest: Prehistoric Terracing in the Rio Bec Region of the Maya Lowlands*. Dellplain Latin American Studies, No. 13. Boulder, Colorado: Westview Press, 1993.

Turner II, B. L. and Peter D. Harrison. *Pulltrouser Swamp: Ancient Maya Habitat, Agriculture, and Settlement in Northern Belize*. Austin: University of Texas Press, 1993.

Turner II, B. L. and Charles H. Miksicek. "Economic Plant Species Associated with Prehistoric Agriculture in the Maya Lowlands." *Economic Botany* 38: 179–193, 1984.

See also: Swidden – Agroforestry

AGRICULTURE IN THE PACIFIC The environments where traditional agriculture was practiced in the Pacific Islands ranged from frost-prone but gardened mountain slopes at 8500 feet (2600 m) in Papua New Guinea to tiny atoll islets lying scarcely above the reach of the waves in the always warm equatorial ocean. Heavy downpours almost every day keep some places in the Pacific Islands permanently humid; short wet seasons followed by long dry spells characterize the rainfall in other places. Still others with almost no rainfall are true deserts. Some single islands contain this whole range — the big island of Hawaii, with its high, massive volcanoes and its sharply contrasting windward and leeward coasts, is a notable example of such climatic variety. A comparable dissimilarity exists in Pacific Island soils, with some young volcanic soils being highly fertile, whereas on atoll islets the only natural soil material may be rough coral rubble, which is alkaline, has a very low water-holding capacity, contains little organic matter, and is unable to supply plants adequately with many of the nutrients required for vigorous growth.

Traditional Pacific Island agriculturalists met this wide range of often challenging conditions with an even wider range of agronomic techniques and crops, which enabled food production on all but the most barren islets or at the highest elevations of the larger volcanic and continental islands. This universe of agro-ecosystems included elaborate terracing and irrigation to grow the water-loving taro, massive drainage works to grow less water-tolerant crops such as sweet potatoes, mulching and composting to enrich the soil and to slow water loss, planting crops in built-up mounds of soil to encourage cold-air drainage and so lessen frost damage, planting in excavated pits to reach the water table on dry atoll islets, and (in systems of shifting cultivation) the use of forest or woodland fallow — planted in some places, spontaneously natural in others — to restore fertility to soils after they had been gardened.

The Origin and Evolution of Pacific Island Agriculture

When, from the sixteenth century onwards, European explorers began to encounter the sophisticated Pacific Island agriculture then practiced, it was not surprising that some of them believed they had sailed to a Garden of Eden, where breadfruit trees and coconut palms provided food without work, and where only a little labor was needed to make irrigated terraces of taro bear heavy harvests of starch-rich corms or tilled beds of sweet potatoes produce many baskets full of nutritious tubers. Initially, Europeans saw this productive agriculture as though it were some sort of a divine gift given to the Pacific Islanders, who had been favored with gardens and orchards that yielded unchangingly through time and that remained continuously in harmony with local environments. It is now clear on the basis of the extensive research into Pacific pre-history carried out only over the past few decades that such a view is far from accurate.

Pacific Island agriculture has never been static. It has been evolving constantly from its beginnings, as migration led people to new environments where there were previously unknown wild plant species, as the agriculture itself changed the environment, as the need for food expanded with growing island populations, as new crop plants were introduced, or as the agriculturalists experimented and introduced productive innovations. The Pacific Island agriculture first seen by Europeans was only an instant snapshot of a long and very dynamic history.

Twenty-five years ago it would have been asserted confidently by many scholars that Pacific Island agriculture had originated in southeast Asia and that Pacific Island cultivated plants had been domesticated in that same hearth. All that the ancestors of Pacific Islanders had done was to carry the Asian techniques and crops with them as they dispersed to the farther oceanic islands. It was recognized that the sweet potato and a few other less important crops did not fit this scenario, having been shown to have originated in the American tropics. However, the presence of these exceptions was explained away by various, often fanciful, theories of migration or simply as the result of introductions by Portuguese or Spanish voyagers during the fifteenth and sixteenth centuries — though recent archaeological research in the Cook Islands (central eastern Polynesia) shows the sweet potato to have been present there by about AD 1000.

We know now that although some plants of southeast Asian origin and domestication were transferred without significant change far into the Pacific (some species of yam, for instance), there is also evidence in support of early indigenous domestication and development of agriculture in the Pacific, specifically in western Melanesia (Solomon Islands and the island of New Guinea). The length of occupation there (at least 40,000 years in New Guinea) is more than sufficient for the experimentation necessary for independent domestication. And chromosomal and paleobotanical studies now indicate that plants that may have been domesticated in this region include sago, one type of *Colocasia* taro, one kind of banana, sugar-cane, *Canarium* (a nut-bearing tree), *Saccharum edule* (a relative of sugar-cane with an edible inflorescence), *kava* (the ritual and social drink still important in many parts of the Pacific Islands), and a variety of other plants, including several fruit trees. This attention to trees and the creation of orchards is a characteristic of food production all across the Pacific and was probably carried by itinerant colonist-cultivators from its place of origin in western Melanesia to Polynesia and Micronesia.

Further, especially strong evidence for the early development of agriculture to the east of southeast Asia comes from an archaeological study in the Papua New Guinean highlands at a place called Kuk. A great deal has been written about Kuk, and the evidence has been interpreted in various and changing ways, but there is general agreement that Kuk demonstrates a long history of agricultural development, beginning about 9000 years ago and involving, among much else, an ever-growing complexity of drainage works and water control in a large swamp, changes over time in cropping combinations from mixed gardens to taro monoculture to sweet-potato dominance, responses to deforestation and land degradation brought about by shifting cultivation on the surrounding slopes, and the development of planted or encouraged tree fallows. Kuk, as well as evidence from prehistory elsewhere in the Pacific, shows that a dynamic agronomic and botanical science has long existed in the Pacific, in terms both of basic understanding and applied techniques. The origins of agriculture in the Pacific can now be said to have a time depth comparable to that of better known sites in southwestern Asia and tropical America.

No less mistaken than the view that traditional agriculture has been static in the Pacific Islands is the view that it has always been in harmony with its environment. Rather, as in the history of any dynamic agriculture, there have been episodes of deforestation, serious soil erosion, land degradation, and crop failure. Pacific Islanders did what all peoples, especially pioneers, do: in their efforts to make a living, they actively manipulated, modified, and, at times, degraded the ecosystems in which they lived, producing environmental changes that in turn required ecological adaptations and social adjustments. Considering the whole landscape, the most widespread of the human-induced changes in the prehistoric Pacific has been deforestation, with the cleared forests replaced by grasslands that required cultivation techniques different from those associated with forest-fallowed gardens. In many places, fire-maintained fern grass savannas underlain by infertile, eroded, or truncated soils came into existence or were extended by agricultural activities. This distinctive plant–soil complex is known as *toafa* in several Polynesian islands and as *talasiga* in Fiji. These dramatic landscape changes resulting from pioneering clearance for agriculture did not, however, bring unmitigated environmental degradation. Rather, in many places, they resulted in what has been termed "landscape enhancement", whereby the eroded soil transported down the slopes filled in the lower valleys and created swampy zones that were ideal sites for what came to be sustained-yield, intensive cultivation of wetland taro. Other responses included the development of dryland cultivation techniques to cope with the changed agricultural environment, irrigated terracing, and an elaborated use of trees.

Traditional Agriculture in the Pacific Islands

The wide range of agricultural systems and techniques devised by Pacific peoples over millennia can be considered as ways of solving the agronomic problems presented by the great variety of island environments. For instance, in forested areas of low population density, soil fertility was maintained by simple no-tillage shifting cultivation wherein natural forest fallow rehabilitated the soil after gardening. Where forest was diminished, leguminous or other nitrogen-fixing trees (such as *Casuarina* spp.) were encouraged or planted. In grasslands a variety of mulching and composting techniques were developed. On atolls where soil was poor or absent and rainfall often low, Islanders created an ingeniously productive and sustainable agricultural environment for the giant swamp taro (*Cyrtosperma chamissonis*) by excavating a pit to reach the water table ("the freshwater lens") below the coral rubble and building up fertile soil in the pit by composting leaves of breadfruit and several other wild or semi-domesticated trees as well as seaweed, pumice, and other materials.

The management of wet and dry conditions by irrigation and drainage was widespread and ranged from very simple ponds and small ditches to elaborate, kilometer-long, stone-lined channels and extensive hillside terracing. Irrigation and drainage were not necessarily spatially exclusive in that the ditches that drained water away from sweet-potato beds were used to provide a moist growing site for taro and other water-loving crops. Water control may also have been used to control insect pests. As population densities increased or as political control expanded enabling greater labor mobilization, some systems of wet cultivation of taro in Polynesia became very intensive, productive of yields as high as from 30 to 60 metric tons per hectare.

The agricultural tool kit of the traditional Pacific was simple, mostly derived from unprocessed natural materials: wood, plant fiber, stone, shell, and bone. Wooden spades were elaborated in places where tillage and swamp cultivation were common. Wooden hoes were made here and there but were rare. The paramount agricultural tool was the digging stick. Used for loosening soil, digging roots and corms, making holes for planting and house posts, or as poles for carrying burdens, digging sticks ranged in size from heavy, two-meter, two-man tools used cooperatively to turn grassland or swampland sod to the light sticks used by girls to open shallow holes in soft forest soil. They remain widely in use today, and modern technology has yet to find a better tool for planting taro.

Before being replaced by steel tools, stone adzes and axes were effective in opening vast areas of forest and were far more sophisticated than they might seem at first glance. For instance, the cutting edges of stone axes might be facetted and asymmetrical to make resharpening more effective and to prevent the blade sticking in the tree during felling. Wooden spades and the way they were handled had similarly subtle attributes.

Organization of agricultural labor varied across the Pacific, with, for example, men doing the clearing but women most of the gardening in Melanesia, whereas in Tonga and Samoa in Polynesia men carried out all the agricultural tasks. Traditional Pacific livestock comprised the pig and chicken. Pigs were of great importance ritually and socially in many places and were a way to "bank" the food produced in tuberous starchy vegetables, which did not store well. During the periods when pigs were being accumulated for ceremonial feasts, considerable land and agricultural labor would be devoted to producing their food.

Like all traditional agriculturalists, Pacific Islanders possessed an enormous store of knowledge about both the domesticated and the wild plants and animals in their environment. This indigenous knowledge was organized by means of complex folk taxonomies that provided a framework for pleasurable intellectual activity as well as serving practical purposes. Similarly, aesthetics was a part of Pacific Island science of agriculture and land use management. Pacific people in traditional landscapes enjoyed the arrangement of productive diversity all around. Medicine here, perfume there, fiber in the hibiscus stem, fruit, timber, edible leaves, and so forth. There is a strong aesthetic pleasure in these observations of utilitarian diversity. As Malinowski wrote about the Trobriand gardens he made famous in Papua New Guinea:

> The gardens are, in a way, a work of art. Exactly as a native will take an artist's delight in constructing a canoe or a house, perfect in shape, decoration and finish, and the whole community will glory in such an achievement, exactly thus will he go about the laying out and developing of his garden. He and his kinsmen and his fellow-villagers as well, will be proud of his labours. ... During all the successive stages of the work, visits are exchanged and mutual admiration and appreciation of the aesthetic qualities of the gardens are a constant feature of village life.

In the Pacific, as elsewhere, the complexity of traditional agriculture has undergone a simplification in modern times. Polycultural gardens of subsistence crops have been replaced by monocultural stands of commercial crops such as coconuts, ginger, coffee, and citrus. Intensive systems of irrigated taro or dryland yam cultivation have fallen out of use, often replaced by the less demanding cassava (manioc). On the other hand, with the current interest in locally based sustainable development, there is a growing concern to revive some of the indigenous traditional systems that served Pacific Islanders well in their past.

WILLIAM C. CLARKE

REFERENCES

Barrau, Jacques. *Subsistence Agriculture in Melanesia*. Honolulu: Bernice P. Bishop Museum, Bulletin 219, 1958.

Barrau, Jacques. *Subsistence Agriculture in Polynesia and Micronesia*. Honolulu: Bernice P. Bishop Museum, Bulletin 223, 1961.

Bellwood, Peter. *Man's Conquest of the Pacific: The Prehistory of Southeast Asia and Oceania*. Auckland: Collins, 1978.

Kirch, Patrick V. *Feathered Gods and Fishhooks: An Introduction to Hawaiian Archaeology and Prehistory*. Honolulu: University of Hawaii Press, 1985.

Kirch, Patrick. V. "Polynesian Agricultural Systems." In *Islands, Plants, and Polynesians: An Introduction to Polynesian Ethnobotany*. Eds. Paul Alan Cox and Sandra Anne Banack. Portland, Oregon: Dioscorides Press, 1991, pp. 113–133.

Klee, Gary A. "Oceania." In *World Systems of Traditional Resource Management*. Ed. Gary A. Klee. London: Edward Arnold, 1980, pp. 245–281.

Malinowski, B. *Coral Gardens and Their Magic*. London: George Allen and Unwin, 2 volumes, 1935.

Oliver, D. *Oceania: The Native Cultures of Australia and the Pacific Islands*. Honolulu: University of Hawaii Press, 2 volumes, 1989.

Yen, D. E. "Polynesian Cultigens and Cultivars: The Question of Origin." In *Islands, Plants, and Polynesians: An Introduction to Polynesian Ethnobotany.* Eds. Paul Alan Cox and Sandra Anne Banack. Portland, Oregon: Dioscorides Press, 1991, pp. 67–95.

Yen, D.E. and Mummery, J.M.J., eds. *Pacific Production Systems: Approaches to Economic Prehistory*. Canberra: Department of Prehistory, Research School of Pacific Studies, The Australian National University, 1990.

See also: Agroforestry in the Pacific

AGRICULTURE IN SOUTH AND CENTRAL AMERICA

Conquest of South and Central America by the Spanish and Portuguese in the sixteenth century was rapidly followed by the introduction of Old World crops. These included both those familiar to European farmers, such as wheat, barley, oats, and many temperate vegetables and fruits catering to European food tastes, as well as tropical crops from Africa and Asia, such as bananas and plantains, sugar-cane, and rice. At the same time many American crops were carried to the Old World — the most important being maize, potatoes, manioc, beans, and squash.

From the time of conquest to the present, agriculture in this region has been dichotomized between small scale subsistence farming and large scale monocrop operations producing for profit. Their development is summarized here.

Some native agricultural methods continued, such as swidden agriculture in temperate and tropical forested regions, field cultivation with the foot plow in the Andes, and intensive chinampa agriculture in central Mexico. Other techniques such as terracing declined, or as with raised fields, disappeared altogether. Some European crops were incorporated as staples by the small farmers — wheat in Mexico (principally among mestizo farmers), barley in the Andes, onions, cabbage or collards almost everywhere. Bananas and plantains spread rapidly throughout the tropics. Agricultural technology continued much as before contact, perhaps because it was appropriate to traditional agriculture which, though small scale, was highly productive.

Although early colonial institutions such as the *en-*

comienda, *corregimiento*, and *repartimiento* were designed to exploit native labor and mineral resources, they also produced surplus foodstuffs to support the European population, mining, and colonial administration. As native populations declined, large land holdings were granted to Spanish and Portuguese immigrants. Some were cultivated dilatorily, but others were transformed into plantation enterprises producing sugar, tobacco, cacao, indigo, or cochineal (also cattle, sheep, and horses) which soon supported a lucrative trade with the home country. Most plantations were dependent on slave labor. Sugar, consumed raw or distilled into aguardiente or rum, became the principal economic enterprise in Brazil and the Caribbean, and was particularly dependent on slave labor.

By the eighteenth century mining became concentrated in a few rich areas. Agriculture acquired more importance in the colonial economy. Land grants were increasingly cultivated to supply colonial needs for food grains and raw materials such as cotton. The *hacienda* became the vehicle for accomplishing this production. Labor was secured through mechanisms such as debt peonage and was administered in such a way that it was almost impossible for the individual laborer to break free.

After independence early in the nineteenth century, slavery was gradually abolished throughout the Americas. The hacienda, however, continued as a major system of ensuring agricultural labor. It only began to disappear with the Mexican Revolution of 1910. Though much less important today, it continues to operate in some Andean countries and in parts of Brazil.

The nineteenth century was marked by widespread expansion of agricultural capitalism in the form of plantation systems producing a wide variety of crops for industrial and consumer markets in Europe and North America. These included henequen from the Yucatan, bananas in Central America, and sugar in the Caribbean and Brazil. Coffee, one of the most profitable crops, was widely introduced, from Mexico to Brazil. In Mexico and parts of Central America it was produced by small farmers, in Costa Rica and Colombia medium sized farmers were the rule; elsewhere large coffee *fincas* predominated. Particularly in the latter case, land was intensively exploited, without efforts at conservation or improvement, leading to rapid deterioration in soil quality and declining production. Where land was abundant, the response was to clear virgin land and plant new coffee groves. Uruguay and Argentina became major world producers of wheat, owing to the rich pampean soil. In both Brazil and the pampas, immigrant labor was employed to clear the land, in exchange for temporary usufruct of part of the cleared area. Evicted after a time in favor of the landowner, some immigrants acquired their own (usually smaller) farms, but most withdrew to the cities, leaving only the larger *fincas* and *estancias* as significant agricultural producers.

Several different systems of agricultural exploitation can be identified in this region today. Yet agriculture has declined in importance in national economies throughout the century. Although in Central America and Bolivia half of the work force is still farming, in most countries only a third or less is employed in agriculture, which represents only 10–11% of the GDP for the region.

The persistent dichotomy between large, extensively cultivated holdings (*latifundio*) and small, intensively cultivated properties (*minifundios*) survives today. Though these are sometimes geographically separated, they frequently occur interspersed with each other interacting as a symbiotic whole. This dichotomy is also at the root of many economic and social problems. Although precise data on distribution of agricultural landholdings have not been reported since the 1970s, the extreme bimodal distribution has probably not changed significantly. In El Salvador, Guatemala, and Peru more than 90 percent of agricultural landholdings were under 10 hectares, yet represented only about 30 percent of the area. In contrast, 26 percent of the area in Guatemala and 61 percent of the area in Peru were larger than 1000 hectares, yet represented less than 0.5 percent of the holdings.

Some small-scale specialized farming techniques survive in scattered locales, such as localized irrigation systems, *chinampas* in the Valley of Mexico, drained fields among the Karinya in eastern Venezuela, or terraces and lazy beds in Mesoamerica and the Andes. Traditional methods of agriculture generally promote ecological stability both because environmental disturbance is minimal and because the agroecosystem is stable. Unfortunately, development, modernization, and institutional pressures from the larger society often lead to abandonment of these techniques, in spite of their suitability to local conditions and often exceptional productivity. Thus most irrigation has been modernized and commercialized and relies on pumping, reservoirs, and lined canals. Yet there are efforts in some places to reintroduce ancient agricultural techniques, like ridged fields in the Titicaca Basin, which have proven superior to current practices. Many traditional practices should be preserved in order to exploit their productive advantages, utilize marginal microenvironments, maintain crop diversity, and minimize production risks.

Swidden agriculture or shifting cultivation is still widely used in forested regions. It may support as many as fifty million people. It is relatively productive and succesfully integrated into the ecological regime of the tropical forest, yet requires less labor than many other methods. It is a technology that does not require capital investment or energy subsidies. When associated with sparse populations its environmental impact is minimal, perhaps even beneficial by stimulating renewed forest growth. When intensified to the point that the forest can no longer replace itself, swidden may produce widespread environmental degradation. Rapid population growth throughout Latin America has spurred migration in search of farmland in lowland tropical forests from the Peten and the Caribbean coast to the Amazon basin. Resulting widespread deforestation has had serious ecological repercussions.

One agriculture technique, the house garden, is nearly ubiquitous among small farmers, even in towns and cities. Their species diversity is high, providing supplementary food, condiments, herbs, medicinal remedies, fuel, fertilizer, and ornamental plants.

Most small-scale agriculture is conducted by peasant farmers. A true peasant owns or controls his land, runs his own operation and makes his own decisions independently. The primary production objective is for subsistence and survival. Peasants do not think in capitalist terms, but instead are oriented around the homestead (land, home, tools), which cannot be converted or exchanged for other means of production. Because cash is limited, the peasant cannot afford to take risks; he often resists trying new crops and techniques; technology remains traditional (paleotechnic).

The strength of peasant farming is its heterogeneity and diversity. Risk is minimized by planting a variety of crops, in several locales, under varying conditions. The most important crops are starchy staples — grains or root crops — and legumes; perhaps supplemented by a high value cash crop — tomatoes, coffee, narcotics.

Large holdings can be roughly divided into haciendas, estancias, plantations, and agroindustrial enterprises. Hacienda organization represents a unique adaptation to abundant land and scarce capital. Labor to work the estate is attracted by offering small subsistence plots to local smallholders or landless laborers (most commonly known as *peons*). Traditionally little or no wages exchanged hands; today real wages may actually be paid, although they are often seriously in arrears because capital remains scarce. For the same reason haciendas try to be self-sufficient in terms of basic needs. They also tend to be inefficient, relying on antiquated technology and cultivating only a small area of the best land, leaving the rest uncultivated or in the hands of the workers. Yet their production of staple cereals and tubers, or milk, meat, and wool may contribute significantly to national economies. Hacienda property may be viewed more as a basis for prestige or a hedge against inflation than as an income producer. For this reason, as well as scarce capital, haciendas also tend to resist innovation. Most surviving haciendas are found in highland areas of the Andes and Mesoamerica, though they have also been the object of land reform from Mexico to Bolivia, Peru and Ecuador.

Estancias are large holdings devoted to livestock production found mostly in the Southern Cone and northern Mexico, although in recent years cattle production has expanded rapidly in the Amazon lowlands and Central America to feed the North American demand for hamburger and processed meats.

The modern plantation contrasts with the traditional hacienda in that it is fully and efficiently operated with large amounts of capital. The plantation generally concentrates on monocrop specialization with the aim of continuous commercial production, usually for export. Although their proportion of total agricultural production has been declining, plantation crops such as sugar, coffee, bananas, pineapples, oil palm, cotton, tobacco, maize, and wheat continue to be the major source of foreign exchange for several countries.

Large farms in central Chile, Northern Mexico, and Central America have also adopted this mode, producing fruits and vegetables for the winter market in North America. Production and marketing of feed products (maize, sorghum and millets, soybeans) and oilseeds (safflower, soybeans, cottonseed) have become important in Brazil, Argentina, Colombia, Guatemala, El Salvador, and Mexico. Commercial farming is almost always profitable, but profits increase exponentially with the size of the operation. They generally practise monocropping with chemical additives such as fertilizers and pesticides. Since larger commercial farmers participate in the international economy, they can survive bad years by falling back on economic institutions (loans, savings, insurance) to carry them through.

Social consequences of this development are that the peasantry is losing control over its productive processes and is being transformed into a rural proletarian underclass that is being exploited as the primary labor force. These factors have deeply transformed production relations within the agricultural sector, resulting in the modern agroindustrial complex. As a result of increasing commercialization of agriculture, the production of cash crops for export and industrial use has expanded at the expense of the production of basic food crops, leading to significant imports in many countries of basic foodstuffs.

Efforts at agricultural modernization vacillate between large, capital intensive projects and smaller "appropriate" technology, designed especially for small farmers. However, the bias is toward large scale projects that are highly visible and politically profitable, but which are often viable only for the short term. In any case they are rarely beneficial to the small farmer.

KARL H. SCHWERIN

REFERENCES

Barkin, David, Rosemary L. Batt, and Billie R. DeWalt. *Food Crops vs. Feed Crops: Global Substitution of Grains in Production.* Boulder, Colorado: Lynne Riener Publishers, 1990.

Cardoso, Eliana and Ann Helwege. *Latin America's Economy: Diversity, Trends and Conflicts.* Cambridge, Massachusetts: MIT Press, 1992.

Crespi, Muriel K. *The Patrons and Peons of Pesillo. A Traditional Hacienda System.* Ph.D. dissertation, University of Illinois, 1968.

Denevan, William M. "Latin America." In *World Systems of Traditional Resource Management.* Ed. Gary A. Klee. New York: Halsted Press, 1980, pp. 217–244.

González, Alfonso. "Latin America: Recent Trends in Population, Agriculture and Food Supply." *Canadian Journal of Latin American and Caribbean Studies* 10(20):3–13, 1985.

Lambert, Jacques. "The Latifundio." In *Latin America. Social Structures and Political Institutions.* Ed. Jacques Lambert. Berkeley: University of California Press, 1967, pp. 59–105.

Rubin, Vera, Julian Steward, et al. *Plantation Systems of the New World.* Washington, D.C.: Pan American Union, 1959.

Teubal, Miguel. "Internationalization of Capital and Agroindustrial Complexes: Their Impact on Latin American Agriculture." *Latin American Perspectives* 14(3): 316–364, 1987.

West, Robert C. and John P. Augelli. *Middle America: Its Lands and Peoples.* Englewood Cliffs, New Jersey: Prentice-Hall, 1989.

Wolf, Eric R. *Peasants.* Englewood Cliffs, New Jersey: Prentice-Hall, 1966.

See also: Potato – Swidden – Food Technology in Central and South America

AGROFORESTRY Indigenous agroforestry is the combination in space or time of trees and food crops such that the trees provide services beneficial to the associated crops. Agroforestry variations practiced by tribal and peasant farmers the world over are historical forerunners to those being developed in modern research centers. While "modern" agroforestry is still largely experimental, indigenous agroforestry is practiced in living agricultural enterprises. The hope of researchers is to convince more farmers to plant trees in their fields, to reverse the declining conditions of the world's soils and groundwater. Not all traditional farmers practice their own agroforestry; the need is to expand the practice. For farmers who do have agroforestry, the concern is to improve it where possible. The first aim, though, is to learn more about models available in indigenous agroforestry.

The major research center devoted to this is the International Council for Research in Agroforestry (ICRAF) in Nairobi, Kenya. Its task is not only to develop new agroforestry systems but also to catalog and describe the world's indigenous ones. They analyze them as to their advantages or disadvantages, and recommend how they can be best improved and expanded. One of its findings is that it is best not to try to institute big changes in indigenous systems but to base small improvements on farmers' knowledge and what they have already developed themselves. Each indigenous agroforestry project is examined for its biological components, how these are arranged in time and space in specific admixtures, the nature of their ecological interactions, how they are managed, what they produce, and what their contribution is to environmental conservation and protection. One question asked is if the indigenous agroforestry project is locked into local conditions as a culturally or biologically unique system, or if it could be extrapolated to neighboring locations.

One type of indigenous agroforestry is interstitial support trees. Trees are planted on field borders, for example, to stop erosion at the edge of a gully or stream. The Indians of arid Sonora, Mexico plant trees along the courses of seasonal flash floods. The torrents are forced to slow down. Particles carried in the water sink to the bottom and are left behind as good soil.

In Costa Rica, living fencerows are trees that can rejuvenate when planted as posts. Small farmers in central Chile plant fruit and nut trees to mark property lines or separate fields of different kinds of crops. They act as habitats for animals, birds, and insects that are actually beneficial to the crops. The trees are capable of coppicing (when cut, they give rise to two or more new stems) and provide fuelwood, poles and fence posts. Grapes, berries, and fiber plants can be interplanted among them. Indigenous agroforestry farmers are aware of the multiple uses of each species of tree, an important aspect of their knowledge.

When trees are brought right into the crop field, it is called agri-silviculture. An important example is the alley cropping of the Cebuano farmers of Naga, Cebu, Philippines. The trees *Leucaena glauca* or *leucocephala* are planted in strips along the contours of hillside fields during resting periods (fallow) when no crops are grown. The alleys are one to two meters wide. Just before crops are planted in the alleys, the trees are cut and their branches piled along the contours to prevent soil erosion. They will coppice and have to be continually cut for firewood so as not to shade out the crops, which are corn in the wet season and tobacco and onions in the dry. After five to six years, the field will be rested again and the trees allowed to flourish. The nitrogen-rich leaves may be periodically stripped and thrown into the alleys to fertilize the soil. The deep roots of the trees are able to save nutrients that have been leached downward out of the topsoil by rain water. This system could be borrowed by other farmers.

There is a remarkable system found among the Fur people of Sudan, Africa. A kind of agrisilvipasture, it includes livestock, trees, staple crops, and pasture, all made possible by the unique legume *Acacia albida*. This is unlike other trees in that it sheds its leaves during the rainy season, rather than in the dry season. Millet or sorghum is planted right underneath the trees, because sunlight can penetrate their crowns. The leaf litter decays rapidly and fertilizes the crops. In the dry season, the trees are in leaf and provide shade, cooling the soil and keeping down the evaporation of water from it. Livestock graze and rest there, and eat the nutritious seed pods of the tree when winds blow them down, or when they and the green leaves are collected to serve as fodder. The livestock manure enriches the soil. This system is limited to where *Acacia albida* can be grown.

Field-and-grove systems have many variations also. They are composed of a single-species crop field and the grove component next to or not far from it in the form of a natural forest, a home garden, or a managed fallow or woodlot.

In field-and-forest, found in Japan, Guatemala, and Mexico, forest floors are harvested of litter that is spread over the fields as a fertilizer. In Japan it is the wet-rice paddies that benefit. Among the Ifugao of Luzon Island, Philippines, the leaves of certain trees are picked and worked into rice field soils.

Where the "forest" component is the multi-storied or mixed home garden, as among the Javanese of Indonesia, it is itself an agri-silviculture and is a good imitator of tropical forest structure. It provides a family with diverse benefits from its fruit trees, vegetables, ornamental plants, and others.

The forest component can be developed out of an old crop field through managed fallow or woodlot. As it is transformed, species composition changes. Among the Bora Indians of Peru, the earliest stage is the cassava field. Part of this is planted to fruit trees, while other portions are allowed to revert directly to natural forest. Then, purposefully planted species provide food, medicine, construction materials, firewood, handicraft materials, and vines for lashings. After all these mature and grow old, high forest dominates. The Ifugao of the Philippines have made such woodlots permanent. They recognize their importance in conserving water for rice fields and preventing land slippage on steep slopes. Among the Huastec Indians of northeastern Mexico, the family-managed woodlots and orchards are not always a later stage of a crop field but may be a site located on a steep slope, ridge, or stream bank where erosion is thereby prevented. These varying methods of agroforestry, many of which have been practiced for years, can be tried out anywhere in the world. They offer a viable alternative to other kinds of forestry practices and are based on the sound scientific ideas of many indigenous cultures.

HAROLD OLOFSON

REFERENCES

Belsky, Jill M. "Household Food Security, Farm Trees, and Agroforestry: A Comparative Study in Indonesia and the Philippines." *Human Organization* 52(2): 130–141, 1993.

Conklin, Harold C. *Ethnographic Atlas of Ifugao: A Study of Environment Culture and Society in Northern Luzon*. New Haven: Yale University Press, 1980.

Jones, Jeffrey, and Norman Price. "Agroforestry: An Application of the Farming Systems Approach to Forestry." *Human Organization* 44(4): 322–330, 1985.

Michon, G., J. Bompard, P. Hecketsweiler and C. Ducatillion. "Tropical Forest Architectural Analysis as Applied to Agroforests in the Humid Tropics: The Example of Traditional Village-Agroforests in West Java." *Agroforestry Systems* 1(2): 117–127, 1983.

Nations, James D. and Ronald B. Nigh. "The Evolutionary Potential of Lacandon Maya Sustained-yield Tropical Forest Agriculture." *Journal of Anthropological Research* 36(1): 1–30, 1980.

Olofson, Harold. "Indigenous Agroforestry Systems." *Philippine Quarterly of Culture and Society* 11: 149–174, 1983.

Terra, G.J.A "Mixed Garden Horticulture in Java." *Malayan Journal of Tropical Geography* 3 (Oct.): 33–43, 1954.

Wilken, Gene. "Integrating Forest and Small-Scale Farm Systems in Middle America." *Agroecosystems* 3: 291–302, 1977.

AGROFORESTRY IN AFRICA Agroforestry is often invoked as a new solution to problems of environmental degradation in the non-Western world, but agroforestry is not a new idea. It is merely a new word used by scientists to describe ancient land use practices by farmers in many parts of the world. Other names stressing different aspects of the same technique include forest farming, forest interculture, layered gardening, and multistory farming.

Professionals working on this topic, notably at the International Center for Research on Agroforestry (ICRAF) in Nairobi, Kenya, have formal definitions which stress that agroforestry is "a holistic approach to land use, based on the combination of trees and shrubs with crops, pastures or animals" (Lundgren, 1982). What this means in practice depends on the climate and environment. There are different traditions of agroforestry in semi-arid tropical regions, rain forests, and other zones.

In the African Sahel, traditional customs of retaining and using sparse tree cover in coordination with livestock raising and cropping tended to moderate the effects of high temperatures and low rainfall and to conserve plant nutrients while making possible a diversity of products, not all of which would suffer when droughts led to crop failure. Whilst many tree and shrub species were involved, one tree in particular, *Acacia albida*, was of such value in the western Sahel that it was often regarded as sacred, and there were severe punishments for its unauthorized felling. Because it is a leguminous, nitrogen-fixing species, yields of millet and sorghum might increase by three or four times if grown in fields which had a scatter of trees of this type. Its leaves and seed-pods were a nutritious fodder for livestock. Materials were obtained from it for the preparation of medicines, and its timber was useful in building.

Another species of importance, particularly in Sudan, was *Acacia senegal*, which produced gum arabic, and was grown in conjunction with millet and other crops in a twenty-year rotation. With the disruption of this and other cropping systems, sometimes in favor of more "modern" methods, the productivity of the land, and the number of people it can support, has been greatly reduced.

In rainforest areas, the retention of tree cover on land where crops are also grown is even more important for protection against erosion and for recycling plant nutrients. Among the many rainforest agroforestry systems known to have existed in the past, those of the Maya culture in Central America used high-yielding nut trees, while in West Africa oil palm was a principal tree species, sometimes grown with bananas, and in Indonesia jackfruit, papaya, and spice-producing trees were grown. Crops grown at ground level under these trees included a great variety of beans, squashes, and vegetables, with cereals in more open areas. Particularly characteristic were cassava (manioc) in Central America, and yams in West Africa.

Many early innovations in agroforestry arose when crops were introduced into new areas. Thus, over the last fifteen hundred years, African agroforestry has incorporated bananas and the Asian yam from Indonesia, then cassava and other crops from the Americas.

Recently, innovations have arisen as scientists have observed and entered into dialogue with farmers and their traditional practices. Thus, a recent innovation is "alley cropping", which is said to have originated when an Indonesian scientist working in Nigeria saw farmers planting a tree species on fallow land to speed the regeneration of the soil, and was led to experiment with trees that could be cut back prior to planting a corn crop but would grow up again quickly after the crop was harvested.

ARNOLD PACEY

REFERENCES

Advisory Committee on the Sahel, U.S. National Research Council. *Agroforestry in the West African Sahel*. Washington, D.C.: National Academy Press, 1984.

Lundgren, B. "Introduction." In *Agroforestry Systems*, vol. 1, part 1, 1982. pp. 3–6.

Niñez, Vera K. *Household Gardens: Theoretical Considerations on an Old Survival Strategy*. Lima, Peru: International Potato Center, 1984.

Raintree, J.B., and A. Young. *Guidelines for Agroforestry Diagnosis and Design*. Nairobi: International Council for Research in Agroforestry, 1983.

Richards, Paul. *Indigenous Agricultural Revolution: Ecology and Food Production in West Africa*. London: Hutchinson, 1985.

Rocheleau, Dianne. "Local Knowledge for Agroforestry and Native Plants." In *Farmer First: Farmer Innovation and Agricultural Research*, Eds. Robert Chambers, Arnold Pacey, and Lori Anne Thrupp. London: Intermediate Technology Publications, 1989.

See also: Agriculture in Africa

AGROFORESTRY IN THE PACIFIC The planting of trees together with the cultivation of annual crops, a combination now generally termed agroforestry, has been strongly promoted in recent years as a way to prevent land degradation and to increase total production of food and useful products from a unit of land. Throughout the Third World, development agencies and government departments of agriculture and forestry have been advocating agroforestry as a way to harmonize forests with farming, or as a way to make up, at least partially, for the destruction of natural forests and their replacement by pasture or by fields of annual crops. In the Pacific these modern, aid-funded attempts to promote agroforestry are ironic, for they take place in a region where agroforestry systems were developed thousands of years ago and where hundreds of species of trees are still used in a bewildering variety of ways.

At least a few trees even have a place in the popular imagination about the Pacific Islands. If asked what particularly characterizes the landscape of the islands, most people would think of a line of coconut palms overhanging a beach beside a coral-reef lagoon. They might also envision the stately and strikingly beautiful breadfruit tree, whose yield of starchy fruit so enchanted Captain Cook and his companions on the first European visit to Tahiti and subsequently led to Captain Bligh's famous voyage to Tahiti in H.M.S. *Bounty* to collect breadfruit cultivars for the West Indies. These conceptions of coconut palms and breadfruit trees in Pacific landscapes are accurate enough, but they only begin to suggest the full significance of trees, both domesticated and wild, in the lives of traditional Pacific peoples.

Recent chromosomal and paleo-botanical studies in the Melanesian islands of the western Pacific reveal that the domestication of plants extends back in time there for thousands of years, thus demonstrating that agriculture evolved endogenously in the Pacific region, rather than being solely or mainly the result of a direct transfer from southeast Asia, as had been believed previously. Plants that may have been domesticated in western Melanesia include — aside from important short-term crops such as *Colocasia* taro and sugar cane — a remarkable number of trees or shrubs. This early emphasis on arboriculture — the cultivation of trees and shrubs — was eventually transported all across the Pacific by the voyaging colonist cultivators, to be incorporated into production systems everywhere and to beget the typical tree-filled environs of human settlements in Polynesia and Micronesia.

Archaeological evidence for a well-developed arboriculture at least 3500 years ago comes from the Mussau Islands, which are now part of the country of Papua New Guinea. Tree species already in use then included: coconut, two or three species of *Pandanus*, *Inocarpus fagifer* (the "Tahitian chestnut", which remains one of the most important Oceanic arboricultural species), *Canarium indicum* (a nutritionally substantial "almond"-producing tree in Melanesia), *Spondias dulcis* (the vi-apple, now of very wide distribution in the tropical Pacific), and other useful trees such as *Pometia* (which provides edible fruit, medicine, and other products), *Pangium* (seeds edible after treatment to remove the poisonous component), *Terminalia* (edible "beach almond", useful timber), *Burckella* (edible fruit), and *Calophyllum* (timber favored for many uses, sticky sap used for caulking canoes). There is also evidence for the early domestication in the Pacific of several species of sago palm (used in some places to produce starch for food, elsewhere its leaves used as long-lasting house thatching), one kind of banana, and kava (a sprawling shrubby plant, the pounded stems and roots of which are used to make the ritual-social drink long important in many parts of the Pacific).

Over 400 species of trees or tree-like plants have been identified as having widespread or localized economic, cultural, and ecological importance in the Pacific Islands. The adoption of these many kinds of trees for human purposes is the cumulative result of a selection process that occurred over thousands of years and that involved both the domestication of previously unknown species encountered when Pacific voyagers landed on uninhabited islands as well as the deliberate transport from island to island of plants already known to be useful. The trees and their products served Pacific peoples in a great variety of ways. For instance, ecologically, trees provided, among other services, shade, erosion control, wind protection, beach stabilization, soil improvement, and frost protection (at high elevations in New Guinea). Cultural and economic uses included, among many others, house timber, firewood, tools, weapons, fishing equipment, abrasives (for example, the "sandpaper" leaves of some fig species), gums and oils, fiber, beverages (for example, the fluid from coconuts, immensely important for drinking on dry atoll islets), caulking, stimulants, medicines, and love potions and perfumes. Many Pacific trees are also of great importance nutritionally. For example, in various places people depend heavily on one or a combination of staple foods from coconut, breadfruit, bananas, sago palm, or several species of *Pandanus*. Many other species provide supplementary and snack foods. Although many tree foods are energy-rich in carbohydrates or vegetable fats or both, it is in other nutritional essentials that they often excel, compared with the starchy root-crop staples. Several fruits are excellent sources of provitamin A; others provide B-complex vitamins or are rich in vitamin C. Most seeds and green leaves from trees (which are widely eaten) are good sources of plant protein and various micronutrients. Spices and sauces derived from tree products can also be of great nutritional and culinary significance. An oily sauce made from the huge red fruit of a *Pandanus* species in highland New Guinea provides a rich, nutritious condiment for many otherwise bland foods. Or coconut milk or cream (squeezed from the coconut flesh) is widely used in cooking in coastal areas, and in places is aged or fermented with sea water and other flavorings to make a tasty sauce that enhances local cuisines.

The multi-purpose nature of many Pacific trees in providing a diversity of different products or services to people is well exemplified by the breadfruit. Its straight trunk is valued for canoe hulls; the inner bark is used to make bark cloth in some areas; the tree's thick, milky sap is used for caulking canoes, as adhesive for bark cloth, and as chewing gum; its large leaves are used as plates and for wrapping food for cooking in earth ovens; the dried inflorescence is burnt as a mosquito repellant; and the fruit is eaten cooked as a staple or important supplementary food in most areas of Polynesia and Micronesia and as a supplementary food in Melanesia, where seed-bearing varieties are often more important than the seedless varieties used as a staple food in Polynesia.

Like most trees, the breadfruit's production of fruit is seasonal, so that people dependent on it for food can expect periods of food shortage recurrently each year. Pacific-Island peoples developed two solutions to this problem. First, high intraspecies diversity had been developed in breadfruit (and most other domesticated crops) by centuries of observation, selection, and transportation of promising varieties from place to place. As only one example of the prolific number of named cultivars that might have been accumulated within a single species, the volcanic island of Pohnpei in Micronesia is reported to have 150 named varieties of breadfruit. Generally, each of the many cultivars followed its own distinct calendar, so that production of breadfruit on atolls and high islands of Polynesia and Micronesia (or, to give another example, the yield of *Pandanus* fruits in highland New Guinea) is staggered over a much longer period than would be available from a single cultivar or individual tree. The second way in which the availability of food from breadfruit was extended over the year was by the pit fermentation of the fruit so that it could be stored. Unlike grains, there were few Pacific-Island indigenous foods except yams (*Dioscorea* spp.) that if unprocessed last long in storage once harvested. Because harvested breadfruit lasts only a few days, a way had to be found whereby the seasonal surpluses could be accumulated for later use. The method developed was pit storage. After the ripe fruit was peeled and cored, it was preserved by a process of semi-anaerobic fermentation, involving intense acidification, which reduces the fruit to a sour paste that lasted in storage for decades. The pits, which served both as fermentation chamber and storage area, were dug in clay soil to prevent water seeping in and then lined with stones, woven mats, and a variety of leaves to keep soil from mixing with the breadfruit paste. Modern food analysis shows that the fermented product contains more carbohydrate, fat, protein, calcium, iron, and B vitamins than the fresh fruit. The pits of breadfruit paste also provided a reserve food supply after tropical cyclones, or hurricanes, devastated gardens and orchards or during times of warfare. Packages of the baked fermented paste wrapped in leaves also provided a portable, long-lasting food for sea voyages.

Although trees had a great significance in people's lives almost everywhere in the traditional Pacific, the hundreds of species utilized were combined in a great variety of unique agroforestry systems, each distinct to particular locales spread over hundreds of islands, each with a unique environment and each occupied by a distinct group of Pacific peoples, with their own particular history and set of agricultural techniques. In forested areas of low population density where shifting cultivation was practiced, a mixture of certain trees might be planted in old gardens, creating orchards that produced food, fiber, and other products for decades while also serving as a kind of fallow for the gardened soil, which eventually would be reused. The spontaneous secondary forest in such areas came to be everywhere dotted with valuable trees, remnants of past orchards and gardens. Elsewhere, in drier, more heavily populated areas where complex irrigation channels had been built to bring water to permanent plots of taro, there might also be permanent and highly diverse tree gardens surrounding the irrigated plots and shading the villages. On atolls, with their severe environmental constraints, a particularly intensive form of agroforestry had been developed to support the often high population densities. Spread through a matrix of planted coconut palms, which were particularly common and immensely useful on atolls, were a variety of other domesticated trees including species of *Pandanus*, breadfruit, and a native fig. As in much of the Pacific, what might look like an untouched natural forest to an uninformed eye was in reality a managed agroforest in which almost every tree was known and owned by an individual or a family and served at least one valuable purpose if not several. Unfortunately, a variety of present-day socioeconomic factors and changes are leading to a decline of traditional agroforestry in the Pacific region.

WILLIAM C. CLARKE

REFERENCES

Budd, W. W. et al., eds. *Planning for Agroforestry*. Amsterdam: Elsevier, 1990.

Clarke, W. C. and R. R. Thaman, eds. *Agroforestry in the Pacific Islands: Systems for Sustainability*. Tokyo: United Nations University Press, 1993.

Kirch, P. V. "Second Millennium BC Arboriculture in Melanesia: Archaeological Evidence from the Mussau Islands." *Economic Botany* 43: 225–240, 1989.

Lebot, V., M. Merlin, and L. Lindstrom. *Kava: The Pacific Drug*. New Haven: Yale University Press, 1992.

Ragone, Diana. "Ethnobotany of Breadfruit in Polynesia." In *Islands Plants, and Polynesians: An Introduction to Polynesian Ethnobotany*. Eds. Paul A. Cox and Sandra Anne Banack. Portland, Oregon: Dioscorides Press, 1991, pp. 203–220.

Vergara, N. T. and P. K. R. Nair. "Agroforestry in the South Pacific Region? An Overview." *Agroforestry Systems* 3: 363–379, 1985.

Yen, D. E. "Arboriculture in the Subsistence of Santa Cruz, Solomon Islands." *Economic Botany* 28: 247–284, 1974.

AIDA YASUAKI Aida Yasuaki, also called Aida Ammei, was born at Yamagata, Japan on February 10, 1747 (March 20, in the present calendar). Aida studied mathematics under Okazaki Yasuyuki, a mathematician of the *Nakanishi-ryu* school, from the time he was 15 years old until he reached the level of the *Tianyuanshu* technique (Chinese Algebra system).

Then he went to Edo (now Tokyo), in September 1769, and became a son-in-law of Suzuki Seizaemon, a Samurai of Shogun. Aida changed his name to Suzuki Hikosuke and worked for Fushin'yaku, a civil engineer. Here he came to know Kamiya Teirei, a student of Fujita Sadasuke (1734–1807). Fujita was a famous mathematician of the *Seki-ryu* school who wrote the *Seiyo Sampo* (Exact Mathematics, 1781) which was one of the best mathematical textbooks at that time.

Aida decided to become a mathematician, retired from his work, changed his name back to Aida Yasuaki, and asked Fujita to teach him mathematics. Fujita did not accept Aida's offer because he was concerned about mistakes in Aida's *Sangaku*. (Sangaku is a votive picture board with mathematical drawings, mostly of geographical problems. They were hung on the walls of shrines and temples for praying for mathematical progress.) Aida grew angry with Fujita and wrote the *Kaisei Sampō* (Counter-argument with Seiyo Sampo, 1785).

He then founded the *Saijō-ryū* school, and both schools disputed mathematics with each other for about twenty years. Through this disputation, Japanese mathematics progressed to a high level. Aida published eight books, nearly 2000 chapters of manuscript. He taught mathematics to Watanabe Hajime (1767–1839), Saito Naonaka (1773–1844), Ichinose Korenaga (fl. 1819), and Kando Seii (nineteenth century).

Aida died at Edo (Tokyo) on October 26, 1817 (December 4 in the present calendar).

The strong point of the *Saijo-ryu* school was in systematic algebraic symbols. Aida created the original symbol for "equal", which was the first notion of equal in Eastern Asia. He wrote the *Sampo Tensei-ho Shinan* (Mathematical Instruction of 'Tensei-ho' (or the Tenzanjutsu technic in the Seki-ryu school)), which is one of the best and most systematic books describing the Japanese algebraic system.

Aida's mathematical technic, the inductive method, was also similar to Seki Kowa's. He also studied the characteristics of irrational numbers. First, Aida computed the approximate value of irrational numbers by a sort of "Horner's method" (Horner, 1819); then he computed the value of the continued fraction. The root 2 is as follows:

approximate value	continued fraction
$\sqrt{2} \approx 1.4142$	$1, 2, 2, 2, 2, 2, 1, \ldots$
$\sqrt{2} \approx 1.414213$	$1, 2, 2, 2, 2, 2, 2, 2, 1, \ldots$
$\sqrt{2} \approx 1.41421356$	$1, 2, 2, 2, 2, 2, 2, 2, 2, 2, 2, 2, 1, \ldots$

Aida then computed all irrational numbers of prime numbers smaller than 100. He was able to do that using the inductive method.

SHIGERU JOCHI

REFERENCES

Aida's publications

1785. *Tosei Jinko-ki* (Today's Jinko-ki). Edo (Tokyo).

1785. *Kaisei Sampo* (Counter-arguements with Seiyo Sampo). Edo.

1787. *Kaisei Sampo Kaisei-ron* (Kaisei Sampo; new edition). Edo.

1788. Kaiwaku Sampo. Edo.

1797. *Sampo Kakujo*. Edo.

1797. *Sampo Kokon Tsuran* (Mathematics for All Ages). Edo.

1801. Sampo Hi-hatsuran. Edo.

1811. *Sampo Tensei-ho Shinan* (Mathematical Introduction of 'Tensei-ho'). Edo.

Secondary sources

Fujiwara Shozaburo. *Meiji-zen Nihon Sugaku-shi* (History of Japanese Mathematics Before the Meiji Era). Tokyo: Iwanami Shoten, 1956.

Hirayama Akira and Matsuoka Motohisa. *Aida Sanzaemon Yasuaki*. Tokyo: Fuji Junior College Press, 1966.

Horner, William G. "A New Method of Solving Numerical Equations of all Orders by Continuous Approximation." *Philosophical Transactions of the Royal Society* 109: 308–335, 1819.

AJIMA NAONOBU Ajima Naonobu (1739–1796), also called Ajima Chokuen, was born at Edo (now Tokyo) in 1732. Ajima's father was eighty *Koku* (a landlord of a village of eighty people), and a samurai warrior of Lord Shinjo, Dewa (now Yamagata prefecture). Ajima studied mathematics first under Irie Masataka, a mathematician of the *Nakanishi-ryu* school, then under Yamaji Nushizumi (also called Yamaji Shuji), who was the third president of the *Seki-ryu* school and an astronomer at *Bakufu Temmongata* (Shogun's Observatory). Then Ajima became an accountant of Lord Shinjo, at the rank of 100 *Koku*.

Yamaji made the *Horeki (Kojutsu) Reki* calendar (Calendar Made in Horeki Era, 1754), which was used from 1755 to 1797; however this calendar was not very accurate. In order to make a new luni-solar calendar, in 1762 he started to observe the sky with Fujita Sadasuke (1734–1807), his assistant. When Fujita was appointed *Sangaku Shihan* (Professor of Mathematics) of Lord Arima (1714–1783) in 1768, he retired from the Shogun's Observatory. After that, Ajima helped Yamaji to observe the sky and taught astronomy at Yamaji's astronomical school. There are four of Ajima's astronomical manuscripts: *Jujireki Bimmo* (Introduction of the 'Works and Days Calendar'), *Anshi Seiyo-reki Koso* (Professor Ajima's Studies for Western Calendars), *Ajima Sensei Bimmo no Jutsu* (Methods of Professor Ajima's 'Bimmo') and *Koshoku Mokyu Zokkai* (Introductions of Eclipses (of the Sun and Moon)). These manuscripts were probably students' notes of Ajima's lectures.

Ajima's works for pure mathematics were studies for logarithm, computing the values of spheres, and series. Ajima and other Japanese mathematicians studied Western mathematics indirectly, that is to say, through Chinese books such as the *Shuli Jingyun* (Mei Juecheng (1681–1763), 1723, China). Ajima studied logarithms from this book, then made a table of logarithms whose values are from 0.9 to 10^{-12}, 108 items. When Ajima used this table and formulae

$$\log XY = \log X + \log Y \quad \text{and} \quad \log 10 = 1,$$

he could compute all logarithmic numbers up to twelve decimal places.

Ajima expanded Japanese mathematicians' traditional method *Tetsu-jutsu*, which is a sort of inductive method. (*Tetsu-jutsu* was created by Takebe Katahiro (1664–1739), second president of the *Seki-ryu* school.) Ajima computed the value of spheres. In this case, he solved a sort of integral equation using *Tetsu-jutsu* twice.

Ajima also wrote the *Sansha San'en Jutsu* (Methods of Three Diagonals and Three Circles), which deals with the same problems as those of Gian Francesco Malfatti (1731–1807).

After Yamaji died, Ajima became the fourth president of the *Seki-ryu* school (or fifth, because Fujita was sometimes counted as the fourth). Ajima died in Tokyo, April 5, 1798 (May 20, 1798, current calendar).

SHIGERU JOCHI

REFERENCES

Fujiwara Matsusaburo. *Meiji-zen Nihon Sugaku-shi.* Tokyo: Iwanami Shoten, 1954.

Hirayama Akira and Matsuoka Motohisa, eds. *Ajima Naonobu Zenshu.* Tokyo: Fuji Junior College Press, 1966.

ALCHEMY IN CHINA In China, as elsewhere, alchemy is a science based on cosmological doctrines, aiming to afford (a) an understanding of the principles governing the formation and functioning of the cosmos, and (b) the transcendence of those very principles. These two facets are complementary and ultimately equivalent: the alchemist rises through the hierarchy of the constituents of being by "exhausting" (*jin* or *liao*, two words also denoting "thorough knowledge") the nature and properties of each previous stage. He thus overcomes the limits of individuality, and ascends to higher states of being; he becomes, in Chinese terms, a *zhenren* or Authentic Man.

While historical and literary sources (including poetry) provide many important details, the bulk of the Chinese alchemical sources is found in the *Daozang* (Daoist Canon), the largest collection of Daoist texts. One fifth of its approximately 1500 texts are closely related to the various alchemical traditions that developed until the fifteenth century, when the extant Canon was compiled and printed. Later texts are included in the *Daozang jiyao* (Essentials of the Daoist Canon) and other minor collections. Modern study of the alchemical literature began in the present century, after the Canon was reprinted and made widely available in 1926. Among the most important contributions in Western languages are those of Joseph Needham, Nathan Sivin, Ho Peng Yoke, Farzeen Baldrian-Hussein, and Isabelle Robinet.

Though the underlying doctrines remained unchanged, Chinese alchemy went through a complex and not yet entirely understood development along its twenty centuries of documented history. The two main traditions are conventionally known as *waidan* or "external alchemy" and *neidan* or "internal alchemy". The former, which arose earlier, is based on the preparation of elixirs through the manipulation of natural substances. Its texts consist of recipes, along with descriptions of ingredients, ritual rules, and passages concerned with the cosmological associations of minerals and metals, instruments, and operations. Internal alchemy developed as an independent discipline around the middle or the late Tang period (618–906). It borrows a substantial part of its vocabulary from its earlier counterpart, but aims to produce an elixir — equated with transcendental knowledge — within the alchemist's person.

At the basis of both traditions are traditional doctrines of metaphysics and cosmology. Chinese alchemy has always been closely related to the teachings that find their classical expression in the early "philosophical" texts of Daoism, especially the *Daode jing* and the *Zhuangzi*. The cosmos as we know it is conceived of as the final stage in a series of spontaneous transmutations stemming from original non-being. This process entails the apparent separation of primeval Unity into the two complementary principles, *Yin* and *Yang*. Once the cosmogonic process is completed, the cosmos is perceived as subject to the laws of cosmology, *wuxing*, the Five Agents or Phases. The alchemist's task is to retrace this process backwards. Alchemy, whether "external" or "internal", provides him with a support, leading him to the point when, as some texts put it, "Heaven spontaneously reveals its secrets." Its practice must be performed under the close supervision of a master, who provides the oral instructions (*koujue*) necessary to an understanding of the processes that the adept performs with minerals and metals, or undergoes within himself.

In order to transcend space and time — the two main

features of the cosmos as it is ordinarily perceived — the alchemist should take extreme care of their correspondences to the work he performs. Space is delimited and protected by talismans (*fu*), and the laboratory (*danwu* or "chamber of the elixirs"), and instruments are properly oriented. According to some texts, heating must conform to minutely defined time cycles. This system, known as "fire times" (*huohou*) and sometimes described in painstaking detail, allows an adept to perform in a relatively short time the same work that Nature would achieve in thousands of years — in other words, to accelerate the rhythms of Nature. Bringing time to its end, or tracing it back to its beginning, is equivalent: in either case time is transcended, and the alchemist gains access to the eternal, constant present that precedes (or follows, though both terms become inadequate) the time of cosmogony and cosmology. The same is true with space: its center, where the alchemist places himself and his work, is a point devoid of dimension. From this point without space and time he is able to move at will along the axis that connects the higher and lower levels of being ("Heaven", *tian*, and the "Abyss", *yuan*).

Among a variety of procedures that the sources describe in an often allusive way, and in a language rich in metaphors and secret names, two stand out for their recurrence and importance. The first is based on lead (*Yin*) and mercury (*Yang*). In external alchemy the two substances are refined and joined in a compound whose properties are likened to the condition of primal Unity. In internal alchemy, lead becomes a cover name for the knowledge of the Dao (Pure Yang, *chunyang*) with which each being is fundamentally endowed, but which is obscured (i.e. transmuted into Yin) in the conditioned state. Mercury, on the other hand, represents the individual mind. The second most important method, which is proper to external alchemy, is centered on cinnabar (Yang). The mercury contained therein (representing as such the Yin principle contained within the Yang) is extracted and newly added to sulphur (Yang). This process, typically performed nine times, finally yields an elixir embodying the luminous qualities of Pure Yang. This Yang is not the complementary opposite of Yin, but, again, represents the One before its apparent separation into the two complementary principles.

The final object of both disciplines is represented as the preparation of an elixir commonly defined as *huandan* (lit., Elixir of Return). This expression, recurring in the whole literature, originally denotes an elixir obtained by bringing the ingredients back to their original condition through repeated cyclical operations — an operation comparable to the process that the adept performs within himself with the support of the alchemical practice. The word *dan* (elixir) also denotes cinnabar, suggesting that the process begins and ends on two corresponding points along an ascensional spiral. This synonymy also shows the centrality of cinnabar in external alchemy, where this substance plays a role comparable to that of gold in the corresponding Western traditions. This role is taken by lead in internal alchemy. Both lead and gold, in their turn, are denoted by the word *jin*. The value of gold, and the word "gold" itself, are therefore mainly symbolic in China: the elixir, whether external or internal, and whatever its ingredients, is often defined as "gold," and Golden Elixir (*jindan*) is a name of the alchemical arts.

The extant *waidan* sources suggest that the two main methods outlined above acquired progressive importance in the history of the discipline. In the *Huangdi jiuding shendan jing* (Book of the Nine Elixirs) and other texts dating from the first centuries AD, cinnabar is never the main ingredient of an elixir, and the lead–mercury compound — sometimes replaced by refined lead alone — is only used as a layer in the crucible together with other ingredients. In these methods, the substances undergo cycles of refining in a hermetically sealed crucible. This process consists of a backward reenactment of cosmogony that brings the ingredients to a state of *prima materia*. The elixir can be finally transmuted into alchemical gold projecting on it a minute quantity of the native metal. Important details on the early phase of Chinese alchemy are also found in portions of the *Baopu zi neipian*, written around AD 320 by Ge Hong. Its descriptions of processes that can be compared with extant sources are, however, often abridged and sometimes inaccurate.

During the Tang dynasty, the *waidan* tradition reached one of its peaks with Chen Shaowei (beginning of the eighth century), whose work describes the preparation of an elixir obtained by the refining of cinnabar. Each cycle yields a "gold" that can be ingested, or used as an ingredient in the next cycle. In the second part of the process, the final product of the first part is used as an ingredient of a *huandan*. Among the representative texts of this period are several collections of recipes, one of the most important of which was compiled by Sun Simo. The first half of the Tang dynasty also marked the climax of contacts between China and the Arabic world. These exchanges may be at the origin of the medieval word *alchymia*, one of whose suggested etymologies is from middle Chinese *kiem-yak* (the approximate pronunciation of the modern *jinye* or "Golden Liquor") with the addition of the Arabic prefix *al-*.

While the Tang period is sometimes defined as the golden age of external alchemy, it also marked the stage of transition to internal alchemy. This shift, sometimes taken to be only due to the multiplication of cases of elixir poisoning, or to the influence of Buddhism, requires further study to be properly evaluated. The very incidence and relevance of cases of accidental poisoning (which claimed their toll even among Emperors) suggests that external alchemy had lost, at least to some extent and in some contexts, its soteriological character, and that its practices had become known outside the legitimate transmission. Some masters may, therefore, have transmitted their doctrine modifying the supports used for the practice. In internal alchemy, the adept's person itself performs the role which natural substances and instruments play in external alchemy. In doing so, this discipline avails itself — in ways and degrees that vary, and which require further study to be correctly understood — of traditional Chinese doctrines based on the analogies between macrocosm and microcosm, of earlier native contemplative and meditative disciplines, and of practices of Buddhist origin (apparently of Tantric character, through the possible medium of the Tiantai school).

Among the forerunners of internal alchemy is the Shangqing (Supreme Purity) tradition of Daoism, as practiced for example by Tao Hongjing. Based on revelations of the late fourth century, this school attributed particular importance to meditation, but also included the compounding of elixirs among its practices. (Shangqing represents in fact the first example of close relations between alchemy and an established movement of "religious" Daoism.) The relevant sources exhibit the earliest traces of the interiorization of alchemy. Among the texts used in this school is the *Huangting jing* (Book of the Yellow Court), a meditation manual often quoted in *neidan* texts.

In Song and Yuan times, the history of *neidan* identifies itself with that of the lines of transmission known as Southern Lineage (*nanzong*) and Northern Lineage (*beizong*). The respective initiators were Zhang Boduan (eleventh century) and Wang Zhe (1112–1170). Both schools placed emphasis on the cultivation of *xing* and *ming*, which constitute two central notions of internal alchemy. *Xing* refers to one's original nature, whose properties, transcending individuality, are identical to those of pure being and, even beyond, non-being. *Ming* denotes the "imprint", as it is, that each individual entity receives upon being generated, and which may or may not be actualized in life (the word also means "destiny" or "life", but neither translation covers all the implications in a *neidan* context). The Northern and Southern lineages, and subtraditions within them, were distinguished by the relative emphasis given to either element. The textual foundation of both lineages was provided by the *Zhouyi cantong qi* of Wei Boyang, and the *Wuzhen pian* (Awakening to Reality), a work in poetry by Zhang Boduan.

During the Ming and Qing dynasties the *neidan* tradition is known to have divided into several schools, but their history and doctrines are still barely appreciated. One of the last greatest known masters of this discipline was Liu Yiming (eighteenth century), who, in his works, propounded an entirely spiritual interpretation of the scriptural sources of his tradition.

FABRIZIO PREGADIO

REFERENCES

Baldrian-Hussein, Farzeen. *Procédés Secrets du Joyau Magique. Traité d'Alchimie Taoïste du XIe siècle*. Paris: Les Deux Océans, 1984.

Eliade, Mircea. *The Forge and the Crucible*. 2nd ed. Chicago: University of Chicago Press, 1978.

Ho Peng Yoke. *Li, Qi and Shu: An Introduction to Science and Civilization in China*. Hong Kong: Hong Kong University Press, 1985.

Needham, Joseph. "Alchemy and Early Chemistry in China." In *The Frontiers of Human Knowledge*. Ed. Torgny T. Segerstedt. Uppsala: Uppsala Universitet, 1978, pp. 171–181.

Needham, Joseph, et al. *Science and Civilisation in China*, vol. V, parts 2–5. Cambridge: Cambridge University Press, 1974, 1976, 1980, 1983.

Pregadio, Fabrizio. "Chinese Alchemy. An Annotated Bibliography of Works in Western Languages." *Monumenta Serica* 44, 1996.

Robinet, Isabelle. "Original Contributions of Neidan to Taoism and Chinese Thought." In *Taoist Meditation and Longevity Techniques*. Ed. Livia Kohn in cooperation with Yoshinobu Sakade. Ann Arbor: Center for Chinese Studies, The University of Michigan, 1989, pp. 297–330.

Schipper, Kristofer, and Wang Hsiu-huei. "Progressive and Regressive Time Cycles in Taoist Ritual." In *Time, Science, and Society in China and the West* (*The Study of Time*, V). Ed. J.T. Fraser, N. Lawrence, and F.C. Haber. Amherst: University of Massachusetts Press, 1986, pp. 185–205.

Sivin, Nathan. *Chinese Alchemy: Preliminary Studies*. Cambridge, Massachusetts: Harvard University Press, 1968.

Sivin, Nathan. "Chinese Alchemy and the Manipulation of Time." *Isis* 67: 513–527, 1976.

Sivin, Nathan. "The Theoretical Background of Elixir Alchemy." In Joseph Needham et al., *Science and Civilisation in China*, vol. V: Chemistry and Chemical Technology, part 4: *Spagyrical Discovery and Invention: Apparatus, Theories and Gifts*. Cambridge: Cambridge University Press, 1980, pp. 210–305.

Sivin, Nathan. "Research on the History of Chinese Alchemy." In *Alchemy Revisited. Proceedings of the International Conference on the History of Alchemy at the University of Groningen, 17–19 April 1989*. Ed. Z.R.W.M. von Martels. Leiden: E.J. Brill, 1990, pp. 3–20.

See also: *Huangdi jiuding shendan jing* – Ge Hong – *Yinyang* – Five Agents or Phases – Sun Simo – Wei Boyang

ALCHEMY IN INDIA Alchemy was an art practiced in ancient India. This is evident from the description in

Artha-śāstra, a monumental work on state craft by Kauṭīlya (400 BC), of a type of gold which was then being prepared by the transmutation of base metals. It was more recently (in 1941–1942) demonstrated in New Delhi in the presence of renowned national leaders. Two marble slabs with the inscription of these two events still adorn the *yajna-vedi* (alter for the Vedic sacrifice) behind the Lakṣmi-Nārāyana temple (popularly known as Birla temple after the name of the donor). The English translation of the first inscription is as follows:

> On the first day of *śukla pakṣa* (bright fortnight) in the month of *Jyeṣṭha* (May–June) of the sambat 1998, i.e. 27th May 1941, Pandit Kṛṣṇalāla Śarmā, in our presence . . . prepared one *tolā* (12 grams approximately) of gold from out of one *tolā* of mercury in Birla house, New Delhi. The mercury was kept inside a fruit of *riṭhā* (bot. *Sapindus trifoliatus Linn*). Inside this, a white powder of some herbs and a yellow powder which were perhaps one and half *ratti* [one *ratti* is equal to approximately 125 milligrams] in weight were added. Thereafter, the fruit of *riṭhā* was smeared with mud and kept over a charcoal fire for about forty-five minutes. When the charcoal became ash, water was sprinkled over it. From inside the fruit which originally contained mercury, gold came out. In weight, the gold was one or two *rattis* less than one *tolā*. It was pure gold. We could not ascertain the mystery behind the performance. The nature as well as the identity of both the powders which were added were not disclosed to us. During the whole experiment, Pandit Kṛṣṇalāla Śarmā was standing ten to fifteen feet away from us. . . .We were all surprised to witness this performance. . ."

The second plaque tells the story of a similar event in 1942, in which mercury was mixed with an unnamed drug, kept over the fire for half an hour, and transformed into gold.

The primary aim of giving the above inscriptions is to show that such alchemical practices are prevalent even now, and they are not mere myths or superstitious beliefs as some people claim.

According to Indian tradition, alchemy is not an end in itself. Mercury, when processed through eighteen different steps which are called *saṃskāras* in Ayurvedic parlance, helps a person to attain positive health and to prevent disease. It also cures several obstinate and otherwise incurable diseases. Prior to administering this processed mercury to human beings, at the seventeenth step, it must be tested on ordinary mercury or other base metals. Depending upon its potency, these base metals become transmutated into gold or silver. Thus alchemy is only a step to test the effectiveness of the recipe before it is administered to human beings to improve their physical and mental health. These are necessary for attaining spiritual perfection in the form of *jīvan-mukti* (salvation while remaining alive). These methods are always kept secret and disclosed only to trusted disciples. It is always ensured that the knowledge does not fall into the hands of unworthy people, who by amassing wealth may create social problems. In fact, acquiring wealth by alchemical methods is considered a great sin. Such wealth should never be used for personal benefit, but for charitable purposes. That is how saints adept in this technique reluctantly agree to demonstrate these methods in public. While doing so, only the end result is shown without disclosing the details of the technique. Pandit Kṛṣṇalāla Śarmā, who demonstrated the method described above, learnt it from a saint of Haradwar named Nārāyaṇa Swamī. However, he did not teach the detailed technique to anybody because he could not find a worthy disciple.

Because of the secrecy involved, many manuscripts describing this alchemical technique have perished. According to anecdotes, Nāgārjuna, the Buddhist monk perfected this technique and wrote several books on it. All these are unfortunately no longer extant. Some of the extant works dealing with both *deha-siddhi* (attaining perfection of the body by rejuvenation) and *lauha-siddhi* (transmutation of base metals into gold, etc.) are as follows:

(1) *Rasahṛdaya-tantra* by Govinda Bhagavat-pāda, fl. ninth century;

(2) *Rasendra-cūdamaṇī* by Somadeva, fl. twelfth century;

(3) *Rasaprakāśa-sudhākara* by Yaśodhara, fl. thirteenth century;

(4) *Rasasāra* by Govindācārya, fl. thirteenth century;

(5) *Rasendra-cintāmanī* by Dhuṇḍukanātha, fl. fourteenth century;

(6) *Rasa-paddhati* by Bindu, fl. fifteenth century;

(7) *Āyurveda-saukhya* by Ṭoḍara Malla, fl. sixteenth century;

(8) *Āyurveda-prakāśa* by Mādhava Upādhyāya, fl. seventeenth century;

(9) *Rasāyana-sāra* by Śyāma Sundarācārya, fl. twentieth century.

In Ayurveda, in addition to mercury, several other metals, minerals, gems, and costly stones are used for therapeutic purposes. Before these ingredients are added to recipes, special processes are required to make them non-toxic and therapeutically potent. The branch of Ayurveda describing such details is called *Rasa-śāstra*. All the books mentioned above belong to this branch. In addition to the processing of mercury both for *deha-siddhi* and *lauha-siddhi*, the technique of processing other metals is described.

Mercury is processed through eighteen different steps both for *deha-siddhi* and *lauha-siddhi*. Although there are some minor differences in different texts, these saṃskāras are (1) *svedana* or fomentation, (2) *mardana* or trituration, (3) *mūrchana* or causing disintegration of particles, (4) *utthāpana* or revival of the natural physical properties of mercury, (5) *pātana* or distillation and sublimation, (6) *bodhana* or potentization, (7) *niyāmana* or regulation of physical properties, (8) *dīpana* or enhancing the power of digestion (of other metals), (9) *grāsa-māna* or determination of the quantity of other metals to be added, (10) *cāraṇa* or impregnation with *bija* (preparations used as seed), (11) *garbhadṛti* or internal digestion, (12) *vāhya-dṛti* or external digestion, (13) *jāraṇa* or assimilation, (14) *rañjana* or coloration, (15) *sāraṇa* or excessive potentization, (16) *krāmaṇa* or enhancing the power of penetration, (17) *vedha* or testing the efficacy and potency of mercury by way of transmutating base metals into gold and silver, and (18) *śarīrayoga* or administration of processed mercury to human beings for the purpose of rejuvenating the body.

Even though the details of all the above mentioned steps are described in books, the description is cryptic and some vital techniques are kept secret. Many people, in their enthusiasm to practice alchemy on the basis of the description in books, have lost lots of energy and money. They are unsuccessful, because the secret codes and hidden techniques can be learned only from a guru or master, and such adept masters disclose these techniques only to worthy disciples who are absolutely free from worldly attachments.

BHAGWAN DASH

REFERENCES

Acarya, Yadavji Trikamji. *Rasāmṛtam*. Banārasa: Motilala Banarasidasa, 1951.

Ārya, Satyendrakumāra. *Āyurveda Rasaśāstra kā udbhava evaṃvikāsa*. Varanasi: Krsnadas Academy, 1984.

Dash, Bhagwan. *Alchemy and Metallic Medicines in Ayurveda*. New Delhi: Concept, 1986.

Dvivedī, Vāsudeva Mūlaśaṅkara. *Pāradavijñānīyam*. Datiyā: Śarmā Āyurveda Mandira, 1969.

Govinda Bhagavatpāda. *Rasahṛdayatantra*. Kaleda: Krsna Gopala Āyurveda Bhavana, 1958.

Kangle, R.P. *The Kautīlīya Arthaśāstra*. Delhi: Motilal Banarasidass, 1992.

Mookerjee, Bhudeb. *Rasa-jala-nidhi*. Vārāṇasi: Śrīgokul Mudranalaya, 1984.

Nityanātha Siddha. *Rasāyanakhaṇḍa of Rasaratnākara*. Varanasi: Chaukhamba Amarabharati Prakasan, 1982.

Panta, Tārādatta, ed. *Rasārṇava or Rasatantram*. Benaras: Chowkhamba Sanskrit Series Office, 1939.

Ray, P.C. *History of Chemistry in Ancient and Medieval India*. Calcutta: Indian Chemical Society, 1956.

Seal, B.N. *Positive Sciences of Ancient Hindus*. Delhi: Motilal Banarasi Dass, 1958.

See also: Āyurveda

ALFONSO X Alfonso X, King of Castile (1252–1284) and a patron of literature and learning, made an important effort to recover Arabic and, very especially, Andalusian astronomical materials by translating them first into Spanish and later into Latin. His collaborators were one Muslim convert into Christianity, eight Christians (of whom four were Spaniards and another four Italians) and a very important group of five Jews. Alfonso failed in his attempt to integrate in his team a Muslim scientist of the importance of Muḥammad al-Riqūṭī but his interest for us here lies in the fact that his translations preserve Andalusian astronomical works which would have been lost otherwise. This is the case, for example, of the *Libro de las Cruzes* (Book of Crosses), a late Latin astrological handbook translated into Arabic in the early ninth century and revised, in the eleventh century, by a certain ʿUbayd Allāh. Other works which are only known through his translations are the *Lapidario* (a book on the magical applications of stones) written by the otherwise unknown Abolays, the two books on the construction of equatoria written by Ibn al-Samḥ (d. 1035) and Ibn al-Zarqāllu (d. 1100), ʿAlī ibn Khalaf's book on the use of the plate for all latitudes (*Lámina Universal*, Toledo, eleventh century) and Ibn al-Zarqāllu's treatise on the construction of the armillary sphere. Other works which are, apparently, original contributed to the European diffusion of Arabic astronomical ideas: the famous *Alfonsine Tables*, extremely popular between the fourteenth and the sixteenth centuries, were strongly influenced by the *Tables* of al-Battānī and marked a turning point in the development of late Medieval European astronomy.

JULIO SAMSÓ

REFERENCES

Comes, Mercè. *Ecuatorios andalusíes. Ibn al-Samḥ al-Zarqāllu y Abū-lṢalt*. Barcelona: Instituto de Cooperación con el Mundo Arabe and Universidad de Barcelona, 1991.

Comes, Mercè, Puig, Roser and Samsó, Julio, eds. *De Astronomia Alphonsi Regis. Proceedings of a Symposium on Alfonsine Astronomy held at Berkeley (August, 1985) and other papers on the same subject*. Barcelona: Instituto "Millás Vallicrosa" de Historia de la Ciencia Arabe, 1987.

Procter, Evelyn S. *Alfonso X of Castile Patron of Literature and Learning*. Oxford: Clarendon Press, 1951.

See also: Ibn al-Zarqāllu

ALGEBRA IN CHINA Arithmetic and geometry are the two oldest branches of mathematics. Algebra has its beginnings from both of them when attempts were made to generalize operations and relationships. Initially, such ideas were expressed in words; in the course of time, they were represented by some form of notation. The symbols facilitated the methods which in turn generated new concepts and methods. For instance, we are now able to solve with ease any arithmetical problem involving what we call a pair of linear equations in two unknowns through the use of the notational equations $ax + by = c$ and $dx + ey = f$. Around AD 825, Muḥammad ibn Mūsā al-Khwārizmī wrote a book expressing equations in words. In order to arrive at a solution, the two sides of an equation were manipulated through two main operations which he called *al-jabr* and *al-muqābala* — the name "algebra" was derived from the first word.

In ancient China, arithmetic developed through the use of the rod numeral system. The development of arithmetic was to the fullest — not only were the methods of addition, subtraction, multiplication, and division known, but the manipulations of the common fraction and the decimal fraction were commonplace, and methods such as those involving proportion and the Rule of Three were widely used. This article will describe very briefly how the mathematicians were able to generalize arithmetical operations and relationships which resulted in general methods of solution. The period covered will be from antiquity to the beginning of the fourteenth century. Traditional mathematics in China was at its height during the thirteenth and early fourteenth centuries; this was a time in Western Europe when the importance of the new arithmetic that evolved from the Hindu–Arabic numeral system was just beginning to be realized.

It might seem strange and improbable that the ancient Chinese were able to find general methods of solving equations since they did not compute through a written system but through the use of rods. However, the rod numeral system was extremely sophisticated and flexible, and the positions where the numerals were placed usually represented certain mathematical concepts. For example, let us consider a set of linear equations in three unknowns which we now write in the following manner:

$$\begin{aligned} a_1x + b_1y + c_1z &= d_1 \\ a_2x + b_2y + c_2z &= d_2 \\ a_3x + b_3y + c_3z &= d_3 \end{aligned}$$

The ancient Chinese would notate such a mathematical concept by placing the numerical values of the as, bs, cs, and ds in rod numerals in the following matrix form:

$$\begin{matrix} a_3 & a_2 & a_1 \\ b_3 & b_2 & b_1 \\ c_3 & c_2 & c_1 \\ d_3 & d_2 & d_1 \end{matrix}$$

The positions occupied by the rod numerals were important — the positions in the first row represented the first unknown, and those in the second and third rows represented the second and third unknowns, respectively.

The aim of the method was to obtain a group of zeros forming a triangle in the top left diagonal half of the matrix through a process of elimination with two columns at a time. Thus the elimination process would result in zeros for the positions which were occupied by a_3, a_2, and b_3 of the above matrix. After this was attained, the third unknown was derived from the third column from the right. This result would be able to derive the second unknown from the second column, and the solutions would in turn derive the first unknown from the remaining column.

The method is called *fang cheng* and can be found in the eighth chapter of *Jiu zhang suan shu* (Nine Chapters on the Mathematical Art). The whole chapter is devoted to the solutions of such equations which include a problem involving five equations in five unknowns and another involving five equations in six unknowns.

The matrix notation was derived through an evolution of a fundamental tradition that used positions occupied by rod numerals to represent concepts or things. Besides the placement of numerals involving the common fraction, problems related to proportion were solved by similar operations. The ancient Chinese were familiar with what is now called the Rule of False Position which, for them, originated from the solution of the concept of a pair of linear equations in two unknowns. The method is called *ying bu zu* and is the precursor of the *fang cheng* method.

The procedure for finding the square root of a number was derived from the geometrical division of a square into smaller areas. This was then arithmetized through the placement of rod numerals to become a general method. The offshoot of this method was the solution of what we call a quadratic equation of the form $x^2 + bx = c$, where b and c are positive. The ninth chapter of *Jiu zhang suan shu* has a problem involving an equation of this form. Besides the method of finding the square root, the book also gives the method of finding the cube root of a number. Knowledge of this method led to the solution of a cubic equation of the form $x^3 + ax^2 + bx = c$, where a, b and c are positive. The seventh-century work *Jigu suanjing* (Continuation of Ancient Mathematics) by Wang Xiaotong has problems involving such equations.

The struggle to solve other types of quadratic equations besides the one mentioned above was depicted by the thirteenth-century mathematician Yang Hui, who quoted from the works of Liu Yi of the eleventh century — Liu Yi's works are now no longer extant. Though the concepts of the different types of quadratic equations were initially derived from a variety of geometrical considerations, the arithmetization of their operations through rod numerals revealed certain patterns and similarities which enabled the emergence of a general method of solution.

In his explanation of the development of the polynomial equation and its solution, Yang Hui quoted another eleventh-century mathematician, Jia Xian. He pointed out that Jia Xian was familiar with the triangular array of numbers now known as the Pascal triangle and was the first to realize its close relationship with the procedure of root extraction. From here, Jia Xian laid the foundation for a ladder or algorithmic method of finding the root of a number of any degree. This eventually provided the breakthrough to finding a solution of any numerical polynomial equation.

It was Qin Jiushao's detailed and systematic methods of explaining the problems in his book, *Shushu jiuzhang* (Mathematical Treatise in Nine Sections), that established beyond doubt the competence of the Chinese mathematicians to solve numerical equations of higher degree. Among the problems in the book, there are three that are involved with equations of the fourth degree and one of the tenth degree. These equations are of the following form:

$$-x^4 + 763200x^2 - 40642560000 = 0$$
$$-x^4 + 15245x^2 - 6262506.25 = 0$$
$$-x^4 + 1534464x^2 - 526727577600 = 0$$
$$x^{10} + 15x^8 + 72x^6 - 864x^4 - 11664x^2 - 34992 = 0$$

Yang Hui, Li Ye and Zhu Shijie were the other thirteenth century mathematicians who were also familiar with the algorithm method of solving a polynomial equation of any degree. This method is now generally accepted as similar to that put forward by Horner in 1819.

The Chinese were able to express the complex concept of a polynomial equation through the placement of rod numerals on the counting board. This notational representation was called *tian yuan shu* (technique of the celestial element) in which an equation was formulated in terms of the unknown called *yuan*. An equation of the form

$$a_0x^n + a_1x^{n-1} + \cdots + a_{n-1}x + a_n = 0$$

was represented in rod numerals in a vertical line as follows:

$$\begin{array}{c} a_n \\ a_{n-1} \\ \vdots \\ a_1 \\ a_0 \end{array}$$

The positions occupied by the rod numerals had meanings—the first row signified that the rod numeral was a constant and the other rows in the downward direction signified that the numerals were the coefficients of the unknown in increasing power.

From here, Zhu Shijie in his *Siyuan yujian* (Precious Mirror of the Four Elements), written in 1303, proceeded to express polynomial equations in two, three and four unknowns with rod numerals. For instance, the equation in two unknowns

$$(-x-2)y^2 + (2x^2+2x)y + x^3 = 0$$

is expressed as follows:

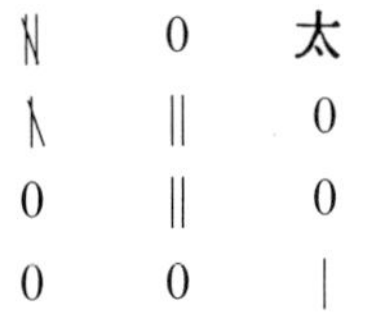

The slanting rod indicates that the numeral is negative. In the first column from the right, the character *tai* indicates the constant of the equation, which in the above case is zero. This column, which is similar to the notation of a polynomial in one unknown, represents $0 + 0x + 0x^2 + x^3$. The second column represents $0y + 2xy + 2x^2y + 0x^3y$, and the last column represents $-2y^2 - xy^2 + 0x^2y^2 + 0x^3y^2$.

Zhu Shijie gave examples to show how a set of simultaneous polynomial equations of varying degrees up to four unknowns could be reduced to an equation in one unknown, and thereby finding the solution. For example, he illustrated how the set of equations of the form

$$-2y + x + z = 0$$
$$-xy^2 + 4y - x^2 + 2x + xz + 4z = 0$$
$$y^2 + x^2 - z^2 = 0$$
$$2y + 2x - u = 0,$$

was reduced to the following single equation in one unknown:

$$4u^2 - 7u - 686 = 0.$$

In *Siyuan yujian*, Zhu Shijie excelled in another area in algebra: he gave correct formulae for the sums of higher order equal difference series. They are of two types which may be described in the following manner:

First type:

$$1+2+3+4+\cdots+n=\frac{1}{2!}n(n+1)$$

$$1+3+6+10+\cdots+\frac{1}{2!}n(n+1)$$
$$=\frac{1}{3!}n(n+1)(n+2)$$

$$1+4+10+20+\cdots+\frac{1}{3!}n(n+1)(n+2)$$
$$=\frac{1}{4!}n(n+1)(n+2)(n+3)$$

$$1+5+15+35+\cdots+\frac{1}{4!}n(n+1)(n+2)(n+3)$$
$$=\frac{1}{5!}n(n+1)(n+2)(n+3)(n+4)$$

$$1+6+21+56+\cdots$$
$$+\frac{1}{5!}n(n+1)(n+2)(n+3)(n+4)$$
$$=\frac{1}{6!}n(n+1)(n+2)(n+3)(n+4)(n+5).$$

Second type:

$$1.1+2.2+3.3+\cdots+n.n=\frac{1}{3!}n(n+1)(2n+1)$$

$$1.1+2.3+3.6+\cdots+n.\frac{1}{2!}n(n+1)$$
$$=\frac{1}{4!}n(n+1)(n+2)(3n+1)$$

$$1.1+2.4+3.10+\cdots+n.\frac{1}{3!}n(n+1)(n+2)$$
$$=\frac{1}{5!}n(n+1)(n+2)(n+3)(4n+1).$$

It was Shen Guo (1032–1095) who initiated the study of this type of series and he was followed by Yang Hui and Zhu Shijie — their basic technique was related to the piling of stacks.

The foundation of algebra, as we know it today, and its impetus for growth arose from the successful development of arithmetic based on the Hindu–Arabic numeral system. In the earlier civilizations, there were various beginnings of algebra which also depicted the struggle to find expressions for arithmetical operations and geometrical relationships. The development of algebra in China has proved to be unique and significant with its growth being maintained continuously until the Ming dynasty (1368–1644).

The essential ingredient that fostered the growth of algebra in traditional China was the rod numeral system. What was extraordinary about it was its notation: the position of each digit of a numeral represented the place value of that digit, such as units, tens, hundreds, and so forth. This notation of a number freed the mind of unnecessary work and enabled arithmetic to be developed to the fullest. The same kind of thinking in the use of the positions of rod numerals to represent concepts or things made possible the subsequent evolution of algebra.

Though the Chinese used rods to develop algebra, the results obtained manifested numerous similarities with our early algebra. In the solution of a set of simultaneous linear equations, the ancient Chinese invented the matrix notation and the *fang cheng* method of elimination. About one thousand five hundred years later, Seki Kowa and Leibniz initiated the study of determinants — Seki Kowa knew of the *fang cheng* method.

In searching for a general solution of a polynomial equation, Jia Xian drew attention to a triangular array of numbers which we now call Pascal's triangle. The method that the Chinese used to solve the polynomial equation was rediscovered by Horner half a century later. The concise notation of expressing the concept of a polynomial equation led Zhu Shijie to invent a notation to express a set of polynomial equations up to four unknowns. He gave examples to show how to solve them. In eighteenth century Europe, it was Étienne Bezout who initiated the study of solving a pair of polynomial equations in two unknowns. Zhu Shijie's formulae for the series of higher order equal difference series also showed that he was a few centuries ahead of his counterparts in Europe.

It is ironical that the use of rods which enabled the expansion and sustenance of algebra in China for over one and a half millennia, was also the reason for its decline. The rods were used not only for the development of mathematics but also for computation. By the Song dynasty (960–1279), a faster paced society could not tolerate the time required for manipulating the rods. The demand for quicker computation led to the invention of the abacus. However, the abacus was only suitable for swift calculations and had neither the potential to foster the growth of mathematics nor the capacity to allow for the conceptual retention of what had already been developed in mathematics. The replacement of the rods by the abacus signalled the demise of traditional mathematics.

Since ancient times, the Chinese mathematicians had been using a base ten place value numeral concept in the rod numerals, and so it would not have been difficult for them to adopt this concept in a written form. If they had made such an adoption during the switch to calculation with the abacus, there would have been a smooth transference of mathematical concepts from the rod medium to the written medium. However, such an adoption was only made when western mathematics entered China beginning with the arrival of Matteo Ricci in 1582. The consolation from this erroneous turn of events was that during the sixteenth and seventh centuries many Chinese were still knowledgeable in traditional mathematics, and they helped greatly to lighten what would have been a tremendous upheaval in the change to the new mathematics.

LAM LAY YONG

REFERENCES

Hoe, J. *Les systèmes d'équations polynômes dans le Siyuan yujian (1303)*. France: Collège de France, Institut des Hautes Études Chinoises, 1977.

Lam Lay Yong. *A Critical Study of the Yang Hui Suan Fa. A Thirteenth-century Chinese Mathematical Treatise*. Singapore: Singapore University Press, 1977.

Lam Lay Yong. "The Chinese Connection between the Pascal Triangle and the Solution of Numerical Equations of any Degree." *Historia Mathematica* 7: 407–424, 1980.

Lam Lay Yong. "Chinese Polynomial Equations in the Thirteenth Century." In *Explorations in the History of Science and Technology in China*. Ed. Li Guohao et al. Shanghai: Shanghai Chinese Classics Publishing House, 1982, pp. 231–272.

Lam Lay Yong. "Jiu Zhang Suanshu (Nine Chapters on the Mathematical Art: An Overview." *Archive for History of Exact Sciences* 47(1): 1–51, 1994.

Lam Lay Yong and Ang Tian Se. *Fleeting Footsteps. Tracing the Conception of Arithmetic and Algebra in Ancient China*. Singapore: World Scientific, 1992.

Libbrecht, Ulrich. *Chinese Mathematics in the Thirteenth Century. The Shu-shu chiu-chang of Ch'in Chiu-shao*. Cambridge, Massachusetts: MIT Press, 1973.

Li Yan and Du Shiran. *Chinese Mathematics. A Concise History*. Trans. J.N. Crossley and A.W.C. Lun. Oxford: Clarendon Press, 1987.

Martzloff, Jean-Claude. *Histoire des Mathématiques Chinoises*. Paris: Masson, 1988.

Needham, Joseph. *Science and Civilisation in China*. Vol. 3: *Mathematics and the Sciences of the Heavens and the Earth*. Cambridge: Cambridge University Press, 1959.

See also: Computation: Chinese Counting Rods – Liu Hui and the *Jiuzhang suanshu* – Wang Xiaotong – Yang Hui – Liu Yi – Jia Xian – Qin Jinshao – Li Ye – Zhu Shijie – Shen Guo – Seki Kowa – Abacus – al-Khwārizmī

ALGEBRA IN INDIA: *BĪJAGAṆITA* *Bījagaṇita*, which literally means "mathematics (*gaṇita*) by means of seeds (*bīja*)", is the name of one of the two main fields of medieval Indian mathematics, the other being *pāṭīgaṇita* or "mathematics by means of algorithms". *Bījagaṇita* is so called because it employs algebraic equations (*samīkaraṇa*) which are compared to seeds (*bīja*) of plants since they have the potentiality to generate solutions to mathematical problems. *Bījagaṇita* deals with unknown numbers expressed by symbols. It is therefore also called *avyaktagaṇita* or "mathematics of invisible (or unknown) [numbers]". Algebraic analyses are also employed for generating algorithms for many types of mathematical problems, and the algorithms obtained are included in a book of *pāṭī*. *Bījagaṇita* therefore also means "mathematics as a seed [that generates *pāṭī* (algorithms)]".

Extant works in *bījagaṇita* include chapter 18 (*kuṭṭaka* only) of Āryabhaṭa's *Mahāsiddhānta* (ca. AD 950 or 1500), chapter 14 (*avyaktagaṇita*) of Śrīpati's *Siddhāntaśekhara* (ca. AD 1050), Bhāskara's *Bījagaṇita* (AD 1150), Nārāyaṇa's *Bījagaṇitāvataṃsa* (before AD 1356, incomplete), and Jñānarāja's *Bījādhyāya* (ca. AD 1500). Śrīdhara's work (ca. AD 750), from which Bhāskara quotes a verse for the solution of quadratic equations, is lost. Chapter 18 (*kuṭṭaka*) of Brahmagupta's *Brāhmasphuṭasiddhānta* (AD 628) has many topics in common with later works of *bījagaṇita*, but the arrangement of its contents is not so systematic as that of the later works, and an unusual stress is placed on *kuṭṭaka* as the title of the chapter suggests. *Kuṭṭaka* (lit. pulverizer) is a solution to the linear indeterminate equation: $y = (ax + c)/b$.

The symbols used for unknown numbers in *bījagaṇita* are the initial letters (syllables) of the word *yāvattāvat* (as much as) and of the color names such as *kālaka* (black), *nīlaka* (blue), *pīta* (yellow), etc. The use of the color names may be related to Āryabhaṭa's *gulikā* (see below). Powers of an unknown number are expressed by combination of the initials of the words *varga* (square), *ghana* (cube), and *ghāta* (product). A coefficient is placed next (right) to the symbol(s) to be affected by it, and the two sides of an equation are placed one below the other. A dot (or a small circle) is placed above negative numbers. Thus, for example, our equation, $5x^5 - 4x^4 + 3x^3 - 2x^2 + x = x^2 + 1$, would be expressed as:

yāvaghaghā 5 *yāvava* $\dot{4}$ *yāgha* 3 *yāva* $\dot{2}$ *yā* 1 *rū* 0
yāvaghaghā 0 *yāvava* 0 *yāgha* 0 *yāva* 1 *yā* 0 *rū* 1

where *rū* is an abbreviation of *rūpa* meaning an integer or an absolute term. The product of two (or more) different unknowns is indicated by the initial letter of the word *bhāvita* (produced): e.g. *yākābhā* 3 for $3xy$.

These tools for algebra had been fully developed by the twelfth century, when Bhāskara wrote a book entitled *Bījagaṇita* (AD 1150), the main topics of which are "four seeds" (*bījacatuṣṭaya*), namely, (1) *ekavarṇasamīkaraṇa* or equations with one color (i.e. in one unknown), (2) *madhyamāharaṇa* or elimination of the middle term (solution of quadratic equations), (3) *anekavarṇasamīkaraṇa* or equations with more than one color, and (4) *bhāvitakasamīkaraṇa* or equations with "the product" (i.e. of the type $ax+by+c = dxy$).

At least part of this algebraic notation was known to Brahmagupta. He uses the words *avyakta* (invisible) and *varṇa* (color) for denoting unknown numbers, when he gives his rules concerning the same four seeds as Bhāskara's, in chapter 18 (*kuṭṭaka*) of his *Brāhmasphuṭasiddhānta*. The details of Brahmagupta's algebraic notation are, however, not known to us.

Bhāskara, a contemporary of Brahmagupta, did know the word *yāvattāvat* meaning an unknown number, but it is not certain if he used it in equations, because he expresses the equation, $7x + 7 = 2x + 12$, without the symbol *yā* as:

7	7
2	12

in his commentary (AD 629) on the *Āryabhaṭīya*. In the same work he refers to four seeds which are said to generate "mathematics of practical problems" (*vyavahāragaṇita*) having eightfold of names beginning with "mixture", but what kinds of seeds he mentioned by the names *yāvattāvat*, *vargāvarga* (square?), *ghanāghana* (cube?), and *viṣama* (odd), are not exactly known. Similar terms (*yāvattāvat, varga, ghana,* and *vargavarga*) occur in a list of ten mathematical topics given in a Jaina canon, *Sthānāṅga* (Sūtra 747), which is ascribed to the third century BC.

Āryabhaṭa used the term *gulikā* (a bead) for an unknown number when he gave his rule for linear equations of the type $ax + b = cx + d$ in his *Āryabhaṭīya* (AD 499). All the equations to which he gave solutions (including *kuṭṭaka*) are linear, although his rules for the interest and for the period of an arithmetical progression presuppose the solution of quadratic equations.

Brahmagupta gave many theorems for *vargaprakṛti* (lit. square nature), that is, the indeterminate equation of the second degree: $Px^2+t = y^2$, but it is Jayadeva (the eleventh century or before) that gave a complete solution for the case $t = 1$ (the so-called Pell's equation).

Bījagaṇita reached its culmination in the twelfth century, when Bhāskara gave solutions to various types of equations of higher degrees by means of *kuṭṭaka* and *vargaprakṛti*. After him significant developments in the field of *bījagaṇita* are not known.

TAKAO HAYASHI

REFERENCES

Bag, A.K. *Mathematics in Ancient and Medieval India*. Varanasi: Chaukhambha Orientalia, 1979.

Colebrooke, H.T. *Algebra with Arithmetic and Mensuration from the Sanscrit of Brahmegupta and Bháscara*. London: Murray, 1817. Reprinted, Wiesbaden: Dr. Martin Söndig oHG, 1973.

Datta, B. and Singh, A.N. *History of Hindu Mathematics*. 2 vols. Lahore: Motilal, 1935/38. Reprinted in one vol., Bombay: Asia Publishing House, 1962.

Ganguli, S. "Indian Contribution to the Theory of Indeterminate Equations of the First Degree." *Journal of the Indian Mathematical Society, Notes and Questions* 19: 110–120, 129–142, 153–168, 1931/32.

Lal, R. "Integral Solutions of the Equation $Nx^2+1 = y^2$ in Ancient Indian Mathematics (*Cakravāla* or the Cyclic Method)." *Gaṇita Bhāratī* 15: 41–54, 1993.

Selenius, C.-O. "Rationale of the Cakravāla Process of Jayadeva and Bhāskara II." *Historia Mathematica* 2: 167–184, 1975.

Sen, S.N. "Mathematics." In *A Concise History of Science in India*. Ed. D.M. Bose et al. New Delhi: Indian National Science Academy, 1971, pp. 136–212.

Shukla, K.S. and Sarma, K.V., eds. and trans. *Āryabhaṭīya of Āryabhaṭa*. New Delhi: Indian National Science Academy, 1976.

Sinha, K.N. "Algebra of Śrīpati: An Eleventh Century Indian Mathematician." *Gaṇita Bhāratī* 8: 27–34, 1986.

Srinivasiengar, C.N. *The History of Ancient Indian Mathematics*. Calcutta: The World Press, 1967.

See also: Arithmetic in India: *Pāṭīgaṇita* – Āryabhaṭa – Śrīpati – Bhāskara – Nārāyaṇa – Brahmagupta – Śrīdhara – Jayadeva

ALGEBRA IN ISLAMIC MATHEMATICS The word algebra is derived from the Arabic *al-jabr*, a term used by its founder, Muḥammad ibn Mūsā al-Khwārizmī, in the title of his book written in the ninth century, *al-Jabr waʾl-muqābalah* (The Science of Equations and Balancing). Algebra is also known as "the science of solving the unknowns in equations."

The simplest equation with one unknown is of the form $ax = b$, with a and b as constants. x here is called *al-jadhr* of the equation. Al-Khwārizmī enumerated six standard second degree equations in his *al-Jabr wa 'l-muqābalah*:

$$ax^2 = bx \qquad ax^2 = b \qquad ax = b$$
$$ax^2 + bx = c \qquad ax^2 + c = bx \qquad ax = bx + c.$$

Also, he provided solutions to these equations using algebraic and geometrical justifications.

The main aim of al-Khwārizmī's algebra was to provide the Muslim community with the necessary arithmetical knowledge essential in their daily calculation needs, such as in matters pertaining to heritage and legacy, transaction, sharing and partnership, loss and profit, irrigation and land-acreage, and geometrical problems. Al-Khwārizmī devoted about half of his *al-Jabr wa 'l-muqābalah* to such problems.

Abū Kāmil Shujāʾ ibn Aslam of Egypt (AD 850–930) gave his treatise on algebra the name *al-Jabr wa 'l-muqābalah*, the same title as al-Khwārizmī's. This treatise gives commentaries on al-Khwārizmī's six standard quadratic equations using Euclid's lemmas in geometry to justify the existence of two roots for a general quadratic equation. In the twelfth century, Abū'l-Fath ʿUmar bin al-Khayyām listed thirty-nine standard cubic equations in his *Risāla fi 'l-barāhīn ʿalā masāʾil al-jabr wa 'l muqābalah,* and solved them by intersecting suitable conic-sections (circle or semi-circle, parabola, hyperbola, and ellipse) using Apollonius's theory of conics. The solutions to the equations were represented by intersections of the curves, but he failed to identify the exact numerical solutions. Many attempts had been made towards finding the geometrical solutions of cubic equations before ʿUmar al-Khayyām mentioned them in his book, such as those by al-Māhānī and Abu'l-Jūd.

Also in the twelfth century, Sharaf al-Dīn al-Ṭūsī examined the cubic equation species classified by ʿUmar al-Khayyām and provided their solutions through a systematic study of the minimum and maximum values of their associated functions. He gave the number of real solutions of a cubic equation in terms of its coefficients.

In the third portion of his *Kitāb fi'l-al-jabr wa'l-muqābalah*, Abū Kāmil discussed indeterminate problems (*mu'ādalah siālah*) of the second degree. Some of these were of Greek origin and could be found in the *Ṣinā'ah al-jabr* by Diophantus, the translation of *Arithmetica* by Qusṭā ibn Lūqā. The problems were then cited by Leonardo Fibonacci in his book *Liber abaci*. Abū Kāmil concentrated on enumerating the possible solutions of simultaneous equations in his *Ṭarāʾif al-Ḥisāb*. They were based on the problem of determining the number of birds that could be purchased with 100 dirhams. Somebody is to buy one hundred fowls, given that, for example, a rooster costs 5 units, a hen 3, and chicks are sold three for one unit. Problems of this type gave rise to a system of equations of the form:

$$x + y + z + u + v = 100 = ax + bx + cz + du + ev.$$

In the case of $a = 2$, $b = \frac{1}{2}$, $c = \frac{1}{3}$, $d = \frac{1}{4}$, $e = 1$, Abū Kāmil gave 2696 possible integer solutions. The analysis marks the birth of a field in algebra which is known today as linear algebra.

Extraction of square and cubic roots became an important subject of discussion in arithmetic and algebra books, or *Ḥisāb al-Hindi* during the heyday of the Muslim mathematicians. The rule for the extraction of roots then was based on binomial expansion of the form $(a + b)^n$. Al-Ṭūsī listed the coefficients in the expansion of $(a + l)^n - a^n$ for some n, in his *Jawāmiʿ* and arranged them in a triangular form which he called *manāzil al-ʿadad*. This triangular arrangement came to be known in the West as Pascal's triangle, after Blaise Pascal, the famous French mathematician who published his *Traité du triangle arithmétique* in 1665. Such an arrangement was also drawn up by al-Karajī in one of his books. This was further mentioned by Samuʿīl (or Samauʿal) ibn ʿAbbas, also called al-Maghribī, (eleventh century) in his *al-Bāhir fi'l-jabr*. ʿUmar al-Khayyām did write a book on the extraction of cubic and fourth roots, but the book is assumed to be lost. The extraction of the fifth root was carried out by al-Kāshī in the fifteenth century in his *Miftāḥ al-Ḥisāb*. He gave a numerical example of extracting the fifth root of 44,240,899,506,197 (order of trillions).

In the expansion of an algebraic term raised to a certain power, the concept of the negative number is extremely important. Muslim mathematicians made a substantial contribution to the development of this concept. Abū'l-Wafā al-Būzjānī in his *Ma Yaḥtāj ilayh al-kuttāb wa'l-ʿummāl min ʿilm al-ḥisāb* considered debts as negative numbers. For example, the calculation for 35 + (–20) = 15 was written as 35 + *dain* 20 = 15 – (*dain* = debt). Abū Kāmil, as a commentator on the *al-Jabr wa'l-muqabalah* of al-Khwārizmī, explained the application of positive and negative signs for the purpose of expanding the multiplication $(a \pm b)(c \pm d)$. This resulted in his rules:

$$(+)(+) = + = (-)(-) \qquad (+)(-) = (-)(+) = -$$

These rules are embodied in his famous work *Kitāb al-jabr wa'l-muqābalah*. Al-Karajī showed clearer examples illustrating operations with negative quantities in his *al-Fakhrī*. These rules are implicitly used throughout the book:

$$a - (-b) = a + b \qquad (-a) + b = -(a - b)$$
$$(-a) - (-b) = -(a - b) \quad (-a) - b = -(a + b).$$

Samuīl and Ibn al-Bannāʾ al-Marrākushī then made some finer rules about calculations involving negative numbers in their works, *al-Bāhir fi'l-jabr* and *Kitāb al-jabr wa'l-muqābalah,* respectively.

The art of proving became an important part of mathematical science. The direct and proof by contradiction methods are two important tools in proving mathematical statements. In some cases, however, they fail to work, especially in proving formulae containing integral terms. In this case, the method of *istiqrāʾ*, or proof by induction, is an appropriate one to use. Al-Karajī wrote an article by that name to explain this method. Samuīl and Ibn al-Haytham used it to prove some formulae on infinite series. A good example of the employment of proof by contradiction was given by Abū Jaʿfar al-Khāzin to establish some properties of right-angled triangles. These can be found in the treatise *Risālah fi'l-muthallathāt al-qāʿimat al-zawāyā* or in the *Tadhkirat al-aḥbāb fī bayān al-tuḥābb* of Kamāl al-Dīn al-Fārisī. The contradiction method based on the logical property "if (statement *a* is true) then it implies (statement *b* is true)" is equivalent to "if (statement *b* is not true) then (statement *a* is not true)". The converse, however, is not always true. This type of reasoning is characteristic of discussions in *manṭiq* or logic.

To explain the method of *istiqrāʾ*, Samuʿīl proved the case $n = k$ through the assumption that the case $n = k - 1$ is true. For example, to prove $a^3b^3 = (ab)^3$, Samuʿīl started with the assumption of $a^2b^2 = (ab)^2$ [which had been proved before], then multiplied both sides by (ab) to obtain $(ab)(a^2b^2) = (ab)(ab)^2 = (ab)^3$. Using the proposition mentioned earlier, that is $(ab)(cd) = (ac)(bd)$, he obtained $(ab)(a^2b^2) = a^3b^3$. Although in this demonstration, Samuʿīl used particular numbers instead of a general k, he successfully showed the method of *istiqrāʾ* correctly as we understand it today. Some writers however, continue to attribute the method to Pascal or Bernoulli in the seventeenth century (Yadegari, 1979).

The inherent idea in the use of logarithm is to expedite the multiplication process by converting it into an addition one. This is done by employing the rules of exponent. Abūʾl-Ḥasan al-Nasawī wrote a book on the idea in Persian which was later translated into Arabic with the title *al-Muqniʿ fi'l-Ḥisāb al-Hindi*. Ibn Yūnus (eleventh century), a well-known Egyptian astronomer, discovered the role of the trigonometric relation

$$\cos(a)\cos(b) = \tfrac{1}{2}[\cos(a+b) + \cos(a-b)]$$

in transforming the process of multiplication into addition. For example, suppose one wishes to obtain the product of 35.84 and 54.46. Since $\cos(69°) = 0.3584$ and $\cos(57°) = 0.5446$, then $(35.84)(54.46) = 1951.5$ by using this identity. This formula had been proven earlier by Abū'l-Wafā al-Būzjānī (d. 998) in his commentary on the *al-Majesṭi* (*Almagest*) of Ptolemy.

The expression of a fraction in the decimal form, based on an extant manuscript, goes back to *Kitāb al-fuṣūl fī al-Ḥisāb al-Hindī* of Abu'l-Ḥasan Aḥmad al-Uqlīdisī. It was written in the year 341H (AD 952). Uqlīdisī operated on the number 19 by consecutive halvings. First, he obtained (using his symbol) 9′5 then 4′75, 2′375, 1′11875 and finally 0′59375 (some commas indicating separations of hierarchy were dropped in the manuscript). Subsequently, Samuʿīl al-Maghribī gave a clearer idea of the notion of decimal fractions in his *al-Qiwāmi fi'l-Ḥisāb al-Hindī*. This book was written in the year AD 1172. The quotient of 210 by 13 was expressed as follows:

Integer	Parts of 10	Parts of 100	Parts of 1000	Parts of 10000	Parts of 100000
16	1	5	3	8	4

The square root of 10 is expressed as 3.16227, a clear definition of a decimal fraction, as the number 16227 is considered part of 1,000,000, with 3 as a whole number. More precisely, 3.16227 means:

$$3 + \frac{1}{10} + \frac{6}{100} + \frac{2}{1000} + \frac{7}{100000}.$$

Jamshīd al-Kāshī (fourteenth century) expressed the decimal fraction in both al-Khwārizmī's and the astronomers' system in the article "*al-Risāla al-muḥīṭīyyah*". He gave the value of 2π as 6.2831853071795865 (in al-Khwārizmī's decimal system) or as 6 – 6, 16, 59, 28, 1, 34, 51, 46, 15, 50 (in the astronomers' sexagesimal system).

The method of writing numbers as decimal fractions later appeared in the West in Stevin's *de Theinde* or its French version, *La Disma*, in 1585. To indicate the integral and fractional portions of the number, Stevin employed a stroke to separate the two. He is considered the founder of the decimal fraction by some writers in the West.

Nicomachus in his *al-Madkhal ilā ʿilm al-ʿadad* (an Arabic translation of *Introductio*) gave the four first perfect numbers: 6, 28, 469 and 8128. Ismail al-Māridīnī (twelfth century) added some others to the list of perfect numbers: 6, 28, 496, 8128, 1130816, (2096128), 33550336, 8589869056, 137438691328, (35184367894528). Actually, the two numbers in brackets are not perfect. This mistake is due to the difficulty in determining the primality of a number.

One can observe from al-Māridīnī's list that it is hard to find an odd perfect number. The numbers seem to be even, and the first digit (remember that Arabs write from the right) of a perfect number obtained by the formula is always 6 or 8. Indeed, the perfect numbers described by the formula $(2^{n-1} - 1)2^n$ are always even, since they are the product of even and odd numbers. However, many think that odd perfect numbers do exist. Euler, centuries later (1849), in his paper in *Tractus de numererum doctrina*, described a necessary condition for the existence of an odd perfect number, but it was not a sufficient condition.

Other types of numbers that became the subject of scrutiny of Muslim mathematicians are the deficient and abundant numbers. The mathematicians supplied some criteria to identify these numbers:

- every odd number less that 945 is deficient;
- every even-times-even number has factors less than itself;
- the first abundant number is 12, and the first odd one in this class is 945; and
- if $2^n S$ is a perfect number, then $2^{n+1} S$ is an abundant number, and $2^{n-1} S$ is a deficient number.

Kamāl al-Dīn al-Fārisī (d. 1320) in his treatise, *Tadhkirah al-aḥbāb fī bayān al-tuḥābb* supplied the rule to find pairs of amicable numbers (*al-a'dād al-mutaḥābbah*) in a systematic way. Thābit ibn Qurra (836–901) developed the theory of amicable numbers and provided a technique to find such pairs in his *Risālah fi'l-a'dād al-mutaḥābah*. Al-Fārisī reached the same conclusions through somewhat different paths. He based his new technique on the systematic knowledge of the divisors of a composite number and their sum. A pair of amicable numbers is defined as a pair of numbers (a, b) with the properties that the sum of all possible proper divisors of a is equal to b and the sum of all possible proper divisors of b is equal to a.

Ibn Ṭāhir al-Baghdādī defined a new variety of numbers known as equivalent numbers or numbers of equal weight (*muta›ādilan*) in his *al-Takmila f i'l -Ḥisāb*. According to him:

> If we have a given number and wish to find two numbers, the parts of which make up this number, we reduce it by one and split the result into two prime numbers, then two others, and so on, as many times as we can. The product of each pair is a number equivalent to the given number. Thus if we are given 57, we split 56 into (3,53), (13,43), etc. The products of 53 by 3 and 4 by 13 are numbers such that the sum of the parts of each is 57. (Saidan, 1977)

Equivalent numbers had not been studied by al-Baghdādī's contemporaries, nor by mathematicians for a few generations after him, until the time of Muḥammad Bāqir al-Yazdī (seventeenth century). Al-Yazdī, in his *‹Uyūn al-Ḥisāb*, considered such numbers and chose evenly-even numbers to be decomposed as al-Baghdādī had done. In this way it was convenient for him to establish some properties of equivalent numbers, such as: if p and q are prime numbers, then p and q are of equal weight. It was felt that numbers of this kind needed more attention and to be examined further as they exhibited interesting unique behaviors.

The problem of finding the sides (x, y, z) of a right angle triangle was studied and addressed by many Muslim mathematicians, including Samu‹īl (twelfth century) and Abū Ja‹far al-Khāzin (tenth century). Al-Khwārizmī considered some basic problems related to a right-angled triangle. These problems became the source of his algebraic problems in his *Kitāb a-jabr wa'l-muqābala*. Samu‹īl in his *al-Bāhir fi'l-Ḥisāb* showed that any triple of the form

$$\left(a, \frac{(a^2 - b^2)}{2b}, \frac{(a^2 + b^2)}{2b}\right)$$

with a, b, c being appropriate positive integers, would describe right-angled triangles. Earlier, al-Khāzin had considered such problems in his *Risālah fī al-muthallathāt al-qā'imat al-zawāyā al-munṭaqat al-aḍlā‹*. He showed that it was not possible that any triple (x, y, z), with x and y being odd (or evenly even) could be the sides of a right-angled triangle with z as an integer. Al-Khāzin then used the results to study the problems of the form $x^m + y^n = z^p$ with m, n, and p some small positive integers. He left out some cases, however, such as $m = n = p = 3$ or 4, for such problems have no solutions. Problems of these types were examined centuries later by de Fermat (1736).

The problem of splitting a cube into three other cubes, that is to find the solution of the equation with three unknowns, $x^3 + y^3 + z^3 = n^3$, was discussed by Ibn Ṭāhir al-Baghdādī (tenth century) in *Takmila fi'l-Ḥisāb*. He gave the answer as

$$(x, y, z) = \left(\frac{n}{2}, \frac{2n}{3}, \frac{5n}{6}\right).$$

This problem then reappeared in the works of E. Barbareta in 1910.

MAT ROFA BIN ISMAIL
(*with the collaboration of*
OSMAN BAKAR *and*
KAMEL ARIFIN MOHD ATAN)

REFERENCES

‹Abd al-Raḥman, Ḥikmat Najīb. *Dirāsāt fī tārīkh al-‹ulūm ‹inda al-‹Arab*. Wizārat al-Ta‹līm al-‹Ālī wa-al Baḥth al ‹Ilmī, Jāmi‹at al-Mawṣil, 1977, p.140

Anbouba, A. "L'Algébra arabe aux IX et X siècles, Aperçu general." *Journal of History of Arabic Science* 2: 66–100, 1978.

Berggren, J.L. *Episodes in the Mathematics of Medieval Islam*. New York: Springer-Verlag, 1986.

Al-Daffa, A.A. and J. Stroyls. *Studies in the Exact Sciences in Medieval Islam*. New York: John Wiley & Sons, 1984.

Al-Daffa, A.A. and J. Stroyls. "Some Myths about Logarithms in Near Eastern Mathematics." In *Studies in the Exact Sciences in Medieval Islam*. New York: John Wiley & Sons, 1984, p. 27.

Fraenkel, Abraham Adolf. "Problems and Methods in Modern Mathematics." *Scripta Mathematica* 9: 162–168, 1943.

Kasir, Daoud S. *The Algebra of ‹Umar al-Khayyām*. Ph.D. Thesis, Columbia University, 1931.

McCarthy, Paul. "Odd Perfect Numbers." *Historia Mathematica* 23: 43–47, 1957.

Naini, Alireza Dja'fari. "A New Type of Numbers in a 17th Century Manuscript: al-Yazdi on Numbers of Equal Weight." *Journal of History of Arabic Science* 7: 125–139, 1983.

Nasr, Seyyed H. *Science and Civilization in Islam*. Cambridge, Massachusetts: Harvard University Press, 1968.

Picutti, E. "Pour l'histoire des sept premiers nombres parfaits." *Historia Mathematica* 16: 123–136, 1989.

Rashed, Roshdi. "Tārīkh al-jabr wa ikhtirā›al-kusūr al-‹usyuriyyah (The History of Algebra and the Discovery of Decimal Fractions)." *Proceedings of the First International Symposium for the History of Arabic Science*. Aleppo: Institute for the History of Arabic Science, 1977, pp. 169–186.

Rashed, Roshdi. *Oeuvres mathématiques*. Paris: Les belles lettres, 1985.

Saidan, A.S. *Tārīkh ‹ilm al-ḥisāb al-‹Arabi*. Amman: al-Lajnah al-Urdunīyah lil-Ta‹rīb wa-al Nashr wa-al Tarjamah, 1973–1984.

Saidan, A.S. "Number Theory and Series Summation in Two Arabic Texts." Proceedings of the First Symposium for the History of Arabic Science, University of Aleppo, 1977, vol. 2, p. 152.

Shawki, J. and al-Daffa, A.A. *Al- ‹Ulūm al-riādiyyāt fi'l-ḥaḍārah al-Islāmiyyah*. New York: John Wiley & Sons, 1985–1986.

Yadegari, Muhammad. "The Use of Mathematical Induction by Abū Kāmil Shujā›ibn Aslam (850–930)." *Isis* 69: 259–262, 1978.

Youschkevitch, A.A.P. *Geschichte der mathematik im mittelalter*. Leipzig: Teubner, 1964.

See also: al-Khwārizmī – Abū Kāmil Shujā› ibn Aslam – ‹Umar al-Khayyām – Sharaf al-Dīn al-Ṭūsī – Qusṭā ibn Lūqā – Samuīl ibn Abbas (al-Maghribī) – Abu'l Wafā – al-Karajī – Ibn al-Bannā›– Ibn Yūnus – al-Uqlīdisī – al-Kāshī – Kamāl al-Dīn al-Fārisī – Thābit ibn Qurra – Sexagesimal System

ALGEBRA, SURVEYORS' Around 1930 a mathematical technique very close to later second-degree algebra was discovered in Babylonian cuneiform tablets, most of them dating from the early second and a few from the late first millennium BC (the "Old Babylonian" and "Seleucid" periods, respectively). Although the texts did not tell, it was supposed that the technique was purely arithmetical, and that its "lengths", "widths" and "areas" were metaphors designating numerical unknowns and their products. The geometry of Euclid's *Elements* II was then believed to represent a Greek geometrical reinterpretation of the arithmetical results of the Babylonians, necessitated by the discovery of irrationality.

A more sophisticated analysis of the Old Babylonian texts shows that the arithmetical interpretation does not hold water. For instance, the texts distinguish sharply between two different concepts which had been understood as one and the same "addition", two different "subtractions", and no less than four different "multiplications". Instead, a non-metaphorical interpretation as "naïve cut-and-paste geometry" imposes itself. A problem where the sum of the area and the side of a square is said to be H is solved as in the illustration shown: The square itself represents the area. From one of its sides a "projection" *1* is drawn, which together with the unknown side contains a rectangle with an area equal to the side – the total area of the square itself and this rectangle is thus $\frac{3}{4}$. This "projection" is bisected, and the outer half is moved so that the two together contain a square of area equal to $\frac{1}{2} \cdot \frac{1}{2} = \frac{1}{4}$. This small square completes the gnomon into which the original area $\frac{3}{4}$ is transformed as a larger square of area $\frac{3}{4} + \frac{1}{4} = 1$. The side of this completed square will then be $\sqrt{1} = 1$. "Tearing out" that rectangular length $\frac{1}{2}$ which was moved around leaves $1 - \frac{1}{2} = \frac{1}{2}$ for the side.

With this technique, the Babylonian scribe school teachers solved problems of much higher complexity than the present one — in non-normalized problems (e.g. if $\frac{2}{3}$ of the area plus $\frac{1}{3}$ of the side equals $\frac{1}{3}$) a change of scale is introduced, but apart from that everything goes by cutting and pasting — the approach is always "naïve" (as opposed to "critical") in the sense that everything can be *seen* immediately to be true. No explicit argument proves that, for example, the two halves of the bisected rectangle really contain a square when arranged as in Figure 1.

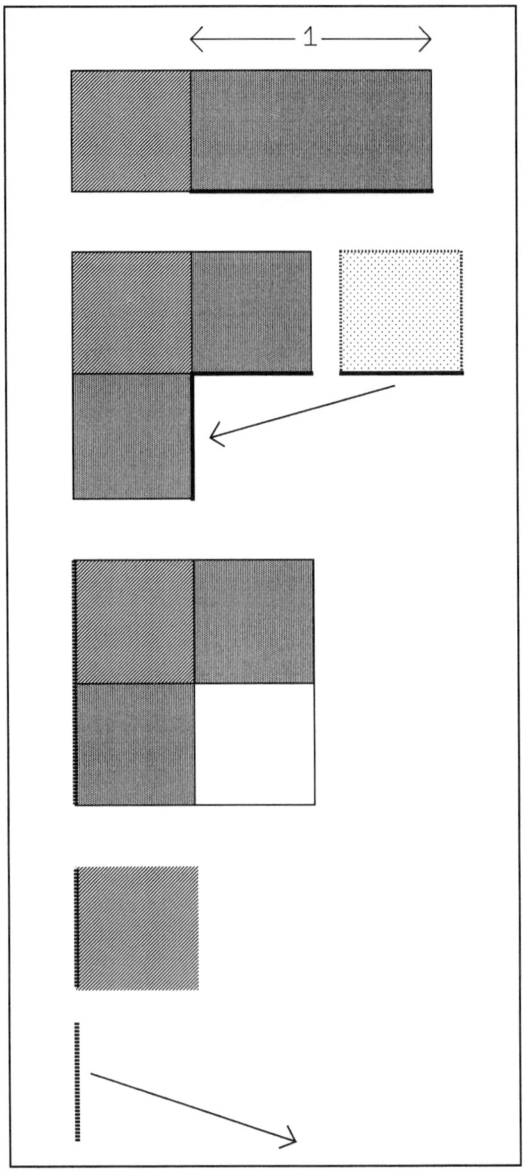

Figure 1: "The sum of the area and the side of a square equals $\frac{3}{4}$".

A few texts, apart from using the terms for "lengths", "widths", and "areas" of fields, hint in other ways at surveying practice. An important example is this problem: "Concerning a field: I have added the four fronts [sides] and the field [the area]". The formulation is unique in mentioning

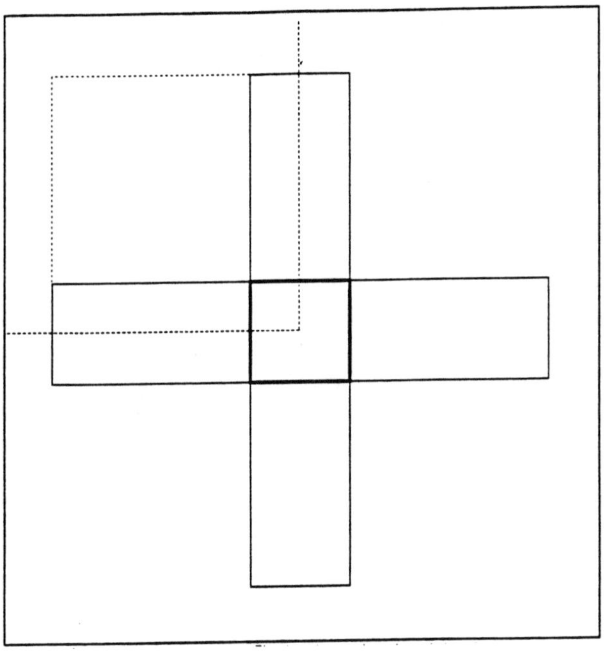

Figure 2: "The four fronts and the area". The dotted lines show the solution: one-fourth of the total area is taken and completed as a square.

the sides before the area, and so is the solution which refers to a configuration where a rectangle with length 1 is glued to each side of the field, see Figure 2.

Certain treatises from the Islamic Middle Ages reveal a close connection to this cut-and-paste tradition. The best known is al-Khwārizmī's *Algebra* (*Kitāb al-mukhtaṣar fī ḥisāb al-jabr wa'l muqābalah* – The Book of Summary Concerning the Process of Calculating, Compulsion, and Equation) from the early ninth century AD. The algebraic technique itself is arithmetical and thus at most obliquely if at all connected to the Babylonian tradition. However, in order to prove the correctness of his algorithms — first the one for "square and ten roots equal 39" — al-Khwārizmī makes use precisely of the diagrams of the figures shown, the second with completion in all four corners instead of quadrisection.

Less widely discussed is a *Book on Mensuration*, written by Abū Bakr. The text is known from a meticulous Latin translation made in the twelfth century by Gerard of Cremona, and contains in its first half a large number of problems similar to those known from the Babylonian texts, both in mathematical substance and method and in the very characteristic grammatical format. However, the differences which are also present and systematic reveal that the Babylonian scribe school algebra is not the source. Instead, both the scribe school and Abū Bakr appear to have drawn on a surveyors' tradition which had been present at least in Northern Iraq, and probably in a wider region, from the early second millennium BC onward.

This tradition nurtured and transmitted a stock of recreational problems with appurtenant techniques for their solution. Those which can be determined with some certainty were of the same quasi-algebraic nature as the Babylonian problems cited above.

If we designate by Q a square area and by s its side, by A a rectangular area and by l, w and d its sides and the diagonal, we may be fairly sure (from what is common to Abū Bakr and the scribe school) that the following problems were present (Greek letters stand for given numbers which cannot be safely determined because the Babylonian and the Medieval tradition give different values — question marks indicate doubt):

$s + Q = 110$	$Q - s = 90$
$4s - Q = \alpha(??)$	$A + (l \pm w) = \alpha,\ l \pm w = \beta$
$4s + Q = 140$	$Q - 4s = 60(?)$
$A = \alpha,\ l \pm w = \beta$	$A = \alpha, d = \beta$

For two squares Q_1 and Q_2, one of which was probably thought of as located concentrically within the other, the four problems $Q_1 + Q_2 = \alpha$, $s_1 \pm s_2 = \beta$ and $Q_1 - Q_2 = \alpha$, $s_1 \pm s_2 = \beta$ were dealt with. All problems are of the second degree (although $Q_1 - Q_2 = \alpha$, $s_1 \pm s_2 = \beta$ reduce trivially to the first degree, since $Q_1 - Q_2$ is easily seen in a diagram to equal $[s_1 + s_2] \times [s_1 - s_2]$).

Since Abū Bakr's treatise contains the problem $d - s = 4$, we may guess that the subscientific tradition solved this problem by equating the diagonal of the 10×10-square with 14. Abū Bakr gives the exact solution $s = 4 + \sqrt{32}$, whereas the Babylonians (who always worked backwards from known solutions in their "algebra") eliminated this "unscientific" problem in the same process as they transformed the restricted stock of surveyors' riddles into a genuine, systematic discipline and into something which can legitimately be regarded as an *algebra*.

The Seleucid cuneiform algebra texts have traditionally been understood as faithful continuations of the Old Babylonian tradition. Close scrutiny of the texts shows even this to be a half-truth. Indeed, the Old Babylonian mathematical tradition lost its higher, algebraic level when the scribe school system collapsed around 1600 BC, and only the directly applicable level survived in an environment whose professional pride was based on other aspects of its practice. The surveyors' tradition, however, survived, and appears to have supplied the material for an algebraic revival in the later first millennium. In the meantime, the surveyors had developed more sophisticated methods — e.g. calculating the heights of triangles and using this for area determination instead of restricting themselves to practically right triangles laid out in the terrain. The most important Seleucid algebraic text also reads as a catalog of new problems and methods for the treatment of rectangles, some of them quite refined (e.g. determining the sides from $l + d$ and $w + d$ or from A and $l + w + d$).

The surveyors' tradition influenced not only Late Babylonian scribal mathematics and Medieval Islamic mensuration but also ancient Greek geometry and arithmetic. What is reflected in *Elements* II is, indeed, not Babylonian "algebra" in general but very precisely *that part of Old Babylonian algebra which it shared with the surveyors' tradition* — whence one may conclude that the real inspiration was *not* Old Babylonian scribal mathematics — long since forgotten — but the still living stock of surveyors' riddles. Proposition 1, it turns out, justifies the geometrical addition of rectangles which have one side in common, whereas propositions 2 and 3 concern the special cases where sides are subtracted from or added to square areas. Propositions 4 and 7 are used, e.g. in two different but equivalent solutions to the problem of finding the sides of a rectangle from the area and the diagonal. Proposition 6 explains the solution of all problems $Q \pm \alpha s = \beta$ (including "the four sides and the area") and $A = \alpha,\ l - w = \beta$, while proposition 5 has a similar relation to rectangular problems $A = \alpha,\ l + w = \beta$ and to $\alpha s - Q = \beta$. Proposition 8, which is used nowhere else in the *Elements*, is associated with the concentric inscription of one square into another, and propositions 9 and 10 are connected to the solution of the problems $Q_1 + Q_2 = \alpha$, $s_1 \pm s_2 = \beta$.

What the Greek text does is not merely repeat the traditional solutions — in fact, what is presented are theorems, not problems to be solved. But the theorems are *critiques* of the traditional "naïve" procedures, showing that what is immediately "seen" is indeed correct and can be proved within the axiomatic framework. The proofs fall into two parts, the first of which establishes the equality of areas and that the quadrangles which are believed to be squares are indeed so. Then on firm ground, the second part goes through the traditional cut-and-paste procedure.

Propositions 11–13 are connected to matters found in Abū Bakr's treatise, and the textual evidence suggests that at least propositions 12–13 (the generalized Pythagorean theorem) were originally developed independently of the Greek theoreticians, as part of the "new" stage of the surveyors' geometry. Propositions 1–10, on the other hand, are completely untouched by the innovations. This critique thus seems to go back to a moment when the "new" development had not yet taken place, or not yet reached the ears of Greek geometers (we may point to the late fifth century BC, the epoch of Hippocrates of Chios and of Theodoros).

Book I of Diophantos' *Arithmetic*, which also contains undressed versions of favorite arithmetical recreational problems, embraces a few problems of the second degree — as it turns out, arithmeticized versions of $A = \alpha,\ l \pm w = \beta$ and $Q_1 + Q_2 = \alpha,\ s_1 \pm s_2 = \beta$ (all four belonging to the original surveyors' stock). As it can be seen from certain passages in Plato, Diophantos' work builds at least in its terminology upon a tradition reaching back to Greek calculators of the fifth century BC or earlier. The present simple problems can be assumed to belong to the early ingredients of this tradition, which then agrees with the chronological conclusions which could be drawn from *Elements* II.

A single Greco–Egyptian papyrus (probably second century AD) shows that the "new" diagonal-centered group of surveyors' problems circulated among the mathematical practitioners of the Greco–Roman orbit at this later moment, without being adopted into the corpus of scientific mathematics. In China, on the other hand, they turn up in the *Shushu jiuzhang* (Nine Chapters on Arithmetic) (one of them in a dress which unambiguously points back to Babylonia and thus shows the Chinese problems to be borrowed). In the later Chinese tradition, however, this interest disappears again.

"Naïve" geometry is also to be found in the Indian *Śulbasūtra* geometry (mid-first millennium BC), and a few commentaries to later Indian algebraic works might suggest a fundament in something similar to our surveyors' tradition. The evidence, however, is all too shaky to allow any conclusion. Apart from its role for the emergence of Babylonian "algebra" and for the Greek metric geometry as found in *Elements* II, the only literate mathematical culture where the anonymous surveyors gained real influence is thus that of medieval Islam. Its reflection in al-Khwārizmī was already mentioned, to which it may be added that al-Khwārizmī's borrowed geometric proofs were taken over by numerous algebraic authors until the mid-sixteenth century — one step in Cardano's solution of the third-degree equation makes use of a three-dimensional generalization. Within the mensuration tradition, the twelfth-century Abraham bar Ḥiyya (Savasorda) betrays familiarity with the surveyors' quasi-algebraic riddles, in a version which shows him to be independent of Abū Bakr.

As the technique was taken over by scientific mathematicians, the reason to uphold a distinct naïve-geometric tradition disappeared. Abū Bakr had already solved many of the problems in two ways, first by naïve geometry and then by *al-jabr* — the originally subscientific discipline committed to writing and made famous by al-Khwārizmī. Abraham bar Ḥiyya, for his part, refers to *Elements* II in his solutions. Nonetheless, the riddles themselves continued to be attractive. They are important in Leonardo Fibonacci's *Pratica geometrie* (1220), which draws on Abraham bar Ḥiyya, on Gerard of Cremona's translation of Abū Bakr, and on at least one other work belonging to the tradition. And they turn up again in Luca Pacioli's *Summa de arithmetica geometria Proportioni: et proportionalita* (1494), which still asks for the side when "the four sides and the area" equals 140, with the same anomalous order of the members as in the Bronze Age. The last shimmering of the tradition is in Pedro Nunez' *Libro de Algebra en Arithmetica y Geometria* (1567), after which every trace disappears, together with the knowledge that such a tradition had ever existed.

From the beginning some 4000 years ago, the surveyors' riddles were algebraic in the sense that the solutions follow steps which correspond to those of a modern solution by means of symbolic algebra. They were also algebraic in the sense that their method was *analytic:* the unknown side of the square is treated as if it were known, and submitted to such operations that at the end it is isolated and thus known — precisely as a modern x.

However, the surveyors' technique lacked two essential characteristics of modern algebra (characteristics which, by the way, are also present in Old Babylonian scribe school algebra). First, it was no general method for finding unknown quantities; the measurable lines and areas which it manipulated were never used as representatives of something else. It was, and remained until it was taken over and reshaped by the written traditions, a collection of sophisticated riddles with no practical purpose beyond itself. Second, it always dealt with the entities which are naturally present in the geometrical configurations of which it spoke: *the* area, *the* side or *the four* sides, etc. It is never, as we may find it even in those of the Old Babylonian scribal texts which otherwise come closest to the background tradition, "one half of the length and one third of the width" of a rectangle. This is also the reason that its problems are — with one very late and thus dubious exception — always normalized, and a supplementary reason that it did not develop into an all-purpose algebra. We may therefore, legitimately, speak of a surveyors' "algebra" — but not of a surveyors' algebra without quotes.

JENS HØYRUP

REFERENCES

Høyrup, Jens. "'The Four Sides and the Area'. Oblique Light on the Prehistory of Algebra." In *Vita Mathematica. Historical Research and Integration with Teaching.* Ed. R. Calinger. Washington, D.C.: The Mathematical Association of America, 1996.

Høyrup, Jens. "Algebra and Naive Geometry. An Investigation of Some Basic Aspects of Old Babylonian Mathematical Thought." *Altorientalische Forschungen* 17: 27–69, 262–354, 1990.

See also: Mathematics, Recreational – al-Khwārizmī – Abraham bar Ḥiyya – Surveying

***ALMAGEST*: ITS RECEPTION AND TRANSMISSION IN THE ISLAMIC WORLD** Around AD 150, Ptolemy wrote his great handbook of astronomy called *Mathematike Syntaxis* (The Mathematical Composition) in the original Greek. Because of its importance, it soon received wide attention throughout the Hellenistic world. Its fame seems to have radiated into the Middle East, because there are hints that the work was known, and perhaps even translated partially or completely, into Middle-Persian (Pahlavi) under the Sassanian ruler Shāhpur I (reigned AD 241–272). A second period of intensive contact of the Persians with Greek science was in the middle of the sixth century, after the closing of the Academy in Athens (AD 529), when several Greek scholars sought refuge in Persia. At this point Ptolemy's work may again have been brought to the attention of Persian scholars. However this was, the first knowledge of the Arabs (who conquered the Middle East and established the Islamic empire of the caliphs around the middle of the seventh century) about Ptolemy and his work betrays Persian influence. The Arabs first knew Ptolemy's astronomical handbook under the title *Kitāb al-majasṭī* (The Book Almagest), which evidently derives from a Greek superlative (*Megiste* [scil. *Syntaxis*], The Greatest Composition). Under this title the book was already known in Arabic–Islamic circles before the translations proper. It would seem that the Arabic spelling *al-majasṭī* is derived from a Middle-Persian form rather than from the Greek directly — a Middle-Persian spelling *mgstyk* is documented, and a form like this may have led to the Arabic spelling *al-majasṭī*. Later, in twelfth-century Spain, the Europeans translated Arabic scientific works into Latin and converted the Arabic *al-majasṭī* into Latin *almagesti*, and henceforth Ptolemy's work has been known in the West under this short title, Latinized from an Arabicized Greek word. It may be added that in the direct transmission, in Greek, the title *Megiste Syntaxis* has not yet been found — here we have only the forms *Mathematike Syntaxis* or *Megale Syntaxis* (The Great Composition).

Direct knowledge of the text of the *Almagest* in Arabic–Islamic science developed through a series of translations from Greek into Arabic. Before the Arabic translations, a translation was made into Syriac, the language common among Christian monks and scholars who later played the main role in the translation of Greek scientific works into Arabic. It is not known when and by whom this Syriac translation was made, but it is probable that the first translators of the work into Arabic knew and used the Syriac version. Still, in the twelfth century Ibn al-Ṣalāḥ had the Syriac version to hand. A first Arabic translation of the *Almagest* was made some time around AD 800 — its text is now lost, but it was often cited by Arabic–Islamic astronomers and bibliographers as "the first" or "the old" or "the ma'mūnian" translation, after the caliph al-Maʾmūn, who patronized translations and scientific work in general. Traces of it are found in the astronomical work of al-Battānī (d. 929) and in Ibn al-Ṣalāḥ. A second translation was made in 827–828 by al-Ḥajjāj ibn Yūsuf ibn Maṭar in cooperation with a Christian, Sergius the son of Elias. Of it two manuscripts have survived until our times, one complete and the other fragmentary. About fifty to sixty years later, a third translation was made by Isḥāq ibn Ḥunayn (830–910). It was soon afterwards revised by Thābit ibn Qurra (836–901) — of this revised version, nine manuscripts are still extant today.

The two versions of al-Ḥajjāj and of Isḥāq revised by Thābit were received in Muslim Spain — they were both used by Gerard of Cremona for his translation of the *Almagest* into Latin (Toledo, ca. 1150–1180). Gerard's Arabo–Latin version then remained the standard version of the *Almagest* in Europe until Renaissance times.

In the following centuries Arabic–Islamic astronomers not only studied the *Almagest* as their main source in astronomy, but they also wrote commentaries on the entire book or on problems of detail. Fuat Sezgin lists more than thirty commentaries, several of which were afterwards commented upon themselves. Among the numerous commentators two shall be named here. Ibn al-Ṣalāḥ (d. 1154) is important, because he had to hand five versions of the *Almagest* (the Syriac version, the "old" version, the version of al-Ḥajjāj, the original of Isḥāq's version, and Isḥāq's version as revised by Thābit) from all of which he tried — by comparing the texts and by observing the stars — to establish the best values in longitude and latitude for a selected number of stars from the star catalog in the *Almagest*. Another famous commentator was Naṣīr al-Dīn al-Ṭūsī (1201–1274) who wrote revisions of many translated Greek mathematical and astronomical texts, which included the *Almagest* — his *taḥrīr* (based on the version of Isḥāq as revised by Thābit) survives in numerous manuscripts, but is still unedited.

The last Oriental translation of the *Almagest*, from al-Ṭūsī's *Taḥrīr* into Sanskrit, was made in 1732, in Jaipur, India, by order of Maharaja Jai Singh II.

The title *al-majasṭī* / *Almagest* became so famous that other authors freely used the name for their own works, both in the East and in Europe, e.g. the *Kitāb al-majasṭī* (The Book Almagest) of the mathematician Abu'l-Wafāʾ al-Būzjānī (940–987 or 998); the *Almagestum novum* of the astronomer Giovanni Riccioli (Bologna, 1651), or even the *Almagestum Botanicum* of Leonard Plukenet (1696).

PAUL KUNITZSCH

REFERENCES

Almagest, Greek text: Ptolemaeus, *Opera*, vol. i, parts i–ii, ed. J.L. Heiberg. Leipzig: Teubner, 1898, 1903 – English translation: G.J. Toomer. *Ptolemy's Almagest*. London: Duckworth, 1984 – the star catalog alone, Arabic and Latin versions: P. Kunitzsch, ed. *Der Sternkatalog des Almagest. Die arabisch-mittelalterliche Tradition*, i–iii. Wiesbaden: Harrassowitz, 1986, 1990, 1991.

Ibn al-Ṣalāḥ. *Zur Kritik der Koordinatenüberlieferung im Sternkatalog des Almagest*. Ed. and trans. P. Kunitzsch. Göttingen: Vandenhoeck & Ruprecht, 1975.

Kunitzsch, P. "New Light on al-Battānī's *Zīj*." *Centaurus* 18:270-274, 1974 – reprinted in P. Kunitzsch, *The Arabs and the Stars*. Northampton: Variorum Reprints, 1989, item v.

Kunitzsch, P. *Der Almagest. Die Syntaxis Mathematica des Claudius Ptolemäus in arabisch-lateinischer Überlieferung*. Wiesbaden: Harrassowitz, 1974.

Pedersen, O. *A Survey of the Almagest*. Odense: Odense University Press, 1974.

Plukenet, L. *Almagestum Botanicum*. London: Sumptibus Autoris, 1696.

Riccioli, G.B. *Almagestum novum*. Bologna: Victorii Benetii, 1651.

Sezgin, F. *Geschichte des arabischen Schrifttums*. vi: *Astronomie bis ca. 430 H*. Leiden: Brill, 1978, pp. 83ff., 88–94.

Toomer, G.J. "Ptolemy." In *Dictionary of Scientific Biography*. vol. XI. New York: Scribner's, 1975, pp. 186–206.

See also: al-Maʾmūn – al-Battānī – Isḥāq ibn Ḥunayn – Thābit ibn Qurra – Naṣīr al-Dīn al-Ṭūsī – Abu'l Wafāʾ al-Būzajānī

ALPHABET As far as we know, literacy goes back to about 3500 BC. The earliest people to have left written symbols are the Sumerians, followed by the Egyptians. The earliest clay tablets unearthed in Mesopotamia are picture writings which have not yet been deciphered, but those written after 3200 BC are clearly Sumerian and their content is well known. Their subject matter includes groups of words, accounts of deeds of sale, and some fragments of early literature.

The Sumerians wrote primarily on clay tablets, producing wedge-shaped characters which became known as cuneiform script. It is quite probable that the idea of writing was introduced into Egypt from Babylon. Soon after the Sumerians, the Egyptians formulated their own system which consists of picture word-signs that we now call hieroglyphs.

By about 2000 BC other forms of writing began to appear in various parts of the Middle East. The Hittites, who lived in modern day Turkey, introduced a new form of hieroglyphs, consisting of some seventy signs, each of which stood for a simple syllable, and about one hundred word-signs. Similar forms of hieroglyphs, consisting of word-signs and syllable-signs, were also used in Crete, Cyprus, Lebanon, Palestine, and Syria. The early script uncovered in Crete is in the form of pictograms and was probably introduced into Crete from Egypt. Although it has not been completely deciphered, it is known that it contains the names of certain gods such as Zeus, Athena, and Poseidon.

All the word-sign methods of writing are complex and difficult to learn, and only professional scribes were able to read and write. To write a letter, keep an account, check a legal document, or read a will, an ordinary person had to rely on the services of trained scribes.

This monopoly on reading and writing persisted until the advent of the alphabet, whose invention is one of the most important and useful inventions of all time. The transformation from word or syllable writing to an alphabetic script is immense — it simplified the process of reading and writing to a degree that enabled a common person to master the art, thus freeing him from the need for professional scribes.

There has always been speculation as to the origin of the alphabet, beginning with the early Greek and Latin authors, who credited the invention of the alphabet to practically every near-eastern country. Diodorus Sciculus states: "The Syrians are the discoverers of the letters, and the Phoenicians having learned them from the Syrians and then passed them to the Greeks". Pliny, on the one hand, claims that: "the invention of the letter is a Phoenician feat", and speaks also of a Mesopotamian origin. Tacitus maintains that the alphabet is of Egyptian origin, and adds that "the Phoenicians took all the credit for what they received before passing it on to others". The various views of the classical authors are mere speculations that demonstrate their awareness of the problem and their cognizance of the importance of the discovery.

There is still controversy about the origin of the alphabet. There are those who still believe that the alphabet is of Mesopotamian origin — they base their opinion on the Ugaritic script which dates back to about 1400 or 1500 BC and which consists of thirty symbols representing thirty consonant letters written in cuneiform script.

The existence of an alphabet in a cuneiform script does not imply, in any way, that the idea of an alphabet originated in Mesopotamia, where there is no evidence that it ever existed. Of the half a million or so cuneiform tablets and fragments of tablets uncovered in Mesopotamia, there is not one that bears a sign of an alphabet.

By about 2000 BC the Mesopotamian cuneiform system of writing was used throughout the Near East and particularly in Canaan, where the rulers were using it in their correspondence even with the pharaohs of Egypt. When the Ugaritic scribes were writing their alphabet using the wedge-shaped cuneiform signs, the Mesopotamians were still content with their word-symbols.

There are also contemporary scholars who believe that the alphabet is of Egyptian origin — they base their views on the yet undeciphered Sinaitic inscriptions which date back to the first half of the second millennium BC The Canaanites, who were working the Egyptian turquoise mines in south western Sinai, left symbols scribbled on rocks and stones, and some scholars believe that the Canaanite workers learned their signs from their Egyptian lords. The Sinaitic pseudo-hieroglyphic and proto-Canaanitic scripts may at best be considered a first attempt at the complex process that led to the alphabetic script.

The champions of an Egyptian origin support their claims by the presence of simple symbols in the Egyptian hieroglyphs — by around 2000 BC. Egyptian scribes, like many others, were introducing symbols which stood for simple sounds or syllables and to a certain degree resemble the symbols of the alphabet. This pseudo-hieroglyph has so many signs that it is considered more syllabic than alphabetic, and instead of simplifying the Egyptian way of writing, it added to its complexity. The Egyptians could not relinquish their complex picture word-signs, and kept using them until about AD 500.

The uncovering of Minoan hieroglyphs and other Minoan pictograms in the beginning of this century suggested to the archaeologist Sir Arthur Evans that the Phoenician alphabet was itself derived from Crete. On the basis of Evans' statement, and in spite of the fact that the excavations on the island revealed no alphabetic signs, Glasgow wrote: "We now know, for instance, that the art of writing came from Crete, Phoenicia being the medium."

Many of the supporters of non-Phoenician origins point out that some of the sounds assigned to the signs that make up the alphabet do not correspond to the Canaanite names represented by the signs. As a matter of fact a few of the signs do not correspond to any known objects, but as we shall see, many of them do. The variations may be attributed to the whim of a scribe or scribes, who forsook fidelity in favor of beauty or simplicity. The alphabet evolved for several centuries before reaching its known Phoenician form.

The most probable origin of the invention of the alphabet is the land of Canaan. It seems that a scribe in Canaan conceived of the idea that a language may be written without the numerous signs used in Babylonian cuneiform or Egyptian hieroglyphs. It took a great deal of effort and insight to realize that complex words may be broken down into simple sounds, and that it does not take many of these simple sounds to write an entire language. This is tantamount to saying that the large number of words that form a language are made up of various combinations of a relatively small number of simple sounds.

Each one of the simple sounds was represented by the picture of a familiar object whose Canaanite name corresponds to the sound. Thus *beit* (house) stood for the *b* sound and *daleth* (door) stood for the *d* sound — the pictures stood for the simple sounds only and not for the word-sign. In a similar fashion all the consonants that form the language were represented. For reasons yet unknown, the vowels were not considered important enough to play a role in this scheme. This is why it is rather difficult to read modern day Arabic and Hebrew scripts, where only the consonants are written.

In addition to the script uncovered in the Sinai desert, there are those unearthed in Byblos (Jbayl) and ascribed to the middle of the second millennium BC. Other proto-Canaanite inscriptions were found in various parts of the Near East, such as Beit Shamesh, Gezer, Lakish, etc. The Gezer calendar is relatively recent — it dates to ca. 950 BC. The alphabetic signs found in various localities are different — they do not resemble one another. This leads to the speculation that once the often competing Canaanite clans knew of the magnificent invention, each group began to formulate its own symbols. Thus began the early stages of the alphabet of which several examples have been recovered, but many aspects of its early development are lost.

Except for the Ugaritic alphabet, which was written on clay tablets, the few early alphabetic signs found in the land of Canaan were scratched on stones or rocks. Toward the middle of the second millennium BC Canaan was under the influence of Egypt, and the Canaanite scribes were writing on Egyptian papyrus. Unfortunately, in the relatively damp soil of Canaan, papyrus did not prove to be a durable material, and the early development of the alphabet written on it is lost — the real origins may never be known.

The earliest known "alphabetic" signs are the proto-Canaanitic scripts found in the Sinai, and they may be considered as first attempts in a long process that led to the alphabet. However, on the whole, they do not suggest true alphabetic designations as a large number of the Sinaitic signs are syllabic rather than alphabetic. Although there are a few similarities between the Sinaitic and the Phoenician signs, many are very different, indicating that radical changes had taken place.

There are indications that the alphabet in its entirety, but not the alphabetic signs, could have been invented around the fourteenth or fifteenth century BC at Ugarit. The Ugaritic literature, written on clay tablets, is the earliest, most comprehensive alphabetic script that has survived. Furthermore, the order of the letters of the alphabet, at least the first few signs, support this possibility — the head of the Ugaritic pantheon is El, whose epithet is *thor* (bull). It is most appropriate to put the head of the pantheon at the head of the alphabet, hence the first letter, *aleph*, meaning bull or ox. The myths of Ugarit often speak of the house of god, they frequently mention the adobe of El, and the construction of a house for Baal receives lengthy, detailed descriptions. Thus the second letter of the alphabet, b, is represented by *beit* (house); and what is a house without a *daleth*, thus *d* is the fourth letter of the alphabet, followed by *hah* (window). There are several other such conjugate pairs in the Phoenician alphabet: *yad* and *kaff* (hand and palm), *myi* and *nahir* (water and river), and *rosh* and *shin* (head and tooth).

The earliest proto-Canaanitic symbols found in Phoenicia belong to the fourteenth century, and the first inscriptions of the Phoenician alphabet, which were left on arrowheads in Jbayl, belong to the twelfth century BC. It is quite probable that the roots of the Phoenician alphabet go back to the Ugaritic alphabet and to the Sinaitic proto-Canaanitic scripts. This hypothesis is strengthened by the few similarities between the Sinaitic and the early Phoenician signs and by the fact that the letters in the Ugaritic and the Phoenician alphabets are listed in somewhat corresponding orders. If this is true, the credit for inventing the alphabet, as such, would go to Ugarit, and the Phoenicians would get the credit for putting the alphabet into its basic easy-to-reproduce form, replacing the Sinaitic syllabic and the Ugaritic cumbersome cuneiform symbols with simple signs. There are indications that the scribes of Ugarit, in their later writings, wrote from right to left and dropped a few of their consonants symbols. The Phoenicians continued the process of converging sounds to accommodate sound variations in their standard language, reducing the total number from thirty characters to twenty-two.

Most recent authors attribute the origin of the alphabet to Phoenicia and more precisely to Jbayl. This is a definite possibility — the invention of the forms of the letters may have taken place in Jbayl. To compare the three scripts, let us reproduce the word 'MLK' meaning king and written without vowels:

Ugaritic  ,

proto-Canaanitic  ,

Phoenician 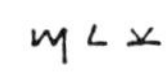 .

Note the similarity between the Phoenician and the English letters. The simplicity of the Phoenician signs rendered them more accessible than all others, and by the end of the second millennium, the Phoenician alphabet became firmly established in the entire land of Canaan.

If doubt persists as to the origin of the invention of the alphabet, it is absolutely clear that the Phoenicians taught the

Table 1: The development of the alphabet

Caananitic (ca. 1500 BC)	Phoenician (ca. 1200 BC)	Picture represents	Meaning	Greek character	Greek name
		aleph	ox	α	alpha
		beit	house	β	beta
		gammaʻ	sickle	γ	gamma
		daleth	door	δ	delta
		hah	window	ϵ	epsilon
		waw	hook	ζ	zeta
		Zayn	–	η	eta
		ḥeth	–	θ	theta
		ṭeh	–	ι	iota
		yad	hand	κ	kappa
		kaff	palm	λ	lambda
		lamed	staff	μ	mu
		myi *or* mem	water	ν	nu
		nahir	river	ξ	xi
		samkeh	fish	o	omicron
		ʻayn	eye	π	pi
		feh or peh	–	ρ	rho
		ṣad	–	σ	sigma
		quof	–	τ	tau
		rosh	head	υ	upsilon
		shin	tooth	ϕ	phi
		tā	cross	χ	chi
				ψ	psi
				ω	omega

letters to the Greeks and then to the rest of the known world. Some authors recognize the importance of this step and do not speak so much about the invention as about the diffusion of the alphabet. Herodotus is one such author and in one often quoted passage, he wrote, "These Phoenicians who came with Cadmus . . . among other kinds of learning, brought into Hellas the alphabet, which had hitherto been unknown, as I think, to the Greeks". The vowels, sometimes erroneously referred to as a Greek contribution, were arrived at either by sheer misunderstanding or because some of the Phoenician consonants were not needed in the Greek language, and the Greeks themselves were having difficulty pronouncing them. Thus the Phoenician aspirate "hah" became a short "e" or epsilon in Greek, the aspirate "heth" became a long "e" or eta, the semi-consonantal "yad" or "yod" became an "i" or iota, and the throaty " 'ayn" became an "o" or omicron.

This alphabet soon spread into the Mediterranean region, or wherever Phoenician and later Greek trading posts were established. This helped in keeping records of commercial transactions and simplified the work of the Phoenicians, the traders of the ancient world.

SEMA'AN I. SALEM

REFERENCES

Abbudi, Henry S. *Encyclopedia of Semitic Civilizations* (in Arabic). Tripoli: Lebanon: Jarrus Press, 1991.

Alexander, Pat, ed. *The Lion Encyclopedia of the Bible.* Tring, England: Lion Publishing Corp., 1986.

Moscati, Sabatino *The World of the Phoenicians*. Trans. Alastair Hamilton. New York: Frederick A. Praeger, 1968.

See also: Cuneiform

AMERICAS: NATIVE AMERICAN SCIENCE It is commonly accepted practice to refer to the peoples and cultures established in the territory now called the Americas as Native Americans. When the expedition of Christopher Columbus reached the Caribbean Islands in 1492, they called the inhabitants "Indians". This is because of the Spaniards' mistaken view that they had arrived in Asia. Although Columbus himself never realized that he had reached another continent and kept referring to the lands as "Cipango", soon the conquerors realized that they were indeed in a continent as far from Asia as from Europe and dealing with a completely different culture. However, the denomination "West Indies" was used by the early chroniclers and up to now all the peoples and the cultures native to these newly encountered lands are called Indians.

The denomination "America" also came as the consequence of a fortuitous event. The navigator Amerigo Vespucci, who had visited and described practically the entire coast of the newly found lands, was consulted by German cartographers who were preparing a revised version of Ptolemy's *Geography*. These cartographers, without a name for the territories added to the original maps, coined the name "Americas" to honor the man who helped them to place these additions to the old maps. This vast territory stretched about seven thousand miles from the North Pole to the South Pole between Europe and Asia, roughly in the middle of the waters surrounding the known world. The peoples and cultures of this territory reached by Columbus, strangers to the conception of the world prevailing in the fifteenth century, are mostly referred to as "American Indians". The name Native Americans is not widely used.

The original occupation of the Americas (North, Central and South) is uncertain. It has been claimed by the colonizers, an obvious political assertion, that the peoples in the Americas were immigrants who arrived from Asia through a glaciation of the Bering Strait no more than 20,000 years ago. Some theories claiming that these peoples are autochthonous were advanced in the early twentieth century, but they were disclaimed on the basis of a lack of scientific support. The best known of these theories is from the Argentinian anthropologist Florentino Ameghino (1854–1911). These claims recur frequently. The theory of a common origin for all of mankind, coming from the heights of Eastern Africa, is the most accepted version for the origin of humans in the Americas. They arrived in migratory fluxes through the Bering Strait, which does not exclude the possibility of ocean access, both through the Atlantic and the Pacific. There is mounting evidence of the presence of humans in these lands since 70,000 BC, but this evidence is very controversial. The most accepted versions place early findings of *Homo sapiens* at about 40,000 BC, without the presence of early hominids.

There is a considerable amount of material relating the impressions of the first Europeans who arrived in the Americas. The first one, the journal of Columbus, is a most important one, but others have also told of the enormous impression made by these peoples and lands on the Europeans. The chroniclers of the conquest reveal their surprise and failure in understanding such different cultures.

Most of the attempts to interpret scientific knowledge in the Americas prior to the arrival of Europeans are limited to a description of what was seen. Buildings, roads, agricultural fields, and artifacts are always listed as evidence of knowledge. Their meaning and cultural context are problematic. An example is the *quipu*, a collection of colored strings with knots, taken as a form of abacus. We now understand these are a very elaborate form of register, both qualitative and quantitative. Also, the wheel appears frequently in pre-Columbian cultures but without the uses so well known in the history of Europe and Asia.

There is a reasonable amount of literature on science and technology in the Americas in the pre-Columbian era written by anthropologists and psychologists. To relate this to the work of historians and philosophers of science and technology is an important step towards a broader understanding of the nature of science and technology and their places in societies. This asks for a clear recognition of the dynamic character of knowledge: from its generation through its organization, both intellectual and social, and through its diffusion. Although these aspects are studied in disciplines such as cognition, epistemology, history, politics, and education, it is practically impossible to understand and explain knowledge in such a fragmented way. This is an important step towards recognizing different modes of thought which lead to different forms of science, which we may call ethnosciences.

Research on the status of knowledge in the Americas in the pre-Columbian period uses methodological instruments and an intellectual posture of classifying it according to accepted, European, views of science. The underlying epistemology is dictated by current scholarship. Thus, research focuses on ethnoastronomy, ethnobotany, ethnochemistry, and so on. The "ethno" prefix stands for "the astronomy of the Aztecs", for example. This reductionist approach does not take into account the meaning of "astronomy" for the Aztecs. This is true for every science. A special situation occurs in the recognition of an ethnomethodology in different cultures. Researchers in this field rely on ethnography, the use of direct observation and extended field research. This produces a naturalistic description of peoples and their culture, uncovering codes, symbols, and categories of analysis which these peoples use to conceptualize, explain, understand, and interpret reality. The generation of these instruments is the essence of studies in cognition and culture. We are thus led to look into the history of science in a broader context, so as to incorporate other possible forms of knowledge, interpretation, and explanation, to deal with nature and natural phenomena.

Generation of knowledge goes on in different environmental settings, according to a multiplicity of stimuli. Practices are created by individuals in response to their immediate needs, motivation, and curiosity, and thus generated, they are intellectually organized as knowledge. A communication system of codes and symbols allows this knowledge to be transmitted informally throughout society and socially organized before it is diffused through education. Its growth into more and deeper knowledge is now a response to further needs and curiosity. These new responses (the *hows* and the *whys*) are incorporated in the pool of common knowledge which keeps a group of individuals together and operational. This is usually called culture.

Cultural forms, such as language, eating and drinking preferences, musical and bodily expression, mathematical practices, religious feelings, familial hierarchy and structure, dressing and behavior patterns, are thus diversified. A larger community is partitioned into several smaller ones, from individuals and their immediate kin to communities and societies, each with distinct cultural variants. The multiplicity of cultural factors is essentially responsible for the dynamic process of the production of new forms of thought and of more sophisticated expressions of the ingenuity of individuals and societies in satisfying their needs of survival and of transcendence. This is the process we call cultural dynamics.

A theory of culture is thus the result of analyzing the intellectual evolution of humankind, focusing on the search for different practices and intellectual tools for explaining, understanding, learning about and managing their natural and socio-cultural environment. In "explaining, understanding, learning about and managing", we recognize the Greek root *mathema*. And a natural and socio-cultural environment is well expressed by the prefix *ethno*. Thus we have coined the word ethnomathematics to describe the mathematical practices of the day-to-day lives of preliterate cultures.

Much has been said about the universality of knowledge, in particular of scientific knowledge. This concept of universality becomes harder to sustain as recent research shows evidence of practices, such as health care and tool use, which are typically scientific, and methodological practices, such as observing, counting, ordering, sorting, measuring, and weighing, all performed in distinctive ways in different cultural environments. This has encouraged further studies on the evolution of scientific concepts and practices within a cultural and anthropological framework. We feel this has been done only to a very limited, and we might say very timid, extent. Indeed, universality of knowledge, more specifically of scientific knowledge, is a fabrication of the colonial ideology.

A new historiography reestablishing authentic universality is needed. There is some agreement about this in several areas of knowledge, but the history of science seems to be immune to this movement. Indeed, the tone of some reviews of recent books and papers in the history of science, in particular mathematics, focusing on this sort of redeeming cultural history or proposing a non-Eurocentric view, is discouraging and sometimes contemptuous. For example, the first non-religious book published in the New World was Juan Diez Freyle's treatise dealing with the arithmetic of the Aztecs, in 1557. Less than a century later, this book had lost its appeal and became completely lost, along with pre-Columbian arithmetic. It was replaced by the arithmetic of the Spaniards. Of course, the suppression of the history of a people involves the removal of all traces of structured knowledge, labeling all forms of knowledge as "popular wisdom", superstition, and folklore.

Early Cultures in the Americas

The early cultures in the Americas are distinct according to broad ranges of latitude. The peoples in the icy northern regions developed into a culture generally known as Inuit, or Eskimos — those in the fertile northern prairies and woods of what is the now the United States and Canada, ranging from the Pacific to the Atlantic, were identified under the generic name of "Indians", organized into several distinct nations. Those in the more desert regions neighboring the Rio Grande were distinct nations with a distinguishable urban organization and rather unique architectural forms. South of the Rio Grande, in the highlands of Mexico and Central and South America there are peoples with cultural styles of a rather different nature. These peoples are now called Aztecs in the region around what is now called the Ciudad de Mexico, Mayas in the Central American jungles, mainly southern Mexico and Guatemala, and Incas, spread throughout the mountainous region called the Andes, from the northern part of South America to the northern part of Chile. Aztecs, Mayas, and Incas are called the "high civilizations" of the Americas. In the "lower lands" — coastal areas, the Amazon Basin and subsidiary basins to the South, and much of the territory east of the Andes — other cultures flourished in a very different style. Apart from the high civilizations, and not mentioning those cultures north of the Rio Grande, we have to mention as best organized those located in the Island of Marajo (in the mouth of the Amazon River), with a considerably developed pottery, the agriculturalist Araucanians in the Southern part of South America, and the Guaranis, the latest major group to organize themselves in an agricultural economy, in central South America.

The Aztecs are distinguished by a monumental religious architecture, with abundant pyramids with precise designs to accommodate astronomical data. The Aztec Calendar reveals these achievements. Sculpture was at a level of sophistication comparable to civilizations in the Old World. Commerce was developed and practiced in large fairs. They used numbers and measures, but had no commerce by weight. The Aztecs were skilled workers in ceramics and in metals (gold, silver, copper, tin, and bronze). They had hieroglyphic writing, using colors which endure to the present day, literature, and a school system.

The Mayas had distinctive urbanization styles. Stone pyramids, temples, and palaces, decorated with very elaborate sculptures, were religious centers, and the population was spread in the surrounding jungles. These centers appear to have been temporary. One hypothesis is that they were abandoned as new ones became ready — another possibility is that the mobility was the result of destruction by wars. These peoples had a very advanced astronomy, with data comparable to those available to the Europeans. From the sixth century, they had developed the zero and positional number system, thus producing an advanced arithmetic. They also had a pictorial writing system which was not recognized as such by Europeans. Some early translators relied on reports of books and documents without recognizing this as reading. Among the classics of their literature we have the elaborated description of the origins of the world and men, the *Popol Vuh*, the books of natural phenomena, the *Chilam Balam*, the *Anals of the Cakchiqueles*, a tribe of Guatemala, and a piece showing the existence of a form of a ritual theater in the high civilizations, the war drama *Rabinal Achi*.

The third high civilization, known as the Incas, took the name after their monarch. The Inca was the highest religious, civilian, and military authority. Although succession was by blood, the successor was not necessarily the eldest son. Like the Aztecs, the religion of the Incas was focused on the divinity of the Sun, and human sacrifices were common. The architecture of the Incas was of a varied nature. Visitors still marvel at what is called cyclopic architecture, buildings and palaces constructed using enormous stones, as in Machu-Pichu, Peru.

The early history of these cultures reveals a great dispersion of nomadic groups, with an economy of hunters and collectors. In general, in all of the Americas, animal domestication was modest. In Mexico, turkeys were domesticated – in Peru the llama and the alpaca, as pack animals and for textiles, were also domesticated, as were dogs and some rodents. Some tribes farmed turtles, some bees and birds, mainly parakeets and parrots. Fish from the coasts of Peru were consumed in the high altitudes, transported and preserved through a sophisticated system of utilization of natural ice.

Agriculture was much more developed. About 6000 BC we see evidence of the domestication of plants, mainly in the northern region (North of Mexico and Southeast United States), among them squash, cotton, and several species of beans. The important development of maize (*Zea mays*) opens up a new era in the development of agriculture and the settlement of people in larger regions, which characterizes the urbanization period. There is much controversy about when this occurred. A recent academic argument about the occurrence of maize in Panama, 5000 versus 3000 BC, is more important for revealing the search for a new historiography for science and technology in pre-Columbian Americas than for the dating question itself.

Intensive Agriculture and Urbanization

Up to the beginning of the Christian era we see the formation of agricultural cultures. They developed different techniques, using fire, intensive irrigation, fertilization, and terrace agriculture. Agriculture was the main subsistence form and the determinant of the political struggles and the development of economy.

As we advance in time, around the second millennium AD we see the development of a variety of urban styles, the weaving of cotton and wool, pottery and ceramic, statuary, metallurgy, especially work in gold, ritual decoration, and architecture. We notice the development of several forms of pyramids, possibly evolving from the common funerary mounds, and other forms of religious architecture. Ceremonial dressing is varied, with feathers, plants, animals, painting, and bright colors. The period also reveals, probably as a consequence of the permanent struggle for land, the development of a variety of languages, with more than two thousand distinct languages or dialects. In the lower lands there is considerable development of agriculture, mainly manioc or cassava (*Manihoc utilissima*) and sweet potatoes (*Ipomoea batatas*).

There is also evidence of an intense coastal commerce. Indeed, navigation in the Pacific seems to be highly developed among the Andean cultures. There is recent evidence of Japanese arrivals in what is now Ecuador about 3000 BC. Also, there is a possibility of commercial exchanges between China and South America in the first century AD, and contacts between Mexico and Africa in the first millennium are most probable. There are no doubts about the practice of considerably advanced navigation among the high civilizations.

There is much controversy about the development of a written/pictorial form of language. As mentioned above, recent findings reveal a written language among the Mayas. They also developed calculations, especially for astronomical and commercial purposes, with a sophisticated vigesimal system. The register of historical and statistical data was rather sophisticated, as with the surprisingly powerful *quipus* among the Incas. Calculating devices, such as the *iupana*, reveal theoretical approaches as yet not sufficiently studied. We are just beginning to recognize the formal knowledge — logic, mathematics, philosophy — possessed by these cultures.

Colonial discourse was absolutely biased in saying that the American peoples were not highly developed. In general, the denomination "New World" finds some support in the evidence of a delay of some two to three thousand years for achievements which might be identified as stages in the cultural evolution of the species as compared with what happened in the Old World. Specifically, agriculture was developed in Asia in the eighth millennium BC, while there is evidence that it was developed in Mexico only in the fifth millennium BC. Ceramics and pottery are seen in Asia about 6000 BC, while they appear in 3000 BC in the northeast of South America. Proto-literacy is recognized in Mesopotamia about 3300 BC, but in Mesoamerica only about 800 BC. Urban cultures are seen in Egypt and Mesopotamia using writing and bronze about 2800 BC, while limited writing and metallurgy are identified in urban complexes of Mexico and Peru only in the beginning of the Christian era. We can recognize that in the first millennium AD, the cultural development of the Americas might be considered parallel to what Europe had gone through centuries before.

What was going on in the centuries preceding the arrival of Columbus was a political process very similar to that which took place in Europe during the Middle Ages, when principalities and other smaller domains were uniting into larger organizations such as kingdoms and nations. The Spanish conquerors were able to take advantage of the local conflicts in destroying the emergent power elites and replacing them with the power of the conqueror, based on a well balanced State–Church structure controlling the means of production. Although effective production was in the hands of natives, the conqueror assumed the role of the native monarchy in controlling the centralized system. Even now production is to a great extent done by impoverished natives. A somewhat recent study (1950) showed that in Bolivia about four thousand native communities were responsible for about 25% of the agricultural production. The conqueror succeeded in deepening these conflicts. Even today different cultures, even different tribes, do not recognize their common history. In the lower lands, East of the Andes and the Amazon Basin and subsidiary basins in the South, and mainly in the Atlantic coast, the strategy was no different.

As we have said before, knowledge is generated as a permanent activity of human beings. It is carried out by individuals and groups of individuals in different environmental settings, according to a multiplicity of stimuli.

Among the several stimuli we have to consider cultural encounters and mutual exposition of different modes of thought. In the evolution of cultural forms subjected to mutual exposition, the possibilities are: first, an absolute domination of one form, either leaving the others in the state of latency or leading them to an eventual total elimination, and second, allowing co-evolution, which eventually leads to new cultural forms. This second possibility occurs in systems which are tolerant of the different, of the stranger. Thus, our analysis of the formation of scientific knowledge and the development of technology among the civilizations of Latin America necessarily goes through an analysis of the occupation of territory and the development of means of production.

UBIRATAN D'AMBROSIO

REFERENCES

Coe, Michael D.. *Breaking the Maya Code*. New York: Thames and Hudson, 1992.

For the Learning of Mathematics vol. 14 no.2, June 1994 [Special issue devoted to ethnomathematics]

Freyle, Juan Diez. *Sumario Compendioso de las quentas de plata y oro...Con algunas reglas tocantes al arithmética*. Mexico: Juan Pablos of Brescia, 1556.

Piperno, Dolores R. "On the Emergence of Agriculture in the New World." *Current Anthropology* 35(5): 637–643, 1994.

Sale, Kirkpatrick. *The Conquest of Paradise. Christopher Columbus and the Columbian Legacy*. New York: Knopf, 1990.

Wright, Ronald. *Stolen Continents. The "New World" Through Indians Eyes*. Boston: Houghton Mifflin Company, 1992.

See also: Metallurgy – Agriculture – Animal Domestication – Navigation – Astronomy – Mathematics – Writing among the Maya – *Quipu* – Ethnomathematics – Knowledge Systems of the Incas – Aztec Science – Calendar – Weights and Measures – City Planning – Stonemasonry

ANIMAL DOMESTICATION Animal domestication is the process by which humans exert direct or indirect, conscious or unconscious influence over the reproduction and evolution of animals that they own or otherwise control. A fully domesticated animal exhibits significant differences from its nearest wild relatives. The domestication of animals and plants is one of the most significant impacts that humans have had on their natural environments, the root of the contemporary global economic system and its marvels and tragedies. For most of human existence, people did not tend crops and care for animals, living instead as gatherers, hunters, and fishers.

The archaeological record indicates that settled societies based on plant and/or animal domestication go back roughly 10,000 years in certain areas of the world. For example, suggestions of sheep domestication reach back nearly 11,000 years in northernmost Iraq. The early domestication of wheat and barley is evidenced in the Levant some 10,000 years ago. About a millennium later, plant and animal domesticates were joined together in agriculture and husbandry complexes, which quickly spread throughout the ancient Near East.

The dog is widely considered the first domesticate. The oldest tentative identification of domestic dog goes back about 12,000 years in northern Iraq. Dog remains are found at sites 10,000 to 12,000 years old in several widely separated areas: Idaho in the United States, Japan, Israel, Russia, England, and Denmark. All dogs are now known to be domesticated wolves, most being descendents of the smaller subtropical subspecies. The earliest dogs may have been pets, self-domesticating scavengers, and sources of meat.

Competing with the dog for honors as the first domesticate is the sheep. Its domestication is suggested some 11,000 years ago, in the hills of Kurdistan, including northernmost Iraq and northwesternmost Iran, and over 10,000 years ago in northern Afghanistan. Goats, too, are ancient domesticates, dating back more than 10,000 years in Iran and northern Afghanistan and more than 9000 years in the Levant.

A second flurry of domestication took place in widely scattered areas between 10,000 and 8000 years ago. Pigs were domesticated some 10,000 years ago, as suggested by their presence in New Guinea; they can only have been brought there by people. They show up about a thousand years later in Jericho near the Dead Sea. Domesticated cattle are evidenced in south central Turkey some 8400 years ago. Chickens turn up in northern China around 8000 years ago.

A third wave of animal domestications occurred roughly 6500 to 5000 years ago, this time involving the New World, too. The species involved included the pigeon (Syria, 6500 years ago), ass (Nile Valley, roughly 6000 years ago), South American camels — llama and alpaca (Peru, approximately 6000 years ago), the horse (Ukraine, southernmost Russia, and westernmost Kazakhstan, about 6000 years ago), dromedary (Arabian Peninsula, probably 5500 years ago), water buffalo (India and southern Iraq, some 5000 years ago), the Bactrian camel (Turkmenia and eastern Iran, perhaps 5000 years ago), the cat (Egypt, perhaps 5000 years ago), and the silkworm (China, perhaps 5000 years ago).

A number of minor domestications have taken place from that time to this, such as the ferret, rabbit, cormorant, turkey, carp, and goldfish. Domestications continue at the present, including parakeets, canaries, mink, laboratory rats, aquarium fish, and experimental domestications of livestock, such as eland and fallow deer.

Why did people give up the hunting of animals for the work entailed in domesticating them? The factors involved differ by species and human circumstance. The simple human tendency to pet keeping no doubt played a role.

Nomadic hunters–gatherers could have domesticated certain herd animals. Such peoples as the reindeer-herding Sami of northernmost Europe sometimes follow migratory herd species all year and influence the direction and spatial cohesion of the animals through driving and herding. They selectively cull the animals, which amounts to the evolutionary pressure associated with domestication. It is not clear, however, whether migratory herding hunters in the past actually undertook independent domestications or adopted the idea from nearby agriculturalists and associated pastoral nomads.

Settled cultivators may have had a variety of reasons to undertake the control and maintenance of animal populations. Their crops, food stores, and even their salty latrine areas attracted animals, inviting reciprocal exploitation. The small southern wolves would steal scraps of food and may have been tolerated as a garbage disposal system. Fields, grain stores, and latrine areas would attract bovine species. The granivorous and cliff-nesting pigeons would descend on early cereal fields and roost in nearby buildings. The opportunities for coevolutionary interactions between people and animals increased once people settled and especially once they took up crop cultivation.

Certain scholars have argued that settled cultivators had no particular economic or ecological need for animals, and animal management does entail work and can prove dangerous. They feel that the earliest animal domestications provided sacrifices for sacred rituals or amusement for gamblers. This argument has not done well in archaeological testing, but the data may not yet be complete enough to dismiss the argument entirely.

In some cases, people may have completely or partially domesticated an animal species and later abandoned it. Gazelles may have been domesticated in the ancient Levant, later being replaced by goats and sheep. These latter may have had some sort of advantage over indigenous domesticates. Military displacement or conquest of one culture by another have replaced one constellation of animals domesticates with another. Religious taboos have caused species to be dropped from husbandry in given areas (e.g. pigs in areas of Jewish or Muslim religion).

To resolve these different issues and disputes over their interpretation will take a lot of work. Much of the world has never had adequate coverage of its archaeological heritage. Archaeological funding has been biased toward those areas regarded rightly or wrongly by Westerners as somehow connected with their own culture histories, such as the Biblical Levant and Mesopotamia, Egypt, ancient Greece and Rome, Europe, and even Mesoamerica and Peru. There are many tantalizing indications that archaeological work in other areas in the non-Western world will transform our present perceptions of the origins of domestication.

CHRISTINE M. RODRIGUE

REFERENCES

Clutton-Brock, Juliet. *Domesticated Animals: From Early Times.* Austin: University of Texas Press – London: British Museum (Natural History), 1981.

Hemmer, Helmut. *Domestication: The Decline of Environmental Appreciation.* Cambridge: Cambridge University Press, 1990.

Isaac, Erich. *Geography of Domestication.* Englewood Cliffs, New Jersey: Prentice-Hall, 1970.

Rodrigue, Christine Mary. *An Evaluation of Ritual Sacrifice as an Explanation for Early Animal Domestications in the Near East.* Ph.D. dissertation, Clark University, Worcester, Massachusetts, 1987.

APPROXIMATION FORMULAE IN CHINESE MATHEMATICS Approximation formulae may be used for various reasons. In some cases, certain computations (such as for example, the extraction of roots digit by digit) are by nature inexact. It may also be that exact solutions of certain problems are unknown or else that such solutions are theoretically known but deemed too complex, so that users prefer elaborating alternative solutions more or less accurate with respect to a certain context of utilization. The history of Chinese mathematics illustrates well these two aspects of the question.

Remarkably, many Chinese approximation formulae are also attested in Babylonian, Greek, Roman, Indian, and Mediaeval European mathematics. For example, the so-called "Heron's formula" for the approximation of square roots

$$\sqrt{a^2+r} \simeq a + \frac{r}{2a} \quad \text{or} \quad \frac{r}{2a+1}$$

is also found in the *Jiuzhang suanshu* (Computational Prescriptions in Nine Chapters, also translated as Nine Chapters on the Mathematical Art) from the Han dynasty (206 BC–AD 220). The approximation formula for the computation of the area of a quadrilateral by taking the product of the half sum of its opposite sides appears not only in the *Wucao suanjing* (Computational Canon of the Five Administrative Services), and the *Xiahou Yang suanjing* (Xiahou Yang's Computational Canon) (fourth century AD) but also in the Babylonian mathematical corpus, in the writings of the Roman agrimensors (surveyors), in Alcuin's *Propositiones ad acuendos juvenes* (Propositions to Sharpen the Minds of the Youth) and numerous other places. The same remark also applies to the formula for the computation of the area of a segment of a circle:

$$A = \frac{h(b+h)}{2},$$

where b is the base of the segment and h its height, a formula which is attested in Heron's *Metrika* as well in the

Jiuzhang suanshu. Such examples are much more numerous than historians of mathematics formerly imagined.

However, Chinese mathematics also contains examples of approximation formulae apparently not attested anywhere else. These all occur in Chinese astronomy and concern astronomical problems such as that of the determination of gnomon shadows or questions of conversions between ecliptic (the band of the zodiac through which the Sun apparently moves in its yearly course) and equatorial coordinates.

The monograph on mathematical astronomy of the Song Dynasty, the *Songshi* (Song History), an official text compiled between 1343 and 1345, for example, reports that between AD 1102 and 1106, Chinese imperial astronomers devised a new theoretical algorithm for the determination of the length of the shadow of the sun cast by a standard gnomon 8 *chi* (feet) long, each day, at noon, during a whole year. If t = the number of days elapsed since the last winter solstice and

$$s_1(t) = 12.83 - \frac{20000t^2}{100\left(100617 + 100t + \dfrac{10000t^2}{725}\right)},$$

$$s_2(t) = 1.56 + \frac{4t^2}{7923 + 9t},$$

$$s_3(t) = 1.56 + \frac{7700t^2}{13584271.78 + 44718t - 100t^2},$$

at Kaifeng, the Song imperial capital whose latitude is 34°48′45″, the length of the shadow of the gnomon is computed as follows (the original text is formulated in words, with no symbols, but its interpretation is straightforward):

if $t < 662.2$	$l(t) = s_1(t)$
if $< 62.2 < t < 91.31$	$l(t) = s_3(182.62 - t)$
if $< 91.31 < t < 182.62$	$l(t) = s_2(182.62 - t)$
if $< 182.62 < t < 273.93$	$l(t) = s_2(t - 182.62)$
if $< 273.93 < t < 303.04$	$l(t) = s_3(t - 182.62)$
if $< 303.04 < t < 365.24$	$l(t) = s_1(365.24 - t)$

(The various numerical coefficients appearing in these formulae all depend on the lengths of the seasons.)

Chen Meidong, a contemporary historian of Chinese astronomy from the Academia Sinica in Beijing, has compared the results provided by these approximate formulae and those of actual observations (simulated for the years 1102–1106 by using the modern astronomical theory of the sun backwards). He has concluded that the error never exceeds 0.02 *chi*, that is much less than a centimeter! Yet Song astronomers were still not satisfied with their approximation and sought new formulae; some of these have reached us. Incidentally, these formulae are vaguely reminiscent of the Indian approximation formula for the sine, cosine, and other trigonometrical functions cited by R.C. Gupta in 1972 — in both cases the approximations use rational fractions.

Another interesting approximation formula occurs in the *Yuanshi* (Yuan History), an official compilation written about AD 1370 but reporting on Guo Shoujing's astronomical reform undertaken one century earlier. (Guo Shoujing (1231–1316) was a specialist on canal draining and astronomy.) Without going into too much detail, we note that Guo's technique was devised to compute certain segments corresponding to given arcs of circles similar to those which arise when converting ecliptic coordinates into equatorial coordinates. For us the question is solved in a relatively simple way by means of spherical trigonometry; but plane and spherical trigonometry were both unknown in China at the time, so that the mathematical techniques of Chinese astronomy had necessarily to rely on approximation formulae. In particular, Guo's technique relied on the three following formulae:

(1) $$p = \sqrt{r^2 - q^2} = \sqrt{(2r - v)v},$$

(2) $$x = p + \frac{v^2}{2r},$$

(3) $$v^4 + (4r^2 - 4rx)v^2 - 8r^3v + 4r^2x^2 = 0,$$

where r represents the radius of the "trigonometrical" circle, and p, q, and v the sine, cosine, and versed sine of the arc x, respectively.

Formula (1) is exact; (2) and (3) are approximate. The polynomial of the fourth degree in v, (3), is the consequence of the elimination of the half chord p between the second part of (1) and (2); (3) serves to compute v given the arc x by finding a root (in fact the smallest positive one) of (3) by means of a technique similar to that which is usually known as "Horner's method" (from a method attributed to the British mathematician William George Horner (1786–1837) who lived five centuries after Guo Shoujing). Lastly we also observe that Guo Shoujing divides the length of the circumference into as many degrees as there are days in a year (365.25 degrees) and computes the radius of the corresponding circle by dividing the circumference by 3 (i.e. by taking π = 3). *A priori*, the reliance on such a value of π would indicate a severe mathematical deficiency. However, when Guo Shoujing devised his techniques, better values of π, such as Zu Chongzhi's (429–500) celebrated approximation π = 355/113, were currently available in China. In fact, according to Toshio Sugimoto, a mathematical analysis of the above formulae shows that when (1), (2) and (3) are computed using approximations of π better than 3, worse results are obtained! The recourse to approximations perturbates the impeccable logic which would hold if exact representations were used.

These various examples show that approximation formulae are well represented everywhere but that some of them are attested only in China. As recent works by historians of Chinese astronomy indicate, the recourse to such formulae seems typical of Chinese astronomy (see the articles in the journal *Ziran Kexue shi yanjiu*, 1982–1994). Are Chinese approximation formulae really original or are they also found elsewhere? Further historical researches not limited to Chinese mathematics will perhaps shed some light on a question which has never really been studied in depth.

JEAN-CLAUDE MARTZLOFF

REFERENCES

Bruins, E. M. and Rutten, M. *Textes mathématiques de Suse*. Paris: P. Geuthner, 1961.

Chen Meidong. "Huangyou, Chongning gui chang jisuan fa zhi yanjiu" (Researches on the length of the shadow of the sun cast by a gnomon during the Huang You and Chongning eras). *Ziran kexue shi yanjiu* 8(1): 17–27, 1989.

Gericke, Helmut. *Mathematik im Abenbland Von den römischen Feldmessern bis zu Descartes*. Berlin: Springer-Verlag, 1990.

Gupta, R. C. "Indian Approximations to Sine, Cosine and Verse Sine." *Mathematics Education* 6(2): 59–60, 1972.

Libbrecht, Ulrich. *Chinese Mathematics in the Thirteenth Century: the Shu-shu chiu-chang of Ch'in Chiu-shao* [i.e. Qin Jiushao]. Cambridge, Massachusetts: The MIT Press, 1973.

Martzloff, Jean-Claude. *Histoire des mathématiques chinoises*. Paris: Masson, 1987. English translation by Stephen S. Wilson. *A History of Chinese Mathematics*. New York: Springer-Verlag, 1997.

Sugimoto, Toshio. "Guanyu Shoushili de Shen Gua de nizhengxian gongshi de jingdu" (On the precision of Shen Gua's arcsine formula in the Shoushi calendar). *Meiji gakuin ronsō* 429: 1–12, 1987.

Waerden, B. L. van der. *Geometry and Algebra in Ancient Civilizations*. Berlin: Springer-Verlag, 1983.

See also: Liu Hui and the *Jiuzhang suanshu* – Gnomon – Guo Shoujing – Zu Chongzhi – *Pi* in Chinese Mathematics – Mathematics

ARITHMETIC IN INDIA: *PĀṬĪGAṆITA* *Pāṭīgaṇita*, which literally means "mathematics (*gaṇita*) by means of algorithms (*pāṭī*)", is the name of one of the two main fields of medieval Indian mathematics, the other being *bījagaṇita* or "mathematics by means of seeds". The two fields roughly correspond respectively to arithmetic (including mensuration) and algebra.

The compound *pāṭīgaṇita* seems to have come into use in relatively later times. In older works, the expressions, *gaṇitapāṭī* and *gaṇitasya pāṭī* (mathematical procedure, i.e. algorithm), are common, and sometimes the word *pāṭī* occurs independently. *Pāṭīgaṇita* is also called *vyaktagaṇita* or "mathematics of visible (or known) [numbers]", while *bījagaṇita* is called *avyaktagaṇita* or "mathematics of invisible (or unknown) [numbers]". Some scholars maintain that the word *pāṭī* originated from the word *paṭṭa* or *paṭa* meaning the calculating board, but its origin seems to be still open to question.

The division of mathematics (*gaṇita*) into those two fields was not practiced in the *Āryabhaṭīya* (AD 499), which has a single chapter called *gaṇita*, but it existed in the seventh century, when Brahmagupta included two chapters on mathematics in his astronomical work, *Brāhmasphuṭasiddhānta* (AD 628). Neither the word *pāṭī* nor *bījagaṇita* occurs in the book, but chapter 12 (simply called *gaṇita*) deals with almost the same topics as later books of *pāṭī*, and chapter 18, though named *kuṭṭaka* or the pulverizer (solution of a linear indeterminate equation), has many topics in common with later books of *bījagaṇita*. Śrīdhara (ca. AD 750) is known to have written several textbooks of *pāṭī* and at least one of *bījagaṇita*.

Extant works of *pāṭī* include Śrīdhara's *Pāṭīgaṇita* (incomplete) and *Triśatikā* (and *Gaṇitapañcaviṃśī*?), Mahāvīra's *Gaṇitasārasaṃgraha* (ca. AD 850), chapter 15 (*pāṭī*) of Āryabhaṭa's *Mahāsiddhānta* (ca. AD 950? or 1500?), Śrīpati's *Gaṇitatilaka* (incomplete) and chapter 13 (*vyaktagaṇita*) of his *Siddhāntaśekhara* (ca. AD 1050), Bhāskara's *Līlāvatī* (AD 1150), and Nārāyaṇa's *Gaṇitakaumudī* (AD 1356).

A book (or chapter) of *pāṭī* consists of two main parts, namely, fundamental operations (*parikarmāṇi*) and "practical problems" (*vyavahārāḥ*). The former usually comprises six or eight arithmetical computations (addition, subtraction, multiplication, division, squaring, extraction of the square root, cubing, and extraction of the cube root) of integers, fractions, and zero, several types of reductions of fractions, and rules concerning proportion including the so-called rule of three (*trairāśika*). The latter originally consisted of eight chapters (or sections), i.e. those on mixture (*miśraka*), mathematical series (*śreḍhī*), plane figures (*kṣetra*), ditches (*khāta*), stacking [of bricks] (*citi*), sawing [of timbers] (*krākacika*), piling [of grain] (*rāśi*), and on the shadow (*chāyā*).

To this list of the practical problems, Śrīdhara added in his *Pāṭīgaṇita* one named "truth of zero" (*śūnyatattva*). A large portion of the work including that chapter is, however, missing in the only extant manuscript. The way the *Gaṇitasārasaṃgraha* of Mahāvīra divides its contents into chapters is unusual, but still it can be characterized as a book of *pāṭī*. It is quite rich in mathematical rules and problems.

In his *Līlāvatī*, Bhāskara separated the rules on proportion from the arithemtical computations, and created with them a new chapter named *prakīrṇaka* (miscellaneous [rules]), in which he also included the *regula falsi*, the rule of inverse operations, the rule of sum and difference, etc. After the ordinary topics of practical problems, he treated *kuṭṭaka* as well as *aṅkapāśa* or the nets of numerical figures (combinatorics). Written in elegant but plain Sanskrit and organized well, the *Līlāvatī* became the most popular textbook of *pāṭī* in India.

In his *Gaṇitakaumudī*, Nārāyaṇa included in the practical problems not only *kuṭṭaka* and *aṇkapāśa*, but also *vargaprakṛti* or the square nature (indeterminate equations of the second degree including the so-called Pell's equation), *bhāgādāna* or the acquisition of parts (factorization), *aṃśāvatāra* or manifestation of fractions (partitioning), and *bhadragaṇita* or mathematics of magic squares. These topics had already been dealt with to a certain extent by his predecessors, but he developed them considerably. He also investigated new mathematical progressions, some of which turned out to be useful when Mādhava (ca. AD 1400) and his successors obtained power series for the circumference of a circle (or π), sine, cosine, arctangent, etc.

TAKAO HAYASHI

REFERENCES

Bag, A.K. *Mathematics in Ancient and Medieval India*. Varanasi: Chaukhambha Orientalia, 1979.

Colebrooke, H.T. *Algebra with Arithmetic and Mensuration from the Sanscrit of Brahmegupta and Bháscara*. London: Murray, 1817. Reprinted, Wiesbaden: Dr Martin Söndig oHG, 1973.

Datta, B. and Singh, A.N. *History of Hindu Mathematics*. 2 vols., Lahore: Motilal, 1935–38. Reprinted in one volume, Bombay: Asia Publishing House, 1962.

Ramanujacharia and Kaye, G.R., trans. "The *Triśatikā* of Śrīdharācārya." *Bibliotheca Mathematica* Series 3, 13: 203–17, 1912–13.

Raṅgācārya, M., ed. and trans. *Gaṇitasārasaṃgraha of Mahāvīra*. Madras: Government Press, 1912.

Sen, S.N. "Mathematics." In *A Concise History of Science in India*. Ed. D.M. Bose et al. New Delhi: Indian National Science Academy, 1971, pp. 136–212.

Shukla, K.S., ed. and trans. *Pāṭīgaṇita of Śrīdhara*. Lucknow: Lucknow University Press, 1959.

Sinha, K.N. "Śrīpati's *Gaṇitatilaka*: English Translation with Introduction". *Gaṇita Bhāratī* 4: 112–133, 1982.

Sinha, K.N. "Śrīpati: an Eleventh-Century Indian Mathematician." *Historia Mathematica* 12: 25–44, 1985.

Sinha, K.N. "Vyaktagaṇitādhyāya of Śrīpati's *Siddhāntaśekhara*." *Gaṇita Bhāratī* 10: 40–50, 1988.

Srinivasiengar, C.N. *The History of Ancient Indian Mathematics*. Calcutta: The World Press, 1967.

See also: Āryabhaṭa – Śrīdhara – Mahāvīra – Bhāskara – Nārāyaṇa – Combinatorics – Magic Squares – Mādhava

ARITHMETIC IN ISLAMIC MATHEMATICS Mathematics flourished during the golden age of Islamic science, which began around the seventh century AD and continued through to about the fourteenth century. Both arithmetic and algebra were advanced dramatically by Muslim mathematicians, who adopted Indian innovations such as decimal numbers and considerably extended them – they also developed earlier Greek concepts of geometry, trigonometry, number theory and the resolution of equations. Islamic mathematicians did far more than just copy Greek and Indian techniques — their additional researches developed and systematized several fields of mathematics. Even modern mathematical language, including terms like "algebra", "root" and "zero", owes an important debt to Arabic scientists. Algebra, for example, comes from the ninth-century Arabic astronomer and mathematician al-Khwārizmī (ca. 780–ca. 850), whose book *Algebra* (*al-Kitāb al-mukhtaṣar fī ḥisāb al-jabr wa'l-muqābala*) described techniques of transposing quantities from one side of an equation to another (*jabr*), then simplifying them (*muqābala*).

Yet while Arabic contributions to algebra have been widely discussed by historians of science and culture, their parallel work in arithmetic has been until recently far less well known. During the early nomadic period of Arabic history, numbers were given names — and around the time of the Prophet Muḥammad (ca. 570–632), the letters of the Arabic alphabet were often used as numerals. However, it was not until the rise of Islam in the seventh and eighth centuries that a recognizably modern system of arithmetic was developed. Muslim arithmetic operations were largely based on ancient Greek definitions, but Islamic scholars pioneered new techniques, including the network or lattice method (*shabaqah*) to multiply numbers, and various techniques of long division.

Muslim mathematicians are best known for their contributions to the pivotal system of modern "Hindu–Arabic numerals" — that is, the technique of expressing all numbers through the repeated use of a few basic symbols. Though originally invented in India, Arab scholars dramatically improved both the writing and manipulation of decimal numerals, and also developed the Hindu idea of positional notation. It is not known exactly when Indian mathematicians began using decimal numbers, but they seem to have been in place by the early sixth century AD — astronomer Āryabhaṭa I (476–ca. 550) did not use them, but they were employed in a limited way by the middle of the sixth century. Decimal numbers were quite popular and spread quickly — by the seventh century, they had reached Iraq and the Middle East, and were praised by Syrian Nestorian bishop Severus Sebokht (fl. AD 630), who considered this new Hindu arithmetic "done with nine signs" even more ingenious than the calculations of Greek mathematicians. Arabs began using Hindu decimal arithmetic around the seventh century, but it was not until the ninth century that Arabic works describing this type of reckoning appear. The earliest known Arabic treatise on decimal arithmetic, *Kitāb al-ḥisāb al-hindī* (Book of Addition and Subtraction According to the Hindu Calculation), was written by Al-Khwārizmī around AD 800 — the Arabic text is lost, but a twelfth century Latin translation is still extant.

Al-Khwārizmī is better known among historians of Muslim mathematics for his many contributions to algebra. Yet his investigations into arithmetic were equally important, and were so widely read in the medieval Latin West that they later gave Europeans one of their early names for arithmetic, the "algorism" or "algorithm". Al-Khwārizmī's book treated all arithmetic operations, spreading knowledge of Hindu techniques throughout the Muslim world. Another important early treatise that publicized decimal numbers was Iranian mathematician and astronomer Kūshyār ibn Labbān's (fl. 1000) *Kitāb fī uṣūl ḥisāb al-hind* (Principles of Hindu Reckoning), a leading arithmetic textbook.

Coupled to the important development of decimal numbers was the equally significant Arabic use of the *ṣifr* (meaning "empty"), or zero, again from Indian roots. An early symbol for zero appears in an AD 876 inscription at Gwalior, India; in the Arab world, Kūshyār introduces the zero in the tenth century as a sign to be placed "where there is no number."

Though scholars concede that much of Arabic arithmetic has its ultimate origins in India, Muslim mathematicians were the first to integrate the various discoveries of Hindu mathematicians into a coherent whole. Historian J.L. Berggren, for example, concludes that while the Hindus were the first to use a "cipherized, decimal positional system", the Arabs pioneered in extending this system to "represent parts of the unit by decimal fractions". In Europe, meanwhile, the zero and decimal system were not widely used until the late twelfth century.

Following al-Khwārizmī, many Arabic mathematicians developed Indian techniques of arithmetic over the next few centuries. The astronomer, translator, and editor al-Kindī (801–ca. 873), for example, wrote several important treatises on arithmetic, including manuscripts on the use of Indian numbers, on lines and multiplication with numbers, on measuring proportions and times, on numerical procedures and cancellation, and many more.

Two centuries later, al-Karajī of Baghdad (fl. 1020) wrote mathematical works that led to his being called "the most scholarly and original writer of arithmetic" by historian Al-Daffa. Al-Karajī's works included a manuscript on the rules of computation entitled *Al-Kāfī fī al-Ḥisāb* (Essentials of Arithmetic), and *al-Fakhrī fī'ljabr wa'l-muqābala*, which was named after his longtime friend, the Baghdad grand vizier.

The depth of early Islamic knowledge of arithmetic is often quite unexpected. Arabic mathematicians were well aware of the existence of irrational numbers, and sometimes developed complex theories to explain their properties. Persian poet and philosopher ʿUmar al-Khayyām, or Omar Khayyam (ca. 1048–ca. 1122) and Persian astronomer Naṣīr al-Dīn al-Ṭūsī (1201–1274) both argued that every ratio of two magnitudes can be considered a number, whether that ratio be commensurable (rational) or incommensurable (irrational). Islamic arithmetic used many of the same Hindu techniques for operating with irrationals as it did with rationals. Also from Indian sources came various operations with numbers, including transformations such as $\sqrt{x^2y} = x\sqrt{y}$ and $\sqrt{xy} = \sqrt{x}\sqrt{y}$.

Islamic arithmeticians did not accept everything offered them by Hindu scholars. For example, negative numbers, long a staple in Indian arithmetic, were transmitted to the Arab world but rejected by it — Arabic mathematicians instead held that negative numbers did not exist.

Modern notation for fractions is also based in part on Muslim arithmetic. Celebrated Hindu mathematicians such as Bhāskara II (1114–ca. 1185) wrote common fractions by just writing a numerator above a denominator, but the idea of a line of separation between the numerator and denominator was an early Islamic development. Decimal fractions, meanwhile, appear in seminal tenth century Arabic texts, such as the *Kitāb al-Fuṣūl fi'l ḥisāb al-hindī* (Book of Chapters on Hindu Arithmetic), written by Damascus mathematician al-Uqlīdisī (fl. 952). In the late twelfth century, al-Samaw'al (fl. 1172) used decimal fractions for division, root extraction, and approximation. By the fifteenth century, decimal fractions had been formally named and systematically developed, but they were not widely used in Europe until the Dutch physicist and engineer Simon Stevin (1548–1620) published *La Thiende* (The Tenth) in 1585, and Scottish mathematician John Napier (1550–1617) reintroduced his decimal point in the early seventeenth century.

While Arabic mathematicians pioneered in decimal arithmetic, they also made considerable contributions to the ancient sexagesimal (base 60) system of arithmetic, which had been developed by the Babylonians in Mesopotamia around 2000 BC. This system was widely used for astronomical calculation throughout the ancient world, particularly in Alexandrian astronomer and geographer Claudius Ptolemy's (ca. AD 100–170) cosmological treatise, *Almagest*, which was later adopted by Islamic scholars as the theoretical base of their astronomy. Sexagesimal addition, subtraction, multiplication, and division became so commonplace among Islamic astronomers it was renamed "the astronomer's arithmetic". Arabic astronomers and mathematicians such as Kūshyār and Samarqand's al-Kāshī (fl. 1406–1429) used sexagesimal numbers to determine approximate roots, extract square roots, and even find the fifth roots of certain numbers.

Islamic arithmetic was often influenced by the needs of astrology, talismans, and sorcery, as in the casting of horoscopes and magic spells. Muslim mathematicians such as Syrian scholar Thābit ibn Qurra (ca. 836–901) and Tunisian historian Ibn Khaldūn (1332–1406) studied amicable numbers, or number pairs where the sum of the factors of each number is equal to the other number. 220 and 284 are amicable numbers, because the sum of the factors of 284 (1+2+4+71+142) equals 220, and vice versa.

Muslim mathematics was also affected by practical considerations such as problems of inheritance and finance, and the need to calculate events in the lunar-based Islamic calendar. For example, al-Khwārizmī devoted the second half of his treatise on algebra to the question of *ʿilm al-farāʾiḍ*, or the calculation of shares of an estate given to various heirs. These problems employed the arithmetic of fractions, and were heavily influenced by religious law and custom. A typical example treated by al-Khwārizmī was to calculate the shares of a dead woman's estate that would accrue to her husband, her son, and her three daughters. As the law required that the husband receive a fourth and each son get twice what a daughter would receive, al-Khwārizmī simply divides the estate into 20 parts, giving 5 to the husband, 6 to the son, and 3 to each daughter. Similar problems involved the topic of *zakāt*, which was the calculation of the share of private wealth that various persons would pay to the community each year.

Islamic arithmetic was often ingenious. Arabic scholars gave us much of the modern system of arithmetic, and while many of its foundations were borrowed from Indian and Greek sources, it is clear that Islamic mathematicians united the various strands of arithmetic into a form recognizable to us today.

JULIAN A. SMITH

REFERENCES

Al-Daffa, Ali Abdullah. *The Muslim Contribution to Mathematics*. London: Croon Helm, 1977.

Berggren, J.L. *Episodes in the Mathematics of Medieval Islam*. New York: Springer Verlag, 1986.

Høyrup, Jens. "Formative Conditions for the Development of Mathematics in Medieval Islam." In *George Sarton Centennial*. Ed. Werner Callebaut et al. Ghent: Communications & Cognition, 1984.

Rashed, Roshdi. *The Development of Arabic Mathematics*. Dordrecht: Kluwer, 1994.

Youschkevitch, Adolf P. *Les Mathématiques Arabes (VIIIe-XVe siècles)*. Paris: Vrin, 1976.

See also: Mathematics – Trigonometry – Number Theory – al-Khwārizmī – al-Kindī – al-Karajī – al-Uqlīdisī – Sexagesimal System – *Almagest* – al-Kāshī – Thābit ibn Qurra – Ibn Khaldūn – Naṣīr al-Dīn al-Ṭūsī

ARMILLARY SPHERES IN CHINA The equatorial armillary sphere was a traditional Chinese astronomical instrument used to observe celestial bodies in an equatorial coordinate system. Its origin is still not very clear. Astronomer Luoxia Hong (ca. 100 BC) of the Western Han Dynasty was probably the first maker of this instrument which possessed a very basic form.

The early equatorial armilla was composed of two layers: the outside layer included a meridian circle, equatorial circle, and vertical circle — all three of these were fixed. The inside layer included a polar axis, right ascension circle, and sighting tube. The right ascension circle could rotate around the polar axis, and the sighting tube could rotate in the right ascension circle freely so it could point to everywhere in the sky.

In the Tang Dynasty (AD 618–907), a third layer was added to the equatorial armillary which included an ecliptic (the band of the zodiac through which the Sun apparently moves in its yearly course) circle and a circle of the moon's path. Astronomers could then measure three coordinate systems with one instrument, but the three-layer armilla was too complex to observe, so from the Northern Song Dynasty (AD 960–1126) a course of simplification was begun. The third layer was canceled, and the so-called "abridged armilla" (*Jian Yi*) appeared. It is in fact two different instruments (one equatorial armilla and one altazimuth) on the same pedestal.

The equatorial armilla was one of the most important astronomical instruments in ancient China. It was the result of the equatorial tradition of Chinese astronomy which lasted more than 2000 years. In ancient China, armillae (and almost all astronomical instruments) were only made by the imperial government, so their size was always very large.

JIANG XIAOYUAN

REFERENCES

The Study Group for the History of Chinese Astronomy. *Zhong Guo Tian Wen Xue Shi* (The History of Chinese Astronomy). Beijing: Science Press, 1981.

Wei Zheng, Linghu Defen. *Sui Shu* (The History of Sui Dynasty). Beijing: Zhonghua Press, 1973, vol. 19.

See also: Luoxia Hong – Astronomy in China

ARMILLARY SPHERES IN INDIA The armillary sphere, known in Hindu astronomy by the terms *Golabandha* and *Gola-yantra* (Globe instrument), was constructed from early times for study, demonstration, and observation. Among texts and commentaries which have either brief mention or detailed treatment of the armillary sphere, the following might be mentioned: *Āryabhaṭīya* of Āryabhaṭa (b. 476), *Pañcasiddhāntikā* of Varāhamihira (505), *Śiṣyadhīvṛddhida* of Lalla (eighth century), *Brāhmasphuṭasiddhānta* of Brahmagupta (b. 628), *Siddhāntaśekhara* of Śripati (1039), *Sūrya-siddhānta*, *Siddhāntairomaṇi* (*Golabandhādhikāra*) of Bhāskara II (b. 1114), and *Goladīpikā* of Parameśvara (1380–1460).

The movable and immovable circles which form parts of the instrument are made out of thin bamboo strips, the earth and the celestial bodies are of wood or clay, and the lines are connected by means of strings. The axis is made of iron and is mounted on two vertical posts, so that it is possible to rotate the sphere as needed.

The *Goladipikā* describes the construction of a simple *Golabandha* with two spheres, the inner one representing the *Bhagola* (Starry sphere) moving inside an outer sphere which represents the *Khagola* (Celestial sphere), both fitted on the same central axis. The movements of the planets, etc. are projected and measured on these spheres. A circular loop made of a thin bamboo strip kept vertically in the north–south direction represents the solsticial colure (*Dakṣiṇottara*). Another similar circle fixed to the former in the east–west direction would be the equinoctial or celestial equator (*Ghaṭikā-maṇḍala*). Still another circle fixed around these, cutting them at right angles and making crosses at the four cardinal points, represents the equinoctial colure. The celestial equator is graduated into 60 equal parts and the other two into 360 equal parts. Another circle is fixed passing through the east and west crosses and inclined at 24° north and south of the zenith and the nadir — this would be the Ecliptic (*Apama-vṛtta*). Several smaller circles are now constructed across the solsticial colure on either side of the celestial equator and parallel to it, at the required declinations — these would be the diurnal circles (*Ahorātra-vṛttas*), of different magnitudes. The orbits of the Moon and other planets are now constructed, crossing the ecliptic (the band of the zodiac through which the Sun apparently moves in its yearly course) at the nodes (*pātas*) of the respective planets and diverging from it, north and south, by their maximum latitudes at 90° from the nodes. A metal rod is inserted through the north and south crosses to form the central axis. This figuration is called the Starry sphere (*Bhagola*).

Three circles are constructed outside the *Kha-gola* (Celestial sphere). The horizontal circle is called the horizon (*Kṣitija*), the east–west circle is called prime vertical (*Samamaṇḍala*), and the north–south circle is called the meridian (*Dakṣiṇottara*). A model of the Earth in spherical form is then fixed to the center of the axis. This would be the figuration at zero latitude.

If the armillary sphere is to be used in any other place, two holes are made in the Celestial sphere at a distance equal to the latitude of the place, below and above the south and north crosses, and the axis of the Starry sphere is made to pass through them. It is also necessary, in this case, to construct a circle called the equinoctial colure (*Unmaṇḍala*), which passes through the two ends of the axis and the east and west crosses. To keep the two sets of circles in position, wooden pieces are fixed to the axis in between them, so that the spheres do not get displaced.

The inner Starry sphere revolves constantly, while the Celestial sphere remains stationary, and directions are reckoned therefrom. A diurnal circle with its radius equal to the sine latitude is also constructed with a point on the central axis as the center, just touching the horizon. The sine and cosine of the place can be measured on this circle.

While the armillary sphere described above is more for study and demonstration, certain other texts speak of more circles and also enjoin observation of celestial bodies. Thus *Brāhmasphuṭasiddhānta* (XXI. 49–69) prescribes the construction of 51 movable circles, *Śiṣyadhīvṛddhida* speaks of diagonal circles (*Koṇa-vṛttas*) and spheres for each planet, and *Siddhānta-śiromaṇi* adds a third circle called *Dṛggola* outside the *Khagola* (Celestial sphere). According to *Śiṣyadhīvṛddhida* the *lagna* (Orient ecliptic point) and time are also found by means of the armillary sphere.

K.V. SARMA

REFERENCES

Dikshit, Sankar Balakrishna. *Bharatiya Jyotish Sastra (History of Indian Astronomy)*, Pt. II. New Delhi : Indian Meteorological Department, 1981, pp. 224–25.

The Goladīpikā by Paramesśvara. Ed. and trans. K.V. Sarma. Madras : Adyar Library and Research Centre, 1957.

"Goḷam Keṭṭal (Golabandham)" (in Malayalam). In *Bhāratīya Śāstra-manjūṣa*. Ed. M.S. Sreedharan. Trivandrum : Bharatiya Sastra-manjusha Publications, 1987. Vol. II, pp. 40–55.

Ohashi, Yukio. *A History of Astronomical Instruments in India*. Ph.D. Thesis, Lucknow University, 1990.

See also: Astronomical Instruments in India

ĀRYABHAṬA Āryabhaṭa (b. AD 476) was a celebrated astronomer and mathematician of the classical period of the Gupta dynasty (AD 320 to ca. 600). This era is called the Golden Age in the history of India, during which Indian intellect reached its high water mark in most branches of art, science, and literature, and Indian culture and civilization reached a unique stage of development which left its deep impression upon succeeding ages. Āryabhaṭa played an important role in shaping scientific astronomy in India. He is designated as Āryabhaṭa I to differentiate him from Āryabhaṭa II, who flourished much later (ca. AD 950–1100) and who wrote the *Mahāsiddhānta*.

Āryabhaṭa I was born in AD 476. This conclusion is reached from his own statement in the *Āryabhaṭīya*: "When sixty times sixty years and three quarters of the *yuga* (now *Mahā*) had elapsed, twenty three years had then passed since my birth" (III, 10).

Since the present *Kaliyuga* (the last quarter of the *Mahāyuga*) started in 3102 BC, Āryabhaṭa was 23 years old in 3600 minus 3101, that is in AD 499. The exact date of birth comes out to be March 21st, when the Mean Sun entered the zodiac sign of Aries in AD 476. The significance of mentioning AD 499 is that the precession of equinoxes was zero at the time, so that the given planetary mean positions did not require any correction. According to some commentators, AD 499 was also the year of compostion of the *Āryabhaṭīya*.

We have no knowledge about his parents or teachers, or even about his native place. Āryabhaṭa composed the *Āryabhaṭīya* while living at Kusumapura, which has been identified as Pāṭaliputra (modern Patna in Bihar State), the imperial capital of the Gupta empire. It is possible that Āryabhaṭa headed an astronomical school there.

The association of Patna, where Āryabhaṭa taught and wrote on mathematics and astronomy, with his professional career, does not settle the question of his birthplace, but it may have been a place where he was educated.

Āryabhaṭa's fame rests mainly on his *Āryabhaṭīya*, but from the writings of Varāhamihira (sixth century AD), Bhāskara I, and Brahmagupta (seventh century), it is clear that earlier he composed an *Āryabhaṭa Siddhānta*. Although voluminous, the *Āryabhaṭa Siddhānta* is not extant. It is also called *Ārdharātrika Tantra,* because in it the civil days were reckoned from one midnight to the next. Its basic parameters are preserved by Bhāskara I in his *Mahābhāskarīya* (Chapter VII). Rāmakṛṣṇa Ārādhya (AD 1472) has quoted 34 verses on astronomical instruments from the *Āryabhaṭa Siddhānta*, of which some were devised by Āryabhaṭa himself.

The *Āryabhaṭīya* is an improved work and the product of a mature intellect. Considering the genius of Āryabhaṭa, it is easy to agree with the view that he composed it at the age of 23. The date is also in fair agreement with the recent research and analysis by Roger Billard. Unlike the *Āryabhaṭa Siddhānta*, the civil days in the *Āryabhaṭīya* are reckoned from one sunrise to the next — a practice which is still prevalent among the followers of the Hindu calendar. The *Āryabhaṭīya* consists of four sections or *pādas* (fourth parts):

1. *Daśagītikā* (10 + 3 couplets in Gīti meter);
2. *Gaṇitapāda* (33 Verses on Mathematics);
3. *Kāla-kriyāpāda* (25 Verses on Time-Reckoning); and
4. *Golapāda* (50 Verses on Spherical Astronomy).

That the *Āryabhaṭīya* was quite popular is shown by the large number of commentaries written on it, from Prabhākara (ca. AD 525) through Nīlakaṇṭha Somayāji (ca. 1502) to Kodaṇḍarāma (ca. 1850).

An Arabic translation of the *Āryabhaṭīya* entitled *Zīj al-Ārjabhar* was made in about 800, possibly by al-Ahwāzī. In spite of the *Āryabhaṭīya*'s popularity, H.T. Colebrooke failed to trace any work of Āryabhaṭa anywhere in India.

The use of modern scientific methodology, as described by Roger Billard in his *L'astronomie indienne*, along with new ephemerides, clearly shows that both of Āryabhaṭa's major works were based on accurate planetary observations. In fact, the use of better planetary parameters, the innovations in astronomical methods, and the concise style of exposition rendered the *Āryabhaṭīya* an excellent textbook in astronomy. In opposition to the earlier geostationary theory, Āryabhaṭa held the view that the earth rotates on its axis. His estimate of the period of the sidereal rotation of earth was 23 hours, 56 minutes, and 4.1 seconds, which is quite close to the actual value.

Āryabhaṭa has been also considered the father of Indian epicyclic astronomy. The resulting new planetary theory enabled Indians to determine more accurately the true positions and distances of the planets (including the sun and the moon). He was the first Indian to provide a method of finding celestial latitudes. He also propounded the true scientific cause of eclipses (instead of crediting the mythlogical demon Rāhu). In fact his new ideas gave rise to the formation of a new school of Indian astronomy: the Āryabhaṭa School or *Āryapakṣa*, for which the basic text was the *Āryabhaṭīya*.

Exposition and computation based on the new astronomical theories were made easy by Āryabhaṭa, because of the development of some mathematical tools. One of them was his own peculiar system of alphabetic numerals. The 33 consonants of the Sanskrit alphabet (Nāgarī script) denoted various numbers in conjunction with vowels which themselves stood for no numerical value. For example, the expression *khyughṛ* (=*khu* + *yu* + *ghṛ*) denoted

$2 \times 100^2 + 30 \times 100^2 + 4 \times 10^3 = 4{,}300{,}000,$

which is the number of revolutions of the sun in a Yuga.

The development of Indian trigonometry (based on sine instead of chord, as the Greeks had done) was another of Āryabhaṭa's achievements which was necessary for astronomical calculations.

Because of his own concise notation, he could express the full sine table in just one couplet, which students could easily remember. For preparing the table of sines, he gave two methods, one of which was based on the property that the second order sine differences were proportional to sines themselves.

Āryabhaṭa seems to have been the first to give a general method for solving indetermininate equations of the first degree. He dealt with the subject in connection with the problem of finding an integral number N which will give a remainder r when divided by an integer a, and s when divided by b. This amounts to solving the equations

$$N = ax + r = by + s.$$

Although at present the topic of indeterminate analysis comes under pure mathematics, in ancient times it arose and was used for practical and astronomical problems. In fact, Āryabhaṭa successfully used his theory of indeterminate analysis to determine a mean conjunction of all planets at the zero mean longitude at the start of the *Kaliyuga* (3102 BC). Recently it has been shown that his algorithm solves more general problems than the Chinese remainder theorem, and works irrespective of the sign of numbers.

The solution of a general quadratic equation and the summation of certain series were some other algebraic topics dealt with by Āryabhaṭa. The methods of adding an arithmetical progression were known in all ancient cultures, but he was perhaps the first to supply a general rule for finding the number of terms (n) when the first term (a), the common difference (d) and the sum (s) were given. His solution is a root of the quadratic equation

$$dn^2 + (2a - d)n = 2s,$$

which comes from the usual formula for the sum of an arithmetical progression.

In geometry, his greatest achievement was an accurate value of π. His rule amounts to the statement

$$\pi = 62832/20000 \text{ nearly}.$$

This implies the approximation 3.1416 which is correct to its last decimal place. How he arrived at this is not known.

From what we know about Āryabhaṭa, it is clear that he was an outstanding astronomer and mathematician. His scientific attitude, rational approach, and mathematical methodology ushered in a new era in the history of the exact sciences in India. It was quite befitting that the first Indian satellite launched on the 19th April, 1975 was named Āryabhaṭa.

R.C. GUPTA

REFERENCES

Ayyangar, A.A. Krishnaswami. "The Mathematics of Āryabhaṭa." *Quarterly Journal of the Mythic Society* 16: 158–179, 1925–26.

Billard, Roger, "Āryabhaṭa and Indian Astronomy: an Outline of an Unexpected Insight." *Indian Journal of History of Science* 12(2): 207–224, 1977.

Elfering, Kurt. *Die Mathematik des Āryabhaṭa I*. Munich: Wilhelm Fink Verlag, 1975.

Jha, Parmeshwar. *Āryabhaṭa I and his Contribution to Mathematics*. Patna: Bihar Research Society, 1988.

Kak, Subhash, "Computation Aspects of the Āryabhaṭa Algorithm." *Indian Journal of History of Science* 21(1): 62–71, 1986.

Sen, S.N. "Āryabhaṭa's Mathematics." *Bulletin of the National Institute of Sciences of India* 21: 27–319, 1963.

Sengupta, P.C. "Āryabhaṭa, the Father of Indian Epicyclic Astronomy." *Journal of the Department of Letters, Calcutta University* 18: 1–56, 1929.

Shukla, K.S., and K.V. Sarma, eds. and trans. *Āryabhaṭiya of Āryabhaṭa*. New Delhi: Indian National Science Academy, 1976.

Shukla, K.S. *Āryabhaṭa: Indian Mathematician and Astronomer*. New Delhi: Indian National Science Academy, 1976.

Shukla, K.S., ed. *Āryabhaṭīya, with the Commentary of Bhāskara I and Someśvara*. New Delhi: Indian National Science Academy, 1976.

See also: Mathematics in India – Astronomy in India – Trigonometry in India

ASADA GORYU Asada Goryu (1734–1799) was a Japanese astronomer who was instrumental in turning Japanese astronomy and calendrical science away from the traditional Chinese style and toward Western models.

In adopting the traditional idea of secular diminution of tropical year length, astronomers at the time were required only to account for the ancient records and modern data of Chinese solstitial observations by a single formula. Classical Western data, such as those listed in the *Almagest* of Ptolemy, became available to Asada through the Jesuit treatises. He endeavored to synthesize Western and Chinese astronomy and to give a numerical explanation, by means of a single principle, of all the observational data available to him — old and new, Eastern and Western.

Copernicus appears in the Sino-Jesuit treatises, not as an advocate of heliocentricism but as an observational astronomer and the inventor of the eighth sphere of trepidation. He is said to have believed that the ancient tropical year was longer than that of the Middle Ages, which in turn was shorter than the contemporary constant. Asada, perhaps struck by this passage, formulated a modified conception in which the length of the ancient tropical year tended to decrease until it reached a minimum in the Middle Ages and to grow longer afterward, varying in a precession cycle of 25,400 years.

It is apparent that what Asada really intended to do was account for the newly acquired Western data. His basic goal, that of "saving the ancient records" by numerical manipulation, differs not at all from that of the traditional approach. His consideration of the precession cycle was theoretical decoration.

NAKAYAMA SHIGERU

REFERENCES

Asada Goryu. *Shocho ho*. Manuscript preserved in Tohoku University Library.

Nakayama, Shigeru. *A History of Japanese Astronomy*. Cambridge, Massachusetts: Harvard University Press, 1969.

ASTROLABE The astrolabe is a portable wooden or metal astronomical instrument which is used to measure the positions and altitudes of celestial bodies, to find the observer's time or latitude, or to solve other mathematical problems. In its complete form, it consists of a main body, or flat plate ("mater" or "mother") to which is attached a series of smaller plates (called climates) engraved with various coordinate lines, according to various latitudes. An alidade, a rotatable straight rule with sights used to find altitudes, is fastened to the back. Attached to the front, above the climates, is a smaller fretted circular plate called the rete — this is a moveable map of the heavens, with pointers indicating various stars. The whole rotating assembly is fastened together by a pin through the center of the mater and climates, and it is secured at the top by a wedge-shaped piece of metal called the horse, after its fanciful resemblance to a horse's head.

To use the astrolabe, an observer would typically rotate the alidade until a star became visible through the sights, and then read its altitude in degrees from a scale on the back of the instrument. Then one would turn the rete until that star corresponded to the almucantar (curves representing parallels to the horizon) for the right altitude. The time could then be determined from the place of the "sun" on the instrument. Astrolabes could measure solar time, sidereal time, and time in unequal hours, depending on how the hour lines were marked.

Although the astrolabe is usually considered an Arabic invention, its true roots go back to the mathematical astronomers of ancient Greece — the word itself comes from the Greek terms for a "star-taking" instrument. Engraving great circles and hour lines demanded considerable skill — in essence, the instrument-maker had to collapse a three-dimensional celestial sphere into the flat, circular plane of the astrolabe. Some historians attribute this discovery to Apponius of Perga (ca. 262–ca. 190 BC), but according to Otto Neugebauer, the planispheric or stereographic projection of the heavens upon a flat surface was first accomplished by Greek astronomer Hipparchus of Nicea/Rhodes (ca. 190–ca. 120 BC). In addition, Hipparchus may have built simple astrolabes, consisting of solid sky maps covered with open networks of lines. Finally, he is credited with developing the projection for an anaphoric clock, an ancestor of the astrolabe. This is an axle with a large circular star map, laid out in a stereographic projection — attached to it is a smaller stationary grill showing the projection of the horizon, and a visible hemisphere for a given latitude. The dial is powered by a clepsydra, and as it rotates, a model of the sun traces out the hours of the day. The anaphoric clock was described by Roman engineer Vitruvius (ca. 25 BC), and may have been built in the famous Athenian "Tower of the Winds" around 50 BC.

The first clear descriptions of the construction of the astrolabe occur in the *Planisphere* of Alexandrian astronomer Claudius Ptolemy (ca. 150 AD). By this time, the complexity of various lines had made the astrolabe's covering bulky, a problem Ptolemy solved by switching the lines to the mater and the star map to the rete. The astrolabe was further developed in the works of Greek mathematician Theon (fl. AD 360–380), now lost, and John Philoponus (fl. 520), both of Alexandria. Syrian Severus Sebokht (fl. 630) also wrote an early treatise on the astrolabe.

Medieval Islamic scientists took the basic planispheric astrolabe of Hellenistic astronomers and improved it dramatically between AD 700–1500, applying it to questions of astronomy, surveying, mathematics, geography, and much more. In AD 843, Baghdad mathematician al-Khwārizmī (fl. AD 810–850) claimed his astrolabe could solve 43 problems — a century later, Persian astronomer al-Ṣūfī (AD 903–986) said that it could answer a thousand astronomical questions. David King divides Arabic astrolabe innovations into five basic categories: the making of tables, non-standard retes, qiblas, multiple climates, and the development of three new forms of the instrument.

Astronomers like al-Farghānī of Baghdad (ca. 830–ca. 861) compiled numerical tables of radii and center distances of both azimuth and almucantar circles for every degree of latitude and azimuth, for every terrestrial latitude. These tables, which exceeded 13,000 entries, were used extensively by Arabic astronomers alongside geometric projections to construct astrolabes for different latitudes. Islamic astronomers also constructed non-standard retes, which would symmetrically represent the otherwise dissimilar northern and southern halves of the ecliptic (the band of the zodiac through which the Sun apparently moves in its yearly course). The Oxford myrtle astrolabe is an example.

Arabic astrolabists inscribed specific markings on their instruments, corresponding to Islamic prayer times and directions. By the thirteenth century, they engraved lists of cities with latitudes, longitudes, and Mecca directions, known as qiblas — this would help observers orient themselves for prayers. Arabic astrolabes also developed multiple plates or climates, giving astronomical tables usually found in handbooks or textbooks. An early example is referred to by Abū Jaʿfar al-Khāzin (d. 961/971).

Finally, Islamic scientists developed at least three new types of instrument: the linear, universal and geared astrolabes. The linear astrolabe consists of a series of scales on a stick which represents the meridian for a given latitude, to which are attached a series of threads through which one can perform all the standard operations of an astrolabe. This was invented by Iranian mathematician Sharaf al-Dīn al-Ṭūsī (d. ca. 1213), and was known as "al-Ṭūsī's cane". The universal astrolabe was developed by Ḥabash of Baghdad (d. ca. 864–874), and Alī ibn-Khalaf (al-Shakkāz) in Toledo in the eleventh century, who devised a special shakkaziya plate for it. Though powerful, it was not widely known, and was reinvented in early fourteenth century Syria. This astrolabe could determine the risings, culminations, and settings of celestial bodies at all latitudes using a single plate. Astronomer al-Zarqāllu (Azarquiel) of Toledo (ca. 1029–1087/1100) simplified this by replacing the rete with an alidade having a movable cursor, and by putting the ecliptic and star pointers on the shakkaziya plate. Geared astrolabes contained complex mechanisms to reproduce the motion of the sun and moon mechanically — their date of invention is unknown, but they were described by Persian astronomer al-Bīrūnī (ca. 973–1048).

The astrolabe was reintroduced into Europe from the Arabs by the tenth century. Gerbert of Aurillac, Pope Sylvester II (ca. 945–1003) imported much astronomical knowledge into the medieval Latin west from Islamic sources, and may have used the astrolabe as a teaching tool. Meanwhile, Hermannus Contractus (1013–1054) transmitted many Arabic concepts in his two influential Latin treatises on the astrolabe: *De Mensura Astrolabii* and *De Utilitatibus Astrolabii*. Ptolemy's *Planisphere* was translated by Hermann the Dalmatian in the twelfth century, and in 1276, an influential Arabic astrolabe treatise by the Egyptian Jew Māshāʾallāh (fl. 762–ca. 815) was translated into Latin, where it formed the basis for the first English book on the astrolabe, by poet Geoffrey Chaucer (ca. 1340–ca. 1400) in 1391.

Though the astrolabe was widely developed in Arab countries, it was virtually ignored in the East. Joseph Needham says that Chinese astronomers made no astrolabes of their own, though they did develop the anaphoric clock independently. The reasons are twofold: Chinese scientists instead developed sophisticated globes and armillary spheres quite early, and they lacked the analytical techniques in their mathematical astronomy which led to the stereographic projection in the West. Needham suggests the astrolabe was imported to China from Persia in 1267 by the Maraghan Observatory astronomer Jamāl al-Dīn ibn Muḥammed al-Najārri, but the mathematical projections remained obscure until the arrival of the Jesuits in the sixteenth century.

Indian scientists also borrowed the astrolabe from Islamic sources. Astronomer Bhāskara II (1114–ca. 1184) used spherical astronomy to construct a wheel-like instrument, called the *phalaka yantra*, which essentially served the same purpose as a primitive astrolabe. However, the true planispheric astrolabe, the *ustaralava*, was first imported by Muslims in the thirteenth century — a Damascus astrolabe of 1204 is still preserved in the Rampur State Library. The astrolabe was described in detail in the *Yantrarāja* of Mahendra Sūri in 1370 – this work is based on Islamic sources. Lahore later became a center for its construction, under families of instrument makers like those of Shaikh Allāh-Dad (fl. 1570–1660). Indian astronomers also developed some of the largest astrolabes in the world. The great brass astrolabe at the Jaipur observatory of Sawai Jai Singh (1686/8–1743) is 3 meters tall, 2.12 meters across, and weighs over 400 kilograms. Jai Singh's *Yantrarājaracanā* also gives instructions for astrolabe construction based on stereographic projections.

The astrolabe was popular as an astronomical instrument until long after the introduction of the telescope in the early seventeenth century. It survived another century, until it was replaced in the 1730s by English astronomer John Hadley's (1682–1744) reflecting quadrant, a precursor to the sextant.

JULIAN A. SMITH

REFERENCES

Abdi, W.H., et al. *Interaction Between Indian and Central Asian Science and Technology in Medieval Times*. New Delhi: Indian National Science Academy, 1990.

Bose, D.M., et al., *A Concise History of Science in India*. New Delhi: Indian National Science Academy, 1971.

Drachman, A.G. "The Plane Astrolabe and the Anaphoric Clock." *Centaurus* 3: 183–9, 1954.

Gunther, R.T. *Astrolabes of the World*. London: Holland Press, 1976.

Jaggi, O.P. *History of Science, Technology and Medicine in India*, esp. vol. 6–7. Delhi: Atma Ram, 1977–1986.

King, David. *Islamic Astronomical Instruments*. London: Variorum, 1987.

Kuppuram, G., and K. Kumudami. *History of Science and Technology in India* 1–12. Delhi: Sundeep Prakashan, 1990.

Michel, H. *Traité de l'Astrolabe*. Paris: Gauthier-Villars, 1947.

Needham, Joseph. *Science and Civilisation in China*. Cambridge: Cambridge University Press, 1954.

Needham, Joseph. *Heavenly Clockwork*. Cambridge: Cambridge University Press, 1960.

Neugebauer, Otto. "The Early History of the Astrolabe." *Studies in Ancient Astronomy* IX, 1949 – reprinted *Isis* 40: 240–56, 1949.

North, J.D. "The Astrolabe." *Scientific American* 230: 96–106, 1974.

Saunders, Harold N. *All the Astrolabes*. Oxford: Senecio, 1984.

See also: Astronomical Instruments – al-Khwārizmī – al-Ṣūfi – *Qibla* and Islamic Prayer Times – al-Farghānī – al-Zarqāllu – Māshā'allāh – Sharaf al-Dīn al-Ṭūsī – Observatories – Jai Singh – Maragha al-Bīrūnī – Bhāskara – Mahendra Sūri – Abū Ja'far al-Khāzin

ASTROLOGY IN CHINA In traditional China there was no distinction between astronomy and astrology. The common word *tianwen* covered both. There was also no distinction between astronomy and astrology in Europe before the end of the seventeenth century. According to the *Oxford English Dictionary*, astrology was of two kinds: (a) natural astrology, which involved the calculation and foretelling of natural phenomena, as the measurement of time, fixing of Easter, predictions of tides and eclipses, and also of meteorological phenomena, and (b) judicial astrology, which was the art of judging the reputed occult and non-physical influences of the stars and planets on human affairs, also known as star-divination or astromancy. Since the end of the seventeenth century the term natural astrology was replaced by astronomy and meteorology, while judicial astrology became the astrology commonly known today.

Traditional Chinese astrology included the two elements of natural and occult science. The latter provided the motivating force that enabled Chinese astronomers to produce the most comprehensive and continuous observational records in the past until the seventeenth century for almost two thousand years. These records are of interest to modern astronomers, but they were never made with such intention. They were meant primarily to enable the emperor to have a foreknowledge of future events concerning himself, his imperial household, his senior officials, his empire, his subjects, and foreign countries. Astronomical observations also played a part in the calculation of calendars that gave auspicious and ominous times and dates for various kinds of events in daily life, ranging from wedding ceremonies to having a bath or a haircut. Traditional Chinese judicial astrology differs from its counterpart in Europe in that it was tailored exclusively to serve the emperor and not the individual. Officially this system is now obsolete among the Chinese, but unofficially there were some whispers linking the event of the demise of their Great Helmsman to the 1976 Tangshan earthquake. What is popularly known today as Chinese astrology is not the traditional official Chinese astrology referred to above. Official and popular astrology are two different entities.

Official astrology found its place in the "Astronomical Chapters" of the Chinese Dynastic Histories, beginning with the *Shiji* (Historical Memoirs) of Sima Qian (145–86 BC). Based on the traditional Chinese belief in the close relationship between heaven (*tian*), earth (*di*) and man (*ren*), the emperor was regarded as the representative of heaven on earth — human actions on the part of the emperor and celestial phenomena had mutual effects. Chinese astrologers divided the whole sky visible to them, making the stars and asterisms correspond to geographical regions. Almost all of them corresponded to China and only a few smaller asterisms corresponded to neighboring countries.

The Polar Star, which was supposed to remain stationary, was regarded as the counterpart of the emperor in heaven. Perhaps because there was seldom any bright star near the North Pole, and the scope for making predictions would be much more limited when a star was far away from the ecliptic (the band of the zodiac through which the Sun apparently moves in its yearly course) beyond the reach of the planets, a number of other stars were also designated to represent the emperor. The asterisms in the circumpolar region, known as the *Ziweiyuan* (Purple Subtlety Enclosure), included those that represented the emperor, the empress, the imperial concubines, and the crown prince. There were stars representing his hierarchy of officials, including ministers and military commanders, and there were stars representing the utilities in the palace, for example the kitchen. The region was enclosed by two chains of stars, representing the walls of the Forbidden City. Many parts of the Forbidden City and the circumpolar region shared the same names. Outside the "walls" of the circumpolar region was the Plough (*Beidou*), an important asterism in both Chinese astrology and Daoism.

Next come the asterisms along the ecliptic. Two other special regions, the *Taiweiyuan* (Great Subtlety Enclosure) and the *Tianshiyuan* (Celestial Market Enclosure), were located there. The former again pertained to the emperor, his household, and his official hierarchy and the latter to the general state of economy in his empire. Distributed along the ecliptic were the twenty-eight lunar mansions. They were used to make a wide range of predictions, from flood and drought in the empire to military activities among the border tribal people. The asterisms near the twenty-eight lunar mansions were also significant. The astrologer could observe, for example, four stars in Pisces, called *Yunyu* (Cloud and Rain) to make a forecast for rain, thus performing the task of the modern meteorologist.

The Sun was the most important astrological object, because it represented the emperor. Solar eclipses and sunspots reflected blemishes on the part of the emperor. Likewise lunar eclipses would refer to those of the empress. The astrologer looked for the presence of planets, comets, and novae near a particular star or asterism to predict an event and where it would happen. Changes in the color or brightness and scintillation due to atmospheric conditions were also noted by the astrologer. He also observed aurora borealis and clouds, noting their color and shape. These were particularly important in the battlefield for gaining advanced information on enemy movements and the outcome of the combat. The astronomical bureau also produced an astrological almanac using the art of *zheri* (calculations of auspicious and inauspicious days) to work out days and times that were auspicious or unlucky for certain events in private and social life, for example having a bath or a haircut, meeting a friend, doing a business transaction, moving house, and holding a wedding ceremony.

There were often occasions when the astrologer was required to give an answer to a specific event, for example when something was lost, when a candidate set out to take the civil examinations, when two armies were facing each other preparing for battle, and so on. There were three sophisticated techniques of divination which fell within the syllabus of candidates taking examinations in the astronomical bureau in the Song Dynasty, namely *taiyi* (Supreme Unity), *dunjia* (Concealing the *jia*s), and *liuren* (Six *ren*s). These three methods did not restrict themselves to the imperial family and the official hierarchy.

Naturally the common people also wished to have a foreknowledge of their individual fate and destiny. The Chinese developed many systems for this purpose, but none of them relied on direct astronomical observation. In the strict sense of the word they hardly qualify to be called astrology. However they generally employed the results of astronomical observations and calculations by using some or all the elements of year, month, day, and time. Furthermore, at least one of the systems contains traces of Greek and Hindu astrology. There are two systems of fate-calculation in general circulation among the Chinese today, namely the *Ziping* method and the *Ziwei doushu* method. These two systems of fate-calculation do not rely on direct observations of the stars and in the *Ziwei doushu* horoscope are worked out without requiring the practitioner to know how to identify the stars that occur in the horoscope.

The history of fate calculation in China is rather obscure as this was not regarded as an orthodox branch of study, and experts writing on this subject often used imprecise language to put off the uninitiated. By the Han period (206 BC–AD 220) Confucian scholars were talking about three types of human fate (*sanming*). One was endowed during birth, and was the only element that could be calculated. One was under the influence of good or evil deeds, and one was governed by catastrophic events that would overrule the first two. It is interesting to draw comparisons with the Han scholars' contemporaries in Europe where the Romans were adopting Stoicism as their state philosophy, believing in the devotion to duty while leaving things to the inevitable. The Chinese had a different belief in life, by talking about three types of life rather than one. To the Chinese it was only the fated life that was predictable, but any predicted event was by no means inevitable. It might be changed according to one's deeds, or by what we nowadays describe as an "act of God". The Chinese system gave encouragement to lead a good life. In this respect it certainly sounded more attractive than Hellenistic astrology.

At first it seems that only the year of birth of the person concerned was taken into account. Even today some Chinese still speak about the "twelve animal cycle" of the years they were born. As time went by, first the month, then the day, and finally the hours (or rather double hour) of birth were gradually included to develop newer systems. The *Ziping* method is one of the most sophisticated systems of fate-calculation. It is attributed to Xu Juyi, said to have lived during the latter half of the tenth century. He was the first to use the time of the day for fate calculation. Many books were also attributed to him, but we do not know exactly which were actually written by him. The most authoritative and comprehensive text that we have on the *Ziping* method is the *Sanming tonghui* (Confluence of the Three Fates) written by Wan Minying during the Ming period (1368–1644). Since then the system has frequently been revised to keep abreast with changes in social structure. This can be testified to by the large number of books written on this subject in China, Hong Kong, Japan, and Korea during the past few decades.

However, back in the third century Indian astrology had made entry into China when the *Sārdulakarnāvadāna* was translated, introducing the names of the "Seven Luminaries" (Sun, Moon, and the five planets) and the *nakṣatras* (Moon-stations). In the year 718 Gautama Siddhārtha translated the *Navagrāha* calendar and introduced the names of two imaginary heavenly bodies, *Rahu* and *Ketu*. Some time afterwards a Nestorian named Adam translated a work called *Simenjing* (lit. Book On Four Departments). This book has long been lost, but it could have been a translation of Ptolemy's *Tetrabiblos* according to recent Japanese authorities on the subject such as Kiyoshi Yabuuchi. Another book with the title *Duliyusijing* is suggested by Michio Yano to carry the name of Ptolemy in a corrupted form. Hellenistic astrology also went from Persia to China through Korea. Another important route taken by Hellenistic astrology was through India, where it was modified under the influence of Hindu astrology. Tantric monks played an important role in the introduction of this form of astrology to China. Amoghavajra (705–774) produced a book known under its abridged title *Xiuyaojing* (Book on the *Nakṣatras* and the Luminaries), in which the twelve signs of the zodiac appeared for the first time in China. Thus by the eighth century imported systems of astrology with Hellenistic and also often Hindu roots had become quite popular. A number of actual horoscopes cast during the Tang period are preserved in the *Qinding gujin tushu jicheng* (Imperial Encyclopedia), edited by Chen Menglei et al. in 1726.

Changes took place when new ideas came into the same melting pot with something that was originally in it. Traditional Chinese star names and astrological terms were adopted by the new imports. Gradually the latter became sinicized. One can hardly notice the Hellenistic and Hindu origin of the *Ziwei doushu system* by looking at its name alone. We do not know when exactly the term "*Ziwei*" was first used here. During the Tang period several names were used; among them was the term "*Taiyi*". The "star" *Taiyi* played the same role as the "star" *Ziwei* in the modern system, and both have somewhat similar reference to the occupant of the most supreme position below or above. The term *Ziwei doushu* first appeared in the title of a book incorporated in the Daoist *Tripitaka*, the *Xu Daozang* in 1607. Similarly traditional fate-calculation methods were at the same time influenced by imported cultures. The Ziping system for example employs a cycle of twelve phases, reminiscent of the Twelve *Nidānas* in Buddhism.

HO PENG YOKE

REFERENCES

Ho Peng Yoke. *The Astronomical Chapters of the Chin Shu*. Paris and the Hague: Mouton & Co., 1966.

Ho Peng Yoke. *Cong li qi shu guandian tan Ziping tuimingfa* (The Ziping Method of Astrology). Hong Kong: Hong Kong University Press, 1988.

Ho Peng Yoke and Ho Koon Piu. *Dunhuang canzhuan Zhan yunqi shu yanjiu* (The Dunhuang MS *Zhan yunqi shu*). Taipei: Yiwen Press, 1985.

Smith, Richard J. *Fortune-Tellers and Philosophy*. Boulder, San Francisco and Oxford: Westview Press, 1991.

Yano, Michio. *Mikkyōno senseijitsu* (Astrology of Esoteric Buddhism). Tokyo: Tokyo Bijitsu, 1986.

See also: Geomancy – Lunar Mansions – Zodiac – Divination in China

ASTROLOGY IN INDIA In India, astrology, *Jyotiṣa*, is defined as *Jyotiṣam sūryādi grahāṇām bodhakamśāstram*, the system which explains the influences of the sun, moon, and planets.

Indian astrology came explicitly to light around 1200 BC, when the monk Lagadha compiled the *Vedānga-Jyotiṣa* on the basis of *Vedas*, in which lunar and solar months are described, with their adjustment by *Adhimāsa* (lunar leap month). *Ṛtus* (seasons), years, and *yugas* (eras) are also described. Twenty-seven constellations, eclipses, seven planets, and twelve signs of the zodiac were also known at that time.

In the period from 500 BC to the beginning of the Christian era some texts were written on the subject of astrology. Nineteen famous sages composed their *Siddhāntas* (texts). *Candra-prajnapti*, *Surya-prajnapti* and *Jyotiṣakaraṇḍaka* were written. The *Sūryasiddhānta*, the ancient text of Indian astrology, was composed around 200 BC.

In the first five centuries of the Christian era, there were some important contributions by Jain writers. *Angavijjā* is a large collection about *Śakuna* (omens). *Kālaka* and *Ṛsiputra* also contributed around this time. At the end of the fifth century, Āryabhaṭa I mentioned in his text *Āryabhaṭīya* that the sun and stars are constant and that day and night are based on the movement of the earth.

The period AD 500–1000 was very productive. Lallācārya, the disciple of Āryabhaṭa, composed two texts — *Śiṣyadhīvṛdhi* and *Ratnakoṣa* — dealing with mathematical theories. The astrologer Varāhamihira composed several texts, and his son Pṛthuyaśā composed a brief horary called *Ṣat-Pañcāśikā*. Bhāskarācārya I wrote a commentary on the *Āryabhaṭīya* in the seventh century, and Brahmagupta composed the *Brāhmasphuṭasiddhānta* and the *Khaṇḍakhādyaka* around AD 635. Other scholars wrote commentaries on the texts of their predecessors and independent texts of their own.

In 1000–1500, there was a great deal of enhancement to the literature concerning the construction of astronomical instruments for observation. In the twelfth century, Bhāskara composed the famous text *Siddhāntaśiromaṇi*. The *Līlāvatī* of Rājāditya is another of the texts of that century. In the fifteenth century, Keśava wrote more than ten books, and his son Gaṇeśa composed the *Grahalāghava* at the age of thirteen.

Many more texts and commentaries were written from the sixteenth century onwards. A few noteworthy ones are: *Tājikanīlakaṇṭhī* of Nīlakaṇṭha (sixteenth century), *Meghamahodaya* by Meghvijayagaṇi (seventeenth century), *Janmapatrīpaddhati* by Lābhacandragaṇi (eighteenth century), and the nineteenth century works of astrologer Bāpūdeva Śāstri.

A knowledge of *pañcānga* is a prerequisite to understanding the subject of astrology. This is the fivefold system of *tithi* (lunar day), *vāra* (weekday), *nakṣatra* (asterism), *yoga* (sum of the solar and lunar longitudes) and *karaṇa* (half lunar day). *Tithi*, the lunar date, is the duration of time in which the Moon moves 12°. The fifteen *tithis* of the white fortnight (from new moon to full moon) are:

1. *Pratipadā*;
2. *Dvitīyā*;
3. *Tṛtīyā*;
4. *Caturthī*;
5. *Pañcamī*;
6. *Ṣaṣṭhī*;
7. *Saptamī*;
8. *Aṣṭamī*;
9. *Navamī*;
10. *Daśamī*;
11. *Ekādaśī*;
12. *Dvādaśī*;
13. *Trayodaśī*;
14. *Caturdaśī*;
15. *Purṇimā* ($15\times 12°=180°$).

In the black fortnight (from full moon to new moon), the fifteenth day is called *Amāvasyā* and the remainder are the same as above. *Tithis* are classified into five groups: *Nandā* (*tithis* 1,6,11), *Bhadrā* (2,7,12), *Jayā* (3,8,13), *Riktā* (4,9,14), and *Pūrṇā* (5,10.15).

The seven *vāras* (weekdays) are based on the names of the *grahas*: Sun, Moon, Mars, Mercury, Jupiter, Venus, and Saturn.

There are twenty-seven *nakṣatras* (asterisms) bifurcating the ecliptic into twenty-seven parts, each of 13.33°. These are mentioned in Table 1.

Table 1: Twenty-seven *nakṣatras* (asterisms)

Kṛttikā	*Rohiṇī*
Mṛgaśiras	*Ārdrā*
Punarvasu	*Puṣya*
Āśleṣā	*Maghā*
Pūrvāphālgunī	*Uttarāphālgunī*
Hasta	*Citrā*
Svātī	*Viśākhā*
Anurādhā	*Jyeṣṭhā*
Mūla	*Pūrvāṣāḍhā*
Uttarāṣāḍhā	*Śroṇā*
Śraviṣṭhā	*Śatabhiṣaj*
Pūrva-Bhādrapada	*Uttara-Bhādrapada*
Revatī	*Aśvinī*
Bharaṇī	

The ecliptic is again bifurcated into twelve parts through *Rāśis* (signs, each of 30°). The twelve signs are equal to twenty-seven *nakṣatras*, or 1 sign = 2.25 constellations. For example, *Aśvīnī*, *Bharaṇī*, and one quarter of *Kṛttikā* make the sign *Meṣa* (Aries). The remaining three quarters of *Kṛttikā*, *Rohiṇī*, and half of *Mṛgaśira* make the sign *Vṛṣa* (Taurus). The same pattern holds true for the other signs: *Mithuna* (Gemini), *Karka* (Cancer), *Singh* (Leo), *Kanyā* (Virgo), *Tulā* (Libra), *Vṛścika* (Scorpio), *Dhanu* (Sagittarius), *Makara* (Capricorn), *Kumbha* (Aquarius), and *Mīna* (Pisces). Thus twenty seven constellations represent twelve signs.

Yoga is the sum of the solar and lunar longitudes. If the sum of their degrees is between 0 and 13.33°, that is called *Viṣkambha Yoga* — from there until 26.66° it is *Prīti* — up to 40° it is *Āyuṣmāna*. The remaining yogas are *Saubhāgya*, *Śobhana*, *Atigaṇḍa*, *Sukarmā*, *Dhṛti*, *Śūla*, *Gaṇḍa*, *Vṛdhi*, *Dhruva*, *Vyāghāta*, *Harṣaṇa*, *Vajra*, *Siddhi*, *Vyatīpāta*, *Varīyāna*, *Parigha*, *Śiva*, *Siddha*, *Sādhya*, *Śubha*, *Śukla*, *Brahma*, *Aindra*, and *Vaidhṛti* ($13.33°\times 27 = 360°$).

Karaṇa (constant or moveable) is the half part of the *tithi*. Constant *Karaṇa Śakuna* belongs to the second half of *caturdaśī Catuṣpada* and *Nāga* to that of *Āmāvasyā* in the black fortnight, while *Kistughna* exists in the first half of the *Pratipada* of the white fortnight in every lunar month. The remaining 14.5 *tithis* of the white and 13.5 *tithis* of the black fortnight contain eight rounds of seven moveable *Karaṇas*: *Bava*, *Bālava*, *Kaulava*, *Taitila*, *Gara*, *Vaṇija*, and *Viṣṭi*.

The subject matter of astrology may be divided into five groups: *Saṃhitā*, *Siddhānta*, *Jātaka*, *Praśāna*, and *Śakuna*. In ancient India, *Saṃhitā* was the miscellaneous collection

Chart 1. Positions of *Grahas* (planets) on 21 March 1994 at 6:02 am at Varanasi

Grahas (Planets)	Sun	Moon	Mars	Mercury	Jupiter	Venus	Saturn	*Rāhu*	*Ketu*
Rāśi (sign)	11	2	10	10	6	11	10	7	1
Anśa (degree)	6	15	16	9	23	22	8	3	3
Kalā	21	7	57	51	19	13	21	3	3
Vikalā	27	45	18	17	40	8	21	44	44

Chart 2. Ascendant as sketched in northern India

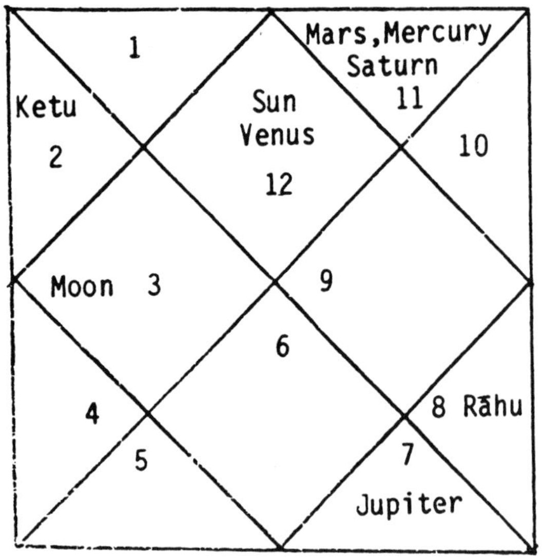

Chart 3. Ascendant as sketched in southern India

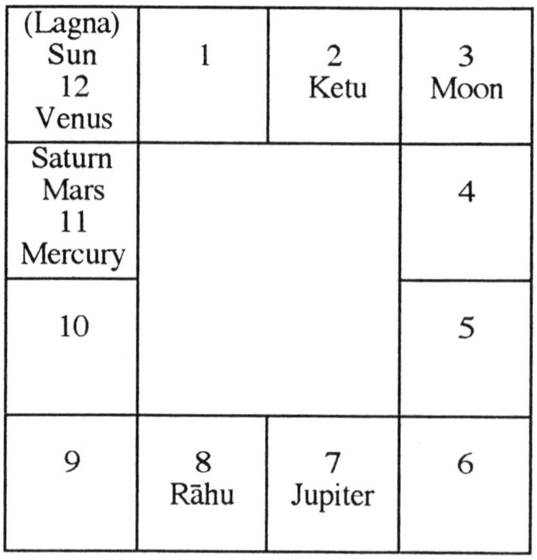

Chart 4. Ascendant as sketched in eastern India (West Bengal and Orissa)

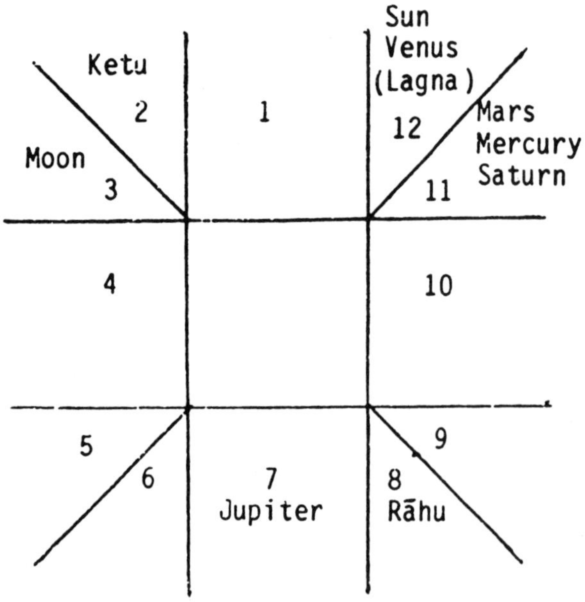

Chart 5. Navamanśa chart

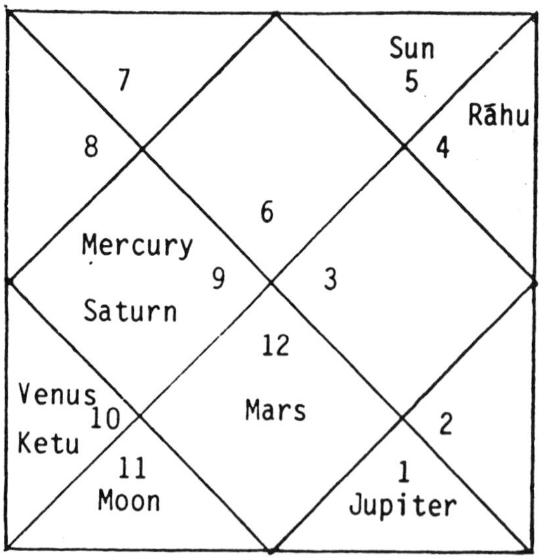

of astrological materials out of which the remaining four grew.

Siddhānta or *Gaṇita* refers to mathematical calculations about time, distance, and position of the planets. On the basis of the proper positions of twelve signs and nine planets, a chart containing twelve chambers may be sketched. In northern, southern, and eastern India, astrologers sketch charts such as numbers 2, 3, and 4 which are called *Janmāṇga* or ascendant.

Jātaka (native) is the person about whom a prediction is made on the basis of a birth chart. Twelve houses represent the health, wealth, brother/sister, mother, offspring, diseases/enemies, wife/husband, death, fate, father, income, and expenses, as in chart number 2. *Daśās* (periods) are defined in numerous ways. The period of any planet becomes favorable or harmful according to its position and power in the horoscope.

There are many other astrological methods in India. As an example, in Kerala-system, numbers are assigned to alphabets, and the astrologer advises the person to say the names of a flower, river, or god on which the calculation depends.

Astrology is applied to many aspects of Indian life. There are rules concerning times for traveling, planting, and building. Favorable times for the preparation of medicines and treatment are also prescribed.

VIJAYA NARAYAN TRIPATHI

REFERENCES

Jagannātha, R. *Principles and Practice of Medical Astrology*. New Delhi: Sagar Publications, 1994.

Lokamani, D. *Bhāratiyajyotiṣaśāstrasyetihāsah* (History of Indian Astrology). Varanasi: Chaukhambha Surabharti, 1990.

Raghunandanaprasāda, G. *Ādhunika-Jyotiṣa* (Modern Astrology). Delhi: Ankur Publications, 1994.

Raman, B.V. *A Manual of Hindu Astrology*. Delhi and London: UBS Publishers, 1992.

Sureṣacandra, M. *Jyotiśa-Sarvasva* (Complete Hindu Astrology). New Delhi: Ranjan Publications, 1994.

ASTROLOGY IN ISLAM A few considerations about the historical development of the term and concept of astrology as an intellectual discipline are in order, so as to avoid the many misconceptions that prevail in this field of historical enquiry.

The first question concerns the terminology applicable in medieval Arabic culture. What we consider astrology in our epistemology has very little connection with its medieval definition. Horoscope-making and interpreting are of course part of the game but with a rather remote bearing on its definition as a science in medieval eyes. In the mind of medieval Arab writers there is but one science of the sky with the moving bodies set in it. It was called *ʿilm an-nujūm* (science of the stars) and it consisted of two distinct treatments of the subject matter of the heavens: a purely mathematical one or *ʿilm al-falak* corresponding to our astronomy, and a humanistic but rather conjectural one which aimed at deducing from the celestial motions their probable significance for the evolution of human affairs, more directly what we now call astrology. The name for this latter discipline was *ʿilm aḥkām an-nujūm* (science of the judgments of the stars): hence the new term *scientia iudiciorum* or judicial astrology in medieval Latin culture. The two methods of treatment were indissolubly linked in the overall picture, and it must be further stressed that the dominating interest of medieval Arabic civilization was the "science of the judgments of the stars". On the practical level, the second portion of this dichotomy was considered an art, and the term *ṣanʿa* applied to it was the equivalent of any other trade or profession.

There were three levels through which Arab authors would approach the study of astrology: a first level through the general science of the stars which could be a predominantly philosophical enquiry bordering on what may be actually labeled cosmology. A second level, purely mathematical, consisted of the consideration of the movements of the spheres, of the celestial bodies they contained, and of phenomena affecting them. This approach corresponded more closely to our astronomy. A third level was in the extension of the above observations to judge their probable impact upon human affairs. The technique of these judgments (*aḥkām*) was determined according to very intricate rules embodied in the age-old lore of astrology proper: but the three levels were considered together to constitute the one science of the stars (*ʿilm an-nujūm*).

This tripartite structure of the science of the heavens among medieval Arab writers was not entirely their creation, for the historical event of early Arab conquests of the entire Middle East had put the young Arab civilization in direct contact with both the still active Hellenistic (Alexandrian) world of thought thriving in Egypt and in the Eastern Mediterranean on the one hand, and the very ancient Babylonian, Persian, and even Hindu cultural traditions on the other. In examining medieval Arabic astrology it is wise to keep in mind these major cultural orientations varying in importance according to the stages of those historical convergences. During the first century of Arab conquests, which corresponds roughly with the rule of the Umayyad Caliphate of Damascus (660–750), the predominant cultural influence came from Syria and Egypt, in which Hellenistic culture and further Christianized Hellenism dominated. Even so, the impact of Greek learning was not directly linked at first with classical Greek science or philosophy, but rather with its late Hellenistic phase heavily marked by neo-Platonic and Alexandrian speculation or mysticism (hermeticism). The full force of the Greek example of learning and of its formative impact on Arabic astrology emerged only under the Abbasid rule (beginning in 750), tentatively at first under al-Mansūr (754–775) and Harūn al-Rashīd (786–809), but dramatically under al-Maʾmūn (813–833) through the direct importation of the works of thinkers and of scientists such as Aristotle in physics and cosmology, and of Ptolemy in astronomy/astrology. Al-Maʾmūn established at Baghdad an astronomical observatory and *Bayt al-Ḥikma* (House of Wisdom) endowed with a great library. It was because of these favorable conditions that the science of astrology, like philosophy and medicine, took its definite hue in Arab literature under the label of *falsafa* (a transliteration of the Greek term philosophy). Maʾmūn's patronage brought scientists and philosophers from all over the Muslim empire. It was likely in the midst of this intellectual fervor that the greatest writer in Arab astrology, Abū Maʿshar, came from his native Balkh in Khurāsān (now Afghanistan) to settle in Baghdad. Although Abū Maʿshar's major work, the *Kitāb al-mudkhal al-kabīr* (Greater Introduction to Astronomy, AD 848) was completed during the generation following the death of al-Maʾmūn (AD 833), its success in molding the framework of Arab astrology that merged the diverse astrological traditions of Greece, Persia, and India must be ascribed to the lively interest in *falsafa* engendered by al-Maʾmūn's sponsorship. With Abū Maʿshar's work, Arab astrology acquired its definitive structure, the result of a syncretism of all Middle East traditions under the umbrella of Aristotelian cosmology and Ptolemaean astronomy/astrology.

Abū Maʿshar came to be quoted as the authority, even by those who criticized him (al-Bīrūnī, Ibn Ridwān). His *Kitāb al-mudkhal al-kabīr* became the "bible" of Arabic astrology because it buttressed the science with a theoretical foundation based on *falsafa*, with Persian and Hindu traditions more or less coherently merged into it. Yet it provided only the introductory theory of astrology as part of the science of Nature. Further extension of the science of judgments of the stars to the full range of human affairs was seen as a kind of adjustment to their inevitable cosmic framework. Arabic astrological science came to include these five principal divisions.

1. A theoretical or introductory part (*mudkhal*) exploring its foundations in physical science and metaphysics. Here Abū Maʿshar's *Kitāb al-mudkhal al-kabīr* shared ultimate authority together with Ptolemy's *Tetrabiblos*.

2. A section dealing with Nativities (*mīlād, mawālīd*) which consisted of drawing up diagrams (horoscopes) of the state of the sky at the time of any beginning. Its most natural occasion was at the time of birth (hence "nativities"), or even of conception when possible, and it would be held as an indication of the probable unfolding of the various life circumstances of the individual person or object for whom it was drawn up. It is not without interest to recall that before the establishment of individual identity status such as birth registers, beginning with the sixteenth century in Europe, natal horoscopes constituted the most reliable record of the chronological span of individual lives. Not every one was born of sufficiently wealthy or honorable stock to be able to afford this luxury. Abū Maʿshar himself lamented the fact that he did not know the exact date of his birth, to compensate for which he had drawn up for himself a "general" (approximative) horoscope.

3. Interrogations (*masāʾīl*) which dealt mostly with enquiries about objects hidden or lost, innermost thoughts or intentions, purposes, etc. A kind of oracle, it aimed to assist individuals in their important decision making or help recover missing objects.

4. Elections or Choices (*ikhtiyārāt*), which were concerned with determining the most favorable moment for starting on important undertakings, such as the construction of cities, opening of hostilities in wartime, investiture or inauguration, or starting on a journey.

5. Weather predictions or meteorology, which were almanacs which astrologers operating for courts, cities, or institutions like universities would issue at the beginning of each new year, as part of their official duties. Weather predictions were of course a prime concern in any predominantly agrarian society.

The art or trade of the astrologer on the other hand was referred to by the term *ṣanʿa* and was treated like any other profession. More specialized applications of astrological science still flourished beside the main stream, particularly in medicine where some Greek treatises of Hippocrates (*On Airs* and *De hebdomadibus* for instance) and Galenic ones about duration of pregnancy were merged into astrological prognostication (*taqdimat al-maʿrifa*) and enjoyed enormous vogue among physicians. Finally all sorts of "predictions" or "interpretation of signs" proliferated in a number of specialized practices of quackery into which some pretense of astrological judgments was introduced. Some of these are chiromancy (interpretation of lines of the hand), spatulomancy (interpretation of form of shoulder blade), and sternutomancy (on sneezing).

An influential sequel to Abū Maʿshar's *Greater Introduction* appeared by Aḥmad ibn Yūsuf, a physician, mathematician, and astrologer of the Tūlūnid era (870–904) in Egypt. Aḥmad wrote a chronicle of this Turkish dynasty and he authored several works of mathematics. He put together an astrological compendium which he entitled *Kitāb ath-Ṭhamara* (Liber Fructus). Since it comprised one hundred short propositions, each one accompanied by a substantial commentary, it came to be designated as *Centiloquium*. In fact its major doctrines are taken straight from Abū Maʿshar's *Kitāb al-mudkhal al-kabīr*. The very passage in Abū Maʿshar's work which probably gave Aḥmad the inspiration for his forgery is met in a special section of the magnum opus (III, 1–2) where Abū Maʿshar enumerates the six benefits to be derived from astrological science, the most alluring of which are the "fructus" (*ṭamara*) to be anticipated from it.

These two works by Abū Maʿshar and by Ibn Yūsuf respectively, the second in the footsteps of the first, influenced the West from the time of their translation into Latin during the twelfth century until the demise of astrology as a science in the Scientific Revolution of early modern times. The nature, influence, and significance of Arabic astrology in the East and the West during the Middle Ages were polarized around the success of these two major works.

RICHARD LEMAY

REFERENCES

Lemay, R. "L'Islam historique et les sciences occultes." *Bulletin d'Études Orientales*, 1992.

Manfred Ullmann. *Die Natur- und Geheimwissenschaften im Islam*. Leiden: E.J. Brill, 1972.

Nallino, C.A. "Astrology" and "Astronomy." In *Encyclopedia of Islam*. Leiden: Brill, 1913–34, pp. 494–501.

Nallino, C.A. *ʿIlm al-falak. Arabian Astronomy. Its History during the Medieval Times*. Badhdad: Maktabat al Mathna, 1960–.

Saliba, G. "Astronomy, Astrology, Islamic." In *Dictionary of the Middle Ages*. vol. 1. Ed. J.R. Strayer. New York: Scribner's, 1978.

Steinschneider, M. *Die Europäischen Übersetzungen aus dem Arabischen bis mitte des 17. Jahrhunderts*. Vienna: C. Gerold's Sohn, 1905–1906.

See also: al-Maʿmūn – Abū Maʿshar – Ibn Ridwān – al-Bīrūnī

ASTRONOMICAL INSTRUMENTS IN INDIA Astronomical knowledge in India can be traced back to the Vedic literature (ca. 1500–500 BC), the earliest literature in India, but no astronomical instrument is mentioned there. Naked eye observations of the sun, moon, and lunar mansions were carried out. It is not clear whether five planets were observed or not.

There is a class of works called *Vedāṅga*, probably composed towards the end of the Vedic period, which is regarded as auxiliary to the *Veda*. It consists of six divisions, including *Jyotiṣa* (astronomy) and *Kalpa* (ceremonial). The *Kalpa* further consists of four divisions, including *Śulba* (method of the construction of the altar). The earliest astronomical instruments in India, the gnomon and the clepsydra, appear in the *Vedāṅga* literature.

The gnomon (*Sanskrit: śaṅku*) is used for the determination of cardinal directions in the *Kātyāyana-śulbasūtra*. A vertical gnomon is erected on a leveled ground, and a circle is drawn with a cord, whose length is equal to the height of the gnomon, with the center the foot of the gnomon. At the two points where the tip of the gnomon-shadow touches the circle, pins are placed, and they are joined by a straight line. This line is the east–west line.

The annual and diurnal variations of the length of the gnomon-shadow are recorded in the political work *Arthaśāstra* of Kauṭilya, the Buddhist work *Śārdūlakarṇa-avadāna*, and Jaina works such as the *Sūrya-prajñapti*. These records seem to be based on observations in North India.

The clepsydra is mentioned in the *Vedāṅga-jyotiṣa*, the *Arthaśāstra*, and the *Śārdūlakarṇ-avadāna*. It was like a water jar with a hole at its bottom from which water flowed out in a *nāḍikā* (one sixtieth of a day).

Towards the end of the *Vedāṅga* astronomy period, certain Greek ideas of astronomy and astrology had some influence in India from the second to the fourth century AD. After that, Hindu astronomy (*Jyotiṣa*) established itself as an independent discipline, and several fundamental texts called *Siddhāntas* were composed. I call this period, from about the fifth to the twelfth centuries AD, the classical Siddhānta period. The main astronomers who described astronomical instruments are Āryabhaṭa (b. AD 476), Varāhamihira (sixth century AD), Brahmagupta (b. AD 598), Lalla (eighth or ninth century AD), Śrīpati (eleventh century AD), Bhāskara II (b. AD 1114), and the anonymous author of the *Sūryasiddhānta*. The *siddhāntas* composed by Brahmagupta, Lalla, Śrīpati, and Bhāskara II contain special chapters on astronomical instruments entitled *Yantra-adhyāya*. The Sanskrit word *yantra* means instrument. No observational data are recorded in the *siddhāntas*, and the extent of actual observations in this period is controversial. Roger Billard maintained that astronomical constants in the *siddhāntas* were determined by actual observations, while David Pingree argued that they were exclusively borrowed from Greek astronomy. In this connection, we should note that the method of determination of astronomical constants by means of observations was correctly explained by Bhāskara II. Let us see the instruments in this period.

The gnomon (*śaṅku*) was continually used in this period. The theory of the gnomon, such as the relationship between the length of gnomon-shadow, the latitude of the observer,

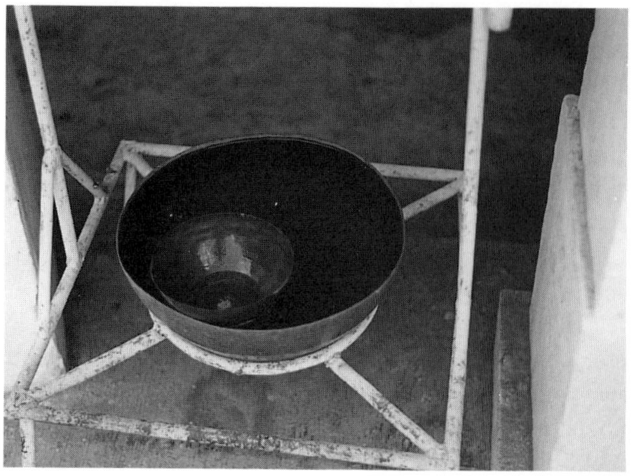

Plate 1. Clepsydra at Rao Madho Singh Museum, Kota, Rajasthan, India. Photograph by the author. Used with his permission.

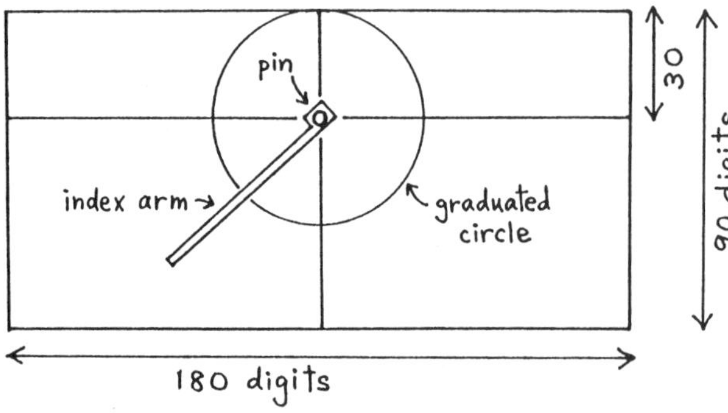

Figure 1: *Phalaka-yantra.*

and time, was developed in this period, and a special chapter called *Tripraśna-adhyāya* in the *Siddhāntas* was devoted to this subject. Trigonometry, invented in India, was fully utilized for this purpose.

The staff (*yaṣṭi-yantra*) is a simple stick, used to sight an object. There are some variations of the staff, such as V-shaped staffs for determining angular distance with the help of a graduated level circle.

The circle-instrument (*cakra-yantra*) is a graduated circular hoop or board suspended vertically. The sun's altitude or zenith distance is determined, and time is roughly calculated from it. Variations of the circle-instrument are the semi-circle instrument (*dhanur-yantra*) and the quadrant (*turya-golaka*).

A circular board kept horizontally with a central rod is the chair-instrument (*pīṭha-yantra*), and a similar semi-circular board is the bowl-instrument (*kapāla-yantra*). The sun's azimuth is determined by them, and time is roughly calculated from them.

A circular board kept in the equatorial plane is the equator-instrument (*nāḍīvalaya-yantra*). It is a kind of equa-

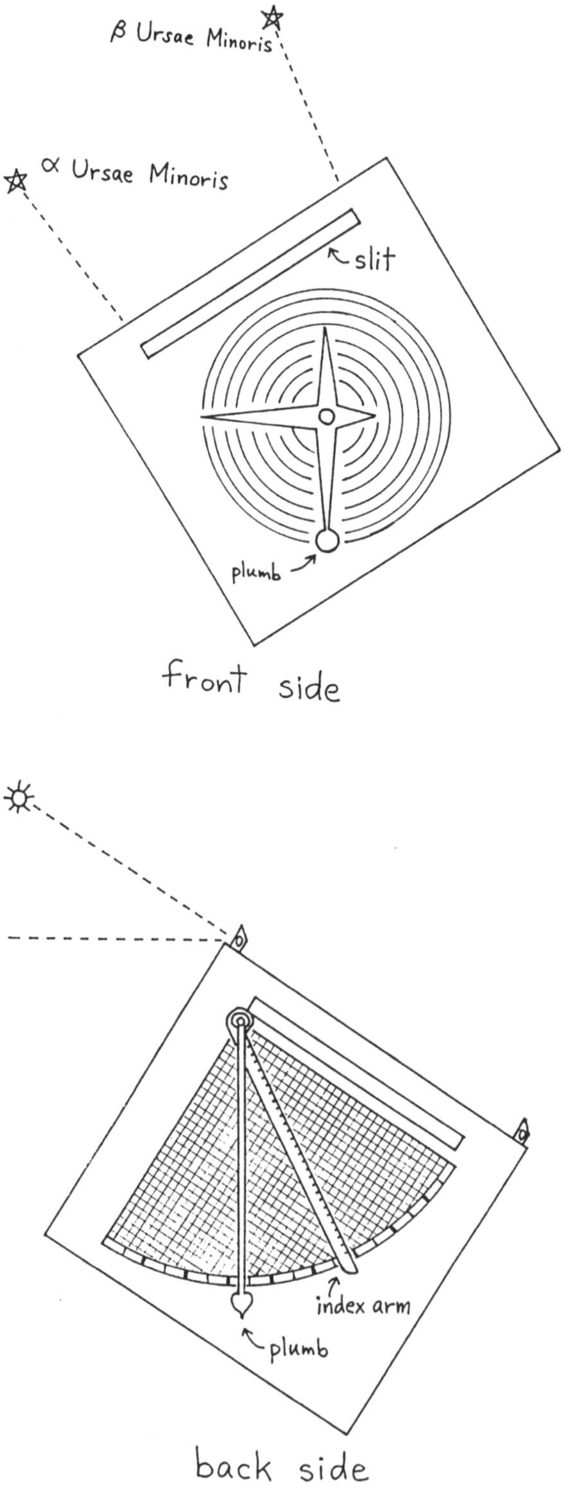

Figure 2: *Dhruva-bhrama-yantra.*

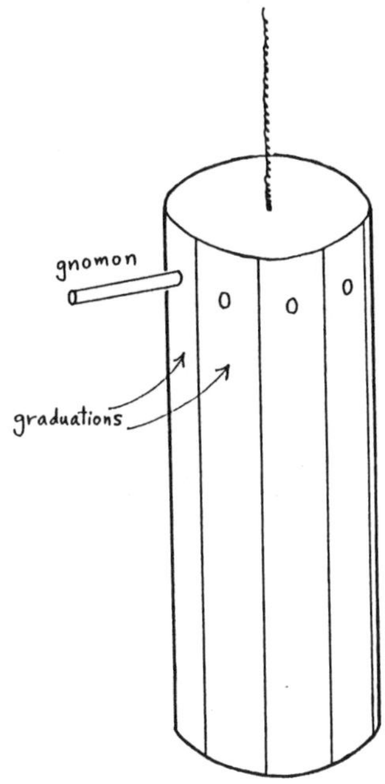

Figure 3: *Pratoda-yantra*.

torial sundial. The combination of two semi-circular boards, one of which is in the equatorial plane, is the scissors-instrument (*kartarī-yantra*). Its simplified version is the semi-circular board in an equatorial plane with a central rod.

The armillary sphere (*gola-yantra*) was unlike the Greek armillary sphere, which was based on ecliptical coordinates, although it also had an ecliptical hoop. Probably, the celestial coordinates of the junction stars of the lunar mansions were determined by the armillary sphere since the seventh century or so. There was also the celestial globe rotated by flowing water.

The clepsydra (*ghaṭī-yantra*) was widely used until recent times. Unlike the clepsydra of the *Vedāṅga* period, the clepsydra of this period is a bowl with a hole at its bottom floating on water. Water flows into the bowl, and it sinks after a certain time interval. The actual use of the clepsydra of this type was recorded by the Chinese Buddhist traveler Yijing (AD 635–713). The clepsydra actually used can be seen in a museum at Kota (Rajasthan) (see Plate 1). Several astronomers also described water-driven instruments such as the model of fighting sheep.

The *phalaka-yantra* (board-instrument) invented by Bhāskara II is a rectangular board with a pin and an index arm, used to determine time graphically from the Sun's altitude (see Fig. 1). This is an ingenious instrument based on the Hindu theory of gnomon.

Plate 2. *Samrāṭ-yantra* at Jantar Mantar, New Delhi, India. Photograph by the author. Used with his permission.

The astrolabe was introduced into India from the Islamic world at the time of Fīrūz Shāh (r. AD 1351–1388) of the *Tughluq* dynasty. Fīrūz Shāh's court astronomer Mahendra Sūri composed a Sanskrit work on the astrolabe entitled *Yantra-rāja* (King of Instruments, the Sanskrit term for the astrolabe) in AD 1370. This is the earliest Sanskrit work on Islamic astronomy. Use of the astrolabe rapidly spread among some Hindu astronomers, and Padmanābha (AD 1423) and Rāmacandra (AD 1428) described the astrolabe in their works.

Some new instruments were made in the Delhi Sultanate and Mughal periods. Padmanābha invented a kind of nocturnal instrument called *dhruva-bhrama-yantra* (polar rotation instrument) (see Figure 2). It was a rectangular board with a slit and a set of pointers with concentric graduated circles. Adjusting the slit to the direction of α and β Ursae Minoris, time and other calculations could be obtained with the help of pointers. Its back side was made as a quadrant with a plumb and an index arm. Thirty parallel lines were drawn inside the quadrant, and trigonometrical calculations were done graphically. After determining the Sun's altitude with the help of the plumb, time was calculated graphically with the help of the index arm.

Later, the quadrant as an independent instrument was described by Cakradhara, and a more exact method to calculate time was explained.

Another new type of instrument in this period was the cylindrical sundial called *kaśāyantra* (whip instrument) by Hema (late fifteenth century AD) or *pratoda-yantra* (whip instrument) by Gaṇeśa (b. AD 1507) (see Figure 3). It is a cylindrical rod having a horizontal gnomon and graduations of time according to the vertical shadow below the gnomon.

The quadrant and the cylindrical sundial exist in the Islamic world also, but the possibility of their influence on these Indian instruments is still to be investigated.

The mahārāja of Jaipur, Sawai Jai Singh (AD 1688–1743) constructed five astronomical observatories at the beginning of the eighteenth century. The observatory in Mathura is not extant, but those in Delhi, Jaipur, Ujjain, and Banaras are. There are several huge instruments based on Hindu and Islamic astronomy. For example, the *samrāṭ-yantra* (emperor instrument) is a huge sundial which consists of a triangular gnomon wall and a pair of quadrants towards the east and west of the gnomon wall. Time has been graduated on the quadrants (see Plate 2).

By this time, European astronomy had begun to be introduced into India, and Jai Singh had certain information about European astronomy. The earliest European style astronomical observatory in India is a private one of William Petrie, an officer of the British East India Company, which was set up in 1786 at his residence in Madras.

YUKIO ŌHASHI

REFERENCES

Billard, Roger. *L'astronomie Indienne.* Paris: École française d'extrème-Orient, 1971.

Kaye, G.R. *The Astronomical Observatories of Jai Singh.* Calcutta: 1918. Reprint: New Delhi: Archaeological Survey of India, 1982.

Kochhar, R.K. "Madras Observatory: the Beginning." *Bulletin of the Astronomical Society of India* 13 (3): 162–168, 1985.

Ōhashi, Yukio. "Sanskrit Texts on Astronomical Instruments during the Delhi Sultanate and Mughal Periods." *Studies in History of Medicine and Science* 10–11: 165–181, 1986–1987.

Ōhashi, Yukio. "Development of Astronomical Observation in Vedic and Post-Vedic India." *Indian Journal of History of Science* 28(3): 185–251, 1993.

Ōhashi, Yukio. "Astronomical Instruments in Classical Siddhāntas." *Indian Journal of History of Science* 29(2): 155–313, 1994.

Pingree, David. "History of Mathematical Astronomy in India." In *Dictionary of Scientific Biography*, vol 15. New York: Scribners, 1978, pp. 533–633.

See also: Observatories – Jai Singh – Astrolabe – Globes – Lunar Mansions – Armillary Sphere – *Śulbasūtras* – Sundials – Gnomon – Clocks and Watches – Āryabhaṭa – Varāhamihira – Brahmagupta – Lalla – Śrīpati – Bhāskara II – Trigonometry – Quadrant – Mahendra Sūri

ASTRONOMICAL INSTRUMENTS IN THE ISLAMIC WORLD Most Islamic observational instruments are lost and known to us only through texts. The state of documentation of the other, smaller Islamic astronomical instruments that do survive leaves much to be desired. Many of the most important instruments are still unpublished, and much that has been written on instruments is on a very amateur level. For these reasons a project is currently underway in Frankfurt to catalogue all Islamic instruments (and European ones) to ca. 1550 as well as various historically significant later Islamic pieces.

Also the most important writings on instruments have not yet received the attention they deserve. For example, a hemispherical observational instrument for a fixed latitude was devised by the tenth-century astronomer al-Khujandī, the leading instrument maker of the early period, and this was modified in the twelfth century to serve all latitudes. There are no surviving examples, and the available manuscripts have yet to be studied. An important work on instruments was compiled in Cairo ca. 1280 by Abū ʿAlī al-Marrākushī — this has yet to be subjected to a detailed analysis. The author simply collected all of the treatises on instruments known to him and incorporated them into his book. An exciting find of the 1980s was a treatise by the fourteenth-century Aleppo astronomer Ibn al-Sarrāj, the leading instrument-maker of the later Islamic period. In this the author described every kind of instrument known to him as well as those invented by himself. This treatise is currently being investigated by François Charette in Frankfurt.

Armillary Spheres and Globes

In the eighth century al-Fazārī wrote a treatise on the armillary sphere, called in Arabic *dhāt al-ḥalaq*, which means "the instrument with the rings". No early Islamic armillary spheres survive, but several other treatises on it were compiled over the centuries. The earliest treatise in Arabic dealing with the celestial globe, called *dhāt al-kursī* (the instrument with the stand) or simply *al-kura* (sphere), was written by Qusṭā ibn Lūqā in the ninth century. This treatise by Qusṭā, who was one of the most important translators of Greek works into Arabic, remained popular for a millennium. Of the various surviving celestial globes, which number over 100, none predates the eleventh century.

The spherical astrolabe, unlike the armillary sphere and the celestial globe, appears to be an Islamic development. Various treatises on it were written from the tenth to the sixteenth century, and only one complete instrument, from the fourteenth, survives. In the ninth century Ḥabash wrote on the spherical astrolabe, the armillary sphere, and the celestial globe, as well as on various kinds of planispheric astrolabes.

Astrolabes

Al-Fazārī also wrote on the use of the astrolabe (Arabic *asṭurlāb*). The tenth-century bibliographer Ibn al-Nadīm states that al-Fazārī was the first Muslim to make such an instrument — he also informs us that, at that time, the construction of astrolabes was centred in Harran and spread from there. Several early astronomers, including Ḥabash, al-Khwārizmī and al-Farghānī, wrote on the astrolabe, and introduced the features not found on earlier Greek instruments, such as the shadow squares and trigonometric grids on the backs and the azimuth curves on the plates for different latitudes, as well as the universal plate of horizons. Also extensive tables were compiled in the ninth century to facilitate the construction of astrolabes.

Another important development to the astrolabe occurred in Andalusia in the eleventh century, when al-Zarqāllu devised the single universal plate (*ṣafīḥa*) called *shakkāziyya* and the related plate called *zarqālliyya* with two sets of *shakkāziyya* markings for both equatorial and ecliptic coordinate systems. The latter was fitted with an alidade bearing a movable perpendicular straight-edge (transversal). Several treatises on these two instruments exist in both Western and Eastern traditions of later Islamic astronomy — the Europeans knew of them as the *saphea*. Ibn al-Zarqāllu's contemporary, ʿAlī ibn Khalaf, wrote a treatise on a universal astrolabe that did not need plates for different latitudes. This treatise exists only in Old Spanish in the *Libros del Saber*, and was apparently not known in the Islamic world outside Andalusia. The instrument was further developed in Syria in the early fourteenth century: Ibn al-Sarrāj devised in Aleppo a remarkable astrolabe that can be used universally in five different ways.

The astrolabes made by Muslim craftsmen show a remarkable variety within each of several clearly-defined regional schools. We may mention the simple, functional astrolabes of the early Baghdad school — the splendid astrolabe of al-Khujandī of the late tenth century, which started a tradition of zoomorphic ornamentation that continued in the Islamic East and in Europe for several centuries — the very different astrolabes of the Andalusian school in the eleventh century and the progressive schools of Iran in the thirteenth and fourteenth centuries — and the remarkable instruments from Mamluk (thirteenth- and fourteenth-century) Egypt and Syria. In the early fourteenth century Ibn al-Sarrāj of Aleppo, a school unto himself, produced the most sophisticated astrolabe ever made. After about 1500 the construction of astrolabes continued in the Maghrib, in Iran, and in India until the end of the nineteenth century. Many of these instruments, especially those from Iran, were beautiful objects of the finest workmanship.

Quadrants

Another category of observational and computational devices to which Muslim astronomers made notable contributions was the quadrant, of which we can distinguish three main varieties. Firstly there is the sine quadrant with an orthogonal grid. This instrument, in a simpler form, had already been described by al-Khwārizmī and was widely used throughout the Islamic period. Some Islamic astrolabes display such a trigonometric grid on the back. The grid can be used together with a thread and movable marker (or the alidade of an astrolabe) to solve all of the standard problems of spherical astronomy for any latitude. Secondly there is the horary quadrant with fixed or movable cursor. This instrument is described already in an anonymous ninth-century Iraqi source and was likewise commonly used for centuries (albeit usually without the cursor, which is not essential to the function of the device). A set of arcs of circles inscribed on the quadrant display graphically the solar altitude at the seasonal hours (approximately, according to an Indian formula). Other Islamic quadrants from the ninth century onwards had markings for the equinoctial hours. The instrument can be aligned towards the sun so that the time can be determined from the observed altitude using the grid. Again this kind of marking was often found on the back of astrolabes. Thirdly there is the astrolabic quadrant displaying one half of the altitude and azimuth circles on an astrolabe plate for a fixed latitude, and a fixed ecliptic (the band of the zodiac through which the Sun apparently moves in its yearly course). The effect of the daily rotation is achieved by a thread and bead attached at the center of the instrument rather than by the movable astrolabe rete. The quadrant with astrolabic markings on one side and a trigonometric grid on the other generally replaced the astrolabe all over the Islamic world (with the notable exceptions of Iran, India, and the Yemen) in the later period of Islamic astronomy.

Sundials

We learn from Islamic tradition that the pious Umayyad Caliph ʿUmar ibn ʿAbd al-ʿAzīz (fl. Damascus, 718) used a sundial to regulate the times of the daytime prayers in terms of the seasonal hours. The earliest sundials described in the Arabic astronomical sources are planar, usually horizontal, but also vertical and polar. The mathematical theory for computing the shadow for the seasonal hours at different times of the year and the corresponding azimuths was available from Indian sources, which seem to have inspired the Islamic tradition more than any of the available Greek works. The treatise on sundial construction by al-Khwārizmī contained extensive tables displaying the polar coordinates of the intersections of the hour lines with the solstitial shadow traces on horizontal sundials for twelve different latitudes. The treatise on sundial theory by Thābit ibn Qurra contains all the necessary mathematical theory for constructing sundials in any plane — likewise impressive from a theoretical point of view is the treatise on gnomonics by his grandson Ibrāhīm.

The earliest surviving Islamic sundial, apparently made in Córdoba about the year 1000 by the Andalusian astronomer Ibn al-Ṣaffār, displays the shadow traces of the equinoxes and solstices, and the lines for the seasonal hours as well as for the times of the two daytime prayers. There is a world of difference between this simple, carelessly-constructed piece and the magnificent sundial made in the late fourteenth century by Ibn al-Shāṭir, so devised that it can be used to measure time with respect to any of the five daily prayers. In the late period of Islamic astronomy a sundial was to be found in most of the major mosques.

Miscellaneous

Several multi-purpose instruments were devised by Muslim astronomers. Notable examples are the rule (*mīzān*) of al-Fazārī, fitted with a variety of non-uniform scales for various astronomical functions, and the compendium of Ibn al-Shāṭir, comprising a magnetic compass and qibla-indicator, a universal polar sundial, and an equatorial sundial. Of particular interest is a circular qibla-indicator made in Isfahan ca. 1700 (but invented much earlier) which consists of a cartographic grid with Mecca at the centre, so devised that the qibla can be read off the outer scale and the distance from Mecca can be read off the non-uniform scale on the diametrical rule.

There are several Islamic treatises on eclipse computers and planetary equatoria for determining the positions of the planets for a given date. With these the standard problems of planetary astronomy dealt with in *zīj*es are resolved mechanically, without calculation. Treatises on eclipse computers are known from the early tenth century, and al-Bīrūnī in the early eleventh describes such an instrument in detail. A newly-discovered manuscript (not yet available for research) contains a treatise by the tenth-century Iranian astronomer Abū Jaʿfar al-Khāzin describing an equatorium called *Zīj al-Ṣafāʾiḥ* (the *Zīj* of Plates). The only known example of this instrument, made in the twelfth century, is incomplete: it is in the form of an astrolabe with tables engraved on the *mater* and additional markings for the foundation of an equatorium. Otherwise the only known early Islamic treatises on planetary equatoria are from eleventh-century Andalusia. The most interesting aspect of the equatorium described by al-Zarqāllu is the ellipse drawn on the plate for the center of the deferent of Mercury — it seems that he was the first to notice this characteristic of Mercury's deferent. Al-Kāshī, the leading astronomer of early-fifteenth-century Samarqand, has left us a description of a planetary equatorium with which not only ecliptic longitudes but also latitudes could be determined and eclipses calculated.

DAVID A. KING

REFERENCES

Gunther, Robert T. *The Astrolabes of the World*. 2 vols., Oxford: Oxford University Press, 1932, reprinted in 1 vol., London: The Holland Press, 1976.

King, David A. *Islamic Astronomical Instruments*. London: Variorum, 1987.

King, David A. "Medieval Astronomical Instruments - A Catalogue in Preparation." *Bulletin of the Scientific Instrument Society* 31: 3–7, 1991.

King, David A. "Some Remarks on Islamic Astronomical Instruments." *Scientiarum Historia* 18: 5–23, 1992.

Mayer, Leo A. *Islamic Astrolabists and their Works*. Geneva: A. Kundig, 1956.

Savage-Smith, Emilie. *Islamicate Celestial Globes – Their History, Construction, and Use*. Washington, D.C.: Smithsonian Institution Press, 1985.

See also: Astronomy – Armillary Spheres – Globes – al-Fazārī – Astrolabes – Quadrants – al-Khwārizmī – al-Farghānī – al-Zarqāllu – Sundials – Thābit ibn Qurra – Ibn al-Shāṭir – Maps and Mapmaking – al-Kāshī – al-Khujandī

ASTRONOMY Astronomy, the study of celestial objects, is a universally human endeavor whose roots lie deeply buried in prehistory. For the sky-watcher devoid of optical aid, the heavens can be thought of as a sort of earth-centered celestial sphere on to which have been sprinkled hundreds of tiny points of light we have called the stars. Half of this inverted bowl of blackness is almost completely dominated by the dazzling presence of the sun, the most prominent and important of the celestial objects. Such is the sun's brilliance that any attempt to view this object directly is to risk serious eye damage or even total blindness. As a result of the earth's spinning motion or rotation, an observer at a given location on the earth sees a sky that alternates between a daytime sky dominated by the sun and a nighttime sky characterized by its absence. As the earth turns on its axis, the sun appears to rise up from a given observer's eastern horizon, pass through a "high noon point" or maximum angle above the horizon, and then descend toward the western horizon.

Approximately one half an earth rotation later, the sun once more rises to repeat the cycle. This rising and falling effect is not limited to the sun. As the earth rotates relative to all celestial objects, they too appear to go through the rising and falling diurnal motions of the sun. Since the rate of the earth's rotation is very nearly constant, this diurnal motion of the sun and stars has long been employed as an important and reliable way of measuring time. The earth's rotation also creates the illusion that the stars of the celestial sphere seem to revolve about two imaginary points located exactly opposite each other. One, the south celestial pole, is visible only from the southern hemisphere of the earth, while the other, the north celestial pole, is visible only from the northern hemisphere. The earth's long term precessional motion carries the locations of these celestial poles along a 47 degree diameter circular path among the stars once every 26,000 years. From time to time, a relatively bright star can be found near the position of one of the celestial poles for a few centuries. Such is the case at present for the north celestial pole, which is currently located near the fairly bright star Polaris, the Pole Star.

In addition to its daily rising and setting, the sun also appears to travel along a great circle on the celestial sphere, which is called the ecliptic. This latter movement is the direct result of the earth's orbital motion about the sun. As the earth arcs along in its orbital path, the apparent position of the sun relative to the more distant background stars appears to change. For an observer on the earth, the sun thus seems to creep gradually from west to east among the stars, completing an entire 360° journey around the ecliptic in exactly the same one year time interval it takes the earth to complete one orbital revolution about the sun. The background stars hence appear to be gradually overtaken in the western sky by the sun as it moves eastward along the ecliptic, engulfed by the solar glare for a month or so, and then reemerge in the predawn sky as the sun leaves them behind in its ongoing easterly movement. The overall result of this annual movement of the sun is a seasonal parade of the heavens in which different stars are visible at different times of the night at different times of the year.

The earth's axis of rotation is also found to be tilted at an angle of $23\frac{1}{2}$ degrees off the vertical to the earth's orbital plane. As the earth orbits the sun, this tilt causes the sun to shine alternately more directly on the northern hemisphere and less on the southern hemisphere and then vice versa over the span of a simple year. This effect is observable as a yearly variation of the sun's highest altitude above the horizon at a given location, and as a change in the time that the sun spends above the horizon. Thus when the sun is shining most directly on the northern hemisphere at the time of the summer solstice, the sun's diurnal motion in the northern hemisphere is characterized by long days and short nights, and in the southern hemisphere by short days and long nights. Half an orbital revolution or six months later at the time of the winter solstice, when the sun is shining more directly on the southern hemisphere, the lengths of night and day are reversed. Halfway between these extremes the sun shines directly down on the earth's equator twice each year. On these dates, the lengths of the days and nights all over the earth are equal, except at the poles, and hence these dates are said to be the equinoxes. It is this combination of the tilt of the earth's axis of rotation and the earth's orbital motion that gives rise to our cycle of seasons here on the earth.

Firmly entrenched in second place in the brightness hierarchy of celestial objects is the moon. Although not as important as the sun, the moon, none the less, exerts several significant influences on the earth, most notably as the chief agent by which tides are produced in the world's oceans. The reflected sunlight we receive from the moon is over one million times fainter than that emanating from the sun, and as a result, the moon can be readily viewed against the backdrop of the stars of the night sky. As the moon orbits the earth in space, it appears to traverse a great circle about the celestial sphere in a fashion not unlike the annual motion displayed by the sun. There are however some important differences between the lunar motion and that of the sun. The moon swings along an apparent path that is tilted at an angle of about five degrees to the ecliptic and takes one-twelfth of the sun's time to make a single journey about the celestial sphere. The moon thus moves at an average rate of about half a degree per hour relative to the background stars, an angular speed easily detectable over the course of a single night by a naked eye observer.

Although the moon's half degree angular diameter is almost exactly the same as that of the sun, the diminished brightness of the moon permits us to look directly upon its face without fear or danger. As a result, the moon presents a number of most interesting and fascinating phenomena to the naked eye observer. Perhaps the most familiar of these is the set of seeming shape changes or phases exhibited by the moon as it journeys about the celestial sphere. These phases arise from the fact that as the moon orbits the earth, the half of the moon's spherical surface which faces the sun, and is hence illuminated by the sun's light, is viewed at different angles by an observer situated on the earth. When the moon is very nearly lined up between the earth and the sun, almost all of the moon's sunlit hemisphere faces away from the earth, and all we see of the moon is a very thin crescent of light. As the moon moves toward progressively larger angular distances from the sun, the thickness of the crescent grows or waxes until the angle between the moon and the sun as seen from the earth is 90°. At this point we see exactly one half of the moon's sunlit surface and the moon appears to have a semi-circular or "quarter-moon" shape. As the sun–moon angular separation increases past 90°, the moon takes on a bulging or gibbous shape whose thickness continues to grow until the moon is very nearly opposite the sun in the sky. When this configuration occurs, the entire sunlit hemisphere of the moon faces the earth, and the now circular-shaped moon is said to be a "full" moon. After passing through the full phase, the moon's shape changes now proceed in reverse order, successively passing through waning gibbous phases, a second or last quarter phase, and finally a waning crescent phase as the angle between the moon and the sun decreases from 180° to nearly zero. The waning crescent moon eventually slides into the predawn solar glare for a few days and then reemerges as a silvery crescent-shaped "new" moon in the post-sunset twilight.

The moon is also unique among celestial objects in that it is the only one for which surface detail can be easily viewed with the unaided human eye. This detail manifests itself in the form of the dark areas on the moon's disk which are called maria or seas and the light areas called continents. This terminology dates back to the Western European Renaissance observers of the moon who imagined the lunar surface to be divided between bright land and dark waters.

In addition to the sun and moon, human beings have recognized since prehistoric times that five other naked eye objects also move about the sky relative to the background stars. These star-like wanderers are called planets, and historically have enjoyed the appellations of the gods and goddesses of ancient Greek and Roman mythology. The five so-called naked eye planets have been named, in order of their increasing distance from the Sun, Mercury, Venus, Mars, Jupiter, and Saturn. A sixth planet, Uranus, possesses a brightness which is just at the limit of naked eye visibility, but the planetary nature of this object does not seem to have been recognized until the English astronomer William Hershel accidently stumbled upon it in 1781.

Two of the planets, Mercury and Venus, have orbits about the sun which are interior to that of the earth. As a result of this orbital geometry and the sun's gravitationally induced faster orbital speeds, these planets exhibit a marked pattern in their appearances in the earth's sky. In a typical cycle, the planet is first visible as an evening "star" in the west after sunset, then appears to move out to a maximum angle of greatest elongation away from the sun before retreating back into the sun's light. After several days or weeks, the planet reemerges from the solar glare, but this time as a morning "star" in the predawn sky. The planet once more moves out to an angle of greatest elongation and then drops back into the solar light. The swiftly moving planet Mercury goes through a complete cycle of appearances or synodic period in about four months, while Venus, whose orbital speed is more closely matched to that of the earth, takes a year and a half for its cycle of appearances. Typically Mercury's appearances as a morning or evening star last about three weeks, while those of Venus extend over several months at a time.

Visually, the plant Mercury appears in the sky as a sparkling object having a somewhat reddish-orange tint. Its apparent brightness is actually comparable to the brightest stars, but because it is almost always observed in twilight, it is usually not as impressive an object as it otherwise might be. The most spectacular of the naked eye planets and the third brightest object in the sky behind only the sun and moon is the planet Venus. The orbital path of Venus can carry it out to an angle of greatest elongation as large as 47 degrees, or about twice that exhibited by the planet Mercury. Thus it is possible to observe this splendid object for as long as four hours after sunset or before sunrise. At its greatest brilliancy, the soft white light of Venus has even been observed to cast very faint shadows as it gleams in the darkness of the pre-dawn or post-sunset night sky.

The three remaining naked eye planets Mars, Jupiter, and Saturn, move in vast orbits about the sun which are exterior to the orbit of the earth. As a result, these planets appear most of the time to move about the celestial sphere in a fashion similar to the west to east movement exhibited by the sun and moon. The times required for each of these planets to make a complete cycle about the celestial sphere, however, are far longer than those for the sun and moon. Mars, for example, completes a single journey around the celestial sphere in just under two years, while Jupiter and Saturn require nearly twelve and thirty years, respectively, to complete similar journeys. As the faster moving earth catches up to and passes one of these exterior planets, the planet exhibits an illusionary phenomenon in the earth's sky called retrograde motion in which the given planet seems to stop its normal west to east motion among the stars, moves "backward" or east to west for several months, stops again, and then resumes its direct or west to east movement. In the midst of its retrograde motion, a given planet will appear to be opposite the sun's position in the celestial sphere as sun from earth. When such a configuration occurs, the planet is said to be "in opposition to the sun," or more simply, "at opposition." At the time of a given planet's opposition, the earth makes its closest approach to the planet, and as a result, the planet shines more brightly than at any other time. Moreover, at opposition the planet rises at sunset, sets at sunrise, and is thus visible throughout the night.

Visually, the planet Mars is perhaps the most remarkable of the exterior planets owing to its distinctly reddish hue. At times of closest approach to the earth, the apparent brightness of this ruddy world is exceeded only by that of the sun, Moon, and Venus. When Mars is not at a close opposition, the fourth brightest object in the sky is the yellowish-white planet Jupiter which shines some ten times more brightly than the average of the brightest of the background stars. The golden-colored planet Saturn is the most distant of the naked eye planets from both the earth and the sun, and thus exhibits a reduced apparent brightness which is comparable to the average of the brightest background stars.

While the paths of the planets about the celestial sphere are not coincident with the ecliptic, they are, none the less, nearly coplanar with it. As a result, the sun, moon, and five naked eye planets move about the celestial sphere in a relatively narrow band of sky centered on the ecliptic which is called the zodiac. Because the sun, moon, and planets move along the zodiac at differing rates, it is possible for objects in the sky to appear to pass close to other objects along the zodiac. When such a passage occurs, the resulting configuration of the two objects is said to be a conjunction. Conjunctions can occur among the sun, moon, and planets, as well as between these moving objects and the bright stationary stars which are to be found along the zodiac. From time to time conjunctions can involve three or more objects, and on rare occasions a conjunction can be so close that the two objects cannot be seen as separate with the unaided eye.

In addition to the imaginary band of planetary paths that is the zodiac, there exists a quite real band of diffuse light, called the Milky Way, which is stationary relative to the stars and girds the celestial sphere like a gigantic faintly glowing heavenly belt. The Milky Way is the naked eye manifestation of the vast galactic system of gas, dust, and stars in which our sun is located. The Milky Way Galaxy, as this system is called, is in the shape of a huge, flat pinwheel which has a substantial bulge at its center. Our sun is situated about two-thirds of the way toward the outer edge of this system, and as a result, our view of the summertime Milky Way in the northern hemisphere is the more prominent one, since at this time we are looking toward the direction in which most of our galaxy is located. On the other hand, in the northern hemisphere winter, our view is now directed away from the galactic center toward the less prominent regions of the galaxy, with the visual result that the wintertime Milky Way is much fainter than its summertime counterpart.

Off the plane of the Milky Way, there exist approximately a dozen or so lesser diffuse objects which are visible to the naked eye and are also set in fixed positions among the stars. Modern telescopic observations reveal that these "fuzzy patches of light" are in reality quite a diverse lot, including clouds of glowing gas, star clusters, and even other galaxies well outside of our Milky Way.

From time to time transitory apparitions and events occur in the sky which can be as awesome as they are spectacular. One such event is a total eclipse of the sun. When the moon passes directly between the earth and the sun, a moving shadow of the moon about 240 kilometers wide is cast upon the surface of the earth. An observer located in the shadow's path will see the sun's disk gradually covered by the dark lunar disk until the sun's light is almost completely blotted out. During this "total" phase of the eclipse, only the light from the sun's outermost atmospheric layers is visible and a darkness comparable to a full moon night descends on the land for a time period ranging from a few seconds to as long as seven minutes. Finally the moon moves out of its direct alignment between the earth and the sun, and the sun reemerges to its full disk and full brightness.

The sun, earth, and moon can also align in such a way that the moon passes into the earth's shadow, thereby producing a total eclipse of the moon. When such an event occurs, an observer on the earth sees a full moon gradually enter the curved shadow of the earth. When the moon is totally immersed in the earth's shadow, it can take on a variety of ruddy hues ranging from an almost totally darkened red to a bright coppery shade of red–orange. This illumination even at the total phase of a lunar eclipse is caused by sunlight being refracted on to the moon's surface by the earth's atmosphere. The variety of colorations exhibited during various lunar eclipses is thus the direct result of the weather conditions in the earth's atmosphere, especially the degree of cloud cover at key locations around the earth. Typically the eclipsed moon spends an hour or so in the total phase before reemerging from the earth's shadow and regaining its full moon brilliance.

Every few years or so the night sky is visited by a strange apparition, a diffuse "long-haired" star-like object called a comet. Comets are known to be collections of ices, dust grains, rocks, and frozen gases which wheel about the sun in huge elongated orbits which alternately carry them from relative proximity to the sun out to the most distant parts of the solar system, thousands of earth–sun distances away. As a comet approaches the sun, the sun's radiant energy causes the ices and frozen gases to evaporate into a glowing coma which surrounds the dust and rocks at the comet's nucleus. As this diffuse, star-like object draws ever closer to the sun, the solar proton wind and radiation pressure drive material out of the diffuse head into a long, streaming tail which can extend over millions of miles of space. For several weeks, like a cosmic messenger, a comet will approach the sun, blossom with a flowing tail, and then fade into the cold blackness that is the periphery of the solar system.

The debris left behind by both these interlopers as well as from the formation process of the solar system permeates the interplanetary medium. As the earth sweeps along its orbit, it is constantly bombarded by objects ranging in size from tiny grains of dust up to small asteroids several kilometers in diameter. Fortunately collisions with the latter are extraordinarily rare! When a given interplanetary particle, called a meteoroid, strikes the earth, it does so at speeds as high as 50 km/sec. At such speeds, friction with the earth's atmosphere causes the object to heat up quickly and glow brilliantly as it falls toward the earth. An observer at the earth's surface sees this event as a "falling star" or "shooting star". Most of the time, such objects disintegrate in the upper layers of the earth's atmosphere, but occasionally a meteoroid is able to traverse the earth's atmosphere and strike the earth's surface. Such an object is then referred to as a meteorite.

Occasionally the earth passes through a large stream of meteoric debris left behind by a comet. Under these conditions, large numbers of meteors can be seen in the form of a meteor "shower". During a typical meteor shower, one can see anywhere from 15–60 meteors per hour above the normal sporadic or background meteor counts of about six meteors per hour. About three or four times per century the earth strikes a particularly large and dense aggregate of meteoroids. Under these circumstances, thousands of meteors per hour flash across the heavens in a display of celestial fireworks which is unmatched anywhere else in the natural world.

From time to time in the remote recesses of interstellar space a star will end its life in a spectacular event called a supernova explosion. For a few days the energy output of this dying object rivals that of all the stars in an entire galaxy. If a supernova detonation occurs at a distance sufficiently close to the earth, the observed result is the transitory appearance of a "new" star in the terrestrial night sky. For time periods ranging from a few days to several months, the star shines at or near its maximum brightness before fading back into naked eye invisibility. One of the more notable of these objects was observed in the year AD 1054. At its maximum brightness, the supernova of 1054 was nearly three times brighter than the planet Venus and could be readily seen in broad daylight. The remains of this stellar blast can be telescopically viewed today as the tattered and twisted gaseous cauldron called the Crab Nebula.

Unlike the mathematical and monolithic universality which characterized the scientific philosophy emergent from the Western European Renaissance, the explanations tendered for the considerable array of celestial phenomena by non-western cultures as well as those of pre-Renaissance Europe and the Mediterranean were far more qualitative in nature and represented a diversity of ingenious viewpoints that were nearly as numerous as the cultures from which they sprang. Generally such explanations appear in a given culture in the form of myths, legends, and folklore, and pay considerable homage to the observed characteristics of the sky and its resident objects. As such, they represent the beginning attempts on the part of human beings to provide rational explanations consistent with observations for the variety of events which occur in the physical world, thereby making that world more comprehensible.

Perhaps the most familiar example of this process in action is to be found in the myths and legends pertaining to the fixed stars. Out of the more or less random distribution of stars in the night sky, one can imagine a variety of figures, shapes, and patterns not unlike the variety of faces and forms that one often fancies in the puffy clouds of a springtime sky. In some instances, a given pattern of stars can bear a striking resemblance to a familiar terrestrial entity. For virtually every culture, such similarities were not fortuitous, but in fact were intrinsic characteristics of the sky which were significant and demanded explanation. The most common approach was to regard the sky as a kind of "Celestial Hall of Fame" into which various legendary characters from a given culture's folklore had been inducted for various reasons. Such "inductees" thus became figures outlined in stars or constellations. The outline of some of the constellations are so compelling in their shapes that a variety of far-flung cultures would often envision very similar portraits for a given star group. Thus, the stars of the highly prominent wintertime constellation of Orion, for example, seem to outline a very fit and trim individual possessed of considerable physical strength. Thus Orion, the mighty hunter of Greek mythology is also al-Babādur (The Strong One) for the Arabs, the great hunter "Bull of the Hills" for the Blackfoot Tribe of the western Canadian plains, and the "Slender First One" to the Navajos of the American southwest. As one might expect, there is also a considerable amount of variation in the sky pictures of various cultures. Even though the J-shaped array of summertime stars which we call Scorpius the Scorpion has been widely depicted as a celestial version of its earthly arachnid namesake, there are many other interpretations of this asterism from other cultures. The Polynesians, for example, saw this star group as the fishhook of their great hero Maui, while the Chinese viewed it as the noble Azure Dragon, the Bringer of Spring. To the Mayas of Central America these stars represented the death god Yalahau, the lord of blackness and waters.

The constellations through which the sun, moon, and planets travel in their respective journeys about the celestial sphere were quite naturally assigned a particularly significant status as the constellations or signs of the zodiac. Traditionally there are twelve such constellations, each of roughly equal extent along the zodiac, and which include Aries the Ram, Taurus the Bull, Gemini the Twins, Cancer the Crab,

Leo the Lion, Virgo the Virgin, Libra the Scales, Scorpius the Scorpion, Sagittarius the Centaur-Archer, Capricornus the Sea-Goat, Aquarius the Water Carrier, and Pisces the Fishes. In addition to the standard twelve constellation zodiacs employed by a majority of the world's cultures, the zodiac has been variously divided throughout human history into as many as 28 constellations by the Chinese and as few as six by the early Euphratean cultures. The denizens of the zodiac exhibit a considerable variation from culture to culture. The Aztec zodiac was graced with the starry presence of a frog, a lizard, a rattlesnake, and a jaguar, while that of the Incas contained a tree, a bearded man, a puma, and the sacred cantua plant.

Numerous explanations were offered for the observed movement of the sun, moon and planets along the zodiac, virtually all of which centered on the basic idea that only gods and goddesses could possess the power to move among the stars. In the case of the sun the concept was further reinforced by the fact that to look directly on the face of the sun's disk was to incur the sun deity's wrath in the form of severe damage to one's eyes. Thus the sun was the sun god Amon-Ra to the Egyptians and the sun goddess Amaterasu to the early Japanese, and so on.

Eclipses and conjunctions in the sky have also inspired a number of mythologically-based explanations. In the Hindu culture, for example, the mortal Rāhu is said long ago to have attempted to partake of the forbidden nectar of immortality. The god Viṣṇu was told of Rāhu's transgression by the sun and moon, and as punishment Viṣṇu proceeded to decapitate Rāhu. Ever since, Rāhu has sought to take vengeance on the sun and moon by pursuing them across the sky in an attempt to eat them. Once in a while, at the time of an eclipse, Rāhu actually catches either the sun or the moon and attempts to devour his prey. As the sun or moon is devoured, it gradually disappears into Rāhu's throat for a time before reappearing at the base of his severed neck as Rāhu attempts to swallow. The entire event is observed here on the earth as an eclipse of the sun or moon.

The sky watchers of antiquity were able to identify a number of basic characteristics relating to the background objects of the celestial sphere. The recognition of the variety of intrinsic colors that characterize the stars, for example, is manifested in names for stars such as the Arabic *Qalb al-ʿAqrab* (Heart of the Scorpion) for the bright red star Antares located at the center of Scorpius and the Hindu *Rohini* (Red Deer) for the ruddy star Alphard in the chest of the constellation of Hydra the Sea Serpent.

Bright stars near the celestial poles have held great meaning and significance to the watchers of the sky. In the third millennium BC the north celestial pole was located in the constellation of Draco the Dragon near a second magnitude star called Eltanin. Because the heavens of the day appeared to rotate about this star, it was quite literally regarded as an object of pivotal importance. As a result, Eltanin was worshipped by a number of cultures, including the Egyptians who used this star to align a number of their important buildings and structures. As the earth's axis of rotation has precessed, other stars have taken on the mantle of Pole Star, most notably by the stars Thuban in the constellation of Draco and Kochab in the constellation of Ursa Minor, the Lesser Bear, and in more recent cultures by the star Polaris at the tip of the tail of Ursa Minor. Both Kochab and Polaris were regarded by the Chinese as Da Di the Great Imperial Ruler of the heavens, about whom the other stars circled in homage. The Pawnee tribe of the American plains named Polaris "The Star That Does Not Walk Around". To the Pawnee this star was related to the god Tirawahat, and as such, was chief over all the other stars. It was this star that saw to it the other stars did not lose their way as they moved across the sky.

Attempts to explain the true nature of the diffuse objects that dot the sky are understandably less prolific in light of the difficulties that are often encountered in observing them. The major exception is, of course, the Milky Way. Of the diffuse objects detectable in the heavens with the unaided eye, the Milky Way is far and away the most extensive and prominent. This delicate band of light which is also highlighted by an array of brighter stars has thus inspired a variety of explanations which include its portrayal as a celestial river by the Chinese and Japanese, as a Path of Souls to an eternal home by the Algonquin tribe of the Lake Ontario region of southern Canada, and as a band of glowing cinders by which one could find one's way home when lost in the darkness by the Bushmen of Africa's Kalahari Desert.

As imaginative and rational as they were, however, the explanations advanced by different cultures for the variety of celestial phenomena observed in the heavens generally became intertwined with the religious beliefs and societal mores of these cultures. As a result, there was a marked tendency for the explanations of celestial phenomena to take on dogmatic qualities in which they were seldom questioned or challenged by alternate points of view. Moreover, the lack of a telescopic astronomy placed severe and fundamental limitations on the level of insight that was possible regarding the nature of celestial objects. Thus the explanations proposed for various celestial phenomena tended to remain largely unchanged in a given culture, and whatever changes that did occur were not so much the result of additional observational insights, but rather due to a gradual evolution brought on as these explanations were passed on from generation to generation or from culture to culture. Even while armed with an impressive instrumental technology, however, human beings still continue to struggle with questions relating to the fundamental nature of what we see in the sky.

Certain observable aspects of the heavens readily lend themselves to practical usage here on the earth, and the greatest levels of achievement enjoyed by non-western astronomers have come in the discovery, recognition, and application of these characteristics. Systematic observations of the sky reveal, for example, that many celestial phenomena, most notably the diurnal and annual motions of the sun and the cycle of lunar phases, occur with precise and predictable regularity. This observable fact of the heavens has thus been employed by cultures worldwide as a method of accurate time keeping.

The diurnal rising and setting of the sun, with its alternating cycle of daylight and darkness, is the shortest and most convenient unit of astronomical timekeeping, and as a result human beings the world over have employed it, quite literally, as an integral part of their daily lives. A second, much longer unit of astronomical time is defined by one complete journey of the sun around the ecliptic. This annual astronomical cycle is of considerable importance owing to the fact that it is intimately related to the cycle of seasons which occur here on the earth. The cycle of seasons, in turn, is virtually identical to the cycle of vegetative growth and those of some animal activity and migrations. Thus agricultural methods and hunting techniques developed by various cultures were necessarily tied deeply to the cycle of seasons and the sun's annual journey along the ecliptic. Intermediate in length between the day and the year is the time interval required for the moon to pass through one complete cycle of its phases. The lunar cycle is particularly attractive as a timing cycle due to the fact that the ever-changing shape of the moon is readily observable on a daily basis. Sequences of shapes inscribed on Cro-Magnon cave walls and artifacts strongly suggest their use as lunar phase timing devices in just this fashion. Similar sequences carved by the inhabitants of Nicobar Island in the Bay of Bengal are known with certainty to be employed for this purpose.

Unfortunately these cycles are not quite numerically compatible with each other. For example, there are about 365 days in a year, but in reality it takes the sun precisely 365.242199 days to complete one cycle around the ecliptic. Similarly there are 29.530588 days to a cycle of lunar phases and 12.36827 cycles of lunar phases in a year. These discrepancies can create difficulties if one wishes to reckon the time of the year, the start of a given season, or the date of an important religious holiday by simply counting the number of days which have elapsed from some defined starting point such as the day of a solstice or equinox. If one counts the number of days as the year progresses, for example, one would find that after 365 days had passed, the sun would not quite yet have completed its journey around the ecliptic, and after 366 days had elapsed, the sun would have moved slightly past one complete cycle. Over several years' time such an effect can add up to a significant discrepancy between the sun's actual position along the ecliptic and the position dictated by the day count. As a result, a variety of schemes, called calendric systems or calendars, have been developed by various cultures around the world which are designed to synchronize two or more astronomical cycles. The most familiar of these is the addition of one day to our calendar every fourth or leap year in order to keep the day count in a given year in agreement with the sun's actual position along the ecliptic.

A number of ingenious techniques were developed by various cultures to monitor the astronomical timekeeping process. The Aztec Temple Mayor, now buried beneath modern Mexico City, was designed in the fifteenth century with two spires that provided a V-shaped notch through which the rays from a sun rising at the time of an equinox shone on to the temple of Quetzalcoatl. At no other times of the year would a rising sun produce this effect. Thus the Aztec temples served quite nicely and deliberately as a device with which the Aztec calendar could be corrected whenever necessary. Similar structural alignments are to be found at Stonehenge in England, in the temples of ancient Egypt, and among the buildings of the ancient peoples of the American Southwest. The Mayas of ancient Mesoamerica developed not only astronomical alignments for many of their structures, but also an incredibly accurate but somewhat complicated astronomical calendar which was based on the annual solar cycle and the synodic period of the planet Venus. The Maya calendar was accurate to within one day every 5000 years. By contrast, the simpler Gregorian calendar used by contemporary society is accurate to within one day in 3300 years.

In addition to the structural alignments, various cultures have also employed natural terrain as calendar correctors. On the top of Fajada Butte at the mouth of Chaco Canyon in the American Southwest, for example, there stand three rock slabs, each of which is about three meters in height. On the rock wall behind these slabs a first millenium AD people called the Anasazi carved a spiral petroglyph in such a way that precisely at noon of the day of the summer solstice, a dagger-shaped beam of sunlight would neatly slice the petroglyph exactly through its center. Through this clever manipulation of sunlight, the Anasazi were able to mark the time of the summer solstice precisely.

The Hopi and Zuni tribes, also of the American Southwest, make use of a so-called sunrise horizon calendar. As the sun moves along the ecliptic, the points of sunrise and sunset along the horizon at a given location exhibit an annual cyclic shift in which the sunrise and sunset points appear to migrate along the horizon from south to north while the sun is moving from the winter solstice to the summer solstice and then north to south along the horizon while the sun is moving from the summer solstice to the winter solstice. As the sunrise and sunset points pass over various key landmarks along the horizon, each passage is taken as a signal to begin the appropriate agricultural activity such as planting various crops, harvesting, etc.

In addition to timekeeping, earth–sky relations can also be employed to find one's way about the surface of the earth. Such techniques are referred to overall as celestial navigation and have been of considerable importance to human cultures, particularly those which are maritime in nature. There are a number of aspects of the heavens which readily lend themselves as navigational aids. As the earth spins on its axis, for example, a star at or near the celestial pole will not appear to change its position in the sky significantly. More importantly, the point on the horizon directly beneath such a star will also remain in a relatively fixed position as well. Thus for observers in the northern hemisphere, the relatively bright star Polaris is located very close to the north celestial pole, and the point on the horizon directly beneath this signpost star has been used for centuries by northern hemisphere peoples to mark the direction we call north.

Other cultures took advantage of the fact that the angular distance of the pole star above the horizon as well as the locations of the rising and setting points of bright stars and constellations along the horizon changed with one's location on the earth's surface. Thus the Caroline Islanders of the central Pacific skillfully navigated by means of this star compass in which 32 points on the horizon were defined by the rising and setting points of bright stars and constellations such as Vega, the Pleiades, Antares, and the Southern Cross. The Polynesians employed a device called the sacred calabash, which was a gourd into which four holes were bored at the same height near the neck. The gourd was then filled with water to the level of the holes. Using the water level as a horizon, altitudes of stars were then measured by sighting through one of the holes over the opposite edge of the gourd. Thus armed with what was in effect the equivalent of our modern sextant, the Polynesians became most adept at deep-water navigation.

Systematic observations of the heavens also reveal that there exist a number of correspondences between celestial events and configurations and natural phenomena here on the earth. For example, the Egyptians recognized that the annual flooding of the all-important Nile River was at hand when the bright star Sirius made its heliacal rising or first appearance out of the pre-dawn solar glare. The Incas of the South American Andes Mountains noticed that the cantua plant blossomed beautifully each year when the sun was located in our zodiacal constellation of Cancer, but which they named appropriately from their observations as the asterism of the sacred cantua plant. The heliacal risings of the bright stars Rigel, Aldebaran, and Sirius served to warn the tribes of the high plains of western America that cold weather was at hand. In light of such readily observable earth–sky correspondences, it was very logical to assume that similar correspondences exist between celestial phenomena and human affairs. Thus evolved the endeavor which we now call astrology.

Whether the astrological leap of logic from earth–sky to human–sky correspondences is a valid one has, of course, been a topic of considerable debate for many centuries, and since the 1600s the premise that such human–sky correlations exist has been emphatically rejected by western science. Nevertheless, astrology, more so than either timekeeping or celestial navigations, demands access to careful and ongoing observations of the entire heavens for the purpose of interpreting the significance here on earth of what is seen to occur in the sky, and whenever possible, to predict future events in the sky as well. Thus a well-developed astrology in China was certainly an important factor in the preparation of the earliest known star catalogue in the fourth century BC, and in the recording of a variety of celestial events, most notably the transitory appearances of sunspots and astrological omens such as comets, which were referred to as *huixing* (broom stars or sweeping stars) and of novae and supernovae explosions, which were called *kexing* (guest stars or visiting stars). So detailed were the records of the Chinese, Japanese, and Korean observations of the supernova event of AD 1054, for example, that modern astronomers were easily able to identify its present remains as the Crab Nebula in the constellation of Taurus, despite the fact that the event went virtually unobserved and unrecorded in Western Europe.

From some cultures, most notably those of the Mayas, Egypt, China, and the Islamic world, careful observations of the sky combined with centuries of relative social and political stability to make possible the discovery of much more subtle and long-term astronomical cycles. The Mayas, for example, were aware of the long-term reappearances of the planet Venus and built the planet's 584-day synodic period into their calendar. Both the Chinese and Islamic observers were aware of the fact that the lunar nodes, or the points on the celestial sphere where the moon's orbit crosses the ecliptic, drift in a westerly direction along the ecliptic at a rate of one complete revolution every 18.6 years and used this knowledge to predict the occurrences of both lunar and solar eclipses.

The Chinese and Islamic observers also recognized that the sun's equinox points drift in a westerly direction along the ecliptic at a rate of nearly one degree per year and made appropriate adjustments in their respective star catalogues and calendric systems in order to account for the protracted effect of this equinotical precession. Awareness of the shifting equinoxes may have also been the province of the Egyptians as well. A number of additions and reconstructions are found to exist in Egyptian temples and other structures which strongly suggest an architectural response to just such long-term changes in the positions of the equinoxes.

ROGER B. CULVER

REFERENCES

Abell, G. *Exploration of the Universe.* New York: Holt, Rinehart, and Winston, 1964.

Allen, R. N. *Star Names.* New York: Dover Publications, 1963.

Culver, R. B. and Ianna, P. A. *The Gemini Syndrome.* Buffalo: Prometheus Books, 1984.

Jobes, G. and Jobes, J. *Outer Space.* New York and London: Scarecrow Press, 1964.

Krupp, E. C., ed. *In Search of Ancient Astronomies.* New York: McGraw-Hill, 1978.

Krupp, E. C. *Beyond the Blue Horizon.* New York and Oxford: Oxford University Press, 1991.

Lockyer, J. N. *The Dawn of Astronomy.* Cambridge, Massachusetts: MIT Press, 1964.

Ruggles, C. and Saunders, N. *Astronomies and Cultures.* Boulder: University of Colorado Press, 1993.

Temple, R. and Needham, J. *The Genius of China.* New York: Simon and Schuster, 1986.

Williamson, R. *Living the Sky.* Norman: University of Oklahoma Press, 1984.

Willis, R. *World Mythology.* New York: Henry Holt and Company, 1993.

Zeilik, M. and Gaustad, J. *Astronomy: the Cosmic Perspective.* New York: Harper and Row, 1983.

See also: Lunar Mansions – Eclipses – Calendars – Time – Navigation – Astrology – Celestial Sphere and Vault

ASTRONOMY IN AFRICA The modern study of sub-Saharan African archaeoastronomy and ancient calendrical reckoning has a varied history. Although investigations have been made for well over a century for some sites, these have sometimes been hampered by ignorance of proper archaeological techniques, highly speculative conjectures about astronomical alignments, and even political policy opposed to the scientific evidence being brought to light, as it was with the Great Zimbabwe Ruin. Thus archaeoastronomy of sub-Saharan Africa can still be said to be in its infancy. We will therefore sample, in this article, a few cases of interest that can point the way for future directions of research. These investigations should serve as good examples of the kinds of research that are presently in progress on the astronomical ideas and methods of the ancient inhabitants of sub-Saharan Africa.

Namoratunga and the Borana Calendar

In 1973 Asmaron Legesse first documented the unique calendrical system of the Borana, a Cushitic-speaking people of southern Ethiopia and northwest Kenya. In this system calendrical time was determined using only seven special stars in conjunction with various phases of the moon. The sun, except indirectly via lunar phases, is ignored. In 1978 B.M. Lynch and Lawrence H. Robbins discovered, near Lake Turkana in Kenya, a site of nineteen stone pillars called by the local people *Namoratunga* (stone people), and suggested that the stone pillars there were used for the alignments needed to derive the Borana calendar. Petroglyphs on these pillars were similar to those on stones at a burial site (also called Namoratunga) 100 kilometers to the south, which was dated at 300 BC. Thus an ancient age was indicated for the astronomical site (called Namoratunga II) because of its association with the burial site (Namoratunga I). The local Turkana people have names for each of these petroglyphs and indicate that they are very old and relate to long-time family names.

Later investigations found the pillars to be magnetic, and they consequently had to be remeasured to see if the astronomical alignments persisted. The new measurement positions are shown. In addition, the working of the Borana calendrical system they were supposed to represent was found to be astronomically inconsistent, requiring a reinterpretation of the original description of its working. Finally, the most recent field work on the Borana calendrical system indicates that the present working of this calendar may differ from the historic one. Let us first examine how the Borana calendrical system may have worked and then see how one might determine if the Namoratunga site was indeed used as an astronomical calendrical marker for this system.

As described by Legesse, the Borana calendar uses six star positions marked by seven stars or star groups. These seven star groups are: Triangulum, Pleiades, Aldebaran, Bellatrix, Orion's Belt, Saiph, and Sirius. New year starts when a new moon is seen in conjunction with Triangulum. Since the lunar phase or synodic cycle is 29.5 days long while the lunar position or sidereal cycle is 27.3 days long, the moon will arrive at the same position on the sky two days before it has completed its phase cycle. The Borana have 27 day names (no weeks) and continue counting days so that at this point the two day names are repeated, starting the second month on a new day name about 29 or 30 days after the start of the first month (depending on the observation). The second month starts when the new phase moon is seen in conjunction with the Pleiades, and the following third through sixth months start when the new moon is seen in conjunction with Aldebaran, Bellatrix, Orion's Belt and Saiph taken together, and finally Sirius. The next six months are defined by observations taken within the counting month. Month seven is defined as the month when a full moon is seen in conjunction with Triangulum, while month eight is defined when a three-quarter moon is in conjunction with Triangulum. The next four months are defined when consequent less-waxed moons are seen in conjunction with Triangulum, and the year begins again when a new moon is finally seen in conjunction with that position again. Allowing for a leap month every three years, as observations dictate, the 354-day lunar (12-month) year can consequently keep up with the 365-day solar cycle.

However, what does "in conjunction" mean? The lunar motion does not allow a "rising with" interpretation of this term since Triangulum is, in general, too faint to be seen next to a rising new moon, which is always seen in twilight. Also, the lunar motion does not correspond to consecutive horizon risings with the given star positions at the start of consequent months at any time within the date of existence of the calendrical system. (Apparent star positions change with the precession of the pointing direction of the Earth's rotation axis.) On the other hand, if "in conjunction" is interpreted as meaning rising at the same position on the horizon, then the 300 BC (but not the present) positions of the Borana calendrical stars do indeed rise at the same horizon rising positions as consequent new moons for the first six months of the calendar. Also, the given consequent phases of the moon rise at the horizon rising position of Triangulum for the next six months, thus defining a workable calendrical system. The 300 BC positions of the Borana calendrical stars are illustrated here; the stars will rise almost vertically this close to the equator. When the Earth's equator is extended on to the sky, the distance a star lies above or below it is called the star's declination measured from +90 to –90 degrees. The distance around this celestial equator starting from the point where the Sun is on the first day of spring (the vernal equinox) is the star's right ascension measured from 0 to 24 hours around the sky or from 0 to 360 degrees.

Additional investigation into the Borana calendrical system has indicated that the present *ayantu* (astronomers) and historians disagree as to what constitutes the Borana calendar stars. In general, the historians call the original stars given by Legesse the Borana stars, while the present *ayantu* use a different overlapping set of stars (as they would have to do since the first set of stars have changed their apparent positions and do not work as a calendrical system as described any more). This present Borana system also has problems since it is apparently based on a "rising with" interpretation of "in conjunction". Most of the stars in this updated system are grouped together, and as a consequence the calendar cannot be used, due to interference from the Sun, for several months out of the year. The system consequently seems a bit ad hoc, and in 1988 Bassi indicated that it might be some sort of remnant of the ancient system suggested by Laurence Doyle in 1986. It is clear, however, that some modification of a stellar–lunar position based calendrical system would have to have taken place if the origin of the Borana calendrical reckoning extends back in time for more than a few hundred years.

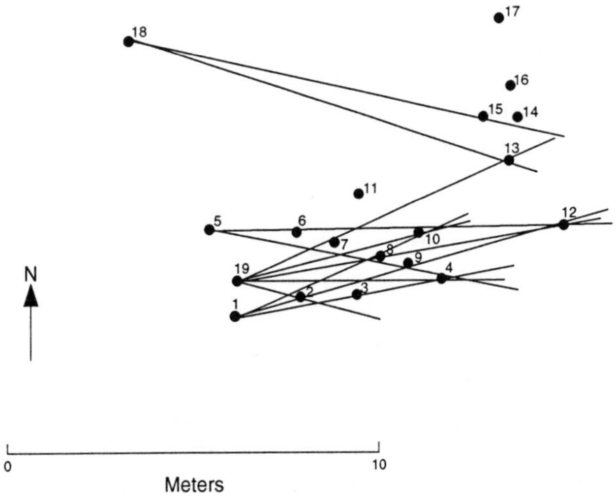

Figure 1: Location of top-center positions of pillars at Namoratunga (exception is Pillar 11, horizontal base position taken).

We next might ask, from the interpretation of the ancient Borana calendrical system, if the nineteen stone pillars at the Namoratunga site could have been used to mark the 300 BC horizon-rising positions of the seven Borana calendar stars. Remeasurement of pillar positions in 1983 produced 25 two-pillar alignments with the 300 BC eastern rising positions of the Borana calendar stars. Some examples of these 300 BC stellar alignments with pillars found are: β Triangulum 1-8-10, Pleiades 1-2-12, Aldebaran 1-3-4 also 19-8-12, Bellatrix 5-6-12, central Orion-Saiph 5-8-9 also 18-15, and Sirius 18-13.

To decide if these alignments are random or likely the result of intentional alignment with these seven specific star positions on the sky, the following test was designed. Seven random star positions were generated and compared with the mapped pillar positions at Namoratunga, giving the number of alignments found with these random star positions on the eastern horizon. This process, generating seven random star positions and comparing them with Namoratunga to determine how many alignments were thus made, was repeated 10,000 times, with the result that the number of times that 25 or more random stellar alignments were found for the Namoratunga pillars was 41. This result would indicate that there seems to be a less than 0.5% probability that the Namoratunga site pillars could have randomly made as many as 25 or more alignments with the 300 BC horizon rising positions of the seven original Borana calendrical stars.

However, several archaeologists, such as D. Stiles, conclude that the stones are instead *soddu* (burial) stones. Few question that they are man-made, as the basalt pillars have a square cross-section, while basaltic material in general naturally cleaves in a hexagonal cross-section. Interpretation of the petroglyphs on the pillars also remains controversial as does the connection with Namoratunga I to the south. Additional dating techniques (desert varnish, for example) may be applicable to constraining the date of the Namoratunga II site, and additional such sites should be locatable if such pillars were in general use for calendrical reckoning. Finally, identification of the petroglyphs with other familial symbols in east Africa (symbols found on the artifacts from the Kushitic pyramids in the Sudan, for example) might prove to be a uniquely African historic tracer supplementing linguistic data on migration patterns and cultural history. In conclusion, the Namoratunga II pillars may have been the site of an ancient calendrical observatory, but much yet remains to be done for this to be convincingly concluded. Perhaps the best possible investigation would be the discovery of another such megalithic site where similar analyses as have been outlined here might be performed.

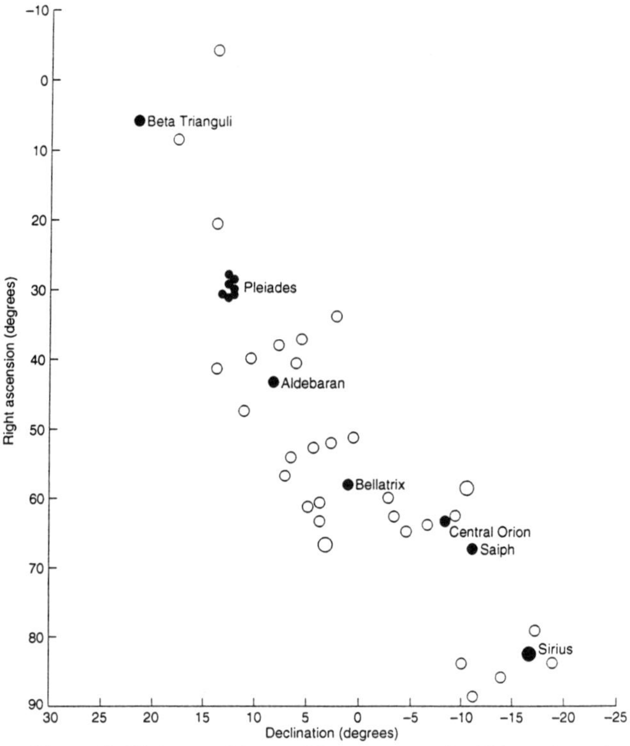

Figure 2: Rising positions of Borana stars for 300 BC (east is at zero degrees declination).

From Kushitic Pyramids to Tanzanian Cave Symbols

Examples of locations of calendrical interest in east Africa that have undergone only preliminary astronomical investigation range from astronomically interesting alignments of the pyramids of Kush to possible lunar symbols found in certain caves in Tanzania. These two cases are presented here as examples of the types of investigations that remain to be done into the astronomical ideas of the varied peoples of this region of the world.

The Kingdom of Kush, in central Sudan, lasted from about 1000 BC to AD 200 around the fifth cataract of the Nile. The capitol was first established at Napata and then relocated to Meroe. While archaeological investigation has revealed an extensive civilization with a perplexing language and writing, the sites of numerous Kushitic pyramids have yet to be thoroughly investigated for astronomical alignments similar to those found in Egyptian pyramids to the north. However, from preliminary investigations of maps of the area around Meroe, it is plain that most of the pyramid entrances face very close to the direction of the eastern rising of the star Sirius. This indicates perhaps an Egyptian calendrical influence brought back from the Kushitic Pharaohs after Kush conquered Egypt in about 800 BC. Any contrasts between calendrical systems of the Kush and the ancient Egyptians could certainly help illuminate some of the cosmology of a still little known civilization that flourished in sub-Saharan Africa almost two millennia ago.

As another example of such research, among ancient cave drawings in Tanzania Mary Leakey has discovered a number of precise circles, drawn with an accuracy that indicated to her that they might have been used for some sort of counting process. She dubbed them “Suns, for want of a better name”. They consist of concentric circles carefully spaced apart and ranging from just one to 29 or 30 circles depending on the location. It is interesting that the number of days in the lunar synodic cycle, 29–30, comes up again. Perhaps they could be called moons more accurately. The number of concentric circles at a particular site may represent, for example, a meeting or gathering cycle, since the travel time between cave sites could be several days. The investigation of these and other symbols to see if they do correlate in a calendrical way could indeed open up new insights into the timekeeping as well as possibly the cosmological ideas of the early inhabitants of these regions.

Great Zimbabwe Ruin and the Pre-Shona Calendar

While it has been estimated that over 1000 megalithic sites can be found in Zimbabwe and surrounding countries, the most famous of these is the Great Zimbabwe Ruin itself in southeastern Zimbabwe. It consists principally of the Great Enclosure with an adjacent area on a nearby hill called the Hill Complex. The pre-Shona people, sometimes called the Karanga, started building here around 400 AD and finished the present structure seen today around the middle of the fourteenth century. Since an initial survey around the turn of the century by Bent and Swan, which was republished in 1969, there have been a number of various astronomical and mathematical features claimed for the Great Enclosure. The site consists of a large oblong wall about 10 meters in height, 3 meters thick, and about 100 meters in length, with various brick altar structures, interior walls, pillars, and stone monoliths inside.

At one end of the enclosure, for example, are two brick towers of about 15 meters in height which were said in early surveys to be related, in that the radius of the larger was the circumference of the smaller, that is, the ratio of their circumferences was the mathematical constant π. Half a decade later some investigators also claimed that the circumference of the whole enclosure was related to certain astronomical distances. Recent investigations by us found no such mathematical relationships inherent in either the internal structures or the surrounding wall structure. Also, by comparing the oldest maps and photographs it was found that the original ruins had been significantly tampered with. Certain smaller towers or pillars had been removed and at least one added. In addition, some of the internal monoliths have been reseated recently. In spite of this, however, preliminary investigations do reveal that the native African peoples that built Great Zimbabwe were aware of the sky and may indeed have marked important astronomical seasonal events.

For example, in a preliminary survey, a chevron pattern on the southeast corner of the large outer wall is bisected by the rising position of the Sun on the summer solstice from inside the enclosure, and aligns with what has been called the altar as well as with an original pillar inside the enclosure. As this large patterning does not appear at any other place on the outer wall it would appear to be a conspicuous candidate for a summer solstice marker built into the Great Enclosure. In addition, a large passageway within the Great Enclosure, about 2 meters in width and 30 or so meters in (curving) length, with 10 meter high brick walls on either side, would allow a limited view of the sky with an angular extent and curvature matching the position and angular extent of the Milky Way overhead on the summer solstice. While the Milky Way was a very important calendrical marker for the Karanga people of this area this observation too must be confirmed with further research. Finally, from a cleared platform at the top of the Hill Complex, two large stones of approximately 5 meters in height, in close proximity to each other, can be seen to form a slit directed precisely east which could have served as a solar marker for the equinoxes. These and other observations are preliminary and a better understanding of the calendrical systems of the early inhabitants of this region would substantially improve further investigations into any astronomical features that may have been built into the ruins at Great Zimbabwe.

The Dogon of Mali and the Star Sirius B

Perhaps the most controversial investigations into the ancient astronomy of sub-Saharan African have come from the historical accounts of the Dogon of Mali, in western Africa. These investigations involve the star Sirius (A) which is the brightest star in the sky — after the Sun — and has played an important historical role in, for example, the Egyptian calendar. There, the helical rising of Sirius always signaled the beginning of the flooding of the Nile in ancient times. However, the fact that Sirius is a double star was not discovered by modern astronomers until 1844 when Friedrich Bessel deduced its existence by the slight wobbling motion observed in Sirius A itself. Named Sirius B, this small companion had to await the invention of larger telescopes in order to be seen, and was first spotted by Alvan Clark in 1862 using one of the largest telescopes at the time (an 18.5 inch aperture refracting telescope). Soon after this the period of the mutual orbit of this binary star was determined to be about fifty years which also enabled a precise determination of the two stars' masses and radii. The mass of Sirius A was found to be about twice that of the Sun, which one would expect from a large, blue star. However, the mass of the much smaller companion was found to be a little larger than the Sun. This came as a surprise since it was only about the size of the Earth. This meant that the density of this star would have to be about one million times that of an average star. However, it was not until 1931 that the astrophysicist S. Chandrasekhar published the first explanation of the structure of Sirius B and several other of these very dense stars, which were to be called white dwarfs.

Also in 1931 the anthropologist and ethnologist Marcel Griaule began collecting the lore of the Dogon. He found that they regard Sirius (which they call *sigi tolo*) as very important, but that its invisible companion *po tolo* (meaning deep beginning) is far more important. *Po tolo* (Sirius B) was said to be closely associated with the *fonio* grain grown in this region (it is sometimes called the star of the *fonio*). They said that because the *fonio* grain is very small and white, as is *po tolo*. They also said that *po tolo* is very heavy — the heaviest of all stars. They said that it is at the center of the sky and its influence makes the stars stay in place. (The node of the Earth's precession plane is very close to Sirius, so that it will not move significantly within several millennia.) Finally they say that *po tolo* circles the star Sirius every fifty years. This is in fact correct, but they apparently have a special ceremony every *sixty* years to celebrate the completion of one cycle associated with *po tolo*. Records of the masks used in this rite apparently indicate that this ceremony, or one like it, has indeed been held since around the thirteenth century. Dogon myths have also indicated, however, that a third star exists in the Sirius system, a star that has not been seen and whose mass influence should perhaps have shown up by now in perturbations of the motions of the other two stars. Nevertheless, from an anthropo-historic viewpoint, the matching of the Dogon star myth with modern astronomical facts continues to be an interesting study. If the authenticity of the investigations can be verified, then much remains to be learned either about the unexpectedly rapid assimilation of modern astronomical fact by a somewhat isolated African culture or about a means in that culture for obtaining direct astronomical information that is beyond their apparent observational abilities. However, the verification of authentic astronomical mythology, in this case, may likely prove to be even more difficult than the verification of authentic archaeoastronomical sites in sub-Saharan Africa have been.

The ancient calendrical systems, archaeoastronomical sites, and astronomical myths of sub-Saharan Africa have just begun to be investigated. While reviewing here some of the previous work that has been done, this article has attempted to point to directions for future research into this fascinating area, while hopefully emphasizing the need for particular scientific restraint when drawing conclusions in this field of academic endeavor. It is indeed a truism in science, however, that the actual scientific facts often turn out to be far more interesting than the speculative hypotheses leading up to them, and this is likely to be the case with the ancient astronomy of sub-Saharan Africa.

LAURANCE R. DOYLE
EDWARD W. FRANK

REFERENCES

Arkell, A.J. *A History of the Sudan – From Earliest Times to 1821*. Westport, Connecticut: Greenwood Press, 1973.

Bassi, M. "On the Borana Calendrical System: A Preliminary Field Report." *Current Anthropology* 29: 619–624, 1988.

Bent, J.T. and R.M.W. Swan. *The Ruined Cities of Mashonaland*. Bulawayo: Books of Rhodesia, 1969.

Doyle, L.R. "The Borana Calendar Reinterpreted." *Current Anthropology* 27: 286–287, 1986.

Doyle, L.R. and T.J. Wilcox. "Statistical Analysis of Namoratunga: An Archaeoastronomical Site in Sub-Saharan Africa?" *Azania: Journal of the British Institute of East Africa* 21: 125–129, 1986.

Garlake, P. *Great Zimbabwe Described and Explained*. Harare: Zimbabwe Publishing House, 1985.

Huffman, T.N. *Symbols in Stone*. Johannesburg: Witwatersrand University Press, 1987.

Krupp, E.C. *Beyond the Blue Horizon*. New York: Harper Collins, 1991.

Leakey, M. *Africa's Vanishing Art: The Rock Paintings of Tanzania*. Garden City, New York: Doubleday, 1983.

Legesse, A. *Gada: Three Approaches to the Study of African Society*. New York: Macmillan, 1973.

Lynch, B.M., and L.H. Robbins. "Namoratunga: The First Archaeoastronomical Evidence In Sub-Saharan Africa." *Science* 200: 766–768, 1978.

McCosh, F.W.J. "The African Sky." *NADA* 12: 30–44, 1979.

Millet, N.B. and A.L. Kelley. *Meroitic Studies — Proceedings of the 3rd International Meroitic Conference, Toronto*. Berlin: Akademie-Verlag, 1977.

Shinnie, P.L. and R.J. Bradley. *The Capital of Kush*. Berlin: Akademie-Verlag, 1980.

Sicard, H.V. "Karanga Stars." *NADA* 2: 42–64, 1969.

Soper, R. "Archaeo-astronomical Cushites: Some Comments." *Azania: Journal of the British Institute of East Africa* 17: 145–162, 1982.

Stiles, D. "The Azanian Civilization and Megalithic Cushites Revisited." *Kenya Past and Present* 16: 20–27, 1983.

Temple, R.K. *The Sirius Mystery*. New York: Inner Traditions International, 1987.

Thomsen, D.E. "What Mean These African Stones?" *Science News* 126: 168–169, 174, 1984.

See also: Namoratunga

ASTRONOMY IN NATIVE NORTH AMERICA The astronomical activities and traditions of the American Indians north of Mexico were based upon practical observation of the sky but were not supported by a written language. For that reason, our knowledge of North American Indian astronomy relies upon the archaeological data, ethnohistoric reports from early encounters between Europeans and the indigenous peoples, and ethnographic information collected more recently by anthropologists. Although this material is distorted and incomplete, it enables us to outline the general character of North American Indian astronomy and to understand some of it in detail. All of these sources confirm that North American Indians farmed, hunted, and gathered by the sky. They developed calendric techniques to order the sacred and ordinary dimensions of their lives. They timed ceremonies by the sky. They extracted symbols from the sky. They told stories about the sky. Throughout all of the cultural territories, physical environments, and linguistic traditions of North America, celestial phenomena were incorporated into ritual, iconography, myth, shamanic activity, and world view.

North American Indians were familiar, of course, with the fundamental cycle of day and night, the daytime path of the sun, and the unmoving pole of the night sky. Cardinal directions, which emerge from the daily rotation of the sky and the circular parade in which the stars march at night, were important to many groups. The moon's phases were monitored, and each monthly cycle was often associated with a seasonal change on earth. The seasonal shift of sunrise, sunset, and the sun's daily path was known. Solstices were recognized, and the rising and setting points of the summer and winter solstice sun sometimes established an alternate directional scheme. Seasonal appearances and disappearances of stars were noted. Constellations were contrived from conspicuous and useful stars. In the historic era, unusual events like eclipses, bright comets, fireballs, and meteor showers attracted notice and sometimes provoked ritual response. Planets were recognized by at least some groups, but explicit evidence of detailed indigenous knowledge of their cyclical behavior has not survived.

Like traditional peoples everywhere, the Indians of North America saw the sky as a realm of power. Access to that power required knowledge of the sky. They acquired that knowledge through careful observation and used that knowledge to order and stabilize their lives. This practical understanding of the sky — the sun, the moon, and the stars — is not the same thing as modern scientific astronomy, for it did not attempt to test and abandon metaphors of nature in the same way science does today. It did, however, help integrate human behavior with nature and consolidate social cohesion. Celestial objects were not just convenient metronomes. They were powerful, supernatural beings, and they revealed the basic structure — the fundamental order of the world. Because cosmic order is, in part, what is meant by the sacred, the Indians' interaction with the sky was an encounter with the sacred.

The earliest account of North American Indian astronomy was reported in 1524 by the Italian explorer Giovanni di Verrazano. He encountered the Narragansett Indians of Rhode Island and mentioned that their seeding and cultivation of legumes were guided by the moon and the rising of the Pleiades. The Pleiades comprise a distinctive cluster of stars. Its value as a signal of seasonal change was recognized throughout the continent. In California, for example, there is explicit evidence of Pleiades lore in all but 12 of the 58 native cultural territories. Those 12 actually correspond to a very small fraction of the entire area and population of the State, and their lack of Pleiades tradition is primarily due to linguistic and cultural extinction. Despite the extraordinary linguistic diversity represented by the 75 mutually unintelligible California Indian languages and 300 different dialects, names for the Pleiades, myths about the Pleiades, and seasonal references to the Pleiades are documented in all five major families of indigenous California languages.

Studies of North American Indian astronomy have emphasized the prehistoric Southwest and the historic Pueblos. This is due to the survival of prehistoric Pueblo architectural monuments and rock art and to the preservation of some information about historic Pueblo astronomical techniques and celestial lore in ethnological reports. Close to the end of the last century, Alexander M. Stephen described in detail the horizon calendar used by the Hopi Sun Chief to establish key dates in the ceremonial and agricultural cycles of the village of Walpi. More recently, historian of science Stephen C. McCluskey demonstrated that Hopi observations of the sun were accurate to 4 arcminutes. This is, however, an average error, and it is difficult to do much better than 30 arcminutes in any single horizon event. McCluskey verifed that the Hopi actually allowed themselves a few days' leeway in scheduling the winter solstice ceremony.

If Alexander Stephen had not seen the Sun Chief perform his duties, we would not know where his sunwatching station was located. Nothing marks the point out at the end of a mesa or on the roof of the highest house. Other references to North American horizon observatories suggest that many were just as subtly blended into the community landscape, and that makes the identification of prehistoric observatories a challenge. Symbolic astronomy was, however, often incorporated into monumental architecture and rock art. By analyzing alignments and iconography, it is sometimes possible to spot the hand of the ancient skywatcher. For example, Pueblo Bonito, an 800-room, five-story, D-shaped apartment-town in Chaco Canyon, New Mexico, makes good use of passive solar heating to stabilize room temperatures in summer and winter. It was completed in the twelfth century AD, and its east–west front wall is oriented cardinally with an accuracy of eight arcminutes. Such accuracy is achievable with simple surveying techniques that rely upon the unaided eye, but it is nevertheless respectable. In addition, archaeologist Jonathan Reyman interpreted corner windows in two rooms as winter solstice sunrise apertures.

The Anasazi built large subterranean community ceremonial chambers known as Great Kivas. Most of these, like

Chaco Canyon's Casa Rinconada, possess cardinal orientation, and the plan of a Great Kiva is thought to mirror the Pueblo concept of the cosmos.

There are many pictographs and petroglyphs in Chaco Canyon, and the star and crescent on a panel near the Penasco Blanco ruin have been promoted by some as a representation of the AD 1054 Crab supernova explosion. That spectacular event very likely was seen by prehistoric Indians in the American Southwest, but there is no way to verify that the Penasco Blanco pictographs are an eyewitness record of it. The supernova interpretation of star/crescent elements in Southwest rock art was first offered in 1955 as an explanation for two sites in northern Arizona. Since then, the number of reported star/crescent groupings has multiplied, but opinion on the supernova is divided. Reyman argues thoughtfully that the Penasco Blanco site is actually a sunwatcher's shrine and implies that the Crab supernova has nothing to do with these star/crescent designs. Even if they do depict the supernova rising with the waning crescent moon on the morning of 5 July 1054, they tell us nothing substantive about prehistoric Pueblo astronomy. In 1990, Robert R. Robbins and Russell B. Westmoreland revived the argument all over again with an analysis of numerical symbolism on prehistoric Mimbres ceramics. One of these has a rabbit in the shape of a crescent moon accompanied by a "star" with 23 rays. This detail is argued to be consistent with the Chinese historical account of the Crab supernova, for the Chinese reported it was visible in the daytime for 23 days.

A spiral petroglyph on Fajada Butte, the most conspicuous landmark in Chaco Canyon, interacts with "daggers" of midday sunlight at the solstices and the equinoxes. These light-and-shadow effects, first reported by Anna Sofaer, Rolf Sinclair, and Volker Zinser, received a great deal of international attention and inspired considerable controversy. Initially, the "Anasazi Sun Dagger" was interpreted as a "precise solar marker", but it may be more correct to regard it as a symbolic seasonal display.

One of the best candidates for a sunwatcher's observing platform is also located in Chaco Canyon, on an upper ledge on a rincon near Wijiji ruin. Rock art on the ledge includes elements as old as the Anasazi and as recent as the Navajo. A line-of-sight to the southeast coincides with a natural rock chimney on the other side of the rincon, and this feature dramatically marks the winter solstice sunrise.

Astronomer Michael Zeilik has emphasized the importance of "anticipatory" observations of the solstices and other astronomical events. The Sun Chief must know ahead of time, with accuracy, when the solstice is due in order to mobilize the community for the ceremonial activity that culminates in the solstice. This is exactly what Walpi's Sun Chief did, and Wijiji has the ability to deliver advance information and confirmation of the solstice. The site is interpreted, then, by extending significant information from the historic period back into the prehistoric context.

Astronomical components have been identified at many other sites in the Southwest, including Hovenweep, Yellow Jacket, Chimney Rock, Casa Grande, and Mesa Verde. Celestial rock art elements are present throughout the Southwest. Few of these involve accurate mapping of constellations, but several seem to invoke magical power attributed to the stars. Von Del Chamberlain has studied Navajo star ceilings in Canyon de Chelly and concluded that most of them are connected with symbolic protection or other celestial magic. Polly Schaafsma interprets war and star imagery in the petroglyphs of New Mexico's Galisteo Basin as part of a tradition of Southern Tewa celestial war magic.

Nine years after astronomer Gerald S. Hawkins rekindled interest in ancient and prehistoric astronomy in 1963, with his studies of solar and lunar astronomical alignments in Stonehenge, another astronomer, John A. Eddy, identified celestial sightlines in the Bighorn Medicine Wheel, a North American antiquity *Time* magazine headlined as "Stonehenge U.S.A." The Bighorn Medicine Wheel is located at an elevation of 9600 feet, above the timberline on Wyoming's Medicine Mountain. It is a ring, 87 feet in diameter, drawn in small rocks. Originally, the Wheel had 27 or 28 spokes of stones that converged on the main cairn at the center. Five other cairns were constructed on the Wheel's rim, and the one spoke that extends beyond the rim, ends in its own cairn. The ring is actually a flattened circle. Its axis of symmetry coincides with the spoke that reaches from the outside cairn, southwest of the rim, to the cairn at the center. Eddy demonstrated that this line also continues to the northeast horizon and the point of summer solstice sunrise. He associated other lines between cairns with the sequential risings of three bright stars in the predawn sky in summer. Although the age of the Bighorn Medicine Wheel is uncertain, a radiocarbon date for a piece of wood retrieved from the central cairn associates it with the seventeenth century. Eddy thought that the Bighorn Medicine Wheel might have been used at that time by historic Plains Indians to make astronomical observations.

It seems likely, however, that the Bighorn Medicine Wheel is much older. It resembles other similar structures, especially in southern Canada, that are known to be thousands of years old. Who built the Bighorn Medicine Wheel and when it was built are still not known with certainty. There is reason to be skeptical about the practical value and validity of the stellar alignments, but the summer solstice sunrise line is congruent with the design. If the solstice alignment were part of the original plan, it may have had as much to do with vision quests and shamanic retreat as with calendric observation.

In the historic era, Plains Indians incorporated the sky into symbols and ritual. The well-known Sun Dance was intended to inspire prayer and visions through self-infliction of pain, fasting, and thirst. Gazing at the sun while suspended from a pole with ropes looped through the flesh of the chest was thought to purify and spiritually strengthen the participant. Although not all Plains tribes practiced this demanding regimen in the Sun Dance, acquisition of sacred power was a common theme. The enclosure in which the ceremony took place is called a Sun Dance Lodge, and it is sometimes built with 28 posts said to represent the days of the lunar month. Originally, the ritual was performed at the time of the summer full moon nearest to the time of the bison hunt.

Another Plains group, the Skidi band of the Pawnee, is known to have possessed a rich and detailed tradition of star lore. This knowledge has been reviewed and analyzed by Von Del Chamberlain in *When the Stars Came Down to Earth*. The Skidi Pawnee Morning Star sacrifice was timed by the movements of planets, especially Mars and Venus, and mythologically, the planets were key players in the Skidi Pawnee Creation myth.

In Nebraska, the Omaha devised a symbol of social cohesion and tribal stability out of the Sacred Pole that was erected ceremonially at tribal gatherings. Omaha myth and ritual allow us to deduce that the Sacred Pole was oriented on the north celestial pole. Its power was linked to the stabilizing character of the hub of the sky.

It is a curious fact that so much public interest in ancient North American Indian astronomy has been directed toward the prehistoric Southwest and the Bighorn Medicine Wheel. Although these sites are valid targets of study, they belong to relatively unpopulated parts of the continent. Even the well-documented traditions of the Plains Sun Dance, the Skidi Pawnee Morning Star Sacrifice, and polar axis symbolism of the Omaha Sacred Pole must be considered marginal traditions of North America.

The mainstream, on the other hand, belongs to the most populated zones of North America. To understand, then, the true character of North American Indian astronomy, it is necessary to look at the Mississippi Valley and California. Nowhere north of Mesoamerica had a comparable population density.

Unfortunately, we know relatively little about the astronomical traditions of the ancient Mississippi Valley. In 1961, however, archaeologist Warren L. Wittry excavated a feature he called the Sun Circle at Cahokia, a great population center and powerhouse of regional trade in central Illinois between AD 700 and 1200. Cahokia is best known for its large mounds, some of which supported temples and residences. Others hosted burials. Monks Mound, the largest prehistoric earthen construction in the world, is the centerpiece of what was the Chicago of the prehistoric Midwest.

Originally, the Sun Circle was a 410-foot-diameter ring of 48 tall posts (perhaps 30 feet high), with a pole in the middle of the circle but offset apparently intentionally five feet from the true center. Three of the posts that once occupied the holes that now remain combined with the center pole to deliver alignments to the rising sun at summer solstice, winter solstice, and the equinoxes. Wittry believed that the Sun Circle's purpose was calendric. It is difficult to understand, however, why a complete ring of posts would be needed and why the posts were so tall. On the other hand, the intentional displacement of the central post makes the astronomical alignments possible.

Despite the ambiguity that persists with Cahokia's Sun Circle, archaeologist Melvin L. Fowler verified the ancient Cahokians' interest in cardinal directions. The city's limits were established by a particular type of earthen mound, and these mounds also defined the primary, and cardinal, axes of the site. In 1994, Fowler described another post circle on the main north–south axis of Cahokia. Its design, size, and astronomical potential mimic the Sun Circle.

The Incinerator Site, a smaller stockaded Mississippian village near Dayton, Ohio, included an arrangement of posts also thought to have astronomical meaning. The largest post, two feet in diameter, was located in a plaza, and it formed a line with a post inside a nearby building. This line pointed to the sunrise on April 24th and August 20th. Both dates potentially have agricultural significance. The April date could mark the end of the frost and the beginning of planting. The August date is linked to the Green Corn Ritual, performed in the historic era when the corn filled the husk but was not yet ripe. Cardinally oriented logs were kindled at this time into the New Fire, which also had solar associations.

European explorers encountered the descendants of the Mississippian mound builders in the southeast United States. The chief of the Natchez was known as the Great Sun. He claimed to be the brother of the sun and greeted the sun each morning, when it first appeared, from the top of his residence mound. We also know the Natchez subdivided the daylight hours into four periods based upon the position of the sun. They measured the year in months based on the observed phases of the moon and named each lunation for an appropriate seasonal phenomenon.

In the Far West, before Columbus, California competed respectably with the Mississippi Valley for the distinction of being the most populous and densely populated zone north of Mesoamerica. Some indigenous astronomical traditions persisted in California until quite late, and valuable information was collected by ethnographers, especially John Peabody Harrington. The revival of interest in California Indian astronomy was largely initiated by D. Travis Hudson and Ernest Underhay, whose book, *Crystals in the Sky*, reconstructed the sky lore of southern California's Chumash Indians. Explicit references to horizon observations of the solstice sun and solstice rituals have been collected from the entire state. The moon's phases were counted, and the lunar months were named by most California groups. They recognized familiar patterns of stars, including the Big Dipper and the Belt of Orion. They named many other stars and used them as seasonal signals. They saw the Milky Way as a route to the sky followed by the souls of the deceased. Elaborate Milky Way ceremonialism is known among the Luiseño, who incorporated it into ritual initiation of the youth, mourning songs for the dead, sacred myth, and rock art.

California possesses many rock art sites and some of the most complex rock art in the world. Some of these sites have been associated with solstitial light-and-shadow events that interact with the rock art. Often, particularly at winter solstice, the light develops into a pointed, knifelike shape that pierces a carved or painted element. These effects appear to be symbolic, for they do not generally pinpoint the solstice with high accuracy. Rather, they "work" over a solstice "season" perhaps two to four weeks long. Although there is no explicit evidence that links solstice effects with California rock art, we do know the names of two Chumash shamans who went into the mountains at the time of winter solstice to paint on the rocks.

In detailed application, Native American astronomy was richly varied in North America, but its purpose and basic character were broadly the same throughout the continent. For that reason, we can rely upon Francisco Patencio, Chief of the Palm Springs Indians of southern California, for words that could apply to nearly all Indians:

> When the sun swung to the north and the moon showed quartered by day overhead, or west, they knew by the signs of the sun and the moon when the seeds of certain plants were ripe, and they got ready to go away and gather the harvest. Every plant that grew, the nesting time of all birds, the time of the young eagles, everything they learned by the signs of the sun and the moon. *Stories and Legends of the Palm Springs Indians* (Los Angeles: Times Mirror, 1943).

E.C. KRUPP

REFERENCES

Ceci, Lynn. "Watchers of the Pleiades: Ethnoastronomy among Native Cultivators in Northeastern North America." *Ethnohistory* 25(4): 301–317, 1978.

Chamberlain, Von Del. "Navajo Indian Star Ceilings." In *World Archaeoastronomy*. Ed. A.F. Aveni. Cambridge: Cambridge University Press, 1989, pp. 331–340.

Chamberlain, Von Del. *When the Stars Came Down to Earth: Cosmology of the Skidi Pawnee Indians of North America*. Los Altos, California: Ballena Press, 1982.

Eddy, John A. "Astronomical Alignment of the Big Horn Medicine Wheel." *Science* 184 : 1035–1043, 1984.

Fowler, Melvin L. "A Pre-Columbian Urban Center on the Mississippi." *Scientific American* 233: 92–101, 1975.

Harrington, John Peabody. *Unpublished Notes on the Chumash and Kitanemuk*. MS 6017, Boxes 2, 3, 5, 9, 12, 538, 543, 545, 550, 651, 705, 745, 747. Washington, D.C.: United States National Anthropological Archives, no date.

Hawkins, Gerald S. "Stonehenge Decoded." *Nature* 200: 306–308, 1963.

Hudson, D. Travis, and Ernest Underhay. *Crystals in the Sky*. Socorro: Ballena Press, 1978.

Krupp, E.C., ed. *In Search of Ancient Astronomies*. Garden City, New York: Doubleday, 1978.

Krupp, E.C. *Echoes of the Ancient Skies*. New York: Harper & Row, 1983.

Krupp, E.C., ed. *Archaeoastronomy and the Roots of Science*. Boulder, Colorado: Westview Press, 1984.

Krupp, E.C. *Beyond the Blue Horizon — Myths and Legends of the Sun, Moon, Stars, and Planets*. New York: HarperCollins, 1991.

Krupp, E.C. *Skywatchers, Shamans & Kings: Astronomy and the Archaeology of Power*. New York: John Wiley & Sons, Inc., 1996.

Malville, J. McKim. "Prehistoric Astronomy in the American Southwest." *The Astronomy Quarterly* 8 (1): 1–36, 1991.

Malville, J. McKim, and Claudia Putnam. *Prehistoric Astronomy in the Southwest*. Boulder, Colorado: Johnson Publishing Company, 1989.

McCluskey, Stephen C. "The Astronomy of the Hopi Indians." *Journal for the History of Astronomy* 8(3): 174–195, 1977.

Robbins, R. Robert, and Russell B. Westmoreland. "Astronomical Imagery and Numbers in Mimbres Pottery." *The Astronomy Quarterly* 8(2): 65–88, 1991.

Schaafsma, Polly. "War Imagery and Magic: Petroglyphs at Comanche Gap, Galisteo Basin, New Mexico." Paper prepared for *Social Implications of Symbolic Expression in the Prehistoric American Southwest*, Symposium at the 55th Annual Meeting of the Society for American Archaeology, Las Vegas, Nevada, April 1990.

Sofaer, Anna, Volker Zinser, and Rolf M. Sinclair. "A Unique Solar Marking Construct." *Science* 206 (4416): 283–291, 1979.

Stephen, Alexander M. *Hopi Journal of Alexander M.Stephen*. Ed. Elsie Clews Parsons. New York: Columbia University Contributions to Anthropology, 23, 2 vols, 1936.

Williamson, Ray A. *Living the Sky: The Cosmos of the American Indian*. Boston: Houghton Mifflin Company, 1984.

Wittry, Warren. "The American Woodhenge." *Explorations into Cahokia Archaeology*. Urbana: Illinois Archaeological Survey, Inc., University of Illinois, 1973, pp. 43–48.

Wroth, L.C. *The Voyages of Giovanni de Verrazzano, 1524–1528*. New Haven: Yale University Press for the Pierpont Morgan Library, 1970.

Zeilik, Michael. "The Ethnoastronomy of the Historic Pueblos, 1: Calendrical Sun Watching." *Archaeoastronomy (Supplement to Journal for the History of Astronomy* 16(8): S1–S24, 1985.

See also: Calendars – Medicine Wheel – Time – Eclipses

ASTRONOMY OF THE AUSTRALIAN ABORIGINAL PEOPLE The Australian Aborigines were almost certainly the world's first astronomers. Their complex systems of knowledge and beliefs about the heavenly bodies have been handed down through song, dance and ritual for some 40,000 years, predating by many millennia those of the Babylonians, the ancient Greeks, the Chinese, or the Incas. The legends which have survived, until very recently, within a virtually unchanged cultural context, show how natural phenomena, including celestial bodies and events, were assimilated by the Australian Aborigines into a holistic value system which was predicated on the close relationship of the individual with the whole natural world. It is significant, in this regard, that theirs was the only known culture with no myth of alienation from Nature, such as the expulsion from Eden of the Judeo-Christian tradition. On the contrary they believed that through their Great Ancestors of the Dreaming they, too, were continuing co-creators of the natural world. Hence they used their knowledge of the stars not only to predict and explain natural occurrences but also to provide celestial parallels with tribal experiences and behavioral codes.

The Aborigines' knowledge of the "crowded" southern sky was probably the most precise possible for people dependent on naked-eye astronomy. They made accurate observations, not only of first- and second-order stars, but even of the more inconspicuous fourth-magnitude stars. Pattern was apparently more important in recognition than brightness, for the Aborigines often identified a small cluster of relatively obscure stars while ignoring more conspicuous single stars in the vicinity. Thus the people of Groote Eylandt named as *Unwala* (the Crab) the small cluster of relatively insignificant (average magnitude 4.4) stars Sigma, Delta, Rho, Zeta and Eta Hydrae, while disregarding the adjacent bright stars Procyon (α Canis Minoris) and Regulus (α Leonis) (magnitude 0.36 and 1.35 respectively) which are not part of an obvious group. Members of the Boorong tribe of the Mallee District of Victoria limited their identification procedures to linear patterns of three or more stars. Unlike the familiar Greek designations, based on a join-the-dot pictorial image, the Aborigines identified a group of stars with the whole cast of characters in a story, the relationship being conceptual rather than visual. Color was also an important factor in the aboriginal designation of stars. The Aranda tribes of Central Australia distinguish red stars from white, blue and yellow stars. They classify the bright star Antares (α Scorpii) as *tataka indora* (very red) while the stars of the V-shaped Hyades cluster are divided into a *tataka* (red) group and a *tjilkera* (white) group. The former are said to be the daughters of the conspicuously red star Aldebaran (α Tauri).

The Aborigines also differentiated between the nightly movement of the stars from east to west and the more gradual annual shift of the constellations. From this latter displacement they devised a complex seasonal calendar based on the location of constellations in the sky, particularly at sunrise or sunset. The Aranda and Luritja tribes around Hermannsburg in Central Australia knew that certain stars lying to the south, namely *Iritjinga* and the Pointers of the Southern Cross, are visible throughout the year, although their position in the sky varies. This amounts to a realization that stars within a certain distance of the South Celestial Pole never fall below the horizon.

Yet what the Aborigines did with their astronomical knowledge was fundamentally different from the procedures of Western science. Tribal Aborigines paid no attention to two of the most basic concepts of western science, numeracy and temporality; they made no measurements of space or time, nor did they engage in even the most elementary mathematical calculations. Their observations of the stars were conducted for essentially pragmatic reasons. Either they were an attempt to discover predictive correlations between the position of the stars and other natural events important to the survival of the tribe — the availability of specific foods or the onset of particular weather conditions, or they provided a system of moral guidance and education in tribal lore — a function regarded as equally necessary to the continuation of tribal identity.

As hunter–gatherers, dependent for their survival on a foreknowledge of environmental changes, the Aborigines noted the correlation between the movements and patterns of stars and changes in the weather or other events related to the seasonal supply of food. Thus the significance attributed to these sidereal occurrences varied with the diet and lifestyle of different tribes. On Groote Eylandt the appearance in the evening sky towards the end of April of Upsilon and Lambda Scorpii indicated that the wet season had ended and that the dry south-easterly wind or *marimariga* would begin to blow, causing changes in climate and animal behavior. At nearby Yirrkalla the appearance of Scorpio in the morning sky in early December heralded the arrival of the Malay fishermen who came in their canoes to collect trepang or *bêche de mer* which they sold to the Chinese. In winter, the most spectacular individual stars in the southern sky are Arcturus (α Bootis) and Vega (α Lyrae). When Arcturus could be seen in the eastern sky at sunrise, the Aborigines of Arnhem Land knew that it was time to harvest the spike-rush or *rakia*, a reed valuable for making fish traps and baskets for carrying food, and a local legend about Arcturus served as an annual reminder of this. On the other hand, amongst the *Boorong* tribe of the Mallee district of western Victoria, Arcturus was personified as *Marpeankurrk* the tribal hero who showed them where to find *bittur*, the pupa of the wood ant, a staple item of diet during August and September. The constellation Lyra represented the spirit of *Neilloan*, or the Mallee-hen who taught the tribe how to finds its eggs, an important source of food in October. Other notable events, like the ripening of tubers and bulbs and the appearance of migratory birds and animals, were correlated with specific positions of Orion, the Pleiades and the Southern Cross at different seasons of the year. For the Pitjantjatjara tribe in the Western Desert, the appearance of the Pleiades in the dawn sky in autumn was the sign that the annual dingo-breeding season had begun. Fertility ceremonies were performed for the dingoes, or native dogs, and some weeks later the tribe raided the lairs, culling and feasting on the young pups. The legends ensured that these nutritional associations were not forgotten and stressed their importance for the continuing survival of the race.

Equally important to the preservation of the tribe was its sense of identity, involving tribal beliefs orally transmitted across generations. These myths outlined the role of the Ancestors of the Dreaming in the scheme of the universe and the behavior appropriate to their descendants. Explanations of natural events which emphasized pattern, order, and laws, rather than unpredictable effects, reinforced the sense of an organic relationship between natural phenomena and social behavior. Many of these legends involved the constellations, so that the night sky provided a periodic reminder of the moral lessons enshrined in the myths.

Like all explanatory systems, including Western science, these legends represented attempts to understand, predict and hence to obtain some control over the natural world. However, unlike the essentially analytical, materialistic, and particularizing approach of Western science, the underlying premise of all the aboriginal myths was a belief in the close spiritual unity of human beings, not only with other species, but also with inanimate objects. Astronomy was an integral part of the Aborigines' total philosophy about the natural world, so the legends emphasized the parallels between the personified heavenly bodies and their earthly counterparts, humanizing and integrating natural phenomena with tribal institutions and customs.

The meaning which a tribe attributed to the celestial bodies was conceptual rather than perceptual. It could not be understood by personal experience or by the intellect, but only through initiation into tribal lore which stressed the intimate, causal association between physical events and the human dramas of good and evil. Lessons about compassion, brotherhood, and respect for the land as Mother, the prohibition of incest and adultery, and taboos on killing or eating totemic animals were nightly reinforced by being enacted in the sky world which thereby established the universal validity of the tribe's ethical laws.

The many and diverse aboriginal myths associated with the heavenly bodies include stories about the Sun, the Moon, the Milky Way, the Magellanic Clouds, Mars, Venus, and the several constellations which form distinctive patterns in the southern sky — notably the Southern Cross and its pointers, the Pleiades, Orion's Belt, Scorpio, Gemini and Aldebaran. The following are a representative selection of these myths.

In most of the Aboriginal creation stories, the Sun is the life-giving spirit. Amongst the Murray River tribes the origin of the Sun is linked to the tossing of a giant emu egg into the sky where it struck a heap of dry wood and burst into flame, bringing light to the hitherto dark world. Thereupon, the Great Spirit *Baiame*, seeing how much the world was improved by sunlight, decided to rekindle the woodpile each day.

In contrast to the ancient Greeks, the Amerind Indians and the Quechua Indians of Peru, all of whom designated the Sun as male and the Moon as female, the Australian Aborigines represented the Sun as female and the Moon as male. In most areas, the Sun is regarded as a woman who daily awakes in her camp in the east and lights a fire to kindle the bark torch she will carry across the sky, thus providing the first light of dawn. She decorates herself with powder made from crushed red ochre, coloring the clouds red in the process. At evening she renews her powder in the western sky before beginning her long passage underground back to her camp in the east. It was probably this underground journey which was instrumental in the classification of the Sun as female, for her torch is thought to bring warmth and fertility to the interior of the Earth, causing the plants to grow. However, in Arnhem Land, where the Sun sets in the sea, she is thought to become a great fish, swimming under the Earth to return in the east next morning while the Moon becomes a fish, passing beneath the earth during the day.

The Moon, being male, is generally accorded greater status, and in many areas powers of death and fertility are accorded to him. An eclipse of the Sun is interpreted as indicating that the Moon Man is uniting with the Sun Woman. Several legends have evolved to account for the Moon's cycles. In coastal areas the correlation between the phases of the Moon and the tides was noted. In Arnhem Land and Groote Eylandt, where high tides occur when the new or full Moon sets at sunset or sunrise respectively, and low tides when the moon is in the zenith at sunrise or sunset, the local Aborigines believe that the high tides, running into the Moon as it sets into the sea, make it fat and round. (Although the new Moon may appear thin, they deduce from the faint outline of the full circle that it too is round and full of water.) Conversely, when the tides are low, the water pours from the full Moon into the sea below and the moon consequently becomes thin.

In most areas the Moon was regarded as more mysterious, and hence more dangerous, than the Sun. Because of the association of the lunar cycle with the menstrual cycle, the Moon was linked with fertility and young girls were warned against gazing at the Moon unless they wished to become pregnant.

The Milky Way was regarded by the Aborigines as a river in the Sky World, the large bright stars being fish and the smaller stars water-lily bulbs. Central Australian tribes believed that the Milky Way divided the sky people into two tribes and thus it served as a perpetual reminder that a similar equitable division of lands should be observed between neighboring tribes. Various legends regarding taboo marriages, adultery, and reminders to celebrate tribal heroes, many of them involving a moral lesson, have evolved in different areas to account for the formation of the Milky Way and the dark region, known to Europeans as the "Coal Sack".

A Queensland version of the origin of the Milky Way associates it with *Priepriggie,* an Orpheus-like hero, as famed for his songs and dances as for his hunting. When he disappeared, his people tried unsuccessfully to perform his dance until they heard singing in the sky. Then the stars, hitherto randomly dispersed, arranged themselves to the rhythm of *Priepriggie's* song. Thus the Milky Way serves as a reminder that the tribal hero should be celebrated with traditional songs and dancing.

Because of its diagrammatic shape, the Southern Cross features in association with various characteristic objects in different areas. Around Caledon Bay on the east coast of Arnhem Land, it is taken to represent a stingray being pursued by a shark — the Pointers. On Groote Eylandt, where fish is the staple diet, the four stars of the Cross represent two brothers, the *Wanamoumitja* (Alpha and Beta Crucis), and their respective camp fires (Delta and Gamma Crucis) where they cook a large black fish (the Coal Sack) which they have caught in the Milky Way. The Pointers are their two friends, the *Meirindilja*, who have just returned from hunting. Desert tribes, on the other hand, see in the kite shape of the Cross the footprint of the wedge-tailed eagle *Waluwara* while the pointers represent his throwing stick and the Coal Sack his nest.

Venus, the Morning Star, was an important sign to the Aborigines, who arose at early dawn to hunt. It, too, was personified and frequently associated with death. Arnhem Land legends identify the home of the morning star, *Barnumbir*, as Bralgu, the Island of the Dead. Afraid of drowning, *Barnumbir* could be persuaded to light her friends across the sea at night only if she were held on a long string by two old women, who at dawn would pull her back to Bralgu and keep her during the day in a basket. Because of this connection, the morning star ceremony is important in the local rituals for the dead since a dead person's spirit is believed to be conducted by the star to *Bralgu*.

One of the most widespread Aboriginal myth cycles concerns the constellations of Orion and the Pleiades and these bear a striking similarity to the Greek story of the seven daughters of Atlas who, when pursued by Orion, flew into the sky as doves to form the constellation of the Pleiades. All identify them with a group of seven young women and nearly all portray them as fleeing from the amorous hunter Orion who, in some versions, is castrated as a punishment and warning to other potential wrongdoers. The whole cluster of Pleiades stories therefore forms part of a much larger group of myths of sexual conquest and submission.

Amongst the Pitjantjatjara tribe, the practical connection between the dingo-breeding season and the appearance of the Pleiades in the dawn sky in autumn, is preserved in a local legend. The *Kungkarungkara* or ancestral women kept dingoes to protect them from a man *Njiru* (Orion), but he succeeded in raping one of the girls, the obscure Pleiad, who died. Even though the women assumed their totemic form of birds and flew into the sky to escape from him, he defies their dingoes and follows the women across the sky, armed with a spear (the stars of Orion's Belt) which has ritual phallic significance. Like the Greek Orion, *Njiru* was also a hunter, and pairs of smaller stars which arise near the constellation of Orion are said to represent his footsteps as he pursues the *Kungkarungkara*.

At Yirrkalla on the coast of Arnhem Land, the constellation of Orion is regarded as a group of fisherman arriving from the east in a canoe with a turtle they have caught, while the Pleiades represent their wives in another canoe with two large fish. As they approached the shore a heavy storm capsized the canoes, drowning the people. All the representative stars are visible in the sky during the wet season — a warning against the dangers of fishing when storms are imminent. In north-eastern Arnhem Land the story carries the added moral that the fishermen drowned as a punishment for catching catfish, forbidden to this tribe by totemic law.

Although relatively insignificant to the naked eye, the Large and Small Magellanic Clouds feature in many aboriginal legends as the camps of sky people. On Groote Eylandt they are believed to be the camps of an old couple, the *Jukara*, grown too feeble to catch their own food. Other star people catch fish and lily bulbs for them in the Milky Way and bring them to the *Jukara* to cook on their fires. The space between the Clouds is their cooking fire, while the bright star Achernar (Alpha Eridani, magnitude 0.49) represents their meal. This story suggests a celestial model of compassion for the aged. At Yirrkalla the Magellanic Clouds are said to be the homes of two sisters. During the middle of the dry season the elder sister (Large Cloud) leaves her younger sister (Small Cloud), but during the wet season she returns so that they can collect yams together. This story reflects the observed fact that at this latitude (12°S) only the Small Cloud is visible during most of the dry season (April to September), whereas both Magellanic Clouds can be seen during the wet season.

Meteors have been variously interpreted by different aboriginal tribes. In north-eastern Arnhem Land, because of their speed and unpredictability, they are believed to be spirit canoes carrying the souls of the dead to their spirit home in the sky. To the Tiwi tribe of Bathurst and Melville Islands, each is the single eye of the one-eyed spirit men, the *Papinjuwari*, who steal bodies and suck the blood of their victims, and their evil eyes are seen blazing as they streak across the sky looking for their prey. In other legends, meteors are associated with fire and linked to the waratah plant, *Telopea speciosissima*, a member of the Protea family, which is resistant to fire and whose brilliant red flowers seemed to the Aborigines like sparks from a fire. This was why, in the early years of white settlement, some Aborigines brought waratahs to the European blacksmiths: they identified the sparks from the anvil with the sparks from meteors and hence with the waratahs.

From this selection of star legends, it will be apparent that, with the possible exception of meteors (and even they can be regarded as recurrent events), the Aborigines' concern was not with extraordinary occurrences, but with the regular patterns of natural phenomena. The star legends served the purpose of integrating a potentially alien universe into the moral and social order of the tribe — by 'humanizing' species and natural objects and ascribing to them behavior patterns and motivations which accorded with those of the tribal unit.

Such a philosophy serves a number of important social functions. In the first place it engenders a level of confidence about Man's place in the universe, not as a superior being but as an equal partner; in this it fulfils a role comparable to that of technology which also offers a level of some control over the environment. Secondly, it cultivates respect for the inanimate as well as the animate, since all partake of the same spiritual identity as Man himself. Thirdly, the legends provide a justification for the customs, rites and morality of the tribe, since these are reflected and enacted in the Skyworld.

The aboriginal myths are not fatalistic as astrology purports to be. Although they link certain natural events with a seasonal configuration of the sky, they make no deterministic predictions about individual lives; the moral values enshrined in the legends are held to be true for the *whole* tribe.

The most radical difference between the vitalistic beliefs which underlie these myths and the materialistic philosophy of western science concerns the relationship of the observer to the observed. In Newtonian science, the observer is assumed to be independent of, and distinct from, the object observed, which, in turn, is regarded as uninfluenced by the observer. Hence, the relationship between physical objects can be validly expressed in mathematical terms which remain true irrespective of the observer. The Aborigines, on the other hand, did not conceive of themselves as observers separated from an objectified Nature, but rather as an integral part of that Nature. The meaning of the celestial bodies, as of everything else in the environment, was neither self-evident nor independent of the observer; rather it depended on the degree of initiation into tribal lore which elucidated the links between tribal customs and natural phenomena. Without this knowledge the individual was disoriented and powerless in an alien universe.

ROSLYNN D. HAYNES

REFERENCES

Elkin, A.P. *The Australian Aborigines*. Sydney: Angus and Robertson, 1964.

Haynes, Roslynn D. "The Astronomy of the Australian Aborigines." *The Astronomy Quarterly* 7 (4): 193–217, 1990.

Isaacs, Jennifer. *Australian Dreaming: 40,000 Years of Aboriginal History*. Sydney: Lansdowne Press, 1980.

Mountford, Charles Percy. *Art, Myth and Symbolism: Records of the American–Australian Expedition to Arnhem Land*, Vol. 1. Melbourne: Melbourne University Press, 1956.

Mountford, Charles Percy. *Nomads of the Australian Desert*. Sydney: Rigby, 1976.

Tindale, Norman. *Aboriginal Tribes of Australia*. Canberra: Australian National University Press, 1974.

See also: Environment and Nature of the Australian Aboriginal People

ASTRONOMY IN CHINA Chinese astronomy became a subject of long debate among historians and astronomers in the first half of the twentieth century. Some based their argument on the *Shujing* (Book of Documents) and concluded that Chinese star clerks had already made astronomical observations between 2000 and 3000 BC. Richard Schlegel even asserted that they knew about the twenty-eight lunar mansions as long ago as the year 16,000 BC. Others doubted the reliability of the records in the *Shujing* claiming that Chinese astronomy could not have originated earlier than 500 or

600 BC. Some said that Chinese astronomy originated from India, others said from Arabia, and there were yet those who said that it was from Mesopotamia. Joseph Needham, for example, favors a Mesopotamian origin.

Recent archaeological studies carried out in China have thrown more light on early Chinese astronomy and enabled scholars to re-examine some of the old interpretations. Archaeologists have been hard at work during the last two decades to establish the Xia dynasty, which, according to tradition lasted from the twenty-first century BC to the sixteenth century BC, and which has until now been regarded as legendary. They have yet to recover written records of that period. Erlitou in Henan province, where bronze vessels dating back to the year 1700 BC were recovered in 1971, is one of their important sites. Even the records in the *Shujing* have recently been re-examined by comparison with computed ancient astronomical phenomena.

In the oracle bone writings we can find records of eclipses, novae, and names of stars and some asterisms. The records indicate that the Yin people between the fourteenth and twelfth centuries BC were already using a luni–solar calendar, where one year consisted of twelve moons or lunar months of either twenty-nine or thirty days each, with an extra month known as the intercalary month added about every three years. From about the sixth century BC until the use of the telescope for astronomical observations in Europe in the seventeenth century, Chinese star clerks kept the most consistent and continuous astronomical records on eclipses, comets, novae, sunspots and aurora borealis in the whole world. Star catalogs were produced during the Warring States period (481–221 BC) by Gan De, Shi Shen and Wu Xian. Chinese astronomical records have found many applications in modern astronomy. One application is in the calculations of the period of Halley's comet. The Crab Nebula has been identified with the supernova observed in China in the year 1054. The first pulsar was discovered in 1967. It was found to be near a site where a "guest star" was recorded by the Chinese. A "guest star" in Chinese astronomy could refer to a supernova or nova, but it could also mean a comet or even a meteor, depending on the context. In this case it refers to a supernova or nova. As a result some astronomers made use of Chinese records in their attempts to discover other pulsars. Chinese records on sunspots have been used to determine sunspot cycles, and recently they were used in the study of the variation of the earth's period of rotation. There are many other applications for Chinese astronomical records, such as in a recent study of the Star of Bethlehem in the Bible.

There has been a prolonged dispute since the last century over the question of the origin of the twenty-eight Chinese lunar mansions. These are twenty-eight asterisms distributed along the ecliptic (the band of the zodiac through which the Sun apparently moves in its yearly course). The moon changes its position among the stars night after night, and appears successively in each of these asterisms, appearing from the earth as if it changes its lodging each night. Hence they were known as lunar mansions or lodges. Chinese astronomers also picked one of the stars in every lunar mansion as a reference point from which distances of other stars in its vicinity were measured. The stars used as reference points were called determinant stars. In ancient Hindu astronomy there were twenty-seven *nakṣatras*, each of which had a principal star (*yogatara).* Nine of the *yogataras* are identical with the Chinese determinant stars. Some thought that the lunar mansions and the *nakṣatras* had a common origin, often with Sinologists and Indianists taking opposite sides. Before a conclusion was reached came a third contender, the *al-manāzil*, the moon-stations in Muslim astronomy. Then it was argued that Muslim astronomy was pre-dated by Hindu astronomy. During the middle of this century the most favored candidate was Mesopotamia, from which ancient Chinese astronomy and others were thought to be originated. This is the view favored by Joseph Needham. The argument against China's favor until then was that all the names of the twenty-eight lunar mansions were not found earlier than the fourth century BC. However, in the year 1978 the names of all the twenty-eight lunar mansions were found inscribed on the lid of a lacquer casket of the early Warring States period, showing that the twenty-eight lunar mansions were already there in China not later than the fifth century BC. Chinese archaeological discoveries in the second half of this century and recent archaeological excavations carried out in India have also shown that both Chinese and Indian civilizations existed much earlier than they were thought. Whether the lunar mansions in Chinese astronomy were influenced by the Hindu *nakṣatras* or whether it was the other way round is an open question. Perhaps there was no influence between the two systems at all. The important thing in the study of history of science and civilization is to learn about mutual exchange of ideas among different cultures rather than to engage in futile disputes, claiming priorities and making scores in meaningless contests, in which there is no prize for the winner other than false pride.

The earliest existing Chinese documents on astronomy are two silk scrolls discovered in the Mawangdui tombs in Changsha, Hunan province in the year 1973. One of them, the *Wuxingzhan* (Astrology of the Five Planets) contains records of Jupiter, Saturn, and Venus, the accuracy of which suggested the use of the armillary sphere for measurement. These records were made between 246 BC and 177 BC. The other, the *Xingxiangyunqitu* (Diagrams of Stellar Objects and Cloud-like Vapors of Various Shapes) illustrates different types of comets. Among the ancient astronomical instruments discovered by archaeologists are the bronze clepsydra from the tomb of Liu Sheng of the second century BC and a bronze sundial of the Eastern Han period (25–220). The title of an early Chinese book on astronomy and mathematics, the *Zhoubi suanjing* (Mathematical Manual of Zhoubi), calls to mind either the Zhou dynasty or the circumference of a circle, and the vertical side of a vertical right-angled triangle. The book contains the so-called *Zhoubi* or *Gaitian* cosmological theory in which the heavens are imagined to cover a flat earth like a tilted umbrella, and shows paths for the sun at different seasons of the year. From the positions of the stars and planets mentioned in the text, Japanese scholars established a period between 575 BC and 450 BC for the observations, and inferred that the book was written within the same period. Christopher Cullen recently showed that the book could well be a work of the early first century.

During the second century Zhang Heng (78–139) constructed an armillary sphere as well as a seismograph for making astronomical observations and for detecting the direction of earthquakes. An armillary sphere consisted of a system of rings corresponding to the great circles of the celestial sphere and a sighting tube mounted in the center. With such an instrument Chinese astronomers could measure the positions of heavenly bodies. The mounting of the Chinese armillary sphere deserves special attention. As pointed out by Needham, it has always been equatorial unlike its counterpart in Europe which was ecliptical, but only changed to equatorial in modern telescopes. The new armillary sphere made by Zhang Heng resulted in more accurate observations and better star maps.

An important role of Chinese astronomy was calendrical calculation. The emperor regarded calendar making as one of his duties associated with the mandate that he received from Heaven. The Chinese calendar took into account the apparent cycle of the sun and the cycle of the moon, both of which, as we know, cannot be expressed in an exact number of days. The astronomer responsible for constructing a lunisolar calendar had to make accurate observations of the sun, the moon, and the planets, but however accurate his observations and his calculations, his calendar would sooner or later, in just a matter of decades, get out of step with observations. Hence throughout Chinese history no less than one hundred calendars had been constructed. Besides, sometimes there were also unofficial calendars adopted in certain regions in China. During the seventh century Indian calendar-making made its presence felt in China. We can read the names of a number of Indian astronomers and calendar experts who lived in Changan, the capital of Tang China.

In the early days of the Tang dynasty the calendar officially adopted was the *Lindeli* calendar constructed by the great early Tang astronomer, mathematician and diviner Li Chunfeng (602–670). By the early eighth century a new calendar became overdue. Eventually the old calendar was replaced by the *Dayanli* calendar constructed by the Tantric monk Yixing (683–727). Yixing's secular name was Zhang Sui. He is regarded as the most outstanding astronomer of his time for his recognition of the proper motion of the stars one thousand years before Edmond Halley in the West. In the year 721 the Tang emperor Xuanzong entrusted Yixing with the task of constructing a new calendar. To do this he constructed new astronomical instruments, including an armillary sphere moved round by wheels driven by water, and he also carried out a large scale research project to measure the length of the earth's meridian. He employed the method of difference, involving equations of the second degree, as well as the method of the remainder theorem in the *Suanzi suanjing* to calculate his new calendar.

From the late seventh century Indian Tantric monks had also come to reside in the Tang capital. Later Yixing made frequent contact with them and assisted them in the translations of some of their *sūtras* into Chinese. Yixing could have acquired from them some knowledge of astrology and mathematics, including the idea of a spherical earth and the sine table. There were then three clans of Indian calendar experts living in the Tang capital, namely the Siddhārtha clan, the Kumara clan and the Kaśyapa clan, of which the Siddhārtha clan was the most active and famous. At least two members of that clan had served as Directors of the astronomical bureau. The most distinguished member was Gautama Siddhārtha, who constructed an iron armillary sphere when he was director and translated the *Navagrāha* calendar into Chinese in the year 718. Most important of all, he compiled the *Da-Tang Kaiyuan zhanjing* (Prognostications Manual of the Kaiyuan Period of Tang Dynasty) between the years 718 and 726. The word *Navagrāha* means the "Nine Luminaries", namely the Sun, the Moon, Mercury, Venus, Mars, Jupiter, Saturn and two imaginary planets *Rahu* and *Ketu*. Although this calendar was never adopted in China, it found its way from there to Korea, where it was used for a period of time.

During the Song dynasty (960–1279) astronomers made more accurate astronomical observations using new and larger armillary spheres. They constructed several such instruments, the most famous of which was that made by Su Song (1020–1101). It was an armillary sphere driven by a water-wheel using the principle of the escapement. A full-scale study of Su Song's instrument is given in Joseph Needham's *Heavenly Clockwork*. Su Song's instrument was both an armillary sphere and a clock, but it was known only as an armillary sphere. Hence until it was pointed out by Needham, Price and Wang Ling, nobody knew that the clock already existed in China when Matteo Ricci introduced the clock in the late sixteenth century.

Astronomy during the Mongol period (1271–1368) was closely connected with the name of Guo Shoujing (1231–1316), the last of the great traditional Chinese astronomers. The *Shoushili* calendar that he constructed was the most advanced and accurate calendar ever produced in traditional China. It was made possible by the precise instruments he built and the method of finite difference he used in his calculations. This was also the time when astronomers from the Arab world came to work in China. The Muslims were still active in the astronomical bureau when the Jesuits came towards the end of the sixteenth century. The Jesuits arrived at a time when Chinese astronomy was in a state of stagnation, when no one with the knowledge and skill of Guo Shoujing could be found. The Chinese learned pre-Copernican astronomy from them. Modern astronomy came to China only around the middle of the nineteenth century. It did not take long for Chinese astronomy to join the mainstream of modern astronomy. Just as Nakayama Shigeru said that the history of Japanese astronomy was the history of Chinese astronomy in Japan, Chinese astronomy has already become modern astronomy.

HO PENG YOKE

REFERENCES

Cullen, Christopher. *Astronomy and Mathematics in Ancient China: The Zhou Bi Suan Jing*. Cambridge: Cambridge University Press, 1996.

Ho Peng Yoke. *Modern Scholarship on the History of Chinese Astronomy*. Canberra: The Australian National University, Faculty of Asian Studies, 1977.

Li Changhao et al. eds. *Zhongguo tianwenxue shi* (History of Chinese Astronomy). Beijing: Kexue chubanshe, 1981.

Nakayama Shigeru. *A History of Japanese Astronomy*. Cambridge, Massachusetts: Harvard University Press, 1969.

Needham, Joseph. *Science and Civilisation in China*. vol.3, Cambridge: Cambridge University Press, 1959.

Needham, Joseph. *Heavenly Clockwork: the Great Astronomical Clocks of Medieval China*. Cambridge: Cambridge University Press, 1986.

Zhang, Yuzhe et al. eds. *Zhongguo dabaikequanshu: tianwenxue* (The Great Chinese Encyclopedia: Astronomy). Beijing & Shanghai: Zhongguo Dabaikequanshu chubanshe, 1980.

See also: Eclipses – Gan De – Lunar Mansions – Yogatara – Armillary Sphere – Li Chunfeng – Su Song – Zhang Heng – Guo Shoujing – Calendars – Clocks and Watches

ASTRONOMY IN EGYPT We have few written records dealing with the heavens, and those that we have are derived from the Greek astronomical tradition and therefore are very late in Egyptian history. Thus, in order to study the astronomy of ancient Egypt we rely on such limited pieces of evidence as "diagonal calendars" carved into coffin lids in the Middle Kingdom (ca. 2150–1780 BC), orientation of tombs and pyramids, and temple decorations (especially during the Ptolemaic period, ca. 300–30 BC).

The primary incentive for studying the heavens seems to have been to establish the civil calendar on a firm foundation. This civil calendar apparently was initiated with the heliacal rising of Sothis (now known as Sirius). The determination of the new year was not dependent on astronomical evidence, however, but on the counting off of 365 days from the previous new year's date. To assist in this process, the heavens were divided into thirty-six *decans*, each characterized by a bright star or a distinctive group of stars, marking thirty-six ten-day periods, to which five epagonal days were added. Because the civil calendar contained exactly 365 days, the civil year and the astronomical year were often out of phase. The two coincided only every 1460 astronomical years (the so-called Sothic cycle).

The calendar for religious festivals, however, seems to have been erected on a lunar model. The Egyptian predicted the phases of the moon based on the observation that there are 309 lunar cycles in 25 Egyptian civil years. In general, two consecutive lunar months were given 59 days, but every four years the last two months were given sixty days. These correlations between the lunar religious festivals and the civil calendar systems allowed the priests to insert the festivals into the state calendar with a fair degree of accuracy.

It has often been pointed out that many pyramids, as well as numerous tombs, are aligned in accord with the cardinal directions, almost always with entrances to the north. The precision of these alignments points to an astronomical determination of the directions, although we do not know what methodology was used. (It is also fair to say that many temples and tombs are not so precisely oriented — many merely face the Nile, which was apparently assumed to flow from south to north.)

The Egyptian observer of the heavens used only a few extremely simple observational tools, and even these were apparently not used extensively before the New Kingdom period (ca. 1550–1085 BC). Nocturnal observations usually involved a *Bay* (notched palm rib) to determine the zenith transit of a celestial object. Daytime observations generally involved determination of the altitude of the sun using length of shadows. These measurements were made with a *merkhet*, a horizontal bar with a vertical block at one end (to cast the shadow) pierced so that a plumb line could be suspended from it. Because these shadows vary so much in length during the course of the day, the merkhet was often constructed so that the shadows fall on an inclined plane, rather than on the bar itself. This modification allowed the instrument to be of a more easily manageable size, but did not necessarily improve its accuracy. Since both day and night were considered to consist of twelve equal hours, the length of the hour varied over the course of the year, complicating any attempt to regulate time by astronomical observations. When celestial observations were not practical, the ancient Egyptian used a water-clock like a Greek clepsydra to determine the hours. A container shaped like a truncated cone was filled with water which was allowed to escape through a small opening in the base. The interior was graduated to show the lapse of time on the basis of how much water had escaped. Here, too, we find that a considerable degree of inaccuracy or indeterminacy seems to have been tolerated.

Egyptian views of the universe were closely tied to their religion. In the beginning, there was only the water, *Nuu*, covered with darkness and containing the germs of the world. Finally, the spirit of the water uttered the word and called the world into existence. The world is a great rectangular box whose longest dimensions run north and south (as does the land in Egypt itself). Earth is the bottom of this box, Egypt its center. The sky stretches over it, supported either by four pillars or four mountain peaks, connected by a chain of mountains. On a ledge just below these mountain peaks a river (Ur-nes) flows around the world. The Nile is a branch of this celestial river.

This river carries a boat on which rides a disk of fire, the sun god, Ra, who is born new every morning, grows and gains strength until noon, when he transfers to another boat that carries him to the northern valley, Dart, which is perpetually covered with night. From here, other boats carry Ra around to the eastern gate once more, from where he is again reborn. On occasion, the solar ship is attacked by a huge serpent and Ra is temporarily swallowed up. Similarly, on the fifteenth of each month the ship of the moon is attacked by a large sow. The moon, wounded in this struggle, enters a period of decline, finally dying some two weeks later. Sometimes in her initial attack, the sow actually manages to swallow the moon temporarily.

During the summer months, the obliquity of Ra's path decreases and the god comes closer to Egypt. During winter he moves farther away as the obliquity increases. This happens because the solar boat always keeps close to the bank of the celestial river that is nearest to the human habitat. When the celestial river overflows in the annual inundation, the solar boat is able to sail closer to the earth. As the water recedes, the boat also descends and draws away from the earth.

GREGG DE YOUNG

REFERENCES

Boker, R. "Über Namen und Identifizierung der ägyptischen Dekane." *Centaurus* 27: 189–217, 1984.

Neugebauer, O. and R. A. Parker. *Egyptian Astronomical Texts*. Providence, Rhode Island: Brown University Press, 1960–1969.

Parker, R. A. "Egyptian Astronomy, Astrology, and Calendrical Reckoning." In *Dictionary of Scientific Biography*. Ed. C. Gillispie. Vol. 16. New York: Scribners, 1971–1980, pp. 706–727.

Slasman, A. *L'astronomie selon les Égyptiens*. Paris: Laffont, 1983.

ASTRONOMY OF THE HEBREW PEOPLE The Hebrew astronomical tradition that shall be surveyed in this essay is that tradition recorded in the Hebrew alphabet dealing with the motions of the heavenly bodies and the structure of the heavens. These writings utilize principally the Hebrew language, but include texts in languages such as Aramaic and Arabic that can be written out in Hebrew script. Topics of specifically Jewish interest, such as calendar computations and doctrinal matters, occupy only a minor portion of this literature. For the most part, the Hebrew astronomical tradition differs little from contemporary writings belonging to the traditions with which Jews found themselves in immediate contact at any given age and place. To the extent that these other traditions may be classified as non-Western, the Hebrew tradition may be by and large considered such.

The earliest substantive materials are found in the *Talmud* and *Midrash*. Most of the discussions center upon the structure and physics of the heavens. Of particular interest are several notices of disagreement between Jewish and non-Jewish experts, for example, on the question of whether it is the star itself, or the spherical shell within which the star was thought to be embedded, which moves around the earth. In other words, we have clear evidence that even at this early date Jewish scholars identified themselves with particular conceptions of the heavens. *Pirqei di-Rebbe Eliezer*, *Baraitha di-Mazalot*, and *Baraitha di-Shmuel* are three post-Talmudic writings whose precise dating is problematic but which certainly precede the flowering of the sciences under the Abbasids. The latter two are the earliest texts which preserve any mathematical astronomy, e.g. a computational scheme for shadows. In addition, al-Khwārizmī's treatise on the Jewish calendar, which belongs to the Arabic tradition but is almost certainly based on Jewish sources, exhibits positional data for the epoch of the Temple. All of this indicates that a Hebrew tradition, drawing upon Indian, Hellenistic, and other sources, had developed by the eighth century.

Without doubt the years spanning the ninth through sixteenth centuries were the most fecund for the Hebrew tradition. During this period, which is more or less commensurate with what Western historians have long called the medieval age, interest in astronomy was especially stimulated in Arabic speaking lands from Spain to Iraq. Contemporary Jewish writings consist for the most part of exposés or translations of the fruits of Arab science. Dozens of works were written, surviving in hundreds of manuscripts; only a few can be surveyed here. Abraham bar Ḥiyya created a Hebrew astronomical vocabulary that endured side by side with that developed by the Tibbons, the famous family of translators of Arabic literature. Abraham ibn Ezra's astrological writings were immensely popular both in the Hebrew original and in Latin translation. Both Abrahams utilize more ancient sources that are no longer extant. Isaac Israeli's *Yesod ʿOlam* is a thorough analysis of all of the astronomical and historical questions connected with the Jewish calendar; his book, written in 1310 in a polished Hebrew style, may be considered the pinnacle of the Spanish Hebrew tradition. All the same, Hispano-Jewish authors continued to produce treatises in Hebrew and Judeo-Arabic through the end of the fifteenth century. For instance, the work of Abraham Zacuto, so important for the Portuguese explorers, appeared around the time of the expulsion of Jews from Spain.

Southern France, Italy, and Byzantium were the other major centers of astronomical activity during this period. Emmanuel ben Jacob, Jacob Anatoli, and Mordecai Comtino are, respectively, perhaps the most important representatives of the Hebrew tradition in those lands. In those areas in particular, Jews drew upon Latin, Romance, and other literatures, as well as Arabic materials.

The two outstanding philosophers of this period, Moses Maimonides and Levi Gersonides, participated strongly in the Hebrew astronomical tradition. Maimonides' wrote a small work on the calendar, and he included in his great law code, the *Mishneh Torah*, a detailed scheme for computing the first visibility of the lunar crescent. However, Maimonides' weightiest contribution is the very high value which he placed on the study of astronomy within the context of his religious philosophy, something which encouraged many Jews to acquaint themselves with, at the very least, non-technical resumés of astronomical knowledge.

Levi Gersonides was without doubt the most creative representative of the Hebrew tradition, indeed one of the greatest scientists of his epoch. He too developed his astronomical views within the framework of a finely tuned and very comprehensive religious philosophy. Levi was both an observer and a theoretician, and, most notable, one of the rare breed who attempted to fit his own original theory, which had to answer to certain philosophical constraints, to his own observations. Among his other major achievements, Levi invented the Jacob's staff, a simple but accurate instrument for measuring the angular distances between stars; studied the errors involved in instrumental measurements; and arrived at a much greater (and hence more realistic) value for the distances of the stars than those accepted by his contemporaries.

Jews living in Islamic lands, most especially the Yemen, continued to study the ancient and medieval texts well into the twentieth century. In those countries the Hebrew tradition was maintained chiefly through the copying of manuscripts, often transcriptions of Arabic texts into the Hebrew alphabet. In European countries the study of Latin texts in Hebrew translation seems to have accelerated in the fifteenth and sixteenth centuries. Some three Hebrew translations of Georg Peurbach's *Theorica* were executed, and several Hebrew commentaries were written, *inter alia* by Moses Isserles of Cracow and Moses Almosnino of Salonika, both of whom were leading rabbis of their times.

There is little in the Hebrew tradition that reflects the great advances associated with the European scientific revolution. Joseph Delmedigo, a widely traveled Cretan who studied under Galileo, published the only lengthy and detailed Hebrew exposition of the new science. Other publications, such as Tuviah Cohen's *Maaseh Tuvia* (Tuvia's Opus), present the new science in somewhat abbreviated form; and yet other writings, such as Solomon Maimon's exposé of Newton's work, remain in manuscript. The Hebrew tradition revived during the nineteenth century, particularly due to the efforts of the *maskilim*, advocates for widening the intellectual horizons of Judaism, who published Hebrew scientific texts in a number of fields. With the re-establishment in the State of Israel of a native Hebrew speaking population, and, no less importantly, institutions interested in teaching and writing about astronomy in the Hebrew language, the quantity and scope of the Hebrew tradition have dramatically increased.

Y. TZVI LANGERMANN

REFERENCES

Beller, E. "Ancient Jewish Mathematical Astronomy." *Archive for History of the Exact Sciences* 38: 51–66, 1988.

Feldman, W.M. *Rabbinic Astronomy and Mathematics*. 2nd ed. New York: Hermon Press, 1965.

Gandz, Solomon. *Studies in Hebrew Astronomy and Mathematics*. Ed. Shlomo Sternberg. New York: Ktav, 1970.

Goldstein, B.R. "The Survival of Arabic Astronomy in Hebrew." *Journal for the History of Arabic Science* 3(1): 31–39, 1979.

Goldstein, B.R. "The Hebrew Astronomical Tradition: New Sources." *Isis* 72: 237–251, 1981.

Goldstein, B.R. "Scientific Traditions in Late Medieval Jewish Communities." In *Les Juifs au regard de l'histoire. Mélanges en l'honneur de Bernhard Blumenkranz*. Ed. G. Dahan. Paris: Picard, 1985, pp. 235–247.

Goldstein, B.R. *The Astronomy of Levi ben Gerson (1288–1344)*. New York: Springer, 1985.

Langermann, Y.T., P. Kuntizsch, and K.A.F. Fischer. "The Hebrew Astronomical Codex Sassoon 823." *Jewish Quarterly Review* 78(3/4): 253–292, 1989.

See also: al-Khwārizmī – Abraham bar Ḥiyya – Abraham ibn Ezra – Isaac Israeli – Levi ben Gerson – Abraham Zacut

ASTRONOMY IN INDIA Astronomy in India, as it was in other ancient civilizations, was interwoven with religion. While the different facets of nature, the shining of the sun, the waxing and waning of the moon, and the alternation of the seasons all excited curiosity and evoked wonder, religious practices conformed to astronomical timings following the seasons, equinoxes, solstices, new and full moons, specific times of the day and the like. In the Vedas, the earliest literature of the Hindus, mention of professions such as *Gaṇaka* (calculator) and *nakṣatra-darśa* (star-gazer), and the mention of a branch of knowledge called *nakṣatra-vidyā* (star science) are illustrative of the fascination that the celestial bodies exerted on the Vedic priests.

The Vedas and their vast ancillary literature are primarily works of a religious nature, and not textbooks on astronomy. Still, they inform about the astronomical knowledge, mainly empirical in nature and often mystic in expression, which Vedic Indians possessed and used in their religious life. One finds in the *Ṛgveda* intelligent speculations about the genesis of the universe from non-existence, the configuration of the universe, the spherical self-supporting earth, and the year of 360 days divided into 12 equal parts of 30 days each with a periodical intercalary month. In the *Aitareya Brāhmaṇa*, we read of the moon's monthly elongation and the cause of day and night. Seasonal and yearly sacrificial sessions helped the priests to ascertain the days of the equinoxes and solstices. The shifting of the equinoxes made the Vedic priests correspondingly shift the year backwards, in tune with the accumulated precession, though the rate thereof was not envisaged. The wish to commence sacrifices at the beginning of specific constellations necessitated the identification of the constellations as fitted on the zodiacal frame. They also noticed eclipses, and identified their causes empirically.

The computational components and work rules for times for Vedic rituals are to be found in the *Vedāṅga Jyotiṣa* (Vedic Astronomical Auxiliary), composed by Lagadha. On the basis of the astronomical configurations given in its epoch, the date of this text is ascertained to be in the twelfth century BC. This work sets out such basic data as time measures, astronomical constants, tables, methodologies, and other matters related to the Vedic ritualistic calendar. It prescribes a five-year lunisolar cycle (*Yuga*) from an epoch when the sun and the moon were in conjunction on the zodiac at the beginning of the bright fortnight, at the commencement of the asterism Delphini (*Śraviṣṭhā*) of the (synodic month of) *Maghā*, and of the (solar month) *Tapas*, when the northward course of the sun began. The constants are contrived to be given in whole numbers for easy memorization. Accordingly, the *Vedāṅga Jyotiṣa* chose a unit of 1830 civil days as its unit, which it divided into five years of 366 days each, the error of the additional 3/4 day in the year being rectified periodically. 1830 days is equal to 62 synodic or lunar months of 29.5 days each, and 60 solar months. The two extra intercalary lunar months are dropped, one at the middle and the other at the end of the cycle, so that the two, the solar and lunar years, commenced together again, at the beginning of each cycle.

During the age that followed, a series of astronomical texts was written, mainly by eighteen astronomers. Passages from some of these texts were quoted in later texts, and five of them were redacted by Varāhamihira (d. 587) in his *Pañcasiddhāntikā*. The texts of this age which are still available are the *Gargasaṃhitā*, and *Parāśarasaṃhitā*, and the Jain texts *Sūryaprajñapti, Candraprajñapti* and *Jambūdvīpa-prajñapti*. The astronomy contained in those is practically the same as that expounded in the *Vedāṅga Jyotiṣa* though with minor differences, including the shifting of the commencement of the year to earlier asterisms due to the precession of the equinoxes.

From about the beginning of the Christian era a number of texts were composed with the generic name *Siddhānta* (established tenet). The scope of these texts was wider, their outlook far-reaching, and their methodology rationalistic. Also, the science began to be studied for its own sake. The stellar zodiac was replaced with the twelve-sign zodiac, and, besides the Sun and the Moon, the planets also began to be reckoned. Their rising and setting, motion in the zodiacal segments, direct and retrograde motion, times of first and last visibilities, the duration of their appearance and disappearance, and mutual occultation began to be computed. These and their synodic motion called *grahacāra* were elaborately recorded. Analyses of these recordings enabled the depiction of empirical formulae for computing their longitudes. While the use of the rule of three (*trai-rāśika*, direct proportion), continued fractions, and indeterminate equations helped computation, the use of trigonometry, both plane and spherical, and geometrical models enabled a realistic understanding of planetary motion, and developed rules, formulae, and tables. This resulted also in fairly accurate prediction of the eclipses and occultation of the celestial bodies. It is to be noted that in Indian astronomical parlance, the word *graha* signifies not only the planets Mars (*Kuja*), Mercury (*Budha*), Jupiter (*Guru*), Venus (*Śukra*) and Saturn (*Śani*), but also the Sun (*Ravi*, *Sūrya*), Moon (*Candra*), the ascending node (*Rāhu*), and the descending node (*Ketu*).

Yuga in Indian astronomy denotes a relatively large number of years during which a celestial body, starting from a specific point on its orbit, made a certain number of revolutions and returned to the same point at the end of the period. In the *siddhānta* texts the starting point is taken as the First point of Aries (*Meṣa-ādi*) (vernal equinox). Through extended observation over long periods, methods were formulated for ascertaining accurately the motion of the planets, as in the *Ārya-bhaṭīya*. Having obtained in this manner the sidereal periods of all the planets, their lowest common multiple was calculated backwards to provide accommodation to the revolutions of the relevant celestial bodies. Thus the *Vedāṅga Jyotiṣa* used a five-year *yuga* to accommodate only the sun and the moon. Later, the planets were included and the *Romakasiddhānta,* redacted by Varāhamihira, used a *yuga* of 2850 lunisolar years, and the *Saurasiddhānta* used a *yuga* of 108,000 solar years. Āryabhaṭa (b. AD 476) formulated a *mahā-yuga* (grand cycle) of 4,320,000 years which was set to commence at the First point of Aries on Wednesday, at sunrise, at Laṅkā, which is a point on the terrestrial horizon of zero longitude, being the Greenwich meridian of Indian astronomy. He devised also a shorter *Yuga*, called *Kali-yuga*, of 432,000 years, which commenced on Friday, February 18, 3102 BC, where, however, the apogee and node were ahead by 90° and behind by 180°, respectively. Pursuing the same principle, other astronomers have devised *yugas* of different lengths, and still others have suggested zero corrections to the mean position of the planets at the beginning of the *yuga*.

Computing the longitude of a planet when the length of the *yuga* in terms of civil days, the *yuga*-revolutions, and the time at which its longitude is required are given, reduces itself to the application of the Rule of Three, if the planetary orbits are perfect circles and the planetary motions uniform. Indian astronomers conceive their motion along elliptical orbits according to the epicyclic model or their own eccentric model. In the former, the planets are envisaged as moving along epicycles which move on the circumference of a circle, and in the latter, the planets are supposed to move along a circle whose center is not at the center of the earth, but on the circumference of a circle whose center is the center of the earth. Both models give the same result. Several sets of sine tables are also derived and several computational steps called *manda* and *śīghra* are enunciated to give the heliocentric positions in place of the geocentric.

When the true longitude of a planet at a particular point of time is to be found, the *ahargaṇa* (count of days) from the epoch up to the sunrise of the day in question is ascertained. To this is added the time elapsed from sunrise on the relevant day to the required point of time. Since the number of days in the *yuga* and the number of the relevant planetary revolutions (which are constants) are known, the revolutions up to the moment in question are calculated by the rule of three, and the completed revolutions discarded. The remainder would be the position of the mean planet at sunrise of the day in question at zero longitude (i.e. Laṅkā or Ujjain meridian). To get the correct longitude, four corrections are applied to the result obtained above. They are (i) *Deśāntara*, the difference in sunrise due to the difference in terrestrial longitude, (ii) *Cara* (ascensional difference), due to the length of the day at the place, (iii) *Bhujāntara*, the equation of time caused by the eccentricity of the earth's orbit, and (iv) *Udayāntara*, the equation of time caused by the obliquity of the ecliptic (the band of the zodiac through which the sun apparently moves in its yearly course) with the celestial equator.

As a striking natural phenomenon, the eclipse had been taken note of and recorded in several early Indian texts. A solar eclipse was recorded in the *Ṛgveda*, where it describes how, when Svarbhānu (the dark planet *Rāhu* of later legends) hid the sun, sage Atri restored it, first as a black form, then as a silvery one, then as a reddish one, and finally in its original bright form. On account of the popular and religious significance associated with eclipses, their prediction assumed great importance in Indian astronomy. In the *Pañcasiddhāntikā*, details of the computation of the lunar eclipse were given in the *Vāsiṣṭha-Pauliśa* and *Saura Siddhāntas* with geometrical diagrammatic representation, and the solar eclipse in the *Pauliśa*, *Romaka* and *Saura Siddhāntas*. The treatment in later *siddhāntas*, like those of Āryabhaṭa, Brahmagupta, Śrīpati and Bhāskara II and also the *Sūryasiddhānta*, is accurate. A number of shorter texts also were written solely on the computation of eclipses, aiming at greater perfection.

The Vedas call the intrinsically dark moon *sūrya-raśmi* (sun's light), and thought that it was born anew every day in different configurations. Each of the fifteen lunar days of the bright and the dark fortnights have high religious significance, and different Hindu rites and rituals are fixed on their basis.

From the early *siddhānta* age, computation of the moon's phases is referred to by the term *Candra-śṛṅga-unnati* (Elevation of the horns of the moon) and is computed elaborately. The first work to give this computation is the *Pauliśa Siddhānta*, which devotes an entire chapter to this subject. Most of the later *siddhāntas* also devote one chapter to the subject, adding corrections and evolving newer methods and also giving graphical representations.

Still another topic which finds computational treatment in

siddhānta texts is the conjunction of planets and stars. Two bodies are said to be in conjunction when their longitudes at any moment are equal. The conjunction of a planet with the sun is called *astamaya* (Setting), with the moon is *samāgama* (Meeting), and with another planet, it is *yuddha* (Encounter). The conjunction of a planet with a star is similar, but with the difference that a star is considered a ray of light and has no *bimba* (orb) or motion. The computational methods followed are similar to those in the case of eclipses, with minor modifications. Full chapters are devoted in *siddhānta* texts to computing this phenomenon.

Although naked eye observations and the star-gazer (*nakṣatradarśa*) find frequent mention in Vedic literature, mechanical instruments are of a later origin. The earliest instruments used were the gnomon (*śaṅku*) for finding the cardinal directions, used in the *śulbasūtras*, and the clepsydra (*nāḍī-yantra*) for measuring time. The *Pañcasiddhāntikā,* devotes one long chapter to "Graphical Methods and Astronomical Instruments". While Āryabhaṭa gives only the underlying principles in his *Āryabhaṭīya*, he has a long section of 31 verses on instruments in his *Āryabhaṭasiddhānta*. Almost all later *siddhāntas* have a full chapter on instruments, which include the armillary sphere, rotating wheels, and shadow, water, circle, semi-circle, scissor, needle, cart, tube, umbrella, and plank instruments. Some of these are used for observation; others are for demonstration. After the advent of Muslim astronomy in India, the astrolabe became common and even Hinduized. A number of texts in Sanskrit also were written on the astrolabe, among which the *Yantrarāja* of Mahendra Sūri and *Yantracintāma* of Viśrāma are important. The pinnacle of this activity came with Sawai Jai Singh, ruler of Jaipur, who patronized a group of astronomers, built five huge observatories, in Delhi, Jaipur, Ujjain, Varanasi, and Mathura, and wrote the work *Yantrarāja-racanā* (Construction of Astrolabes).

Towards the early centuries of the Christian era, texts with the generic name *siddhānta* (tenet), in contrast to the earliest astronomical texts, were composed. These were mathematically based, rationalistic, and adumbrated by geometric models and diagrammatic representations of astronomical phenomena. While Varāhamihira selectively redacted in his *Pañcasiddhāntikā* five of the early *siddhāntas*, his elder contemporary, Āryabhaṭa (b. AD 499) produced a systematic *siddhānta* treatise entitled *Āryabhaṭīya*, in which he speaks also of the diurnal rotation of the earth. The work is divided into four sections, covering the following subjects: (i) planetary parameters and the sine table (*Gītikā-pāda*), (ii) mathematics (*Gaṇitapāda*), (iii) time reckoning and planetary positions (*Kālakriyā-pāda*), and (iv) astronomical spherics (*Gola-pāda*). This *siddhānta*, which started the Āryabhaṭan school of astronomy, was followed by advanced astronomical treatises written by a great number of astronomers and was very popular in the south of India, where astronomers wrote a number of commentaries and secondary works based on it.

Brahmagupta (b. AD 598) started another school through his voluminous work *Brāhmasphuṭasiddhānta* in twenty chapters. This work shows Brahmagupta as an astute mathematician who made several new enunciations. He also wrote a work by the name of *Khaṇḍakhādyaka* in which he revised some of Āryabhaṭa's views. Among the things that he formulated might be mentioned a method for calculating the instantaneous motion of a planet, correct equations for parallax, and certain nuances related to the computation of eclipses. Brahmagupta's works are also significant for their having introduced Indian mathematics-based astronomy to the Arab world through his two works which were translated into Arabic in about AD 800.

Bhāskara I (ca. AD 628), who followed in Āryabhaṭa's tradition, wrote a detailed commentary on *Āryabhaṭīya* and two original treatises, the *Laghubhāskarīya* and the *Mahābhāskarīya*. Vaṭeśvara (b. A.D. 880) wrote his erudite work *Vaṭeśvarasiddhānta* in eight chapters, in which he devised precise methods for finding the parallax in longitude (*lambana*) directly, the motion of the equinoxes and the solstices, and the quadrant of the sun at any given time. He wrote also a work on spherics. Śrīpati, who came later (ca. AD 999), wrote an extensive work, the *Siddhāntaśekhara*, in twenty chapters, introducing several new enunciations, including the moon's second inequality.

The *Sūryasiddhānta*, the most popular work on astronomy in North India, is attributed to divine authorship but seems to have been composed about A.D. 800. It adopts the midnight epoch and certain elements from the old *Saurasiddhānta*, but differs from it in other respects. It promulgates its own division of time-cycle (*yuga*) and evinces some acquaintance with *Brāhamasphuṭasiddhānta*.

The *Siddhāntaśiromaṇi* of Bhāskara II (b. AD 1114) comprises four books: *Līlāvatī* dealing with mathematics, *Bījagaṇita* with algebra, *Gaṇitādhyāya* with practical astronomy, and *Golādhyāya* with theoretical astronomy. The author's gloss on the work which he calls *Vāsanā* (Fragrance) is not only explanatory but also illustrative and highly instructive. Extremely popular and widely studied in North India, the work has been commented on by generations of scholars.

The *siddhānta* texts composed later, when Muslim astronomy had been introduced into India, bear its influence in the matter of parameters and models though the general set up remains the same. The more important among these are the *Siddhāntaśindhu* (1628) and *Siddhāntarāja* (1639) of Nityānanda, *Siddhāntasārvabhauma* (1646) of

Munīśvara, *Siddhāntatattvaviveka* (1658) of Kamalākara and the *Siddhāntasārakaustubha* (1732) of Samrāṭ Jagannātha.

In order to relieve the tedium in working with the very large numbers involved when the *mahāyuga* or *yuga* is taken as the epoch, a genre of texts called *karaṇas* was evolved which adopted a convenient contemporary date as the epoch. The mean longitudes of the planets at the new epoch were computed using the *siddhāntas* and revised by observation, and the resulting longitudes were used as zero corrections at the epoch for further computations. In order to make computations still easier, planetary mean motions were calculated for blocks of years or of days, and depicted in the form of tables. Each school of astronomy had a number of *karaṇa* texts, produced at different dates and often exhibiting novel shortcuts and methodologies. While the North Indian texts had tables with numerals, South India, particularly Kerala, had its own traditions, and depicted the daily motions of the planets, sines of their equations, and other matters in the form of verses with meaningful words and sentences, employing the facile *kaṭapayādi* notation of numerals. The earliest *karaṇa* texts are the redactions by Varāhamihira in his *Pañcasiddhāntika* of the *Romaka, Pauliśa* and *Saura Siddhāntas*, the epoch of all the three being 21st March, 505 AD.

Hindu religious life, which served as the incentive to the development of astronomy in India in its beginnings, has continued to be so even today for the orthodox Indian. The *pañcāṅga* (five-limbed) almanac which is primarily a record of the *tithi* (lunar day), *vāra* (weekday), *nakṣatra* (asterism), *yoga* (sum of the solar and lunar longitudes) and *karaṇa* (half lunar day), can be said to direct and regulate the entire social and religious life of the orthodox Hindu. Though the primary elements of the almanac are identical throughout India, other matters like the sacred days, festivals, personal and community worship, fasts and feasts, and social celebrations, customs, and conventions differ from region to region. These matters are also recorded in the almanacs of the respective regions. In order to bring about some uniformity in the matter, the Government of India appointed a Calendar Reform Committee in 1952 which made several recommendations, but conditions have not changed much.

In astronomy, as in other scientific disciplines, the Indian ethos had been to depict the formulae and procedures, but refrain from giving out the rationale, though much rationalistic work would have gone before formulation. This position is relieved by a few commentators, such as Mallāri in his commentary on the *Grahalāghava* of Gaṇeśa Daivajña and Śaṅkara Vāriyar in his commentaries on the *Līlāvatī* of Bhāskara II and the *Tantrasaṅgraha* of Nīlakaṇṭha Somayāji. It is also interesting that there are texts wholly devoted to setting out rationales, like the *Yuktibhāṣā* of Jyeṣṭhadeva. Even more interesting are texts like *Jyotirmīmāṁsā* (Investigations on Astronomical Theories) by Nīlakaṇṭha Somayāji, which open up a very instructive chapter of Indian astronomy.

K.V. SARMA

REFERENCES

Āryabhaṭīya of Āryabhaṭa. Ed. and trans. by K.S. Shukla and K.V. Sarma. New Delhi: Indian National Science Academy, 1976.

Billard, Roger. *L'Astronomie Indienne*. Paris: École Française d'extrème Orient, 1971.

Bose, D.M. et al. *A Concise History of Science in India*. New Delhi: Indian National Science Academy, 1971.

Dikshit, S.B. *Bhāratīya Jyotiśh śāstra* (History of Indian Astronomy). New Delhi: Director General of Meteorology, 2 pts., 1968, 1981.

Jyotirmīmāṁsā of Nīlakaṇṭha Somayāji. Ed. K.V. Sarma. Hoshiarpur: Visveshvaranand Institute, 1971.

Pañcasiddhāntikā of Varāhamihira. Trans and intro. by T.S.K. Sastry and K.V. Sarma. Madras: K.V. Sarma, P.P.S.T Foundation, Adyar, 1993.

Pingree, David. *Jyotiḥśāstra*. Wiesbaden: Otto Harrassowitz, 1981.

Report of the Calendar Reform Committee, Government of India. New Delhi: Council of Scientific and Industrial Research, 1956.

Sarma K.V. *A History of the Kerala School of Hindu Astronomy*. Hoshiarpur: Vishveshvaranand Institute, 1972.

Sastry, T.S.K. *Collected Papers on Jyotisha*. Tirupati: Kendriya Sanskrit Vidyapeetha, 1989.

Sen, S.N. and K.S. Shukla. *History of Astronomy in India*. New Delhi: Indian National Science Academy, 1985.

Subbarayappa, B.V., and K.V. Sarma. *Indian Astronomy: A Sourcebook*. Bombay : Nehru Centre, Worli, 1985.

Tantrasangraha of Nīlakaṇṭha Somayāji, with Yuktidīpikā and Laghuvivṛti. Ed. K.V. Sarma. Hoshiarpur: Vishveshvaranand Institute, 1977.

See also: Lunar Mansions – Eclipses – Gnomon – Clocks and Watches – Time – Astronomical Instruments – Observatories – Armillary Sphere – Algebra in India: *Bijagaṇita* – Precession of the Equinoxes – Religion and Science

ASTRONOMY IN THE INDO–MALAY ARCHIPELAGO

"... in recent years anthropology has begun to face up to the implications of the truism that people do not respond directly to their environment but rather to the environment *as they conceive of it*: e.g. to animals and plants as conceptualized in *their* minds and labeled by *their* language. In other words, understanding human ecology requires understanding human conceptual systems. It is not enough to understand the role of a type of organism in economics and/or ritual. One must also understand how the people involved classify and think about it." (Dentan, 1970)

All societies have their own body of knowledge through which they seek to understand the natural environment and their relationship to it. Thus we may be able to understand a society better by going beyond the categories of Western science and beginning to consider the interrelationship of a society with its environment from the viewpoint of its members. This attempt to understand how members of a society themselves conceive of their environment has come to be known as ethnoecology. Ethnoastronomy, the subject of this article, may be seen as a branch of ethnoecology wherein the interrelationship of human populations with their celestial environment is the focus of interest.

The modern nations of Indonesia and Malaysia, with a combined population of nearly two hundred million, encompass the homelands of well over two hundred distinct ethnic groups whose cultures and languages form part of a common Austronesian heritage, a heritage they share with Polynesians and Micronesians, among others. Inhabiting mountain sides, river valleys, and coastal plains and faced with a rather unpredictable tropical monsoon climate, the peoples of Indonesia and Malaysia have developed a diverse agriculture which includes both inundated rice farming and the shifting cultivation of rice and other food crops. Spread across an archipelago of over 13,000 islands, they have also developed sophisticated systems of navigation. These indigenous agricultural and navigational practices have been found to be informed by an astronomical tradition that is, at once, unique to this cultural area and richly diverse in its local variation. This article describes several of the many techniques of astronomical observation employed by the peoples of Indonesia and Malaysia to help regulate their agricultural cycles and navigate their ships.

The Celestial Landscape

The passage of time is mirrored in all of nature: in the light of day and dark of night, the flowering of plants, the mating and migratory behavior of animals, the changes in weather, the ebb and flood of the tides, and in the recurring cycles within cycles of the sun, moon, planets, and stars. Of these cycles, perhaps the most obvious is the diurnal rising and setting of the sun, moon, planets and stars, as well as the synodic, or cyclic changes in the phases of the moon and in the time of day that it rises and sets. More subtle than these might be the annual changes in the sun and stars: the north–south shift in the path of the sun across the sky (including its rising and setting points and its relative distance above the horizon at noon) and the appearance, disappearance, and reappearance of familiar patterns of stars at various times of night. As an integral part of the natural landscape, these recurring celestial phenomena have long provided farmers and sailors with dependable markers against which agricultural and navigational operations can be timed.

Orientation in space and the art of wayfinding have often relied upon knowledge of these same celestial phenomena. The English term "orient" is derived from the Latin for "rise" and later became associated with the East as the direction in which the sun appears at dawn. Although the times that individual stars rise and set shifts gradually throughout the year, as viewed from a given latitude, the azimuths at which a star rises and sets vary only slightly over a lifetime, thereby providing a reliable "star compass" by which to determine direction. The sun, moon, and planets also rise and set generally east and west, depending upon their individual cycles, affording additional guides by which people are able to orient themselves on land and sea.

Agricultural Time-keeping

Many traditional desert and plains cultures use the shift in the rising and/or setting points of the sun along the horizon both to mark important dates and seasons and to commemorate the passage of years. These environments are conducive to the use of these horizon-based solar calendars: a permanent location from which to make sightings, a series of permanent distant horizon markers and a clear view are all that is needed for such a calendar. England's Stonehenge and the Big Horn Medicine Wheel of Wyoming are striking examples. Such conditions are not common in Indonesia and Malaysia. Here the landscape may consist of anything from a nearby or distant mountain to, more often, some nearby trees; the horizon is therefore a rather undependable device against which to sight and measure the rising and setting positions of the sun. However, in cultivated areas of the region, even from swiddens located deep within the rainforest, one can usually find a field or homesite from which much of the sky is visible.

There are several types of observations of annually recurring celestial phenomena that can be made where permanent, distant horizon markers are not commonly available. These include cyclic changes in the phases of the moon, the annual changes in the appearances of familiar groups of stars, and the annual changes in the altitude of the sun at noon. Variations of these types of observations as practiced by traditional Indonesian and Malaysian farmers will now be presented. For the sake of clarity, they are grouped using Western astronomical categories.

Solar Gnomons. Most often seen on sundials in western cultures, a solar gnomon is simply a vertical pole or other similar device that is used to cast a shadow. The altitude of

Figure 1: Solar gnomon.

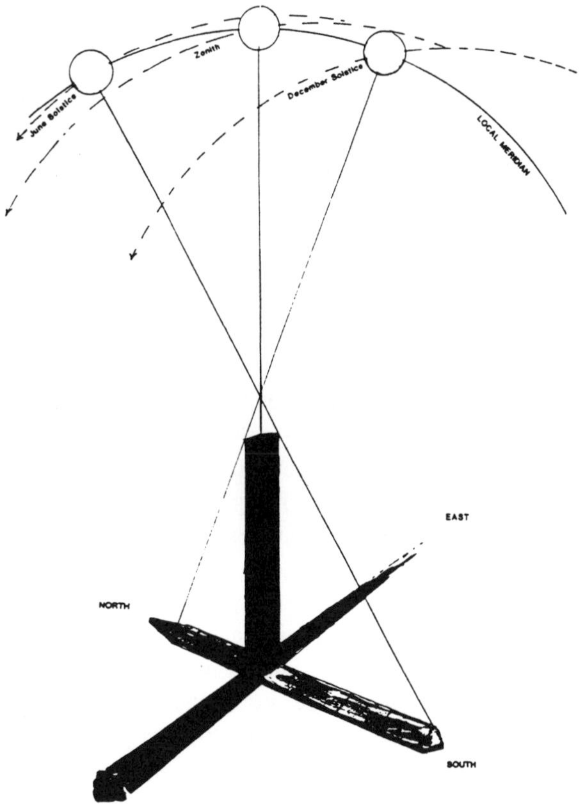

Figure 2: Javanese *bencet*.

the sun above the horizon varies not only through the day, but through the year as well. By measuring the relative length of this shadow each day at local solar noon, one can observe and more or less accurately measure the changing altitude of the sun above the horizon (or, reciprocally, from the zenith) through the year and thereby determine the approximate date.

Two distinct types of solar gnomons have been reported. Both measure the altitude sun at local solar noon to determine the date. One type has been attributed to various groups of the Kenyah and may still be in use.

It consists of a precisely measured (= span of maker's outstretched arms + span from tip of thumb to tip of first finger), permanently secured, plumbed, and decorated vertical hardwood pole (*tukar do*) and a neatly worked, flat measuring stick (*aso do*), marked with two sets of notches (see Figure 1).

The first set corresponds to specific parts of the maker's arm and ornaments worn upon it, measured by laying the stick along the radial side of the arm, the butt end against the inside of the armpit. To mark the date, the measuring stick is placed at the base of the vertical pole, butt end against the pole and extending southward. This is done at the time of day that the shadows are shortest, local solar noon. On the day that the pole's noontime shadow is longest (the June solstice), a notch is carved on the other edge of the stick to mark the extent of the shadow made by the pole. This observation indicates that the agricultural season is at hand. From then on, the extent of the noonday shadow is recorded every three days as a record-keeping device. Dates, both favorable and unfavorable, for various operations in rice cultivation such as clearing, burning, and planting are determined by the length of the shadow relative to the marks on the stick that correspond to parts of the arm and to the marks made every three days.

On Java a highly accurate gnomon, called a *bencet*, was in use from about AD 1600 until 1855 (see Figure 2). A smaller, more portable device than that employed by the Kayan and Kenyah, the *bencet* divides the year into twelve unequal periods, called *mangsa*, two of which begin on the days of the zenith sun, when the sun casts no shadow at local solar noon, and another two of which begin on the two solstices, when the sun casts its longest midday shadows. At the latitude of Central Java, 7 degrees south, a unique condition exists which is reflected in the *bencet*. As the illustration shows, when, on the June solstice, the sun stands on the meridian (that is, at local solar noon) and to the north of the zenith, the shadow length, measured to the south of the base of the vertical pole, is precisely double the length of the shadow, measured to the north, which is cast when the sun, on the December solstice, stands on the meridian (at noon) south of the zenith. By simply halving the shorter segment and quartering the longer, the Javanese produced a calendar with twelve divisions which are spatially equal but which range in duration from 23 to 43 days. The twelve *mangsa* with their starting dates and numbers of days are shown in Table 1.

Apparitions of Stars at Dawn and Dusk. The second category of observational techniques regularly employed by traditional farmers of the region include all of those which involve apparitions of stars or recognized groups of stars (asterisms) at last gleam at dawn or first gleam at dusk. Because of the earth's orbital motion about the sun, it can be observed that each star rises and sets approximately four minutes earlier each night; similarly, each star appears to have moved about 1 degree west, when viewed at the same time each night. As a result, around the time of its conjunction with the sun, any given star becomes lost in the sun's glare and is therefore not visible for approximately one month each year. It also means that the altitudes of stars above the horizon vary as a function of both the time of night and the day of the year, such that a certain star or group of stars, when observed at the same time each night (in this case near dusk or dawn) will appear at a given altitude above the eastern or western horizon on one and only one night of the year. Hence the use of any technique or device to measure the altitude of a star or group of stars as it appears at first or last gleam can provide the observer with the approximate date.

In general the term "acronical" refers to any stellar apparition which occurs at first gleam while "cosmical" is used to describe stars at last gleam. A subset of these, "heliacal" (Greek *helios*: sun) apparitions of stars are those which occur just prior to and shortly after conjunction with the sun. The heliacal setting of a star or group of stars occurs on the date that the star or stars are last observed before conjunction, just above the western horizon at dusk. Likewise, the heliacal rising of a star or group of stars occurs on the date that the star or stars are first observed, after several weeks' absence, above the eastern horizon at dawn.

Just about three months after its heliacal rise, the star, now about ninety degrees west of the sun, appears on the meridian at first gleam, an event which may be termed a "cosmical culmination" of the star. Two months later the star is nearing opposition with the sun and undergoes an "acronical rising" in the east at first gleam. In less than one month, after opposition, the "cosmical setting" of the star is seen in the western sky. After two more months the star

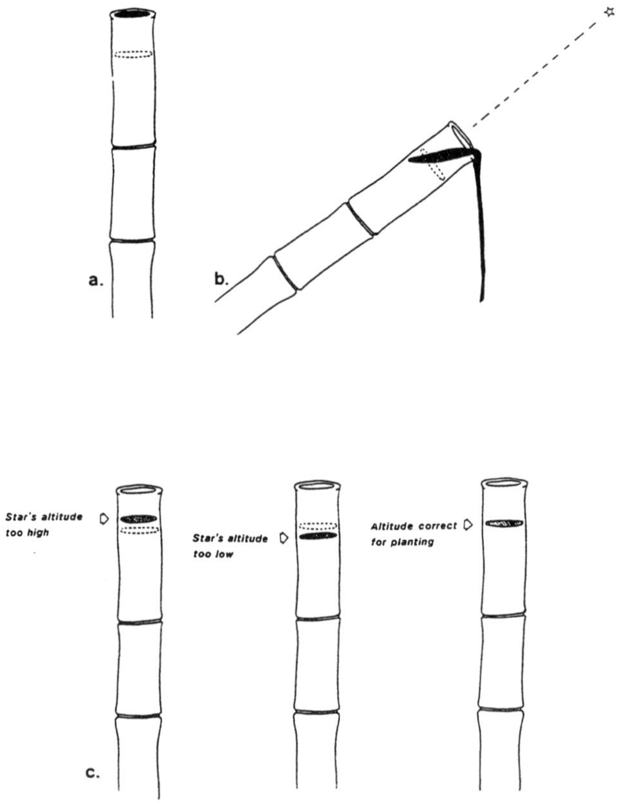

Figure 3: Bamboo device for measuring the altitude of a star.

may be seen on the meridian at first gleam, accomplishing its "acronical culmination". Finally, about three months later, the star undergoes "heliacal setting" as it is once again outshone by the sun.

These categories are taken from Western mathematical astronomy and are used here as a way of organizing and presenting this material to the Western scholar. Actual observations made by local Indonesian and Malaysian farmers may not always be as precise as their assigned astronomical categories might imply. The demand for such precision varies greatly between and within cultures and with local environmental conditions. It would not be unusual to find a local farmer, for example, first noting the heliacal rising of the Pleiades at dawn several days or more after its mathematically calculated reappearance.

The apparitions of stars at first and last gleam have been systematically observed by traditional cultures everywhere. From Indonesia and Malaysia there are references in the literature, too numerous to describe in detail here, to the calendrical use at both dusk and dawn of the stars we know as the Pleiades, Orion and, to a lesser extent, Antares, Scorpius, and Crux. Culminations at both first and last gleam of the Pleiades, Orion, and Sirius are noted in the literature. Interestingly, the observation for calendrical purposes

Table 1: The Pranatamangsa calendar

Ordinal number	Name(s) of mangsa	Duration in days	First day(s)	Civil calendar
Ka-1	Kasa	41	22 June	21 June
Ka-2	Karo or Kalih	23	2 August	1 August
Ka-3	Katelu or Katiga	24	25 August	24 August
Ka-4	Kapator Kasakawan	25	18 September	17 September
Ka-5	Kalima or Gangsal	27	13 October	12 October
Ka-6	Kanem	43	9 November	8 November
Ka-7	Kapitu	43	22 December	21 December
Ka-8	Kawolu	26/27	3 February	2 February
Ka-9	Kasanga	25	1 March	ult. February
Ka-10	Kasepuluh or Kasadasa	24	26 March	25 March
Ka-11	Desta	23	19 April	18 April
Ka-12	Sada	41	12May	11 May

of such culminations seems to be unique to peoples of the Indo-Malay Archipelago.

The first example in this category was practiced by a small Dayak group related to the Kenyah–Kayan complex mentioned earlier. Like their neighbors, they were swidden rice farmers. Unlike their neighbors who tracked the sun, they depended upon the stars to fix the date of planting. To do so, they nightly poured water into the end of a vertical piece of bamboo in which a line had been inscribed at a certain distance from the open end. The bamboo pole was then tilted until it pointed toward a certain star (unrecorded) at a certain time of night (also unrecorded), causing some of the water to pour out (see Figure 3). It was then made vertical again and the level of the remaining water noted. When the level coincided with the mark, it was time to plant.

Near Yogyakarta, Central Java, the ritual practitioner raised his hand toward the East in the direction of Orion at dusk, rice seed in his open palm. On the night that kernels of rice rolled off his palm, it was time to sow seed in the nursery. Using a planetarium star projector, the author has fixed the date of this at about 4 January.

Lunar Calendars. Lunar calendars comprise the third category. These calendars are based upon the 29.5 day synodic period, usually measured from new moon to new moon and often subdivided by phase. Because there is not an even number of lunar months in a solar year and because agricultural cycles are, after all, tied to the solar year, simple lunar calendars alone are of little use in farming. However, when it is somehow pegged to the solar year by reference to the apparent annual changes in the positions of the sun or stars or to other phenomena in nature that regularly recur on an annual basis, a lunar calendar can be of use to the farmer.

Indigenous lunar calendars fall into two general categories: lunar–solar and sidereal–lunar. Examples of the lunar–solar calendar include the Balinese ceremonial calendar, still in use, and the old Javanese *Saka* calendar, used from the eighth through sixteenth centuries. Both are apparently of a common Hindu origin and are primarily lunar; both employ complex mathematical techniques to provide the intercalary days which periodically synchronize the lunar with the solar year.

A second type of lunar calendar, best described as sidereal–lunar, uses the apparitions at dusk and dawn of stars and asterisms as well as the appearance of other signs in nature (such as winds, birds, and flowers) to determine which month is current. In these cases, it is only important to know which month it is for a few months each year (that is, during the agricultural season), thereby obviating the need for codified schemes for realigning the lunar with the solar/stellar year. Such "short" lunar calendars are found spread throughout the region. The Iban calendar provides a good example.

The Iban are a riverine people practicing shifting agriculture in the vicinity of their longhouses, situated in low hills of Sarawak and West Kalimantan. The stars play a central role in Iban mythology and agricultural practices. Several Iban stories tell how their knowledge of the stars was handed down to them by their deities and according to one village headman, "If there were no stars we Iban would be lost, not knowing when to plant; we live by the stars" (Freeman, 1970). The Iban lunar calendar was annually adjusted to the cosmical apparitions of two groups of stars: the Pleiades and the three stars of Orion's belt.

The first observation is probably the most difficult. It is the reappearance of the Pleiades on the eastern horizon just before dawn after two month's absence from the night sky.

This heliacal sighting, around June 5th of the civil calendar, informs the observer that the month, taken from new moon to new moon, that is current is the fifth lunar month. It is during this month that two members of the longhouse go into the forest to seek favorable omens so that the land selected will yield a good crop. This may take from two days to a month, but once the omens appear, they return to the longhouse and work clearing the forest begins. If it takes so long for the omens to appear that Orion's belt rises before daybreak (heliacal rising around June 25th), the people "must make every effort to regain lost time or the crop will be poor." This reappearance of Orion at dawn occurs during the next or sixth lunar month, the time to begin clearing the land.

The remaining observations of the stars are more easily accomplished. They are all cosmical culminations, occurring "overhead" at last gleam, and are seen to be approaching for several weeks. When the Pleiades undergoes its cosmical culmination (September 3rd) and the stars of Orion's belt are about to do so (September 26–30), it is the eighth month and time to burn and plant. For good yields the burn should occur between the time that the two asterisms culminate at first gleam, usually when the two are in balance or equidistant from the meridian (September 16th). Rice seed sown after the star Sirius has completed its cosmical culmination, (October 15th) will not mature properly. Planting may carry into the tenth lunar month (October/November), but it must be completed before the moon is full or the crop will fail. At this point the lunar calendar ends: only months five through ten are numbered and fixed while the remaining months vary according to how quickly the crop matures (e.g. *bulan mantun*, the "weeding month"). The lunar months from November–April are simply not numbered; it is difficult to see the stars during the rainy season and unimportant in any case.

Navigation

As we have just seen, the rice farmers of Indonesia and Malaysia have long noted correlations between celestial and terrestrial cycles and incorporated periodic changes in the sky into their agricultural calendars. Meanwhile, neighboring seafaring societies have used many of these same phenomena to orient themselves in both space and time for the purpose of navigation. Of these societies, the Bugis of South Sulawesi are perhaps the best known. Maintaining a tradition of seafaring and trade that spans at least four centuries, the Bugis are reputed to have established and periodically dominated strategic trade routes across Southeast Asia, stretching northeast from North Sumatra to Cambodia, north to Sulu and Ternate, and east to Aru and Timor. Their maritime prowess notwithstanding, Bugis systems of navigational knowledge and practice have only recently come under study.

Bugis navigators employ a system of dead reckoning which depends upon the knowledge of a variety of features of the natural environment to negotiate the seas in their tall ships. Although these features include land forms, sea marks, currents, tides, wave patterns and shapes, and the habits of birds and fish, navigators rely most heavily upon the prevailing wind directions, guide stars, waves, swells, and, increasingly, the magnetic compass.

The major wind patterns across Island Southeast Asia are governed by the monsoons. From approximately May through October, winds from the east and southeast bring generally fair weather and steady breezes; from November through April, the west monsoon brings first calm air, then rain and squalls. For Bugis seafarers these winds are of such fundamental importance that they have named the two monsoons for their respective prevailing directions in the area of the Flores Sea: *bare*' (west) and *timo*' (east).

For the Bugis of the small coral islet of Balobaloang, located midway between Ujung Pandang (formerly Makasar) on Sulawesi and Bima on Sumbawa, trade routes are generally north and south across the Flores Sea. In principle, this allows them to take advantage of both easterly and westerly winds by reaching in either direction, although storms and heavy seas usually confine them to port during the west monsoon.

Bugis navigators have long relied upon the stars and star patterns to set and maintain course. Although most sailors seem to know a few star patterns and their use, the navigators know many more. These star patterns or asterisms are known to rise, stand and/or set above certain islands or ports when viewed from others and thereby pinpoint the direction of one's destination forming a "star compass". For example, in late July and early August *bintoéng balué* (Alpha and Beta Centauri) (Asterism A, Figure 4) is known to make its nightly appearance at dusk in the direction of Bima as viewed from Balobaloang, that is, to the south.

As the night passes and a given asterism is no longer positioned over the point of destination, the navigator's thorough familiarity with the sky allows him to adjust mentally to the new conditions: he can derive through visualization the points on the horizon at which the stars rise or set relative to wherever they currently appear. When a certain asterism simply is not visible, other associated but unnamed stars may be used to remind the navigator where the original asterism set or is about to rise or they may be used instead of the missing asterism. This, by the way, appears to be analogous to the "star path, the succession of rising or setting guiding stars down which one steers" described by David Lewis for several cultures in Polynesia and Micronesia. The

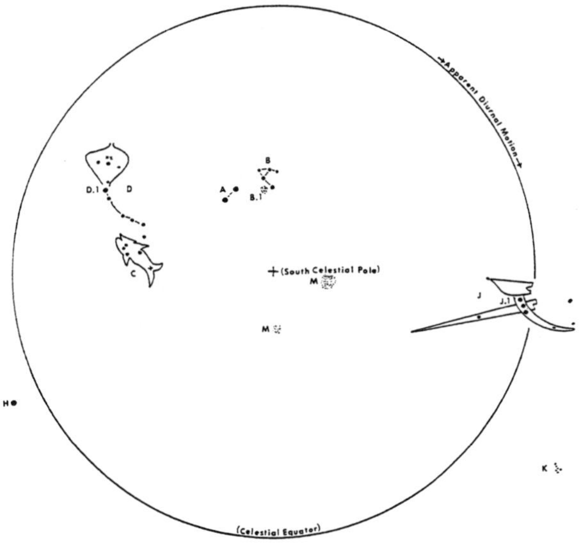

Figure 4: Bugis asterisms: southern sky.

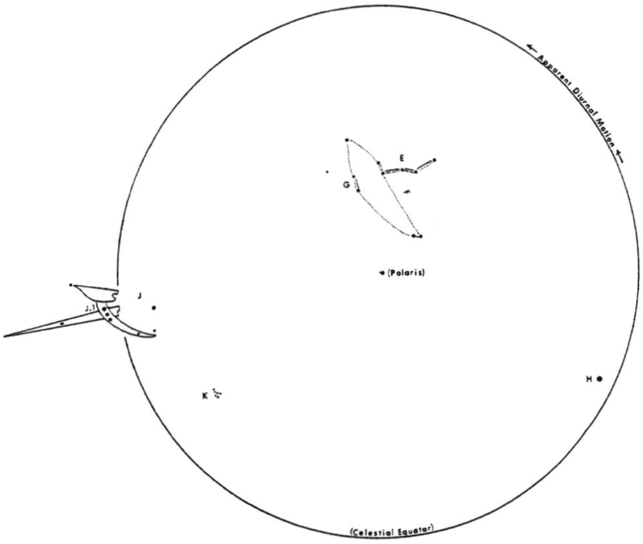

Figure 5: Bugis asterisms: northern sky.

stars identified by Bugis navigators are listed in Table 2 and illustrated in Figures 4 and 5; those which are most relied upon will now be described.

Perhaps the most frequently used asterism among the Bugis of Balobaloang is that *bintoéng balué*, mentioned above. These two bright stars are used to locate Balobaloang from Ujung Pandang and Bima from Ujung Pandang or Balobaloang. With regard to their rise/set points, navigators observe that they appear "in the south" at dusk during the middle of the east monsoon, the peak period for sailing; they further note that they rise southeast and set southwest. Their brightness makes them visible even through clouds. The name *balué* is derived from *balu* which means "widow from death of the betrothed before marriage" with the affix é forming the definite article. Hence: "the one widowed before marriage". No graphical figure is attributed to this asterism.

Just to the west of Alpha and Beta Centauri is *bintoéng bola képpang* (Crux), visualized as an "incomplete house of which one post is shorter than the other and, therefore, appears to be limping" (Asterism B, Figure 4). Crux is used in conjunction with Alpha and Beta Centauri to navigate along southerly routes; like them, it is known to set southwest. Interestingly, it was emphasized that Crux is also used to help in predicting the weather. This asterism is located in the Milky Way which is known to the Bugis as *bintoéng nagaé* (the dragon), whose head is in the south and whose tail wraps all around the sky. As such, Crux is surrounded by a bright haze of starlight. On the eastern side of the house, however, there is a small dark patch totally devoid of light which is seen as a *bembé'* (goat) (B.1, Figure 4). Between the squall clouds of the rainy season the goat in the sky may be seen standing outside the house trying to get in out of the rain. There are nights, however, when the goat is gone from the protection of the house. Hidden by haze, the missing goat portends a period of calm air and little rain.

In the northern sky the asterism which figures most prominently in Bugis navigation is *bintoéng kappala'é*, (the ship stars) in the Western constellation of Ursa Majoris (Figure 5: asterisms E and F = two versions). The "ship" is used when traveling north. In particular, it rises northeast and sets northwest over Kalimantan from Ujung Pandang and Balobaloang. Associations of this group of stars with the hull of a boat or ship appear to be common throughout Indonesia.

Adjoining the ship and likewise used to navigate northward is *balu Mandara'* (widow of the Mandar) (Asterism G, Figure 5). These two stars, Alpha and Beta Ursa Majoris, remind the Bugis of Alpha and Beta Centauri (thus the name *balu*), while "Mandar" recalls their northern seafaring neighbors.

Several asterisms and a planet are used for sailing east and west: *bintoéng timo'* (eastern star) (Altair; H, Figure

Table 2: Bugis stars and asterisms familiar to navigators

Asterism	Bugis name[a]	English gloss	International designation
A	*bintoéng balué*	widow-before-marriage	Alpha & Beta CENTAURI
B	*bintoéng bola képpang*	incomplete house	Alpha-Delta, Mu CRUCIS
B.1	*bembé'*	goat	Coal Sack Nebula in CRUX
C	*bintoéng balé mangngiweng*	shark	SCORPIUS (south)
D	*bintoéng lambarué*	ray fish, skate	SCORPIUS (north)
D.1	(identified w/o name)	lost Pleiad	Alpha SCORPII (Antares)
E	*bintoéng kappala'é*	ship	Alpha-Eta URSA MAJORIS
F	*bintoéng kappala'é*	ship	Alpha-Eta UMA; Beta, Gamma UMI
G	*bintoéng balu Mandara'*	Mandar widow	Alpha, Beta URSA MAJORIS
H	*bintoéng timo'*	eastern star	Alpha AQUILAE (Altair)
J	*pajjékoé* (Mak.)[b] or *bintoéng rakkalaé*	plough stars	Alpha-Eta ORIONIS
J.1	*tanra tellué*	sign of three	Delta, Epsilon, Zeta ORI
K	*worong-porongngé* or *bintoéng pitu*	cluster seven stars	M45 in TAURUS (Pleides)
M	*tanra Bajoé*	sign of the Bajau	Large and Small Magellanic Clouds
[]	*wari-warié*	(no gloss)	Venus: morning
[]	*bintoéng bawi*	pig star	Venus: evening
[]	*bintoéng nagaé*	dragon stars	Milky Way

[a] Note that the Bugis term for 'star(s)' is *bintoéng*; the suffix *é* may be translated as the definite article 'the' in English.
[b] Although *pajjékoé* is a Makasar term, it is most commonly used on Balobaloang to indicate this asterism.

5), *pajjékoé*, Makassar for "the plough" and also known as *bintoéng rakkalaé*, Bugis for "the plough stars" (Orion; Figures 4 and 5), *tanra tellué* (the sign of three) (the belt of Orion; Figures 4 and 5), *wari-warié* [no gloss] or *bintoéng élé'* (morning star; Venus as morning star), and *bintoéng bawi* (pig star; Venus as evening star), so named since it is believed that wild pigs will enter and destroy a garden or orchard when this object shines brightly in the west. Both the pig star and the plough, by the way, speak of an agrarian lifestyle not practiced by Bugis seafarers but culturally shared with their kin who farm the lands of Sulawesi as well as other islands of the archipelago.

Although the stars are useful guides, they are not always visible. Because it is possible to see landfall during the day, navigators appear to plan their voyages so as to maximize its usage. On a voyage from Balobaloang to Bima the captain scheduled departure in mid-afternoon, allowing him to back-sight on Balobaloang and other islands of the atoll and observe the sun as it set in the west until dusk when Alpha and Beta Centauri appeared in the sky. There was, in fact, a period of about thirty minutes where *both* the receding island and the stars could be seen, providing a good opportunity to maintain course as attention was shifted from land forms to the stars.

Except during the height of the west monsoon, it is uncommon to experience extended periods of totally overcast skies. Should the primary guiding asterism be concealed by clouds, the navigator depends upon his knowledge of other asterisms or unnamed stars to fix his direction. If it is very cloudy, day or night, the navigator turns to wave directions and the magnetic compass to maintain course.

Courses are committed to memory in terms of destinations and their required compass headings under various winds. That is, certain points of the compass are associated with certain destinations from various ports. For example, it is known that to sail from Balobaloang to Bima during the east monsoon, one must head due south, while during the west monsoon one heads south-southwest to southwest, depending on the strength of the wind and current. Likewise, to reach Ujung Pandang from the island during the east monsoon one travels somewhat east of north, while during the west monsoon a heading to the north-northwest is preferred. This difference takes into account drift from wind and currents, while true directions are also known.

With regard to the study of indigenous astronomical systems, there appear in regional agricultural calendars two types of celestial observations that may be unique to this cultural region. They are (1) observations of "cosmical" and "acronical" culminations — meridian transits at last and first gleam — of groups of stars, and (2) observations of the lunar month for a limited number of months each year, creating discontinuous sidereal-lunar calendars. The use of stars by the Bugis navigators of South Sulawesi likewise appears to represent a system which, along with its transformations used by Oceanic mariners, may be unique to the Austronesian world.

Implicit in the discussion of agricultural calendrical and navigation systems is the understanding that celestial observations do not stand alone. That is, many other environmental markers — changes in wind and weather and the appearances of flora and fauna also inform agricultural and navigational decision-making. It is suggested that by noting more carefully these and other signs in nature to which members of non-Western societies attend, we may gain a deeper appreciation of the true richness of human knowledge across cultures.

GENE AMMARELL

REFERENCES

Ammarell, Gene. "Sky Calendars of the Indo-Malay Archipelago: Regional Diversity/Local Knowledge." *Indonesia* 45:84–104, 1988.

Ammarell, Gene. "The Planetarium and the Plough: Interpreting Star Calendars of Java." *Yale Graduate Journal of Anthropology* 3:11–25, 1991.

Ammarell, Gene. *Bugis Navigation*. Doctoral Dissertation, Yale University, 1995.

Casiño, Eric S. "Jama Mapun Ethnoecology: Economic and Symbolic." *Asian Studies* 5:1–3, 1967.

Covarrubias, Miguel. *Island of Bali*. New York: Knopf, 1937.

Dentan, Robert K. "An Appeal to Members of the Society from an Anthropologist." *Malayan Nature* 23:121–122, 1970.

Freeman, J.D. *Report on the Iban*. London School of Economics Monographs on Social Anthropology no. 41. New York: Humanities Press Inc., 1970.

Hose, Charles. "Various Methods of Computing Time among the Races of Borneo." *Journal of the Straits Branch of the Royal Asiatic Society* 42:4–5, 209–210, 1905.

Hose, Charles and W. McDougall. *The Pagan Tribes of Borneo*. London: Macmillan, 1912.

Lewis, David. *We, The Navigators*. Honolulu: University of Hawaii Press, 1972.

Pelras, Christian. "Le Ciel et Les Jours. Constellations et Calendriers Agraires Chez Les Bugis (Celebes, Indonesie)." In *De La Voute Celeste Au Terroir, Du Jardin Au Foyer*. Paris: Éditions de l'École des Hautes Études en Sciences Sociales, pp. 19–39, 1987.

van den Bosch, F. "Der Javanische Mangsakalender." *Bijdragen tot de taal-, land- en volkenkunde* 136: 251–252, 1980.

See also: Agriculture – Ethnobotany – Navigation – Colonialism and Science – Malay Science – Celestial Sphere – Medicine Wheel – Swidden – Sundials – Gnomon – Calendar

ASTRONOMY IN THE ISLAMIC WORLD From the ninth to the fifteenth century, Muslim scholars excelled in every branch of scientific knowledge; their contributions in astronomy and mathematics are particularly impressive. Even though there are an estimated 10,000 Islamic astronomical manuscripts and close to 1000 Islamic astronomical instruments preserved in libraries and museums, and even if all of them were properly catalogued and indexed, the picture that we could reconstruct of Islamic astronomy, especially for the eighth, ninth, and tenth centuries, would be quite deficient. Most of the available manuscripts and instruments date from the later period of Islamic astronomy, that is, from the fifteenth to the nineteenth century, and although some of these are based or modelled on earlier works, many of the early works are extant in unique copies and others have been lost almost without trace; that is, we know only of their titles. The thirteenth-century Syrian scientific biographer Ibn al-Qifṭī relates that the eleventh-century Egyptian astronomer Ibn al-Sanbadī heard that the manuscripts in the library in Cairo were being catalogued and so he went to have a look at the works relating to his field. He found 6500 manuscripts relating to astronomy, mathematics, and philosophy. Not one of these survives amongst the 2500 scientific manuscripts preserved in Cairo today.

The surviving manuscripts thus constitute but a small fraction of those that were actually copied; nevertheless they preserve a substantial part of the Islamic scientific heritage, certainly enough of it for us to judge its level of sophistication. Only in the past few decades has the scope of the activity and achievements of Muslim scientists become apparent, and the days are long past when they were regarded merely as transmitters of superior ancient knowledge to ignorant but eager Europeans. Islamic astronomy is to be viewed on its own terms. The fact that only a small part of the available material, mainly Greek and Indian material in Arabic garb, was indeed transmitted to Europe is to be viewed as an accident of Islamic history. There is no need to apologize for using the expression "Islamic astronomy". Within a few decades of the death of the Prophet Muḥammad in the year 632, the Muslims had established a commonwealth stretching from Spain to Central Asia and India. They brought with them their own folk astronomy, which was then mingled with local traditions, and they discovered the mathematical traditions of the Indians, Persians, and Greeks, which they mastered and adapted to their needs. Early Islamic astronomy was thus a pot-pourri of pre-Islamic Arabian starlore and Indian, Persian, and Hellenistic astronomy, but by the tenth century Islamic astronomy had acquired very distinctive characteristics of its own. A.I. Sabra labels this process "appropriation and naturalization".

Astronomy flourished in Islamic society on two different

levels: folk astronomy, devoid of theory and based solely on what one could see in the sky, and mathematical astronomy, involving systematic observations and mathematical calculations and predictions. Folk astronomy was favored by the scholars of the sacred law (*fuqahā*ʾ), not least because of various religious obligations that demanded a basic knowledge of the subject; these legal scholars generally had no time (or need) for mathematical astronomy. That discipline was fostered by a select group of scholars, most of whose activities and pronouncements were, except in the case of astrological predictions, of little interest to society at large.

The astronomers also played their part in applying their discipline to certain aspects of Islamic religious practice. It was not Islam that encouraged the development of astronomy but the richness of Islamic society, a highly-literate, tolerant, multi-racial society with a predominant cultural language, Arabic: but neither did Islam, the religion, stand in the way of scientific progress. The Prophet had said: "Seek knowledge, even as far as China". To be sure, over-zealous orthodox rulers occasionally pursued, killed, or otherwise attacked "scientists" or destroyed or burnt their libraries, but these were exceptions. The scholars of the religious law, who saw themselves as the representatives of Islam, generally ignored the pronouncements of the scientists, even on matters relating to religious practice. Astronomy was the most important of the Islamic sciences, as we can judge by the volume of the associated textual tradition, but a discussion of it in the broader context of the various branches of knowledge, which has been attempted several times elsewhere, is beyond the scope of this essay.

Arab Starlore

The Arabs of the Arabian peninsula before Islam possessed a simple yet developed astronomical folklore of a practical nature. This involved a knowledge of the risings and settings of the stars, associated in particular with the acronychal settings of groups of stars and simultaneous heliacal risings of others, which marked the beginning of periods called *naw*ʾ, plural *anwā*ʾ. These *anwā*ʾ eventually became associated with the 28 lunar mansions, a concept apparently of Indian origin. A knowledge of the passage of the sun through the twelve signs of the zodiac, associated meteorological and agricultural phenomena, the phases of the moon, as well as simple time-reckoning using shadows by day and the lunar mansions by night, formed the basis of later Islamic folk astronomy.

More than twenty compilations on the pre-Islamic Arabian knowledge of celestial and meteorological phenomena as found in the earliest Arabic sources of folklore, poetry, and literature, are known to have been compiled during the first four centuries of Islam. The best known is that of Ibn Qutayba, written in Baghdad about the year 860. Almanacs enumerating agricultural, meteorological, and astronomical events of significance to local farmers were also compiled; several examples of these survive from the medieval Islamic period, one such being for Cordoba in the year 961. The Yemen possessed a particularly rich tradition of folk astronomy, and numerous almanacs were compiled there.

Since the sun, moon, and stars are mentioned in the *Qur*ʾ*ān*, an extensive literature dealing with what may well be labeled Islamic folk cosmology arose. This was inevitably unrelated to the more "scientific" Islamic tradition based first on Indian sources and then predominantly on Greek ones. Since it is also stated in the *Qur*ʾ*ān* that man should use these celestial bodies to guide him, the scholars of the religious law occupied themselves with folk astronomy.

Persian and Indian Sources

The earliest astronomical texts in Arabic seem to have been written in Sind and Afghanistan, areas already conquered by the Muslims in the seventh century. Our knowledge of these early works is based entirely on citations from them in later works. They consisted of texts and tables and were labelled *zīj* after a Persian word meaning "cord" or "thread" and by extension "the warp of a fabric", which the tables vaguely resemble. The Sasanian *Shahriyārān Zīj* in the version of Yazdigird III was translated from Pahlavi into Arabic as the *Shāh Zīj*, and the astronomers of al-Manṣūr chose an auspicious moment to found his new capital Baghdad using probably an earlier Pahlavi version of this *zīj*. The various horoscopes computed by Māshāʾallāh (Baghdad, ca. 800) in his astrological world history are based on it.

Significant for the subsequent influence of Indian astronomy in the Islamic tradition was the arrival of an embassy sent to the court of the Caliph al-Manṣūr from Sind ca. 772. This embassy included an Indian well versed in astronomy and bearing a Sanskrit astronomical text apparently entitled the *Māhasiddhānta* and based partly on the *Brāhmasphuṭasiddhānta*. The Caliph ordered al-Fazārī to translate this text into Arabic with the help of the Indian. The resulting *Zīj al-Sindhind al-kabīr* was the basis of a series of *zīj*es by such astronomers as al-Fazārī, Yaʿqūb ibn Ṭāriq, al-Khwārizmī, and others, all prepared in Iraq before the end of the tenth century. The *Sindhind* tradition flourished in Andalusia, mainly through the influence there of the *Zīj* of al-Khwārizmī (see below). As a result, the influence of Indian astronomy is attested from Morocco to England in the late Middle Ages.

Greek Sources

The *Almagest* of Ptolemy (Alexandria, ca. 125) was translated at least five times in the late eighth and ninth centuries. The first was a translation into Syriac and the others were into Arabic, the first two under al-Maʾmūn in the middle of the first half of the ninth century, and the other two (the second being an improvement of the first) towards the end of that century. All of these were still available in the twelfth century, when they were used by Ibn al-Ṣalāḥ for his critique of Ptolemy's star catalogue. The translations gave rise to a series of commentaries on the whole text or parts of it, many of them critical, and one, by Ibn al-Haytham (ca. 1025), actually entitled *al-Shukūk fī Baṭlamiyūs* (Doubts about Ptolemy). The most commonly-used version in the later period was the recension of the late-ninth-century version by the polymath Naṣīr al-Dīn al-Ṭūsī in the mid-thirteenth century. Various other works by Ptolemy, notably the *Planetary Hypotheses* and the *Planisphaerium*, and other Greek works, including the short treatises by Autolycos, Aristarchos, Hypsicles, and Theodosios, and works on the construction known as the analemma for reducing problems in three dimensions to a plane, were also translated into Arabic; most of these too were later edited by al-Ṭūsī. In this way Greek planetary models, uranometry, and mathematical methods came to the attention of the Muslims. Their redactions of the *Almagest* not only reformulated and paraphrased its contents but also corrected, completed, criticized, and brought the contents up to date both theoretically and practically.

Theoretical Astronomy

The geometrical structure of the universe conceived by Muslim astronomers of the early Islamic period (ca. 800–1050) is more or less that expounded in Ptolemy's *Almagest*, with the system of eight spheres being regarded essentially as mathematical models. However, in Ptolemy's *Planetary Hypotheses* these models are already taken as representing physical reality; this text also became available in Arabic. Several early Muslim scholars wrote on the sizes and relative distances of the planets, and one who proposed a physical model for the universe was Ibn al-Haytham (*fl.* Cairo, ca. 1025). In order to separate the two motions of the eighth sphere, the motion of the fixed stars due to the precession of the equinoxes, and the motion of the fixed stars due to the apparent daily rotation, he proposed a ninth sphere to impress the apparent daily rotation on the others.

Of considerable historical interest are various Arabic treatises on the notion of the trepidation of the equinoxes. This theory, developed from Greek sources, found followers who believed that it corresponded better to the observed phenomena than a simple theory of uniform precession. The mathematical models proposed were complex and have only recently been studied properly (notably those of Pseudo-Thābit (date unknown) and Ibn al-Zarqāllu (Andalusia, ca. 1070), who seems to have relied on his predecessor Ṣāʿid al-Andalusī). The theory of trepidation continued to occupy certain Muslim scholars (in the late period mainly in the Maghrib), as it did European scholars well into the Renaissance.

Other significant Islamic modifications to Ptolemaic planetary models, devised to overcome the philosophical objections to the notion of an equant and the problem of the variation in lunar distance inherent in Ptolemy's lunar model, belong to the later period of Islamic astronomy. There were two main schools, one of which reached its fullest expression in Maragha in north western Iran in the thirteenth century (notably with al-Ṭūsī and his colleagues) and Damascus in the fourteenth (with Ibn al-Shāṭir), and the other developed in Andalusia in the late twelfth century (with al-Biṭrūjī). The latter tradition was doomed from the outset by a slavish adherence to (false) Aristotelian tenets and by mathematical incompetence. The former was based on sophisticated modifications to Ptolemy's models, partly inspired by new observations; Ptolemy himself would have been impressed by it, as have been modern investigators, for the tradition has been rediscovered and studied only in the latter half of this century. In the 1950s E.S. Kennedy discovered that the solar, lunar, and planetary models proposed by Ibn al-Shāṭir in his book *Nihāyat al-suʾl* (The Final Quest Concerning the Rectification of Principles) were different from those of Ptolemy; indeed they were mathematically identical to those of Copernicus some 150 years later. In this work Ibn al-Shāṭir laid down the details of what he considered to be a true theoretical formulation of a set of planetary models describing planetary motions, and actually intended as alternatives to the Ptolemaic models. He maintained the geocentric system, whereas Copernicus proposed a hypothesis, which he was unable to prove, that the sun was at the centre of things. Nevertheless this important discovery raised the interesting question of whether Copernicus might have known of the works of the Damascene astronomer. Since the 1950s we have progressed to a new stage of inquiry: we now know that there was a succession of Muslim astronomers from the eleventh century to the sixteenth who concerned themselves with models different from those of Ptolemy, all designed to overcome what were seen as flaws in them. The question we may now ask is: was Copernicus influenced by any of these Muslim works? The answer is unsatisfactory, namely, that he must have been; definitive proof is, however, still lacking.

Mathematical Astronomy — The Tradition of the Zījes

The Islamic *zīj*es constitute an important category of astronomical literature for the historian of science, by virtue of the diversity of the topics dealt with, and the information that can be obtained from the tables. In 1956, E.S. Kennedy published a survey of about 125 Islamic *zīj*es. We now know of close to 200, and material is available for a revised version of Kennedy's *zīj* survey. To be sure, many of these works are lost, and many of the extant ones are derived from other *zīj*es by modification, borrowing, or outright plagiarism. Nevertheless, there are enough *zīj*es available in manuscript form to reconstruct a reasonably accurate picture of Muslim activity in this field.

Most *zīj*es consist of several hundred pages of text and tables; the treatment of the material presented may vary considerably from one *zīj* to another. The following topics are handled in a typical *zīj*:

(1) chronology;
(2) trigonometry;
(3) spherical astronomy;
(4) solar, lunar, and planetary mean motions;
(5) solar, lunar, and planetary equations;
(6) lunar and planetary latitudes;
(7) planetary stations;
(8) parallax;
(9) solar and lunar eclipses;
(10) lunar and planetary visibility;
(11) mathematical geography (lists of cities with geographical coordinates), determination of the direction of Mecca;
(12) uranometry (tables of fixed stars with coordinates);
(13) mathematical astrology.

As noted above, by the eighth century a number of Arabic *zīj*es had been compiled in India and Afghanistan. These earliest examples, based on Indian and Sasanian works, are lost, as are the earliest examples compiled at Baghdad in the eighth century. With the *zīj*es compiled in Baghdad and Damascus in the early ninth century under the patronage of the Caliph al-Maʾmūn, we are on somewhat firmer ground. These follow either the tradition of Ptolemy's *Almagest* and *Handy Tables* or the Indian tradition. Manuscripts exist of the *Mumtaḥan Zīj* of Yaḥyā ibn Abī Manṣūr and the Damascus *Zīj* of Ḥabash, each of which was based on essentially Ptolemaic theory rather than Indian. The *Zīj* of al-Khwārizmī, based mainly on the Persian and Indian traditions, has survived only in a Latin translation of an Andalusian recension. Amongst the most important and influential later works of this genre, are: the *Ṣābiʾ Zīj* of al-Battānī of Raqqa, ca. 910; the *Ḥākimī Zīj* of Ibn Yūnus, compiled in Cairo at the end of the tenth century; the *zīj* called *al-Qānūn al-Masʿūdī* by al-Bīrūnī, compiled in Ghazna about 1025; the *Zīj* of Ibn Isḥāq, compiled in Tunis, ca. 1195; the *Īlkhānī Zīj* of Naṣīr al-Dīn al-Ṭūsī, prepared in Maragha in north-western Persia in the mid-thirteenth century; and the *Sulṭānī Zīj* of Ulugh Beg from early-fifteenth-century Samarqand.

Although the *zīj*es are amongst the most important sources for our knowledge of Islamic mathematical astronomy, it is important to observe that they generally contain extensive tables and explanatory text relating to mathematical astrology as well. Islamic astrological texts form an independent corpus of literature, mainly untouched by modern scholarship. Often highly sophisticated mathematical procedures are involved. It should also be pointed out that in spite of the fact that astrology was anathema to Muslim orthodoxy, it has always been (and still is) widely practiced in Islamic society.

All early Islamic astronomical tables have entries written in Arabic alphanumerical notation (*abjad*) and expressed sexagesimally, that is, to base 60. A number written in letters equivalent to "23 30 17 seconds" (Ulugh Beg's value for the obliquity) stands for 23 + 30/60 + 17/3600 degrees, that is, $23°30'17''$. In sexagesimal arithmetic, more so than in decimal arithmetic, it is useful to have a multiplication table at hand, and such tables, with 3600 or even 216,000 entries, were available.

Already in the early ninth century Muslim astronomers had restyled the cumbersome Indian sine function using the Greek base 60 (which the Greeks had used for their even more cumbersome chord function). Likewise the Indian shadow functions, unknown in Greek astronomy, were adopted with different bases (12, 6, $6\frac{1}{2}$ and 7, and also 60, and occasionally 1). Most *zīj*es contain tables of the sine and (co)tangent function for each whole, or half, or quarter degree of arc. Entries are generally given to three sexagesimal digits, corresponding roughly to five decimal digits. However, certain Muslim scholars compiled more extensive sets of trigonometric tables that were not included in *zīj*es. In the early tenth century al-Samarqandī prepared a set of tables of the tangent function with entries to three sexagesimal digits for each minute of arc. Later in the same century Ibn Yūnus tabulated the sine function to five sexagesimal digits, equivalent to about nine decimal digits, for each minute of arc, also giving the differences for each second. He also tabulated the tangent function for each minute of arc, and the solar declination for each minute of solar longitude. His trigonometric tables were not sufficiently accurate to warrant this number of significant figures, and indeed over four centuries were to elapse before the compilation in Samarqand of the magnificent trigonometric tables in the *Sulṭānī Zīj* of

Ulugh Beg, which display the values of the sine and tangent to five sexagesimal digits for each minute of argument and are generally accurate in the last digit.

Planetary Tables and Ephemerides

Given the Ptolemaic models and tables of the mean motion and equations of the sun, moon, and planets such as were available to Muslim astronomers in the *Almagest* and *Handy Tables*, or the corresponding tables based on Indian models that exemplify the *Sindhind* tradition, Muslim astronomers from the ninth to the sixteenth century sought to improve the numerical parameters on which these tables were based. Most of the leading Muslim astronomers of the early period made solar observations and computed new solar equation tables. Ibn Yūnus is the only astronomer from the first four centuries of Islam known to have compiled a new set of lunar equation tables. The majority of Islamic planetary equation tables are Ptolemaic, and where exceptions do occur, such as in the tables of Ibn al-Aʿlam and Ibn Yūnus for Mercury, we find that they are based on a Sasanian parameter rather than on any new observations.

Ptolemy used the same data as Hipparchus for his determination of the solar apogee and hence obtained the same result. The Muslims thus inherited the notion that the solar apogee is fixed with respect to the fixed stars (although the planetary apogees move with the motion of precession), and it is to their credit that their earliest observations established that the solar apogee had moved about 15° since the time of Hipparchus. Most early Muslim astronomers accepted the *Mumtaḥan* value of 1° in 66 Persian years (actually a parameter attested in earlier Persian sources) for both precession and the motion of the apogees. Ibn Yūnus possessed all the necessary data that could be used to demonstrate that the motion of the solar apogee is not the same as the motion due to precession, but he chose to use the same value for both, 1° in $70\frac{1}{4}$ Persian years, which happens to be remarkably close to the actual rate of precession. Al-Bīrūnī (Central Asia, ca. 1025) seems to have been the first to distinguish the proper motion of the solar apogee from the motion of precession (this discovery is sometimes erroneously attributed to al-Battānī). It was Ibn al-Zarqāllu (Andalusia, ca. 1070) who was the first to assign a numerical value to both motions, although he also subscribed to the theory of trepidation.

All Islamic *zīj*es contained tables of mean motions and equations for computing solar, lunar, and planetary positions for a given time. Some of the equation tables are arranged in a form more convenient for the user (so that one simply has to enter the mean motions, and calculations are avoided). Auxiliary tables were sometimes available for generating ephemerides without the tedious computation of daily positions from mean-motion and equation tables. From the ninth to the nineteenth centuries Muslim astronomers compiled ephemerides displaying solar, lunar, and planetary positions of each day of the year, as well as information on the new moons and astrological predictions resulting from the position of the moon relative to the planets. Al-Bīrūnī described in detail how to compile ephemerides in his astronomical and astrological handbook *al-Tafhīm fī ṣināʿat al-tanjīm* (Instruction in the Art of Astrology). Manuscripts of ephemerides had a high rate of attrition since the tables could be dispensed with at the end of the year: the earliest complete extant examples are from fourteenth-century Yemen, discovered in Cairo in the 1970s and still unpublished; on the other hand, literally hundreds of ephemerides survive from the late Ottoman period.

Stellar Coordinates and Uranography

Most *zīj*es contain lists of stellar coordinates in either the ecliptic (the band of the zodiac through which the Sun apparently moves in its yearly course) or the equatorial systems, or occasionally in both. A survey of the stellar coordinates in Islamic *zīj*es would be a valuable contribution to the history of Islamic astronomy, and could help determine the extent to which original observations were made by Muslim astronomers. An impressive amount of research on Arabic star names and their later influence in Europe has been conducted in the last few years by Paul Kunitzsch.

In his *Ṣuwar al-kawākib* (Book of Constellation Figures) the tenth-century Shiraz astronomer al-Ṣūfī presented lists of stellar coordinates as well as illustrations of the constellation figures from the Hellenistic tradition and also information on the lunar mansions following the Arab tradition. Later Islamic works on uranography are mostly restricted to Persian and Turkish translations of al-Ṣūfī, although some astrological works also contain illustrations of the constellations that have recently attracted the attention of historians of Islamic art.

Spherical Astronomy and Spherical Trigonometry

Most *zīj*es contain in their introductory text the solutions of the standard problems of spherical astronomy, such as, to give only one example, the determination of time from solar and stellar altitude. Rarely is any explanation given of how the formulae outlined in words in the text were derived. There were two main traditions. In the first, the problems relating to the celestial sphere are reduced to geometric or trigonometric problems on a plane. The construction known as the analemma was a singularly powerful tool for solutions of this kind. In the second, the problems are solved by applications of rules of spherical trigonometry. Both techniques are ultimately of Greek origin, and Muslim scholars made substantial contributions to each.

There is some confusion about these contributions in the modern literature. It has been assumed by modern writers that when a medieval writer used a medieval formula that is mathematically equivalent to the modern formula derived by a specific rule of spherical trigonometry, the medieval scholar must have known the equivalent of the modern rule of spherical trigonometry. In fact, however, the medieval formula may have been derived without using spherical trigonometry at all. The first known Islamic treatise dealing with spherical trigonometry independently of astronomy is by the eleventh-century Andalusian Ibn Muʿādh. The contributions to spherical astronomy by scholars such as Thābit ibn Qurra, al-Nayrīzī, Abu'l-Wafāʾ al-Būzajānī, al-Khujandī, Kūshyār ibn Labbān, al-Sijzī, and Abū Naṣr, are outlined in the recently-rediscovered *Maqālīd fī ʿilm al-hayʾa* (Keys to Astronomy) of al-Bīrūnī, also from the eleventh century.

Already in the work of Ḥabash in the mid-ninth century we find a Muslim astronomer at ease with both spherical trigonometrical methods and analemma constructions for solving problems of spherical astronomy. In the *zīj*es of scholars of the caliber of Ibn Yūnus and al-Bīrūnī we find various methods for solving each of the standard problems of medieval spherical astronomy. The auxiliary trigonometric tables compiled by such scholars as Ḥabash, Abū Naṣr (Khwārizm, ca. 1000), and al-Khalīlī (Damascus, ca. 1360) for solving all of the problems of spherical astronomy for any latitude are a remarkable testimony to their mastery of the subject.

Observation Programmes and Regional Schools of Astronomy: al-Maʾmūn's Circle

In the early ninth century the Abbasid Caliph al-Maʾmūn patronized observations first in Baghdad and then in Damascus, gathering the best available astronomers to conduct observations of the sun and moon. Some of the results were incorporated into a *zīj* called *al-Mumtaḥan*, "tested", although the details of the activities at the two observation posts are somewhat confusing. The *Mumtaḥan Zīj* was apparently compiled in Baghdad by Yaḥyā ibn Abī Manṣūr, but upon his death, according to Ḥabash, the Caliph ordered his colleague Khālid al-Marwarrūdhī to prepare some new instruments and conduct a one-year programme of solar and lunar observations in Damascus in order to compile a new *zīj*. According to Ḥabash this was done, but no such *zīj* is otherwise known to have been prepared before Ḥabash's own *Damascus Zīj*. Also simultaneous observations of a lunar eclipse were conducted at Baghdad and Mecca, and the longitude difference used together with the newly-measured latitudes of the two localities to find the *qibla* at Baghdad.

These observations, like later ones, were mainly directed towards determining the local latitude and current value of the obliquity, and towards deriving improved parameters for the Ptolemaic planetary models and more accurate star positions. The armillary sphere, the meridian quadrant, and the parallactic ruler were known to the Muslims from the *Almagest*, and they added new scales and other modifications, often building larger instruments even when smaller ones would have sufficed. Our knowledge of the instruments used by al-Maʾmūn's astronomers is meager. An armillary sphere used by Yaḥyā in Baghdad was said to display markings for each ten minutes of arc, but even contemporary astronomers were not impressed by the precision of the results obtained using it. A mural quadrant made of marble with a radius of about five meters was used in Damascus, as well as a vertical gnomon made of iron standing about five meters high. Al-Maʾmūn also patronized measurements of the longitude difference between Baghdad and Mecca in order to establish the *qibla* at Baghdad properly, as well as measurements of the length of one degree of terrestrial latitude.

Other Observational Programmes

Besides the officially-sponsored observations conducted in Baghdad and Damascus in the early ninth century, there are numerous instances of other series of observations conducted in different parts of the Muslim world.

The two brothers called Banū Mūsā made observations in their own house in Baghdad and also in nearby Samarra about thirty years after the *Mumtaḥan* observations. They also arranged for simultaneous observations of a lunar eclipse in Samarra and Nishapur in order to determine the difference in longitude between the two cities. In view of their proficiency in mathematics, it is most unfortunate that neither of the two *zīj*es compiled by them has survived.

Al-Battānī carried out observations during the period 887 to 918 in Raqqa in northern Syria. He appears to have financed his observational activity himself, and although we have no description of the site where he made his observations, the instruments mentioned in the *zīj* based on his observations include an armillary sphere and mural quadrant, as well as a parallactic ruler, an astrolabe, a gnomon, and a horizontal sundial.

The observational activities of the Baghdad family known as the Banū Amājūr were almost contemporaneous with those of al-Battānī in Raqqa. Father and two sons, and also a freed family slave, all made observations and each compiled a *zīj*, none of which survives. In the accounts of their eclipse observations recorded by Ibn Yūnus it appears that the place where they conducted their observations had some kind of a balcony fitted with slits for observation, but the details are obscure. A particularly interesting account of a solar eclipse in the year 928 that they observed by reflection in water includes a remark that the altitude of the sun was measured on an instrument marked for each third of a degree.

A large mural quadrant was erected at Rayy (near modern Tehran) about the year 950, but we have information only on its use to establish the local latitude and obliquity of the ecliptic. In Shiraz not long thereafter, al-Ṣūfī used an armillary sphere with a diameter of about five meters to derive the same parameters and to "observe" equinoxes and solstices. Al-Ṣūfī is best known for his work on the fixed stars, but it seems that this was based more on "observation", looking at the heavens with the naked eye, than on "measurement", looking at the heavens with precision instruments and making estimates of positions. Another contemporary astronomer who conducted observations on which we have no information other than the main parameters of his *zīj* was Ibn al-Aʿlam.

In the late tenth century the distinguished mathematician and astronomer Abuʾl-Wafāʾ al-Būzajānī made observations in Baghdad. Most of these appear to have been directed towards the determination of the solar parameters, and the obliquity of the ecliptic and the latitude of Baghdad, although Abuʾl-Wafāʾ also collaborated with al-Bīrūnī in Khwārizm (modern Khiva in Turkmenistan) on the simultaneous observation of a lunar eclipse in the year 997. We have no information on the nature of the site where Abu'l-Wafāʾ made his observations, other than its location in a specific quarter of Baghdad.

Contemporaneous with the activity of Abu'l-Wafāʾ was the establishment in 988 of an observatory in the garden of the Baghdad residence of the Buwayhid ruler Sharaf al-Dawla. The organization of a building and programme of observations was entrusted to Abū Sahl al-Qūhī, a mathematician of considerable standing. We know from contemporary historical records that a special building was erected for the observations, which in turn were witnessed by "judges, scientists and scholars of note, astronomers, and engineers". In view of the favourable conditions under which this observatory was established, and the competence of its director, it is somewhat surprising that the two recorded "observations" that were witnessed by so many dignitaries were the entry of the sun into Cancer and Libra in the year 988. Al-Bīrūnī describes the main instrument that was constructed as a hemisphere of radius 12.5 meters on which the solar image was projected through an aperture at the centre of the hemisphere. Activity at the observatory stopped in 989 with the death of Sharaf al-Dawla, so that the institution lasted not much more than a year.

In 994 Abū Maḥmūd al-Khujandī made a measurement of the obliquity using a meridian sextant of about twenty meters radius. This instrument was erected in Rayy, but al-Khujandī confessed to al-Bīrūnī that it was so large that the centre of the sextant had become displaced from its intended position.

The Egyptian astronomer Ibn Yūnus made a series of observations of eclipses, conjunctions, and occultations, as well as equinoctial and solstitial observations. We are extremely fortunate to have not only his reports of these observations but also his citations of earlier observations of the same kind made by individuals such as Ḥabash and the Banū Amājūr. Ibn Yūnus' purpose in making these observations and recording them in the introduction to his *Zīj* is somewhat obscured by the fact that he does not list those observations or present those calculations with which he derived his new solar, lunar, and planetary parameters. Neither does he mention any locations for his observations other than his grandfather's house in Fustat and a nearby mosque in al-Qarāfa. The popular association of Ibn Yūnus with an observatory on the Muqaṭṭam Hills outside Cairo is, as Aydın Sayılı has shown, a myth. Nevertheless, Ibn Yūnus mentions at least one instrument, probably a meridian ring, that was provided by the Fatimid Caliphs al-ʿAzīz and al-Ḥākim. In a later medieval Egyptian source Ibn Yūnus is reported to have received 100 dinars a day from al-Ḥākim, and it may be that such extremely high payments were made to Ibn Yūnus when he was making satisfactory astrological predictions for the Caliph. Al-Ḥākim made an abortive attempt to found an observatory in Cairo, but this was after the death of Ibn Yūnus in 1009. At some time during his reign there was an armillary sphere in Cairo with nine rings, each large enough that a man could ride through them on horseback.

The observations of al-Bīrūnī were conducted between 990 and ca. 1025 in several localities between Khwārizm and Kabul. His recorded observations include determinations of equinoxes and solstices, eclipses, and determinations of the obliquity and local latitude.

The corpus of tables known as the *Toledan Tables* was compiled in the eleventh century, based on observations directed by Ṣāʿid al-Andalusī and continued by Ibn al-Zarqāllu. Only the mean motion tables in this corpus of tables are original; most of the remainder were lifted from the *zīj*es of al-Khwārizmī and al-Battānī.

In the thirteenth century there was a substantial observational program at Maragha. The results are impressive only in so far as theoretical astronomy is concerned. Otherwise the trigonometric and planetary tables in the major production of the Maragha astronomers were modified or lifted *in toto* from earlier sources. This is not a happy outcome for a generously endowed observatory fitted with the latest observational instruments, known to us from texts. In the early fifteenth century the scene had moved to Samarqand in Central Asia: there, a group of astronomers, directed by the astronomer-prince Ulugh Beg, did impressive work. Only the 40-meter meridian sextant survives from the observatory. These men produced a set of tables which it would be foolish to judge before they have been properly studied. The same is true for the short-lived observatory in Istanbul under the direction of Taqiʾl-Dīn (1577).

Regional Schools of Astronomy

After the tenth century there developed regional schools of astronomy in the Islamic world, with different interests and concentrations. They also had different authorities (for example, in the furthest East al-Bīrūnī and al-Ṭūsī, and in Egypt Ibn Yūnus). The main regions were Iraq, Iran and Central Asia, Muslim Spain, Egypt and Syria, the Yemen, the Maghrib, and later also the Ottoman lands. Only recently have the complex tradition of Muslim Spain (tenth to fourteenth centuries), the colourful tradition of Mamluk Egypt and Syria (thirteenth to early sixteenth centuries), the distinctive tradition of Rasulid Yemen (thirteenth to sixteenth centuries), and the staid tradition of the Maghrib (twelfth to nineteenth centuries) been studied. The traditions of Ottoman Turkey and Mogul India are currently being researched.

Transmission to Europe

The Europeans learned of Islamic astronomy mainly through Spain, a region where, because of political problems and the difficulty of communications, the most up-to-date writings were not always available. This explains, for example, how it came to pass that the Europeans came across two major works of Muslim astronomers from the East, al-Khwārizmī and al-Battānī, at a time when these works were no longer widely used in the Islamic East. It also explains why so few Eastern Islamic works became known in Europe. None of the Eastern Islamic developments to Ptolemy's planetary theory was known in Andalusia or in medieval Europe. Al-Biṭrūjī's unhappy attempt to develop planetary models confused Europeans for several centuries; he must be worth reading, they naïvely thought, because he was trying to reconcile Ptolemy with Aristotle. As far as astronomical timekeeping was concerned, this does not seem to have been of much concern to the Muslims in Spain; hence nothing of consequence was transmitted.

On the other hand, some early Eastern Islamic contributions, later forgotten in the Islamic East, were transmitted to Spain and thence to Europe; they have been considered European developments because evidence to the contrary has seemed to be lacking. A good example is the horary quadrant with movable cursor (the so-called *quadrans vetus*), which was invented in Baghdad in the ninth century and (at least in the version with the cursor) virtually forgotten in the Islamic East thereafter; it came to be the favourite quadrant in medieval Europe. What, if any, astronomical knowledge was transmitted through Islamic Sicily remains a mystery, and nothing of consequence is known to have been learned about the subject by the Crusaders.

In the European Renaissance there was no access to the latest Islamic works. So the Europeans contented themselves with new editions of the ancient Greek works, with occasional, almost nostalgic, references to Albategnius (al-Battānī), Azarquiel (al-Zarqāllu), Alpetragius (al-Biṭrūjī), and the like. A few technical terms derived from the Arabic, such as alidade, azimuth, almucantar, nadir, saphea, and zenith, and a few star-names such as Aldebaran, Algol, Altair, and Vega, survived. When the Europeans did come to learn of some of the major Islamic works and to try to come to terms with them it was as orientalists and historians of astronomy, for by this time the Islamic materials other than observation accounts were of historical rather than scientific interest. Thanks to orientalists like the Sédillots in Paris, works that had been completely unknown to Europeans and mainly forgotten by Muslims were published, translated, and analyzed. Islamic astronomy was highly respected by such scholars and others, like the historian of astronomy J.-B. Delambre, who, innocent of Arabic, took the trouble to read what his colleagues had written about the subject. However, Islamic astronomy, indeed Islamic science in general, received a blow below the belt from P. Duhem, a physicist and philosopher ignorant of Arabic who simply ignored what scholars like the Sédillots had written. His thesis, that the Arabs were incapable of scientific thought and that whatever merits their science may have had were due to the intellectually superior Greeks, still has many followers, but only amongst those ignorant of the research of the past 150 years.

In the period after ca. 1500 Islamic astronomy declined. All of the problems had been solved, some many times over. Much of the innovative activity had led into a cul-de-sac, from which it would not emerge until modern times, thanks to investigations of manuscripts and instruments. Not that interest in astronomy died out. From Morocco to India the same old texts were copied and studied, recopied and restudied, usually different texts in each of the main regions, but there was no new input or output of any consequence. Astronomy continued to be used as the handmaiden of astrology, and for the regulation of the calendar and the prayer times. Where there was innovation — such as, for example, in the remarkable device made in Isfahan ca. 1700 that correctly displays the direction and distance of Mecca for any locality — one must suspect the existence of an earlier tradition. However, the old traditions died hard, and Muslim astronomers for several centuries spent more time copying old treatises and tables than compiling new ones.

During the millennium beginning ca. 750 and especially in the period up to ca. 1050, although also in the period up to ca. 1500, Muslim astronomers did first-rate work, most of which was not known in medieval Europe at all. Those few Islamic works from the early period that were transmitted, notably the *zīj*es of al-Khwārizmī and al-Battānī (especially through the *Toledan Tables*) and the banal summary of the *Almagest* by al-Farghānī, convey only an impression of classical astronomy in Arabic garb. However, they were in no way representative of contemporary Islamic astronomy in the East, and whilst the Europeans laboured for centuries to come to terms with them, Muslim astronomers were making substantial contributions to their subject that have only been revealed by modern scholarship.

There is a wealth of material relating to this subject that remains untouched. Very few Islamic astronomical works have been published or have received the attention they merit. Three out of close to 200 Islamic *zīj*es have been published in the optimum way (text, translation, and commentary). We have no published edition of the Arabic versions of the *Almagest* (except for the star catalogue), or of any Arabic recensions or commentaries. Many of the published Arabic scientific texts were printed in Hyderabad, most with no critical apparatus. Likewise most of the historically-important Islamic astronomical instruments are still unpublished, although the catalogue currently in preparation in Frankfurt promises to make them better known.

In 1845 L. A. Sédillot, whose privilege it was to have access to the rich collection of Arabic and Persian scientific manuscripts in the Bibliothèque Nationale in Paris, wrote: "Each day brings some new discovery and illustrates the extreme importance of a thorough study of the manuscripts of the East". Sédillot also realized the importance of Islamic astronomical instruments. Given the vast number of manuscripts and instruments now available in libraries and museums elsewhere in Europe, the United States, and the Near East, and the rather small number of people currently working in this field, Sédillot's statement is no less true today than it was a century and a half ago.

DAVID A. KING

REFERENCES

Berggren, J. Lennart. *Episodes in the Mathematics of Medieval Islam*. New York: Springer, 1986.

Goldstein, Bernard R. *Theory and Observation in Ancient and Medieval Astronomy*. London: Variorum, 1985.

Heinen, Anton. *Islamic Cosmology*. Beirut and Wiesbaden: Franz Steiner, 1982.

Kennedy, Edward S. "A Survey of Islamic Astronomical Tables." In *Transactions of the American Philosophical Society*, n.s., 46 (2): 121–177, 1956.

Kennedy, Edward S., et al. *Studies in the Islamic Exact Sciences*. Beirut: American University of Beirut Press, 1983.

King, David A. *A Survey of the Scientific Manuscripts in the Egyptian National Library*. Winona Lake, IN: Eisenbrauns, 1986.

King, David A. *Islamic Mathematical Astronomy*. London: Variorum, 1986. 2nd rev. ed., Aldershot: Variorum, 1993.

King, David A. "Some Remarks on Islamic Scientific Manuscripts and Instruments and Past, Present and Future Research." In *The Significance of Islamic Manuscripts*. Ed. John Cooper. London: Al-Furqan Islamic Heritage Foundation, 1992, pp. 115–144.

King, David A. *Astronomy in the Service of Islam*. Aldershot: Variorum, 1993.

King, David A. "Islamic Astronomy." In *Astronomy before the Telescope*. Ed. Christopher Walker. London: British Museum Press, 1996, pp. 143–174.

King, David A., and George Saliba, eds. *From Deferent to Equant: Studies in the History of Science in the Ancient and Medieval Near East in Honor of E.S. Kennedy*. (Annals of the New York Academy of Sciences 500.) New York: New York Academy of Sciences, 1986.

Kunitzsch, Paul. *Untersuchungen zur Sternnomenklatur der Araber*. Wiesbaden: Otto Harrassowitz, 1961.

Kunitzsch, Paul. *The Arabs and the Stars*. Northampton: Variorum, 1989.

Lorch, Richard P. "Arabic Mathematical Sciences" Instruments, texts, transmission. Aldershot: Variorum, 1995.

Ragep, F. Jamil. *Naṣīr al-Dīn al-Ṭūsī's Memoir on Astronomy (al-Tadhkira fī ʿilm al-hayʾa)*. 2 vols, New York: Springer, 1993.

Sabra, Abdelhamid I. "The Appropriation and Subsequent Naturalization of Greek Science in Medieval Islam." *History of Science* 25: 223–243, 1987.

Saliba, George. "The Astronomical Tradition of Maragha: A Historical Survey and Prospects for Future Research." *Arabic Science and Philosophy* 1: 67–99, 1991.

Samsó, Julio. *Las ciencias de los antiguos en al Andalus*. Madrid: Mapfre, 1992.

Samsó, Julio. *Islamic Astronomy and Medieval Spain*. Aldershot: Variorum, 1994.

Sayılı, Aydın. *The Observatory in Islam*. Ankara: Turkish Historical Society, 1960, reprinted New York: Arno Press, 1981.

Sezgin, Fuat. *Geschichte des arabischen Schrifttums*. V: Mathematik, VI: Astronomie, VII: Astrologie, Meteorologie und Verwandtes. Leiden: E.J. Brill, 1974, 1978, and 1979.

Suter, Heinrich. "Die Mathematiker und Astronomen der Araber und ihre Werke." *Abhandlungen zur Geschichte der mathematischen Wissenschaften* 10, 1900, and 14: 157–185, 1902, both reprinted Amsterdam: Oriental Press, 1982.

van Dalen, Benno. *Ancient and Mediaeval Astronomical Tables — Mathematical Structure and Parameter Values*. Utrecht: Utrecht University, 1993.

Varisco, Daniel M. *Medieval Agriculture and Islamic Science — The Almanac of a Yemeni Sultan*. Seattle: University of Washington Press, 1993.

Wright, Ramsey R. *The Book of Instruction in the Elements of Astrology by ... al-Bīrūnī*. London: Luzac, 1934.

See also: Astronomical Instruments in the Islamic World – Religion and Science in Islam – Stars in Islamic Science – Lunar Mansions in Islam – Ibn Qutayba – Time – *Zīj* – Māshā'allāh – al-Khwārizmī – *Almagest* – al-Ma'mūn – Ibn al-Haytham – Naṣīr al-Dīn al-Ṭūsī – Precession of the Equinoxes – Ṣā'id al-Andalusī – al-Zarqāllu – Maragha – al-Biṭrūjī – Astrology – Ulugh Beg – al-Bīrūnī – al-Battānī – al-Ṣūfī – Ibn Mu'ādh – Observatories in Islam – Armillary Sphere – Qusṭā ibn Lūqā

ASTRONOMY IN MESOAMERICA Mesoamerica is an area of cultural interaction extending from central Mexico south into Nicaragua. Much of the indigenous knowledge of the peoples in this region did not survive the initial years of violence and disease followed by centuries of less dramatic, but equally destructive, political oppression. Nevertheless, among many indigenous groups, portions of that ancient knowledge is retained today. Added to the knowledge of contemporary peoples are data available through comparison of ethnographic, historical, and archaeological records. Perhaps the most amazing resources are the written records of events which were tied to astronomical phenomena. Some are as much as 2000 years old, from Epi-Olmec, Maya, Mixtec, Zapotec, and Aztec cultures. Although hundreds of distinct ethnic groups exist within Mesoamerica, the presence of the shared calendars and a body of shared astronomical knowledge suggest that communication between these groups reaches back even farther than two millennia.

The set of twenty day names combined with the numbers from one to thirteen (260 day calendar), and the year of eighteen months of twenty days with five added days (365 day calendar) are common throughout the region. The two calendars combine to form a cycle of 52 years, known to more than fifty linguistic groups including languages of the Maya, the Mixe-Zoque, the Zapotec, Mixtec, and Aztec families. These cultures shared ceremonies, mythic traditions, and a common system of knowledge which included the prediction of expected solar and lunar eclipses, observation of the seasonal rising and setting of the Pleiades, and awareness of the periods of Venus, Mars, Jupiter, and Saturn. Dates of significant conjunctions of planets, the moon, and bright stars and constellations were recorded on monuments, from at least as early as AD 156.

The movement of the sun, moon, stars, and planets seems to have been measured in part by observing the important points of rising or setting on the horizon and noting the alignment with hills, mountains, and buildings. Unusual shaped structures such as Building J at Monte Alban and the Caracol at Chichen Itza may have had observation as their primary purpose. The year sign used in Mexican codices is thought by some to represent some sort of sundial used to measure the movement of the sun. Some modern Maya daykeepers have calendar boards on which they mark the passing of the days. The ancient peoples of Mesoamerica also produced mirrors which may have been employed in observation, though such use has not been substantiated.

The degree of precision required to note astronomical phenomena varies greatly depending upon the event. Phases of the moon are easily observed changing daily. Solar events such as the equinoxes and solstices are measurable within a range of several days. First perceptible movement of the outer planets (Mars, Jupiter, and Saturn) from stationary points (the change from forward to retrograde motion and vice versa) may not occur for a week or more. Some planetary conjunctions may be visually notable for weeks, but conjunctions involving the moon last for only a day or two. Heliacal risings and eclipses are measured by hours. An awareness of these variations is necessary to evaluate supposed records of such events.

Scattered throughout Mesoamerica are structures oriented so that the solstices and equinoxes could be observed by persons stationed on their summits or looking through their doors. These include Group E at Uaxactun, buildings from the Puuc region in the northern Yucatan peninsula, the Templo Mayor in Mexico City, and Alta Vista near Zacatecas on the Tropic of Cancer. Temple 22 at Copan, the Governor's Palace at Uxmal, and the Caracol at Chichen Itza were built with architectural features marking the rising and setting points of Venus. At Yaxchilan summer ceremonial activities recorded in hieroglyphic texts on stelae and lintels coincided with the summer solstice.

Cycles composed of fractions of days were counted in multiples of the cycle in order to preserve whole numbers. For example, the lunar sidereal period (the time required for the moon to return to the same position against the background of the stars), was counted in sets of three. This cycle is observed today by the Quiche of Momostenango as an 82 day period equal to three sidereal periods of 27.3 days. This same 82 day cycle, and multiples of it, was recorded by Classic Period Maya in inscriptions at Palenque and Calakmul. The synodic period of the moon (the time required for the moon to return to the same position relative to the sun), 29.53 days, was frequently recorded in Classic Maya inscriptions. The lunar data recorded the number of days since the first visible crescent, the number of the current lunation within a set of six months, the name or patron of the month, and whether a month contained 29 or 30 days.

Although debated since the turn of the century, it now seems certain that the ancient Mesoamericans recognized thirteen zodiacal constellations. Animals and other figures representing the constellations have been identified in the murals of Bonampak, in the sky band on Las Monjas at Chichen, the Acanceh mural in Yucatan, the Maya Dresden, Paris, and Madrid codices, as well as the Borgia, Cospi, the Vaticanus codices, and the colonial Florentine Codex.

The changing positions of Pleiades throughout the year continues to be important in Mesoamerica. Various single stars are named by modern peoples; undoubtedly more were named in the past. Barbara and Dennis Tedlock, for example, have recorded that Alnitak, Rigel, and Saiph in the constellation Orion are known as the *three hearth stones* to modern Quiche.

The undivided segment of the Milky Way is identified by modern Quiche as the white road, and the half of the Milky Way which contains the divided segment is referred to as the road of the underworld. Linda Schele associates various ancient iconographic representations with the changing position of the Milky Way in the night sky. She suggests that when it crosses the ecliptic (the band of the zodiac through which the sun apparently moves in its yearly course), it represents the place of creation, when it rims the horizon the portal to the underworld is overhead, in its east–west direction it becomes the Cosmic Monster, and when it "sinks" from a horizontal to a vertical position relative to the ecliptic it is depicted as a canoe.

Obsidian is thought to be the result of meteors hitting the earth. In numerous cultures throughout Mesoamerica meteorites are called excrement of the stars. Several Maya groups refer to meteorites as the cigar butts of the gods. As in the Old World, comets are generally considered bad omens. Both meteors and comets are illustrated repeatedly in the codices of central Mexico.

The planet Mercury has been associated with the Venus Almanac in the Dresden Codex, and with a passage in the Codex Fejervary of the Borgia group of codices. It was in a line with Jupiter, and Venus near the Pleiades in the western sky at twilight in AD 162, on the Initial Series date on the Tuxtla Statuette. Twelve days later Mercury had reached maximum elongation and was heading downward toward the horizon, and Jupiter and Venus, still near the Pleiades, were less than 1.3 degrees apart. If the two numbers given in the text of that inscription, 4 and 8, signify days — and there is no persuasive evidence that they do — this conjunction may have been recorded.

Venus remains perhaps the most important planet throughout Mesoamerica. Five sidereal periods are equivalent to eight of earth's solar years, a ratio that was recorded frequently. Maximum elongation of Venus as evening and morning star is recorded on the two Long Count dates on La Mojarra Stela 1 from Veracruz (AD 143 and AD 156). The Venus almanac in the Dresden Codex follows Venus through 65 synodical revolutions of 584 days. Heliacal rising after inferior conjunction is repeatedly recorded with war events among the Classic Period Maya.

The second date on La Mojarra Stela 1 is the day of the first rising of Mars in the east at sunset. At Palenque during the reign of Kan Balam, a conjunction of Mars, Jupiter, Saturn, and the moon was recorded on AD July 19, 690. The Dresden Codex contains a 780-day Mars almanac.

The two Long Count dates on the La Mojarra stela both record first stationary points of Jupiter. Conjunctions of Saturn and Jupiter of less than two degrees' separation occur regularly at slightly less than twenty years (also slightly less than a *katun* of 20 periods of 360 days). These conjunctions and attendant rituals are recorded at Palenque and Yaxchilan. Jupiter is associated with several Maya rulers including Kan Balam II at Palenque.

In ancient times, as now, astronomical knowledge was maintained by priest scholars who were distinct from political rulers. Astronomical knowledge was used, however, by political leaders to reinforce and legitimize their authority. Written records include descriptions of the sky on dates which are significant because they are births or period endings. Dates for other events, such as accessions, wars, and heir designations, seem to have been chosen to coincide with upcoming planetary, lunar, and solar phenomena. A few dates seem to have been recorded primarily to commemorate an eclipse or some dramatic planetary conjunction.

In other parts of the world, the invention and spread of writing is associated with the necessity of keeping records of commercial transactions. In Mesoamerica, however, writing was stimulated by, if not invented for, the desire to record and predict astronomical phenoma. Even with the loss of literacy, astronomical knowledge has endured through oral tradition, supported by persistent ceremonial practices.

MARTHA J. MACRI

REFERENCES

Aveni, Anthony. *Skywatchers of Ancient Mexico.* Austin: University of Texas Press, 1980.

Bricker, Victoria R. and Harvey M. Bricker. "The Mars Table in the Dresden Codex". In *Research and Reflections in Archaeology and History: Essays in Honor of Doris Stone.* Ed. E. W. Andrews V. New Orleans: Tulane University, Middle American Research Institute, Publication 57, 1986, pp. 51–80.

Freidel, David, Linda Schele, and Joy Parker. *Maya Cosmos*. New York: Morrow Press, 1993.

Closs, Michael. "Venus and the Maya World: Glyphs, Gods, and Associated Phenomena." In *Tercera Mesa Redonda*. Ed. M. Greene Robertson. Palenque, Chiapas, Mexico: Pre-Columbian Art Research Center, 1979, pp. 147–166.

Justeson, John. "Ancient Maya Ethnoastronomy: An Overview of Hieroglyphic Sources." In *World Archaeoastronomy: Selected Papers from the Second Oxford International Conference on Archaeoastronomy* Held at Merida, Yucatan, Mexico 13–17 January 1986. Ed. Anthony F. Aveni. Cambridge: Cambridge University Press, 1989, pp. 76–129.

Lounsbury, Floyd. "A Palenque King and the Planet Jupiter." In *World Archaeoastronomy: Selected Papers from the Second Oxford International Conference on Archaeoastronomy* Held at Merida, Yucatan, Mexico 13–17 January 1986. Ed. Anthony F. Aveni. Cambridge: Cambridge University Press, 1989, pp. 246–259.

Tedlock, Barbara. *Time and the Highland Maya*. Revised Edition. Albuquerque: University of New Mexico Press, 1992.

See also: Calendar – Long Count

ASTRONOMY IN TIBET Tibetan astronomy is a living form of traditional astronomy, and is the basis of the Tibetan calendar which is used in Tibet and in Tibetan communities in India, etc.

There are four branches of Tibetan astronomical science (*rtsis*). The most important branch is *skar-rtsis* (star calculation) which is based on the *Kālacakra* astronomy of India. Another branch is *dbyaṅs-'char* which is based on Indian divination, *svarodaya*. Another is *nag-rtsis* (black calculation), based on Chinese astrology and natural philosophy, and the lastly introduced branch is *rgya-rtsis* (Chinese calculation), based on the Shixian calendar of China.

Indo–Tibetan Astronomy

The Tibetan *Tripitaka* is a collection of Tibetan translations of Buddhist works, some of which include astronomical information. There is a Tibetan translation of the *Śārdūlakarṇa-avadāna*, which is a Buddhist work in which early Indian astronomy and astrology of the Vedāṅga period, the post-Vedic period before Greek influence, are mentioned. There is also a Tibetan translation of an early Indian astrological text ascribed to Sage Garga. Astronomical knowledge in these texts is from an early period, and not of *skar-rtsis*. The most important texts from an astronomical point of view are the Tibetan translation of the *Kālacakra-tantra* and its commentary *Vimalaprabhā*.

Kālacakra-yāna Buddhism is the last stage of Esoteric Buddhism in India. Its most fundamental text is the *Kālacakra-tantra*. It is not known when and where it was composed. Some people say it was introduced into Tibet in AD 1027, and was introduced into India from Central Asia sixty years before. I believe that it was composed in the eleventh century, because the year 1027 is used as the beginning of the sixty-year cycle (*bṛhaspaticakra*) in the text itself. I also believe that it was composed in India, because it adopts the Hindu system of astronomy without any apparent influence of Chinese or other astronomy.

According to the commentary *Vimalaprabhā*, there was the original text or *Mūlatantra*, where the *Siddhānta* system of astronomy was explained, and the text on which it comments is the abridged text or *Laghu-tantra*, where the *Karaṇa* system of astronomy is explained. The *Mūla-tantra* is not extant, and it is difficult to say whether it actually existed as a whole or not, but some fragments are quoted in the *Vimalaprabhā*. The *Siddhānta* system of astronomy is called *grub-rtsis* in Tibetan, and the *Karaṇa* system is called *byed-rtsis*. In Tibetan astronomy, these two systems are basically the same, and only the length of a year and a month are different. In the *Siddhānta* system, one sidereal year = 365.270645 days and one synodic month = 29.530587 days. In the *Karaṇa* system, a sidereal year = 365.258675 days, while a synodic month = 29.530556 days.

In the Tibetan calendar, there are two intercalary months for 65 ordinary months. This is harmonious with the *Siddhānta* system, but not with the *Karaṇa* system. The *grub-rtsis* is usually followed now.

From about the twelfth century, the *Kālacakra* calendar has been followed in Tibet. In the fourteenth century, a comprehensive treatise of *Kālacakra* astronomy entitled *mKhas-pa-dga'-byed* (AD 1326) was composed by an encyclopaedic scholar Bu-ston Rin-chen-grub (1290–1364). After Bu-ston, lHun-grub-rgya-mtsho wrote the *Pad-dkar-żal-luṅ* (AD 1447), and his system was developed as the Phug school. The most famous work of this school is the *Vaiḍūrya dkar-po* (AD 1683) written by Saṅs-rgyas-rgya-mtsho, who was the regent of the fifth Dalai Lama. Another famous work is the *Ñin-byed-snaṇ-ba* (AD 1714) of Dharmaśrī.

There is another school, mTshur-phu, whose fundamental text is the *Ñer-mkho-bum-bzan* (AD 1732) written by Karma Nes-legs-bstan-'dzin.

Let us use astronomical calculation in the *mKhas-pa-dga'-byed* as a case study. It is one of the earliest treatises of the *skar-rtsis* branch of Tibetan astronomy, and will give a general idea of *skar-rtsis*. As *skar-rtsis* is based on *Kālacakra* astronomy, it is similar to Hindu astronomy. First, mean motions of the planets are calculated, and then the equation of the center and the epicyclic correction are applied. The operation of the equation of the center (Sanskrit: *manda-karman*) is called *dal-ba'i-las* in Tibetan, and the operation of the epicyclic correction (Sanskrit: *śīghra-karman*) is called *myur-ba'i-las*.

Three kinds of days are used. They are *ñin-żag* (Sanskrit: *sāvana-dina*), *tshes-żag* (Sanskrit: *tithi*), and *khyim-żag* (Sanskrit: *saura-dina*). A *ñin-żag* is a civil day measured from sunrise to sunrise. A mean *tshes-żag* is a thirtieth part of a synodic month. The equation of the center of the sun and moon are applied so as to make a *tshes-żag* correspond to the change of twelve degrees of the longitudinal difference between the sun and moon. A mean *khyim-żag* is a 360th part of one year.

The ecliptic is divided into twelve *khyim* (Sanskrit: *rāśi*) or zodiacal signs, and also into twenty-seven *rgyu-skar* (Sanskrit: *nakṣatra*) or lunar mansions. Each day as well as *rgyu-skar* is divided into sixty *chu-tshod* (Sanskrit: *nāḍī*). One *chu-tshod* is further divided into sixty *chu-sran* (Sanskrit: *vināḍī*). One *chu-sran* is divided into six *dbugs* (Sanskrit: *prāṇa*).

The mean motion of the sun and moon is calculated from the following simple formulae, which correspond to the *grub-rtsis* system:

$$\text{length of a } \textit{khyim-żag} = \text{length of a } \textit{tshes-żag} \times \left(1 + \tfrac{2}{65}\right);$$

$$\text{length of a } \textit{tshes-żag} = \text{length of a } \textit{ñin-żag} \times \left(1 - \frac{1 + \frac{1}{707}}{64}\right).$$

The equation of the center of the sun is given for each zodiacal sign. Twelve zodiacal signs are divided into four quadrants. Then 6/135, 4/135, and 1/135 of the mean daily motion of the sun are subtracted from or added to the mean daily motion of the sun in each sign. The variables 6, 4, and 1 are called *dal-rkaṅ* (slow step). The ecliptic is divided into the first half (*rim-pa*) and the second half (*rim-min*). The first as well as the second half is further divided into the first part (*sṅa-rkaṅ*) and the second part (*phyi-rkaṅ*). So, one part consists of three signs. The first point of the first half is the apogee.

This *dal- rkaṅ* is, in fact, the difference between the mean motion and the true motion of the sun during one zodiacal sign's movement of the mean sun in terms of *chu-tshod*. Hence, the maximum equation is the total of the variables, that is eleven *chu-tshod* or 2°26′40″. The solar apogee is located at the first point of Cancer in this system.

One anomalistic month is roughly considered as 28 *tshes-żag*, and a correction is applied to the length of each *tshes-żag*. This correction is called *zla-ba'i-myur-rkaṅ* (fast step of the moon). The word *myur* (fast) shows that it was considered to be the epicyclic correction rather than the equation of center. Since the period of 28 *tshes-żag* is a little longer than the actual anomalistic month, a special correction is also applied so as to diminish the period of 28 *tshes-żag* at the rate of one *tshes-żag* per 3780 *tshes-żag*. So, one anomalistic month becomes about 27.55459 civil days.

One anomalistic month is divided into four quadrants, each of which consists of seven *tshes-żag*. Then 5, 5, 5, 4, 3, 2, and 1 *chu-tshod* are added to or subtracted from the length of each *tshes-żag*. These values were probably originally meant to be the difference between the mean motion and the true motion of the moon during one *tshes-żag* in terms of *chu-tshod*. Since the time interval during which the moon moves the arc of one *chu-tshod* is about 1.01 *chu-tshod*, this was considered to be one *chu-tshod*, and the same value was used for the correction of the length of a *tshes-żag*. The maximum equation is the total of the variables, that is 25 *chu-tshod* or 5°23′20″.

Five planets are divided into *żi-ba'i-gza'* which corresponds to inner planets, and *drag-gza'* which corresponds to outer planets. The sidereal period *(dkyil-'khor)* of each planet is given as follows:

Mercury (lhag-pa): 87 days 58 *chu-tshod* 12 *chu-sraṅ*;
Venus (pa-saṅs): 224 days 42 *chu-tshod*;
Mars (*mig-dmar*): 687 days;
Jupiter (*phur-bu*): 4332 days;
Saturn (*spen-pa*): 10766 days.

Just like the case of the sun, *dal-rkaṅ* (slow step) for each zodiacal sign is given for each planet. The mean daily motion of each planet in the case of outer planets, or of the sun in the case of inner planets, is corrected as follows.

Corrected daily motion = $A \mp A(d/135)$, where A is the mean daily motion, and d is *dal-rkaṅ*. The value of *dal-rkaṅ* for each planet is:

Mars: 25, 18, and 7.
Mercury: 10, 7, and 3.
Jupiter: 11, 9, and 3.
Venus: 5, 4, and 1.
Saturn: 22, 15, and 6.

The total of the value of *dal-rkaṅ* is the maximum equation in terms of *chu-tshod*. The maximum equation of each planet is:

Mars: 50 *chu-tshod* or 11° 6′ 40″,
Mercury: 20 *chu-tshod* or 4° 26′ 40″,
Jupiter: 23 *chu-tshod* or 5° 6′ 40″,
Venus: 10 *chu-tshod* or 2° 13′ 20″,
Saturn: 43 *chu-tshod* or 9° 33′ 20″.

The longitude of the apogee of each planet in this system is Mars: 126°40′, Mercury: 220°, Jupiter: 160°, Venus: 80°, and Saturn: 240°.

The "parameter of step" (*rkaṅ-'dzin*) is used to count steps of epicyclic correction. The period of sixty *chu-tshod*'s change of the "parameter of step" is considered one step.

Table 1: The values of *myur-rkaṅ* for each planet

Planet	First part	Second part
Mars	24, 23, 23, 23, 21, 21, 18, 15, 11, and 3	11, 38, 80 and 53
Mercury	16, 16, 15, 14, 13, 11, 7, 5, and 0	4, 11, 20, 28 and 34
Jupiter	10, 10, 9, 8, 6, 6, 2, and 1	3, 6, 9, 11, 16, and 7
Venus	25, 25, 25, 24, 24, 22, 22, 18, 15, and 8	6, 30, 99, and 73
Saturn	6, 5, 5, 4, 4, 2, 2, and 0	2, 4, 5, 6, 8, and 3

Sixty *chu-tshod* correspond to one lunar mansion, and there are 27 lunar mansions, so one cycle consists of 1620 *chu-tshod*. One cycle is divided into two halves, and each consists of fourteen steps. The fourteenth step of the first half and the first step of the second half consist of thirty *chu-tshod* only.

In the case of the outer planets, daily motion of the "parameter of step" is the mean daily motion of the sun minus the daily motion of the planet which has been corrected by its equation of center. In the case of the inner planets, the daily motion of the "parameter of step" is the daily motion of "parameter of fast step" (*myur-rkaṅ-'dzin*) minus the true daily motion of the sun. The "parameter of fast step" is, in fact, the daily motion of the planet's revolution, because it is defined as the quotient of 1620 *chu-tshod* divided by the planet's sidereal period. The variable of the epicycle correction is given as *myur-rkaṅ* (fast step) for each step.

The method of the correction is as follows: Let M be the true daily motion of the planet, D the daily motion of the planet in the case of an outer planet and the daily motion of the sun in the case of an inner planet, both of which have been corrected by the equation of center of the planet itself, K the daily motion of the "parameter of step", and m the *myur-rkaṅ*. Then, for the steps from the first step to the thirteenth step of the first half and from the second step to the fourteenth step of the second half, the following equation gives the true daily motion of the planet

$$M = D \pm K\frac{m}{60}.$$

For the fourteenth step of the first half and the first step of the second half, the following equation is used

$$M = D \pm K\frac{m}{30}.$$

The first half (*rim-pa*) as well as the second half (*rim-min*) are further divided into the first part (*sṅa-rkaṅ*) and the second part (*phyi-rkaṅ*). The correction is plus in the first part of the first half and the second part of the second half, and minus in the second part of the first half and the first part of the second half.

The values of *myur-rkaṅ* for each planet are shown in Table 1. The values are arranged for the first half. In the second half, the same value is used in reverse order.

The total of the value of *myur-rkaṅ* in one part is the maximum epicyclic correction in terms of *chu-tshod*. The maximum correction of each planet is:

Mars: 182 *chu-tshod* or 40° 26′ 40″,
Mercury: 97 *chu-tshod* or 21° 33′ 20″,
Jupiter: 52 *chu-tshod* or 11° 33′ 20″,
Venus: 208 *chu-tshod* or 46° 13′ 20″,
Saturn: 28 *chu-tshod* or 6° 13′ 20″.

In a paper written in 1986, I compared the astronomical constants used by Bu-ston with those of some schools of Hindu astronomy, and pointed out that they are close to those of the Ārdharātrika school of Hindu astronomy.

Sino–Tibetan Astronomical Science

The *nag-rtsis* was said to be introduced into Tibet from China in the seventh century. It is based on Chinese astrology and natural philosophy. According to the *Zla-ba'i-'od-zer* (AD 1684) of Dharmaśrī, a popular work of *nag-rtsis*, the most fundamental elements of the *nag-rtsis* are as follows.

(1) *Khams* (also called *'byuṅ-ba*), which are the five elements of Chinese natural philosophy: wood, fire, earth, metal, and water.

(2) *Lo-'gros*, which are twelve animals used to name each year of a twelve-year cycle: rat, ox, tiger, rabbit, dragon, snake, horse, sheep, monkey, bird, dog, and boar. This twelve-year system is the Chinese system, which is widely used in East Asia. The combination of the *khams* and *lo-'gros* is used to name each year of a sixty-year cycle. This is also the Chinese system.

(3) *Sme-ba*, which is the Chinese "nine stars" used for astrological purposes.

(4) *sPar-kha*, which is eight symbols of Chinese natural philosophy, of which the most fundamental text is the famous *Yijing* (I Ching, Book of Changes).

(5) *Zla-ba*, which is twelve months for each of which twelve animals are attributed. The first month of spring is a tiger, and so on.

(6) *Tshes*, which is the date of the month.

(7) *Dus-tshod*, which is a twelfth part of a day, for each of which twelve animals are attributed. The midnight is rat, and so on.

(8) *gZa'*, which is eight planets: the sun, moon, five planets, and *rāhu*. The *rāhu* (dragon's head, or the ascending node of the lunar orbit) is not Chinese, but of Indian origin.

(9) *sKar-ma*, which is the Chinese 28 lunar mansions or lodges.

The *rgya-rtsis* is based on the Tibetan version (AD 1725) of the Mongolian translation (AD 1711) of the Chinese astronomical work *Xiyang xinfa suanshu* (AD 1669) compiled in the Qing dinasty, which is the theoretical text on the Shixian calendar. The Shixian calendar is the last luni-solar calendar in China.

The traditional Tibetan calendar, which is based on the Phug school of *skar-rtsis*, is still used by Tibetan people. Also, several Tibetan astronomical texts are extant, and the process of astronomical calculation is explained in detail in these texts. More extensive study of Tibetan astronomy by historians of astronomy will be fruitful. In 1987, Chinese scholars Huang Mingxin and Chen Jiujin published a *skar-rtsis* text of epoch AD 1827, the *Rigs-ldan-sñiṅ-thig* of Phyag-mdzod gsuṅ-rab, with Chinese translation and astronomical commentary. This is a good introduction to Tibetan astronomy.

YUKIO ŌHASHI

REFERENCES

Primary sources: Tibetan classical texts

Bu-ston Rin-chen-grub. *Bu-ston skar-rtsis phyogs-bsgrigs*. Lanzhou: Gansu minzu chubanshe, 1992.

Bu-ston Rin-chen-grub. *rTsis-kyi bstan-bcos mkhas-pa dga'-byed*. Xizang: Xizang renmin chubanshe, 1987.

Chandra, Lokesh, ed. *The Collected Works of Bu-ston*, part 1. New Delhi: International Academy of Indian Culture, 1965.

Dharma-śrī. *rTsis-kyi man-ṅag ñin-byed snaṅ-ba'i rnam-'grel gser-gyi śiṅ-rta*. Xizang: Xizang renmin chubanshe, 1983.

Dharma-śrī. *rTsis-kyi man-ṅag ñin-mor byed-pa'i snaṅ-ba*. Xizang: Xizang renmin chubanshe, 1990.

Karma Ṅes-legs-bstan-'dzin. *rTsis-kyi lag-len ñer-mkho'i bum-bzaṅ*. New Delhi: T.G. Dhongthog Rinpoche, 1977.

mKhyen-rab-nor-bu. *rTsis-kyi man-ṅag rigs-ldan sñiṅ-thig*. Leh: O-rgyan-rnam-rgyal, 1976.

Saṅs-rgyas-rgya-mtsho. *Baiḍūr-dkar-po*, 2 vols. Dehra Dun: Sakya Centre, 1978.

Saṅs-rgyas-rgya-mtsho. *Vaiḍūrya-g.ya'-sel*, 2 vols. New Delhi: T.Tsepal Taikhang, 1971.

sMin-gliṅ Lo-chen Chos-dpal. *'Byuṅ-rtsis man-ṅag zla-ba'i 'od-zer*. Dharamsala: Bod-gźuṅ chos-don lhan-khaṅ, 1968.

sMin-gliṅ Lo-chen Dharmaśrī. *rTsis-kyi man-ṅag ñin-mor byed-pa'i snaṅ-ba*. Leh: Dondup Tashi, 1976.

Primary sources: Tibetan modern texts

Bod-ljoṅs groṅ-khyer lHa-sa sman-rtsis-khaṅ-gi *Bod-gi lo-tho* rtsom-sgrig-khaṅ. *Bod-ljoṅs skar-rtsis gnam-rig-gi dus-rtags gnam-dpyad*. Xizang: Xizang renmin chubanshe, 1978.

bSam-'grub-rgya-mtsho. *sKar-rtsis kun-'dus nor-bu*, 2 vols. Qinghai: Zhongguo zangxue chubanshe, 1990.

Tshe-tan-źabs-druṅ. *Mi'u-thuṅ sgyu-ma'i gom-'gros*. Xining: Qinghai minzu chubanshe, 1985.

Tshe-tan-źabs-druṅ. *rGya-rtsis-gi-don*. Lanzhou: Gansu minzu chubanshe, 1980.

Secondary sources

Huang, Mingxin and Chen, Jiujin. (Tibetan title:) *Bod-kyi rtsis-rig-gi go-don daṅ lag-len*; (Chinese title:) *Zangli de yuanli yu shijian*. (The Principle and Practice of Tibetan Calendar, in Tibetan and Chinese). Beijing: Minzu chubanshe, 1987.

Ōhashi, Yukio. "Putun no tenmon-rekihō." (Mathematical Astronomy of Bu-ston, in Japanese). In *Chibetto no Bukkyō to Shakai*. Ed. Yamaguchi Zuiho. Tokyo: Shunjusha, 1986, pp. 629–646.

Petri, Winifred. "Tibetan Astronomy." *Vistas in Astronomy* 9: 159–164, 1968.

Schuh, Dieter. *Untersuchungen zur Geschichte der Tibetischen Kalenderrechnung*. Wiesbaden: Franz Steiner Verlag, 1973.

Tshul-khrims-chos-sbyor. *sKar-nag-rtsis-kyi lo-rgyus phyogs-bsdud mu-tig phreṅ-ba* (A Concise History of *skar-rtsis* and *nag-rtsis*, in Tibetan). Lhasa: Xizang-renmin-chubanshe, 1983.

Yamaguchi, Zuihō. "Chibetto no rekigaku" (Tibetan Calendrical Science, in Japanese). *Suzuki-gakujutsu-zaidan-kenkyu-nenpo* 10: 77–94, 1973.

See also: Lunar Lodges – Calendars

ATOMISM IN ISLAMIC THOUGHT Atomism, the view that there are discrete irreducible elements of finite spatial or temporal space, played a significant role in Islamic intellectual history. It was upheld by most practitioners of the uniquely Islamic discipline of *kalām*. However, some practitioners of *kalām* (i.e. *mutakallimūn*) as well as all but one of the practitioners of *falsafa* (i.e. *falāsifa* — those engaged in the Neoplatonized Peripatetic philosophy of medieval Islam) were anti-atomists. There was thus a lively debate between atomists and anti-atomists, regarding not only matter theory, but also other areas of natural philosophy and cosmology, particularly theories of space, time, void, motion, and causality.

Kalām has no counterpart in the Western tradition. Even though primarily theological in orientation, it is not equivalent to theology. The subject matter of *kalām* includes not only theological topics, e.g. the nature and attributes of God, prophecy, and revelation, but also philosophical problems of cosmology, logic, anthropology, psychology, etc. The origins of *kalām* are obscure and a subject of debate. Suffice it to say that *kalām* arose in the mid-eighth century, and that during the later half of that century, questions about the nature and attributes of objects were being discussed. In their discussion of such questions among themselves and with others from the various religious and intellectual traditions of the Hellenized Near East, the *mutakallimūn* had access to views and theories propounded by the intellectual, doctrinal, and sectarian movements of Late Antiquity: Neoplatonism, Stoicism, Manicheanism, Dualism, Bārdaiṣānism, etc. Little is known about the *mutakallimūn*'s manner of access to these views. It was probably oral and through personal contact. Such a transmission is in sharp contrast with the large-scale translation of Greek philosophical and scientific texts during the late eighth and ninth centuries which gave rise to *falsafa*.

The *mutakallimūn* of the late eighth century held three theories of matter and its attributes. In the first, bodies are the only constituents of the world. All secondary qualities like sound, taste, color, etc. are thus corporeal. It follows that perceptible objects consist of several interpenetrating bodies. This view, whose origins are Stoic, was held by some Dualists. Its early *kalām* subscribers were Ḥishām ibn al-Ḥakam (d. ca. 795), and al-Aṣamm (d. 815). Later, the anti-atomist Ibrāhīm ibn Sayyār al-Naẓẓām (d. 835–845) advocated it, albeit holding that motion was the sole accident.

In the second theory, unextended accidents are the only constituents of the world. Extended bodies result from a combination of accidents, namely, color, taste, hot/cold, rough/smooth. The origins of this view lie in Neoplatonism and Christian theology. Its *kalām* subscribers were Ḍirār ibn ʿAmr (d. 815), Ḥafṣ al-Fard (fl. 810), and Ḥusayn al-Najjār (d. ca. 835–845).

The third theory holds that accidents and bodies constitute the world, and that bodies are constituted from atoms. This view had its partisans among the Dualists and the Bārdaiṣānites. It was appropriated into *kalām* by Abū al-Hudhayl al-ʿAllāf (d. 841) of the Basrian Muʿtazilī school of *kalām*; the Baghdadī Muʿtazilī Bishr ibn al-Muʿtamir (d. 825–840); and Muʿammar ibn ʿAbbād al-Sulamī (d. 830). Towards the mid-ninth century, the atomic theory displaced its rivals to become the dominant physical theory of *kalām*. Atomism was upheld by the Ashʿarī *kalām* school formed in the tenth century in opposition to Muʿtazilīs, particularly regarding questions of man's free will and God's absolute power. However, atomism was attacked by the *falāsifa* who upheld Aristotelian arguments. In the eleventh century, the Basrian Muʿtazilī Abū al-Ḥusayn al-Baṣrī (d. 1044) embraced *falsafa* physical theory and abandoned atomism. Atomism declined further in the twelfth and later centuries with the growing influence of Ibn Sīnā's (d. 1037) philosophy. Even though atomism was never actually abandoned, it was no longer central to post-twelfth century *kalām*.

As none of the writings on physical theory of the eighth and ninth century *mutakallimūn* has survived, their theories must be reconstructed from extant fragments. The principal source has been the doxography, *Maqālāt al-islāmīyīn* (The Doctrines of Muslims) by the former Muʿtazilī and founder of Ashʿarī *kalām*, Abū al-Ḥasan al-Ashʿarī (d. 935). The following account of early atomism may be drawn: Early *kalām* atomists distinguished between the atom (denoted by *jawhar* (atom), *juz*ʾ (part), *al-juzʾalladhī lā yatajazza*ʾ (the indivisible part) and the body. The body has length, breadth, and depth, while the atom lacks these dimensions (however al-Ṣāliḥī [fl. end of ninth/early tenth century] held that the atom was a body). Rather, they believed that dimensions are produced by combinations of atoms. Hence, a minimal length arises when two atoms combine (or, in the formulation of the *mutakallimūn*, a line is formed by the combination of two atoms). There were different views on the minimal number of atoms which constitute a body having length, breadth, and depth: Abū al-Hudhayl held that it was six; Muʿammar held that eight were required; and Abū al-Qāsim al-Balkhī (d. 931) held that four sufficed.

There are obvious parallels between *kalām* and Greek atomism regarding proofs for the existence of atoms, as well as terms which denote the atom. Yet the concept of dimensionless atoms combining to form bodies with dimension is unique to *kalām*. New light has been shed on this puzzling view, and on *kalām* physical theory in general, by the rediscovery of eleventh century sources. Of particular importance are the Basrian Muʿtazilī texts of Ibn Mattawayh and Abū Rashīd al-Nīsābūrī (both fl. first half of eleventh century), as well as Ashʿarī texts by Abū al-Maʿalī al-Juwaynī (d. 1085) and Ibn Fūrak (d. 1015). These sources reveal that, in his reformulation of *kalām*, the Basrian Muʿtazilī Abū Hāshim al-Jubbāʾī (d. 933) had redefined the atom as "that which occupies space (*mutaḥayyiz*)", a designation which was also applied to the body. From this designation, as well as arguments advanced in support of the theses that the atom has magnitude (*miṣāḥa*) and its shape resembles a cube, it is clear that the atom must somehow be extended (the Ashʿarī *mutakallim* Abū Bakr al-Fūrakī (d. 1085) states: the atom is the smallest of what is small with respect to volume). Paradoxically, these *mutakallimūn* insisted that despite its magnitude the atom lacked length, breadth, and depth. Like their earlier colleagues, they continued to hold that dimensions were produced by combinations of atoms. Moreover, they considered any difference between their view of the atom and the earlier view as marginal; it was partially conceptual but partially a result of the manner of expression.

These texts suggest the interpretation that a geometry of discrete space underlies *kalām* atomism (as in Epicurean atomism). In ancient and medieval thought, any distinction between geometrical and physical space was inconceivable. If physical space was continuous, then so was geometrical space. Likewise, discrete physical space meant discrete geometrical space. Hence, *kalām* formulations of 'indivisible', 'dimension', 'magnitude', and 'body' need to be analyzed within discrete geometry. Here, a point, which is defined as that which has no parts, is equivalent to the indivisible magnitude (i.e. atom). Next, a line, which is terminated by two end points, must consist of at least two indivisibles. It follows that two indivisibles constitute the least line and are the least to constitute the dimension of length. The *kalām* atom cannot, thus, have length, breadth, or depth, but it has magnitude for it is an indivisible of discrete (and not continuous) geometry. The combination of atoms to form linear dimensions is no longer problematical; unlike points of continuous geometry which lack both dimension and magnitude, atoms/indivisibles of discrete geometry have minimal magnitude yet lack dimension. Such atoms combine to form objects with larger magnitudes and dimension. In a discrete geometry whose indivisibles are square-shaped (as seen from a continuous geometry, for the *mutakallimūn* state that the atom *resembles* a square; having no dimensions it cannot be a square!), one may configure minimal bodies from four, six, or eight atoms.

In his critique of atomism, Aristotle argued that atomism entails indivisible parts of space, time, and motion. Accordingly, most eighth and ninth century atomist *mutakallimūn*, and many later *mutakallimūn* upheld the minimal parts of space, time, and motion. In the tenth century, Abū Hāshim al-Jubbā'ī abandoned the minimal parts of space (and consequently minimal parts of time and motion) in response to difficulties raised by anti-atomists. His successors did not all adopt this view; some continued to say that space, time, motion and matter were constituted out of indivisibles. We may also note that the atomist *mutakallimūn* upheld the existence of void spaces.

Atomism posed several conceptual and geometrical difficulties, some of which are traceable to Aristotelian arguments against atomism. It may be relevant to mention some Islamic contributions. One argument formulated by Abū al-Hudhayl, which is based on Zeno's dichotomy paradox, was the ant and sandal argument. Imagine an ant creeping over a sandal. In order to traverse the sandal, the ant must first traverse half the sandal; but to traverse half, the ant must first traverse half of this, and so on. Hence the traversal cannot commence unless the division terminates at an indivisible (i.e. atom). Abū al-Hudhayl's student, the antiatomist al-Naẓẓām, responded with his theory of leap (*ṭafra*), saying the ant does not traverse through all points on the path of traversal, but it traverses through some and leaps over others. Hence, al-Naẓẓām claimed, one may traverse from one location to another without traversing all intervening points. The theory of leaps played an important role in discussions of physical theory, if only to illustrate the absurdity of the actually infinite division of matter which was attributed to al-Naẓẓām. Al-Naẓẓām also formulated a clever argument against atomism. Imagine a rotating millstone. Since both an inner circle and the millstone's circumference must complete a rotation in equal time, when an inner circle is ten atoms in length, and the circumference is a hundred atoms, for each unit of space traversed by an atom on the inner circle, an atom on the circumference would have to traverse ten units. Explaining this, al-Naẓẓām resorted to his theory of leaps: When an atom on the inner circle moves one unit, an atom on the circumference moves one unit and leaps nine units. Abū al-Hudhayl, however, responded that the atom on the inner circle moves for one time unit and rests for nine units, while the atom on the circumference moves for all ten units. Al-Naẓẓām objected that this entails that particles of a solid body cannot adhere to each other but must be set loose to allow for such moments of motion and rest. Hence, a rotating solid body must disintegrate (*tafakkuk*) and its internal configuration of atoms be modified. The atomist *mutakallimūn*, unable to answer al-Naẓẓām's challenge, accepted the internal disintegration of a rotating body. However, they considered it to be analogous to a salt shaker where salt particles move freely within the confines of the shaker.

Why did the *mutakallimūn* embrace atomism? This question has puzzled researchers, particularly given the difficulties atomism raised. The thesis that atomism was theologically more acceptable than the continuously divisible matter theory of the *falāsifa* has been widely accepted. Continuous divisibility raises problems of inifinity, and the *mutakallimūn* were mindful of the relationship between ending infinite regress in the argument for the temporal creation of the world and the argument for the divisibility of matter. Yet the question still remains as to why three theories of matter were considered theologically sound in early *kalām*, and why early atomists did not claim that their theory was, theologically, the most sound. The assertion that anti-atomists were heretics is only found after the tenth century when atomism had triumphed. The question of the affinity of atomism to occasionalism has also been raised, but this needs to be reexamined in the light of new sources.

Atomism was also upheld by the famous physician and *faylasūf* Abū Bakr Muḥammad ibn Zakariyāʾ al-Rāzī (d. 925). However, surviving accounts are scanty. We only know that his atoms were extended (like Democritus' atoms). Atomism is part of al-Rāzī's cosmology of the five eternals: God, Soul, Space, Time, and Matter upheld by the Ṣābians of Ḥarrān. Some of al-Rāzī's views may derive from Irānshahrī (fl. late ninth century) about whom very little is known.

ALNOOR DHANANI

REFERENCES

Baffioni, Carmela. *Atomismo e Antiatomismo nel Pensiero Islamico*. Naples: Instituto Universitario Orientale, 1982.

Dhanani, Alnoor. *The Physical Theory of Kalām: Atoms, Space, and Void in Basrian Muʿtazilī Cosmology*. Leiden: Brill, 1994.

Frank, Richard. "Bodies and Atoms: The Ashʿarite Analysis." In *Islamic Theology and Philosophy: Studies in Honor of George F. Hourani*. Ed. Michael E. Marmura. Albany: State University of New York Press, 1984, pp. 39–53.

Pines, Shlomo. *Beiträge zur islamischen Atomenlehre*. Berlin: Heine, 1936.

Sorabji, Richard. *Time, Creation, and the Continuum: Theories in Antiquity and the Early Middle Ages*. Ithaca: Cornell University Press, 1983.

Sorabji, Richard. *Matter, Space, and Motion: Theories in Antiquity and their Sequel*. Ithaca: Cornell University Press, 1988.

van Ess, Joseph. *Theology and Science: The case of Abū Isḥāq al-Naẓẓām*. Ann Arbor: Center for Near Eastern and North African Studies, University of Michigan, 1978.

van Ess, Joseph. *Theologie und Gesellschaft im 2. und 3. Jahrhundert Hidschra*. 6 vols to date. Berlin and New York: Walter De Gruyter, 1991.

White, Michael. *The Continuous and the Discrete: Ancient Physical Theories from a Contemporary Perspective*. Oxford: Oxford University Press, 1992.

Wolfson, Harry A. *The Philosophy of the Kalam*. Cambridge: Harvard University Press, 1977.

See also: Ibn Sīnā – al-Rāzī

ĀTREYA Ātreya was the teacher of Agniveśa, renowned as a writer on medicine. He is popularly known as Punarvasu Ātreya or only Punarvasu. The term Ātreya implies a person who is either a disciple or a descendant of Atri, the Vedic sage. On the basis of his teachings, Agniveśa and five other disciples, Bhela, Jatukarṇa, Parāśara, Kṣārapāṇi and Hārita, composed separate works. Agniveśa's work is still available today. *Bhela-saṃhitā* is available only in a mutilated form. *Hārita-saṃhitā*, which is now in print, appears to be the work of a later author who was also known as Hārita. Works of the other three disciples, though quoted profusely by later commentators, are no longer extant.

In the past, there were several other medical authorities with the term Ātreya suffixed to their names. For example, there is Bhikṣu-Ātreya and Kṛṣṇa-Ātreya. Another Ātreya was the head of the Ayurvedic faculty at Taxila University (600 BC), and his disciple Jīvaka was the physician of Lord Buddha. These should not be confused with Punarvasu Ātreya.

Since Agniveśa's work was reprinted in a period prior to 600 BC, he and his teacher Ātreya must have flourished much before this time. The *Bhela-saṃhitā* states that Candrabhāgā was the mother of Punarvasu Ātreya, and he was the physician of King Nagnajit of Gāndhāra. This king is mentioned in *Śatapatha-Brāhmaṇa* and *Aitareya-Brāhmaṇa* which were composed prior to 3000 BC. Therefore, Punarvasu Ātreya must have flourished during or prior to this period.

Though he was the court physician of Gandhāra, evidence in *Caraka-saṃhitā* indicates that he traveled with his disciples in Pāñcāla-kṣetra, the forests of Caitraratha, Pañcagaṇga, Dhaneśāyatana, the northern side of the Himalayas, and Triviṣṭap.

Apart from his teachings, which are codified in *Caraka-saṃhitā*, manuscripts of an independent work called *Ātreya-saṃhitā* are also available. This work has four sections and deals with both the theory and practice of Ayurveda in detail.

BHAGWAN DASH

REFERENCES

Agniveśa. *Carakasaṃhitā*. Bombay: Nirnayasagara Press, 1941.

Bhela. *Bhelasaṃhitā*. Delhi: Central Council for Research in Indian Medicine, 1977.

Mukhopadhyaya, Girindranath. *History of Indian Medicine*. New Delhi: Oriental Books Reprint Corporation, 1974.

Śarmā, Priyavrata. *Āyurveda kā vaijñānika itihāsa*. Varanasi: Caukhambha Oriyantaliya, 1981.

Vāgbhaṭa. *Aṣṭāṅgahṛdaya*. Varanasi: Krishnadas Academy, 1982.

Vṛddhajīvaka. *Kāśyapasaṃhitā*. Varanasi: Chaukhambha Sanskrit Series Office, 1953.

AZTEC SCIENCE The name Aztec most commonly refers to a group of people who dominated the Valley of Mexico, and indeed much of central and southern Mexico, in the fifteenth and early sixteenth centuries. These people, who called themselves Mexica, settled their city of Tenochtitlan (today Mexico City) in the mid-1300s, in the midst of a large number of already-settled cities. The Mexica, as hunter and gatherer immigrants from the northern deserts, were latecomers to the Valley of Mexico and combined their nomadic-style culture with the ways of life of long-settled villagers and urban dwellers. In the fifteenth and early sixteenth centuries numerous different ethnic groups coexisted in the Valley of Mexico; these groups were politically organized in city-states and exhibited emblems of their specific cultural identities (such as patron gods, clothing styles, and distinctive dialects). Three of these groups, the Mexica of Tenochtitlan, the Acolhua of Texcoco, and the Tepaneca of Tlacopan, joined in a military alliance which conquered much of Mexico and created the Aztec empire.

Mexica society was hierarchical and highly specialized. In general, pronounced distinctions between nobles and commoners and between members of different ethnic groups also meant differences in access to certain scientific knowledge and specialized training. Noble boys were educated in priestly schools where the curriculum included literacy skills, astronomically-based calendrics, and the learning of histories, orations, and songs. Some highly-placed individuals, such as the sixteenth-century Acolhua king of Texcoco, Nezahualpilli, actively pursued scientific knowledge and were renowned as philosophers. This particular king spent endless nocturnal hours on his palace roof studying the movements of heavenly bodies, and consulted extensively with other learned individuals, undoubtedly priestly specialists. Medicinal knowledge and skills were in the hands of highly trained physicians, who appear in the documents as men or women skilled in herbal remedies and treatments of injuries and afflictions. Specific forms of scientific knowledge, embedded in industrial arts such as stone or featherworking, were generally the special province of defined ethnic groups, who passed on their craft from parent to child. Engineering and architecture (manifest, for instance, in the creation of buildings, dikes, and the development of elaborate irrigation works) would have required special training, perhaps in an apprentice-style setting. However, the documents are silent on this. Much knowledge of agronomics, as in genetic engineering of food crops, was probably developed by farmers themselves, who slowly but persistently developed increasingly productive strains of maize and other crops. While documentary evidence is scanty, it can be concluded that some scientific knowledge was developed and passed on in a formal, literary context while other knowledge (such as seed selection and midwifery) was maintained in a more informal "folk" realm.

The Mexica were the last of a long succession of complex states and civilizations in central Mexico. As such, they inherited many cultural traditions from prior civilizations, which included a great deal of scientific knowledge. Urban planning had a long history in Mesoamerica, and several cities in central Mexico display a consistency in their alignment since at least Teotihuacan times (ca. AD 1–750). While still a controversial subject, alignments of cities and specific structures within those cities were probably linked to astronomical phenomena. The Mexica were devout admirers of their predecessors the Toltecs (ca. AD 960–1160), to whom they attributed much of their scientific knowledge. This included medicine, geology and mining, astronomy and calendrics, architecture, and fine technical arts (especially featherworking, metalworking, and stoneworking). While the Mexica honored the Toltecs with these inventions, most of these skills and surely much of this knowledge clearly predated the rise of the Toltec civilization in central Mexico.

Mexica scientific knowledge was empirically based, but also closely linked to religious beliefs. To the Mexica and their neighbors, the natural and supernatural worlds shaded into one another, and their practical, scientific inquiries cannot be understood apart from their religious concepts and cultural symbolism. So empirical astronomy was intertwined with astrology; the involved system of calendrics was largely based on prolonged astronomical observations, but the calendars themselves were applied to ritual as well as practical ends. Medicine combined pragmatic remedies with shamanism, divination, and magical cures. Glyphic writing, sculpture, architecture, and the luxury crafts of stone-, feather-, and metal-working all relied on sophisticated and well-honed practical technologies; the resulting works served secular goals and/or displayed a complex religious symbolism. Animals and plants resided in the everyday, visible natural world, but also carried a heavy load of abstract meaning in the less tangible world of mythology and religious symbolism. To the Mexica and their neighbors, then, an empirical, scientific realm of understanding and inquiry was not readily separable from a more abstract, religious realm.

Mexica scientific concepts and knowledge are understood only imperfectly today, mainly due to the paucity of primary source materials on the subject. The ancient peoples of central Mexico, including the Mexica, maintained extensive libraries of pictorial books (codices), but virtually all of these were destroyed during or shortly after the Spanish conquest in 1521. Some of the information contained in these books was reconstructed following the Conquest and set down in written and/or pictorial form, usually under the supervision of Spanish friars. The most famous of these, and one which includes considerable information on astronomical knowledge, natural history, and medicinal practices is *Historia general de las cosas de Nueva España* [The General History of the Things of New Spain (Florentine Codex)], derived from native informants and compiled in the Narhuatl (Aztec) language by the Franciscan friar Bernardino de Sahagūn. Codices based on native glyphic writing traditions were also produced in early colonial times. Some of these contain interesting scientific details, such as images of plants and animals in place name glyphs and a star-gazing priest in the Codex Mendoza; or herbal remedies pictured and described in the Badianus Herbal. Also available, and providing variable enlightenment on native scientific concepts, are Spanish-language histories and descriptions of the Mexica and their neighbors by Spanish secular and religious officials. Among the most revealing of these are the natural histories of Francisco Hernández and Gonzalo Fernández de Oviedo y Valdés, although the latter focuses primarily on areas to the south and east of the Aztec imperial domain. These post-Conquest sources blend native concepts and information with European understandings and conventions, and must be read in that light. Written sources are augmented by the material remains of the people themselves, discovered and interpreted archaeologically. The remains of structures provide clues to architectural and engineering skills; urban layouts suggest detailed understandings of the movements of celestial bodies; artifacts in metal, stone, and feathers (and the tools that produced them) reveal sophisticated industrial technologies; and ancient food remnants, such as corn cobs, demonstrate a steady enhancement of crop productivity through hybridization and selective breeding.

To illustrate the extent and goals of Mexica scientific inquiry, a closer look at astronomy, natural history, and medicine is presented below.

Cosmology, Astronomy and Astrology

Like their forebears and contemporary neighbors, the Mexica were sophisticated observers of astronomical phenomena. Systematic celestial observations and studies were reportedly the domain of the elite. Imperial rulers such as Motecuhzoma (r. Mexica 1502–1520) had duties which included observing star groups in the night sky, as well as carefully following the morning star, Venus. Nezahualpilli, sixteenth-century ruler of neighboring Texcoco, could observe and record (and probably measure) the movements of celestial bodies from the roof of his royal palace. This "observatory" was structured so a man could just lie down and contemplate the night sky through small perforations through which were placed lances with cotton spheres atop. While this description of observing techniques is vague and not entirely clear, it does indicate that careful, rigorous, naked-eye techniques were employed to follow changes in the celestial realm. While rulers may have spent some of their nocturnal hours studying the heavens, this activity was more commonly performed by priests, who spent many waking hours at night. The Codex Mendoza shows one such priest engaged in observing the heavens for the purpose of marking the passage of time. Some temples may have served as "sighting stations" or observatories, from which pairs of crossed sticks could have been aligned to gain accurate lines-of-sight to celestial phenomena. Lines-of-sight were also established between temples and recognizable points on the horizon, and between specific urban structures. The passage of the sun through the seasons could be readily charted in this fashion, with solstices and equinoxes especially marked. For instance, solar equinox sightings in Tenochtitlan were made from the Temple of Quetzalcoatl through an opening between the twin temples of the Great Temple. These various devices and techniques also allowed the ancient Mexicans to track the phases of the moon, record the arrangements of star groups (constellations), follow the movements of the Pleiades, calculate the rotation of Venus, and predict eclipses.

The Mexica conception of the universe and its heavenly bodies was a combination of scientific observations and ideological constructs. Celestial and terrestrial space was united in a hierarchical scheme, with thirteen heavenly layers and nine layers of the underworld. Heavens and underworld were linked by earth, which was counted in each. The world above the earth combined visible phenomena with invisible gods and goddesses; for instance, the moon occupied the layer above the earth, followed by the clouds in the next tier, then the sun, Venus, and the Fire Sticks constellation (perhaps the belt and sword of Orion) in successive levels. It should be kept in mind that certain celestial bodies, especially the sun, moon, and Venus, were not only visible phenomena but were also accorded divine status. Heavenly layers also contained invisible deities, with the male/female creator god at the apex. This whole arrangement is quite different from the heliocentric (sun-centered) and geocentric (earth-centered) concepts developed in the Eastern Hemisphere.

The movements of the sun, stars, moon, and planets had important practical and ritual applications for the Mexica and their neighbors. They were especially concerned with the passage of time, and developed complex and accurate calendars. These included a calendar of 365 days (based on the sun's passage) and a ritual calendar of 260 days (of uncertain derivation). The former was especially used in seasonal agricultural planning, while the latter was the basis for divination and astrological determinations. Combined, these two calendars yielded a 52-year cycle which carried a heavy load of ritual and symbolism. To the Mexica, time was cyclical and repetitive, and much as the individual seasons and years came and went with marked similarity, so also did the 52-year "centuries". However, the fate-oriented

Mexica also saw cycles in the creation and destruction of the universe; legends told of four prior worlds and their destructions, the Mexica living in the "fifth sun". A great deal of ritual and sacrifice was required of the central Mexicans to assure the continuance of this world, which could end at the closing of each 52-year unit. Thus a great deal of cultural interpretation was lent to the systematic visual astronomical observations of priests and rulers.

Similarly, ancient Mexicans frequently oriented entire cities along astronomical lines, especially 15–20 degrees east of north. The basis for this common spatial arrangement is still unclear, although it may relate to the movements of the Pleiades. Still, in orienting their centers along consistent, meaningful lines, the earthly, sacred, and scientific were meshed into a single cultural realm.

Natural History and Ecology

Like the celestial bodies above the earth, the natural phenomena on and in the earth were viewed both scientifically and symbolically. In this close perceived link between natural and supernatural, the creatures of the earth performed a variety of functions in both visible and invisible cultural realms. Birds and beasts, for instance, were valued as providers of food, fur, and feathers, but were also frequent subjects of metaphors and players in myths and legends. However, the roles that creatures played in myths, the meanings they carried in the ritual calendar, or the messages they conveyed in metaphors were based on empirical observations of behavior, life cycles, and anatomy. For example, a human fugitive was likened to a fleet deer, the patting of tortillas compared to the flapping wings of a butterfly, and an eavesdropper described as the little mouse which inhabited every nook and cranny of a house. Close, systematic observations also led to an understanding of the transformations of the hummingbird, which seemingly died during nighttime or winter periods and came to life in the warmth of day or springtime.

Much of the study of wild creatures undoubtedly took place in the animals' natural habitat, so that, for instance, the wiley hunting techniques of the bobcat or the nesting of certain birds in "inaccessible places" are recorded. However, the Mexica also maintained a large zoo and aviary in the city of Tenochtitlan, and such a setting would have provided ample opportunity for observation and study. This was particularly significant for understanding birds and beasts from distant parts of the empire; these were especially valued for their fine feathers or precious pelts.

Habitats of various creatures were meticulously recorded, surely from direct attention to nature: the grey fox was a cave-dweller, the tadpole lived in fresh water among algae and waterlilies, the raccoon preferred forests and crags, and so on. The many parts of nature were understood as interconnected: the American Bittern's enthusiastic nighttime singing predicted heavy rains and abundant lacustrine fish; the Ruddy Duck's evening antics signaled rain at dawn; the song of another bird heralded the onset of frost.

Relationships among the various creatures themselves, including humans, were readily understood by the Mexica, whose hunter-gatherer experience was combined with knowledge acquired in agrarian and urban contexts. Human involvement in ecological dynamics at times had significant consequences. The Mexica were aware of the rarity of certain creatures, such as the wood ibis or certain serpents; indeed, they considered it a bad omen to catch the wood ibis, a cultural constraint which would have served to preserve this rare avian. In another example, the ruler Ahuitzotl (1486–1502) was reportedly responsible for bringing Great-tailed Grackles to Tenochtitlan from eastern lands and ordering their protection. However, these birds thrived to such an extent that they became an ecological burden and lost their popularity. In these contexts, the Mexica were active participants in the continuous ecological drama, engineering changes with often-unexpected consequences.

Medicine and Health

The Mexica were scientific and empirical in their discovery and use of medical cures for a large number of illnesses and injuries. The specialized doctor, or *ticitl*, used practical approaches to his/her profession, examining and diagnosing physical problems and prescribing appropriate cures. Physicians could call on a large pharmacopoeia derived from herbs, roots, animals, and minerals to relieve symptoms, heal injuries, and restore health. They were skilled in soothing burns, setting broken bones, and suturing lacerations, many of these latter injuries undoubtedly suffered in the frequent battles fought in this militaristic society.

Procedures could be multi-staged and quite complicated. Broken bones, for instance, were first set, then splinted with poultices of specified ground roots or herbs. There was some variety in these usages, as the curative ground roots or herbs could be spread on the injury, drunk with pulque, or enjoyed in a bath. A bitten tongue was first subjected to a mixture of chile cooked with salt, followed by a more comforting application of bee or maguey honey. The Mexica considered at least 132 herbs to have curative properties, and these were applied to at least forty ailments other than injuries. Curable ailments included nosebleed, pimples, headache, diarrhea, fatigue, coughs, chest pains, nausea, and many fevers and infections. Some infirmities could be served by a variety of cures — for instance, forty-five different herbs could be used to relieve fevers and eighteen could be applied to festered skin. On the other hand, some remedies had multiple uses — one herb, for instance, could reportedly slow bleeding, inhibit vomiting, relieve side and chest pains, and in general restore a person's strength.

Medicinal aids were often mixed into complex curative potions. Additionally, cures were at times combined with other types of remedies. One of the most popular of these was the sweatbath, often recommended as helpful in childbirth problems, skin festering, and traumas. Practical remedies were also at times joined with divinitory and magical cures; magic and divination were used in both diagnosis and curing of certain illnesses. The Mexica believed that some illnesses could have supernatural causes; these included soul loss and the intrusion of unpleasant supernatural substances.

In another vein, some illnesses required adherence to strict taboos: a patient with a fever was not to eat hot tortillas or chiles, and one with a head wound was to refrain from eating fish or meat. A preventative element is also evident in Mexica medicine. For instance, one designated herb was said to aid digestion and prevent fevers, a nursing mother could prevent diarrhea in her infant by shunning avocados, and stammering or lisping could be avoided by weaning children at a young age. Thus medicine and the maintenance of health in Mexica society were complex matters, relying on a blending of practical, scientific knowledge and strong religious beliefs.

Applications of Mexica Science

What did the Mexica do with the vast amount of knowledge they and their forebears had acquired about the world around them? What was the purpose of the many systematic and rigorous studies they made of the heavens and the earth?

While some priests and rulers may have enjoyed the process of discovery for its own sake, it is clear that scientific knowledge among the Mexica was preeminently geared toward practical and religious applications. Long-term, rigorous observations of the heavens, and the recording of celestial movements, furnished the Mexica with temporal order. These scientific inquiries provided the basis for a sophisticated calendrical system, the ability to make appropriate seasonal preparations, and the capacity to predict and prepare for extraordinary and sometimes fearful events (such as eclipses).

The Mexica were hunter-gatherers long before they entered the Valley of Mexico and there adopted a well-developed agrarian lifestyle. Bolstered by both of these traditions, the Mexica were keen observers of nature and its processes. They understood the anatomy, behavior, life cycles, and ecology of wild plants and animals. They also had a sophisticated knowledge of cultigens, and drew on a long agrarian tradition of seed selection, agricultural technology, and even the ability to predict certain pertinent weather conditions from animal behavior. They applied this extensive and varied knowledge first and foremost to increase and protect their food supply. However, natural resources also enhanced their life conditions by providing fibers, furs, and feathers for clothing and adornment; building materials for domestic and godly shelter; and precious stones and metals for a show of status.

Nature also provided a wealth of medicinal cures, discovered, combined, and applied successfully by specialized physicians. Administering the proper remedy to the specific ailment undoubtedly required considerable experimentation, but unfortunately the documents are silent on this process.

We therefore have quite a bit of information on the Mexica's scientific results, but little understanding of their methods. Those methods, such as rigorous naked-eye astronomy, repeated observations of earthly phenomena, and centuries-long seed selection, yielded a significant body of sophisticated scientific knowledge, applied by the Mexica to cultural ends. The Mexica consistently linked the scientific realm to the religious and utilized empirical discoveries to enhance and embellish their everyday lives.

FRANCES F. BERDAN

REFERENCES

Aveni, Anthony. *Skywatchers of Ancient Mexico.* Austin: University of Texas Press, 1980.

Caso, Alfonso. "Calendrical Systems of Central Mexico." In *Handbook of Middle American Indians,* vol. 10. Ed. Robert Wauchope. Austin: University of Texas Press, 1971, pp. 333–348.

Hernández, Francisco. *Historia Natural de Nueva Espana.* 2 vols. Mexico: Universidad Nacional de Mexico, 1959 (orig. written 1571–1577).

Ortiz de Montellano, Bernard R. *Aztec Medicine, Health, and Nutrition.* New Brunswick: Rutgers University Press, 1990.

Sahagún, Bernardino de. *Florentine Codex: General History of the Things of New Spain.* 12 vols. Trans. and Ed. Arthur J. O. Anderson and Charles E. Dibble. Salt Lake City: University of Utah Press, 1950–1982 (originally written 1570s).

See also: City Planning – Agriculture – Architecture – Calendars – Astronomy – Medicine – Metallurgy – Stonemasonry – Astrology – Magic and Science – Environment and Nature – Religion and Science – Crops

B

BAKHSHĀLĪ MANUSCRIPT The Bakhshālī Manuscript is the name given to the oldest extant manuscript in Indian mathematics. It is so called because it was discovered by a peasant in 1881 at a small village called Bakhshālī, about eighty kilometers northeast of Peshawar (now in Pakistan). It is preserved in the Bodleian Library at Oxford University.

The extant portion of the manuscript consists of seventy fragmentary leaves of birchbark. The original size of a leaf is estimated to be about 17 cm wide and 13.5 cm high.The original order of the leaves can only be conjectured on the bases of rather unsound criteria, such as the logical sequence of contents, the order of the leaves in which they reached A.F.R. Hoernle, who did the first research on the manuscript, physical appearance such as the size, shape, degree of damage, and knots, and the partially preserved serial numbers of mathematical rules (9–11, 13–29, and 50–58).

The script is the earlier type of the Śāradā script, which was in use in the northwestern part of India, namely in Kashmir and the neighboring districts, from the eighth to the twelfth centuries. G.R. Kaye, who succeeded Hoernle, has shown that the writing of the manuscript can be classified into at least two styles, one of which covers about one fifth of the work. There is, however, no definitive reason to think that the present manuscript consists of two different works.

The information contained in the manuscript, the title of which is not known, is a loose compilation of mathematical rules and examples collected from different works. It consists of versified rules, examples, most of which are versified, and prose commentaries on the examples. A rule is followed by an example or examples, and under each one the commentary gives a "statement", "computation", and a "verification" or verifications. The statement is a tabular presentation of the numerical information given in the example, and the computation works out the problem by following, and often citing, the rule step by step.

Thus, the most typical pattern of exposition in the Bakhshālī Manuscript is:

- rule (*sūtra*);
- example (*udāharaṇa*)
  - statement (*nyāsa/sthāpanā*);
  - computation (*karaṇa*);
  - verification(s) (*pratyaya/pratyānayana*).

A decimal place-value notation of numerals with zero (expressed by a dot) is employed in the Bakhshālī Manuscript. The terms for mathematical operations are often abbreviated, especially in tabular presentations of computations. Thus we have: *yu* for *yuta* (increased), *gu* for *guṇa* or *guṇita* (multiplied), *bhā* for *bhājita* (divided) or *bhāgahāra* (divisor or division), *che* for *cheda* (divisor), and *mū* for *mūla* (square root). For subtraction, the Bakhshālī Manuscript puts the symbol, + (similar to the modern symbol for addition), next (right) to the number to be affected. It was originally the initial letter of the word *ṛṇa*, meaning a debt or a negative quantity in the Kuṣāṇa or the Gupta script (employed in the second to the sixth centuries). The same symbol is also used in an old anonymous commentary on Śrīdhara's *Pāṭīgaṇita*, which is uniquely written in the later type of the Śāradā script (after the thirteenth century). Most works on mathematics, on the other hand, put a dot above a negative number.

The problems treated in the extant portion of the Bakhshālī work involve five kinds of equations, namely (1) simple equations with one unknown (fifteen types of problems), (2) systems of linear equations with more than one unknown (fourteen types), (3) quadratic equations (two types, both of which involve an arithmetical progression), (4) indeterminate systems of linear equations (three types, including the so-called Hundred Fowls Problem, in which somebody is to buy one hundred fowls for one hundred monetary units of several kinds, and (5) indeterminate equations of the second degree (two types: $\sqrt{x+a} = u$ and $\sqrt{x-b} = v$, where u and v are rational numbers; and $xy = ax + by$).

The rules of the Bakhshālī work may be classified as follows:

(1) fundamental operations, such as addition and subtraction of negative quantities, addition, multiplication, and division of fractions, reduction of measures, and a root-approximation formula,

$$\sqrt{a^2 + r} \approx a + \frac{r}{2a} - \frac{(r/2a)^2}{2(a + r/2a)};$$

(2) general rules applicable to different kinds of problems: *regula falsi*, rule of inversion, rule of three, proportional distribution, and partial addition and subtraction;

(3) rules for purely numerical problems: simple equations with one unknown, systems of linear equations with more than one unknown, indeterminate equations, and period of an arithmetical progression;

(4) rules for problems of money: equations of properties, wages, earnings, donations, etc., consumption of income and savings, buying and selling, purchase in proportion, purchase of the same number of articles, price of a jewel, prices of living creatures, mutual exchange of commodities, installments, a sales tax paid both in cash and in kind, and a bill of exchange;

(5) rules for problems of travelers: equations of journeys, meeting of two travelers, and a chariot and horses;

(6) rules for problems of impurities of gold; and

(7) rules for geometrical problems: volume of an irregular solid and proportionate division of a triangle.

All the rules of the first category, namely the fundamental operations, occur only as quotations in the computations of examples. Many of the other rules could belong to either *miśraka-vyavahāra* (the practical problems on mixture) or *śreḍhī* (the practical problems of series) in a book of *pāṭī* (algorithms) such as Śrīdhara's *Pāṭīgaṇita* and *Triśatikā* (eighth century), etc., but they have not been arranged according to the ordinary categories of *vyavahāra* (practical problems).

We apparently owe the present manuscript to four types of persons: the authors of the original rules and examples, the compiler, the commentator, and the scribe. Possibly, however, the commentator was the compiler himself, and "the son of Chajaka" (his name is unknown), by whom the Bakhshālī Manuscript, or at least part of it, was "written," was the commentator, or one of the commentators. The colophon to the section that deals exclusively with the *trairāśika* (rule of three) reads:

> This has been written by the son of Chajaka, a brāhmaṇa and king of mathematicians, for the sake of Hasika, son of Vasiṣṭha, in order that it may be used (also) by his descendents.

Immediately before this statement occurs a fragmentary word *-rtikāvati*, which is probably the same as the country of Mārtikāvata mentioned by Varāhamihira (ca. AD 550) among other localities of northwestern India such as Takṣaśilā (Taxila), Gandhāra, etc. (*Bṛhatsaṃhitā* 16.25). It may be the place where the Bakhshālī work was composed.

A style of exposition similar to that of the Bakhshālī work ("statement," etc.) is found in Bhāskara I's commentary (AD 629) on the second chapter called *gaṇita* (mathematics) of the *Āryabhaṭīya* (AD 499). Both Bhāskara I's commentary and the Bakhshālī work attach much importance to the verification; it became obsolete in later times. The unusual word *yāva* (*yāvakaraṇa* in Bhāskara I's commentary) meaning the square power, and the apparently contradictory meanings of the word *karaṇī*, the square number and the square root, occur in both works.

Bhāskara I does not use the symbol *yā* (the initial letter of *yāvattāvat* or "as much as") for unknown numbers in algebraic equations even when it is naturally expected, while he employs the original word *yāvattāvat* itself in the sense of unknown quantities (in his commentary on *Āryabhaṭīya* 2.30). This probably implies that he did not know the symbol. The symbol is, on the other hand, utilized once in the Bakhshālī work in order to reduce the conditions given in an example to a form to which the prescribed rule is easily applicable; after the reduction, the symbol is discarded and the rule is, so to speak, applied mechanically (fol. 54^{v}). This restricted usage of the symbol seems to indicate that the work belongs to a period when the symbol was already invented, but not very popular yet.

There has been quite a bit of dispute over the dates of the manuscript. Hoernle assigned the work to the third or the fourth century AD, Kaye to the twelfth century, Datta to "the early centuries of the Christian era," and Ayyangar to the eighth or the ninth century. The above points suggest that the Bakhshālī work (commentary) was composed not much later than Bhāskara I (the seventh century).

TAKAO HAYASHI

REFERENCES

Ayyangar, A.A.K. "The Bakhshālī Manuscript." *Mathematics Student* 7: 1–16, 1939.

Channabasappa, M.N. "On the Square Root Formula in the Bakhshālī Manuscript." *Indian Journal of History of Science* 2: 112–124, 1976.

Channabasappa, M.N. "The Bakhshālī Square-root Formula and High Speed Computation." *Gaṇita Bhāratī* 1: 25–27, 1979.

Channabasappa, M.N. "Mathematical Terminology Peculiar to the Bakhshālī Manuscript." *Gaṇita Bhāratī* 6: 13–18, 1984.

Datta, B. "The Bakhshālī Mathematics." *Bulletin of the Calcutta Mathematical Society* 21: 1–60, 1929.

Gupta, R.C. "Centenary of Bakhshālī Manuscript's Discovery." *Gaṇita Bhāratī* 3: 103–105, 1981.

Gupta, R.C. "Some Equalijection Problems from the *Bakhshālī Manuscript*." *Indian Journal of History of Science* 21: 51–61, 1986.

Hayashi, T. *The Bakhshālī Manuscript: An Ancient Indian Mathematical Treatise*. Groningen: Egbert Forsten, 1995.

Hoernle, A.F.R. "On the Bakhshālī Manuscript." *Verhandlungen des vii Internationalen Orientalisten Congresses* (Vienna 1886), Arische Section, 1888, pp. 127–147.

Hoernle, A.F.R. "The Bakhshālī Manuscript." *The Indian Antiquary* 17: 33–48 and 275–279, 1888.

Kaye, G.R. "The Bakhshālī Manuscript." *Journal of the Asiatic Society of Bengal* NS 8: 349–361, 1912.

Kaye, G.R. *The Bakhshālī Manuscript: A Study in Medieval Mathematics*. Archaeological Survey of India, New Imperial Series 43. Parts 1 and 2: Calcutta, 1927. Part 3: Delhi, 1933. Reprinted, New Delhi: Cosmo Publications, 1981.

Pingree, D. *Census of the Exact Sciences in Sanskrit*. Series A (in progress), vols. 1–4. Philadelphia: American Philosophical Society, 1970–81.

Pingree, D. *Jyotiḥśāstra: Astral and Mathematical Literature*. Wiesbaden: Harrassowitz, 1981.

Sarkar, R. "The Bakhshālī Manuscript." *Gaṇita Bhāratī* 4: 50–55, 1982.

See also: Zero – Śrīdhara – Bhāskara I

THE BALKHĪ SCHOOL OF ARAB GEOGRAPHERS The Balkhī School refers to a group of four authors who recognize the fact that their geographic work is interlinked. It is also known to scholars as the Classical School of Arab Geography or the Islamatlas. Abū Zaid Aḥmad ibn Sahl al-Balkhī (d. 322/834), who wrote *Ṣuwar al-aqālīm*, is the earliest of them and presumably the originator. The other three authors are Abū Isḥāq Ibrāhīm ibn Muḥammad al-Iṣṭakhrī (ca. AD 950) (*Al-masālik wa'l-mamālik*), Muḥammad ibn Ḥawqal (d. between 350/861 and 360/972) (*Ṣūrat al-arḍ*) and Abu ʿAbdallāh Muḥammad ibn Aḥmad al-Muqaddasī (d. ca. 390/1000) (*Aḥsān al-taqāsim*).

Their work is based on a series of maps covering the Islamic Empire together with a text which consists mainly of notes on the maps. Many copies of these works survive; the earliest surviving manuscript being a version by Ibn Ḥawqal dated AH 479/AD 1086. This is the earliest Arabic manuscript to contain a map. Yet, copies of al-Iṣṭakhrī's book were still being produced as recently as the middle of the nineteenth century AD. There is so much material available that scholars have identified two separate editions of al-Iṣṭakhrī, three of Ibn Ḥawqal and two of al-Muqaddasī, although one of the versions of Ibn Ḥawqal does not contain maps. These different texts can be associated with similar sets of maps, and these maps can be compared and relationships established which enable us to trace the development of "Balkhī" cartography. The standard set of maps consists of a world map, two oceans (Indian and Mediterranean), four Roman provinces (i.e. areas which were originally Byzantine) and fourteen Persian provinces.

It is obvious that the maps are conceived as a set covering the Muslim Empire with reasonable detail, and there is no attempt to cover non-Islamic areas in the same way. It has been suggested that this policy of including only Islamic regions is deliberate. Each map is given a page or so of textual description, and each of these descriptions is planned in such a way that lists of routes, towns, mountains, rivers, etc., are given for each province. Thus they bear a certain resemblance to the work of Ibn Khurdādhbih, although the latter's work was not accompanied by maps nor did he limit himself only to the Dar al-Islam. There is a likelihood that the Balkhī School material and the work of Ibn Khurdādhbih are based on Persian (Sassanid) materials which survived the Islamicization of the Persian home area.

The Balkhi maps cannot be connected together like the sectional maps of al-Idrīsī to form one large map of the known world. Al-Iṣṭakhrī and Ibn Ḥawqal show no interest in projection, scale, or mathematical geography and do not mention latitudes and longitudes at all. The only form of measurement given is that of days' journeys (*marḥala*). The maps are very geometrical in design. Lines are straight or arced, rivers are parallel lines, lakes often perfect circles. Towns can be circles, squares, four pointed stars or something similar. Stopping places on routes resemble small tents or caravanserais.

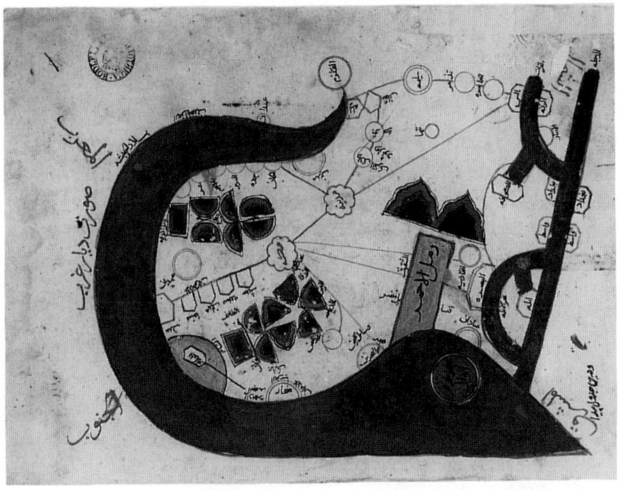

Figure 1: Map of Arabia from al-Iṣṭakhrī's *Ṣuwar al-buldān*. From the Bodleian Library, Oxford. MS Ouseley 373, folio 9v. Used with permission of the library.

al-Muqaddasī's text is based on the same principles but is a considerable improvement over that of his predecessors. He also includes a section on astronomical geography and geodesy including a note on the Greek system of climates. Both Ibn Ḥawqal and al-Muqaddasī are more up-to-date and are more at home in Europe and North Africa, having a preference for the western part of the Empire rather than the Persian speaking areas. al-Muqaddasī's maps however have a closer affinity with those of al-Iṣṭakhrī, whereas we would expect, from the nature of his text, something much more advanced. He has however a different selection of maps, there being no world map nor one of the Caspian Sea and a completely new map of the Arabian desert.

The works of these authors were reproduced continually throughout the centuries not only in Arabic but also in Persian or Turkish translation. Other writers occasionally borrow a selection of the maps or an individual map. Some of the later versions are very corrupt and hardly recognizable. A world map derived from this school appears regularly in the works of Ibn al-Wardī and often in texts of al-Qazwīnī's cosmography showing how popular these maps were in the Muslim world.

GERALD R. TIBBETTS

REFERENCES

Ahmad, S. Maqbul. "Djughrafīya" and "Kharīṭa." In *Encyclopedia of Islam*. Leiden: E.J. Brill, 1960.

al-Muqaddasī. *Aḥsan al-taqāsim fī maʿrifat al-aqālim*. Trans. André Miquel. Damascus: Institut français, 1963.

Dunlop, D.M. "al-Balkhī". In *Encyclopedia of Islam*. Leiden: E.J. Brill, 1960.

Miller, Konrad. *Mappae arabicae: arabische Welt- und Landkarte des 9–13 Jahrhunderts*. 6v. Stuttgart, 1926–31.

Miquel, André. "Ibn Ḥauqal" and "al-Iṣṭakhrī." In *Encyclopedia of Islam*. Leiden: E.J. Brill, 1960.

Tibbetts, G.R. "The Balkhī School of Geographers." In *History of Cartography*. vol. II, book 1. *Cartography in the Traditional Islamic and South Asian Societies*. Ed. J.B. Harley and David Woodward. Chicago: University of Chicago Press, 1992, pp. 108–136.

See also: al-Muqaddasī – Ibn Ḥawqal – Ibn Khurdadhbīh – al-Idrīsī – Maps and Mapmaking in the Islamic World – al-Qazwīnī

BAMBOO Few other plants or materials have been so pleasing to the eye, inspired so many artists, poets and philosophers, or contributed so materially to the development of civilization as bamboo has.

Bamboo (*Arundinaria*) is a grass that grows with great rapidity in a wide variety of climates with over 1500 species world wide. It can be used in a huge variety of ways. It is both immensely strong and very light. It is easily worked with simple tools and is strikingly beautiful both in the economy of its natural form and in its finished state. Yet not only has its key role in the development of science and technology been largely overlooked by Western historians of science, but its contemporary uses tend to be contemptuously undervalued as simple and traditional. It was dismissed as "poor man's timber". The reality is that bamboo has been one of the prime nutrients of Asian culture and may yet turn out to have a significant role to play in civilizations around the world.

The uses of bamboo are so great as to be almost limitless. It has been used to create and preserve knowledge by inspiring metaphysics and geometry and by providing pen, paper, and books. It has sustained life by providing housing, food, and medicine. It has been used to make a profusion of tools and domestic essentials from baskets to scaffolding, cables, and steel reinforcing, from musical instruments to windmills, bridges, and airplanes.

As a raw material for papermaking bamboo has, perhaps, made its greatest contribution in the past, whilst also having the capacity to revolutionize paper production in the future. Bamboo is a super producer yielding two to six times more cellulose per acre than pine which can be cropped annually, but its capacity to provide paper pulp has gone largely unrecognized outside India and China. In India alone more than forty factories make 600,000 tons of paper annually from bamboo. If this use of bamboo were extended world wide, the saving of old growth forest currently used in wood chip production would be greatly reduced. However, it is in its transition from being virtually a free good to being a valued resource that potential problems lie. Demand is already exceeding supply in India, and while government economists in Delhi argue that bamboo can become a cash crop that will improve the lot of the poor, it is possible that the landless peasants will end up being unable to afford even poor man's timber. In Colombia and Costa Rica, bamboo, when coated with concrete in a form of lath and plaster construction, has proved to be a cheap and effective way of building earthquake-proof housing. Similar techniques are under development in the Philippines and Hawaii to use bamboo in cyclone-proof buildings.

Bamboo has proved extremely effective in stabilizing soil and preventing erosion. It also has a great capacity to extract nitrates and phosphates from effluent. These characteristics make it an ideal plant for handling two of the world's most significant problems, soil degradation and water quality. If employed in either or both these ways it can become a virtually free and renewable resource for paper pulp, building and furniture material, steel reinforcing, and a profusion of decorative items. As Western civilization gradually overtakes the developing world, research on the more industrial uses has begun, and it is being used in making parquet flooring and laminated beams. It is to be hoped that, as it becomes a first world resource, it does not become just another opportunity for exploitation. If the developing economies leave the knowledge of its uses and the control of production in the hands of the local people, bamboo has the capacity to benefit all of us.

DAVID TURNBULL

REFERENCES

Farrelly, David. *The Book of Bamboo*. San Francisco: Sierra Club Books, 1984.

BANŪ MŪSĀ The Banū Mūsā were three brothers, Muḥammad, Aḥmad, and al-Ḥasan, who were amongst the most important figures in the intellectual life of Baghdad in the ninth century. We do not know their dates of birth, but Muḥammad died in 873 and could hardly then have been less than 70 years old, because the youngest brother al-Ḥasan was already a brilliant geometrician in the reign of al-Ma᾿mūn (813–833). Their father, Mūsā ibn Shākir, was a noted astronomer and a close companion of al-Ma᾿mūn when the latter was residing at Marw in Khurasan before he became caliph. When Mūsā died, al-Ma᾿mūn became the guardian of his sons, who were given a good education in Baghdad, becoming skilled in geometry, mechanics, music, mathematics, and astronomy.

Under the successors of al-Ma᾿mūn, the brothers became rich and influential. They devoted much of their wealth and energy to the quest for the works of their predecessors, especially in Greek and Syriac, and sent missions to the lands of the Byzantine Empire to seek out manuscripts and bring them to Baghdad. Muḥammad is said to have made a journey to Byzantium in person. The brothers acted as sponsors to a group of scientists and translators, to whom they paid about 500 *dinars* a month. The most outstanding of these scholars were Thābit ibn Qurra and Ḥunayn ibn Isḥāq, who rendered numerous works, many of which would otherwise have been lost, from Greek and Syriac into Arabic. Muḥammad was on friendly terms with this group of scholars, particularly with Ḥunayn, who translated and composed books at the request of his patron. The brothers therefore played a leading role in the transmission of Greek works into Arabic, and in the foundation of the long and important contribution of the Islamic world to the sciences. Some twenty works are attributed to the Banū Mūsā, of which three have survived. The best-known and most important of the brothers' books, which was largely the work of Aḥmad, is their *Kitāb al-ḥiyal* (Book of Ingenious Devices). The work comprises descriptions of some 100 small machines, including alternating fountains, self-filling and self-trimming lamps, and a clamshell grab. About eighty of the devices, however, are trick vessels of various kinds that exhibit an astonishing mastery of automatic controls. The inspiration for Aḥmad's work is undoubtedly to be found in the machine treatises of the Hellenistic writers, particularly the *Pneumatics* of Philo of Byzantium (mid-third century BC) and the *Pneumatics* of Hero of Alexandria (fl. ca. AD 60). *The Kitāb al-ḥiyal*, although some of the devices in it come directly from Philo or Hero, goes well beyond its Greek predecessors, particularly in the use of small pressure variations and conical valves and other components in automatic controls. Indeed, the work of the Banū Mūsā in the variety and ingenuity of their control systems was unsurpassed until quite recent times. There may have been some didactic intention in their writing, but most of their constructions are quite trivial to our eyes. Nevertheless, many of the ideas, techniques, and components that they used were to be of considerable importance in the development of machine technology.

DONALD R. HILL

REFERENCES

The English edition of *Kitāb al-ḥiyal* is: *The Book of Ingenious Devices by the Banū Mūsā*. Translated and annotated by Donald R. Hill. Dordrecht: Reidel, 1979. The Arabic edition is: *Kitāb al-ḥiyal*. Ed. Ahmad Y. al-Hassan. Aleppo: Institute for the History of Arabic Science, 1981.

Hill, Donald R. "Arabic Mechanical Engineering: Survey of the Historical Sources." *Arabic Sciences and Philosophy* 1(2): 167–186, 1991.

See also: Thābit ibn Qurra – Ḥunayn ibn Isḥāq – al-Ma᾿mūn

ABRAHAM BAR ḤIYYA (SAVASORDA) Abraham bar Ḥiyya, also called Savasorda (latinized from the Arabic *ṣāḥib al-shurṭa* = Chief of the guard), flourished in Barcelona, in Christian Spain, but was probably educated in the kingdom (*tā᾿ifa*) of Saragossa, during the period in which it was ruled by the Arabic dynasty of the Banū Hūd. Thus his scientific education could be related to the well known scientific talents of some of the Banū Hūd kings.

Having mastered the Arabic language and culture, he was a pioneer in the use of the Hebrew language in various fields. He wrote on philosophy, ethics, astronomy, astrology, mathematics, and calendrial calculations. He clearly indicated that his Hebrew compositions were written for Jews living in southern France, who were unacquainted with Arabic culture and unable to read Arabic texts.

Two mathematical compositions by Abraham Bar Ḥiyya, and four astronomical ones are known.

• *Yesodey ha-tevuna u-migdal ha-emuna* (The Foundations of Science and the Tower of the Faith) was supposed to be a scientific encyclopedia, of which only the mathematical sections survived. Presumably an adaptation from some unknown Arabic composition, it dealt with basic definitions and knowledge in arithmetic, geometry, and optics.

• *Ḥibbur ha-meshiḥa we ha-īshboret* (The Composition on Geometrical Measures) dealt with practical geometry. This book enjoyed a very large diffusion in medieval Europe in its Latin version, the *Liber embadorum*, translated by Plato of Tivoli (1145), who was assisted by the author himself. The importance of this text for the development of practical geometry in Europe has been noted by ancient and modern scholars.

• *Sefer ṣurat ha-areṣ we-tavnit kaddurey ha-raqi ʿa* (Book on the Form of the Earth and the Figure of Celestial Spheres), together with *Ḥeshbon mahalakhot ha-kokhavim* (Calculations of the Courses of the Stars) and The *Luḥot* (The Astronomical Tables), offered a basic astronomical knowledge founded on Arabic sources such as the works of al-Farghānī and al-Battānī.

• *Sefer ha-'Ibbur* (The Book of Intercalation) dealt with calendrial calculations and aimed "to enable the Jews to observe the Holy Days on the correct dates".

Bar Ḥiyya can rightly be considered the founder of Hebrew scientific culture and language.

TONY LEVY

REFERENCES

"Ḥeshbon mahalakhot ha-kokhavim." In *La obra Sefer Heshbon Mahlekoth ha-Kokabin de R. Abraham bar Ḥiyya ha-Bargeloni*. Ed. Jose-Maria Millas-Vallicrosa. Barcelona: Consejo Superior de Investigaciones Cientificas, 1959.

"Ḥibbur ha-meshiḥa we ha-tishboret." In *Chibbur ha- Meschicha we ha-Tishboret*. Ed. Michael Guttmann. Berlin, 1912.

Millas-Vallicrosa, Jose-Maria. "La obra enciclopedica de R. Abraham bar Ḥiyya." In *Estudios sobre historia de la ciencia espanola*. Barcelona, 1949, pp. 219–62.

Sarfatti, Gad. *Mathematical Terminology in Hebrew Scientific Literature of the Middle Ages*. Jerusalem: The Magnes Press, 1968, pp. 61–129 (in Hebrew).

Steinschneider, Moritz. "Abraham Judaeus, Savasorda und Ibn Ezra." *Zeitschrift für Mathematik und Physik* 12: 1–44, 1867. Reprinted in *Gesammelte Schriften*. Berlin: M. Poppelauer, 1925, pp. 388–406.

"Yesodey ha-tevuna u-migdal ha-emuna." In *La obra enciclopedica Yesode ha-tebuna u-migdal ha-emuna de R. Abraham bar Ḥiyya ha-Bargeloni*. Ed. Jose-Maria Millas-Vallicrosa. Madrid/Barcelona: Consejo Superior de Investigaciones Cientificas, 1952.

See also: al-Battānī

AL-BATTĀNĪ Abū ʿAbd Allāh Muḥammad ibn Jābir ibn Sinān al-Raqqī al-Ḥarrānī al-Ṣābiʾ was an extremely important Islamic astronomer of the ninth to tenth centuries. He was probably born in Ḥarrān before 858, and had Sabian ancestors. He lived most of his life in Raqqa (Syria) where he made most of his observations, but there is also evidence that he visited Baghdad and Antioch.

Apart from a few astrological tracts which have not been studied so far, he compiled (after 901) his *al-Zīj* (astronomical handbook with tables) also called *al-Zīj al-Ṣābi'* (Sabian *Zīj*), a work which marks the stage of full assimilation of Ptolemaic astronomy in Islam. This process had produced its first results ca. 830 with the *zīj*es which were the consequence of the program of observations undertaken in Baghdad and Damascus under the patronage of Caliph al-Maʾmūn. Al-Battānī's *zīj* contains a set of instructions for the use of the numerical tables which have an essentially practical character. We do not find in them careful descriptions of the Ptolemaic models implied in the tables, and the author makes surprising simplifications, such as not describing Ptolemy's model for Mercury or not mentioning the equant point around which the mean motion of the center of the epicycle takes place. Nevertheless he describes, sometimes very carefully, the observations he made in Raqqa between 887 and 918, which allowed him to establish new and more precise mean motion parameters, a new eccentricity (0;2,4,45°) for the Sun and Venus, the longitude of the apogee (82;17°) of these two celestial bodies, a very accurate determination of the obliquity of the ecliptic (23;35°) (the band of the zodiac through which the Sun apparently moves in its yearly course), measurements of the apparent diameters of the Sun and the Moon, and their variation in a solar year and anomalistic month respectively. These new parameters show a clear improvement over those of Ptolemy and led al-Battānī to establish some important corrections on Ptolemaic theory such as the mobility of the solar apogee, the fact that the obliquity of the ecliptic is not a fixed value, and the possibility of solar annular eclipses.

Apart from the *Almagest* and the *Planetary Hypotheses* (used by Battānī to determine the geocentric distances of the planets), Theon's *Handy Tables* constitute a major Ptolemaic influence in the *Zīj*: the planetary equation tables (with the obvious exception of those for the equation of the center of Venus), for example, derive from Theon, and al-Battānī's work constitutes one of the important instruments for the diffusion of the *Handy Tables* during the Middle Ages. The *Zīj* was translated twice into Latin (by Robert of Ketton and Plato of Tivoli) in the twelfth century, as well as into Spanish (thirteenth century) under the patronage of Alfonso X. It influenced strongly the Latin version of the *Alfonsine Tables*, was known in Jewish circles through the summary made in Hebrew by Abraham bar Ḥiyya (d. ca. 1136) and was quoted by European astronomers until the seventeenth century. Al-Battānī died in 929.

JULIO SAMSÓ

REFERENCES

Bossong, Georg. *Los Canones de Albateni. Herausgegeben sowie mit Einleitung, Anmerkungen und Glossar versehen*. Tübingen: Max Niemeyer Verlag, 1978.

Hartner, Willy. "Al-Battānī." In *Dictionary of Scientific Biography*, vol. I. New York: Charles Scribner's Sons, 1970, pp. 507–516.

Kennedy, Edward S. "A Survey of Islamic Astronomical Tables." *Transactions of the American Philosophical Society* (Philadelphia) 46 (2): 32–34, 1956.

Nallino, Carlo A. *Al-Battānī sive Albatenii Opus Astronomicum*. 3 vols. Milan: Publ. Reale Osservatorio di Brera, Milano, 1899–1907. Reprint Frankfurt: Minerva G.M.B.H., 1969.

See also: al-Maʾmūn – Abraham bar Ḥiyya – *Zīj* – *Almagest*

BAUDHĀYANA India's most ancient written works are the four *Vedas*, namely *Ṛgveda*, *Yajur-veda*, *Sāma-veda*, and the *Atharva-veda*. There are different schools which are represented by various *Saṁhitās* or recensions of the *Vedas*. To assist their proper study, there are six *Vedāṅgas* (limbs or part of the *Veda*), namely *Śikṣā* (phonetics), *Kalpa* (ritualistics), *Vyākaraṇa* (grammar), *Nirukta* (etymology), *Chandas* (prosody and metrics), and *Jyotiṣa* (astronomy, including mathematics and astrology). These auxiliary Vedic works (except the last) are written in *sūtra* or aphoristic style.

The *Kalpa Sūtras* deal with the rules and methods for performing Vedic rituals, sacrifices, and ceremonies, and are divided into three categories: *Śrauta*, *Gṛhya*, and *Dharma*. The *Śrauta Sūtras* are more specifically concerned with the sacrificial ritual and allied ecclesiastical matters. They often include tracts which give rules concerning the measurements and constructions of *agnis* (fireplaces), *citis* (mounds or altars), and *vedis* (sacrificial grounds). Such tracts are also found as separate works and are called *Śulba Sūtras* or *Śulbas*. They are the oldest geometrical treatises which represent in coded form the much older and traditional Indian mathematics. The root *śulb* (or *śulv*) means "to measure" or "to mete out".

The names of about a dozen *Śulba Sūtras* are known. The oldest of them is the *Baudhāyana Śulba Sūtra*. It belongs to the *Taittirīya Saṁhitā* of the *Black Yajur-veda*, and is the thirtieth *Praśna* or chapter of the *Baudhāyana Śrauta Sūtra*. The title *Baudhāyana Śulba Sūtra* shows that its author or compiler was Baudhāyana, or perhaps more correctly, it belonged to the school of Bodhāyana. Other Vedic works bearing the same name are known. It is more proper to consider these works as belonging to the Baudhāyana school than to regard them as authored by the same person.

Here we are concerned with Baudhāyana the Śulbakāra (that is, the author of the *Baudhāyana Śulba Sūtra*), or the śulbavid (expert in śulba mathematics and constructions). We do not know his biographical details. Georg Bühler believed that he hailed from the Andhra region, but a recent study by Ram Gopal shows that he probably came from northern India. His dates are also uncertain, being any time between 800 and 400 BC. Taking into account the views of A.B. Keith, W. Caland, David Pingree, and Ram Gopal, he may be placed about 500 BC or earlier. However, it must be noted that much of the material in *Baudhāyana Śulba Sūtra* is traditional and, thus, still older than its date of compilation and coding.

The *Baudhāyana Śulba Sūtra* is not only the earliest but also the most extensive among the *Śulbas*. The subject matter is presented in a systematic and logical manner. Of course, the language is somewhat archaic, and due to the aphoristic style, the rules are highly condensed. The *Baudhāyana Śulba Sūtra* was commented on by Dvārakānātha Yajva (ca. seventeenth century). His Sanskrit commentary called the Ś*ulbadīpikā* was published more than a century ago by G.F.W. Thibaut (1848–1914) in his edition of the *Baudhāyana Śulba Sūtra*. A recent edition (Varanasi, 1979) of the *Baudhāyana Śulba Sūtra* also contains the above commentary as well as another called *Bodhāyana śulbamīmāṁsā*, which was written by Vyaṅkateśvara (or Veṅkateśvara) Dīkṣita who lived during the Vijaya-nagaram kingdom (ca. 1600).

The text of the *Baudhāyana Śulba Sūtra* is divided into three chapters which comprise a total of 272 passages or 519 aphorisms. The main subject is the measurement, construction, and transformation of various altars and fireplaces. The forms of the three obligatory *agnis* (whose tradition was older than even the *Ṛgveda*) were square, circle, and semicircle, but those of optional *citis* involved all sorts of plane figures, including the above three and also the rectangle, rhombus, triangle, trapezium, pentagon, and some complicated shapes.

Some mathematical topics covered in the *Baudhāyana Śulba Sūtra* are a fine approximation for the square root of two and the so-called Pythagorean theorem. In the latter, the sides of a right-angled triangle obey the rule $a^2 + b^2 = c^2$. This rule is generally called the Pythagorean theorem, although it was known in Babylonia and China, as well as in India, earlier than the time of the Greek philosopher and mathematician Pythagoras.

The *Baudhāyana Śulba Sūtra* also contains formulas for circling a square and squaring a circle, and provides rules for basic simple geometric constructions, such as drawing a perpendicular on a given line or drawing the right bisection of a line. Some other elementary plane figures such as the isosceles triangle, trapezium, and rhombus are also covered for construction.

The *Baudhāyana Śulba Sūtra* deals with the measurements and constructions of a large number of fire-altars (*Kāmya agnis*). These were needed as part of rituals performed to attain certain desired objects according to the religious beliefs of the Vedic people. It was essential that the shape, size, area, and orientation of the relevant altar be according to the prescribed instructions. Otherwise, there was a risk of divine wrath.

The standard forms of some of the optional altars were those which resembled certain birds. The most significant and perhaps the oldest of such altars was the *śyenaciti* (falcon-shaped altar). It was to be constructed when one desired heaven (after death) because "the falcon is the best flyer among the birds." The spatial dimensions, the number of bricks (of prescribed shapes and sizes), the number of layers, etc., are all given. It is interesting to note that the archaeological remains of this most striking and complicated *śyenaciti* reported to be built in Kausambi in the second century BC, still survive.

Early Indian geometry developed because of the need for accurate altar constructions and transformations which often required quite advanced mathematical knowledge.

R.C. GUPTA

REFERENCES

Bhattacarya, V., ed. *Baudhāyana Śulba Sūtra with Commentaries of Vyaṅkateśvara and Dvārakānātha*. Varanasi: Sampurnanand Sanskrit University, 1979.

Datta, B. *The Science of the Śulba*. Calcutta: University of Calcutta, 1932. Reprinted 1991.

Ganguli, S.K. "On the Indian Discovery of the Irrational at the Time of Śulba Sūtras." *Scripta Mathematica* I: 135–141, 1932.

Gupta, R.C. "Baudhāyana's Value of $\sqrt{2}$." *Mathematics Education* 6(3): 77–79, 1972.

Gupta, R.C. "Vedic Mathematics From the Sulba Sutras." *Indian Journal of Mathematics Education* 9(2): 1–10, 1989.

Gupta, R.C.. "Sundararāja's Improvements of Vedic Circle-Square Conversions." *Indian Journal of History of Science* 28(2): 81–101, 1993.

Kashikar, C.G. "Baudhāyana Śyenaciti: A Study in the Piling Up of Bricks". In *Proceedings of 29th All-India Oriental Conference*. Poona: Bhandarkar Institute, 1980, pp. 191–199.

Kulkarni, R.P. *Geometry According to Śulba Sūtra*. Pune: Vaidika Samsodhana Mandala, 1983.

Prakash, Satya, and R.S. Sharma, eds. *Baudhāyana Śulba Sūtra*. New Delhi: Research Institute of Ancient Scientific Studies, 1968.

Seidenberg, A. "The Origin of Mathematics." *Archive for History of Exact Sciences* 18(4): 301–342, 1978.

Sen, S.N., and A.K. Bag, eds. and trans. *The Śulba Sūtras*. New Delhi: Indian National Science Academy, 1983.

Thibaut, G. *Mathematics in the Making in Ancient India*. Calcutta: Bagchi, 1984.

See also: *Śulbasūtras*

BEN CAO GANG MU The application of herbal, mineral, animal, and man-made substances in medical therapy has been recorded in China since antiquity. These pharmaceutical compilations are called *Ben Cao*.

In the Ming dynasty, a masterpiece, *Ben Cao Gang Mu* (Compendium of Materia Medica), was published in AD 1590. In China, as well as in other parts of the world, this work is the best-known and most respected description of traditional pharmaceutics. The value of this enormous achievement, written by one person, goes far beyond the scope of a pharmaceutical-medical drug work. In fact, it constitutes an extensive encyclopedia of knowledge concerning nature and the technology required for the medicinal use of nature.

The compilation of *Ben Cao Gang Mu* was based on the *Zheng Lei Ben Cao* (Classified Materia Medica, compiled at the end of the eleventh century AD). Li Shizhen, the author, took from it 1479 drugs. He supplemented these with 39 drugs that had been included in the drug compendia of the Jin Yuan period, as well as with 374 drugs that he himself described for the first time, so that the *Ben Cao Gang Mu* contained a total of 1892 drugs. This number is confirmed by the table of contents; the actual number should be 1898. To continue the statistics: the number of drug illustrations is 1160 and that of the recorded prescriptions is 1196, of which Li Shizhen collected or newly compiled 8161 himself.

The entire work was divided into two chapters of illustrations and another fifty-two chapters containing the texts. The first four text chapters are of a general nature; chapters five through fifty-two contain the monographs. The total number of drugs was divided into sixteen groups, and then sixty subgroups. Thus, for example, the group of herbs was divided into the subgroups of mountain herbs, aromatic herbs, swamp herbs, poisonous herbs, convolvulous herbs, water herbs, stone herbs, moss and lichen herbs, and finally, various herbs.

For each drug the following ten items were discussed. As they were used only when the information was necessary, not all are contained in each monograph:

(1) information concerning a previously false classification of the drug;
(2) information on secondary names of the drug, including the sources of these names;
(3) collected explanations;
(4) the pharmaceutical-technological preparation of the drug;
(5) explanation of doubtful points;
(6) correction of mistakes;
(7) information on the thermo-influence and taste of the drug;
(8) enumeration of the main indications of the drug;
(9) explanations concerning the effects of the drug;
(10) enumeration of prescriptions in which the drug is used.

As for the author, Li Shizhen was born into a medical practitioner's family. When he was unable to pass the official career exams after three attempts, he was determined to devote himself to the study of medicine. Through thirty years' clinical practice and learning from many experts, he, after innumerable hardships, completed the manuscript in

1578, but the first edition was not finished until after his death.

LIAO YUQUN

REFERENCES

Li Shizhen. *Ben Cao Gang Mu*. Beijing: Jen min wei sheng chu pan she, 1975.

Needham, Joseph. *Science and Civilisation in China*, vol. V. Cambridge: Cambridge University Press, 1974.

The Collected Works of Study on Li Shizhen. Hubei: Science and Technology Press, 1985.

BHĀSKARA II Undoubtedly, the greatest name in the history of ancient and medieval Indian astronomy and mathematics is that of Bhāskarācārya (b. AD 1114). His *Līlāvatī* is the most popular book of traditional Indian mathematics. He is usually designated as Bhāskara II in order to differentiate him from his earlier namesake who flourished in the early part of the seventh century.

According to Bhāskara's own statement towards the end of his *Golādhyāya*, he was born in Śaka AD 1036 or AD 1114. He also adds that he came from Vijjaḍaviḍa near the Sahya mountain. This place is usually identified with the modern Bijapur in Mysore. S.B. Dikshit is of the opinion that Bhāskara's original home was Pāṭaṇa (in Khandesh), where a relevant inscription was discovered by Bhau Daji in 1865. According to the inscription, Manoratha, Maheśvara, Lakṣmīdhara, and Caṅgadeva were the names of the grandfather, father, son, and grandson, respectively, of Bhāskara. Caṅgadeva was the chief astronomer in the court of the king Siṅghaṇa and had established in AD 1207 a *maṭha* (residential institution) for the study of the works of his grandfather. Bhāskara's father, Maheśvara, was also his teacher.

Bhāskara's *Līlāvatī* (The Beautiful) is a standard work of Hindu mathematics. It belongs to the class of works called *pāṭī* or *pāṭīgaṇita*; that is, elementary mathematics covering arithmetic, algebra, geometry, and mensuration. Its popularity is shown by the fact that it is still used as a textbook in the Sanskrit medium institutions throughout India. It provides the basic mathematics necessary for the study of almost all practical problems, including astronomy. The subject matter is presented through rules and examples in the form of about 270 verses which can be easily remembered. There is a story that the author named the work to console his daughter Līlāvatī, who could not be married due to some unfortunate circumstances; but the truth of the story cannot be ascertained. Bhāskara addressed the problems to a charming female Līlāvatī, who, according to some scholars was his wife (and not his daughter).

The great popularity of *Līlāvatī* is illustrated by the large number of commentaries written on it since it was composed about AD 1150. Some of the Sanskrit commentators were: Parameśvara (about 1430), Gaṇeśa (1545), Munīśvara (about 1635), and Rāmakṛṣṇa (1687). Only a few of these have been published. Gaṇeśa's gives a good exposition of the text with a demonstration of the rules. However, the best traditional commentary is the *Kriyākramakarī* (ca. 1534), which is a joint work of Śaṅkara Vāriyar and Mahiṣamaṅgala Nārāyaṇa (who completed it after the demise of Śaṅkara).

There are a number of commentaries and versions in regional Indian languages. Quite a few modern scholars have edited, commented on, and translated *Līlāvatī*. At least three Persian translations are known, the earliest being by Abū al-Fayḍ Fayḍī (AD 1587). The English translation by H.T. Colebrooke (London, 1817) was based on the original Sanskrit text and commentaries.

Bhāskara's *Bījagaṇita* (Algebra) is a standard treatise on Hindu algebra. It served as a textbook for Sanskrit medium courses in higher mathematics. In it the author included an exposition of the subject based on earlier works. Among the sources named were the algebraic works of Śrīdhara and Padmanābha. Besides operations with various types of numbers (positive, negative, zero, and surds), it deals with algebraic, simultaneous, and indeterminate equations. There is a separate chapter on the Indian cyclic method called *cakravāla*. He attributes the method to earlier teachers but does not specify any name. Due to the difficult nature of some of the topics, the *Bījagaṇita* was not as popular as the *Līlāvatī*.

Bhāskara's *Siddhānta-śiromaṇi* (AD 1150) is an equally standard textbook on Hindu astronomy. It has two sections: *Grahaṇita* (Planetary Mathematics) and the *Golādhyāya* (Spherics). Often these two sections appear as independent works. There is a lucid commentary on the whole work by the author himself. It is called *Mitākṣarā* or *Vāsanābhaṣya*. Other commentators include Lakṣmīdāsa, Nṛsiṁha, Munīśvara (1638/1645), and Rāmakṛṣṇa. The fourteenth chapter of the *Golādhyāya* is the *Jyotpatti*, which may be regarded as a small tract on Hindu trigonometry.

Usually it is customary to regard *Līlāvatī*, *Bījagaṇita*, *Grahaṇita*, and the *Golādhyāya* as the four parts of the *Siddhānta Siromaṇi* to make it a comprehensive treatise of Bhāskarācārya's Hindu mathematical sciences. His two other works are the *Karaṇa-kutūhala* (whose epoch is 1183), a handbook of astronomy, and a commentary on Lalla's *Śiṣyadhīvṛddhida-tantra* (eighth century AD). Some other works have also been attributed to him, but his authorship of *Bījopanaya* is questionable.

Bhāskara introduced a simple concept of arithmetical infinity through what he calls a *khahara*, which is defined by a positive quantity divided by zero, e.g. 3/0. His arithmetical and algebraic works are full of recreational problems to provide interesting pedagogical examples.

Perhaps the most important part of Bhāskara's *Algebra* is his exposition of the Indian cyclic method. We now know that the method was already known to Jayadeva (eleventh century AD or earlier). A modern expert, the late C.-O. Selenius, praised it by remarking that "no European performance in the whole field of algebra at a time much later than Bhāskara II, nay nearly up to our times, equalled the marvelous complexity and ingenuity of cakravāla method." Fermat proposed the equation $61x^2 + 1 = y^2$ in 1657 to Frénicle as a challenge problem. However, by applying the above method, Bhāskara had already solved the problem five centuries earlier. Bhāskara's solution (which he got just in a few lines) in its smallest integers was $x = 226,153,980$, $y = 1,766,319,049$.

The feat was possible not only due to the technique but also because of a well-developed symbolic notation. Colebrook remarks

> Had an earlier translation of Hindu mathematical treatises been made and given to the public, especially to the early mathematicians in Europe, the progress of mathematics would have been much more rapid, since algebraic symbolism would have reached its perfection long before the days of Descartes, Pascal, and Newton.

Another gem from Bhāskara's *Algebra* is a very short proof of the so-called Pythagorean theorem.

The geometrical portion of *Līlāvatī* covers mensuration regarding triangles, quadrilaterals, circles, and spheres. A special rule gives the numerical lengths of the sides of regular polygons (from three to nine sides) in a circle of radius 60,000. The last chapter entitled *Aṅka-pāśa* is devoted to combinatorics.

Bhāskara's *Jyopatti* contains many trigonometrical novelties which appear in India first in this tract.

Although an equivalent of the differential calculus formula

$$\Delta \sin \theta = \cos \theta \, \Delta \theta$$

already appeared in Muñjāla's *Laghumānasa* (AD 932), Bhāskara II gave its geometrical demonstration. He knew that when a variable attains an extremum, its differential vanishes. He is credited even with a knowledge of the mean value theorem and the Rolle's theorem of differential calculus. A crude method of infinitesimal integration is implied in his derivation of the formula for the surface of a sphere. This he gave in the *Vāsanābhāṣya* on the *Golādhyāya* (chapter III).

Bhāskara's *Siddhānta-śiromaṇi* is one of the most celebrated works of Hindu astronomy. It is a comprehensive work of Brahma-pakṣa. He praised Brahmagupta, who belonged to the same school, but his own astronomical work became more famous.

Although based on the works of predecessors, rather than on any fresh astronomical observations, the *Siddhānta-śiromaṇi* served as an excellent textbook. Systematic presentation of the subject matter, lucidity of style, and simple rationales of the formulas made it quite popular. It also contained some improved methods and new examples. For instance, he gave a very ingenious method of finding the altitude of the Sun in any desired direction in his *Golādhyāya* (III, 46). His professional expertise and all-round knowledge made him a truly great and revered *ācārya* (professor) of Hindu astronomy and mathematics for generations.

R.C. GUPTA

REFERENCES

Colebrooke, H.T. *Algebra with Arithmetic and Mensuration from the Sanscrit of Brahmegupta and Bháscara.* London: Murray, 1817. Reprinted Wiesbaden: Martin Sandig, 1973.

Datta, B. and A.N. Singh. "Use of Calculus in Hindu Mathematics." *Indian Journal of History of Sciences* 19(2): 95–104, 1984.

Gupta, R.C. "Bhāskara II's Derivation for the Surface of a Sphere." *Mathematics Education* 7(2): 49–52, 1973.

Gupta, R.C. "Addition and Subtraction Theorems for the Sine and their Use in Computing Tabular Sines." *Mathematics Education* 8(3B): 43–46, 1974.

Gupta, R.C. "The *Līlāvatī* Rule for Computing Sides of Regular Polygons." *Mathematics Education* 9(2B): 25–29, 1975.

Jha, A., ed. *Bīja Gaṇita of Bhāskarācārya.* Banaras: Chowkhamba, 1949.

Kunoff, Sharon. "A Curious Counting/Summation Formula from the Ancient Hindus." In *Proceedings of the Sixteenth Annual Meeting of the CSPM.* Edited by F. Abeles et al. Toronto: CSHPM, 1991, pp. 101–107.

Sastri, Bapu Deva, ed. *The Siddhānta-śiromaṇi by Bhāskarācārya with His own Vasanābhāṣya.* Benares: Chowkhamba, 1929.

Selenius, Clas-Olof. "Rationale of the Chakravāla Process of Jayadeva and Bhāskara II." *Historia Mathematica* 2(2): 167–184, 1975.

Srinivasiengar, C.N. *The History of Ancient Indian Mathematics.* Calcutta: World Press, 1967.

See also: Algebra in India – Mathematics in India – Geometry in India – Parameśvara – Munīśvara – Trigomometry in India – Combinatorics – Muñjāla

BIAN QUE Bian Que, one of the most famous medical men in ancient China, lived during the time of the Zhou dynasty, that is between 500 and 600 BC He was good at various subjects in medicine and skilled in diagnosis and treatment, especially in pulse-taking and acupuncture.

Si Maqian (about 145–96 BC), a well-known historian, wrote a biography of Bian Que in his work *Shi Ji* (Historical Record). In it, he discussed Bian Que's practicing medicine, wandering from town to town and effecting miraculous cures, even bringing the dying back to life. In the end Bian Que was killed by a jealous commissioner of the Imperial Academy of Medicine. Si Maqian says: "Bian Que expounded medicine as the guiding principle of (medical) technique; the later generations follow it and could not change any more." He was regarded as the founder of traditional Chinese medicine, but according to the historical materials in the biography, Bian Que lived and flourished for hundreds of years. That is because its author confused him with another famous medical man, Qin Yueren. Probably Qin Yueren had a close relationship with Bian Que's teachings.

In the earliest bibliography *Qi Lue*, written in the first century BC, Bian Que was assumed to be the author of some medical works such as *Bian Que Nei Jing* (The Internal Classic of Bian Que) and *Bian Que Wai Jing* (The External Classic of Bian Que). Although all of them have been lost, the contents can be partly found in other extant medical works, which reflect what he contributed to traditional Chinese medicine.

Bian Que said that the blood and *qi* (vital energy) move along the channels of the body, making fifty circuits per day. Medical historians consider this the first intimation of the circulatory system of the blood. Considering the limited knowledge of anatomy by the ancient Chinese, this statement seems astonishingly prophetic. Actually this view was constructed on the principle of cyclicism, a very common mode of thought for the Chinese.

It was he, too, who founded the extremely complicated old Chinese system of pulse-taking, according to which a doctor was supposed to be able to diagnose an illness solely by the condition of the pulse. It was not just a question of whether the beat was strong or weak, regular or uneven. Pulses could also be distinguished as "sounding like a sickle, first exuberant then dying away"; "flowing along quietly like flying hair or feathers"; "beating deep and strong like a thrown stone"; "sounding delicate like the string of an instrument" or "slipping along like a fish or a piece of wood on the waves" and so on. From the above selection of possible pulse beats it is easy to see with what extraordinarily sharpened sensitivity the Chinese doctor examined his patient. Bian Que also discussed how to diagnose illness by the condition of the complexion. For example, a yellow complexion and nails were an indication of jaundice; red dealt with the heart; white was always due to the lungs, etc.

As for therapy, Bian Que used the traditional techniques such as acupuncture, moxibustion, and drugs to cure various diseases. We do not know who invented those treatments; however, all of them were improved owing to Bian Que's practice and teaching. Indeed, these are still followed in clinical practice and diagnosis today.

LIAO YUQUN

REFERENCES

Li Bocong. *Study of Bian Que and his School*. Shanxi: Science and Technology Press, 1990.

Liao Yuqun. "Study of Bian Que's Sphygmology." *Chinese Journal of Medical History* 18 (2): 65, 1988.

Sima Qian. "Biography of Bian Que." In *The Historical Record*. Beijing: Zhonghua Press, 1959.

Wang Shuhe. *Mai Jing* (Classic of the Pulse). Shanghai: Shang-wei sheng chu pan she, 1957.

AL-BĪRŪNĪ (Part 1) Abū Rayḥān Muḥammad ibn Aḥmad al-Bīrūnī was born on Thursday, 3rd of Dhū al-Hijjah, 362 H (4th September, AD 973) at "Madīnah Khwārizm". His exact birth place is still a matter of dispute. It is conjectured that he was born in the outskirts *(bīrūn)* of Kāth, at al-Jurjāniyah, Khwārizm or at a place called Bīrūn, as implied by his nickname al-Bīrūnī. The only clue given by al-Bīrūnī was that he was born in a city in Khwārizm. The name Abū Rayḥān (perfume or herb) was given to him because of his love for sweet fragrances. Al-Bīrūnī died in 443 H (AD 1051).

Al-Bīrūnī was a devout Muslim, yet there was no conclusive evidence of his adhering to any particular *madhḥab* (denomination) throughout his life. His native language was the Khwarizmian dialect. He knew Persian but preferred Arabic because he believed that Arabic was more suitable for academic pursuit. He received some of his early education under the tutelage of the astro-mathematician Abū Naṣr al-ʿIrāq and Abd al-Samad from Khwārizm in addition to his formal education at the *madrasah*, an institution where Islamic sciences are studied.

Al-Bīrūnī's first patron was the Sāmānid Sultan Abū Sālih Manṣūr II who reigned in Bukhara until the city was invaded by the Ghaznavid Sultan Maḥmūd in 389 H (AD 999). Later al-Bīrūnī went to Jurjān to the court of Abu'l Haṣṣan Qābūs ibn Washmjīr Shamṣ al-Ma ʿālī (r. AD 998–1012), under whose patronage he wrote *alāthār al-bāqiya min al-Qurūn al-khāliya* (Chronology) which was completed in 390 H (AD 1000). Al-Bīrūnī found the Sultan indiscriminate and harsh.

His next sojourn was in Khwārizm and Jurjāniyah, under the service of the Sāmānid Prince Abu'l Abbas al-Maʾmūn ibn Muḥammad II. He was the best patron and al-Bīrūnī received the respect he very much deserved. It was during this period that al-Bīrūnī met the physician al-Jurjāni. His *Taḥdīd Nihāyāt al-amākin li-taṣḥiḥ masāfāt alMasākin* (Determination of the Coordinates of Cities) was completed in AD 1025, and his *Kitāb fī taḥqīq ma li 'l Hind* (Book on India) was finally published in 421 H (AD 1030). Al-Bīrūnī's other book, *Kitāb altafhīm li-awāʾil sinā ʿat al-tanjīm* (The Book of Instruction in the Art of Astrology) which was dedicated to Rayḥānah, daughter of al-Hassan, was written in Ghaznah, AD 1029. Al-Bīrūnī's *magnum opus* was an astronomical encyclopedia, *al-Qānūn al-Masūdī fī al-hay'ah wa 'ltanjīm* (Canon Masudicus) which comprises eleven treatises divided into 143 chapters. It was completed in 427 H (AD 1035). Apart from emphasizing the importance of astronomy, he gives accurate latitudes and longitudes and also geodetic measurements. His *Kitāb al-jamāhir fī maʿrifat al-jawāhir* (Mineralogy) was completed less than a decade later (435 H/ AD 1043). Al-Bīrūnī's *Kitāb al-ṣaydanah fiʿl-ṭibb* (Materia Medica, or Pharmacology) was completed only in the form of a rough draft before he died. The list of books mentioned thus far is not exhaustive; altogether he wrote about 146 treatises.

Al-Bīrūnī was both a scientist and a philosopher, but he never used the word "science" in the sense the word is understood today — that knowledge which is popularly thought to be exact, objective, veritable, deductive, and systematic. He used the Arabic word *ʿilm*, which means knowledge.

Solving scientific problems, which to al-Bīrūnī is analogous to "untying knots", (*Kitāb fī ifrād al-maqāl fī umr al-ẓlāl* [The Exhaustive Treatise On Shadows]) is the main activity of scientists. That science to al-Bīrūnī was a problem-solving activity and a scientific problem was a problem circumscribed by the Holy *Quʾrān* and *Sunnah* can be discerned by examining, in particular, the Introductions of his major books. *India*, for example, was written by al-Bīrūnī primarily because in his opinion, "while Muslims had been able to produce fairly objective works on such religions as Judaism and Christianity, they had been unable to do so with regard to Hinduism". In *Taḥdīd*, he clearly states another aspect of a scientific problem. He says that "geography is very essential for a Muslim for knowing the right direction of Mecca (*qibla*)".

Al-Bīrūnī was certainly a very prolific, multidimensional scholar. He did serious work in almost all branches of science in his time and his 146 treatises range from 10 to 700 pages each.

ABDUL LATIF SAMIAN

REFERENCES

Primary sources

al-Āthār al-bāqiya min al-gurūn al-khāliya. English trans. by Edward Sachau, *The Chronology of Ancient Nations*. London: Minerva GMBH, 1879.

al-Qānūn al- Masʿūdī (Canon Macudicus). Hyderabad: Osmania Oriental Publications Bureau, 1954–1956.

Kitāb al-jamāhir fī marifat al-jawāhir. Ed. by F. Krendow. Hyderabad-Dn: Osmania Oriental Publications Bureau, 1936.

Kitāb al-ṣaydanah fiʿl-ṭibb. Ed. and trans. by Hakid Mohammed Said et al. Pakistan: Hamdard Academy, 1973.

Kitāb al-tafhīm li-awāʿil ṣinā ʿat al-tanjīm. Tehran: Jalal Humāʿl, 1940. English trans. by R. Ramsay Wright, *The Book of Instruction in the Art of Astrology*. London: Luzac, 1934.

Kitāb fī ifrād al-maqāl fī amr al-ẓilāl. Hyderabad-Dn: Osmania Oriental Publications Bureau, 1948. Ed. and trans. as *The Exhaustive Treatise on Shadows*. E.S. Kennedy. Syria: University of Aleppo. 1976.

Kitāb fī taḥqīq ma liʿl Hind. English trans. by Edward Sachau, *Alberuni's India*. London: Trubner, 1888.

Taḥdīd nihāyat al-amākin li-taṣḥiḥ. English trans. by Jamil Ali, *The Determination of the Coordinates of Cities, al-Biruni's Tahdid al-Amakin*. Beirut: American University of Beirut, 1967.

Secondary sources

Boilot, D.J. "L'oeuvre d' al-Bīrūnī: essai bibliographique." *Mélanges de l' Institut dominicain d'études orientales* 2:161-256, 1955.

Kennedy, E.S. "al-Bīrūnī." In *Dictionary of Scientific Biography*, vol. II. New York; Charles Scribner's Sons, 1970.

Khan, Ahmad Saeed. *A Bibliography of The Works Of Abū 'l Rāiḥān Al-Bīrūnī*. New Delhi: Indian National Science Academy, 1982.

Nasr, Seyyed Hossein and William C. Chittick. *al-Bīrūnī: An Annotated Bibliography of Islamic Science*. Tehran: Imperial Iranian Academy of Philosophy, 1975.

Nasr, Seyyed Hossein. *An Introduction to Islamic Cosmological Doctrines*. Revised edition. Albany: State University of New York Press, 1993.

Said, Hakim Mohammed and Ansar Zahid Khan. *Al-Bīrūnī– His Times, Life and Works*. Karachi: Hamdard Foundation, 1979.

See also: *Qibla* and Islamic Prayer Times – Geography in Islam – Religion and Science

AL-BĪRŪNĪ (Part 2: Geographical Contributions) The geographical knowledge of the Muslims, in part derived from the Greeks and others, and contemporaneously developed and advanced by themselves, had reached a very high level of development by the tenth century. It is in this development that the work of al-Bīrūnī is significant. Al-Bīrūnī presented a critical summary of the total geographical knowledge up to his own time. He made some remarkable theoretical advances in general, physical, and human geography. Al-Bīrūnī did not confine himself to a simple description of the subject matter with which he was concerned. He compared it with relevant materials and evidence, and evaluated it critically, offering alternative solutions.

George Sarton identifies al-Bīrūnī as one of the great leaders of this period because of his relative freedom from prejudice and his intellectual curiosity. Although his interests ranged from mathematics, astronomy, physics and the history of science to moral philosophy, comparative religion and civilization, al-Bīrūnī became interested in geography at a young age. He is considered to be the greatest geographer of his time.

In the area of physical geography, he discussed physical laws in analyzing meteorology and climatology. He wrote of the process of streams development and landscape evolution. He introduced geomorphological enquiry to elucidate a history of landscape. He developed the mathematical side of geography, making geodetic measurements and determining with remarkable precision the coordinates of a number of places. Some of his noteworthy contributions to geography include: a theory of landform building processes (erosion, transportation, and deposition); proofs that light travels faster than sound; explanations of the force of gravity; determination of the sun's declination and zenithal movement; and discussion of whether the Earth rotates on its axis. He described various concepts of the limits for which he seems to have had recourse to contemporary sources not available to earlier geographers. He made original contributions to the regional geography of India.

In the study of physical phenomena, including landforms, weather, and geology, al-Bīrūnī adopted the methods of the physical sciences and drew conclusions with scientific precision. Long before Bernhard Varenius (AD 1622–50), al-Bīrūnī developed a schema for physical geography: (a) terrestrial conditions, describing the shape and size of the Earth; (b) cosmic concepts, dealing with the measurement of the circumference of the Earth and the establishment of the exact location of places; (c) classification of natural phenomena either in accordance with their nature or with their position in time and space. He studied phenomena in time (chronological science) and also tried to study them in space. In his view, geography was an empirical science.

Based on available knowledge concerning the surface of the Earth, he deduced and described the shape and forms of land surface. Al-Bīrūnī examined questions concerning the Earth's shape, size, and movement.

He explained running water as the most effective agent by which the surface of the land is sculpted. He further asserted that as the rivers of the plains of India approached the sea they gradually lost their velocity and their power of transportation, while the deposition process along their beds increased proportionately. Al-Bīrūnī considered the changes in the course of a river a universal phenomenon. He also recognized the influence of the sun upon the tide and suggested that heavenly bodies exert a gravitational effect on the tides.

Al-Bīrūnī recognized that the heat of the atmosphere and the Earth's surface is derived from the sun through the transfer of energy by rays, and that it varies with the length of time that the Earth is exposed to the rays. He recognized the wind's force and velocity and argued that the wind, in all its phases, it determined by certain causes.

Al-Bīrūnī noticed the peculiarities of the Indian monsoon, observed the time of its breaking, and described its westward and northward movements and the unequal distribution of rain in different areas of India.

Finally, he added that the habitable world does not reach the north on account of the cold, except in certain places where it penetrates into the north in the shape, as it were, of tongues and bays. In the south it reaches as far as the coast of the ocean, which in the east and west is connected with what he calls the comprehending ocean (*India*, Vol. I, p. 196).

In short, al-Bīrūnī recognized geography as an empirical science, and he dealt with the terrestrial globe as a whole. He stressed its nature and properties. He also tried to investigate the causes of global phenomena and described them as they exist.

AKHTAR H. SIDDIQI

REFERENCES

Ahmad, S. Maqbul. "Djughrafia." (Geography) In *The Encyclopedia of Islam*, Vol. II. Leiden: E.J. Brill, 1965, pp. 575–87.

Barani, S. Hasan. "Muslim Researches in Geodesy." In *Al-Bīrūnī Commemorative Volume*. Calcutta: Iran Society, 1951, pp. 1–52.

Kazmi, Hassen Askari. "Al-Bīrūnī's Longitudes and their Conversion into Modern Values." *Islamic Culture* 49: 165–76, 1974.

Kazmi, Hassen Askari. "Al-Bīrūnī on the Shifting of the Bed Amu Darya." *Islamic Culture* 50: 201–11, 1975.

Kramer, J.H.. "Al-Bīrūnī's Determination of Geographical Longitude by Measuring the Distance." In *Al-Bīrūnī Commemorative Volume*. Calcutta: Iran Society, 1951, pp. 177–93.

Memon, M.M. "Al-Bīrūnī and his Contribution to Medieval Muslim Geography." *Islamic Culture* 33: 213–18, 1959.

Sachau, Edward C. *The Chronology of Ancient Nations*. London: W.H. Allen and Co., 1879.

Sachau, Edward C. *Alberuni's India*, Vols I and II. London: Kegan Paul, Trench, Trubner and Co., 1910.

Sarton, George. *Introduction to the History of Science*, Vol. I. Baltimore: Williams and Wilkins, 1927, pp. 693–737.

Sayılı, Aydın. "Al-Bīrūnī and the History of Science." In *Al-Bīrūnī Commemorative Volume*, ed. M.M. Said. Karachi: Times Press, 1979, pp. 706–12.

See also: Geography in Ancient India – Maps and Mapmaking in India

AL-BIṬRŪJĪ Al-Biṭrūjī (fl. 1185–1192) was an Andalusian astronomer whose complete name seems to be Abū Isḥāq [ibn?] al-Biṭrūjī, Nūr al-Dīn. Nothing is known about his biography apart from the fact that his name probably derives from the region of Los Pedroches, near Cordoba, and that he was a disciple of the philosopher Ibn Ṭufayl (ca. 1110–1185). The latter was already dead when al-Biṭrūjī wrote his only known work, the *Kitāb fī-l-hay'a* (Book on Cosmology), which seems to have been read by the anonymous author of a book on tides dated in 1192. On the other hand, al-Biṭrūjī's treatise was translated into Latin by Michael Scott in Toledo in 1217.

Al-Biṭrūjī was a member of the Andalusian school of Aristotelian philosophers composed of Ibn Bājja (1070?–1138), the aforementioned Ibn Ṭufayl, Ibn Rushd (Averroes, 1126–1198) and Mūsā ibn Maymūn (Maimonides, 1135–1204). All these authors criticized Ptolemaic astronomy due to its mathematical character which did not agree with Aristotelian physics. Al-Biṭrūjī was, however, the only one who made a serious, although unsuccessful, attempt to create an astronomical system which could have a physical reality. In it he uses homocentric spheres in the Eudoxan tradition which he combines with materials derived from the Toledan astronomer Ibn al-Zarqāllu (Azarquiel, d. 1100). It is interesting to remark that al-Biṭrūjī's Dynamics are not exclusively Aristotelian but also use Neoplatonic concepts such as the *impetus* theory which he seems to have been the first to introduce into Western Islam. Al-Biṭrūjī's system was soon known in Western Europe and became very popular among Scholastic philosophers of the thirteenth century, who considered it a serious alternative to Ptolemy.

JULIO SAMSÓ

REFERENCES

Carmody, Francis J. *Al-Biṭrūjī. De motibus celorum. Critical edition of the Latin translation of Michael Scott*. Berkeley: University of California Press, 1952.

Goldstein, Bernard R. *Al-Biṭrūjī: On the Principles of Astronomy*. 2 vols. New Haven: Yale University Press, 1971.

Samsó, Julio. *Las Ciencias de los Antiguos en al-Andalus*. Madrid: Mapfre, 1992, pp. 330-356.

See also: *Hay'a* – al-Zarqāllu

BITUMEN IN PREMODERN CHINA Reports of bitumen seepages and production concentrate on four provinces: northern Shaanxi and eastern Gansu from the Han period onwards, Sichuan from the Ming onwards, and Xinjiang which has records of bitumen seepages during the periods of the Northern Dynasties (386–581), the Tang, and the Qing. In addition, during the Ming and Qing periods, minor production sites existed in Guangdong, Taiwan, Zhejiang, Anhui, and Liaoning (Liaodung).

The earliest record of bitumen seepages refers to northern Shensi, that is the oil fields of Yanchang and Yongping of today. The *Hanshu*, first century AD, states that in Gaonu district (modern-day Yan'an) "there is the Wei River; it can burn". This statement is elucidated by later sources, which say that on the water of the Wei River "fat" (*fei*) is floating which can burn and which can be taken for greasing cart axles or as fuel for lamps.

Sources from the Song onwards provide more details on the locations of Shaanxi petroleum seepages and production methods. In his famous account of about 1070 on using petroleum carbon for ink, Shen Gua (1031–1095) mentions the production of "stone oil" (*shiyou*) in the two prefectures Fu and Yan and of "stone juice" (*shiye*) in Yanchuan. He wrote that stone oil grew in the sand and stones of the waterfront, was mixed with spring water, and, with a view to ink manufacture, was produced inexhaustibly within the earth, in contrast to the limited supply of pine wood. In spring, local people collected it with pheasant tail brushes and put it into pots. The method of collection described by Shen Gua is similar to that described in Agricola's *De re metallica* of the mid-sixteenth century.

Song and Yuan sources report that near Yan'an bitumen was used to make so-called "stone candles" (*shizhu*). Lu You (1125–1210) says in his *Laoxuean biji* of 1190 that they are solid like stone and give a very bright light. Moreover, "they also gutter like wax [candles]. Their smoke, however, is thick and may fumigate and defile curtains and clothes. This is why people in the western [part of the empire] do not esteem them". The stone cave supplying the raw material for the production of the stone candles was, however, closed in 1270 by government order.

Perhaps the first notion of a well intentionally dug for getting bitumen can be found in the *Da Yuan da yitongzhi*, the comprehensive administrative and geographical description of the Yuan period:

> South of Yanchang district, at Yinghe, there is a stone oil well which has been cut open. Its oil can burn and also cures itches and ringworms of the six domestic animals.

> Annually, 110 *jin* [pounds] are handed in. Moreover, in Yongping village, 80 *li* north-west of Yanchuan district, there is another well, which annually procures 400 *jin*. They are handed over to the [Shaanxi] Route (*lu*).

As in many other cases, the output of oil seepages was low. The record also shows that the government was heavily interested in this resource, probably because of its importance for warfare.

The use of bitumen in warfare from the Yumen oil field in western Gansu is reported by an eighth-century source, stating that, in 578, the Emperor Wu of the Northern Zhou, when being besieged by the Turks at Jinquan, took this "fat" and set it afire, thus burning the enemies' assault weapons. When brought into contact with water, it burnt even more, with the result that Jinquan was saved.

With the development of deep drilling, Sichuan became an important producer of bitumen, being a by-product of the search for brine and, later, natural gas. Sixteenth-century accounts report that the sources of the fire wells of Jiading and Qianwei are all "oily". Because of its odor, locals call the stone oil "realgar oil" (*xionghuangyou*) and "sulphur oil" (*liuhuangyou*). It was thought that the sources and veins of stone fat, realgar, and sulphur were interconnected. Officials filled bamboo tubes with oil and ignited it for lighting the way. Moreover, special iron lamps were constructed which fitted the use of bitumen as fuel. An indigenous terminology developed which was based on the colors of bitumen. White bitumen was called "rice-soup oil" (*mitangyou*), green "green-bean oil" (*lüdooyou*), yellow "gardenia-nut oil" (*zhiziyou*), and black "black-lacquer oil" (*moqiyou*). Petroleum occurred at depths of 180–220 meters, ca. 360 meters, and 950–990 meters, and there were experts who knew exactly to which depths the hoisting buckets should be sent to have them filled with petroleum exclusively.

The most important application of bitumen in crude and distilled form was certainly in the military field, where it was used for incendiary weapons. The sources do not inform us about details of the processing and distilling methods, but this may be due to the fact that these were treated as military secrets. The second most important use of bitumen was for lighting purposes, for which numerous references, particularly for regions near bitumen seepages or wells, can be found. In Sichuan, bitumen appears to have served sometimes as fuel for the evaporation of brine. Bituminous products were also used for greasing cart-axles and the bearings of water power trip-hammers, manufacturing ink, sealing leather wine sacks, covering floors, and tempering iron.

Moreover, bitumen was often applied for medical, pharmaceutical, and alchemical purposes. For instance, Li Shizhen's *Bencao gangmu*, published in 1596, states that it is smeared on ulcers, ringworms, worms, and scabies, and is used for drugs which heal flesh wounds caused by iron arrows. A source of the seventh century tells us that when swallowing the bitumen of Jiuzi in Xinjiang, fallen out teeth and hair can be induced to grow again. Later sources are less sanguine about internal application and sometimes stress that bitumen is poisonous.

Although, in China, bitumen seepages and deposits were discovered and used from an early period onwards, the overall economic, technical, and scientific role of bitumen remained small and did not extend much beyond the limits of the small number of production areas. The only exemption was in the military field, where testimonies of fire and flame throwers using distilled petroleum are numerous. The revolutionary impact of the Drake Well in mid-nineteenth century Pennsylvania was not due to the depth reached in deep drilling, but to the fact that beforehand a thorough scientific investigation of the properties, refining methods, and possible applications of petroleum had taken place. It was this combination of scientific investigation, technical achievement, establishment of production plants, and creation of a huge demand that was responsible for the rise of the petroleum industry in the West.

HANS ULRICH VOGEL

REFERENCES

Anon. "Lüelun wo guo shiyou, tianranqi de kaifa jiqi lishi jiaoxun" (A Brief Account of the Development of Petroleum and Natural Gas in our Country and its Historical Lessons). *Wen-wu* 6: 47–53, 1975.

Kuo, Sampson Hsiang-chang. *Drilling Oil in Taiwan: A Case Study of Two American Technicians' Contribution to Modernization in Late Nineteenth-Century China*. Ph.D. dissertation, Georgetown University, Washington, D.C., 1981.

Li Jung. "An Account of the Salt Industry at Tzu-liu-ching. Tzu-liu-ching-chi." *Isis* 39: 228–234, 1948.

Needham, Joseph. *Science and Civilisation in China*. Vol. III, vol. IV, part 2, and vol. V, part 7. Cambridge: Cambridge University Press, 1954.

Pi Xiaozhong et al. "Wo guo gudai laodung renmin kaifa tianranqi he shiyou de bufen lishi ziliao" (Some Historical Material Concerning the Development of Natural Gas and Petroleum in Ancient Times by the Working People of our Country). *Jingyanshi tongxun* 1: 36–44, 1977.

Shen Lisheng. *Zhongguo shiyou gongye fazhanshi; Diyi juan: Gudai de shiyou yu tianranqi* (A History of the Development of the Petroleum Industry in China; Part 1: Petroleum and Natural Gas in Ancient Times). Rev. ed. Beijing: Shiyou gongye chubanshe, 1984.

Vogel, Hans Ulrich, Joseph Needham et al. "The Salt Industry." Section 37 of *Science and Civilisation in China*. Ed. Joseph Needham. Cambridge: Cambridge University Press, forthcoming.

Xing Runchuan. "Wo guo lishi shang guanyu shiyou de yixie jizai" (Some Records on Petroleum in the History of our Country). *Huaxue tongbao* 4: 63, 1976.

See also: Salt (China) – Tribology (China) – *Ben Cao Gang Mu*

BRAHMAGUPTA "Brahmagupta holds a remarkable place in the history of Eastern civilization. It was he who taught the Arabs astronomy before they became aqcuainted with Ptolemy" (Sachau, 1971). Bhāskara II described Brahmagupta as *Gaṇakacakracūdāmaṇi* (Jewel among the circle of mathematicians).

Brahmagupta was born in AD 598 accordining to his own statement: "... when 550 years of the Śaka era had elapsed, Brahmagupta, son of Jisṇu, at the age of 30, composed the *Brāhmasphuṭasiddhānta* for the pleasure of good mathematicians and astronomers". Thus he was 30 years old in Śaka 550 or AD 628 when he wrote the *Brāhmasphuṭasiddhānta*. That he was still active in old age is clear from the title epoch of AD 665 used in another of his works called *Khaṇḍa-khādyaka*. Pṛthūdaka Svāmin, an ancient commentator on Brahmagupta, calls him Bhillamālācārya, which shows that he came from Bhillamāla. This place has been identified with the modern village Bhinmal near Mount Abu close to the Rajasthan–Gujarat border.

We have no knowledge of Brahmagupta's teachers, or of his education, but we know he studied the five traditional *Siddhāntas* on Indian astronomy. His sources also included the works of Āryabhaṭa I, Lāṭadeva, Pradyumna, Varāhamihira, Siṃha, Śrīṣeṇa, Vijayanandin, and Viṣṇucandra. He was, however, quite critical of most of these authors.

The *Brāhmasphuṭasiddhānta* is Brahmagupta's most important work. It is a standard treatise on ancient Indian astronomy, containing twenty-four chapters and a total of 1008 verses in *āryā* meter. The *Brāhmasphuṭasiddhānta* claimed to be an improvement over the ancient work of the Brahmapakṣa, which did not yield accurate results. Brahmagupta used a great deal of originality in his revision. He examined and criticized the views of his predecessors, especially Āryabhaṭa I, and devoted two chapters to mathematics. There have been many commentators on this work. The earliest known was Balabhadra (eighth century AD), but his commentary is not extant.

Chapter 7 is on *Gaṇita* (Mathematics). It deals with elementary arithmetic, algebra, and geometry. The subject is presented under twenty-eight topics of logistics (arithmetical operations) and determinations, including problems related to mixtures, plane figures, shadows, series, piles, and excavations. He wrote in a concise and understandable style, whether dealing with simple mathematics or complex geometry. In the treatment of surds, Brahamagupta is remarkably modern in outlook. The *Brāhmasphuṭasiddhānta* includes formulas for the rationalization of the denominator, as well as a marvelous piece of pure mathematics in the rule for the extraction of the square root of a surd. Still more remarkable algebraic contributions are contained in a chapter entitled *Kuṭṭaka*, which is a traditional name for indeterminate analysis of the first degree. The second order indeterminate equation

$$Nx^2 + c = y^2 \tag{1}$$

is called *varga-prakṛti* (square nature). An important step towards the integral solutions of such equations is what is called Brahmagupta's *Lemma* in the history of mathematics. In modern symbology the Lemma is as follows:

> If (α, β) is a solution of (1) with $c = k$, and $(\alpha' \ \beta')$ is its solution with $c = k'$, then $(\alpha\beta' \pm \alpha'\beta, \beta\beta' \pm N\alpha\alpha')$ will be its solution with $c = kk'$.

This lemma not only helps in finding any number of solutions from just one solution, but it also helps in solving the most popular case of $c = 1$, provided we know a solution for $c = -1$, or ± 2 or ± 4. It was rediscovered in Europe by Euler in 1764.

In geometry, Brahamagupta's achievements were equally praiseworthy. He wrote a fine symmetric formula for the area of a cyclic quadrilateral, which appeared for the first time in the history of mathematics. Even more important are his expressions for the diagonals of a cyclic quadrilateral.

Brahmagupta's name has been immortalized by yet another achievement. A "Brahmaguptan quadrilateral" is a cyclic quadrilateral whose sides and diagonals are integral (or rational) and whose diagonals intersect orthogonally. He gave a simple rule for forming such figures in the *Brāhmasphuṭasiddhānta* (chapter 7, verse 38): If a, b, c and α, β, γ are the sides (integral or rational) of two right-angled triangles (c and γ being hypotenuses), then $a\gamma$, $b\gamma$, $c\alpha$ and $c\beta$ are the required sides of a Brahamaguptan quadrilateral.

Prior to Brahmagupta, the usual method for computing the functional value intermediary between tabulated values was that of linear interpolation, which was based on the rule of proportions. He was the first to give second order interpolation formulas for equal as well as unequal tabulated argumental intervals. Mathematically, his rule is equivalent to the modern Newton–Stirling interpolation formula up to the second order.

The *Khaṇḍa-khādyaka* is a practical manual of Indian astronomy of the *Karaṇa* category. The author claims that it gives results useful in everyday life, birth, marriage, etc. quickly and simply, and is written for the benefit of students. The work consists of two parts called the *Pūrva* and the *Uttara*. The former comprises the first nine chapters and expounds the midnight system. The latter six chapters provide corrections and additions. This work has been studied by a great number of commentators, from Lalla in the eighth century to Āmarāja in the twelfth. It was translated into Arabic first by al-Fazārī (eighth century) and then by al-Bīrūnī.

Brahmagupta's genius made use of mathematics (traditional as well as that which he developed) in providing better astronomical methods. He used the theory of quadratic equations to solve problems in astronomy. He knew the sine and cosine rule of trigonometry for both plane and spherical triangles. He supplied standard tables of Sines and Versed Sines.

The historian of science George Sarton called him "one of the greatest scientists of his race and the greatest of his time".

R.C. GUPTA

REFERENCES

Primary sources

Sharma, R.S. et al., eds. *The Brāhma-sphuṭa-siddhānta* (with Hindi translation). 4 volumes. New Delhi: Indian Institute of Astronomical and Sanskrit Research, 1966.

Chatterjee, Bina, ed. and trans. *Khaṇḍakhādyaka with the commentary of Bhaṭṭotpala*. Calcutta: World Press, 1970.

Sengupta, P.C., trans. *The Khaṇḍakhādyaka of Brahmagupta*. Calcutta: Calcutta University, 1934.

Secondary sources

Colebrooke, H.T. *Algebra with Arithmetic and Mesuration from the Sanscrit of Brahmegupta and Bhāskara*. London: Murray, 1817.

Gupta, R.C. "Second Order Interpolation in Indian Mathematics." *Indian Journal of History of Science* 4: 86–98, 1969.

Gupta, R.C. "Brahamagupta's Rule for the Volume of Frustum-like Solids." *Mathematics Education* 6(4B): 117–120, 1972.

Gupta, R.C. "Brahmagupta's Formulas for the Area and Diagonals of a Cyclic Quadrilateral." *Mathematics Education* 8(2B): 33–36, 1974.

Kak, Subhash. "The Brahmagupta Algorithm for Square-Rooting." *Gaṇita Bhāratī* 11: 27–29, 1989.

Kusuba, Takanori. "Brahmagupta's Sūtras on Tri- and Quadrilaterals." *Historia Scientarum* 21: 43–55, 1981.

Pottage, John. "The Mensuration of Quadrilaterals and the Generation of Pythagorean Triads etc." *Archive for History of Exact Sciences* 12: 299–354,

Sachau, Edward, trans. *Alberuni's India*. New York: Norton, 1971.

Sarton, George. *Introduction to the History of Science*. Baltimore: Williams and Wilkins, 1947.

Venkutschaliyenger, K. "The Development of Mathematics in Ancient India: The Role of Brahmagupta." In *Scientific Heritage of India*. Ed. B.V. Subbarayappa and S.R.N. Murthy. Bangalore: Mythic Society, 1988, pp. 36–47.

See also: Geometry in India – Arithmetic in India: *Pāṭīgaṇita* – Algebra in India: *Bījagaṇita* – Astronomy in India

C

CALCULUS Calculus, both integral and differential, was known *to a limited extent* and used for specific purposes in India during early and medieval times, and is attested to by the works of Āryabhaṭa (b. 476), Brahmagupta (b. 598), Muñjāla (b. 932), Bhāskarācārya (b. 1114), Nārayāṇa Paṇḍita (1356), Nīlakaṇṭha Somayāji (b. 1444), and Jyeṣṭhadeva (ca. 1500–1610). In the same manner as Hindu geometry grew in response to the religious needs of designing sacrificial altars, it would seem that calculus too evolved when there was the need to ascertain favorable and unfavorable times for religious rites and rituals on the basis of the moments of the eclipses, conjunction of planets, and conjunction of planets and stars. Calculus did not evince further growth in India.

In order to determine the accurate motion of a planet at a particular moment (*tātkālika-gati*), Bhāskarācārya divides the day into a very large number of intervals of time units called *truṭi*, equal to 1/33750 of a second, and compares the successive positions of the planet, the motion during that very small unit of time being considered constant. He also suggests that the differential coefficient vanishes when the interval is diminished to the absolute minimum. He illustrates this by three examples.

Hindu ideas on infinitesimal integral calculus are shown in the methods employed and formulae arrived at for the calculation of the area of a circle, and the surface area and volume of a sphere, all of which are virtually the same as these derived by modern mathematics.

Two methods are adopted for finding the surface area of a sphere. One of them is to draw, from a point on the sphere taken as the pivot, a very large number of circles, parallel to each other and of small and equal interstices, and to slice the sphere at these circles. The sum of the areas of the strips peeled off each slice, the longest of them being at the equator, and the smallest, equal to zero, at the pivot and the bottom, would be equal to the surface area of the sphere. It is explained that when the breadth of the strips tends to zero, the total surface area could be obtained by the formula $4\pi^2$.

In the same manner, the volume of a sphere is equal to the sum of the numerous cones that would be formed with their bases on the surface of the sphere and their apexes being at the center of the sphere. When the number of cones is infinite, their bases could be taken as flat, and the volume of the cones could be calculated by the usual formula. From this postulate, the formula for the volume of the sphere is identified as $\frac{4}{3}\pi^3$. The several methods enunciated for the determination of the value of the circumference of a circle, involving the irrational quantity, π, as exposited in several works from Kerala, like the *Yuktibhāsā* of *Jyeṣṭhadeva*, also concern integral calculus.

K.V. SARMA

REFERENCES

Āryabhaṭīya of Āryabhata, Critical Edition. Ed. K.S. Shukla and K.V. Sarma. New Delhi: Indian National Science Academy, 1976.

Balagangadharan, K. "Mathematical Analysis in Medieval Kerala." In *Scientific Heritage of India: Mathematics*. Ed. K.G. Poulose. Tripunithura: Government Sanskrit College, 1991, pp. 29–42.

Datta, Bibhutibhusan, and Awadesh Narayan Singh. "Bhaskara and the Differential Calculus." In *The History of Ancient Indian Mathematics*. Ed. C.N. Srinivasiangar. Calcutta: The World Press, 1967, pp. 91–93.

Datta, Bibhutibhusan, and Awadesh Narayan Singh. "Infinitesimal Calculus." In *Mathematics in Ancient and Medieval India*. Ed. A.K. Bag. Varanasi-Delhi: Chaukhamba Orientalia, 1979, pp. 286–98.

Datta, Bibhutibhusan, and Awadesh Narayan Singh. "Use of Calculus in Hindu Mathematics." *Indian Journal of History of Science* 19(2): 95–104, 1984.

Sengupta, P.C. "Infinitesimal Calculus in Indian Mathematics." *Journal of the Department of Letters, Calcutta University* 22: 1–17, 1932.

Siddhāntaśiromaṇi of Bhāskarācārya. Ed. Murali Dhara Chaturvedi. Varanasi, 1981, *Spaṣṭādhikāra*, 36–38: pp. 119–21: *Buvanakośa*, 58–61: pp. 363–67.

Yuktibhāṣā (of Jyeṣṭhadeva) (in Malayalam). Ed. Rama Varma (Maru) Thampuran and A.R. Akhileswara Iyer. Trissivaperur: Mangalodayam Limited, 1948. Trans. K.V. Sarma (Ms).

See also: Āryabhaṭa – Brahmagupta – Muñjāla – Bhāskarā – Nārayāṇa Paṇḍita – Nīlakaṇṭha Somayāji – Jyeṣṭhadeva

CALENDARS IN EAST ASIA Successive attempts to improve a typical luni-solar calendrical system for reconciling two fundamentally incommensurable periods — the tropical year and the synodic month — were made throughout the history of Chinese and Japanese calendars until their replacement by the Gregorian solar calendar in modern times, the Japanese in 1883 and the Chinese in 1912.

The length of a synodic month varies between 29.0 and 30.1 days. The luni-solar calendar provided for "short" months of twenty-nine days and "long" months of thirty days. Calendrical scientists attempted to arrange short and long months so that the moon's conjunction would take place on the first day of every month. The day notation of the lunar month represented the phase of the moon; for instance, the fifteenth day of the month was always a full moon, while the first day was a new moon.

In addition, the Chinese had an independent system of solar intervals (*qi*) for indicating seasonal changes, the most important phenomena in the regulation of agriculture. The tropical year was divided into twelve equal intervals of time, the element of a purely solar calendar. The middle point of each interval was called its *zhongqi* or interval center. The synodic month was always slightly shorter than an equal interval, and thus an interval center did not occur in certain months. Such months were designated as intercalary months, and in this way the sequence of synodic months was reconciled with the seasons of the tropical year. The year that included an intercalary month had thirteen synodic months. This occurred roughly once every three years.

The Chinese calendar calculators were, however, not satisfied with providing a conventional calendar, in which the courses of the sun and moon were reconciled, merely for civil use; they also tried to include the anomalistic motions of the sun and moon.

Prior to the Tang period (roughly seventh to ninth centuries), a mean synodic month of approximately 29.5306 days was used for this purpose. Shortly before the Tang, however, a proposal was made to take anomalistic motions of the sun and moon into consideration in calendar making. Liu Zhuo (AD 544–610) used the term *dingshuo* (true synodic month) as opposed to the older concept of *pingshuo* (mean synodic month). During the Tang period the idea of the true synodic month was adopted in the official calendar in order to attain better agreement with the actual phases of the moon, and also for convenience in eclipse-prediction calculations.

Liu Zhuo also made the first recorded distinction between *dingqi* and *pingqi*. The former is the period of time for the sun to move through a thirty-degree angle of the ecliptic (the band of the zodiac through which the sun apparently moves in its yearly course), which therefore varies in length in accordance with the anomalistic movement of the sun. The latter is the average length of time to move through a thirty-degree angle. This concept, *dingqi*, however, was not actually adopted in the official calendar until the Qing period (the seventeenth century on).

Daily life is never affected by such small discrepancies in the calendar as those due to the anomalistic motions of the sun and moon. The astronomical attainments of the Chinese calendrical schemes went far beyond the concern of the common people. They included not only luni-solar phenomena but also the periodic motions of the five planets. They were called *li*, which corresponds to "ephemeris" as well as "calendar" in the narrow sense.

In the case of the Julian calendar reform data were doubtless gathered from prior calendars and not from actual observations. One recalls in this connection Lillius' effort in the Gregorian calendar reform to reduce the toil of observation and computation so as to provide an artificial system independent of astronomical tables and indeed of any link with actual celestial phenomena. The Western calendar was not, in terms of astronomical precision, overwhelmingly rigorous.

The difference of purpose between calendar-making in the West and East is clear. Whereas the Western calendar was divorced from the development of astronomical science, and aimed at schematic convenience for civil and religious proposes, the East Asian calendar, disregarding civil convenience, was an attempt to represent faithfully the movements of the heavenly bodies. It thus became much more complicated than its practical applications warranted. A sharp separation of scientific astronomy and civil calendar-making did not take place in traditional East Asia. Its scope was confined to the composition of a luni-solar ephemeris which stood or fell on the accuracy with which it could predict eclipses, which is the best way to check the validity of any luni-solar calendar. Analysis of planetary motions was rather ancillary to the main stream of Chinese calendar making.

Cosmological discussions often took place during the Han and Six Dynasties periods (second century BC to sixth century AD), although they were discontinued soon after. In the Tang period the Chinese astronomers were no longer interested in this particular speculative pursuit. While occasional cosmological debates were found among philosophers, scientific cosmology was long set aside and forgotten in professional calendrical science circles until the time of the Jesuits' impact in the seventeenth century.

Perhaps the most striking thing about the recastings of the civil calendar in China was their frequency. In two thousand years there were more than fifty revisions of the Chinese calendar, and then of the Japanese calendar, whereas in the West there was only one major reform, the Gregorian. Why were the Chinese so preoccupied with the calendar that they were driven to such repeated efforts? There were two major reasons, one political and the other technical.

Among the ancient Chinese the idea prevailed that a ruler received his mandate from heaven. In the early period, therefore, after important changes within individual reigns and always after important changes of dynasty, the new emperor was prompted to reform key institutions, especially the official calendar, in order to confirm the establishment of a "new order" which a new mandate implied. A new mandate meant a new disposition of celestial influences. The Han Emperor

Wu (r. 1480–1487 BC) was advised by his councilor to revise the calendar and change the color of ceremonial vestments, in order to make it clear that his rule was based on a true mandate. The result was the imperial edict by which the *Taichu* calendar was adopted.

This notion was responsible for the subsequent course of development of Chinese calendrical science. Calendrical science enjoyed governmental sponsorship throughout its history, and had more prestige than other branches of science; it was China's most genuine contribution to the exact sciences. The history of Chinese and traditional East Asian astronomy is, for the most part, the history of calendar-calculations.

In the course of time, however, the political importance of calendar reform dwindled. The restriction of calendar reform to change of dynasty was not strictly observed by the Northern Wei (fifth century AD) and succeeding northern dynasties. Some rulers unduly took advantage of calendar reform to gain or regain popular support. The calendrical significance and authority of calendar revision were thus largely lost. In the Tang dynasty, the motive for calendar reform became simply to correct disagreements of the calendar with observed celestial phenomena. Hence, reforms were carried out whenever a small error was found. This accounts for the frequent revisions in later phases.

Beyond the concern with timekeeping, the Chinese astronomers dealt mainly with apparent solar and lunar movements and eclipse prediction. The Chinese luni–solar calendar required revision whenever the discrepancy between calculation and observation became noticeable. For this purpose, the government founded the Office of Astronomy, and incumbent astronomers were responsible for the improvement of the calendar. The government enjoyed a monopoly on the regulation of time, which contributed to imperial control of daily life and which strengthened the central power and prestige at a sacrifice of civil convenience. People were forced, in matters of everyday life, to follow a calendar based on a complicated, "astronomically rigorous" and "state-authorized" ephemeris, which, unlike our solar calendar, could never be prepared by laymen, but depended on the skill of official astronomers

Despite early recognition that a single perpetually valid calendrical scheme was impossible, the calendar calculators never tired of proposing revisions. Their incentive must be ultimately attributed to the rational desire to conform to the celestial order, a philosophical compulsion to accept the phenomena of the material sky as the most authentic reality. It was because of this orientation that calendar-making maintained its status as an integral part of Chinese science and learning for such a long time.

Modernization and Westernization

As the official calendar was quite an important part of the social system, westernization in modern times was first infiltrated through calendrical science.

Western astronomy had arrived in China with the Jesuits in the sixteenth century. Calendar-making was a vital function of the Chinese bureaucracy, and the Jesuits hoped that by demonstrating the superiority of Western astronomy they would succeed in persuading the bureaucratic elite that Western cultures and Christianity were superior. Astronomy was well suited to their purposes, not only because it was so important to the Chinese bureaucracy but because the phenomena with which it dealt possessed a universality that transcended East and West, and the objectivity of celestial phenomena did not permit much human manipulation. Furthermore, it was quantitatively precise. Thus, when the tally sheet was in, Chinese astronomers had to acknowledge that Western astronomy had superior features. All that the Jesuits made available from Western tradition, however, were peripheral data and methods of calculation. The structure, styles, and purposes of Chinese calendrical astronomy remained unchanged. As Xu Guangqi, a high Chinese official who collaborated with the Jesuit Matteo Ricci on several projects, put it, "We melted down their materials and poured them into the *Datong* (the traditional Chinese calendar then in use) mold."

In Japan, the eighth Shogun Yoshimune (1644–1751) attempted to use this refurbished Chinese astronomy as the basis for calendrical reform, but he was effectively opposed by conservative court circles in Kyoto. The revision of Horeki that came into effect in 1754 was still modeled upon the traditional *Shoushi* calendar. Yoshimune's desire was fully realized, however, in the Kansei revision of 1797.

After that, the scientific quality of the traditional luni-solar calendar was improved and perfected, but still the traditional mold was maintained throughout the Tokugawa period (1603–1867).

Despite the earlier recognition of Western superiority in science, a radical change of political institutions in Japan had to precede the official adoption of the Western Gregorian calendar in 1873. The decisive factor in the adoption of the Gregorian calendar was the high value the government placed on westernization. Many Japanese proposed a more reasonable solar calendar at the time of the reform, but the government preferred the Gregorian, despite its obvious shortcomings, because of diplomatic relations with the West.

With the establishment of the Republic of China in 1912, the Chinese luni-solar calendar was officially abolished.

To what extent the Gregorian calendar was observed among the populace is questionable, but after the second World War, it became firmly established in everyday life in East Asia.

NAKAYAMA SHIGERU

REFERENCES

Yabuuchi, Kiyoshi. "Astronomical Tables in China, From the Han to the T'ang Dynasties." English section of *Chugoku chusei kagaku gijutsu shi no kenkyu* (Researches on the History of Science and Technology of Medieval China). Ed. Yabuuti Kiyoshi. Tokyo: Kadokawa Shoten, 1963.

Yabuuchi, Kiyoshi. *Chugoku no tenmon rekiho* (Chinese Astronomy and Calendrical Science). Tokyo: Heibonsha, 1969.

See also: Time – Astronomy

CALENDARS IN EGYPT Otto Neugebauer described the ancient Egyptian civil calendar as "the only intelligent calendar which ever existed in human history". His statement can be appreciated by tracing the ancient Egyptians' gradually developing sense of time from an initial awareness of the day to the full realization that the year had a constant length of 365 days. The introduction of the civil calendar, based on solar years as opposed to lunar years, resulted in the development of our present-day Gregorian calendar.

Humans first became conscious of the existence of the day through the regular reappearance of the sun after its disappearance the evening before. Observing the regular change of one crescent moon to the next, they became aware of the longer time unit of the month. The practical demands of temple administration led them to develop their ability to count and calculate and they realized that one lunar month was made up of 29 or 30 days.

With the advent of the agricultural civilization in the Nile Valley, the season, rather than the month, gained importance as a time unit, since the Nile flood was a regularly recurring phenomenon that had to be predicted with a certain degree of accuracy. The ancient Egyptians must soon have realized that a purely lunar calendar was not capable of adequately predicting such annual events. They hit upon another device by which the onset of the inundation could be determined. This was the heliacal rising of Sothis (Sirius, the brightest star in the sky) which coincided closely with the beginning of the annual flooding of the Nile. They observed that after a 70-day period of invisibility, Sothis could be seen again just before sunrise, and this first appearance was taken by the Egyptians to signal the beginning of the new year.

The Egyptians, having reckoned that there was a 365 day span between two successive heliacal risings of Sothis, initially resorted to a luni-stellar calendar which consisted of twelve lunar months totalling 354 days, to which an additional thirteenth intercalary month was added every few years to make up for the cumulative discrepancy arising from the 11 day difference between the lunar and stellar years. The rule of thumb used was that whenever Sothis rose during the last eleven days of the fourth month of the last season, the intercalary month should be added.

Such a solution was somewhat cumbersome, and although the luni-stellar calendar was used to determine religious festivals, a more practical calendar had to be introduced to meet the demands of the state administrative system. Around 4200 BC, the Egyptians adopted a 365 day Civil Calendar which was divided into three seasons, each of four thirty-day months. The year consisted of twelve months, each of which would count thirty days plus "five days upon the year". Events were dated as occurring on the first, second, third ... day of the first, second, third ... month of the specific season.

The 365 day common year of the ancient Egyptians has also been called the "vague year" because its length fell short of the astronomical year by just over a quarter of a day each year. In Greek documents, the twelve months of this vague year are referred to by Greek names which originated from a cult connected with these months. These months are: (1) Thoth, (2) Phasphi, (3) Altyr, (4) Choriak, (5) Tybi, (6) Mechir, (7) Phamenoth, (8) Pharmuthi, (9) Pachon, (10) Payni, (11) Epiphi, (12) Mesori. These months are still preserved today in Egyptian folk tradition.

Each of these 30-day months was divided into three ten-day "weeks" or decans. Thus, there were thirty-six decans each year. The Egyptians defined each decan to begin when a particular star or constellation appeared on the horizon. The period of time between the appearance of one decan on the horizon and the next was defined to be an hour. From the surviving "diagonal calendar" examples, it is clear that the night was considered to contain only twelve hours, rather than the eighteen which might have been expected when we consider that we see approximately half the celestial sphere on any given night. The discrepancy seems to arise because the Egyptians did not include the period of twilight and the darkening of the sky as part of the night. Only the period of totally darkened sky corresponded to the term "night".

The idea that the day also included twelve hours seems to be developed merely in analogy with the night. These hours might be measured on a shadow clock, since the length of shadow cast by any object changes relative to the position of the sun in the heavens. The position does not just change over the course of the day, however, but also over the course of the year as the sun travels between the tropics. As a result, these shadow clocks do not give more than approximate seasonal hours. The use of clepsydra (water clocks), in which water escapes from a graduated container at a regular rate, may be closer to our modern concept of equal hours.

The Egyptian calendar constructors and time keepers did not go beyond the hour in measuring time. It is to the Babylonians, with their sexagesimal mathematical system, that we owe our modern concepts of sixty minutes in an hour and sixty seconds in a minute.

In time, the Egyptian 365 day calendar became completely out of step with the real tropical year, as it fell behind the latter by one day every four years. Thus, in 4 × 365 years, i.e. in 1460 years, the civil year would have lost a complete solar year and would again correspond with the seasons. This 1460 year period is generally referred to as the "Sothic Cycle" and reflects the fact that every 1460 years the so-called "heliacal rising of Sothis" fell on New Year's Day and thus marked the onset of a new Sothic period. The Egyptians seem not to have noticed the discrepancy between their civil calendar and the true solar year. If, at some point they did become aware of this difference, the strong conservative element in the culture apparently made them reluctant to make any amendment to their calendar.

In 238 BC, King Ptolemy III made an unsuccessful attempt to correct the "vague year" by adding an extra day every four years. Modification of the ancient calendar only ensued two centuries later when, after Julius Caesar's conquest of Egypt in 46 BC, he followed the advice of the famous Alexandrian astronomer, Sosigenes, and adopted a modification of the ancient civil calendar for the Roman Republic. Sosigenes suggested that the length of the year should be $365\frac{1}{4}$ days; that there should be a four-year cycle of three years of 365 days and a fourth year of 366 days; and that the year should begin on the first of January.

The Julian method of intercalation was not only convenient, it also provided a calendar whose year closely approximated the tropical year. Nevertheless, its deficiency appeared in time, making further reform imperative. The Julian Calendar, as we now call it, assumed the length of the year to be equal to 365.25 days, whereas the tropical year (defined as the passage of the sun from one equinox to the next) is nearly 365.2422 days. Thus, the error amounted to a gain of one whole day in 128 years. To deal with this problem, Pope Gregory XIII decreed that the eleven days, that had accumulated over the sixteen centuries since the Julian Calendar had been introduced, should be removed from the calendar and that henceforward no century should be considered a leap year unless divisible by 400. This Gregorian Calendar, which we use today, can thus be seen to be based on the principle of the Civil Calendar of ancient Egypt.

JEHANE RAGAI
GREGG DE YOUNG

REFERENCES

Neugebauer, O. *The Exact Sciences in Antiquity*, 2nd ed. New York: Dover, 1969.

Parker, R. A. *The Calendars of Ancient Egypt.* Chicago: University of Chicago Press, 1950.

Parker, R. A. "Calendars and Chronology." In *The Legacy of Egypt*. 2nd ed. Ed. J. R. Harris. Oxford: Clarendon Press, 1971, pp. 13–26.

See also: Clocks and Watches – Sexagesimal System

CALENDARS IN INDIA Calendrical science developed in India in three distinct phases, and each phase influenced the succeeding one. The first phase produced the Vedic calendar, covering a period from an unknown antiquity to the Mauryan emperor Asoka (ca. 300 BC); the second had the Greco-Indian calendar for a short period between the post-Asokan and post-Kuṣāṇa period (ca. AD 300). And the third produced the Siddhāntic calendar from the Gupta period (AD 319) which is still used in India today. All three phases are characterized by formulations of fitting a lunar year into a sidereal solar one by suitably intercalating lunations.

References to calendrical elements like lunation, intercalation, year, and solstice occur even in the earliest part of Vedic literature. A separate calendrical literature, deemed a part of the Vedas, was developed in a small text, *Vedāṅga Jyotiṣa*, but this text has not been fully deciphered yet. There are some astronomical statements in the text, and also in the earliest part of Vedic literature, which do not hold true in the latitude belt and in the time period when the Vedic Aryans are believed to have settled in India.

It is a separate course of study as to how and when the Vedic Indians obtained this calendar, but they adopted it without verifying it.

This calendar is based on the following imperfect parameters :

5 sidereal years = 1830 days.
62 lunations = 1830 days.
67 sidereal lunar months = 1830 days.

The star *beta Delphini* indicates the winter solstice (which was indeed the case around 1500 BC) which is apparently a fixed point. The lunar orbit is divided into twenty-seven equal arcs from *beta Delphini* each equal to 13°20′, called *nakṣatra*, the first arc named *Dhaniṣṭhā*. The arcs are named after a prominent star of the division. The theory is that if, in five years, two lunations are intercalated, then the lunar year will remain tied with the solar one. This period is called a 5-yearly cycle or *yuga*.

In practice, a cycle begins when the sun and the moon are in conjunction at *beta Delphini* (i.e. from the winter solstice). The first, second and fourth year each contain twelve new moon ending lunations. One lunation is intercalated at the middle of the third year and another at the end of the fifth. The second cycle then again begins from a similar new moon at winter solstice.

This imperfect calendar was destined to collapse, as the succeeding cycles will begin from new moons shifted by some 4.5 days from the winter solstice at *beta Delphini* per cycle, accumulating to one lunation in six or seven cycles. However, the calendar did not collapse, and so the accumulated lunation was extracalated, but we are not told in the text how this was done. Perhaps the rule is hidden in an obscure part of the text, and so we have to guess it.

A lunation is divided into thirty parts called *tithis*. Accordingly, 30 *tithis* = 1830/62 days. The excess 30/62 of a *tithi* over a day is called an omitted *tithi*. A month's days are designated by the *tithi* of the day and not by ordinal numbers. The Jainas used this cycle in their calendar with a marginal change: that months were full-moon-ending and a cycle began from a full moon at summer solstice in the middle of the *Āśleṣā* division (180° away from *beta Delphini*).

Kautilya wrote a book entitled *Arthaśāstra* which is believed to reflect the social and administrative conventions of the Mauryan empire. This book, on the subject of measuring time, only reproduced the Jain school of the 5-yearly cycle. We are thus assured of its use until then.

The second phase of the Indian calendar began by the first century BC after Śaka penetration into northwestern India.

The Śakas, a central Asian people, conquered Bactria in 123 BC and established an era with epoch at 123 BC, now called the Old Śaka Era. This epoch has variously been fixed at 88 BC (Konow) and 110 BC (Herzfeld), but, as M.N. Saha has shown, 123 BC fits into all the circumstantial evidence.

The Śakas used the Greek calendar based on the Metonic cycle in their homeland. This was a nineteen year cycle at whose beginning and end the sun and moon are in the same relative position to each other. When they reached India they adopted Indian culture and Aryanized or Indianized the Greek calendar, producing a Greco-Indian calendar. It replaced the classical 5-yearly cycle and it circulated in India up to ca. AD 200. The Kuṣāṇas, another Śaka tribe, penetrated further inside India, adopting Indian culture and using this calendar. Kings of the Śaka and Kuṣāṇa dynasties left behind some inscriptions bearing dates. These always refer to an era, as in era of king Kaṇiṣka. Further, they contain a mysterious expression — *etaye purvam* — whose meaning has not been deciphered.

The Indian form of the calendar is that the months are full moon ending; the 19-year cycle begins from a full moon at the autumnal equinox in the month Kārtika; months are assigned Sanskrit or Sanskritized Greek names; and, in special cases, the day is designated by the moon's *nakṣatra* on that day. The Greek form is that days are designated by ordinal numbers and not by *tithis*, and intercalation is made at the middle of the year in Chaitra.

All these features are fully reflected in the inscriptions of the Kuṣāṇa kings. For example, present researches have shown that Kaṇiṣka established his era in AD 78 after deleting the hundredth place in the old Śaka era for that year.

Now, 123 + 78 = 201 = 1, deleting the hundredth place.
Hence, AD 78 = 1 Kaṇiṣka era (current) and
0 Kaṇiṣka era (elapsed).

We interpret *etaye purvam* as elapsed year, and accordingly the Julian equivalent of Kaṇiṣka era in these inscriptions is:

AD 78 = 0 Kaṇiṣka era.

The day of the month can also be identified with the *tithi* with tolerable accuracy. We have seen that the moon gains 12° over the sun per *tithi*. In case of new-moon-ending months, we should expect, on 20th Āṣāḍha:

Moon − sun = 20 × 12° = 240° (near about)
or sun = moon − 240°.

If the moon is near *Uttara Phalguni* (tropical longitude in AD 90 = 144°46′), we should have: sun = 144°46′ − 240° = 265° (say), so that the sun's position comes near the winter solstice, which cannot happen in Āṣāḍha. However, if the month is full-moon-ending, then the moon − sun = 5 × 12° = 60°, or sun = moon − 60° = 144°46′-60° = 85° (say), i.e. near the summer solstice, which indeed happens in Āṣāḍha.

If we compute the mean tropical longitude of the sun (L) and mean elongation of the moon in days (D), we get:

Date	L	D
June 21, AD 90	87°.16	4.81

These lunisolar positions are in complete agreement with the inscriptional dates. We may perhaps identify these dates as Julian equivalents of the inscriptional dates. Using this scale we can decipher all the other dates in the inscriptions of the Kusana kings and use the same method to decipher the dates of the old Śaka Era.

Table 1: In a period of 4,320,000 years

	Āryabhaṭa	*Sūryasiddānta*
Sidereal revolution of the sun	4,320,000	4,320,000
Sidereal revolution of the moon	57,753,336	57,753,336
Sidereal revolution of Jupiter	364,224	364,220
Total civil days	1,577,917,500	1,577,917,828
Intercalary lunations	1,593,336	1,593,336
Omitted *tithis*	25,082,580	25,082,252

This second phase was short-lived, but had far-reaching effects on the later Indian calendar. In fact, this was perhaps the most productive period in Indian calendrical history.

This Greco–Indian calendar was discontinued after fall of the Kusana empire.

In the third phase, the Indian calendar was thoroughly revised. New calendrical elements were introduced, new techniques in computational works were formulated, and the application of cycles for intercalation was abandoned. The new calendar that emerged is called the Siddhāntic calendar. This calendar was perfected by later Indian astronomers and is still used all over India with some marginal changes in some regions. In this phase the solar zodiac with twelve signs was introduced. The *nakṣatra* division was recast so that the first arc-division, renamed *Aśvinī*, began from the first sign *Meṣa* (Aries). This common point was the vernal equinox. By AD 550, the vernal equinox shifted to *zeta Piscium*, and this star (or a point very close to it) has since been the beginning of the Indian zodiac.

The new technique for computations was that an epoch, *Kaliyuga*, was formulated such that all the luminaries, like the sun, moon, and planets, were assumed to be in conjunction at the initial point of the zodiac at this epoch. Sidereal motions of the luminaries were so assigned that each of them made an integral number of revolutions in a period of 4,320,000 years (see Table 1).

Astronomical parameters in this phase were not uniform in different schools. We cite below some calendrical parameters from the Āryabhaṭian school (AD 476) and the Sūryasiddhānta school of the tenth century.

For any day in question, elapsed days from *Kali* are computed from the above figures, and from the rates of sidereal motion of the luminaries their positions are found in the zodiac. The epoch of *Kali* was computed by Bentley at 18th February, 3102 BC midnight at Ujjain (longitude 75°45′ east). This date has been generally accepted by all astronomers.

Āryabhaṭa formulated corrections of mean motions of the luminaries, but such formulations are not found in earlier works. Our presumption is that only mean motions were considered in the pre-Āryabhaṭian period.

In this revised calendar, the luni-solar year begins from the new moon just preceeding the sun's entry to the initial point in the spring month *Chaitra*. Now, twelve lunations fall short of 10.8 days from a sidereal year, and this accumulates to one lunation in two or three years. Whenever such an extra lunation forms, it is intercalated in the year; there is no role of any cycle here. This luni-solar year, with days designated by the *tithi* of the day, is used all over India for religious purposes.

A solar year is also devised beginning from the sun's entry into the first sign *Meṣa*, and is divided into twelve solar months, each month being the period the sun stays in a sign. Days of solar months are designated by ordinal numbers. The first month is *Vaiśākha*. Datings are recorded by luni-solar year in some regions, and by solar year elsewhere. The era generally used is the Kaṇiṣka era renamed the Śaka era.

In a year containing twelve lunations, the twelve new moons will, most generally, be distributed over the twelve signs, i.e. each sign will contain a new moon. However, in an intercalary year when thirteen lunations are distributed over twelve signs, two new moons must occur in one sign, and in extreme cases, one sign may not contain any new moon at all. Such situations are natural phenomena and are no problem for astronomers.

Out of the thirteen lunations occurring in an intercalated year, any one may be earmarked as the extra month. However, metaphysics has taken a toll here. As one sign will contain two new moons, so two lunations, one from each new moon, will occur in that sign. One lunation out of these two has to be selected as the intercalary one. Volumes of religious literature have been written, perhaps more voluminous than their astronomical counterparts, on which lunation is the proper one and which is the intercalary one. Similarly, when a sign becomes void of a new moon, religious literature is controversial as to which lunation is to be assigned to that sign. Almost every region of India has its own convention on these points.

A.K. CHAKRAVARTY

REFERENCES

Chakravarty, A.K. *Origin and Development of Indian Calendrical Science*. Calcutta: Indian Studies, 1975.

Chatterjee, S.K. and Chakravarty, A.K. "Indian Calendar from Post Vedic Period to 1900 AD." In *History of Astronomy in India*. New Delhi: Indian National Science Academy,1985, pp. 252–307.

Chattopadhyaya, Debiprasad. *History of Science and Technology in Ancient India*, vol 1, *The Beginnings*. Calcutta: Firma KLM Private Ltd., 1966.

Konow, Sten. *Corpus Incriptionum Indicarum* vol ll, part l. Calcutta: Government of India Central Publication Branch,1929.

Report of the Calendar Reform Committee. New Delhi: Government of India, Council of Scientific and Industrial Research,1955.

Report of the State Almanac Committee. Alipur: Government of West Bengal, 1963.

Shamasastry, R., ed. *Vedāṅga jyotiṣa*. Mysore: Government Press, 1936.

Tilak, B.G. *Vedic Chronology and Vedāṅga Jyotiṣa*. Poona: Tilak Brothers, Gaikwar Wada, 1925.

See also: Astronomy – Lunar Mansions – Astrology

CALENDARS IN ISLAM Even the most ancient peoples considered the day the earliest unit of time-telling. However, the day is not convenient for long intervals of time. Even allowing for a primitive life-span of thirty six years, a person would live some 13,500 days, and it is very easy to lose track of some of those.

It seems natural to turn to the Earth's next most prominent body, the moon, for another unit. Here, the ready-made period of the lunar phases provides the natural unit called the lunar month. The lunar month is approximately 29.5 days (more exactly: 29.5306 or 29:12:2.8). In pre-agricultural times, the month provided a convenient and simple method for measuring moderately long periods of time. A lifetime of something like 400 months is a more convenient figure than 13,500 days. Lunar months untied to the seasons, as they would have been initially, would produce some problems for a settled agricultural society, which is not the case for a hunting or herding society of nomads, who could wander more freely in search of grain or grass. Farmers not only had to stay where they were, but needed to increase their chances of a good harvest by sowing at a proper time to take advantage of seasonal rains and seasonal warmth. Therefore, to keep the lunar months approximately seasonal, an intercalary month needs to be added in seven of each 19-year cycle (19 solar years = 235 lunar months, i.e. 19 lunar years of 12 months each, + 7 lunar months). In due course, the practice of intercalation was gradually perfected; the intercalary months came to be added in the third, sixth, eighth, eleventh, fourteenth, seventeenth, and nineteenth year of each cycle, in a popular scheme which transformed the lunar calendar into a seasonal, or luni-solar calendar, in which a month remained in pace with the sun to within 20 days.

The luni-solar calendar practice was adopted by the Hebrews, Greeks, and Romans and remains in use in the Jewish, Hindu, and Chinese calendars of today. The early Christians continued to use the Jewish (luni-solar) calendar tied to the new moon for more than three centuries and established the day of Easter on that basis. On the other hand, the Roman Republic, which had also used a luni-solar calendar for a long period, moved to a purely solar calendar in 46 BC as a result of abuse of power by Pontiffs in the matter of intercalation.

Julius Caesar decided to put an end to this chaotic situation by introducing the solar calendar system. In this transformation, 46 BC was allowed to continue for 445 days (!) and came to be known as 'The Year of Confusion', in effect the last one. The adopted solar calendar with a year length of 365.25 days became known as the Julian calendar.

The month, being tied to the moon, begins at a fixed phase in the lunar calendrical system. Since any phase will do, the month can begin at each full moon, the first quarter, etc. However, the obvious way is to begin each month with the new moon — on the evening when the growing crescent first becomes visible immediately after sunset. Practically all lunar calendars began their new months with the new visible crescent. Some modern users of the lunar calendar (the Chinese and Jews) have changed this to the astronomical new moon, i.e. the conjunction, while Muslims maintain the first visible phase of the new crescent as the beginning of a new month.

The Islamic calendar was both a religious as well as a civil calendar, necessary to determine periods of the annual month-long fast, pilgrimages, times of marriage and divorce, years of taxes on wealth for the needy, religious observances, and general and historical timekeeping.

In Arabia, the use of a lunar calendar is known to have existed from very early times. The original practice is believed to have been to use twelve lunar months to a year. On the Pre-Islamic Arab calendar, the annual Pilgrimage (*Ḥajj*) to *Ka'bah* in Mecca was a most significant and important event. Although the event of the pilgrimage to *Ka'bah* was basically religious, it was also important for trade and business. When the *Ḥajj* was out of season, it created difficulties in procuring the crops and sacrificial animals for trade and use. To overcome this, the Arabs in Mecca are believed to have introduced a system of intercalation known as *al-Naasi'*. A Meccan by the name of Qalmas is reported to be the first person assigned this task who, during each pilgrimage, would announce the dates for the coming years' *Ḥajj* and inform whether the intercalation was due for the coming *Ḥajj* season.

In addition to the month of *Dhu 'l-Hijjah* (in which the pilgrimage is held), three other months (*Rajab*, *Shawwal*, and *Thu 'l Qa'dah*) were sacred to the pre-Islamic Arabs. During these months, certain things, such as wars with opponents, were forbidden. In time, the practice of intercalation was much abused (as with the Roman calendar) thus affecting the sacred months and the related prohibitions by changing a sacred month into a non-sacred one.

The calendar in use in Medina (North of Mecca) remained in the original 'twelve months to a year' form. The early Muslims continued to use the Meccan calendar while in Mecca, but shifted to the *Medinan* calendar after the Prophet Muḥammad and his companions migrated to Medina in AD 622. Following the conquest of Mecca in the eighth year of *Hijrah* (December AD 629), the Muslims continued to use the Medinan calendar, and the *Meccan* calendar ran parallel.

However, during the Prophet Muḥammad's last pilgrimage in the tenth year of *Hijrah* (AD 632), the Meccan practice of intercalation was abolished through a Quranic injunction, thus reverting the Arab-Islamic calendar to the simple 'twelve months to a year' practice.

The Islamic calendar also required that the beginning of a month be based on the first sighting. This was particularly important for religious events like the beginning and end of the fasting month and the date of the pilgrimage to the *Kaʿbah*.

There was no commonly accepted permanent calendar in pre-Islamic Arabia. Nevertheless, the custom of counting years was, in one form or another, prevalent among the Arabs and their neighbours. Opinions are divided as to the origin of this practice. According to Ibn al-Jazwī, it dates from the time when the children of the prophet Adam multiplied and spread on earth. In South Arabia, a calendar system originated in 115 or 109 BC when the Himyarites adopted one using the reigns of the Tubba as the epoch years of their era. The inhabitants of Sanaa (Yemen) also adopted a calendar, using the victory over the Yemen by the Abyssinians, and later, the Persian conquest.

The origin of a chronology of events in North Arabia may be traced as far back as the construction of the *Kaʿbah* by the prophet Abraham and his son the prophet Ismail. The northern Arabians are also reported to have their famous battle days as marking epochs of eras. A local calendar is supposed to have existed in Medina also. Al-Masudi maintains that the people of Medina used the dates of their castles and palaces as their local calendar at the time of the prophet Muḥammad's migration, but others reject this, claiming the adoption of an era by the people of Medina to a month or two after the prophet's arrival, which continued until his death in AD 632. According to a widely held view, the *Hijrah* era was set up at the time of Caliph ʿUmar ibn Khattab in AD 637/638, i.e. about five or six years after the introduction of the purely lunar calendar. After many consultations, the year of the prophet Muḥammad's *Hijrah* was accepted for the beginning of the Islamic (or *Hijrah*) calendar.

Based on scientific understanding, certain ground rules were laid down which form the part of *sharīʿah* (the Islamic legal system) governing the calendar. Some of these were:

(1) the length of a month should be 29 or 30 days;

(2) the length of a year should be 354 or 355 days;

(3) the maximum number of consecutive 30 day-months is 4; the maximum number of consecutive 29 day-months is 3;

(4a) each new month begins with the first light of the new crescent moon visible on the western horizon after sunset;

(4b) one should try sighting the new moon on the 29th day of the month, but if it cannot be seen (even due to clouds), complete the month as of 30 days;

(4c) the visual sighting report must be corroborated by a witness;

(4d) the persons involved in the reporting must be reliable, adult, truthful, and sane, with good eyesight (implied), who are punished if proved to be purposely misleading;

(4e) the visual sighting report should not conflict with basic scientific understanding and natural laws. Indeed, professional scientists' involvement is essential to ascertain the reliablity of the reported sighting, and the scientific test would include a check on related parameters (e.g. the shape of the crescent, its position and altitude, time of observation, sky conditions); and finally

(4f) the sighting must be carried out in an organized way every month.

Sharīʿah also allows for the correction of a mistake. If on the 28th day of an Islamic month, the new moon has been sighted, a correction will be made to the beginning of the month, since a month should have only 29 or 30 days. If the month concerned is *Ramaḍan* then an extra fast would have to be completed after Eid celebrations.

Clearly, earliest visibility of the new moon plays a very critical role in Islamic calendar regulation. The Islamic 'State' placed special emphasis on astronomy research, and it became a standard element in formal religious and legal education. The early Muslim community, based under the clear skies of Arabia and assisted by State-sponsored research, had no serious problems in following these injunctions to regulate their calendar. Muslims contributed enormously to the development of the science of the new Moon's earliest visibility and its advance prediction for greatly varying geo-environmental situations. In this, they built upon the work of earlier researchers as well (e.g. Babylonians, Hindus, Jews). Besides the rigorous science, simple schemes were devised to construct long-term calendars especially to interconvert Islamic dates with Christian and other luni-solar calendrical dates. Based on technical information, one of the simple schemes (known as the schematic or *Istalahi* system) involved a cycle of thirty lunar years in which months approximately alternate between 29 and 30 days and 11 years consist of 355 days, the rest 354.

Backed by extensive research, Muslim scientists developed visibility tables and produced many reference works. All was well when astronomy and scientific endeavor were at their zenith in the Muslim lands.

Scientific interest in predicting the time of the first (earliest) new Moon sighting (for a clear sky) goes back to a period at least as early as the Babylonian era. Based on careful observational data, a simple criterion was developed and passed on to the Muslims through the Hindus, apparently with very little further improvement. This problem was thoroughly investigated by the early Muslim astronomers in the eighth to tenth centuries AD and included such notable persons as Ḥabash al-Ḥāsib and al-Battānī. The physics based system(s), developed up to the eleventh century, saw very little further development until more recent times, when improvements were made. There has also been a series of simpler criteria involving time difference (lag) between moonset and sunset and the 'age' of the moon.

MOHAMMAD ILYAS

REFERENCES

Ahmad, M. "Origin of the Hijra Calendar." *Al-Nahdah*: 53–54, 1991.

Faruqi, N.A. *Early Muslim Historiography*. Delhi: Idarah-i Adabiyat-i Delhi, 1979.

Hashmi, A.Q. *Taqweem Tareekhi*. Islamabad: Islamic Research Institute, 1987 (in Urdu).

Ilyas, M. *A Modern Guide to Astronomical Calculations of Islamic Calendar, Times & Qibla*. Kuala Lumpur: Berita Publishing, 1984.

Ilyas, M. *Astronomy of Islamic Times for the Twenty-first Century*. London: Mansell, 1989.

Ilyas, M. *New Moon's Visibility and International Islamic Calendar for the Asia-Pacific Region, 1407–1421H*. Penang/Islamabad: COMSTECH/RISEAP/USM, 1994.

Ilyas, M. *Islamic Astronomy and Science Development: Glorious Past, Challenging Future*. Kuala Lumpur: Pelanchuk, 1996.

O'Neal, W.M. *Time and the Calendars*. Sydney: Sydney University Press, 1975.

Whitrow, G.J. *The Nature of Time*. New York: Holt, Rinehart and Winston, 1965.

CALENDARS IN MESOAMERICA For many reasonably educated persons, the greatest single achievement of ancient Mesoamerica was the Maya calendar. In fact, a calendar was in use throughout Middle America, but it is its Maya form that has captured the world's imagination, the claim often being made that its accuracy exceeded that of the calendar in use in Europe at the time of the Conquest and, by some, even of that which we use today. While most of their neighbors kept account of time with three periods of differing numbers of days, the combinations of which formed a repeating calendar of 18,980 days, the Maya employed a mechanism that synchronized these same periods and others of their own making with a system that tracked cumulative time.

This cumulative record, or Long Count, expressed historical dates with five-place numbers that recorded days accumulated from an arbitrary zero-point which has been determined to correspond to August 13, 3114 BC in our present Gregorian calendar. (The correlation of Maya dates to European are to the no-zero-year Gregorian calendar, employing J. Eric S. Thompson's 1935 correlation of 584,285 days.) Accumulating 1-day (*k'in*), 20-day (*winal*), and 360-day (*tun*) periods, the Long Count employed a qualified vigesimal (20-base) place-number system — the qualification being an 18-base for the third place. The names and values of the Long Count places, from the unit up, were expressed in the writing system with distinct hieroglyphs for each order of magnitude (Figure 1).

Thus, for example, a Long Count date with the numerical value of *9 bak'tunob*, *8 k'atunob*, *9 tunob*, *13 winalob*, and *0 k'inob* records a day that was 1,357,100 days after August 13, 3114 BC, which is to say March 26, 603. These values were communicated by positioning the numbers as prefixes or superfixes of the glyphs for the periods. The *-ob* inflection of Mayan nouns indicates plural. Because double digits are often required to express decimal equivalents of numbers in the places of the Maya system, Mayanists have adopted the convention of inserting periods between the places of Long Count dates. The above date, for example, is written 9.8.9.13.0. Though Maya historical dates that employed the Long Count were generally limited to five places, the calendrical system itself was open-ended and, when called upon to do so (as to meet mythological needs) could accommodate any number of places and express any full number of days. The greatest known to have survived is recorded at Coba as 13.13.13.13.13.13.13.13.13.13.13.13.13.13.13.13.13.13.13.0.0.0.0,

PLACE	NAME	DAYS	CALC	VALUE	GLYPHS
1	*K'in*	1 day	(1 X 1)	1 *K'in*	
2	*Winal*	20 days	(20 X 1)	20 *K'inob*	
3	*Tun*	360 days	(18 X 20)	18 *Winalob*	
4	*K'atun*	7200 days	(20 X 360)	20 *K'atunob*	
5	*Bak'tun*	144,000 days	(20 X 7200)	20 *Bak'tunob*	

Figure 1: The names and values of the Long Count places with their respective glyphs. [Drawings by the author.]

which expressed in our decimal-system amounts to almost 22 quintillion tropical years!

Long Count dates were carved on monuments in the Maya heartland for over 600 years, the earliest known being 8.12.14.8.15 (July 8, 292), the latest 10.4.0.0.0. (January 20, 909). Moreover, the *Dresden Codex* records the much later date, 10.17.13.12.12 (October 29, 1178). Though extensively used by the Maya, the Long Count was probably not their invention. Earlier Long Count dates have survived on several monuments on the southwestern Maya periphery — the earliest, 7.16.3.2.12 (December 7, 36 BC), at Chiapa de Corzo — in what was probably a proto-Zoquean (non-Mayan) speaking area.

The Maya coordinated this cumulative Long Count with the pan-Mesoamerican repeating calendar of 18,980 days. Known generally as the Calendar Round and called the *xiuhmolpilli* by the Aztecs and *hunab* by the Maya, this period consisted of three component cycles of 13, 20, and 365 days. The first two of these were a set of consecutive day-numbers (1 to 13) and a set of twenty day-names specific to each of some sixty language-groups within Mesoamerica (Figure 2). The permutations of the thirteen numbers and twenty names designated each passing day with a combination that repeated every 260 (13×20) days. This resulting 260-day period was termed the *tonalpohualli* by the Aztecs and is called the *tsolk'in* by Mayanists. Mesoamericans visualized this period as a four-sided cosmogram (Figure 3) that can be described as a 'Maltese' cross (the broad arms of which are identified in a counterclockwise order as East, North, West, and South) superimposed over a 'St. Andrew's' cross (the narrow arms of which define the diagonal directions Northeast, Northwest, Southeast and Southwest). The eight projecting arms of this scheme are each comprised of thirteen days proceeding out from the center and thirteen more returning, with the four outer edges of the 'Maltese' arms also each contributing thirteen to complete the round of 260 days. This 260-day *tonalpohualli* was the most significant aspect of the calendar for the daily life of the typical Mesoamerican. It formed the basis of numerous rituals, character evaluations, omens, and prognostications, traced out in Aztec, Mixtec, and Maya books.

While the 260-day ritual period seems to have been the result of the combination of a cultural fascination with the number 13 and a vigesimal counting system derived from the number of digits of the hands and feet, the 365-day period — the Maya *haab* or Aztec *xihuitl* — to which we now turn, was an attempt at a full-number approximation of the tropical year (of 365.2422 days, as calculated today). This "vague year" was achieved by a set of eighteen named periods of twenty consecutively-numbered days each, and five extra consecutively-numbered terminal days (*Nemontemi* in Aztec, *Wayeb* in Maya) to total 365 days (18×20 + 5) (Figure 4). Every day, then, could be expressed with a day-number, a day-name and a numbered position in one of the periods of the "vague year". The workings of this Calendar Round system are well illustrated by comparing the modes of expressing succeeding dates that include the five terminal days of the vague year in the Aztec *xiuhmolpilli*

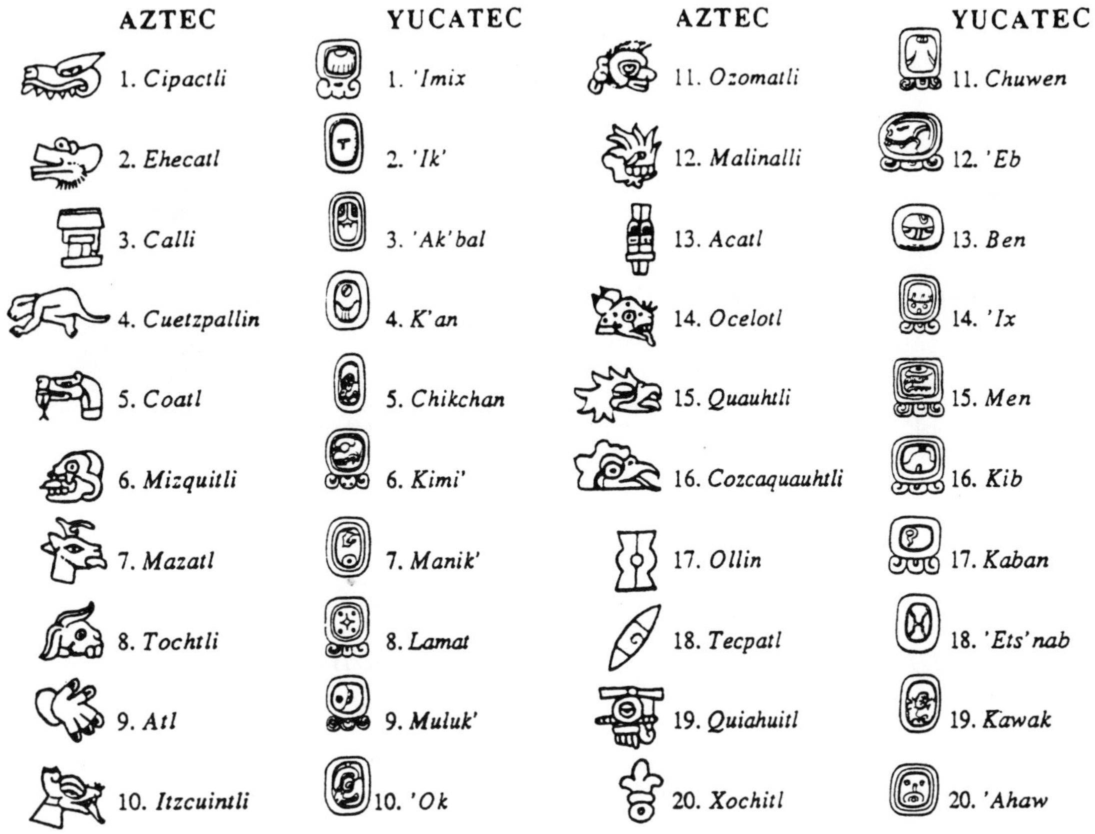

Figure 2: The twenty Aztec (Nahuatl) and Yucatec (Mayan) day-names and signs. [Drawings by the author.]

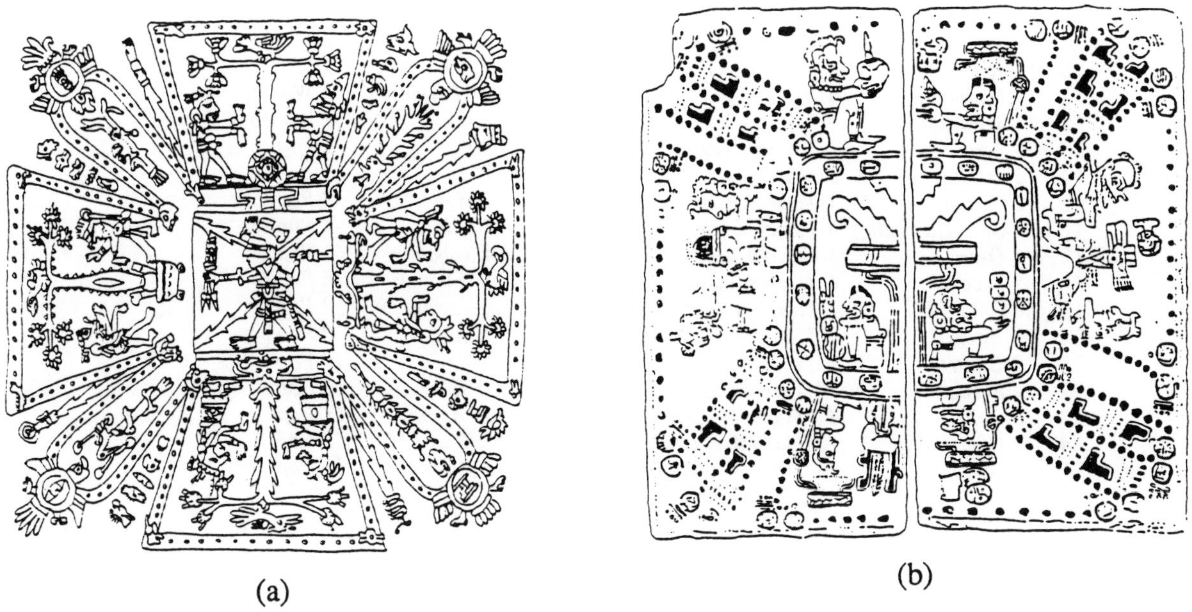

Figure 3: The 260-day period as portrayed by the (a) Aztecs in the *Fejervary-Mayer Codex* and (b) Maya in the *Madrid Codex*.

YUCATEC	AZTEC	YUCATEC	AZTEC
1. *Pop*	1. *Tlaxochimaco*	10. *Yax*	10. *Izcalli*
2. *Wo'*	2. *Xocotlhuetzi*	11. *Sak*	11. *Atlcahualo*
3. *Sip*	3. *Ochpaniztli*	12. *Keh*	12. *Tlacaxipehualiztli*
4. *Sots*	4. *Teotleco*	13. *Mak*	13. *Tozoztontli*
5. *Sek*	5. *Tepeilhuitl*	14. *K'ank'in*	14. *Hueytozoztl*
6. *Xul*	6. *Quecholli*	15. *Muwan*	15. *Toxcatli*
7. *Yaxk'in*	7. *Penquetzaliztli*	16. *Pax*	16. *Etzalcualiztl*
8. *Mol*	8. *Atemoztli*	17. *K'ayab*	17. *Tecuilhuitontlii*
9. *Ch'en*	9. *Tititl*	18. *Kumk'uh*	18. *Hueytecuilhuitl*
		Wayeb	*Nemontemi*

Figure 4: The eighteen 20-day periods and final five days of the Aztec *xihuitl* and Maya *haab*.

AZTEC			MAYA		
XIUHMOLPILLI			*HUNAB*		
	TONALPOHUALLI	***XIHUITL***		***TSOL K'IN***	***HAAB***
1.	*1 Cipactli*	*14 Huetecuilhuitl*	1.	*1 'Imix*	*14 Kumk'u*
2.	*2 Ehecatl*	*15 Huetecuilhuitl*	2.	*2 'Ik'*	*15 Kumk'u*
3.	*3 Calli*	*16 Huetecuilhuitl*	3.	*3 'Ak'bal*	*16 Kumk'u*
4.	*4 Cuetzpallin*	*17 Huetecuilhuitl*	4.	*4 K'an*	*17 Kumk'u*
5.	*5 Coatl*	*18 Huetecuilhuitl*	5.	*5 Chikchan*	*18 Kumk'u*
6.	*6 Mizquitli*	*19 Huetecuilhuitl*	6.	*6 Kimi*	*19 Kumk'u*
7.	*7 Mazatl*	*20 Huetecuilhuitl*	7.	*7 Manik'*	*20 Kumk'u (0 Wayeb)*
8.	*8 Tochtli*	*1 Nemontemi*	8.	*8 Lamat*	*1 Wayeb*
9.	*9 Atl*	*2 Nemontemi*	9.	*9 Muluk*	*2 Wayeb*
10.	*10 Itzcuintli*	*3 Nemontemi*	10.	*10 'Ok*	*3 Wayeb*
11.	*11 Ozomatli*	*4 Nemontemi*	11.	*11 Chuwen*	*4 Wayeb*
12.	*12 Malinalli*	*5 Nemontemi*	12.	*12 'Eb*	*5 Wayeb (0 Pop)*
13.	*13 Acatl*	*1 Tlaxochimaco*	13.	*13 Ben*	*1 Pop*
14.	*1 Ocelotl*	*2 Tlaxochimaco*	14.	*1 'Ix*	*2 Pop*
15.	*2 Quauhtli*	*3 Tlaxochimaco*	15.	*2 Men*	*3 Pop*
16.	*3 Cozcaquauhtli*	*4 Tlaxochimaco*	16.	*3 Kib*	*4 Pop*
17.	*4 Ollin*	*5 Tlaxochimaco*	17.	*4 Kaban*	*5 Pop*
18.	*5 Tecpatl*	*6 Tlaxochimaco*	18.	*5 'Ets'nab*	*6 Pop*
19.	*6 Quiahuitl*	*7 Tlaxochimaco*	19.	*6 Kawak*	*7 Pop*
20.	*7 Xochitl*	*8 Tlaxochimaco*	20.	*7 'Ahaw*	*8 Pop*
21.	*8 Cipactli*	*9 Tlaxochimaco*	21.	*8 'Imix*	*9 Pop*
22.	*9 Ehecatl*	*10 Tlaxochimaco*	22.	*9 'Ik'*	*10 Pop*
23.	*10 Calli*	*11 Tlaxochimaco*	23.	*10 'Ak'bal*	*11 Pop*
24.	*11 Cuetzpallin*	*12 Tlaxochimaco*	24.	*11 K'an*	*12 Pop*
25.	*12 Coatl*	*13 Tlaxochimaco*	25.	*12 Chikchan*	*13 Pop*

Figure 5: The succession of 25 Calendar Round dates that pass through the five terminal days of the *xihuitl* of the Aztec *xiuhmolpilli* and the *haab* of the Maya *hunab*.

and Maya *hunab* (Figure 5). The presence of these five days causes the position of any given *tsolk'in* day to drop back by five days with each successive vague year. Because 260 and 365 share a common denominator of 5, any specific combination of day-number, day-name, and vague year position repeats every 18,980 days ($260 \times 365 \div 5 = 18,980$), or about 13 days short of 52 tropical years. As with the Long Count, evidence for the origins of the Calendar Round lies beyond the western pale of the Maya, dating from as early as the sixth century BC in the Zapotec region.

The *xiuhmolpilli* was of particular importance to the Aztecs, whose mythology claimed that the age in which they lived and in which the current Sun-God, *Tonatiuh*, reigned was preceded by four earlier ages, each with its own sun and each of which had been destroyed in its own way — the first by devouring jaguars, the second by winds, the third by a rain of fire and the fourth by a world-flood. During this fifth era, under *Tonatiuh*, the future seems to have been doled out to the Aztecs one *xiuhmolpilli* at a time. Each successive *xiuhmolpilli* had to be freshly solicited by a great deal of bloodletting and an elaborate "New Fire" ceremony that was carried out during the five *Nemontemi* days of each fifty-second *xihuitl*. If, following the ceremony, the stars continued on their steady march across the midnight-sky of the last day of the *xiuhmolpilli* — presumably because the gods had found the ritual to their satisfaction — the people could be assured that another fifty-two "years" had been granted them.

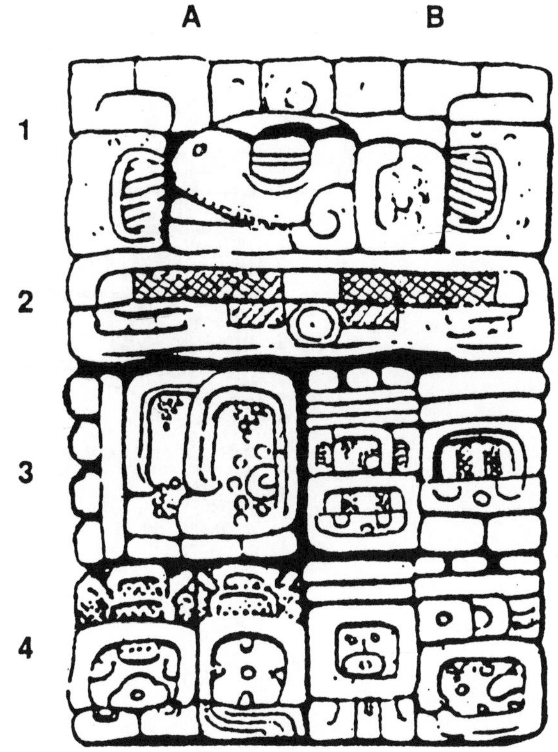

Figure 6: Quirigua, Stela I: hieroglyphic text with Initial Series date. [Drawing after Annie Hunter.] From *Biologia Centrali-Americana*, by Alfred P. Maudlsay, reprinted in 1974 by Milpatron Publishing Corp. Used with permission of the publisher.

In the Maya region, this endlessly repeating pan-Mesoamerican cycle was synchronized with the accumulated days of the Long Count as the mathematical consequence of the arbitrary assignment of the Calendar Round date *4 'Ahaw 8 Kumk'u* to the Long Count zero-date (August 13, 3114 BC). Written out in full and preceded by a distinctive hieroglyph that served to introduce them, dates so synchronized stood at the very beginning of the hieroglyphic texts on the carved monuments of Mayadom — whence their designation, Initial Series dates. To take an example, the Initial Series date on Quirigua Stela I (Figure 6) — which reads in typical Maya fashion, from top-left to bottom-right, two columns at a time — displays a large Initial Series Introductory Glyph (ISIG) in the initial position (A1–B2) followed by the Long Count *9.18.10.0.0* (A3–A4) and the Calendar Round *10 'Ahaw 8 Sak* (B4), establishing a date corresponding to August 19, 800. (Glyphs are conventionally located by assigning upper-case letters to the columns of a text, numbers to its rows, and lower-case letters to divisions within the glyph-blocks and designating them in that order, as A3, C2a, F7b, etc. Number-values are dots for 1, bars for 5, and various special symbols for 0.)

At times, supplementary to the Long Count and Calendar Round and inserted parenthetically between the *tsolk'in* and *haab* entries, were glyphs that recorded additional cycles of 7 or 9 days duration, as well as certain lunar information. Supplementing the cycles mentioned, the Maya tracked one of 819 days — a product of three basic periods ($7 \times 9 \times 13$) — relating it to a 4-day cycle (which the Aztecs also recognized) that linked each day with one of the four cardinal directions and associated colors (East, red; North, white; West, black; South, white). On Piedras Negras Stela 1 (Figure 7), for example, the ISIG (A1) is followed by the Long Count *9.12.2.0.16* (A2–A6) and the *tsolk'in* position *5 Kib* (A7), while the *haab* position, *14 Yaxk'in*, does not appear until eight glyph-blocks later (C2). The intervening blocks include two that record the seventh position of the nine-day cycle (A8–A9), one that records the moon's age as eight days (A10), another the moon's number as third in a set of six (A11), two that name the moon (B1–C1), and one that identifies it as a thirty-day month (B2). Following *14 Yaxk'in* are four non-calendrical glyphs (B3–E1) describing the event of the Initial Series date (July 7, 674).

When there was a need to record multiple dates, each

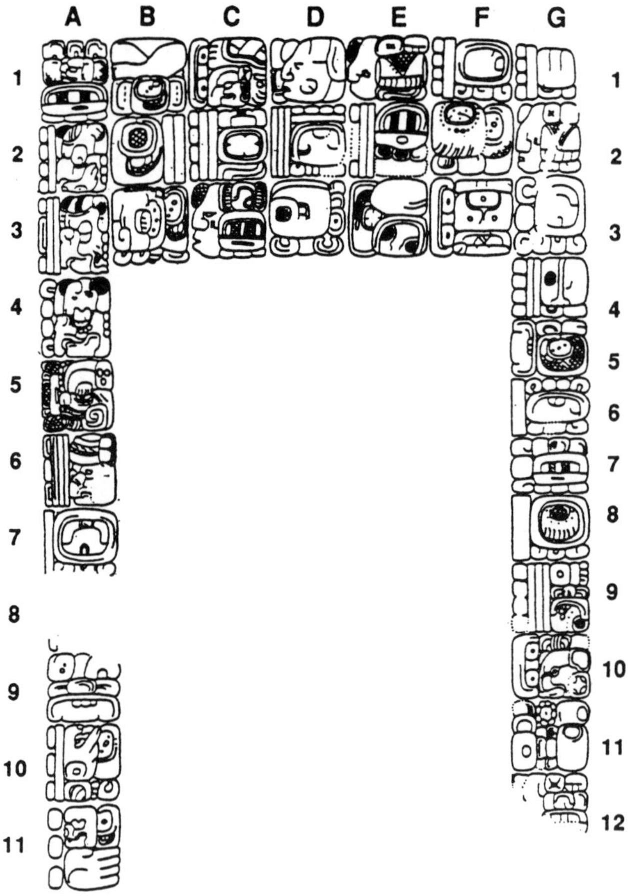

Figure 7: Piedras Negras, Stela 1: hieroglyphic text. [Drawing by John Montgomery.]

new one (expressed as a Calendar Round date) was typically linked to its predecessor by what scholars call a Distance Number. The Distance Number employed the same place system as the Long Count, but because the number of days to be recorded usually did not require all five places, they were entered in the reverse order of the Long Count — commencing with *k'inob* and proceeding only through those higher places necessary to record the distance. Thus on the Piedras Negras stela, following the initial event, come two glyphs that record the Distance Number *15 k'inob*, *9 winalob*, and *12 tunob* (D2–E2), then two more that indicate an advance of that distance forward in time (D3–E3), and finally, two glyphs that record the Calendar Round date, *9 Chuwen 9 K'ank'in*. The calendrical mathematics are:

9. 12. 2. 0. 16	*5 Kib*	*14 Yaxk'in*	(July 7, 674)
+ 12. 9. 15			
9. 12. 14. 10. 11	*9 Chuwen*	*9 K'ank'in*	(November 16, 686)

After two non-calendrical glyphs (F2–G2) that describe the event of the date so reached, come a second distance number, *5 k'inob* (F3), and the resulting third Calendar Round date, *1 Kib 14 K'ank'in* (G3–G4). The mathematics are:

9. 12. 14. 10. 11	*9 Chuwen*	*9 K'ank'in*	(November 16, 686)
+ 5			
9. 12. 14. 10. 16	*1 Kib*	*14 K'ank'in*	(November 21, 686)

This new date is followed by the single glyph that describes the event (G5). Then come the Distance Number, *5 k'inob*, *2 winalob*, and *1 k'atun* (G6–G7), a fourth Calendar Round date, *5 Imix 19 Sak* (G8–G9), and fourth event (G10–G12). This final calculation is:

9. 12. 14. 10. 16	*1 Kib 14 K'ank'in*	(November 21, 686)
+ 1. 0. 2. 5		
9. 13. 14. 13. 1	*1 Imix 19 Sak*	(September 23, 706)

As was the case for the second of these two monuments, the Maya often accompanied their calendrical entries with lunar data. From this propensity and from the structure of a Lunar Table in the *Dresden Codex*, it is clear that their concern was to anticipate eclipse-possible dates. Because eclipses (both lunar and solar) can only take place during a period of eighteen days on either side of node-passage of the Moon over the path of the Sun, which occurs twice during a tropical year, the Maya grouped the months in semesters of six (and for corrective purposes, occasionally of five). In an attempt to approximate the Moon's 29.53059-day synodic period (as calculated today), they ascribed twenty-nine and thirty days, alternately, to successive months, compensating for the gradual accumulation of excess time by introducing an extra thirty-day month at predetermined intervals. One result was the *Dresden Codex* Lunar Table with its eclipse-period of 11,960 days, or just over 405 complete lunations. However, that excess time would be allowed to accumulate until after eight repetitions of the period (95,680 days), when the table could be adjusted to correspond more closely to the moon's position by subtracting one day. The mathematics behind this adjustment are: $29.53059 \times 405 = 11{,}959.88895$; $11{,}960 - 11{,}959.88895 = 0.11105$; $11{,}960 \times 8 = 95{,}680$; $11{,}959.88895 \times 8 = 95{,}679.116$; $95{,}680 - 95{,}679.116 = 0.884$, or 21.216 hours. To further refine the Lunar Table's use, follow-up adjustments had to be made with succeeding repetitions of the cycle.

Though not central to the calendar itself, the planet Venus was nevertheless also a concern of Maya calendrical savants. Observations and calculations recorded in the inscribed monuments show that the Maya employed a 584-day period to predict culturally important points in the planet's synodic period. Both the Maya and the Aztecs assigned numerological significance to the correspondence between five of these Venus-periods and eight *haabs* ($5 \times 584 = 2{,}920$; $8 \times 365 = 2{,}920$). Also, in the *Dresden Codex* is an elaborate table that consists of thirteen sets of five periods of 584 days, or 37,960 days. Here again, because of the discrepancy between the full number 584 and the actual synodic period of Venus (583.97 days, by modern measurements) and because of the added cultural requirement that the first heliacal rising of the planet should occur on the *tsolk'in* date *1 'Ahaw*, an adjustment like that in the Lunar Table had to be made after several runs of the Venus Table.

The Mesoamerican calendar, if by no means invented by the Maya, nevertheless reached its most elaborate form at their hands. Tracking basic periods of 4, 7, 9, 13, 20, 365, and 584 days and greater periods that were the products of various combinations or lowest common multiples thereof — 260 (13×20), 819 ($7 \times 9 \times 13$), 2,920 (8×365; 5×584), 3,276 ($4 \times 7 \times 9 \times 13$), 18,980 ($13 \times 20 \times 365$) and 37,960 ($13 \times 20 \times 365 \times 2 = 13 \times 5 \times 584$) days — and groupings of months of alternately twenty-nine and thirty days in duration, they synchronized them all with the cumulative and open-ended Long Count that recorded every passing day from its original *4 'Ahaw 8 Kumk'u* zero point. Where the European calendar intercalated extra days as a device for keeping its dates in close harmony with the seasons (the vernal equinox, for example, occurring always on or about March 22), the Mesoamerican calendar, even in its most refined (Maya) mechanism, simply ground out the one-day intervals of its several parts, accumulating or repeating them endlessly and without interruption, all of its components creeping gradually backwards through the seasons. It was thus impossible to identify a position in the tropical year by knowing a Calendar Round date. The first *4 'Ahaw 8 Kumk'u* fell on August 13, but the second fell twelve days earlier on August 1 and the third, twelve more on July 19. In the year 265, near the beginning of the Classic Period, after having returned to its original August 13 date 113 times, *4 'Ahaw 8 Kumk'u* fell on May 16, and by the close of the Classic in the late ninth century, it had fallen back to December 15.

However skilfully the Maya may have combined their astronomical observations with calendrical calculations to predict lunar and solar eclipse stations, heliacal risings and settings, conjunctions, stationary points, and retrograde motions of the five visible planets, the fact remains that, for all its reputation, the calendar itself was no more accurate than the 365-day calendar the ancient Egyptians adopted around 4250 BC which was the product of twelve thirty-day months ($12 \times 30 = 360$), plus — like the Mesoamerican *xihuitl* or *haab* — five extra days. Indeed, it was an Egyptian mathematician, Sosigenes, who, recognizing the faults of the rigid 365-day year, persuaded Julius Caesar to adopt his scheme for reforming the Roman calendar in 45 BC, by intercalating an extra day every fourth year into an otherwise 365-day year. However, it would be unhistorical to judge the Maya calendar by a standard set by the Egyptian, the Julian, or the later Gregorian calendar. The cultural need of the Egyptians was to anticipate (for agricultural reasons) the annual flooding of the Nile. That of the Julian reform was to ensure that the Roman solsticial festival *fors fortuna* would fall regularly on June 22. The Christians, in turn, adopted the Julian calendar in 325 at the Council of Nicaea from a need to establish predictable dates for Easter, which — for religious reasons — had to fall on the first Sunday following the first full moon after the vernal equinox. It was this same need that led Pope Gregory XIII to authorize a further reform of the Julian calendar in 1582.

However, the cultural needs met by the Mesoamerican calendar were of a numerological–astrological nature. Quite apart from the divinations of the purely numerical *tsolk'in*, the Maya assigned significance to astronomical predictions based upon complex multiples of their several whole-number cycles which were often linked to the actions of remote or fabled ancestors or mythological supernaturals or to multiples of the never whole-numbered synodic periods of the heavenly bodies. Contrived calendrical calculations into the remote past and distant future were made to establish mythological beginnings, regenerations, and calamities to come. Day-to-day events, even to the timing of military attacks or battles among themselves or surrender to the Spanish, seem to have fallen under this numerological spell. However closely the *Dresden* Venus Table may appear to approach a modern astronomical ephemeris of Venus, its purpose was to return the latter's heliacal rising to *1 'Ahaw*, the birth date of the mythological Venus-associated hero, *Hunahpu* — irrespective of any temporary losses in accuracy its mechanisms may have produced. If the calendrical efforts of Mesoamericans can be described as an attempt to bring about a marriage between their whole-number numerology and their perceived rhythms of the Sun, Moon, planets, and stars, it was one in which it was the latter who were to love, honor, and obey. However, if the test of a good marriage is its durability, then the Mesoamerican calendar — particularly in its Maya form — was not only mathematically sophisticated, conceptually elegant, and visually beautiful, but a highly successful one as well.

TOM JONES

REFERENCES

Closs, Michael P. "The Nature of the Maya Chronological Count." *American Antiquity* 42 (1): 18–27, 1977.

Edmonson, Munro S. *The Book of the Year, Middle American Calendrical Systems*. Salt Lake City: University of Utah Press, 1988.

Edmonson, Munro S. "The Middle American Calendar Round." In *Epigraphy. Supplement to the Handbook of Middle American Indians*, volume 5. Ed. Victoria Reifler Bricker. Austin: University of Texas Press, 1992, pp.154–167.

León-Portilla, Miguel. *Time and Reality in the Thought of the Maya*. Trans. Charles L. Boiles and Fernando Horcasitas. Boston: Beacon Press, 1973.

Morley, Sylvanus Griswold. *An Introduction to the Study of the Maya Hieroglyphs*. Bureau of American Ethnology, Bulletin no. 57. Washington: Smithsonian Institute, 1915; reprinted New York: Dover, 1975.

Thompson, J. Eric S. *Maya Hieroglyphic Writing: an Introduction*. Norman: University of Oklahoma Press, 1960.

See also: Time – Long Count

CALENDARS IN SOUTH AMERICA At the time of the Spanish conquest around the year 1530, the Incas had integrated different cultures, all belonging to Andean civilization, into a large empire, stretching from southern Colombia through Ecuador, Peru, and Bolivia, into northern Chile and Argentina. Much can be learned from the art of the Incas and their predecessors, including architecture, stone sculpture, pottery, and textiles. However, we have no access to written records as defined by western culture; understanding intellectual aspects of this civilization is almost exclusively dependent on what informants from Cuzco, the Inca capital in southern Peru, told the Spanish chroniclers. These were most impressed by Inca statecraft, although they were never present when it was still practiced. Their interpretation of Inca religion was heavily influenced by Christian opinions. Religion to them included memories of the Inca past and of state and private rituals related to, for instance, the calendrical organization of agriculture and llama husbandry. As Inca history is discussed in the article on Knowledge Systems of the Incas, this one concentrates on the calendar as it functioned in Cuzco.

Most of our information on the Inca calendar and related astronomical practices derives from only a few chroniclers: Juan de Betanzos [1551],[1] a soldier and the first chronicler to write on this subject in the context of a full account of Inca culture; an anonymous author [ca. 1565] with precise information on the integration of astronomical observations into the calendar; Cristóbal de Molina [1574], a priest with great knowledge of Inca rituals but excluding those for one third of the year; and Bernabé Cobo [1653], a late Jesuit author who gave our most comprehensive description of Inca culture including the calendar for the whole year. His information probably derives from a now lost account by Juan Polo de Ondegardo [1559], whose own description of the calendar only survives in a much revised short version [1585] for use by Catholic priests. Later chroniclers and dictionaries mostly follow Polo's version and add little information towards a reconstruction of the original Inca calendar. This reconstruction provides very different results from the general picture that the chroniclers present us with.

All chroniclers conceive the Inca calendar as consisting of twelve months. Notwithstanding this agreement, there is

[1] In case of an old text, square brackets indicate the date of writing or first publication.

much confusion about the nature of the months, their order, and their correspondence to Christian months. Some of this confusion may derive from much local variation among calendars in the empire, but most of the blame is due to Spanish misreporting on the Cuzco calendar itself. Betanzos, Polo, and Cobo describe a system of twelve "months" with a fixed position in the solar year. Molina speaks of "months" *and* "moons" and refers to rituals according to the phases of the moon for three times of the year, roughly corresponding to June, September, and January. Some modern scholars therefore argue that the whole calendar was lunar, but this was not the case (Ziólkowski and Sadowski, 1992). On each of the three occasions important lunar rituals were celebrated after an independent solar observation had been made for that time of the year. We have no information on lunar rituals in Cuzco for other times of the year; we can only conclude that the integration of solar and lunar rituals in Peru was done in a way different from Western (Christian, Jewish) presuppositions in this respect.

There are various difficulties in accepting the Inca organization of months according to the Polo version of 1585. For instance, two names are taken together for one month that in other versions refer to separate months; in other calendars a name may be used as an alternative for either one or a next month. In some versions, a month name is suppressed for no apparent reason at all; months were said to be of rather different lengths, either longer or shorter than 30 or 31 days. And the anonymous chronicler mentions for two double-months that the moon observed in one month (either the first or the second) would always reach the other, leaving us to conclude that the fixed double-month was shorter than two movable synodic (lunar) months ($2 \times 29\frac{1}{2} = 59$ days). Comparing the various descriptions of the Inca year, it becomes clear that originally it contained thirteen and not twelve months and that each version suppressed one or another month name to accomodate this calendar to the Spanish one. It may have been a goal of Catholic missionaries fighting against "pagan" rituals, but there was also Inca consent, realizing the advantages of combining Inca and Catholic celebrations, first openly and later secretly.

From indications by Molina and Cobo we can conclude that the Inca calendar placed one month around the June solstice (21 June), celebrating *Inti Raymi* (feast of the sun), and two months, *Capac Raymi* (royal feast) and *Capac Raymi Camay Quilla* (royal feast, feast of the moon) around the December solstice, which honored the fact that in the southern hemisphere the sun is strongest then. For practical reasons, *Inti Raymi* may have been celebrated not on the days of the June solstice but within its month during the full moon either before or after in a given year. People took advantage of the dry season's clear skies when the moon shone the whole night.

During *Capac Raymi*, considerations were different. While the June solstice marked the middle of a month, the December solstice, six and a half Inca months apart, marked the date separating the two months of *Capac Raymi*. Moreover, these months fell in the middle of the rainy season when normally no good observations of the sun and the moon could be made. The month of *Capac Raymi Camay Quilla* was celebrated, first awaiting the new moon with some celebrations. The major celebrations came with the next full moon and took some seven days. If the new moon arrived early after the solstice (December 21), all celebrations took some 22 (15+7) days; but if it was late, they could take up some 52 (30+15+7) days. Then the full moon celebrations would not occur in *Camay Quilla* but extend into the next month. Although after those days the moon would still be visible late at night, no more attention would be paid. However, the day after these 52 days (February 13) was important for another reason. Then the sun in Cuzco — this city being $13\frac{1}{2}°$ south of the equator and within the tropics — passes the zenith point in the sky at noon. Although the chroniclers were only vaguely aware of the importance of this event, Polo indicates that then a new year began, with people preparing for the coming harvest. The sun also passes the zenith 53 days before the December solstice, and we can assume that a similar calendrical calculation was made for this period. The anonymous chronicler considered it as one in which the moon of one month would always reach the other. Its beginning was celebrated with a feast, the *Itu*, which only Inca nobility could attend, and the last six days before the December solstice were also important for celebration. These days could begin with a full moon if the new moon had arrived late in this 53 day-period.

Apparently, the Incas divided the time from the first to the second passage of the sun through the zenith (107 days) into four months, of which the central ones of *Capac Raymi* were most important. The average length of the thirteen months was about 28 days and as the connection to the length of a synodic month (29 days) was lost, there is no reason that the months could not be of various lengths as reported by the chroniclers. Having established the framework of the Inca calendar, an outline can be given of its ritual organization serving various, often contrasting, social needs.

The calendar organized, first, a multitude of sacrifices (llamas, guinea-pigs, birds, sometimes children) and offerings (mostly cultivated plants). The months were organized according to four seasons of three months and an independent month after the sun's second zenith passage. Sacrifices and offerings expressed the different seasonal concerns.

A quite different organization of the months attended more political interests in terms of tribute due by Inca and non-Inca subjects and of the large feasts of state presented by the king, queen, and high nobility. In this organization, months were brought together in groups of two and then of four. This organization cross cut the seasonal one but kept the same independent month. Our first chroniclers mention how tribute was brought to Cuzco every four months and how non-Inca nobles accompanying these deliveries stayed for the same time in town. Betanzos adds to this that services of labor and distribution of food, especially meat and cloth, followed the same schedule. At first, the colonial administration showed interest in adopting this schedule for its own needs, but after about 1570 it changed to the European custom of having tribute paid twice a year, around Christmas and St. John (24 June), and chroniclers no longer paid attention to the earlier system.

I mentioned above the most important four month-period around the December solstice. In between the times of planting and harvest, this was a season when the fields needed less attention and when heavy rainfall made travel difficult. *Capac Raymi* and its preceding month were dedicated to the initiation rituals of noble boys, these being assisted by noble girls who had had their own rituals of first menstruation at other times of the year. In the lunar rituals of *Camay Quilla* and the following month, noble men and women pledged their allegiance to the king, and rains were ritually dismissed awaiting the ripening of the fruits. In the four months around planting, land was first prepared for cultivation, ploughed and planted, and then rituals followed to help the budding plants grow and attract rains. Again, the two central months were dedicated to state rituals. First, the king himself helped with the planting. Keen observations were made of sunset throughout the first month to have the various timings for planting and irrigation of different sorts of plants, at lower, warmer and higher, colder places well orchestrated. In the second month, dedicated to the moon, the queen, and women in general, "evil" and "illnesses" were ritually driven out of town in support of peace among people and of the growth of plants. Finally, the four months of harvest included the early harvest, the general harvest around Cuzco, the reception of foreign lords with their presents and the tribute of their people, and storage. Again the most important ceremonies occurred in the central months when first harvest was ritually brought into town and then the Inca himself ploughed, opening a new agricultural year, and ate with all nobles and foreign lords in the plaza.

Good descriptions of Andean rituals were also made for the central provinces around Lima, the vice royal capital in colonial times, and some of these ceremonies can be compared with Incaic ones in Cuzco. But when those other records became available in the early seventeenth century, more than eighty years had already passed after the Spaniards had arrived in the Andes. Much research still has to be carried out making those comparisons fruitful for a reconstruction of prehispanic calendars in general.

R. TOM ZUIDEMA

REFERENCES

Duviols, Pierre. *Cultura Andina y Represion. Procesos y Visitas de Idolatrias y Hechcerias Cajatambo, Siglo XVII*. Cusco: Centro de Estudios Rurales Andinos "Bartolome de las Casas", 1986.

Moseley, Michael E. *The Incas and Their Ancestors. The Archaeology of Peru*. London: Thames and Hudson, 1992.

Salomon, Frank, and George L. Urioste. *The Huarochirí Manuscript. A Testament of Ancient and Colonial Andean Religion*. Austin: University of Texas Press, 1991.

Urton, Gary. *At the Crossroads of the Earth and the Sky. An Andean Cosmology*. Austin: University of Texas Press, 1981.

Ziólkowski, Mariusz S., and Robert M. Sadowski. *La Arqueoastronomía en las Investigaciones de Las Culturas Andinas*. Quito: Banco Central del Ecuador/Instituto Otavaleno de Antropologia, 1992.

Zuidema, R. Tom. "The Moieties of Cuzco." In *The Attraction of Opposites. Thought and Society in the Dualistic Mode*. Ed. Uri Almagor and David Maybury-Lewis. Ann Arbor: The University of Michigan Press, 1989, pp. 255–75.

Zuidema, R. Tom. *Inca Civilization in Cuzco*. Austin: University of Texas Press, 1990.

Zuidema, R. Tom. "At the King's Table. Inca Concepts of Sacred Kingship in Cuzco." In *Kingship and the Kings*. Ed. Jean-Claude Galey. London: Harwood Academic Publishers, 1990, pp. 253–78.

Zuidema, R. Tom. "Inca Cosmos in Andean Context from the Perspective of the Capac Raymi Camay Quilla Feast Celebrating the December Solstice in Cuzco." In *Andean Cosmologies Through Time: Persistence and Emergence*. Ed. Robert V.H. Dover, Katherine Seibold and John H. McDowell. Bloomington: Indiana University Press, 1992, pp. 17–45.

Zuidema, R. Tom. "El Encuentro de los Calendarios Andino y Español." In *Los Conquistados. 1492 y la Población Indígena de las Américas*. Ed. Heraclio Bonilla. Quito, Ecuador: Tercer Mundo editores, FLACSO, 1992, pp. 287–316.

Zuidema, R. Tom. "De la Tarasca a Mama Huaco. La Historia de un Mito y Rito Cuzqueño." In *Religions Des Andes et Langues Indigènes Equateur - Pérou - Bolivie avant et après la Conquête Espagnole. Actes Du Colloque III D'Études Andines*. Ed. Pierre Duviols. Aix-en-Provence: Publications de l'Université de Provence, 1993, pp. 331–381.

CANDRAŚEKHARA SĀMANTA Candraśekhara Sāmanta (1835–1904) was a self-made astronomer who had the distinction of revising the traditional calendar of Orissa during the nineteenth century. He was a scion of a junior branch of the chiefs of the small estate of Khandapara and bore the title *Sāmanta* (Feudatory) on that account. His full traditional name was Sāmanta Śrī Candraśekhara Singh Harichandan Mohāpatra, and he was called locally Pathani Sāmanta. Young Candraśekhara had little modern education, but learned Sanskrit and astrology from his uncle, which he further developed by an intensive study of two of the authoritative texts of the times, the *Sūryasiddhānta*, and the *Siddhāntaśiromaṇi* of Bhāskarācārya. Exhibiting an uncanny interest in watching the skies, he became aware that the times of the rising and setting of the sun and moon and of the other celestial bodies were at variance with the times indicated in local almanacs arrived at by computation based on traditional texts. Often, when almanacs differed in their indication of times, and even of dates (*tithis*), there was confusion in the matter of fixing the days for domestic and social festivals and of sacred worship at temples like that of Lord Jagannātha at Puri.

Deeply religious and equipped with the fundamentals of traditional astronomy, Candraśekhara took it upon himself to remedy the prevailing state of affairs, and from the age of 23 he commenced watching the transit, conjunction, rising and setting of the celestial bodies, and recorded his observations consistently and systematically. As aids to accuracy in observation, he designed, all by himself, several astronomical instruments which had been described in the *Sūryasiddhānta* and the *Siddhāntaśiromaṇi* and their commentaries. Among the instruments were an armillary sphere and a vertical wheel, which in modern terms served the purpose of the transit and alta-azimuth instruments. He also used the clepsydra (water clock) for measuring sidereal time. An instrument which he designed and used constantly for celestial observation in place of a telescope was a T-square frame which he called *mānayantra* (measuring instrument) with the main limb twenty-four digits and the crosspiece marked with notches and having holes at distances equal to the tangents of the angles formed at the free end of the main piece. He also designed an automatically revolving wheel (*svayam-vāhaka*), with spokes partly filled with mercury, which he also used to measure time. He effectively used the gnomon, which when fitted with a small mirror could be used at night for measuring time and angular distances.

Candraśekhara's observations, experiments, and recordings, on the basis of which he coined corrections to earlier enunciations and innovated new methodologies and practices, lasted thirty-four years. This enabled him to introduce reforms to all aspects of astronomical computation, which he set out in an extensive work *Siddhāntadarpaṇa*, in 2500 verses. This is an astronomical manual in five chapters, devoted, respectively, to Mean planets (*Madhyama-adhikāra*), True planets (*Sphuṭa-adhikāra*), Problems of Direction, Time and Place (*Tripraśna adhikāra*), Spherics (*Gola-adhikāra*), and Time and Appendices (*Kāla adhikāra* and *Pariśiṣṭa*). The main contributions of Candraśekhara include corrections to the sidereal periods of the star planets and to the main inclinations of the planetary orbits to the ecliptic (the band of the zodiac through which the sun apparently moves in its yearly course), identification, and evaluation of evection, variation, and annual equation among the moon's inequalities, and horizontal parallax of the sun and the moon. Recognition and honors were late in coming when the Government of India conferred on him the title of *Mahāmahopādhyāya* (Scholar of scholars), the highest title for a Sanskritist.

K.V. SARMA

REFERENCES

Bandopadhyay, Amalendu. "Astronomical Works of Samanta Chandrasekhar." *Journal of the Asiatic Society* 30: 7–12, 1989.

Baral, Haris Chandra. "Chandra Sekhar Samanta." *Indian Review* 23: 459–61, 1922.

Ray, Joges Chandra. "Centenary of Chandra-sekhara and a Reformed Hindu Almanac." *Modern Review* 1936, pp. 56–60.

Sarma, S.R. "Perpetual Motion Machines and Their Design in Ancient India." *Physics* 29(3): 665–76, 1992.

Sengupta, P.C. "Hindu Luni-solar Astronomy." *Bulletin of the Calcutta Mathematical Society* 24: 1–18, 1932.

Siddhāntadarpaṇa by Candraśekhara Sinha. Ed. Joges Chandra Ray. Calcutta: Indian Depository, 1897.

See also: Armillary Sphere – *Sūryasiddhānta*

CARAKA The oldest of the Ayurvedic classical works still in use by Ayurvedic physicians is the *Carakasaṃhitā*, and its celebrated editor was Caraka. Caraka's contribution was so significant that the work bears his name, rather than that of Agniveśa, the original author.

Caraka was a popular name in ancient India from the Vedic period up to the time of Kaniṣka (AD 1), but the editor of the *Carakasaṃhitā* belongs to a period prior to 600 BC, the pre-Buddhist period. We know this because the prose style of this work resembles the Brāhmaṇas and Upaniṣads, because there are no references to Buddha or Buddhist philosophy, because there are references to the Vedas and Vedic gods, and because there is no Puranic mythology in the text.

The *Carakasaṃhitā* consists of 120 chapters which are divided into eight sections. Unfortunately, the original complete work is no longer extant. Seventeen chapters of the sixth section, and the entire seventh and eighth sections were later supplemented by another scholar-physician called Dṛḍhabala (AD 12).

Out of the eight specialized branches of Ayurveda, Caraka's work primarily deals with six: (1) internal medicine, (2) toxicology, (3)psychiatry, (4) pediatrics, (5) rejuvenation therapy, and (6) sexology, including the use of aphrodisiacs. The other two, surgery and the treatment of eye, ear, nose, and throat diseases, are scantily described. The important topics dealt with in this text are anatomy, physiology, etiology (disease causation), pathology, therapeutics, form and time of treatment, conduct of the physicians, medicaments and appliances, and rules for diet and drugs. In addition, it contains a detailed classification and nomenclature for diseases, as well as information on embryology, obstetrics, personal hygiene, sanitation, and the training of physicians. It also describes the cosmological, biological, physico-chemical, metaphysical, ethical, and philosophical ideas prevalent in ancient India.

BHAGWAN DASH

REFERENCES

Agniveśa. *Carakasaṃhitā.* Bombay: Nirnayasagara Press, 1941.

Dash, Bhagwan. "Caraka." In *Cultural Leaders of India: Scientists.* Ed. V. Raghavan. New Delhi: Publication Division, Ministry of Information and Broadcasting, Government of India, 1976, pp. 24–43.

Ray, Priyadaranjan, and Hirendra Nath Gupta. *Caraka saṃhitā: a scientific synposis.* New Delhi: National Institute of Sciences of India, 1965.

Śarmā, Priyavrata. *Caraka-cintana.* Varanasi: Chowkhamba Vidyabhawan, 1970.

Śarmā, R.K., and Bhagwan Dash. *Carakasaṃhitā: text with English translation and critical exposition based on Cakrapāṇidatta's Āyurveda dīpikā.* Varanasi: Choukhambha Sanskrit Series Office, 1976.

CELESTIAL VAULT AND SPHERE The celestial vault is the apparent surface of the sky — both the night sky against which the stars appear to be placed and the blue expanse of the daytime sky.

The idea that the celestial vault, and particularly the daytime sky, is a physical cover of some kind is found in the cosmologies and myths of many cultures. Ethnic groups in northern Eurasia, for example, thought of the sky as a roof to protect the earth and life. The Lapps, Finns, Yakuts, Japanese, and Hebrews described it more specifically as a tent roof, resting on a central pillar. Similar ideas were applied to the night sky. The Milky Way appeared to some as a stitched seam, and the stars as tiny holes in the fabric of the celestial vault.

In the eleventh century the Islamic physicist Ibn al-Haytham noted that the celestial vault usually appears somewhat flattened. He explained this as an illusion created by our perception of distant objects towards the horizon, and the lack of such indicators of distance towards the sky. He furthermore used the flattened appearance of the celestial vault to explain the illusion that the sun and moon are enlarged near the horizon — an explanation that survived well into the twentieth century.

Astronomers in different cultures have for many centuries assumed that the celestial vault represents the upper half of an imaginary (or sometimes real) spherical surface, called the celestial sphere. The other half of this sphere is below the horizon. The apparent positions, sizes, and movements of the sun, moon, planets, comets, and stars were described as if these bodies were situated on the inner surface of the celestial sphere. Estimates of the sizes and separations between the celestial bodies were therefore expressed as lengths or distances on the celestial sphere, rather than in terms of angles. For example, Chinese observations of comets from the second century BC to the nineteenth century were expressed in terms of the *chi*, a length standardized as 311 mm in the ninth century. The *chi* appears to have corresponded to an angle of a little less than 1°. Similarly, the Babylonian *ammat* (about 524 mm) represented either 2° or 2.5° at different times. Also, the Islamic *fitr* and *shibr* (spans later standardized as 152 mm and 178 mm respectively) were equivalent to about 0.6° and 0.8° respectively in the astronomical work of Ibn Yūnus during the eleventh century. These lengths and angles imply that the astronomers in each of the three cultures assumed that the imaginary sphere used to express their estimates had a radius of only some tens of meters.

CORNELIS PLUG

REFERENCES

Gray, L.H., ed. *The Mythology of All Races.* New York: Cooper Square, 1964.

Ho Peng Yoke. "Ancient and Medieval Observations of Comets and Novae in Chinese Sources". In *Vistas in Astronomy.* Ed. A. Beer. New York: Macmillan, 1962, vol. 5, pp. 127–225.

Neugebauer, Otto. *A History of Ancient Mathematical Astromomy.* Berlin: Springer-Verlag, 1975.

Newton, Robert. *Ancient Astronomical Observations and the Accelerations of the Earth and Moon.* Baltimore: Johns Hopkins, 1970.

Plug, Cornelis. "The Registered Distance of the Celestial Sphere: Some Historical Cross-cultural Data". *Perceptual and Motor Skills* 68: 211–217, 1989.

See also: Astronomy – Ibn al-Haytham – Ibn Yūnus

CHAO YUANFANG Virtually the only known detail of the life of Chao Yuanfang, Chinese physician and medical author, is that he was Medical Erudite (*taiyi boshi*) at the court of the Sui dynasty sometime between 605 and 616. In that position he supervised the compilation of the *Zhubing yuanhou lun* (Treatise on the Origin and Symptoms of Diseases), the first Chinese text comprehensively devoted to etiology and symptomatology. The work, which is integrally preserved, was submitted to the throne in 610.

The fifty chapters of the *Zhubing yuanhou lun* discuss more than 1700 syndromes, classified into sixty-seven categories of internal and external diseases. The final sections are concerned with gynecology, obstetrics, and pediatrics. A special feature of the text lies in its therapeutic methods. Rather than ingestion of drugs or acupuncture, Chao advocated a therapy based on such practices as diet and *daoyin* (a set of bodily postures, similar in some respects to Indian *hatha yoga*). The *Huangdi neijing* (Inner Canon of the Yellow Emperor) provides the broad theoretical foundation to Chao's work.

The *Zhubing yuanhou lun* exerted a remarkable influence on the history of Chinese medicine, as reflected for example in the *Qianjin yaofang*, written by Sun Simo. It was highly esteemed in Japan as well, where it inspired the classification of diseases in the earliest extant medical compilation, the *Ishinpō*.

FABRIZIO PREGADIO

REFERENCES

Rall, Jutta. "Über das Ch'ao-Shih Chu-Ping Yüan-Hou Lun, ein Werk der chinesischen Medizin aus dem 7. Jahrhundert." *Oriens Extremus* 14: 143–178, 1967.

Unschuld, Paul. *Medicine in China; A History of Ideas*. Berkeley: University of California Press, 1985.

See also: Acupuncture – *Huangdi Neijing* – Sun Simo

CHEMISTRY IN CHINA Of all the sciences in antiquity chemistry was one of the least understood. Until people had some comprehension of chemical elements, compounds, atoms, molecules, ions, and bonding, they found it difficult to make much sense of the chemical processes occurring naturally. This does not mean that there is no real history of chemistry; it means only that we should assess the early chemical practitioners on the accuracy of their observations rather than that of their interpretations. The modern Chinese term for chemistry is *huaxue*, the study of change, and what the ancient Chinese did well was to note and describe the chemical changes which were observed when substances were mixed, heated, dissolved or treated in some way. The fact that their interpretation may seem a little abstruse by modern standards does not detract from the worth of their observations.

Before chronicling chemistry in ancient China it is important to understand exactly what had stimulated an interest in chemical processes. The Daoist religion sought to lead its adepts into such a harmonious relationship with the world that they would escape the horrors of disease and the tragedy of death. It was not life after death but life without death, although the precise notion varied from age to age. Those who had discovered the *dao*, through meditation, study, forms of exercise, appropriate sexual activity, and purity of life, would retire to an isolated mountain retreat and there live in complete harmony with nature. They were called *xian* or "immortals", nearer to gods then men. The idea of longevity as a reward for the godly life is not unique to Daoism; similar notions occur in Judaism. In the genealogy of the patriarchs we read, "And Enoch walked with God after he begat Methuselah three hundred years and begat sons and daughters" (Gen. 5^{22}). However, what is special to Daoism is the use of drugs to acquire immortality.

Until the Song dynasty, Chinese drugs were grouped into three classes: superior drugs which allowed one to realize vital powers, medium drugs to enrich one's nature, and inferior drugs which cure disease. This particular classification is given in *Zhenghe bencao* (Herbal of the Zhenghe Era) and it is the superior drugs that became associated with the attainment of immortality. Initially the drugs were quite crude materials: minerals straight from the ground or dried herbs, but subsequent developments lead to the compounding of these "simples" to give "elixirs of immortality". It was in the preparation of elixirs that the ancient Chinese made their first chemical observations.

The preparation of elixirs was not an activity associated with a popular or debased version of Daoism and abhorred by the more sophisticated Daoist adepts. Although the importance of elixirs in Daoist practice varied somewhat from dynasty to dynasty it always played some part in the pursuit of the Daoist goal. Several emperors of ancient China expended considerable effort to obtain the elixir of life. ixShihuangdi (259–210 BC), of the Qin dynasty, equipped large expeditions under the command of Xu Fu and others to go abroad to uncover arcane secrets of preparing elixirs. Emperor Wu (156–87 BC) appointed numerous magic specialists to high positions.

Two prominent components of Daoist elixirs were cinnabar (mercuric sulfide) and gold. The reasons for this selection are not difficult to appreciate. Cinnabar is bright red, reminiscent of healthy blood, and it was thought that the infusion into the body of extra healthy blood would enhance health and life span. The absence of understanding of chemical composition among the ancient Chinese meant that the physical appearance (i.e. the color) had greater significance than the obviously different chemical properties of cinnabar and blood. Gold was probably the most prized component of an elixir because it was obtained from the ground in an uncombined (and thus, uncorrupted) form and in the air it did not tarnish. If these attributes could be acquired by the human body then it would, indeed, last for ever. One difficulty recognized by Chinese alchemists was that both cinnabar and gold are insoluble in water and, therefore, unlikely to be absorbed in the body. Much effort, therefore, was expended in devising method of solubilizing both cinnabar and gold.

A manual in the Daoist Patrology (*Dao Cang*) entitled *Sanshi liu Shui Fa* describes the use of niter (potassium nitrate) in vinegar (dilute ethanoic acid) to bring numerous minerals, including cinnabar, into solution. In some cases there appears to be some confusion between solution and suspension; in other cases solution would occur in any aqueous solvent. However, in a few cases, particularly in the solubilization of cinnabar, some interesting chemistry is involved. A solution of pure niter in pure aqueous ethanoic acid has no effect on cinnabar but solubilization does occur if chloride ion is present as an impurity. This is probable as niter was obtained from marine sources. Thus the fanciful descriptions of Chinese alchemy may have a sound basis in chemistry if allowance is made for probable impurities. The same applies to an account in the *Sanshi liu Shui Fa* for the solubilization of gold where the required contaminant is iodate, a common impurity in crude niter deposits. The fact that these accounts of Chinese alchemical activity can be explained in modern chemical terms is evidence of the accuracy of the original observations.

Gold is not common in China and it was not regularly used as a form of currency. Because of the shortage of gold, much alchemical activity was directed at making gold from more commonly available materials. This appears, at first sight, to be identical with European alchemical activity but the motivation was completely different. In Europe it was, until fairly late in the thirteenth century, the pursuit of wealth; in China it was from earliest times the pursuit of immortality. Cinnabar was a popular material upon which to work (gold and cinnabar often occur together in the earth) and the *Gui Jia Wen* contains the lines

> The span of life is up to me, not heaven.
> The reverted cinnabar becomes gold, and millions of years are mine.

In the most famous alchemical manual (the *Neipian* of the *Bao puzi*) the author, Ge Hong, gives a recipe for preparing gold from tin by heating with various forms of alum (potassium aluminum sulfate) on a horse dung fire. What was made was tin sulfide or mosaic gold, a yellow material which glitters just like gold. Present-day repetition of the procedure, using an electric furnace rather than a horse dung fire, confirms the production of tin sulfide.

The elixirs made from simples like cinnabar and gold contain, in general, so many components, some of which will contain impurities, that it is difficult to hazard a guess at what chemical changes are occurring. Picturesque titles were given to many elixirs. A mixture of calomel (mercuric chloride), cinnabar, sulfur, and realgar (arsenic sulfide) was known, according to the *Danjing Yaojue* of Sun Simo of the Tang dynasty, as Grand Unity Three Envoys Elixir. Needless to say, the elixirs did not have the required effect and their use in Daoist practice gradually died out, leaving only a memory of sustained but flawed chemical activity. There are just a few exceptions to this generalization of futility in Chinese alchemical activity, the most famous of which is the production of gunpowder, known originally as *huoyao* or "fire drug". Niter and sulfur are mentioned in early *materia medicas*, including *Shennong Bencao Jing* from the Han. Possibly in pursuit of elixirs, they were mixed and inadvertently contaminated with carbonaceous materials, giving a material which burned very brightly. Careful adjustment of the proportions gave an explosive, used in China for firearms and civil engineering and for fireworks. The first printed recipe for gunpowder is given in a military manual *Wujing zongyao* by Zeng Gongliang, printed in AD 1044. It was the optimized formula which traveled to Europe and had such a dramatic impact on the conduct of warfare. It is ironic that the most significant achievement of Chinese alchemists in their efforts to make elixirs of immortality was an instrument of warfare.

With a decline in interest in alchemy during the Yuan, Meng, and Qing dynasties there was no emergence of modern chemistry from alchemical activity as there was in Europe. The advanced state of modern Chinese chemistry (epitomized by the laboratory synthesis of insulin in 1965) is due to the introduction of chemistry from the West during the nineteenth century, partly by missionaries and partly by young Chinese students studying in Europe and America. Among missionaries (although the term is not entirely appropriate) was John Fryer, the son of a poor English clergyman, who worked at the Kiangnan Arsenal translating manuals and textbooks for the Chinese government. Between 1867 and 1897 he acted as focal point for the propagation of western scientific ideas in China. Even during the political upheavals of the first part of this century, there was a general increase in Chinese scientific education, and a number of Western style scientific institutions, like the Peking Union Medical College, were established. In chemical research the dominant figure was Wu Xian. He completed his Ph.D. at

Harvard Medical School in 1920 and returned to the Peking Union Medical School. By 1925 he was a full professor and initiated an extensive program of research into blood chemistry. Chemical research started elsewhere and, fueled by the innate talent of the Chinese people, it has grown, in spite of all the political upheavals of the Communist revolution and the downfall of Mao, to make China one of the leading nations in the world for chemical research.

ANTHONY R. BUTLER
CHRISTOPHER GLIDEWELL

REFERENCES

Needham, Joseph. *Science and Civilisation in China*. Cambridge: Cambridge University Press, Vol. 5, Parts 2–5, 1974.

Reardon-Anderson, James. *Chemistry in China, 1840–1949*. Cambridge: Cambridge University Press, 1991.

Sivin, Nathan. *Chinese Alchemy: Preliminary Studies*. Cambridge, Massachusetts: Harvard University Press, 1968.

See also: Alchemy – Ge Hong

CHEN YAN Chen Yan, a famous physician of the Southern Song Dynasty in the twelfth century, was born in Qingtian (now in Zhejiang Province), with Wuzhe as his surname and Hexi as his nickname. He was extraordinarily clever and proficient in prescription and pulse-taking, as well as in treatment. It was said that he could prophesy the prognosis of incurable diseases. In 1161, he assembled a six-volume work of prescriptions with eighty-one categories which deals with the pulse and diagnosis of *Yang* diseases, followed by their etiology, and the variorum of the *Canon of Pulsology*, all with relevant recipes. This work is entitled *Yi Yuan Zhi Zhi* (Treatment on the Basis of Etiology), which was never published.

Chen was especially interested in the study of pathogenesis. He claimed that the causes of all diseases could be divided into three categories; this was based on Zhang Zhongjing's hypothesis. Eventually, he compiled the *San Yin Ji Yi Bing Zheng Fang Lun* (Discourses on the Manifestations and Recipes of Diseases of Three Pathogenic Categories) in 1174, in eighteen volumes. He summarized all pathogens into three categories:

- endogenous emotional changes, the Seven Emotions (joy, anger, sadness, ratiocination, grief, apprehension, and fear), which originate from the internal viscera and affect the extremities;
- exogenous pathogenic factors, the Six Excesses (wind, cold, heat, damp, dry, and fire), which originate from the channels and affect the internal viscera; and
- causes neither exogenous nor endogenous, which derive from immoderate food and drink, wounds, ulcers, and insect and beast bites. The etiological work exerted some influence on later generations.

Chen also summarized medical science in four Chinese characters, reading "organic name, functional property" claiming that all new commentaries on ancient classics could be categorized under four branches: pulsology, manifestation, disease, and treatment. Hence, all medical students should recognize a disease on the basis of pulsology and make a diagnosis on the basis of manifestations.

In the years 1165–1173, he compiled the *Herbology of Compiled Categories* with subheadings based on the abovementioned ideas. Though it was an anonymous compilation with a "Preface by Master Hexi", it was commonly believed to be Chen's.

HONG WULI

REFERENCES

Chen, Yan. *On the Three Causes Epitomized and Unified: Disorders and Recipes* (Reprint). Beijing: People's Health Publishing House, 1957.

Liu, Shijue. "Chen Wuze's Native Home and his Medical Activities." *TCM Information* 6: 40–41, 1986.

Wang, Miqu. "Exploration on Chen Wuze's Medical Psychological Ideas." *Xinjiang Journal of TCM* 2 :17–18, 1982.

See also: *Yinyang* – Medicine in China – Zhang Zhongjing

CHILDBIRTH The pelvic changes that evolved when hominids became upright walking creatures resulted ultimately in what is known as the "obstetrical dilemma" for humans. The woman's pelvic bones must not obstruct safe passage of the infant's head, and her soft vaginal tissue must be able to stretch to the maximum. Meanwhile, the baby must be mature enough to survive outside the woman's body but small enough to pass through her narrowed birth canal. Managing childbirth within these parameters developed throughout human history by means of observation, trial, and errors accommodated, however, by high fertility. Obstetrics conducted in hospitals is a recent biomedical specialty, and in the past decades, Western obstetrics has developed still more technologies. However, for non-western mothers, little has changed in 2500 years of written record. From that record, we can draw one of two opposing conclusions: first that childbirth customs are appropriate adaptations to biological needs; or, second, that high technology is always better than childbirth in a natural setting.

Some writers portrayed childbirth practices in non-Western societies as ideal, resulting in a quick and painless process. Men who were not allowed to observe births were told about them by informants who themselves also had been excluded from the birth scene. These writers believed that the non-Western woman just goes in the fields, gives birth, and returns to work in a few hours. Later, women anthropologists who observed non-Western births gave very different reports. Management of childbirth anywhere requires attention to the problems of safety, pain, fatigue, protracted labor, malposition of the fetus, delivery, expulsion of the placenta, and healing. Solutions to problems are mandated by the universal physiology of childbirth. In Western obstetrics, these problems of childbirth are managed with technological equipment; in non-Western societies, the same problems are managed with hands.

Childbirth is not only a biological occurrence; it is also a sociocultural event. Differences in the management of birth reflect different values and beliefs. In some cultures, myth is used to explain pregnancy and birth. Sometimes birthing is a private process, sometimes a communal event. How a society reacts to physiological processes and the extent to which there is contact with other cultures accounts in part for diversity in the ways of managing childbirth. Each birthing system relates to the place women hold in a community, the value given individual children in that community's social structure, how much medical care is available and how it is distributed, and the religious belief system that is in place. Thus childbirth practices develop within the context of each particular society.

Such cultural strictures determine, among other things, how the pregnant woman should behave, who should take care of her, how abnormalities are to be explained and how to prevent them, what, it is believed, causes labor, how it should be stimulated, where the birth should take place with which persons in attendance, and what should be done after birthing for mother and newborn. Descriptions of diverse birthing practices can be found in the writings of George Engelmann and Hermann Ploss, Max Bartels, and Paul Bartels. Birth customs are catalogued in the Outline of World Cultures. MacCormack and Kay edited anthologies detailing birth practices in mostly non-Western cultures. Following are illustrations of some of this diversity from East India, Japan, Malaysia, the Navajo Nation, Seriland of the northwest coast of Mexico, Guatemala, Mexico, Ireland, and Ibo and Bariba tribal homelands in Africa.

Many "normal" activities of women are proscribed during pregnancy with the view of preventing childbirth difficulties. Some of these pre-birth customs, are considered adaptive by Western medicine. For example, pregnant American Indian women are urged to keep moving and working during pregnancy and to limit their intake of food. The goal is not to have too large an infant for easy delivery. In West Africa, on the contrary, women are fattened in order to make them desirable for marriage and childbearing. Each custom is beneficial: genetically, American Indian women tend to have very large infants, whereas the babies of African and African-American women are likely to be quite small.

The women themselves explain their customs only as "we have always done it that way". The outsider sees the rationale for many proscriptions deriving from principles of sympathetic magic, like affecting like. Ancient Mexicans and recent Navajo, for example, believe that weaving is dangerous during pregnancy because the umbilical cord will tie up the baby. This belief has a very long history. It is recorded as fact by ancient Greek authorities of Aristotle, Galen, and Soranus. Across Western and non-Western cultures, the woman and sometimes her mate are told to avoid certain foods and actions to prevent problems in birthing or abnormalities in the baby. Conversely, a woman's cravings for exotic foods must be satisfied or the baby will be marked. Some of the rules placed to solve birth problems may be seen in the belief system of the larger culture; for example, the *bidan kampung* (traditional midwives) of Muslim Malaysia consider the humoral system of Islam in selecting a western-facing direction (with modification for days of the week) to align the woman for birthing; west faces Mecca. For Navajo, on the other hand, it is vital that the mother's head **never** face west, for that is their direction for death. Women who are taught that congenital malformations are caused by unnatural phenomena wear amulets to protect the unborn child. Charms include medals, pictures and prayers that represent their religion, iron filings, and pieces of crystal, coral or jet. It could be said that magical thinking can also characterize high technological Western cultures, for childbirth outcomes cannot always be predicted. Rituals of testing are enacted to resolve anxiety of the unknown. Childbirth is feared to a certain extent by all people.

Non-Western women are given prenatal massages in many cultures, notably Japan and Mexico. The purpose of the massage is to fix the fetus in a position favorable for delivery. In some societies, the fetal position is also assisted by a belt around the woman's abdomen. The Japanese apply the maternal *obi* in a special ceremony in the fifth month of pregnancy. Mexican women wear a narrow red cord, which shows the ascent of the uterus. Traditional Navajo women maintain their custom of wearing a red belt woven with green figures, five to six inches in width, throughout pregnancy. This belt is later used during labor and delivery.

When a woman believes that her childbirth labor is starting, she takes herbal teas. Mexican women drink *té de manzanilla*, camomile tea, believing that if it is false labor, the tea will make the pains go away, and if it is true labor, the pains will come stronger and harder. This belief has pharmacological support. Navajo women use tea from a penstemon for the same purpose, or one from rabbitbrush. Ideally, the birth should be fast. Mexican and Guatemalan women are given the ancient Aztec herb *cihuapatli*, which has strong effects on the uterus. They are permitted little sleep but are encouraged to walk with their contractions to hasten delivery. When in pain they are allowed to scream, unlike West African Bariba and Ibo. The proud Bariba women attempt to deliver alone to conquer their fear. Traditional Navajo women were trained to make no sound in order to preserve the secrecy of the process and today are silent and motionless during labor.

Because childbirth is the obvious result of sex, women in societies that value female purity, innocence, and virginity have particular difficulties. Irish peasants hide their pregnancies as long as possible, and do not publicize births. Hindu women also do not advertize their condition, resulting in little attention to their prenatal health, and distressing labor. These problems occurred in Western culture, too. Following Biblical injunction, the laboring woman was expected to suffer, until in the mid-nineteenth century, Queen Victoria accepted the newly-discovered anesthesia for her birthings.

In non-Western childbirth, a midwife conducts the birth process. She uses many ways to stimulate labor, from placing an amulet on the laboring woman or an axe under her bed, to giving herbal teas or some of the husband's semen for her to swallow. To lubricate the birth canal, she massages with whatever is typically available, such as lard, clarified butter *(ghee)* or cooking oil.

Posture during labor also varies. In many non-Western societies, the laboring woman is encouraged to keep an upright position, utilizing gravity to move the fetus in the birth canal. Anderson and Staugård summarized traditional birth positions of twenty-two tribes in Africa and found kneeling to be used in twelve, squatting in ten, holding rope in seven, lying in six, and half sitting in five. These postures may relate to different pelvic dimensions; for example, women of Biafra commonly used the squatting position, which works best for women who have a short pelvic outlet. The eighty thousand women who have had traditional genital surgery (female circumcision) present special challenges for safe delivery, because scar tissue does not stretch.

A common practice of American Indians has the woman in either a kneeling or squatting position, with one attendant supporting the woman from behind, pushing on the abdomen above the uterus, while another attendant is in front, pressing from below. The mother might be pulling on a rope during contractions. Or she might pull on a hammock, which is suspended above where she stands, as has been recorded for the Kalopalo Indians of Central Brazil. In the latter case, four attendants help the laboring woman, two pressing down on each shoulder, and two pressing on her feet to direct force during a contraction. Navajo women pull on the special red belt that they wore during pregnancy, looped over a beam in the ceiling. For the Seri, a hunting–gathering people of Sonora, Mexico, two women who are usually close relatives help with the birth. One assistant sits on the ground, her legs spread but her knees doubled under her, to make a lap for the parturient to sit on. The second assistant, also sitting on the ground but facing the laboring woman, holds her hands in position to receive the infant. The parturient and midwives often wait for hours in the described positions. To speed the labor, the parturient may be given tea from the roots of indigenous plants, or a concoction of alcohol, limes, cinnamon, and aspirin. The women often say that they do not want to marry because childbirth hurts too much. Women of the nearby agricultural Yaquis drink an infusion of the narcotic plant Datura for childbirth pain.

For difficult deliveries, Punjabi midwives straddle the chest of the laboring woman, pressing hard on the abdomen with their hands; or they push with the heels of their feet on the pubic bone.

In matrilineal societies such as the Navajo, the attendants are commonly women from the maternal side of the family. Patrilineal societies such as Egyptian or Japanese, where descent is traced through the father's line, require the presence of the husband's kin, to assure continuity of the line. The birthing woman is accompanied by her husband's mother, and goes to her mother-in-law's house to recuperate from the birth.

Delivery of the placenta (afterbirth) is a grave concern in all societies. The medical ethnobotany of every group contains a variety of plants to give the woman whose placenta is slow to separate. According to humoral doctrines, which are prevalent in many non-Western cultures including India, China, Middle America, and Latin America, those herbs that acted on the uterus were categorized as "hot" and "dry", and many of the same plant genera are used cross-culturally. A retained placenta is a most dreaded complication of birth: "anyone can deliver a baby but it is the placenta which kills" (Bariba). Once the midwife delivers the placenta, she disposes of it according to its meaning in her culture, respect as a spiritual entity or fear of its power. In Malaysia, the placenta is considered to be the fetus's semi-human sibling, and thus is washed and ceremonially buried, which is similarly done by Guatemalans and Seri.

Some of the materials used to act on the uterus, either to bring on menstruation, stimulate labor, hasten birth, or separate the placenta, have been subjected to phytochemical analysis. *Artemisia* species, the botanical name first given by the Greeks for their goddess of women, and used world-wide in Europe, the Middle East, China, Latin America, North America, and Mexico, contain estrogens, female hormones. *Matricaria*, named since Roman times from the Latin word for womb, relieves painful uterine cramps through its spasmolytic compounds. Analysis of other materials, such as the husband's semen, given to the laboring woman in some African cultures, shows that they are high in prostaglandins, chemical compounds used in biomedicine to initiate labor by causing the mouth of the uterus to begin opening.

Indian peoples of Mexico use plants passed down in oral history from Uto-Aztecan forbearers (such as *yyahautli*, marigold, to stimulate labor, and *toloache*, jimson weed, for labor pain) and also those learned from Europeans who came in the colonial period, *coriander* and *rosemary*, for their actions on the uterus. Hindu mothers are given Ayurvedic herbs, while Chinese herbs are used all over eastern Asia. Childbearing women take herbs for prenatal care, labor strength, and post partum healing. The woman may drink the plant material in teas; it may be applied to her abdomen, inserted in her vagina, inhaled, or she may sit over infusions in vapor baths.

After the baby is born, the umbilical cord must be tied and cut, and again there are many practices to be followed. The cord is cut with a piece of sharp bamboo in Malaysia, a razor blade or broken glass in Ghana. It may be tied with hair, string, or animal tendon. The variety of dressings include pounded pottery, herbs, salt, ground shells, and ash, materials that can harbor bacilli causing neonatal tetanus. The cord remains may be buried in a place that is associated with the anticipated sex role behavior of the child. For example, a Navajo woman may bury her infant son's cord near the sheep corral and her infant daughter's cord near the weaving room. Mexican, Malaysian, and Japanese mothers kept each child's cord for the child's future welfare.

The immediate postpartum period lasts for three days in cultures predominately Christian, and four days in American Indian societies, the respective magic numbers for each. After giving birth, and for a period of time prescribed by the culture, twenty days for many American Indians, forty days for traditional Jews, Christians, and Muslims, the woman is supposed to be confined to her home. It is believed that the woman is vulnerable, requiring protection during the entire post partum period. Some Mexican patients still observe *la dieta* during which time baths, walking barefoot, and otherwise being subjected to cold are proscribed. This includes avoiding "cold" foods, foods that are believed to have a quality of chilling the body.

The newly delivered woman is healed with heat in many cultures. The Navajo woman may have heat applied by warming green cedar and holding it to the abdomen with her red sash. She is bathed on the fourth day after delivery, especially if she had a ceremonial Sing during her labor. Her lacerations are treated with a salve of baked cactus, or a fumigant treatment may be prepared by digging a hole and heating stones therein. Soranus described a similar fumigant two thousand years ago. "Mother roasting", as it is called, is a therapy used by many societies in Southeast Asia, including Malaysian. A fire is maintained beside the delivered woman for a prescribed number of days. This "dries out" the womb. A ritual fire is kept burning in the dwelling of the Seri woman for four days following the birth, in this case for the infant's sake. Hot leaves are pressed on the Ibo mother's abdomen, immediately after birth. The *temascal* or sudatory was used by ancient Aztecs after childbirth, as well as by ancient Romans and is used by present day orthodox Jewish women after menstruation as well as after childbirth.

Sexual intercourse is not permitted anywhere until the puerperium, however defined, is over. Mexicans, like so many others, believe that intercourse when there is vaginal bleeding is abhorrent. The Navajo believe that even indirect contact with vaginal blood is very dangerous to the husband and anyone else. In many cultures, contact with blood was considered polluting to the individual and to the community. The woman can now be purified, by bath and by religious ceremony. Leviticus XII, 2 provides the Laws of Purification and Atonement for Jews, Christians and Muslims. The postpartum bath is a Shinto rite of purification in Japan.

Non-Western Birth Today

It is estimated that worldwide, eighty percent of births are attended by Traditional Midwives, Traditional Birth Attendants, or simply female relatives. The status of the traditional midwife has varied through time and place. For some non-Western cultures, her status has been very low, for it is women's work, and deals with blood. With many ethnic groups in India, midwifery was looked upon as a most degrading occupation.

The Traditional Midwife is typically a woman who is respected in her community, nonliterate, postmenopausal, has herself borne children, and learned her craft as an apprentice to another woman. She is looked down on by the Western medical community. Traditional midwives may get Western training to meet difficult births, although they may not always find such instruction useful. When Western technology is not available, the traditional midwife's birth skills may prove to be superior. It may be hoped that the Certified Nurse Midwives of each culture may bridge the birth customs of the non-Western traditional midwife with the practices of Western biomedicine.

MARGARITA A. KAY

REFERENCES

Anderson, Sandra and Frants Staugård. *Traditional Midwives.* Gabarone, Botswana: Ipelegeng Publishers, 1986.

Basso, Ellen B. *The Kalopalo Indians of Central Brazil.* New York: Holt, Rinehart & Winston, 1973.

Engelman, George Julius. [1882] *Labor Among Primitive Peoples, showing the development of the obstetric science of to-day, from the natural and instinctive customs of all races, civilized and savage, past and present.* New York: AMS Press, 1977.

Frazer, Sir James George. *The Golden Bough.* New York: Macmillan Publishing Company, 1922.

Hahn, Robert A. and Marjorie Muecke. "The Anthropology of Birth in Five U.S. Ethnic Populations: Implications for Obstetrical Practice." *Current Problems in Obstetrics, Gynecology and Fertility* 10 (4): 133–171, 1987.

Kay, Margarita, ed. *Anthropology of Human Birth.* Philadelphia: F.A. Davis, 1982.

MacCormack, Carol P., ed. *Ethnography of Fertility and Birth.* London: Academic Press, 1982.

Mead, Margaret and Niles Newton. "Cultural Patterning and Perinatal Behavior." In *Childbearing: Its Social and Psychological Implications.* Ed. S. A. Richardson and Alan F. Guttmacher. Baltimore: Williams and Wilkins, 1967, pp. 142–244.

Murdock, G.P. *Outline of World Cultures.* New Haven: Human Relations Area Files, 1963.

Ploss, Hermann Heinrich, Max Bartels and Paul Bartels. *Das Weib in der Natur- und Völkerkunde: anthropologische Studien.* Vol II. Leipzig: T. Grieben (L. Fernau), 1902.

Population Information Program. "Traditional Midwives and Family Planning." In *Population Reports.* Baltimore: Johns Hopkins University Press, 1980, pp. 437–488.

Shaw, E. "Female Circumcision." *American Journal of Nursing* 85: 684–687, 1985.

See also: Magic and Science – Ethnobotany

CHINA Scholars in the past found it difficult to focus their attention on Chinese science because in traditional China knowledge did not come under the same groupings as in the West. For example the Chinese pharmacoepias included knowledge on natural history; the official dynastic histories, some Daoist writings, and the works of some great poets contain information on astronomy and alchemy, and compendia on military science mention meteorology and firearms together with magic and divination. It is true that there are some Chinese monographs on science and technology, such as the *Tiangong kaiwu* (Exploitations of the Work of Nature) and the *Jiuzhang suanshu* (Mathematical Manual of the Nine Sections). Nevertheless knowledge of traditional Chinese science exists mainly in the official dynastic histories, the compendia (*leishu*), the gazetteers, religious works like the *Daozang* (Daoist *Tripitaka*), the numerous book collections (*congshu*), and the general literature.

In sixteenth-century Europe, Francis Bacon (1561–1626), Jean Fernal (1497–1558), and Jerome Cardano (1501–1576) all made reference to three great inventions, namely the compass, the art of printing, and gunpowder. In the minds of many there was only some vague connection between China and these three inventions. Indeed, prior to the middle of the twentieth century the history of Chinese science received little attention both from inside and outside China. It gained recognition in the middle of the twentieth century when Joseph Needham launched his monumental work, *Science and Civilisation in China*. Needham did not pioneer the study of the history of Chinese science. Sinologists and scientists before him had written on acoustics, alchemy, architecture, astronomy, gunpowder, hydraulics, mathematics, and medicine in traditional China. However, Needham demonstrated the originality of many Chinese discoveries and inventions, including the "three great inventions" of Renaissance Europe, and made Chinese science a new discipline of study. Modern scholarship and recent archaeological discoveries have increased our understanding of Chinese science and helped in the reappraisal of some old interpretations.

Excavations carried out first at Anyang between 1928 and 1937 and again after 1950, followed by those done at Zhengzhou in Henan province, substantiated the existence of the Shang dynasty (sixteenth–eleventh centuries BC) and brought to light many bronze artifacts, foundry sites, and proto-porcelain wares. At the ruins in Anyang the remains of the capital of the last Shang kings were found. This later period of the Shang dynasty is known as Yin (fourteenth–eleventh centuries BC). Besides bronze artifacts, archaeologists uncovered gold, jade, pottery, and shell objects, wooden artifacts with traces of lacquer, traces of silk fabric, and even a chariot. The most important find, however, was the oracle bones, which were carapaces of a specie of tortoise or shoulder blades of buffaloes that bear inscriptions written by the people of Yin for divination purposes. They contain records of eclipses and novae, as well as names of stars and asterisms. They also show that the Yin people were already using a luni-solar calendar with twelve moons or lunar months in one year. Each month consisted of either twenty-nine or thirty days each, and every two or three years there was one extra month, known as the intercalary month, added to the year. The numerals in the oracle bones were the earliest known Chinese numerals until the discovery of the inscriptions on the pottery at the ruins of Banpo, which dated back to about the year 3000 BC. From the oracle bone numerals we know that the Yin people were already using a decimal system. We also find that the Yin people already had some knowledge of irrigation, agriculture, sericulture, and wine-making. For example, for tooth decay they used a character that included the character for "worms", suggesting that they had made an effort to attribute the problem to a cause. Chinese archaeologists have been quite active in recent years trying to re-establish the Xia dynasty. According to their tradition this dynasty existed for four centuries immediately before Shang, which itself lasted six centuries. Erlitou in Henan province, where bronze vessels dating back to the year 1700 BC were recovered in 1971, is one of the more promising sites where they hope to discover epigraphical evidence to confirm the existence of that ancient dynasty,

Bronze vessels and pottery are some of the works of art characteristic of the Chinese. Shang bronzes show that the technique of bronze casting had already reached an advanced level. It was also in Shang China that the earliest proto-porcelain ware was discovered. It took the form of a wine container (*cun*), made of kaolin clay, with a yellowish green glaze on the surface about the mouth and a translucent deep green glaze on the inner and outer surfaces. The Western Zhou period (eleventh century–771 BC) produced beautiful bronze vessels with inscriptions and proto-porcelain wares, some of which can be seen in many museums today.

In the spring-and-autumn period (722–481 BC) two early texts, which Confucius (551–479 BC) himself referred to, namely the *Shujing* (Book of Documents) and the *Shijing* (Book of Odes), were written. Both contain astronomical material from about the tenth to the fifth century BC. The earliest sighting of Halley's comet in the year 613 BC is recorded in the *Chunqiu* (Spring-and-Autumn Annals), which also contains references to solar and lunar eclipses, meteor streams, and other comets. The *Zhouli* (Records on the Rites of Zhou) informs us that the Zhou kings employed star clerks to observe astronomical and meteorological phenomena and to make astrological prognostications therefrom. The earliest star catalog is said to have been made by a court astronomer Wu Xian, of whom little is known. Two other catalogs were produced by Gan De and Shi Shen between 370 BC and 270 BC. The three original catalogs have long been lost. Some fragments claimed to be from the originals are quoted in many old astronomical writings. In the early fourth century Chen Zhuo (fl. ca. 310) constructed a star map supposedly based on such information.

Two weapons with iron plates taken from meteorite sources and dating back to the Western Zhou period were discovered in 1949. In 1976 a steel weapon of the sixth century BC was excavated in Changsha, Hunan province, showing that by then China had already entered the Iron Age. Subsequently more iron artifacts of the fifth century BC were found in Jiangsu and Henan provinces. It is interesting that although Europe knew about iron earlier, the Chinese came to know about the making of cast iron not later than the fifth century BC, soon after iron was known to them, in contrast to the much later development of cast iron technology in Europe. Another interesting technology developed by the Chinese of about that period is the breast-strap harness seen on the horses among the terra cotta warriors guarding the tomb of Qin Shihuangdi in Xian. Showing good understanding of the horse, this method was far superior to the throat-and-girth harness of the Roman chariot adopted in Europe in the early days. Some time between the years 600 and 1000 the breast-strap technique went to Europe and evolved into the modern harness.

Between the sixth and the fourth centuries BC the "Hundred Schools of Philosophical Teachings" flourished in China. Some of these schools played an important role in the development of science and technology in China. The philosopher Mozi (b. ca. 479 BC) is remembered for his techniques of defense and for his knowledge of mechanics and optics embodied in his *Mojing* (The Canon of Mozi). Confucius is said to have edited the *Yijing* (Book of Changes), which contains one of the most subtle and influential methods of Chinese divination. In the past, especially in China, it was difficult to draw a sharp line of demarcation between science and divination. These two subjects were included within the term *shuxue* (mathematics). The writings of the Daoist philosophers and the Naturalists exerted a great influence on early Chinese science, and indeed they provided a set of natural laws that was supposed to be "universal", even more so than what modern scientists believe. Indeed the attempt to be too universal made it difficult for science in traditional China to separate itself from philosophy and to develop into modern science. The greatest name among the Naturalists was Zou Yan (fl. ca. 300 BC) who lived in the eastern seaboard state of Qi (in modern Hebei province). He was the greatest exponent of the *yin* and *yang* theory and that of the *wuxing* in ancient China.

Traditional Chinese science was based on the philosophy of Harmony of Nature, in which heaven (*tian*), earth (*di*), and human beings (*ren*) were all mutually related. Attempts were made to develop a philosophy to explain everything — from astronomy to astrology, from mathematics to fortune-telling, from alchemy to magic, from philosophy to ethics, from politics to fine arts, and from medicine to music — in a common concept based on *qi*. The word *qi* has a wide range of meanings, which cannot be adequately covered by a single term in translation. Its modern meaning includes "air", "gas", "vapor", "steam", "weather", "trend", "demeanor", "manner", "temper", and a sort of life-giving force, reminding us of the Greek concepts of *pneuma* and *psyche*, the Hindu idea of *prāṇa*, and what modern scientists call "matter-energy".

From *qi* the two opposite and yet complementary cosmological forces *yin* and *yang* and the five *xings* were derived. *Yin* conveys the idea of darkness, the shady part of a mountain, coldness, cloudiness, rain, anything that is feminine, and so on. *Yang*, on the other hand, refers to brightness, the sunny side of a mountain, warmth, clear sky, sunshine, anything that is masculine, etc. The term "*wuxing*" was generally translated as "Five Elements". This gave rise to some confusion. When the Jesuits rendered the Four Elements of the ancient Greeks into Chinese they used the term "*siyuan*", (four *yuans*), but when they translated the Chinese term *wuxing* they called it Five Elements. Unlike the Greek Elements (earth, air, fire, and water), the Chinese *xing*'s were not stationary but were in a state of constant motion rather than rest. They are Fire, Water, Wood, Metal, and Earth. There is an "order of production or generation" in which Fire produces Earth, Earth produces Metal, Metal produces Water, Water produces Wood, and Wood produces Fire. Then there is "an order of conquest or destruction" in which Fire conquers Metal, Metal conquers Wood, Wood conquers Earth, Earth conquers Water, and Water conquers Fire. In Chinese medicine for example, the *Huangdi neijing* divides different parts of the body into *yin* and *yang* and associates them with the five *xings*. For example, the heart, the small intestines and the tongue are associated with Fire, the kidneys and the ears with Water, the gall-bladder and the eyes with Wood, the lungs, the large intestines and the nose with Metal, and the stomach, the spleen and the mouth with Earth. The foundation of health, tranquillity, and well-being rests on the perfect equilibrium and harmony of the two cosmological forces *yin* and *yang*, which continually ebb and flow. Everything is produced by union and perishes by decomposition. From the two orders of mutual production and mutual conquest the Chinese derived two principles, namely the "principle of control" and the "principle of masking". Any of the five *xings* that conquers another *xing* is controlled by the *xing* that conquers it. For example, Fire conquers Metal, but the process is controlled by Water. Any *xing* that conquers another *xing* is masked by a *xing* that produces the conquered *xing*, i.e. Fire conquers Metal, but Earth masks the effect.

Then comes the system of the *Yijing* (Book of Changes) with its sixty-four hexagrams said to be derived from the theory of *yin* and *yang* and that of the *wuxing*, although the system was sometimes used to support those two theories instead. Traditionally the origin of the system traced back to the eleventh century BC when the sage father of the founder of the Zhou dynasty got inspiration from two mystical diagrams, the *Hetu* (River Diagram) and the *Luoshu* (*Luo* Chart). These two diagrams supposedly held the secrets of *yin* and *yang* as well as the *wuxing*. In the *Yijing* heaven and earth, thus *yang* and *yin* originated from *Dao*. This term probably refers to natural order in this context but is opened to a broad range of diverse interpretations. The system of the *Yijing* was mainly employed for the purpose of divination. Besides divination it also found its use in the "explanation" of scientific phenomena, such as in astronomy and alchemy.

In traditional Chinese thought, divination and mathematics were inseparable. The word *shuxue*, the modern term for mathematics, was only first used in the modern sense in the last century when Li Shanlan (1811–1882) translated Western mathematical works into Chinese. Before then the same term in traditional China referred not only to divination and mathematics but also to astronomy and music. One is reminded of the word mathematics as defined by Boethius (480–524) to include arithmetic, geometry, astronomy, and music, which became the *quadrivium* in medieval Europe. In the eyes of traditional Chinese mathematicians, such as Liu Hui (fl. 263) and Qin Jiushao (1202–1262), divination was the loftiest form of *shuxue* while mathematics (in its modern sense) was only its common form. *Shu* was something that could be predicted by calculations, whether by means recognized by modern scholars as mathematics, such as finding the unknown number in a mathematical equation, or predicting rain in weather forecasts and telling the future using the system of the *Yijing*.

Buddhism and Daoism gradually gained ground over the Confucianists after Han China in the third century. A revival of learning during the time of Song China (960–1279) saw the neo-Confucianists attempting to cover science, particularly cosmology, within their schools of philosophy. The most celebrated among them was Zhu Xi (1130–1200), who sought to explain everything, both natural and human, with *li* (Nature's pattern), *qi*, and *shu*, while Zhang Zai (1020–1077) applied the concept of *qi* to explain natural phenomena and moral issues. Then there was the *xiangshu* (numbers of the symbols of the *Yijing*) school that focused its attention on the elucidation, verification, and application of the system of the *Yijing*. The most famous member of the school was Shao Yong (1011–1077), who used the system to explain natural phenomena and human affairs and to work out past and future events. His new order of the sixty-four hexagrams was shown by Leibniz (1646–1716) to be similar to the arrangements using the binary notation. Although Zhu Xi's philosophy was adopted as state orthodoxy, it does not imply that it met no opposition. For example, a contemporary, Lu Xiangshan (1139–1192) developed the neo-Confucian school of Idealism or of the Mind, which was later further developed by the Ming philosopher Wang Shouren (1472–1529), who had a large following.

The earliest text on Chinese astronomy and mathematics is the *Zhoubi suanjing* (Mathematical Manual of Zhoubi), which Christopher Cullen showed recently to be a compilation of the first century. In 1973 many important discoveries were made in the excavation of a Western Han (206 BC–AD 9) tomb in Mawangdui, Hunan province. They included a manuscript recording the ephemerides of Jupiter, Saturn, and Venus between the years 246 BC and 177 BC and a silk manuscript illustrating comets in various shapes. Astronomical instruments such as bronze clepsydrae and bronze sundials have also been discovered in other excavations.

The Commentary to the *Jiuzhang suanshu* (Nine Chapters on Mathematical Arts) by Liu Hui (fl. 263) was the most influential text in the history of Chinese mathematics. In 1983 the *Suanshushu* manuscript (Mathematics Book), written on bamboo strips, was discovered in an ancient tomb dating back to the second century BC at an excavation carried out in Hubei province. A preliminary study shows some similarities between this manuscript and the *Jiuzhang suanshu*. Liu Hui also wrote another, but much shorter text, the *Haidao suanjing* (Sea Island Mathematical Manual), using the right-angled triangle to measure distance and height. Before the time of Liu Hui, around the year AD 190, Xu Yue wrote his *Shushu jiyi* (Recording Omitted Items in Mathematics). Between AD 280 and 473 the *Sunzi suanjing* (Mathematical Manual of Sunzi) appeared. Of all the problems it contains, the one that has aroused most attention among modern scholars is that of the Remainder Theorem. The problem is to find a number, which when divided by 3 leaves a remainder 2, when divided by 5 leaves a remainder 3, and when divided by 7 leaves a remainder 2. The required answer is 23. Actually this problem involves three indefinite simultaneous linear equations of one unknown, and there is an infinite number of answers of which the required answer is the smallest. Historians of mathematics study this purely as a case of the Remainder Theorem. However, a more important application of this problem was recently revealed. It was found to be the method adopted by Chinese astronomers up to the eighth century to calculate and construct new calendars.

Several other mathematical texts were written in the following three centuries. One must note the evaluation of the ratio of the circumference to the diameter of a circle made by Zu Congzhi (429–500) and his son Zu Keng. They gave three different values, of which the most accurate was $3.1415925<\pi<3.1415926$. It was not until the sixteenth century that Viéte in Europe was able to match this accuracy. The method used by Zu Congzhi was described in his book, the *Zuishu*, which unfortunately is already lost. Chinese mathematics reached its golden age of development in the twelfth and the thirteenth centuries. During the twelfth century, mathematicians like Liu Yi and others could solve cubic equations numerically. Their writings are no longer extant, but fortunately we have the works of four great thirteenth-century mathematicians, namely Qin Jiushao, Li Zhi (1192–1279), Yang Hui (fl. 1261–1275) and Zhu Shijie (fl. ca. 1280–1303). The study of numerical solutions of equations of higher degrees is yet another of the characteristics of Chinese mathematics. Li Zhi also had a great influence in the development of mathematics in Japan.

The Chinese emperor regarded calendar-making as one of his duties that came with the mandate bestowed upon him from Heaven. The Chinese calendar took into account the apparent cycle of the sun and the cycle of the moon, both of which cannot be expressed in an exact number of days. The astronomer responsible for constructing a luni-solar calendar had to make accurate observations of the sun, the moon, and the planets, but however accurate his observations, his calendar would eventually, in just a matter of decades, go out of step with observations. In 1972 archeologists working at the site of a Han tomb at Linyi in modern Shandong province, discovered a calendar for the year 134 BC. Throughout the history of China no less than one hundred calendars had been constructed, not to mention other unofficial calendars sometimes used in certain regions. The most renowned among the Chinese calendars were the *Dayanli* calendar completed by the Tang Tantric monk Yixing (683–727) in 727 and the *Shoushili* calendar prepared under the Mongols by Guo Shoujing (1231–1316) in 1280. They used the method of differences in mathematics to make their calculations. Accurate and new astronomical observations had to be made for the purpose of calendar making. New astronomical instruments were constructed for this purpose. Yixing, for example had to construct his own armillary sphere. The Song dynasty (960–1279) is remembered for the several large armillary spheres made for this purpose. The most famous was that made by Su Song (1020–1101). It was an armillary sphere driven by a water wheel using the principle of the escapement.

In the early stages medicine and magic were indistinguishable from each other, as both were practiced by the shamans (*wu*). Shamans and doctors were together referred to as *wuyi* (shamans and doctors). During the Spring-and-Autumn Period (722–480 BC) there were signs that physicians and shamans had already parted company. There was the celebrated physician Qin Yueren, better known as Bian Que (fl. 501 BC), who was the counterpart of Hippocrates (465–370 BC) in China, but lived at least one generation earlier. He was already acquainted with the four important diagnostic procedures used in Chinese medicine: observing external signs (*wang*), listening to sounds (*wen*), asking the patient's history (*wen*), and feeling the pulse (*qie*). These are sometimes referred to as looking, listening, asking, and touching. The earliest Chinese medical writing known to us until recently is the *Huangdi neijing* (The Yellow Emperor's Manual of Corporeal Medicine). Consisting of two parts, the book appears to have been written by several earlier unknown authors but took its final form during the second century BC. Between the winter of 1973 and the spring of 1974 some valuable medical writings were discovered during the excavations at Mawangdui near the city of Changsha in modern Hunan province. The tomb where these writings were found dated back to the year 168 BC and hence the medical writings concerned must have belonged to an earlier date. These are the two most ancient Chinese medical writings extant.

The earliest Chinese pharmacopoeia that we have is the *Shennong bencaojing* (Pharmacopoeia of the Heavenly Husbandman). We do not know its exact authorship and neither are we certain about when it was written, although the date could not be later than the second century. It sets a tradition followed by a long series of succeeding pharmacopoeias over a period of more than a thousand years. It mentions the *yin* or *yang* property and the indications of each medicinal substance, noting that some combinations of two or more of them are either beneficial or counter-indicative. It divides all the 365 items of medicine that it contains into three categories. Those in the first category have the efficacy of nourishing or prolonging life; those in the second are, as a rule, non-toxic and can be used to restore the constitution of the patient; and those in the third are usually toxic or have side effects. The last category is used for combating diseases. Thus the *Shennong bencaojing* catered to the needs of both the physicians and the aspirants of longevity.

While the physicians had parted company with the shamans, some of the latter devoted their attention to prolonging the human life span. Some of the substances they ingested, such as mercuric sulfide, mica, licorice, and asparagus, are listed in the *Shennong bencaojing* pharmacoepia. The search for the way to physical immortality probably began in China at least 2500 years before our time. It is recorded that in the Warring States Period (480–221 BC) someone presented an elixir to the prince Jingxiang wang in the State of Qu. Some modern writers attribute the first practice of alchemy in China to Zou Yan. Later on the story of the elixir told by the shamans got the fancy of the emperor Qin Shihuangdi, who made several attempts to procure an elixir for himself so that he could live and rule his empire to eternity. Another patron of the shamans was the emperor Han Wudi, who reigned about 133 BC. It was then that the shamans talked about making gold as the first step towards the elixir. At about the same time, Liu An, the Prince of Huainan, compiled, with the help of a group of shamans and alchemists retained by him, the *Huainan wanbishu* (The Ten Thousand Infallible Arts of the Prince of Huainan). Tradition says that this book dealt mainly with alchemy, especially the making of gold, but unfortunately we no longer have the complete text, and the fragments of the book now left are only concerned with magic. In the following century Liu Xiang, who was said to have inherited this book, was commissioned by the emperor to make gold for the Treasury. His failure landed him in prison.

The earliest Daoist alchemical treatise that is still extant today was written by the alchemist Wei Boyang in the second century AD. Entitled *Cantongqi* (The Kinship of the Three), the text is very obscure, containing a number of alchemical terms with "hidden" meanings. The use of the system of the *Yijing* (Book of Changes) has contributed directly to the obscurity of the text of the *Cantongqi*. Alchemy gained popularity following the work of Wei Boyang. During the fourth century Ge Hong wrote his *Baopuzi neipian* (The Esoteric Chapters of the Solidarity Master), showing the great advancement made in alchemy since the time of Wei Boyang. Ge Hong was at the same time an accomplished physician. Other famous physicians who were also great alchemists after Ge Hong were Tao Hongjing (456–536), Sun Simo (?581–?682), and his disciple Meng Shen (621–718). For the next eight hundred years or so after Ge Hong, alchemy continued to flourish and many alchemical works were written. Most of these works are lost; those that survived are included in the present version of the *Daozang* (Daoist *Tripitaka*), a collection of 1464 Daoist works, only a small fraction of which deal with the subject of alchemy.

In many of their experiments the Chinese alchemists used the process of sublimation and of distillation. They often made use of the reaction-vessel, called *yaofu*. Many of the elixir recipes contained toxic ingredients such as mercury, arsenic, and lead. Hence such elixirs would be quite poisonous. Quite a number of Chinese emperors showed great interest in the elixir and some of them unwittingly perished. For example, the Jin emperor Aidi died in his very prime, aged only twenty-five, as a result of his attempt to avoid growing old. The emperor Wenxuandi of Northern Qi, on the other hand, exercised more caution. When presented with the elixir he decided that the most opportune moment to test it would be on his death bed. At least three Tang emperors died as a result of taking the elixir. Political motives might be behind the early demise of some Chinese emperors, but the alchemists responsible were quickly punished for their failure. With the evidence so efficiently destroyed the full case is difficult to investigate.

There were probably many charlatans among the alchemists, who, when their elixirs brought about the early demise of unfortunate emperors, managed to escape before it became too late. However, there were also alchemists sincerely interested in their work who believed in the elixirs they made. So strong was their faith that many alchemists must themselves have fallen martyr to their own beliefs, or become victims to mistakes in following the obscure and contorted instructions of their predecessors. The most experienced or industrious experimenters were often the most enthusiastic believers, and in the end the surest victims. In this respect the elixir mania must have acted as an inhibiting factor to the progress of chemical knowledge in China.

Many Chinese alchemists were aware of the toxicity of their products. Some tried to neutralize the poison; some recommended only symptomatic treatment; some believed that the ill effects were only side-effects associated with the elixir and as such should be completely ignored; some turned to the vegetable kingdom to look for suitable ingredients; and some turned away from the material elixir and practiced "physiological alchemy" (*neidan)* following a regime of meditation and breathing exercise. By the ninth century alchemy in China had already seen its best days. The alchemists gradually relied more and more on the vegetable kingdom for their raw material. Hence there was an alchemical work that included the term *bencao* (pharmacopoeia) in its title, namely the *Waidan bencao* (Pharmacoepia of Operative Alchemy) by Cui Fang (fl. eleventh century). After the turn of the fourteenth century several books on elixir plants made their appearance. The Chinese alchemists seemed to have gone round one full circle back to the days of the ancient shamans. Alchemy merged again with the tradition of Chinese medicine set by the *Shennong bencaojing* pharmacopoeia.

Hence traditional Chinese medicine and alchemy arose from the same source. At first the shamans practiced both magic and medicine. About the sixth century BC medicine and magic took different courses. Some shamans developed the art of prolonging human life and reached the conclusion that one had to make gold by artificial means as a first step to physical immortality. That was the beginning of alchemy. Alchemy maintained a close link with medicine, dealing only with different aspects of human life. Quite a number of alchemists were at the same time famous physicians, as in the case of Ge Hong, Tao Hongjing and Sun Simo. A preliminary study shows a close connection between alchemy and Chinese medicine in the similarities between the basic principles used in medical and elixir prescriptions. Medicine and alchemy borrowed from each other. For example, alchemical works are liberally quoted in the more important pharmacopoeias of later time such as in the *Bencao gangmu* (the Great Pharmacoepia) written in 1596. Although the Chinese alchemists did not succeed in their quest for the elixir of immortality, they played a part as the iaotrochemist in Chinese medicine. Another by-product of their experiments was that they stumbled upon gunpowder when some of them caused an accident when they did not exercise sufficient caution in using saltpeter and sulfur, allowing some carbon impurities to get in from the charcoal or wood that they used as fuel.

There was early inter-cultural transmission of scientific and technological knowledge between China and her neighbors, especially west Asia, before the Christian era. Buddhism first came to China in the second century. Scientific knowledge followed the wakes of missionary and pilgrimage activities. During the eighth century Indian monks and calendar experts, Nestorians, Arab merchants, Korean and Japanese students, and others lived together in the Tang capital Changan (modern Xian) which had a population estimated to be over one million. Mutual exchange of knowledge was inevitable. Muslim astronomers found employment in the astronomical bureau in thirteenth-century China and some Chinese astronomers could have been sent by Hulagu Khan to work in the observatory in Maragha under its director Naṣir al-Dīn al-Ṭūsī (1201–1273). The Renaissance in Europe was influenced in no small measure by Arabic and Indian learning, but Europe also acquired knowledge of Chinese science and technology through the Arabs and the invasion of the Mongols during the thirteenth century. Since the seventeenth century science in Europe has advanced by leaps and bounds while Chinese traditional science remained in a state of stagnation. Towards the end of the sixteenth century the Jesuits arrived in China, using science as a tool to promote their mission. Science from Europe, particularly mathematics and astronomy, demonstrated its superiority to traditional science. In the latter half of the nineteenth century modern science came to China from Europe and North America, and by the twentieth century China had joined the world in the common enterprise of modern science.

HO PENG YOKE

REFERENCES

Cullen, Christopher. *Astronomy and Mathematics in Ancient China: The Zhou Bi Suan Jing*. Cambridge: Cambridge University Press, 1996.

Du Shiran et al. *Zhongguo kexuejishushi gao* (Draft Version of the History of Science and Technology in China). 2 volumes, Beijing: Kexue chubanshe, 1982.

Ho, Peng Yoke. *Li, Qi and Shu: An Introduction to Science and Civilization in China*. Hong Kong: Hong Kong University Press, 1985.

Needham, Joseph. *Science and Civilisation in China*. 7 vols. in estimated 30 parts, 15 parts published; others pending. Cambridge: Cambridge University Press, 1954.

Ronan, Colin A. *The Shorter Science and Civilisation in China* (abridgment of Needham's *Science and Civilisation in China*). Cambridge: Cambridge University Press, 1978.

See also: *Yinyang* – Astronomy – Metallurgy – Calendars – Armillary Sphere – Alchemy – Ethnobotany – Medicine – East and West – Magic and Science

CHINESE MINORITIES In China there are now fifty-six nationalities, of which the Han is the majority. Each nationality and its ancestors has made its contribution to Chinese civilization.

Many nationalities have lived in China since ancient times. From 2100 BC to 476 BC, the Xia, Shang, and Zhou established their states successively, but they had small territories. During the Warring States (475 BC–221 BC), feudalism was developed and the Huaxia nationality grew out of the Xia, Shang, and Zhou nationalities in the middle and upper reaches of the Yellow River. The Han evolved from the Huaxia. Minorities surrounding Huaxia were generally called Man, Yi, Rong, Qiang, and Di in ancient books. Since the Warring States, the main minority groups have been the Xiongnu, Eastern Hu, Turk, Shusheng, Di qiang, Baiyue, Miao and Man. From them, many other minorities evolved, such as Xiongnu, Wuhuan, Xianbei, Rouran, Qidan (Khitan), Mongul, Dingling, Chile, Tiele, Turk, Huihe, Uyguy, Shusheng, Nüzhen, Manchu, Wusun, Rouzhi, Shache, Bai Yue, Baipu, Li, Liao, Baiman, Wuman, Zhuang, Dai, Yi, Miao, Tibetan, and Dangxiang. Fusion, disintegration, and evolution are the essentials of the development of Chinese minorities. During the course of history, many minorities changed. Some emerged, some evolved into new nationalities, some fused into the Han or the other nationalities. In the 1950s, fifty-five minorities were identified. Historically, dozens of states were established by Chinese minorities. Some of the States were powerful and important in the history of China. The Northern Dynasties (386–581), namely, the Northern Wei Dynasty (535–556), the Northern Qi Dynasty (550–577), the Northern Zhou Dynasty (577–581), the Xixia Dynasty (1032–1234), the Liao Dynasty (907–1125), the Jin Dynasty (1115–1234), the Yuan Dynasty (1279–1368), and the Qing Dynasty (1644–1911) were established respectively by the Xianbei, Dangxiang, Qidan, Nüzhen, Mongol, and the Manchu.

The Chinese made great contributions to the development of sciences; many of these were made by minorities. Numerous scientific works were written in various languages. Many cultural relics of minorities were preserved or unearthed throughout China. Some imperial courts of minorities paid great attention to science and technology. They formulated policies for promoting the development of science and organized scientific projects. Such cases were astronomy and national surveying at the beginning of the Yuan Dynasty, the projects to find the sources of the Yellow River, and the compilation of scientific works in the early Qing Dynasty.

Some achievements in science, technology, and medicine will be mentioned here.

Astronomy

Astronomy is an ancient natural science in China. It was related to politics and philosophy and played an important role in Chinese culture. Both the Han and minorities devoted their attention to astronomy, especially to calendars. Most minorities had their own calendars. Some of them, such as the calendars of Tibetan, Yi, and Hui nationalities, had a great impact on Chinese astronomy. Some are still used by minorities today. The Yi calendar has had a long history. It can be traced back to the Warring States Period. Historians of astronomy, such as Chen Jiujin, believe that the calendar is a ten-month solar calendar, which means that a year has ten months, and a month has thirty-six days. The other five days are added at the end of a year for holidays. Furthermore, they believed that the *Xia Xiao Zheng* (Lesser Annuary of the Xia Dynasty) the oldest extant calendar in China, was a ten-month one. But Luo Jiaxiu, a Yi scholar, disagrees with Chen's views. Luo argued that the Yi calendar was a twelve-month one. The Tibetan calendar was a lunisolar one and was formulated during the seventh or eighth century. It grew out of Indian calendars and was influenced by Chinese ones. Since the thirteenth century, Tibetan scholars wrote many astronomical works, most of which are preserved today. Tables and illustrations of solar and lunar eclipses from AD 1387 to 1687 were compiled. Today's Tibetan calendar is based on one compiled in 1927 which was based on two astronomical works written in 1687 and 1714 respectively, which were the most important works in the history of Tibetan calendars.

The Dai nationality has a luni-solar calendar. Its epoch was in AD 638. In their calendar a year had 365.25875 days and a lunar month 29.530583 days. The first almanac in the Dai language was *Su-Ding*. About ten calendar books are preserved at the Central Institute for Nationalities in Beijing. The Hui nationality grew out of the Arabians and Persians who came to China. Their calendar, which is the Arabic calendar, had great influence in China then and was introduced by central Asian scholars in the Yuan Dynasty. During the Yuan and the Ming Dynasties, a department of Hui calendar was established in the Royal Observatory. A Qidan scientist, Yelü Chucai, proposed the concept of *li-cha* in AD 1180. This corresponded to the concept of latitude. In AD 412, Hulan, a Xianbei technician, and Cao Chun made the only iron armillary sphere in China. In the eighth century, four instruments for determining the length of the sun's shadow at the equinoxes and the solstices were made by a Tibetan scientist. At the beginning of the Yuan Dynasty, an astronomer Jamal al-Dīn, who came from Persia and became a Hui, later made seven pieces of astronomical instruments by order of the Emperor Hubilie, including the first Chinese terrestrial globe.

From the fourth century on, many astronomers from minority groups were in royal observatories. There are many astronomical relics of minorities in China. Here are some examples:

- the only Mongolian astronomical map (in Huhehot, eighteenth century);
- a color picture of the twelve signs of the zodiac in AD 1116 on the top of a Qidan's tomb located at Xuanhua in Hebei Province;
- a Dai language astronomical map on the stone tablet in AD 1801.

There are hundreds of astronomical books in Tibet, Inner Mongolia, Xin-jiang, Qinghai, Beijin, Liaonin, and other places which have not been studied.

Mathematics

Arabic numerals were introduced into China by Islamic believers in the thirteenth century. Five magic squares of the sixth order in Medieval Arabic numerals were unearthed in 1956. Some minorities, such as Tibetan, Mongol, Nahxi, and Dai, have their own numerals. Tibetan numerals are similar to Indian ones; some in fact are the same. Mongolian numerals were influenced by Tibetan ones. The Uygur, Kazak, and Hui nationalities adopted Arabic numerals earlier than the Han nationality. A kind of sand or ash plate for calculation was used by Tibetan, Mongolian, Uygur, and other minorities in the north. The *Zhuerhai* was an old Mongolian activity in which mathematical knowledge is related. Euclid's *Elements* was introduced into China in the thirteenth century. Menge, a grandson of Ghenghis Khan, was the first person to study the *Elements* in China. It is said that an Arabic language text of the *Elements* in fifteen volumes was preserved in the Royal Observatory of the Yuan Dynasty. Some mathematical treatises in both Chinese and minorities' languages are preserved now. Among them, the most important is Mingantu's *Ge Yuan Mi Lu Jie Fa* (Quick Method for Determining Segment Areas, 1774) written in Chinese. Mingantu (ca. 1692–1763) was a Mongolian mathematician and astronomer. He was the head of the Royal Observatory for about fifty years in the Qing Dynasty. He played an important role in national scientific projects. For example, he participated in compiling three series of scientific books and went to Xinjiang twice for surveying. His mathematical works dealt with the series of trigonometric functions and π. The series of $\sin\alpha$, π, and $\operatorname{vers}\alpha$ was introduced into China by a Frenchman, Petrus Jartoux (1668–1720). Mingantu spent thirty years studying these series and discovered six new series of trigonometric functions. He laid the foundation for the study of series expansions and their operations. In addition, he derived the so-called Catalan numbers before L. Euler (1758) and E.C. Catalan (1838) and also used the method of reversion of a series and found two kinds of counting functions.

Geography, Water Conservancy and Agriculture

Most minorities lived in peripheral regions. Many were explorers and travelers. Uygur Yiheimishi traveled to the western regions to explore its geography by order of the Emperor Hubilie of the Yuan Dynasty in 1272 and 1275. Nüzhen Dushi explored in Tibet three times in the 1280s; he also explored the sources of the Yellow River by order of Hubilie. When he returned, he reported his story and drew a map which was the first map of the sources of the river in China. A Tibetan book about the sources of the Yellow River was quoted by the geographer Zhu Siben in 1320. A Manchu, Shulan, and a Mongolian, Laxi, organized an expedition to explore the sources of the Yellow River by order of the Emperor Kangxi in 1704. They verified that there are three sources and reported details of the region, including animals, plants, climate, and geography.

One of the most important navigators in China was Zhen He of Hui nationality in the Ming Dynasty. He travelled by boat seven times from 1405 to 1433, reaching many Asian and African countries. Zhen He also drew a nautical chart. Ma Huan, who was a Hui and Zhen He's companion, recorded the geographical and local conditions and customs for more than thirty countries in his travelog. A Qidan scientist, Yelü Chucai, described the geographies of the western region of China and some Asian countries at the beginning of the Yuan Dynasty. A geographical work with one thousand three hundred volumes was compiled and a color national map was drawn by the Yuan court from 1285 to 1303. Many minoritiy scholars participated in the project. Among them, Jamal al-Dīn was the leader. Another minority scholar, Samus (1278–1351), wrote at least four geographical books. A complete geographical survey of the empire had been organized by the Qing court. Some Mongolian, Manchu, and Tibetan scholars participated in the project. Mount Qomolangma was mapped on the national map in 1719.

A series of hydraulic engineering projects was accomplished during the Warring States Period; many of them were related to minorities. For example, Dujiangyan was built by people of Qiang nationality under the leadership of Li Bin of the Han nationality in BC 250. Some scholars pointed out that Karez wells, which are irrigation systems of wells connected by underground channels, were dug by the minorities in Xinjiang two thousand years ago. However, others think that Karez wells appeared in the Qing Dynasty. In the Tang Dynasty an irrigation system which could irrigate thousands of *qing* of fields was built by the Bai nationality in Dali of Yunnan. Sibo and Uygur built some irrigation systems in Xinjing in the nineteenth century. Treatises on water conservancy were written. Among them, *Revision and Enlargement of a General Discussion of the Protection Works Along the Yellow River* (1321) by Samus is the oldest extant monograph on water conservancy in China. Linqin (1791–1864) of Manchu wrote a book entitled *Illustrations and Explanations of the Techniques of Water Conservancy and Civil Engineering* that is the first monograph on tools of water conservancy in China.

Minorities also made great contributions to agriculture and animal husbandry. For example, many breeds of animals were improved and disseminated to the central plains of China by minorities. The Xiongnu cultivated the mule by hybridizing between a horse and an ass two thousand years ago. Many crops, such as garlic, pea, sesame, and watermelon were introduced into the central plains of China from the western region by minorities. Rice was cultivated first in the southern and southwestern regions. Some books on agriculture and horticulture were written by minorities. For example, there is *Selected Clothing and Food*, written by Uyguy Lu Mingshan in 1314.

Medicine

Every minority group had its medical system. Many medical books written by minorities both in minority languages and Chinese are extant today. Hundreds of Tibetan medical books have been written. The most important work was *rGuyd-bzhi* (Four Medical Tantras) written by the medical sage gYu-thog Yon-Tan MGon-po (708–833) in the eighth century. Almost a hundred texts of the book with various commentaries have been produced. Also, about a hundred painted scrolls were attached in the seventeenth century, which formed a book entitled *Tibetan Medical Thangka*. Thousands of drugs were recorded in these medical works. Tibetan medicine influenced Mongolian medicine which was formed in the thirteenth century. Dozens of Mongolian medical works were created in the Ming and Qing Dynasties. Mongolians were expert in surgery and orthopedics. A method for rescuing a patient with shock therapy was used in the twelfth century. The Hui and Uygur nationalities' medical systems were influenced by Arabic and Chinese ones. Arabic medicine was introduced into China in the Tang Dynasty and developed into Hui Hui medicine. In the Yuan Dynasty, institutes for Hui Hui medicine were set up, and many physicians who came from central Asia studied in the institutes. Freezing anaesthesia was used by Qidan nationality in the Liao Dynasty.

Some anatomy was also studied. Tibetans studied the body's organization by means of dissection. Some illustrations of dissections were drawn in the eighth century. A well-made diagram of the human body was made in the seventeenth century in which knowledge of anatomy was used correctly. In the 1040s, autopsies were made in the region of the Zhuang nationality, where they created a diagram named Ou Xifan's Organs in which some pathological changes of the modes dissected were described. Two vivisections were made by Mongolians in 1263.

Minorities also made contributions to nutrition. Husihui, who was a Hui or Mongolian and a royal physician in the Yuan Dynasty, wrote the first monograph on nutrition in China in 1330.

Technology and Architecture

The technique of iron smelting was spread to minorities' regions during the Qin and the Han Dynasties. In the Northern Dynasty, Qimu Huaiwen, a descendant of the Xiongnu, invented a new steel smelting technique called "cofusion steel", used for knife blades. The knife he made was quenched with animal fats and urine; it could cut thirty suits of armor. His method was still used in the Sui Dynasty (589–618). Wrought iron was introduced into China from Persia by minorities in Xinjiang in the Tang Dynasty. Mongolians studied the technique from Qidan, and wrought iron workshops were set up. Most minorities had their own techniques of making arms such as sabers and armor. For example, before the eleventh century, a branch of the Qiang nationality made a kind of armor known as Monkey Armor which was harder than others at that time. In the thirteenth century, the Qidan, the Nüzhen, and the Mongolian made firearms. Also, the oldest grenade was made and used by the Yi in the 1850s.

Spinning and weaving had been studied by the minorities in Xinjing and southern China since the Han Dynasty. They produced excellent cotton goods, some of which were given in tribute to imperial courts. Silk technology was also studied. They embroidered some kinds of brocade, such as the Zhuang and Dong brocade, which are both well known. Silk knitting by minorities in the Northern Dynasty has been unearthed. Some minorities were good at making woollen knitwear. Various technologies of spinning and weaving were invented. The technique of wax printing was an important one which was invented by minorities in the Han Dynasty.

The architectural arts of minority groups are splendid. The yurt was invented by the Xiongnu nationality. It was used by many minorities in Northern China. Today's Mongolian yurt is derived from it. The minorities in the south who invented the technology built the Gan Lan building, which was a kind of wooden building with two or more stories. Heated *kang* or a heated brick bed was first used by the Nüzhen nationality. It then spread throughout northern China.

Minorities also made their contributions to urban construction. Many cities and towns were built, such as Chengde, Kunming, Beijing, Shenyang, Datong, and Huhehot. Many religious buildings were constructed by minorities. Temples, mosques, towers, and grottoes built by them are standing in great numbers in China. The Yongningsi temple in Luoyang built in AD 516, the Dunhuang grotto, Tri-towers in Dali (built in the Tang Dynasty), the Yonghegong Palace in Beijing, the Ta'ersi Temple in Xinning, and the Batala Palace in Lhasa are some examples.

Minorities built great numbers of bridges, many of which are still standing. They invented the chain bridge, and designed and built the Lanjiqiao Bridge, the first iron chain bridge in China, in the fifth century.

Chinese minorities have made their contribution to science. It should be mentioned that the studies of the history of science of Chinese minorities have just begun. There is much work to do.

GUO SHIRONG
FENG LISHENG

REFERENCES

Chen Jiujin. *A History of the Yi Nationality's Astronomy*.

Li Di. *A History of Science and Technology of Chinese Minorities*, vol. 1. Nanning: Guangxi Science and Technology Press, 1996.

Luo Jiaxiu. *Studies on the Calendars of the Yi Nationality from Ancient to Today*. Chendu: Sichuan Nationalities Press, 1993.

CITY PLANNING: AZTEC CITY PLANNING The Aztecs were the most urbanized of the ancient civilizations of Mesoamerica. The last in a long line of urban societies, they selected principles of city planning from an ancient Mesoamerican heritage and adapted these to their needs. Most Aztec urban centers were modest settlements best called towns, but the central capital Tenochtitlan was a huge metropolis of a different order.

Most Aztec towns were founded between AD 1150 and 1350 when the Aztec peoples first immigrated into the central Mexican highlands. They established new settlements and dynasties leading to a system of autonomous city-states. The construction of a royal palace marked the official founding of a new city or town, most of them city-state capitals. In 1430, two Aztec peoples, the Mexica and Acolhua, formed a tributary empire, and their capitals, Tenochtitlan and Texcoco, became the preeminent cities of the Valley of Mexico. By the time Spanish conquerors arrived in 1519 this empire had conquered much of Mesoamerica, and Tenochtitlan had grown into a city of 200,000.

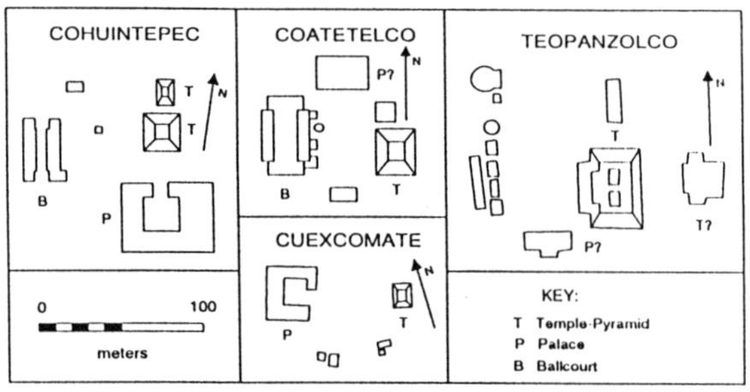

Figure 1: Plans of the ceremonial precincts of four Aztec towns in Morelos, Mexico.

Some Aztec towns were destroyed in the Spanish conquest, but most were settled by Spaniards in the colonial period and continue to be occupied today. Underneath almost any modern town in central Mexico are the disturbed ruins of an Aztec town, making archaeological fieldwork difficult or impossible. Adding to the problems of studying these towns is a rarity of documentary data. Available evidence suggests that towns outside of Tenochtitlan conformed to a general Mesoamerican urban pattern in which a central ceremonial precinct exhibits formal planning while a surrounding residential area does not. A common feature of Aztec towns is the temple-pyramid facing west on to a public plaza, as in Figure 1. Other buildings constructed around the plaza include a palace and often a ballcourt (used to play a game that combined ritual and politics).

Outside the precinct, there is little evidence for formal planning in Aztec towns. Only a handful of examples have been mapped in detail, but all show a dispersed residential pattern. Houses were spread widely across the settlement, with much of the town area taken up by horticultural gardens. Most towns were small; those in the Valley of Mexico averaged 9000 inhabitants and those in Morelos 3000.

Tenochtitlan

According to native historical accounts, the Mexica people founded Tenochtitlan in AD 1325 at a place designated by their god Huitzilopochtli, which was a desolate swampy island in Lake Texcoco (the god's sign, an eagle holding a snake while sitting on a cactus, is now the national symbol of Mexico). Tenochtitlan grew rapidly into the largest city in Mesoamerica, a thriving imperial capital without parallel in the ancient New World. Among the social changes responsible for its explosive growth were rapid population growth, the proliferation of craft production and market trade, and the expansion of the Aztec empire with increasing tribute flow into the capital.

Although no explicit articulations of urban planning concepts have survived, three primary principles were responsible for creating the form of Tenochtitlan: the city's island location, the grid principle, and imperial ideology. Most of Tenochtitlan's 13.5 square kilometers were reclaimed from Lake Texcoco. Tenochtitlan was connected to the shore by causeways, one of which supported an aqueduct that brought fresh water to the city. Spanish observers were struck by the great number of canals in the city, which they likened to Venice. The canals were used as transportation arteries and for agricultural purposes. Raised fields or *chinampas*, an extremely productive method of farming, were built to cultivate reclaimed swampy land in the outer neighborhoods of the city. These fields were worked by families living on their individual, small plots.

Tenochtitlan was laid out following a grid pattern with major avenues, along the cardinal directions, radiating out from a central ceremonial precinct. The avenues divided Tenochtitlan into four quarters, each with its own smaller ceremonial precinct. These quarters were in turn divided into neighborhoods, many of which had ethnic or occupational significance. Outside of the chinampa areas, houses were packed tightly together. A northern area called Tlatelolco, the setting of the famous central marketplace, was originally a separate town but was later incorporated into Tenochtitlan.

The Mexica rulers borrowed the grid principle from the ancient imperial capitals Teotihuacan and Tula to help establish the city's legitimacy as the imperial capital of central Mexico. The imposition of a common grid expressed the power of the rulers to shape their city and differentiate it from other Aztec cities, and it was also convenient for reclaiming and subdividing new land from the lake.

Imperial ideology also contributed to the form and layout of the capital. At the center of Tenochtitlan was a large walled compound of temple-pyramids, altars, priests' residences, and other sacred buildings. The central temple of the Aztecs, the so-called *Templo Mayor*, had two stairways

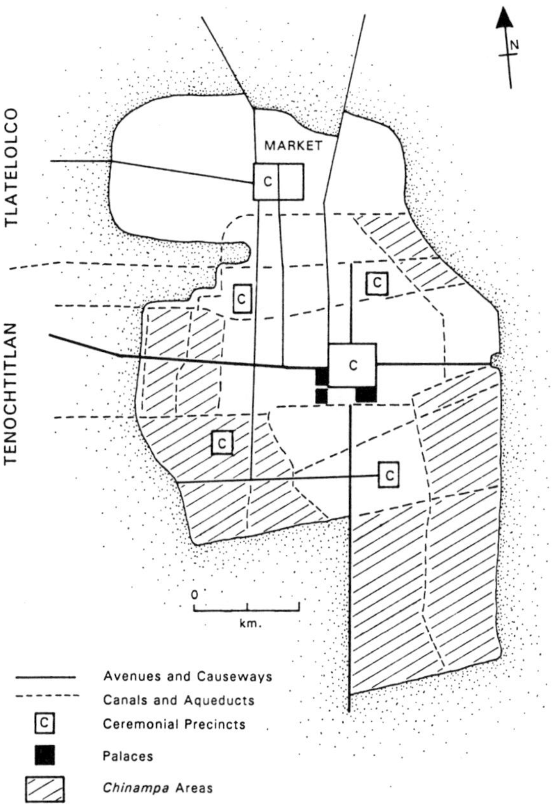

Figure 2: Plan of Tenochtitlan, the Aztec imperial capital (redrawn from Calnek 1972).

and two temples — one devoted to the rain god Tlaloc and the other to Huitzilopochtli, the god of warfare and patron deity of the Mexica. Excavations at the Templo Mayor in downtown Mexico City have uncovered the spectacular remains of several construction stages and numerous buried offerings. The Templo Mayor, viewed as the symbolic center of the Aztec empire, was the setting for elaborate state ceremonies including human sacrifices. The palaces of the Mexica emperors were built around the outside of the sacred precinct.

All Aztec urban centers exhibited established Mesoamerican principles of planning in which cities and towns were laid out around a central ceremonial precinct containing a temple-pyramid, palace, and other buildings that were formally arranged around a public plaza. Tenochtitlan also drew on central Mexican principles of imperial city planning that emphasized a grid layout and densely occupied residential zones. This city, the largest ever built in the pre-Columbian New World, combined these ancient principles with new ideas derived from its island location and imperial status. When destroyed in the Spanish conquest of 1519, Tenochtitlan and other Aztec urban centers were at the height of their glory. Only traces of the cities still survive.

MICHAEL E. SMITH

REFERENCES

Broda, Johanna, David Carrasco, and Eduardo Matos Moctezuma. *The Great Temple of Tenochtitlan: Center and Periphery in the Aztec World*. Berkeley: University of California Press, 1987.

Calnek, Edward E. "The Internal Structure of Tenochtitlan." In *The Valley of Mexico*. Ed. Eric R. Wolf. Albuquerque: University of New Mexico Press, 1976, pp. 287–302.

Marcus, Joyce. "On the Nature of the Mesoamerican City." In *Prehistoric Settlement Patterns*. Ed. Evon Z. Vogt and Richard M.Leventhal. Albuquerque: University of New Mexico Press, 1983, pp. 195–242.

Rojas, José Luis de. *México Tenochtitlan: Economía e Sociedad en el Siglo XVI*. Mexico City: Fondo de Cultura Económica, 1986.

Smith, Michael E. *The Aztecs*. Oxford: Blackwell, 1996.

CITY PLANNING: INCA CITY PLANNING Although the Incas were not great city builders, they redesigned their capital, Cuzco, on a grand scale, and founded numerous new settlements. Ollantaytambo is one such new town. Because it has what "may be the oldest continuously occupied dwellings in South America" (Kubler, 1975), and because many parts of the town are well preserved, it would seem the perfect object for the study of the town planning principles of the Incas.

Ollantaytambo is located about 90 km to the northwest of Cuzco at the confluence of the Urubamba and the Patakancha rivers. It is built on a narrow, gently sloping bench of artificially leveled ground squeezed in between Cerro Pinkuylluna to the east and the Patakancha river to the west. It is carefully sited so as not to occupy prime agricultural land, yet to provide easy access to the terraced fields to the north and the south. The glacier-fed Patakancha provides an ample water supply for both agricultural and domestic uses. Tucked in between the high mountains at the mouth of the Patakancha valley, the town is well protected from the fierce afternoon winds which often sweep through the broader Urubamba valley.

The town was laid out on a regular grid, trapezoidal in shape, of four longitudinal and seven transversal streets. The transverse streets, oriented at 110.5° east of magnetic north, are perfectly parallel to each other, suggesting that the Inca were knowledgeable about geometry, and that they had a method of surveying to lay out the streets.

In the time of the Incas, there was a large plaza in the middle of the town. On its north and south sides it was bordered by great halls, long buildings with many openings towards the plaza. If what Garcilaso de la Vega tells us about Cuzco holds for Ollantaytambo, then the plaza was the theater for ceremonies and festivities, and the great halls were the place where the revelers withdrew to pursue their activities on rainy days (Garcilaso de la Vega lib. VII, cap. X; 1976:II,108).

The fifth transverse street manifestly divided Ollantaytambo into two parts with distinct architectural features: the street facing walls in the southern half of town were built of cut and fitted stones, whereas in the northern half they were built of unworked field stones. The blocks in the southern half show a rigorously repetitive block design of two walled-in habitation compounds, called *kancha*, arranged back to back, that is not reflected in the northern half. Cobo stated:

> The Incas imposed in all their kingdom the same division in which the city of Cuzco was divided, Hanan Cuzco and Hurin Cuzco; cutting each town and lordship into two parts or factions called hanansaya and hurinsaya, which means "the upper district" and "the lower district", or the part or faction from above and the faction from below; . . . (Cobo lib. XII, cap. XXIV; 1964:II,112).

This division, which is a social one and probably has ancient Andean roots, was used by the Incas to control, administer, and inspire their subjects (Cobo lib. XII, cap. XXIV; 1964:II,112). The two moieties were essentially equal, and each was governed by its own leader, with the *hanan* leader having first rank. It is possible, but not demonstrated, that the architectural division observed at Ollantaytambo is a referent to this social division.

Because of the rigor and clarity with which the town of Ollantaytambo has been laid out, one might be tempted to interpret it as an exemplar of Inca town planning. A comparison from a formal point of view with other Inca sites, however, does not sustain this proposition. Outside Ollantaytambo regular grids are only found at Chucuito, on Lake Titicaca, and Calca. The Chucuito grid too is trapezoidal in shape, and that of Calca appears to be strictly orthogonal. It is not known with certainty whether the Chucuito grid is of Inca origin. Calca, on the other hand, is known to have been established by *Waskhar*, the twelfth Inca. Whether or not the blocks in these two sites were occupied by *kancha* is not known. Cuzco, the navel of the Inca empire, shows an ordered street pattern, but the grid and the block sizes are not as regular and uniform as those of Ollantaytambo. At least some blocks in Cuzco were built up with compounds very similar to the *kancha* at Ollantaytambo. Huánuco Pampa, a large administrative center in the central highlands, lacks a regular street pattern, and although there are discernible town blocks, they are quite irregular in size and shape. Many *kancha*-like compounds are scattered throughout the site, but their design is quite varied and irregular, and the compounds do not define blocks or streets. At Patallaqta, some twenty kilometers downstream from Ollantaytambo, the *kancha* design is prominently in view. There are two arrangements of *kancha*; one consists of two *kancha*, back to back, similar to Ollantaytambo, except that it is not walled in; the other is made of four smaller *kancha*. Even though the *kancha* are not walled-in, their groupings nevertheless define town blocks and streets. Tambo Colorado, an administrative center in the coastal valley of Pisco, has neatly rectangular compounds, but the center is not built on a grid pattern of streets, and the compounds only remotely recall the *kancha* design. The only feature common to the sites discussed above appears to be a plaza. At Huánuco Pampa the plaza is huge; it measures about 540 by 370 meters, is surrounded by buildings on all sides, and is dominated by what must be the largest remaining *usnu*, a ceremonial platform. Two great halls face the plaza on its east side. The plaza, *Hawkaypata*, in Cuzco was also quite large, about 200 by 250 meters. It too had an *usnu*, and was bordered with great halls on its north and east sides. Tambo Colorado's trapezoidal plaza is surrounded by buildings on three sides and has an *usnu* on its fourth side, but no great halls face it. Whether the plaza at Ollantaytambo had an *usnu* is not established. Thus, if the plaza is a recurrent element of Inca settlements, its configuration is by no means standard.

If no specific town planning rules seem to emerge from the comparison of different Inca settlements with Ollantaytambo, it stands to reason that Ollantaytambo is not representative of Inca town planning. It has been said that the Incas laid out their new settlement in the image of their capital, Cuzco. However, a comparison of the various plans analyzed above does not fare much better; there are no more similarities between Cuzco and the other sites than there are between them and Ollantaytambo. Rowe suggested that Cuzco was laid out in the shape of a Puma. Neither of the plans reviewed can be made to fit this shape. Perhaps similarities between Cuzco and other sites should not be searched for in the physical form, but instead, as Gasparini and Margolies suggested, "in the meaning and the functions of the form". They argue that, if one makes abstraction of the physical form, one will note that certain elements are "repeated with considerable insistence". They list the division into *hanan* and *hurin*, a principal and a secondary plaza, the great halls on the plaza, the *usnu*, the *inkawasi* (house of the Inca), the *aqllawasi* (house of the Chosen Women), the temple of the sun, and the storehouses. In light of the difficulties of attributing specific functions to particular ruins, and the uncertainties regarding a physical referent for the division into *hanan* and *hurin* discussed above, the effort to establish similarities among settlement patterns, and to derive planning canons, is considerably weakened.

The Spanish chronicler Bernabe Cobo described an elaborate set of about forty lines, called *zeq'e*, radiating out in all directions from the Qorikancha, the holiest of all Inca shrines, in Cuzco. Along each line were arranged a number of sacred places and objects. Each line was the responsibility of a specific royal family of Cuzco. The families were in charge of officiating at the shrines in their care and of providing the appropriate sacrifices on the designated days. Although Tom Zuidema argued that the *zeq'e* influenced the plan of Cuzco, and perhaps of other Inca settlements, it is obvious that the radiality of the *zeq'e* did not affect the city's layout which is orthogonal in its core. As John Hyslop noted "it is still uncertain to what degree the Incas are actually portraying zeque systems [even] in radial layouts, or just the principles and spatial relationships found in the system".

At least in part, discovering general planning principles may be complicated because the principles, whatever they were, may have been altered to fit the specific features of a site and its topography.

In Cuzco, in spite of colonial modifications, the major features of the Inca plan can still be appreciated. The grid-like street pattern was squeezed and bent into the neck of land between the Watanay and the Tulumayo rivers to conform to the Puma shape. At Patallaqta the street pattern also is grid-like, yet it is shifted and broken to fit the crescent-shaped plateau on which it is built. At Machu Picchu, where the terrain is much more accidented, or broken, than at Patallaqta, the intent of an orthogonal layout can still be detected. Alleyways and staircases generally cross at right angles. The modifications brought to the layout are subtle adaptations to the specific topography. Streets and alleyways, like the terraces, follow the contour lines, and staircases follow the terrain's fall lines. The town occupies two natural ridges, and is completely split into two sectors by a plaza which is molded into the saddle between the ridges. The *kancha* are squeezed, stretched, and distorted to fit the available space on the terraces, as they do at Wiñay Wayna, about an hour from Machu Picchu.

One of the most striking site planning efforts, and some of the most spectacular adaptations of building forms to suit the particular terrain, are seen at Phuyupatamarka. The buildings literally grow out of the bedrock and espouse the terrain to become one with the site; the man-made world again is inseparable from the natural world.

Today's tourists, arriving at Machu Picchu by train and bus, miss one of the most important features of Inca site planning; they are deprived of the dramatic encounter with the site the ancient traveler experienced when coming through the gate of *Intipunku*. That encounter involves a succession of open spaces that are scaled from the vast horizons of the Andes to the intimate courtyard in town, strung together by narrow passages over precipitous cliffs, steep staircases, alleyways, and gates. Along this pathway the site is veiled and unveiled in a sequence of vistas that reveal the lay of the land and the setting of the town, and attract attention to details of nature or architecture framed by narrow passages or gates.

The gate of Intipunku affords a distant view of Machu Picchu, still too far away to discern individual architectural features, but close enough to grasp its general layout, and to appreciate its scenic setting and subtle incorporation into the majestic landscape. When one descends the trail from Intipunku, Machu Picchu disappears from view completely only to reappear in a succession of mere glimpses. Only when one rounds the bend in the trail, just below the so-called Watchman's House, does the site reveal itself again in its entire splendor, spread out immediately at one's feet with the sugar loaf peak of Wayna Picchu looming over it. Continuing down the stairway toward the main gate, most of the site vanishes again, and it is not until the very last step before the gate that Wayna Picchu is unveiled again in a dramatic view, perfectly framed by the gate.

Moving on through the site, one's view is alternatingly constrained by high walls bordering the alleyways and broadened as one emerges onto a staircase, a terrace, or an open space. There is a progression form the open and public spaces through narrow alleyways and gates to the enclosed and private spaces. At many junctures, one's attention is drawn to important features in the landscape. Sometimes this is achieved by narrowing one's field of vision, sometimes by widening it. The windows of the Temple of the Three Windows concentrate the view upon the half dome of Putukusi to the east and to the west the plaza, defined by the temple and the two buildings which flank it, opens wide to a sweeping view of the snow-capped Cordillera Vilcabamba and its many sacred peaks.

Machu Picchu is not unique in this; orientations upon features in the landscape are rather common in Inca architecture and can be observed at many other sites.

The rigid street grid of Ollantaytambo was by no means the norm for Inca settlements. Where regularity in the layout of the streets is found, it is probably a by-product of the rectangular design of the *kancha*, and not so much a reflection of the intent to parcel out the land in a grid. In accidented terrain the *kancha* and the building forms were "deformed" and modified to adapt to the topography, rather than made to conform to an abstract principle. While some commonalities and the repetition of certain elements can be observed at various sites, there is simply not enough empirical evidence available to induce a set of formal town planning rules by which, if they existed, the Incas laid out their settlements. It may be that the incorporation of views and the orientation on architectural and landscape features was the all-pervasive site and town planning criterion. To support this hypothesis, if it is at all tenable, many more sites will need to be analyzed dynamically; that is, as one moves through the sites rather than by merely investigating their plans.

JEAN-PIERRE PROTZEN

REFERENCES

Cobo, Bernabé. *Historia del Nuevo Mundo* [1653]. Ed. P. Francisco Mateos. Madrid: Ediciones Atlas, 1964.

Garcilaso de la Vega, Inca. *Comentarios reales de los Incas* [1604]. Ed. Aurelio Miro Quesada. Caracas: Bibliotheca Ayacucho, Vols 5 and 6, 1976.

Gasparini, Graziano, and Luise Margolies. *Inca Architecture*. Trans. Patricia J. Lion. Bloomington: Indiana University Press, 1980.

Hyslop, John. *Inca Settlement Planning*. Austin: University of Texas Press, 1990.

Kubler, George. *The Art and Architecture of Ancient America: the Mexican, Maya, and Andean Peoples*. 2nd ed. Harmondsworth, Middlesex: Penguin Books, 1975.

Morris, [Edward] Craig, and Donald E. Thompson. *Huánuco Pampa: An Inca City and its Hinterland*. London: Thames and Hudson, 1985.

Niles, Susan. "Looking for 'Lost' Inca Palaces." *Expedition* 30 (3): 56–64, 1988.

Protzen, Jean-Pierre. *Inca Architecture and Construction at Ollantaytambo*. New York: Oxford University Press, 1993.

Rowe, John Howland. "Inca Culture at the Time of the Spanish Conquest." In *Handbook of South American Indians*, vol. 2, Ed. Julian H. Steward. Washington, D.C.: Bureau of American Ethnology, 1946, pp. 183–330.

Rowe, John Howland. "What Kind of Settlement was Inca Cuzco?" *Ñawpa Pacha* 5:59–76, 1967.

Rowe, John Howland. "An Account of the Shrines of Cuzco." *Ñawpa Pacha* 17:1–80, 1979.

Zuidema, Reiner Tom. *The Ceque System of Cuzco: The Social Organization of the Capital of the Inca*. Leiden: E.J. Brill, 1964.

CITY PLANNING IN INDIA As the birthplace of Indian culture, the towns of the Indus Valley represent an important and rich source of information concerning urban development in the Indian subcontinent. The Indus River is one of the largest and most important rivers in south central Asia. From its source in the Himalayan Mountains to its terminus in the Indian Ocean it traverses a course of over two thousand miles. For millennia it has been an essential route for travel, trade, and communication and has been the source of much of India's agricultural production. The valley which surrounds the Indus River has witnessed the birth, growth, and death of many cities.

Excavations of Indus towns have demonstrated the most ancient town planning in the world. The grid pattern (straight streets intersecting other straight streets at right angles) is among the most common and universal types of town planning. The discovery and excavation of the most famous of these sites, Mohenjo-daro, by Sir John Marshall (1931) and Ernst Mackey (1938) supported the contention of Dan Stanislovski (1946) that the development of the grid pattern in western societies — Greece, Rome, and Europe in the Middle Ages — can be linked directly to the development of town planning in the Indus valley.

A recent excavation on the Western plains of the Indus — the Rahman Dheri site — reveals a town plan from the early Indus period. This site comprises an area of roughly 22 hectares and could have been home to ten to fourteen thousand people. Radio carbon-dating of artifacts places development somewhere in the first half of the fourth millennium BC. Rahman Dheri is built in the classic grid pattern. It is rectangular in shape surrounded by an immense wall and bisected by a major traffic artery which runs roughly southeast to northwest. Perpendicular to this road exists a pattern of regularly occurring laneways which appear to create individual dwelling lots.

The successor to the plan of Rhaman Dheri is thought to be the Surkotada site in Kutch which dates to ca. 2500–2000 BC. Though smaller in scale than its predecessor the Surkotada site is built along the same grid pattern as Rahman Dheri. This settlement was enclosed by a stone rubble and brick fortifications. This produced two separate areas each of roughly sixty square meters.

Similar to Surkotada, only on a larger scale, the Kalibangan site was completely surrounded by an enormous rampart. This site, about 200 kilometers to the southeast of Hirappa on the banks of the (now dry) Ghaggar river, was composed of two mounds. The smaller mound, named the citadel, was to the west and was roughly 240 meters by 120 meters. The larger mound, named the lower city, was to the east and measured 360 meters by 240 meters. The lower city demonstrates the grid plan divided into blocks of which the east–west side was roughly 40 meters in width. The width of lanes and streets in the lower city ranges from 1.8 to 7.2 m. Interestingly each lane or street is some multiple of 1.8 meters. This site is thought to date to the early Indus period, roughly 3000 BC.

The evolution of Indus town planning reached its zenith at Mohenjo-daro. The town is geographically located on a floodplain of the Indus river; the city occupies about one square kilometer — over five times larger than Kalabangan. The basic grid of Mohenjo-daro is about 180 meters square, subdivided into 16 sections of about 40 square meters. Mohenjo-daro existed from ca. 2500 to 2000 BC. During this time it bore the brunt of often severe flooding, and as a result the original grid plan was often modified and transformed. The city itself was originally planned in the same way as Kalabangan: an oblong walled city divided into a citadel and a downtown area divided by an open space. Similar to its predecessor communities, the town plan of Mohenjo-daro indicates the same parallel grid street structure that has come to characterize urban planning in this region at this time.

As this brief history of the evolution of town planning in the Indus river valley demonstrates the grid pattern was commonplace. The importance of the Indus river valley communities to the historical development of culture and civilizaton in the Indian subcontinent and the East is well known. Stanisloski has suggested that the grid pattern form of town planning has its roots in the Indus valley. Parallel developments in Nepal, Sri Lanka (formerly Ceylon), Burma, Korea, Vietnam, and China also indicate how widespread the influence of the Indus river valleys communities were. It is suggested that this form of town planning was taken up by the Greeks, Romans, and other Western Europeans and eventually became a standard form of urban organization in Europe and the New World.

Excavations at and along the Indus River valley reveal cities strikingly similar in topography to contemporary Western European and North American cities. While some may suggest that the grid pattern evolved in a happenstance fashion without benefit of central planning or administration, substantial evidence exists to support the opposite point of view. First, the grid pattern can be seen in a variety of towns all along the Indus river which existed and flourished over a several thousand year period. Second, the similarity between these towns and the general division between a lower city area suggests the development of town planning conventions as early as 2500 BC. Third, there is a the remarkable mathematical symmetry in Kalabangan where lanes and streets (which range from 1.8 to 7.2 m in width) all occur in multiples of 1.8 m.

The contribution of this historic period to the development of Western cities should not be underestimated. Much of what we take to be enlightened urban planning was considered over four thousand years ago by planners in south eastern Asia.

PAUL GREGORY

CITY PLANNING: MAYA CITY PLANNING Much of ancient Maya civilization flourished in the period from AD 250 to 900 in the area encompassing the whole of Mexico's Yucatan Peninsula, the Mexican states of Chiapas and Tabasco, and the countries of Belize, Guatemala, western Honduras, and El Salvador. During the course of nearly twenty centuries of development, the distinctive architectural styles of the Maya heartland evolved and collapsed. As with other aspects of the Native American legacy, ancient Maya architecture was once seen as far too sophisticated to have been the product of aboriginal genius, and was long credited to Old World voyagers ranging from Egyptians to Israelites, and from white men or Danes to "great white Jewish Toltec Vikings" (Silverberg, 1968).

As scholarly knowledge of Maya architecture, settlement patterns, and city planning has grown exponentially over the course of the last sixty years, so too has our respect and admiration for this ancient American tradition. Until quite recently, scholars perpetuated the belief that the Maya had neither a literate tradition or history, nor true cities or political economies. They often typecast Maya architecture as inferior, because it did not employ the "true" arch in the Roman sense. And they were unable to demonstrate that Maya centers were planned and constructed under the direction of professional architects with drafted plans. Given the geomantic principles, astronomical alignments, and geometric sophistication inherent in Maya cities, it would be more difficult to demonstrate that the Maya had no such architects or recorded plans. This review will begin with an overview of the role of astronomy, cosmology, and geomancy in Maya architectonic arrangements, and then move to a broader consideration of the question of site planning in the ancient Maya heartland.

Maya Architecture, Sculptural Programs, and Cosmology

Archaeologist and architectural historian H.E.D. Pollock has defined ten primary architectural and planning styles for the lowland Maya heartland. To these may be added another four highland Maya areas that I believe may be distinguished for the Quichean highlands of southern Guatemala, the Cotzumalhuapan Pacific coastal tradition, the Middle Classic (AD 400–650) styles of Kaminaljuyu and vicinity, and the hybrid regional developments of the present-day state of El Salvador.

Where the Maya lowlands are concerned, Pollock describes the influence of environment and topography on planning and architectural layout.

> The flood plains of Copan and Quirigua; the flat or gently sloping banks of the Pasion River; the steeply rising hills along the Usumacinta River; the mountainous shelf of Palenque and alluvial plain of Comalcalco; the low, wet, but often sharply broken terrain of the Peten; southern Campeche; and southern Quintana Roo; the savannas, valleys, and hills of the Chenes and Puuc regions; and the flat northern Yucatan plain all offered different opportunities and challenges to the ancient builder.

Custom was mediated by local and regional environmental conditions and the adaptive traditions which arose from them. Clearly, civil planning developed well beyond the purely "organic" (Andrews, 1975) organization of monuments oriented to topographic and natural features of the immediate environment. Even in the earliest periods of Maya city planning, cosmological, astronomical, ritual, genealogical, and dynastic variables were integrated into site plans and construction programs. Plans were formulated in terms of multiple variables, not the least of which were the mechanistic dimensions of astronomical and cosmological cycles and dimensions, and the organic or topographic, and natural or practical realities of urban planning.

The period immediately preceding the Classic period was a time of great dynamism. According to archaeologist C. Bruce Hunter, "between 600 BC and AD 250 great activity in city planning was taking place all over Mesoamerica. Cultural areas were defined, ceremonial complexes in the heart of the cities were constructed, and trade routes and luxury goods fostered trade competition between regions." Many of the greatest urban centers were built at this time. By the end of the Preclassic era — ca. AD 100 — Tikal and Uaxactun in the Peten lowlands and Cerros and Cuello in Belize had added to the architectural traditions of the era. The ancient cities of Dzibilchaltun in the northern lowlands, El Mirador in the southern lowlands, and Kaminaljuyu in the Guatemalan highlands defined the ceremonial construction for each of these subregions.

On the basis of recent glyphic transcriptions, we can now say that the iconography and architectural order inherent in the earliest city planning incorporated cosmological, dynastic, ritual, architectonic, and geomantic principles used in the organization of sacred geography and the built environment. My own study of the sculptural program embodied in the architecture and planning of the rock-cut temples of Malinalco, Mexico, and Carolyn Tate's more recent architectural survey of the ancient Maya center of Yaxchilan provide compelling evidence for Mesoamerican site plans based on pre-Columbian solar configurations or cosmograms, as well as the use and deployment of solstice-based axial alignments in city planning (Mendoza, 1975; Tate, 1992).

Geomancy and Maya City Planning

Geomancy, referred to as "mystical ecology" and as cosmological ecology or sacred geography, was the ancient Chinese practice of situating burials, monuments, and entire towns on the basis of cosmological principles. Because ancient Chinese scholars documented this practice so extensively, it is largely identified with Chinese civilization, although recent studies make clear the use of geomancy in ancient Mesoamerica.

Much of the divinatory practice embodied in geomancy is particularly relevant to Mesoamerican, and specifically Maya, city plans. According to John Carlson, geomancy is "a divinatory art involving the interpretation of local topographical features for the purpose of properly locating and orienting the constructions of man — be they graves, houses, or entire cities". However, Carlson has redefined geomancy as a "profound system of thought, deeply rooted in ancient and fundamental oriental philosophy and involving perception of the dynamic balance of the controlling forces of nature. It is the system in which man divines his own place in relation to the play of these forces, and it is the mechanism by which he can influence them in order to re-establish equilibrium when imbalance is perceived" (Carlson, 1981).

Maya city centers all provide archaeological evidence for the interpretation that plans were based on cosmological and geomantic models; however, few such sites have been examined with such principles in mind. Recent studies, such as those of William Rust, Jeremy Sabloff and Gair Tourtellot, Wendy Ashmore, Robert Sharer, and Carolyn Tate, make clear the ancient Olmec connection to Maya ceremonial and settlement planning, the influence of changing economic and demographic patterns on planning, the icons of power, divinity, and divine mandates invoked through creation myths and expressed through architectural vocabularies and sculptural programs, the concentric patterning of civil monuments; and the deployment of solar cosmograms in site planning, respectively. We shall examine one site in detail.

Yaxchilan

The site of Yaxchilan, Chiapas, which overlooks the Usumacinta River separating Mexico from Guatemala, is one site that has been examined from the perspective of world plan as embodied in city plan. The site underwent dramatic transformations in construction during the reigns of Shield Jaguar, his father Bird Jaguar III, and his grandfather. The site was altered to establish "specific axes of solstitial orientation for architecture depending on the function of the building or space or the message of its sculptural program, and was conceived to define ritual areas used by particular rulers" (Tate, 1992). Yaxchilan was constructed on an elongated shelf and used terraces, tiers, and elevated plaza-platforms within the context of a larger acropolis-centered pattern. Yaxchilan also incorporated a series of hills into the larger plan of the site, and was situated in a position providing the only unobstructed vantage point from which to observe solstice alignments between major buildings and prominent mountain peaks — serving as clefts in the horizon and as points of astronomical reference. Other recent observations document that major buildings at Yaxchilan devoted to rites of passage, particularly those pertaining to dynastic accession, succession, calendrically-based Period Endings, and to the capture and sacrifice of living kings, are oriented with respect to the eastern horizon and the summer solstice, whereas funerary chambers and obituary monuments and stelae are oriented along solstitial axes pertaining to the winter solstice at 118 degrees east of magnetic north. Several of the obituary monuments and funerary shrines in question are aligned to 115–116 degrees, as opposed to the more exacting 118 degrees east of magnetic north; apparently so as to allow for the use of doorways with half-quatrefoil overhead jambs to provide for architectural-solar hierophanies, or light and shadow patterns serving to record the passage of summer and winter solstice events. Such information provides a direct correspondence with ancient Maya identifications of the summer solstice with royal and or divine renewal and the planet Venus, and the cosmological association of the southern horizon with caverns as the portals to the underworld, the moon, and the land of the dead. The fact that Yaxchilan's southern precinct skirts the mouths of two prominent caves is cosmologically significant. According to Carolyn Tate, the caves "must have played a role in Bird Jaguar's decisions on how to shape the ritual landscape" (Tate, 1992).

Major buildings, monumental stelae, altars, and ritual markers from Yaxchilan all bear orientations that speak to political and ritual events. Buildings were aligned horizontally, and "situated from high to low vertically" (Tate, 1992). By employing "alignments of stelae that conceptually linked the reigns of successive rulers" the architects and planners coordinated the organization of the royal house of that era.

Recent studies at the sites of Tikal and Quirigua (Guatemala), Cerros (Belize), and Kohunlich (Mexico), provide dramatic examples of the means by which the Maya world view and the realm of the supernatural were incorporated into city planning in a manner reminiscent of the Christian, Muslim, Buddhist and Hindu towns and cities of the Old World. While the sites of Cerros and Kohunlich provide architectural arrangements centered on monumental depictions of solar, lunar, terrestrial, and venusian cosmograms, the sites of Tikal and Quirigua incorporate these same dimensions in addition to those iconographic vocabularies and Twin Pyramid groups and precincts intended to symbolize the Classic Maya conception of the cosmos. Maya city planning has only recently undergone scrutiny with an eye to the larger mechanistic and organic dimensions and concepts of time and space inherent in Maya planning and design of the built environment. Wendy Ashmore has recently identified what she deems the five principle components of the Maya pattern pertaining to Tikal and related sites. These are (1) north–south axial arrangements, (2) complementary and paired functions of buildings centered along the main axes that serve to demarcate supernatural and celestial from underworld or terrestrial dimensions, (3) subsidiary eastern and western monumental arrangements within precincts providing for a triangulation with monuments located on the northern perimeter of individual precincts — and identified with the celestial vault and the cosmic tree of the north, (4) ball courts as transitional zones between north and south; and (5) the construction of causeways and paved roads linking precincts into a symbolically coherent cosmogram. To this listing I would add specific buildings that embody solar cosmograms and cosmic templates, as well as sculptural programs and iconographic vocabularies deployed across whole sites to imbue them with a sacred geography and cosmological ecology. We can no longer assume that the Maya built in a random and largely organic fashion — by accretion and accomodation as opposed to design and structure. Primitivist and eurocentric portrayals of the ancient Maya have eroded in recent years under the onslaught of a new world order reinterpreted on the basis of surviving hieroglyphic stairways, monumental texts, commemorative stelae, funerary shrines, frescoed murals, painted ceramics replete with textual information, and pre-Columbian and contact period codical documents and screenfold books. These sources are now providing a revolutionary perspective on the Maya.

RUBEN G. MENDOZA

REFERENCES

Ashmore, Wendy. "Deciphering Maya Architectural Plans." In *New Theories on the Ancient Maya*. Ed. Elin C. Danien and Robert J. Sharer. Philadelphia: The University Museum, University of Pennsylvania, 1992, pp. 173–184.

Carlson, John B. "A Geomantic Model for the Interpretation of Mesoamerican Sites: An Essay in Cross-cultural Comparison." In *Mesoamerican Sites and World-Views*. Ed. Elizabeth P. Benson. Washington, D.C.: Dumbarton Oaks Research Library and Collections, 1981, pp. 143–211.

Freedman, Maurice. "Geomancy: Presidential Address 1968." *Proceedings of the Royal Anthropological Institute of Great Britain and Ireland for 1968*. London: Royal Anthropological Institute of Great Britain and Ireland, 1969, pp. 5–15.

Hunter, C. Bruce. *A Guide to Ancient Maya Ruins*. 2nd ed. Norman: University of Oklahoma Press, 1986.

Mendoza, Ruben G. "The Nahuatl Temples and Their Relationship to Cosmology." *Southwestern Anthropological Association Newsletter* 15(1): 3–7, 1975.

Morley, Sylvanus G., George W. Brainerd and Robert J. Sharer. *The Ancient Maya*. 4th ed. Stanford: Stanford Univ. Press, 1983.

Pollock, H. E. D. "Architecture of the Maya Lowlands." In *Handbook of Middle American Indians,* Volume 2, *Archaeology of Southern Mesoamerica*. Austin: University of Texas Press, 1965, pp. 378–440.

Rust, III, William F. "New Ceremonial and Settlement Evidence at La Venta, and its Relation to Preclassic Maya Cultures." In *New Theories on the Ancient Maya*. Ed. Elin C. Danien and Robert J. Sharer. Philadelphia: The University Museum, University of Pennsylvania, 1992, pp. 123–129.

Sabloff, Jeremy A., and Gair Tourtellot. "Beyond Temples and Palaces: Recent Settlement Pattern Research at the Ancient Maya City of Sayil (1983–1985)." In *New Theories on the Ancient Maya*. Ed. Elin C. Danien and Robert J. Sharer. Philadelphia: The University Museum, University of Pennsylvania, 1992, pp. 155–160.

Schele, Linda and David Freidel. *A Forest of Kings: The Untold Story of the Ancient Maya*. New York: William Morrow, 1990.

Sharer, Robert J. *Quirigua: A Classic Maya Center & Its Sculptures*. Durham, North Carolina: Carolina Academic Press, 1990.

Sharer, Robert J. "The Preclassic Origin of Lowland Maya Sites." In *New Theories on the Ancient Maya*. Ed. Elin C. Danien and Robert J. Sharer. Philadelphia: The University Museum, University of Pennsylvania, 1992, pp. 131–136.

Silverberg, Robert. *Mound Builders of Ancient America: The Archaeology of a Myth*. New York: Greenwich, 1968.

Tate, Carolyn E. *Yaxchilan: The Design of a Maya Ceremonial City*. Austin: University of Texas Press, 1992.

See also: Geomancy in China

CLOCKS AND WATCHES There was a long tradition in the Hellenistic world for the construction of large waterclocks. Vitruvius, writing in the first century BC, mentions the waterclocks constructed by an Alexandrian engineer called Ctesibius (ca. 300 BC) which incorporated gearing, automata and audible time signals. The well-known Alexandrian writer Hero, who flourished around AD 60, is known to have written a book on waterclocks. The Byzantine historian Procopius described a monumental waterclock constructed at Gaza in the sixth century AD. It is almost certain that the inspiration for the construction of waterclocks in the Islamic world came from this Hellenistic tradition.

Islam

Monumental waterclocks are described in detail in two Arabic treatises. Al-Jazarī in his machine book *Kitāb fī maʿrifat al-ḥiyal al-handasiyya* (The Book of Knowledge of Ingenious Mechanical Devices), completed in Diyar Bakr in 1206 describes two such machines. Riḍwān ibn al-Saʿātī, in his treatise *Kitāb ʿamal al-sāʿāt wa'l-ʿamal biha* (Book on the Construction of Clocks and on Their Use) dated 1203, describes the waterclock built by his father Muhammad at the

Jayrun gate in Damascus. It fell into disrepair after Muhammad's death and was restored to working condition under his son's supervision. It was a large construction, having a timber working face about 4.73 meters wide by 2.78 high, built into the front of a masonry structure. The clock had several design defects which undoubtedly caused the breakdown that Riḍwān undertook to repair. Moreover, Riḍwān himself was not an engineer and his description, though containing some valuable information, omits to deal with some important constructional details.

Al-Jazarī's two clocks, on the other hand, were manufactured and constructed in a very workmanlike manner. Although very similar in principle to al-Saʿati's they did not incorporate any design defects. The first and larger of the two was described in such careful detail that it was possible to construct a full-size working facsimile from al-Jazarī's instructions and illustrations for the World of Islam Festival, in the London Science Museum in 1976.

The working face of the clock consisted of a screen of bronze or wood about 225 cm high by 135 cm wide set in the front wall of a roofless wooden house which contained the machinery. At the top of the screen was a zodiac circle about 120 cm in diameter, its rim divided into the twelve signs. It rotated at constant speed throughout the day. Below this circle were the time-signalling automata which were activated at each hour. (The clock worked on 'unequal' hours: i.e. the hours of daylight or darkness were divided by twelve to give hours that varied in length from day to day.) These included doors that opened, falcons that dropped balls on to cymbals and the figures of five musicians — two drummers, two trumpeters and a cymbalist. The musicians were operated by the discharge of water from an orifice, whereas all the other automata were operated by a heavy float that descended at constant speed in a reservoir. A cord tied to a ring at the top of the float led to a system of pulleys that activated various tripping mechanisms.

The speed of descent of the float was controlled by very ingenious water machinery that included a feedback control system and a flow regulator, the latter for varying the rate of discharge daily in order to produce the 'unequal' hours. The same system was used by Riḍwān and both writers attribute its invention to Archimedes. There is a treatise that exists only in Arabic and is attributed to Archimedes. The treatise almost certainly contains Hellenistic, Byzantine, and Islamic material but its first two chapters describe water machinery that is essentially the same as that used by Riḍwān and al-Jazarī. There is every likelihood that these chapters were indeed the work of Archimedes.

Al-Jazarī's book also contains descriptions of four other waterclocks, two of which embody the principle of the closed-loop, and four candle-clocks which, on a small scale, are as impressive from an engineering point of view as the waterclocks.

Other Arabic works add to our knowledge of Islamic hydraulic timekeeping. A certain Ibn Jalaf or Ibn Khalaf al-Murādī worked in Andalusia in the eleventh century. Unfortunately, the unique manuscript of his treatise on machines is badly defaced, but it is possible to determine the essential details of the automata and waterclocks that are described in it. The most important feature that they incorporate is complex gear-trains which include segmental gears (i.e. gears in which one of the wheels has teeth only on part of its perimeter, a device that makes intermittent action possible).

Al-Khāzini's justly famous book on physics, *Kitāb Mīzān al-Ḥikma* (Book of the Balance of Wisdom) was completed in 1121–1122. In the eighth treatise two steelyard clepsydras are described. On the short arm of the beam was a vessel that discharged water at a constant speed from a narrow orifice. Two sliding weights were suspended to the long arm, which was graduated into scales. At a given moment the weights could be moved to bring the beam into balance and the time could then be read off from the scales.

A number of noteworthy references to water clocks can be found scattered among the works of Arabic writers. For example, we read of one such device in the works of the Anadusian poet ʿAbbās ibn Firnās (d. 887). Two water clocks were constructed by the famous astronomer al-Zarqallu in 1085 on the banks of the river Tagus at Toledo. The remains of two large waterclocks still exist in Fez, Morocco. One of these is in the street opposite the Bu'ananiyye Madrasa and was built in 1357. According to the historian of Fez, Al-Jazna'ī in *Zahrat al-Ās*, at each hour pellets fell on to a gong and a pair of door leaves opened. The door openings and the brackets for the gongs still exist. A second clock was constructed in 1361. It had an astrolabic dial and it also released pellets and had doors that opened. It is in the minaret of the Qarawiyyīn mosque. The doors and the dial still survive, but in neither of the Fez clocks does any trace of the water-machinery remain. Another North African waterclock, in Tlemcen, was mentioned by Abū Zakariyyā Yaḥyā ibn Khaldūn in *Bughyat al-ruwwād*.

In 1276–1277 a work entitled *Libros del Saber de Astronomia* was produced in Castilian under the sponsorship of Alfonso X of Castile. This consists of various works that are either translations or paraphrases of Arabic originals. It includes five timepieces, one of which is of significance in the history of horology. This consisted of a large drum made of walnut or jujube wood tightly assembled and sealed with wax or resin. The interior of the drum was divided into twelve compartments, with small holes between the compartments through which mercury flowed. Enough mercury was enclosed to fill just half the compartments. The drum was mounted on the same axle as a large wheel powered by a weight-drive wound around the wheel. Also on the axle was a pinion with six teeth that meshed with thirty six oaken teeth on the rim of an astrolabe dial. The mercury drum and pinion made a complete revolution every four hours and the astrolabe dial made a complete revolution in twenty four hours.

This type of timepiece had been known in Islam since the eleventh century.

China

The commonest method of timekeeping in China, for many centuries, was the inflow clepsydra. In its simplest form this consisted of two vessels: a reservoir and a receiver below it. The reservoir had an overflow pipe near its top and an orifice in its underside. The water supply ran into the reservoir and the overflow pipe ensured that the water-level remained constant. The rate of flow from the orifice into the receiver was therefore also constant as was the rise of the water in the receiver. A float in the receiver was used to activate various time-recording mechanisms.

The inflow clepsydra was used extensively in China from the Han period (202 BC to AD 221) onwards. Two main types were developed. The first type involved the introduction of one or more compensating tanks between reservoir and receiver. At each successive stage the retardation of flow due to diminishing pressure head was more and more fully compensated. The introduction of at least one compensating tank can be dated to the second century AD. As many as six tanks are known to have been used. In the second type an overflow or constant head tank was placed in the series, a practice that began in the middle of the sixth century AD. There can be no doubt that this type of waterclock was widely used in China throughout the medieval period.

The most notable achievement of Chinese horologists was the clock described by Su Song in *Xin yixiang fa yao* (New Design for an Armillary Clock or Description of a New Astronomical Instrument) begun in 1088 and completed in AD 1094. This was a monumental clock, 30 to 40 feet in height, having a rotating armillary sphere and celestial globe, together with numerous jack-work figures with both audible and visible effects. These devices were driven, through a complex system of gearing, by a water-wheel 11 feet in diameter carrying 36 scoops on its perimeter. Water stored in an upper reservoir was delivered to a constant-level tank by a siphon, whence it was discharged on to the scoops of the wheel, each of which had a capacity of 0.2 cubic feet. The wheel was provided with a very ingenious escapement system, in essence two steelyards upon each of which the scoops acted in turn. When a scoop was full, its weight overcame the balancing system and it fell freely for a given distance until checked, without recoil, by a locking device. The next scoop then came under the delivery jet and the cycle repeated itself. There is no evidence that the system of escapement used in Su Song's clock was ever transmitted westward.

The mechanical clock was invented in western Europe towards the end of the thirteenth century. Almost certainly its inventor came from the ranks of the makers of water-clocks. The verge escapement made the mechanical clock possible, but all its other features — weight-drive, automata, gear-trains and segmental gears — were present in Islamic waterclocks. It is highly probable that these ideas were transmitted from Islam to the European makers of waterclocks. An Islamic influence on the genesis of the mechanical clock may therefore be postulated.

Several of Taqī al-Dīn's writings are concerned with timekeeping, and one of these, *Al-kawākib al-darriyya fī al-binkāmāt al-dawriyya* (The Brightest Stars for the Construction of Mechanical Clocks), written about 973/1565, has been edited with Turkish and English translations.

In this he described the construction of a weight-driven clock with verge-and-foliot escapement, a striking train of gears, an alarm and a representation of the moon's phases. He also described the manufacture of a spring-driven clock with a fusee escapement. He mentions several mechanisms invented by himself, including, for example, a new system for the striking train of a clock. He is known to have constructed an observatory clock and mentions elsewhere in his writings the use of the pocket watch in Turkey. This is a surprisingly early reference to the use of watches in Turkey; the manufacture of watches began in Germany about 1525 and in England about 1580.

Taqī al-Dīn's descriptions are lucid with clear illustrations, showing that he had mastered the art of horology. Clockmaking did not, however, become a viable indigenous industry and Turkey was soon being supplied with cheap clocks from Europe. Taqī al-Dīn himself commented on the low price of these European clocks, which entered Turkey, he said, from Holland, France, Hungary and Germany.

DONALD R. HILL

REFERENCES

Al-Jazarī. *The Book of Knowledge of Ingenious Mechanical Devices*. Ed. and trans. D.R. Hill. Dordrecht and Boston: Reidel, 1974.

Al-Jazarī. *Kitāb fī maʿrifat al-ḥiyal al-handasiyya*. Ed. Ahmad Y. Al-Hassan. Aleppo, Syria: University of Aleppo, 1979.

Al-Khāzinī, ʿAbd al-Raḥmān. *Kitāb mīzān al-ḥikma* (Book of the Balance of Wisdom). ed. Hashim al-Nadwa. Hyderabad: 1940.

'Archimedes', *Fī ʿamal al-binkāmāt* (On the Construction of Waterclocks) Trans. Donald R. Hill. London: Turner & Devereux, 1976.

Bedini, Silvio A. "The Compartmented Cylindrical Clepsydra." *Technology and Culture* 3: 115–41, 1963.

Hill, Donald R. *Arabic Water Clocks*. Aleppo: Institute for the History of Arabic Science, 1981.

Ibn Khaldūn, Yaḥyā ibn Muḥammad. *Bughyat al-ruwwād*. Algiers: Imprimerie Orientale Pierre Fontana, 1904–13.

Needham, Joseph, Wang Ling and Derek de Solla Price. *Heavenly Clockwork: The Great Astronomical Clocks of Mediaeval China*. Cambridge: Cambridge University Press, 1960.

Needham, Joseph. *Science and Civilisation in China,* Vol. 4, Part II. Cambridge: Cambridge University Press, 1965.

Price, Derek de Solla. "Mechanical Water Clocks of the 14th Century in Fez, Morroco." *Ithaca*: 599–602, 1962.

Tekeli, Sevim. *The Clocks in the Ottoman Empire in the 16th Century and Taqi al-Din's: The Brightest Stars for the Construction of Mechanical Clocks*. Ankara: Ankara University, 1966.

See also: Al-Jazarī – Al-Khāzini – al-Zarqallu – Ibn Khaldūn – Alfonso X – Su Song – Taqī al-Dīn

CLOCKS: ASTRONOMICAL CLOCKS IN CHINA Timekeeping was a necessary preoccupation of the Chinese imperial state from its inception in the third century BC. It was essential to the effective performance of ritual, which required the selection of auspicious hours as well as days, and to the proper interpretation of the astrological portents carefully recorded by the Astronomical Bureau. In addition it underpinned the fine tuning of increasingly elaborate systems of mathematical astronomy. The earliest devices used by astronomers were simple outflow or inflow waterclocks (clepsydras), which depended on the flow of water from a vessel through a small orifice. A graduated float indicated the time elapsed. Several later sources mention the Han dynasty scholar Zhang Heng (AD 78 to 139) as having succeeded in rotating an armillary sphere by means of "flowing water" so that it kept time with the heavens.

More detailed accounts survive of a similar project under the Tang dynasty in the eighth century AD, when the Buddhist monk Yixing and his colleague Liang Lingzan made a water-powered armillary sphere, which incorporated jackwork to announce the hours and quarters by bells and drums. In neither case do we have enough data to begin to reconstruct the mechanism involved. However, in AD 1086 the high official and scholar Su Song was commissioned to construct an astronomical timekeeper which was to be much more complex than any of its predecessors, and a detailed illustrated description of this device has come down to us in his monograph entitled *Xin yixiang fa yao* (Description of a New Astronomical Instrument or New Design for an Armillary Clock).

From this book we learn that the device Su Song constructed was in effect a clock-tower more than ten meters high, surmounted by a great bronze armillary sphere automatically rotated, with a rotating celestial globe in an inner chamber. On the front of the tower was a complex array of jacks and annunciators to mark the passage of time. The mechanism which turned the shafts was based on a great water-driven scoop wheel, with a weighbridge and trip-lever movement which allowed the wheel to advance one step each time a standard weight of water had run from a constant-head tank. This happened once every hundreth of a day, about 15 minutes, so that the correspondence with the heavens was somewhat jerkily maintained. While the utility of this device to astronomers can only have been moderate, its possession must have been seen as a considerable enhancement to imperial prestige. Su Song's tower fell out of use when the capital fell to invaders in AD 1126.

CHRISTOPHER CULLEN

REFERENCES

Needham, Joseph, Wang Ling and Derek J.de Solla Price. *Heavenly Clockwork*. Cambridge: Cambridge University Press, 1960.

See also: Zhang Heng – Divination in China – Calendars in China

COLONIALISM AND MEDICINE IN MALAYSIA The history of British medicine in the Malay Peninsula is synonymous with the history of tropical medicine. It is also a history of colonial medicine, which is a chapter of the history of British colonialism and imperialism in the East.

Three factors were crucial in determining the outcome of the struggles by colonial powers to expand and consolidate their empires in the East: sea-power, colonial settlers, and diseases. The nation that controlled the sea could occupy distant lands with ease. If there were conflict, sea-power enabled a colonial nation to intercept and prevent vital supplies from reaching rival colonies. Colonial settlers were important for political and economic reasons. A claim for the motherland could easily be made over an area well settled with colonial immigrants. Economic exploitation and control could be established by a network of colonial settlements, which also functioned as buffers against economic challenge and security threats from natives or rival powers. Finally, disease was of immense importance in moderating colonial expansion. Diseases could exact a heavy toll on the army, officials, colonial settlers, and their laborers or slaves. Malaria and yellow fever reportedly wiped out some of the British colonies in the tropics and Africa.

British colonial rule over the Malay Peninsula and Borneo Island (now Malaysia) lasted for more than a century. Their use of medicine as a strategic, diplomatic, and political tool to enhance their grip on the indigenous people and resources of the Malay Peninsula is well-documented. Medicine was also an instrument of colonial conscience, one aspect of which was the call to discharge "the white man's burden," including the task of "civilizing" the natives. In the Peninsula this entailed, among other things, the aggressive promotion of tropical medicine, and it contributed to Malayan negative perceptions of colonial medicine. Indeed, colonial medicine came to be perceived as a ruthless campaign to stamp out their traditional medicine.

In spite of that, colonial medicine could still claim success among the colonial laborers and urban dwellers, although among the natives and villagers the campaign seemed to have failed. There are several reasons for this. One was the pre-existence of an elaborate traditional medical system. Second, the natives enjoyed better health than the imported colonial laborers, and even the colonial masters, many of whom were afflicted either with beriberi or malaria. Third, local traditional medicine was far more effective in treating some forms of tropical diseases.

The effectiveness of Malay medicine, however, was generally overlooked. It was difficult for colonials to accept that a primitive medical tradition could be equal to or better than colonial medicine. The official attitude further institutionalized such prejudices, as Malay medicine was officially perceived with contempt.

Since the beginning of British colonialism in the Malay Peninsula, combating tropical diseases was a priority. Initially, it was important to the officials and indentured colonial laborers, and it assumed a strategic and political dimension.

If one were to put a date on the genesis of British colonialism in the Malay Peninsula, it is likely to be 1786. This was when the East India Company took possession of Penang from the Sultan of Kedah. By 1830 the Company had added Singapore and Malacca to its inventory list, and by 1895, four of the richest Malay states in the Peninsula were under British influence. Throughout the period, the number of Britons sent to the area increased tremendously. The Peninsula's climate and geography, however, were harsh, and many of the early colonialists succumbed either to the heat or diseases. The large-scale importation of foreign indentured laborers, known then as *coolies*, was a significant feature of the period. They were crucial for economic and political reasons, but they were also decimated by diseases, especially malaria and beriberi. The result was a loss of productivity and a forced suspension of many economic projects. Thus, disease, or its absence, among officials, settlers, and coolies acquired an economic imperative; while health or the lack of it among natives acquired a strategic and political dimension.

During the colonial period, the course of events in the Peninsula was inevitably intertwined with that in England. The 1870s saw the emergence of a new political figure in Britain, Joseph Chamberlain (1836–1914). In 1893 he was appointed Secretary of State for the Colonies. He reorganized the Office from an ill-equipped, almost forgotten, department to one of the best around. Indeed, colonial appointments before he took office were seen as more akin to polite exile than glamorous jobs. Chamberlain changed that. He was able to infuse a new attitude, as well as a new policy, so that the colonial jobs gained the same respectability as home appointments.

As far as the history of colonial medicine is concerned, Chamberlain's most significant political act was his appointment of Dr Patrick Manson (1844–1922) as the Medical Advisor to the Colonial Office. Dr Manson was then the most distinguished authority in tropical medicine, which was then a marginal subject within mainstream medicine. The main concern of British medicine was the diseases of Europe, not of the tropics, and health policy was geared to serve home needs and rarely those of the colonies. Under Chamberlain, with Manson as the medical advisor, there was a reorientation of policy and priorities. The medical problems and health policies of the colonies were given serious attention. Soon Chamberlain's name became synonymous not only with the development of British colonialism but also with the development of tropical medicine. Indeed, under his stewardship, Britain's colonial fortune took a positive turn; so too did the health of Britons and the state of medicine in the colonies.

Chamberlain was forthright about the fact that British lives were more important to him than native lives. The participation of his ministry in the development of British tropical medicine was solely to ensure their welfare. The welfare of the natives was looked upon only in terms of Britain colonial interests, if ever. This policy was reflected in the establishment of the London School of Tropical Medicine in 1899. Its purpose was to foster research and development of the "new" British tropical medicine. The school was backed up by a number of research stations in the colonies, the biggest and most important of which was the Kuala Lumpur Pathological Institute, located in the Federated Malay States (British Malaya). The Institute was formed a year after the London School. Indeed, the rapid development of British tropical medicine in the early part of the century owed much to the linkages of the London School with the Kuala Lumpur Institute.

The Institute's founder was Sir Frank Swettenham (1851–1945), a charismatic colonial official who knew a lot about colonial and local politics but next to nothing about medicine. Swettenham's tenure coincided with the period of intensive colonial economic activities in the Malay States. These were sustained mainly by the extensive importation of coolies from abroad, particularly from China and India. As indicated earlier, this importation was important for both economic and political reasons. It showed the native Malays that the British could easily handle any boycott by local labor; but the importation soon hit a snag. Many of them were decimated by unknown diseases, one of which was later identified as beriberi.

Owing to the epidemic, the local colonial administration was burdened by thousands of sickly and unproductive coolies. A search for a beriberi cure was imperative. Swettenham had come to know about the interest at the Colonial Office regarding tropical diseases, and about the plan to set up a school of tropical medicine in London, but he also knew that the School's priority was research on tropical diseases afflicting European communities in the tropics. In the Peninsula, however, beriberi was a non-European disease. Swettenham was aware that it was extremely difficult, if not impossible, to persuade the Colonial Office to allow it to devote a portion of its research time and money to beriberi. It was at this stage that he toyed with the idea of a local institution modeled after the school, which would facilitate the search for beriberi's cure.

In London, luck seemed to be on his side. Swettenham's plan for a local medical research institute was received positively. It was seen as simply the compliance of a colonial official to London's policy on tropical medicine. Such plans from distant colonies normally received an impersonal go-ahead signed routinely by nameless individuals on behalf of the Secretary. Instead, Swettenham's plan came to the attention of Chamberlain, who personally signed the authorization papers for the Institute. At home, Swettenham did not bother to wait for the official go-ahead; he had made arrangements to set up the Institute even before Chamberlain had signed the order.

Perhaps owing to the excellent staff, the Institute lived up to its expectation. It was able to come up with the cure for several tropical diseases, including beriberi. The riddle to the latter was conclusively resolved exactly nine years after the founding of the Institute. Thus it succeeded in alleviating the colonial authorities from one of the major causes of the local economic and social problems. The Institute was a bastion of colonial medicine, and far removed from the practices of traditional medicine. Notwithstanding that, some of the major breakthroughs of the Institute were affected directly or indirectly by the latter. For instance, in the search for the cure of beriberi the Institute's scientists looked to traditional cures for clues, and indeed found a lead in Malay medical practices. Colonial pharmacologists examined and benefited greatly from the Malay materia medica, particularly in the search for new drugs and alkaloids. Institute authorities still were extremely reluctant to acknowledge any hint of contribution by native medicine, especially in advancing tropical medicine.

To understand the events that followed, it is necessary to take a brief look at the prevailing colonial health policy, particularly with regard to the natives. There were several discernible phases. The initial phase placed the emphasis on the health of the officials and laborers with no regard to the natives. The second phase involved the missionary zeal to promote colonial medicine among the natives and other non-European inhabitants of the colonies. The *Ordinance to Provide for the Registration of Medical Practitioners in the Colony* (1905). was passed to control the practice of medicine, by stipulating minimum qualifications for those who wanted to practice medicine in the Malay States. On the face of it, the Ordinance seemed to spell a blow to traditional medical practitioners, particularly the natives and the Chinese, but interestingly, the Ordinance carried with it an exclusion clause:

> Nothing contained in this Ordinance shall be construed to prohibit or prevent the practice of *native* systems of therapeutics according to Indian, Chinese or other Asiatic method.

Throughout the British rule the native Malays relied extensively on traditional medical systems for their medical needs. The backbone of the traditional system was the medicine man, *bomor* or *pawang*. They did almost everything, from preparing herbal medicines, setting dislocated bones, and delivering babies, to variolation and minor surgery like circumcision. Occasionally, Europeans sought treatment from the *bomor* and claimed that they were cured of their ailments by the Malay medicine man.

Evidently, colonial medical officials were intrigued by the ability of the native Malays to live a healthy life in the tropical climate. Many noticed that the native Malay view of nature was totally different from theirs. Nature or the land was treated by the Malays with respect, regarded more as a living or spiritual being than a thing, and more like a member of their family than an artifact of convenience. The Malays maintained a relationship with their land in what one would describe now as balanced ecological relationship. In fact, the need to seek harmony and equilibrium was central to their view of disease and health. Nature was multidimensional and hierarchical, having both a metaphysical and physical dimension — reflecting the spiritual and physical constitution of man. Their doctrine of diseases and health reflected the belief that health was a manifestation of the harmony of human's inner constitution and nature's constitution; disease was the result of a lack of harmony.

The complexities and intricacies of the Malay medical doctrines were left much to themselves. Occasionally they became objects of anthropological curiosity by colonial scholars. However, this attention was academic; no medical officials or policy makers seemed interested in knowing more about Malay medicine or the Malay medical system. It was not until the later stage, the missionary zeal phase, that they began to take notice of Malay medicine. At this stage, however, Malay medicine was seen more as a rival than a complement.

The effort to introduce colonial medicine to the Malays was apparently a frustrating task to the majority of the colonial officials. Frequently, they expressed their frustrations at what they alleged was the Malays' indifference to "superior medical treatments". For instance, the Colonial Resident Councillor of a Malay State remarked:

> If a Malay thinks he is going to be kept in a hospital, he will not apply for medicine as an outdoor patient and will prefer to remain in his own house and be treated by a Malay doctor on simple herbs, together with a judicious supply of still simpler incantations. (Harun, 1989)

The Colonial authorities also took active steps to restrict the practice of Malay medicine men. This perhaps can be best seen in the checking of the traditional inoculation or variolation practice. Variolation was widely practiced whenever smallpox broke out in a village. It involved the direct inoculation of smallpox virus treated with certain medicinal herbs. Colonial doctors described it as "a barbarous practice", and the practice was targeted for curtailment. In 1905, when smallpox broke out in the eastern Malay state of Kelantan, the Government enforced a law banning variolations. In their place, they introduced large-scale vaccinations. The authorities also recruited some traditional inoculators as "assistant vaccinators", although the actual task of vaccination was undertaken by the colonial medical department.

To the Malays, the inoculation rituals sanctioned by their traditional culture were much preferable to vaccination. Their acceptance of colonial medicine lagged behind official expectation. In fact, the authorities noted a decrease in the number of natives seeking medical treatment at colonial clinics. They realized that the campaign to promote colonial medicine among the Malays would fail unless their confidence was won over.

The new campaign strategy perhaps can be best seen in the work of Dr J.D. Gimlette. He was a colonial medical officer best known for his sympathetic and detailed study of Malay medicine which was originally part of the new campaign. By the time he finished, it had taken an unexpected form. For all practical purposes, it had become a crusade for the preservation of Malay medicine. Gimlette, probably to the displeasure of his superiors, openly urged the colonial authorities to soften their attitudes toward the *bomors*. Indeed he argued that "it would be unfair to damn him [the *bomor*] as an accursed sorcerer who poisons honest folk to gain his private ends." Gimlette even drew parallels between the practice of colonial doctors and that of the medicine man:

> It is too much to say that the work of the bomor in clinical medicine is merely fanciful; he endeavors to prepare a *penawar*, that is to say a "neutraliser," for every kind of poisonous principle; this idea of neutralization distinctly anticipates modern science. His knowledge of local materia medica is often profound, and after all, some of his theories as to the etiology of tropical diseases are conceptions now known to modern science in the form of animal parasites (protozoa, spirochaetes, etc.), which are invisible except under high powers of microscope.

Gimlette also claimed that Malays knew mosquitoes were the cause of malaria, evidently earlier than Manson. According to Gimlette, Malays' apparent resistance to malaria was largely owing to their knowledge of mosquitoes. For instance, the Malay traditional ritual of smoking their homes and villages through the burning of *padi* (paddy) straw during harvest season effectively expelled mosquitoes in their homes or in the surrounding area. Malays also bred fish in their *padi* fields during the flooding period. This practice was akin to a form of biological control of pests, since the fish preyed on small organisms and thus destroyed all the mosquito larvae breeding in the *padi* fields. The Malays also practiced selective clearing of jungles as another form of mosquito control. Gimlette was convinced that the secret of Malays' resistance to malaria or "harmony" with mosquitoes, lay in their practice of such rituals.

It is not known what the actual reaction to Gimlette's views was, particularly within the colonial medical establishment. It is likely that his views would have caused some uneasiness among officialdom. Gimlette assigned a common level of legitimacy to Malay medicine and colonial medicine. It is interesting to note that soon after the publication of his book on *Malay Poison*, Gimlette retired in 1921 from the Colonial medical service. He returned to England and embarked on a project to compile a dictionary of Malay folk medicine. Because of ill health, the project was not completed. However, his friends, H.W. Thomson and W.W. Skeat, took over the project and completed the manuscript, which was posthumously published in 1934 as the *Dictionary of Malayan Medicine.*

The official version of the development of colonial or tropical medicine was no different from the history of medicine. The motives were Hippocratic idealism and the fanatical dictum that there should be no mention of non-scientific or non-modern contributions to the development of modern medicine. Understandably, the Malay's contribution to the development of colonial medicine in the Peninsula was largely ignored.

The prevailing colonial view stereotyped native medicine as nothing more than primitive panaceas or "snake-oil cures." However, studies like Dr Gimlette's challenged established prejudices and institutionalized misconceptions. They established that native medicine was a storehouse of perennial wisdom, and collection of empirical observations of scientific significance. Still, Malay medicine was often referred to contemptuously in colonial discourse.

No branch of modern science can be said to have developed almost exclusively under colonial ideological and political impetus than tropical medicine.

HAIRUDIN BIN HARUN

REFERENCES

Amery, Julian. *The Life of Joseph Chamberlain*. London: Macmillan, 1951.

Gimlette, J.D. "Some Superstitious Beliefs Occurring in the Theory and Practice of Malay Medicine." *Journal of the British Royal Asiatic Society* 65: 29–35, 1913.

Gimlette, J.D. *Malay Poisons and Charms Cures*. London: J. Churchill, 1915.

Gimlette, J.D. *A Dictionary of Malayan Medicine*. Completed by H.W. Thomson. Oxford: Oxford University Press, 1934.

Gullick, J.M. *The Malay Society in the Late Nineteenth Century: The Beginnings of Change*. Oxford: Oxford Univ. Press, 1987.

Harun, Hairudin. "Medicine and Imperialism: A Study of British Colonial Medical Establishment, Health Policy and Medical Research in the Malay Peninsula, 1786–1918." Ph.D. dissertation, University of London, 1988.

Harun, Hairudin. "Science and Colonialism: A Study of the Wellcome's Malay Manuscripts." *Sarjana* 4: 1–20, June 1989.

Harun, Hairudin. "Colonial Strategy and the Kuala Lumpur Institute for Medical Research." Departmental Seminar Paper No. 1, Faculty of Science, University of Malaya, 1989.

Harun, Hairudin. "Galenism: A Study in the Philosophy of Malay Medicine." *Sarjana* 6: 81–96, 1990.

Manson-Bahr, P. *Patrick Manson: The Father of Tropical Medicine*. London: Nelson, 1962.

Scott, H. Harold. *A History of Tropical Medicine*. London: Edward Arnold & Co., 1939.

See also: Colonialism and Science

COLONIALISM AND SCIENCE From the very first decades, scientific advances played essential roles in European overseas expansion. Improved astronomical instruments and calculations and an array of new navigational devices and cartographic techniques made the Portuguese voyages into the uncharted South Atlantic possible. Subsequent voyages of exploration and (for the Europeans) discovery led to further refinements of both instruments and navigational data and stimulated the introduction of ever more sophisticated ways of measuring time, distance, and location. Scientific curiosity was among the main motives of Europeans who planned or led expeditions of discovery and conquest. Astronomers and cartographers often sailed with explorers and merchants for the express purposes of testing new instruments, taking astronomical or meteorological readings from distant latitudes, and charting regions that had hitherto been unknown to the Europeans. The application of the data gathered on these voyages to improving maps, to compiling more accurate navigational charts, and to devising new instruments greatly enhanced European advantages over other peoples in navigation, trade, and warfare by sea.

As European armies and administrators advanced inland, first in the Americas and the outlying island of Africa and Asia and later throughout the globe, botanists and geologists followed, collecting "exotic" specimens to carry home to the gardens and cabinets of their rulers and national scientific societies. European savants examined and classified the bewildering mass of information and objects that flowed in from the rest of the globe, and struggled to fit it into the preexpansion world picture; or, more commonly, used it to build a new vision of both the earth and the cosmos. A host of self-appointed, usually amateur, ethnologists also reported extensively on the diverse peoples and cultures they encountered overseas. Their observations became the basis for elaborate, allegedly scientific, and invariably hierarchical classifications of human types. These classifications in turn provided evidence to support highly contentious theories regarding the relationships among human groups, usually termed races, and between humans and other species, particularly different varieties of apes.

Both in Europe itself and increasingly (but on a much reduced scale) in colonial enclaves overseas, scientific institutions — such as the Hortus Botanicus in Leiden, Kew in London, and the Sibpur botanical gardens in Calcutta — served as clearing houses for flora and fauna collected from around the globe and focal points for taxonomic inquiries and experimentation in procreation and hybridization. The elected members of Europe's scientific societies, most notably London's Royal Institution and the Académie des Sciences in Paris, provided the backing to mount scientific expeditions, certified that what were regarded as scientific procedures had been followed by the explorer or naturalist in question, and made the findings of overseas investigations available to the educated public of their respective nations and the European scientific community as a whole.

From the outset the scientific side of the process of European overseas expansion was all but monopolized by one of a number of approaches to the natural world that had jostled for supremacy in the medieval era and coexisted in the West since ancient times. The predominance of what recent scholars have characterized as the Baconian or mechanistic strain of scientific thinking in European colonial enterprises was of special importance. It meant that other options, such as the organic approach to nature and the cosmos, which might have been more compatible with and accommodating towards non-Western epistemologies, were excluded or relegated to marginal roles in this critical area of interaction between European and non-Western peoples and cultures.

The key attributes of the dominant Baconian or mechanistic strain of Western science rendered European colonial observers and policy-makers particularly unreceptive to non-Western ways of thinking about and interacting with the natural world. The underlying assumptions of the Baconian approach also strengthened the resolve of those among the colonizers who were ironically the most concerned with the well-being of the subjugated peoples to impose Western epistemological presuppositions and standards and procedures of investigation on colonized peoples. Though there were important differences among those scientists who adhered to the Baconian mode, it is possible to distil a number of key attributes that most colonial theorists and administrators would have agreed characterized the proper, or Western, approach to scientific thinking. Though developed in the West and shown by recent research to have been vitally affected by the social and cultural milieux in which it arose, the science carried overseas by European colonizers was seen to be value neutral, objective in its procedures, privileging abstraction and reason, empirically grounded, and somehow transcending time and space and thus universally valid. These attributes gave the practitioners and advocates of Western science confidence that the spread of this epistemology — and the institutions and procedures associated with it — to the rest of the peoples of the globe was both beneficial and inevitable.

This view is, of course, still held today by many scientists as well as development specialists, social theorists, and assorted intellectuals with an interest in the so-called Third World. Its champions among the colonizers viewed the diffusion of Western science as part of a larger campaign to rationalize the world and conversely to banish what they viewed as superstitious or subjective, intuitively-oriented epistemologies. Although only the most reflective and outspoken colonial observers would have acknowledged it, Western science was also aggressively expansive and intolerant of rival non-Western epistemologies — just as it had been of alternatives in Europe itself. By the early nineteenth century, it was also inextricably linked to the great technological advances that drove the Industrial Revolution, and the market-oriented, fossil fuel intensive, interdependent global order that was associated with the resultant transformation of the West. Thus, the Western scientific approach proved to be quite compatible with prevailing Western hegemonic ideologies, such as the *mission civilisatrice*.

Despite attempts to develop general paradigms for the diffusion under the aegis of European colonial domination of Western science, its role in the process of colonization and its impact on non-Western peoples and cultures varied widely by time and place. Important differences can been seen, for example, in the scientific fields and types of investigations that were prominently associated with overseas colonization before and after the Industrial Revolution. In the early centuries of exploration European scientists overseas were mainly preoccupied with astronomy, zoology, and botany. As western Europe's need for markets and raw materials grew with the spread of industrialization, fields like meteorology, geology, chemistry, and applied mathematics became increasingly integral to the colonial enterprise. As European colonizers came to administer much of Africa and Asia directly, more sophisticated techniques of surveying, codifying, and statistical gathering became essential for activities as diverse as revenue collection and the formulation of legal codes. Recent scholarship has also shown that the relationships between colony and metropolis in terms of scientific investigation and exchange also shifted dramatically after World War I. These transformations were accelerated in the interwar years and after World War II by the process of decolonization, and a concomitant shift from an emphasis on extraction to a fixation with the rhetoric of development, which became enshrined in modernization theory after 1945.

The process by which Western science was diffused and its impact on overseas societies also differed greatly depending on the timing and nature of European interaction with non-Western cultures, the colonizers' assumptions about the level of sophistication of indigenous epistemologies and material cultures, and the actual attainments of colonized peoples in science and technology. The process of Western scientific diffusion, for example, in settlement colonies such as Australia and Canada, that had relatively small and scattered autochthonous populations, followed a very different course than in densely populated areas, such as South Asia and China, where small communities of European administrators, missionaries, and merchants encountered highly developed scientific and technological traditions with ancient links to their own.

Even within the different types of colonial societies there

were critical variations. Thus, for example, the New World colonies in both Latin and North America, which were the first areas of extensive European conquest and large-scale settlement, were the sites of significant scientific investigations and developed their own scientific societies as early as the eighteenth century. By contrast, Australia and New Zealand were not settled until the following century and were not integrated into the scientific networks of their European metropolis, Great Britain, until the late 1800s. In addition, from as early as the late seventeenth century, the pursuit of science was stunted in much of Latin America, as compared to Canada and the United States, due to Spain's long political and economic decline and its increasing marginalization as a center of scientific learning.

None the less, in all of the settlement colonies, indigenous learning and epistemologies were pushed to the periphery. Ethnologists studied "aboriginal" beliefs and traditions for their folkloric or antiquarian value, not because they had anything to teach the colonizers about the ecology or topography of the lands they occupied, much less because they might stimulate major modifications in European understandings of the natural world. The research agendas of all of the settlement colonies were also largely dictated by the savants of scientific societies and institutions of higher learning in Europe itself. Though the size and the close links of the elite in the United States to Europe made original work by colonial scientists, such as Benjamin Franklin, more likely than in other settlement areas, until the early twentieth century settler colonists were primarily engaged in applied field research, heavily focused on the collection of local flora and fauna samples. Much of this work was directly linked to efforts to uncover and extract the great natural wealth of settlement areas. The empirical findings of New World botanists, geologists, or ethnologists, who were often merely temporarily transplanted Europeans, were mainly funneled to scientists in European centers for interpretation and theoretical elaboration.

After about 1900, scientific connections between European centers and settlement areas — both those still formally colonized and those, such as the United States, long independent — were radically transformed. Dominions, such as Canada and Australia, were accepted as partners, rather than assistants, in international scientific inquiry. The settlement areas more generally emerged as important centers of scientific research and theoretical innovation in their own right–as evidenced, for example, in Frederick Banting's discovery of insulin in Canada and Macfarlane Burnet's work on immunology in Australia. This was particularly true of the United States, where the already substantial growth of a internationally recognized scientific community early in the twentieth century was greatly accelerated in the 1930s by the forced migration of leading scientists from Europe, such as Albert Einstein. By the 1940s, this former colony and settlement area had taken the lead in many areas of scientific endeavor, from the most abstract theory to applied research.

In comparison with the tropical dependencies, where small numbers of Europeans ruled large and ancient African or Asian societies, the diffusion of Western science to the settlement areas met with little serious resistance. To begin with, the transfer was overwhelmingly to settler populations of European origins and heritage. This meant, on the one hand, that questions of racial ability with regard to the development of scientific communities and institutions in colonized areas did not arise, even though the arrogance of metropolitan savants was often a source of considerable discomfort for the colonials. On the other hand, because the dominant settlers despised and marginalized or discarded altogether the epistemologies of the indigenous peoples, there were no alternative thought systems or ways of doing to rival or provide foci of resistance to hegemonic Western scientific approaches.

Scientific friction between metropolis and settlement colony arose from personal and institutional rivalries, but scientists in both locales shared epistemologies, methodologies, and presuppositions about the place of humans in the natural order. Local resistance came not from subjugated indigenous peoples defending their own, very different traditions, but from the poorly (or un)educated majority of the settler population. The latter were not so much opposed as indifferent to the endeavors of a rather isolated scientific elite, whose labors appeared to be wasted unless they could be turned to some immediate practical advantage. Thus, the key challenge for the scientific communities in the settler colonies was expanding popular understanding of and approval for scientific research, and winning government funding for scientific institutions and projects. Evidence of practical application was key to success in these endeavors, and thus a premium was placed on applied research, at least until the middle of the twentieth century.

The linkages between science and colonization in non-settler areas of Africa, Asia, and Oceania, where small numbers of Europeans dominated large and diverse colonized populations, were markedly different than in Australia or the United States. Here it is important to distinguish between areas that the Europeans conquered outright and ruled directly, such as India and much of sub-Saharan Africa and Southeast Asia, and those they dominated indirectly or "informally" through military threats and periodic interventions, economic influence, and the manipulation of indigenous leaders. The latter areas would include most of the Islamic Middle East, Persia/Iran, China, Siam/Thailand, and Japan — until it became a Western-style scientific and technological power in its own right in the early twentieth century. Even within these two general categories of non-settlers, critical distinctions need to be taken into account. Colonial policies regarding the diffusion of scientific learning and investigation, for example, were quite different in British India or French Indochina than in most of sub-Saharan Africa or the islands of the Pacific, even though all of these areas were formally colonized. Within the indirectly controlled areas in the informal spheres of influence, the European approach to Islamic societies in scientific matters differed in important respects from their interaction with China, Japan, or Thailand.

In some ways the diffusion of Western science in some formally colonized areas like India shared more with informally controlled areas, particularly China and the Islamic Middle East, than with Africa or the Pacific, which were ruled directly. In India, China, and the Middle East, the European colonizers recognized not only that the indigenous peoples had produced ancient and sophisticated civilizations — which usually meant that they had certain key attributes, including writing, specialized elites and cities — but that they had long nurtured scientific traditions of their own. Informed European observers even acknowledged that the West owed a large debt to Islamic peoples both for the recovery of much of classical Greek learning that had been "lost" to European civilization after the fall of Rome, and for original scientific advances made by Muslim peoples and transmitted to Europe through Arab and Jewish scholars in Spain and Italy. A handful of European observers also recognized that Islamic civilization had served as a vital conduit of scientific learning and inventions from civilizations further east, particularly India and China.

Recognition of the earlier scientific achievements of selected non-Western peoples informed the attitudes of European colonizers towards the diffusion of science in the societies so designated in important ways. In the early centuries of European overseas expansion, it fed mutual curiosity and ongoing interchange. The first Jesuit missionaries in China, for example, evinced a strong interest in Chinese astronomy, chemistry, and medical techniques. In India, Portuguese and later Dutch merchants and officials routinely consulted Indian physicians well into the eighteenth century. European physicians not only studied the prescriptions and techniques of Indian *vaidas* and *hakims*; they readily conceded the superiority of indigenous specialists in treating tropical diseases.

By the early decades of the eighteenth century, most European observers had concluded that even the most advanced, non-Western scientific traditions were hopelessly outdated and encrusted with superstition and quackery. If it survived at all, recognition of non-Western achievements was focused on the distant past. In the accounts of European authors who concerned themselves with matters scientific or technological, the stagnation of Indian or Chinese learning, in stark contrast to the ever-changing and highly progressive advance of Western knowledge, was a constant refrain. In areas that escaped direct colonization, particularly China and Japan, scientific knowledge and technological expertise had been and remained the most effective ways for gaining access to Asian rulers and impressing Asian elites with the efficacy of Western ways, which seemed otherwise dubious, if not crude. None the less, the sense that peoples like the Chinese and Japanese had long been civilized and had once excelled in science and technology deeply affected the educational policies pursued by Christian missionaries, when they were permitted to proselytize in these areas, and the trading policies and diplomatic exchanges of the great powers. Japan as a whole closed in on itself after 1600, but Dutch traders at Deshima continued to make the latest European scientific treatises available to an emerging Japanese literati. Christian educators in China frequently pushed advanced training in the sciences (particularly medical) and mathematics. By the late-nineteenth century, regionally-based segments of the Chinese scholar-gentry were avidly importing Western technologies in "self-strengthening" campaigns designed to ward off the imperialists' partition of the Middle Kingdom.

From the mid-nineteenth century similar patterns of diffusion can be found in other non-Western areas, which the Europeans regarded as long civilized but where they increasingly exercised informal sway. French missionaries strove to impress their Lebanese students with the superiority of Western scientific learning, while Siamese rulers strove to educate an elite corps in the engineering skills of the West. Because their own scientific traditions shared much with those of the Europeans — not the least a common, extensive borrowing from the ancient Greeks — the approach of the Muslim peoples of the eastern Mediterranean to these interchanges was somewhat different from that of the Siamese or, most critically, the Chinese. Having given so much to the West in the sciences, Arab and Turkic reformers could claim that it was in no way demeaning for Muslim peoples to borrow back in order to catch up to their European rivals. Though not all accepted this argument, it made the introduction of Western scientific learning palatable for those who supported advocates of forced "modernization", such as Muḥammad ʿAlī in Egypt and the Ottoman ruler Maḥmūd II. Unable to make a convincing argument for such historical precedents, the Confucian literati of China vehemently resisted the introduction of Western learning or reluctantly conceded its necessity in the name of China's survival.

Although similar tensions were present in formally colonized areas that had longstanding scientific traditions, such as India, decisions about how much and what sort of Western scientific and technological diffusion should occur were taken out of the hands of the indigenous peoples. Colonization resulted in a growing influx of European scientists, engineers, and medical practitioners, and a great expansion in their activities and influence. The collection of flora and fauna that antedated colonial rule continued and in some respects expanded, but trained scientists and technicians were now employed by colonial administrations for massive and systematic geological and land tenure surveying operations. Statisticians oversaw census counts; botanists tested new plant varieties in state supported gardens; and hydrologists directed the overhaul and expansion of irrigation systems. As these activities suggest and scholars from formerly colonized areas have recently argued, science was a central pillar of the European colonial order.

Colonial science was also field-oriented, overwhelmingly applied, and devoted primarily to enhancing efforts to maximize the extractive potential of conquered areas. Its diffusion to the colonized peoples was consciously channeled and often constricted out of security concerns or fears of future economic competition. Rather than an abstract, objective and dispassionate concomitant of colonial expansion, Western science was the linchpin of rationalizing offensives that sought to reshape the thinking of colonized peoples, reconfigure the spatial relationships in colonized areas, and remake institutional frameworks and social relations in ways that accorded with Western epistemologies and presuppositions about the logical order of things.

Much of the literature on the diffusion of Western science in the colonial empires has concentrated on the biographies and activities of European scientists and on the policies pursued by European colonial administrators. As a consequence the varied and critical roles played by the colonized peoples in the dissemination of Western scientific learning and techniques have been obscured. At a range of levels, their contributions were essential. In the era when naturalists were dominant, the collection and identification of specimens would have been difficult, if not impossible, without Javanese or Indian guides and assistants, and villagers who were vital repositories of local knowledge about animal life or the rain forest. Working in the celebrated miniaturist tradition that peaked under the Mughals, Indian artists sketched flora and fauna in the field and often illustrated the published collections on these subjects that remain standard references to the present day.

As the grip of colonial administrations tightened from the middle of the nineteenth century, subjugated peoples found new ways to become involved in the scientific enterprises that buttressed the new order. Indian surveyors, engineers, and medical practitioners served in colonial administrations both in South Asia and British colonies in Africa and Southeast Asia. Indian chemists, physicists and biologists worked for British firms, traveled to Europe for advanced study or to take up research positions, and increasingly taught in one of a growing number of universities in India itself. Indian doctors and surveyors set up private practice in colonial towns.

The assumption of many of these roles depended on the spread of higher educational opportunities open to Indians in the sciences and mathematics. The very existence and steady expansion of these opportunities underscores a vital difference between India and most of the other colonized areas. Here again, the colonizers' recognition of precolonial scientific attainments proved critical. The work of Sir William Jones and other "Orientalists" in the late-eighteenth and early nineteenth centuries left little doubt that ancient Indian civilizations had excelled in many of the sciences and mathematics. Though increasingly disdainful of these achievements and impressed with contemporary Indian backwardness in matters scientific and technological, key colonial officials who strove to reform Indian society premised their educational initiatives on the assumption that Indians could master the most advanced Western scientific learning and mathematical techniques. This assumption and the policy decisions regarding higher education that followed gave Western-educated Indians much greater agency in the process of the diffusion of Western science than was enjoyed by any other colonized people. By the last decades of the nineteenth century, racist objections to these policies were raised by both prominent officials and private individuals. However, the pathbreaking work done by such Indians scientists as P.C. Ray and J.C. Bose made a shambles of the racists' aspersions, and bolstered the demands of Indian nationalists for greater opportunities in higher education, particularly in the Western sciences.

With the partial exception of colonies like French Indochina and the Netherlands Indies, opportunities for advanced training in the Western sciences in much of the rest of Asia, sub-Saharan Africa, and the Pacific islands were minimal or non-existent. In these areas, scientific work was the monopoly of the European colonizers. As in settlement areas, such as Australia and the United States, indigenous epistemologies and local knowledge were dismissed as irrational, ignorant, and superstition-bound. These might be studied in order to understand the "native mind" to control the colonial populace better. However, they had nothing to contribute to Western scientific learning or methodology. Thus, in most non-settler colonies, which the Europeans deemed devoid of scientific traditions of their own, racist assumptions regarding the mental capacity of the indigenous peoples insured that little or no advanced training in science or mathematics was available. These policies in turn left most colonized peoples ill-prepared for a post-colonial world where Western scientific knowledge and skills have proved vital for successful economic competition, development planning, and many aspects of intellectual discourse.

The colonizers' disdain for indigenous epistemologies and approaches to the natural world has also obscured a number of other important patterns involving the interaction of science and colonialism. However extensive or limited the volume of Western science diffused to different colonies, in all there was a persistence of indigenous scientific thinking and practices. In some cases the colonizers themselves were complicit in this process, as when the British made use of Indian *hakims* and *vaidas* in their campaigns to eradicate epidemic diseases, such as smallpox and cholera in India, or when merchants or officials in Africa resorted to local cures for snakebite or dysentery. In some colonies, most notably India but also in Kenya, Senegal, Indonesia, and other areas, the defense of indigenous epistemologies and procedures was explicitly linked to the nationalist assault on the European colonizers. In the thinking of visionary leaders, such as Mohandas Gandhi, the restoration and reworking of these traditions posed what well may be viable alternatives to the approaches to economic and social development championed by the industrial West.

In virtually all of the formerly colonized societies, non-Western approaches to science in areas as diverse as medical practice, preservation of soil fertility, and care for the emotionally disturbed have persisted. In view of the clear limits in terms of resource exhaustion, environmental degradation, and social inequity of both major Western development alternatives — market-capitalist and socialist command economy — this persistence may ultimately prove critical. It may be that the predominant direction of scientific diffusion will be reversed; that approaches to work, resource use, social organization, and economic well-being pioneered by Gandhi and other non-Western thinkers will have much to offer the beleaguered societies of the post-industrial world.

MICHAEL ADAS

REFERENCES

Adas, Michael. *Machines as the Measure of Men: Science, Technology and Ideologies of Western Dominance*. Ithaca: Cornell University Press, 1989.

Basalla, George. "The Spread of Western Science." *Science* 156: 611–22, 1967.

Cohen, I. B. "The New World as a Source of Science for Europe." In *Actes du IX Congrès International d'Histoire des Sciences, Madrid 1959*. Barcelona: Asociación para la historia de la ciencia española, 1960, pp. 96–130.

Fleming, Donald. "Science in Australia, Canada and the United States: Some Comparative Remarks." In *Proceedings of the 10th International Congress on the History of Science*. Ithaca, New York, 1962, pp. 180–96.

Inkster, Ian. "Scientific Enterprise and the Colonial 'Model': Observations on the Australian Experience in Historical Context." *Social Studies of Science* 15: 677–704, 1985.

Kumar, Depak, ed. *Science and Empire: Essays in Indian Context*. Delhi: Anamika Prakashan, 1991.

Kwok, D.W.K. *Scientism in Chinese Thought, 1900–1950*. New Haven: Yale University Press, 1965.

McLeod, Roy. "On Visiting the 'Moving Metropolis': Reflections on the Architecture of Imperial Science." *Historical Records of Australian Science* 5(3): 1–16, 1982.

Petitjean, Patrick, et al. *Science and Empires: Historical Studies about Scientific Development and European Expansion*. Dordrecht: Kluwer Academic Publishers, 1992.

Vaughn, Megan. *Curing Their Ills: Colonial Power and African Illness*. Cambridge, England: Polity Press, 1991.

See also: Knowledge Systems: Local Knowledge – Ethnobotany

COLONIALISM AND SCIENCE IN AFRICA By 1914, most of the African continent had been subordinated to colonial rule, with 36% of African territory seized by the French, 30% by the British, and the remaining 34% dominated by Belgium, Germany, Italy and Portugal. This of course was not done without bloodshed. Millions of Africans died in the process of resistance against the invasion of the continent, waging guerrilla warfare for as many as thirty years in some cases. Colonialism was essentially a process of administration which denied Africans political and economic autonomy while subordinating them to the dictates of an evolving global capitalist ethic and world market controlled by a Euro-American elite. Indigenous science and technology were adversely affected by it. Ironically though, institutions such as the British Imperial Institute, through its Scientific and Technical Department, actually aided in the transfer of existing scientific and technical know-how from the colonies to the metropole.

One of the major effects of colonialism was the subordination of science and science education to the logic of colonial production and class structures. Science and technology development in most cases ceased to emanate from the womb of African civilization and indigenous problem-solving and experimentation. A new set of norms, values and relations of production was introduced and superimposed on existing pre-capitalist structures. In so far as colonialism was an exploitative system geared towards the redeployment of resources in the form of mineral and agricultural wealth, from the periphery to the center, it necessarily

ed to the destabilization of existing processes of accumulation of knowledge and technique. Colonial capitalism was geared primarily towards the export of surplus from the continent, and colonial education was to focus primarily on the creation of a new collaborative, non-threatening elite supportive of the colonial agenda and committed to the running of the colonial (and neo-colonial) machine. It was less geared to innovation and creativity than to servile complicity.

References to glass-making and iron fabrication may illustrate the extent to which indigenous capability was undermined. As a consequence of the relatively cheap glass products from Europe dumped on the Nigerian market, indigenous glass producers ceased to produce glass from local raw materials such as silica and potash and proceeded to simply re-smelt and refashion imported European glass objects. In the case of iron there was a deliberate attempt to undermine indigenous initiative by the introduction of the "scrap iron policy" whereby metal designed as "scrap" was imported into the region from Britain. It has been argued that this policy was a deliberate attempt on the part of the British to replace iron products with those from European factories. The policy succeeded in strangulating the local metallurgical industry and arrested the further development of indigenous technology in this field.

The Scientific and Technical Department of the British Imperial Institute, located in London, was aimed at the acquisition of knowledge with respect to various indigenous technologies through experimental research, technical trials, and commercial valuation from within the vast colonial domain, including African colonies. The Institute consisted of a series of well-equipped research laboratories, and a supporting staff of trained scientists and researchers kept active in the analysis of natural and manufactured products from the colonies. Specific attention was placed on medicinal plants and it was explicitly stated that the numerous plants held in high repute as medicinal agents in the colonies should be investigated both as regards their active ingredients and their medicinal value. The above guidelines led to experimentation on *Khaya senegalensis* and a variety of Guinea corn, amongst others, plants held in high repute by indigenous experts. Traditional medicinal knowledge, integrated into the British scientific laboratory, was in fact a contributing factor in pharmaceutical development in the metropolitan center in the first three decades of the twentieth century. Research activity was not confined to medicinal plants, however, and one of the areas of investigative research was textile, specifically silk. Loni or Boko silk from Bauchi, Nigeria was one of those products analyzed. An up-to-date display of the various research findings was kept in the Galleries of the British Imperial Institute, London, and was of particular interest to merchants and industrialists on the look out for new products. Details were provided on application to the Commercial Information Office of the Institute or other authorized officers.

It is clear therefore that colonialism had far reaching implications for both the colony and the metropole. It weakened indigenous African capability with respect to experimentation, problem-solving and the creation of indigenous utilitarian objects and processes. At the same time though it facilitated the transfer of knowledge from Africa and other colonial dependencies while leaving in return an education system more geared to the reproduction of Christian religious norms and values and alienation than the further development of scientific and technological capability.

GLORIA T. EMEAGWALI

REFERENCES

Andah, Bassey. *Nigerian's Indigenous Technology*. Ibadan: Ibadan University Press, 1992.

Khapoya, Vincent. *The African Experience*. New Jersey: Prentice-Hall, 1993.

Odhiambo, T. and T. Isoun. *Science for Development in Africa*. Nairobi: ICIPE Science Press and Academy Science Publishers, 1988.

Prah, Kwesi Kwaa. *Culture, Gender, Science and Technology in Africa*. Windhoek: Harp Publications, 1991.

Thomas-Emeagwali, Gloria. *Science and Technology in African History with case studies from Nigeria, Sierra Leone, Zimbabwe and Zambia*. New York: Edwin Mellen, 1992.

Thomas-Emeagwali, Gloria. *African Systems of Science, Technology and Art: The Nigerian Experience*. London: Karnak, 1993.

COLONIALISM AND SCIENCE IN THE AMERICAS

Probably more than any other cultural form, the set of explanations and dealings with natural realities that constituted the scientific knowledge of the peoples of the newly found lands were absolutely novel to the Europeans arriving with Christopher Columbus. They marveled at the constructions, urban organization, clothing, and ornamentation of these peoples, mistakenly called Indians. To explain and understand what they were witnessing, European knowledge was not appropriate. The surprise was total.

Viking traders, and probably some other European, Asian, and African navigators, had previously visited these lands. There is evidence of this in or about the tenth century, but the records of these visits did not reach what we call the official literature.

The Portuguese were the most advanced people in Europe in acquiring navigational knowledge. People interested in navigation would study at the court of King Dom João II, in

Lisbon, and frequent the entourage of the Infante Henrique, his son in charge of the navigation projects. Astronomers, ship builders, and pilots from all over the world would be there. Among them was Martin Behaim, who, upon returning home to Nürnberg in 1490, was asked by his city authorities to produce a report on his studies abroad. He presented this in the form of a model of the Earth: a globe which he called the *Erdapfel*. There, between Europe, Asia, and Africa, shown in a detailed picture, lay a vast mass of water.

Surely views like this convinced Christopher Columbus, also a student in the court of Dom João II, of the possibility of reaching Cipangu and the domains of Grand Kahn. He convinced the King and Queen of Spain, Fernando and Isabel, who financed a fleet of three ships, which left Palos, near Seville, on August 3rd, 1492. On October 12, 1492 they landed on an island called Guanahani in the language of its inhabitants. They did not recognize where they had landed; Columbus and the others believed they had reached Asia. The early days of the conquest and first encounters with the natives are related in descriptions of the conquest, by Columbus himself and by other chroniclers.

What the Spaniards encountered, as far as science and technology are concerned, is described in the article on "Americas: Native American Science" in this Encyclopaedia. The centuries preceding the arrival of Columbus saw a political process very similar to the one which took place in Europe in the Middle Ages, when principalities and other smaller domains were uniting into larger organizations such as kingdoms and nations. The Spanish conquerors were able to destroy the emergent power elites and replace them with the power of the conqueror, based on a well balanced State-Church structure controlling the means of production.

In Central America there was a tax system collecting products, while in the Inca empire tax was paid in labor. We can identify something that approaches private property and commerce in the Aztec model, while among the Incas the State is all powerful in the economy. Recent research reveals important commercial development in the coastal area of Peru and the beginnings of economic uses of sea resources. Although by and large the economic practices are similar to those in Europe, there is no indication at all of something that resembles capitalism. Maybe these are examples of non-capitalist economies. It is important to note the relatively low use of technology in agriculture. For example, they had no plowing devices and practically no fertilizers. Thus, production relied more on human factors, such as a more attractive division of lands with well coordinated forms of cooperation and control of distribution, than on the intensive use of technology. At the same time, a considerable intervention of the State in major infrastructures such as irrigation and roads, associated with a rather efficient tributary system, indicate economic models which differ from those in the Old World (Europe and Asia). This is an important area of scholarship.

This was the situation the conquerors encountered when they met the highland civilizations. Before that a number of colonial settlements had been created in the Dominican Republic: Isabela in 1494, and Santo Domingo in 1496. In the beginnings of this process, with agriculture depending on Indian slave labor, local tribes were exhausted. This caused slave hunting expeditions to the other islands of the Caribbean. To the South, the arrival of the Portuguese Pedro Alvares Cabral in Brazil led to an exploratory mission by Amerigo Vespucci, in 1501, under the Portuguese crown. The establishment of colonial settlements waited until 1532, when Martim Affonso de Sousa founded São Vicente in coastal Brazil. In 1534 the Portuguese king divided Brazil into large areas, the *capitanias*, mostly coinciding with the current coastal states of Brazil, and with the right of property to be inherited. This family distribution of the lands of Brazil established much of the style of land possession in the Portuguese colony.

Meanwhile, the search for slaves lead the Spaniards to found Puerto Rico in 1508, Jamaica in 1509 and Havana in 1511. From there to Florida was a small step for Juan Ponce de León in 1513. The exploration of the coast of the Gulf of Mexico by Alonso de Pineda was completed in 1519. Also, the remnants of the failed expeditions of Alonso de Ojeda and Diego de Nicuesa to Colombia and Panama were led by Vasco Nuñez de Balboa to the Pacific Ocean and the founding of Panama in 1519. From there the exploration of the Central America and Mexico and incursions to South America were obvious steps. Founding several cities, the conquerors approached the heart of the Aztec and Inca empires.

The recognition of high civilizations in the region approached by the Spaniards led to political, religious, and philosophical disputes. Under instructions by the King of Spain to establish trading relations, Hernán Cortés contacted the Aztec King Moctezuma. Attracted by the richness of the Aztecs, Cortés joined with peoples conquered by the Aztecs against Moctezuma, and ingeniously used Aztec mythology as a war strategy to defeat him. Cortés completed the conquest of the region and established the Kingdom of New Spain. In 1539, he was recalled to Spain, leaving behind an empire stretching from the northern part of South America to the southern part of the United States, up to the Colorado River.

The successes in Central America led to efforts to conquer South America. In 1530 Francisco Pizarro reached Peru and won over the Inca Atahuallpa, who was killed after being baptized. Resistance to the conquerors lasted until 1572, with a prolonged siege of Cuzco, the old Inca capital, ending with the capture and beheading of the last Inca, Topa Amaru. From Peru, the conquest was carried on to the South and North. Founding of the cities of Quito (1534), Lima (1535), Buenos Aires (1536), Asunción (1537), Santa Fé de Bogotá (1538), Santiago de Chile (1541), Potosí (1545), La Paz (1549), and Caracas (1562) completed the period called "the Conquest". All of Latin America, with the exception of the Amazon Basin, was in the hands of the Spanish and the Portuguese. Attempts by the French and the Dutch to conquer regions in Brazil did not succeed.

In the colonial times that followed, we see the exploitation of the lands and resources and of the peoples of the conquered regions. The colonizers brought with them traditional European agricultural and mining techniques with which they exploited the native production, mainly in metallurgy. The means of production were changed by the colonizers. This was done to a great extent indirectly. The native religions were simply destroyed. Food habits were also considerably modified. Wheat, rice, coffee, citrus, sugar cane, and bananas were all grown for export. This had severe consequences for the nutrition of the natives. Famine, never known in pre-Columbian times, became a major concern for Latin America. Urbanization, again following the styles of Europe, led to the development of architectural styles unsuitable to the weather and geological conditions of the conquered lands. This can be traced to the current inadequacy of the cities in the face of natural disasters such as earthquakes, floods, and high winds.

In the sciences in general, Latin America was the recipient and not the producer of scientific advances. The peripheral status of the countries was maintained throughout independence. Indeed, the colonial style and submission of the native population, with land distribution determined by the conquerors and colonizers, was kept after independence. Independence was for the *criollos*, never for the native populations, throughout the three Americas. Even after independence, education was modeled on the former imperial system.

Colonial science, considered as contributing to the mainstream of scientific development, is at best very modest. It is very important to recognize isolated cases, but they do not add to universal scientific development. Native contributions are almost exclusively in medicine, particularly in pharmacology.

The development of Native American science and philosophy was interrupted with the conquest and colonization. Remnants of pre-Columbian forms of knowledge are limited, and research today only begins to reveal their depths.

UBIRATAN D'AMBROSIO

COLONIALISM AND SCIENCE IN INDIA Colonization was not merely a political phenomenon; it had far-reaching economic and cultural ramifications. It was an exercise in power, control, and domination. Scientific and technological changes greatly facilitated this progress. Technoscience and colonialism are closely linked and to some extent share a cause and effect relationship. In recent years a good deal of work has been done on the nature, course, and consequence of this relationship in different geographical and culture areas. Some scholars see in it utilitarian and developmental images; many others find it utterly exploitative, while some prefer to opt for a middle path and emphasize both the regenerative and retrogressive aspects of the science and colonization nexus. So the debate continues, and several works have appeared with case studies on Africa, Latin America, and Asia. India, being a prime example of classic colonization, has also received considerable attention.

The fact that India has a very long techno-scientific tradition and a rich cultural heritage is fairly well known. It was, however, during the seventeenth to eighteenth centuries, the post-Renaissance epoch (that of Descartes and Newton) that Europe began to outdistance India in scientific and material advancement. The rise of modern science in Europe profoundly disturbed the balance of scientific development among traditional societies. It is also possible that the various sciences and technologies were on a decline in India around 1790. There was definitely no "conscious" spirit of technological innovation and scientific enquiry to match the spirit of Europe. The result was colonization.

The advancing European trading companies became deeply involved in political and military rivalries, culminating in the establishment of the British paramountcy over the Indian sub-continent. A new empire was in the making. The colonizers were out to collect the maximum possible information about India, its people and resources. They reported what was best in India's technological traditions, what was best in India's natural resources, and what could be most advantageous for their employers. The English East India Company, for its part, was quick to realize that the whole physical basis of its governance was dependent upon the geographical, geological, and botanical knowledge of the areas being conquered. The colonizers fully recognized the role and importance of science in empire building.

The most interesting feature of this early phase of colonial science lies in its highly individualistic character. State followed the trade, and certain individuals on the spot would largely determine what was advantageous for both. These colonial scientists would try their hands at several fields simultaneously, and were in fact botanist, geologist, zoologist, physicist, chemist, geographer, and educator — all rolled into one. As data gatherers they had no peers, but for analysis and recognition, they had to depend upon the metropolitan scientific culture whose offshoots they were and from which they drew sustenance. The colonial government quickly patronized geographical, geological, and botanical surveys; after all these were of direct and substantial economic and military advantage. Medical or zoological research did not hold such promises. Research in physics and chemistry was simply out of the question, for there were no laboratories, equipment, or specialized training. The reigning spirit remained that of exploration. Systematization or analysis of its results had to wait for some time, but then, even disjointed and often haphazard studies served some purpose.

Other positive achievements were the establishment of scientific bodies and museums. Pre-British India never had anything like as scientific society, not to say a journal, which could provide some sort of a platform for scientific workers. William Jones was the first to realize this and founded the Asiatic Society in Calcutta in 1784. This society soon became the focal point of all scientific activities in India. It was followed by the Madras Literary and Scientific Society (1805), the Agricultural and Horticultural Society of India (1817), Calcutta Medical and Physical Society (1823), and the Bombay Branch of the Royal Asiatic Society (1829). Trigonometrical and topographical surveys were organized under the Great Trigonometrical Survey of India (1818), and a Geological Survey was established in 1851. Scientific research thus for a long time remained an exclusive governmental exercise, and this largely determined the nature and scope of scientific research in India. Colonial science primarily implied "natural history" and its star (if not sole) attraction was the exploitation of natural resources. It was basically plantation research with emphasis on experimental farms, the introduction of new varieties, and the various problems of cash crops. Next came surveys in geology and meteorology. Another major area of concern was health. The survival of the army, the planters, and other colonizers was at stake. The importance of medical research was always recognized, but the quantum of emphasis varied from time to time. In any case, however, research was not to be a curiosity-oriented affair. Financial considerations were invariably there. The colonial administrators consistently held that scientists in India should leave pure science to Britain and apply themselves only to the applications of science. They would goad the various organizations to work along only economically beneficial lines. Colonial researchers often found themselves unable to distinguish between basic and applied research. This was particularly true of the geologists and the botanists. Their problem was how to discover "the profitable mean course" in which scientific research, having a general bearing, would at the same time solve the local problems of immediate economic value. The dilemma was fairly acute.

In the field of education, science was unfortunately never given a high priority. In 1835 Thomas Macaulay not only succeeded in making a foreign language, English, the medium of instruction, but his personal distaste for science led to a curriculum which was purely literary. A few medical and engineering colleges were opened, but they were meant largely to supply assistant surgeons, hospital assistants, overseers, etc. What India got was some sort of a hybrid emerging out of a careless fusion between literary and technical education. What is more, adoption of English as the sole medium of instruction in science rather hampered its percolation to the lower classes. Colonial education widened the social gulf and accentuated the age-old divide. Even in government institutions, growth was kept under a self-regulatory check. The Tokyo Engineering College was established in 1873, much later than the Engineering College at Roorkee, and by 1903 it had a staff of 24 professors, 24 assistant professors, and 22 lecturers. The Massachusetts Institute of Technology was established in 1865 and by 1908 it had 306 teachers. And Roorkee, even after 100 years of its existence (i.e. in 1947), had only three professors, six assistant professors, and twelve lecturers.

Colonialism involved not only exploration and classification but also coding and decoding cultures. Its cultural projects showed deeper penetration and greater resilience than its economic forays. With the help of schools, universities, textbooks, museums, exhibitions, newspapers, etc. local discourse was influenced and colonized. Modernity was presented as a colonial import and not something intrinsic to humanity's rational nature. Colonial rule, with its sharp tools, dissected and bared differences and was not inclined to synthesize. Colonialism usually stalls the possibilities of exchange and prefers one-way traffic. One may talk of transfer — transfer of knowledge, systems, or technologies — but it was a transfer restricted or guided to achieve certain determined objectives. As education and awareness grew, several Indians participated in the official scientific associations and institutions, but very often they searched for a distinct identity and established institutions, scholarships, and facilities of their own.

In the first half of the nineteenth century, Bal Gangadhar Shastri and Hari Keshavji Pathare in Bombay, Master Ramchandra in Delhi, Shamhaji Bapu and Onkar Bhatt Joshi in Central Provinces, and Aukhoy Kumar Dutt in Calcutta worked for the popularization of modern science in Indian languages. In 1864 Syed Ahmed founded the Aligarh scientific Society and called for the introduction of technology to industrial and agricultural production. Four years later Syed Imdad Ali founded the Bihar Scientific Society. These societies did not live long. In 1876 Mahendra Lal Sarkar established the Indian Association for Cultivation of Science. This was completely under Indian management and without any government aid or patronage. Sarkar's scheme was very ambitious. It aimed not only at original investigations but at science popularization as well. It gradually developed into an important center for research in optics, acoustics, scattering of light, magnetism, etc. In Bombay, Jamshedji Tata drew up a similar scheme for higher scientific education and research. This was opposed by the then Governer-General, Lord Curzon. Yet it finally led to the establishment of the Indian Institute of Science at Bangalore in 1909. There was thus greater awareness by the turn of the century.

In the first quarter of the twentieth century, those who put India on the scientific map of the world were J.C. Bose, who studied the molecular phenomenon produced by electricity on living and non-living substances, Ramanujan, a mathematical genius, and P.C. Ray who analyzed a number of rare Indian minerals to discover in them some of the missing elements in Mendeleef's Periodic Table. C.V. Raman's research on the scattering of light later won him the Nobel Prize in 1930 and gave the name to Raman spectroscopy. Meghnad Saha pioneered the field of astrophysics, while S.N. Bose's collaboration with Einstein led to what is known as the Bose–Einstein equation. These were great sparks, individual and sometimes lonely, yet imbued with both scientific and national spirit. They thought over what role science and technology would play in building modern India, and they dreamt of freedom. The colonial government was aware of its limitations and discomfiture, and gradually permitted greater indigenization of its scientific institutions and cadre. In the wake of the First World War the government realized that India must become more self-reliant scientifically and industrially. It appointed an industrial commission in 1916 to examine steps that might be taken to lessen India's scientific and industrial dependence on Britain. However, few of the Commission's recommendations were actually implemented. Discontent continued to grow. India's national leaders appreciated the importance of science and technology in national reconstruction and worked closely with the scientific talent of the time. A National Planning Committee was formed in 1937 for this purpose, in which several leading Indian scientists and technologists participated. In 1942, the Council for Scientific and Industrial Research was established. The end of colonial science was pretty near. With the A.V. Hill Report in 1944 on Scientific Research in India, the curtain dropped.

The foregoing analysis illustrates that in a colonial situation field sciences may have been developed through imported scientists as an economic necessity, but little fundamental research was possible. A few colonial scientists made important contributions that no doubt enriched science in general, but their activities hardly succeeded in introducing science to the Indian people or in ameliorating their condition. Colonial science did, on the whole, support and help sustain exploitation and underdevelopment.

DEEPAK KUMAR

REFERENCES

Adas, Michael. *Machines as the Measure of Men: Science, Technology and Ideologies of Western Dominance*. Ithaca: Cornell University Press, 1989.

Alvarez, Claude. *Homo Faber: Technology and Culture in India, China and the West, 1500 to Present Day*. The Hague: Nijhoff, 1980.

Arnold, David. ed. *Imperial Medicine and Indigenous Societies*. Manchester: Manchester University Press, 1988.

Bala, Poonam. *Imperialism and Medicine in Bengal*. New Delhi: Sage Publications, 1991.

Brockway, Lucile H. *Science and Colonial Expansion: The Role of the British Botanic Gardens*. New York: Academy Press, 1978.

Chakravarty, Suhash. *The Raj Syndrome: A Study in the Imperial Perceptions*. New Delhi: Chanakya Publications, 1991.

Cipolla, Carlo M. *Guns and Sails in the Early Phase of European Expansion, 1400–1700*. London: Collins, 1965.

Dharampal. *Indian Science and Technology in the Eighteenth Century: Some Commentary European Accounts*. Delhi: Impex India, 1971.

Ellsworth, Edward W. *Science and Social Science Research in British India, 1788–1880*. Westport, Connecticut: Greenwood Press, 1991.

Gaeffke, Peter and Utz, David A., eds. *Science and Technology in South Asia*. Philadelphia: Department of South Asia Regional Studies, University of Pennsylvania, 1985.

Goonatilake, Susantha. *Crippled Minds: An Exploration into Colonial Culture*. New Delhi: Vikas, 1982.

Grove, Richard. *Green Imperialism*. Delhi: Oxford University Press, 1995.

Headrick, Daniel R. *The Tools of Empire: Technology and European Imperialism in the Nineteenth Century*. New York: Oxford University Press, 1981.

Hoodbhoy, Pervez. *Islam and Science*. London: Zed Books, 1991.

Hutchins, Francis G. *The Illusion of Permanence: British Imperialism in India*. Princeton: Princeton University Press, 1967.

Kumar, Deepak, ed. *Science and Empire: Essays in Indian Context*. Delhi: Anamika Prakashan, 1991.

Kumar, Deepak. *Science and the Raj 1857–1905*. Delhi: Oxford University Press, 1995.

Kumar, Krishna. *Political Agenda of Education: A Study of Colonialist and Nationalist Ideas*. New Delhi: Sage Publications, 1991.

Leslie, Charles, ed. *Asian Medical Systems*. Berkeley: University of California Press, 1977.

MacKenzie, John M., ed. *Imperialism and the Natural World*. Manchester, Manchester University Press, 1990.

MacLeod, Roy and Kumar, Deepak, eds. *Technology and the Raj*. New Delhi: Sage Publications, 1995.

MacLeod, Roy and Lewis, M., eds. *Disease, Medicine, and Empire*. London: Routledge, 1988.

McClellan, J.E. *Colonialism and Science: Saint Domingue in the Old Regime*. Baltimore, Johns Hopkins University Press, 1992.

Meade, Teresa, and Walker, Mark, eds. *Science, Medicine, and Cultural Imperialism*. New York: St. Martin's, 1991.

Mendelssohn, K., *Science and Western Domination*. London: Thames and Hudson, 1976.

Nandy, Ashis. *Alternative Sciences*. New Delhi: Allied, 1980.

Nandy, Ashis. *The Intimate Enemy: Loss and Recovery of Self Under Colonialism*. Delhi: Oxford University Press, 1983.

Petitjean, Patrick et al., eds. *Science and Empires*. Dordrecht: Kluwer, 1992.

Pyenson, Lewis. *Cultural Imperialism and Exact Sciences: German Expansion Overseas, 1900–1930*. New York: P. Lang, 1985.

Pyenson, Lewis. *The Empire of Reason: Exact Science in Indonesia, 1850-1950*. Leiden: Brill.

Reingold, Nathan and Rothenberg, Marc, eds. *Scientific Colonialism*. Washington: Smithsonian Institution Press, 1986.

Sangwan, Satpal. *Science, Technology and Colonialization: An Indian Experience, 1757–1857*. Delhi: Anamika Prakashan, 1991.

Zimmerman, F. *The Jungle and the Aroma of Meats: An Ecological Theme in the Hindu Medicine*. Berkeley: University of California Press, 1987.

See also: Western Dominance – Ramanujan

COLONIALISM AND SCIENCE IN THE MALAY WORLD

By the sixteenth century, the legends of the Golden Chersonese, about the distant land having more gold and exotic treasures than people, were well established among the sea going nations of Europe. Legend had it that the Golden Chersonese was geographically located midway between India and China. The Golden Chersonese description apparently fit the Malay Peninsula very well. Indeed, it came to be known among the early Europeans by that very name until it was replaced by another, Malacca. By whatever name the Peninsula was known, it was a major focal point of Eastern trade. Indeed, Malacca — actually a tiny coastal city-port situated in the middle of the Malay Peninsula — was not only a commercial Mecca but also a cultural center and meeting point for the major civilizations of the time.

In 1575, Louis De Roy, writing on the virtues of the new knowledge and inventions spurned by the Renaissance spirit, made a special note about the meeting of Magellan and the King of Malacca. According to De Roy, the meeting was made possible by the new knowledge of the seas and the use of the magnetic compass for navigation. Malacca, to the Europeans then, was not just a reference to the city-port but rather to the whole Malay Archipelago, including the Peninsula and Islands of Indonesia, that were under Malaccan rule. By the time De Roy's article was published, Malacca had actually been under the annexation of Portugal for 65 years, since the Portuguese defeated the Malays in 1511. Some scholars have suggested that the downfall of Malacca also marked the decline of Eastern scientific and technological superiority *vis-à-vis* the European. Others have cautioned against making simplistic conclusions, pointing out that the reasons for the downfall of Malacca were far more complex than those that have been suggested.

The Portuguese ruled Malacca until the mid-seventeenth century. In 1614 the Dutch successfully wrested Malacca from the Portuguese. By the mid-eighteenth century Dutch power in the Malay Archipelago declined, giving way to the British. The British were relatively latecomers to the region, but they were none the less regarded as the best prepared — intellectually, militarily and economically — among the colonizers of the region. They were poised for a long stay.

There were two major factors that were crucial to the expansion and consolidation of the Crown's influence: the colonial economy and the health of officials, colonialists, and *coolies* (colonial laborers). In the early phase of British colonialism, spices, timber and other agricultural products were the backbone of the colonial economy. Knowledge of economically and medicinally useful plants of the East was indispensable both for the sustenance of the economy and the maintenance of health. This knowledge was to be found in the new discipline of plant science, later known as economic botany, and much later as agricultural science.

The cultivation of spice trees, and the transfer of economic plants like rubber trees from the Amazon to the Malay Peninsula, were some of the colonial activities made possible by the new plant science and its scientific institutions. Indeed, apart from medicine, no branch of science was regarded as so crucial to the sustenance of colonialism as economic botany. Before the advent of synthetic drugs, herbs and plants were the main source of medicines. For instance, quinine, an extract from the chichona tree that grew wildly in the Peruvian jungle, was extensively used to cure malarial fever. Hence, knowledge of the indigenous plants was important not only for economic reasons but for health as well. Health and diseases also assumed an economic and political significance. They were moderating factors in the colonial expansion and exploitation of the East, especially in the tropics. Tropical diseases like malaria were a menace to the health of colonial officials and their laborers. Indeed, malaria reportedly wiped out some of the remote British colonies and military bases and was blamed for slowing down colonial campaigns in the Far East and Africa. In the following discussion we will see how the development and history of British colonialism and science in the Malay Archipelago became intertwined.

Botany and British Colonialism

As early as the seventeenth century there were attempts to grow exotic and foreign food crops away from their natural habitats. The first attempt to grow pineapples in England was reportedly made during the reign of Charles II (1660–1685). The colonization of the New World and later, Asia, led to the introduction of many new crops to Britain. Maize, a staple crop of the native Americans, which yields ten times more than wheat, was among the crops that revolutionized European tastebuds. Another imported high yielding crop, that reportedly saved Ireland from starvation, was the potato. However, there were still many food crops that resisted local cultivation and therefore had to be brought over from the colonies. Supply was notoriously irregular and rarely adequate to meet demands. It was not until the nineteenth century that supply became more reliable. This reliability owed much to the active colonization of new fertile lands in the East and the new discipline of economic botany.

In the nineteenth century, the political ideology of the times, the botanic gardens, and the colonial office functioned as an integrated framework, enforcing an economic structure whereby the East became the producer of raw materials for the West's industries. Indeed, this framework was a characteristic description of colonialism which involved, among other things, the establishment of plantations and the large scale cultivation of economic plants, and the systematic transfer of economic plants from an area less productive or less conducive (either politically, economically, or scientifically) to a more productive and conducive one. Colonialism also entailed the need to coordinate and transfer labor from one colony to another in order to ensure the viability of a plantation or an agro-economic project. Underwriting the success of the colonial plantations and agriculture based economic projects were the many research oriented botanic gardens established throughout the colonies. The most prominent of these botanic gardens were the Calcutta Garden in India, the Peradeniya Gardens in Ceylon (Sri Lanka), and the Penang, Malacca, and Singapore Gardens in the Malay Peninsula, known then as British Malaya.

The Royal Botanic Gardens at Kew in London was the center for scientific research and coordination of the work of the botanic stations in the colonies. Kew's role was very well illustrated in the case of the discovery and cultivation of the cinchona tree — the medicinal tree from which the malarial medicine, quinine, was extracted. The discovery of cinchona involved a large army of field workers and researchers collecting and gathering seemingly irrelevant bits of information and materials, from every remote corner of the British Empire and beyond. The information and materials were sent to Kew for analysis. The discovery and eventual successful extraction of quinine, which turned out to be the most potent antidote for malaria, was a credit to Kew and its network of botanic gardens in the colonies. Indeed, quinine was not just an important scientific discovery; it also had important demographic and political effects. Through the control of malaria it saved the lives of countless colonial officials and their *coolies* (laborers) and made possible the large scale colonial exploration and exploitation of Africa and the Far East.

Colonial Botany in the Malay Peninsula

Colonial botany came to the shores of the Malay Peninsula in the year 1822 with the establishment of the botanic garden in Singapore by Sir Stamford Raffles. The staff of the botanic garden was instrumental in overseeing the introduction and improvement of spice cultivation in the Straits Settlements. Indeed, in the Malay Peninsula *botany* was almost indistinguishable from *economy*. Whether the relationship between botany and colonial economy in the colonies had any influence in the formalization of economic botany back in the motherland is a difficult question to answer with certainty; but what is certain is that the colonial economy benefited much from the knowledge and expertise provided by economic botany.

Back in Britain, botany was given a boost following several developments, one of which was the appointment in 1841 of Joseph Hooker (1817–1911), far more widely known as a colonialist than as a botanist, as head of the Royal Botanic Garden at Kew. Kew was responsible for coordinating and supervising all the scientific research on economic plants and agriculture within the British colonial empire. It was under Kew's supervision that in 1858 new botanic gardens were established in Singapore superseding the one established by Raffles. In 1887 botanic gardens were also established in Penang followed soon after by another in Malacca.

Nowhere is the importance of Kew to British colonialism so clear as in the case of the transfer of the wild Brazilian rubber plant from Latin America to the Malay Peninsula. Initially, various species of wild rubber seeds were collected from the Amazon jungle and sent to Kew Gardens for study and analysis. Eventually a species, *Hevea brasiliensis*, was identified as promising and was transferred to the botanic gardens in Singapore. Under the supervision of the Singapore gardens which H.N. Ridley headed, politically backed by the colonial office, and heavily financed by the Crown companies, experimental plantations were set up in various parts of the Malay Peninsula. The plantations turned out to be experimentally successful and, as anticipated, destined to become commercially successful too. At the turn of the twentieth century, with the advent of motor vehicles industry in Europe and America, rubber became one of the most important raw materials for the new industry. Rubber planting indeed became one of the most profitable colonial commercial ventures. Most importantly, rubber could only be found in abundant quantity and of high quality within the British Empire.

The transfer, identification and eventual cultivation of economic plants would not have been successful without colonial botany equally recording a success in another related activity: taxonomical works. Indeed, one of the main tasks of the botanic gardens was taxonomical work closely associated with identifying economically and medically useful plants. The varieties of plant life in the tropics became a testing ground for the new system of classification. In the Malay Peninsula, the colonial taxonomical campaigns of identifying and naming tropical plants soon turned out to be an enormous task that almost overwhelmed the colonial scientists and their machinery. The variation of plant life in the tropics rendered it practically impossible for their taxonomists and the new taxonomy to start fresh from scratch. However, starting from scratch was the last thing that they had to do. Long before the arrival of the British, the natives had already classified most of the plants, especially the economically useful ones. More often than not, all the colonial taxonomist had to do was to refer to the local taxonomy and be creative. They eventually discovered that the local folk taxonomy was almost encyclopedic, encompassing the encounter of many generations of indigenous wisdom with the local flora.

Colonial Botany and Malay Ethnobotany

The encounter of the colonialists with Malay ethnobotany probably went as far back as the arrival of the first Europeans to the area. Tome' Pires, a Portuguese scholar and explorer, who visited Malacca during the Portuguese rule, noted the existence of a botanical tradition among the Malays. The Malay botanical tradition was described as being composed of a complex system of plant taxonomy and knowledge as well as the cultural–educational rituals associated with teaching and imparting the knowledge to successive generations.

The majority of the colonial botanists, with very few exceptions, were genuinely intrigued by the sophistication and mechanics of Malay plant taxonomy. They seemed to agree that the Malay taxonomy was much more than a fanciful accidental uttering of a primitive tribe oblivious to any sense of rationality or systematic thinking. In fact, in may ways it was comparable to the modern colonial system.

In the Colonial scientific taxonomy, plants having the same kind of flower and fruit but which differ in details of size, shape, and color of the flowers, fruits or leaves, constituted a *genus*. A specific name came after the generic name either as an adjective qualifying it or as noun in the genitive case if the species was in honor of some person.

Colonial botanists noted that Malay folk taxonomy, on the other hand, was based on either the use or morphological characteristics of the plants, or at times both. Most economically useful plants were classified by their use, whilst medically useful plants and others were often classified by the morphology of their niche, leaves, fruits, or flowers. What was found as a pleasant surprise to many of the colonial botanists was the discovery that in many instances the colonial scientific taxonomy and Malay taxonomy shared common grounds. There were also considerable numbers of Malay taxonomical names that found no equivalents whatsoever with the colonial names. Names like *kedundung*, *Tinjau Belukar*, and *Puding* were morphological names referring to the chief characteristics of the respective plant. They could be either a reference to their growth habit, shape of leaves, flowers, fruits, or combinations of any of those. They did not correspond neatly to the modern colonial taxonomy. However, in spite of that, Malay ethnotaxonomy was a tremendous aid to the colonial reclassification of plants in the tropical Peninsula to modern taxonomy.

Many of the taxonomical and other scientific findings were published in the *Kew Bulletin*. The *Bulletin* was not only important academically but it was also a medium of communication and instruction between Kew and the imperial network of botanic stations. It played the dual role of a scientific medium and a channel for colonial interests, reflecting faithfully the nature of colonial scientific activities.

Colonial scientific activities in the Malay Peninsula, especially the botanical and the taxonomical exercises, were not just scientific *per se*; they were first and foremost an integral part of the colonial activity to sustain British colonialism. In short, they were by and for colonialism.

HAIRUDIN B. HARUN

REFERENCES

Allan, N. *The Oriental Collections of the Wellcome Institute for the History of Medicine*. London: Wellcome Institute of Medicine, 1984.

"An Outline of Malayan Agriculture." *Malayan Planting Manual No.2*. Dept. of Agriculture, Federation of Malaya, 1950.

Bird, I.L. *The Golden Chersonese and the Way Thither*. London: Oxford University Press, 1976, first published 1883.

Brockway, L.H. *Science and Colonial Expansion: The Role of British Botanical Gardens*. London: Academic Press, 1979.

Burkill, I.H. *A Dictionary of the Economic Products of the Malay Peninsula*. 2 vols. London: Government Printers, 1935.

Cromer, J.A. *Wayside Trees of Malaya*. Vol.1. Singapore: n.p., 1940.

Desmond, Ray. *Dictionary of British and Irish Botanists and Horticulturists*. London: Taylor and Francis, 1979.

Emerson, R. *Malaysia: A Study of Direct and Indirect Rule*. Kuala Lumpur: University of Malaya Press, 1964.

Greham, G.S. *A Concise History of the British Empire*. London: University Press, 1970.

Green, J.R. *A History of Botany in the U.K. from the Earliest Times to the End of the 19th Century*. London: J.M. Dent & Sons, 1914.

Greene, E.L. *Landmarks of Botanical History*. Stanford, California: Stanford University Press, 1983, first published 1909.

Gimlette, J. D. *Malay Poisons and Charm Cures*. London: J & Churchill, 1915. Reprinted Kuala Lumpur: Oxford University Press, 1971.

Hervey, D.F.A. "Malay Games." *Journal of the Royal Anthropological Society* 33, July–December, 1903.

The Hervey Malay Collection. The Wellcome Institute for the History of Medicine, London.

Harun, Hairudin B. *Medicine and Imperialism: A Study of British Colonial Medical Establishments, Health Policy and Medical Research in the Malay Peninsula, 1786–1918*. Ph.D. dissertation, University of London, 1988.

Harun, Hairudin B. "Science and Colonialism: A Study of Wellcome's Malay Manuscripts." *SARJANA: Journal of the Faculty of Arts and Social Sciences, University of Malaya* 4: 1–20, June, 1989.

Morton, A.G. *History of Botanical Science: An Account of the Development of Botany from the Ancient Times to the the Present Day*. London: Academic Press, 1981.

Oliver, F.W. *Makers of British Botany*. Cambridge: Cambridge University Press, 1913.

Ridley, H.N. *The Flora of the Malay Peninsula*. 5 vols. London: L Reeve & Co., 1925.

Thomson, H.W. and J.D. Gimlette. *A Dictionary of Malayan Medicine*. London: Oxford University Press, 1934.

Turill, W.B. *The Royal Botanic Gardens Kew: Past and Present*. London: Herbert Jenkins, 1959.

See also: Ethnobotany

COMBINATORICS IN INDIAN MATHEMATICS Having prescribed the rule

$$C_n^r = \prod_{k=1}^{r} \frac{n-k+1}{k}$$

for the number of combinations of r things taken at a time from n things, Bhāskara II (AD 1150) remarked: "This [rule] has been handed down [to us] as a general [method], being employed [for their own purposes] by the experts [of specific fields of study], namely, for the tabular presentation of possible meters in metrics, for the number of ways of opening ventilating holes, etc., and the diagram called Partial Meru in arts and crafts, and for the varieties of tastes in medicine. [But], for fear of prolixity, they are not explained here" [*Līlāvatī* 113-114]. It is impossible to tell exactly when and by whom this rule was formulated, but from ancient times Indian peoples have had a keen interest in arranging things in order in various aspects of human life, and in theorizing in various areas of study.

The *Bhagavatī*, one of the twelve canonical books of the Jainas, counts by enumeration the number of all the possible cases when one person, two persons, three persons, etc. are distributed in the seven nether worlds.

Most books of Sanskrit and Prakrit prosody, such as the *Chandaḥsūtra* of Piṅgala (ca. AD 200?), *Vṛttajātisamuccaya* of Virahāṅka (between the sixth and the eighth centuries), *Jayadevacchandaḥ* of Jayadeva (before AD 900), *Chando'nuśāsana* of Jayakīrti (ca. AD 1000), *Vṛttaratnākara* of Kedāra (before AD 1000), *Chando'nuśāsana* of Hemacandra (ca. AD 1150), etc., devote one of their chapters (usually the last one) to "six kinds of ascertainment" (*ṣaṭpratyaya*), namely:

(1) spread (*prastāra*), to spread a list of all the possible variations of a given type of meter consisting of short and long syllables according to a certain method;
(2) lost (*naṣṭa*), to find out a lost variation when its serial number in the list is given;
(3) mentioned (*uddiṣṭa*), to calculate the serial number in the list when a particular variation is mentioned;
(4) short- (and-long-) calculation (*laghukriyā/galakriyā*), to calculate the number of the variations in the list that have a given number of short or long syllables (for this purpose, a diagram called the Mount Meru-like spread (*meruprastāra*), equivalent to the so-called Pascal's triangle, is constructed);
(5) number (*saṃkhyā*), to calculate the number of all the variations in the list; and
(6) way (*adhvan/adhvayoga*), to calculate the space of writing materials required for writing down the list.

A chapter on prosody in the *Nāṭyaśāstra* of Bharata (before AD 600), a dramaturgical work, and one in the *Agnipurāṇa* (ca. AD 800), a work on sacred traditions, also contains a section on these topics.

Four out of the six kinds of ascertainment, namely, spread, lost, mentioned, and number, were applied to the melody (combinations of seven notes) and the rhythm (combinations of a half, one, two, and three beats) in Indian music [Śārṅgadeva's *Saṅgītaratnākara*, ca. AD 1250], for which a diagram called Partial Meru (*khaṇḍameru*) was employed. This was also the case with the combinations of five sub-categories taken severally from five categories of "carelessness" (*pramāda*) when they contain different numbers of sub-categories, in Jaina philosophy [Nemicandra's *Gommaṭasāra*, ca. AD 980].

Medical treatises such as the *Carakasaṃhitā* (ca. AD 100) and *Suśrutasaṃhitā* (ca. AD 200) treated combinations of the six basic tastes (*rasa*), and of the three humors (*doṣa*) of the body. Varāhamihira dealt with the problem of combinations of nine and sixteen basic perfumes in the *Bṛhatsaṃhitā* (ca. AD 550).

Indian mathematicians prior to Bhāskara, too, are known to have been interested in these problems of combinatorics. Brahmagupta devoted one whole chapter consisting of nineteen (or twenty) stanzas of his astronomical work, *Brāhmasphuṭasiddhānta* (AD 628), for combinatorics related to Sanskrit prosody. The problem of tastes was taken up by the mathematician Śrīdhara in his *Pāṭīgaṇita* (ca. AD 750). Mahāvīra treated the six kinds of ascertainment of Sanskrit prosody as well as tastes in his mathematical treatise, *Gaṇitasārasaṃgraha* (ca. AD 850).

In addition to the traditional rule for combination mentioned above, Bhāskara gave six rules for permutations concerning sequences of numerical figures in a chapter called "nets of numbers" (*aṅkapāśa*) of his famous *Līlāvatī*. However, it is Nārāyaṇa who for the first time treated various problems of permutation and combination, including the system of "spread-lost-mentioned", of different areas systematically from the viewpoint of mathematics with the help of variously defined sequences of numerals. Chapter 13 entitled "nets of numbers" of his mathematical treatise, *Gaṇitakaumudī* (AD 1356), consists of about one hundred stanzas for rules of combinatorics and forty-five for examples.

TAKAO HAYASHI

REFERENCES

Alsdorf, L. "Die Pratyayas. Ein Beitrag zur indischen Mathematik." *Zeitschrift für Indologie und Iranistik* 9: 97–153, 1933. Translated into English with notes by S.R. Sarma: "The Pratyayas: Indian Contribution to Combinatorics." *Indian Journal of History of Science* 26: 17–61, 1991.

Bag, A.K. "Binomial Theorem in Ancient India." *Indian Journal of History of Science* 1: 68–74, 1966.

Chakravarti, G. "Growth and Development of Permutations and Combinations in India." *Bulletin of the Calcutta Mathematical Society* 24 (2): 79–88, 1932.

Das, L.R. "La théorie des permutations et des combinations dans le gaṇita hindou." *Periodico di matimatiche* 12: 133–140, 1932.

Datta, B. "Mathematics of Nemicandra." *The Jaina Antiquary* 1(2): 25–44, 1935.

Datta, B. and Singh, A.N. "Use of Permutations and Combinations in India." Revised by K.S. Shukla. *Indian Journal of History of Science* 27: 231–249, 1992.

Gupta, R.C. "Varāhamihira's Calculation of ^nC_r and Pascal's Triangle." *Gaṇita Bhāratī* 14: 45–49, 1992.

Hayashi, T. "Permutations, Combinations, and Enumerations in Ancient India" (in Japanese). *Kagakusi Kenkyu*, Ser. II 18(131): 158–171, 1979.

Kusuba, T. "Combinatorics and Magic Squares in India. A Study of Nārāyaṇa Paṇḍita's *Gaṇitakaumudī*, Chaps 13–14." Ph.D. Dissertation, Brown University, 1993.

Singh, P. "Nārāyaṇa's Treatment of Net of Numbers." *Gaṇita Bhāratī* 3: 16–31, 1981.

Singh, P. "Contributions of Some Jain Ācāryas to Combinatorics." *Vaishali Institute Research Bulletin* 4: 90–110, 1983.

Singh, P. "The So-called Fibonacci Numbers in Ancient and Medieval India." *Historia Mathematica* 12: 229–244, 1985.

See also: Bhāskara – Brahmagupta – Śrīdhara – Varāhamihira – Mahāvīra – Nārāyaṇa

COMBINATORICS IN ISLAMIC MATHEMATICS

Among the factors which influenced mathematics in Arab-Islamic civilization was a cultural life which fostered, particularly between the ninth and twelfth centuries, rich interaction between the different intellectual activities of the time.

Taken in a general sense as a study of configurations in one-, two-, or three-dimensional spaces, combinatory analysis had its beginnings in early studies in music, chemistry, astrology, and especially in metrics and Arabic linguistics. From the ninth century on combinatory processes asserted themselves in truly mathematical disciplines like algebra, astronomy, and trigonometry.

Combinatory Practices in Non-mathematical Fields

In music, many works contain a study of the combinations of sounds and rhythms, for example, the *Rasāʾil* (Epistles) of the Ikhwān al-ṣafāʾ(Brethren of Purity, tenth century) and the *Kitāb al-mūsīqā al-kabīr* (Great Book on Music) of al-Fārābī (ninth century). In the field of chemistry, as early as the eighth century, Jābīr ibn Ḥayyān introduced combinatory arguments in his theory of balance, which was based on the principle that everything is a combination of numbers. Jābir goes so far as to identify the language, i.e. the combination of words, in chemistry which he considers to be a morphology of metals.

In astronomical astrology, the conjunctions of the seven planets known at that time played an important role in predicting events. One needed to know the number of conjunctions which corresponded with the number of combinations of seven elements two times two, three times three ··· seven times seven. There was a great need to construct configurations by manipulating numbers or letters. Many mathematicians worked on this problem trying to invent processes of construction and new types of magic squares, for example, Thābit ibn Qurra (d. 901), Abū'l-Wafā᾽ (d. 997), Ibn al-Haytham, Ibn Munʿim (d. 1228), and Ibn al-Bannā᾽ (d. 1321).

In the field of Arabic poetry, it was al-Khalīl ibn Aḥmad (d. 791) who originated research on meter. He began by deriving a superficial structure of Arabic verse (which also serves for prose). Then he derived an underlying structure based on ten measures which, in combination, provided the five fundamental meters. In turn, these meters provided other meters to Arabic poetry thanks to a purely combinatory operation, circular permutation.

In the field of Arabic prose, there is even more interesting information about combinatory practices. As early as the eighth century, al-Khalīl undertook an analysis of the structures of the Arabic language and attempted to develop a theory based on the results of his investigations by studying the constitution of the roots of the language from the twenty-eight letters of the alphabet, and then the composition of words from those roots. To that end, he calculated combinations, without repetitions, of letters two by two, three by three, four by four, and five by five, in order to determine the numbers of roots of two to five letters in the language. The results of his calculations, without any explanation, appear in his book *Kitab al-ʿayn*. After him, a long tradition was established in this field, but the combinatory aspects have not been systematically studied.

That being said, the examples cited above are really too sparse and isolated to prove any continuity in combinatory practices. In particular, we have not yet found anything which demonstrates the beginning of the mathematization of these combinatory practices in the Muslim East or in Andalusia before the twelfth century.

Combinatory Practices in Mathematics

Combinatory practices in the field of mathematics prior to the twelfth century were relatively modest and concerned only two disciplines, astronomy and algebra.

In astronomy, for example, there is the book of Thābit ibn Qurra, *Al-shakl al-Qaṭṭāʿ*, in which combinations occur, two by two, in six sizes occurring in the formula fixed for the secant figure. In his book, *Maqālīd ʿilm al-hay'a* (Keys to the Science of Astronomy), al-Bīrūnī also uses small combinatory results for the purpose of determining the unknown elements of a spheric triangle as a function of the known elements.

In algebra, there are two works which contain certain aspects of combinatorics. The first, the *Kitāb al-ṭarā᾽if fi'l-ḥisāb* (Book of Rare Things in the Art of Calculation) by Abū Kāmil (d. 930), deals with the resolution of certain systems of indeterminate equations, stated in the form of problems of birds, the aim being to enumerate, for each system, all the possible solutions. The second work is the *Kitāb al-Bāhir* by as-Samaw'al (d. 1175) on algebra, its objects and its instruments.

We have, as yet, no way of proving that these combinatory practices from within the mathematical tradition gave rise to any real research in the field between the ninth and the twelfth centuries, and in instances where research did occur, we do not know the results. In any case, it was not the combinatory elements encountered in mathematical problems which inspired the mathematicians of the Maghreb who became involved in combinatorics. It was the tradition of linguistics and lexicography which was the basis of their research.

Ibn Munʿim

The oldest mathematical work from the Maghreb which deals with combinatory problems is the *Fiqh al-ḥisāb* of Ibn Munʿim, a scholar originally from Andalusia, living in Marrakesh in the Almohad era. To our knowledge, his book was the first in the entire history of mathematics to have devoted a whole chapter to this type of problem and to have stated them and solved them according to a common procedure.

Combinatorics After the Thirteenth Century

In the second half of the thirteenth century or the beginning of the fourteenth, another Maghrebin mathematician, Ibn al-Bannā᾽, took up some of the results of Ibn Munʿim in at least two of his works, the *Rafʿal-ḥijāb* and the *Tanbīh al-albāb*. In the latter work, he notes only a table of combinations from Ibn Munʿim without giving exact references, but he explicitly lays claim to the demonstration of a result long attributed to Pascal (d. 1662) which gives C_n^p as a function of C_n^{p-1}, thus allowing the calculation of the combinations 2 times 2, 3 times 3, etc., of a certain number of objects, using the arithmetical formula:

$$C_n^p = \frac{n(n-1)\cdots(n-p+1)}{1\cdot 2\cdot 3\cdots p\cdot(p-1)},$$

which today takes the form of

$$C_n^p = \frac{n!}{p!(n-p)!}.$$

In the same spirit, he deduced the expression of arrangements without repetition of a certain number of objects, after having calculated the number of their permutations.

It is important to point out that, at the same time, in the East, the mathematician al-Fārisī (d. 1321) identified the values of the different combinations of the elements of the arithmetic triangle of al-Karajī and used them to establish a theorem of the theory of numbers such that one can determine the number of divisors belonging to a certain whole. Unfortunately, we do not know if this contribution of al-Fārisī spread throughout the East nor can we affirm that it was known in the Muslim West.

However that may be, we can discern two significant processes from Maghrebian writings after Ibn Munʿim. First, there occurred the extension of the field of application of formulas and of combinatory reasoning. This field was no longer concerned just with the elements of an alphabet, since that became an abstract model operating in different mathematical fields: in astronomy, with the reformulation and enumeration of compound ratios occurring in spheric trigonometry; in geometry, with the classification of figures and the enumeration of all the equations one can deduce; and finally in algebra, with the enumeration of systems of equations (the work of Ibn Haydūr (d. 1413) and of equations of a certain degree greater than three (the work of the Egyptian Ibn alMajdī, d. 1447, in his commentary on the *Talkhīṣ* (Summary) of Ibn al-Bannāʾ).

In the second place, there was a taking account of enumeration in general, in very different fields not always related to mathematics. The most typical case is that of Ibn al-Bannāʾ, who devoted part of his work *Tanbīh al-albāb* to these types of problems: the enumeration of different cases of possible inheritances when the heirs are n boys and p girls, the enumeration of prayers to be said according to the exigencies of Malekite ritual in order to compensate for having forgotten other prayers, and the enumeration of all the possible readings of the same sentence according to the rules of Arabic grammar.

The presence of the same combinatory vocabulary in the works of different authors and the fact that none of them laid explicit claim to their results reinforces the characters of continuity of combinatory practices, a continuity made possible, probably, by the persistence of this new chapter in the teaching of certain professors of that time.

However, if to our eyes this new material and these new instruments seem objectively to be of a piece, we do not know to what degree those who contributed to their formulation were conscious of them or what importance they held in *their* eyes.

The reasons for this are to be found in several places at once: the state of the society itself and the nature of its activities and preoccupations which did not allow for the significant development of combinatorics; the absence of local or regional institutions responsible for renewing programs and imposing, and then perpetuating, the teaching of new ideas; and the imprint of certain specialists whose authority influenced at certain moments the content of scientific teaching, fixing it or simplifying (or deleting) certain theoretical developments and certain chapters. At least that is what Ibn al-Bannāʾ implies in his *Rafʿal-ḥijāb* and what Ibn Khaldūn confirms in his *Muqaddima* (Prolegomena), writing about the abandonment of theoretical problems and the perpetration of a more utilitarian mathematics.

AHMED DJEBBAR

REFERENCES

Al-Fārābī. *Kitāb al-mūsiqā al-kabīr*. al-Qāhirah: Dār al-Kātib, 1967.

Al-Maghribī. *Al-bāhir fī al-jabr d'as-Samaw'al*. Ed. S. Ahmad and R. Rashed. Damascus: Imprimerie de l'Université de Damas, 1972.

Djebbar, A. *Enseignement et recherche mathématiques dans le Maghreb des XIIe–XIVe siècles*. Paris: Publications Mathématiques d'Orsay, 1981.

Djebbar, A. *L'analyse combinatoire au Maghreb: l'exemple d'Ibn Munʿim (XIIe–XIIIe siècles)*. Paris: Publications Mathématiques d'Orsay, 1985.

Ikhwān al-ṣafāʾ. *Rasāʾil*, vol. 1. Beirut: Dār ṣādir, 1957, pp. 109-113.

Kraus, P. *Jābir Ibn Ḥayyān, contribution à l'histoire des idées scientifiques dans l'Islam*. Paris: Les Belles Lettres, 1986.

Rashed, R. *The Development of Arabic Mathematics: Between Arithmetic and Algebra*. Dordrecht and Boston: Kluwer Academic Publ., 1994.

See also: Ibn Munʿim – Magic Squares – Algebra – al-Karajī – Number Systems – Ibn al-Bannāʾ– Ikhwān al-ṣafāʾ– Ibn Khaldūn

COMPASS The origin of the magnetic compass is obscure. Medieval and Renaissance European compasses have been analyzed in detail, but until recently little work has been done on their history in non-Western cultures. Most modern accounts place the birth of the magnetic compass in ancient China, but its inventors are unknown, its development unclear, and historians are divided over whether its later appearance in the West was independent, or borrowed from China through overland or Arabic maritime intermediaries.

The Chinese compass seems to have been derived from a "south controlling spoon" (*si nan shao*) carved from lodestone (magnetite) and used in the early diviners' boards the Han Dynasty (202 BC–AD 220). Just as in ancient Greece, Chinese philosophers were first aware of magnetic attraction; the ability of the lodestone to pick up iron was mentioned in the *Lü Shi Chun Qiu* (Master Lu's Spring and Autumn Annals, third century BC), the *Lun Heng* (Discourses Weighed in the Balance, AD 83), and a host of other Chinese annals between the third century BC and the sixth century AD. The discovery of magnetic directivity (the tendency of magnets to point north and south) made the lodestone particularly important in geomancy and divination. Historian Joseph Needham argued that this was known publicly by the first century AD, and may have been a secret of court magicians as early as the second century BC.

Between the first and sixth centuries AD, Chinese scholars discovered that magnetic directivity could be induced in small iron needles by stroking them on lodestones. These needles, when floated in bowls of water (sometimes hidden inside carved wooden fish), would then guide Chinese navigators by pointing to the north and south; these early "wet compasses" were the first type. The second variety, a "dry compass" consisting of a needle mounted on a pivot (often concealed inside a wooden turtle), was developed much later (early twelfth century). Chinese encyclopedias such as the *Taiping Yulan* (Taiping Reign-Period Imperial Encyclopedia) also mention "dry" compasses made by suspending a magnetic needle on a silk thread.

Chinese physicists may also have been first in discovering magnetic declination. The fact that the compass does not point exactly to the geographic north, but rather to the magnetic north a few degrees away, was mentioned in philosopher Shen Gua's *Meng Qi Bi Than* (Dream Pool Essays) in 1088 AD. He says the compass needle "always inclines slightly to the east". Knowledge of this effect was extremely important for ocean navigation. Chinese knowledge of declination goes back to the late Tang period (eighth–ninth centuries AD).

The development of the magnetic compass in India is highly uncertain. According to some scholars, the compass is mentioned in fourth century AD Tamil nautical books; moreover, its early name of *maccha-yantra* (fish machine) suggests a possible Chinese origin. In its Indian form, the wet compass often consisted of a fish-shaped magnet, floated in a bowl filled with oil.

The earliest references to the magnetic compass in Europe appear in the grammatical and philosophical treatises *De Nominibus Utensilium* and *De Naturis Rerum* of English monk Alexander Neckam (1157–1217), French poet Guyot de Provins' (fl. 1184–1210) *La Bible*, and the *Historia Orientalis seu Hierosolymitana* of French preacher Jacques de Vitry (ca. 1165–1240). Based on these sources, historians have dated the European appearance of both dry and wet compasses to the middle of the twelfth century.

Among Arabic sources, the earliest descriptions of the magnetic compass occur in the thirteenth century. A Persian collection of stories, the *Jāmiʿal-Ḥikāyāt* (ca. 1232) of Muḥammad al-ʿAwfī says that sailors navigate using a piece of iron rubbed by a magnet, and the 1282 lapidary of Bailak al-Qabajaqi, the *Kanz al-Tijar*, mentions a wet compass seen in 1242. Some of these early treatises discuss the use of fish-shaped compass needles among Islamic navigators in the Indian Ocean, suggesting a possible borrowing from Indian sailors; but it is also possible that Arabs got the compass from Europeans (scholars arguing this hypothesis often point out the Arabic word for compass, *al-kunbas*, appears to come from Italian roots). Much more research needs to be done before this argument can be settled.

JULIAN A. SMITH

REFERENCES

Fleury, Paul Mottelay. *Bibliographic History of Electricity and Magnetism*. London: Charles Griffin, 1922.

Mitchell, A. Chrichton. "Chapters in the History of Terrestrial Magnetism." *Terrestrial Magnetism and Atmospheric Electricity* 37: 105–146, 1932; 42: 241–280, 1937, and 44: 77–80, 1939.

Needham, Joseph. *Science and Civilisation in China*, vol. 4, pts. 1 and 3. Cambridge: Cambridge University Press, 1962 and 1971.

Smith, Julian. "Precursors to Peregrinus." *Journal of Medieval History* 18: 21–74, 1992.

COMPUTATION: CHINESE COUNTING RODS The basic questions in the study of arithmetic in any civilization are: how did the people add, subtract, multiply, and divide, and what were the notations and media that were used to perform these operations.

The ancient Chinese used bamboo sticks or animal bones to count. The first five numbers were tallies in the following form:

𝍠 𝍡 𝍢 𝍣 𝍤

The representations for six, seven, eight, and nine were as follows:

𝍥 𝍦 𝍧 𝍨

The horizontal rod represented the quantity five.

It is easy to imagine how these rod numerals were handled for the addition or subtraction of small numbers such as two plus five or nine minus two. In the case of the addition of, for example, six and seven, the three vertical rods were grouped together and the two horizontal rods representing two fives were replaced by a single rod positioned horizontally to the left of the three vertical rods. The sum, thirteen, was expressed in the following form:

𝍩𝍢

In this manner, a place value notation was created; the tens' place was to the left of the units' place. The nine numerals that could occupy this place were similar in idea to the nine numerals shown above but had a significant difference: the vertical rods were turned into horizontal ones and the horizontal rods were turned into vertical ones. They were of the following form:

𝍩 𝍪 𝍫 𝍬 𝍭 𝍮 𝍯 𝍰 𝍱

The notion of place values was then extended to express larger numbers. The first set of nine numerals occupied positions whose place values were units, hundreds, ten thousands, and so forth; the second set of nine numerals occupied positions whose place values were tens, thousands, hundred thousands, and so on. For example, the rod numeral for thirty six thousand one hundred and eighty seven appeared as

𝍢𝍮𝍠𝍰𝍦

and the rod numeral for thirty thousand one hundred and eighty seven appeared as

𝍢 𝍠𝍰𝍦

which revealed that the thousands' place was empty.

Any number, however large, thus could be represented by the rod numeral system, which used a place value notation with ten as base. The rod numeral system was essentially a computing mechanism; the results of the calculation were recorded in the written numbers.

The rods were used for reckoning in the fifth century BC. According to *Qian Han shu* (Standard History of the Western Han Dynasty), they were cylindrical bamboo sticks, 0.1 *cun* (approximately 0.231 cm) in diameter and 6 *cun* long. They were carried in a hexagonal bundle made up of two hundred and seventy one pieces. Computation was performed on a flat surface such as a table or a mat. By the sixth century AD, the rods were shorter in length and square in cross-section. During the Tang dynasty (AD 618–907), it was quite common for officials, astronomers, engineers, mathematicians, and others to carry their bundles of rods wherever they went. Besides animal bones and wood, the rods could be carved from horn, ivory, or jade.

Sun Zi suanjing (The Mathematical Classic of Sun Zi) written around AD 400 is the earliest existing work to have a description of the rod numerals. It also gives detailed descriptions of how multiplication and division were performed. The step-by-step procedures are the same as the earliest known methods of multiplication and division using Hindu–Arabic numerals, which were described by Muḥammad ibn Mūsā al-Khwārizmī in his book on arithmetic and by Abū al-Ḥasan Aḥmad ibn Ibrāhīm al-Uqlīdisī in *Kitāb al-Fuṣul fī al-Ḥisāb al-Hindī*. This fact together with other evidence supports the thesis that the Hindu–Arabic numeral system has its origins in the Chinese rod numeral system.

Just as the Hindu–Arabic numeral system provided the mainstay and impetus for the growth and development of our arithmetic and algebra, the rod numeral system provided the same support for the development of mathematics in ancient and medieval China.

The process of division with rod numerals resulted in an important notation for the expression of a fraction. For example, the notation for four-fifths is

𝍣
𝍤

This is the same as the notation for the common fraction in Hindu–Arabic numerals which also originated from the earliest method of division with numerals. The use of this notation enabled an extensive study on operations with fractions — a subject with which other ancient civilizations had difficulties that were frequently insurmountable.

With the concept of the rod numeral system ingrained in their minds and using the rods as the medium of expression, the ancient Chinese developed arithmetic and algebra. The problems in *Jiu zhang suan shu* (Nine Chapters on the Mathematical Art) (ca. AD 100) show that they were able to solve problems involving fractions, the Rule of Three, proportion, areas and volumes, square roots and cube roots, the Rule of False Position, sets of simultaneous linear equations, and negative numbers.

For numerous centuries the rod numeral system played a dual role: in calculation and in the development of mathematics. It aided the progress of the Chinese civilization till the Ming dynasty (1368–1644) when it was gradually replaced by the abacus as an instrument for computation. Mathematics flourished from the firm foundation depicted in *Jiu zhang suan shu* and reached its peak in the thirteenth century. When the rod numerals became outmoded and fell into disuse, mathematics also underwent its period of decline and transition.

LAM LAY YONG

REFERENCES

Crossley, John N. and Alan S. Henry. "Thus Spake al-Khwārizimī: A Translation of the Text of Cambridge University Library Ms. Ii.vi.5." *Historia Mathematica* 17:103–131, 1990.

Lam Lay Yong. "Linkages: Exploring the Similarities Between the Chinese Rod Numeral System and our Numeral System." *Archive for History of Exact Sciences* 37: 365–392, 1987.

Lam Lay Yong. "A Chinese Genesis: Rewriting the History of our Numeral System." *Archive for History of Exact Sciences* 38: 101–108, 1988.

Lam Lay Yong and Ang Tian Se. *Fleeting Footsteps. Tracing the Conception of Arithmetic and Algebra in Ancient China*. Singapore: World Scientific, 1992.

Li Yan and Du Shiran. *Chinese Mathematics. A Concise History.* Trans. by J.N. Crossley and A.W.C. Lun. Oxford: Clarendon Press, 1987.

Mei Rongzhao. "The Decimal Place-Value Numeration and the Rod and Bead Arithmetics." In *Ancient China's Technology and Science.* Beijing: Foreign Languages Press, 1983, pp. 57–65.

Needham, Joseph. *Science and Civilisation in China*. Vol. 3: "Mathematics and the Sciences of the Heavens and the Earth." Cambridge: Cambridge University Press, 1959.

Qian Baocong, ed. *Suan jing shi shu* (Ten Mathematical Classics). Beijing: Zhong hua shu ju, 1963.

Saidan, A.S. *The Arithmetic of Al-Uqlīdisī*. Dordrecht: D. Reidel, 1978.

See also: Sun Zi – al-Khwārizmī – al-Uqlidīsī – *Jiuzhang Suanshu* – Abacus – Algebra in China

CONICS The theory of conic sections was proposed in Greece around 350 BC. The most comprehensive ancient treatise on the subject was the *Conics* of Apollonius of Perga (fl. ca. 200 BC). This treatise consisted of eight books, of which only the first four have come down to us in Greek. Apollonius considered an arbitrary cone with a circular base, whose apex is an arbitrary point not in the plane of the base. He obtained the conic sections as intersections of this cone by a plane not through the apex. Apollonius distinguished three types of conic sections: the ellipse, the parabola, and the hyperbola.

The story of the translation of the *Conics* into Arabic is interesting, because it shows the difficulty of translating a technical mathematical text. The first Arabic geometers with an interest in the *Conics* were the three Banū Mūsā ("sons of Mūsā", fl. ca. AD 830). At first they possessed only one poor Greek manuscript, containing the first seven Books, and they were unable to understand the very technical contents. Then one of the brothers, al-Ḥasan, decided to develop the theory of the ellipse as the intersection of a cylinder and a plane. He believed that the cylinder was easier than the cone, and he hoped that the theory might be more accessible this way. al-Ḥasan wrote a book on his discoveries, entitled *Al-shakl al-mudawwar al-mustaṭīl* (The Rounded Elongated Figure). This work is lost.

After the death of al-Ḥasan, his brother Aḥmad obtained a second manuscript of the first four Books of the *Conics* in Syria, together with the commentary of Eutocius of Ascalon (ca. AD 500). This manuscript was much better than the other one, and with the help of the book of his deceased brother, Aḥmad was able to make sense of the contents of the Greek text. He then inserted cross-references in the Greek in order to make the text more comprehensible. Now that he understood the first four Books, Aḥmad was able to make sense of Books V–VII on the basis of the other manuscript. He then had the Books translated: Books I–IV by Hilāl al-Ḥimṣī, Books V–VII by Thābit ibn Qurra. Aḥmad supervised and corrected the translation. Thus Aḥmad ibn Mūsā and Thābit rescued one of the most interesting mathematical works of antiquity, namely Books V–VII of the *Conics*. Book VIII of the *Conics* seems to have been lost altogether. It was reconstructed around 1000 by Ibn al-Haytham, and around 1700 by Edmund Halley.

By the ninth century, geometers in the Arabic tradition had already written treatises on conic sections. Thābit ibn Qurra authored a work on plane sections of a cylinder, which is extant (and which may well resemble the above mentioned lost treatise on the same subject by al-Ḥasan ibn Mūsā). A popular subject was the determination of the surface areas of segments of conic sections and the volumes of solids obtained by revolving a segment of a conic section about one line (such as the axis or the base). The Arabic geometers knew that Archimedes had solved this problem for the parabola, but they did not know his method of solution. The treatises on this subject by Thābit ibn Qurra, and by the tenth-century mathematicians al-Qūhi and Ibn al-Haytham, are among the highlights of Arabic mathematics.

The Arabic geometers applied conic sections in the solution of a group of problems (trisection of the angle, construction of two mean proportionals, construction of the regular heptagon, etc.) which could not be solved by means of ruler and compass. In the tenth century AD, most geometers in Iraq and Iran considered straight lines, circles, and conic sections the only legitimate means of construction in geometry. The reason was that these objects belonged to fixed (immovable) geometry (*al-handasa al-thābita*), whereas all other curves and instruments were believed to be created by motions and instruments, which did not belong to mathematics but to mechanics.

Conics were also used in the solution of the "Problem of Alhazen" (named after Al-Ḥasan ibn al-Haytham, mentioned above). In this problem, the positions of the eye, the object, and a convex or concave spherical, cylindrical, or conical mirror are given. One asks for the points of the mirror at which the object is reflected to the eye. Ibn al-Haytham constructed these points of reflection, and he showed that the object is reflected at no more than four points on the mirror. His solution is very intelligent but also extremely complicated, and several seventeenth-century European mathematicians complained about its difficulty. Earlier, the solution had been simplified by al-Muʾtaman ibn Hūd, the king of Zaragoza (d. 1085), but this simplification was unknown in Christian Europe.

In the tenth century, al-ʿAlāʾ ibn Sahl studied reflection in parabolic lenses (which had been studied before) and refraction in hyperbolic lenses (this topic was new). He showed that rays parallel to the axis are refracted to one of the foci of the hyperbola. The reasoning implies knowledge of what we now call Snell's law of refraction, but it is unclear how al-ʿAlā discovered this law.

An important field of application of conic sections was the theory of cubic equations. A cubic equation is an equation of the form $x^3 + ax^2 + bx + c = 0$, for a, b, c given numbers. The Arabic geometers were unable to solve this equation algebraically, that is to say, by expressing the root x in terms of a, b, and c. However, they learned to solve it geometrically. The history of the theory is connected with a certain division of a line segment mentioned by Archimedes in Proposition 4 of Book II of *On the Sphere and Cylinder*. Archimedes used the divided line segment but he did not show how it could be divided. In the ninth century, al-Māhānī showed that the division of the line segment, which Archimedes uses, depends on a cubic equation of the form $x^3 + c = ax^2$ (a and c are positive). However, he was unable to solve this equation. Abū Jaʿfar al-Khāzin (tenth century) knew the work of al-Māhānī and also a commentary by Eutocius of Ascalon (ca. AD 500), who shows how the line segment can be divided by means of conic sections. Apparently, Abū Jaʿfar realized that one could use Eutocius' construction using conic sections to find the root x of the equation $x^3 + c = ax^2$ to which al-Māhānī had reduced the problem. Soon other Arabic geometers discovered that conic sections could also be used to solve other "types" of cubic equations. (Because they only worked with positive coefficents, they distinguished different types of equations, e.g. $x^3 + c = ax^2$ is of a type different from $x^3 + ax^2 = c$, etc.) In the eleventh century, ʿUmar al-Khayyām gave in his *Algebra* a geometric solution (by means of conic sections) of all types of cubic equations. His theory was considerably improved by Sharaf al-Dīn al-Ṭūsī around 1200.

The study of conic sections was of theoretical interest only. The Arabic mathematicians developed an instrument, the perfect compass, with which conic sections could be drawn, but this does not seem to have been much used and no examples have survived.

JAN P. HOGENDIJK

REFERENCES

Hogendijk, Jan P. *Ibn al-Haytham's Completion of the Conics*. New York: Springer, 1984.

Hogendijk, Jan P. "Greek and Arabic Constructions of the Regular Heptagon." *Archive for History of Exact Sciences* 33: 197–330, 1984.

Knorr, W. *Textual Studies in Ancient and Medieval Geometry*. Boston: Birkhäuser, 1989.

Rashed, R. "A Pioneer in Anaclastics. Ibn Sahl on Burning Mirrors and Lenses." *Isis* 81: 464–491, 1990.

Toomer, Gerald. *Apollonius, Conics, Books V to VII. The Arabic translation of the lost Greek original in the version of the Banū Mūsā*. New York: Springer, 1990.

See also: al-Muʾtaman ibn Hūd – Ibn al-Haytham – Thābit ibn Qurra – Banū Mūsā – al-Qūhi – ʿUmar al-Khayyām – Sharaf al-Dīn al-Ṭūsī – al-Māhānī – Ibn Sahl

CONSTRUCTION TECHNIQUES IN AFRICA Architecture is a deliberate attempt to construct through the collective will. African architecture is therefore a product of the organization of society involving the effective use of materials to express the people's building desires. In this way, it forms an integral part of their history, culture, and of their responses and adaptations to the African environment.

The cultural diversity of African peoples, which is further reinforced by ecological differences, fostered diversity in the available technology and architectural styles. Such a wide diversity only allows for the broadest of generalizations.

African architecture is of great antiquity, as we know from the many places where architectural remains exist. The Great Pyramids of Egypt, the minarets and stelae of Ethiopia, the mud city walls of Hausaland, Yoruba, and Edo, the palaces and the mosques and terraces of eastern Africa, and the Great Zimbabwe of central Africa, are all evidence of such architectural remains. Africans applied technologies in their respective environments relying on available materials. For example, people in rocky areas exploited the local stones and built either dry or mud structures. Those in riverine areas raised their dwellings on stilts, while those in dry environments built bentwood frame structures covered with thatch or leaves or used sun-dried brick walls with reinforced mud arches to support a flat or domed mud roof.

A study of the different structures of Africa helps to reveal the level at which Africans were able to master their environment and how they effectively mobilized and utilized labor. It also helps show the levels of political centralization and standards of living attained by the various states. Unfortunately, all this was negatively affected by the slave trade, colonialism, and the activities of the new elite.

Among the architectural forms of Africa are: military architecture, funerary and royal architecture, and religious and settlement architecture.

Settlement Architecture

Settlements in Africa were in most cases systematically planned around a central market, palace, and temple complex with structures showing the social standing of their owners. Modest houses were built of mud brick made from clay and straw. Others were storied houses built from stone blocks afforded only by the rich. For example, JenneJeno, a city by the Niger's inland delta in what is today Mali, built in the third century B.C, had by 800 A.D developed into a full-fledged city with diverse economic activities and sociopolitical stratification. Most people employed wood in rafters, floor supports, stair treads, cupboards, doorways, lintels, and windows. This, along with the thatched roof, increased the risk of fire in such buildings. The vernacular houses had in most cases a courtyard for domestic activities like animal rearing, where pens, feeding troughs, and mangers were kept. In some houses, storage jars for water or grain were half buried in the courtyard floor. For example, the Kalenjin people of Kenya built sunken cattle kraals enclosed with banked earth and fences near their houses, while in Zimbabwe, the Nilotes lived in walled villages. In Tanzania, the Iraqw constructed the sunken flat-roofed dwelling known as *tembe*, a form of habitation for defense against enemies. *Tembe* is a pit covered with a flat roof of mud and dung rarely above the ground level.

Religious Architecture

Religion encouraged the development of architecture. Towns and temples were built for specific gods. Each god had his own domain which was jealously protected. The king, the heir and successor to the god, had to maintain order to avoid the wrath of the gods, and ensure agricultural fertility and the happiness of society. Architects had to make the temple a scale model of the universe. All portions of the temple building reflected the aspirations of the god. Such temples were separated by high brick perimeter walls from other buildings regarded as impure. The city of Oxyrhynchus on the banks of a tributary of the Nile River in Egypt was specifically devoted to some Egyptian gods, each with his own temple. One of them, the temple of Sarapis, performed religious, economic, and social functions. Each of these functions had its cluster of buildings with workshops and living units connected with the cult. There was a center of banking activity, and a public market for traders who paid a tax for selling there. There was also accommodation for visitors seeking oracles or interpretations of dreams or for the sick. In Zimbabwe, skilled stonemasonry was employed for temple construction and defense works. These were built of plain split blocks, untrimmed, varying in thickness and built without mortar laid on rocky foundations. Some of the temples date back to 1100 AD. This stone masonry was also employed in royal and funerary architecture.

Royal and Funerary Architecture

Architecture was used as an emblem of power and prestige by the king through the control of the mason's guild. The king's residence was built and renovated in accordance with his power and wealth. For example, murals of animals connected with the divinity of Yoruba Obas are often depicted on palaces. Elephants, ostriches, and lions signifying majesty, and monkeys and snakes symbolizing wisdom, are colored by using local materials such as charcoal, indigo, clay, and snail shells. While the structure and materials used in the palace indicated power and status, funerary architecture was also a symbol of power, representing the final abode for the dead. These included the Egyptian pyramids and stelae of Aksum. Pyramids, built by Egyptian Pharaohs, are huge structures indicating the power and desire of the builder to leave behind a big palace and to make his remains totally secure. They are therefore the most indestructible of architectural structures, built of limestone blocks faced with granite which was also used for flooring. The step pyramid at Sakkara built ca. 2650 BC during the Third Dynasty was constructed completely in stone, where small limestone blocks like imitation sun-dried bricks were used. The imbedded columns and ceiling joists were stone and were direct copies of bundles of plants used in earlier construction, where papyrus stems were tied together to serve as pillars. The Great Pyramid of Cheops contains two and half million blocks of stone and weighs 4,883,000 tons in which every necessity of life, ranging from food to vehicles, weapons and cosmetics, was provided inside the tomb. This is the first evidence in recorded history where hewn stone was first used in coursed work. Columns were also an architectural innovation from Egypt, where the environment provided reeds and papyrus which enabled builders to cut capitals of columns into the shapes of lotus flowers. Egyptians also used the vault by the Second Dynasty, using bricks which were replaced with stone by the Sixth Dynasty.

The giant monolithic stelae of the Ethiopians of Aksum are also original and beautiful. Their method of construction is the dry wall type in which blocks of stone were accurately dressed without the use of mortar. These had doors, windows, and the butt-end of beams all carved in hard stone. These 33 meter high stelae were at times dwellings and at others tombs. This technique was later developed into the rock-cut churches of Lalibela of AD 1000–1270, succeeded by the Portuguese styles of the sixteenth century at Gondar.

Military Architecture

The need for security as a result of internecine wars and the slave raids which intensified in the eighteenth and nineteenth centuries led to the construction of imposing defensive structures. Forts and city walls were built of mud or stones and at times topped with thorn thickets with cunningly concealed entrances to stop enfilade fire. The Kano city wall, which by the nineteenth century measured 30 feet high, 40 feet thick at the base, and 16 kilometers in circumference, is one such defensive mechanism. Only one third of this walled area was built up. The rest was used for farming and cattle rearing to provide for their needs in time of siege and for temporary dwellings for rural refugees and immigrants especially in the dry season. The gate, the weakest point of the defense, received the greatest attention by convoluting it to force attackers to adopt a zig-zag course, thereby exposing them to fire from the defenders. Gates also had riveted iron doors. The balcony or battlement was crenellated, giving defenders not only security but also a height advantage over their assailants. The battlement supported enough defenders and enabled support teams to bring weapons, pass information, and replace the wounded. Other structures like worship centers and ordinary houses were also constructed with this security consciousness in mind, especially by using non-combustible materials.

The minarets constructed on mosques served as lookout posts for sentries on duty to provide advanced warning of assaults. This outpost of defense was also a safe place for retreat. The minaret of Bamara Mosque in the Malindi quarters of Zanzibar for example, was built in the form of a tower with staircases leading to the roof from where the call to prayer was made. It was built of stone set on a square base, while the Gobirau minaret of Katsina was built of sun-dried bricks. The Mihrab, the prayer niche, received the greatest attention in terms of decoration. It had a fluted round classical arch. Then there was the trefoil or poly-lobed arch designed with a groove in the middle. The Mombasa minaret-bearing Mosques are dated to the seventeenth century. The construction of circular houses was also to ensure security. Each compound was made a self-contained spatial unity where entry into and departure from it was regulated through a single gate. The circular shape gave centrality inside. For example, among the Atyap (Kataf) of Nigeria, each living structure was not only circular in shape but was also provided with a *Lebut* (battlement) reached through earthen steps. The *Lebut* had openings for viewing assailants and dropping arrows and firing spears. The *Wa'ad* or store room protected grains against fire and termites, while the sleeping apartment was provided with mud beds and a fireplace under the bed to reduce dampness. Others who could not build such impregnable defensive structures resorted to naturally defensive sites like forests or inaccessible hills as settlements.

It is often held that advanced architecture in Africa is of Hamitic origin which came together with iron smelting from other civilizations from across the Sahara. This tendency to attribute to alien influences whatever creativeness belongs to the African led Paul Oliver to also assert that African building types do not merit the term architecture since, to him, the materials used in their construction are of short durability as they do not include stone, marble, metal, or brick. He therefore refers to them as Shelter. R.L.B. Maiden, in supporting this view in an effort to disprove African architectural ability, added that the exquisitely refined workmanship of the Arabs in stone, wood, metals, and mud never appear to have become a legacy of the peoples of the Sudan. He added that the arts and architectural forms of the peoples of Western and Central Sudan have seldom developed to any great degree beyond crude and primitive designs. The primitive factor attributed to Africa and her heritage has been allowed to prejudice objective evaluation of African creative works. African architecture developed from a conscious effort to create functional as well as psychological space and to combine these in an aesthetically satisfying building form. It is therefore mainly in Africa that mud, stone, wood, and grass have been developed into the fullest potential both in structural terms and as a means of aesthetic expression. However, these have hardly received the attention of scholars, because their experience has been in a culture that recognizes concrete and steel as the major authentic building materials; but to fully understand the relevance of African architecture, it is not just enough to look at the materials, walls, and roofs that enclose the space. One must go beyond mud, stone, and wood and also look at the meanings that wall decorations convey to observers.

Materials and Techniques of Construction

The use of mud as a basic building material was common in Africa. Mud is easy to shape though poor in stability. It can hardly withstand the lateral thrust of a battlement and it is easily saturated during rainfall. Thus, mud walls, if not carefully built, easily crack and buckle. Africans stabilized mud buildings by constructing thicker walls and battering them towards the top, thereby diverting the weight downwards. It was also treated by molding it into bricks which were dried in the sun or fired and used in building. Mud was molded into egg-shape bricks by mixing it with clay and grass. It was trodden with bare feet and molded into bricks of rectangular or circular shapes of uniform size for ease of handling and for a uniform structure, before baking in fire or drying under the sun. In this way it could withstand fire, termites, mold, or rainfall. Walls were also made waterproof through smoothing by using locust bean pericarp soaked in water, boiled into a brown paste, and applied to the walls.

Patterns of sweeping arm movements on mud plaster give a characteristic identity to African mud walls. The mud vaults crossing each other or meeting at a terminal create a loftiness and an ecological unity comparable with that created by the Gothic arches. The structural azara (deleb palm) composition that supports in part the heavy mud roof provides a beautiful soffit, and when rendered, takes on an interplay of colors. These decorations have meanings rooted in religious beliefs and practices. For example, Ife art and the sculptured supports in the roof eaves along the verandah enclosing the Yoruba compound are images of revered personalities or gods, religious or cultural symbols, or even pictorial representations of a human experience. They give form and character to the building and create a different dimension to the language of its architecture.

Another form of aesthetics on the walls is painting. Painting helps promote psychological feeling in the built environment and increases the architectural scale and proportion in that space. Aesthetics of the roof in African traditional architecture embodied an architectural idea. For example, the development of Hausa vaults and domes was instigated and encouraged by the desire to simulate the formal and visual imagery of Islam in northeastern Africa. The structural principle of the Hausa vault and dome is completely different from the North African Roman derived stone dome. While the latter depends on the strength of stone in compression, the former incorporates in nascent form the same structural principles that govern pre-stressed concrete design.

The Yoruba roof is in the form of a saddle where one side slants outwards from the courtyard and the other inward. This is associated with Yoruba vegetal environment in roofing and thatching materials. The Edo impluvium (roof) surrounding a closed courtyard on all four sides, thus letting in air and light without the need of windows, reflects the indoor–outdoor nature of African social life and characterizes the architecture of hot countries.

Africans also explored and exploited stonemasonry. Huge blocks of granite, limestone, basalt, and diorite were transformed from raw material into well shaped, polished masonry for various architectural designs. Each stone type had its quarrying technique. For example, galleries were hollowed out of limestone, where blocks of fine stone were extracted and used to construct the Great Pyramids faced with granite blocks. Sandstone deposits like those of El-Kob of Upper Egypt and of Nubia were mined by open face techniques. For hard stone, quarries first had to cut out a groove and make deep notches into which they inserted wooden wedges which they soaked. The resultant swelling of the wood helped in splitting the block along the groove.

Wooden mallets and copper chisels were used to work stone like limestone and sandstone, while the pick, chisel, and hard stone hammer were used in working metamorphic rocks like granite, gneiss, diorite, and basalt. Iron metal tools like the pick and adz were used to break boulders.

Roofs were reinforced by raffia, while deleb palm was used at panels. Bamboo sticks and grass used for the ceiling were arranged in a conical shape to give it an arch shape or in straight lines for the balcony. Joists were let into the masonry to strengthen it, while beams supported floors of rooms in upper stories or roofs. Basements of large buildings were made solid by laying large blocks of hewn stones at the corners. Tree trunks were firmly fixed at the two ends of each entrance of a door/gate. One side of a bamboo door was permanently tied to one end of the trunk using ropes of raffia frond. On closing the door, the other end of the bamboo was tied to the trunk. The door was further supported by using large props of wood or boulders slopingly placed against the door from inside. Wooden or riveted iron doors were also used.

The consideration given to proportion, balance, and gravity of African structures was to ensure their stability. Integrating architects with mathematicians, artists, stone workers, wood workers, priests, and other related specialists brought refinements into building engineering. The guild of building specialists checked labor quality and productivity, and fostered the spirit of competition and excellence. Labor incentives through the award of titles also increased output. This grew out of the realization of the importance of labor. Digging the earth, mixing it, carrying water, molding bricks, breaking and transporting boulders, laying the foundation, building the superstructure and thatching the roof, were all done by hand. This required a well organized society. Thus, because Africans did not use animal, water, or mechanical devices except the lever and wooden cross bar, they had to employ much human strength. It was therefore not easy to get the cooperation of all the operatives in the building construction. Building is a hazardous task, and force had to be used at times to recruit labor.

In Egypt for example, the Memphis troops had to participate in operations of economic relevance and in the great building operations. Teams of young literate recruits who served as the king's bodyguards supervised the breaking and transportation of stones for the pyramids and expeditions to mines and quarries. Paramilitary corps, called Sementi, prospected and exploited the gold mines. Labor was divided into porters and quarrymen. It is, however, wrong to assume, as others have done, that African massive structures were the handwork of ceaseless slave labor overseen by cane wielding supervisors (Larymore, 1908). Though force was sometimes used, its continuous application could lead to labor instability through desertion and migration to distant areas as a form of protest. What was required to mobilize, retain, and utilize labor was therefore not force but political centralization and organization. This is why non-centralized and weak states could not build and maintain sophisticated structures. Architectural excellence required effective societal organization, a high level of buoyancy, and mobilization of some of its labor force to the building sector.

Colonialism changed both indigenous architecture and Africans' attitudes towards their own culture. Imported products and practices seemed to have more value than traditional ones. This encouraged the erroneous notion that buildings of traditional materials or locally available resources are substandard. Part of reclaiming African heritage includes appreciating the complexity and skill of African architecture and construction.

BALA ACHI

REFERENCES

Bivar, A.H.D. and P.L Shinnie. "Old Kanuri Capital." *Journal of African History* 111(1): 4–6, 1962.

Davies, N.M. *Ancient Egyptian Paintings*. 3 vols. Chicago: University of Chicago Press, 1936.

Denyer, Susan. *African Traditional Architecture*. London: Heineman Educational Books, 1978.

Dodo, B. *The Traditional Architecture of Kataf Land*. B.Sc. Thesis, A.B.U Zaria, 1988.

Edwards, I.E. *The Pyramids of Egypt*. London: Penguin, 1970.

Foyle, A.M. *The Development of Architecture in West Africa*. 2 vols. Ph.D. Dissertation, University of London, 1959.

Freeman-Grenville, P. "Some Preliminary Observations on Medieval Mosques Near Dar es Salaam." *Tanzania Notes and Records* 36: 64–65, 1954.

Garlake, P.S. *The Early Islamic Architecture of the East African Coast*. Nairobi: Oxford University Press, 1966.

Larymore, Constance Belcher. *Resident's Wife in Nigeria*. New York: E.P. Dutton, 1908.

McIntosh, Susan and Roderick. "Finding West Africa's Oldest City." *National Geographic* 162 (3): 396–418, 1982.

Maiden, R.L.B. *Historical Sketches*. Zaria, Nigeria: Gaskiya Corporation, 1965.

Morgan, P. "The Walls of Kano City." *Journal of the Royal Institute of British Architects* 1929, p. 402.

Moughtin, J. "The Friday Mosque, Zaria City." *Savanna* 1(1): 143–166, 1972.

Oliver, Paul, ed. *Shelter in Africa*. New York: Praeger, 1977.

Williams, P. "An Outline History of Tropical African Art." In *Africa in the Nineteenth and Twentieth Centuries*. Ed. J.C Anene and G.N Brown. Ibadan: University Press, 1966, pp. 59–91.

ḤASDAI CRESCAS Ḥasdai ben Abraham Crescas (1340–1412) was one of the most influential personalities of Spanish jewry in the end of the fourteenth century. As chief rabbi of the Aragonian Jewish communities during the persecutions of 1391, he was responsible for their reconstruction. Although his literary work was not very large, it constitutes an important contribution to the history of philosophical and scientific ideas, mainly as a critique of medieval Aristotelianism.

Under the influence of Maimonides' *Guide for the Perplexed* (translated from Arabic into Hebrew in the beginning of the thirteenth century), Aristotelianism had spread in Jewish circles, raising many controversies. Besides those who opposed the study of philosophy only on the ground that it undermined religious beliefs, others invoked scientific and philosophic arguments to give support to their own critique of philosophy. Crescas is certainly the most important of the latter critics.

Apart from a short composition, *Biṭṭul ʿIqqarey ha-Noṣrim* (Refutation of the Principles of Christianity), Crescas' main work is his *Or Adonay* (Light of the Lord) completed in the last years of his life. In the first part of this book the fundamental concepts of Aristotle's physics and metaphysics, as adapted by Jewish philosophers, were analysed and criticized.

Crescas' main purpose was to establish that the existence of God, His unity and His incorporeality could not be fully proved by demonstrative reasoning, as Maimonides had argued: "it is impossible to arrive at a perfect understanding of these principles except by way of prophecy, (...), and yet it will be shown that reason agrees with the teachings thus arrived at". In this sense, one could say that Crescas did not criticize philosophy as such, but rather the intellectualization of religion by Aristotelian scholastics. Actually, in many respects, Crescas opposed to Aristotelian philosophical notions other philosophical notions, more relevant — according to him — to the purposes of philosophy itself.

The main importance of Crescas' contribution lies in his

critique of Aristotle's concept of the infinite, and its applications to the notions of space, number and time.

There are three main conclusions of Crescas' analysis which are relevant to history of science. The first is in opposition to Aristotle who had argued that the place of a body (i.e. space) was the inner limit of the enclosing body, Crescas asserted that space was prior to bodies and that the place of a body was equivalent to its extension. Thus the existence of a void was not self-contradictory. Secondly, again in opposition to Aristotle who had argued that the universe was finite, because nothing could be infinite, neither a body nor a substance, Crescas claimed that there was no absolute limit to space. Space was an infinite receptacle, which could contain many other worlds. By analogy, space could be thought of as God's omnipresence. Finally, Aristotle and aristotelians argued that an infinite quantity was impossible, because it would lead to the absurd conclusion of one infinite quantity being greater than another. Crescas' reply was of importance since it could be viewed as anticipating some modern concepts: the relations "equal to", "greater than" and "smaller than" do not apply to the infinite, in the same manner that they apply to finite quantities. Thus, the notion of an actual infinity is not self-contradictory.

TONY LÉVY

REFERENCES

Copenhaver, Brian. "Jewish Theologies of Space in the Scientific Revolution: Henri More, Joseph Raphson, Isaac Newton and their Predecessors." *Annals of Science* 37: 489–548, 1980.

Lévy, Tony. "Monde juif et science gréco-arabe, avant la Renaissance." In *Figures de l'infini. Les mathématiques au miroir des cultures*. Paris: Éditions du Seuil, 1987.

Or Adonay. Hebrew text in *Sefer Or ha-Shem* by Rabbi Hasdai Crescas. 1st edition, Ferrara, 1555. Reprint, with an introduction in Hebrew by E. Shweid. Jerusalem: Makor Publishing, n.d.

Pines, Shlomo. "Scholasticism after Thomas Aquinas and the Teachings of Hasdai Crescas and his Predecessors." Jerusalem: *Proceedings of the Israel Academy of Sciences and Humanities* 1(10): 1–101, 1967.

Wolfson, Harry Austryn. *Crescas' Critique of Aristotle. Problems of Aristotle's Physics in Jewish and Arabic Philosophy*. Cambridge, Massachusetts: Harvard University Press, 1929.

CROPS IN PRE-COLUMBIAN AGRICULTURE Before the European invasion of the Americas, the native peoples harvested a large variety of crops under many types of agricultural and gathering regimes.

Dense concentrations of people had existed for centuries in Mesoamerica and Andean America, and had evolved complex societies of agricultural peasantry and stratified urban elites. These peoples, in fact most Native Americans, had selected and adopted a miracle grass, maize (*Zea mays*), which reached yields of over one hundred to one in regions of good soils, intensive irrigation, and monoculture such as the *chinampas* or "floating gardens" of the central valley of Mexico. These high yields led to maize dominance in many diets, and the societies which evolved around this cereal may have resembled those of paddy rice China, which had highly intensified, quasi-gardening systems, rather than those of Europe or Africa, which had lower yielding but more diversified systems associated with wheat, barley, millet, and sorghum.

Maize was often associated, especially in Mesoamerica, with various varieties of American beans, especially *frijoles* of the genus *Phaseolus*, which climbed the maize stalks; with squashes (*Cucurbitae*) and their relatives; and with chilli peppers (*Capsicum*), often planted between the maize stalks. The insufficiencies of this basic diet led to heavy foraging and to the harvesting of unoccupied areas and forests. In fact, even in settled societies there appears to have been a harvesting continuum from irrigated maize gardens to collecting, often in forests which were themselves results of centuries of human selection and propagation. Thus, the modern student finds, paradoxically, intensive monoculture, plus slash and burn, plus areas of great biodiversity, all used by the same population.

Tubers of many kinds were widely cultivated but were somewhat more geographically restricted than the maize, beans, squash, and chili complex. Yams (genus *Dioscorea*), cassavas, or maniocs (*Euphorbiacae*), some of highland origin, were intensely planted, and eaten after conversion to flours and breads, in tropical zones such as the Caribbean and the Amazon Basin. In many such regions, cassava was the staple diet. Musae (bananas and plantains) also appear to have been or become a staple in parts of the tropical lowlands, but scholars dispute their origin, some claiming that they are not native to the Americas.

In the Andes, tubers were also of great importance. Varieties of potatoes (*Solanum*) have been cultivated there for about eight thousand years, selected by humans for the many microenvironments of the mountain tropics. Andean farmers sometimes plant as many as two hundred varieties in a single field, although most are being phased out in favor of the modern selected potatoes. Several other tubers were cultivated in highland America, such as achira (*Canna edulis*); arracacha (*Arracacia xanthorrhiza*); jicama (*Pachyrhizus erosus*); machua (*Tropaeolum tuberosum*); ova (*Oxalis tuberosa*); and ulloco (*Ullucus tuberosus*), still a staple in parts of the Andes. This area also produced tubers and grains which could be grown at great altitudes in poor climatic and soil conditions. Maca (*Lepidium meyenii*), a tuber, can be grown just below the snowline, as can kaniwa (*Chenopodium pallidicanle*), and the better known quinoa (*Chenopodium quinoa*), both cereal grains rich in proteins and other nutrients.

Several other crops in America were traded widely before the European arrival. Cacao (*Theobroma cacao*), probably of Amazon origin, was known in the regions today called coastal Ecuador and Venezuela, but it was in Mesoamerica that cacao reached its greatest prominence. Planted along the Pacific coast from Colima to El Salvador, and in parts of Honduras, Tabasco, and Veracruz, cacao was important in trade, tribute, and ritual. In some parts of the region it was used as coinage, and there are also reports that in some states its consumption was limited to the aristocracy. Tobacco (*Nicotiana tabacum*) was somewhat different. Although traded in many regions of America, its use may have been limited to ceremonial occasions.

Mesoamerica also produced vegetable and insect dyes, especially blues from indigo (*Indigofera tinctoria*) and reds from cochineal (*Opuntia cochinellifera*). Both items were traded, paid over as tribute, and used for ceremonial purposes. American cotton and thread from cactus were parts of textile trades and tribute.

European invasion and its consequences had widespread effects on American crops and their diffusion. The catastrophic fall of the native population and the introduction of large and medium-sized domesticated animals, such as horses, cattle, sheep, goats, and pigs, caused abandonment of fields, destruction of crops, and the disappearance of many forms of ground cover. European need for wood for construction, shipbuilding, and fuel led to the cutting down of forests and the crops harvested in them. Colonial demands for uniformity in taxation caused peasants to give up the growing of marginal cultigens, or ones unfamiliar, distasteful, or useless to the dominant culture. The introduction of animals, and of plants such as wheat, citrus trees, and sugarcane, all cultural preferences of Europe, took up agricultural lands, and drove out native species.

Many scholars have stressed the more positive side of these intrusions and exchanges. Introduced crops such as wheat and animals such as cattle transformed the American diet, and the vast quantities of wheat grown on the prairies of the United States and Canada, and on the Pampas of Argentina, have fed large numbers of people in many parts of the world.

The more positive view also emphasizes the diffusion of American plants throughout the world, while admitting that the impact of some of these exports has been noxious (tobacco) or ambivalent. The potato, for example, the introduction of which to northwestern Europe met some resistance at first, was blamed by some for the Irish famine. Its high yields and ease of cultivation led to a monoculture, rapid population growth, then famine and mass emigration — ironically to the Americas — when the potato crop failed. The advantages brought, nevertheless, by this productive, nutritious, human selection to the everyday diet of peoples living in the temperate zones are, of course, immense.

Some American crops, while having a generally beneficial impact when exported, had their "booms" modified or cut short by new technologies or crops. The two great American dyes, indigo and cochineal, after enjoying a century or two of flourishing trade, were destroyed by the invention of aniline dyes in the mid-nineteenth century. Sugarcane, not originally an American plant, nevertheless achieved great importance only when exported to the Americas. Plantation sugar, always prized by Europe but historically in short supply because of poor access to tropical areas and labor, boomed when northeast Brazil and the colonial islands of the Caribbean combined tropical climates, African slavery, intensified production, and reasonably close markets, and began to export huge shipments. Sugarcane transformed much of the world's diet, and was an important fuel for the industrial and commercial revolutions of the nineteenth century. Although it continued to be of importance its monopoly was destroyed by the rise of beet sugar.

There can be little doubt, however, about the continuing beneficial impact of the diffusion beyond the Americas of many of the native crops. Maize, with rice one of the world's two "miracle" cereals as far as yields are concerned, has spread, now largely as a hybrid, throughout the world, and its productivity has increased since the "Green Revolution". It has become a staple in parts of the Mediterranean, the Balkans, Africa, and South Asia, and continues to be the largest food item for the masses in many parts of Latin America.

Some American crops required modification or additions before their use expanded. Cacao did not suit European tastes when first encountered; indeed it was repugnant to many, until people learned to add sugar and vanilla. After Swiss and Dutch houses learned the process of making milk chocolate, cacao became an item of world diet, and West Africa is the main producer today.

Indeed, so pervasive today are some of the high-yielding American crops that their pre-Columbian American origins have been forgotten by the general public. Indian curry and Hungarian paprika, both based on American chilli peppers, Italian tomato sauce, Irish potatoes, West Africa ogi and kenkey (maize doughs), and Swahili posho (maize meal), Turkish tobacco, and Swiss chocolate, are now thought of as age-old, integrated parts of regional and national cultures far from the Americas.

MURDO J. MACLEOD

REFERENCES

Alcorn, Janis B. "Huastec Noncrop Resource Management. Implications for Prehistoric Rain Forest Management." *Human Ecology* 9(4): 395–417, 1981.

Andrews, Jean. *Peppers: The Domesticated Capsicums.* Austin: University of Texas Press, 1984.

Bergman, John F. "The Distribution of Cacao Cultivation in Pre-Columbian America." *Annals of the Association of American Geographers* 59: 85–96, 1969.

Crosby, Alfred W., Jr. *Ecological Imperialism: The Biological Expansion of Europe.* Cambridge: Cambridge University Press, 1993.

National Research Council. *Lost Crops of the Incas: Little-Known Plants of the Andes with Promise for World-Wide Cultivation.* Washington, D.C.: National Academy Press, 1989.

Sauer, Carl O. "Cultivated Plants of South and Central America." In *Handbook of South American Indians*, Vol.6: *Agricultural Origins and Dispersals*. New York: Bureau of American Ethnology, 1952, pp. 487–543.

Sauer, Jonathan D. *Historical Geography of Crop Plants: A Select Roster.* Boca Raton, Florida: CRC Press, 1993.

Stone, Doris, ed. *Pre-Columbian Plant Migration.* Cambridge, Massachusetts: Harvard University Press, 1984.

Warman, Arturo. *La historia de un bastardo: maíz y capitalismo.* Mexico: Instituto de Investigaciones Sociales. UNAM/ Fondo de Cultura Económica, 1988.

Wood, R. *Quinoa the Super Grain.* New York: Japan Publications, 1989.

See also: Potato – Agriculture – Food Technology – Sugar

CUNEIFORM Cuneiform was the script used throughout the Ancient Near East from the fourth to the first millennium BC. Although a hieroglyphic script came to predominate very early in Egypt, the international second millennium correspondence of the Egyptian Sun King, Iknaton, used the cuneiform script. In the last millennium BC Mesopotamian art depicts a set of scribes, one writing with a stylus on clay tablets, another on parchment. This later overlap between the use of parchment and clay as writing material foreshadowed the demise of cuneiform, the script by means of which literacy as a technological advance spread throughout the ancient world from early Mesopotamian sites such as Uruk in the lower Euphrates River Valley and Susa to the east.

The Script Form

Cuneiform (from Latin *cuneus* 'wedge'; cf. German *Keilschrift* 'wedge-script') is so named because of the wedge-shaped signs which make up the characters of the script. A stylus used in writing on clay tablets created wedge-shaped signs that resemble the head of a nail. Pressing one corner of the stylus into the damp clay of the tablet made a *Winkelhaken* or corner wedge (). This corner wedge could be elaborated by adding a vertical () or horizontal tail (), or by tilting the tail at an angle. These basic shapes combine to produce the more than 700 signs of the cuneiform writing system. A corner wedge alone could, as a syllabic unit, represent the syllable sound of 'u' as in 'U-gan-da' or, as a logogram, refer to the word for the numeral 'ten'. Two corner wedges combined with a vertical wedge () form the phonetic syllable 'ud' which also has the logographic meaning 'day'.

Alphabets used to write modern Western languages reduce those languages to writing on the principle of one sound to one written symbol. The principle underlying the cuneiform system was by no means so simple. A single sign might have phonetic or logographic values. While vowel sounds had a one-to-one correspondence with a sign as in 'a, i, u': 'a' (), 'i' ( ), and 'u' (), there were no signs for consonants like 'p', 't', 'k', or 'm'. Cuneiform signs had phonetic values as syllables, so there are signs instead for 'pa, ta, ka, ma, pi, ti, ki, mi' and so on. Often a sign also had a logographic value as in the case of *ud*. This logographic-syllabic principle underlying the cuneiform script, unlike that underlying an alphabet, thus used signs corresponding either to entire words (logograms) or to phonetic (syllabic) units.

The syllable sign 'ud' is typical of signs that refer to vowel plus consonant sequences. Cuneiform also had syllabic signs for consonant–vowel sequences, e.g.

 'di', 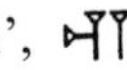 'gi', or  'ni'

and, less frequently, for consonant–vowel–consonant sequences:

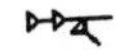 'tar' or  'kur'.

This system of writing thus easily recorded a language with consonant-vowel sound patterns and consonant clusters in the middle of a word, but not words which begin or end with consonant clusters like 'st', or 'sk' as in English 'stop, skate, nest' or 'desk'.

While cuneiform syllabic signs represented speech sounds, the same sign might also have non-phonetic, even non-word, uses as well. In addition to syllabic and logographic uses, signs might function as determinatives. As a determinative, a cuneiform sign preceded (or followed) one or more other signs without any phonetic value. The non-phonetic determinative merely signaled to the reader that a word in question was of a particular semantic type. Typically, the determinative for 'female' preceded a woman's name, the determinative for 'male' a man's, or the determinative for 'deity' the name of a god or goddess. Used as a logogram, however, a cuneiform sign stood for an entire word. Thus, either the single sign ( ) or the syllabic sequence *ni-in-da* might stand for Sumerian 'bread'. Likewise, 'deity' might be written logographically in a Sumerian text using the 'an' sign ( ) or syllabically with the sequence *di-in-gi-ir*. Such dual mechanisms illustrate not only the relation between logographic and phonetic representation of words but also something of the syllabic structure of the Sumerian language which had word-medial (*-nd-* and *-ng-* in *ninda* and *dingir*) but not word-initial or word-final consonant clusters.

Scribes often used signs logographically for the lexical content of common words and syllabically for grammatical markings such as the possessive of a noun. Thus, 'of god' in Sumerian might be written phonetically as *di-in-gi-ir-ra* (*dingir* plus the possessive suffix *a*) or logographically with the 'god' sign plus the syllabic sign, *-ra*. Transliteration conventions using the Roman alphabet distinguish among determinative, logographic, or syllabic uses of cuneiform signs on a tablet by using lower case, upper case, and raised letters. Signs used phonetically are transliterated with lower case letters; signs used logographically but interpretable on the basis of a Sumerian word are transliterated using capital letters for the Sumerian word form. The Sumerian scribe's writing of the logogram for 'god' plus syllabic *-ra* would thus be transliterated DINGIR-ra to distinguish it from a scribe's entirely syllabic rendition, *di-in-gi-ir-ra*.

Determinatives are transliterated as raised sequences, either before or after the 'determined' word. Characteristically the plural determinative follows, while others often precede. Thus, 'gods', written with the logogram for 'god' ( ) plus a plural determinative whose Sumerian phonetic value was 'mesh' ( ), might be transliterated in Roman letters DINGIR^{MESH} and translated 'gods'. Similarly, the Sumerian sungod, Utu, would begin with the determinative for 'deity' ( ) and continue with the name, Utu, written syllabically ^{DINGIR}U-tu. However, the sungod's name was usually written logographically using the logogram for 'day', ^{DINGIR}UD ( ), transliterated ^{DINGIR}UTU (Sungod). To bring the use of cuneiform signs full circle, one might point out that the 'god' sign ( ), besides its logographic reference to the Sumerian word, *dingir* 'god', also had a syllabic reading, *an*. For a scribe, the same sign thus had multiple values. The sign ( ) had a phonetic (syllabic) value *an*, a logographic value 'god' (Sumerian DINGIR), and a non-phonetic use as the determinative for 'deity' to which modern scholars assign the phonetic value, *dingir*, for purposes of transcription.

Like determinatives, which had no phonetic counterpart in the speech represented by cuneiform signs, horizontal lines were often drawn across the tablet as a kind of punctuation aid. Because the structure of the languages written in the cuneiform script was very different from English, punctuation principles might also be expected to vary. In fact the scribes' horizontal lines across the tablet probably marked something corresponding in part to our sentences and partly to our paragraphs. These scribal lines occur from earliest Sumerian tablets down through adaptations of cuneiform to write Semitic and Indo-European languages. Cuneiform signs are read left to right, as are western European alphabetic symbols.

Languages Written in Cuneiform

Questions about the range of sounds that written symbols might need to distinguish, and indeed which units of language should be represented by individual signs, have concerned linguists seeking to reduce languages to writing. Western European languages often solved these problems by borrowing the Roman alphabet. Cultures of the Ancient Near East borrowed the cuneiform script, as the alphabet did not yet exist in the third millennium BC. Cuneiform scribes thus dealt with issues of defining the linguistic units to be represented by altering a system of cuneiform signs already in use to write Sumerian. Scribes in fact used the cuneiform script to write languages with sets of sounds and structures varying widely from those of Sumerian. They then had to adapt the script to fit the needs of the languages they spoke. The words of some of the languages had long series of suffixes, for example, while others used word-internal vowel changes much as English does in 'sing, sang, sung' or 'stand, stood', and Indo–European languages had word-initial and word-final consonant clusters.

A major adaptation of cuneiform was for the East Semitic Akkadian that served as the basis for the spread of cuneiform throughout much of the Near East. Besides Old Akkadian, which had its own conventions for phonetic representations using the cuneiform script, major East Semitic languages include the Old Babylonian in which the southern Mesopotamian Code of Hammurapi was written and Old Assyrian, which records the northern merchant trade between Assur and Anatolian Kanesh even before the time of Hammurapi. Babylonian and Assyrian continue in ever-changing regional forms for nearly 2000 years as Middle Babylonian, Middle Assyrian, Neo-Babylonian, and Neo-Assyrian. To the east of Mesopotamia the ancient language of Susa, Elamite, also used cuneiform, while Hurrians penetrated scribal schools in Northern Mesopotamia, Syria, and Anatolia at different times, and adapted cuneiform for their language. In second millennium Anatolia, the scribes of Hattusa used cuneiform to write Hittite, Luwian, and Palaic, Indo-European languages related to Greek, Latin, and the Germanic language family from which English is descended.

When scribes adapted cuneiform for use in writing other languages, some things changed more than others. While the signs often came to have peculiarly local values in different scribal schools, the most constant aspect of the cuneiform script was its core repertoire of determinatives and logograms. Where cuneiform was used, determinatives such as those for male, female, deity, land, and plural remained the same. Logograms for basic vocabulary, for god, ox, sheep, child, king, mountain, and river, also remained constant. The Akkadian scribe, for example, now had all the options available that the Sumerian scribe had, plus those that involved the phonetic representation of Akkadian. Akkadian nouns had three different forms depending on whether they were used as subject or object in the sentence, or as a possessive. The Akkadian word for 'god' was thus *ilum* (nominative), *ilam* (accusative), or *ilim* (genitive), written as phonetic *i-lu-um, i-la-am, i-li-im* or logographic DINGIR-lum, DINGIR-lam, DINGIR-lim where syllabic signs differentiate case, or simply DINGIR, if the context was clear without the case ending.

When cuneiform spread via an Akkadian-based scribal school, scribal practices also spread. Thus, sign sequences were now often interpretable, not only on the basis of recognizing Sumerian, but also Akkadian, words. This created a further complication for scholars transliterating the tablets into a Roman alphabet. Instead of signs interpretable as Sumerian logograms (Sumerograms in capital letters), there are now signs interpretable as Akkadian logograms (Akkadograms). Scholars use italic capitals to transliterate Akkadograms. Since the Hittites borrowed the script as a form of Akkadian cuneiform, Hittite scribes had still more options for writing 'of god'. While the Sumerian scribe wrote 'of god' as phonetic *di-in-gi-ir-ra* or logographic DINGIR-ra and the Akkadian scribe used phonetic *i-li-im*, logographic DINGIR-lim or DINGIR-RA-lim, the Hittite scribe could write phonetic *si-u-ni-ya-as* (Hittite 'god': nominative *sius*, genitive *siunias*), Sumerographic DINGIR-(RA), Akkadographic *I-LIM*, or some combination, DINGIR-(RA-)*LIM* or DINGIR-(RA-)*LIM*-as.

The Origin of the Cuneiform Script

The earliest attested use of cuneiform as a developed system of symbols to write a language was its use to write Sumerian ca. 3000. Modern Japanese, which has borrowed Chinese (logographic) kanji characters as a basis for reducing Japanese to writing, likewise uses syllabic symbols to augment logographic symbols. Because the Japanese syllabic symbols (*hiragana*) are an independent Japanese invention to adapt the Chinese script to a language structure for which the kanji characters were inadequate, scholars have hypothesized that Sumerian, with its logographic-syllabic script, represented an adaptation of an earlier "Proto-Tigridian" script of which we have no record, one used for a pre-Sumerian language of the Tigris river valley. Studies in the origin of writing, however, suggest that writing emerged as a new technology among people who spoke Sumerian.

The forerunners of the earliest cuneiform tablets were "impressed" tablets, tablets impressed first, not by the stylus which came to be used to write cuneiform, but by small clay artifacts known as tokens. The token-impressed tablets were economic documents recording amounts of various commodities that the society stored or transferred. Discrete token symbols for commodities such as sheep, barley, or jugs of wine and oil were impressed on to the tablet. At first repetitions of the sheep symbol, for example, referred to the quantity of sheep in question and repetitions of the 'oil jug' impression designated the quantity of oil. As the technology for writing and enumeration matured, forms for the commodity became stylized. At first, instead of impressing the token into the clay, the scribe used a stylus to draw the shape of the token. However, soon the shape drawn by the stylus was replaced by an abstraction made by impressing the corner of the stylus into the clay, by abstract cuneiform signs. Over the millennia in which cuneiform was used, the shapes of the signs evolved from more pictographic forms to highly stylized symbols. Later Ugaritic use of wedge-shaped signs as an alphabet broke with the tradition in assigning new values to arbitrary sign forms.

Early Non-cuneiform Scripts

The existence of other early scripts in the Ancient Near East attests to the rapid spread of literacy and numeracy. While Egyptian hieroglyphs are clearly independent script forms, the basis for reducing the early Egyptian language to writing is probably not independent of the new technology which arose in Mesopotamia as a response to economic and lifestyle changes involving the storage and transfer of surpluses of essential commodities. In the second millennium too a hieroglyphic script, Hieroglyphic Luwian (at first known as Hieroglyphic Hittite) developed independently of the Egyptian hieroglyphs (in Anatolia and North Syria). This Luwian Hieroglyphic and the cuneiform script coexisted in different functions. The local hieroglyphic script recorded royal inscriptions in the Luwian language on stone monuments but cuneiform was the script of clay tablets. The fact that Hittite cuneiform tablets of Anatolia refer to 'old

Table 1: Cuneiform signs and their sign list keys (cf. Borger or Labat)

Numerical key	Usual value	Sign shape
411	u	
480	dish	
1	ash	
381	ud	
579	a	
142	i	
457	di	
85	gi	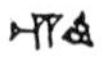
231	ni	
12	tar	
366	kur	
597	NINDA	
13	an	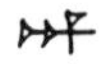
533	mesh	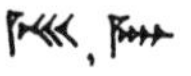

wooden tablets' has led to speculation that hieroglyphic may have been inscribed into a wax coating on wooden tablets. Since none of these wooden tablets has survived, such inferences, of course, remain speculative. The forms of the hieroglyphs are unrelated to the forms of cuneiform signs, but again the two scripts share a common conceptual basis for reducing language to writing. Hieroglyphic Luwian, like cuneiform and Egyptian hieroglyphics, is based on logographic-syllabic principles for reducing the spoken word to writing.

Decipherment

The cuneiform script was deciphered in the nineteenth century as a result of a trilingual inscription found at Behistun in Iran, not far from the Persian capital, Persepolis. The languages of the inscription were Elamite, Old Babylonian, and Old Persian, all written in a form of cuneiform. Both Elamite and Old Babylonian were unknown languages at the time, but Old Persian was known from Zoroastrian religious texts. The Old Persian inscription was first deciphered on the basis of hypotheses about the recurring symbols in Old Persian royal names. From there, hypotheses related the signs used to write the Old Persian with signs in the unknown scripts. Little by little, Old Babylonian and Elamite began to be deciphered. With the decipherment of the script came the decipherment and study of the different languages and cultures that used cuneiform to record their history and socioeconomic structure. A not insignificant aid in the subsequent decipherment of Sumerian turned out to be the discovery of grammatical texts that Akkadian scribes had developed as pedagogical devices for teaching the learned language of the second millennium BC, Sumerian, thereby training scribes to support the ever-growing bureaucracy of the ancient world.

CAROL F. JUSTUS

REFERENCES

Borger, Rylke. *Akkadische Zeichenliste.* (Alter Orient und Altes-Testament, Sonderreihe. Veröffentlichungen zur Kultur und Geschichte des Alten Orients, 6.) Neukirchen-Vluyn: Butzon & Bercker Kevelaer, 1971 (with updates).

Gelb, Ignace J. *A Study of Writing.* Chicago: University of Chicago Press, 1963.

Gordon, Cyrus H. *Forgotten Scripts: Their Ongoing Discovery and Decipherment.* New York: Basic Books, 1968.

Hawkins, J. D., and Anna Morpurgo-Davies. *Hittite Hieroglyphs and Luwian: New Evidence for the Connection.* Göttingen: Vandenhoeck & Ruprecht, 1973.

Kramer, Samuel Noah. *The Sumerians: Their History, Culture, and Character.* Chicago: University of Chicago Press, 1963.

Labat, Réné. *Manuel d'épigraphie akkadienne (Signes, Syllabaire, Idéogrammes).* New revised and corrected edition. Paris: Librairie Orientaliste Paul Geuthner, 1976.

Nissen, Hans J., Peter Damerow, & Robert K. Englund. *Early Writing and Techniques of Economic Administration in the Ancient Near East.* Translated by Paul Larsen. Chicago: University of Chicago Press, 1993.

Powell, Marvin A., ed. "Aspects of Cuneiform Writing." *Visible Language* 15: 4, 1981.

Rüster, Christel, & Erich Neu. *Hethitisches Zeichenlexikon. Inventar und Interpretation der Keilschriftzeichen aus Boghazköy-Texten.* (Studien zu den Boghazköy-Texten. Kommission für den Alten Orient der Akademie der Wissenschaften und der Litertur, Beiheft 2.) Wiesbaden: Otto Harrassowitz, 1989.

Schmandt-Besserat, Denise. *Before Writing.* Volume 1: *From Counting to Cuneiform.* Austin: University of Texas Press, 1992.

Steve, M.-J. *Syllabaire élamite: histoire et paléographie.* (Civilisations du Proche-Orient. Archéologie et Environment–Philologie et Cultures–Histoire. Série II Philologie, Volume 1.) Neuchâtel–Paris: Recherches et Publications, 1992.

Walker, C. B. F. *Cuneiform.* Berkeley & Los Angeles: University of California Press, 1987.

D

AL-DAMĪRĪ Al-Damīrī, Muḥammad ibn Mūsa ibn ʿĪsā was born in Cairo, Egypt in AD 1341. Although he began work as a tailor, he soon decided to study with the leading teachers of the time such as Bahāʾ al-Dīn al-Subkī, Jamāl al-Dīn al-Asnāwī, Ibn ʿAqīl, and others. He became a professional theologian and taught in different centers such as al-Azhar University, achieving a great recognition for his preaching and his ascetic life. A very religious man, he made the pilgrimage to Mecca six times between AD 1361 and AD 1367 and died in Cairo in AD 1405.

The majority of al-Damīrī's works are conventional commentaries and epitomes of earlier works such as the one on al-Nawawī's *Minhāj* (a manual of Islamic law). He also wrote sermons and treatises on canon law. Most of these works seem to be lost. His most famous work is *Ḥayāt al-ḥayawān* (Life of the Animals), a zoological dictionary which contains information on the animals mentioned in the *Quʾrān* and in the Arabic literature. It includes not only the zoological aspects but everything related to the animals mentioned.

The work contains 1069 articles describing a lesser number of animals because the same animal is occasionally described twice using two different names. The animals are described in alphabetical order and usually contain seven sections: (1) grammatical and lexicographical peculiarities of the name; (2) a description of the animal according to the leading authorities; (3) Muslim traditions in which the animal is mentioned; (4) juridico-theological considerations regarding the animal; (5) proverbs about the animal; (6) the medicinal properties of the products derived from the animal; and (7) rules for the interpretation of dreams in which the animal appears.

There are three versions of this work: the large (*al-kubrā*), the medium (*al-wusṭā*), and the small (*al-ṣugrā*) and it has been republished several times and translated into Persian and Turkish.

EMILIA CALVO

REFERENCES

Kopf, L. "al-Damīrī" In *Encyclopédie de l'Islam*, 2nd ed., vol. II, Leiden: E.J. Brill; Paris: G.P. Maisonneuve, 1965, pp. 109–110.

Sarton, George. *Introduction to the History of Science*. Baltimore: Williams & Wilkins, 1948, pp. 1214,1326, 1639–1641.

Vernet, Juan. "Al-Damīrī." In *Dictionary of Scientific Biography*, vol. III. New York: Charles Scribner's Sons, 1971, pp. 548–549.

DECIMAL NOTATION Decimal notation is a system which imparts to nine figures (digits) an absolute numerical value and also a positional value which latter increases their value ten times by being shifted by one place to the left. Thus, the digits: 1, 2, 3, 4, 5, 6, 7, 8, and 9, coupled with the figure '0' which stands for zero or *śūnya* (nothing, empty), while expressing just their individual values when standing alone, can express also any quantity of any magnitude by their repeated use in the same number, and shifting of places, as needed. The importance of this contrivance is apparent from the words of the great French mathematician P.S. Laplace, when he says: "The idea of expressing all quantities by nine figures whereby both an absolute value and one by position is imparted to them is so simple that this very simplicity is the reason for our not being sufficiently aware how much admiration it deserves" (Srinivasiaingar, 1967). Halstead observes: " The importance of the creation of the zero mark can never be exaggerated. This giving to airy nothing, not merely a local habitation and a name, a picture, a symbol, but helpful power, is the characteristic of the Hindu race from whence it sprang. It is like coining the *Nirvāṇa* into dynamos. No single mathematical creation has been more potent for the general on-go of intelligence and power".

In Indian tradition, the need for enumeration and decimal notation stemmed from the adoration of gods and for ritualistic purposes. From Vedic times, the Hindus used the decimal notation for numeration. The *Ṛgveda* (ca. 2000 BC) groups gods into three (1.105.5); there are three dawns (8.41.3); there were seven rays of the Sun-god (1.105.9); there were seven sages (4.42.8), and seven seas (8.40.5). There were 180 Marut-gods, or three times sixty (8.96.8); the God Śyāvā gave as gifts cows numbering 210 or three times seventy (8.19.37). There were 21 followers of Indra, or three times seven (1.133.6), and the number of horses prayed for was thrice seven times seventy or $3\times7\times70$ (8.46.26). In Vedic literature, besides the primary numbers, one to nine, expressed by the terms, *eka, dvi, tri, catur, pañca, ṣaṭ, sapta, aṣṭa*, and *nava,* the decuple terms from ten to ninety, expressed by *daśa, viṃśati, triṃśat, catvārimśat, pañcāśat, ṣaṣṭi, saptati, aśīti* and *navati* are found. These are then sequentially multiplied by ten, taking terms from 100 to 10 to the power of 12, the terms being *śata, sahasra, ayuta, niyuta, prayuta, koṭi, arbuda, nyarbuda, samudra, madhya, anta,* and *parārdha*. In the matter of the arrangement of decuples in compound number-names, the practice generally followed in Vedic literature was to put the term of higher denomination first, except in the case of the two lowest denominations, where the reverse method was followed. See, for example, *sapta śātāni viṃśatiḥ* (seven hundreds and twenty, *Ṛgveda* 1.164.11), *sahasrāṇi śata daśa* (thousands hundred, and ten, *Ṛgveda* 2.1.8) and *ṣaṣṭ, sahasra navatim nava* (sixty thousands, ninety and nine, 60,099 *Ṛgveda* 1.53.9).

With respect to written symbols for numbers, since no palaeographic records of the Vedic age have been preserved, little can be said. However, a few Vedic passages occur where written numerical symbols are mentioned. In the *Ṛgveda* (10.62.7) certain cows with the mark of '8' *(aṣṭa-karṇī)* are referred to, and *Yajurveda-Kāṭhaka Saṃhitā* makes mention of pieces of gold with the mark '8' imprinted on them *(aṣṭa-pruddhiraṇyam, aṣṭāmṛdam hiraṇyam*, 13.10). Inscriptions and manuscripts, of later ages, all over India, use numerical symbols profusely. The tendency had been, from early ages, to spell out the numbers or make use of things permanently associated with a number to represent that number. For instance, eyes, hands, etc. were used for 2; moon, sky, etc. were used for 1, seasons for 6, and week for 7. Another method was to attribute specific numerical values for the letters of the alphabet and use those letters to indicate the specified numbers, a method which was mentioned by the Sanskrit grammarian Pāṇini of the fourth century BC. These methods were very popular in the classical age in India, especially with mathematicians and astronomers.

K.V. SARMA

REFERENCES

Datta, Bibhutibhusan, and Avadhesh Narayan Singh. *History of Hindu Mathematics: A Source Book.* Bombay: Asia Publishing House, 1962.

Gupta, R.N. "Decimal Denominational Terms in Ancient and Medieval India." *Ganita Bharati* 5(1–4): 8–15, 1986.

Halstead, G.B. *On the Foundation and Technic of Arithmetic.* Chicago: Open Court, 1912.

Ray, Priyadranjan, and S.N. Sen. *The Cultural Heritage of India. vol. VI: Science and Technology.* Calcutta: Ramakrishna Mission Institute of Culture, 1986.

Ṛgveda: *The Hymns of the Ṛgveda.* Trans. R.Th. Griffith. Delhi : Motilal Banarasidass, 1986.

Sharma, Mukesh Dutt. "Indian Invention of Decimal System and Number Zero." *Vedic Path* 44 (1): 32–37, 1982.

Srinivasaingar, C.N. *The History of Ancient Indian Mathematics.* Calcutta: World Press, 1967.

See also: Mathematics in India – Sexagesimal System – Zero

DEŚĀNTARA In Indian astronomy, the *Deśāntara* of a place is its terrestrial longitude, i.e. the 'distance of the place' from a universally accepted zero meridian. In modern times, the meridian at Greenwich in England is accepted as the zero meridian, and the longitude is expressed in terms of the angle subtended, at the pole, by the Greenwich meridian and the meridian of the place in question. Indian astronomy had, from early times, taken as the zero meridian the meridian passing through the ancient city of Ujjain in Central India, cutting the equator at an imaginary city called Laṅkā and passing through the south and north poles. Again, in order to facilitate the conversion of the local time to that of the zero meridian and vice versa, the *Deśāntara* was expressed in terms of time-measures like *nāḍī* (or *ghaṭī*), equal to 24 minutes, as converted from the corresponding degrees. Since the earth completes one eastward rotation of 360° in 24 hours, it is 15° an hour or 1 degree in 4 minutes. In terms of Indian measures, since 60 *nāḍīs* are equal to 24 hours or two and half *nāḍīs* make one hour, the rotation of 15° corresponds to a period of two and a half *nāḍīs.* Since 1 *nāḍī*= 60 *vināḍīs*, and 1 *vināḍī*= 6 *prāṇas*, the rotation will be 1 degree in 10 *vināḍīs,* or 1 minute in 10 *prāṇas*. The *deśāntara* which is expressed in terms of time measure is done through either *nāḍīs*, *vināḍīs* or *prāṇas.*

Since the planetary positions derived by Indian astronomical computation are all related to the mean sunrise at the zero meridian, viz., the Ujjain meridian, to arrive at the positions at local places, a longitude correction or *deśāntara-saṃskāra* is called for. It is calculated in time-measures, as above, and is subtracted if the place in question is east of Ujjain and added if it is west of Ujjain.

The *deśāntara-saṃskāra* is expressed also in terms of the distance, i.e. in *yojanas*, of the desired place from the Ujjain meridian in the same latitude, at the rate of 55 *yojanas* for 10 *vināḍīs* or 1 minute.

K.V. SARMA

See also: Astronomy in India

DEVĀCĀRYA Devācārya, son of Gojanma and author of the astronomical manual *Karaṇaratna* (lit. Gem of a Manual), hailed from Kerala in South India. The epoch of *Karaṇaratna*, i.e. the date from which planetary computations were instructed to be commenced in that work, is the first day of the year 611 in the Śaka era, which corresponds to February 26 of AD 689. This places Devācārya in the latter half of the seventh century. We know he came from Kerala because he used the *Kaṭapayādi* system of letter numerals and the *Śakābdasaṃskāra*, which is a correction applied to the mean longitudes of the planets from the Śaka year 444 or AD 521, and a unique method of computing the solar eclipse, all of which are peculiar to Kerala, and because his work is popular in that part of the land.

Devācārya uses the elements of the Āryabhaṭan school of astronomy as the basis of his work, as he himself states towards the commencement of *Karaṇaratna*. His work is based both on the *Āryabhaṭīya* and on the second work of Āryabhaṭa, the *Āryabhaṭa-siddhānta*, as abridged in the *Khaṇḍakhādyaka* of Brahmagupta. The influence of the *Sūryasiddhānta* and of Varāhamihira are also apparent. It is also noteworthy that Devācārya himself innovated several thitherto unknown methodologies and techniques.

In eight chapters, the *Karaṇaratna* encompasses almost all the generally accepted aspects of Hindu astronomical manuals. Chapter I of the work is concerned with the computation of the longitudes of the sun and moon, and also the five basic elements of the Hindu calendar (*pañcāṅga*). Computation and graphical representation of the lunar and solar eclipses are the subjects of chapters II and III. Chapter IV deals with problems related to the gnomonic shadow, and chapter V with the calculation of the time of the moonrise and allied matters. In chapter VI, heliacal rising of the moon and elevation of the moon's horns are dealt with. The last two chapters are concerned with the derivation of the longitudes of the planets, planetary motion, and planetary conjunctions.

Several peculiarities characterize Devācārya's work. Among these are the computation of the sun, moon, moon's apogee, and moon's ascending node using the 'omitted' lunar days, the *Śakābda* correction, and the application of a third visibility correction for the moon. However, what is most significant in the work is the recognition of the precession of the equinoxes and the rule that he gives for its determination on any date. Devācārya's measure of the rate of precession is 47 seconds per annum, its modern value being 50 seconds. Devācārya's importance lies in the fact that his work formed a record of the astronomical practices and methodologies for the quick derivation of astronomical data that prevailed in India during the seventh century AD.

K.V. SARMA

REFERENCES

The Karaṇa-ratna of Devācārya. Ed. Kripa Shankar Shukla. Lucknow: Department of Mathematics and Astronomy, Lucknow University, 1979.

Pingree, David. *Census of the Exact Sciences in Sanskrit*. Philadelphia: American Philosophical Soc. Ser. A, vol. 3, 1976, p. 121.

See also: Āryabhaṭa – *Sūryasiddhānta* – Varāhamihira – Precession of the Equinoxes

DIVINATION IN CHINA From neolithic times into the twentieth century, divination occupied a prominent place in Chinese culture. By the third millennium BC at the latest, specialists in reading stress cracks in animal bones had already emerged as a distinct occupational group in north China. During the Shang dynasty (ca. 1500–ca. 1100 BC) the interpretation of oracle bones (primarily the scapulae of cattle and the plastrons of turtles) reached a high degree of sophistication. Royal diviners sought spiritual advice on a wide range of practical problems, from the weather, agriculture and hunting to administrative affairs, warfare, sacrifices, and medical matters. Shang oracle bone inscriptions also indicate an interest in dream divination and "spirit-possession" (shamanism), both of which persisted in China for about three thousand years.

Another form of divination dating from the Shang dynasty involved calculations made with the stalks of milfoil plants (*shi*). By the Zhou dynasty (ca. 1100–256 BC) milfoil calculations were used primarily to create six-line figures called hexagrams (*gua*). These figures, which represented symbolically the images or structures of change in the universe, became the core of a mantic work known as the *Zhouyi* (Zhou Changes), later called the *Yijing* (*I Ching*, Classic of Changes). The premise of the *Yijing*, as articulated in its "Great Commentary," was that the future was knowable, and that heavenly patterns of change had discernable earthly manifestations. Over time, virtually all Chinese divination techniques came to be associated with, and in some sense validated by, this enormously influential Confucian classic.

During the late Zhou period, a great number of mantic systems developed, including various "schools" of astrology (*zhanxing*, *xingming*, etc.), geomancy (*dixing*, *xingfa*, *kanyu*, *fengshui*), physiognomy (*xiangren*, *kanxiang*), and the numerological arts (*shushu*). This growing interest in divination was part of a more general burst of philosophical creativity in the so-called Warring States period (453–221 BC) — an era marked by social and geographical mobility, the exchange of new ideas, increased professional specialization, and the introduction of new technologies, including advanced techniques of astronomical and calendrical calculation. From the late Zhou into the twentieth century, almanacs or "day books" (*rishu*, *tongshu*, etc.), which specified propitious and impropitious times for a wide range of activities (bathing, beginning projects, getting married, etc.), became indispensable guides for daily behavior in China.

During the Han dynasty (206 BC to AD 222), the correlative cosmology of Dong Zhongshu (ca. 179–ca. 104 BC) — based on the pervasive ideas of *yin* and *yang* and the so-called "five elements" (*wuxing*; also known as the "five agents" or "five phases") — influenced virtually all forms of Chinese divination, not to mention the related realms of medicine and natural science. One striking and enduring feature of this cosmology was that *yinyang/wuxing* correlations often counted for more than empirical observation, even when, as often proved to be the case, the correspondences and analogies were internally inconsistent or mutually incompatible.

Another enduring feature of Han thought was the notion of "cosmological kingship" — the idea that the emperor was Heaven's agent, charged with maintaining cosmic harmony. Divination helped Chinese rulers to understand unfolding patterns of change and thus facilitated their effort to assure that the social order and nature's Way (*dao*) were fully congruent. Not surprisingly, the Han emperors, like the rest of society, employed occult specialists (*fangshi*) of various sorts to tell the future, to heal illnesses, and to engage in other forms of cosmic "magic". So influential were such specialists that from Han times onward, every official dynastic history, as well as many local gazetteers and other historical sources, contained special sections devoted to the biographies of *fangshi*, or more broadly, "technicians" (*fangji*).

The Han period witnessed the invention or refinement of almost every major system of Chinese divination. Han *fangshi* specialized not only in time-honored mantic techniques, but also in relatively new types of prognostication, such as the evaluation of wind and weather (*fengjiao*, *wangqi*, etc.) and the analysis of written characters (*xiangzi*). Throughout the remainder of China's imperial history, until 1912, the introduction or emergence of new philosophical systems such as Buddhism, Religious Daoism and Neo-Confucianism had little effect on Chinese divination. The only major innovations of later eras were the widespread use of "spiritual sticks" (*lingqian*) and divining blocks (*jiao*) by individuals, and the group-oriented activity known as spirit-writing (*fuji*).

Divination played an important role in Chinese society at all levels. In addition to helping people to "resolve doubts" (*jueyi*) about the future, divination categorized and explained the workings of the world in culturally significant ways. The elaborate schemes used by fortune-tellers to analyze heavenly phenomena, earthly forms, physical objects, personality types, and medical problems were undoubtedly more well known and widely understood than other forms of scientific explanation available in imperial China. To the extent that science can be viewed as an ordering device for managing data, divination served Chinese purposes nicely.

Although the mantic arts in China were practiced openly by a wide variety of private individuals, both professionals and amateurs, the emperor and his officials also made regular use of fortune-tellers. Throughout the imperial era, from 221 BC to 1912, government diviners chose auspicious dates and times for all kinds of ritual events, and from the fourteenth century onward, the official state calendar, in imitation of popular almanacs, included tables of lucky and unlucky days. Experts in wind, rain, and cloud divination served Chinese bureaucrats as meteorologists; geomancers, with their special knowledge of landforms and waterways, assisted in government public works projects and military affairs; and physiognomers served as personnel specialists, mediators, and even interrogators — the functional equivalent of social psychologists. Diviners also dispensed medical advice, since many of them were actually doctors and all shared the same social status and cosmological outlook, based on the idea of "responsive correspondence" (*ying* or *xiangying*).

During the seventeenth and eighteenth centuries, exponents of the so-called School of Evidential Research (*kaozheng xue*), sparked by a fervent desire to "seek truth from facts" (*shishi qiu shi*), brought a host of technical and empirical studies (including mathematical astronomy and physical geography) into the mainstream of Confucian scholarship. In the minds of some, this development had "cosmologically subversive" implications for China. In fact, however, *kaozheng* scholarship did not succeed in effecting the kind of radical transformation of consciousness that occurred in Europe about the same time. Only in the late nineteenth century, when nationalistic Chinese intellectuals began to criticize the concept of cosmological kingship, in order to pry power from the throne, did the old world view begin to crumble.

After the fall of the Qing dynasty (1644–1912), the new Chinese Republic launched a systematic campaign to eradicate all forms of "superstition" (*mixin*). This effort included a vigorous assault on the inherited cosmology, a denunciation of all forms of divination, and an attack on traditional Chinese medicine. Intellectuals during the New Culture Movement (ca. 1915–1925) enthusiastically embraced Western science and railed against such "outmoded" ideas as *yinyang* and *wuxing*. Yet, in Chinese popular culture, these concepts continued to have remarkable staying power, expressed not only in almanacs and the predictions of diviners but also in the diagnoses and prescriptions of doctors. One reason for this persistence, of course, is that the traditional mantic and medical arts continued to perform important roles in Chinese society, providing mental and physical assistance to individuals in a wide variety of circumstances. They do so to this day, not only in Taiwan, Hong Kong, and overseas Chinese communities, but also in the People's Republic of China.

RICHARD J. SMITH

REFERENCES

Henderson, John. *The Development and Decline of Chinese Cosmology*. New York: Columbia University Press, 1984.

Ho Peng Yoke. *Li, Qi and Shu: An Introduction to Science and Civilization in China*. Hong Kong: The Chinese University Press, 1985.

Loewe, Michael, and Carmen Blacker, eds. *Oracles and Divination*. Boulder: Shambala, 1981.

Morgan, Carole. *Le tableau du boeuf du printemps: Étude d'une page de l'almanach chinois*. Paris, Institut des hautes études chinoises, 1980.

Smith, Richard J. *Fortune-tellers and Philosophers: Divination in Traditional Chinese Society*. Boulder, Colorado: Westview Press, 1991.

Smith, Richard J. *Chinese Almanacs*. Hong Kong: Oxford University Press, 1992.

Smith, Richard J. and Kwok, D. W. Y., eds. *Cosmology, Ontology and Human Efficacy: Essays in Chinese Thought*. Honolulu: University of Hawaii Press, 1993.

Yuan Shushan. *Zhongguo lidai buren zhuan* (Biographies of Diviners in China by Period). Shanghai: n.p., 1948.

Zhang Yaowen. *Wushu zhanbu quanshu* (Complete Book of Five Arts Prognostication. Taibei: Dahua Publishers, n.d.

See also: *Yijing* – Geomancy in China – Maps, Geomantic – *Yingyang* – Five Phases – Magic and Science

DYES The human urge to paint the body with symbolic, warlike, or identifying colors may have been among the earliest impulses which led to the discovery of color yielding clays and plants. The dyeing of human clothing followed. Encounters, both warlike and peaceful, between groups then led to the identification of certain colors with specific regions or groups of producers, and exchanges began. This specialization and trade eliminated many of the poorer dyes of prehistoric times, which may have numbered many hundreds, and by the time recorded history took note of dyes only a relative few still saw widespread use.

Dye exchanges at first were very local. The major ones in ancient times were usually confined to the great areas of early culture and urbanization such as China, northern India, and the eastern Mediterranean. Later, trade in dyestuffs became more long distance, and, with European intrusions into Asia and invasion of the Americas, transoceanic and worldwide. This huge trade in natural dyes was destroyed and again reduced to a local level by the invention of coal tar or aniline dyes and other chemical dyes, in the mid-nineteenth century.

The production and exchange of colorants was always associated with certain other industries. Cloth manufacture and weaving were certainly the main ones. Others arose out of the limitations of some natural dyes. Many were not "fast"; that is they tended to fade or discolor when exposed to light, frequent washing, or wear and tear. Accordingly, much effort was historically expended in the search for mordants (dye fixers). Many of these were readily available, such as blood, dung, or urine. Various acids, alkalis, and wetting agents were widely used. Alum, an astringent, was probably the most common, and large cargoes were mined and shipped to dye factories and market towns. Minerals such as copper, tin, and iron also came to be used for mordanting, and thus added to the importance of mining for these minerals. Tree-borne dyes led to forest cultivation and above all, to extensive and destructive logging industries. The dye-yielding attributes of plants such as indigo stimulated the creation of plantation complexes.

The production of natural dyes and their use in dyeing textiles were often complicated skills requiring apprenticeships and years of training. Certain towns became known as dye centers or markets, such as ancient Tyre or medieval Venice in the Mediterranean, or Oaxaca in colonial Mexico. Guilds of dyers and castes devoted to weaving and dyeing were of considerable importance in European and Indian societies from medieval times.

Before European intrusion reached sub-Saharan Africa, Asia, Australia, and the Americas, there were several dyes which were produced and distributed over large areas. Perhaps the best known of the early root dyes was madder (from the Rubiaceae, of which there are over thirty species). *Rubia tinctorum* roots were used to obtain red dye in India, and were also known to the ancient Persians, Egyptians, Greeks, and Romans. For centuries Baghdad was the center of the madder trade, and it was cultivated extensively — rather than gathered as elsewhere — in Mesopotamia. After its use spread to Western Europe, the Dutch became the most systematic and scientific madder growers, combining it with their leading role in cloth making (wool) and cloth importing (silks and linens). The French began to compete in the eighteenth century, but the French revolution damaged the industry and it never revived.

Before the European expansion, the most widely used dye made from plant leaves was woad (*Isatis tinctoria* of the Cruciferae family). It yielded various blues and grays. Although easy to produce in the temperate areas of Europe and Asia, it was not a very brilliant or fast dye. Woad in medieval and early modern Europe was manufactured and marketed by powerful guilds which, by monopolies, boycotts, and powerful legislation, managed to prevent largescale intrusions of other blues, especially indigos, from America and southeast Asia, until the seventeenth century.

The leading dye made from woad has always come from brazilwood (*Caesalpinia echinata*), although many other trees such as lima, sapan, and peachwood, all soluble redwoods, are often lumped together as brazilwoods. These medium-sized trees have been widely used for dyestuffs in many parts of the world. They are cut into small logs, ground to powder, then soaked and fermented, often with an aluminum or other metal ore mordant. They yield reds and browns, except on silks. Before the sixteenth century, India, Sumatra, and Ceylon (Sri Lanka) were the main producers. Other woods, such as camwood from India, which imparted a rough feeling to cloth because of its resins, were also traded. Cutch or kutch, a brown dye, has been manufactured in India for over two thousand years and comes from the leaves and twigs of various acacia and mimosa trees.

Humans have elaborated dyes from the bodies of insects and animals for millennia. The most expensive, prestigious dye of ancient times was Tyrian purple. It was manufactured in Crete by 1600 BC, but is usually associated with the Phoenicians. Tyre became the great market center for this dye until captured by the Arabs in AD 638. Phoenician traders had spread its use all over the Mediterranean. It was so rare and costly that in many areas this fast, blue to purple dye was restricted to the clothing of high ecclesiastics, the aristocracy, or simply the ruling family. This colorant is extracted from a gland found in several shellfish, most notably *Murex trunculus*. A few similar dyes were used in the Americas before and after the European invasions, especially in Nicoya (Costa Rica) and on the Peruvian coast, but the American sources never produced enough dye to be of importance beyond local markets.

Insects have also been significant to the natural dye industry. Kermes is the oldest and most widespread of these dyestuffs. The insects harvested, *Coccus arborum* and *Coccus ilicis*, live on the holm oak (*Quercus ilex*), the shrub oak (*Quercus coccifera*) and a few other trees. Kermes is Armenian for "little worm" and is a scarlet dye. It is mentioned in the Old Testament and in the writings of ancient Greece, but its origins are probably Asiatic and it was much used in India.

The European invasions of south and southeast Asia and of the Americas greatly changed dye usage by bringing new and better colorants into the international markets. Indigo, which produces a range of blues, became for about two centuries the most important of all dyestuffs. It is a vegetable dye of considerable fastness, and has been known in parts of Asia for over four thousand years. *Indigofera tinctoria*, of which only the leaves bear dye, is of the order leguminosae, and belongs to the pea family. It was found in India, southeast Asia, Africa and America. It has been discovered in both Egyptian and Inca tombs, and its continent of origin is obscure.

Dutch ships carried indigo throughout the Indian Ocean, and then, in the seventeenth century, to Europe, where, despite a struggle, it eventually displaced woad. Bengal supplied large quantities in the late eighteenth century to the British textile industry. When American indigo, mostly produced in Central America, the Carolinas, and Georgia, began to flood the market in the eighteenth century, the woad industry collapsed in the face of this cheaper and better dyestuff.

Cochineal, a scarcer and more expensive dye, had a similar history. It is made from an American insect, *Coccus cacti*, which feeds on a cactus (*Nopalia* or *Opuntia cochinellifera*), and was used in Mesoamerica, especially in the Oaxaca region, long before the Europeans arrived. While its production is elaborate and costly, the result is a superior scarlet or crimson dye, and it soon replaced kermes in Asian and European markets and dyeworks. After bullion it was the most expensive item carried by the Spanish treasure fleets.

The brazilwood industry grew after the products of the American continents began to enter world commerce. Vast new stands were found in Brazil — hence the name — and to this was added logwood or campeachy wood, a large, tropical American tree (*Haematoxylon campeacheanum*) found especially in Campeche, Tabasco, and Belize. It yields black or blue dyes, plus edible seeds called allspice. By the seventeenth century it was in use in Africa and Europe.

One crop from America moved to Asia. Annato (*Bixa orellana*) was used by Mesoamerican peoples largely as a food additive and colorant, but when taken to southeast Asia and India this dye, which is a poor, fugitive yellow to red, was used for cloth, especially monks' robes. It is so culturally accepted in these regions today that many writers describe it as indigenous.

The new dominance achieved by Asian and above all American dyes such as indigo and cochineal, both of which were spread by European expansion, lasted less than two centuries. They, in their turn, were overwhelmed by the new coal tar or aniline dyes invented in the mid-nineteenth century. Many of the natural dyes, however, remain in use in local and peasant economies.

MURDO J. MACLEOD

REFERENCES

Cannon, John. *Dye Plants and Dyeing*. Portland, Oregon: Timber Press, 1994.

Donkin, R.A. *Spanish Red: An Ethnographic Study of Cochineal and the Opuntia Cactus*. Philadelphia: American Philosophical Society, 1977.

Leggett, William Ferguson. *Ancient and Medieval Dyes*. Brooklyn: Chemical Publishing Co., 1944.

Robinson, Stuart. *A History of Dyed Textiles*. Cambridge, Massachusetts: MIT Press, 1969.

Zanoni, Thomas A., and Eileen K. Schofield. *Dyes From Plants: An Annotated List of References*. New York: The New York Botanical Garden, 1983.

See also: Textiles

E

EAST AND WEST As traditionally used in the West, the terms East and West imply that the two are somehow of equal importance. While that might be arguable in the nineteenth and twentieth centuries it was certainly not true during the long reaches of human history prior to the nineteenth century. By 500 BC the globe supported four major centers of civilization: the Chinese, the Indian, the Near Eastern, and the Western, considering Greek culture antecedent to what eventually became the West. Of the four the West was probably the least impressive in terms of territory, military power, wealth, and perhaps even traditional culture. Certainly this was the case after the fall of the Roman Empire in the fifth century AD. From that time until about AD 1500 the West probably should be regarded as a frontier region compared to the other centers of civilization.

From roughly 500 BC to AD 1500 a cultural balance was obtained between the four major centers of civilization. During these millennia each center continued to develop its peculiar style of civilized life, and each continued to spread its culture and often its control to peoples and lands on the periphery. While the inhabitants of each center of civilization were aware of the other centers, sometimes traded with them, and occasionally borrowed from them, the contacts were sufficiently thin so that no one center threatened — commercially, militarily, politically, or culturally — the existence of the others. During this long period, that is through much of civilized human history, there was no question of Western superiority or hegemony. No visitor from Mars would likely have predicted that the West would eventually dominate the globe.

Obviously the Greeks and Romans knew quite a bit about the Near-Eastern world, especially about the Persian empire, that of Alexander the Great, and the successor states formed after its collapse. About India and China they knew much less, and what they knew was much less accurate. Although Herodotus (ca. 484–425 BC) reported some things about India, most of the information available to the Greeks came from the writers who described Alexander's campaigns in the Indus Valley (326–234 BC) and from Megasthenes. They described India as fabulously rich, the source of much gold and precious stones. It was hot; the sun stood directly overhead at midday and cast shadows toward the south in summer and toward the north in winter. They described huge rivers, monsoons, tame peacocks and pheasants, polygamy, and the practice of *suttee* (widow burning). However, they also reported fantastic things such as gold-digging ants, cannibals, dog-headed people, and people with feet so large that they served as sun shades when sitting. The Romans knew that India was the source of spices and that China, which they called Serica and well as Sinae, was the source of silk. There are even possible traces of Asian influence in some Roman silver and ivory work and perhaps even some influence of Buddhism on Neo-Platonism and Manichaeanism.

From about the fourth century, even before the fall of the Roman empire, until the return of Marco Polo from China in the late thirteenth century, Europe or the West added little factual information to its understanding of India and China. After the fall of the Roman empire there was no direct trade between Europe and Asia, and thus there were no opportunities to test the stories by observation. The rise of Islam in the seventh and eighth centuries completed Europe's isolation. During these centuries the old stories inherited from the Greeks were retold and embellished with little effort to distinguish fact from myth. To these were added three legends of more recent origin: the stories celebrating the heroic exploits of the mythical Alexander; those rehearsing Saint Thomas the Apostle's missionary journey to India and his subsequent martyrdom; and those describing the rich, powerful, Christian kingdom of Prester John somewhere to the east of the Islamic world with which European rulers dreamed of allying against the Muslims. Even the trickle of precious Asian products brought to Europe by intermediaries seemed only to confirm the image of Asia as an exotic and mysterious world, exceedingly rich and exceedingly distant.

The rise of the Mongol empire in Asia during the thirteenth century resulted in direct overland travel between Europe or the West, and China. The Mongols' success also revived hopes among European rulers of finding a powerful ally to the east of the Muslims. Even the devastating Mongol incursions into Poland and Hungary in 1240 and 1241 scarcely dampened their enthusiasm. Already in 1245 the pope sent an embassy led by John of Plano Carpini to the Mongol headquarters near Karakorum. He was followed during the ensuing century by a fairly large number of envoys, missionaries, and merchants, several of whom wrote reports of what they saw and did in Eastern Asia. Marco Polo's was the most comprehensive and reliable, and the most widely distributed of the medieval reports. By the time the Polos first arrived at Kublai Khan's court in 1264 it was newly established at Cambaluc (Beijing), from which the khan ruled the newly conquered Cathay (China). Like many other foreigners during the Mongol period (the Yuan Dynasty in China, 1260–1368) the Polos were taken into the khan's service. They were employed in the Mongol administration for seventeen years during which time Marco traveled extensively throughout China. On his return to Europe he produced the first detailed description of China in the West based primarily on first-hand observation and experience. No better account of China appeared in Europe before the middle of the sixteenth century. Marco Polo described China as the wealthiest, largest, and most populous land in the thirteenth-century world. While his understanding of Chinese culture was minimal he accurately and admiringly described cities, canals, ships, crafts, industries, and products. He noted the routes, topography, and people encountered in his travels, including his voyage home through Southeast Asia to Sumatra, Ceylon and along the west coast of India.

The decline of the Mongol empire and the establishment of the Ming Dynasty in China in 1368 severed the direct connection between Europe and China. The fall of Constantinople in 1454 and establishment of the Turkish empire in the Near East disrupted the older connections between Europe and the near East and India. Europe's isolation from the outside world was complete, not to be restored until the opening of the sea route around the tip of Africa in the waning years of the fifteenth century. During this period no European appears to have traveled to China. Some few travel reports refer to India and Southeast Asia. Of them only that written by the humanist Poggio Bracciolini in 1441 and based on Nicolò de' Conti's travels added to the West's store of knowledge about India and confirmed some of the more accurate of the ancient Greek reports. In fact amid the Renaissance humanists' enthusiasm for the rediscovery of ancient Greek literature, the Greek reports of India received new respect and attention.

During the long era of cultural balance before AD 1500 many important technological and scientific inventions and innovations appear to have migrated to the West from the other centers of civilization, more often from China than from the others. The migration of technology was usually gradual, involving one or more intermediaries, the inventions usually being established in the West without any clear ideas about their origins. Much of the basic technology that enabled the Europeans to sail directly to Asia in 1500 and later to begin their march towards global domination, was known earlier in the Asian centers and only later adopted or separately invented in Europe.

Among the more important technological borrowings were gunpowder, the magnetic compass, printing, and paper, all apparently originating in China. For none of them is the path of migration entirely clear, and thus for none of them can the possibility of independent invention be entirely ruled out. Gunpowder, for example, was known in China by 1040 and did not appear in Europe until the middle of the thirteenth century. The magnetic compass was fully described in an eleventh-century Chinese book, *Meng Qi Bi Tan* (Dream Pool Essays), written by Shen Gua in 1088. It began to be used in Europe during the late twelfth or early thirteenth century. Most likely Europeans learned about it from the Arabs. The case for moveable-type printing having been borrowed from the Chinese is more hotly debated than that for gunpowder or the compass. Wood-block printing was used in China by the seventh century, and paper was invented much earlier; the first printed books appeared there during the ninth century, six centuries before the invention of printing in the West by Johannes Gutenberg in 1445. Block printing probably became known in Europe through the introduction of printed playing cards and paper money during the Mongol period; medieval travelers frequently mentioned these. Because of the large number of Chinese characters the Chinese continued to prefer printing from page-sized blocks of wood carved as a single unit; European printing almost immediately employed moveable type, thus convincing some scholars that it was a separate invention. However, while they may have preferred block printing the Chinese also developed moveable type as early as the eleventh century. For none of these basic inventions taken separately — gunpowder, the compass, and printing — is the case for its diffusion from China to Europe indisputably demonstrated. Taken together, however, along with a rather large number of other technological and scientific innovations such as paper, the stern-post rudder, the segmented-arch bridge, canal lock-gates, and the wheelbarrow, which all appear to have migrated from China to the West, it becomes apparent that the general flow of technology and science in pre-modern times was from East to West. This would seem to increase the likelihood that these basic innovations also migrated to Europe from Asia. Those who used the technological innovations probably cared little about their ultimate origin and apparently did not seek it out. Nevertheless European mariners after 1500, confronted first hand with evidence that printing, gunpowder, the mariner's compass, and the like had been in use much longer in Asia than in Europe, frequently suggested that they had been borrowed from the Asians.

Even before gunpowder a group of military innovations found their way to Europe from China and India, again through intermediaries and apparently without Europeans' being aware of their origins. The Chinese form of the Indic stirrup was the most important of these and may have been as important to military development in the eighth century as gunpowder was later. The Javan fiddle bow and the Indian Buddhist pointed arch and vault were acclimated in Europe before 1100. The traction trebuchet along with the compass and paper appeared in the twelfth century. Still more important for subsequent Western scientific achievements was the adoption of Hindu–Arabic mathematics in the twelfth and thirteenth centuries: the Indian system of arithmetical notation, trigonometry, and the system of calculating with nine Arabic numbers and a zero were all practiced in India as early as AD 270. Some components of Indian mathematics may have come from Babylonia or China. They came to Europe, however, through the translation of Arabic writings, and the European borrowers usually credited India rather than China, Babylonia, or the Arabic intermediaries as the source of the new mathematics. Along with Indian mathematics, Europeans learned some elements of Indian astronomy and also became fascinated with the Indian idea of perpetual motion.

Also before 1500, Europeans sometimes attempted to imitate desirable Asian products, not always successfully. Already in the sixth century the Byzantine emperor Justinian monopolized the silk trade in his realm and expressed his determination to learn the secret of its manufacture. In 553 a monk supposedly smuggled some silkworm eggs into Constantinople carrying them in a hollow stick, perhaps of bamboo. Nothing is said in the story about the importation of silk technology, but less than a century later sericulture had obviously taken root in Syria. From there it spread to Greece, Sicily, Spain, Italy, and France. In Italy during the fourteenth century, water power was used in silk spinning, as it had been used in China much earlier. Attempts to imitate Chinese porcelain, however, were unsuccessful. The best attempts to do so were made in northern Italian cities during the fifteenth century. None, however, approached Chinese porcelain in composition, color, or texture. Nor did the Dutch Delftware of the seventeenth century, another attempt to imitate the Chinese product. Not until the eighteenth century were European craftsmen able to produce a hard-paste porcelain to rival that of China. Also appearing in Europe before 1500 were less important devices or techniques such as the Malay blowgun, playing cards, the Chinese helicopter top, the Chinese water-powered trip-hammer, the ball and chain governor, and maybe Chinese techniques of anatomical dissection.

While impressive, the West's importation of Asian science and technology before 1500 in no way deflected Western culture from its traditional paths. Seldom was the provenance of the new inventions or techniques known, and they could all rather easily be incorporated into the traditional Christian European world-view. They did not provoke any serious questions about the European way of life, its religious basis, its artistic and cultural traditions, or even its traditional scientific views. This is also true for the artistic and cultural borrowings from Asia prior to 1500. They too were often unconscious, and even when they were not they were regarded as embellishments or decoration, rather than in any way a challenge to traditional themes. For example, the incorporation into the Christian calendar of Saints Baarlam and Josephat, derived as they were from stories of the life of Buddha, resulted not in a Buddhist challenge to the Christian faith but simply in the addition of two new saints to the growing Christian pantheon.

Nevertheless, the borrowing and adaptation of Asian science and technology by the West before 1500 was indispensable in making the long overseas voyages to Asia and the protection of the ships and shore installations possible. Without gunpowder, cannons, the compass, the stern-post rudder, etc. there would have been no European expansion. While they had lagged well behind the other centers of civilization through most of civilized history, by 1500, European marine and military technology were beginning to equal that of the Near East, India, and China, and were obviously superior to that of the peripheral areas of Africa, Southeast Asia, and the Americas. The Portuguese voyages down the coast of Africa and Columbus' voyage across the Atlantic attest to Europe's rapidly improving technology.

A small Portuguese fleet under Vasco da Gama reached Calicut on the Malabar Coast of India in 1498, thus establishing direct contact between Europe and Asia by sea and also inaugurating a new era in the relationships between the four major centers of civilization on the globe. Soon after 1500 the Europeans began to dominate the seas of Asia, the Portuguese in the Indian Ocean being the first. The Portuguese, and later the Dutch and the English, moved along sealanes, visited seaports, and fitted into a trading world which had been developed earlier by Muslim traders and which stretched from eastern Africa to the Philippines. As the Europeans at first tried to compete in that world and later tried to dominate and even to monopolize it, they found Muslim merchants and merchant-princes to be their most formidable opposition. That they were able to move so rapidly and effectively into that international trading system was due not so much to their superior technology and firepower as to the fact that the great Muslim empires of the sixteenth century — the Ottoman Turkish, the Safavid Persian, and the Mughul in northern India — seem to have been too busy consolidating their newly won empires to contest the European intrusion. These Muslim empires, as well as the Southeast and East Asian empires of Siam, Vietnam, China, and Japan had all as a matter of governmental policy turned away from the sea and looked inward to control of their land empires and to land taxes as the source of their wealth and power. Had any or all of these major Asian and Near-Eastern states seriously resisted the western incursion the story might have had a different conclusion. Apart from Muslim traders in the Indian Ocean those who formidably opposed European power in the sixteenth and seventeenth centuries were small Muslim commercial port-city states like Makassar on Celebes and Aceh on Sumatra. The Mughul emperors in northern India usually allowed the Europeans to trade freely in their ports, often exempting them from customs duties. Apart from the illegal but locally tolerated Portuguese settlement in Macao the Europeans were not permitted to trade in China at all. China briefly contested Dutch maritime power only in 1624 after the Dutch had spent two years raiding the coast of Fukien Province, burning villages, seizing junks, enslaving their crews, and constructing a fort on one of the Pescadores Islands; all in an effort to force the Chinese to allow them to trade freely at some port along the Chinese coast. Confronted with a full scale Chinese war fleet the Dutch commander hastily sued for peace and gratefully accepted the Chinese admiral's offer to let the Dutch trade on Formosa, which was not yet considered Chinese territory. From 1500 until about the middle of the eighteenth century the Europeans were able to carve out empires in the Americas, insular Southeast Asia, and the Pacific Islands, but they did not threaten the major centers of civilization in Asia.

During the first three centuries of direct maritime contact between Europe and Asia (1500–1800) the Westerners showed little sense of cultural superiority toward the high cultures of Asia. If anything they tended to exaggerate the wealth, power, and sophistication of the other centers of civilization. Most Europeans were confident that Christianity was indeed the true religion, and they quickly began to send missionaries to convert Asian peoples. By 1600, if not earlier, they were also justifiably confident that European mathematics, science, and technology were superior to that of the Asians. Those areas aside, however, Europeans were endlessly fascinated with what they discovered beyond the line and realized that they still had much to learn from the high cultures of Asia. Between 1500 and 1800 the currents of cultural influence continued to flow mainly from east to west, but during this era the impact of Asia on the West was usually more conscious and deliberate than previously, and it was primarily in areas other than science and technology.

The new seaborne commerce with Asia almost immediately brought greatly increased quantities and varieties of Asian products into Europe. Pepper and fine spices were the first to appear on the docks in European ports, but they were soon followed by such goods as Chinese porcelain and lacquerware, tea, silks, Indian cotton cloth, and cinnamon. Following these staples of the trade came also more exotic products: Japanese swords, Sumatran or Javanese krisses, jewelry, camphor, rhubarb, and the like; the list is very long. Some of these products, such as tea, provoked striking social changes in Europe. Attempts to imitate others resulted in new industries and in new manufacturing techniques, Delft pottery in imitation of Ming porcelain, for example. Attempts to compete with cheap Indian cotton cloth seem to have touched off a technical revolution in the British textile industry which we customarily regard as the beginning of the industrial revolution.

Along with the Asian products came descriptions of the places and peoples who had produced the products. The earliest sixteenth-century descriptions usually seem designed to inform other Asia-bound fleets about the conditions of trade. After Christian missions were established they were often intended to elicit support for the missionaries. However, before long the travel tales and descriptions became popular in their own right and profitable to publish. During the seventeenth century what had been a sizeable stream of literature about Asia became a veritable torrent. Hundreds of books about the various parts of Asia, written by missionaries, merchants, mariners, physicians, soldiers, and independent travelers were published during the period. For example, during the seventeenth century alone there appeared at least twenty-five major descriptions of South Asia, another fifteen devoted to mainland Southeast Asia, about twenty to the Southeast Asian archipelagoes, and sixty or more to East Asia. Alongside these major independent contributions stood scores of Jesuit letterbooks, derivative accounts, travel accounts with brief descriptions of many Asian places, pamphlets, newssheets, and the like. Many of the accounts were collected into the several large multivolume compilations of travel literature published during the period. In addition to the missionaries accounts, travel tales, and composite encyclopedic descriptions, several important scholarly studies pertaining to Asia were published during the seventeenth century: studies of Asian medicine, botany, religion, and history; and translations of important Chinese and Sanskrit literature.

The published accounts range in size from small pamphlets to lavishly illustrated folio volumes. They were published in Latin and in almost all of the vernaculars, and what was published in one language was soon translated into several others, so that a determined enthusiast could probably have read most of them in his own language. They were frequently reprinted in press runs which ranged from 250 to 1000 copies. Five to ten editions were not at all uncommon, and some of the more popular accounts would rival modern "best sellers". In short the Early-Modern image of Asia was channeled to Europe in a huge corpus of publications which was widely distributed in all European lands and languages. Few literate Europeans could have been completely untouched by it, and it would be surprising if its effects could not have been seen in contemporary European literature, art, learning, and culture.

From this literature European readers could have learned a great deal about Asia and its various parts. Perhaps most obviously their geographic horizons would have been continually expanded. Gradually Europeans gained accurate knowledge about the size and shape of India, China, and Southeast Asia. During the seventeenth century, for example, several puzzles which had plagued earlier geographers were solved: for example, the identification of China with Marco Polo's Cathay, the discovery that Korea was a peninsula and Hokkaido an island. By the end of the century Europeans had charted most of the coasts of a real Australia to replace the imagined antipodes as well as those of New Guinea, the Papuas, numerous Pacific islands, and parts of New Zealand. Interior Ceylon and Java, as well as Tibet were visited and accurately described by Europeans before the end of the century. By 1700 only areas of continental Asia north of India and China, the interior of Australia, New Zealand, and New Guinea and parts of their coastlines remained unknown to the Europeans. Most of these lacunae were filled during the eighteenth century. Even more impressive than the greatly expanded geographic knowledge available to European readers in early modern times was the rapidly increasing and increasingly detailed information about the interiors, societies, cultures, and even histories of Asia's high cultures. Already during the seventeenth century European readers could have read detailed descriptions and even viewed printed cityscapes and street scenes of scores of Asian cities, interior provincial cities as well as capitals and seaports, and they could have learned countless details about Asia's various peoples, their occupations, appearance, social customs, class structures, education, ways of rearing children, religious beliefs, and the like. Details regarding Asia's abundant natural resources, crafts, and arts were described as well as its commercial practices and patterns of trade. Asian governments were described in exceedingly close detail, especially for major powerful states such as China, the Mughul Empire, Siam, and Japan. Jesuit missionaries in China, for example, described the awesome power of the emperor, his elaborate court, the complex imperial bureaucracy and its selection through competitive written examinations, and the Confucian moral philosophy on which it was all based. They also described the frequently less orderly and less savory practice of Chinese government, complete with detailed examples of officials' abuse of power and competing factions within the administration. Similar details were reported for the governments of all the major states as well as for countless smaller states. By the end of the seventeenth century European observers had published many sophisticated accounts of Asian religions and philosophies; not only the frequently deplored Hindu "idolatry" and widow burning, but also the Hindu world view which lay beneath the panoply of deities and temples, the various schools and sects of Hinduism, and the ancient texts of Hindu religion. Similarly sophisticated and detailed accounts of Confucianism and Buddhism were available, as well as descriptions of the beliefs of peoples like the Formosan aborigines, the Ainu of Hokkaido, and the inner Asian and Manchurian tribes. Seventeenth and eighteenth-century readers could also have learned much about Asian history; especially, but not exclusively, that of Asia's high cultures. By the mid-seventeenth century, for example, a very detailed sketch of China's long dynastic history culled from official Confucian histories by Jesuit missionaries had been published. Martino Martini wrote the *Sinicae historiae decas prima*, which was published in Munich in 1658. During the eighteenth century an important Chinese history, the *Tongjian Gangmu* (Outline and Details of the Comprehensive Mirror[for aid in government]) by Zhu Xi, was translated into French in its entirety by Joseph-Anne-Marie de Mailla and published in Paris (*Histoire générale de la Chine*, 1777). Not only history, however, but also news was reported to Early-Modern readers. Their image of Asia was surely not that of a static world far away. Among the more important events reported in almost newspaper-like detail during the seventeenth century alone were the Mughul emperor's successful campaigns in the south of India, the Maratha challenge to Mughul supremacy, the fall of the Indian states of Golconda and Vijayanagar, the Manchu Conquest of China in 1644, the feudal wars and the establishment of the Tokugawa shogunate in Japan (1600), and the internal rivalries and wars in Siam and Vietnam. Natural disasters such as earthquakes, fires, volcanic eruptions, and the appearance of comets were also regularly reported. Readers of this richly detailed, voluminous, and widely distributed literature may well have known relatively more about Asia and its various parts than do most educated westerners today.

The post-1500 literature on Asia also contains a great many descriptions of Asian science, technology, and crafts: such things as weaving, printing, papermaking, binding, measuring devices, porcelain manufacture, pumps, watermills, hammocks, palanquins, speaking tubes, sailing chariots, timekeepers, astronomical instruments, agriculture techniques and tools, bamboo and other reeds for carrying water, as well as products such as musical instruments, wax, resin, caulking, tung-oil varnish, elephant hooks and bells, folding screens, and parasols. Some, such as Chinese-style ship's caulking, leeboards, and strake layers on hulls, lug sails, mat and batten sails, chain pumps for emptying bilges, paddle-wheel boats, wheelchairs, and sulphur matches were quickly employed or imitated by Europeans. Some provoked documented experimentation and invention: Della Porta's kite in 1589, and Simon Stevin's sailing chariot in 1600. The effects of the new information on the sciences of cartography and geography are obvious and profound. Simon Stevin, whose sailing chariot was inspired by descriptions of similar Chinese devices, also introduced decimal fractions and a method of calculating an equally-tempered musical scale; both of which might also have been inspired by Chinese examples. The sixteenth-century mariners' cross staff may have been inspired by the Arab navigators' *kamals*. The Western science of botany was profoundly influenced by the descriptions of Asian flora and by the specimens taken back to Europe and successfully grown in European experimental gardens; rice, oranges, lemons, limes, ginger, pepper, and rhubarb were among the most useful. More important for botany, however, the Asian plants provoked comparisons with familiar plants, the development of comprehensive classification schemes, and thus the beginnings of modern plant taxonomy. Some Asian cures (herbs and drugs) were borrowed, especially for tropical medicine. Chinese acupuncture, moxibustion, and methods of diagnosis by taking the pulse were minutely described and much admired by European scholars, but it is not yet clear to what extent they were actually used in Europe.

The flow of cultural influence was not exclusively from East to West between 1500 and 1800. Many Asians became Christian. Able Jesuit missionaries translated scriptures and wrote theological works in Chinese and other Asian languages; they also translated European mathematical and scientific treatises into Chinese. The Kangxi emperor himself studied Euclidian geometry, Western astronomy, geography, the harpsichord, and painting under Jesuit guidance. Like many Chinese he was fascinated by European clocks. Many Asians, including the Chinese, learned how to use western firearms and cast western-style cannon. However, the cultural consequences of these efforts were disappointingly small and of short duration. Before the end of the eighteenth century they seem largely to have disappeared along with the Christian mission. The Japanese, however, even during the closed-country period after 1640, were far more curious about Western science. Samurai scholars, for example, studied Dutch medicine and science and in the eighteenth century repeated Benjamin Franklin's kite-flying experiment. Nevertheless through most of the early modern period the Europeans were far more curious about and more open to influence from Asia than were any of the high cultures of Asia to influence from the West

During the seventeenth and eighteenth centuries the new information about Asia influenced Western culture primarily in areas other than science and technology. The extent of this impact remains to be comprehensively studied, and even of that which is known only a few examples can be mentioned here. Asian events and themes entered European literature in scores of instances from Lope de Vega, Ariosto, Rabelais, and More in the sixteenth century to the several Dutch, German, and English plays and novels depicting the fall of the Ming Dynasty and the triumph of the Manchus in the seventeenth century, to Voltaire's literary and philosophical works in the eighteenth. Even popular literature, seventeenth-century Dutch plays and pious tracts, for examples, shows surprising familiarity with the new information about Asia. Asian influences in European art, architecture, garden architecture and the decorative arts also began in the sixteenth century and culminated in the *chinoiserie* of the eighteenth century. Confrontation with China's ancient history challenged the traditional European Four-Monarchies framework of universal history and touched off a controversy among European scholars which by the mid-eighteenth century resulted in an entirely new conception of ancient world history. Some scholars, beginning with Pierre Bayle in the late seventeenth century, have detected a Neo-Confucian influence in the thought of Spinoza and in some aspects of Leibniz's philosophy. Chinese government and especially the examination system were frequently held up for emulation by European states during the seventeenth and eighteenth centuries. It might well be that the institution of written civil-service exams in Western states, beginning with eighteenth-century Prussia, was inspired by the Chinese example. Of more general importance that any single instance of influence, however, was the challenge presented by long-enduring, sophisticated, and successful Asian cultures to traditional European assumptions about the universality of their own. Perhaps here can be found the beginnings of cultural relativism in the West.

By 1600 European science and technology generally and especially marine and military technology outstripped that of any Asian society, whatever had been borrowed earlier from them. Also, while the fascination with and appreciation of the high cultures of Asia continued through most of the eighteenth century, the West between 1600 and 1800 experienced the radical transmutation of its traditional culture which resulted in the development of a rational, scientific approach to the use of nature, and society — to agriculture, business, industry, politics, and above all warfare — which we have traditionally associated with the Scientific Revolution, Enlightenment, and Industrial Revolution. This transmutation has enabled Western nation-states during the past two centuries to establish the world-wide competitive empires which came to dominate all the other centers of civilization and has resulted in the global dominance of Western culture. The triumph of this transmuted Western culture has reversed the centuries-long East-to-West flow of cultural influence and threatens all of the world's traditional cultures. It should be remembered, however, that this rational, scientific, industrialized Western culture was not an obviously natural outgrowth of traditional Western culture, that in its early development it received important basic components from Asia, and that its triumph threatens traditional Christian Western culture almost as seriously as it threatens those of Asia.

EDWIN J. VAN KLEY

REFERENCES

Chaudhuri, K. N. *Trade and Civilization in the Indian Ocean: An Economic History from the Rise of Islam to 1750*. Cambridge: Cambridge University Press, 1985.

Cipolla, Carlo M. *Guns, Sails, and Empires: Technological Innovation and the Early Phases of European Expansion, 1400–1700*. New York: Pantheon, 1965.

Hodgson, Marshall. *The Venture of Islam: Conscience and History in a World Civilization*. Chicago: University of Chicago Press, 1974.

Lach, Donald F. *Asia in the Making of Europe*. Vol. I: *The Century of Discovery*. Bks. 1–2. Chicago: University of Chicago Press, 1965. Vol. II: *A Century of Wonder*. Bks. 1–3. Chicago: University of Chicago Press, 1970, 1977.

Lach, Donald F. and Van Kley, Edwin J. *Asia in the Making of Europe*. Vol. III: *A Century of Advance*. Bks. 1–4. Chicago: University of Chicago Press, 1993.

McNeill, William H. *The Rise of the West: A History of the Human Community*. Chicago: University of Chicago Press, 1963.

McNeill, William H. *The Pursuit of Power: Technology, Armed Force, and Society Since* AD *1000*. Chicago: University of Chicago Press, 1982.

Needham, Joseph. *Science and Civilisation in China*. 9+ vols. Cambridge: Cambridge University Press, 1954.

Parker, Geoffrey. *The Military Revolution: Military Innovation and the Rise of the West, 1500–1800*. Cambridge: Cambridge University Press, 1988.

Wolf, Eric R. *Europe and the People without History*. Berkeley: University of California Press, 1982.

See also: Navigation – Compass – Magnetism – Gunpowder – Paper and Papermaking – Shen Gua – Military Technology – Mathematics – Technology – Zero – Colonialism and Science – Medicine

EAST AND WEST: AFRICA IN THE TRANSMISSION OF KNOWLEDGE FROM EAST TO WEST Africa's role in the transmission of knowledge from the East to the West has been strongly conditioned and circumscribed by its geographical location and relative isolation from European and Asian centers of civilization. Africa is to the south of Europe, home of the "West"; it is also positioned to the west and south of Asia, the historical and cultural "East" of Europe, from which it is separated by distance, the Sahara desert, and the Indian Ocean. The contiguity of the Eurasian landmass makes it easy for East–West communications to bypass Africa. Nevertheless, at various periods of pre-modern history Africa transmitted systems or elements of knowledge and culture to Europe, some generated in Africa, others developed in Asia in the Near, Middle, or Far East. The parts of Africa most actively involved in such exchanges are delimited by the waters of the Mediterranean and Red Seas, the latter guiding to Arabia and the Indian Ocean. The precise role played in these exchanges by Egypt, Ethiopia, and North Africa across time was subject to fluctuations in the environment, growth of early civilizations, long-distance trade, and the development of cities and elites.

Historically, the intellectual contribution of African societies to world civilization has been overshadowed by the magnitude of African economic and cultural inputs. In addition, numerous direct contacts between Africa and Asia (especially in the northern part of the continent and across the Indian Ocean) by-passed Europe for simple geographical reasons, or failed to have an impact on Western civilization because of circumstances of political, cultural, or religious nature (for example, the rise of Islam). This lack, inconsistency, and sporadic nature of intellectual interaction between Africa and the West are overlooked in some enthusiastic attempts to assert African primacy in world civilization based on incorrect interpretation of evidence. Reconstruction of the African cultures of the past faces serious problems because of the lack of writing systems among the majority of pre-modern African societies. In the absence of a written record, the historical existence of knowledge is often inferred from contemporary anthropological evidence or archaeological data. To complicate things further, the secret of some African scripts that had existed in the past has been lost, as is the case with Libyan, Meroitic, and Tuareg. Thus the internal written evidence for the ancient period is currently limited to texts recorded in Egyptian hieroglyphics and the Ethiopic script. However, few of the surviving texts deal with scientific knowledge or formal disciplines of learning.

African elements in the culture of Ancient Egypt came primarily from the south (the Nile Valley above Aswan, where ancient Nubia and Kush were located) and the West (Libyan desert, Saharan oases and possibly Lake Chad region, and the rim of the Mediterranean). The prosperity and cultural splendor of Egypt during the Old and Middle Kingdoms (ca. 2600–1780 BC) were second to none. Both depended on a flourishing agriculture which in turn reflected and supported the high development of astronomy, mathematics, hydrology, and technology. The texts and practices of ancient Egyptians provide evidence of elaborate knowledge of medicine, human anatomy, and chemistry. About the time of the transition from the Middle to New Kingdom (ca. 1500 BC) Egyptian influence reached the Greek periphery. The early cultures of Crete and Archaic Greece show traces of Egyptian and North African influence. Later, Phoenician colonies in North Africa (most prominently Carthage) became a bridge between Africa and the northern Mediterranean. Although there is no record of formal transmission of knowledge from the Punic roots to Rome, some elements of Punic culture spread through the expanse of the Roman Empire as far north as the British Isles. Beginning with Herodotus (fifth century BC), early Greek descriptions of Egypt and Africa make clear the Greek admiration for Egypt and vaguely acknowledge a certain cultural continuity with Egyptian learning and philosophy. Greek mythology hints at North African (Libyan) religious influence, and Greek folklore (e.g. Aesop's fables) reflects a cultural–philosophical connection with sub-Saharan Africa ("Ethiopia" of the Ancients). Egypt was recognized by the Greeks as both an integral part of Africa and a crossroads on the way east. After the conquest by Alexander the Great, Egypt (especially Alexandria) became a major center of cosmopolitan Hellenistic culture and long-distance trade connecting the Mediterranean with Arabia, Ethiopia, India, and East Africa. Late Greek and Greco–Roman astronomy, mathematics, geography, medicine, and philosophy drew upon the knowledge and thinking of Hellenized Egyptians. The geographical and astronomical works of the Librarian of Alexandria, Claudius Ptolemy (second century AD), continued to influence Western and Islamic scholarship until the sixteenth century.

With the rise of Christianity, Egypt, Ethiopia and North Africa became a major stage for religious rather than scientific developments. Some of them involved religious controversies with Constantinople, Rome, and Near Eastern neighbors, and resulted in cultural and spiritual separation of the Asian and African Christian communities from mainstream Christian societies even where geographical isolation and remoteness (as in Ethiopia) were not a factor. A further transformation of cultural networks occurred after the Arab conquest of Egypt and North Africa in the seventh-early eighth century AD. Arabic became the language of learning and Islam the dominant ideology. A fragile cultural connection between Ethiopia, Egypt, and Byzantium survived for a few more centuries in the religious sphere; it channeled limited influences in theology, art, and hagiographic literature. One unrecognized consequence of this link is the use of Coptic (originally Egyptian hieroglyphic) elements in the Glagolitic (later Cyrillic) script invented in the ninth century for the Slavic vernaculars by the Byzantine monks Cyril and Methodios.

The early Islamic science was a synthesis of Hellenistic, Middle Eastern, and Indian sciences, to which was added a considerable body of original knowledge, especially in mathematics, optics, geography, and medicine. Egypt and North Africa became a bridge by which these sciences were then transmitted from the Middle Eastern centers, especially Baghdad, to the rest of the Islamic world and to Europe. Islamicized societies of tropical Africa were most influenced by the legal, theological, and philosophical thought. This Islamic contact stimulated the adoption in the Middle Ages of the Arabic alphabet for writing in some African languages, such as Swahili in East Africa and Hausa in West Africa. Europe, on the other hand, primarily borrowed Islamic knowledge in the fields of natural and applied sciences and philosophy. The Iberian peninsula was involved in this process in a number of ways: first, through the spread of Asian food and technical crops and cultivation techniques as a result of Arabo–Berber occupation in the eighth century; then in the eleventh to thirteenth centuries under the Almoravid and Almohad empires when Spain was united with northwest Africa as far south as the Senegal River; third, through the Christian translation activities of the late eleventh to sixteenth centuries, and finally as a transit point for transmission to the New World.

From the period of the later Crusades and under the Mamluk dynasties Egypt became the dominant Islamic partner of European traders on the Mediterranean and gained exclusive control of the Red Sea and thus access to the trade of the Indian Ocean. By the thirteenth century the transmission to the Mediterranean of the magnetic compass and the lateen sail (both of them products of the Indian Ocean societies) had been accomplished. After the Mongol conquest of Baghdad in 1258 Cairo became the foremost center of culture and learning for the Near East and Africa. In addition, it was a major transit point on the pilgrimage route to Mecca, visited by many Africans for purposes of both ritual and learning. Cairo kept this role despite the ravages of the Black Death in the mid-fourteenth century and the Ottoman Turkish conquest in 1516, and has retained it to a certain extent to this day. However, even before the fall of Byzantium in 1453 the balance of secular learning in the eastern Mediterranean had shifted, and Muslims became eager students of the early modern West. In the western Mediterranean, the European and Islamic powers interacted in the hostile atmosphere of the Christian Reconquista and piracy; by the sixteenth century the technological balance of power had tilted in Europe's favor, and the geographical arena of progress moved to the Atlantic.

In modern times, the European discovery of the New World and the expansion of the West to overseas colonial empires led to new patterns of communication which could connect Africa with Asia and the West Indies or the Americas without involving Europe. The new global networks served very different purposes and produced dramatically different results (one such network was the Portuguese sea-borne empire spread over three continents; its shipping lanes led from

Goa in India east to Macao or west to Mozambique and Angola to Brazil, the vessels often completely by-passing Portugal itself). The new societies developing overseas were molded by the West but no longer European; in the Americas they experienced heavy influxes of Africans capable of preserving in their midst some of the living African culture even under the conditions of slave plantation economy. Traditional knowledge of African medicine, philosophy, and folklore, including occasional Islamic elements, colored the social, cultural, and spiritual life of African immigrants in the New World and fed into the culture of the larger society despite class barriers and frequent misperception as "magic" or "superstition". These influences originated mostly in the tropical and equatorial areas of western Africa. Another channel of unstructured cultural transmission to early Latin America was the importation, within Iberian culture, of "Moorish" (that is Islamic or North African) elements, for example in architecture. A special, modern example of cross-cultural interaction in the "Old World" is the South African colonial society where European farmers benefited from observing the herding practices and environmental adaptation of the indigenous groups (the Khoi-, San-, and Bantu-speakers). The Dutch-based Afrikaans language of white South Africans contains African loan-words illustrative of such informal, unrecorded, and often unrecognized transmission of African knowledge to Western societies.

M.A. TOLMACHEVA

REFERENCES

Bentley, Jerry H. *Old World Encounters: Cross-Cultural Contacts and Exchanges in Pre-Modern Times*. New York and Oxford: Oxford University Press, 1993.

Bernal, Martin. *Black Athena: The Afroasiatic Roots of Classical Civilization*. 3 vols. London: Free Association Books, 1987.

Davidson, Basil. *Africa in History*. New York: Collier Books, 1974.

Diop, Cheikh Anta. *The African Origin of Civilization: Myth or Reality*. New York: Lawrence Hill and Company, 1974.

James, George G. M. *Stolen Legacy: the Greeks Were Not the Authors of Greek Philosophy, but the People of North Africa, Commonly called the Egyptians*. New York: Philosophical Library, 1954 (Reprint: San Francisco: Richardson Associates, 1976.)

McNeill, William H. *The Rise of the West: A History of the Human Community*. Chicago: University of Chicago Press, 1963, esp. ch. I–VI.

Mazrui, Ali A. *World Culture and the Black Experience*. Seattle: University of Washington Press, 1974.

Meier, August and Elliott Rudwick, eds. *The Making of Black America*. Vol. I. New York: Atheneum, 1969.

Moreno Fraginals, Manuel. *Africa in Latin America*. London: Holmes & Meier, 1984.

Petry, Carl F. *The Civilian Elite of Cairo in the Later Middle Ages*. Princeton: Princeton University Press, 1981.

EAST AND WEST: CHINA IN THE TRANSMISSION OF KNOWLEDGE FROM EAST TO WEST Despite the distances between them and their totally different cultures, there has been more or less continuous communication between the West and China since classical Greek times. Although the connection was indirect and limited to trade in luxury goods, there were even in ancient times marvelous resemblances between Western and Chinese inventions in technology and engineering. Therefore it is natural to believe that a transmission of ideas and knowledge must have occurred, even if it is impossible to describe the exact exchange of any particular scientific achievement.

In the thirteenth century there was some considerable personal contact between the West and China, of which that of William of Ruisbroek (1210–1270), Marco Polo (1254–1324), and Odoric of Pordenone (1286–1331) are but the most famous examples. By the end of the sixteenth century in China an era of isolation ended and a new stage of intercultural exchange began. In this period the first Jesuit missionary fathers entered China, bringing with them knowledge of the sciences of Renaissance Europe. The Jesuits' scientific instruction was intended to aid their religious teaching by adding to the prestige of the culture they represented. Although in many respects Chinese technology was more advanced than European technology until the Renaissance, it lost ground in subsequent centuries. Francis Bacon (1561–1626) saw Chinese inventions, namely the magnetic compass, gunpowder, and the printing press, as crucial for the transformation of European society. However, after the generation of Leibniz (1646–1716), Wolff (1679–1754), and Voltaire (1694–1778) with the end of the Enlightenment movement, European philosophers and historians began to speak in a disparaging way of Chinese culture. From the late eighteenth century on, the myth of a closed China was born, and Europeans saw the scientific exchange between China and the outside world solely in terms of what the Chinese borrowed from the West. Though Chinese technological advances were old and undeniable, many authors of textbooks on the history of science, technology, and medicine still assured their readers that China never created science as a persisting institution and never consciously developed technology on the sound basis of a theory for applied science.

The British historian of science Joseph Needham (1900–1995) with his work *Science and Civilisation in China*, began to restore the reputation of China as a cradle of scientific inventions. He showed the patterns of transmission of these inventions and enriched the knowledge of Chinese science and technology. Nathan Sivin, an American historian, introduced a different angle. He presented the argument that it is impossible to equate divisions of modern science and engineering with premodern Chinese divisions. Chinese science was not integrated under the authority of philosophy, as schools and universities merged them in European and Islamic cultures. The Chinese had sciences but not "science", not a single conception or word for the sum of all scientific divisions. Traditional Chinese terms for science might exclude empirical methods but include ethical or religious principles discovered through reflection on authoritative texts.

Since there is neither sound evidence for early adoption of Chinese science in the West, nor agreement as to whether Chinese science is distinct from sciences in the West, we have to answer each complex of questions separately. First of all, when immediate contact between China and the West was too limited to be noteworthy, we have to determine whether the Western world owed much to foreign influences which in turn made use of knowledge from China. Second, we must ask whether the conception of Chinese sciences inevitably leads to disparaging treatment by those scholars who are measuring two radically different things by the same standard. To draw a more balanced picture of Chinese scientific thought, we have to look for autochthonous features of ancient Chinese sciences and explain their characteristics.

To solve the first part of the puzzle, we have to consult Chinese historical records which contained much information on foreign relationships. Although there is clear evidence for Arabic, Babylonian, and Indian relations with China, recent archaeological finds demonstrate the originality of Chinese culture in many ways. We still do not know enough about the routes and intermediaries by which neighboring states imported cultural artifacts from China. Many transmissions which finally reached the West transformed Chinese, Indian, Iranian, and Greek contributions simultaneously.

Under the Han rule (221 BC–AD 220), as a result of military campaigns and diplomatic activities, China's immediate contacts with other cultures grew to a degree so far unrivaled in Chinese history. In this period China spread her concepts and skills around all of Asia. As early as 130 BC, Chinese government officials set out to explore the routes into Central Asia and North China. The success of these missions is very difficult to reexamine, and yet it is probable, because in the official documents of these times we find many references to China's intercultural exchange. The first record of official visitors arriving at the Han court from Japan is for the year AD 57, and we can furnish proof of a mission from Rome which had reached China by ship in AD 166.

With the downfall of the Han Dynasty in AD 220 China entered a period of decline and disunity with respect to the transmission of ideas. After a period of domestication and growth under the later Han, Buddhism came from India into China. The spread of Indian thought was attended by a dispersion of Chinese thought westwards. For the next three hundred years the Silk Road which linked China with the West was the most important connection between the different culture groups. On this trade route goods and ideas were carried between the three great civilizations of Rome, India, and China.

Under the Sui reign (589–618), when China was consolidated again, grandiose plans aiming to unify the empire culminated in projects for new canal systems. As a result a great canal linking central and southern China was constructed and long campaigns in Manchuria and on the Korean frontier were prepared. There was a direct relationship between waterway construction works and active foreign policy. Using the new canals for logistical support, the Sui realm was able to establish sovereignty over old Chinese settlements in the south, and extend its influence to other territories, especially in central Vietnam. In addition to these military operations expeditions were sent to Taiwan, and relations with Japan were opened. Sui colonies were established along the great western trade routes, and rulers of several minor local states of Central Asia became tributaries. This was the time when China was in contact with the eastern Turks, who occupied most of the Chinese northern frontier, and the even more powerful western Turks, whose dominions stretched westward to the north of the Tarim Basin as far as Sassanid Persia and Afghanistan. We can learn from the writings of the Arabian scientist and philosopher Al-Bīrūnī (973–1048) that the Turkish sphere of influence was a fertile soil for intercultural exchange. As a consequence of the history of Turkish engagement in Inner Asia remarkable transmissions of technological know-how took place and helped to spread, among other things, Chinese defense engineering and mechanical skills.

Various imports and influences into Arabic empires originated essentially from China, or were at least transmitted through the intermediary of the Turks. As we can prove by discoveries of imported pottery and textiles the links between Iran in the Sassanid period (AD 226–651) and China were extremely close. With the decay of Turkish power the new Tang dynasty (618–906) extended its power all over East Asia. Chinese western dominions extended even farther than in the great days of the Han, and trade developed with the West, with Central Asia, and with India. The Chinese capital was thronged with foreign merchants and monks. Every great city contained a variety of non-Chinese communities and had Zoroastrian, Mazdean, and Nestorian temples, along with Buddhist monasteries. This set the stage for even more transfers of knowledge and inventions.

Perhaps it is not exaggerated to call China under the Song rule (960–1279) the most advanced civilization at that time. During the Song dynasty an agricultural revolution produced plentiful supplies for a population of more than one hundred million. Acreage under cultivation multiplied in all directions, and a variety of early ripening rice, imported during the eleventh century from regions in modern Vietnam and Cambodia, shortened farming time to below one hundred days and made two crops a year the norm and three crops possible in the warm South. Among other new crops the most important was cotton, which provided clothing for rich and poor alike; silk and hemp were also important. Improved tools, new implements, and mechanical devices that raised manpower efficiency were widely used. Although advanced skills were guarded as trade secrets, many technical inventions of these times found their way into printed manuals used at home and abroad. Productivity of such minerals as lead, tin, silver, and gold increased tremendously. In manufacturing the Chinese improved processing in skill-intensive patterns; they began with mass production, and a division of labor as well, while skills and products entered into diversified specialization. High quality earthenware progressed to genuine porcelain, which attained international fame.

Despite the fact that until the early eleventh century Chinese maritime trade had been dominated by foreigners, Chinese artisans developed a new type of ship at the end of the century, helping the Song empire to take control over the transport of merchandise and passengers across the waters of East Asia. The Chinese ocean-going junks with large proportions and tonnage were bigger, more solid, but also more comfortable than the Arabian and Indian ships. From the twelfth to the fourteenth centuries the Chinese fleet was at the peak of its power, at least on the routes linking China with the ports of south India. Chinese trade with Southeast Asia increased for the very last time at the end of the fourteenth century when China flooded Sumatra and Borneo with ceramics, and also coins, in exchange for spices, aromatics, medicinal drugs, and precious woods. Chinese ceramics of this period have been found in quite large quantities not only along the silk road, but also on the sites of ancient ports or depots along the navigation routes towards Indonesia, and in the great commercial cities of the Middle East.

In the thirteenth century after the Mongol conquest of China and the founding of the Yuan dynasty (1280–1367) Europeans developed a lively interest in Chinese affairs. The Roman Catholic Church also looked for potential converts among the non-Muslim people of Asia. After Franciscan envoys brought back information on what was known as China in the mid-thirteenth century, Pope Nicholas IV dispatched a mission to the court of the Grand Khan. The Mongol capital Khanbaliq (Beijing) became the seat of an archbishopric, and in 1323 a bishopric was established. From then on missionaries traveled to China and brought back first-hand information to medieval Europe. These reports inaugurated an era of discoveries and created a new vision of the world, with China as a part. Furthermore, shortly after direct contact with China was established, European philosophers and clerics speculated upon Chinese modes of thinking. Chinese medical treatises were translated into Persian, and Persian pottery techniques show some influences of Chinese handicraft. In addition, Chinese-type administration and chancellery practices were adopted by various Mongol dominions in Central Asia and the Near East. Some scholars suggested that the invention of gunpowder and printing in Europe was due to a sort of stimulus diffusion from China, although there is no sound evidence for a direct influence from China via the Near East.

During the Ming reign (1368–1644) Europe was unable to maintain its contacts with the Far East, partially because the Black Death broke off many overseas trade relations. Toward the end of the fourteenth century China was almost forgotten. Meanwhile Ming dynasty bureaucracy reoriented foreign policy, which made foreign contacts much more difficult. Ming China's influence in Southern Asia reached its climax during the early fifteenth century when official exploratory voyages brought most important South Asian states into the Ming political sphere. Besides protecting China's southern borders, these voyages were undertaken to monopolize the overseas trade by preventing private individuals from taking control of seafaring activities. Foreign states responded to these overtures not only because they feared military reprisals if they refused, but also because they saw great commercial benefits in relations with China. In these years Chinese missions established contacts with most of the important countries from the Philippines to the Indian Ocean, the Persian Gulf, and the east coast of Africa.

The Ming rulers also maintained China's traditional relationships with foreign peoples; they took for granted that the Chinese emperor was everyone's overlord and that other rulers of non-Chinese states were in a strict sense nothing but feudatories. Foreign rulers were expected to acknowledge the supremacy of the Ming emperor and to send periodic missions to the Ming capital to demonstrate their fealty and present tribute of local commodities. Tributary envoys from continental neighbors were received in selected frontier zones while those from overseas states were only accepted at three key ports on the southeast and south coasts. All envoys received valuable gifts in acknowledgment of the tribute they presented and also were permitted to buy and sell private trade goods at officially supervised markets. Luxury goods flowed out of China and some rarities flowed in. In order to preserve the government's monopolistic control of foreign contacts and trade, and to keep the Chinese people from being tainted by so-called barbarian customs, the Ming rulers prohibited private dealings between Chinese and foreigners and forbade any private voyaging abroad.

In the sixteenth century China came into contact with Jesuit missionaries who impressed the Chinese with the superiority of Western astronomy. The most famous of them, Adam Schall von Bell (1591–1666), was trained in Rome in the astronomical system of Galileo. After curing the Empress Dowager of a strange illness, Schall became an important adviser to the first emperor of the Qing dynasty (1644–1911/12). He was soon given an official post and he also translated Western astronomical books and reformed the old Chinese calendar. Matteo Ricci (1552–1610), another Jesuit, produced the first edition of his remarkable map of the world, the Great Map of Ten Thousand Countries, which showed China's geographical relation to the rest of the world. Moreover Ricci taught the rudiments of mathematics and translated many mathematical treatises on Western science and engineering into Chinese, notably the first six books of Euclid. At the same time, many books and correspondences of Catholics like Ricci were published in Europe and caused an interest in Chinese culture. This period of intensive cultural exchange came to an end with the imperial decree of 1717 which prohibited the preaching of Christianity and ordered the deportation of all missionaries from the empire with the exception of those working at the court.

From this time on, for more than one hundred years China reduced her contact with the West to a minimum. Trade was rigidly limited to a few ports where officials regulated it strictly and taxed all merchandise excessively. Chinese attitudes towards foreign relations clashed with those of the rising Western powers, especially after the newly expanding states of Britain, France, and Holland all began to develop major overseas empires. In these times, during the last decades of the eighteenth century, China's image changed, and Chinese affairs were viewed in a rather negative way. After the turn of the century China no longer received favorable attention in the West, but metamorphosed into the very archetype of a backward country. By teaching the Chinese the science and technology of the West, Europeans believed they had to stimulate the scientific development of a culture without any noteworthy tradition of its own. For a very long time it was quite inconceivable to Europeans that China had anything to offer in return.

As mentioned above, to draw a more balanced picture of Chinese scientific thought, we have to look for special features of ancient Chinese sciences and explain their singularities. It is easy to understand why traditional Chinese sciences are hard to describe in the modern terms of Western science or engineering. For example the Chinese science of geomancy (*fengshui*, or wind and water) cannot be assigned to any department of modern science. Since the concepts of geomancy differ significantly from premodern European natural philosophy, even recourse to ancient traditions of the West does not help. The geomancer aimed to adapt the dwelling places of the living and the dead in a suitable way to arrange them in harmony with the energy balances existing in a region. In Chinese thought every place has its peculiar topographical characteristics which can alter the local energies of nature. Directions of watercourses, shapes of wooded areas, or forms of hills are treated as important aspects for everything living in this region, and the forms and structure of objects built by humans are believed to be significant factors too. As the art of geomancy was dedicated to exploring the relations between the landscape and the living conditions of its inhabitants, it was an official state science, directed by the Board of Rites in the capital and patronized by the emperor himself.

Western scientists who have written about geomancy agree that this science recognizes certain types of energy which permeate the earth and atmosphere and animate the forms of nature. Further understanding has been obstructed by the impossibility of equating these energies with phenomena recognized by modern physics. In one aspect they evidently correspond to the emanations from the earth which are detected by water-diviners, or to the earth's magnetic currents. In another aspect they correlate to traditional Chinese medicine and acupuncture. Just as Chinese doctors commanded a great diversity of therapies and techniques, which they generally used in combination, geomancers did the same, and so traditionally both techniques are described as departments of the same science. Acupuncture is based on the same principles as geomancy, being concerned with the flow of subtle energies in the human body which correspond to those perceived by geomancy in the body of the earth. These remarkably abstract and comprehensive systems of acupuncture and geomancy are based on concepts describing the relations of the body, the mind, the immediate physical surroundings of the body, the earth, and the cosmos which are very different concepts from those in the West. What makes geomancy even more complicated are the differences between special schools of this science. One school believed that natural shapes in the landscape tended to affect the characters and destinies of those living within sight of them, while another school paid more attention to astronomical factors, horoscopes, and reading of the geomancer's compass. In Western terms geomancy seems like a mixture of obscure medicine, applied proto-physics, esoteric speculations, superstition, or even swindling.

Another example of the interrelationship of different branches of traditional Chinese learning are the alchemical sciences (*fulian*). We can distinguish two major divisions of alchemy, internal (*neidan*) and external (*waidan*). Internal alchemy was concerned with longevity practices and interpreted immortality as the highest kind of health. A special branch also existed for the alchemical process of internal transformations. Adepts of this school considered the interior of the human body as a laboratory in which elixirs could grow by meditation, breath control, or sexual gymnastics.

In theories of external alchemy, the transmutation of metals into gold and the production of universal remedies for diseases were the focus of interest. Chinese adepts of external alchemy by and large were not interested in exploring chemical reactions, but in simulating cosmic series of transformation and creation. In the laboratories alchemists intended to produce new substances or to convert given materials into new substances by means of allegorical imitations of natural phenomena rather than by controlled chemical experiments.

It is important to outline some Chinese cosmological ideas briefly. This will help us to relate theories of creation and change to patterns of Chinese mathematical astronomy (*lifa*) and astrology (*tianwen*). In traditional Chinese thought, ideas about the origin of the world do not involve any concept of creation by an almighty creator but only by impersonal processes of spontaneous self-creation. The first fully developed cosmological idea is the *gaitian* (heaven as cover, or umbrella heaven) theory, whose origins are around the first century AD, although its first traces are to be found as early as 239 BC. According to this theory, heaven and earth are flat and parallel planes. A variation of this theory depicts heaven and earth as bodies having a mild curvature very similar to the curvature of an umbrella. In both theories heaven is thought to rotate once daily about an imaginary axis, normally held to be vertical, and carry with it all the stars and heavenly bodies. Since the observer is some distance from the vertical axis, it is no contradiction that the polestar is not overhead. Rising and setting of the sun, moon, and stars are optical illusions caused by their entering and leaving the observer's pretended narrowed range of vision. Advanced variations of this theory initiated the idea of a spherical or complete heaven (*huntian*) which was thought to surround the earth and rotate daily about an axis inclined to the horizontal. In this theory heaven and earth are compared with the shell and yolk of an egg, e.g. the earth is said to be completely enclosed by heaven, rather than merely covered from above. Chinese astronomers continued to think in flat-earth terms until the seventeenth century, when Jesuit missionaries introduced Western theories.

Traditional Chinese astronomy does not use a zodiac of twelve signs laid out along the sun's annual path through the constellations. The celestial sphere is sliced into twenty-eight unequal segments, which are said to radiate from the north celestial pole in the same way in which lines of longitude radiate from the poles of a terrestrial globe. These slices are called lunar mansions or lodges (*xiu*). Each mansion bears the name of the constellation found in it. By means of this system, astronomers were able to follow the progress of the stars in the sky, but the mansions also have an astrological function. Each mansion has a corresponding terrestrial territory to which the predictions based on phenomena observed in the sky are applied. The appearance of comets, haloes, or clouds guided the actions of the governors of the states associated with the mansions where those phenomena were observed. Along with the mansion system, twenty-four guiding stars were chosen by which the position of the other stars in the sky could be determined. In this, Chinese astronomy differs greatly from Greek helical astronomy, which is based on the observation of the rising and setting of stars just before dawn and just after dusk. In China, where the celestial Pole symbolized the Emperor, astronomers studied the circumpolar movements of the constellations around the Pole.

Another science very closely related to mathematical astronomy and in the same way entirely different from its Western correspondent is Chinese astrology. To distinguish mathematical astronomy from astrology we have to remember that from the very beginning astronomy was designed to make celestial phenomena predictable, whereas astrology served as an aid for interpretations of those phenomena which were unpredictable. In the West astrology is in a way the same as horoscopy, but in the Chinese context astrology and horoscopy differ widely. In traditional China the appearance of celestial phenomena guided the actions of the governors of the states. The Chinese astrologer observed and interpreted anomalous celestial or meteorological phenomena to reveal faults and shortcomings in the political order. There was a close correspondence between the cosmic and the political domains. For instance, from the second century BC there has been the theory of *fenye* (field allocation). The sky was mapped upon political segments, so that strange phenomena discovered in a particular segment could be related directly to the corresponding political realm existing on earth. Throughout Chinese history the mediator between celestial circumstances and mundane affairs was the Emperor, who was responsible for the undisturbed course of all regularities on heaven and earth. Therefore, astrologers interpreted celestial omens as indications of imperial negligence or correctness. This attitude towards celestial phenomena also influenced calendrical and planetary astronomy. As a result of this attitude towards omens, almost every government aimed to control and sponsor mathematical astronomy and astrology.

Since changes in the heavens predicted important changes on the Earth, Chinese astronomy and astrology were incorporated into the system of government from the dawn of the Chinese state in the second millennium BC. The result was a system of astronomical observations and records, thanks to which star catalogs and observations of eclipses and novae that go back for millennia survived. Chinese records, therefore, are still of value to every student of the history of astronomy. In our times Western astronomers have identified ancient Chinese observations of the sudden appearance of bright stars with the supernova explosions whose remainders have been detected by radio astronomy. Moreover, observations of sunspots made from the first century BC onwards helped to solve some problems of the variation of solar activity over the centuries.

Astronomy and astrology had no real effect on Chinese mathematics (*suanxue*). On the whole Chinese mathematics was algebraic and numerical in its approach rather than geometric. Since ancient Chinese mathematics was primarily oriented toward practical application, any search for the hypothetical meaning of numbers was rather a system of occultism built around numbers than an exploration into the realm of abstract mathematics. Historians of mathematics have claimed that only the Greeks produced an abstract, logical mathematics that could function as the language of science. Chinese mathematics, however, consisted of reckoning rules, and, in spite of their great sophistication, these were only intended for practical use.

Perhaps mathematics serves best to demonstrate the whole problem of contingent affection of China by the West. In the first phase this Chinese science was disregarded, because it was inconceivable that a non-European culture was endowed with an efficient mathematical system. In the second phase, a few Western scholars made an attempt to understand such things as an abacus or counting-rods, but still theoretical primary sources were not used. In the next phase mathematical treatises were carefully studied and summarized in a reliable form. Today we can find several first hand characterizations of the subject by experts in this field, but it is very difficult to obtain a balanced picture. It is clear that ignorance of Chinese sciences was mainly caused by misunderstanding.

CHRISTOPH KOERBS

REFERENCES

Elvin, Mark, et al. "Symposium: The Work of Joseph Needham." *Past and Present* 87: 17–53, 1980.

Ho, Peng Yoke. *Li, Qi and Shu: An Introduction to Science and Civilization in China*. Seattle: University of Washington Press, 1987.

Kuhn, Dieter. "Wissenschaft und Technik." In *Das alte China: Geschichte und Kultur des Reiches in der Mitte*. Ed. Roger Goepper. München: Bertelsmann, 1988, pp. 247–279.

Li, Guohao, et al., eds. *Explorations in the History of Science and Technology in China*: A Special Number of the "Collections of Essays on Chinese Literature and History." Shanghai: Shanghai Chinese Classics Publishing House, 1980.

Nakayama Shigeru, and Nathan Sivin, eds. *Chinese Science: Explorations of an Ancient Tradition*. Cambridge, Massachusetts: MIT Press, 1973.

Nakayama Shigeru. *Academic and Scientific Traditions in China, Japan, and the West*. Tokyo: University of Tokyo Press, 1974.

Needham, Joseph. *Science and Civilisation in China*. (7 vols. projected). Cambridge: Cambridge University Press, 1954.

Sivin, Nathan. *Science and Technology in East Asia*. New York: Science History Publications, 1977.

Sivin, Nathan. "Science and Medicine in Imperial China – The State of the Field." *Journal of Asian Studies* 47(1): 41–90, 1988.

Zurndorfer, Harriet T. "Comment la science et la technologie se vendaient à la Chine au XVIIIe siècle: Essai d'analyse interne." *Études chinoises* 7(2): 59–90, 1988.

See also: Gunpowder – Compass – Metallurgy – Agriculture – Chemistry – Navigation – Astronomy – Astrology – Calendar – Geomancy – Mathematics – Chinese Science – Alchemy – Metallurgy – Cosmology – *Gaitian* – *Huntian* – Lunar Mansions – Stars – Eclipses

EAST AND WEST: INDIA IN THE TRANSMISSION OF KNOWLEDGE FROM EAST TO WEST The exhange of ideas was more balanced in the time before the European Scientific Revolution, after which the rapid growth of knowledge in the West dwarfed the interregional traffic that had taken place earlier. The Western tradition of the last few centuries has become the only system studied in universities and practiced in centers of science and technology worldwide. Quite often there is no interaction between this new tradition and the earlier knowledge from regions such as South Asia, even though there are many areas of learning that could enrich the Western tradition.

A study of the growth of the European and the South Asian scientific traditions shows considerable areas of overlap and mutual influence from very early times. When Europeans in the Renaissance looked back to Greek sources for new inspiration, they were in fact looking to Greek sources partly influenced by the South Asians.

Generally speaking, India was outside the world of shared ideas and values of pre-classical Greece. After the wars with the Persian empire the myth of a division into East and West was born, as was the concept of Europe. The conditions for a large scale traffic of culture and ideas between Greece and Asia were created when the Persian Empire became a bridge from the Mediterranean to the Indus. One sees South Asian concepts that arose between 700 BC and 500 BC in the later Vedic hymns, the *Upaniṣads*, and among the Buddhists and the Jains, being echoed in Greek thought.

There are striking parallels between the two traditions. The *Upaniṣads* seek one reality; this has its echoes in Xenophanes, Parmenides, and Zeno. Pythagoras is thought to have been influenced by the Egyptians, Assyrians, and Indians. He believed in the possibility of recalling previous lives, which is also typical of South Asian philosophy. Pythagoreans abstained from destroying life and eating meat, as do Jains and Buddhists. They expounded many theories in the religious, philosophical, and mathematical sphere that were known in sixth century BC India.

In Plato's philosophy, the 'cycle of necessity', a concept similar to Karma, was central. Humans were reborn as animals or other humans. The Indian elements *pṛthvī* (earth), *ap* (water), *tejas* (fire or heat), *vāyu* (air), and *ākāśa* (ether, or a non-material substance) have their counterpart in Empedocles, who belived that matter had four elements: earth, water, air, and fire.

After Alexander's encounter there was explicit dialogue with India. Several who traveled with Alexander are said to have met with Indian sages. In late antiquity India was seen in some debates as the origin of philosophy and religion. In the second century AD Lucianus stated that before philosophy came to the Greeks, Indians had developed it. It has also been suggested that Gnostic thought was influenced by Buddhist literature. Gnostic Carpocratians strongly supported the idea of transmigration. At least one Gnostic philosopher, Bardesanes of Edessa (ca. AD 200) had traveled extensively in India. Mani, the Persian Gnostic of the third century AD, incorporated several Buddhist ideas into Manichaeism. By the second century AD India had almost replaced Egypt as the presumed origin of Greek thought and learning.

At a later period, Plotinus, the father of the Neoplatonic school, took part in the military campaign against the King of Persia. Neoplatonism recommended abstention from sacrifice and meat eating. Neoplatonism, *vedānta*, yoga systems, and Buddhism all have strong similarities. In the second century AD, Clement of Alexandria spoke often about the existence in Alexandria of Buddhists, being the first Greek to refer to the Buddha by name. He was aware of the belief in transmigration and the worshipping of *stupas*.

During the Roman Empire, contacts between the two places continued. There was heavy trade in luxuries with South India and Sri Lanka by the Romans, and ambassadors were sent to Rome. An Indian delegation visited Europe in Emperor Antoninus Pius' reign. In the reverse direction, Apollonius of Tyana traveled to India. These repeated interactions between the two regions probably resulted in the exchange of ideas from South Asia to Greece. The Buddhists had sophisticated discussions prior to Heraclitus around the concept of being in a state of flux. Buddhists and Ājīvakas added joy and sorrow to the five elements, which precedes Empedocles's views that love and hate acted mechanically on the elements. The Buddhists and others taught a doctrine of the mean several centuries earlier than Aristotle (340 BC). In medicine, the Hippocratic treatise *On Breath* deals in much the same way with the pneumatic system as we find in the Indian concept of *vāyu* or *prāṇa*. In his *Timaeus*, Plato discussed pathology in a similar way to the doctrine of *tridoṣa*.

The above examples should not suggest that there were no transmissions in the opposite direction. The ancient world had much cross flow of intellectual traffic. A well known example from Greece to South Asia concerns ideas on geometry and astronomy.

When the Classical age collapsed, European and South Asian contacts continued in the Middle Ages through Arab intermediaries. The Arabs performed the functions earlier performed by the Persian, Alexandrian, and Greek empires which brought together the ideas of East and West. It is useful to trace the transmission of Indian sciences to Europe as well as trace those that were not transmitted but remained in the region only to be rediscovered much later. This is done to some extent in the article on Indian mathematics in this Encyclopaedia.

The European Renaissance and Scientific Revolution brought about many changes that have been considered unique. However, the evidence indicates that many of the results were known, some albeit in an incipient form, in South Asia.

Alchemy was an important precursor to the development of chemistry. Greek alchemical texts do not show an interest in pharmaceutical chemistry, a marked contrast with China and India. In the *Atharva Veda* (eighth century BC) there are references to the use of gold for preserving life. The transmutation to gold of base metals is discussed in the Buddhist texts of the second to fifth centuries AD by concoctions using vegetables and minerals.

In the West, iatrochemists, especially Paracelsus, were of the view that the human body consisted of a chemical system of mercury, sulfur, and salt. Sulfur and mercury were already known to the alchemists; salt was introduced by Paracelsus. This theory differed from the four humors theory of the Greeks advocated by Galen (AD 129–200). An Indian alchemist by the name of Ramadevar taught a salt-based alchemy in Saudi Arabia in the twelfth century.

In medicine, the work of Suśruta laid the foundations for the art of surgery. Suśruta emphasized observation and dissection, and described many instruments like those used in modern surgery, listed several kinds of sutures and needles, and classified operations into types. The operations described included those for hydrocele, dropsy, fistula, abscess, tooth extraction, and the removal of stones and foreign matter. The ancient Indian surgeons practiced laparotomy and lithotomy, plastic surgery, and perineal extraction of stones from the bladder. The region had considerable knowledge in dentistry including artificial teeth making. In AD 1194 the king Jai Chandra when beaten in battle was recognized by his false teeth. Suśruta describes details of operations for the conditions of obstructions in the rectum and for removal of a dead fetus without killing the mother, considered a very difficult procedure. He describes plastic surgery of the nose and cataract operations on the eye.

At the end of the eighteenth century the British studied Indian surgical procedures for skin grafting to correct for deformities of the face, which became the starting point for the modern specialty of plastic surgery. Dharmapal collected several illuminating accounts by Britons on Indian medical practices in the eighteenth century. This included one by J.Z. Holwell, who gave a detailed report on the practice of inoculation against the smallpox. The smallpox epidemics in the nineteenth and early twentieth centuries have been attributed to the cessation of this practice before the vaccination system could become widespread.

In the West after Democritus, atomic theories were further expanded by Lucretius in the first century BC but then virtually vanished from intellectual view for 1600 years. In the seventeenth century Gassendi, Boyle, Newton, and Huygens revived the atomic perspective. Atomic views of several schools such as the Buddhists, the Jains, and the Vaiśeṣika persisted. The Vaiśeṣika's theory of atomism considered atoms as eternal and spherical in form. The disintegration of a body results in its breaking down to constituent atoms. A solid block like ice or butter melts, and this is explained as a loosening of the atoms, giving rise to fluidity.

Evolution is one other element in the modern phalanx of scientific ideas. Evolutionary ideas had existed among pre-Socratic Greek and Indian thinkers. However, evolutionary thinking in Greek tradition was brought to a sudden end by the ideas of Plato and Aristotle. Plato viewed the real world as consisting of unchanging forms or archetypes; Aristotle viewed the physical world as a hierarchy consisting of kinds of things. For Aristotle the universe was unchanging and eternal. The idea of evolution is found in the *Upaniṣads*, the writings of the Buddhists, and others.

The *Encyclopaedia Britannica* lists three major innovative transformations in British agriculture in the Era of Improvement in the eighteenth century. They were the invention by Jethro Tull in 1731 of the drill plough, "whereby the turnips could be sown in rows and kept free from weeds by hoeing thus much increasing their yields"; the introduction of rotation of crops in 1730–38 by Lord Townshend, and the selective breeding of cattle introduced by Robert Bakewell (1725–95). There is evidence that all three were in existence in India, as reported by British scientists working there.

Roxburgh, generally recognized to be the "father of Indian botany" in the contemporary tradition, put this as follows: "the Western World is to be indebted to India for this system of sowing", meaning the implicit rotation of crops in the Vedic period where rice was sown in summer and pulses in winter in the same field. Other British works have attested to the use of "careful breeding of cattle", various kinds of drill ploughs, and rotation of crops and mixed cropping.

The Scientific Revolution had a deep impact on the philosophical underpinnings of Europe. There was for example the dethroning of human exclusivity as the special creation of God, as exemplified by the trials of Copernicus or the criticisms of Darwin's evolutionary theory. The discovery of the unconscious by Freud and others is also in this class. None of these events would have had the same impact on South Asia, whose cosmology allowed for a large number of worlds, for evolution and change, for humans as part of a larger living world, and for a subconscious.

Aside from these historical examples, are there innovations occurring even now which are drawing sustenance from the earlier South Asian tradition?

Helmut von Glasenapp observed that ancient South Asian ideas on fundamental issues had several parallels with those in modern science. Some of these concepts were: (1) an infinite number of worlds exists apart from our own; (2) worlds exist even in an atom; (3) the universe is enormously old; (4) there are infinitely small living beings parallel to bacteria; (5) the subconscious is important in psychology; (6) doctrines of matter in both Sāṃkhya and Buddhism are similar to modern systems; (7) the world that presents itself to the senses is not the most real; and (8) truth manifests itself differently in different minds giving the possibilty of a multiplicity of valid truths.

Following are examples of innovations based on the past taken from a few disciplines. The first is medicine. A recent study has documented the use of honey and sugar as treatment for wounds and ulcers in both Āyurvedic and contemporary biomedicine. The tranquilizer Reserpine was based on an ancient āyurvedic medicine. Hoechst, the West German pharmaceutical company, used Āyurvedic literature to help identify useful medicinal plants. By the early 1980s, over two hundred Indian medicinal plants were being tested every year in this program.

A recent study has evaluated the effect of *Rasayana* therapy which aims at promotiong strength and vitality. The study covered six drugs from classical Āyurvedic literature. The clinical studies indicated that the drugs toned up the cardiovascular and respiratory systems and improved physical stamina. On the biochemical side, a significant drop in lipids was noticed.

References to curative plants in the Indian tradition go back to the Ṛgvedic period (3500–1800 BC). The *Suśrutasaṃhitā* and *Carakasaṃhitā*, two compendia which are summaries of earlier works, dealt with about seven hundred drugs, some of them outside the subcontinental region. Clearly, a vast reservoir of explorable scientific knowledge exists.

One of the areas of study with a very long tradition in South Asia is psychology. There is also a very long tradition of sophisticated discussions on epistemology. There is potential for a fruitful interaction between these and the contemporary study of the mind, including the philosophy of language, methodology, ontology, and metaphysics.

Memory, motivation, and the unconscious are shown to have parallels in the theories of Freud and Jung, as well as in Patañjali. Similar parallels have been noted between the psychoanalytical theorists Heinz Hartmann and Erik Erikson, and the Hindu theory on the stages of life, as well as between Buddhism and early twentieth century analytical thought. Strong parallels between the concept of self-realization used in subcontinental traditions such as Vedantic Hinduism, Theravāda, and Mahāyāna Buddhism and the concept of self-actualization as developed in humanistic psychology by Arthur Maslow and Carl Rogers have been demonstrated.

Francisco Varela, a theoretical biologist and student of cognitive science and artificial intelligence, and co-workers have used Buddhist insights in extending the limitations of both the neo-Darwinian adaptation in biological evolution and of the current paradigm in cognitive sciences. Having noted that in Buddhist discourse, classical Western dichotomies like subject and object, mind and body, organism and environment vanish, Varela applies these discourses to several areas where these dichotomies had traditionally appeared. These include cognitive psychology, evolutionary theory, linguistics, neuroscience, artificial intelligence, and immunology.

Their position is that if cognitive science is to incorporate human experience, then it must have a means of exploring the dimension which is provided by Buddhist practice. Buddhist experiences of observing the mind are in thc tradition of scientific observation. They can lead to discoveries about the behavior and nature of the mind, a bridge between human experience and cognitive science.

Another area of interest is that of adaptation in evolutionary biology. In the conventional view it is assumed that the environment exists prior to the organism, into which the latter fits. This is not so. Living beings and the environment are linked together in a process of codetermination or mutual specification. In this light, environmental features are not simply external features that have to be internalized by the organism; they are themselves results of a long history of codetermination. The organism is both the subject and object of evolution. The processes of coevolution result in the environment's being brought to life through a process of coupling. Taking the world as pregiven and the organism as adapting can be categorized as dualism. Buddhism transcends this duality in its codeterminative perspective.

The standard arithmetic that we use today, based on the decimal place system and the use of zero, was transmitted through the Arabs from South Asia. It entails certain standard procedures, algorithms, to perform various operations. But are these the only such operations that exist and are these the ones that are computationally the most efficient? Could there be algorithms that did not get transmitted from India through the Arabs, or those that were developed after the transmission?

Indeed there are many such, as Ashok Jhunjhunwala, a professor of electrical engineering in the Institute of Technology, Madras, has discovered recently. He has examined everyday practices in arithmetic in areas not yet influenced by European techniques such as those used by artisans and businessmen in the non-Europeanized sector. He came across simple but fast methods of calculation. He described eight of these methods which are faster than conventional methods. These included means of finding area, multiplication, squaring, division, evaluation of powers, square roots of numbers, divisibility of numbers, and factorization. They also included methods to catch errors. Jhunjhunwala has compared the speed of some of these old approaches with contemporary ones and found that some are faster. He is now applying these general methods to speed up calculations in computers.

Jhunjhunwala's collection of mathematics at the local level shows the proliferation of methods possible once the decimal system is understood. Local groups discovered new tricks, a process of grass roots creativity very much like the different responses to changing agroclimatic conditions across the world and the resultant variations in agricultural practices.

Time is yet another area to explore. There are many different philosophical discussions on *Saṁsāra* concerning what could be termed the nature of long duration processes. According to some Jain views, time was one of the causal factors in the evolution of nature; and Buddhism alone has a very large tapestry of conceptions of time.

One of these approaches developed an elaborate theory

not only of atoms but also of moments, with some schools recognizing four types of moments and others three. Other theories were also proposed by different schools to relate the theory of moments to the fact of continuity of temporal events.

Virtual Reality brings into question the constructor and the constructed. These types of questions are regularly dealt with in Buddhist and other South Asian philosophies. In the virtual realities that use visual representations, parallels also exist with visualization techniques in certain branches of Buddhism. The author of a text on the topic, Howard Rheinhold, says that the Virtual Reality "experience is destined to transform us because it's an external mirror of something that Buddhists have always said, which is that the world we think we see 'out there' is an illusion."

The ethical and conceptual questions of the future brought about by modern science and technology could have many uses for South Asian perspectives.

There are many stores of valid information still to draw from. One authoritative estimate of manuscripts, roughly covering the areas of mathematics and astronomy, is about 100,000. Yet the recently published book *Source Book of Indian Astronomy* lists only 285. Of these only very few have been studied. They are mines of mathematical ideas and applications that have hardly been touched.

A passage from Suśruta stimulated the growth of modern plastic surgery in the nineteenth century in Europe. However, as Krishnamurty points out, that was only a stray reference in the many procedures described. It had the fortune of catching the imagination of a western expert. There could very well be many other descriptions that could be rediscovered for modern medicine. Under treatment for mental diseases Suśruta gives a very large list of plants. It is possible that screening of these plants could give rise to a much larger set of useful remedies.

Varela stated that the infusion of Eastern ideas into the sciences of the West would have as much an impact as did the Renaissance rediscovery of Greek thought. How far this may be true is for the future to decide. However, this would help make the present Western knowledge system more universal while still maintaining the rigor developed in the last few centuries. It would help both to enlarge the knowledge terrain covered by the present system as well as retrieve what is relevant from other traditions.

SUSANTHA GOONATILAKE

REFERENCES

Bose, D. M., S.N. Sen, and B.V. Subarayappa. *A Concise History of Science in India*. New Delhi: Indian National Science Academy, 1971.

Halbfass, Wilhelm. *India and Europe: An Essay in Understanding* Albany: State University of New York Press, 1988.

Jain, S.K. *Medicinal Plants*. Delhi: National Book Trust of India, 1968.

Jhunjhunwala, Ashok. *Indian Mathematics: An Introduction.* New Delhi: Wiley Eastern Limited, 1993.

Krishnamurty, K.H. *A Source Book of Indian Medicine: An Anthology*. Delhi: B.R. Publishing Corporation, 1991.

Lach, Donald F. *Asia in the Making of Europe,* vol. 2. Chicago: University of Chicago Press, 1977.

Mohanty, J. N. "Consciousness and Knowledge in Indian Philosophy." *Philosophy East and West* 29(1): 3–11, 1979.

Rawlinson, H.G. "Early Contacts Between India and Europe." In *Cultural History of India*. Ed. A.L. Basham. Oxford: Clarendon Press, 1975.

Sangwan, Satpal "European Impressions of Science and Technology in India (1650–1850)." In *History of Science and Technology in India,* vol 5. Ed. G. Kuppuram and K. Kumudamani. Delhi: Sundeep Prakashan, 1991.

Varela, Francisco J., Evan Thompson and Eleanor Rosch. *The Embodied Mind: Cognitive Science and Human Experience*. Cambridge, Massachusetts: MIT Press, 1991.

von Glasenapp, Helmuth. "Indian and Western Metaphysics." *Philosophy East and West* 111(3): 1953.

See also: Mathematics in India – Knowledge Systems: Local Knowledge – Time

EAST AND WEST: ISLAM IN THE TRANSMISSION OF KNOWLEDGE FROM EAST TO WEST The Arabic term for knowledge or science, *ʿilm*, had a developing history steeped in religious reflection and writing as a consequence of the extensive use of its Semitic root *ʿ-l-m* in the *Qurʾān*. In contrast to the pre-Islamic "Time of Ignorance" (*jāhilīyah*), the advent of Muḥammad as Prophet conveying the words of God in the *Qurʾān* marked the presence of a new understanding of how human beings are to submit (*islām*) to the will of God in all aspects of their lives. Knowledge on the part of humans was viewed as having its source without exception in the Divine. For the Muslim, as a result, the place of primacy goes to religious sciences directly (study of the *Qurʾān*, the *Ḥadīth*, commentaries on these, etc.) or indirectly (grammatical studies, law, etc.) which deal with what has been divinely revealed. Knowledge of the sciences were frequently divided into the religious and the foreign (al-Khwārizmī, ca. AD 976; this is the philosopher, who should not be confused with the mathematician al-Khwārizmī of ca. AD 830), the religious and non-religious or rational (al-Ghazālī, 1058–1111) or the traditional religious and the philosophical (Ibn Khaldūn, 1332–1406), although the philosopher al-Fārābī (870–950) gave a different account classifying the sciences into those of language, logic, mathematics, natural philosophy, metaphysics, politics, law, and theology. While such distinctions frequently reflected believers' deep distrust of foreign learning as superfluous to the guidance and injunctions of the *Qur'ān*, the perceived value of medicine, mathematics, astronomy, and astrology led ultimately both to the full incorporation of Greek scientific and philosophical learning into Muslim intellectual and cultural life and to its enhancement and development at the hands of Islamic scholars.

In the course of its expansion, Islam came to have dominance over populations of Christians, Jews, Zoroastrians, and others in Egypt and the greater Middle East who were already possessed of Greek medicine, technology, and scientific learning and who were active participants in the construction of long traditions of intellectual reflection on God, nature, and human learning. In addition to interest in medicine universal to all peoples, Arabian Muslims and converted native populations had practical religious interests in astronomy and geography because of Islamic precepts calling for prayer by the faithful in the direction of Mecca at five separate times daily and because of the need for exact calculations for holy days and for the proper arrangement and placement of mosques and their interiors. For those in positions of power, there was also the need to discern matters related to quality and character of life present and to come as read in the stars by astrologists. Another sort of practical need which contributed to the demand for Greek scientific texts was that posed by theological debate both internal and external to Islam. Christians as well as Jews and their Muslim students marshalled powerful arguments with enviable sophistication thanks to Syrian and Greek intellectual roots in a Hellenistic cultural tradition which had a long theological interest in Aristotelian logic due to controversies among Nestorians, Monophysites, and others. In these contexts there arose a demand for deeper understanding of the conceptual foundations of Greek knowledge incompletely conveyed in medicine and other vital areas with a resulting need for translations of works which set out the theoretical and philosophical structures for medical, astronomical, astrological, and other studies. This gave rise to an extraordinary translation movement which can be divided into three major periods by reference to the work of the Nestorian Christian Ḥunayn ibn Isḥāq al-ʿIbādī (808/9–873) and his assistants whose achievement and influence in the transmission of knowledge was without equal in the entire Middle Ages. Thanks in great measure to the *Fihrist* or *Catalogue* of the Baghdad bookseller Ibn al-Nadīm (ca. 935–990), our knowledge of works translated and of works available in his day is substantial.

Transmission of Knowledge to Islam

The transmission of works of science, philosophy, and other intellectual endeavours was well underway even prior to the establishment in Baghdad of the *Bayt al-Ḥikmah* (House of Wisdom) as an institute dedicated to translation during the rule of al-Ma'mūn (813–833). An Arabic rendering from Sanskrit of the *Zīj al-Arkand* derivative upon the *Khaṇḍakhādyaka* of Brahmagupta which was available around 738 provided important astronomical tables employed by al-Khwārizmī (d. 850) whose work was so important to the Latin West. He also made use of Ptolemy's *Geographia* and *Megale Syntaxis*, the latter available in Arabic as *al-Majistī* by about 796 and known in the Latin West as the *Almagest*. As early as the late seventh century the Jewish physician Māsarjawayh had rendered the medical *Panducts* of Ahrūn of Alexandria from Syriac into Arabic perhaps for use at Jundīshāpūr, the famous medical center. Organized and sponsored translations from Indian, Iranian, and Greek sources into Syriac and Arabic began to be produced in earnest under the ʿAbbasids, particularly during the reigns of al-Manṣūr (754–775) and Hārūn al-Rashīd (786–809) at Baghdad, and under the influence of the Barmakid family of physicians and administrators.

The account of Ḥunayn's role in the transmission of knowledge is a story of personal achievement and high scientific and scholarly standards. Removed from medical studies by his teacher at Baghdad, the physician, translator and scholar Yuhannā ibn Māsawayh of the *Bayt al-Ḥikma*, Ḥunayn disappeared into Byzantium (*bilād al-Rūm*) to reappear two years later with a mastery of Greek and a penchant for quoting Homer in the original. The now trilingual (Syriac, Arabic, and Greek) Ḥunayn established his expertise with a translation of a Galenic work and attained a funded place among his colleagues. Sometimes translating into Arabic directly from Greek or Syriac, other times translating from Greek into Syriac and then from Syriac into Arabic, Ḥunayn and his assembled team of translators, ʿĪsā ibn Yaḥyā, Mūsā ibn Khālid, Yaḥyā ibn Hārūn, his son Isḥāq, his nephew Ḥubaysh ibn al-Ḥasan and others, set a high standard of quality for their work. They sought out Greek manuscripts to collate into a single version, worked to standardize Arabic technical vocabulary, and rendered phrases and sentences, not translating word for word, in a sophisticated effort to capture the sense of the texts. Ḥunayn was particularly interested in the work of the physician and epitomizer Galen (b. 129) and translated over a hundred of his works and studied the corpus throughout his lifetime. Other medical works translated included works of Hippocrates and the *Materia Medica* of Dioscorides which constituted the Arabic pharmacopoeia, this latter possible only because of the physician Ḥunayn's extensive Greek vocabulary on herbs and drugs. They went on during or shortly after his lifetime to translate into Arabic or Syriac, and to comment upon or abridge many of Plato's and Aristotle's works. This second phase of the translations also includes the work of the mathematician Thābit ibn Qurra, a Sabian from Ḥarrān and contemporary and collaborator with Isḥāq ibn Ḥunayn. Thābit worked with a group of astrologers and mathematicians in the Ḥarrānian tradition and translated Nicomachus of Gerasa's *Introduction to Arithmetic*, improved existing translations by Isḥāq of Euclid's *Elements* and Ptolemy's *Almagest*, commented on Aristotle's *Physics*, and worked to carry on the Greek Neopythagorean metaphysics of numbers and astrology. Also active in the same era were the Muslim Abū ʿUthmān al-Dimashqī who translated Aristotle's *Topics* and Porphyry's *Isagoge* as well as texts of Alexander and medical and mathematical works and the Christian Qusṭā ibn Lūqā. Many of the works were translated more than once.

The third and final phase of translation began with the work of the Syrian Greek Christian Abū Bishr Mattā ibn Yūnus (d. 940) who had Muslim, Christian, and Jewish teachers, was an accomplished trilingual writer, and had as his student the renowned philosopher al-Fārābī. His student Yaḥyā ibn ʿĀdī (893/4–974) was a philosopher, theologian, and logician as well as a translator who was frequently called upon to check the work of his master and to render Syriac translations by Ḥunayn, Isḥāq, and others into Arabic. He also wrote commentaries on most of Aristotle's logical works while writing works of his own on logic, mathematics, natural philosophy, metaphysics, ethics, theology, scripture, and medicine. Of the same school were Ibn Zurʿah (942–1008) and Ibn al-Khammār (942–1017). Their work brought to a close for the most part the great translation movement which had some beginnings during the reign of the Ummayads but was for the most part a phenomenon of the era of the Abbasids. A massive portion of Greek scientific, philosophical and other literature had been translated and was now in the process of being assimilated.

Yet the story of the transmission of knowledge is not merely one of texts moving from one language to another. It is, in an equally important way, a story of the movement of ideas and their reception and development in a new language and cultural framework. This is all the more so in the case of knowledge and its movement from East to West via the Islamic route, for what was transmitted was often not only texts but methodological thought and intellectual advances and achievements made on the basis of wisdom garnered from foreign sources. Nearly all of the translators mentioned above were also physicians, scientists, or philosophers in their own rights. It was primarily out of their desire for knowledge that they studied foreign works. Translation was merely a consequence of a thirst for knowledge on the part of the translators, their patrons, and Muslim intellectuals. Thinkers such as the Arab al-Kindī and the Turk al-Fārābī worked closely with those who were studying foreign texts and preparing translations. By the time of the Persian Ibn Sīnā (980–1037), the cultivation of foreign knowledge within the context of Islam was yielding new fruits on Islamic soil. Ibn Sīnā (Avicenna) himself developed a philosophical system with a new approach to metaphysics and other philosophical matters which had a powerful impact on the Latin West and which has come down in a continuous tradition in present-day Iran. Al-Bīrūnī (973–1043), a contemporary and correspondent of Ibn Sīnā, wrote extensively on mathematics, astronomy, astrology, the astrolabe, pharmacology, gems, metals, and other matters in addition to his work on Indian thought. Working with translated Greek texts on mathematics and optics as well as more physical acccounts of visual perception, Ibn al-Haytham (d. 1039), a physician as well as astronomer and mathematician, developed a theory of optics (*perspectiva* in Latin) and intromissive visual perception which became dominant in Islam and later in the Latin West, where Roger Bacon championed his work. In the area of alchemy, the names of Jābir ibn Ḥayyān (d. ca. 815) and of the physician and scholar al-Rāzī (d. ca. 925) stand out, although contemporary scholars raise questions about Jābir's contributions to the corpus of works traditionally attributed to him. While there remained vestiges of distrust of writings inspired by foreign sciences (e.g. of the metaphysical and psychological teachings of the Andalusian philosopher Ibn Rushd [d. 1198] whose philosophical thought was recognized to have religious ramifications), the advancement of science broadly defined continued in Islam until the fifteenth century.

Transmission of Knowledge to the Latin West

In the Latin West science and advanced learning were actively sought out from Greek and Arabic sources and then translated into Latin and carried north for an economically and politically expanding and intellectually developing Europe. The West did possess some remnants of earlier Greek learning including translations of portions of Aristotle's *Organon* by Boethius (d. 524). Moreover, for the most part, Greek science and philosophy, while recognized as arising from a pagan culture, were not feared as foreign in the way that they were in Islam; rather, they were recognized as intellectual treasures to be recovered and put to use in the service of Latin Christianity. Individual scholars began to seek out Greek texts and learning and came to the frontiers of Islam in search of knowledge and more sophisticated intellectual understanding from a culture which had benefited from and added to the Hellenistic learning which it had received. Spain was by far the most productive geographic area for translations from Arabic, although some important work was done in Sicily with a few other works of significance being rendered into Latin in the Near East. The Latin movement can also be distinguished into three periods during each of which translations were being made from Arabic as well as Greek sources.

While an important Barcelona manuscript containing Latin translations and other materials from the late tenth century provides important evidence of early interest in Arabic scientific knowledge about the astrolabe, the first translations of significance from Arabic began to appear in the eleventh century. Constantine the African (d. ca. 1087), apparently a widely traveled and well-educated North African Muslim, converted to Christianity and became a Benedictine monk at Monte Cassino. According to one source, he undertook the task of translating and composing works on medicine when he had learned of the poverty of Latin Medieval medicine. His contribution to the Latin corpus was substantial and included works by Galen and Hippocrates, as well as work by ʿAlī ibn al-ʿAbbās al-Majūsī/Haly Abbas, Ḥunayn ibn Isḥāq/Johannitius, and others. These translations and other work by Constantine played a substantial role in forming the foundations of medicine and its study by Europeans. However, it was the twelfth and thirteenth centuries which saw floodgates open and a wealth of learning pour forth into Latin Europe.

In the twelfth century, scholars engaged in translation and transmission of Arabic texts were, like their counterparts working in the Islamic milieu, scientists and scholars first and translators second. The great majority of them found that Spain provided the needed environment for their work. Some of them were natives to the area but a large number came from afar in search of scientific learning. An exception to this was Adelard of Bath (d. ca. 1142) who studied in France and then traveled widely in Italy and the Near East before returning to his native England. Among his translations were two of monumental importance to the Latin West: in astronomy al-Khwārizmī's *Zīj* (astronomical handbook with tables) and in mathematics Euclid's *Elements*. In 1126 he translated Abū Maʿshar's astrological *Introductorium in astronomiam*. John of Seville (fl. ca. 1130–42) translated Qusṭā ibn Lūqā's *De differentia spiritus et anime* and a large number of astrological texts by Māshāʾallāh, Abū Maʿshar and al-Farghānī. Plato of Tivoli was also interested in astrology and translated Ptolemy's *Quadripartitium* and a work by al-Bāttanī, an astronomer and disciple of Thābit ibn Qurra the Ḥarrānian astrologer and mathematician.

At Toledo, the great center for translation and study, Dominicus Gundissalinus (Domingo Gundisalvi, ca. 1110–90) devoted himself to the study of Arabic and Jewish thought, translating a number of important philosophical works. Gundissalinus apparently worked frequently with an assistant. The nature of their collaboration is far from clear even though an account of it is given in the dedication of his translation of Ibn Sīnā's *On the Soul*. It may have involved an intermediate step of translation into the local vernacular or a reading of the text out loud. He is also thought to have translated Isaac Israeli's *Book of Definitions* and al-Kindī's *De intellectu* as well as al-Fārābī's *De intellectu, De scientiis, De ortu scientiarum*, and *Liber excitativus ad viam felicitatis*. Regardless of the precise nature of his collaborations, it is clear that his study and translations had substantial influence on the thought of Gundissalinus himself as evidenced in his own philosophical lectures which in turn together with the translations played an important role in the formation of the thought of the great philosophers of the thirteen century and later such as William of Auvergne, Albertus Magnus, Thomas Aquinas, Duns Scotus, and others who made extensive use of these works.

At Toledo at the same time was the great scholar, lecturer, and translator par excellence of the age, Gerard of Cremona (ca. 1114–87), who had come from Italy to Spain in search of Ptolemy's *Almagest* which he then proceeded to translate from Arabic into Latin. The list of Gerard's translations is long and detailed and includes works on mathematics by Euclid (*Elements*), Thābit ibn Qurra, al-Khwārizmī, al-Kindī and others; works on astronomy by Thābit and al-Farghānī; a medical work by Ibn Sīnā, the great *Qānūn fī al-ṭibb* (Canon of Medicine), which was in use into and beyond the Renaissance; philosophical works of Aristotle; and a vast array of other works of Islamic philosophical and scientific learning. Gerard used the common style following the work of Boethius in translating from Greek into Latin in the early sixth century: *verbum ex verbo*. The attempt was made to give translations which reflected as precisely as possible the original text in the grammatical structure of the original language's sentences and in word-for-word translation. In the case of Indo-European Greek texts, this was a somewhat reasonable way of proceeding. However, in the case of Latin translations of Semitic Arabic works, some of which had already gone through translation from Greek into Syriac and then into Arabic, the medieval Latin reader was left with an intellectually challenging exercise, with the result that later translations from Greek were often preferred. Nevertheless, the translations from Arabic by Gerard (and Gundissalinus) frequently conveyed well and clearly the sense and letter of the Arabic.

Translators of the later twelfth and early thirteenth centuries include the Englishman Alfred of Sareshal who translated the pseudo-Aristotelian *De plantis*, on which he wrote a commentary, and Ibn Sīnā's *De mineralibus*. Most accomplished of the Latin translators from Arabic of this era was Michael Scot who was in Toledo in 1215, Bologna in 1220, and Sicily in 1227 as court astrologer to Frederick II, where he died in about 1236. Michael completed his translation of al-Biṭrūjī's *De spheris* in 1217 at Toledo where he also translated Aristotle's *On Animals* in part. His greatest achievement, however, was responsibilty for making works of Ibn Rushd/Averroes available to the Latin West where they continued to be read through the time of the Renaissance. Michael is likely the translator of Ibn Rushd's great or long commentaries on Aristotle's *On the Heavens*, *On the Soul*, *Physics*, and *Metaphysics*, and *On Animals* in Ibn Rushd epitomes. Hermannus Alemannus, William of Luna, Petrus Gallegus, and Philp of Tripoli working in the mid-thirteenth century complete the list of well-known translators from Arabic. Herman is responsible for the Latin of Ibn Rushd's middle commentaries on the *Ethics* and *Poetics* as well as for the widely circulated ethical epitome, *Summa Alexandrinorum*. William also translated some of Ibn Rushd's texts. Petrus rendered Ibn Rushd's epitome on Aristotle's *On the Parts of Animals*.

While this enormous wealth of Greek and Islamic learning was being rendered into Latin from Arabic sources in Spain, Sicily, and the Near East, twelfth century Europe was also the recipient of a vast array of translations directly from the Greek by James of Venice, Henricus Aristippus, a certain Ioannes, and others. In the thirteenth century there were substantial efforts of translation from Greek by Robert Grosseteste, William of Moerbeke, and others. The transmission of knowledge into the Latin West came along two different roads, each of which had its orgins in Greek learning. Via the Islamic route, however, the Latin West also received the sophisticated reflections and intellectual advancements of Islamic thinkers whose religiously oriented philosophical and theological reflections on the world and its Creator were often much closer to the mentality of Christian Europe than were the texts of Ancient pagan authors. The philosophical texts of the Islamic philosophers Ibn Sīnā and Ibn Rushd, sophisticated interpretations and complex new intellectual syntheses, early on "explained" Aristotelian works to Medieval Latins and thereby had a profound influence on the methodologies and analyses worked and developed by Christian theological and philosophical thinkers from the twelfth and thirteenth centuries and beyond.

RICHARD C. TAYLOR

REFERENCES

Alverny, Marie-Thérèse d'. *La transmission des textes philosophiques et scientifiques au Moyen Âge*. Ed. Charles Burnett. Aldershot, Hampshire, U.K.; Brookfield, Vermont, U.S.A.: Variorum, 1994.

Butterworth, Charles E. and Blake Andree Kessel, eds. *The Introduction of Arabic Philosophy into Europe*. Leiden: E.J. Brill, 1994.

Colloque international de Cassino (1989: Cassino, Italy). Rencontres de cultures dans la philosophie médiévale: traductions et traducteurs de l'antiquité tardive au XIVe siècle. Ed. Jacqueline Hamesse and Marta Fattori. Louvain-la-Neuve: Université catholique de Louvain; Cassino, Italy: Università degli studi di Cassino, 1990.

Dictionary of the Middle Ages. Ed. Joseph R. Strayer. New York: Scribner, 1982–1989.

The Encyclopaedia of Islam. Ed. H. A. R. Gibb et al. New ed. Leiden: Brill, 1960.

Gardet, Louis and M. M. Anawati. *Introduction à la théologie musulmane; essai de théologie comparée*. Paris: J. Vrin, 1948.

Ibn al-Nadīm, Muḥammad ibn Isḥāq. *The Fihrist of al-Nadim*. Ed. and trans. Bayard Dodge. New York: Columbia University Press, 1970.

Jayyusi, Salma Khadra, ed. *The Legacy of Muslim Spain*. Leiden: E.J. Brill, 1992.

Kretzmann, Norman, ed. *The Cambridge History of Later Medieval Philosophy: From the Rediscovery of Aristotle to the Disintegration of Scholasticism, 1100–1600*. Cambridge: Cambridge University Press, 1982.

La Diffusione delle Scienze Islamische nel Medio Evo Europeo. Rome: Accademia nazionale dei Lincei. Fondazione Leone Caetani, 1987.

Lindberg, David C., ed. *Science in the Middle Ages*. Chicago: University of Chicago Press, 1978.

Lindberg, David C. *The Beginnings of Western Science: the European Scientific Tradition in Philosophical, Religious, and Institutional Context, 600* BC *to* AD *1450*. Chicago: University of Chicago Press, 1992.

Nasr, Seyyed Hossein. *An Introduction to Islamic Cosmological Doctrines: Conceptions of Nature and Methods Used for its Study by the Ikhwan al-Safa', al-Biruni, and Ibn Sina*. Revised ed. London: Thames and Hudson, 1978.

O'Leary, De Lacy. *How Greek Science Passed to the Arabs*. London: Routledge and K. Paul, 1949.

Peters, F. E. *Aristotle and the Arabs; the Aristotelian Tradition in Islam*. New York: New York University Press, 1968.

Peters, F. E. *Aristoteles Arabus. The Oriental Translations and Commentaries of the Aristotelian Corpus*. Leiden: E. J. Brill, 1968.

Rosenthal, Franz. *Knowledge Triumphant; the Concept of Knowledge in Medieval Islam*. Leiden: Brill, 1970.

Young, M.J.L. et al., eds. *Religion, Learning, and Science in the Abbasid Period*. Cambridge: Cambridge University Press, 1990.

See also: Qibla – Religion and Science – Astronomy – Astrology – *Almagest* – Ḥunayn ibn Isḥāq – Thābit ibn Qurra – *Elements* – Isḥāq ibn Ḥunayn – al-Bāttanī – al-Rāzī – Ibn al-Haytham – Jābir ibn Ḥayyān – Ibn Sīnā – Ibn Rushd

ECLIPSES Eclipses of both the moon and sun are frequently recorded in the history of non-Western cultures throughout ancient and medieval times. These records originate almost exclusively from Babylonia and China in ancient times, and from East Asia (China, Korea, and Japan) and the Islamic world in the medieval period. As yet, virtually no eclipse observations have been uncovered from other major civilizations such as ancient Egypt, India, and Central America.

Eclipses were often noted on account of their spectacular nature, or because they were regarded as omens. However, early records provide many examples of a more scientific attitude, resulting in the careful measurement of times and other details. Ancient references to eclipses often prove of value in dating historical events, while the more careful observations have played a major role in present-day knowledge of variations in the rate of rotation of the Earth. Eclipses are thus among the most interesting of all celestial phenomena mentioned in history.

The oldest known allusions to eclipses in any civilization are recorded on a series of astrological tablets (known as *Enuma Anu Enlil*) from the city of Ur, in southern Iraq. The earliest of these accounts may be translated as follows:

> "If in the month Simanu an eclipse occurs on day 14, the [Moon-] god in his eclipse is obscured on the east side above and clears on the west side below, the north wind blows, [the eclipse] commences in the first watch of the night and it touches the middle watch... The king of Ur will be wronged by his son, the Sun-god will catch him and he will die at the death of his father...".

Professor P.J. Huber of the University of Bayreuth, who provided the above translation, regards the historical details as relating to the murder of King Shulgi of the third dynasty of Ur by his son. Huber calculates the date of the eclipse as April 2 in 2094 BC (Julian Calendar).

One of the earliest surviving reports of a solar eclipse is from Assyria. This is found in the *Eponym Canon*, which lists the names of the annual magistrates (*limmu*) who gave their names to individual years (similar to the Athenian archons or Roman consuls) under a year which is equivalent to 763–762 BC.

> Revolt in the citadel. In (the month) Sivan the Sun was eclipsed.

Sivan corresponded to May–June. The only large eclipse visible in Assyria for many years occurred on June 13 in 763 BC.

Between about 750 and 50 BC (or possibly down to the first century AD), Babylonian astronomers systematically maintained a watch for eclipses (as well as other celestial phenomena) and carefully estimated the time of occurrence and magnitude (the maximum degree of obscuration of the moon or sun). Their aim was to use these observations to enable better predictions of future eclipses to be made, although the ultimate goal was astrological. Both genuine observations and failed predictions (the latter which were often described as eclipses which "passed by") are recorded on the late Babylonian astronomical texts which were recovered from the site of Babylon rather more than a century ago. Most of these clay tablets, many of which are badly damaged, are now in the British Museum. Dates are expressed in terms of a luni-solar calendar. Although earlier years were counted from the accession of each king, they began to be continuously numbered from the Seleucid Era (311 BC) onwards.

Among Babylonian records of eclipses, the best known is the solar obscuration of April 15 in 136 BC. This is reported on two separate British Museum tablets, which give overlapping details. A composite translation — based on work by Prof. H. Hunger of the University of Vienna — is as follows:

> "Year 175 (Seleucid), intercalary 12th month, day 29. At 24 degrees after sunrise, solar eclipse. When it began on the south-west side, in 18 degrees of day in the morning it became entirely total. Venus, Mercury and the Normal Stars were visible. Jupiter and Mars, which were in their period of disappearance, became visible in its eclipse [...] It threw off [the shadow] from south-east to north-west. [Time-interval of] 35 degrees for onset, maximal phase and clearing".

Jupiter and Mars were too close to the sun to be seen under normal circumstances. The unit of time here translated as "degree" (i.e. *us*) was equivalent to 4 minutes. Measurements were presumably made with a water clock, although direct evidence is lacking. The "Normal Stars" were certain reference stars in the zodiac belt. Many other detailed descriptions of both lunar and solar eclipses are preserved on the Late Babylonian astronomical texts.

Occasional allusions to eclipses are found in the Old Testament, notably in the Book of Joel (II, 31). However, the most direct reference to an eclipse in the history of ancient Palestine is recorded by Flavius Josephus, a Jewish historian of the first century AD. This event occurred only a few days before the death of Herod the Great and is thus of importance in dating the birth of Jesus Christ. Josephus gives the following account:

> "As for the other Matthias who had stirred up the sedition, he [Herod] had him burnt alive along with some of his companions. And on the same night there was an eclipse of the Moon. But Herod's illness became more and more severe... "

The events described occurred shortly before the Passover. Calculation shows that between 17 BC and AD 3, the only springtime eclipses visible in Palestine occurred on March 25 in 5 BC (a total obscuration) and March 13 in the following year (a partial obscuration). The latter date is usually preferred by historians, although the two dates are conveniently close together.

For several centuries after this period, scarcely any eclipses are recorded outside East Asia. However, in the ninth century both Muslim astronomers and chroniclers began independently to note the occurrence of eclipses of both moon and sun. Between AD 830 and 1020, Muslim astronomers, largely based in either Baghdad or Cairo, regularly made careful observations of both local times and magnitudes. The intention was to test the reliability of contemporary eclipse tables and sometimes also to determine the difference in longitude between selected cities. Astronomers would first roughly predict when an eclipse was to occur and then assemble one or more parties of observers. Our principal source for such measurements is the handbook by the Cairo astronomer Ibn Yūnus, who died in AD 1009. This work, dedicated to Caliph al-Ḥakīm, is entitled *al-Zīj al-Kabīr al-Ḥakīmī*.

The following account of the lunar eclipse of AD 927 by Ibn Amājūr of Baghdad is fairly typical. As is customary, years are counted from the Hijra, which occurred in AD 622.

> (315 AH, a Friday — month and day of month not stated). "This eclipse was observed by my son Abū al-Ḥasan. Altitude of the star al-Shi'rā alYamāniya [i.e. Sirius] at the start, 31 deg in the east; revolution of the sphere between sunset and the start of the eclipse, approximately 148 degrees or 9 hours and 52 minutes equal hours, which is 10 unequal hours. Estimated magnitude of the eclipse, more than one-quarter and less than a third, approximately $3\frac{1}{2}$ digits".

The only feasible date for the eclipse was AD 927 September 14, which was indeed a Friday in the year 315 A.H. Medieval Muslim astronomers were in the habit of measuring eclipse times indirectly by determining the altitude of the sun, moon or a bright star (probably with a sextant or astrolabe) and then converting to local time with the aid of an astrolabe or tables. In estimating magnitude, they followed the practice of the ancient Babylonians and Greeks of using digits, each equal to one-twelfth of the diameter of the luminary.

After AD 1020, virtually no eclipse observations made by Muslim astronomers have survived. However, chroniclers continued to report eclipses — largely on account of their spectacular nature — for many centuries. An appealing eyewitness account of the total solar eclipse of AD 1176 Apr 11 was penned by Ibn al-Athīr, who was aged 16 when it occurred:

> (571 AH). "In this year the Sun was eclipsed totally and the Earth was in darkness so that it was like a dark night and the stars appeared. That was the forenoon of Friday, the 29th of the month Ramaḍān at Ibn 'Umar's island, when I was young and in the company of my Arithmetic teacher. When I saw it, I was much afraid; I held onto him and my heart was strengthened. My teacher was learned about the stars and told me, 'Now you will see all this go away', and it went quickly".

The island mentioned, now named Cizre, lies on the frontier between Turkey and Syria. This same eclipse also alarmed Saladin's army as the soldiers were crossing the Orontes River in Syria.

The very earliest observations of eclipses from China are recorded on inscribed bones dating from some time between 1350 and 1050 BC. These are known as "oracle bones" since they are mainly concerned with divination. The following example provides a useful illustration:

> "The divination on day *guiwei* was performed by Zheng: 'Will there be no disaster in the next ten days?' On the third day *yiyu* an eclipse of the Moon was reported". (These events occurred) in the 8th month".

Guiwei and *yiyu* are the 20th and 22nd days of a 60-day cycle.

Since the year is missing, the date of this event is uncertain; few bone inscriptions are intact, leading to considerable difficulties in dating.

Over the next 1500 years, little interest was shown in lunar eclipses in China, since they were regarded as of minor astrological consequence. By contrast, solar obscurations were considered to be unfavorable omens, especially for the ruler, and thus the records are much more complete. As many as thirty-six eclipses of the Sun are reported in a single ancient work, the *Chunqiu* — a chronicle covering the period from 722 to 481 BC. This reports total eclipses in 709, 601 and 549 BC, but without any descriptive details. Three further accounts (from 669, 664 and 612 BC) which describe eclipse ceremonies in which drums were beaten and oxen were sacrificed underline the importance attached to solar obscurations at this early period.

Ancient records contain few mentions of eclipse times but from the fifth century AD, eclipses of both the moon and sun began to be carefully timed using water clocks. The main purpose was to verify the accuracy of astronomical tables. Solar eclipses were usually timed to the nearest *ke* (mark) — 1/100 of a day and night — roughly equal to 15 minutes. For lunar eclipses the time was often expressed to the nearest fifth of a night watch instead. The period from dusk to dawn was divided into five equal *geng* (night watches), each of which was in turn divided into five equal intervals, often termed "calls". Hence the night watches and their subdivisions varied in length with the seasons.

The earliest eclipse whose time is carefully recorded is noted in the *Songshu*, the official history of the Liu-Song Dynasty (AD 420–479); its date corresponds to AD 434 September 5:

> "Yuanjia reign period, 11th year, 7th month, 16th day, at full Moon... The Moon began to be eclipsed at the 2nd call of the 4th watch, in the initial half of the hour of *chou*. The eclipse was total at the 4th call".

The initial half of the *chou* hour was between 1 and 2 a.m. Many similar accounts are to be found elsewhere in the *Songshu* and in later official histories.

A few total solar eclipses are reported in vivid detail in Chinese history, and the onset of darkness and appearance of stars is described. Examples are found in AD 120, 454, 761 and 1275.

The very earliest Korean records of eclipses appear to be merely copied from Chinese history and probably do not become independent until around AD 700. Later accounts tend to be very brief, seldom giving more than the day of occurrence. These are mainly found in the *Koryo-sa*, the official history of Korea from AD 936 to 1392, and in later chronicles. The Chinese luni-solar calendar was closely followed in Korea, although years were numbered from the accession of Korean monarchs. Around AD 1200 there are occasional references to the king, dressed in white robes and accompanied by his closest ministers, attempting to rescue the moon from its eclipse, but these may have been no more than ceremonies.

Japanese descriptions of eclipses tend to be more detailed than those of other East Asian countries and historical sources are much more diverse. The following account of the total solar eclipse of AD 975 August 9 is found in a privately compiled history known as the *Nihon Kiryaku*:

> "Ten-en reign period, 7th month, day *xinwei*, the first day of the month. The Sun was eclipsed. Some people say that it was entirely total. During the hours *mao* and *chen* (between 5 and 9 a.m.) it was all gone. It was the color of ink and without light. All the birds flew about in confusion and the various stars were all visible. There was a general amnesty (on account of the eclipse)".

In this brief article it has only been possible to cite a minute proportion of the available records of eclipses from the non-Western world. However, it should be clear from the selection offered here that many detailed and important descriptions are available. Ancient and medieval observations often reveal a high level of sophistication, particularly in techniques of measurement. In particular, medieval Arab astronomers consistently determined the times of eclipses to within about five minutes while contemporary Chinese measurements were only a little inferior.

F. RICHARD STEPHENSON

REFERENCES

Beijing Observatory. *Zhongguo Gudai Tianxiang Jilu Zongji* (A Union Table of Ancient Chinese Records of Celestial Phenomena). Kiangxu: Kexue Jishi Chubanshe, 1988, sections 4 and 5.

Newton, Robert R. *Ancient Astronomical Observations and the Accelerations of the Earth and Moon*. Baltimore: Johns Hopkins University Press, 1970.

Said, Said S., F. Richard Stephenson, and Wafiq S. Rada. "Records of Solar Eclipses in Arabic Chronicles." *Bulletin of the School of Oriental and African Studies* 274(4): 170–183, 1989.

Stephenson, F. Richard. "Historical Eclipses." *Scientific American* 274(4): 170–183, 1982.

Stephenson, F. Richard. "Eclipses in History." In *Encyclopedia Britannica* vol. 16, 1993, pp. 872b–877b.

Stephenson. F. Richard. *Historical Eclipses and Earth's Rotation*. Cambridge: Cambridge University Press, 1996.

See also: Ibn Yūnus – Calendars – Korea

***ELEMENTS* – RECEPTION OF EUCLID'S *ELEMENTS* IN THE ARABIC WORLD** Euclid (ca. 300 BC) was a Greek mathematician who worked in Alexandria. His name and his major work, the *Elements*, are famous not only among mathematicians, but also to students of other disciplines.

The *Elements* were sent by the Byzantine emperor to the Abbasid caliph al-Manṣūr (r. 754–775) into the Islamic civilization. We do not know if any of the scholars of his time prepared a translation or at least a summary for him or the scholarly community.

The earliest translation, according to medieval historical sources, was carried out by al-Ḥajjāj Ibn Yūsuf Ibn Maṭar (fl. between 786 and 833) for the famous Barmekide Yaḥyā Ibn Khālid (d. 805), wazir of caliph al-Hārūn al-Rashīd (r. 786–809) or, perhaps, the caliph himself. This translation seems to be lost in Arabic, except for very short quotations in writings of later authors, primarily about non-mathematical subjects.

Some twenty years later, al-Ḥajjāj composed a second Arabic version of the *Elements*, which he dedicated to caliph al-Maʾmūn (r. 813–833). One of those sources reports he did this to gain the caliph's support, because al-Maʾmūn's love for knowledge and his generous attitude towards scholars were well known (Codex Leidensis 399,1, 1897). To achieve this goal al-Ḥajjāj is said to have rewritten the *Elements* extensively with respect to their language, methods, and didactical features for people with knowledge in and love for this discipline, without changing the contents itself.

Some fragments of his version 2 seem to have been preserved in different forms:

(1) as part of a collection of mathematical texts put together in about the sixteenth or seventeenth century in Persia;

(2) as footnotes in Arabic translation manuscripts in those parts which transmit by and large the edition of Isḥāq Ibn Ḥunayn's (d. 910/911) translation by Thābit Ibn Qurra;

(3) in parts of Book III and the complete Book IV of the preserved Arabic translation manuscripts;

(4) perhaps as a variant of books XI-XIII in two of the translation manuscripts;

and in a barely recognizable form

(5) as parts of a commentary written by al-Nayrīzī probably at the beginning of the tenth century; or

(6) as a paraphrase composed around a hundred years later by Ibn Sīnā (d. 1036).

Those fragments, although they differ in some essential aspects, are characterized by a technical language using expressions from algebra, arithmetic, early logical or theological writings, as well as some words borrowed from Syriac, characteristic of the early translations into Arabic.

Furthermore, two of the Latin translations of the twelfth century, one by Adelard of Bath (so-called version 1), the other ascribed by the editor to Hermann of Carinthia (Busard 1968, 1977), contain at least extracts, possibly also more, of one of the Ḥajjāj versions. It is still an unsettled question as to how close these translations are to an original Ḥajjāj text and to which Ḥajjāj version they are related. A third possible, but also still disputed, source for the Ḥajjāj tradition is the only surviving Syriac fragment of Book I. Finally, some of the Hebrew translations of the *Elements* contain footnotes which are related to some of the footnotes ascribed to al-Ḥajjāj in the above mentioned Arabic manuscripts.

Al-Ḥajjāj's work, evidently, was crowned by success. In 828/29 he was ordered to translate the *Almagest*. The scholarly community eagerly accepted his translation and edition. They organized meetings discussing the *Elements*, the *Almagest*, and, probably other subjects. They composed editions and commentaries. Among the earliest works of this type are the editions made by al-Jawharī and al-Kindī (d. ca. 873). Except for short fragments, their texts are lost.

A fresh start was made in the second half of the ninth century. Perhaps because of the shortcomings ascribed to the earlier translation or for some other reasons unknown to us so far, Isḥāq Ibn Ḥunayn prepared a completely new translation of the *Elements*, probably before 872. Together with his translations of other important mathematical and astronomical texts, it was revised by Thābit ibn Qurra at about the same time. We do not know yet the reasons for nor the extent of those revisions.

The version contained in nearly all preserved Arabic translation manuscripts seems to be this edition — at least in the Books I, II, V and VII–IX. Other books, however, seem to be based on al-Hajjāj's work (see above). Most of the Hebrew translation manuscripts as well as the Latin translation by Gerard of Cremona are also derived to a great extent from Thābit's edition. This so-called Isḥāq/Thābit tradition exercised the greatest influence upon mathematical research and teaching in the medieval Muslim world until the late thirteenth century. It is characterized by a very literal translation of a text connected with the Greek edition made by Theon of Alexandria (fourth century). It differs in several aspects from the Greek critical edition made by J. Heiberg (1883). In some cases, Thābit is ascribed variant readings or even whole proofs taken from newly available Greek sources or other Arabic manuscripts, probably derived from the Ḥajjāj tradition. In other cases, the differences in text, diagrams, and proofs might well have been already present in the Greek manuscript translated by Isḥāq Ibn Ḥunayn. This leads to two questions that we are unable to answer at the moment:

(1) Did Isḥāq follow the methodological rules devised by his father Ḥunayn Ibn Isḥāq and collate manuscripts to establish a reliable Greek text or not?

(2) Which criteria motivated him in his choice between different variants of the same theorem?

Until the end of the thirteenth century several new Arabic editions based on the two ninth century traditions were produced. The most successful in terms of later influence were the two ascribed to Naṣīr al-Dīn al-Ṭūsī. Only one, however, consisting of fifteen books, was composed by him. The other one, which contains only the genuine thirteen books of Euclid, was finished in 1298, after al-Ṭūsī's death. Both versions replaced the earlier traditions, judging by the number of preserved copies, marginal notes in manuscripts, and references to the *Elements* in mathematical and non-mathematical works.

While the two ninth century traditions follow the Greek text relatively closely, the two so-called al-Ṭūsī revisions are a free rewriting of the *Elements* with respect to language, didactics, proofs, and variations. Proofs are generally abbreviated and condensed, further definitions or lemmata have been added, and for some difficult points new proofs have been provided. Such a free attitude towards a received text by an accepted authority deviated from values characteristic of or defended by the scholarly community of the medieval Islamic civilization. It can be observed, nevertheless, since the beginning of the ninth century in nearly all Arabic editions and commentaries on the *Elements*. This is one of the reasons that contemporary research about textual history is so difficult and the study of the mathematical results of the transmission so fruitful.

Although all fifteen books of the *Elements* attracted the attention of Muslim scholars, the central points of interest were the parallel postulate (Book I), the definitions of ratio and proportion (Book V), and the geometrical theory of quadratic irrationalities (Book X).

Efforts to prove the parallel postulate started with Greek mathematicians and philosophers, since the form given by Euclid did not fit into their ideas about what an axiom or a postulate should be. Parts of this discussion were translated into Arabic as Simplikios' (sixth century) commentary on the definitions of Book I and, perhaps, Heron's (second century?) commentary on the *Elements*. Both commentaries are lost in Greek and Arabic. Extracts from both are preserved in a revised form in al-Nayrīzī's commentary on the *Elements*. Extracts from Heron's commentary can also be found in Ibn al-Haytham's great commentary. The Arabic analysis of the parallel postulate started at the beginning of the ninth century with al-Jawharī and other scholars.

The geometers working on this postulate followed two different approaches. One approach, represented by ʿUmar al-Khayyām (d. 1131), sought to establish a theory of parallels independent of the Euclidean postulate which would allow them to prove theorem I,29, the first which Euclid proves on the basis of the parallel postulate, without using this postulate. In modern terms one could call this an effort to establish an absolute geometry. The second approach, represented by al-Jawharī, Ibn al-Haytham, Naṣīr al-Dīn al-Ṭūsī, and other scholars, was concentrated solely on proving the parallel postulate.

With respect to Book V the main result of the debate was, however cautious and reluctantly formulated, the introduction of a new number concept, which enlarged — speaking in modern notions — the Greek concept of numbers as natural numbers (without one) to the positive real numbers (without zero) by ʿUmar al-Khayyām and Naṣīr al-Dīn al-Ṭūsī. It was developed discussing compound ratios, i.e. the multiplication of ratios, already dealt with by Greek mathematicians. One of the reasons stimulating this research was its use in astronomy. Closely connected was the debate about the proportion theory for geometrical magnitudes. Although Arabic commentators did not reject this theory as wrong, they were dissatisfied with the fundamental definitions V,5 and V,7 about identity of ratios and inequality of ratios. They argued that the equimultiple concept used in those definitions, for instance definition V,5: $a : b = c : d \Leftrightarrow (na \lesseqgtr mb \Leftrightarrow nc \lesseqgtr md$, for all positive integers n, m), did not express the essence of a ratio clearly, which in their opinion was the measuring of one quantity by another one. They replaced it by a definition built upon the so-called Euclidean algorithm or, speaking in modern terms, on the comparison of the expansion of each ratio into a continued fraction. Among the medieval scholars who wrote treatises on it were al-Jawharī, al-Māhānī (ninth century), Thābit ibn Qurra, and ʿUmar al-Khayyām. The latter thought he had proven the equivalence between both theories of proportion, which allowed the use of all Euclidean theorems in Books V and VI in the second theory. Since proportions were of primordial importance for deriving and proving new results in geometry and astronomy, this was an essential project. His proof was based on a theorem that to three proportional magnitudes a fourth proportional always exists. The fundament of his proof for that theorem was a principle attributed to Aristotle about the infinite divisibility of (geometrical) magnitudes. Lacking the notion, not to mention a strict definition, of continuity, this principle, however, was insufficient for his purpose.

The main issue discussed with respect to Book X — which was perceived as the most difficult book of the *Elements* — was the question of how the geometrical magnitudes treated in it were related to the roots of non-square numbers used by algebraists and arithmeticians. To answer this question in a meaningful way meant developing a new conceptual understanding of the objects involved and creating a technical language which made it possible to speak about geometrical and algebraic-arithmetical irrationals on the same mathematical level. After several incomplete trials during the ninth and tenth centuries the equivalence of the different kinds of geometrical irrationals of Book X and simple as well as compound algebraic-arithmetical irrationals was established in the sense that one type could be translated into the other one. This debate not only brought the heirs of theoretical Greek mathematics together with practitioners trained in Eastern traditions, it also attracted laymen from other fields, who ordered — at least until the fourteenth century — copies of such commentaries on Book X, asked for explanations of their contents, and were proud to possess them.

SONJA BRENTJES

REFERENCES

Primary sources

Busard, Hubertus L. L. *The Translation of the "Elements" of Euclid from the Arabic into Latin by Hermann of Carinthia(?)*. Leiden: E. J. Brill, 1968; Amsterdam: Mathematisch Centrum, 1977.

Busard, Hubertus L. L. *The First Translation of Euclid's "Elements" Commonly Ascribed to Adelard of Bath*. Toronto: Pontifical Institute of Medieval Studies, 1983.

Busard, Hubertus L. L. *The Latin Translation of the Arabic Version of Euclid's "Elements" Commonly Ascribed to Gerard of Cremona*. Leiden: E. J. Brill, 1983.

Heiberg, I. L., ed. *Euclidis Elementa*. Euclidis Opera Omnia. Leipzig: Teubner, 1883.

Ibn al-Nadīm. *Kitāb al-Fihrist*. Leipzig: Harrassowitz, 1871/1872.

Ibn al-Qifṭī. *Taʾrīkh al-ḥukamāʾ*. Ed. Julius Lippert, Leipzig: Harrassowitz, 1903.

Secondary sources

Brentjes, Sonja. "Textzeugen und Hypothesen zum arabischen Euklid in der Überlieferung von al-Ḥaǧǧāǧ Ibn Yūsuf Ibn Maṭar (zwischen 786 und 833)." *Archive for History of Exact Sciences* 47(1): 53–92, 1994.

Dodge, Bayard, ed. and trans. *The Fihrist of al-Nadīm. A Tenth-Century Survey of Muslim Culture*. New York: Columbia University Press, 1970.

Dold-Samplonius, Yvonne. "al-Māhānī." In *Dictionary of Scientific Biography* vol. X. New York: Charles Scribner's Sons, 1974, pp. 21–22.

Jaouiche, Khalil. *La théorie des parallèles en pays d'Islam: contribution à la préhistoire des géometries non-euclidiennes*. Paris: J. Vrin, 1986.

Matvievskaia, Galina Pavlovna. *Učenie o čisle*. Moscow: Izdatel'stvo Nauka, 1967.

Murdoch, J. E. "The Transmission of the *Elements*." In *Dictionary of Scientific Biography* vol. IV. New York: Charles Scribner's Sons, 1971, pp. 437–459.

Rosenfeld, Boris A. *A History of Non-Euclidean Geometry. Evolution of the Concept of a Geometric Space*. New York: Springer, 1988.

Sabra, A.S. "Jauharī." In *Dictionary of Scientific Biography* vol. VII. New York: Charles Scribner's Sons, 1973, pp. 79–80.

Sezgin, Fuat. *Geschichte des Arabischen Schrifttums*. vol. V. Leiden: Brill, 1974.

Youshkevitch, Adolph Pavlovitch and Boris Rosenfeld. "Khayyāmī." In *Dictionary of Scientific Biography* vol. VII. New York: Charles Scribner's Sons, 1973, pp. 323–334.

See also: Isḥāq ibn Ḥunayn – Thābit ibn Qurra – al-Nayrīzī – Ibn Sīnā – *Almagest* – al-Jawharī – al-Kindī – Ḥunayn ibn Isḥāq – Naṣīr al-Dīn al-Ṭūsī – Ibn al-Haytham – ʿUmar al-Khayyām – al-Māhānī

ENGINEERING Engineering has become the paradigm or model for almost all technology in the West, hence terms such as "genetic engineering", or "systems engineering". Applied to the non-Western world, this use of language can be grossly distorting, preventing recognition of other paradigms. Thus the word "engineering" is best limited to *construction works* and related *mechanical devices*. Moreover, the smallest of such works, as carried out by individual house builders or farmers practising irrigation, may often be better considered in other contexts, such as "building" or "agriculture".

Three kinds of construction works need to be considered, namely those related to fortifications, monuments, and water management. Methods of construction were nearly always labor-intensive, so various ways of organizing a labor force are important for understanding what was involved. In Southeast Asia, for example, irrigation systems for rice culture were often built by groups of farmers working cooperatively. Thus on the island of Bali, there was a social unit known as a *subak* comprising all the farmers whose fields were watered from the same source, who had to come to a mutual arrangement about regulation of water flow, repairs and new construction.

By contrast, very large construction works were usually carried out by order of a king or emperor, supervised by state officials, and using conscript or corvée labor. An example is the New Bian Canal in China, completed in AD 635, on which five million people were said to have worked at different times.

The organization of labor, supervision of construction, and payment for materials might take other forms if the work was being done for a rich villager or other notable, or if it had been initiated by a religious community based at a temple. The latter were often very important, and in China, Buddhists helped pioneer the use of iron in bridge building as well as in the architecture of temples between AD 500 and 1200. Buildings for religious or funerary purposes were also built at the instigation of rulers, often on a very ambitious scale. One thinks of the pyramids of Egypt, Maya temples in Central America, and numerous temples in India and Southeast Asia.

Such buildings could be linked to apparently utilitarian engineering works. The temples of Angkor in Cambodia, founded about AD 880, were built on mounds created by the excavation of large reservoirs or tanks that were part of a system for supplying water for rice culture, and were also linked to a system of transport canals. Stone for building the temples was brought to the site along these canals. The irrigation works reached their fullest extent by about 1150, after which environmental damage — the formation of a ferrous hard pan — seems to have contributed to their decline.

Other remarkable works of hydraulic engineering include large reservoirs with elaborate systems of feeder canals in Sri Lanka, where some canals pass through tunnels from one catchment to another. These systems evolved slowly from about 200 BC to AD 1300, and like the works at Angkor, may have had monumental as well as utilitarian significance. There were also early canal irrigation systems in India, and in Iran there was a specialized technique for tapping groundwater known as the *qanat*.

In the Americas, there were the impressive Chicoma canals in Peru and the dikes in the former Lake Texcoco built by the Aztecs to separate artificial cultivated islands (*chinampas*) from the salty water of the main body of the lake. The Aztec capital of Tenochtitlan adjoined this system of islands and dikes.

Reviewing early engineering along these lines leaves one with a strong impression of its connections with agriculture, especially where irrigation or other water control was necessary, and indicates the frequent role of state or government authorities with their interest in using engineering also for religious or monumental constructions. The latter were important in demonstrating and legitimizing the authority of the state and its ruler.

This view of engineering can be interpreted in terms of the theory of "hydraulic civilization" associated particularly with the work of Karl Wittfogel. According to him, the food requirements of growing populations made it essential to develop centralized states capable of organizing large-scale irrigation works. Thus hydraulic engineering is seen as primary, with many other forms of large-scale engineering developing from it. The importance of irrigation in much of Asia, in Egypt, and in parts of Central America seems obvious, and it is at least plausible to think that the necessity for such irrigation works was a driving force for innovation in engineering techniques and for growth of government organizations capable of managing them.

An alternative interpretation of world history arises from the argument that military power emerged for the first time as a distinct force in human affairs — as opposed to an occasional and *ad hoc* assembly of fighting men — toward the end of the Neolithic, not long before 3000 BC. Proponents of this view point out that it was in states that expanded as a result of the growth of organized military power that engineering developed most rapidly. The earliest examples were in Mesopotamia and Egypt, but there were parallel developments in many other parts of Asia, notably with the rise of the first Chinese dynasties. Later on, the emergence of empires in Central America and Peru can be seen as having a roughly similar technological outcome. By contrast, kingdoms that remained small, tribal societies, and isolated communities of cultivators and pastoralists had different kinds of technology that hardly involved engineering at all. What may have made the difference was that the experience of creating and commanding armies provided a model for recruiting, organizing, and disciplining the labor force needed to construct fortifications and other large-scale works. In this view, then, military engineering was primary, and monument building and hydraulic engineering were extensions of the same basic techniques into other areas that interested rulers and their officials.

Military technology, of greater or lesser significance depending on which of these interpretations is adopted, is the subject of a separate article in this volume. Looking in more detail at the civil branches of engineering, we may note that state officials often played a part in surveying and supervising works, and typically had a good knowledge of the mathematics needed for laying out sites or estimating quantities. The first pyramids in Egypt were accurately square and were built on carefully leveled foundations. Centuries later, Chinese bureaucrats studied books on arithmetic that told them how to calculate quantities and estimate volumes of earth-moving needed on construction sites — but books on machines were rare.

Much early engineering involved the excavation of canals or the building of earthworks in the form of fortifications, dikes, irrigation terraces, or dams, with few striking innovations beyond increases in scale, surveying methods, or the use of gabions (wickerwork containers of stones) in China. Where irrigation canals needed gauging weirs or distribution boxes to divide the flow equally between several farmers, some intriguing designs were developed in Iran, many of them made of stone.

The earliest engineers used stone in construction long before iron tools were available, so that quarrying and dressing the material presented considerable problems. In Egypt, the pyramid builders used wedges and dolerite hammers to work limestone, and the architect and minister of state Imhotep, who was active around 2660–2650 BC, was credited with the invention of building in stone. Transport was by barge where possible and then on sledges hauled along lubricated wooden tracks. In the absence of pulleys or any other lifting gear, stones were raised by hauling up ramps.

In the Americas, the most remarkable engineering in stone was the terracing and building accomplished by the Incas in Peru. As in Egypt, there were no iron tools, but stones of quite hard rock, including granite, were shaped by grinding until they fitted together very precisely. In the Americas, as in ancient Egypt and many other parts of the world, sun-dried mud brick provided a further option for the construction of buildings and fortifications. A notable example is the fortress of adobe brick at Paramonga, in the lowlands of Peru. In West Africa, mud or clay was sometimes used rather as modern architects use concrete, to form domes and shell-vaults (for example, at Zaria in northern Nigeria). City walls of mud brick (adobe) date back to the twelfth century AD at Kano (also Nigeria). They were extended twice in later centuries to encircle a 17-kilometer perimeter, and in 1903 successfully withstood cannon fire from the British army.

While much hydraulic engineering depended on earthworks, spillways needed to be faced in stone to prevent erosion, and a dam designed so that overflow water could pass over its crest would usually need to be stone-faced throughout. A fine example was the long, low Cauvery River dam, built to serve an irrigation system in the Chola Kingdom of

South India. It was built of masonry laid in clay around AD 200. At about the same date, Persian engineers were using iron dowels set in lead to fix together the massive stones they used in facing some of their dams, and also poured molten lead into the space between stones.

Looking back to an earlier age to consider mechanical aids for construction, one should note that the movement of large blocks of stone along the ground on rollers was not a simple procedure, and was rarely practiced in Egypt. However, the consensus remains that the invention of the wheel derived from experience with rollers, perhaps in Mesopotamia. There, at any rate, wheeled vehicles were in use by 2500 BC. Moreover, in conformity with the view that military technology was the leading sector in engineering, the first known examples were on chariots and army supply vehicles.

The development of mechanical engineering from that time on is hard to document in detail. Some commentators stress the emergence of mechanisms involving wheels or rotary motion, such as the pulley (by 800 BC) or the first toothed gears, probably made in the eastern Mediterranean regions soon after 500 BC. Another approach is to notice the importance of three main kinds of mechanical development — firstly, a slow evolution of levers, pulleys and hoists for lifting stone on construction sites; secondly, the development of siege engines, catapults and other military hardware, and thirdly, the evolution of water-raising devices, sometimes for drawing water from wells, but sometimes also to supplement irrigation from canals or rivers.

One kind of mechanical water-raising device was the chain-of-pots. The simplest form of this may have originated before 100 BC as a "necklace" of pots hanging over a roller or pulley which was hauled around manually. The bottom of the necklace dipped in water, and as it was pulled round, pots filled at the bottom, and then were tipped over the roller at the top to empty into a trough. An animal-power version, driven through wooden gears, and with a large wheel replacing the roller or pulley, developed five or six centuries later. Known as the *saqiya* or "Persian wheel", it came to be widely used in North Africa, the Islamic countries, India, and China. For example, as recently as 1904 some 4000 were recorded in the Dongola province of Sudan, each irrigating about 15 acres (6 hectares).

Other mechanical water-raising devices include the pedal-driven "dragon spine" pump used in Chinese rice culture, and the *noria*, a water wheel with containers attached to its rim, powered by the flow of a river. The latter was known in western Asia by 100 BC and also has a long history in India and China extending into the twentieth century. Other water raising mechanisms used in modern peasant cultures have included a form of Archimedean screw in Vietnamese rice paddies and lever-based devices in Indonesia.

The first water-powered mills for grinding corn were adapted to mountainous areas with streams rushing down steep gullies, and their design was perhaps suggested as much by experience of the potter's wheel as also undoubtedly by the *saqiya* and *noria*. Thus we might expect the first water mills to have been made in a mountainous district where there were people with experience of all these devices and also where there were quarries producing millstones for rotary hand-mills. These considerations support evidence indicating that Iran or Iraq is the place of origin of the water mill some time prior to 350 BC. There were quarries known for their millstones both in Iran, and on the upper Tigris, in a region now in Turkey.

The mills invented at this date had horizontal, propeller-like water wheels that drove the millstones directly, and it appears that they were so effective that they were soon adopted in southern Europe and also China. Later, this type of corn mill was used in northern European countries (where it became known as the "Norse mill"), and most notably in the foothills of the Himalayas, especially in Nepal.

The alternative type of water mill which has a vertical water wheel driving the millstones through the wooden gears developed later, and had a more restricted distribution. However, it is noteworthy that it used the same kind of gears as the more widespread animal-powered *saqiya* or Persian wheel, namely "peg and lantern" gears. That is, a wooden crown wheel with projecting peg teeth on the same horizontal axle as the water wheel meshed with a lantern pinion on the vertical axle.

Almost the only non-Western countries in which this type of gearing was developed beyond the basic corn mill or Persian wheel format were Iran, Iraq, and Syria in western Asia and China in the East. It is striking how little mechanical devices of this kind were used elsewhere. However, in western Asia and China, a considerable variety of water-powered (and animal-powered) machinery was developed, including machinery associated with papermaking and sugar refining, and windmills were used also, all before AD 950. In preparing pulp for papermaking, for example, water wheels turned cam shafts which operated trip hammers. On a smaller scale, experiments were also made with geared astrolabes and clock mechanisms, and in China, an astronomical clock powered by a slow-running water wheel was made by Su Song in the 1080s. At about the same time in Chinese industry, water-powered bellows provided the draught for a few of the many iron furnaces, and water-powered spinning mills were evolving in districts that produced the coarser kinds of textiles.

One distinctive feature of Chinese machinery (as compared with practice in the Islamic world) in the period up to AD 1300 was a preference for using an endless rope running over pulleys for transmission of power from water wheels, instead of gears and shafts. A comparable contrast in style and practice in the West occurred in the nineteenth century when American designers of New England textile mills favored rope drives from water wheels or turbines to different parts of a mill, whereas European millwrights still preferred gears and shafting.

In conclusion, it is worth observing that although a steady development of engineering can be discerned in several parts of Asia up to about AD 1200, it is hard to identify major innovations dating from after that time. Engineering was never the most important branch of technology over much of that continent, but even where it had developed strongly there was little further progress. One factor was undoubtedly the devastating Mongol conquests of Iran, Iraq, and China, completed by 1260, and then the European conquests that were beginning in 1500. The Mongols destroyed irrigation systems in parts of northern Iran, and allowed others to decay, thus undermining the basis of much engineering in that region. However, it is difficult to believe that conquest and political disruption are the whole story, and one must also look to institutional and economic factors that are beyond the scope of this article.

Another general point is that by concentrating on the civilian uses of technology and leaving the military aspects for separate treatment, this article inevitably demonstrates the great importance of hydraulic engineering. This may seem to justify the idea that many non-western societies were "hydraulic civilizations". However, we probably need to be more aware of the many mechanical and structural ideas originating from the experience of building siege engines, chariots, city defenses, moats and forts. Maybe, too, we ought to appreciate that military engineers traveled more than their purely civilian counterparts and probably contributed more to the spread of ideas. In the third century AD Persian rulers employed a captured Roman army on a dam-building project, presumably learning a little of Roman engineering in the process. In AD 751 a battle between Chinese and Islamic forces in Central Asia led to the employment of Chinese experts on papermaking in Samarqand, and resulted in the transfer of papermaking technology into the Islamic world.

However, the most important point we tend to neglect is that the organization and discipline of armies seems likely to have been a model, at many points in Western, as well as non-Western history, for the organization and management of civilian technology, especially large scale construction projects.

ARNOLD PACEY

REFERENCES

Bray, Francesca. *The Rice Economies: Technology and Development in Asian Societies*. Oxford: Blackwell, 1986.

Coe, Michael D. "Chinampas of Mexico." *Scientific American* 211: 90–98, July 1964.

Denyer, Susan. *African Traditional Architecture*. London: Heineman, 1978.

Forbes, R.J. *Studies in Ancient Technology*. 9 vols. Leiden: Brill, 1955–64.

Needham, Joseph, with Wang Ling. *Science and Civilisation in China*, vol. 4, part 2, *Mechanical Engineering*. Cambridge: Cambridge University Press, 1965.

Robertson, Donald. *Pre-Columbian Architecture*. New York: George Braziller, 1963.

Sleeswyk, A.W. "The Development of the Earliest Wheels." *Polhem: Tidskrift för Teknikhistoria* 10: 109–130, 1992.

Smith, Norman. *A History of Dams*. London: Peter Davies, 1971.

Wittfogel, Karl A. *Oriental Despotism*. New Haven, Connecticut: Yale University Press, 1957.

Wulff, Hans E. *The Traditional Crafts of Persia*. Cambridge, Massachusetts: MIT Press, 1966.

See also: Military Technology – Technology in the Islamic World – *Qanat* – Irrigation – Agriculture – Stonemasonry – Road Building – Pyramids – Clocks and Watches – Construction

ENVIRONMENT AND NATURE: AFRICA Africa's location astride the equator subjects the continent's environment to tropical and subtropical conditions. Mean temperatures range above 15°C in the winter and above 25°C in the summer. During the northern summer the weather is hot and wet over Central Africa and the Guinea coast, arid conditions prevail over northern and most of southern Africa, and the east coast is affected by the southwest monsoon. During the northern winter the Sahara and the Mediterranean are affected by the westward flows of warm dry air, and the northeast monsoon regulates the weather over the east coast; warm rainy conditions prevail elsewhere except along the southwest coast. The surrounding waters have mean sea-surface temperatures above 20°C, but regional peculiarities distinguish the coastal zones: the Guinea coast with its abnormal sub-humid conditions; the hyperarid southwest coast with the cool and foggy Namib desert; the temperate Cape zone; the east coast affected by the monsoon winds and rains, the tropical easterlies, and the semi-diurnal tides up to six meters; and the Red Sea region with its extremely hot and arid climate and the highest seawater salinity in the world.

Only 9% of Africa's total land surface is well suited to agriculture. Most of the continent's ecological systems occupy latitudinal belts with the climate varying from sub-humid to semiarid. The dense tropical forest extends several hundred miles on either side of the equator and along the Guinea coast where precipitation may reach 3000 mm a year. To the north and south it thins out to woodland savanna and drier forest or bush. The southern edge of the Sahara, called the Sahel (Arabic for "coast, edge"), forms a transitional zone between savanna and desert. The Horn of Africa is persistently dry, but the Kalahari desert is mostly grassland and scrub, with waterholes and even some marshes. The Ethiopian highlands form the largest single area on the continent where fertile soils, numerous rivers, and abundant rainfall allow for productive agriculture. The width of the coastal zone varies from less than 1 km to over 300 km (200 miles). Most African countries have mixed farming, and most land by area serves as pasture. The rainfall has had a profound influence on the distribution of population. Three rainfall thresholds mark the risk margin in traditional economies: (1) where average annual precipitation is below 500 mm agriculture cannot be sustained and total dependence on cattle is necessary; (2) where rainfall varies between 500–750 mm per year it is impossible to subsist on agriculture alone, and supplementary stock-rearing is the customary solution; (3) rainfall in excess of 750 mm per year is generally more predictable and more evenly distributed. Even then, the actual potential for agriculture and livestock raising depends on soil fertility, water evaporation, and ground water retention. African soils are generally poor, and agriculture requires large labor investment for subsistence maintenance. Only where soil fertility is combined with adequate water resources is surplus production possible.

Precarious existence in marginal environments makes for particularly strong connections between ecology and society. The harsh lands where water is scarce and grazing restricted historically produced fierce competition within a society and between neighboring societies. Cultivators and herdsmen displaced hunters and gatherers; possession of iron for weapons and implements played a key role in this process. For insurance against famine, in addition to stock-keeping, farmers cultivate different plots, if possible in more than one ecological zone to take advantage of variations in the seasonal rainfall patterns and temperatures either using the same crop (e.g. maize or bananas) or different crops. Usually, two harvests are possible. Cultivation became linked with the idea of civilization, and the conceptual opposition between culture and nature found expression in the dychotomies of farm/bush, coast/hinterland, civilization/barbarism noted in various settled societies. This polarity was also reflected by the classification of animals into wild or carnivorous, therefore "evil" and unfit for human consumption, and domesticated or herbivorous, suitable for food.

Subsistence patterns affect residential mobility, division of labor, dominant forms of property and ownership patterns, land tenure, social networks, typical marriage age, and marriage transactions. The transhuman cycle depends on the particular botanical and hydrological environment and may vary from regular moves between wet-season villages and dry-season camps to a continual migration of over 1000 km throughout the year. In mixed-economy societies, women often provide the bulk of the wet-season horticultural labor and men the bulk of the pastoral activities. Traditional forms of wealth associated with production and sustenance, such as iron hoes or cattle, are still included in bridewealth payments. In sub-Saharan Africa, land was traditionally held in tenure by the producers rather than owned. Towns grew in ecological oases or on trade routes. Markets developed between two ecological zones whose growing and harvest seasons differed, to trade for seed, food (in case of famine), and local specialties. There were seasonal markets, like those on the east coast regulated by the monsoon calendar, and "weekly" markets (e.g. held every fifth day as in the Shambaa country).

There is no formalized or universal world view in African systems of thought. Man is the center of the universe but he is not master of it. If harmony between man and nature is lost, the ensuing chaos harms man the most. In most cosmologies, the earth exists for man's sake but he must show it respect. Nature is seen as animated by spirits, and people depend on it for survival. Nature rituals, especially agricultural rites, punctuate the annual religious cycle: planting, first-fruits, and harvest ceremonies, etc. Special rituals relate to stock-keeping. Rain-making rituals are the most important. They are performed by specialists — priests, witchdoctors, or royalty (such as the Lovedu rain-queen in southern Africa fictionalized by Ryder Haggard as "She Who Must be Obeyed"). Shaka, the founder of the Zulu warrior state, declared himself the greatest of all rain-makers. Such specialists are not "mere" witchdoctors but well-trained in weather lore: people are aware that there is order in the laws of nature which normally do not change. They do not ask for rain in the middle of the dry season; the rituals are performed before the rains to ensure their proper and timely coming, if the rains fail to arrive, or to stop the rain when too much water may damage the crops and fields. In Zambia, when the rains start, people wait for a few days before working in the fields. An Ashanti farmer addresses his ancestors and the earth at the beginning of cultivation, saying: "The yearly cycle has come round and I am going to cultivate. When I work, let a fruitful year come upon me, do not let a knife cut me, do not let a tree break and fall upon me, do not let a snake bite me". At planting, the Venda of South Africa distribute grain symbolically cooked in the field over a grass fire. At first-fruits ceremonies, grains of the new crop may be sprinkled in the field for the spirits or the remains of last-year's crop cooked and eaten. Among the Basuto the first gourds are taboo until the chief tastes them and offers a cooked dish of the new food to the ancestors.

Traditional African views of nature are thoroughly socialized. West African cultivators hold that while gold is rare and expensive, iron was given to man by God for universal use as tools. The Zulu believe that the sun, moon, and stars were given by God so men could see, and that cattle were created by God to be man's food. The Masai think that cattle were given by God exclusively to them, and so consider raiding the herds of their neighbors legitimate. A chief's fields must be cultivated first because their fertility, as his virility, guarantees the whole community's well-being. At the Yoruba harvest festivals, gifts of yam tubers due the chief from each household serve as a tool of unwritten population census. When women pound meal, their long-stemmed pestles are thought to knock against the sky. The stars of the Pleiades are thought to follow one another "like cattle returning home from pasture". Water is sacred, and boiling it "kills" the water spirit. Many plants and seeds are used for medicinal purposes. Sacrifices at religious festivals, rainmaking ceremonies, and seances of divination and spirit-possession involve agricultural produce (e.g. yam, various cereals) and cooked foods, and domestic animals and fowl (especially chickens, sheep among Muslims, and on very special occasions cattle).

The universe is unending in space and time. The past and present matter greatly; the future matters only within the next growing cycle. The major rhythms of time are the seasons of rain and dry weather, migrations of birds and animals, flowering of certain plants, lunar months, and day-and-night sequences. The minor rhythms are in the lives of humans, animals, and plants as they move from birth to growth, then procreation and death. The calendars of African peoples are punctuated by references to natural phenomena and production activities. Since there is little temperature variation, the difference is not between the hot and cold seasons, but wet and dry. Usually, the two major seasons are the Greater Rains and the Lesser Rains. The Bantu word for "year", *mwaka,* literally means the rains at the beginning of the Greater Rains. On the East African coast they come in March; the Lesser Rains beginning in July or August are not a good planting season, but about October come the Latter Rains, the second cultivating season. Despite long-term Islamic influence and many Arabic loan-words, the Swahili language has preserved Bantu names for the cardinal points, dominant winds, days of the week, and times of day which reflect the sailing, agricultural, and stock-rearing aspects of Swahili culture. Some coastal cities created an office of "Master of the Sea" or "Master of the Beach". The word "Swahili" itself is derived from the Arabic for "coasts" or "coastal settlements". The name "Benadir" for the coast of southern Somalia means "harbors".

The pre-Islamic Bantu calendar calculated the annual cycle in twelve months of three ten-day "weeks" (decades, as in ancient Egypt) but within the year it distinguished lunar months in two phases. Different African peoples have "weeks" of four, five, seven, eight, nine, and ten days. Traditional African time-reckoning is not mathematical but event-based. For example, the historical chronology of the Shambaa people of Tanzania is based on the memory of famines. The daytime computation of the Ankole of Uganda begins with the morning milking (about 6 a.m.) and follows the sun with references to times of rest, drawing water for people, driving cattle to drink, returning them to grazing, etc. The lunar months may be called "hot month", "month of rains", "weeding month", "bean harvest month", "hunting month", and so on.

The dwelling and settlement patterns too are affected by reference to the universe and nature. The Dogon of West Africa orientate their village north to south and visualize its plan as a human body prostrated on the ground. The smithy is the head, the family houses are in the center, and separate women's and men's houses, like hands, are on the east and west. Millstones and a foundation altar are lower down like sexual organs, and other altars are at the feet. The shape and orientation of the house mimics the Dogon vision of the earth and the sky as a couple in union. The rooms, the roofs, and the supporting posts indicate the male and female parts; the placement of the hearth, door, and stones supporting cooking pots, even the sleeping arrangements are regulated by traditional spatial orientation.

Environmental factors have conditioned some major processes in African history, such as population movements and settlement, agricultural production, state formation, and long-distance trade. For example, the arid expanses of the Sahara desert both protected and isolated West Africa from more extensive contacts with North Africa, the Mediterranean, and the Nile Valley, and the remoteness and paucity of resources of the equatorial forest led to historically low population density and lack of contacts with the outside world. The highlands of Ethiopia kept the country independent and self-sufficient but also isolated for much of its history. By contrast, the Swahili culture of the East African coast was totally predicated on coastal location with easy and regular contacts with the Indian-Ocean world, its prosperity fed by fishing, agriculture, and trade. In the West Sudan, the rise of the medieval empires of Ghana, Mali, and Songhay was supported by efficient agriculture, access to gold and control of the rivers. However, most African rivers, including the Nile, are not navigable throughout their complete course or through the year; they are thus unable to serve as major communication arterials. Material culture of different localities still demonstrates its dependence on availability and qualities of timber, fiber, clay, and pigments. Traditional architectural forms developed in West Africa reflect the needs of mud construction in the rainy climate, while in East Africa stone architecture was limited to the coast because of reliance on coral rock.

The numerous migration legends of African peoples are misleading in that they highlight the movement of small segments of population while masking a general cultural stability reflective of ecological zones rather than specific ethnic groups. In areas suited to agriculture, shifting cultivation was a usual response to soil exhaustion or population growth until this century. In drier areas with sparse vegetation and meager topsoil, ecological changes could lead to long-distance migration, especially by herding populations. The West African *jihads* of the eighteenth and nineteenth centuries were spearheaded by mobile, pastoralist communities such as the Fulani, and affected mostly the Sahel and savanna zones of the Sudanic belt where both cattle and horses (for cavalry) could live. In the sixteenth century, agricultural Ethiopia suffered major invasions first by the nomadic Somalis and then by the cattle-herding Galla (Oromo). In a later expansion of the Amhara Ethiopian empire, the Amhara armies invading Oromo territories could not feed themselves because the Oromo pastoralists did not grow crops and could evacuate their herds from the path of an advancing army. Once some Oromo settled and became cultivators, they avoided stock-rearing for fear of cattle-raiding Somalis or Masai. In the nineteenth century, Southern Africa became the theater of two major late migrations based on cattle-herding and a search for pasture: the southbound Mfecane led by the Zulu and the northbound Great Trek of the Dutch Boer farmer immigrants.

The wagons used by the Boers were the first wheeled transport in sub-Saharan Africa, where trade was moved by head porterage in consequence of the tsetse fly infestation and lack of salt in animal diets, which caused the absence of draft and pack animals. Growing contacts with the Western world led to the development of new trade patterns and regional economies: new trade routes connected the forest societies with the Atlantic coast, and some mobile groups (e.g. with access to rivers in the west or sea in the east) became the major agents of change. In the interior, populations seeking safety from slave-raiding moved their villages into inaccessible hills while those seeking trade moved their towns from mountain slopes into plains. By the mid-nineteenth century a transcontinental system of routes was in place.

To this day, Africa's food production and raw material bases are still traditional, small-scale and rural; food productivity is the lowest in the world. Sixty-five per cent of the continent's population derives its livelihood from agriculture. However, the picture of a static self-subsistence economy is false. African societies have shown both adaptability to change and readiness to experiment. The early spread of three Asian crops — yam, taro, and banana — revolutionized the economy of the humid regions. American food crops (especially cassava, sweet potato, maize, and groundnuts), introduced early by the Portuguese, spread quickly across the continent and still continue to displace indigenous crops. Former hunters and fishermen, like the Bakuba of the Congo basin (Zaire), once they were able to produce agricultural surplus, underwent an economic revolution which allowed them to develop handicrafts and create a kingdom in the seventeenth century. The risk and responsibility for trying the new crops dwelled with the chiefs or kings. In the first half of the nineteenth century the sultan of Zanzibar became a major innovator and plantation owner when he transformed the island into a clove-producing garden; coconut plantations were developed, some numbering hundreds of thousands of trees. While for almost two thousand years the main articles of African trade with the outside world had been gold, slaves, and ivory, agricultural diversification in the nineteenth and twentieth centuries has led to increasingly varied international trade.

European settlement succeeded only in temperate zones free of tropical fevers. The emphasis on cash crops and the spread of perennial crops (oil palm, cocoa, coconuts, coffee, cotton, etc.) have reduced shifting cultivation. Application of Western methods of agriculture has been successful in cultivating sisal, tea, and pyrethrum but has shown no advantage in the production of rubber, groundnuts (peanuts), oil palms, cocoa, coffee, or sugarcane. Indigenous cattle, although considered poor by Western standards, have shown greater resistance to epidemics of rinderpest, lung sickness, etc., several of which devastated regions of eastern Africa in the nineteenth century, facilitating European colonization. Numerous attempts to introduce Western agricultural machinery have failed because of their unsuitability to weak soils, traditional multi-crop multi-story cultivation, and the socially-determined patterns of labor and land tenure. Today, the hoe remains the major farm implement and women still grow 80% of the food consumed by their families and 50% of the cash crops. The lack of protein, especially in the children's diet, persists as a major obstacle to the development of a productive and healthy African society.

Africa is a continent rich in resources which faces enormous environmental challenges, including deforestation, desertification, loss of biodiversity, and degradation of water resources. In 1960 Africa was a net food exporter but since the 1970s it has developed a dependence on food imports and aid. The droughts and famines of the 1970s, 1980s and 1990s have been exacerbated by civil wars and government policies, in addition to poor communications and transportation. Deforestation, inefficient cultivation, and destruction of the wildlife are tolerated to feed people quickly. Draining marshes helps reduce the incidence of malaria, typhoid, river blindness, and cholera, but creation of new reservoirs increases bilharzia (schistosomiasis). Hydroelectric power dams built in recent decades have eased Africa's energy dependence, but during droughts the water supply is sometimes insufficient for electricity production. Nevertheless, in Africa the potential for preserving and successfully exploiting the environment is higher than in any other developing areas of the world. Planning, changes in mentalities of both rural and urban populations, and wide dissemination of appropriate information are keys to the solution of the many current problems.

M.A. TOLMACHEVA

REFERENCES

Allen, James deVere. *Swahili Origins: Swahili Culture & the Shungwaya Phenomenon*. London: James Currey, 1993.

Bassett, Thomas J. and Donald E. Crummey, eds. *Land in African Agrarian Systems*. Madison: University of Wisconsin Press, 1993.

Church, R. J. H. *West Africa: A Study of the Environment and of Man's Use of It*. 8th ed. London and New York: Longman, 1980.

Clark, J. Desmond and Steven A. Brandt. *From Hunters to Farmers: The Causes and Consequences of Food Production in Africa*. Berkeley: University of California Press, 1984.

Connah, Graham. *Three Thousand Years in Africa: Man and his Environment in the Lake Chad Region of Nigeria*. Cambridge and New York: Cambridge University Press, 1981.

Croll, Elisabeth, and David Parkin, eds. *Bush Base: Forest Farm; Culture, Environment, and Development*. London and New York: Routledge, 1992.

Feierman, Steven. *Peasant Intellectuals: Anthropology and History in Tanzania*. Madison: University of Wisconsin Press, 1990.

Fitzgerald, W. W. A. *Travels in the Coastlands of British East Africa and the Islands of Zanzibar and Pemba: Their Agricultural Resources and General Characteristics*. London: Chapman and Hall, 1898. (Reprint: Folkestone and London: Dawsons of Pall Mall, 1970).

Forde, Daryll, ed. *African Worlds; Studies in the Cosmological Ideas and Social Values of African Peoples*. London and New York: Oxford University Press, 1960.

Griaule, Marcel. *Conversations with Ogotemmeli: an Introduction to Dogon Religious Ideas*. London: Oxford University Press for the International African Institute, 1965.

Gulliver, Philip. *The Family Herds: A Study of Two Pastoral Tribes in East Africa, the Jie and the Turkana*. London: Routledge and Kegan Paul, 1955.

Herskovits, Melville. "The Cattle-Complex in East Africa." *American Anthropologist* 28: 230–72, 362–80, 949, 1926; 29: 528, 633–64, 1927.

Karp, Ivan and C. S. Bird, eds. *Explorations in African Systems of Thought*. Bloomington: Indiana University Press, 1980. (Reprint: Washington, DC: Smithsonian Institution, 1987).

Koponen, Juhani. *People and Production in Late Precolonial Tanzania: History and Structures*. Helsinki: Scandinavian Institute of African Studies, 1988.

Mbiti, John S. *African Religions and Philosophy*. 2nd ed. Oxford: Heineman, 1991 (orig. 1969).

McCann, James C. *People of the Plow: A History of Ethiopian Agriculture, 1800–1990*. Madison: University of Wisconsin, 1995.

McLoughlin, P. F. M., ed. *African Food Production Systems*. London and Baltimore: Johns Hopkins Press, 1970.

Parrinder, Geoffrey. *African Traditional Religions*. 3rd ed. New York: Harper, 1976.

Pollock, N. C. *Animals, Environment, and Man in Africa*. Farnborough: Saxon House, 1974.

Pottier, Johan, ed. *Food Systems in Central and Southern Africa*. London: School of Oriental and African Studies, 1985.

Saha, Santosh C. *A History of Agriculture in West Africa: A Guide to Information Sources*. Lewiston, New York: Edwin Mellen Press, 1990.

Thomas, M. F. and G. W. Whittington, eds. *Environment and Land Use in Africa*. London: Methuen, 1969.

Thornton, Robert. *Space, Time and Culture among the Iraqw of Tanzania*. New York: Academic Press, 1980.

Tubiana, M.-J. and J. *The Zaghawa from an Ecological Perspective: Foodgathering, the Pastoral System, Tradition and Development of the Zaghawa of the Sudan and the Chad*. Rotterdam: Balkema, 1977.

Webb, James L. A., Jr. *Desert Frontier: Ecological and Economic Change along the Western Sahel, 1600-1850*. Madison: University of Wisconsin Press, 1994.

See also: Agriculture – Medicine – Construction Techniques

ENVIRONMENT AND NATURE: THE AUSTRALIAN ABORIGINAL PEOPLE Australian Aboriginal people's traditional relationship to their land from a philosophical, economic, and spiritual viewpoint was quite different from that of the Europeans who arrived towards the end of the eighteenth century. Aboriginal environmental philosophy was related to their being observers, knowers, and users, rather than managers and interferers. This approach was the essence of their genius, enabling them to survive successfully for 40,000 years — perhaps much longer. Their philosophy could be described as non-materialistic ecocentrism, expressed through totemism, Dreaming, and the law, contrasting markedly with European materialistic anthropocentrism.

During the millennia before European occupation the continent had undergone big climatic changes followed by enormous environmental ones. Aboriginal people experienced the gradual extinction of the megafauna, commencing from the earlier part of their occupation, and, between 15,000 and 6000 years ago, the decrease in the size of the continent as the sea level rose, mostly imperceptibly.

One of the significant technological innovations affecting Aboriginal use of the environment was the grinding stone or "grindstone". Dating from around 18,000 years ago, it gave impetus to settlement in the arid center where there was heavy dependence on seed foods. A lack of suitable stones in some areas, e.g. southwest Queensland's Channel Country, brought about trade in the course of which grindstones were carried very long distances. Aboriginal mythology underlines the importance of the grindstone right up to the time of European occupation.

As the sea level rose, mainland Australia became separated from Tasmania as well as from New Guinea. As the new conditions became stabilized, an extremely rich and diverse flora and fauna developed. Then, in the period from 1500 BC to the beginning of the Christian era, colder and drier conditions emerged — perhaps the reason that around this time Aboriginal people made major changes in their methods of exploiting resources. Technological innovation included the emergence of the small tool tradition, and in Victoria, the introduction of water control systems. The building of brush and stone fence traps and pitfalls may also have been developed during this period to allow bigger catches of wallabies and kangaroos. There is some evidence of the reorganization of society at this time, perhaps to enable better exchange and sharing of resources in the face of new conditions.

Economic Use of the Environment

Although the archaeological record gives a general picture of Aboriginal life, the most detailed knowledge comes from the ethnographic record made since European occupation. Aboriginal cultures and lifestyles varied considerably throughout the continent, according to specific needs, beliefs, knowledge, and availability of resources.

Aboriginal people were hunter–gatherers who mostly operated freely within a limited range — their own 'country' — in small groups or bands who moved according to the season and the availability of resources. They were not acquisitive, often leaving behind all but the most needed tools as they moved around. With few (arid zone) exceptions, they did not store food. Included in the diet was a wide variety of flora and fauna — fish, land and sea mammals, some insects, birds, lizards, and snakes, depending on where people lived — providing a healthy, mostly nutritious diet, low in fat and sugar. Generally people did not take out more plants or animals than were needed for day to day survival.

Housing, implements, weapons, clothing, and other utilitarian, ritual, and decorative items were fashioned from materials taken from the environment, e.g. stone, plant material (wood, fibers, thorns, leaves, twigs, branches, gum), animal products (bone, hair, skin, feathers), and soil (mud, ochre).

Examples of crafted wooden instruments and weapons include the spear, waddy or club, and boomerang (each used in fighting and hunting), the scoop-shaped dish (a carrying bowl, ash cooking bowl, or shovel), digging stick, shield, playstick, message stick, and spear-thrower. Wood was also the basic construction material for boats. Shelter in most areas was constructed of wood and branches, and of bark; they were quickly made and easily abandoned. Wood could be selected for its shape, requiring little or no working for some implements such as building poles, witchetty grub hooks, skewers, brooms, walking sticks, axehandles, spindles, firesticks, fire-making sticks, laddersticks, and implements of secret rituals such as droughtsticks for causing drought. Other plant materials were used in making the tools of daily life, e.g. plant fibers for making string to be woven into bags, nets, headbands, etc., and for making baskets, thorns for tattooing, leaves for bedding, branches for brush fences, and gum for glueing.

Animal skins were used against cold winters for clothing, and animal hair was woven for string from which such items as headbands, phallocrypts, belts, net traps, and bags were made. Bone items included awls and the bones used in 'boning' rituals.

Stone items included hafted hatchets and adzes, unhafted cutting tools chipped from hard stone or quartz, clubs (hand held and thrown), and grindstones. These items were often traded by those communities who had the best supply of raw material. In places where stone abounded, e.g. the Flinders Ranges, stones were also utilized as meat chopping blocks, throwing weapons for killing, gravestones and other sorts of site markers and stone arrangements.

Modification of the Environment?

Debate continues regarding the extent to which Aboriginal people modified the environment in times past. Many scholars believe that although they utilized materials from the environment, they did not *substantially* alter the environment itself. Others differ. Two issues often debated are the extinction of the megafauna and the use of fire.

It has only recently been confirmed that in antiquity Aboriginal people killed megafauna species at all. As to fire, the fact that they used it extensively is not disputed; what is debated is whether or not they used it deliberately to manipulate and alter the environment substantially. There is still no conclusive evidence that they did this to a degree that made the Australian environment vastly different from the way it would have been had they never inhabited the continent. From everything known about Aboriginal environmental philosophy and practice, large scale intervention in the environment to change the way things were would seem unlikely. Aboriginal people's skill lay in their observation and utilization of nature as it was, including the natural potential fire regime.

The concept of Aboriginal "firestick farming" arose in the 1950s as a reaction to the somewhat idealistic view of Aboriginal people's "[having lived in] harmony with the environment". This notion is part of a general view that in recent millennia Aboriginal people modified their environment to the point of being the most significant agents of change within it. More likely, however, is the view presented here, that as change occurred — mostly imperceptible at any given time — Aboriginal people gradually adapted to the new conditions.

The "firestick farming" model has become popular because of its idealistic portrayal of "traditional" Aboriginal people in European terms, firstly as more like (superior) European farmers, and more recently as "resource managers". This is also often the case with new views of Native Americans and other indigenous people.

Traditional Ways of Looking After the Land

Aboriginal people, like all hunter-gatherers, were above all observers of their environment. Their traditional concept of "looking after the land" was diametrically opposed to a European view. First and foremost it was a question of knowing, and to some extent, of not doing rather than of doing. Every individual had both the need and the right to know his own country to which they were attached through totemic affiliation and the ties of kinship, and to be informed on its terrestrial and celestial natural phenomena.

Dreaming history (mythology) was largely the encyclopedia of knowledge. Here were "stored" maps, vital environmental information, and laws concerning people's relationship to the land. Knowing the land meant above all maintaining the Dreaming. The Dreaming had an ecological rationale, the actions and habitat of the totemic ancestral species mostly coinciding with those species' typical behavior and habitats. Place names reflected the Dreaming; they remain today as windows on Australia's environmental past, many incorporating names of species of flora and fauna once prevalent but now extinct.

Knowing the rituals and accompanying songs was also considered crucial in looking after the country. Recently Aboriginal people have even blamed their own failure to maintain the songs and rituals for the dramatic loss of species following European occupation: "We can't look after the country now; no one knows the songs any more."

Totemism, whereby an individual or a clan was in a special relationship with a species or phenomenon in the natural world, was a powerful link between the people and the environment. Maintaining a proper ritual attitude towards one's totemic species was an essential part of looking after the country. In some places this meant maintaining a taboo on killing one of one's own or a near neighbor's totem species — an act akin to fratricide. In some regions, individuals had only one personal totem; elsewhere, through clan membership individuals had a number of totems.

In parts of Australia the lands belonging to a totem clan were located in that totem species' quintessential habitat. This meant that in those places species were protected from human predation in their best refuges. Taboos on hunting and food gathering also applied around certain sacred sites, sometimes ensuring that the best waterholes were completely protected. There are examples also of taboos on killing female kangaroos, which would help to ensure the continuity of the species, and on interfering with certain plants because these were a specific animal's food. These examples, however, do not represent large scale strategic planning in conservation, even though their evolution would to some extent have had that effect.

Over all, then, Aboriginal people's "looking after the land" (rather than "land management" — a very European concept) was based on using knowledge skilfully and performing the associated rituals. Indeed, it was the very fact of *not* having management schemes that interfered unduly with the ecosystem, which was the secret of Aboriginal survival. The "strategy" for success was knowing.

Environmental Dispossession

According to the philosophy which European colonizers took with them to Australia from 1788 onwards, it was the duty of individuals to give value to the land — otherwise seen as valueless — by going beyond the laws of nature through work. As the Australian environment was not "worked" in their understanding, they regarded it as completely empty — *terra nullius* — when they arrived. They would give it value by working it and extracting profit from it. Aboriginal non-materialistic ecocentrism, valuing both the human and non-human world in a holistic way, had nothing in common with European anthropocentrism and materialism. The philosophy of the dominant group soon prevailed, and Aboriginal people were dispossessed of their land and their traditional lifestyle, if not of their lives. The change to the environment as Aboriginal people knew it was irrevocable. Indeed, the speed and scale of faunal loss may have been unprecedented in the history of the world. For many communities today, the only records of the species which sustained their forefathers are sub-fossil material, old songs, and Dreamings.

Aboriginal People and the Environment Today

Dispossession has not occurred to the same degree for all. In the process of adapting to their changed environment as Aboriginal people have always done, some desert communities have in recent times introduced new ways of looking after the land, while not abandoning old ways of maintaining the Dreaming and the rituals which reinforce knowledge. Also, there is increasing interest in traditional Aboriginal environmental knowledge, and here and there attempts are being made to apply it with Aboriginal help, to "bring back" the country.

Traditional ways of caring for the country, however, operated within a supporting philosophical, religious, and economic framework — on a very different landscape, generally for different goals, and above all, on a continent wide basis. It is therefore unlikely that the traditional package of environmental nurture will be reintroduced on a large scale.

Perhaps more significantly, Aboriginal people are seeking — and getting — more involvement in environmental management. Some see it as one way of getting back the country. While dispossession may have left many with limited traditional knowledge to apply, it has not removed their sense of belonging to the land. Moreover, 1993 native title legislation (*Mabo*) is motivating a new appreciation of Aboriginal people's long relationship to the unique Australian environment.

DOROTHY TUNBRIDGE

REFERENCES

Hallam, Sylvia J. *Fire and Hearth: a Study of Aboriginal Usage and European Usurpation in South-western Australia*. Canberra: Australian Institute of Aboriginal Studies, 1975.

Hetzel, B.S. and H.J. Frith, eds. *The Nutrition of Aborigines in Relation to the Ecosystem of Central Australia*. Melbourne: CSIRO, 1978.

Horton, D.R. "The Burning Question: Aborigines, Fire and Australian Ecosystems." *Mankind* 13: 237–51, 1982.

Mulvaney, D.J. and J. Golson, eds. *Aboriginal Man and Environment in Australia*. Canberra: Australian National University Press, 1971.

Mulvaney, D.J. and J. Peter White, eds. *Australians to 1788*. Broadway NSW: Fairfax, Syme & Weldon Associates, 1987.

Newsome, A.E. "The Eco-mythology of the Red Kangaroo in Central Australia." *Mankind* 12: 327–33, 1980.

Rose, F.G.G. *The Traditional Mode of Production of the Australian Aboriginals*. Sydney: Angus & Robertson, 1987.

Rowland, M.J. "Aborigines and Environment in Holocene Australia: Changing Paradigms." *Australian Aboriginal Studies* 2: 62–76, 1983.

Strehlow, T.G.H. "Personal Monototemism in a Polytotemic Community." In Eike Haberland, Meinhard Schuster and Helmut Straube, eds. *Festschrift für AD. E. Jensen*. München: Klaus Renner Verlag, 1964.

Tunbridge, Dorothy. *Flinders Ranges Dreaming*. Canberra: Aboriginal Studies Press, 1988.

Tunbridge, Dorothy. *The Story of the Flinders Ranges Mammals*. Sydney: Kangaroo Press, 1991.

Waddy, J.A. *Classification of Plants & Animals from a Groote Eylandt Aboriginal Point of View*. Vol I. Australian National University North Australian Research Unit Monograph, 1988.

Williams, Nancy M. and Eugene S. Hunn, eds. *Resource Managers: North American and Australian Hunter Gatherers*. Boulder, Colorado: Westview Press for the American Association for the Advancement of Science, 1982.

Young, Elspeth, et al. *Caring for the Country: Aborigines and Land Management*. Canberra: Australian National Parks & Wildlife Service, 1991.

ENVIRONMENT AND NATURE: BUDDHISM Buddhism persists in much of Asia after some 2500 years, and is spreading into other parts of the world as well. This persistence reflects the continuing validity and utility of its philosophical and moral principles. For example, many of the basic principles of the modern science of ecology are reflected in Buddhist thinking. Buddhism has an inherent environmental ethic which could contribute to the reduction of environmental problems and the development of a more ecologically sustainable society. Bhikkhu Bodhi states the matter succinctly:

> With its philosophic insight into the interconnectedness and thoroughgoing independence of all conditioned things, with its thesis that happiness is to be found through the restraint of desire in a life of contentment rather than through the proliferation of desire, with its goal of enlightenment through renunciation and contemplation and its ethic of non-injury and boundless loving-kindness for all beings, Buddhism provides all the essential elements for a relationship to the natural world characterized by respect, care, and compassion.

At first glance, the ecology of Western science and the philosophy and religion of Buddhism may appear to be quite different, but closer inspection reveals some remarkable parallels. Both Buddhism and ecology view all life as subject to the elemental laws of nature. Moreover, the Buddha's teaching, called *Dhamma*, means the discovery of the nature of things, and this includes the character and processes of the natural environment. Biologist D.P. Barash asserts that as ecology continues to develop it approaches the principles of Buddhism. Holistic concepts in ecology, like nutrient cycling, ecosystems, and Gaia, are similar to Buddhist notions regarding the unity, wholeness, integrity, interrelatedness, and interdependence of all life. Buddhism does not segregate organism and environment, human and nature, and related phenomena into dichotomies or antitheses, in contrast to the dualistic thinking common in the West.

Buddhism does not stop with an analysis of nature, nor with mere respect for it. It teaches reverence and compassion. The Buddha told a monk who had built a shelter of baked mud to tear it down. He pointed out that in the process of constructing such a shelter, the life in the corresponding soil had been destroyed. Modern conservationists have often appealed to public sentiment by pointing to endangered cute and furry creatures like the panda, koala, or chimpanzee, but they seldom point to invertebrates and microorganisms, although ecologists realize that these smaller life forms are also indispensable to the functioning of ecosystems. In contrast, as the above anecdote illustrates, the Buddha recognized and respected the unity of all life.

An ethic for a sustainable society is inherent in Buddhism. In a work on the conservation of biological diversity, Jeffrey McNeely and his co-authors identify the basic ingredients for a sound environmental ethic. These include recognizing the unity and interdependence of nature, the limits of resources and the need for sustainability, compassion for life in all of its diversity, the right to exist and other natural rights, thought and action in relation to responsibility, and the dependence of the individual on nature. All of these principles are inherent in Buddhist thought: simplicity, moderation, nonviolence, respect, compassion, love, and identity in the interactions with nature of the individual and society as a whole.

LESLIE E. SPONSEL
PORANEE NATADECHA-SPONSEL

REFERENCES

Badiner, A.H., ed. *Dharma Gaia: A Harvest of Essays in Buddhism and Ecology*. Berkeley, California: Parallax Press, 1990.

Barash, D.P. "The Ecologist as Zen Master." *American Midland Naturalist* 89: 214–217, 1973.

Batchelor, Martine, and Kerry Brown, eds. *Buddhism and Ecology*. London: Cassell Publishers, Ltd., 1992.

Bodhi, Bhikkhu. "Foreword." In *Buddhist Perspectives on the Ecocrisis*. Ed. Klas Sandell. Kandy, Sri Lanka: Buddhist Publication Society, 1987, pp. v–viii.

Callicott, J. Baird, and Rogert T. Ames. "Epilogue: On the Relation of Idea and Action." In *Nature in Asian Traditions of Thought: Essays on Environmental Philosophy*. Ed. J.B. Callicott and R.T. Ames. Albany: State University of New York Press, 1989, pp. 279–289.

Kabilsingh, C. "Early Buddhist Views on Nature." In *Dharma Gaia: A Harvest of Essays in Buddhism and Ecology*. Ed. A.H. Badiner. Berkeley, California: Parallax Press, 1990, pp. 8–13.

McNeely, Jeffrey A., Kenton R. Miller, Walter V. Reid, Russell A. Mittermeier, and Timothy B. Werner. *Conserving the World's Biological Diversity*. Washington, D.C.: World Resources Institute, 1990.

Sponsel, Leslie E., and Poranee Natadecha-Sponsel. "The Relevance of Buddhism for the Development of an Environmental Ethic for the Conservation of Biodiversity." In *Ethics, Religion and Biodiversity: Relation Between Conservation and Cultural Values*. Ed. Lawrence S. Hamilton. Cambridge, England: White Horse Press, 1993, pp. 75–97.

ENVIRONMENT AND NATURE: CHINA China represents one of the four ancient civilizations of the world, the others being Egypt, Mesopotamia, and India. The one unifying feature of these civilizations is that they all developed along the flood plains of major rivers. The Chinese civilization started along the valleys of the Yellow and Wei Rivers and gradually expanded to the middle and lower parts of the Yellow River and eventually across the North China Plain. It was not by accident that this occurred. Flood plains are well-suited to agriculture and China's greatest natural resource is her agricultural land. The river waters also served other purposes such as domestic (drinking, cooking, washing), aesthetic, recreational, irrigation and fishing.

The early Chinese knew that their livelihood was dependent on nature and in turn believed that their fate and nature were intertwined. This belief caused them to hold the soil in reverential regard and, in some areas, even to consider the rain to be the life-giving seed of the Supreme Ruler in heaven. Humans were considered equals to all forms of life. Prosperity and happiness depended on an ability to adjust successfully to the various forces of nature. To go against nature or to tamper with it might disrupt its equilibrium and ultimately prove to be harmful.

This cooperation with nature, however, did not mean that the early Chinese submitted passively to life. Instead they felt that humanity was the means through which nature's full potential could be realized. In fact, according to some historians, the Chinese have had more of an effect on their environment than any other people (Keswick, 1986). Nature, in turn, has endowed the Chinese with some of the world's most productive agricultural areas. The Chinese regarded the alteration of their environment as a type of adornment and not as a form of mastery and control. Through their diligence and devotion, they were able to develop a natural system to explain nature and the environment. Their understanding was further modified according to ancient traditions and customs, which preserved in the Chinese a spirit of sacred reverence for the divine powers of nature and a sense of community with it. They looked upon nature and the environment not as a dead or inanimate thing but rather as a living breathing organism, in a fashion somewhat similar to the present-day Gaia Hypothesis. The Chinese saw a chain of spiritual life running through every form of existence which bound together, as in one living body, all things existing in heaven above and earth below. They had an aesthetic, poetic, emotional, and reverential way of looking at natural objects.

The fourth and third centuries BC marked a transition period during which ancient Chinese natural philosophy developed into a more sophisticated and systematized theory. Out of their observation, identification, and dependency on nature came Daoism, a philosophy based on nature. The Daoists championed the independence of each individual, and maintained that the only concern should be to fit into the great order of nature, i.e. the *Dao*. They believed that if left to itself, the universe proceeds smoothly according to its own harmonies. They emphasized the unity and spontaneity of nature. Mankind's efforts to change or improve nature only destroy these harmonies and produce chaos. Humanity's place in nature was to be in harmony with it since both humans and nature obey the same laws. Thus, they are in a constant dynamic relationship with one another, each able to affect the other's flow. As Laozi (ca. 561–467 BC) stated in the *Dao De Jing*, "Man follows the way of earth, earth follows the way of heaven, heaven follows the rules. The rules are the *Dao*".

At the same time that Daoism was flourishing, another school of philosophic thought was also flourishing, known as the Naturalist School. This attempted to explain nature's workings on the basis of certain cosmic principles, one of which was the basic dualism of nature, the *Yinyang* Principle. *Yang* represents masculinity, light, hotness, dryness, hardness, roundness, activity, heaven, the sun, etc. *Yin* represents femininity, dark, coldness, wetness, softness, squareness, passivity, earth, the moon, etc. Rather than being in perpetual conflict, *yinyang* are mutually complementery and balancing. The greater *yang* grows, the sooner it will yield to *yin*; likewise the greater *yin* grows, the sooner it will yield to *yang*. In addition, *yang* always contains some *yin* and *yin* always contains some *yang*. *Yin* and *yang* continually interact, creating cyclical change. For example, *Yinyang* helped explain the seasons and the proper actions which should occur, in spring: plow, in summer: weed, in autumn: harvest, in winter: store. The interdependence of the two principles was well symbolized by an interlocking figure, which today is used as the central design element in the flag of South Korea. This *yinyang* concept was eventually incorporated into Daoism, with *Dao* being the natural process that unites the two.

When perfect balance and harmony exist between the *yin* and *yang* elements and qualities, growth of all living things flourishes and *qi*, the cosmic breath, enhances the environment. *Qi* is believed to exist in all living and nonliving things. This concept of *qi* is pervasive throughout Chinese customs and traditions, from acupuncture to kung fu. It is through *qi* that all things in the universe are related and attached to each other and integrated in the united whole.

The early understanding of the forces of nature which was accumulated over the years, together with the concepts of *yinyang*, *qi*, and later concepts, eventually developed into a complex science known as *feng shui*. Simply put, *feng shui* is an early form of ecology, conservation, and environmental science. It unites heaven, earth, and humans in such a way that a respect for nature is developed along with the ideas of renewability and sustainability of resources.

The most famous of the Chinese philosophers was Confucius (551–479 BC). He was primarily concerned with society and interpersonal relationships. As such, the Confucian view of nature is only relevant with respect to how it affects human relationships and interactions. However, he did recognize the importance of the human–nature relationship. This can be seen in the Classics, a set of books associated with Confucianism.

In the *Yijing* (Book of Changes), it is stated that "The great man is the one who unites his morals with heaven and earth". *Li Ji* (Book of Rites) in turn states, "Man is the heart of the heaven and the earth. . ., so a holy man must treat heaven and earth as the root of behavior". It also says that "to chop down a tree or kill an animal at the improper time is unfilial". The *Doctrine of the Mean* states "Let the states of equilibrium and harmony exist in perfection, and a happy order will prevail throughout heaven and earth and all things will be nourished and flourish". Therefore, in order to unify with nature, we should obey the regularity of nature and take care of it. According to the *Yijing*, also known as the *I Ching*, the Chinese concepts are revere nature, unify with nature, adapt to nature, spare nature, learn from nature, and play in nature. The Confucian thus asks people to act in a proper manner and to take care of nature; so by leading a moral life, people can assist heaven and earth to maintain a harmonious balance.

Mencius (371–289 BC), a disciple of Confucius, said "If you do not interfere with the seasons of husbandry, the grain will be more than can be eaten. If you do not use a net in a fish pond, the fish and turtles will be more than can be consumed. When cutting wood only go into the forest and hills when it is the proper time, and the wood will be more than can be used. Not allowing the exhaustive consumption of fish and turtles, nor the exhaustive use of the woods, enables the people to live and die generation after generation without interruption". In other words, resources are for the perpetual use of mankind.

The last major influence on the Chinese concept of nature and the environment is that of Buddhism, which came to China from India in the first century AD. Buddhism gave to the Chinese a respect for the lives of all living things, because it is believed that all things have the potential to achieve Buddhahood and that animals are but a form in which something has returned in its new incarnation. Everything in the world has a relationship with everything else, be it animate or inanimate. The only difference is in the closeness of the relationship.

The concepts of nature and the environment in ancient China are basically a group of general principles, such as the cyclic nature of the natural world, the equality of the natural laws, the creativity of nature as expressed in its biodiversity and abundance, and the belief that everything contains both a positive and negative part. Through Confucianism, these concepts became part of morality; through Daoism, they became part of philosophy; through Naturalism, they became part of cosmology; through *feng shui*, they became a part of science; through Buddhism, they became part of religion. The concepts were easily understood and incorporated into an understanding as to how humans should live their lives, i.e. in harmony with nature.

KENNETH J. E. BERGER

REFERENCES

Keswick, Maggie. *The Chinese Garden: History, Art and Architecture*. 2nd ed. New York: St. Martin's Press, 1986.

Lau, D. C. *Lao Tzu*. Harmondsworth: Penguin Classics, 1963.

Legge, J. *The Chinese Classics*, vols. 1, 2, and 3, 2nd ed. Hong Kong: Hong Kong University Press, 1960.

Legge, J., trans. *Yi-Jing: The Book of Changes*. Oxford: Clarendon Press, 1979.

Needham, J. *Science and Civilisation in China*, vols. 1 and 2. Cambridge: Cambridge University Press, 1954 and 1956.

Rossbach, S. *Feng Shui: The Chinese Art of Placement*. New York: E. P. Dutton, 1983.

Skinner, S. *The Living Earth Manual of Feng Shui*. London: Routledge & Kegan Paul Ltd., 1982.

See also: *Yinyang* – Geomancy – Divination

ENVIRONMENT AND NATURE: THE HEBREW PEOPLE

It is perhaps inappropriate to include a section on Hebrew thought in an encyclopaedia on non-Western cultures. When the Hebrew Bible was adopted in translation as the Christian Old Testament, it became one of the intellectual cornerstones of western civilization. The two principal divisions of Jewish culture and ethnicity, central and eastern European (*Ashkenazi*) and Mediterranean and Near Eastern (*Sephardi*) both have extensive European roots. Nevertheless Christianity and Judaism diverged in critical ways over the interpretation of their shared scriptures.

The Hebrew Bible and Jewish Law

The essential core of Jewish belief is the first five books of the Bible, notably the set of 613 commandments (*mitzvoth*) handed down to Moses on Mount Sinai. The rabbinical interpretations of Mosaic law (*halakhah*) and explanations of biblical narratives, codified principally in Babylon and Palestine during the first centuries of the Common Era as the Talmud, form the essence of traditional Jews' understanding of the cosmos and permissible activities within it. While "Hebrew thought" as a body of literature in the Hebrew (or Aramaic) language is by no means identical with "Jewish thought" the two will be treated here together.

In the Hebrew Bible, the one God created the heavens, the earth, and their creatures, and pronounced them "good". God placed the first humans in the beautiful Garden of Eden "to dress it and keep it". Humans were given control over soil and biota: "the fish of the sea, the fowl of the air", but the Bible explicitly reserves control over physical forces of the environment, notably climate, to God. The Bible links humans' ability to shape nature to their liking to their obedience to God's will. In the biblical narratives, few humans are righteous enough to accomplish dominion over nature.

God commanded the patriarch Abraham and his descendants, the Israelites, to worship God exclusively and to obey His commandments. The books of Numbers and Deuteronomy explain environmental change in the form of divine blessings and curses for human good deeds and misdeeds. The Israelites are to receive a homeland, abundant rainfall, and crop yields if they obey the *mitzvoth*, but drought, insect plagues, and the "scorched earth" resulting from warfare if they disobey.

The Hebrew Bible and Talmud have little explicitly to say about environmental conservation in the modern sense. There are proscriptions in Jewish law against cruelty to animals, destruction of fruit trees during warfare, tilling the soil during the sabbatical year, and working (and hence altering nature) on the Sabbath day. There is little evidence in the Bible that Middle Eastern, Iron Age farmers and pastoralists even understood that humans were capable of causing extensive environmental degradation (with the possible exception of overgrazing,) and considerable evidence that they attributed environmental deterioration to God's wrath.

These cast considerable doubt on some scholars' assertions that the "Judeo–Christian tradition" is the root of the West's callous treatment of nature, simply on the basis of a few verses in Genesis mandating human use and domestication of plants and animals. Needless to say, both Christian and Israeli settlers in recent centuries have subsequently cited the Genesis injunction for human dominion over nature and Isaiah's prophecy that the desert shall "bloom like the rose" as a license for resource exploitation.

Today when many environmentalists seek to undo the dualism between humanity and nature, it is worth noting that the Hebrew Bible indeed has no words for "environment" or "nature" that distinguish these concepts from culture: only the universal and integrative concepts of "everything" (*olam*) or Creation. In the Bible's original Hebrew (though not in most English translations) animals have souls (*nefesh*, *ruach*) just as people do. Trees and stars praise God and shout for joy. The Bible is full of evocative nature poetry, notably in the prophets, Job and the Psalms. Eagles, lions, the cedar forests of the mountains of Lebanon, the desert after a rain shower, lush pastures: these seemed especially inspirational to the ancient Hebrew mind.

The Bible also describes an annual cycle of holy days that were based on a lunar calendar and changing seasons as they would have affected ancient farmers and herders. The festival of *Shavuot* (Weeks) for example, celebrates the gathering of first fruits of the land. *Sukkot* (Tabernacles) in autumn culminates a series of days set aside for repentance and atonement with prayers for the winter rains so essential for agriculture in the Near East. While such environmental connotations were weakened among urban Jews of the Diaspora (who nevertheless continued to observe the holy days), they are quite obvious to visitors of rural Israel today.

Middle Ages and Early Modern Times

With the post-biblical Diaspora or exile from Palestine following Roman destruction of Judaism's institutions and sacred sites, Jews literally scattered to the four corners of the earth. Often forbidden by host governments from attaining the most basic of human rights, owning land, living outside of urban ghettoes, or working at most occupations, many Jews were prevented by anti-semitism from forming a close relationship with environments where they settled. Preserving their faith and studying Torah and Talmud were largely indoor activities.

Although development of Jewish environmental thought was thus limited, medieval Jewish philosophers and mystics who developed the Kabbalah teachings believed that all of nature was imbued with divine emanations, or even with divine substance. Some leaders of the eighteenth century pietist Chassidic movement of eastern Europe were noted for retreating alone to the fields and forests for prayer and mystical insight. The agrarian Zionist movement of the late nineteenth and early twentieth centuries advocated a return of Jews to Israel, in the belief that working the soil of their ancient homeland would redeem and ennoble oppressed European Jewry.

Recent Past and Present

The recent history of Hebrew thought on the environment and nature is best exemplified by the modern state of Israel. Since its inception in 1947, it has experienced many of the same environmental problems as other industrialized countries, with particular concerns about water scarcity and reforestation. Israel's push for development of new settlements and industries despite a hostile international political scene, and its land use needs for military security often overrule environmental preservation priorities. Hostilities between Arabs and Jews clearly have had an environmental impact, notably in the security landscape and in tensions between Arab range lands and Jewish reforestation projects.

Nevertheless Israel is one of the few nations in the Near East with a variety of environmental organizations, such as the Society for the Protection of Nature in Israel (founded in 1953,) a Ministry of the Environment, and a system of nature preserves and national parks. Two noteworthy parks, Chai-Bar and Neot Kedumim, are maintained to preserve specimens of vanishing desert fauna and flora, respectively, mentioned in the Bible.

The Israeli response to nature has not lost its ancient religious roots. Some Orthodox Jews in Israel use traditional prayers for rain as a response to drought, and halakhic rulings about agricultural fields together with scientific agronomy practice.

Many Jewish environmentalists throughout North America today seek to integrate their faith and environmental concern. Some are re-examining the rabbinical concept of *Tikkun Olam* (healing everything) as the performance of *mitzvoth* necessary to restore planetary health. Vegetarianism for some fits in well both with Jewish dietary laws and environmentalist beliefs about the sanctity of life. Clearly the synthesis of ancient traditional texts, religious practice, and recent environmental awareness will lead to a rapid evolution of environment and nature in Jewish thought in the near future.

JEANNE KAY

REFERENCES

Feliks, Yehuda. *Nature & Man in the Bible: Chapters in Biblical Ecology*. London: Soncino Press, 1981.

Hareuveni, Nogah. *Tree and Shrub in Our Biblical Heritage*. Kiryat Ono, Israel: Neot Kedumim Ltd. 1984.

Kay, Jeanne. "Human Dominion over Nature in the Hebrew Bible." *Annals of the Association of American Geographers* 79 (2): 224–232, 1989.

The Melton Journal: Issues and Themes in Jewish Education no. 24, 1991 and no. 25, 1992.

Roth, Cecil and Wigoder, Geoffrey, eds. *Encyclopaedia Judaica*. 17 vols. Jerusalem: Keter Publishing House, Ltd. 1971.

ENVIRONMENT AND NATURE: INDIA The philosophical and scientific ideas developed in India over the centuries are, on analysis, found to be deeply related to ecological issues, both generally and specifically. Like the ancient ideas of China and of the Hellenic world, the ancient Indian ideas of the comparable period are cosmological and comprehensive in character. The ideas presented in the *Vedas* and *Upaniṣads*, or in Laozi's *Dao* or in Parmenides' *Nature of Being* all were engaged in search of the first principle, One. They were all obliged to relate it to Many – many individual objects of knowledge. The One–Many relationship is pregnant with both cosmological and ecological implications.

Broadly speaking, the objects we know around us are biotic (living) or abiotic. However, to many thinkers, especially to the pluralists and evolutionists, this two-fold classification is simplistic and inadequate. They try to draw our attention to different grades of being or Reality — physical, chemical, paleontological, botanical, biological, psychological, and spiritual. The scientific philosophers of the Vedic insights, of the *Samkhya* persuasion, and also of latter times could discern different subgrades within each of these grades. These graded characteristics of different living and non-living Beings have to be recognized if we are to understand the complex and the interactive character of our environment. This insightful approach to ecology as an integral part of cosmology is evident in the tradition of Indian thought. It is very clearly available, for example in *Caraka-Saṁhitā*.

In the ancient India the good of human life used to be thought of in the context of life's environment. The right relation between the individual and his environment received serious attention even at the levels of primary and secondary education. The aim was to impart basic knowledge of personal hygiene and medicine. This education was meant for all. In a way medical education was universal. It is interesting to note the five compulsory subjects for all high school students: Grammar (*Śabdavidyā*), Art (*Śilpasthānavidyā*), Medicine (*Cikitsāvidyā*), Logic (*Hetuvidyā)*, and Science of Spiritual philosophy (*Adhyātmavidyā*).

Both abstract and concrete areas of knowledge, understood in their interconnection, are all relevant to the exercise of adjusting to our environment. Linguistic communication, artistic articulation, logical ratiocination, and medication are in different ways intended to awaken the best in us and to strike a balance with the large world around us. From the strength of body to the span of life, from good health to peace of mind, all are a unified function of food, personal hygiene, right actions, and character.

Life has a rhythm of its own, which is not necessarily manifest. The *Dharmaśāstra* are full of injunctions on how to attain that rhythm in terms of purity, diet, regulation, ablutions, behavior, and physical and mental disciplines. These have to be followed as part of daily (*dinacaryā*) as well as seasonal routines (*ṛtucaryā*). Food and drink habits and fulfilment of natural urges, avoidance and indulgence in sexual acts, and eating some ordinary things like curds, buttermilk, and honey are all necessary to define our correct relation with the environment. However, this normative rhythm can hardly be generalized, for it is integrally related to the individual constitution (*prakṛti and svāsthya-vṛtti*).

The traditional Indian medical system (*Āyurveda*) takes a comprehensive view of a person. It neither encourages asceticism and mortification of flesh nor promotes unregulated sensualism. It highlights the importance of a "sound mind in a sound body". It points out that *svāsthya-vṛtta* or philosophy of hygiene is to be supplemented by *sadvṛtta* or the right life. Rightly understood, the philosophy of hygiene is a way of life containing in it the principles of eugenics, ethics, and healing.

In ancient Indian thought the basic principles of ecology are ontologically oriented and cosmological in implication. The three main concepts of the Vedas which are evident in *Āyurveda* are *satya* (truth, right, reality, or being), *svadhā* (self-position, self-power, or spontaneity) and *ṛta* (proper, suitable, and settled order). The basic point which is being emphasized here is that the true nature of reality cannot be changed by human will which is not informed of that reality. No technological skill, no arbitrary will of this or that person, can go indefinitely against the true nature of reality without causing harm to those who try to follow this wrong path. Even the most sophistical biotechnology cannot tamper with the essential nature of reality, the true Nature or *Prakṛti*. This role of the concept of truth has both ontological and axiological implications. Humans are required both to know reality in its true nature and to live and shape their lives accordingly.

Humans are also aided by the principle of becoming or the dynamic nature of reality in shaping their best possible life. This possibility is contained in the concept of *svadhā*. Reality has within itself the impetus or power for self-unfolding or gradual disclosure. It is not static, fixed, or self-enclosed.

The world as reality and as a whole is perpetually expanding. This macrocosmic truth of *svadhā* is at work also in human nature. Consequently, human freedom knows no bounds. Nature and its laws are not antagonistic to the human will to be free. When we can discover the laws governing the true nature of the relation between individuals and the world, our knowledge of the world helps us to live in harmony with it. That harmony creates a favorable environment which not only nourishes our bodies but also enables us to be free from social conflict and tension. When we are free from ill health, we are better placed in our relation to nature. When we fail to strike the balance with nature, we are not only likely to be poorer in health but also less capable of getting the best out of our environment.

Finally, besides the true nature of reality (*satya*) and the power of self-disclosure (*svadhā*) what helps us to have a right environment is suggested by the concept of *ṛta*. It is orderliness or the law-governed character of reality or nature. From the change of seasons to the changing periods of life, *ṛta* is clearly perceptible. Even human cultures are found to exhibit certain rhythms. Cultures are characterized simultaneously by fragile and stable features. The rhythm of nature is not antagonistic to the spirit of human freedom. Our lives, both individual and collective, are marked by a sort of dynamic equilibrium.

The above three concepts embodying certain abstract principles may appear irrelevant to our actual lives. However, to the thinkers of *Āyurveda* the principles of philosophy and ethics, or those of good living, are inseparable from the basic characteristics of reality.

The same principle is illustrated in defining the ideal principles of town-planning and village-planning. Aśoka, the Buddhist emperor of ancient India, said that trees, plants, and shrubs of medicinal value had to be planted around every village and along the roadsides. The people were allowed to use the leaves, fruits, and bark of the trees. This practice survives today. Like geography, history, science, and arithmetic, the general principles of hygiene and physiology and simple methods of curing cuts, wounds, and everyday ailments were prescribed for everyone's general education. The underlying belief was that individuals had to be able to take care of their elementary medical needs.

The Indians believed that every village should be so constructed that its population must have the professional service of a medical practitioner (*vaidya*). People were advised to reside in a place with plenty of water, herbs, sacrificial sticks, flowers, grass, and firewood, and which yielded abundant food. They were also advised to live where there was safety of property and person, where the outskirts were beautiful and pleasing, and where there was a strong presence of learned people.

Ancient Indian thinkers paid attention to the development and preservation of the right type of environment for human life. The state was assigned a very important role for the purpose. It was expected to lay down rules and regulations for rubbish disposal and drainage systems. The state also determined where and how the quality of food and drink should be preserved. Those who violated these rules and thereby polluted the environment were punished. Equally conscious were the decision-makers in charge of public health. The state had a very important role to perform for preservation and promotion of a healthy environment. Various types of punishment were prescribed for cutting for the tender sprouts of trees.

Kautilya mentioned elaborate laws for the promotion of agricultural activities. For example, he prohibited high agricultural taxes and bonded agricultural labor and attached much importance to animal husbandry and soil preservation.

D.P. CHATTOPADHYAYA

REFERENCES

Kauṭilya. *Kauṭilya's Arthaśāstra*. Trans. R. Shamasastry. Mysore: Mysore Print and Publishing House, 1961.

See also: Medicine in India: Āyurveda

ENVIRONMENT AND NATURE: ISLAM The sacred text of Islam, the *Qur'ān*, contains a theology of ecology, for nature and ethics are at the very core of its moral worldview. Indeed, so central is the theme of the affinity of nature and ethics that even outside observers, such as Marshall G. Hodgson, have epitomized the dictates of the Islamic commitment as "the demand for personal responsiblity for the moral ordering of the natural world".

The creation of humanity is one of the grandest themes of the *Qur'ān*. It is alluded to either philosophically in a symbolic language or biologically, employing the medium of natural science. Philosophically, the first assertion is that of the purposefulness and meaningfulness of human life. The Quranic teleology is preeminently moral: humans are to execute the will of God, but it is an undertaking which they have imposed upon themselves. It is a pledge made by man and woman to God. Hence, God, for His part, has endowed them with all the faculties essential to undertaking this august moral mission. Nature is the testing ground of this moral responsibility. Men and women are thus enjoined to read its "signs". For this purpose, nature has been created both orderly and predictable. The creation of human beings and the creation of nature are thus two chapters of the single theme of moral responsibility and trust that is the *sine qua non* of Islamic commitment.

Nature, therefore, is a trust or *amāna* and a theatre for a Muslim's moral struggle. According to the *Qur'ān*, heavens, earth, and mountains refused to assume this responsibility which humans took upon themselves voluntarily. By doing this, no doubt, humans showed ignorance and hubris — but also their willingness to serve God's purpose. As trust is a mutual commitment, it may also be surmised that God, by entrusting people with this responsibility, expressed confidence in their ability. No wonder that in the Quranic worldview and Islamic tradition the individual is known as the trustee, or *khalīfa*, of God.

The Islamic rationale for an ecological ethics rests on the Quranic notions of *khalīfa* and *amāna*. Nature, being the gift of God to man, is accepted in Islam as an estate over which we have temporary control but no sovereign authority. Our relationship with nature thus can never be ethically neutral. Islam views nature essentially in a teleological perspective and therefore the claims of man's dominion over her have no resonance in Islam.

While Islam is a monotheistic faith belonging to the Arabrahamic tradition, its teachings on enviornment and nature contrast sharply with its sister religion of Christianity. To give an example: the Hebrew story of creation is transformed in Christianity into the doctrine of the fall. Creation thus appears to the Christian mind as "fallen", and nature is viewed as opposed to grace. St Augustine, to take one example, believed that nature was "unredeemed", just as many Christian theologians maintain that nature cannot teach man anything about God and is therefore of no theological and spiritual interest. Salvation is the humbling of nature by the miraculous, the intrusion of the supernatural into history. The nearest thing in the physical universe that reflects the miraculous is man. Holiness exists only in a man-made environment. Thus, nature, so devoid of God's presence and grace, may be "tortured"; it may be justifiably subjected to scientific experimentation. The Islamic view is very different. Creation (nature) in the Quranic view always bears the "signs of God" and is necessary for man's salvation. It is in accordance with this that Islam holds that there is no such thing as a profane world. All the immensity of matter constitutes a scope for the self-realization of the spirit. All is holy ground; or as the Prophet Muḥammad said, "the whole of this earth is a mosque". Earth, creation and nature thus have a sacramental efficacy in Islam which can be ill-accommodated with the perverse applications of the "dominion ethics". The claim for nature's "salvational worth", however, may never be construed as a token of its autonomy. In fact, Muslim theologians have always claimed that nature has no meaning without reference to God; without Divine purpose it simply does not exist. (Hence, nature is simply known as the created order.)

The ethical link between faith in God and love for His creation is fully demonstrated in the life of the Prophet Muḥammad himself who declared that whoever is kind to the creatures of God is kind to himself. One authentic tradition narrates that a man once came to the Prophet with a bundle and said, "I passed through a wood and heard the voice of the young of birds, and I took them and put them in my carpet and their mother came fluttering around my head". And the Prophet said, "Put them down". And when he had put them down, the mother joined the young. And the Prophet said, "Do you wonder at the affection of the mother towards her young? I swear by Him who has sent me, verily, God is more loving to His servants than the mother of these birds. Return them to the place from where you took them, and let their mother be with them".

Such environmental teachings were actually translated into envrionmental policies and legislation in the classical Muslim civilization. For example, the Muslims developed the notion of *ḥaram* — inviolate zones — outsides towns, near water-courses and other areas where development was forbidden. A second type of inviolate zone was *ḥima* which applied to forests, woods, and wild habitation and was desiged to conserve wildlife. Ibn ʿabd as-Salam (fl. thirteenth century) formulated the first statements of animal rights. Muslims were also concerned about the protection of the urban environment; Islamic town planning and architecture provide ample demonstrations of this. Many classic cities, like Fez, were built with the full understanding of carrying capacity and were designed so that the city's population would not increase beyond a critical limit. The debate about conservation, protection of animals and their habitats, and the Islamic teachings on environment can be clearly seen in such classics as *Disputes Between Animals and Man*, which is a part of the *Rasā'īl Ikhwān al-ṣafā'* (The Epistles of the Brethren of Purity) written in the tenth century. The deep respect for nature and environment is also evident in Sufism, the mystical strand of Islam, both in its thought and practice. The titles of some of the classic Sufi works reflect their concern with nature: *Gulistān* (The Rose Garden) and *Bustān* (The Fruit Garden) by Saʿdī of Shīrāz, and Farīd al-Dīn ʿAṭṭār's *Manṭiq al-ṭayr* (The Conference of the Birds) provide good illustrations.

Contemporary Muslim socieites have lost much of their traditional consciousness and concerns. Both colonialism and the mad rush for modernization have played their part in this oversight, but today we can detect a minor resurgence in Islamic environmental consciousness.This is evident both in the intellectual and academic efforts to shape a contemporary Islamic environment theory — for example in the works of Gulzar Haider, Seyyed Hossein Nasr, and Othman Llewellyn — and the frequent use of traditional Islamic concepts and technologies (such as *qanat*, the ingenious system of wells drained through a network of tunnels) in developing environmentally sound practices in the Middle East and Iran, Pakistan, and Malaysia.

Equipped with the ethical insights of *khalīfa* and *amāna*, and impelled by the Quranic dictates to assume personal moral responsibility in the world of nature, Muslims have a responsibility to meet the challenge of ecology to religious consciousness and provide mankind with a healing vision of the harmony of man and nature under God. Like everything else of value in Islam, its ecological insight can be summed up under the seminal concept of *tawḥīd* (unity). *Tawḥīd*, Islam's eternal quest for the unity of life and purpose, spirit and matter, human beings and nature, law and ethics, faith and morality, implies that man not dominate the earth or commit violence in any form.

PARVEZ MANZOOR
ZIAUDDIN SARDAR

REFERENCES

Ateshin, Hussain Mehmet. "Urbanisaton and Environment: An Islamic Perspective." In *An Early Crescent: The Future of Knowledge and Environment in Islam*. Ed. Ziauddin Sardar. London: Mansell, 1989, pp.163–194.

ʿAṭṭār, Farīd al-Dīn. *The Conference of the Birds*. Harmondsworth, England: Penguin, 1984.

Haider, Gulzar. "Habitat and Values in Islam: A Conceptual Formulation of an Islamic City." In *The Touch of Midas: Knowledge, Values and Environment in Islam and the West*. Ed. Ziauddin Sardar. Manchester: Manchester University Press, 1984, pp. 170–210.

Haider, Gulzar. "Man and Nature." *Inquiry* 2(8): 47–50, 1985.

Haider, Gulzar. "The City Never Lies." *Inquiry* 2(5): 38–44, 1985.

Haider, Gulzar. "The City of Learning." *Inquiry* 2(7): 45–51, 1985.

Hakim, B. *Arabic Islamic Cities: Building and Planning Principles*. London: KPI Publishers, 1986.

Hodgson, Marshall G. S. *The Venture of Islam: Conscience and History in a World Civilization*. Chicago: University of Chicago Press, 1974.

Johnson-Davies, Denys, trans. *The Island of Animals* (Translation of *The Disputes Between Animals and Man of The Epistles of the Brethern of Purity*). London: Quartet Books, 1994.

Khalid, Fazlun and Joanne O'Brien, ed. *Islam and Ecology*. London: Cassell, 1992.

Llewellyn, Othman. "Desert Reclamation and Conservation in Islamic Law." *The Muslim Scientist* 11: 9–29, 1982.

Llewellyn, Othman. "Sharian Values Pertaining to Landscape Architecture." In *Islamic Architecture and Urbanism*. Ed. Aydin Germen. Dammam, Saudi Arabia: King Faisal Univesity, 1983.

Manzoor, S. Parvez. "Environment and Values: The Islamic Perspective." In *The Touch of Midas: Knowledge, Values and Environment in Islam and the West*. Ed. Ziauddin Sardar. Manchester: Manchester University Press, 1984, pp. 150–169.

Manzoor, S. Parvez. "Islam and the Challenge of Ecology." *Inquiry* 2(2): 32–38, 1985.

Nasr, Seyyed Hossein. *Encounter of Man and Nature*. London: Allen and Unwin, 1978.

See also: Religion and Science

ENVIRONMENT AND NATURE: JAPAN In traditional Japan, the word *shizen*, also pronounced *jinen*, meant naturalness, or the mode of being which is natural. Its literal meaning is "from itself (*shi/ji*) thus it is (*zen/nen*)". In modern Japanese *shizen* by extension came to refer to nature, or the environment encompassing all between heaven and earth, especially the earth, oceans, mountains, rivers, flora, and fauna.

Premodern Japanese had no single word signifying nature as a unified entity. Nevertheless passages about nature abound in their ancient literature, philosophy, and religion. Words like *ten*, literally meaning heaven, and *tenka*, meaning heaven and earth, meant something like nature. An understanding of traditional conceptions of nature can be garnered by examining Japanese thoughts about aspects of nature such as heaven, earth, mountains, rivers, trees, flowers, and fields.

Ancient Japanese evinced an unabashed intimacy with the natural world in their earliest poetry as compiled in the eighth-century anthology, the *Man'yōshū* (Collection of Myriad Leaves). Its poems spoke of the world of mountains, rivers, flora and fauna in anthropomorphic, animistic terms, investing each entity with a living personality infused with *kami*, or mysterious spiritual energy. Mountains were deemed most divine. Indeed, in their verses the ancient poets immortalized Mt. Fuji as a peerless deity. Rivers were considered living forces, manifesting immense spirituality in their rushing flows of clean, life-giving water. The sea was viewed as a more awe-inspiring, fearful spiritual force, one supplying sustenance but also destruction. References to birds and beasts fill the poems, evoking a sense of the four seasons and the human feelings linked to them. The poems themselves were likened to the natural world: the *Man'yōshū* included myriad *yō*, or "leaves" of poetry.

The earliest histories of Japan, the *Kojiki* (Records of Antiquity, 712) and the *Nihon shoki* (Chronicles of Ancient Japan, 720), open with myths relating the Shintō cosmogony. They give the *Man'yōshū* vision of nature a strongly religious, spiritual grounding, leaving no doubt that the world of nature was the world of creative religious spirit. After heaven and earth congealed out of an undifferentiated, egg-like mass, successive generations of personified *kami* begot one another. Finally the *kami* pair, Izanagi and Izanami, created the first island, Onogorojima, by letting the brine drip off a spear they had plunged into the ocean's depths. On Onogorojima, Izanami gave birth to the elements of nature including the rivers, mountains, birds, beasts, and flowers and trees, and the forces of nature such as fire. Because of the divinity of the progenitors, Japanese traditionally considered their archipelago and all of nature within it as sacred.

According to one account, Izanagi and Izanami also gave birth to Amaterasu the Sun Goddess and her brother Susanoō the impetuous god of storms. Because both Amaterasu, a benevolent *kami*, and Susanoō, a mischievous if not malevolent one, are deemed divine beings, Japanese have not dismissed destructive forces of nature as evil. Nor have they branded acts against nature as necessarily wrong. Indeed, Shintō myths relate that Susanoō wreaked havoc in his sister's rice fields, implying that similar actions contrary to the general good of the world of nature might still have some divine sanction via Susanoō's example. Perhaps this partly explains why many Japanese, despite their close religious, poetic, and mythic ties to nature, have at times tolerated abuses of it.

Pollution, however, was considered anathema, and was dealt with via Shintō purification rites. Pollution did not necessarily mean physical dirtiness or noxious environmental conditions; however, those forms were recognized among the more spiritual nuances associated with the Shintō notion of pollution. Traditional abhorrence of physical and spiritual pollution perhaps explains the energetic opposition of many Japanese, though certainly not all, to environmental pollution.

Amaterasu and Susanoō parented the imperial line and the ancestral stock of all its human subjects. Belief in the divinity of their islands, their imperial family, and everything within their natural environment thus results from the supposed ancestry of Japanese in Izanagi and Izanami. This became a basic tenet of Shintō, one regularly repeated throughout Japanese history. In the *Engi shiki* (Religious Regulations of the Engi Era, 927) ritual prayers record a similar vision of nature as fully infused with divine spirituality. Kitabatake Chikafusa's (1293–1354) political tract, the *Jinnō shōtōki* (The Legitimate Succession of Divine Emperors, ca. 1340), opened with the declaration, "Japan is a sacred land (*shinkoku*)". Japanese Buddhists and Confucians almost unanimously have accepted the same, essentially Shintō doctrine.

Belief in the divinity of the archipelago encouraged some Japanese to xenophobia: foreigners, typically viewed as barbarians, were feared as potential agents of pollution. Japanese soil, they felt, would be violated if the barbarians were allowed on it. Since 1945, the Japanese have become more accustomed to the presence of foreigners in their country. Yet still their concern for nature often seems Japan-specific: some southeast Asian forests have been depleted to accommodate the Japanese preference for disposable, wooden chopsticks, even though forest conservation practices are followed within Japan.

Introduced to Japan in the mid-sixth century, Buddhism advanced various attitudes towards nature. The Four Noble Truths, the original teachings of the historical Buddha, Siddhārtha Gautama (563–483 BC), characterize existence as suffering, implying that *samsara*, or the environment of reincarnation, was similar. Their religious solution, *nirvana*, which literally means "putting out the flame" (of existence in this world), seems to offer an escapist otherworldliness which might have permitted a relative disengagement from nature.

The Mahāyāna Buddhist ideas of the Indian thinker Nāgārjuna (ca. AD 150–250), which were accepted by most Japanese Buddhists, affirmed the natural order. Nāgārjuna equated *samsara* with *nirvana*, disallowing otherworldliness. Nāgārjuna's view was based on the doctrine that everything is empty (*sunya*). Empty here means empty of self-sustaining substance, i.e. a substance existing in and of itself. Nāgārjuna insisted that everything that exists does so through spatial, temporal, and causal relations with the remainder of the universe. Nothing exists independently of everything else. His ideas could be construed as anticipating the ecologist's belief that all life is interrelated, and that destruction of small niches endanger the entire ecosystem. Nāgārjuna's ideas surely facilitated a more positive appraisal of nature by Japanese Buddhists.

Traditional Buddhist cosmology, however, claims that the world is subject to creation and disintegration just as humans experience cycles of death and rebirth. Arguably this view could allow a cavalier attitude towards the natural environment since regardless of one's efforts the world will inevitably disintegrate and then begin anew. The Buddhist belief that attachment to things leads to suffering also might vitiate wholehearted involvement in an environmental ethic geared toward conservation of nature. The Buddhist two-level theory of truth, assigning ultimate status to *sunya*, and relegating common sense to secondary validity, allows for concern for nature but not in a primary way.

Otherworldly tendencies appeared in the popular Jōdo, or "Pure Land", School. Based on the Indian writing *Sukhāvatīvyūha-sūtra* (Discourse on Paradise), Jōdo posits both a heaven called the Pure Land presided over by Amida Buddha, and a multi-leveled hell where sinners suffer eternally. Yet some theorists claim that the Pure Land and hell are merely "expedients" meant to motivate non-believers to meditate on Amida Buddha as the way to salvation. If so, then the Pure Land becomes a symbol of *nirvana*, or existential extinction, while hell becomes a hyperbole of the Buddha's claim that life in this world, the world of nature, entails suffering.

Yet the ideas of many Japanese Buddhists evinced a religiously based concern for nature. The Kegon, or Flower Garland, school asserted that every particle of existence was infused with Buddha-nature, making the natural universe a spiritual one as well, one to be saved from suffering. The *Konkō kyō* (Sutra of the Golden Light), an important text in early Japanese Buddhism, claimed that rulers who promoted the Buddha's teachings would be protected, as would be their domains, by the Four Deva Kings, tutelary divinities who protected Buddhism throughout the universe. Compassion for all sentient beings, the core ethic inculcated by Māhāyana Buddhists, instilled in some a concern for the natural world.

Buddhist poets like Saigyō (1118–1190) even extolled the world of nature as the primary arena of Buddhist values. Others debated whether plants and trees, though nonsentient beings, could actually attain Buddhahood. Many Japanese Buddhists argued that they could. Some also claimed that nature possessed a healing and even soteriological capability. Probably influenced by Shintō beliefs, the Shugendō (Order of Mountain Ascetics) school had its practitioners make pilgrimages to sacred mountains to glimpse scenery foreshadowing the Pure Land. Buddhist temples aesthetically enhanced the environment. With their rock gardens, moss gardens, and vegetable gardens, temples were practically involved in local environmental improvement as a way of meditation. Zen Buddhists see enlightenment as an experience to be had in this world and in this body. Disregard of nature, therefore, cannot be allowed. Yet Buddhists are not known for authoring agricultural or environmental tracts. It was the Confucian and Neo-Confucian scholars who, in addition to admiring nature's beauty and revering it religiously, made the world of nature the focus of proto-scientific research designed to conserve the environment for future generations.

Neo-Confucianism, the last major philosophical force to emerge in traditional Japan, encouraged scientific interest in and ethical concern for the natural world. Prompted by the sophisticated Buddhist metaphysics with its (to Neo-Confucians) repulsive doctrine of emptiness, Confucians reformulated their originally socio-political thought along novel metaphysical lines so as to refute the Buddhist challenge. Originally a Chinese movement of the Song dynasty (967–1279), Neo-Confucianism ultimately became a pan-Asian force decisively affecting China, Korea, Japan, Vietnam, and other East Asian areas up until the modern period. In Japan, it was a dominant force during the Tokugawa period (1600–1867).

Neo-Confucians endorsed common sense, declaring that the natural environment was both substantial and fully real. They believed that there was no other world. Rejecting emptiness, they asserted that everything consisted of a quasi-material, psycho-physical energy called *ki*. Giving "matter" its rationale was another ontological element, *ri*, or "principle". The fusion of *ki* and *ri* accounted for the diversity within the natural world. The latter was created by heaven and earth, which engaged in constant production and reproduction as its Way. Most Neo-Confucians recognized the complementary forces of yin and yang as the cardinal modes of material being. They also admitted the five elements of earth, wood, fire, water, and metal, as essential processes defining all development within nature. Because nature and man were created by the same elements and by the same forces, Neo-Confucians reinterpreted their earlier socio-political ethic of humaneness in mystical terms of forming one body with the universe. This mysticism identified the human body with all that existed. Some even spoke of heaven and earth as their parents, and the myriad entities of nature, organic and inorganic, as their companions. Most Neo-Confucians declared human nature (*sei*), or the original psycho-physical disposition of human beings, to be good, not empty as the Buddhists claimed. Furthermore Neo-Confucians claimed that the nature (*sei*) of the universe, i.e. its moral character, was originally good. The human project, as defined by Neo-Confucians, was to preserve this original goodness by moral self-cultivation and by moral action in the world.

Confucius (551 BC–479 BC) too had respected the natural world. The *Analects*, the most authentic record of his thought, states that the wise person loves water (Chinese: *shui*; Japanese: *sui*), while the humane person loves mountains (Chinese: *shan*; Japanese: *san*). This reveals that early Confucianism linked moral concerns to ecological ones. It is also artistically significant for the Chinese word for landscape painting is *shansui*, and the Japanese is *sansui*, denoting a combination of bodies of water with mountains or hills, which were the constituent elements in a landscape painting. The latter were not just idealized depictions of nature; they were equally reflections of the moral consciousness of the artist. Furthermore Confucians and Neo-Confucians, following Confucius' views about mountains and bodies of water, believed that morality involved right behavior towards nature and humanity.

Evidence abounds revealing a practical interest in the world of nature by Confucians and Neo-Confucians in Japan. By the late-seventeenth century, forests throughout the archipelago had been depleted due to an overexploitation as a result of a boom in the construction of castles and urban residences. Confucian scholars diagnosed the environmental crisis and called for its cure. Yamaga Sokō (1622–1685), for example, argued that forests should be conserved to ensure future productivity. Sokō admonished loggers to harvest lumber only in the proper season, not to overcut, and to reforest areas they had cut. One of Sokō's disciples, Tsugaru Nobumasa, daimyo of Hirosaki domain in northeastern Honshū, claimed that the three fundamental concerns of a feudal lord were for (1) his family line, (2) his heir, and (3) his mountain forests.

In his *Daigaku wakumon* (Dialogues on the Great Learning), Kumazawa Banzan (1619–1691) argued that humane government involved afforestation, river dike repair, and other conservation practices that would maximize agricultural productivity. Noting the crisis at hand, his disciples declared that "mountains and rivers are the foundations of a country". Full of hope, despite the depleted forests that were all too evident to him, Kumazawa claimed even that bald mountains could be covered with trees again if oats and other cover crops were planted so as to retain a thin layer of soil long enough for a forest ecosystem to reappear. Ultimately, the *Dialogues on the Great Learning* advocated harnessing samurai energy for the sake of the agricultural ecosystem: Kumazawa argued that samurai should be allowed to return to the countryside to labor as farmers rather than required to live in castle towns where they fell prey to urban vices. Kumazawa's proposals were so radical, however, that they were not heeded. Generations after his death, however, his more environmentally-oriented ideas, as advocated by others, did find favor.

Kaibara Ekken (1630–1714), influenced by the Neo-Confucian call to investigate the principles of things, authored the first systematic botanical study in Japan, the *Yamato honzō* (Flora and Fauna of Japan, 1709), describing and classifying over 1550 trees, plants, flowers, birds, fish, seashells, etc., into some thirty-seven categories. Ekken's "Preface" declares that because heaven and earth produce and reproduce myriad lifeforms, scholars must study their principles. Many later studies in herbology, botany, pharmocology, and zoology were influenced by Ekken's *Flora and Fauna*. He also encouraged the research of Miyazaki Yasusada (1623–1697) which culminated in the conservation-minded *Nōgyō zensho* (Agricultural Encyclopedia, 1696). In his preface, Ekken observed that a sage government, which nourished and educated the people, was facilitated by studying the *Agricultural Encyclopedia* because through it people could understand how to assist heaven and earth by cultivating the productive forces of nature.

Some late-Tokugawa intellectuals were influenced by notions of Western science as introduced to Japan via Dutch traders permitted at Nagasaki. Satō Nobuhiro (1769–1850), exposed to Western science but also influenced by Shintō, advocated techniques of agricultural management, based on the scientific study of natural law, to improve agriculture. He argued that this would realize the divine aim inherent in creation. Rejecting such clever short cuts, Ninomiya Sontoku (1787–1856) insisted instead on following the natural, creative cycles of heaven in agriculture, and repaying the virtue of heaven with conservation techniques which would rejuvenate nature. Miura Baien (1723–89), influenced by Western science, slightly modified the Neo-Confucian project of investigating things by advocating the investigation of the rational order (*jōri*) of heaven and earth in a disinterested, objective way.

Given the dense population which appeared during the Tokugawa, and its heavy taxation of the fragile ecosystem of the archipelago, one extremely poor in natural resources, Japan could easily have become, as Conrad Totman has suggested, an eroded moonscape rather than a verdant archipelago. In the seventeenth century, Japan was well on its way to a state of deforestation. It was saved largely by Confucian and Neo-Confucian scholars who called for systematic, scientific action. Yet it would be far-fetched to deny credit to Shintō and Buddhism, though admittedly they were less conspicuous in enunciating a conservation program saving Japan's forests, and thus its mountains, rivers, and fields.

Following the Meiji Restoration of 1868, when Japan embarked upon a course of rapid Westernization, concern for the natural environment lessened as the new Meiji state presided over the beginnings of industrialization by fostering polluting industries such as railroads. Though anathema to many, the goal of the Meiji state was to match, if not surpass, the industrial prowess of the Western nations which had imposed unequal treaties on Japan during the nineteenth century. Voices of protest against the noxious side effects were either ignored or muffled until Japan had effectively modernized. Widespread water pollution caused by the Ashio Copper Mine in Tochigi Prefecture from the late-1870s resulted in a major ecological disaster and a political scandal which simmered for nearly a century, pitting farmers against private industry and the government. Though notions of "the supremacy of agriculture" (*nohon shugi*) circulated in the late-Meiji, these were frequently no more than ideological currents.

The leading twentieth-century philosopher of nature was Watsuji Tetsurō (1889–1960). Watsuji's *Fūdo* (Environment, 1935) criticized Heidegger's *Being and Time*, arguing that space, i.e. natural setting, and not just time, was crucial to human culture. Watsuji correlated the latter with three environmental zones: monsoon, desert, and pastoral. Japanese culture, he claimed, emerged from a monsoon zone where climatic vagaries produced passive, forbearing, inconsistent, sentimental, intuitive, and temperamental traits.

Since 1945, a decreasing number of Japanese have chosen to remain in the countryside close to nature. Just as the Meiji state rushed to modernize, so did postwar Japan race to renew its industrial sector oblivious to air and water pollution during the 1950s and early-1960s. In Minamata, Kyūshū, a chemical factory was finally held responsible for mercury poisoning, fifteen years after the poisoning first occurred in 1953. By the 1960s, Tokyo became internationally notorious for its air pollution. Citizen's protest groups rallied in the 1970s and 1980s, forcing the government to regulate pollution. Since the 1970s, environmentalists have argued for "environmental rights" (*kankyōken*), basing their claims on the 1947 Constitution which guarantees Japanese a "wholesome and cultured life".

JOHN A.TUCKER

REFERENCES

Primary sources

Analects. D. C. Lau, trans. *Confucius: The Analects*. New York: Penguin Books, 1979.

Nihongi: Chronicles of Japan from the Earliest Times to AD *697*. New York: Paragon Book Reprint Corporation, 1956.

Collection of Myriad Leaves (*Man'yōshū*). Ian Hideo Levy, trans. *The Ten Thousand Leaves: A Translation of the Man'yōshū*. Vol. 1. Princeton: Princeton University Press, 1981.

Discourse on Paradise (*Sukhāvatīvyūha-sūtra)*. In E. B. Cowell, et al. *Buddhist Mahayana Texts*. Sacred Books of the East, vol. 49. Oxford: Clarendon Press, 1890. Reprint, New York: Dover Press, 1969.

Kaibara, Ekken. *Flora and Fauna of Japan* (*Yamato honzō*). In Araki Kengo and Inoue Tadashi, eds. *Kaibara Ekken/Muro Kyūsō*. Nihon shisō taikei, vol. 34. Tokyo: Iwanami shoten, 1985.

Kitabatake, Chikafusa. *Legitimate Succession of Divine Emperors (Jinnō shōtōki)*. H. Paul Varley, trans. *A Chronicle of Gods and Sovereigns: Jinnō shōtōki of Kitabatake Chikafusa*. New York: Columbia University Press, 1980.

Kumazawa, Banzan. *Dialogues on the Great Learning (Daigaku wakumon)*. Galen M. Fisher, trans. *Dai Gaku Wakumon: A Discussion of Public Questions in Light of the Great Learning*. Transactions of the Asiatic Society of Japan. Second Series. Vol. XVI, May 1938.

Miyazawa, Yasusada. *Agricultural Encyclopedia* (*Nōgyō zensho*). In Furushima Toshio and Aki Kōichi, eds. *Kinsei kagaku shisō*, Nihon shisō taikei, Vol. 62. Tokyo: Iwanami shoten, 1972.

Records of Antiquity (*Kojiki*). Donald L. Philippi, trans. *Kojiki*. Tokyo: University of Tokyo Press, 1969.

Religious Regulations of the Engi Period (*Engi shiki*). Felicia Gressitt Bock, trans. *Engi shiki. Procedures of the Engi Era, Books I–V*. Monumenta Nipponica monograph. Tokyo: Sophia University, 1970.

Sutra of the Golden Light (*Konkō kyō*). Excerpts in Wm. Theodore de Bary, editor. *The Buddhist Tradition*. New York: Vintage Books, 1972.

Watsuji, Tetsurō. *Fūdo* (*Environment*). Leopold G. Scheidl, trans. *Die Geograhischen Grundlagen des Japanischen Wesens*. Tokyo: Kokusai bunka shinkōkai, series B, no. 35, 1937.

Secondary sources

Ames, Roger T., and J. Baird Callicott, eds. *Nature in Asian Traditions of Thought: Essays in Environmental Philosophy*. Albany, New York: State University of New York Press, 1989.

Earhart, Byron H. "Nature in Japanese Religion." *Kodansha Encyclopedia of Japan*, vol. 5. Tokyo: Kodansha, Ltd., 1983, pp. 357–358.

Imamichi, Tomonobu. "Concept of Nature." In *Kodansha Encyclopedia of Japan*, vol. 5. Tokyo: Kodansha, Ltd., 1983, p. 358.

Sakamoto, Yukio. "On the 'Attainment of Buddhahood' by Trees and Plants." In *Proceedings of the IXth International Congress for the History of Religions* (1958). Tokyo: Maruzen, 1960, pp. 415–22.

Totman, Conrad. *The Green Archipelago: Forestry in Preindustrial Japan*. Berkeley: University of California Press, 1989.

Tsuda, Sōkichi. "Outlook on Nature." In *An Inquiry into the Japanese Mind as Mirrored in Literature*. Trans. Matsuda Fukumatsu. Tokyo: Japan Society for the Promotion of Science, 1970.

See also: Agriculture in Japan

ENVIRONMENT AND NATURE: NATIVE NORTH AMERICA An old Pueblo song says:

I add my breath to your breath
That our days may be long on earth
That the days of our people may be long
That we may be one person
That we may finish our roads together
May our mother bless you with life
May our Life Paths be fulfilled.

The reciprocal relationship embodied in this song is one of the central insights into Native American understandings of the natural world. One does not act in isolation, but must act with respect towards the others with whom one shares the universe. This understanding has implications for the way cultures understand their world and the way in which they interact with that world.

This discussion treats Native Americans as if they were one culture. However, it is estimated that in 1492 there were at least 400 distinct linguistic groups on the continent. These were, and are, very different cultural and political nations.

Consider the natural world from an indigenous perspective. The universe is alive. Involved with a living universe, the Native American is engaged in a constant, vital dialogue with a network of relations: human and non-human, and natural and supernatural. A member of the Pueblo culture, Paula Gunn Allen states that the Indian lives within "a circular, dynamic universe: where all things are related, are of one family, then what attributes man possesses are naturally going to be attributes of all beings".

Native Americans practice a "person to person" relationship with the environment as the power of life itself is personified and inextricably linked with, and identical to, the natural world. This sense of relationship is explicit in every aspect of life. Consider Ojibwa Winona LaDuke's comments on the ordinary, mundane activity of hunting and gathering:

> Whether it is wild rice, whether it is fish, whether it is deer or turtles, when you go and take something from the land, you pray before you take it. You offer tobacco, you offer a prayer to that spirit and to the creation of a part of that. You take those things because you have a relationship with all the other parts of the creation. That is why you are allowed to take those things. You take that and you give something back as a reciprocal arrangement, because that is how you maintain your relationship.

As countless sacred rituals, ceremonies, songs and teaching stories make clear, maintaining the relationship between human life and non-human life is the heart of Native American spirituality.

The vital importance of maintaining essential relationships has serious implications for how Native Americans deal with their fur-clad relatives, the animals. As Allen describes the relationship, "All are seen to be brothers or relatives (and in tribal systems relationship is central), all are offspring of the Great Mystery, children of our mother, and necessary parts of an ordered, balanced and living whole."

Anthropologist Joseph Epes Brown comments that nonhumans are the links between humans and the Great Mystery. To realize the self, kinship with all beings must be realized. To gain knowledge, humans must humble themselves before all creation, down to and including the lowliest ant. Knowledge may come through vision quests, and this knowledge is transmitted by animals. Nature is a mirror which reflects all things, including that which it is important to learn about, understand and value throughout life. Further, many tribes acknowledge that humans found a world to come into and a way of making a living as the result of conscious, caring gifts from the animals. Sacred objects are donated, as when the eagle grants the use of his feathers, and food is found because holy animals offer their flesh as a gift.

Most Native American legends speak of other species as beings who could shed their fur masks and look human. They once shared a common language with humans, and continued to understand after humans lost their ability to speak to the animals. The animals partook of the sacredness of life, and were often the descendants of the powerful beings who had lived on the earth before humans. The southwestern Chemehuevis, writes anthropologist Carobeth Laird, see animals this way:

> It is to be remembered that the pre-human Immortals (the gods, if you will) were Animals Who Were People. These Forerunners . . . have long since taken their final departure; yet they remain as the visible animals of this everyday world. . . Mythic Coyote, supertrickster and pattern-setter for mankind, is not [coyote], raiding fields and howling on the hills before dawn — and yet in a certain mystical sense, he is. This is true of every animal from the largest mammal to the smallest insect.

Even now, the Native Americans respect their relationship with animals and believe that the respect and caring go both ways. Thus, communication is possible, as Yukon Indian Irene Isaacs explains, "You can tell them and they do it, that wolf. You tell them, 'Kill something for me!' And then the wolf will. Then you come, and they are going to kill a moose and you're going to find it and have moose meat. They do that, wolves. So they understand Indians. That's what my dad and my mother and my grandfather told me" (McClellan, 1987).

According to anthropologist Richard Nelson, the Alaskan Koyukon believe that animals and humans are distinct beings, their souls being quite different, but that animals are powerful beings in their own right. Consequently a complex collection of rules and respectful activities surrounds everyday life and assists humans in remaining within the moral codes that bind all life. Hunting, therefore, is conducted with respect and with ritual from the moment the hunt is conceptualized until the animal's remains are properly disposed of. At all times, there is a mutual obligation to be courteous. Animals are not offended at being killed for use, but killing must be done humanely, and there should be no suggestion of waste. Nor can the body be mistreated: irreverent, insulting or wasteful behavior could mean that the animal no longer makes itself available for killing.

Even gathering must be done respectfully. Yukon women gathering roots are pleased by the find of caches of roots already collected by mice. However, they are careful to leave some for the mice. Otherwise the mice are likely to come raiding Indian supplies during the winter.

For most Native Americans, the land and its residents have more shades of meaning than non-Native societies comprehend. The land is a source of sustenance, and is exploited for that purpose. However, the land also holds other meanings for past and present Indian cultures. Nelson remarks that to the Koyukon, the land "is permeated with different levels of meaning — personal, historical, and spiritual. It is known in its finest details, each place unique, each endowed with that rich further dimension that emerges from the Koyukon mind".

The practical consequences of Native American relationships with their lands and fellow beings are profound. Historian J. Donald Hughes argues that there were significant differences between the ecological modifications made by Native American cultures and those made by the European invaders. To picture Native Americans as people who passed through the land leaving no trace, as some do, is to deny an ecological reality, and, in passing, to damn the Native American with faint praise. Instead, Hughes argues that Native Americans had the capacity to do considerable ecological damage, and most certainly changed the ecosystems.

> Like all human cultures, the forest Indians were agents of change in nature. For at least 10,000 years, and perhaps much longer, they had lived within the forest ecosystems, hunting, fishing and gathering with skill and experience. Their land was not a wilderness, but a woodland park which had known expert hunters for millennia. The deer behaved differently, in a sense they were different animals, because they were subjected to the selective pressures of hunting. The forest itself reflected the presence and character of its human inhabitants.

Hughes also points out, however, that when European immigrants arrived in North America, they found what they took to be a wilderness, complete with extensive tree cover, predatory animals, and abundant game.

That the lands remained productive was in part due to Native American lifestyles that were mobile (even the horticulturalists moved to new areas when the soil was exhausted). It is arguable that it was also the result of deliberate stewardship. Many tribes deliberately took (and still take) steps to ensure that ecological changes did not damage their source of livelihood. Among the Yukon Indians, these steps took the form of a belief in ownership of certain areas. "This kind of ownership meant that a headman and the people who travelled with him had the first right to use the products of the local land and water, but they also had the duty of taking care of these resources... It was part of their job to see that people did not overhunt, overtrap or overfish the area, as well as to see that as far as possible everybody got enough to eat..." (McClellan, 1987).

Of great concern to Native Americans was the conservation of game animals. Anthropologist Howard Harrod notes the presence of several Trickster stories amongst the Plains Indians, in which the Trickster character exploits trusting animals to satisfy an insatiable appetite. In this pursuit he is foolish enough almost to eliminate all the animals, usually missing just one. "That Trickster is foolish and even a dangerous hunter appears in the theme of the 'last animal' ... As a consequence of escape, often brought on by a momentary lapse in Trickster's character, an animal survives to continue the species. Clearly the hunting techniques of Trickster are not affirmed; rather, the foolishness and potentially disastrous consequences of such activities are underlined in these stories."

Recent anthropological work among the twentieth century Cree and Koyukon has uncovered what researchers feel are deliberate conservation practices. Adrian Tanner observes that the Cree are constantly assessing plant and animal population levels. Long and short term changes in populations are understood both in spiritual terms (the activities of the Animal Masters) and in terms of environmental factors (food supplies, water, weather patterns, forest fires, and hunter activities). The Cree govern their hunting activity on the basis of these observations:

> An area which appears short of an adequate game population is accepted as such, and avoided for several years until the adverse conditions (either natural or religious) are resolved through time... Where the animal population of a territory is adequate, various survey methods are used in order to plan the winter's activities in conformity with the observed balance between the various available species.

Among the Koyukon, conservation activities are also based on their keen observation of the ecological dynamics of their lands. The people regulate their harvests in a number of ways to ensure that plant and animal populations remain healthy. They may avoid taking more individuals than they believe can be naturally replaced, or they may take special measures to enhance the productivity of a species, such as not killing female waterfowl, bears, and moose during spring breeding. Hunting activities are spread over as wide an area as possible. Young plants and animals are not harvested, but are allowed to mature. Trappers are very cautious about where and how many animals are harvested.

While such practices are noteworthy, the anthropologists are careful to point out that there are no objective data to prove that these practices actually achieve their conservation goals. Richard Nelson believes, however, that the Koyukon at least could be considered to be practicing a "conservation ethic".

In the end, however, practicing, for Native Americans, means more than just taking a living from the land. It involves certain recognitions of the nature of life, and in living a certain way, as a Navajo prayer for the good life suggests:

With beauty before me may I walk
With beauty behind me may I walk
With beauty above me may I walk
With beauty below me may I walk
With beauty all around me may I walk
As one who is long in life and happiness may I walk
In beauty it is finished.

ANNIE L. BOOTH

REFERENCES

Allen, Paula Gunn. "The Sacred Hoop: A Contemporary Indian Perspective on American Literature." In *The Remembered Earth*. Ed. Geary Hobson. Albuquerque: Red Earth Press, 1979. pp. 222–239.

Allen, Paula Gunn. *The Sacred Hoop: Recovering the Feminine in American Indian Traditions*. Boston: Beacon, 1986.

Booth, Annie L., and Harvey M. Jacobs. "Ties That Bind: Native American Beliefs as a Foundation for Environmental Consciousness." *Environmental Ethics* 12(1): 27–43, 1990.

Brown, Joseph Epes. *The Spiritual Legacy of the American Indian*. New York: Crossroad Publishing Co., 1985.

Capps, Walter H., ed. *Seeing With a Native Eye*. New York: Harper and Row, 1976.

Cronon, William. *Changes in the Land: Indians, Colonists and the Ecology of New England*. New York: Hill and Wang, 1983.

Dooling, D.M. and Paul Jordan-Smith, eds. *I Become Part Of It: Sacred Dimensions in Native American Life*. New York: Parabola Books, 1989.

Harrod, Howard L. *Renewing The World: Plains Indian Religion and Morality*. Tucson: University of Arizona Press, 1987.

Hughes, J. Donald. "Forest Indians: The Holy Occupation." *Environmental Review* 2: 2–13, 1977.

Hughes, J. Donald. *American Indian Ecology*. El Paso: Texas University Press, 1983.

LaDuke, Winona. "Environmentalism, Racism, and the New Age Movement: The Expropriation of Indigenous Cultures." *Left Green Notes* 4: 15–18, 32–34, 1990.

Laird, Carobeth. *The Chemehuevis*. Banning: Malki Museum Press, 1976.

McClellan, Catherine. *Part of the Land, Part of the Water*. Vancouver: Douglas and McIntyre, 1987.

Nelson, Richard K. *Make Prayers to the Raven*. Chicago: University of Chicago Press, 1983.

Tanner, Adrian. *Bringing Home Animals: Religious Ideology and Mode of Production of the Mistassini Cree Hunters*. New York: St Martin's Press, 1979.

ENVIRONMENT AND NATURE: SOUTH AMERICA – THE AMAZON Half of all life on Earth, as estimated either by total weight or number of species, is concentrated in the rain forests of the equatorial zone, even though they comprise only about 6% of the terrestrial surface of the planet. Half of the world's tropical rain forests are in the Amazon of lowland South America where there are an estimated 80,000 plant species. In a tropical rain forest there are up to ten times as many species as would be found in the same area of forest in a temperate region. Moreover, not only are there many more species, but there are also many more possibilities for interactions among species. Thus, the primary characteristic of tropical rain forests is biological diversity.

In recent decades, ecologically oriented anthropologists have been accumulating data from field research which demonstrates that traditional indigenous societies possess detailed and accurate knowledge of the ecology of the tropical rain forests and associated ecosystems such as rivers and lakes. This should be intuitively obvious because indigenous societies have survived successfully for millennia in the numerous and diverse ecosystems of the Amazon. However, because of the stigma of labels such as "primitive" and "tribal," and the associated racist and ethnocentric prejudices, Westerners have seldom recognized let alone begun to appreciate indigenous intelligence, philosophy, science, technology, and creativity in the Amazon. This Western mythology has been reinforced by the absence of an accumulated written record of the knowledge of Amazon societies because they are based on an oral rather than a literate tradition. It has also been reinforced by the fact that in indigenous cultures science, philosophy, religion, and education are usually integrated with other aspects of their lives, in contrast to Western society's segregation of these aspects. Most important of all, this Western mythology has served along with other prejudices to rationalize the colonization, exploitation, oppression, and destruction of the indigenous inhabitants of the Amazon, a process that continues to this day.

What is the evidence for indigenous knowledge, science, and technology in the Amazon? We can find the answer within the three related lines of evidence in anthropology: cultural ecology, ethnoecology, and historical ecology. Each of these is briefly discussed below.

Cultural ecology and its more recent rubric, ecological anthropology, focus on how culture influences the interaction between a human population and the ecosystems within its habitat. Since the 1950s studies of cultural ecology have accumulated which demonstrate that most traditional indigenous populations of the Amazon enjoy relatively good nutrition and health and a relatively high quality of life. The descendants of the original colonizers of the Amazon created societies which usually were to some degree ecologically and socially sustainable. These societies, especially in the interior forests away from major rivers, were most often characterized by a population with low density and high mobility combined with a rotational system of land and resource use for hunting, gathering, and fishing as well as gardening. This rotational system of resource use and management usually effectively avoided irreversible resource depletion and environmental degradation. For example, the Yanomami, whose ancient national territory overlaps the modern borders of the states of Brazil and Venezuela, traditionally live in communities of some 40 to 150 people with an average size of only 60 individuals. Moreover, the Yanomami spend as much as 60% of the year away from their village on treks which involve camping and foraging (hunting, fishing, and gathering) for weeks at various places in the forest. While the Yanomami also farm in the vicinity of their village, this is secondary to their foraging activities deep in the surrounding forest.

The relative degree of equilibrium between the human population and the natural resources of its habitat in many (but not all) of these indigenous societies was usually reinforced by a world view, cosmology, and values emphasizing harmony between humans and nature. For instance, the Desana, of the Colombian portion of the northwest Amazon, created an elaborate system of myths, rituals, and symbols which appear to contribute to the regulation of the dynamics of their own human population in relation to the prey species they depended on for subsistence. The shaman, a part-time religious practitioner in each community, monitored the condition of the human predators and their animal prey to maintain balance between the society and nature.

Ethnoecology focuses on the world view, knowledge, taxonomy, and values of indigenous societies, usually concentrating on a particular domain of nature such as garden soils, medicinal plants, or mammalian prey. The accumulating research in ethnoecology demonstrates that indigenes, both individually and collectively within a particular society, possess an impressive wealth of accumulated data on the ecology of the various ecosystems in their habitat. This knowledge has been tested through trial and error by daily interaction with their ecosystems, and also much has been passed through many generations for centuries or even millennia. As a specific example, the Ka'apor recognize at least 768 species of plants, and they know each species from the stages of seed through reproductive adult. Beyond mere taxonomic identification, the Ka'apor have the intelligence, knowledge, technology, skills, and creativity to use these plants for their subsistence, economic, social, medicinal, ritual, and aesthetic activities. Moreover, at least 112 of the plants known to the Ka'apor have various medicinal uses.

Historical ecology focuses on the dialectical processes in the mutual influences of culture and ecosystem. The Amazon is not all pristine, primary, or virgin rain forest. Although on the surface it appears that the immediate environmental impact of indigenous societies is negligible because of their limited population and technology as well as their world view and values, the net result of centuries or even millennia of accumulated use and management of the land and natural resources in their habitat can be a significant environmental impact. However, most if not all of this impact is within the levels and processes of natural disturbances. For example, a swidden garden plot is fairly comparable to the clearing and vegetation dynamics created by a natural tree fall. A swidden is made by cutting the trees and brush in a small circular area of forest, usually under one to two hectares. The slash is allowed to dry out during one or more months of the dry season and then burned. The fire transforms the nutrients which are otherwise locked in the vegetation into fertilizer on the surface of the soil. Then as the rainy season begins the crops are planted and within a few months harvesting begins. The garden is only intensively used for two to three years, and then it is gradually abandoned, although it may still occasionally be visited even decades later to harvest fruit or other useful plants and to hunt. Under the traditional conditions of a subsistence economy, low population density, and plenty of forest in reserve for future gardens, usually this rotational system of farming is sustainable and avoids irreversible resource depletion and environmental degradation.

Through time swiddening creates a mosaic of plant and associated animal communities at different stages of development or succession. Indigenes often concentrate plant species which are attractive to game for subsequent hunting. Moreover, animals who temporarily visit swiddens transport seeds from the surrounding forest which in turn increases the biological diversity of the garden plots. In these and other ways swiddens may enhance the biodiversity of the Amazon. Indeed, it has been estimated that more than ten percent of the Amazon forest is anthropogenic, the product of human influence and manipulation of the species of plants and animals in the ecosystems.

The indigenous peoples of the Amazon and elsewhere have contributed much to the rest of the world, although this is seldom recognized, let alone appreciated. For example, manioc or cassava (*Manihot esculenta*, a root crop), was probably domesticated originally in the northwest Amazon by about 2000 BC. It grows well in poor soils and is resistant to drought and pests. Conveniently the starchy roots are naturally stored in the ground until there is a need to harvest them for food. Since European colonization, manioc has spread throughout the tropics of the Old World until today it ranks as the fourth most important source of food energy for the world.

The knowledge behind these and many other indigenous practices is encyclopedic in scope. However, with contact this and other aspects of culture are usually degraded or even destroyed as the people are subjected to ecocide, ethnocide, and genocide. This ongoing holocaust in the Amazon is not only a terrible human tragedy, but a great loss for the human species and the ecosystems and biodiversity of the Amazon in general. For over five hundred years, experiments in "economic development" by Western "civilization" in the Amazon have repeatedly failed with grave economic, social, and ecological consequences. The accumulated historical record of these many embarrassing failures clearly demonstrates that to this day Western society does not yet know how to develop the Amazon without destroying it. Actually Western eyes have not recognized that, prior to European "discovery," the Amazon had already been discovered, colonized, and developed by indigenes for millennia, usually in a manner which was sustainable ecologically and socially. If Westerners would recognize and appreciate the intelligence, philosophy, science, technology, and creativity of the traditional indigenous societies of the Amazon, then they might learn sustainable, non-consumptive uses of the forest and associated ecosystems. Otherwise, the indigenous and natural worlds of the Amazon will be severely degraded if not destroyed, and likewise eventually humanity as a whole may be degraded or even destroyed by Western ignorance, arrogance, carelessness, and greed.

LESLIE E. SPONSEL

REFERENCES

Balée, William. "The Culture of Amazonian Forests." *Advances in Economic Botany* 7: 1–21, 1989.

Balée, William. *Footprints of the Forest: Ka'apor Ethnobotany — the Historical Ecology of Plant Utilization by an Amazonian People*. New York: Columbia University Press, 1994.

Bodley, John H. *Victims of Progress*. Mountain View, California: Mayfield Publishing Co., 1992.

Bodley, John H. "Native Amazonians: Villagers of the Rainforest." In *Cultural Anthropology: Tribes, States, and the Global System*. John H. Bodley. Mountain View, California: Mayfield Publishing Co, 1994, pp. 45–74.

Clay, Jason W. *Indigenous Peoples and Tropical Forests: Models of Land Use and Management from Latin America*. Cambridge, Massachusetts: Cultural Survival, Inc., 1988.

Crosby, Alfred W. *The Columbian Exchange: The Biological and Cultural Consequences of 1492*. Westport, Connecticut: Greenwood Press, 1972.

Descola, Philippe. *In the Society of Nature: A Native Ecology in Amazonia*. New York: Cambridge University Press, 1994.

Good, Kenneth. "Yanomami of Venezuela: Foragers or Farmers, Which Came First?" In *Indigenous Peoples and the Future of Amazonia: An Ecological Anthropology of an Endangered World*. Ed. Leslie E. Sponsel. Tucson: University of Arizona Press, 1995, pp. 113–120.

Hames, Raymond B., and William T. Vickers, eds. *Adaptive Responses of Native Amazonians*. New York: Academic Press, 1983.

Hecht, Susanna B., and Alexander Cockburn. *The Fate of the Forest: Developers, Destroyers and Defenders of the Amazon*. New York: Verso, 1989.

Moran, Emilio F. *Through Amazonian Eyes: The Human Ecology of Amazonian Populations*. Iowa City: University of Iowa Press, 1993.

Plotkin, Mark J. *Tales of a Shaman's Apprentice: An Ethnobotanist Searches for New Medicines in the Amazonian Rain Forest*. New York: Viking, 1993.

Prance, Ghillean T., and Thomas E. Lovejoy, eds. *Amazonia*. New York: Pergamon Press, 1985.

Redford, Kent H., and Christine Padoch, eds. *Conservation of Neotropical Forests: Working from Traditional Resource Use*. New York: Columbia University Press, 1992.

Reichel-Dolmatoff, Gerardo. *Amazonian Cosmos: The Sexual and Religous Symbolism of the Tukano Indians*. Chicago: University of Chicago Press, 1971.

Roe, Peter G. *The Cosmic Zygote: Cosmology in the Amazon Basin*. New Brunswick, New Jersey: Rutgers University Press, 1982.

Roosevelt, Anna, ed. *Amazonian Indians From Prehistory to the Present: Anthropological Perspectives*. Tucson: University of Arizona Press, 1994.

Sponsel, Leslie E. "Amazon Ecology and Adaptation." *Annual Review of Anthropology* 15: 67–97, 1986.

Sponsel, Leslie E. "The Environmental History of Amazonia: Natural and Human Disturbances, and the Ecological Transition." In *Changing Tropical Forests: Historical Perspectives on Today's Challenges in Central & South America*. Ed. Harold K. Steen and Richard P. Tucker. Durham, North Carolina: Forest History Society, 1992, pp. 233–251.

Sponsel, Leslie E., ed. *Indigenous Peoples and the Future of Amazonia: An Ecological Anthropology of an Endangered World*. Tucson: University of Arizona Press, 1995.

Stone, Roger D. *Dreams of Amazonia*. New York: Penguin, 1993.

Sullivan, Lawrence E. *Icanchu's Drum: An Orientation to Meaning in South American Religions*. New York: Macmillan, 1988.

Weatherford, Jack. *Indian Givers: How the Indians of the Americas Transformed the World*. New York: Fawcett Columbine, 1988.

Wilson, Edward O. *The Diversity of Life*. Cambridge, Massachusetts: Belknap Press of Harvard University Press, 1992.

ENVIRONMENT AND NATURE: SOUTH AMERICA – THE ANDES The Andes were the home of the Incas, an ancient civilization with an empire larger than the Romans; only peoples of the Himalayas have adapted as well to high mountain environments. The ancestors of the Incas arrived as nomadic hunters and gatherers roughly 20,000 years ago. They began camelid pastoralism and plant agriculture around 8000 years ago; many of the contemporary man-nature relationships can be identified as early as 4000 years ago.

A high plateau outcrops along the Andean chain, and supports an essentially treeless grassland — the Ecuadorian *paramo*, the Peruvian *puna*, the Bolivian *altiplano*, and the Argentine salt *puna* — which is the home of the native herders and sierra farmers. Because similar elevations support trees elsewhere, one theory which has been proposed independently by historians and ecologists is that somehow the native people destroyed the trees through mismanagement. However archaeologists have shown this idea to be wrong; pollen samples prove that these areas were grasslands long before any substantial human settlements existed. They supported large herds of wild camelids, guanaco, and vicuña, whose subsequent domestic relatives, llama and alpaca, were the basis of the first Andean civilizations and continue to be economically important today.

The llama and alpaca were critical for survival of several highland groups. Not only did they provide wool for clothing, meat for food, dung for fertilizer, and labor power as caravan animals carrying trade goods, but they also served as "banks on the hoof", resources which could be mobilized to bail out the family in the irregularly occurring but not infrequent environmental catastrophes (sustained droughts, unseasonal frosts, etc.). The gods responsible for protecting the flocks and for insuring their increase were believed to dwell among the surrounding peaks. Annually ceremonies (*pagos*) were held to pay the gods and to petition for the increase of flocks. A male and female animal were selected from the herds, fed the local corn beer, and "married" to begin the next season's offspring; the llamas and alpacas were treated and thought of as kinfolk of the herders. Today these ancient fertility ceremonies have been conflated with Catholic rituals, and are held during Carnival. Often small *illa* (animal figures) of the desired type of animal are blessed, and buried with offerings. In Inca times these were sometimes gold and silver, while today they are ceramic, stone, or tin.

Plant agriculture was an exceedingly risky undertaking at this elevation; folklore suggests that crop failure occurred as often as 30–40% of the years. Andean people thus sought to reduce risk through a variety of strategies. Because rains could fall in one field but not another a kilometer away, or because early frosts or late freezes could hit a valley bottom one time, but a high slope another time, the farmer might own two dozen or more small plots, scattered over several elevations, in several micro-environmental areas. Each farmer maintained a dozen or so varieties of potatoes and other seed plants, because in one year a tuber which thrived in dry soils would be the only one to produce, in another year a variant which thrived in moist soils excelled, and in a third year a variety which did well in saltier soils yielded best. Elaborate freeze-drying techniques were employed to produce dehydrated potatoes (*chuño*) and other tubers, a technology which served two purposes: many of the high elevation plants had significant quantities of phytotoxins, which were removed by the dehydration process, and the resulting detoxified tubers could be stored up to twenty years. Herders provided a necessary component of this process; in following their herds, they located "salt" licks with comestible earths. They collected these geophageous clays and traded them with agriculturalists. These clays were employed as sauces to consume with meals; residual phytotoxins in the tubers which had not been removed by freeze-drying or by cooking became bound with the clays, and bioavailable minerals were exchanged, thus doubly enhancing the quality of the food.

Pachamama (earth mother) was responsible for providing plant foods for the humans. Proper behavior involved offering a few coca leafs or other items before planting the field, much more substantial rituals at the harvest, and practices in daily life such as always decanting a few drops "for Pachamama" each time one took a drink (thus attempting ritually to insure adequate moisture for the crops).

Andean farmers developed a series of strategies to modify the landscape to control nature, particularly moisture and temperature parameters. In some of the mountain valleys, only the valley bottoms had adequate moisture for crops, but as cold air settles, the valley bottoms had the greatest risk from frosts. Elaborate terrace systems were constructed, with water provided by extensive canal networks. Thus hill slopes initially too steep to be plowed were remade into a myriad of small, level, irrigated plots. Because they were along the lower valley flanks, they were too low for high elevation freezes but above the pools of supercooled area that collected on the valley bottoms, and thus often frost-free during most of the growing season.

On the plateaus, another strategy was employed, that of constructing raised, ridged fields. The flat plains usually were poorly drained, so suffered from too much water in the rainy season, but too little in the dry season. Construction of raised, ridged fields, with swales or canals between the ridges, resulted in ridge tops above the waterlogged soils in the rainy season; the intervening canals trapped water which was curated through the dry season to provide a continuing water supply. In addition to moisture, temperature had to be modified. Thus the field patterns were constructed either parallel to, or perpendicular to, the path of the sun, an orientation which permitted maximum solar energy capture by the water. This water kept the fields slightly warmer at night, and often radiated enough heat to prevent frost damage while the surrounding unmodified grasslands suffered heavy freezes. This technology was employed through the Inca period. European invaders brought the horse and plow (and in this century, the tractor). Raised fields could not be plowed efficiently by European technology, and hence they were leveled. Although Western technology permits much less labor to be employed per unit area, and thus is more efficient in that sense, the productivity per unit area has fallen off. Recently raised fields have been reconstructed by anthropologists in the Titicaca basin. These fields provide both increased yields and decreased environmental risk: local seed growing without any high-tech inputs have produced two to four times more per unit than crops growing on the European-style fields; and on more than one occasion, unusually severe, out-of-season frosts have destroyed all the crops in the plowed flat fields, but have at worse only inflicted a little freeze damage on plant tops in the raised fields. While more labor intensive, the original indigenous technology has proved superior to western technology in these environmental circumstances.

The indigenous concept known (in Aymara) as *taypi* structures part of the human relationship with nature. *Taypi* physically is the center between two opposing concepts, the point of necessary convergence where the cosmological centrifugal forces that permit differentiation exists simultaneously with the centripetal force that ensures their mediation. It is thus both an integrating and separating center. The human body is divided into three components, with the *taypi* integrating center the heart and stomach. The cosmos is divided into an upper world, this world, and a lower world, with the earth as the mediating point between the upper and lower world. Hence humans, as residents of this earth, arbitrate not only between forces of good and evil, but also serve to mediate between the natural environment and the gods. Just as the human body is animated and integrated as a whole by an exchange of fluid elements (blood circulating, water drunk and expelled, air inhaled and exhaled), so too is nature: analogies are made between the circulation of fluids to animate humans and the circulation of fluids to animate nature.

The human imagery overlay upon nature is even extended to the organization of some of the Andean tribes. One of the mechanisms employed to deal with the environmental risks is to spread humans over the landscape, in a pattern similar to the spreading of fields discussed above. A village might send family members up to a new hamlet in the herding zones, or down to lower elevation corn growing areas, spreading production risk so that in case there was complete failure in one zone, the community would have direct access to resources in other environmental zones because they had residents living there. In the case of one group from the Qollawaya, the settlements in the lower, middle, and upper ecological zones were conceptualized and referred to by terms of the human body, with the middle elevation home village thus seen as the heart and vitalizing component of the group, and the landscape itself as replicating the human body. Andeans hence perceive their lifeways as essentially harmonious with and replicating nature.

DAVID L. BROWMAN

REFERENCES

Bastien, Joseph. *Mountain of the Condor: Metaphor and Ritual in an Andean Ayllu*. St. Paul, Minnesota: West Publishing Co., 1978.

Browman, David L., ed. *Arid Land Use Strategies and Risk Management in the Andes*. Boulder: Westview Press, 1987.

Brush, Stephen B. *Mountain, Field and Family: The Economy and Human Ecology of an Andean Valley*. Phildelphia: University of Pennsylvania Press, 1977.

Classen, Constance. *Inca Cosmology and the Human Body*. Salt Lake City: University of Utah Press, 1993.

Erickson, Clark L. "Prehistoric Landscape Management in the Andean Highlands: Raised Field Agriculture and its Environmental Impact." *Population and Environment* 13(4): 285–300, 1992.

Gade, Daniel W. Plants, *Man and the Land in the Vilcanota Valley of Peru*. The Hague: W. Junk Publishers, 1975.

Isbell, Billie Jean. *To Defend Ourselves: Ecology and Ritual in an Andean Village*. Austin: University of Texas Press, 1978.

Orlove, Benjamin S. *Land and Power in Latin America: Agrarian Economies and Social Process in the Andes*. New York: Holmes and Meier Publishers, 1980.

Winterhalder, Bruce, and R. Brooke Thomas. *Geo-ecology of Southern Highland Peru: A Human Adaptation Perspective*. Boulder: University of Colorado Institute of Arctic and Alpine Research, 1978.

See also: Potato – Crops in Pre-Columbian Agriculture

EPILEPSY IN CHINESE MEDICINE A description of epilepsy in ancient Chinese medicine first appeared in *Huangdi Neijing* (The Yellow Emperor's Classic of Internal Medicine). It is believed to be a collective work of a group of Chinese physicians, written around the period of Warring States (770–221 BC). In this book, there is a vivid description of a grand mal seizure (*Ling-Shu* volume) and a theory that emotional shock of the pregnant mother is a cause of her child's epilepsy (*Shu-Wen* volume). Since that time, the literature elaborates on the varieties of seizure phenomena and speculates on the causes of epilepsy. In Chinese traditional medicine, epilepsy sometimes tends to be confused with psychosis or mania. This ambiguity stems from the fact that the word *Dian* (falling sickness) sometimes refers to epilepsy, and at other times refers to insanity. The word *Dian* is used sometimes with the word *Kuang* (mania) or *Feng* (psychosis), to form the compound word *Dian-Kuang*, or *Feng-Dian*, a practice that further blurs the distinction between epilepsy and psychosis.

There are several different classifications of epilepsy in Chinese traditional medicine, some of which are based on clinical observation of the symptoms and/or the presumed causes. The most fascinating classification is based on the noises the patient makes during epileptic attacks. Sun Simo, in *Qianjin Yaofang* (AD 682, Tang Dynasty), named epileptic attacks after various animals whose cry the epileptic cry resembles. He divided epileptic attacks into six types; *Yang Dian* (goat epilepsy), *Ma Dian* (horse epilepsy), *Zhu Dian* (pig epilepsy), *Niu Dian* (cow epilepsy), *Qi Dian* (chicken epilepsy), and *Gou Dian* (dog epilepsy). The terminology used in this classification probably has had an adverse effect on the way Chinese society perceives epilepsy, as shown in the colloquial expression of epilepsy as *Yang-Dian-Feng*, which can be translated literally as "goat-falling sickness-psychosis." Therefore, the confusion between epilepsy and psychosis, and the comparison of epileptic attacks to animal behavior may have contributed to the misunderstanding and discrimination against epilepsy in Chinese culture.

The philosophy of treatment of diseases in Chinese traditional medicine is different from that of Western medicine. In general, traditional doctors believe that life is the result of a combination, in specific proportions, of *yang* energy from the sun and *yin* energy from the earth. It is believed that when the balance between yin and yang is disturbed, diseases will follow. Another doctrine, *Wuxing* (Five Elements), consists of the belief that the human body is composed of a harmonious mixture of five primordial substances; metal, wood, water, fire, and earth. The treatments for diseases are designed to restore the balance of the energy and these elements by a variety of means. The therapy for epilepsy includes herbs, acupuncture, *Mai Yao* (injection of herbs into the acupuncture points), *Mai Xien* (burying a piece of goat intestine into the acupuncture points), or massage.

CHI-WAN LAI
YEN-HUEI C. LAI

REFERENCES

Huangdi Neijing. Shu-Wen (around 770–221 BC). Henan, China: Henan Ke Xue Ji Shu Chu Ban She, 1962.

Huangdi Neijng. Ling Shu (around 770–221 BC). Beijing, China: Ren Min Wei Sheng Chu Ban She, 1964.

Lai, C.W., and Y.H. Lai. "History of Epilepsy in Chinese Traditional Medicine." *Epilepsia* 32(3): 299–302, 1991.

Sun Simo. *Qianjin Yaofang.* (AD 652). Beijing, China: Lin Shui Wen Ming Shu Ju, 1944.

See also: *Huangdi Neijing* – Sun Simo (Sun Simiao) – Medicine in China – Five Elements – *Yinyang* – Acupuncture

EPILEPSY IN INDIAN MEDICINE Āyurveda is the ancient Indian medical system (4500–1500 BC). Its name is derived from the Sanskrit *Ayu*, meaning life-combined state of body, senses, mind and soul, and *veda*, meaning science. Āyurveda is therefore the science of life. It is considered by many to be the oldest system of medicine in the world. Epilepsy was referred to as *Apasmara* in Āyurveda. The prefix *Apa* means negation or loss of, and *smara* means consciousness or memory. Epilepsy was considered to be caused by both endogenous and exogenous factors. Among the exogenous factors were internal hemorrhage, high fever, excessive sexual intercourse, disturbances of the body due to fast running, swimming, jumping, or leaping, eating of foods that are contaminated, nonhygienic practices, and extreme mental agitation caused by anger, fear, lust, or anxiety. An endogenous disturbance refers to a metabolic derangement in the form of a disturbance of *doṣas* (humors) which are aggravated and lodged in the channels of *hrt* (brain). It was also recognized that epilepsy could arise secondary to other diseases.

Aura was recognized and was called *Apasmara Poorva Roopa*. An actual attack of seizures was said to occur when the patient saw non-existent objects (visual hallucinations), fell down, and had twitching in the tongue, eyes, and eyebrows with jerky movements in the hands and feet. In addition, there was excessive salivation. After the paroxysm was over, the patient awakened as if from sleep.

Apasmara was classified into four types. The *Vatika* type was characterized by frequent fits with uncontrollable crying, unconsciousness, trembling, teeth gnashing, and rapid breathing. Upon regaining consciousness, the patient had a headache. In the *Pattika* type, the patient became agitated, had sensations of heat and extreme thirst and an aura of the environment being on fire followed by frequent fits accompanied by groaning and frothing at the mouth and finally falling on the ground. In the *Kaphaja* type the onset of convulsions was delayed and was preceded by an aura during which the patient felt cold and heavy and saw objects as white. The seizure was accompanied by falling and frothing at the mouth. The fourth type, *Sannipatika*, was due to a combination of all of the above. This type was considered incurable, occurred in older people, and resulted in emaciation.

The first step in the treatment was the "awakening of the heart" (meaning to wake the patient from unconsciousness) by using drastic measures to clear those doṣas that block the channels of the mind. After the patient was awake, drug formulations to alleviate epilepsy were administered. The ingredients of these preparations included sulfur, aged ghee (butter fat), and many herbs such as *Achyranthes aspena*, *Holanthena antidysenterica*, *Alstonia scholaris*, and *Ficus carica*. Blends of herbal formulations such as Pancamula and Triphala were also used. Bhagwan Dash describes pharmaceutical processes and preparations which involved fermenting, extracting, preparing inhalable substances, filtrating, heating in a closed cavity, purifying, and pill-making. General measures to correct exogenous factors, such as proper hygiene and a balanced diet were recommended. Epileptics like the insane were kept away from dangerous situations, namely, water, fire, treetops, and hills.

BALA V. MANYAM

REFERENCES

Dash, Bhagwan. *Materia Medica of Ayurveda*. New Delhi: Concept Publishing, 1980.

Kurup, P.N.V. *Birds-Eye-View on Indigenous Systems of Medicine in India*. Delhi: Depak, 1977.

Sharma, Priyavrat. *Caraka-saṃhitā* text with English translation, vols. I and II. Varanasi: Chaukhambha Orientalia 1981.

Singhal, G.D., S.N. Tripathi, and K.R. Sharma. *Ayurvedic Clinical Diagnosis Based on Madhava- Nidana, Part 1*. Varanasi: Singhal Publications, 1985.

See also: Āyurveda – Caraka

ETHNOBOTANY There are numerous definitions of ethnobotany. The simplest is that it concerns the study of the uses of plants in societies. The term was first employed by John Harschberger in 1895. It was used narrowly in reference to the use of plants by aboriginal people. More recent authors believe that ethnobotany should consider not only the uses of plants but the entire range of relations between humans and plants.

Ethnobotany is a distinct field of research with a strongly interdisciplinary outlook. A number of subdivisions have developed, such as archaeoethnobotany (often called paleoethnobotany), ethnopharmacology, ethnoecology, and ethnomycology. There is even a possible subdivision which could be called "horticultural ethnobotany". With this rapid proliferation of interests, ethnobotanists have widened the definition to encompass the study of uses in aboriginal societies, native technological manipulation, classifications of the plants involved, indigenous nomenclature, agricultural systems, magico-religious and mythological concepts connected with plant uses, and the general sociological importance of the flora in indigenous societies.

Ethnobotany itself is certainly not new. The earliest humans must have been incipient ethnobotanists. They must have classified plants in their surroundings — those of little or no utility, those which were useful in many practical ways, those which alleviated pain or otherwise ameliorated illness and those that caused illness or even killed outright. They must have wondered at the unwieldy effects of the few hallucinogenic or psychoactive species, and they could explain their extraordinary properties by assuming that they were endowed with spiritual power from supernatural forces.

It was not long before the knowledge and manipulation of the properties of plants became associated mainly with certain individuals, and the early medicine men or shamans ultimately acquired great powers over many aspects of the life and beliefs of the general population. Members of the general population, often especially women, were likewise conversant with the properties and uses of their food, medicinal, and other economic plants of daily use. Many peoples around the world are very knowledgeable about their ambient vegetation as a result of inherited knowledge, the result of hundreds of years of experimentation.

Modern ethnobotany can be of great value to many fields. Human health, new and better medicinally valuable phytochemicals, the domestication of new crop plants for food and industrial purposes, better understanding and use of plant biodiversity, and general environmental conservation — all might well reap significant advantages from proliferation of ethnobotanical investigations.

The world's flora, estimated to contain some half million species of flowering plants, have been analyzed with modern chemical techniques only on a limited scale. Tropical rain forests, which contain a greater part of the world's plant species, have been the least chemically studied even though they are extremely rich laboratories. The Amazon Valley, for example, has an estimated population of 80,000 species of angiosperms. When this wealth of species in other rain forest areas of the world is considered the opportunities for great advances in our knowledge of phytochemistry are obvious. If we consider only alkaloids, for example, the increase in our knowledge is significant: In 1950 we were aware of 2000; in 1970 the number had increased to 4000, and in 1990 it had reached 10,000.

If enough material of 80,000 species must be collected for chemists from such a difficult and extensive area for travel, transportation, and availability of resources, it is obvious that the task can never be completed, especially by random collecting. Ethnobotany can be a significant help.

More and more tribal peoples are succumbing to acculturation or westernization as a result of road and airport building, dam construction, increased missionary pressure, heightened tourism, or disruption from their centuries-old dwelling sites. In many areas of the great native Amazon Basin knowledge is disappearing even faster than the forests. When an Indian can get from a missionary, tourist, or commercial agent our effective and easily used drugs, this native knowledge disappears rapidly. Of the numerous subdivisions of the field of ethnobotany, none is more important than ethnopharmacology — the search for new medically valuable plants.

In the small area of the northwestern Amazonia (Colombia, Ecuador, and a small part of western Brazil), more than

1600 species of biodynamic plants used by the Indians as medicines are considered to be an adjunct of routine cures by medicine men, or poisons (as in the manufacture of curare or arrow poisons). If chemists concentrate on those plants which Indian use has demonstrated to be in some way bioactive, understanding of the chemistry of the flora of the region can be accomplished more efficiently.

Another example of the value of ethnobotanical research lies in the Indians' familiarity with minor variations in plant species. This may be of great help in one phase of the study of biodiversity. One of the enigmas that botanists and other specialists have not had much success in understanding concerns the native's easy recognition of varieties of many of the plant species of their forests. These variants are so well established in the Indians' classifications that they usually have distinct names, in spite of the botanists' inability to see any distinguishing morphological characteristics. This skill is manifest not only in the few cultivated economic plants but also in their classification of many wild plants of no utilitarian, ceremonial, magical, or mythological importance. The Indians can usually tell at once and frequently on sight and often at a significant distance, without feeding, tasting, smelling, crushing, tearing, or other physical manipulation, to which category a plant belongs. While it is of interest to the anthropologist and psychologist, it is of extremely practical importance to chemists and botanists.

Yoco (*Paullinia yoco*, Sapindaceae) is a gigantic liana from the bark of which numerous tribes of the Colombian Amazonia prepare a very strong stimulating drink taken in the early morning before eating solids. The drink has 3% caffeine. There are native names for at least thirteen of these lianas. While several give a creamy white drink, others a slightly reddish brown product, there are no discernable morphological differences in them. Often several of the named "kinds" grow in relatively close proximity, so it cannot be due to ecological differences. Nor is it due to different parts of the liana.

Another example is ayahuasca or caapi (*Banisteriopsis caapi*, Malpighiaceae), a forest liana the bark of which is the source of an interesting vision-producing hallucinogenic drink employed widely by many tribes in the western Amazonia. Since the source plant was identified and botanically described as a species new to science in 1853 by the British plant explorer and ethnobotanist, Richard Spruce, many specialists and amateurs have written about it. We are certain now that one species of *Banisteriopsis* is used. There is now no doubt also that Indians can visually identify at a distance different "kinds" of the *caapi* liana. There is a long list of these variants, and the natives maintain that they are employed to prepare drinks of different strengths or in connection with various ceremonies or dances because of longer or shorter intoxications or because of special magico-religious needs. There is one of the named variants from which a drink may be prepared according to the kind of animals the hunter wants to find and kill.

Could the ability of the Indian to recognize various types of *Banisteriopsis caapi*, often at a distance, be due to slight differences in the bark of the liana, presumed chemical diversity (which could not be seen), variation in leaf shape, ecological factors (growing in dense or open forests), or other dissimilarities? None of these can be considered logical explanations of the uncanny ability consistently to identify the named type of the liana. It remains an enigma which deserves further study.

Whatever the explanation of this perspicacity, specimens of eighteen "kinds" of the ayahuasca vine were collected by Dr E. Jean Langdon, an anthropologist who worked amongst the Siona in the Colombian Amazonia. They were submitted for botanical identification by a botanical specialist, Dr Timothy Plowman, and almost all were referred to the single species, *Banisteriopsis caapi*. As Dr Langdon states: "Further exploration between the conjunction of botany–chemistry–culture warrants further investigation".

This example of the unexplored horizon awaiting scientific investigation is typical of many other unexplained examples of the wealth of aboriginal concepts of natural phenomena awaiting technical solution around the world.

Ethnobotany, if intensified, can contribute extraordinarily to many fields of modern science, technology and sociology, but time is of the essence in view of the worldwide uncontrolled acculturation of indigenous societies. As Prince Philip, former president of World Wildlife Fund, has pointed out: "The tropical forests and their traditional inhabitants are under very severe pressure from human encroachment, and it is sadly inevitable that many of these plant species and the tribes which understand their use are rapidly disappearing forever."

RICHARD EVANS SCHULTES

REFERENCES

H.R.H. Prince Philip. In the foreword to *The Healing Forest, Medicinal and Toxic Plants of the Northwest Amazonia*. Portland, Oregon: Dioscorides Press, 1990.

Schultes, Richard Evans. "Ethnobotany." In *The New Royal Horticultural Society Dictionary of Gardening*. London: Macmillan, 1992, pp. 213–216.

ETHNOBOTANY IN CHINA Ethnobotany is the study of the links that human beings, as individuals or societies, have maintained and maintain with plants. China is a country with more than fifty different national minorities, with very different traditional cultures, languages, food habits, etc. Chinese ethnobotany could thus be the juxtaposition of the different ethnobotanies of all the ethnic groups living in China. On the other hand, one generally understands "Han" when one says "Chinese". This term, which is the name of the second dynasty after unification of the empire (206 BC–AD 220) is also the name of the main ethnic group in China, with about 96% of the population. It is the ethnnobotany of the Han Chinese that we are going to present here.

With 3,692,000 sq. miles, China is about the same size as the United States. Its relief can be roughly compared to stairs going down from the western plateaux to the plains of the east and the south. Its climate is under the double influence of a huge continental mass and the strong impact of the summer monsoon: generally the winters are cold to very cold — except for the extreme south-east — and dry. The summers are very warm with high rainfall and moisture. The diversity of climates and reliefs has probably favored a particular richness in vegetation. With about 25,000 species of higher plants, 2600 species of ferns, 2100 species of mosses and liverworts, and some 5000 species of fungi, China has the third richest flora of the world after Malaya and Brazil.

In Chinese mythical history, the beginning of the relation between man and plants is so strong that it is considered the first step into "civilization". Speaking of Shennong, the Divine farmer, one of the three mythical emperor creators of the Chinese civilization, one says that "before him, men did not know cereal grains, they ate animals and drank their blood. Then Shennong tasted the herbs, sorted cereals, and taught the art of tilling to the people. Then he tasted herbs again and sorted simples to save the people from illness". Eventually, the Yellow Emperor, successor of Shennong, taught cooking and medicinal prescriptions. In the Chinese tradition, the access of mankind to civilization was achieved when man became a tiller and a gatherer; agriculture supplied what was needed for basic food, and gathering gave access to the complementary plants, medicinal, and pot herbs. Actually, the local biodiversity appears to have been rather well appreciated and exploited all through Chinese history, because China is one of the main centers of domestication of plants and possesses a very rich *materia medica*. The local richness did not prevent a great interest in plant introduction. However, introduced plants, like corn, sweet potato, tomato, and potato, have been used mainly as complementary food and never caused any fundamental change in food habits, as happened in Europe, for instance, with American plants. Among the 116 botanical species cultivated as vegetables today in China, at least 37 would have been locally domesticated. Also, considering the cultivars — which are the form under which these species are cultivated — their number amounts to several thousands. As for trees and shrubs, more than 300 species are exploited today — even if they are not all cultivated — for their fruits, eaten fresh or, more frequently, after transformation or preparation.

There is a traditional opposition between crops cultivated on dry lands — millets, wheat, barley, sorghum — in Central and Northern China, and rice and various tubers cultivated in irrigated fields of the south. Nowadays, thanks to agricultural hydraulic development, rice is also cultivated in northern parts of the country, where climatic conditions permit it. On the other hand, corn, following irrigation development, replaced traditional millet crops. The ordinary diet consists of large amounts of a basic food, called *fan*, made from steamed or boiled grains, such as rice, millet, or sorghum, or various kinds of noodles made from wheat, pulses, rice, etc. It is accompanied by *cai* (dishes), generally made from vegetables and soya bean in the form of *tofu*. Among these dishes, fish or meat, important for special occasions, are still not very frequent in everyday life, especially in the countryside.

As for clothing, even if today synthetic fibers, wool and silk have become more and more important, in the sixties, in mainland China, cotton was the only material used for common suits and even shoes. In the countryside, some work clothes, sun hats, and raincoats are still frequently made of bamboo or palm leaves. They combine several advantages: since they are not heavy, they provide good shelter and allow the air to circulate freely. The use of wood for buildings is entirely limited by the resources of the environment: it is a wooden framework which bears the roof of the traditional standard Chinese house, the non weight-bearing walls being made of brick or tamped earth. However, from its shoots eaten as a delicacy, to the chopsticks used to eat, from the scaffolding, drill-hafts and brine conduits of the salt-fields in Sichuan to irrigation flumes, and from ordinary furniture to the most exquisite pieces of wickerwork, bamboo is probably the member of the vegetable kingdom which plays the most important part in the material life of the Chinese.

Traditional Chinese medicine uses as *materia medica* some 4773 products of plant origin, 740 of animal origin, and 82 minerals.

Chinese people are also very interested in the beauty of plants. Varieties of some twelve ornamentals are particularly valued: tree-peony (*Paeonia suffruticosa* Andr.), Japanese apricot (*Prunus mume* Sieb. et Zucc.), hybrid roses, camellias, rhododendrons, Wintersweet (*Chimonanthus praecox* Link.), Chrysanthemums [*Dendranthema morifolium* (Ramat.) Tzvel.], orchids (*Cymbidium* sp.), daffodils (*Narcissus tazetta* var. *chinensis* Roem.), hybrid gladioli, pinks (*Dianthus caryophyllus* L.) and Indian lotus (*Nelumbo nucifera*

Gaertn.). Besides a great number of other "traditional" ornamentals like bamboos or cockscombs (*Celosia argentea cristata*), new varieties appear frequently. Grafted succulents became very popular and, more recently, Kaffir lilies (*Clivia* sp.). Just for the pleasure of watching new flowers or to be photographed in front of a blossoming magnolia or plum, crowds of people will visit public parks in the spring, while in autumn it would be unthinkable not to go and admire the red leaves of maples in the nearby hills of the city of Nanking, for instance.

Plants obviously play a very important part in the life of the Chinese people, but one may wonder whether there is any particular kind of relation which would define a "Chinese ethnobotany". A first striking point, which can be well appreciated in agriculture and horticulture, is the closeness of the human-to-plant relation. This is obvious with a crop like rice where, after the nursery stage, every seedling in the paddy is transplanted to the field where it will ripen. Propagating other plants like yams or taros also necessitates the individual handling of every cutting. The sowing of all the plants cultivated in gardens, like cucurbits and beans, is individual, and even for crop plants like millets or wheat it is remarkable that a sowing machine allowing seed by seed control has existed at least since the thirteenth century. Grafting seems to have begun during the first centuries AD but it became very important in horticulture as early as the sixth century and was much used for fruit and flower production. Professional horticulturists have used grafting since the Song dynasty (960–1279) when it was believed it could create new varieties of flowers (tree-peonies and chrysanthemums) for amateurs who would pay fortunes just to watch the most beautiful. Another example of individual treatment of a plant is the technique of tree-potting (*penzai*, *bonsai* in Japanese): to keep the tree in small proportions, it is necessary to prune its branches and roots every year and to look after it day after day.

In ancient China, scholars also developed a very exclusive relation with plants. They would rank them from the most precious to the lowest. Following their personal taste, men of letters would develop real "friendships" with the ones they apreciated most and which would live in their studios, along with "the books and the cithar (zither)". Joseph Needham and Li Hui-Lin have shown that this interest in special plants led to various monographs on them. Plants have also been a favorite theme for literature and poetry. Since the thirteenth century several encyclopedic books have been composed entirely with quotations from texts dealing with plants. One famous book, published in 1231, *Portraits of the Plumflower*, is devoted only to the Japanese apricot (*Prunus mume* Sieb. et Zucc.). The author, Sun Boren, through a hundred ink drawings, each accompanied by a poem, evokes the feelings that he experienced watching the flower, from its bud stage to the fall of the last petal. Actually, in the painting of flowers, Chinese artists were more concerned with emotion than representation. Using a brush and black ink, they were looking for the essence, more than for the true form of, bamboos, pine trees, orchids, or plum-flowers. Zou Yigui (1696–1772), a famous painter, criticized the Western paintings he had seen at the imperial court as being too realistic, and he considered their authors not painters, although they possessed skill, but simple artisans.

In ancient China, plants as medicine were not distinct from minerals or animals and belonged to a system of classification where the whole *materia medica* was separated into three ranks, *san pin*, for non-poisonous products, good for the health and used to "nourish life", to very active products, generally highly toxic and used only under very strict conditions. During the sixth century, Tao Hongjing, a famous physician, developed this classification: within each of the three grades, plants were subdivided into herbs, trees, grains, potherbs, and fruits. At the end of the sixteenth century, another physician, Li Shizhen (1518–1593) in a famous book called *Bencao gangmu* (Classified Pharmacopaeia, 1596), changed the basis of this classification. Considering almost obsolete the three ranks system, he reorganized the whole materia medica "from the lowest to the most precious", from minerals to man. He kept the five categories mentioned above for plants but in a new order — herbs, grains, vegetables, fruits, and trees — and divided those into thirty-three subgroups following various criteria like ecology, taste, and toxicity. Besides this learned system, created and used by physician-scholars, there was a folk taxonomy not exclusive from the previous classification, where the ethnobotanical data were distributed into six ranks. Following Brent Berlin's vocabulary, they are:

- **kingdom** marked by two words meaning plant, *caomu* (herb tree) and *zhiwu* (planted thing);
- **life-form** with three main taxa, *cao* (herb), *mu* (tree) and *teng* (vine) and the ambiguous statute of *chu* (bamboo) "neither herb nor tree";
- **intermediate** where groups are named by juxtaposition of terms used for taxa of generic rank, like *tao-li* (peach-prune), *song-bai* (pine-cypress);
- **generic** marked by monosyllabic terms like *tao* (peach), *xing* (apricot), *zao* (jujub);
- **specific** where one finds some of the monosyllabic terms also used for generic taxa but with a narrower meaning like *xing* (apricot: *Prunus armeniaca* L.). However, terms of this rank are generally polysyllabic. They may be **secondary plant names**, lexical extensions on the basis of a generic term, like *yinxing* (silver apricot: *Ginkgo*

biloba L.), or *jiazhutao* (narrow bamboo peach: oleander, *Nerium odorum* Soland.). They may also be **primary plant names** like *Lu meiren* (the beautiful lady of Lu: poppy, *Papaver rhoeas* L.), *mudan* (male cinnabar: tree peony, *Paeonia suffruticosa* Andr.);

- **varietal** a system, already described in the first encyclopedia *Erya* (third–second centuries BC) is still effective today.

The long tradition of observation, culture, and use of plants along with a fair knowledge of plant life never led to a systematic approach. Among the rich literature about plants one cannot find anything like a primer of botany. Up to the middle of the nineteenth century, plants, in China, were considered in their cultural environment. Besides horticulture, agriculture, or art, concern with plants was philological: considering plant names with obscure meanings occuring in ancient texts, literati would try to find which plants would fit these names. To do so, they compared textual evidence or interpretation with the results of inquiries they made among their contemporary countrymen. Chinese traditional scholars' plant knowledge can be described as cultural botany.

GEORGES MÉTAILIÉ

REFERENCES

Anderson. E.N. *The Food of China*. New York and London: Yale University Press, 1988.

Berlin, Brent. *Ethnobiological Classification*. Princeton, New Jersey: Princeton University Press, 1992.

Bickford, Mary, ed. *Bones of Jade, Soul on Ice — The Flowering Plum in Chinese Art*. New Haven, Connecticut: Yale University Art Gallery, 1985.

Bray, Francesca. *Science and Civilisation in China*. Vol. 6:2. *Agriculture*. Cambridge: Cambridge University Press, 1986.

Bretschneider, Emil. "Botanicon Sinicum." *Journal of the North-China Branch of the Royal Asiatic Society*, New Series 16(2): 1–467, 1881. (Facsimile edition, Nendeln, Liechenstein: Kraus Reprint Ltd., 1967, vol. 1–2).

Goody, Jack. *The Culture of Flowers*. Cambridge: Cambridge University Press, 1993.

Li, Hui-Lin. *The Garden Flowers of China*. New York: The Ronald Press Company, 1959.

Needham, Joseph. *Science and Civilisation in China*. Vol. 6:1. *Botany*. Cambridge: Cambridge University Press, 1986.

Sirén, Osvald. *Chinese Painting: Leading Masteers and Principles*. London and New York: Lund Humphries and the Ronald Press Company, 1955–1958, 7 vols.

Sivin, Nathan. "Li Shih-chen." In *Dictionary of Scientific Biography*. Ed. Charles G. Gillispie. New York: Charles Scribner's Sons, 1973, pp. 390–398.

See also: Food Technology in China – Agriculture in China – Bamboo – Li Shizhen

ETHNOBOTANY IN INDIA Ethnobotany deals with the relationship between humans and plants. Archaeological or paleobotanical evidence about the use and cultivation of plants for food, medicine, housebuilding, etc. suggests a long history, yet the word 'ethnobotany' was coined by J.W. Harshberger only in 1895. Human–plant relationships can be classified into two topics, material and abstract. Material relationships include use in food, medicine, building, painting and sculpture, and in acts of domestication, conservation, and improvement or destruction of plants. Abstract relationships include faith in the good or evil powers of plants, sacred plants, worship, taboos, and folklore. Ethnobotanical enquiry extends beyond botany, and has significant input of other branches of science such as archaeology or medicine.

The tribal people of India mostly live in forests, hills, and relatively isolated regions. They are variously termed *Ādivāsī* (original settlers), *Ādim Niwāsī* (oldest ethnological sector of the population), *Ādimjāti* (primitive caste), Aboriginal (indigenous), *Girijan* (hillsmen), *Vanyajāti* (forest caste), *Vanavasi* (forest inhabitants), *Janajāti* (folk communities), *Anusuchit Janjati* (scheduled tribes), and other such names signifying their ecological, historical, or cultural characteristics. While the constitutional term is *Anusūchit Janajāti* (Scheduled Tribes), the most popular term is *Ādivāsī*.

Certain characteristics make India ethnobotanically rich. The region supports very varied and rich flora, from desert to tropical rain forests. Of deeper significance is the tradition of the Vedas and other ancient literature, an ethnobotanical continuum that enables contemporary investigators to delve into the distant past and often to link modern folklore with that of ancient cultures.

Ethnobotanical work in India can be broadly classified. First there is the ethnobotany of distinct ethnic groups such as the Mikir of Assam, Bhils of Rajasthan, and Thārūs of Uttar Pradesh. Next there is the ethnobotany of a specific region. Most ethnobotanical research is aimed at one utility group, e.g. the study of famine foods, special diets for festivals, beverages, spices, medicines in general, medicines for particular diseases, perfumes, oils, dyes, narcotics, and plants for tools, worship, ceremonies, or personal adornment.

People in remote areas cultivate numerous vegetables representing distinct genetic stocks adapted to local conditions. Some examples are *Piper peepuloides* (a condiment in the Khasi hills), *Digitaria cruciata* var *esculenta*, (a minor millet of Khasi Hills), and species of *Alocasia*, *Amorphophalus, Colocasia* and *Dioscorea* (cultivated by tribes in the northeast). In remote mountains or desert regions, wild plants (tubers, leaves, flowers, fruits, seeds, and grains) still provide considerable quantities of food. Examples of notable and less known plants in ethnomedicine are Yarrow (*Achillaea millefolium*), an anthelmintic; Cutch (*Acacia catechu*), and Indian Stinging Nettle (*Tragia involucrata*) for intestinal diseases; and Prickly Chaff Flower (*Achyranthes porphyristachya*) and Nutgrass (*Cyperus rotundus*) for liver complaints.

Many ethnobotanists have given attention to the diversity of traditional tools, gadgets, and articles of personal adornment. Good examples are the single-pan balance among the Mikirs of Assam, the variety of cattle-traps among the Goṅḍs and Bhils, and containers, utensils, bridegroom's hats, and other articles of personal adornment and agricultural or other tools. Researchers have found much ingenuity in the choice of raw materials and in the design or art of decorating these articles.

Vegetable dyes for coloring hair, teeth, palms, and other parts of the body have been used since ancient times. Hair, palms, and feet are usually dyed with Henna (*Lawsonia inermis*). Teeth are blackened by chewing walnut bark. The Mikirs of Assam burn green stems of *Murraya koenigii* and certain other plants and apply the gum exudate to their teeth. Tattooing on arms, legs, chest, cheeks, etc. is still popular in some societies; common designs include figures of gods and goddesses, flowers, or animals. Mikir women (Assam) make a dye from the juice of *Baphiacanthus cusia* with which they draw a perpendicular line from their foreheads, down over their noses to their chins. The aboriginals make indigenous musical instruments from naturally available materials. The Santals in Bihar and the Gonds in central India make drums from the wood of mango, Toddy Fish-tail Palm (*Caryota urens*), Bengal Padauk (*Pterocarpus marsupium*), Malay Bushbeech (*Gmelina arborea*), and Little Flower Crepe Myrtle (*Lagerstroemia parviflora*) trees. Stringed instruments are common. Flutes are made from bamboo, and are played singly or in accompaniment with drums.

The role of ethnobotany in conserving plant resources has been studied in sacred groves and traditional agriculture.

Many cultivators have not taken to using improved varieties of crops; they continue to use traditional land races or wild relatives of crops, thereby maintaining their genetic material. Some characteristics such as hardiness, disease resistance, and adaptability to waterlogging, drought, or cold in land races have been utilized by breeders; these have averted famines.

The study of mythological associations with plants like tree worship, plants in offerings, carvings in temples, and plants in Indian epics has been fascinating. Work has been done on plants associated with deities, sages, origins from the bodies of gods, and planets; with taboos on sowing, plucking flowers or fruits, and eating wild fruits in certain seasons; with woods suitable (or unsuitable) for making idols of gods, and with driving away the evil eye.

Detailed interdisciplinary studies on the ethnobotany of any particular plant include people's first association with that plant, the impact of selective use on the biology of the plant or on the ecosystem, and the implications of usage on the life and culture of people. Studies have been done on *Bauhinia*, *Coix, Coptis*, *Ficus*, *Saussurea*, and *Selaginella*.

Among the chief tools of research in ethnobotany in India, mention may be made of the following:

- *Literature*. Ancient or unnoticed, published or unpublished literature is an important source of information. Though the identity of a number of plants referred to in ancient works has sometimes been in doubt, Vedic literature has been a valuable resource.
- *Field work*. Information is collected from local informants on the local name, parts used, processing, and dosage of the drug. Voucher specimens are collected for recording in an herbarium.
- *Herbaria and Museums*. Notes on herbarium and museum specimens are a good source of data. They are attached to an actual specimen, and identification of the plant is precise.
- *Archaeological remains*. Sometimes, archaeological remains provide useful data. In India, an attempt was made to describe plants from reliefs on the Great Stupa of Sanchi, and the Stupa of the first and second centuries BC respectively. Plant remains from the neolithic period apparently used as food, fodder, and shelter, were also studied.

Ethnobotany has been recognized as a priority area of research by the government.

S.K. JAIN

REFERENCES

Harshberger, J.W. "Some New Ideas." *Philadelphia Evening News*, 1895.

Jain, S.K. "Ethnobotany." *Interdisciplinary Science Reviews* 11(3): 285–292, 1986.

Jain, S.K., ed. *A Manual of Ethnobotany*. Jodhpur: Scientific Publishers, 1987.

Jain, S.K. *Ethnobotany — Its Concepts and Relevance*. Presidential Address. 10th All India Botanical Conference. Patna: 1987.

Jain, S.K., ed. *Methods and Approaches in Ethnobotany*. Lucknow: Society of Ethnobotanists, 1989.

Jain, S.K., ed. *Contributions to Indian Ethnobotany*. Jodhpur: Scientific Publishers, 1990.

Jain, S.K., ed. *Dictionary of Indian Folkmedicine and Ethnobotany*. New Delhi: Deep Publications, 1991.

Jain, S.K. and Roma Mitra. "Ethnobotany in India". In *Contributions to Indian Ethnobotany*. Ed. S.K. Jain. Jodhpur: Scientific Publishers, 1990, pp. 1–17.

Jain, S.K., et al. *Bibliography of Ethnobotany*. Howrah: Botanical Survey of India, 1984.

Janaki-Ammal, E.K. "Introduction to Subsistence Economy of India." In *Man's Role in Changing the Face of the Earth*. Ed. L.T. William Jr. Chicago: University of Chicago Press, 1956, pp. 324–335.

ETHNOBOTANY IN MESOAMERICA The ethnobotany of Mesoamerica is extremely rich. The region is the original home of many of our most useful plants. Amongst the numerous species, maize (*Zea mays*), avocado (*Persea americana*), henequen (*Agave fourcroydes*), maguey (*Agave cantala*) and chili pepper (*Capsicum frutescens*) immediately come to mind. All of these are inherited from aboriginal peoples who domesticated them long before the arrival of Europeans.

Perhaps the ethnobotanical aspect most typical of the region was the medicinal and ceremonial use of psychoactive or hallucinogenic plants. The Indian populations of Mesoamerica discovered and still employ in their magic and medicine many species with psychophysical properties. Even more significant is the evidence that from ancient times these plants have been considered sacred. That explained why the few plants with these unworldly effects amongst the half-million species in the world have such weird effects when ingested; they must be endowed with spiritual power, according to the belief of the aborigines. They are capable, through visual, auditory, or other hallucinations and related bioactivity, of allowing the medicine men or even ordinary individuals to communicate with ancestors or the spiritual forces who govern the affairs of the world below and who are able, if not propitiated, to inflect sickness, suffering, death, or calamities on people or on whole tribal groups.

We know much today about the use of bioactive plants for magico-religious and medicinal purposes from archaeological finds, ancient monuments, petroglyphs, codices, missionary reports and other early publications. It is obvious from all of these and other sources that the use of plants capable of producing physiological and psychological alterations has had deep significance amongst native peoples in many parts of the world, especially in Mesoamerica.

In view of the great importance of psychoactive plants in the life of Indians of Mesoamerica, a brief discussion of the cultural importance of several of the numerous psychoactive plants employed by the Indians of Mexico and Guatemala in ancient and modern times will best indicate this particularly significant aspect of Mesoamerican civilizations.

PEYOTE (Lophophora williamsii)

The hallucinogenic peyote cactus has a long history in Mexico as a sacred plant employed in numerous medicinal and ceremonial ways. Peyote is a low, spineless cactus. It is still used ceremonially by Indians in central and northern Mexico, particularly the Huichol, Tarahumare, and Cora. The Huichols each year make a sacred pilgrimage to the peyote deserts and cut off the tops of the peyote plants, leaving the roots in the ground to regenerate tops. The tops are dried and taken back for use throughout the year. It is the dried top, known as the "peyote button", that is eaten during their ceremonies and dances.

Early reports, mostly by post-Conquest ecclesiastical writers, usually condemned the use of this innocuous hallucinogen as the work of the devil, and they tried, unsuccessfully, to exterminate this pagan religious ceremony. One report equated the eating of peyote to cannibalism. One of the early writers intimated that the Chichimecas and Toltecs were acquainted with peyote as early as 300 BC, but the accuracy of this dating depends on interpretation of native calendars.

Recently, however, excellently preserved archaeological findings in Trans-Pecos, Texas, and Coahuila, Mexico, have confirmed ceremonial use of peyote. Identifiable remains of peyote buttons in great abundance and in a context suggestive of ritual use, have been dated by Carbon 14, and they span 8000 years of intermittent human occupation. In Coahuila, ceramic bowls dated from 100 BC to AD 200–300 depict significant use of peyote; four peyote-like ornaments and a hunch-backed man holding a pair of peyotes are depicted in these bowls. There can be little doubt that peyote is the most important psychoactive plant still utilized in the dry regions of Mexico

Of the many alkaloids (30 or more) in the peyote cactus, only one, mescaline, is responsible for the richly colored, kaleidoscopically moving visions induced when peyote buttons are ingested.

RED BEAN or Mescal Bean (Sophora secundiflora)

Together with the archaeological cache of peyote in Texas, two other plants were discovered. One is the red bean or coral bean (*Sophora secundiflora*); the other is a suspected psychotropic plant known today as the Mexican buckeye (*Ungnadia speciosa*).

An early Spanish explorer of Texas, Cabeza de Vaca, mentioned mescal beans as an article of trade amongst Indians of the southern regions of North America in 1539.

Sophora secundiflora was employed in northern Mexico until recently in certain ceremonies, but, as in the southwestern United States its use as an intoxicant has disappeared. According to the Stephen Long expedition of 1820, the Arapaho and Iowa tribes were using the large red beans as medicine and a narcotic. A well developed mescal bean cult existed amongst at least twelve tribes of the United States. There are so many parallels between the peyote cult and the former Red Bean Dance that the origin of the ceremony must have had a southern or Mexican origin.

The active principal of *Sophora secundiflora* is cytisine which is common in the legume family. This alkaloid belongs to the same group as nicotine; it is a strong poison, attacking the phrenic nerve controlling the diaphragm. Death can occur from asphyxiation. It may possibly be because of the great danger in cases of overdosing with the red bean that its ceremonial use has disappeared. It is of interest, however, that the "road-man" or leader of the peyote ceremony today always wears a necklace of the red beans during the peyote ceremony, undoubtedly as a reminder of a once sacred plant.

OLOLIUQUI and BADOH NEGRO (Turbina corymbosa and Ipomoea violacea)

Early Spanish ecclesiastical chroniclers, writing at the time of the conquest of Mexico or slightly later, described the medico-religious use of a brown lentil-like seed which the Aztecs called *ololiuqui*, coming from a plant known as *coatlxoxouhqui* (snake plant), indicating a vine with heart-shaped leaves and a tuberous root, clearly a member of the Morning Glory family. Its present botanical name is *Turbina corymbosa*, but it has been known as *Ipomoea sidaefolia* and *Rivea corymbosa*.

We know much about the pre-conquest use of this hallucinogenic plant. Furthermore, there are clear, although crude, illustrations of ololiuqui in the codices. The plant is depicted in mural paintings showing the water goddess with a stylized vine reminiscent of this sacred vine.

But even more scientifically meaningful is the report and illustration of ololiuqui in a most extraordinary book: *Rerum Medicarum Novae Hispaniae Thesaurus, seu Plantarum Animalium, Mineralium Mexicanorum Historia* (Medical Thesaurus of New Spain, or the Story of the Plants, Animals and Minerals of Mexico). Shortly following the conquest, the King of Spain sent his personal physician, Dr Francisco Hernández, to live with the Aztecs and study their medicines. His notes did not appear as a book, however, until 1651, although some of Hernández' notes were published as early as 1615 by Ximénez, who wrote of ololiuqui, without identifying it, because, he stated "it will not be wrong to refrain from telling where it grows, for it matters little that this plant be here described or that Spaniards be made acquainted with it". Another report, dated 1629, stated that "...when drunk, the seed deprives of his senses him who has taken it, for it is very powerful". Still another source said that "it deprives those who use it of their reason... The natives communicate with the devil... and they are deceived by the various hallucinations which they attribute to the deity which they say resides in the seeds". (Ruiz de Alarcón, 1629)

Besides being an hallucinogen, ololiuqui was used as a magic potion with reputedly analgesic properties. Aztec priests, before making human sacrifices, rubbed the victims with an ointment of the ashes of insects, tobacco and ololiuqui to benumb the flesh and lose all fear. The seeds were venerated and placed in idols of the ancestors.

Mexican botanists identified ololiuqui correctly in the last century, but doubts arose, since intoxicating chemical constituents were unknown in the Morning Glory Family. In 1911, a German specialist, Christian Hartwich, suggested that ololiuqui might be a species of the Solanaceae, and in 1915, an American ethnobotanist, William Safford, assuming that the chroniclers had been misled by the Indians, definitely identified it as a species of the solanaceous genus Datura. There were voices of protest. Not until 1939, however, were actual specimens collected in Oaxaca and used in Mazatec rituals.

It is now employed by more than six tribes, and in many Oaxacan villages the morning glory is planted "as an ever present help in time of trouble". When employed ceremonially as a medicine, the patient must himself collect the seeds, and they are ground and made into a beverage by a virgin, usually a child.

In 1960, the seeds of another Morning Glory, *Ipomoea violacea* (formerly called *I. tricolor*), was identified as a sacred hallucinogen amongst the Zapotecs of Oaxaca. These seeds are black — thus the local name *badoh negro* — and larger than those of *Turbina corymbosa*, with which the badoh negro, usually employed alone, may be mixed. It has been suggested that these black seeds represent the ancient Aztec narcotic seed called *tlitliltzin*, a term in Nahuatl derived from the word for black with a reverential suffix. An early chronicler, Pedro Ponce, had written of "ololiuqui, peyote and tlitliltzin", indicating that three different inebriating agents were represented (Wasson, 1963).

The seeds of both ololiuqui and badoh negro contain a number of ergoline alkaloids, chemically related to the synthetic d-lysergic acid diethylamide (LSD).

Figure 1: Two species of Datura: *D. Inoxia* and *D. ceratocaula*. From *The Badianus Manuscript; an Aztec Herbal of 1552*. Edited and translated by Emily Walcott Emmart, 1940. Reprinted by permission of the Johns Hopkins University Press.

TOLOACHE (Datura inoxia, D. discolor, D. Wrightii, D. kymatocarpa, D. ceratocaula)

Species of Datura have a long history in both hemispheres as a genus valued as medicines and hallucinogens. The main center of ceremonial use of Datura, however, lies in the Southwest of the United States and Mexico where a number of species are employed. All are known as toloache in Mexico.

Today, many Indians believe that they can acquire supernatural helpers through the drug, and much secret knowledge is thought to be gained during the ceremonial intoxication. In tribes of northern Mexico boys who are studying to be medicine men must undergo Datura intoxication once a year. Modern Tarahumare add *D. inoxia* to *tesquino* (a drink prepared from maize) to make it strong and to induce visions. Some Mexican Indians consider toloache an hallucinogen inhabited by a malevolent spirit, unlike peyote.

All species of Datura owe their activity mainly to their active constituent scopolamine.

PUFFBALLS (Lycoperdon marginatum, L. mixtecorum)

More ethnobotanical research must be carried out on the use of two species of puffballs amongst the Mixtec Indians of Oaxaca. The first of these two species is known by the Mixtec name *gi-i-wa* (fungus of first quality), the second as *gi-i-sa-wa* (fungus of second quality).

Lycoperdon mixtecorum causes a state of half sleep in which voices and echoes are heard and, according to the natives, voices respond to questions posed to the spirits.

There is a puffball, still unidentified, reportedly employed by the shamans amongst the Tarahumare.

The active chemical in these fungi with auditory activity has not been identified. There are, however, medical records attesting to this physiological activity amongst patients who ate puffballs in North America and were rushed to hospitals hearing voices.

MUSHROOMS (Teonanacatl) (Conocybe, Panaeolus, Psilocybe and Stropharia species)

Undoubtedly the most important hallucinogenic plants employed from ancient to modern times from central and southern Mexico and Guatemala involve a number of species of several genera of mushrooms.

The Spanish conquerors of Mexico were disturbed by the religious and ceremonial use of "diabolic mushrooms" known by the Nahuatl name *teonanacatl* (divine flesh). As with other native religious rites, the early clerics tried to stamp out such a loathsome religion that venerated mushrooms. It has been said that there was little that Christianity could offer comparable to the supernatural power of teonanacatl. One of the early Spanish writers reported: "They possessed another method of intoxication which sharpened their cruelty; for it they used certain small toadstools . . . they would see a thousand visions and especially snakes . . . and in this wise with that bitter victual by their cruel god were they hassled" (Sahagun, 1829). The King of Spain's personal physician, who studied the medicines of the Aztecs shortly after the conquest, reported in a more scientific vein three intoxicating mushrooms that "cause not death but madness that occasionally is lasting, of which the symptom is uncontrolled laughter. . . There are others which, without inducing laughter, bring before the eyes all kinds of things, such as wars and the likeness of demons".

There is plentiful archaeological evidence indicating a long use of teonanacatl. Frescoes from central Mexico made at least 1700 years ago unmistakenly refer to these intoxicating mushrooms. Clay figurines from Jalisco about 1800 years old have mushroom effigy "horns". In Colima, a clay artifact dated between AD 100 and 300 illustrates figures

Figure 2: Teonanacatl ("flesh of the gods"): the Aztec sacred, hallucinogenic mushrooms. Painting by a Mexican artist of the sixteenth century, showing an Aztec-looking devil encouraging an Indian to eat the fungus. From the Magliabecchiano Codex, Biblioteca Nazionale di Firenze, Italy. Used with permission.

dancing around a mushroom. Stylish mushrooms decorate the pedestal of the statue of Xochipili, the Aztec god of sacred flowers whose body is decorated with a number of intoxicating plants. This statue was discovered on Mt. Popocatepetl and dated about AD 1450.

In Guatemala, however, the deep roots of mushroom use in ceremony goes back to the famous "mushroom stones" dated between 300 and 900 BC. Consisting of an upright stem with either an anthropomorphic or animal figure and crowned with an umbrella-like top, they are now thought to have been associated with the sacred Mesoamerican ball game ritual.

It was nearly five centuries after the arrival of Europeans that serious ethnobotanical research on the sacred mushrooms of Mexico began. Since in this period of time the ceremonial use of these fungi had not been seen, W.E. Safford, an American ethnobotanist suggested, in 1915, that the Aztecs had misled the chroniclers and that, since he assumed that the dry brown top of the peyote cactus superficially resembled dried mushrooms that *teonanacatl* and *peyote* were synonymous and referred to the same plant. Unfortunately, this "identification" received some acceptance, despite voices of protest. It was not until specimens of the mushrooms were collected in the Mazatec country of Oaxaca that the first of many species — *Panaeolus sphinctrinus* — was botanically identified. Later ethnobotanical and anthropological research in a series of expeditions to the same

Figure 3: Animal effigy mushroon stone. Highland Guatemala, Protoclassic Period (100 BC–AD 300). Collection of the Museo Nacional de Guatemala. Photograph by G. Kalivoda, courtesy of R.M. Rose.

region has led to the identification of more than thirty species of Psilocybe, Stropharia, and Conocybe ceremonially employed by at least nine tribes in southern Mexico.

Intensive research on these fungi has led to a tremendous increase in ethnobotanical, mycological, chemical, and anthropological knowledge and to an extensive bibliography. Not only have many new species of mushrooms been described, but chemical investigation of these plants has led to the discovery of the active alkaloids, psilocybine and psilocine. Psilocybine is unique in being an indole alkaloid with phosphorus in its formula — a curious structure known chemically as a phosphoric hydroxyl radical.

This relatively recent discovery and research on what perhaps is a most significant aspect of Mesoamerican ethnobotany has been certainly an important factor in stimulating ethnomycological and even general ethnobotanical investigation in Mesoamerica as well as in other areas in both hemispheres.

RICHARD EVANS SCHULTES

REFERENCES

De la Cruz, Martinus. *The Badianus Manuscript; an Aztec Herbal of 1552*. Ed. and trans. Emily Walcott Emmart. Baltimore: Johns Hopkins University Press, 1940.

Hernández, Francisco. *Nova Plantarum Animalium et Mineralium Mexicanorum Historia*. Rome: Deversini, 1651.

Ruiz de Alarcón, Hernando. *Tratado de las supersticiones y costumbres gentílicas...entre dos Indios...desta Nueva España* (Treatise on the Heathen Superstitions that Today Live Among the Indians Native to this New Spain, 1629). Trans. and ed. J. Richard Andrews and Ross Hassig. Norman: University of Oklahoma Press, 1984.

Sahagun, Bernardino de. *Historia General de las Cosas de Nueva España* (General History of the Things of New Spain). Santa Fe, New Mexico: School of American Research, 1953–1983.

Wasson, R. Gordon. *The Hallucinogenic Mushrooms of Mexico and Psilocybin*. Cambridge, Massachusetts: Botanical Museum, Harvard University, 1963.

See also: Crops

ETHNOBOTANY IN NATIVE NORTH AMERICA Native American peoples developed a sophisticated plant-based medical system in the ten millennia before the European conquest of America. Although there were significant differences between the systems developed by the many native groups, about which many fine works have been written, there were also many broad similarities which will be detailed here. There are approximately 20,000 species of plants in North America. Native Americans used about 2500 of them medicinally (Moerman, 1986, 1991). The utilized portion (the medicinal flora) is a distinctly non-random assortment of the plants available. The richest sources of medicines are the sunflower family (Asteraceae), the rose family (Rosaceae) and the mint family (Menthaceae). By contrast, the grass family (Poaceae) and the rush family (Juncaceae) produce practically no medicinal species. This remarkable volume and extraordinary selectivity demonstrate without any doubt the falseness of demeaning claims which suggest that native American medicines were chosen at random, that they "just used everything and stumbled on something useful once in a while".

Health and Disease

To understand the character and effectiveness of a medical system, one must understand the health status of the people who use it. Native American peoples generally did not suffer from the heart and circulatory diseases of modern times; their diets were rich in fiber and carbohydrates and low in fats. They lived vigorous lives which provided hearty exercise on a daily basis. They experienced little cancer. Cancer is largely a modern disease of civilization. Although the situation is obviously very complex, an apparently necessary condition for cancer is carcinogens which are largely products manufactured by industrial societies (organic chemicals and dyes, nuclear radiation, etc.). Even into the current day there is evidence that the Navajo have lower rates of cancer than surrounding people.

In addition, they suffered little from the classic infectious diseases which ravaged European society over the past two millennia. In large part, this seems to be due to the fact that most such diseases (plague, typhoid, smallpox, cholera, etc.) are zoonoses, diseases of animals which, under conditions of domestication, underwent massive evolutionary change and subsequently affected the human keepers of these animals. Native Americans never domesticated animals to any significant degree (the guinea-pig and llama of Peru were apparently only coming under domestication in the few hundred years before European contact). Once these diseases were introduced into North America, they devastated native populations which had no immunity to them. However, until the sixteenth century, while Europeans underwent successive epidemics which regularly killed a quarter or half of the population, native Americans were spared this devastation.

What medical problems *did* native Americans face? In the southeast and southwest, there is evidence of a decline in health status after the invention of agriculture. Then the diet became simpler, which apparently led to some deficiency diseases. Hunting and gathering peoples avoided that problem, but they, like Europeans, may have experienced some zoonotic infections particularly from beaver, and some trichinosis from bears. These would have been direct zoonoses which individuals contracted directly from the infected animal, not remote zoonoses which, once passed to one human being, were subsequently passed from person to person. Like rabies, a terrible disease for the individual who contracts it, these direct zoonoses are not serious threats to a whole society because they are not contagious in the ordinary sense of the term.

Native Americans probably paid a price for the vigorous life they led. Accidents, sprains, broken bones, cuts, lacerations and the like were common. There was a range of arthritic conditions, some probably the result of injury like those just mentioned, and perhaps some similar to rheumatoid arthritis. There is ample evidence that native peoples engaged in warfare; this would have been a source of serious medical problems. There was a range of occasional problems associated with menstruation, pregnancy, childbirth, and lactation which required attention. Living in smoky houses, it is not surprising that they had a broad range of treatments for irritated eyes; they also treated colds, headaches, cold sores and bruises, the normal insults of daily life everywhere (Vogel, 1970).

Drugs

To address this range of problems, native Americans inevitably resorted to drugs based on various plants. There were some non-plant substances used medicinally. Castorecum from beaver was utilized for various conditions, and some minerals and clays were used as well, but by far the preponderance of medicinal substances came from plants. Every native American group for which we have any information had a botanical pharmacy. While some were quite small (the Inuit had few plant resources on which to rely) most were quite elaborate with hundreds of plant drugs used for a broad range of conditions.

This straightforward proposition raises a number of much more challenging questions. Native American healers, even into the early twentieth century, regularly knew the identity of 200 or 300 medicinal plants which they could readily distinguish from the 3000 to 5000 species which grow in any particular area. Among 100 sophisticated and well-educated modern Americans, it seems unlikely that very many could identify 200 species of plants of any kind unless they were professional botanists. How did non-literate people, without reference to botanical keys or floras compiled by professionals, maintain this extraordinary amount of knowledge (Berlin, 1992)?

Another challenging question is this: Why is it that plants might be of medicinal value in any case? Why do poppies (*Papaver)* or nightshade (*Solanum)* induce unconsciousness? Why does willow (*Salix)* bark tea relieve headaches? Why does wormseed (*Chenopodium*) kill intestinal parasites? Why does milkweed (*Asclepias)* cause vomiting? Although there are many obstinate details yet to understand, the broad outlines of an answer to this question may be sketched. Plants often produce substances of a variety of sorts to protect themselves against browsers (most often insects or worms, but also vertebrates), to defend their space against other plant competitors, or for a wide variety of other particular purposes. These substances share one character: they are somehow biologically active. In using them as medicine, people appropriate these (usually) defensive chemicals to induce reactions which they desire. They are usually toxic; they are therefore customarily used in moderation (for example, by "prescription").

Sometimes the answer to particular questions seems fairly straightforward: the toxic cardiac glycosides in milkweeds (*Asclepias* spp.) discourage browsing by insects, worms, deer or other vertebrates. However, for people who want to induce vomiting, to clear their systems of foul humors or the like, such plants are ideal medicines. Similarly, the substances in wormseed which deter various worms from eating the roots also kill intestinal parasites in human beings. There are non-human analogues to these uses of drugs which are of great interest. In the case of milkweeds, an insect, the monarch butterfly, might be said to use the plant medicinally as well. By virtue of a complex evolutionary adaptation, they manage to tolerate the milkweed toxins, secreting them throughout their bodies, making themselves in turn unpalatable to predatory birds, notably blue jays (Brower et al., 1988). In a case similar to the human use of goosefoot as an anthelmintic, there is evidence to indicate that occasionally chimpanzees, when they appear not to feel well, seek out the leaves of a particular species of *Aspillia* which has anthelmintic properties. Otherwise, the species is ignored (Wrangham and Goodall, 1987).

Some cases are less clear cut. Salicin is a substance found in the leaves of most species of willow; the chemical is named after *Salix*, the genus of willows. This water-soluble substance washes off the leaves during rain, and acts as an herbicide on plants growing around the tree. It is also the chemical precursor of acetylsalic acid, a synthetic drug which we know as aspirin. Various naturally occurring salicylates (also found in certain birches, in the common ornamental *Spiraea*, and in wintergreen) have the same general biological effects as aspirin (named after *Spiraea)* relieving headache and reducing inflammation and fever. The advantage of aspirin is that it is less toxic than the natural chemicals. All of the salicylates seem to work as a result of their inhibiting effect on a class of substances known as the prostaglandins which are involved in all these biological processes. The prostaglandins are also involved in the maintenance of the mucous layer in the stomach and intestine; this is why aspirin (and the other salicylates) can cause stomach upset. What is not clear is why a herbicide should have this effect on mammalian biochemistry. Many similar cases remain to be understood.

Psychological Drugs

There is little evidence of native American use of drugs for recreational purposes. In addition, there is only very little evidence indicating the use of drugs in religious or other ritual. This differentiates native North American peoples from indigenous peoples of Mexico, Central and South America, and many other places as well. The one significant native American use of a consciousness altering substance is the role of peyote in the Native American Church. This, however, is a recent development of the twentieth century with no obvious precursors. There are a number of highly active drugs available for such uses, among them various members of the Solanaceae family (particularly nightshade and jimsonweed), as well as a number of hallucinogenic mushrooms, most notably *Amanita muscaria*. The one drug of this sort which was widely used was tobacco (*Nicotinia* spp.). Uncured native tobacco is a much more powerful drug than the highly processed modern variation. In Mexico, powerful tobacco concoctions were ingested in a number of ways to induce substantial transformations of consciousness. This seems not to have happened north of the Rio Grande. All accounts of its use in North America indicate that it was utilized very sparingly: individuals sat in a circle and shared the smoke from an ounce or so of tobacco. Its function was clearly more symbolic than biological. There may be a few exceptions to this generalization, but they are all either controversial or very poorly documented. When native American peoples wanted to transform consciousness — in the vision quest, for example — they did so with disciplines like starvation, concentration, isolation and sleep deprivation.

The medicinal knowledge of native North American peoples is extraordinary. Just how this knowledge was developed remains a mystery. Native American peoples came from Asia; the flora of Asia is in many ways similar to that of North America. It is quite likely that the first migrants to the New World brought with them detailed knowledge of medical botany, much of which was applicable to this new flora.

Most remarkable, however, may be this: I am unaware of any significant medicinal use of any indigenous American plant species which was not used medicinally by one or another native American group. An interesting example involves recent research on taxol, a substance of great potential medical value found in the common yew, *Taxus brevifolia*, and the Canadian yew, *Taxus canadensis*. Taxol has shown substantial effect in the destruction of tumors in a number of forms of cancer, particularly ovarian cancer, until now a highly refractory form of the disease. Native Americans did not use yew to treat cancer (see above), but they did use it for a variety of other conditions, among them skin problems, wounds, rheumatism, and colds. In general, if one is interested in finding potentially useful botanical chemicals from the North American flora, it would clearly be wise to focus first on that portion of the flora which had been used by native Americans. Their experience and knowledge can yet guide our scientific efforts to enhance human health.

DANIEL E. MOERMAN

REFERENCES

Berlin, Brendt. *Ethnobiological Classification. Principles of Categorization of Plants and Animals in Traditional Societies*. Princeton, New Jersey: Princeton University Press, 1992.

Brower, Lincoln P., et al. "Exaptation as an Alternative to Coevolution in the Cardinolide-Based Chemical Devense of Monarch Butterflies (*Danaus plexippus* L.) against Avian Predators." In *Chemical Mediation of Coevolution*. Ed. Kevin C. Spencer. San Diego: Academic Press, 1988, pp. 447–476.

Csordas, Thomas. "The Sore That Does Not Heal: Cause and Concept in the Navajo Experience of Cancer." *Journal of Anthropological Research* 45(4):457–485.

Duke, James A. and Edward S. Ayensu. *Medicinal Plants of China*. Algonac, Michigan: Reference Publications Inc., 1985.

Moerman, Daniel E. *Medicinal Plants of Native America*. University of Michigan Museum of Anthropology, Technical Reports, Number 19. 2 vols. 1986.

Moerman, Daniel E. "The Medicinal Flora of Native North America: An Analysis." *Journal of Ethnopharmacology* 31:1–42, 1991.

Vogel, Virgil J. *American Indian Medicine*. Norman: University of Oklahoma Press, 1970.

Wrangham, R. W. and Jane Goodall. "Chimpanzee Use of Medicinal Leaves." In *Understanding Chimpanzees*. Ed. Paul G. Heltne and Linda Marquardt. Chicago: University of Chicago Press, 1987.

See also: Medicine of the Native North Americans

ETHNOBOTANY IN THE PACIFIC The island communities of the Pacific, sometimes collectively known as Oceania, are inhabited by indigenous peoples from three major cultural or ethnic regions: Polynesia, Melanesia, and Micronesia. Because they lacked a written language of their own at the time of first contact with Europeans in the eighteenth century, much of the earlier information on the ethnobotany of the area has been constructed from the journals and other records of explorers, missionaries, and anthropologists. The increasing westernization of the islands has led to a greater reliance on imported food, clothing, fibers, and other materials, resulting in a decline in the use and specialized knowledge of native plants and agriculture. Fortunately, this influence is greatest in the urban areas and many of the traditional practices continue to flourish in the more rural and isolated communities. Some practices or items of material culture have lost their original roles, but instead have since acquired social, cultural, or historical significance, and consequently are being perpetuated by the present generations. A considerable body of plant lore still exists among the aged chiefs and inhabitants, and several anthropologists and ethnobotanists have strived to document this information. Because of space shortage, only a brief synopsis is presented and the reader is directed to the bibliography for a more complete treatment.

One popular theory on the origins of the Pacific islanders contends that their navigator ancestors migrated eastward from the Indo-Malay area, carrying with them from island to island the plants they needed for food, social, and cultural purposes. The food plants include all of the starchy perennial plants which are propagated vegetatively, including the taro, *Colocasia esculenta*, and other nourishing *Araceae* of the genera *Alocasia, Amorphophallus*, and *Cyrtosperma*; the various yams, *Dioscorea* spp., whose numbers of cultivated species decrease from west to east; the breadfruit, *Artocarpus altilis*, of Captain Bligh and the mutiny on HMS *Bounty* fame, and for which the greatest number of cultivated varieties are found in eastern Polynesia; the principal bananas, *Musa* spp.; also the sugar cane, *Saccharum officinarum*, which provides a nourishing drink; and the kava plant, *Piper methysticum*, the source of the kava beverage which is used in many different activities in the islands.

Throughout its long history, the Pacific most certainly witnessed modification of its vegetation as its inhabitants adapted to the new environment and improved their techniques, and as waves of new migrations of humans arrived with new methods, ideas, and plants. Two important plants which fall into this category are the sweet potato, *Ipomoea batatas*, and the manioc or cassava, *Manihot esculenta*, being cultigens which were introduced from the American tropics in the last century, and which have become important components of the local diet.

The most widely reported uses of plants in the Pacific are for food and utensils, medicine, ceremony and rituals, general construction, fuelwood, boat and canoe building, cordage and fiber, fish poisons, woodcarving, tools, weapons and traps, mats, clothing, dyes and pigments, and perfumes and oils.

The dominant food economy is based on the cultivation of the small number of starchy perennial plants mentioned above, which are propagated by vegetative means and which provide tubers and starchy fruits. They are usually embellished for taste, flavor, and variety by various condiments and complimented by seafoods and other plants and shrubs. They include the coconut tree, *Cocos nucifera*, which is unquestionably the most utilized of all plants in the Pacific. By one count as many as 125 different uses have been reported. Food uses include eating the soft flesh of immature nuts as an important weaning and adult food; drinking the juice of immature nuts as a nutritious and refreshing local beverage which is often sold and is considered a sacred offering to visitors in Kiribati; using the kernel of the endosperm of the mature nut raw, cooked, or fermented in a variety of ways as a staple food and as a major food for chickens and pigs, as well as an ingredient in locally produced commercial livestock feeds. It is also used for fish and rat bait. The kernel is dried to make copra from which coconut oil is obtained for use in cooking, soap making, as scented oil and in perfumery, and in medicinal potions. The sap from the flower spathe is used to make fermented and unfermented toddy and syrup, which are of considerable nutritional importance in Micronesia and on the atolls. Also, the husks of some cultivars of green nuts are eaten in atoll Polynesia and Micronesia. Other parts of the plant serve as a major source of fuel, for construction and fiber and cordage. Young and mature leaves are used for weaving baskets, food containers and parcels, table mats for feasts, to beat water during fish drives, and for pounding and stabilizing banks of taro beds. The shells are used to make drinking cups for water and kava, small bowls, cooking vessels, funnels, storage utensils, fish hooks, lures, and various types of body ornaments.

Other supplementary food sources include using the pith of the trunk of the sago palm, *Metroxylon* spp., Polynesian arrowroot, *Tacca leontopetaloides*, mature seeds or drupes, aerial root tips, and hearts of meristem of the screwpine, *Pandanus* spp., seeds and flesh of ripe fruit of the tropical or beach almond, *Terminalia catappa*, seeds of the cycad, *Cycas circinalis*, and the young fronds of many ferns and tender shoots of a wide range of species.

The beverage called kava, prepared as an infusion from macerated stems and rhizomes of the kava plant, *Piper methysticum*, is probably the best known and most distinctive item of the material culture of the Pacific. It continues to occupy a central place in everyday life in the islands, although its role has been somewhat diminished by time and outside influences. Besides being the social beverage of chiefs, noblemen, and more recently of commoners, it has also been used to welcome distinguished visitors at formal gatherings, at initiation and completion of work, in reconciling with enemies, in preparing for a journey or an ocean voyage, for installation in office, validation of titles, ratification of agreements, celebration of important births, marriages, and deaths, as a libation to the gods, to cure illnesses and to remove curses, as a prelude to tribal wars — in fact, in almost all phases of life in the islands.

Despite the gradually declining reliance on traditional medicine, over 120 different plant species are reportedly used for medicinal purposes. Herbal preparations and potions are used to treat a variety of ailments, although not always identified with the equivalent Western disease names, including cuts and bruises, bleeding, skin disorders, pain, fever and inflammation, vomiting and nausea, gastrointestinal and kidney disorders, cardiovascular disorders, muscular diseases, parasitic and microbial infections, arthritis and joint pains, bone fractures, diabetes, pulmonary disorders like asthma, obstetric and gynecological conditions including infertility, menstruation and pregnancy, postpartum care and nursing, general incontinence, etc. Among the plant species of most widespread medicinal importance are the beach almond, coconut, pandanus, *Vigna marina, Centella asiatica, Premna taitensis, Ageratum conyzoides, Hernandia nymphaeifolia, Entada phaseoloides, Triumfetta procumbens, Morinda citrifolia, Guettarda speciosa*, and *Terminalia catappa.* Traditional practitioners and midwives continue to play an important role in healthcare delivery and, with the rapidly rising costs of Western medicine, there is renewed interest in herbal cures and practices. Presently, an extensive literature on the past and current medicinal practices is being assembled by anthropologists and ethnobotanists. However, the active chemical components have been identified for only a small number of the herbal potions.

Also of medicinal value are species used as fumigants or insect repellants, soap substitutes, shampoos or hair conditioners, and antitoxins. Of the soap substitutes, *Colubrina asiatica* is used almost universally for this purpose. Among those used as antitoxins, or to treat puncture wounds from marine animals or jellyfish stings are *Avicenna maritima, Cassytha filiformis, Excoecaria agallocha, Sophora tomentosa,* and *Tournefortia argentea.* The seeds of *Calophyllum inophyllum*, the heartwood of *Santalum yasi*, and the leaves of *Vitex* spp. are burned as mosquito repellants.

Tree trunks and timber from about eighty different species have found multiple uses, including for house, canoe, and boat building, tool and weapon making, woodcarving, fishing equipment, food utensils, games, toys, and musical instruments. General construction purposes utilize sawn or hewn timber for house poles, beams, rafters, flooring, walls, pilings, bridges, wharves, etc. Some of the more commonly used species include *Bruguiera gymnorhiza, Calophyllum inophyllum, Casuarina equisetifolia, Cocos nucifera, Guettarda speciosa, Hibiscus tiliaceus, Inocarpus fagifer, Intsia bijuga, Pandanus tectorius, Rhizophora* spp., *Terminalia catappa,* and *Thespesia populnea.* For thatching purposes, various grasses, and the leaves and fronds of coconut palms and pandanus are extensively used.

The timber and woody parts of many of the same species are used to construct warclubs, spears and spearpoints, bows and arrows, fishing poles, fish traps, fish hooks, floats, paddles, digging sticks, needles and awls, adzes and tool handles, tapa cloth beaters, bowls, including kava bowls, ladles, spoons, stirrers, mortars and pestles, coconut huskers, breadfruit splitters, and food containers. For toys and games, the seeds or fruit of *Abrus precatorius, Barringtonia asiatica, Caesalpinia bonduc, Erythrina variegata,* and others, are used for small balls, marbles and lagging pieces, while the wood of *Gardenia taitensis* is carved into marbles and cricket balls in some Polynesian islands. Pandanus leaves are made into kites and whistles while parts of the coconut palm are made into toy windmills, toy boats, rattles, sledges and clappers. The thin epidermis of pandanus leaves serves as a substitute for cigarette paper for handrolled cigarettes. Musical drums (like the *lali* of Fiji) and the slit-gongs (of Vanuatu), beside being musical instruments, are still extensively used for summoning people to village meetings or to church services, the most favored species for this purpose being *Cocos nucifera, Guettarda speciosa, Pemphis acidula, Terminalia catappa,* and *Thespesia populnea.*

Ocean going crafts of diverse structures and sophistication were developed to serve specific functions such as for transportation between the islands, for fishing, as racing outrigger or war canoes, and chiefly for voyaging. The components of such crafts, most of which were obtained from plants, include hulls, keel and prow pieces, outriggers or floats, booms, ribs and spreaders, planking, platforms, shelters, masts and mastheads, paddles, steering oars, and sails, bound together with fiber cordage. The hull and keel pieces were composed of extremely strong and durable timber. *Calophyllum inophyllum* has been favored for the hulls of the larger Polynesian and voyaging canoes. Other species also used for this purpose include *Cordia subcordata, Intsia bijuga,* and *Hernandia nymphaeifolia.* Sails were plaited together from pandanus or coconut leaves and other species were used as adhesives and for caulking.

Cordage and fiber for lashings on crafts, housing, weapons, fishing lines, stringing fish nets, and in various handicrafts are obtained from the husk fiber or coir of the coconut, the bast fiber of *Hibiscus tiliaceus,* and the leaves of the pandanus. These three materials are also used for straining coconut cream, kava, and other liquids, and for stuffing and caulking. The dried fibrous pandanus drupes, coconut coir and *Hibiscus tiliaceus* bast fiber are used as brushes for painting the tapa cloth, other ceremonial clothing, and handicrafts.

Of the various species used in the production of clothing and handicrafts, the most important are the coconut palm, pandanus, breadfruit tree, *Hibiscus tiliaceus,* and the paper mulberry, *Broussonetia papyrifera.* The leaves of the first two and the bast fiber of *Hibiscus tiliaceus* are used throughout the Pacific for making hats, ordinary and ceremonial mats and garments, fans, and a wide array of handicrafts. The bark of *Broussonetia papyrifera* and the breadfruit tree are processed to make the cloth called tapa which is best known in Fiji, Tonga, and Samoa. Dyes, pigments, and preservatives, obtained from a number of species, in particular *Ficus tinctoria, Bruguiera gymnorrhiza, Morinda citrifolia* and *Rhizophora* spp., are used for dying, painting, or strengthening and preserving bark cloth, mats, baskets, and other plaited ware, hats, breechcloths, grass skirts, canoe sails, and for decorating parts of the human body on ceremonial occasions. The dyes and pigments are obtained from the bark and, less commonly, from leaves, roots, flowers, sap, fruit, and seeds.

Although most wood species are used as firewood, some are particularly favored because of their high heat content (*Pemphis acidula, Suriana maritima, Premna serratifolia*), ability to cook slowly and produce slow-burning charcoal (*Bruguiera gymnorhiza, Rhizophora* spp., *Casuarina equisetifolia*, coconut shells), or provide a desired taste to foods. All parts of the coconut palm are used for fuel, and it is by far the main source of fuel in the atoll countries, as well as in some of the larger low-lying islands.

Sophisticated fishing gear used by coastal communities has been complemented by fish poisons or stupificants, with some ten species being used for this purpose. The seeds or roots are crushed, and the juice is scattered in a tidal pool or the lagoon. The suffocated fish rise to the surface and are gathered in baskets woven from coconut leaves. The poisons do not affect the edibility of the fish. The most commonly used species are *Barringtonia asiatica* (seeds), *Tephrosia piscatoria* (roots), and *Derris* spp. (roots). The use of these fish poisons has recently been outlawed in some countries.

YADHU N. SINGH

REFERENCES

Barrau, J. "The Oceanians and their Food Plants." In *Man and His Foods*. Ed. C.E. Smith, Jr. University, Alabama: University of Alabama Press, 1973, pp. 87–105.

Buck, P.H. (Te Rangi Hiroa). *Samoan Material Culture*. Honolulu: Bernice P. Bishop Museum Bulletin 75, 1930.

Cox, P.A. and S.A. Banack, eds. *Islands, Plants, and Polynesians*. Portland, Oregon: Dioscorides Press, 1991.

Parham, J.W. *Plants of the Fiji Islands*. Suva: The Government Printer, 1972.

Singh, Y.N. "Kava: An Overview." *Journal of Ethnopharmacology* 37: 13–45, 1992.

Thaman, R.R. "Batiri kei Baravi: The Ethnobotany of Pacific Island Coastal Plants." *Atoll Research Bulletin* 361: 1–62, 1992.

ETHNOMATHEMATICS Ethnomathematics, as a field of inquiry, began in about 1970, although the term itself did not come into use until about ten years later. Its basic tenet is that mathematical ideas are cultural expressions embedded within cultural contexts. The emergence or elaboration of mathematical ideas follows no necessary or universal path. The ideas that are stressed, their expressions, and their applications vary depending on the culture. Whether an idea arises within a culture or is stimulated by contact with another culture, it becomes enmeshed in the complex of ideas particular to the culture. This perspective is of particular importance because mathematics had long been viewed as culture-free or culture-neutral.

Ethnomathematics calls for a definition of mathematical ideas that is broader in scope than just those associated with modern mathematics. By modern mathematics we mean the category so designated by professional mathematicians worldwide and spread through Western-style schooling. Modern mathematics itself is the confluence of ideas from people in many cultures which became merged through translation, media, and standardization of expression. Mathematical ideas, however, whether or not they fed this stream, are those ideas involving number, logic, spatial configuration, and more important, the organization of these into systems and structures. To fully appreciate the ideas, they must be viewed in their cultural and ideational contexts.

The number of different cultures existing during the past 300 years, using the criterion of mutually exclusive speech communities, is in the range of 5000–6000. Although today there is an overlay of a few dominant cultures, traditional cultures still exist, in some cases blended with, or within, the dominant cultures. Moreover, there are, particularly within large or small industrialized nation-states, subcultures, part cultures, and composite cultures which have developed shared ideas and particular ways of doing things. To learn about the mathematical ideas of cultures that had no writing systems or whose traditions are no longer extant, we must depend on information that can be gleaned from artifacts or from reports of observations left by others. Even where the ideas are current, they are often implicit rather than explicit, and so must be obtained through the interpretation of observations and conversations. Thus, the study of mathematical ideas in their cultural contexts often interacts with or draws upon such fields as archaeology, ethnology, linguistics, culture history, and cognitive studies.

For some, the primary goal of ethnomathematics is to broaden the history of mathematics to one that is global and humanistic. For others, the pedagogical implications and uses are paramount.

Through a discussion of planar graphs and concepts of space/time, the sense and perspective of ethnomathematics will be made more specific.

Planar Graphs

The mathematical idea of tracing figures continuously is found in several diverse cultures. Just as the contexts for the tracings differ from culture to culture, so do the associated geometric or topological ideas.

In modern mathematics, the concept of continuous figure

tracing falls within graph theory. Described geometrically, graph theory is concerned with arrays of points (called vertices) interconnected by lines (called edges). The question said to have inspired the founding of graph theory by the mathematician Euler was: "For a graph, can a continuous path be found that covers every edge once and only once? Also, if such a path exists, can the path end at the point it started?" According to the story, there were seven bridges in Königsberg, Prussia where Euler lived. The townspeople wondered if, on their Sunday walks, they could start from home, cross each bridge once and only once, and end at home. Between Euler in 1736 and Hierholzer some 130 years later, a complete answer was found. The answer depends on the degrees of the vertices — the degree of a vertex is the number of edges emanating from it. First of all, not all graphs can be traced continuously covering every edge once and only once. Such a path exists if the graph has one pair of vertices of odd degree, provided that you start at one of them and end at the other. Also, if all vertices have even degree, such a path can be traced starting anywhere and ending where you began. There can be no such path when a graph has more than one pair of vertices of odd degree.

Much the same question was of concern to the Malekula who live in Oceania in what is now the Republic of Vanuatu. There, however, the issue is getting to the Land of the Dead. According to the Malekula, when a man dies, in order to get to the Land of the Dead, his ghost must pass a spider-like ogre who challenges him to trace a figure in the sand. He must trace the entire figure without lifting his finger or backtracking and, if possible, ending at the point he started. If he cannot meet the challenge, he cannot proceed to the Land of the Dead.

From the ethnographic literature about 100 figures and the exact tracing paths used by the Malekula are known. Analysis of the tracing courses corroborate the Malekula's concern for the problem and their adherence to its stipulations and solution. However, in addition, the tracing courses demonstrate the use of systematic procedures involving general systems that extended beyond individual figures to groups of figures. There are three or four of these extended systems, one of which is briefly described here.

For each figure in the group, there is an initial procedure, namely some ordered sequence of motions (call it A). This is followed by the same procedure modified by formal transformations. Call a transformed procedure A_T. For the group of figures, only a particular set of transformations is used: rotation through 90°, 180°, 270°; horizontal reflection; vertical reflection; each alone or in combination with inversion. Inversion is the reversal of the order of the procedure. One figure and its initial procedure, A, are shown in Figure 1.

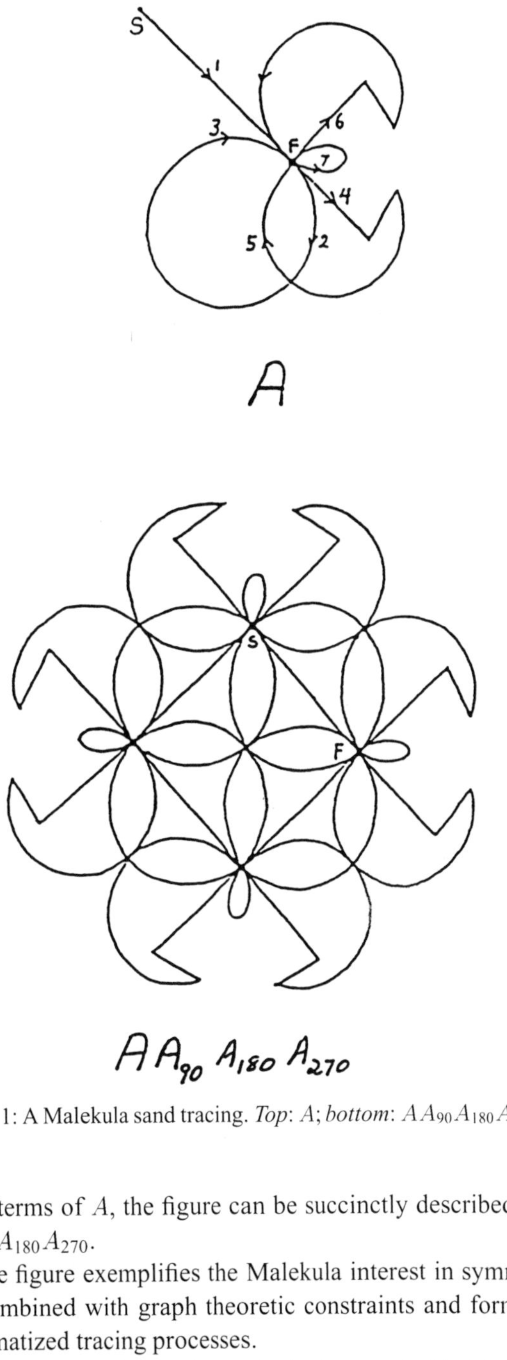

Figure 1: A Malekula sand tracing. *Top*: A; *bottom*: $AA_{90}A_{180}A_{270}$.

In terms of A, the figure can be succinctly described as $AA_{90}A_{180}A_{270}$.

The figure exemplifies the Malekula interest in symmetry combined with graph theoretic constraints and formal, systematized tracing processes.

The study of the mathematical ideas embedded in the Malekula sand tracing tradition leads to an appreciation of this as an intellectual endeavor. However, the global history of mathematics is also enriched. The question of continuous figure tracing is seen as one that arises in different human settings and that has intrigued and challenged quite diverse people. As such, it has been used in the teaching of modern mathematics to create a more inclusive and humanistic view of mathematics.

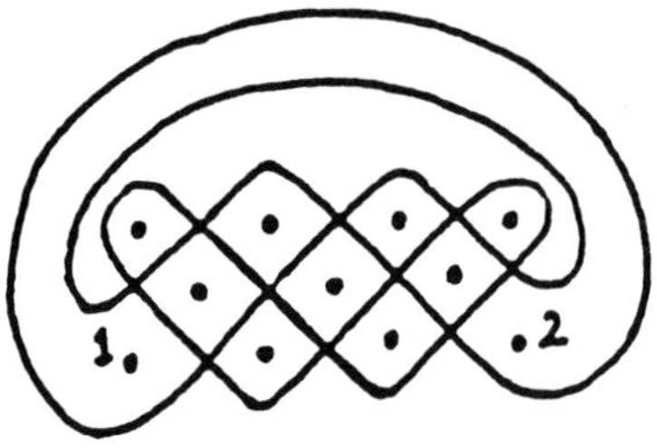

Figure 2: A Tshokwe sand tracing.

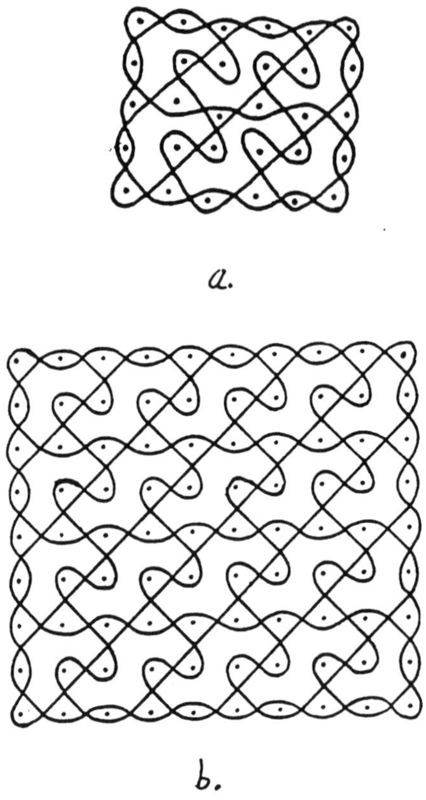

Figure 3: Tshokwe sand tracings. *Top*: (a); *bottom*: (b).

Another, quite different, sand tracing tradition is found among the Tshokwe of what is now the Angola/Zaire region of Africa. In this tradition, a rectangular array of dots is first constructed. Then a skilful storyteller intrigues his audience by drawing a continuous figure around the dots as the story related to the figure emerges. Some of the stories highlight the topological fact that the resultant figure defines regions in which certain dots are isolated from others.

In Figure 2, for example, dots 1 and 2 represent a husband and wife; the other dots are their neighbors. The husband built barriers to keep his wife from the neighbors so that she would attend to her chores instead of visiting. A large collection of these figures and their stories have been made during the past fifty years. The collection is rich and varied. There are, for example, sets of figures sharing general characteristics of shape but differing in a construction parameter. For example, compare Figures 3a and 3b.

Again, exploration of this tradition has increased our appreciation and enriched the global history of mathematics. In addition, however, now having several instances of continuous figure tracing, we see that a particular mathematical idea can lead to different elaborations through combination with a different assortment of mathematical ideas.

Concepts of Space/Time

All cultures define time and the space around them by the physical and mental imposition of order. Because these orderings play such a significant role in how experience is perceived and interpreted, it is extremely difficult to comprehend that others may define them differently. Western concepts of time and space are an intimate part of modern mathematics. We briefly describe another view, that of the Navajo of North America, with particular emphasis on points of contrast with the Western view.

In Western culture, until the late nineteenth century, Euclidean geometry was believed to describe truths about the physical world. Basic to Euclidean geometry are points, lines, surfaces, and solids, and the belief that they can be used to separate space into parts. For example, a line can be separated into two parts by a point, or a surface can be separated into parts by lines. It is also assumed that space has three dimensions, it has no gaps (continuous), it extends in all directions without bound (infinite), it has zero curvature, and neither size nor shape are changed if something is in one place rather than another (uniform). At the end of the seventeenth century — particularly because of the work of Isaac Newton — the three spatial dimensions were augmented by time as a fourth. This time dimension, however, is distinct from the space dimensions. That is, a configuration in space may change with the passage of time but spatial properties are absolute and not affected by time. Mathematicians now understand that Euclidean geometry is a mental construct and that, under other assumptions, there are other geometries. Also, for the physical universe, time and space have become interrelated by Einstein's theories of relativity and by the cosmological theory that space itself has been undergoing expansion since the universe began as a single point. Nevertheless, the Euclidean model (with an augmented time dimension) still underpins the world view incorporated in modern mathematics and science.

For the Navajo, space and time are so inextricably interwoven that one cannot be discussed without the other. They see the universe as dynamic, made up of processes rather than objects and situations. They do not conceptualize things as wholes made up of clearly distinguishable static parts. We, for example, see the body as a physical unit which has distinct parts with specific locations and specific boundaries. For us, arms, legs, teeth, and eyes are part of the body but, say blood pressure, is not. To the Navajo, on the other hand, the body is a dynamic whole, that is, a *system* of interrelated parts. To be a part of the body means to be involved in making the body work. Blood pressure then, without a static specific place, is as much a part of the body as an arm or leg.

For the Navajo, of course, specific locations and spatial boundaries do exist. However, while we view a location as where something *is*, the Navajo view it in terms of process — an object is in process of being in a specific place as the result of the withdrawal of motion. Spatial boundaries, as well, have dynamic components: some interrupt action but the action can continue once the boundary is surmounted; others require that actions be modified.

Another contrast is the description of two overlapping surfaces. We see as significant that the surfaces have a region in common. Since our focus is on the region as a set of spatial points, we can describe the region with no concern for time or motion. The Navajo see the overlap as part of an active, ongoing process. What is of primary significance is whether the same or different elements are in contact and, hence, are defining the region. If, for example, the surfaces that overlap are a snake and a rock, the overlap is different in kind if the snake is sleeping on the rock or if the snake is slithering over the rock.

As contrasted to focusing on when and where something is, the Navajo focus on its motion — whether it is coming or going, getting faster or slower, or moving purposefully or aimlessly. Distance, too, is conceptualized in terms of movement with respect to markers.

In the Navajo world view, space is continuous, has three dimensions, and is finite in that the universe is expanding outward but will eventually shrink back to its starting point. However, above all, interrelatedness and motion are ever present, incorporating and subsuming both space and time.

Differences in world view have ramifications for approaches to solutions of problems as well as to their contents. The analytic approach is fundamental to mathematics and its teaching. Problems are broken up into subparts in the belief that the solution is the sum of the solutions of the subparts. Furthermore, the steps used in mathematical problem solving superimpose processes on what are viewed as static entities and fixed relationships, in the belief that these processes have no effects on the entities or relationships. By contrast, Navajo problem solving is holistic and focuses on the problems' dynamic interrelationships.

The Navajo culture and world view were discussed by ethnologists and linguists in the first half of the twentieth century. However, only much more recently have studies concentrated on their mathematical ideas and on the ramifications of their world view for school learning of mathematics (Moore, 1993; Pinxten, 1983). From these studies we have gained insight into how very deeply mathematical ideas are embedded in culture. The ethnomathematical perspective, which views mathematical ideas as cultural products, provides an enlarged framework within which there can be more than one world view, making clearer contrasting underlying assumptions and enabling diverse contributions.

The ethnomathematics endeavor has drawn together researchers and educators from many parts of the world. For example, Claudia Zaslavsky's seminal book on African mathematical ideas and practices inspired further investigations by many African scholars. Notices of their work appear in an ongoing newsletter published by the African Mathematical Union's Commission on the History of Mathematics in Africa. And, in Mozambique, Paulus Gerdes, a mathematics educator has made extensive use of the Tshokwe figures discussed earlier, using them as a basis for introducing students to a variety of numerical, topological, and algorithmic ideas. The association of these ideas with an indigenous tradition is being used by him (Gerdes, 1993) to underscore the premise that mathematics is not the exclusive product or province of an outside, dominant culture. More anthropologically based studies continue to enlarge the global history of mathematics and mathematical ideas. See, for example, Ascher (1991), Washburn and Crowe (1988) and Frank (1992).

MARCIA ASCHER

REFERENCES

Ascher, Marcia. "Graphs in Culture: a Study in Ethnomathematics." *Historia Mathematica* 15(3): 201–227, 1988.

Ascher, Marcia. *Ethnomathematics*. Belmont, California: Brooks/Cole, 1991.

D'Ambrosio, Ubiratan. "Ethnomathematics: a Research Program in the History of Ideas and in Cognition." *International Study Group on Ethnomathematics Newsletter* 8(2): 5–8, 1988.

Frank, Rosyln M., and Jon D. Patrick. "The Geometry of Pastoral Stone Octagons: the Basque Sarobe." In *Proceedings of Oxford 3: The Third International Conference on Archaeoastronomy*. London: Group D Publishers, 1992.

Gerdes, Paulus. *SONA Geometry: Reflections on a Drawing Tradition Among Peoples in Africa South of the Equator*. Maputo, Mozambique: Mozambique's Higher Pedagogical Institute, 1993.

Keitel, Christine, Peter Damerow, Alan Bishop, and Paulus Gerdes. *Mathematics, Education and Society*. Paris: UNESCO, 1989.

Moore, Charles G. "Research in Native American Mathematics Education." *For the Learning of Mathematics*, 14(2):9–14, 1994.

Pinxten, Rik, Ingrid van Dooren, and Frank Harvey. *The Anthropology of Space*. Philadelphia: University of Pennsylvania Press, 1983.

Washburn, Dorothy K., and Donald W. Crowe. *Symmetries of Culture*. Seattle: University of Washington Press, 1988.

Zaslavsky, Claudia. *Africa Counts: Number and Pattern in African Culture*. Boston: Prindle, Weber and Schmidt, 1973.

See also: *Quipu* – Calendars in Mesoamerica – Mathematics, Aztec – Mathematics, Maya – Mathematics, Native American – Geometry in Africa: Sona Geometry – Mathematics in Africa – Mathematics: West African Mathematical Games

F

AL-FARGHĀNĪ Al-Farghānī, Abu-l-ʿAbbas Aḥmad ibn Muḥammad ibn Kathīr was born in Farghana, Transoxania and died in Egypt, ca. 850. He was a famous astronomer during the time of the ʿAbbasid caliph al-Maʾmūn and a contemporary of al-Khwārizmī, al-Marwarudhī, al-Jawharī, and Yaḥya ibn Abi-Manṣūr. He was well known in the Latin Middle Ages under the name of Alfraganus, thanks principally to his widely read book, *Compilatio astronomica* (also called *Liber 30 differentiarum*, Book of the 30 Chapters), which is a summary of Ptolemy's *Almagest*. The work still survives in Arabic under the following titles: *Jawāmiʿ ʿilm al-nujūm wa'l-ḥarakāt al-samāwiyya, Uṣūl ʿilm al-nujūm, ʿIlal al-aflāk*, and *Kitāb al-fuṣūl al-thalāthīn*. The *Jawāmiʿ* (sometimes translated as *Elements*) provided the medieval reader with a rather comprehensive account of Ptolemy's astronomy through a well-organized, accessible, and non-mathematical presentation. The work was translated into Latin at least twice in the twelfth century: by John of Spain (John of Seville) in 1135, and by Gerard of Cremona before 1175. The *Jawāmiʿ* was also translated into Hebrew during the thirteenth century by Jacob Anatoli. Copies of this translation exist today in Berlin, Munich, Vienna, and Oxford, among other places. In 1590, drawing from Anatoli's translation, Jacob Christmann published the third Latin version of the book in Frankfurt-am-Main. A later Latin translation of the text, along with al-Farghānī's original Arabic, was published in 1669 by Jacob Golius. Widely circulated in the West during the Middle Ages, the *Jawāmiʿ* was frequently referenced by medieval writers, and it is generally accredited today for having contributed considerably to the propagation of knowledge on the Ptolemaic system. In addition to the *Jawāmiʿ*, al-Farghānī wrote on the astrolabe. A number of his manuscripts on the subject survive under the following titles: *Fī ṣanʿ at al-asṭurlāb, al-Kāmil fi'l-asṭurlāb*, and *Kitāb ʿamal al-asṭurlāb.*

AHMED BOUZID

REFERENCES

Carmody, F. J. *Arabic Astronomical and Astrological Sciences in Latin Translation.* Berkeley: University of California Press, 1956.

Duhem, P. *Le système du monde: histoire des doctrines cosmologiques de Platon á Copernic.* Paris: A. Hermann, 1959. vol. ii, pp. 204–214.

Sabra, A. I. "Al-Farghānī." In *Dictionary of Scientific Biography.* Ed. C. C. Gillispie. New York: Charles Scribner's Sons, 1981, vol. IV, pp. 541–45.

Saliba, G. *A History of Arabic Astronomy: Planetary Theories during the Golden Age of Islam.* New York: New York University Press, 1994.

See also: Astrolabe – *Almagest*

AL-FAZĀRĪ Abū Isḥāq Ibrāhīm ibn Ḥabīb ibn Sulaymān ibn Samura ibn Jundab al-Fazārī (d. ca. 777) was a Muslim astronomer and the first Muslim constructor of astrolabes. He was the author of many scientific works whose manuscripts are not extant, but the Arabic historians Abū'l-Faraj Muḥammad ibn Nadīm al-Warrāq al-Baghdādī (d. 993) in his *Kitāb al-fihrist al-ʿulūm* (Bibliography of Sciences) and Jamāl al-Dīn ʿAlī ibn al-Qifṭī (1173–1248) in his *Taʾrīkh al-ḥukamā* (History of Sages) mention one mathematical and five astronomical works by al-Fazārī:

(1) *Kitāb fī tasṭīḥ al-kura* (Book on the Projection of a Sphere onto a Plane),
(2) *al-Zīj ʿalā sinī al-ʿarab* (Astronomical Tables According to Arabic Years),
(3) *Kitāb al-ʿamal bi'l-asṭurlāb al-musaṭṭaḥ* (Book on the Use of the Plane Astrolabe),
(4) *Kitāb al-ʿamal bi'l-asṭurlābāt dhawāt al-ḥalaq* (Book on the Use of Astrolabes with Rings),
(5) *Kitāb al-miqyās li'l-zawāl* (Book on the Gnomon for the Noon), and
(6) *Qaṣīda fī ʿilm al-nujūm* (Poem on the Science of Stars).

His son Muḥammad translated an astronomical work *Brāhma-sphuṭasiddhānta* by the sixth century Indian astronomer and mathematician Brahmagupta from Sanskrit into Arabic. The Arabic name of this work, *Sindhind*, came from the word *siddhānta*, astronomical texts, and the Arabic name for India, *Hind*. The extant fragments of *Sindhind* have been translated into English by David Pingree.

BORIS ROSENFELD

REFERENCES

Matvievskaya, Galina P. and Boris A. Rosenfeld. *Mathematicians and Astronomers of Medieval Islam and Their Works (8–17th c.)*, vol.2. Moscow: Nauka, 1983, p. 29 (in Russian).

Pingree, David. "The Fragments of the Works of al-Fazārī." *Journal of Near Eastern Studies* 27 (2): 103–123, 1970.

Sarton, George. *Introduction to the History of Science*, vol. 1. Baltimore: Williams & Wilkins, 1927, p. 530.

Suter, Heinrich. *Mathematiker ind Astronomen der Araber und ihre Werke*. Leipzig: Teubner, 1900, pp. 3–5.

See also: Astrolabe – Brahmagupta

FIVE PHASES (*WUXING*) Five phases, or *Wuxing*, is one of the basic concepts used by the ancient Chinese along with *Qi*, *Yin*, and *Yang*, to explain natural phenomena. The term *Wuxing*, formerly translated as "Five Elements" is now rendered as "Five Phases". The reason for such a rendering is simply that the term *Wuxing* in Chinese did not necessarily mean "five elements" as we understand them today but is a term which implies something dynamic, ever moving, and transforming in a regular pattern through the operation of *Qi* in nature.

Before we trace the origins and development of the *Wuxing* concept, we should bear in mind that early Chinese thought was perennially involved with the relationship of humanity and nature. Humans were seen to hold an integral but not an assertive place in nature. They had to understand nature and live harmoniously with it. To provide a rationale for it, *Zou Zhuan* (Master Zou's Enlargement of the Spring and Autumn Annals) of the fifth century BC says:

> There are Six *Qi* in nature. When they descend, they give rise to the Five Tastes; display themselves in the Five Colors, and are evidenced by the Five Sounds. When they are in excess, they generate the Six Diseases. The Six *Qi* are *yin* and *yang*, wind and rain, dark and light. They divide to form the Four Seasons, showing the Five Periods in sequence. When they are in excess, they bring about calamities. Excess in *yin* results in cold diseases; excess in *yang*, hot diseases; excess in wind, the diseases of the extremities; excess in rain, the diseases of the stomach; excess in dark, delusions; excess in light, diseases of the heart.

Realizing that nature is vicissitudinary, the early Chinese classified, through their observations and experiences, all the natural phenomena as well as mundane affairs in groups of fives. Classification as such was based on their perception of nature and the society in which they lived. Hence, in several pre-Qin (third century BC) texts we come across such terms as *Wufang* (Five Directions), *Wushi* or *Wuchen* (Five Periods/Seasons), *Wude* (Five Powers), *Wucai* (Five Materials), and other fivefold categorizations. The oldest reference to the term *Wuxing* without the nomenclature for the set of five is found in the *Guo Yu* (Discussion on the Ancient Feudal States) of Western Zhou in the eighth century BC.

The person purported to have systematized and stabilized these ideas of categorization was Zou Yan who flourished between 350 BC and 270 BC. He referred to the set of Water, Fire, Metal, Wood, and Earth as "Five Powers". He cited the rise and fall of past dynasties to put forward a cosmological theory of monarchy. As a reflection of the order on Nature, the political order was subject to the Five Powers which conquer each other. This sequence later came to be known as the Mutual Conquest Order (*Xiang Ke* or *Xiang Sheng*), wherein Water conquers Fire by extinguishing it, Fire conquers Metal by melting it, Metal conquers Wood by cutting and carving it, Wood conquers Earth either by digging it up or growing out of it, and Earth conquers Water by damming it up and constraining it. Because the Five Powers dominate alternately, it is seen that the process of domination is itself "controlled" by the Power which conquers the conqueror. For example, Wood conquers Earth, but Metal controls the process. There is yet another sequence which is found in the *Chun Qiu Fan Lu* (String of Pearls on the Spring and Autumn Annals) by Dong Zhongshu of the second century BC. The sequence gives rise to the Mutual Production Order (*Xiang Sheng*), in which the Five Powers generate each other. The sequence begins with Wood which produces Fire (being consumed as fuel). Fire produces Earth (by forming ash); Earth produces Metal (by fostering the formation of metallic ores); Metal produces Water (by liquefying itself when heated), and Water produces Wood (by nourishing the plants). If both the orders of Mutual Conquest and Mutual Production are taken together, it is seen that in the process of destruction of one Power by another, the process of change is somehow "masked" by some other process which produces more of the substrate, or produces it faster than it can be destroyed by the primary process.

Another *locus classicus* of the pre-Qin period which describes the Five Powers as Five Processes is found in the *Hong Fan* (Great Plan) chapter of the *Shu Jing* (Historical Classic). The text gives the names of *Wuxing* in numerical order, followed immediately by their respective descriptions. Thus, we have

- Water: wetting, descending.
- Fire: flaming, ascending.
- Wood: (allowing to be) carved, straightened.
- Metal: (allowing to be) molded.
- Earth: sowing and harvesting.

The description suggests five sorts of fundamental processes characterized by their respective quality in nature.

In their effort to understand the physical changes in Nature, the early Chinese tried to establish a viable system in terms of abstract ideas. In the *Wuxing* chapter of *Guan Zi* of the late fourth century BC, the set of Wood, Fire, Earth, Metal, and Water is fitted to 72-day divisions of the 360-day year in order to show the change of seasons in an evolving manner. The set of five is associated with five types of *Qi* that bring about changes in nature. Then, in another *Wuxing* chapter in *Bai Hu Tong* (Comprehensive Discussion in the White Tiger Hall) of the first century, the word *Xing* is specially used to mean "the activity of *Qi* by the natural order". In the *Chun Qiu Fan Lu*, too, *Xing* means "activity". It is clear that *Wuxing* is no longer seen as a set of chemically or physically distinct substances, nor is it a force itself which is capable of performing actions. The *Wuxing*, then, are phases of change brought about by *Qi* in Nature.

ANG TIAN SE

REFERENCES

Chen, Jiujin. "Yin-Yang, Wuxing, Bagua Qiyuan Xin Shuo." (A New Theory of Yin Yang, the Five Phases, and the Eight Trigrams) *Studies in the History of Natural Sciences*. 5(2): 97–112, 1986.

Chen, Mengjia. "Wuxing Zhi Qiyuan." (Origin of the Five Phases) *Yanjing Xuebao* 24: 35–53, 1938.

Graham, A.C. *Yin-Yang and the Nature of Correlative Thinking*. Singapore: The Institute of East Asian Philosophies, Occasional Paper and Monograph Series No.6, 1986.

Hu, Weijia. *Zhongguo Gudai Kesue Sixiangzhong Yin-Yang, Wuxing, Qide Qiyuan Yanjiu*. (A Study of the Origins of Yin-Yang, Five Phases, and Qi in the Scientific Thought in Ancient China) Unpublished M.A. Thesis. Beijing: Chinese Mining University, 1989.

Hu, Weijia. "Yin-Yang, Wuxing, Qi Guanniande Xingcheng Ji Qi Yiyi." (The Origins of Yin-Yang, Five Phases and Qi, and Their Significance: A Study of the Scientific Thought System of the Pre-Qin Period) *Studies in the History of Natural Sciences* 12(1): 16–28, 1993.

Needham, Joseph. *Science and Civilisation in China*. Vol.2. Cambridge: Cambridge University Press, 1969.

Ren, Yingqiu. *Yin-Yang Wuxing* (Yin Yang and the Five Phases). Shanghai: Kexue Jishu, 1960.

Yosida, Mitzukuni. "The Chinese Concept of Nature." In *Chinese Science*. Ed. Nakayama Shigeru and Nathan Sivin. Cambridge, Massachusetts: MIT Press, 1973.

See also: *Yinyang*

FOOD TECHNOLOGY IN AFRICA Food processing embodies all treatments applied to foodstuffs after harvest, capture or slaughter to prepare them for consumption or preservation. Treatments may be single or suitably combined physical, mechanical, chemical, and biological procedures which modify foodstuffs aesthetically, nutritionally, texturally, and organoleptically.

In traditional Africa, food processing is a daily domestic or village-level activity. Women constitute the main workforce and appropriate skills are acquired informally during the process of acculturation. Male professional blacksmiths, stone-cutters and wood carvers, basket makers, and others produce the grindstones, earthenware pots, pans and plates, wooden pestles, mortars and spoons, baskets, knives, calabashes, and gourds used in the processing.

In addition to familiar food plants and animal resources some uncommon products have food uses in certain areas of Africa, depending on their availability and on the prevailing culture and religion. For example, toxic castor oil seeds (*Ricinus communis*) and leaves of the toxic legume *Cassia obtusifolia* are processed into food condiments in Nigeria and Sudan respectively. *Spirulina*, a "vegetable" microbe, is consumed in Chad, and caterpillars called mopane worms constitute human food in Zimbabwe. Other items include dog meat, cow and goat blood in parts of Nigeria, Kenya, and Zimbabwe, and gall bladder juice, hides, and skins in Sudan and Nigeria respectively.

First stage processing is a preliminary treatment used to release valuable seeds and fruits from heavy pods, heads, or husks. Melon pods (*Citrullus vulgaris*), breadfruit heads (*Treculia africana*), and oil palm husks (*Elaeis guineense*) are examples, and they are generally allowed to ret or soften over a period of time. The process may occur in the farm or bush to circumvent transportation and pollution problems.

Retting occurs naturally, but it is aided traditionally by hitting the structure on a hard surface or with a machete in order to rupture it. After about a week, the rotten seedcontaining pulp is scooped into a basket. Loose material leaches out, leaving the seeds which are washed and sun-dried.

Sundrying precedes other steps for treating cereal heads to release grains and for shelling legumes and other fruits like the castor seed. Dried millet, sorghum, and rice heads or husks are threshed on hard ground or pounded lightly in a mortar to detach the grains; rice is invariably pounded. Mixtures of grains or seeds are winnowed with flat winnowing baskets. Then manual screening separates the food material from the chaff.

On a large scale sun-dried pods of legumes such as the cowpea (*Vigna* sp.) are packed in jute bags, and then shelled by being beaten by sticks or trampled upon. One method of separating mixtures of seeds and chaffs is to spread them on a mat across a distance of 4–5 meters. The lighter chaffs fall off during the throwing, and denser seeds are retained on the mat.

Overall, first stage processing eliminates bulky wastes from harvests and prepares foodstuffs for markets and for further processing. The products mostly remain inedible or unpalatable.

In second stage processing, foodstuffs undergo textural and organoleptic modification to yield palatable products or flours and oils. Two categories of second-stage processing will be treated: those not involving fermentation and those which are fermented.

Cooking with water, frying in oil, and roasting are commonly used methods. Individual cooked items or suitable mixtures may be pounded into sticky doughs and eaten with relishes; pounded yam is popular in West Africa, and small amounts are also used for thickening soups or relishes. Ordinary cooking suffices for preparing some cocoyam (*Colocasia esculenta*) varieties but at least one toxic variety is detoxified by prolonged boiling.

Leguminous seeds are generally cooked and consumed with their skins (testas). However, groundnuts (*Arachis hypogaea*) and the bambara nuts (*Voandzeia subterranea*) are often cooked in the pod and later extracted manually.

Smoking is rarely used for preserving plant-based unfermented foods. In contrast, sundrying is widely practiced. Raw or lightly cooked thin slices of sweet potatoes may be so preserved in Zimbabwe, but thorough cooking and washing of thinly sliced cassava is adopted before sundrying, to remove the cyanide and guarantee a safe product. In the Chad Republic, mats of *Spirulina*, a microscopic alga which grows in ponds around Lake Chad, are sun-dried for use as a vegetable in local sauces.

Roasting is traditionally carried out with a burning or glowing fire and is commonly applied to tubers and plantains. Among the legumes, only groundnut seeds are roasted. Cowpea seeds are dehulled, without roasting, by first soaking them in water overnight. Placing the container on the surface of water in a deep bucket or pan makes the testas float, and they are skimmed off. Wet dehulled cowpea seeds are pounded further and ground into a slurry which is seasoned and fried in hot palm oil to form palatable balls called *akara* in Nigeria.

Many kinds of dried foodstuffs are processed into flours. In Southern and Eastern Africa, flours from grains are called meals, as in maize meal, and production begins with pounding and winnowing to remove coarse chaffs. The grains or the mortar are then moistened to prevent scattering during a second pounding. The material is then sun-dried for a few hours and finally ground on a grindstone. The finished meal is sun-dried on a mat before storage or use.

Yam and sweet potato tubers are peeled, chipped, and blanched by soaking the chips overnight in hot water before sundrying, pounding, and grinding. In processing cassava flour, chips of peeled and sun-dried tubers are ground.

Many other types of foodstuffs are processed into flour. Shelled groundnuts are roasted and ground for meat seasoning in Nigeria and as a source of butter in Zimbabwe. Various dried spices are also processed into powders.

Cooking oil is extracted from raw or cooked palm fruits and from roasted groundnut seeds. By pounding oil palm fruits in a deep mortar the oil-laden fibrous pericarps are abraded from the kernels which are sorted manually. Batches of the fibrous mass are subsequently warmed in a wide-mouthed earthenware pot to melt the oil. Then suitable portions are pressed between the palms of the hands and the oil flows into a bowl or other container.

Food processing by fermentation is widespread in Africa and various cultures have developed methods of fermenting selected food resources for different purposes. Virtually every kind of foodstuff can be fermented, and suitable processes produce heavy main course meals, condiments and flavor enhancers, non-alcoholic and alcoholic beverages, meat substitutes, and foods for the elderly, infants, and the convalescent.

Cassava tuber is processed mainly by fermentation to detoxify it by eliminating the cyanogenic glucosides linamarin and lotaustralin and to impart desirable flavors. Important fermented products from grated material include *gari* (West Africa), *attieke* and *plakali* (Ivory Coast), and *Oyoko* (Zaire).

Major products of fermenting cassava are *chikwange* (Central Africa), *ntuka* (Zaire) and *fufu* (Zaire and Nigeria). The fermentations are retting processes because they soften the tubers. For *chikwange* and one form of Nigerian *fufu* the fermented pulp is mashed and sieved through a basket into a cloth bag; the solids are concentrated by pressing the water out. Meals are prepared by steaming portions of the processed material, and then kneading, boiling, and pounding into a gel. In Zaire the fermented pulp is steamed (*ntuka*), or sun-dried, milled, and cooked (*fufu*).

African cereal-based fermentations result in diverse foods and beverages which satisfy different nutritional needs. Maize (*Zea mays*), sorghum (*sorghum bicolor*) and various millets: pearl millet (*Pennisetum glaucum*), and finger millet (*Eleusine coracana*) are the fermentation substrates and products include acidic porridges, non-alcoholic beverages, and opaque beers.

The first step in preparation of acidic non-alcoholic beverages and porridges is soaking the cereals. Subsequent sieving, sedimentation of solids, and decantation of liquor leaves a product which can be molded into various sizes for sale. They are particularly useful for infant weaning and convalescent feeding.

Examples of acidic non-alcoholic drinks are Southern African *mahewu*, Nigerian *kununzaki* and Sudanese *hulu mur*. In *mahewu* preparation, a suspension of maize meal in boiled water (1:9) is cooked and cooled. Some wheat flour is added and the drink is ready after a day's fermentation. Nigerian *kununzaki* uses flour produced by grinding grains soaked in water for one or two days. Sudanese *hulu mur* is prepared by flavoring cold water with flat sheets of fermented and baked sorghum dough prepared with equal amounts of flours from malted and unmalted sorghum.

African acidic alcoholic beverages are produced from malted grains. Malting involves germination of the appropriate grain followed by malt grinding and mashing in warm water to saccharify the malt. Unmalted grain or another carbohydrate source may be added to the mash before fermentation, which may proceed naturally or by induction with a starter from a previous brew.

Clear beers are relatively rare in Africa. Nigerian *otika*, Cameroonian *amgba* and Sudanese *assaliya* are exceptions. *Talla* is an Ethiopian home-brewed beer which differs from the others in some respects. First, it is brewed with barley or wheat, hops, or spices. Secondly, it has a smokey flavor due to the addition of bread darkened by baking and use of a fermentation vessel which has been smoked by inversion over smoldering wood.

Sugary plant juices and saps and honey are the substrates for wine and spirit production. West African palm wines are produced by fermenting saps from oil palm trees (*Elaeis guineense*) and from Raphia palm trees (*Raphia* spp.). In Zimbabwe sap from the ilala palm tree (*Hyphaene benguellensis*) is fermented into wine, while in Northwest Africa a wine called *lagmi* is produced from the sap of the date palm. Over-ripe bananas and plantains are also used for wine production in Southwestern Nigeria. For West African palm wines, the best known, fresh juice tapped into gourds contains up to 10% sugar and has a neutral pH. The fermentation occurs naturally and in twenty-four hours lowers the sugar content considerably, acidity develops, and alcohol builds up so that the wine becomes intoxicating.

Spirits are processed in Africa by distillation of fermented sugary substrates. Zimbabwean *uchema*, Nigerian *ogogoro* and Ghanaian *ekpeteshi* are distilled from Zimbabwean illala palm wine and from West African palm wines. Kenyan *chang'aa* and Nigerian *kai-kai* are distilled from fermented cane sugar juice. Generally the alcoholic content of African spirits ranges from 20–30%.

Food fermentations based on oilseed and leguminous seeds are sometimes called vegetable protein fermentations. West African *iru* or *dawadawa* (from locust bean), *ugba* (from oil bean) various *ogiris* (from melon, fluted pumpkin: *Telfaria occidentals* castor bean and sesame) and Sudanese *sigda* (from sesame oilseed press cake) are fermented vegetable protein foods, and they are flavor-enhancers.

Sudanese *kawal* produced by fermentations of leaves of a toxic legume *Cassia obtusifolia* and Nigerian *ule* are examples of African food products of leaf fermentation. *Kawal* and *ule* are flavoring materials for stews and other relishes.

Meat fermentation is uncommon in West Africa. However in Sudan virtually every part of a carcass including bones, hide, and gall bladder juice is fermented for food purposes. Fermented gall bladder juice is called *itaga*. It is prepared by adding some sorghum flour or grains to the juice which is then hung up to dry slowly. *Itaga* is pounded and used as a kind of spice for fatty meat dishes.

Fish is processed like meat, and fermentation is also uncommon. It appears to be a last resort for preserving a day's unsold catch. *Bonome* and *guedj* are fermented dried products of Ghana and Senegal respectively. Fresh fish is allowed to undergo putrefactive fermentation in the open air for twenty hours. It is then eviscerated, soaked in salty sea water, and dried in the sun for two to four days with or without filletting. Sudan has a variety of fermented fish products including pastes and sauces prepared from fresh water Nile fish.

Many insects, especially gregarious and seasonal types, were traditionally acceptable as food in Africa. Examples are locusts, swarming winged reproductive termites, the large cricket (*Brachytrypes*), and caterpillars of the African silkworm (*Anaphe* spp.). Traditional processing of insects included cleaning and salting, then roasting or tying in parcels and boiling. Although the practice is now restricted to isolated areas of the continent it is noteworthy that in Zimbabwe the commercial product *madora,* also called mopane worms, consists of edible caterpillars. *Madora* is available in markets or in supermarkets. The caterpillars are generally squeezed after capture to press out the digestive fluid, then salted and preserved by sundrying.

The art of fermenting milk is widespread in Africa. Cattle and camels are important dairy animals, and milk fermentation products vary widely. For example, *jben* and *ayib* are Moroccan and Ethiopian cheeses, *nono* is a Nigerian yoghurt-like product, and Zimbabwe and Sudan also have fermented milk products. African milk fermentations occur naturally with attendant souring or coagulation, the milk having usually been collected into clay or earthenware pots or animal skin bags, gourds, or calabashes. They are often churned to produce butter. Moroccan *jben* is prepared by placing the coagulated milk (*raib*) in a cloth at room temperature and draining the whey. In Kenya various additives such as wood ash, animal blood, urine, and sometimes leafy vegetables may be added to preserve a fermented sour milk called *maziwa lala*.

Scientific Bases of Some Traditional Practices

African traditional food processing is a cultural activity which evolved independently from modern science and technology. Nevertheless valid scientific explanations can be offered for many food practices.

In many traditional fermentations, starters consisting of

small amounts of previously fermented materials are introduced into new processes. Scientifically speaking, the starter contains a dense population of relevant microbes which accelerate the new process. Also, most fermentations are carried out in warm conditions, or fermentation time is prolonged in cold weather; these practices are consistent with scientific knowledge of beneficial and adverse effects of warmth and cold on the growth and activities of microbes.

Traditional food preservation with plant extracts is exemplified in the processing of Nigerian palm wine and *nono* with *nche* and *kuka*. Both preservatives inhibit spoilage microorganisms by means of chemicals: *nche* contains phenolics and alkaloids, and *kuka* contains tartarates. Again, the use of small amounts of urine, among other things, for improving the shelf life of Kenyan sour milk has been explained in the context of the lactoperoxidase system which inhibits Gram negative, catalase positive bacteria in milk. Traces of hydrogen peroxide and thiocyanate are known to be present in urine and these may exhibit antimicrobial effects together with lactoperoxidase which normally occurs in bovine milk.

Traditional food softening or flavoring with extracts from ashed monocot heads or husks is exemplified by the use of Sudanese *kambo* from sorghum, Nigerian *ngu* from palm oil husk and Zimbabwean products from yellow nut grass (*Cyperus esculentus*) and maize cobs. Ashing concentrates inorganic elements, and high levels of potash in the leachates influence flavor and soften foods by hydrolysis.

African traditional food processing involves diverse food resources, equipment, practices, and products. Productivity is low, and final products often lack good hygienic quality, uniform composition, and predictable shelflives.

Research efforts to modernize food processing systems have been intensified in African institutions. Production of Southern African opaque beers and *mahewu*, Nigerian *gari*, *dawadawa* and *poundo* yam have become industrialized, and microbiologically safe processes have been developed for production of Moroccan and Ethiopian dairy products. These developments are encouraging and continuing research efforts will probably assist future generations in preserving Africa's rich traditional food heritage.

RICHARD NNAMDI OKAGBUE

REFERENCES

Applications of Biotechnology to Traditional Fermented Foods. Washington D.C: National Academy Press, 1992.

Dirar, Hamid A. "Traditional Fermentation Technologies and Food Policy in Africa." *Appropriate Technology* 19(3): 21–23, 1992.

Emeagwali, Gloria Thomas, and Rashid O. Lasisi. "Changes in Cassava Processing Technology in Nigeria." In *The Historical Development of Science and Technology in Nigeria*. Ed. Gloria Thomas-Emeagwali. New York: Edwin Mellen Press, 1992, pp. 47–61.

Ene, J.C. *Insects and Man in West Africa*. Ibadan: Ibadan University Press, 1963.

Mahungu, N.M., Y. Yamaguchi, A.M. Almazan, and S.K. Hahn. "Reduction of Cyanide During Processing of Cassava into Some Traditional African Foods." *Journal of Food and Agriculture* 1(1): 11–15, 1987.

Odunfa, S.A. "African Fermented Foods." In *Microbiology of Fermented Foods*. Ed. B.J. Wood. London: Elsevier Publications, 1985, pp. 155–191.

Okagbue, R.N. "The Scientific Basis of Food Processing in Nigerian Communities." In *African Systems of Science, Technology and Art*. Ed. Gloria Thomas-Emeagwali. London: Karnak House, 1993.

Olatunji, F.O. "Appropriate Technologies in the Processing of Foods in Nigeria." *Nigerian Food Journal* 2(1): 61–64, 1984.

Tregold, M.H. *Food Plants of Zimbabwe*. Gweru: Mambo Press, 1990.

FOOD TECHNOLOGY IN CHINA Agriculture started in China in about 5000 BC. The earliest crops cultivated were millet in the north and rice in both the north and south. Barley, soybean and wheat came later. In the classical period (1100–200 BC), the principal grains of the realm were millet (both *panicum* and *setaria*), rice, wheat, barley, and soybean. Much ingenuity was expended by the Chinese to develop methods for processing these crops into attractive and nutritious articles of food and drink.

The kernels of both millet and rice are relatively soft. They were steamed to produce tender, palatable granules called *fan*. Pottery steamers of great antiquity (4000–5000 BC) have been found in the neolithic sites in Banpo near Xian in the north, and at Hemudu near Hangzhou in the Yangzi delta. It was probably the use of steaming that fortuitously led to the discovery of a distinctive technology for the conversion of cereal grains into alcoholic drinks.

The Discovery of Qu (Chinese Ferment)

The fragrant, fluffy *fan* granules obtained by steaming millet or rice grains were not only appetizing to humans, but also highly attractive as a medium for the growth of airborne fungi, such as those of the genus *Aspergillus* or *Rhizopus*. As these fungi proliferate they produce enzymes which hydrolyze the starch in the granules into fermentable sugars. Yeasts then multiply on the granules and convert the sugars to alcohol. This mixed culture (or ferment) could be used to inoculate larger amounts of *fan* to produce an alcoholic drink, or dried and stored as a stable solid product known as *qu* for use in future fermentations. As we shall see, in addition to its role in alcoholic fermentations, *qu* ferment was the foundation upon which a rich array of fermented foods were later developed.

Alcoholic Drinks

The use of *qu* to make an alcoholic drink or wine called *jiu* was probably practiced in China as early as 2000 BC. The technology for *qu* and *jiu* reached a mature level by the Han Dynasty (206 BC–AD 220). Extensive recipes for preparing varieties of *qu* and *jiu* (as well as vinegar from it) are given in the *Qi Min Yao Shu* (Important Arts for the Peoples' Welfare), of AD 540. In making wine, *fan* (steamed millet or rice) was mixed with *qu* and the semi-solid medium allowed to ferment until the substrate was depleted. The semi-solid mash was then pressed and clarified to give the finished wine. One interesting innovation was the cumulative addition of fresh *fan* to the medium before the substrate was spent. This technique made it possible for skilled brewers in the sixth century to produce wines with alcoholic contents comparable to those of grape wines in the West.

During the Song Dynasty (AD 960–1280) the art of heating the wine before storage, later called pasteurization, was introduced, and a red *qu* based on strains of *Monascus* sp. was developed, leading to the production of the so called yellow or red wine. Although bronze stills were known as early as the Han, distilled spirit did not become commercial until after the Song dynasty. The fermented mash was placed directly on a grid in a boiler and the alcohol carried over by steam distillation. The result was a spirit with a strikingly rich flavor and an alcohol content of greater than 50%.

Malt Sugar

Sprouted barley, *nie* (malt) was of equal antiquity as *qu* in China and it has been suggested that some of the earliest wines might have been made with malt as the saccharifying agent. But soon *qu* displaced *nie*, and the *nie*-based fermentation died out. *Nie* itself, however, lived on as the agent for converting steamed millet or rice into malt sugar, the most important sweetener in ancient China. Although its importance declined somewhat when the refining of sugar from sugar canes was introduced from India in the Tang Dynasty (AD 618–905) malt sugar continued to be a popular treat in the Chinese diet until the present day.

Products from Wheat Flour

Since the kernels of wheat are hard, they cannot readily be made palatable by cooking. Thus, wheat was less popular than millet or rice in ancient China. However, the situation changed during the Han when mills for grinding wheat into flour or *mian* became widely available. The dough made from *mian* could be processed into a dazzling variety of delicious pastas, cakes and breads (including leavened breads). Most of these products were cooked by steaming or boiling. As a result, wheat flour soon became the staple food in the North while rice remained the staple food in the South.

Probably the best known Chinese food processed from wheat flour is the filamentous noodle, *mian tiao*, which was first made in the second century AD. There is, however, no evidence that Marco Polo ever brought filamentous noodles from China to Italy. Filamentous pasta was already known in Italy before Marco Polo was born. Thus, whether there is a connection between the filamentous noodle of China and the spaghetti and macaroni of Italy remains an intriguing problem in the history of food technology.

Another important product processed from flour is wheat gluten *mian jin*, an valuable source of protein for vegetarians. It is obtained by the continuous washing of wheat dough with water. The phenomenon was first described in the *Qi Min Yao Shu* (AD 540), but gluten did not become a popular component of Chinese foods until the Song.

Soybean Processing

Although soybean is an excellent source of protein, it is hard to cook and difficult to digest. Three methods have been developed to process soybeans into wholesome, attractive, and nutritious products. The first is simply to allow the beans to sprout in the dark. Soybean sprouts were known before the Han as a medicament; they did not become popular as food until the Song. Since then sprouts from both soybean and mung bean have remained standard fare in the Chinese diet.

The second method is to process the beans into *tou fu* (bean curd). The steps involved are: (1) soaking the beans in water overnight, (2) grinding the beans to a puree, (3) filtering the puree through cloth, (4) heating the milk to near boiling, (5) adding a coagulant, such as bittern, to the cooled milk, and (6) pressing the soft curd in a wooden frame. The procedure is depicted in a mural in a late Han tomb (second century AD), although the product did not become commercialized until the Sung. Since then *tou fu* has been the most important vegetarian source of protein in the Chinese food system.

The third method is to ferment the cooked beans with *qu* in a high salt medium to give a triad of relishes and condiments; they are *shi*, fermented bean relish, *jiang*, fermented bean paste, and *jiang you*, soy sauce, probably the best known processed soybean product in the West. When cooked soybeans were incubated with *qu* and salt, the product was *shi*. When they were incubated with *qu*, wheat flour and salt, the product was *jiang*. Both *shi* and *jiang* were already made on a large scale in early Han. Recipes for their preparation can be found in the *Qi Min Yao Shu*. Soy sauce was simply the liquid drained from *shi* or *jiang*. It was known during the late Han, but the name *chiang you* did not come into common usage until the Song.

Much of this distinctive food technology was transmitted to Japan during the Tang and later dynasties. Thus, *qu* and *jiu* are known as *koji* and *saké* in Japan; *shi*, *jiang* and *jiang you* evolve into *natto*, *miso* and *shoyu*; while *tou fu* remains as *tofu*. All these products are now integral parts of the food system in Japan and other countries of East Asia.

However, the best known dietary product from China is actually a drink. Tea was first prepared by steaming, crushing and drying the leaves of *Camellia sinensis* during the Han. It became widely popular during the Tang, and was disseminated to the West in the seventeenth century. Tea and coffee are now the two major beverages consumed throughout the world.

H.T. HUANG

REFERENCES

Anderson, E. N. *The Food of China*. New Haven, Connecticut: Yale University Press, 1988.

Bray, Francesca. *Science and Civilisation in China*, Vol. VI, Part 2, *Agriculture*. Ed. Joseph Needham. Cambridge: Cambridge University Press, 1984.

Huang, H. T. "Han Gastronomy: Chinese Cuisine *in statu nascendi*." *Interdisciplinary Science Reviews* 15(2):139–152, 1990.

Shih Sheng-Han. *Ch'i Min Yao Shu: An Agricultural Encyclopaedia of the 6th Century*. Beijing: Science Press, 1962.

FOOD TECHNOLOGY IN LATIN AMERICA About 5000 years BC, native populations in the Caribbean and Latin America developed food technologies which enabled them to exploit their natural biota and become settled. By the time of the Columbian encounter, Indians had evolved unique agricultural techniques suitable for efficient food production in various terrains and climates. Their indigenous staples were cassava (*Manihot esculenta)*, maize *(Zea mays)*, potatoes *(Solanum tuberosum)*, and yams (*Ipomea batatas*), beans, and squash supplemented by fruits and vegetables, fish, turkey, deer, dogmeat, and guinea-pigs, as well as assorted beverages and flavorants.

Food technology began with soil preparation and with the selection of seeds, roots, tubers, or cuttings. The ground was cleared through slash-and-burn and girdling, which is removing a ring of tree bark. Farmers then worked the earth into numerous knee-high mounds several feet in diameter (*montones*), collectively called *conucos*, among the stumps in irregularly shaped fields, or *milpas*, fertilized by ashes and decaying vegetation. These mounds produced less water runoff, and hence less erosion, than traditional European fields plowed in rows.

In the Caribbean, Middle America, and South America, *conucos* were planted with bitter cassava, with a vegetable triad such as maize-beans-squash, or with potatoes in the Andean highlands. In the lake regions of Mexico, floating garden islets *(chinampas)* formed of dense intertwining roots covered with soil and plants could be paddled over the water. In the Andes, irrigated stone-walled terraces (*andenes*) allowed farming steep grades in narrow strips at higher altitudes dominated by llamas, quinoa, and potatoes, descending in ever broadening tiers to cacao, cacti, and tropical fruits at lower levels, each with its own biota, creating a diverse diet in a small vertical space. Where the parched soil of valleys was unyielding, natives removed the earth's surface in expanses as large as an acre to form *hoyas* (pits) which reached natural moisture and permitted cultivation.

Indians learned to cultivate strains of indigenous flora by taking cuttings, by planting tubers, or by carefully selecting seeds which they planted individually in small holes poked by a planting stick called a *macana* in the Caribbean and a *taclla* in the Andes. These methods produced more uniform quality than European broadcast seeding and resulted in higher caloric production per acre than Old World counterparts of wheat, rice, millet, and barley.

The bitter cassava (also known as *yuca brava,* manioc, mandioca, tapioca, and *farinha*) roots from two to six inches in diameter were dug up, peeled, and grated. The wet pulp was then stuffed into long mesh baskets and hung from a beam which allowed for twisting to express the poisonous prussic acid. When the toxic juice was removed, the pulp was forced through basket sieves or pounded, and the resulting flour was spread in flat three-foot cakes on hot stone grills or dried and stored for later use. The poisonous liquid was safe to drink after boiling. From Colombia southward down the Pacific coast of South America, a non-toxic sweet cassava was similarly processed and often made into preserves.

The true staff of life for most of Latin America was maize. This Indian corn occurred in hundreds of varieties and colors. Indigenous cultures hybridized the silky tassels by hand to breed the most desired traits. Some kernels were so hard they had to be soaked or chewed before boiling gruels. Women pounded grilled kernels with stones to produce a flour meal for tortillas or other dishes. Dried grain was also stored indefinitely in elevated, aerated storage silos.

Indians of the upper Andes invented freeze-dried potatoes by alternately exposing the tubers to night frost and sunshine. Within a few days all the water was removed, leaving a potato preserve called *chuñu* which could be further whitened or reconstituted by steeping in water. Lightweight and easily transported, the *chuñu* could be preserved indefinitely like cassava or maize. There were over three thousand types of potatoes, sweet potatoes, ocas, and other tubers. Bean and seed pulses were also important foods throughout Latin America. They too were dried for preservation if not cooked immediately.

In addition to drying, there were several other technologies for preserving foods. These included salting, pickling, fermenting, toasting, and ensiling. Fish was salted as was meat cut into strips. Brine solutions and lime mixtures were also used. Vegetables and fruits were pickled in a variety of herbal vinaigrettes. Fruit and vegetable juices were fermented and distilled into alcoholic beverages which were also used as preserving agents.

Among native beverages, cacao *(Theobroma cacaoyer)* was the most lucrative exchange commodity in New Spain and parts of Peru. Cacao trees were grown by Indians in orchards shaded by large "mother" trees. When the large red pods ripened, they were picked, and hundreds of bitter-tasting seeds were removed by chewing the luscious fruit. The sun-dried seeds were used as money or ground and roasted for a bitter brew restricted to nobles and ceremonial rites. Under Spanish colonization, the chocolate beverage gained widespread popularity in Mesoamerica.

In the Pampas, another stimulating beverage came from a native tea known as *maté*, steeped in boiling water like East Indian varieties. When coffee was introduced to tropical Latin America, it was prepared much like cacao and was widely consumed. All three of these nonalcoholic beverages grew in popularity with the increasing availability of sugar and vanilla which replaced pepper as the preferred seasoning.

In addition to flavor enhancement, many herbs and spices were used to preserve foods and to add medicinal benefits. Seaweed supplemented natural salt sources in Andean diets to protect natives from goiter. Tropical fruits added vitamin C. Cinnamon, cloves, and capsicum peppers preserved as well as flavored foods. Sarsaparilla was a refreshing beverage which soothed gastrointestinal ailments. Another vegetable product, *guayacán* treated syphyllis while *cuasi,* a valuable insecticide, also eased fever symptoms. The most important treatment for malaria came from quinine or *chinchona* which remains a viable cure as well as a popular tonic water.

Alcoholic beverages were consumed in moderation before conquest. In Mexico, the agave or maguey was fermented and distilled into pulque, mezcal, and tequila. Maize was fermented into a beerlike beverage called *chicha* which was popular throughout Middle America and the Andes. *Chicha* could also be fermented from a variety of fruits. Other alcoholic beverages came from fermentation of potatoes, cassava, and tropical fruits such as pineapples and bananas. These were sometimes mixed into brandy called *aguardiente*. None of these beverages became popular in Europe, but rum — fermented from cane sugar — became commercially successful throughout North America and Europe.

Prior to European contact, indigenous populations had few meat sources and no beasts of burden other than the llama and the dog. With the introduction of horses, cattle, oxen, sheep, swine, goats, mules, and donkeys, they gained transportation, meat, and power to plow, haul, and operate mills. They also received protein-rich meat and dairy supplements to their diets. By-products from domesticated and feral herds included fibers, leather, tallow, bones, and horns.

Among the food cultigens brought to the New World, the grape–grain–olive culture remained heavily European. Sugar-cane and coffee, however, dominated enormous regions where native labor consumed the stimulants while working plantations. Both commodities were labor intensive requiring hand planting and harvesting. Sugar works (*trapiches* or *engenhos*) used human, animal, or water power for the crushing mills which released the sugary syrup. Slaves operated a train of boiling vats to purify and clarify the liquid for evaporation. The resulting muscovado sugar was 96% pure and was packaged in cones or loaves for shipment to European refineries. Hand-picked coffee beans were spread on patios to dry before roasting and shipping.

Many of the technologies used by indigenous populations are practiced today throughout rural areas of Latin America. Scientists are studying their techniques and lore for insights into environmental protection, expanded food production, and medicinal applications. Global dietary diversity has been accompanied by language enrichment as a result of food exchange. As a result of Latin American food technology, words such as barbecue, olla, mezquite, cassava, tapioca, quinine, banana, palm, tomato, etc. have joined the alimentary menu.

The Indians practiced ecological efficiency and perspicacity unrivaled by the colonial powers who imposed a wheat-wine-oil-meat agriculture in lands more suited to manioc, maize, and potatoes. None the less, dietary diversity has been enhanced throughout the world as a result of native Latin American technologies which first unlocked the secrets of preparing and preserving exotic New World foods which have become staples for subsistence.

JUDITH VIDAL

REFERENCES

Crosby, Alfred W. Jr. *The Columbian Exchange: Biological and Cultural Consequences of 1492*. Westport: Greenwood, 1972.

Sauer, Carl Ortwin. *The Early Spanish Main*. Berkeley: University of California, 1992.

Super, John C. "Nutritional Regimes in Colonial Latin America." In *Food, Politics, and Society in Latin America*. Ed. John C. Super and Thomas C. Wright. Lincoln: University of Nebraska, 1985.

Weatherford, Jack. *Indian Givers: How the Indians of the Americas Transformed the World*. New York: Fawcett, 1988.

See also: Potato – Ethnobotany – Sugar – Crops in Pre-Columbian Agriculture – Agriculture

FORESTRY IN INDIA The history of forestry in India is related to the history of civilization. If we look at the development of human kind we find that with the evolution of the human race various concepts evolved, such as those of family, tribe, etc. as well as developmental sciences like forestry.

One of the earliest civilizations of the Indian subcontinent was that of the Indus Valley (third or fourth millennium BC). It was here that cedar (*Cedrus deodara*) and rosewood (*Dalbergia latifolia*) used for coffins was found. A wooden mortar of ber (*Zizyphus mauritiana*) for pounding grain and charred timber of *Acacia* spp., *Albizzia* spp., teak (*Tectona grandis*), haldu (*Adina cardifolia*), and *Soyamida febrifuge* were also found. This shows that neolithic people not only made extensive use of wood but also understood its particular characteristics for different purposes. This concept today is known as forest utilization.

Evidence of tree worship during the Indus Valley civilization is exhibited by various seals of the Harappan culture which depict the pipal (*Ficus*) and weeping willow (*Salix*) trees. Trees were an essential and integral part of the life support system and considered existing agencies of the creator or God.

Just as forests played an important part in Vedic India, tree worship was also practiced by the Aryans. Because of the human dependence on trees, they were venerated and protected by religious injunctions; their planting was encouraged by a promise of eternal bliss in future life. *Ṛgv̄eḍa* and *Aṭharvav̄eḍa*, basic texts of the Vedic period, contain several hymns praising and endowing trees, plants, and vegetation with various divine qualities, highlighting their medicinal significance, and enunciating the policy of conservation or sustainable management. Various *Upaniṣads*, a later group of philosophical treatises that explain the theology of Hinduism, like *Bṛhaḍāruṇyaka, Chāndogya, Chulikā,* and *Muṇdakopniṣaḍa*, conclude, that trees had life akin to human life. *Maṭsyapurāṇa* and *Varāhapurāṇa* describe the benevolence of trees along with the rituals of tree planting. *Skanḍapurāṇa* contains a long list of trees which should not be cut except for the purpose of *yajnas* (holy rituals). Whereas planting of trees led to heavenly comforts (*Agnipurāṇa*) indiscreet felling of trees meant torture in hell.

Vedas also contain valuable information about various species of birds. In *Ṛgv̄eḍa* for instance there is a mention of *Garuda* (eagle), *Mayūra* (pea fowl) along with various Himalayan pheasants, partridges, and other species of birds. Surprisingly *Ṛgv̄eḍa* also mentions not only anatomical details of some common birds but details about their staple food too. For example *Vartika* (partridges) are said to have well developed bills, legs, and rounded wings; their food consisted of grain, grass, weed, seed, tender shoots, insects, and even white ants. These kindled the sparks of wildlife management during the later period of civilization.

Great saints and sages who understood these texts, passed on their knowledge and wisdom to their disciples. Such education was given in the *Gurukuls* (schools) which were located in the forests. The students lived in the forests and continued their studies. Thus a sense of tolerance and coexistence with various forms of plant and animal life was infused in early childhood. This *Aranya* (forest) culture provided early exposure to nature study and ecology, as well as the policy of development without destruction in current parlance. Forest and environment consciousness was thus ingrained into the educational system from the very onset.

Even epics and religious texts written later on describe the protective role of trees and their unlimited usefulness. *Rāmāyaṇa* and *Mahābhāraṭa* describe the rich biodiversity and multiplicity of flora and fauna. In *Bhagwaṭgīṭā* Lord Krishna compared himself to the *Aśvaṭha* (ficus) tree in order to emphasize the importance of trees.

Other religions of that time also mentioned the importance of forests and forestry. According to Lord Buddha it was the obligation of every good Buddhist to plant and nurture at least one tree every five years.

Moving along chronologically we find that forest management practices were well documented during the reign of Chandra Gupta Maurya (321–296 BC). Reliable historical documents, such as the *Indika* of the Greek ambassador Magasthenese, the *Arṭhaśāsṭra* of Kautilya and the *Mudrārākśasa* of Viśākhaḍaṭṭa vividly depict various aspects of forestry and wildlife. Kautilya's *Arṭhaśāsṭra*, recognized as the pioneer work in economics in India, indicates the existence of a regular Forest Department headed by *Kūpyādhakśa* with definite duties and responsibilities for various officers of the department. Some of the important duties of *Kūpyādhakśa* were to increase the productivity of forests, classification, price fixation, and disposal of various types of forest produce, raising block plantations of important species (e.g. sandalwood), pasture development in saline-alkaline waste lands, and joint management of forests with people dependent on forests.

Forests were legally classified into three main classes: reserved forests, forests donated to eminent Brahmans, and forests for public use. They were classified into six categories on the basis of crown density, luxuriance, growth, and origin. Planted forests were called *Upvana*.

Forest and game laws were stringent and draconian, containing corporal punishment. The death sentence was envisaged for poaching of elephants. Awards were provided to a person who collected elephant tusks from dead elephants and deposited them with government officials. Kautilya's treatise also dealt at great length not only with the use of various trees and shrubs but also with the specific use of various parts. For example it indicated that flowers of *Palāś, Kusuma,* and saffron are used as dyes; *Munja* and *Love* grass are used for making ropes, and fruits of *Aonla, Harara,* and *Bahera* are used as medicines.

Emperor Ashoka adopted and improved these practices. Plantation of fruit-bearing shade trees for the benefit of travelers and common people was started on an ambitious scale. Ashoka was the founder of *Abhyāraṇya,* the *sanctum sanctorum* of wildlife, now known as National Parks and sanctuaries. Ashoka's edicts at Sarnath Varanasi bear ample testimony to the above.

Forestry flourished during the Gupta period (320–800). *Śukranīti*, a well known work of that time, throws light on the improvements in forest management practices. Seeds of "Social Forestry" were sown during this era. *Śukranīti* dwells on the concept of village forests, choice of species, afforestation and maintenance techniques, fertilization procedures, irrigation schedules, and measures to increase flowering and fruiting in trees. Names of various multipurpose trees such as *Kaḍamba, Sīśama, Peepal, Mango, Nīm, Coconut*, and *Imlī* are listed in this policy document for planting near villages. Incidentally many of these are also listed in the latest I.C.R.A.F. (International Council for Research in Agroforestry, in Kenya) booklets. This shows the worth of these ancient publications.

It is thus clear that in ancient India trees were the best friend of people in a hostile environment. They were held sacred, worshipped, and studied in great detail for service to humanity. Forestry practices evolved gradually, in a scientific and rational manner, and are important even today, although in the modern parlance new names have been coined for them. As a matter of fact some of the knowledge which originated in India on various aspects of forestry later spread to various parts of the world.

DEEPA PANDE

REFERENCES

Bhagwat Geeta. Gorakhpur: Geeta Press, 1992.

Chattopadhyay, Debiprasad. *Science and Society in Ancient India.* Calcutta: Research India Publications, 1979.

Ghildiyal Vineet and Sharma R.C. "Some Himalayan Birds and their Conservation in Rigvedic India." *Himalayan Research and Development* 4(2): 24–25, 1985.

Kautilya's Arthaśāstra. Mysore: Wesleyan Mission Press, 1923.

Pande, Deepa and I.D. Pande. "Forestry in India through the Ages." In *History of Forestry in India*. Ed. Ajay S. Rawat. New Delhi: Indus Publishing Company, 1992, pp. 151–162.

Sinha, B.C. *Tree Worship in Ancient India*. New Delhi: Books Today, 1979.

G

GAITIAN *Gaitian* is the Chinese name for a scheme of cosmography, i.e. a description of the overall layout of heaven and earth. The term may be interpreted as "umbrella [-like] heaven".

The classical description of the *gaitian* view is found in the *Zhoubi suanjing* (Mathematical Manual of Zhoubi), dating from around the beginning of the Christian era. It seems likely however that it is a systematization of what was the common view of the shape of the heavens at least as early as 250 BC. Heaven and earth are more or less flat and parallel planes, although they may sometimes be described as gently curved like the cross-section of an umbrella. Earth is stationary, while heaven rotates once daily about an imaginary vertical axis through the north celestial pole, carrying the heavenly bodies with it. Day and night occur because this rotation carries the sun beyond the range of the observer's sight and back again. The rising and setting heavenly bodies are an optical illusion. In winter the sun is further away from the celestial pole than in summer, and is hence more distant from the observer and lower in his sky. The *Zhoubi suanjing* gives the height of heaven above earth as 80,000 *li*. The Chinese observer is said to be 103,000 *li* from the subpolar point, while the greatest and least radii of the sun's daily orbit round the pole are 119,000 *li* (summer solstice) and 238,000 *li* (winter solstice). The scheme gives a fairly good qualitative explanation of obvious phenomena, including the six-month alternation of day and night at the earth's north pole. By the second century AD it becomes clear that mathematical astronomers preferred not to think in *gaitian* terms, but were using the so-called *huntian* (continuous heaven) scheme.

CHRISTOPHER CULLEN

REFERENCES

Cullen, Christopher. *Astronomy and Mathematics in Ancient China, the Zhou Bi Suan Jung*. Cambridge: Cambridge University Press, 1996.

See also: *Zhoubi Suanjing – Huntian*

GAN DE As one of the earliest astronomers in China, Gan De, in the fourth century BC, made many observations of the heavenly bodies, especially Jupiter, which was then called *Suixing* (the year-star). He wrote two books: *Suixing Jing* (Treatise on Jupiter) and *Tianwen Xingzhan* (Astrological Prognostications). Unfortunately these books were lost long ago, and only some quotations from them are extant in the *Kaiyuan Zhanjing* (The Kaiyuan Treatise on Astrology), compiled between AD 718 and 726.

In the *Kaiyuan Zhanjing*, Gan De is quoted as saying: "Jupiter was very large and bright. Apparently, there was a small reddish star appended to its side. This is called 'an alliance'." Included in the quotation were a date and rough coordinates, both in the ancient Chinese system.

Xi Zezong, of the Institute for the History of Natural Sciences in the Academia Sinica, Beijing, claims that this record is evidence of the earliest discovery of the brightest moon of Jupiter, Ganymede, in the summer of 365 BC.

Although Ganymede's magnitude of 4.6 is somewhat brighter than that of the naked eye's limit, Jupiter is 760 times brighter and located less than 5.9 arcmin away from this moon. Only people possessing eyes of extraordinary power would be able to see the satellite. Could Gan De have been one of these extraordinary people? We will never know.

Some people claim, on the basis of experiments, that under good observational conditions, the normal naked eye can determine Jupiter and its brighter satellites, especially Ganymede. However, Gan De's reference to the reddish color continues to be mystifying, because Ganymede is too faint for its color to be perceived with the naked eye.

HUANG YI-LONG

REFERENCES

David W. Hughes. "Was Galileo 2,000 Years Too Late?" *Nature* 296: 199, 1982.

Qutan Xida. *Kaiyuan Zhanjing* (The Kaiyuan Treatise on Astrology). Taipei: Taiwan Commercial Press, 1985 reprint.

Xi Zezong. "The Discovery of Jupiter's Satellite Made by Gan De 2,000 Years Before Galileo." *Chinese Physics* 2(3): 664–667, 1982.

Yang Zhengzong, Jiang Shiyang, and Hao Xiangliang. "Experimental Test for Jupiter's Satellites to the Naked Eye." *Kexue Tongbao* 28 (7): 927–929, 1983.

GAS: EXPLOITATION AND USE OF NATURAL GAS IN PREMODERN CHINA Ancient Chinese sources give many references to strange fiery phenomena burning on water or rising out of the earth. For instance, the *Hanshu* says that in 61 BC the emperor sacrificed at the "fire well" (*huojing*) of Hongmen (Shaanxi). We do not know the cause of this fiery discharge, but it is clear that this fire well was an

Figure 1: Fire wells of different quality, as shown on a mid-eighteenth century scroll. The scroll was photographed by Rewy Alley in Beijing in 1954. The location of the original is unknown. Photograph used with the permission of the Needham Research Institute, University of Cambridge.

object of religious worship, because a fire well temple had been erected there. No mention was made of any use for industrial purposes.

The first reports on the industrial use of a fire well come from the middle and late third century AD. One of these reports state that there is a fire well in Linqiong, Sichuan province. Once Zhuge Liang (181–234), counselor-in-chief of Liu Bei (162–223), came to see it, whereafter the fire turned stronger. People put pans on the well to boil salt. When brands of common hearths were entered into the well, it at once extinguished and did not burn again. Thus, it is clear that the prosperity and decline of the well was seen as corresponding to political developments. Another source, for instance, interpreted the extinction of the well as an omen for the annexation of Shu by Wei.

We cannot ascertain whether the Linqiong fire well was really a natural gas well, because later texts appear to suggest that petroleum was involved. Whatever the case, there is no doubt that it was productive for only a short period. That later authors referred to this well time and again is because it constituted a strange and interesting historical phenomenon. It is therefore hardly justified to speak of the "systematic use" of natural gas for evaporating brine "on an industrial scale" starting in the second or even fourth century BC, as Joseph Needham claimed. Moreover, Needham also assumed that deep drilling, which was one of the preconditions for a systematic exploitation and use of natural gas, had already been invented during the first century BC or AD. Recent research showed, however, that deep drilling originated in the middle of the eleventh century.

After the Linqiong fire well had become extinguished, probably in the late third century, Chinese sources do not mention the use of fire wells for industrial purposes until the sixteenth century. Although they still report on the outflow of mist (*yanqi*) and hidden vapors (*yinqi*) from wells, it is clear that these discharges were conceived as threats to well salt production. For instance, in the tenth century explosive "hidden gas" was reported from the Ling well, a shaft well in the Lingjing industrial prefecture. Moreover, it is said that in the eleventh century, when workers had been let down into the well for repair work, they were killed. Only by the installation of a so-called "rain basin" (*yupan*) over the top of the well could these difficulties be overcome. From the rain basin water sprinkled down like rain, which carried the gas down into the well. This device was derived from the observation that gas discharge was restrained during rainy days.

Systematic use of natural gas began in the sixteenth century. In the beginning, the number and output of industrially used fire wells appear to have been still limited. Until the end of the eighteenth century a fire well usually could supply fuel for only one to four pans, though exceptionally productive wells like the Yongtong well existed, which is said to have fed sixteen pans. A scroll of the mid-eighteenth century shows interesting details of fire wells of differing productivity probably in northern Sichuan.

The standard situation appears to have been that one well was feeding one pan which was placed directly over the well. A bamboo tube inserted into the well served as burner. It was covered by a stone, in case the fire was to be extinguished. The scroll also shows a very productive fire well supplying two pans. The gas distribution device consisted of a large bamboo tube on which a porcelain vessel filled with water was put. The gas was distributed to the pans by two small pipes connecting the burners with the large bamboo tube. Finally, less productive fire wells not only could not feed more than one pan, but had also to be supported with firewood.

The situation changed dramatically with the development of deep drilling techniques in the nineteenth century, when the high pressure gas deposits in the deeper strata of the Ziliujing gas field were tapped. For instance, the French missionary Imbert reported in September of 1827 that a single fire well of Ziliujing could feed more than 300 pans and that the wells were so productive that a part of the gas could not be used and thus had to be burnt off. A new generation of productive fire wells supplying 400 to 700 pans is mentioned by Li Rong in his famous account of Ziliujing from the end of the nineteenth century. The use of the resources of such highly productive wells necessitated the invention of a number of devices for the control and distribution of the gas flow and for coping with variations in the well's pressure. The utilization of gas was a highly dangerous undertaking which could lead to explosions. That was also the reason why at least in Ziliujing gas was only rarely used for household purposes.

Natural gas was an amazing phenomenon which required explanation. Song Yingxing, in 1637, said that the fire wells contained simply cold water, but not the slightest evidence of the *qi* of fire. One can only see the notion (*yi*) of fire which is bursting forth from the burner's pipe, so that the brine will boil violently. Yet if the bamboo pipelines for conducting the gas are opened and examined, no sign of charring or burning can be seen. "To use the spirit (*shen*) of fire without seeing the solid form (*xing*) of fire — this is indeed one of the strangest things in the world" (Sun and Sun, 1966). Later Chinese authors had less strange ideas of the phenomenon of fire wells. For instance, in 1791, Xu Deqing came to the conclusion that the existence of fire and wind in the earth of high-lying terrains should be regarded as something quite normal, similar to the existence of fresh water sources in marshy and low-lying regions.

HANS ULRICH VOGEL

REFERENCES

Imbert, [Laurent-Joseph-Marius]. "[Letters of Sept. 1826 and later from Wutongqiao, Jiadingfu, and Ziliujing in Sichuan]." *Annales de l'association de la propagation de la foi* 3: 369–381, 1828, and 4: 414–418, 1830.

Li Jung. "An Account of the Salt Industry at Tzu-liu-ching. Tzu-liu-ching-chi." Trans. Lien-Che Tu-Fang. *Isis* 39: 228–234, 1948.

Needham, Joseph. *Science and Civilisation in China*. Cambridge: Cambridge University Press, 1954.

Sun, E-Tu Zen, and Shiou-Chuan Sun, trans. *T'ien-kung k'ai-wu: Chinese Technology in the Seventeenth Century, by Sung Ying-Hsing*. University Park and London: Pennsylvania State University Press, 1966.

Vogel, Hans Ulrich. "Feuerbrunnen in China und ihre Bedeutung für die Technikgeschichte." *Heidelberger Jahrbücher* 35: 199–218, 1991.

Vogel, Hans Ulrich. *Naturwissenschaften und Technik im vormodernen China: Historie, Historiographie und Ideologie, Kurseinheit 3: Geschichte der Tiefbohrtechnik im vormodernen China: Invention, Innovation, Diffusion, Transmission*. Hagen: Fern Universität Gesamthochschule - in Hagen, 1993.

Vogel, Hans Ulrich. "The Great Well of China." *Scientific American* 268 (6): 116–121, 1993.

Vogel, Hans Ulrich. "Chinese and Western Scientific Explanations of Sichuan Brine and Natural Gas Deposits prior to 1900." To be published in *T'oung Pao*.

Vogel, Hans Ulrich, Joseph Needham et al. "The Salt Industry". Section 37 of *Science and Civilisation in China*. Ed. Joseph Needham. Cambridge: Cambridge University Press, forthcoming.

See also: Salt in China

GE HONG Ge Hong (ca. 283–ca. 343), alchemist, physician, astronomer and government officer, was the greatest alchemist and physician in fourth-century China. We also know him under several other names, such as Ge Zhiquan, Zhiquan Zhenren, Baopuzi, and Xiao Ge Xianweng. His exact dates are not known with certainty. Some say that he was born between 280 and 286. Other dates given for his birth are ca. 253, ca. 280, ca. 281, 283, and 284. The year of his death has been variously given as ca. 333, 340, 343, 361, and 364. Chen Guofu's elaborate study suggests the period 283 to 343 for him. Ge Hong's autobiography is contained in his *Baopuzi waipian* (Exoteric Chapters of the Preservation-of-Solidarity Master), but it says precious little about the author's scientific achievements. His biography is given in the *Jinshu* (Official History of the Jin Dynasty). Since he is regarded by the Daoists as having attained physical immortality, his hagiography is found in abundance in Daoist literature.

When he was young Ge Hong studied alchemy under Zheng Yin and Bao Jing, whose daughter he later married. He also learned much about medicine. He joined the government side as an army officer in a military campaign to suppress an uprising in the year 303. After peace was restored he left without seeking reward and traveled widely in search of books. In the year 306 he accepted an invitation from his friend Qi Han to become his military adviser when the latter was appointed Governor of Guangzhou. Qi Han, the author of the *Nanfang caomuzhuang* (Records of Plants and Trees in the Southern Region) was one of the greatest botanists in traditional China, but unfortunately he was soon assassinated. Ge Hong remained for some time in the South and then returned to his place of birth, Nanyang, in Jiangsu province. A belated award came for his past service in the army. He was given the title Marquis of Guannei, and was recommended to the emperor for appointment as a member of the bureau of historiography. He declined the offer, but requested that the emperor make him magistrate of Goulou in south China, which was a rich source of cinnabar that he needed to prepare the elixir of life. Eventually the emperor granted him his wish. Ge Hong then went to live in the Luofoushan Mountain where he carried out his alchemical and proto-chemical experiments. It was probably there that he completed his famous alchemical work, the *Baopuzi*. Some say that the year was 317.

The *Baopuzi* consists of a *Neipian* (Esoteric Chapters) in twenty chapters and a *Waipian* (Exoteric Chapters) in thirty-two chapters. Among those in the *Neipian*, chapters 4, 11, and 16 are of special interest to the alchemists, while the other chapters also contain bits of information on alchemy scattered here and there. Chapter 16 specifically describes the transmutation of base metal into gold or silver. Chapter 1 of the *Waipian* contains Ge Hong's autobiography. The rest of the *Waipian* says nothing about science or alchemy. The relevant sections in the *Neipian* contain names of more than fifty elixirs supposed to have various efficacies — some could only prolong the human life span, while others could transform the aspirant to a holy immortal varying from three days to three years depending on the elixir itself. Ge Hong seemed to be successful in aurifiction, the process of imparting an appearance of gold to base metals by artificial means. It is interesting that the attributes of the elixir of being a panacea, of being able to translate base metals into gold or silver by means of projection, and of being able to impart longevity and perpetual youth, were already mentioned by Ge Hong in the case of some of the elixirs in his *Baopuzi*. Sulfur and mercury were used in many of his elixir recipes.

Ge Hong was an eminent physician in his own right. His *Baopuzi neipian* describes the medicinal values of many plants and minerals. During his stay at the Luofoushan Mountain he also wrote the *Jingui yaofang* (Prescriptions in the Treasury of Medicine), in one hundred chapters. Later he added an abridged version, entitled *Zhouhou jiuzufang* (Handbook of Medicine for Emergencies) in three chapters. Ge Hong's writings contain prescriptions for various types of diseases, including eye trouble, ailments of women and children, and infectious diseases such as smallpox and tuberculosis. They prescribed soya beans, cow's milk and goat's milk for the treatment of beri-beri, and for treating bites from mad dogs the recommendation was to apply the brain of the culprit over the wound.

Ge Hong also showed considerable interest in astronomy. The *Jinshu* (Official History of the Jin Dynasty) quotes from a piece of his lost writings on the construction of astronomical instruments by Zhang Heng in the second century. It also narrates Ge Hong's participation in the great cosmological debate of his time. There were then two rival schools, the *Gaitian* (Canopy Heaven) and the *Huntian* (Spherical Heaven). The former was more ancient and pictured a canopy heaven covering a square earth like a tilted umbrella. The latter visualized the earth as the yolk of an egg floating on water at the center of a spherical heaven that was itself supported by water. The argument used against it was that the sun rising from and setting into water would be quenched of its fire and heat. In supporting the *Huntian* theory, Ge Hong skilfully quoted the hexagrams and passages from the *Yijing* (I Ching, Book of Changes) to silence its critics. Ge Hong also observed the effect of the moon on the water in the sea and on the tides. In his *Baopuzi* he referred to the waves of the sea heaving up and down with the waxing and waning of the moon and the increase in magnitude of the tide when the moon was full.

HO PENG YOKE

REFERENCES

Needham, Joseph, with Ho, Peng Yoke, Lu Gwei-Djen and Nathan Sivin. *Science and Civilisation in China*, vol 5 part 3. Cambridge: Cambridge University Press, 1976.

Ware, James R. *Alchemy, Medicine and Religion in the China of* AD*320: the Nei P'ien of Ko Hung, (Pao p'u tzu)*. Cambridge, Massachusetts: MIT Press, 1966.

GENDER AND TECHNOLOGY The history of science and technology is often portrayed exclusively in terms of men's achievements. Yet it is also often said that women were most probably responsible for domesticating plants and sowing the first crops, devising food processing and cooking methods, and inventing pottery and textiles.

Much of this refers to prehistory and, hence, to a period for which there are no records. Even in historical times, however, it is difficult to find documentation for what women actually did. Hardly any woman innovator is recorded by name, although in China a shrine or monument was erected in 1337 to commemorate a woman named Huang Daopo, who visited the lower Yangtse in 1295–1297 and introduced improved implements for spinning and weaving.

Reports by anthropologists which comment on grain milling show that in 95 percent of communities where this job was done by hand, it was done by women. But as soon as watermills (or windmills) were introduced and the operation could be run as a business, milling became an exclusive preserve of men. Similarly, pottery was originally molded by hand, without a wheel, and this is thought to have been done mostly by women — but:

> The earliest surviving potter's wheel was found at Ur . . . on it rested the large-scale production of pottery which made it a man's trade and not, like earlier pottery, a woman's. (Roberts)

Again, women may have pioneered the domestication of plants, and in recent years women have been observed bringing threatened wild plants into cultivation. They may have been the first farmers, therefore, and in Africa are still farmers to a greater extent than men. However, in cultures where that is true, it remains true only as long as hand hoe cultivation persists. As ox-drawn plows replaced hand hoes, primary tillage became a masculine responsibility, and other operations were taken over by men as they, too, were mechanized.

We can see now that the history of technology, as conventionally represented, mentions few women because historians do not start at the beginning, when processes were first invented. They start halfway through, when the processes were first mechanized and men took over.

It is possible to think of this development as the result of a rational division of labor between the sexes. Women invented processes related to the domestic sphere of life. Men took over when it became possible to extend the scope of a process and move it into the public sphere as a business enterprise — or as a means of exercising power, but one suspects deeper reasons than this. Men seem always to have been especially fascinated and proprietorial in their attitudes to sources of power, such as the ox pulling the plow, or the water wheel driving the mill. Then, also, Lewis Mumford sees tools and machines as symbolizing male attributes — a hammer parallels a man's fist, for example. He correspondingly sees slow, organic, nurturing processes as more characteristic of women.

A parallel contrast is expressed by two schools of thought in Chinese hydraulic engineering that go back more than two thousand years. Engineers who favored control of river flooding by erecting high dikes, weirs, and dams were felt to be challenging and taming nature, thereby expressing masculine qualities. However, other engineers favored "the deepening of river beds by excavation of 'feminine' concavities", and that was a matter of "going along with Nature", as approved by the best Daoist archetypes.

ARNOLD PACEY

REFERENCES

Goodrich, L.C. "Cotton in China." *Isis* 34: 408–410, 1942–43.

Mumford, Lewis. *The Myth of the Machine: Technics and Human Development*. New York: Harcourt, Brace, Jovanovich, 1966.

Needham, Joseph, Wang Ling, and Lu Gwei-Djen. *Science and Civilisation in China*, Vol. 4, part 3, *Civil Engineering and Nautics*. Cambridge: Cambridge University Press, 1971, p. 249.

Roberts, J.M. *The Pelican History of the World*. Harmondsworth: Penguin, 1983, p. 73 (reprinted from *History of the World*. New York: Knopf, 1976).

See also: Textiles – Food Technology

GEODESY Geodesy, the measurement of the earth, is an essential component of both astronomy and geography. The spherical shape of the earth was a concept developed in the early stages of Greek science, and one finds this concept wherever the legacy of Greek science was taken seriously. Figures for the radius of the earth were reported by Aristotle and a number of other Greek scientists, although a judgment of their accuracy is hindered by our ignorance of an exact measure in modern terms of the units in which the radius was expressed. The coordinates in Ptolemy's geography were fixed by him from various distances measured in terms of *stadia*, converted on the assumption that one degree equaled 500 *stadia*; for Eratosthenes, the degree equaled about 700 *stadia*. On the other hand sources in Latin and Syriac rather later than Ptolemy worked on the assumption that one degree equaled 75 Roman miles, which we can evaluate because the length of the Roman mile is well established, so we know that it is very precise. The radius and circumference of the earth follow immediately from the length of one degree.

In the early stages of Arabic science, especially the work heavily patronized in the early ninth century by the Caliph al-Mam'ūn, various Greek and Syriac sources were consulted with a view to fixing the earth's dimensions. Here the intention was to evaluate the degree in terms of the Arabic mile, which we know to be 4/3 Roman miles, 1972 meters. The Arab scientists of the time, however, were uncertain as to the interpretation of the various figures given in their sources, and they were certainly confused about the true length of the Roman mile. While they were aware that the Roman and Arabic miles were respectively 3000 and 4000 cubits, they were ready to assume that it was the cubits and not the miles which differed.

Against this background they set about to measure the length of the degree by direct observation of the latitude variations along a north-south traverse in the desert of Iraq. They chose a region south of Sinjār, which is about 100 km west of Mosul. According to extant reports two expeditions were made, one led by al-Marwarrūdhī, the other by ʿAlī ibn ʿIsā al-Asṭurlābī. In one account it is said that they agreed on the ratio 56 miles per degree, but other accounts of the expeditions report $56\frac{2}{3}$, or $56\frac{1}{4}$ miles. There is probably a blending here of observation together with a conversion of the late Roman figure of 75, since $\frac{3}{4}\times75=56\frac{1}{4}$. The accounts are somewhat lacking in circumstantial detail, and probably the methods employed, which must have involved a sufficiently portable quadrant for the measurements of latitude along the route, would not have sufficed to provide more than a rough verification of the received figures, sufficient at least to distinguish between the two different interpretations of the mile.

The great scientist al-Bīrūnī (fl. AD 1030) carried out some work of great value in his determination of a series of longitude differences for stages leading from Baghdad to Ghazna. These were reported in both his treatise on geodesy *Taḥdid nihāyāt al-amākan li-tasḥīḥ masāfāt al-masākin* (The Determination of the Coordinates of Cities), and in his astronomical treatise the *Qānūn al-Masʿūdī.* He also reported his own attempt at a determination of the length of the degree. The latter took place in the Punjab, where the plain to the south may be seen from a peak (479 m) in the Salt Range, some 50 km South West of Jhelum. Al-Bīrūnī says that he measured the dip of the distant horizon as 34 minutes of arc, from which according to his calculation, he finds slightly more than 56 miles to the degree. He considers however that this should be regarded not as a new result but as a confirmation of the earlier result fixed at the time of al-Mam'ūn. In fact his procedure was flawed in various ways, for his measurement of the height of the mountain was considerably in error, and a correct interpretation would have required that refraction be taken into account. Moreover we may calculate, taking refraction into account, that he would have obtained a dip nearer to 29 minutes, so that his whole account is quite suspect. The method has been suggested independently by a number of scientists, but even when carried out with the greatest care, it is doomed to failure because the degree of refraction is unpredictable for a ray grazing the earth's surface.

Al-Bīrūnī's determination of the longitude differences was much more successful. The tables of mean motion in his astronomical treatise were referred to the meridian through Ghazna (near Kabul), and he was naturally obliged to determine the time/longitude difference between that place and Baghdad. The primary information available to him, as to Ptolemy before him, was provided by travelers who would have converted a traveler's time to a distance. The region between Baghdad and Ghazna is divided up by al-Bīrūnī into shorter stages along two different routes, one passing North through Rayy and Jurjaniyya, and the other South through Shiraz. For each short stage, he calculated the difference of longitude, given the latitudes at each stage, and the distance from one stage to the next. In every case the travelers' distance is discounted to allow for the fact that the real path deviates from a perfect great circle. The problem of fixing the longitude difference is posed and solved correctly in terms of spherical trigonometry, and he gives all the numerical details, so that one can judge the accuracy of his work. In spite of occasional lapses in the calculation the results represent a very considerable improvement over that available in Ptolemy or al-Khwārizmī; the final longitude difference between Baghdad and Ghazna is only some 15 minutes of arc short of the true value. People would not be able to improve on that until entirely new methods were discovered in the late seventeenth century, such as the use of Jupiter's satellites.

RAYMOND MERCIER

REFERENCES

Mercier, Raymond P. "Geodesy." In *Cartography in the Traditional Islamic and South Asian Societies*. In *The History of Cartography*, Vol. 2, Book 1. Ed. J.B. Harley and David Woodward. Chicago: University of Chicago Press, 1992, pp. 175–188.

See also: al-Bīrūnī

GEOGRAPHICAL KNOWLEDGE Cultures differ in their assumptions about the nature of life, the place of their existence, and the place of humanity within their environment. Places, like space and time, are constructed socially and culturally and as such can provide ways to understand the literal and metaphorical geographies of humans. The specificity of place derives from the fact that each place is the focus of unique mixtures of wider and more local relations between social groups, and it is this mixture which accumulates continually to create the distinct place. While distinguishing labels, for people involved in these social relations, such as woman, Canadian, Maori, can be starting points, no one person or one group is exclusively one complete identity, Western, or non-Western, in the end of the twentieth century.

Human cultures are neither essentially coherent nor wholly homogeneous. The borders between nations, classes, and cultures have become more fluid. The very notion that the majority of the world's population is defined by not being something (in this case non-Western) infers the continuing power of the past legacy of European and more recently American colonialism or imperialism. The majority of the world's population are non-Western. They did not originate from Europe's colonial expansion into the Americas, Australia, and New Zealand.

This geographical inquiry into the historical experience of non-Western cultures is made from the perspective of a Western woman, a cultural/historical geographer, born in Canada, now residing in Australia. A renewed concept of culture refers not to any unified entity. In a post-colonial world, the idea of an authentic culture as an internally coherent world is no longer tenable. Increasing global interdependence has made it clear that neither "we" nor "they", Western nor non-Western cultures, are neatly separated and mutually exclusive. All of us inhabit an interconnected, late twentieth century world, marked by porous national and cultural boundaries that are enmeshed with power, domination, and inequality.

It is however, impossible to deny persisting continuities of long standing traditions, literate or oral, sustained habitation, and cultural geographies of non-Western traditions. Human history is rooted in the earth. To understand this history, it is necessary to consider human dwelling and also to realize that European expansion meant the taking of land which was already owned by indigenous residents. With the beginning of European expansion in 1500, most indigenous peoples lived on their lands practicing agriculture, hunting, and gathering. While all of these people have expanded geographically into adjacent territories, or into lands near where they once lived, there is no comparison with the great spread of people from European nations. It has been this modern colonialism which forced many indigenous peoples off their own land: Aborigines of Australia, Maoris of New Zealand, the indigenous peoples of North and South America. This compulsion of the Western world to explore, conquer, occupy, and own other lands has involved such a long continuous process that it has been called the distinctive Western characteristic.

In contrast, a deep sense of belonging to the land, is central to the identity of indigenous peoples throughout the world, as exemplified by this Maori statement, that while "people disappear, land remains; land doesn't belong to the people, people belong to the land" (Yoon, 1986). The names many of these First Peoples call themselves, such as Inuit, Maori, Saami, mean "people" and the names of their places mean "land". They literally know themselves as people of the land. "We have opted to become part of the environment", explains Burnum Burnum of Australia. Although differing widely in their customs, culture, economic status, physical surroundings, and impact on the land, many indigenous societies often share a set of values that are in marked contrast to Western priorities. The Cree of Quebec, San of Kalahari, and Tukano of Brazil are all connected through a profound relationship and spiritual attachment with the land.

Human beings have created various cultural space and time systems in order to maintain their unique spiritual relationship with the universe. Before the dominance of Western ideas and colonial empires throughout the world, literate societies of Japan, China, and India were eloquent in describing their human relationship with non-human species and their place within the greater environment. Oral traditional societies had highly developed constructions of their environments which were transmitted through stories, dance, and by paintings to succesive generations. For non-Western peoples, although widely different in customs, culture, and impact on the land, there was a shared belief in a deep bond with nature, an awareness that all life, rivers, skies, people, ants, and rocks were inseparably interconnected. Physical and spiritual worlds were woven together in one web. Humans were not separate from nature but were a part of the nature of the world. Geographical knowledge concerning human and environmental nature is always socially and culturally construed. Through specific examples from non-Western cultures of both literate and oral traditions, an understanding of the diversity of specific geographical knowledges can be explored.

The Mythic Creation of the World

Each culture possesses creation myths which speak of the creative beginnings, an energetic phase prior to the appearance of physical matter and life. Marriage between the Ancestral couple is common to creation myths of India, Egypt, Mesopotamia, China, Central America, and New Zealand. In all these myths, creation is sparked by a cosmic copulation. The intercourse of Geb and Nut in Egyptian mythology and that of Shiva and Paravati in the Indian myth, and Ranginui and Papatuanuku, the primeval pair in Maori myth, reflect creation themes. The vision of the sacred earth as a ledger of cosmology is unique to Australian Aborigines with their creation story written like a book in the earth's topography.

There is evidence of great length and complexity in Australian Aboriginal culture, the oldest continuing society in the world, reaching back to the Pleistocene Age which is based on the remembrance of the origin of life. Aborigines speak of the forces and powers which created their world as Creative Ancestors. During this epoch of world creation called the Dreaming, the Ancestors travelled across a barren flat terrain. Their hunting, loving, and fighting shaped the land, deeply marking the surface of the earth and leaving their imprints forming the topographical landscape, making riverbeds, and forming rocks and deserts. Before their travels, the Ancestors would sleep and dream the adventures of the following day. In this way, moving from dreaming to action, they had the ability to alternate between pure spiritual powers and an apparition of other forms. In Western scientific language, the earth evolved through a phase of powerful geological and climatic forces which shaped the earth's crust, creating mountains, oceans, and rock formations. The main difference between these two myths is that Western scientific description acknowledges only physical forces, while Aboriginal explanation attributes consciousness to the creative forces and to everything created.

The Dreamtime epoch concluded with three fundamental conditions making possible the embodiment of conscious life: the earth's distinct topography, all variety of species of life, and patterns for social relationships. With the completion of this task, the Ancestors grew weary as the world was shaped and filled with species and varieties of ancestral transformations and they retired into the earth, the sky, the clouds, and the creatures to exist like a potency within all that they had created. Their journeys were preserved in stories, paintings, ceremonies, and patterns of living and extended a universal psychic consciousness to every living being as well as to the earth and the primary forces and elements.

The Western tendency to view land as material object, existing separately from humanity, allowed for it to be bought, sold, and exploited. In contrast, Aboriginal culture shared the identity of the forms, principles, and activities by which the natural world was created and saw life and sustenance embedded in the landscape which must be cared for as a friend and provider in continuing reciprocity. "With the birds, the fish, the animals, the plants, and all of nature we are still Yorro Yorro, standing up in accordance with the blueprint of Creation", explains Mowaljarlai, concerning the role of men in maintaining the connection between cosmic regions and earthly community life in the Kimberley, Western Australia.

Understanding Nature and Humanity

Generally, the Japanese philosophical and religious traditions represent a holistic ecocentric view characterized by an aesthetic and material oneness with all things. In pre-industrial Japan, the intimate relationship between people and their surroundings can be grasped by the Japanese word *fudo*, which translates into the natural environment. While nature and humanity have been viewed as separate but connected in Western thought, fudo refers to the idea that nature and society are inseparable. *Shi-zen* or *Ji-nen* is the Japanese word which most closely corresponds to the European term nature. The connotations of this term signify a concept of nature in which nature is not objectified. Rather, *Shi-zen/Ji-nen* represents the manner of being and becoming of nature, human and environment. *Shi-zen*, originally a Chinese word, was adopted by the Japanese about 1500 years ago. In ancient Chinese culture, the term had fundamental significance. Lao-Ze wrote: Man is based on earth, earth is based on heaven, heaven is based on the Way (*Dao*) and the Way is based on nature (*Zu-ran*, or *Shi-zen* in Japanese).

The contrast between European and Japanese garden design illustrates the difference in perspectives of nature. In the English garden, a person seeks escape from humanity into a natural space free of artificiality. In the Japanese garden, one tries to express, within a defined space, the totality of mountains and rivers. This is an artistic, human act representing the highest meaning of *Shi-zen*. The English garden is a public place, open to all. The Japanese garden is of a private nature created solely for the few who are artistically gifted. Nature relates equally towards all in the English garden. The *Shi-zen* of the Japanese garden requires those who are able to experience *Shi-zen*. This implies that *Shi-zen* can never be separate, but within the person.

It is the same for Ikebana art and the *Shi-zen* of the tea-ceremony. In the Japanese tea ceremony, the everyday routine becomes an art form. Sitting in a quiet, small room allows for the mind and body to become alert to the slightest sound, movement of the wind, perfume of flowers, changes in the intensity of light. This time passed within a small room heightens the sense of place by creating a moment of spiritual connection with the environment.

Sense of Place

The sense of place exists in the area where nature and humanity interact and shape each other. Within cultures, information concerning the geographical environment is expressed in certain ways, such as the making of maps, conceptualizing landscapes, and giving places names. In the Pintupi language of Central Australia, a place, or a country, or a camp is called *ngurra*. *Ngurra* signifies both the physical place where people share food, sleep, and dance, and also the metaphysical act of dreaming the country into existence. In this way, the question of Aboriginal identity is obtained by knowing the place. As Kuningga, the ancestral cat, traveled north he crossed plains, identified water sources and, after reaching a ceremonial site, continued north through a gap in the range. Kuningga's dreaming track weaves north, describing the particular geography of the land and of the natural resources contained within through a song cycle. Thus Kuningga's song can be understood as an oral map of the country. Places where significant events happened, where power was stored, or where the Ancestors had retired into the ground and still remain today are known as *Yarta Yarta* or special sites because of their concentrated ancestral potency.

Certain landscapes create a sense of awe by their presence. In Japan, Mt. Fuji, rising 13,000 feet above sea level, has been depicted in poems, paintings, and stories from ancient times. The sheer physical presence of mountains which appeared to connect the earth with the sky allowed for them to be considered Japanese shrines. The word *oyama* sometimes means shrine and is derived from *yama* (mountain). Tadahiko Higuchi in his book *The Visual and the Spatial Structure of Landscape*, described seven protoype shapes for shrine and temple complexes which were translated by Kazuo Matsubayashi. They are:

1. Water-distributing: where water flowed out of the mountain creating a perfect rice irrigation place;
2. Bowl: a small, peaceful, plain surrounded on all sides by green mountains;
3. Lotus Flower Calyx: the site of a mountain temple surrounded by eight other mountains, which resembled the calyx of a lotus;
4. Wind and Water (*fusui*): usually having mountains to the north, hills to the east and west, and a sea or river to the south, which was the ideal place to build a city, house or tomb;
5. Valley: enclosed by mountains on both sides, which was thought to be the home of the spirits of the dead;
6. Thickly-forested: a small mountain covered in thick foliage in sharp contrast to its situation near a flat plain, which was believed to be the sacred dwelling of a god;
7. Country-viewing: a low hill or mountain standing in the middle of a plain, commanding a view of the surrounding land.

The close relationship between the people and their mountainous environment, with each of the types serving an unique purpose, demonstrates the vital reciprocity between humanity and nature.

In Maori tradition, the special relationship between people and their land is revealed, first by the myths explaining the origin of the universe, in which the material world proceeds from the spiritual and the spiritual world interpenetrates the material, physical world of Te Ao Marama, and secondly by the use of dominant landmarks, mountains, or water sources, as symbols of tribal identity. While many indigenous peoples live intimately with nature, the Maori appear to be unique in their formulation of their *pepeha* or popular sayings of tribal identity. Every established Maori community may apply their standard *pepeha* or introduction by naming their mountains, river, and tribe at intertribal meetings, now as in the past, actively maintaining the oral tradition and identity of place.

> Makeo is the mountain – Ko Makeo te manunga
> Waiaua is the river – Ko Waiauna te awa
> Tutamure is the Ancestor – Ko Tutamure te tangata

Hong-Key Yoon has described the pattern of structure of these sayings which includes the first two phrases for natural identity in terms of landscape and the third for cultural identity in terms of people. The first phrase mentions the most significant landform, almost always a *Maunga* or mountain followed by the second phrase giving the name of a prominent body of water or *te wai*, *te awa* or river, *te awauni* or big river, and *te moana* or sea. The third phrase names the tribe, giving social and cultural identity. Maoris associated intense feelings with the mountains, considering them to be the place of buried ancestral bones, while the valleys and rivers, often serving as tribal boundaries, did not assume as intense a sanctity.

These *pepeha* revealed the symbolic relationship of the people within their place and were combined with the functional territorial boundary delineating home territory, while also perpetuating cultural heritage. Today, the Maori traditions of respect for place are represented by the Maori Secretary in the Ministry of the Environment. The name of their movement, *Maruwhenua* comes from the understanding of the human responsibility for protecting the land. Mmaruwhenua reflects the saying, "People perish, but the land endures".

Sustaining Natural and Human Place

Indigenous knowledge of nature has ensured the survival of many peoples living in fragile habitats. For example, the Tuareg of West Africa practice nomadic pastoralism in a land so arid that it cannot sustain continual habitation, and the Inuit of the frigid Arctic regions have traditionally depended on hunting and fishing. Modern hunting and gathering depends on cooperation betwen group members and the formation and maintenance of economic relations with other groups. In Ituri, the tropical rainforest of northeastern Zaire, a mutually supportive, economic relationship has been created between the Efe, who practice a semi-nomadic lifestyle, and the Lese, who live as shifting cultivators. The Efe trade forest products, nuts, construction materials, meat, and fish for produce grown by the Lese, and for their pottery, tools, and arrows.

In many areas of the world, indigenous societies have classified soils, climate, and animal species. Many of these words and descriptions for insects and plants are yet to be known by botanists or entomologists. The Hanunoo people of the Philippines distinguish 400 more plant species in their forests than the 1200 ones identified by the scientists working in the same location. The Kayapo of Brazil gather about 250 types of wild fruit, hundreds of tubers, leaves, and nuts, and at least 650 plants which are used for medicinal purposes. This vast, complex, indigenous knowledge is now viewed as being important not only to the sustenance of the traditional owners, but to the very survival of the world.

The human relationship to the natural environment has been conceptualized differently over time and through space. At the end of the twentieth century, understanding of the human experience of nature, from both a Western and non-Western cultural tradition, has become an urgent imperative. The challenge now is to regard humanity and earth in global terms while also seeking an understanding of the social and ecological implications of the whole of humanity planetized. There has been much discussion concerning the damaging perspectives of the Western, anthropocentric approach to nature which has led to actions based on a premise of the separation between mind/body, human/nature, and individual/society. This view has been blamed for much of the environmental destruction and exploitation of the world's human and natural resources. At the same time, there have been appeals to consider alternative modes of conceptualizing humanity and nature, such as developing a reverential attitude toward nature and acting in a socially cooperative manner, based on a non-Western, ecocentric perspective.

Cultural expressions, Western or non-Western however, show a universality only at the broad structural level. To go beyond this, it is necessary to explore the specific historical contexts of human habitation in place and through time. Non-Western and Western cultures are historically dynamic. It is important therefore to emphasize the particular and convey the diversity of traditions within and between groups, in order to create mutual understanding.

NANCY HUDSON-RODD

REFERENCES

Burger, Julian. *The Gaia Atlas of First Peoples*. London: Gaia Books, 1990.

Buttimer, Anne. *Geography and the Human Spirit*. Baltimore and London: John Hopkins University Press, 1993.

Lawlor, Robert. *Voices of the First Day (Awakenings in Aboriginal Dreamtime)*. Vermont: Inner Traditions International, 1991.

Marsden, Maori. "God, Man and Universe: A Maori View." In *Te Ao Hurihuri (Aspects of Maoritanga)*. Ed. Michael King. Auckland: Reed Books, 1992, pp. 118–139.

Matsubayashi, Kazuo. "Spirit of Place: The Modern Relevance of an Ancient Concept." In *The Power of Place (Sacred Ground in Natural and Human Environments)*. Ed. James Swan. Madras: Quest Books, 1991, pp. 334–346.

Mowaljarlai, David and Jutta Malnic. *Yorro Yorro: Everything Standing up Alive, Spirit of the Kimberley*. Broome, Western Australia: Magabala Books Aboriginal Corporation, 1993.

Myers, Fred. *Pintupi Country, Pintupi Self*. Washington, D.C.: Smithsonian Institution Press, 1986.

Said, Edward. *Culture and Imperialism*. London: Chatto and Windus; New York: Knopf, 1993.

Tellenbach, Hubertus and Bin Kimura. "The Japanese Concept of Nature." In *Nature in Asian Traditions of Thought (Essays in Environmental Philosophy)*. Ed. Baird Callicott and Roger Ames. New York: State University of New York Press, 1989, pp.153–163.

Yoon, Hong-Key. *Maori Mind, Maori Land (Essays on the Cultural Geography of the Maori People from an Outsider's Perspective)*. Berne: Peter Lang, 1986.

See also: Knowledge Systems – Ethnobotany – Maps and Mapmaking – Environment and Nature

GEOGRAPHY IN CHINA The term geography (*di li*) which first appeared in Chinese documents during the Spring–Autumn period (770–476 BC) literally meant the study of the order and morphology of the land. In the beginning the purpose of such a study was for survival; then later it was used to facilitate the production of food and gathering materials for shelter. Agriculture was central to the economy, hence information was collected on climate, land and water supply, and for crop cultivation and water control, e.g. to construct dikes. Because of these needs, studies in "geography of production" increased; thus the Chinese have accumulated much geographic knowledge since ancient times.

The two frequently cited geographic classics are *Yu Gong* and *Di Li Zhi*. They demonstrate the importance of geography in delineating major regions in the country and in collecting locational information; these geographical applications facilitated the governing of the ancient Chinese empire. Mapping of large regions and local areas followed.

Yu Gong (The Tribute of Yu, who was a semi-legendary

emperor,) is a short but brilliant work, written in the Warring States period (475–221 BC). It presents chapters on mountains and rivers, and on two methods of organizing geographical space: delineating nine physical regions, and allocating five political regions in China. Thus this work presents the earliest examples of regional and systematic geographies.

The term geography was first used as a title of a work in *Di Li Zhi* (Treatise of Geography), a chapter in the *Han Shu* (History of Han Dynasty), by Ban Gu (AD 32–92). It includes three sections, of which the first and third were reprints of earlier works, including *Yu Gong*.

The second section is devoted to the geography of 103 prefectures (*jun/guo*) and their subunits, the 1587 counties. Within each administrative unit, some physical, cultural, economic, and demographic features are described. Cities are mentioned. Interestingly, it also describes petroleum as some kind of liquid that can be burnt. However, the major focus of this work is on the history and changes of boundaries of administrative divisions and settlements; such information was important to the governing of a large empire. Unfortunately, too often this particular focus was emphasized at the expense of a more thorough treatment of the physical characteristics of an area, and their effects on the inhabitants and the economy.

The publication of *Di Li Zhi* marked the beginning of a chapter on geography to be included in a dynastic history; this chapter appeared in the second of the twenty-four such histories. Thus the content and the methodology presented therein became a model that was followed in the later dynastic histories.

Ban Gu's work also planted the seeds of a later development: the publication of local gazetteers (*di fang zhi*) with maps included in these volumes. Beginning in the fifth century, the production of these gazetteers increased in number and gradually became the most prevalent form of geographic writing and compilation in pre-modern China. During the Tang Dynasty (AD 618–907), the scope of some gazetteers began to expand to cover a province and even the whole country; some of these compilations were multi-volume and were really geographical encyclopaedias. Today they constitute a voluminous geographical and historical literature.

Beginning in the Tang Dynasty, field studies became more important; perhaps, by this time, long distance travel in China had become more manageable. Noted scholars spent many years away from home traveling and studying the country. They left with us their vivid descriptions and insightful interpretations of the geography of China. Among these books, the best known is the *Xu Xiage Yuji* (Dairy of The Travels of Hsu), written in the seventeenth century. Thanks to the efforts of these scholars, accurate geographical knowledge was accumulated, e.g. of the source of the Huang River, the process of erosion, and the retreat of the sea in the area of today's eastern lowlands. The last of these was studied by Shen Gua in the eleventh century; similar observations to that of Shen's were made in Europe by J. Hutton in the nineteenth century.

Despite the richness of the pre-modern literature, geography was not formally included in the civil examination in imperial China. Only during the late nineteenth century, as the modern school system was developed, was geography incorporated as an instructional subject in schools. Today, in Taiwan and on the Chinese mainland, geography courses, which vary in area coverage and degree of difficulty, are taught in every grade in the secondary schools. The two major purposes of this instruction are to provide students with basic knowledge of China and of other countries in the world and to foster the spirit of patriotism.

In historical China, geography and cartography played important roles in scientific inquiry and applications. In this century, the subject geography is taught in schools and colleges, and mapping activities are routinely carried out by government agencies and educational institutions.

MEI-LING HSU

REFERENCES

Cao, Wanru. "Some Questions concerning the History of Chinese Geography." *Zi Ran Ke Xue Shi Yan Jiu* 1(30): 242–250, 1982.

Fung, Yee-wang. "The Development of Geographical Education in Hong Kong and China (1949–1988): A Comparative Study." In *International Perspectives on Geographic Education*. Ed. A. David Hill. Skokie, Illinois: Rand McNally, 1992, pp. 51–61.

Hou, Renzhi. *Selected Readings of Famous Geographical Writings in Ancient China (in Chinese)*. Beijing: Kexue Chubanshe, 1959.

Hsu, Mei-Ling. "The Han Maps and Early Chinese Cartography." *Annals of the Association of American Geographers* 68: 45–60, 1978.

Hsu, Mei-Ling. "The Qin Maps: A Clue to Later Chinese Cartographic Development." *Imago Mundi* 45: 90–100, 1993.

Lu, Jonathan J. "China's Geographic Education." In *International Perspectives on Geographic Education*. Ed. A. David Hill. Skokie, Illinois: Rand McNally, 1992, pp. 63–76.

Needham, Joseph and Ling Wang. *Science and Civilisation in China*. Volume 3, Chapter 23. Cambridge: Cambridge University Press, 1959.

See also: Maps and Mapmaking in China – Petroleum

GEOGRAPHY IN INDIA The study of geography (*Bhūgol*) as a systematic literature was not in use in ancient India. But various names, such as *Bhuvanakośa* (Terrestrial Treasure), *Trilokya Darpaṇa* (World's Mirror) and *Kṣetrasamāsa* (Combination of Countries), were used to denote geographic phenomena. Beginning with the *Ṛgveda* (1500 BC), tribes, rivers, and mountains of India were mentioned. As the Aryans, who brought the *Ṛgveda* with them, first settled in northwestern India (Punjab and the Indus plain) only the related regional rivers like Sarasvatī (extinct), Sarayu (Sutlej) and Sindhu (Indus) were mentioned. Later, by the beginning of the Christian era when the three other *Vedas* were developed and the *Purāṇas* were written, a more extensive description of Indian geography evolved. The Aryan settlements advanced not only eastward to Bengal and Assam, but also to South India, which though Dravidian in terms of linguistic and physiognomic characteristics, became Hinduized. Kauṭilya's *Arthaśāstra*, written around 300 BC, prescribed a state-planned colonization policy for undeveloped (primarily natural vegetation covered) areas with people from foreign lands and kingdoms' surplus labor. Each village was to have five hundred families, mostly *śūdra* cultivators or laborers ; villages were to be grouped in a hierarchy of settlements of eight hundred, four hundred, two hundred, and ten units with a large town (*sthānīya*), small town (*droṇamukha*) smaller town (*kārvaṭika*), and large village (*saṃgrāhara*) to serve each group respectively.

The river Ganges (*Gaṅgā*) is mentioned in *Vāyupurāṇa* as a purifier of sinners, passing through thousands of mountains and irrigating hundreds of valleys. It was in *Kauṣītaki* that the southern mountain — Vindhyas — was first mentioned, while both *Vāsiṣtha Dharmasūtra* and the *Code of Manu* refer to Vindhyas by name. The two great epics — *Rāmāyaṇa* and *Mahābhārata* — give extensive descriptions of geographic features of both north and south India. When the king Daśaratha, father of Rāma, performed *Aśvamedhayajña* (Horse Sacrifice), to establish his supremacy over the world, both north and south Indian kings were referred to. In the *Bhīṣmaparva* section of *Mahābhārata*, Sañjaya, the chariot driver of the blind king Dhṛitarāṣtra, identified the nations, mountains, and rivers of India.

Beginning with the *Ṛgveda* through the *Purāṇas*, cosmological and cosmogonic interpretations of causes of wind, precipitation, day and night, seasons, and planetary movement have been made. Explanation of artistic, mechanical, instrumental, and philosophical origins of the universe has also been given. The artistic origin refers to god as an artist who skillfully constructed the universe. The mechanical origin is conceived as a sacrifice of the primeval body (*Ādipuruṣa*) who not only had the soul and the nucleus of the universe, but also embodied the Supreme Spirit, resulting in the formation of the earth, sky, wind, the moon, the sun, and other terrestrial elements. The philosophical concept considered the beginning as an empty space with no atmosphere or sky, and the universe was born out of its own nature, possibly by its own inherent heat. *Māhāpurāṇa*, composed by a Jain teacher in the ninth century AD, further crystalizes this idea that the world endures under its own nature and is divided into hell, earth, and heaven. The instrumental origin idea is reflected in the union of heaven and earth caused by the action of different gods, such as *Agni* (fire), *Sūrya* (sun), and *Indra*. "The central idea of various cosmogonic theories of the Vedic and post-Vedic period appears to be (1) the existence of water in the beginning, and (2) the creation of a cosmic nucleus — *Prajāpati*" (Ali, 1966). *Prajāpati* is an embodiment of propagation and hence maker of the universe.

The ancient literature described a cyclical human development on the earth. The Hindu view, expressed in the *Code of Manu*, speaks of four ages: Kṛta, Tritā, Dvāpra, and Kali, with a sharp break at the end of each age; physical and spiritual deterioration occurs at the completion of the four-age period with the universe coming to an end and then beginning a new cycle. The Jains also believed in the cosmic cycles, but unlike Hindus they did not foresee any sharp break at the ends of the periods (ages). The Jains divided a full cycle into six periods, each with ascending and descending halves. At the end of the sixth period, designated as "very wretched" (*Duḥṣama-Duḥṣama*), human deterioration reaches its peak with a fierce storm wiping out almost all inhabitants. After this a new six-period cycle starts again.

In a general view of explaining the distribution of continents and oceans the world was divided into seven regions or islands (*dvīpas*). S.M. Ali identifies those dvīpas as Śālmali (East Africa), Kuśa (Middle East), Polaksa (Mediterranean), Krauñcha (Europe), Pushkara (East Asia), Śaka (Southeast Asia), and Jaṃbū (Northern, Central, and Southern Asia). The southern part of the Jaṃbūdvīpa was inhabited by Hindus in the land called Bharata. At the center of the Jaṃbūdvīpa was the Meru identified with Pamir knots, and at the northern extremity adjoining the Arctic ocean was the Uttarakuru. Jaṃbūdvīpa was surrounded by an "ocean", meaning a physical barrier. Yet another Puranic conception compares the earth with a lotus. The pericap of the earth-lotus was the Meru; each petal represented a continent of the *Māhādvīpa* (great island), situated equidistant from each other and surrounded by oceans. Jaṃbūdvīpa was the largest of the continents.

Bhāratavarsha (India) has been described in the *Purāṇas* as having various shapes: a half moon, a triangle, etc. It was given the shape of a rhomboid by Eratosthenes in BC 320 and that of four equal triangles in the *Mahābhārata* composed around BC 100. Ptolemy (AD 100–170) in mapping India ignored the peninsula and drew an almost straight line from the Gangetic delta to the Makran coast in Baluchistan. Varāhamihira (AD 550) identified seven divisions of India surrounding the central division of Pañchāla (Punjab/Haryana area); they were Magadha (east), Kaliṅga (southeast), Avanta (south), Sindhu-Sauvīra (west), Ānarta (southwest), Madra (north), and Kauṇḍa (northeast).

Though only the southern part of Jaṃbūdvīpa was considered Bhāratavarsha, the Indians referred to the latter as the whole earth because the landmass of South Asia surrounded by mountains in the north and seas in the south was not only a physical–cultural cul-de-sac, secluded from other peoples but was so productive and large that to the people of India it constituted the world. Puranic legends claim that the king Bharata, whose name was identified with the country, ruled the entire earth, meaning all of India. Such was the case with the emperor Aśoka (272–232 BC) as inscribed in his Fifth Rock Edict.

Geopolitics took a special form in ancient India. The *Code of Manu* and the *Arthaśāstra* developed a set of stratagems and practices based on the position of states which were parts of a circle (*Mandala*). A kingdom was considered to be surrounded by four circles of states. The first circle consisted of friends and friends' friends. The three outer circles represented the states that belonged to the enemy, the middle king, and the neutral king. A sixfold interstate policy — peace, war, marking time, attack, seeking refuge, and duplicity — was evolved to maintain a balance of power among the circle of states. Policies related to extension or defense of a state depended on one's power in relation to the nearness and distance from other states. As in the Puranic *Bhāratavarsha*, *Arthaśāstra* defined the sphere of influence of an imperial ruler (*Cakravartin*) by the conquest of all land between the Himalāyas and the southern sea, meaning Cape Comorin. Bāṇa's *Kādambarī* and *Harshacharita* while conforming with *Arthaśāstra*'s boundary specified the eastern limit as the mythical sunrise or Udaya mountains (possibly the Blue Mountains or Pawnpuri of Mizoram) and the western limit by another mythical mountain-sunset or Mandara (possibly the Western Ghats or Sahayadri).

ASHOK K. DUTT

REFERENCES

Ali, S. M. *The Geography of The Purāṇas.* New Delhi: People's Publishing House, 1966.

Cunningham, Alexander. *The Ancient Geography of India.* Varanasi: Indological Book House, 1979.

De Bary, W. M. Theodore, ed. *Introduction to Oriental Civilizations: Sources of Indian Tradition.* New York and London: Columbia University Press, 1967.

Goyal, S.R. *Kautilya and Megasthenes.* Meerut: Kusumanjali Prakashan, 1985.

Majumdar, R. C. and A.D. Pusalker, eds. *The History and Culture of the Indian People: The Vedic Age.* London: George Allen and Unwin Ltd, 1951.

Majumdar, R.C. and A.D. Pusalker, eds. *The History and Culture of the Indian People: The Age of Imperial Unity.* Bombay: Bharatiya Vidya Bhavan, 1953.

Majumdar R.C. and A.D. Pusalker, eds. *The History and Culture of the Indian People: The Classical Age.* Bombay: Bharatiya Vidya Bhavan, 1954.

Sircar, D. C. *Studies in the Geography of Ancient and Medieval India.* Delhi: Motilal Banarsidass, 1971.

Thapar, Romila. *A History of India 1.* Middlesex, England: Penguin Books, 1966.

Tripathi, Maya Prasad. *Development of Geographic Knowledge in Ancient India.* Varanasi: Bharatiya Vidya Prakashan, 1969.

GEOGRAPHY IN THE ISLAMIC WORLD Geography is the study of the earth's surface as the space within which the human population lives. The internal logic of this study has tended to split modern geography into two parts: physical and human. Such a division was inapplicable in the geography of the Middle Ages, the golden age of scientific inquiry in the Islamic civilization. Nevertheless, if we confine the meaning of science and technology to the natural and exact sciences, then a survey of geography in Islam may be divided into three categories: exploration and navigation; physical geography; and cartography and mathematical geography.

Prompted by the sense of Islamic brotherhood, and the quest for knowledge and piety, Muslim scholars engaged in many exploration and navigational activities between the ninth and twelfth centuries. The journeys were not confined to the political boundaries of the Islamic empire but extended to distant regions such as China, Southeast Asia, southern Africa, and Russia. In northern Europe, the extent of the Muslims' travels may be gauged from the fact that some ten million pieces of the Islamic state's coins have been found around the Baltic. In Sweden alone twelve thousand such coins were discovered from 169 sites.

From extant documents, it may be deduced that no less than twenty geographers were involved in navigation and exploration, resulting in a massive store of information and descriptions of the *terra-cognita.* Out of these, five may be regarded as outstanding. These geographers are: al-Muqaddasī (b. tenth century), Ibn Jubayr (b. AD 1145), Yāqūt (b. AD 1179), Ibn Baṭṭūṭa (b. AD 1304), and Ibn Mājid (b. AD 1400). The contributions of Ibn Faḍlān (tenth century), Ibn Ḥawqal (tenth century), Masʿūdī (d. AD 456), al-Bīrūnī (b. AD 972), al-Idrīsī (b. AD 1099), Ibn Rushd (AD 1300), Jawainī (ca. AD 1000), Abdur Razzāq (AD 1000), Yaʿqūbī (d. AD 897), al-Marwazī (d. AD 887), Nāṣir-i-Khusraw (b. AD 1003), al-

Māzinī (b. AD 1080), al-Maghribī (d. AD 1274), and Ibn Khurdādhbih (ca. AD 900) are also noteworthy.

Al-Muqaddasī or al-Maqdisī spent twenty years traveling the length and breadth of the Islamic empire. He perceived that it was important to be able to substitute hearsay with personal observations, and set out on a series of field trips to gather information and knowledge about places. One of his major works was *Kitāb aḥsan al-taqāsim fī maʿrifat al-aqālīm* (The Best Climatic Divisions), which was regarded as accurately informative of areal differences in climate in the Islamic world. Ibn Jubayr was meticulous in his description of places such as Mecca and Medina in his *The Travels of Ibn Jubayr*, which became a basis for historical comparison with later writings. Yāqūt 's *Muʿjam al-buldān* (Dictionary of Geography) was another major contribution, and his extensive traveling within the Islamic empire resulted in a vivid account of the Islamic world before its fall to the Mongols. Ibn Baṭṭūṭa certainly stands out as the most seasoned traveler. In twenty years he journeyed as far as 120,675 km (75,000 miles) covering countries like China, Southern Ukraine, Sumatra, the Malay Peninsula, Kampuchea, Maldive and the Volga. His detailed accounts of life in these places contributed further to the geographical preoccupation with different areas. However, Ibn Mājid and Sulaymān al-Mahrī were better informed about the South-east regions of the world. The former's *Kitāb al-fawāʾid fī uṣūl ʿilm al-baḥr waʾl-qawāʿid* (Principles of Navigation) and the latter's *ʿUmda* and *Minhāj* became important references for those who wanted first-hand information on the Malay Archipelago.

The fact that collectively the Muslim travelers and navigators covered more than two thirds of the earth's surface illustrates the Muslim scholars' disdain for armchair scholarship. That distant places were visited for collecting information at a time when vehicles, vessels and instruments were underdeveloped, and when both land and sea routes were more unsafe than otherwise, serves to prove the inquiring spirit of the Muslim geographers.

In physical geography the contributions made by Muslim geographers in the Middle Ages pertain to their theoretical speculations about the lithosphere, atmosphere, hydrosphere and biosphere. In cosmology, they came out with new notions regarding the relationship between cosmological cycles and geographic changes. For instance, The Sincere Brotherhood (*Ikhwān al-ṣafā*), an association of scholars whose treatises were considered the standard for scientific knowledge of their day, conceived of the influence of the angle of declination on insolation, the influence of relief on precipitation, and the relationship of precipitation with the origin of streams and springs.

With regard to the lithosphere, the solid part of the earth, Muslim geographers thought of landforms as being the results of erosion, deposition and tectonic activities. Soils were considered the results of organic decomposition, and winds (gases in modern usage) trapped underground and seeking outlets were held responsible for earthquakes.

Their theories on atmosphere embraced both local atmospheric phenomena and the distribution of macro-climatic elements. Thus, by envisaging atmospheric processes, such as insolation, convection and evaporation, they were able to account for the formation of clouds, rainfall, snow, fog, and dew as different forms of evaporation due to differences in temperatures. They also divided the world into threee climatic zones: hot, cold, and temperate, based on their understanding of the changing positioning of the sun.

In terms of the hydrosphere, most of the Muslim geographers' works were focused on understanding the causes of tides, and the origin of oceans and reasons for their salinity. In accounting for tides, the geographers were divided. Some pointed to the influence of heat generated by the sun and moon; others attributed the tides to the influence of wind on the movement of sea water. Muslim geographers, in general, concurred with their Greek counterparts in thinking the oceans had originated from the remnants of primitive precipitation which did not get dried when matter was formed in the universe. They, however, differed again with respect to marine salinity. Some ascribed it to the transformation of calm water to salty hard water after prolonged gestation.

Regarding the biosphere, geography in the Islamic Middle Ages saw the categorization of flora and fauna according to the grade of their creation. Thus, al-Jāḥiẓ wrote of four classes of animals based on their locomotion: those that creep, walk, swim, and fly.

By contrast, Ibn Maskwaih's division was based on the creatures' stage of evolution, with the lowest stage being occupied by vegetation which did not require seeds to grow, and humans occupying the highest stage of evolution. The categorization by the *Ikhwān al-Ṣafā* was based on senses, whereby creatures with one sense occupied the lowest level and those with five senses the highest. In spite of these diferences, the Muslim biogeographers agreed that the spatial distribution of organisms was governed by climate, topography, and soils.

Yaʿqūbī was noted for his *Kitāb al-buldān* (The Book of Countries) written in AD 891, in which he described in detail the physical characteristics of Baghdad, Samarra, Iran, Turan, and the present Afghanistan. Ibn Rustah discussed the importance of locational and physical environmental factors on the development of cities, such as Medina and Mecca, in the seventh volume of his *al-Aʿlāq al-nafīsah* (The Book of Precious Things), an encyclopaedia completed in AD 903. Al-Maqdisī described an imperfect spherical earth, the northern hemisphere of which contained more land than water. He also gave a detailed account of the world climate in terms of zones. Yet the most daring speculation may be ascribed to al-Bīrūnī, who theorized that, based on empirical evidence of its physical environment, the Indian subcontinent might have been a continental shelf before depositions raised it into a landmass. Lastly, al-Qazwīnī (b. AD 1203) produced two books, *ʿAjāʾib al-buldān* (The Wonders of the Lands, or Geography) and *ʿAjāʾib al-makhlūqāt* (The Wonders of Creation, or Cosmology), which were notable for their reference to more than fifty authors. The significance of this fact is that the large number of practicing geographers that were cited indicates how active and prolific physical geography was in the medieval Islamic civilization.

The achievement of Muslim geographers in mathematical geography may be described in various phases of development. An important phase was the establishment of the school of mathematical geography in Shiraz, Persia after AD 950. This school produced important scholars such as ʿAbd al Raḥmān al-Ṣūfī and Ibn al-Aʿlam. ʿAbd al Raḥmān reviewed Ptolemy's Star catalogue, while Ibn al-Aʿlam compiled an astronomical table which became a main source of reference for centuries.

Important developments took place in the school in Cairo towards the end of the tenth century and the early eleventh century. Important scholars from this school included Ibn Yūnus, who prepared astronomical tables and corrected several errors committed by the Greeks Ptolemy and Hipparchus. For instance, according to Ptolemy, the precession of equinoxes occurs at 1° in 100 years, but the correct measure was 1° in 70 years, as calculated by Ibn Yūnus. His book entitled *Al-zīj al-ḥakimī* (Tables of Wisdom) explained, among other mathematical ideas, the method of calculating longitutes and latitudes.

Among the notable works written in Central Asia of this school was al-Bīrūnī's book entitled *Taḥdīd nihāyāt alamākin* (The Determination of the Coordinates of Cities).

Other scholars of high reputation were al-Zarqāllu, Naṣīr al-Dīn al-Ṭūsī and Ulugh Beg. Al-Zarqāllu produced several books on astronomy including the *Toledan Tables*. His other work entitled *Qānūn* (Canon) was used by European scholars during the Renaissance. Al-Ṭūsī is remembered for his book *Zij-i-īlkhānī,* which recorded all the notes and observations he made from the observatory of Maragha. His other important writing was on the astrolabe and the problems related to the measurement of time, celestial altitudes, meridians and the like. Ulugh Beg was a Tartar prince (AD 1393–1449) who was also noted for the observations he conducted from the observatory in Samarkand. He managed to point out several errors committed by Ptolemy in the latter's tables and made new tables of latitudes and longitudes.

With respect to cartography, Muslim scholars followed either one of two schools: one based on the cartographical method of Ptolemy, and the other on that pioneered by al-Balkhī. Ptolemy's teachings were followed by al-Idrīsī, and al-Balkhī's were continued by al-Istakhrī, Ibn Ḥawqal and al-Maqdisī. The Muslims' maps were drawn and utilized in conjunction with countries and regions. Idrīsī became famous in the West later, because of his maps and cartographical work. His maps were valued not so much because of their accuracy but because of the symbols he used to distinguish various physical and cultural forms.

A prominent feature of the achievement of Muslim scholars in mathematical geography and cartography was the invention of scientific instruments of measurement. Among these were the *astrolab* (astrolabe), the *ruba* (quadrant), the gnomon, the celestial sphere, the sundial, and the compass. The portable astrolabe was used by navigators for measuring altitudes up to the seventeenth century. The *ruba* was used for measuring the value of angles. The gnomon was also used for measuring altitudes of the sun and other planets. The use of celestial sphere was to explain celestial movements, the sundial for calculating daily time and the azimuth (compass bearing) of Mecca, and the compass for finding direction during navigation.

AMRIAH BUANG

REFERENCES

Alavi, S.M.Z. *Geography in the Middle Ages.* Delhi: Sterling Publishers Ltd, 1966.

Arnold, T.W. "Arab Travellers and Merchants, AD 1000–5000." In Newton, A.P. *Travel and Travellers of the Middle Ages.* London: Kegan Paul, 1926.

Institut für Geschichte der Arabisch-Islamischen Wissenschaften. *Islamic Geography.* Frankfurt: Strauss Offsetdruck, 1992 (in Arabic).

Kennedy, M.H. and E.S. Kennedy. *Geographical Coordinates of Localities from Islamic Sources.* Frankfurt: Institut für Geschichte der Arabisch–Islamischen Wissenschaften, 1987.

Nafis, Ahmad. *Muslim Contribution to Geography.* Lahore: Muhammad Ashraf, 1972.

Tibetts, G.R. *A Study of the Arabic Texts Containing Material on South-east Asia.* London: Royal Asiatic Society, 1979.

Tozer, H.F. *A History of Ancient Geography,* 2nd ed. New York: Biblio and Tannen, 1964.

See also: Maps and Mapmaking – al-Muqadasī – Ibn Jubayr – Yāqūt – Ibn Baṭṭūṭa – Ibn Mājid – Ibn Ḥawqal – Masʿūdī – al-Bīrūnī – al-Idrīsī – Ibn Rushd – Yaʿqūbī – al-Marwazī – Nāṣir-i-Khusraw – al-Maghribī – Ibn Khurdādhbih – Ikhwān al-ṣafā – al-Maghribī – Abū Maʿshar – Ibn al-Aʿlam – Ibn Yūnus – al-Zarqāllu – Naṣīr al-Dīn al-Ṭūsī – Ulugh Beg – Maragha – Astrolable – Balkhī – Sundials

GEOGRAPHY IN MESOAMERICA The most original intellectual creation to emanate from Mesoamerica was the 260-day sacred almanac, known variously as the *tzolkin* among the Maya and as the *tonalpohualli* among the Aztecs. Pre-dating the 365-day secular calendar, it ran concurrently with the latter to produce a never-ending series of 52-year cycles ($365\times52 = 260\times73 = 18{,}980$ days), giving rise to the Mesoamerican belief that history repeated itself every 52 years.

Though the origin of the 260-day calendar has long been debated, the most convincing explanation for its astronomic underpinnings was first given by Zelia Nuttall in 1928, who argued that it represented the interval between zenithal sun positions at Copán, the major Maya astronomical center located in the mountains of western Honduras. At the latitude of Copán (14.8°N), the sun passes overhead at noon on August 13 on its apparent southward journey to the Tropic of Capricorn and again on April 30 on its apparent northward journey to the Tropic of Cancer. In 1945, R. Merrill called attention to the "coincidence" of the August 13 date with the start of the so-called Maya Long Count, which in the Goodman–Martínez–Thompson correlation between the Maya and Christian calendars fixes the beginning of the present cycle of the world as August 13, 3114 BC. (The Long Count represented a meshing of the 260-day sacred almanac and the 365-day secular calendar in such a way that each day was as uniquely and precisely identified as they are in the Julian Day system employed by modern astronomers.)

In 1948 a Guatemalan scholar, R. Girard, contended that the calendar's birthplace lay along the same parallel of latitude but in the mountains of his country instead. While concurring in the astronomical importance of the 14.8° parallel of latitude, the present writer was forced to reject both Copán and the highlands of Guatemala as the calendar's cradle, for reasons both of history and of geography. No pre-Columbian site situated along that parallel in the highlands of Central America predates the fifth century AD, whereas the sacred calendar is known to have been in use for several centuries before the birth of Christ. Furthermore, many of the day-names used in the sacred almanac commemorate lowland tropical animals. The only place where the requisite astronomy, history, and geography come together is at Izapa, on the Pacific coastal plain of Mexico, where a large ceremonial center of Formative age (1500 BC–AD 300) is found amidst a tropical rainforest ecological niche.

Field work at Izapa in 1974 revealed not only that the entire ceremonial center is oriented to Tacaná, a commanding volcano of 13,428 feet elevation on the northern horizon, but also that the highest volcano in Central America, Tajumulco (13,845′) marks the azimuth of the rising sun on the summer solstice (June 22) as seen from Izapa. This use of prominent topographic features to serve as calendrical markers was subsequently traced to the oldest Olmec ceremonial centers of San Lorenzo (1200 BC, oriented to Zempoaltepec (11,138 ft.) at the winter solstice sunset), La Venta (1000 BC, oriented to Cerro Santa Martha (4600 ft.) at the summer solstice sunset), and Tres Zapotes (800 BC, oriented to Cerro San Martín (4600 ft.) at the summer solstice sunrise), but also to the earliest ceremonial centers of the Mexican plateau, including Cholula, Cuicuilco, Tlatilco, and Tlapacoya. Significantly, only two of the earliest Maya sites, Uaxactun and Tikal, have mountains within view, but both of them are likewise oriented to the winter solstice sunrise over the highest peak within sight — the first to Baldy Beacon (3346 ft.) and the second to Victoria Peak (3680 ft.). In each instance, the local site-factor which pinpointed the ceremonial center's location was the availability of water, both for domestic uses and, in the early Olmec centers, for transport, but only when the situational factor of the proper solsticial orientation coincided with it.

Interestingly, where prominent mountains are not visible on the horizon, the early Mesoamericans substituted architectural alignments for solsticial orientations. The greatest of pre-Columbian cities, Teotihuacán, about thirty miles northeast of present-day Mexico City, combines an intriguing blend of both principles: it is located precisely in line with the winter solstice sunrise over the highest mountain in Mexico — Orizaba, 18,700 ft. — but the peak itself is obscured by a low ridge of hills on the southeastern horizon. It would appear that a "relay station" was built there to allow the priests of Teotihuacán to calibrate their calendar with the southernmost position of the sun, whereas the entire city itself was meticulously gridded to the sunset position on August 13 — the day the present cycle of the world was believed to have begun.

The commemoration of the August 13 sunset is also found in the layout of other ceremonial centers on the Mexican plateau of later, Toltec vintage, but its most widespread use in city planning was amongst the Maya. The location of the earliest Maya ceremonial center at Edzná (dating to 150 BC) was primarily dictated by the presence of the largest *aguada*, or temporary lake, in all of the Yucatán, but its internal layout clearly reflects their religious preoccupation with the August 13 sunset. Similarly, numerous structures throughout the Maya regions of Yucatán and Petén reflect the August 13 alignment, among them the Codz Pop at Kabah, the Pyramid of the Magician at Uxmal, and El Caracol at Chichén Itzá, to name but a few. The crowning blend of celestial mechanics, time, and space is to be found in the Maya capital of Tikal, where five sky-scraper pyramids, all over 200 feet in height and all constructed in the eighth century, mark the alignments of the August 13 sunset (Temple I to Temple IV), the equinoxes (sunrise Temple III to Temple I; sunset Temple I to Temple III), and the winter solstice sunrise (Temple IV to Temple III) in a sophisticated astronomical matrix. Further, an alignment from Temple V to Temple I not only marks a perfect right angle to the August 13 alignment but one from Temple V to Temple II likewise pinpoints the westernmost point in the rotation of Polaris, which at that time was the closest thing to a polestar which existed, even though it was a full 8° away from its present position. (Clearly, the repeated use in Mesoamerican ceremonial centers of the August 13 sunset orientation not only forcefully argues for the sacred almanac's astronomical origin but it also makes any alternative explanation, such as the length of the human gestation period or the simple permutation of the numbers 13 and 20, untenable.)

In the fifth century, the Mesoamericans appear to have carried out two remarkable geographic expeditions. Although they both had ultimate religious goals in mind, these endeavors probably reflected the closest approach the pre-Columbian civilizations ever made to what we regard as scientific inquiry. The priests of Teotihuacán dispatched an expedition into the northern desert to determine where the "sun stopped" in its annual migration, and in consequence of this, the astronomical center of Chalchihuites was founded on the Tropic of Cancer. Lacking a better means of recording key alignments, they dug trenches through the earth and plastered them with adobe to mark the summer solstice sunrise. (In the tenth century the Toltecs added a trench to mark the beginning of their New Year, February 12.) Perhaps under Teotihuacán's sponsorship, the Maya carried out a similar expedition at about the same time to locate the place where the 260-day sacred almanac could be calibrated. The result of this venture was the founding of Copán, in the western mountains of Honduras, whose oldest recorded Long Count date is AD 426. Thanks to its key geographic location, Copán was ultimately to become the Mayas' principal center for astronomic studies — the late and distant heir of Izapa where the first intellectual stirrings had begun nearly two millennia earlier.

VINCENT H. MALMSTRÖM

REFERENCES

Aveni, Anthony F., Horst Hartung, and J. Charles Kelley. "Alta Vista (Chalchihuites): Astronomical Implications of a Mesoamerican Ceremonial Outpost at the Tropic of Cancer." *American Antiquity* 47: 316–335, 1982.

Malmström, Vincent H. "Origin of the Mesoamerican 260-day Calendar." *Science* 181: 759–60, 1973.

Malmström, Vincent H. "A Reconstruction of the Chronology of Mesoamerican Calendrical Systems." *Journal for the History of Astronomy* 9: 105–116, 1978.

Malmström, Vincent H. "Architecture, Astronomy, and Calendrics in PreColumbian Mesoamerica." In *Archaeoastronomy in the Americas* Ed. Ray A. Williamson. Los Altos, California: Ballena Press/Center for Archaeoastronomy, 1981.

Merrill, R. H. "Maya Sun Calendar Dictum Disproved." *American Antiquity* 10: 307–11, 1945.

Nuttall, Zelia. "Nouvelles lumières sur les civilisations américaines et le système du calendrier." *Proceedings of the Twenty-second International Congress of Americanists*. Rome, 1928, pp. 119–48.

See also: Long Count – Calendars – Time – City Planning

GEOGRAPHY OF NATIVE NORTH AMERICANS Geographical inquiry focuses on the creation and recreation of space and place. These two dominating themes of geography are useful organizers for discussing Native American geography especially when spatial conditions and landscapes are investigated within the context of human–land relations and environmental perception. Native American geography can be described on the basis of size of area, territoriality, material culture, and decision-making with regard to natural resources. Contrary to Western belief systems, Native American–land relations are primarily structured around a set of beliefs that recognize Earth as the spiritual and material center of life.

Generally, Native Americans do not divide their world between the real and the spiritual. The world includes every existing entity and those spirits who occupy them. All aspects of nature are sacred; there is no separation between the great universal and human actions. The Native American world is based on recognition, worship, and action that expresses a oneness with nature. The Ojibway of the Upper Great Lakes region view of Earth as "spirit garden" dramatizes the relationship between spirituality and reality. European concepts such as "wilderness" and "resource maximization" stand in sharp contrast to Native American environmental perception.

Geography requires an ability to recognize and reconnoiter within an area. While Euro-Americans often view area in terms of ownership and control, based on politically structured space, dominance, and hierarchies, Native Americans generally view area based on concepts of territoriality. Territoriality implies a recognition of social and physical landscapes. Acute observation skills, familiarity with local and regional physical characteristics, and recognition of other social groups within the area are consistent attributes of Native American geography. For example, the Apache utilized an area of the American Southwest generally between the Rio Grande River and the Great Plains. Primarily a mountain people dependent on species found at various elevations during changing seasons, the Apache also developed hunting skills that allowed them to take advantage of grasslands species such as buffalo. They shared buffalo hunting with distant Apache relatives and permitted other Apache to hunt and gather resources in their mountain territory when grasslands resources were scarce. While immediate territories were defined by occupancy, use of distant territories for resource gathering was permissible. The Mescalero Apache had a sense of resource-holding, that is they held all resources as a common good and shared vital sources as other Apache made similar demands.

An important aspect of territoriality is the ability to communicate conditions of space and place. The Inuit are superb cartographers, able to produce maps of territory with accurate scale, symbology, and direction. The maps also convey a sense of space and place where value is placed on animal trails, natural features, and settlement locations. An Inuit map of the north therefore demonstrates Inuit reality as opposed to another culture's maps of northern places.

Places usually recognize the spiritual integrity of natural object or events that magnify the wishes of a creator. The Hopi of the American Southwest named each of their mesa homelands with respect to physical features and conditions of Hopi creation beliefs. Hopi place names not only imply the physical but also the spiritual conditions of existence. Prior to contact with non-Hopi groups, especially non-Indian groups, each mesa was a world unto itself which anchored spiritual and material existence. Recent events within Hopi mesas, such as surface coal mining by western industrialists, have wounded sacred ground.

The use of natural resources for survival is generally framed within the context of small Indian populations, exploitation to meet nothing in excess of needs, and returning something to nature for the use of an item. Outside the framework of capitalism and Newtonianism, Native Americans developed a resource use paradigm that focused on the unity of all beings and their systemic value rather than individual worth. For example, individual animals within a forest ecosystem were recognized in relationship to other plants, soil, water, and air. The use of a specific animal was always observed with respect to other elements of the ecosystem with an attitude of thankfulness and recognition of the animal as a gift. The Ojibway practiced this respectfulness of animals taken in a hunt.

This is not to suggest that specific species were not singularly important to American Indian groups. The buffalo provided nearly all of the needs of Plains Indians while Northwest Coast Indians were nearly exclusively dependent upon salmon. Given the low population numbers of Indians, neither species were threatened with extinction due to overkill. It is reasonable to believe that resource demands and human populations would have remained static if Indian societies had not been impacted by European explorers with different technologies and valued resources.

Reliance upon specific species developed over centuries of adaptation. As American Indian groups entered new areas of North America and as North American physical environments were transformed by primarily climatic events, groups had to adjust, migrate or face decline. Many groups did adjust to what may appear to non-Indian eyes to be a landscape of harsh conditions. However, as the Piute and Shoshone of the Great Basin discovered, evolving desert conditions were not insurmountable. There were more than ample resources to support Piute and Shoshone before modern pollution deteriorated resource habitats and species.

The American Indian concept of spirituality in nature has historically predicated their geography. Since contact with Euro-Americans, population dynamics, resource use, and creation of places have been modified. Traditional views including geography are being challenged by modern internal and external conditions of tribal groups. Within the current era of Indian self-determination a balance between traditional beliefs and progressive ideas is being sought. Human-land relationships and environmental perception may be redefined, thus causing a new geography of Native Americans.

MARTHA L. HENDERSON

REFERENCES

Albers, Patricia and Jeanne Kay. "Sharing the Lands: A Study in American Indian Territoriality." In *A Cultural Geography of North American Indians*. Ed. Thomas E. Ross and Tyrel G. Moore. Boulder, Colorado: Westview Press, 1987.

Basehart, Harry. "The Resources Holding Corporation Among the Mescalero Apache." *Southwestern Journal of Anthropology* 23: 277–291, 1967.

Churchill, Ward. "The Earth is Our Mother: Struggles for American Indian Land and Liberation in the Contemporary United States." In *The State of Native America*. Ed. Annette M. James. Boston: South End Press, 1992.

Jackson, J.B. "The Order of a Landscape: Reason and Religion in Newtonian America." In *The Interpretation of Ordinary Landscapes: Geographical Essays*. Ed. D.W. Meining. New York: Oxford University Press, 1976.

Johnston, Basil. *Ojibway Heritage*. New York: Columbia University Press, 1976.

Matthiessen, Peter. *Indian Country*. New York: Penguin Books, 1979.

Rundstrom, Robert A. "A Cultural Interpretation of Inuit Map Accuracy." *Geographical Review* 80 (2): 155–168, 1990.

Sack, Robert David. *Conceptions of Space in Social Throught: A Geographical Perspective*. Minneapolis: University of Minnesota Press, 1980.

Tuan, Yi Fu. *Topophilia: A Study of Environmental Perception, Attitudes and Values*. Minneapolis: University of Minnesota Press, 1976.

See also: Environment and Nature

GEOMANCY IN CHINA The Chinese word for geomancy is *feng-shui*, which means "wind and water" or *ti-li* which means "the principles (patterns) of the land". The term *ti-li* is also the Chinese word for geography, which suggests an intimate relationship between Chinese geomancy and geography.

In Chinese geomancy, a place having certain landforms and orientations is believed to be more auspicious than others. An auspicious place is where vital energy (*sheng-ji*) is accumulated and available to humans who occupy the site. The function of vital energy is to give birth to and support all living things, and it is stored in certain places meeting geomantic requirements, blessing the people who use the site in harmony with the surrounding landscape.

The flow of vital energy underground (often through mountain ranges) is analogous to that of blood through the veins of the human body. Therefore, a geomancer's job is to find a spot where vital energy is accumulated, in a similar way that an acupuncturist finds a critical spot of the body where a needle can be planted.

An auspicious site (geomancy cave) is sometimes compared with a melon on a vine. The nutrients from the soil are taken through the roots, and transported through vines to be stored in the melon before being consumed by humans. In a comparable manner, the vital energy flows underground through the veins and is deposited in the geomancy cave to be available for humans who use the place appropriately by building, for example, a house or a grave.

The most important geomantic condition of an auspicious site is that it be sheltered by a surrounding range of hills on three sides, with one side open. The range on the left (normally East) side is called 'azure dragon'; the right (normally West) range, 'white tiger'; the hill behind an auspicious site is called the main mountain or 'black warrior'. A horseshoe shaped basin is an auspicious site. The most desired orientation of the site is to be facing south. The land also needs to have an access to a watercourse; the most desirable ones flow slowly with many bends.

In geomancy, the people who occupy auspicious sites are believed to benefit from the land by enjoying longevity, accumulating wealth, or achieving fame in the world. Traditionally, geomancers were normally consulted when building cities, temples, graves and other human settlements. Professional geomancers are supposed to know how to choose an auspicious site by applying geomantic principles, and can be seen as the traditional Chinese version of modern geomorphologists, location analysts or landscape and architectural planning consultants.

It is virtually impossible to understand the East Asian cultural landscape without having knowledge of geomancy. For instance, important cities like Beijing and Nanjing in China, Seoul and Kaesong in Korea, and Kyoto and Nara in Japan all were chosen and planned geomantically. Thus, Chinese geomancy is defined as "a unique and comprehensive system of conceptualising the physical environment which regulates human ecology by influencing man to select auspicious environments and to build harmonious structures (i.e. graves, houses, and sites) on them" (Yoon, 1976).

In the evaluation of the quality of the landscape (the environment, place) in geomancy, the following three images are important. First, the landscape is perceived as a magical being which can influence people mysteriously, either auspiciously or inauspiciously. An auspicious site can bless people with a happy life, while an inauspicious site can cause people to suffer misfortunes, including disease, bankruptcy, death, and infertility.

Secondly, the landscape is personified and regarded as a system of either a living organism or an inanimate object. Either an element of a local landscape such as a hill or stream, or the entire landscape itself may be treated as a functioning system of an object such as a cow, boat or a flower. Personification of a landscape as a beautiful, balanced and peaceful object normally indicates an auspicious site, while personification as an ugly, dangerous and unbalanced object suggests an inauspicious site. Depending on how a geomancer perceives a local landscape, it may be compared to (personified as) any object in the world.

Thirdly, the landscape is seen as a vulnerable being which can easily be hurt or remedied by human interference, because the vital energy which flows beneath the surface is extremely vulnerable. Vital energy is only one particular phase of the *Yinyang* energy according to a geomantic classic called *Zang-shu*. When *Yinyang* energy belches out, it becomes the wind. When the wind ascends, it becomes a cloud. When it descends, it becomes rain. When it flows under the ground, it becomes vital energy, but when it emerges out of the ground, it is no longer vital energy (*Guo Bu*). That is why the vital energy of an auspicious site should be utilized without disturbing it, by building a house or a grave in harmony with the surrounding landscape.

In the history of China, geomancy was often more popular for choosing an auspicious site for a grave rather than for a house, although the same geomantic principles were applied for both cases. All the important principles concerning an auspicious place are in fact about ideal conditions

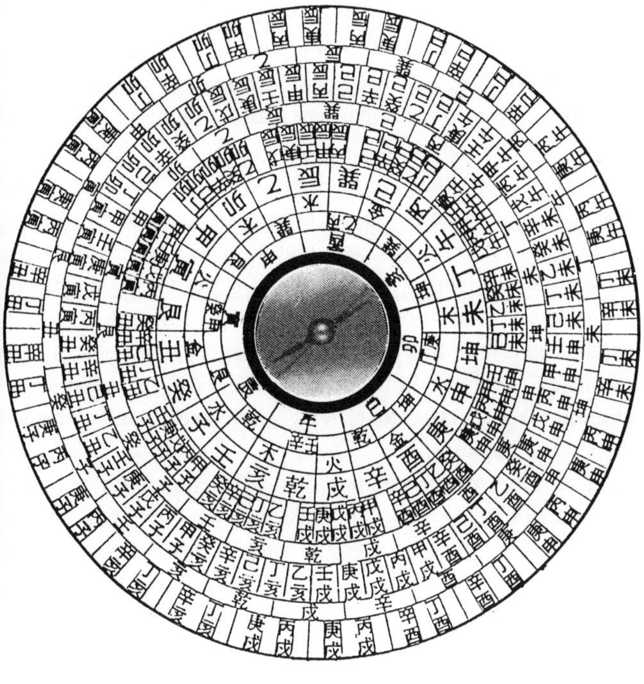

Figure 1: A geomantic compass currently in use by professional geomancers.

relating to a living person's dwelling site. The analysis of geomantic principles concerning auspicious locations suggests that Chinese geomancy began as an art of selecting a comfortable place to live, and then came to be applied to selecting grave sites as well.

By the sixth and seventh centuries, geomantic principles as we know them today were probably well established in China. Ever since then, geomancy in China has been a powerful part of the art of environmental planning, landscape design, and determining various types of settlement location.

This art was probably diffused to Korea with an early wave of Chinese cultural diffusion. Koreans in turn introduced this art to Japan. Judging from ancient capital sites in Korea and Japan, this art was introduced and practiced in those countries by the seventh century. Geomancy is still practiced in China, Taiwan, Hong Kong, Singapore, Korea, and Japan. In fact, in Hong Kong, Singapore, and Korea, a number of geomantic institutes were established, and professional geomancers practice their art in planning building structures and in the selection of residential and grave sites.

Geomancy has been adapted to the modern urban environment, and even many Chinese living in cities in America apply geomantic principles in selecting their house sites and designing house structures. In city situations, the main points of geomantic considerations are, in place of the landform conditions, street patterns, roof lines of neighbors, the shape and situation of a house section, and the floor plan of a house. In some cases, the furniture arrangement inside a house is decided by applying Chinese geomancy to present day city dwellings.

There have been numerous geomantic textbooks and manuals throughout the history of China, and even now there are more than a dozen printed editions of geomantic books available for purchase in the Chinese language alone.

HONG-KEY YOON

REFERENCES

De Groot, J.J.M. *The Religious System of China*. Vol. 3. Leiden: Librairie et Imprimeries, 1897.

Feuchtwang, Stephan D.R. *An Anthropological Analysis of Chinese Geomancy*. Vientianne: Editions Vithagna, 1974.

Hsu Shan-chi and Hsu Shan-shu (Ming Dynasty). *Jen-tsu Hsu-chih*. Taipei: Chulin Shu-chu, Hsin-chu, 1969.

Kuo Pu. "Tsang-shu." In *Tili Chen-tsung* (commentary by Chiang Kuo) Shin-chu: Chu-lin Shu-chu, 1967.

March, Andrew. "An Appreciation of Chinese Geomancy." *Journal of Asian Studies*. 27 (2): 252–267, 1968.

Rossbach, Sarah. *Feng shui: The Chinese Art of Placement*. New York: E.P. Dutton, 1983.

Yoon, Hong-key. *Geomantic Relationships Between Culture and Nature in Korea*. Taipei: Oriental Culture Service, 1976.

Yoon, Hong-key. "The Image of Nature in Geomancy." *GeoJournal*. 4 (4): 341–348, 1980.

Yoon, Hong-key. "An Early Chinese Idea of a Dynamic Environmental Cycle." *GeoJournal* 10 (2): 211–212, 1980.

Yoon, Hong-key. "Environmental Determinism and Geomancy: Two Cultures, Two Concepts." *GeoJournal*. 6 (1): 77–80, 1982.

See also: Geography in China – Maps, Geomantic – Divination

GEOMANCY IN THE ISLAMIC WORLD The term geomancy comes from the Latin *geomantia*, first used in Spain in the twelfth century as a translation of the Arabic *ʿilm al-raml* (the science of sand), the most common name for this type of divination. The practice is to be distinguished from a totally unrelated Chinese form of prognostication based on land forms, unfortunately also called geomancy in English. The origin of this distinctly Islamic art is a matter of speculation, but it appears to have been a well established practice in North Africa, Egypt, and Syria by the twelfth century.

The divination is accomplished by forming and then interpreting a design, called a geomantic tableau, consisting of sixteen positions, each of which is occupied by a geomantic figure. The figures occupying the first four positions are determined by marking sixteen horizontal lines of dots on a piece of paper or a dust board. Each row of dots is examined to determine if it is odd or even and is then represented by one or two dots accordingly. Each figure is then formed of a vertical column of four marks, each of which is either one or two dots. The first four figures, generated by lines made while the questioner concentrates upon the question, are placed side by side in a row from right to left. From these four figures the remaining twelve positions in the tableau are produced according to set procedures. Various interpretative methods are advocated by geomancers for reading the tableau, often depending upon the nature of the question asked. The course and seriousness of an illness, the outcome of pregnancy, the location of lost or buried objects, and the fate of a distant relative are among the most popular questions addressed to a geomancer.

The acknowledged master of geomancy was Abū ʿAbdallāh Muḥammad ibn ʿUthmān al-Zanātī, who lived before AD 1230. Virtually nothing is known of his life, though his name suggests that he was from the North African Berber tribe of Zanāta that was known for practicing other forms of fortune telling, particularly scapulomancy (divination by inspection of shoulder blades). One of the great codifiers of geomancy was ʿAbdallāh ibn Maḥfūf who lived, probably in Syria or Egypt, before AD 1265 and whose treatise is preserved today in several Arabic manuscript copies. Brief discussions of geomancy are included in a Persian encyclopedia composed at the end of the twelfth century by the celebrated theologian Fakr al-Dīn al-Rāzī and in short Persian tracts by the mathematician, philosopher, and founder of the observatory at Maragha in northwest Iran, Naṣīr al-Dīn al-Ṭūsī (d. AD 1275).

The majority of existing treatises on the subject are from the fourteenth century and later, with numerous ones still being written in the nineteenth and twentieth centuries. In view of the relatively few written sources on the topic before the fourteenth century, an intricate metal geomantic tablet, now in the collections of the British Museum, is of considerable importance. It was made in AD 1241–1242 (H. 639) by the metalworker Muḥammad ibn Khutlukh al-Mawṣilī, who also made an incense burner in Damascus about AD 1230–1240. This unique device is of a brass alloy inlaid with gold and silver, with a front plate carrying twenty dials and four sliding arcs and a back plate engraved with inscriptions, both plates held in a rectangular frame which has a triangular suspensory device on top similar to that commonly found on astrolabes. No writings before or after its construction mention such a mechanical contrivance for establishing a geomantic reading, and there is no other known geomantic device from any culture remotely similar to it. It is evident that the designer of this elaborate device was well-versed in the geomantic literature of his day.

In Iran the term *raml* is applied to two types of divination. One type, frequently described by travelers, employed the throwing of brass dice that were strung together in groups of four. Although these dice are commonly referred to as geomantic dice, they are not marked so as to produce a geomantic figure, and thus the divination using such dice is a form of lot casting or sortilege different from true geomancy. *Raml* is also used in Iran for the traditional form of geomancy, and in modern Persian writings the art often attains an astounding degree of complexity, with successive tableaux generated from previous ones. A large number of lithographed Persian texts were published in India in the nineteenth century.

From the twelfth century until the seventeenth century, geomancy, in a slightly altered form, was very popular in Europe, where only astrology seems to have outranked it in popularity. The Spanish translator Hugh of Santalla working at Tarazona in Aragon appears to have been the first to prepare a Latin paraphrase of an Arabic treatise on the subject, with his slightly younger contemporary working in Toledo, Gerard of Cremona (d. AD 1187), translating another Arabic tract. After the seventeenth century, interest in geomancy faded abruptly in the West, and today it remains relatively unknown there.

There are many non-Western areas where geomancy and derivative methods of divination are still practiced. Geomancy in sub-Saharan Africa and Madagascar, which employs simplified but clearly derivative versions of classical Islamic geomancy, has been the subject of several anthropological studies. The mathematical structure of the practice has also received some scholarly attention. In nearly all Islamic lands geomancy and related methods of divination are still practiced, in forms varying from the simple casting of a favorable or unfavorable geomantic figure to the complex interpretation of tableaux employing a large number of procedures.

EMILIE SAVAGE-SMITH

REFERENCES

Bascom, William. *Ifa Divination: Communication Between Gods and Men in West Africa.* Bloomington, Indiana: Indiana University Press, 1969.

Charmasson, Thérèse. *Recherches sur une technique divinatoire: la géomancie dans l'occident médiéval.* (Centre de Recherches d'Histoire et de Philologie, V, Hautes Études Médiévales et Modernes, 44.) Geneva: Librairie Droz and Paris: Librairie H. Champion, 1980.

Fahd, Toufic. *La divination arabe: études religieuses, sociologiques et folkloriques sur le milieu natif de l'Islam*. Leiden: Brill, 1966.

Jaulin, Robert. *La Géomancie: analyse formelle*. (École Pratique des Hautes Études, Sorbonne, Cahiers de l'Homme, n.s., t. 4.) Paris: Mouton & Co., 1966.

Pedrazzi, Maino. "Le Figure della Geomanzie: Un Gruppo Finito Abeliano." *Physis* 14(2): 146–161, 1972.

Peek, Philip M., ed. *African Divination Systems — Ways of Knowing*. Bloomington, Indiana: Indiana University Press, 1991.

Savage-Smith, Emilie and Marion B. Smith. *Islamic Geomancy and a Thirteenth-Century Divinatory Device*. (UCLA Von Grunebaum Center, Studies in Near Eastern Culture and Society, 2.) Malibu: Undena Publications, 1980.

Savage-Smith, Emilie. "Geomancy." In *The Oxford Encyclopedia of the Modern Islamic World*. Ed. John L. Esposito. New York: Oxford University Press, 1995, pp. 53–55.

Skinner, Stephen. *Terrestrial Astrology: Divination by Geomancy*. London: Routledge & Kegan Paul, 1980.

Smith, Marion B. "The Nature of Islamic Geomancy with a Critique of a Structuralist's Approach." *Studia Islamica* 49: 5–38, 1979.

GEOMETRY Geometry is the branch of mathematics that deals with the partitioning of a physical or abstract space and the relationships induced by that partitioning. Historically, as a human activity, it is said by the Roman historian Herodotus (d. 425 BC) to have begun in Egypt where the annual inundation of the Nile River obliterated field boundaries. In order to preserve the royal system of land distribution and taxation, official priest-surveyors measured the land and reestablished boundary markers. Their land-measuring techniques were imported to Greece and given the collective name "geometry" (*geo*, "earth"; *metrein*, "to measure"). Thus the practices of land measurement became the science of geometry. However, all peoples and societies possess and use some geometrical knowledge. For example, decorative pottery motifs and textile or basket weaving patterns demonstrate a geometric knowledge and facility on the part of their creators. Certainly the blanket patterns of the Ibans of Borneo and the rafter carvings of New Zealand's Maori people reflect an appreciation of symmetry; a complexity of line relationships is evident in the *sona* tracings of the Tshokwe culture in southern Africa, and a mastery of geometrical figures is illustrated in the sand drawings of North America's Navajos. All these non-Western cultures possess "a geometry" but to survey these traditional geometries would present a monumental task and be beyond the scope of this present survey. Therefore the discussion below will view geometry in a narrower sense, that is, as an exact science and as a subject of acknowledged societal concern either documented by existing historical records or the evidence of archaeological excavations.

Egypt and Babylonia

Egyptian geometry was practical in its conception and tied to measurement and the bureaucratic needs of the state. A small collection of surviving papyrus problems provided a limited glimpse into the workings of Egyptian geometry. The ancient Egyptians possessed correct computation formulas for determining the areas and volumes for a variety of plane and solid figures. Apparently they deduced such formulas by techniques of "dissection and rearrangement". Correct formulas for the areas of rectangles, triangles and isosceles trapezia were employed. Problem 50 of the Ahmes Papyrus (1650 BC) gives the area of a circle with known diameter d as $A = (\frac{8}{9}d)^2$ from which it is found that the Egyptians of this time used a value $\pi \approx 3.1605$. Problem 48 of the same sequence provides an illustration that has been interpreted as a polygonal approximation for the area of a circle and from which the formula was derived. In total, the Ahmes Papyrus contains 19 problems of a geometrical nature. The Moscow Papyrus (1850 BC), in its 14th problem, offers a correct computational procedure for determining the volume of a truncated square pyramid of given height h and with bases whose sides measure a and b in length; $V = h/3(a^2 + ab + b^2)$. Problem 10 of the Moscow Papyrus requires the surface area of a "basket" with known diameter and has resulted in a yet unresolved controversy as to whether the Egyptians were considering the basket as a hemisphere or as a semicylinder. If the former interpretation is accepted this would indicate that the Egyptians possessed rather sophisticated geometrical insights at a very early date. Egyptian geometry remained utilitarian in nature and was never advanced to the status of an abstract science.

In a similar manner, Babylonian geometry was also of a computational nature and devoted to solving practical problems. In our context, the term "Babylonian" designates the civilization that occupied the Tigris–Euphrates region in the period 3500 BC–539 BC and includes the Sumerian, Akkadian, Chaldean and Assyrian peoples. Knowledge of Babylonian geometry has been obtained from a limited examination of cuneiform texts and mathematical tables. This examination reveals that their authors possessed computation procedures for obtaining the areas of rectangles, right-angled triangles, isosceles triangles and trapezia with one side perpendicular to the parallel sides. They knew several numerical properties of a circle and approximated the area of a given circle radius r by $A = 3r^2$. A clay tablet from the Old Babylonian Period (1900–1650 BC) supplies a more accurate value of π as 3.125. The Babylonians could determine the length of a chord l, given the diameter of a circle d and the length of the sagitta in question, a: $l = \sqrt{d^2 - (d - 2a)^2}$. Tablet no. 322 in the Plimpton Collection at Columbia University, whose origin is traced to 1800–1650 BC, provides a list of Pythagorean triples indicating that the Babylonians of this period were familiar with the "Pythagorean Theorem". A clay table excavated at Tell Harmal in Iraq and dated to about 2000 BC indicates that its users knew some properties of similar triangles and could employ these properties in solving numerical problems involving triangles. Babylonian scribes also could compute the volumes for simple solids and possessed an approximate formula for the volume of a truncated square pyramid of height h and bases of length a and b, i.e. $V = h/2(a^2 + b^2)$. Although several excavated clay tablets bear geometric diagrams, these diagrams serve as illustrations for relevant concrete problem situations and are not theoretical constructions. Apparently for the Babylonians, geometry remained an adjunct to numerical problem solving and did not evolve into a separate discipline.

Ancient India

Geometric activity on the Indian subcontinent can be traced back to the Indus Valley Civilization (ca. 3000 BC). Excavations at this civilization's urban centers of Harappa, Lothal and Mohenjo-Daro reveal the existence of a baked-brick technology allowing for the erection of numerous structures: houses, baths, and market places. This construction entailed large-scale city planning and the use of geometry. Archaeological evidence indicates that the Indus peoples were familiar with the basic properties of rectangles, triangles and circles and could employ right triangle principles for planning and structural purposes.

With the arrival and dominance of the Aryan culture in the region (1500–800 BC), there developed a sacred Vedic literature of rituals and customs. Part of this literature concerned the construction of sacrificial fire altars of specific shapes and dimensions. Cord stretching techniques were described to obtain the required results. This collection of construction procedures and techniques became known as the *Śulbasūtras* (800 BC) or "the rules of the cord". Using cords and pegs, squares, circles, rectangles and trapezia of specified dimensions and areas could be laid out and constructed by the Vedic priests. Although outwardly religious texts, the *Śulbasūtras* presented a codified system of geometry. Besides obvious geometric constructions, the texts supplied a technical vocabulary for geometrical principles and concepts including a theory of similar triangles. Techniques for transforming shapes while preserving areas considered the quadrature of rectangles and circles, and procedures were also given for doubling the area of a circle and transforming a square into a circle. The principle of constructing a square on the diagonal of a rectangle ("Pythagorean Theorem") was also known to the writers of *Śulbasūtras*. No formal "proofs" were given but some results were justified by arguments.

The subsequent influence of the *Śulbasūtras* on the latter development of Indian geometry is controversial: early researchers such as G.R. Kaye could discern no influence, whereas more modern writers such as T.A. Sarasvati Amma believe that the *Śulbasūtras* established the foundation for all Indian geometry. Whatever the influence of these Veda texts on geometrical thinking, later developments in this field were spotty and isolated. Geometry became closely associated with astronomy and cosmography. It was dominated by numerical problem solving based primarily on the use of right-triangle principles and was mainly concerned with finding the length of circular chords. The post-Vedic period saw the rise of the Jaina school of philosophical and scientific thought. The Jainas, in their cosmological considerations, held two geometric figures, the circle and trapezium, in high regard. Much of their geometry focused on mensuration problems involving these figures. In general, they considered geometry important denoting it as "the lotus of mathematics". The Jainas employed the approximation $\pi = \sqrt{10}$.

The next Indian mathematical text of note to appear was the *Āryabhaṭīya* (ca. 499) of Āryabhaṭa I (476–550). While many of its problems concern the applications of geometry to astronomy, its author also considered the computation of area for triangles, trapezia and circles and the volumes of spheres and triangular pyramids. Āryabhaṭa's formula for the area of a circle is correct; his approximation for π is 62832/20000 or 3.1416. However, his results for the volume of a sphere are $V = \pi^{3/2}r^3$. Subsequent mathematical authors would refine and extend Āryabhaṭa's work. Brahmagupta (ca. AD 628) in his *Brahmasphutasiddhānta* (Correct Astronomical System of Brahma) provided sections on: *Kṣetra*, plane figures; *Khāta*, cubic figures; *Citi*, solids composed of piles of bricks; *Krakaca*, truncated solids and *Chāyā*, plane figures resulting from shadows. Much of Brahmagupta's results are limited to triangles and quadrilaterals inscribable in a circle. His most notable findings include:

- the correct determination of the area of a cyclic quadrilateral of sides a, b, c, d, with s as its semi-perimeter i.e.

$$A = \sqrt{(s-a)(s-b)(s-c)(s-d)};$$

- the length of the quadrilateral's diagonals d_1 and d_2:

$$d_1 = \sqrt{\frac{(ac+bd)(ad+bc)}{(ab+cd)}}, \quad d_2 = \sqrt{\frac{(ac+bd)(ab+cd)}{(ad+bc)}},$$

- and the volume of a cone as $\frac{1}{3}$ the product of its base area and its height.

Mahāvīra (ca. 850), a Jaina mathematician, improved upon Brahmagupta's theories in his *Ganitasārasamgraha* (Summary Compendium of Mathematical Astronomy), greatly extending geometric terminology and classification systems. For example, he divided quadrilaterals into five different subcategories: equal sides, opposite sides equal, etc. and triangles into three categories: equilateral, isosceles and scalene. Bhāskara II (b. 1114), author of one of the most illustrious of Indian mathematical classics, *Līlāvatī* (The Beautiful), obtained accurate formulas for the volume and surface area of a sphere and obtained close approximations for the length of a circular arc in terms of its known chord. Mādhava of Sangamagrāmma (ca. 1340–1425), a Kerala astronomer, is noted for his work on the circle as well as spherical geometry.

Euclid's *Elements* were known and studied in India in the fourteenth century, but the first complete Sanskrit translation of this work appeared in 1718. The translation was performed by Jagannātha Samrāt at the request of his patron king astronomer Jayasimka of Jaipur.

The Islamic World

The founding of Islam in AD 622 was marked by the flight of Muḥammad from Mecca to Medina. Initially experiencing a period of rapid expansion, it was not until the founding of the Abbasid Caliphate, with the establishment of Baghdad as an intellectual and political center (AD 726) that Islamic civilization could begin to express its own intellectual and scientific traditions. The early Abbasid caliphs were patrons of the collection and translation of foreign scientific texts into Arabic, thus establishing a basis for Islamic science. Although Muslim scholars became the heirs of existing western (and eastern) scientific theories and traditions, they did not remain mere translators and passive communicators of this knowledge. In many fields they became true innovators and amplifiers of their mathematical and scientific inheritance. Geometry was one such field that was enriched by Islamic contributions.

The Greek sources that most influence Islamic geometric thinking were Euclid's *Elements*, Archimedes' *On the Sphere and Cylinder* and Apollonios of Perga's *Conics*, all of which found their way into Arabic during the eighth to the ninth centuries. Muslim geometers pursued the theoretical problems poised in these texts and used their principles in a wide variety of new applications, particularly the study of optics and the design of mathematical instruments. The earliest documented Islamic geometry appears as a separate section of al-Khwārizmī's *Kitāb al-mukhtaṣar fī ḥisāb al-jabr wa'l-muqābala* (Compendious Book on Calculation by Completion and Balancing), where rules for mensuration and geometric computations involving areas and volumes are stated. The value of π is given as $3\frac{1}{7}$. Within fifty years of al-Khwārizmī's work, his successors were involved with far more complex geometric theories.

Thābit ibn Qurra (830–890) discussed the geometric verification of algebraic results in his *Qawl fī taṣḥīḥ masā'il al-jabr bi'l-barāhīn al-handasīya* (On the Verification of Problems of Algebra by Geometrical Proofs). In a private correspondence with a colleague, he expressed dissatisfaction with the existing Socratic proof of the "Pythagorean theorem" and went on to derive three new proofs, one of which provides a generalization of the theorem to all triangles. Thābit also wrote on the trisection of an angle problem. His grandson, Ibrāhīm ibn Sinan (d. 946), continued researching this problem and provided for "ruler and compass" constructions of the conic sections in his *Rasm al-quṭūʿ al-talātha* (Outline of Three Sections). In approximately AD 950, Abū'l Wafā' wrote a book on applied geometry for craftsman, *Kitāb fī mā yaḥtāj ilayh al-ṣāniʿ min al-aʿmāl al-handasiyya* (Book on Necessary Geometric Construction for the Artisan), in which he presented several original constructions for conic sections. Around the year 1000 Abū ʿAbdullāh al-Ḥasan ibn al-Baghdādī published a comprehensive treatment of incommensurables improving on the theory given in the *Elements*.

One of the recurrent themes in Islamic geometry is the concept of parallel lines and the provability of Euclid's fifth postulate. Ibn al-Haytham (ca. 965–1039), known in the West as Alhazen, attempted to reformulate Euclid's theory of parallel lines by assuming the constructibility of the lines. He presented his theory in a work entitled *Maqāla fī sharḥ muṣādarāt kitāb Uqlīdis* (Commentary on the Premises of Euclid's Elements). ʿUmar ibn Ibrāhīm al-Khayyāmī, or ʿUmar al-Khayyām (ca. 1048–1126), in investigating the concept of parallelism, developed a series of eight propositions eventually leading to the establishment of Euclid's fifth postulate. ʿUmar al-Khayyām's work was based on the possible relationships of the angles of a quadrilateral. In turn, this work was expanded by Naṣīr al-Dīn al-Ṭūsī (1201–1274). He also explored parallelism with the aid of a quadrilateral, but allowed for the existence of acute and obtuse angles within the quadrilateral seeking to obtain a contradiction to his premises concurring the nature of parallel lines. He published his work on the fifth postulate in 1250; it was entitled *Al-risāla al-shāfiya'an al-shakk fī al-khuṭūṭ al-mutawāziya* (Discussions which Remove Doubts about Parallel Lines).

Eventually the theories of Islamic scholars such as Ibn al-Haytham and al-Ṭūsī reached Europe; there they profoundly affected the nature of geometric thinking.

China and the Far East

Most geometrical considerations in early China were empirically based. However, during the Mohist school of activity (ca. 300 BC) a deductive approach to understanding geometry was undertaken. Mohist philosophy and scientific theory was based on the use of logic. The Mohist Canon (*Mozi*) (ca. 330 BC) develops a formalist approach to plane geometry building upon a concept of points and lines. Mohist theories did not become popular, and the development of a theoretical geometry progressed no further.

The earliest extant references specifically on Chinese mathematics are the *Zhoubi suanjing* (Arithmetical Classic of the Gnomon and the Circular Paths of Heaven) and *Jiuzhang suanshu* (Nine Chapters on the Mathematical Art, ca. 100 BC). A discussion of right-triangle relationships is given in *Zhoubi,* along with a dissection proof demonstrating that the sum of the squares of the sides adjacent to the right angle in a right triangle is equal to the square of the hypotenuse. Four chapters of the nine chapters of *Jiuzhang* are specifically devoted to geometrical computations involving areas, volumes and work with right triangles. Correct procedures for finding a variety of areas and volumes are given; the area of a circle of radii r is noted as $A = \pi r^2$ with $\pi = 3$; but the volume of a sphere with diameter d is incorrectly found, $V = \frac{9}{16}d^3$. An approximation for the area of a circular segment with chord C and sagitta S is given as $A = S/2(S + C)$. Similarity among right triangles is employed in several problem situations.

In the third century, the scholar Liu Hui wrote a commentary on the *Jiuzhang* in which he provided dissection proofs for many of its geometrical formulas. Liu also employed a circle dissection technique involving a 192-sided polygon to estimate π as 3.141024. He extended the right triangle theory of the *Juizhang* to include more complex problem situations involving pairs of similar right triangles. Using a technique called *chong-cha*, Liu provided proofs for solution procedures and published his findings in an appendix to *Juizhang*. Eventually, this appendix became a separate geometrical classic called *Haidao suanjing* (Sea Island Mathematical Manual). Zu Chongzhi (429–500) refined Liu's approximation for π obtaining a value of 355/113 or 3.1415929. Further, using a geometric slicing technique, Zu found the volume of a sphere, radii r, correctly as $V = 4/3\pi r^3$.

In 656, the geometric theory contained in *Zhoubi suanjing*, *Juizhang suanshu* and *Haidao suanjing* was sanctioned by the Tang dynasty as a formal curriculum for its scholar officials. This curriculum was also eventually adopted in Japan and Korea, establishing a limited geometrical outlook based on empirical problem solving. This outlook would remain unaltered until the sixteenthth century when a translation of Euclid's *Elements* appeared in the Chinese language. This translation of the first six books of Euclid was made by the Jesuit missionary Matteo Ricci and the Chinese scholar Xu Guanggi.

Pre-Columbian America

Although no written records exist to document the geometric knowledge of the early native American civilizations, archaeological sites testify to their use and understanding of geometry. In particular, the city planning and construction techniques employed by the Olmec, Maya, Teotihuacan, Toltec and Aztec peoples of South and Central America and the Anasazi of the North American Southwest indicate that these peoples utilized the properties of circles, squares and rectangles and employed right triangle theory.

FRANK J. SWETZ

REFERENCES

Amma, T.A. Sarasvati. *Geometry in Ancient and Medieval India*. Delhi: Motilal Banarsidass, 1979.

Berggren, J.L. *Episodes in the Mathematics of Medieval Islam*. New York: Springer-Verlag, 1986.

Closs, Michael P., ed. *Native American Mathematics*. Austin: University of Texas Press, 1986.

Gillings, Richard J. *Mathematics in the Time of the Pharaohs*. Cambridge, Massachusetts: MIT Press, 1972.

Joseph, George Gheverghese. *The Crest of the Peacock: Non-European Roots of Mathematics*. London: I.B. Tauris, 1991.

Kaye, G.R. *Indian Mathematics*. Calcutta: Thacker, Spink and Co., 1915.

Kulkarni, R. P. *Geometry According to Sulba Sutra*. Pune: Vaidika Samsodhana, 1983.

Li Yan and Du Shiran. *Chinese Mathematics: A Concise History*. Trans. John N. Crossley and Anthony W. C. Lun. Oxford: Clarendon Press, 1987.

Seidenberg, Abraham. "The Ritual Origin of Geometry." *Archive for History of Exact Sciences* 1: 488–527, 1962.

Swetz, Frank J. *The Sea Island Mathematical Manual: Surveying and Mathematics in Ancient China*. University Park, Pennsylvania.: The Pennsylvania State University Press, 1992.

Youschkevitch, A.P. *Les Mathématiques Arabes (VIIe–XVe Siècles)*. Paris: J. Vrin, 1976.

See also: Liu Hui – *Śulbasūtras* – Āryabhaṭa – Brahmagupta – Mahāvīra – *Elements* – Bhāskara – Mādhava – al-Khwārizmī – Thābit ibn Qurra – Ibrāhīm ibn Sinān – Abū'l Wafā' – Ibn al-Haytham – Naṣīr al-Dīn al-Ṭūsī – ʿUmar al-Khayyām – Liu Hui and the *Jiuzhang suanshu* – *Zhoubi suanjing*

GEOMETRY IN AFRICA: *SONA* GEOMETRY The *sona* tradition is a part of the heritage of the Tchokwe, Lunda, Lwena, Xinge and Minungo peoples that inhabit the northeastern part of Angola, and of the Ngangela and Luchazi peoples of southeastern Angola and Western Zambia. When the Tchokwe met at their central village places or at their hunting camps, they usually sat around a fire or in the shadow of leafy trees spending their time in conversation, illustrated by drawings in the sand. These drawings are called *lusona* (singular) or *sona* (plural).

Most of these drawings belonged to an old tradition. They referred to proverbs, fables, riddles, animals, etc. and played an important role in the transmission of knowledge and wisdom from one generation to the next.

Every boy learned the meaning and execution of the easier *sona* during the intensive schooling phase of the circumcision and initiation rites. The significance and creation of more difficult *sona* were known only by specialists, the *akwa kuta sona* (those who know how to draw), who transmitted their knowledge to their sons.

The designs have to be executed smoothly and continuously. In order to facilitate the memorization of their

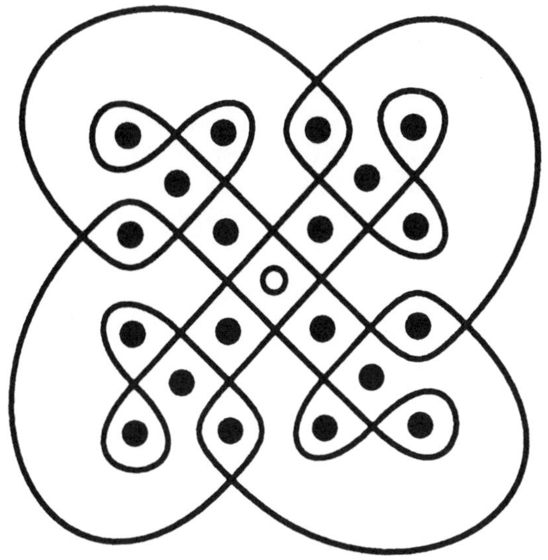

Figure 1: Monolinear *lusona* with rotational symmetry.

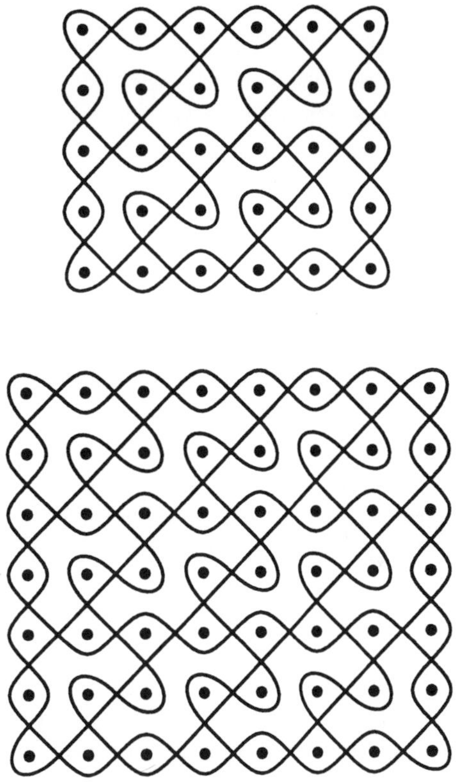

Figure 2: Geometrical algorithm for 'chased chicken' patterns.

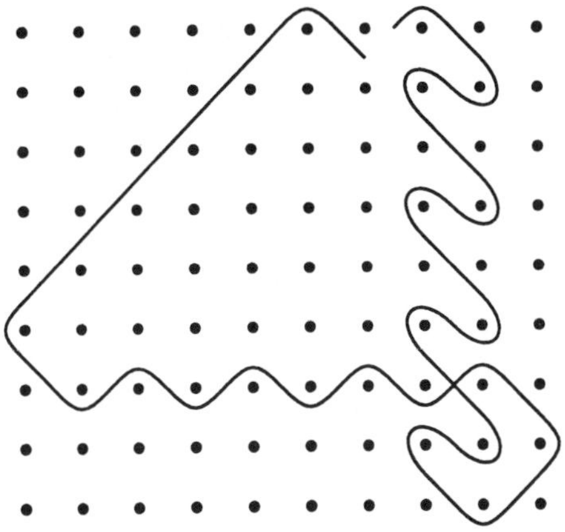

Figure 3: Two instances of 'chased chicken' *sona*.

Figure 5: Representation of a leopard with five cubs.

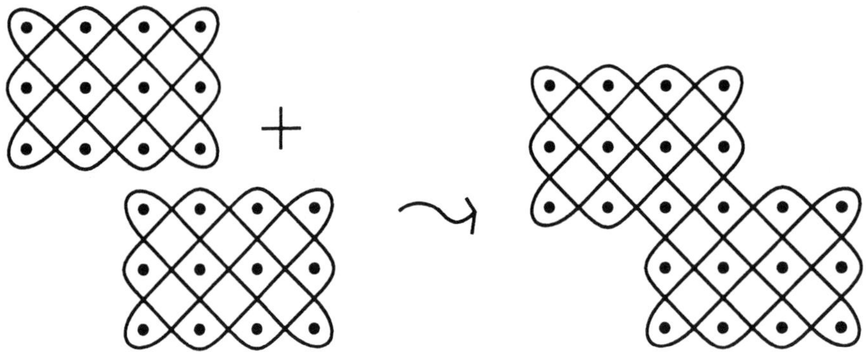

Figure 4: Example of a composition rule.

standardized 'sona', the drawing experts used the following mnemonic device. After cleaning and smoothing the ground, they first set out with their fingertips an orthogonal net of equidistant points. Then one or more lines were drawn that 'embraced' the points of the reference frame. By applying their method, the drawing experts reduced the memorization of a whole drawing to that of mostly two numbers (the dimensions of the reference frame) and a geometric algorithm (the rule of how to draw the embracing line(s)).

The sona tradition vanished almost completely. Tentative designs have been made to try to reconstruct the mathematical knowledge of the drawing experts.

They considered symmetry and monolinearity (i.e. a whole figure made up of only one line) important. Note the example of a *lusona* that is monolinear and that displays a rotational symmetry of order 4 (Figure 1).

They invented geometrical algorithms for the construction of classes of 'sona'. Figure 2 displays two monolinear drawings belonging to the same class and constructed in agreement with the same algorithm (Figure 3). Both drawings represent the marks left on the ground by a chicken when it is chased.

The drawing experts also invented a series of rules for the systematic construction of monolinear sona. They probably knew why the rules were valid: they could prove in one way or another the truth of the theorems that these rules expressed.

Figure 4 illustrates one such a rule. This rule serves for chaining monolinear patterns and has been applied four times in the Tchokwe representation of a leopard with five cubs (see Figure 5, previous page).

PAULUS GERDES

REFERENCES

Ascher, Marcia. "Graphs in Cultures (II): a Study in Ethnomathematics." *Archive for History of Exact Sciences* 39(1): 75–95, 1988.

Fontinha, Mário. *Desenhos na areia dos Quiocos do Nordeste de Angola*. Lisbon: Instituto de Investigação Científica Tropical, 1983.

Gerdes, Paulus. "On Mathematical Elements in the Tchokwe 'Sona' Tradition". *For the Learning of Mathematics* 10(1): 31–34, 1990.

Gerdes, Paulus. *Lusona: Geometrical Recreations of Africa*. Maputo: Mozambique's Higher Pedagogical Institute, 1991.

Gerdes, Paulus. *Geometria sona: Reflexões sobre uma tradição de desenho em povos da África ao sul do Equador*. Maputo: Mozambique's Higher Pedagogical Institute, 1993 (3 vols).

GEOMETRY IN CHINA The earliest evidence of a systematic organization of regular shapes reproduced on material objects found in China dates from the third millennium BC or even earlier. Painted motifs displaying geometrical patterns such as symmetrical arrangements of triangles, lozenges, or circles have been found on pottery pieces unearthed at Banpo (near present day Xi'an) and other archaeological sites. These designs demonstrate an early interest in spatial ordering and are perhaps at the origin of subsequent developments even though we are now unable to establish any continuity between prehistoric and historic Chinese mathematics. None the less, several Chinese myths and legends attest that the plumb line, the compass, the carpenter square, and the gnomon (a post of standard height) were commonly used in Zhou China (1121–256 BC). This last instrument, in particular, was considered so important that the Chinese began founding their astronomical and calendrical conceptions on the determination of the length of gnomon shadows.

Chinese mathematical computations based on the gnomon have been preserved in several important sources from Chinese antiquity such as the *Huai Nan Zi* (The Book of the Prince of Huai Nan) and the *Zhoubi suanjing* (The Canon of Gnomon Computations from the Zhou Dynasty). In these sources, gnomons and their shadows are considered as sides of similar right-angled triangles and properties of similarity are applied to the determination of the height of the sun, at noon, above the flat earth. Later on, the same technique was generalized and used for the determination of earthly distances not directly measurable such as the depth of a well, the height of a mountain or the distance of an island from the shore. Towards the end of the third century AD, these techniques were edited in algorithmic form in a book entitled *Haidao suanjing* (The Sea Island Computational Canon).

In a quite different spirit, the Mohists — a sect of preaching friars from the fourth and third centuries BC — approached geometry from a logical point of view and attempted to define the point, the circle, the square, space, and even parallelism (this last point is controversial). Certain of these definitions bear a striking resemblance to those found in Greek mathematics, particularly in Euclid's *Elements*. Still, there are important differences between both approaches: the Mohist definitions do not concern a single science, geometry, but include optics, mechanics, and economics. Moreover, the Mohists allow their definitions to incorporate familiar objects from the immediate human environment such as the sun, the bolt of a door, and other very concrete notions. More importantly, the grammatical structure of the Chinese language is such that objects and properties of objects are not distinguished in any way in the Mohist definitions.

Mohist logic did not arouse much interest and was never again studied in China; Chinese geometry developed without concern for geometrical definitions, let alone axiomatic-deductive constructions. But during the Han dynasty (206 BC–AD 220) Chinese mathematical knowledge was for the first time separated from other domains and recorded in the *Jiuzhang suanshu* (Computational Prescriptions in Nine Chapters, also translated as Nine Chapters on the Mathematical Art). Composed of 243 problems and almost as many prescriptive solutions (*shu* = method, device, prescription, algorithm), this very influential manual was eventually revered as a kind of mathematical Bible, so that as late as the seventeenth century certain respected Chinese mathematicians considered that all imaginable problems and computational methods would necessarily be interpretable in terms of the traditional nine chapters of the *Jiuzhang suanshu*.

In fact, the geometry of the *Jiuzhang suanshu* is not precisely an autonomous body of knowledge and consists essentially in a threefold compilation of problems pertaining to planimetry, stereometry, and right-angled triangles. Planimetry is included in the first chapter of the *Jiuzhang suanshu* and comprises a collection of formulae needed to compute the area of fields in the form of squares, rectangles, triangles, trapezia, circles, rings, and segments of circles. Stereometry is found in the fourth and fifth chapters and concerns the computation of volumes or capacities of cubes, parallelepipeds, pyramids, spheres, as well as that of various prisms, dikes, moat walls, ditches, excavations, and cylindrical or prismatic grain silos. Problems on right-angled triangles are found in the last chapter of the *Jiuzhang suanshu* and bear upon Pythagoras' theorem, where some dimension of a right-angled triangle is sought given simple algebraic combinations of other dimensions (such as the sum of its base and hypotenuse, its area, and so on). Lastly, the determination of the side and diameter of a square and a circle respectively inscribed in the same right-angled triangle are also considered. Non-geometrical subjects are also dealt with in the same chapters (for example, fractions in the first chapter, square and cube roots in the fourth). In addition, various concerns (units of measurement for example) frequently interfere with the geometrical aspect of problems, so that they seem highly realistic at first sight. In reality, most of them are artificial, since the situations they describe are often the inverse of what would naturally occur in real life: one is often asked to determine the dimensions of a figure given its area or its volume.

Given that problems on planimetry, stereometry, and right-angled triangles quite similar to the above ones are also well represented in the known corpus of Babylonian mathematics, it would seem that there is no essential difference between Chinese and Babylonian geometry. That is not the case, however, for on the one hand the solutions of Chinese problems are practically never stated using particular numbers but rather general formulations and on the other hand, certain Chinese mathematicians have felt the need to justify geometrical results. Remarkably, these justifications would appear quite correct and even assimilable to informal proofs in the eyes of modern mathematicians not at all acquainted with Chinese mathematics.

Chinese proofs of geometrical results have been preserved in various commentaries on the *Jiuzhang Suanshu* composed between the third and the seventh century AD by Liu Hui (fl. ca. end of the third century), the author of most of them, Zu Chongzhi (429–500), a famous astronomer and mathematician known for his approximation $\pi = 355/113$, and Li Shunfeng (602–670), an astronomer and compiler of the *Suanjing shi shu* (Ten Computational Canons), a collection of mathematical manuals representative of Chinese mathematical knowledge from the origins to the seventh century. As may be expected, the commentaries do not focus on mathematics alone but focus also on many others aspects of problems needed to understand their original context and pertaining to history, geography, philology, pedagogy, and philosophy. The commentaries are thus intrinsically non-homogeneous. In their turn, the proofs themselves are often interspersed with numerous digressions and are no more homogeneous that the rest of the commentaries they are imbedded in. Moreover, they frequently rely on all sorts of results (Pythagoras' theorem, for example) not yet proved when they were first used. Reasoning is thus neither constrained by formal modes of presentation nor by restrictions on what is admissible and what is not. In particular, numerical computations are liberally used, even when strictly speaking they would not necessarily be needed. More surprisingly however, but probably for mnemotechnical reasons, most proofs are formulated in a terse language; deductions are only suggested and based on a limited number of heuristic principles whose formulation is also extremely concise and which are used over and again.

One of these heuristic principles can be described as a generalized form of the so-called "Cavalieri's principle" (named by analogy with a famous proposition found in Bonaventura Cavalieri's *Geometria indivisibilibus continuorum* (Bologna, 1635)).

In its Chinese version, this principle is limited to solids and states that when the areas of the plane sections of two solids always have the same proportion between them, their volumes also have the same proportion. According to another heuristic principle, when a figure is dissected into several pieces as if it were a puzzle, and when the pieces are adequately reassembled, the area (or the volume) of the initial and final figure are equal. This principle is used with standard colored pieces (*qi*) but these are not necessarily the equivalent of concrete pieces of some puzzle since their number is sometimes infinite as is the case in the derivation of the volume of the pyramid. This dissection technique is so productive that its usage extends beyond the bounds of geometry. Very often, properties that we would call "algebraic" (especially when the equivalent of equations comes into play) are manipulated figuratively by means of dissections so that algebraic relations are directly visualized without any recourse to discursive reasoning or even to particular computations.

As numerous quotations from classical texts (such as Confucius's *Analects*, the *Yijing* [I Ching, Book of Changes] or Zhuangzi's works) found in the commentaries on the *Jiuzhang suanshu* suggest, this approach largely depends on Confucian and Daoist ideas. According to the Confucian conception, when presenting new knowledge, all the details should not immediately be revealed to students in order to oblige them to make efforts which would result in a deeper understanding. Similarly, Zhuangzi favored conciseness of expression, but for a very different reason: he believed that, as sophisms show, discursive reasoning is intrinsically limited and therefore any efficient access to knowledge should necessarily include all imaginable modes of apprehension of reality and not only discursive reasoning. On the whole, these conceptions induced a defiance towards modes of reasoning based on language alone; hence the recourse to non-linguistic modes of communication such as those based on computations or figurative techniques. Still it remains possible that the *Jiuzhang suanshu* and its commentaries are also the result of various other influences which did not necessarily originate in the Chinese world. In this respect it should be noted, for example, that equivalents of Cavalieri's principle were already used by Archimedes and other Greek mathematicians many centuries before the advent of the *Jiuzhang suanshu* and its commentaries. Dissections are also attested in antiquity and even more strikingly, the Chinese proof relating to the volume of the sphere relies, among other things, on the knowledge of the fact that the volume of a solid defined by the intersection of two orthogonal cylinders inscribed in the same cube is equal to two-thirds of the volume of the cube, a result explicitly stated in Archimedes's *Method of Mechanical Theorems for Eratosthenes* and in Heron of Alexandria's *Metrica*.

Anyway, after the seventh century, the Chinese repertory of planimetrical and stereometrical formulae was continuously enriched. Among the most interesting results from this period, one must cite Hero's formula for the computation of the area of a triangle given its sides, which was published in Qin Jiushao's *Shushu jiuzhang* (Mathematical Works in Nine Chapters, 1247). Despite its title, this work is very different from the *Jiuzhang suanshu* and is better known for its study of the "Chinese remainder problem" (simultaneous congruences) and its algebraic developments. After the seventh century however, and until the seventeenth, geometrical proofs were never again recorded in Chinese mathematical books. This fact perhaps has something to do with the remarkable development of Chinese algebra which took place during the Song and Yuan dynasties, for algebraic computations have the capacity to generate new results by means of mere computations.

At the beginning of the seventeenth century, the Chinese became aware of axiomatic-deductive modes of reasoning when, in 1607, the first six books of Christopher Clavius's commentary on Euclid's *Elements* were translated into Chinese by the Jesuit missionary Matteo Ricci (1552–1610) and Xu Guangqi (1562–1633) under the title *Jihe yuanben* (*jihe* = geometry or quantity and *yuanben* = elements, so that *Jihe yuanben* means either "Elements of Geometry" or "Elements of the Measure of Quantities"). Xu Guangqi was a Christian convert and influential high official responsible for the reform of Chinese astronomy undertaken from 1630 onwards on the basis of imported European knowledge. It has often been noted that some Persian or Arabic version of Euclid's *Elements* had already reached the Chinese imperial library three centuries earlier, during the Mongol domination of China, but there exists no evidence of any Chinese translation made during that period, and no trace of influence on subsequent Chinese mathematics has ever been detected.

At first, the new geometry was not much studied save by a few converts and Chinese students of Western astronomy who served the Jesuits at the imperial bureau of astronomy. However, as time went on, many Chinese scholars impressed by the successes of European sciences (especially mathematical predictions of celestial phenomena such as eclipses of the sun and calendrical computations) began to study the *Elements* seriously. The majority of these scholars considered that Euclid stood at the very basis of the new Western knowledge. At the same time they also believed that once properly reinterpreted, the results of the *Elements* were the same as those of their nine mathematical chapters; they also thought it desirable to dissociate the content of the *Elements* from its form. Far from representing the very model of logical clarity and the universal basis of all true knowledge, the formal structure of argumentation was judged inappropriate and even noxious. Such reactions against the *Elements* were also fairly common in Europe during the seventeenth century, but, in sharp contrast with what happened in China, these European reactions (which often went hand in hand with the rejection of Aristotelianism and scholasticism) were never powerful enough to dominate the mathematical scene, except perhaps in elementary education. In fact, during the seventeenth and eighteenth centuries, China was dominated by a strong reaction against all sorts of speculative reasoning prevalent during the preceding centuries which was designated as the source of Chinese decline. Consequently, numerous scholars advocated the development of concrete sciences (*shixue*), that is of applied sciences whose usefulness could be tangibly demonstrated by an increased social welfare. Thus, if mathematics were to be used at all, computational results had to be expressed in the form of easily understandable instructions. In this respect, syllogisms like those of the *Elements* were particularly difficult to use, inasmuch as they also involved numerous repetitions, a characteristic which was also contrary to the canons of Chinese literary composition. In addition, certain critics also wrote that Euclid's *Elements* were written in a cryptic language devised purposefully to obscure simple mathematical results. Many scholars found still another reason to defy geometry; they believed that they had detected an essential similarity between formal discourses typical of the rhetoric of geometrical demonstrations and the scholasticism of the theological speculations widely diffused by the European missionaries. Hence there was a rejection of forms of expression which tended to speculative form.

These ideas remained unchallenged in China until the end of the nineteenth century, but if axiomatic-deductive reasoning was never studied as such during that period, geometrical results were nevertheless taken seriously as early as the second half of the seventeenth century. In particular, a famous mathematician, Mei Wending (1633–1721) developed at length the computational aspect of the stereometry of regular and semi-regular polyhedrons on the basis of a very incomplete description of these that he had found in the *Celiang quanyi* (Complete Treatise on Measurements), a manual based on Clavius's *Geometria practica* and translated into Chinese ca. 1635. Others tried to recast to content of the *Elements* into the mold of the nine chapters of the *Jiuzhang suanshu*. From the end of the seventeenth century, new Chinese and Manchu translations of the *Elements* were realized under the patronage of the Kangxi emperor. These were all given the same title as the former translation of the *Elements* published a century earlier. However, they were very different from Clavius's commentary since they were all essentially based on Father Gaston-Ignace Pardies's (1636–1673) *Elements de géométrie*, a French manual very different from Euclid's initial text and intended for the teaching of geometry in Jesuit colleges. Pardies's manual (first published in 1671) was so popular that it remained in usage in Europe during a whole century and was translated into several European languages. The new translations of the *Elements* all remained in manuscript form, but these eventually gave birth to a final text which was eventually incorporated into a famous mathematical encyclopedia published by imperial order at the end of Kangxi's reign, the *Shuli jingyun* (Collected Essential Principles of Mathematics, 1723). Whereas this encyclopedia remained in use in China for two centuries, the not yet translated part of the first translation of the *Elements* was carried out between 1852 and 1856 by Alexander Wylie (1815–1887), a British Protestant missionary, and Li Shanlan (1811–1882), a renowned translator of Western manuals on astronomy, mathematics, botany, and other scientific subjects. This time, the translation was not based on Clavius' commentary but on a still unidentified edition (probably of British origin) of the *Elements* in seventeen books. After 1760, other elementary Western textbooks on geometry, which have been forgotten since, were again translated into Chinese.

After the 1911 revolution, Chinese mathematics was never again studied for scientific purposes (except by historians) and China gradually made its way into the international mathematical community. Nowadays, there is of course no distinction between geometry developed in China and in other countries.

JEAN-CLAUDE MARTZLOFF

REFERENCES

Ang Tian Se. "Chinese Interest in Right-angled Triangles." *Historia Mathematica* 5(3): 253–266, 1978.

Ang Tian-Se, and F.J. Swetz. "A Chinese Mathematical Classic of the Third Century: the *Sea Island Manual* of Liu Hui." *Historia Mathematica* 13: 99–117, 1986.

Dijksterhuis, E.J. *Archimedes*. Trans. C. Dikshoorn. Princeton, New Jersey: Princeton University Press, 1987.

Elia, Pasquale d'. "Prezentazione della Prima Traduzione Cinese di Euclide." *Monumenta Serica* 15: 161–202, 1956.

Graham, A. C. *Later Mohist Logic Ethics and Science*. London and Hong Kong: The Chinese University Press and the School of Oriental and African Studies, 1978.

Ho Peng-Yoke. "Liu Hui." In *Dictionary of Scientific Biography*, vol. 8. Ed. Charles C. Gillipsie. New York: Scribners, 1973, pp. 418–425.

Kokomoor, F.W. "The Distinctive Features of Seventeenth Century Geometry." *Isis* 10 (34): 367–415, 1928.

Lam Lay Yong and Shen Kangshen. "Right-angled Triangles in Ancient China." *Archive for the History of Exact Sciences* 30(2): 87–112, 1984.

Lam Lay Yong and Ang Tian Se. "Circle Measurements in Ancient China." *Historia Mathematica* 13: 325–340, 1986.

Li Di. *Zhongguo shuxue shi jianbian* (A Concise History of Chinese Mathematics). Shenyang: Liaoning Renmin Chubanshe, 1984.

Libbrecht, Ulrich. *Chinese Mathematics in the Thirteenth Century: the Shu-shu chiu-chang of Ch'in Chiu-shao (i.e. Qin Jiushao)*. Cambridge, Massachusetts: MIT Press, 1973.

Martzloff, Jean-Claude. *Histoire des mathématiques chinoises*. Paris: Masson, 1987. English translation by Stephen S. Wilson: *History of Chinese Mathematics*. New York: Springer-Verlag, 1995.

Martzloff, Jean-Claude. "Eléments de réflexion sur les réactions chinoises à la géométrie euclidienne à la fin du XVIIe siècle—le *Jihe lunyue* de Du Zhigeng vu principalement à partir de la préface de l'auteur et de deux notices bibliographiques rédigées par des lettres illustrées." *Historia Mathematica* 20(2): 160–179 and 20(3): 460–463, 1993.

Waerden, B. L. van der. *Geometry and Algebra in Ancient Civilizations*. Berlin: Springer-Verlag, 1983.

Wagner, D.B. "Liu Hui and Tsu Keng-chih on the Volume of a Sphere." *Chinese Science* 3: 59–79, 1978.

Wagner, D.B. "An Early Derivation of the Volume of a Pyramid: Liu Hui, Third Century AD" *Historia Mathematica* 6: 164–188, 1979.

Wu Wenjun, ed. *Jiuzhang suanshu yu Liu Hui* (Liu Hui and the *Jiuzhang suanshu*). Beijing: Beijing Shifan Daxue Chubanshe, 1982.

Ziggelaar, August. *Le physicien Ignace-Gaston Pardies S.J.* Odense: Odense University Press, 1971.

See also: Liu Hui and the *Jiuzhang suanshu* – *Elements*: Reception in the Islamic World – *Pi* in Chinese Mathematics – Liu Hui – Zu Chongshi – Li Shunfeng – Qin Jiushao – Algebra

GEOMETRY IN INDIA An examination of the earliest known geometry in India, Vedic geometry, involves a study of the *Śulbasūtras*, conservatively dated as recorded between 800 and 500 BC, though they contain knowledge from earlier times. Before what is conventionally known as the Vedic period (ca. 1500–500 BC), there was the Harappan civilization dating back to the beginning of the third millennium BC. Even a superficial study of the Harappan cities show its builders as extremely capable town planners and engineers requiring fairly sophisticated knowledge of practical geometry. An interesting conjecture has been suggested by a drawing on a seal found from Harappa (ca. 2500 BC): was there an awareness then that the area of a polygon inscribed in a circle approaches the area of the circle as the number of sides of the polygon keeps increasing? This is the basic idea behind techniques that were developed for the mensuration of the circle in a number of mathematical traditions including Indian.

The *Śulbasūtras* are instructions for the construction of sacrificial altars (*vedi*) and the location of sacred fires (*agni*) which had to conform to clearly laid down instructions about their shapes and areas if they were to be effective instruments of sacrifice. There were two main types of ritual, one for worship at home and the other for communal worship. Square and circular altars were sufficient for household rituals, while more elaborate altars whose shapes were combinations of rectangles, triangles, and trapezia were required for public worship. One of the most elaborate of the public altars was shaped like a giant falcon just about to take flight (*Vakrapraksa-śyena*). It was believed that offering a sacrifice on such an altar would enable the soul of the supplicant to be conveyed by a falcon straight to heaven.

It is clear that if in the construction of larger altars they had to conform to certain basic shapes and prescribed areas or perimeters, two geometrical problems would soon arise. One is the problem of finding a square equal in area to two or more given squares; the other is the problem of converting other shapes (for example, a circle or a trapezium or a rectangle) into a square of equal area or vice versa. The constructions were achieved through a judicious combination of concrete geometry (in particular what would be known today as the principle of dissection and reassembly), ingenious algorithms, and the application of the so-called Pythagorean theorem. The essence of the dissection and reassembly method involves two commonsense assumptions.

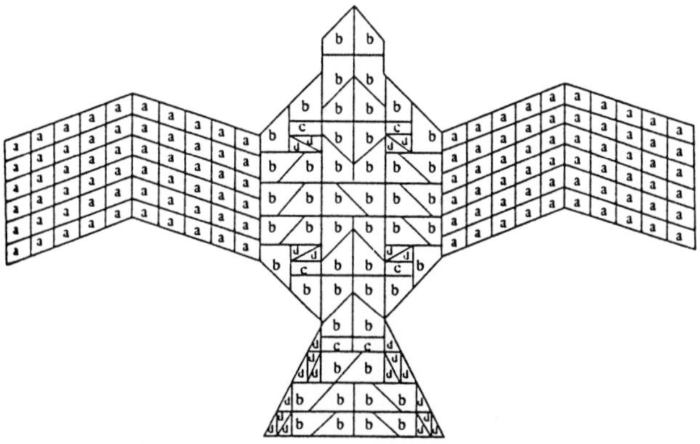

Figure 1: The first layer of a Vakrapaksa-syena altar. The wings are made from 60 bricks of type 'a', and the body, head and tail from 50 type 'b', 6 of type 'c' and 24 type 'd' bricks. Each subsequent layer was laid out using different patterns of bricks with the total number of bricks equalling 200.

The first is that both the area of a plane figure and the volume of a solid remain the same under rigid translation to another place. The second says that if a plane figure or solid is cut into several sections, the sum of the areas or volumes of the sections is equal to the area or volume of the original figure or solid. The reasoning behind this approach was very different from that behind Euclidean geometry, but the method was often just as effective, as shown in the Indian (and Chinese) "proofs" of the Pythagorean theorem.

In the *Kātyāyana Śulbasūtra* (named after one of the authors) the following proposition appears: "The rope (stretched along the length) of the diagonal of a rectangle makes an (area) which the vertical and horizontal sides make together." (2.11). Using this version of the Pythagorean theorem, the *Śulbasūtras* show how to construct both a square equal to the sum of two given squares and a square equal to the difference of two given squares. Further constructions include the transformation of a rectangle (square) to a square (rectangle) of equal area and of square (circle) to a circle (square) of approximately equal area. The constructions "doubling the square" and "squaring the circle" lead naturally to devising algorithms for the square root of 2 and other numbers, for implicit estimates of π, and for constructing similar figures in required proportions of a given figure.

The composers of the *Śulbasūtras* made it clear that their work was not original but could be traced to earlier texts, notably the *Saṃhitās* and the *Brāhmaṇas* of which the most relevant text, *Śatapatha Brāhmaṇa*, is at least three thousand years old. In spite of its obscurities and archaic character the text is valuable for an early discussion of the technical aspects of altar construction. The instructions given in *Śatapatha Brāhmaṇa* (X.2.3.11–14) for constructing a falcon-shaped altar consisting of 95 layers of bricks are as follows:

Area of the body (Atman) = $56 + \frac{12}{7}\sqrt{56}$;

Area of two wings = $2(14) + \frac{3}{7}\sqrt{14} + (\frac{1}{5})(\frac{1}{7})(3)(\sqrt{14})$;

Area of tail = $14 + \frac{3}{7}\sqrt{14} + (\frac{1}{10})(\frac{1}{7})(3)(\sqrt{14})$.

The total area is about 116 square *purushas*, which is an over-estimate of the required 101.5 square *purushas*, arising in part from a rounding off error involved in taking 14 rather than 13 + 8/15.

A major strand running through the history of Indian geometry and also providing the main motivation for the development of the subject was a recognition of the impossibility of arriving at an exact value for the circumference of a circle given the diameter (i.e. the incommensurability of π). A passage in Āryabhaṭa's *Āryabhaṭīya* (AD 499) — Verse 10 of the section on *Gaṇita* — reads:

> Add 4 to 100, multiply by 8, and add 62,000. The result is *approximately* the circumference of a circle whose diameter is 20,000. (Giving an implicit value of 3.1416 for π. This was the most accurate estimate for π known at that time. About six hundred years earlier (ca. 150 BC), there was an implicit estimate of π as the square root of 10 in a Jaina text called *Anuyoga Dwāra Sūtra*.)

It was the word "approximately" that gave food for thought to commentators of Āryabhaṭa's work from Bhāskara I (ca. AD 600) to Nīlakaṇṭha (b. AD 1445). The first formal proof of the transcendental nature of π was given by the Swiss mathematician Lambert in a paper to the Berlin Academy in 1671. However, about one hundred and fifty years earlier, Nīlakaṇṭha's commentary on *Āryabhaṭīya* contained the following statement:

> Why is only the approximate value (of circumference) given here? Let me explain. Because the real value cannot be obtained. If the diameter can be measured without a remainder, the circumference measured by the same unit (of measurement) will leave a remainder. Similarly the unit which measures the circumference without a remainder will leave a remainder when used for measuring the diameter. Hence the two measured by the same unit will never be without a remainder. Though we try very hard we can reduce the remainder to a small quantity but never achieve the state of 'remainderlessness'. This is the problem. (Adapted from Sarasvati Amma, 1979)

Once the incommensurability of π was accepted, the approach of the Indian mathematician was to obtain as accurate a value of this quantity as possible, and the strategy to be followed was expressed thus by Śaṅkara Variyār and Nārāyana Kriyākramakarī (ca. 1550)]:

> Thus even by computing the results progressively, it is impossible theoretically to come to a final value. So, one has to stop computation at that stage of accuracy that one wants and take the final result arrived at ignoring the previous results. (Adapted from Sarma, 1975)

The major breakthrough came from the revolutionary idea, most probably that of Mādhava (ca. 1340–1425), that it was possible to obtain an infinite series whose sum would be exactly equal to π and that an increasingly close rational approximation of the quantity could be obtained by taking partial sums successively of higher order. While the question of the slow convergence of this series was not explicitly discussed, the need for increasing rapidity of convergence was recognized and some remarkable corrections to be applied to truncated series were deduced. The work on infinite series for circular functions provided an impetus to derivation of other infinite series for trigonometric functions, namely arctangent, sine, and cosine series. Often inductive reasoning (and intuition) built upon geometrical representation helped them to discover these results, but the proof of these results can withstand any rigorous criterion applied to it today. An implicit estimate for π based on infinite-series expansion given by Mādhava around 1400 is correct to eleven decimal places.

Another area of geometry in which the Indian contribution was significant was in the study of the properties of a cyclic quadrilateral. In the *Brāhma Sputa Siddhānta*, Brahmagupta (b. AD 598) gives these results:

(1) The area of a cyclic quadrilateral is given by the product of half the sums of the opposite sides, or by the square root of the product of four sets of half the sum of the sides (respectively), diminished by the sides.

(2) The sums of the products of the sides about the diagonal should be divided by each other and multiplied by the sum of the opposite sides. The square roots of the quotients give the diagonals of a cyclical quadrilateral.

The derivations of these results are first referred to in a tenth-century commentary on Brahmagupta's work, but find their full expression in the sixteenth century Kerala text *Yuktibhāṣā* by Jyeṣṭhadeva. This makes use of Ptolemy's theorem that in a cyclic quadrilateral the product of the diagonals is equal to the sum of the products of two pairs of opposite sides.

Notable extensions in this area are contained in Nārāyana Pandita's *Gaṇita Kaumudī* in the fourteenth century and Parameśvara's *Lilāvati Bhāsya*, a detailed commentary on Bhāskaracharya's *Lilāvati*. In the latter is found a new rule for obtaining the radius of the circle in which a cyclic quadrilateral is inscribed. The great interest in the cyclic quadrilateral in Indian mathematics arose from the fact that it was an important device for deriving a number of important trigonometric results which were in almost all cases used in astronomy.

In India geometry never became idealized in the way it did in Greece. Geometry was largely concrete and empirical in character. It did, however, have an algebraic character which is best seen in the genesis of trigonometry there. Because of their geometric emphasis, the Greeks used chords in their astronomical calculations, whereas the Indians developed the notion of sines and versines (i.e. 1 − cosine of an angle) as early as AD 500. Āryabhaṭa was perhaps the first Indian astronomer to give a special name to these functions and draw up a table of sines for each degree. Approximation formulae were developed for these functions, culminating in the construction of sine tables in Kerala during the fifteenth century where the values in almost all cases are correct to the eight or ninth decimal place, a remarkable degree of accuracy.

GEORGE GHEVERGHESE JOSEPH

REFERENCES

Colebrook, H.T. *Algebra with Arithmetic and Mensuration from the Sanscrit of Brahmegupta and Bhascara*. London: John Murray, 1817.

Datta, B. *The Science of the Śulbas: A Study in Early Hindu Geometry*. Calcutta: Calcutta University Press, 1932.

Gupta, R.C. "Paramesvara's Rule for the Circumradius of a Cyclic Quadrilateral." *Historia Mathematica* 4: 67–74, 1977.

Joseph, G.G. *The Crest of the Peacock: Non-European Roots of Mathematics*. London: Penguin, 1992.

Joseph, G.G. "What is a Square Root ? A Study of Geometrical Representation in Different Mathematical Traditions." In *Proceedings of 1993 Annual Meeting of the Canadian Mathematics Education Study Group*. Ed. M.Quigley, University of Calgary, 1994, pp.1–14.

Kulkarni, R.P. *Geometry according to Śulba Sūtra*. Pune: Vadika Saṁśodhana Maṇḍala, 1983.

Varma, T. Rama and A. Aiyer, eds. *Yuktibhāṣā*. (in Malayalam). Trichur: Mangalodayam, 1948.

Sarasvati Amma, T.A. *Geometry in Ancient and Medieval India*. Delhi: Motilal Banarisidass, 1979.

Sarma, K.V., ed. *Lilavati of Bhaskaracara with the Commentary Kriyākramkarī*. Hoshiarpur: Vishveshvaranand Institute, 1975.

Staal, F. *Agni: The Vedic Ritual of the Fire Altar*. Berkeley: Asian Humanities Press, 1983.

See also: *Śulbasūtras* – Geometry, Practical – Nīlakantha – Āryabhaṭa – Nārāyana – Mādhava – Bhāskara – Jyeṣṭhadeva – Parameśvara

GEOMETRY IN THE ISLAMIC WORLD Many original Muslim geometric works are "books" or chapters of great mathematical, astronomical, and encyclopedic works. Geometrical parts are contained in the algebraic treatise of al-Khwārizmī (ca. 780–ca. 850), in the *Miftāḥ al-ḥisāb* (Key of Arithmetic) of Jamshīd al-Kāshī (d. ca. 1430), in the astronomical *al-Qānūn al-Mas'ūdī* (Masudic Canon) of Abū'l Rayḥān al-Bīrūnī (973–1048) and in his treatises on astrolabes, as well as in the encyclopedic *Kitāb al-shifā'* (Book of Healing) of Abū 'Alī ibn Sīnā (Avicenna, 980–1037). There are also special geometric treatises, commentaries on Euclid, and treatises on geometrical theorems, calculations, constructions, and the foundations of geometry.

Calculations of areas of plane figures and of volumes of solids were 1040 considered in the chapter "On mensuration" of the algebraic treatise of al-Khwārizmī. Many rules in this chapter coincide with the rules of Hero's *Metrics* (first century AD) and of Indian mathematicians of the fifth and sixth centuries. For the area S of a circle with radius r, he supplies the rule $S = \pi r^2$, where $\pi = 3\frac{1}{7}, \sqrt{10}, 62832/20000$.

The contents of this chapter are similar to the Hebrew *Mishnat ha-middot* (Treatise on Measuring) which was believed to be part of the *Talmud* written in the second century AD. Recently Gad Zarfatti showed that this book was written after al-Khwārizmī's treatise and under its influence.

Many treatises on geometric calculation were written by Thābit ibn Qurra. In his *Kitāb fī misāḥa al-ashkāl al-musaṭṭaḥa wa 'l-mujassama* (Book on Mensuration of Plane and Solid Figures), he provides many rules for measuring area and volume. For example, there is his general rule for the calculation of the volume V of a right round cylinder, cone, and truncated cone expressed by one formula: $V = \frac{1}{3}h(S_1 + S_2 + \sqrt{S_1 S_2})$.

In his *Kitāb fī misāḥa qaṭ 'al-makhrūṭ alladhī yusammā al-mukāfī* (Book on Mensuration of a Conic Section Named Parabola) and other works, Thābit tackles many geometric problems. Among these are the problems of calculating the area of a segment of a parabola, the volume of segments of solids obtained by the revolution of segments of a parabola, and of the area of a surface of an oblique round cylinder. Thābit calculated the area of segments of an ellipse by means of an equi-affine transformation of the ellipse onto a circle with equal area. His grandson Ibrāhīm ibn Sinān (908–946) in his *Kitāb fī misāḥa al-qaṭ 'al-mukāfī* (Book on Mensuration of a Parabola) calculated the area of a segment of a parabola by means of a general affine transformation. Wayjan al-Qūhī (tenth-eleventh centuries) and Ibn al-Haytham (965–ca. 1040) calculated the volumes of segments of solids obtained by the revolution of segments of parabolas that Thābit had not considered.

Al-Bīrūnī, in the mathematical part of his *Masudic Canon,* calculated the sides of a regular triangle, square, pentagon, hexagon, octagon, and decagon inscribed in a given circle. Also, he provided the length $C = 2\pi r$ as the circumference of a circle where π = 3.7 30 59 10 in sexagesimal fractions and 1628631471/518400000 in decimal. Al-Kāshī in his *al-Risāla al-muḥīṭiyya* (Treatise on Circumference) calculated the length $C = 2\pi r$ by means of the calculation of the perimeter of regular inscribed and circumscribed polygons with $3 \cdot 2^{28}$ sides. The number of sides was chosen on the condition that the difference between these perimeters for the great circle of the sphere of fixed stars must be less than the "width of a horse hair". The result was expressed in sexagesimal and decimal fractions as 3.8 29 44 0 47 25 53 7 25 and 3.14 159 265 358 979 325. In the last case the calculation is correct up to 17 digits.

The teachers of Thābit ibn Qurra, the brothers Muḥammad (d. 872), Aḥmad, and al-Ḥasan Banū Mūsā ibn Shākir, in their *Kitāb al-shakl al-mudawwar al-mustaṭīl* (Book on an Oblong Round Figure) proposed the construction of an ellipse by means of a string attached to its foci (the so-called "Gardiner's construction"). Thābit himself solved two classical problems of antiquity: the construction of two mean proportional magnitudes between two given ones ($a : x : y : b$), which is the Delic problem of the duplication of a cube (if $b = 2a$), and the trisection of an angle. He also proposed the spatial construction of a semiregular polyhedron with 14 faces, which some now say was discovered by Archimedes.

The great philosopher Abū Naṣr al-Fārābī (ca. 870–950) proposed many geometric constructions of parabolas, regular polygons, squares equal to three given equal squares, constructions with one opening of the compass, and constructions on the sphere. These were also studied by Thābit's pupil's pupil Abū'l-Wafā al-Būzjānī (940–998) in the *Kitāb fī mā yakhtāju ilayhi al-ṣāni' min a'māl al-handasiyya* (Book on What is Necessary from Geometric Construction for the Artisan).

Ibrāhīm ibn Sinān ibn Thābit proposed many ways to construct a parabola (including al-Fārābī's one), ellipse, and hyperbola. One of the constructions of an ellipse is based on the compression of a circle; one of an hyperbola is based on a projective transformation of a circle.

Al-Qūhī described the instrument for the construction of conics, which he invented. It is a compass in which one leg varies in length and the second leg can change the angle of its slope to the plane. Many methods for the construction of conics were described by al-Bīrūnī in *Isti'āb al-wujūh mumkina fī ṣan'a al-asṭurlāb* (The Exhaustion of Possible Ways of Constructing an Astrolabe). In particular he mentions the "Gardiner's construction" of Banū Mūsā, the perfect compass of al-Qūhī, and constructions of all three kinds of conics based on projective transformations.

Al-Fārābī revised the order of the first definitions of Euclid. The first definition of Book One of the *Elements* was that of a point as "that which has no parts". Later Euclid defined a line as "a breadthless length" and a surface as "that which has length and breadth only". In Book 11 he defined a solid as "that which has length, breadth, and depth". Al-Fārābī in following Aristotle defined a solid as the first abstraction from a physical body, a surface as the abstraction from a thin solid, a line as the abstraction from a narrow surface, and a point as the abstraction from a short line. The same order of geometric definitions was used by al-Bīrūnī in the geometric part of his *Kitāb al-tafhīm li-awā'il ṣinā'a al-tanjīm* (Book of Instruction in the Elements of the Art of Astrology).

In his *Taḥrīr Uqlīdis* (Exposition of Euclid) Naṣīr al-Dīn al-Ṭūsī added to Euclid's axioms that of the existence of points, lines, and surfaces, and the axiom of possibility of choice of a point on an arbitrary line and surface and of a line on an arbitrary surface and through an arbitrary point. There are also analogous axioms ascribed to al-Ṭūsī's version of the exposition of Euclid's *Elements* and in the geometric part of the encyclopedic work of his pupil Quṭb al-Dīn al-Shīrāzī (1236–1311). In both these works there are also attempts to prove Euclid's first three postulates about the ideal ruler and compass.

In al-Shīrāzī's commentaries to al-Ṭūsī, the rules of the theory of syllogisms from Aristotle's *Analytics* are systematically used.

In many Islamic treatises the problem of mathematical atomism was discussed. Aristotle's ideas, on which Euclid's *Elements* is based, ousted the mathematical atomism of Pythagoras and Democritus. According to them, finite solids, surfaces, and lines consist of a finite number of "atoms". Pythagoras said these atoms were points without sizes, and the sizes of finite figures were created by the distances between them. According to Democritus these atoms were particles with very small but finite sizes. In medieval Islam the ideas of mathematical atomism were held by Muslim scholastics, the *mu'tazila* and *mutakallimūn*, who wanted to find a rational explanation for Islam. They believed that time consists of a separate "now" and explained by means of this view that God creates the world anew every instant. Among the *mu'tazila* there were adherents of both types of mathematical atomism. The first school was headed by Abū'l-Qāsim al-Ka'bī (d. 932) whose nickname al-Ka'bī means "cubical"; the second was headed by Abū'l-Ḥāshim al-Jubbā'ī (820–933). Democritus' ideas were also held by the philosopher and alchemist Abū Bakr al-Rāzī (865–952). The idea of mathematical atomism was also mentioned by al-Shīrāzī in his commentaries on a treatise by al-Ṭūsī.

Islamic mathematicians often attempted to prove Euclid's fifth postulate on which the theory of parallel lines in his *Elements* is based. This postulate is much more complicated than the first four. It asserts that, if a straight line falling on two straight lines makes the interior angles on the same side together less than two right angles, the two straight lines, if produced indefinitely, meet on that side in which the angles together are less than two right angles. Attempts to prove this axiom as a theorem were made by Archimedes (third century BC), Posidonius (second–first centuries BC), Ptolemy (second century AD), Proclus (fifth century), and Simplicius (sixth century). In their proofs there was the logical error of *petitio principii*, that is the implicit use of an assertion equivalent to the one being proved.

The first Islamic mathematician who attempted to prove the fifth postulate was al-'Abbās al-Jawharī (first half of the ninth century). In his *Iṣlāḥ li-kitāb al-Uṣūl* (Improvement of the Book "Elements") he proposed a proof based on the assertion that a straight line intersecting both sides of an angle can always be drawn through any point inside this angle.

Thābit ibn Qurra made two attempts to prove the fifth postulate. His two treatises on parallel lines were written in Syriac and then translated by him into Arabic. The first proof is based on the assertion of the existence of "a simple motion"; that is a parallel translation equivalent to the existence of equidistant straight lines and to the existence of a rectangle. The second proof is based on the assertion that, if two straight lines diverge on one side of a line falling on them, they necessarily converge on the other side. From this follows the existence of a parallelogram. An analogous error was made by al-Faḍl al-Nayrīzī (d. 922) in his *Risāla fī bayān al-muṣādara al-mashhūra min Uqlīdis* (Treatise on the Proof of the Known Postulate of Euclid).

Ibn al-Haytham also made two attempts to prove the fifth postulate. One attempt was based on the same error as in Thābit's first treatise. In this proof Ibn al-Haytham first considered quadrangles with three right angles ("the Lambert quadrangles") and three hypotheses about their fourth angle. The hypotheses of acute and obtuse angles were refuted by means of the existence of a rectangle. The second attempt was written in his last years. Here was the first proof of the fifth postulate which was free from logical error. At first the postulate was replaced by another more evident one and was proved by means of it. This new postulate was: two intersecting straight lines cannot be parallel to a third line. This postulate in the form "if the point A and a straight line 'a' are given, it is impossible to draw more than one not intersecting line through A in their plane". This is now known as "the Playfair postulate" or "the strong Hilbert axiom of parallels", after Europeans who later came up with the same axiom.

Also free from logical error was ʿUmar al-Khayyām's proof of the fifth postulate in his *Sharḥ mā ashkala min muṣādarāt kitāb Uqlīdis* (Commentaries on Difficulties in Introductions of Euclid's Book). In this work Khayyām proved the postulate on the basis of a postulate which he called the "fourth principle of the philosopher (Aristotle)". "Two converging lines intersect and it is impossible for them to converge in the direction of divergence". In this proof he first considered equilateral quadrangles with two right angles at their low bases (which we now call "the Saccheri quadrangles") and three hypotheses about their equal upper angles. The hypotheses of acute and obtuse angles are refuted by means of this fourth principle. These two hypotheses, which are analogous to those of Ibn al-Haytham (his quadrangles are halves of Khayyām's), are fulfilled in hyperbolic and elliptic non-Euclidean geometries respectively. Ibn al-Haytham and Khayyām actually proved some theorems of these geometries first.

Naṣīr al-Dīn al-Ṭūsī considered the theory of parallel lines. In the treatise, al-Ṭūsī explained the proofs of al-Jawharī, Ibn al-Haytham (only the second) and Khayyām (not completely) and considered the same quadrangles and the same three hypotheses as Khayyām. However, he made the same logical error *petitio principii*. In the book written after the critical letters of Qayṣar al-Ḥanafī (ca. 1170–1251) he explained the same proof, but before it he formulated the postulate similar to Khayyām's, and thus does not make a logical error.

Al-Shīrāzī also made an attempt to prove the fifth postulate based on a logical error: he supposed that the distance between lines which do not intersect is constant.

We saw that in the proofs of the fifth postulate by Thābit ibn Qurra and Ibn al-Haytham "a simple motion" was used. The application of motion to geometry was used by Pythagoreans who considered lines as traces of moving points and surfaces as traces of moving lines. However, this application was rejected by Aristotle, who considered geometric solids as abstractions from physical bodies, surfaces as abstractions from thin solids, lines as abstractions from narrow surfaces, and points as abstractions from short lines. Following Aristotle, Euclid tried in his *Elements* to avoid the application of motion. Khayyām criticized the application of motion by Ibn al-Haytham.

We also saw that Thābit ibn Qurra and his grandson Ibrāhim ibn Sinān used affine transformations in their geometric treatises, and al-Bīrūnī also used projective transformations.

An astronomical instrument called an astrolabe was very popular in the Middle Ages. It was based on the projection of the celestial sphere onto its plane. Many treatises on the construction of astrolabes contain descriptions of different projections. In the *Kitāb ṣanʿa al-asṭurlāb* (Book of the Construction of the Astrolabe), Aḥmad al-Farghānī (d. 861) described the stereographic projection. That and others were described by al-Bīrūni in his *Exhaustive Ways of Constructing an Astrolabe*. Al-Bīrūnī applied a stereographic projection for determining the azimuth of *qibla* (the direction to Mecca). He also used the stereographic and some other projections of a sphere on to a plane in his *Risāla fī tasṭiḥ al-ṣuwar wa tabṭiḥ al-kuwar* (Treatise on the Projection of Designs on to a Plane and on the Map of Spheres on a Plane), which was devoted to problems of cartography.

In the *Elements*, Euclid used the plane geometrical algebra inherited from the Pythagoreans. In this, he represents the products $a \times b$ as rectangles with sides a and b and provides a geometric representation of algebraic equalities. For instance, there is a representation of the formula $(a + b)^2 = a^2 + 2ab + b^2$ by decomposing the square $(a+b)^2$ on two squares a^2 and b^2 and two rectangles $a \times b$. In Archimedes' *Lemmas* there is an analogous decomposition of a circle with diameter $a+b$ on two circles with diameters a and b and two "arbelons". Analogous representations of the formula $(a+b)^3 = a^3+3a^2b+3ab^2+b^3$ by decompositions of the cube $(a + b)^3$ and of the sphere with diameter $a + b$ were proposed by Abū Saʿīd al-Sijzī (ca. 950–ca. 1025).

Apparently the idea of "geometric" names for powers: square-square for x^4, square-cube for x^5, cubo-cube for x^6 and so on was borrowed from Diophantus' *Arithmetics*.

Islamic mathematicians used rectangular and oblique coordinates connected with conics: the axes of these coordinates were one of the diameters of a conic and the tangent line at one of its ends. Each point of a conic was determined by half of its chord conjugate with this diameter from a given point to the diameter and by the segment of this diameter from its end (vertex) to the meeting point of the chord with the diameter. They called these coordinates "ordered lines" and "line cut from the vertex" respectively (Our terms "ordinate" and "abscissa" came from Latin translations of these expressions). These coordinates were used by Khayyām in his *Risāla fī'l-barāhīn ʿalā masā'il al-jabr wa'l-muqābala* (Treatise on Demonstrations of Problems of Algebra), which was devoted to the solution of cubic equations by means of the intersection of conics.

Islamic astronomers and geographers following Ptolemy used the geographical coordinates longitude and latitude for the surface of the Earth, and analogous ecliptical coordinates with the same names, the horizontal coordinates azimuth and altitude, and the equatorial coordinates right ascension (or horary angle) and declination for the celestial sphere.

Thābit ibn Qurra determined the positions of the ends of the shadows of the gnomon on the plane of a plane sundial by rectangular coordinates which he called, by analogy with the geographical coordinates, longitude and latitude, and by polar coordinates which he called azimuth and length of shadow.

Muḥammad al-Khwārizmī in his geographical treatise *Kitāb ṣūra al-arḍ* (Book of the Picture of the Earth) proposed the classification of curved lines and used the names of different kinds of lines in his descriptions of rivers, seacoasts, and islands between points with given geographical coordinates.

In addition to their original contributions, Muslims made many translations of Greek works. In many cases, the Arabic version of the work is the only one extant today. The most important Greek geometric work, Euclid's *Elements,* was translated into Arabic in the ninth century AD by al-Ḥajjāj ibn Yūsuf ibn Maṭar and Isḥāq ibn Ḥunayn al-ʿIbādī. The last translation was edited and supplemented by Thābit ibn Qurra (836–901), who was also the translator of other works of Euclid, of many works of Archimedes, of the *Conics* of Apollonius, and of many Greek mathematical works. These, together with a geometric treatise of his teachers the brethren Banū Mūsā, and his *Kitāb al-mafrūḍāt* (Book of Assumptions) constituted the so called "middle books" which were studied between Euclid's *Elements* and Ptolemy's *Almagest*. Later many Muslim revisions of other Greek mathematical works appeared.

BORIS ROSENFELD

REFERENCES

Rosenfeld, Boris. *A History of Non-Euclidean Geometry.* New York: Springer-Verlag, 1988.

Youschkevitch, Adolphe. *Les mathématiques arabes* (VIII–XV siècles). Paris: Vrin, 1976.

Zarfatti, Gad B. *Munḥe ha-mathematikah ba-sifrut ha-madaʿit ha-ʿivrit shel yeme habenayim* (Mathematical Terminology in the Hebrew Scientific Literature of the Middle Ages). Jerusalem: Hebrew University, 1968.

See also: al-Shīrāzī – Sundials – Gnomon – ʿUmar al-Khayyām – Naṣīr al-Dīn al-Ṭūsī – Astrolabe – Conics – Sexagesimal system – Mathematics in Islam – Thābit ibn Qurra – al-Bīrūnī – Banū Mūsā – al-Khwārizmī – Maps and Mapmaking in Islam – Geography in Islam – Atomism in Islam – al-Farghānī – al-Fārābī – al-Qūhī – Ibn al-Haytham – al-Rāzī – al-Kāshī – *Elements* – Ibrāhīm ibn Sinān – al-Sijzī – *Almagest* – al-Māhānī – al-Khujandī

GEOMETRY IN JAPAN Western mathematics is concerned very much with logic based on proof and demonstration. Japanese mathematics in the Edo period (1603–1867), which is called *Wasan*, has its origin in China, and from the beginning put its emphasis on application, mainly on application in daily life. At the beginning of *Wasan*, experts studied practical aspects of mathematics, such as deriving the areas of figures and the volumes of solid bodies. However, in the developing period, they began to devise questions which provided conclusions based on the beauty of figures and complicated calculation. They especially liked questions on geometrical figures, and they had their own ways of studying. When they studied circles, they tried to increase their number, change their positions, and consider them as ellipses. They also tried to expand into three dimensions. Because of the national isolation policy of Japan, *Wasan* experts were kept in guild-like groups. The study of *Wasan* became very complicated; practitioners could not concentrate on the main themes but had to act as if their studies were rather like a hobby.

Wasan was formulated in the latter half of the seventeenth century by a Japanese mathematician, Takakazu Seki (or Kowa Seki) with the text *Jinko-ki* as its basis. The interest of *Wasan* experts in those days was to solve the questions left unsolved in *Jinko-ki* and other mathematical texts: that is, to calculate the areas and volumes of planes and solid figures, and also the length of chords and arcs and the area surrounded by them. Takakazu Seki, before he succeeded in systematizing *Wasan*, tried hard to solve those questions, and in the process studied parabolas, hyperbolas, and also the spirals of Archimedes. And he partially succeeded in measuring ellipses.

In the field of algebra, Seki used a determinant of five dimensions in his work *Kaifukudai-no-ho* published in 1683. In the area of geometry, he derived a relative equation between a side of a regular polygon and radii of an inscribed circle and a circumscribed circle. This arose from a theory on regular polygons. He also made great accomplishments in the field of *Enri*: this word may be translated as "circle principle" or "circle theory," being derived from the fact that the mensuration of the circle is the first subject that it treats. He repeated doubling the numbers of the side of an inscribed, regular polygon, and achieved the correct number for π. He devised a calculation method for circles — integral calculation — studying the relation among the chord, the arrow, and the arc. He also worked in other fields, such as the calendar.

After the publication of such works as *Seiyo Sampō*, *Shinpeki Sampō*, *Sampō Kokon Tsuran*, and *Sampō Tensei-ho Shinan*, interest in *Wasan* was very much aroused. For example, *Sangaku* (Mathematical Tablets) were dedicated to shrines and Buddhist temples. Some of these are beautiful, with figures of red, blue, and yellow. The definitions of the figures are not written on the tablets, but the drawn figures (questions) themselves illustrate the things necessary to solve the questions. These questions posed by *Wasan* experts are combined questions, such as a square and a square, a straight line and a circle, a circle and plural circles (or ellipses). There are questions on the length of a square side, a straight line, and the radius of a circle. There are also questions on the number and areas of circles put into one circle. These questions are expanded from "plane" to "solid", and from "limited" to "unlimited".

Wasan experts tried to derive the theorems to solve questions, and in the process they reached general laws by something like an inductive method. They sought a general equation with some sort of inspiration. One of these *Wasan* experts, Nushizumi Yamaji, stated in his *Sampo Ruiju* (vol. 3) the same ideas as in Descartes' circle theorem but in a different way. A theorem like Descartes' was introduced in the mathematical texts of the first half of the nineteenth century, such as *Kokon Sankan* by Gokan Uchida.

Wasan in the Bunka and Bunsei period (1804–1818–1829) was at its peak. In *Suri Mujinzo* by Nagatada Shiraishi, one finds a technique or theorem called *Bosha Jutsu*, which has exactly the same idea as Casey's. This theorem is considered an extension of Ptolemy's.

Shingen Takeda wrote *Shingen Sampō*. In the first volume of this work, there is a problem called Clever Ways to Cross the Twenty-eight Bridges in Naniwa (now Osaka). This is the same kind as Euler's Königsberg problem. Takeda's Twenty-eight Bridges marked the beginning of topology.

Another *Wasan* expert, Zen Hodoji, learned many things through visiting various districts and teaching students in each district. In one of his works, *Kan-shinko Sanhen*, there are questions which are close to the present "inversion" formula. We are not sure how he reached the inversion formula, but the result was correct.

As was stated before, *Wasan* experts were inspired to find a solving technique. *Kyokkei Jutsu* and *Henkei Jutsu* should be introduced here. *Kyokkei Jutsu* is explained in *Sampo Kyokkei Shinan* written by Hiroshi Hasegawa and his student, Giichi Akita. Special figures — *Kyokkei* — are used in this method. Beside this, Hasegawa wrote *Sampo Henkei Shinan* with his student, Teishin Heinouchi. In this, you find a sentence to the effect that "Questions with reasonable conditions, that is, questions with conditions enough to solve them are called "complete questions" while others with too many or too few conditions are called "incomplete questions". Some of these questions can be solved with insufficient conditions, and they are called *Kyoku Dai* (special questions). This book deals mainly with variations of *Kyoku Dai*. Both *Kyokkei Jutsu* and *Henkei Jutsu* deal with figures in special cases instead of general figures, but each technique is different. *Kyokkei Jutsu* has a tendency to make a mistake by trying to get an answer; *Henkei Jutsu* has no such tendency, but you find it difficult to get a *Kyoku Dai*. Each method is unique as a method of geometrical studies.

In the Meiji period (1868–1911), the new education system established in 1872 eliminated *Wasan* studies from public education, and adopted western mathematics.

However, in the Taisho and Showa periods (1912–1928–1988) mathematical leaders such as Tsuruichi Hayashi and Matsusaburo Fujiwara collected materials and wrote theses on various themes of *Wasan*, and its popularity was revived. The History of Science Society of Japan was established in 1941 and Kin-nosuke Ogura, counselor of the society, insisted that Japanese mathematics history be carefully considered. The Mathematics Club, later The History of Mathematics Society of Japan, was established in 1958 mainly to study *Wasan*. Seminar groups appeared in many districts, and magazines and theses were published. In 1960, the *History of Japanese Mathematics before the Meiji Period* (5 vols.) edited by the Japan Academy was complete. All of the works mentioned in this essay can be found in this set.

YOSHIMASA MICHIWAKI

REFERENCES

Hayashi, Tsuruichi. *Wasan Kenkyu Shuroku* (Collected Papers on Old Japanese Mathematics), vols 1 and 2. Tokyo: Kaiseikan, 1913.

Meiji Zen Nippon Sugakushi (History of Japanese Mathematics before the Meiji Period, 5 vols) Tokyo: Noma Kagaku Igaku Kenkyū Shiryō, 1979.

Michiwaki, Yoshimasa, Makoto Oyama, and Toshio Hamada. "An Invariant Relation in Chains of Tangent Circles." *Mathematics Magazine* 48(2): 80–87, 1975.

Michiwaki, Yoshimasa, and Noriko Kimura. " On the Relevancy of the Descartes Circle Theorem and Links of Six Spheres Theorem." *Journal of History of Science, Japan*, Series 2, 22(147), 1983 (in Japanese).

Michiwaki, Yoshimasa. "An Investigation of Mathematics Education." *2nd International Congress on the Cultural History of Mathematics*, 1993, Japan.

Ogura, Kin-osuke. *Nihon no Sugaku*, pp. 120–142, Tokyo: Iwanami Shoten, 1964.

Smith, David and Yoshio Mikami: *A History of Japanese Mathematics*. Chicago: Open Court, 1914.

See also: Seki Kowa

GEOMETRY IN THE NEAR AND MIDDLE EAST This article concentrates on ancient Mesopotamia and Egypt. Within this area, it concentrates on the geometrical knowledge of the scribal traditions, not on what *could* have served for the construction of buildings. It must be observed that high technical precision is often achieved through the combination of quite elementary mathematics and sophisticated non-mathematical techniques. Anybody who has played with rulers and compasses knows that a regular pentagon can easily be made more precise through trial and error than by "exact" construction with all its accumulated errors. It is therefore next to impossible to extrapolate from the high precision of for example prestige buildings to the particular kind of geometry used in their design, unless some kind of blueprint has survived. Even in such cases, when the tracing of a ground plan has subsisted or the building can be seen to be planned around base lines geared to the standard brick, all that may be concluded is often that *some kind* of geometrical knowledge is involved.

On the other hand, in Mesopotamia as well as Egypt, written texts with geometrical contents have survived, most of which are scribe school problems. Even when unambiguous (which is not always the case), they confront us with a problem of bias: scribal mathematics consisted in calculation; the texts always aim at the determination of a number (an area, a volume, the number of man-days needed to dig a ditch, etc.). Geometrical construction was not the concern of the scribes.

Mesopotamia

The sources for our knowledge of Mesopotamian geometry reach back to the later fourth millennium BC. The earliest sources are not very informative, but they demonstrate that the length and area metrology, together with the basic techniques for area computation, go back to the first phase of writing. Sources from the third millennium become more copious, in particular towards its end. Sophisticated scribe school problems dealing with complex volumes only turn up in the Old Babylonian period (early second millennium), and disappear again from the horizon after 1600 BC. Finally, some interesting texts have survived from the later half of the first millennium BC.

Length metrology may have been created in the same process as writing and written administration. Already in the earliest written documents, area metrology is keyed to the length system — but some of its units appear to be old "natural" (irrigation, ploughing, or seed) measures which have been normalized so as to fit, while others have been created anew. These systems survive until the end of the Old Babylonian period. In the late period, new quasi-natural measures (seed measures) are created, which allow easier practical computation.

The basic volume measures coincide with the area measures, which are tacitly imagined as provided with the height 1 cubit. The system of hollow measures, on the other hand, is perplexingly complex, involving both traditional measures that have been normalized so as to fit the length system, and corrections introduced in order to facilitate practical computation of rations, cylindrical volumes, etc. A Mesopotamian specialty is the multiplicity of "brick measures", using the terminology of volume measures but referring in fact to the number of bricks, for which several standard dimensions coexisted.

The Old Babylonian texts show us that square and rectangular areas as well as right triangles and trapezia were calculated as we would do it. No unambiguous standard term for the height of a triangle or a trapezium is present, and field plans show us that complex areas, including skew triangles and trapezia, were subdivided into "practically rectangular" quadrangles and "practically right" triangles. No concept of a quantified angle is present, the field plans seem simply to distinguish "right" from "wrong" angles.

In practical mensuration, working in an always slightly uneven terrain and not in an abstract Euclidean plane, what is laid out as a rectangle with traditional methods mostly turns out to have opposing sides slightly different. In this situation, the Mesopotamian scribes calculated the area according to the "surveyors' formula": average length times average width. The result always exceeds the true area, but not significantly if the angles of the quadrangle are approximately right. In school problems, where the formula served as a pretext for calculation, it was occasionally used in situations where the outcome was absurd. In practical field measurement, a single instance is known where the error is at least 100%; whether this reflects general carelessness or the ignorance of a single scribe is not clear.

Most often, the circular area was determined as 1/12 of the square on the circumference (which is often the dimension most easily determined in practice by means of a piece of string). Translated into our idiom, this means that 4π was approximated as 12, or that $\pi = 3$. Other computations, such as the determination of the diameter from the circumference, agree with this; the Babylonians were thus aware that the circular area was the semi-product of the radius and the circumference.

Simple volumes (prisms, cylinders) were determined as the product of the base and the height — the formulation being that the base was "raised" to the height. Raising is a general term for multiplication based on considerations of proportionality — derived originally from volume determinations but in general use. The idea which underlies volume determination is indeed one of proportionality, since the base was understood as already provided with a standard height of 1 cubit. Similarly, a length could be understood as provided with a standard width of one length unit, and area computations (when not implicit in the construction of a rectangle) were hence also done by raising.

Complex volumes, such as truncated pyramids and cones, were determined by *ad hoc* methods, not always correctly or consistently. Several texts determine the volume of a truncated cone as that of a cylinder whose diameter is the average of the diameters of top and bottom. One of them determines the volume of a truncated square pyramid in the same way, but adds a correction term that makes the computation accurate (it is not quite sure whether the underlying reasoning is also correct). Other texts determine the volumes of truncated cones and pyramids by raising the average between top and bottom area to the height.

No later than the Old Babylonian period, the Pythagorean theorem was known and used for diagonal calculations. The purpose of a table text making use of Pythagorean triples is not fully decided, but it seems to be arithmetical rather than geometrical.

Already in the twenty-third century BC, it was known that the square on the parallel transversal which bisects a trapezium is the half-sum of the squares on the parallel sides. No text tells us how this formula was reached, but it follows simply from consideration of two squares of which one is embedded concentrically in the other. Similar naïve-geometric reflections may have led to the discovery of the Pythagorean theorem; their presence is well attested in the so-called "algebra" texts.

A few texts testify to an interest in geometrical arrangements and in regular polygons; in the actual cases, however, the texts always calculate something (mostly an area). A text with a regular hexagon draws a height in one of the six isosceles triangles from which it is composed and computes its length, and uses this in the determination of the area.

The few late Babylonian texts exhibit some changes in the area techniques, such as a more general use of heights in area determination.

Egypt

Third-millennium sources for Egyptian geometry are rare, and even though the precision of the great pyramids demonstrates both architectural and geometrical skill, they tell us little about the geometrical techniques that were in use.

The important mathematical texts — the *Rhind Mathematical Papyrus* and the *Moscow Papyrus* — date from the early second millennium BC. Together with some minor texts and fragments and with what can be grasped from administrative documents, they present us with a coherent picture. In spite of some unsolved problems of interpretation and the presence of still open questions, nothing suggests that new sources might change our understanding of Egyptian mathematics radically. Nor does Egyptian technology ask for *mathematical* knowledge beyond what we know from the papyri.

The evidence for the length and area metrologies of the third millennium is meager, but sufficient to show that they were intricate; the metrologies of the early second millennium derive from and expand a subset of the third-millennium systems.

In this "classical" period, the most widely used length unit was a "royal cubit" of seven palms. Also in use was a "short cubit" of 6 palms, together with units equal to the diagonal and semi-diagonal of a square with side equal to the royal cubit.

Area units were derived from the length units. The basic unit for land measurement was $(100 \text{ cubit})^2$; for practical purposes, however, the Egyptians would often think in terms of strips, as the Babylonians might also do.

Volume and capacity units were also keyed to the length system, but in a way that shows them to have originated independently — how else are we to explain that the basic volume unit was not 1 cubit^3 but $\frac{2}{3}$ cubit^3?

The areas of squares, rectangles and right triangles were calculated from length and width. The areas of trapezia and isosceles triangles were found as the product of height and average width, and the Rhind Mathematical Papyrus explains that this calculation of the triangular area means that the triangle is transformed into a rectangle. It should be noted that the height is a given or measured number, not the outcome of a computation.

The area of the circle was found as $(\frac{8}{9}d^2)$, where d is the diameter. Translated into modern terms this means that the interesting constant was not π but $\sqrt{\pi/4}$, which was approximated as 0.888... (where the true value is 0.8862...).

One problem (concerned with a "basket") has been interpreted as the correct determination of the surface of a semi-sphere (equal to twice the cross section of the sphere). Another interpretation is that the text wants to determine the curved surface of a semi-cylinder, which will then also be correct, and would imply that the semi-circumference of the circle was found in agreement with the determination of the area.

Volumes of prisms and cylinders were found as the product of base and height (multiplied with a metrological conversion factor $1\frac{1}{2}$). The *Moscow Papyrus* gives a correct determination of the truncated pyramid as $(a^2 + ab + b^2)(h/3)$ (where a and b are the sides of the upper and lower square and h the height).

Pyramid slopes were measured as horizontal recess, measured in palms, per cubit increased height. Most often, in texts as well as practice, the value is $5\frac{1}{4}$, corresponding to a ratio 3:4 in pure numbers. Countless attempts have been made to demonstrate that the Cheops pyramid hides knowledge of either π or the Golden section (or both!) in its proportions. The proximity of certain ratios to these "mystical" numbers is, however, a mere consequence of the simple determination of the slope. Moreover, as we have seen, the Egyptians were not interested in the ratio π but in the equivalent number $\sqrt{\pi/4}$. There is no reason to believe that the pyramids testify to occult mathematical knowledge beyond the level revealed in the mathematical papyri. As Otto Neugebauer has pointed out, nothing is less secret than the "secret" knowledge of the Egyptians, the Greek magicians, the Gnostics, etc., all of whom have left an immense number of texts on occult topics, exceeding by far anything these cultures ever wrote on mathematics. Had the Egyptians possessed secret mathematical knowledge, we would certainly have been informed, as we are about the secret spells to be used in the Netherworld.

Even the oft-claimed Egyptian knowledge of the Pythagorean theorem (or at least of the fact that a 3-4-5-triangle is right) should be treated with circumspection. No single text suggests so. One of the arguments in favor of the assumption has been that this knowledge *might* have been used to construct right angles. This trivially correct observation made by Moritz Cantor has since then been quoted as information that they did so. The other is that the pyramid recess of 3:4 could have been chosen for this reason. As we have seen, however, the actual number used by the Egyptians was $5\frac{1}{4}$; if the vertical distance was measured in cubits and the horizontal in palms, what was then to be done with the hypotenuse?

All of this concerned the mathematics of the scribes. Egypt, however, offers an exception to the rule that non-scribal geometrical techniques cannot be decoded. This exception regards the use of square grids, first of all in the pictorial arts but also in architectural planning. In the pictorial arts the grids were connected to a so-called "canonical system" for the proportions of the human body. That is, whereas scribal geometry aimed at finding numbers from geometrical configurations, the artists determined geometrical configurations from numbers.

General Observations

Histories of mathematics tend to distinguish "exact" from "approximate" formulae. In so far as we are dealing with school traditions built on some kind of didactical argument, this distinction may be defended. Both Babylonian and Egyptian scribal training certainly involved argument, even though it was not deductive, but it forgets that formulae can only be exact with regard to ideal mathematical objects; when it comes to applying the formulae in real-life practice, *all* formulae become approximate.

Histories of mathematics — in particular general histories — also tend to translate into familiar terms. To a certain extent this is necessary, but it hides from view the fact that other cultures may use conceptualizations which are different from ours but just as good and consistent. Speaking of a Babylonian and an Egyptian "value for π" suggests that the Babylonians, as we, looked for the ratio between the circumference and the diameter of a circle. However, to the Babylonians, the fundamental ratio was the ratio between the area of the circle and the area of the square on the circumference, and the terminology they used demonstrates that they really thought of this as a constructed geometrical square. The Egyptians, for their part, were interested in the ratio between the sides of the squared circle and the circumscribed square. Both conceptualizations are fully legitimate, also according to the gauge of mathematical theory, but they are certainly different from ours. If we neglect this, we only use the foreign culture as a mirror where we see — ourselves.

Later Developments

In the first millennium BC, Egypt was conquered first by the Assyrians, then subdued by the Persian armies, and finally — after Alexander — ruled by a Greek dynasty. We cannot follow the phases of this in detail in the rare mathematical texts, but we observe that the Assyrian or Persian administrators and tax collectors brought some of their own methods to Egypt, such as, for instance, the "surveyors' formula". We also see that the practical geometries of the Babylonian and Egyptian cultures served as inspiration for the Greeks when they developed their theoretical geometry. It is obvious, however, that the Greek theoretical development did not influence geometrical practice significantly, in spite of the attempts of certain Alexandrian mathematicians to educate the practitioners (in particular Hero). The ancient mathematical theoreticians were never sincerely interested in the problems encountered when geometry was to be practised — how to make sure that a ruler is straight, what to do if the compass opening is insufficiently large, how to prevent that the unavoidable imprecision in geometrical construction sums up to something unacceptable (in Euclid's *Elements*, a circle is never drawn; it is presupposed to have been drawn by an anonymous "helping hand", as pointed out by C. M. Taisbak). The synthesis between geometrical theory and geometrical practice was only achieved by the mathematicians of the Islamic Middle Ages, from al-Khwārizmī to al-Kāšī.

JENS HØYRUP

REFERENCES

Primary sources

Chace, Arnold Buffum. Ludlow Bull, and Henry Parker Manning. *The Rhind Mathematical Papyrus*. Oberlin, Ohio: Mathematical Association of America, 1929.

Struve, W.W. *Mathematischer Papyrus des Staatlichen Museums der Schönen Künste in Moskau*. Berlin: Julius Springer, 1930.

Babylonia

Friberg, Jöran. "Mathematik." In *Reallexikon der Assyriologie und Vorderasiatischen Archäologie* VII. Berlin and New York: de Gruyter, 1990, pp. 531–585.

Vogel, Kurt. *Vorgriechische Mathematik*. II. *Die Mathematik der Babylonier*. Hannover: Hermann Schroedel; Paderborn: Ferdinand Schöningh, 1959.

Egypt

Gillings, Richard J. *Mathematics in the Time of the Pharaohs*. Cambridge, Massachusetts: MIT Press, 1972.

Iversen, Erik. *Canon and Proportion in Egyptian Art*. 2nd ed. Warminster, England: Aris & Phillips, 1975.

Høyrup, Jens. "The Formation of 'Islamic Mathematics'. Sources and Conditions." *Science in Context* 1: 281–329, 1987.

See also: Metrology – Surveying – Algebra: Surveyor's Algebra – Geometry in Islam – Mathematics in Egypt

GLOBES The celestial globe is the oldest form of celestial mapping, for its origins can be traced to Greece in the sixth century BC. The stars were perceived as though attached to the inside of a hollow sphere enclosing and rotating about the earth. The earth, known from early classical antiquity to be spherical, was imagined at the center of the globe, while the stars were placed on its surface. Since this three-dimensional model of the skies presented the stars as seen by an observer outside the sphere of stars, the relative positions of the stars on a celestial globe are the reverse, east to west (or right to left), of their appearance when viewed from the surface of the earth. The sequence of the zodiacal constellations will be counterclockwise when the globe is viewed from above the north pole.

No celestial globes from antiquity have survived, but the basic principles of their design were maintained, with some modifications and elaborations, in the Islamic world, where the earliest preserved celestial globes were made. By the ninth century celestial globes were being made in the Arabic-speaking world. The most important early Islamic center of globe-making was the city of Harran, between the northern reaches of the Euphrates and Tigris rivers. In the ninth and tenth centuries it was an important town at the intersection of major caravan routes and had a prominant Sabian community whose pagan religious interest in the stars and sun was perhaps conducive to the study of astronomy. Many of the early astronomical instrument-makers were members of the Sabian sect at Harran. Several Arabic treatises were composed in the ninth century on the design and use of celestial globes, including one by the famous astronomer of Harran, al-Battānī who was known in the Latin world as Albategni, and others by Ḥabash al-Ḥāsib and Qusṭā ibn Lūqā in Baghdad. In the following centuries, additional treatises on the subject were composed, including one by ʿAbd al-Raḥmān ibn ʿUmar al-Ṣūfī (d. 983) whose treatise on constellations became the model for constellation iconography in the Islamic world.

Over 180 Islamic celestial globes are known to be preserved today. The earliest was made in AD 1080 in Valencia, Spain, and the most recent in Ottoman Turkey in 1882. Regardless of date, the stars represented on Islamic globes are those listed in the medieval star catalogs, and only the forty-eight constellation outlines recognized in antiquity are indicated. When constellation outlines are drawn around the stars, the clothing and faces of the human figures, such as Orion or Virgo, reflect the artistic conventions common in the artisan's day. Since the positions of the stars change over time with the precession of the equinoxes, the star positions on a globe, correct when the globe was made, remain valid for only three-quarters of a century.

In addition to the celestial equator and the ecliptic (the band of the zodiac through which the sun apparently moves in its yearly course), the Tropic of Cancer and the Tropic of Capricorn were also frequently shown on Islamic globes, as well as the north and south equatorial polar circles. On some later globes the ecliptic equivalents of tropic and polar circles were indicated, apparently in an attempt to complete the symmetry. On every Islamic globe preserved today there is also a set of six great circles at right angles to the ecliptic — six ecliptic latitude-measuring circles, reflecting the common use of ecliptic-based coordinates for measuring star positions. To function as an instrument, the sphere needed to be placed in a ring assembly, allowing for adjustment to a particular location. If supplemented by a gnomon or quadrant providing the altitude of the sun, the globe could then be used by an astronomer or astrologer to determine a range of astronomical data, including the length of the unequal daytime hour for a given day and location, or the time elapsed on a certain day, or data for a horoscope. It is questionable, however, whether many of the globes preserved today were of more than didactic or artistic value.

In terms of design, Islamic celestial globes fall into several distinct categories. The first includes the largest and the most elaborate artifacts, all of which display the forty-eight constellation outlines and approximately 1022 stars. Those in the second category do not have constellation outlines. Only a selection of the most prominent stars, usually between 20 and 60, are shown.

The third type of design is one in which the globe has neither constellation outlines nor any stars. In general these globes are the smallest. They have on them only the great and lesser circles (ecliptic, equator, tropic, and polar circles), all of which are labeled. This design is not mentioned in any of the written sources, and evidence so far available suggests it originated in Iran in the late seventeenth or early eighteenth century.

Only a few painted wood or papiermâché Islamic globes have survived, all of them hand drawn or painted. The method that dominated globe making in Europe — namely, laying printed paper gores over a wood or fiber core — seems not to have been practiced in the Islamic world. The vast majority of Islamic globes are hollow metal spheres and were made in two ways: from two hemispheres of cast or raised metal, or cast in one piece by the lost wax process. While globes made of wood or papiermâché or with metal hemispheres are of considerable antiquity, seamless globes, on the basis of evidence so far available, appear to have originated in northwestern India toward the end of the sixteenth century, the earliest confirmed date for one being AD 1589–90. They became the hallmark of all workshops in the Punjab and Kashmir areas of India through the nineteenth century. The workshop that excelled in this technique was a four-generation family of instrument makers in Lahore (in modern Pakistan). During more than a century, this remarkable workshop produced numerous astronomical instruments, including twenty-one signed globes (the earliest made in AD 1622). The technique of making seamless globes continued to be practiced in India after this workshop ceased to make them.

Throughout the ten centuries of their production in the Islamic world, celestial globes maintained the medieval tradition of displaying only the classical constellations and stars. On metal globes the stars were usually indicated by inlaid silver points. None of the surviving Islamic celestial globes records the stars and constellations of the southern hemisphere first mapped by Europeans during explorations of the sixteenth century.

While the production of celestial globes was widely and continuously practiced in the Islamic world, there was no comparable tradition of terrestrial globe making. There seems to have been little interest in terrestrial globes until the sixteenth century, when early modern European terrestrial globes became known to Ottoman Turkish astronomers and in the next century to those at the Mughal Indian court.

EMILIE SAVAGE-SMITH

REFERENCES

Brieux, Alain, and Francis R. Maddison. *Répertoire des facteurs d'astrolabes et de leurs oeuvres: Première partie, Islam*. Paris: Centre National des Recherches Scientifiques. In press.

Maddison, Francis R., and Emilie Savage-Smith. *Science, Tools and Magic*. (The Nasser D. Khalili Collection of Islamic Art, Vol. XII.) London: The Nour Foundation in association with Azimuth Editions and Oxford University Press, 1996.

Savage-Smith, Emilie. *Islamicate Celestial Globes: Their History, Construction, and Use*. (Smithsonian Studies in History and Technology 46.) Washington, D.C.: Smithsonian Institution Press, 1985.

Savage-Smith, Emilie. "The Classification of Islamic Celestial Globes in the Light of Recent Evidence." *Der Globusfreund* 38/39: 23–35 and plates 2–6, 1990.

Savage-Smith, Emilie. "Celestial Mapping." In *The History of Cartography*. Volume Two, Book One: *Cartography in the Traditional Islamic and South Asian Societies*. Ed. J. B. Harley and David Woodward. Chicago: University of Chicago Press, 1992, pp. 12–70.

Savage-Smith, Emilie. "The Islamic Tradition of Celestial Mapping." *Asian Art* 5(4): 5–27, 1992.

See also: al-Battānī – Precession of the Equinoxes – Qusṭā ibn Lūqā – al-Ṣūfī – Gnomon – Sundials – Quadrants – Ḥabash al-Ḥāsib

GNOMON IN INDIA The gnomon is an instrument used widely in early astronomy. The shadow of a vertical rod on a horizontal plane determines the cardinal directions, the latitude of the place of observation, the celestial coordinates of the sun, and the time of the observation.

A fairly complete account of its use in India was given by Varāhamihira in the *Pañcasiddhāntikā*. This was written in AD 505 and summarized the astronomical information current in India at that time. The *Āryabhaṭīya* of Āryabhaṭa also provides the main results of the theory of the gnomon, and these features appear again in the works of Bhāskara, Brahmagupta, and many later astronomers.

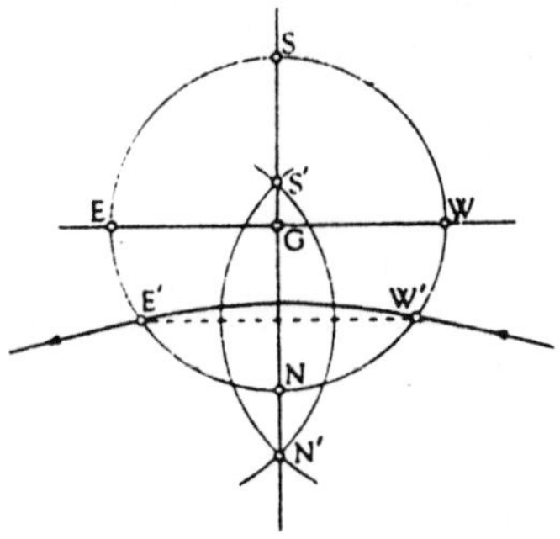

Figure 1: Finding the cardinal direction (Neugebauer, 1971).

The Cardinal Directions

The procedure is illustrated in Figure 1. G is the foot of the gnomon. The path of the end of the shadow enters and leaves a circle, center G, at W′ and E′. Then the line E′W′ is in the east–west direction. With E′, W′ as centers, circular arcs are drawn intersecting at N′, S′. Then N′S′, the perpendicular bisector of E′W′, is in the north–south direction and intersects the circle at N and S, the north and south points. The east and west points, E and W, can be found by the same procedure since they are on the perpendicular bisector of NS.

This method depends on the symmetry of the shadow path about the north-south line. It does not take into account the small change in the declination of the sun during the day. Brahmagupta prescribed a correction for this error in the *Mahābhāskarīya*. This method of finding the cardinal directions, described in the *Pañcasiddhāntikā*, is found in a much earlier treatise, the *Śulbasūtra*, which contains mathematical topics related to the construction of sacrificial altars.

The Noon Shadow

Trigonometric formulas enable us to find the latitude of the observer and the sun's declination from the shadow of the gnomon at noon, when the sun is on the local meridian. In Figure 2, g is the height of the gnomon, s the length of the shadow, Z the zenith, S the sun, and z the zenith distance of the sun. On any day, the zenith distance at noon is $z = \phi \pm \delta$, where ϕ is the observer's latitude and δ the declination of the sun. The Indian formula is

$$\operatorname{Sin} z = \frac{Rs}{\sqrt{s^2 + g^2}},$$

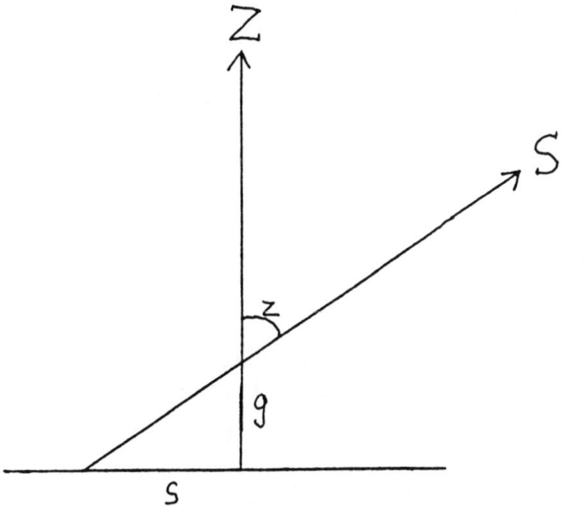

Figure 2: The noon shadow.

where the Sine of an angle is defined as R times the modern sine function, R being a constant angle, taken to be 120 minutes in the *Pañcasiddhāntikā*. When the sun is on the equator, $\delta = 0$, and the formula above gives us the latitude in terms of the length of the noon shadow. On other days, the formula yields the change $\pm\delta$ in the zenith distance, as a function of the noon shadow. On the days of the solstices, the declination has the maximum value ε, the obliquity of the ecliptic (the band of the zodiac through which the sun apparently moves in its yearly course).

When the sun is on the prime vertical (the great circle on the celestial sphere through the east and west points and the zenith), let z_1 be the zenith distance, and $a_1 = 90 - z_1$, the altitude of the sun, and λ, the sun's longitude. Then the *Pañcasiddhāntikā* formulae are

$$\operatorname{Sin} a_1 = \frac{R \operatorname{Sin} \delta}{\operatorname{Sin} \phi} = \frac{\operatorname{Sin} \lambda \operatorname{Sin} \varepsilon}{\operatorname{Sin} \phi}.$$

With these two formulae, we can find the declination and longitude of the sun from the shadow length, when the sun is on the prime vertical.

With the gnomon, it was also possible to find the time after sunrise, from the length of the shadow, using formulae which are equivalent to those used in modern spherical astronomy.

The second formula above gives the sun's longitude λ and declination δ, when it is on the prime vertical. λ and δ can be determined at any time, from the length of the gnomon's shadow and the distance of its endpoint from the east–west line. The *Pañcasiddhāntikā* also gives an approx-

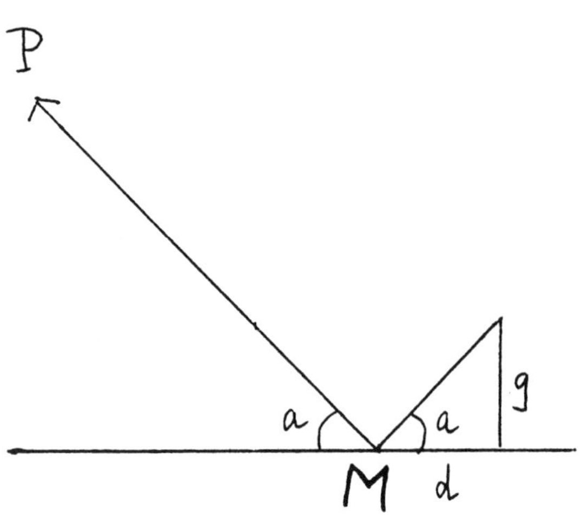

Figure 3: Determining the altitude of the moon or planets.

imate empirical algebraic formula which would have been useful in very early astronomy:

$$\frac{d}{2t} = \frac{s - s_0}{g} + 1.$$

This gives t, the time after sunrise in the morning or the time before sunset in the afternoon. s_0 is the noon shadow and d the length of daylight. This formula is derived from the following considerations:

(a) there is a linear relation between s and $1/t$,
(b) at noon $2t = d$,
(c) $s - s_0 = g$ at $4t = d$.

The *Yavanajātaka* of Sphujidhvaja also has this formula. Chapter 20 of Kauṭilya's *Arthaśāstra* gives the relation between the time after sunrise and the gnomon shadow. The *Arthaśāstra* also gives the rule for the uniform variation of the noon shadow from zero at the summer solstice to g, the gnomon height, at the winter solstice, a reasonable approximation for an observer on or near the Tropic of Cancer, for example at Ujjain. However, the rule for the uniform variation of the length of daylight from 12 to 18 *muhurtas*, also found in the two books above, implies a latitude of about 35 degrees, which suggests a Babylonian origin.

The theory of the gnomon presented above can be applied to the moon and planets also. The altitude of the moon or planet is determined in the following manner, illustrated in Figure 3.

The moon or planet (P) is seen reflected in a mirror M in the same horizontal plane as the foot G of the gnomon of height g at a distance d from the gnomon. Then the altitude, a, is given by the formula

$$\text{Sin}\, a = \frac{Rg}{\sqrt{g^2 + d^2}};$$

d is called the reversed shadow and takes the place of the shadow length s in the case of the sun.

The eleventh-century Arabic scholar al-Bīrūnī wrote *Kitāb fī ifrād al-maqāl fī umr al-ẓilā* (The Exhaustive Treatise on Shadows), which contains a comprehensive account of the theory and applications of the gnomon shadow. Al-Bīrūnī refers to many Indian sources, for example:

(a) the method described above for finding the cardinal directions. He calls it the method of "the Indian circle";
(b) the algebraic formula; and
(c) the time from the shadow. Al-Bīrūnī follows the procedure of Brahmagupta.

GEORGE ABRAHAM

REFERENCES

Primary sources

Burgess, E. *Sūrya-Siddhānta*. Calcutta: University of Calcutta, 1935.

Kennedy, E.S. *The Exhaustive Treatise on Shadows by al-Biruni*, 2 vols. Syria: University of Aleppo, 1976.

Neugebauer, O., and D. Pingree. *The Pañcasiddāntikā of Varāhamihira*. Copenhagen: Munksgaard, 1970–72.

Pingree, David. *The Yayanajātaka of Sphujidhvaja*. Cambridge, Massachusetts: Harvard University Press, 1978.

Sastri, T.S. Kuppanna. *Mahābhāskarīya*. Madras: Government Oriental Manuscripts Library, 1957.

Sastri, Ganapati. *The Arthśāstra of Kautilya*. Madras: Kuppuswami Sastri Research Institute, 1958.

Sen, S.N., and A.K. Bag. *The Śulbasūtras*. New Delhi: Indian National Science Academy, 1983.

Sharma, R.S. *Brāhmasphuṭa Siddhānta*. New Delhi: Indian Institute of Astronomical and Sanskrit Research, 1966.

Shukla, K.S., and K.V. Sarma. *Āryabhaṭīya of Āryabhaṭa*. New Delhi: Indian National Science Academy, 1976.

Shukla, K.S. *Mahābhāskarīya*. Lucknow University, 1960.

Thibaut, G., and M.S. Dvivedi. *The Pañcasiddāntikā*. Benares: Medical Hall Press, 1889.

Secondary sources

Abraham, George. "The Gnomon in Early Indian Astronomy." *Indian Journal of History of Science* 16(2): 215–218, 1981.

Pingree, David. "History of Mathematical Astronomy in India." In *Dictionary of Scientific Biography*, vol. 15. New York: Scribners, 1978, pp. 533–633.

Smart, W.M. *Text Book on Spherical Astronomy*. Cambridge: Cambridge University Press, 1977.

Somayaji, D.A. *Ancient Hindu Astronomy*. Dharwar: Karnataka University, 1971.

See also: Astronomy in India – Astronomical Instruments in India – Varāhamihira – Āryabhaṭa – Bhāskara – Brahmagupta – *Śulbasūtras* – Sphujidhvaja – al-Bīrūnī

THE *GOU-GU* THEOREM One of the basic theorems of geometry in both the East and West concerns the relationship between the sides of a right triangle and their squares, known in the West as the "Pythagorean" theorem, but understood in an equivalent form as the *Gou-Gu* theorem in China. The ancient Egyptians, Babylonians and Chinese probably discovered this remarkable property of right triangles by empirically examining the simplest case of 3-4-5 triangles. Whether in its geometric form or more familiar algebraic expression, $3^2 + 4^2 = 5^2$, the theorem concludes that the sum of the squares on either "side" of the right angle is equal to the square on the hypotenuse (*Xian*). In China, this was established for right triangles in general, i.e. not just for the 3-4-5 triangle, or for those with sides of integer lengths. The Greek made this discovery as well, but proved it rather differently in the argument presented at the end of the first book of Euclid's *Elements*, Proposition I-47.

One of the great treasures of Chinese mathematics is the *Jinzhang Suanshu* (Nine Chapters on the Art of Mathematics). It is the final chapter that is the most famous, in which the *Gou-Gu* theorem is introduced. Even before the *Nine Chapters* was written, results dealing with right triangles had been presented in an earlier, astronomical–mathematical work, the *Zhoubi Suanjing* (*The Arithmetic Classic of the Zhou Gnomon*). The *Nine Chapters*, however, goes well beyond the simple applications found in the *Zhou Gnomon* text, which is concerned primarily with applications in astronomy. In fact, Chapter Nine contains twenty-four problems, each of which deals primarily with right triangles and solutions of quadratic equations. One of these is a variation on one of the oldest of China's mathematical problems (this is Problem 6 in Chapter 9 of the *Nine Chapters*) (see Figure 2):

> In the middle of a pond that is ten "*chi*" in diameter, a reed grows one "*chi*" above the surface of the water. When pulled toward the edge of the pond, the reed just reaches the perimeter. How long is the reed?

The solution to this problem is a straightforward application of the *Gou-Gu* theorem.

In commenting on a passage from the *Nine Chapters* that reads, "Combining each square of *Gou* and *Gu*, taking the square root will be *Xian* (the hypotenuse)," Liu Hui explains the *Gou-Gu* theorem as follows:

> The *Gou*-square is the red square (*Zhu fang*), the *Gu*-square is the blue square (*Qing fang*). Putting pieces inside and outside according to their type will complement each other, then the rest (of the pieces) do not move. Composing the *Xian*-square, taking the square root will be *Xian* (the hypotenuse) (Guo, 1990).

The reference to moving pieces inside and outside is related to a diagram, no longer extant, and makes use of the so-

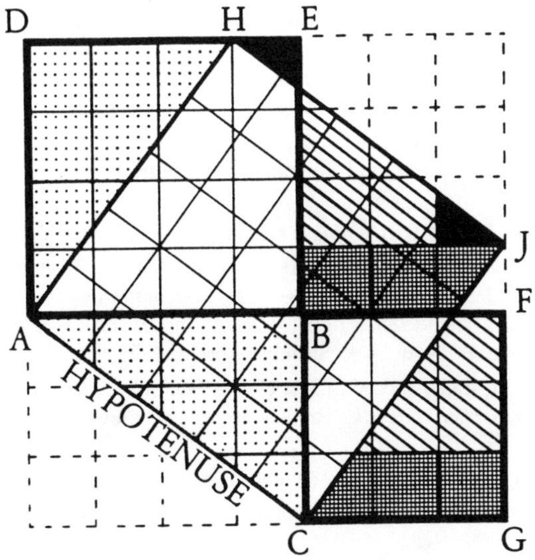

Figure 1: The *Gou-Gu* theorem based upon the *Xian Tu* = *Xian* diagram, meant to accompany the *Zhoubi Suanjing* (see Figure 3). Note that the square AHJC on the hypotenuse of the right triangle ABC is comprised of 25 unit squares, equal to the sum of the two squares on each side of the triangle, ADEH (16 unit squares) and CBFG (9 unit squares). *Drawing by the author.*

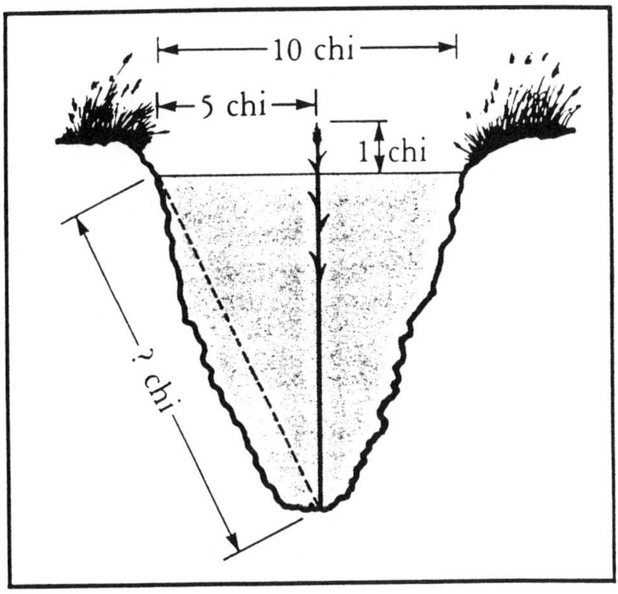

Figure 2: A problem that can be solved by applying the Gou-Gu theorem (see text: How long is the reed?). *Drawing by the author.*

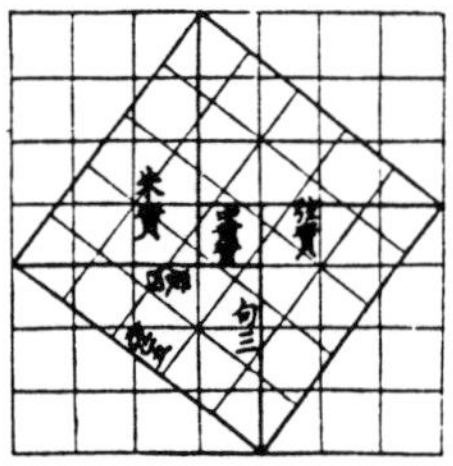

Figure 3: The *Xian* figure.

called "Out-In" method which was taken as an axiom by ancient Chinese mathematicians. The power of this axiom can be seen, however, from the following illustration, where the two triangles ABC and ACD are equal:

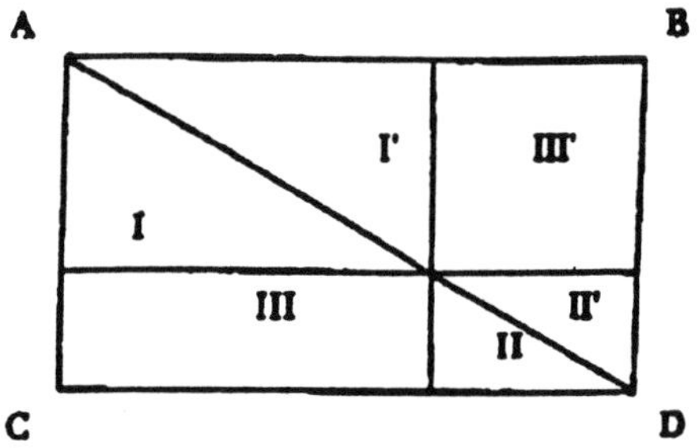

Since the areas I and I′, II and II′ are equal, then it follows from the "Out-In Complementary Principle" that III must be equal to III (Wu, 1983).

This now helps to explain Liu Hui's commentary on the *Gou-Gu* theorem. First of all, his references to colored squares are similar to colors mentioned in the diagrams illustrating the *Gou-Gu* theorem in the *Zhoubi Suanjing*. There, in the "*Xian* figure", the central square is yellow, while the squares of the *Xian* square are red. Often the *Gou* square is blue.

Applying the "Out-In" complementary principle to the *Xian* figure, and following Liu Hui's commentary on the *Gou-Gu* theorem, the sum of the squares based on each leg of the right triangle ABC, namely the squares ADEB and BFGC, is equal to the square of the hypotenuse AC, namely the square AHJC (see Figure 1 above). In accordance with the "Out-In" principle, if we move those parts of the two small squares (ADEB and BFGC) that are on the outside of the large square (AHJC) to its inside, we can see that they fill the inside exactly and that the combined areas of the two small squares equal that of the larger one. Since the areas are analogous to the squared sides of the triangle, the sum of the squared legs equals the squared hypotenuse.

It has been argued that geometry never developed further in China than it did with Liu Hui's commentary because this was sufficient for Chinese needs. After Liu Hui, Chinese geometry does not seem to have made much further progress. Although some authors suggest that this was due primarily to the practical orientation of ancient Chinese mathematics, it may have been its actual success, its comprehensiveness, that caused the stagnation of any further development. As D.B. Wagner has suggested:

> Liu Hui's conceptual framework was adequate, for example, to deal with a much broader range of geometric solids than those which he actually considers in his commentary. Had he felt a need to push his methods to their inherent limits, he would surely have contributed a great deal more to the mathematical tradition. Here we can see the double influence of the enormous prestige of the *Chiu-chang suan-shu:* it provided a challenge and an inspiration; but it was often a strait jacket which confined the interests of mathematicians to certain specific problems.

Like Euclid, Liu Hui summarized his art so successfully that his successors may have felt little need, or room, for improvement.

JOSEPH W. DAUBEN

REFERENCES

Dauben, Joseph W. "The 'Pythagorean Theorem' and Chinese Mathematics. Liu Hui's Commentary on the *Gou-Gu* Theorem in Chapter Nine of the *Jiu Jang Suan Shu*." In *Amphora. Festschrift in Honor of Hans Wussing*. Leipzig: B.G. Teubner, 1992, pp. 133–155.

Guo Shuchun. *Jiu Zhang Suan Shu*. Shenyang: Liaoning Educational Press, 1990, and Taipei: Taiwan Nine Chapters Press, 1990.

Li Yan and Du Shi-Ran. *Chinese Mathematics. A Concise History*. Oxford: Clarendon Press, 1987.

Martzloff, Jean-Claude. *Histoire des mathématiques chinoises*. Paris: Masson, 1988. English translation: *History of Chinese Mathematics*. New York: Springer-Verlag, 1995.

Vogel, Kurt, trans. and ed., *Chiu Chang Suan Shu. Neun Bücher arithmetischer Technik*. Braunschweig: F. Vieweg & Sohn, 1968.

Wagner, Donald B. "An Early Chinese Derivation of the Volume of a Pyramid: Liu Hui, Third Century AD." *Historia Mathematica* 6: 164–188, 1979.

Wu Wenchun. "The Out-In Complementary Principle." In *Ancient China's Technology and Science*. Beijing: Foreign Languages Press, 1983, pp. 66–89.

See also: Liu Hui and the *Jiuzhang Suanshu* – *Zhoubi Suanjing*

GUNPOWDER Gunpowder was probably discovered in China accidentally. Since the Han period (202 BC–AD 220) Chinese alchemists attempted to make gold or to prepare an elixir of immortality. Sulfur and saltpeter were among the raw materials used for their experiments, and charcoal was among the different types of fuel used in their laboratories. The *Zhenyuan miaodao yaolue* (Classified Essentials of the Mysterious Dao of the True Nature of Things), a book of the late Tang but probably containing material from much earlier dates, carries a note of caution to the alchemists warning them to exercise due care when dealing with sulfur, saltpeter and charcoal because there were cases where the operators had their hands scorched or their thatched huts set on fire.

In China gunpowder found its early use in amusements, in religious and ceremonial functions, taking the form of fireworks and rockets, and in construction works, such as blasting rocks in the opening of waterways and roads. For example explosives were used by Gao Ping to open a rocky water route linking Guangdong province and Annam in the ninth century. There are no records that show the use of gunpowder in the battlefield before the tenth century. An early fire weapon that the Chinese used was the "incendiary arrow" (*huojian*) to send a variety of ignited substances to the enemy camp. At first naphtha was one of the combustion agents used. By 919 there appeared the force pump that could throw flaming petrol, or "Greek Fire", directed towards the enemy. Both naphtha and "Greek Fire" were imports from the West.

The *Songshi* (Official History of the Song Dynasty) mentions that in 970 Feng Jisheng submitted to the emperor a report on the manufacture of "incendiary arrows" and that in 1000 Tang Fu made "incendiary arrows", "fire balls" (*huoqiu*) and "thorny fire balls" (*huojili*). These were bombs and grenades containing gunpowder for hurling over enemy walls and camps by means of trebuchets. Later on the bombs progressed from those with weak casings to those with stronger ones.

In the middle of the tenth century the "fire-lance" (*huoqiang*) was invented. At first it consisted of a bamboo tube with its septa cleared. Then it was filled with rocket composition and capable of shooting out flames horizontally. According to Needham the "fire-lance" had enormous repercussions in China for some seven hundred years from the middle of the tenth century, playing prominent roles in many battles. It also marked a transition from the bulky petrol flame-thrower to a much lighter and portable weapon using gunpowder, and heralded the use of the principle of the tube in military technology. Bamboo tubes gradually gave way to barrels made of metal. The "fire-lance" took a big step forward in 1259 when the "fire-emitting lance" (*tuhuoqiang*) was developed. This type of "fire-lance" could fire scrap metal and broken porcelain, if not also small darts or arrows, on which poison was sometimes applied. The range of the projectiles was short compared to a single bullet from a gun barrel. The Chinese later developed a similar firearm with a large barrel mounted on a carriage, rather like a field gun but which emitted a large quantity of projectiles instead of a single shell. Again these projectiles could be scrap metal, stone, broken porcelain, arrows, and poisonous or noxious substances.

The earliest document giving the gunpowder formula and describing its applications in various forms of firearms is the Song military compendium, *Wujing zongyao* (Collection of the Most Important Military Techniques), written by Zeng Gongliang in 1044. It describes with illustrations a number of weapons using gunpowder, such as the grenade, the bomb, etc. and gives several gunpowder recipes. The compositions suggest that the earlier Chinese gunpowder was more of the deflagratory type. A much higher ratio of saltpeter for explosives was used later.

Around the year 1000, gunpowder weapons were deployed on the battlefield in China. Large quantities of gunpowder were prepared and stored up during the Northern Song (960–1126). In 1232 the "thunderclap bombs" were brought into action. In the early twelfth century knowledge of gunpowder was acquired by the Jurchen people who eventually occupied northern China and established the Jurchen empire (1115–1234). In 1232 they used the "thunder-crash bomb" and the "flying fire lance" (*feihuoqiang*) to defend their capital Bianjing against the Mongol invaders. Bianjing fell the next year. Gunpowder and firearms technicians were captured by the Mongols and taken into their service. The Mongols invaded Europe from the year 1236, deploying firearms of the offensive type. For example trebuchets were used on at least two occasions, in July 1237 at the battle of Ryazau, and in July 1241 "incendiary arrows" were used at the battle of Wahlstadt. The Mongols also used their firearms against the Arabs. In 1258 they used "fire-jars" to attack Baghdad, probably a reference to the thunder-crash bombs, which the Mongols also used against the Japanese in a sea battle near Kyushu in 1272. The next important city to fall after Baghdad was Damascus. The Muslims set up the Mamluk caliphate in Cairo and in 1260 defeated the Mongols in a battle in Syria. From then on, the two sides made little headway against each other for some years, during which the Mamluk caliphate turned its attention to military affairs and weaponry. Some Mongolian soldiers surrendered and some firearms and technicians fell into the hands of the Muslims. The Muslims thus acquired the technique of making firearms. They used them successfully against the Sixth Crusade (1248–1254).

By the latter part of the thirteenth century the narrow

cylindrical tube of the fire-lance had already developed into one with much wider bore for the cannon (*huopao*). For example cannons were deployed when the Mongolian fleet went to Java in 1292 and 1293. Many examples of Chinese cannon, both of bronze and iron, are known from 1330 onwards. Several types of cannons are described in the fourteenth-century military book, the *Huolongjing* (Fire Dragon Manual).

During the second half of the fourteenth century Chinese firearms were unmatched anywhere in East Asia. They were used by Zhu Yuanzhang, the founder of the Ming dynasty, to overthrow the Mongols and to suppress local rebellions. After his victory he kept his firearms in secret arsenals. There was no need to improve these weapons as they met no challenge, so that he only saw the necessity of guarding both the weapons and the knowledge from falling into undesirable hands. This happened during a period when Europe was undergoing great social change and when European firearms began to come to ascendancy. Traditional Chinese weaponry is now something of the past, but firework displays that we see on auspicious occasions around the world today remind us of the original role played by gunpowder in China.

HO PENG YOKE

REFERENCES

Needham, Joseph. *Gunpowder as the Fourth Power, East and West.* Hong Kong: Hong Kong University Press, 1985.

Needham, Joseph, et al. *Science and Civilisation in China.* Cambridge: Cambridge University Press, 1986.

See also: Military Technology

GUO SHOUJING Guo Shoujing (1231–1316), an astronomer, mathematician, and hydraulic engineer, was born in Xintai (in modern Hebei province). The names of his parents are not known, although records show that his grandfather, Guo Yong, was knowledgeable in the classics, in mathematics, and in water works. As a boy of fourteen Guo Shoujing was able to construct an advanced type of clepsydra, or water clock, for his time: the *lianhualou* (lotus clepsydra), so named because the top of the receiver was in the shape of a lotus flower. In about the year 1251 he restored an old bridge across the Dahuoquan River north of Xinzhou, also in Hebei.

In 1260 Guo accompanied his friend Zhang Wenqian (1217–1283), who was sent by Qubilai Qan to Daming to pacify the local population. There he began to construct astronomical instruments, such as a bronze clepsydra and a bamboo armillary sphere. The same year Zhang Wenqian recommended him to Qubilai for his expertise on irrigation. Guo was soon commissioned to improve irrigation and water communication within the region south of the capital Dadu (modern Beijing) and north of the Yellow River. The mission was successfully accomplished. In the year 1264 Guo again accompanied Zhang to the Circuit of Xixia (in modern Gansu province) to restore the irrigation canals that had been blocked or damaged during the years of war. He reported on what he saw during his travels and made many recommendations to Qubilai on the improvement of irrigation and water communication systems.

There was an urgent need to reform the existing calendar, the *Damingli* In 1276 Qubilai captured the Southern Song capital, Linan (modern Hangzhou) and thought that the time was ripe to promulgate a new calendar. Guo and Wang Xun (ca. 1235–1281) were commissioned to lead a special bureau established for the project. Guo Shoujing built seventeen new astronomical instruments to obtain more accurate astronomical observations on which the accuracy of the new calendar would depend. Thirteen of the instruments were used in the capital Dadu, and four were used for field work. In 1279 Qubilai established an Astronomical Bureau at the capital. Zhang Wenqian was appointed Director, while Guo together with Wang Xun were made the two co-Directors. A new observatory was built at the capital along with new bronze astronomical instruments. Guo organized a large scale measurement of the length of the shadow of the gnomon cast by the sun in different latitudes, from the capital in the north to Nanhai (modern Guangzhou) in the south, to determine the length of the meridian.

Guo used spherical trigonometry and the method of finite difference involving a cubic equation to do his calculations. The new calendar, the *Shoushili*, was completed in 1280. This was the most accurate calendar ever made in traditional China. It was promulgated the next year. Guo Shoujing was promoted to the Directorship of the Astronomical Bureau.

In 1292 Guo held a joint appointment as head of the water works bureau in the capital. The same year he constructed water locks and canals linking various cities to the capital. The Qan was greatly pleased on completion of the project. While he was in the capital Guo made a clepsydra whose bells and drums would chime and sound on the hours for his master. After the death of Qubilai, Guo's advice was also sought by Timur. Guo died in 1316. The Jesuit Adam Schall von Bell (1591–1666) was correct up to a certain point in calling Guo Shoujing "China's Tycho Brahé".

HO PENG YOKE

REFERENCES

"Biography of Guo Shoujing." In Ke Shaomin, *Xin Yuanshi* (New Official History of the Yuan Dynasty) Ch.171, Biographical chapter 68. First published 1922, reprinted Taipei: Yiwen Press (no date).

Ho Peng Yoke and M. Wang. "Kuo Shou-ching." In *In the Service of the Khan. Eminent Personalities of the Early Mongolian Period (1200–1300)*. Ed. I. de Rachewiltz et al. Wiesbaden: Asiatische Forschungen,1993, pp. 282–299.

Needham, Joseph. *Science and Civilisation in China.*, vol. 3. Cambridge: Cambridge University Press, 1959.

See also: Armillary Spheres – Clocks and Watches – Bamboo – Astronomical Instruments – Irrigation – Calendars – Gnomon

H

ḤABASH AL-ḤĀSIB Ḥabash al-Ḥāsib was one of the earliest Muslim astronomers and a major contributor to the development of trigonometry. He was born in Marw, Turkestan (modern Mary, Turkmenistan) and died between AD 864 and 874. Ḥabash lived during the Abbasid empire, when the caliphs became the stewards of civilization while Europe languished in the Dark Ages. This empire preserved ancient science and philosophy by translating ancient Greek, Syriac, Sanskrit, and Persian texts into Arabic. Ḥabash himself based his work and methods on Ptolemy's *Almagest*. Ḥabash held Ptolemy in high regard, calling him "the wise Ptolemy" and describing his work as having the "utmost in research and precision" (Langermann, 1985). Nevertheless, Ḥabash led the way in the Arabic development of astronomy and computational techniques that far surpassed the Ptolemaic system in accuracy, efficiency, and elegance.

As an astronomer Ḥabash worked at calculating more precise celestial distances and at developing more accurate ways of calculating these distances. He calculated such values as the circumference of the earth, the diameter of the moon, and the distance of one minute along the orbit of the sun. Ḥabash's work in trigonometry consisted of calculating tables of trigonometric values and developing new trigonometric functions. He calculated tables (called *zīj*es) of sine values at one-degree intervals to three places, and he is considered to be the first to construct a table of tangent values.

In Ḥabash's time much of Muslim science served the needs of the religion. For example, Islam requires the faithful to face Mecca when they pray. To this end Ḥabash developed an analemma, or graphical, method for finding the azimuth to Mecca, called the *qibla*. Muslim calendars were dependent upon the appearance of the new crescent of the moon. Ḥabash is thought to be the first astronomer to calculate when the new crescent appears. Islam also requires the faithful to make a pilgrimage to Mecca. Ḥabash once calculated what he called the distance "by the straight arrow," or the great-circle distance between Baghdad and Mecca to be 677 miles, while the actual overland distance was known to be about 712 miles. In one of his few surviving works, *Kitāb al-ajrām wa-l-ab ʿād* (The Book of Bodies and Distances), Ḥabash reported that al-Maʾmūn, the caliph at the time and Ḥabash's patron, was pleased with the small difference between the two distances considering the uneven terrain between the two cities.

LAWRENCE SOUDER

REFERENCES

Berggren, J.L. "A Comparison of Four Analemmas for Determining the Azimuth of the Qibla." *Journal for the History of Arabic Science* 4(1): 49–65, 1980.

Berggren, J.L. *Episodes in the Mathematics of Medieval Islam*. New York: Springer-Verlag, 1986.

Langermann, Y. Tzvi. "The Book of Bodies and Distances of Ḥabash al-Ḥāsib." *Centaurus* 28: 108–128, 1985.

Tekeli, S. "Ḥabash al-Ḥāsib, Ahmad Ibn ʿAbdallah al-Marwazi." In *Dictionary of Scientific Biography*. Ed. Charles C. Gillispie. New York: Charles Scribner's Sons, 1972, pp. 612–620.

See also: *Almagest* – Trigonometry – *Qibla* and Islamic Prayer Times – Calendars

AL-ḤAJJĀJ Al-Ḥajjāj ibn Yūsuf ibn Maṭar was the earliest translator of Greek mathematical and astronomical texts into Arabic. Few details of his life are known. According to a report of the bio-bibliographer Ibn al-Nadīm, al-Ḥajjāj prepared an Arabic version of Euclid's *Elements* under the sponsorship of the ʿAbbasid caliph Hārūn al-Rashīd (170–193 AH/AD 786–809). Later, under caliph al-Maʾmūn (198–218 AH/AD 813–833), he prepared a second (and improved) version. There is some indication that the mathematician Thābit ibn Qurra may have helped to revise the latter version.

Neither version exists today. Quotations from what purports to be the second version serve as the basis for a commentary by al-Nayrīzī. This treatise exists in a unique but incomplete manuscript containing Books I–VI and a few lines from Book VII. Recent studies indicate that the quotations found in this manuscript have been heavily edited by al-Nayrīzī.

Although definitive studies have not yet been completed, it appears that one of the Ḥajjāj versions may have served as the basis for the summary of the *Elements* prepared by Ibn Sīnā for inclusion in his philosophical encyclopedia, *Kitāb al-Shifāʾ* (The Cure [of Ignorance]). It is commonly believed that the Latin translation by Adelard of Bath is also derived from the Arabic version of al-Ḥajjāj.

Al-Ḥajjāj is also credited with producing an Arabic version of Ptolemy's *Almagest*, perhaps by way of a Syriac intermediary. The relation of this translation to the extensive tradition of Arabic commentaries and to the Latin translations is not yet established.

GREGG DE YOUNG

REFERENCES

Brentjes, S. "Varianten einer Ḥaǧǧāǧ-Version von Buch II der *Elemente*." In *Vestigia Mathematica: Studies in Medieval and Early Modern Mathematics in Honor of H.L.L. Busard*. Ed. M. Folkerts and J.P. Hogendijk. Amsterdam: Editions Rodopi, 1993, pp. 47–67.

De Young, G. "New Traces of the Lost Arabic Translations of Euclid's *Elements*." *Physis* 28: 647–666, 1991.

Sezgin, F. "Al-Ḥaǧǧāǧ b. Yūsuf." *Geschichte des arabischen Schriftums* Band V: *Mathematik*. Leiden: Brill, 1974, pp. 225–226.

See also: al-Maʾmūn – Thābit ibn Qurra – al-Nayrīzī – Ibn Sīnā – *Almagest*

AL-HAMDĀNĪ Al-Hamdānī, Abū Muḥammad al-Ḥasan ibn Aḥmad ibn Yaʿqūb, was born in Ṣanʿāʾ, Yemen, AD ca. 893, and died after AD 951.

Al-Hamdānī's family belonged to the tribe Hamdān, which played an important role in the pre-Islamic and Islamic history of the Yemen. This might have contributed to his interest in South Arabian history and geography as well as to his political activities as a South Arabian nationalist, particularly in his poems against the North Arabian tribes. After a learned education, he travelled extensively, for example to Iraq to study grammar and lexicography, and lived for a long time in Mecca. He corresponded with learned men such as the philologists of Kūfa. Living in the Yemenite towns Rayda and Ṣaʿda he wrote many books, when he was not in prison for political reasons.

Ṣāʿid ibn Ṣāʿid al-Qurṭubī, a famous *qāḍī* (judge) of Toledo, said in 1068, in a work entitled *Ṭabaqāt al-umam* (Book of the Categories of Nations), that al-Hamdānī and al-Kindī were the only two Arab philosophers (most philosophers were Iranians, Turks, Spaniards, or Jews). As such al-Hamdānī was well aquainted with Aristotle and other Greek philosophers and with their theories on the generation of substances from the four elements: fire, air, water, and earth. He was also familiar with the ideas about the cardinal humors — blood, choler (yellow bile), phlegm, and melancholy (black bile) — and their relations to the planets and the seasons and maybe also of atoms. Al-Hamdānī was a defender of alchemy, considering the methods of metallurgy imitations of the processes in nature, but his approach was practical without magic procedures. Philosophy was the foundation of astronomy and astrology, chemistry and medicine. Of his astronomical work *Sarāʾir al-ḥikma* (The Secrets of Wisdom), only one chapter has been recovered, in which several Greek scholars and an Indian work are cited. Al-Hamdānī is also said to have compiled astronomical tables and a book on horoscopes and the projection of rays, as well as a medicinal work, all of which are lost. Astronomy and geography are related, and one of al-Hamdānī's two famous works is his *Ṣifat jazīrat al-ʿarab* (Description of the Arab Peninsula), which also contains sections on meteorology, botany, agriculture, and mining. It is based on his own observations supplemented by information from Greek and Arab scholars.

His other famous work is a history of South Arabia, *Al-Iklīl* (The Crown), which is only partly preserved. Al-Hamdānī had a certain knowledge of the epigraphic South Arabic script but did not understand the inscriptions; his sources were oral and written information from people and archives. He also wrote a trilogy on the South Arabic economy, on agriculture, camel breeding, and gold and silver, of which only the last part has been preserved. *Kitāb al-Jawharatayn al-ʿatīqatayn* (The Book on Gold and Silver) treats the technology and chemistry of these metals from their extractions from the mines to the fabrication of coins with sections on steel technology, gilding, soldering, and other pertinent matters.

Al-Hamdānī is South Arabia's best representative of mediaeval Islamic learning based on observation and a critical use of the Greek, Iranian, and Indian heritage.

CHRISTOPHER TOLL

REFERENCES

Primary sources

Al-Iklīl (The Crown). Part 1–2: facs. ed. Berlin 1943; part 1: ed. O. Löfgren, Uppsala 1954, 1965 (Bibliotheca Ekmaniana 58:1–2); part 8: ed. and trans. N. A. Faris. Princeton, New Jersey: Princeton University Press, 1940, 1938; part 10: ed. M. al-D. al-Khaṭīb, Cairo 1949.

Kitāb al-Jawharatayn al-ʿatīqatayn (The Book on Gold and Silver). Ed. and (German) trans. Christopher Toll. Uppsala: Almqvist & Wiksell, 1968.

Sarā'ir al-ḥikma (The Secrets of Wisdom). Ed. M. ʿA. al-Akwaʿal-Ḥiwālī, Ṣanʿāʾ, 1978.

Ṣifat jazīrat al-ʿarab (Description of the Arab Peninsula). Ed. D. H. Müller, Leiden: Brill, 1884, 1891; reprinted Amsterdam: Oriental Press, 1968; (partial) trans. L. Forrer, Leipzig 1942 (Abhandlungen für die Kunde des Morgenlandes 27:3); reprinted Nendeln: Kraus Reprint Ltd, 1966.

Secondary sources

Abdallah, Yusuf, ed. *Al-Hamdani, a Great Yemeni Scholar*. Ṣanʿā, 1986.

Löfgren, O. "al-Hamdānī." In *Encyclopaedia of Islam,* 2d ed., vol. 3. Leiden: Brill, 1971, pp. 124–125.

Toll, Christopher. "al-Hamdānī." In *Dictionary of Scientific Biography*, vol. 6. New York: Scribners, 1972, pp.79–80.

Toll, Christopher. "al-Hamdānī as a Scholar." *Arabica* 31: 307–317, 1984.

Toll, Christopher. "The 10th Maqāla of al-Hamdānī's Sarā'ir al-ḥikma." In *On Both Sides of al-Mandab. Ethiopian, South-Arabic and Islamic Studies Presented to Oscar Löfgren*. Ed. U. Ehrensvärd and Christopher Toll. Stockholm: Svenska Forskningsinstitutet i Istanbul, 1989.

HARIDATTA Haridatta was the promulgator of the Parahita system of astronomical computation widely used in Kerala in South India, from where it spread to the neighboring state of Tamilnadu. There are two basic texts of the system, the *Grahacāranibandhana* and the *Mahāmārganibandhana*, the latter of which is no longer extant. Haridatta inaugurated the system, as the legend goes, on the occasion of the 12-yearly religious festival held at the temple town of Tirunāvāy on the banks of the Bhāratappuzha River in AD 683. The system was called *Parahita* (suitable to the common man), because it simplified astronomical computation and made it accessible for practice even by ordinary people.

Haridatta based his system on the *Āryabhaṭīya* of Āryabhaṭa (b. 476), but made it simpler in several ways. First, he dispensed with the rather cumbersome and terse numerical symbolism used by Āryabhaṭa and substituted the facile, easily manipulated *ka-ṭa-pa-yā-di* system of notation. In this system, specific letters were used for representing digits, which could be arranged to form meaningful words and even sentences, which could be remembered with much less possibility of error.

Computations in Indian astronomy involved long numbers for planetary revolutions and other parameters for the aeon. To avoid these long numbers in multiplication and division, Haridatta ingeniously introduced a sub-aeon of 576 years or 210,389 days, and accurately determined the zero-correction for this sub-aeon for the mean motion of the several planets. In actual practice, Haridatta directed the aeonary days for any current day being divided by the sub-aeonary days. The quotient would then give the number of completed sub-aeons, and the remainder the days in the current sub-aeon. The above quotient multiplied by the sub-aeonary zero-corrections of the several planets would give the mean planets at the commencement of the current sub-aeon. In order to make calculation easier, when a large number of years passed, Haridatta gave these zero-corrections for chunks of six sub-aeons. To find the mean motion of the several planets for the completed days in the current sub-aeon, Haridatta gave certain sets of simple multipliers and divisors. The mean motion arrived at using these, when added to the mean position at the commencement of the current sub-aeon, would give the mean planet for the current date. In order to obviate the large numbers that would accumulate to mean planets when the Āryabhaṭan parameters were used, Haridatta prescribed a corrective called *Vāgbhāva* (Corrections to be Applied to the Different Planets from AD 523).

The Parahita system promulgated by Haridatta has been extremely popular in Kerala, and a very large number of texts and tracts based on the system have been produced in that part of India, both in Sanskrit and in the local language Malayalam. The system has also been regularly used for the computation of the daily almanac.

K.V. SARMA

REFERENCES

Datta, B.B., and A.N. Singh. "Kaṭapayādi System." In *History of Hindu Mathematics*. Bombay: Asia Publishing House, 1962, pt. I, pp. 69–70.

Sarma, K.V. *Grahacāranibandhana: A Parahita Manual by Haridatta*. Madras: Kuppuswami Sastri Research Institute, 1954.

Sarma, K.V. *A Bibliography of Kerala and Kerala-based Astronomy and Astrology*. Hoshiarpur: Vishveshvaranand Institute, 1972.

Sarma K.V. "Parahita System of Astronomy." In *A History of the Kerala School of Hindu Astronomy*. Hoshiarpur: Vishveshvaranand Institute, 1972, pp. 7–9.

See also: Astronomy in India – Āryabhaṭa

AL-HĀSHIMĪ ʿAlī ibn Sulaymān al-Hāshimī flourished some time in the second half of the ninth century, probably somewhere in the central lands of Islam. Virtually nothing is known about him other than the fact that he wrote a rather uncritical work on *zīj*es (astronomical handbooks) that nevertheless preserves a great deal of otherwise unknown or little known information. This book, *Kitāb fī ʿilal al-zījāt* (Explanation of Zījes), was written at a time before Ptolemaic astronomy had become the dominant astronomical tradition in Eastern Islam. As such, it contains considerable material about the Indian and Persian astronomical traditions, at least insofar as they were received and preserved during this early period of Islamic science.

Hāshimī mentions some sixteen *zīj*es, of which two are Greek (Ptolemy's *Almagest* and Theon's *Canon*, which is based upon it); seven are Indian or derived mainly from Indian sources (the *Arjabhar*, the *Zīj al-Arkand*, the *Zīj al-Jāmiʿ*, the *Zīj al-Hazūr*, the *Zīj al-Sindhind* of al-Fazārī [eighth century] as well as a second *zīj* by him, and the

Zīj of Ya‛qūb ibn Ṭāriq [eighth century]); two are Persian but mainly Indian in inspiration (the *Zīj al-Shāh* of Khusro Anūshirwān and the *Zīj al-Shāh* of Yazdigird III); and five are from the ninth century and use material from these three traditions in varying degrees (the *Zīj al-Sindhind* of al-Khwārizmī, the *Zīj al-Mumtaḥan* of Yaḥyā ibn abī Manṣūr, the first Arabic *zīj* that was principally Ptolemaic, two *Zīj*es of Ḥabash, one mainly Indian, the other mainly Ptolemaic, and the *Zīj al-Hazārāt* of Abū Ma‛shar).

Besides its importance as a historical resource, Hāshimī's work provides some valuable clues about the state of Islamic science during this early period. It is clear that the impressive number of astronomical works floating about made a work explaining them desirable, giving evidence for the vitality of science during this period. The influx of "foreign" knowledge had its detractors, though, and Hāshimī felt compelled, in a passing comment, to affirm that the cycles of Indian astronomy were not from their prophets, whatever might be claimed; nor were they for the purpose of "soothsaying", something he knows would be unIslamic. Rather he asserts that they are mathematically derivable and as such safe, an interesting and not atypical way of handling religious opposition to astronomy in Islam.

This makes the subsequent history of Islamic astronomy all the more remarkable. Indian astronomy with its cycles and computational tradition, but lacking a full-blown cosmology would, at first glance, seem to be a much more congenial tradition for Islam, which had its own religious cosmology and metaphysics. In fact Hāshimī implies as much in introducing the *Sindhind*; he also reports that Shāh Anūshirwān [sixth century] preferred Indian to Ptolemaic astronomy. So the subsequent predominance and triumph in Islam of Ptolemaic astronomy, based ultimately on Aristotelian physical principles and cosmology, was a remarkable occurrence indeed.

F. JAMIL RAGEP

REFERENCES

al-Hāshimī, ‛Alī ibn Sulaymān. *The Book of the Reasons Behind Astronomical Tables (Kitāb fī ‛ilal al-zījāt)*. A Facsimile Reproduction of the Unique Arabic Text Contained in the Bodleian MS Arch. Seld. A.11 with a Translation by Fuad I. Haddad and E. S. Kennedy and a Commentary by David Pingree and E. S. Kennedy. Delmar, New York: Scholars' Facsimiles and Reprints, 1981.

See also: al-Khwārizmī – Yaḥyā ibn abī Manṣūr – Abū Ma‛shar – *Almagest* – *Zīj*

HAY'A The Arabic term *hay'a* had several distinct significations when used in the medieval astronomical context. Its basic meaning is "structure" or "configuration"; it was used as such to connote the physical structure of the universe as a whole (*hay'at al-‛ālam*) or a distinct part of that structure, as in the *hay'a* of the Earth or of Venus. It also referred to those works of theoretical astronomy in which that structure is dealt with in detail. As part of the phrase *‛ilm al-hay'a* (the science of *hay'a*), it came to refer, especially after the eleventh century, to astronomy in a general sense. This is an indication of thc importance given to physical cosmography in Islamic astronomy. Here the main focus will be on *hay'a* in the sense of the Islamic astronomical genre devoted to cosmography.

Before proceeding, it is important to note two points. Although the term *hay'a* was occasionally used to indicate a religious cosmology, works on *hay'a* are mainly cosmological in the scientific and not the religious sense. Second, this tradition, though departing from it in numerous ways, owes its origins and inspiration to the Hellenistic scientific tradition.

The main ancient sources for the Islamic *hay'a* tradition were the cosmological writings of Aristotle, in particular his *De Caelo* and his attempt in the *Metaphysics* to give a coherent structure to the models of Eudoxus, and the astronomical writings of Ptolemy, especially his *Planetary Hypotheses*. These works provided the fundamental assumption of Islamic theoretical astronomy, namely that the subject matter of astronomy was the simple *physical* bodies, both celestial and sublunar. An astronomer writing in the *hay'a* tradition was charged with transforming mathematical models of celestial motion, usually those of Ptolemy's *Almagest*, into physical bodies that could be nested, along with the sublunar levels of the four elements, into a coherent cosmography (*hay'a*). Despite the obvious connection and dependence on Greek natural philosophy, this "astronomical" (or external) cosmology was often contrasted with both the tradition of Aristotle's *De Caelo* and the astrological corpus, both of which were held to deal with the essential (or internal) aspects of the bodies. Note that because *hay'a* was intended to give a general picture of the universe, it also dealt with the "configuration" of the four sublunar elements; thus most general treatments of astronomy after the eleventh century came to have a section on the "configuration of the Earth" that included general discussions of geography.

The prevalent *hay'a* was that the universe was a plenum composed of nine contiguous, solid, spherical bodies called orbs, all concentric with an immobile, spherical Earth. The lowest of these, that of the moon, enclosed the four sublunar levels of the four elements: fire, air, water, and earth. Each of the "concentrics" had embedded within it additional, non-concentric, spherical orbs called eccentrics and epicycles, the former hollowed-out spheres surrounding the Earth, the latter full spheres within the eccentrics. There was one concentric each for the seven "planets" in the following order—moon, Mercury, Venus, sun, Mars, Jupiter, Saturn—another for the fixed stars, and a starless ninth that was the source of the daily motion. Except for the sun, which was embedded within an eccentric, the planets were situated on the inside surface of their epicycles. Opinions varied as to how motion occurred but generally this was held to result from a combination of the proper motion effected by the orb's own soul and the accidental motion that was a consequence of being contained inside other orbs. All the celestial orbs were composed of a special fifth element called "aether"; unlike the four sublunar elements, it could only rotate with uniform motion. Among the most famous and influential works on *hay'a* were the *Hay'at al-ʿālam* (Configuration of the World) by Ibn al-Haytham (d. ca. 1040), the *Tadhkira fī ʿilm al-hay'a* (Memoir on Astronomy) by Naṣīr al-Dīn al-Ṭūsī (d. 1274), and *Al-mulakhkhaṣ fī al-hay'a al-basīṭa* (Epitome of Plain *hay'a*) by al-Jaghmīnī (thirteenth or fourteenth century).

Although the above cosmography derived for the most part from Ptolemy's *Planetary Hypotheses*, Islamic theoreticians found much to complain about and proposed numerous alternatives that were meant to reform or supersede his models. In his *Doubts About Ptolemy*, Ibn al-Haytham pointed to certain devices that Ptolemy had introduced in the *Almagest*, such as the equant, that violated the principle of uniform circular motion; as such, it was not possible to physicalize such models in any straightforward way. In the thirteenth century, a number of writers, beginning with Naṣīr al-Dīn al-Ṭūsī, who was the first director of the Mongol-sponsored Marāgha observatory, proposed alternative models that were meant to reform the Ptolemaic system. These efforts continued for at least three more centuries and included the work of his colleague Mu'ayyad al-Dīn al-ʿUrḍī, his student Quṭb al-Dīn al-Shīrāzī, Ibn al-Shāṭir, the timekeeper of the Umayyad Mosque in Damascus during the fourteenth century who proposed an astronomy without eccentrics, and a number of people associated with the Samarqand observatory in the fifteenth century.

A much more radical attempt to transform the Ptolemaic system occurred in twelfth-century Islamic Spain. A number of Aristotelians, including the famous philosopher Averroes (Ibn Rushd), sought a radical reform of Ptolemaic astronomy that would rid it of epicycles and eccentrics altogether. The motivation was to return to a purer astronomy in which there was only a single center of celestial motion. The end product of this movement was the work of al-Biṭrūjī. His system, reminiscent of that of Eudoxus, was not very successful from a mathematical point of view, but it was quite influential in the Latin West.

This was not the only case in which Islamic *hay'a* influenced science in other cultures. Byzantine astronomers were certainly aware of this tradition; we also know of several *hay'a* works that were translated into Sanskrit in the eighteenth century. Given that the *Planetary Hypotheses* was unknown during the Latin Middle Ages, it is clear that the main source for the European *theorica* tradition was Islamic *hay'a*; Ibn al-Haytham's *Configuration of the World* was certainly quite influential. In recent years, very strong similarities between the models and motivations of several Renaissance astronomers (including Copernicus) and the late *hay'a* tradition have been discovered, and it now seems quite likely that the former were influenced by Ṭūsī, Ibn al-Shāṭir, and other post twelfth-century writers on *hay'a*. However, because their works seem not to have been translated into Latin, the most likely means of transmission is through Byzantine sources that made their way to Italy (and perhaps Vienna) in the fifteenth century.

F. JAMIL RAGEP

REFERENCES

al-Biṭrūjī, Nūr al-Dīn abū Isḥāq. *On the Principles of Astronomy*. Edition, translation, and commentary by Bernard Goldstein. 2 vols. New Haven: Yale University Press, 1971.

Chittick, William C. "Islamic Cosmology." In *Encyclopedia of Cosmology*. Ed. Norriss Hetherington. New York and London: Garland, 1993, pp. 322–329.

Heinen, Anton. *A Study of As-Suyūṭī's "al-Hay'a as-sanīya fī l-hay'a as-sunnīya."* Beirut: Orient-Institut der Deutschen Morgenländischen Gesellschaft, 1982.

Langermann, Y. Tzvi. *Ibn al-Haytham's "On the Configuration of the World."* New York: Garland (Harvard Dissertations in the History of Science), 1990.

Nasr, Seyyed Hossein. *An Introduction to Islamic Cosmological Doctrines: Conceptions of Nature and Methods Used for Its Study by the Ikhwān al-Ṣafā', al-Bīrūnī, and Ibn Sīnā*. Revised edition. Albany, N.Y.: State University of New York Press, 1993.

Ragep, F. Jamil. *Naṣīr al-Dīn al-Ṭūsī's Memoir on Astronomy (al-Tadhkira fī ʿilm al-hay'a)*. 2 vols. New York: Springer-Verlag, 1993.

Sabra, A. I. "An Eleventh-Century Refutation of Ptolemy's Planetary Theory." In *Studia Copernicana* XVI. Warsaw: Ossolineum, 1978, pp. 117–131.

Sabra, A. I. "The Andalusian Revolt Against Ptolemaic Astronomy: Averroes and al-Biṭrūjī." In *Transformation and Tradition in the Sciences*. Ed. E. Mendelsohn. Cambridge: Cambridge University Press, 1984, pp. 133–153.

Saliba, George. *A History of Arabic Astronomy: Planetary Theories during the Golden Age of Islam*. New York: New York University Press, 1994.

Swerdlow, N. M., and O. Neugebauer. *Mathematical Astronomy in Copernicus's De Revolutionibus*. 2 parts. New York: Springer-Verlag, 1984.

See also: Astronomy in the Islamic World – *Almagest* – Geography in the Islamic World – Ibn al-Haytham – Naṣīr al-Dīn al-Ṭūsī – Marāgha – al-ʿUrḍī – Quṭb al-Dīn al-Shīrāzī – Ibn al-Shāṭir – al-Biṭrūjī

HUANGDI JIUDING SHENDAN JING The *Huangdi jiuding shendan jing* (Book of the Divine Elixirs of the Nine Tripods of the Yellow Emperor) also known as *Jiudan jing* (Book of the Nine Elixirs) is one of the earliest extant Chinese alchemical texts. Quotations and other references in Ge Hong's *Baopu zi neipian* suggest that the text circulated in a form at least substantially similar to the current one by the end of the third century. The extant version, dating from the latter half of the seventh century, is found in the *Daozang* (Daoist Canon). It is followed by an extended commentary, that represents in its own right a compendium of early medieval Chinese alchemy.

Besides its early date, the *Book of the Nine Elixirs* is significant as it provides one of the most complete descriptions of the alchemical practice in China, from the preliminary rituals to the compounding and ingestion of the elixirs. The ritual features include purifications, interdictions, secrecy, retirement, choice of time, delimitation of space, and ceremonies for transmitting the text, for starting the fire, and for ingesting the elixirs.

The compounding begins after the preliminary ritual practices. The first step consists in the preparation of an amalgam of refined lead and mercury, said to contain the essences of Heaven and Earth, or *Yin* and *Yang*. This compound is used as a layer in the crucible together with the ingredients of the elixirs. Then one prepares a mud used for sealing the crucible and prevent the loss of *qi* (energy), thus recreating inside the vessel conditions similar to those of primeval chaos.

The methods of the Nine Elixirs proper are based on processes of sublimation (*fei*). The substances most commonly used are mercury, orpiment, realgar, malachite, arsenolite, magnetite, cinnabar, alum, and hematite. The ingredients are first placed in a tightly sealed crucible, and heated on a fire progressively made more intense. After the required number of days, the crucible is left to cool and is opened. The essence is carefully collected from the higher part of the crucible, and mixed with a moistening substance. The elixirs should be ingested at dawn, facing the sun. Their ingestion is said to bestow eternal life, command over gods and spirits, protection from calamities, and sometimes magical powers. Some of the Nine Elixirs can be transmuted into alchemical gold. The text emphasizes that the purpose of this final transmutation is to make sure that the compounds have been correctly prepared.

FABRIZIO PREGADIO

REFERENCES

Needham, Joseph, et al. *Science and Civilisation in China*, vol. V: *Chemistry and Chemical Technology*, parts 2–5. Cambridge: Cambridge University Press, 1974.

Pregadio, Fabrizio. "The *Book of the Nine Elixirs* and Its Tradition." In *Chûgoku kodai kagakushi ron* [Studies on the History of Ancient Chinese science]. vol. II. Ed. Yamada Keiji and Tanaka Tan. Kyoto: Kyōto Daigaku Jinbun Kagaku Kenkyūjo, 1991, pp. 543–639. See also the Italian translation of the *Book of the Nine Elixirs* in *Cina* (Rome) 23: 15–79, 1991.

See also: Alchemy in China – Ge Hong – *Yinyang*

HUANGDI NEIJING The *Huangdi Neijing* (The Yellow Emperor's Medical Canon) is the most famous and oldest medical classic of China, being complied by various unknown authors from the period of the Warring States (475–221 BC) onwards. The book consists of two main sections, each of eighty-one chapters: the *Suwen* (Plain Questions) and *Ling Shu* (Miraculous Pivot), the latter being earlier entitled the *Zhen Jing* (Needling Classic).

These texts are the very earliest to be written on Chinese medical theory. However Keegan has shown conclusively that not one of the *Neijing*s now extant are the original. Lu and Needham also summarized the evidence and considered that the corpus was probably already well-formed by the first century BC.

The Canon achieved something close to its known form with the recension (ca. AD 762) which came about with the editorship of Wang Ping during the Tang dynasty. Another edition, the *Huangdi Neijing Taisu* (Great Simplicity), recently discovered in Japan, may be considered closer to the original text. The insertion of chapters into the text, including those on *wuyun liuqi* (the theory of the five elemental cycles and six climates), undoubtedly took place some time close to the mid-tenth century (Lu and Needham, 1980). Since this date, however, the contents have been fixed.

The book takes on a question and answer form, in which the legendary Yellow Emperor questions his ministers about the need for a medical reformation, in accord with the prevailing tone of Han thought. In the opening paragraphs the Emperor questions his minster-physician Qi Bo:

> The Yellow Emperor asked: "I have heard that the men of ancient times lived through a hundred springs and autumns, and remained active and did not reach senility. Now our men only reach half that age and they are senile. Are we so very different? Or have we missed something?" Qi Bo replied: "The men of ancient times, they understood the Way, they modelled themselves on Yin and Yang and were at peace with the arts of guiding their destiny. They ate and drank in moderation, they rose and retired at regular hours, neither made wild schemes, nor wearied their bodies at work. . . . The men of the present day are not the same. They drink liquor to excess, they take wild thinking as the norm. When drunk they perform the act of love, seeking to exhaust their energies and waste their lives. They do not know how to be content. They have no time to control the passage of their thoughts, looking only for what cheers – not for the joys of health" (Ch. 1).

This sets the tone for the preventative and rehabilitative character which dominates Chinese medicine. Later in the chapter, the same need for ataraxia (peace of mind) is stressed: "Detach and be at peace, humble and empty — and the true energies will follow. The vital spirit, guard it within! How then can disease arrive?"

Among the Canon's other ideas are the well developed tract system of twelve main channels and numerous collaterals (the *jingluo*); and the science of the five elements (*wuxing*): fire, earth, metal, water, and wood. It also introduces the classificatory dialectical tool of *Yin/Yang* (earth/heaven, dark/light, night/day, solid/hollow, etc.). Many chapters also describe the relationship between human beings and their environment, and refer in great detail to circadian, or internal, rhythms within the body. These biological clocks foreshadowed the modern science of chronobiology.

Surgery is also mentioned, and the *bian* stone needle as the forerunner of acupuncture. The Canon accurately identifies gangrene in the feet and says it is more common in the toes, where it advises amputation to save life. This is just one of the many ideas prevalent in the book which accords with modern medical theory.

RICHARD BERTSCHINGER

REFERENCES

Keegan, David Joseph. *The Huang Ti Nei Ching: the Structure of the Compilation: the Significance of the Structure.* Ph.D Dissertation, University of California at Berkeley, 1988.

Lu, Gweidjen and Joseph Needham. *Celestial Lancets.* Cambridge: Cambridge University Press, 1980.

Veith, Ilza, trans. *The Yellow Emperor's Classic of Internal Medicine.* Berkeley: University of California Press, 1966.

See also: Acupuncture – *Yinyang* – Five Phases

HUANGFU MI Huangfu Mi was born in 215 in Anding Chao'na (now Pingliang County, Gansu Province) and died in 282. He was born to a poor family and did not receive any education until he was twenty years of age, after which he devoted all his time to reading, extensively and intensively, all kinds of ancient literature. He eventually became a great master in the art of classics.

During the period 256–260, he suffered *Feng Bi* or apoplexy and became hemiplegic. From then on, he began to study medicine. In view of the overlapping of many texts in several medical classics, he decided to compile a new book on the art of acupuncture and moxibustion under the principle of "sorting out the similarities, excluding the meaningless and useless words and sentences, deleting the unnecessary repetition and extracting the essence". He did this by incorporating the relevant portions from the three classics, namely *Huangdi Neijing Suwen* (Plain Question of Huangdi's Inner Canon), *Zhenjing* (Classic of Acupuncture), and *Mingtang Kongxue Zhenjiu Zhiyao* (Essentials of Treatment for Acupuncture and Moxibustion with Anatomical Charts and Acupoints). A twelve-volume *Zhenjiu Jiayi Jing* (A–B of Acupuncture and Moxibustion) was written and divided into two parts. One of them deals with the physiology of the human body, indications for each acupoint, diagnostic measures, way of puncturing, and pathophysiology. The other part is devoted to clinical knowledge, including internal medicine, surgery, pediatrics, and gynecology, with special reference to internal medical science. Emphasis is laid on the unification of channels and acupoints and on exploration of indications and contraindications for acu-moxibustion therapy. This work, from which most of the works on acupuncture and moxibustion of later ages were derived, exerted great influence on the development of the art of acupuncture and moxibustion in successive generations. Erudite and learned, Huangfu Mi was very productive in his writings, including works on history and classical science.

During his time, it was a custom to swallow processed stone remedies for seeking longevity and as an aphrodisiac. Huangfu Mi also practiced this behavior and was brought to the brink of death after a critical illness. After he fortunately recovered, he wrote a book, *Hanshi San Lun* (On Stone Powder To Be Swallowed Cold) to advise people to avoid such harmful and potentially fatal behavior.

HONG WULI

REFERENCES

Fang, Xuanling, et al. "Biography of Huangfu Mi." In *Book of Jin Dynasty* (Reprinted). Beijing: Zhonghua Book Company, 1974, pp. 1409–1418.

Huangfu, Mi. *A–B Classic of Acupuncture-Moxibustion* (Reprinted) (revised and annotated by Shandong College of Traditional Chinese Medicine). Beijing: People's Health Publishing House, 1979.

Wu, Shijian, and Liu Changgan. *Annotations on Book of Jin Dynasty*. Peking: Jinshi's Xylographic Edition, 1928.

See also: Medical texts in China

ḤUNAYN IBN ISḤĀQ Abū Zayd Ḥunayn Ibn Isḥāq al-ʿIbādī (Johannitius, AD 808–873), a physician, philosopher, and translator, was born in al-Ḥīrah (Hira), now southeast of al-Najaf (Iraq). He and his ancestors were Syrians who belonged to the Nestorian church. The family nickname, al-ʿIbādī, is derived from "al-ʿIbād," a Christian Arab tribe.

Ḥunayn studied medicine in his youth in Baghdad. The thirteenth-century medical historian, Ibn Abī Uṣaybiʿa, author of the book *ʿUyūn al-Anbāʾ fī Ṭabaqāt al-Aṭibbāʾ* (Sources of Information about the Classes of Physicians), gives a lively account of the difficulties that confronted Ḥunayn as a student of medicine. Yūḥannā Ibn Māsawayh (Mesue, d. AD 857) was a famous physician and a highly respected teacher of medicine. He conducted classes in a vast room in his large house. Questions from young Ḥunayn, who was very keen on amassing knowledge, interrupted the serenity of Mesue's lessons. Mesue reprimanded Ḥunayn several times without effect, then finally ordered him out of the classroom and told him to give up medicine. Mesue advised him to trade in coins instead, which he believed would be financially more rewarding than the practice of medicine. Mesue had in mind the fact that most of the residents of Hira were merchants and money changers. So far as one can tell, Ḥunayn abandoned Mesue's tutorship and decided on self-education.

Bakhtīshūʿ Ibn Jibrāʾīl, a contemporary of Ḥunayn, held regular meetings at which physicians and philosophers debated questions and problems raised by Caliph al-Mutawakkil (d. AD 861). Ḥunayn was invited to attend these meetings at which he showed medical acumen. Al-Mutawakkil is reported to have endowed a school of translation, and appointed Ḥunayn as its head. Among the members of this school were Ḥunayn's son, Isḥāq Ibn Ḥunayn (d. AD 910), who translated from Greek into Syriac and Arabic. He also took charge of revising, against the original Greek, translations by his colleagues. It is worth noting that Isḥāq's translations were carefully revised and corrected, whenever necessary, by Ḥunayn himself. Towards the end of his life, Ḥunayn embarked on an Arabic translation of Galen's *De partibus artis medicativae* (On the Parts of Medicine; Fī Ajzāʾal-Ṭibb), of which he had earlier rendered the Greek text into Syriac. Death prevented him from completing his translation. Isḥāq attended to Ḥunayn's unfinished work.

Another eminent translator in Ḥunayn's school was his nephew, Ḥubaysh Ibn al-Ḥasan al-Aʿsam, who was involved in translating Galen's (d. ca. AD 200) *De anatomicis administrationibus* (On Anatomical Procedures; *Fī ʿAmal al-Tashrīḥ*) into Arabic. This work consists of fifteen books. It was translated into Syriac by Ayyūb al-Ruhāwī (Job of Edessa) for Jibrāʾīl Ibn Bakhtīshūʿ (d. AD 828–829). A revised Syriac translation was made by Ḥunayn himself for Yūḥannā Ibn Māsawayh. It is generally accepted that Ḥubaysh produced an Arabic version about the end of Ḥunayn's life, and with his active collaboration.

Ḥunayn's school of translation helped to establish a firm foundation upon which medieval Arabic medicine and allied sciences were securely built. Renowned Arabic-speaking physician-philosophers, like al-Rāzī (Rhazes, d. AD 925) and Ibn Sīnā (Avicenna, d. AD 1037), and many others, owe Ḥunayn and members of his school a great debt for rendering fundamental Greek texts of antiquity into Arabic, the language of learned men at the time.

Al-Mutawakkil elevated Ḥunayn to the post of Chief Court Physician and dismissed Bakhtīshūʿ Ibn Jibrāʾīl. The latter conspired against Ḥunayn, and, unfortunately, Ḥunayn fell from grace. Al-Mutawakkil ordered the sequestration of Ḥunayn's entire library.

Ḥunayn was forty-eight when he compiled the first draft of his *Risālat Ḥunayn Ibn Isḥāq ilā ʿAlī Ibn Yaḥyā fī Dhikr mā Turjima min Kutub Jālīnūs bi-ʿIlmih wa-baʿḍ mā lam Yutarjam* (Missive to ʿAlī Ibn Yaḥyā on Galen's Books Which, so far as Ḥunayn Knows, Have Been Translated, and Some of Those Books Which have not been Translated). The fact that Ḥunayn's *Missive*, which presents impressive information about some of his predecessors and contemporaries, was compiled, he says, after the loss of his library, indicates his complete dedication to his work. ʿAlī Ibn Yaḥyā, to whom Ḥunayn's *Missive* was dedicated, was a powerful friend and a private confidant of Caliph al-Mutawakkil. Through his intercession, Ḥunayn later recovered his own library.

The following are a few examples to shed light on the depth of Ḥunayn's involvement in translating Galen:

- The *Pinax* (Index; Fīnaks) was translated into Syriac for a doctor named Dāʾūd and into Arabic for Muḥammad Ibn Mūsā;
- *De ordine librorum suorum* (On the Order of his Books; *Fī Marātib Qirāʿ at Kutubih*), into Arabic for Aḥmad Ibn Mūsā;
- *De sectis ad eos, qui introducuntur* (On Sects; Fī'al–Firaq, into Syriac when Ḥunayn was twenty years of age, and another translation for Ḥubaysh when Ḥunayn was forty;
- *De febrium differentiis* (On the Types of Fevers; Fī

Aṣnāf al-Ḥummayāt), the first book to be translated by Ḥunayn, into Syriac, and another translation for his son when he discovered textual gaps which he filled in by collating against other manuscripts, and again into Arabic for Abu'l-Ḥasan Aḥmad Ibn Mūsā;

• *De crisibus* (On Crisis; Fī'al-Buḥrān). First Ḥunayn corrected a Syriac translation by Sergius of Resaina (d. AD 536), then produced an Arabic version for Muḥammad Ibn Mūsā. All these and many more are discussed in detail in Ḥunayn's *Missive*.

In spite of his fame as a translator, he was scoffed at by doctors when he wished to practice the art of physic. They admitted that he was an excellent translator, yet he was not a doctor, in the same way that a blacksmith could forge a beautiful sword without himself being a swordsman.

Ḥunayn was also a medical author in his own right. His books are mainly based on Greek sources, such as: *Fī Awjāʿ al-Maʿidah* (On Stomach Ailments) and *al-Masāʾil fi 'l-Ṭibb li 'l-Mutaʿallimīn* (Questions on Medicine for Students). His book entitled *al-ʿAshar Maqālāt fi 'l-ʿAyn* (The Ten Treatises on the Eye), covering both the theory and practice of medicine, ranks highly among medieval books of ophthalmology.

Historians of medicine and science are greatly indebted to Ḥunayn and his school of translation. Without their efforts, many of the works of antiquity would have probably been lost forever. Two examples will suffice. Galen's *On Anatomical Procedures* (*Fī ʿAmal al-Tashrīḥ*) of which Books IX (in part) and X–XV (inclusive) are lost in the original Greek, is preserved in its entirety in Arabic manuscripts. Again, Galen's book *De optimo medico cognoscendo* (On Examinations by Which the Best Physicians are Recognized) is entirely lost in the original Greek. Ḥunayn's Arabic version has survived in two manuscripts.

ALBERT Z. ISKANDAR

REFERENCES

Anawati, George Chehata. "Ḥunayn Ibn Isḥāq." In *Dictionary of Scientific Biography*. Ed. Charles C. Gillispie. New York: Scribner's, 1970–1980, vol. 15, pp. 230–234.

Bergsträsser, G. *Ḥunain ibn Isḥāq über die syrischen und arabischen Galen-übersetzungen*. Leipzig: F.Z. Brockhaus, 1925.

Flügel, G. *Ibn al-Nadīm, Kitāb al-Fihrist*. Leipzig: F.C. W. Vogel, 1871–1872, vol. 1, pp. 294–295.

Ghalioungui, Paul. *Questions on Medicine for Scholars by Ḥunayn Ibn Isḥāq*. Trans. Galal M. Moussa. Cairo: Al-Ahram Center for Scientific Translations, 1980.

Iskandar, Albert Zaki. "Ḥunayn the Translator." In *Dictionary of Scientific Biography*. Ed. Charles C. Gillispie. New York: Scribner's, 1970–1980, vol. 15, pp. 234–249.

Lippert, J. *Ibn al-Qifṭi's Taʾrīḥ al-Ḥukamāʾ*. Leipzig: Dieterich, 1903, pp. 171–177.

Meyerhof, Max. *The Book of the Ten Treatises on the Eye Ascribed to Hunain ibn Isḥāq*. Cairo: Government Press, 1928.

Müller, A. *Ibn Abī Uṣaybiʿa. ʿUyūn al-Anbāʾ fī Ṭabaqāt al-Aṭibbāʾ*. Cairo and Königsberg: Bulaq, 1882–1884.

Mūsā, Muḥammad Jalāl, ed. *Al-Masāʾil fi al-Ṭibb li al-Mutaʿallimīn li-Ḥunayn Ibn Isḥāq*, [Cairo]: n.d.

See also: Isḥāq Ibn Ḥunayn – al-Rāzī – Ibn Sīnā – Ibn Māsawayh

HUNTIAN *Huntian* is the Chinese name for a scheme of cosmography, often mentioned as in opposition to the *gaitian*. According to explanations of this scheme, heaven is a sphere surrounding a flat earth whose upper surface lies across its diametral plane. Heaven is therefore continuous (*hun hun ran*), unlike the situation described by the *gaitian*, under which heaven has a boundary. The heavenly sphere rotates once daily about an axis inclined at 35 degrees to the horizontal, carrying with it the heavenly bodies, whose risings and settings occur as they move above and below the plane of the earth's surface. The Chinese observer is imagined to be at the center of the earth, and hence at the center of the heavenly sphere. Since the latitude of most ancient Chinese capitals was about 35 degrees north, the *huntian* universe acts as a sort of cosmic planetarium for the official astronomers at those capitals. Observers near the edge of the earth would see a very odd and asymmetrical series of celestial phenomena, but the diameter of the sphere is so large (one version gives a figure equivalent to over 140,000 miles) that the entire inhabited world is effectively at the centre of the universe.

The first full account of the *huntian* cosmography is that of Zhang Heng, written ca. AD 117. Thereafter it is the usual scheme referred to by those concerned with mathematical astronomy, and a number of texts point out its advantages compared with the *gaitian*. The innovation of the *huntian* was apparently connected with the introduction of the armillary sphere in China at some time after 100 BC. The nested rings of such an instrument are highly suggestive of the skeleton of the celestial sphere. It would be a mistake, however, to see debates on cosmography as a crucial concern for astronomers in China, since the algebraic/arithmetic character of Chinese mathematical astronomy meant that the detailed geometry of the cosmos was much less of an issue than it was in ancient Greece and Rennaissance Europe.

CHRISTOPHER CULLEN

REFERENCES

Cullen, Christopher. *Astronomy and Mathematics in Ancient China*. Cambridge: Cambridge University Press, 1996.

I

IBN ʿABBĀD Abū al-Qāsim Ismāʿīl Ibn ʿAbbād Ibn al-ʿAbbās Ibn ʿAbbād Ibn Aḥmad Ibn Idrīs, also known as Kāfī al-Kufāt, and al-Ṣāḥib, was a famous vizier and man of letters of the Buwayhid period. There is disagreement about his place and date of birth, but he was probably born at Iṣṭakhr on 16 Dhū'l-qaʿda 326/14 September 938. His family included high dignitaries, and his own father had been a *kātib* (clerk) and then vizier, or minister of state, under the Buwayhid Prince Rukn al-Dawla (r. 335/946–366/976).

Ibn ʿAbbād himself became the disciple and secretary of Abū'l-Faḍl Ibn al-ʿAmīd (d. 360/970), the father of the Buwayhid vizier Abū'l-Fatḥ Ibn al-ʿAmīd (b. 337/948–9). His close relationship to the Buwayhid *amīrs* (princes) is said to have begun in 347/958, when he accompanied Mu'ayyid al-Dawla (reigned 366/976–373/983) to Baghdad as a clerk. He was later confirmed in this office when Mu'ayyid al-Dawla became governor of Isfahan. Ibn ʿAbbād's career took a more significant turn when he was appointed vizier, replacing Abū'l-Fatḥ Ibn al-ʿAmīd.

Ibn ʿAbbād served two rulers: Mu'ayyid al-Dawla, and Fakhr al-Dawla (reigned 373/983–387/997). The main source for the study of his vizierate remains the volume of the *Rasāʾil* (ʿAzzām, 1946), which is a collection of administrative pieces, appointment and other official letters by Ibn ʿAbbād. According to the sources, the personality of Fakhr al-Dawla was not really compatible with that of Ibn ʿAbbād, although he recognized the latter's administrative skills and talents. When Ibn ʿAbbād died on 24 Ṣafar 385 / 30 March 995, Fakhr al-Dawla confiscated his property, and from that time onward, no other member of his family was to be appointed to a high official position.

Ibn ʿAbbād was not only a statesman and a politician; he was also a talented writer whose works cover a very wide spectrum. In their article in the *Encyclopedia of Islam*, Claude Cahen and Charles Pellat give a classification of his works: dogmatic theology, history, grammar and lexicography, literary criticism, poetry, and belles-lettres.

In Abū Ḥayyān al-Tawḥīdī's *Mathālib al-Wazīrayn*, a comprehensive list of Ibn ʿAbbād's works is given. Some of the most important are:

Kitāb al-Muḥīṭ bi-ʿl-lugha (The Comprehensive Treatise About Language) in 10 vols.
Kitāb Dīwān rasā'ilihi (Collection of Letters) in 10 vols.
Kitāb al-Kāfī. Rasāʾil (The Book of al-Kafi [that which is sufficient]. Letters [or Correspondence])
Kitāb al-Aʿyād wa-faḍā'il al-Nowrūz (The Book of Feasts and the Excellent Qualities of New Year's Day).
Kitāb al-Wuzarāʾ (The Book of Viziers).
Kitāb Dīwān Shiʿr(ihi) (Collection of Poetry).

In the *Yatīmat al-dahr* (III, 204), al-Thaʿālibī commented on a *Risāla* on medicine said to have been written by Ibn ʿAbbād: "I found that it combined beauty of style, elegance of expression, and mastery of the subtleties and particularities of medicine, and it showed that he was thoroughly familiar with this science, and had a penetrating knowledge of its intricacies".

Ibn ʿAbbād often inspired very contradictory opinions and feelings as is shown by Abū Ḥayyān al-Tawḥīdī's hostility on the one hand, and by al-Thaʿālibī's praise and admiration on the other. However, regardless of personal like or dislike, he certainly was a highly exceptional personality in Muslim history. Maybe one can refer to him as a "patron-vizier", a talented individual who had the ability to mix politics and literature, and a poet whose court once counted as many as 23 poets.

ZEINA MATAR

REFERENCES

Abū Ḥayyān al-Tawḥīdī. *Mathālib al-Wazīrayn*. Ed. I. al-Kaylānī. Damascus: Dār al-Fikr, 1961.

al-Thaʿālibī. *Yatīmat al-Dahr*. Ed. M. Muḥiyyaddīn ʿAbd al-Ḥamīd, 4 vols. Miṣr: Dār al-Māmūn, 1956–1958.

ʿAzzam, A.A. and S. Ḍayf, eds. *Rasāʾil al-Ṣāḥib Ibn ʿAbbād*. Cairo: Dār al-Fikr alʿArabī, 1946.

Cahen, C. "Buwayhids." In *Encyclopaedia of Islam II*, Vol. 1. Leiden: Brill, 1960, pp. 1350–1357.

Cahen, C. and Pellat, C. "Ibn ʿAbbād." In *Encyclopaedia of Islam II*, Vol. 2. Leiden: Brill, 1971, pp. 671–673.

Kahl, O. and Z. Matar. "The Horoscope of as-Ṣāḥib Ibn ʿAbbād." *Zeitschrift der Deutschen Morgenländischen Gesellschaft* 140(1): 28–31, 1990.

Matar, Z. and A.Vincent. "A Little-Known Note (Ruqʾā) Attributed to the Buyid Vizier al-Ṣāḥib Ismā'īl Ibn ʿAbbād." In *Occasional Papers of the School of Abbasid Studies*, No. 2, 1988, pp. 46–56.

Yāqūt al-Rūmī. *Muʿjam al-Udabāʾ*. Ed. A.F. Rifāʿī, 20 vols. Miṣr: Dār al-Māmūn, 1936–1938.

IBN AL-AʿLAM Ibn al-Aʿlam, Abū'l-Qasim ʿAlī ibn ʿIsa al-Husain, al-ʿAlawī, al-Sharīf was a tenth century astronomer, apparently established in Baghdad. The year of his death is recorded by Ibn al-Qifṭī as 375/985. The *Zīj* (astronomical handbook with tables) which he wrote is lost, but substantial information about it may be gleaned from notices in the work of other astronomers. His work was patronized by the Būyid ruler ʿAḍūd al-Dawla but suffered from lack of support in the disturbed period which followed his death in 372/982.

A near contemporary, the great astronomer Ibn Yūnus of Cairo, reports that Ibn al-Aʿlam had fixed the length of the year as 365; 45,40,20 days, determined the position of Regulus (α Leonis) in the year 365/975–6 as 15; 6 Leo, and also fixed the rate of precession, one degree in 70 Persian years. He remarks that Ibn al-Aʿlam was known everywhere for the exactitude of his observations and the extent of his geometrical knowledge.

Al-Bīrūnī in his work *Tamhīd al-mustaqarr li-taḥqīq maʿnā al-mamarr* (On Transits) gives the name of his *Zīj* as *al-ʿAdūdī*, and incidentally provides values for the radius of the epicycles for each of the planets.

In spite of the loss of the *Zīj* full information about its parameters may be obtained from two sources compiled in the fourteenth century, the Persian *Zīj-i Ashrafī*, and an anonymous Arabic collection known as the *Dastūr al-Munajjimīn*. Information is also available from Greek sources, in which Ibn al-Aʿlam is referred to as Alim (*'Αλημ*). The Greek manuscripts are of the fourteenth and fifteenth centuries, but they preserve texts older than the Persian and Arabic compilations. These include *scholia* to the *Almagest* datable to AD 1032 which refer to tables composed for the Greek calendar from the work of Ibn al-Aʿlam, as well as two horoscopes for the years AD 1153 and AD 1162 which have been calculated from such tables, apparently for the Emperor Manuel Comnenus. The tabulation of the equations indicates that the technique of 'displacements' was used in order to provide values of the equations which were always positive.

RAYMOND MERCIER

REFERENCES

Kennedy, E.S. *Al-Bīrūnī on Transits*. Beirut: American University of Beirut, 1959.

Mercier, Raymond. "The Parameters the Zīj of Ibn al-Aʿlam." *Archives Internationales d'Histoire des Sciences* 39: 22–50, 1989.

Tihon, Anne. "Sur l'identité de l'astronome Alim." *Archives Internationales d'Histoire des Sciences* 39: 3–21, 1989.

See also: *Zīj* – Ibn Yūnus

IBN AL-ʿARABĪ Muḥyī al-Dīn ibn al-ʿArabī is one of the most influential Muslim thinkers of the past seven hundred years. Born in Murcia in present-day Spain in 1165, he set out for the western lands of Islam in 1200, traveled in the Arab countries and Turkey, and, in 1223, settled in Damascus, where he lived until his death in 1240. He wrote voluminously and attracted the attention of scholars and kings during his own lifetime. His magnum opus, *al-Futūḥāt al-makkiyya* (The Meccan Openings) — inspired sciences that were "opened" up to his soul during his pilgrimage to Mecca — will fill some 15,000 pages in its new edition. His most widely studied work, *Fuṣūṣ al-ḥikam* (The Bezels of Wisdom), is a short explication of the various modalities of wisdom embodied by twenty-eight of God's prophets, from Adam to Muḥammad.

Ibn al-ʿArabī's writings investigate every dimension of Islamic learning, from the *Quʾrān* and the *ḥadīth* (the sayings of Muḥammad) to grammar, law, philosophy, psychology, and metaphysics. His basic intellectual project was to illustrate the unity of all human endeavors and the underlying, interrelated functions of all human thinking. He cannot be classified as a philosopher, theologian, scientist, or jurist, though his works address most of the basic epistemological issues of these disciplines. He saw himself as an inheritor of the wisdom of the prophets, but one who was given the duty of explaining this wisdom in the subtlest intellectual discourse of the day — at a period that is looked back upon as the high point of Islamic learning. He provides no system, but rather a unified vision that is capable of spinning off innumerable systems, each of them appropriate to a given field of learning or level of understanding. He offers many ways of approaching the basic questions of human existence, such as the nature of reality itself, the role of God, the structure of the cosmos and the human psyche, the goal of human life, and the relationship of minerals, plants, and animals to other creatures. In short, he provides basic patterns for establishing complex systems of metaphysics, theology, cosmology, psychology, and ethics.

Ibn al-ʿArabī was followed by a number of major thinkers who systematized his "openings" in various ways, depending upon their own orientations and intellectual contexts. The diverse interpretations given to his works are especially obvious in a series of over one hundred commentaries that have been written on his *Bezels of Wisdom* from the thirteenth century down to modern times. His stepson Ṣadr al-Dīn Qūnawī (d. 1274) had probably the keenest philosophical mind among Ibn al-ʿArabī's followers. Qūnawī in turn trained many disciples, several of whom wrote widely influential works. Saʿīd al-Dīn Farghānī (d. 1296) provided systematic expositions of the teachings of both Qūnawī and Ibn al-ʿArabī in Arabic and Persian. Fakhr al-Dīn ʿIrāqī (d. 1289) was a poet who wrote a delightful summary of Qūnawī's teachings in mixed Persian prose and poetry that helped popularize Ibn al-ʿArabī's teachings. Muʾayyid al-Dīn Jandī (d. ca. 1300) wrote in Arabic the first detailed commentary on Ibn al-ʿArabī's *Bezels of Wisdom*. The intellectual tradition established by Ibn al-ʿArabī and Qūnawī gradually merged with various branches of Islamic philosophy, yielding a wide range of intellectual perspectives that dominated the Islamic wisdom tradition down to the coming of Western colonialism.

In order to grasp Ibn al-ʿArabī's importance for the history of scientific thought in Islam, one needs to understand his basic accomplishment, which was to establish an honored place in the Islamic intellectual tradition for supra-rational knowledge. From their inception in the eighth and ninth centuries, the mainline schools of theology and philosophy in Islam had endeavored to understand the Quranic revelation on the basis of rational modes of investigation taken over from the Greek heritage. Parallel to this, there developed a more practical, existential approach that found the goal of human life in direct experience of the presence of God. This second approach, which came to be called by the umbrella term "Sufism", laid stress upon supra-rational modes of knowledge that are collectively known as *kashf* (unveiling), that is, the lifting of the veils that separate the human soul from God. Unlike some Sufis, Ibn al-ʿArabī was not opposed to acknowledging the authority of reason. However, he maintained that unveiling was a higher form of knowledge, because it grows out of the unmediated perception of God's actuality. In Ibn al-ʿArabī's way of looking at things, reason tends innately to divide and discern. It eliminates connections between God and the cosmos and understands God as distant and transcendent. In contrast, unveiling works by seeing sameness and presence; hence God is perceived as near and immanent. Perfect knowledge of God and of reality as a whole depends upon a happy balance of reason and unveiling. Only through this balance can God be perceived in appropriate modes as both absent and present, near and far, transcendent and immanent, wrathful and merciful. Ibn al-ʿArabī's works are devoted largely to explaining the vast range of these appropriate modes.

The long term effect of the marriage between reason and unveiling effectuated by Ibn al-ʿArabī is symbolized by his meeting when still a boy — of perhaps fifteen years — with the philosopher Ibn Rushd (Averroes, d. 1198). Ibn al-ʿArabī had already experienced the opening of the unseen worlds, and Ibn Rushd, who was a friend of his father, had asked to meet him. In the brief exchange that took place, Ibn Rushd asked if unveiling and reason achieved the same goals. Ibn al-ʿArabī replied, "Yes and no". Then, in cryptic language, he affirmed that reason was a valid route to achieve knowledge of the nature of things, but he denied that it exhausted the possibilities of human knowing. In the West, the teachings of Ibn Rushd were employed to help establish nature as an autonomous realm of intellectual endeavor. Under the discerning eye of reason, God was abstracted from perceived reality, eventually becoming a hypothesis that could be dispensed with. The world of nature was now the proper site for rational analysis and dissection, and the result has been the ever-increasing fragmentation of human knowledge, with a total divorce between science and ethics. In contrast, Ibn Rushd was forgotten in the Islamic world, but Ibn al-ʿArabī and his followers succeeded in establishing a harmony between reason and unveiling. Hence Muslim intellectuals were never able to conceive of nature as a realm cut off from God. If God is present in all things, then the ethical and moral strictures that he establishes through revelation need to be observed at every level. It becomes impossible to investigate the natural world without at the same time investigating its relationship with God and recognizing the moral and ethical demands that this relationship entails.

Ibn al-ʿArabī's career and teachings exemplify the dimensions of Islamic learning. The worldview to which he gave detailed expression provided a perspective from within which Muslim intellectuals were able to answer the deepest questions of the human mind. Ibn al-ʿArabī's achievements contributed to an intellectual equilibrium that refused to subordinate the spiritual demands of human beings to corporeal demands and that gradually established a vast framework within the context of which the intellectual disciplines came to be ever more united and interrelated. This holistic perspective on knowledge in turn prevented the fragmentation of the Islamic worldview and allowed no room for "declarations of independence" by specific schools of science or philosophy. Given that ethics, morality, and spiritual development lay at the heart of this perspective, it was impossible to divorce any branch of science or learning from these concerns.

WILLIAM C. CHITTICK

REFERENCES

Addas, Claude. *Quest for the Red Sulphur: The Life of Ibn ʿArabī.* Cambridge: The Islamic Texts Society, 1993

Chittick, William C. *The Sufi Path of Knowledge: Ibn al-ʿArabī's Metaphysics of Imagination.* Albany: State University of New York Press, 1989.

Chittick, William C. "Ebn al-ʿArabī." In *Encyclopaedia Iranica*, Vol. VII. Costa Mesa: Mazda, 1996, pp. 664–670.

Chodkiewicz, Michel. *An Ocean Without Shore: Ibn ʿArabī, The Book, and the Law.* Albany: State University of New York Press, 1993

Corbin, Henry. *Creative Imagination in the Sūfism of Ibn ʿArabī.* Princeton: Princeton University Press, 1969

Hirtenstein, S., and M. Tiernan, eds. *Muhyiddin Ibn ʿArabī: A Commemorative Volume.* Shaftesbury: Element, 1993

IBN AL-BANNĀ᾽ Ibn al-Bannā᾽ al-Marrākushī, Abū-l-ʿabbās Aḥmad ibn Muḥammad ibn ʿUthmān al-Azdī was born in Marrakesh, Morocco on 29 December 1256 and died, probably in Marrakesh, in 1321. He studied the Arabic language, grammar, the *Qu᾽rān*, *ḥadīth* (commentaries of the prophet), and also mathematics, astronomy, and medicine; but his fame is due to his knowledge of mathematics. His teachers in this field were Muḥammad ibn Yaḥyā al-Sharīf, Abū Bakr al-Qallūsī, and Abū ʿAbd Allah ibn Makhlūf al-Sijilmāsī. He also studied medicine with al-Mirrīkh.

Among his disciples were al-Lajā᾽ī, teacher of Ibn Qunfūdh, Muḥammad ibn Ibraḥim al-Abūlī, Abū-l-Barakāt al-Balāfiqī and Ibn al-Najjār al-Tilimsānī.

He is credited with having written more than eighty works. Among them are an introduction to Euclid, a treatise on areas, an algebra text, a *Kitāb al-anwā᾽* (about asterisms and stars used in meteorology and navigation), an almanac, two abridgements of treatises by Ibn al-Zarqāllu on the use of the *ṣafīḥas* (astrolabes) *zarqāliyya* and *shakkāziyya* entitled, respectively, *Risālat al-ṣafīḥa al-mushtaraka ʿalā al-shakkāziyya* (Epistle on the Shakkāziyya Plate), in 23 chapters, and *Taqbīl ʿalā risālat al-ṣafīḥa al-zarqāliyya* (Epistle on al-Zarqālī's Plate). However, his two most important works are the *Minhāj* and the *Ṭalkhīṣ*.

The *Kitāb minhāj al-ṭālib li taʿdīl al-kawākib* (The Way of Him Who Seeks the Equation of the Planets) is a very practical book for calculating astronomical ephemerides. This *zīj* (astronomical handbook with tables) is based on the one by Ibn Isḥāq. The two of them are highly dependent on Ibn al-Zarqāllu's astronomical theories.

The *Talkhīṣ aʿmāl al-ḥisāb* (Summary of Arithmetical Operations) was widely diffused in the Arabic world because of its conciseness, which makes it easy to memorize. Al-Qalaṣādī, among others, wrote an important commentary on it. A different version of this work, more extensive and complete, has been edited recently by A. S. Saidan under the title *Al-maqālāt fī ʿilm al-ḥisāb li-Ibn al-Bannā᾽* (The Treatises on the Science of Computation by Ibn al-Bannā᾽). The procedures found in the *Talkhīṣ* are studied here in a more detailed way.

EMILIA CALVO

REFERENCES

Al-Fāsī, M. "Ibn al-Bannā᾽ al-ʿadadī al-Marrākussī." *Revista del Instituto Egipcio de Estudios Islámicos* 6: 1–10, 1958.

Calvo, Emilia. "La Risālat al-ṣafīḥa al-mušraka ʿalà al-šakkāziyya de Ibn al-Bannā᾽ de Marrākuš." *Al-Qanṭara* 10: 21–50, 1989.

Ibn al-Bannā᾽. *Taljīṣ aʿmāl al-ḥisāb*. Edition and French translation by Muhammad Souissi. Tunis: Université de Tunis, 1964.

Puig, Roser. "El Taqbīl ʿalà risālat al-ṣafīḥa alzarqāliyya de Ibn al-Bannā᾽de Marrākuš." *Al-Qanṭara* 8: 45–64, 1987.

Renaud, H.P.J. "Notes critiques d'histoire des sciences chez les musulmans. II: Ibn al-Bannā᾽ de Marrakech, ṣūfī et mathématicien (XIII–XIVs. J.C.)." *Hespéris* 25: 13–42, 1938.

Saidan, A.S. *Al-maqālāt fī ʿilm al-ḥisāb li-Ibn al-Bannā᾽*. Ammān: Dār al-Furqān, 1984.

Sarton, George. *Introduction to the History of Science*. vol. II. Baltimore: Williams & Wilkins, 1931, pp. 998–1000.

Suter, H. and M. Ben Cheneb. "Ibn al-Bannā᾽." In *Encyclopédie de l'Islam*, 2nd ed., vol. III. Leiden: E.J. Brill, 1971, pp. 753–754.

Vernet, Juan. *Contribución al estudio de la labor astronómica de Ibn al-Bannā᾽*. Tetuán: Editoria Marroquí, 1951.

Vernet, Juan. "Ibn al-Bannā᾽ al-Marrākushī." In *Dictionary of Scientific Biography*, vol. I. New York: Charles Scribner's Sons, 1970, pp. 437–438.

See also: Ibn al-Zarqāllu – Ibn Qunfūdh

IBN AL-BAYṬĀR Ibn al-Bayṭār al-Mālaqī, Ḍiyā᾽ al-Dīn Abū Muḥammad ʿAbdallāh ibn Aḥmad was a pharmacologist born in Málaga, Spain at the end of the twelfth century (ca. AD 1190–1248). He studied in Seville with Abū-l-ʿAbbās al-Nabātī, ʿAbdallāh ibn Ṣāliḥ, and Abū-l-Hajjāj. He was interested in the works of al-Ghāfiqī, al-Zahrāwī, al-Idrīsī, Dioscorides, and Galen.

Around AD 1220 he migrated to the East and, in 1224, arrived in Cairo where he was named chief herbalist by the Ayyūbid Sultan al-Kāmil. He traveled through Arabia, Palestine, Syria, and Iraq. Ibn Abī Uṣaybiʿa was one of his followers and left a mention of his teacher full of praise in his *ʿUyūn*.

Among Ibn al-Bayṭār's works we can mention *Al-Mughnī fī'l-adwiya al-mufrada* (The Complete Book on Simple Drugs), dedicated to al-Kamil's son, Sultan al-Ṣāliḥ and dealing with the simple medicines, and *Al-Jāmi᾽ li-mufradāt al-adwiya wa-l-aghdhiya* (Compendium of Simple Drugs and Food), which enumerates alphabetically some 1400 animal, vegetable, and mineral medicines, as well as some 150 authorities including al-Rāzī and Ibn Sīnā.

Ibn al-Bayṭār's main contribution is the systematization of the discoveries made by the Arabs during the Middle Ages in this field. He was also concerned with synonimy, finding the technical equivalents between the Arabic and Persian, Berber, Greek, Latin, and Romance languages. The *Jāmi᾽* had great influence in the Near East, but less in the West. Andrea Alpago used it in his works on Ibn Sīnā.

Other works of Ibn al-Bayṭār, less known than the aforementioned two, are *Mizān al-ṭabīb* (The Measure of the Physician), *Risāla fī'l-aghdhiya wa'l-adwiya* (Treatise on Food and Medicines), *Maqāla fī'l-laymūn* (Treatise on the Lemon), and *Tafsīr kitāb Diyusqūrīdis* (Explanation of Dioscorides' Book) in which he inventories 550 medicines found in the first four books of Dioscorides.

EMILIA CALVO

REFERENCES

Dubler, C.E. "Ibn al-Bayṭār en Armenio." *Al-Andalus* 21: 125–130, 1956.

Meyerhof, Max. "Esquisse d'histoire de la pharmacologie et botanique chez les musulmans d'Espagne." *Al-Andalus* 3: 31–33, 1935.

Samsó, Julio. *Las ciencias de los Antiguos en al-Andalus*. Madrid: Mapfre, 1992.

Sarton, George. *Introduction to the History of Science*, vol. II, pt. 2. Baltimore: Williams & Wilkins, 1931.

Torres, M.P. "Autores y plantas andalusíes en el Kitāb al-Yāmiʾde Ibn al-Bayṭār." In *Actas del XII Congreso de la U.E.A.I. (Málaga, 1984)*. Madrid, 1986, pp. 697–712.

Vernet, Juan. "Ibn al-Bayṭār." In *Dictionary of Scientific Biography*, vol. I. New York: Charles Scribner's Sons, 1970, pp. 538–539.

Vernet, Juan. "Ibn al-Bayṭār." In *Encyclopédie de l'Islam*, 2nd ed., vol. III. Leiden: E.J. Brill, 1971, pp. 759–760.

IBN AL-HĀʾIM ʿAbd al-Ḥaqq al-Ghāfiqī al-Ishbīlī is known as Ibn al-Hāʾim. He was an Andalusian astronomer, probably from Seville. He dedicated (ca. 1204) his very important *al-Zīj al-Kāmil fī'l-Taʿālīm* (The Perfect Handbook on Mathematical Astronomy) to the Almohad Caliph Abū ʿAbd Allāh Muḥammad al-Nāṣir (1199–1213). This work, which is extant in a unique and incomplete manuscript in the Bodleian Library at Oxford University, was influential in the Maghreb and contributed to the development there of a new kind of astronomy in the Andalusian tradition. As a *zīj* (astronomical handbook with tables), it is exceptional in Western Islam, because it contains a highly technical and complete exposition of Ptolemaic astronomy with geometrical demonstrations, and not a simple set of instructions for the use of the tables. It also conveys new information on the astronomical works of the Toledan school of the eleventh century, the main representative of which was Ibn al-Zarqāllu/Azarquiel (d. 1100), as well as new planetary parameters. Ibn al-Hāʾim appears as a defender of Zarqāllian orthodoxy and harshly criticises Ibn al-Kammād (fl. beginning of the twelfth century) for his modifications of Ibn al-Zarqāllu's doctrines. Ibn al-Hāʾim's *zīj* contained numerical tables (three have been preserved in the *zīj* of Ibn Isḥāq, who flourished in Tunis at the beginning of the thirteenth century) but an incomplete copy must have circulated early for Ibn al-Raqqām (d. 1315) states that Ibn al-Hāʾim did not include any tables in his work.

JULIO SAMSÓ

REFERENCES

Samsó, Julio. *Las Ciencias de los Antiguos en al-Andalus*. Madrid: Mapfre, 1992, pp. 320–326.

See also: Ibn al-Zarqāllu – Ibn al-Raqqām

IBN AL-HAYTHAM (ALHAZEN) Among the mathematicians of classical Islam, few are as famous as al-Ḥasan ibn al-Ḥasan ibn al-Haytham (Alhazen in the Latin West). A physicist and astronomer as well as mathematician, he quickly gained a wide reputation, first in Arabic, in the Islamic East as well as the Islamic West, and then from the translations of his works in optics and astronomy into Latin, Hebrew, and Italian.

But his renown, completely justified by the importance of his contributions and especially of the scientific reforms accomplished in them, contrasts singularly with the paucity of information we have on the man, his teachers, or his scientific milieu. Also, the significance of his works surrounded the man with the aura of a legend. Sources available to us consist of narratives recounted by ancient bibliographers where legend becomes mixed up with the rare historical evidence. These same narratives are precisely what modern bibliographers continue to reproduce partially or totally until today. After a critical reading of these sources, very little information remains: born in Iraq, most likely in Bassorah, sometime in the second half of the tenth century, Ibn al-Haytham arrived in Cairo, under the reign of Faṭimid al-Ḥākim. He proposed a hydraulic project to control the waters of the Nile, but it was rejected by the caliph. He continued to live in Cairo until his death, after 432/1040.

From the thirteenth century until today, biographers have confused al-Ḥasan ibn al-Haytham with Muḥammad ibn al-Haytham, a philosopher and theorist of medicine who lived in Baghdād at the same time. This confusion, due undoubtedly to similarity of the two names of these contemparary authors, is serious as it brings into question the authenticity of certain writings attributed to al-Ḥasan ibn al-Haytham.

Biobibliographers, notably al-Qifṭī, cite 96 titles of Ibn al-Haytham, not all of which have survived. Half of his writings are in the field of mathematics, fourteen on optics, including the authoritative and voluminous *Kitāb al-Manāẓir* (Book of Optics), twenty-three on astronomy, two in philosophy (one on the *Place* and the other on the *Indivisible Part*), three on statics and hydrostatics, two on astrology, and four on various other topics. This accounting shows clearly that Ibn al-Haytham grappled with all the mathematical sciences of that time, or at least the most advance part of this discipline. We will see that he was always at the leading edge of research or at the culmination of one tradition and the beginning of a new period. It is precisely this quality which distinguishes his contributions. Ibn al-Haytham lived at a privileged time, his work following a century of intense research in these fields by eminent scholars such as the Banū Mūsā, Thābit ibn Qurra and his grandson Ibrāhīm ibn Sinān, al-Qūhī, and Ibn Sahl, to name a few. We will now briefly examine the principal aspects of his research.

Mathematics

Ibn al-Haytham's mathematical research was particularly in the fields of geometry and not of algebra. Ancient biobibliographers attributed a book on algebra to him, but it has not survived. From the outset, geometers wanted to combine closely the study of the positions of figures and their metric properties: in other words, to combine the geometry of Apollonius with that of Archimedes. This combination is not a static synthesis, but a new organization of geometry which possessed a real heuristic value. Already initiated by al-Ḥasan ibn Mūsā and followed by Thābit, this work led to the study of geometric transformations and of projective methods. It was this work which Ibn al-Haytham developed further in his own geometrical studies.

The contributions of Ibn al-Haytham in geometry can be divided into several groups, the most important of which are in infinitesimal mathematics and the theory of conic sections and their applications. He composed twelve treatises on infinitesimal mathematics and then on conic theory. To those can be added a third area in which Ibn al-Haytham takes up several problems relating to the foundations of mathematics and their methods in his treatise *Maqāla fī 'l-taḥlīl wa 'l-tarkīb* (On Analysis and Synthesis), his *Kitāb fī al-ma 'lūmāt* (On the Known Things), his *Sharḥ Uṣūl Uqlīdis fī 'l-handasa wa 'l-ʿadad wa talkhīṣuhu* (Commentary on the *Elements* of Euclid), and his *Kitāb fī Ḥall shukūk Kitāb Uqlīdis fī 'l-uṣūl wa-sharḥ maʿānīh* (Solutions to Doubts) again concerning Euclid. In these books, he deals as much with the constitution of a new discipline, a kind of proto-topology, as with the theory of the demonstration within the difficulties raised by the fifth postulate of Euclid, or with the theory of parallels. Ibn al-Haytham also edited an important paper on number theory, four treatises on arithmetic, and the same number on practical geometry.

Of the twelve papers on infinitesimal mathematics, only seven have survived. The first three are devoted to the study of lunes and the quadrature of a circle. Note that the calculation of the area of lunes involves the calculation of sums or differences of areas of sectors or of triangles, the comparison of which has recourse to that of the ratio of angles or of the ratio of segments. In the most important of the three papers, Ibn al-Haytham begins by setting up four lemmas, the results of which demonstrate the role of the function f, defined as

$$f(x) = \frac{\sin^2 x}{x}$$

in the study of lunes.

In his study *Tarbīʿ al-dāʾira* (On the Quadrature of a Circle), he examines the relationship between proving the existence of a magnitude or a property and the question of effectiveness of its construction.

Other treatises on infinitesimal mathematics deal with the volume of a solid curve: *Misāḥat al-mujassam al-mukāfiʾ* (The Measurement of a Paraboloidal Solid) and *Misāḥat al-kura* (The Measurement of a Sphere). In calculating the volume of a paraboloid, Ibn al-Haytham deals rapidly with the volume of a revolving paraboloid, which had already been studied by Thābit ibn Qurra and al-Qūhī. He then moves on to his own invention: how to calculate the volume of a paraboloid obtained from the rotation of a parabola around its ordinate. He shows that this volume is 8/15 of the volume of the circumscribed cylinder. His calculation is equivalent to that of the integral.

$$\pi \int_a^b k^2(b^2 - 2b^2y^2 + y^4)\,dy = \frac{8}{15}\pi k^2 b^5 = \frac{8}{15}V,$$

with V being the volume of the circumscribed cylinder.

Ibn al-Haytham proceeded in this study with the help of the method of integral sums, which he also applied in calculating the volume of a sphere. In order to do this calculation, Ibn al-Haytham generalized the proposition X-1 of Euclid's *Elements*. He devoted the seventh paper in this group to that. This group includes finally an important treatise devoted to isoperimetric and isepiphane problems. It was the most advanced mathematical work of its time, and for several centuries following. In it, to study these *extrema*, he had to undertake the first substantial research on the theory of a solid angle. Moreover he combined both a projective method and an infinitesimal method.

Ibn al-Haytham's second group of mathematical writings dealt with the theory of conic sections. He was well acquainted with the *Conics* of Apollonius and had copied them in his own hand, so he knew that, in Greek, the eighth and last book was lost. He tried to reconstruct the book according to the indications of Apollonius. In addition to his writings on conics, he applied the theory of the intersection of conics to the resolution of problems which cannot be constructed with a compass or ruler, problems either passed down (for example, the regular heptagon) or posed by him (for example, the solution of a solid arithmetic problem). Ibn al-Haytham was one of the first mathematicians who insisted on demonstrating the existence of the point of intersection of two conics in these last examples.

It is impossible in this space to explicate the mathematical results of Ibn al-Haytham's work. But let us simply note his expression of what is called Wilson's theorem, and the converse of Euclid's theorem for perfect numbers.

Indeed, in the course of solving the problem called the Chinese remainder, he stated Wilson's theorem, which can be written as:

n is prime

$(n-1)! \equiv -1 (\mathrm{mod}\, n)$.

As for the converse of Euclid's theorem of perfect numbers, he tried to show that any even perfect number is in Euclidean form, in other words in the form $2^p(2^{p+1}-1)$ with $(2^{p+1}-1)$ prime.

Optics

A brief look at the work of Ibn al-Haytham on optics reveals not only its revolutionary nature but also its comprehensiveness, touching all the known branches of optics: optics in its proper sense in his *Book on Optics* and his *Discourse on Light*; catoptrics, notably burning mirrors (parabolic and spherical burning mirrors); dioptrics, in *al-Kura al-muḥriqa* (The Burning Sphere); and meteorological optics in *Ḍawʾ al-qamar* (The Light of the Moon), *Aḍwāʾ al-kawākib* (The Light of the Stars), *Fī ṣūrat al-kusūf* (On the Shape of the Eclipse) and *al-Hāla wa-qaws quzaḥ* (The Halo and the Rainbow). With this extension, Ibn al-Haytham modified the meaning of optics. Optics is not any more reduced to a theory of direct vision, a geometry of the gaze with which a theory of vision is associated, but also bears significantly on the theory of light, its propagation, and its effects as a material agent. This leads us to the revolution accomplished by Ibn al-Haytham in optics and more generally in physics.

Ibn al-Haytham sought to bring about a program of reform, which led him to take up a whole range of different problems. The basic aspect of this reform was to clarify the difference between the conditions of the propagation of light and the conditions of the vision of objects. This led, on the one hand, to giving physical support to the rules of propagation — making a firm mathematical analogy between a mechanical model of the movement of a solid ball thrown against an obstacle and that of light — and on the other hand to proceeding geometrically at all times, both by observation and by experimentation. Optics consisted henceforth of two parts: one, a theory of vision and the associated physiology of the eye and psychology of perception, and the other, the theory of light to which are linked geometric optics and physical optics. The organization of the *Optics* reflects this new situation: there are chapters devoted entirely to propagation, such as the third chapter of the first book and books IV to VII; others deal with vision and related problems. This reform also resulted in the emergence of new problems, never before posed, such as the famous Alhazen's problem in catoptrics, the examination of the spherical lens and the spherical diopter, not only as burning instruments but also as optical instruments in dioptrics, and to experimental control, viewed as much as a general practice of investigation as the norm of a proof in optics, and more generally in physics. Let us now take a quick look at how this reform in optics was carried out.

Ibn al-Haytham rejected any doctrine of a ray stemming out from the eye, called a visual ray, in order to defend the intromissionist theory of visible forms. But unlike the intromissionists of antiquity, he did not believe that objects sent off "forms" or totalities which emanated from the visible under the effect of light. He saw them rather as forms reducible to their elements: a ray emanating towards the eye from every point of a visible object. Looked at thus, the eye becomes a simple optical instrument. Ibn al-Haytham then explained how the eye perceives the visible with the help of its rays emitted from all points. In the *Optics* he devotes the first three chapters to the foundations of this theory. In the three following chapters, he deals with catoptrics. The seventh and last chapter is devoted to dioptrics. His theories rest on two qualitative laws of refraction and on several quantitative rules, all controlled experimentally with the help of an instrument which he designed and built himself. The two qualitative laws, known to his predecessors Ptolemy and Ibn Sahl, can be stated as follows:

1. The incident ray, the normal at the point of refraction, and the refracted ray are in the same plane; the refracted ray gets closer (respectively far away) from the normal, if light passes from a milieu less (respectively more) refringent to a milieu more (respectively less) refringent.

2. The principle of inverse return.

Instead of pursuing the path opened up by his predecessor Ibn Sahl, who discovered Snell's law, Ibn al-Haytham returned to a study of angles in order to establish the quantitative rules. He devoted a substantial part of the seventh book to a study of the refracted images of an object, notably if the surface of separation of two milieux is either planar or spheric. It was in the course of this study that he fixed his attention on the spherical diopter and the spherical lens. He returned to the spherical lens in his treatise *On the Burning Sphere*, one of the high points of research in classical optics, in order to improve upon certain results that he had already obtained in his *Optics*. This treatise was the first deliberate study on the spherical aberration for parallel rays falling on a glass sphere and giving off two refractions.

Astronomy

By their number, their thematic variety, the power of the analysis they show, and by their results, the works of Ibn al-Haytham in astronomy yield nothing to his works in mathematics or optics. It should be noted only that an elementary treatise of Muḥammad ibn al-Haytham, the *Commentary on the Almagest* is often erroneously attributed to Ibn al-Haytham. The attribution of the book *On the Configuration of the Universe* to him is also doubtful. The authentic works of Ibn al-Haytham have not yet been seriously studied, *a fortiori*, apart from a few rare and particular contributions, such as the *Samt al-qibla bi-al-ḥisāb* (Determination of the Direction of Mecca by Calculation). It remained for subsequent astronomers, notably al-ʿUrḍī, one of the founders of the school of Marāgha, to recognize their debt to Ibn al-Haytham's book *al-Shukūk ʿalā Baṭlamyūs* (Doubts on Ptolemy). Before assessing his contribution in astronomy, we must wait until his books have received the editing and study that they deserve.

The impact of the work of Ibn al-Haytham varies according to the field. In mathematics, his influence can be seen in the works of Ibn Hūd, al-Khayyām, Sharaf al-Dīn al-Ṭūsī, and al-Samaw'al, among others. But we do not know anything of successors who might have tried to follow up on his research on lunes, the solid angle, or the measurement of figures and solid curves. In optics, the Latin translation of his *Optics* (under the title *Perspectiva* or *De Aspectibus*, reedited in 1572 under the title *Opticae Thesaurus*) and his treatise *On Parabolic Burning Mirrors* provided a basis of research for centuries of scholars such as Witelo, Roger Bacon, J. Peccham, Frederick of Fribourg, Kepler, and Snell, among many others. In Arabic, there is the commentary of Kamāl al-Dīn al-Fārisī. Finally in astronomy, there is the work of al-ʿUrḍī which shows the influence of his work. It is too soon to measure the impact of the writings of Ibn al-Haytham on his successors in this field also, but they appear to be immense.

ROSHDI RASHED

REFERENCES

Nazīf, M. *Al-Ḥasan ibn al-Haytham, buḥūthuhu wa kushūfuhu al-baṣariyya*. Cairo, 1942–1943, 2 vols.

Rashed, R. "Analysis and Synthesis According to Ibn al-Haytham." In *Artifacts, Representations and Social Practice*. Ed. C.C. Gould and R.S. Cohen. Boston: Kluwer Academic, 1994, pp. 121–140.

Rashed, R. *The Development of Arabic Mathematics: Between Arithmetic and Algebra*. Boston: Kluwer Academic, 1994.

Rashed, R. *Géométrie et dioptrique au Xᵉ siècle: Ibn Sahl, al-Qūhī et Ibn al-Haytham*. Paris: Les Belles Lettres, 1993.

Rashed, R. *Mathématiques infinitésimales du IXᵉ au XIᵉ siècle*. Vol. II. *Ibn al-Haytham*. London: al-Furqān Islamic Heritage Foundation, 1993.

Rashed, R. *Optique et mathématiques: recherches sur l'histoire de la pensée scientifique en arabe*. Aldershot: Variorum, 1992.

Rashed, R. (ed. and co-author), *Encyclopedia of the History of Arabic Science*. London: Routledge, 1996, Routledge, 3 vols.

Sabra, A.I. *The Optics of Ibn al-Haytham: Books I–III on Direct Vision*. London: Warburg Institute, 1989.

See also: Geometry – Physics – Optics – Ibn Hūd – al-Khayyām – Sharaf al-Dīn al-Ṭūsī – al-Samaw'al – *Almagest*

IBN AL-KAMMĀD Abū Jaʿfar Ahmad ibn Yūsuf ibn al-Kammād was an Andalusian astronomer who flourished in Cordoba towards the end of the eleventh century and the first half of the twelfth. He was probably a direct disciple of Ibn al-Zarqāllu / Azarquiel (d. 1100) as well as the student who helped him in observations made in Cordoba during his last years. He compiled three sets of astronomical tables (*zīj*es) of which only one (*al-Muqtabis*) is extant in a Latin translation made by Johannes of Dumpno in Palermo (1262). In it he appears as a follower of the Zarqallian tradition, although he often corrects his master's parameters. Towards 1204, Ibn al-Hāʾim makes a strong criticism of one of his other two *zīj*es (*al-Amad ʿalāʾl-abad*, Valid for all Eternity) because of his departures from Zarqallian orthodoxy. Also extant, in Arabic, is a small astrological work in which Ibn al-Kammād studies the problem of the length of human pregnancy and the determination of the exact moment of the conception.

His work was influential in late Maghribean and Egyptian astronomy, for quotations and tabular materials from his *zīj*es appear in the thirteenth century *zīj* of Ibn Isḥāq al-Tūnisī in Abūʾl-Ḥasan ʿAlī al-Marrākushī's treatise on *mīqāt* (astronomy applied to Muslim worship) and in a fourteenth century anonymous Egyptian treatise on the same topic.

JULIO SAMSÓ

REFERENCES

Josep Chab'as and Bernard R. Goldstein, "Andalusian Astronomy: al-Zîj al-Muqtabis of Ibn al-Kammâd." *Archive for History of Exact Sciences* 48: 1–41, 1994.

Millás-Vallicrosa, José M. *Las traducciones orientales en los manuscritos de la Biblioteca Catedral de Toledo*. Madrid: Consejo Superior de Investigaciones Científicas, 1942, pp. 231–247.

Samsó, Julio. *Las Ciencias de los Antiguos en al-Andalus*. Madrid: Mapfre, 1992, pp. 320–324.

Vernet, Juan. "Un tractat d'obstetricia astrològica." In *Estudios sobre Historia de la Ciencia Medieval*. Ed. J. Vernet. Barcelona-Bellaterra: Universidad de Barcelona-Universidad Autónoma de Barcelona, 1979, pp. 273–300.

See also: *Zīj* – Ibn al-Hā᾿im

IBN AL-MAJŪSĪ Abū'l-Ḥasan ʿAlī ibn al-ʿAbbās ibn al-Majūsī was considered one of the leading Muslim physicians of his time. He was of Zoroastrian ancestry, and therefore called the Magian, and he is known in Latin as Haly ʿAbbās. He was born in the old Persian city of Arrajān, in the southwest of Iran, at the end of the first quarter of the tenth century. He studied medicine under a leading medical tutor (*shaykh*) named Abū Māhir Mūsā ibn Yūsuf ibn Sayyār, who died ca. AD 983.

When he had developed a reputation for excellence and skill, Ibn al-Majūsī was invited to become a physician-in-ordinary at the palace of King ʿAḍud al-Dawlah Fannā Khusraw, who reigned from AD 949–983 in Shīrāz. He was the Buwayhid Shāh who founded al-ʿAḍudī Bīmāristān, the famous hospital in Baghdad that lasted for almost three centuries. In that period, the Buwayid Dynasty's power and glory reached their apex.

In recognition of the King's generosity and patronage to the sciences and the arts, Ibn al-Majūsī dedicated his medical encyclopedia, *Kāmil al-Ṣināʿah al-Ṭibbīyah*, known also as *Kitāb al-Malikī* (Latin *Liber regius*, presented to the King's royal library) in Shīrāz.

The book comprises two parts, the theoretical and the practical, each of which has ten treatises. It brought together original contributions on public health, preventive medicine, dietetics, materia medica, therapy, surgery, clinical observations, and practical medico-ethical procedures. *Kitāb al-Malikī* was first rendered into Latin by Constantine Africanus (AD 1020–1087) under the title *Pantegni*, without giving credit to its original author. Al-Majūsī was fully recognized when the book was translated from Arabic into Latin by Stephen of Antioch in 1127. Since then, many editions in Latin and Arabic have appeared, and a facsimile edition was published in 1985 in two volumes.

Among the Arabic compendia on the theme of the healing arts in Islam, Majūsī's *Kitāb al-Malikī* stands as one of the leading texts in its style, systematization, and precision. It was well read among students and practitioners alike. The judge and historian Ibn al-Qifṭī praised Majūsī by saying that he "excelled in the study of medicine, [and] worked hard to comprehend its doctrines and laws from the original sources".

Ibn al-Majūsī died in Shīrāz in AH 384/AD 994, leaving behind an essential document in the Arabic legacy to the history of the medical sciences in this golden age.

SAMI KHALAF HAMARNEH

REFERENCES

Browne, Edward Granville. *Arabian Medicine*. Cambridge: Cambridge University Press, 1921.

Gurlt, Ernst. *Geschichte der Chirurgie und ihrer Ausübung Volkschirurgie, Altterthum, Mittelalter, Renaissance*. Vol. 1, Berlin: Hirschwald, 1898.

Hamarneh, Sami Khalaf. "Al-Majūsī's Observations and Instructions on Medicine and Public Health." *Hamdard Medicus* 23(1/2): 3–36, 1980.

Ibn al-Majūsī, ʿAlī ibn al-ʿAbbās. *Kamil al-Ṣināʿah* or *Kitāb al-Malikī*. Arabic ed. in Būlāq, Cairo: 2 vols, AH 1294/AD 1877. Facsimilie edition: Frankfurt: Institute for the History of Arabic-Islamic Science, 1985.

Ibn al-Majūsī. *Discourses 2 and 3 on Anatomy and Physiology, Liber regius*. French trans. P. de Koning. Leiden: Brill, 1903.

Leclerc, Lucien. *Histoire de la médecine arabe*. Paris: Leroux, 1876 and Rabat, Morocco: Ministry of Islamic Affairs, 1980.

Al-Qifti, Jamāl al-Dīn ʿAlī ibn Yūsuf. *Tārīkh al-Ḥukamā᾿*. Ed. Julius Lippert. Leipzig: Weicher, 1903.

Richter, P. "Al-Majūsī 'On Dermatology.'" *Archiv fur Dermatologie und Syphilis* 113: 849–64, 1912, and 118: 199–208, 1913.

Sezgin, Fuat. *Geschichte der arabischen Schrifttums*. Leiden: Brill, 1967.

IBN AL-NAFĪS Ibn al-Nafīs, ʿAlā᾿ al-Dīn Abu'l-Ḥasan ʿAlī Ibn Abī al-Ḥazm al-Qurashī was born in a village near Damascus (Syria). He studied medicine at the Great Nūrī Hospital (*al-Bīmāristān al-Nūrī*) in Damascus, founded by the Turkish ruler Nūr al-Dīn Maḥmūd Ibn Zankī (d. AD 1174). He chose to live, practice, and teach medicine in Egypt, where he eventually became a Chief of Physicians, and was also a personal doctor to the then-ruler al-Ẓāhir Baybars (r. ca. AD 1260–1277).

In addition to practicing medicine, he was a Shāfiʿī jurist, thoroughly educated in Islamic theology and jurisprudence at the Masrūriyya School (*Madrasa*) in Cairo. Hence, he is classed as a "jurist physician," thus deviating from the traditional image of the physician-philosopher Galen (d. ca. AD 200), which was so faithfully emulated by many medieval Arabic-speaking doctors, as for example al-Rāzī (Rhazes, d. AD 925) and Ibn Sīnā (Avicenna, d. AD 1037).

Ibn al-Nafīs was a prolific author. Among his medical works is *Kitāb al-Shāmil fi 'l-ṣinā^ca al-Ṭibbiyya* (Comprehensive Book on the Art of Medicine). He jotted down preparatory notes for this voluminous book in three hundred volumes, of which he managed to publish only eighty. Its manuscripts, so far unpublished, are to be found in Cambridge University Library (Cambridge, England), the Bodleian Library (Oxford, England), and al-Mutḥaf al-ʿIrāqī (Iraq). In 1960, three autographed manuscripts of this book were discovered in the Lane Medical Library (Stanford University). The first, referred to by Ibn al-Nafīs himself as the thirty-third *mujallad* (bound volume), is dated AH 641/AD 1243–1244. According to the author, the two other manuscripts are its forty-second and forty-third volumes.

It is of historical significance that Ibn al-Nafīs, in *Kitāb al-Shāmil*, divides the procedure to be followed by doctors in surgical operations into three stages: first, the "stage of presentation for clinical diagnosis"; second, the "operative stage"; and third, the "postoperative period", during which the patient remains under the doctor's supervision until full recovery is achieved.

Another important work is *Sharḥ Tashrīḥ Kitāb al-Qānūn fi 'l-Ṭibb li-Ibn Sīnā* (Commentary on Anatomy in Avicenna's Canon of Medicine). In it he gives the earliest known account of pulmonary circulation: "... This is the right cavity of the two cavities of the heart. When the blood in this cavity has become thin, it must be transferred into the left cavity, where the pneuma is generated. But there is no passage between these two cavities, the substance of the heart there being impermeable. It neither contains a visible passage, as some people have thought, nor does it contain an invisible passage which would permit the entry of blood, as Galen thought. The pores of the heart there are compact and the substance of the heart is thick. It must, therefore, be that when the blood has become thin, it is passed into the arterial vein (pulmonary artery) to the lung, in order to be dispersed inside the substance of the lung, and to mix with the air. The finest parts of the blood are then strained, passing into the venous artery (pulmonary vein) reaching the left of the two cavities of the heart, after mixing with the air and becoming fit for the generation of pneuma ...".

Ibn al-Nafīs' discovery of pulmonary circulation antedates the accounts mentioned by Michael Servetus (AD 1553), Realdo Colombo (AD 1559), and Andrea Cesalpino (d. AD 1603). Andreas Vesalius (d. AD 1564), who does not mention blood circulation, refutes Galen in his statement: "... I do not see how even the smallest amount of blood could pass from the right ventricle to the left ventricle, through the interventricular septum ..." We now say it was William Harvey (d. AD 1657) who, through experimentation which lasted almost twenty years, discovered the entire path of blood circulation, although some Chinese historians of medicine say the Chinese described this centuries earlier.

Ibn al-Nafīs wrote a large textbook, *Al-Muhadhdhab fi 'l-Kuḥl al-Mujarrab* (The Polished Book on Experimental Ophthalmology), divided into two sections: "On the Theory of Ophthalmology," followed by detailed accounts of "Simple and Compounded Ophthalmic Drugs."

A very popular and concise book was Ibn al-Nafīs' *Mūjiz al-Qānūn fi 'l-Ṭibb* (Abstract of Ibn Sīnā's Canon of Medicine). It has been claimed — probably unfairly — that the tedious prolixity of Ibn Sīnā's *Canon of Medicine*, together with the incomprehensibility of some of its statements, and Ibn al-Nafīs' *Abstract of Avicenna's Canon of Medicine*, with its undue brevity and popularity among Arabic-speaking students of medicine, led to the decline of late medieval medical education.

On the philosophy of religion, Ibn al-Nafīs wrote *Al-Risāla al-Kāmiliyya fi 'l-Sīra al-Nabawiyya* (Missive on the Complete Prophetic Conduct), also known by the title *Fāḍil Ibn Nāṭiq*. It was written along the lines of Ibn Ṭufayl's (d. AD 1185) *Ḥayy Ibn Yaqẓān*. In it, Ibn al-Nafīs imagines the generation — inside a cave on a deserted island — of a human being. The author guides the reader to the way in which this lone human being would arrive at the discovery of science and philosophy, then the knowledge of prophecies, and particularly the *Sīra* (conduct) of the Prophet Muḥammad, and the legal doctrines of Islam.

In his old age, Ibn al-Nafīs bequeathed his own house, including an extensive private library, to the Qalāwūn Hospital, founded in AD 1284 by Sultan Qalāwūn (r. AD 1279–1290). Ibn al-Nafīs died in his eighties, a bachelor who devoted all his time to the practice of medicine on which he wrote several books. He died on 17th December 1288 (11th Dhu'l-Qaʿdah 687).

ALBERT Z. ISKANDAR

REFERENCES

ʿAlāʾ al-Dīn Ibn al-Nafīs. *Mūjiz al-Qānūn*. Lucknow, AH 1324/AD 1906.

Al-Wafāʾī, Muḥammad Ẓāfir and Qalʿajī, Muḥammad Ruwās, eds. *Al-Muhadhdhab fi 'l-Kuḥl al-Mujarrab li-ʿAlī (Ibn al-Nafīs)*. Casablanca: Al-Najāḥ al-Jadīdah Press, Manshūrāt al-Munaẓẓama al-Islāmiyya li-l-Tarbiya wa'l-ʿUlūm wa'l-Thaqāfa (Isīkū), 1988.

Iskandar, Albert Zaki. "Ibn al-Nafīs." In *Dictionary of Scientific Biography*. New York: Scribner's, 1970–1980, vol. 9, pp. 602–606.

See also: Ibn Sīnā – al-Rāzī

IBN AL-QUFF (AL-KARAKĪ) Abū'l-Faraj ibn Yaʿqūb ibn Isḥāq Ibn al-Quff al-Karakī was born on the 22nd of August, 1233, in the city of Karak (hence the name al-Karakī) in the district of Transjordan in larger Syria (*Bilād al-Shām*). The first and most intimate biography, by a friend of al-Karakī's family, was by Abū'l-Faraj's first teacher in the healing art — the prominent physician and historian of medicine Ibn Abī Uṣaybiʿah (d. 1270). The meeting between al-Karakī's family and Uṣaybiʿah took place in Ṣarkhad, Syria. As Uṣaybiʿah was the physician-counselor to the governor (*wālī*) of Ṣarkhad and the entire province, al-Karakī's family moved from Karak north to Ṣarkhad. The father was summoned by the governor to serve as a secretary to the Department of Welfare (*dīwān al-birr*). A very close friendship developed between the physician as a leading and resourceful practitioner, and Yaʿqūb al-Karakī as an able secretary, adviser, historian, linguist, and man of letters.

At that time, the son Abū'l Faraj was 11 or 12 years of age, a bright fellow who had acquired the basics in education in Karak. The father asked the physician if he would teach his son the healing art, and Uṣaybiʿah willingly accepted the challenge. He soon taught him the basics of the field: the theory and practical aspects of medicine, methods of treatment and the identification of causes and symptoms of diseases.

The major texts which were studied included some of the writings of Hippocrates (known as the Hippocratic *corpus*, such as the *Aphorisms* and *Prognoses*) and some of the medically important catechisms, such as *Al-Masāʾil* by Ḥunayn ibn Isḥāq (809–873), and the major writings of al-Rāzī (Latin Rhazes, 865–925), particularly the clinical and therapeutic texts.

After the fall of the Ṣarkhad province into the hands of the Ayyūbid King, al-Ṣāliḥ Najm al-Dīn (1245–49), al-Karakī's family moved to Damascus (the Syrian capital of the Ayyūbids) where the father was promoted to a higher position. The son continued his study under some of the most illustrious physicians at the time. Damascus had great hospitals, including the hospital located in the Citadel for the Royal family, civil servants, and army personnel. There Ibn al-Quff had his training to become a physician.

This medieval period had witnessed great cultural, political, and techno-scientific changes. There was the fall of the prestigious Fātimiḍ (Shīʿite) dynasty in Cairo, and the rise of the Ayyūbid (Sunnite) dynasty in Cairo and Damascus under the leadership of Sultān Ṣalāḥ al-Dīn (Saladin, r. 1171–93) and his successors. There was also the rise of the *Bahri Mamluks* (slave sultans 1250–1381). However, the true founder of the "Slave" dynasty was Sultān al-Ẓahir Baybars (d. 1277), who was first purchased to serve in the Ayyūbid army and then became their ruler. During this time, the healing arts reached new heights.

Having had excellent training and having become a worthy practitioner-surgeon, Ibn al-Quff was summoned in about 1262 (at the age of 29) to be the physician-surgeon in ʿAjlūn, his home country in Transjordan. There he served the profession for a decade, at the end of which he published his first medical encyclopedia *Al-Shāfī-al-Ṭibb* (The Comprehensive of the Healing Arts), completed in early 1272. It is composed of twelve treatises encompassing the entire medical field. It was a significant contribution to the field at the time and contained many timely observations and innovations.

From ʿAjlūn, al-Karakī was summoned to the Syrian capital, Damascus. There he served at its Citadel and hospital from 1272 until his untimely death in early July 1286 at the age of 52. We know that many medical students came to hear his lectures and listen to his eloquence. He also continued to fulfill his duty in caring for the sick and the wounded and continued his research and publications. Among these are two commentaries: one on the Hippocratic *Aphorisms*, entitled *al-Uṣūl*, edited in Cairo, and the other a commentary on the generalities of the *Qānūn* of Ibn Sīnā (d. 1037). The last *Sharh* was complete about 1274.

His other work is *Jamiʿ al-Gharaḍ*, on preventive medicine and the preservation of health in 60 chapters, completed about 1275. It is extant in several manuscripts. This is possibly the finest work of its kind in medieval times, and it needs translation into English for wider audiences.

Al-Karakī's best and most renowned manual is *al-ʿUmdah* on surgery. It was published in Hyderabad, India in 1356 AH/1937 and ranks second after *al-Taṣrīf* (The Thirtieth Treatise) by al-Zahrāwī (Abulcasis, ca. 939–1013).

Only recently have al-Karakī's literary contributions begun to draw wider recognition. His writings should place him among the greatest physician-surgeons and public health experts during the Arab-Islamic Golden Age.

SAMI K. HAMARNEH

REFERENCES

Brockelmann, Carl. *Geschichte der arabischen Litteratur,* vol. 1. Leiden: E.J. Brill, 1943, p. 649; and *Supplement* 1, 1937, p. 899.

Clot, Antoine B. *Compte rendu des travaux de l'école de médecine d'A. Zabel* (Egypt). Paris: Cavellin, 1833, pp. 117–78.

Hamarneh, Sami K. *The Physician, Therapist and Surgeon — Ibn al-Quff (al-Karakī 1233–1286), an Introductory Survey of his Time, Life and Works*. Cairo: The Atlas Press, 1974.

Hamarneh, Sami K. "Ibn al-Quff's Contributions to Arab-Islamic Medical Sciences." *Hamdard* 3(1): 27–36, 1991.

Hamarneh, Sami K. "Nutritions and Dietetics in Ibn al-Quff al-Karakī's Writings." *Hamdard* 3(4): 23–37, 1990.

Ibn Abī Uṣaybiʿah. *ʿUyūn al-Anbāʾ fī Ṭabaqāt al-ʿAṭibbāʾ*. Cairo: Būlāq edition, 1299 AH/1882, vol. 2, pp. 273–4.

Ibn al-Quff al-Karakī's *Book on Preventive Medicine and the Preservation of Health*. Ed. S.K. Hamarneh. Amman: University of Jordan Press, 1989.

Leclerc, Lucien. *Histoire de la médecine Arabe*, vol. 2. Paris: Leroux, 1876, pp. 203–4.

Sarton, George. *Introduction to the History of Science*. Baltimore: Williams and Wilkins, 1927.

Sobhy, Georgy. "The Arabian Surgeon Ibn al-Quff." *Medical Association of Egypt* 20: 349–57, July 1937.

See also: Ibn al-Majūsī – Abu'l-Ḥasan

IBN AL-RAQQĀM Muḥammad ibn al-Raqqām al-Andalusī (d. 1315) was an Andalusian astronomer, mathematician, and physician. He was probably born in Murcia and left the city when it was conquered by Alfonso X in 1266. He lived in Bejaia (Algeria), Tunis and, after 1280, accepted the invitation of Muḥammad II (1273–1302) and established himself in Naṣrid Granada where he taught mathematics and astronomy (as well as medicine and law). Among his students in astronomy we find king Naṣr (1309–1314). His son, Ibrāhīm ibn Muḥammad ibn al-Raqqām, was an astrolabe maker: one of his instruments (made in Guadix in 1320) is still preserved in Madrid.

Among his extant works we should mention his *Risāla fī ʿIlm al-Ẓilāl* (Treatise on the Science of Shadows) as well as two sets of astronomical tables (*zīj*es) the first of which was probably compiled in Bejaia in 1280–1281, while the second was made in Tunis and later adapted to the coordinates of Granada. His book on *Shadows* is a brilliant treatise on Gnomonics where Ibn al-Raqqām explains how to build all kinds of sundials (among which we find a portable sundial which includes a compass) using projections on a plane which, ultimately, derive from Ptolemy's *Analemma*. The first of his *zīj*es, entitled *al-Zīj al-Shāmil fī Tahdhīb al-Kāmil* (A General Set of Astronomical Tables in which [Ibn al-Hā'im's] Kāmil Zīj is Corrected) follows narrowly the theoretical contents of Ibn al-Hāʾim's *al-Zīj al-Kāmil fī'l-Taʿālīm* to which Ibn al-Raqqām adds the numerical tables which had been lost in the manuscript of Ibn al-Hā'im's work. His second *zīj, al-Zīj al-Qawīm fī Funūn al-Taʿdīl wa-l-Taqwīm* (The Solid Handbook to Calculate Equations and Planetary Positions) is a summary and adaptation of the first one to the coordinates of Tunis and Granada. Both works bear witness to the diffusion of Ibn al-Zarqāllu's astronomical ideas in the Maghreb and Andalusia.

JULIO SAMSÓ

REFERENCES

Carandell, Joan. "An Analemma for the Determination of the Azimuth of the Qibla in the *Risāla fī ʿIlm al-Ẓilāl* of Ibn al-Raqqām." *Zeitschrift für Geschichte der Arabisch-Islamischen Wissenschaften* 1: 61–72, 1984.

Carandell, Joan. *Risāla fī ʿilm al-Ẓilāl de Muḥammad ibn al-Raqqām al-Andalusī*. Barcelona: Instituto Millás-Vallicrosa de Historia de la Ciencia Arabe, 1988.

Samsó, Julio. *Las Ciencias de los Antiguos en al-Andalus*. Madrid: Mapfre, 1992, pp. 414–415, 421–427.

See also: Ibn al-Zarqāllu – Alfonso X – *Zīj* – Sundials

IBN AL-SHĀṬIR Ibn al-Shāṭir, ʿAlāʾ al-Dīn ʿAlī ibn Ibrāhīm was born in Damascus *ca.* 1305. He was the most distinguished Muslim astronomer of the fourteenth century. Although he was head *muwaqqit* at the Umayyad mosque in Damascus, responsible for the regulation of the astronomically defined times of prayer, his works on astronomical timekeeping are considerably less significant than those of his colleague al-Khalīlī. On the other hand, Ibn al-Shāṭir made substantial advances in the design of astronomical instruments. Nevertheless, his most significant contribution to astronomy was his planetary theory.

In his planetary models Ibn al-Shāṭir incorporated various ingenious modifications of those of Ptolemy. Also, with the reservation that they are geocentric, his models are the same as those of Copernicus. Ibn al-Shāṭir's planetary theory was investigated for the first time in the 1950s, and the discovery that his models were mathematically identical to those of Copernicus raised the very interesting question of a possible transmission of his planetary theory to Europe. This question has since been the subject of a number of investigations, but research on the astronomy of Ibn al-Shāṭir and his sources, let alone on the later influence of his planetary theory in the Islamic world or Europe, is still at a preliminary stage.

Ibn al-Shāṭir appears to have begun his work on planetary astronomy by preparing a *zīj*, an astronomical handbook with tables. This work, which was based on strictly Ptolemaic planetary theory, has not survived. In a later treatise entitled *Taʿlīq al-arṣād* (Comments on Observations), he described the observations and procedures with which he had constructed his new planetary models and derived new parameters. No copy of this treatise is known to exist in the manuscript sources. Later, in *Nihāyat al suʾl fī taṣḥīḥ al-uṣūl*

(A Final Inquiry Concerning the Rectification of Planetary Theory), Ibn al Shāṭir presented the reasoning behind his new planetary models. This work has survived. Finally, Ibn al-Shāṭir's *al-Zīj al-jadīd* (The New Astronomical Handbook), extant in several manuscript copies, contains a new set of planetary tables based on his new theory and parameters.

Several works by the scholars of the mid-thirteenth century observatory at Maragha are mentioned in Ibn al-Shāṭir's introduction to this treatise, and it is clear that these were the main sources of inspiration for his own non-Ptolemaic planetary models.

The essence of Ibn al-Shāṭir's planetary theory is the apparent removal of the eccentric deferent and equant of the Ptolemaic models, with secondary epicycles used instead. The motivation for this was at first sight aesthetic rather than scientific, but his major work on observations is not available to us, so this is not really verifiable. In any case, the ultimate object was to produce a planetary theory composed of uniform motions in circular orbits rather than to improve the bases of practical astronomy. In the case of the sun, no apparent advantage was gained by the additional epicycle. In the case of the moon, the new configuration to some extent corrected the major defect of the Ptolemaic lunar theory, since it considerably reduced the variation of the lunar distance. In the case of the planets, the relative sizes of the primary and secondary epicycles were chosen so that the models were mathematically equivalent to those of Ptolemy.

Ibn al-Shāṭir also compiled a set of tables displaying the values of certain spherical astronomical functions relating to the times of prayer. The latitude used for these tables was 34°, corresponding to an unspecified locality just north of Damascus. These tables display such functions as the duration of morning and evening twilight and the time of the afternoon prayer, as well as standard spherical astronomical functions.

Ibn al-Shāṭir designed and constructed a magnificent horizontal sundial that was erected on the northern minaret of the Umayyad Mosque in Damascus. The instrument now on the minaret is an exact copy made in the late nineteenth century. Fragments of the original instrument are preserved in the garden of the National Museum, Damascus. Ibn al-Shāṭir's sundial, made of marble and a monumental 2×1 m in size, bore a complex system of curves engraved on the marble which enabled the *muwaqqit* to read the time of day in equinoctial hours since sunrise or before sunset or with respect to either midday or the time of the afternoon prayer, as well as with respect to daybreak and nightfall. The gnomon is aligned towards the celestial pole, a development in gnomonics usually ascribed to European astronomers.

A much smaller sundial forms part of a compendium made by Ibn al-Shāṭir, now preserved in Aleppo. It is contained in a box called *ṣandūq al-yawāqīt* (jewel box), measuring $12\times 12\times 3$ cm. It could be used to find the times (*al-mawāqīt*) of the midday and afternoon prayers, as well as to establish the local meridian and the direction of Mecca.

Ibn al-Shāṭir wrote on the ordinary planispheric astrolabe and designed an astrolabe that he called *al-āla al-jāmiʿa* (the universal instrument). He also wrote on the two most commonly used quadrants, the astrolabic and the trigonometric varieties. Two special quadrants which he designed were modifications of the simpler and ultimately more useful sine quadrant. One astrolabe and one universal instrument actually made by Ibn al-Shāṭir survive.

A contemporary historian reported that he visited Ibn al-Shāṭir in 1343 and inspected an "astrolabe" that the latter had constructed. His account is difficult to understand, but it appears that the instrument was shaped like an arch, measured three-quarters of a cubit in length, and was fixed perpendicular to a wall. Part of the instrument rotated once in twenty-four hours and somehow displayed both the equinoctial and the seasonal hours. The driving mechanism was not visible and probably was built into the wall. Apart from this obscure reference we have no contemporary record of any continuation of the sophisticated tradition of mechanical devices that flourished in Syria some two hundred years before his time.

Ibn al-Shāṭir died in Damascus ca. 1375. Later astronomers in Damascus and Cairo, none of whom appears to have been particularly interested in his non-Ptolemaic models, prepared commentaries on, and new versions of, his *zīj*. In its original form and in various recensions this work was used in both cities for several centuries. His principal treatises on instruments remained popular for several centuries in Syria, Egypt, and Turkey, the three centres of astronomical timekeeping in the Islamic world. Thus his influence in later Islamic astronomy was widespread but, as far as we can tell, unfruitful. On the other hand, the reappearance of his planetary models in the writings of Copernicus strongly suggests the possibility of the transmission of some details of these models beyond the frontiers of Islam.

DAVID A. KING

REFERENCES

Primary sources

Zīj of Ibn al-Shāṭir. MS Oxford Bodleian A30.

Secondary sources

Brockelmann, Carl. *Geschichte der arabischen Litteratur*. 2 vols., 2nd ed., Leiden: E. J. Brill, 1943–1949, and 3 suppl. vols. Leiden: E. J. Brill, 1937–1942, II, p. 156, and suppl. II, p. 157.

Hartner, Willy. "Trepidation and Planetary Theories: Common Features in Late Islamic and Early Renaissance Astronomy." *Accademia Nazionale dei Lincei, 13° Convegno Volta*, 1971, pp. 609–629.

Janin, Louis. "Le cadran solaire de la Mosquée Umayyade à Damas." *Centaurus* 16: 285–298, 1971.

Janin, Louis, and D. A. King, "Ibn al-Shāṭir's *ṣandūq alyawāqīt*: An Astronomical 'Compendium'." *Journal for the History of Arabic Science* 1: 187–256, 1977, reprinted in D.A. King, *Islamic Astronomical Instruments*, London: Variorum, 1987, XII.

Kennedy, Edward S. "A Survey of Islamic Astronomical Tables." *Transactions of the American Philosophical Society*, n.s. 46 (2): 121–177, 1956.

Kennedy, Edward S. and Imad Ghanem. *The Life and Work of Ibn al-Shāṭir, an Arab Astronomer of the Fourteenth Century*. Aleppo: Institute for History of Arabic Science, 1976.

Kennedy, Edward S. et al. *Studies in the Islamic Exact Sciences*. Beirut: American University of Beirut, 1983.

King, David A. "Ibn al-Shāṭir." In *Dictionary of Scientific Biography*, vol. 12. New York: Charles Scribner's Sons, 1975, pp. 357–364.

King, David A. *A Survey of the Scientific Manuscripts in the Egyptian National Library*. Winona Lake, Indiana: Eisenbrauns, 1986, no. C30.

King, David A. "The Astronomy of the Mamluks." *Isis* 74: 531–555, 1983, reprinted in D.A. King, *Islamic Mathematical Astronomy*, London: Variorum; 2nd rev. ed., Aldershot: Variorum, 1993, III.

King, David A. "L'astronomie en Syrie à l'époque islamique." In *Syrie - mémoire et civilisation*. Ed. S. Cluzan et al. Paris: Institut du Monde Arabe, 1993, pp. 391–392, 435 and 439.

Saliba, George. "Theory and Observation in Islamic Astronomy: The Work of Ibn al-Shāṭir of Damascus (1375)." *Journal for the History of Astronomy* 18: 35–43, 1987.

Schmalzl, Peter. *Zur Geschichte des Quadranten bei den Arabern*. Munich: Salesianische Offizin, 1929.

Suter, Heinrich. *Die Mathematiker und Astronomen der Araber und ihre Werke*. Leipzig, 1900, reprinted Amsterdam: Oriental Press, 1982, no. 416.

See also: al-Khalīlī – *Zīj* – al-Jazarī – Astronomical Instruments

IBN AL-YĀSAMĪN The name of this mathematician is Abū Muḥammad ʿAbdallāh ibn Muḥammad ibn Ḥajjāj al-Adrīnī, more commonly known as Ibn al-Yāsamīn. As his name indicates, he was originally from a Berber tribe from the Maghreb (North Africa), and, according to Ibn Saʿīd, he was black like his mother. We know nothing about the exact date of his birth, but can reasonably place it in the second half of the twelfth century. We also know nothing specific about his place of birth, which could have been in the Andalus (Spain) or the Maghreb. But since some historians have given him the surname al-Ishbīlī, he may have been born or grown up in Seville. In any case, Ibn Saʿīd states that his formative education occurred in Seville.

This education was not restricted to mathematics, since we know he also became famous in the fields of law and literature, particularly in the Andalusian poetry of the Muwashshaḥāt. That being said, however, we know nothing of the context in which he received this rich education nor of his teachers. The only information we have is from Ibn al-Yāsamīn himself about one of his professors, Abū ʿAbdallāh Muḥammad ibn Qāsim al-Shalūbīn, who taught him algebra and the science of calculation.

We also do not know exactly when Ibn al-Yāsamīn began to publish his mathematical writings. Ibn al-Abbār tells us only that Ibn al-Yāsamīn's famous algebraic poem was drafted in Seville and that, in 1190 also in Seville, he was using it in his teaching.

Concomitant with his mathematical activities was Ibn al-Yāsamīn's dedication to poetry, and according to Ibn Saʿīd, some of his poems had even been set to music and sung at this time. It may have been his literary success which led to his frequenting the court of the third Almohad caliph Abū Yūsuf Yaʿqūb (1184–1199) and of his successor Muḥammad al-Nāṣir (1199–1213). These frequent court visits and the fame of his literary and mathematical publications probably gained Ibn al-Yāsamīn some enemies. Also some of his contemporaries accused him of leading a dissolute life. But none of these can be seen as the cause of his assassination in 1204 in Marrakesh.

The Mathematical Writings of Ibn al-Yāsamīn

The best-known work of Ibn al-Yāsamīn is a poem of fifty-three verses in rajaz meter entitled *al-Urjūza al-yāsmīnīyya fī al-jabr wa'l-muqābala* (Poem on Algebra and Restoration). In it the author defines the algebra known in his time — number, root, and sequence, then the six canonical equations of al-Khwārizmī with the processes of solving them, and finally, the operations of algebra — the restoration, comparison, multiplication, and division of monomials.

This poem has been widely read throughout the centuries both within the Maghreb and beyond. Thus, there are many commentaries on it by other famous mathematicians such as Ibn Qunfudh (d. 1407) and al-Qalaṣādī (d. 1486) in the Maghreb, Ibn al-Hāʾim (d. 1423), and Sibṭ al-Māradīnī (d. 1501) in Egypt and elsewhere.

For a long time, the contribution of Ibn al-Yāsamīn to mathematics was known only through his *Urjūza* in algebra. It is quite possible, moreover, that it was the success of this poem which led him to write a second one on irrational quadratic numbers and maybe a third on the method of false position. But the distribution of these last two poems was relatively modest, compared to that of the first one, and aside from the rare copies that exist, we have not yet found any explicit reference to the contents of the two other poems in any works on calculation written after the twelfth century.

The same situation exists for Ibn al-Yāsamīn's fourth written work on mathematics, entitled *Talqīḥ al-afkār bi rushūm ḥurūf al-ghubār* (Fertilization of Thoughts with the Help of Dust Letters [Hindu Numerals]). This work is much more important than his poems, as much in quantity as in quality. Indeed, it is a book of two hundred folios which contains classic chapters on the science of calculation and on geometry. Among the works of the Muslim West which have come down to us, it is the only one which consolidates these two disciplines. Its importance is also due the nature of its materials and its mathematical tools which make it an original book and also one which is totally representative of this period of transition in which three mathematical traditions were juxtaposed — of the East, the Andalus, and the Maghreb — before they became blended in the same mold.

For example, the following elements contribute both to the originality of his work and to its being anchored in the great Arab mathematical tradition of the ninth to eleventh centuries:

In arithmetic, contrary to the Maghrebian tradition which prevailed from the fourteenth century on, Ibn al-Yāsamīn treats multiplication and division first, before addition and subtraction. This approach, which can be found again later in the work of Ibn al-Zakariyā' al-Gharnāṭī, seems to be based on Andalusian mathematical practice.

For fractions, the remarks and suggestions of Ibn al-Yāsamīn concerning the reading of certain expressions demonstrate that the symbolism of fractions had not been established definitively in his time. This was not the case, it appears, for the symbolism of equations which had been established relatively early, since there is no difference between the symbols used in the *Talqīḥ al-afkār* of Ibn al-Yāsamīn and those used in the *Bughyat aṭ-ṭullāb* (The Hope of Students) of Ibn Ghāzī al-Miknāsī (d. 1513).

As for the presence of geometry in a work on the science of calculation, this is not exceptional in regard to the Arab mathematical tradition, viewed in its entirely, in so far as similar chapters (that is, chapters which deal with problems in metric geometry) had already been inserted in works edited in the East, such as the *Takmila fī l-ḥisāb* (Complement to Calculus) of ʿAbd al-Qāhir al-Baghdādī or the *Kitāb al-Kāfī* (The Sufficient Book) by al-Karajī.

In spite of what we have noted about the contents of *Talqīḥ al-afkār*, there has not been any explicit reference to the book in Maghrebian mathematical writing. There could be at least two possible explanations for this: first, a break in the tradition whose cause is to be found outside the scientific milieu of that time. This hypothesis is not implausible if one takes into account the personality of Ibn al-Yāsamīn and his controversial behavior and also his close ties to Almohad power.

The second reason, which is also plausible and which can be added to the first, can be found in mathematical practice after Ibn al-Yāsamīn, a practice which bore the strong imprint of mathematicians from Marrakesh, like Ibn Munʿim (d. 1228), al-Qāḍi al-Sharīf (d. 1282–1283), and Ibn al-Bannāʾ (d. 1321). We have observed this same phenomenon of the absorption of a mathematical tradition first in the East with the first written Arab arithmetical works from the ninth century and, in particular, with the work of al-Khwārizmī, and then in the Andalus with writings of the tenth century, like those of al-Majrīṭī and his pupils.

AHMED DJEBBAR

REFERENCES

Djebbar, A. "Quelques aspects de l'algèbre dans la tradition mathématique arabe de l'Occident musulman." *Actes du Premier Colloque Maghrébin d'Alger sur l'Histoire des mathématiques arabes* (Alger, 1–3 December, 1986). Alger: Maison du Livre, 1988, pp. 99–123.

Guergour, Y. "Étude comparative entre deux commentaires du *Talkhīṣ* d'Ibn al-Bannāʾ: celui d'Ibn Qunfudh et celui d'Ibn Zakariyā al-Gharnāṭī. *5e Symposium International sur les Sciences Arabes*. Grenada: March–April, 1992. Proceedings of the symposium forthcoming.

Ibn al-Abbār. *At-Takmila li kitāb aṣ-Ṣila*. Ed. ʿIzzat al-ʿAṭṭār al-Ḥusaynī. Cairo: Maṭbaʿat as-saʿāda, 1956, p. 43.

Ibn Saʿid. *Al-Ghuṣūn al-yāniʿa fī maḥāsin shuʿarā' al-mi'a as-sābiʿa*. Ed. Ibrāhīm al-Ibyārī. Cairo: Dār al-maʿārif, 1945, pp. 42 and 47.

Jalāl Shawqī. *Manẓūmāt Ibn al-Yāsamīn fī aʿmāl al-jabr wa l-ḥisāb*. Kuwait: Mu'assassat al-Kuwayt li t-taquaddum al-ʿilmī, 1988.

Souissi, M. *Al-lumʿa al-māradīniyya fī sharḥ al-Yāsamīniyya*. Kuwait, 1983.

Zemouli, T. *Mu'allafāt Ibn al-Yāsamīn ar-riyāḍiyya*. Master's thesis. E.N.S. d'Alger, 1993.

See also: Ibn al-Bannaʾ– al-Majrīṭī – al-Karajī

IBN AL-ZARQĀLLU Ibn al-Zarqāllu, Abū Isḥāq Ibrāhīm ibn Yaḥyā al-Naqqāsh, sometimes known as Azarquiel, was born in the first quarter of the eleventh century to a family of artisans. He entered the service of the *qāḍī* (judge) Ṣāʿid of Toledo first as an artisan, and after as the director of a group which carried out astronomical observations. Ibn al-Zarqāllu lived in Toledo until ca. AD 1078, when he moved to Cordoba where he composed his last works under the patronage of the king of Sevilla al-Muʿtamid ibn ʿAbbād and where he died in AD 1100. His work exerted considerable influence on later authors such as Ibn al-Kammād, Abū'l-Ḥasan ʿAlī, Ibn al-Bannāʾ, Abraham ibn ʿEzra, Ibn al-Hāʾim, Ibn Isḥāq, and Ibn Bāṣo.

Ibn al-Zarqāllu's works are basically astronomical, although an astrological treatise by him is also extant. Other works of Ibn al-Zarqāllu are described below:

The *Almanac* is a reelaboration by Ibn al-Zarqāllu of the work of an unknown author called Awmātiyūs. It allows the determination of planetary longitudes, without computation, by combining Ptolemaic parameters with the Babylonian doctrine of the goal-years. The goal-years consisted of cycles peculiar to each planet. These cycles included an entire number of solar years which, in turn, comprised an exact number of synodic and zodiacal revolutions. These cycles were known by the Babylonian astronomers, and they are also found in Ptolemy's *Almagest*. The advantage of these cycles for astronomers is that the positions of the planets can be calculated for a complete cycle which will be repeated, so these positions will always be the same for a given date within the cycle.

Treatise on the Motion of the Fixed Stars is preserved in a Hebrew translation. It is probably the most complete medieval text on the trepidation theory. Trepidation consists of a back-and-forth vibration within fixed limits, which is supposed to account for the variation in velocity of the slow eastward motion observed in the fixed stars. The treatise proposed three different geometrical models to demonstrate this theory.

Fī sanat al-šams (On the Solar Year) was probably written between AD 1075 and 1080 and based on 25 years of solar observations. Here Ibn al-Zarqāllu established the proper motion of the solar apogee to be of 1° in 279 Julian years. The text is lost but it can be reconstructed from the works of later astronomers like Ibn al-Kammād, Ibn al-Hāʾim, Ibn Isḥāq, Ibn al-Raqqām and Ibn al-Bannāʾ.

Risālat al-ṣafīḥa al-zarqāliyya (Treatise on the Zarqaliyya Plate) and *Risālat al-ṣafīḥa al-shakkāziyya* (Treatise on the Shakkāziyya Plate) are two treatises on the use of a universal astrolabe called *ṣafīḥa* (plate). There is an Alphonsine translation in the *Libros del Saber de Astronomía* of the treatise on the use of the *ṣafīḥa zarqāliyya*. This instrument offers the possibility of making calculations for any given latitude by means of only one plate.

Kitāb al-ʿamal bi 'l-ṣafīḥa al-zījiyya (Treatise on the Plate for the Seven Planets) describes the equatorium, an instrument consisting of the representation, drawn to scale, of a planetary model, which is used to determine the position of a planet at a given moment. Both his treatise on its construction and his treatise on its use are extant.

He also took an active part in the elaboration of the *Toledan Tables* which seem to have been the result of the work of a group of astronomers directed by the *qāḍī* Ṣāʿid.

Finally, he was probably the author of a treatise on the construction of the armillary sphere, translated or adapted by Isḥāq ibn Sīd, which was incorporated into the Alphonsine *Libro de las Armellas*.

EMILIA CALVO

REFERENCES

Boutelle, M. "The Almanac of Azarquiel." *Centaurus* 12: 12–19, 1967. Reprinted in *Studies in the Islamic Exact Sciences*. Beirut: American University of Beirut, 1983, pp. 502–510.

Comes, Mercè. *Ecuatorios andalusíes. Ibn al-Samḥ, al-Zarqāllu y Abū'l Ṣalt*. Barcelona: Instituto de Cooperación con el Mundo Arabe-Universidad de Barcelona, 1991.

Millás Vallicrosa, José María. *Estudios sobre Azarquiel*. Madrid-Granada: Consejo Superior de Investigaciones Científicas, 1943–1950.

Pedersen, F.S. "*Canones Azarchelis*. Some Versions and a Text" *Cahiers de l'Institut du Moyen-Age Grec et Latin*, 54. Université de Copenhague, 1987.

Puig, R. *Los tratados de construcción y uso de la azafea de Azarquiel*. Madrid: Instituto Hispano-Arabe de Cultura, 1987.

Puig, R., *Al-ṣakkāziyya. Ibn al-Naqqāš al-Zarqāllu*. Barcelona: Universidad de Barcelona, 1986.

Samsó, Julio. *Las ciencias de los Antiguos en al-Andalus*. Madrid: Mapfre, 1992.

Soomer, G.J. "The Solar Theory of al-Zarqāl. A History of Errors." *Centaurus* 14: 306–336, 1969.

Vernet, J. "Al-Zarqāllī." In *Dictionary of Scientific Biography*, vol. XIV. New York: Charles Scribner's Sons, 1976, pp. 592–595.

See also: Ibn al-Kammād – Ibn al-Bannāʾ– Abraham ibn ʿEzra – Ibn al-Hāʾim – Astrolabe – Armillary Sphere

IBN BAṬṬŪṬA Ibn Baṭṭūṭa (1304–1369) was the greatest Muslim traveler of his time. He was born in Tangier to a well-educated Moroccan family that produced many judges. After receiving basic education in his home town, at age 21 he headed toward Mecca both to make a pilgrimage and to study under some notable Muslim scholars in Egypt, Syria, and Hejaz. After reaching Egypt via Tunis and Tripoli, he decided to become a traveler to gain firsthand knowledge about as many parts of the world as possible. Ibn Baṭṭūṭa's fascination for travel took him to Syria, Iraq, southern Iran, and Azerbaijan. He then decided to spend the next three years (1327–1330) in the holy towns of Mecca and Medina in Hejaz. Ibn Baṭṭūṭa's next expedition (1330–1332) started from the seaport of Jidda. He sailed across the Red Sea to

Yemen, traveled across Yemen by land, and from Aden he sailed along the coast to various trading ports of East Africa. On his way back, he turned his boat to the Persian Gulf area and concluded this trip by another pilgrimage to Mecca.

Ibn Baṭṭūṭa's travels were supported by the contributions of the rulers, governors, and other prominent residents of the places that he visited. After hearing of the benevolence of Muḥammad bin Tughlaq — the ruler of Delhi, India — Ibn Baṭṭūṭa decided to go to India. This time he adopted a very unusual route: by moving northward, he first passed through Egypt and Syria; he then traveled extensively in Anatolia (Asia Minor)and the territories of the Golden Horde where he was well received by the local sultans and other prominent people. Ibn Baṭṭūṭa's *Riḥlah* (Book of Travels) provides a clear and lucid picture of Constantinople, Saray (the capital of the Khan of the Golden Horde), and other Sultanates. After crossing the Steppes with a caravan, Ibn Baṭṭūṭa passed through the towns of Bukhara, Samarkand, Balkh, and Herat. He then crossed the Hindu Kush Mountains and visited many important towns and cities in the Indus Valley — particularly Sukkur, Multan, and Lahore — before he finally reached Delhi. Sultan Muhammad bin Tughluq received him with respect and presents. The Sultan also appointed him the chief justice of Delhi. Ibn Baṭṭūṭa enjoyed the patronage of the Sultan for several years.

In 1342, he appointed Ibn Baṭṭūṭa the ambassador to China. After his ship wrecked near Calicut on the Malabar Coast of India, he decided to go to the Maldives where he married into the royal family. In the following years, Ibn Baṭṭūṭa visited Sri Lanka (Ceylon), supported the Sultan of the Maldives in a war, went again to the Maldives, and visited Bengal, Assam, and Sumatra. The Sultan of Sumatra provided him with a new ship to go to China. He arrived at the Chinese port of Zaytun and reached Beijing via the inland waterways. On his return journey, he eventually reached Mecca via Sumatra, Calicut, the Persian Gulf, Baghdad, Syria, and Egypt. Ibn Baṭṭūṭa's narrative for this entire journey is sketchy and some scholars doubt that he ever reached Beijing.

In April–May of 1349, Ibn Baṭṭūṭa embarked for his home from Alexandria via Tunis, Sardinia, and Algeria. He eventually reached Fez (Morocco), from where he went to the Kingdom of Granada, in Spain. His next destination was western Sudan. After traveling across the Sahara, he visited the Empire of Mali and returned to retire in Fez.

Ibn Baṭṭūṭa was an outstanding adventurer. His education and experience earned him numerous honors and awards, including the position of judge in many parts of the Muslim world. After his retirement, he again held the office of *qāḍī* (judge) in a Moroccan town and dictated his recollections to Ibn Juzayy — a royal poet. Ibn Baṭṭūṭa's *Riḥlah* is a valuable document for understanding the ways of life in the fourteenth century Muslim world.

BILAL AHMAD

REFERENCES

Beazley, C. R. *The Dawn of Modern Geography*. Vol. III. Oxford: Clarendon Press, 1906.

Dunn, R. E. *The Adventures of Ibn Baṭṭūṭa: A Muslim Traveller of the Fourteenth Century*. London: Croom Helm, 1986.

Dunn, R. E. "International Migrations of Literate Muslims in the Later Middle Period: The Case of Ibn Baṭṭūṭa." In *Golden Roads: Migration, Pilgrimage and Travel in Mediaeval and Modern Islam*. Ed. I.R. Netton. Richmond: Curzon Press, 1993, pp. 75–85.

Gibb, H. A. R. *The Travels of Ibn Baṭṭūṭa,* AD *1325–1354*. 3 vols. Cambridge: Cambridge University Press for the Hakluyt Society, 1958–1971.

Ibn Baṭṭūṭa. *Riḥlat*. Beirut: Dar Sadir, 1964.

Newton, A. P. *Travel and Travellers of the Middle Ages*. London: Kegan Paul, 1930.

IBN BUṬLĀN Abū'l-Ḥasan al-Mukhtār ibn ʿAbdūn ibn Saʿdūn ibn Buṭlān was a physician, philosopher and Christian theologian from Baghdad (eleventh century). He had for a teacher a Nestorian priest, Abū'l-Faraj ibn al-Ṭayyib, a commentator on Aristotle, Hippocrates, and Galen, who was interested in botany and wrote on the humors, wine, and natural qualities. Teaching at the hospital founded in Baghdad by ʿAḍud al-Dawla, he directed his pupil in reading and the use of several medical works. He taught him so well that later, Ibn Buṭlān, in a controversy with the physician Ibn Riḍwān, maintained that one could understand the basic precepts of medicine by simply reading books.

On the subject of the *al-Masāʾil fi 'l-Ṭibb fi 'l-Mutaʿallimīn* (Questions on Medicine for Students) by Ḥunayn ibn Isḥāq, he believed that Ibn Riḍwān, who had refuted it, had not understood it at all, "because he didn't study it under the direction of masters in this art". That is why, in spite of his great knowledge of the works of the Ancients, he refused to conform blindly and to the letter. Also he asked why clear-sighted doctors had lost the habit of caring for certain maladies, as the Ancients had done, with warm medicines, and preferred to use cold ones. It seems also that Ibn Buṭlān was instructed in the practice of medicine by Abū'l-Ḥasan Thābit ibn Ibrāhīm al-Ḥarrānī, about whom many praises were said.

Ibn Buṭlān left Baghdad, crossed Syria, and arrived in Egypt, where he undertook several polemics with Ibn Riḍwān. These covered diverse subjects, touching in particular the philosophical questions and ideas of Aristotle concerning place, movement, and the soul on which Ibn Riḍwān had commented before. He returned to Constantinople, where he arrived in 446/1054, at the time of the schism which would eventually separate the Greek from the Latin church. The patriarch Michel Cérulaire asked him to edit a treatise on the Eucharist and the use of bread without leavening. That was also the year when a terrible epidemic of the plague broke out in the capital of the Byzantine Empire. Ibn Buṭlān kept a diary and cited the names of several savants who had succumbed to it. After this he went to Antioch where he directed the establishment of a hospital. Eventually he retired from traveling, and died in 460/1068.

His work is very diverse, covering theology, philosophy, logic, and medicine. His interpretation of the work of the Ancients is original and animated by a critical spirit. He relied on logic and the grammar of languages, and attempted to explain apparent opposition by showing that they had very different points of view. Thus Aristotle studied organic forces relative to their nature, while Galen did so relative to their perceptible action in the organs which were their instruments. Aristotle divided the organs according to their physical constitution, Galen in relation to the illnesses which affected them. In the same way, with regard to the "egg yolk" color of bile, Ibn Buṭlān tended to agree with Galen who explained it by the predominant action of heat, which rends the bile hotter and lighter. On the other hand, Ḥunayn ibn Isḥāq explained it as the result of a mixture of bile and phlegm. These differences can be explained by the ambiguity of the term *muḥḥ,* which signifies at the same time both the egg yolk alone and the entire interior of the egg, both the yellow and the white.

Ibn al-Qifṭī retained for us a list of the problems that Ibn Buṭlān posed. For example, he wondered about the chemical nature of the force of physical attraction between lovers. In the field of physiology, he questioned why it is that when men dream they are urinating, they wake themselves up without urinating in their beds, while when they dream of a sexual encounter, there is an emission of sperm. He wondered why this was so, when you consider how much easier it is to urinate than to ejaculate when one is awake.

Ibn Buṭlān's most important medical work is the *Taqwīm al-ṣiḥḥa* (Strengthening of Health), a treatise on hygiene dedicated to general questions on the four elements, the humors, and the temperaments. The author studied the nature and value of nutrition, as well as the influence of the environment, water, climate, and housing on health. The originality of the work lies in its form: it is presented in small tableaux. It was translated into Latin and German. Another noteworthy work is *Da ʿat al-Aṭibbā* (The Physician's Banquet), on the subject of medical ethics, which included a satire on charlatans and ignorant physicians. There is also a treatise on the maladies caused by food, with recommended remedies used by monks. Finally, there is a treatise devoted to a discussion of whether a chicken is hotter than a smaller bird.

ROGER ARNALDEZ

REFERENCES

Ibn Abī Uṣaybiʿa. *ʿUyūn al-anbāʾ fī Ṭabaqāt al-Aṭibbā* (Sources of Information on the Classes of Physicians). Bayrūt: Dār Maktabat al-Hayāh, 1965.

Ibn al-Qifṭī. *Taʾrīkh al-Ḥukamā* (History of Philosophy). Ed. J. Lippert. Leipzig: Dieterichische Verlagsbuchshandlung, 1903.

Leclerc, L. *Histoire de la médecine arabe.* Paris: E. Leroux, 1876. Reprinted Rabat: Ministère des habous et des affaires islamiques, Royaume de Maroc, 1980 and New York: B. Franklin, 1960.

Sarton, George. *Introduction to the History of Science.* Baltimore: Williams & Wilkins, 1927.

Schact, J. "Ibn Buṭlān." In *Encyclopedia of Islam.* Leiden: Brill, 1960+.

Schact, J. and Meyerhof, M. *The Medico-Philosophical Controversy between Ibn Buṭlān of Bagdadh and Ibn Riḍwān of Cairo.* Cairo: Egyptian University Faculty of Arts, 1937.

Usāma ibn Munqidh. *Kitāb al-Iʿibar.* Al-Qāhirah: Maktabat al-Thaqāfah al-Dīnīyah, 1980.

See also: Ḥunayn ibn Isḥāq – Ibn Riḍwān

ABRAHAM IBN EZRA Abraham ibn Ezra was born in Toledo, Spain in 1089. In his youth, he studied all the branches of knowledge that Arabic and Jewish gifted (and well to do) youngsters could master, and was mainly known as a poet. Around 1140, he left Spain and wandered through Italy, southern France, and England. Also, legend says that in his old age he traveled to the Holy Land. During his itinerant life, Ibn Ezra met scores of scholars and wrote a number of works, of which his commentary on the Pentateuch and the Prophets in the most widely known.

He was a real polymath, who wrote on Hebrew philology (*Moznei Leshon ha-Kodesh*), translated several works on grammar from Arabic into Hebrew, and wrote on the calendar, mathematics (*Sefer ha-Mispar*, Book of the Number), and philosophy and ethics (*Yesod Mora* on the meaning of the commandments). He is considered one of the Jewish Neoplatonists, in particular regarding his description of the soul. In his view, intellectual perfection is the only way to enjoy a relationship with the divine Providence. As a scientist, Ibn Ezra (also known by the name of Avenezra, sometimes misspelled Avenaris) is mainly known for his works on astronomy (*Sefer ha-ʿIbbur*: *Ta'amei ha-Luḥot*, Book on Intercalation) and for his treatise on mathematics mentioned above. He also composed a number of brief astrological works, most of them still unpublished. It is not known whether Ibn Ezra ever practiced medicine. He certainly showed in his biblical commentary a fair degree of knowledge in medicine and biology.

It has been said that Ibn Ezra wrote over one hundred works, which seems rather exaggerated; certainly many fewer have survived. Ibn Ezra was the Paracelsian type of scholar, learning from each new experience, from each encounter with other scholars, living a simple life and despising wealth. It is particularly striking that he wrote only in Hebrew, contrary to nearly all his contemporaries in Spain who wrote in Arabic. This is mainly due to the fact that he wandered throughout Europe and North Africa, using the language that was common to all his coreligionists. He may be considered an ambassador of Spanish scholarship to the Jewish Diaspora at large. He died in 1164.

SAMUEL S. KOTTEK

REFERENCES

Akabia, A.A. "Sefer ha-'Ibbur le-Rabbi Abraham ibn Ezra; Biurim, he'arot vehagahot." (The Book of the Leap Year; Explanations, Remarks, and Proofs.) *Tarbiz* 26 (3): 304–316, 1957.

Ben Menaḥem, N. "Iyyunim be-mishnat Rabbi Abraham ibn Ezra." (Research in the Teachings of Rabbi Abraham ibn Ezra) *Tarbiz* 27(4): 508–520, 1958.

Heller-Wilensky, S.A. "Li-sheelat meḥabero shel Sefer Sha'ar Hashamayim hameyuḥas le-Abraham ibn Ezra." (To the Question of the Author of the Book Sha'ar Hashamayim Attributed to Abraham ibn Ezra) *Tarbiz*: 32(3): 277–295, 1963.

Leibowitz, Joshua O., and S. Marcus. *Sefer Hanisyonot, The Book of Medical Experiences, attributed to Abraham ibn Ezra*. Jerusalem: Magnes, 1984.

Levy, Raphael. *The Astrological Works of Abraham ibn Ezra*. Baltimore: Johns Hopkins University Press, 1927.

Millás-Vallicrosa, José M. "Avodato shel Rabbi Abraham ibn Ezra beḥokhmat ha-tekhunah." (The Work of Rabbi Abraham ibn Ezra in Astronomical Science) *Tarbiz* 9: 306–322, 1938

Sarton, George. *Introduction to the History of Science*. Baltimore: Williams & Wilkins, 1931, Vol. II, Part 1, pp. 187–189.

Steinschneider, M. "Abraham ibn Ezra." In *Zur Geschichte der mathematischen Wissenschaften im XII. Jahrhundert*. Supplement to *Zeitschrift für Mathematik & Physik* 25: 59–128, 1880.

IBN ḤAWQAL Ibn Ḥawqal al-Nasibī, Abū'l-Qāsim Muḥammad ibn ʿAlī was born in Nisibis (now Nusaybin, Turkey) in the second half of the tenth century. He worked as a merchant and traveled, beginning in AD 943, through the Muslim world, visiting the Maghreb and Andalusia between 947 and 951; Egypt, Armenia, and Azerbaijan around AD 955; Iraq, Persia, Transoxiana, and Khwarazm between 961 and 969. In AD 973, he was in Sicily.

Ibn Ḥawqal is the author of a book on geography entitled *Kitāb al-masālik wa'l-mamālik* (Book on the Routes and Kingdoms), also known as *Kitāb ṣūrat al-ʿarḍ*, which belongs to the category of the so-called *Atlas of Islam*. It consists of a description of the Islamic countries, although some non-Islamic regions of Sudan, Turkey, Nubia, and Sicily are also described.

Ibn Ḥawqal based his work on al-Iṣṭakhrī's book and incorporated new material from his travels which led to three succesive revisions of his *Kitāb al-masālik*: the first one in AD 967, dedicated to Sayf al-Dawla; the second ca. AD 977, and the third ca. 988. The final result was a book whose descriptive part surpassed the works of earlier authors.

From the contents of his work, his sympathy with the Fāṭimid movement can be deduced. He showed a certain interest towards Fāṭimid politics, although he cannot be considered a Fāṭimid *dāʿī* (propagandist). He also gives economic information. His interests are focused not on rare or precious goods but on basic agricultural and artisanal products.

Ibn Ḥawqal's *Kitāb al-masālik* influenced the work of later geographers such as Abū'l-Fidā. He is also the author of a book on Sicily which is not preserved.

EMILIA CALVO

REFERENCES

Ibn Ḥawqal. *Kitāb al-masālik wa'l-mamālik*. Ed. M.J. de Goeje in "Bibliotheca Geographorum Artabicorum" vol. II. Leiden: Brill, 1873.

Ibn Ḥawqal. *Kitāb ṣūrat al-ʿarḍ*. Ed. J.H. Kramers in "Bibliotheca Geographorum Artabicorum" vol. II, 2nd ed. Leiden: Brill, 1938. Reprinted 1967.

Miquel, André. *La géographie humaine du monde musulmane jusqu'au milieu du XIe siècle*. Paris: Mouton, 1967.

Miquel, André. "Ibn Ḥawkal." In *Encyclopédie de l'Islam*, 2nd ed., vol. III. Leiden: E.J. Brill; Paris: G.P. Maisonneuve, 1971, p. 810–811.

Vernet, Juan. "Ibn Ḥawqal." In *Dictionary of Scientific Biography*, vol. VI. New York: Charles Scribner's Sons, 1972, p. 186.

Wiet, G. *Configuration de la Terre*. Paris and Beirut: Commission Internationale pour la Traduction des Chefs d'oeuvres, 1964.

See also: Balkhī School – Geography

IBN HUBAL Muhadhdhib al-Dīn Abū'l-Ḥasan ʿAlī ibn Aḥmad ibn ʿAli ibn Hubal al-Baghdādī was a famous physician, medical authority, and accomplished poet. Born in Baghdad in 1121 (AH 515), he migrated to Khilat (modern

Ahlat, on the shore of Lake Van, Turkey), and became very prosperous in the service of the local ruler. He later moved to Mardin to serve another lord, and died at Mosul (in modern Iraq) in 1213 (AH 610). His chief work, *Kitāb al-Mukhtār fī al-Ṭibb* (The Choice Book of Medicine) was written in about 1165. It resembles the medical encyclopedias of Ibn Sīnā and al-Rāzī in that it is a compendium of Galenic medical knowledge, supplemented by personal clinical practice. *The Choice Book* is divided into three main parts, comprising anatomy and general principles, a pharmacopoeia, and a list of maladies arranged according to the affected organs, running from head to foot. Although this large book does not seem to have been translated into Latin during the Middle Ages, the number, diffusion, and varying ages of the manuscript copies attest to its popularity. No copy appears to survive of Ibn Hubal's *Kitāb al-Ṭibb al-Jamāli* (Book of Medicine for Jamal al-Din al-Wazir). A short work on logic, *al-Ārā' wa'l-mushāwarāt*, in manuscript in Paris (B.N. MS 2348) is ascribed to him.

The manuscripts of *The Choice Book* in Leiden, Paris, Cairo, and India are listed in Brockelmann; there are also several copies in Turkey, and fragmentary ones in Princeton and the British Library in London (the London text starts with the diseases of the brain). The whole Arabic text was edited in Hyderabad 1943–44, but Albert Dietrich maintains that a proper critical edition is still badly needed. Two chapters of Ibn Hubal's medical encyclopedia were published from the Leyden manuscript with an accompanying French translation; they describe the causes, symptoms, and treatment of stones in the kidneys and bladder. Ibn Hubal employs the usual medieval medical terminology derived from earlier Greek physicians: four fluids, or humors, within the body (blood, phlegm, choler [yellow bile] and melancholy [black bile]) are held to constitute in their balance and proportions the essential foundations of good health; pain and disease reveal an evil condition or imbalance among the fluids. Most remedies and treatments consist of trying to correct the malfunctioning fluid through diet, bleeding, or alteration of the patient's physical surroundings. In spite of what we might today consider the erroneous basis of his theoretical approach, Ibn Hubal shows an impressive concern for personal observations and clearly relies on extensive clinical practice. In discussing stones, for instance, he attributes the condition to excessive bodily heat which causes phlegm, a "thick fluid", to form deposits in the kidneys and bladder. He prescribes medicines such as horseradish, ginger, and chicken soup to break up the stones or cause them to be passed. He also cites his own personal experience of a more desperate remedy which he witnessed: a surgical operation to remove a bladder stone from a boy. He cites al-Rāzī and Rufus of Ephesus, but conveys the impression of a knowledgeable practical physician, mindful of the agonizing pain caused by the condition he is discussing.

E. RUTH HARVEY

REFERENCES

Dietrich, Albert. *Medicinalia Arabica.* Göttingen: Vandenhoeck u. Reprecht, 1966.

Ibn Abī Uṣaybi'ah. *'Uyūn al-Anbā fī ṭabaqāt al-aṭibbā'*. Beirut: Dār Maktabat al-Ḥayāh, 1965 pp. 471–2.

Ibn Hubal. *Traité sur le calcul dans les reins et dans la vessie.* Ed. P. De Koning. Leiden: E.J. Brill, 1896.

Ibn Hubal. *al-Mukhtār fī al-Ṭibb*. 4 vols. Hyderabad: 1943–44.

Vernet, J. "Ibn Hubal." In *The Encyclopaedia of Islam,* new edition. Leiden: Brill, 1968–1971, vol. III p. 802.

IBN ISḤĀQ AL-TŪNISĪ Abū-l-'Abbās Aḥmad ibn 'Alī ibn Isḥāq al-Tamīmī al-Tūnisī was a Tunisian astronomer of the early thirteenth century. He compiled an impressive astronomical handbook with tables (*zīj*) a manuscript of which (copied ca. 1400), has been recently discovered by David A. King. It contains an important set of tables (completed ca. 1218) which mark the starting point of a Maghribian astronomical school. Parts of these tables seem original and are based, according to the famous historian Ibn Khaldūn, on observations made by a Sicilian Jew. The rest is a miscellaneous collection of materials which derive from Andalusian sources, many of which seem lost, by Ibn al-Zarqāllu (d. 1100), Ibn Mu'ādh al-Jayyānī (d. 1093), Ibn al-Kammād (fl. ca. 1125), Ibn al-Hā'im (fl. ca. 1204) as well as others. The canons (instructions for the use of the numerical tables) were not written by Ibn Isḥāq but by some later author who also used materials derived from the aforementioned Andalusian *zīj*es. Ibn Isḥāq's *zīj* is, therefore, a first rate new source for the study of both Andalusian and Maghribian astronomy.

JULIO SAMSÓ

REFERENCES

Kennedy, E.S. and David A. King. "Indian Astronomy in Fourteenth Century Fez: the Versified Zīj of al-Qusunṭīnī." *Journal for the History of Arabic Science* 6: 3–45, 1982. Reprinted in King, D.A. *Islamic Mathematical Astronomy.* London: Variorum Reprints, 1986.

King, David A. "An Overview of the Sources for the History of Astronomy in the Medieval Maghrib." *Deuxième Colloque Maghrebin sur l'Histoire des Mathématiques Arabes.* Tunis: Université de Tunis, 1988, pp. 125–157.

Samsó, Julio and Honorino Mielgo. "Ibn Isḥāq al-Tūnisī and Ibn Mu'ādh al-Jayyānī on the Qibla." In *Islamic Astronomy and*

Medieval Spain. Ed. J. Samsó. Aldershot: Variorum Reprints, 1994, no. VI

Samsó, Julio and Eduardo Millás. "Ibn al-Bannā᾿, Ibn Isḥāq and Ibn al-Zarqālluh's Solar Theory." In *Islamic Astronomy and Medieval Spain*. Ed. J. Samsó. Aldershot: Variorum Reprints, 1994, no. X.

IBN JULJUL Ibn Juljul al-Andalusī, Sulaymān ibn Ḥasan was born in Córdoba in AD 943 and died ca. 994. He studied medicine with a group of Hellenists presided over by Ḥasdāy ibn Shaprūṭ, a Jewish physician and vizier of the Caliph ʿAbd al-Raḥmān III, and later became the personal physician of Caliph Hishām II (976–1009).

Ibn Juljul is the author of *Ṭabaqāt al-aṭibbā᾿ wa'l-ḥukamā᾿* (Generations of Physicians and Wise Men), the oldest extant summary in Arabic on the history of medicine (it was finished in AD 987), after Isḥāq ibn Ḥunayn's *Ta᾿rīj al-aṭibbā᾿* (History of the Physicians). It contains fifty-seven biographies grouped into nine generations. Thirty-one of them concern Asian authors, and the rest refer to African and Andalusian scholars.

Ibn Juljul used Eastern sources (Hippocrates, Galen, Dioscorides) and Western ones (Orosius, Isidore) and established the chronological limits of the Latin influence on medicine in Andalusia. The work has chronological errors but provides interesting information about the oldest translations into Arabic, in the time of the Caliph ʿUmar II (AD 717–719).

Other works of Ibn Juljul are *Tafsīr asmā᾿ al-adwiya al-mufrada min kitāb Diyusqūridūs* (Explanation of the Names of the Simple Drugs from Dioscorides' Book), written in 982, from which only a fragment is preserved, containing the transcription of the Greek names of 317 simple medicines, their translation into Arabic and their identification; *Maqāla fī dhikr al-adwiya al-mufrada lam yadhkurha Diyusqūridūs* (Treatise on the Simples not Mentioned by Dioscorides), which includes sixty-two simple medicines not mentioned in Dioscorides' *Materia Medica; Maqāla fī adwiyat al-tiryāq* describing the components of the theriaca; and *Risālat al-tabyīn fī-mā ghalaṭa fīhi baʿḍ al-mutaṭabbibīn* (Treatise on the Explanation of the Errors of some Physicians).

EMILIA CALVO

REFERENCES

Dietrich, A. "Ibn Djuldjul." In *Encyclopédie de l'Islam*, 2nd ed., vol. III. Leiden: E.J. Brill; Paris: G.P. Maisonneuve, 1971, p. 778–779.

Garijo, Ildefonso. "El tratado de Ibn Ŷulŷul sobre los medicamentos que no mencionó Dioscórides." In *Ciencias de la Naturaleza en Andalusia. Textos y Estudios I*. Ed. Expiración García Sànchez. Granada: Consejo Superior de Investigaciones Científicas, 1990, pp. 57–70.

Ibn Juljul. *Kitāb ṭabaqāt al-aṭibbā᾿ wa'l-ḥukamā᾿*. Ed. F. Sayyid. Cairo: Maṭbaʿat al-Maʿhad al-ʿIlmī al-Faransī lil-Āthār al-Sharqīyah, 1955.

Ibn Juljul. *Ibn Ŷulŷul, tratado octavo*. Edition and Spanish translation by Ildefonso Garijo. Córdoba: Universidad de Córdoba, 1992.

Ibn Juljul. *Ibn Ŷulŷul, tratado sobre los medicamentos de la tríaca*. Edition and Spanish translation by Ildefonso Garijo. Córdoba: Universidad de Córdoba, 1992.

Samsó, Julio. *Las Ciencias de los Antiguos en al-Andalus*. Madrid: Mapfre, 1992.

Vernet, Juan. "Ibn Juljul." In *Dictionary of Scientific Biography*, vol. 7. New York: Charles Scribner's Sons, 1973, pp. 187–188.

Vernet, Juan. "Los médicos andaluces en el "Libro de las generaciones de los médicos" de Ibn Ŷulŷul." *Anuario de Estudios Medievales* 5: 445–462, 1968. Reprinted in *Estudios de Historia de la Ciencia Medieval*. Barcelona: Universidad de Barcelona, 1979, pp. 469–486.

IBN JUMAY᾿ A contemporary of the great Jewish doctor and philosopher Moses Maimonides, Ibn Jumay᾿ was one of the physicians in the service of Ṣalāh ad-Dīn. He was born of a Jewish family in Fustat (Egypt) and studied with another physician of some renown, ʿAdnān ibn al-ʿAynzarbī (d. 548/1153). The relevant biographical dictionaries mention Ibn Jumay᾿'s talents in medicine as well as his highly developed linguistic consciousness, inducing him always to carry al-Jawharī's *Kitāb aṣ-ṣaḥāḥ* (The Truthful Guide) to class so he could check words of which he was uncertain. Ibn Jumay᾿ died in 594/1198.

In Ibn Abī Uṣaybiʿa's dictionary of medical doctors, *ʿUyūn al-anbā᾿ fī ṭabaqāt al-aṭibbā᾿* (Sources of Information about the Classes of Physicians) Ibn Jumay᾿ is presented as the author of eight works on medical or medicine-related subjects, the most important of them being *Kitāb al-irshād li-maṣāliḥ an-nufūs wa'l-ajsād* (Guide to the Welfare of Souls and Bodies), a compendium of the different fields of the art of medicine. Others also deal with practical questions of the doctor's craft such as first aid or nutritive advice.

The only one of Ibn Jumay᾿'s works published to date, *al-Maqāla aṣ-ṣalāḥīya fī iḥyā᾿ aṣ-ṣinā᾿a aṭ-ṭibbīya* (Treatise to Ṣalāḥ ad-Dīn/Saladin on the Revival of the Art of Medicine), is not mentioned by Ibn Abī Uṣaybiʿa. It is a deontological work, that is, it deals with the doctor's profession on a more theoretical level.

This treatise, as Ibn Jumay᾿ mentions in the introduction, owes its composition to a conversation he had with his sovereign on the deplorable state of medicine in his time, the reasons for this, and ways to ameliorate the situation.

Thus, formally the work stands in the literary tradition of the epistle, a genre frequently employed by Ibn Jumay› in other works and by Arabic medical authors in general. Its contents — the complaints about the declining state of the art and considerations about its improvement — were not unknown in his time either. The theme goes back to Galen or even Hippocrates.

The treatise falls into three chapters. The first concerns the presentation of medicine, including the qualities of medicine and the need for it, as well as the difficulties of medicine and their consequences. Chapter two deals with the reasons for the decline of medicine, including a brief presentation of its history, and chapter three suggests ways to revive the art of medicine.

Whereas compendia of a more technical presentation of medicine in medieval Arabic literature are comparatively numerous, the same cannot be said about this kind of work, with its introductory and deontological character. In that lies the importance of this doctor of the twelfth century AD, Ibn Jumay›.

HARTMUT FÄHNDRICH

REFERENCES

Fenton, Paul B. "The State of Arabic Medicine at the Time of Maimonides According to Ibn Ĝumay›'s Treatise on the Revival of the Art of Medicine." In *Moses Maimonides – Physician, Scientist, and Philosopher*. Ed. Fred Rosner and Samuel S. Kottek. Northvale, New Jersey and London: Jason Aronson Inc., 1993, pp. 215–229, 270f.

Meyerhof, Max. "Sultan Saladin's Physician on the Transmission of Greek Medicine to the Arabs." *Bulletin of the History of Medicine* 18: 169–178, 1945.

Treatise to Ṣalāḥ ad-Dīn on the Revival of the Art of Medicine by Ibn Jumay›. Ed. and trans. Hartmut Fähndrich. Wiesbaden: Kommissionsverlag Franz Steiner, 1983.

Ullmann, Manfred. *Die Medizin im Islam*. Leiden: Brill, 1970.

See also: Moses Maimonides – Medicine in Islam

IBN KHALDŪN ʿAbd al-Raḥmān ibn Khaldūn (1332/732–1406/808) spent the first two-thirds of his life in North Africa and Muslim Spain, fleeing in 1382/784 to Egypt, where he remained until his death. Though he is best known for the lengthy Introduction (*Muqaddima*) to his massive philosophical history of civilization (*Kitāb al-ʿIbar*), Ibn Khaldūn spent much of his life in political activities. Born and raised in Tunis, he read the *Qur'ān* and studied the religious sciences as well as Arabic and poetry, then was educated in logic, mathematics, natural science, and metaphysics. He also received specialized training in court correspondence and administrative matters, subjects that allowed him to become a court secretary to the Marinid ruler Abū ʿInān in Fez at about the age of twenty-two.

After some vicissitudes, including almost two years of prison, Ibn Khaldūn went to Grenada in 1362 to become an advisor and tutor to Muḥammad V. That position lasted only about two years, no longer than his subseqent position as prime minister or *ḥājib* to Prince Abū ʿAbd Allāh of Bougie. Following these forays into practical politics, Ibn Khaldūn endured several years of upheaval (1366/766–1375/776), settled for about four years in Qalʿat Ibn Salāma near Oran and began work on his history, then moved to Tunis under the patronage of Abū al-ʿAbbās in order to have access to documents and libraries. After a few years there, court intrigues led him to seek tranquility in Egypt.

During the next quarter of a century he served the Mamlūk Sulṭān Barqūq as judge (*qāḍī*) and chief judge (qāḍī al-quḍā), professor at various universities (including the prestigious al-Azhar), and one time university president. A few years before his death he met with the famous Mongol chieftain Tamerlane. But the period in Egypt was, above all, a time for revising his *Kitāb al-ʿIbar* and working on the Introduction (*Muqaddima*) to it.

The *Kitāb al-ʿIbar* is a multi-volume effort that, in his words, sets forth "the record of the beginning and the suite of the days of the Arabs, Persians, Berbers, and the most powerful of their contemporaries". Its Introduction consists of six very long chapters that explore the character of human civilization in general and Bedouin civilization in particular, as well as the basic kinds of political associations, and then the characteristics of settled civilization, the arts and crafts by which humans gain their livelihoods, and, finally, the different human sciences. Ibn Khaldūn starts by explaining the merit of history and how to go about writing it. Properly speaking, the reason to write history or the "inner meaning of history" is, by means of reflection, to get "at the truth, subtle explanation of the causes and origins of existing things, and deep knowledge of the how and why of events". Though his enterprise is therefore "rooted in philosophy" and to be considered a branch of it, Ibn Khaldūn acknowledges a problem with the way history has come down. Many unqualified people have trammeled with the books of history written by competent Muslim historians; they have introduced tales of gossip imagined by themselves as well as false reports. Moreover, other historians have compiled partial reports of particular dynasties and events without looking to the way things have changed over time, without looking at natural conditions and human customs. Consequently, Ibn Khaldūn considers his task to be that of showing the merit of writing history, investigating the various ways it has been done, and showing the errors of previous historians. What needs to be known, and thus what he sets out to make known, are "the principles of politics, the nature of existent things, and the differences among nations, places and periods with regard to ways of life, character, qualities, customs, sects, schools, and everything else . . . plus a comprehensive knowledge of present conditions in all these respects . . . complete knowledge of the reasons for every happening and . . . [acquaintance] with the origin of every event". Yet in the end Ibn Khaldūn hints that he has almost digressed in the whole undertaking. What he wanted to do was to explain the nature of civilization and its accompanying accidents, but he fears he has strayed from his basic point.

CHARLES E. BUTTERWORTH

REFERENCES

Ibn Khaldūn. *The Muqaddimah, An Introduction to History*. Trans. Franz Rosenthal. New York: Pantheon Books, 1958.

Mahdi, Muhsin. *Ibn Khaldūn's Philosophy of History*. London: George Allen and Unwin, 1957.

IBN KHURDĀDHBIH Abu'l-Qāsim ʿUbayd Allāh ibn Khurdādhbih (also spelled: Ibn Khurradādhbih), the first scholar to write on world geography in Arabic, was born in ca. AH 205/AD 820 (or AH 211/AD 825), and died in ca. AH 300/AD 912. Probably born in Khurāsān, he was brought up in Baghdad. His grandfather, Khurdādhbih, was a Zoroastrian, latter converted to Islam, and his father was the governor of Ṭabaristān. When he grew up, he became the director of posts and information in Jibāl (Media) and subsequently became director-general of the same department in Baghdad and later in Sāmarrā (Iraq). He became a companion of the ʿAbbāsid Caliph al-Muʿtamid (AH 256–279/AD 870–892).

Ibn Khurdādhbih was a versatile writer; besides writing on geography, he wrote on history, genealogy, music, wines, and even on the culinary art. Al-Nadīm, in his *al-Fihrist*, lists at least eight works to his credit. The Arab historian Abu'l-Ḥasan ʿAlī ibn al-Ḥusayn al-Masʿūdī (d. AD 956) considered him an *imām* (leader) in authorship and mentions his voluminous historical work dealing with the ancient kings and peoples of Iran (*Murūj*, 1965). Ibn Khurdādhbih also claimed to have translated into Arabic the geographical treatise of Claudius Ptolemy (ca. AD 90–168) from a "foreign language" (probably Syriac or Greek), but the translation is not extant.

However, Ibn Khurdādhbih's major work on geography, entitled *Al-Masālik al-Mamālik* (Roads and Kingdoms), was published by M.J. De Goeje in 1889. In fact, this work is an abridgement (prepared not later than AD 885–886) of his larger work (not extant) written in ca. AD 846–847. Considering the vast amount of information contained in the work and the early date of its compilation, it may be said that Ibn Khurdādhbih was the father of Arab-Islamic geography; no work of such magnitude existed before him. *Al-Masālik al-Mamālik* deals briefly with mathematical and physical geography, but the major portion of the work is devoted to descriptions of land and sea routes in the four directions emerging from al-Sawād. Then, it deals with marvels of the world, seas and mountains, sources of the rivers, and reports on countries like India and Central Asia.

It is not unlikely that Ibn Khurdādhbih relied heavily for his information on the ancient Sassanian government records which must have become available to him as the person in charge of the department of posts and information in Baghdad and elsewhere. Again, in his methodology and arrangement of the material and in the use of geographical terms and Persian couplets, a distinct Persian influence is discernible. The ancient Persians used to divide the known world into seven circular regions called *kishvars* (kingdoms) with Irānshahr at the center and the remaining six circles drawn around it. Such an arrangement is observable in his descriptions of the various routes emerging from al-Sawād, which, he says, was called *dil-i Irānshahr* (the heart of Iraq).

Ibn Khurdādhbih was not only the first to write on geography in Arabic; he also set the style for writing on descriptive geography. Several later Arab-Islamic geographers utilized his work as a major source of information.

SAYYID MAQBUL AHMAD

REFERENCES

Ahmad, S. Maqbul. *Arabic Classical Accounts of India and China*. Shimla: Indian Institute of Advanced Study, 1989.

Ahmad, S. Maqbul. "Djughrāfiyā." In *Encyclopedia of Islam*. Leiden: Brill, 1960.

Al-Masʿūdī. *Murūj al-Dhahab wa Maʿādin al-Jawhar*. (Meadows of Gold and Mines of Gems). Trans. Aloys Sprenger. London: Allen, 1841.

Ibn Khurdādhbih. *Al-Masālik al-Mamālik*. Ed. M.J. De Goeje. Leiden: E.J. Brill, 1889.

Krachkovsky, I.I. *Istoria Arabskoi Geograficeskoi Literatury*, Moscow and Leningrad 1957; Arabic translation: *Tārīkh al-adab al-jughrāfī al-ʿArabī* by Ṣālāḥ al-Dīn ʿUthman Hāshim, vol. I. Beirut: Dār al-Gharb al-Lubnān, 1987.

See also: Geography in Islam – Navigation

IBN MĀJID Shihāb al-Dīn Aḥmad ibn Mājid ibn Muḥammad ibn ʿAmr ibn Faḍl ibn Duwayk ibn Yūsuf ibn Ḥasan ibn Ḥusayn ibn Abī Maʿlaq al-Saʿdī ibn Abu'l-Rakāʾib al-Najdī was the greatest Arab navigator of the fifteenth century AD and one of the greatest of the Middle Ages. We do not know the date of his birth or death, but he must have died at an advanced age sometime in the first decade of the sixteenth century. Born in Julfār (Oman), he belonged to an illustrious family of navigators. Both his father and grandfather were *muʿallims* (masters of navigation) of repute.

Ibn Mājid wrote a number of works, both in prose and poetry, on nautical theory and on describing the seas (mainly the Indian Ocean) which served as guides for the Arab navigators of later periods. Among his important works in prose is the *Kitāb al-fawāʾid fī ʿuṣūl ʿilm al-baḥr wa'l-qawāʿid* (The Book of Benefits on the Principles of the Science of Navigation), dated AH 895/AD 1489–90. This book and many others have been reproduced in the editions mentioned in the bibliography.

Ibn Mājid considered himself the fourth of the great Arab navigators of the Middle Ages; the other three were Muḥammad ibn Shādān, Sahl ibn Abān, and Layṯ ibn Kahlān, who belonged to the Abbāsid period. He thought his own works were more accurate than theirs, since they were more current and since many of the ports mentioned in the older works no longer existed. Apart from his practical experiences as a navigator, he had studied and improved upon the work of his father (*al-Ḥijāzīya*) and had studied a number of earlier Arabic works on astronomy and geography.

On the practical side of navigation, it is not unlikely that Ibn Mājid was in contact with the Indian navigators of his time, the *Ṣūliyān* (Cholas of Tamil Nadu, India) and with the Gujarati and the Konkani (Maharashtra, India) navigators, whose *qiyāsāt* (readings of the ports and harbors) he seems to have known. He was particularly knowledgeable about Siam and Bengal, and these navigators frequented these regions more than the Arab navigators did.

In his works Ibn Mājid covered a number of subjects relating to navigation, nautical astronomy, oceanography, and geography. In his *Kitāb al-fawāʾid*, he pays special attention to subjects of a more general nature, like guidelines to navigators such as the prerequisites for sailing on the sea, and he describes the lunar mansions, the stars corresponding to the thirty-two divisions (*aqnān*) of the compass card, the winds and seasons of the seas, nautical instruments, and the essentials required by captains of boats. These include knowledge of the rising and corresponding settings of the stars (*al-anwāʾ*), latitudes and longitudes, landfalls, and tides. He emphasizes that before sailing, captains should see that their instruments are in perfect order, the sailors obedient, and the seasons suitable. They should be patient and soft-spoken, should not deprive merchants of their rights, and they should be courageous, literate, and well-behaved. He also presents a systematic description of the sea coasts of the Oikumene (the known world at that time), which no geographer had done before. From his writings, it appears that he did not conceive that a *terra incognita* existed in the southern quarter of the earth, as other geographers of his time believed. He thought that the Indian Ocean was connected with the Atlantic through a sea channel, which he calls *al-madqal* (place of entry).

Ibn Mājid claimed to have made several contributions to navigation and to determining the direction of the *qibla* (Mecca) from different positions of the earth with the help of the compass card. He also claimed to have fixed a magnetized needle on the mariners' compass (probably on a day box). Ibn Mājid had met Vasco da Gama, and had guided him to Calicut, India. Although the Portuguese sources do not mention him by name in this regard, we know from an Arabic work, *al-Barq al-Yamānī fi'l-fatḥ al-ʿUṯmānī* (The Yemenite Lightning on the Ottoman Conquest) by al-Nahrawālī that it was he who directed Vasco da Gama from Malindi (East Africa) to India.

SAYYID MAQBUL AHMAD

REFERENCES

Ahmad, S. Maqbul. "Ibn Mājid." In *Encyclopaedia of Islam.* Leiden: E.J. Brill, 1960, pp. 856–869.

Ahmad, S. Maqbul. "Ibn Mājid." In *A History of Arab-Islamic Geography 9th–16th Century* AD. Ed. Sayyid Maqbul Ahmad. Mafraq, Jordan, Al al-Buyt University, 1995, in press.

Ferrand, Gabriel. *Instructions nautiques et routiers arabe et portugais des XVe et XVIe siècles.* Paris: Geuthner, 1921.

Khūrī, Ibrāhīm, and ʿIzzat Ḥasan. *Kitāb al-fawāʾid fī ʿuṣūl ʿilm al-baḥr wa'l-qawāʿid.* Damascus: al-Maṭbaʿah al-Taʿāwunīyah, 1390/1971.

Shumovsky, T.A. *Ṯalāṯa rahmānajāt al-majhūla li Aḥmad b. Mājid.* Moscow and Leningrad: Nauka, 1957.

Tibbetts, Gerald R. *Arab Navigation in the Indian Ocean Before the Coming of the Portuguese; being a translation of Kitāb al-Fawāʾid fī sūl al-baḥr wa'l-qawāʾid of Aḥmad b. Mājid al-Najdī.* London: Royal Asiatic Society of Great Britain and Ireland, 1971.

See also: Navigation – *Qibla* – Lunar Mansions – Compass

IBN MĀSAWAYH, YŪḤANNĀ Abū Zakariyyāʾ Yūḥannā ibn Māsawayh was born in Baghdad during the caliphate of Hārūn ar-Rashid (786–809), and not in 777 as was stated by

Leo the African. His father, Māsawayh, was a pharmacist in the service of the physician Jibrāʾīl ibn Baḫtishūʿ, with whom he came from Jundīshāpūr (in Persia) to Baghdad. His mother was a slave, named Risala, whom Māsawayh bought from the physician Dawūd ibn Sarābiyūn. Thus, Yūḥannā ibn Māsawayh belonged to the milieu of Christian Nestorian physicians, who played an important part during the eighth and ninth centuries. He married the daughter of his colleague Abdallāh at-Tayfūrī and had a son of poor intelligence. He was very famous as a teacher and practitioner; he became the personal physician of four successive caliphs from al-Maʾmūn to al-Mutawakkil. He died in Samarra in 857.

As for many other authors of this period, it is difficult to distinguish legend and history in his biography. It was said that he did translations from Greek into Arabic, but none is extant under his name; most probably, he only commissioned some of them. For instance, it is well attested that Ḥunayn ibn Isḥāq undertook the translation of Galen's *Methodus medendi* (Arabic *Kitāb hīlat al-burʾ*; Methods of Healing) at Ibn Māsawayh's request. The relationship between both physicians was nevertheless strained, at least at the beginning: Ibn Māsawayh is supposed to have driven Ḥunayn out of his teaching position, who then went traveling in order to learn Greek and purchase manuscripts. It was also reported by Arabic medieval historians that Ibn Māsawayh had the opportunity of dissecting an ape, which had been given to the caliph by the prince of Nubia in 836 as a present. Following in Galen's footsteps Ibn Māsawayh afterwards wrote an anatomical monograph, which can perhaps be identified with his *Kitāb at-tashrīḥ* (Book of Anatomy). Apart from his knowledge of Greek medicine, Ibn Māsawayh had access to some Indian works: he quotes, for instance, Āryabhaṭa in his ophthalmological treatise *Kitāb dafal al-ʿain* (Book on the Defectiveness of the Eye).

Ibn Abī Uṣaybiʿa listed 42 works; some others were quoted by al-Rāzī and al-Bīrūnī. But only 31 are extant, as far as we know, and very few have been edited or studied. Those that have been edited — and sometimes translated into English or French — are the *Aphorisms*, the works on barley-water, simple drugs, and perfumes, as well as the medical calendar; the ophthalmological treatises have also been analysed. It seems that among the works which were translated into Latin during the Middle Ages under the names of "Mesuë" or "Johannes Damascenus", only the *Aphorismi Johannis Damasceni* and some ophthalmological fragments can be attributed to Ibn Māsawayh; the other ones, mainly pharmacological, are probably apocryphal for the most part. The *Aphorismi* — a faithful translation of *Nawādir aṭ-ṭibb* — were largely diffused from the twelfth century; a second translation appeared as the sixth book of Rāzī's *Secrets of Medicine:* it was done during the thirteenth century by the Dominican Giles of Santarem.

Some original features of Ibn Māsawayh's medicine can be drawn from his *Nawādir aṭ-ṭibb.* Dedicated to Ḥunayn ibn Isḥāq and modeled on the Hippocratic *Aphorisms*, this short work was intended to give practical advice. The eight first aphorisms can be considered as a kind of commentary on the first Hippocratic aphorism: "Life is short, art long, opportunity fleeting, experiment dangerous, judgment difficult". They stressed the necessity for physicians to be both learned and skilled. Throughout the remaining 124 aphorisms the following topics are covered: the link between body and soul, the observance of astrological and climatological rules, the attention that physicians have to pay to the healthy nature of their patients, the numerical ratios which rule human temperaments, as well as natural substances used for treatment. It has to be noted that this last idea was deeply developed by Ibn Māsawayh's contemporary, al-Kindī. Ibn Māsawayh seems also to have been very much attached to the idea that physicians must mainly reinforce nature by using drugs similar to it; medical treatment with substances contrary to disease had to be prescribed cautiously and their sole goal was to purge. Physicians had to be cautious not to alter nature too much. For example, Ibn Māsawayh stated: "It is important that, against diseases, the strongest contrary is not introduced into the body, since this would be very harmful; it can be compared with a very cold wind which, during the same day, blows after a very hot one" (aph. 60). In the same manner, treatment by diet was preferred to pharmacopoeia: "If the physician can treat with food, to the exclusion of drugs, he will be very successful" (aph. 108). In addition to several pharmacological treatises, Ibn Māsawayh composed an important work on dietetics.

Ibn Māsawayh's works remain very little known, despite their importance in the history of Arabic medicine. Detailed studies would shed light on the decisive stage constituted by the beginning of the ninth century, before the spread of Ḥunayn's translations.

DANIELLE JACQUART

REFERENCES

Ibn Māsawayh, Yūḥannā. *Le livre des axiomes médicaux*. Ed. Danielle Jacquart, Danielle et Gérard Troupeau. Genève: Droz; Paris: Champion, 1980.

Levey, Martin. "Ibn Māsawaih and his Treatise on Simple Aromatic Substances." *Journal of the History of Medicine* 16: 394–410, 1961.

Prüfer, Carl and Meyerhof, Max. "Die Augenheilkunde des Jūḥannā b. Māsawaih." *Der Islam* 6: 217–256, 1916.

Sbath, Paul. "Le livre des temps d' Ibn Massawaih." *Bulletin de l'Institut d'Egypte* 15: 235–257, 1933.

Sbath, Paul. "Traité sur les substances simples aromatiques par Yohanna ben Massawaih." *Bulletin de l'Institut d'Egypte* 19: 5–27, 1937.

Sbath, Paul. "Le livre sur l'eau d'orge de Youhanna ben Massawaih." *Bulletin de l'Institut d'Egypte* 21: 13–24, 1939.

Sezgin, Fuat. *Geschichte des arabischen Schrifttums* III. Leiden: Brill, 1970, pp. 231–236.

Sournia, Jean-Charles and Troupeau, Gérard. "Médecine arabe: biographes critiques de Jean Mésué (VIIIe siècle) et du prétendu 'Mésué le Jeune' (Xe siècle)." *Clio Medica* 3:109–117, 1968.

Ullmann, Manfred. *Die Medizin im Islam*. Leiden: Brill, 1970.

See also: al-Kindī – Leo the African

IBN MUʿĀDH Ibn Muʿādh al-Jayyānī, Abū ʿAbd Allāh Muḥammad (d. ca. 1093) has traditionally been assigned a birthdate of 989. However, recent scholarship suggests that Ibn Muʿādh was born somewhat later, in the early eleventh century. The only secure date we have for him is 1079, the year of a solar eclipse he describes from first-hand observation. The ending "al-Jayyānī" to his name indicates that he was from Jaén in Andalusia, where he evidently served as a *qāḍī* (judge) for much of his life. Among his few surviving astronomical and mathematical works are included the *Tabulae Jahen,* a set of astronomical tables based on those of al-Khwārizmī and probably translated into Latin by Gerard of Cremona; *Maqāla fi sharḥ al nisba* (On Ratio), a commentary on Book 5 of Euclid's *Elements*; and *Kitāb majhūlāt qisiyy al-kura* (Determinations of the Magnitudes of the Arcs on the Surface of a Sphere), a work on trigonometry. Certainly the most original, and perhaps the most historically significant of his extant works, is the brief treatise *On Twilight and the Rising of Clouds*.

While no Arabic exemplar has yet come to light (the original title was probably *Ma 'l-fajr wa 'l-shafaq*), this work has none the less reached us in three other linguistic forms: a fourteenth-century Hebrew translation from the Arabic (represented by one manuscript), a late twelfth-century Latin translation from the Arabic, probably by Gerard of Cremona (represented by 25 manuscripts), and a fourteenth-century Italian translation from the Latin (represented by one manuscript). The sheer number of Latin manuscripts — plus the Italian translation — indicates the seriousness with which this treatise was received and disseminated in the medieval West. That it was commonly misattributed to Ibn al-Haytham, author of the magisterial *Kitāb al-manāẓir* (*De aspectibus*), may have had something to do with this. There is also clear, albeit indirect, evidence that *On Twilight* exerted its share of influence in the medieval East as well.

Ibn Muʿādh's purpose in *On Twilight* is to determine the height of the atmosphere under the assumption that the first light of dawn is produced when rays from the rising sun tinge vapors at the very upper edge of the atmosphere. Although he offers no practical justification for this inquiry, he does, in a couple of querulous asides, berate those (religious conservatives?) who would squelch rational inquiry out of mere ignorance. The determination itself depends upon four basic parameters: the depression of the sun below the horizon at first light (18°); the mean distance between earth and sun (1110 terrestrial radii); the relative size of sun and earth (5.5:1 in terrestrial radii); and the circumference of the earth (24,000 miles). On the basis of these parameters and using simple trigonometric functions, Ibn Muʿādh calculates the atmosphere to be around 52 miles high. This figure remained canonical in the Latin West until the end of the sixteenth century, when the issue of atmospheric refraction was raised to prominence by Tycho Brahe. Within this context, it soon became clear that Ibn Muʿādh's calculation was useless because it failed utterly to take atmospheric refraction into account. Consequently, his figure of 52 miles was drastically reduced by Johan Kepler and succeeding astronomers.

A. MARK SMITH

REFERENCES

Hermelink, H. "Tabulae Jahen." *Archive for History of Exact Sciences* 2: 108–12, 1964.

Plooj, Edward B. *Euclid's Conception of Ratio and his Definition of Proportional Magnitude*. Rotterdam: van Hengel, 1950.

Sabra, A.I. "The Authorship of the *Liber de crepusculis*." *Isis* 58: 77–85, 1967.

Saliba, George. "The Height of the Atmosphere According to Muʿayyad al-Dīn al ʿUrḍī, Quṭb al-Dīn Al Shīrāzī and Ibn Muʿadh." In *From Deferent to Equant*. Ed. David A King and George Saliba. New York: New York Academy of Sciences, 1987, pp. 445–65.

Smith, A. Mark. "The Latin Version of Ibn Muʿādh's Treatise 'On Twilight and the Rising of Clouds'." *Arabic Sciences and Philosophy* 2: 83–132, 1992.

Smith, A. Mark and Goldstein, Bernard R. "The Medieval Hebrew and Italian Versions of Ibn Muʿāhd's 'On Twilight and the Rising of Clouds'." *Nuncius* 8: 611–643, 1993.

Villuendas, M.V. *La trigonometría europea en el siglo XI: Estudio de la obra de Ibn Muʿad, El Kitāb mayhūlāt*. Barcelona, 1979.

See also: al-Khwārizmī – Ibn al-Haytham – Astronomy in the Islamic World

IBN MUNʿIM The oldest mathematical work from the Maghreb which deals with combinatory problems is the *Fiqh al-ḥisāb* of Ibn Munʿim, a scholar originally from Andalusia, living in Marrakesh in the Almohad era. To our knowledge, his book was the first in the entire history of mathematics to have devoted a whole chapter to these types of problems and to have stated them and solved them according to a common procedure.

Aside from Ibn ʿAbd al-Malik, the biographers of the Maghreb do not mention Ibn Munʿim, even though they write at length about mathematicians of lesser significance, and they use the contents of his book. The little information we do have on Ibn Munʿim comes from his own introduction to his mathematical text, cited above, and from the book of Ibn ʿAbd al-Malik.

According to the latter source, Ibn Munʿim's full name was Aḥmad ibn Ibrāhīm ibn ʿAli Ibn Munʿim al-ʿAbdarī. He was originally from the town of Denia on the east coast of Spain, near Valencia, and he lived in Marrakesh where he taught and where he died in 626H/1228. He was known as one of the best scholars of his era in geometry and number theory. At the age of thirty, he began to study medicine which he practiced successfully at the same time as his mathematical activities.

Only three of Ibn Munʿim's numerous mathematical texts and letters are known today: one on magic squares, another on geometry, and the third on the science of calculation. And of these, only the last, *Fiqh al-ḥisāb*, is extant. Ibn Munʿim wrote it under the reign of the fourth Almohad caliph, al-Nāṣir (1199–1213). During his reign celebrated scholars like the grammarian Abū Mūsā al-Jāzūlī, the algebraist Ibn al-Yāsamīn, and the doctors of the Ibn Zuhr family all lived in the Almohad capital, or even within the court itself. This leads one to believe that there was a variety of scientific activity, thanks to a generous and often enlightened patronage.

Combinatory analysis is taken up the eleventh section of the first chapter of *Fiqh al-ḥisāb*, entitled "an accounting of words which are such that human beings can express them only by one of them". This section is not, however, in the eyes of this author, a complete overview of practical calculations. He takes care to explain, in the course of his exposition, that he proposes first to treat the problem in a general manner, even though he is obliged, in order to make his ideas clear, to formulate it in specific terms using the Arabic alphabet. In fact, this study goes beyond the linguistic framework in which it is formulated, as much by the way of posing the problems and linking them to each other, by the methods of reasoning used, as by the established results.

Ibn Munʿim begins by setting out the problem as a mathematician; he defines precisely the framework in which he is stating the chosen hypotheses and the degree of generality researched. Then he establishes, using a set of colors of silk as an abstract model, a rule which enables one to determine all the possible combinations of n colors p times p. In order to do that, he constructs, in accordance with an inductive method, a triangular numeric table, identifies its elements, with the desired combinations, and deduces the relationships:

$$C_n^p = C_{n-1}^{p-1} + C_{n-2}^{p-1} + \cdots + C_{p-1}^{p-1}.$$

Thus, he presents, to our knowledge for the first time, the famous arithmetical triangle which algebraists from the Muslim East like al-Karajī (d. 1029) had already constructed but for other purposes and using another procedure.

Ibn Munʿim's study continues by establishing, using induction, relative formulas with permutation, with or without repetitions, of a group of letters such that they give, by recurrence, the number of possible readings of a word of n letters, taking into account all the signs (vowels and sukūns for Arabic) used by a given language. This is the content of problems 2, 3, and 4. In problem 5, the author concludes the first part by establishing a formula of arrangements, without repetition, of n objects p times p, which take into account the vowels and the sukūn accompanying the letters.

The second part, much longer, seeks to enumerate the combinations with repetitions, adopting a method analogous to the preceding one and which necessitates recourse to the table of numbers. It was moreover this same method that the French scholar Mersenne rediscovered and applied to his work in the seventeenth century.

To set up problem 6, Ibn Munʿim goes back to his model of bunches and proposes to solve a difficult dilemma, apparently removed from the initial problem and stated thus: being given threads of silk in n colors, we want to determine the number of bunches it is possible to make with p threads of k colors, so that $p_1, p_2, \ldots, p_k$ threads are, respectively, of the same color.

All these propositions, and even more the techniques, allow the resolution of the problem which Ibn Munʿim formulates as follows: to determine the number of words of one to ten letters which it is possible to make with the letters of the Arabic alphabet, including all possible repetitions of letters in a words, and including vowels and the sukūn which can appear on letters.

The third part of his study includes, along with several applications, a series of tables which enables the determination, more and more closely, of all the elements $(P_n, A_n^p, C_n^p, \ldots)$ which occur in the counting of words that it is possible to pronounce in a given language.

In addition to the results included in this chapter of *Fiqh al-ḥisāb*, the way in which Ibn Munʿim established his results is also notable. In fact, he uses two types of reasoning which can be called inductive and combinatorial. While inductive reasoning is a traditional tool of Islamic mathematics with its privileged domains and unique stature, one cannot say the same for combinatorial reasoning which appeared, to our knowledge for the first time, in the *Fiqh al-ḥisāb*. His systematic use of combinatorial reasoning for establishing general propositions appeared as a clear acknowledgment of its mathematical character.

One can even suppose that if combinatorial reasoning had enjoyed a quantitative development in the field of application, it would have resulted in its explicit recognition as a process of reasoning beside analysis, synthesis, induction, and reasoning *ad absurdum*.

The elaborate nature of his procedures and results which appear in the *Fiqh al-ḥisāb*, as well as the spirit of the method which emerged from them leads to the theory that the beginning of the mathematization of combinatorial problems within the framework of Arabic science occurred prior to the work of Ibn Munʿim.

While we await the confirmation, or correction, of this conjecture, the *Fiqh al-ḥisāb* remains the oldest known Arabic work from the Muslim West in which an autonomous chapter on combinatorial analysis appeared. But its importance does not end there: with regard to the linguistic tradition of the Maghreb, the book was a culmination in that it laid out a general solution to a given problem. Also, for mathematics, the work represents an important link, marking the end of one stage in the progress of combinatorics, that of calculation using tables, and the beginning of another stage, that of the extension of formulas and their use in solving problems.

AHMED DJEBBAR

REFERENCES

Djebbar, A. *Enseignement et recherche mathématiques dans le Maghreb des XIIᵉ–XIVᵉ siècles*. Paris: Publications Mathématiques d'Orsay, 1981.

Djebbar, A. *L'analyse combinatoire au Maghreb: l'exemple d'Ibn Munʿim (XIIᵉ–XIIIᵉ siècles)*. Orsay, France: Université de Paris-Sud, 1985.

Rashed, R. *The Development of Arabic Mathematics: Between Arithmetic and Algebra*. Dordrecht and Boston: Kluwer Academic, 1994.

See also: Combinatorics in Islamic Mathematics

IBN QUNFUDH His name is Abū'lʾ Abbās Aḥmed Ibn al-Ḥasan Ibn ʾAli Ibn al-Khaṭīb, and he is known under the two names of Ibn Qunfudh and Ibn al-Khaṭīb. He was born in 710 AH (AD 1339) in Constantine, Algeria and came from an old family which was cultured and well to do. We know that Ibn Qunfudh began his studies with his father and his maternal grandfather, in order to follow them eventually under the direction of other professors from his same town.

After his elementary education, he returned to Fez, in Morocco, where he remained for eighteen years. There he studied with several professors, covering different scientific themes, and teaching and publishing some of his own work. We know that it is in this city that he edited in 771 (1370) his most important mathematical work, the *Ḥaṭṭ an-niqāb ʾan wujūh ʾamāl al-ḥisāb*, which is a commentary on the *Talkhīṣ aʾmāl al-ḥisāb* of Ibn al-Bannāʾ, written in 721 (1321). We think that he acquired his advanced education either completely or partially during his stay in Fez.

During the period of famine which raged in all of Maghreb in 776 (1374), Ibn Qunfudh went back to Constantine in order to take up duties as a *khaṭīb* (preacher), *mufti* (jurist), and *qāḍī* (judge). At the same time, he devoted himself to the teachng and editing of scholarly work. No information has come down to us on the subject of the contents of this teaching, and we do not even know if he taught the contents of his own mathematical writings at Constantine or in the cities where he was posted, or even if he was content only to edit them, following in that the tradition of several older authors, who wrote on subjects which were very different from each other without having taught them.

What follows is an attempt to give a glimpse of Ibn Qunfudh's contributions to mathematics, of which we know only four titles:

1. *Mabād'as-sālkin fī sharh rajz Ibn al-Yāsamīn* (Principles for Those who are Concerned with the Commentary on the Poem of Ibn al-Yāsamīn). Ibn Qunfudh began this poem according to the traditional procedure of commentators of the Middle Ages. What is not traditional is his use of mathematical symbolism to resolve equations and to represent polynomials. This was to become normal practice during this period, since this commentary was adopted by his students. The symbolism was universally used in mathematical works from Maghreb. This hypothesis is reinforced by the presence of this same symbolism in the book *Ḥaṭṭ an-niqāb* and in the book written by Yaʾqūb al-Muwāḥidī *Tahṣīl al-munā fī sharh Talkhīṣ Ibn al-Bannā.*

2. *Bughyat al-fāriḍ min al-ḥisāb wa l-farā'id* (The Desire of the Genealogical Specialist to Know Arithmetic and Successional Division). To this day, no copy of this book has been found.

3. *at-Talkhīṣ fī sharh at-Talkhīṣ* (Abridged Commentary on *Talkhīṣ*). This is a resumc of *Ḥaṭṭ an-niqāb,* of which two copies are extant today.

4. *Ḥaṭṭ an-niqāb ʾan wujūh ʾamāl al-ḥisāb* (Lifting the Veil on the Operations of Calculation). Five copies of this work are extant.

Since this is Ibn Qunfudh's most important work, it is useful to describe certain aspects of its contents. Reading it permits us to make several important observations on mathematical writing in this era. The author begins his book with some advice and directions designed to facilitate the reading of any book; then he provides a detailed list of the writings of Ibn al-Bannāʾ. In the beginning of each chapter, he gives an abstract of its contents, something which one finds nowhere else in the literature of other mathematicians of this time. It is also important to note again the existence of Ibn Qunfudh's mathematical symbolism, particularly in the chapters on roots and on algebraic equations. However, recent research reveals that this symbolism had been used previously at the end of the twelfth or beginning of the thirteenth century, especially in Ibn Yāsamīn's book *Talqīḥ al-afkār*. Elsewhere, Ibn Qunfudh supplies arithmetic rules which one doesn't find in Ibn al-Bannāʾ's work, notably in the chapter on the product. Ibn Qunfudh's terminology also differs significant from Ibn al-Bannāʾ's.

Finally, one finds in the work of Ibn Qunfudh an equation of which the second number is zero, an equation which had formerly been studied by Ibn Badr. His originality lies in the symbolic way in which he expressed it:

ش

0 ل 7 ﻻ 8,

or, in modern symbols, $8x - 7 = 0$.

Ibn Qunfudh died in 810 AH (AD 1407)

Y. GUERGOUR

REFERENCES

Djebbar, A. *L'analyse combinatoire au Maghreb: l'exemple d'Ibn Munʿin*. Paris: Publications Mathématiques d'Orsay, Number 85–01, 1985.

Djebbar, A. *Enseignement et recherche mathématiques dans le Maghreb des XIIIe–XIVe siècles*. Paris: Publications Mathématiques d'Orsay, Number 81-02, 1981.

Djebbar, A. "Quelques apects de l'algèbre dans la tradition mathématique Arabe de l'occident musulman." *Colloque Maghrébin sur l'histoire des mathématiques Arabes*, 1–3 décembre 1986. Alger: la Maison des livres, 1988.

Guergour, Y. *Dirāsa ʾan Ibn Qunfudh was taḥqīq sharhihi li urjuzat Ibn Yāsamīn al-jabria*. Mémoire de D.E.A. en Histoire des Mathématiques E.N.S. d'Alger, 1986.

Guergour, Y. *Les écrits Mathématiques d'Ibn Qunfudh al-Qasanṭīnī (810/1406)*. Doctoral thesis, Algeria, 1990.

Ibn Qunfudh. *Uns al-faqīr wa ʾizz al-ḥaqīr*. Ed. Muhamed al-Fasi and Adolphe Faure. Rabat: Université de Recherches Scientifiques, 1965.

Zemouli, T. *Les écrits mathématiques d'Ibn al-Yāsamīn*. Doctoral thesis, Algeria, 1992.

See also: Ibn al-Bannā

IBN QUTAYBA Ibn Qutayba, Abū Muḥammad ʿAbdallāh ibn Muslim, was born in 828 in Kufa or Baghdad, and died in 884 or 889 in Baghdad.

Ibn Qutayba was a scholar of typical Arabic-Islamic education. His studies included all branches of the traditional Arabic and Islamic knowledge of his time: religion, history, biography, philology, lexicography, literature, and some science. He left twenty works of varying lengths covering all the fields mentioned. In his career, for some years he was *qāḍī* (a judge according to Islamic rules) in Dinawar (northern Iran); later he lived and taught in Baghdad, where he also died. Of particular scientific interest is his *Kitāb al-anwāʾ* (*Book on the anwāʾ*). *Anwāʾ* are asterisms and stars used by the Arabs in pre-Islamic and early Islamic times to determine seasons, predict weather — especially rain, and guide them in their nightly desert travels. Many Arabic philologists and lexicographers wrote books of this type, but most of these did not survive. Therefore Ibn Qutayba's *Book on the anwāʾ*, which was printed in Arabic, in Hyderabad/Deccan (India) in 1956, is of great interest. In it, the author has assembled information on the popular astronomical and meteorological knowledge of the old Arabs, from the time before their acquaintance with Greek, Persian, and Indian astronomy. Facts, traditions, terminology, and nomenclature are amply described. Much of this material continued to be used later in the most active period of Arabic–Islamic astronomy. And some of it even lived on into our time, as, for example, the star name Aldebaran (for α Tauri).

PAUL KUNITZSCH

REFERENCES

Brockelmann, Carl. *Geschichte der arabischen Litteratur*. Leiden: Brill, I, 1943, pp. 120–123; Supplement I, 1937, pp. 184–187.

Huseini, I.M. *The Life and Works of Ibn Qutayba*. Beirut: American Press, 1950 .

Lecomte, Guy. *Ibn Qutayba, l'homme, son oeuvre, ses idées*. Damascus: Institut français de Damas, 1965.

IBN RIḌWĀN Abu'l-Ḥasan ʿAlī Ibn Riḍwān ʿAlī Ibn Jaʿfar, a self-educated physician philosopher, was born in AD 998 at Gīzah, a suburb south of Cairo (al-Fusṭāṭ). He was the son of a poor baker (*farrān*) and had to earn money in his youth by practicing medicine, teaching, and telling people's fortunes from astrological signs. When he was fifty-nine, he wrote a book entitled *Fī Sīratihī* (On His Own Conduct). The work is now lost, but its text is partly preserved in Ibn Abī Uṣaybiʿa's *ʿUyūn al-Anbāʾ fī Ṭabaqāt al-Aṭibbāʾ* (Sources of Information about the Classes of Physicians).

In his youth, he believed in astrology (*ʿilm al-nujum*) and was convinced that the stars at the time of his birth indicated a prosperous medical career. To realize his ambition, he first sought out a popular teacher in Cairo, who instructed Ibn Riḍwān to memorize Ḥunayn Ibn Isḥāq's (d. AD 873) *Kitāb al-Masāʾil fi 'l-Ṭibb li 'l-Mutaʿallimīn* (Questions on Medicine for Students). Ibn Riḍwān watched him teaching: pupils read the text; the teacher listened but did not utter one word of explanation and did not even bother to correct their errors. Ibn Riḍwān sought to find other teachers and put questions to each of them, based on the writings of Hippocrates and Galen, which they displayed on shelves in their private libraries. He concluded that they merely knew the titles of books but were ignorant of their contents. He pondered over traveling to Iraq for further education, but financial difficulties prevented his going there. He decided on self-education at the tender age of fifteen. From perusing Galen's *On the Doctrines of Hippocrates and Plato*, he concluded that he should first study geometry and logic. He studied the well-known books on these two subjects, then proceeded to read textbooks on medicine proper. In this way he reached an understanding of the principles of the art of healing.

Ibn Riḍwān worked very hard as a practicing doctor until he reached the age of thirty-two, when he became well known and earned enough money to build his own residential Palace of Candles (*Qaṣr al-Shamʿ*). At the age of fifty-nine he divided his time between practicing medicine, daily physical exercise (*al-riyāḍah*), and reading books on literature, Islamic law, and medicine. His fame reached the Fātimid Caliph al-Mustanṣir (r. AD 1036–1094), who appointed him Chief of Physicians (*Raʾīs al-Aṭibbāʾ*) in Egypt. He was kind-hearted, treated the poor for free, and was always willing to extend a helping hand to the needy.

In old age, Ibn Riḍwān's mind became disturbed (*taghayyar ʿaqluhu*) as a result of being robbed of all his cherished possessions. He had adopted an orphan young girl whom he brought up in his own house. She absconded with all the precious items he had accumulated and twenty thousand gold dīnārs which he had kept in the house. An extensive search for the girl was abortive. Ibn Riḍwān died in AD 1067.

Among Ibn Riḍwān's extensive bibliography are his treatise *Fī Dafʿ Maḍārr al-Abdān fī Arḍ Miṣr* (On the Prevention of Bodily Ills in Egypt) and the *Al-Nāfiʿ fī Kayfiyyat Taʿlīm ṣināʿ at al-Ṭibb* (Useful Book on the Quality of Medical Education).

In *On the Prevention of Bodily Ills*, he describes the Nile and the Muqaṭṭam Hills on both sides of the river. The hills in the east hold back the "hot and humid" winds, which are most favorable for the temperament (*mizāj*) of animals and which do not reach *al-Fusṭāṭ*. Ibn Riḍwān gives an account of "the six non-natural causes" (*al-asbāb al-sitta al-ḍarūriyya*) which determine health and sickness: the air surrounding the body, food and drink, movement and rest, sleep and wakefulness, retention and evacuation, and psychic events.

Ibn Riḍwān was discourteous in criticizing members of the medical profession. For example, chapter five of this treatise is entitled "On the Incorrectness of Most of Ibn al-Jazzār's [d. AD 1009] Reasons for the Unhealthy Air in Egypt". Furthermore, disagreements arose with Ibn Buṭlān of Baghdad. These are well-documented by Meyerhof and Schacht.

In his *Useful Book*, Ibn Riḍwān preserved for posterity a unique document, the late Alexandrian medical curriculum (sixth–seventh century AD). In it he attributes the decline of medicine in his time to the popularity of poor-quality compendia (*kanānīsh*) and summaries and commentaries of the books of Hippocrates and Galen, compiled by incompetent physicians. The Alexandrian curriculum specifies the titles of textbooks of logic, medicine proper, and mathematics (including astronomy). It consists of preparatory (introductory) courses and main courses. The preparatory courses contain optional subjects, including language and grammar, and compulsory subjects: physics, arithmetic, numerals, measurement, geometry, the compounding of drugs, astrology, and ethics.

The main courses include logic and medicine proper. Sixteen of Galen's books were to be studied in seven grades: Grade one included Galen's *On Sects*, *On the Art of Physics*, *On the Pulse, to Teuthras*, and *To Glaucon on Therapy*. In the final and seventh grade, students studied Galen's *On the Method of the Preservation of Health*.

ALBERT Z. ISKANDAR

REFERENCES

Dols, Michael W. *Medieval Islamic Medicine. Ibn Riḍwān's Treatise "On the Prevention of Bodily Ills in Egypt"*. Berkeley: University of California Press, 1984.

Ibn Abī Uṣaybiʿa. *ʿUyūn al-Anbāʾ fī Ṭabaqāt al-Aṭibbāʾ*. Ed. August Müller. Königsberg: Selbstverlag, 1884, vol. 2, pp. 99–105.

Iskandar, Albert Z. "An Attempted Reconstruction of the Late Alexandrian Medical Curriculum." *Medical History* 20(3): 235–258, 1976.

Meyerhof, Max, and J. Schacht. *The Medico-Philosophical Controversy Between Ibn Buṭlān of Baghdad and Ibn Riḍwān of Cairo. A Contribution to the History of Greek Learning Among the Arabs*. Faculty of Arts, Publication no. 13. Cairo: The Egyptian University, 1937.

See also: Ibn Buṭlān

IBN RUSHD (AVERROËS) Abu'l-Walīd Muḥammad ibn Aḥmad ibn Muḥammad ibn Rushd (Averroës, AD 1126–1198), a native of Cordoba, Spain, was the namesake of his famous grandfather. Later, to avoid any confusion, he was nicknamed *al-ḥafīd* (grandson). Like his father and grandfather, he was a well-known jurist. By profession, following in the footsteps of his father, he became a *qāḍī* (judge), who specialized in religious matters, and at one time was the Imam of the great mosque of Cordoba. He adhered to the Mālikī sect, one of the four great sects of Islam, and wrote *Kitāb al-Muqaddimāt al-Mumahhidāt* (A Book of Introductions that Pave the Way), for the followers of the sect.

At his father's insistence, Ibn Rushd studied Islamic law (*Sharī'a*) under the teacher, al-Ḥāfiẓ Abū Muḥammad Ibn Rizq. Ibn Rushd also studied the science of Tradition (*Ḥadīth*), but he is known to have been more interested in Islamic law. He memorized the *Muwaṭṭa'* of Imam Mālik and was also greatly impressed by the Ash'arite science of *Kalām* (Theology). Later, he turned against the Ash'arī school of thought and attacked its proponent, the Imam al-Ghazālī (d. AD 1111). Ibn Rushd was also acquainted with the doctrines of the Mu'tazila theology.

His teachers of medicine were Abū Marwān Ibn Jurrayūl, a first-class practitioner, and Abū Ja'far Hārūn al-Tarjālī, a well-known physician philosopher in Seville, who was knowledgeable about Aristotelian philosophy, as well as the medical writings of the Ancients. Al-Tarjālī was employed by the Almohad (*al-Muwaḥḥid*) ruler Abū Ya'qūb Yūsuf (r. AD 1163–1184). Abū Yūsuf Ya'qūb al-Manṣūr (r. AD 1184–1199), before succeeding his father, attended meetings in Seville, to which notable philosophers, physicians, and poets were invited. Abū Bakr Ibn Ṭufayl (Abubacer), Ibn Zuhr (Avenzoar), and Ibn Rushd were regular attendants at these meetings. Ibn Rushd's education under al-Tarjālī qualified him as a physician-philosopher; his studies in religious law, under al-Ḥāfiẓ, entitled him to be considered a jurist (*faqīh*).

In 1153, when he was in Marrakech (Morocco), Ibn Rushd supported the Almohad ruler 'Abd al-Mu'min (d. AD 1163) in furthering education by founding colleges. From his commentary on Aristotle's *De Caelo*, one learns that Ibn Rushd conducted astronomical observations when in Marrakech, and in his commentary on the *Metaphysics*, he mentions his yearning for his early studies in astronomy. Ibn Rushd studied the writings of Arabic-speaking astronomers and expressed his own opinion regarding the three kinds of planetary motions: those that can be detected by the naked eye, those that can be seen by instruments of observation (remarking that some occurred over long periods of time that exceeded the lifetime of observers), and those planetary movements whose existence can only be surmised by reasoning. Between AD 1169 and 1179, Ibn Rushd visited many places in the Almohad realm. He was actively researching matters pertaining to astronomy when he met Ibn Ṭufayl, a philosopher and astronomer in his own right. Ibn Ṭufayl enhanced Ibn Rushd's career by introducing him to Abū Ya'qūb Yūsuf, who appointed Ibn Rushd as *qāḍī* of Seville in AD 1169. Ibn Rushd returned to Cordoba in AD 1171, at which time he was still holding the office of *qāḍī*. In AD 1182, he succeeded Ibn Ṭufayl as a personal doctor to Abū Ya'qūb Yūsuf, and was elevated to the office of grand *qāḍī* of Cordoba.

For ten years Ibn Rushd enjoyed the patronage of Abū Ya'qūb Yūsuf. Out of jealousy, jurists (*al-fuqahā'*) of the Mālikī sect, whose influence had grown as a result of their religious zeal during the period of the Crusades, successfully conspired against him for his free-minded views. He was summoned to appear in court, and his philosophical writings were deemed contrary to the teachings of Islam. Ibn Rushd fell from grace in AD 1195 and was banished to Lucena, a province of Cordoba.

Very little is known about Ibn Rushd as a teacher of medicine. Nevertheless, the names of two of his students are known: 'Abd Allāh al-Nadrūlī and 'Īsā ibn Aḥmad ibn Muḥammad ibn Qādir. The latter transcribed the whole text of Ibn Rushd's book *Al-Kulliyyāt* (Generalities) from the author's own autograph, during his lifetime, and with his approval.

Kitāb al-Kulliyyāt (Latin, *Colliget*), consists of seven books: *Tashrīḥ al-A'ḍā'* (The Anatomy of Organs), *al-ṣiḥḥa* (Health), *al-Maraḍ* (Disease), *al-'Alāmāt* (Symptoms), *al-Adwiya wa'l-Aghdhiya* (Drugs and Foods), *Ḥif ẓ-al-ṣiḥḥa* (Hygiene), and *Shifā' al-Amrāḍ* (Recovery from Disease). The purpose of the book, according to the author, is to provide medical men with an introduction to concise accounts of the different parts of medicine (*ajzā' al-ṭibb*).

Scientific collaboration existed between the eminent practitioner Ibn Zuhr (Avenzoar, d. AD 1162) and Ibn Rushd. Ibn Rushd asked Ibn Zuhr to write a book on therapy, which he did. It was called *al-Taysīr fī'l-mudāwāt wa'l-tadbīr* (An Aid to Therapy and Regimen). *Al-Taysīr* and the *Kulliyyāt* together were meant to cover the whole science of medicine, possibly instead of Ibn Sīnā's (Avicenna, d. AD 1037) *Kitāb al-Qānūn fi'l-Ṭibb* (Canon of Medicine), which Ibn Zuhr severely criticized.

A merchant from Baghdad presented Ibn Zuhr with a beautifully transcribed and ornamented copy of Ibn Sīnā's *Kitāb al-Qānūn*. After reading it for the first time, Ibn Zuhr condemned the book and kept tearing off the margins of its leaves, which he used for jotting down prescriptions for his patients. It took almost a century before a copy of *Kitāb al-Qānūn*, completed by its author in Hamadan (Persia), actually reached Cordoba, Ibn Rushd's home town.

ALBERT Z. ISKANDAR

REFERENCES

Arnaldez, Roger, and Albert Zaki Iskandar. "Ibn Rushd." In *Dictionary of Scientific Biography*, vol. 12. New York: Scribner's, 1970–1980, pp. 1–9.

Ibn Abī Uṣaybiʿa. *ʿUyūn al-Anbāʾ fī Ṭabaqāt al-Aṭibbāʾ*. Ed. August Müller. Königsberg: Selbstverlag, 1884, volume 2, pp. 75–78.

Ibn Rushd. *Kitāb al-Kulliyyāt*. MS No. 14524.b.61. London: The British Library.

See also: Ibn Ṭufayl – Ibn Sīnā

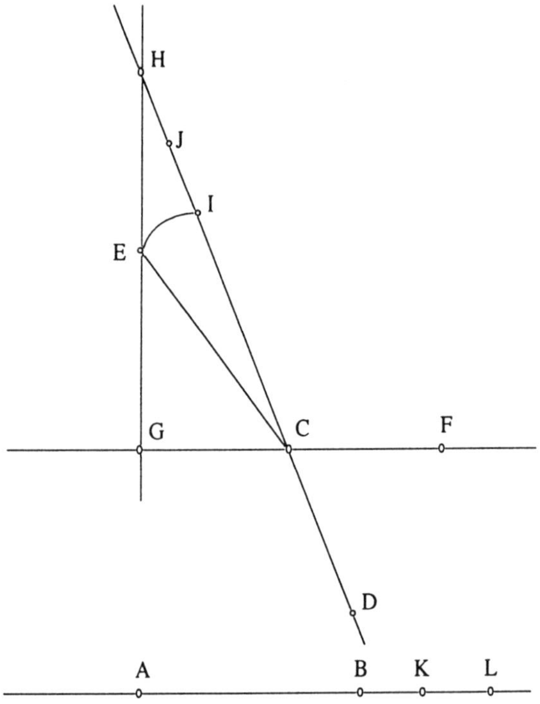

Figure 1: Refraction in lenses.

IBN SAHL Ibn Sahl, Abū Saʿd al-ʿAlāʾ, was a first-class mathematician. From his correspondence as well as from diverse information transmitted by mathematicians of the second half of the tenth century, we can deduce that he flourished under the Buwayhid Dynasty, and probably in Baghdād, between 970 and 990. Many of Ibn Sahl's important writings have been lost, namely two treatises, *On the Measurement of the Parabola* and *On the Centers of Gravity*, and a kind of anthology of problems about which we have no direct information. His book *Fī al-Ḥarrāqāt* (On Burning Instruments) was also on the point of being lost. This book, written in Baghdād around 984, is the first known contribution on the geometric theory of lenses.

Let us begin with the contribution of Ibn Sahl to optics. He wrote a memoir on the transparency of the celestial sphere, which was commented on by Ibn al-Haytham. The memoir and the commentary have come down to us. In this, composed in the course of his reading of Book V of Ptolemy's *Optics*, Ibn Sahl takes up not only the rules of refraction set out by his predecessor, but also demonstrates that every medium, including the celestial sphere, is invested with a certain opacity which defines it. Ibn al-Haytham had already captured this idea perfectly when, on reading this same memoir of Ibn Sahl, he wrote that his predecessor wanted to demonstrate that "there is no limit to transparency, and, for each transparent body, another always exists which is more so". That is to say that the mathematician better understands the notion of a medium and its definition by a certain characteristic opacity.

But Ibn Sahl's real discovery takes place when he poses the still unthought-of question of burning by refraction. He no longer defines the medium by a certain opacity, but characterizes it by a constant ratio. It is this concept of constant ratio distinguishing the medium which is the masterpiece of his study of refraction in lenses. This ratio, postulated by Ibn Sahl but never calculated, is nothing but the inverse of the refraction index n of the medium in relation to air. It therefore deals with Snellius' law of refraction, its formulation being very close to Snellius' own formulation some six centuries later. Let us look again at *On Burning Instruments*.

At the beginning of the study of refraction in lenses, Ibn Sahl considers a plane surface GF limiting a piece of crystal. He considers the straight line DC following which light propagates in the crystal, the straight line CE according to which it refracts in the air, and the perpendicular at G to the surface GF which cuts the straight line CD at H and the refracted ray at E.

Here Ibn Sahl clearly applies the law by which the ray CD in the crystal, the ray CE in the air, and the perpendicular GE to the surface plane of the crystal, are all on the same plane. As usual, with no conceptual comment, he writes, "the straight line CE is therefore smaller than CH. We separate from the straight line CH the straight line CI equal to the straight line CE; we divide HI in half at point J; we set the ratio of the straight line AK to the straight line AB equal to the ratio of the straight line CI to the straight line CJ; we extend the straight line BL along AB and set it equal to the straight line BK".

In these sentences, Ibn Sahl concludes that CE/CH is less than 1, which he will use throughout his research on lenses constructed in the same crystal. He will not fail to give this same ratio again, nor to reproduce this same figure, each time he discusses refraction in the crystal.

This ratio is nothing but the inverse of the index of refraction in the crystal in relation to air. In fact, let us consider i_1 and i_2, the angles formed respectively by CD and CE with the perpendicular GH. One has

$$\frac{1}{n} = \frac{\sin i_1}{\sin i_2} = \frac{\mathrm{CG}}{\mathrm{CH}} \cdot \frac{\mathrm{CE}}{\mathrm{CG}} = \frac{\mathrm{CE}}{\mathrm{CH}}.$$

Ibn Sahl takes on segment CH point I such that CI=CE, and point J at the middle of IH. One has

$$\frac{\mathrm{CI}}{\mathrm{CH}} = \frac{1}{n}.$$

The division CIJH henceforth characterizes the crystal for all refraction.

Ibn Sahl made innovations not only in optics, but also in mathematics. Among his mathematical inventions, I shall mention only one paper, set down as a follow-up to his contemporary, the mathematician al-Qūhī: his study of the method of projection of the sphere. Al-Qūhī is, in fact, the author of a treatise on *The Art of the Astrolabe by Demonstration*. This treatise is composed of two books, of which the first opens with an introductory chapter on the theory of projections. The propositions which al-Qūhī set forth seemed "difficult to understand" to one of his contemporaries, who turned to Ibn Sahl for clarification of the notions found therein and for the proof by synthesis of what al-Qūhī showed by analysis. We have here a situation of privileged history: we are witness to the elaboration of a new chapter of geometry by two contemporary mathematicians.

In conclusion, I remark upon the emergence of Ibn Sahl, a figure until recently almost unknown, and two chapters: anaclastics and the study of projection of the sphere. Ibn Sahl also made a substantial contribution in infinitesimal mathematics as well as in construction problems — the regular heptagon, for instance.

ROSHDI RASHED

REFERENCES

Rashed, R. "A Pioneer in Anaclastics: Ibn Sahl on Burning Mirrors and Lenses." *Isis* 81: 464–491, 1990. Reprinted in *Optique et mathématiques: recherches sur l'histoire de la pensée scientifique en arabe*. Aldershot: Variorum, 1992.

Rashed, R. *Géométrie et dioptrique au Xᵉ siècle: Ibn Sahl, al-Qūhī et Ibn al-Haytham.* Paris: Les Belles Lettres, 1993.

Rashed, R. (ed. and co-author). *Encyclopedia of the History of Arabic Science*. 3 vols. London: Routledge, 1996.

IBN SARABI (SERAPION) There are two different Arab authors known to the medieval 'Latin west as 'Serapion', and the confusion between them in all of the authorities seems inextricable. The elder of the two authors was a Syrian, Yuḥānnā ibn Sarabiyun, who lived in the ninth century. He wrote two books in Syriac, which were translated into Arabic: they are called the *Large Kunnāsh* (it was in twelve parts) and the *Small Kunnāsh* (in seven parts). Manuscript copies of the Arabic versions exist in European libraries; what sounds like a whole copy of the *Large Kunnāsh* is in Istanbul. Both works deal with medicine and diet, but not surgery; al-Rāzī (Razes), who may have been a younger contemporary, cites Serapion, naming the *Kunnāsh*, and ʿAlī ibn ʿAbbās al-Majūsī (d. 994) criticizes Serapion for ignoring surgery. The *Small Kunnāsh* was translated into Latin by Gerard of Cremona (d. 1187) and entitled *Practica Joannis Serapionis dicta Breviarium*. There are many manuscript copies of this version and some early printed texts. Moses ben Mazliach translated Gerard's version into Hebrew, and a manuscript of this translation still survives.

The younger Serapion is an even more obscure figure. He was the author of a treatise on drugs called *Kitāb al-adwiya al-mufrada,* which was translated into Latin as *Liber de medicamentis simplicibus* by Simon de Cordo of Genoa and Abraham of Tortosa in about 1290. The younger Serapion cannot have lived earlier than the eleventh century, because he cites among his authorities not only al-Rāzī, but also Ibn al-Wāfid (997–1075). There are many manuscript copies of the Latin translation, and several early prints. Usually the works of both Serapions appear together. In the Latin text, the author explains that he intends to collect and reconcile the views of Dioscorides and Galen on a wide variety of medicinal substances, and then to add in the opinions of others. In his chapter on pearls, for instance, he describes the generally accepted power of pearls to strengthen the heart; he then goes on to cite four Arab authorities on the effects of pearls on the eyesight, the blood, the nerves, melancholia, and as a dentifrice.

E. RUTH HARVEY

REFERENCES

Campbell, Donald. *Arabian Medicine and its Influence on the Middle Ages*. 2 vols. London: Kegan Paul, Trench, Turbner & Co., 1926.

Peters, Curt. "Johannan b. Serapion." *Le Muséon* 55:139–42, 1942.

Sezgin, Fuat. *Geschichte des Arabischen Schrifttums III: Medizin, Pharmacie, Zoologie, Tierheilkunde, bis ca. 430 H.* Leiden: Brill, 1970, pp. 228, 240.

See also: al-Rāzī – al-Majūsī

IBN SĪNĀ (AVICENNA) Abū ʿAlī al-Ḥusain ibn ʿAbdallāh ibn Sīnā (980–1037), also known as Avicenna, was born in the city of Būkhārā in the Eastern part of Persia into an Ismaʾili family. He demonstrated an incredible genius for learning, and having mastered the *Quʾrān* and the sciences of his time, became a physician at the age sixteen. Ibn Sīnā gained favor with the Sāmānid dynasty for having cured Prince Nūḥ ibn Manṣūr and was thus allowed to use the royal library. He mastered other sciences such as psychology, astronomy, chemistry, and pharmacology by age eighteen, and towards the end of his life said he had learned everything he knew by then.

Ibn Sīnā lived in a tumultuous time when different princes were engaged in a power struggle, resulting in his traveling from city to city. He first went to Gorgān, an area close to the Caspian Sea, crossing the desert on foot, and after a while traveled to Khorāsān, Rayy, and Qazwīn, until he finally settled in Hamadān at the request of prince Shams al-Dawlah. In 1022, following the death of the prince, Ibn Sīnā went to Iṣfahān where he found the peace and serenity that the intellectual life demands. In 1037 while traveling with ʿAlāʾ al-Dawlah he became ill and died as the result of colic. He is buried in Hamedan.

Ibn Sīnā gained the title of Shaykh al-Raʾīs (Master of the Wise) because he composed numerous treatises, 276 of which have been alluded to by his commentators. It is noteworthy that he wrote a great number of these works on horseback while fleeing from one city to another.

Ibn Sīnā's most extensive and elaborate work is *Kitāb al-shifāʾ* (The Book of Healing) which itself consists of four segments: *al-Manṭiq* (Logic), *Tabīʿiyyāt* (Natural Philosophy), *Riyāḍāyāt* (Mathematics), and *Ilāhāyyāt* (Metaphysics). To this monumental work should be added *Kitāb al-nijāt* (The Book of Deliverance), a synopsis of the *Shifāʾ*, and his last major work *al-Ishārāt wa 'l-tanbīhāt* (The Book of Directives and Treatments). Among his other philosophical works are *ʿUyūn al-ḥikmah* (Fountain of Wisdom), *Kitāb al-hidāyah* (The Book of Guidance), *al-Mabdaʾ wa 'l-maʿād* (The Book of Origin and End), and the first major work written in Persian for ʿAlāʾ al-Dawlah, *Dānishnāmah-yi ʿAlāʾī* (The Book of Knowledge). Finally, there are a few works which are mystical and visionary, such as *Ḥayy ibn Yaqẓān* (Son of the Living Awake), and *Risālat al-ṭaīr* (Treatise of the Bird).

Ibn Sīnā was a great synthesizer in that he not only incorporated the ideas of some of his predecessors such as al-Kindī and al-Fārābī, but made extensive use of Greek philosophy, in particular Neoplatonism, Plato and Aristotle. From Plotinus he adopted the emanation scheme, from Plato his theory of archetypes and from Aristotle his logic, physics, and psychology. The genius of Ibn Sīnā was in interpreting these different thoughts within the unitary matrix of Islam.

Ibn Sīnā has been called "philosopher of being" for he brought the question of being and the study of ontology to the forefront of philosophical debates and thereby corrected a deficiency that he saw in Greek philosophy, in particular in Aristotle. Ibn Sīnā divided all beings into three categories: necessary, contingent, and impossible. Furthermore, he argued that existent beings were made up of an essence and existence. This distinction became the central theme of medieval ontological discussions as well as of the subsequent debates within the Islamic philosophical tradition.

Ibn Sīnā, who had studied the Ptolemaic and Aristotelian systems, adopted the Islamic astronomical view based on the nine spheres and interpreted them within the scheme of the emanation of intellects. Each order of intellect accordingly corresponds to one of the heavens. For example, the second intellect corresponds to the highest heaven which is located above the fixed stars, and the tenth intellect corresponds to the moon below which the corporeal domain of our world is located.

Medicine

Ibn Sīnā was in the tradition of such grand Muslim physicians as al-Rāzī and al-Majūsī. By the age of twenty, he had already become an accomplished physician and gained the respect of the court for having cured several members of the royal family. Eventually he came to be known as the "prince of physicians". He is the author of the *Qānūn al-Ṭibb* (The Canon of Medicine), which became a standard text both in the Christian West and the Islamic world. In addition to his *magnum opus*, Ibn Sīnā wrote a number of treatises on medicine both in Arabic and Persian some of which deal with particular diseases. He also wrote a book in which principles of medicine are written as poetry in order to facilitate their memorization by medical students. This tradition continues among medical students today.

In medicine, Ibn Sīnā was in many ways a follower of Greek ideas on medicine. However, he made the following revisions:

1. Unified the central principles of Greek medicine and interpreted them within an Islamic framework;

2. Made the theoretical and practical aspects of medicine to represent a unified whole;

3. Documented the effects of medications on the body and commented on the necessity of a methodology in the study of pharmacological issues; and

4. Introduced a number of medications and treatments for illnesses unknown to Greeks.

Ibn Sīnā believed that the science of medicine was divided into two categories, theoretical and practical. Illnesses were brought about either by internal or external causes. Internal illnesses were caused by an imbalance of temperament in the human body. These causes could also be the result of psychological elements or what is now called "psychosomatic medicine".

Ibn Sīnā is known to have prescribed a variety of treatments for ailments, ranging from psychotherapy and diet to exercising. In the first volume of the *Canon of Medicine,* he states that medicine is the science of knowing the structure of the human body, and he goes on to discuss the fundamental principles necessary to maintain human health. He discusses surgery as an independent branch of medicine and elaborates on the science of anatomy. In the second volume, Ibn Sīnā explains various healing properties of simple and compound drugs and their various applications. In the third and fourth volumes, he presents his diagnosis of a variety of illnesses known to him, ranging from meningitis and cancer to tuberculosis and gastrointestinal illnesses. In the fifth volume of this work, Ibn Sīnā offers an extensive discussion of pharmacology, providing a detailed description of the effects of medications on body.

Some of his observations are extraordinarily advanced for his time. For example, he attributed the cause of plague to mice, commented on a number of contagious diseases, used single hair lines of horses for stitching after surgery, experimented with various drugs as anesthetics when operating on patients, described surgical processes for many operations such as gall bladder, used spices to stop bleeding after surgery, and understood the relationship between temperature and the spread of diseases. He even experimented on animals using different drugs.

Ibn Sīnā's treatment of the subject of psychology as a science which is both independent and part of the field of medicine is extensive. He first argues for the existence of a soul which is rather similar to the Aristotelian *psyche* and then shows how this incorporeal part of man interacts with his corporeal dimension. Ibn Sīnā then goes on to say that the health of the body to a large extent depends on the health of the soul and he offers much elaboration on that point throughout his works.

Perhaps one of the most important contributions of Ibn Sīnā to the field of medicine is the technical medical terminology which he introduced and which is still used by physicians in the Islamic countries.

Psychology

In *The Book of Healing*, Ibn Sīnā treats the subject of the "three kingdoms" (minerals, plants, and animals). He extends the hierarchy of the incorporeal world to the material world, concluding that whereas each domain has its soul it is only in humans that these souls reach their completeness. Ibn Sīnā's treatment of the human soul, its relationship with the body and the active intellect, is extensive. He introduces five internal senses and their equivalent five external senses.

The scope of Ibn Sīnā's writings on sciences includes such topics as motion, the process of sedimentation, and classification of minerals in the *Shifā'*, which came to be known in the West as *De mineralibus*. In a famous debate with al-Bīrūnī, Ibn Sīnā responded to ten questions on Aristotle's *De caelo* and eight other questions such as how vision is possible, the relationship between water and light, the nature of a vacuum, heat, cooling, and how it is that ice floats on water.

Ibn Sīnā's philosophy remained the dominant philosophical school in the medieval west where he became known through translations of his works in Islamic Spain by Avendeuth, who was a Jewish philosopher, Gerard of Cremona. The Sicillian school also benefited much from Ibn Sīnā who was translated by Michael Scot. Ibn Sīnā's works continued to be translated through the sixteenth century influencing such western philosophers as the Augustinian Gundisalvo, Alberttus Magnus, William of Auvergne, Alexander of Hales, Roger Bacon, and Duns Scotus.

Ibn Sīnā left an indelible mark on the history of science, medicine, and philosophy.

MEHDI AMINRAZAVI

REFERENCES

Anwaīi, G.C. *Essai de bibliographie avicennienne*. Cairo, 1950.

Arberry, Arthur J. *Avicenna on Theology*. London: J. Murray, 1951.

Avicenna Commemoration Volume. V. Courtois, ed. Calcutta: Iran Society, 1956.

Carra de Vaux, Bernard. *Avicenne*. Amsterdam: Philo Press, 1974.

Corbin, Henry. *Avicenna and the Visionary Recital*. Williard Trask, trans. New York: Pantheon Books, 1960.

Gohlman, William E. *The Life of Ibn Sīnā*. Albany: State University of New York Press, 1974.

Gruner, O. Cameron. *A Treatise on the Canon of Medicine of Avicenna*. London: Luzac, 1930

Krueger, Haven C. *Avicenna's Poem on Medicine*. Illinois, C.C. Thomas, 1963.

Le Livre de Science. Trans. Mohammad Achena and Henri Masse, 2 vols. Paris: Les Belles Lettres, 1955–1958.

Le Livre des directives et remarques, A. M. Goichon, trans. Beirut: Commission internationale pour la traduction des chefs d'oeuvre, 1951.

Morewedge, Parviz. *The Metaphysica of Avicenna ibn Sīnā*. New York: Columbia University Press, 1973.

Nasr, Seyyed H. *An Introduction to Islamic Cosmological Doctrines*. New York: Random House, 1978.

Rahman, F. *Avicenna 's Psychology*. London: Oxford University Press, 1952.

Shah, Mazhar H. *The General Principles of Avicenna's Canon of Medicine*. Karachi: Naveed Clinic, 1966.

See also: al-Kindī

IBN ṬĀWŪS Raḍī al-Dīn Abū-l-Qāsim ʿAlī ibn Mūsā ibn Jaʿfar ibn Muḥammad ibn Aḥmad ibn Muḥammad ibn Ṭāwūs was a religious scholar, historian and astrologer. He was born in al-Ḥilla (Iraq) on 15 Muḥarram 589/21 January 1193. He belonged to a distinguished family of scholars, the Banū Ṭāwūs, whose ancestry can be traced back to ʿAlī ibn Abī Ṭālib.

Ibn Ṭāwūs grew up in the town of al-Ḥilla, which had been established by the famous scholar Abū Jaʿfar al-Ṭūsī (385–458/995–1066) as a center of learning boasting five *madrasas* (institutions for the study of the Islamic sciences). Ibn Ṭāwūs spend his formative years in his native town, and in Baghdad where he lived for fifteen years. He made pilgrimages to the holy Shīʿī shrines of Najaf, Kerbela, and al-Kazimayn. He had a reputation for saintly powers, and concerned himself extensively with popular religious practices.

Ibn Ṭāwūs concentrated mainly on devotional literature, but he also wrote in the fields of history and astrology. (See the listing of primary sources in the bibliography.) In AH 650/AD 1252, he owned over 1500 volumes in his private library. Professor E. Kohlberg of the Hebrew University of Jerusalem has recently contributed a superb reconstruction of this library, and a detailed critical analysis of the works it contained.

Ibn Ṭāwūs was one of the most prolific medieval Shīʿī scholars, who aimed at providing the believer with a set of guiding moral and ethical principles. The sources stress his "erudition, knowledge, asceticism, devoutness, trustworthiness, understanding of *fiqh* (science, or knowledge), loftiness and godfearingness [along with his talents as] poet, man of letters, writer, and eloquent speaker". (Āghā Buzurg Tihrānī, 1983.) A man of deep faith, concerned with religious matters, Ibn Ṭāwūs could also display wisdom in sensitive political situations, as in this apocryphal story:

> When Sultan Hulagu conquered Baghdad in 656/1258, he ordered that the scholars give an opinion on which was preferable: the unbelieving but just sultan, or the Muslim despotic one. The scholars were then gathered in al-Mustanṣiriyya for this purpose. They refrained from answering. Raḍī al-Dīn ʿAlī Ibn Ṭāwūs was present at this Council. When he saw their hesitation, he took the *futya* (*fatwa*, legal opinion), and wrote that the unbelieving but just ruler was preferable to a Muslim despot.

Although Ibn Ṭāwūs declined the office of Naqābat al-Ashrāf under the Caliph al-Mustanṣir (d. 640/1242), he apparently changed his mind during the Mongol Period, and received the title from Naṣīr al-Dīn al-Ṭūsī (d. 672/1273). This important office, which had originated under the Buwayhids (also Buyids, a dynasty in Persia and Iraq, 320–454/932–1062), enabled the Sayyid families to represent their community in political, religious, and scholarly matters.

The date on which Ibn Ṭāwūs died in Baghdad is generally accepted as 5 Dhū-l-qa ʿda 664/9 August 1266, and he was probably buried in Najaf.

ZEINA MATAR

REFERENCES

Primary sources

Ibn Ṭāwūs. *Kitāb al-Luhūf fī qatlā al-ṭufūf*. Tehran: 1904.

Ibn Ṭāwūs. *Faraj al-mahmūm fī-maʿrifat nahj al-ḥalāl wa-l-ḥarām min ʿilm al-nujūm*. Qum: Manshūrāt al-Raḍī, 1944.

Ibn Ṭāwūs. *al-Malāḥim wa-l-fitan*, 1963.

Ibn Ṭāwūs. *Kashf al-muḥajja*. Tehran: Dafteri Nashri Farhangi Islāmī, 1989.

Ibn Ṭāwūs. *Kitab al-Yaqīn bi-ikhtiṣāṣ mawlānā ʿAlī bi-imārat al-muʿminīn*. Beirut: Tawzīʿ Dār al al-ʿUlūm, 1989.

Secondary sources

Kohlberg, E. *A Medieval Muslim Scholar at Work: Ibn Ṭāwūs and his Library*. Leiden: Brill, 1992.

Matar, Zeina. *The Faraj al-Mahmūm of Ibn Ṭāwūs: A Thirteenth Century Work on Astrology and Astrologers*. Ph.D. Dissertation, New York University, 1987.

Matar, Zeina. "Some Additions to the Bibliography of Mediaeval Islamic Astronomy from the *Faraj al-Mahmūm* of Ibn Ṭāwūs." *Archiv Orientalni* 57:319–322, 1989.

Matar, Zeina. "Dreams and Dream-Interpretation in the *Faraj al-Mahmūm* of Ibn Ṭāwūs." *The Muslim World* 80 (3–4): 165–175, 1990.

Momen, M. *An Introduction to Shiʿi Islam*. New Haven: Yale University Press, 1985.

Strothmann, R. *Die Zwölferschia: Zwei Religionsgeschichtliche Charakterbilder aus der Mongolenzeit*. Leipzig: Harrassowitz, 1926.

Millás, José María. *Tractat de l'assafea d'Azarquiel*. Barcelona: Universitat de Barcelona, 1933.

Poulle, Emmanuel. "Le quadrant nouveau médiéval." *Journal des savants* 148167, and 182–214, April–June 1964.

Sarton, George. *Introduction to the History of Science*, vol. 2. Baltimore: Williams & Wilkins, 1931.

Vernet, Juan. "Ibn Tibbon." In *Dictionary of Scientific Biography*, vol. 8. New York: Charles Scribner's Sons, 1976, pp. 400–401.

IBN TIBBON Ibn Tibbon, Jacob Ben Machir is known as Profeit in Romance languages and as Prophatius Judaeus in Latin. These names come from the translation of *mehir* (prophet) into the languages of Southern France.

Ibn Tibbon was probably born in Marseille ca. 1236. His family was originally from Granada and, for four generations, had been devoted to the translation of Arabic religious, philosophical, and scientific texts. Through these translations, Arabic learning and therefore Greek scientific traditions, were made available to the scholars of medieval Europe. Ibn Tibbon studied medicine at Montpellier and probably lived in Gerona, Spain between AD 1266 and 1267. He spent most of his life in Lunel, where his great-grandfather had established and practiced medicine at the beginning of the twelfth century, and Montpellier where he died in AD 1305.

Ibn Tibbon translated works by Autolycus of Pitane, Euclid, Menelaus of Alexandria, Qusṭā ibn Luqa, Ibn al-Haytham, Ibn al-Ṣaffīr, Ibn al-Zarqāllu, al-Ghazālī, Jābir ibn Aflaḥ, and Ibn Rushd from Arabic into Hebrew.

He is also the author of some works dealing with astronomy:

- *Robaʿ Yisrael* (Quadrant of Israel), written between AD 1288 and 1293, in which an astronomical instrument, called the *quadrans novus*, is described. It consists of a simplification of the face of the astrolabe. Some examples of this instrument have been preserved.
- The *Almanac*, calculated for Montpellier and dated 1 March 1300. It is based on Ibn al-Zarqāllu's reelaboration of Awmātiyūs' almanac, although the author says that he has used the Toledan Tables, which is incorrect.

Ibn Tibbon is also the author of the Prologue to Abraham bar Ḥiyya's *Sefer Hesbon Mahlekot ha-Kokabim* (Calculation of the Courses of the Stars).

EMILIA CALVO

REFERENCES

Ibn Tibbon. *Il quadrante d'Israel*. Ed. and trans. Guiseppe Bofitto and C. Melzi d'Eril. Florence: Libreria Internazionale, 1922.

IBN ṬUFAYL Abū Bakr Muḥammad ibn ʿAbd al-Malik ibn Muḥammad ibn Muḥammad ibn Ṭufayl al-Qaisi was an Arab physician famous for his encyclopedic learning. He was born at Guadix in Granada (modern Spain) in the first decade of the twelfth century. He served local rulers as physician and diplomat, and eventually became the court physician to Abū Yaʿqūb Yūsuf, the Almohad prince who became the most powerful Muslim ruler of his day. Ibn Ṭufayl enjoyed a close friendship with Abū Yaʿqūb until the latter's death in 1184; he was esteemed and cherished by Abū Yaʿqūb's successor, Abū Yūsuf Yaʿqūb, until his own death in 1185–6.

Ibn Ṭufayl's importance for history is twofold: the influence of his unique work, *Ḥayy ibn Yaqẓān*, and his patronage of one of the most important Muslim philosophers of the Middle Ages, Ibn Rushd (known to the west as Averroes). The medieval Arab historian, ʿAbd al-Wāḥid al-Marrākushi, recorded that Ibn Ṭufayl would bring all sorts of learned men to his sovereign's notice. It was in this way that Ibn Rushd was given audience with Abū Yaʿqūb Yūsuf, and there was led on by the praise and encouragement of Ibn Ṭufayl to demonstrate his familiarity with Aristotelian philosophical traditions. The Sultan himself joined in the discussion, and Ibn Rushd was commissioned to compose his great commentaries. It was through the mediation of Ibn Rushd's works that Aristotle was reintroduced to the Latin West. The astronomer al-Bīṭrūjī (known in the West as Alpetragius), was a disciple and friend of Ibn Ṭufayl; he mentions Ibn Ṭufayl's criticisms of the traditional astronomical theories on epicycles and eccentrics.

Ibn Ṭufayl was said to have written poetry as well as works on medicine and astronomy, but nothing seems to survive except *Ḥayy ibn Yaqẓān* and (possibly) a poem on medicine recently rediscovered (see Russell below). The highly unusual *Ḥayy ibn Yaqẓān* (Life, Son of Awareness) is a philosophical tale in the form of a fable: it tells how Ḥayy, cast up as a baby or spontaneously generated on an island, is fostered by a doe and grows to adulthood without any human contact whatsoever. Ibn Ṭufayl's interest lies in positing the natural unfolding of the human reason from the blank of infancy to the highest possible intellectual development;

Ḥayy reaches the summit of human possibility in a mystical experience of the godhead. Eventually Ḥayy meets another human being who is so filled with awe at Ḥayy's intellectual and spiritual attainments that he persuades him to make an attempt to convert human society to wisdom. Humanity proves to be unworthy of Ḥayy's teaching, and Ḥayy and his new-found friend retreat back to the island again to spend the rest of their lives in contemplation.

Ḥayy ibn Yaqẓān draws on the philosophy of Ibn Sīnā. It was edited as early as 1671 by Edward Pocock, who provided a Latin translation, thus inserting Ibn Ṭufayl's ideas into the mainstream of European culture. The *Philosophus Autodidactus*, as Pocock named it, was sent to Huygens, Locke, and Leibniz; it was possibly an influence on Defoe's *Robinson Crusoe*, and almost certainly one on Rousseau's *Émile*. English translations have been appearing since 1686; there are Dutch, German, French, Catalan, Hebrew, Persian, and Russian versions. It is still stirring up scholarly debate and controversy.

E. RUTH HARVEY

REFERENCES

De Vaux, Carra. "Ḥayy b. Yakẓan" and "Ibn Ṭufayl." In *Encyclopedia of Islam*. New edition, vol. 3. Leiden: Brill, 1971.

Gauthier, Léon. *Ibn Thofail: Sa Vie, ses Oeuvres*. Paris: Leroux, 1909.

Hawi, Sami S. *Islamic Naturalism and Mysticism: A philosophical Study of Ibn Ṭufayl's Ḥayy bin Yaqẓān*. Leiden: Brill, 1974.

Ibn Ṭufayl. *Ḥayy ben Yaqdhan, roman philosophique d'Ibn Thofail: traduction française*, 2nd ed. Paris: Vrin, 1983.

Ibn Ṭufayl. *Le Philosophe sans Maître*. Presentation de Georges Labica, traduction de Léon Gauthier. Algiers: Société Nationale d'Édition et de Diffusion [1969?]; reprinted 1988.

Nasri, Hani. "The 'Mystic' and Society According to Ibn Bajjah and Ibn Ṭufayl." *International Philosophical Quarterly* 26: 223–227, 1986.

Russell, G.A. "The Role of Ibn Ṭufayl, a Moorish Physician, in the Discovery of Childhood in Seventeenth-Century England." In *Child Care through the Centuries*. Ed. John Cule and Terry Turner. Cardiff: British Society for the History of Medicine, 1986, pp. 166–177.

See also: Ibn Rushd – al-Bītrūjī – Ibn Sīnā

IBN WĀFID Ibn Wāfid, Abū'l-Muṭarrif ʿAbd al-Raḥmān ibn Muḥammad was a pharmacologist and physicist who lived and worked in Toledo during the first half of the eleventh century (ca. 1008–1075). He studied the works of Aristotle, Dioscorides, and Galen in Cordoba but then he moved to Toledo where he planted a botanical garden for the king of this city, al-Maʾmūn.

Ibn Wāfid is the author of a book entitled *Kitāb fī-l-adwiya al-mufrada* (Book on the Simple Medicines) which is a synthesis of Dioscorides and Galen. It is an extensive work to which the author devoted twenty years. It was abridged and translated into Latin by Gerard of Cremona. Translations into Catalan and Hebrew are also extant.

Ibn Wāfid is also the author of a pharmacopeia and manual of therapeutics entitled *Kitāb al-wisād fī'l-ṭibb* (Book of the Pillow on Medicine) which, according to Juan Vernet, could be a misreading of the Arabic title *Kitāb al-rashshād fī'l-ṭibb* (Guide to Medicine). This work can be considered complementary to the preceding one because Ibn Wāfid describes compound medicines in it, and it is a practical book: the information given is based on experience. Ibn Abī Uṣaybiʿa attributes to Ibn Wāfid a work entitled *Mujarrabāt fī-l-ṭibb* (Medical Experiences) which could probably be identified with this Book of the Pillow.

Ibn Wāfid is also the author of two works entitled *Tadqīq al-naẓar fī ʿilal ḥāssat al-baṣar* (Observations on the Treatment of Illness of the Eyes) and *Kitāb al-Mughīth* (Book on Assistance) which are not preserved, and of a treatise on balneology which is preserved in a Latin version entitled *De balneis sermo* (Venice, 1553).

Ibn al-Abbār attributes to Ibn Wāfid a book entitled *Majmuʿ al-Filāḥa* (Compendium of Agriculture) although his authorship is now being questioned.

EMILIA CALVO

REFERENCES

Hopkins, J.F.P. "Ibn Wāfid." In *Encyclopédie de l'Islam*, 2nd ed., vol. 3, Leiden: E.J. Brill; Paris: G.P. Maisonneuve, 1971, p. 987.

Ibn Wāfid. *El libro de la almohada de Ibn Wāfid de Toledo (Recetario medico árabe del siglo XI)*. Ed. Camilo Alvarez de Morales. Toledo: Instituto Provincial de Investigaciones y Estudíos Toledanos, 1980.

Ibn Wāfid. *Kitāb al-adwiya al-mufrada*. Ed. Luisa Fernanda Aguirre de Cárcer. Madrid: C.S.I.C-I.C.M.A., 1993.

Samsó, Julio. *Las Ciencias de los Antiguos en al-Andalus*. Madrid: Mapfre, 1992, pp. 267–270 and 281.

Vernet, Juan. "Ibn Wāfid." In *Dictionary of Scientific Biography*, vol. 14. New York: Charles Scribner's Sons, 1976, pp. 112–113.

IBN YŪNUS Ibn Yūnus, Abu'l-Ḥasan ʿAlī ibn ʿAbd al-Raḥmān ibn Aḥmad ibn Yūnus al-Ṣadafī was one of the greatest astronomers of medieval Islam. Unfortunately nothing of consequence is known about his early life or education. As a young man he witnessed the Fatimid conquest of Egypt and the founding of Cairo in 969. In the period up to the reign of Caliph al-ʿAzīz (975–996), he made astronomical observations that were renewed by order of the Caliph al-Ḥākim, who succeeded al-ʿAzīz in 996 at the age of eleven and was much interested in astrology. Ibn Yūnus' recorded observations continued until 1003.

Ibn Yūnus' major work was a monumental *zīj* or astronomical handbook with tables. The *Ḥākimī Zīj*, dedicated to the Caliph, is distinguished from all other extant *zīj*es by beginning with a list of observations made by Ibn Yūnus and others made by some of his predecessors. Despite his critical attitude toward these earlier scholars and his careful recording of their observations and some of his own, he completely neglects to describe the observations that he used in establishing his own planetary parameters — nor does he indicate whether he used any instruments for these observations. Indeed, the *Ḥākimī Zīj* is a poor source of information about the instruments he used. In view of the paucity of this information, it is remarkable that the statement that Ibn Yūnus worked in a "well-equipped observatory" is often found in popular accounts of Islamic astronomy. A. Sayılı has shown how this notion gained acceptance in Western literature.

Ibn Yūnus' *Zīj* was intended to replace the *Mumtaḥan Zīj* of Yaḥyā ibn Abī Manṣūr, prepared for the Abbasid Caliph al-Maʾmūn in Baghdad almost 200 years earlier. When reporting his own observations, Ibn Yūnus often compared what he observed with what he had computed using the *Mumtaḥan* tables.

The observations he described are of conjunctions of planets with each other and with Regulus, solar and lunar eclipses, and equinoxes; he also records measurements of the obliquity of the ecliptic (chapter 11) and of the maximum lunar latitude (chapter 38).

In spherical astronomy (chapters 12–54) Ibn Yūnus reached a very high level of sophistication. Although none of the several hundred formulae that he presents is explained, it seems probable that most of them were derived by means of orthogonal projections and analemma constructions, rather than by the application of the rules of spherical trigonometry that were being developed by Muslim scholars in Iraq and Iran.

The chapters of the *Zīj* dealing with astrological calculations (77–81), although partially extant in the anonymous abridgment of the work, have never been studied. Ibn Yūnus was famous as an astrologer and, according to his biographers, devoted much time to making astrological predictions.

Ibn Yūnus' second major work was part of the corpus of spherical astronomical tables for timekeeping used in Cairo until the nineteenth century. It is difficult to ascertain precisely how many tables in this corpus were actually computed by Ibn Yūnus. Some appear to have been added in the thirteenth and fourteenth centuries. The corpus exists in numerous manuscript sources, each containing different arrangements of the tables or only selected sets of tables. In its entirety the corpus consists of about 200 pages of tables, most of which contain 180 entries each. The tables are generally rather accurately computed and are all based on Ibn Yūnus' values of 30°0′ for the latitude of Cairo and 23°35′ for the obliquity of the ecliptic. The main tables in the corpus display the time since sunrise, the time remaining to midday, and the solar azimuth as functions of the solar altitude and solar longitude. Entries are tabulated for each degree of both arguments, and each of the three sets contains over 10,000 entries. The remaining tables in the corpus are of spherical astronomical functions, some of which relate to the determination of the five daily prayers of Islam. The impressive developments in astronomical timekeeping in fourteenth-century Syria, particularly the tables of al-Khalīlī for Damascus, also owe their inspiration to the main Cairo corpus.

It is clear from a contemporary biography of Ibn Yūnus that he was an eccentric, careless, and absent-minded man who dressed shabbily and had a comic appearance. One day in the year 1009, when he was in good health, he predicted his own death in seven days. He attended to his personal business, locked himself in his house, and washed the ink off his manuscripts. He then recited the *Qurʾān* until he died — on the day he had predicted. According to his biographer, Ibn Yūnus' son was so stupid that he sold his father's papers by the pound in the soap-market.

DAVID A. KING

REFERENCES

Primary sources

MS Leiden Cod. Or. 143 and MS Oxford Hunt. 331 (major fragments of *al-Zīj al-kabīr al-Ḥākimī*). MS Paris B.N. ar. 2496 (anonymous abridgment constituting the sole source for some additional chapters).

MS Dublin Chester Beatty 3673 and MS Cairo Dār al-Kutub *mīqāt* 108 (complete copies of the Cairo corpus of tables for timekeeping).

Secondary sources

Caussin de Perceval, A. P. "Le livre de la grande table Hakémite." *Notices et extraits des manuscrits de la Bibliothèque Nationale* 7: 16–240, 1804.

King, David A. "The Astronomical Works of Ibn Yūnus." Ph.D. dissertation, Yale University, 1972.

King, David A. "Ibn Yūnus". In *Dictionary of Scientific Biography*. vol. 14. New York: Charles Scribner's Sons, 1976, pp. 574–580.

King, David A. "Ibn Yūnus' *Very Useful Tables* for Reckoning Time by the Sun." *Archive for History of Exact Sciences* 10: 342–394, 1973, reprinted in D.A. King. *Islamic Mathematical Astronomy*. London, 1986, 2nd rev. ed., Aldershot: Variorum, 1993, IX.

Sayılı, Aydın. *The Observatory in Islam*. Ankara, 1960, reprinted New York: Arno Press, 1981, pp. 130–156, 167–175.

Stevenson, F. R., and S. S. Said. "Precision of Medieval Islamic Eclipse Measurements." *Journal for the History of Astronomy* 22: 195–207, 1991.

See also: Yaḥyā ibn Abī Manṣūr – al-Maʾmūn – al-Khalīlī

IBN ẒUHR Abū Marwān ʿAbd al-Mālik ibn Ẓuhr (Latin: Avenzoar; 484–557/1092–1162) belonged to the Arabian tribe of Iyād, Banū Ẓuhr. His father, Abū al-ʿAlā ibn Ẓuhr (d. 525/1131), was a respected physician in the courts of the Murābiṭ dynasty (482–541/1090–1147). He trained his son in the art and craft of medicine.

Like his father, Ibn Ẓuhr started his career in the service of the Murābiṭ dynasty and earned a good reputation. During the reign of ʿAli ibn Tashfīn (499–537/1106–1143), he served at the palace in Marrakesh, Morocco, where his life was full of trials and tribulations. In ca. 535/1141, he was stripped of his official position and imprisoned. Although he was pardoned and released, he endured a hard life in prison and the experience left a resentment in his heart against the ruling dynasty.

The beginning of the reign of al-Muwāḥḥidūn proved to be a blessing for Ibn Ẓuhr. He was not only appointed as an official physician but also became a *wazīr* in the court of Abū Muḥammad ʿAbd al-Mūmin (d. 558/1163). During this period, Ibn Ẓuhr transmitted his knowledge and skills to his children with meticulous attention. His son and daughter both became famous physicians. One of Ibn Ẓuhr's treatises, *al-Tadhkirah* (The Remembrance), was dedicated to his son in appreciation of his achievements.

At least nine of Ibn Ẓuhr's works are known, but only a few are extant. He is said to have written two works and dedicated them to ʿAbd alʾ-Mūmin. *Al-Aghdhīyyah* (On Dietetics) is a text on the therapeutic properties of selected foods. *Al-Tiryāq al-Sabʿīnī* is a book of antidotes against poisoning by enemies. This work, perhaps, was Ibn Ẓuhr's way of expressing gratitude to his benevolent patron.

Al-Taysīr fī al-Mudāwāt wa al-Tadbīr (On Preventive Regimen and Treatment), in thirty treatises, is considered to be his monumental work. His friend Ibn Rushd considered it a great source of medical knowledge and therapeutic advice. In its Latin version, *Alteisir scilicet regiminis et medelae*, the treatise remained in wide circulation across Europe for centuries. Both Arabic and Latin copies of this work are extant True to the medical norms of his days, Ibn Ẓuhr presented in this work a mix of astrology and folklore blended with therapeutics and pharmacology. *Al-Kulliyāt* (The Collection) a general compilation on medical practice, appears as an appendix.

The great historiographer Ibn Abi ʿUṣaybiʿah, in his encyclopedic work *ʿUyūn al-Anbā fī Ṭabaqāt al-Aṭibbā*, has listed only seven of Ibn Ẓuhr's works:

1. *Fī al-Zīnah* (On Beautification)
2. *Al-Tiryāq al-Sabʿīnī* (On Antidotes)
3. *Fī ʿIlāl al-Kilā* (On Diseases of the Kidney)
4. *Fī ʿIllat al-Baraṣ wa al-ʾBahaq* (On Leprosy and Vitiligo)
5. *Al-Aghdiyyah* (On Dietetics)
6. *Al-Tadhkirah* (The Remembrance)
7. *Al-Taysīr fī al-Mudāwāt wa al-Tadbīr* (On Preventive Regimen and Treatment)

He does not mention two of his treatises. One of the missing ones is *Al-Iqtiṣād fī Iṣlāḥ al-Anfūs wa al-Ajsād* (Treatment and Healing of the Soul and Body). Addressed to the general reader, it focuses on problems of hygiene and therapeutics The second is *Jāmiʿ al-Asrār al-Ṭibb* (Compendium of Medical Wisdom). In addition to a discussion on dietetics, this work describes the physiology of several organs including the spleen, liver, and bladder.

Ibn Ẓuhr was a prolific writer and a highly successful medical practitioner. Some of his work showed the influence of the Hippocratic and Galenic traditions, but his original insights came from a rich family heritage in medical practice. Thus, going against the Greco-Roman medical dictates, Ibn Ẓuhr engaged in experimental work and recorded his observations. His detailed study of the pericardial abscess, pharyngeal paralysis, intestinal and mediastinal tumors, and inflammation of the middle ear were great improvements over the work of his predecessors.

Ibn Ẓuhr was one of the great Muslim clinicians and therapists. Through numerous translations of his works, he remained highly influential in the medical academies of the West. In his later age, he suffered from a malignant tumor that ultimately caused his death. He was buried in Seville, Spain.

MUNAWAR AHMAD ANEES

REFERENCES

Arnaldez, Roger. "Ibn Ẓuhr." In *Encyclopedia of Islam*, new edition. Leiden: Brill, 1969, pp. 976–979.

Hamarneh, Sami K. "Ibn Ẓuhr." In *Dictionary of Scientific Biography*. Ed. Charles C. Gillispie. New York: Scribners, 1976, pp. 637–639. Reprinted in *Health Sciences In Early Islam*. Ed. Munawar Ahmad Anees. San Antonio, Texas: Zahra Publications, 1984.

Wüstenfeld, F. *Geschichte der Arabischen Aertze und Naturforscher*. Göttingen, 1840. Reprinted Hildesheim: G. Olms, 1963.

See also: Ibn Rushd

IBRĀHĪM IBN SINĀN Ibrāhīm ibn Sinān ibn Thābit ibn Qurra was born in Baghdad in 296/909 where he died 37 years later in 335/946. Grandson of the famous mathematician Thābit ibn Qurra, cousin of the famous literary figure Hilāl al-Ṣābiʾ, son of a great physician and mathematician Sinān ibn Thābit, he thus belonged to an intellectual aristocracy whose members frequented the corridors of powers and the upper levels of the worlds of science and medicine. Ibrāhīm was born into and raised in this world, before being the object of a short-lived persecution to which he makes allusion but explains neither the cause nor the duration. Ibrāhīm ibn Sinān was not only the "heir" to great tradition, but a mathematician of genius in his own right who made his own mark on the mathematics of his era.

Ibrāhīm ibn Sinān was also heir to a historical tradition. He belonged to a privileged generation, the fourth since the Banū Mūsā. The translation of the major mathematical texts had been essentially completed and the great traditions of research had already been well established: that of the algebraists, beginning with al-Khwārizmī and extended by Abū Kāmil; that of the geometers, al-Jawharī, al-Nayrīzī, etc. who followed the work of Euclid; and the tradition of the Banū Mūsā which, thanks to mathematicians like Thābit ibn Qurra, had already gathered considerable results, developed new methods, and elaborated theories. Ibrāhīm ibn Sinān clearly was a part of this tradition in which Archimedean geometry, a geometry of measurement, and the geometry of Apollonius, which was concerned with the properties of positions, were all combined. Profiting from scholarly works, especially those of his grandfather Thābit ibn Qurra, Ibrāhīm ibn Sinān developed the study of geometric transformations and their applications to conic sections, as well as to the measurement of the area of a portion of a parabola. He extended their work on sundials in developing a theory of a whole class of instruments. Finally the questions posed by his predecessors about analysis and synthesis prompted him to write the first treatise worthy of the name on the subject.

Ibrāhīm ibn Sinān explained the composition of his work in his own autobiography which happily still exists. He published the autobiography after his twenty-fifth year, in 934. According to his own account, he began his research at age fifteen; at sixteen or seventeen he had written a first version of his book *Fī Ālāt al-aẓlāl* (On Instruments of Shadows), which he revised at the age of twenty-five. A year later he was discussing and criticizing Ptolemy's views on The Determination of the Anomalies of Saturn, Mars, and Jupiter (*Fī istikhrāj ikhtilāfāt Zuḥal wa al-Mirrīkh wa al-Mushtarī*) in a treatise that he completed six years later, at the age of twenty-four. In geometry, Ibn Sinān wrote *Fī al-dawāʾir al-mutamāssa* (The Tangent Circles), *al-Taḥlīl wa'l-Tarkīb* (Analysis and Synthesis), *Fī al-masaāʾil al-mukhtāra* (Selected Problems), *Fī misāḥat qiṭʿ al-maḥrūṭ al-mukāfī'* (The Measurement of the Parabola), and the *Fī rasm al-quṭūʿ al-thalātha* (Drawing of the Three Conic Sections). All these works had been published and revised before Ibn Sinān was twenty-five.

To illustrate the approach of Ibn Sinān let us look briefly at his *Measurement of the Parabola*, beginning with what he wrote about his own studies:

> My grandfather had determined the measurement of this section. But several contemporary geometers led me to understand that a work of al-Māhānī on the same subject, which they presented to me, was easier than my grandfather's. I was not pleased that al-Māhānī's work was more advanced than my grandfather's without there being one among us who surpassed his work. My grandfather had determined his result in twenty propositions. He proceeded from several arithmetic lemmas included in the twenty propositions. The question of the measurement of the section appeared to him clearly through the method of *reductio ad absurdum*. Al-Māhānī also proceeded from arithmetic lemmas. He then demonstrated his demand by the method of *reductio ad absurdum* in five or six propositions, which involved lengthy discussions. I myself then determined the measurement in three geometric propositions without recourse to any arithmetic lemma. I demonstrated the surface area of this same section using the method of direct proof, and I did not need the method of *reductio ad absurdum*.

In addition to expressing pride in his heritage and the certainty of an exceptional scholar, these remarks also reflect the qualities of Ibn Sinān as a mathematician: brevity, ease, and elegance.

Ibn Sinān's approach is the following: in a first proposition, he demonstrates that the affine transformation conserves the ratios of the areas in the case of triangles and polygons; he then demonstrates in a second proposition, that it is the same for the ratio of the area of a portion of a parabola to that of an associated triangle.

Ibn Sinān again had recourse to geometric transformations in his *Drawing of the Three Conic Sections*. In this treatise, he constructed an ellipse by transforming the circle by an orthogonal affinity — a process already found in the works of the Banū Mūsā. It was also by means of an affinity that he deduced all the hyperbolas of a particular hyperbola for which the latus rectum is equal to the transverse axis.

The intensity of mathematical activity, the large amount of work done, the new demands of brevity, elegance, and rigor in demonstration, as well as the general interest in geometric transformations, all led Ibn Sinān to take up the theory of analysis and synthesis again. Thus he published the first treatise devoted to this subject. He wrote: "I found that the geometers of this time had neglected Apollonius's method of analysis and synthesis, as was the case for the majority of things I brought up, and that they had limited themselves just to analysis, limiting themselves to such a degree that they led analysis to the point of letting people think that this analysis was not the synthesis that they were doing".

In this treatise, Ibn Sinān undertook two tasks at once: one didactic, the other logical. On the one hand, he proposed a method (*ṭarīq*) for geometry students which allowed them to solve geometry problems; on the other hand, he reflected on geometric analysis itself, proposing a classification of geometry problems according to their numbers and the hypothesis that they are to verify, and explaining for each class of problems the respective parts of analysis and synthesis.

There is not enough space here to take up, even very briefly, the works of Ibn Sinān. But the examples discussed above do show how this eminent geometer left his mark on all the fields in which he worked, including mathematics and philosophy. The import of his work can be discovered in the work of Ibn al-Haytham who followed up on the research of Ibn Sinān in his own treatise on analysis and synthesis and in his writings on sundials.

ROSHDI RASHED

REFERENCES

Bellosta, H. "Ibrāhīm ibn Sinān: On Analysis and Synthesis." *Arabic Sciences and Philosophy* 1(2): 211–232, 1991.

Rashed, R. "Ibrāhīm ibn Sinān ibn Thābit ibn Qurra." In *Dictionary of Scientific Biography*, vol. VII. New York: Charles Scribners' Sons, 1973, pp. 2–3.

Rashed, R. *Mathématiques infinitésimales du IXᵉ au XIᵉ siècle*. Vol. I. *Fondateurs et commentateurs: Banū Mūsā, Ibn Qurra, Ibn Sinān, al-Khāzin, al-Qūhī, Ibn al-Samḥ, Ibn Hūd*. London: al-Furqān Islamic Heritage Foundation, 1995.

Rashed, R. R. Morelon, and H. Bellosta. *Astronomie, Géométrie et Philosophie d'Ibn Sinān (Oeuvres Complètes)*, forthcoming.

Saidan, A.S. *Rasā'il Ibn Sinān*. (The Works of Ibrāhīm ibn Sinān. In Arabic). Kuwait: Qism al-Turāth al-ʿArabī, 1983.

Suter, H. "Abhandlung über die Ausmessung der Parabel von Ibrāhīm b. Sinān b. Thābit." In *Vierteljahrsschrift der Naturforschenden Gesellschaft in Zürich*. vol. 63, 1918, pp. 214ff.

Youschkevitch, A.P. *Les mathématiques arabes*. Paris: Vrin, 1976.

See also: Banū Mūsā – Thābit ibn Qurra – Sinān ibn Thābit

AL-IDRĪSĪ Abū ʿAbd Allāh Muḥammad ibn Muḥammad ibn ʿAbd Allāh ibn Idrīs, known as al-Sharīf al-Idrīsī, was one of the great Arab geographers of medieval Islam. He was born in Ceuta (Morocco) in AH493/AD1100 and died there in AH 560/AD1166. Al-Idrīsī belonged to a ruling family, the Alavī Idrīsīs, who were claimants to the caliphate and had ruled in the region around Ceuta from AD 789–985. His ancestors were the nobles of Malaga, Spain who, unable to maintain their authority there, migrated to Ceuta in the eleventh century. Al-Idrīsī was educated in Cordoba and began traveling at a very early age. At the age of 16, he visited Asia Minor and then traveled in southern France, England, Spain, and Morocco.

This was a period of the growing power of the Normans in Europe and the Mediterranean region. It is said that Roger II (1097–1154), the Norman ruler, invited al-Idrīsī to come and stay with him in Palermo (Sicily), saying he would be safe from Muslim kings who were trying to murder him. Al-Idrīsī accepted the king's invitation and went to Palermo sometime in AD 1138 and stayed there until after the death of Roger in AD 1154. Then he returned to Ceuta.

Sicily at this time was an important center, where Arab-Islamic and Western European cultures intermingled. Roger himself was very interested in the promotion of arts and sciences, and we learn from al-Idrīsī that he was also interested in geography and astronomy. Roger gave al-Idrīsī an important task, probably from political motives, which was to construct a world map. Roger was still engaged in expanding his empire in North Africa, and thought al-Idrīsī would be a suitable person for this task. Although he was well-traveled in North Africa and Europe, al-Idrīsī was not a geographer in the true sense of the word at this stage. But he began constructing the map under the patronage and supervision of Roger. Finding that the courtiers at his palace did not possess sufficient knowledge of the geography of the world, Roger sent envoys to different regions to collect fresh data. After the information was acquired, al-Idrīsī utilized only such data on which there was unanimity; the rest was discarded. He also used the Arabic geographical works which were available to him, such as the Arabic version of Claudius Ptolemy's (ca. AD 90–186) *Geography* (Arabic *Jughrāfiya*), *Kitāb Ṣūrat al-Arḍ* (Book on Routes and Kingdoms) by Ibn Hawqal, *Al-Masālik wa 'l Mamālik* by Ibn Khurdādhbih (AH 232/AD 846), and the lost work of al-Jayhānī entitled *Kitāb al-Masālik wa'l Mamālik*. Thus it seems that al-Idrīsī had a vast amount of geographical material on the known world at his disposal.

He worked on the map for fifteen years, basing it primarily on Ptolemy's map, with some modifications. When it was completed, Roger asked him to have it carved on a round silver disc, which al-Idrīsī did with great skill with the help of artisans. The silver map had all the physical features and names of places drawn on it. Roger was so pleased with al-Idrīsī's performance that he asked him to write a book on world geography. He produced the voluminous compendium *Nuzhat al-Muṣtāq fi ikhtirāq al-ʾĀfāq*, the full Arabic text of which has now been published as *Opus Geographicum* under the auspices of the Italian Institute of the Middle and Far East in Rome and the Institute of Oriental Studies at Naples University. Although the silver map has not survived, the original world map has. The *Nuzhat* is supposed to be a description of this world map. Al-Idrīsī divided the known world into seven climes (*iqlīm*) running parallel to the equator up to 64° N. latitude and divided each of the climes into ten longitudinal sections. Thus there are seventy odd sectional maps, and the arrangement of the book follows the seventy divisions. In many cases, there is more information in the book than is depicted on the maps. Al-Idrīsī's world map is the most detailed and largest map drawn by any Muslim cartographer. His book is indeed a mine of information on physical, topographical, human, cultural, and political geography. From the world map and the book, one can see that his knowledge of Europe, North Africa, and West and Central Asia was much deeper and more correct and extensive than it was of South Asia or the Far East.

Al-Idrīsī had a mathematical basis for his map, but he did not provide latitudes or longitudes as al-Maghribī did, who followed the pattern of al-Idrīsī. As he tried to include all the information at his disposal, some distortions were bound to take place, especially in the northern and southern regions. For example, the shapes and positions of the islands in the Indian Ocean were distorted. Even with these errors, his book is an encyclopedia of geographical knowledge of medieval times, and was an important geographical textbook for a long time.

SAYYID MAQBUL AHMAD

REFERENCES

Ahmad, S. Maqbul. *India as Described by al-Sharīf al-Idrīsī in his 'Kitāb Nuzhat al-Muṣtāq fi ikhtirāq al-ʾĀfāq'*. Leiden: E.J. Brill, 1960.

Ahmad, S. Maqbul. "al-Idrīsī." In *Dictionary of Scientific Biography*, vol. 7. Ed. Charles C. Gillispie. New York: Charles Scribner's Sons, 1970–80, pp. 7–9.

Ahmad, S. Maqbul. "Cartography of al-Sharīf al-Idrīsī." In *The History of Cartography*, vol. 2, Book 1, *Cartography in the Traditional Islamic and South Asian Societies*. Chicago: University of Chicago Press, 1992, pp. 156–176.

Ahmad, S. Maqbul. *A History of Arab-Islamic Geography*. Mafraq, Jordan: Al al-bayt University, 1995.

al-Idrīsī. *Opus Geographicum*. Ed. E. Cerulli and A. Bombaci. Naples: Istituto Universitario Orientale di Napoli, 1970.

Oman, Giovanni. "al-Idrīsī." In *Encyclopaedia of Islam*, vol. 3. Leiden: E.J. Brill, 1960, pp. 1032–35.

Oman, Giovanni. "Notizie bibiliografiche sul geografo arabo al-Idrīsī (XII secolo) e sulle sue opere." *Annali dell'Istituto Universitario Orientale di Napoli* 11: 2561, 1961.

IKHWĀN AL-ṢAFĀʾ The forty-eight treatises composed by the Ikhwān al-Ṣafāʾ (Brethren of Purity) constitute an encyclopedia of the sciences which has had a long and influential career in the Arabic/Islamic world. There has been considerable debate among historians over the identity of the authors of this collection. Current scholarship places the date of composition in the last half of the fourth/tenth century and the place of composition in Basra.

These treatises contain a strong gnostic element, an emphasis on the esoteric knowledge that exists within the exoteric or phenomenal features of everyday life but is open only to the initiated. This interest in the esoteric or inner knowledge, however, does not imply that they were not interested in the physical world. It is necessary to have a thorough understanding of nature in order to penetrate to the inner knowledge that it expresses. An appreciation of nature, therefore, is one of the first steps toward union with the Divine Knowledge.

This gnostic element can be seen already in the classification of knowledge proposed by the Ikhwān. They begin with introductory or preparatory topics (*riyāḍiyya*), advance to religious sciences (*sharʿiyya*), and finally reach the highest, or philosophic (*falsafiyya*) sciences. This system is different from many Arabic/Islamic discussions of the structure of the sciences in that it places religious sciences in a position subordinate to philosophic or intellectual sciences. These higher sciences include arithmetic (subsuming also geometry, astronomy, and music), logic, natural history (including physical principles, cosmology, generation and corruption, meteorology, mineralogy, botany, and zoology), and theology (the science of the Creator, of spiritual beings, and of psychical beings, as well as politics and political discipline, in addition to eschatology). When arranging the treatises in this encyclopedia, however, the Ikhwān adopt a slightly different arrangement in four sections: propaedeutic studies, physical sciences, psychical sciences, and the metaphysical and revealed sciences.

The gnostic character of these treatises is evident in the tendency toward Neoplatonic and Neo-Pythagorean formulations of philosophic problems. The cosmological scheme of the Ikhwān clearly illustrates this facet of their thought. God originally created the universe in a series of emanations from Himself. First in this series was Intellect, followed by its Archetypes and the World Soul. (Intellect, the Ikhwān tell us, instructed the World Soul through the Archetypes.) The World Soul, through a further series of emanations, produced individual souls or faculties that individuate or give form to Prime Matter. The first of these individual emanations were the nine celestial spheres (the invisible *primum mobile*, the sphere of the fixed stars, and the spheres of the seven planets: Saturn, Jupiter, Mars, the Sun, Venus, Mercury, and the Moon), followed by the four terrestrial elements (fire, air, water, earth). This represents the farthest separation of the soul from God. Drawn by an irresistible desire for union with God, it now commences a return in which it mounts higher and higher through the mineral, plant, animal, and, finally, human realms. (Man is seen as straddling the boundary between the earthly and the celestial.) Soul rises through these realms by a process of purification from the contamination with Matter by means of personal morality and knowledge. Just as there is a hierarchy of soul in the natural world, so there is a hierarchy among human souls, expressed in the metaphor of the Spiritual city (a Neoplatonic interpretation of Plato's Republic within the context of Islamic culture): the enlightened souls form a Spiritual City under the leadership of imams and prophets — chief of whom is the Prophet Muḥammad — whom God has sent to instruct humankind in the process of freeing itself from the constraints of matter.

Some scholars have claimed that these treatises represent a deliberate attempt to formulate an Ismā‘īlī position. (The Ismā‘īlīs represent one branch of the Shī‘i community. The Shī‘ites believe that the rightful line of leadership in the Islamic community should have passed to ‘Alī, the son-in-law of Muḥammad, and thence to his sons, rather than to Mu‘āwiyah, who was not a close relative of Muḥammad but only a senior member in his tribe. The Ismā‘īlīs split with the Shī‘i community over the line of succession from ‘Ali's son, al-Ḥusayn. The sixth imam, Ja‘far al-Ṣādiq, deposed his eldest son, Ismā‘īl, as his successor and named instead his son Mūsa al-Kāẓim. The Ismā‘īlī sect traces its beginning to the refusal of some to accept this decision on the part of the imam Ja‘far.) Whether or not that is the case, these treatises were early adopted by the Ismā‘īlī sect of Islam as a statement of their beliefs. They continue to play an important role in Islamic intellectual life even today.

GREGG DEYOUNG

REFERENCES

Bausani, A. "Scientific Elements in Ismā‘īlī Thought: The Epistles of the Brethren of Purity." In *Ismā‘īlī Contributions to Islamic Culture*. Ed. S. H. Nasr. Tehran: Imperial Iranian Academy of Philosophy, 1977, pp. 121–140.

Dieterici, F. *Die Abhandlungen der Ichwan aṣ-ṣafa in Auswahl zum ersten Mal aus arabischen Handschriften herausgegeben*. Leipzig: J.C. Hinrichs, 1886.

Diwald, S. *Arabische Philosophie und Wissenschaft in der Enzyklopädie Kitāb Ihwan aṣ-Ṣafa›*. 3 vols. Wiesbaden: Harrassowitz, 1975.

Marquet, Y. *La philosophie des Ihwan al-Safa›*. Algiers: Société nationale d'édition et de diffusion, 1975.

Nasr, S. H. *An Introduction to Islamic Cosmological Doctrines*. Cambridge, Massachusetts: Harvard University Press, 1964.

Stern, S. "New Information About the Authors of the ‹Epistles of the Sincere Brethren'." *Islamic Studies* 3: 405–428, 1964.

Ullmann, M. *Die Natur- und Geheimwissenschaften im Islam*. Leiden: Brill, 1972.

INDIA: MEDIEVAL SCIENCE AND TECHNOLOGY This article deals not with the ancient period of Indian history, but with the time after the conquest of northern India by the Turko–Afghāns between AD 1192 and 1206. Its focus is on Medieval India.

India had already had commercial and cultural relations with Greece, Afghanistan, and Central Asia since prehistoric times. When the Muslims arrived in India a cross fertilization of scientific and intellectual ideas began. Muslims migrated to India after AD 1258 because of the Mongol invasion of Iran, Iraq, and Central Asia; some of them were astronomers, mathematicians, and physicians.

"With the establishment of the Ghaznavid and Mughal rule in India the Greek or rather more advanced Ptolemaic astronomy in an Arabic version reached India and began to be studied and taught among the Muslim and Hindu astronomers who appreciated its merit." (Sen, 1971). Al-Bīrūnī (d. AD 1050) claims that he translated Euclid's *Elements* and Ptolemy's *Almagest* into Sanskrit, but these translations are not available. In any case, the Arabic versions of these two books were introduced by the Muslims, and towards the close of the twelfth century AD mathematical books in Arabic also began to trickle into India.

Some of the Muslim mathematicians and astronomers who settled in India knew these Arabic translations and translated them into Persian. Thus Greco–Arab astronomy and mathematics were introduced. Some of this material was incorporated into textbooks used in the educational institutions. Akbar (d. AD 1605) made those subjects compulsory.

Both mathematics and astronomy were used by the Hindus and Muslims for religious purposes. Mathematics was also studied for its practical utility, as in the construction of buildngs and monuments. Later, geometry was further developed for this purpose.

The outstanding mathematician-astronomers of this period were Śrīdharācārya (ca. AD 991), Śrīpati (AD 1039–1056), Śatānanda (fl. AD 1099), and Bhāskarāchārya II (AD 1150).

Al-Bīrūnī studied the Indian sciences in the original Sanskrit texts and translated some of them into Arabic. His book *Kitāb al-Hind* (Book on India) gives valuable information about astronomical methods and Indian astronomers, some of whom were personally known to him. His *Ras'āil* (Treatise), the *al-Qānūn al-Mas'ūdī* (Masudic Canon) contains useful source material for the history of science in India in the first half of the eleventh century.

A new development in astronomy was the introduction of Arabo–Persian–Greek *Zīj* literature (astronomical tables). Abu'l-Faḍl mentions eighty-six *Zīj*es in his *Ā'īn-i Akbarī*. There were several *Zīj*es prepared in pre-Mughal India.

Astrolabes were used in India after the Muslims arrived. In 1370, during the reign of Sultan Fīrōze Shāhā Tughluq, an astrolabe was constructed and named *Asṭurlāb-i Fīrōze Shāhī*. This testifies to the fact that Indians possessed adequate knowledge of applied technology during the fourteenth century.

Ḍiyā' ad-Dīn Baranī records the names of the astronomers and astrologers who flourished during this period, and adds that other sciences such as *ramal* (geomancy) and *al-Kīmyā* (alchemy) also flourished.

Another fact worth mentioning is that the first substantial contact between the Āyurvedic (Indian) and Ūnānī (Greco–Arab) systems of medicine began during this period, which resulted in mutual enrichment in therapeutics, materia medica, and pharmacology. There were seventy hospitals in Delhi alone under Sulṭān Muḥammad ibn Tughluq (AD 1297–1348) having 12,000 *Vaidyas* (Hindu physicians) and *Ṭabībs* (Arab physicians) paid by the state.

Medicinal herbs and plants were widely cultivated. Al-Bīrūnī's *Kitāb al-Saydanah fi't-Ṭibb* (Book on Pharmacy and Materia Medica) contains useful information for the pharmacographia Indica. Compiled in the middle of the eleventh century, it is an encyclopedia of simple drugs containing medicinal herbs and minerals used in the Ūnānī system, arranged in alphabetical order. There are names of hundreds of Indian medicinal herbs and plants.

The arrival of the Turko–Afghāns in India also brought some changes in the area of technology. In the field of metallurgy, iron, copper, brass, gold, silver, and other minerals were extracted from the mines in different parts of India by a simple technology with the help of a clay furnace in which wood and charcoal were burnt. As the army was important, iron was much in use for the manufacture of arms and armor. For the minting of coins, mostly gold, silver, lead, bronze, brass, and copper were used.

The premier industry in India was cotton textile; the dyeing, printing, and painting of clothes had been done since ancient times. In connection with the innovation in textile technology, the general use of the spinning wheel should be mentioned. There is a lack of positive evidence ascribing its origin to India. Moreover, the word for it is *charkha* which is Persian. In the absence of better evidence, literary ones from Indo-Persian works can be put forward to show that the spinning wheel in which the belt-drive technique was employed was in use at the end of the twelfth century. Another piece of equipment which was much used in this period was the bow-string (*Tant* and *Kamān*) for cleaning the cotton and separating the seeds (ginning). This technique is still used in India today.

Al-Bīrūnī gives evidence that around AD 1030 the materials used for writing in India were mainly black tablets, palm leaves, the bark of the Tūz tree called *Bhūrja*, and silk. The manufacture of white paper started in India in the thirteenth century. Amīr Khosrāw mentions paper (*Kāghaz*) several times in his work. He states that paper was made with cotton, linen, silk (*Qash, Ḥarīr*), and reed (*Kilk*). They were soaked in water, then pounded and turned into pulp, and dried. After that, they were cut with sharp scissors. He adds that this light paper was quite costly. As regards the preparation of ink, the earliest recipes for hair dye in India are found in the *Navanītaka* (ca. second century AD). From this Nityanath Siddha (AD 1200) derived his recipes for ink as recorded in his *Rasaratnākara*. The ingredients used were metallic or herbal substances such as lamp-black charcoal, gum, burnt almond husk, or gold and silver powder.

In the field of irrigation, the large Perumamilla tank bears an inscription dated AD 1291. There is a difference of opinion about the device for water raising used in ancient India called *Araghaṭa* or *Ghati-Yantra* which are Sanskrit words. Several scholars have argued that it was the same as the Persian wheel or *sāqiyah* and it was not introduced into India by the Turko–Afghans. But these two words are not clearly mentioned in any early source. It was actually the simple *noria,* a wheel carrying pots and buckets on its rim which did not involve gearing. One view is that the Persian wheel was introduced by the Turko–Afghāns in India in the thirteenth century. It was used to raise well water with the help of a chain of buckets driven by animal power using pin-drum gearing.

The planks of the hulls of Indian ships were sewn together by coir and generally not nailed. G.F. Ḥourānī says that al-Maʿsūdī states "...in ships of the Indian Ocean iron nails do not last because the sea-water corrodes the iron and the nails grow soft and weak in the sea and therefore the people on its shores have taken to threading cords of fibre instead and these are coated with grease and tar". Ibn Baṭṭūṭa, who arrived in India in September 1333, gave a detailed account of the technique used in the ships' manufacture.

Indian builders were employed by the Muslim rulers for the construction of monuments, and several expert masons and craftsmen were also brought from Persia and Afghanistan, introducing the Arabo-Persian architectural and decorative traditions. They used the lettering of the Quranic verses as well as leaves, flowers, buds, and geometrical designs (*arabesque*). This brought about great changes in Indian architectural style. The imported style consisted of domes, portals, minarets (*mīnār*), and pendentive and squinch arches. In so far as building materials are concerned, stones and bricks were used, and lime-mortar was employed perhaps for the first time in India.

So many articles made of glass have been discovered, especially at Kopia in Uttar Pradesh, and the Taxila and Satavahana sites, that it cannot be argued that all of them were imported goods. Most of the articles are beads, bangles, bowls, slags, small vessels, tiles, and flasks. But very little information is available concerning the technique of glass-making, including the raw materials, furnaces, or tools used.

In the field of military technology, it is generally asserted that the stirrup was used in ancient India, and some evidence shows that the flat iron stirrup was introduced by Turkish conquerors in the early thirteenth century. Nailed horseshoes were used after the Turko-Afghān conquest.

To manufacture a sword the two ingots were covered with soft earth in order to prevent carbon loss; then both the ingots were forge-welded by hammering when the steel was softened. The process of hardening steel by quenching the red hot metal into water or oil is still employed today. The first hammering was for preparing a blank, while the second was to give it shape. The blunt sword was sharpened on an abrasive wheel and polished later with oxymel.

The question arises whether artillery and gunpowder were used before Babar's time (AD 1530) or not. This is a controversial issue. Some historians state that mechanical artillery was in use in India as early as the first quarter of the thirteenth century; others are of the view that it was not used until AD 1365–1366 (Makhdoomee, 1936). Naptha or Greek fire was used by the Turko–Afghān invaders, but this also may mean gunpowder.

One of the difficulties in writing this history is that the source materials, especially the manuscripts in Sanskrit, Arabic, and Persian, have not been edited and published. Some of them are uncatalogued so that they are unknown even to those who do research on the history of science in India. Therefore, it is evident that a thorough assessment of India's contributions to scientific and technological development is not possible now, and the history of science and technology in late ancient and early medieval India presented here is neither thorough nor complete.

M.S. KHAN

REFERENCES

al-Bīrūnī. *Kitāb al-Hind* (Alberuni's India). Trans. E.C. Sachau. New Delhi: Oriental Reprint, 1983.

Bag, A.K. "Al-Bīrūnī on Indian Arithmetic." *Indian Journal of the History of Science* 10: 174–186, 1975.

Chowdhury, Mamata. "The Technique of Preparing Writing Materials in Early India with Special Reference to al-Bīrūnī's Observations." *Islamic Cultures* 48(1): 33–38, 1974.

Chowdhury, Mamata. "The Technique of Glass Making in India." Paper presented at *National Seminar on Technology and Science in India during 1400–1800* AD. New Delhi: Indian National Science Academy, 1978.

Gode, P.K. "Recipes for Hair Dye in the *Navanītaka* and Their Close Affinity with the Recipe for Ink Manufacture after AD 1000." *Studies in Indian Cultural History* 1: 101–110, 1961.

Ḥourānī, George Fadlo. *Arab Seafaring in the Indian Ocean in Ancient and Early Medieval Times*. Princeton: Princeton University Press, 1951.

Khan, M.S. *Al-Biruni and Indian Sciences*, forthcoming.

Khan, M.S. "Arabic and Persian Source Materials for the History of Science in Medieval India." *Islamic Culture* April–July, 1988, pp. 113–139.

Khan, M.S. "The Teaching of Mathematics and Astronomy in the Educational Institutions of Medieval India." *Muslim Education Quarterly* 6, 1989.

Khan Ghori, S.A. "Development of *Zīj* Literature." In *History of Astronomy in India.* Ed. S.N. Sen and K.S. Shukla. New Delhi: Indian National Science Academy, 1985, pp. 21–48.

Makhdoomee, M.A. "Mechanical Artillery in Medieval India." *Journal of Indian History* 11: 189–195, 1936.

Mukherjee, B.N. "Technology of Indian Coinage, Ancient and Medieval Period." In *Technology in Ancient and Medieval India.* Ed. Aniruddha Roy and S.K. Bagchi. Delhi: Sundeep Prakashan, 1986, pp. 47–70.

Ray, Priyadaranjin. *History of Chemistry in Ancient and Medieval India*. Calcutta: Indian Chemical Society, 1956.

Ray, Priyadaranjan. "Medicine as it Evolved in Ancient and Medieval India." *Indian Journal of the History of Science* 6: 86–100, 1970.

Sen, S.N., D.M. Bose and B.V. Subbarayappa. *A Concise History of Science in India*. New Delhi: Indian National Science Academy, 1971.

White, Lynn. "Tibet, India and Malaya as Sources of Western Medieval Technology". *American Historical Review* 65(3): 517, 1960.

See also: al-Bīrūnī – Astronomy in India – *Zīj* – Āyurveda – Agriculture – Technology – Metallurgy – Textiles – Paper and Papermaking – Irrigation – Navigation – Architecture – Military Technology

INO TADATAKA Ino Tadataka, 1745–1818, was a Japanese cartographer and an energetic field observer. A major astronomical and geodetic problem of the time in Japan was the finding of the length of a meridian by Japanese measure. Since Sino-Jesuit works had set zero longitude at Beijing, that of Japan had to be accurately measured so that, in predicting a solar eclipse, the Sino-Jesuit method could be employed for the Japanese longitude. After making over 2000 measurements of latitude, Ino calculated the length of a meridian which agreed (within several tenths of a second of a degree) with the figure given in the Dutch translation of Lalande's *Astronomie* which had been imported into Japan.

He did not excel in devising new methods or new theories in either astronomy or geodesy. While he was active, knowledge of Western astronomy was available through Dutch translations and Sino-Jesuit works and after, through the works of Lalande; but Ino had no knowledge of Dutch or dynamics and little understanding of astronomical theories. When calculating the length of the meridian, he considered the earth as a perfect sphere rather than a spheroid. Moreover, when observing the position of fixed stars, he did not take into account the effects of refraction, parallax, or nutation.

In his surveying, Ino did not use modern triangulation but relied upon the old traverse method. His mapmaking approach resembled the Sanson–Flamsteed method (it is presumed that his method was developed independently), which is appropriate only for small areas. Ino none the less used the method for an area as large as all of Japan. Despite Ino's scientific failings, his map of Japan, based upon surveys covering the length and breadth of the land, has an important place in geographical history.

NAKAYAMA SHIGERU

REFERENCES

Most of Ino's works, consisting mainly of maps, observations, records of his surveys, field notes, and diaries are preserved in the Ino Memorial Hall in Sahara City.

Otani, Ryokichi. *Ino Tadataka*. Tokyo: Iwanami, 1932.

IRRIGATION IN INDIA AND SRI LANKA Assessments by historians of Asia's irrigation systems and irrigation-related civil engineering techniques have been based on the scantiest of historical or empirical data. Naturally, they have ranged from one extreme to the other. Of these, the one most easily recognized and debated was provided by Karl Wittfogel whose theories led to the idea of "hydraulic civilizations".

A diametrically opposite assessment has been provided by some Indian historians who have concluded that there was no significant irrigation technology in use at all. Symbolic of this view is R. Majumdar and H.C. Raychaudhuri's *An Advanced History of India*, in which the authors make a categorical statement on the "comparative absence of artificial irrigation" in eighteenth century India.

However both views — fairly representative of the historiographical terrain — have had to be revised considerably because of the emergence of new historical materials and investigations during the last two decades. These are reflected in new literature specifically devoted to the subject. Illustrative of these materials is the report of Alexander Walker, an English specialist who toured India in the eighteenth century. Walker produced an elaborate treatise on Indian agriculture in which he drew the conclusion that "the practice of watering and irrigation is not peculiar to the husbandry of India, but it has probably been carried there to a greater extent, and more laborious ingenuity displayed in it than in any other country".

This display of ingenuity, however, is not restricted to eighteenth century India and has indeed found expression in a plethora of irrigation systems in Asia each designed to appropriate its own specific ecosystem potential. There is evidence of large scale irrigation works in several Asian countries including China and Sri Lanka. The systems studied on the sub-continent include gigantic artificial lakes, large-scale and small-scale embankments, and diversion channels. They include schemes for taking water up a hill against gravity, elaborate canal distribution networks, innovations like the *khazans* on the west coast of India, where unmanned wooden sluice gates control the sway of salt and sweet water in low lying paddy fields adjoining the coastal or tidal rivers, and storage tanks with a bewildering variety of names.

"The irrigation history of Indian has been studied only in fragments" (Sengupta, 1993). But even this admittedly fragmentary picture that is emerging is far more fascinating than the simplistic or impressionistic scenarios of Wittfogel or later historians like Raychaudhuri and Majumdar. The most fundamental aspect of irrigation technology and civil engineering works to be noted is that almost all of them are related to monsoon precipitation in one way or another. Over ninety per cent of the annual run-off in the peninsular rivers and eighty per cent in the Himalayan rivers occurs during the four months of the monsoon. Thus, unlike the case of temperate ecosystems, irrigation becomes extremely crucial: in the wet season which stretches approximately four months, in several places less, there can be too much precipitation over intense bursts. This is followed by a dry season during which there is no precipitation at all. The result is predictable: periods of excess water followed by drought.

In this context, the basic design of irrigation technology is intimately related to precipitation: how to save it, store it, and divert it, so that spatially it reaches areas where there is no water (diversion techniques) or temporally makes it available during the dry months (storage techniques). Thus, rain-fed rivers are diverted into channels, or river basins are interconnected. Or the rainfall is directly collected in huge storage facilities on the land.

If this is the scenario, (and it is as valid today as it was in 3000 BC), one would normally expect a much richer history of irrigation techniques in Asian conditions — where rice is a basic crop adapted to growing largely in water — than in any other part of the planet, particularly the temperate zones. It would also follow that the irrigation designs evolved for coping with such situations would not readily be available in other ecosystems. For this reason, it has taken some time for engineers and historians trained in other culture areas to appreciate their worth and function.

The irrigation experience of China is documented in Joseph Needham's *magnum opus*, *Science and Civilisation in China*, and will not concern us here. We shall restrict ourselves in this essay to a consideration of the irrigation and civil engineering techniques that arose in the Indian subcontinent including Sri Lanka and which were the result of a close interaction and adaptation between overall environmental situations and human ingenuity.

Irrigation Technologies

In the circumstances related above, it stands to reason that the primary design objective of irrigation engineers would be predominantly in the direction of a water storage system. The following listing is given by Shankari and Shah.

There were storage systems designed purely for drinking water: *nadi*, tanks, *bowari*, *jhalara*, and *pokhar*. Some were reserved only for human beings; others were for human beings and animals. These we shall ignore here.

The second category of storage technologies relates to irrigation, and there is considerable evidence of the spread of such technologies through the length and breadth of the subcontinent. Though the structures here were all designed for irrigation, they also provided other useful functions of soil conservation and ground water percolation and recharge.

Irrigation water stored in such storages was conveyed through two methods: first, under the force of gravity, or gravity irrigation, and second, through extractive or lift techniques or devices of some kind, including for instance the Persian wheel.

There were three main classes of such irrigation-related storage systems. The first comprised tank and pond irrigation systems. These in turn were of two types: *above surface* storage works, where a reservoir was created above the ground through a fairly long embankment. The corresponding structures were called *Keri*, *Eri*, *Cheruvu*, *Kalvai* and *Kunta* in south India and *Ahar* in Bihar (northern areas). The second category involved *below surface* storage works, which included dug ponds from which water was lifted by some means manual or mechanical: *pokhar*, *talab*, *jhil*, *beel*, and *sagar*.

The second class of irrigation systems comprised land inundation systems: the land was flooded and saturated before cultivation and then drained off prior to planting (another term used is flood irrigation). These were primarily above surface types or referred to (in India) as submergence tanks. Important variations of these are the *khadin* and the *johad* in Rajasthan and the *bundhies* of Madhya Pradesh.

The *bundhies* were built generally in a series and therefore captured every possible drop of rainwater that fell. If there were a surplus, a waste weir was provided. There was generally a sluice at the deepest part of the storage reservoir. A *stambh* or pillar would indicate the location of the sluice. Sluices could be of different types: pipe sluices or sluices made of masonry for the larger tanks. (Some sluices open, as in the *khazans*, with the pressure of the incoming tides and discharge water automatically when the tide has fallen; these are made from wood.) The crop which was grown in the *bundhies* after the water was drained did not require any irrigation until harvest.

The third class comprised *in situ* techniques through which storage facilities were created to retain precipitation and ground water infiltration. The difference between class two and three was that in the former, cultivation followed drainage; in the latter it occurred simultaneously.

The sizes of these storage tanks varied and the tanks themselves were generally known from the command areas they irrigated: the smallest ones irrigated around fifty acres, the medium ones, about a hundred and the major ones five hundred and above. The large tanks were clearly impressive in scale. The Veeranam Tank in Tamilnadu has a bund (embankment) 16 km in length. The Gangaikonda Cholapuram tank, also in Tamilnadu, was constructed by a Chola king from AD 1012 to 1044 and survives even today with a 25 km embankment.

The construction of tanks was a widely dispersed skill.

To create a tank reservoir, an earthen embankment, usually curved, was erected in a concave form across the flow of water. The water was retained in the belly of the curve from where it was drawn and directed through channels to irrigate plots at lower levels through gravity irrigation. After the tank was emptied, the tank bed itself could be used temporarily for raising a crop utilizing residual moisture. Many of the tanks in an area were interlinked and functioned as parts of an integrated system. The tank at the higher level released its surplus water as runoff to tanks at a lower level and the next in turn. These were called chain tanks. There were also tanks which were fed by canals from a river. Chain tanks were generally created in the upper reaches of the river basin and river fed tanks in the lower reaches. Their construction would have required detailed cooperation among several communities in the region.

In Mysore in 1966 Major R.H. Sankey, Chief Engineer, wrote, "Of the 27,269 square miles covered by Mysore, nearly 60% has, by the patient industry of its inhabitants been brought under the tank system. Unless under exceptional circumstances, none of the drainage of these 16,287 square miles is allowed to escape. To such an extent the principle of storage has been followed, that it would now require some ingenuity to discover a site within this great area suitable for a new tank." The profusion of such tanks was not a feature of Karnataka alone. Experience was similar in Tamilnadu, Goa, or Bihar. The area north of the Vindhya mountain range in middle India for instance had more than eight thousand submergence tanks. In one district of Rajasthan alone, there were more than five hundred *khadins*.

In Sri Lanka, dry areas were populated with what are known as "tank villages". "The one-mile to an inch topographical maps of the island", writes D.L.O Mendis, "show nearly 15,000 of these, of which over 8000 are in working condition today. Tradition has it that some 30,000 of these small tanks had been constructed down the ages and there is a reference in the chronicles to 20,000 in the ancient province alone in the 12th century."

Water Conveyance

Apart from the storage works, the subcontinent witnessed the emergence of competent and impressive water conveyance systems designed to divert waters of rivers and flowing streams. Some diversions were accomplished without a check or embankment across the river; in such instances, the flood waters of the river were drained through a natural diversion. The *Kuhls* or *Guls* in the Himalayan areas, the *dongs* of the Northeastern states, and the *pynes* of Bihar all reflect this feature.

The second category involved check dams as a basic feature: the river bed was first raised and the resulting raised water diverted into a channel as was the case with the *bandharas*. Some of these schemes were fairly small, like those in the hilly areas. Others could be extremely large scale and it is reports of the latter that probably gave Wittfogel material for his speculations. According to Major T. Greenway, these were works of "truly gigantic magnitude, vast embankments and drainage channels equal to ordinary English rivers in capacity."

Historical Development of Irrigation Techniques

The interesting question that is now being asked is whether one can talk in terms of an evolution of irrigation technologies and civil engineering techniques from the earliest times to the present? The question is important in view of the fact that many of these storage systems, diversion channels, and embankments are largely intact and still in use. There are still parts of the country where Persian wheels are operated. Tanks and storage vessels are once again being made functional and weirs continue to be constructed.

The answer seems obvious: while more complex technological mechanisms did emerge as time passed, the earlier and the later techniques have continued to co-exist. The only major new innovation seems to be the idea of dams; these are new in terms of function, since the generation of hydropower was not intended in earlier times. The idea of large dams, once considered the temples of modern India, has taken a severe beating in recent years. They are now considered unsuited to tropical ecosystems, since the reservoirs invariably lead to displacement of large numbers of people, submergence of forests, and destruction of wildlife, and in places like Sri Lanka, submergence of smaller functioning reservoirs.

This being said, it is possible still to identify certain periods as distinct historical events in a possible history of irrigation and civil engineering on the sub-continent. There is archaeological evidence of artificial irrigation from pre-Harappan and Harappan times (ca. 5500–3500 BC): this took the form of a large number of wells. One well found had a brick lining going down twelve meters. Post-Harappa, the major irrigation find is Inamgaon on the west coast of India where under the influence of the Jorwa culture (ca. 3400–3000 BC), one finds evidence of a major diversion scheme reflected in a massive embankment 240 meters by 2.2 meters wide to divert the river into a channel. The channel itself was 200 meters long and 4 meters wide.

The first storage tanks in their rudimentary form appear around 2500 BC — also in Sri Lanka — with the invention of iron tools. Hereafter, there are increasing references in literary works of both Sanskrit and Tamil up to the Gupta

(AD 350) period. There are tank related inscriptions which give details of tank construction, maintenance, sources of funds for maintenance, and so on. The word *eri* also comes into circulation by the seventh century as a term for tanks.

The *anicut* (weir) technique is probably older than the *eri* or tank. The most famous of the *anicuts*, the Kaveri Anicut on the river Kaveri, is linked to a Chola king of the second century AD. It involved a dam on the river Kaveri 300 meters long and 12–18 meters wide and 5 meters deep. There is a dispute about the age of the *anicut*, since the *anicut* technique itself bears a strong resemblance to the Sri Lankan technique of massive stone dams and sluices, a technique which developed, according to Sri Lankan historian R.A.L.H. Gunawardene, only in the seventh century AD.

Sir Arthur Cotton paid eloquent testimony to the engineering talent involved in the large scale irrigation works. He wrote: "There are a multitude of old native works in various parts of India... These are noble works, and show both boldness and engineering talent. They have stood for hundreds of years . . . it was from them that we learnt how to secure a foundation in loose sand of unmeasured depth. In fact, what we learnt from them made the difference between financial success and failure, for the Madras river irrigations executed by our engineers have been from the first the greatest financial successes of any engineering works in the world, solely because we learnt from them... With this lesson about foundations, we built bridges, weirs, aqueducts, and every kind of hydraulic work . . . we are thus deeply indebted to the native engineers...".

Social Arrangements/Religious Sanction

A significant feature of these irrigation works related to their construction and maintenance. Since water availability could be problematic with monsoon failure, those associated with the emergence of these works could gain religious merit for their deed. Though large systems were often sponsored by the state — to include kings, queens, local chieftains, *zamindars* (landowners) — village communities, temples, and even individuals are associated with their construction. Thus a public park in Pondicherry bears an inscription recording a tank built by a *dasi* — a temple dancer/courtesan, while another inscription in Karnataka (AD 1100) records a tank and shrine constructed by a village watchman.

All the major dynasties including the Mauryas (Sudarshan Lake near Kathiawar), the Cholas, the Hoysalas and the Vijayanagar Kings and Muslim Sultans were associated with irrigation works. Of these, the most impressive schemes are associated with the Cholas. However, these kings depended upon a cadre of skilled hydraulic engineers. Dikshit records the performance of one such engineer in the fourteenth century:

"The Kalludi (Gauribidanur taluk) inscription of 1388 AD is well known. According to it, when Vira Harihara Raya's son Sri Pratapa Bukkaraya was in Penugonda city in order that all the subjects might be in happiness — water being the life of the living beings — Bukkaraya in open court gave an order to the master of ten sciences, the hydraulic engineer (Jalasutra) Singayya Bhatta that he must bring the Henne (Pennar) river to Penugonda. Accordingly Singayya Bhatta conducted a channel to the Siruvara tank and gave the channel the name Pratapa Bukka Raya Mandalanda Kaluve".

The day to day operation and maintenance of both large and small works were mostly in the hands of local communities and of special professionals like the *nirkattis* of Tamilnadu. In many areas, produce from certain lands was set aside specifically to meet the maintenance costs of tanks. During the installation of the colonial regime, these revenues were appropriated by the colonial power and consequently the maintenance of such irrigation works fell into bad times leading to declines in efficiency.

CLAUDE ALVARES

REFERENCES

Dharampal. *Indian Science and Technology in the 18th Century*. Delhi: Impex India, 1983.

Dikshit, G.S., G.R. Kuppuswamy, and S.K. Mohan. *Tank Irrigation in Karnataka: A Historical Survey*. Bangalore: Gandhi Sahitya Sangha, 1993.

Majumdar, R.C. and H.C. Raychaudhuri. *An Advanced History of India*. 3rd edition. London: Macmillan, 1967.

Mendis, D.L.O. "Lessons from Traditional Irrigation and Ecosystems." In *The Revenge of Athena: Science, Exploitation and The Third World*. Ed. Ziauddin Sardar. London and New York: Mansell, 1988, p. 317.

Needham, Joseph. *Science and Civilisation in China*. Cambridge: Cambridge University Press, 1954.

Sengupta, N. *User-Friendly Irrigation Designs*. New Delhi and Newbury Park: Sage Publications, 1993.

Shankari, U. and E. Shah. *Water Management Traditions in India*. Madras: PPST Foundation, 1993.

Somashekhar, R. *Forfeited Treasure: A Study on the Status of Irrigation Tanks in Karnataka*. Bangalore: Prarambha, 1991.

Wittfogel, Karl A. *Oriental Despotism: a Comparative Study of Total Power*. New Haven, Connecticut: Yale University Press, 1957.

IRRIGATION IN THE ISLAMIC WORLD From Andalusia to the Far East, from the Sudan to Afghanistan, irrigation in Islamic countries is the basis of all agriculture and the source of all life. After the Roman empire, the classical Islamic empires relied on the great cities like Damascus, Baghdad, Cairo, Cordoba, and Fez. In the countryside, all the small villages, made up of groups of rush huts or houses of stone, wood, or concrete were organized around a water source: a mountain fed by a living spring, a cistern of rain water, or wells bringing water up to the surface from deep beds. The Arabic word for water, *Mā'*, is also the word for center. The water for religious ablutions at the center of the courtyard facing every mosque is the symbolic echo of a physical fact.

Islamic irrigation recovered pre-Islamic purviews while at the same time developing and expanding them. The historical record describes a vast network of canals from hot climatic zones to cold, rainy ones. The first were developed in the river valleys of semi-arid and arid regions. The Nile, the Tigris and Euphrates, and the Guadalquivir have offered rural and urban communities from earliest times both water and silt, a means of providing refreshment and nourishment. In short, they provided for the development of civilization itself.

Where no great rivers existed, human societies dug wells and prospected deep lying aquifer beds, in the south of the Arabian peninsula, in Yemen, and in Hadramawt, as well as in the Sudan. There, where the effects of Indian monsoons are often felt, a tropical agriculture (coffee trees, date palms, banana and tamarind trees) served as a point of departure for those tropical plant species to Mediterranean regions.

Lastly, on the borders of agricultural zones stretched vast steppes with winter rainfalls varying from 50 to 150 millimeters per year, semi-nomadic lands which had been used for thousands of years. Animal husbandry and intensive agriculture at oasis sites have for centuries contributed to the econimic base of Yemenites and Maghreb tribes. This association between semi-nomadic grazing and intensive irrigated agriculture is a distinctive characteristic of Islamic irrigation, a system of a complementary nature between field, pasture, and natural resources in the environment.

For classical Islam, written documentation is inseparable from the latest results of rural archeology; all of the *Kitāb-al-Filāḥat* (Books of Agriculture) — Maghrebian, Andalusian, Egyptian, Iraqi, Persian, or Yemenite — insist meticulously on the deployment of equipment and on the control of water. What are examined are the means of water distribution, following the season and the species being cultivated. A Tribunal of Waters looks at legal cases, and that of the Islamic writer Valence has lasted from the time of the Christian *Reconquista* in 1248 until today. Irrigation was linked to a social time, following necessities of nature but also following the social rank of the users.

In the current state of research, one of the oldest calendars of irrigation was found in the *Filāḥat al-Nabaṭiyyah* (Nabataean Agriculture, by Ibn Waḥshiyyah), which in the 1000s gave information on pre-Islamic Mesopotamia and Abbasidian Iraq. For western Islam, jurisprudence provided historians with precise details of daily life and ecological behavior, and explained the elements relating to the history of that environment, such as torrential rains in the autumn and the spring, the shifting of waterholes, and complaints from the owners of cultivated estates which had been deprived of water from one day to another. The *fiqh* had to regulate any unexpected events from case to case. The joint purchase of land and water, inherent to the *fiqh*, confronted the fact of the capriciousness of nature: the agrarian weather of micro-climates really was the final law. In Islamic law ownership of land is always linked to ownership of waters.

Lastly, financial writings integrate the agrarian dependence on water into the framework of the whole political and social evolution of Islamic history. Irrigation is the crucial element of agriculture in Islamic lands.

The rains follow two dominant climatological patterns: one Mediterranean, the other tropical and sub-tropical. The Mediterranean basin, along with southern Africa and California, has the only climate which has rain in the winter and maximum heat in the summer. The vegetation there needs a hot, humid season in its tillage phase. In other respects, the torrential spring and autumn rains create a major risk for the thin, mountainous soil which is predominant in the Mediterranean basin. A negative quality can thus indeed have positive effects.

This link between the climate and the soil in Andalusia entailed three inseparable factors: an extremely fragile ecology at risk of erosion, a highly organized and detailed system of irrigation which was adapted to regional conditions, and a stable political and administrative system (except in time of war).

In certain historical cases, like that of Andalusia, in spite of the fluctuating tides of conquest and reconquest, there existed a true ecological pattern which lasted for several centuries and which served as a real ecological laboratory for eastern and western Islamic statutes, based on a high level of natural science, climatology, and botany.

The extensive spread of theoretical and technical knowledge made of the Andalusian model a precedent which could be applied to all societies, ethnic and religious differences aside, which were trying to understand the environment in its historical dimension.

It was water which governed the classification of land types. The rich agronomical and administrative literature of Arabic Islam divided cultivatable land into rainy land and irrigated land. The former were those lands which received enough winter rain to permit the cultivation of olive trees, grape vines, and citrus trees, the Mediterranean trilogy.

Irrigated plantings used water stored in deep beds; the surface of the land was hot and irrigated and was worked in squares irrigated by ditches or *acequias*. The device which allowed the flowing of water into the ditches across fields, squares, or gardens was a wheel of variable dimension called a *noria*. More simply, on river banks and along the shore a system of scales and counterweights called *shaduf* was used.

The plan of irrigated water brought about a kind of intensive horticulture comparable to that achieved in the great Asiatic deltas. In the development of the countryside irrigation allowed plant species which became the basis of the popular diet — e.g. rice and legumes rich in protein — to become acclimated to the Muslim west. In Andalusia, sugar cane, rice, and cotton were cultivated.

In plateau regions (Meseta, Morocco, Iran) where aquifer beds were deep, the most profitable procedure was the traditional system of *qanats*. The *qanats* were deep drainage tunnels which directed water to springs or artesian wells. Some were over 16 kilometers long, and their deep tunnel was dug from the outlet up to the mother-well. The gradient incline of the aquifer bed can range from almost zero to very steep. Well sites were marked and were used for ventilation and disposition of debris from land clearing. The science of the *qanats*, traditionally, was based on empirical knowledge and was passed down by gesture or through oral tradition. The earliest descriptions are Iranian. Later they were developed on the Castilian Meseta (Madrid) and in Morocco (in Marrakesh). Today, one can see them in Central America; they are the oil-qanats.

In between the canals of flowing water and the system of *qanats* the Islamic world watered its gardens and fields, its city courtyards and domestic patios, and its mosques, with water from wells whose volume was increased through a process described by Ibn al-ʿAwwām in the *Kitāb al-Filāthat* and in *Nabatean Agriculture*. Chapters relevant to finding the underground water level and sinking wells are followed by considerations on ways to increase the volume of water stored, ways to raise water from wells that are too deep, and ways to modify the taste of brackish or salty water.

The detailed observation of the earth or of its vegetation, in order to identify the presence of water, as well as the empirical nature of the investigation to sample the earth, in order to define various qualities of water, such as neutral, salty, or brackish and bitter, systematically mobilized the senses. Empiricism was able to bring about the creation of an irrigated, productive agriculture area. This success was built on the very controlled work of the *fellahs*, a political encouragement of individual appropriation, and on a high level of applied knowledge.

This multi-secular, integrated system was the reason behind the development of numerous plant species crucial to the existence of civilization.

LUCIE BOLENS

REFERENCES

al-Hassan, A.Y. and Hill, D.R. *Islamic Technology*. Paris: UNESCO, 1986.

Bolens, L. *Agronomes andalous du Moyen Age*. Genève: Droz, 1981.

Bolens, L. "L'irrigation en Andalusia." In *El Agua en Zonas Aridas: Arqueologia e Historia*. Almeria: Instituto de Estudios Almerienses de la Diputaciòn de Almeria, 1989, pp. 69–95.

Fahd, T. "Un traité des eaux dans *al-Filāḥa an-Nabaṭiyya*." In *La Persia nel Medioevo*. Roma: Academia dei Lincei, 1971, pp. 277–326.

Glick, T.F., ed. *Irrigation and Society in Medieval Valencia*. Cambridge, Massachusetts: Harvard University Press, 1970.

Watson, A. *Agricultural Innovation in the Early Islamic World (700–1100)*. Cambridge: Cambridge University Press, 1983.

See also: Technology in the Islamic World – Agriculture in the Islamic World – *Qanat*

IRRIGATION IN SOUTH AMERICA Large-scale irrigation systems (canal irrigation) went hand-in-hand with the rise of cities and truly complex societies on the coast of Peru during pre-Hispanic times. Such systems delivered water to hundreds of hectares of potentially fertile land along a dry desert coast. Relatively late in prehistory, from 300 BC onwards, a number of well-known urban cultures (e.g. Moche, Lima) undertook the construction and expansion of irrigation works, which supported dense populations. This process culminated around AD 1000 in the construction of exceptionally large and sophisticated works by the Chimu, including an intervalley canal of impressive proportions from the Rio Chicama. These works made it possible for a population of some 25,000 to reside at the capital city of Chan Chan. Irrigation agriculture played a primary role in sustaining urban communities near the coast, where large irrigable plains of potentially fertile land were brought into production.

Similar irrigation systems in the Andean highlands are also of note. Some researchers, such as Michael Moseley, believe that irrigation agriculture developed here, as an extension of hunter-gatherer practices. Because of the steep slope, short canals would have been sufficient to water desired areas, and over the course of a millennium or more, a sophisticated irrigation technology developed. By AD 600, canal irrigation brought hundreds of hectares of otherwise unusable land into production in the Ayacucho valley of south-central Peru. These networks helped support a population of some 20,000 or more at the urban site of Huari, which is situated at an elevation of some 3000 meters, above any substantial acreage of arable land.

Other agricultural systems are also known for the Andes, and for elsewhere in South America. These include: (a) sunken gardens on the desert coast, where plots of land were dug down to the water table and then planted, (b) terraces along the highland slopes, which served to retain moisture and enhance production, and (c) raised or drained fields along lake margins and in the Amazon basin, which served to reclaim inundated wetland. Aspects of irrigation technology theoretically come into play with each. Upland irrigation tends to create a higher water table downslope, which can make sunken fields possible in otherwise very dry areas. Slope irrigation can deliver a supply of water to terraced fields, and such systems are documented for the Inca. During the dry season, spring water can be run into the swales of raised field systems, which can allow for double cropping in some settings, e.g. Lake Titicaca. Irrigation plays a role in other agricultural systems of the Andes. All of these agricultural systems were geared towards intensive production, as a means for dense human adaptation to the Andean environment.

In general, the construction and maintenance of irrigation systems is portrayed in the archaeological literature as a powerful force leading to the rise of civilization and urban life. This theory is based upon the view expressed by Karl Wittfogel, and subsequently anthropologists such as Julian Steward, that water management plays a key role in crystallizing political authority; it is also referred to as the "Hydraulic Hypothesis". According to this theory, highland peoples came to control a scarce resource of high value – water — which led to a hierarchical system for its management and distribution. Such a scenario is thought to have led inevitably to the formation of stratified society and subsequently state bureaucracy. Political power is seen as a direct outgrowth of the struggle for water, from which centralized authority and state-level government emerge. Not everyone agrees, and some scholars, such as Robert Adams, have argued forcefully that irrigation is the consequence of political power, rather than its cause.

The Peruvian data do not support the "Hydraulic Hypothesis" of civilization's emergence. A strong maritime economy typified subsistence strategy along the coast during pre-ceramic times, and large agglutinated settlements that housed hundreds of people were built along the coast between 2500 and 1800 BC, e.g. Huaca Prieta, El Paraíso, Río Seco, Playa Culebras, and others. There is an active debate on the role that crops, including maize, had in this developmental process, but it seems quite clear that irrigation agriculture played little part and certainly not to the scale of later times. According to Michael Moseley, the socio-political organization that accompanied these settlements was highly evolved, incorporating the management of labor mobilization (a form of taxation) on a regional basis. Irrigation agriculture took hold around 1800 to 900 BC in the prehistoric sequence for Peruvian cultures. By this time, authority figures and labor taxation systems had already emerged. Irrigation technology did not bring them about, but vice versa.

Still, the matter of irrigation's impact on society is a topic of central concern. For the north coast of Peru, it has been argued that irrigation technology went hand-in-hand with hierarchical social organization during late pre-Hispanic times. According to Patricia Netherly, water was a valuable commodity, and groups benefited from a hierarchically organized social structure, capable of resolving conflict when it occurred. Michael Moseley adds the perspective that large-scale irrigation projects were designed and engineered by rulers in these settings. Together both views provide a clear understanding of how irrigation was carried out, on the one hand through kinship organization, and on the other through government intervention in labor-intensive projects. Irrigation may not have given rise to civilization and urban life in the Andes, but it clearly evolved within the power structure of subsequent developments.

Today, systems of native irrigation management are under study, as are the impacts of modern attempts to improve them (Mitchell, 1991; Bolin, 1990). It is noted that highland communities often attempt to maintain their rights to water use and management, although in Peru such rights no longer exist in the legal sense. All water belongs to the Peruvian state, and there are national laws and regulations that govern its use. Highlanders, especially those who regard themselves as "Owners of the Water", fall into conflict with people in the lowland valleys. According to Bolin, there is a great need for agricultural and government agencies to investigate the indigenous patterns of water use, and the competition between and within villages for access to irrigation water. As revealed by Mitchell and Netherly, such patterns of water rights are likely to be deeply embedded within kinship organization and indigenous social structure.

GRAY GRAFFAM

REFERENCES

Adams, Robert M. *The Evolution of Urban Society.* Chicago: Aldine, 1966.

Bolin, Inge. "Upsetting the Power Balance: Cooperation, Competition, and Conflict along an Andean Irrigation System." *Human Organization* 49(2): 140–148, 1990.

Hastorf, Christine A. *Agriculture and the Onset of Political Inequality Before the Inca*. Cambridge: Cambridge University Press, 1993.

Isbell, William H, and Gordon F. McEwen. *Huari Administrative Structure: Prehistoric Monumental Architecture and State Government*. Washington, D.C.: Dumbarton Oaks Research Library and Collection, 1991.

Mitchell, William P. *Peasants on the Edge: Crop, Cult, and Crisis in the Andes*. Austin: University of Texas Press, 1991.

Moseley, Michael E. *The Incas and Their Ancestors: The Archaeology of Peru*. London: Thames and Hudson, 1992.

Netherly, Patricia J. "The Management of Late Andean Irrigation Systems on the North Coast of Peru." *American Antiquity* 49(2): 227–254, 1984.

ISA TARJAMAN During the Yuan dynasty (1271–1368) the Arabs played a role in Chinese science and technology similar to that of the Indians in the Tang, bringing in stimulating outside influences which were then incorporated and synthesized into Chinese mathematics, astronomy, and medicine. Isa Tarjaman (1227–1308) or Aixue in Chinese was a remarkable example. Isa Tarjaman, sometimes called Isa the Interpreter or Isa the Mongol, was a Nestorian Arab from Syria. He was skilled as a mathematician and astronomer as well as in medicine and pharmacy. He came to China in about 1247, then worked for the Mongol Khans till his death in 1308. In 1263 Kublai Khan appointed Isa the director of the Muslim Astronomical Bureau and Medical and Pharmaceutical Bureau. During this time he suggested that Kublai prepare a new calendar in the Arabian style which he finished in 1267 in cooperation with a Persian astronomer Zhama Ludin or Jamāl al-Dīn, who was working in China too. This was called the *Wan-nian* (ten thousand years) calendar. Meanwhile, they made seven kinds of Arabian astronomical instruments for the Huihui (Muslim) Observatory. All of these exerted some influence on the Chinese astronomer Guo Shoujing (1231–1316) and his work.

From 1283 to 1286, Isa was sent to Il-khan as a member of a delegation. So he visited the famous Maraghā Observatory and worked together with some Arabian and even Chinese astronomers who were working there for a time. Then he brought some of these astronomical and mathematical works back to China. These were studied by staff members at the Muslim Astronomical Bureau in Beijing.

As the director of the Medical and Pharmaceutical Bureau, Isa Tarjaman established a Capital Hospital to introduce Arabian medicine to China; his wife Sara also worked there. An important Arabian medical work entitled *Huihui Yaofang* (Collection of Muslim Prescriptions) was compiled under his leadership. It is interesting that some of the contents of this work were taken from Ibn Sīnā's *Canon*. Therefore, Isa made a great contribution to the history of the Sino-Arabian scientific exchange, and he thus was praised by the Mongol Khans. After returning from Il-Khan he rose to be a Hanlin Academician, then the minister of State in 1297. After his death the Mongol court made him the Fuolin Prince. He was the only Arab to attain this highest official position in China.

SUN XIAOLI

REFERENCES

Needham, Joseph. *Science and Civilization in China*, vol. 3. Cambridge: Cambridge University Press, 1959.

Shen, Fuwei. *Zong Xi Wenhua Jiaoliu Shi* (History of Sino-Western Cultural Exchange). Shanghai: People's Publishing House, 1985.

Song, Lian. *Yuan Shi* (History of the Yuan Dynasty, 1370), chap. 134, "Biography of Aixue (Isa Tarjaman)", book 11. Beijing: Zhonghua, 1979.

See also: Jamāl al-Dīn – Guo Shoujing – Maraghā

ISḤĀQ IBN ḤUNAYN Isḥāq ibn Ḥunayn (215–298 AH/AD 830–910) is best known for his role in the translation of Greek texts, both philosophical and mathematical, into Arabic. In addition to his work in the translation institute founded by his father, Ḥunayn ibn Isḥāq (193–263 AH/AD 809–877), he served as a physician in the court of the caliphs al-Muʿatamid and al-Muʿtadid. Ḥunayn was a Nestorian Christian, but some authorities report that Isḥāq converted to Islam.

As a translator, Isḥāq followed the scientific approach developed and applied by his father with such success to the Galenic corpus. Rather than translate mechanically word-by-word, they attempted to render the meaning of each Greek thought unit into an appropriate form in the new language. (Both Isḥāq and his father preferred to translate first from Greek into their native Syriac before producing an Arabic version. Given this methodology, the paucity of Syriac translations among the extant manuscript remains has long puzzled historians.)

Isḥāq's greatest contributions lay in his translations of Greek philosophical texts, especially the works of Aristotle. He produced Syriac versions of a part of the *Prior Analytics* and all of the *Posterior Analytics* and the *Topics*. He also rendered the *Categories*, *On Interpretation*, *Physics*,

On Generation and Corruption, *On the Soul*, parts of the *Metaphysics*, and the *Nichomachean Ethics* into Arabic. He may also have translated the *Rhetoric* and the *Poetics*. In addition to Aristotle's works, he translated Galen's *Number of the Syllogism* and part of his *On Demonstration*, as well as some logical and philosophical works by Alexander of Aphrodisias, Porphyry, Themistius, and Proclus.

Isḥāq was also instrumental in translating several important Greek mathematical treatises into Arabic. These include Euclid's *Elements*, his *Optics*, and his *Data*, as well as the *Almagest* of Claudius Ptolemy, *On the Sphere and the Cylinder* by Archimedes, the *Spherics* of Menelaus, and minor works by Autolycus and Hypsicles. Isḥāq's translations of Euclid and Ptolemy were revised by the mathematician, Thābit ibn Qurra, who compared the translation with additional Greek manuscripts that he had at his disposal and noted differences between the Arabic and Greek versions. All extant manuscripts seem to reflect some aspects of this editing process, although not all manuscripts contain all the comments attributed to Thābit.

As a leading physician, Isḥāq also produced a number of original works on medicine. Unfortunately, these seem not to have survived. His *Tārīkh al-Aṭibbā'* (History of Physicians), an extended version of a book of the same title by John Philoponus, is extant. Isḥāq has added the names of philosophers active during the lifetimes of each physician mentioned. This work has been helpful to historians of both medicine and philosophy.

GREGG DE YOUNG

REFERENCES

De Young, G. "The Arabic Textual Traditions of Euclid's *Elements*". *Historia Mathematica* 11: 147–160, 1984.

De Young, G. "Isḥāq ibn Ḥunayn, Ḥunayn ibn Isḥāq, and the Third Arabic Translation of Euclid's *Elements*." *Historia Mathematica* 19: 188–199, 1992.

Shehaby, N. "Isḥāq ibn Hunayn, Abū Ya'qūb." In *Dictionary of Scientific Biography*. Ed. C. Gillispie. New York: Scribner, 1970–1981, vol. 7, pp. 24–26.

See also: Ḥunayn ibn Isḥāq – *Almagest* – Thābit ibn Qurra

ISLAMIC SCIENCE: THE CONTEMPORARY DEBATE

The debate on the meaning, nature, and characteristics of a contemporary Islamic science first emerged during the late seventies. The rise of OPEC, the Iranian revolution, and a growing consciousness in Muslim societies of their cultural identity led many scientists and academics as well as institutions to emphasize the distinctive scientific heritage of Islam. The reflections on the history of Islamic science generated a question of contemporary relevance: how could modern Muslim societies rediscover the spirit of Islamic science as it was practiced and developed in history? A number of international seminars and conferences — most notably the series of seminars on "Science and Technology in Islam and the West: A Synthesis" held under the auspices of the International Federation of Institutes of Advanced Studies in Stockholm in 1981 and Granada in 1982; the "International Conference on Science in Islamic Polity — Its Past, Present and Future" backed by the Organization of Islamic Conference and held in Islamabad, Pakistan in 1983; and "The Quest for a New Science" Conference organized by the Muslim Association for the Advancement of Science (MAAS), in Aligarh, India in 1984 — were held to explore the question. The conferences provide a launchpad for the debate and revealed a great deal of confusion around the whole notion of Islamic science and its meaning and relevance to contemporary times.

During the last decade, as the debate spread to all parts of the Muslim world, a number of different and distinct approaches to Islamic science have come to the fore. A considerable amount of literature, ranging from the sublime to the mediocre, has been produced. So, what we now have is an embryonic, diffused, sociology-of-knowledge type of discipline which roughly defines a vigorously contested territory. The proponents of various schools of thought, each deeply entrenched in its position, regularly publish in the discipline's core journal: *MAAS Journal of Islamic Science*.

The discussions have centered around two basic questions: what is Islamic science? Also, how does it differ from the practice of conventional science? Five points of view can be distinguished on these questions. The first is what we may call western-type *scientific fundamentalism*: it totally rejects the whole idea of Islamic science. The second position represents the other end of the spectrum and takes us towards deep subjectivity: *mystical fundamentalism*. The third and fourth viewpoints maintain that Islam has a great deal to say about science, but are divided on where exactly Islam enters the picture. The third position is mainly concerned in reading the verses of the *Qur'ān* in the theories, advances, and discoveries of modern science. This tendency has been labeled Bucaillism, after the French surgeon who initiated the movement. The fourth, *science in Islamic polity*, school is the official, establishment position in that it is backed and supported by the Islamic Secretariat of the Organization of Islamic Conference, the international body that brings the Muslim world together on a political platform. The fifth viewpoint argues that Islamic science is something quite different from science as it is practiced today. It is attributed largely to the *Ijmalis*, a group of independent scholars and thinkers who have championed a future oriented critique of contemporary Muslim thought; and to the *Aligarh* school, which has evolved around the Center for Studies on Science in Aligarh, India.

Scientific Fundamentalism

Those who reject the notion of Islamic science argue from the conventional, positivist perspective on science. Science, it is argued, is neither Western nor Eastern, Christian nor Islamic. It is neutral, value-free, and universal. While advances in science may raise ethical issues, scientific knowledge itself has nothing to do with values. At best, the proponents of this position argue, value judgment can be applied in giving emphasis and pursuing a particular area of scientific research. However, basic scientific research in all areas must be pursued for its own sake; religion should be kept firmly away from science.

Initially, this approach, which ignores recent advances in the philosophy of science and sociology of knowledge, was the dominant view amongst rank and file Muslim scientists. However, recent debates on eugenics, efforts in theoretical physics to produce a Theory of Everything that can be proudly displayed on a T-shirt, and the emergence of complexity, has seriously undermined the position of scientific fundamentalists. The positivist view of science is being increasingly questioned in the Muslim world.

Mystical Fundamentalism

In this perspective, Islamic science becomes the study of the nature of things in an ontological sense. The material universe is studied as an integral and subordinate part of the higher levels of existence, consciousness, and modes of knowing. Thus we are talking about science not as a problem solving enterprise, but more as a mystical quest for understanding the Absolute. In this universe, conjecture and hypothesis have no real place; all inquiry must be subordinate to the mystical experience. This school of thought has a very specific position on Islamic science in history. All science in the Muslim civilization was "sacred science", a product of a particular mystical tradition — namely the tradition of gnosis, stripped of its sectarian connotations and going back to the Greek neoplatonists. "Traditional science" does not necessarily mean science as it has existed in Muslim tradition and history, but products produced within the tradition of Islamic mysticism or Sufism. Traditional science is *science sacra*, the Science of Ultimate Reality, as taught by Sufi masters and mystics of other traditions. The goal of Islamic science today, they argue, is to rediscover the classical Islamic traditions and their sacred nature. This perspective is represented by a small and influential group.

Bucaillism

This is a combination of religious and scientific fundamentalism. Bucaillists try to legitimize modern science by equating it with the *Qurān* or to prove the divine origins of the *Qur'ān* by showing that it contains scientifically valid facts. Bucaillism grew out of *The Bible, the Qur'ān and Science* by Maurice Bucaille published in 1976. Bucaille, a French surgeon, examines the holy scriptures in the light of modern science to discover what they have to say about astronomy, the earth, and the animal and vegetable kingdoms. He finds that the Bible does not meet the stringent criteria of modern knowledge. The *Qur'ān*, on the other hand, does not contain a single proposition at variance with the most firmly established modern knowledge, nor does it contain any of the ideas current at the time on the subjects it describes. Furthermore, the *Qur'ān* contains a large number of facts which were not discovered until modern times. The book, translated into almost every Muslim language from the original French, has spouted a whole genre of literature looking at the scientific content of the *Qur'ān*. Subjects ranging from relativity, quantum mechanics, and the big bang theory to the entire field of embryology and much of modern geology have been discovered in the *Qur'ān*. Conversely, experiments have been devised to discover what is mentioned in the *Qur'ān* but not known to science — for example, the program to harness the energy of the jinn! Bucaillism takes the reverence of science to a new level: Bucaillists do not just accept all science as Good and True, but attack anyone who shows a critical or skeptical attitude towards science and defend their own faith as "scientific", "objective", and "rational". This is the most popular version of Islamic science.

Science in Islamic Polity

In this perspective, science is seen in similar terms to those of the Western paradigm as neutral and universal. But, it is argued, we can approach science with a secular or an Islamic attitude. The Islamic approach to science is to recognize the limitations of human reason and acknowledge that all knowledge comes from God. Within an Islamic polity — that is, an idealized "Islamic state" — the principles and injunctions of Islam which are the basis of the state would automatically guide science in the direction of Islamic values. The individual Muslim scientists would also bring their own values to bear on their work. The "Statement on Scientific Knowledge seen from Islamic Perspective", issued after the first International Conference on Science in Islamic Polity in 1983, states that science is one way humanity seeks "to serve the Supreme Being by studying, knowing, preserving and beautifying His creation". The Islamic framework seeks a "unifying perspective, combining the pursuit of science and the pursuit of virtue in one and the same individual". The "Islamabad Declaration" also called for the creation of "the Islamic science and technology system" by the end of the century. However, the emphasis on Islamic values in this perspective has remained largely at the level of rhetoric. Much of the work done at the national and international level within the framework of the "Islamic Conference Standing Committee on Scientific and Technological Cooperation" (COMSTECH) has been very conventional and concerned largely with nuclear physics, biotechnology, and electronics.

The Ijmali Position and the Aligarh School

Both the Ijmalis and the Aligarh school argue that while Western science claims the norm of neutrality for itself, it is in fact a value-laden and culturally biased enterprise. The Ijmalis emphasize the "repulsive facade" of the metaphysical trappings of Western science, the arrogance and violence inherent in its methodology, and the ideology of domination and control which has become its hallmark. These things are inherent both in the assumptions of Western science as well as its methodology. Thus attempts to rediscover Islamic science must begin by a rejection of both the axioms about nature, universe, time, and humanity as well as the goals and direction of Western science and the methodology which has made meaningless reductionism, objectification of nature, and torture of animals its basic approach. But science in this framework is not an attempt to reinvent the wheel; it amounts to a careful delineation of norms and values within which scientific research and activity is undertaken. At the Stockholm Seminar in 1981, Muslim scientists identified a set of fundamental concepts of Islam which should shape the science policies and scientific activity of Muslim societies. The concepts generate the basic values of Islamic culture and form a parameter within which an ideal Islamic society progresses. There are ten such concepts, four standing alone and three opposing pairs: *tawḥīd* (unity), *khalīfa* (trusteeship), *ʿibādat* (worship), *ʿilm* (knowledge), *halal* (praiseworthy) and *haram* (blameworthy), *adl* (social justice) and *zulm* (tyranny) and *istislah* (public interest) and *dhiya* (waste). When translated into values, this system of concepts embraces the nature of scientific inquiry in its totality: it integrates facts and values and institutionalizes a system of knowing that is based on accountability and social responsibility. How do these values shape scientific and technological activity? Usually, the concept of *tawḥīd* is translated as unity of God. It becomes an all-embracing value when this unity is asserted in the unity of humanity, unity of person and nature, and the unity of knowledge and values. From *tawḥīd* emerges the concept of *khalīfa*: that mortals are not independent of God but are responsible and accountable to God for their scientific and technological activities. The trusteeship implies that "man" has no exclusive right to anything and that he is responsible for maintaining and preserving the integrity of the abode of his terrestrial journey. But just because knowledge cannot be sought for the outright exploitation of nature, one is not reduced to being a passive observer. On the contrary, contemplation (*ʿibādat*) is an obligation, for it leads to an awareness of *tawḥīd* and *khalīfa*, and it is this contemplation that serves as an integrating factor for scientific activity and a system of Islamic values. *ʿIbādat*, or the contemplation of the unity of God, has many manifestations, of which the pursuit of knowledge is the major one. If scientific enterprise is an act of contemplation, a form of worship, it goes without saying that it cannot involve any acts of violence towards nature or the creation nor, indeed, could it lead to waste (*dhiya*), any form of violence, oppression, or tyranny (*zulm*) or be pursued for unworthy goals (*haram*); it could only be based on praiseworthy goals (*halal*) on behalf of public good (*istislah*) and overall promotion of social, economic, and cultural justice (*adl*). Such a framework, argue the Ijmalis, propelled Islamic science in history towards its zenith without restricting freedom of inquiry or producing adverse effects on society. When scientific activity was guided by the conceptual matrix of Islam, it generated a unique blend of ethics and knowledge. It is this blend — which produces a distinctive philosophy and methodology of science — that distinguishes Islamic science from other scientific endeavors. Rediscovering a contemporary Islamic science, argue the Ijmalis, requires using the conceptual framework to shape science policies, develop methodologies, and identify and prioritize areas for research and development.

While accepting the position of the Ijmalis, the Aligarh School has added a number of other concepts to their framework of inquiry into Islamic science. However, both schools maintain that the practice of science within the conceptual framework of Islam involves a change in methods and direction from the dominant style and practice of science. The difference in emphasis and priorities of Islamic science, together with its assumptions and methods, generates a scientific enterprise that is radically different from Western science. Imagine a biology without vivisection or animal experimentation, or physics based on synthesis rather than reduction. The Aligarh school focuses on such methodological issues in an attempt to bring Islamic values to the level of the laboratory. The Ijmalis, on the other hand, are interested in developing a futuristic framework within which science policies and research and development activities in the Muslim World could be shifted gradually towards a generally accepted, and consensus oriented, framework of Islamic science. The two approaches are complementary and attempt to produce a viable alternative while recognizing the complexity of the issues and the difficulties involved in solving the problems generated by Western science.

Outlook

Given the fragmentary nature of the debate on Islamic science, little of pragmatic and practical value has emerged so far. This debate, it appears, has largely followed the same process as the attempts to develop a "science for the people" in the early 1970s. The rhetoric has become more and more extreme while the practical dimension has, on the whole, been ignored. However, unlike the "science for the people" movement, the Islamic science debate has produced a theoretical framework which is evolving towards a consensus. While explorations of the theoretical framework for Islamic science will continue, it will be the pragmatic policy and methodological work that will determine and shape its future.

ZIAUDDIN SARDAR

REFERENCES

Ahmad, Rais and S. Naseem Ahmad. *Quest for A New Science*. Aligarh: Centre for Studies on Science, 1984.

Al-Attas, Syed Muhammad Naquib. "Islam and Philosophy of Science." *MAAS Journal of Islamic Science* 6(1): 59–78, 1990.

Anees, Munawar Ahmad. "What Islamic Science is Not." *MAAS Journal of Islamic Science* 2(1): 9–20, 1986.

Anees, Munawar Ahmad. *Islam and Biological Futures*. London: Mansell, 1989.

Anees, Munawar Ahmad and Merryl Wyn Davies. "Islamic Science: Current Thinking and Future Directions." In *The Revenge of Athena: Science, Exploitation and the Third World*. Ed. Ziauddin Sardar. London: Mansell, 1988, pp. 249–260.

Bakr, Osman. *Taweed and Science*. Kuala Lumpur: Secretariat for Islamic Philosophy and Science, 1991.

Bucaille, Maurice. *The Bible, the Qur'ān and Science*. Paris: Seghers, 1976.

Butt, Nasim. *Science and Muslim Societies*. London: Grey Seal, 1991.

Hoodbhoy, Pervez. *Islam and Science*. London: Zed Books, 1991.

Jamison, Andrew. "Western Science in Perspective and the Search for Alternatives." In *The Uncertain Quest: Science, Technology and Development*. Ed. Jean-Jacques Saloman et al. Tokyo: United Nations Press, 1994, pp 131–167.

Kirmani, Zaki. "Islamic Science: Moving Towards a New Paradigm." In *An Early Crescent: The Future of Knowledge and the Environment in Islam*. Ed. Ziauddin Sardar. London: Mansell, 1989, pp 140–162.

Kirmani, Zaki. "An Outline of an Islamic Framework for a Contemporary Science." *MAAS Journal of Islamic Science* 8(2): 55–76, 1992.

Manzoor, S.P. "The Unthought of Islamic Science." *MAAS Journal of Islamic Science* 5(2): 49–64, 1989.

Nasr, Seyyed Hossein. *The Need for A Sacred Science*. Richmond, Surrey: Curzon Press, 1993.

Ravetz, J.R. "Prospects for an Islamic Science." *Futures* 23(3): 262–272, 1991.

Salam, Abdus. "Islam and Science." *MAAS Journal of Islamic Science* 2(1): 21–46, 1986.

Sardar, Ziauddin. "A Revival for Islam, A Boost for Science." *Nature* 282: 354–357, 1979.

Sardar, Ziauddin. "Can Science Come Back to Islam?" *New Scientist* 88: 212–216, 1980.

Sardar, Ziauddin, ed. *The Touch of Midas: Science, Values and the Environment in Islam and the West*. Manchester: Manchester University Press, 1982.

Sardar, Ziauddin. *Explorations in Islamic Science*. London: Mansell, 1989.

J

JĀBIR IBN AFLAḤ Almost nothing is known of Jābir's life, but remarks by Maimonides (d. 1204), e.g. that he knew Jābir's son, place Jābir probably in the first half of the twelfth century, and the name "al-Ishbīlī" establishes a connection with Seville. Indeed, legend associates his name (wrongly) with the building of the Torre del Oro and of the tower now belonging to the cathedral in Seville. Jābir's principal work was a commentary (or correction, *iṣlāḥ*) on Ptolemy's *Almagest*, the standard textbook on mathematical astronomy, which he had seen in two translations from the Greek. In this treatise he not only simplified the mathematics and separated theory from calculation (there are no tables in the book), but also indulged in violent criticisms of Ptolemy. The introduction to the commentary contains a list of Ptolemy's "errors", which are considered in detail in the body of the book. His best known astronomical claim of this kind was his assertion, against Ptolemy, that Venus and Mercury must lie above the Sun because of their lack of parallax.

Jābir's lasting contribution was his statement of the requisite theorems in trigonometry. The essence of his plane trigonometry is to be found in the *Almagest* itself, but his clear enunciations of his results for triangles obviated the need for construction lines that cluttered so many diagrams. Again, his theorems on spherical triangles, which replaced Ptolemy's theorems involving six quantities by proportions involving four, were clearly taken over from a group of scholars, such as Abū 'l-Wafā' and Abū Naṣr ibn ʿIrāq, who worked in Baghdad and elsewhere in the eastern provinces of Islam about AD 1000. Curiously, Jābir quotes no Arabic author in his work, not even Ibn Muʿādh, who had lived and worked in eleventh-century Seville.

The text of the *Iṣlāḥ al-majisṭī*, at least in the trigonometrical part, was revised, perhaps by the author. Ibn Rushd (d. 1198) and al-Biṭrūjī in Spain were both influenced by the work. It was also known in the East, for al-Shīrāzī (d. 1311), one of the Marāghā astronomers, made a compendium of it. There were two translations into Hebrew, by Moses ben Tibbon (1274) and by Jakob ben Makhir (revised by Samuel ben Yehuda of Marseilles, 1335), and Jābir appears to have had considerable influence in Hebrew astronomy. But the most lasting influence of the work was through the translation into Latin by Gerard of Cremona (d. 1187). For the Latins not only was "Geber" a vigorous critic of Ptolemaic astronomy, but his treatise established trigonometry in the West. This can be seen in an anonymous commentary, in an anonymous compilation of the plane trigonometry, *De tribus notis* (On Three Known [quantities]), and in the works of Simon Bredon, Richard of Wallingford (both fourteenth century), and others. Finally, "Geber" was the source of much of Regiomontanus' *De triangulis* (On Triangles), perhaps the source for the trigonometrical section of Copernicus' *De revolutionibus* (On the Revolutions [of the Heavenly Spheres]).

RICHARD P. LORCH

REFERENCES

Al-Bīrūnī. *Kitāb maqālīd ʿIlm al-hayʾa* (Book of the Keys of Astronomy). Ed. M.-Th. Debarnot. Damascus: Institut français de Damas, 1985.

Benjamin, F. S. and G. J. Toomer, eds. *Campanus of Novara and Medieval Planetary Theory, Theorica Planetarum.* Madison: University of Wisconsin Press, 1971.

Lorch, Richard P. "The Astronomy of Jābir ibn Aflaḥ." *Centaurus* 19: 85–107, 1975.

Lorch, Richard P. "The Astronomical Instruments of Jābir ibn Aflaḥ and the Torquetum." *Centaurus* 20:11–34, 1976

Lorch, Richard P. "Jābir ibn Aflaḥ and the Establishment of Trigonometry in the West", item VIII of Richard Lorch, *Arabic Mathematical Sciences: Instruments, Texts, Transmission.* Aldershot, 1995, 42 pp.

JĀBIR IBN ḤAYYĀN Abū Mūsā/Abū ʿAbd Allāh Jābir ibn Ḥayyān, for a long time the reigning alchemical authority both in Islam and the Latin West, is at the same time among the most important and most enigmatic figures of the history of Islamic science. Doubts already existed in the medieval Arabic tradition as to whether the large corpus of alchemical, philosophical, and religious texts attributed to Jābir were authentic. Scholars of modern times have shared these doubts, and some have gone as far as to conclude that Jābir may never have existed at all, and that contrary to the received view, the Jābirian corpus is not the work of a disciple of the sixth Shīʿī Imām Jaʿfar al-Ṣādiq (d. 765); rather, it was produced piecemeal by several generations of Ismāʿīlī authors, the oldest of whom lived no earlier than the second half of the ninth century.

This essentially is the widely accepted position of Paul Kraus who still remains the greatest Jābirian scholar of modern times. Some historians have disagreed with this position, and others have tended to revise it. At this juncture, then, very little about Jābir can be claimed with certainty. It might be safe to say on the basis of the most recent evidence that Jābir was a historical Muslim figure of the eighth century, that there was a small authentic core of Jābirian writings, and that out of this core developed the largely apocryphal and grand Jābirian corpus as we know it today.

Jābir is generally referred to as an alchemist. But if turning base metals into gold is the essential preoccupation of alchemy, then Jābir is hardly an alchemist, for transmutation of metals was only a minor part of his concerns. He was concerned, rather, with developing an all-embracing metaphysical and natural scientific system based upon immutable universal principles. It is this search for universal principles that led him to the study of language, music, and numbers. The Pythagoreans said that things are numbers; Jābir says that things are the names that designate them. An analysis of the name of a thing is for Jābir an analysis of the thing itself; this daring ontological claim of an equivalence between language and reality sounds more metaphysical than alchemical.

Jābir did write extensively about chemical processes and techniques, and in this field he made some highly original and historic contributions. For example we find in his treatises the theory that all metals are composed of sulphur and mercury existing in various proportions. It was this idea that led to the phlogiston theory of modern chemistry. Another Jābirian contribution is the introduction of sal ammoniac in the repertoire of chemistry. Two varieties of sal ammoniac were distinguished: natural (ammonium chloride) called *al-ḥajar,* and derived (ammonium carbonate), called *al-mustanbaṭ*. The latter was obtained by the dry distillation of hair and other animal substances. Again, the use of organic materials, both plant and animal, in addition to the inorganic, is a monumental Jābirian innovation.

Jābir's chemical processes are never carried out in a theoretical vacuum; we find in his writings both a developed theory of matter and a sophisticated cosmology. He believed that matter was ultimately composed of four "natures" — hot, cold, moist, and dry. But unlike the familiar Aristotelian qualities, Jābir's natures were not abstractions; rather, they were real, material, and independently existing entities; hot, cold, moist, and dry were the "first elements" out of which were born the "second elements": air, water, earth, and fire. The former were simple, the latter were compound; the former were primary, the latter were derived; the second elements could be resolved into natures, but natures were immutable. Jābirian natures were, then, the ultimate building blocks of the world.

A striking aspect of Jābir's cosmology is his parallel idea of the "first creation" and the "second creation". The former was an act of God, the latter an act of man. The difference between these two is that God acts in a timeless fashion, whereas man effects his creation in a temporal domain. Thus man can imitate God's work, but he requires time to accomplish it. From this emerges the Jābirian idea of artificial generation. Birds, for example, are found in nature, but these creatures can also be produced over a period of time in a laboratory; so can human babies. Through a manipulation of natures man can generate even such living beings as are not found to exist naturally — monsters, dwarfs, giants, freaks, and so on. Indeed, Jābir's idea of artificial generation sometimes strikes one as thoroughly modern.

If Jābir is the first alchemist of Islam then he is the pioneer of all that is important and characteristic of Islamic alchemy: the sulphur–mercury theory, the introduction of sal ammoniac, the use of organic substances, the idea of artificial generation of life, the production (though not recognition) of mineral acids, and the conceptual distinction between heat and temperature. He was widely known in Medieval Europe, and at least three of his treatises were translated into Latin.

S. NOMANUL HAQ

REFERENCES

Haq, S. Nomanul. *Names, Natures and Things: The Alchemist Jābir ibn Ḥayyān and his Kitāb al-Aḥjār (Book of Stones)*. Boston: Kluwer Academic Publishers, 1993.

Holmyard, E. J. *Alchemy*. London: Penguin Books, 1957.

Kraus, Paul. "Jābir ibn Ḥayyān; Contributions à l'Histoire des Ideés Scientifiques dans l'Islam II: Jābir et la Science Grecque." *Mémoires de l'Institut d'Égypte* 45(1), 1942.

Kraus, Paul. "Jābir ibn Ḥayyān: Contributions à l'Histoire des Ideés Scientifiques dans l'Islam I: Le Corpus des Écrits Jābiriens." *Mémoires de l'Institut d'Égypte* 44(1), 1943.

Needham, J. "Arabic Alchemy in Rise and Fall." In *Science and Civilisation in China,* vol. 5, pt. 4. Cambridge: Cambridge University Press, 1980.

JAGANNĀTHA SAMRĀṬ Paṇḍita Jagannātha (1652–1744), who bore the title "Samrāṭ", writer on astronomy and mathematics, and designer of astronomical instruments, was the religious preceptor and collaborator in astronomical pursuits of Jai Singh Sawai (1688–1744), the astronomer-prince of Jaipur in Rajasthan. Born into a Vedic family, the son of Gaṇeśa, Jagannātha was attached to the court of Jai Singh from an early age and assisted his patron in all his social, religious, and scientific activities.

At Jai Singh's behest, Jagannātha mastered Arabic and Persian, the two foreign languages prevalent in the Mughal court, which he utilized in the study of Islamic astronomy and put to beneficial use translating into Sanskrit texts in those languages for his patron. In this way, Jagannātha produced his *Rekhāgaṇita* and *Siddhāntasārakaustubha*, which are translations of Euclid's *Elements of Geometry* and Ptolemy's *Almagest*, respectively, from their Arabic versions by Naṣīr al-Dīn al-Ṭūsī. It is interesting that in the case of *Rekhāgaṇita*, Jagannātha himself coined more than a hundred Sanskrit equivalents of technical terms.

In his original work *Siddhānta-samrāṭ*, composed at the behest of Jai Singh, Jagannātha described the construction and application of a number of astronomical instruments. He also mentioned the reasons that his patron, who was obsessed with metallic instruments like the astrolabe for making celestial observations and reading out the results, later opted for huge outdoor observatories with stone and mortar. The reason was that Jai Singh found that the metallic instruments did not give minute readings, and were also susceptible to wear and tear, and to climatic conditions, which brick observatories were not. His *Yantraprakāra* is a more elaborate work on the subject of astronomical instruments, which includes descriptions of some more instruments, computations, a number of tables, and allied data.

Jagannāha's *Siddhānta-samrāṭ* and *Yantraprakāra* carry a number of recordings of celestial observations of different types, for periods short or long, which demonstrate how the instruments designed, and observatories constructed, as above, had been put to use for correcting parameters, preparing almanacs and the like. The part played by Jagannātha in these endeavors was considerable and significant.

K.V. SARMA

REFERENCES

Pingree, David. *Census of the Exact Sciences in Sanskrit*. Philadelphia: American Philosophical Society, 1981. Series A, vol. 3, pp. 86–88; vol. 4, p. 95.

The Rekhāgaiṇta or Geometry in Sanskrit composed by Samrāṭ Jagannātha. Ed. K.P. Trivedi. Bombay: Nirnaya Sagar Press, 1901.

Samrāṭ-siddhānta. Ed. Ram Swarup Sharma. New Delhi: Indian Institute of Astronomical and Sanskrit Research, 1967.

Sharma, M.L. "Jagannāth Samrāṭ's Outstanding Contribution to Indian Astronomy in Eighteenth Century AD." *Indian Journal of History of Science* 17 (2): 244–51, 1982.

Sharma, Virendra Nath. "Sawai Jai Singh's Hindu Astronomers." *Indian Journal of History of Science* 28 (2): 131–55, 1993.

Siddhāntasamrāṭ of Jagannāthasamrāṭ. Ed. Muralidhar Chaturveda. Sagar: Samskrita Parishat, Saugar University. 1976.

Upadhyaya, B.L. *Prācīna Bhāratīya Gaṇita*. New Delhi: Vijnana Bharati, 1971.

Yantraprakāra of Sawai Jai Singh. Ed. and Trans. S.R. Sarma. New Delhi: Department of History of Medicine and Science, Jamia Hamdard, 1987.

See also: Jai Singh – *Almagest* – Naṣīr-al-Dīn Ṭūsī – *Elements* – Astronomical Instruments in India

JAI SINGH Jai Singh, or Jai Singh Sawai (Jaya Siṃha Savā'ī), the eighteenth century statesman-astronomer of India, was born on November 3, 1688 to the royal house of Amber, in the present state of Rajasthan, India. His ancestors were semi-autonomous rulers of their princely state under the Mughals and occupied important posts at the Mughal court. Jai Singh lived his life during one of the most troubled, uncertain, and critical periods of Indian history, and he was involved directly or indirectly in just about every political or military conflict of his time. With his diplomacy and political maneuverings, he acquired a great deal of authority and influence throughout the Mughal empire. Making full use of his prestige and power, Jai Singh embarked upon a program of reviving astronomy in his country.

Jai Singh displayed an early inclination towards astronomy and mathematics and soon acquired mastery over these two subjects. He realized that the astronomical predictions based on the Hindu, Islamic, or European books (which were available to him) did not agree with actual observations. He reasoned that the disagreements between predictions and observations were primarily due to the outdated parameters found in the astronomical books, and would not be alleviated until new parameters based on careful observations were made available. Consequently, he decided to obtain new parameters.

With the blessings of the reigning emperor, Muḥammad Shāh, Jai Singh initiated a multifaceted program in astronomy. He designed instruments, built observatories, compiled an excellent library, assembled competent astronomers of different scientific backgrounds, and sent a fact-finding scientific mission to Europe. His scientific career lasted for more than 20 years. He died in 1743, at the age of 54.

Jai Singh started out first with traditional instruments of brass built according to the designs given in the texts of the Islamic school of astronomy. However, the metal instruments did not measure up to his expectations; he discovered with disappointment that the instruments gave inaccurate results once their axes wore down, displacing their centers. The instruments were also unsteady during observing because of their portable nature. He discarded these instruments, therefore, in favor of the instruments of masonry and stone of his own design and tried to achieve the desired precision from their large sizes and steadiness from relatively inflexible structures. His instruments range anywhere from 1 m to 25 m in height.

Jai Singh built five observatories in cities of north India, at Delhi, Jaipur, Varanasi, Ujjain, and Mathura and equipped them with instruments of his own design. His observatory in Delhi was completed in 1724 and the others within a decade. His observatories, all except that of Mathura, are still extant in good to fair states of preservation. His observatory at

Table 1: Major instruments of masonry and stone of Jai Singh

1.	Jaya Prakāśa (Hemispherical dial I)	2	Delhi, Jaipur
2.	Samrāṭ yantra (Equinoctial sundial)	6	Delhi, Jaipur, Ujjain, Varanasi
3.	Rāma yantra (Cylindrical dial)	2	Delhi, Jaipur
4.	Rāśi valaya (Ecliptic dial)	12	Jaipur
5.	Ṣasthāṃs'a yantra (Sixty-degree instrument)	5	Delhi, Jaipur
6.	Dakṣinottara Bhitti (Meridian dial)	6	Delhi, Jaipur, Ujjain, Varanasi, Mathura
7.	Digaṃs'a yantra (Azimuth circles)	3	Jaipur, Ujjain, Varanasi
8.	Nāḍīvalaya (Equinoctial dial)	5	Jaipur, Ujjain, Varanasi, Mathura
9.	Kapāla A (Hemispherical dial II)	1	Jaipur
10.	Kapāla B (Hemispherical dial III)	1	Jaipur

Jaipur has the largest number of instruments and is in the best preserved state. The observatories of Delhi and Jaipur are big tourist attractions these days and visited by hundreds of thousands of people each year.

An inventory of Jai Singh's major instruments of stone and masonry is presented in Table 1.

The Samrāṭ, Ṣaṣthāṃs'a, and Dakṣinottara Bhitti are Jai Singh's high precision instruments. With these instruments, he extended precision to the very limit of naked eye observing, i.e. $1'$ of arc.

Although the telescope had become common with European astronomers and had acquired refinements such as the micrometer and cross-hair, there is no evidence that Jai Singh benefited from it. His instruments do not use a telescopic sight, and with all their ingenuity of concept and design are no more than what may be called "naked eye tools" somewhat in the tradition of the medieval astronomers such as Ulugh Beg of Samarkand. It is reasonable to believe that the invention of the telescopic sight which had come into vogue with European astronomers only a few decades earlier did not come to his attention early enough. It should be pointed out, however, that Jai Singh was familiar with the telescope and had observed with it. His personal library inventory lists a telescope bought for him at a cost of 100 rupees.

Jai Singh's early training as an astronomer had been under Hindu *pundits*, and they remained the mainstay of his program until the very end. At one time there were at least twenty-two astronomers working at the observatory of Jaipur alone. Jagannātha Samrāṭ, Kevalarāma, and Nayanasukhopādhyāya were his principal astronomers. These astronomers constructed instruments, collected data, translated books and compiled original works in astronomy. The translated works included Ptolemy's *Almagest*, Euclid's *Elements*, and De La Hire's *Tabulae Astronomicae*.

Jai Singh was equally interested in the Islamic and the European traditions of astronomy. He collected astronomical works in Persian and Arabic and patronized Muslim astronomers of the Persian-Arabic school. The Muslim astronomers included Muḥammad Ābid, Sheikh Asad Ullah, Sheikh Muḥammad Shafī, and Dayānat Khān. These astronomers procured astronomical books for the royal library, constructed instruments, helped with the translations, collected data at the observatories of Delhi and Jaipur, and traveled to distant lands at the command of their patron.

Figure 1: Sawai Jai Singh (1688–1743). Courtesy of the Sawai Man Singh II Museum, Jaipur (used with permission of the author).

By 1725, the involvement of the Muslim *nujūmīs* or astronomers in Jai Singh's astronomical program began to taper off and, in its place, the involvement of Europeans, primarily Jesuit priests, increased. The European astronomers included De Bois, Figuerado, Boudier, Gabelsberger, and Strobl. The Europeans played the role of conveyors of

European knowledge to the Raja. Accordingly, they led a delegation to Europe, procured texts and instruments, translated De La Hire's tables, and carried out mathematical computations. However, the knowledge these Europeans brought to Jai Singh and his astronomers had already become outdated in Europe, for it did not include the theories of Galileo, Kepler, or Newton; nor did it include observational techniques such as those employed by Flamsteed in England.

In 1727, Jai Singh dispatched a scientific delegation to Europe after learning that "the business of the observatory was being carried out there." The delegation, first of its kind from the East, was led by Figuerado, and it reached Portugal in January 1729. The delegation stayed on in Portugal for over a year. It did not travel to Paris or London, however, where the most advanced work in astronomy was being done. In 1730, the delegation returned to Jaipur, the capital of Jai Singh'state at the time, with some instruments, books on mathematics, and the tables of De La Hire. The delegation did not bring any books elaborating the heliocentric world view, such as proposed by Newton, Kepler or Copernicus, since these publications were prohibited by the Catholic Church.

After collecting data for nearly a decade, Jai Singh succeeded in obtaining new astronomical parameters. With these parameters, he prepared a set of astronomical tables called a *Zīj*. The *Zīj*, completed sometime between 1731 and 1732, was dedicated to the reigning monarch, Muḥammad Shāh and is, therefore, called *Zīj-i Muḥammad Shāhī*. *Zīj-i Muḥammad Shāhī* is a 400-page long traditional work of astronomy similar to the *Zīj-i Sulṭānī* of Ulugh Beg. *Zīj-i Muḥammad Shāhī* may be considered Jai Singh's most important contribution to the astronomy of India. The *Zīj* remained a valuable resource for traditional astronomers of the country for nearly 150 years.

For the sake of rejuvenating astronomy in his country, Jai Singh expended a great deal of energy as well as his personal fortune, but he failed to initiate the new age of astronomy in India. He himself remained unaware of the Copernican revolution that had swept the intellectual circles of Europe. Lack of good communication systems, and a complex interaction of intellectual stagnation, religious taboos, theological beliefs, national rivalries, and the simple human failings of his associates share the blame for it. Jai Singh's scientific accomplishments were medieval in retrospect, but his scientific outlook was quite modern.

VIRENDRA NATH SHARMA

REFERENCES

Bhatnagar. V.A. *Life and Times of Sawai Jai Singh*. Delhi: Impex India, 1974.

Garrett, A. ff. and Guleri, Chandradhar. *The Jaipur Observatory and Its Builder*. Allahabad: Pioneer Press, 1902.

Kaye, G.R. *The Astronomical Observatories of Jai Singh*. New Delhi: Archaeological Survey of India, reprint, 1982.

Sarma, Sreeramula Rajeswara. "Yantraprakāra of Sawai Jai Singh." Supplement to *Studies in History of Medicine and Science*, Vols. X and XI, New Delhi, 1986 and 1987.

Sharma, Virendra Nath. *Sawai Jai Singh and His Astronomy*. New Delhi: Motilal Banarasidass, 1995.

See also: Jagannātha Samrāṭ – *Zīj* – Observatories in India – Astronomical Instruments in India

JAMU *Jamu* is the Indonesian and Malay term for indigenous pharmaceuticals made from fresh or dried medicinal herbs. These medicines are popular among the ethnic Malays in the whole Malayan Archipelago (Malaysia, Singapore, Brunei Darussalam, Indonesia). It is assumed that *jamu* originates from the Javanese principal courts. On the other hand there are different local recipes and ways of preparation with an old tradition, which contribute to the idea of a common popular knowledge.

Jamu can be an infusion of herbs, a mixture of fresh, dried or dried and powdered medicinal plants, and even an extract of herbs. It can be made of a single plant, but mostly it is a composition of up to forty different ones. Today *jamu* is sold in the form of powder, tablets, pills, tonics and capsules for internal use; and in the form of ointments, oils, tonics, or compresses for external use. The powdered *jamu* are packed in individual portions, which are then sold in nearly every small shop across Indonesia. Today these *jamu* as well as pills and capsules are available in Malaysian and Singaporean supermarkets, too.

In the traditional way, *jamu* is produced at home. An experienced family member provides friends and relatives with the homemade version of the product. To supplement family income, female family members sell the surplus to neighbours or villagers. These "*jamu* women" are still a familiar sight in Indonesia, even in the big cities. As a kind of peddlar, the *jamu* woman wanders through villages or capital streets, offering the contents of her *jamu gendong* (*jamu* being carried on the back) to a regular clientele.

Indonesia was, and is, the trendsetter in the production, development and marketing of *jamu*. At the beginning of the twentieth century her *jamu* production reached a turning point. Some innovative entrepreneurs began to produce home made *jamu* for their sale, thereby founding a flourishing cottage industry. The profits which accrued in this fashion kicked off a process which led to the modern *jamu* industry, in which more than 350 factories are registered in Indonesia. Their product is exported to several countries, the largest consumers being Malaysia and Singapore. The industrialization of *jamu* manufacture had severe consequences. It altered the outer appearance of *jamu*, and demanded more and more modernizing and profitable sale strategies. It was necessary to change the image from an old fashioned herbal remedy to a proudly promoted "ancestral heritage", which fit better into the Western back-to-nature trend.

Jamu concentrates on aspects and interrelationships which Western medicine does not take as seriously. The *jamu* idea of medication is based on a broad conception of body, health, and sickness; the notion of care and therapy, for example, tend to be kept less separate. *Jamu* is more than a pharmaceutical in the Western sense; it is food supplement, prophylaxis, and specific remedy all at once. Keeping healthy also necessitates staying beautiful and attractive as long as possible. Cosmetics, general tonics and, above all, aphrodisiacs are very popular, and have considerable stature in *jamu* medicine.

Culturally different world views and concepts of knowledge find their expression in all aspects of culture, including the "hardware" of medicine, the pharmaceuticals. If we characterize the Western world view roughly as individualistic, bound to linear thinking, the ideas of cause and effect and the explanation of isolated phenomena on a microlevel, the Malay world view can be seen as sociocentric, bound to the mutual dependency of microcosmos and macrocosmos, and to the maintenance of balance and harmony on all levels of existence.

Medicines reflect these different aspects. Western pharmaceuticals consist of molecular defined substances or mixtures of substances, which are considered to be active. Any function of inactive substances is denied. The number of ingredients tends to be kept low in order to avoid synergistic effects. In contrast *jamu* are preparations of dozens of entire plants, with all active, inactive, and unknown substances plus the fibrous material. *Jamu* is a whole cosmos of ingredients, in which each part is supposed to play its proper role. *Jamu* has to be taken regularly for some weeks or months, and helps, in a long run, to balance the state of health between the poles of hot and cold.

Jamu cannot be completely analyzed. Pharmacologists are able to screen only some of the active ingredients, not to mention those which are called inactive. Synergistic effects are a matter of course and are considered to be of special importance in the efficacy and safety of *jamu*. The discussion of synergism and risk reduction started in 1976, when the Indonesian scientist Sutrisno formulated his SEES (Side effect eliminating/Secondary effectiveness enhancing substance) Theory which postulates fewer side effects and a well balanced total efficacy of entire plant extracts. Western science considered the SEES-Theory not sufficiently proven. Much research has been carried out to prove that some of the active *jamu* ingredients have curative value. This is an attempt to translate a non-Western medical concept into a Western one, which soon reaches methodological limits.

These facts are not perceived as contradictions by the Malay — or, in a melting pot like Singapore also Indian and Chinese — clients. As a vivid element of Malay culture *jamu* is consumed in any form, in tiny villages and big capital cities, and plays a constant role in everyday health care.

CHRISTINE TUSCHINSKY

REFERENCES

Afdhal, A.F., and R.L. Welsch. "The Rise of the Modern Jamu Industry in Indonesia." In *The Context of Medicines in Developing Countries*. Eds. S.v.d. Geest and S. Reynolds Whyte. Dordrecht: Kluwer, 1988, pp. 149–172.

Jordaan, R.E. "On Traditional and Modern Jamu in Indonesia and Malaysia: A Review Article." *rima — Review of Indonesian and Malaysian Affairs* 22(1):150–163, 1988.

Rehm, K.D. "Jamu–die traditionellen Heilmittel Indonesiens." *Curare* 3: 403–410, 1985.

Sastroamidjojo, S. *Obat Asli Indonesia*. Jakarta: Dian Rakyat, 1948/1988.

Simandjuntak, E.S. "Meningkatan Pemasaran Jamu, Menjual Gairah Seks." *Prisma* 2: 74–84, 1984.

Tuschinsky, C. *Produktion, Handel und Konsumtion nicht-westliche Medikamente in Südost-Asien: Malaiische jamu in Singapore*. Hamburg: Lit, 1992.

See also: Medicine in the Malay Peninsula

JAPANESE SCIENCE The history of Japanese science, technology and medicine is divided into two distinctly different periods, traditional and modern, the dividing line being the late nineteenth century when Japan started wholesale Westernization. Since the subject matter of modern Japanese development is not much different from Western development, this article deals only with the traditional and transitional periods.

The Traditional Period

The written history of Japan started approximately in the sixth century when Chinese writing, culture, and institutions were brought in by Korean immigrants. Technical professions began in Japan with the immigration of Korean experts in the sixth to eighth centuries. The immigrants slowly settled into Japanese society, and the Chinese view of nature gradually began to take root in Japan. Thus, traditional Japanese science, technology, and medicine is basically a ramification of Chinese paradigms. In the following, we shall elucidate the points of Japanese deviation from the Chinese prototype.

Whereas the Western world of learning had an integral structure of trivium, quadrivium, scientia, or Wissenschaft, the corresponding Chinese world of learning was thoroughly structured with Confucian classics at the center of a structured bureaucracy. Scientific subjects were of peripheral importance, with mathematics and calligraphy being necessary for the civil service examination system, and astronomy, medicine, and architecture being fields with their own branch offices in the bureaucracy. The small replica of the Chinese world of learning and bureaucracy was created during the eighth century in Japan.

In principle, the Japanese Astronomical Bureau was a miniature reproduction of the Chinese bureaucracy, but a closer look reveals significant remodeling to meet local requirements. While in China the astronomical organizations that computed the ephemerides and observed celestial phenomena had important astronomical bureaux, in Japan all of these activities were subsumed under the single *Yinyang* Board, the name of which indicates a clear priority for divination. After that time, the Japanese *Yinyang* art developed into a complex of divination techniques.

Given a Chinese manual of calendrical science from which to determine basic parameters and computational procedures, there was little that local talent or preference could add. On the other hand, when unforeseen and ominous celestial phenomena were observed they had to be interpreted without delay. Thus in astrology the Japanese were thrown upon their own resources, and the practical application of imported knowledge was valued over basic theory.

It was not the regularities of the eternal truths of mathematical astronomy but the unforeseeable omens of the astrologers that attracted attention in Japan. Whereas abstraction and involved theoretical argument were by no means rare in Chinese science, they were vastly less important in Japan.

As the Chinese influence became less noticeable from the tenth century on, the bureaucracy also diminished in strength. Throughout the medieval period, it was in family businesses rather than the government bureaucracy where those with talent enough to pass civil service examinations were employed. Kinship ties returned to their former importance.

Because they were separated from the Asian continent, the Japanese were not totally overwhelmed by Chinese influence. Japanese society was much less bureaucratic than that of China or Korea, and as a result was more able to respond to the incoming intellectual flux. The Japanese received Chinese influence selectively to meet local needs.

Meanwhile, Buddhism reached Japan from India by way of China and Korea. Although enthusiasm for Buddhism waned somewhat in China and Korea, the Japanese have retained it throughout history. In medieval times, Buddhist monks competed with official astronomers in predicting eclipses by use of their *Fu-tian* calendar. In the paradigmatic treatise of Japanese divination art, *Sukuyōkyō*, excerpts from classics of Tantric Buddhism were mixed side by side with Chinese Daoist texts. In Japanese medical classics, certain Buddhist recipes were mixed and preserved. This reflects Japanese intellectual pluralism.

The Platonic conviction that eternal patterns underlie the flux of nature is so central to the Western tradition that it might seem that no science is possible without it. Nevertheless, although Chinese science assumed that regularities were there for the finding they believed that the ultimate texture of reality was too subtle to be fully measured or comprehended by empirical investigation. The Japanese paid even less attention to the general while showing an even keener curiosity about the particular and the evanescent.

This attitude may be seen in the career of Shibukawa Harumi (1639–1715), the first official astronomer to the Japanese Shogun. In the preface to his early treatise *Shunju jutsureki* (Discussions on the Calendar Reflected in the Spring and Autumn Annals, the oldest Chinese chronicle), he stated that astronomers had rigidly maintained that when Confucius dated the events in his *Annals of the Spring and Autumn Era* he made conventional use of the current calendar with little care for its astronomical meaning, so that the dates were not very reliable. "This error is due to the Chinese commitment to mathematical astronomy, so that they do not admit that extraordinary events happened in the heavens. Extraordinary phenomena do in fact take place in the heavens. We should therefore not doubt the authenticity of [Confucius'] sacred writing brush."

Once admitting, as Shibukawa did, that regular motion was too limited an assumption, one could easily conceive such notions as that astronomical parameters could vary from century to century. In the official Chinese calendar in the thirteenth century and earlier, the discrepancy between ancient records and recent observations was explained by a secular variation in tropical year length. Shibukawa revived this variation in the Japanese calendar, and Asada Goryu (1734–1799) extended it to other basic parameters to account for Western as well as Eastern observations then available to him.

The variation terms used in Japanese astronomy were too large to survive empirical testing, and were eventually discarded. It is, however, noteworthy that their cosmological outlook was so flexible that they accepted the notion that all astronomical parameters were subject to change and the whole universe was changeable, while in the West the Aristotelian notion of an unalterable universe was followed rigorously, and irregular motions in the sky were inconceivable.

In early Chinese history, higher mathematics was always associated with calendrical science, in terms of solving indeterminate equations for ascertaining the time of grand conjunction. In Japan, interest in mathematics was also raised in connection with calendrical science in the seventeenth century, and was soon developed as mathematical puzzles to be enjoyed by hobbyists, in the same way as there were enthusiasts of flower arrangement and the tea ceremony. This form of Japanese mathematics called *wasan* was mostly viewed as a popular art. It was not considered a scholarly activity.

I shall not attempt a conceptually rigorous definition of scholarship and hobby here, but merely note that scholarship is usually thought to have some public function while hobbies are regarded as private indulgences which may or may not have significant social value. While art does not require legitimation, scholarship does, and there has to be some basis on which to differentiate one from the other. Some kind of legitimation was found to be necessary for mathematicians and one way of doing so was to invite Confucian scholars to write prefaces to their books, in effect borrowing Confucianism's prestige.

The Japanese mathematicians were generally socially marginal curiosity seekers. During the Tokugawa period, mathematics was not formally recognized by the government. The Tokugawa mathematical tradition existed entirely in the private sector and had no official support. Whenever scholars demand legitimacy from society, they display a sense of mission which reinforces their commitment. This sense of mission is associated with the rise of professions, but the Japanese mathematicians cannot be called professionals.

With the use of mathematical rods, Chinese mathematics developed quite sophisticated algebraic formulae. Building on the Chinese tradition, a significant development was made by Japanese mathematicians to develop numerical algebra into symbolic algebra by writing on paper and making it possible to obtain a more general treatment called *wasan*. Curiously enough, the formulas of *wasan* resemble modern algebraic formulae but they are written vertically. However, due to stylistic differences, there was no intercourse between Japanese and Western developments in mathematics.

Wasan mathematics developed a unique custom in the formative period consisting of the posting of mathematical problems called *idai* (bequeathed questions). A mathematician would pose scores of problems of several kinds at the end of a book and then publish it. Another mathematician would post answers to these problems and present his own in the same manner. According to convention, yet a third mathematician would post answers to these problems and issue his own in relay fashion. This mechanism of *idai* caused a chain reaction and greatly stimulated the formation of *wasan* groups. The tradition began with twelve problems from the *Shinpen Jinkoki* published in 1641. A succession of mathematical lineages soon developed and reached their peak during the life of Seki Kowa (Takakazu). Practically all of the important problems in the history of *wasan* date from the period above.

Another feature of *wasan* development was the *sangaku* (mathematical tablet) form that came in the later phase. These tablets included both problems and answers and were dedicated at shrines. The best mathematicians made their accomplishments known through books, but it was largely the custom of *sangaku* which supported the activities of the wasan enthusiasts. That *wasan* was a hobby which cost money to pursue is best shown by the elegant diagrams which embellished such work. In fact, the offering of *sangaku* had a strong attraction as it showed a desire to keep their work on more or less permanent view.

Chinese natural philosophy developed certain abstract concepts in ancient times. The *qi* concept, one of the basic ideas which appeared in the *Huangdii Neijing* (The Yellow Emperor's Internal Classic), the earliest Chinese medical compilation, is today interpreted as something akin to energy which is imponderable but permeates and circulates in the macro- as well as the microcosmos. When the *qi* concept was introduced into Japan, medical doctors tended to interpret it as materialistic, unable to comprehend such a highly abstract concept of imponderable energy. Given the difficulty of understanding the subtleties of Chinese philosophical speculation, the Japanese physicians adopted a more simplistic, pragmatic, and empirical approach.

For instance, certain tumors and internal swellings were thought to be stagnated or congealed *qi*. Indeed, the seventeenth century Japanese physician, Goto Gonzan, attempted to explain the cause of all medical disorders by stagnation of this kind.

Traditional Chinese medicine was concerned primarily with function and only secondarily with tissues and organs. Internal disorders were never local in Chinese medicine. The major physiological and pathological theory was to regulate energy flow internally and externally between the microcosmos and macrocosmos, so that holistic diagnoses could be applied. They depended heavily on pulse diagnosis as the indication of energy flow. More materialistic Japanese physicians such as Yoshimasu Todo (1701–1773), however, developed stomach and abdomen diagnosis by touching the location of disease, rejecting the traditional pulse diagnosis. Yoshimasu, the foremost figure of the radically critical school declared that the *qi* of the universe had nothing to do with medicine. This group was prepared to take a position much closer to that of the solidists in the West than had been possible in Japan at an earlier time. Functional analysis lost its importance, and the physical organs could be studied for their own sake. This was nothing less than a gestalt change.

In accordance with the extension of the materialistic inclination of Japanese physicians, they were interested in anatomy and found that Western schema were a great deal more accurate than the Chinese anatomical charts. In mathematical astronomy and calendrical science, Western superiority in accuracy was quickly recognized. In medicine, there is good reason to doubt that there was any difference in therapeutic efficacy before the late nineteenth century. It is above all in the comparison of anatomical charts that the strength of Western medicine would be apparent.

In the second half of the eighteenth century, Sugita Genpaku took up the study of anatomy because it seemed the most tangible, and therefore the most comprehensible, part of Dutch medicine. Following the solidist breakthrough, Sugita's successors in medicine studied physics and chemistry and opened up the world of modern science. The Copernican influence was minor by comparison, because the Japanese cosmos had not been defined by religious authority. The impact of anatomy challenged the energetic and functional commitments not only of medicine but of natural philosophy. Its effect was bound to be revolutionary.

The learned treatises of the Chinese and Japanese medical traditions lacked terminology not only for brain function but also for mental processes. It was only later scholars with considerable knowledge of Western anatomy, such as Sugita Genpaku, who could abandon the Chinese tradition entirely and display as much anatomical interest in the brain as in the viscera.

Materia medica and natural history books were brought first from China and later from Holland. The first thing for a Japanese natural historian to do was to identify local species with imported Chinese books. This identification work endowed Japanese scholars with new insights into comparative study. While the original Chinese texts were intended for the collection of medicinal herbs for pharmaceutical purposes, the Japanese developed this study into a natural history of visual observations for identification.

In feudal Japan, all occupations were hereditary. The post of official astronomer to the Shogun was created to recognize the personal achievement of Shibukawa Harumi, and was passed down to his descendants. It had no significance beyond the technical, and thus was of no interest to the generalists. Those who held technical posts of this kind, such as medical doctors and Confucian scholars, were from the beginning separate from the general samurai bureaucracy.

When the official astronomer and his subordinates were compiling the ephemerides, Confucian scholars were not consulted. The astronomers had inherited the tradition through their family line since medieval times. At the time of a calendar reform, once every hundred years for a Chinese-type luni-solar calendar, the descendants were usually incapable of innovative work, but able to perform the routine calculations necessary for the distribution of the yearly calendar according to a given formula. In most such cases, talented astronomers in the private sector were appointed to the post of official astronomer. Beginning with Shibukawa Harumi, the number of such hereditary official astronomers amounted to seven families toward the end of shogunate rule.

Often professionals were encouraged to adopt talented young boys rather than their own sons. Under the rules of samurai society in the Tokugawa period, only the first son was entitled to inherit the family stipend. Later born sons were adopted by other sonless samurai families or left to find ways of obtaining their livelihood themselves. For either, they had to show their ability by studying hard. These younger sons would provide Japan with a pool of talented professionals.

In Japan, there was no social or political reason for a single philosophical orthodoxy. Although nominally based on the centralized Chinese model, Japanese feudal society remained multifocal, consisting of the Shogunate and a hundred fiefs. Thus freedom to choose between several paradigms seems to have been as desirable as the search for a unitary principle was in China.

In China and Korea, the official character of science and learning was rigidly maintained by adherence to the civil service examination system. The Japanese imitated the Chinese bureaucracy for only a short time in the seventh and eighth centuries, and subsequently this failed to take root in Japanese society. It was only in the last quarter of the nineteenth century with the establishment of a modern government system that a centralized bureaucracy complete with civil service examinations emerged. During the Tokugawa period, it was centralized feudalism, in which the Tokugawa government had overwhelming power but in which each fief government was able to maintain a degree of autonomy by having their own independent schools, bureaucracy, and policies, in competition with other fiefs.

The Tokugawa shogunate wanted to instigate centralized civil and medical service examinations but was unable to impose such a system on the fief governments. As a result, there was no rigidly maintained state orthodoxy of science and learning, although the exclusion of Christian tenets was executed with considerable harshness. In short, the Japanese have enjoyed the merits of multi-culturalism. It is not the kind of multi-culturalism which involves the peaceful coexistence of different races in one society but rather signifies the acceptance of different cultures by largely one single group.

Science and learning had been substantially privatized and able to respond to imported foreign ideas by the late eighteenth and early nineteenth centuries. Even during the semi-seclusion period under the Tokugawa shogunate, the activity of Dutch learning was practiced by city intellectuals, medical doctors, and merchants in the private sector. Only after the Western threat of invasion became substantially felt in the nineteenth century, particularly after the news of the Opium War and the arrival of Commodore Perry's gunboats in 1853, did the knowledge of Western learning become the business of the public sector. Samurai became interested in Western military technology, finally overthrowing the feudal government and replacing it with a centralized state.

From the seventeenth century onwards, when Western knowledge began to exert claims of its own against the backdrop of Chinese learning, Japanese thinkers were critically attentive. As soon as they were convinced that European technical knowledge was superior, the Japanese switched to the new paradigm with remarkable speed. They quickly modified their attitudes and oriented themselves towards achieving certain desirable goals that had been reached in the West. For the Chinese, the encounter with European ideas was traumatic; to accept them was to reject traditional values, and to reject them would leave them with no defence against dismemberment by Western powers. For the Japanese it was merely the appearance of another paradigm to be accepted in their eclectic manner. The Japanese were flexible enough to accept the Western paradigm without strong resistance.

The Transitional Period

Let us now turn to the occupational support groups who successively played a leading role in the introduction of Western science to Japan

Astronomers and Interpreters. The adoption of Western 'barbarian' astronomy in the seventeenth century by the Chinese dismayed conservative elements in Japan. However, it did not promote the desire in others for a similar change in the Japanese calendar, which was modeled on that of the ancient Chinese. As has been mentioned, the eighth Shogun Yoshimune and his court astronomers recognized the superiority of Western over Chinese astronomy, but astronomers were primarily government bureaucrats or technicians whose scope remained limited to their assigned duties of drawing up the official calendar. Their interest in Western science was limited to the precision of astronomical data and methods of calculation, and they made no attempt to jeopardize their hereditary posts by entertaining revolutionary paradigms such as were developing in Europe at that time.

Professional interpreters at Nagasaki were the most well-versed in the Dutch language, through which they must have become acquainted with the concepts of Western science. But again, they were also hereditary government officials who remained within the boundaries of their duty of faithful translation and nothing more. Neither official astronomers nor interpreters published their work for general audiences.

Independent Scholars and Physicians. From the late eighteenth century on, a sizeable number of Dutch books containing the term *Natuurkunde* (study of nature) found their way into the country and aroused the interest of various independent scholars in the private sector who set about translating them, although their foreign language skills were much inferior to that of the Nagasaki interpreters. The majority of these "Dutch scholars", as they came to be called, were avant-garde physicians who were primarily freelancers, with no strict subordinative links to the governing elite or subsequent interest in maintaining an existing *status quo*. Thus they were not inhibited in stepping out of their line of work and were free to extend their interests to anything Western except perhaps the Christian doctrine. Astronomy was the first area in which people sensed that the West was superior, but the notion that the West was superior in other fields of scientific endeavor first spread among these independent physicians.

As they inched their way through Dutch texts, they realized that Western science was more than a variant of the natural history line of their own tradition. As they saw it, its essence could be translated as *kyuri* (literally, investigating the principles of things, a new Confucian term, or "natural philosophy", being a systematic and fundamental investigation and consideration of the nature of things). Thus, although at first there was no clearly established tradition for the term, *kyuri* was later to become the most common, and given the vocabulary of the period, it was an informative translation. In recognizing an enquiry into principles at the bottom of such traditional practical studies as medicine and calendar-making, in becoming aware, that is, of natural philosophy or physics, they also grasped the hierarchical structure of modern science, from basic to applied. Above and beyond the culturally-bound achievements of Western science, they seemed to have sensed that it contained a revolutionary paradigm, since the belief in underlying laws in Nature which could be formulated was weak in Japanese traditional thought. This was in marked contrast to the official astronomers who looked upon science as something that could be handled at the technical level and assigned it only a supplementary role.

The physicians recognized the importance of physics and chemistry as the basis of medical work, and several founded schools for the teaching of Western medicine. Their students, during their apprenticeship and internship days, moved from one school to another in major medical centers like Nagasaki, Osaka, Kyoto, and Edo (Tokyo), and played a role of diffusing knowledge of Western culture as well as medicine. While some physicians remained in the practice of medicine, others turned to the teaching of the Dutch language and even the discussion of international knowledge and politics and also founded schools for such. Thus, the cultural influence of the physicians was more considerable than that of the astronomers who limited their activities to the translation of technical works.

But these support groups were by no means adequate preparation for the reception of contemporary science. The Dutch scholars were still only amateur supporters of uninstitutionalized paradigms, comparable in this respect to gentleman members of the Royal Society. Their schools were few and regarded on the whole as avant-garde. Had they flourished in the seventeenth and eighteenth centuries their work might have had greater consequences. By the mid-nineteenth century, however, modern science in the West had entered the systematically structured world of the university and reached new advanced levels. Reception of learning that was now systematically pursued in Europe within the structures of the university required the creation of an institutional system. It had to be met not by self-taught amateurs but with professional scientists who had received a modern systematic education. These early scholars were not aware of the institutionalized aspect of Western science up until the mid-nineteenth century, remaining bookish translators of the *kyuri*, paradigmatic aspect of Western science.

Samurai. The Opium War between the Chinese and Western colonialists created concern among the samurai cognoscenti, but Commodore Perry's visit to Japan in 1853 and the subsequent threat of war caused all samurai to realize that the Westerners' science was needed for national defense. As a result, young samurai flocked to the schools of Western learning established by physicians. As they were traditionally the ruling class, and in time of war the warriors, it was natural that samurai interest in Western science and technology training was based on more real and pressing needs than reception of and support for particular scholarly paradigms.

First, they tried to learn the Western art of manufacturing firearms, but soon realized that it was impossible to catch up and compete with Western forces by a crash project of manufacturing cannons. They could purchase hardware, but what was really needed was the software of Western military training and tactics. Moreover, the acquisition of Western military discipline was one of the contributing factors in the overthrow of the ruling Tokugawa family by anti-Tokugawa samurai and the subsequent establishment of the Meiji government in 1868. The samurai recognition of Western science was therefore political rather than cultural.

After the revolution, convinced that Western scientific learning was essential, the ruling class set out to disseminate their conviction through the establishment of an educational infrastructure. With Meiji effort and initiative, modern science was fully assimilated in a wholesale introduction through governmental institutions rather than a piecemeal cultural infiltration, as in the previous era, and modern scientific and technological professions became the artificial creation of the new Western-oriented government.

The main practitioners of these new professions were former samurai. In the past they had received hereditary family stipends in exchange for their loyalty to the feudal powers, the Shogunate or local feudatories. But in the 1870s, efforts were made by the Meiji government to curtail the inherited family stipends of the samurai class as a step towards social modernization. While other classes, farmers, artisans and merchants, could continue to be engaged in their inherited vocations, samurai completely lost their guaranteed source of income. Consequently, they had to find new ways of living independently. Since the samurai could not compete with other classes in the fields of traditional occupations like agriculture and medicine, science and technology was one of these new fields into which jobless samurai were attracted and invited. Almost all of the early graduates of the engineering colleges were samurai. Even as late as 1890, the percentages of Imperial University samurai graduates in engineering and science were 86 and 80 respectively.

Thus, Japanese modern science and technology professions were, in the beginning of their formation, very much "samurai-spirited". The samurai were the class long accustomed by mental habit to think in terms of public affairs and by behavior patterns to play the game of public office. Unlike the European pattern in which science and technology training was one experience of the rising bourgeoisie, the new Japanese scientific and technological professions in the last quarter of the nineteenth century were dominated by the proud old samurai class, comprising the top five per cent of the total population.

Table 1 summarizes the sequence of support groups' thrust toward Westernization in Japan.

In the first stage, Western science was confined merely to technical knowledge by technicians in the government sector. In the second, Western knowledge in the form of its rev-

Table 1: The sequence of support groups' thrust toward Westernization in Japan

Occupation	Astronomers and interpreters	Physicians	Samurai
Leading century period	18th century	Late 18th–early 19th century	Mid-19th century
Interest	Technical	Cultural	Political
Status	Technicians	Freelancers	Planners and administrators
Role	Referees of superiority	Diffusion and popularization	Institutionalization

olutionary paradigms was garnered by vanguard physicians but they were alienated from the central power structure. Only in the third stage was Western science fully admitted into the power structure, and after the Meiji Restoration, through institutionalization, it was recognized as a most legitimate study for the youth of the samurai class.

The modernist samurai leadership soon established a policy of Westernization in which the establishment of Western-style educational institutions occupied a central position. During the 1870s and 1880s greater emphasis was placed on science and technology in the Japanese educational curriculum, from elementary school to university level, than in any other nation. For instance, mathematics and science occupied about one-third of the school curriculum at the lower grades (first four years) and two-thirds at the upper grades of the eight-year elementary education, though due to the shortage of qualified teachers available, it is somewhat questionable to what extent these ideal plans were put into practice. At the tertiary level too, the emphasis on science and technology was evident in the high percentages of graduates in scientific disciplines of Tokyo University (85 percent in the 1880s). In Europe in the nineteenth century, the voluntary activities of scientific or professional societies usually preceded their inclusion into the university curriculum. In the case of Meiji Japan, however, and as is generally the case when a foreign discipline is artificially transplanted under state sponsorship, this process was reversed. Tokyo University was established as an indispensable part of the modern technocratic bureaucracy. Graduates in science and technology were mostly samurai descendants who were accustomed to working in the public sector and inspired in technocratic policy-making, rather than by engineering apprenticeship of the traditional artisan class. They were adaptable in "public science" initiated by the government, which was indispensable for the operation of a modern state, in areas such as geographical and geological surveys, weights and measures, meteorological observations, sanitation, printing, telegraph and telephone system, military works, railways , and surveys of national resources.

In addition, the government participated in private entrepreneurial activities, constructed and managed pilot plants, and guided and subsidized new kinds of industries. Their enterprises were, however, from the beginning exposed to financial risk. Many of the projects of the Ministry of Technology eventually proved to be too far ahead of their times, as they intended to introduce the technology of an industrialized society into a pre-industrial environment. For instance, their railway construction and textile-mill enterprises were economically unsuccessful at the time.

Some of the samurai experiments in the public sector were successfully transferred into the private sector, where they were transformed to the market-oriented "private science" of the traditional artisan, such as Toyota Sakichi, the founder of Toyota Motors. The privatization of science phase was finally completed after World War II.

NAKAYAMA SHIGERU

REFERENCES

Nakayama, Shigeru. "Japanese Scientific Thought." In *Dictionary of Scientific Biography*, vol. 15, Supplement 1. Ed. Charles C. Gillispie. New York: Charles Scribner's Sons, 1978, pp. 728–58.

See also: Seki Kowa – Shibukawa Harumi – Environment and Nature – Astronomy – *Yinyang* – Computation: Chinese Counting Rods – Asada Goryu – Divination in China

AL-JAWHARĪ Probably of Iranian origin, al-ʿAbbās Ibn Saʿīd al-Jawharī was one of the court astronomers/astrologers of caliph al-Maʾmūn (r. 813–833) in charge of the construction of astronomical instruments. He participated in astronomical observations carried out in Baghdad in 829–830 and in Damascus in 832–833. He is said to have composed an astronomical handbook (*zīj*), which is lost, except for indirect references (Sezgin, 1973). In his house in Baghdad, meetings were held at which the participants discussed Ptolemy's *Almagest*, Euclid's *Elements*, and problems derived from the two books. A not yet studied

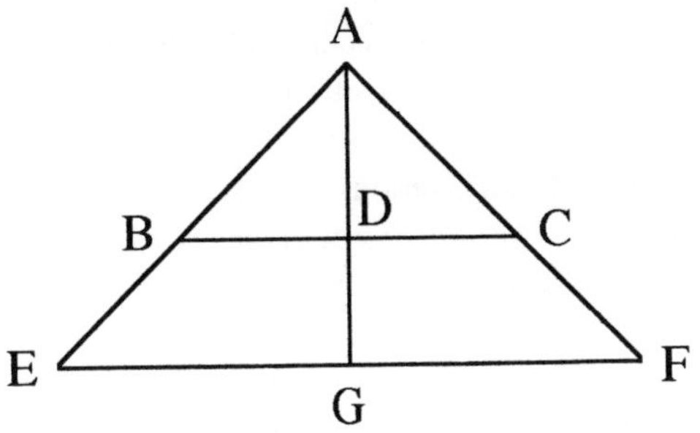

Figure 1: Jawharī's theorem.

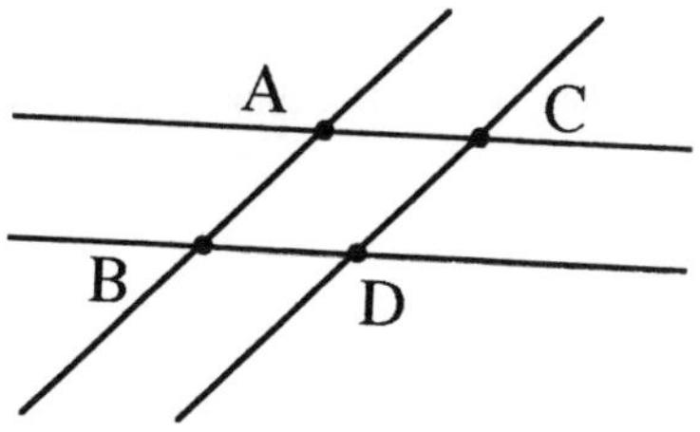

Figure 2: The distance between one point of a line and its corresponding point on a second, parallel line always equals the distance between a second point on the first line and its corresponding point on the second.

manuscript, *Kalām fī maʿrifat buʿd al-shams ʿan markaz al-arḍ* (Speech about the Knowledge of the Distance between the Sun and the Center of the Earth), might be an extract of his *Zīj* or an independent astronomical treatise. In astrology, he was considered an expert at horoscopes determining an individual's length of life.

His main achievements in the mathematical sciences are in geometry. He edited or commented on the *Elements*. Extracts of this work are preserved as independent manuscripts containing fragments of his *Ziyādāt fī'l-maqāla al-khāmisa min kitāb Uqlīdis* (Additions to Book V of Euclid's Book) and as quotations of his attempted proof of the parallel postulate (Book I) in Naṣīr al-Dīn al-Ṭūsī's *al-Risāla al-shāfiya ʿan al-shakk fī'l-khuṭūṭ al-mutawāziya* (The Healing Treatise on the Doubt Concerning Parallel Lines). In this treatise, al-Ṭūsī also cites one of al-Jawharī's additions to Book I, namely that the angles contained by three lines drawn from one point in different directions equal four right angles.

Al-Jawharī's attempted proof of the postulate was contained in the same edition of the *Elements* as the aforementioned additional proposition. It evidently was inspired by the proofs of Simplikios (sixth century) and Aghānīs/Aghānyūs (Agapios?, fl. ca. 511), since he used the same variant of the so-called Eudoxos-Archimedes axiom as Aghānīs and at least two propositions attributed by ʿAlam al-Dīn Qayṣar (d. 1251) to Simplikios. Since both proofs were contained in Simplikios' commentary on the definitions, postulates, and axioms, its Arabic translation obviously was made in the first half of the ninth century.

The theorem upon which al-Jawharī's proof is built states principally that if a triangle ABC is divided by a line AD and this line and the two sides AB and AC are extended and cut by a line EF such that AB=BE and AC=CF, then AD=DG with G being the point of intersection between line EF and the extension of line AD (see Figure 1).

The proof of this proposition and consequently the proof of the postulate ultimately depend upon two propositions possibly introduced by al-Jawharī himself. The problem of the whole proof lies in the incomplete proof of the second part of the first of the two theorems, which al-Ṭūsī had already discovered. This second part states principally that the distance between one point of a line and its "corresponding" point on a second, parallel line always equals the distance between a second point on the first line and its "corresponding" point on the second line, AC=BD. Al-Jawharī's proof of this part not only treats a special case, but also fails to prove the equality of the two joining lines.

The fragment of al-Jawharī's additions to Book V contains three propositions which try to prove Euclid's definitions V, 5 (identity of ratios), V, 7 (one ratio > a second ratio), and the negation of definition V, 5. This illustrates how difficult those definitions were. Usually, a definition is not proven, since it was regarded as an evident assumption or a statement agreed upon between scholars. Al-Jawharī's explanations dressed as formal proofs do not really clarify those definitions, since all he did was simply to repeat them for special objects, namely, natural numbers.

Al-Jawharī is also credited with the translation from Persian into Arabic of a book about poison of supposed Indian origin, the so-called *Kitāb al-Shānāk* (The Book of Shānāk), for al-Maʾmūn.

SONJA BRENTJES

REFERENCES

Primary sources

al-Ṭūsī, Naṣīr al-Dīn. *al-Risāla al-shāfiya ʿan al-shakk fī l-khuṭūṭ al-mutawāziya.* Hyderabad, 1359h, pp. 17–24.

Ibn al-Qifṭī. *Taʾrīkh al-ḥukamāʾ*. Ed. Julius Lippert. Leipzig: Harrassowitz, 1903.

Kalām fī maʿrifat buʿd al-shams ʿan al-arḍ. MS Bairūt, American University 223.

Zīyādāt fī l-maqāla al-khāmisa min kitāb Uqlīdis. MS Tunis, Aḥmadīya 16167; MS Istanbul, Feyzullāh 1359; Teheran, Dānishkada-i Adab. Ǧ 284; Hyderabad, Osmaniya University Library A 510.

Secondary sources

Jaouiche, Khalil. *La théorie des parallèles en pays d'Islam: contribution à la préhistoire des géometries non-euclidiennes.* Paris: J. Vrin, 1986.

Kennedy, E.S. "A Survey of Islamic Astronomical Tables." *Transactions of the American Philosophical Society*, n.s. 46(2): 123–177, 1956.

Rosenfeld, Boris A. *A History of Non-Euclidean Geometry. Evolution of the Concept of a Geometric Space.* New York: Springer, 1988.

Sabra, A.S. "Simplicius's Proof of Euclid's Parallels Postulate". *Journal of the Warburg and Courtauld Institutes* 32: 1–24, 1969.

Sabra, A.S. "al-Jawharī." *Dictionary of Scientific Biography,* vol. VII. Ed. Charles C. Gillispie. New York: Charles Scribner's Sons, 1973, pp. 79–80.

Sezgin, Fuat. *Geschichte des Arabischen Schrifttums.* Leiden: Brill, 1970, vols. III, V, and VI, 1970, 1973, and 1974.

See also: al-Maʾmūn – Astronomical Instruments – *Almagest* – *Elements* – *Zīj* – Naṣīr al-Dīn al-Ṭūsī

JAYADEVA Ācārya Jayadeva is an early Indian mathematician, known only through a long aphoristic (*Sūtra*) quotation in twenty verses from an unknown work of his. In this passage, Jayadeva sets out, step by step, an ingenious method for solving the indeterminate equation of the form $Nx^2 \pm C = y^2$. This quotation was extracted by Udayadivākara, an astronomer of Kerala, in his commentary called *Sundarī* on the *Laghu-bhāskārīya* of Bhāskara I (b. 629). Udayadivākara flourished about AD 1073. This means that Jayadeva lived before that date.

The above extract was made in the context of an astronomical problem involving two simultaneous equations: (1) $7y^2 + 1 = z^2$, and (2) $8x + 1 = y^2$. Here, Udayadivākara states that the value of y in the first equation can be found by an ingenious method called *varga-prakṛti* (lit. 'square-nature') enunciated by Ācārya Jayadeva, and the value of x in the second equation by the method of inversion.

The extract from Jayadeva forms an account of *varga-prakṛti* as he conceives it and solves problems through it. First he defines the term: "When (in an equation of the type $Ax^2 \pm C = y^2$) the square of an optional number is multiplied by a given number and then the product is increased or decreased by another number, and the result is in the nature of a square, such an equation is called *varga-prakṛti*." He then goes on to explain the technical terms which would be used in the course of his exposition, such as *Kaniṣṭha-mūlam* (lesser root), *jyeṣṭha-mūlam* (greater root), *kṣepa* (interpolator), *bhāvanā* (visualization), and its two forms, *Samāsa-bhāvanā,* and *viśesa-bhāvanā* or *tulya-bhāvanā* and *atultya-bhāvanā*, all of which form the step by step processes for the solution of the indeterminate equation envisaged. Working through these processes is termed the *cakravāla* (cyclic method) through which any number of solutions can be found. The actual method of solving the equation is given in the last five verses, towards the close of which Jayadeva quips, "Thus have we identified a very ingenious method for solving the problem which is as difficult as it is for a flea to fly against the wind". Jayadeva is perhaps justified in making such a comparison, for his cyclic method was set out later by other authors like Bhāskara II (b. 1114) and Nārāyaṇa (1356). The historian of mathematics Hermann Hankel has remarked: "It is above all praise; it is certainly the finest thing which was achieved in the theory of numbers before Lagrange."

K.V. SARMA

REFERENCES

Hankel, H. *Zur Geschichte der Mathematik in Alterum und Mittelalter.* Leipzig: Teubner, 1874, pp. 203–04.

Selenius, Clas-Olaf. "Rationale of the Chakravāla Process of Jayadeva and Bhāskara II." *Historia Mathematica* 2 : 167–84, 1975.

Shukla, Kripa Shankar. "Ācārya Jayadeva, The Mathematician." *Gaṇita* 5: 1–20, 1954.

Udayadivākara's Commentary, Sundarī, on the Laghu-bhāskarīya of Bhāskara I. Mss. available in the Kerala University Oriental Research Institute and Manuscripts Library, Trivandrum, and a modern copy in the Tagore Library, Lucknow University.

AL-JAZARĪ Al-Jazarī, Badīʾ al-Zamān Abū'l- 'Izz Ismaʿil ibn Al-Razzāz, was an engineer who worked in al-Jazira during the latter part of the twelfth century. His reputation rests upon his book, *Kitāb fī maʿrifat al-ḥiyal alhandasiyya* (The Book of Knowledge of Ingenious Mechanical Devices), which he composed in 1206 on the orders of his master Nāsir al-Dīn Maḥmūd, a prince of the Artuqid dynasty of Diyar Bakr. All that we know of his life is what he tells us in the introduction to his book, namely that at the time of writing he had been in the service of the ruling family for twenty-five years. The book is divided into six categories (*nawʿ*). Each of the first four contains ten chapters (*shakl*), and the last two consist of only five each. The categories are as follows:

1. water clocks and candle clocks;
2. vessels and pitchers for use in carousals or celebrations;
3. vessels and basins for hand washing and phlebotomy;
4. fountains and musical automata;
5. water lifting machines; and
6. miscellaneous.

There are many illustrations, both of general arrangements and detailed drawings, and these are of considerable assistance in understanding the text, which contains many technical expressions that have since fallen into disuse. Some thirteen manuscript copies, made between the thirteenth and the eighteenth centuries, are extant to bear witness to the widespread appreciation of the book in the Islamic world.

There are, however, no references to al-Jazarī in the standard Arabic biographical works of the Middle Ages, and there is no known translation into a European language before the twentieth century.

Only one of the complete machines, a twin cylinder pump driven by a paddle-wheel, can be said to have direct relevance to the development of mechanical technology. Many of the devices, however, embody techniques and mechanisms that are of great significance, since a number of them entered the general vocabulary of European engineering at various times from the thirteenth century onwards. Some of these ideas may have been received directly from al-Jazarī's work, but evidence is lacking. Indeed, it seems probable that a large part of the Islamic mechanical tradition — especially water clocks and their associated mechanisms and automata — had been transmitted to Europe before al-Jazarī's book was composed. Even leaving aside the question of direct transmission, we still have a document of the greatest historical importance. First, it confirms the existence of a tradition of mechanical engineering in the Eastern Mediterranean and the Middle East from Hellenistic times up to the thirteenth century. Al-Jazarī was well aware that he was continuing this tradition and was scrupulous in acknowledging the work of his predecessors, including Apollonius of Byzantium, the Pseudo-Archimedes, the Banū Mūsā (ninth century), Hibat Allah ibn al-Ḥusayn (d. 1139–40), and a certain Yūnus al-Asṭurlābī.

Other writings and constructions, whose originators were unknown to al-Jazarī, are also mentioned.

Second, his use of and improvement upon the earlier works, together with his meticulous descriptions of the construction and operation of each device, enables us to make an accurate assessment of mechanical technology by the close of the twelfth century.

DONALD R. HILL

REFERENCES

Hill, Donald R. "Arabic Mechanical Engineering: Survey of the Historical Sources." *Arabic Sciences and Philosophy* 1(2): 167–186, 1991.

The Arabic edition is: *Kitāb al-ḥiyal*. Ed. Ahmad Y. al-Hassan. Aleppo: Institute for the History of Arabic Science, 1981.

The English edition of *Kitāb al-ḥiyal* is: *The Book of Ingenious Devices by the Banū Mūsā*. Trans. Donald R Hill. Dordrecht: Reidel, 1979.

See also: Banū Mūsa – Engineering – Technology in the Islamic World

JIA XIAN Little is known of Jia Xian's life except that he served as a minor functionary during the reign of Emperor Renzong of Northern Song dynasty in the first half of the eleventh century. He learnt mathematics from Chu Yan, a famous astronomer and mathematician (fl. AD 1022–1054). Jia Xian was said to have written two books, *Huangdi Jiuzhang Suanjing Xicao* (Detailed Workings of the Nine Chapters on Mathematical Art), and *Suanfa Xuegu Ji* (A Collection of Ancient Mathematical Rules). While the latter was irretrievably lost, the former was largely found in Yang Hui's *Xiangjie Jiuzhang Suanfa* (A Detailed Analysis of the Mathematical Rules in the Nine Chapters) of AD 1261 preserved in Ms. form in chapter 16,344 of the *Yonglo Dadian* encyclopedia compiled by Xie Jin in 1407 during the Ming dynasty. In his preface, Yang Hui explicitly stated that his text was an attempt to expound and preserve the works of his predecessors, notably those of Jia Xian.

Jia Xian's main contribution to mathematics lies in his innovative method for the solution of numerical equations of higher degrees. The processes for square and cube root extractions began with *Jiuzhang Suanshu* (Nine Chapters on the Mathematical Art) at the beginning of the Christian era, and continued to appear in all later mathematical books without many improvements. It was Jia Xian who came out with a method called *Zengchang Kaifang* (Method of Multiplying and Adding for Root Extractions) which could be extended to the solution of numerical higher equations for approximate values of roots. This method is similar to that developed by P. Ruffini (1765–1822) and W.G. Horner (1786–1837) in 1802 and 1819 respectively.

Accompanying the method was an array of binomial coefficients up to the sixth power tabulated in the form of a triangle similar to that by Blaise Pascal (1623–1662) in 1665, now commonly known as the Pascal Triangle. This triangle of binomial coefficients was subsequently copied and expanded by Zhu Shijie in the thirteenth century. Thus, Jia Xian's mathematical knowledge to some extent laid the foundation for the rapid development of mathematics during the twelfth and thirteenth centuries.

ANG TIAN SE

REFERENCES

Guo, Shuchun. "Jia Xian Huangdi Jiuzhang Suanjing Chu Tan." (An Initial Inquiry into *Jia Xian's Huangdi Jiuzhang Suanjing Xicao*). *Studies in the History of Natural Sciences* 7(4): 328–334, 1988.

Guo, Shunchun. "Jia Xian." *Zhongguo Gudai Kexuejia Zhuanji* (Biographies of Ancient Chinese Scientists). Ed. Du Shiran. Beijing: Kexue Chubanshe, 1992, pp. 472–479.

Mei, Rongzhao. "Jia Xiande Zengcheng Kaifang Fa." (Jia Xian's Zengcheng Method for the Extraction of Roots). *Studies in the History of Natural Sciences* 8(1): 1–8, 1989.

Qian, Baozong. *Qian Baozong Kexueshi Lunwen Xuanji* (Collection of Qian Baozong's Essays on the History of Science). Beijing: Kexue Chubanshe, 1983.

See also: Liu Hui and the *Jiuzhang Suanshu* – Gou-gu Theorem

AL-JURJĀNĪ Abū'l-Faḍā'il Ismā'īl ibn al-Ḥusayn al-Jurjānī, Zayn al-Dīn, sometimes called Sayyid Ismā'īl, was the most eminent Persian physician after Ibn Sīnā (Avicenna), and the author of the first great medical compilation written in Persian. Born at Jurjan, 50 miles east of modern Gurgan, east of the Caspian, he was a pupil of Ibn Abī Ṣādiq (d. 1066–77), who had himself been a pupil of Ibn Sīnā. In 1110 al-Jurjānī entered the service of the ruler of Khwarizm (modern Khiva), the Khwārizmshāh Quṭb al-Dīn Muhammad (d. 1127) and his son Atsiz; later he moved to Marw (Merv) and served the rival sultan Sanjar. Al-Jurjānī died at Marw in about 1136 (AH 531).

Al-Jurjānī's great work is entitled *Dhakhīra-i-Khwārazmshāhī* (The Thesaurus of the King of Kharazm). Comparable in size and scope to Avicenna's *Canon*, the *Thesaurus* is a compendium of medical knowledge and clinical practice. It rapidly became a classic and established the medical and scientific vocabulary in Persian; its influence was extensive and long-lasting. It was translated into Hebrew, Urdu, and Turkish, and remained in use until the nineteenth century. There are very many manuscripts, both complete and fragmentary, but no available modern edition (although Keshavarz records an Indian edition of 1865–6, and one produced in Tehran in 1965).

In addition to the *Thesaurus*, al-Jurjānī composed other works in Persian which appear to be, in the main, abridgements of it. Chief among these is the condensed version called *Mukhtaṣar-i Khuffī-i 'Alā'ī* (Abridgement for the Boots of Alā') in two long volumes for Shah Atsiz to carry in his tall riding boots. This work was printed in Agra (1852) and Cawnpore (1891), but neither edition is readily available. There are several manuscripts in Turkey, one in the British Library, and an Arabic translation in Paris and at the University of California.

The *al-Agrāḍ al-ṭibbīya* (Aims of Medicine), composed for the vizier of Atsiz, is partly another abridgement of the *Thesaurus* combined with an account of the symptoms and treatment of local diseases. The *Yadgar i-ṭibb* or (Remembrancer) is largely on pharmacology.

Al-Jurjānī's works have not been much studied, and the relationships between the various texts and translations have not been worked out. A work in Arabic, *Zubdat al-ṭibb* (Essence of Medicine) is frequently assigned to him; this is presumably a translation of one of the Persian treatises mentioned above, but it is not clear which one. Some of the other manuscript works ascribed to al-Jurjānī are probably excerpts from his lengthy books.

E. RUTH HARVEY

REFERENCES

Browne, Edward G. *Arabian Medicine*. Cambridge: Cambridge University Press, 1921.

Elgood, Cyril. *A Medical History of Persia and the Eastern Caliphate*. Cambridge: Cambridge University Press, 1951.

Elgood, Cyril. *Safavid Medical Practice*. London: Luzac, 1970.

Ibn Abī Uṣaybi'ah. *'Uyūn al-Anbā fī ṭabaqāt al-aṭibbā'*. Beirut: Dār Maktabat al-Ḥayāh, 1965.

Naficy, Abbas. *La Médecine en Perse, des origins à nos jours; ses fondements théoriques d'apres l'encyclopédie medicale de Gorgani*. Paris: Éditions Vega, 1933.

Schacht, J. "al-Djurdjani." *The Encyclopaedia of Islam*, new edition. Leiden: Brill, 1965, vol.II, p.603.

See also: Ibn Sīnā

K

KAMALĀKARA Kamalākara was one of the most erudite and forward-looking Indian astronomers who flourished in Varanasi during the seventeenth century. Belonging to Maharashtrian stock, and born in about 1610, Kamalākara came from a long unbroken line of astronomers, originally settled at the village of Godā on the northern banks of the river Godāvarī. Towards AD 1500, the family migrated to Varanasi and came to be regarded as reputed astronomers and astrologers. Kamalākara studied traditional Hindu astronomy under his elder brother Divākara, but extended the range of his studies to Islamic astronomy, particularly to the school of Ulugh Beg of Samarkand. He also studied Greek astronomy in Arabic and Persian translations, particularly with reference to the elements of physics from Aristotle, geometry from Euclid, and astronomy from Ptolemy. He wrote both original treatises and commentaries on his own works and those of others.

Kamalākara's most important work is the *Siddhānta-Tattvaviveka*, written in AD 1658. The work which is divided into fifteen chapters and contains over 3000 verses, faithfully follows the *Sūryasiddhānta* in the matter of parameters, general theories, and astronomical computation. However, in certain matters Kamalākara made original contributions and offered new ideas. Though he accepted the planetary parameters of *Sūryasiddhānta*, he agreed with Ptolemaic notions in the matter of the planetary system. He presented geometrical optics, and was perhaps the only traditional author to do so. He described the quadrant and its application. He proposed a new Prime Meridian, which is the longitude passing through an imaginary city called Khalādātta, and provided a table of latitudes and longitudes for twenty important cities, in and outside India, on this basis. Kamalākara was an ardent advocate of the precession of the equinoxes and argued that the pole star also does not remain fixed, on account of precession. Kamalākara wrote two other works related to the *Siddhānta-Tattvaviveka*, one a regular commentary on the work, called *Tattvavivekodāharaṇa,* and the other a supplement to that work, called *Śeṣavāsanā,* in which he supplied elucidations and new material for a proper understanding of his main work. He held the *Sūryasiddhānta* in great esteem and also wrote a commentary on that work.

Kamalākara was a critic of Bhāskara and his *Siddhāntaśiromaṇi*, and an arch-rival of Munīśvara, a close follower of Bhāskara. This rivalry erupted into bitter critiques on the astronomical front. Thus Ranganātha, younger brother of Kamalākara, wrote, at the insistence of the latter, a critique on Munīśvara's *Bhaṅgī* method (Winding method) of true planets, entitled *Bhaṅgī-vibhaṅgī* (Defacement of the *Bhaṅgi*), to which Munīśvara replied with a *Khaṇḍana* (Counter). Munīśvara attacked the theory of precession advocated by Kamalākara, and Ranganātha refuted the criticisms of his brother in his *Loha-gola-khaṇḍana* (Counter to the Iron Sphere). That in turn was refuted by Munīśvara's cousin Gadādhara in his *Loha-gola-samarthana* (Justification of the Iron Sphere). These kinds of astronomical and intellectual battles were typical of the philosophical and religious disputes which were common in ancient India.

K.V. SARMA

REFERENCES

Dikshit, S.B. *Bhāratiya Jyotish Sastra (History of Indian Astronomy)*. Trans. by R.V. Vaidya. Pt. II. *History of Astronomy during the Scientific and Modern Periods*. Calcutta: Positional Astronomy Centre, India Meteorological Department. 1981.

Dvivedi, Sudhakara. *Gaṇaka Taraṅgiṇi: Lives of Hindu Astronomers*. Benares: Jyotish Prakash Press, 1933.

Loha-gola-khaṇḍana of Ranganātha and *Loha-gola-samarthana* by Gadādhara. Ed. Mithalala Himmatarama Ojha. Varanasi: Sañcālaka, Anusandhāna Saṃsthāna, 1963.

Pingree, David. *Jyotiḥśāstra – Astral and Mathematical Literature*. vol. VI, fasc. 4 of *A History of Indian Literature*. Ed. Jan Gonda. Wiesbaden: Otto Harrassowitz. 1981.

Siddhānta-Tattvaviveka. A Treatise on Astronomy by Bhaṭṭa Kamalākara, with *Śeṣavāsanā* by the same author. Ed. Sudhakara Dube. Benares: Benares Sanskrit Series, 5 vols., 1880–85 : Revised by Muralidhara Jha, Benares: Krishna Das Gupta for Braj Bhusan Das & Co, 1924–35.

See also: Astronomy in India – Ulugh Beg – Astronomy in the Islamic World – Ranganātha – *Sūryasiddhāntha* – Precession of the Equinoxes – Bhāskara – Munīśvara

AL-KARAJĪ Al-Karajī, Abū Bakr Muḥammad, was a Persian mathematician and engineer. He held (ca. 1010–1015) an official position in Baghdad during which time he wrote his three main works.

1. His *al-Fakhrī fī 'l-jabr wa 'l-muqābala* (Glorious on Algebra) contains an exposition of algebra and auxiliary topics (e.g. summing series of integers). The next and larger part is a collection of problems: some are commercial and recreational, but for the most part they belong to indeterminate algebra. All of the latter sort are taken from Diophantus or Abū Kāmil.

2. The *Badīʿ fī'l-ḥisāb* (Wonderful on Calculation) is more original. After restating many of Euclid's theorems, al-Karajī shows how to calculate with square, cubic, and biquadratic roots and with their sums, then how to extract the root of a polynomial in *x* (this is the earliest known mention of the procedure). A second part contains problems on indeterminate algebra. Here, al-Karajī has taken the basic methods used by Diophantus in his problems and classified them, the result being a useful introduction to Diophantus's algebra.

3. The composition of the *Kāfī fī'l-ḥisāb* (Sufficient on Calculation) belongs to the usual duty of a mathematician holding an official position: to write a simple textbook in a way accessible to civil servants and containing the elements of arithmetic with integers and fractions (common and sexagesimal), the extraction of square roots, the determination of areas and volumes, and elementary algebra. All this is illustrated by numerous examples.

4. The *ʿIlal al-jabr wa'l-muqābala* (Grounds of Algebra) is a small compendium on basic algebra (reckoning with roots, solving the basic six forms of the equations of the first two degrees), without resort to any geometrical demonstrations.

5. The *Inbāṭ al-miyāh al-khafīyah* (Locating Hidden Waters) was written after al-Karajī's return to Persia. It is concerned with finding subterranean water, extracting it, and transporting it in accordance with the soil's configuration.

Other, lost works of al-Karajī's are known to have dealt with indeterminate algebra, arithmetic, inheritance algebra, and the construction of buildings. Another contained the first known explanation of the arithmetical (Pascal's) triangle; the passage in question survived through al-Samaʾwal's *Bāhir* (twelfth century), which heavily drew from the *Badīʿ*.

Although much of his work was taken from other's writings, there is no doubt that al-Karajī was quite an able and influential mathematician. However, the quality of his writings is uneven, as he seems to have worked hastily sometimes.

JACQUES SESIANO

REFERENCES

al-Karajī. *Inbāṭ al-miyāh al-khafīyah*. Hyderabad: Osmania University, 1941.

al-Karajī. *ʿIlal al-jabr wa'l-muqābala*. Ankara (*Atatürk kültür merkezi yayını*, sayı 40), 1991.

Anbouba, Adel. *L'Algèbre al-Badīʿ d'al-Karagī*. Beirut: Université libanaise, 1964.

Hochheim, Adolf. *Kâfî fîl Ḥisâb* (Genügendes über Arithmetik). Halle: Nebert, 1878–1880.

Sesiano, Jacques. "Le traitement des équations indéterminées dans le *Badīʿ fī'l ḥisāb* d'Abū Bakr al-Karajī." *Archive for History of Exact Sciences* 17: 297–379, 1977.

Woepcke, Franz. *Extrait du Fakhrî*. Paris: Duprat, 1853.

AL-KĀSHĪ Al-Kāshī, or al-Kāshānī (Ghiyāth al-Dīn Jamshīd ibn Masʿud al-Kashī (al-Kashanī)), was a Persian mathematician and astronomer. He was born in Kāshān in northern Iran, and worked at first in Herat (now in Afghanistan) at the court of *khāqān* ("khān of khāns" Shāhrukh, the son of Tīmur). In 1417 he was invited by Ulugh Beg, the son of Shāhrukh, the ruler of Samarqand (now in Uzbekistan) to become the director of his astronomical observatory. His scientific treatises were mostly written in Arabic, and partly in Persian. Al-Kāshī died in Samarqand in about 1430.

Al-Kāshī's main mathematical work is *Miftāḥ al-ḥisāb* (Key of Arithmetic). This work, written in the tradition of Arabic mathematical texts, contains five books: (1) Arithmetic of integers, (2) Arithmetic of fractions, (3) Arithmetic of astronomers, (4) Geometry, and (5) Algebra.

In Book One, al-Kāshī considers duplication (multiplication by two), mediation (division into two), addition, subtraction, multiplication, and division. As Naṣīr al-Dīn al-Ṭūsī (1201–1274) had done in his arithmetic treatise, he also considers extraction of roots of arbitrary integer powers by means of what we now call the Ruffini–Horner method and expresses the approximate value of the root by means of "Newton's binomial formula" for $(a + b)^n$. In Book Two, al-Kāshī introduces decimal fractions and calls the digits of these fractions "decimal minutes", "decimal seconds", "decimal thirds", etc. Arithmetic of astronomers considered in Book Three is arithmetic of sexagesimal fractions borrowed by Islamic astronomers from Ptolemy, who in turn had borrowed them from Babylonian astronomers. Islamic astronomers designated figures from 1 to 59 by means of letters, and zero by ō. Like the ancient Babylonians, al-Kāshī extended the sexagesimal system on to integers and called the sexagesimal digits "raised", "twice raised", "three times raised", etc. In Book Four, the rules of mensuration of many plane and solid figures, including buildings, cupolas, and stalactite surfaces, are formulated. In Book Five, besides explaining the algebraic solution of quadratic equations and equations reducible to them, al-Kāshī discusses the solutions of equations by means of the rule of two errors, actions with roots, algebraic identities, and Thābit ibn Qurra's rule for the determination of amicable numbers. Al-Kāshī claims that he wrote an algebraic treatise with the classification of equations of the fourth degree, and that for every type he proposed the solution by means of the intersection of conics more general than the conics by means of which ʿUmar al-Khayyām solved cubic equations. This algebraic treatise is not extant and since the number of these equations does not coincide with those mentioned by al-Kāshī, this treatise was not finished.

The *al-Risāla al-muḥīṭiyya* (Treatise on Circumference) is devoted to the calculation of the ratio of circumferences of circles to their radii (now this ratio is designated by 2π). Al-Kāshī calculates the perimeters of regular inscribed and circumscribed polygons with $3{\cdot}2^{28}$ sides. The number of sides is chosen on the condition that the difference between these polygons for the great circle of the sphere of fixed stars must be less than the "width of a horse hair". The result is expressed in sexagesimal and decimal fractions.

In the *Risāla al-watar wa'l-jayb* (Treatise on Chord and Sine) al-Kāshī calculates the sin 1° according to known sin 3°. This problem was necessary for composition of the tables of sines with five sexagesimal digits in Ulugh Beg's *Zīj-i Ulugh Beg* (Ulugh Beg's Astronomical Tables). This problem was reduced to the cubic equation $4x^3 + q = 3x$ ($x = \sin 1°$, $q = \sin 3°$) and was solved by the method of successive approximations

$$x_1 = \frac{q}{3}, \quad x_2 = \left(\frac{q + 4x_1^3}{3}\right), \quad \ldots .$$

In the field of astronomy, al-Kāshī was author of *Zīj-i Khāqānī* (Khāqān Astronomical Tables) written in 1413–1414 in Herat. He also composed a treatise on astronomical instruments written in 1416 dedicated to the ruler of Iṣfāhān Iskandar, who was the nephew of Shāhrukh, and was one of the authors of Ulugh Beg's *Astronomical Tables*. He was also the translator of the last tables from Persian into Arabic. In the treatise *Nuzha al-ḥadāiq* (Delight of Gardens), al-Kāshī describes an instrument for representation of the movements of planets which he invented. He was also the author of numerous other astronomical treatises.

BORIS ROSENFELD

REFERENCES

Ğamšīd b. Masʿūd al-Kāšī. *Der Lehrbrief über der Kreisumfang (ar-Risāla al-Muḥīṭiyyaa).* Übersetzt und erläutert von P. Luckey. Herausg. von A. Siggel. Berlin: Akademie-Verlag, 1953.

Jamshīd Ghiyāth al-Dīn al-Kāshī. *Key of Arithmetics. Treatise on Circumference.* Moscow: Gostehizdat, 1956 (Arabic with Russian translations by B.A. Rosenfeld).

Kennedy, Edward S. *The Planetary Equatorium of Jamshīd Ghiyāth al-Dīn al-Kāshī.* Princeton: Princeton University Press, 1960.

Luckey, Paul. "Die Rechenkunst bei Ğamšīd b. Masʿūd al-Kāšī mit Rückblicken auf die ältere Geschichte des Rechnens." *Abhandlungen für die Kunde des Morgenlandes*, Bd. 31, Wiesbaden: F. Steiner, 1951.

Matvievskaya, Galina P. and Boris A. Rosenfeld. *Mathematicians and Astronomers of Medieval Islam and their Works (8–17th centuries)* (in Russian). Moscow: Nauka, 1983, pp. 480–486.

Youschkevitch, Adolphe, and Boris Rosenfeld. "Al-Kāshī (or al-Kāshānī)." In *Dictionary of Scientific Biography.* Ed. Charles Gillispie. New York: Scribner, 1973, vol. 7, pp. 255–262.

AL-KHALĪLĪ Al-Khalīlī, Shams al-Dīn Abū ʿAbdallāh Muḥammad ibn Muḥammad, lived in Damascus, ca. 1365. He was an astronomer associated with the Umayyad Mosque in the centre of Damascus. A colleague of the astronomer Ibn al-Shāṭir, he was also a *muwaqqit* — that is, an astronomer concerned with *ʿilm al-mīqāt*, the science of timekeeping by the sun and stars and regulating the astronomically-defined times of Muslim prayer. Al-Khalīlī's major work, which represents the culmination of the medieval Islamic achievement in the mathematical solution of the problems of spherical astronomy, was a set of tables for astronomical timekeeping. Some of these tables were used in Damascus until the nineteenth century, and they were also used in Cairo and Istanbul for several centuries. The main sets of tables survive in numerous manuscripts, but they were not investigated until the 1970s.

Al-Khalīlī's tables can be categorized as follows:

- tables for reckoning time by the sun, for the latitude of Damascus;
- tables for regulating the times of Muslim prayer, for the latitude of Damascus;
- tables of auxiliary mathematical functions for timekeeping by the sun for all latitudes;
- tables of auxiliary mathematical functions for solving the problems of spherical astronomy for all latitudes;
- a table displaying the *qibla*, that is, the direction of Mecca, as a function of terrestrial latitude and longitude; and
- tables for converting lunar ecliptic coordinates to equatorial coordinates.

The first two sets of tables correspond to those in the large corpus of spherical astronomical tables computed for Cairo that are generally attributed to the tenth-century Egyptian astronomer Ibn Yūnus.

Al-Khalīlī's fourth set of tables was designed to solve all the standard problems of spherical astronomy, and they are particularly useful for those problems that, in modern terms, involve the use of the cosine rule for spherical triangles. Al-Khalīlī tabulated three functions and gave detailed instructions for their application. The functions are as follows:

$$f_\phi(\theta) = \frac{\sin\theta}{\cos\phi} \quad \text{and} \quad g_\phi(\theta) = \sin\theta\tan\phi,$$

$$K(x, y) = \arccos\left(\frac{x}{R\cos y}\right),$$

computed for appropriate domains. The entries in these tables, which number over 13,000, were computed to two sexagesimal digits and are invariably accurate. An example of the use of these functions is the rule outlined by al-Khalīlī for finding the hour-angle t for given solar or stellar altitude h, declination δ, and terrestrial latitude ϕ. This may be represented as:

$$t(h, \delta, \phi) = K\{(f_\phi(h) - g_\phi(\delta)), \delta\},$$

and it is not difficult to show the equivalence of al-Khalīlī's rule to the modern formula

$$t = \arccos\left(\frac{\sin h - \sin\delta\sin\phi}{\cos\delta\cos\phi}\right).$$

These auxiliary tables were used for several centuries in Damascus, Cairo, and Istanbul, the three main centres of astronomical timekeeping in the Muslim world.

Al-Khalīlī's computational ability is best revealed by his *qibla* table. The determination of the *qibla* for a given locality is one of the most complicated problems of medieval Islamic trigonometry. If (L, ϕ) and (L_M, ϕ_M) represent the longitude and latitude of a given locality and of Mecca, respectively, and $\Delta L = |L - L_M|$, then the modern formula for $q(L, \phi)$, the direction of Mecca for the locality, measured from the south, is

$$q = \arccos\left(\frac{\sin\phi\cos\Delta L - \cos\phi\tan\phi_M}{\sin\Delta L}\right).$$

Al-Khalīlī computed $q(\phi, L)$ to two sexagesimal digits for the domains $\phi = 10°, 11°, \ldots, 56°$ and $\Delta L = 1°, 2°, \ldots, 60°$; the vast majority of the 2880 entries are either accurately computed or in error by $\pm 1'$ or $\pm 2'$. Several other *qibla* tables based on approximate formulas are known from the medieval period. Al-Khalīlī's splendid qibla table does not appear to have been widely used by later Muslim astronomers.

DAVID A. KING

REFERENCES

Primary sources

MS Paris B.N. ar. 2558, copied in 1408 (contains all of the tables in al-Khalīlī's major set).

Secondary sources

King, David A. "al-Khalīlī." In *Dictionary of Scientific Biography*, vol. 15. Ed. Charles C. Gillispie. New York: Charles Scribner's Sons, 1978, pp. 259–261.

King, David A. "Al-Khalīlī's Auxiliary Tables for Solving Problems of Spherical Astronomy." *Journal for the History of Astronomy* 4: 99–110, 1973, reprinted in David A. King. *Islamic Mathematical Astronomy*. London: Variorum, 1986, 2nd rev. ed., Aldershot: Variorum, 1993, XI.

King, David A. "Al-Khalīlī's Qibla Table." *Journal of Near Eastern Studies* 34: 81–122, 1975, reprinted in *Islamic Mathematical Astronomy*. London: Variorum, 1986, 2nd rev. ed., Aldershot: Variorum, 1993, XIII.

King, David A. "The Astronomy of the Mamluks." *Isis* 74: 531–555, 1983, reprinted in *Islamic Mathematical Astronomy*. London: Variorum, 2nd rev. ed., Aldershot: Variorum, 1993, III.

King, David A. "L'astronomie en Syrie à l'époque islamique." In *Syrie – mémoire et civilisation*. Ed. S. Cluzan et al. Paris: Institut du Monde Arabe, 1993, pp. 392–394 and 440.

King, David A. *Shams al-Dīn al-Khalīlī and the Culmination of the Islamic Science of Astronomical Timekeeping*, forthcoming.

Van Brummelen, Glen. "The Numerical Structure of al-Khalīlī's Auxiliary Tables." *Physis – Rivista Internazionale di Storia della Scienza*, N.S., 28: 667–697, 1991.

See also: Ibn al-Shāṭir – Religion and Science in Islam – *Qibla* and Islamic Prayer Times – Ibn Yūnus – Sexagesimal System

AL-KHARAQĪ Abū Muḥammad ʿAbd al-Jabbār ibn ʿAbd al-Jabbār al-Kharaqī was a Persian astronomer, mathematicians and geographer. He was born at Kharaq near Marw and worked at Marw at the court of Saljuq Sultan Sanjar (1118–1157). Al-Kharaqī wrote in Arabic.

His main work was *Muntahā al-idrāk fī taqāsīm al-aflāk* (The Highest Understanding of the Divisions of Celestial Spheres). The work consists of three books: astronomical, geographical, and chronological. It is written in the tradition of Arabic astronomical textbooks initiated by al-Farghānī, continued by Ibn al-Haytham, and widely spread after al-Kharaqī in the form of Arabic and Persian books on the science of cosmography (*ʿilm al-hayʾa*). In these books the planets are considered not as supported by imaginary circles according to Ptolemy's *Almagest*, but as supported by massive material spheres in which they move in tubes.

His *Kitāb al-tabṣira fī ʿilm al-hayʾa* (Introduction to the Science of Cosmography), written in the same tradition, is the shortened version of the first work, and contains two separate books, "On the Heavens" and "On the Earth".

Al-Kharaqī was also the author of two lost mathematical treatises, *al-Risala al-shāmila* (Comprehensive Treatise), devoted to arithmetic, and *al-Risala al-maghribiyya* (The North African Treatise), on the "calculus of dirham and dinar".

BORIS ROSENFELD

REFERENCES

Matvievskaya, Galina P. and B.A. Rosenfeld. *Mathematicians and Astronomers of Medieval Islam and Their Works* (8th–17th c.). Moscow: Nauka, 1983, pp. 325–326. (in Russian)

Sarton, George. *Introduction to the History of Science*, Vol. 2. Baltimore: Williams & Wilkins, 1927, pp. 204–205.

Wiedemann, Eilhard, and K. Kohl. "Einleitung zu Werken von al-Charaqi." In *Aufsätze zur arabischen Wissenschaftsgeschichte*, vol. 2. Ed. E. Wiedemann. Hildesheim, New York: Olns, 1970, pp. 628–643.

See also: Ibn al-Haytham – *Almagest* – *Hayʾa*

UMAR AL-KHAYYĀM ʿUmar al-Khayyām (Ghiyāth al-Dīn Abū'l-Fatḥ ʿUmar ibn Ibrāhīm Khayyām (al-Khayyāmī) al-Naysābūrī (Nīshāpūrī), 1048–1131) was a Persian mathematician, astronomer, philosopher, and poet. His scientific treatises were written primarily in Arabic; his poems were mostly written in Persian. The name Khayyām means "the tentmaker" — probably it was the profession of his father or grandfather. He was born and died in Nīshāpūr. Khayyām was a student in Balkh (now in Afghanistan), and worked at first in Bukhārā and Samarqand (now in Uzbekistan). In 1074 he was invited by Saljūq Sultan Mālikshāh Jalāl al-Dīn to the capital Iṣfahān to participate in the reform of the solar Persian calendar and to organize an astronomical observatory. The calendar reform was completed in 1079; the new calendar era was named according to the names of the sultan Mālikī or Jalālī. In 1092, after Mālikshāh's death Khayyām fell into disgrace, and his observatory was closed. This was because of suspicions that some of his verses were antiIslamic. To allay these suspicions, he made a pilgrimage to Mecca. Later he worked in Marw (now Mary in Turkmenistan) which was the new capital of the Saljūq sultans.

Two of Khayyām's treatises are very important for the history of mathematics. One is the *Risāla fī'l-barāhīn ʿala masāʾil al-jabr wa'l-muqābala* (Treatise of Demonstrations of Problems of Algebra), written in Samarqand and dedicated to the judge Abū Ṭāhir who had given him the opportunity to devote himself to the science. The topic of this work was the theory of cubic equations. Here the complete classification of cubic equations with positive roots was given, and for every type of equation the solution by means of the intersection of circumferences, equilateral hyperbolas, and parabolas was also given.

The second work is the *Sharḥ mā ashkala min muṣādarāt kitāb Uqlīdis* (Commentaries on Difficulties in Introductions of Euclid's book). It was written in Iṣfahān in 1077 and contains the commentaries to Euclid's *Elements*. The treatise consists of three *maqālas*: (1) on the theory of parallels, (2) on the theory of ratios, and (3) on the theory of compound ratios. The ideas of this work were developed further by Naṣīr al-Dīn al-Ṭūsī (1201–1274), John Wallis (1616–1703), and Girolamo Saccheri (1667–1733) (Khayyām's quadrangle is also called Saccheri's quadrangle).

Before the *Risāla fī'l-barāhīn* Khayyām wrote the *Risāla fī taqsīm rubʿ al-dāʾira* (Treatise on Division of a Quarter of a Circle) and *Mushkilāt al-ḥisāb* (Problems of Arithmetic). In the first, a geometric problem is considered. It is reduced to a cubic equation whose solution by means of the intersection of the circumference and equilateral hyperbola and an approximate numerical solution are found. Here the problem of classifications of cubic equations is formulated. The second work is not extant, but is mentioned in the former as "the treatise on the demonstration of methods of Indians of extraction of square and cube roots and on generalization of these methods for determination of bases of square-squares, square-cubes, cube-cubes and so on as many as you like, what was not earlier". These words show that this work was devoted to the extraction of roots of nth degree, which was explained later in the arithmetic treatise of Naṣīr al-Dīn al-Ṭūsī. Probably in this work, as in al-Ṭūsī's treatise, "Newton's binomial formula" for $(a+b)^n$ for an arbitrary natural n was explained.

Khayyām also made many contributions to the study of astronomy and the calendar. His *al-Zīj Mālikshāhī* (Malikshah Astronomical Tables) is not extant, except for the catalog of fixed stars from this work. In Khayyām's calendar there were eight intercalary days in 33 years. The error is one day in five thousand years (in the Gregorian calendar the error is one day in 3330 years). After the closing of his observatory Khayyām wrote the Persian historical treatise *Nowrūz-nāmeh* (Book on the New Year) on reforms of the solar Persian calendar and on the Persian New Year holiday Nowrūz in pre-Islamic Iran.

In the field of physics, Khayyām wrote *Fī ikhtiyāl maʿrifa miqdāray al-dhahab wa'l-fiḍḍa fī jism murakkab minhumā* (On the Art of Determination of Quantities of Gold and Silver in a Body Consisting of Them) and *Fī'l-qustāṣ al-mustaqīm* (On the Right Balance). Both of these works are included in the *Kitāb mīzān al-ḥikma* (Book of the Balance of Wisdom) of Khayyām's student ʿAbd al-Raḥmān al-Khāzinī, who worked in Marw. In this he explains a method to determine specific gravity.

Khayyām also made significant contributions to music and philosophy. He wrote a *Qaul ʿalā ajnās allatī bi'l-arbaʿa* (Speech on Genera Which are [formed] by a Quarter) on the division of a quarter into three intervals in accordance with three genera of tonality — diatonic, chromatic, and enharmonic.

In philosophy, Khayyām was a follower of Aristotle and Ibn Sīnā (Avicenna). On the question of universalia he was near to Peter Abelard. Many philosophical questions were considered in Khayyām's poetry, particularly in his well known *Rubāʿīyāt* (Quatrains).

BORIS A. ROSENFELD

REFERENCES

Daoud Kasir. *The Algebra of Omar Khayyam*. New York: Columbia University Press, 1931. 2nd ed., 1972.

Matvievskaya, Galina P. and Boris A. Rosenfeld. *Mathematicians and Astronomers of Medieval Islam and their Works (8–17th centuries)* (in Russian). Moscow: Nauka, 1983, vol. 2, pp. 314–319.

Rosenfeld, Boris, and Adolphe Youschkevitch. *Omar Khayyam (1048–1131)*. Moscow: Nauka, 1965 (in Russian).

Swami Govinda, Tirtha. *The Nectar of Grace, ʿOmar Khayyam's Life and Works*. Allahabad: Kitabistan, 1941.

ʿUmar al-Khayyāmī. "Discussion on Difficulties in Euclid." *Scripta Mathematica* 24 (4): 275–303, 1959.

ʿUmar al-Khayyām. *Treatises*. Moscow: Nauka, 1962 (Arabic and Persian with Russian translations by B.A.Rosenfeld).

Youschkevitch, Adolphe, and Boris Rosenfeld. "Al-Khayyāmī (or Khayyām)." In *Dictionary of Scientific Biography*. Ed. Charles Gillispie. New York: Scribners, 1973, vol. 7, pp. 323–334.

See also: *Elements* – Naṣīr al-Dīn al-Ṭūsī

AL-KHĀZINĪ Abū'l-Fatḥ (or Abū Manṣūr) ʿAbd al-Raḥmān (also called in some sources ʿAbd al-Raḥmān Manṣūr) was an astronomer and expert in mechanics and scientific instruments. He lived in Marw (Khurāsān) ca. 1115–ca. 1131. A slave (and later a freedman) of Byzantine origin, he was bought by a treasurer (*khāzin*) of the Seljuk court at Marw, called Abū'l-Ḥusayn (or Abū'l-Ḥasan) ʿAlī ibn Muḥammad al-Khāzin al-Marwāzī, who gave him a good scientific education.

As an astronomer his main work is his *al-Zīj al-Sanjarī*, an astronomical handbook with tables, compiled between ca. 1118–ca. 1131 and dedicated to the Seljuk Sultan Sanjar ibn Malikshāh (1118–1157). This *zīj* seems to be influenced by the work of Thābit ibn Qurra, al-Battānī and al-Bīrūnī, but parts of it seem to have been checked against a limited number of his own observations (planetary, solar, and lunar in moments of conjunctions and eclipses) made in Marw. He is credited with a careful determination of the obliquity of the ecliptic (the band of the zodiac through which the sun apparently moves in its yearly course), but he adopts the Battanian value of 23; 35° and concludes (against al-Battānī and most of the successive Islamic astronomical tradition) that this parameter is a constant. His *zīj* includes a rich chronological section and a lot of materials related to the theory of Indian cycles. It also contains very important developments on the theory of planetary visibility as well as a very elaborate set of eclipse tables. He seems to have been interested by the problem of *qibla* determination (attempting to figure out the direction to Mecca) and the canons of his *zīj* mention a double entry *qibla* table computed for each integer degree of difference in latitude (from 1° to 30°) and in longitude (from 1° to 60°) between Mecca and other localities. This *qibla* table seems to be lost and the one ascribed to al-Khāzinī by the fourteenth century author al-Mustawfī does not seem to have anything to do with our author. Al-Khāzinī's *Zīj* was used in Byzantium by Georges Chrysococces (fl. ca. 1335–1346) and Theodore Meliteniotes (fl. ca. 1360–1388).

Among his minor astronomical works we find his treatise on instruments (*Risāla fī-l-ālāt*) which deals with observational instruments as well as with analog computers and simple helps for the naked eye. Furthermore, his short treatise on "the sphere that rotates by itself with a motion equal to the motion of the celestial sphere" (*Maqāla li'l-Khāzinī fī ittiḥād kura tadūru bi-dhātihā bi-ḥaraka musāwiya li-ḥarakat al-falak wa maʿrifat al-ʿamal bihā sākina wa mutaḥarrika*) — probably the earliest of all his extant works — shows the link between his interest both in astronomy and in applied mechanics. It describes a solid sphere, marked with the stars and the standard celestial circles and half sunk in a box, the rotation of which is propelled by a weight falling in a leaking reservoir of sand. Similar devices were known and built in Classical Antiquity.

The most important of al-Khāzinī's works is probably his *Kitāb mīzān al-ḥikma* (Book of the Balance of Wisdom), a treatise on the physical principles that underlie the hydrostatic balance as well as the construction and use of the instrument. This work was written in 1121–22 and dedicated to Sultan Sanjar: the instrument it describes was meant for Sanjar's treasure, for its main application was to discriminate accurately between pure and adulterated metals as well as between real gems and fakes. A similar balance, also called *mīzān al-ḥikma*, had been built by al-Khāzinī's predecessor al-Asfizārī also for Sanjar, but its scales had been destroyed. Al-Khāzinī's book is a long work, divided into eight books and eighty chapters, and its quotations bear witness to the Classical sources on pure and applied mechanics that reached the Islamic lands (Aristotle and the pseudo-Aristotelian treatise on mechanical problems, Euclid, Archimedes, Menelaus) as well as to the development of the discipline in Islamic civilization (Muḥammad ibn Zakariyyā' al-Rāzī, al-Bīrūnī, Ibn al-Haytham, Abū Sahl al-Qūhī, ʿUmar al-Khayyām). There is nothing specifically new in al-Khāzinī's physical ideas which derive mainly from Aristotle and Archimedes, but his treatise has the obvious interest of its very careful description of a highly precise instrument with which he has been able to calculate tables of the specific gravities of many substances, both metallic and non metallic, attaining, in some cases, results which are correct to within one percent.

JULIO SAMSÓ

REFERENCES

Hall, Robert E. "Al-Khāzinī." In *Dictionary of Scientific Biography*, vol. VII. New York: Charles Scribner's Sons, 1973, pp. 335–351.

Kennedy, Edward S. "A Survey of Islamic Astronomical Tables." *Transactions of the American Philosophical Society* (Philadelphia) 46(2): 37–39, 1956.

King, David A. "The Earliest Islamic Mathematical Methods and Tables for Finding the Direction of Mecca." *Zeitschrift für Geschichte der Arabisch-Islamischen Wissenschaften* 3: 134–138, 1986. A corrected version of this paper appears in D.A. King, *Astronomy in the Service of Islam*. Aldershot: Variorum Reprints, 1993, no. XIV.

Lorch, Richard. "The *Qibla*-Table Attributed to al-Khāzinī." *Journal for the History of Arabic Science* 4(2): 259–264, 1980.

Lorch, Richard. "Al-Khāzinī's Sphere That Rotates by Itself." *Journal for the History of Arabic Science* 4(2): 287–329, 1980.

See also: Thābit ibn Qurra – *Zīj*

AL-KHUJANDĪ Abū Maḥmūd Ḥāmid ibn al-Khiḍr al Khujandī (ca. 990) was a mathematician and astronomer who lived and worked under the patronage of the Buwayhid ruler, Fakhr al-Dawla (978–997).

In mathematics he worked on equations of the third degree and tried to show, although in an imperfect manner, that the sum of two cubed numbers cannot be another cube ($a^3 + b^3 \neq c^3$). His name is included among others such as Abū'l-Wafā al-Buzjanī (940–97), Abū Naṣr Manṣūr ibn ʿAlī ibn Irāq (ca. 1000) and Naṣīr al-Dīn al-Ṭūsī (1201–1274) to whom the discovery of the sine law is attributed. The sine law says that if ABC is any triangle, then

$$\frac{c}{b} = \frac{\text{Sin C}}{\text{Sin B}}$$

Although we do not have any reliable information on whether al-Khujandī founded an observatory on Jabal Tabrūk, in the vicinity of Rayy, he tells us that he made observations on planets for Fakhr al-Dawla with armillary spheres and other instruments and prepared a *zīj* (astronomical handbook with tables) entitled *al-Fakhrī* based on these observations. According to E.S. Kennedy an incomplete copy of a *zīj* in the Library of the Iranian Parliament (Tehran Ms. 181) may be attributed to al-Khujandī.

His fame depends on an important instrument constructed for the measurement of the obliquity of the ecliptic (the band of the zodiac through which the Sun apparently moves in its yearly course). This instrument surpassed all previous ones in size. Al-Khujandī praises his instrument and says that it is of his own invention. It is a sixty degree of meridian arc and is called *al-suds al-Fakhrī*, after Fakhr al-Dawla. Although *suds* means the sixth part of a circle (that is to say sextant), in reality it is constructed in place of a mural quadrant. As the distance between the summer solstice and the winter solstice is about 47°, a sixty degree of meridian arc is enough for observations of the sun. It has a radius of about twenty meters, and each degree was subdivided into 360 equal parts. Therefore each ten second portion was distinguished on the scale, and with it the limit of precision had been pushed to seconds.

This arc was constructed between two walls erected on the meridian. On the arched ceiling there was a hole through which the sun's rays passed and there were projections on the divisions of the arc. Al-Bīrūnī tells us that the aperture sank by about a span, and caused a slight displacement of the center of the arc.

In 384 H. (AD 994), al-Khujandī observed the summer solstice in the presence of a group of scientists. They gave their written testimony concerning the observations. Although the cloudy weather prevented his observing the winter solstice, he calculated the position of the sun from the observations made preceding the solstice. The result was 23°32′19″. According to al-Bīrūnī, displacement of the center of the arc also affected the result. By comparing this result to the Indian astronomers' and Ptolemy's he deduced that the obliquity of the ecliptic was decreasing. He also fixed the latitude of Rayy (35°34′39″).

He describes a kind of astrolabe which is called *shāmila* (universal instrument) in his *Risāla fī Aʿmāl al-ʿāmma* (Treatise on the Construction of the Universal Instrument). The astrolabe, which carried a stereographic projection of the heavens on a circular plate, is a portable instrument. As we know from existing specimens, they ranged in size from two inches to two feet in diameter. They were used to measure the altitudes of the sun and to show the places of the stars in certain latitudes. Khujandī's *shāmila* could be used over a larger territory than former ones. He also provides methods of projection concerning intersections of circles of the azimuth by the equator and *muqanṭarāt*. Abū Naṣr Manṣūr mentions al-Khujandī's two methods on this topic. Al-Birūnī, in his *Tahdīd Nihāyāt al-Amākin* (Limit of Ends of Places) states that al-Khujandī was unique in his age for his constructions of astrolabes and other astronomical instruments.

SEVIM TEKELI

REFERENCES

Kennedy, E. S. "A Survey of Islamic Astronomical Tables." *Transactions of the American Philosophical Society* 46 pt. 2:133, 1956.

Samsó, J. "Al-Khudjandī." In *Encyclopedia of Islam*. Vol.V. Leiden: E.J. Brill, 1986, pp. 46–47.

Samsó, J. *Estudios Sobre Abū Naṣr Manṣūr ibn Alī ibn Iraq*. Barcelona: Asociación para la Historia de la Ciencia Española, 1969, pp. 89–93.

Sarton, George. *Introduction to the History of Science*. Baltimore: Williams and Wilkins, 1927.

Sayılı, Aydın. *The Observatory in Islam*. Ankara: Turkish Historical Society, 1960.

Tekeli, Sevim. "Al-Khujandī." In *Dictionary of Scientific Biography*, Vol. VII. New York: Scribner's, 1973, pp. 352–54.

Tekeli, Sevim. "Nasirüddin, Takiyüddin ve Tycho Brahe'nin Rasat Aletlerinin Mukayesesi." Ankara Üniversitesi, *Dil ve Tarih-Coğrafya Fakültesi Dergisi* 16 (3–4): 301–393, 1958.

Wiedemen, E. "Über den Sextant des al-Chogendī." *Archive für Geschichte der Naturwissenschaften und der Technik* 2: 148–51, 1919.

AL-KHWĀRIZMĪ Al-Khwārizmī, Muḥammad ibn Mūsā, is the earliest Islamic mathematician and astronomer of fame, and his works had considerable influence in the medieval world. His name suggests that he was of Persian origin, but his treatises were written in Baghdād during the caliphat of al-Maʾmūn (813–833).

The *Algebra* (possibly *al-Kitāb al-mukhtaṣar fī ḥisāb al-jabr wa 'l muqābala*), al-Khwārizmī's best known work, consists of four parts of very unequal length. The first part explains the fundamentals of algebra: the resolution of the six basic types of equations of first and second degree (with positive coefficients and at least one positive solution), then basic algebraic reckoning (with expressions involving an unknown or square roots); then, six examples of problems each ending with one of the six equations, and lastly various other problems of the same kind. This structure as well as some characteristic features (no symbolism, numbers written in words, illustrations of algebraic rules by geometrical figures) remained customary in early textbooks on algebra. The second part contains a few considerations on the application of the rule of three to commercial transactions. Part Three covered surfaces and volumes of elementary plane and solid figures, mostly without any use of algebra. The lengthy Part Four is devoted to the application of algebraic methods (or simple arithmetic) to the sharing out of estates in accordance with wills and Islamic legal requirements regarding the parts due to heirs.

Several Latin translations of Parts I and II were made in the Middle Ages.

The *Arithmetic* (possibly *Kitāb al-ḥisāb al-hindī*) exposes the newly introduced Indian positional system of numerals and its use in arithmetical operations and (square) root extraction, for integers and fractions, both common and sexagesimal. This work is lost in Arabic, but has survived in two Latin translations.

The *Zīj al-sindhind* (Astronomical Tables), al-Khwārizmī's best known work in astronomy, is a set of tables based mainly on Indian material. Instructions on the use of the various tables are provided; with their help, one can determine the mean motion of the seven known celestial bodies, also daily motions, sizes, and eclipses of the sun and moon; calendric and trigonometrical tables are included. We do not know the original text, since only a Latin translation of a later reworking has come down to us.

The *Kitāb ṣūrat al-arḍ* (Geography) is a list of longitudes and latitudes of cities, mountains, (geographical points along) sea coasts, islands, center points of regions and countries, and detailed courses of rivers. The information is drawn from Ptolemy's *Geography*, but improvements are found, mostly, of course, for the regions under Islamic rule or travelled through by Arabian merchants.

The *Istikhrāj taʾrikh al-yahūd* (Calendar of the Jews) is the oldest extant description of the modern Jewish calendar. After reporting that God inspired it to Moses (its Mesopotamian origin was unknown to al-Khwārizmī), we are told about the year and the intercalation of seven intercalary months within the nineteen year (Metonic) cycle and how to determine on which day *rosh ha-shana* falls. Next we are given the positions of the celestial bodies at the beginning of three eras (Adam, Temple of Jerusalem, and Alexander). Finally, we are taught how to determine the mean positions of the sun and moon for any given date and the time elapsed since their last conjunction.

The *Kitāb al-ʿamal bi 'l-asṭurlāb* (On the Use of the Astrolabe) contains short instructions, but perhaps not in their original form. We are taught how to find altitudes of celestial bodies, and how to determine time or latitudes from celestial observations.

The purpose of another minor work, *Kitāb al-rukhāma* (On the Sundial), is to explain the construction of a plane sundial; the appended tables are given for various latitudes besides that of Baghdād.

A *Kitāb al-taʾrīkh* (Chronicle) is known by quotations from other historians; it reported purely historical events and covered at least the years 632 to 826.

As to lost works, a short treatise *Kitāb ʿamal al-asṭurlāb* (On the Construction of the Astrolabe) is attributed by Arabian bibliographers to al-Khwārizmī (it may be extant as an anonymous work). At the end of his *Algebra*, Abū Kāmil explains a short way for calculating the result of the duplication on the chessboard's cells ($1+2+2^2+\cdots+2^{63} = 2^{64} - 1$), which he attributes to al-Khwārizmī but which appears in no known work of his. Another lost work was concerned with the rule of false position.

Al-Khwārizmī has enjoyed a great reputation, particularly as the first algebraist. Although he seems not to have made a great number of original contributions, he was a learned man of great versatility and didactical ability. His role was primarily that of a disseminator of science in early Islamic times. As some of his works were studied in eleventh century Spain, he played the same role for the Christian world through Latin translations. The latinization of his misread Arabic name as *algorizmus*, later misread as *algoritmus*, later misinterpreted as *algorithmus*, has kept his name if not his fame alive until the present time.

JACQUES SESIANO

REFERENCES

King, David. *Al-Khwārizmī and New Trends in Mathematical Astronomy in the Ninth Century*. New York: Hagop Kevorkian Center for Near Eastern Studies, 1983.

Mukhammed ibn Musa al Khorezmi, k 1200-letiyu so dnia roždieniya. Moscow: Nauka, 1983.

Toomer, Gerald. "Al-Khwārizmī." In *Dictionary of Scientific Biography*, vol. VII. Ed. Charles C. Gillispie, pp. 358–365. New York: Charles Scribner's Sons, 1973.

See also: Algebra – Sundials – Astrolabe

AL-KINDĪ Al-Kindī, Abū Yūsuf Yaʿqūb ibn Isḥāq was born ca. 801 in Kūfah and died ca. 866 in Baghdad.

Surnamed the Philosopher of the Arabs, little is known with certainty about the life of this early Muslim philosopher and scientist. He flourished in Baghdad, then capital of the Abbasid Empire and center of its intellectual life. Greek, Persian and Indian works were being translated into Arabic and a multitude of religious and other thinkers were developing new and sophisticated schools of thought and literature. The *Mutakallimūn*, Islamic theologians, were actively engaged in controversies over God's attributes and freedom of the will as well as over the methodology of knowledge. During the life of al-Kindī, Baghdad experienced the political ascendency and fall of the dialectical theological movement of the *Muʿtazilah,* a movement that used a rationalistic approach to defend the religious dogmas of Islam. Al-Kindī sympathized with some ideas of this movement such as the uniqueness of God, his justice, and the use of a rational approach in the defense of Islam against its opponents. Also like *al-Muʿtazilah*, al-Kindī looked at God as the ultimate source of all being, yet believed his creations were independent in their daily function.

Most of the works of al-Kindī have, unfortunately, disappeared. Ibn al-Nadīm (d. ca. 987), one of the earliest bibliographers of Islam, lists 242 works, mostly essays and epistles, dealing with a wide range of sciences: logic, metaphysics, mathematics, spherics, music, astronomy, geometry, medicine, astrology, theology, psychology, politics, meteorology, prognostics, and alchemy.

The scientific works of al-Kindī are by far the most numerous. In fact there are among the early Arabic authors many who considered him a mere scientist and not a philosopher. Z. al-Bayhaqī (d. 1169), who is often cited, referred to him as an engineer, and Ibn Khaldūn (d. 1406) did not list him among the philosophers. Whatever the case may be, one can find in al-Kindī's presently known works the outline of a philosophical system. Based on his treatise *Fī al-Falsafeh al-ula* (On First Philosophy), al-Kindī defined philosophy as the "knowledge of the realities of things according to human capacity." He stressed the cumulative character of philosophy and the duty to receive the truth gratefully from whatever source it comes. He distinguished between philosophy and theology proper as two different disciplines, both pursuing and reaching the same goal, which is truth. The first pursues it through arduous research and effort, the second through prophetic knowledge which is granted to certain individuals immediately and without any effort. He established the various divisions of philosophy on the basis of the different channels of human knowledge, the sense experience of material entities and that of rational cognition of immaterial ones. First philosophy or metaphysics is concerned with the First Principle, the True One, the Necessary and Uncaused Being who is eternal and infinite. Probably drawing upon the ideas of the pre-Islamic John Philoponus who championed the concept of creation *ex nihilo*, he opposed the Aristotelian theory of the eternity of the universe. The chief attribute of God, argued al-Kindī, is unity. God is the only real agent or cause in the world. Unlike the Muslim theologians, he did not ignore the role of secondary agents in the process of nature. He established the premise that the causes of generation and corruption are the heavenly bodies, which are superior to the physical bodies and possess intelligence.

The writing of al-Kindī on the soul and the intellect are neither numerous nor comprehensive. Besides his important short treatise *Risālah fī māʿiyat al-ʿaql* (On the Intellect), he wrote *Kalām fī al-nafs mukhtaṣar wajīz* (A Discourse on the Soul Abridged and Concise) where he explains the nature of the union between the soul and the body as different from the union between the elements. Inspired by Aristotelian, Plantonic, and Pythagorean sources, he considered the soul as the principle of life and only accidentally related to the body, being nobler in nature. In his *On the Intellect*, al-Kindī presents a Platonic interpretation of Aristotle's noetics. He defines the intellect as "a simple essence cognizant of things in their true realities". There are four intellects, the first of which is separate and seems to be construed in the image of Aristotle's active intellect. It exists outside the human soul and is the cause of "all intelligible thoughts and secondary intellects". The four intellects: the active, the potential, the habitual, and the manifest or acquired, played an important role in the history of medieval discussions on the nature of the intellect. In his *Risālah fī dafʿal-aḥzān* (On the Means to Drive Away Sadness), al-Kindī offers practical advice to overcome sorrow based on the principle that true riches are not material by nature. What causes sadness is the loss of externals and the failure to attain what we highly cherish. By despising external things and educating our souls to seek what is spiritual, we avoid the miseries of sorrow.

With many of his scientific works not available, it is difficult to offer a systematic exposition of al-Kindī's scientific works and contributions. They covered almost all fields of the physical sciences and went beyond the mere transmission of Hellenic scientific data by adding to it through observations of his own and some rudimentary experimentation. Al-Kindī took over the Greek scientific heritage which was available in translations or which he had helped to translate, and tried to assimilate, summarize, and at times develop it and experiment with it. In his classification of the sciences, he starts from the idea of a hierarchy of beings according to which the lower sciences deal with sensible beings and higher sciences with the nobler and intelligible beings. There is a close relationship, however, between science and philosophy, especially mathematics, which he considers preparatory and essential to the study of philosophy. He uses mathematical concepts and arguments when discussing infinity and plurality in proving his theory of the unity of God.

Al-Kindī was not satisfied with the role of the transmitter and commentator. In many of his scientific works we feel the surge of an investigative spirit. In works on optics, pharmacology and music for example, he provided new data and approaches, thus enhancing our knowledge of these subjects, among many others. At times, he sought to verify some statements through experimentation, as when he shot an arrow in the air to verify Aristotle's statement that substances expand when heated. In his work *Risālah fī ikhtilāf al-manaṣir* (*De Aspectus)*, which is based on Euclid and Heron of Alexandria's optics, he rejected the Euclidean theory of the emission of light, and using geometrical arguments, he offered amendments so it conformed with observable data. Furthermore, al-Kindī gave an original interpretation independent from that of Aristotle of the azure blue we see in the sky. He argued that the color we actually see is the reflection of light from vapors and particles of dense bodies carried up into the atmosphere.

Likewise, al-Kindī, in his medical and pharmaceutical works, was a contributor of new ideas trying to improve upon the knowledge of antiquity. In his work on pharmacology *Risālah fī maʿrifat qūwah al-adwīyah al-murakkabah* (translated into Latin as *De medicinarum compositorum grabidus investigandis libellus)*, he applied the principles of posology. He based the efficacy (*qūwah*) of compound medicine upon geometrical progression. He linked the degree of intensity with the numerical changes in the qualitative forces that produce them. The efficacy corresponds to the proportion of the sensible qualities: warm, cold, dry, and humid. If a compound medicine was to be warm in the first degree, it had to possess double the equable mixture. If it was to be warm in the second degree, it had to possess four times as much, etc.

The same tendency to improve on the ancient sciences is seen in his works on music. For example, he used the letters of the alphabet to designate the notes of the scale, a procedure used in Europe a century later. Al-Kindī's musical works in Arabic on the theory of music paved the way for the *Kitāb al-musīqi al-kabīr* (Great Book on Music) by al-Fārābī.

Al-Kindī's influence in medieval Europe might not be as great as the other Arabic philosophers such as Ibn Sīnā and Ibn Rushd; he was nevertheless a courageous intellectual and scientist. Whether he was the first philosopher in Islam is still not ascertainable in the absence of many of his philosophical works. There is no doubt, however, that he was the first in Islam to bridge the gap between philosophy and the Islamic dogma, thus establishing the conceptual framework that became characteristic of philosophy in Islam.

GEORGE N. ATIYEH

REFERENCES

ʿAbd al-Hādī Abū Ridah. *Rasāʿil al-Kindī l-falsafīyah*. 2 vols. Cairo: Dār al-Fikr alʿArabī, 1950–53.

Atiyeh, George N. *al-Kindī, the Philosopher of the Arabs*. Islamabad, Pakistan: Islamic Research Institute, 1967.

Cortabarria, A. "La classification des sciences chez Al-Kindi." *Mélanges de L'Institut dominican d'études Orientales* 11: 49–76, 1972.

Cortabarria, A. "Al-Kindi Vu par Albert Le Grand." *Mélanges de L'Institut dominican d'études Orientales* 13: 117–146, 1977.

d'Averny, M.T. and F. Hurdy, "Al-Kindi, De radiis." *Archives d' histoire doctrinale et litteraire du moyen-age*. 41: 139–260, 1974.

Fakhri, Majid. *History of Islamic Philosophy*. New York: Columbia University Press, 1970.

Ivry, Alfred L. *Al-Kindī's Metaphysics*. Albany: State University of New York Press, 1974.

Jolivet, J. *L'intellect selon Kindi*. Leiden: Brill, 1971.

Levey, M., *The Medical Formulary or Aqrābadhin of al-Kindī*. Madison: University of Wisconsin Press, 1966.

Rashed, Roshdi, "Al-Kindī's Commentary on Archimedes' The Measurement of the Circle." *Arabic Science and Philosophy* 3: 7–53, 1993.

Rescher, Nicholas, *Al-Kindī, an Annotated Bibliography*. Pittsburgh: University of Pittsburgh Press, 1964.

KNOWLEDGE SYSTEMS: LOCAL KNOWLEDGE The concept of local knowledge has recently come to the fore in the field of the sociology of scientific knowledge, where it is a common empirical finding that knowledge production is an essentially local process. Knowledge claims are not adjudicated by absolute standards; rather their authority is established through the workings of *local* negotiations and judgments in particular contexts. This focus on the localness of knowledge production provides the condition for the possibility for a fully-fledged comparison between the ways in which understandings of the natural world have been produced by different cultures and at different times. Such cross-cultural comparisons of knowledge production systems have hitherto been largely absent from the sociology of science. A necessary condition for fully equitable comparisons is that Western contemporary technosciences, rather than being taken as definitional of knowledge, rationality, or objectivity, should be treated as varieties of such knowledge systems. Though knowledge systems may differ in their epistemologies, methodologies, logics, cognitive structures, or in their socio-economic contexts, a characteristic that they all share is their localness. Hence, in so far as they are collective bodies of knowledge, many of their small but significant differences lie in the work involved in creating assemblages from the "motley" of differing practices, instrumentation, theories, and people (Hacking, 1992). Much of that work can be seen as strategies and techniques for creating the equivalences and connections whereby otherwise heterogeneous and isolated knowledges are enabled to move in space and time from the local site and moment of their production and application to other places and times.

In this view, all knowledge systems from whatever culture or time, including the contemporary technosciences, are based on local knowledge. However within the master narrative of modernism, local knowledge is an oxymoron. Exploring this contradiction and the manifold meanings of local requires a brief excursion into postmodernism as well as some of the arguments underpinning the sociology of scientific knowledge.

Though postmodernism eludes definition and is more likely a stage of modernism than a marked epistemological break, there has been a recent coalescence of strands of thought in a wide variety of areas that have questioned the assumptions underlying modernism. Postmodernism is most frequently equated with the collapse of the concepts of rationality and progress held to accompany the emergence of the post-industrial society and is consequently concerned with the rejection of universal explanations and totalizing theories. But perhaps the strand that is most truly pervasive in the constellation of reformulated approaches to understanding the human condition is the emphasis on the local.

In physics, ecology, history, feminist theory, literary theory, anthropology, geography, economics, politics, and sociology of science, the focus of attention has become the specific, the contingent, the particular. This is the case whether it is a text, a reading, a culture, a population, a site, a region, an electron, or a laboratory. Within this diversity of uses of local there seem to be two broad and rather different senses being used. On the one hand there is the notion of a voice or a reading. The voice may be purely individual and subjective or may be a collection of voices belonging to a group, class, gender, or culture, but in all cases the notion captures one of the basic characteristic elements of postmodernism, courtesy of deconstruction, that all texts or cultures are multivocal and polysemous. That is they have a multiplicity of meanings, readings and voices and are hence subject to "interpretive flexibility" (Collins, 1985). On the other hand, local is used both in the more explicitly geopolitical sense of place and in the experiential sense of contextual, embodied, partial, or individual. A range of disciplines from meteorology to medicine now recognize the necessity of focusing on the particular conditions at specific sites and times rather than losing that specificity in unlocalized generalizations.

The sociology of scientific knowledge is one of the most classically post of all modernisms and is therefore an area in which the local is also a thematic presence which is only now coming into focus. Some philosophers of science have come to re-evaluate the role of theory and argue that scientists practicing in the real world do not deduce their explanations from universal laws but rather make do with rules of thumb derived from the way the phenomena present themselves in the operation of instruments and devices. Similarly philosophers and sociologists of science alike have recognized for some time the lack of absolute standards and the role of tacit knowledge in technoscientific practice, and have sought to display the context in which the practice of science is manifested as craft skills and collective work. However the recognition of the social and material embodiment of skills and work in the cultural practice of individuals and groups has only recently coalesced into the general claim that all knowledge is local. Knowledge, from this constructivist perspective can be local in a range of different senses. "It is knowledge produced and reproduced in *mutual interaction* that relies on the *presence* of other human beings on a direct, face-to-face basis." (Thrift, 1985). It is knowledge that is produced in contingent, site, discipline or culture specific circumstances (Rouse, 1987). It is the product of open systems with heterogeneous and asynchronous inputs "that stand in no necessary relationship to one another" (Pickering, 1992). In sum scientific knowledge is "situated knowledge" (Haraway, 1991).

Perhaps the most important consequence of the recognition of the localness of scientific knowledge is that it permits a parity in the comparison of the production of contemporary technoscientific knowledge with knowledge production in other cultures. Previously the possibility of a truly equitable comparison was negated by the assumption that indigenous knowledge systems were merely local and were to be evaluated for the extent to which they had scientific characteristics. Localness essentially subsumes many of the supposed limitations of other knowledge systems compared with western science. So-called traditional knowledge systems have frequently been portrayed as closed, pragmatic, utilitarian, value laden, indexical, context dependent, and so on. All of which was held to imply that they cannot have the same authority and credibility as science because their localness restricts them to the social and cultural circumstances of their production. Science by contrast was held to be universal, nonindexical, value free, and as a consequence floating, in some mysterious way, above culture. Treating science as local simultaneously puts all knowledge systems on a par and renders vacuous discussion of their degree of fit with transcendental criteria of scientificity, rationality, and logicality. Now the multidisciplinary approaches to understanding the technosciences which together constitute the sociology of scientific knowledge can be made more fully anthropological by the addition of a new sub-discipline called comparative technoscientific traditions.

Emphasizing the local in this way necessitates a re-evaluation of the role of theory which is typically held by philosophers and physicists to provide the main dynamic and rationale of science as well as being the source of its universality. Karl Popper claims that all science is cosmology and Gerald Holton sees physics as "a quest for the Holy Grail", which is no less than the "mastery of the whole world of experience, by subsuming it under one unified theoretical structure" (Allport, 1991). It is this claim to be able to produce universal theory that Western culture has used simultaneously to promote and reinforce its own stability and to justify the dispossession of other peoples. It constitutes part of the ideological justification of scientific objectivity, the "god-trick" as Donna Haraway calls it; the illusion that there can be a positionless vision of everything. The allegiance to mimesis has been severely undermined by analysts like Richard Rorty, but theory has also been found wanting at the level of practice, where analytical and empirical studies have shown it cannot provide the sole guide to experimental research and on occasion has little or no role at all. The conception of grand unified theories guiding research is also incompatible with a key finding in the sociology of science: "consensus is not necessary for cooperation nor for the successful conduct of work". This sociological perspective is succinctly captured in Leigh Star's description:

> Scientific theory building is deeply heterogeneous: different viewpoints are constantly being adduced and reconciled... Each actor, site, or node of a scientific community has a viewpoint, a partial truth consisting of local beliefs, local practices, local constants, and resources, none of which are fully verifiable across all sites. The aggregation of all viewpoints is the source of the robustness of science (Star, 1989, 46).

Theories from this perspective have the characteristics of what Star calls "boundary objects"; that is they are "objects which are both plastic enough to adapt to local needs and constraints of the several parties employing them, yet robust enough to maintain a common identity across sites". Thus theorizing is itself an assemblage of heterogeneous local practices.

If knowledge is local we are faced with a problem: how are the universality and connectedness that typify technoscientific knowledges achieved? Given all these discrete knowledge/practices, imbued with their concrete specificities, how can they be assembled into fields or knowledge systems; or in Star's terms "how is the robustness of findings and decision making achieved?" Ophir and Shapin ask, "How is it, if knowledge is indeed local, that certain forms of it appear global in domain of application?" The answers, considered here, lie in a variety of social strategies and technical devices that provide for treating instances of knowledge/practice as similar or equivalent and for making connections, that is in enabling local knowledge/practices to move and to be assembled.

Research fields or bodies of technoscientific knowledge/practice are assemblages whose otherwise disparate elements are rendered equivalent, general, and cohesive through processes that have been called "heterogeneous engineering" (Law, 1987). Among the many social strategies that enable the possibility of equivalence are processes of standardization and collective work to produce agreements about what counts as an appropriate form of ordering, what counts as evidence, etc. Technical devices that provide for connections and mobility are also essential. Such devices may be material or conceptual and may include maps, calendars, theories, books, lists, and systems of recursion, but their common function is to enable otherwise incommensurable and isolated knowledges to move in space and time from the local site and moment of their production to other places and times.

Some of these devices have been revealed relatively unproblematically through direct observation. Others are less susceptible to investigation and analysis, being embodied in our forms of life. One way to catch a glimpse of these hidden presuppositions and taken for granted ways of thinking, seeing, and acting, is to misperceive, to be jolted out of our habitual modes of understanding through allowing a process of interrogation between our knowledge system and others. Such an interrogative process of mutual inter-translation can enable us to catch sight of the cultural glasses we wear instead of looking through them as if they were transparent.

This challenging of the totalizing discourses of science by other knowledge systems is what Foucault had in mind when he claimed that we are "witnessing an insurrection of subjugated knowledges" and corresponds to an emphasis on the local that has emerged in anthropology at least since Clifford Geertz's *Interpretation of Cultures*. In his critique of global theories and in his emphasis on "thick description" Geertz pointed out that cultural meanings cannot be understood at the general level because they result from complex organizations of signs in a particular local context and that the way to reveal the structures of power attached to the global discourse is to set the local knowledge in contrast with it.

Equally there is the pervasive recognition characterized as post-colonialism that the West has structured the intellectual agenda and has hidden its own presuppositions from view through the construction of the other. Nowhere is this more acute than in the assumption of science as a foil against which all other knowledge should be contrasted. In the view of Marcus and Fischer we are at an experimental moment where totalizing styles of knowledge have been suspended "in favour of a close consideration of such issues as contextuality, the meaning of social life to those who enact it and the explanation of exceptions and indeterminants". In this emphasis on the local we are postparadigm.

However we should not be too easily seduced by the apparently liberating effects of celebrating the local since it is all too easy to allow the local to become a "new kind of globalizing imperative" (Hayles, 1990). In order for all knowledge systems to have a voice and in order to allow for the possibility of inter-cultural comparison and critique, we have to be able to maintain the local and the global in dialectical opposition to one another. This dilemma is the most profound difficulty facing liberal democracies now that they have lost the convenient foil of communism and the world has Balkanized into special interest groups by genders, race, nationality, or whatever. By moving into a comparatist mode there is a grave danger of the subsumption of the other into the hegemony of western rationality, but conversely unbridled cultural relativism can only lead to the proliferation of ghettos and dogmatic nationalisms. We cannot abandon the strength of generalizations and theories, particularly their capacity for making connections and for providing the possibility of criticism. At the same time we need to recognize reflexively that theory and practice are not distinct. Theorizing is also a local practice. If we do not recognize this joint dialectic of theory and practice, the local and the global, we will not be able to understand and establish the conditions for the possibility of directing the circulation and structure of power in knowledge systems. It is in the light of this recognition that I want to consider the ways in which the movement of local knowledge is accomplished in different knowledge systems and their consequent effects on the ways in which people and objects are constituted and linked together; that is their effects on power. The essential strength of the sociology of scientific knowledge is its claim to show that what we accept as science and technology could be other than it is. The great weakness of the sociology of scientific knowledge is the general failure to grasp the political nature of the enterprise and to work towards change. With some exceptions it has had a quietist tendency to adopt the neutral analyst's stance that it devotes so much time to criticizing in scientists. One way of capitalizing on the sociology of science's strength and avoiding the reflexive dilemma is to devise ways in which alternative knowledge systems can be made to interrogate each other.

Considerable advances in understanding the movement of local knowledge have been made possible through Bruno Latour's insightful analysis. For Latour the most successful devices in the agonistic struggle are those which are mobile and also "immutable, presentable, readable and combinable with one another". These immutable mobiles are the kinds of texts and images that the printing press and distant point perspective have made possible. These small and unexpected differences in the technology of representation are on his account the causes of the large and powerful effects of science. That which was previously completely indexical, having meaning only in the context of the site of production, and having no means of moving beyond that site, is standardized and made commensurable and re-presentable within the common framework provided by distant point perspective. Hence that which has been observed, created, or recorded at one site can be moved without distortion to another site. At centers of calculation such mobile representations can be accumulated, analyzed, and iterated in a cascade of subsequent calculations and analyses.

Latour's account has been augmented by the work of Steven Shapin and Simon Schaffer in the *Leviathan and The Air Pump* (1985). They have shown that experimental practice in science is sustained by a range of social, literary, and technical devices and spaces that we take for granted but which had to be created deliberately to overcome the fundamentally local and hence defeasible character of experimentally derived knowledge claims. In the seventeenth century, the problem for Robert Boyle, one of the earliest experimentalists, was to counter the arguments of his opponent Thomas Hobbes about the grounding of true and certain knowledge which they both agreed was essential in a country riven by dissent and conflicting opinion. Reliable knowledge of the world, for Hobbes, was to be derived from self-evident first principles, and anything that was produced experimentally was inevitably doomed to reflect its artifactual nature and the contingencies of its production; its localness would deny it the status of fact or law. Boyle recognized the cogency of these arguments and set out to create the forms of life within which the knowledge created at one site could be relayed to and replicated at other sites. In order for an empirical fact to be accepted as such it had to be witnessable by all, but the very nature of an experimental laboratory restricted the audience of witnesses to a very few. Boyle, therefore, had to create the technology of what Shapin calls virtual witnessing. For this to be possible three general sorts of devices or technologies had to be developed. Socially groups of reliable witnesses had to be formed. Naturally in the seventeenth century they were gentlemen. These gentlemen witnesses had to be able to communicate their observations to other groups of gentlemen so that they too might witness the phenomena. This required the establishment of journals using clear and unadorned prose that could carry the immutable mobiles, experimental accounts, and diagrams. The apparatus had to be made technically reliable and reproducible, but perhaps most importantly the physical space for such empirical knowledge had to be created.

Hobbes, of course, was right: experimental knowledge is artifactual. It is the product of human labor, of craft and skill, and necessarily reflects the contingencies of the circumstances. It is because craft or tacit knowledge is such a fundamental component of knowledge production that accounts of its generation, transmission, acceptance, and application cannot be given solely in terms of texts and inscriptions. A vital component of local knowledge is moved by people in their heads and hands. Harry Collins, a sociologist of science, has argued that this ineradicable craft component in science is ultimately what makes science a social practice. Because knowledge claims about the world are based on the skilled performance of experiments their acceptance is a judgment of competence not of truth. An example of the centrality of craft skill is the TEA laser, invented in Canada by Bob Harrison in the late sixties, which British scientists attempted to replicate in the early seventies. "No scientist succeeded in building a laser using only information found in published or other written sources" and furthermore the people who did succeed in building one were only able to do so after extended personal contact with somebody who had himself built one. Now TEA lasers are blackboxed and their production is routine and algorithmic. But in order to become routinized, Harrison's local knowledge had to be moved literally by hand.

Joseph Rouse (1987, 72) in considering the contemporary production process of scientific knowledge has summarized the implications of this understanding of science:

> Science is first and foremost knowing one's way about in the laboratory (or clinic, field site etc.). Such knowledge is of course transferable outside the laboratory site into a variety of other situations. But the transfer is not to be understood in terms of the instantiation of universally applied knowledge claims in different particular settings by applying bridge principles and plugging in particular local values for theoretical variables. It must be understood in terms of the adaptation of one local knowledge to create another. We go from one local knowledge to another rather than from universal theories to their particular instantiations.

According to the historian of science Thomas Kuhn the way a scientist learns to solve problems is not by applying theory deductively but by learning to apply theory through recognizing situations as similar. Hence theories are models or tools whose application results from situations being conceived as or actually being made equivalent. This point is implicit in the recognition that knowledge produced in a laboratory does not simply reflect nature because nature as such is seldom available in a form that can be considered directly in the lab. Specially simplified and purified artifacts are the typical subject of instrumental analysis in scientific laboratories. For the results of such an artificial process to have any efficacy in the world beyond the lab, the world itself has to be modified to conform with the rigors of science. A wide variety of institutional structures have to be put in place to achieve the equivalences needed between the microworld created inside the lab and the macroworld outside in order for the knowledge to be transmittable. The largest and most expensive example of this is the Bureau of Standards, a massive bureaucracy costing six times the R&D budget. Without such social institutions the results of scientific research are mere artifacts. They gain their truth, efficacy, and accuracy not through a passive mirroring of reality but through an active social process that brings our understandings and reality into conformity with each other.

The result of the work of Latour, Collins, Shapin, Star, Hacking, Rouse, and others has been to show that the kind of knowledge system we call Western science depends on a variety of social, technical, and literary devices and strategies for moving and applying local knowledge. It is having the capacity for movement that enables local knowledge to constitute part of a knowledge system. This mobility requires devices and strategies that enable connectivity and equivalence, that is the linking of disparate or new knowledge and the rendering of knowledge and context sufficiently similar as to make the knowledge applicable. Connectivity and equivalence are prerequisites of a knowledge system but they are not characteristics of knowledge itself. They are produced by collective work and are facilitated by technical devices and social strategies. Differing devices and strategies produce differing assemblages and are the source of the differences in power between knowledge systems.

In conclusion, it has been argued that Western science, like all knowledge in all societies, is inherently local, and furthermore other non-Western societies have developed a variety of social and technical devices for coping with that localness and enabling it to move. Some of them are technical devices of representation like the mason's templates, and the Incan *ceques* and *quipus*. Some of them are abstract cognitive constructs, like the Anasazi and Incan calendars, and the Micronesian navigation system. All of them also require social organization, rituals, and ceremonies. All of them have proved capable of producing complex bodies of knowledge and in many cases have been accompanied by substantial transformations of the environment. The major difference between Western science and other knowledge systems lies in the question of power. Western science has succeeded in transforming the world and our lives in ways that no other system has. The source of the power of science on this account lies not in the nature of scientific knowledge but in its greater ability to move and apply the knowledge it produces beyond the site of its production. However at the end of the twentieth century we can now perceive that there is a high cost to pay for science's hegemony. Much of that cost in terms of environmental degradation and ethnocide is due not so much to the totalizing nature of scientific theories but to the social strategies and technical devices science has developed in eliminating the local.

The task of resisting and criticizing science may now be addressed by reconsidering the causes of its dominating effects. Without the kinds of connections and patterns that theories make possible we will never be able to perceive the interconnectedness of all things. Without the awareness of local differences we will lose the diversity and particularity of the things themselves. Thus, rather than rejecting universalizing explanations what is needed is a new understanding of the dialectical tension between the local and the global. We need to focus on the ways in which science creates and solves problems through its treatment of the local. Science gains its truthlike character through suppressing or denying the circumstances of its production and through the social mechanisms for the transmission and authorization of the knowledge by the scientific community. Both of these devices have the effect of rendering scientific knowledge autonomous, above culture, and hence beyond criticism. Equally problematic is the establishment of the standardization and equivalences required in order that the knowledge produced in the lab works in the world. The joint processes of making the world fit the knowledge instead of the other way round and immunizing scientific knowledge from criticism are best resisted by developing forms of understanding in which the local, the particular, the specific, and the individual are not homogenized but are enabled to talk back.

DAVID TURNBULL

REFERENCES

Allport, P. "Still Searching for the Holy Grail." *New Scientist* 132: 51–52, Oct 5, 1991.

Collins, Harry. *Changing Order: Replication and Induction in Scientific Practice*. London: Sage, 1985

Foucault, Michel. *Power/Knowledge: Selected Interviews and Other Writings 1972–77.* New York: Pantheon Books, 1980.

Geertz, Clifford. *The Interpretation of Cultures: Selected Essays.* New York: Basic Books, 1973.

Hacking, Ian. "The Self-Vindication of the Laboratory Sciences." In *Science as Practice and Culture.* Ed. Andrew Pickering. Chicago: University of Chicago Press, 1992, pp. 29–64.

Haraway, Donna. *Symians, Cyborgs and Women: The Reinvention of Nature*. London: Free Association Books, 1991.

Hayles, Katherine. *Chaos Bound: Orderly Disorder in Contemporary Literature and Science.* Ithaca: Cornell University Press, 1990.

Latour, Bruno. "Visualisation and Cognition: Thinking With Eyes and Hands." *Knowledge and Society* 6: 1–40, 1986.

Law, John. "Technology and Heterogeneous Engineering: The Case of Portuguese Expansion." In *The Social Construction of Technological Systems: New Directions in the Sociology and History of Technology.* Ed. W. Bijker, T. Hughes, and T. Pinch. Cambridge, Massachusetts: MIT Press, 1987, pp. 111–34.

Marcus, G.E. and M.M.J Fischer. *Anthropology as Cultural Critique: An Experimental Moment in the Human Sciences.* Chicago: University of Chicago Press, 1986.

Ophir, Adi and Steven Shapin. "The Place of Knowledge: A Methodological Survey." *Science In Context* 4: 3–21, 1991.

Pickering, Andrew, ed. *Science as Practice and Culture.* Chicago: University of Chicago Press, 1992.

Rouse, Joseph. *Knowledge and Power: Towards a Political Philosophy of Science.* Ithaca: Cornell University Press, 1987.

Shapin, Steven and Simon Schaffer. *Leviathan and the Air Pump: Hobbes, Boyle and the Experimental Life.* Princeton, New Jersey: Princeton University Press, 1985.

Star, Susan Leigh. "The Structure of Ill-Structured Solutions: Boundary Objects and Heterogeneous Distributed Problem Solving." In *Distributed Artificial Intelligence.* Ed. L. Gasser and N. Huhns. New York: Morgan Kauffman Publications, 1989, pp. 37–54.

Thrift, Nigel. "Flies and Germs: a Geography of Knowledge." In *Social Relations and Spatial Structures*. Ed. Derek Gregory and John Urry. London: Macmillan, 1985, pp. 366–403.

KNOWLEDGE SYSTEMS OF THE AUSTRALIAN ABORIGINAL PEOPLE Is there only one legitimate knowledge production system? Most say yes — science. Others admit different systems of knowledge which exist as separate, sealed, and static domains. In fact there is not only a multiplicity of knowledge production systems, but they interact and cross-fertilize each other. This is especially unwelcome to science which holds itself to be compromised by involvement with 'less rigorous' systems of knowledge production. It is a radical idea to claim that the various knowledge production systems are all legitimate and none may claim to be hermetic or exclusive. This claim is not idle theoretical posturing. It is intimately tied up with possibilities for social justice for non-Western peoples engaged in confrontations with polities with Western scientific traditions, and with possibilities for transforming science and other knowledge production systems to render them more sensitive to changing human aspirations.

In Australia societies working with radically different knowledge systems have confronted each other for two hundred years. In this confrontation many Aboriginal Australian peoples, their political systems, and their knowledge traditions have been systematically destroyed; a few have survived. Many Australians, Aboriginal and non-Aboriginal alike, feel that reconciliation and reconstruction is needed. This is related to claims that the surviving Aboriginal knowledge production systems can be useful in finding new directions for Australian life and demands for social justice for Aboriginal Australians.

Recent Yolngu history provides an example. In 1970 the Yolngu people challenged invasion of their lands by a European mining company in the High Court of Australia. This court case was one of the earliest attempts by Aboriginal Australian people to regain control over their land. The Yolngu claimed the mining that had begun was wrongful in the light of their own particular relationship with the land. The Commonwealth of Australia had granted the company mineral leases so they too became defendants in the case. Impressed as the presiding judge was with the evidence presented by the Yolngu, he decided that the Yolngu people were governed by a law which did not deal with land as property; the Yolngu lost the case. The finding of the court was that the Yolngu feeling of obligation to the land was stronger than a feeling of ownership. "...it seems easier, on the evidence, to say that the clan belongs to the land, than the land belongs to the clan" (Williams, 1986). The Australian legal system constructed Aboriginal Australia as "nature" rather than "culture" in primarily serving the colonial enterprise. Some five years later however Yolngu people were granted their lands with the passage of the 1976 Land Rights Bill. The Yirrkala Land Case had been instrumental in changing the political climate for Aboriginal land rights.

Yolngu were shocked by the Court's finding in 1970, and still are. As they see it the court proceedings made a judgment on the rationality of the concepts which underlie the ordering of Yolngu society. There was never any determined effort to mediate between the knowledge worlds during the court case — that, of course, is not the function of Western courts.

Both knowledge systems, the scientific and the Yolngu, have conventional accounts of how knowledge production occurs in their communities. Let us briefly consider the orthodox ways that the two knowledge systems are understood. The orthodox view of scientific knowledge production has it that nature is represented in scientific knowledge which, because it offers a true account of the way things are, gives those who know in this way power in the remaking of society. Science is seen as a process which employs, in the words of Barry Jones, until recently Minister for Science in the Australian Government, "collection, organization and dissemination of data that is then converted to 'information' and then further refined to 'knowledge'". Power, in this account, while necessarily extrinsic to knowledge, flows from knowledge; particular kinds of power are more readily achieved or deployed if one can accurately represent one's situation.

Yolngu Aboriginal Australians likewise have a strong sense of how their society is reproduced through the functioning of their knowledge production system. Previous existing situations are the background against which new knowledge is produced. Yolngu society is remade as each new generation reconstitutes the balance of Yolngu life.

The possibility of reframing Yolngu concepts within Western knowledge and vice versa requires that each understand the other. To enable understanding between knowledge systems there must be the possibility both for collective solidarity and for plurality and difference. On the Yolngu side two metaphors have been developed which originate in the natural process of the Yolngu lands. On other hand Balanda researchers couch their framework in terms of metaphors of construction. (The term *Balanda* is a Yolngu term for non-Aboriginal Australian. It predates British invasion of the continent, deriving from the Macassan word "Hollander". This word is borrowed from the Macassan traders with whom Yolngu had substantial dealings until the beginning of this century.)

What is felt of Yolngu knowledge in the Western world might at first be far from the way Yolngu people experience their knowledge. Also, what is felt about Western knowledge in the Yolngu world often at the beginning can seem far from the way that Balanda people experience their own understandings.

In the practices of everyday life the Yolngu cosmos is acknowledged as one which has been created. A far back reality, apart from the time of ordinary human life, is a necessary and inevitable component of everyday living. The explanation that "ancestral beings long ago created the world we share in their actions of social living" underlies every aspect of ordinary Yolngu life. This contrasts with everyday living in the Western cosmos where hints of a transcendental "other" world rarely intrude. The explanation which underlies everyday living in the Western world is: "the world is the way it is and that is the way we know it": an empiricist's explanation.

The process of making knowledge of one world available in another is a familiar practice for Yolngu, because their world exists in two mutually exclusive constituents: the *Dhuwa* and the *Yirritja*. These categories each have *rom* (objects natural and manufactured, places and people, words, knowledge, and texts). Dhuwa *rom* is made available to Yirritja clans for their use and vice versa. In particular contexts Dhuwa understandings can be made use of by Yirritja, and likewise there are accepted ways of representing Yirritja in the Dhuwa world.

A metaphor which derives from the Yirritja side of the Yolngu world, *Ganma*, is the dialectic of the meeting and continual mutual engulfing of two rivers. The rivers have different sources, and as they flow into each other their separate linear forces become the force of a vortex. This vortical flow gives deeper penetration into understanding and knowledge. In terms of understanding different ways of thinking, *ganma* is taken as describing the situation where a river of water from the sea (Western knowledge) and a river of water from the land (Yolngu knowledge) continually engulf and re-engulf each other on flowing into a common lagoon. In coming together the streams of water mix across the interface of the two currents and foam is created at the surface, so that the process of ganma is marked by lines of foam marking the interface of the two currents.

On the Dhuwa side the *Milngurr* metaphor is applied. This sees the dynamic interaction of knowledge traditions as the interaction of fresh water from the land bubbling up in fresh water springs to make waterholes, and salt water moving to fill the holes under the influence of the tides. Salt water from the sea and fresh water from the land are eternally balancing and re-balancing each other. When the tide is high the salt water rises to its full. When the tide goes out fresh water begins to occupy the waterhole. *Milngurr* is ebb and flow. In this way the Dhuwa and Yirritja sides of Yolngu life work together.

In the Western scientific tradition, nature is taken as something quite different from knowledge. Scientific knowledge sees itself (and all other knowledge systems) as a representation of reality. This is in stark contrast to the Yolngu system which does not see nature and knowledge as distinct and different sorts of things. What is taken as important in scientific knowledge production is adjudication over true and good representations. While many involved with science adhere to representationalism, when we look at what scientists actually do in producing scientific knowledge, we see that this refers only to the end products, not the process of production of scientific knowledge. Here there is a robust reliance on material practices as constituents of knowledge.

In constructing their world people bring the material world into the social using practices which employ a wide variety of signing systems. There is a deconstructive moment which gives way to a reconstructive task within the practices of the production systems involved. Understanding the specific traditions of discourse and practice in which we are embedded as both the source of our frustrations and our aspirations invites us to engage our own and other's self-understandings critically.

On the Yolngu side it is seen that knowledge was in its 'architectural' elements laid down in the far past and that succeeding generations of Yolngu have been engaged in striving to formulate balance within this general framework. On the Western side renewal is seen as the task of understanding our traditions of discourse and practice as a source of both frustration and aspiration, and continual reformulation of the foundation upon which social action might be realized. Each envisages a continuing striving with ends as uncertain.

In both the Yolngu and scientific understandings of our practices of translation there is an assumption that human nature is unchanging, or changing in a way that points to a stable universal state. It is too limited a view of Yolngu knowledge production which sees it as essentially retrogressive, aiming to re-establish a time, commonly called 'The Dreaming' in English. This view mistakenly identifies 'The Dreaming' as a stable set of relations ideally suited to a supposed 'true human nature'. Similarly it is an unwarrantedly short-sighted view of scientific knowledge which sees it as progressing to the establishment of a perfect representation of nature which will enable the establishment of a perfected set of human relations. Compared to science we might want to characterize Yolngu knowledge as idealist (as distinct from empiricist), but it is also profoundly anti-representationalist, recognizing knowledge as constituted in linguistic and material practices.

This view specifically rejects representationalism, but this is not to reject science as a knowledge production system. On the contrary it is to focus on those very elements of science which are most important in the practical business of scientific knowledge production. Practicing scientists for the most part admit the role that their craft skills play; their work is painstakingly practical both in its abstract and material aspects. Fingers, numbers, minds, solids, liquids, gases, paper, pens, machines, theories, words, students, and technicians all mix together in laboratory practices, albeit in established, regular, and ordered ways. In the model of scientific knowledge production based on a non-representationalist view of knowledge, science does not start with nature but in practices with nature in laboratories.

Knowledge is a presentation of the life-world. In this sense it can be regarded as both a continually reconstituted framework through which the lifeworld is organized, structured, and configured and the (likewise continually reconstituted) techniques by which this is achieved.

The material practices which constitute a knowledge production system can be in forms which are institutionalized or non-institutionalized, and engage discourses which are narrative or argumentative, mythical or non-mythical, inscriptional or non-inscriptional, numeric or non-numeric, oral or non-oral methods of translation and re-translation, and presentation and re-presentation. Science has particular forms of material practice; other knowledge production systems exemplify others. As David Turnbull asserts, the significant differences between knowledge production systems lie in "the ways in which differing knowledge systems have developed techniques for enabling knowledge to move in space and time from the local site and moment of its production to other sites and generations".

In Bruno Latour's terms, in the processes of knowledge production allies are recruited and a knowledge claim becomes dehistoricized and decontextualized. It becomes established as a universal fact when people begin to treat it as a fact.

In science we understand the microworld as the laboratory where 'nature' is interrogated by 'society'. Here 'society' has agency, but not 'nature' which is passive, awaiting the informed gaze of scientists. Laboratories are institutionalized microworlds and signs elicited are inscribed through all manner of technology and incorporated into sub-texts and texts as inscriptions (graphs, lists of numbers etc.). It is a simplified context set apart from the bewildering complexity of everyday life.

The analogy in Yolngu knowledge production is *bunggul*, a time and place agreed upon by both Yirritja and Dhuwa agents (the fundamental categories of Yolngu life). Signs are elicited through the mutual interrogation of Yirritja and Dhuwa. These fundamental categories of Yolngu life are constituted by people and places, flora and fauna, words and songs, stories and metaphors, dances and graphic symbols. Everything, every person, and every place that matters in the Yolngu world is either Yirritja or Dhuwa, in much the same way as the Chinese characterize everything as *yin* or *yang*. It is roughly equivalent to assigning gender to nouns, as many languages do. Yirritja and Dhuwa contain entities which science would classify as natural, supernatural, and social.

I have previously compared English — a common language of scientific knowledge production — and the Yolngu language in terms of the types of entities on which sentences pivot. The rather startling result is that we see that the sorts of entities which Yolngu language use emphasizes are fundamentally different from the sorts of entities that English language has its users continually conjuring into existence.

This account of languages as systems of differentiation is different from the more common notions of language either as a neutral mirroring of the world or as a functionally derived social image. Languages necessarily adopt particular categories in making these representations. In this sense language can be regarded as material practice that pre-exists experience and a translation of experience into forms that actively shape the contexts that render experience meaningful.

Forms of rationalization and standardization by which conformity is wrought can generally be understood as taking the form of patterns which seem to arise 'naturally' from the world. They can be linguistic patterns or metaphors in the form of idealized narratives or visual images. In comparing scientific and Yolngu knowledge production systems we find that fundamentally different forms of rationalization are employed. In the knowledge production system of science the base ten number system, which is used in conjunction with a system of visual metaphors, provides a fundamental

Table 1: A comparative summary of contemporary scientific and contemporary Yolngu knowledge production systems

	Interrogative microworlds	Systems of differentiation logic in language	Forms of rationalization	Means of codification
Contemporary European derived technoscience	Institutional labs where 'the social' interrogates 'the natural'; signs inscribed graphically and numerically into written texts	A system which differentiates spatio-temporal objects	Base ten tallying recursion; visual metaphors (extensions of qualities); standard units for measurement	Numeric value; co-ordinated space numbered time
Contemporary Yolngu Aboriginal knowledge	Institutional *bunggul* where there is mutual interrogation of Yirritja and Dhuwa; signs are incorporated into written, oral, graphic, song and dance texts	A system which differentiates vectors within a matrix	Narrative metaphors of idealized journeys (*djalkiri*) and kinship recursion (*gurrutu*)	Balanced mesh of relatedness

form of rationalization. Using these forms quantification of everything can be achieved; numeric value can be ascribed. In contrast the forms of rationalization and standardization employed in Yolngu Aboriginal knowledge production use fundamentally different types of patterns. Here the pattern of kinship is idealized in a dual binary recursion. Just like the number system, the *gurrutu* system of Yolngu Aboriginal Australia is a pattern of names. This pattern is worked with a series of narrative metaphors which record idealized journeys across the landscape. These are the so-called song-lines or myth tracks of Aboriginal Australia, the *djalkiri*.

The Yolngu knowledge production world is characterized by a balanced mesh of relatedness. In this matrix (Table 1) the relation of every element to every other element of the Yolngu world can be precisely and economically elaborated. The elements of this code are provided by *gurrutu* and *djalkiri*.

Considering the model of technoscientific knowledge production first: in the interrogative microworld of the laboratory 'nature' is interrogated by 'society', and signs elicited. The signs are rationalized and codified using the common forms of numeric value, coordinated space, and numbered time. 'Society' and 'nature', encoded as value and arraigned in time and space, are the products of this knowledge production system, as well as the entities which constitute the system.

In Yolngu knowledge production, during *bunggul* there is a structured mutual interrogation of things *yirritja* by *dhuwa* agents, and of things *dhuwa* by *yirritja* agents. The signs elicited are constituted as entities by the pervasive ontology implicit in Yolngu language. Both sides identify and agree upon the incorporation of the signs into texts using *gurrutu* and *djalkiri*. Within the general framework of understanding the world as a balanced mesh of relationships, new texts are elaborated by appointed agents. While the new understandings begin to influence the balance between *yirritja* and *dhuwa* , and the internal workings of both of these halves of the world, there is always the possibility of re-negotiation. *Yirritja* and *dhuwa* as balanced halves of the world are both the products of the knowledge production system and its constituting elements.

HELEN WATSON-VERRAN

REFERENCES

Latour, Bruno. "Give Me a Laboratory and I will Raise the World." In *Science Observed*. London: Sage, 1983, p. 141–170.

Latour, Bruno. *Science in Action*. Milton Keynes: Open University Press, 1986.

Latour, Bruno and Steve Woolgar. *Laboratory Life: The Construction of Scientific Facts*. Princeton, New Jersey: Princeton University Press, 1989.

Turnbull, David. "Moving Local Knowledge." Paper presented at the conference *Understanding the Natural World: Science Cross-Culturally Considered*, Amherst, Massachusetts, June, 1991.

Watson, Helen, et al. "Australian Aboriginal Maps." In *Maps are Territories*. Ed. David Turbull. Geelong: Deakin University Press, 1989, p. 30.

Watson, Helen with the Yolngu community at Yirrkala, and D.W. Chambers. *Singing the Land, Signing the Land*. Geelong: Deakin University Press, 1989, p. 57.

Watson, Helen. "Investigating the Social Foundations of Mathematics: Natural Number in Culturally Diverse Forms of Life." *Social Studies of Science* 20: 238–312, 1990.

KNOWLEDGE SYSTEMS IN CHINA The science and technology developed in China often led the world before the fifteenth century. Statistics of significant discoveries in the world from the sixth century BC to the nineteenth century AD show that before the year 1500, discoveries made in China comprised more than half of the total (Guo and Guo, 1987). Then, the percentage dropped rapidly, and in the nineteenth century it became less than 1%.

The mode of development of science in China is, roughly speaking, a slowly progressed "mode of accumulation". It is not full of ups and downs like the Western saddle-shaped "mode of revolution".

In contrast to the West, Chinese scientific and technological achievements mainly belonged to the technique and empirical type. The Chinese were less inclined to use theoretical and experimental methods. The four best known inventions: compass, gunpowder, paper, and printing, were the most important technological achievements, which, when they came to Europe, exerted a great influence on the West.

The most developed sciences in ancient China were astronomy, mathematics, medicine, agronomy, and other related branches. This was determined by the needs of the agricultural Chinese society.

The characteristics of the development of science and technology in China were formed by factors that existed before the Christian era. At that time Chinese culture was already very different from Greek culture. There were both a different value-orientation and a different way of thinking, which had long-term influences over the directions of the development of Chinese society and Chinese science and technology.

In regard to value-orientation, the highest goal for Chinese ancient thinkers was in searching for the harmony and balance in one's mind, and in the relationship between Man and Man, Man and Society, and Man and Nature. The core of the Confucian thought is *jen* (loving people). Therefore, the only important criterion for learning is usefulness for humanity's purpose; other factors do not matter at all. This is clearly utilitarianism. In this regard, it is very different from ancient Greece, where study was for knowledge's sake only and without any practical intention.

Huishi was the only exception among Chinese ancient thinkers. But he was uniformly criticized by thinkers from all the other schools such as Confucians, Daoists, Mohists, and Legalists, and was suppressed by them. This contributed to the fact that the ancient Chinese were strong in technological achievements and relatively weak in theoretical ones. Even though the Mohists discussed almost all the problems contained in classical logic, they never made a single step towards formalization or axiomization. Therefore, they never came up with their own axiomatic system. The main reason for this was that logic, at that time, didn't seem to have any practical use.

The Chinese also developed mathematics to a very high level. For example, in the *Shushu jiuzhang* (Mathematical Treatise in Nine Sections) compiled by Qin Jiushao in 1247, there was a method of solving high degree equations which was the same as that developed by W. G. Horner (1786–1837). Qin was five centuries earlier than Horner. However, the Chinese never developed axiomatic systems similar to Euclidean geometry. Therefore, despite the fact that China had the most complete, most systematic astronomical records in the world at that time, and that she was rich in cosmological constructions, such as the *gaitian* (hemispherical dome), the *huntian* (celestial sphere), and the *xuan ye* (infinite empty space) hypotheses, she never produced any geometric cosmological models similar to Ptolemy's or Copernicus'. Chinese mathematics characteristically took practical use as its aim, and centered around calculation. Its earliest representative work is *Jiuzhang Suanshu* (Nine Chapters on the Mathematical Arts) written in the Western Han dynasty. Ancient Western mathematics was mainly geometry; ancient Chinese mathematics was mainly algebra. Today, when computing technique once again is the primary concern of so many mathematicians, the practical value of ancient Chinese mathematical inclination becomes more and more obvious.

The development of modern science and technology makes people realize that it is a double-edged sword; it can bring benefits as well as destruction. Therefore, its development must be controlled ethically and politically, under strict management and control. In this respect, Chinese value orientation is a very significant model.

In regard to the Way of thinking, Chinese traditional philosophy has always inclined to an organismic view of nature. Chinese philosophers regarded heaven, earth, and man as a united whole. This was fundamentally different from other ways of thinking, as there was no distinction between Object and Self. When French thinker L. Levi-Brühe (1857–1939) considered the Chinese traditional way of thinking primitive, he completely misunderstood. The Chinese idea of the unification of heaven and man mainly emphasizes that humans are only a part of heaven (nature); the Way of Heaven and the Way of Man are one and the same. Humans should be in a harmonious existence with nature, and these two should not be opposed to each other.

Chinese philosophy also claims that heaven, earth and

everything else are constituted by a pair of contradictions. Chinese philosophers use two symbols, *yin* and *yang,* to represent this pair of contradictions. *Yinyang* is a unity of contradictions. There is an antagonistic as well as a complementary and containing relationship between them. There is yang in yin and there is yin in yang. They can convert to each other under certain conditions. Therefore, in order to achieve harmony in nature, the stability of a society, and a long rule with eternal peace, as well as the health of individuals, "the doctrine of the mean" has to be applied to keep a dynamic balance between yin and yang. The philosophy advocates that nothing should be overdone, and all actions should be done appropriately. This way of thinking is different from the dichotomy in two-valued logic, in which every proposition is either ture or false. Its symbolic representation is the yin-yang fish taiji pattern.

A mechanistic mode of thought has never been dominant throughout Chinese history. This organismic way of thinking affected the development of Chinese science to a very large degree. It can be found in ancient Chinese cosmology, astronomy, medicine, agronomy, and social theories. It contributed tremendously to keeping peace and stability during the Middle Ages. Some believe that Chinese society in the Middle Ages was the most mature, most developed society in the world at that time. It lasted twice as long as that of the Europeans. However, at the same time it restrained the growth of capitalism, and served as one of the most important reasons that prevented China from having a scientific revolution.

According to the Chinese traditional view, the origin of the universe is the result of the changes of yin and yang. Imbalances of yinyang caused natural disasters like earthquakes, as well as social disturbance. The yinyang theory is also the theoretical foundation of Chinese medicine. It considers the human body as a whole. If a body loses its balance of yin and yang, the person becomes sick. The task of a medical doctor is to use every method to help the patient to resume this organismic dynamic balance. However, Chinese medicine is not a two-value system; it is a multiple value system, totally different from the Western medical system. But these two can go hand in hand and complement each other, thereby enriching the treasury of our medical knowledge.

The Chinese traditional way of thinking is an organismic one which emphasizes integrity and perceivability, and allows fuzziness. It is, relatively speaking, weak at analytical thinking and it does not value axiomatic systems. Generally it is not good for macrocosmic descriptions in the area of physics, especially mechanics. However, it is closer to the microcosmic world of modern science, such as the area of quantum mechanics, or to descriptions of cosmological, artificial intelligence, and social systems. The return from accuracy back to fuzziness is one of the main indications of contemporary science. Nowadays many scientists realize that this Chinese way of thinking could indeed bring helpful inspirations to the progress of modern science.

YANG DI-SHENG

REFERENCES

Guo Jianron and Guo Guangyin. "On the Effects of Chinese National Psychology and Tradition on the Cultural Development of Scientific Technology." *Central Nationality Institution Journal*, 1987.

History of Chinese Scientific Thinking. Zhejiang, China: Education Publishing House, 1992.

Needham, Joseph. *The History of Chinese Scientific Technology. Selected Works of J. Needham* (Chinese Edition). Liaoning, China: Science and Technology Publishing House, 1986.

Scientific Traditions and Culture; on the Causes for the Falling Back of Modern Chinese Scientific Technology. Proceedings of a conference. Shanxi, China: Science and Technology Publishing House, 1983.

Wu Wenjun. "On the Study of the History and the Present Situation of Chinese Mathematics." *Journal of Natural Dialectics* 4, 1990.

See also: Mathematics in China – Algebra in China – *Yinyang* – Medicine in China – Calculation: Chinese Counting Rods – Liu Hui and the *Jiuzhang Suanshu* – Environment and Nature in Chinese Thought

KNOWLEDGE SYSTEMS OF THE INCAS On the fourteenth of November, 1533, Francisco Pizarro and a small Spanish army entered the town of Cajamarca. On the next day they took Atahuallpa, the last Inca king, prisoner after he had come to their encounter with a large army. It was the first and the last time that the Spaniards received a glimpse of the independent Inca state which had conquered an empire into southern Colombia and northern Argentina and Chile. Perhaps the empire had already been weakened by the civil war that Atahuallpa had won over his brother Huascar, the crowned king. While in prison, Atahuallpa had Huascar killed, and after some months in Cajamarca the Spaniards executed Atahuallpa. But more than these events, it was the possession of superior arms, including horses, and the help of native troops choosing their side, that allowed the Spanish army to cross the country almost without resistance and to enter the capital of Cuzco a year later. Here they set up Manco Inca as a puppet king. Less than two years later Manco Inca fled Cuzco and withdrew to the eastern slopes of the Andean mountains where heavy forest made it difficult to defeat him and his successors. In 1572, thirty-six years later, the last Inca king, Tupac Amaru, was captured and executed in Cuzco. It was the end of Inca resistance, and the viceroy Francisco de Toledo could consolidate a colonial government that ruled until the early nineteenth century when the Andean countries of Colombia, Ecuador, Peru, Bolivia, Argentina and Chile declared their independence.

Some of the first Spaniards that described their participation in the conquest of the Inca empire included in their accounts valuable information about its people. But only after some twenty years did the first chronicles appear, by Juan de Betanzos [1551],[1] one of the first Spaniards to learn the Inca language of Quechua well, and Pedro Cieza de Leon [1551]. These give an integrated view of Inca culture, referring to aspects of history, social and political organization, agriculture and husbandry, and religion. Although much of this information derived from the memory of Inca nobles in Cuzco, it was not based on direct observation of a thriving culture with all its rituals and feasts. Moreover, Andean civilization never developed a system of writing, recognized by the chroniclers as such, through which we would be able to hear the independent voices of its peoples on what they thought about their cosmos, their gods and goddesses, and their myths and rituals, without being directed by the questions of representatives (administrators, lawyers, priests) of Spanish domination. Almost all the documentation needed for Inca administration was recorded on knotted cords with numerical information organized in bundles, called *quipus*. Hundreds of these *quipus* have been found in graves on the desert coast of Peru. Most are from Inca times but some are from the earlier Wari culture, demonstrating a different system of knots and chords. We can study the sophistication of Inca mathematics, analyze some of the information, mostly on bureaucratic and economic but also on ritual matters, as released by *quipu* specialists through Spanish prodding, and be aware of additional information on the Inca past and culture memorized by the specialists. But most of this research is still in its infancy (Ascher, 1981).

Other promising sources of information on Inca culture are the many textiles that survived in coastal graves, were conserved in heirlooms from colonial times, and were represented in colonial art. The Incas inherited sophisticated techniques of weaving from earlier cultures, but the style of their art was very much their own. Certain types of male tunics show highly standardized geometric patterns. Others include a wide variety of square designs, *tucapus*, that in their distribution as possible signs with meaning are reminiscent of a writing system, although only for a few *tucapus* are there clues to their interpretation. None the less, a careful comparison with written sources on the use and iconography of Inca textiles can help us to establish this art as an independent, prehispanic source of documents on aspects of Inca culture.

The colonial chroniclers provide us with the only written knowledge of Andean culture at the time of the Spanish conquest. Beginning with Betanzos and Cieza, they organize their material in a historical framework. Thus they tell how Manco Capac, the first Inca king, came out of a cave some fifty kilometers south of Cuzco with three brothers and four sisters and founded the future capital. At the time of his writing, the conquered people in the valley of Cuzco still remembered their pre-Inca past. From here the Incas first conquered surrounding valleys and gave their kingdom an administrative structure recognizing the old local lords as lower-ranked members of their own nobility. This nobility consisted of different ranks defined according to criteria of kinship to the reigning king and queen; thus the local lords were defined by the chroniclers as "Incas by privilege". Like the Inca nobility, they themselves were exempt from contributing labor to Cuzco but their subjects were not. Moreover, they and their subjects participated in the ritual organization of the town. From this base the empire was conquered.

When the Spaniards were still in Cajamarca, they had some inkling about the enormous extent of the Inca empire but no idea how it was obtained and who had been the successive conquerors. On their arrival in Cuzco, they reconstructed the last two royal successions before the Spanish conquest. Twenty years later a certain consensus was arrived at of the earlier dynasty. But as this list of kings was not based on historical records, its reconstruction conformed more to a pattern of western than of indigenous ideas about royalty, succession, conquest, and history. One of the most arduous tasks of research deals therefore with the questions of evaluating how the Spaniards reconstructed the conquest of the Inca empire, how their Inca informants presented this conquest to them, and how the conquered peoples, especially those outside the original kingdom, understood how this Inca domination had occurred.

Betanzos, Cieza, and others after them, especially Pedro Sarmiento de Gamboa [1572], placed as the central person in the epic of Inca conquests the ninth king, called Pachacuti Inca. Cuzco had been attacked by the kingdom of the Chancas during the reign of his father, Viracocha Inca. As the latter fled from town, Pachacuti Inca defeated the enemies and was crowned king even though he had not been designated as crown prince. He rebuilt the city, reorganized the people who lived in its valley, elevating in rank those

[1] Square brackets after the name of an old author indicate the date of writing or of first publication.

who had helped him in the defense and lowering others, and he established different institutions of government. Thus he divided the valley into districts based on the system of agricultural irrigation; he organized a system of worship of those ancestral mummies that he decided to recognize as such; he assigned administrative posts to relatives and other nobles; and he established the calendar of state celebrations.

The story of Pachacuti Inca has all the characteristics of a foundation myth and with its help we can get an understanding of how the histories of the Incas before and after this king were constructed in colonial times. Memories of earlier times were organized according to the ancestral system, wherein a ranking order of mummies was used to establish a dynastic order. But the dynastic sequence after Pachacuti Inca cannot be taken without critical examination either. The many conquests of the Inca empire towards Colombia and Chile and Argentina were said to be accomplished mostly by Pachacuti Inca's son, Tupa Yupanqui, and grandson, Huayna Capac. (The latter died some seven years before the Spaniards arrived). But the Inca informants to Spanish questioning recognized that some rulers after Pachacuti Inca had been suppressed from the official list of kings. Thus it is very possible that the history of Inca conquest was not as short as suggested. Here the memories of the conquered peoples can help us to arrive at a more realistic picture of Andean history.

Intensive documentary research beginning some fifty years ago on various local kingdoms integrated into the Inca empire allows us to get some idea of their cultural and institutional differences and their ties to earlier peoples whose art and archaeology we study. Perhaps the most important kingdoms that the Incas conquered were those of the Chimu on the north coast of Peru and around Lake Titicaca. All had had a prestigious history of their own. The Chimu had built a kingdom along the coast from about the valley of Lima to near the present-day border of Peru with Ecuador. Their capital of Chanchan had been the largest city in South America. They were the descendants of the Mochica people (ca. AD 0 to 700) whose pottery with realistic painted scenes established one of the high points of Andean arts. The Chimu are the only kingdom on which we have some dynastic information from before the Spanish and even Incaic conquests. Recently, much work has been accomplished reconstructing their culture and political organization (Moseley and Cordy-Collins, 1990).

Around Lake Titicaca the cultures of Pucara (ca. 200 BC–AD 400), north of the lake, and Tiahuanaco (ca. AD 200–900) south of it once flourished. The impressive stone ruins of their ceremonial centers played an important role in Inca mythology. Chroniclers constructed the myth of a creator god coming from Lake Titicaca or Tiahuanaco, called Tunupa, around the lake and Viracocha in Cuzco. In the sixteenth century three different languages were spoken: Aymara, that extended to the west of Cuzco and near Lima, Puquina, and Uru. Puquina was probably the most important language at the time of Tiahuanaco, and Uru may have been there and in the high plains of Bolivia even earlier. But since colonial times Uru was retained mostly by people with a fishing and hunting economy, and Puquina died out. Today, Aymara is mostly spoken by people living south and east of Lake Titicaca.

Most of our historical information on Aymara culture derives from colonial administrative documents. Among these are, however, two of the most important ones for research on local government and economic organization in the Andes at the time of Spanish conquest. The first, from 1567, describes the former kingdom of the Lupaca south of Lake Titicaca. Its economy was based not only on highland crops grown there but also on crops from distant valleys down on the Pacific coast. Instead of obtaining these lowland crops through trade it had sent its own people to grow them. The second document, from 1568–70, concerns a former kingdom north of the lake; here cultivation of coca in the lower valleys east of the Andes had been important.

In conquering Chimu and Lake Titicaca, the Incas had concentrated their rule near former centers of power. In other cases they established new provincial capitals away from such centers. The most important and best researched example of this kind is the city of Huanucopampa in central Peru near the eastern slopes of the Andes. It formed part of a chain of provincial capitals going north from Cuzco, including Vilcas Huaman, Bombon, Huanucopampa, Cajamarca, Tumibamba (present day Cuenca in Ecuador) and Quito. Huanucopampa was perhaps the largest of all. It is also best preserved, as its location on a high plateau did not attract later Spanish settlement. Huanucopampa expresses in an admirable way the basic ideas of Inca town planning. Around a huge rectangular plaza the four quarters of town were organized, each with its own pattern of social activities. In the center of the plaza stood the *Ushnu*. Chroniclers describe the *Ushnu* in the plaza of Inca towns as a platform or pyramid near a small round structure of stone where offerings of corn beer were pledged to the Sun god, the ancestors, and the forces of the underworld. Cuzco had such a ritual complex but the platform was probably a temporary structure used only during special feasts. But in Huanucopampa the large platform was built of stone and could have served also during military parades. On the east side of the plaza and aligned with the *Ushnu* stood the palace of the Inca governor. The king probably resided here when passing through town. It contained three courtyards, as one chronicler also describes for a royal palace in Cuzco. Thus we can assume that Inca and non-Inca subjects could enter the first and largest courtyard, that only nobility entered the second courtyard, and that the family of the governor or the king had exclusive access to the third and smallest courtyard to which were attached their living quarters. There was a straight road leading from the plaza to the inner court and the various gates through which visitors had to pass still stand, each crowned by two stone lions facing one another like on the lion gate of Mycenae. Above the town, on its outskirts, can still be seen the hundreds of storehouses for the tribute brought by the people of the province of Huanucopampa. Two extensive documents from 1562 describe in detail the demographic composition of this province and the contributions that its villages had to make to the state. Huanucopampa and its province have provided us with the best opportunity for research on the organization of an Inca province and its economy.

Many documents now being studied allow us to reconstruct the local socio-political situations in Inca times. Regular forms of organization, like those consisting of moities and of three or four local groups, reveal patterns of regional variation which have survived in many places. Although the social functions of these organizations may have changed after almost half a millennium, they retain their ritual importance. A form of state organization that could be combined with the ones already mentioned consisted of bringing together families in numbers of 10, 50, 100, 500, 1000, 5000, 10,000, and finally 40,000, the latter number expressing an ideal way of organizing an Inca province. In a similar way, the population of Cuzco itself consisted of ten *panacas* and ten *ayllus*, the first referring to the relatives of the king who administered these groups. *Panacas* and *ayllus* played a crucial role in the calendrical rituals of the capital.

With the Inca conquests of the empire, the administrative model of Cuzco was replicated in the organization of the whole empire. For instance, the four quarters of town, the *suyus*, were extended into the four provinces of the empire. Inca nobles and lords of the Incas-by-privilege became imperial administrators (*tucricuc*) supervising local lords (*curacas*) who themselves governed their territories like the king in Cuzco. But *curacas*, or their representatives or sons, also had to visit Cuzco. Periods of four months' or a year's residence in town are mentioned, but we do not known how regularly they were repeated. A specially important time for such visits was after harvest, when lords came with their tribute and presents for the Inca king. He consulted with them about their governments and discovered who did not want to comply with his obligations. Presents that the king received from one province were awarded by him to the lords of another.

Of particular interest is the imperial (re-)organization of the *acllas*, the "chosen ones". Girls were selected in local organizations according to set criteria of beauty and assigned tasks, like weaving and making corn beer, in houses built for that purpose. From there they could be reselected and sent to provincial capitals and even to Cuzco. Their organization touched on almost all aspects of Inca culture. Ranking was expressed in an idiom of age-classes, and these applied also to men from their time of initiation and marriage to the time that they turned over their social obligations to their sons. Some *acllas* were married off by their *curaca*, an Inca governor or the king, either as a principal or as a secondary wife, or were dedicated to a role in Inca religion. When, with the Spanish conquest this organization fell apart, one way *acllas* were dealt with was by converting them into nuns. But their roles were not really comparable to those of nuns, because this female organization with the queen at its head in most respects paralleled the male hierarchy. The concept of *aclla* also played a central role in the imperial organization of sacrifice, the *capac hucha*, as this practice had grown out of the calendrical organization of local rituals in Cuzco. Children were selected as *acllas* and dedicated to various but always specifically recorded purposes. Thus they were sacrificed, including couples of a boy and a girl. In their own locality, sacrifice might be for the the purpose for obtaining good pottery clay. In far-away places sacrifices commemorated important events, such as winning a battle. In Cuzco it was part of a great state ritual, either for the time of planting, of the December solstice or of the harvest. *Aclla* sacrifice was the culminating act in a organization of rituals that crystalized the cosmological values of Inca government.

Inca religion forms an integral part of Andean religion in general as observed in various parts of the empire. Because of the importance of Cuzco as imperial capital and the intellectual interest of the Spanish in its history, we are also well-informed about Inca religion. However, for reasons of colonial history we also have extensive accounts of myths, rituals, and religion in the mountain provinces of Huarochiri and Cajatambo, central Peru, where in the early seventeenth century the Spaniards believed they observed a resurgence of indigenous beliefs that they tried to eradicate. While the original belief systems in Cuzco and the two other provinces may not have been that wide apart, the circumstances of their recording were very different. Chroniclers in Cuzco, among whom there were well-educated people, came to the Inca past through their own curiosity and were informed by Inca nobles. Initially, both parties may have looked for a common intellectual basis. However, the religious expressions in Huarochiri and Cajatambo were seen as a rebellious return to pagan beliefs in villages after more than eighty years of teaching that Christianity was the only true religion. I will deal here with the problem of studying Inca religion, although many myths and rituals may have been recorded in Huarochiri and Cajatambo in a form closer to prehispanic reality.

Unlike the situation of extensive pantheons in Aztec and Maya religions and of illustrations of gods, goddesses, and priests dressed as such in prehispanic and early colonial codices and chronicles, the Spaniards described only a few gods and goddesses for the Incas, and even those were not visualized in any way in Inca art. Nevertheless, documents and chronicles are replete with names of sacred places in the form of mountains and rocks, lakes and springs, and all other kinds of natural and man-made objects with ritual significance. Foremost were three male gods, the Sun, the Thunder, and Viracocha and one female god, Pachamama the "mother earth". The ruling king was considered to be the son of the Sun, but from Cuzco we have no mythical description of the latter's actions. The Thunder god was seen in various parts of the country as an active mountain god, but in Cuzco only indirect references to his deeds occur. The god Viracocha seems to have corresponded to a type of mythical figure, known elsewhere in the Andes under various names. The myth of the god Coniraya in Huarochiri, there compared to Viracocha, has perhaps best preserved this prehispanic image. He pursues a woman along a river down to the Pacific Ocean. Here she escapes as he cannot follow her any further. Along the way and through various acts that present him as a trickster, he defines the interests of people in their land for cultivation, water for irrigation, and wild animals for use in rituals. An original way of representing Viracocha in Cuzco, although found in a late chronicle, is as a force of nature, a giant, who during a month of heavy rain comes down the Villcanota river near Cuzco, flowing from the southeast to the northwest, threatening to destroy it. But Betanzos and Cieza describe Viracocha as a god who brought forth from the island of Titicaca (the rock of the cat), in the lake of the same name southeast of Cuzco, the sun, the moon and the ancestors of different peoples. Viracocha sends these underground to their local places of origin, from where they reemerge to establish local government. Viracocha himself also travels northwest, following the Villcanota river near Cuzco, but continues until he arrives at the sea in Ecuador. There he disappears, not into the sea but over it towards the horizon. This version of the myth was well adapted to the imperial interests of the Incas and may have also received a colonial reinterpretation because of the Spanish interests of including Peru in a universal empire. Soon the exploits of Viracocha were also phrased in terms of those of two early apostles, Saint Bartholomew or Saint Thomas, who had been said to have traveled through the country bringing Christianity long before the Spaniards did. According to this colonial reinterpretation, Viracocha was the creator god of the Incas. But his mythology became embedded also, in fragmentary form like that of other gods, in the legendary history of kings. Stories were told of the eighth king, Viracocha Inca, that were more fitting for his namesake the god Viracocha, and in a similar vein his son Pachacuti Inca was associated with the Thunder god. In fact, one chronicler recognizes these relationships as such when he says that each king took as his name of nobility that of his god. Pachamama was the great goddess of the earth, but again no stories of her exploits are told. The myths of some Inca queens are more interesting, considered in their roles as ancestral deities. Probably, the attention to Pachamama was mostly developed in early colonial times.

Of more immediate interest for the study of Andean religion are the numerous sacred places, the *huacas*, mentioned in various local documents. Sometimes myths are told about them as actors like humans. Two indigenous chroniclers from the early seventeenth century give us extensive hierarchies of *huacas* of more than local importance, although our sources tell us little about their possible organization. In about 1560, however, the Spaniards had become aware that in Cuzco the cult of the *huacas* was organized according to a highly sophisticated scheme of directions as seen from the temple of the Sun in the town. Families each took care of the cult of a particular *huaca* and larger social groups were associated with the *huacas* along one direction, the *ceque*, or with groups of *ceques*. The organization of the *ceques* served various social purposes. For instance, the topographical description of the valley was of interest for land distribution in agriculture, irrigation, husbandry, and mining, especially in terms of quarries of stone used for building purposes. Through the *ceque* system a formal description can be given of Cuzco's political organization including its calendrical organization of state rituals as defined in terms of its system of astronomical observation. Polo de Ondegardo, the chronicler who discovered the *ceque* system in Cuzco, mentions that other villages, towns, and provinces of southern Peru and Bolivia also had their *ceque* systems. Some interregional *ceques* are described, of use in the imperial system of *capac hucha* sacrifice, that suggests a hierarchy and network of *ceque* systems. *Ceque* systems were recorded on *quipus*. They supported an Andean way of reflecting in an abstract way on cultural values.

Andean civilization did not direct its interests towards the use of writing like Mesoamerican civilization did. Thus the intellectual aspects of Inca culture are difficult to grasp. But so much can still be studied of early Andean practices through those of their descendants, as in techniques of agriculture and weaving, that it will become possible to define the originality of Andean civilization.

R. TOM ZUIDEMA

REFERENCES

Ascher, Marcia and Robert Ascher. *Code of the Quipu. A Study in Media, Mathematics, and Culture*. Ann Arbor: University of Michigan Press, 1981.

Hemming, John. *The Conquest of the Incas*. London: Abacus, 1972.

Kendall, Ann. *Everyday Life of the Incas*. New York: G. P. Putnam's Sons, 1973.

MacCormack, Sabine. *Religion in the Andes. Vision and Imagination in Early Colonial Peru*. Princeton, New Jersey: Princeton University Press, 1991.

Masuda, Shozo, Izumi Shimada, and Craig Morris, eds. *Andean Ecology and Civilization. An Interdisciplinary Perspective on Andean Ecological Complementarity*. Tokyo: University of Tokyo Press, 1985.

Morris, Craig, and Donald E. Thompson. *Huánuco Pampa. An Inca City and Its Hinterland*. London: Thames and Hudson, 1985.

Moseley, Michael E. and Alana Cordy-Collins, eds. *The Northern Dynasties. Kingship and Statecraft in Chimor*. Washington, D.C.: Dumbarton Oaks Research Library and Collection, 1990.

Pärssinen, Martti. *Tawantinsuyu. The Inca State and Its Political Organization*. Helsinki: Suomen Historiallinen Seura, 1992.

Urton, Gary. *The History of a Myth. Pacariqtambo and the Origin of the Incas*. Austin: University of Texas Press, 1990.

Zuidema, R. Tom. *Reyes y Guerreros. Ensayos de Cultura Andina*. Ed. Manuel Burga. Lima: Fomciencias, 1989.

Zuidema, R. Tom. *Inca Civilization in Cuzco*. Austin: University of Texas Press, 1990.

Zuidema, R. Tom. "At the King's Table. Inca Concepts of Sacred Kingship in Cuzco." In *Kingship and the Kings*. Ed. Jean-Claude Galey. London: Harwood Academic Publishers, 1990, pp. 253–78.

Zuidema, R. Tom. "Guaman Poma and the Art of Empire: Toward an Iconography of Inca Royal Dress." In *Transatlantic Encounters. Europeans and Andeans in the Sixteenth Century*. Ed. Kenneth J. Andrien and Rolena Adorno. Berkeley and Los Angeles: University of California Press, 1991, pp. 151–202.

See also: *Quipu* – Textiles – Mummies – Calendar – City Planning – Stonemasonry – Irrigation – Time

KNOWLEDGE SYSTEMS IN INDIA Traditionally all knowledge in India has been traced to the Vedas. The Vedas are considered to be divine revelation. They were organized into four major branches: *Ṛgveda*, *Yajurveda*, *Sāmaveda*, and the *Atharvaveda*. Various other branches of knowledge grew up as auxiliaries that were to be developed in order to interpret and put to practical use the material of the Vedas.

There were a total of fourteen *Śāstras* or branches of knowledge: the four Vedas, the four Upavedas (auxiliary Vedas), and the six Vedāṅgas (parts of the Vedas). The four Upavedas were (1) *Āyurveda*, literally "The Science of Life", which constituted the medical system; (2) *Arthaśāstra*, which constituted state craft and political theory; (3) *Dhanurveda*, literally, archery, but practically constituting the art of warfare in its varied aspects, and (4) *Gāndharvaveda*, constituting music, drama, and the fine arts.

Similarly, the knowledge systems required for understanding, interpreting, and applying the Vedas were organized into six branches called *Vedāṅgas*, literally " the limbs of the Vedas", with the Vedas personified in a human form. The six *Vedāṅgas* are *Vyākaraṇa* (Grammar), *Chandas* (Metrics), *Śikṣā* (Phonetics), *Nirukṭa* (Etymology), *Kalpa* (Ritual), and *Jyotiṣa* (Astronomy and Mathematics).

These *Vedāṅgas* were essential, since the Vedas had to be understood correctly (needing etymology and grammar), pronounced and chanted accurately (needing metrics and phonetics), and used properly in various contexts (needing ritual), and the times for these performances had to be computed correctly, requiring the knowledge of computation of the flow of time and of planetary movements (needing astronomy and mathematics). Even though the *Śāstras* originally evolved in the context of the Vedas, they also developed an independent identity and their own corpus of literature and applications that extended well outside the originally formulated requirements of the Vedic context.

In later periods the list of *Śāstras* became much larger and the area covered was much wider. For example, in his famous text *Kámasūtra*, the author Vātsyāyana provides a list of sixty-four arts with which any scholar should be familiar.

Since the various branches of Indian knowledge systems are extremely diverse, we will focus upon a few that can best illustrate some characteristic different features. These are: (1) the fact that linguistics occupied a seminal place even for exact sciences (unlike Western knowledge systems); (2) the nature of theorization and theory building in Indian tradition; (3) the algorithmic nature of Indian computation, and (4) the sociology of organization of knowledge — the "classical" and the "folk" streams and their interrelation.

In any scientific discourse it is essential to achieve precision and rigor. In the Western tradition, the geometry of Euclid is considered the paradigm of an ideal theory, and various other branches of knowledge tried to emulate Euclid by setting out their knowledge on the basis of a formal axiomatic system. In contrast, in Indian tradition, an attempt was made to use natural language and to refine and sharpen its potential by technical operations so that precise discourse was possible even in natural language. This is so, particularly in Sanskrit, where we find that even the most abstract and metaphysical discussions regarding grammar, mathematics, or logic are still written in natural language. In Indian knowledge systems, it is the science of linguistics that occupied the central place which, in the West, was occupied by mathematics.

Linguistics

Linguistics is the earliest of Indian sciences to have been rigorously systematized. This set an example for all the other Indian sciences. Linguistics is systematized in *Aṣṭādhyāyī* — the text of Sanskrit grammar by Pāṇini. The date of this text is yet to be settled with any certainty. However, it is not later than 500 BC. (The dates mentioned here are those based on Western scholarship. An indigenous Indian dating and chronology in this matter have yet to be established.) In the *Aṣṭādhyāyī*, Pāṇini achieves a complete characterization of the Sanskrit language as spoken at his time, and also manages to specify the way it deviated from the Sanskrit of the Vedas. Given a list of the root words of the Sanskrit language (*dhātupatha*) and using the aphorisms of Pāṇini, it is possible to generate all the possible correct utterances in Sanskrit. This is the main thrust of the generative grammars of today that seek to achieve a purely grammatical description of language through a formalized set of derivational strings. It is understandable that until such attempts were made in the West in the recent past, to the Western scholars the Paninian aphorisms (*sūtras*) looked like nothing but some artificial and abstruse formulations with little content.

Science in India seems to start with the assumption that truth resides in the real world with all of its diversity and complexity. Thus for the linguist what is ultimately true is the language as spoken by the people. As Patañjali, a famous grammarian who wrote a commentary on Pāṇini's *Grammar* emphasizes, valid utterances are not manufactured by the linguist, but are already established by practice in the world. Nobody goes to a linguist asking for valid utterances, the way one goes to a potter asking for pots. Linguists do make generalizations about language as spoken in the world, but these are not the truth behind or above the reality. They are not the idealization according to which reality is tailored. On the other hand, what is ideal is real, and some part of the real always escapes our idealization of it. It is the business of the scientist to formulate these generalizations, but also at the same time to be attuned to the reality, to be conscious if the exceptional nature of each specific instance. This attitude seems to permeate all Indian science and makes it an exercise quite different from the scientific enterprise of the West.

Astronomy and Mathematics

Indian mathematics finds its beginning in the *Śulbasūtras* of the Vedic times. Purportedly written to facilitate the accurate construction of various types of sacrificial altars of the Vedic ritual, these *sūtras* lay down the basic geometrical properties of plane figures like the triangle, the rectangle, the rhombus, and the circle. Basic categories of the Indian astronomical tradition were also established in the various *Vedāṅga Jyotiṣa* texts.

Rigorous systemization of Indian astronomy begins with Āryabhaṭa (b. AD 476). His work *Āryabhaṭīya* is a concise text of 121 aphoristic verses containing separate sections on basic astronomical definitions and parameters; basic mathematical procedures in arithmetic, geometry, algebra, and trigonometry; methods of determining the mean and true positions of the planets at any given time; and descriptions of the motions of the sun, moon, and planets along with computation of the solar and lunar eclipses. After Āryabhaṭa there followed a long series of illustrious astronomers with their equally illustrious texts, many of which gave rise to a host of commentaries and refinements by later astronomers and became the cornerstones for flourishing schools of astronomy and mathematics. Some of the well known names belonging to the Indian tradition are: Varāhamihira (d. AD 578), Brahmagupta (b. AD 598), Bhāskara I (b. AD 629), Lalla (eighth century AD), Muñjāla (AD 932), Śrīpati (AD 1039), Bhāskara II (b. AD 1115), Mādhava (fourteenth century AD), Parameśvara (ca. AD 1380), Nīlakaṇṭa (ca. AD 1444), Jyeṣṭhadeva (sixteenth century AD), Gaṇeśa, and Daivajña (sixteenth century AD). The tradition continued right up to the late eighteenth century, and in regions like Kerala, original work continued to appear until much later.

The most striking feature of this tradition is the efficacy with which the Indians handled and solved rather complicated problems. Thus the *Śulbasūtras* contain all the basic theorems of plane geometry. Around this time Indians also developed a sophisticated theory of numbers including the concepts of zero and negative numbers. They also arrived at simple algorithms for basic arithmetical operations by using the place-value notation. The reason for the success of the Indian mathematician lies perhaps in the explicitly algorithmic and computational nature of Indian mathematics. The objective of the Indian mathematician was not to find "ultimate axiomatic truths" in mathematics, but to find methods of solving specific problems that might arise in astronomical or other contexts. The Indian mathematicians were prepared to set up algorithms that might give only approximate solutions to the problems at hand, and to evolve theories of error and recursive procedures so that the approximations might be kept in check. This algorithmic methodology persisted in the Indian mathematical consciousness until recently, so that Ramanujan in the 20th century might have made his impressive mathematical discoveries through its use.

Āyurveda: The Science of Life

The third major science of the classical tradition is *Āyurveda*, the science of life. Like linguistics and astronomy this finds its early expression in the Vedas, especially the *Atharvaveda*, in which a large amount of early medicinal lore is collected. Systemization of *Āyurveda* takes place during the period from the fifth century BC to the fifth century AD in the *Caraka Saṃhitā*, *Suśruta Saṃhitā* and the *Aṣṭāṅga Saṅgraha*, the so-called *Bṛhat-trayee* texts which are still popular today. This is followed by a long period of intense activity during which attempts are made to refine the theory and practice of medicine. This process of accretion of information and refinement of practice continued right up to the beginning of the nineteenth century.

Folk and Classical Traditions

There exists a vast amount of knowledge which represents the wisdom of thousands of years of observation and experience. While in any given area (such as medicine) there may be a body of experts or learned professionals, knowledge also prevails in more diffuse or scattered forms. In Indian tradition, it seems to be a general principle running through all types of learning, that knowledge can and does prevail in various forms and also gets communicated in many ways.

The general picture that emerges seems to be that the "classical texts" in any area of learning may set out broad general principles as well as their application in a given context, say a particular region of the country. But in various different contexts or regions, knowledge gets expressed based on the given situation, and the generalities get adopted, modified, or even overridden sometimes based on the specificity. This can perhaps best be illustrated in the case of medicine, where classical medical texts themselves deal with this issue. A classical text of *Āyurveda* such as *Caraka Samḥitā* expounds general principles of drug action on the six factors: *Dravya*, *Guna*, *Rasa*, *Vīrya*, *Vipāka*, and *Prabhava*. It also discusses remedies for several diseases and lists specific drugs. These may get modified to suit local conditions. In any recipe for a drug, one can substitute a non-principal component with an equivalent, which may be listed in the text or selected on the basis of the principle of *Rasa*, *Vīrya*, etc. From time to time traditional physicians produce texts and manuals which set out prescriptions for drugs in any given area based on what is available and suitable to the requirements of that area. For example, the text *Rājamṛgaṅka* lists 129 recipes, and in his foreword the editor states that it is a compilation that must have been made by a *vaidya* (physician) from Tamil Nadu, since it contains recipes based on herbs readily available in Tamil Nadu. Such recipes are not only easier to formulate, but they are also more suited to the area, in accordance with Caraka's dictum "For a person who belongs to a particular country or a region, herbs from the same region are most wholesome".

The fact that it is the particularity of the context that is the overriding consideration and that Sastric (i.e. scientific) principles are to be considered as precepts and guidelines and not applied in a mechanical or legalistic manner is clearly stated in many classical texts. "A Vaidya who comprehends the principles of *Rasa*, etc. would discard treatment if not wholesome to the patient in a given situation, even if it is prescribed in the texts; on the contrary he would adopt treatments that are helpful to the patient, even if they do not find a mention in the texts".

It is also interesting to note what the texts of *Āyurveda* say about folk knowledge. The *Caraka Saṃhitā* states that "the goatherds, shepherds, cowherds and other forest dwellers know the drugs by name and form." Similarly, the *Suśruta Saṃhitā* states "one can know about the drugs from the cowherds, tapasvis, hunters, those who live in the forest and those who live by eating roots and tubers".

This is an overview of Indian knowledge systems and does not go into the details of achievement in a variety of areas, particularly those pertaining to material sciences. Our attempt is to highlight basic characteristics of these knowledge systems, particularly in those respects where they differ from their modern counterparts.

A.V. BALASUBRAMANIAN

REFERENCES

Balasubramanian, A.V. and M. Radhika. *Local Health Traditions: An Introduction.* Madras: Lok Swasthya Parampara Samvardhan Samithi, 1989.

Cardona, G. *Pāṇini: A Survey of Research.* The Hague: Mouton, 1976.

Srinivas, M.D. "The Methodology of Indian Mathematics and Its Contemporary Relevance." *PPST Bulletin-S* 12: 1–35, 1987.

See also: Āyurveda – Mathematics in India – Astronomy in India

KOREAN SCIENCE Korea is a nation with five thousand years of history. Located on the periphery of East Asian civilization, it was greatly influenced by China. However, Koreans have developed a culture and tradition independent from that of the Chinese. At the same time, the history of Korean science can be regarded as a branch or an adaptation of Chinese traditional sciences, although, in most cases, Chinese sciences and technology were not adopted in their original form.

Korea had its own unique Paleolithic Age some five hundred thousand years ago. Some of the tools of the Paleolithic Age found in Korea are not found in any other parts of East

Asia. The Korean people of the Neolithic Age differed from the Chinese people in that they belonged to what is often called the northern race. Unlike the lineage from the Paleolithic Age, the lineage from the Neolithic Age has continued to survive down to the formation of the Korean people. Elements of the Neolithic Age combined with each other and then with elements of the Bronze Age in the formation of the Korean people.

The highly developed Bronze Age culture in Korea around the year 1000 BC was influenced by a northern culture that was different from the Chinese scientific one. Chinese science and technology were adopted on the basis of an indigenous technological tradition. Koreans have always attempted to re-shape Chinese technology whenever it was introduced into Korea to make it uniquely Korean and to develop new technology.

Typical examples of this kind of endeavor are found in the bronze sword and mirror. The lute-shaped bronze sword and Korean-style sword and the bronze mirror which has a thick lined design with two knobs, and the thin lined mirror, are uniquely Korean bronze instruments. These instruments reveal unique design and casting skills indicating the highly developed state of bronze-making technology. This technology in Korea has a different lineage from that of the Chinese. This is found in the chemical make-up of the bronze instruments themselves. Even the earliest bronzes found in Korea there are alloys of zinc and bronze.

Koreans demonstrated a special creativity by developing an advanced skill in bronze production and alloy making. Bronze makers in Korea used both mud molds and stone molds to produce bronze items. Stone molds, which are frequently found in Korea, are very seldom discovered in China.

Casting technology utilizing stone molds led to the development of a unique iron ax casting system that developed from the third century BC in Korea. In this period iron axes were produced in massive quantities by using stone molds. It is a commonly held belief that the Iron Age in Korea started in the fifth to fourth centuries BC with the introduction of Chinese iron culture. The mass production of iron axes can be explained as the creation of a new technology, which would not have been possible if there had not been an autonomous tradition of iron technology in Korea.

The development in iron-casting technology brought about a dramatic increase in agricultural production on the basis of the mass production of farm instruments. Iron ingots of the late Iron Age found in the southern regions of the Korean peninsula are unique symbols of power and wealth. Iron ingots developed by Koreans spread to Japan, setting in motion the development of an iron culture there.

The widespread use of iron brought about new kinds of earthenware. Kaya culture (or Kimhae culture) of the region produced hardened quality earthenware produced through the combination of the technology of the indigenous people of the area (who were known for their designs which lacked ideogrammatic characters) and Chinese ashen-earthenware skills.

Other regions on the Korean peninsula, such as Koguryo, Paekche, and Silla, were also developing a technological culture. The people of Koguryo, Silla, and Paekche each constructed their tombs in unique ways. The painters of the Koguryo tombs painted their walls with individual and powerful lines and colors. The many kinds of pure gold objects discovered in kings' tombs, with their distinctive designs and refined inscriptions, testify to what seems to have been a highly refined manufacturing technology.

Besides those kinds of technical cultural traditions, several examples of scientific inquiry are known to have been written in the Three Kingdoms period (first century BC–seventh century AD). Among the most outstanding are the developments in astronomy and medicine. Records indicate that Koguryo had maps of stars in the sky inscribed on stones, and a special astronomic observatory. The paintings of the stars found in a mural of Koguryo support this likelihood. Paekche also had astronomical observatories, and special calendrical specialists were appointed in the sixth century.

The Ch'omsongdae observatory in Kyongju is symbolic of the astronomical enterprise of the Three Kingdom period of Korea. Ch'omsongdae, a stone brick edifice, built in AD 647, is the oldest surviving astronomical observatory. The delicate aesthetic lines that characterize Chomsongdae are a symbol of the Korean sense of beauty. Records of astronomic observations dating from the Silla period are indications of very active institutionalized astronomical observation in that period. Astronomers in Silla learned calendrical studies from the Chinese, but were not content with just learning or copying from them. They endeavored to develop their own kind of astronomy. Ch'omsongdae is a valuable part of the heritage of their efforts.

Although actual records of medical prescriptions for long life and for the curing of diseases are scanty, they enable us to picture the systematic herbal medicine practiced in the period.

The period of Unified Silla is best represented by the great quantity of surviving remnants and records. Korea achieved a high level of creative development in science and technology in this period.

The Sokkuram grotto and the Darani scripture, among others, are representative of the achievements of Silla science and technology. The grotto, the most highly regarded edifice left by the craftsmen of Silla in the eighth century, is seen as the highest point in traditional architecture because of its geometric design and daring construction skills. There is no doubt that the Sokkuram grotto was a copy of a rock cave temple in China. However, while the Chinese edifice was built into a natural rock cave, the Sokkuram grotto is an artificial stone cave created on the basis of a harmony of various compositions including circles, spheres, triangles, hexagons, and octagons.

Silla master craftsmen also produced a number of beautiful bells. They successfully combined ancient Chinese bells and tinkle-bells in creating a unique Silla style. A special feature of the Silla bells is found in the dragon shaped hooks for hanging them. Historical records assert that Silla's craftsmen produced an alloy of bronze combined with copper, tin, and lead for a better sound. This was found to be true in a chemical analysis. The Chinese naturalist, Li Shizhen, wrote in his *Bencao Gangmu* that "Persian bronze is good for making mirrors and Silla bronze for making bells". The bell preserved in the Kyongju National Museum was made in 771. It has a height of 3.3 meters. Its solemnity and the arrangement of its design patterns are typical of Silla bells.

Master craftsmen of Silla began developing wood print type in the early part of the eighth century. The scroll of Darani scripture found in 1966 was assessed to be a wood type print on white hammered paper produced between 705 and 751. This makes it the oldest surviving printed material in the world. On the basis of surviving evidence, this would prove that printing technology was invented in Korea prior to its development in China. Whatever the merits of this claim are, the fact that Silla technology had developed, in the eighth century, to the level of producing wood type print, is a significant demonstration of its technology. It indicates that the gap between the level of technology in China and Silla was not very wide.

The science and technology of Koryo developed on the basis of the achievements of the sciences and technology in Silla. They were also greatly influenced by the Sung and Yuan dynasties in China, and there are signs of the indirect influence of Islamic scientific and technological culture. The representative achievements of science and technology in Koryo are the development of wood block printing, the invention of bronze type printing technology, and the development of Koryo celadon.

Wood block printing in Koryo developed as a result of the aristocratic interest in print copies of calligraphy in pine wood blocks. It was also stimulated by a religious motivation. Buddhist scripts were made into wood type in order to draw on the power of Buddha in the struggle against the Mongol invasion. Eighty Thousand Scriptures, the world's largest and oldest surviving wood block printing set known, was also produced out of this motivation. The printing of the Eighty Thousand Scriptures is regarded as the highest level of technology in wood block printing.

The invention of a bronze type printing system was created from a totally different motivation. Koryo, having less demand for books than China, could not maintain the enormous amount of wood blocks, time, and labor to produce a variety of kinds of books. Bronze type was invented as a solution to this problem.

Ceramic type printing was invented by Bi Sheng in the China of the eleventh century. However, a metal type printing system was not established even by the fourteenth-century because of the difficulties in casting and the ink and paper needed for metal type printing. However, all these were possible with the science and technology available in Koryo. There, craftsmen knew of the technique of producing sand molds to make bronze print type, and were already producing oil ink suitable for printing with metal type and good quality paper. Because of this, the master craftsmen of Koryo were able to invent bronze type from wood type.

The central element in making bronze type was making sand molds. This was one of the greatest contributions in the development of printing technology. It may have been developed from the accumulated knowledge from production of various bronze vessels and casting of the grand bells.

The craftsmen of Koryo were successful in developing their skills for making celadon. They also adopted the inlaying method, which until that time had only been used in metal ornament making, for celadon making. This illustrates that, while Koryo celadon skills were adopted from Sung China they were never a simple copying of the Sung products.

Astronomy and herbal medicine were also two of the central pillars of science in Koryo as in the previous dynasties. Koryo scientists also developed geography. Major efforts in astronomy in the Koryo period were in observational astronomy and in calendar making. Official records of astronomical observations contain records of observation over 475 years. Among the records, there is a record of 132 occasions of an eclipse of the sun. This is on a par with the astronomic records of Islamic astronomers. There are also notable observations of sunspots.

A systematic foundation of herbal medicine in Korea was established from the sixth to the seventh centuries at the height of the Three Kingdom period. It came as a combination of traditional medicinal prescription and the influence of medical theories from China. A system was developed for utilizing herbs found in Korea. *Bencao jingjizhu* (The Shennong Pharmacopoeia with Collected Annotations) written by Dao Hungzhing contains eleven medicinal herbs that originate form Korea. And *Ishimpō* (Tamba no Yasayori's

Collected Prescriptions), a famous medicinal book in tenth century Japan, contains various quotations of prescriptions from medical texts from Paekche and Silla. By the ninth century, twenty-two kinds of medicinal herbs originating in Korea were recorded in the medicinal texts used in China and Japan. A national medical school was established in the tenth century, and national examinations for medical practitioners were introduced. In the twelfth and thirteenth centuries, Sung medical science was introduced into Korea, while native prescriptions of local medicines emerged in Koryo in the form of three volumes entitled *Hyangyak kuguppang* (First-aid Measures with Local Medicines). This represented the first medical text describing herbal medicinal practice in the Koryo period.

In the early period of the Choson dynasty, efforts to create an independent culture brought about a forceful development of science and technology. In 1402, King Taejong proceeded with the development of bronze type print despite strong opposition from high ranking government officials. However, the books printed with the newly developed type (*kemi*) were not of better quality than the books printed with wood type. Furthermore, it resulted in lower efficiency in printing one kind of book, and it did not bring about any improvements in cost and labor productivity. It was a low efficiency technology. However, this project provided the basis for greater improvements when the succeeding king adopted it as a state-funded program. As a result Choson type printing developed to a state of perfection. An improvement in technology was achieved which was not seen in any other part of East Asia.

Achievements in science and technology reached their highest point during the King Sejong period, which is regarded as the golden age of Korean traditional science. The invention of the rain gauge is one of its achievements. Rain measuring instruments and watermarks were developed between 1441 and 1442, enabling the scientific measurement of rainfall. The invention of a cylindrical rain gauge resulted from efforts to obtain a precise measure. As a result the scientists in the court of King Sejong were able to develop a scientific method for the quantitative measurement of natural phenomenon. Government officials maintained the measurement and recording of rainfall throughout the country, utilizing a highly systematic method, for over four hundred years. Korea was the only country in the fifteenth century which undertook meteorological observation through the use of a quantitative measuring device.

New astronomical observatories were built during the reign of King Sejong. The great equatorial torquetrum observation platform built in Kyongbok Palace contains a torquetrum, armillary clock, armillary sphere, gnomon, direction markers, an automatic striking clepsydra, a jade clepsydra, and various other kinds of sundials. In order to construct these astronomical observatories, which took seven years to complete, King Sejong sent mathematicians, astronomers, and technicians to China to study astronomical observational instruments. Scientists designed an astronomical instrument that was modeled on the Guo Shoujing system of the Yuan dynasty in China. However, major aspects of the instrument were modified to reflect Korean characteristics. The astronomical observatory built through such a process was the largest of its kind and among the best equipped in the fifteenth century.

Astronomers of the King Sejong period developed an independent calendar on the basis of their observations and calculations. The publication of the calculation of the motions of the seven governors, the inner and outer parts (*Ch'ilchongsan Naep'yon* and *Ch'ilchongsan Oep'yon*), is the product of such efforts. The outer part is regarded as one of the most authoritative texts written in Chinese characters on Islamic astronomy and calendar science.

The scientific achievements of the King Sejong period were found in all fields of endeavor. Koryo celadon was developed into Punch'ong ware by early Choson craftsmen to produce unique Choson celadons that differed in style and quality from Koryo celadons. A blue and white porcelain was first imported from China during the King Sejong period and began to be produced in Korea from the middle of the fifteenth century. However the geometric shape and design of the porcelain were entirely different from the Chinese originals. Choson white porcelain was able to stand equally in terms of quality and quantity.

Military technology also developed unique characteristics in the early period of Choson. The development of Choson style firearms and turtle ships are representative examples. Firearms began to be used widely from the late Koryo period after introduction of the skill from China. However, by the King Sejong period, firearms production in Choson had abandoned a great many of the Chinese features leading to the development of unique firearms. By totally recasting the firearms, which were improved to strengthen the national system, the Choson court, they were modified and standardized until they had their own style.

There were also notable achievements in geography. The map of the world drawn by Choson geographers in 1402, while based on some of the central maps in China, was a more complete one. Although it did not overcome the Chinese world view, it contained a depiction of Europe and Africa and the Far East which can be said to make a truly world map with the most up-to-date geographical knowledge.

In the King Sejong period, actual national measurements were undertaken to produce a complete map of the country.

The map of Choson produced in the fifteenth century, currently preserved in Japan, is a map of the highest quality in comparison with others produced in the same period. Geographical scientists in the King Sejong period endeavored, in addition to making maps from thorough field surveys and study of literature, to do proper geographic work that included national and provincial maps.

The field of medicine experienced a total systematization and concentration of Korean herbal medicine and Oriental medicine. Study of local medicinal herbs was collated into a basic foundation for herbal medicine in Korea. It was developed into a comprehensive system for prescription. The *Hyangyak chipsongbang* (Great Collection of Native Korean Prescriptions), in which a total of 703 Korean native medicines were included, was completed in 1433. This development provided an impetus for breaking away from a dependence on Chinese medicine. Concurrent with these efforts was the editing and publishing of *Uibang yuch'ui* (Classified Collection of Medicinal Prescriptions). This was a medical encyclopedia, completed in 1445, which incorporated 153 different Korean and Chinese texts, and was regarded as one of the greatest medical texts of the fifteenth century. Also, the *Tongui pogam* (Precious Mirror of Eastern Medicine) was completed from 1556–1610 by Hochun.

Agricultural technology also experienced great development in the King Sejong period. Much of the agricultural technology in Korea until this period was based on Chinese agricultural texts. However, they could not provide appropriate guidelines for farming in Korea. To provide these, an agricultural technology text, *Nongsa chiksol* (Theories and Practice of Farming), was edited and published in 1429. This book surveyed the various farming methods in different fields and summarized the most developed and practical methods. This book contributed greatly to the improvement in agriculture in Choson and became the basic farming textbook.

However, Korean traditional sciences suffered a series of ruptures along with their creative development. A number of foreign invasions which decimated the entire territory disrupted the creative tradition in Korean scientific endeavors. Each time Koreans labored to overcome these disruptions and ruptures. The history of Korean sciences and technology is at the same time a history of these efforts. The introduction of Western science and the efforts to systematize the traditional sciences and technologies by "Sirhak" scholars from the seventeenth to the eighteenth centuries is just one example. These scholars advocated practical learning (*sirhak*) under the slogan *silsa kusi* (verification of truth on the basis of factual studies), and accepted some of the little European science that came their way. They thus began what might have become a scientific reformation because they were influenced by modern science and technology in Europe through Qing China, where the same kind of movement had been going on for some time, partly as a reaction to mystical tendencies in late Confucian philosophy. Sirhak scholars pioneered new frontiers in scientific and technological theory and scientific philosophy. However, their efforts were frustrated because of the onset of another round of ruptures.

JEON SANG-WOON

REFERENCES

Jeon, Sang-woon. *Science and Technology in Korea: Traditional Instruments and Techniques*. Cambridge, Massachusetts: MIT Press, 1974.

Needham, Joseph et al. *The Hall of Heavenly Records: Korean Astronomical Instruments and Clocks 1380 – 1780*. Cambridge: Cambridge University Press, 1986.

See also: Maps and Mapmaking in Korea – Calendars – Eclipses – Sundials – Mathematics in Korea

KŪSHYĀR IBN LABBĀN In his book *al-Madkhal fī ṣināʿat aḥkām al-nujūm* (Introduction to the Art of Astrology) the author calls himself Kūshyār ibn Labbān ibn Bāshahrī al-Jīlī. This name indicates that Kūshyār was a son of Labbān who was a son of Bāshahrī and that he hailed from Jīlān, a region of modern Iran south of the Caspian Sea. The date of the book is some time around AD 992, the year for which positions of the fixed stars are given. In the same book he refers to his two earlier books on astronomical tables (*zīj*es), *al-Zīj al-jāmiʿ* (Comprehensive) and *al-Zīj al-bāligh* (Far-reaching). In some manuscripts 'Abū al-Ḥasan' (the father of al-Ḥasan) is added at the top of his name. No further information is available about his family and life.

One of the most famous of his books is the *Kitāb fī uṣūl ḥisāb al-hind* (Book on the Principles of Hindu Reckoning) which is known as the oldest surviving Arabic book on arithmetic using Hindu numerals. The Arabic text is divided into two parts. In the first section of the first part Indian numerals and the decimal system of notation are introduced. In the following sections are (2) addition, (3) subtraction, (4) multiplication, (5) results of multiplication, (6) division, (7) results of division, (8) square root, and (9) arithmetic checks. The second part comprising sixteen sections is devoted to sexagesimal computations using sexagesimal tables.

According to E.S. Kennedy's classification of the subject matter of Islamic *zīj*es, Kūshyār's *Comprehensive astronomical table* covers the following subjects: chronology, trigonometric functions, spherical astronomical functions, equations of time, mean motions, planetary equations, planetary latitutdes, stations and retrogradations, parallax, eclipse theory, visibility conditions, geographical locations, star tables, and astrological tables.

Kūshyār's book on astrology seems to have been one of the most popular handbooks on this subject, especially in the eastern half of the Muslim world, as is witnessed by the abundance of surviving Arabic manuscripts and translations into Persian, Turkish, and Chinese. The book consists of four books, following the model of Ptolemy's *Tetrabiblos*. Almost all the chapters in the first book have corresponding ones in the *Tetrabiblos*. The second book deals with so-called judicial astrology where Kūshyār shows his knowledge of topics of Persian and Indian origin. Most of the subjects in the third and fourth books are found in Book III and IV of the *Tetrabiblos*. The last two chapters of the third book are devoted to a subject which was not unknown to Ptolemy but which found a significant development in Persian astrology, namely, the rules for computing the so-called *tasyīr* arc for determining the length of an individual's life.

In the introduction of this book he clearly distinguishes between two branches of the science of stars: astronomy and astrology in modern terms. The former, dealing with spheres of planets, their motion, and the computation of their positions, is more fundamental and is grasped by instruments and observation, and is to be proved by geometry. The latter branch concerns the knowledge of human deeds which is derived from the planets, their power, and their influence upon whatever is below the sphere of the moon. This is grasped by experience and analogy (*giyās*).

MICHIO YANO

REFERENCES

Kennedy, E.S. "A Survey of Islamic Astronomical Tables." *Transactions of the American Philosophical Society*, vol. 46, pt 2, 1956.

Levey, Martin, and Marvin Petruck. *Kūshyār ibn Labbān: Principles of Hindu Reckoning*. Madison: University of Wisconsin Press, 1965.

Saidan, A.S. "Kūshyār ibn Labbān ibn Bāshahrī, Abū'l-Ḥasan, al-Jīlī." In *Dictonary of Scientific Biography*. Ed. Charles C. Gillispie. New York: Charles Scribner's Sons, vol. 7, 1973, pp. 531–533.

Sezgin, F. *Geschichte des Arabischen Schrifttums*. Band V (Mathematik), 1974, Band VI (Astronomie), 1978, and Band VII (Astrologie), 1979. Leiden: Brill.

Yano, Michio, and Mercè Viladrich. "Tasyīr computation of Kūshyār ibn Labbān." *Historia Scientiarum* 41: 1–16, 1991.

L

LALLA Lalla, an eighth century Indian astronomer, was an exponent of the school of astronomy founded by Āryabhaṭa (b. AD 476). He was the son of Tāladhvaja and grandson of Sāmba alias Trivikrama, and hailed from Daśapura in Mālava in Western India.

Lalla was a popular astronomer who wrote both on astronomy and astrology. His most important work is the *Śiṣyadhīvṛddhida* (Treatise Which Expands the Intellect of Students), which, as he says, was composed to expatiate astronomy as set out by Āryabhaṭa. He uses the parameters enunciated in the *Āryabhaṭīya*, but propounds corrections to them every 250 years commencing from AD 498, the time of Āryabhaṭa. The first such correction falls in AD 748, which gives an indication of Lalla's date. The *Śiṣyadhīvṛddhida* is in two sections, entitled *Grahādhyāya*, dealing with planetary computations, and *Golādhyāya*, dealing with spherics, and theoretical and cosmological material. The first section, which is comprised of chapters I–XIII, treats of the mean and true planets, the three problems relating to diurnal motion, eclipses, rising and setting of the planets, the moon's cusps, planetary and astral conjunctions, and complementary situations of the sun and the moon. The second section (chapters XIV–XXII), deals with the graphical representation of the motion of the planets, the rationale of the rules enunciated earlier, rejection of popular false notions on astronomy, and astronomical instruments. Another work of Lalla, known from quotations by later authors on astronomy, is *Siddhāntatilaka*.

Lalla wrote a work on natural astrology, entitled *Ratnakośa*, which is still in manuscript form. Lalla's verses on mathematical topics are frequently quoted by later writers, but the complete text from which these verses are taken has yet to be found. This provides the justification for Lalla's being referred to in later works as *Tri-skandhavidyākuśalaikamalla*, "the one stalwart versed in all three branches", that is, mathematics, astronomy, and astrology.

Though Lalla follows Āryabhaṭa in certain aspects, he follows Brahmagupta (b. AD 598), and Bhāskara I (fl. AD 629), in certain others. It is also interesting that some of his innovations are followed by later astronomers like Śrīpati (tenth century), Vaṭeśvara (ca. AD 900), and Bhāskara II (b. AD 1114). This makes Lalla an important link in Indian astronomical tradition.

K.V. SARMA

REFERENCES

Chatterji, Bina. *Śiṣyadhīvṛddhida Tantra of Lalla with the commentary of Mallikārjuna*. 2 vols. New Delhi: Indian National Science Academy, 1981.

See also: Astronomy in India – Āryabhaṭa – Brahmagupta – Bhāskara – Astronomical Instruments in India – Śrīpati

LEO THE AFRICAN Leo the African was born in Islamic Granada as Al-Ḥasan ibn Muḥammad al-Wazzān al-Zayyātī around 1485 and was educated in Fez, where his family settled several years before the conquest of Granada. He is known primarily for his geographical writings based on four trips. The first, from Fez to Constantinople, probably took place in 1507 and 1508. The second, sometime between 1509–1511, was his first to Timbuctu following the caravan route south, probably as far as the Niger river. The third trip (1512–1514) took him across the Sahara via Lake Chad to Egypt, providing important information on the Sudan. The fourth trip (1515–1518) took him to Constantinople as ambassador of the Moroccan sultan, Egypt, and Arabia; on his return he was captured by the Italian pirate Pietro Bovadiglia. In Italy he was given as a slave to Pope Leo X whose name he took upon his conversion to Christianity.

Although he wrote an Arabic–Hebrew–Latin vocabulary in 1524, he is best known for his geographical treatise written and published in Italian as *Della descrittione dell'Africa* (Venice, 1550), although it was no doubt based on an Arabic draft. The treatise is divided into nine books, whose principal significance was to provide European cartographers with detailed information about areas of sub-Saharan Africa practically unknown in the West. Leo also made a substantial contribution to natural history: In his description of African plants, animals, and minerals he points out errors made by ancient writers such as Pliny. Leo's *Description* also contains invaluable information on Islamic institutions such as, for example, his description of the *muhtasib*, or market inspector, of Fez. Leo returned to North Africa in 1529, reconverted to Islam, and died in Tunis some time after 1554.

THOMAS F. GLICK

REFERENCES

Mauny, Raymond. "Note sur les 'Grands Voyages' de Léon l'Africain." *Hésperis* 41: 379–394, 1954.

The History and Decription of Africa. 3 vols. London: The Hakluyt Society, 1963.

LEVI BEN GERSON Levi Ben Gerson (1288–1344), also known as Gersonides or Leo de Balneolis, his Provençal name, was one of the most original Jewish thinkers of the Middle Ages, and he wrote on logic, philosophy, biblical exegesis, mathematics, and astronomy. He lived in Orange, where the de Balneolis family was prominent, and occasionally in Avignon (France). During the last years of his life he maintained relations with the papal court of Clement VI (1342–1352), to whom he dedicated a Latin version of his work on trigonometry and on the Jacob Staff (*Tractatus instrumenti astronomie*, 1342).

On mathematics, his *Maʿaseh Ḥoshev* (Work of the Computer, 1321) deals with arithmetic, summations of series, algebra, and combinatorial analysis. At the request of the French musical scholar Philippe de Vitry, he composed his *De numeris harmonicis* (On Harmonic Numbers, 1343, extant only in Latin) to demonstrate that numbers belonging to geometrical progressions of ratio 2 or 3 and first term 1, or generated by product of terms in these progressions, differ by a number greater than 1, excepting pairs 1–2, 2–3, 3–4, and 8–9. On geometry, he wrote a commentary on Books I–V of Euclid's *Elements*, a treatise on Euclid's parallel postulate, preserved incomplete, and commentaries apparently lost on Menelaus' *Sphaerica* and Thābit ibn Qurra's *Risāla fī Shakl al-qaṭṭāʿ* (On the Secant Figure).

Levi's most important scientific achievements are contained in the *Sefer Tekhunah* (Book of Astronomy), in fact, part 1 of the fifth book of his main philosophical work, *Milḥamot Adonai* (Wars of the Lord). Preserved in Hebrew and Latin versions, it is a lengthy work, divided into 136 chapters, which contains planetary observations and research from 1321 to 1340, and is based on a profound understanding of the astronomical tradition as well as on a sound criticism of some of his predecessors, mainly Ptolemy and al-Biṭrūjī. Levi's purpose was to construct a true astronomy, able to satisfy at the same time the requirements of experience, natural philosophy, and metaphysics. The crucial role attributed to observation led him to construct the Jacob Staff (an instrument for determining the angular distance between two stars or planets widely used in navigation until the eighteenth century, whose precision he increased by inventing a transversal scale for linear measurement, later also applied to the astrolabe), and to investigate successfully the theory of the *camera obscura*, which he used for observing eclipses. Levi employed his own observations both for deriving parameters for his new solar and lunar models and for testing them. Of special interest is his lunar model, which avoided the inadequacy of Ptolemy's model at half-moon phase (when it ought to appear twice as large as at opposition) and eliminated the use of the Ptolemaic epicycle (which would allow us to see both sides of the moon, again contrary to appearances). Using his own solar and lunar models, Levi composed ca. 1335 a set of tables and canons for computing eclipses, later partially modified when it was incorporated into the *Sefer Tekhunah*, which includes tables for the sine function, spherical astronomy, and solar and lunar mean motions and corrections. The chapters of the work dealing with planetary models are apparently unfinished, but a measure of Levi's originality is his theory of planetary distances and sizes: although Levi considered as not fully decided the problem of the position of Mercury and Venus with respect to the sun, assuming that these planets were placed above it, he computed the distance of the sphere of the fixed stars to be $159 \times 10^{12} + 6.515 \times 10^{8} + 1.338 \times 10^{4} + 944$ earth radii (instead of Ptolemy's 20,000 earth radii, generally accepted in the Middle Ages).

J.L. MANCHA

REFERENCES

Freudenthal, G., ed. *Studies on Gersonides. A Fourteenth-Century Jewish Philosopher-Scientist.* Leiden: E. J. Brill, 1992.

Goldstein, B. R. *The Astronomical Tables of Levi ben Gerson.* Hamden, Connecticut: Archon Books, 1974.

Goldstein, B. R. "Medieval Observations of Solar and Lunar Eclipses." *Archives Internationales d'Histoire des Sciences* 29:101–156, 1979.

Goldstein, B. R. *The Astronomy of Levi ben Gerson (1288–1344).* New York: Springer, 1985.

Goldstein, B. R. "A New Set of Fourteenth Century Planetary Observations." *Proceedings of the American Philosophical Society* 132:371–399, 1988.

Lange, G. *Sefer Maassei Choscheb. Die Praxis des Rechners.* Frankfurt: Louis Golde, 1909.

Rabinovitch, N. L. "Rabbi Levi ben Gershon and the Origins of Mathematical Induction." *Archive for History of Exact Sciences* 6:237–248, 1970.

See also: Thābit ibn Qurra – al-Biṭrūjī – Astrolabe

LI BING Li Bing (fl. 322–247 BC) is famous in Chinese history as an expert in irrigation works. He had a good knowledge of astronomy and geography as well as of engineering and technology. In 316 BC Shu (Sichuan province) was conquered by the State of Qin. In order to establish the province as an important base and to harness the Minjiang River floods, King Huizhao (r. 306–251 BC) appointed Li Bing governor of Shu.

During 277–250 BC Li decided to build the large waterworks at Guanxian county, from where the river flows into the plain of Sichuan. The whole project is known as Dujiang Yan. Li decided to divide the river into two great feeder

Canals, the Neijiang (Inner Canal) on the east and Waijiang (Outer Canal) on the west. This was done by means of piled stones, known as *Yuzui* (Fish Snout). The inner canal was used for irrigation, while the outer one, the mainstream, acted as a flood channel, and also carried some boat traffic. In order to construct the inner canal Li made a great rock cut through the end of a ridge of hills. This is known as the Bapoing Kou (Cornucopia Channel). Between the primary division-head and the rock cutting the channels were separated by the Feisha Yan (Flying Sands Spillway) which was adjusted to regulate the volume of flow into the inner canal.

In order to measure the water level so as to control it, three stone figures were put at three different places at the canal intake. With the help of observing the water level at the inner canal intake the feeding capacity into that canal was controlled by division dams of the Yuzui, Feisha Yan and Baoping Kou. The whole engineering operation made it possible to supply water for an area of 4.4 million acres to support a population of five million people, most of them engaged in farming, while at the same time remaining free from drought and floods. It can be compared with the ancient works of the Nile, and it is still in use today. Apart from this great contribution, Li also developed the deep-drilling technology for producing well-salt and built six bridges near Guangdu (Chengdu). The exploitation of water resources and the manufacture of well-salt greatly promoted the development of agriculture, industry, and commerce in the Sichuan area and made it become a famous 'Land of Abundance' in China since Li's time.

SUN XIAOLI

REFERENCES

Chang, Ju. *Huayang Guoji* (Historical Geography of Sichuan). Shanghai: Commercial Press, 1939.

Needham, Joseph. *Science and Civilisation in China*, vol. 4, pt. 3. Cambridge: Cambridge University Press, 1971, pp. 288–289.

Sima, Qian. *Shiji* (Historical Records, 90 BC), chapter 29. Beijing: Chung Hua Shu Chu, 1959. English edition New York: Columbia University Press, 1961.

LI CHUNFENG Li Chunfeng (602–670) was a Chinese mathematician, astronomer, and early historian of science. A native of Shaanxi province in northwest China and influenced by his Daoist father, Li Chunfeng served as a high-ranking court astronomer and historian for several decades in the early Tang dynasty (618–906). Both Li Chunfeng's son and grandson successively occupied the position of court astronomer.

As a mathematician Li Chunfeng is not considered to be original, but he played a role in indeterminate analysis, or the Chinese remainder theorem. He is credited, along with Zu Chongzhi in the fifth century, with developing the *zhaocha* method of finite or divided differences in mathematical astronomy. Using algebra to deal with the *ping* (floating differences) and *ding* (fixed differences) Li Chunfeng calculated the "angular speed of the sun's apparent motion" (Needham, 1959).

This method of finite differences was applied by Li Chunfeng in designing the Linde (Unicorn Virtue) calendar, which was adopted in 665. The new calendar was an improvement over the previous ones in predicting the movements of the planets and in the placing of the long (30 days) and short months (29 days) as well as the intercalary months. To compensate for the extra days in a solar year of 365.2422 days and a lunar month of 29.5306 days, one intercalary month needs to be added every three or four years.

Li Chunfeng also earned a place in the history of *hunyi* (armillary spheres), predecessor to the telescope. His armillary sphere had three nests of concentric rings: the inner sighting tube ring, the intermediate nest, and the outer stationary nest. Li's innovation lies in the intermediate ring, which improved the observations of celestial bodies and allowed for finer accuracy.

Li Chunfeng wrote the treatises on astronomy, calendar, and portents in the official dynastic histories, the *Jinshu* (History of the Jin Dynasty) and *Suishu* (History of the Sui Dynasty), in addition to authoring another extant work, *Yisi zhan* (Omen-taking in 645). Typical of astronomers in his time, Li Chunfeng could not avoid the moral and political responsibility of interpreting omens to the country, and included portent astrology and omen lore in his writings.

Li Chunfeng commented on the ten mathematical manuals that were officially designated for curriculum use both in Tang China and in Korea and Japan. These appear below with translated titles (Ho, 1985): *Zhoubi suanjing* (Arithmetical Classic of the Gnomon and the Circular Paths of Heaven); *Jiujing suanjing* (Nine Chapters on the Mathematical Art); *Sunzi suanjing* (Mathematical Manual of Sunzi); *Haidao suanjing* (Sea Island Mathematical Manual); *Wucao suanjing* (Mathematical Manual of the Five Government Departments); *Xiahou Yang suanjing* (Mathematical Manual of Xiahou Yang); *Zhuishu* ('Stitching' Method); *Zhang Qiujian suanjing* (Mathematical Manual of Zhang Qiujian); *Wujing suanshu* (Arithmetic in the Five Classics), and *Jigu suanjing* (Continuation of Ancient Mathematics). *Zhuishu* is not extant and was substituted by *Shushu jiyi* (Memoir on Some Traditions of Mathematical Art).

JENNIFER W. JAY

REFERENCES

Ho, Peng Yoke. *Li, Qi and Shu: An Introduction to Science and Civilization in China.* Hong Kong: Hong Kong University Press, 1985.

Ho, Peng Yoke. *The Astronomical Chapters of the Chin Shu, with Amendments, Full Translation and Annotations.* Paris: Mouton, 1966.

Li, Chunfeng. "Lüli" (Calendar), "Tianwen" (Astronomy), "Wuxing" (Five Elements), in *Suishu* (History of the Sui Dynasty). Compiled by Wei Zheng et al. Beijing: Zhonghua shuju, 1962.

Li, Chunfeng. "Lüli" (Calendar), "Tianwen" (Astronomy), "Wuxing" (Five Elements), in *Jinshu* (History of the Jin Dynasty). Compiled by Fang Xuanling et al. Beijing: Zhonghua shuju, 1974.

Li, Chunfeng. *Yisi zhan* (Omen-taking in 645). Beijing: Zhonghua shuju, Congshu jicheng ed., 1085.

Li, Chunfeng, annot. *Sunzi suanjing, Wucao suanjing, Shushu jiyi, Xiahou Yang suanjing, Jiuzhang suanshu, Zhoubi suanjing, Haidao suanjing.* Beijing: Zhonghua shuju, Congshu jicheng ed., 1985.

Li, Yan and Du, Shiran. *Chinese Mathematics: a Concise History.* Oxford: Clarendon Press, 1986.

Libbrecht, Ulrich. *Chinese Mathematics in the Thirteenth Century: The Shu-shu chiu-chang of Ch'iu Chiu-shao.* Cambridge, Massachusetts: MIT Press, 1973.

Liu, Xu et als. *Jiu Tangshu* (Old History of the Tang Dynasty). Beijing: Zhonghua shuju, 1975. Chapter 79.2717–19.

Needham, Joseph with Wang Ling. *Science and Civilisation in China.* Volume 3: *Mathematics and the Sciences of the Heavens and the Earth.* Cambridge: Cambridge University Press, 1959.

See also: Liu Hui – Armillary Spheres – Zu Chongzhi

LI GAO Li Gao was born in 1180 in Zhending (now Zhengding County of Hebei Province) and died in 1251. His surname was Mingzhi, and Dongyuan, the Old Man, was his nickname. He was an exponent of one of the four major schools of the Jin-Yuan period.

After his mother died of an unknown disorder which was treated to no avail by unskilled practitioners, Li made up his mind to pursue medical studies. His tutor Zhang Yuansu, who advocated the idea that "modern diseases can't be cured by ancient prescriptions", taught him to probe for new ideas and prescriptions so as to free himself from the yoke of the ancient art of prescribing. He was conversant in the nature and properties of drugs and in their beneficial potentials. The recipes he formulated were all organized according to the theory of "king, minister, assistant, and attendant", commonly composed of ten to twenty kinds of herbal drugs. In line with the classical theory, all the ingredients were closely related with both complementary and opposing relationships.

He emphasized the role of stomach *qi* during all seasons. *Qi* is the root of all body functions. The spleen and stomach are the roots of growth and metabolism; various disorders arise when their functions are jeopardized. He expounded the etiology and mechanism of disease due to endogenous pathogens (those coming from within), which are different from those caused by exogenous ones, and he stressed the importance of the spleen and stomach in the course of endogenous disorders. He criticized quacks that mix up disorders of different etiology and apply methods for exogenous ailments to tackle the problems of endogenous ones, with ensuing complications. Clinically, he stressed that the spleen and stomach should always be nourished so as to renew the body. He formulated a famous recipe, the *Buzhong Yiqi* (Benefiting-Interior Reinforcing-Qi) decoction, and the *Shengyang Yiwei* (Ascending Yang Replenishing-Stomach) decoction, which are both common remedies applied today. Due to his emphasis on spleen and stomach, he was given the title of the head of an academic school, the Earth-Replenishing School. He was also a prolific writer. Among his works, the most famous include *Nei Wai Shang Bien Huo Lun* (Differentiation for Endogenous and Exogenous Disorders), *Pi Wei Lun* (On Spleen and Stomach), *Yi Xue Fa Ming* (Medical Inventions), *Dong Yuan Shi Xiao Fang* (Dongyuan's Trial Effective Recipes), *Lan Shi Mi Cang* (Clandestine Collections in Orchid Mansion), *Huo Fa Ji Yao* (Essentials of Flexible Methods), *Yao Lei Fa Xiang* (Pharmaceutical Normalcy), *Yong Yao Xin Fa* (Mastery of Drug Application), and *Shang Han Hui Yao* (Collected Essentials of Disease Due to Cold Evil). In his later years he passed his academic ideas and works on to his disciple Luo Tianyi.

HONG WULI

REFERENCES

Jiang, Jingbo. "Mr. Li Dongyuan's Theory and His Works." *Traditional Chinese Medicine of Guangdong* 2: 11–13, 1963.

Li, Gao. *On Spleen-Stomach* (Reprinted). Shanghai: Shougu Bookstore's Lithographic Edition, 1913.

Taki, Mototane. *Textual Research on Chinese Medical Works* (Reprint). Beijing: People's Health Publishing House, 1956.

See also: Medicine in China

LI SHANLAN Li Shanlan (1811–1882) was a native of Haining, Zhejiang. Although he belonged to a moderately fortunate family and was given formal training in the classics, he never passed government examinations beyond the first level, and had to give up the dream of entering officialdom. Under difficult circumstances, he took refuge in

Shanghai, a city newly opened to foreign trade as a consequence of the Opium War.

From 1852 to 1859, Li went into service with the London Missionary Society who employed him as a co-translator of all sorts of Western scientific works. Missionaries had an insufficient mastery of literary Chinese and for that reason they had to rely on Chinese co-workers, even though these, like Li Shanlan, generally had no knowledge of foreign languages. During this period, Li translated with Alexander Wylie (1815–1887), Joseph Edkins (1829–1890) and others the part of Euclid's *Elements* not yet translated into Chinese (Books 7–15), Augustus de Morgan's *Elements of Algebra* (1835), Elias Loomis's *Elements of Analytical Geometry and of Differential and Integral Calculus* (1851), John F. W. Herschel's *Outlines of Astronomy* (1849), and other manuals on mechanics and botany. But what made Li famous in the eyes of his contemporaries was not so much the numerous translations he was responsible for but his own mathematical works.

Contrary to what might be expected, Li's works were not based on new Western mathematics but on ancient Chinese conceptions, especially the positional algebra *tianyuan shu* developed five centuries earlier during the Song and Yuan dynasties. Li however did not stick slavishly to the venerable discoveries of his ancestors. Rather he used much ingenuity emulating Westerners and trying to beat them at their own game. Using ancient tools, he thus conducted research of his own into logarithms, infinite series, and combinatorics. In particular, he made extensive use of algebraic computations, analogical reasoning, and inductive generalizations built on the knowledge of a few particular cases. But he never relied on hypothetico-deductive reasoning. One of his original results, which may be expressed as follows using modern symbolism

$$\sum_{j=0}^{k} \binom{k}{j}^2 \binom{n+2k-j}{2k} = \binom{n+k}{k}^2,$$

where the $\binom{n}{p}$ represent the usual binomial coefficients occasionally appears in modern books.

Owing to the generosity of Zeng Guofan (1811–1872), a famous general responsible for the suppression of the Taiping revolt, Li Shanlan's collected mathematical works were published in Nanking in 1867 under the title *Zeguxi zhai suanxue* (Mathematics from the Studio "Devoted to the Imitation of the Ancient Chinese Tradition").

In 1869, Li was appointed Professor of Mathematics at the newly created *Tongwen guan*, a college in Beijing devoted to the teaching of foreign languages, technology, and science. He based his lectures partly on Western mathematics and partly on Chinese mathematics, and remained in that post until his death. His works were much admired in China by mathematicians and non-mathematicians alike. They can be considered at the same time the most creative mathematics written in China during the whole of the nineteenth century and the swan song of Chinese traditional mathematics.

JEAN-CLAUDE MARTZLOFF

REFERENCES

Fang, Chao-ying. "Li Shanlan." In *Eminent Chinese of the Ch'ing Period* (1644–1912). Ed. Arthur W. Hummel. Washington, D.C. Government Printing Office, 1943.

Horng, Wann-sheng. *Li Shanlan, the Impact of Mathematics in China During the Late Nineteenth Century.* Ph.D. dissertation, City University of New York, 1991.

Martzloff, Jean-Claude. "Li Shanlan (1811–1882) and Chinese Traditional Mathematics." *The Mathematical Intelligencer* 14(4): 32–37, 1992.

LI SHIZHEN The great Chinese naturalist and pharmacologist Li Shizhen was born in 1518 near modern Qichun (Hubei). Born into a family of doctors, he concentrated on the study of medicine at an early age under the guidance of his father. His reputation as a physician soon reached the prince of Chu, who in 1543 entrusted him with medical and administrative responsibilities at his court in Wuchang. From 1544 to 1549 Li practiced at the Imperial Academy of Medicine (Taiyi Yuan) in Beijing. After that, he returned to his native village, where he worked as a doctor until his death in 1593.

The experience as a physician suggested to Li the need of a thorough revision of the traditional repositories of pharmaceutical knowledge (*bencao*). Earlier standard sources such as the work by Tang Shenwei had been made obsolete by the adoption of new drugs and prescriptions, along with errors in the identification and classification of some substances. Li apparently began to conceive the compilation of a definitive pharmacopoeia around 1552, and worked on this project for more than twenty-five years. He spent much of that time reviewing a massive amount of literature, and traveling extensively to collect specimens and recipes.

The result of these efforts represents the culmination of the literary tradition of Chinese pharmacology. Li Shizhen's work, entitled *Bencao gangmu* (The Pharmacopoeia Arranged into Headings and Sub-headings), contains 1892 entries, classified into sixteen main sections (e.g. Minerals, Trees, Reptiles) and sixty-two categories. The entries describe 275 minerals, 1094 plants, 444 animals, and 79 miscellaneous substances (those in the sections Water, Fuel, and Earth). Li Shizhen himself contributed no less than 374 new entries.

Li classified the *materia medica* according to an essentially binomial system, describing species as variants of a genus. His work deals with mineral first, followed by plants, invertebrates, vertebrates, mammals, and man. A typical entry is divided into sections concerned with nomenclature, places of occurrence, varieties, problems of identification, medical properties, uses, and prescriptions. The *Bencao gangmu* contains more than eleven thousand recipes, about eight thousand of which were collected or devised by the author, and more than one thousand illustrations. The sources, numbering about one thousand, consist of medical and pharmacological works, along with historical, literary, philosophical, and other texts. While Li — like most traditional pharmacologists before him — accepted the authority of the early, pre-Tang pharmacopoeias, he did not refrain from criticizing and correcting the views of later authors.

The manuscript of the *Bencao gangmu*, completed in 1578, went through several revisions until 1590, when Li took it to a Nanjing publisher for publication. It is uncertain whether the author was able to see his work printed before he died in 1593: recent research has suggested that printing may have not been completed until 1596 rather than 1593 as often indicated. The *Bencao gangmu* was re-edited several times in China and Japan. References in early sinological works reached the attention of eighteenth- and nineteenth-century Western scientists, including Carl von Linné (Linnaeus) and Charles Darwin, who refer to it in their own works.

FABRIZIO PREGADIO

REFERENCES

Lu, Gwei-Djen. "China's Greatest Naturalist; A Brief Biography of Li Shih-chen." *Physis* 8: 383–392, 1966.

Needham, Joseph, et al. *Science and Civilisation in China*, vol. VI: *Biology and Botanical Technology*, part 1: *Botany*. Cambrdge: Cambridge University Press, 1986.

Pan Jixing. "Charles Darwin's Chinese Sources." *Isis* 75: 530–534, 1985.

Sivin, Nathan. "Li Shih-chen." In *Dictionary of Scientific Biography*. New York: Charles Scribner's Sons, 1973, vol. VIII, pp. 390–398.

Unschuld, Paul. *Medicine in China; A History of Pharmaceutics*. Berkeley: University of California Press, 1986.

See also: Medicine in China – *Bencao gangmu* – Ethnobotany in China – Tang Shenwei

LI ZHI Considered by George Sarton as essentially an algebraist, Li Zhi (1192–1279) was one of the greatest mathematicians of his time. He is known by some as Li Ye, a name that he was supposed to have taken at the later stage of his life to avoid the word *zhi* which was adopted by a member of the royalty. He used an algebraic process known as *tianyuanshu* (method of the celestial element) to set up equations of any degree. He was not the originator of this method, but it was through his writings that the *tianyuanshu* was handed down to later generations. It was introduced to Japan and had a profound influence on Japanese mathematics. Known as *tengenjitsu* to Japanese mathematicians it enabled Seki Takakazu (also known as Seki Kowa, ca. 1642–1708) to develop a formula for infinite expansion, which is now arrived at using infinitesimal calculus.

Li Zhi was born in Luancheng in modern Hebei province. He served the Jurchen government as a magistrate in Junzhou (in modern Henan province). In 1232 Junzhou was captured by the Mongols and Li Zhi made his escape taking refuge in various places and living as a recluse. He became a noted scholar in north China. Qubilai Qan intended to give him an official appointment, but he declined the offer. He read widely and was a voluminous writer. Unfortunately most of his books are lost. According to one source, before Li Zhi died he told his son Li Kexiu to burn all his writings except the *Ceyuan haijing* (Sea Mirror of Circular Measurements) which he wrote in 1248, saying that posterity might find it useful. Somehow, another mathematical text, the *Yigu yanduan* (New Steps in Calculation) which he wrote in 1259, also escaped the fire. It is only through these two works that we know about Li Zhi's contributions to mathematics.

Consisting of one hundred and seventy problems, the *Ceyuan haijing* begins with a diagram of a circle inscribed in a right-angled triangle. Diameters of the circle parallel to the vertical and horizontal sides of the triangle, and tangents to the circle parallel to these two sides of the triangle produced fifteen different right-angled triangles of different sizes. Li Zhi worked out the relations between the sides of these triangles, and between the diameter of the circle and the parallel sides of two of the trapezia formed in the diagram. His understanding of the properties of the circle and the right angled triangles enabled him to work out many of his problems with great ease. The same problem that his contemporary Qin Jiushao (ca. 1202–ca. 1261) solved with a tenth degree equation was handled by Li Zhi simply using a quartic equation instead.

The *Yigu yanduan* contains sixty-four problems involving mainly a square and a circle, with a few cases on two different circles, two different squares, and a circle together with either a rectangle or a trapezium. Again equations of higher degrees are used. This text stabilized the terminology used in Chinese mathematics about equations of higher degrees.

HO PENG YOKE

REFERENCES

"Biography of Li Zhi." In Ke Shaomin, *Xin Yuanshi* (New Official History of the Yuan Dynasty) Ch. 171, Biographical Chapter 68, first published 1922, reprinted Taipei, Yiwen Press (no date).

Chemla, Karine. *Étude de livre "Reflets des mesures du cercle sur la mur" de Li Ye.* Doctoral dissertation, University of Paris, 1982.

Ho Peng Yoke. "Li Chih, Thirteenth-Century Chinese Mathematician." In *Dictionary of Scientific Biography*, vol.8, Ed. Charles C. Gillispie. New York: Scribner, 1973, pp. 313–320.

Ho Peng Yoke and Chan Hock-lam. "Li Chih." In *In the Service of the Khan. Eminent Personalities of the Early Mongolian (1200–1300).* Ed. I.de Rachewiltz *et al.* Wiesbaden: Harrassowitz, 1993, pp. 316–335.

See also: Mathematics in Japan – Seki Kowa

LIU HONG Liu Hong was a Chinese astronomer (AD 129–210), descended from the imperial family of the Eastern Han Dynasty. Liu Hong was very interested in astronomy from his childhood. About AD 160, he was assigned to be an officer of the Imperial Observatory. In this position, he participated in determining a series of astronomical dates. In AD 174–175, he offered *Qi Yao Shu* (The Art of Seven Planets) and its reformulated edition *Ba Yuan Shu* (The Art of Eight Elements) to the imperial government. The two works have been lost, but there are indications that they related to Buddhist astrology in middle Asia. In A.D. 187–188, he composed the *Qian Xiang Li* (Qian Xiang Calendar), a work of traditional Chinese mathematical astronomy. The method to describe lunar motion in this work was so advanced that the imperial government adopted it immediately. In AD 206, he examined and approved the formal edition of *Qian Xiang Li*. This is one of the best calendars in ancient China. It used new data for the tropical year ($365\frac{145}{589}$ days) and the synodic month ($29\frac{773}{1457}$ days), calculated the advance of the lunar perigee, established the concept of the moon's path, and calculated the regression of node, established the first table of the lunar apparent motion in ancient China, and also made progress in the calculation of eclipses and planetary motion.

JIANG XINOYUAN

REFERENCES

Chen Meidong. "The Life, Thinking and Astronomical Achievements of Liu Hong." *Studies in the History of Natural Sciences* 5(2): 129–142, 1986.

Fan Ye. *Hou Han Shu (The History of the Eastern Han Dynasty* vol. 12. Beijing: Zhonghua Press, 1965.

Fang Xuanlin. *Jin Shu* (The History of Jin Dynasty), vol. 17. Beijing: Zhonghua Press, 1974.

See also: Calendars in East Asia

LIU HUI AND THE *JIUZHANG SUANSHU* Liu Hui (fl. 263) reminds us of Euclid in that we know nothing much about their lives, but their works had a great impact on mathematics. Mathematics had many practical applications in ancient China, such as in the surveying of farm land, in the distribution of grains, in taxation, in civil engineering, and so on. Before the second century BC there were separate writings by different authors on each of the ancient applications, giving a total number of nine. The term *jiushu* (Nine Mathematics) probably referred to the nine mathematical applications. The syllabus for the education of royal princes in the Eastern Zhou period (771–256 BC) included *jiushu*. All those writings have been lost, but it is believed that they formed the original source from which the *Jiuzhang suanshu* (Nine Chapters on the Mathematical Arts) was later derived. The Han mathematicians Zhang Cang (fl. 165–142 BC) and Geng Shouchang (fl. 75–49) both had a hand in the arrangement of the *Jiuzhang suanshu* and in adding their own commentaries. Both their works were also lost. Another mathematician, Xu Shang (fl. 26–27 BC) might have something to do with the *Jiuzhang suanshu*, but again none of his writings remained. Our modern version of its text is based on the *Commentaries* of Liu Hui (fl. 263), who also rearranged the subject matter in the main text.

The Nine Chapters or Nine Sections in the book include: (1) *fangtian* (square farm), referring to land survey; (2) *sumi* (millet and rice), on proportions and percentages; (3) *shuaifen* (distribution by progression), on partnership and the rule of three, including taxation; (4) *shaoguang* (diminishing breadth), on finding an unknown side of a figure given the area and other sides; (5) *shanggong* (consultations on engineering works), on the volume of a solid; (6) *junshu* (impartial taxation), on the time taken to carry a contribution of grains to the capital and allocation of tax burdens according to population; (7) *yingbuzu* or *yingnu* (excess and deficiency), using a typical Chinese method of "rule of false position" to solve problems of the type $x = b$; (8) *fangcheng* (calculation by tabulation), on simultaneous linear equations; and (9) *gougu* (base and height of a right triangle), dealing with right-angled triangles involving quadratic equations.

Liu Hui also wrote another book, the *Haidao suanjing* (Sea-Island Mathematical Manual), applying the principle of the right-angled triangle to the measurement of heights and distances. The size of this book is approximately that of one chapter in the *Jiuzhang suanshu*. It has been suggested that it was meant to be a supplement to the latter. Also known as *Jiuzhang suanjing* (Mathematical Classics of the Chapters), the *Jiuzhang suanshu* became the most influential text in the history of mathematics in China. It was studied by mathematicians throughout the centuries, and it appeared in many book titles on mathematics. Together with its commentator Liu Hui, it is the most popular topic of research among historians of mathematics in China today.

In 1983 an ancient mathematical text written on bamboo strips was excavated from a tomb dated in the second century BC at the Zhangjiashan hill in Jiangling, Hubei province. Preliminary studies of this bamboo manuscsript, entitled *Suanshushu* (Book on the Art of Arithmetic), suggested a certain resemblance to Liu Hui's version of the *Jiuzhang suanshu* text. A full study of this ancient relic may answer some questions on the origin of this important ancient Chinese mathematics text.

HO PENG YOKE

REFERENCES

Ho Peng Yoke. "Liu Hui." In *Dictionary of Scientific Biography*. Ed. Charles Gillispie, vol. 8. New York: Scribner, 1973, pp. 418–24.

Jiuzhang suanshu yu Liu Hui (*Jiuzhang suanshu* and Liu Hui). Ed. Wu Wenjun. Beijing: Beijing Normal University Press, 1982.

Needham, Joseph. *Science and Civilisation in China*. vol. 3. Cambridge: Cambridge University Press, 1959.

See also: Mathematics in China – Gou-gu Theorem

LONG COUNT One of the most original creations of pre-Columbian Mesoamerica was the so-called Long Count, a system of recording events in time with unique precision. No parallel system existed in any other society until the Julian Day count was begun in Western Europe in 1582. Because it was most widely used by the Maya, its development was first credited to them, but subsequent research has shown that the earliest inscriptions which employed the Long Count not only lay well outside of the Maya core-area geographically but also pre-dated the rise of the Maya historically.

The Mesoamerican Long Count was a meshing of the two time-reckoning systems which had been already been in use in the region for about a millennium: a sacred almanac of 260 days and a secular calendar of 365 days. In both counts a vigesimal system (based on the number twenty) was used to group the days into 'bundles', for a total of thirteen in the sacred almanac and a total of eighteen in the secular calendar; the five remaining days in the latter count were considered "unlucky" or "worthless". The basic unit of the count was the day, or kin, in Maya. Twenty days constituted one uinal, and eighteen uinals made up one tun, for a total of 360 days. (The tun was the only deviation the Maya allowed from a strictly vigesimal system, and was only employed in their count of days.) Twenty tuns comprised a katun (7200 days) and twenty katuns constituted a baktun (144,000 days).

The fundamental premise of the Long Count was that it recorded the number of days that had elapsed from a date in the distant past which marked the beginning of the present cycle of the world. Because the Maya recorded their Long Count dates by enumerating, in order, the number of baktuns, katuns, tuns, uinals, and kins which had elapsed from that date, together with the number and name of the day in both the sacred almanac and secular calendar which it represented, any given day was identified by a minimum of nine elements. Thus, in the present convention for transcribing Long Count days, the first day of the Long Count was designated as 13.0.0.0.0. 4 Ahau 8 Cumku.

Although many attempts were made to reconcile the Long Count with the Christian calendar, one of the first credible explanations was that offered by the American newspaper editor, Joseph T. Goodman, in 1905. According to his calculations, the Maya had fixed the beginning of the present cycle of world as August 11, 3114 BC—a date which accords with Julian Day number 584,283; hence, his correlation has been identified by that value. However, because Goodman was not an academic, little attention was paid to his findings until 1926 when a Mexican astronomer named Juan Martínez Hernández re-calculated the beginning date of the Long Count as August 12, 3114 BC, yielding an equivalent Julian Day number of 584,284. The following year, when the British archaeologist J. Eric Sydney Thompson made his own attempt to reconcile the two calendars, he determined that the correct date should be August 13, 3114 BC, for an equivalent Julian Day number of 584,285. Because all three of these correlations produced results within two days of each other and have subsequently been shown to accord most closely with known astronomical events, the so-called Goodman-Martínez-Thompson is now the accepted means of equating the Long Count with our own calendar. It should be noted, however, that in 1935 Thompson revised his calculations and adopted Goodman's original formula of 584,283 instead; unfortunately, this 'correction' served to confuse the correlation issue for an entire generation because it rendered all astronomical equivalencies impossible. Subsequently, I and others demonstrated that the 584,285 correlation is the only one which has astronomical validity.

As early as 1930, the American mathematician John Teeple concluded that the Long Count had been devised on a date which the Maya recorded as 7.6.0.0.0. 11 Ahau 8 Cumku. Teeple had been struck by the fact that seventy-three cycles of the 260-day sacred almanac equated with fifty-two cycles of the 365-day secular calendar and that the day-numbers and names of each count then came back into phase with each other. He believed that the Maya had also employed seventy-three cycles of larger intervals, such as katuns, in order to define longer spans of time, and that to fix the beginning of the present era of the world at an appropriately early date, they had projected their count backwards for a total of 146 katuns, or 7 baktuns and 6 katun. This meant that the Long Count had been originated on September 18, 236 BC, a date which Thompson rejected as being too early because it assigned its creation to a people other than the Maya, who he championed.

In a paper published in 1978, using a different premise — namely that the originator of the Long Count was bound by the recognition of the astronomical importance of August 13th as its initiating impulse — I reached the same starting date for its origin as Teeple had done. Also in this paper I argued that, by using internal evidence contained in the Long Count, it is possible to establish both the beginning dates of the secular 365-day calendar and the sacred 260-day almanac as well. The former was set in motion at the summer solstice around the year 1320 BC, whereas the latter was initiated with the zenithal passage of the sun over Izapa in southernmost Mexico on August 13 about the year 1358 BC. (Each of these dates may vary by as much as four years, due to the fact that the indigenous Mesoamerican calendars did not correct for leap years.) All of the earliest, or Baktun 7, Long Count inscriptions have been discovered along an axis which extends from the Soconusco region of southern Mexico and adjacent Guatemala, through central Chiapas into the Gulf coastal plain of Veracruz, an area which constituted the original homeland of the Zoque-speaking "Olmec" people.

VINCENT H. MALMSTRÖM

REFERENCES

Goodman, Joseph T. "Maya Dates." *American Anthropologist*, N.S. 7: 642–647, 1905.

Malmström, Vincent H. "A Reconstruction of the Chronology of Mesoamerican Calendrical Systems." *Journal for the History of Astronomy* 9: 105–116, 1978.

Teeple, John E. *Maya Astronomy*. Washington, D.C.: Carnegie Institution, Publication 403, Contribution 2, 1930.

LUNAR MANSIONS IN CHINESE ASTRONOMY Early Chinese astronomers grouped the stars visible from their country into more than 280 small constellations, each containing an average of about five stars. Among these star groups, twenty-eight asterisms encircling the sky in the vicinity of the celestial equator had special significance. Although their general appearance was much the same as that of any other constellation, these select star groups, known as *xiu* (lunar mansions) were of great importance in both positional astronomy and astrology. The expression *xiu* is a derivative of the term *su*, meaning "to stay the night"; at an archaic period the lunar mansions represented the nightly resting places of the Moon in its monthly circuit around the Earth.

Just when the stars were first divided into groups by the Chinese is not known and this is also true of the *xiu*. In past centuries, the mean great circle through the *xiu* was a better approximation to the celestial equator than at present although it should be emphasized that individual lunar mansions often lie at a considerable distance from the mean circle. By applying precession of the equinoxes, it can be shown that towards the middle of the third millennium BC the circle of the lunar mansions attained its best match to the celestial equator. Although this suggests a possible date of origin for the *xiu*, there is no evidence of an advanced culture in China at such a remote period.

Direct historical evidence on the origin of the lunar mansions is lacking until about three thousand years ago. In several of the ancient folk songs assembled in the *Shijing* (Book of Odes) — some probably dating from around 1000 BC — there are allusions to about ten star groups. Most of these are identifiable with the names of lunar mansions as listed in later texts. Again, several bronze vessels which were cast at much the same period as the folk songs were composed are inscribed with the names of a few lunar mansions. This suggests that at least some of the *xiu* had acquired special significance by early in the first millennium BC. However, it is not until several centuries later that the earliest preserved list of all twenty-eight names is encountered.

In 1978, archaeologists unearthed a lacquer chest from a tomb in Hubei province. This chest had belonged to a nobleman by the name of Yi who had died about 433 BC during the Warring States Period. On the lid of this chest are engraved the names of twenty-eight constellations in a roughly circular pattern. In general, the various ideographs are identifiable with the names of the lunar mansions as found in later lists. Further, they are cited in a sequence which characterizes more recent texts.

Table 1 gives a standardized list (together with translations) of all twenty-eight lunar mansions as found in texts from the third century BC onwards. In this table, the *xiu*

Table 1: The twenty-eight lunar mansions

Number	Name	Translation	Determinative star	Width
1	*Jue*	Horn	α Vir	12°
2	*Kang*	Neck	κ Vir	9°
3	*Di*	Base	α Lib	15°
4	*Fang*	Chamber	π Sco	5°
5	*Xin*	Heart	σ Sco	5°
6	*Wei*	Tail	μ Sco	19°
7	*Ji*	Basket	γ Sgr	11°
8	*Nandou*	Southern Dipper	φ Sgr	27°
9	*Niu*	Ox	β Cap	8°
10	*Xunu*	Maid	ϵ Aqr	12°
11	*Xu*	Emptiness	β Aqr	10°
12	*Wei*	Rooftop	α Aqr	17°
13	*Yingshi*	Encampment	α Peg	17°
14	*Dongbi*	Eastern Wall	γ Peg	9°
15	*Kui*	Stride	ζ And	16°
16	*Lou*	Harvester	β Ari	11°
17	*Wei*	Stomach	35 Ari	15°
18	*Mao*	Mane?	17 Tau	11°
19	*Bi*	Net	ε Tau	18°
20	*Zuixi*	Turtle Beak	ϕ Ori	1°
21	*Shen*	Triad	δ Ori	8°
22	*Dongjing*	Eastern Well	μ Gem	33°
23	*Yugui*	Ghost Vehicle	θ Cnc	4°
24	*Liu*	Willow	δ Hya	15°
25	*Qixing*	Seven Stars	α Hya	7°
26	*Zhang*	Extended net	ν Hya	17°
27	*Yi*	Wings	α Crt	18°
28	*Zhen*	Axletree	γ Crt	17°

are numbered in their traditional order, commencing with *Xue* (in Virgo). Among the first seven lunar mansions, a few names relate to various features of the Azure Dragon. This is one of the four mythical creatures (the others being the Dark Warrior, White Tiger and Red Bird) denoting the cardinal directions. The remaining *xiu* designations are little more than a random assemblage — e.g. Basket, Ox, Rooftop — and in general they are much more mundane than those of the Western zodiacal signs.

An early almanac known as the *Yueling* (Monthly Observances), probably dating from the Warring States Period (481–256 BC), marks the passing months by the position of the sun in various *xiu*, as well as the culmination of certain lunar mansions at dawn and dusk. By at least the Former Han Dynasty (202 BC–AD 9), the lunar mansions were used to specify the locations of other star groups as well as the changing positions of the sun, moon and planets — and also temporary objects such as comets and supernovae. A series of "determinative stars" (*chuxing*: one for each *xiu*) was selected to specify right-ascension. This coordinate was known as *ruxiudu* (degrees within a mansion). The meridian through each determinative star was adopted as the eastern boundary of a zone of right-ascension bearing the same name as the lunar mansion itself. As a result, the term *xiu* came to mean both the asterism and the range of right-ascension which it defined.

Commencing in the late second century BC, several preserved lists give the equatorial extension of these zones in *du* ("degrees", of 365.25 to a circle). The widths of individual zones were far from regular, ranging from only one or two degrees to as much as 33 degrees, see Table 1, which gives the calculated widths during the Han Dynasty. As yet, the reason for this uneven spacing has not been fully explained. In some regards, the *xiu* resembled the Western zodiacal signs, which from ancient Babylonian times were used for defining celestial longitude. However, the signs of the zodiac were of regular extent (each thirty degrees) and — unlike the lunar mansions — the effect of precession caused a gradual easterly displacement relative to the twelve constellations after which they were named.

For almost every solar eclipse occurring during the Former Han Dynasty, the official history of the period, the *Han-shu*, gives an estimate of the right-ascension of the sun to the nearest degree, followed by an astrological prognostication. Several later histories follow a similar practice. For instance, on a date equivalent to March 4 in 181 BC, the following entry is recorded in the *Han-shu*:

> The Sun was eclipsed and it was total; it was 9 degrees in *Yingshi* (which represents) the interior of the Palace chambers. At that time the (Dowager) Empress of Kao [-tzu] was upset by it and said, 'This is on my account.' The next year it was fulfilled.

The Empress Dowager died some eighteen months after the eclipse.

Down the centuries, similar measurements of right-ascension for comets and supernovae have proved of great importance in modern astronomy. Using such observations, it has proved possible to investigate in detail the past orbit of Halley's Comet and also identify the present-day remnants of several supernovae which occurred in our own galaxy before the invention of the telescope.

The *Xingjing* (Star Manual), a star catalog which has been dated by recent astronomical computation to around 70 BC, gives the earliest preserved measurements of north polar distance for the determinative stars of the *xiu* — as well as for many other stars. Using these and later measurements, and several carefully drawn star maps from around AD 1000 onwards, it is possible to identify the individual constituents of the *xiu* with stars listed in Western catalogs with considerable confidence. Certain lunar mansions, such as *Shen* (equivalent to the central portion of Orion) contain some of the brightest stars in the sky. Others, such as *Yugui* (part of Cancer) consist of stars only barely visible to the unaided eye.

The earliest chart of the lunar mansion asterisms dates from around 25 BC. This is preserved on the ceiling of a tomb in the city of Xian and came to light in 1987. Many later maps are extant — some displaying considerable detail. These often depict the night sky on a circular plan with the lunar mansion boundaries shown as straight lines radiating from the celestial pole at the center of the chart. Illustrations of the *xiu* are frequently found in tombs. It is interesting to note that whereas occidental representations of the zodiacal signs almost always use the various symbols — e.g. a crab for Cancer, two fish for Pisces — Chinese illustrations of the lunar mansions typically portray the outlines of the constellation patterns instead. These are often highly schematized.

Individual determinative stars cover a fairly wide range in declination. As a result, with the passage of time the effect of precession causes the widths of the *xiu* in right ascension to change somewhat. In particular, measurements around AD 1280 by the great Yuan Dynasty astronomer Guo Shoujing reveal that by his time the width of *Zuixi* — the narrowest lunar mansion — had decreased to almost zero. Over the next few centuries, *Zuixi* was tacitly assumed to exist with zero width until around 1650 when the Jesuit Adam Schall von Bell, who was then Astronomer Royal of China, rectified the situation. He took the bold step of reversing the order of *Zuixi* and the adjacent mansion *Shen*, thus assigning a finite width to *Zuixi* once more.

Jesuit missionaries, who made such a profound impression on Chinese astronomy in the seventeenth and eighteenth centuries, introduced improved techniques of measurement as well as knowledge of the far southern stars which were invisible from China. However, they did not attempt to supplant the indigenous lunar mansion system (which was fundamentally equatorial) with the zodiacal scheme which was standard in the West.

During the first millennium AD, the Chinese system of mapping the stars and its associated lunar mansions spread to other parts of East Asia, namely Korea, Japan and Vietnam. However, this appears to have been the limit of its sphere of influence. Although a parallel scheme of *nakṣatra* (star groups in the path of the Moon) — complete with determinative stars (*yogatārā*) — developed in India, measurements of the positions of these stars cannot be traced until the fifth century AD. By this time the *xiu* had already been firmly established in China for many centuries. It is evident that about one-quarter of the *yogatārā* coincide with the *chuxing*, but there are also major differences between the two systems. The origin of the various similarities is presently unknown.

Today, the lunar mansion system has been abandoned in East Asia — except at the popular level — in favor of Western techniques of specifying star positions.

F. RICHARD STEPHENSON

REFERENCES

Stephenson, Richard F. "Chinese and Korean Star Maps and Catalogs." In *The History of Cartography*. Eds. J.B. Harley and D. Woodward, Chicago: University of Chicago Press, vol. 2, part 2, pp. 511–578.

See also: Stars in Chinese Astronomy

LUNAR MANSIONS IN INDIAN ASTRONOMY In Indian astronomy the 27 or 28 *nakṣatras* (constellations) with their *yogatārās* (junction stars), all situated in the zodiac, correspond to the lunar mansions called *xius* in the East Asian, and *manāzil* in the Islamic tradition. *Nakṣatra* (lit. *na-kṣatra*, non-moving, fixed; *nakta-tra*, 'guardian of the night'), meaning 'star', refers to the 27 asterisms or star groups that occur on or on the sides of the zodiac. It refers also to the 27 equal spaces into which the zodiac can be divided, each space being equal to: $360°/27 = 13°20'$ or $800'$, commencing from the First point of Meṣa, which is the starting point of the zodiac in Indian astronomy. Now, the zodiac is divided into 12 equal parts, each being 30°, called 'sign' or *rāśi,* to accommodate the 12 solar months. Hence, each *rāśi* holds, inside it, two and a quarter *nakṣatra*-spaces, each distinguished by a prominent star which is called *yoga-tārā*.

Since the performance of rituals and sacrifices at specified times, days, seasons, and years was obligatory for the Vedic Indians, they were interested in the preparation of a workable calendar to be able to ascertain the specified times. This required the study of the motions of the sun and the moon, which moved along or near the zodiac. The fixed stars and constellations provided the astronomers with a stellar frame of reference against which they could follow and measure the movements of the sun, the moon, and the planets. Hence Indian astronomy identified and concentrated on the study, from very early times, of these stars only, to the exclusion of the general array of stars that stud the heavens. Thus the *nakṣatra* system of the Hindus came into being.

The *nakṣatras* had been identified even during the time of the *Ṛgveda*, though only those which were relevant to specific Vedic prayers were mentioned therein. In the *Yajurveda* and the *Atharvaveda*, however, all the *nakṣatras* were mentioned in the order in which they appear on the zodiac, since in those texts contexts required their mention in a row. In the *Yajurveda* literature, the several *nakṣatras* were assigned presiding deities, pictured as male or female or neuter, and the plurality specified. Legends have also been narrated to explain some of the characteristics of the *nakṣatras*, besides

Table 1: Constellations of Hindu astronomy

No.	Nakṣatra	Presiding deity	Gender	Plurality	Yogatārā
1.	Kṛttikā	Agni	Feminine	Plural	η Tauri
2.	Rohiṇī	Prajāpati	Feminine	Singular	α Tauri
3.	Mṛgaśiras	Soma	Neuter	Singular	λ Orionis
4.	Ārdrā	Rudra	Feminine	Singular	α Orionis
5.	Punarvasu	Aditi	Masculine	Dual	β Geminorum
6.	Puṣya	Bṛhaspati	Masculine	Singular	δ Caneri
7.	Āśleṣā	Sarpa	Feminine	Plural	α Caneri
8.	Maghā	Pitṛ	Feminine	Plural	α Leonis
9.	Pūrvaphalgunī	Aryamā	Feminine	Dual	δ Leonis
10.	Uttaraphalgunī	Bhaga	Feminine	Dual	β Leonis
11.	Hasta	Savitā	Masculine	Singular	δ Corvi
12.	Citrā	Indra	Feminine	Singular	α Virginis
13.	Svātī	Vāyu	Feminine	Singular	α Bootis
14.	Viśākhā	Indrāgnī	Feminine	Dual	α Librae
15.	Anurādhā	Mitra	Feminine	Plural	δ Scorpii
16.	Jyeṣṭhā	Indra	Feminine	Singular	α Scorpii
17.	Mūlā	Nirṛti	Feminine	Singular	λ Scorpii
18.	Pūrvāṣāḍhā	Āpaḥ	Feminine	Plural	δ Sagittarii
19.	Uttarāṣāḍhā	Viṣvedevāḥ	Feminine	Plural	α Sagittarii
20.	Śroṇā	Viṣṇu	Feminine	Singular	α Aquilae
21.	Śraviṣṭhā	Vasu	Feminine	Plural	β Delphini
22.	Śatabhiṣaj	Indra	Masculine	Singular	λ Aquarii
23.	PūrvaProṣṭhapada	Ajekapād	Masculine	Plural	α Pegasi
24.	UttaraProṣṭhapada	Ahirbudhnya	Masculine	Plural	γ Pegasi
25.	Revatī	Pūṣā	Feminine	Singular	ζ Piscium
26.	Aśvinī	Aśvin	Feminine	Dual	β Arietis
27.	Bharaṇī	Yama	Feminine	Plural	41 Arietis

specifying them as benefic or malefic, which aspect was elaborated in later astrological literature.

Yogatārā (Junction star) is the cardinal star in a *nakṣatra* which is made up of several stars. Normally, the *Yogatārā* would be the brightest star in the group, and the zodiacal signs would mostly be named after that star. The several constellations of Hindu astronomy, the details regarding them, and their *yogatārās* are listed in Table 1.

It is interesting that besides the above details, the work *Nakṣatra-kalpa*, an ancillary text of the *Atharvaveda*, also provides, among other things, the number of stars making up each constellation.

K.V. SARMA

REFERENCES

Bose, D.M., S.N. Sen, and B.V. Subbarayappa. *A Concise History of Science in India.* New Delhi: Indian National Science Academy, 1971.

Chakravarty, S.K. "The Asterisms." In *History of Oriental Astronomy, IAU Colloquium, 91.* Ed. G. Swarup et al. Cambridge: Cambridge University Press, 1987, pp. 23–28.

Dikshit, S.B. *Bhāratīya Jyotish Śāstra (History of Indian Astronomy)* Trans. R.V. Vaidya. pt. I. *History of Astronomy During the Vedic and Vedāṇga Periods.* Delhi: Manager of Publications, Civil Lines, 1969.

Modak, B.R. "Nakṣatra-Kalpa." In *Proceedings of the Twenty-sixth International Congress of Orientalists*, New Delhi, 1964, vol. III, pt. I. Poona: Bhandarkar Oriental Research Institute, 1969, pp. 119–122.

Report of the Calendar Reform Committee, Government of India. New Delhi: Council of Scientific and Industrial Research, 1955.

Yano, M. "The Hsiu-yao Ching and its Sanskrit Sources." In *History of Oriental Astronomy, IAU Colloquium 91.* Ed. G. Swarup et al. Cambridge: Cambridge University Press, 1987, pp. 125–134.

See also: Astrology in India – Astronomy in India

LUNAR MANSIONS IN ISLAMIC ASTRONOMY The old Arabs, before their acquaintance with Greek-based astronomy, had their own folk astronomy. They knew the fixed stars and used a number of stars and asterisms, the so-called *anwā'*, for orientation in nightly desert travels and for fixing seasons and predicting weather, especially rain. At an unknown time, and through unknown channels, they received from India the system of the 28 lunar mansions (*manāzil al-qamar*), stars or asterisms or spots in the sky, roughly along the ecliptic (the band of the zodiac through which the sun apparently moves in its yearly course), near which the moon was seen in the sky during its monthly revolution. Each mansion was identified with one of their *anwā'*-stars or asterisms. The complete list is presented in Table 1 (opposite).

Later, after the spread of Greek-based "scientific" astronomy in the Arabic–Islamic civilization, i.e. from the eighth century onwards, astronomers knew about this system and gave exact identifications of the mansions' stars from among the 1025 stars fixed up in the star catalogue in Ptolemy's *Almagest*. But the lunar mansions were not actually used by the Arabic–Islamic astronomers in their work. Their place was mostly in astrology. Here they were included in the numerous systems of divination, and in this context they were also borrowed into the Latin translations of Arabic astrological works, in Spain, in the late tenth century and, more often, in the twelfth and thirteenth centuries. Thus their corrupted, Latinized names are found throughout European astrological writings from that time on.

PAUL KUNITZSCH

REFERENCES

Kunitzsch, P. "al-Manāzil." In *Encyclopaedia of Islam*, new edition, vol. vi. Leiden: Brill, 1991, pp. 374–376. Reprinted in Kunitzsch, P. *The Arabs and the Stars*, Northampton: Variorum, 1989, item XX.

LUOXIA HONG Luoxia Hong was a Chinese astronomer active in the Western Han Dynasty about 100 BC He came from southwest China as a folk astronomer. When the Emperor Wu (140–87 BC) decided to draw up a new calendar, Luoxia Hong and more than twenty other astronomers were called together in the capital Chang'an. These folk and imperial astronomers put forward their plans for a new calendar. After comparing eighteen plans, the emperor believed that the best one was put forward by Luoxia Hong and another astronomer Deng Ping. The new calendar was applied to the whole country in 104 BC and called *Tai Chu Li* (Tai Chu Calendar). This calendar became the standard model for nearly two thousand years. It was a luni-solar calendar with the 19-year and 7-leap cycle, and included the calculation for the motion of the sun, moon, planets, and for eclipses. In fact it was a mathematical astronomy system. The emperor intended to confer an official position on Luoxia Hong to cite his achievements, but Luoxia refused it and preferred to live in seclusion.

Luoxia Hong made an equatorial armillary sphere to measure the data of his new calendar. He is one of the candidates for the inventor of this instrument.

JIANG XIAOYUAN

Table 1: Lunar mansions and their modern identification (see LUNAR MANSIONS IN ISLAMIC ASTRONOMY)

1.	*al-sharaṭān* (also *al-naṭḥ*)	$\beta\gamma$ or $\beta\alpha$ Arietis
2.	*al-buṭayn*	$\epsilon\delta\rho$ Arietis
3.	*al-thurayyā*	the Pleiades
4.	*al-dabarān*	α Tauri
5.	*al-haqʿa*	$\lambda\phi^{1,2}$ Orionis
6.	*al-hanʿa*	$\gamma\xi$ Geminorum; also *al-taḥāyī*, $\eta\mu\nu$ Geminorum (to which also $\gamma\xi$ Geminorum are sometimes added)
7.	*al-dhirāʿ*	$\alpha\beta$ Geminorum
8.	*al-nathra*	ϵ Cancri or M 44
9.	*al-ṭarf*	κ Cancri + λ Leonis
10.	*al-jabha*	$\zeta\gamma\eta\alpha$ Leonis
11.	*al-zubra* (also *al-kharātān*)	$\delta\theta$ Leonis
12.	*al-ṣarfa*	β Leonis
13.	*al-ʿawwāʾ*	$\beta\eta\gamma\epsilon$ Virginis (to which δ Virginis is sometimes added)
14.	*al-simāk*	α Virginis
15.	*al-ghafr*	$\iota\kappa\lambda$ Virginis
16.	*al-zubānā*	$\alpha\beta$ Librae
17.	*al-iklīl*	$\beta\delta\pi$ Scorpii
18.	*al-qalb*	α Scorpii
19.	*al-shawla*	$\lambda\upsilon$ Scorpii; sometimes instead, *al-ibra*, for the same stars or for M 7
20.	*al-naʿāʾim*	$\gamma\delta\epsilon\eta\sigma\phi\tau\zeta$ Sagittarii
21.	*al-balda*	a starless region between nos. 20 and 22
22.	*saʿd al-dhābiḥ*	$\alpha\nu\beta$ Capricorni
23.	*saʿd bulaʿ*	$\mu\epsilon$ Aquarii (sometimes FL, 7 or ν Aquarii are also included)
24.	*saʿd al-suʿūd*	$\beta\xi$ Aquarii + c^1 Capricorni
25.	*saʿd al-akhbiya*	$\gamma\pi\zeta\eta$ Aquarii
26.	*al-fargh al-muqaddam* (or *al-fargh al-awwal*)	$\alpha\beta$ Pegasi
27.	*al-fargh al-muʾakhkhaṛ* (or *al-fargh al-thānī*)	γ Pegasi + α Andromedae
28.	*baṭn al-ḥūt* (also *al-rishāʾ*)	β Andromedae

REFERENCES

Ban Gu. *Han Shu* (The History of the Han Dynasty), vol. 21. Beijing: Zhonghua Press, 1962.

Sima Qian. *Shi Ji* (The Historical Record), vol. 26. Beijing: Zhonghua Press, 1959.

The Study Group for the History of Chinese Astronomy. *Zhong Guo Tian Wen Xue Shi* (The History of Chinese Astronomy). Beijing: Science Press, 1981.

See also: Calendars in East Asia – Armillary Sphere

M

MĀDHAVA OF SAṄGAMAGRĀMA During the Muslim rule in north India, there was a decline in Hindu culture. This adversely affected the creative spirit in indigenous art, literature, and science. Southern India was comparatively less affected, and traditional culture and the sciences flourished there. The followers of Āryabhaṭa I made enormous contributions to the development of mathematical sciences. It was a golden age of Indian mathematics.

Mādhava of Saṅgamagrāma, who flourished about AD 1400, was the first great astronomer and mathematician of the Late Āryabhaṭa school, which he in fact founded. Saṅgamagrāma has been identified as the modern Irinjalakkuda, a town near Cochin in Kerala State. Mādhava belonged to the Emprantiri subcaste group of Kerala Brahmins. We have no knowledge about his parents and teachers, or of the exact dates of his birth and death. Various dates ranging from AD 1336 to 1418 are used in his works. Hence the period of activity of his life has been roughly fixed from AD 1340 to 1425.

There is no doubt that Mādhava was an extraordinarily brilliant man. He used his talent and sharp intelligence to acquire knowledge by private study, and could thus overcome the difficulty of finding a good Guru because of his inferior status in the dominant Namputiri Brahminic community. He was a self-taught genius and not a gifted pupil. He was generally referred to as *Golavid* (Master of Spherics) by subsequent scholars and followers of his School, such as Nīlakaṅṭa Somayāji (AD 1444–1545) and Acyuta Piṣāraṭi (AD 1550–1621).

Mādhava wrote all his works in Sanskrit, the classical language of India. One of his earliest works is the *Candra-Vākyāni* (Moon Sentences). This was composed as a revision of Kerala's ancient traditional astronomical work attributed to Vararuci, who lived a thousand years earlier. The *Candra-Vākyāni* gives 248 mnemonic phrases regarding the longitudinal position of the moon for each of the 248 days which comprise a period of nine anomalistic months.

The *Sphuṭacandrāpti* (Computation of True Moon) contains 51 verses and is a work of the *Karaṇa* category. A *Karaṇa* is a handbook on practical astronomy. In this one, he provides an ingenious method for finding the true position of the moon.

Mādhava's *Veṇvāroha* (Bamboo Climbing) is an elaboration of his *Sphuṭacandrāpti* and consists of 74 verses. In this work the author created a facile procedure to find the true lunar positions at intervals of about half an hour. It is dated as AD 1403, and is the most popular astronomical work of Mādhava. Acyuta Piṣāraṭi wrote a Malayalam commentary on it.

A recently identified astronomical work of Mādhava is *Agaṇita-grahacāra*. It is an extensive work on planetary computations using somewhat novel methodologies. It is a treatise of the *Karaṇa* category and must have been composed just after AD 1418 which is the latest date mentioned.

Among other unpublished works of Mādhava there are two short astronomical tracts. One is the *Madhyamānayana-prakāra* (Method for Computing Mean Positions) which is extant in a unique manuscript at the India Office Library. The other is *Lagnaprakaraṇa*, of which at least three manuscripts exist in South India. This work deals with computations of ascendents.

It is possible that Mādhava composed a work on *Golavāda* (Spherics) which earned him the appellation *Golavid*. But the reported manuscript from a private collection has been eaten by white ants. Mādhava was also the author of a number of stray or free verses which have been cited by later authors and commentators.

Scientific Contributions

The traditional "moon sentences" of Vararuci (fourth century), used in Kerala, gave daily longitudes of the moon only up to minutes of the arc or angle. Mādhava computed more sophisticated moon sentences which expressed the longitudes correctly up to seconds. By making use of the popular system of alphabetic numerals, called the *Kaṭapayādi Nyāya*, these mnemonic phrases were made short and aphoristic. Mādhava also provided a value of π using a system of word numerals (*bhūta-saṁkhyās*).

The knowledge of an accurate value of π enabled Mādhava to obtain a better value of the traditional *Sinus Totus* (Total Sine, or radius). Mādhava's sine table is quite precise and accurate. He may have used traditional methods for getting the table or the newly discovered power-series expanion of sine (see below). For computing sine for the argument intermediary between any two tabulated angles, he knew a formula which is equivalent to the modern Taylor series approximation up to the second order. Higher interpolation based on second order finite differences had been known in India since the time of Brahmagupta (seventh century AD).

For computing π to any desired degree of accuracy, Mādhava discovered a number of series including the one $\pi/4 = 1 - \frac{1}{3} + \frac{1}{5} - \frac{1}{7} \cdots$, often called the Leibniz series after the German mathematician G.W. Leibniz, who rediscovered it in 1673 or so.

Another formula perhaps known to Mādhava is now called the Gregory Series, after the Scottish mathematician James Gregory (1638–1675). The Indian proof is found in the *Yuktibhāṣā* (AD sixteenth century) and other works.

One of Mādhava's major achievements was the discovery of the power-series expansions of sine and cosine, which were rediscovered in Europe at the time of Newton, and which are equivalent to

$$\sin x = x - \frac{x^3}{3!} + \frac{x^5}{5!} - \frac{x^7}{7!} + \cdots,$$

$$\cos x = 1 - \frac{x^2}{2!} + \frac{x^4}{4!} - \frac{x^6}{6!} + \cdots.$$

The two Sanskrit verses which embody the method of computing sine based on power series up to x^{11} are quoted by Nīlakaṇṭha in his commentary on the *Āryabhaṭīya* (II, 17b).

In this connection it is relevant to discuss a small tract called *Mahājyānayana-prakāra* (Method for the Computation of the Great Sines). It gives the power-series methods for computing *Mahājyās* (Great Sines). Unfortunately there is no mention of an author's name in it, although Mādhava's rule for computing sines up to x^{11} is mentioned. K.V. Sarma attributed the tract to Mādhava, but Gold and Pingree consider it to be the work of his follower(s). Perhaps Mādhava explained his theory during lectures to his pupils, whose lecture notes may be the basis of the above tract.

R.C. GUPTA

REFERENCES

Primary sources

Candra-Vākyāni. Ed. K.V. Sarma as the appendix to his edition of *Sphuṭacandrāpti* as well as of *Veṇvāroha* (see below).

Mahājyānayana-prakāra. Ed. and trans. D. Gold and D. Pingree. *Historia Scientiarum* 42: 49–65, 1991.

Sphuṭacandrāpti. Ed. K.V. Sarma. Hoshiarpur: Vishveshvaranand Institute, 1973.

Veṇvāroha. Ed. K.V. Sarma, with a Malayalam commentary of Acyuta Piṣāraṭi. Tripunithura: Sanskirt College, 1956.

Secondary sources

Gupta, R.C. "Mādhava's Power Series Computation of the Sine." *Gaṇita* 27: 19–24, 1976.

Gupta, R.C. "South Indian Achievements in Medieval Mathematics." *Gaṇita Bhāratī* 9: 15–40, 1987.

Gupta, R.C. "The Mādhava-Gregory Series for $\tan^{-1} x$." *Indian Journal of Mathematics Education* 11(3): 107–110, 1991.

Gupta, R.C. "On the Remainder Term in the Mādhava-Leibniz's series." *Gaṇita Bhāratī* 14: 68–71, 1992.

Hayashi, T. et al. "The Correction of the Mādhava Series for the Circumference of a Circle." *Centaurus* 33: 149–174, 1990.

Pingree, D. *Census of the Exact Sciences in Sanskrit*, Series A. vol. 4. Philadelphia: American Philosophical Society, 1981.

Sarma, K.V. "Date of Mādhava, a Little-known Indian Astronomer." *Quarterly Journal of the Mythic Society* 49(3): 183–186, 1958.

Sarma, K.V. *A History of the Kerala School of Hindu Astronomy*. Hoshiarpur: Vishveshvaranand Institute, 1972.

MAGIC AND SCIENCE The concepts "magic" and "science" are products of Euro-American history; thus their use in other regions of the globe, from the colonial era to modern anthropological studies, is intertwined with Western intellectual history. Consequently, understanding the meaning of, and relationship between, magic and science in the context of western notions of rationality is essential when examining phenomena in non-Western societies placed in these categories. Magic and science are, in essence, labels used to exclude and include according to an intellectual value system rooted in European history. They are part of the cultural baggage taken abroad by Euro–American travelers, and used to identify "otherness" in foreign cultures. In non-Western cultures, similar practices of "magic" were not necessarily excluded or marginalized by the growth of science as they were in the West.

The European evolution of these two words from the classical (Greco-Roman) era to the twentieth century shows a growing gap between magic as occult or hidden knowledge on the one hand, and science and religion as public knowledge on the other. Increasingly in the European intellectual tradition, science was defined in narrower ways while pushing magic out of the realm of knowledge. This contributed to an evolutionary paradigm applied by Westerners to non-Western societies, of progress from magic to religion to science, a model now called into question by modern anthropologists. Because of this conceptual evolution in the intellectual history of Europe, "magic" and "science" can be used in a number of senses in modern English usage (see the *Oxford English Dictionary*). The gradual transformation of the word science as a distinctive rationality valued above and against magic is part of a uniquely European duality not generally found in non-Western societies, where magic can exist side-by-side with science and religion.

History of the Terms in the Western Intellectual Tradition

European intellectual history is full of self-imposed oppositions: temporal–spiritual, natural–supernatural, pagan–Christian, devil–God, magic–religion, magic–science. All of these are subject to a moral scale of Good versus Evil, a distinction in Western thought that has its roots in the Judeo–Christian monotheistic system positing a single, omnipotent, all-good Deity. This way of thinking is very different from, for example, the Chinese world view embodied in *Yin* and *Yang*, opposites that create balance (positive–negative, active–passive forces) without the identification of Evil versus Good. Thus, the Western dualities are not, as some westerners visiting other cultures have assumed, universals found in all cultures; rather they are a particular product of the belief systems and intellectual history of Europe.

None the less, despite these polar oppositions in European thought, changes in definitions over time caused overlaps and gray areas. Thus magic, as the opposite of religion or science, has a history that complicates the way the word is used at different times and places by different classes of people. The self-defined shape of European history is one of progress from magic to religion to science: from root definitions in classical culture (Greco–Roman), through the medieval magical and religious mentality, through major religious and intellectual changes (renaissances) in the medieval and early modern world, to the development of science as a separate discipline in the modern world. At the same time, this chronological picture is muddied by the slippery definition of magic as it changes in relation to the growth of religion and science. Throughout the development of these distinctions, from the fourteenth through the nineteenth and twentieth centuries, Europeans and Americans went abroad and applied these differing notions to the peoples they met.

The root meaning of magic contains a sense of exclusion found throughout the history of the term: in the Roman world, the Latin *magia*, derived from *magi* (Persian astrologers like the Magi of the Christmas story), implied a foreigner, even when the practitioner was a Roman. This was someone who possessed secret and powerful knowledge both feared and respected, displayed in the ability to manipulate unseen or spiritual agencies, in such arts as divination, astrology, curses, oracles, and amulets to ward off evil (Luck, 1985). Thus the word magic has at its root a sense of marginality in its otherness and its paranormal, unknown, and supernatural associations, but also a strong sense of power held exclusively and secretly by the *magus*. The root word for science, on the other hand, is more normative: *scientia* includes knowledge, art, or skill. It derives from the Greek heritage a strong sense of human rationality, but comes to include divinely-revealed knowledge as well in the Christian era (from ca. 200).

In the progress model, the magic of the European past is associated with the medieval period (ca. 500–1350), in contrast to the rationality of the Renaissance (fourteenth and fifteenth centuries) and Enlightenment (eighteenth century) intellectual revolutions. "Medieval" thinking has earned the label "magic" from later generations on several grounds. The medieval otherworldly emphasis and reliance on divine revelation led to a lack of distinction between natural and supernatural and contributed to an allegorical way of thinking about nature, so that objects such as a flower and events such as storms or illnesses were read as divine messages. Medieval thinking rested on a belief in a wonder-working, ever-present God and also a magic-working evil presence in the devil; sometimes the miracles of God's saints and the tricks of the devil appear similar in method in the eyes of later thinkers (Flint, 1991; Thomas, 1971). Specifically, the belief in supernatural powers in words is found in both Germanic animism (worship of nature spirits) and the Christian tradition: in Germanic animism charms (ritual words and actions) invoke the inherent virtues of a plant. In the Christian eucharist, bread and wine are changed into flesh and blood through prayer. The two were joined in Christianized folk medicine in the production of charms using Christian prayers as the powerful words (Jolly in Neusner et al., 1989). To Protestants (after the sixteenth century) and anthropologists (late nineteenth century), these practices are all magical in their manipulation of nature through word-magic (as opposed to true religious prayer as supplicatory). Yet to the medieval mind, magic was defined by *who* — God or the Devil — not by *how* — supplication versus manipulation. Thus medieval thinking was rejected as backward by later rationalists and was used to describe cultures that had not advanced out of magical or superstitious thinking.

Religion, in the history of European "progress", moves away from magical thinking and opens the door for rationalism. The tradition of logic and deductive reasoning dates back to the Greeks and partially survived into the Middle Ages through Roman-Christian church leaders and thinkers; the recovery of Aristotle through Arab sources in the twelfth century helped spur a renaissance in learning among medieval scholars so that human reason was placed alongside divine revelation as a way of knowing truth. This interest in the potential of human reason to understand things in conjunction with divinely-revealed knowledge cleared a space for reason to function independently over the succeeding centuries. The separation of natural from supernatural, and reason from revelation, allowed thinkers to focus on the human study of natural phenomena. Magic in this context became things not in the category of the divine (miracles) and not subject to human reason either: black magic associated with the devil and witches, such as curses and evil spells, or the low magic of ignorant persons based on false reasoning, such as herbal charms and love potions. High, white magic was associated with the intelligentsia of the high Middle Ages and Renaissance (twelfth through fifteenth centuries).

These early scientists dabbled in the occult, a gray area in between divine knowledge and human reason: occult phenomena were insensible (not subject to human sense perception), such as magnetism, gravity, or the pull of the stars, but might be intelligible (something to be reasoned about); these occult phenomena, classified as "natural magic", became the sciences of astrology, astral medicine, and alchemy, for example.

In the Scientific Revolution, science became a separate, and increasingly higher, discipline from religion; it came to mean exclusively the human (versus divine) study of natural (versus supernatural) phenomena. This secularization of knowledge was the product of the Italian Renaissance of the fourteenth and fifteenth centuries and the Enlightenment of the seventeenth and eighteenth centuries. Simultaneously, Protestant ethical values contributed to this process a utilitarian view of the created order that effectively circumscribed religion into a rational system: God made the world to work by certain laws that humans could understand and systematize (Tambiah, 1990). Human study could produce a true understanding of reality independent of divine insight. Magic was now clearly marked as something not rational: it could be proven to be a hoax (prestidigitation or sleight-of-hand), and was relegated to the entertainment industry where it could be enjoyed as an illusion, a deception that could be scientifically explained. As science became the religion of the modern west, magic was being exorcised from modern consciousness not as demonic but as irrational and backwards.

This simple pattern of progress from magic to religion to science is misleading in two ways: it is anachronistic in applying later definitions of magic back on to earlier periods where the word had different meanings, and it does not take into account the overlaps and continuities whereby magic survives alongside religion and well into science. For example, the Scientific Revolution is compromised by intellectual dabbling in the occult: the great shift from a geocentric worldview (earth at the center) to heliocentric (sun at the center) was founded not just on forward-looking developments in mathematics and astronomy but was motivated by a backward-looking interest in a supposedly Egyptian magical tradition, Hermeticism (Tambiah, 1990). Differences of class in relation to conceptions of magic further complicate this picture of Euro-American intellectual history: the older ways of folk belief, in medicine for example, as a viable, not magical, method of manipulating nature or spiritual agencies was retained among many classes long after the religious authorities or the intellectual elite had dispensed with it, and in some cases had begun investigating it as witchcraft (anti-religion) or fraud (unscientific). All of these divergent attitudes toward magic were carried abroad by Europeans and Americans: magic as demonic, evil, and fearful; magic as medieval or backward; magic as unscientific, irrational or uncivilized; but always as something "other."

Modern ethnography, the study of cultures, is a product of this western history and its intellectual legacy of magic versus religion or science. The earliest ethnographers were explorers and missionaries, some of whom made an effort to observe and document these "new" peoples. The paradigm of progress some missionaries used in meeting non-urban, preliterate peoples was to categorize them as children needing to be fostered into adulthood; other colonizers used a model to exploit the "Indians" as sub-human slaves. For the missionaries, conversion was one step in the maturing process necessary for the native peoples to reach the "level" of civilization mastered by the Europeans.

Modern anthropology attempted to break with the religiously-biased view of these missionaries and take an objective observer stance which was, none the less, still colored by an evolutionary model of progress from magic to religion to science (Herbert, 1991). This model is clearly evident in the works of early nineteenth-century founders Edward Tylor and Sir James Frazer and into the twentieth century in Jacob Bronowski and Bronislaw Malinowski (Tambiah, 1990). The anthropological definition focused on magic as unscientific manipulations of nature or supernatural forces and classified it according to its false premises (imitative magic, contagious magic, sympathetic magic). These notions of magic were assumed as a universally valid construct applicable cross-culturally. Consequently, observed peoples were placed into the spectrum of development from magic to science. This model is the subject of debate in late twentieth-century anthropological scholarship, by such authors as Francis Hsu and Stanley Tambiah, who question whether the European concept of magic can be used accurately to classify a set of phenomena in a non-European culture.

Indigenous Views in Non-Western Societies

In Western thought, then, magic has become something marginal, separate from or opposite to a mainstream tradition of religion or science. Non-Western practices of magic seen in their own cultural context are not the opposite of religion or science, but are complementary to their political, social, religious orders; magic is not the "other" in their worldview, but is part of the norm. Magical practices in non-Western societies can function as part of their cultural identity, alongside scientific development or as a subgroup of religion. In many parts of the world, syncretism is more prominent as a response to alternate worldviews, resulting in coexisting modes of rationality rather than competing ones.

The ancient civilizations of Asia offer examples of traditions developing their own modes of rationality with different dynamics than the European model, between the spheres of religion, science, and magic. In China, the traditions are as complex and overlapping as they are in Europe: ancestor-worship, Confucianism, Daoism, and Buddhism as belief-systems evolved and interacted amid the simultaneous development of science and technology. These belief-systems cannot be easily categorized as religion, philosophy, or magic along Western lines. The Buddhist emphasis on the world as illusory and the Daoist focus on metaphysics lent themselves more readily to practices resembling magic in a Western sense (appeals to supernatural aid, fortune-telling), as did ancestor worship. Confucianism, on the other hand, has both religious and philosophical elements; its concern with the social and political world resembles more the secular humanism of the western tradition. All of these coexist as alternate, and sometimes complementary, modes of rationality in China.

Similarly in the Chinese world, science and technology do not necessarily replace magico-religious belief: villagers' responses to crisis (for example, a plague) incorporate both medicinal remedies such as serums proven through experimentation in the Western scientific tradition, and rituals seeking to appease the gods. While Westerners would be under some pressure to justify such magical practices with some rational or pseudo-scientific explanation, the Chinese do not feel compelled to argue about where or how the practice fits into some duality of true or false, natural or supernatural, orthodox or heretical, scientific or magical. Ancestor worship and recourse to geomancers (practitioners of *feng shui*, the art of finding spiritually-correct locations for buildings) and fortune tellers (Daoist priests or other spiritualists) continue in twentieth-century China and in Chinese communities in the West without shame or apology (Hsu, 1983).

Elsewhere in Asia, "magic" continues as part of mainstream culture because it is closely linked with cultural identity. In Korea, female shamans perform ritual cures and exorcisms (*kut*) as part of the uniquely Korean fabric of life, alongside imported traditions from China and the West; in this way shaman practice is part of a unique Korean identity (Kendall, 1985). Shintoism in Japan, like ancestor-worship in China, is part of the Japanese cultural heritage embedded in everyday life, so much so that it easily accommodated an incoming religious system such as Buddhism or the development of science and technology, with which it has no reason to quarrel. Thus, syncretism, rather than opposing dualities, is a common pattern in the dynamics of magic, science, and religion in Asian cultures.

In many of the ancient near eastern and south Asian cultures, practices resembling magic (fortune-telling, amulets, sorcery) form a subgroup of either religion or science/medicine. Magicians in India, for example, are closely linked with the religious traditions of Hinduism, Buddhism, and Islam. Indian thought emphasizes the illusory nature of the physical world; religious masters (gurus, yogi, and other mendicants) are able to manipulate the natural world precisely because they have reached a state of enlightenment where the world is truly an illusion to them. Likewise street magicians and stage entertainers practice "deceptions" (sawing people, producing trees from a basket, pulling an egg or a bird out of a bag, making ropes rise) that echo Hindu or Islamic stories and thus embody certain truths about the world as a wondrous, deceptive, and illusory place that one looks *through* to find meaning. Although such street magicians are a separate, low (Muslim) caste in India, they are a prominent part of India's cultural landscape, popular as reflections of Indian values (Seigel, 1991).

The interconnection of belief, science, and magic is also seen in Islamic culture. Medieval Islam fostered scientific, technological, and medical research because of their belief in a monotheistic Deity who made all things rational for humans to study (a view that eventually sparked medieval European science). Yet Islam also has magical–mystical subgroups, some of whom were condemned by Islamic law: their explorations of astronomy have astrological connections, as in Europe; their mathematical concepts have occult meaning to some; Sufi mystics distanced themselves from the Islamic intellectual heritage and sought knowledge through other, spiritual or interior, means. While these branches of Hinduism, Buddhism, or Islam may be minority groups, they are not marginal to their own cultures. Rather, they form an important extension of mainstream ideas that influences the whole culture, although not without conflict.

Societies in Africa, the Pacific, and Southeast Asia that did not have the urban, literate characteristics of these older civilizations provide clear examples of how poorly the conceptual dichotomy of magic versus science works as a model for understanding cultures where so-called magical practices are part of the norm. In Mali (Africa), the *Sundiata*, a twentieth-century version of an oral tradition dating back to the thirteenth century, speaks of the war magic used by Malian sorcerers to make a tribe successful so that others feared them; their magic oracles dispensed wisdom for successful living, a combination of prophecy and character insight. Likewise, many Pacific island cultures practiced a kind of potent magic in love, war, and healing that relied on a sorcerer's ability to conjure nature and spirits.

These practices encompass both white and black magic in the European paradigm, to both heal and curse; such mastery is always dangerous to the practitioner because of the power of the sources he or she is using. In New Caledonia (Melanesia), sorcerers concoct love potions; in Samoa (Polynesia), chants ward off a headache caused by a god. In Pohnpei (Micronesia), tribal groups employ sorcery to win a war, calling on ghosts or natural forces such as tides; during the Spanish occupation, they used both the borrowed technology of guns and their own tradition of sorcery to hold off attacks (Hanlon, 1992). Marquesans (in French Polynesia) also practiced a kind of war "magic" closely linked to their tribal identity: chants of power specific to their people and location used words ritually to invoke natural/supernatural forces to aid them in battle. The concept of *mana* in Polynesian cultures embodies this sense of powerful forces that can be channeled through words and actions, and is a more authentic construct for understanding their practices than is "magic".

Such beliefs and practices, prior to European contact, were part of political and social value systems; the practitioners were feared for their power, but were not marginal to an intellectual belief system, as they became in European culture. The retention of these beliefs in the power of the old ways after contact, sometimes in defiance of, and sometimes integrated with, Western beliefs, is a form of cultural resistance and identity.

Many non-Western cultures, confronted with the European intruder, perceived the actions of the newcomer in terms of their own construct of supernatural or occult power. For example, in the Americas, native American Indians identified literate Europeans (mostly missionaries) as shamans, and their books as tools for manipulating natural/supernatural forces. Literacy was a powerful form of knowledge to acquire, and therefore was classified with other powerful forms of knowledge in their culture held by their shamans (magic and magicians to the Europeans). Thus literacy, and the Christian religion wound around it, was incorporated into the native belief system (Axtell, 1988). Indeed, in the same way that the taking of photographs was feared by some groups as a type of sorcery for capturing their souls, so too the writing down of history and ethnographic observations (by Westerners) is sometimes perceived as a kind of sorcery: the power of shaping and defining cultural identity past and present.

Just as the magic–religion divide is indistinct, so too the magic-science line is easily crossed. Westerners cast doubt on the ancient Polynesians' ability to navigate the Pacific to reach new islands such as the Hawaii chain without the technological tools used by Europeans to accomplish such tasks. New research and recreated voyages, however, confirm the knowledge of wind, sky, and water, and the skills of navigation contained in the remnants of Polynesian chants and other oral traditions, usually categorized as magico-religious rituals. This ambiguous line between magic and science is also visible in modern globalized medical practices that incorporate both Western medical techniques and traditional medicine from non-Western societies. The Chinese practice of acupuncture, once viewed in the United States as magic or pseudo-science, is now being mainstreamed into American medicine in lieu of drugs. In Java, a doctor claims the ability to produce heat and electricity from his own body for healing purposes and he documents this ability using the Western scientific mode of proof (experimentation, repeatability), performing spontaneous combustion on video (*Ring of Fire*). Such global syncretism is increasing rather than decreasing, blurring the distinctions between magic and science as defined in European intellectual history.

Traditionally, then, magic has been defined in opposition to science or religion in Western intellectual development. However, such practices in non-Western cultures that appear to resemble this category or are similar to practices Westerners label magic, may not in those cultures have been defined as a class in opposition to some concept of religion or science. The realization that "one man's religion (or science) is another man's magic" is leading to redefinitions of these terms and the development of meanings and categories unique to each cultural context.

KAREN LOUISE JOLLY

REFERENCES

Axtell, James. *After Columbus: Essays in the Ethnohistory of Colonial North America.* New York: Oxford, 1988.

Flint, Valerie I.J. *The Rise of Magic in Early Medieval Europe.* Princeton: Princeton University Press, 1991.

Hanlon, David. "Sorcery, 'Savage Memories,' and the Edge of Commensurability for History in the Pacific." In *Pacific Islands History: Journeys and Transformations.* Ed. Brij Lal. Canberra: The Journal of Pacific History, 1992.

Herbert, Christopher. *Culture and Anomie: Ethnographic Imagination in the Nineteenth Century.* Chicago: University of Chicago Press, 1991.

Hsu, Francis L.K. *Exorcising the Trouble Makers: Magic, Science and Culture.* Westport, Connecticut: Greenwood Press, 1983.

Kendall, Laurel. *Shamans, Housewives, and Other Restless Spirits: Women in Korean Ritual Life.* Honolulu: University of Hawaii Press, 1985.

Kieckhefer, Richard. *Magic in the Middle Ages.* Cambridge: Cambridge University Press, 1990.

Luck, Georg. *Arcana Mundi: Magic and the Occult in the Greek and Roman Worlds.* Baltimore: Johns Hopkins University Press, 1985.

Malinowski, Bronislaw. *Magic, Science and Other Essays.* Boston: Beacon Press, 1948.

Neusner, Jacob, Ernest Frerichs, and Paul Virgil McCracken Flesher, eds. *Religion, Science, and Magic: In Concert and In Conflict.* New York: Oxford University Press, 1989.

Seigel, Lee. *Net of Magic: Wonders and Deceptions in India.* Chicago: University of Chicago Press, 1991.

Tambiah, Stanley. *Magic, Science, Religion, and the Scope of Rationality*. Cambridge: Cambridge University Press, 1990.

Thomas, Keith. *Religion and the Decline of Magic.* New York: Charles Scribner's Sons, 1971.

See also: Geomancy – Navigation in Polynesia – Navigation in the Pacific

MAGIC SQUARES IN CHINESE MATHEMATICS By modern definition a magic square is an arrangement of numbers in a square whereby the sum of the numbers in every individual row, in every individual column and in each of the two diagonals of the square is identical. Figures 1 and 2 are examples of magic squares. In Figure 1 all rows, columns and diagonals add up to the sum 15 and in Figure 2 they add up to 34.

Each number occupies a cell. There are 9 cells in Figure 1 and 16 cells in Figure 2. Figure 1 is known as a 3×3 magic square or a magic square of the order 3; similarly Figure 2 is a 4×4 magic square or a magic square of the order 4. Methods for constructing these two magic squares as well as some magic squares of higher orders are given in the *Xugu zheqi suanfa* (Continuation of Ancient Mathematical Methods for Elucidating the Strange), composed in 1275 by Yang Hui. Numbers can also be arranged to form a circle or even a cube that shares some of the properties of the magic square. Nowadays all these come under the heading "number theory" or even "mathematical recreation" in libraries. However, the word "magic" in the magic square has lost some of its original meaning.

Chinese magic squares have attracted the attention of modern mathematicians and historians of science. Studies by modern scholars have brought forth many interesting results, but how and why certain Chinese magic squares were constructed cannot be fully understood if one approaches the subject entirely from the standpoint of modern mathematics. The modern Chinese equivalent of the term "mathematics" is *shuxue*, but prior to the middle of the nineteenth century the same term had a much wider meaning, embracing mathematics, philosophy, astrology, and divination. Studies from the angle of modern mathematics have recently been supplemented by investigations from a different perspective.

The first-century book *Da-Dai Liji* (Record of Rites by the Elder Dai), contains the following arrangement of numbers:

2, 9, 4   7, 5, 3   6, 1, 8.

This set of numbers also occurs in a probably earlier mathematical text, the *Shushu jiyi* (Memoir on Some Traditions of Mathematical Art), said to be written in the year 190 BC by Xu Yue. This is the earliest magic square on record. It was used mainly in astrology and divination. Then during the twelfth century one of Zhu Xi's (1130–1220) disciples Cai Yuanding (1145–1198) identified the Bright Hall and the Nine-palace arrangement as the legendary *Luoshu* chart mentioned in ancient texts. Together with another legendary chart, the *Hetu* River Diagram, the riddles of the universe were supposed to be embodied therein. Hence the two charts were used to interpret not only natural phenomena, but also philosophy and human behavior. They even found their use in Daoist ceremony and magic. Numerology played a profound role in Chinese magic squares. The occurrence of certain numbers, e.g. 5, 9, 25, 49, 50, and 64 in Chinese magic squares is significant. All these were closely associated with the *Hetu* and *Luoshu* charts and with the *Yijing* (I Ching, The Book of Changes).

Yang Hui's *Xugu zheqi suanfa* is the earliest and best source for Chinese magic squares. The methods of construction of some of the larger magic squares remind us of feet movements of the Daoists performing ceremonies and magic. About three decades later Ding Yidong wrote his *Dayan suoyin* which contains a number of magic squares and magic circles constructed on a similar basis. Then in about the year 1593 Chen Dawei collected a number of magic squares in his *Suanfa tongzong* (Systematic Treatise on Arithmetic) without involving himself with the theoretical aspect of the subject. In 1661 Fang Zhongtong (1633–1698) wrote the *Shuduyan*, which contains magic circles, cubes, and spheres, besides magic squares. Next comes Zhang Chao's (b. 1650) *Xinzhai zazu* (Miscellanea of Zhang Xinzhai), which includes a Supplement to the magic

4	9	2
3	5	7
8	1	6

Figure 1

4	9	5	16
14	7	11	2
15	6	10	3
1	12	8	13

Figure 2

squares of Cheng Dawei's work. The last description of magic squares by a traditional Chinese scholar came in the latter part of the nineteenth century when Bao Qishou wrote his *Binaishanfang ji* (Collections of Writings in the Binai Mountain Studies).

The *Luoshu* chart magic square probably first went to India and then to the Arab countries, where other magic squares were later developed in their individual ways. From the Arab countries magic squares were said to be first brought to Europe by a Byzantine, Manuel Moschopoulos (fl. ca. 1295–1316). Transmission of knowledge between China and her western neighbors was seldom uni-directional. In 1956 a thirteenth-century 6 × 6 Muslim magic square was excavated near the city of Xi'an. Magic squares even had a role to play in trade between China and the Arab countries in the past. In 1906 when Queen Mary of the British Empire visited Hyderabad in India she was presented with a Chinese porcelain plate decorated with Arabic inscriptions and a magic square. This plate is now preserved at the Victoria and Albert Museum in South Kensington. It was manufactured in the eighteenth century at the world renowned Chinese kiln center, Jingdezhen. It was used originally by the Muslims as a medicine bowl, so that the inscriptions were taken from the *Quʾrān*, and the magic square was believed to possess powerful virtues for protecting life, healing the sick, and bringing about a comfortable delivery when a pregnant woman sat on it. Other specimens of Chinese porcelain plates bearing magic squares in a corrupted form are among the collections of museums and private collectors today.

HO PENG YOKE

REFERENCES

Andrews, W.S. *Magic Squares and Cubes*. Chicago: Open Court Publishing, 1908.

Cammann Schuyler. "Islamic and Indian Magic Squares." *History of Religion* 8(1):181–209, 8(2): 271–299, 1969.

Lam, Lay Yong. *A Critical Study of the Yang Hui Suan Fa*. Singapore: University of Singapore Press, 1977.

Needham, Joseph. *Science and Civilisation in China* vol.3. Cambridge: Cambridge University Press, 1959.

See also: Yang Hui – Xu Yue – Astrology – Zhu Xi – *Yijing* – Divination

MAGIC SQUARES IN INDIAN MATHEMATICS The oldest datable magic square in India occurs in Varāhamihira's encyclopedic work on divination, *Bṛhatsaṃhitā* (ca. AD 550). He utilized a modified magic square of order four in order to prescribe combinations and quantities of ingredients of perfume. It consists of two sets of the natural numbers 1 to 8, and its constant sum (p) is 18. It is, so to speak, pan-diagonal, that is, not only the two main diagonals but also all "broken" diagonals have the same constant sum. Utpala, the commentator (AD 967), also points out many other quadruplets that have the same sum.

One of the four candidates for Varāhamihira's original square, with a rotation of 90°, coincides with the famous Islamic square, which al-Bīrūnī and al-Zinjānī frequently used as a basic pattern for talismans.

Varāhamihira called his square *kacchapuṭa* (the carapace of a turtle?), which reminds one of the title of a book on magic, *Kakṣapuṭa* (date unknown). The book contains a method for constructing a magic square of order four when a constant sum (p) is given. It also contains a square having the sum 100, which is attributed to Nāgārjuna.

In his medical work, *Siddhayoga* (ca. AD 900), Vṛnda prescribed a magic square of order three to be employed by a woman in labor in order to have an easy delivery. Its sum is thirty. This is the first datable instance of a magic square of order three in India, although there is a legend that a Garga, who may or may not be the author of the *Gargasaṃhitā* (ca. first century BC or AD), recommended magic squares of order three in order to pacify the *navagraha* (nine planets).

The famous Jaina magic square, which is incised on the entrance of a Jaina temple, Jinanātha, in Khajuraho, is assignable to the twelfth or the thirteenth century on a palaeographical basis. It is pan-diagonal. Several Jaina hymns that teach how to make magic squares have been handed down, but their dates are uncertain.

As far as is known, Ṭhakkura Pherū, a Jaina scholar, is the first in India who treated magic squares in a mathematical work. His *Gaṇitasāra* (ca. AD 1315) contains a small section on magic squares that consists of nine verses. He gives a square of order four, and alludes to its rearrangement; classifies magic squares (*jaṃta* = Sanskrit *yantra*) into three (odd, even, and evenly odd) according to the order (n), i.e. the number of cells (*kuṭṭha* = Sanskrit *koṣṭha*) on a side of the square, gives a square of order six and prescribes one method each for constructing even and odd squares.

The method for even squares divides the square into component squares of order four, and puts the numbers into cells according to the pattern of a standard sqaure of order four. That for odd squares first places in the central column the arithmetical progression whose first term and common difference are unity and $(n + 1)$ respectively; and then, starting from the numbers in the central column and proceeding by knight move, successively increases the number by n. The square thus obtained is the same as the one obtained by the so-called diagonal method (cf. Figure 21).

Nārāyaṇa wrote a comprehensive work on mathematics

2	3	5	8
5	8	2	3
4	1	7	6
7	6	4	1

Figure 1: Varāhamihira's magic square ($p = 18$).

10	3	13	8
5	16	2	11
4	9	7	14
15	6	12	1

(a)

2	11	5	16
13	8	10	3
12	1	15	6
7	14	4	9

(b)

10	3	5	16
13	8	2	11
4	9	15	6
7	14	12	1

(c)

2	11	13	8
5	16	10	3
12	1	7	14
15	6	4	9

(d)

Figure 2: Magic squares reconstructed from Varāhamihira's square ($p = 34$).

8	11	14	1
13	2	7	12
3	16	9	6
10	5	4	15

Figure 3: The Islamic square of order four ($p = 34$).

$a-3$	1	$a-6$	8
$a-7$	9	$a-4$	2
6	$a-8$	3	$a-1$
4	$a-2$	7	$a-9$

(a) $a = \frac{p}{2}$ (p: even)

$b-3$	1	$a-6$	8
$a-7$	9	$b-4$	2
6	$a-8$	3	$b-1$
4	$b-2$	7	$a-9$

(b) $a = \frac{p+1}{2}$, $b = \frac{p-1}{2}$ (p: odd)

Figure 4: Patterns for magic squares of order four given in the *Kakṣapuṭa*.

30	16	18	36
10	44	22	24
32	14	20	34
28	26	40	6

Figure 5: Nāgārjuna's magic square ($p = 100$).

16	6	8
2	10	18
12	14	4

Figure 6: Vṛnda's magic square of order three ($p = 30$).

7	12	1	14
2	13	8	11
16	3	10	5
9	6	15	4

Figure 7: Jaina magic square of Khajuraho ($p = 34$).

12	3	6	13
14	5	4	11
7	16	9	2
1	10	15	8

(a) Original Square

1	7	12	14
10	16	3	5
8	2	13	11
15	9	6	4

(b) Rearranged

Figure 8: Pherū's square of four and its rearrangement ($p = 34$).

1	32	34	33	5	6
30	8	28	27	11	7
24	23	15	16	14	19
13	20	21	22	17	18
12	26	9	10	29	25
31	2	4	3	35	36

Figure 9: Pherū's square of order six ($p = 111$).

15	9	6	4
8	2	13	11
1	7	12	14
10	16	3	5

(a) Standard Square

			4				8
	2				6		
1				5			
		3				7	
			12				16
	10				14		
9				13			
		11				15	

(b) First Stage

63	33	30	4	59	37	26	8
32	2	61	35	28	6	57	39
1	31	36	62	5	27	40	58
34	64	3	29	38	60	7	25
55	41	22	12	51	45	18	16
24	10	53	43	20	14	49	47
9	23	44	54	13	19	48	50
42	56	11	21	46	52	15	17

(c) Magic Square Obtained ($p = 260$)

Figure 10: Pherū's construction method for 'even'-order magic squares.

	n^2	
	·	
	·	
	·	
	·	
	·	
	$2n+3$	
	$n+2$	
	1	

(a) Central Column

37	48	59	70	81	2	13	24	35
36	38	49	60	71	73	3	14	25
26	28	39	50	61	72	74	4	15
16	27	29	40	51	62	64	75	5
6	17	19	30	41	52	63	65	76
77	7	18	20	31	42	53	55	66
67	78	8	10	21	32	43	54	56
57	68	79	9	11	22	33	44	46
47	58	69	80	1	12	23	34	45

(b) Square of Order Nine Obtained ($p = 369$)

Figure 11: Pherū's construction method for 'odd'-order magic squares.

1	8	13	12
14	11	2	7
4	5	16	9
15	10	3	6

(a)

1	14	4	15
8	11	5	10
13	2	16	3
12	7	9	6

(b)

Figure 12: Nārāyaṇa's square of four made by 'horse-move' ($p = 34$).

−5	9	19	17
21	15	−3	7
1	3	25	11
23	13	−1	5

(a) $a = -5, d = 2$

$p = 40$

−14	14	34	30
38	26	−10	10
−2	2	46	18
42	22	−6	6

(b) $a = -14, d = 4$

$p = 64$

Figure 13: Nārāyaṇa's square of four made by an arithmetical progression.

7	15	22	20
23	19	8	14
10	12	25	17
24	18	9	13

(a) $p = 64$

Cf. Fig. 12a

16	14	7	30	23
24	17	10	8	31
32	25	18	11	4
5	28	26	19	12
13	6	29	22	20

(b) $p = 90$

Cf. Fig. 21

Figure 14: Nārāyaṇa's square of four made by n sets of arithmetical progressions.

$\frac{35}{2}$	$\frac{49}{2}$	$\frac{59}{2}$	$\frac{57}{2}$
$\frac{61}{2}$	$\frac{55}{2}$	$\frac{37}{2}$	$\frac{47}{2}$
$\frac{41}{2}$	$\frac{43}{2}$	$\frac{65}{2}$	$\frac{51}{2}$
$\frac{63}{2}$	$\frac{53}{2}$	$\frac{39}{2}$	$\frac{45}{2}$

$t = \frac{33}{2}, p = 100$

Based on Fig. 12a

Figure 15: Nārāyaṇa's square of four made by addition of a number (t).

2	3	2	3
1	4	1	4
3	2	3	2
4	1	4	1

(a) Prelim-A

5	0	10	15
10	15	5	0
5	0	10	15
10	15	5	0

(b) Prelim-B

17	13	2	8
1	9	16	14
18	12	3	7
4	6	19	11

(c) B over A

$p = 40$

8	2	13	17
14	16	9	1
7	3	12	18
11	19	6	4

(d) A over B

$p = 40$

Figure 16: Nārāyaṇa's method for 'even-womb' squares (I): folding method.

1				2			
		8				7	
4				3			
		5				6	

(a) First Stage

1	32	49	48	2	31	50	47
56	41	8	25	55	42	7	26
16	17	64	33	15	18	63	34
57	40	9	24	58	39	10	23
4	29	52	45	3	30	51	46
53	44	5	28	54	43	6	27
13	20	61	36	14	19	62	35
60	37	12	21	59	38	11	22

(b) Magic Square Obtained ($p = 260$)

Figure 17: Nārāyaṇa's method for 'even-womb' squares (II): according to the pattern of standard square (Fig. 12a).

entitled *Gaṇitakaumudī* (AD 1356). Its last chapter, called *bhadra-gaṇita* (Mathematics of Magic Squares), comprises fifty-five verses for rules and seventeen verses for examples, and is devoted exclusively to magic squares and derivative magic figures (*upabhadra*) of various shapes.

The topics treated are: definitions of technical terms; determination of the mathematical progressions to be used in magic squares by means of *kuṭṭaka*, i.e. indeterminate equations of the first degree; how to make a square of order four by *turagagati* (horse move); the number of pan-diagonal magic squares of order four, 384, including every variation made by rotation and inversion.

Then Nārāyaṇa gives three general methods for constructing a square having any optional order (n) and constant sum (p) when a standard square of the same order is known — (1) by means of an arithmetical progression having an appropriate first term (a) and common difference (d), (2) by means of n sets of arithmetical progressions whose common differences are all unity, and (3) by adding an appropriate number (t) to every term of the standard square.

Nārāyaṇa next explains two methods each for constructing *sama-garbha* (even-womb) or evenly even, *viṣama* (odd-womb) or evenly odd, and odd squares when the sum is given. The two methods for the first kind are: (1) by folding two preliminary squares *karasaṃpuṭa-vat* (just like the folding of two hands), and (2) by arranging numbers in the component squares of order four according to the pattern of a standard square. For the second kind they are: (1) by putting numbers zigzag in the square, and (2) by tranposing certain numbers in the natural square. These two methods are not completely mechanical, and require *mati* (intelligence). The methods for the third kind are: (1) by folding two preliminary squares just as in the case of the first kind, and (2) by starting from the central cell of any side of the square and proceeding diagonally.

In the last section Nārāyaṇa gives a number of examples for two kinds of derivative magic figures, *saṃkīrṇa* (miscellaneous) and maṇḍala (circular). Both kinds are made by rearranging ordinary magic squares.

In his encyclopedic work on Hindu Law, *Smṛtitattva* (ca. AD 1500), Laghunandana gives a method for making squares of order four having any optional sum that is determined according to the purpose (see Table 1).

Significant scholarly work has been done on the importance of magic squares both mathematically and philosophically. Cammann and Roşu both discuss the significance of

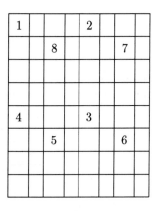

1	195	194	193	5	6	190	7	9	10	186	185	184	14
169	27	171	25	173	23	175	22	20	178	18	180	16	182
168	167	166	32	33	34	162	35	37	38	39	157	156	155
141	142	143	53	52	51	147	50	48	47	46	152	153	154
140	139	138	60	61	62	134	63	65	66	67	129	128	127
113	114	115	81	80	79	119	78	76	75	74	124	125	126
112	111	92	88	89	90	98	110	93	94	95	101	100	106
85	86	105	109	108	107	99	87	104	103	102	96	97	91
84	83	82	116	117	118	77	120	121	122	123	73	72	71
57	58	59	137	136	135	64	133	132	131	130	68	69	70
56	55	54	144	145	146	49	148	149	150	151	45	44	43
29	30	31	165	164	163	36	161	160	159	158	40	41	42
28	170	26	172	24	174	21	176	177	19	179	17	181	15
196	2	3	4	192	191	8	189	188	187	11	12	13	183

$n = 14, p = 1379$

Figure 18: Nārāyaṇa's method for 'odd-womb' squares (I): zigzag method.

100	92	93	94	5	6	7	8	9	91
20	89	83	84	16	15	87	18	82	11
30	29	78	77	75	26	74	73	22	21
40	39	38	67	65	66	64	63	32	31
41	52	43	44	56	55	47	48	59	60
51	42	58	57	46	45	54	53	49	50
61	69	68	37	35	36	34	33	62	70
71	72	28	27	25	76	24	23	79	80
81	19	13	14	86	85	17	88	12	90
10	2	3	4	96	95	97	98	99	1

$n = 10, p = 505$

Figure 19: Nārāyaṇa's method for 'odd-womb' squares (II): transposing method (conjectural reconstruction).

4	5	1	2	3
5	1	2	3	4
1	2	3	4	5
2	3	4	5	1
3	4	5	1	2

(a) Prelim-A

20	25	5	10	15
25	5	10	15	20
5	10	15	20	25
10	15	20	25	5
15	20	25	5	10

(b) Prelim-B

19	15	6	27	23
25	16	12	8	29
26	22	18	14	10
7	28	24	20	11
13	9	30	21	17

(c) B over A: $p = 90$

Figure 20: Nārāyaṇa's method for odd squares (I): folding method.

22	21	13	5	46	38	30
31	23	15	14	6	47	39
40	32	24	16	8	7	48
49	41	33	25	17	9	1
2	43	42	34	26	18	10
11	3	44	36	35	27	19
20	12	4	45	37	29	28

$n = 7, p = 175$

Figure 21: Nārāyaṇa's method for odd squares (II): diagonal method.

1	24	37	36	2	23	38	35	3	22	39	34
42	31	6	19	41	32	5	20	40	33	4	21
12	13	48	25	11	14	47	26	10	15	46	27
43	30	7	18	44	29	8	17	45	28	9	16

(a) Preliminary Magic Oblong

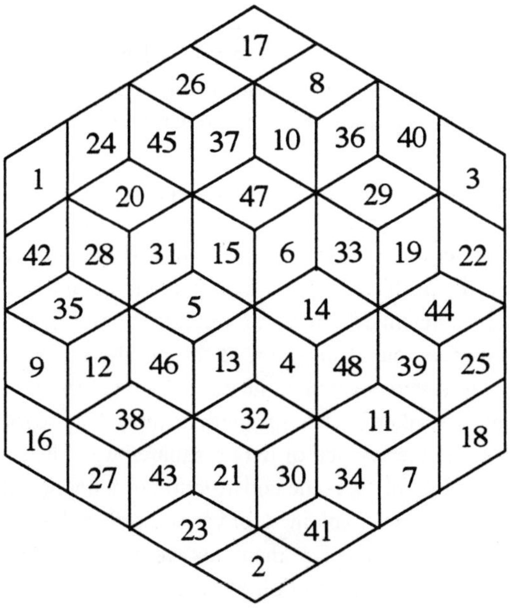

(b) Magic Lotus: $p = 294$

Figure 22: (a) Nārāyaṇa's magic lotus with six petals: preli minary magic oblong. (b) Nārāyaṇa's magic lotus with six petals ($p = 294$).

1	16	25	24	2	15	26	23
28	21	4	13	27	22	3	14
8	9	32	17	7	10	31	18
29	20	5	12	30	19	6	11

(a) Preliminary Magic Oblong

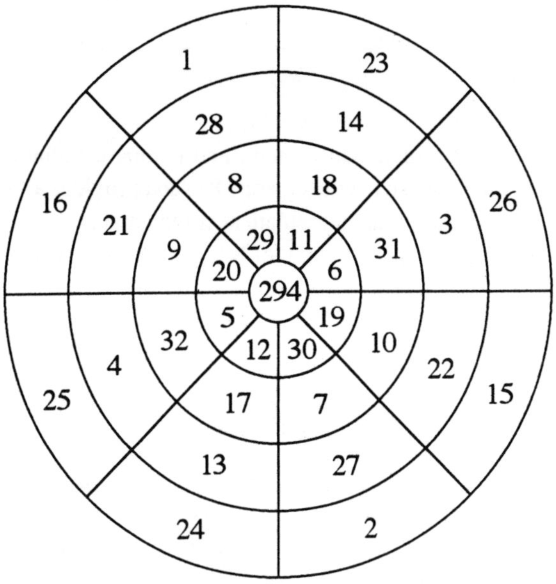

(b) Magic Circle: $p = 360$

Figure 23: (a) Nārāyaṇa's magic circle: preliminary magic oblong. (b) Nārāyaṇa's magic circle ($p = 360$).

1	8	$a-7$	$a-2$
$a-5$	$a-4$	3	6
7	2	$a-1$	$a-8$
$a-3$	$a-6$	5	4

$a = \frac{p}{2}$

Figure 24: Pattern for magic square of four by Laghunandana.

Table 1: Purposes of magic squares of order four

Sum (p)	Purpose
20	To neutralize poison
28	To protect crops from insects
32	To accelerate delivery
34	To protect travelers
50	To exorcise evil spirits
64	To protect warriors
72	For women having no children
84	To soothe crying children

magic squares in Indian thought and compare Indian, Islamic, and Chinese magic squares. Kusuba provides an English translation with mathematical commentary, as well as an edition of the Sanskrit text, of Nārāyaṇa's work on magic squares.

TAKAO HAYASHI

REFERENCES

Cammann, S. "Islamic and Indian Magic Squares." *History of Religions* 8: 181–209 and 271—299, 1968 and 1969.

Datta, B. and A.N. Singh. "Magic Squares in India." *Indian Journal of History of Science* 27:51–120, 1992.

Goonetilleke, W. "The American Puzzle." *The Indian Antiquary* 11: 83–83, 1882.

Grierson, G.A. "An American Puzzle."*The Indian Antiquary* 10: 89–90, 1881.

Hayashi, T. "Hōjinzan (A Japanese translation with mathematical commentary of Nārāyaṇa's 'Mathematics of Magic Squares')." *Epistēmē* (Tokyo: Asahi Press), Series II, 3: i–xxxiv, 1986.

Hayashi, T. "Varāhamihira's Pandiagonal Magic Square of the Order Four." *Historia Mathematica* 14: 159–166, 1987.

Kapadia, H.R. "A Note on Jaina Hymns and Magic Squares." *Indian Historical Quarterly* 10: 148–153, 1934.

Kusuba, T. "*Combinatorics and Magic Squares in India. A Study of Nārāyaṇa Paṇḍita's Gaṇitakaumudī, Chaps. 13–14.*" Ph.D. Dissertation, Brown University, 1993.

Ojha, G.K. *Aṅkavidyā*. Delhi: Motilal Banarsidass, 1982.

Roşu, A. "Études ayurvediques III:Les carrés magiques dans la médicine indienne." In *Studies on Indian Medical History*. Ed. G.J. Meulenbeld and D. Wujastyk. Groningen: Egbert Forsten, 1987, pp. 103–112.

Roşu, A. "Les carrés magiques indiens et l'histoire des idées en Asie." *Zeitschrift der Morgenländischen Gesellschaft* 139: 120–158, 1989.

Singh, P. "Total Number of Perfect Magic Squares: Nārāyaṇa's Rule." *Mathematics Education* 16A: 32–37, 1982.

Singh, P. "Nārāyaṇa's Treatment of Magic Squares." *Indian Journal of History of Science* 21:123–130, 1986.

Vijayaraghavan, T. "On Jaina Magic Squares." *Mathematics Student* 9: 97–102, 1941.

See also: Varāhamihira – Nārāyaṇa

MAGIC SQUARES IN ISLAMIC MATHEMATICS One of the most impressive achievements in Islamic mathematics is the development of general methods for constructing magic squares. A magic square of order n is a square divided into n^2 cells in which different natural numbers (mostly the n^2 first naturals) must be arranged in such a way that the same sum appears in each of the rows, columns, and two main diagonals. If, in addition to this basic property, the square remains magic when the borders are successively removed, we speak of a "bordered square". If the sum in any pair of complementary diagonals (i.e. pairs of parallel diagonals lying on each side of a main diagonal and having together n cells) shows the constant sum, the square is called "pandiagonal".

Squares are usually divided into three categories: odd-order squares ($n = 2k + 1$, k natural); evenly-even squares ($n = 4k$), and evenly-odd squares ($n = 4k + 2$). There are general methods for constructing squares of any order from one of these three categories. Except for the smallest possible order, $n = 3$, there are numerous possibilities of forming magic squares of any given order. There may be, however, some limitations concerning additional magical properties; for instance, there are no pandiagonal squares of evenly-odd order.

Information about the beginning of Islamic research on magic squares is lacking. It may have been connected with the introduction of chess into Persia in early Islamic times. Initially, the problem was a purely mathematical one; thus, the Arabic ancient designation for magic squares is *wafq al-aʿdād* (harmonious disposition of the numbers). We know that treatises were written in the ninth century, but the earliest extant ones date back to the tenth century. It appears that, by that time, the science of magic squares was already established. Not only was the construction of a magic square, simple or bordered, explained for various orders, but several additional conditions or refinements were also considered. For example, the construction of "composite squares" was well-known: if $n = r \cdot s$, with $r, s \geq 3$, one may fill successively, according to a magical arrangement for the order r, r^2 subsquares of order s (Figure 1). Another contemporary achievement was the construction of bordered squares in which the even numbers are separated from the odd ones (Figure 2).

Treatises explaining general constructions were common in the eleventh and twelfth centuries. They also explain how

31	36	29	76	81	74	13	18	11
30	32	34	75	77	79	12	14	16
35	28	33	80	73	78	17	10	15
22	27	20	40	45	38	58	63	56
21	23	25	39	41	43	57	59	61
26	19	24	44	37	42	62	55	60
67	72	65	4	9	2	49	54	47
66	68	70	3	5	7	48	50	52
71	64	69	8	1	6	53	46	51

Figure 1: A composite square.

36	16	108	110	10	113	8	116	118	2	34
50	48	24	100	107	97	7	102	18	46	72
52	56	60	103	91	89	23	3	58	66	70
96	54	17	47	51	81	83	43	105	68	26
94	13	29	49	59	57	67	73	93	109	28
11	27	35	45	69	61	53	77	87	95	111
92	117	101	85	55	65	63	37	21	5	30
32	80	121	79	71	41	39	75	1	42	90
38	78	64	19	31	33	99	119	62	44	84
82	76	98	22	15	25	115	20	104	74	40
88	106	14	12	112	9	114	6	4	120	86

Figure 2: A bordered square separating the numbers according to parity.

81	7	116	66	80
79	78	68	64	61
63	56	70	84	77
67	76	72	62	73
60	133	24	74	59

Figure 3: Magic square from a given first row.

to construct a magic square for a given sum, or for a word of n letters or n words to be put in the first row, since a number among the units, the tens, the hundreds, and one thousand is associated with each of the twenty-eight Arabic letters. The words occurring in the first row are either proper names or words of a religious nature (Figure 3; the five words correspond to the numbers 81, 7, 116, 66, and 80). Pandiagonal squares are constructed in various ways for evenly-even orders, but incompletely for odd orders (the method fails for certain orders).

From the thirteenth century on, magic squares become more and more associated with magic and divinatory purposes. Consequently, some texts merely picture squares and mention their attributes. Some others, though, keep the general theory alive, mostly to enable the reader to construct amulets by himself.

Interest in magic squares in Europe first arose towards the end of the Middle Ages, when two sets of squares associated with the seven planets were learned of through astrological and magic texts (whence the name), but without any information on their construction. Thus, the entire theory had to be built anew, and it is only in very recent times that the extent of Islamic research has come to light. It should also be remarked that the methods of construction spread eastwards around the twelfth century towards India and China.

JACQUES SESIANO

REFERENCES

Sesiano, Jacques. "An Arabic Treatise on the Construction of Bordered Magic Squares." *Historia Scientiarum* 42: 13–31, 1991.

Sesiano, Jacques. "Quelques méthodes arabes de construction des carrés magiques impairs." *Bulletin de la Société vaudoise des Sciences naturelles* 83 (1): 51–76, 1994.

Sesiano, Jacques. "Herstellungsverfahren magischer Quadrate aus islamischer Zeit, I-III." *Sudhoffs Archiv* 64 (2): 187–196, 1980;

65 (3): 251–265, 1981 & 71 (1): 78–89, 1987; 79 (2): 193–226, 1995.

Sesiano, Jacques. *Un traité médiéval sur les carrés magiques*. Lausanne: Presses polytechniques et universitaires romandes, 1996.

MAGIC SQUARES IN JAPANESE MATHEMATICS The study of magic squares in Japan began in the beginning of the Kan-ei period (1624–1643) with the import of *Suan fa tong zog* published in 1592 by Chen Dawei of China. Almost all the famous *wasan* (Japanese mathematics) experts, such as Takakazu Seki (or Seki Kowa, ca. 1642–1708) studied squares. On a smaller scale, the study has continued, and we have gained some new insights lately.

The study of squares concerns itself with some aspects of combinatorial mathematics, so some of its general methods are applied as well. For example, some parts of perfect 5×5 squares are made by the "Knight Jump Arrangement". The study of compositions, which was started by a Russian scholar, is limited to perfect 4×4 squares, but the method was also applied to 8×8 squares, as in the study by Motoaki Abe. Rakuho Abe obtained a fantastic result of irregular, perfect 7×7 squares. This is shown below:

16	24	20	17	37	14	47
23	28	48	1	38	6	31
39	7	45	9	42	18	15
40	4	29	25	21	46	10
35	32	8	41	5	43	11
19	44	12	49	2	22	27
3	36	13	33	30	26	34

An irregular, perfect 7×7 square.

In this the total of the diagonal or pan-diagonal numbers is 175; the total of the four corner numbers (16+47+34+3) is 100; the total of the numbers around the centers (9+21+41+29) is 100; and the total of the middle numbers of the four sides (17+10+33+40) is also 100.

In the middle of the Heian period, that is, in 970, Tamenori Minamoto edited *Kuchi-zusami* for the education of young nobility. In the twelfth chapter we find this sentence: 4 and 2 makes a shoulder, left 3 and right 7, legs are 6 and 8, head 9, body 5, tail is 1.

4	9	2
3	5	7
8	1	6

The legs, head, and tail seem to be those of some animal (a tortoise?), and maybe they suggest the arrangement of numerals. We are not sure how Minamoto came by this knowledge of squares, but the 3×3 square in *Kuchi-zusami* is the oldest record as far as we know.

Table 1: A general way to make squares

A

7	4	1
8	5	2
9	6	3

B

2	9	6
3	5	7
4	1	8

C

4	9	2
3	5	7
8	1	6

D

12	19	8
9	13	17
18	7	14

A book by Yūeki Andō (1624–1704), a clansman of Aizu, is the first in the world in which the author explains a general way to make squares by increasing from 3×3 to 30×30 squares (see Table 1).

For example, if we exchange positions in square A like 1 ↔ 6, 3 ↔ 8, 9 ↔ 4, and 2 ↔ 7, we get square B. Then the four corners are all even numbers. If we then turn these four corner numbers 90 degrees clockwise, we get square C.

21	16	11	6	1
22	17	12	7	2
23	18	13	8	3
24	19	14	9	4
25	20	15	10	5

5×5 square, with a 3×3 square inside.

In the 5×5 square like the figure shown, the thick-lined square is a 3×3 square. Change it into D of Table 1. Next, work on the four corners of the outside square, and exchange positions in the following manner: 1↔15, 5↔23, 25↔11, and 21↔3. Do not turn the four corner numbers, because they are odd numbers. You will complete the work by putting D, the 3×3 square, inside it.

Takakazu Seki commented on the general way of making squares in the books, *Hōjin-no-hō* and *Ensan-no-hō*, written in the third year of the Tenwa period (1683). He divided them into odd-celled squares and two types of even-celled squares. Squares by Katahiro Takebe (1664–1739) are introduced in Volume 4 of the *Ichigen Kappō*, written by Shukei Irie. They are not found in the West.

In 3×3 squares, put the numerals in order, turn the center joined by both diagonals, lines, and rows 45 degrees leftward. Then turn the center of the lines and rows 180 degrees:

7	4	1
8	5	2
9	6	3

⟶

4	1	2
7	5	3
8	9	6

⟶

4	9	2
3	5	7
8	1	6

Yoshisuke Matsunaga (1692–1744), a pupil of Seki's pupil, wrote *Hōjin-Shin-jutsu*. There are also studies by Yoshihiro Kurushima (d. 1757), and by Naonobu Ajima (1732–1798). The question on step children is in *Jingōki* by Mitsuyoshi Yoshida. The question is close to that of Joseph's problem in the West.

Shūtarō Teramura (1902–1980) studied how many "parent–child" squares could be made in all, and in 1926 he made 605 squares. Following are some examples.

In the first, the parent square is a perfect 8×8 square. The child one (beteween thick-lines) is a perfect 4×4 square.

2	29	51	48	1	30	52	47
56	43	5	26	55	44	6	25
15	20	62	33	16	19	61	34
57	38	12	23	58	37	11	24
4	31	49	46	3	32	50	45
54	41	7	28	53	42	8	27
13	18	64	35	14	17	63	36
59	40	10	21	60	39	9	22

A parent–child square, 8×8, with a 4×4 square inside.

In another example, the parent square is also a perfect 8×8 square, and the child one (upper-middle part) is a perfect 4×4 square.

2	29	51	48	1	30	52	47
56	43	5	26	55	44	6	25
13	18	64	35	14	17	63	36
59	40	10	21	60	39	9	22
4	31	49	46	3	32	50	45
54	41	7	28	53	42	8	27
15	20	60	33	16	19	61	34
57	38	12	23	58	37	11	24

A parent–child square.

To make a perfect 4×4 square, choose optional numbers for x, y, z, u, and t. $X = x + y + z + u$. Make the square as follows:

x	y	z	u
$z - t$	$u + t$	$x - t$	$y + t$
$\frac{1}{2}X - z$	$\frac{1}{2}X - u$	$\frac{1}{2}X - x$	$\frac{1}{2}X - y$
$\frac{1}{2}X - x + t$	$\frac{1}{2}X - y - t$	$\frac{1}{2}X - z + t$	$\frac{1}{2}X - u - t$

To make perfect 8×8 squares, choose optional numbers for a, b, and c, and construct the square as in Table 2.

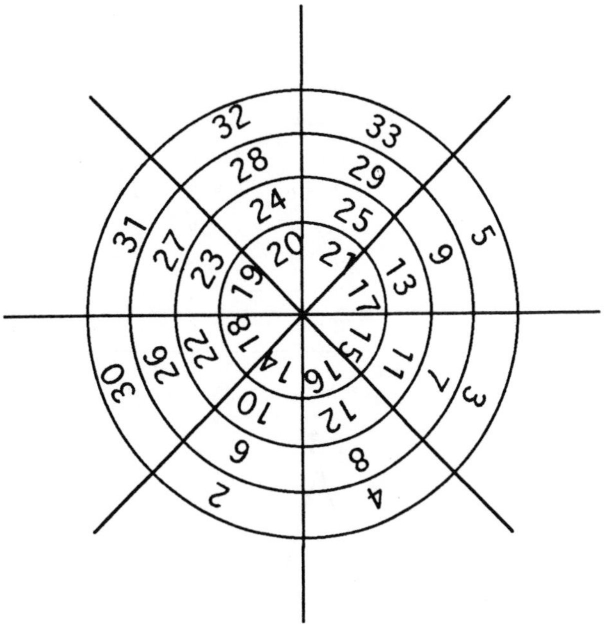

Figure 1: Circular square.

In an N square, in which a line has n divisions, and which consists of n^2 numerals, the total of the vertical and horizontal lines should be the same as that of both diagonals. This square is a Magic Square, and if each sum of the numbers of all the parallel lines is the same, the square is a Perfect Magic Square. The study of Michiwaki and Moriyama (1963) tells how to make it.

There are also many special squares. The best examples of Circular Squares are introduced in Takakazu Seki's *Hōjin-no-hō* and *Ensan-no-hō* (see Figure 1).

Star Squares were studied by Shigematsu Urata and Rakuhō Abe in 1955 (see Figure 2).

The first cubic square in Japan was introduced as "cubic design" in *Rakusho-kikan* by Yoshizane Tanaka (1651–1719). In it, he piled up the following 4×4 squares in order:

(1)

14	54	43	19
59	3	30	38
22	46	51	11
35	27	6	62

(2)

20	44	53	13
37	29	4	60
12	52	45	21
61	5	28	36

(3)

33	25	8	64
24	48	49	9
57	1	32	40
16	56	41	17

(4)

63	7	26	34
10	50	47	23
39	31	2	58
18	42	55	15

Arata Sakai started the study of perfect 4×4 cubic squares in 1938, and Shigematsu Urata completed it in 1948. Motoaki Abe made a 5×5 cubic square in the latter part of the

Table 2: Making perfect 8×8 squares.

a	b	c	d	e	f	g	h
$c+x$	$d-x$	$e+x$	$f-x$	$g+x$	$h-x$	$a+x$	$b-x$
$e+y$	$f-y$	$g+y$	$h-y$	$a+y$	$b-y$	$c+y$	$d-y$
$f-y$	$h+x$	$a-x$	$b+x$	$c-x$	$d+x$	$e-x$	$f+x$
$-e-y$	$-f+y$	$-g-y$	$-h-y$	$-a-y$	$-b+y$	$-c-y$	$-d-y$
$-g+z$	$-h-z$	$-a+z$	$-b-z$	$-c+z$	$-d-z$	$-e+z$	$-f-z$
$-a$	$-b$	$-c$	$-d$	$-e$	$-f$	$-g$	$-h$
$-c-z$	$-d+z$	$-e-z$	$-f+z$	$-g-z$	$-h+z$	$-a-z$	$-b+z$

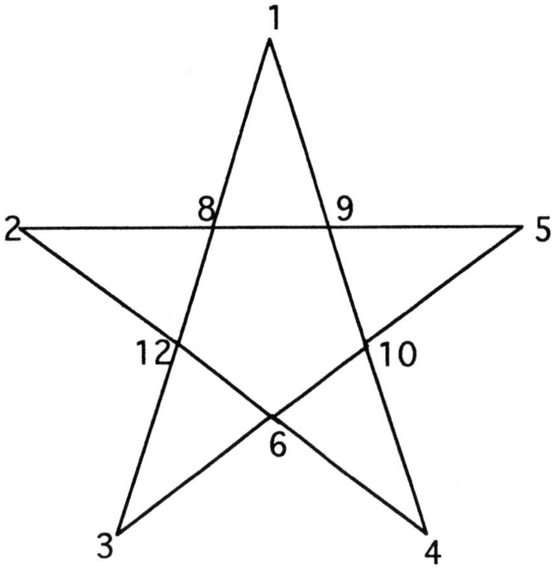

Figure 2: Star square.

Taishō period. A 7×7 cubic square was made by Motoaki Abe and Shōji Shimada.

In 1948 Rakuhō Abe made a perfect 6×6 cubic square with a plane 6×6 square as its base, but the diagonal total was not regular. Michiwaki and Moriyama studied the way to make special perfect 9×9 cubic squares, and the way to count their numbers.

YOSHIMASA MICHIWAKI

REFERENCES

Abe, Rakuhō. "Three Kinds of Excellent Squares–Difficult Puzzling Since Two Hundred Years and Its Solutions." In *Mathematical Sciences*, 1977, pp. 39–41.

Abe, Rakuhō. "Square of Wasan–History, Characteristics, Works." *BUT Bulletin, Science University of Tokyo*, 1987, pp. 2–5.

Hayashi, Takao. "Square in Kuchi-zusami." *Journal of History of Mathematics, Japan* 131:34–37, 1991.

Hirayama, Akira and Abe, Rakuhō. *Study of Squares*. Osaka: Kyoiku Tosho, 1983.

Katō, Heizaemon. "Study of Wasan." *Zatsuron* 3. Japan Association for the Advancement of Science, 1956.

Michiwaki, Yoshimasa and Moriyama Yoshio. "A Note on Perfect and Magic Squares 1." *Research Reports of Nagaoka Technical College* 1(2):79–98, 1963.

Michiwaki, Yoshimasa and Moriyama Yoshio. "A Note on Perfect and Magic Squares 2." *Research Reports of Nagaoka Technical College* 1(3):57–69, 1964.

Michiwaki, Yoshimasa and Moriyama Yoshio. "A Note on Perfect and Magic Squares 3." *Research Reports of Nagaoka Technical College* 2(4):69–76, 1966.

Michiwaki, Yoshimasa and Moriyama Yoshio. "A Note on Perfect and Magic Squares 4." *Research Reports of Nagaoka Technical College* 3(1):19–30, 1967.

Michiwaki, Yoshimasa and Moriyama Yoshio. "A Note on Perfect and Magic Squares 5." *Research Reports of Nagaoka Technical College* 4(2):69–80, 1968.

Michiwaki, Yoshimasa and Moriyama Yoshio. "A Note on Perfect and Magic Squares 6." *Research Reports of Nagaoka Technical College* 5(2):77–84, 1969.

See also: Seki Kowa – Takebe Katahiro – Ajima Naonobu

MAGNETISM IN CHINA In Chinese civilization, the *ci-shi* (lodestones) and their ability to attract iron are mentioned in a number of Zhou period Classics. It is stated in the *Shanhai Jing* (Classic of the Mountains and Rivers) that lodestones are found at *ao-ze*. A remark in the *Guan Zi* (Book of Master Guan) indicates that the finding of lodestones was once viewed by ancient prospectors as an indication for the possible presence of other mineral deposits. Both the *Gui Gu Zi* (Book of the Devil Valley Master) of the fourth century BC and *Lushi Chun Qiu* (Master Lu's Spring–Autumn Annals) of 239 BC mention that the lodestone attacts iron.

The attractive interaction between the lodestone and iron was interpreted as being caused by *xiang gan* (mutual influence), a sympathetic response between two interacting entities. Wang Chong, in the *Lun Heng* (Discourse Weighed in Balance) of AD 82 offers the interpretation that some items attract others, while some cannot, because their *qi* is different and consequently they cannot mutually influence one another. The character *qi* literally means vapor or breath. In the course of time, it took on a number of proto-scientific significances: the basic constituent entity of all things in nature and the media of the *yinyang* interactions. There are a number of scholars who translate *qi* as 'energy', 'field', or 'matter-energy', but in ancient times, the concepts of energy and field were not yet explicitly developed. *Qi*, as it is given in the Chinese texts, is not identified by its intrinsic properties, but by its general relationship with others. Though the pattern relationship is identifiable with certain aspects of the modern concept of energy, field, or matter-energy, it is better to preserve the term *qi* and not to translate it.

A later account (AD 300) by Guo Pu also comments on the role of *qi* in such interactions.

> The lodestone draws iron, and amber collects mustard-seeds. [Since] their *qi* are invisibly interconnected and their measures are silently met, the mutual influence between them consequently occurred.

Such a view of interaction at a distance involving *qi* is compatible with modern views.

The strength of such attractive interactions was later discussed in terms of the weight of iron pieces or the number of needles that the lodestone was capable of supporting.

Tao Hong Jing stated in the *Ming Yi Bie Lu* (Informal Records of Famous Physicians) that some "[lodestones] could suspend a chain of more than ten needles".

Based on such quantitative measures of the attractive power, lodestones were graded as follows:

> A piece of genuine lodestone is called the *yan nian sha* if it is capable of attracting, on its four sides, one catty of iron piece (the equal weight of the lodestone); called the *ji cai shi* if it is capable of attracting, on its four sides, eight ounces weight of iron piece; and called simply the *ci shi* if it is capable of attracting, on its four sides, four ounces of iron piece (*Lei Gong Pao Zhilun*).

The directional property of lodestones was probably first used in the *sinan* ceremony, a practice evolved directly from the tradition of using a gnomon and the positions of the sun and stars to determine the time and directions. The term *sinan* means 'south verification' or 'south controller'. The discovery of lodestone's south pointing property led naturally to its association with the term *sinan* and the *sinan* (south verification) devices. However, early mentions of the *sinan* devices provide little information on their construction. In the *Gui Gu Zi* there is mention of the fact that when the Zheng people engage in collecting jade, they always carry with them a *sinan* to avoid being lost.

The device described by Wang Chong has been widely considered to be the earliest form of the magnetic compass.

The Chinese literature of the period ranging from the Han to the Tang dynasties provides a number of hints that, by the first century AD, the fact that the directive property of the lodestone could be transferred to small pieces of iron was discovered, and by the seventhth century AD, the magnetized needle began to appear in certain compasses, replacing lodestones for greater precision. But the records are not sufficiently specific to provide a definitive account of these developments. Explicit descriptions of magnetic properties and compass construction are found in the work of the Song dynasty.

Properties of magnetized needles are discussed in the work of Shen Gua (1029–1093). We have from his *Meng Xi Bi Tan* (Meng Xi Essays) of 1086, the following description of declination:

> *Fang Jia uses the lodestone to rub the point of a needle;*
> *The needle is then able to point to the south.*
> *It does not, however, point directly at the south,*
> *It always inclines slightly to the east.*

Following this description is the passage on the needle supporting methods:

> [The needle] can be supported by making it float on the surface of water, but it is rather unsteady. It may be balanced on the finger-nail, or on the rim of a cup, so that it can turn more easily, but such supports being hard and smooth is liable to fall off. The best way of supporting the needle is by suspension. The method uses a single cocoon fiber of new silk to suspend the needle. By attaching silk to the center of needle with a piece of wax the size of a mustard-seed and hanging in a windless place, the needle will then always point to the south undisturbed.

He discusses three types of needle supporting methods: floating, pivot, and suspension. For the study of magnetic properties, Shen Gua preferred the suspension method using a silk thread. The other two methods of supporting the needle, by floating it on the water surface and by pivoting it on a hard smooth surface, correspond to those used in the wet and dry compass, respectively.

Shen Gua's work also discusses magnetic polarity. He noticed that some needles pointed north and some south. He believed that different natures caused this, just as animals shed at diferent seasons.

An important discussion of the magnetic compass is found in the *Wujing Zongyao* (Collection of the Most Important Military Techniques) of AD 1044.

> When troops encountered gloomy weather or dark night, and could not distinguish the directions, they would let an old horse in the front to lead them, or else they would make use of the south-pointing carriage, or the south-pointing fish to identify the directions. Now the carriage method has not been handed down. In the (south-pointing) fish method, a thin leaf of iron is cut into the shape of a fish two inches long and half an inch broad, having a pointed head and tail. It is then heated in a charcoal fire until it becomes thoroughly red-hot, taken out by the head with iron tongs, and placed so that its tail is in the *zi* direction (due north). In this position, it is quenched with water in a basin, so that its tail is submerged for several tenths of a inch. It is then kept in a *miqi* (tightly closed box). To use it, a small bowl filled with water is set up in a windless place, and the fish is laid as flat as possible upon the watersurface so that it floats, whereupon its head will point in the *wu* direction (south).

The passage not only provides a clear discussion on the construction of a compass, but also reveals a different method for magnetizing iron pieces. Thus no later than the early Song dynasty (960–1279), two methods of magnetizing were known. Other than rubbing with the lodestone, an iron piece could also be magnetized by quenching it from red heat through the Curie point, held in a north–south direction [in the magnetic field] of the earth. The discovery of the thermoremanence phenomena was of great scientific significance. Since the earth's magnetic field is relatively weak and the soft iron does not retain its magnetism long, questions were raised as to whether such magnetized iron could function satisfactorily as a compass. Recent archaeological evidence revealed a long history of steel development in China, beginning from the later part of the spring–autumn period of the Zhou dynasty. By the second century BC, two methods for steel production from pig iron were developed, one by the puddling of molten iron and the other by decarbonization of cast iron in the solid state without the formation of graphite. Thus, good steel and steel needles were certainly available in the early Song dynasty.

An important use of the compass is in navigation. Clear and accurately datable statements on such a use are found in the *Pingzhou KeTan* (Pingzhou Table Talk) of 1119, when the author says that ship's pilots look at a south-pointing needle or sample of the mud collected from the sea-bottom to determine their whereabouts in dark weather. This is confirmed in other books of the period.

Based on these statements and other sources, Needham has concluded that "description of the use of the compass for navigation on Chinese ships antedates the first knowledge of this technique in Europe by just under a century, but there are indications that it was used for this purpose in China somewhat earlier."

Needham is also of the view that "Chinese sailors remained faithful to the floating-compass for many centuries. Although the dry pivoted compass had been described early in the twelfth century AD, it did not become common on Chinese vessels until it was reintroduced from the West in the sixteenth century by the Dutch and Portuguese by way of Japan. Associated then with it was the compass-card (the windrose attached to the magnet) which had probably been an Italian invention at the beginning of the fourteenth century". The European mariner's compass is characterized by sixteen compass points in contrast to the twenty-four of the Chinese compass. The discovery of the clay depiction of a dry-pivoted compass in the tomb of Zhu Jinan (1140–1197) in 1985 raised a number of pertinent questions. The compass depicted in the arm of the two clay figurines unearthed from the tomb has sixteen, not twenty-four, compass-points. Since the tomb was sealed in 1198, the compass depicted here could not have been influenced by those reintroduced from the West.

Figure 1: Clay figurine with a dry-pivoted compass in its arm unearthed in 1985 from the tomb of Zhu Jinan.

Both the dry-pivoted and floating compass were used in Chinese ships. The choice was often based on accuracy. It should be noted that, though the gyroscope was invented in China around the first century BC by Ding Huan, no known reference to its use in connection with the mariner's compass is available.

CHEN CHENG-YIH (JOSEPH)

REFERENCES

Chen Ding-Rong and Xu Jian-Chng. "Jiangxi Linchuan-Xian Song Mu (A Song Tomb in the Lin-chuan County, Jinagx Province)." *Kao-Gu* (*Archaeology*) 4: 329–334, 1990.

de Saussure, L. "L'Origine de la Rose des Vents et l'Invention de la Boussole." *Archives des Sciences Physiques et Naturelles* 5(2 and 4), 1923.

Needham, Joseph. *Science and Civilisation in China*. Cambridge: Cambridge University Press, 1962, vol. 4, part I, pp. 229–334.

Tsun Ko. "The Development of Metal Technology in Ancient China." In *Science and Technology in Chinese Civilization*. Ed. Cheng-Yih Chen . Singapore: World Scientific, 1987, pp. 225–243.

Wang Zhen-Duo. "Sinan Zhinan Zhen yu Luo-Jing-Pan (Discovery and Application of Magnetic Phenomena in China)." *Zhong Guo Kao Gu Xue Bao (Chinese Journal of Archaeology*) 3: 119–260, 1948.

Wenren Jun. "Nan-Song Kan-Yu Han-Luo-Pan de Fa-Ming zhi Fa-Xian (The Discovery of a South Song Pivoted Compass)." *Kao-Gu* (*Archaeology*) 12: 1127–1131, 1981.

See also: Navigation – Metallurgy – Compass – Shen Gua

MAGNETISM IN MESOAMERICA That the pre-Columbian peoples of Mesoamerica were familiar with the property of magnetism has been suggested by several researchers, among them the geographer Robert Fuson and the anthropologist Michael Coe. Indeed, a flattened oblong piece of hematite discovered during Coe's excavation of the Olmec site of San Lorenzo in southern Veracruz state in 1966 has been thoroughly examined by John Carlson, who suggests that it probably was fashioned for use as a compass. In 1975, a basaltic sculpture at the site of Izapa, on the Pacific coastal plain of Mexico near the Guatemalan border, was found to possess a strong magnetic field. Variously described as being the representation of either a frog (Norman, 1976) or a turtle's head (Malmström, 1976), it has a north-seeking pole in its snout and a south-seeking pole at the back of its head. The discovery led the latter researcher to speculate that the stone's carver may have associated the property of magnetism with the homing instinct of the turtle. Because it was the only magnetic object found at the site, critics of the notion that it was a human artifact argued that it may have been struck by lightning and its magnetic field had been induced in that manner.

However, in 1979, several additional magnetic sculptures were discovered in the Pacific coastal plain of Guatemala, including seven which now repose in the central plaza of the town of La Democracia, and two more which were identified at the nearby sugar plantation of El Baúl. Those at La Democracia are extremely crude depictions of human beings, and, because of their rotundity, have been termed the "Fat Boys" by archaeologists. When an entire body is depicted, the two magnetics poles are usually found on either side of the navel; when only a head is portrayed, the two magnetic poles are almost invariably centered on the right temple. The sculptures at El Baúl include a rampant jaguar, with magnetic poles in each upraised paw, and a tablet showing two men seated on a bench with their arms folded over their chests. This single block of stone has four magnetic poles, one north-seeking pole between each of the men's folded arms and one south-seeking pole below each man in the space beneath the bench. In 1983 a small humanoid sculpture in the plaza of Tuxtla Chica, Mexico, just back of Izapa, was found to be magnetic in the right side of its head. Clearly, the patterns of polarity discerned suggest a conscious intent on the part of the sculptors to fashion their carvings around a known center of magnetic attraction, for in none of the stones has any inset of foreign material been made. That such recurring patterns could have been the result of random lightning strikes must also be ruled out. Because the "Fat Boys" are considered to date from 1500–2000 BC, it is possible that these sculptures represent the oldest known magnetic artifacts in the world. But to what use, other than art and magic, this knowledge was put, we have no answer as yet.

VINCENT H. MALMSTRÖM

REFERENCES

Carlson, John B. "Lodestone Compass: Chinese or Olmec Primacy?" *Science* 189: 753–760, 1975.

Fuson, Robert H. "The Orientation of Mayan Ceremonial Centers." *Annals of the Association of American Geographers* 59: 508–510, 1969.

Malmström, V.H. "Knowledge of Magnetism in Pre-Columbian Mesoamerica." *Nature* 259(5542): 390–391, 1976.

Norman, Garth. "Izapa Sculpture." *Papers of the New World Archaeological Foundation*, no. 30, 1976.

See also: Compass

MAHĀDEVA Mahādeva (ca. 1275–1350) composed the extensive set of planetary tables, *Mahādevī*, named after him and dated 28th March, 1316. He belonged to a family of astronomers and in his work *Grahasiddhi* describes himself as the son of the astrologer Paraśurāma, son of Padmanābha, son of Mādhava, son of Bhogadeva of the Gautamagotra, a follower of *Sāmaveda* and a performer of sacrifices. The planetary tables contained in the *Mahādevī*, prepared for facilitating the computation of the daily almanac, were extremely popular in the Gujarat and Rajasthan regions, and numerous manuscripts of the work have been located in these places. While the basic text was restricted to 43 verses, the author himself wrote a set of instructions for using the tables, called *Grahasiddhi*, which also were extensively used. The popularity of *Mahādevī* is also attested to by the several commentaries that were written on the work, including that of Nṛsiṃha (1528), Dhanarāja (1635) and Mādhava, and a host of anonymous commentators.

Mahādeva is a synonym of Śiva, one of the trinities of Hinduism, and so formed one of the words commonly used to name Hindus. There are several astronomers of medieval India who bore the name Mahādeva. Among these are: Mahādeva, younger brother of Viṭṭhala, from Gujarat, author of *Tithicakranirṇaya*, also called *Tithinirṇaya, Tithiratna*, and *Mahādevasiddhānta*; Mahādeva, author of *Jātakāpaddhati*, called also *Mahādevapaddhati* after his name; Mahādeva, son of Luṇiga and author of commentaries on the *Cintāmaṇisāraṇikā* of Daśabala and the *Jyotiṣaratnamālā* of Śrīpati; Mahādeva of the Kauṇḍinyagotra, son of Bopadeva, and author of *Kāmadhenu* called also *Tithikāmadhenu*; Mahādeva, son of Kahnaji Vaidya, author of *Muhūrtadīpaka* in 57 verses, written in 1640, *Praśnapradīpa* called also *Praśnaratna* (1647), *Bhāveśaphalapradīpa* (1647), *Kālanirṇayasiddhānta* (1652), and a commentary on his own *Muhūrtadīpaka* (1661); Mahādeva Pāṭhaka (1842–1899), son of Revāśaṅkara, and author of *Varṣadīpaka*, called also *Varṣadīpikā* (1861), *Jātakatattva*, written in 1872, *Pitṛmārgapradīpa* in 57 verses (1874), *Varṣapaddhati*, a compilation (1874), and *Āśubodhajyotiṣa*.

K.V. SARMA

REFERENCES

Dvivedi, Sudhakara. *Gaṇaka Taraṅgiṇī or Lives of Hindu Astronomers*. Ed. Padmakara Dvivedi. Benares: Jyotish Prakash Press, 1933.

Pingree, David. "Sanskrit Astronomical Tables in the United States." *Transactions of the American Philosophical Society* 58 (3): 1–77, 1968.

Pingree, David. *Census of the Exact Sciences in Sanskrit, Ser. A, Vol, 4*. Philadelphia: American Philosophical Society, 1981.

AL-MĀHĀNĪ al-Māhānī, Abū ‹Abd Allāh Muḥammad ibn ‹Īsā was born in Māhān, Kerman, Iran. He lived in Baghdad, ca. 860 and died ca. 880.

Little is known about al-Māhānī's life, and few of his works are extant. In the *Ḥākimite Tables* Ibn Yūnus cites observations of conjunctions and lunar and solar eclipses made by al-Māhānī between 853 and 866. In the only extant astronomical work, *On the Determination of the Azimuth for an Arbitrary Time and an Arbitrary Place* (Maqāla fī Ma'rifat as-samt li-aiy sā'a aradta wa-fī aiy mauḍi' aradta), al-Māhānī added arithmetical solutions to two of the graphic ones. His method corresponds to the cosine formula in spherical trigonometry, and is later applied by al-Battānī.

Al-Māhānī worked on the fundamental problems of mathematics of his time and is especially known for his commentaries to Euclid's *Elements*, to Archimedes', *De Sphaera et Cylindro* (On Spheres and Cylinders), and to the *Sphaerica* by Menelaus. In the last treatise, now lost, he inserted explanatory remarks, modernized the language, especially the technical terms, and remodeled or replaced obscure proofs. It was revised and finished by Aḥmad ibn Abī Sa‹īd al-Harawī in the tenth century. Al-Ṭūsī considered al-Māhānī's and al-Harawī's improvements valueless and used the edition by Abū Naṣr Manṣūr ibn ‹Irāq. This redaction, the most widely known Arabic edition, is included in the collection of the *Intermediate Books*. These were the books read between Euclid's *Elements* and Ptolemy's *Almagest*.

Of the commentaries to the *Elements* only those to Book V and to Book X are extant. In the former al-Māhānī compared magnitudes by comparing their expansion in continued fractions, referring to Thābit ibn Qurra. Ratio is defined as "the mutual behavior of two magnitudes when compared with one another by means of the Euclidian process of finding the greatest common measure". Two pairs of magnitudes were for him proportional when "the two series of quotients appearing in that process are identical". Essentially the same theory was worked out later by al-Nayrīzī. Neither established a connection with Euclid's definition, which was first done by Ibn al-Haytham. In the commentary to Book X al-Māhānī examined and classified not only quadratic irrationalities but also those of the third order. In contrast with Euclid, for whom magnitudes were only lines, he considered integers and fractions alike as rational magnitudes, while regarding square and cube roots as irrational ones. Al-Māhānī then explicated the contents of Book X using rational and irrational numbers instead of geometric magnitudes.

According to al-Khayyāmī, al-Māhānī was the first to

attempt an algebraic solution of the Archimedean problem of dividing a sphere by a plane into segments the volumes of which are in a given ratio (*De Sphaera et Cylindro* II, 4). He expressed this problem in a cubic equation of the form $x^3 + a = cx^2$, but he could not proceed further. Al-Khayyāmī relates that the problem was thought unsolvable until al-Khāzin succeeded by using conic sections.

YVONNE DOLD-SAMPLONIUS

REFERENCES

Dold-Samplonius, Yvonne. "al-Māhānī." In *Dictionary of Scientific Biography* vol. IX. New York: Charles Scribner's Sons, 1970, pp. 21–22.

Matvievskaya, Galina. "The Theory of Quadratic Irrationals in Medieval Oriental Mathematics." In *From Deferent to Equant.* Ed. David A King and George Saliba. New York: New York Academy of Sciences, 1987, pp. 253–277.

See also: Ibn Yūnus – al-Battānī – al-Ṭūsī – Thābit ibn Qurra – al-Nairīzī – Ibn al-Haytham – ‘Umar al-Khayyām – *Elements*

MAHĀVĪRA The Rāṣtrakūṭa dynasty of the medieval period was founded in the Deccan, South India, by Dantidurga about the middle of the eighth century AD. A king of this dynasty named Amoghavarṣa ruled from AD 815 to 877. The long period of his rule in known for its material prosperity, political stability and academic fertility. He was rich, powerful, and peace-loving, and he patronized art and learning.

In the later part of Amoghavarṣa's reign there lived a great mathematician named Mahāvīrācārya. Mahāvīra was a Digambara Jaina and wrote an extensive Sanskrit treatise called *Gaṇitasāra saṅgraha* (Compendium of the Essence of Mathematics) about AD 850. It is devoted to elementary topics in arithmetic, algebra, geometry, mensuration, etc. The work is important because it is a collection summarizing elementary mathematics of his time and providing a rich source of information on ancient Indian mathematics. It is written in the style of a textbook and was used as one for centuries in all of South India. Its importance is greater still because the *Pāṭīgaṇita* of Śrīdhara (ca. AD 750), written in the same style as the *Gaṇitasāra saṅgraha*, is not extant in full.

Mahāvīra shows sufficient originality not only in presenting older material lucidly, but also in introducing several new topics. A commentary called *Bālabodha* in Kannada was written by Daivajña Vallabha. A Sanskrit commentary was composed by Varadarāja. Dates for these two commentators are not known, nor are their works available in print. There were other translations in the eleventh century and in 1842.

The nine chapters of the *Gaṇitasāra saṅgraha* are as follows:

1. Terminology (70 verses);
2. Arithmetical operations (115 verses);
3. Operations involving fractions (140 verses);
4. Miscellaneous operations (72 verses);
5. Rule of three (43 verses);
6. Mixed operations ($337\frac{1}{2}$ verses);
7. Geometry and mensuration ($232\frac{1}{2}$ verses);
8. Excavations ($68\frac{1}{2}$ verses);
9. Shadows ($52\frac{1}{2}$ verses);

The total number of verses, 1131, shows that the book is quite comprehensive. Another noteworthy feature is that the Jaina tradition of Indian mathematics is also preserved within the scope of the *Gaṇitasāra saṅgraha.*

The authorship of the astronomical work *Jyotiṣapaṭala* is also ascribed to Mahāvīra. Some manuscripts of this title are mentioned in the *Jina-ratna-kośa* and the *New Catalogus Catalogorum*, but no author is mentioned. Another work attributed to him is *Chattisu*, but this is also a sort of elaboration of part of the *Gaṇitasāra saṅgraha* made by Mādhavacandra Traividya (about AD 1000). Whatever the case, the reputation of Mahāvīra relies solely on his *magnum opus.*

During ancient times, unit fractions were considered quite important. There are some interesting results in the *Gaṇitasāra saṅgraha* on this topic. One rule gives a practical method for expressing any given fraction as a sum of unit fractions. Let p/q be the given fraction (p being less than q). We add a suitable integer x to q such that $(q + x)$ become exactly divisible by p, say, r times. Then Mahāvīra's rule is

$$\frac{p}{q} = \frac{1}{r} + \frac{x}{(r \cdot q)}.$$

Mahāvīra made a very significant remark in connection with the square-root of a negative number. He said "A negative number is non-square by its nature, whence there is no (real) square-root from it."

This remark is the first clear recognition of the imaginary quantities in mathematics which had to wait for several more centuries for their formal definition.

Mahāvīra was one of the earliest Indian mathematicians to deal with the lowest common multiple which he calls *niruddha*. It was evolved to simplify operations with fractions.

Arithmetical and geometrical progressions were already handled earlier. An extensive treatment is available in the *Gaṇitasāra saṅgraha*. In the absence of modern theories of logarithms and equations, problems were solved by methods of trial and repetition.

Mahāvīra seemed to be expert in handling all sorts of equations reducible to quadratic forms and gave a variety of examples of them.

In mensurational problems in geometry, Mahāvīra usually gave two rules: one for rough and the other for better or accurate results. He dealt with all the usual plane figures. For π, he conformed to the Jaina values 3 (rough), and $\sqrt{10}$ (better). He was the first Indian to deal with mensuration related to an ellipse which he calls *āyata-vṛtta* (elongated circle), but his rules are approximate.

For an ellipse of semi major and minor axes his "accurate" results are:

$$\text{Area} = b\sqrt{(4a^2 + 6b^2)};$$

$$\text{Perimeter} = \sqrt{(16a^2 + 24b^2)}.$$

For the exact rectification of the ellipse, one had to wait for about eight hundred years to acquire the powerful tool of calculus. In this situation Mahāvīra's first attempt is to be appreciated.

Regarding the volume of a sphere of radius r, Mahāvīra gave the formula:

$$v = \frac{9}{2}r^3,$$

which was the practical rule of Jaina tradition. He gave another rule for the purpose, but it gives a better result only with an emendment of the text. For the curved surface of a spherical segment, his rule has been newly interpreted to yield the formula

$$S = \pi r^2\theta \sin\theta,$$

where θ is the semi-angle subtended by a diameter of the base of the segment at the center of the sphere. This peculiar formula gives quite a good result in all practical cases (i.e. for θ up to $60°$). The modern exact formula is $2\pi r^2(1 - \cos\theta)$. For the volume of frustum-like solids, Mahāvīra gave a generalization of Brahmagupta's rule based on the theory of averages.

In the end we mention Mahāvīra's extensive contribution to the formation of rational figures. He calls a triangle or quadrilateral *janya* (generated) when its sides, altitudes, and other important measures can be expressed in terms of rational numbers.

R.C. GUPTA

REFERENCES

Datta, B. "On Mahāvīra's Solution of Rational Triangles and Quadrilaterals." *Bulletin of the Calcutta Mathematical Society* 20: 267–294, 1928–29.

Dube, Mahesh. "Poet and Mathematician Mahāvīrācārya" (in Hindi). *Arhat Vacana* 3(1): 1–26, 1991.

Gupta, R.C. "Mahāvīrācārya on the Perimeter and Area of an Ellipse." *Mathematics Education* 8(1B): 17–19, 1974.

Gupta, R.C. "Mahāvīrācārya Rule for the Surface-Area of a Spherical Segment." *Tulasī Praj?ñā* 1(2): 63–66, 1975.

Gupta, R.C. "Mahāvīrācārya's Rule for Volume of Frustum-like Solids." *Aligarh Journal of Oriental Studies* 3(1): 31–38, 1986.

Gupta, R.C. "The Mahāvīrā-Fibonacci Device to Reduce p/q to Unit Fractions." *HPM Newsletter* 29: 10–12, 1993.

Jain, Anupam, and S.C. Agrawal. *Mahāvīrācārya: A Critical Study* (in Hindi). Hastinapur (Meerut): Digambar Jain Cosmographical Research Institute, 1985.

Jain, L.C., ed. *The Gaṇitasāra saṅgraha* (with Hindi translation). Sholapur: Jain Sanskriti Sanraksaka Sangh, 1963.

Rangacharya, M., ed. and trans. *Gaṇitasāra saṅgraha of Mahāvīrācārya.* Madras: Government Press, 1912.

Sarasvati, T.A. "Mahāvīra's Treatment of Series." *Journal of the Ranchi University* I: 39–50, 1962.

See also: Śrīdhara

MAHENDRA SŪRI Mahendra Sūri, Jain astronomer, and pupil of Madana Sūri, was a protégé of the progressive minded Sultan Fīrūz Shāh Tuglaq, who ruled in Delhi from AD 1351 to 1388. The Sultan was one of the pioneers of the cultural exchange between Hindus and Muslims and was much interested in astronomy. His most important contribution in this field was the introduction of the astrolabe into India from the Islamic world. He induced Mahendra Sūri to study the astrolabe and familiarize Indian astronomers with the instrument through the Sanskrit language. This persuasion resulted in Mahendra Sūri's writing the *Yantrarāja* (King of Instruments), in 1370, the first work written in Sanskrit on the astrolabe.

The *Yantrarāja* sets out in five chapters the theory of the astrolabe, the construction of the instrument with its several planes and designs, and lines, circles, and other markings to be made on the planes while making the instrument and graduating it for making observations and recordings. It is to be noted that Mahendra Sūri first describes the ordinary astrolabe, which he calls *saumya-yantra* (northern instrument), wherein the astrolabe is projected from the south pole, and then the *yāmya-yantra* (southern instrument), where the instrument is projected from the north pole. He then introduces a *miśra* (mixed) instrument, which he calls *phaoṅīndra-yantra* (the serpentine instrument), wherein the two types are combined.

The commentary on the *Yantrarāja* by the author's pupil, Malayendu Sūri (fl. 1377) explains the practical application of the instrument for taking readings. He also provides tables of the latitudes of about 75 cities in and outside India, and also one for 32 fixed stars. In this connection the commentator says that the Muslims have recorded more than 1022 stars, but that he has selected only 32, being those required for practical use in Indian astronomy and astrology. There is still another commentary on the work, which is more elaborate than that of Malayendu Sūri, written by Gopīrāja about AD 1540, which is still in manuscript form.

K.V. SARMA

REFERENCES

Dvivedi, Sudhakara. *Gaṇaka Taraṅgiṇī or Lives of Hindu Astronomers*. Ed. Padmakara Dvivedi. Benares: Jyotish Prakash Press, 1933.

Pingree, David. *Census of the Exact Sciences in Sanskrit. Series A, vol. 4*. Philadelphia: American Philosophical Society, 1981.

Yantrarāja of Mahendra-guru with the commentary of Malayendu Sūri, along with the *Yantraśiromai* of Viśrāma. Ed. K.K. Raikwa. Bombay: Nirnaya Sagar Press, 1936.

Yantrarāja of Mahendra Sūri, with the commentary of Malayendu Sūri. Jodhpur: Rajasthan Prachyavidya Shodh Samsthan, 1936.

See also: Astronomy in India – Astronomical Instruments in India – Astrolabe

AL-MAJRĪṬĪ Al-Majrīṭī, Abū-l-Qāsim Maslama ibn Aḥmad al-Faraḍī was born in Madrid, Spain in the second half of the tenth century and died in Cordoba ca. AD 1007. He settled early in Cordoba where he studied with Abū Ayyūb ibn ʿAbd al-Gāfir ibn Muḥammad and Abū Bakr ibn Abī ʿĪsà. He was engaged in making astronomical observations in about AD 979; he may have served as court astrologer.

He had a number of disciples who made his work known throughout the provinces of Spain. Among them were al-Kirmānī (d. AD 1066); Abū-l-Qāsim Aṣbagh, known as Ibn al-Samḥ (d. AD 1035) who is the author of a treatise on the construction and use of the astrolabe in 130 chapters and of the *Book of the Plates of the Seven Planets*, of which the original Arabic version is lost. However, it was translated into Spanish and included in the *Libros del Saber de Astronomía*; Abū-l-Qāsim Aḥmad known as Ibn al-Ṣaffār (d. AD 1034), who is the author of a treatise on the astrolabe attributed in its Latin version to al-Majrīṭī; Ibn al-Khayyāṭ; al-Zahrāwī; and Abū Muslim ibn Khaldūn of Seville.

Maslama's most important work is the adaptation of al-Khwārizmī's astronomical tables (*zīj*) which were elaborated ca. 830. This adaptation is not preserved in his original Arabic form but in a Latin translation made by Adelard of Bath (fl. 1116–1142) and revised by Robert of Chester. The work done by Maslama in this adaptation illustrated his high degree of astronomical and mathematical knowledge.

Another work is an Arabic commentary on Ptolemy's *Planispherium*, entitled *Tasṭīḥ basīṭ al-kura* (Projecting the Sphere onto a Plane), which deals with the stereographic projection on which the conventional astrolabe is based. This adaptation is only preserved in a Latin version by Hermann of Dalmatia (1143) and in a Hebrew abridgement. This work was the point of departure of a long series of Andalusian treatises on this topic. Maslama also knew the *Almagest* as well as al-Battānī's tables.

Maslama also wrote a treatise on *muʿāmalāt* (commercial arithmetic) which probably dealt with sales, cadastre (an official record of property ownership and value), and taxes using arithmetical, geometrical, and algebraic operations.

Some other works are attributed to him, such as *Rutbat al-ḥakīm* (The Rank of the Sage), composed after AD 1009, and *Ghāyat al-ḥakīm* (The Aim of the Sage), translated into Spanish in AD 1256 by order of Alfonso el Sabio and distributed throughout Europe under the name of *Picatrix*.

EMILIA CALVO

REFERENCES

Neugebauer, O. *The Astronomical Tables of al-Khwārizmī. Translation with Commentaries of the Latin Version edited by H. Suter supplemented by Corpus Christi College Ms 283*. Copenhagen: I kommission hos Munksgaard, 1962.

Samsó, Julio. *Las ciencias de los Antiguos en al-Andalus*. Madrid: Mapfre, 1992, pp. 84–98.

Suter, H. *Die Astronomischen Tafeln des Muḥammed ibn Mūṣa al-Khwārizmī in der bearbeitung des Maslama ibn Aḥmed al-Madjrīṭī und der latein. Uebersetzung des Athelard von Bath*. Copenhagen: A.F. Høst & Søn, 1914.

Vernet, Juan, and M.A. Catalá. "Las obras matemáticas de Maslama de Madrid." *Al-Andalus* 30: 15–45, 1965.

Vernet, Juan. "Al-Majrīṭī." In *Dictionary of Scientific Biography*, vol. IX. New York: Charles Scribner's Sons, 1974, pp. 39–40.

Vernet, Juan. "Al-Madjrīṭī." In *Encyclopédie de l'Islam*, 2nd ed., vol. V. Leiden: E.J. Brill; Paris: G.P. Maisonneuve, 1986, p. 1105.

MAKARANDA Makaranda was a resident of Kāśī (or Benares, Varanasi). In AD 1478 he wrote an extensive astronomical manual with the title *Makaranda*. This was doubly significant, first because the title was reminiscent of his own name, and, second because it called the computed result obtained *makaranda* (honey), and gave the several astronomical terms names of parts of plants, such as *guccha* (flower cluster), *kanda* (bulb), *vallī* and *latā* (creeper), and the like. Makaranda based his work on the parameters of and practices prescribed by the modern *Sūryasiddhānta*, to which he added certain corrections to insure greater accuracy, and provided a number of astronomical tables for ease in computing the daily almanac.

Makaranda's tables, which are often long and extend to several centuries, involved much labor and ingenuity in their preparation. They cover such subjects as *tithi* (lunar day, 5 tables), *nakṣatras* (asterisms, 4 tables), *yogas* (complementary positions of the sun and the moon, 3 tables), *saṅkrāntis* (entry of the sun into the zodiacal signs, 3 tables), mean motion of planets and their anomalies (11 tables), length of daylight on different days (1 table), weekdays (2 tables), and eclipses and allied matter (10 tables).

In order to render the work of the user easier, Makaranda provides, in certain cases, two sets of tables, one for single years and the other for groups of years, all from the epoch date of the commencement of Śaka year 1400 (AD 1478). Thus, when calculations are made for a date which is several years after the epoch, multiples of the group-years can be skipped, just taking note of the readings for the number of group-years skipped and applying the same to the relevant year of the current group. In the case of *tithis* the group is taken as 16 years; for *nakṣatras* and *yogas*, it is nearly 600 years, from AD 1478 to 2054, in 24 year periods. For the precession of the equinoxes, it is nearly 100 years, from AD 1758 to 1838, in 20-year periods, and for *saṅkrāntis* for 400 years, from AD 1478 to 1877, in 57-year periods.

Makaranda's tables were widely used in the entire northern belt of India, comprised of Gujarat, Rajasthan, Uttar Pradesh, Bihar, and Bengal, as attested by the profusion of manuscripts of the work found in this region. The work has also been commented on by several authors, from Ḍhuṇḍirāja (fl. 1590), through Puruṣottama Bhaṭṭa (fl. 1610), Divākara (fl. 1606), and Kṛpakara Miśra (fl. 1815), to Nīlāmbara Jhā (b. 1823). The *Siddhāntasudhā* of Paramānanda Ṭhakkura is based on the work of Makaranda.

K.V. SARMA

REFERENCES

Dvivedi, Sudhākara. *Gaṇaka Taraṅgiṇī or Lives of Hindu Astronomers*. Ed. Padmakara Dvivedi. Benares: Jyotish Prakash Press, 1933.

Makaranda with the Ṭīkā of Gokulanātha, Divākara and Viśvanātha. Kasi: Benarsi Press, 1884.

Pingree, David. "Makaranda." *Transactions of the American Philosophical Society*. 58(3): 39–46, 1968.

Pingree, David. *Census of the Exact Sciences in Sanskrit, Series A, Vol. 4*. Philadelphia: American Philosophical Society, 1981.

See also: *Sūryasiddhānta* – Lunar Mansions – Astronomy in India

AL-MA᾽MŪN Al-Ma᾽mūn, Abul'-ʿAbbās ʿAbdallāh ibn Hārūn, was born in 786 in Baghdad, and died in 833 near Tarsus, in a campaign against the Byzantines.

Al-Ma᾽mūn was not himself a scientist. He was the seventh caliph of the dynasty of the Abbassids, son and second successor of the famous caliph Hārūn al-Rashīd (well known from the tales of the Arabian Nights). He ruled the Islamic empire from 813 to 833, at first from Marw (in the Eastern province of Khurāsān, where he was based before his accession), and from 819 from the capital of Baghdad (founded in 762 by the caliph al-Manṣūr). In the intellectual history of the Islamic world and in the history of science, al-Ma᾽mūn played an important role as an instigator and patron of many important activities. He was a firm adherent of the *Muʿtazila*, a rational school of Islamic theology which was strongly influenced by Greek philosophy; in 827 he declared Muʿtazilism the official doctrine for the whole empire. His interest in philosophy and the sciences manifested itself in many ways. He initiated and patronized the translation of scientific works, mostly from Greek, but also from Persian and Syriac, into Arabic. One translation (of several successive versions) of Ptolemy's *Almagest*, and two translations (also of several successive versions) of Euclid's *Elements* were distinctly called, after him, "the Ma᾽mūnian version(s)." In the *Elements*, the theorem I 5 was given, after him, the nickname *al-ma᾽mūnī*, "the Ma᾽mūnian [theorem]." Later, in the medieval Latin translation, this degenerated into *elnefea, id est fuga* and, more contracted, *eleufuga* or *elefuga*. In 832 he founded the *Bayt al-Ḥikma* (House of Wisdom, in continuation of a similar institution established earlier by his father Hārūn), which was established for collecting scientific texts, translation, and teaching. Further, on his order astronomers carried out new measurements of many of the astronomical parameters transmitted in Greek texts, such as precession of the equinoxes, inclination of the ecliptic, length of the year, length of a degree of geographical latitude, geographical coordinates, etc. For many of these data, they arrived at remarkably better values. The results of the observations made by al-Ma᾽mūn's astronomers were laid down in a work called *al-Zīj al-mumtaḥan* (The Revised Tables, dated 829–830; in medieval Latin, *Tabulae probatae*); it is only preserved in a reworked form. These values were afterwards widely quoted and used by other Islamic and also medieval Western astronomers. From quotations in the *Elementa astronomica* of al-Farghānī (ninth century), al-Ma᾽mūn's name entered the West in the forms *Almeon* (Johannes Hispalensis's Latin translation of al-Farghānī's *Elementa*) and *Maimon* (Gerard of Cremona's translation of the same). On Riccioli's map of the moon (1651) a crater was called Almaeon (i.e. the aforementioned Almeon); as Almanon the name survives on modern charts, in permanent memory of this remarkable Eastern ruler.

PAUL KUNITZSCH

REFERENCES

Èche, Yousuf. *Les bibliothèques arabes publiques et semi-publiques en Mesopotamie, en Syrie et en Egypte au moyen âge*. Damascus: Institut francais de Damas, 1967.

Kennedy, Edward S. "A Survey of Islamic Astronomical Tables." *Transactions of the American Philosophical Society*, N.S. 46: 123–177, 1956.

Rekaya, M. "al-Ma᾽mūn b. Hārūn al-Rashīd." In *Encyclopaedia of Islam*, new edition, vol. VI. Leiden: Brill, 1991, pp. 331–339.

Sourdel, Dominique. "Bayt al-Ḥikma." In *Encyclopaedia of Islam*, new edition, vol.I. Leiden: Brill, 1960, p. 1141.

Vernet, Juan. "Las 'Tabulae Probatae'." In *Homenaje a J.M. Millás Vallicrosa*, II, Barcelona, 1956, pp. 501–522.

Yaḥyā ibn Abī Manṣūr. *The Verified Astronomical Tables for the Caliph al-Ma᾽mūn* (facsimile edition of the Arabic text, from MS Escorial 927). Frankfurt am Main: Institute for the History of Arabic-Islamic Science, 1986.

See also: Precession of the Equinoxes–*Elements*–*Almagest*

MAPS AND MAPMAKING A map is an artifact, typically printed on paper, that selectively links places in the world to what comes with them — rents or taxes, voting rights or military obligations, species abundance, or incidence of rainfall. That is, maps are permanent, graphic objects which are a very recent phenomenon with relatively shallow roots in human history. Almost all maps have been created in this century, most in the past few decades.

In the Western world serious interest in map history is datable in its current form only to the 1960s. This history initially married the interests of Euro-American academic cartographers, who were strongly committed to Western positivism, to a pre-existing European antiquarianism which was dominated by a passion for decorative printed maps of the fifteenth to eighteenth centuries. This spawned a hero saga involving such men as Eratosthenes, Ptolemy, Mercator, and the Cassinis, that tracked cartographic progress from humble origins in Mesopotamia to the putative accomplishments of the Greeks and Romans. The rediscovery of these in the European Renaissance led directly to the development of the triumphant scientific cartography that swept the world in the wake of Western colonialism.

As we now acknowledge, this story is disingenuous at best, and perhaps false, in almost every particular. Although the oldest surviving uncontested map is Babylonian, in no way is this map "the" origin of mapmaking, which was originated around the world again and again. Such maps as the Babylonians and Egyptians did make were not "built on" by Greek, Roman, or subsequent European mapmaking. Almost all of those seem to have been invented independently. Indeed Greco-Roman contributions to the history of mapmaking have been greatly exaggerated; if ancient Greeks actually made any maps they left none behind. In any case later European mapmaking was not much indebted to them, nor was this ever the "scientific" enterprise it was claimed, but rather a profoundly ideological activity serving colonial and other interests of the United States and Western Europe, the Soviet Union and Japan.

At the heart of this canonical history lies a family tree, which, like any family tree, forces its conclusion by selective pruning. The lineage Babylon – Egypt – Greece – Rome – Western Europe – United States is tracked only by invoking a sort of magic "moving spotlight of history" that after shining on Babylon for a while capriciously shifts to the Nile, then darts to Greece and thence to Rome, before lighting up Western Europe on its way across the Atlantic. This ethnocentric lineage is then reified as "Europe" or, less implausibly, "the West", while the rest of the world is then constructed as "the East".

While this reification is self-contradictory (on its own account the "Western" tradition is Asian–African–European–North American), it is most thoroughly discredited by its inability to account for the historical record, which it constructs by inventing what is lacking and suppressing or marginalizing what is unwelcome, such as evidence of an independent tradition of mapmaking in India dating to the first millennium BC. This canonical history is not only racist, it is profoundly elitist. It therefore requires explanations rooted in race, environment, or genius that are incapable either of accounting for the universal ability of humans to orient themselves and navigate large environments, or of explaining what it is that propels the manufacture of the millions of maps produced today. By confining its attention to a single line of descent, the history overlooks the evolution of mapmaking that leads to tourist placemat maps, world globe socks, and the Flat Earth Society as inexorably as it does to geological mapping, world globes, and the National Geographic Society.

The interesting division in the history of mapmaking lies between the mapmaking activities of the world's people who live (or lived) as relatively small societies, and those of the world's people who live (or lived) as relatively large groups dominated by centralized bureaucracies and extensively mediated relationships (by, for example, money, scripted languages, recorded legal codes, books, audio recordings, telephones, television). To overdraw this distinction does little justice to the complexity of our data, for many peoples live in both kinds of societies, and there are numerous variations and degrees of difference, but where speech serves to connect people, maps are less common than where the reach of speech is inadequate. Examples of small societies include the Kung of the Kalahari, the Burusho of the Karakoram, the Yolgnu of Arnhem Land, the Navajo of the American Southwest, and the Sambia of New Guinea. These societies, from all over the world and from every historical period, have rarely produced, as a tradition, the cultural artifacts recognized as maps today, though when they have, the maps have varied as markedly from one group to another as have any other attribute. Contrast a *rebbelith* constructed by a Marshall Islander out of cowry shells and palm fronds, depicting Pacific Ocean island groups and sea swells, and a *dhulang* painted by an Aboriginal Australian on bark, representing the "footprints of the Ancestors".

Examples of large societies with centralized bureaucracies include dynastic Egypt, imperial Persia, the Indic societies centered on Harappa and Mohenjo-Daro, dynastic China in both the Hwang and Yangtze valleys, the Islamic empires of the Umayyads and the Abbasids, Heian Japan, Ghana, Zimbabwe, and Aztec Mexico. At one time or another, most of these societies made, or wrote about making, maps, however sporadically. Taking other cultural differences into account, these maps were very similar, consisting of large-scale cadasters or plans, very small scale cosmographical diagrams, or both; and typically the writing about mapmaking described comprehensive small scale surveys that as far as is known were never carried out. For instance, we can compare the cuneiform tablet from Nuzi in contemporary Iraq (ca. 2300 BC), the estate plan of the Todai-ji temple precincts in Japan (AD 756), and the plan of Inclesmoor, England (ca. AD 1405) Though they derive from distinct, probably unrelated traditions, all depict landed property at large scales in a mix of plan and elevation (for what have frequently been characterized as legal purposes), are oriented to the cardinal directions, and display watercourses (the Nuzi and Todai-ji plans also display horizon-closing hills or mountains depicted in profile). Or we can compare the Babylonian "world map" of 600 BC, a European *mappaemundi* of the Isidorus type from the thirteenth century, and a Jain cosmographical diagram of *Adhai-dvipa* (the "two and a half continents") from sixteenth century India. They again derive from distinct traditions widely scattered in space and time, yet all portray, in plan, at very small scale, an ocean-girt cosmos of great similarity. Finally, compare Julius Caesar's purported project to map the known world (ca. 44 BC) with the equally unrealized imperial proposal of AD 646 for the mapping of Japan; or the Alexandrian Ptolemy's (AD 90–168) project for mapping the world with that of the Chinese Pei Xiu's (AD 223–271). In none of these cases are maps known to have been made; certainly none survive.

Thus there would seem to be three distinct functions related to one another only in hindsight: (1) a large scale graphic property control function, (2) a graphic cosmological speculation function, and (3) a proto-bureaucratic inventory and control function that never quite got off the ground. These functions emerge in so many different societies because ancient civilizations all evolved broadly similar social structures. Furthermore, none of these maps, whether made by the smaller or larger societies, seems to be causally connected to the maps that begin to emerge with the development in the current millennium of new forms of social organization including capitalism and the modern nation-state. Those maps are best characterized by the small-scale topographic surveying that comes to be taken as the conceptual model for all mapmaking, so realizing the dreams of Caesar, the emperor of Japan, Ptolemy, and Pei Xiu, and in the light of which the canonical history has rewritten the past.

In general, then, humans have generated maps in three distinct social settings: (1) sporadically at the very large and very small scales in the great "historical" civilizations, (2) with still increasing fervor at every scale in the modern societies of the past five hundred years; and (3) rarely, in the relatively small societies about which much is known with certainty, and with comparative idiosyncracy. These settings are not historically unrelated.

Looking at the historical record now, we greatly exaggerate the importance of maps for the administration of the "great" civilizations by assuming they must have done things the way we do them. This has led historians not only to assume that when people wrote about mapping they must have made maps, and that where one map survives a hundred must have been made, but to postulate mapmaking traditions where instead, as we have seen, there were probably traditions of cosmological speculation, property control or centralized management. Other maps appearing in the historical record may have played no part in any of these traditions, but instead have arisen from isolated efforts to address individual problems, such as laying new drains or defending property in a law suit. That is, they may have been based on no prior model, and left no progeny, and so be akin to what geneticists call a sport. In any event, as their existence and the rest of the record attest, mapmaking was a marginal activity for all these peoples, among whom the functions served by mapmaking today were served by other (usually scripted) forms of inventory and control. This is to say that the historical record is spotty not because survival rates were low, but because maps were actually infrequently made.

This is also why uncontested maps more than five hundred years old are rare at any scale from anywhere in the world. Cosmographical diagrams are more common, although still extremely rare, and large scale plans more common still (though again the numbers are absolutely tiny). But prior to the fifteenth century small-scale geographic maps are rare to the point of non-existence in any cultural tradition. But then no unquestioned map of any kind predates the second millennium BC. Whether prehistoric humans made maps is uncertain, because the interpretation of their artifacts is mired in controversy. If they did not it was not because they were unable, but because the function served by maps either was not called for, or was fused with other functions in a synthesis not recognized as maplike today. Reputable scholars assert with conviction the maplike qualities of the wall painting at Çatal Hüyük (6200 BC), and a case has been made for the petroglyphs at Valcamonica (2500 BC) and elsewhere. But if prehistoric humans did make maps, they were neither made often nor in very many places. They likely served broadly pictorial, religious or ritual functions, and their production was discontinuous with the practice of mapmaking encountered in historic populations.

The oldest extant maps about which there is scholarly consensus are, as noted, Babylonian. Dozens of large scale Babylonian cuneiform maps and plans survive from the second and third millennium BC, but only a couple of small scale maps survive, and these from the first millennium BC. The existence of the so-called Turin gold mining map from about 1150 BC is the sole survivor of an Egyptian mapmaking tradition of roughly similar age that otherwise is represented only by cosmographical diagrams and pictures of gardens, canals, and other features. Recent scholarship posits an Indic tradition of mapmaking stretching back to the first millennium, but the earliest extant artifacts are an allegorical wall sculpture from about AD 400, and a Jain cosmographical diagram of the thirteenth century AD. There is textual evidence of a Hindu tradition of cosmographical globe construction dating from the first millennium BC, but again no actual globes predate the fifteenth century AD.In China three maps survive from the second century BC, but no others until the twelfth century AD, and maps do not become common at even this date. Evidence also suggests a Tibetan mapmaking tradition rooted in the first millennium BC, though again, with the exception of a mandala transmitted to Japan in the ninth century AD, no survivals predate the eighteenth century. Textual evidence also supports a Hellenistic mapmaking tradition, but no maps survive of any character. Except for Medieval European copies of Roman itineraries, no small scale Roman maps survive, despite the elaborate instructions for producing them in Ptolemy's *Geography*, and even large-scale survey and property maps do not exist in abundance. That is, with respect to the ancient world there are many more textual suggestions that something like mapmaking was carried out than there are surviving artifacts, the numbers of which, with the exception of Babylonian and Roman plats and surveys, may be counted on the fingers of two hands. That is, mapmaking was relatively widespread but everywhere relatively uncommon.

The record is not much different for the medieval world. Islamic scholars elaborated sophisticated theoretical schemes for the construction of maps from the seventh century on, but if any were made, none survives from periods prior to the tenth century, and maps remain rare until the fifteenth and sixteenth centuries. In Medieval Europe cosmographical diagrams and large scale plans are extant from the seventh century, but with the exception of the late medieval portolan charts, maps were otherwise practically unknown. There is textual evidence of relatively small-scale mapmaking in Japan as long ago as the seventh century AD, but again, nothing survives. Property maps of paddy fields are extant from the eighth century, but the earliest surviving maps of Japan or the world date from the fourteenth century. Textual evidence supports a mapmaking tradition in Vietnam as early as the eleventh century, but no artifactual maps predate the fifteenth century. No Mesoamerican maps predate the Conquest, though there is ample reason to assume a pre-existent tradition of cosmographical diagrams and property (or "community") mapping among the Nahua, Mixtec, Otomi, Zapotec, Totonac, Huastec, Chinantec, Cuicatec, and Mazatec. The oldest surviving Malay maps are from the sixteenth century. No indubitable maps made prior to the fifteenth century survive from Sub-Saharan Africa, South America, Australia, Oceania, or North America, though in many places the record was systematically destroyed, and historical research may yet uncover evidence of mapmaking traditions unknown today. Despite these lacunae, the record suggests that large scale property maps and small scale cosmographical diagrams were made rarely, but with increasing frequency, everywhere in the world since the third millennium BC. Mini-traditions seem to develop often, only to die.

The significance of these data is obvious. Human societies did not need maps and got on without them for hundreds of thousands of years. But during the last two or three millennia BC, larger, more complicated societies including Babylonia, Egypt, perhaps the Indic societies centered on Mohenjo-Daro and Harappa, and China began to articulate, sporadically and apparently independently, graphic notation systems linking location with rights and obligations, as in the large scale property maps, and with speculative attributes of the larger environment, as in cosmographical diagrams. Similar graphic notation systems filling broadly similar social functions emerged fitfully in other ancient civilizations around the world, again apparently independently, although extensive trade and other connections among these groups are acknowledged and cross-fertilization undoubtedly took place. The articulation of such similar notation systems in so many of these societies strongly supports the notion that maps of this character inevitably emerge in societies whose increasing size and complexity call for them. But the sporadic nature of this articulation no less strongly suggests that at the size and degree of complexity reached by ancient civilizations, the map function could be satisfied by other, better established functions, such as writing, and so failed to establish itself no matter how many times it was seeded. Mapping is nowhere well-rooted until the rise of the modern nation-state, with which it co-evolves as an instrument of polity, to assess taxes, wage war, facilitate communications, and exploit strategic resources.

While it is not "wrong" to refer to these early graphic notation systems as maps, it is critical to accept that they were not referred to as such by those who made them. For example, early maps were generally free of the spatiality so characteristic of what we think of as maps, and were probably not discriminated from other graphic-textual productions. Until modern times no society distinguished — or made — such maps as distinct from religious icons, landscape painting, construction drawings, itineraries, and so on. For example, the Chinese word *tu*, frequently translated "map", can also be translated "picture", "diagram", or "chart". *Tu* of geographical subjects frequently had poems painted on them as was common on paintings of other subjects. This not only reflects the conceptual continuity that tied together the Chinese practice of what today the Chinese themselves think of as discrete genres (painting, mapping, drawing), but that synthesis of painting, calligraphy, and poetry that so effectively distinguishes, say, Ming painting from that of the European Renaissance. This synthesis lent Chinese *tu* an explicitly expressive character inconceivable in twentieth century conceptualizations of cartography.

Such inclusiveness characterizes other words frequently translated "map," including the Arabic *naqshah* (painting, any kind of visual representation), its Indian derivation *nakṣa* (picture, plan, general description, official report), the Sanskrit *chitra* or *alekhya* (painting, picture, delineation), the Latin *mappa* (cloth) and *carta* (formal document), the Mexican *lienzo* (linen, cloth, canvas), and the Yolgnu-Australian *dhulang* (painting, map, diagram, graphic representation). Mesoamerican *lienzos* in their community maps drew history and territory together (or, perhaps, from their perspective did not rip history and territory apart). Where the Mixtec made do with one, contemporary societies are obliged to use three or four discrete forms, such as plat, deed, title search, or genealogy. The *lienzos* served the Mixtec, as cosmographical diagrams did the Jains and medieval Christians, every bit as effectively as topographic surveys serve the interests of the modern state. The functions a society evolves depend on what kind of society it is. But indeed few of the graphic notations produced in ancient or medieval civilizations would be considered maps today. Maps construed as topographic surveys gained currency only in the last five or so hundred years, and within this period only in relatively stable states with entrenched centralized bureaucracies and wellestablished academies. Though few people used maps in 1400, by 1600, people around the world found maps indispensable. There is a divide here that is impossible to evade. Recall the dates at which maps really begin to appear in the historical record: Islamic artifacts may date to the tenth century, but maps do not become common until the fifteenth and sixteenth centuries; the oldest surviving map of China may be from the second century BC, but maps do not become abundant until the seventeenth century; large scale Japanese maps may survive from the eighth century, but the oldest Japanese world map is from the fourteenth century; the oldest surviving Hindu globe is from the fifteenth century; Vietnamese and European maps become plentiful only in the fifteenth and sixteenth centuries; Mesoamerican maps survive largely from the sixteenth century, Malay maps from the sixteenth century. Again and again we find large, centralized societies, from everywhere in the world, inaugurating mapmaking traditions during the transition to the early modern period.

For mapmaking this transition has been well studied only in Europe, but there is reason to believe that common processes were at work in all the societies struggling with similar socioeconomic transformations. Certainly the nascent European mapmaking tradition was transported around the globe, but the ability it demonstrated to import material from other traditions (well documented, for example, in the cases of Islamic, Burmese, Chinese, and Japanese mapmaking) and the apparent ease of its adoption actually suggest a merging of mapmaking traditions into a kind of transnational or worldwide tradition that less differentiated West from East.

Intriguingly, the functions the new maps initially served were not those that would strike us as obvious (for instance neither ownership of open-field strips nor routes were mapped, at least in Europe); nor were the state functions they did serve newly created in the fifteenth century. That is, at least in the case of early modern Europe, people did not adopt maps because of unique characteristics of either maps or state, notwithstanding testimony of participants in this process of adoption to the contrary. Thus, in 1602 the duc de Lesdiguières complained to Henry IV of France that, "Your majesty will understand much better than I can set it out in writing, if [you] will look at the map of Dauphiné with the Piedmont border," and Michelangelo complained about the Habsburg emperor, Charles V, that, "If only the emperor ... had ordered a drawing to be made of the course of the river Rhône, he would not have met with losses so severe, nor retired with his army so disarrayed." In fact, Charles did use maps, extensively. About the very battle to which Michelangelo referred, Martin de Bellay wrote of seeing Charles, "Studying the maps of the Alps and the lower region of Provence so enthusiastically that the emperor had convinced himself that he already possessed the land in the same way he owned the map." The most general admonition seems to have been Castiglione's of the 1520s to the effect that there were, "matters, the which though a manne were liable to keep in mynde (and that is a harde matter to doe) yet can he not shew them to others" without a map or painting. These anecdotes reveal the growing currency of maps in the early modern period, but they also make very clear their novelty. Accompanying these novel attitudes was a steep increase in the use of maps for military, administrative, and speculative humanistic purposes.

Why? Why after 1400 did people suddenly start using maps if they were ignored in what would become their essential applications, and if the functions they served were not new? The canonical answers — profoundly Eurocentric — include: the Renaissance rediscovery of Ptolemy and the consequent revolution in world view; the Scientific Revolution and a shift to quantitative ways of thinking about locations; the encouragement the painterly "realism" of the Van Eycks and the Limbourgs gave to the emergence of topographical views (and thus maps of town and country); the estate mapping called for by rationalized land management policies; and the new role of the emerging nation-state. Unfortunately these have limited explanatory power.

It is as unlikely that any of these "led to" mapmaking as it is that mapmaking "led to" them. Indeed the record suggests that the administrators promoting map use were sufficiently sophisticated to recognize Ptolemy for the ancient he was; that any interest they might have had in quantification was driven by the pragmatics of fortification; and that realism was less important than information (the many inelegant sketches made for military lodging-masters are exemplary in this regard). Projections, quantification and realism — in a word the scientific cartography that is so central a subject of the canonical history — seem to be, if not beside the point, certainly secondary. As one expert on early modern cartography cautions, "The sort of precision to be found in scale maps was often not required by decision-makers, who could make do perfectly adequately on most occasions with rough-and-ready picture or position maps, lacking scale or standardized conventional signs."

Then what did "lead" to the increase in mapmaking? Any answer to this question is conjectural given the state of our knowledge, but the implication that mapmaking emerges as a rationalizing tool of control during periods of relative prosperity in capitalist state economies is supported by evidence from the Habsburg, Bourbon, and Tudor realms as well. Certainly the aggressive use of maps in this period as tools for control of land for taxation, resource exploitation, and other purposes was limited neither to Europe nor its colonies, but was a feature of administration in the Ottoman Empire, southeast Asian states, China, and Japan as well.

In order to expand their economies, these emergent nation-states all jostled for access to territories not already under the control of other centralized bureaucracies. There they encountered large numbers of relatively small societies. The record is unequivocal. Only a handful of these societies had established mapmaking traditions, though many more used sign systems of various kinds. Yet all of these peoples were intimately acquainted with the territories they inhabited and, of course capable of making maps in any medium when requested by traders, soldiers, missionaries, explorers, or others from expanding states.

There are many instances of the latter. For instance, the 1743 *Carte de l'Amérique Septentrionale* of Jacques-Nicolas Bellin explicitly acknowledges its Indian sources, now known to be those compiled by Pierre La Vérendrye from oral, gestural, and graphic information provided in 1729 by the Indians Tacchigis, Auchagah, and others. In a later example, Peter Fidler, working for the Hudson's Bay Company in 1801, asked the Blackfoot Indian Chief, Ac Ko Mok Ki, to describe the Rocky Mountains and adjacent plains between southern Alberta and central Wyoming. Ac Ko Mok Ki drew a map embracing some 200,000 square miles, depicting the salient topography, hydrography, and culture (including population estimates of indigenous peoples) of an area only partly known to him first-hand. The information on this map was immediately compiled onto Aaron Arrowsmith's map of North America, which in turn provided the foundation for Lewis and Clark's mapping of the Missouri basin. In a still more recent example, maps collected in the 1920s from Canadian Inuits played an important role in the Euro-American mapping of the North American Arctic. One expert on this subject has pointed out that, "Knud Rasmussen and other members of the Fifth Thule Expedition were 'astonished' at the ability of the Caribou Inuit to draw accurate maps without previous experience with the instruments or media with which they were supplied." Throughout the history of contact between North American indigenous peoples and Europeans, the latter have solicited maps from the former with every expectation of acquiring useful information, and this practice has been continuous from early contact times through the very recent present. The history of contact between indigenous peoples and colonizers elsewhere in the world has taken a similar form. Critical in this process, however, was the solicitation of the maps, which as far as we know were not naturally occurring products of a pre-existing cultural tradition.

But not all maps were explicitly solicited. Some were created to fulfill novel functions called into being by the new social relations instituted by colonial powers (for example, relations involving private property, individual ownership, exclusive use-rights and so on). Examples of such maps, among many, include that of the upper Mississippi drawn by Non Chi Ning Ga, an Iowa Indian chief, as part of a land claim made in Washington, D.C., in 1837; or that drawn to support the petition of the Chippewa to their lands in 1849, which combines totems, myth, history, and geography in a way that resembles the *lienzos* drawn by the Nahua and Mixtec. For that matter many of the *lienzos* were drawn in response to similar calls for legal documentation of land claims in Spanish and Mexican courts, where they are still used for legal purposes today. Indeed contemporary societies of the kind under consideration everywhere continue to create maps to establish, preserve, or reclaim territorial rights, as the example of the Yalata map Kinsley Palmer made with the Southern Pitjanjatjara in Australia illustrates so well.

None the less, only in a few isolated cases is there evidence of a prior tradition of making maps in anything like the sense we understand today. Among these some scholars number the Luba of Zaire, whose *lukasa* memory devices were used to pass on mythological and moral lessons to initiates in a once powerful secret society, and the Dakota of North America, whose winter counts comprised a kind of community annals; and others. In all these cases the map is so fused with other functions as to call the ascription map into doubt. Mnemonic device or annals would serve just as well. Among those about whom there is greater scholarly consensus may be numbered the Semang of West Malaysia, the Bataks of Sumatra, and the Dayaks of Borneo, all of whom etched maps in bamboo and sometimes on paper provided by missionaries. These too fuse the map function with others; they were divinatory as well as cosmographical, and sometimes charted routes to other worlds as had ancient Egyptian coffin paintings. Among traditions producing maps with the heightened spatiality associated with survey maps are those of the Inuit and the Marshall Islanders. The map-like quality of these artifacts — those of the Inuit are often carved or sculpted in wood, while those of the Marshall Islanders are constructed out of palm fronds and shells — has long been acknowledged because their explicitly navigational and pedagogic functions are so similar to our maps. The even richer tradition of the aboriginal Australians provides another, well-studied, example. Their *dhulang* (bark paintings) embody an understanding of the world other than that of the contemporary academic cartographer, but there is no doubt that they are both religious icons as well as objects linking territory to what comes with it for the purpose of supporting the reproduction of the social relations of power.

Because maps often get confused with fundamental human abilities like wayfinding and spatial intelligence, to deny that a people makes maps is seen by many as attacking them in some fundamental way. This is illusory. Just as people lived (and continue to live) without writing, carrying on a rich human life in the absence of scripts, so people lived (and continue to live) without maps. Maps are brought into being only when the social relations in a society call for them. Mapmaking was carried out in ancient civilizations around the world, but rarely, and usually only at very large and very small scales.

DENIS WOOD

REFERENCES

Buisseret, David, ed. *Monarchs, Ministers and Maps: The Emergence of Cartography as a Tool of Government in Early Modern Europe*. Chicago: University of Chicago Press, 1992.

Harley, J.B. and David Woodward, eds. *The History of Cartography*. Chicago: University of Chicago Press, 1992.

Kain, Roger J.P. and Elizabeth Baigent. *The Cadastral Map in the Service of the State: A History of Property Mapping*. Chicago: University of Chicago Press, 1992.

Turnbull, David. *Maps Are Territories: Science Is an Atlas*. Geelong, Victoria: Deakin University, 1989.

Wood, Denis. *The Power of Maps*. New York: Guilford Press, 1992.

MAPS AND MAPMAKING IN AFRICA With the exception of medieval Islamic mapmaking, the corpus of precolonial African maps is too small for us to generalize about distinctive cartographic traditions. B.F. Adler (1910) provides the only summary of sub-Saharan mapping, and most of his examples are maps solicited from Africans by European explorers. The paucity of extant maps may be explained by a number of factors. First, among literate cultures like the Muslim Hausa and Jula of West Africa, there existed effective substitutes such as travel guides written in Arabic script. These commonly took the form of itineraries that listed the names of towns between a starting point and destination (e.g. the road from Bornu to Mecca). In 1824 Joseph

Dupuis provided examples of these "native charts" kept by merchants and pilgrims which he used to construct his own maps of West Africa. Second, the scarcity of maps in non-literate societies may be explained by a common recourse to drawing maps on the ground. European explorers witnessed this indigenous form of mapmaking and were frequently impressed by its accuracy. However, the ephemeral nature of ground maps has left us with few traces of this apparently widespread practice (see below). Third, the demand for maps was probably very limited because of the hazards of traveling beyond one's territory. Even merchants risked being captured by neighboring groups if they did not travel in caravans, carry letters of introduction, or have contacts in distant communities. Under these circumstances, there was little demand for maps as conveyors of geographical knowledge to outsiders. Finally, in searching for African maps, we should be wary of looking for Western forms of mapping in societies whose spatial concepts and relationships to land are fundamentally different. Western maps are constructed upon culturally specific notions of property, territory, and political authority over bounded areas. One should not assume *a priori* that these concepts are held by African peoples. Moreover, as Paul Bohannan shows in his discussion of the "genealogical map" of the Tiv of central Nigeria and the "rain shrine neighborhoods" of the Tonga of Zambia, there is tremendous variety within Africa itself in the conceptualization of space. Rarely does one find the expression of socio-spatial relationships in two-dimensional maps. Despite this restricted development of mapmaking in precolonial Africa, there are a number of interesting maps to consider.

One of the earliest examples of African mapmaking is an Egyptian map dating from about 1150 BC. Fragments of the picture map depicting the Wadi Hammamat area between Thebes and the Red Sea port of Quseir are preserved on papyrus in the Muzeo Egizio at Turin. Other ancient Egyptian examples include plans of gardens and maps of the afterlife painted on stone during the second millennium BC.

One of the foremost cartographers of the Middle Ages was al-Sharīf al-Idrīsī. He was born in Ceuta, Morocco, in 493 AH/AD 1100 and is believed to have died there in 560/1165. He was the court geographer of the Norman king of Sicily, Roger II, for whom he wrote his celebrated *Nuzhat al-mushtāq fī' khtirāq al-āfāq* (The Book of Pleasant Journeys into Faraway Lands, also known as the Book of Roger). This work contains a world map and seventy sectional maps that build upon both the Balkhī and Ptolemaic cartographic traditions. Although written in the twelfth century, al-Idrīsī's work was still influential five centuries later among North African chartmakers. The portolan charts of the al-Sharafī al-Ṣifāqsī family that thrived in the Tunisian town of Sfax for over eight generations are compilations based on al-Idrīsī and Catalan sea charts.

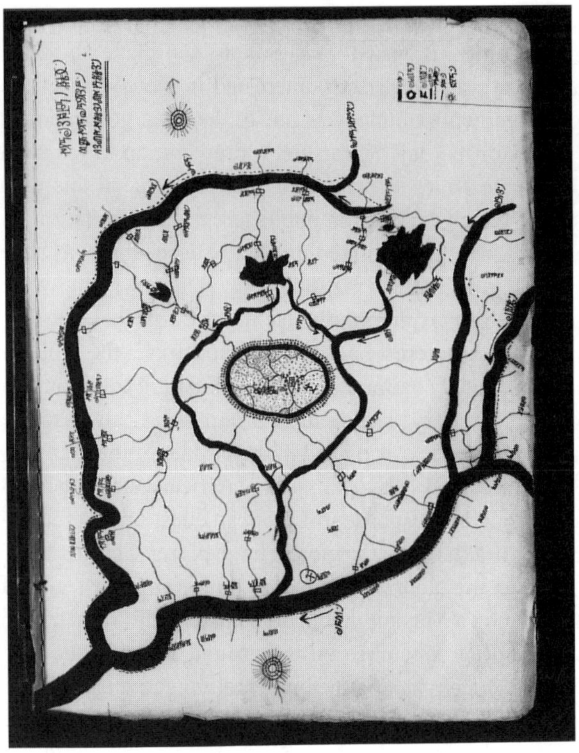

Figure 1: Ethiopian map of Tigre. From the Bibliothèque Nationale de France, Collection Antoine d'Abbadie 225 fol. 3-cliché A85/489.

A distinctive form of schematic mapping dating from the eighteenth century is represented in Ethiopian manuscript maps of Tigre. These maps consist of three concentric circles in which Aksum, the center of Ethiopian Christiandom, is situated in a box in the innermost circle, as in the figure here. The outer circles are divided into segments that contain the cardinal directions and the names of outlying districts. At least five versions of this map are known to exist, two of which are found in the manuscript titled *Kebrä Nägäst* (The Glory of Kings). Below the circle map is a "wheel of wind" or "wind rose" in which the cardinal directions are again shown with east at the top. More research is needed to examine the relationships between these circle maps and the texts in which they are found. Evidence of African maps and mapmaking in the nineteenth century is largely found in European accounts of exploration and travel. In many instances, African mapmaking was stimulated by European interest in the geography of unexplored areas. A well-known example is the map drawn by Sultan Bello for Hugh Clapperton during his visit to the Sokoto Caliphate in 1824. Clapperton was particularly interested in the course of the Niger River whose outlet was one of the great geographical mysteries of the day. Sultan Bello drew a map in the sand showing the Niger's course and later reproduced this map on paper which Clapperton published in the account of his journey. Although dismissed by some Europeans as a "rude representation", Bello's map and geographical writings were valued by later explorers like Heinrich Barth in 1859. His map is also of interest because it demonstrates the "rule of ethnocentricity" common to most mapmaking traditions, in which the territory of a cultural group — in this case the Sokoto Caliphate — is placed in the middle of the map.

There are many other examples of Africans drawing maps on the ground in response to European questions on the geography of a particular region. The explorer Charles Beke was shown the incorrect course of the Gojab river south of Abyssinia by a Muslim merchant named Hádji Mohammed Núr who drew its course on the ground with his stick. In 1881, the Bohemian doctor Emile Holub recounts a similar experience when travelling in the mid-1870s in the Marutse empire of the upper Zambezi River. Before leaving to explore the headwaters of the Zambezi, Holub asked the Maurtse chief, Sepopo, to suggest a good travel route. To Holub's great interest, Sepopo drew a map in the sand whose accuracy was confirmed by two other persons familiar with the area. While on a frontier reconnaissance mission in the dense tropical forests of southeastern Liberia in 1899, the French explorer Capt. Henri d'Ollone asked a person named Tooulou to draw on the ground with a piece of charcoal the distribution of the different ethnic groups in the region. After a moment's reflection and to d'Ollone's great surprise, Toolou drew a detailed ground map showing the location of villages, rivers, and mountains, as well as ethnic groups. The information gained from solicited maps was occasionally incorporated into European maps. For example, sections of the map of the Sahara produced by the French geographer Henri Duveyrier in 1864 were based on maps drawn in the sand by "Cheikh-'Othmân". The German geographer Karl Weule preserved the maps he solicited by requesting that they be drawn on paper that he provided. Three of these solicited maps are found in Adler's 1910 book.

Among the Luba of central Africa, ephemeral maps were a common feature of initiation ceremonies. During one stage, the initiate is taken into a meeting house where elders have chalked maps on the wall showing major lakes and rivers, the location of various chiefdoms, and the dwelling places of spirits. While facing the map, initiates are quizzed about the residence of certain chiefs and spirits within the Luba kingdom. In the final stage of initiation, elders teach initiates about the origins of Luba kingship and customary taboos. In recounting the origin myths, elders use memory boards (*lukasa*) as mnemonic devices to aid their storytelling. These small rectangular boards are covered with beads and shells that map the migration history of the founding royal family. Rivers and villages are represented by the patterning of beads and shells into configurations that are recognizable to initiates.

An example of cosmographical mapping is found among the Bakongo of Zaïre in their initiation and funerary art. The Kongo cosmos is pictured ideographically as the sun moving through four phases: dawn, noon, sunset, and midnight. The cosmogram is composed of a cross representing the cardinal directions with a small circle at each end point. The horizontal line (*kalunga*) divides the realms of the living and the dead through which all persons travel. Only the most courageous and generous in life return as immortal spirits in natural forms and forces in the landscape.

A more secular and ambitious mapmaking took place in the Kingdom of Bamum in contemporary western Cameroun under the leadership of King Njoya. Njoya stands out as a highly creative and politically astute individual who collaborated with a succession of German, British, and French authorities to consolidate his rule. He was responsible for developing an alphabet which enabled him to write the history of his kingdom in his own language. He also appears to have been a self-taught mapmaker whose earliest preserved work (1906) is composed of a plan of his farm and a route map between Fumban, the capital of Bamum, and his fields. Njoya further honed his mapmaking skills when the German cartographer Moisel visited Bamum in 1908. One of Njoya's most impressive mapping projects was a topographic survey of his kingdom that employed up to sixty individuals. When the British took control of western Cameroun from the Germans in 1916, Njoya displayed his political skills by presenting a map of his kingdom to the British political officer stationed in Foumban. This map with its southerly orientation shows a well-defined territory in which all roads lead to the royal capital and historic center of political authority. In his letter to the King of England that accompanies the map, Njoya seeks British protection against the Germans. In this

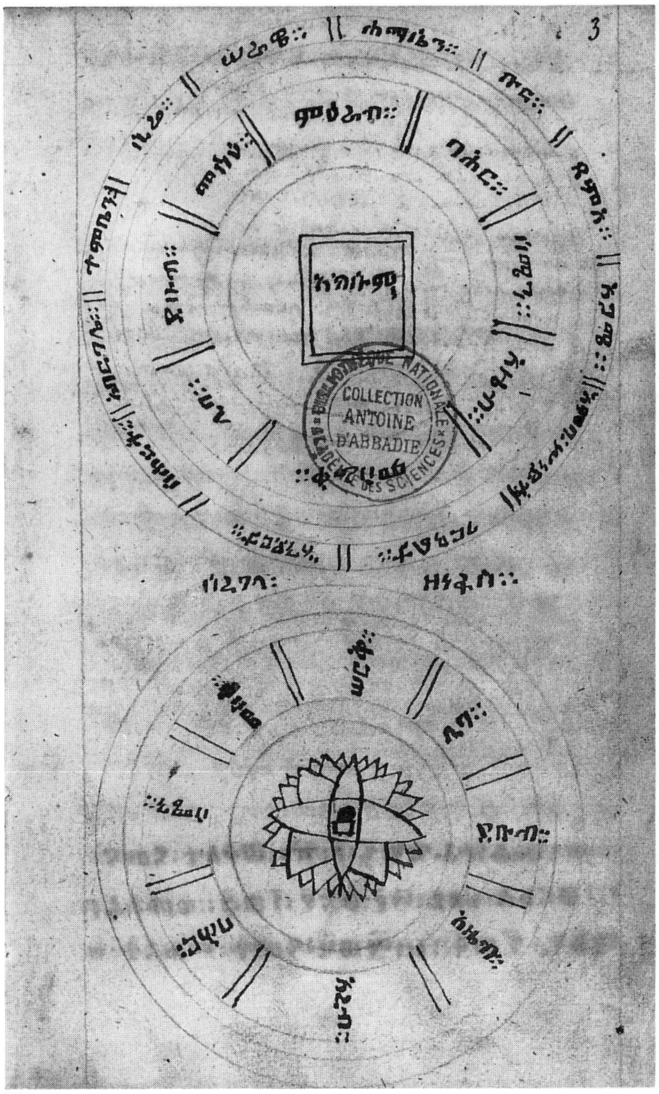

Figure 2: King Njoya's map. From the Public Record Office, Kew, CO 649/7 (Crown copyright reserved).

context, Njoya's map becomes an instrument of power that legitimates his (contested) claim to be the ruler of Bamum and symbolizes his willingness to collaborate with colonial authorities.

In conclusion, the dearth of African maps seems to imply that mapmaking was not a common means of expressing spatial information. One could even argue that the maps solicited by explorers reflected European mapping traditions rather than African custom. However, the ability of individuals from across the continent to produce consistently accurate ground maps suggests that this was an indigenous practice. Ironically, these ephemeral maps led to the drawing of new and improved maps of Africa by Europeans who ultimately employed them in their partition of the continent into colonies.

THOMAS J. BASSETT

REFERENCES

Adler, B.F. *Karty Piervobytnyh Narodov, Izviestia Impieratorskavo Obshchestva Lubitielei Estiestvoznania, Antropologii i Etnografi, sostoyaszchavo pri Impieratorskom Moskovskom Universitietie*, Tom 119. St. Petersburg, Trudy Geograficheskavo Otdielienia, Vypusk II, 1910.

Ahmad, S. Maqbul. "Cartography of al-Sharif al-Idrisi." In *The History of Cartography*, Vol. 2, Book 1, *Cartography in the Tra-*

ditional Islamic and South Asian Societies. Ed. J. Brian Harley and David Woodward. Chicago: University of Chicago Press, 1992, pp. 156–174.

Bohannan, Paul. "'Land', 'Tenure' and Land Tenure." In *African Agrarian Systems*. Ed. Daniel Biebuyck. Oxford: Oxford University Press, 1963, pp. 101–112.

Denham, Major, Captain Clapperton, and Dr. Oudney. *Narrative of Travels and Discoveries in Northern Central Africa in the Years 1822, 1823 and 1824*. London: John Murray, 1826.

Dupuis, Joseph. *Journal of a Residence in Ashantee*, London: Frank Cass & Co., 1966.

Pankhurst, Alula. "An Early Ethiopian Manuscript Map of Tegré." In *Proceedings of the 8th International Conference of Ethiopian Studies, University of Addis Ababa, 1984*. Ed. Tadesse Beyene. Addis Ababa: Institute of Ethiopian Studies, 1989, vol. 2, pp. 73–88.

Reefe, Thomas. "Lukasa: A Luba Memory Device." *African Arts* 10 (4): 48–50, 88, 1977.

Shore, A.F. "Egyptian Cartography." In *The History of Cartography*, Vol. 1: *Cartography in Prehistoric, Ancient, and Medieval Europe and the Mediterranean*, Ed. J. Brian Harley and David Woodward. Chicago: University of Chicago Press, 1987, pp. 117–129.

Struck, Bernhard. "Köing Ndschoya von Bamum als Topograph." *Globus* 94: 206–209, 1908.

Thompson, Robert F. and Joseph Cornet. *The Four Moments of the Sun: Kongo Art in Two Worlds*. Washington: National Gallery of Art, 1981.

See also: al-Idrīsī – Balkhī School

MAPS AND MAPMAKING IN ASIA (PREHISTORIC)

There is plenty in Asian prehistoric art to interest the historian of cartography. It is sometimes difficult to draw a line between a picture of a place and a map of a place and it is, of course, impossible to know the original meaning of any graphic representation made in the period before writing. Nevertheless, there is a great deal of evidence that prehistoric people had both the mental capacity and the communicative and graphic skills to make maps. Such depictions were made on a variety of surfaces and in various ways but most commonly as paintings or engravings (petroglyphs) on boulders, rocks, cave walls, and cliff faces as well as on artifacts (bone, pottery, bronzes etc).

The earliest recognizable representation of spatial relationships involves, in Asia as in Europe, a continuous line — representing a boundary or enclosure — within which an event (animal trapping or herding, people standing or dancing) is taking place. For example, a painting on the wall of a Mesolithic rock shelter in Bhimbetka (Madhya Prades, India) shows, in profile, a family group mourning a dead child. The scene is placed in a hut, or some special place, which is represented by the encircling line. Similarly, a rock-carved enclosure from Mongolia (Bayan Khongor Province) provides a close parallel with enclosures in the Palaeolithic art of the Franco-Cantabrian region of western Europe that have been identified as 'hut or game enclosures'. Yet another Asian parallel comes from Armenia. In this case, the enclosure depicted, possibly as early as the third millennium BC, bears a marked resemblance to an an early historic petroglyph from Jordan carved into a large stone with a text on another side referring to the enclosure as an animal pen.

Picture Maps

In all these examples, only one element of the composition, the enclosure, is rendered in plan (i.e. viewed from above); all other figures — persons, huts, animals — are shown in profile (i.e. as seen from ground level). It is the defined space of the enclosure, however, that provides the cartographic key, the intended spatial relationships of the figures. These spatial representations are very simple as graphic compositions. Nevertheless, they contain the essence of a map and can be thought of as simple picture maps. In contrast, one of the most detailed prehistoric depictions of a village fails to meet the criteria for even a picture map, for there is nothing in it to suggest the critical *spatial* element. The petroglyphs in question, the Boyar *pisanitsas*, are found high on a cliff near Minusinsk, overlooking the Yenesei River in Siberia. They are thought to date from the last millennium BC. The main assemblage is nearly ten meters long but nowhere in it is any feature or figure shown in plan. Nor is it easy to be certain, especially in the absence of an enclosure or frame, that the array of huts and scatter of human figures were created as an assemblage and are not merely a palimpsest of engravings made on quite different occasions, perhaps separated by long intervals of time.

However, another highly detailed rock painting, one of many on the cliffs at Cangyuan (Yunnan Province, China) dating from the last millennium BC, can be thought of as a picture map. Not only are the huts neatly and obviously deliberately positioned along the perimeter of a circular enclosure but their arrangement is topologically consistent: the huts on the far side of the enclosure from the viewer have been drawn upside down in order to maintain the relationship beteen the piles supporting each hut and the line that represents the village boundary or fence. Other lines project from the central enclosure like paths and indeed there are figures, human and animal, shown in profile, walking along them.

Plan Maps

The identification of objects and landscape portrayed from above is always difficult and the more so in rock art where all external evidence is lacking and diagnosis has to rest on intrinsic visual characteristics. However, the visual similarity between certain landscape features, when viewed from high above (i.e. in plan) and some of the rock art found in Asia suggests that such examples were intended to be mimetic. Thus, included in the assorted petroglyphs of Mugur Surgol (upper Yenesei) are representations of the local herders' huts (yurts) and their surrounding stockyards. Four different sets of rock markings — 'map signs' — are involved: solid outlines (squares or rectangles), rectangular outlines with internal subdivisions or compartments, rectangular outlines filled with stippling, and empty outlines, also rectangular or sub-rectangular. Each of the petroglyphs in question usually comprises one solid or compartmentalized shape and one or more stippled, or empty, rectangular or subrectangular shapes. These assemblages have been interpretated by archaeologists as plans of the Mongolian-type yurts found among the local Tuva, together with the yards and enclosures around them. Similar 'hut and yard' markings are found in the Altai Mountains. All these are strikingly like those dating from the end of the Neolithic and, especially, the middle Bronze Age (third and second millennia BC) found in the western Alps, as for example in Mont Bégo, France. Another category of plan map is more typical of Mongolian rock paintings. In these, assemblages comprising relatively large rectangles, usually stippled within, associated with human or human-like figures and the outline of a bird (eagle) with wings outstretched, have been interpreted as representations, rich in religious symbolism, of local graves, again as seen from above. This style of burial is known to be ancient and both graves and grave plans may date from the late Bronze Age.

Celestial Maps

Given the long history of astronomy in many parts of Asia, prehistoric representations of at least the major constellations might be expected. Certainly by the end of the last millennium BC, the period of the Han Dynasty in China, the use of 'ball and chain' patterns to represent groups of stars was evidently an already established tradition. Fully-fledged prehistorical celestial maps, however, have yet to be discovered, in the literature or in the field. There are indications of what is to come. In 1990, the discovery of a very early historical map painted on the vault of a tomb was reported (*The Times*, 1 February 1990). It shows the heavens divided into twenty-eight lunar mansions, seven for each of the cardinal points, which are personified as a Daoist deity. The map, executed in pastel polychrome, matches the description given by the Han historian Sima Qian (ca.145–87 BC). Elsewhere in Asia, representations of constellations, as well as 'ingenious calendars' and the solar and lunar motifs associated with cosmologial myths, are said to be common in the rock art of Armenia. This rock art dates largely from the third milliennium BC, the date also of an astronomical observatory that was excavated at Metsamor (Armenia). Similar motifs may also be found in the pottery decoration of the period.

Cosmological Maps

It would be surprising if cosmological maps were not proven to be by far the most important category of Asian mapping in prehistoric times, just as they have been throughout the historic period (prior to European involvement). A preoccupation with the origin and structure of the universe, and above all with the afterlife and its location, is a fundamental attribute of human life — a manifestation of humanity's 'cosmic anguish'. Most, if not all, prehistoric rock art would have been associated with religion, and much of it must have reflected various aspects of local belief. Those responsible for the paintings and peckings, often made in virtually inaccessible places, would have been shamans or members of a priestly élite. One notable characterisitic of prehistoric rock art, in Asia as in Europe, is its concentration into what are best seen as former ritualistic or holy places, on or at the base of a particular mountain peak, in or within a close radius of a high mountain pass, for instance, or associated with burial places. Thus the 'hut and yard' maps should be seen as a record of a fossilized prayer, perhaps for the safety or prosperity of the homestead and its inhabitants, and represented as an icon rather than as an exact configuration of a nearby homestead, yet to be uncoverd by archaeologists. As the Chinese tomb painting already described demonstrates, celestial cartography and cosmological cartography often overlap in their religious significance. Individual cosmological motifs are widespread in Asian rock art, such as those relating to the structure of the cosmos (the Tree of Life or *axis mundi*) or to access to the next world (boats, ladders). They are also found in certain types of decoration, such as the bronze drums from Borneo and other parts of Indonesia. However, examples of cosmological maps from the prehistoric period, as opposed to these individual motifs, are few. One Mesolithic rock painting from Madhya Pradesh, India, is thought to represent, with its three bands, the three parts of the cosmos (water, air, and earth).

At present, information about rock art in general in Asia is both chronologically and geographically patchy. Some localities, for example in central and southern India, the peripheral provinces of China and Mongolia, the upper reaches of the great rivers of Siberia (Ob, Yenesei, Lena, Amur), around the high passes of the Pamirs, and the mountains of Armenia in the vicinity of Uchtasar, are comparatively well known, but such regions are separated by vast areas of ignorance, either through lack of exploration or through lack of reporting, especially in literature accessible to western scholars. Undoubtedly, though, future research and the wider reporting of new discoveries will yield a wealth of prehistoric maps and map-like representations throughout most of Asia, characterized by strong regional traditions, most of which will have a close counterpart elsewhere in the world.

CATHERINE DELANO SMITH

REFERENCES

Devlet, M.A. *Petroglify Mugur Sargola.* Moscow: Nauka, 1980.

Harley, J.B. and David Woodward, eds. *The History of Cartography.* Vol. 1 *Cartography in Prehistoric, Ancient, and Medieval Europe and the Mediterranean.* Chicago: University of Chicago Press, 1987. Vol. 2 Book 1, *Cartography in the Traditional Islamic and South Asian Societies,* 1992; Vol. 2 Book 2, *Cartography in the Traditional East and Southeast Asian Societies*, 1994.

Wang Ningsheng. *Yunnan Cangyuan bihua di faxian yu yanjiu* (The Rock Paintings of Cangyuan County, Yunnan: Their Discovery and Research). Beijing: Wenwu chubanshe, 1985.

See also: Maps and Mapmaking: Celestial East Asian Maps – Lunar Mansions in East Asian Astronomy

MAPS AND MAPMAKING OF THE AUSTRALIAN ABORIGINAL PEOPLE One of the most common forms of representation in Australian Aboriginal culture is the map. Bark paintings are often maps, as are sand sculptures, body painting, and rock art. Spear throwers and log coffins may be decorated with maps. In 1957 the anthropologist Donald Thomson visited the central Australian desert country belonging to the Pintupi people and described this experience.

> I was able to ... live and hunt with a group of desert-dwelling aborigines who still followed the life of their ancestors ... On the eve of our going (return) Tjappanongo produced spear throwers, on the backs of which were designs deeply incised, more or less geometric in form. Sometimes with a stick or with his finger, he would point to each well or rock hole in turn and recite its name, waiting for me to repeat it after him ... I realized that here was the most important discovery of the expeditition — that what Tjappanongo and the old men had shown me was really a map, highly conventionalized, like the marks on a "message" or "letter" stick of the aborigines, of the waters of the vast terrain over which the Bindubu hunted.

Message sticks and *Toas* (clubs) may incorporate geographical information. Aboriginal maps in whatever form are typically landscape maps depicting known places in the geographical environment. Such maps are relatively well recognized. Somewhat more controversially it has been claimed that some rock art may be read as celestial maps. Why is it that, counter to the orthodoxy, maps are so ubiquitous in a culture that has no written language and, seemingly, has little of the social complexity held to characterize contemporary Western culture? The answer is that Aboriginal culture is far from simple, having as it does one of the richest religious systems, and that its central values are embodied as knowledge, knowledge that is spatially organized because the land and relationships to it underpin everything.

Aboriginal culture is spatialized linguistically, socially, religiously, artistically, and epistemologically. Aboriginal ontology is one of spatialized activities, of events and processes, people, and places. To talk of things is to speak of the relationships of processes at named sites. It is to consider the connections between actions of the Ancestral beings and humans. Every moment of daily life is replete with spatial references; asking someone to move over may be phrased as 'move northwards please'. Dreams and narratives are cast in a framework of spatial coordinates. Visiting groups at ceremonial gatherings distribute themselves in a spatial replication of the location of their homelands. Ceremonial and intiation grounds are spatially constructed and oriented either to other sacred sites or to the sun. The pervasiveness of spatiality in Aboriginal daily life jointly derives from the semantic structure of the langauge in which the subjects of sentences are not things but relations and from the centrality of the land in Aboriginal cosmology.

It is the land that is the source of value and meaning, of rights and obligations. Everywhere is sacred since all the land was created in the Dreaming by the activities of Ancestral Beings as they moved across the landscape. These journeys left Dreaming tracks, knowledge of which is recreated in song, story, and ceremony. Everyone has a spiritual linkage to the land by virtue of birth such that they *are* the land. Knowledge of the Dreaming tracks, of the activites that created the land of one's birth, is therefore evidence of possession of the land and by the land. Continued prosperity of the land depends on the fulfilment of the ceremonials and rituals which are in effect both a celebration of ownership and a continuation of the act of creation. The landscape is the source of meaning and value and the repository of history and events and can be read as a map of itself and its own creation.

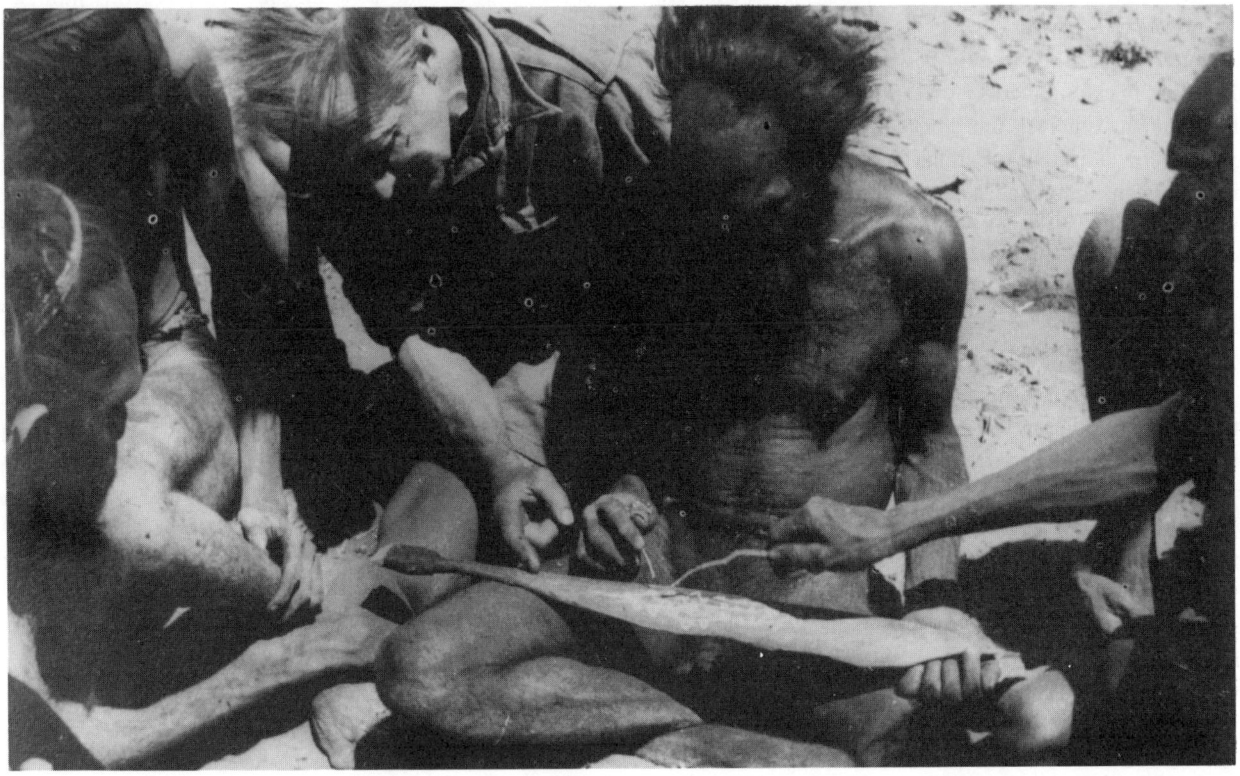

Figure 1: Tjappanongo, a Pintupi elder, explaining the incisions on the spear thrower to Donald Thomson.

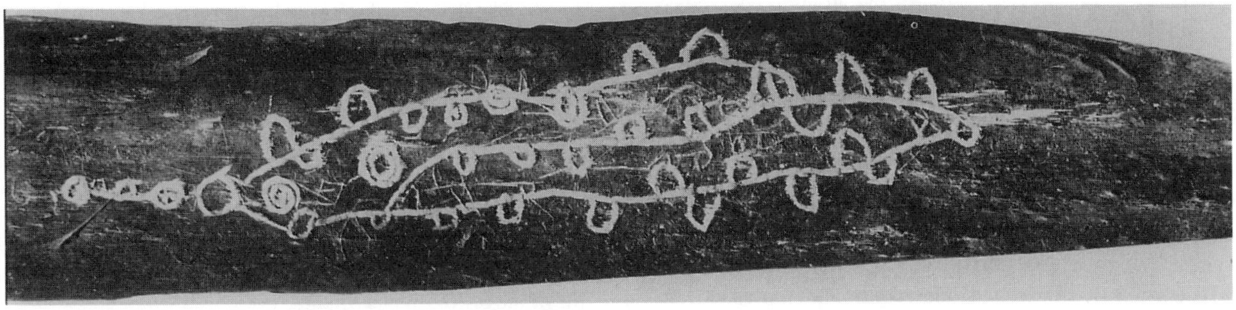

Figure 2: Spear thrower map. Photographs by Donald Thomson, on loan to the Museum of Victoria. Courtesy of Mrs D.M. Thomson.

However, it is knowledge which is the primary marker of status and item of exchange. Surface knowledge is the outside knowledge that anyone can speak of; inside knowledge is that which only the initiated can speak of and which is gradually revealed through life as maturity is attained. The way the Yolgnu of Eastern Arnhemland structure their knowledge system is typical of the ways in which it is possible for Aboriginal groups to have a detailed understanding of their environment. Their knowledge system is dependent on the joint articulation of two modes of patterning. One is genealogical — *gurrutu* the kinship system; the other is spatial — *djalkiri* the footsteps of the Ancestors or the dreaming tracks. The kinship system provides an unlimited process of recursion that enables all things to be named and related and thus imposes an order on the social and natural world that gives it coherence and value. It provides the framework within which social obligations with regard to life, death, marriage, and land can be negotiated. The other mode of patterning is provided by the stories, myths, or dreamings that relate the travels and activities of the ancestors in creating the landscape in the form of tracks or songlines that traverse the whole country. The kinship system and the songlines together form a knowledge network that allows for everything to be connected. The concept of connectedness is an extremely powerful one in Aboriginal culture and is exemplified by the Yolgnu term *likan* ,which in the mundane sphere means elbow — the connection of the upper and lower arm — and in the spiritual sphere connotes the connections among ancestors, persons, places, and ceremonies. A wide variety of Australian Aboriginal paintings have been interpreted as being simultaneously geographic and social; they represent both the tracks of the Ancestors and detailed maps of places. Hence bark paintings are encoded knowledge of connections.

The Kunwinjku people of Western Arnhemland paint both bark and bodies at the Mardayin ceremony in the 'x-ray' style that shows internal body parts. In the Mardayin ceremony the bodies of the initiates are painted so that in effect their own body parts are mapped with a design that represents the body parts of the Ancestral beings and features of the landscape. These paintings can be read on one level as maps of the way Kunwinjku "conceive of the spatial organization of sites in their land in terms of an abstract model of the divided yet organically related body parts of the ancestral beings that created those lands. Such sites are described as transformations of the actual body parts of the ancestral being, and all the sites thus created are considered to be intrinsically connected" (Taylor, 1989).

The connective function of bark paintings like this helps children to learn the shape of the *wanga* and to have respect for it and the animals in it by integrating the activities of the ancestors, people, and places. Ownership of the land thus means having the right speak of it, to have the knowledge of it, and also to have responsibility for it.

While the land may have boundaries which can be known with precision, it is not good custom to display them, because they are permeable rather than fixed entities with rites of access being required and most frequently granted. Areas can be owned by more than one group, and routes can be common property. Boundaries are more properly the subject of negotiation and exchange in ceremony and ritual. Moreover Yolgnu conceptions of place do not correspond to Western legal notions of enclosure but are more typically open and extendable "strings" of connectedness.

Consequently, while Australian Aboriginal groups constantly map their land, this is a very different process from that of the dominant white society. Being mapped in the white manner may have advantages, for example, in making land claims. In fact it is standard procedure for anthropologists to record Dreaming tracks on western topographical maps as evidence that this knowledge is the property of the claimants. However, this is a very diffferent process from that of mapping the "precise boundaries of all Aboriginal groups in Australia" as is the aim of the recently published map *Australia's Extant and Imputed Traditional Aboriginal Territories* (1993). Aboriginal conceptions of identity with the land do not equate with the notions of boundary precision, exclusion, and individual property rights and the linkages to the state implicit in western maps. Aboriginal maps celebrate that identity by providing connectedness leaving their permeability, imprecision, and inclusion open to negotiation in ceremony and ritual.

DAVID TURNBULL

REFERENCES

Keen, Ian. "Metaphor and the Meta-Language: 'Groups' " in Northeast Arnhemland." *American Ethnologist* 22: 502–27, 1995.

Morphy, Howard. "'Now You Understand' — An Analysis of the Way Yolgnu Have Used Sacred Knowledge to Retain Their Autonomy." In *Aborigines, Land and Land Rights*. Ed. Nicolas Peterson and Marcia Langton. Canberra: Australian Institute of Aboriginal Studies, 1983, pp. 110–133.

Morphy, Howard. *Ancestral Connections: Art and an Aboriginal System of Knowledge*. Chicago: University of Chicago Press, 1991.

Myers, Fred R. *Pintupi Country, Pintupi Self: Sentiment, Place and Politics among Western Desert Aborigines*. Washington, London, and Canberra: Smithsonian Institution Press and Australian Institute of Aboriginal Studies, 1986.

Sutton, Peter. *Dreamings: The Art of Aboriginal Australia*. Ringwood: Viking, 1988.

Taylor, Luke. "Seeing the 'Inside': Kunwinjku Paintings and the Symbol of the Divided Body." In *Animals into Art* . Ed. Howard Morphy. London: Unwin Hyman, 1989, pp. 371–389.

Turnbull, David. *Maps Are Territories; Science is an Atlas*. Geelong: Deakin University Press, 1989.

Watson, Helen, the Yolngu Community at Yirrkala, and David Wade Chambers. *Singing The Land, Signing The Land*. Geelong: Deakin University Press, 1989.

Williams, Nancy M. *The Yolgnu and Their Land: A System of Land Tenure and the Fight for its Recognition*. Canberra: Australian Institute of Aboriginal Studies, 1986.

Wood, Denis. "Maps and Mapmaking." *Cartographica* 30(1): 1–9, 1993.

MAPS AND MAPMAKING: CELESTIAL EAST ASIAN MAPS For the purpose of this article, East Asia will be understood to comprise the three countries of China, Korea, and Japan. In common with many other aspects of culture, mapping of the night sky in both Korea and Japan began much later than in China and furthermore closely followed Chinese tradition. Hence the main emphasis in this article will be on Chinese achievements.

The origins of celestial mapping in China are lost in the mists of time. Occasional star names — such as *Huo* (the Fire

Star, identified with the bright red star Antares) — are found on bone inscriptions dating from a little before 1000 BC. Over the succeeding millennium, there is a gradual increase in the number of star groups cited in the available literature. However, no astral map is known to be extant until as late as the first century BC. From then onwards, the development of celestial mapping in China can be traced with more confidence, although it is known that many important star charts and celestial globes have failed to survive.

Since the stars appear to us as scattered points of light, any attempts to divide them into groups must necessarily be artificial. This is illustrated by the great diversity between the constellation patterns as depicted on Eastern and Western charts of the night sky. Chinese stellar maps further differ from their Western counterparts in the following ways: there is virtually no symbolic representation of the constellations by figures of men, animals, etc.; star groups tend to be much smaller, typically containing only about five stars; individual stars are represented by dots of almost equal size — regardless of brightness; an equatorial (rather than ecliptic) co-ordinate system is standard, with division of the sky into twenty-eight unequal sectors (the *xiu* or "lunar lodges").

Chinese celestial mapping reveals negligible traces of Western influence (whether of European, Arab, or Indian origin) until as late as the seventeenth century when the Jesuits introduced improved techniques as well as knowledge of the far southern constellations which are invisible from China.

Whereas many occidental constellation names are derived from Babylonian and Greek mythology, Chinese names tend to be much more ordinary. From an early period, the celestial vault came to be viewed as a direct counterpart of the Chinese empire. Stars or star groups were regarded as representing members of the imperial family, courtiers and other officials — as well as domestic animals, crops and important buildings (from palaces to prisons). Any event occurring in or near a particular constellation — such as the close passage of a planet or comet — was regarded as an omen of change affecting the terrestrial equivalent.

Archaeological investigations during the last twenty years have brought to light a number of important Chinese celestial maps, mostly adorning the ceilings of tombs. However, the existence of other works has been known for centuries. The earliest extant astral map from China depicts only twenty-eight star groups — the lunar mansions. This rather crude but colorful artifact was discovered as recently as 1987. It is painted on the ceiling of a tomb dating from around 25 BC which is located in the ancient capital of Xi'an. Individual constellations are depicted in a ring of approximate diameter 2.5 meters. No further chart of the night sky is preserved until the sixth century AD, but there is documentary evidence that a number of accurate stellar charts and celestial globes were produced by leading astronomers in the intervening time. Their purpose was to predict the risings and settings of the constellations as well as to follow the movements of wandering celestial bodies such as planets and comets. None of these works survived for more than about three centuries.

A star map dating from AD 526 was discovered during excavations at Luoyang — another ancient capital of China — in 1973. This is also painted on a tomb ceiling. Although the constellations are very sketchily depicted, the Milky Way (*Tianhe*, the "Celestial River") features very prominently.

Clearly, there is no particular reason why stellar maps painted on tomb ceilings should compare in quality with those produced by contemporary astronomers. Nevertheless, with a single notable exception, little else of significance is accessible until well after AD 1000. A paper celestial map measuring approximately 110×25 cm was among the vast number of manuscripts uncovered by Sir Aurel Stein in a grotto at Dunhuang (Xinjiang province) in 1907. This chart, which is now in the British Library, portrays in 13 sections the whole of the night sky visible from China. Stars are depicted in three colors (red, black, and yellow), reflecting an ancient tradition. A suggested date for this very crude artifact is around AD 700.

Two huge star maps dating from AD 941 and 950 were uncovered from royal tombs near Hangzhou between 1958 and 1975. These charts are engraved on stone slabs which formed the ceilings of the tombs. Both charts depict with fair accuracy the twenty-eight lunar lodge constellations and a few polar star groups — notably *Beidou* (Northern Dipper, identical with the Big Dipper). The celestial equator and polar circle (the circle of constant visibility) are also carefully positioned.

Many astral maps and celestial globes are known to have been produced during the highly advanced Song Dynasty (AD 960–1275) but most have long since disappeared. Only one of the original artifacts, dating from AD 1247, is known to exist today, but late replicas of a further chart, first printed in AD 1094, are still available. The great astronomer Su Song is known to have produced a celestial globe 1.7 meters in diameter in AD 1092. This was said to represent 1464 stars in 283 constellations — standardized figures from ancient times. Although the globe was destroyed, star charts in copies of the *Xinyi xiang fayao* by Su Song are believed to represent accurately the configurations on this globe. However, today the earliest extant version dates from AD 1781. This reproduction, now in the National Library at Beijing, portrays the entire night sky as seen from Central China in five sections. The equator, ecliptic, circle of constant visibility, and boundaries of the lunar lodges are all marked.

One of the sections, showing the southern hemisphere, has a central void which corresponds to the region of sky near the south pole which is permanently below the Chinese horizon. Apart from isolated constellations such as *Denglonggu* (Frame of the Lantern, identical with the Southern Cross), no far southern stars appear to have been mapped by the Chinese until the Jesuit era.

Careful modern measurements show that typical positional errors on the Su Song star maps — or at least the extant copies — are fairly large, some 4 degrees. A much superior map, engraved on stone in AD 1247, is exhibited at the Suzhou Museum in Jiangsu province. This circular chart, 1.05 meters in diameter, is engraved on a rectangular stele measuring 2.2×1.1 meters. The surface is still in good condition and rubbings are occasionally taken. Entitled *Tianwentu* (Astronomical Chart), the map portrays the whole of the visible night sky on a polar projection. It depicts 1436 stars in 277 constellations. The various reference circles and lines are shown, and the Milky Way is well defined. Most star positions prove to be accurate to within about 2 degrees, which is tolerably good precision.

In recent years, several well-preserved celestial maps have been uncovered during excavations of Buddhist mausoleums of the Liao Dynasty, which flourished in Northern China between AD 916 and 1125. Two of these colorful maps, painted on the ceilings of the tombs, although having no pretensions to accuracy, are particularly interesting since they show both the lunar lodges and symbols (rather sinified) of the 12 signs of the Western zodiac. The signs of the zodiac were first introduced to China around AD 600, with the translations of certain Buddhist *sūtras*. Not long afterwards, horoscope astrology — based on Western methods — became popular in China although it had negligible influence on official practice. However, not for several centuries do we find significant pictorial relicts. Apart from the Liao tombs, a large bell, which was cast in AD 1174, is also adorned with the signs of the zodiac.

A number of substantial star maps is preserved from the Ming Dynasty (AD 1368–1644), but apart from those revealing Jesuit influence (after about 1630) none is of the caliber of the Suzhou chart of AD 1247. The Ming was a period of decline in both astronomy and mathematics. Many Jesuit celestial maps and globes — notably those produced in China by Adam Schall von Bell in 1634, Ferdinand Verbiest in 1673, and August von Hallerstein in 1757 — still survive. These are superior to even the best Chinese efforts. The Jesuit astronomers did not try to supplant the Chinese constellations with those of Western origin, but they measured the positions of the various stars with previously unrivaled precision and in some cases added further stars to individual groups. All significant representations of the night sky throughout the Qing Dynasty (1644–1911) either directly or indirectly reveal Western influence in these ways and are thus outside the scope of this article.

The earliest surviving astral map produced in Korea dates from AD 1395. Entitled *Ch'onsang yolch'a punyajido* (Chart of the regular division of the celestial bodies), this left a profound impression on later celestial mapping. Indeed, until the end of the last dynasty (1910) virtually all surviving star charts produced in Korea which do not show Western influence appear to be copied from it. History records that the original celestial map, which is circular and engraved on a marble slab measuring 2.1×1.2 meters, is an accurate reproduction of a much older map which had been presented to one of the three early kingdoms of Korea by a Chinese emperor. Although this stele was lost in a river during a battle in the Korean peninsula in AD 670, a rubbing still survived in 1395, and this was used to make a new engraving. Both the 1395 chart and a careful stone replica produced in 1687 are still preserved in Seoul museums. They are in good condition, although the earlier artifact has suffered damage down the centuries.

Each of the stone engravings depicts the usual circles and lunar lodge boundaries, as well as the Milky Way. However, several constellation patterns differ considerably from medieval Chinese representations, while measurements of star positions best fit a very early date-perhaps around 30 BC. Hence the two steles may well preserve traditions of celestial mapping which are far older than the detailed Chinese charts which are still accessible.

In 1973 a crude map showing the twenty-eight lunar lodges in a square measuring 80×80 cm was found in a Japanese tomb dating from around AD 700. This work, the earliest Japanese celestial chart so far discovered, clearly shows Chinese influence. Detailed maps and globes of the night sky based on Chinese originals are known to have been produced at various stages in later Japanese history. However, none has survived before the sixteenth century. A chart compiled by Abe Yasuyo around AD 1315 was the oldest extant Japanese celestial map until it was destroyed during the Second World War. Fortunately some replicas still exist.

The two oldest surviving star maps from Japan both date from around AD 1540. Each is circular, extending to the edge of the region of perpetual invisibility (the area of sky always below the horizon), and shows the traditional Chinese constellations and co-ordinates. Both the Suzhou celestial map of AD 1247 and the Korean chart of 1395 have had an important effect on celestial cartography in Japan and many Japanese copies of each survive from the seventeenth century onwards.

F. RICHARD STEPHENSON

REFERENCES

Jeon Sang-woon. *Science and Technology in Korea: Traditional Instruments and Techniques*. Cambridge, Massachusetts: MIT Press, 1974.

Miyajima Kazuhiko: "Japanese Celestial Cartography before the Meiji Period." In *History of Cartography*. Ed. J.B. Harley and David Woodward. Chicago: University of Chicago Press, 1994, vol. II, part 2, chapter 14.

Needham, Joseph and Wang Ling. *Science and Civilisation in China*. Cambridge: Cambridge University Press, 1959, vol. III.

Pan Nai. *Zhongguo hengxing guance shi* (History of Stellar Observations in China). Shanghai: Shelin Chubanshe, 1989.

Stephenson, F. Richard: "Mappe celesti nell'antica Cina." *l'Astronomia* no 98:1827, 1990.

Stephenson, F. Richard: "Chinese and Korean Star Maps and Catalogs." In *The History of Cartography*. Vol. II, part 2, *Cartography in the Traditional East and Southeast Asian Societies*. Ed. J.B. Harley and David Woodward. Chicago: University of Chicago Press, 1994, pp. 511–578.

Zhongguo Shehui Kexueyuan Kaogu Yanjiusuo (Archaeological Research Institute, Chinese Academy of Social Science). *Zhongguo Gudai Tianwen Wenwu Tuji* (Album of Ancient Chinese Astronomical Relics). Beijing: Wenwu Chubanshe, 1979.

See also: Lunar Lodges – Celestial Vault and Sphere – Su Song – Zodiac – Stars in Chinese Astronomy

MAPS AND MAPMAKING: CELESTIAL ISLAMIC MAPS

The earliest evidence of Islamic interest in celestial mapping is a vaulted ceiling in a small eighth-century provincial palace, known as Quṣayr ʿAmrah, in the desert of present-day Jordan. One of the bathrooms in this palace has a domed ceiling painted to resemble the vault of the heavens. It is the oldest astronomical dome of heaven preserved today. Though the ceiling has badly deteriorated, enough remains to ascertain that the artist was influenced by late Antique and Byzantine two-dimensional flat maps of the skies. The iconography of the constellation figures, the lack of stars, and the method of projection are in keeping with classical and early Western medieval maps of the heavens. The sequence of the constellations on the domed ceiling, however, is not as one would see it when looking up into the sky, but rather as it would appear when looking down onto a celestial globe.

The celestial globe is the oldest form of celestial mapping, for its origins can be traced to Greece in the sixth century BC, though the earliest preserved example is an eleventh-century Islamic globe made in Valencia. This three-dimensional model of the skies presented the stars as seen by an observer *outside* the sphere of stars, so that the relative positions of the stars are the reverse, east to west (or right to left) of their appearance when viewed from the surface of the earth. Islamic celestial globes were made from the ninth through the nineteenth centuries, steadfastly maintaining the basic classical design and encouraging the concept of a spherical universe rotating around the earth.

No flat two-dimensional star maps on paper or parchment from Islamic lands have survived, if indeed any were made, although planispheric maps of the skies drawn on parchment exist today from the Roman and Byzantine worlds. Displaying the entire surface of a sphere on a flat surface — such as parchment or paper or a metal plate — requires a system of mathematical projection. In addition to the method called stereographic projection known in late antiquity, other methods for flat mapping were described in the early eleventh century by the versatile scholar Abū al-Rayḥān Muḥammad ibn Aḥmad al-Bīrūnī (b. 973) working in Iran. Three of his proposed methods correspond in modern terms to orthographic projection, azimuthal equidistant polar projection, and globular projection. His ideas, however, appear to have had no direct effect upon subsequent celestial mapmaking in the Islamic world, and no maps employing these novel methods are known today.

The evidence for Islamic interest in flat mapping of the entire sky is found only in instrument design and production. The flat, planispheric astrolabe was, in fact, the most commonly used form of celestial map in the Islamic world, for it consisted of a pierced star map placed over a projection of the celestial coordinate system as it related to the observer's position on earth. The resulting representation of the positions of the stars with respect to the local horizon forms a two-dimensional model of the heavens. The method of stereographic projection required for the astrolabe's construction was certainly described by the Greek astronomer Ptolemy in the first century AD, though not the instrument itself. It was also Ptolemy who compiled a star catalog giving coordinates for 1022 stars, with descriptions of forty-eight constellation outlines based mostly on Greek mythological characters which served as mnemonic devices for mapping the skies. This star catalog was the basis for all the medieval Islamic star catalogs as well as instruments employing stars. It seems certain that the astrolabe was a Greek invention, but its design and production were perfected in the Islamic world, where it was manufactured in many variations from Spain to India from the early ninth through the nineteenth century.

The influence of early modern European celestial mapping is evident in an astrolabe plate engraved in Iran in 1654–1655 by the instrument maker Muḥammad Mahdī of Yazd. His metal plate reproduces the northern and southern hemispheric maps from a planispheric celestial map printed about 1650 by the Parisian engraver Melchior Tavernier, whose brother Jean-Baptist Tavernier had made six trips to the Near East before his death in 1689 and probably served as the conduit by which the map reached Iran. Tavernier's star map included the new chartings of the southern skies at the end of the sixteenth century, and these were carefully rendered by Muḥammad Mahdī, who changed the labels of the Ptolemaic constellations into Arabic, but did not attempt to give Arabic names to the newer non-Ptolemaic constellations. Muḥammad Mahdī made at least two additional copies of these plates, but they appear to have had no subsequent influence on Islamic celestial cartography.

Celestial space was also often represented by schematic diagrams that did not involve the mathematical determination of coordinates (necessary for globes) or methods of projection (required for planispheric maps or astrolabes). In such diagrams, concentric circles were often used to indicate in general terms the orbits of the planets and the sphere of the stars about the earth, which was viewed as being at the center of the universe. More complex and abstract diagrams illustrating the orbit of an individual planet employed concentric and eccentric circles to explain the peculiarities of the planet's path.

Mapping the entire sky or heavens was not the only form of celestial mapping to occupy Islamic thinkers. Maps of individual constellations rather than the entire sky had the advantage of not employing a coordinate system or requiring knowledge of projection methods. The most important guide to constellation diagrams in the Islamic world was undoubtedly an Arabic treatise written in the tenth century by ʿAbd al-Raḥmān ibn ʿUmar al-Ṣūfī, a court astronomer in Isfahan. He provided two drawings of each of the forty-eight classical constellations, one as it would be seen in the sky by an observer on earth and one as it appears on a celestial globe. According to one account, al-Ṣūfī obtained his images by laying very thin paper over a celestial globe and tracing the constellation outlines and individual stars. While taking pains to make clear to the reader the mirror-image relationship between constellations as imagined in the sky and those on a celestial globe, by treating each one individually al-Ṣūfī ignored the spatial interrelationships of the constellations to each other. Al-Ṣūfī also preceded the discussion of each constellation with a survey of traditonal Bedouin constellations visualized in the same area of the sky, such as the constellation of a lion much larger than Leo with gazelles running before the large lion. Numerous copies of this popular treatise exist today, with dress and general presentation of the figures changed to reflect local artistic fashions and conventions, and many later writers incorporated the constellation images into their treatises. Sometimes in later works the constellations were illustrated without stars, with only the animal or human mythological form that gave rise to the constellation outline, and occasionally even the understanding of the mythological figure was lost or confused. The diagrams of individual constellations of stars yielded an easily understood non-mathematical guide to portions of the skies, while giving wide scope to the artist in interpreting the animal and human outlines.

Another form of Islamic celestial mapping was the emblematic or symbolic representation of celestial bodies (planets as well as constellations of stars) and their spatial relationships. The twelve zodiacal signs were often represented as emblematic motifs rather than as constellation diagrams. No attempt was made to represent the stars forming the asterism or even the basic outline of the constellation. Rather, each constellation was represented by a commonly accepted convention, such as a two-headed man sitting cross-legged for Gemini, or for Libra a squatting man with scales over his shoulders or held overhead. The seven classical planets (Moon, Mercury, Venus, Sun, Mars, Jupiter, and Saturn) were designated by human personifications. Venus, for example, was portrayed as a woman playing a lutelike instrument or Jupiter as a man reading a book. In astrological writings, zodiacal signs were related to the planets by a series of "domiciles". Thus the moon was most frequently associated with, or domiciled in, Cancer, and the sun in Leo. The remaining five planets were each assigned two zodiacal signs as their domiciles; Venus, for example, was assigned to both Libra and Taurus. Artisans working in metal or with manuscript painting would represent these spatial relationships by depicting Taurus as a bull ridden by a lute player (Venus), or Cancer with a lunar disk, or Leo as a lion surmounted by the radiant disk of the sun. In a second system, the zodiacal signs were combined with the "exaltation" or "dejection" of a planet, whch could also be represented symbolically. The lunar nodes (the northern and southern intersections of the moon's orbit with the ecliptic) were referred to as the head and tail of a dragon, which came to be regarded as a pseudo-planet associated with Sagittarius and Gemini and often represented graphically. Such symbolic representations of celestial bodies and their spatial relations formed an important part of medieval Islamic understanding and graphic interpretation of the heavens. They did not require the difficult technical knowledge necessary for geometrically projected mappings and they were an attractive subject matter for artists to interpret flexibly.

Each type of Islamic celestial mapping was directed at a different audience. The fairly educated audience who could interpret the symbolic and allegorical representations of celestial bodies was probably more select than those who could appreciate a constellation diagram. Different still would be those who could appreciate as a scientific instrument the celestial globe with its stars positioned by coordinates or the astrolabe with the stellar coordinates projected geometrically onto a flat plate. The Islamic world's apparent lack of flat stellar maps of the entire visible sky drawn on paper or parchment has yet to be fully explained.

EMILIE SAVAGE-SMITH

REFERENCES

Kunitzsch, Paul and J. Knappert. "al-Nudjūm" (The Stars). In *Encyclopaedia of Islam*, 2nd ed., vol. 8. Leiden: Brill, 1991, pp. 97–105.

Savage-Smith, Emilie. "Celestial Mapping." In *The History of Cartography. Volume Two, Book One: Cartography in the Traditional Islamic and South Asian Societies.* Ed. J. B. Harley and David Woodward. Chicago: University of Chicago Press, 1992, pp. 12–70.

Savage-Smith, Emilie. "The Islamic Tradition of Celestial Mapping." *Asian Art* 5(4): 5–27, 1992.

Savage-Smith, Emilie and Colin Wakefield. "Jacob Golius and Celestial Cartography." In *Learning, Language and Invention: Essays Presented to Francis Maddison.* Ed. W. D. Hackmann and A. J. Turner. London: Variorum, 1994, pp. 238–260.

Wellesz, Emmy. "An Early al-Sufi Manuscript in the Bodleian Library in Oxford: A Study in Islamic Constellation Images." *Ars Orientalis* 3: 1–26, 1959.

See also: Globes – al-Bīrūnī – Astrolabe – Stars in Islamic Astronomy – al-Ṣūfī

MAPS AND MAPMAKING IN CHINA The Chinese have one of the world's longest histories of mapmaking — more than two thousand years. An adequate account of this history, however, has yet to be written. For some time spans, the first to the tenth centuries, for example, losses from warfare and neglect have probably been great. Almost no examples of maps remain, so that one must make inferences on the basis of textual sources and other evidence. For other periods, especially from about the seventeenth century on, so many maps, as well as supplementary textual sources, survive that no one has adequately surveyed them.

Despite these difficulties, it is still possible to make some broad generalizations. Before the twentieth century, mapmaking in China was an activity of the educated elite, those who formed the pool from which posts in the government bureaucracy were filled. In the course of their duties, these scholar-officials developed mathematical techniques and instruments needed to produce measured maps of high accuracy. Such maps have features familiar to users of modern maps — scalar indications, directional markers, conventional signs to represent topographic features, and a lack of perspective. Their look is planimetric, all features being represented as if lying on the same plane and as if viewed from above. Maps like these form only a portion of the body of surviving works, and partly for this reason it would be misleading to characterize Chinese mapmaking before the twentieth century as what we in the West now call science. As will be discussed later, it was a broader activity than the word "science" usually connotes, one that often involved measurement.

The foundations for quantitative mapping were established early in China's cartographic history. The magnetic properties of lodestones, or south-pointers, were known from about the third century BC. There is textual evidence that during the second century BC the idea of map scale was understood. Around this time Chinese surveyors and mapmakers already had considerable technical resources available to them for producing maps drawn to scale: graduated rods, carpenter's squares, plumb lines, compasses for drawing circles, and even sighting tubes that could be used for measuring inclination. In addition, a reference frame suggestive of a coordinate system for identifying locations had been hinted at by astronomers who divided the heavens into sectors, or lunar lodges (*xiu*).

A few hundred years after these beginnings, Pei Xiu (223–271), a mapmaker known for large-area maps drawn to scale, formulated a set of principles necessary to produce accurate maps. These principles stressed the importance of consistent scale, directional measurements, and adjustments in land measurements to correct for irregularities in the terrain being mapped. The principles had some influence on later mapmakers. Jia Dan (730–805) drew a large area scale map following Pei Xiu's principles, as did Shen Kuo (or Gua, 1031–1095). But none of the works by these mapmakers survives, so that it is impossible to judge how well they followed Pei's principles. Moreover, none of Pei's own work survives, so it is not known how well Pei followed his own ideas.

Some researchers have claimed that maps dating from the fourth to the second century BC are evidence that mensurational techniques were being applied to mapmaking. One of these, discovered in Hebei Province, is a bronze architectural plan of a mausoleum. Perhaps better known examples are two silk maps discovered at Mawangdui, near Changsha, Hunan. Other early maps are a set of maps drawn on wood discovered at Fangmatan in Gansu. There is some disagreement over what areas the maps found at Mawangdui and Fangmatan represent and how they should be reconstructed. Such questions need to be resolved before it can be determined whether the maps were drawn to scale. Pei Xiu, in articulating his principles of mapmaking, had complained about the lack of accuracy in maps before his time. It is not yet clear whether his assessment of his predecessors was justified.

In any case, a map produced within a half century of Shen Kuo's death suggests that mapmakers were capable of following Pei's principles quite rigorously: this is the much celebrated *Yu ji tu* (Map of the Tracks of Yu [the Great, legendary emperor]). It was carved in stone in 1136 and measures 80×79 cm. It represents all of China, and is impressive for the accuracy of its depiction of China's coastline and the courses of the Yellow and Yangtze Rivers. The map is also notable for the square grid imposed on its surface. According to a note on the map, each grid increment represents 100 Chinese miles. The grid is thus a scaling device, not a coordinate system like latitude and longitude. After the *Yu ji tu* the best known examples of grid maps are printed in the *Guang yutu* (Enlarged Terrestrial Atlas, ca. 1555) by Luo Hongxian (1504–1564). This atlas is a revision of a map no longer extant, the *Yutu* (Terrestrial map, 1320) by Zhu Siben (1273–1337). It contains a general map of China and individual maps of provinces, all with grids. According to Luo, the grid, in addition to serving as a scaling device, served as an aid in aligning sections of maps drawn or printed on different sheets.

The origins of the Chinese cartographic grid are unknown. Its use is certainly consistent with Pei Xiu's call for scale and attention to directions, and it has some similarities with the lunar lodges long used in Chinese astronomy. The polymath Zhang Heng (78–139), credited with an armillary sphere, a seismograph, and a topographic map, has been suspected of using a grid. But there is no direct evidence that any mapmaker before the *Yu ji tu* employed the device.

On other Chinese maps made at roughly the same time as the *Yu ji tu*, the grid is conspicuous by its absence, perhaps an indication that their makers were not particularly concerned with measurement and scale. A number of these maps are, like the *Yu ji tu*, engraved in stone, but are not considered as accurate as the *Yu ji tu*. For example, the *Hua yi tu* (Map of Chinese and Foreign Lands, 1136), engraved on the other side of the stone on which the *Yu ji tu* appears, and the *Jiu yu shouling tu* (Map of the Prefectures and Counties of the Nine Districts [the empire], 1121). As their titles suggest, these maps, like the *Yu ji tu*, represent all of China. An example of a stone map depicting a smaller area is the *Pingjiang tu* (Map of Pingjiang Prefecture, 1229). It, too, lacks a grid, and its scale has been found to vary from about 1 : 1300 to 1 : 2800 in the central portion and from about 1 : 10,000 to 1 : 77,000 at the periphery.

The evidence thus suggests that by the twelfth century Chinese mapmakers had the resources to produce measured maps of high quality, but in many cases did not make full use of those resources. This is not necessarily a defect. In general Chinese mapmakers seem to have regarded their task as encompassing more than representation according to a consistent scale. The Chinese word *tu*, commonly translated as map, also means picture, diagram, or chart. As the range of meaning suggests, the forms of Chinese mapmaking extend beyond what are easily recognizable as forerunners of modern measured maps.

Maps served a variety of functions for which uniform scale might be necessary or desirable: navigation, water conservancy, public works, defense and military planning, government administration, and record keeping for land tax accounting. But they also served purposes for which attention to scale might not be so important: they might be used to symbolize political power, to represent unseen other worlds, or to illustrate configurations of energy (or *qi*), knowledge of which was useful in "siting" (*dili* = land patterns or geonomy, or more popularly, *fengshui* = wind and water or geomancy).

The media used for mapmaking also show considerable diversity. Already mentioned have been flat maps engraved on bronze and stone, drawn on silk, wood, and paper, and printed on paper. Maps were also painted on walls of caves and tombs. Three-dimensional relief models were also made. One of the largest of these, described in the *Shi ji* (Records of the Grand Historian, ca. 91 BC, by Sima Qian), is contained in the tomb of Qin Shihuang (d. 210 BC), founding emperor of the Qin Dynasty (221–207 BC). It supposedly consists of representations of the heavens above and the empire below, with mercury-filled streams representing rivers and the sea.

Mapmaking in pre-twentieth-century China did not develop into a distinct specialty of learning as it did in Europe. As the variety of map functions suggests, mapmaking was located at the intersection of a number of activities and traditions of learning. This is consistent with the educational background of the intellectual elite who made maps. Their educations generally emphasized broad learning, frequently encompassed humanistic and scientific disciplines, and often centered around texts. Not surprisingly the activity of mapmaking often involved textual study. This was the case with Pei Xiu, Jia Dan, Shen Kuo, and Luo Hongxian, those who are credited with advancing quantitative mapmaking techniques.

Among the major sources of geographic information a mapmaker might consult were the dynastic histories, which conventionally included geographic records, and local gazetteers, compendia of historical and geographical information focusing on China or one of its various subdivisions. Gazetteers often included maps, usually at the head of the geographic section. These maps often lack scalar indications and are frequently pictorial. In such instances, they do not seem to have been intended to be used or studied in isolation, but were meant to be complemented by the accompanying text. The maps give an idea of the spatial relationships between geographic features, while the texts provide detailed information on distances and directions. This relationship between image and text does not seem to have been restricted to maps in gazetteers. Jia Dan and Shen Kuo, for example, say that their maps were accompanied by extensive notes, and Luo Hongxian's *Guangyu tu* consists mostly of text.

The relationship between map and text seems to be related to the close ties among painting, calligraphy, and poetry. From about the tenth century on, many Chinese artists, themselves often members of the bureaucratic elite, conceived the highest work of visual art as a combination of all three arts, each contributing in different ways to the aesthetic effect of the entire artifact. *Tu* (maps or pictures) of geographic subjects often had poems inscribed on them, suggesting that maps were valued like other forms of visual art for their emotional and expressive effects. In addition to employing quantitative techniques and devising means to present the information gleaned from those techniques, Chinese mapmakers drew upon the resources of the visual arts to express their responses to the land. The "language" of Chinese maps thus seems to have been more than one of denotation, of correspondence to material realities. This seems to be true even with maps of utilitarian function. Scholars have long noted, for example, that some of the representations on the nautical chart in the *Wubei zhi* (Treatise on Military Preparations, ca. 1621, comp. Mao Yuanyi) resemble elements often seen in Chinese landscape paintings.

One result of the interactions between mapmaking and other areas of learning seems to be that grid maps in particular and measured maps in general constitute a small proportion of the surviving corpus of Chinese maps made before the twentieth century. The disparity in the numbers of measured maps and more pictorial maps seems especially pronounced during the seventeenth through the nineteenth centuries. This disparity has fostered the conclusion that after the sixteenth century Chinese mapmaking declined, as mapmakers seemed to pay less attention to quantitative techniques and accuracy.

It seems true that there was little innovation in measured mapping in China after the twelfth century. Improvements in the mariner's compass and measurements of celestial latitude, for example, do not seem to have influenced Chinese mapmakers, as such developments affected European mapmaking. In addition, Chinese mapmakers did not develop projection techniques for transferring points from a spherical to a plane surface. Mapmakers seem to have believed the earth's surface was generally flat, so that, in their minds, drawing maps on plane surfaces would not result in appreciable distortion.

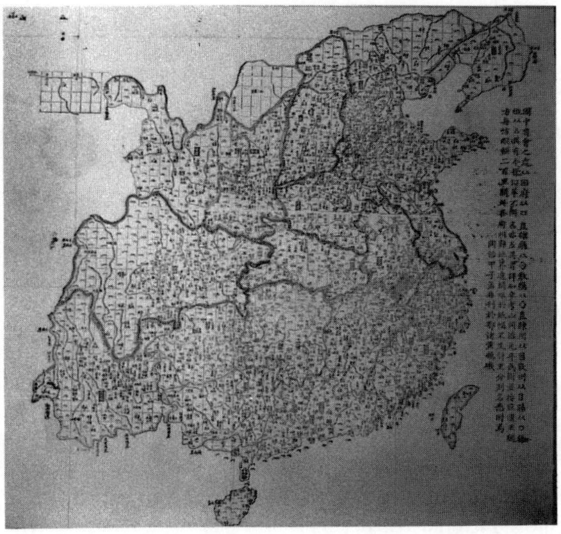

Figure 1: Grid map of China dating from 1864. Each grid increment equals 200 Chinese miles. A legend explaining the various signs on the map appears on the right. 57×58 cm. Courtesy of the Library of Congress (G7820 1864.H8 Vault).

In 1267 a Persian astronomer presented the Chinese imperial court with a "geographic record" that represented the earth as round, but the representation seems to have had no effect on Chinese mapmaking. The Chinese did not begin to adopt techniques of spherical projection until the late sixteenth century when Jesuit missionaries, notably Matteo Ricci (1552–1610), introduced Ptolemaic cartography with its coordinate system into China. Some Chinese copied Ricci's maps, and in the eighteenth century, Jesuits were commissioned by Chinese emperors to survey and map the entire empire. In carrying out this work, the Jesuits employed Chinese assistants. But even so, Chinese adoption of European techniques was slow, since the Jesuits were limited in their access to China, and many Chinese intellectuals resisted a view of the earth as round. The bulk of maps made in China continued to be made in the traditional manner up to the late nineteenth century. There was little uniformity in mapmaking, even within the government. Maps might be planimetric; they might be pictorial; they might be drawn to scale; they might not be; they might have grids; they might lack them; and late in the nineteenth century they might have square grids as well as lines of latitude and longitude (see Figures 1 and 2).

Circumstances in the nineteenth century, however, fostered change in this situation. As China weakened as a result of domestic problems and encroachments by European powers, many Chinese intellectuals began to believe that China needed reform. Some pointed out that European

Figure 2: Pictorial map of Hainan island dating from the nineteenth century. Notes on the map describe marriage and other ceremonies of the Li people. The mapmaker has distorted the shape of the island so that it resembles a rectangle, an auspicious form. North is at the bottom. 184×93 cm. Courtesy of the Library of Congress (G7822.H3E62 18–.H3 Vault).

maps were superior to those of the Chinese. In response, the government tried to standardize mapmaking practices, stipulating that standard projections and scales be used. Progress in instituting these changes was difficult since there was a shortage of personnel capable of employing the necessary techniques. Not until after the collapse of the empire in the twentieth century did the European conception of cartography as primarily quantitative fully supplant the traditional Chinese idea of map as an intersection of various lines of learning including the literary and visual arts.

In the first half of the twentieth century the quality of measured maps improved as more Chinese were trained in the earth sciences. During the Second World War, however, accurate measured maps were still not in good supply — a situation that hampered the Chinese military. To remedy that situation, the Chinese government took steps to promote cartography. In 1956 it established a State Bureau of Surveying and Mapping which undertook to produce a series of topographic maps for the whole country. Since then more institutions have been established to make maps, and others, such as the Wuhan Technical University of Surveying and Mapping, have been set up to teach cartography. As part of China's effort to modernize, Chinese cartography has employed the latest technological advances: remote sensing, satellite photography, and geographic information systems.

CORDELL D.K. YEE

REFERENCES

Cao Wanru et al., eds. *Zhongguo gudai ditu ji* (An Atlas of Ancient Maps in China). 1 vol. to date. Beijing: Wenwu chubanshe, 1990-.

"Cartography in China." Chapters 3–9 of *The History of Cartography*. Vol. 2, book 2, *Cartography in the Traditional East and Southeast Asian Societies*. Ed. J.B. Harley and David Woodward. Chicago: University of Chicago Press, 1994.

Chen Cheng-siang. "The Historical Development of Cartography in China." *Progress in Human Geography* 2: 101–120, 1978.

Mills, J.V.G., trans. *Ying-yai sheng-lan: "The Overall Survey of the Ocean's Shores"* [1433], by Ma Huan. Cambridge: Cambridge University Press, 1970.

Needham, Joseph, et al. *Science and Civilisation in China.* Vol. 3, *Mathematics and the Sciences of the Heavens and the Earth*, 1959. Vol. 4, *Physics and Physical Technology*, pt. 1, *Physics*, 1962; pt. 3, *Civil Engineering and Nautics*, 1971. Cambridge: Cambridge University Press, 1954.

Taylor, D.R.F. "Recent Developments in Cartography in the People's Republic of China." *Cartographica* 24 (3): 1–22, 1987.

Unno, Kazutaka. "Tōyō chizugakushi" (A Bibliographical Review of the History of Asian Cartography: from the End of the Second World War). *Kagakushi kenkyū* (Journal of History of Science, Japan), 2d series, 30: 1–14, 1991.

See also: Zhang Heng – Armillary Spheres – Geomancy – Maps and Mapmaking: Chinese Geomantic Maps – Shen Gua (or Kuo) – Luoxia Hong

MAPS AND MAPMAKING: CHINESE GEOMANTIC MAPS The Chinese geomantic map is a type of cartographic expression of a landform that portrays a place as geomantically auspicious or inauspicious. In search of an auspicious site for a grave, settlement, temple, or house, geomancers often make extensive field surveys using compasses. Determining an auspicious site, which is often found on a foothill, is a complex business achieved by considering landforms, water courses, and the direction that the locality faces. The results of important surveys are often recorded in maps. Although the geomantic sketch map is a large scale map, the scale is not accurate beyond the immediate area of the auspicious site, because the map is normally based on eye measurements and impressionistic descriptions of a locality.

Chinese geomantic maps have basically the same surveying and cartographic methods and identical or similar map symbols. A geomantic map may represent an auspicious site (the geomancy cave: normally designated as a grave or a house site) by the symbol of a small single circle or an even smaller circle inside a small circle. A geomantic map is a topographic map featuring mountain ranges most prominently, because the auspicious vital energy is said to flow through these ranges before accumulating in a geomancy cave. Mountain ranges are expressed in stylized but realistic shapes, using different symbols for different parts of a mountain: a single black line marks the end of foothills and the beginning of flat land; a serrated demarcation line shows slopes along the edge of mountain ridges; the mountain ridges and peaks are colored black. A smaller water course is symbolized by broken lines. A bigger river may be represented by a broken line with a solid line on one side. A large lake, river or sea may adopt the symbol of scales, which are also used in other types of maps in East Asia.

The first main characteristic of a geomantic map is that it is a center oriented map. The focus of the map is on the auspicious site of a given landscape which normally occupies the central position of the map. In the depiction of the topographic formations, only the mountains facing the auspicious site are presented in the map.

The second interesting point concerns the perspectives on the map. The point of perspective on the modern topographic map of contour lines is in the sky directly above the concerned landform. However, that for the relief formations of the mountain slopes on the geomantic map is on the ground at the auspicious site, while that for watercourses on the map is in the sky directly above the waters concerned.

The third point is that there is a lack of map symbols indicating cardinal direction. Unlike modern maps, the top of the geomantic map is assumed to be southward, because an auspicious site is normally facing a southern direction. Although geomancy maps did not adopt a symbol of direction, the exact direction that an auspicious site faces is often noted with Chinese characters which enable a map reader to tell directions.

The fourth point is that the geomantic maps are generally large scale maps focusing on a small catchment area or a small basin. Other non-geomantic traditional Chinese maps are normally small scale maps.

The fifth point concerns the emphasis on mountain ranges in the map. Especially emphasized are the relationships between the key mountain ranges encircling an auspicious site because mountain ranges are believed to be the route of the flow of *qi* (vital energy) which influences people auspiciously.

The origin of cartographic skills used in the geomantic map is not clearly understood yet by modern scholars. Fully developed geomantic maps had already appeared in a Ming Dynasty (1368–1644) geomantic manual, *Renze Shuji*. The technique of presenting relief formations in Chinese geomantic maps is perhaps more sophisticated than that on any other type of map in traditional China. However, their cartographic contribution is yet to be evaluated.

HONG-KEY YOON

REFERENCES

Xu Shanji, and Xu Shanshu. *Renze Shuji*. Xinzhu, Taiwan: Zhulin Shuzhu, 1969.

Yoon, H. "The Image of Nature in Geomancy." *GeoJournal* 4(4): 341–348, 1980.

Yoon, H. "The Expression of Landforms in Chinese Geomantic Maps." *Cartographic Journal* 29: 12–15, 1992.

MAPS AND MAPMAKING IN INDIA In comparison to Europe, the Islamic world, and East Asia, the cartographic achievements of South Asia appear modest and until the 1980s were scarcely recognized by historians of science. In recent years, new archaeological evidence, surviving sacred and secular texts, and other evidence have led to a change in scholarly opinion. We now believe that some form of mapping was practiced in what is now India as early as the Mesolithic period, that surveying dates as far back as the Indus Civilization (ca. 2500–1900 BC), and that the construction of large-scale plans, cosmographic maps, and other cartographic works has occurred continuously at least since the late Vedic age (first millennium BC). Because of the ravages of climate and vermin, surviving maps from prior to the eighteenth century are rare and are largely in stone, metal, or ceramic. However, cosmographies painted on cloth date back at least to the fifteenth century, while palm leaf architectural plans from seventeenth-century Orissa are believed to be copies of manuscripts originally prepared as early as the twelfth century. Few surviving maps on paper predate the eighteenth century.

Though not numerous, a number of map-like graffiti appear among the thousands of Stone Age Indian cave paintings; and at least one complex Mesolithic diagram is believed to be a representation of the cosmos. Other map-like grafitti continued to be produced by tribal Indians over most of the historic period. The principal reason for supposing that surveying was a feature of the Indus civilization is that the regularity of its planned grid-pattern urban settlements could not easily be achieved without it. Moreover, excavations have uncovered several objects that appear to have been simple surveying instruments and measuring rods. The uniformity and modular dimensions of the bricks for so much of the architecture over the vast extent of the Indus civilization are also noteworthy. During the ensuing period of the Vedic Aryans, the building of enormous sacrificial altars was an important religious activity. Texts known as *Śulbasūtras* set forth in great detail how these altars were to be constructed and called for drawing on the ground a plan prefiguring each altar. Again, a system of modular measures was employed. The ancient practice of building gigantic altars, once widespread, has died out over most of India, but survived into the latter half of the twentieth century in the Indian state of Kerala. Hindu temples were also built according to ancient detailed textual prescriptions, known as *Śilpaśāstras*, which also specified drawing on the ground, at full scale, the outline of the structure to be. That practice is still followed. Comparable, though simpler, rules applied, at least in theory, to the building of houses. The *Śilpaśāstras* also contained a variety of models for laying out towns and cities; relatively few present-day settlements in India suggest that such models were actually followed.

For the historic period, one of the earliest surviving artifacts that clearly embodies recognizable map symbolizations is an allegorical wall sculpture from Udayagiri in Madhya Pradesh, ca. AD 400, which shows the confluence of the Ganges and Yamuna Rivers in Madhyadesa (the sacred Central Region of India) over which the then Gupta Empire held sway. Other early datable works are sculpted bas relief cosmographies, some of which were quite elaborate. The earliest of these, depicting *Nandīśvaradvīpa*, the eighth continent of the Jain cosmos, was carved in Śaka 1256 (AD 1199–1200). It may safely be assumed that cosmographic paintings adorned the walls and portals of many ancient temples and monasteries of India's main religious groups — Hindus, Jains and Buddhists — just as they still do in Jain religious edifices and in Buddhist establishments in other parts of the world. However, largely because of the ravages of time and partially because of the iconoclasm of Muslim invaders, virtually none of these survive from prior to the twelfth century, apart from fragmentary remains in the Buddhist caves at Ajanta.

Astronomy and its handmaiden astrology were well developed sciences in ancient India. The major texts provided detailed observations that enabled latter-day scholars to prepare elaborate celestial diagrams. Despite these, there is no unequivocal evidence that astronomical charts accompanied those early works or were otherwise drawn. Horoscopes were prepared to show the positions of major heavenly bodies at particular times (especially at times of birth), and iconic representations of those bodies as deities often appeared in sculpture and paintings, as did the signs of the zodiac. But none of these were formed into assemblages that one would readily designate as maps. During the Mughal period (1526–1857), however, planar astrolabes and celestial globes were manufactured in northwestern India. One Muslim family practiced the trade in Lahore over a period of several generations. Though some of these works used Sanskrit, rather than Persian, as was the norm, in naming various heavenly bodies, the tradition in which all were made has been dubbed "Islamicate". A variety of related objects in the form of giant masonry instruments appeared in the astronomical observatories constructed during the period ca. 1722–39 by the Rajput king, Sawai Jai Singh, in his capital at Jaipur and in Delhi, Varanasi, Ujjain, and Mathura. Many of these are still usable.

At least five Hindu cosmographic globes are known, all based largely on *Purāṇic* texts from the mid-first millennium BC to the mid-first millennium AD. The oldest of these is a brass globe, probably from Gujarat, dated Śaka 1493 (AD 1571). The largest (diameter ca. 45 cm) and most elaborate, and the only one to contain a substantial component of geographic information along with its mainly mythic elements, is a papier-maché globe, probably from eastern India, that appears to date from the mid-eighteenth century. One of the other globes is a painted wooden production, also thought to be from eastern India, probably from the mid-nineteenth century. Two are late nineteenth century bronze creations of unknown provenance. Unlike most Western globes, none of these was constructed with the use of gores, the triangular or moon-shaped pieces that form the surface of modern globes.

A number of planispheric world maps also survive. All but one of these, a crude Marathi map on paper, probably from the mid-eighteenth century, are essentially Islamicate productions, in which mythic elements (e.g. the Land of Gog and Magog) coexist with known geographic places. The largest and most ornate of the world maps is a richly illuminated eclectic painting on cloth, with text in Arabic, Hindustani, and Persian; it probably dates from the eighteenth century and may be of either Rajasthani or Deccani provenance. Additionally, a 32-sheet atlas of the "Inhabited Quarter" (i.e. the part of the world suitable for human life), forms part of a 1647 encyclopedic work by Sadiq Isfahani of Jaunpur in what is now Uttar Pradesh. The orientation on maps made by Muslims is typically toward the south.

Indigenous regional maps, mostly from the eighteenth and early nineteenth centuries, derive mainly form Rajasthan, Kashmir, Maharashtra, and Gujarat, probably in that order of frequency. These range in size from very large (several meters on a side) to page-size productions, the larger works being almost always painted on cloth. No clear schools of cartography emerge, though one can usually distinguish among regions of origin. Map symbols and orientation vary markedly, though some regional tendencies may be noted. No map has a consistent scale, though some contain textual notes on distances between places.

Route maps most commonly appear in strip form, occasionally as lengthy scrolls, and typically show the places and physical and man-made features encountered between two given points. Some route maps, largely relating to pilgrimages, which frequently take the form of circuits, are likely to be more complex. Surviving navigational charts are few in number, entirely from Gujarat, and in a tradition presumably derived from the Middle East. The oldest such known work is dated AD 1710.

The most common genre of maps are those that relate to relatively small localities, especially individual towns, as well as plans of specific forts, palaces, temples, gardens, and tombs. Such maps served many purposes: guides to pilgrims, aids for engineering projects, plans or documents for military activities, commemorations of historical events, text illustrations, interior adornments, and so forth. Locality maps typically combine a largely pictorial rendition of specific structures, drawn in either an oblique perspective or in frontal elevation, with an essentially planimetric rendition of the encompassing space. Hill features on such maps (as well as on regional maps) are characteristically shown in frontal perspective. Colors for rendering hills, water features, and vegetation are naturalistic and not very different from what one would encounter on Western maps. The largest known Indian map, depicting the former Rajput capital at Amber in remarkable house-by-house detail, measures 661 × 645 cm. (260 × 254 inches, or approximately 20 × 21 feet).

Although hundreds of Indian maps have now been studied and described, hundreds of additional recently discovered works await analysis; and it may be safely predicted in light of the interest aroused by recent research that many more maps will soon come to light.

JOSEPH E. SCHWARTZBERG

REFERENCES

Arunachalam, B. "The Haven-Finding Art in Indian Navigational Traditions and Cartography." In *The Indian Ocean: Explorations in Exploration, Commerce, and Politics.* Ed. Satish Chandra. New Delhi: Sage Publications, 1987, pp. 191–221.

Bahura, Gopal Narayan and Singh, Chandramani. Catalogue of Historical Documents in Kapad Dwara, Jaipur; Part II: Maps and Plans. Jaipur: published with the permission of the Maharajah of Jaipur, 1990.

Caillat, Collette and Ravi Kumar. *The Jain Cosmology.* Basel: Ravi Kumar, 1981.

Digby, Simon. "The Bhugola of Ksema Karna: a Dated Sixteenth Century Piece of Indian Metalware." *AARP (Art and Archaeology Research Papers)* 4: 10–31, 1973.

Gole, Susan. *Indian Maps and Plans: From Earliest Times to the Advent of European Surveys.* New Delhi: Manohar, 1989.

Habib, Irfan. "Cartography in Mughal India." *Medieval India, a Miscellany* 4: 122–34, 1977; also published in *Indian Archives* 28: 88–105, 1979.

Savage-Smith, Emilie. *Islamicate Celestial Globes: Their History, Construction, and Use.* Washington, D.C.: Smithsonian Institution Press, 1985.

Schwartzberg, Joseph E. "South Asian Cartography." Part 2 of *Cartography in the Traditional Islamic and South Asian Societies*, Vol 2, Book 1 of *The History of Cartography*. Ed. J.B. Harley and David Woodward. Chicago: University of Chicago Press, 1992, pp. 295–509.

Tripathi, Maya Prasad. *Development of Geographic Knowledge in Ancient India.* Varanasi: Bharatiya Vidya Prakashan: 1969.

See also: Astrology in India – Astronomy in India

MAPS AND MAPMAKING: ISLAMIC TERRESTRIAL MAPS The cultural boundaries of premodern Islamic civilization (ca. 700–1850) extended from the Atlantic shores of Africa to the Pacific Ocean and from the steppes of Siberia to the islands of South Asia. Widely different traditions of empirical and theoretical cartography developed and coexisted within this cultural sphere. The academic study of these traditions is at its preliminary stages, and it is likely that further research will unveil hitherto unknown aspects of the cultural history of maps in the Islamic world.

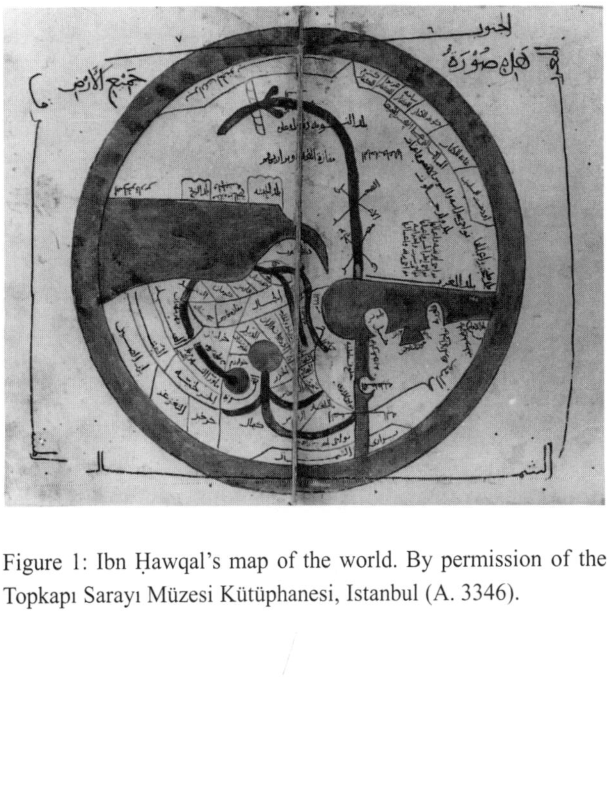

Figure 1: Ibn Ḥawqal's map of the world. By permission of the Topkapı Sarayı Müzesi Kütüphanesi, Istanbul (A. 3346).

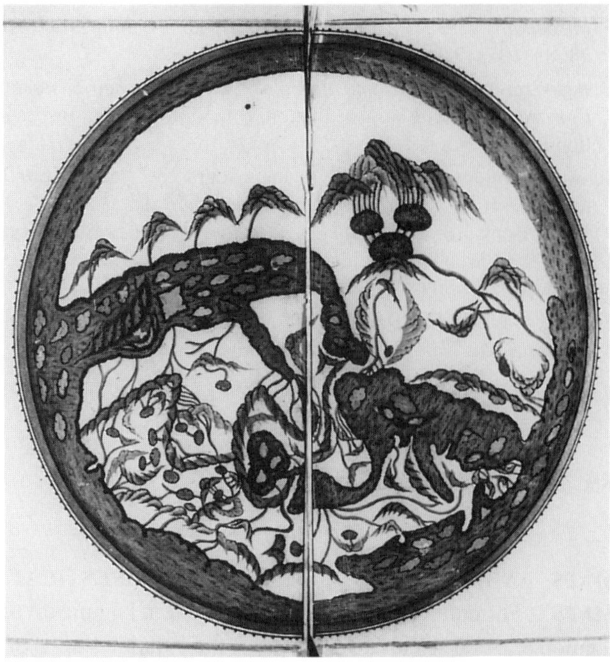

Figure 2: al-Idrīsī's world map from the Sofia manuscript. By permission of the Cyril and Methodius National Library, Sofia (Or. 3198, fols. 4v-5r).

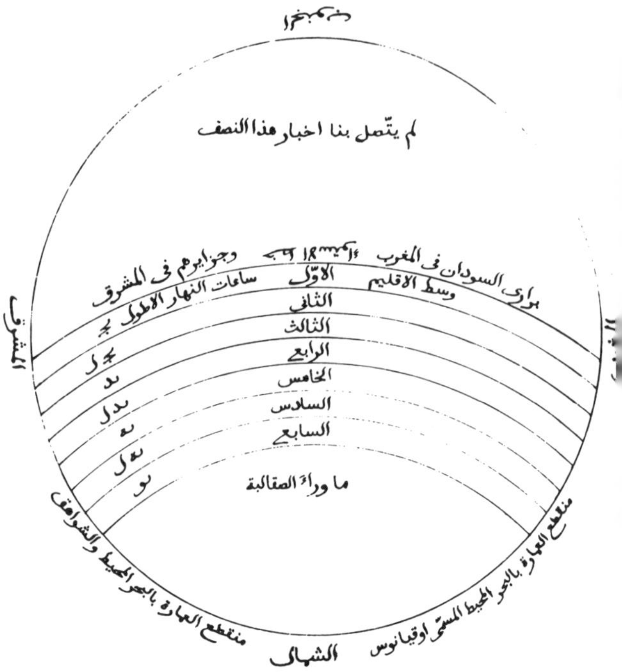

Figure 3: Yāqūt's climate map. From *Jacut's geographisches Wörterbuch*, 6 vols, ed. Ferdinand Wüstenfeld (Leipzig: F.A. Brockhaus, 1866–73), vol. 1, between pp. 28 and 29 (fig. 4).

Practice

If one leaves aside purely literary references to maps and map use in historical sources for the first three centuries of Islamic history, the first significant corpus of Islamic terrestrial maps that survive are those that accompany texts written by the Balkhī school of geographers (al-Balkhī, al-Iṣṭakhrī, Ibn Ḥawqal, and al-Muqaddasī) during the tenth century. This set of maps normally comprised a world map, maps of the three seas (the Mediterranean, the Indian Ocean, and the Caspian Sea), and maps of seventeen regions of the Islamic world. Not based on any projection and lacking a scale, it is possible that these maps were based on geographical writings that described the postal routes and administrative divisions of the Islamic states.

The other major cartographic school to develop in this period of Islamic history was the Ptolemaic. No early specimen of this school survived — the most spectacular of these seems to have been a world map produced for the ʿAbbāsid caliph al-Maʾmūn (r. 813–33) — and one has to turn to the celebrated geographical compendium of al-Sharīf al-Idrīsī, entitled *Nuzhat al-mushtāq fīʾkhtirāq al-āfāq* (The Book of Amusement for Those Yearning to Penetrate the Horizons) to appreciate fully the strength of the Ptolemaic cartographic tradition. Al-Idrīsī's work, completed in 1154 and accompanied by a large world map engraved on silver (no longer extant), contained a small world map and seventy sectional maps that collectively represented the whole of the known world. Latitudes and longitudes, not shown on the maps, were given in the text.

Cartographic representation of the globe during the High Caliphal and Early Middle Periods (ca. 700–1250) was not confined to geographical mapping of the Balkhī and Ptolemaic schools. In their attempts to depict the world Muslims also resorted to geographical diagrams. At least three different traditions of diagrammatization were used: the seven-*climata* scheme, the seven-*kishvar* system, and the *qibla* charts. The seven-*climata* scheme, Ptolemaic in origin, divided the inhabited portion of the earth into seven zones (Arabic *iqlīm*) based on latitude calculations. In the seven-*kishvar* system, of Persian origin, the inhabited portion was represented in seven circular regions (Persian *kishvar*), arranged so that six of the regions totally engulfed the seventh central one. The *qibla* charts, occasioned by the religious prescription to perform various ritual acts in the direction of the Kaʿba in Mecca, divided the world into four, eight, eleven, twelve, or more sectors around the Kaʿba.

These cartographic traditions, originally developed during the tenth and eleventh centuries, formed the basis of further cartographic activity during the Later Middle Period and the Period of the Great Regional Empires (ca. 1250–1850). There took place a certain degree of interaction between the

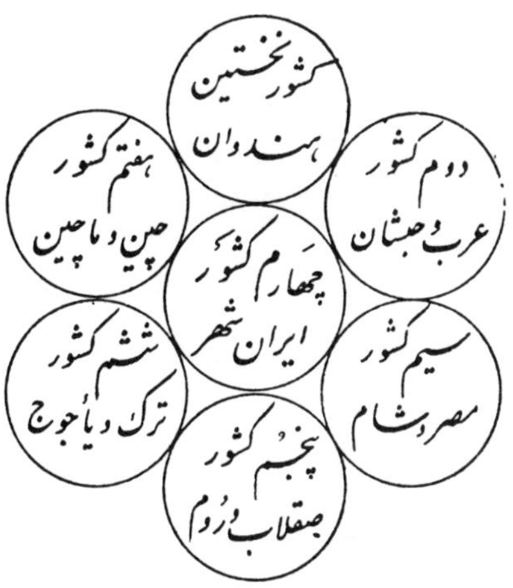

Figure 4: al-Bīrūnī's seven *kishvars*. From al-Bīrūnī, *Kitāb al-tafhīm li-avāʾil ṣināʿat al-tanjīm*, ed. Jalāl al-Dīn Humāʾī (Tehran, 1974), 196.

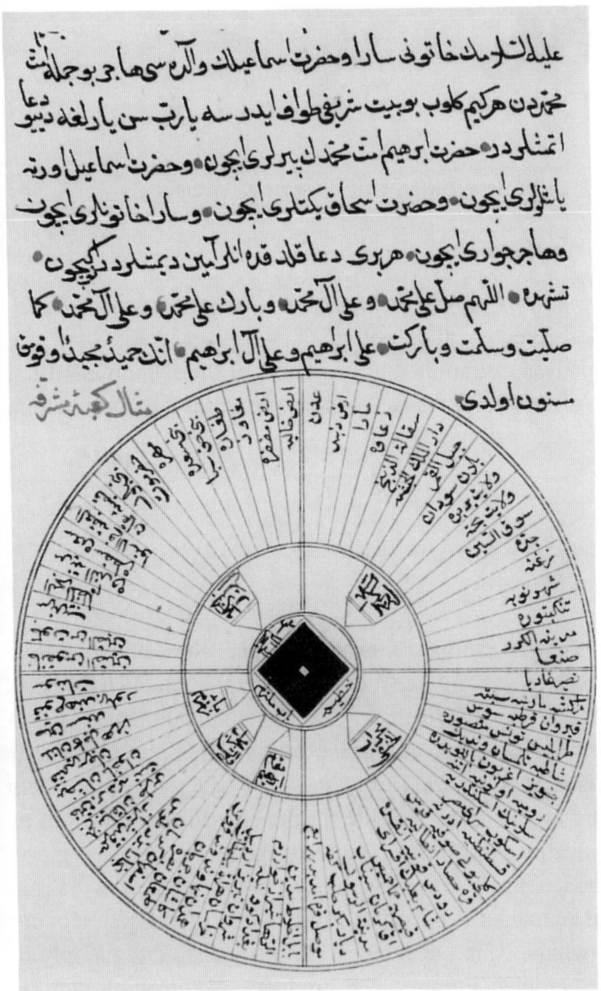

Figure 5: Seventy-two section scheme of sacred geography. By permission of the Topkapı Sarayı Müzesi Kütüphanesi, Istanbul (B. 179, fol. 52r).

Figure 6: Plan of the fortress of Van. By permission of the Topkapı Sarayı Müzesi Arşivi, Istanbul (E. 9487).

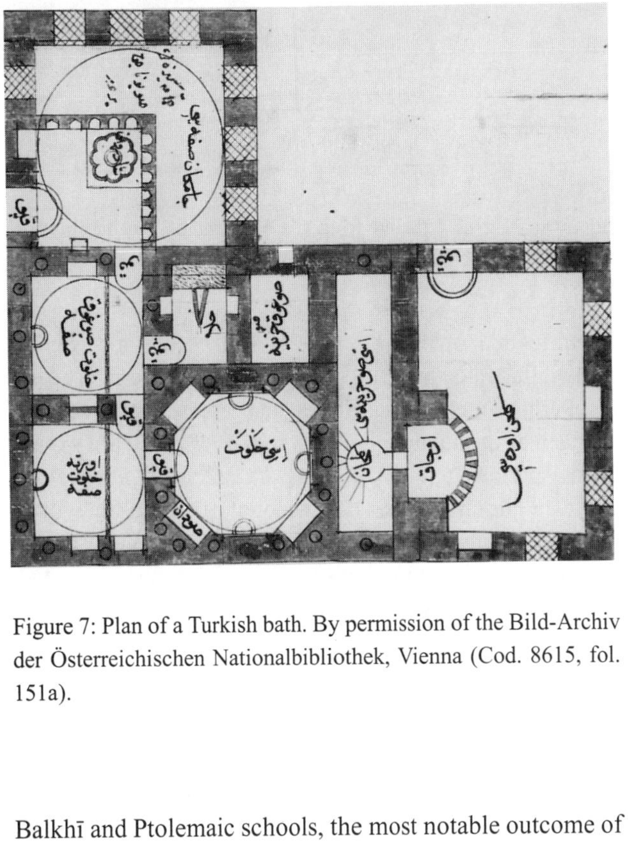

Figure 7: Plan of a Turkish bath. By permission of the Bild-Archiv der Österreichischen Nationalbibliothek, Vienna (Cod. 8615, fol. 151a).

Balkhī and Ptolemaic schools, the most notable outcome of which was the attempt to place a graticule on the circular world map by Ḥamd Allāh Mustawfī (d. 1339) and Ḥāfiẓ-i Abrū (d. 1430). The formation of the Gunpowder Empires during the sixteenth century opened up new directions in Islamic mapmaking. We are particularly well informed about Ottoman maps and mapmaking. Graphic representation of space was used systematically for administrative purposes in the spheres of military operation and state-sponsored architectural construction. At least two new genres, visual

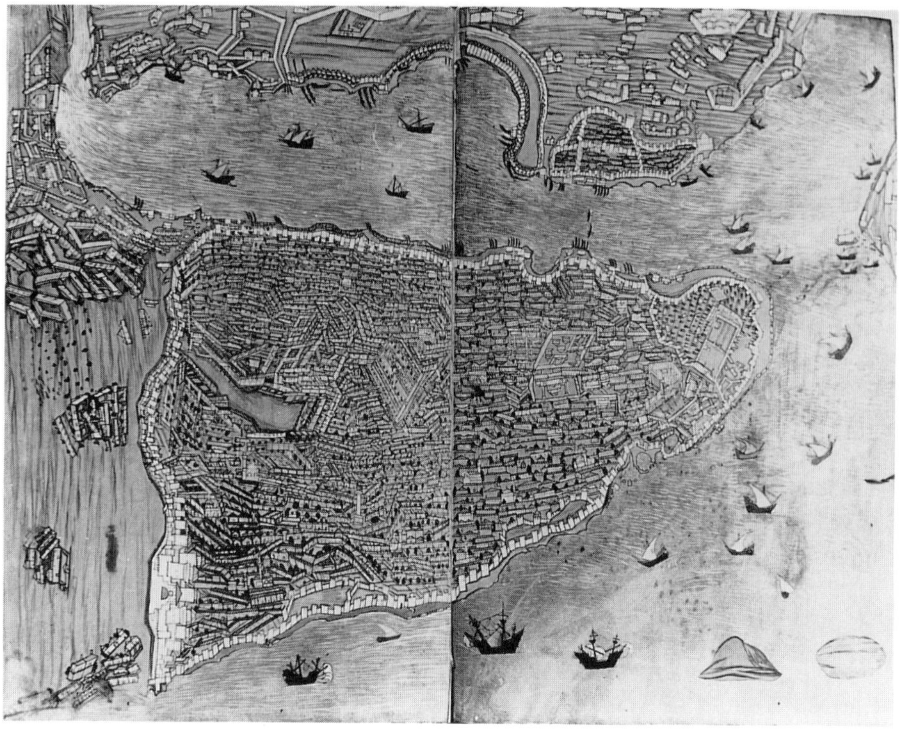

Figure 8: Plan of Istanbul from the Hünernāme. By permission of the Topkapı Sarayı Müzesi Kütüphanesi, Istanbul (H. 1523, fols. 158b–159a.).

itineraries and town views, were also introduced and heavily used in illustrated histories produced under imperial patronage. Perhaps the most significant and representative of such spectacular productions — of which over thirty are extant for the period 1537–1630 — is the *Mecmūʿ-i menāzil* of Maṭrakçı Naṣūḥ (d. 1564), an account of Sultan Süleymān I's campaign into eastern Anatolia, Persia, and Iraq in 1533–35. Not much is known on cartographic production in the Safavid and Mughal Empires.

Theory

Quite apart from cartographic practice and only tenuously related to it, a strong tradition of rigorous investigation of the earth was maintained in Islamic civilization from at least the ninth century onward. In addition to purely narrative geographical accounts, serious attention was paid to geodesy — the measurement of distances, or the determination of exact points, on the curved surface of the earth. As a result, a considerable number of geographical tables exist in Arabic and Persian, normally included in astronomical works. These either list places under climates, with no longitudes given and only the latitude of the climates specified, or assign longitude and latitude values for each place individually. The question of projection was also addressed in some detail. These theoretical aspects of Islamic cartography are perhaps best exemplified in the works of the scholar al-Bīrūnī (d. after 1050), whose output constitutes the culmination point of Islamic geodesy.

Map Use

The extant corpus of Islamic maps, when coupled with purely literary references to cartographic practice, suggest that the mapping instinct in Islamic civilization was by and large harnessed to the cause of scholarship and imperial artistic production. The majority of Islamic map artifacts were produced by the elite for the elite for the purposes of edification, illustration, and propagation of imperial glory. Practical application was the exception rather than the rule and remained confined to military maps and architectural drawings, which developed into distinct traditions only in the Ottoman Empire. The products of elite high culture, however, were infinitely more likely to be preserved, if only in literary form, than their popular or folk counterparts, so that the surviving cartographic record of premodern Islamic societies is only partially reflective of everyday mapping practices.

AHMET T. KARAMUSTAFA

REFERENCES

Harley, J. B., and David Woodward, eds. *The History of Cartography*, Volume 2, Book 1, *Cartography in the Traditional Islamic and South Asian Societies*. Chicago: University of Chicago Press, 1992.

Kamal, Youssouf. *Monumenta Cartographica Africae et Aegypti*. 5 vols. in 16 pts. Cairo, 1926–51. Facsimile reprint, 6 vols., ed. Fuat Sezgin. Frankfurt: Institut für Geschichte der Arabisch-Islamischen Wissenschaften, 1987 [maps].

Miller, Konrad. *Mappae arabicae: Arabische Welt- und Länderkarten des 9.–13. Jahrhunderts*. 6 vols. Stuttgart, 1926–31 [maps].

See also: Balkhī – Ibn Ḥawqal – al-Muqqadasī – al-Idrīsī – *Qibla* – Ottoman Science – Geodesy – al-Bīrūnī

MAPS AND MAPMAKING: ISLAMIC WORLD MAPS CENTERED ON MECCA Two remarkable world maps came to light in 1989 and 1995, respectively. The maps have Mecca at the center and are so devised that the direction of Mecca (*qibla*) and distance to Mecca can be read directly for any locality in the Islamic world between Andalusia and China (see the frontispiece for an illustration of the first world map). Both maps are engraved on circular brass plates of diameter 22.5 cm and clearly hail from the same workshop. From considerations of the calligraphy and decoration they can be associated with Isfahan and dated ca. 1700. But both are clearly copied from different originals, and so they are from a series of such world maps about which until recently we knew nothing.

The highly sophisticated cartographical grid (see Figure 1) is to be used in conjunction with the scale around the circumference, on which the *qibla* can be read, and with the non-uniform scale on the diametral rule, on which the distance to Mecca in *farsakh*s can be read. (A *farsakh* is equivalent to 3 miles) There are some 150 localities marked on each of the maps, but the selection is not identical: their coordinates are at first sight based on those in the *Sulṭānī Zīj* of Ulugh Beg (Samarqand, ca. 1430), based in turn on those in the *Īlkhānī Zīj* of Naṣīr al-Dīn al-Ṭūsī (Maragha, ca. 1250), but there are several localities whose coordinates testify to the fact that they were taken from the common source of both the *Sulṭānī* and *Īlkhānī Zīj*es, namely, the mysterious anonymous *Kitāb al-Aṭwāl wa-l-ʿurūḍ li-l-Furs*, a source known to us only from citations by the astronomer-prince Abu'l-Fidāʾ (Hama, ca. 1325). The *Kitāb al-Aṭwāl wa-l-ʿurūḍ* may be as early as the twelfth century. On the other hand several of the localities in India featured on the maps were not founded until the early fifteenth century. In fact, the earliest known geographical table featuring all of the localities on the two maps is found in a treatise on the astrolabe compiled in Najaf in 1702/03 by one ʿAbd al-Raḥīm ibn Muḥammad, who wrote that he compiled his table from al-Ṭūsī, Ulugh Beg, "and others". But he also presented the *qiblas* and distances to Mecca for the 274 localities in his list, and at least some of these feature on various Persian astrolabes from the mid-seventeenth century, that is, some fifty years before he compiled his treatise. In fact ʿAbd al Raḥīm simply copied a table compiled in Kish near Samarqand in the first half of the fifteenth century. This was, in fact, the actual source of the geographical data on the two world maps. It seems probable that the mapping goes back to al-Bīrūnī in the early eleventh century if not even Ḥabash in the mid-ninth century.

Although the origin and development of these Mecca-centered world maps is still obscure, it is clear that they are entirely Islamic in their conception. Indeed they represent the culmination of Islamic mathematical cartography, and have no parallel in sophistication between Antiquity (the

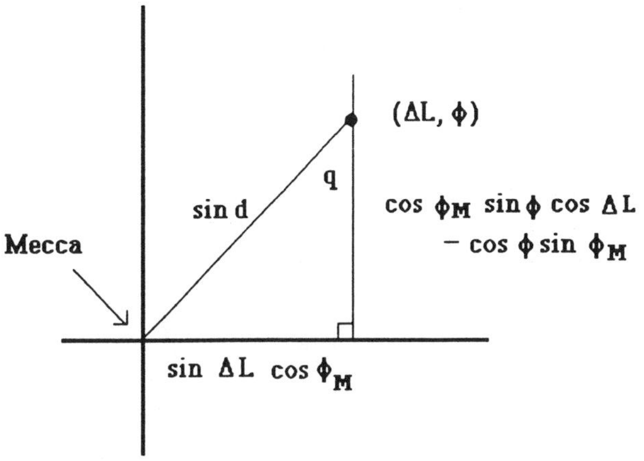

Figure 1: The mathematics underlying the theory of the grid on the Isfahan world maps, enabling the user to read the *qibla* on the circumferential scale and the distance on the diametral scale. An approximation has been used on the world map so that the latitude curves are arcs of circles; this produces slight inaccuracies noticeable only on the edges of the map (that is, in Andalusia and China). Drawn by the author.

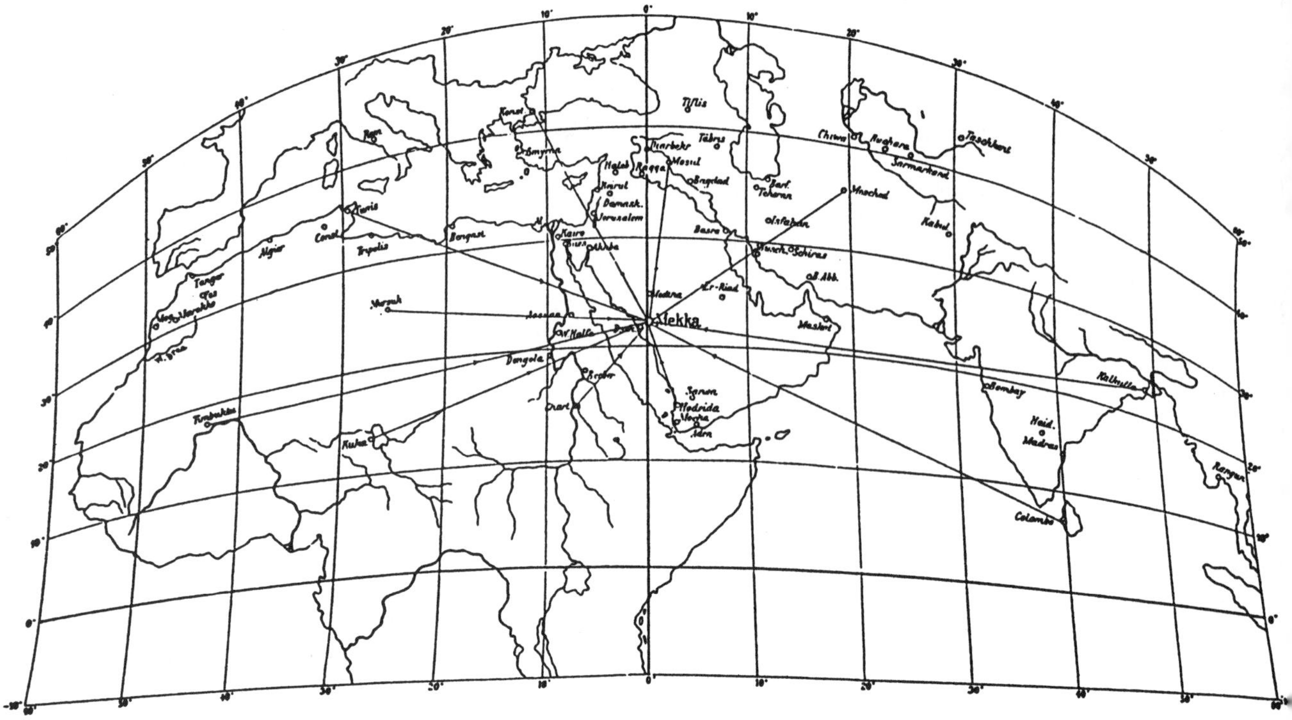

Figure 2: The world map proposed by Carl Schoy ca. 1920 for preserving direction and distance to Mecca. There are major cartographic distortions for regions on the other side of the world. On the Isfahan world map, even though it was made ca. 1700, we are dealing essentially with the world as known to Ptolemy. (From Carl Schoy, *Gnomonik der Araber*. Berlin and Leipzig: Walter de Gruyter, 1923, between pp. 44 and 45.)

world map of Ptolemy, ca. 125) and European cartography of the seventeenth century. Prior to their rediscovery it was thought that the first person to construct a world map centered on Mecca from which one could read off the *qibla* and distance to Mecca was the German historian of Islamic science Carl Schoy, who published such a map ca. 1920 (see Figure 2).

DAVID A. KING

REFERENCES

King, D. A. "Weltkarten zur Ermittlung der Richtung nach Mekka." In *Focus Behaim-Globus*. Ed. G. Bott. 2 vols. Nuremberg: Germanisches Nationalmuseum, 1992, vol. 1, pp. 167–171, and vol. 2, pp. 686–691.

King, D.A. "Two Iranian World Maps for Finding the Direction and Distance to Mecca." *Imago Mundi* 49 (1997), pp. 1–20.

Lorch, Richard P. and David A King. "Qibla Charts, Qibla Maps, and Related Instruments." In *The History of Cartography*, vol. 2, book 1: *Cartography in the Traditional Islamic and South Asian Societies*. Ed. J.B. Harley and David Woodward. Chicago: University of Chicago Press, 1992, pp. 189–205.

See also: *Qibla* – Ulugh Beg – *Zīj* – Astrolabe – Naṣīr al-Dīn al-Ṭusī – al-Bīrūnī – Astronomy in the Islamic World – Religion and Science in Islam I

MAPS AND MAPMAKING IN JAPAN The first record of Japanese mapmaking is an Imperial edict in AD 646 included in the *Nihon Shoki* (Chronicles of Japan), 720, ordering that each province reports its territorial range to the central government by means of a map. However, no fragment of such a map is extant now. The earliest extant maps relate to land ownership and date from the eighth century. These maps are almost all preserved in the Shōsōin, Nara, and consist of a map of Tōdai-ji temple precincts, 756, and over twenty maps of paddy fields. On these maps we can observe the traditional Chinese grid system. The majority of them are drawn on hemp.

The extant early general maps of Japan include: one dated the twelfth month, the third year of the Imperial era known as Kagen (1305/06), now owned by the Ninna-ji temple, Kyoto; *Dai-Nihonkoku Zu* (Map of Great Japan) in the 1548 codex of an encyclopedia, *Shūgaishō* (Collection of Oddments); and *Nansembushū Dai-Nihonkoku Shōtō Zu* (Orthodox Map of Great Japan in Jambudvīpa) drawn in the mid-sixteenth century (Tōshōdai-ji temple, Nara). These maps show the coastline and the boundary lines between provinces by means of smooth curves, and have the common characteristic of showing the main routes from the province of Yamashiro, where the capital Kyoto had been located since 794, through almost all the provinces. Maps with such characteristics are called "Gyōki-type" maps, because they have an inscription indicating that the author was Gyōki (668–749), a revered Buddhist priest. But no one believes today that they were actually the work of Gyōki, for the capital in his time was not Kyoto but Nara, Yamato Province. Later, in the ritual of *Tsuina* (which had been instituted on Gyōki's advice in 706 in order to offer prayers for national peace and the health of the people), a map of Japan came to be used to provide a concrete image of the country. It may be imagined that by association maps so used came to be called by his name.

The earliest extant map of the world is the *Go-Tenjiku Zu* (Map of the Five Indies), drawn by Jūkai, a Buddhist priest, in 1364, and now owned by the Hōryū-ji temple in Nara. The actually existing Buddhist continent called Jambudvīpa is drawn; India occupies a great part of the continent, and China, Persia, Japan, and many countries of Central Asia are shown. There are ten other extant maps which are copies made in later periods, and which belong to the same group. Incidentally, the above-mentioned *Shūgaishō* also contains a rough map of Jambudvīpa entitled *Tenjiku Zu* (Map of India).

European cartographical works were introduced to Japan in the late sixteenth century, and the coastlines of the Gyōki-type maps came to be drawn in detail, especially in the area of Kyūshū. This revision seems to have been done by the European pilots and by the Portuguese Ignacio Moreira, who resided in Japan between 1590 and 1592.

Exact maps of provinces were needed for state administration. But there are no known records about the compilation of national maps after the year 646 mentioned above (except for government orders of 738 and 796) until 1591, when the Toyotomi government embarked on such a project. The extant maps of two counties in Echigo Province are rare examples of the results of this project. The Tokugawa Shogunate, which succeeded the Toyotomi government and held its hegemony for two and a half centuries until 1867, gave orders for cartographical records of all provinces on five occasions during its reign.

These orders were issued in the tenth year of the Imperial era of Keichō (1605); about the tenth year of Kan'ei (ca. 1633); the twelfth month of the first year of Shōhō (1645); the tenth year of Genroku (1697); and the sixth year of Tempō (1835). No detailed records remain as regards the first and second projects. After the third project, prescriptions aiming at some regularization of form were issued: for example, the scale was fixed at six *sun* to one *ri* (1 : 21,600). For the third project, the Shogunate also required the province to submit the plans of cities where the clan offices were located. These maps and plans are huge and beautifully colored. All of the maps of the fifth project, and some earlier ones, are still extant in the National Archives in Tokyo.

General maps of Japan based on the provincial maps were compiled on each occasion except the fifth project. The earliest extant official map of Japan is the so-called Keichō map of Japan, compiled in the mid-seventeenth century and now in the National Diet Library, Tokyo. On this map the eastern half of Honshū is completely curved to the north, and depicted as smaller than it actually is. The island of Shikoku, which actually has projections in all four directions of the compass, is drawn in a rectangular form.

The best of the official maps of Japan compiled during the Edo era is the so-called Shōhō map of Japan, based on the results of the third cartographical project, and completed around 1670. Compiled by the famous surveyor Hōjō Ujinaga, this map shows the shape of the Japanese archipelago which is very close to actuality.

These maps strongly influenced private works of cartography. *Nihon Bunkei Zu* (Map of Japan Divided into Parts), published in 1666, and the series of works by the painter Ishikawa Ryūsen beginning in 1687, were based on the Keichō map of Japan; the former was the first printed atlas of Japan. Seki Sokō's *Nihon Bun'iki Shishō Zu* (Easily Understandable Atlas of the Regions of Japan), 1698, and Nagakubo Sekisui's works (first edition, 1779) were also based on this "Shōhō map of Japan".

In traditional cartography, little attention was paid to the existence of spherical coordinates, but latitudes are shown in the portolan charts of Japan (ca. 1670, Mitsui Bunko Library, Tokyo; National Museum, Tokyo). The first map of Japan made under the influence of these charts and having a network of parallels and meridians was Mori Kōan's *Nihon Bun'ya Zu* (Astronomical Map of Japan), 1754. In this map, however, parallels and meridians were simply added to the already existing official map. We also find such a network on Nagakubo's maps of Japan, but with only the degrees of latitude. It is evident that they imitated Mori's idea, and moreover had a greater social influence, because these latter maps were often printed, while Mori's remained only a manuscript.

The first map of Japan that was based on actual observation of degrees of longitude and latitude was completed by officials of the Shogunate astronomical observatory in 1821. The surveying began in 1800 and mainly concentrated on the coastlines, as the title *Dai-Nihon Enkai Yochi Zenzu* (Maps of the Coastlines of Great Japan) shows. The supervisor of the surveying throughout was the astronomer Inō Tadataka, and all the maps thus made in this project are called "Inō's maps". The 1821 maps consisted of 214 sheets on a scale of 1 : 36,000, 8 sheets of 1 : 216,000, and 3 sheets of 1 : 432,000.

The making of topographical maps by triangulation began in 1781, and in 1944 the entire country was depicted on maps with a scale of 1 : 50,000. Western style charts began to be systematically executed by the navy in 1871.

KAZUTAKA UNNO

REFERENCES

Cortazzi, Hugh. *Isles of Gold: Antique Maps of Japan.* New York: Weatherhill, 1983.

Namba, Matsutaro, Muroga, Nobuo, and Kazutaka Unno. *Old Maps in Japan.* Osaka: Sōgensha, 1973.

Oda, Takeo, Nobuo Muroga, and Kazutaka Unno. *Nihon Kochizu Taisei Sekaizu Hen* (The World in Japanese Maps until the Mid-Nineteenth Century). Tokyo: Kōdansha, 1975. (in Japanese).

Unno, Kazutaka. "Japanese Cartography." In *The History of Cartography*, Vol. 2, book 2, *Cartography in the Traditional East and Southeast Asian Societies.* Ed. David Woodward. Chicago: University of Chicago Press, 1994, pp. 346–477, pls. 22–29.

Unno, Kazutaka, Takeo, Oda, and Nobuo Muroga. *Nihon Kochizu Taisei* (Monumenta Cartographica Japonica). Tokyo: Kōdansha, 1972 (in Japanese).

See also: Ino Tadataka

MAPS AND MAPMAKING IN KOREA Korea has a long history, but evidence of mapmaking comes only from about the fifth century AD, when a nobleman's survivors painted a map of Liaodong city (now in Liaoning Province, China, but then within the borders of the Korean Koguryŏ kingdom [37 BC–668]) on a wall of his tomb. Since the tomb is in northern Korea, the map probably was meant to comfort him with an image of his far-off hometown. Koguryŏ's southern neighbors, the kingdoms of Paekche (18 BC–663) and Silla (57 BC–935) are known through written historical sources to have used maps in local administration, but no examples have survived. The same is true for the long Koryŏ dynasty (918–1392). The people of Koryŏ times had a sure sense of the peninsular outline of their country and are said to have commemorated it in the shape of the kingdom's standard silver bullion unit. Outlines of Korea also appear on a few Chinese maps of the twelfth century. Toward the end of the dynasty an enterprising scholar composed an elaborate map of the world — probably just the China–Korea region — and this work was probably connected in spirit to the *Kangnido* of 1402, which, in a copy of about 1470, is the oldest Korean-made map to survive.

The Artifactual Record

The succeeding Chosŏn dynasty built a much stronger and more durable state than had previously been known in the Korean peninsula. A vast amount of mapping was carried out in the fifteenth century, but aside from some prominent exceptions there is little remaining trace of this early production. Most surviving Korean maps are of the seventeenth century or later, and are most commonly found in albums, often hand-drawn but occasionally including wood-block prints. Larger maps are usually found preserved in silk or paper scrolls or on screens. A large sample of this material is now available in published color illustrations.

From the seventh century on, Korean kings presided over a strongly centralized, bureaucratic state. Provincial governors and district magistrates were responsible for producing maps of their areas in response to requests from central authorities for geographical, demographic, and fiscal information. Thus the central government maintained and frequently updated a broad range of data from which provincial and national maps could be compiled. The most complete such compilation to survive in its entirety is the *Sinjŭng Tongguk yŏji sŭngnam* (Complete Conspectus of Korean Territory, newly expanded; called the *Sŭngnam* for short) of 1531, descended from earlier editions going back to 1481.

Beginning in the late fifteenth century, the government generally limited publication and dissemination of maps for national security reasons, and the maps that were available, such as those in the *Sŭngnam*, were rather spare in detail. But security concerns also spawned the defense map (*kwanbangdo*) genre, which was rich in information relating to logistics and communciation.

Geomancy

In the late ninth century, Koreans began to articulate a national geomantic structure for their country. In its elementary form, geomancy (*p'ungsu*, Chinese *fengshui*) is an originally Chinese science employed in siting graves, human habitations, temples and shrines, and towns and cities. The landscape is the locus of forces which deliver energy and power through networks of montane "arteries" (*maek*) or riparian "veins" (*p'a*), which will be determined to be either favorable or unfavorable for the use envisaged for a particular site. Even today, specialists in this kind of knowledge can be found in most Korean communities, and indeed all over East Asia. But Korea is unique in developing this science into a nationwide framework for understanding and defining the physical aspects of the nation itself. From an early date the source of Korea's geomantic forces was seen to be located in Mount Paektu (2744 m), which through Korea's mountain ranges and water sheds distributed vital forces to the entire country. It was thus of crucial importance to determine and map the mountain ranges of the land, so that the relationship of any particular spot to the overall network of "shapes and forces" (*hyŏngse*) could be clearly understood. Whether on the national or local level, a shapes-and-forces map (*hyŏngsedo*) would clearly configure the mountain ranges and water sheds, so that the physical lay of the land was easily preceived.

In the seventeenth century, following the Japanese (1592–1598) and Manchu invasions (1627, 1637), Korea went through a series of reforms which resulted in enhanced revenues and a more developed military. This was accompanied by economic growth and increased trade with China and Japan. In society at large these changes produced greater occupational diversity, a certain degree of social mobility, and a broader access to education. An intellectual movement in pursuit of "practical studies" (*sirhak*) extended scholarship into new areas of science and statecraft research. Thus it is not surprising that we see a development which might be called the privatization of cartography, leading to works by individual mapmakers that steadily grew in quality and diversity. Maps were no longer restricted to government offices but were much more broadly distributed throughout society. This resulted in a higher rate of preservation of maps into modern times.

In tracking this greater patronage we can notice two different kinds of map collections, administrative and popular atlases. The administrative atlas typically would consist of a set of maps of Korea's eight provinces. Each provincial map would show the districts of the province, often with a shapes-and-forces accent that clarified the geographical character of the area. It would indicate principal roads and bridges, post stations and military bases, temples and schools, and other sites of official or civic importance. Distances between towns would be indicated either by notations on the map or in a table in the atlas. Accompanying essays would give a general historical and administrative overview of the province, and lists would be provided showing the rank of each magistracy, with statistics on its tax revenues, military reserves, and various other categories.

The popular atlas often had a similar organization, but it was distinguished by the addition of a general map of the world, or *ch'ŏnhado* (map of "all-under-heaven"), a general map of China, sketchier maps of Japan and the Ryukyu kingdom, and a map of Korea as a whole.

Korean Knowledge of Western Cartography

Jesuit missionaries, led by Matteo Ricci, began to introduce western maps of the world to China in the last years of the sixteenth century, and copies of printed editions of Ricci's world maps of 1602 and 1603 each reached Korea within a year of their publication. In 1708, a royal order was given to copy a variant version of the Ricci map, and this occasioned a Korean essay on Western cartography. Western maps continued to be imported throughout the eighteenth and early nineteenth centuries. A Jesuit-directed summation of mathematics and astronomy published in Chinese in 1723 updated Korean understanding of geographic latitude and explained methods of calculating longitude through the observation of eclipses, and Jesuit determinations of latitude and longitude for major Chinese centers and even for Seoul itself were found in the same source.

In spite of this more than passing exposure to western maps and mapmaking, one can find no visual trace of western influence on Korean maps themselves. An apparently Anglo-Chinese map from the 1790s was even printed by woodblock in Korea, in 1834, by the greatest Korean cartographer of all, Kim Chŏngho, yet Kim's own maps stay completely within the Korean tradition. On the other hand, there is abundant evidence that by the eighteenth century, Koreans were systematically compiling latitude and longitude data for their country, and although no indication of it is found on any Korean map of Korea, a kind of geodetic measurement did play a role in the accuracy of the data used to make the maps.

The Kangnido

The first rulers of the Chosŏn dynasty sponsored a number of projects designed to strengthen their legitimacy. Among these were a recarving on stone of a seventh century Korean star map and a terrestial map of the world. The star map reflected a very ancient Chinese astrographical tradition long superseded in both China and Korea, but the terrestial map, the *Honil kangni yŏktae kukto chi to* (Map of Integrated Lands and Regions and Historical Countries and Capitals) was based on two fourteenth-century Chinese maps. In 1402, Kwŏn Kŭn (1352–1409), a Confucian scholar and royal advisor, was ordered to supervise the compilation of a world map based on these two sources. The actual drawing was the work of Yi Hoe. To his images of China and the greater world, Kwŏn and Yi added Yi's own map of Korea and a re-

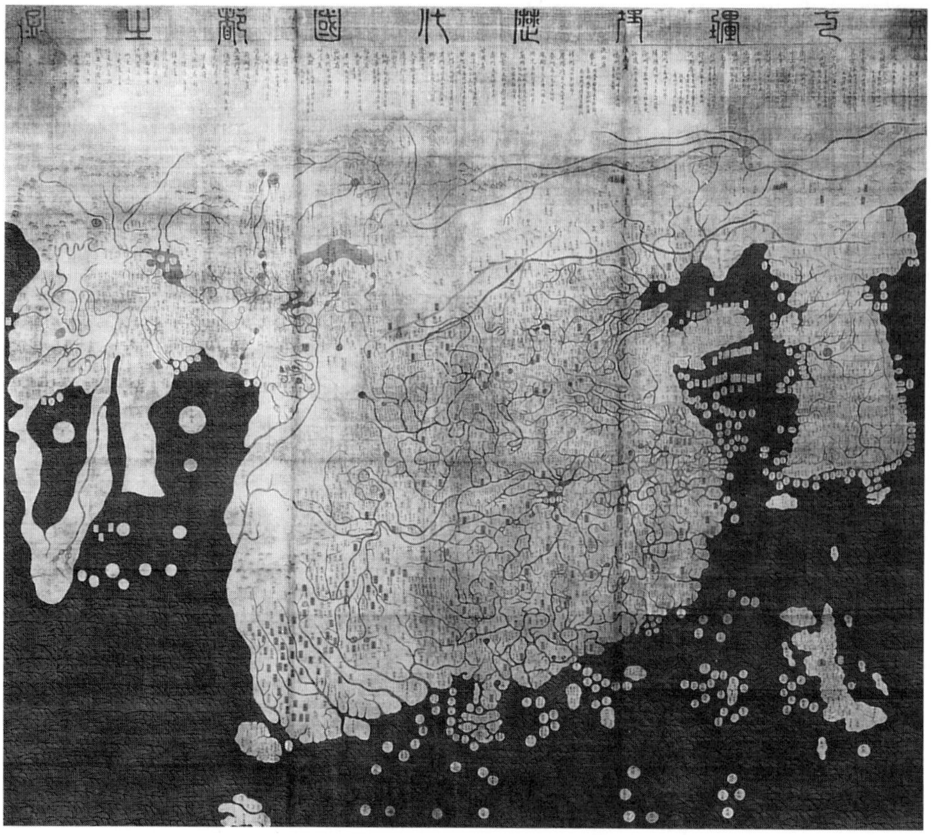

Figure 1: *Honil kangni yŏktae kukto chi to* (Map of Integrated Lands and Regions and Historical Countries and Capitals). By permission of the Ryukoku University Library, Kyoto, Japan.

cently imported image of Japan, thus, in their understanding, bringing the map to completion.

While this original *Kangnido* was lost long ago, a Korean copy made around 1470 and seized by Japanese invaders in the 1590s has survived in Japan. This copy contains Kwŏn Kŭn's original preface, which details the source materials.

At first glance, the *Kangnido* seems a collection of distortions. China and India make up a dominating, undivided mass in the center, while Europe and Africa hang from the western side and Korea hangs from the eastern as if these two masses were geographically equal. An upended Japan floats uncertainly in the East China Sea. The *Kangnido* was a conflation of other maps, with considerable relative distortion of the land masses, rather than cartography in the strict sense. As the first East Asian map to include Europe and Africa, and such features as the Mediterranean and Black Seas, the Arabian peninsula and the Persian Gulf, the Nile and the Red Sea, as well as to add Korea and Japan on the east, the *Kangnido* is an epochal achievment in Korean as well as world mapmaking.

Popular Cosmography: the Ch'ŏnhado

The *Kangnido* was in Korea too exotic to pass unrevised into the mainstream of Korean mapmaking. The unpronounceable names, in many cases Arabic originals filtered through Chinese transcriptions; the unimaginable distances; strange countries of which in many cases nothing was known — very little of this related to the concept of the world known in Korea. This world was essentially of Chinese definition. It was the *tianxia* (*ch'ŏnha* in Korean pronunciation) or "all under Heaven" that was at peace during the reigns of sage emperors, or that fell into disorder during the rule of bad ones. The geography of this world could be known through the Chinese classics and through the geographical sections and accounts of foreign countries in the long chain of Chinese dynastic histories that began to be compiled in the first century BC. In addition there was the more fantastic and whimsical geography recorded in the *Shanhai jing* (Classic of Mountains and Seas), the older parts of which date from around the second century BC but which evoke even

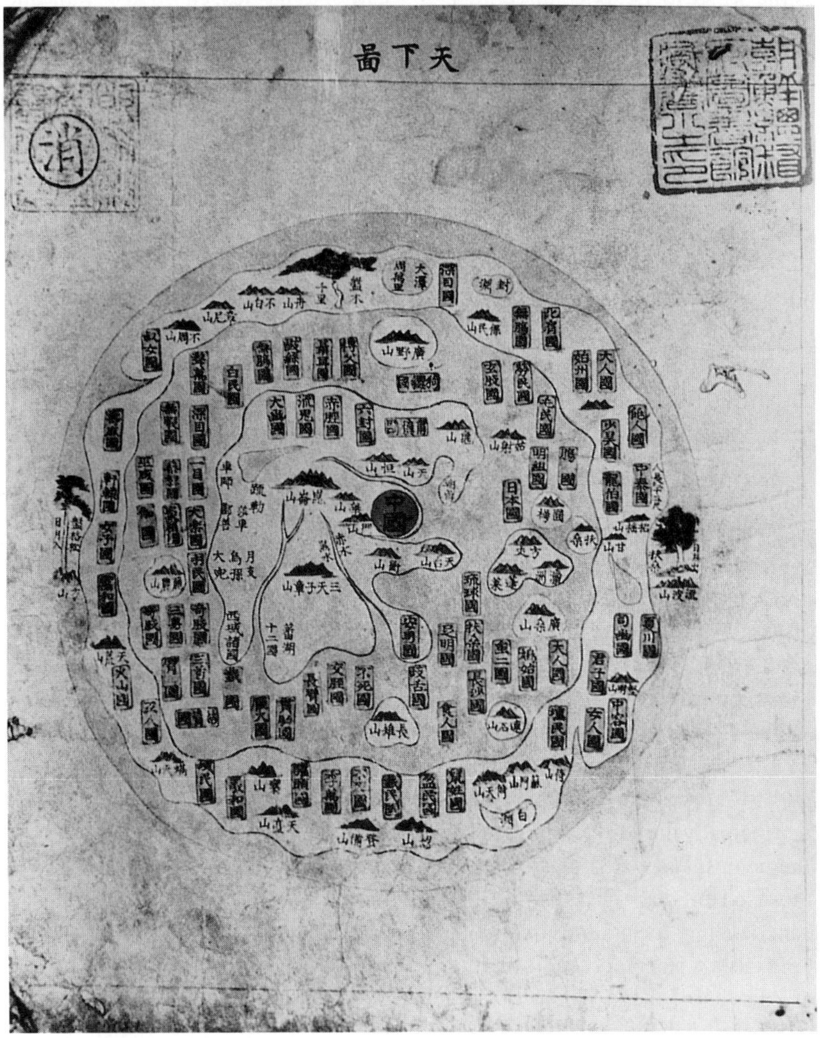

Figure 2: *Ch'ŏnhado* (Map of the World). By permission of the National Central Library, Seoul.

more ancient myths and traditions. This source added to the classical geography such imagined places as the "Land of the One-armed", the "Country of Women", "Mount Incomplete", and many more. This mix of ancient historical places and imagined fantastic ones constituted the world that was mapped in the *ch'ŏnhado* (world maps).

Although based almost completely on Chinese source material, the numerous examples of *ch'ŏnhado* constitute a completely Korean genre of map. An inner continent dominated by China, with the Korean peninsula always clearly depending from the eastern side, is surrounded by a sea full of kingdoms, which itself is surrounded by a circular land strip with even more amazing places. The eastern and western sides of this ring feature trees near which the sun rises and sets.

The origins of the *ch'ŏnhado* genre are unclear. A number of scholars have believed that these maps were Buddhist in origin, and some have thought that the basic shape of the map could have been derived from geographic theories popularized by the Chinese naturalist philosopher Zou Yan (fl. 3rd century BC). While morphologically plausible, it is impossible to find any cartographic link over the nearly two millennia that separate Zou Yan from the emergence of the *ch'ŏnhado* in the sixteenth century.

I believe that the core of the *ch'ŏnhado* — its inner continent — evolved from the *Kangnido*. In addition to many geographical and morphological considerations, the Tenri *Kangnido* is also chronologically congenial to this theory, since it contains placenames that fix its copying no earlier than 1568, making a plausible bridge to the sixteenth century *ch'ŏnhado*.

Korean National Maps

Under King Sejong (r. 1418–1450), the Korean court pursued a wide range of technical and scientific projects in astronomy, calendrical science, horology, musicology, pharmacology, agronomy, and geography and cartography. Distances between Seoul and the seat of each district, and between the districts themselves, were carefully measured and recorded. Polar altitude observations determined the latitude of the capital and of the extreme northern and southern frontiers, permitting an accurate estimate of the length of the country. Sejong also gave serious attention to the montane structure of the nation, collecting and studying data on the nation's mountain "arteries" and river "veins". King Sejo (r. 1455–1468), Sejong's son and follower ordered the complete mapping of the country, including not only a national map but maps of each of Korea's eight provinces and approximately 330 counties.

The national map that resulted from this preparation was the *Tongguk chido* (Map of the Eastern Country), which was presented to King Sejo in 1463 by his two chief cartographers, Chŏng Ch'ŏk (1390–1475) and Yang Sŏngji (1415–1482), but is thought to be principally the work of the former. The general outlines of the peninsula's coastline are astonishingly suggestive of those on modern maps, while the shapes-and-forces treatment detailing the nation's mountain network and watersheds provides a rich appreciation of its physical geography. Every district of the kingdom is indicated, along with its distance from Seoul and its provincial affiliation. Equally distinctive is the map's chief flaw, an unduly flattened northern frontier. Early Korean mapmakers appear to have had considerable difficulty in grasping the outline of the northern frontier, which was defined by the Amnok and Tuman (internationally, Yalu and Tumen) rivers that respectively flowed off the western and eastern slopes of Mount Paektu. It was not until the eighteenth century that this problem was essentially solved. So for about three centuries the conventional shape of the country was associated with Chŏng Ch'ŏk's outline, and cartographers now refer to such maps as in the Chŏng Ch'ŏk style.

The great summation of the administrative geography promoted by the fifteenth century Korean courts was the *Sinjŭng Tongguk yŏji sŭngnam* (Complete Conspectus of Korean Territory, newly expanded), which was first drafted in 1481, and went through several revisions before the final one of 1531. Called the *Sŭngnam* for short, this work detailed the administrative history of Korea's provinces and districts. It connected each district to its earliest known organization in earlier dynasties; clarified its rank and position in the administrative and military chains of regional governance; listed its schools, monasteries, Confucian shrines, post stations, signal-fire stations, natural and economic resources, famous native sons and virtuous women; and concluded with a sampling of literature associated with its history and public and private institutions. The final edition of the *Sŭngnam* was so thorough and so well done that it was never supplemented or re-edited.

We know that the cartographer Yang Sŏngji, whose work over the years for Kings Sejo and Sŏngjong (1469–1494) had laid the foundation for the *Sŭngnam*, was well known for his strict views on defense and national security. He argued that maps should not circulate outside of a few designated government offices. These views may have been a factor in the spareness of detail on the *Sŭngnam* maps that appeared later in Sŏngjong's reign. Another important change on the *Sŭngnam* maps concerned the shapes-and-forces treatment. Even the *Kangnido*'s representation of Korea had displayed the principal mountain ranges of the kingdom. But on the *Sŭngnam* maps, shapes-and-forces indications completely disappeared. The reasons for this change are not clear.

During the first half of the eighteenth century there were new developments in the mathematical and observational sciences which had a strong impact on the maps of the later Chosŏn period. Much of this was connected with western knowledge, which had began to flow steadily into Korea from Jesuit sources in China. Jesuit participation in the national mapping project of the Kangxi Emperor (1661–1722), which took place from 1709 to 1716, brought Jesuit cartographers who were mapping Manchuria to Korea's borders, and in 1713, Jesuit-trained Manchu and Chinese specialists made surveys and observations within Korea itself. The concepts of latitude and longitude, and related observational and mensurational techniques, were clearly described in Sino-Jesuit manuals which were introduced into Korea at least by 1715. A dramatic improvement in spatial representation is found on Korean maps that were developed during the following thirty or forty years.

The man responsible for this cartographic revolution was Chŏng Sanggi (1678–1752), a brilliant scholar who labored for decades in the privacy of his home. He developed what he called the "hundred-li foot" (*paengni ch'ŏk*), a 100-*li* visual scale bar (metaphorically a "foot;" I will call his unit a "scale-foot") calibrated in 10-*li* (4.3 km) intervals, which he inserted just after his introduction to his maps. Chinese books of that period specified a ratio of 200 *li* to one terrestial degree of latitude or longitude. This ratio was cited and used by late-eighteenth century Korean mapmakers (although the Korean *li* was considerably shorter than the Chinese), and calculations based on their maps show that they probably followed such a formula to determine geodetic distance. Perhaps Ch'ong Sanggi had used the same procedure.

Shapes-and-forces cartography made a strong return on Chŏng Sanggi's maps, and the vernacular painting style of the day depicted the mountain ranges in an attractive manner that created both cartographic clarity and an aesthetic dimension. A variety of symbols marked military bases, post stations, and other facilities Some copies of Chŏng's map's are works of art as well as cartographic masterpieces. His sons and grandsons continued to be active in cartography throughout the eighteenth century.

The Maps and Writings of Kim Chŏngho

In 1791, the Korean government sponsored the creation of a national grid for the purpose of complete cartographic standardization. While details of this project are uncertain, several albums of local maps copied on standardized *sŏnp'yo* (line guides) survive. Evidently the grid coordinates of each district seat were determined with reference to copies of Chŏng Sanggi maps owned by the government, and then marked on uniform line guides which were sent to district magistrates for development into standard district maps. The detail of these is astonishing. Every river and stream, every principal road and bridge, every public facility and many private ones such as schools, monasteries, and shrines, are indicated.

In 1834, an obscure cartographic genius named Kim Chŏngho (fl. 1834–1864) developed one of these national collections of district maps into a uniform national album, in which he re-edited hundreds of local maps into standard rectangular sheets. This national grid map was called the *Ch'ŏnggudo*, or Map of the Blue Hills, after an ancient poetic sobriquet for Korea.

Kim Chŏngho was himself a professional woodblock carver. It is not known whether he became one in order to market his maps, or began as a blockcarver and branched out into mapmaking. But it is as a blockcarver that he first comes to notice, in 1834, preparing for a famous scholar, Ch'oe Han'gi (1803–1875), a woodblock of a Chinese copy of an English hemispheric map of the world. Ch'oe obligingly returned the favor, writing an enthusiastic preface for the *Ch'ŏnggudo*. But this contact represents the only instance when the Korean world is known to have given any documentary notice to Kim's activities. This is in spite of the fact that in 1861, Kim would produce an even more remarkable national grid map that would bring him enduring fame. He must have been of very humble social status. We have no indication of his ancestry or native place. Traditions speak of endless trips throughout the country to check details and redo existing maps, and of a faithful daughter who took care of him and helped with the blockcarving. The traditional story holds that he was arrested in 1864, supposedly for endangering national security by revealing the nation's geography to potential enemies. But scholars doubt this, reasoning that such a grave incident would surely be reflected in official records, and noting that too many copies of the 1861 map (and of an 1864 recut edition) survive to permit belief that such a thing could have occurred. The 1860s were a time of great tension, with foreign incursions and a large-scale persecution of the country's Catholic community, which had been growing since 1784 in spite of many purges and constant suppression. Could Kim's printing of a foreign map of the world in 1834 have been taken against him? Ch'oe Han'gi, the actual patron of that project, had no problem with this, but then he was an upper-class scholar of influence and repute. Kim had no such insulation, and could have been the victim of petty policemen far down in the official structure. Whatever the reason, he disappeared in 1864 without a trace.

However humble Kim's status may have been, his writing shows that he was an accomplished scholar and a respectable writer of classical Chinese. From 1834 to 1861 his doings and whereabouts are completely unknown. In 1861, under the pseudonym Kosanja (The Master of Old Mountains), a second grid map appeared entitled *Taedong yŏjido* (Terrestial Map of the Great East [Korea]). Kim's short preface deals mainly with the importance of maps for military affairs.

The *Taedong yŏjido* was a completely reconsidered cartographic image of the country. The shapes-and-forces treatment received here its greatest representation. Mountain ranges were now represented by a solid black line, thinner for lower ranges, thicker for higher ones, with special jagged teeth on the line to represent particularly rugged stretches, or white peaks to indicate snowy heights. The clarity of the overall effect is impressive. With the *Taedong yŏjido*, a complete union of cartographic display and woodblock publishing technique was achieved. Kim seems to have wished to present simply the earth and its natural and human features, with the cartographic structure left, so to speak, underground.

The complete printed *Taedong yŏjido* is likely to have been very expensive. To make his vision of Korea more cheaply available, Kim produced a single sheet version which put the whole nation on a rectangle of about 115×76 cm, giving it the title of *Taedong yŏji chŏndo* (Complete Terrestial Map of the Great East [Korea]). The short text that filled the Bay of Wonsan gave the map something of the quality of a patriotic morale poster. After introducing Mount Paektu as "the grandfather of Korea's mountain arteries," giving the dimensions of the seacoasts and northern rivers, and proclaiming the greatness of the nation's legendary founders Tan'gun and Kija, he concluded with his ringing climax: "'Tis a storehouse of Heaven, a golden city! Truly, may it enjoy endless bliss for a hundred million myriad generations! Oh, how great it is!"

The District Map

We have seen that in the late eighteenth century, the Korean government had taken special measures to see that all the districts of the country were uniformly mapped. Although the purpose of this program was to create a national standard, in fact local maps reflected the distinctive features of the district. These highly skilled, uniform maps of districts could not displace the traditional local maps in popularity.

The traditional local map must properly be called a map-painting. It was composed and executed by a practiced painter rather than by a mapmaker, and it set the community into the surrounding landscape in the manner of a landscape painting. The village was seen from a fixed orientation, most commonly with north at the top, and usually in bird's-eye view; roads, rivers, and walls divided up the space in realistic proportions. Distances were indicated by short notes inserted at a focal point in the road. Houses were nestled together in homey elevation, smoke rising from the chimney, a chicken pecking at the ground by the wall. The village well was in its proper place, but visited by ladies and waterboys. The school house with its surrounding pines was a miniature all by itself. In the distance, tucked into the hills, were the familiar shrines and monasteries. In Korean, as in the other East Asian languages, the word for "picture" and "map" was the same, and traditional Korean map-paintings exactly reflected that ambivalence.

City Maps

Seoul was the nation's first city, and was the home of virtually all of the nation's prominent people. Those who lived there seem to have been very fond of maps of the city mounted on folding screens in one of the principal rooms of the house. Although these maps often showed the close urban detail of streets and buildings, one does not get the feeling that those who bought them did so in order to find their way around. The city was large enough so that a bird's-eye view would be unable to reveal the order and scale of its streets and alleys. Thus most maps of Seoul are executed in aerial plan, although the surrounding mountains were characteristically drawn in pictorial elevation.

P'yŏngyang is Korea's oldest city. Once a capital of the ancient kingdom of Koguryŏ, it was always an important regional center in later dynastic days. As the capital of P'yŏng'an Province on Korea's northwestern border facing China, it had great strategic and commercial significance. Maps of the city are commonly encountered and must have always been a popular souvenir. Although maps in aerial plan are known, the favored orientation for its mappers was a bird's-eye view looking toward the city over the Taedong River from the east. More than maps of Seoul, those of P'yŏngyang took on many of the qualities of the map-paintings of smaller towns.

Defense Maps

After the invasions of Korea by the Japanese in 1592–98 and the Manchus in 1627 and 1637, Korean statesmen adopted policies that put a high emphasis on defense and military preparedness. Thus for the years of the seventeenth through the nineteenth centuries, a great variety of defense maps (*kwanbangdo*) were produced. Most of them are unique and few were ever copied, accounting for the fact that most of those surviving today are original works. Some, such as the *Yogye kwanbang chido* (Map of the Defenses of the Liaoji Area) are vast panoramas that stretch from the northeastern coast of Korea northward toward the Amur River and westward to Beijing itself. Others show particular sections of the northern frontier. There are many having the character of charts that map Korea's coastal areas. There are detailed paintings of mountain fortresses. One sees in these works a great variety of styles, media, and painting skills. These maps, which in their day were in the category of classified information, are now prized for their unique approach to national and international cartography, and in many instances also for their highly stylized and artistic manner of execution.

Though one of East Asia's smaller countries, and always the object of Chinese cultural influences, Korea has had a proud and distinguished tradition of mapmaking all its own, and not a small number of unique cartographic achievements. The *Kangnido* is East Asia's oldest surviving world map, and by world standards at the time of its composition in 1402, one of the best realizations of world geography known from that time. The *Ch'ŏnhado*, though it reflects only Chinese geographic views and source material, is a map that China itself never produced. Chŏng Sanggi's maps of Korean provinces achieved a high standard of cartographic excellence and artistic distinction. Kim Chŏngho produced two great grid maps of Korea, and achieved a standard of cartographic imagination and excellence perhaps unknown elsewhere in East Asia. Korea's great corpus of maps, now in Yi Ch'an's magnificent album, are deserving of serious

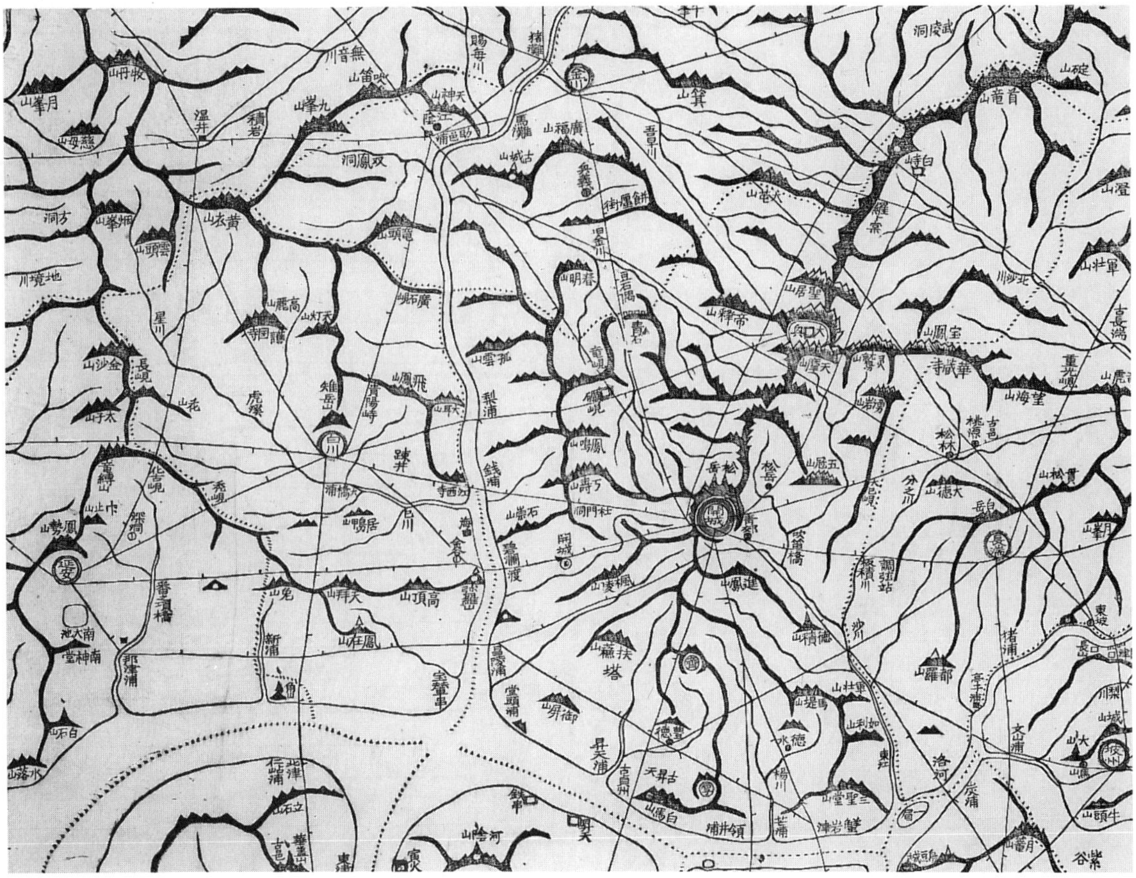

Figure 3: *Taedong yŏjido* [Terrestial Map of the Great East (Korea)]. Courtesy of the East Asian Library, University of California at Berkeley.

attention and the further research efforts of the world's cartographic historians.

GARI LEDYARD

REFERENCES

Ledyard, Gari. "Cartography in Korea." In *The History of Cartography,* Volume 2, Book 2, *Cartography in the Traditional East and Southeast Asian Societies.* Ed. J. B. Harley and David Woodward. Chicago: University of Chicago Press, 1994, pp. 235–345.

Lee, Chan (Yi Ch'an). *Han'guk ŭi ko chido* (Old Maps of Korea). Seoul: Pŏm'u Publishers, 1991.

MAPS AND MAPMAKING: MARSHALL ISLAND STICK CHARTS The so-called stick charts of the Marshall Islands in the Pacific Ocean are a rare instance in which a kind of cartography was developed in a non-literate culture. Their designation comes from the fact that they are constructed of slender sticks and twigs, and sticks lashed together into complex patterns. These patterns represent what occurs when mature ocean swells sweep past one or more of the coral atolls that make up the Marshall Island chain. The charts are not carried to sea, but are used as illustrative devices to train young men in the skills of piloting canoes through the archipelago, out of site of land, by noting various swell phenomena.

The basic concepts that are illustrated on the charts are refraction and reflection of ocean swells. When well-defined swells approach a shore they are bent according to the angle the swell line encounters when it reaches shallow water and the shore. This occurs because the portion of a swell slows down as it encounters shallow water, while the offshore, deep water part continues unaffected. This is wave refraction.

Refraction occurs as long as a swell is in contact with shallow water, and with a small roughly circular atoll this may be entirely around it. As a result, the surface waters off the protected side of the island will be confused by the intersections of opposed arms of the refracted swells. Further off the protected shore the swells reform as they continue on their course.

Wave energy is also reflected from a shore line which sends a smaller reflected wave back at a complementary angle, just as light striking a reflective surface or a ball striking a hard surface bounce back.

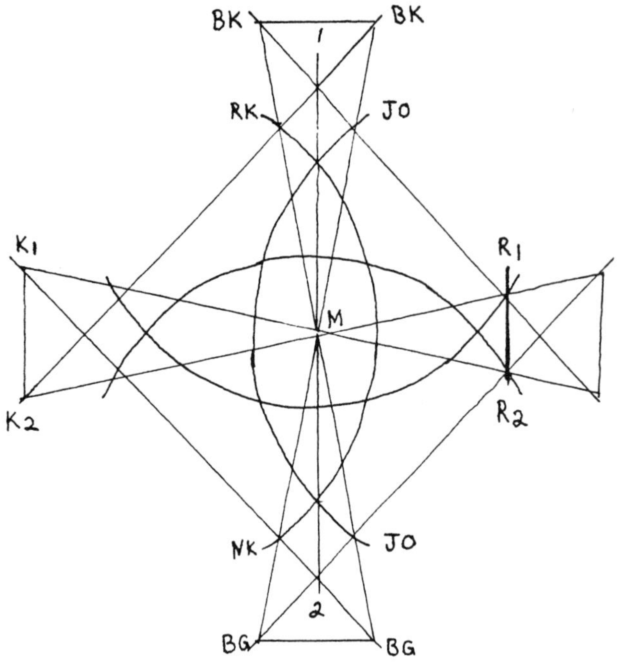

Figure 1: Schematic diagram of a mattang chart.

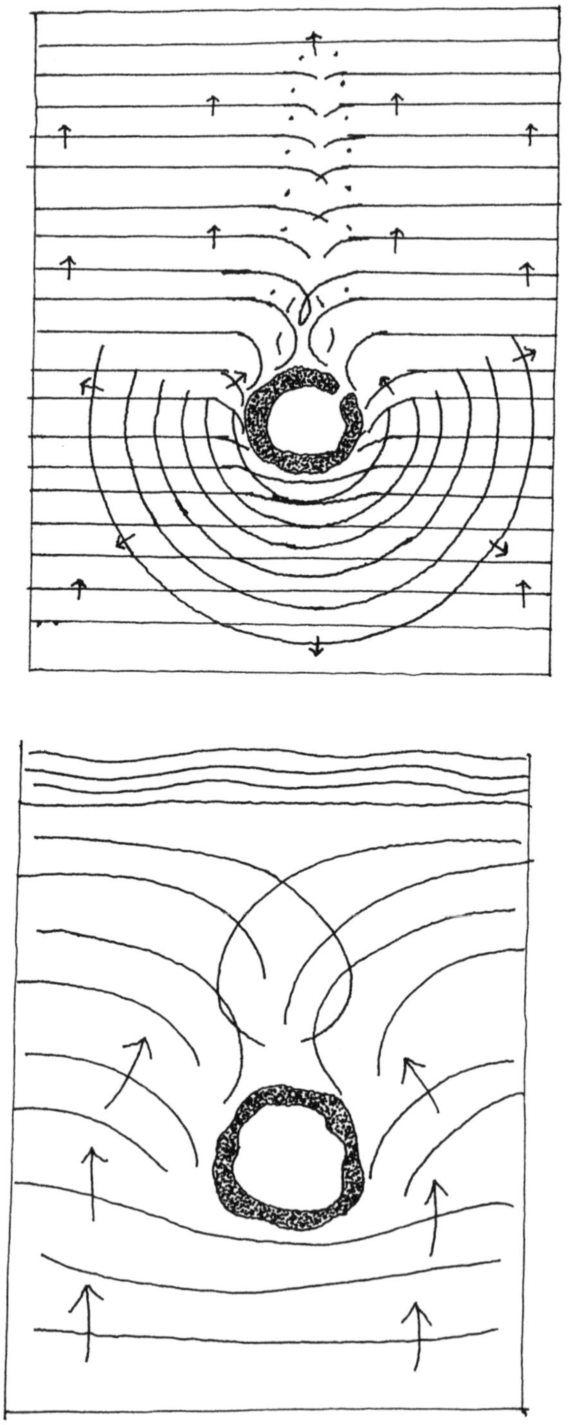

Figure 2: (a, *top*) Wave reflection and refraction around an atoll. (b, *bottom*) Wave refraction around an atoll.

The Marshallese stick charts represent these phenomena: one or several lines of ocean swells approaching an island, their refracted and reflected swells, and the interactions of these with each other. The stick patterns on a chart can be quite intricate when more than one island and more than one swell system is depicted. In such cases a cardinal direction (rear) is indicated, which is the one generated by the north Pacific tradewind system, *rilib* (backbone), that strikes the Marshall Islands from a northeasterly or easterly direction.

There are three kinds of charts: the *mattang* which illustrates the general principles of swell refraction and the intersection of swell lines; the *meddo* (sea) which depicts the relative positions of more than one island, wave data, and sometimes other pertinent hydrographic information; and the *rebbelith* which is like the *meddo* but includes many islands or most of the group, and has less detailed swell information. Sometimes shells are used to designate islands. There are some indications that the *rebbelith* type may have developed after Marshallese sailors learned about European nautical charts.

A partial interpretation of the schematic diagram shown here (Figure 1), taken from an actual mattang, will illustrate most of the wave phenomena used by Marshallese navigators. The center intersection of the chart (M) represents any island; at the right the vertical line R/1–R/2 represents

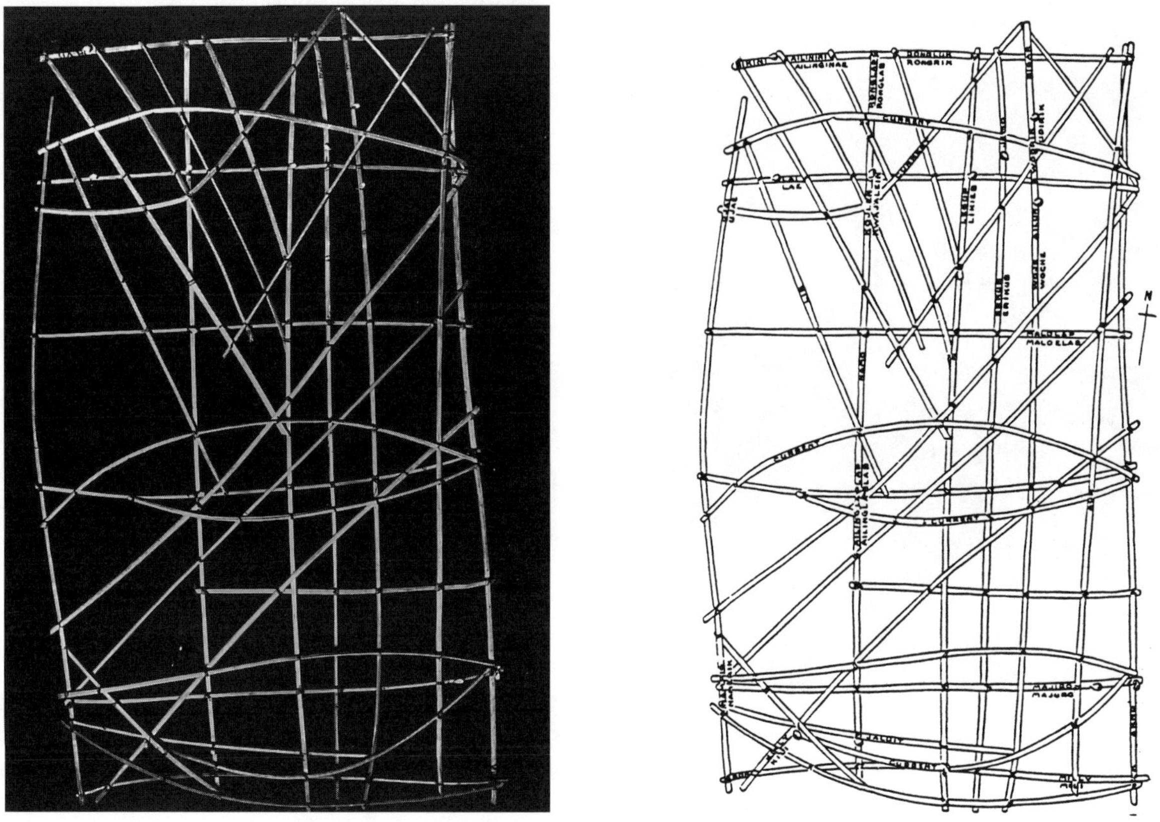

Left. Meddo Chart, collected by Robert Louis Stevenson. From the University of Pennsylvania Museum of Archeology and Anthropology (negative number S8-140153). Used with their permission. *Right*. Key to the Robert Louis Stevenson Chart.

the cardinal direction (rear) and dominant east-to-west swell (*rilib*). As *rilib* encounters island M it is refracted, the upper (northern) refracted arm (RK) is called *rolok*, the lower (southern) refracted arm (NK) is termed *nit in kot*. Opposing *rilib* is a weaker west-to-east swell called *kaelib*, both refracted arms of which are termed *jur in okme* (JO). Two other, usually weak, swell systems, one from the north called *bundokerik* (BK), and its opposed system. called *bundokeing* (BG) are also indicated as they are refracted by island M.

At sea where the *rolok* (RK) or *nit in kot* (NK) arms intersect at a certain angle with *jur in okme* (JO) as well as at other similar intersections of reflected and refracted swells their combined energies cause the water surface to peak up in a characteristic way and briefly break, producing a distinctive kind of white cap. This visible sign is called a *bot* (node). The narrow sector of sea in which *bot* are visible is termed the *okar* (root), because it leads to a tree, that is, to land where trees grow. In the diagram, the line 1–M–2 represents the northern and southern *okar* of the *rilib* and *kaelib* swell systems. In following the *okar*, the navigator notes the change in angle of the intersecting swells, if the angle increases (because the refraction is greater) he is getting closer to land, and the converse. Usually, however, dead reckoning has been good enough for the navigator to know which side of an island he is on.

It is well to keep in mind that a coral atoll is a very low-lying island, and from a distance the trees growing on it are its most conspicuous feature. Even so, from a canoe, atolls are not visible for more than a few miles away. Actually, there are other useful signs that indicate an atoll over the horizon. Among them are flight directions and patterns of certain birds, high accumulations of overhanging clouds, colored reflections from lagoons on the undersides of cumulus clouds, and floating vegetation, and navigator's rely heavily upon them. However, at night and in situations of poor visibility the information derived from swells can be critical. The Marshallese navigator does not use the information from swells only for making direct landfalls. Rather, on a voyage which passes in the vicinities of intervening

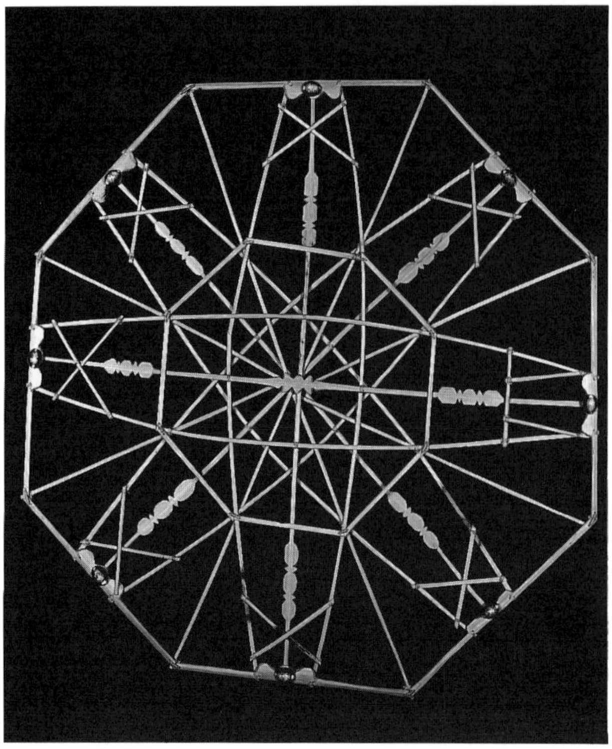

Figure 3: Stick Chart from a Marshall Islands *Mattang* Chart. From the University of Pennsylvania Museum of Archaeology and Anthropology (negative number S8-79948). Used with their permission.

islands he can mark his progress toward his destination by noting the reflected and refracted wave signatures of unseen islands as he passes by them.

WILLIAM H. DAVENPORT

REFERENCES

Davenport, William H. "Marshall Islands Navigational Charts." *Imago Mundi* 20: 28–36, 1960.

See also: Navigation in the Pacific – Navigation in Polynesia

MAPS AND MAPMAKING IN MESOAMERICA When Spanish conquistadores first set foot in Mexico in the early sixteenth century, they found that many of the indigenous peoples they encountered made maps — graphic records of space. Since Mesoamericans had no previous contact with civilizations outside the New World, their mapping traditions developed independently and thus were distinct from those of Europe and Asia. Maps from central Mexico depended upon pictographic writing to convey place-names and geographic information. Frequently territorial maps would include historical narratives, also written with pictographic symbols. With the Spanish conquest, the isolation of Mesoamerican mapping came to an end as it came under the sway of European forms and style. However, native peoples have continued using and making maps that are indebted to the native tradition.

Because preconquest Mesoamerican maps were frequently made on perishable paper or cloth, few such territorial maps survive today. Scholars therefore depend upon eyewitness Spanish accounts of native mapping in the decades after the Conquest as well as "prehispanic-style" maps — those made in the decades after the Spanish Conquest that show little European influence — to reconstruct Mesoamerican mapping in the century or so before the Conquest. Unfortunately, the maps and mapping of earlier Mesoamerican civilizations — the Olmec, the Classic Zapotec, the Classic Maya — are lost to time.

Nahua, or Aztec, maps are best known because Nahuatl speakers dominated central Mexico at the time of the Conquest. Nahuas also had the greatest degree of interaction with colonizing Spaniards. Many prehispanic-style maps from the Nahua heartland in the Valley of Mexico (the site of the present-day Mexico City) survive. These maps had practical uses within Nahua society. For individuals, Nahua mapmakers carefully measured house lots, orchards, and gardens to make large-scale maps that documented their property. For neighborhoods, they drew up maps so that ward leaders could apportion lands and collect tribute. For towns, they created cadastral records that included individual maps of each family's plot. Not all Nahua maps showed such familiar territory: Nahua military spies mapped the layouts of foreign cities to help army commanders in planning their battles of conquest.

In the densely settled Valley of Mexico, scale-model maps were of some importance. Here, precise ways of measuring plots of land using ropes and measuring sticks were in evidence in the years after the Conquest. Two basic units of measurement coexisted in the Valley, the *quahuitl,* measuring about 2.5 meters, and the *cemmatl* (which ranged from 2.5 meters to 1.77 meters). Lines, representing one linear unit, and dots, representing twenty, were used as counters. Fractions of the basic units could be shown with glyphs representing arrows (*cemmitl*), hearts (*cenyollotli*), bones (*omitl*), or hands (*cemmatl*). In some maps, area is also calculated and recorded in square *quahuitl*. Such measured precision in maps and specialization among maps may have been unique to the urbanized and hierarchical Nahua.

The Nahua were, however, very much like other Mesoamericans in making community maps, which were perhaps the most ubiquitous maps in the prehispanic world.

They are sometimes called *lienzos*, after the Spanish word for "linen", since many are painted upon cloth. These maps often showed the boundaries or extent of territory of a native city-state that the Nahua called an *altepetl*. To make community maps, mapmakers used pictographs arranged on a paper or cloth sheet to mimic the distribution of places and geographic features in space. Mapmakers did not carry out specific measurements of the contours of the landscape, but rather used symbols standing for the names of both places and physical landmarks, placing them relative to each other on the sheet. Thus community maps depended upon a mapmaker's knowledge of names — rather than absolute geography — to define the landscape. These community maps aimed not only to map territory, but to record history. The historical narrative would be written in pictures and in pictographs upon the surface of the map and would often tell of a community's travels to reach its present territory. It would show the battles a community fought and alliances it struck in order to cement its rights to lands. The Codex Xolotl of ca. AD 1540 is perhaps the earliest of such map-histories known from the Valley of Mexico; the Mapas de Cuauhtinchan nos 1, 2, 3, and 4 are notable map histories made outside of the Valley near the important prehispanic center of Cholula.

The mapping by the Mixtec, who live in the modern state of Oaxaca, centered on community maps similar to those made by the Nahua, their northern neighbors. Since Mixtec communities tended to be small and independent, never coalescing into the complex hierarchical states of the Nahuas in the Valley, they had little need for the same kind of property, ward, and war maps. The community maps they made in abundance brought together spatial records with historical narratives, again recorded with pictures and pictographs. While Nahua community maps focused on peregrination and conquest, Mixtec ones emphasized genealogy — specifically the genealogy of each community's ruling family. In a characteristic community map, the Lienzo of Zacatepec of ca. AD 1540–1560, the boundary markers of Zacatepec's lands are written down with pictographic place-names to create an inner frame within a rectangular sheet of cloth. Within this boundary map, the important ancestors of Zacatepec's rulers are shown with figures and named with pictographs.

While the Spanish conquest inexorably changed Mesoamerican culture, native communities continued to make traditional maps to document their boundaries vis-à-vis those of adjacent communities and to keep records of community history. While these maps are best known from Nahuatl- and Mixtec-speaking communities, other ethnic groups in Mexico had their versions of community maps. Among them were the Otomí, Zapotec, Totonac, Huastec, Chinantec, Cuicatec, and Mazatec. Today, many towns and villages in Mexico still hold community maps. While these maps may have been redrawn and reinterpreted in the past five centuries, their roots remain in the native traditions of the preconquest period.

Less is known about the history of Maya mapping, even though the Classic Maya of AD 300–900 left us a rich record, written in a partly phonetic script, of their dynastic histories. Existing Maya maps from the post-conquest period show the heavy influence of European forms and convention. These Maya maps use the European, not the Maya, alphabet to write place-names. And other Maya records of territory take the form of written records rather than maps. Such quick conversion of Maya maps to European forms is perfectly understandable: having a writing system of their own, the Maya could easily adapt a foreign one to create territorial records that took a written, rather than a map form. In addition, the few Maya maps that we know of were probably not created to be used among the Maya, but to be presented to Spaniards in courts. Thus the effectiveness of these maps in proving territorial claims was only increased by shedding native conventions and adopting European ones.

BARBARA E. MUNDY

REFERENCES

Burland, C.A. "The Map as a Vehicle of Mexican History." *Imago Mundi* 15: 11–18, 1960.

Castillo F., Victor M. "Unidades nahuas de medida." *Estudios de Cultura Náhuatl* 10: 195–223, 1972.

Glass, John B. "A Survey of Native Middle American Pictorial Manuscripts." In *Handbook of Middle American Indians*, vol.14. Ed. Howard F. Cline. Austin: University of Texas Press, 1975, pp. 3–80.

Glass, John B., with Donald Robertson. "A Census of Native Middle American Pictorial Manuscripts." In *Handbook of Middle American Indians*, vol. 14. Ed. Howard F. Cline. Austin: University of Texas Press, 1975, pp. 81–252.

Glass, John B. "Annotated References." In *Handbook of Middle American Indians*, vol. 15. Ed. Howard F. Cline. Austin: University of Texas Press, 1975, pp. 537–724.

Mundy, Barbara E. "Mesoamerican Maps" In *The History of Cartography*. Chicago: University of Chicago Press, forthcoming.

Oettinger, Marion, Jr. *Lienzos Coloniales: una exposición de pinturas de terrenos communales de México, siglos* XVI-XIX. Mexico: Universidad Nacional Autónoma de México, Instituto de Investigaciones Antropológicas, 1983.

Robertson, Donald. *Mexican Manuscript Painting of the Early Colonial Period: The Metropolitan Schools*. New Haven: Yale University Press, 1959.

Smith, Mary Elizabeth. *Picture Writing from Ancient Southern Mexico: Mixtec Place Signs and Maps*. Norman: University of Oklahoma Press, 1973.

MAPS AND MAPMAKING IN NATIVE NORTH AMERICA At or soon after their first encounters, Europeans reported that native North Americans made maps. Most of the early reports were of maps made for Europeans to communicate geographical information, as for example a map of the lower Colorado River made by a Halchidhoma Indian in 1540. Ever since, Europeans and non-native North Americans have continued to solicit and receive geographical intelligence in this way, whether they were explorers, traders, soldiers, settlers, missionaries, government agents, field scientists, or cultural anthropologists. It was, and until very recently remained, an important aspect of frontier information exchange. Natives communicating in this way included men of every social status, less frequently women, and occasionally even children. It occurred in every culture area and in most, if not all, tribes. Because of the publicity given to nineteenth-century Arctic exploration, it has sometimes been assumed that the Inuit were particularly skilled map-makers. There is, however, no evidence that the maps they made for Europeans were either better or significantly different from those made by Indians.

In the encounter process, most maps were made with pencil on paper. Occasionally, however, more traditional materials were used e.g. bark, textile, skin, hard animal tissue, three-dimensional models, and even rock. The distinctive characteristics of these encounter maps were, however, neither the media nor the mapmaking processes. Although Europeans were slow to recognize it, the natives' maps differed fundamentally from their own in three important ways. The geometry was topological, a product of cultural tradition, individual experience, the shape and size of the medium, and the purpose for which the map was made. Unlike European maps, it was not a consequence of plotting locations on a graticule selected to conserve particular properties at the expense of others. Neither scale, direction, or shape were conserved, except for very small areas. The second important difference was in the categorization and magnitudinal ordering of the phenomena represented. Whereas most European maps were general, serving a range of functions and a diversity of users, each native map was made for a specific purpose and audience. Content and emphasis were determined by these and only the essentials were represented. Hence, large physical features occurring within the area mapped were frequently either omitted, diminutive, or highly generalized, whereas small but contextually significant features were included and perhaps exaggerated. Thus, a large, complexly-shaped lake might be represented by a simple circle, ornamented perhaps by a detailed representation of one culturally or contextually significant peninsula. Similarly, a long, complex, maritime coastline might be represented by an essentially straight line. The third important difference was in the natives' use of pictographs as the approximate equivalent of the Europeans' combination of words, toponyms, and conventional map symbols. Pictographs were constructed according to culturally established principles to communicate complex mixes of information about size, importance, number, relationships, time, distance, direction, and events, as well as material and organic phenomena. Richer and far more flexible than the symbols on European maps, their information content was rarely intelligible to the aliens. Indeed, when, as they so frequently did, European's copied natives' maps, they often omitted or generalized the pictographic content. Regrettably, most extant examples of encounter maps are contemporary transcripts or, worse still, printed engravings: a small, debased sample of the many that were made, now scattered in archives, museums, libraries, government departments, and private collections in Europe as well as North America. Nevertheless, in their time, this type of native map served the aliens well; helping them to open up their *terrae incognitae*, locate resources, plan military campaigns, etc. Though usually unacknowledged, they were frequently incorporated into the first generation of maps made by the aliens. Sometimes, misinterpretation and insensitive incorporation resulted in gross errors on maps made by Europeans, as with the Great River of the West, the Long River, and the gross westerly displacement of the mouth of the Mississippi River on printed eighteenth-century maps.

For some parts of the non-Eurasian world there is uncertainty as to whether pre-contact natives did or did not make maps. Each culture had a strong oral tradition, an important part of which was concerned with landscape, place, and spatial relations. Place — and feature — names were the "survey pegs" of spatial memory. This was almost certainly so among pre-Columbian North Americans, but there is considerable direct and indirect evidence that they also expressed their oral maps cartographically. For example, in 1540, Francisco Vásquez de Coronado discovered a painted skin in an abandoned Zuni pueblo. The pueblo was the first of the Seven Cities of Cibola ever to be seen by Europeans, and the painting represented their relative sizes and a route, probably the one linking them. Although difficult to date, much, if not most, North American rock art pre-dates contact with Europeans. A very small proportion of these works have map-like appearances, at least when viewed from the European cultural perspective. Lacking texts, however, and in the light of the characteristics of native maps as now known, proof is difficult. Establishing similitude between a topological representation and its referent is much more difficult than for a projectively constructed map. Some celestial paintings on the roofs of rock overhangs in the Southwest undoubtedly represent exceptional astronomical juxtapositions known to have occurred at precisely established dates in pre-Columbian times. Others, like Map Rock, Idaho (a name given by the earliest white settlers in the lower Snake valley, because the engraving on it looked to them like a map) do appear to represent topologically and pictographically the pattern of geographical features in the regions in which they are sited. These, however, are exceptional cases and irrefutable evidence for terrestrial maps in pre-contact rock art has still to be established, though the case for plans of small areas and features is much stronger.

Ironically, the strongest evidence for mapmaking in pre-contact times is afforded by very early post-contact accounts by Europeans of the indigenous use of maps, usage of a kind that could not have been derived from European practices. It is debatable whether at the time of first contact any of the native languages contained nouns equivalent to map, though at later stages many of them certainly did. Nevertheless, some Indians certainly possessed an ability to read network patterns as maps. Montagnais and Naskapi divination by scapulimancy involved inducing random patterns of cracks by heat or percussion on mammalian bones, and these were sometimes interpreted to be maps of actual river systems. Likewise, the women of these tribes sometimes read as trail networks patterns made by biting folded pieces of birchbark, especially when the patterns had emerged in mistake for something else intended. Maps inscribed or painted on the inner surface of birchbark (or on growing trees from which the bark had been conspicuously stripped) were frequently positioned at conspicuous or strategic sites in the Northeastern forests to convey to friends — and sometimes enemies — spatially-organized information about recent events or planned activities. There are accounts of maps being made by older men to instruct young braves about to undertake long journeys for the first time. Map modeling with whatever materials were at hand was sometimes a cooperative activity, at its best leading to a geographical consensus prior to military plannings and strategic briefings. Cosmological maps were common and usually combined geographical with mythical elements, the former most frequently near the center. Where geographical features were endowed with religious meaning, they might be topologically displaced. For example, on Lakota maps of the Black Hills, the Devil's Tower, located sixty miles to the northwest, was placed within the so-called Race Track that circumscribed them. Southern Ojibway birchbark midé-migration scrolls preserve the tradition as to how the tribe received its religion: from the east via a great waterway. At the left end of some of the scrolls, the waterways of part of Minnesota are readily interpretable. A little to the right, Lake Superior is highly stylized, but further to the right (east) it becomes increasingly difficult to relate the linework to the Great Lakes–St. Lawrence River waterway.

Most native maps represented relatively small areas and/or were linear, but a few embraced areas as large as one third of a million square miles. The information upon which a map of a very large area was based could not have been derived from the direct experiences of an individual or even the cumulative experience of one group. Most of the content must have been based on information received by gesture and speech, perhaps over several generations.

Potentially, at least, maps made by native North Americans are significant in four broad contexts: legal, historical, anthropological, and scientific. Though lawyers have been slow to recognize it, an understanding of the fundamental differences between native and Euro-American maps helps, in some cases, to explain disputes between the two cultures concerning treaty and land sale agreements. Likewise, but with a few notable exceptions, archaeologists and historians have been slow to recognize the significance of maps made by natives in such contexts as exploration, mapping, locating resources, place — and feature — naming, etc. Similarly, cultural anthropologists have been slow to explore the relations between maps and other graphical forms or to use them as evidence in establishing world views, reconstructing lifestyles, etc. Scientists, including linguists, have not even begun to recognize their potential significance in such contexts as human cognition, genetic epistemology, information exchange, language, categorization of phenomena, etc. These failures stem, in part, from the absence of a cartobibliography to the primary materials. Furthermore, scientists remain unaware of the now considerable body of secondary publications that, for the most part, is confined to the literature of the humanities. Most serious in this respect is the absence of a global survey of maps and mapmaking in traditional cultures worldwide. Such a survey is vital in order to resolve some important questions. Did all peoples make maps? If not, was the spatial organization of information part of an older oral tradition? If the latter, why did it find expression in mapmaking at different stages in cultural development? Hopefully, it will become possible to address these and similar questions once impending scholarly compendia are published.

G. MALCOLM LEWIS

REFERENCES

De Vorsey, Jr., Louis. "Silent Witnesses: Native American Maps." *Georgia Review* 46(4): 709–726, 1992.

Lewis, G. Malcolm. "Indian Maps: Their Place in the History of Plains Cartography." In *Mapping the North American Plains*.

Ed. Frederick C. Luebke et al. Norman: University of Oklahoma Press, 1987, pp. 63–80.

Lewis, G. Malcolm. "Metrics, Geometries, Signs, and Language: Sources of Cartographic Miscommunication between Native and Euro-American Cultures in North America." *Cartographica* 30(1): 98–106, 1993.

Lews, G. Malcolm ed., *Cartographic Encounters*. Chicago: University of Chicago Press, forthcoming.

Lewis, G. Malcolm and David Woodward eds., "Cartography in the Traditional African, American, Arctic, Australian, and Pacific Societies", *The History of Cartography*, Vol. 2, Book 3. Chicago: University of Chicago Press, forthcoming.

Ruggles, Richard I. *A Country So Interesting: The Hudson's Bay Company and Two Centuries of Mapping, 1670–1870*. Montreal and Kingston: McGill-Queen's University Press, 1991.

Rundstrom, Robert A. "A Cultural Interpretation of Inuit Map Accuracy." *Geographical Review* 80(2): 155–168, 1990.

Vollmar, Rainer. *Indianische Karten Nordamerikas*. Berlin: Dietrich Reimer Verlag, 1981.

Waselkov, Gregory A. "Indian Maps in the Colonial Southeast." In *Powhatan's Mantle: Indians in the Colonial Southwest*. Ed. Peter H. Wood et al. Lincoln: University of Nebraska Press, 1989, pp. 292–343.

MAPS AND MAPMAKING IN SOUTHEAST ASIA Indigenous maps from Southeast Asia were drawn for a multitude of secular and religious purposes and exhibit remarkable diversity in respect to medium, style, and content. Even within individual countries and cultures the surviving cartographic corpus is likely to be quite varied, though Burma and Vietnam were moving toward a national style in recent centuries. Broadly speaking, maps fall within four main traditions:

- tribal, for groups that never came under the cultural dominance of the great traditions of Hinduism, Buddhism, or Islam;
- Hindu–Buddhist, for cultures primarily shaped by influences emanating from India;
- Sinic, for the Vietnamese; and
- Malay, for Islamicized peoples in what are now Indonesia and Malaysia.

The number of surviving maps is great for Burma and Vietnam, and few for Thailand (largely because of the Siamese tradition of purging documents that were no longer of current utility), Cambodia, Indonesia, and Malaysia. For Laos and the Philippines, indigenous maps are not known to exist. The corpus of known indigenous maps, however, is not a reliable guide to past cartographic output, since the accidents of preservation and documentation depend in large measure on the attitudes of colonial powers and individual scholars with respect to indigenous culture. It is likely that many maps still await discovery, especially in Buddhist temples and monasteries on the Southeast Asian mainland.

Tribal maps are documented mainly for the Sakai, an isolated Negrito people of West Malaysia, the Bataks of Sumatra, and various Bornean groups collectively known as Dayaks. For all of these the traditional medium was bamboo, into which designs were etched to create the map image. For the Sakai the mapped images served as magical charms to help insure success in hunting, fishing, and other activities in the places depicted or to ward off diseases and natural hazards associated with those places. Batak maps were mainly cosmographic or divinatory. Dayak maps, the most recent of which are exceedingly vivid designs on paper (supplied and preserved by missionaries), relate mainly to mortuary cults and relate to routes taken by the dead to reach the netherworld or the upper world and to the activities that take place therein. Dayaks also had a system of reading pigs' livers as if they were maps to discern auguries of good or bad outcomes for contemplated actions in particular places.There are no reliable guides to when mapping commenced among Southeast Asian tribal groups.

Hindu–Buddhist maps were drawn on palm leaf, indigenous bark paper (often assembled accordian-style in long folding manuscripts), cloth, and other media. In dealing with maps in this tradition it is necessary to distinguish between essentially religious and secular productions. The former are largely cosmographic, most depicting all or part of the Buddhist cosmos. Maps of the terrestrial plane, centered on Mount Meru, the *axis mundi*, generally utilized a planimetric perspective. Those that also showed the manifold heavens and hells normally depicted the universe as if seen mainly in frontal elevation, but sometimes rotated the horizontal terrestrial plane ninety degrees so as to maintain the traditional planimetric view for that region. Other Meru-centered maps show that mountain and its seven rings of surrounding mountains in vertical profile. One such depiction, more than ten meters in height, is carved into a cliff in central Myanmar (Burma). A sumptuously illuminated cosmographic narrative, quite popular in Thailand, is the *Trai Phum* (Story of Three Worlds, *Traībhūmikathā* in Sanskrit). This work, compiled in ca. 1345, is a variant on the Jataka tales, relating to the past lives and wanderings of the Buddha. It has gone through a number of recensions and exists also in modified form in Myanmar and Cambodia. The longest manuscript, dating from B.E. 2319 (AD 1776), provides a continuous picture on 272 folding panels with an overall length of 32 meters. Six of these panels depict Asia from the Arabian Sea coast to Korea. No surviving manuscript predates the sixteenth century. Hindu cosmographies were probably also painted in Southeast Asia, but apart from one attributed to Cambodia, none is known to survive on the mainland. Complex cosmographic works, however, continue to be made in various media in Bali.

The oldest known geographic map in the Hindu–Buddhist tradition, dating from the late seventeenth or early eighteenth century, is a long folded-panel work showing ecclesiastical and other property holdings in what is now southern Thailand. Thereafter, there is a long hiatus in Siamese secular cartography, the next known work being a military map from the late eighteenth or early nineteenth century. Known Burmese secular maps do not predate the latter half of the eighteenth century. But one may safely assume that the mapmaking was by then a well established art. Numerous maps, many covering rather large areas, were drawn for Francis Hamilton during his sojourn in Burma in 1795 and attest to the geographic sophistication of educated Burmese at that time. During the nineteenth century, if not earlier, the Burmese state engaged official surveyors and cartographers. Maps on European paper and cloth became increasingly common with the passage of time. Burmese maps were drawn largely for military and political intelligence purposes, sometimes to plan or to document specific campaigns. Others, showing property holdings and types of land use, were used for purposes of taxation. Many were used as aids for laying out new cities, monastic complexes, irrigation systems, and other public works. A common feature of Burmese maps was the use of a regularly ruled grid, very likely to transfer a sketch from one scale to another, though maps were never drawn to a uniform scale. Many of the maps from what is now Myanmar were actually drawn by ethnic Shans (a Thai people) and other ethnic groups within the Burmese political orbit. From insular Southeast Asia, only one presumably geographic map in an essentially Hindu tradition is known. Drawn on a batik shawl, this ornate and enigmatically patterned work probably dates from the early nineteeenth century. It has no text, but is believed to relate to the former princely states of Surakarta and Yogyakarta, which, though ruled by Muslim sultans, maintained a largely Hindu court style.

Vietnamese cartography was mainly secular and, as in Burma, drawn for the military, administrative, and other political needs of the government. (Cosmographic maps did exist, but these were mainly in the Hindu–Buddhist tradition previously described.) The oldest textual references indicate that secular maps existed as early as the late eleventh century, though mapping appears not to have become a serious concern until the government was reorganized along Sinic bureaucratic lines some four centuries later. Thereafter, the map style was a regional variant of a general Sinic tradition, though changes were evident from one period to another. Many of the maps were bound, after the Chinese fashion, into atlases, which, with their abundant text, served also as gazetteers. The oldest atlas, made in the 1270s, related to a royal inspection tour of the then relatively circumscribed Vietnamese state, but it does not survive and its form is not known. In addition to atlases and regional maps, the Vietnamese drew route maps, maps of river systems and coastal maps to facilitate navigation; city plans, some of which were remarkably detailed, maps of individual forts and other edifices, and maps documenting tribute missions to the Chinese court in Beijing.

Surviving maps from the Malay cultural realm, though few in number, are quite diverse. The oldest, dating from the late sixteenth century, is a large cloth map covering the western third of Java. Most of the map space is given over to the depiction of the small chiefdom of Timbanganten, whose prince commissioned the work. The symbolization on this map is particularly distinctive and has no close analog in any other known work. All the remaining maps show varying degrees of European influence. A map of the Bornean Sultanate of Pontianak and another of part of east central Java apparently related to administration and/or taxation. Others related to navigation and trade. One shows ports and commercial products of the Malay Peninsula. The others, covering most of Southeast Asia, are detailed hydrographic charts, copied on to cowhide, from an early nineteenth century Dutch original, but with adaptations indicating independent depth soundings and new coastal observations. These maps were drawn and used by wide-ranging Bugi pirates and contain abundant text written in the Bugi script, though Roman numerals were used to indicate ocean depths. Apart from the surviving nautical maps, there are textual and other grounds for supposing that Indonesians made use of maps in long distance navigation prior to the advent of Europeans in the region.

JOSEPH E. SCHWARTZBERG

REFERENCES

Hamilton, Francis (formerly Francis Buchanan). "An Account of a Map of the Countries Subject to the King of Ava, Drawn by a Slave of the King's Eldest Son." *Edinburgh Philosophical Journal* 2: 262–71, 1820.

Kennedy, Victor. "An Indigenous Early Nineteenth Century Map of Central and Northeast Thailand." In *Memoriam Phya Anuman Rajadhon: Contributions in Memory of the Late President of the Siam Society*. Ed. Tej Bunnag and Michael Smithies. Bangkok: The Siam Society, 1970, pp. 315–48 and eleven appended map plates.

Phillimore, Reginald Henry. "An Early Map of the Malay Peninsula." *Imago Mundi 13:* 175–79, 1956.

Reynolds, Frank E. and Mani B. Reynolds. *Three Worlds According to King Ruang: A Thai Buddhist Cosmology*. Berkeley, California: Asian Humanities Press, 1982.

Schärer, Hans. *Ngaju Religion: The Concept of God Among a South Borneo People.* Trans. Rodney Needham. The Hague: Martinus Nijhoff, 1963.

Schwartzberg, Joseph E. "Cartography in Southeast Asia." In *Cartography in the Traditional East and Southeast Asian Societies*, Vol. 2, Book 2 of *The History of Cartography,* Ed. J.B. Harley and David Woodward. Chicago: University of Chicago Press, 1994, pp. 689–842.

U Maung Maung Tin and Thomas Owen Morris. "Mindon Min's Development Plan for the Mandalay Area." *Journal of the Burma Research Society* 49(1): 29–34, plus two maps, 1966.

Whitmore, John K. "Cartography in Vietnam." In *Cartography in the Traditional East and Southeast Asian Societies*, Vol. 2, Book 2 of *The History of Cartography*. Eds. J.B. Harley and David Woodward. Chicago: University of Chicago Press, 1994, pp. 478–508.

MAPS AND MAPMAKING IN TIBET For purposes of this article Greater Tibet includes not only the present Autonomous Region of Tibet, within China, but also the Chinese provinces of Qinghai and Sikang, the Indian regions of Ladakh and Sikkim, much of Nepal, and Bhutan, which have all been greatly influenced by Tibetan Buddhism and other aspects of Tibetan culture. Despite its small population, this region has given rise to a remarkably rich and varied cartographic tradition. Moreover, that anciently rooted tradition still survives, despite attempts by the West, China, and India to impose political and cultural hegemony over the lands of its development.

Given the pervasive role of religion in Tibetan culture, it is hardly surprising that maps serving a variety of religious functions form the greater part of the region's cartographic corpus. These are commonly painted on cloth *thaṇkas*, which frequently occupy a focal position in family and monastic altars. The most abstract of Tibetan religious maps are *maṇḍalas*, which are regarded as representations of the cosmos and serve as objects of meditation for clergy and laity alike. These assume many forms, in both two and three dimensions, and utilize a wide range of media. Elaborate *maṇḍalas* fashioned from colored sand or carved in yak butter may be created over a period of weeks, but are kept for even less time than it took to construct them, thereby demonstrating the transience of all earthly existence. Many cosmographic maps were divinatory. Others were didactic, for example, paintings of the *bhāvacakra* (wheel of life) depicting the various realms of existence through which souls may transmigrate on the path to *nirvāna* (the ultimate aim of existence in which the soul is liberated from the painful round of rebirth). Other forms of didactic painting are essentially hagiographic, showing various incarnations of the Buddha, *bodhisattvas*, or revered lamas surrounded by collages of sacred landscapes depicting particular places associated with their existence, both on earth and in non-terrestrial portions of the Buddhist cosmos. Though largely mythic, such places are often shown with a considerable sense of verisimilitude. Maps showing the cosmos as a whole or major portions of it are numerous. Some of these are large mural paintings. A particularly vivid set of fresco murals in Bhutan shows the evolution of the cosmos as a succession of disaggregations of terrestrial and extra-terrestrial elements in space-time as described in various Indian and Tibetan Buddhist texts dating back to the fifth century. In these and many other cosmographic maps Mount Meru, the *axis mundi*, and four principal surrounding continents in the terrestrial plane figure prominently. Maps of other continents are also painted, two of particular note being Sukhāvatī (the Western Paradise) and the mountain-girt utopian realm of Śambhala (whence James Hilton's "Shangri-la"). Bards used to wander about Tibet illustrating stories of the way to Śambhala with map scrolls on which routes began in recognizable places, but led ultimately to mythic space. To devout Tibetans, however, the distinction between the real and the mythic would seem a false dichotomy.

Geographically identifiable locales do figure prominently in Tibetan religious maps. Many show individual religious edifices and centers such as the Potala and Tashilunpo, the residences of the Dalai and Panchen Lamas respectively; Lhasa and Shigatse, in which those two residences are respectively located; or Samye, Tibet's oldest monastic complex. Others show sacred landscapes in which groups of such centers, especially those proximate to Lhasa, are clustered. Some maps focused on Mount Kailas, in western Tibet, viewed as the earthly manifestation of Mount Meru, around which a complex of sacred pilgrimage places developed. Large sacred maps were displayed along the walls of the monasteries where they were held during certain holidays, when pilgrims would flock to major religious centers. One such painting, displayed annually in the Nepali town of Patan, has a length of more than twenty-five meters and provides a rich panoramic view of the Vale of Kathmandu. Maps were made not only to serve the needs of pilgrims, but were also commissioned to commemorate pilgrimages completed. A particularly beautiful example, dedicated in AD 1802, depicts the patron and his entourage at many of the major pilgrimage places in central Nepal.

When Tibetans began to make maps is not known, but the roots of Tibetan cartography run deep. One of the most remarkable maps, depicting the anciently known world, appears in a modern recension of a dictionary of the Zang Zung language (probably Indo-European), formerly spoken in western Tibet. Though the places shown, reaching as far west as Egypt and Greece, were compressed into the form of a rectangular *maṇḍala,* thereby losing much of their geographic logic, most could be related to the period of Alexander the Great (late fourth century BC). The map itself focuses on Parsogard, the Achaemenid capital from 550 to 522 BC. There are grounds to believe that some version of this map was transmitted from antiquity to the present, despite the fact that no pre-modern version is known to survive. Another simpler, but equally cryptic, surviving world map was brought to Japan in AD 891.

Apart from the two world maps just noted, one map of Nepali provenance (but stylistically akin to Tibetan maps) covers a very large region of Central and Southwest Asia, with western limits at Baghdad and the Russian city of Saratov, on the Volga River. It has been argued that this map was commissioned ca. 1860 in relation to a contemplated, but never consummated, grand alliance to drive the British from Asia. This richly detailed map, despite its late date, includes numerous features that are essentially mythic.

A remarkable series of maps is that of the so-called Wise Collection in London's India Office Library. Wise's identity has not been established; but there are grounds to believe that the maps were prepared by Buddhist recruits (later dubbed *pundits*) from India's Himalayan territories who traveled over much of Tibet in disguise to gather intelligence. British training notwithstanding, the style of these maps is distinctly Tibetan.

With the aforementioned exceptions, regional maps from Tibet seldom cover more than a few thousand square kilometers. None is known to predate the late eighteenth century. As a group, the maps in question are remarkably detailed and utilize a well developed set of symbolic and color conventions. They typically combine diverse perspectives: planimetric, high oblique, and frontal, depending on the features shown. Important buildings are sometimes shown as if seen from several sides simultaneously. There is no consistent orientation. Regrettably, only a few Tibetan maps have yet been studied as such by scholars with the needed cultural sensitivity and linguistic skills.

JOSEPH E. SCHWARTZBERG

REFERENCES

Banerjee, N.R. and O.P. Sharma. "A Note on a Painted Map of the Kathmandu Valley at the National Museum, New Delhi." *Marg, A Magazine of the Arts* 38 (3): 77–80, n.d. [ca.1986].

Gole, Susan. "A Nepali Map of Central Asia." *South Asian Studies* 8: 81–89, 1992.

Gumilev, L.N. and B.I. Kuznetsov. "Two Traditions of Ancient Tibetan Cartography." *Soviet Geography: Review and Translation* 11(7): 565–79, 1970.

Huber, Toni. "A Tibetan Map of lHo-kha in the South-Eastern Himalayan Borderlands of Tibet." *Imago Mundi* 44: 8–23, 1992.

Schwartzberg, Joseph E. "Maps of Greater Tibet." In *Cartography in the Traditional East and Southeast Asian Societies*, Vol. 2, Book 2 of *The History of Cartography*, Ed. J.B. Harley and David Woodward. Chicago: University of Chicago Press, 1994, pp. 607–681.

Slusser, Mary Shepherd. "The Cultural Aspects of Newar Painting." In *Heritage of the Kathmandu Valley: Proceedings of a Conference in Lübeck, June 1985*. Ed. Niels Gutschow and Alex Michaels. Sankt Augustin, Germany: Hans Richarz Publikations-Service, 1987, pp. 3–27.

MAPS AND MAPMAKING IN VIETNAM The tradition of mapping began in Vietnam just over half a millennium ago. As the state of Dai Viet, led by the young king Le Thanh-tong (1460–1497), adopted the administrative model of the Ming dynasty in China and became more centralized, mapping served as one of the procedures to that end. The new bureaucratic structure placed officials in the provinces and districts and ordered that these officials gather and relay information to the capital of Thang-long (now Hanoi) on the land and people within their jurisdictions. Maps were part of the required information, and they led eventually to an atlas covering the entire country. This atlas and its maps in the Chinese mode set the pattern of administrative cartography well into the nineteenth century.

The same reign also saw the extension of Vietnamese territory to the south as Thanh-tong crushed the forces of Champa in 1471. A map showing the stages of the expedition became the model for military and commercial cartography in the following three centuries. As these centuries were a time of military and commercial expansion, particularly of the *Nam Thien* (Southern Push) further down the east coast of mainland Southeast Asia through Champa and into Cambodia, this style of map came to reflect the dynamics of the Vietnamese population.

International contacts in the late eighteenth and early nineteenth centuries led to a new style of administrative mapping which combined a more realistic Western (particularly French) style with stronger Chinese elements. Yet, by the end of the nineteenth century, as French colonialism controlled the Vietnamese state, there appeared a shift towards a more complete sinic mode of illustrating maps.

No technical study of Vietnamese maps based on the original documents exists. Looking at reproductions, mainly in black and white, yields the following observations. The maps available to us are almost entirely hand drawn in black ink, some with added color to emphasize their features. Only in the mid-nineteenth century did printing begin to play a major role in cartography here.

The earliest extant map, most likely from the sixteenth century, is quite simple in form and style. Lines were drawn to show the rivers and the land separating them. Nothing indicates water in the streams, and the standard Chinese three-ridge pattern for mountains served to suggest the highland regions above the plains. Names (in Chinese characters) marked location with no other symbol employed.

This changed in the seventeenth century as the earliest existing major corpus of indigenous maps appeared. They had become more artistic. Water was drawn in the streams and off the coast (river currents and roiling sea waves). Mountains retained the same style, but more accurately designated Vietnam's uplands. In addition, palaces, temples, and walls appeared (in frontal elevation), while jurisdictional locations were marked by written characters in rectangular boxes with little sense of hierarchy among them. Other characters showed the locations of natural features and additional human constructions.

In the eighteenth century, Vietnamese cartography took on in places a more artistic and sinic mode. This change appeared particularly in the style of portraying the mountains. In place of the simpler three-ridge style, there came to be a more naturalistic landscape mode which well fit the vertical karst topography of northern Vietnam (and southern China). In addition, waves lapped on the shore, and temples and walls received a more sinic treatment. The maps were drawn in black ink with colors (red, grey, and blue, for example) used to highlight roads, buildings, and mountains.

The nineteenth century saw more sophisticated international influences appear in Vietnamese mapping. Southern political forces had contact with French military figures and adopted the Vauban style of fortification. Simultaneously, the new Nguyen dynasty chose to hew more closely to the contemporary Qing dynasty of China. Though elements of the old Le style would continue, more realistic elements appeared on Vietnamese maps within the context of Chinese forms. A regional division of the country led the north to maintain the old style as French forms appeared in the south. The central region around the capital of Hue gradually forged a national style which, by 1840, showed the country of Dai Nam and its hydrographic complexities in a European perspective.

While becoming more realistic in outline, the artistic style shifted from the earlier sketchiness to a more specifically Chinese mode. This change appeared particularly in the mountains with their vertical and naturalistic mode, continuing to match the karst topography of the north. The new international style developed through the nineteenth century until the French conquest in the 1880s. Thereafter, in the early years of the twentieth century, the reaction against French rule among Vietnamese literati led to the greater emphasis on the Chinese style over the realistic mode.

The first centuries of Vietnamese mapping (fifteenth to eighteenth) produced two different genres, one administrative, the other military/commercial. The first, the atlas, originated in Le Thanh-tong's governmental transformation of the 1460s. In it, each of the thirteen provinces had a map showing the location of its prefectures and districts and a list of the types and numbers of villages contained therein. In addition, the atlas included a map of the country, noting the location of each province, and a map of the capital Thang-long. Despite originating in the fifteenth century, the earliest extant atlas, the *Hong Duc Ban Do* (Maps of the Hongduc Era), comes from the mid-late seventeenth century. As noted above, a single map of the country, showing its administrative units, does seem to be from the sixteenth century. In the atlases, natural features, particularly rivers, and some human construction like city walls and temples were also noted. The maps generally are seen from the sea (the east) and hence have a western orientation. Since the Le dynasty (1428–1787) rose from the then southern provinces of Thanh-hoa and Nghe-an, these two maps came first after those of the country and the capital.

The other genre involved itineraries of different kinds, each a series of maps showing the way (usually from the capital region around present-day Hanoi) out to a distant location. The genre seems to have originated in Le Thanh-tong's major campaign south against Champa in 1470–1471. Again, the earliest extant example comes from the seventeenth century, the *Thien Nam Tu Chi Lo Do Thu* (Book of the Major Routes of Thien Nam). At that time, the southern route was joined to three shorter routes to China: northeast, northwest, and north (the standard route to Beijing). These maps, and particularly those of the southern route, provide much more detail of daily life than does the atlas. They illustrated features of the land, the rivers, and the sea as well as elements of Vietnamese common and commercial life (markets, inns, temples, military installations, etc.). Another seventeenth century itinerary, the *Binh Nam Do* (Map of the Conquest of the South), comes from the southern regime of the Nguyen and takes us through their territory of central and southern Vietnam to Cambodia. It shows the path of the Southern Expansion (*Nam Thien*) of the Vietnamese people then taking place.

The eighteenth century produced new, more artistic versions of the seventeenth century atlas and itineraries. In addition, a new itinerary appeared in various forms — that of the Vietnamese embassy north to the Qing capital of Beijing in northern China. Once across the Chinese border, the delegation traveled mainly by river and canal, and these sets of maps were drawn from that perspective, showing the mountains and villages stretching away from both sides of the route. Finally, at the end of the century, there appeared the only known map of the brief Tay-son dynasty (1788–1802), a single page showing Siam (Thailand) and the routes through it from north to south. This map, like all other pre-1800 ones, had very little sense of the river systems to the south, particularly the Mekong Delta and the Great Lake of Cambodia.

Three major developments led to cartographic change in the nineteenth century: the international context, the initial regional control of the country (north, center, and south), and the borrowing of the Chinese geography form. The combination of these three brought the gradual emergence of a new form of mapping. This single form, the geography with maps accompanying the text, replaced the Le atlas as an administrative tool. Curiously, the genre of the itinerary disappeared, continuing only in the reproduction of earlier texts.

Just as the northern and southern regional autonomy was finally integrated into the central government during the 1830s, so their styles merged with that of the center and its usage of European and Chinese elements, as seen in the *Dai Nam Toan Do* (Complete Map of Dai Nam). The map of the entire country, now stretching from China to Cambodia, in particular took on a more Western form. The coast and the southern rivers, even the Great Lake of Cambodia, appear much more familiar, and the provincial maps are more realistic in our eyes, showing rivers in a dendritic manner for example.

Blending the new maps with the geography form of the Chinese *yi-tong-zhi* (Vietnamese *nhat-thong-chi*), Nguyen cartography reached its peak first in the *Dai Nam Nhat Thong Du Do* (Maps of the Unity of Dai Nam) of 1861 and finally in the imperial geography, the *Dai Nam Nhat Thong Chi* (Record of the Unity of Dai Nam). The latter, compiled and printed between 1865 and 1882, is a text describing the country of Dai Nam, its capital of Hue, and the 29 provinces and presenting maps of each of them. Even though the French had already seized the six provinces of the south, the Tu-duc Emperor kept this section in the geography together with those of the center and north as a symbolic representation of the whole.

Mapping of the nineteenth century was thus almost entirely involved with administrative matters, like the old Le atlas. The only itineraries were copies of the seventeenth and eighteenth century maps. The country map in the administrative collections did, however, provide a broader international setting than had occurred under the Le. This was done in a form similar to the European mapping of the time.

The major new form of map in the nineteenth century was part of the *dia-bo,* the land registers. Recording the land plots of each village, these documents show them in outline and provide their dimensions and type. The government produced these maps in a number of survey (and resurvey) campaigns through the first half of the century. Another new form of map was the symbolic diagram of the imperial tombs. The resulting drawings emphasized the symbolic relationships of mountains and water and show the strongly Chinese style adopted by the Nguyen dynasty.

Vietnamese mapping was mainly the result of four hundred years of administrative effort. Beginning with Le Thanh-tong's reforms in the 1460s, the need to know the bureaucratic jurisdictions and their villages led to the Le atlases and the Nguyen geographies. The fifteenth century military campaign began the Vietnamese itinerary which to all appearances became commercial. The Vietnamese cartographic tradition was almost entirely internal, not external. Until the 1830s maps rarely showed much beyond their borders, the major exception being the eighteenth century embassies to China. In addition, all the routes illustrated were by land, not by sea.

JOHN K. WHITMORE

REFERENCES

Truong Buu Lam, ed. *Hong Duc Ban Do* (Maps of the Hong Duc Era). Saigon: Bo Quoc-gia Giao-duc, 1962.

Whitmore, John K. "Cartography in Vietnam." In *The History of Cartography* . Vol. II, Book 2, *Cartography in the Traditional East and Southeast Asian Societies*. Ed. J.B. Harley and David Woodward. Chicago: University of Chicago Press, 1994, pp. 478–511.

See also: Geography in China

MARĀGHA The observatory located just outside Marāgha, the Ilkhanid capital, represents the highest development of astronomical research institutions within the context of Islamic cultures. It is reported, on the authority of Naṣīr al-Dīn al-Ṭūsī, its first director, that construction of the observatory began under the patronage of Hūlāgū, Mongol conqueror of the region, in Jumādī I, 657 (April/May, 1259). The remains of the observatory buildings still occupy a hilltop position near the city in Azerbayjan.

Construction at the site was overseen by Mu'ayyad al-Dīn al-ʿUrḍī, who wrote a treatise describing his efforts. In addition to the main observatory building, he constructed a mosque and a residence for Hūlāgū, who took an active part in the work of the astronomers. The main building was described by contemporary witnesses as "huge" and there are reports that it contained a library of over 400,000 treatises. There is also a report of a "high tower" as well as a domed structure, which may or may not have been the main observatory structure. This dome had a hole in the top to allow the sun's rays to enter. It was used to determine the equinoxes and to measure the mean motion of the sun and its changing elevation. There are still traces of a large masonry mural quadrant outside one of the main buildings.

Al-ʿUrḍī was also in charge of constructing the instruments for the observatory (although it seems he did not construct all of them personally). He reports that this aspect of the construction was completed during 660 (1261/1262). In addition to the previously mentioned mural quadrant (whose radius must have been more than four meters), which was used to determine the latitude of the observatory as well as the obliquity of the ecliptic, the observatory included the following instruments: (1) an armillary sphere of five rings and an alidade (sighting device), maximum radius about 1.6 meters; (2) a solstitial armilla, a circle of 1.25 meters placed in the plane of the ecliptic and equipped with an alidade for determining the solstice points; (3) an equinoctial armilla, similar to the previous instrument but with a second ring paralleling the plane of the equator, for determining the equinoxes accurately; (4) an instrument with two holes on a sighting rod, used to determine the apparent diameters of sun and moon, as well as for observing eclipses; (5) an azimuthal ring, equipped with two quadrants and two alidades, to measure angles of elevation; (6) a parallactic ruler whose measurements were comparable in accuracy to those made on a circle of radius 2.5 meters; (7) an instrument for measuring azimuthal altitude and the sine of the complement of that angle of elevation; (8) a similar instrument that measures the azimuthal altitude and the sine of the angle of elevation; (9) an instrument similar to a parallactic ruler but not fixed in the meridian plane. In addition to such large-scale instruments, there were many smaller ones, such as planispheric astrolabes and quadrants, terrestrial and celestial globes, time keeping devices, star charts, maps, representations of the heavenly spheres, and many others.

In addition to being the best-equipped observatory of the medieval period, the Marāgha institution was also the best-staffed. In addition to Naṣīr al-Dīn al-Ṭūsī, the director, such outstanding astronomer/mathematicians as ʿAlī ibn ʿUmar al-Qazwīnī, Mu'ayyad al-Dīn al-ʿUrḍī (already mentioned in conjunction with his role as architectural supervisor and chief instrument maker at the site), Fakhr al-Dīn al-Akhlātī, Fakhr al-Dīn al-Marāghī, Muḥyi al-Dīn al-Maghribī, Quṭb al-Dīn al-Shīrāzī, Shams al-Dīn al-Shirwānī, Najm al-Dīn al-Qazwīnī, ʿAbd al-Razāk ibn al-Fuwwatī (the librarian), Kamāl al-Dīn al-Ayta, Athīr al-Dīn al-Abhārī, the two sons of Naṣīr al-Dīn al-Ṭūsī — Aṣīl al-Dīn and Ṣadr al-Dīn — and even a Chinese astronomer, Fao Mun Ji. (The presence of the latter indicates the growing interaction between the Arabic/Islamic and the Chinese astronomical traditions made possible and encouraged during the period of Mongol domination.) The activities of these scholars were supported by a full complement of technicians, instrument makers, and administrators.

It was generally argued by astronomers that in order to produce a full set of astronomical tables, observations had to be made over a thirty year period (the time for Saturn to complete one revolution around the sky). Most observatories prior to this had been forced to make do with less ambitious research programs, for they could not survive the death of their patrons. The Marāgha Observatory is a notable exception to this rule. Hūlāgū, the royal patron, died in 663 AH/AD 1265 and Naṣīr al-Dīn al-Ṭūsī, the first director, died in 672 AH/AD 1274. After al-Ṭūsī's death, the observational activity continued at a somewhat slower pace under the leadership of his son, Ṣadr al-Dīn al-Ṭūsī. And thirty years later, al-Ṭūsī's younger son, Aṣīl, was named director of the institution. This is the last recorded activity on the site. Thus it seems that the observatory continued to function for nearly half a century before finally being abandoned and falling into ruins.

One possible reason for the extraordinary longevity of this institution is the use of *waqf* funds to support the operation of the observatory. (A waqf is a grant of property in perpetuity for the support of religious or charitable institutions. Thus, not only are many mosques supported by income derived from waqf properties, but also orphanages, soup kitchens, hospices, schools, libraries, hospitals, and other public service institutions. It is, thus, surprising to find waqf funds devoted to the operation of an observatory.) Several sources indicate that al-Ṭūsī, in addition to directing this observatory, was also charged with administering waqf funds throughout the state. These sources indicate that about a tenth of all waqf revenues was earmarked for the expenses of the observatory. It is not clear whether this unusual use of waqf funds generated any sort of public outcry. If nothing else, it seems to indicate that the work of the observatory was intended to be ongoing, not merely an intense short term effort.

Perhaps the use of waqf funds to support the observatory was easier to justify in light of the extensive instructional activity carried on there. It was not uncommon to endow a *madrasa* (an institution of higher learning usually focused on religious and legal sciences) with waqf funds. Our sources tell us that the observatory was a center of teaching in mathematical, astronomical, and related sciences. The educational program here was apparently organized in a way analogous to that of a madrasa, although the details given in the sources are too vague to be certain.

In addition to teaching the classics of the ancient mathematical and astronomical tradition, the Maragha Observatory was also the seat of a major movement to criticize and reform Ptolemaic (Greek) astronomical theories. The problem stemmed from Ptolemy's perceived disregard for the essential sphericity of the celestial motions. Ptolemaic theory described each planet as carried daily from east to west across the vault of the sky by the motion of the sphere of the fixed stars. This diurnal motion is complicated by the fact that each planet sets a bit late when compared to the fixed stars and moves across the heavens in a path not parallel to the fixed stars. To accommodate these observations, the Greeks assumed that each planet must be fixed to a sphere (epicycle) whose center was carried by a second sphere, the deferent, whose axis is not parallel to the axis of the sphere of the fixed stars and whose proper motion of rotation (zodiacal motion) is against that of the rotation of the fixed stars which produced the diurnal motion. The combined rotations of this epicycle-deferent system also neatly accounted for the occasions when the planets appeared to reverse their proper west-to-east motion and temporarily moved east-to-west before slowing to a stop and resuming their proper zodiacal motion (retrogade motion).

The motion of the epicycle center, as observed from the earth, is not uniform, however. Ptolemy assumed, then, that the earth must not be at the center of the deferent orb. The distance (eccentricity) of the center of the deferent from the earth was fixed by observational considerations. However, when this distance is determined, it is observed that the calculated variation in velocities is only half what it should be. To obtain the observed fluctuations in angular velocity, he was forced to postulate that the uniform angular motion took place about a point (equant) located at twice the value of the eccentricity from the earth. This postulate allows uniform motion in the sense of sweeping out equal angles in equal increments of time to occur about the equant sphere but not uniform motion in the sense of equal arcs traversed in equal increments of time. Herein lies the central problem that the Marāgha astronomers tried to resolve: how to construct a model that fits the assumption of uniform circular motion in both senses of that term and that gives results as accurately as did Ptolemy's equant model.

Al-Ṭusi, assuming that Ptolemy's circular motions must be performed by real physical spheres, developed a system of solid spheres rolling inside one another to account for the observed zodiacal motions of the planets. The net result of these motions (each truly uniform and circular) is that the planet follows the path predicted by Ptolemy with the observed variations in velocity, thus overcoming the fundamental conceptual problem of the Ptolemaic system.

Al-Ṭusi is most famous among modern historians of science for his modification of a theorem by Proclus demonstrating that when a circle, whose diameter is equal to the radius of a second circle, rolls inside the circumference of larger while rotating about its own axis in the opposite direction with the same speed, a point on the circumference of the smaller circle will execute a simple harmonic oscillation along a diameter of the larger circle. Al-Ṭusi used his device, now known as a "Tusi Couple", to describe the anomalous motion of the equant point in Ptolemy's description of Mercury's motion. (Al-Ṭusi's idea was developed and further applied by his pupil Ibn al-Shatir.) An essentially similar geometric device was employed by Copernicus more than two centuries later. This striking similarity has led to question the originality of Copernicus. However, whether or not he was aware of, or even copied, al-Ṭusi's earlier model (Arabic manuscripts describing this model appear to have been present in Rome while Copernicus was studying there, and there may also have been Byzantine Greek translations of some of these treatises), Copernicus can still claim a place in history for his daring switch from a geostatic to a heliostatic cosmological model.

GREGG DEYOUNG

REFERENCES

Kennedy, E. S. "Late Medieval Planetary Theory." *Isis* 57(3): 365–378, 1966.

Saliba, G. "The Role of Maragha in the Development of Islamic Astronomy: A Scientific Revolution Before the Renaissance." *Revue de Synthèse* 108: 361–373, 1987.

Saliba, G. "The Astronomical Tradition of Maragha: A Historical Survey and Prospects for Future Research." *Arabic Sciences and Philosophy* 1: 67–99, 1991.

Sayılı, Aıydın. *The Observatory in Islam and its Place in the General History of the Observatory.* 2nd ed. Ankara: Turk Tarih Kurumu Basimevi, 1988.

Seemann, Hugo J. "Die Instrumente der Sternwarte zu Maragha nach den Mitteilungen von al-ʿUrdi." *Sitzungsberichte der physikalisch-medizinischen Sozietät zu Erlangen* 60: 15–126, 1928.

Varjāvand, Parvīz. *Kāvush-i Raṣd-khāna-i Marāgha* [Excavation of the Marāgha Observatory]. Tehran: Amīr Kabīr, 1987.

See also: Naṣīr al-Dīn al-Ṭūsī – al-ʿUrḍī – al-Shīrāzī – Armillary Sphere – Quadrant – Maps, Celestial – Astrolabe – Clocks and Watches – al-Qazwīnī – Astronomy in China – Astronomy in Islam – Astronomical Instruments in Islam

AL-MĀRIDĪNĪ, JAMĀL AL-DĪN AND BADR AL-DĪN

Jamāl al-Dīn al-Māridīnī was a competent astronomer who lived in Damascus or Cairo ca. 1400. He authored numerous short treatises, mainly dealing with instruments. One of the most remarkable of these instruments is a universal quadrant, the only known example of which is from sixteenth-century Flanders (now in the Adler Planetarium, Chicago). His name indicates that he or his family came from Mardin, now in southern Turkey, but biographical information is lacking; indeed it is not even clear where he worked. He is often confused with his grandson, Sibṭ al-Māridīnī.

Badr al-Dīn al-Māridīnī, known as Sibṭ al-Māridīnī, lived in Cairo, ca. 1460. He was a grandson (Arabic *sibṭ*) of Jamāl al-Māridīnī and was one of the leading astronomers in Cairo. He compiled a large number of treatises, including many short works on the standard instruments of his time, the trigonometric and astrolabic quadrant and sundials. These treatises became extremely popular and exist in hundreds of copies. In fact they were studied by everyone in Egypt who was interested in astronomy in the succeeding centuries. As a result of this, more significant works were forgotten and the level of astronomy declined.

DAVID A. KING

REFERENCES

King, David A. "An Analog Computer for Solving Problems of Spherical Astronomy: The Shakkāziyya Quadrant of Jamāl al-Dīn al-Māridīnī." *Archives internationales d'histoire des sciences* 24: 219–242, 1974, reprinted in David A. King. *Islamic Astronomical Instruments*. London: Variorum, 1987, Repr. Aldershot: Variorum, 1995, X.

King, David A. *A Survey of the Scientific Manuscripts in the Egyptian National Library*. Winona Lake, Indiana: Eisenbrauns, 1986, nos. C47 (Jamāl al-Dīn) and C97 (Badr al-Dīn).

King, David A. "The Astronomy of the Mamluks." *Isis* 74: 531–555, 1986, reprinted in David A. King, *Islamic Mathematical Astronomy*. London: Variorum, 2nd rev. ed., Aldershot: Variorum, 1993, III.

Suter, Heinrich. *Die Mathematiker und Astronomen der Araber und ihre Werke*. Leipzig, 1900, reprinted Amsterdam: Oriental Press, 1982, nos. 421 (Jamāl al-Dīn) and 445 (Badr al-Dīn).

MĀSHĀ>ALLĀH Māshā>allāh ibn Atharī was a Jewish astrologer who was born in the southern city of Basra ca. 730 in the later Umayyad caliphate and survived till ca. 815. If he was ever converted to Islam, as his name Mā>shā>allāh seems to imply, it is not made clear in any of the original sources such as the *Fihrist*. His respectful tone toward the Caliphs in his *Fī qiyām al-Khulafā>* would certainly befit a Muslim more than a Jew. At any rate he wrote extensively in Arabic, and he was called upon by the second Abbasid Caliph al-Manṣūr (754–775) to assist in the laying of the foundations of Baghdad in 762. Hence by that time he was surely a well known astrologer active in the Babylonian region, representative perhaps of the Jewish culture long settled in the area.

A celebrated astrologer in his time, Māshā>allāh was either the founder or a leader of a school of Jewish astrologers in Mesopotamia. Apart from his participation in the laying of the foundations of Baghdad, we know rather little of the circumstances of his life. Though sparse on biographical details, the principal Arab bibliographers al-Nadīm (*Fihrist*, AD 987) and Ibn al-Qifṭi (*Tā>rīkhal-ḥukamā>*, ca. AD 1248) credit Mā>shā>allāh with a sizeable number of works. On the other hand, the *Jewish Encyclopedia* confuses him in part with Maslama al-Majrīṭī, probably because of the similarity of names. Paul Kunitzsch has recently established that the Latin treatise on the astrolabe long ascribed to Māshā>allāh and translated by John of Seville is in fact by Ibn al-Ṣaffār, a disciple of Maslama al-Majrīṭī.

In his astrological doctrine Mā>shā>allāh seems to have depended principally on the Persian tradition, though a late Byzantine translation of one of his works shows that he could also quote from Greek authors such as Plato, Aristotle, and Hermes. Mā>shā>allāh's works have not survived extensively in Arabic and it is thanks to the Latin translations, mostly by John of Seville in twelfth century Spain, that we have access to a large number of those ascribed to him.

Māshā>allāh's contribution in Arabic astrology must be linked with the widespread use of astrological prognostication in the Mesopotamian area since very ancient times. The thirteenth century bibliographer Ibn Khallikān recalls the extensive use of this "projective tool" in the politics and social life of the early Arab empire. What has survived of Māshā>allāh's works shows that the substance of his astrological knowledge owes but little to Greek philosophy proper. Assuredly, there flourished in Sassanian times a trend to borrow from Greek (Alexandrian) astrology, apparently only to consolidate a native tradition of learning. And it is this Sassanian lore above all that nourishes Māshā>allāh's production. On the other hand, it is historically confirmed that the passion for Greek learning among the Arabs was engendered by the caliph al-Ma>mūn's (813–833) enthusiasm for the philosophy of Aristotle, too late for Māshā>allāh to share significantly in this outburst. The fresh start taken by "scientific" astrology under al-Ma>mūn's prodding is noticeable in the works of al-Kindī, of Abū Ma<shar, and the members of the *Bayt al-Ḥikma* (House of Wisdom) founded by al-Ma>mūn. Compared with their works, those of Māshā>allāh and of his generation show no clear and systematic use of the axioms of peripatetic philosophy to buttress the principles of astrological science as was done so decidedly a generation later.

Māshā>allāh's reputation remained high in the following generation as witnessed by Abū Ma<shar's praise of him, according to the *Mudhākarāt* (Memorabilia) of Ṣa>id ibn Shaḏān. Yet, it is mostly for one single astrological topic, the "projection of rays" (*matraḥ al-shuʿāʿāt*), that Abū Maʾshar lavishes praise on Māshāʾallāh.

An important historical factor affecting the preservation of Māʾshāʾallāh's works, and consequently of his reputation as a leader in Arab astrology, must be stressed at this point. As stated above, his works were virtually superseded in Arabic astrology by the works of the generation which followed his death. Why then do we possess so many of his manuscripts in Latin translation? This seems to have resulted from the special interest manifested in Māshāʾallāh's works by John of Seville (Johannes Avendauth, or Ibn Dawūd). Having set about to translate for the benefit of Latin scholars the works of the "prince" of Arab astrologers Abū Maʿshar (no Arab himself but Persian/Afghan since he came from the "Bactrian" city of Balkh), John enlarged his program of translation to include the corpus of Jewish astrologers of the early Abbasid era.

Within the scope of Arab astrological doctrines there developed a side which stemmed from the application of the scheme of planetary conjunctions in relation to the zodiacal signs and various triplicities in which they might occur. But it was Jupiter's special prerogative to foster religious life. The conjunction theory animating astrological history turned principally around the conjunctions of Saturn and Jupiter to signal the emergence, duration, and waning of sects (Arabic *milal*) when referred to religions, or of dynasties and empires (*duwal*) when applied to military domination. A basic element of the scheme was the association of each planet with what was classified as "great civilizations", mostly religions with cultures that set religion as their guiding light. Saturn as the most distant planet was appropriated to the Judaic faith, the Sun to the Roman Empire, and Venus to Islam. This was in a nutshell the astrological exercise of the "horoscope of religions" as it came to be known in the Latin West.

The most famous of such interpretations was embodied in Abū Maʿshar's *Great Conjunctions*. Associating the spread of the Judaic religion with the rule of the planet Saturn, the sway of the Roman Empire with that of the Sun, and of the Arab Empire with that of Venus, etc., the scheme thereby stressed the primeval character of the Jewish faith in accordance with the status of Saturn, the highest and most distant planet. The vicissitudes of planetary conjunctions thereafter unfolding along the variety of signs in which they occurred successively, rendered Judaic monotheism anachronistic. It was replaced in time by the Roman Empire, which in turn was displaced by "manicheism" when the prophet Mani appeared, until the rise of Islam.

In ninth-century Baghdad there was great concern among the Abbasid rulers and the Arab elites who were confronted with widespread social unrest among non-Arab subjects. One wave of unrest was further encouraged in *shuʿūbīya* (ethnic) groups by astrological predictions hinting at the impending end of Arab domination foreboded by planetary conjunctions. This was the occasion for the consolatory *Risāla fī Mulk al-ʿarab wa-kammiyatihi* which al-Kindī wrote for the Caliph, a consolatory Epistle in which he calculated that the conjunctions of Saturn with Jupiter held forth a total duration of 693 years for the Muslim Empire, which meant at that time four more centuries of Arab rule, whereas Abū Maʿshar's similar calculations in his *Great Conjunctions* foresaw only 310 years or so for it.

The association of Judaism with Saturn is not present in Māshāʾallāh's scheme of astrological history, although his interpretation of planetary conjunctions was extended to Islam and in particular to the rule of the Caliphs. Yet the status assigned to Judaic culture in the scheme of planetary conjunctions of Arab astrology stirred echoes in Jewish communities in Spain, inspiring their leaders in astrological science to use the scheme of conjunctions to confirm the unique "vocation" of Israel. This endeavor is at the core of the *Megillat ha-Megalle* by Abraham bar Ḥiyya of Barcelona. This book purports to illustrate the fate of Israel in the cosmic framework of universal history by referring to Holy Scriptures, Rabbinic literature, and philosophy. Yet the entire last chapter is devoted to the scheme of planetary conjunctions and its significance for the fate of Israel, confirming by "science" the data of sacred literature. The enterprise ostensibly aimed to counteract the "scientific" demoting of Judaic civilization to a lower historical role by Arab astrological writers. Another Jewish astronomer astrologer in Spain, Abraham ibn Ezra, also critiqued the theory of planetary conjunctions.

The medieval Anglo-Irish scientific literature used to include two versions of some work by Māshāʾallāh' in the form of Chaucer's *Treatise on the Astrolabe*, and an Irish tract following closely the text published by J. Heller in Nurnberg in 1549 under the title *De elementis et orbibus coelestibus*. Since the publication of Paul Kunitzsch's thesis, however, which shifts the authorship of the Latin source used by Chaucer from Māʾshāʾallāh the Jewish astrologer from Iraq, to Maslama, the eleventh century Arab astronomer from Spain, it would appear that this part of Māshāʾallāh's claimed influence upon Western literary tradition must be abandoned.

A compilation of Arab "authorities" in astrology called *Liber Novem Iudicum* (Book of the Nine Judges) included Māshāʾallāh as one of its principal authorities. The compilation of the Nine Judges did much to carry Māshāʾallāh's astrological rules and reputation to the four corners of medieval Europe.

RICHARD LEMAY

REFERENCES

Goldstein, B. "The Book on Eclipses of Māshāʾallāh." *Physis* 6: 205–213, 1964.

Kunitzsch, P. "On the Authenticity of the Treatise on the Composition and Use of the Astrolabe Ascribed to Messahalla." *Archives Internationales d'Histoire des Sciences* 31: 42–62, 1981.

Pingree, D. *The Astrological History of Māshāʾallāh.* Cambridge, Massachussetts: Harvard University Press, 1971, pp. 129–143.

Pingree, D. "Māʾshāʾallāh." In *Dictionary of Scientific Biography.* Vol. IX. Ed. Charles C. Gillispie. New York: Scribners, 1974, pp. 159–162

Pingree, D. "Māshāʾallāh: Some Sasanian and Syriac sources." In *Essays on Islamic Philosophy and Science.* Ed. F.G. Hourani. New York: 1975, pp. 5–14.

Samsó, J. "Māʾshāʾallāh." In *Encyclopedia of Islam* VI. Leiden: Brill, 1991, pp. 710–712.

See also: al-Maʾmūn – al-Kindī – Abū Maʿshar – Abraham bar Ḥiyya – Ibn Ezra

AL-MASʿŪDĪ (Abuʾʿl Ḥasan ʿAlī Ibn al-Ḥusayn al-Masʿūdī) In the tenth century AD Muslims were arguably the leaders of world sciences, including geography. Many Muslim geographers, including al-Masʿūdī, established the principles of science and research in geography. Al-Masʿūdī's geographic curiosity took him as far as China and Madagascar. His conceptual orientation, which combined geography and history to explain cultural history, made him an outstanding scholar.

Not much is known about al-Masʿūdī's early life and education, except that he was born in Baghdad in AD 893 and died in Old Cairo, Egypt, in AD 956. His two surviving books, *Murūj al-dhahab wa-Maʿādin al-Jawhar* (Meadows of Gold) and *Kitāb al-tanbīh waʾl Ishraf* (Book of Indication and Revision) provide no biographical information. Presumably, his early education was with historians, philosophers, scientists, *ḥadīth* specialists (those who study the words of the Prophet Muḥammad), grammarians, and literary critics. He began his travels at the age of twenty-three to seek geographic knowledge and ended them when he was fifty-six years old. He lived and traveled during a period of Islamic Renaissance in many fields including geography.

In his book, al-Masʿūdī described the shape of the earth, seas and their depths, islands, mountains and rivers, mines, marshes, and lakes. He also provided information on inhabited areas and explained why land and sea changed their forms.

As a historian and geographer, al-Masʿūdī broadened the concept of geography by including all the known branches of geography in his discussions. In doing so he benefited equally from the scholarly works of both Muslims and non-Muslims. He quoted quite extensively from Greek, Persian, and Indian sources. Al-Masʿūdī also studied people within their own habitats. To him, "nature" (*ṭabīʿā*) meant the processes of the external physical world. He related those processes to humans in the universe, the activity of God in history, and the growth and development of societies.

An aspect of the development of Arabic geographic literature of his period was the production of maritime literature and travel accounts. Al-Masʿūdī also made important contributions to the field of oceanography, but his reputation as a scholar and scientist is based on his two surviving books on history and natural history.

Al-Masʿūdī had a universal outlook. He focused on a wide range of topics dealing with Islamic and non-Islamic cultural histories and provided elaborate cultural, historical, geographical, ethnological, climatic, and maritime descriptions of the known world of his time. He was a cultural geographer of the first order, and far ahead of his time.

MUSHTAQUR RAHMAN

REFERENCES

Khalidi, Tarif. *Masʿūdī's Theory and Practice of History*. Ph. D. Dissertation, University of Chicago, 1970.

Nasr, Syed Hossein. *Science and Civilization in Islam*. Cambridge, Massachusetts: Harvard University Press, 1968, pp. 47–48.

Sarton, G. "The Times of al-Masʿūdī." In *Introduction to the History of Science*, vol. I. Baltimore: Williams and Wilkins, 1927, pp. 619–620.

Shoul Ahmad, M.H. *Al-Masʿūdī and His World*. London: Ithaca Press, 1979, pp. 1–17.

Siddiqi, Akhtar Hussain. "Al-Masʿūdī's Geographic Concept." *International Journal of Islamic and Arabic Studies* 7(2): 43–71, 1990.

Sprenger, Aloys. trans. *Al-Masʿūdī's Historical Encyclopedia.Meadows of Gold and Mines and Gems*. Vol I. London: W. H. Allen and Company, 1941.

MATHEMATICS A concise and meaningful definition of mathematics is virtually impossible. Mathematics has developed into a worldwide language with a particular kind of logical structure. It contains a body of knowledge relating to number and space, and prescribes a set of methods for reaching conclusions about the physical world. Also, it is an intellectual activity which calls for both intuition and imagination in deriving 'proofs' and reaching conclusions. Often it rewards the creator with a strong sense of aesthetic satisfaction.

Mathematics initially arose from a need to count and record numbers. As far as we know there has never been a society without some form of counting or tallying, i.e. matching a collection of objects with some easily handled set of markers, whether it be stones, knots or inscriptions such as notches on wood or bone. Also, it is precisely such an artifact that helps us to locate the early beginnings of mathematics.

High in the mountains of Central Equatorial Africa, on the borders of Uganda and Zaire, lies Lake Edward, one of the furthest sources of the River Nile. Though this area, Ishango, is remote and sparsely populated today, about twenty thousand years ago a small community lived by the shores of the lake that fished, gathered food or grew crops depending on the season of the year. The settlement had a relatively short life span of a few hundred years before being buried in a volcanic eruption. Archaeological excavations at Ishango unearthed a bone tool handle which is now on display at the Musée d'Histoire Naturelle in Brussels. The original bone may have petrified or undergone chemical change through the action of water and other elements. What remains is a dark brown object on which some markings are clearly visible. At one end is a sharp, firmly fixed piece of quartz which may have been used for engraving, tattooing, or even writing of some kind. Along the Ishango Bone, as it is now called, is a series of notches arranged in three distinct columns. One column contains four groups of notches with 9, 19, 21, and 11 markings. On a second column are four groups of 19, 17, 13, and 11 markings. And the third column has eight groups of notches in the following order: 7, 5, 5, 10, 8, 4, 6, 3, with the last pair (6,3) being spaced together, as are (8,4) and (5,5,10), suggesting a deliberate arrangement in distinct sub-groups.

Conjectures based on underlying numerical patterns of the notches are well summed up by de Heinzelin, the archaeologist who helped to excavate the Ishango Bone. The Bone "may represent an arithmetical game of some sort devised by a person who had a number system based on ten as well as a knowledge of duplication (or multiplying by 2) and of prime numbers". Further, from the existing evidence of the Ishango tools, notably harpoon heads, northwards up to the frontiers of Egypt, de Heinzelin supports the possibility that the Ishango numeration system may have traveled as far as Egypt to influence the development of its number system, the earliest decimal-based system in the world.

There is, however, another explanation which highlights a link that has played a crucial part in the historical development of mathematics. The close link between mathematics and astronomy has a long history and is tied up with the need for societies to record the passage of time, both out of curiosity as well as practical necessity. The alternative explanation is that the bone markings constitute a system of sequential notation — a record of different phases of the moon. Whether this is a convincing explanation would depend in part on establishing the importance of lunar observations in the Ishango culture and in part on how closely the series of notches on the bone matches the number of days contained in the successive phases of the moon. Marshak attempted to do both and concluded that the Ishango Bone, with its markings of different indentations, shapes, and sizes, was a lunar calendar where the different types of engraving indicate that it was also a calendar of events, probably of a ceremonial or ritual nature. There is still a good deal of controversy and interest around this bone.

One of the most ingenious methods of recording and storing information before the emergence of the computer is a *quipu*. The *quipu* (a Quechua word from the language of the Incas of South America) is an arrangement of colored wool or cotton cords with clusters of knots tied in the cords to represent numerical magnitude. In essence, it resembles a mop which has seen better days. It was widely used as a device for storing numerical and other information before the Spanish conquest in an area which would today include all of Peru, parts of Bolivia, Chile, Ecuador, and Argentina. A vast amount of information can be kept on a quipu using different colored cords for summation between and within categories and relative placement of the knots for indicating different numerical magnitudes. A quipu has been discovered which is a record of a census taken in 1587 of the Andean population of Lupaqa disaggregated by province, ethnic groups, size, and age/sex distribution of households, using cords of different thickness, colour, and configuration. Altogether forty-six different items of information were kept on this recording device no larger than an ordinary kitchen mop.

For a highly centralized society such as the Inca, an essential prerequisite for maintaining good order and efficient organization was the existence of detailed and up-to-date information (or government statistics, as we would describe such information today). A whole inventory of resources which included agricultural produce, livestock, and weaponry — as well as people — was maintained and updated by a group of specially trained officials known as *quipucamayus* (quipu-keepers). Each village under the rule of the Inca had its own specially trained *quipucamayu* and certain larger villages had as many as thirty. One of the main tasks of the *quipucamayu* was to devise efficient and economical methods of storing information. The more the *quipucamayu* considered the pattern of distribution, taking account of the relative sizes and positions of different cords, the better the logical structure of the final representation. Cord placement, color coding, and number representation were the basic design features, repeated and recombined to define a format and convey a logical structure. This search for a coherent numerical/logical structure involves mathematical thinking reminiscent of data base management in a modern computer and a study of spatial configuration structures of different quipus is a good introduction to a field of modern mathematics known as graph theory.

It is possible to view the appearance of a written number (or numeral) system as a culmination of earlier developments. First was the recognition of the distinction between more and less (a capacity we share with certain other animals). From this developed first simple counting, then the different methods of recording the counts as tally marks, of which the Ishango Bone is one example. This progression continued with the emergence of more and more complex means of recording information, culminating in the construction of devices such as the quipu. Before the appearance of such devices, there must have existed an efficient system of spoken numbers founded on the idea of a base to enable numbers to be arranged into convenient groups.

There is ample historical and anthropological evidence indicating that a variety of bases for counting systems have been used over the ages and around the world. The base of a counting system tells how numbers are grouped and constructed. Our system, a decimal or base ten system, has words for numbers from one to ten displaying no common root. Beyond ten, the number words generally show a variation of the unit word for the multiples of ten.

To illustrate, take the case of a few counting systems with different bases chosen from round the world. For an indigenous Australian group, the Gumulgal, counting proceeds as *urapon* (1), *ukasar* (2), *ukasar-urapon* (3), *ukasar-ukasar* (4), *ukasar-ukasar-urapon* (5), ... to indicate counting by twos (or a base-2 system). In a Melanesian language, Sesake, counting proceeds as *sekai* (1), *dua* (2), *dolu* (3), *pati* (4), *lima* (5), *la-tesa* (6), *la-dua* (7) ... *dua-lima* (10), ... *dua-lima-dua* (20). This suggests a base-5 counting system where words for numbers six or seven use the roots for words one and two; ten is literally two-fives and twenty is two-fives twice. In a Micronesian language, Kiribati, counting proceeds as: *tenuana* (1), *uoua* (2), *tenua* (3), *aua* (4), *nimua* (5), *onoua* (6), *itiua* (7), *wanua* (8), *ruainua* (9), *tebwina* (10), *tebwina-ma-tenuana* (11), ... *vabui* (20), vabui-ma-tenuana (21), ... *tebubua* (100), ... *tenga* (1000), ... *tebina-tenga* (10,000), ... *tebubuna-tenga* (100,000). Here we have a straightforward base-10 counting system. In Kiribiti, number words also vary according to the object being counted. Thus the number word for 9 is *ruaman* when counting animals, *ruakai* when counting plants, *ruai* when counting knives, *ruakora* when counting baskets, and *rauawa* when counting boats. In certain other systems, such as the Aztec number system, the choice of the bases would depend on the type of objects being counted. Cloths or tortillas would be counted in twenties, while round objects such as eggs or oranges would be counted in tens.

Counting by tens has been the most widespread system probably because it is associated with the use of fingers on both hands. The other scale (base-20) had its most celebrated development as a written number system during the first millennium of the present era among the Maya of Central America. There have also been other base-20 systems, of which the Yoruba system from Nigeria and the Welsh system from Britain are two well-known examples.

An unusual feature of the Yoruba system is its heavy reliance on subtraction. The subtraction principle operates in the following way. As in our number system, there are different names for the number one (*okan*) to ten (*eewa*). The numbers eleven (*ookanla*) to fourteen (*eerinla*) are expressed as compound words which may be translated into 'one more than ten' to 'four more than ten'. But once fifteen (*aarundinlogun*) is reached the convention changes, so that fifteen to ninteen (*ookandinlogun*) are expressed as 'twenty less five' to 'twenty less one' respectively where twenty is known as *oogun*. Similarly, the numbers twenty one to twenty four are expressed as additions to twenty, and twenty five to twenty-nine as deduction from thirty (*ogbon*). At thirty-five (*aarundinglogoji*), however, there is a change in the way the first multiple to twenty is referred to: forty is expressed as 'two twenties' (*ogoji*) while higher multiples are named *ogata* ('three twenties'), *ogerin* ('four twenties'), and so on to 'ten twenties', for which a new word *igba* is used. It is in the naming of some of the intermediate numbers that the subtraction principle comes into its own. To take a few examples, the following numbers are given names which indicate the decomposition shown on the right:

$$45 = (20 \times 3) - 10 - 5$$
$$108 = (20 \times 6) - 10 - 2$$
$$300 = (20) \times (20 - 5)$$
$$318 = 400 - (20 \times 4) - 2$$
$$525 = (200 \times 3) - (20 \times 4) + 5.$$

All the numbers from 200 to 2000 (except those that can be directly related to 400 or *iriniwa*) are reckoned as multiples of 200. From the name *egebewa* for 2000, compound names are constructed for number in excess of this figure, using subtraction and addition wherever appropriate similar to the above examples.

The origin of this unusual counting system is uncertain. One conjecture is that it grew out of the widespread practice of using cowrie shells for counting and computation. A description of the cowrie shell counting procedure given by Mann in 1887 is interesting. From a bag containing a large number of shells the counter draws four lots of five to make twenty. Five twenties are then combined to form a single pile of one hundred. The merging of two piles of one hundred shells gives the next important unit of Yoruba numeration, two hundred. As a direct result of counting in fives, the subtraction principles comes into operation. Take the decomposition of 525, given earlier, as an illustration. We begin with three piles of *igba* (200), remove four smaller piles of *oogun* (20), and then add five (*aarun*) cowrie shells to make up the necessary number.

This complicated system of numeration in which the expression of certain numbers involves considerable feats of arithmetical manipulation has certain advantages for computation. As an example of a calculation which exploits the strength of the Yoruba numeration to the full, consider the multiplication: 19×17.

The cowrie calculator begins with twenty piles of twenty shells each. From each pile, one shell is removed (-20). Then three of the piles now containing nineteen shells each are also removed. The three piles are adjusted by taking two shells from one of them, and adding one each to the other two piles to bring them back to twenty $(-20 \times 2 - (20 - 3))$. At the end of these operations we have

$$400 - 20 - (20 \times 2) - (20 - 3) = 323.$$

While the Yoruba system shows what is possible in arithmetic without a written number system, it is clearly impractical for more difficult multiplications.

However, there are other mechanical aids which are more versatile, such as the Chinese or Japanese abacus (*soroban*). In the speed of addition and subtraction, the abacus still holds its own against the electronic calculator. In a test problem consisting of, first, adding a column of eleven numbers and then subtracting four numbers from the result where there was no number of less than three digits and most contained four or five digits, a third grade abacus operator in China or Japan (i.e. an operator possessing the minimum level of competence acceptable for employment in a bank or similar institution) performed this computation in thirty seconds. An operator using the calculator took ninety seconds. A first-grade abacus operator would have been expected to have finished the computation in twenty seconds. There is even the well-attested claim of an expert operator taking less than fifteen seconds to add ten numbers each of ten digits!

There are two modes of performing arithmetical operations. The first arose in cultures which, either because of scarcity of writing materials or because of the limitation of their written number systems resorted to physical devices such cowrie shells or an abacus to carry out multiplication. The second mode involves 'paper-and-pencil' methods of written numbers. The origins of many of these paper-and pencil algorithms are found in parts of the non-Western world which gave birth to different place value number systems: Babylonia during the third millennium BC, China around the third century BC, the Maya Empire around the beginning of the Common Era, and in India a few centuries later. And in the process of transmitting the Indian numerals westwards some of the efficient algorithms, initially devised for these numerals, were lost.

To illustrate, consider multiplying 97 by 93. The method favored today involves two sets of multiplication, one requiring the "9-times table", then adding the results of the two sets of multiplication on paper to get the final answer: 90/21. Of course these days, one would use an electronic calculator but feel rather lost if the calculator was not readily at hand . But there is a startling simple yet mathematically profound method available from the distant past.

Take the problem: 97×93. 97 is 3 less than 100; 93 is 7 less than 100. Multiply the deficiencies 3 and 7 to get 21. Subtract from 97 the deficiency corresponding to the other number, 93, which is 7 *or* the other way round. Both subtractions will give 90. The final answer is the merging of the two parts: 90/21. This procedure can be mathematically 'productive' if one then proceeds to answer the following "what ifs":

- What if both numbers to be multiplied are just over 1000?
- Both just under 1000?
- One just over and one just under 1000?
- Both just over or just under 100?
- Or one just over and one just under 100?
- What about numbers near 50?
- Or near 20?
- What if they are not both within the range of the same base?

In all these cases, the procedures devised involve using the principal strength of our number system — the place value principle. An examination of why the method works leads us quite naturally to the interconnection of arithmetic, algebra, and geometry and also a reminder that algebra is but a generalization of arithmetic. A method claimed to be as old as the Vedas (ca. 500 B.C), it is found in works of later Indian mathematicians, notably Brahmagupta (ca. AD 600) and Bhāskarāchārya (ca. AD 1100). It was known to the Arabs through whom our number system, which began in India, spread westwards into Europe. Yet this method was lost and what we have in place are rather cumbersome procedures, some more suitable perhaps for multiplication with number systems without place value like the Roman numerals.

The rules devised by mathematicians for solving problems about numbers of one kind or another may be classified into three types. In the early stages of mathematical development, these rules were expressed verbally, and consisted of detailed instructions about what was to be done to obtain a solution to a problem, for which reasons this approach is referred to as rhetorical algebra. In time, the prose form of rhetorical algebra gave way to the use of abbreviations for recurring quantities and operations, heralding the appearance of syncopated algebra. Traces of such algebra are to be found in the works of the Alexandrian mathematician Diophantus (ca. AD 250), but it achieved its fullest development in the work of Indian and Arab mathematicians during the first millennium AD. During the past five hundred years symbolic algebra has developed so that, with the aid of letters and signs of operation and relation ($+, -, \times, /, =$), problems are stated in such a form that the rules of solution may be applied consistently and systematically. The transformation from rhetorical to symbolic algebra has been a long one and marks one of the most important advances in mathematics. It had to await the development of a positional number system (i.e. the Indian numerals) which allowed numbers to be expressed concisely and with which operations could be carried out efficiently.

As early as 1800 BC, the Babylonians had developed sophisticated methods of solving equations, building on their invention of a positional number system. A four thousand year old Babylonian clay tablet, now kept in a Berlin museum, gives the value of $n^3 + n^2$ for $n = 1, 2, \ldots, 10, 20, 30, 40, 50$, from which it is deduced that the Babylonians may have used these values in solving cubic equations after reducing them to the form $x^3 + x^2 = c$. Linear and non-linear equations in two and three unknowns were also solved correctly within the framework of a rhetorical algebra.

Early Indian algebra also contained solutions of linear, simultaneous, and even indeterminate equations. An example of an indeterminate equation in two unknowns (x and y) is $3x+4y = 50$, which has a number of positive whole-number solutions for (x, y). For example $x = 14$, $y = 2$ satisfies the equation, as do the solution sets (10,5), (6,8) and (2,11). But it is only from AD 500 that there emerged a distinctive feature — the use of symbols, such as a dot or the letters of the alphabet, to denote unknown quantities. In fact it is this very feature of algebra that one immediately associates with the subject today. A general term in Indian algebra for any unknown was *yāvat tāvat*, which was shortened to the algebraic symbol *yā*, In Brahmagupta's work, Sanskrit letters appear, which are the abbreviations of names of different colors which he used to represent several unknown quantities. Thus the letter *kā* stood for *kālaka*, meaning 'black', and the letter *nī* for *nīlaka* meaning 'blue'.

The word *al-jabr* appears frequently in Arab mathematical texts that followed al-Khwārizmī's influential *Ḥisāb al-jabr w'al-muqābala* written in the first half of the ninth century AD. There were two meanings associated with *al-jabr*. The more common one was "restoration" as applied to the operation of adding equal terms to both sides of an equation, so as to remove quantities, or to restore a quantity which is subtracted from one side by adding it to the other. Thus an operation on the equation $2x + 5 = 8 - 3x$ which leads to $5x - 5 = 8$ would be an illustration of *al-jabr*. There was also another, less common meaning: multiplying both sides of an equation by a certain number to eliminate fractions. Thus, if both sides of the equation $\frac{9}{4}x + \frac{1}{8} = 3 + \frac{5}{8}x$ were multiplied by 8 to give the new equation $18x + 1 = 24 + 5x$, this too would be an instance of *al-jabr*. The common meaning of *al-muqābala* is the "reduction" of positive quantities in an equation by subtracting equal quantities from both sides. For the second equation above, applying *al-muqābala* would give $18x - 5x + 1 - 1 = 24 - 1 + 5x - 5x$ or $13x = 23$. The words *al-jabr* and *al-muqābala*, linked by *wa* (and) came to be used for any algebraic operation, and eventually for the subject itself. Since the algebra of the time was almost wholly confined to the solution of equations the phrase meant exactly that.

Apart from giving the name to the subject, the great contribution of the Arabs to algebra was to devise an efficient system of classifying equations. Starting with al-Khwārizmī, they reduced all equations to six main types. For each type they offered solutions and when possible a geometric rationale. Their work culminated in ʿUmar al-Khāyyam's geometric solution of the cubic equation in the middle of the eleventh century AD. The Arab work on equations is one illustration of their ability to bring together two strands of mathematical thinking — the geometric approach which had been carefully cultivated by the Greeks and the algebraic /algorithmic methods which had been used to such effect by the Babylonians, Indians, and Chinese.

The development of Chinese algebra was a direct result of their number system — the rod numeral system. Apart from its notational facility, the rod numeral system was helpful in suggesting new approaches to algebraic problems, notably the use of a variant of the modern matrix method for solving simultaneous equations about one thousand five hundred years before its so-called discovery in Europe and the "method of double false" which constitutes an important algorithm for solving higher order equations in the modern subject of numerical analysis.

The Chinese work on solutions of numerical equations of higher order has its origins in the method of extracting square and cube roots found in their premier text, *Jiuzhang suanshu* (Nine Chapters on the Mathematical Arts), written around the beginning of the Common Era. The combination of the root extraction procedure with the use of what we know as the Pascal Triangle (although it was already known to the Chinese about 500 years before the seventeenth century French mathematician, Pascal), meant that the Chinese were solving equations of the ninth degree (i.e. equations involving the term x^9) around AD 1250, using a variant of the Horner–Ruffini method which only came into modern mathematics at the beginning of the nineteenth century.

Any definition of the subject matter of mathematics would include activities that relate to spatial configurations or geometry. There are two theories regarding the origins of geometry. Herodotus, the Greek historian who lived in the fifth century BC, wrote that geometry arose in ancient Egypt from the need to parcel out, in an equitable fashion, precious agricultural land whose boundaries were annually obliterated by the overflow of the River Nile. The ideas of the Egyptian surveyors (or 'rope-stretchers') were eventually passed on to the Greeks who proceeded to build that most impressive edifice known as Greek geometry. However, Egyptian geometry had shown the way, for in the Moscow Papyrus, an important source of Egyptian mathematics from about 1850 BC, the correct rule for calculating the volume of a truncated pyramid appears.

An alternative explanation sees the origins of geometry in religion and ritual. A good illustration would be a class of ritual literature from ancient India, known as the *Śulbasūtras* dealing with the measurement and construction of various sacrificial altars. They also happen to be the earliest text of Indian geometry dated around 800–500 BC. They provided instructions for two types of rituals, one for worship at home and the other for communal worship. Square and circular altars were sufficient for household rituals while more elaborate altars involved combinations of rectangles, triangles, and trapezia for public worship. The geometry of the *Śulbasūtras* grew out of the need to ensure strict conformation of the orientation, shape, and area of altars to the prescriptions laid down in the scriptures. They include a general statement of the so-called Pythagorean theorem, an approximation procedure for obtaining the square root of 2 correct to five decimal places and a number of accurate geometric constructions including ones for 'squaring the circle' (approximately) and constructing rectilinear shapes whose area was equal to the sum or difference of areas of other shapes. The earliest known demonstration of Pythagoras' theorem is found in an ancient Chinese text, *Zhoubi suanjing*, at least three hundred years before Pythagoras (ca. 500 BC). Over a thousand years before Pythagoras, the Babylonians knew and used the result now known under his name.

To most of us geometry deals with lines, angles, circles, and polygons. These are the central concepts that appeared in the best-known text in geometry, Euclid's *Elements* (ca. 300 BC). To these were subsequently added subjects such as symmetry, coordinates, vectors, and other curves. Many of these concepts appear in different cultures in a variety of contexts: in architecture, drawings, decorations, etc. Do such examples constitute mathematics or can they at best be used as no more than peripheral illustrations of certain geometrical notions? In 1986, Paulus Gerdes posed the following 'non-standard' problems to a workshop of mathematics educators who had some difficulty working them out, although artisans in Mozambique, some of them illiterate, solved them as a matter of course:

- Construct a circle given only its circumference: a problem in laying out a circular floor for a traditional Mozambican house.
- Construct angles that measure 90°, 60°, or 45° with only strips of straw: a problem in basket weaving.
- Fold an equilateral triangle out of a square: a problem in making a straw hat.
- Construct a regular hexagon out of straw: a problem in making a fish trap.

Does solving these problems involve mathematical thinking? It could be argued that the artisans who first discovered the optimal solutions to these problems were engaged in "creative" geometrical thinking. Mary Harris has extended this argument to tasks which are usually seen as women's work. "Many of the male teachers are so unfamiliar with the construction and shape and size of their own garments that they cannot at first perceive that all you need to make a sweater (apart from technology and tools) is an understanding of ratio and all you need to make a shirt is an understanding of right-angled and parallel lines, the idea of area, some symmetry, some optimisation, and the ability to work from 2-dimensional plans to 3-dimensional forms . . . It is interesting to take Gerdes' analyses . . . and apply them . . . in the different context of women's culture."

All that is needed in many non-Western cultures is to "defrost" the frozen mathematics of the cultures contained in useful objects such as baskets, mats, pots, houses, sand-drawings, sculptures, fish-traps, etc.

There is, however, a major stumbling block: the wide spread acceptance of the hegemony of a Western version of mathematics, following from the assumption that mathematics is largely a European creation. Two tactics have been used to propagate this Eurocentric myth.

The first is Omission and Appropriation. Prior to the "Renaissance", European acknowledgment of the debt it owed to Arab mathematics and its antecedents was fulsome both in words and deeds. Indeed the course of European cultural history and the history of European thought are inseparably tied up with the activities of Arab scholars during the Middle Ages and their seminal contributions to mathematics, the natural sciences, medicine, and philosophy. By the seventeenth century, however, the perception concerning the origins of mathematical knowledge had begun to change, due to the workings of a number of forces. With the European expansion in the American continents, the development of the slave trade, and the imposition of colonial rule in many parts of the world, the assumption of white superiority became dominant over a wide range of social and economic activities, including the writing of the history of mathematics. Moreover, the rise of nationalism in nineteenth and twentieth century Europe and the consequent search for the roots of European civilization led to an obsession with Greece and the myth of Greek culture as the cradle of all knowledge and values. This was despite ample evidence of significant mathematical developments in Mesopotamia, Egypt, China, pre-Columbian America, India, and the Arab world, showing that Greek mathematics owed a significant debt to most of these cultures. In recent years a grudging recognition of the debt owed by Greece to earlier civilizations and the important contribution of Arab mathematicians has led to some revisions to a 'purely' Eurocentric trajectory of the historical development of mathematics. But the modifications still ignore for the most part the routes through which Hellenistic and Arabic mathematics entered Europe and take little account of the mathematical knowledge produced by India, China, and other cultures. Even those texts which include the Indian and Chinese mathematics often confine their discussion to a single chapter which may go under the misleading title of "Oriental" or "Eastern" mathematics. That these cultures contributed to the mainstream development of mathematics is rarely recognized, and little consideration is given to the mathematical research that is currently taking place in these and other non-Western regions.

The second tactic is Exclusion by Definition. A Eurocentric approach to the history of mathematics is intimately connected with the dominant view of mathematics as a socio-historical practice and intellectual activity. Despite the development of contrary trends in the last two centuries, the standard textbook approach sees mathematics as a deductive system, ideally proceeding from axiomatic foundations and revealing, by the 'necessary' unfolding of its pure abstract forms, the eternal/universal laws of the 'Mind'.

The Indian and Chinese concepts of mathematics were very different. Their aim was not to build an imposing edifice on a few self-evident axioms, but to validate a result by any method. Some of the most impressive works in Indian and Chinese mathematics (the summations of complex mathematical series, the use of Pascal's triangle in solutions of higher order numerical equations, the derivations of infinite series, and the 'proofs' of the Pythagorean theorem) involve the use of visual demonstrations that are not formulated with reference to any formal deductive system. The Indian view of the nature of mathematical objects, like numbers, is also based on a framework developed by Indian logicians and linguists which differs at the foundational level from the set theory universe of modern mathematics.

The view that mathematics is a system of axiomatic/deductive truths inherited by the Greeks and enthroned by Descartes has been traditionally accompanied by a cluster of values that reflect the social context in which it originated:

(1) an idealist rejection of any practical, material(ist) basis for mathematics : hence the tendency to view mathematics as value-free and detached from social and political concerns;

(2) an elitist perspective that sees mathematical work as the exclusive preserve of a pure, high-minded and almost priestly caste, removed from mundane preoccupations and operating in a superior intellectual sphere.

Non-Western mathematical traditions have therefore been dismissed on the grounds that they are dictated by utilitarian considerations with little notion of rigor in proof. Any attempt at excavation and restoration of non-Western mathematics is a multi-faceted task: confront historical bias, question the social and political values shaping the mathematics curriculum, and search for different ways of knowing or establishing mathematical truths found in various traditions.

GEORGE GHEVERGHESE JOSEPH

REFERENCES

Ascher, Marcia. *Ethnomathematics: A Multicultural View of Mathematical Ideas.* Pacific Grove, California: Brooks/Cole Publishing Co., 1991.

Bishop, Alan. "Western Mathematics: The Secret Weapon of Cultural Imperialism." *Race and Class* 32 (2): 51–65, 1990.

D'Ambrosio, Ubiritan. "Ethnomathematics and its Place in the History and Pedagogy of Mathematics." *For the Learning of Mathematics* 5 (1): 44–48, 1985.

Gerdes, Paulus. "How to Recognize Hidden Geometrical Thinking: a Contribution to the Development of Anthropological Mathematics." *For the Learning of Mathematics* 6 (2): 10–17, 1986.

Harris, Mary. "An Example of Traditional Women's Work as a Mathematical Resource." *For the Learning of Mathematics* 7 (3): 26–28, 1987.

de Heinzelin, J. "Ishango." *Scientific American* 102: 105–116, June 1962.

Ifrah, George. *From One to Zero: A Universal History of Numbers.* New York: Viking, 1985.

Joseph, George G. *The Crest of the Peacock: Non-European Roots of Mathematics.* London: Penguin Books, 1993.

Marshack, A. *The Roots of Civilisation.* Mount Kisco, New York: Moyer Bell, 1991.

Nelson, David, George G. Joseph, and Julian Williams. *Multicultural Mathematics.* Oxford: Oxford University Press, 1993.

Zaslavsky, Claudia. *Africa Counts.* New York: Lawrence Hill Books, 1979.

See also: Algebra – Geometry – Computation – Gou-gu Theorem – *Zhoubi Suanjing* – *Śulbasūtras* – Liu Hui and the *Jiuzhang Suanshu* – al-Khwārizmī – ʿUmar al-Khāyyam – Brahmagupta – Bhāskara – Abacus – Decimal System – *Quipu*

MATHEMATICS IN AFRICA SOUTH OF THE SAHARA

Most books on the history of mathematics devote only a few pages to Africa, and even then only to Ancient Egypt and to northern Africa during the Middle Ages. Generally they ignore the existence of mathematics in Africa south of the Sahara. They often deny that Egyptian mathematics is African. With the publication of Claudia Zaslavsky's *Africa Counts: Number and Pattern in African Culture* in 1973 this dominant Eurocentric view of the history of mathematics in Africa became challenged. When one uses a broad definition of mathematics — including counting, locating, measuring, designing, playing, explaining, classifying, sorting, etc. — it becomes clear that mathematics is a pan-cultural phenomenon manifesting itself in many ways. In African history, we have evidence of counting and numeration systems, games and puzzles, geometry, graphs, record-keeping, money, weights and measures, etc. Mathematics in Africa may not be considered in isolation either from the development of culture and economy in general, or from the evolution of art, cosmology, education, philosophy, natural sciences, medicine, logic and language, graphic systems, and technology in particular. The application of historical and ethnomathematical research methods in recent years has contributed to a better knowledge and understanding of the history of mathematics in Africa.

In this article evidence for early mathematical activity in Africa will be given, followed by examples from geometry, games, riddles, and puzzles with mathematical 'ingredients'. Some topics for future research will be indicated and some comments about the development of mathematics south of the Sahara and in other regions, in particular northern Africa, will be presented.

A small piece of the fibula of a baboon, marked with 29 clearly defined notches, may be one of the oldest mathematical artifacts known in the world. Discovered in the early seventies during an excavation of a cave in the Lebombo Mountains between South Africa and Swaziland, the bone has been dated to approximately 35,000 BC. This bone resembles calendar sticks still in use today by the San (Bushmen) in Namibia. The San hunters developed very good visual discrimination and visual memory for survival in the harsh environment of the Kalahari desert. From the San in Botswana, information has been collected on their counting, measurement, time-reckoning, classification, and tracking. Well-known as early evidence for mathematical activity in Africa is a bone now dated from about 8000 BC to 20,000 BC, dug up at Ishango (Zaire). The bone has what appear to be tallying marks on it, notches carved in groups that have been explained as early lunar phase count or as an arithmetical game of some sort.

Georges Njock has characterized the relationship between African art and mathematics as follows: "Pure mathematics is the art of creating and imagining. In this sense black art is mathematics."

Mathematicians have primarily analyzed symmetries in African art. Symmetries of repeated patterns may be classified on the basis of the twenty-four different possible types of patterns which can be used to cover a plane surface (these are the twenty-four plane groups attributed to Federov). Of these, seven admit translations in only one direction and are called strip patterns. This classification has been applied to decorative patterns that appear on the raffia pile cloths of the Bakuba (Zaire), on Benin bronzes, and on Yoruba adire cloths (Nigeria), showing that all seven strip patterns and many of the plane patterns occur. The use of group theory in the analysis of symmetries in African art attests to and underlines the creative imagination of the artists and artisans involved and their capacity for abstraction. These studies do not focus, however, on how and why the artists and artisans themselves classify and analyze their symmetries. This is a field open for further research.

Why do symmetries appear in human culture in general, and in African craftwork and art, in particular? Paulus Gerdes analyzed the origin of axial, double axial, and rotational symmetry in African basketry. He showed how six

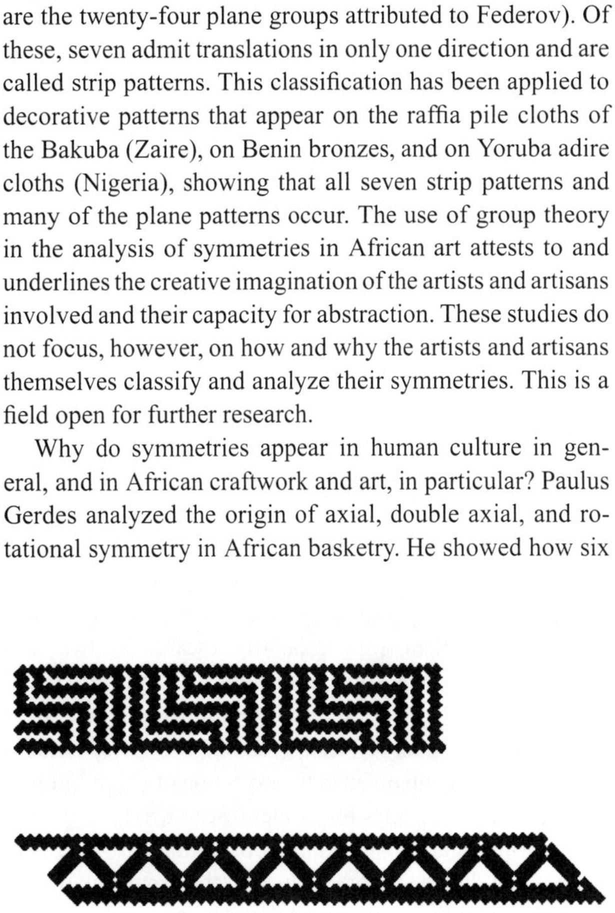

Figure 1: Two examples of woven strip patterns on baskets (Mozambique).

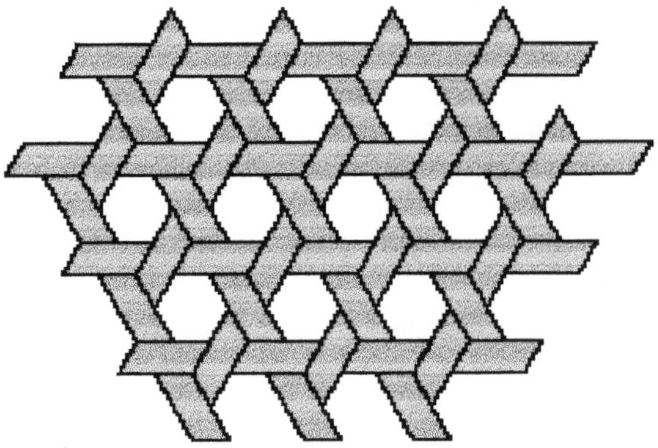

Figure 2: Hexagonal weaving pattern (Congo, Kenya, Madagascar, Mozambique).

and fivefold symmetry emerged quite naturally when artisans were solving some problems in (basket)weaving (see Figure 2).

Beehive, conical, cylindrical, and rectangular shapes are common in African architecture. In West Africa the mathematician-scholar and the architectural designer-builder was often the same person. An example of geometrical know-how used in laying out the rectangular house plans in Mozambique is the following. Two ropes of equal length are tied together at their midpoints. A bamboo stick, whose length is equal to the desired breadth of the house, is laid down on the floor, and at its endpoints pins are hit into the ground. An endpoint of each of the ropes is tied to one of the pins. Then the ropes are stretched and at the remaining two endpoints of the ropes, new pins are hit into the ground. These four pins determine the four vertices of the house to be built.

The geometric shapes of pots, baskets, fishtraps, houses, etc. generally represent many practical advantages and are frequently the only possible or the optimal solutions of a production problem. Some scholarly research has been undertaken concerning knowledge about the properties and relations of circles, angles, rectangles, squares, regular pentagons and hexagons, cones, pyramids, cylinders, symmetry, etc. probably involved in the invention of the techniques (Gerdes, 1992). Themes for further research are the geometry of string figures, the geometry of settlement patterns in Africa, and the geometry involved in the ornamentation of the walls of buildings all over Africa (e.g. the attractive colorful patterns of the Ndebele in South Africa and the *litema* patterns drawn by Basotho women in Lesotho).

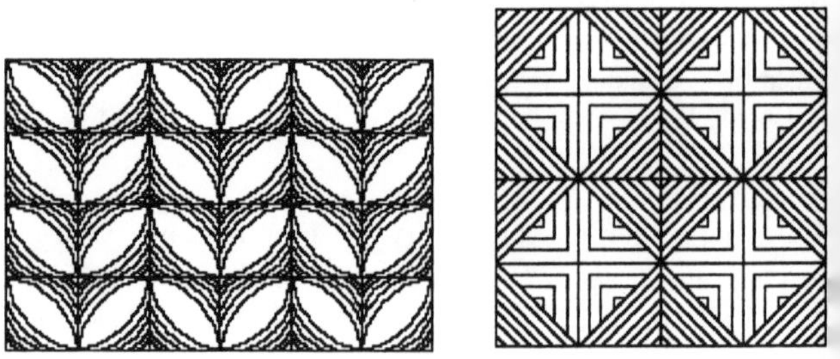

Figure 3: Examples of *litema* wall patterns (Lesotho).

Games

Among the games with mathematical ingredients are counting rhymes and rhythms, arrangements, three-in-a-row-games like *Shisima* (Kenya), *Achi* (Ghana), *Murabaraba* (Lesotho), *Muravarava* (Mozambique), and games of chance. Board games like Mancala games, both two-row versions such as *Oware* (Ghana), *Awélé* (Côte d'Ivoire), *Ayo* and *Okwe* (Nigeria), and four-row versions such as *Omweso* (Uganda), *Tshisolo* (Zaire), and *Ntchuva* (Mozambique) also exhibit use of mathematical knowledge.

Recent research in Côte d'Ivoire showed that the rules of some games, like *Nigbé Alladian*, reveal a traditional and empirical knowledge of probabilities.

Riddles and Puzzles

From the Kpelle (Liberia) a riddle has been reported about a man who has a leopard, a goat, and a pile of cassava leaves to be transported across a river, whereby certain conditions have to be satisfied: the boat can carry no more than one at a time, besides the man himself; the goat cannot be left alone with the leopard, and the goat will eat the cassava leaves if it is not guarded. How can he take them across the river? This type of problem is also known from Ethiopia, Liberia, Tanzania, Uganda, and Zambia. More difficult to solve is the following puzzle from the Valuchazi (eastern Angola and northwestern Zambia) about three women and three men who want to cross a river in order to attend a dance on the other side. With the river between them there is a boat with the capacity for taking only two people at one time. However, each of the men wishes to be the only husband of all the three women. Regarding the crossing they would like to cross in pairs, each man with his female partner, but failing that any of the other men could claim all the women for himself. How can they cross? In order to solve the problem or to explain the solution, the Valuchazi make auxiliary drawings in the sand.

The relationships between the development of mathematics in Africa south of the Sahara and the development of mathematics in Ancient Egypt, in both Hellenistic and Islamic northern Africa, and across the Indian and Atlantic Oceans, deserve further study.

Throughout history there have been many and varied contacts between Africa south of the Sahara and North Africa. Since the birth and spread of Islam, relations have been intensified and/or extended. *ʿIlm al-Ḥisāb* (arithmetic), as part of the Islamic sciences was introduced some time after the eleventh century in Nigeria, first in Kanem-Borno and later, probably in the fifteenth century in Hausaland. Arithmetic was taught in both secular and Islamiyya schools, was used in the courts for the calculation of inheritance, and for collecting and distributing poor dues, business, and land surveying. A famous mathematician was Muḥammed ibn Muḥammed from Katsina (now northern Nigeria), who worked on chronograms and magic squares. He had been a pupil of Muḥammed Alwali of Bagirmi. He made a pilgrimage to Mecca in 1730, and he died in Cairo in 1741. Recently a manuscript of his was found in Marrakesh, Morocco. Magic squares were used in amulets among the Fulbe, and in Niger, Benin and Timbuktu (Mali). Formal logic was one of the fields of study in the economic and educational-scientific centre of Timbuktu, where recently three Arabic mathematical manuscripts were found in the Aḥmad Baba Library. One of the three manuscripts, whose calligraphy is typical for Africa south of the Sahara, seems to have been written by a mathematician from Mali, al-Arwani. The other two contain references to medieval mathematicians from the Maghreb. Systematic research in libraries and archives will probably lead to the discovery of more mathematical manuscripts from Muslim scholars south of the Sahara.

What mathematics was brought to the Americas by the slaves? Which mathematical ideas have survived in one way or another? Mancala, and maybe other games with mathematical components, are played in the Caribbean and may be compared with their African 'ancestors'. In Africa the slave trade was extremely destructive of the existing mathematical traditions. This is because of breaking professional continuity and depriving Africa of bearers of mathematical knowledge and skills such as that of the drawing experts from Angola and Zambia and that of calculators like Thomas Fuller (1719–1790). Recent ethnomathematical research in Nigeria and Mozambique shows the survival, none the less, of a rich tradition of mental calculations among illiterate people.

The destructive impact of colonialism and the slave trade on Africa is one of the principal reasons Georges Njock gives to explain why mathematics in Africa has had a slower development in the last five centuries than in Europe. Other reasons he gives deal with the geography of the continent (migratory movement) and wars. This also constitutes an important research area that deserves further study.

PAULUS GERDES

REFERENCES

African Mathematical Union. Commission on the History of Mathematics in Africa (AMUCHMA). *Newsletter*, 1986.

Gerdes, Paulus. "On the History of Mathematics in Africa South of the Sahara." *AMUCHMA Newsletter* 9: 3–32, 1992.

Gerdes, Paulus. *Ethnogeometrie. Kulturanthropologische Beitrage zur Genese und Didaktik der Geometrie*. Bad Salzdetfurth/Hildesheim: Verlag Barbara Franzbecker, 1992.

Gerdes, Paulus. "Fivefold Symmetry and (Basket)weaving in Various Cultures." In *Fivefold Symmetry in a Cultural Context* Ed. I. Hargittai. Singapore: World Scientific Publishing, 1991, pp. 243–259.

Gerdes, Paulus. "On Ethnomathematical Research and Symmetry." *Symmetry: Culture and Science* 1(2): 154–170.

Kani, Ahmad Mohammad. "Arithmetic in the Pre-Colonial Central Sudan." In *Science and Technology in African History with Case Sudies from Nigeria, Sierra Leone, Zimbabwe, and Zambia*. Ed. G.Thomas-Emeagwali. Lewiston: Edwin Mellen Press, 1992, pp. 33–39.

Lumpkin, Beatrice. "Africa in the Mainstream of Mathematics History." In *Blacks in Science*. Ed. I. Van Sertima. New Brunswick, New Jersey: Transaction Books, 1983, pp. 100–109.

Njock, Georges E.. "Mathématiques et environnement socioculturel en Afrique Noire." In *Présence Africaine, New Bilingual Series*. 135(3): 3–21, 1985.

Washburn, Dorothy, and Donald Crowe. *Symmetries of Culture: Theory and Practice of Plane Pattern Analysis*. Seattle: University of Washington Press, 1988.

Zaslavsky, Claudia. *Africa Counts: Number and Pattern in African Culture*. Brooklyn, New York: Lawrence Hill Books, 1979.

See also: Number Systems in Africa – Geometry in Africa: Sona Geometry – Arithmetic – Magic Squares

MATHEMATICS OF AFRICA: THE MAGHREB In discussing the mathematical activity which occurred within the framework of Arab-Islamic civilization, it is not possible to speak of one specific tradition from the medieval Maghreb, nor still less of specific traditions from each of the three regions of the Maghreb–Ifriqya, the central Maghreb and the far Maghreb. Indeed, there has been one Arab mathematical tradition, born and developed in the East, partially transmitted to the Muslim cities of the west of Central Asia and then, later, to southern Europe by the intermediary of Latin and Hebrew translations. This tradition was assimilated, revived, and enriched by different scientific milieux.

Mathematics in the Maghreb from the Ninth to the Eleventh Centuries

Considering the close economic, political, and cultural ties which linked the Maghreb with the Andalus, it is not possible to separate the mathematical activities which took place in these two regions of the Muslim West. Indeed, the period which extends from the end of the eighth century to the end of the eleventh was characterized by the development of two scientific traditions linked to each other and brought to life by scholars who, apart from social divisions and differences in laws and religion, were united by the cultural and scientific environment which was made up of exchanges between the Andalus and the Maghreb, and of frequent contacts with the cities of the east — Baghdad, Damascus, and, later, Cairo.

That being said, the beginnings of scientific activities in the Andalus and the Maghreb are not well known. Indeed, information on the earliest mathematical activities of the region is rare and not very specific. But it is reasonable to state that, during the period of installation and consolidation of Arab-Islamic power, the development of the study and teaching of the Arabic language and of the different religious sciences of the period probably favored the teaching of sciences like medicine and arithmetic, which answered the needs of certain fringes of urban society (in particular, caring for the wealthy and administrating the distribution of inheritances).

In the Andalus, it is likely that from the beginning of the ninth century, the first translations of Greek and Indian works appeared in Cordoba from the centers of the empire and served to bring scientific learning to everyone from children to rulers. This was the case, for example, for ʿAbd ar-Raḥmān II (822—852) who benefited from this education and, in turn, actively supported further scientific activity by financing the establishment of an important library with works bought in the East. But it was not until the middle of the ninth century that scientific centers began to exist independently, outside the walls of palaces and princely mansions, in Cordoba, Toledo, Seville, and Saragossa.

During the last third of the ninth century and throughout the tenth the activities of teaching and research in the different branches of mathematics were greatly stimulated by the patronage of two great Umayyad caliphs, ʿAbd ar-Raḥmān III (912–961) and his son al-Ḥakam II (961–976). These activities continued in the eleventh century with the blossoming of important scientific centers in capital cities of principal states which, in turn, stemmed from the brilliance of the caliphate of Cordoba.

The information which exists today about scientific activities in the Maghreb leads one to believe that the beginnings of mathematics, on the scale of all of the Maghreb, occurred in Ifrīqyā (present day Tunisia) as early as the end of the eighth century. Among the scholars whose names are still known is Yaḥyā al-Kharrāz, who had as a student Yaḥyā al-Kinānī (828–901), the author of the first book of *ḥis'* (control of weights and measures) written in the Maghreb.

From the ninth century, the name of only one mathematician remains, Abū Sahl al-Qayrawānī, of Iraqi origin. He is also the first Maghrebian mathematician whose book is known: *Kitāb fi al-ḥisāb al-hindī* (Book of Indian Mathematics). As the title indicates, the book was in the new Arabic arithmetical tradition, of Indian origin, which had been inaugurated by mathematicians in Baghdad at the end of the eighth century and the beginning of the ninth.

As in other regions of the Islamic world, patronage of scientific activity was prevalent in the Maghreb and functioned as it did in the great cities of the East, with the purchase of books, the financing of manuscript copies, grants to scholars, and the construction of schools and institutions. The only precise information we have is on the *Bayt al-ḥikma* (House of Wisdom) founded by Ibrāhīm II (875–902) in Raqqāda. This institution, which survived as a scientific center until the Fatimid dynasty, welcomed mathematicians, astronomers, and astrologers.

Maghrebian mathematical activity of the tenth century is not well known. It seems that the patronage initiated by the Aghlabids in the ninth century continued and that the study of mathematics and astronomy flourished, particularly in the course of the first two decades of the reign of the Fatimid Caliph al-Muʿizz (953–975). But biographers have given us only a few names of people who were known for their mathematical activities or their interest in the discipline, such as Yaʿqūb ibn Killīs (d. 990) and al-Huwarī (d. 1023). To these names must be added the names of scholars who came under the influence of the intellectual centers of the Umayyad Andalus or of Fatimid Egypt, like Ibn Yāsīn and al-ʿUtaqī.

We are somewhat better informed about mathematical activity in the eleventh century, but information is still fragmentary. Some scholars of this period are well known, for example Ibn Abir-Rijāl (d. 1034–1035) and Abū'l-Ṣalt (1067–1134). They wrote on mathematics, astronomy, and astrology, but only certain of their works in the latter two disciplines survive.

Mathematics in the Maghreb in the Almohad Period (Twelfth–Thirteenth Centuries)

We already know about the importance of the twelfth century in the political and economic history of the Maghreb. But the cultural and scientific history of that period remains a vast, unexplored field. For example, in the field of mathematics, only three scholars of this period have been the subject of research studies. They are Abū Bakr al-Ḥaṣṣār (twelfth century), Ibn al-Yāsamīn (d. 1204), and Ibn Munʿim (d. 1228). These three scholars are important because of several factors. The first is that they were the first mathematicians in the Maghreb whose writings have survived.

The full name of al-Ḥaṣṣār is Abū Bakr Muḥammad ibn ʿAbdallāh ibn ʿAyyāsh al-Ḥaṣṣār. It appears that in addition to his mathematical activities, he was also a reader of the *Quʾrān* and a specialist in inheritances. It is also probable that he lived a long time and that he taught in Sebta, since he seems to have had ties with other mathematicians in that city. Only two of his writings have survived. The first, entitled *Kitāb al-bayān wat-tadhkār* (Book of Demonstration and Memorization) is a manual of calculation dealing basically with arithmetical operations on whole numbers and fractions. His second book is entitled *Kitāb al-kāmil fī ṣināʿat al-ʿadad* (Complete Book on the Art of Numbers). It was in two volumes, but only the first volume is extant. It takes up the themes of the first book, with new chapters on the breakdown of a number into prime factors, on common divisors, and on common multiples.

Reference to al-Ḥaṣṣār in two Andalusian works which are now lost leads us to the conclusion that, in one way or another, the Andalusian arithmetical tradition was present in the Maghreb in the twelfth century. This presence was reinforced, moreover, both by the direct diffusion of Andalusian works on algebra, geometry, and astronomy and by the utilization of the contents of the works of al-Ḥaṣṣār by later Maghrebian mathematicians like Ibn Munʿim, Ibn al-Bannāʾ (d. 1321), and Ibn Ghāzī (d. 1513).

It appears that al-Ḥaṣṣār lived before Ibn al-Yāsamīn, but there is no way to verify that fact. However, the contents of their mathematical writings are quite similar insofar as they are written in the Andalusian tradition of the twelfth century, and they expand that tradition by the introduction of symbolism for certain objects and certain arithmetical operations and by the important development of a chapter on fractions.

We can say nothing about the works of Ibn Munʿim on geometry or on the construction of magic squares, as they have not yet been discovered. But we are better informed about his writings on combinatory analysis, arithmetic, and number theory. The *Fiqh al-ḥisāb* (Science of Calculation), his only surviving text, contains the usual chapters of calculation manuals, dealing with the four arithmetic operations applied to whole numbers and to fractions, but also chapters at a higher level which deal with extraction of the exact or approximate root of a number, the properties of figurative numbers, the sum of series of whole numbers, and the determination of amicable numbers.

It is not known what the importance of mathematical activity in Marrakesh was at the time that Ibn Munʿim settled there. He himself is not specific about the Maghrebian mathematicians of his time or their predecessors. On the other hand, he refers very specifically to Andalusian scholars, citing their names, titles of their works, and even passages of their writings. This only serves to confirm the important scientific ties between the Maghreb and the Andalus, ties which were considerably strengthened in the fields of science and philosophy during the reigns of the first four Almohād caliphs, from 1130 to 1213.

It is important to note that the writings of al-Ḥaṣṣār, Ibn al-Yāsimīn, and Ibn Munʿim were not the only writings which were studied. Indeed, one can cite the names of several whose work may have been just as important: Abū'l Qāsim al-Qurashī (d. 1184) who taught algebra in Bougie, al-Qāḍī al-Sharīf (d. 1283) who was a student of Ibn Munʿim in Marrakesh, al-Qalʿī (d. 1271) who also lived in Bougie and who taught the science of inheritances, and Ibn Isḥaq al-Tūnusī (d. after 1218), who was known for important work in astronomy. Unfortunately, none of the mathematical writings of these scholars remains, and we cannot even speculate on their contents.

The Mathematics of Ibn al-Bannāʾ and Commentaries on His Work (Fourteenth–Fifteenth Centuries)

In the history of scientific activity in the Maghreb, the fourteenth century is a special time, not only because of the quantity of mathematical work but also because of the content of that work and its influence on the teaching of mathematics in the Maghreb in the centuries to follow.

The majority of mathematical work done in this century was reviews, in the form of commentaries and summaries of work which had already been discovered or assimilated in the course of previous centuries. New contributions were the exception. The work of the mathematician Ibn al-Bannāʾ (1256–1321) becomes even more important in this light, since he was both one of the last innovators in the great Arab mathematical tradition and also one of the initiators of a new tradition in the teaching of mathematics.

Ibn al-Bannāʾ was born and raised in Marrakesh where he also died, but he also lived and taught in Fez for a period. He seems to have been one of the last scholars to have engaged in research activity in that he grappled with problems which were new at that time and to which he found original solutions, for example in combinatory analysis. He also introduced original ideas or processes in algebra (on the existence of solutions to a second degree equation) and in calculus (on non-decimal bases).

Ibn al-Bannāʾ was also notable for the richness and diversity of his work. In the inventory which Ibn Haydūr (d. 1413) made of his writings, he records ninety-eight titles, thirty-two of which deal with mathematics and astronomy. This may have been the reason for the high social status he enjoyed, being honored by the highest authorities in the far Maghreb, which led him to leave Marrakesh and settle in Fez for a period of time.

Ibn al-Bannā᾽ wrote on geometry and an important work on algebra which was used as a teaching text in the Maghreb until the fifteenth century. But it is his writings on the science of calculation which have made him famous. Three of them remain in existence: the *Talkhīṣ aʿmāl al-ḥisāb* (Summary of the Operations of Computation), *al-Qānūn fi l-ḥisāb* (Canon of Mathematics), and the *Rafʿal-ḥijāb* (Lifting of the Veil).

While the third title is the most important in mathematical terms, it is the first which is best known. A manual on the operations of calculation, it is characterized by great conciseness, rigor in its formulation, and especially, by a total absence of mathematical symbolism. For these reasons, many mathematicians after Ibn al-Bannā᾽ have published commentaries on this manual.

A comparative study of the most important chapters of these commentaries leads to the following observations on their content and on the general level of mathematics at that time: first, that the level of mathematics was not lower than that of earlier periods but that certain themes which had been taught since the tenth century no longer appear. Second, there are no new contributions in the commentaries, neither on the theoretical level nor on the level of the applications of ideas or earlier techniques. The most significant innovation in the commentaries is in the progressive utilization of a relatively elaborate symbolism. This symbolism had already appeared in the writings of al-Ḥaṣṣār and Ibn al-Yāsamīn, but its use seems to have taken hold throughout the thirteenth century and the beginning of the fourteenth. It reappeared in certain commentaries in the fourteenth and fifteenth centuries, both in arithmetic and in algebra.

Mathematics in the Maghreb After the Fifteenth Century

The activities of mathematicians from the Maghreb who lived between the sixteenth and nineteenth centuries are not clearly known, but it is possible to get an idea of their contributions from the titles of writings which still exist. The mathematical disciplines which were taught or treated in the writings are metric geometry and calculation. It is possible to affirm that the content of these works differs from that of earlier mathematical writing both in form and level. There are poems, for example those of al-Akhḍarī (d. 1575), and al-Wansharīsī (d. 1548), glosses and commentaries like those of Ibn al-Qāḍī (d. 1616) and Muḥammad Bannīs (d. 1798), and summaries like those of Ibrāhīm al-Ribāṭī (d 1926). But the level of these works is inferior to those of the fifteenth century, which are themselves much poorer than works of the thirteenth and fourteenth centuries, both in ideas and techniques. The same is true in other sectors of intellectual activity in the Maghreb which were characterized by a narrowing of the respective domains of investigation and an impoverishment of their contents. This situation, the result of a long process of decline, had an indirect effect in mathematics in a gradual reduction both of content and in the field of application. Thus, the only activities left to mathematicians, aside from teaching and editing manuals, were mathematical practices linked directly to activities or preoccupations of a religious nature, like the distribution of inheritance, and donations to rightful claimants, the determination of times for fixing moments of prayer, or the construction and use of astronomical instruments.

AHMED DJEBBAR

REFERENCES

Aballagh, M. Rafʿal-ḥijāb d'Ibn al-Banna᾽. Thèse de Doctorat. Paris I-Sorbonne, 1988.

Djebbar, A. Mathématiques et Máthématiciens du Maghreb, Médiéval (IXe–XVIe siècles): Contribution à létude des activités scientifiques de l'Occident musulman. Thèse de Doctorat. Université de Nantes (France). 1990.

Souissi, M. *Talkhīṣ aʿmāl al-ḥisāb*. Tunis, 1969.

See also: Ibn al-Yāsimin – Ibn Murʿim – Ibn al-Bannā᾽

MATHEMATICS IN WEST AFRICA: TRADITIONAL MATHEMATICAL GAMES Exploring the socio-cultural environment of Africa is an interesting way of learning about concrete mathematics. Games, always a popular childhood activity, should be studied with care in as much as they are a reflection of the society and its fundamental values. Today we are witnessing the introduction of "educational games" to the African continent, as if African games did not exist. In this article, we will describe several African games and the mathematics behind them.

Exhibition Games

In Africa there are games which have hidden underlying algorithms. The person who knows the algorithm is assured of winning, which makes him a "magician" or "sorcerer" to those who do not know it. The "sorcerer" often chants incantations or performs other magic tricks to show his power. We will study two of these games.

I. *The game of the sorcerer*. The game of the sorcerer is

very old. Sorcerers perform it to demonstrate their supernatural powers. Forms of the game are played throughout Africa–in Senegal (Game of the Devil), in the bend of the Niger River (the Sorcerer's Apprentice), in the Ivory Coast (*Lokoto*), and in Mali (*Gamma*).

There are two players: the sorcerer and his victim. The game is played on a board of twelve compartments set up in two rows of 6 and 48 beads.

Before beginning play, the players place four beads in each compartment. The sorcerer and the victim each take two stones from eleven compartments. From the twelfth, the sorcerer takes three beads, the victim one. (The sorcerer thus has two extra beads). He must now "magically" pass the two extra beads to the victim. He rubs the beads, chants incantations, blows on his thumbs, and then instructs the victim to fill the compartments with four beads each as he does the same. When the beads are redistributed, three beads are left in the victim's hand, one in the sorcerer's, and the trick is done.

If we analyze the mathematical aspects of the game, we see that it has two parts:

1. The taking of the stones (beads).
   The sorcerer: $2 \times 11 + 3 = 25$.
   The victim: $2 \times 11 + 1 = 23$.
2. The replacing of the stones.
   Each player must put four beads in each of the compartments, filling as many as he can with the beads in his hand.
   The sorcerer: $25 = 4 \times 6 + 1$.
   The victim: $23 = 4 \times 5 + 3$.

After redistributing the beads, the victim has three beads in his hand and the sorcerer, one. The game works whenever the number of compartments is even.

II. *The game of the unknown stone.* The player asks a spectator to select in his mind one of sixteen different stones. The player must discover the unknown, selected stone. First he arranges the stones in two rows of eight. Then he asks the spectator to indicate the row that the unknown stone is in. Mentally the player makes two groups of four stones each from the stones in the indicated row. He rearranges them (or distributes each group on different lines) and asks the spectator again to indicate which row has the unknown stone. Upon the indication of the row he makes two groups of two stones each from one of the groups of four stones. Then he rearranges them again.

Again, the spectator indicates the row with the unknown stone. The player then redistributes the only two stones left, which could possibly be the unknown stone, into two different rows. The spectator again indicates the row with the unknown stone. Having identified the stone, the player then collects all the stones together, mixes them up, pulls out the "unknown" stone, and the game is over.

The mathematics of this game shows that the player must divide the stones into two rows in order to locate the unknown stone.

$16 = 2^4$ where 4 is the number of times that the player must divide the sixteen stones, two being the number of rows. More generally, if n is the total number of stones, and k is the number of times the player must divide the stones into two rows, then the following equation holds true: $n = 2^k$. Inversely, if one knows n, one can calculate k from the formula $k = \log_2 n$.

Games of Chance

Men have always played games of chance. In Africa, there are many situations in which chance is called into play: when one needs to make a hard choice (heads or tails, the short straw), or when one needs to be sure of winning. Games of chance are based on theories of probability.

The cowrie is a sea shell with two sides: a front and a back. In Africa, cowries were considered rare and valuable, and so played an important role in African life. Its form, similar to an aura according to the mystics, makes it an instrument of divination. Cowries also play a role in initiations, funerals, and engagements, and in certain regions of Africa, they were — and still are — used as a medium of exchange.

All the cowrie games are collective games. The number of players is two or more. The principle of the game is simple. Two players (or two representatives of two teams) throw the cowries at the same time. The cowries fall on the ground, landing on one of two sides. There are usually four shells. The configurations formed by the cowries are examined so that the results can be read. The results are interpreted according to pre-established rules: Heads is given a $+$; tails is given a $-$.

Thus, when the four stones are thrown, there are five possible configurations:

Heads (+)	Tails (−)			
+ + + +	− − − −	10 points	− − − −	win
+ + + −	+ + + +	5 points	+ + + +	win
+ + − −	+ + − −	2 points	+ + − −	win
+ − − −	+ + + −	0 points	+ + + −	lose
− − − −	+ − − −	0 points	+ − − −	lose
Configurations		Example 1		Example 2

The calculation of points varies. There are two ways to compute points, either numerically (Example 1) or not (Ex-

Figure 1: Game of the Sorcerer. Photograph by the author.

ample 2). Generally, if points are calculated numerically, the winner is declared when a player reaches a certain total number of points, that number being fixed before play begins.

Word Games and Traditional Learning in Africa

In many African countries, traditional learning is oral. Information and wisdom are transmitted from generation to generation in the form of proverbs, stories, chants, riddles, and games. This provides real lessons in language, history, geography, natural science, arithmetic, measurement, cosmography, etc. Teaching is seen as the natural and traditional way of communicating the secrets of the adult world to a child. Following are some examples of counting games.

Cumulative chants.

Memory Games: The Yé Gonan. The Yé Gonan is a children's game from the Ivory Coast. There are two players, one who asks questions and one who answers. The players use eight stones and eight holes dug in a line in the sand, each one containing a stone. The player who is answering looks at the game and makes himself/herself acquainted with the rules. One of the rules is to retain correctly the sense of the course imposed by the questioning player. Then the answering player turns his back on the game. For each turn, the questioner taps on the hole and always asks the same question: "Am I taking a stone?" The answerer, who has his back to the game, can only answer "Yes" or "No, it's empty." Each time that he has taken a stone, the questioner comes back to the first hole. The answerer who reaches the eighth hole without making a mistake, wins the game. The players then change roles.

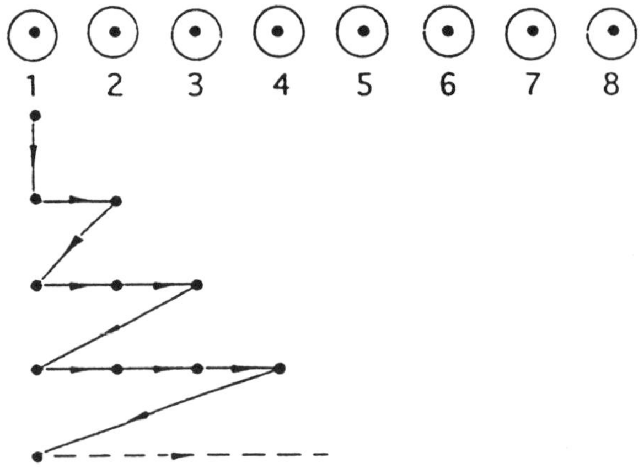

The mathematical problem posed by the game is the following: when the questioner is at hole number n, which contains a stone, which is the total number Z_n of questions posed from the beginning of the game to the taking of that stone?

The answer is $Z_n = n(n+1)/2$.

It is a triangular number. For $n = 8$, one finds that: $Z_8 = 9 \times \frac{8}{2} = 36 = 6^2$. The number Z_8 is therefore a squared number: it is the smallest number that is, at the same time, both triangular and squared. The Pythagorean tradition considered the number $n(n+1)/2$ as a secret value. It is interesting to find this same representation in the Pythagorean tradition both among the moors in Mauritania and the Wolofs in Senegal.

$Z_1 = 1$	•	1
$Z_2 = Z_1 + 2 = 1 + 2$	• •	2
$Z_3 = Z_2 + 3 = 1 + 2 + 3$	• • •	3
.	• • • •	4
.	• • • • •	5
.	• • • • • •	6
.	• • • • • • •	7
.	• • • • • • • •	8
$Z_n = Z_{n-1} + n$		
$= 1 + 2 + \cdots + (n-1) + n$	$\cdots$	n

Cumulative chants diagram.

Riddle problems. In The Vultures, a shepherd, spending the night under a baobab tree, heard an old vulture pose the following riddle to some children: "There are 33 baobab trees; on each baobab there are 33 vultures; each vulture had laid 33 eggs; each egg yields 33 chicks; and each chick has 33 barbed feathers — How many barbed feathers are there altogether?" The shepherd, wanting to answer, fell dead. This is why, they say, the Fulani do not want to answer.

The solution to the problem lies in the calculation of powers. There are 33^6 vultures: $33^6 = 1\,291\,467\,969$.

Magic squares. The Fulani are familiar with simple magic squares. They appear drawn with Arabic numerals as amulets. They are also used as a kind of game. Someone who knows the magic square proposes it to a group of people who then try to figure it out by putting stones, pebbles, or pieces of dung in a square drawn on the ground.

2	9	4
7	5	3
6	1	8

Saturn square (sum = 15).

21	26	19
20	22	24
25	18	23

Number of Allah (sum = 66).

23134	23137	23143	23127
23142	23128	23133	23138
23129	23145	23135	23132
23136	23131	23130	23144

The Four Angles (sum = 92541).

In a magic square, the sum of the numbers in each line, horizontal, vertical or diagonal, is the same. The number of Allah is 5 + 30 + 30 + 1 = 66. In the Square of Saturn, the total is 15, and in the Four Angels, the total is 92, 541. Magic squares are well-known in Muslim countries, as they are in China and Japan.

SALIMATA DOUMBIA

REFERENCES

Béart, C. *Jeux et Jouets de l'Ouest Africain.* Dakar: IFAN, 1955.

Comoe, Krou. *Ludistique Mathématique.* Abidjan: Université d'Abidjan, 1978.

Deledicq, A. and A. Popova. "Wari et Solo." Supplement to the *Bulletin de liaison des Professeurs de Mathématiques*, no. 14, December 1977.

Doumbia, S. and J.C. Pil. *Les Jeux de cauris.* Abidjan: CEDA, 1992.

I.R.M.A. *Mathématiques dans l'Environnement Socio-Culturel Africain*, 1984.

Lombard, C. *Les jouets des enfants Baoulé.* Paris: Quatre vents, 1978.

Neveu, J. *Mathematical Foundations of the Calculus of Probabilities.* San Francisco: Holden-Day, 1965.

Plot. Dossier jeux, maths et sociétés. Décembre 1994.

See also: Mathematics, Recreational – Magic Squares

MATHEMATICS OF THE AUSTRALIAN ABORIGINAL PEOPLE To consider notions of Aboriginal mathematics usefully in a short space we need to avoid the Western controversies over what mathematics is about and how we know it, while still demonstrating something held in common between the two discourses we identify as Western and Aboriginal mathematics. Before we can say anything about notions we might understand as mathematical in the life of Aboriginal Australian communities it will help to look briefly at the cluster of meanings associated with the term 'mathematics' in Western life.

Bowing to the antiquity of the notion of mathematics we can note the Greek origins of the words mathematics and mathesis/mathetic. This cluster of words coming to us in English through Latin carries original ideas of learning, seeing, and mental discipline. To these are added overtones of systematicity: arithmetic, geometry, and reasoning.

Western mathematics involves rigorous and systematic ways of seeing and working things out, and is intimately tied up with the constitution of the social order at both the macro and the micro level. That much is uncontroversial. But asking about the objects which constitute mathematics brings us to less solid ground. Most people want to answer that mathematics is constituted of abstract objects, and, constrained to ask what abstract objects are, we find ourselves amongst the disagreeing philosophers and mathematicians.

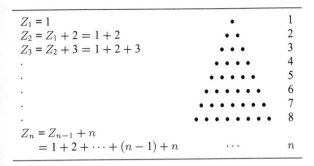

A common sense way of dealing with this is to see numbers as representative of the abstract objects we find in mathematics. Many of us are comfortable with numbers, seeing them as having their origins in very practical matters, like tallying, though we do not expect that they should remain tied to these practical origins. Conceptually we happily deal with numbers and by extension other abstract objects, by analogy to the ways we understand ourselves as dealing with fully fledged material objects.

A minimalist basis for identifying an Aboriginal mathematics is to identify a discourse with the characteristics of rigor and systematicity, which is tied up with constituting the social order, and in which abstract objects feature.

On this basis we could point to any Aboriginal community and identify an "Aboriginal mathematics", although among the many different Aboriginal Australian communities such a discourse would go under different names. This account will confine itself to looking at *gurrutu*, a mathematical-like discourse and set of practices, expressed in the community life of Yolngu Aboriginal Australians who own lands in the northeast section of Australia's Northern Territory. *Gurrutu* here is treated as representative of a cluster of discourses in a wide variety of Aboriginal languages, which in some respects are mathematical.

To render *gurrutu* in English is just as difficult as expressing 'mathematics' in the cluster of Yolngu languages. But just as Western mathematics can be represented, inadequately, as the rigorous discourse about abstract entities, which has grown out of the material practices of tallying and land division (arithmetic and geometry), Yolngu Aboriginal *gurrutu* can be inadequately represented as the rigorous discourse about abstract entities which has grown out of the material practices of kinship and land ownership. Here the practical modeling processes of tallying and mapping of kin relations provide patterns within which abstract objects come to life.

This short essay attempts to reveal some of the flavor of *gurrutu* by considering the nature of the abstract entities which constitute it and by talking of some of the rules with which these entities are manipulated. We can begin by looking at the origins of sets of abstract objects in Western mathematics and *gurrutu*.

A story about origins of one familiar set of abstract objects in Western mathematics — numbers — sees them as originating in the practices of material tallying. This is modeling an event or episode in the world with a material encoding process. We can see this as using a finger to encode the passing of a sheep through a gate, the placing of a pebble to encode the pointing at a soldier, or the engraving of a line on a bone or a piece of wood to record the filling of a vessel with grain.

We can easily imagine the involvement of fingers and toes in tally keeping in a nonlinguistic way. One separated digit codes for one separated item. But if we then extend the coding operation and say a word which codes for the finger or toe, we have done something much more complex, and ended up with a code which is much more useful than the material code of fingers and toes or display of stones/shells.

In saying a word as a finger is held up to code for an item involved in some event we understand that the word we say does not name either the item or the finger. It names a position in a progression. Numerals are words that code for a position in a series.

A numeral system is characterized by having a sequential base pattern and recursivity. Numerals constitute an infinite series by having a base about which repetition occurs, and a set of rules by which new elements are generated. Fingers and toes are used to encode events involving itemized matter, and in talk language is used to encode signs created in both these material signing systems.

The contemporary numeral system which has developed in association with Indo-European languages like English, has ten as its base; in other words ten is the point in the series which marks the end of the basic set of numerals. As each ten is reached, the basic series is started again, each time recording in the numeral how many tens have been passed. The rule by which new elements are devised is addition of single units and base ten units.

A similar story can be told of origins of one set of abstract objects in *gurrutu*. We can imagine Yolngu plotting the spatial pattern of land sites owned by related groups of people. Stones and shells are taken to represent positions in the mesh of family relations and the sites owned. The shells/stones symbolize the links between people and land, and placed on sites through which land is owned they show the pattern of the network of linkages between related groups of people. And just as we find our bodies embedded in Western mathematics, so too we find bodies embedded in *gurrutu*. On biological grounds of life expectancy we could expect a manageable such representation of family-land links to recognize the presence of three generations, differentiated along matrilineal and patrilineal lines.

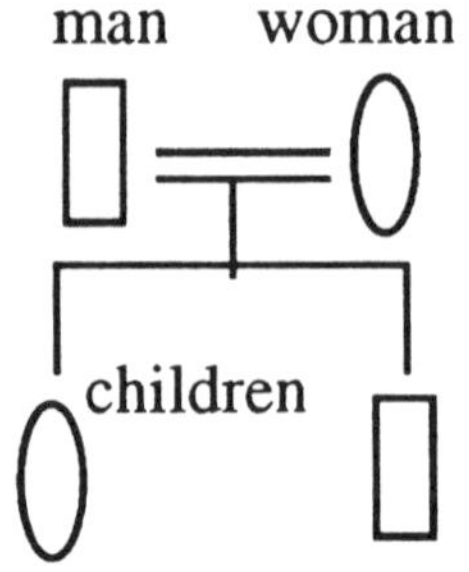

Figure 1: Hypothetical ideal family tree.

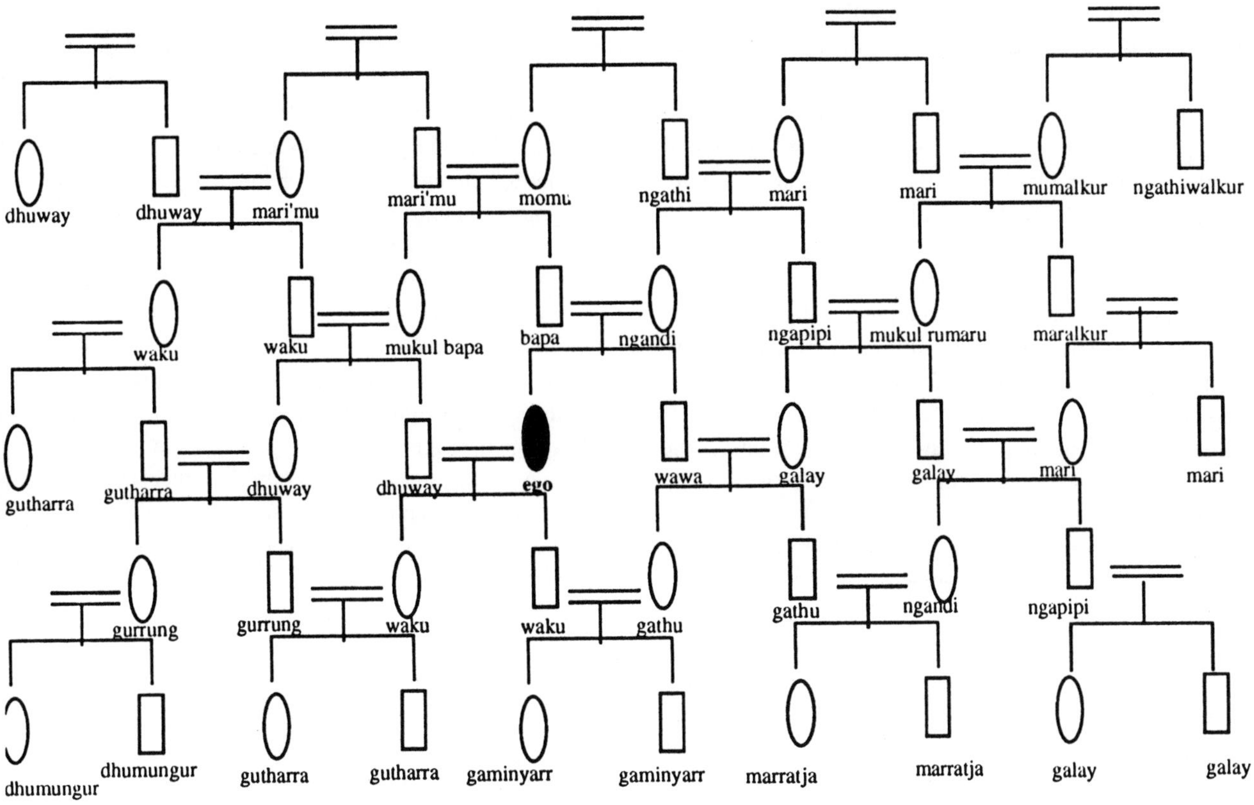

Figure 2: Diagram of kin relations.

If we extended the operations and name relations between the positions plotted with shells/stones, we would have eight pairs of reciprocal names naming across generations or across matrilineal or patrilineal lines (two parental lines raised to the power of three generations, give eight reciprocal pairs). The brother/sister reciprocal pair completes the set of named relations. These reciprocal naming pairs name relative positions in the kinship mesh. They do not name clans, they do not name individual people, and they do not name the sites of land owned by those clans and their members. They are abstract objects that are named by these names: *Gurrutu* names are names of positions in a formal series of relations; positions in a matrix which has a material form in the spatial pattern of sacred sites through which land is owned.

We can list these eight reciprocal pairs which constitute the base of the *gurrutu* recursion

ngandi (woman); ngapipi (man)	waku
bapa (man); mukul bapa (woman)	gathu
mari	gutharra
ngathi (man); momu (woman)	gaminyarr
mari'mu	marratja
mukul rumaru (woman); maralkur (man)	gurrung
galay	dhuway
mumalkur (woman); ngathiwalkur (man)	dhumungur

The dual names on the left hand side of the pairs represent male and female holders of a position, necessarily brother and sister. Each of those pairs have the reciprocal relation of *yapa* and *wawa*. Elaborating this list of names is analogous to listing the set of numerals from one to ten. Zero holds an odd position, enabling the number pattern to proceed in both directions (this makes negative numbers possible). Zero is not itself a member of the base set of names. Similarly the reciprocal pair *yapa and wawa* enable the pattern to work both backwards and forwards from any position; they are not themselves members of the base set of reciprocal name pairs.

Perhaps the best way to indicate the structure of the system of terms is to map them in the conventional form that genealogies are mapped. The basic unit in this map is the hypothetical ideal family tree as in Figure 1.

This 'ideal unit' illustrates the fundamental contrast between the two different sorts of relations which constitute the *gurrutu* system of relations: that between wife and husband and brother and sister.

When we map the series of eight reciprocal kin relation pairs with this unit we get the diagram, shown in Figure 2, of the relations named. The name series of *gurrutu* constitutes a recursion where elements, i.e. positions or abstract objects, constituted in the series provide the basis for further constitution of elements. Together the names and rules of generation form an infinite series.

Turning now from the "architecture" of the primary set of abstract objects constituting *gurrutu*, which we have understood by juxtaposing them with the "architecture" of numbers, the primary set of abstract objects constituting mathematics, we now look at the primary principle or rule for manipulating these primary abstract objects in the systems. The primary principle for manipulating the abstract objects constituted in the pattern derives ultimately from the practical, material system which has formed the template for the recursion of abstract objects.

Going back to numbers we can see that the primary rule of that recursion is a version of the process of holding up another finger as another sheep leaps through the gateway. The important thing about this operation from the point of view of recursion of abstract objects is that each finger is exactly equivalent to each other finger. Also, any collected set of fingers has a precise and identifiable relation to any other collections of fingers; they relate to each other as a specific ratio. The primary principle of the number system is the principle of ratio: the ratio of any term to another is determined by the number of times one contains the other.

The primary principle of working the primary set of abstract objects in *gurrutu* is quite different. Going back to the mapping exercise which underlies *gurrutu* (in the same way that we went back to the modeling exercise that underlies number) we can see that putting down a stone/shell to mark each new position in the matrix immediately sets up a two way naming process with every position in the matrix which immediately sets up a two way naming possibility with every position so far mapped. The important thing about the process from the point of view of the recursion of abstract objects is that the entity — the relation between two positions — must be specified from both ends before it comes to life as an entity. This is the principle of reciprocity — two opposing elements constituting unity. A reciprocal is an expression so related to another that their product is unity. It is the inherent duality constituting the unity of relation, which is the fundamental principle of Yolngu Aboriginal mathematics.

Intuitively reciprocity can be understood as two sides of a coin. In the old Australian penny (which had the head of the current king or queen of England on one side and a kangaroo on the other), 'head' and 'tail' constitute reciprocals, and together constitute a unity. The notion of reciprocal is often difficult for learners of Western mathematics centered around numbers to understand; they need to understand fractions before it is possible to understand reciprocals in a rigorous manner. The Aboriginal system of mathematics makes the rigorous constitution of reciprocity easy to comprehend. If you take any base family relation as a unit, for example *mari-gutharra* — the universal matrilineal relation across two generations (mother's mother or mother's mother's brother) you can understand it as constituted by the mutual engagement by both sides; the descendant and the grandmother/great uncle are equally important in constituting the unity described by *mari-gutharra*. Each side is necessary to constitute the unity held between them.

Just as reciprocity can be secondarily derived in the system of numbers pivoting around ratio, similarly the notion of ratio can be secondarily derived in *gurrutu*. *Milmarra* is a name given to this way of rendering *gurrutu*. One sort of ratio involves juxtaposing a *yothu–yindi* (mother–child) relation with a *mari-gutharra* (child mother's mother) relation. For any given clan or person, given any two of these positions the third can be derived as the one correct answer to complete the expression.

HELEN WATSON-VERRAN

MATHEMATICS OF THE AZTEC PEOPLE The tribal records of the Aztecs indicate that they left their legendary homeland in AD 1168 and founded their capital Tenochtitlan (present day Mexico City) in 1325. By the fifteenth century, their capital had become the center of an expansionist empire. When Cortés arrived in 1519, Tenochtitlan dominated all other cities and had reached the height of its power and magnificence. The language of the Aztecs, Nahuatl, is still spoken today in Central Mexico. An overview of the Nahuatl number sequence is given in Table 1.

The term for five, *macuilli*, derives from *maitl* 'hand', *cui* 'to take' and *pilli* 'fingers'. It means something like 'fingers taken with the hand'. The term for ten, *matlactli*, comes from *maitl* 'hand' and *tlactli* 'torso'. The term for fifteen is a new basic word for which there is no known etymology. The vigesimal nature of the number system clearly shows up in the introduction of new basic terms for 400 and 8000. The word for 400, *tzontli*, means 'hair' or 'growth of garden herbs'. In either case, it signifies multitude or abundance. The word for 8000, *xiquipilli*, refers to a 'bag of cacao beans'.

The Aztec had written numerals for the first four vigesimal powers: $20^0 = 1, 20^1 = 20, 20^2 = 400$ and $20^3 = 8000$. These symbols are shown in Figure 1: (a) a dot represents the unit 1; (b) a flag represents 20; (c) a hank of hair or garden herb represents 400; (d) a bag of cacao beans is used for 8000. It can be seen that the symbol for 400 reflects the word for that number, *tzontli,* 'hair' or 'growth of garden

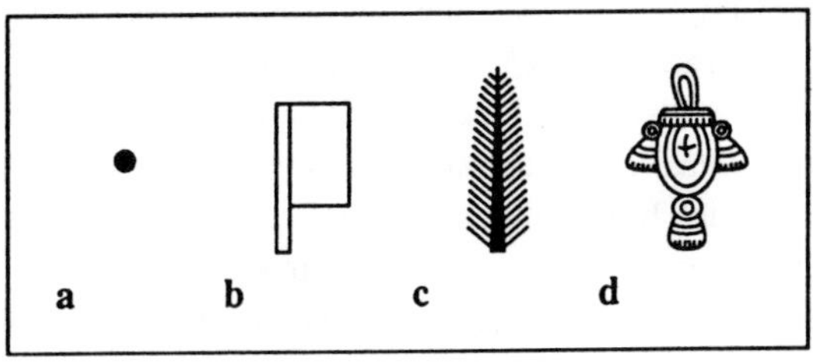

Figure 1: Aztec numerals: (a) 1; (b) 20; (c) 400; (d) 8000. (Drawing by M.P. Closs).

herbs'. Similarly, the sign for 8000 reflects the word for that number, *xiquipilli,* 'bag of cacao beans'.

A tally of these four numerals was used to represent other numbers. Thus, quantities from 1 to 19 were represented by the appropriate number of dots or circles. In the same way, multiples of 20 less than 400 were represented by repeating the sign for 20 as many times as necessary. Similarly, the symbols for 400 and 8000 were repeated to form multiples of 400 and multiples of 8000, respectively. The largest number I have seen recorded with these numerals occurs in the Vatican Codex. It is composed of two signs for 8000 and 9 signs for 400 yielding $(2 \times 8000) + (9 \times 400) = 19,600$.

Table 1: The Nahuatl number sequence

1	ce	
2	ome	
3	ei, yei	
4	nahui	
5	macuilli	
6	chicuace	(5) + 1
7	chicome	(5) + 2
8	chicuei	(5) + 3
9	chiconahui	(5) + 4
10	matlactli	
11	matlactli once	10 + 1
12	matlactli omome	10 + 2
13	matlactli omei	10 + 3
14	matlactli onnahui	10 + 4
15	caxtolli	
16	caxtolli once	15 + 1
17	caxtolli omome	15 + 2
18	caxtolli omei	15 + 3
19	caxtolli onnahui	15 + 4
20	cempoalli	one score
30	cempoalli ommatlactli	20 + 10
37	cempoalli oncaxtolli omome	20 + 15 + 2
40	ompoalli	2×20
60	eipoalli	3×20
100	macuilpoalli	5×20
400	tzontli	
401	centzontli once	(1×400) + 1
405	centzontli onmacuilli	(1×400) + 5
500	centzontli ipan macuilpoalli	(1×400) + (5×20)
8000	cenxiquipilli	1×8000

Numerals are found in a variety of contexts. One of the most common of these is as numerical coefficients in calendar dates. The Aztecs, along with other Mesoamerican cultures, employed two major calendrical cycles, a sacred almanac of 260 days called the *tonalpohualli* and an annual calendar of 365 days. The *tonalpohualli* was constructed from a sequence of numbers from 1 to 13 paired with a sequence of twenty day names. The 260-day and the 365-day calendars were combined so that each day could be specified by both a sacred date and an annual date. Since the least common multiple of 260 and 365 is 18,980 (= 52×365), the combined cycle of the two calendars repeats after 52 years of 365 days each. This 52-year period, known as the *xiuhmolpilli* 'sacred bundle', played a significant role in Aztec religious life.

A given 365-day year was designated by the sacred almanac name of its 360th day. We refer to this name day of the year as the year bearer. Because of the structure of the calendar, only four day names could function as year bearers. The thirteen numerical coefficients with the four day names yield 52 year bearers, one for each year in the 52 year period. In the Aztec codices and stone monuments, such years are frequently named. In these cases, the numerical coefficients are represented by a tally of from 1 to 13 small circles. Examples of the year bearer dates, 13 Rabbit and 8 Reed, are shown in Figures 2a–2b.

The Aztecs also used numerals in chronological counts which varied from a few days to thousands of years. One section of the Mendocino Codex gives the different stages in the education that a boy or girl receives from its parents. The tasks which children must learn are depicted and their daily food ration is shown by a tally of tortillas that appears over their heads. The ages are indicated by a simple tally of blue disks representing the number of years. These run from three up to fourteen as one progresses through the manuscript. The corresponding daily ration rises from half a maize cake to two per day. The largest chronological interval of this type is depicted in another portion of the Mendocino Codex where an elderly male and female are shown drinking a fermented beverage denied to those who are younger. Both persons are seventy years old and their age is indicated by a tally of 10 blue disks together with three disks marked with the flag symbol for 20 as shown in Figure 2c.

The Vatican Codex contains other chronological counts which measure far larger intervals. A count which refers to the years of the second age of the world is shown in Figure 2d. Each of the cross-hatched disks with hair on top is blue in the original and represents a period of 400 years,

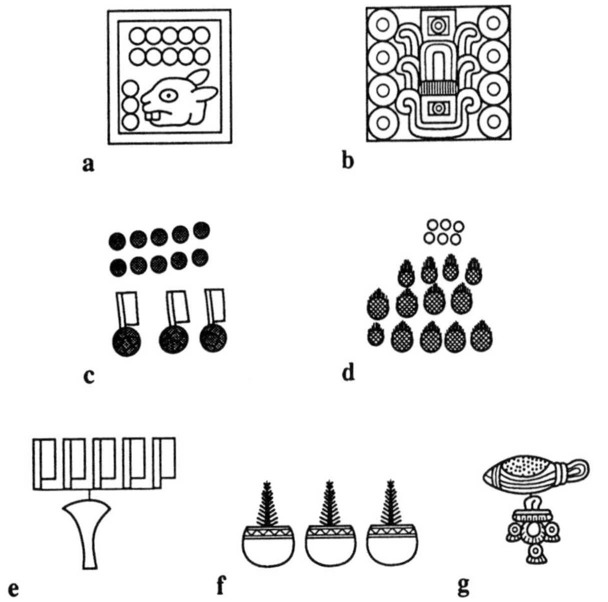

Figure 2: Contexts of Aztec numeral usage. *Calendar dates (year bearers)*: (a) 13 Rabbit; (b) 8 Reed. *Chronological counts*: (c) 70 years; (d) 5206 years. *Tribute records*: (e) 100 copper hatchets; (f) 1200 coarse clay pots; (g) 8000 balls of unrefined copal (incense) wrapped in palm leaves. (Drawing by M.P. Closs).

being a conflation of the standard blue disk for the 365-day period and the numeral for 400. (In this regard, recall that the Aztec term for 400 is *tzontli*, 'hair'). In the upper row is a tally of six smaller disks marking 6 years. The total count measures a period of $(13 \times 400) + 6 = 5206$ years.

Some of the Aztec codices contain lists in which the quantities of the various tribute items to be received from conquered towns are recorded. In these documents, the towns are represented by hieroglyphic toponyms, the tribute items by hieroglyphic signs, and the quantities by numerals or by a simple tally of the items. If the total is less than 20, the item is often represented by a simple tally of the sign for the object itself or by a single depiction of the tribute item with an attached tally of numerals. For larger totals, a combination of the two types of tallies is often used. Some examples of tribute records are shown in Figures 2e–2g.

In the land documents of the Texcocan Aztecs a more concise and sophisticated system of numerical notation was employed. It was used for recording the measurements of perimeters and areas of land holdings in at least two locations in the Valley of Mexico. The system made use of only four symbols — a vertical tally stroke, a bundle of five strokes linked at the top, a dot and a corn glyph (*cintli*) — and position to indicate the value of measurements (Harvey and Williams, 1980, 1986).

An example from the Códice de Santa Maria Asunción, dating from around 1545, is illustrated in Figure 3. The land record is divided into three parts. The first section (*tlacatlacuiloli, tlacanyotl*) contains a census by household. The name for each head of household is written in glyphic form beside a conventional symbol for household head. In the present example, the tlacatlacuiloli shows the head Pedro, his wife Juana, their two young daughters Ana and Martha, and the head's younger brother, Juan Pantli, his wife Maria and their son Balthasar. The shaded heads indicate that the individual is deceased.

The second section (*milcocoli*) consists of a record of land parcels associated with each household. In this section, the scribe drew the approximate shape of each field. The measurement of each side was recorded using lines and dots, a line equal to one linear unit and a dot equal to twenty. In the Texcocan area, the usual unit of linear measure was the *quahuitl* of approximately 2.5 meters. Units less than one quahuitl were indicated by symbols such as a hand, an arrow, or a heart. The modern equivalents of these signs can only be estimated at present. In addition to recording linear measurements around the field perimeters, each field contains a hieroglyph in the center which indicates the type of soil. In particular, the milcocoli section for our example records the approximate shape, perimeter measurements, and soil type of four fields belonging to the household, two to the household head and two to his brother. The hand glyph in the first field indicates a fraction of a quahuitl.

The third section (*tlahuelmantli*) is a second record of the same lands as in the milcocoli section. However, in this case, the lands are shown in a stylized form as rectangles of the same size. Interestingly, the nahuatl term tlahuelmantli literally means 'smoothed, leveled, equalized'. The majority of these fields have a protuberance in the upper right hand corner. In addition to the standardized field shapes, the placement of numbers in the tlahuelmantli is different from in the milcocoli. The numerical quantities using lines and dots are entered either in the center or on the bottom line of the rectangle and in the protuberance. When numbers, never exceeding 19, are entered on the bottom line, a cintli glyph occurs near the top border of the rectangle. In addition, most fields contain a number ranging from 1 to 19 in the upper right corner of the protuberance. It has been determined that these numbers report the area of the field in square quahuitl by use of positional notation.

The system of recording area is dependent on three separate registers. The first register is located in the upper right protuberance and is used to record a unit count from 0 to 19. When the value of this register is 0, the protuberance is left empty or is not drawn. Otherwise, 1 to 19 strokes are recorded with groups of five being bundled together by a

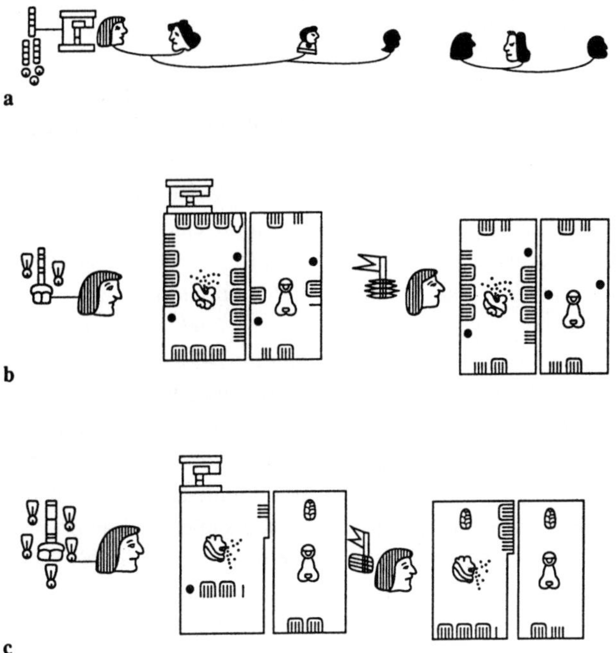

a

b

c

Figure 3: A land record from the Codice de Santa Maria Asuncion. (Drawing by M.P. Closs from Harvey and Williams, 1986).

connecting line. The second register is the bottom line of the rectangle. It is used to record from 1 to 19 multiples of 20 (the vigesimal numbers from 20 to 380) by using a simple tally of 1 to 19 strokes. The third register is in the central portion of the rectangle. It is used to express higher multiples of 20, that is multiples of 20 which are greater than or equal to 400. In this register, a dot has value 400 and a stroke has value 20. The second and third registers are never used concurrently. If there is no entry in the third register, the cintli glyph is drawn towards the top of the rectangle. This may signify 0 in that register. In order to compute the area, the number in the second or third register is multiplied by 20 and added to the number in the first register.

Table 2: The length of the sides, in quahuitl, of the four fields in the milcocoli section and the area, in square quahuitl, of the corresponding fields in the tlahuelmantli section

	Sides	Area
Field 1	39, 15, 39, 15+	$(31\times20)+4=624$
Field 2	25, 8, 26, 8	$(10\times20)+0=200$
Field 3	38, 9, 39, 8	$(16\times20)+13=333$
Field 4	20, 9, 20, 8	$(9\times20)+0=180$

In the example under consideration, the length of the sides, in quahuitl, of the four fields in the milcocoli section and the area, in square quahuitl, of the corresponding fields in the tlahuelmantli section is shown in Table 2.

The problem of how the areas in the tlahuelmantli were determined remains to be resolved. In general, the area of a quadrilateral cannot be determined from the lengths of its sides. As a result, the information in the milcocoli section is not sufficient to obtain the true areas of the fields.

MICHAEL P. CLOSS

REFERENCES

Harvey, Herbert R. and Barbara J. Williams. "Aztec Arithmetic: Positional Notation and Area Calculation." *Science* 210: 499–505, 1980.

Harvey, Herbert R. and Barbara J. Williams. "Decipherment and Some Implications of Aztec Numerical Glyphs." In *Native American Mathematics*. Ed. Michael P. Closs. Austin: University of Texas Press, 1986, pp. 237–259.

Payne, Stanley E. and Michael P. Closs. "A Survey of Aztec Numbers and Their Uses." In *Native American Mathematics*. Ed. Michael P. Closs. Austin: University of Texas Press, 1986, pp. 213–235.

See also: Calendars

MATHEMATICS IN CHINA Chinese mathematics, called *suanshu* or *shushu*, was considered an autonomous science separate from calendar computations and divination. Traditionally, it was subdivided into nine chapters — as in the *Jiuzhang suanshu* (Nine Chapters of the Mathematical Art) — but these were not different mathematical methods, but rather fields of application. The algorithms were given mostly in short, mnemotechnic verses, from which it becomes clear that mathematical handbooks were made for reckoning clerks. Whereas Greek mathematical handbooks were geometrical and deductive in nature, Chinese mathematics always had a kind of "algebraic" outlook, not in the modern sense (since Viète), but in the algorithmic sense of all medieval arithmetic books. The Chinese knew general rules, but these were only written down in the form of numerical problems.

Chinese mathematical handbooks were written for pedagogical purposes, i.e. for training some minor officials in all kinds of calculations. This means that they never give the rationale of a method, but only instructions for use. As a consequence we mostly do not know how mathematicians developed their sometimes very complicated procedures. This is a disadvantage for a correct interpretation of mathematical knowledge as such, although we have the same situation in India, the Islamic countries, and pre-Renaissance Europe. We know Chinese mathematics only through its application to economic or astronomic problems. Mathematicians were technicians, only trained in applied mathematics, who used the methods perhaps without actually understanding the logic of the procedures. This means that we are almost wholly uninformed about the "pure" mathematicians who invented these procedures. It is true that some applications for daily life problems did not require great mathematical skill, but this cannot be said about certain astronomical problems. With the exception of Qin Jiushao's *ta-yen* rule and the interpolation formula, we have not much information about astronomical mathematics.

The Chinese way of calculating was, from very early times, decimal. This had an important influence on the use of the counting board (this is not the same as the abacus). This counting board in turn influenced mathematical notation. In 'literary' style the number 35,672 was written as "san *wan* wu *qian* liu *bai* qi *shi* er" (3 ten thousands 5 thousands 6 hundreds 7 tens 2). But on the counting board, rods were used in a series ABABAB, according to the diagram in Table 1.

The zero was a blank space on the counting board, but from 1247, after Qin Jiushao, it was printed as 0. Positive numbers (*zheng*) were probably represented by red counting rods, and negative numbers (*fu*) by black ones. In the 'literary' representation, characters for positive and negative were added, e.g. *fu* 525 = –525. This cumbersome notation had no influence on the calculating work that was performed on the counting board.

The calculating board was a kind of chess board on which numbers were represented by counting rods. Complicated calculations could be done in a mechanical way, according to prescriptions in handbooks. The abacus is actually a simplified device, very suitable for practical use, but restricted to simple calculations for daily use. When European mathematical methods were introduced, they were called "Pen Calculations" (*bisuan*).

The *Jiuzhang suanshu* was a textbook on arithmetic, containing 246 problems. It was used by reckoning clerks for more than a thousand years, for it contained solutions to all the problems the bureaucracy was confronted with. Mathematics of a simple kind was one of the essential accomplishments of the Confucian gentleman, and was considered on the same level as propriety, music, archery, and calligraphy. It belonged to the so-called minor arts. However, the astronomical and calendrical sciences were of a special kind, and were the task of the Bureau of Astronomy. In general, mathematicians had nothing to do with this Bureau, the only exception being Qin Jiushao. The more creative mathematicians of the Song-Yuan period were not official mathematicians: Qin Jiushao and Yang Hui were civil servants, Zhu Shijie was a wandering teacher, and Le Zhi (Li Ye) a reclusive scholar. That those men were more creative mathematicians seems to be due to the fact that they were not official reckoning clerks.

History of Chinese Mathematics

The history of Chinese mathematics can be divided into several periods. The most important feature of Chinese arithmetic before the Qin dynasty (221 BC) was undoubtedly the invention of the decimal notation, which is found on the oracle bones of the Shang dynasty. This notation lies at the basis of the calculation with counting rods.

The oldest mathematical handbook from the Han dynasty

Table 1: Series of rods on a counting board

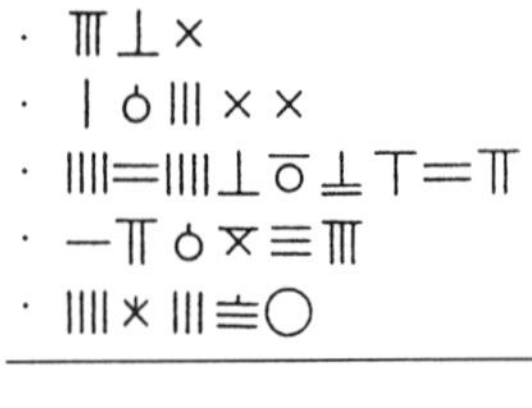

864
15,344
424,657,627
175,938
49,380

The number 35, 672 is written ··· 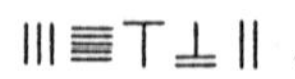 .

(206 BC–220) is the *Zhoubi suanjing* (Arithmetic Classic of the Gnomon and the Circular Paths of Heaven), the contents of which date from the Zhou and Qin periods. The work is also important for the history of astronomy. It contains the *Gou-Gu* theorem, later known as the Pythagorean theorem. In the commentary written by Zhao Shuang, several formulae for the resolution of right-angled triangles are given.

The *Jiuzhang suanshu* is undoubtedly the most influential Chinese handbook. As usual, it does not contain proofs. In general, the work is on a high level and all later works are tributary to this mathematical classic. But it has all the disadvantages of a classic: its methods were applied blindly, without any interest in the rationale. In this way it obstructed further developments. Problems are classified according to the field of application, such as *fang tien,* the calculation of surfaces, and *su mi*, commercial calculations. As usual for classics, several commentaries on the book were written.

In the period from the Han to the Tang dynasties (221–960 AD), the *Haidao suanjing* (Sea Island Mathematical Manual) was written by Liu Hui. It contains calculations of heights and distances of inaccessible points. It can be considered a kind of proto-trigonometry, relying on the proportions in right triangles. Another important text, the *Sunzi suanjing* (Mathematical Manual of Sunzi) contains the first problem using indeterminate equations.

All these works, together with several others, were brought together in a collection called *Shibu suanjing* (The Ten Mathematical Classics), an official compilation made during the Tang at about 750, as a textbook for the imperial examinations.

The Song and Yuan dynasties (960–1368) were the culminating point in the development of Chinese mathematics. The works of this period that have survived intact are:

Qin Jiushao's *Shushu jiuzhang* (Mathematical Treatise in Nine Sections), 1247; Li Zhi's *Ceyuan haijing* (Sea Mirror of Circle Measurements), 1248 and his *Yigu yanduan* (New Steps in Computation), 1259; Yang Hui's *Xiangjie jiuzhang suanfa* (Compendium of Analyzed Mathematical Methods in the 'Nine Chapters'), 1261, *Riyong suanfa* (Mathematical Methods for Daily Use), 1262 (incomplete), and *Yang Hui suanfa* (Yang Hui's Mathematical Methods), 1274–1275, and Zhu Shijie's *Suanxue qimeng* (Introduction to Mathematical Studies), 1299, and *Siyuan yujian* (Precious Mirror of the Four Elements),1303.

These works contain the following new algebraic methods:

1. *Indeterminate analysis* (the Chinese remainder problem). Qin Jiushao gives a fully elaborated method of solution, even for the cases where the moduli are not relatively prime. It may have been derived from calendrical computations, but undeniably he was the first to see its full mathematical significance. This problem was again solved in Europe in the eighteenth century, with the work of Euler and Gauss.

2. *Numerical higher equations.* These occur for the first time in the works of Qin Jiushao, as well as in Li Zhi and Zhu Shijie's works. Such equations were solved in Europe in the beginning of the nineteenth century by Ruffini and Horner.

3. The *tien-yuan notation for nonlinear equations* was first used by Li Zhi and Zhu Shijie. It was peculiar to Chinese mathematics.

4. The *Pascal triangle* was first given by Yang Hui and Zhu Shijie. Its European date is the sixteenth century.

5. *Cubic interpolation formulae* were used by Guo Shoujing (1231–1316) and are identical with the Newton-Sterling formulae (1711/1730).

6. A considerable development in the field of *series* beyond the studies of Shen Gua appeared in the works of Qin Jiushao, Yang Hui, and in particular of Zhu Shijie.

7. Guo Shoujing's *prototrigonometry* developed out of the arc-sagitta method of Shen Gua.

Many changes took place during the Ming and Qing dynasties (1368–1911). After the occupation of China by the Mongols (the Yuan dynasty), the mathematical tradition was lost, and there was a kind of regression to pre-Song mathematics. The only valuable work of the Ming was the *Suanfa tongzong* (Systematic Treatise on Arithmetic) written by Cheng Dawei in 1593. It is actually a compilation of older works, but it was reprinted many times. It was in this period of low-tide mathematics that European knowledge was introduced by the Jesuits. The most important translation was that of Euclid's *Elements of Geometry* by Matteo Ricci and Xu Guangqi (1607). With the assistance of Li Zhizao, Ricci also translated the *Epitome of Practical Arithmetic* under the title *Tongwen suanzhi* (Treatise on European Arithmetic, 1631). The first result of the influence of Western upon Chinese mathematics was Mei Wending's mathematical work. Important in the field was also the *Shuli jingyun* (Collected Basic Principles of Mathematics) compiled under emperor Kang Xi in 1712.

During the Qing dynasty a lot of research was done concerning old autochthonous mathematics: various collections of books were republished, among them the important *Suanjing shishu* (Ten Mathematical Manuals), Song and Yuan texts were collated, and Ruan Yuan published his *Chouren zhuan* (Biographies of Mathematicians and Astronomers, 1799). Also in the same period, there were studies of Pierre Jartoux's formulae on trigonometry and research on the theory of equations and on summing finite series.

After the Opium War (1840), several Western mathematical handbooks were introduced by Protestant missionaries, and several of them were translated by Li Shanlan and Hua Hengfang. At the same time new schools were founded, helping to spread modern mathematical knowledge.

Mathematical Methods

In discussing mathematical methods, we have to restrict ourselves to a description of the contents of mathematical writings in general, before they were influenced by western methods. In the field of arithmetic, numbers were classified into four groups: (1) *Yuanshu* ("Original Numbers"), whole numbers; (2) *shoushu*, decimal fractions; (3) *tongshu*, fractions; and (4) *fushu*, "multiples of 10"

Proportions are very important; they are linear (*fanzhui che,* inverted-wedge difference) or square (*fangzhui cha,* squared-wedge difference). Calculations of interest were an important application in arithmetic.

Chinese geometry cannot be compared with Greek geometry, because the Chinese did not have the concept of deductive systems. What we find in their mathematical handbooks are some practical geometrical problems concerning plane areas and solid figures. The more theoretical geometry had not developed beyond its embryonic form in the propositions of the Mohist Canon (*Mojing*, fourth century, BC), and, as Needham pointed out, " their deductive geometry remained the mystery of a particular school and had little or no influence on the main current of Chinese mathematics." The mathematical genius of the Chinese was pragmatic in nature, and for this reason their geometry was restricted to problems of land surveying and the capacity of various vessels. Moreover, Chinese geometry was algebraic in approach, and many problems were reduced to algebraic equations.

The Pythagorean theorem was known in China from early times. Similar triangles provide a basis for the solution of several geometrical problems, and in fact for the solution of all "proto-trigonometric" problems. The value of π was 3.14 (Liu Hui, ca. 250), $\sqrt{10}$ (Zhang Heng, 78–139) and even 3.1415926... (Zu Chongzhi, fifth century). For practical calculations the values 3, $\frac{22}{7}$ and $\sqrt{10}$ were mostly used.

As for planimetry, in addition to the usual surface calculations, some more complicated formulae for the area of all kinds of triangles derived from the sides, for the area of quadrangles, circles, annuli and segments, are known: in stereometry, the formulae of a fustrum of a pyramid and cone, or of a "wedge", belong to the more special applications. Calculations at a distance (proto-trigonometry) are applied to measurements of height and distance of a far-off mountain or wall, also starting from a slanting basis, and to measurements of a square or round tower, all by means of a gnomon.

Algebra

Algebra is undoubtedly the most important field of Chinese mathematics. When we call it "algebra" we do not mean the modern algebraic notation, but the kinds of problems which were extended in later times to algebraic procedures. Although the Chinese did not know these general formulations, they were well acquainted with the general procedures.

Simultaneous linear equations were thoroughly known to the authors of the *Jiuzhang suanshu*. This method is called *fangcheng*. One of the problems solved is:

$$3x + 2y + z = 39$$
$$2x + 3y + z = 34$$
$$x + 2y + z = 26$$

The method is almost perfect; it makes use of multiplication and elimination by subtraction. In the *Nine Chapters* no diagrams are given, but we find them fully elaborated in Qin Jiushao's work. For instance he solves:

$$64x + 30y + 75z = 29,400$$
$$792x + 568y + 815z = 392,000$$
$$140x + 88y + 15z = 58,000$$

Study of series and progressions started with the "technique of small increments" — the problem of piling up stacks — in Shen Gua's *Mengqi bitan* (ca. 1078). In Qin Jiushao's work there are also a couple of problems on arithmetic progressions and their summation. The same holds true for Yang Hui. The "technique of third order differences, linear, square, and cube" was developed in Guo Shoujing's work on calendrical computations, *Shou shili* (1280) which was "adopting the principle of third-order interpolation to calculate the tables for the positions of the sun and moon" (Li and Du, 1987). The most important work on stacking and finite differences was done by Zhu Shijie in his *Siyuan yujian*, where he discusses 32 problems in the field. In modern algebraic notation, his procedure is:

$$\sum_{r-1}^{n} \frac{1}{p!} r(r+1)(r+2)\cdots(r+p-1)$$
$$= n(n+1)(n+2)\cdots(r+p+1)(r+p).$$

As Zhu Shijie understood the formula for series, he was able to invent an accurate formula for finite differences, up to the fourth order. His formula

$$f(n) = n\Delta + \frac{1}{2!}n(n-1)\Delta^2 + \frac{1}{3!}n(n-1)(n-2)\Delta^3 + \frac{1}{4!}n(n-1)(n-2)(n-3)\Delta^4$$

is the same as Newton's interpolation formula.

Further investigations on the summing of finite series were done by Li Shanlan (1811–1882).

Undoubtedly the most important contribution of the Chinese to the development of algebra was the solution of numerical equations of higher degree. In the West this method is known as the Horner–Ruffini procedure (1819). It is an extension of the method for square and cube root extraction already known to the authors of the *Jiuzhang suanshu*. In Yang Hui's work, the equation $-5x^4+52x^3+128x^2=4096$ is solved. In Qin Jiushao's work an equation of the tenth degree is solved. Representing the applied procedure in modern algebraic language gives a wrong idea of the way these equations were solved on the counting board.

The *tianyun shu* is an extension of the solution of equations of higher degree with one unknown to equations of higher degree with four unknowns. The technique reached its culminating point in Zhu Shijie's *Siyuan yujian*. An example with three unknowns is:

$$-x-y-xy^2-z+xyz=0$$
$$x-x^2-y-z+xz=0$$
$$x^2+y^2=z^2$$

Elimination of x and y results in $z^4-6z^3+4z^2+6z-5=0$, which can be solved with the Horner–Ruffini procedure.

In the field of indeterminate analysis, there are two kinds of indeterminate problems. One is the hundred fowls problem, of the type $ax+by=c$. The other is the Remainder Problem of the type $N \equiv r_1(\mathrm{mod}\, m_1) \equiv r_2(\mathrm{mod}\, m_2)\cdots$. It appears for the first time in the *Sunzi suanjing* (fourth century), but is fully elaborated in Qin Jiushao's work.

Chinese mathematics is part of medieval mathematics, i.e. a collection of algorithms, given without proof. There was no systematic deductive algebra anywhere in the world before modern times. This means that Chinese arithmetic must not be evaluated by a modern standard.

ULRICH LIBBRECHT

REFERENCES

Li Di. *Zhongguo shuxue shi jianbian* (Concise History of Chinese Mathematics), 1984.

Li Yan and Du Shiran. *Chinese Mathematics: A Concise History.* Oxford: Clarendon Press, 1987.

Libbrecht, Ulrich. *Chinese Mathematics in the Thirteenth Century,* Cambridge, Massachusetts: MIT Press, 1973.

Martzloff, Jean-Claude. *Histoire des Mathématiques Chinoises.* Paris: Masson, 1988.

Zhongguo Shuxue jianshi (Concise History of Mathematics), Chinan: Shantung Chiao yü chubanshe, 1986.

See also: *Pi* in Chinese Mathematics – Algebra in Chinese Mathematics – Computation: Chinese Counting Rods – Liu Hui and the *Jiuzhang Suanshu* – Sundials – Xu Guangqi – Li Shanlan – Guo Shoujing – *Zhoubi Suanjing* – Li Zhi – Yang Hui – Decimal System – Trigonometry – Gou-Gu Theorem – Shen Gua – Zhu Zhijie – Calendar – Geomancy – Divination in China – Qin Jiushao – Abacus

MATHEMATICS IN EGYPT By 'Egyptian mathematics', we understand the mathematics developed in Ancient Egypt between the end of the fourth millennium BC, when a centralized state came into being in the Nile Valley and a system of writing was invented, and 332 BC, when Egypt was conquered by Alexander the Great and Greek cultural elements entered into the hitherto fairly autonomous development of Egyptian civilization. During these three thousand years, the role of mathematics remained of central importance; the Egyptian educational system was geared principally to produce scribes who could count and calculate the work, rations, land, and grain of the State and the private landowner.

The judgments of Egyptian mathematics ordinarily to be found in histories of science fall into two main schools: either they are dismissive, claiming that mathematics in Egypt was merely of a 'practical' or 'trial and error' nature; or they are laudatory in the extreme, the Egyptians being seen as the precursors and inventors of all important concepts in mathematics. In fact, these two judgements share an identical misconception about mathematics itself, namely that it really exists only insofar as it corresponds to a particular Western image of theorem and proof. However recent interest in alternative (algorithmic and effective) methods in mathematics, as well as a better understanding of the complex and sophisticated nature of Egyptian civilization, has opened the way for a deeper comprehension of mathematics in the Nile Valley.

The real question is what mathematics represented for the Egyptians themselves, what they aimed at, what structural, intellectual, and sociological activities were connected with their mathematics. The Egyptians left us no philosophical or self-reflexive works commenting on their own activity; the nature and role of Egyptian mathematics must be pieced together from such texts as administrative and accounting documents, historical annals, and, above all, school texts.

We have, in fact, astonishingly few sources: the choice of excavation sites and the fragility of papyrus under humid conditions means that most written documents come, almost exclusively, from cemeteries and temples in the desert fringe along the Nile valley and only rarely from towns or cities in the fertile band around the Nile and its Delta. In addition, mathematical texts are particularly rare — some five papyri, mostly fragmentary, a pair of wooden exercise tablets, and a small inscribed stone flake. Moreover, with the exception of the last, all the texts are originals or copies dating from the same period, the Middle Kingdom between 2000 and 1700 BC. The accounting documents are a little more plentiful and from a more balanced time span, but not all have been published and those that have still await a study of the mathematical techniques used in their fabrication.

But there are some things that we do know. In Egypt, there were no centers of mathematical research in the modern sense, no journals or books exposing new results. There is not even an Egyptian word for 'mathematics' in our sense: the only two Egyptian titles of mathematical school books that have come down to us are "The right method for entering into things, for knowing everything that is, every obscurity, . . ., every secret" (Rhind Mathematical Papyrus) and "The right method in matters of writing (?)" (Kahun Papyrus LV.4).

However, Egypt, like almost all societies, needed to create computational skills, to count and keep track of the collection and distribution of material wealth. Indeed, writing in Egypt was created precisely to make permanent records of these numerical data; the earliest signs were used to indicate metrological values along with the titles and proper names of those from whom came and to whom were to be delivered this wealth; mathematics and writing were linked from the very start and were the staple of Egyptian educational training. In particular, the mathematical texts we possess — including the Rhind Mathematical Papyrus — are school textbooks. So it is to the world of apprentice scribes and their teachers that we must look for an understanding of Egyptian mathematics.

Numbers

The third millennium had seen the development of some eight or nine distinct metrological systems — a discrete system, used for counting individual objects; a length system; an area system, used for measuring field surfaces; a distinct system for surfaces of linen; a capacity system for grain, and a weight system for measuring precious metals. The signs used to write the units of each system differed among themselves, as did the ratio between successive units in each system. In all cases though the Egyptians used an additive system of notation, in which a unit was repeated as many times as necessary to express the value desired.

For example, the discrete system is decimal in structure while the length system has 4 fingers to the palm and 7 palms to the cubit. Thus whereas fourteen goats would be written with 1 ten and 4 ones — written ∩|||| — a length of fourteen fingers would have 3 palms and 2 fingers:

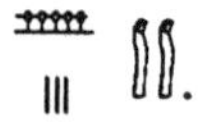

By the time of the Middle Kingdom, there is a distinct rationalization of these systems. Some, such as that for linen, simply disappear; some, such as that for lengths, are simplified. In the mathematical texts, all arithmetical operations are performed in the discrete system, now employed as a universal calculational system; an abstract concept of number is born. The discrete system being decimal, the base 10 became the standard. A significant part of mathematical training was devoted not only to learning how to perform mathematical calculations in the decimal system but also to mastering the conversions into and out of the surviving, non-decimal metrological systems.

The Mathematical Texts

The careful organization of the mathematical papyri reflect their pedagogic nature. They are constituted of numerical tables (to which we shall return) and solved problems: each problem is expressed in ordinary language with no symbols aside from the numbers themselves. The data and the results are given as concrete numerical values, and the solutions presented in the form of algorithms, that is a specific sequence of steps leading to the solution, generally followed by a numerical verification of the result obtained.

The subjects of the problems in the mathematical papyri touch on arithmetic calculations involving distribution of rations, work assignments, and the conversion of quantities of grain into beer and bread; mensuration problems involving the determination of volumes of circular and rectangular containers, areas of triangular, rectangular, and circular fields, slopes of pyramidal and conic constructions, and parts of ships, as well as others. The problems are roughly grouped by subject but, more precisely, by the type of algorithm used in their solution. Each group of problems is a series of exercises in the learning and use of an algorithm.

A typical example is given by the following Problem 26 of the Rhind Mathematical Papyrus; the words in red ink in the original are given in bold-face; the numbering of the steps of the algorithm and phrases in square brackets are my additions:

A quantity; its $\frac{1}{4}$ is added to it. It becomes 15.

(1) Calculate starting from 4; you will make its $\frac{1}{4}$: 1.

(2) [Add them together.] Total: 5.

(3) Calculate starting from 5, to find 15.

\1 5

\2 10

3 is the result.

(4) Calculate starting from 3, 4 times.

1 3
2 6
\4 12
12 is the result.

The quantity: 12. Its $\frac{1}{4}$: 3. Total: 15.

[Verification]

1 12
[1/2 6]
1/4 3
Total: 15.

It is important, in order to understand Egyptian mathematics, that each problem in a mathematical text operates on three distinct levels. The first is the global strategy used, embodied by the *algorithm* and is the same for all problems in a group. In our example it is the common Egyptian method, now called "false position". A (false) solution is supposed, in our example 4; the result of calculation with this value (steps 1 and 2) is compared to the true answer (step 3) and the original choice corrected by the necessary factor (step 4).

The second level is that of arithmetical *operations*: addition, subtraction, multiplication, division, square root extraction, etc., each of which corresponds to one step of the algorithm. It is introduced by a standardized formulation and includes a numerical computation. In our example, step (4) is a multiplication: as always indicated by "Calculate starting from N, M times"; likewise, step (3) is a division.

The third level is that of *techniques*, that is the set of means for carrying out the operation. Here the Egyptian scribe had a wide range of choices. For the simple operations of addition and subtraction (see step 2 in the example), no details were shown; the scribe was presumed to be able to do these without such aids. In the case of multiplication and division, the computations show what was done.

The multiplication of step (4), in our example, 3×4, is carried out in the following manner: starting with the first multiplicand, 3, in the right-hand column and 1 in the left, the scribe seeks the value of the second multiplicand, 4, in the left-hand column (since this is multiplication); in this case, the scribe obtains the columns directly by two successive uses of the same technique, namely *doubling*. The 4 is marked and the answer, 12, read off in the opposite column.

The division of step (3), 15+5, is carried out by putting the divisor 5 in the right-hand column, facing 1, and then, after a single doubling, finding the dividend, 15, in the right-hand column (since this is division), the sum of the two entries. Both are checked and the answer, $3 = 1 + 2$, read off as the sum of the corresponding entries in the left-hand column.

Note that though the two operations, in the particular case treated here, are the inverse of each other, the only technique used to effect both of them is doubling. However, the Egyptian scribe was not limited to this single technique; others, such as halving (illustrated in the verification step of our problem), multiplying or dividing by 10, multiplying by $\frac{2}{3}$, inversing, etc. were equally available. Given a particular arithmetic operation, the scribe chose the set of techniques appropriate to the numerical values in play.

Tables and Fractions

Fractions played a particular role in Egyptian mathematics. A fraction was viewed as the inverse of an integer; there could thus be only one fraction associated with a given integer N: namely, what we call the 'unit fraction' $1/N$ — there being a single exception, the fraction $\frac{2}{3}$. These fractions were manipulated in precisely the same way as integers, using the same panoply of techniques.

Only one fraction of a given kind could be used in the writing of a given number. Thus, for example, the double of the number $1+\frac{1}{5}$ was not written $2+\frac{1}{5}+\frac{1}{5}$ but rather $2+\frac{1}{3}+\frac{1}{15}$. Finding such expressions is a centrally difficult problem: it occurs each time it is necessary to double an 'odd' fraction, calculate two-thirds of a fraction, or add fractions together. It constitutes the core area of Egyptian mathematics, and furnishes the content of Egyptian mathematical *tables*. Such difficult calculations could be done once and for all and the results simply copied and looked up when the need arose in the solution of a specific problem.

The problems in Egyptian mathematical texts were chosen in order to cover the domain of the possible by a *network of typical examples*, a process which permitted the student (and later the practicing scribe) to place any new problem in this framework. The Egyptian approach to the question of generalization was not the discovery and application of a 'general formula' in which each case might be enveloped, but rather through interpolation in a pattern of known results — a method equally used in some branches of mathematics today.

If Egyptian mathematics is viewed not as a poor simulacrum of proof-oriented mathematics, but on its own ground, it will be seen as an adequate and rational response to the socio-economic and educational needs of Egyptian society. Mathematics even provided a model of a rational practice, equally applicable to other domains, wherever an efficient mode of action on the world was needed. Egyptian medical texts, for instance, were constructed in the same manner, with tables of remedies and a systematic network of procedural prescriptions.

This approach was not unique to Egypt; one finds analogous underlying principles at work in the mathematics of other civilizations, such as Mesopotamia, China, and India. It is not a question of borrowing; the mathematical traditions in question developed too early and were too widely geographically spread for this. Rather it shows the success of this algorithmic approach to the solutions of the problems posed by sufficiently advanced societies.

These different civilizations differed in their choices of number system, of modes of writing, of arithmetical operations, as well as the techniques used to carry these out. And the differences determined in turn the very different paths of development that mathematics knew in each society. If similar needs led to similar responses on the fundamental level, the different ways of implementing them created very different mathematics.

JAMES RITTER

REFERENCES

Caveing, Maurice. "Le statut arithmétique du quartième égyptien." In *Histoire de Fractions, Fractions d'Histoire*, Ed. Paul Benoit, Karine Chemla, Jim Ritter. Basel: Birkhäuser, 1992.

Chace, Arnold, *et al.*. *The Rhind Mathematical Papyrus*. 2 vols. Oberlin: Mathematical Association of America, 1927–1929.

Gillings, Richard. *Mathematics in the Time of the Pharaohs*. Cambridge, Massachusetts: MIT Press, 1972. Reprint New York: Dover, 1982.

Peet, T. Eric. *The Rhind Mathematical Papyrus, British Museum 10057 and 10058*. London: Hodder and Stoughton, 1923. Reprint: Liechtenstein: Kraus, 1970.

Ritter, James. *Pratiques de la raison en Mésopotamie et en Égypte aux IIIe et IIe millénaires*. Thesis, University of Paris XIII, 1993.

Ritter, James. "Measure for Measure." In *Introduction to the History of Science*. Ed. Michel Serres. London: Blackwell, 1995.

Struve, W. W. *Mathematischer Papyrus des Staatlichen Museums der schönen Kunst in Moskau*. Berlin: Springer, 1930. Reprint: Würzburg: JAL, 1973.

See also: Weights and Measures in Egypt

MATHEMATICS OF THE HEBREW PEOPLE We shall call a "Hebrew mathematical text" any text or work whose language is Hebrew (usually written in Hebrew characters), and whose content is mathematical in a narrow sense, that is, does not include astronomy (apart from relevant mathematical sections), astrology, or calendar calculations. We shall provide a general survey, both chronological and geographical, of mathematical activity among Jewish scholars, as attested by Hebrew mathematical texts, up to the sixteenth century.

Apart from a few passages that are to be found in biblical and postbiblical (rabbinical) literature and which are relevant to the history of mathematics (number words and fraction words, practical rules of geometry), the oldest mathematical tract in Hebrew is the *Mishnat ha-Middot*, by an unknown author. This tract gives practical rules for the measurement of areas and volumes, and then deals with the measurements (*middot*) of the Tabernacle erected by the Jews in the desert. It has been recently shown that its composition was probably influenced by the geometrical part of al-Khwārizmī's *Algebra*. This tract remained unknown to most medieval Jewish scholars and its Hebrew mathematical terminology was of no consequence.

Two famous scholars are central in the eleventh and twelfth centuries in Spain: Abraham bar Ḥiyya (ca. 1065–ca. 1145) and Abraham ibn ʿEzra (1092–1167). This period saw the actual birth of Hebrew mathematics.

Abraham bar Ḥiyya, also called Savasorda (latinized from the Arabic *Ṣāḥib al-shurṭa*), flourished in Barcelona in Christian Spain, but was probably educated in the Arabic kingdom of Saragossa. Bar Ḥiyya wrote books in Hebrew on mathemathics, astronomy, astrology and philosophy. He clearly indicated that his Hebrew compositions were written for Jews living in southern France (in Hebrew, *Ereṣ Ṣarfat*) who were unacquainted with Arabic scientific culture and unable to read Arabic texts. Bar Ḥiyya can thus rightly be considered the founder of Hebrew scientific culture and language, and specifically the father of Hebrew mathematics. We know of two mathematical compositions by Bar Ḥiyya: the extant parts of a scientific encyclopedia and a geometrical compilation.

The first of these books, *Yesodey ha-Tevuna u-Migdal ha-Emuna* (The Foundations of Science and the Tower of Faith), is presumably an adaptation from some unknown Arabic composition; the geometrical and arithmetical parts are extant. The study of its content sheds some light on eleventh century mathematical literature in Western Islamic lands and its diffusion.

The *Ḥibbur ha-Meshiḥa we ha-Tishboret* (The Composition on Geometrical Measures) enjoyed a very large diffusion as the *Liber embadorum* in its Latin translation by Plato of Tivoli (1145). Much has been said by ancient and modern scholars concerning the importance of this text for traditions of practical geometry in Europe.

Abraham ibn ʿEzra was born in Tudela in Aragon. Raised in Arabic culture, and probably knowing Latin, Ibn ʿEzra spent the last part of his life traveling over Europe. He was a poet, a grammarian, an astronomer-astrologer, and a biblical commentator of great reputation. Ibn ʿEzra, one generation after Bar Ḥiyya, also created a new scientific language, different from that of his predecessor, closer to biblical Hebrew and less influenced by Arabic. His mathematical work consists essentially of the important book *Sefer ha-Mispar* (The Book of Number), but numerous arithmetical and numerological remarks are scattered in his others works, including his biblical commentaries.

His *Book of Number*, written probably before 1160, expounds for the first time the decimal positional notation including zero. The foundations of arithmetic are then discussed in the following order: multiplication, division, addition, substraction, fractions, proportions, and square roots.

In the thirteenth and fourteenth centuries, the works of both Abrahams opened up to the Jewish communities unfamiliar with Arabic access to basic mathematical knowledge. They obviously mention the names of classical authors, Greek or Arabic, and of their works, but these works themselves only became partly available in Hebrew during the thirteenth century. This first major period of translations continued until the first third of the fourteenth century.

Several of the translators were members of a single family of scholars. The first, Yehuda ibn Tibbon, left Granada in the middle of the twelfth century, at the time of the arrival of the Almohads in Al-Andalus, and settled in Lunel in the south of France. There, he undertook the translation into Hebrew of ethical and philosophical works written in Arabic by Jewish scholars from Spain. His son, Samuel ben Yehuda, translated the famous *Guide for the Perplexed* of Maimonides from Arabic into Hebrew. The third and fourth generations of this family include translators of scientific works.

Thus, Jacob Anatoli (1194?–1256?), working in Italy under the patronage of Frederic II of Sicily, translated Ptolemy's *Almagest* and Ibn Rushd (Averroes)' *Compendium on Astronomy*, lost in the Arabic original. According to recent findings, he is perhaps the first translator into Hebrew of Euclid's *Elements*. Moses ben Samuel ben Yehuda of Montpellier (active between 1240 and 1283) translated, among others things, works by Euclid (*The Elements*), Theodosius (*Sphaerics*), al-Fārābī (The Commentary on *The Elements*), Ibn al-Haytham (The Commentary on *The Elements*), and al-Ḥaṣṣār (The so-called *Arithmetic*) Lastly, Jacob ben Makhir of Montpellier (1236?–1305?) also translated works by Euclid (*The Elements*, the *Data*, and perhaps the *Optics*), Autolycus of Pitane (*On the Moving Sphere*), Menelaus (*Sphaerics*), Jābir ibn Aflaḥ (*Astronomy*), and Ibn al-Haytham (*Astronomy*).

To this period belongs a Jewish scholar from Toledo, who also had contacts with Frederic II of Sicily during the first half of the thirteenth century, Yehuda ben Solomon ha-Kohen (probably b. 1215) wrote in Arabic a (lost) encyclopedia of sciences, which he translated himself into Hebrew, *Midrash ha-ḥokhma* (The Learning of Wisdom). This book includes a redaction, or free translation, of Euclid's *Elements* (Books I–VI and XI–XIII).

A generation after Jacob ben Makhir, Qalonymos ben Qalonymos (Maestro Calo) of Arles (b. 1287) translated Archimedes (*The Sphere and the Cylinder* and the commentary upon it by Eutocius, perhaps also *The Measurement of the Circle*), Nicomachus of Gerasa (a lost Arabic paraphrase of the *Introduction to Arithmetics*), Thābit ibn Qurra (on the *Figura Sector* of Menelaus), Jabir ibn Aflaḥ (on the same problem, a text apparently lost in Arabic), and Ibn al-Haytham (a part of the commentary on *The Elements*, already translated in part by Moses ibn Tibbon). Qalonymos also translated several texts for which the original Arabic has not yet been identified, or is now lost: thus we have important fragments of a *Treatise on Cylinders and Cones* by Ibn al-Samḥ of Granada (eleventh century), perhaps extracted from a larger book of this Andalusian scholar. Qalonymos is also probably the author of an important mathematical and philosophical composition on the nature of numbers, the *Sefer ha- Melakhim* (The Book of Kings).

Thus, we possess more or less complete Hebrew versions of the treatises of the following classical authors: Euclid, Archimedes, Autolycus, Theodosius, Menelaus, Ptolemy, Nichomacus. The conspicuous absence of the name of Apollonius does not mean that he was not known to the Jewish scholars.

The diffusion of the Hebrew versions of important mathematical works created the conditions for the composition of original works by Jewish scholars who were also excellent astronomers. This, at least, is the case of two scholars that we will mention here, as much for the depth of their talent as for the importance of their works. Levi ben Gershom, also called Gersonides (1288–1344) is the author of an important mathematical work. His writings on harmonic numbers (extant only in a Latin translation), on arithmetic and combinatorics (*Ma'aseh Ḥoshev*, Work of the Reckoner), and on the geometry of Euclid have been recently studied from the point of view of their contents, of their Arabic and perhaps Latin sources, and their possible influence on later works.

Immanuel ben Jacob Bonfils of Tarascon, one generation after Gersonides, must be mentioned in connection with two mathematical areas: that of the introduction of decimal fractions, and that of calculations related to the measurement of the circle. The sources and the scope of his developments in both areas have not yet been fully investigated.

In 15th-century Italy, there were a number of translators and commentators of note. The movement toward Italy is not only geographical: the Hebrew texts which we possess, and which have been only partially studied, show an increasing influence of Latin and also the vernacular languages.

Mordekhai (Angelo) Finzi of Mantua (active between 1441 and 1473) is without a doubt the best known of these translator-scholars. He translated several compositions of Abū Kamīl (a tenth-century Arabic scholar); he seems also to have been the translator (or perhaps the author) of a compendium on geometry in eleven chapters. Finzi is the author of a *Ma'amar be-Ḥeshbon Medidut ha-Gigiyyot we ha-Ḥaviyyot* (Treatise on the Measurement of Buckets and Barrels) in which he quotes Bar Ḥiyya and "the masters of the abacus", and whose analysis should yield useful clues on the history of stereometry before Kepler. Finally, we must emphasize Finzi's role in the diffusion of algebraic knowledge: in addition to Abū Kamīl's *Algebra*, Finzi translated from Italian or Latin into Hebrew the noteworthy algebraic composition of a certain Maestro Dardi of Pisa (who wrote in the fourteenth century), dealing with complex equations involving powers up to the twelfth degree.

The contacts of Jewish scholars in Italy with those in Constantinople, especially after the arrival of the Ottomans, constitute perhaps one of the channels by which ideas or texts of the Arabic East diffused.

There is also some mathematics of note in the Judeo-Byzantine scholarly world of the fifteenth to sixteenth centuries. The specific interest in studying this scholarly milieu — besides the talents of its representatives — lies in the use made therein of a triple heritage: that of texts available in Hebrew (Bar Ḥiyya, Ibn ʿEzra, the Greek and Arabic classics in translation, Gersonides, etc.), Arabic texts (sometimes not extant in Arabic) not translated into Hebrew but quoted in Hebrew compositions, and finally the Greek-Byzantine texts.

From this point of view, attention of scholars must be drawn to the mathematical works of Mordekhai Komtino (1402–1482) and of his two students Kalev Afendopulo (1460?–1525?) and Eliahu Mizraḥi (1455?–1526).

The *Sefer ha- Ḥeshbon we- ha- Middot* (Book on Reckoning and Measurements) of Komtino shows a direct knowledge of ancient Greek sources such as Hero, and probably also of Byzantine-Greek texts. The *Sefer ha-Mispar* (The Book of Number) by Mizraḥi, of which extracts were published in Latin in the sixteenth century, should be analyzed in the light of recent studies on the history of Arabic and Greek-Byzantine arithmetic. Finally, the commentary by Afendopulo of the Hebrew version of Nicomachus' *Introduction to Arithmetics* is a long text which includes several excursuses on philosophy, astronomy and astrology.

In this very brief — and far from exhaustive — survey of Hebrew mathematics, an attempt has been made to indicate the importance of these texts and to describe their transmission as a general set of mathematical ideas of Hellenistic origin between the eleventh and the sixteenth centuries, that is, before the rediscovery of Greek original texts and the relaunching of activity in the Renaissance.

TONY LÉVY

REFERENCES

Lévy, Tony. "Hebrew Mathematics in the Middle Ages: An Assessment". In *Tradition, Transmission, Transformation. Ancient Mathematics in Islamic and Occidental Cultures*. Ed. F. Jamil Ragep and Sally P. Ragep with Steven J. Livesy. Leiden: E.J. Brill, 1996.

Lévy, Tony. "Gersonide, le Pseudo-Tusi, et le postulat des parallèles. Les mathématiques en hébreu et leurs sources arabes." *Arabic Sciences and Philosophy* 2(1): 39–82, 1992.

Lévy, Tony. "Gersonide, commentateur d'Euclide: traduction annotée de ses gloses sur les *Eléments*." In *Studies on Gersonides*. Ed. Gad Freudenthal. Leiden: E.J. Brill, 1992, pp. 83–147.

Lévy, Tony. "Les Eléments d'Euclide en hébreu (XIIIe–XVIe siècles)." In *Perspectives médiévales (arabes, latines, hébraïques) sur la tradition scientifique et philosophique grecques*. Ed. M. Aouad, A. Elamrani-Jamal, A. Hasnaoui. Paris–Louvain: Peeters, 1996, pp. 79–93.

Sarfatti, Gad. *Mathematical Terminology in Hebrew Scientific Literature of the Middle Ages*. Jerusalem: The Magnes Press, 1968 (Hebrew, with English summary).

Steinschneider, Moritz. *Mathematik bei den Juden* (1893–1901). Hildesheim: Olms, 1964.

See also: al-Khwārizmī – Bar Ḥiyya – Decimal System – Ibn Tibbon – Moses Maimonides – Ibn Ezra

MATHEMATICS IN INDIA A widely held view is that Indian mathematics originated in the service of religion. Support for this view is sought in the complexity of motives behind the recording of the *Śulbasūtras*, the first written mathematical source dated around 800–500 BC, dealing with the measurement and construction of sacrificial altars. This view ignores the skills in mensuration and practical arithmetic that existed in the Harappan (or Indus Valley) culture which dates back to 3000 BC. Archaeological remains indicate a long established centralized system of weights and measures. A number of different plumb-bobs of uniform size and weights have been found that could be classified as decimal, i.e. if we take a plumb-bob weighing approximately 27.534 grams as a standard representing 1 unit, the other weights form a series with values of 0.05, 0.1, 0.2, 0.5, 2, 5, 10, 20, 50, 100, 200 and 500 units. Also, scales and instruments for measuring length have been discovered, including one from Mohenjo-Daro, one of the two largest urban centers, consisting of a fragment of shell 66.2 mm long, with nine carefully sawn, equally spaced parallel lines, on average 6.7056 mm apart. The accuracy of the graduation is remarkably high, with a mean error of only 0.075 mm.

A notable feature of the Harappan culture was its extensive use of kiln-fired bricks and the advanced level of its brickmaking technology. While fifteen different sizes of Harappan bricks have been identified, the standard ratio of the three dimensions—the length, breadth and thickness–is always 4:2:1, considered even today as the optimal ratio for efficient bonding. A close correspondence exists between the standard unit of measurement (the "Indus inch" (33.5 mm) and brick sizes, in that the latter are integral multiples of the former. (An Indus inch is exactly twice a Sumerian unit of length (*sushi*). 25 Indus inches make a Megalithic yard, a measure probably in use north-west Europe around 2000 BC. These links have led to the conjecture that a decimal scale of measurement originated somewhere in western Asia and then spread as far as Britain, Egypt, Mesopotamia and the Indus Valley.)

This relationship between brick-making technology and metrology was to reappear 1500 years later during the Vedic period in the construction of sacrificial altars of bricks. However, a most intriguing suggestion of B.V. Subbarayappa is that the Harappan numeration system contains certain similarities with the Kharoṣṭhī and the 'Ashokan' variant of the *Brāhmi* numeration systems which emerged in India about two thousand years later. He notes the following similiarities:

- There are identical symbols for the numbers one to four and for a hundred in all three numeration systems, and
- All three were ciphered systems employing a decimal base.

He suggests that deciphering the inscriptions on the large number of excavated seals and other artifacts require that they be recognized as numerical records rather than as literary passages. Given the failure so far to decipher the Harappan script, this approach is certainly worth further examination.

The earliest written evidence in India of a recognizable antecedent of our numeral system is found in an inscription from Gwalior dated 'Samvat 933' (AD 876) where the numbers 50 and 270 are given as  and . Notice the close similarity with our notation for 270 showing in both an understanding of the place value principle as well as the use of zero. There is earlier evidence of the use of Indian system of numeration in South East Asia in areas covered by present-day countries such as Malaysia, Cambodia and Indonesia, all of whom were under the cultural influence of India. Also, as early as AD 662, a Syrian bishop, Severus Sebokt, comments on the Indians carrying out computations by means of nine signs by methods which "surpass description" (Joseph, 1993).

The spread of these numerals westwards is a fascinating story. The Arabs were the leading actors in this drama. Indian numerals probably arrived at Baghdad in 773 AD with the diplomatic mission from Sind to the court of Caliph al-Manṣūr. In about 820, al-Khwārizmī wrote his famous *Kitāb ḥisāb al-adad al-hindī* (Book of Addition and Subtraction According to the Hindu Calculation, also called just Arithmetic), the first Arab text to deal with the new numerals. The text contains a detailed exposition of both the representation of numbers and operations using Indian numerals. Al-Khwārizmī was at pains to point out the usefulness of a place-value system incorporating zero, particularly for writing large numbers. Texts on Indian reckoning continued to be written, and by the end of the eleventh century this method of representation and computation was widespread from the borders of Central Asia to the southern reaches of the Islamic world in North Africa and Egypt.

In the transmission of Indian numerals to Europe, as with almost all knowledge from the Islamic world, Spain and (to a lesser extent) Sicily played the role of intermediaries, being the areas in Europe which had been under Muslim rule for many years. Documents from Spain and coins from Sicily show the spread and the slow evolution of the numerals, with a landmark for its spread being its appearance in an influential mathematical text of medieval Europe, *Liber Abaci* (Book of Computation), written by Fibonacci (1170–1250), who learnt to work with Indian numerals during his extensive travels in North Africa, Egypt, Syria, and Sicily. The spread westwards continued slowly, displacing Roman numerals, and eventually, once the contest between the abacists (those in favor of the use of the abacus or some mechanical device for calculation) and the algorists (those who favored the use of the new numerals) had been won by the latter, it was only a matter of time before the final triumph of the new numerals occurred, with bankers, traders, and merchants adopting the system for their daily calculations.

The beginnings of Indian algebra may be traced to the *Śulbasūtras* and later Bakhshālī Manuscript, for both contain simple examples involving the solution of linear, simultaneous and even indeterminate equations. An example of an indeterminate equation in two unknowns (x and y) is $3x+4y = 50$, which has a number of positive whole-number (or integer) solutions for (x, y). For example, $x = 14$, $y = 12$ satisfies the equation as do the solution sets (10, 5), (6, 8) and (2, 11).

But it was only from the time of Āryabhaṭa I (b. AD 476) that algebra grew into a distinct branch of mathematics. Brahmagupta (b. AD 598) called it *kuṭṭaka gaṇitā*, or simply *kuṭṭaka*, which later came to refer to a particular branch of algebra dealing with methods of solving indeterminate equations to which the Indians made significant contributions.

An important feature of early Indian algebra which distinguishes it from other mathematical traditions was the use of symbols such as the letters of the alphabet to denote unknown quantities. It is this very feature of algebra that one immediately associates with the subject today. The Indians were probably the first to make systematic use of this method of representing unknown quantities. A general term for the unknown was *yāvat tāvat*, shortened to the algebraic symbol *yā*. In Brahmaguptas's work Sanskrit letters appear, which are the abbreviations of names of different colors, which he used to represent several unknown quantities. The letter *kā* stood for *kālaka*, meaning 'black', and the letter *nī* for *nīlaka* meaning 'blue'. With an efficient numeral system and the beginnings of symbolic algebra, the Indians solved determinate and indeterminate equations of first and second degrees and involving in certain cases more than one unknown. It is likely that a number of these methods reached the Islamic world before being transmitted further westwards by a similar process and often involving the same actors as the ones that we discussed earlier in the spread of Indian numerals.

The beginnings of a systematic study of trigonometry are found in the works of the Alexandrians, Hipparchus (ca. 150 BC), Menelaus (ca. AD 100) and Ptolemy (ca. AD 150). However, from about the time of Āryabhaṭa I, the character of the subject changed to resemble its modern form. Later, it was transmitted to the Arabs who introduced further refinements. The knowledge then spread to Europe, where the first detailed account of trigonometry is contained in a book entitled *De triangulis omni modis* (On All Classes of Triangles), by Regiomontanus (1464).

In early Indian mathematics, trigonometry formed an integral part of astronomy. References to trigonometric concepts and relations are found in astronomical texts such as *Sūryasiddhānta* (ca. AD 400), Varāhamihira's *Pancha Siddhāntạ* (ca. AD 500), Brahmagupta's *Brāhma Sputạ Siddhānta* (AD 628) and the great work of Bhāskara II called *Siddhānta Śiromaṇi* (AD 1150). Infinite expansion of trigonometric functions, building on Bhāskara's work, formed the basis of the development of mathematical analysis — a precursor to modern calculus to be discussed later.

Basic to modern trigonometry is the sine function. It was introduced into the Islamic world from India, probably through the astronomical text, *Sūryasiddhānta*, brought to Baghdad during the eighth century. There were two types of trigonometry available then: one based on the geometry of chords and best exemplified in Ptolemy's *Almagest*, and the other based on the geometry of semi-chords which was an Indian invention. The Arabs chose the Indian version which prevailed in the development of the subject. It is quite likely that two other trigonometric functions — the cosine and versine functions — were also obtained from the Indians.

One of the most important problems of ancient astronomy was the accurate prediction of eclipses. In India, as in many other countries, the occasion of an eclipse had great religious significance, and rites and sacrifices were performed. It was a matter of considerable prestige for an astronomer to demonstrate his skills dramatically by predicting precisely when the eclipse would occur.

In order to find the precise time at which a lunar eclipse occurs, it is necessary first to determine the true instantaneous motion of the moon at a particular point in time. The concept of instantaneous motion and the method of measuring that quantity is found in the works of Āryabhaṭa I, Brahmagupta and Muñjāla (ca. AD 930). However, it was in Bhāskara II's attempt to work out the position angle of the ecliptic, a quantity required for predicting the time of an eclipse, that we have early notions of differential calculus. He mentions the concept of an "infinitesimal" unit of time, an awareness that when a variable attains the maximum value its differential vanishes, and also traces of the "mean value theorem" of differential calculus, the last of which was explicitly stated by Parameśvara (1360–1455) in his commentary on Bhāskara's *Līlāvatī*. Others from Kerala (South India) continued this work with Nīlakaṇṭha (1443–1543) deriving an expression for the differential of an inverse sine function and Acyuta Piṣāraṭi (ca. 1550–1621) giving the rule for finding the differential of the ratio of two cosine functions.

However, the main contribution of the Kerala school of mathematician-astronomers was in the study of infinite-series expansions of trigonometric and circular functions and finite approximations for some of these functions. The motivation for this work was the necessity for accuracy in astronomical calculations. The Kerala discoveries include the Gregory and Liebniz series for the inverse tangent, the Liebniz power series for π, the Newton power series for the sine and cosine, as well as certain remarkable rational approximations of trigonometric functions, including the well-known Taylor series approximations for the sine and cosine functions. And these results had been obtained about three hundred years earlier than the mathematicians after whom they are now named. Referring to the most notable mathematician of this group, Mādhava (ca. 1340–1425), Rajagopal and Rangachari (1978) wrote: "(It was Mādhava who) took the decisive step onwards from the finite procedures of ancient mathematics to treat their limit-passage to infinity, which is the kernel of modern classical analysis". The growing volume of research into medieval Indian mathematics, particularly from Kerala, has refuted a common perception that mathematics in India after Bhāskara II made "only spotty progress until modern times" (Eves, 1983).

GEORGE GHEVERGHESE JOSEPH

REFERENCES

Primary sources

Algebra with Arithmetic and Mensuration from the Sanscrit of Brahmegupta and Bhāscara. Ed. H.T. Colebrooke. London: Murray, 1817; reprinted Wiesbaden: Sandig, 1973.

Āryabhaṭiya of Āryabhaṭa. Ed. and trans. K.S. Shukla and K.V. Sarma. New Delhi, Indian National Science Academy, 1976.

Gaṇitasāra-saṃgraha of Mahāvīrācārya. Ed. and trans. M. Rangacharya. Madras: Government Press, 1912.

Pañcasiddhāntikā of Varāhamihira. Ed. and trans. T.S.K. Sastry and K.V. Sarma. Madras : PPST Foundation, 1993.

Secondary sources

Bag, A. K. *Mathematics in Ancient and Medieval India*. Varanasi: Chaukhambha Orientalia, 1979.

Datta, B. and A.N. Singh. *History of Hindu Mathematics*, 2 vols. Bombay: Asia Publishing House, 1962.

Eves, H. *An Introduction to the History of Mathematics*, 5th ed. Philadelphia: Saunders, 1983.

Joseph, George G. *The Crest of the Peacock: Non-European Roots of Mathematics*. London: Penguin, 1993.

Rajagopal, C.T. and M.S. Rangachari. "On an untapped source of medieval Keralese Mathematics." Archives for History of Exact Sciences, 18, 1978, 89–108.

Sarma, K.V. *A History of the Kerala School of Hindu Astronomy*. Hoshiarpur: Vishveshvaranand Institute, 1972.

Srinivasiyengar, C. N. *History of Indian Mathematics*. Calcutta: World Press, 1967.

Subbarayappa, B.V. *Numerical System of the Indus Valley Civilisation*. Mimeo, 1993.

See also: *Śulbasūtras* – Weights and Measures in India – Surveying – Technology and Culture – Geometry in India – Arithmetic in India – *Bakhshālī* Manuscript – *Almagest* – Eclipses – al-Khwārizmī – Brahmagupta – Āryabhaṭa – Bhāskara – Parameśvara – Nīlakaṇṭha Somayāji – Acyuta Piṣāraṭi - Mādhava

MATHEMATICS IN ISLAM Roughly speaking, one can distinguish two types of mathematics in Antiquity and the Middle Ages: One is "practical" (or subscientific) mathematics, concerned with practical calculation, administration, trade, land measurement, tax collecting, etc. The other is "theoretical" mathematics, studied either for its own sake, or connected with astronomy, philosophy, or religion.

The following survey of the history of mathematics in Islamic civilization will concentrate on "theoretical" mathematics. "Practical" mathematics in Islamic civilization is of great historical interest, but until recently the subject has received less attention than it deserves, perhaps because this kind of mathematics is less exciting from an intellectual point of view. Therefore it is not yet possible to give an adequate survey of its history.

The history of theoretical mathematics in Islamic civilization is intimately connected with astronomy. In Iran there was a pre-Islamic astronomical tradition which simply continued after the country was conquered by the Muslims. When the Abbasids came to power in 750, an interest in astronomy, and hence mathematics, developed in the new capital Baghdad. Iranian astronomy was influenced by Indian astronomy, and around 775 Indian astronomers and mathematicians were received at the court of the Caliph, and Sanskrit astronomical works were translated into Arabic. After 800, the Greek mathematicial and astronomical works became increasingly popular, and many of these works were translated into Arabic, including the *Elements* of Euclid and the *Almagest* of Ptolemy, an astronomical compendium full of very complicated mathematics. In the rest of this article, the term "Arabic" will be used in a linguistic sense, referring to Arabic as a scientific language. Thus "Arabic mathematics" is mathematics written in the Arabic language; an "Arabic mathematician" is a mathematician writing in Arabic. The reader should bear in mind that a large number of these "Arabic mathematicians" were actually Persians.

The most important traces of the Indian heritage in the Arabic tradition were the system of numbers and the sine function (which replaced the Greek chord in trigonometry). It is possible that early Arabic algebra was influenced by India to a much greater extent than is recognized today by most modern historians, many of whom assume a Babylonian influence, transmitted in some unknown way. However, the Greek influence was predominant in Arabic mathematics. It was a living continuation of Greek mathematics. New areas of mathematics were developed (algebra, trigonometry), new problems were solved (such as the *qibla* problem, determining the direction of Mecca), and a lot of creative work was done. However, there was no "revolution" in Arabic mathematics comparable with, for example, the development of analytic geometry and calculus in seventeenth-century Europe.

Arithmetic

Various systems of numeration were used by the Arabic mathematicians and astronomers. Many mathematicians and astronomers used the *abjad* (alphabetical) numeration. In this system, the letters of the alphabet have numerical values 1,2, ..., 9, 10, 20, ..., 90, 100, 200, ..., 900, 1000. Almost all astronomical calculations were performed in a sexagesimal system in which the sexagesimals were denoted in the abjad system. The Hindu–Arabic system, consisting of nine symbols and a zero, was introduced into the Arabic world by al-Khwārizmī around AD 830, but it was not received with much enthusiasm. It was only used for large numbers, for example the tangents of angles near 90 degrees. In algebraical works, al-Khwārizmī and others did not use numbers but words, for example "three squares and four is equal to seven things" (meaning: $3x^2 + 4 = 7x$ in modern notation). Fractions were denoted in various complicated ways, often without any symbolism. In the tenth century, al-Uqlīdisī used decimal fractions and a symbol equivalent to our decimal point.

Algebra

The name of this part of mathematics is derived from al-Khwārizmī's *Kitāb fī'l-jabr wa'l-muqābala* (Book on Restoration and Confrontation). In this treatise al-Khwārizmī gives a systematic treatment of linear and quadratic equations. The contents were not new, because the Babylonians in Iraq were able to solve quadratic equations two thousand years before al-Khwārizmī. However, the treatise is very well written and al-Khwārizmī gives many worked examples. He treats the reduction of any linear or quadratic equation to one of six standard forms $ax = b$, $ax^2 = c$, $ax^2 = bx$, $ax^2 + bx = c$, $ax^2 + c = bx$, $ax^2 = bx + c$, and he then gives the solution of each of these forms: first the procedure for finding x, and in the case of the last three forms, also a geometrical motivation using rectangles and squares. The operation *al-jabr*, which gave its name to the whole field of solving equations, means the "restoration" of a negative term (example: $3x^2 - 7x = 4$ is "restored" to $3x^2 = 4 + 7x$). After al-Khwārizmī, Abū Kāmil solved complicated quadratic equations with irrational coefficients. In the tenth century, al-Karajī treated various properties of quadratic irrationals, which Euclid had proved geometrically. Al-Karajī also discussed cubic irrationalities, and he explained the extraction of the root of a polynomial (which is assumed to be a perfect square). In the tenth and eleventh centuries, cubic equations were solved geometrically by the intersection of conic sections, for example by ʿUmar al-Khayyām, and the theory was perfected by Sharaf al-Dīn al-Ṭūsī. A few mathematicians worked on equations of degree higher than three, but most of this work seems to be lost.

Theory of Numbers

Some Arabic mathematicians were interested in indeterminate equations, that is, equations with an infinite number of solutions. The solutions had to be rational numbers (in the way of Diophantus of Alexandria, whose *Arithmetica* was translated into Arabic) or integers. Examples of the last kind are: find integers x, y and z such that $x^2 + y^2 = z^2$ (such numbers x, y and z are the sides of a right-angled triangle); or the famous problem of "congruent numbers": find numbers which are the surface area of a right-angled triangle whose sides are all integers. In the tenth century it was believed that no numbers x, y and z such that $x^3 + y^3 = z^3$ existed, but it seems that this fact could not be proven by means of the mathematical methods then available.

Some progress was made in the theory of perfect and amicable numbers. A number is perfect if it is equal to the sum of its own divisors. Euclid had proved a formula for even perfect numbers, and four such numbers were known in antiquity: 6, 28, 496, 8128. In or before the thirteenth century, three more were found by Arabic mathematicians: 33,500,336; 8,589,869,056 and 137,438,691,328. Two numbers are called amicable if they are the sum of the divisors of each other. The example 220, 284 was known from antiquity. Thābit ibn Qurra discovered and proved that if p, q and r are three prime numbers of the form $p = 3 \cdot 2^{n-1} - 1$, $q = 3 \cdot 2^n - 1$, and $r = 9 \cdot 2^{n-1}$, then the numbers $2^n \cdot pq$ and $2^n \cdot r$ are amicable. Perhaps the subject of amicable numbers was popular because of its magical applications, and such applications may also explain the interest in magic squares, which continued throughout the Arabic tradition.

The reader should bear in mind that the ancient and medieval concept of "number" is not necessarily the same as the modern one. In ancient Greek mathematics, a "number" was always a positive integer number, or a fraction (that is, a ratio between two positive integer numbers). The Greeks knew that there were proportions in geometry which could not be expressed by this limited concept of number (such as the ratio between the side and diagonal of a square, that is $1 : \sqrt{2}$). Nevertheless, they did not extend their concept of number to include the modern idea of an irrational or real number. The reason is that this involves difficulties with the infinite, which they would have avoided (and which are carefully hidden in naïve presentations of the concept of real number in school mathematics). Instead of working with real numbers, the Greeks developed a theory of proportions of geometrical magnitudes (for example, line segments). This theory is found in Book V of Euclid's *Elements*, which was widely studied by the Arabic mathematicians. In astronomy one has to calculate the (approximate) length of segments which cannot be represented as a number according to the orthodox view. Thus it is understandable that the Arabic mathematicians, most of whom were also astronomers, came closer to the naive concept of real number, as it is nowadays taught in schools. Many mathematicians, such as Sharaf al-Dīn al-Ṭūsī, continued to believe that any rigorous treatment of numbers and algebra had to be based on the Euclidean theory of proportions.

Geometry

The basic geometrical work in Arabic mathematics was the *Elements* of Euclid (ca. 300 BC). This work was very thoroughly studied, and more than fifty commentaries in Arabic were written, most of which have not been studied in modern times. Euclid's parallel postulate was one of the main points of attention. Most Arabic mathematicians found this unsatisfactory, and attemps were made, by Ibn al-Haytham, Naṣīr al-Dīn al-Ṭūsī, and others, to replace it by a more appropriate axiom. Many other Greek geometrical works were translated into Arabic, so the Arabic literature is an important source of new information on Greek geometrical works which are now lost.

The Arabic mathematicians were very interested in geometrical constructions. Such constructions should preferably be made by ruler and compass, that is to say, by successive intersections of straight lines through two known points and circles with known centers and radii (the intersection of such known figures produces new known points). The most interesting problems from Greek geometry (trisection of the angle, construction of two mean proportionals, etc.) cannot be constructed in this way. The Arabic mathematicians also used conic sections as means of construction.

Some of the mathematicians were clearly aware that they progressed beyond the ancient Greeks, and there was a feeling that the authorship of a geometrical construction added to one's prestige as a mathematician. Some mathematicians even plagiarized others' works, as in the case of the regular heptagon. This figure cannot be constructed by means of ruler and compass, but the Arabic geometers knew an ancient construction (attributed to Archimedes) by means of a straight line which had to be moved in an unclear way until two triangles are equal. Around 968 Abū'l-Jūd proposed a ruler- and compass construction of the heptagon. The elementary error in this construction was soon discovered by al-Sijzī. The missing link was filled by al-ʿAlā' ibn Sahl by means of conic sections, and this was then plagiarized by al-Sijzī (who repented later) and also by Abū'l-Jūd. Other constructions of the heptagon were found by al-Qūhī (ca. 970) and Ibn al-Haytham (ca. 1000), who were both very proud of their achievement. Most mathematicians felt that the heptagon had not really been constructed by the ancients, and thus the "moderns" had gone beyond the level reached by the Greeks. A similar problem was the inscription of an equilateral pentagon in a given square. This was solved by al-Qūhī, who stressed that this problem had not been solved by the ancient geometers.

An important branch of geometry was spherical trigonometry. In the eighth and ninth centuries, spherical trigonometry developed rather chaotically, in the context of astronomy and on the basis of a mixture of Indian and Greek methods. In the tenth century the field became an independent subject of study, and special treatises were devoted to it. The Arabic mathematicians studied the spherical triangle, that is a triangle on a sphere, such that the "sides" of the triangle are arcs of great circles (i.e. circles whose center is the center of the sphere). In the tenth century, "angles" were also defined in a spherical triangle, so that a spherical triangle has six elements, three angles and three sides, like a plane triangle. The main problem of spherical trigonometry could now be phrased thus: if three elements are given, how do we compute the rest? Treatises on this problem were written by Naṣīr al-Dīn al-Ṭūsī and the eleventh century Spanish mathematician Ibn Muʿādh.

The determination of the *qibla* (direction of Mecca) from a locality with given geographical coordinates can also be regarded as a problem of spherical trigonometry. In the eighth and ninth centuries only approximate solutions of the problem were known, but exact solutions were found in the tenth century, and tables were computed in the twelth century and later.

The Arabic mathematicians devoted much time and energy to the computation of trigonometrical tables (sines and tangents) with ever increasing accuracy. The main difficulty is the determination of the sine of one degree; since this quantity cannot be expressed in square roots of rational numbers, the standard approximation methods fail. Around 1420, al-Kāshī found a method to approximate the sine of 1 degree with any degree of accuracy, using algebra. Al-Kāshī was able to express this as a root of a cubic equation with known coefficients, and he developed a very fast algorithm to approximate the root. He also computed the number π to 16 decimals.

Mathematics and Astronomy

It is historically impossible to separate Arabic mathematics from Arabic astronomy. Astronomy was the main field of application of mathematics, and calculations that were done by the astronomers vastly surpassed those of "practical mathematics" (commerce, administration, etc.). Many people studied mathematics in order to become astronomers (or astrologers). One finds a lot of very interesting mathematics in astronomical handbooks (*zījes*). The astronomical problems were sometimes too difficult to be solved by the methods of medieval mathematics, or, in cases where the problems could be solved, the computations were sometimes too complicated to be performed in a reasonable amount of time. Therefore, such computations were often simplified by clever approximation devices, which presuppose considerable numerical insight.

Mathematics was also important for astronomy on a more philosophical level. The philosophical foundation of Ptolemy's astronomical models was unacceptable for some Arabic mathematicians (such as Naṣīr al-Dīn al-Ṭūsī) because some of the circular motions in his theories of the moon and the planets were non-uniform.

The later Arabic mathematicians sought to remove these flaws by adding one or more circles to these models, producing the same effects as Ptolemy in a philosophically acceptable way. The most famous device is the so-called Ṭūsī couple, consisting of a circle of radius r, which rolls along the interior of a second circle with the double radius $2r$. As a result of this motion, any point on the first circle oscillates on a fixed straight line.

Transmission

Arabic mathematics influenced the development of mathematics in medieval Europe in various ways. In the twelfth century, some Arabic mathematical works (such as the *Arithmetic* and the *Algebra* of al-Khwārizmī) and Arabic versions of Greek mathematical and astronomical works (such as Euclid's *Elements* and Ptolemy's *Almagest*) were translated into Latin, mainly in Spain. These translations were the beginning of the development of mathematics in Christian Europe. The thirteenth century European mathematician Leonardo Fibonacci learned mathematics in the city of Bougie in Algeria and during his travels to other Islamic countries, and he then wrote various influential mathematical works. The transmission of mathematics from Arabic to Latin was far from complete, and until 1450 the level of mathematics in Europe was below that in the Islamic world.

JAN P. HOGENDIJK

REFERENCES

Berggren, J.L. *Episodes in the Mathematics of Medieval Islam*. New York: Springer, 1986.

Juschkevitch, A.P., 1961. *Istoria Matematiki b Srednie Veka*. (History of Mathematics in the Middle Ages.) Translated into German as: *Geschichte der Mathematik im Mittelalter*. Leipzig: Teubner, 1964; the section on Arabic mathematics was translated into French as: Youschkevitch A.P., *Les mathématiques Arabes (VIIe–XVe siècles)*. Paris: Vrin, 1976.

Kennedy, E.S. *Studies in the Islamic Exact Sciences*. Beirut: American University, 1983.

Matvievskaya, G.P. and B. Rozenfeld. *Matematiki i Astronomy Musul'manskogo Srednevekovya i ikh Trudy* (Mathematicians and Asronomers of Medieval Islam and Their Works). Moscow: Nauka, 1983 (in Russian).

Rosenfeld, B. *A History of non-Euclidean Geometry*. New York: Springer, 1988.

Samsó, J. *Las Ciencias de los Antiguos en Al-Andalus*. Madrid: MAPFRE, 1992.

Woepcke, F. *Études sur les mathématiques arabo-islamiques*. 2 vols, Frankfurt: Institut für Geschichte der arabisch-islamischen Wissenschaften, 1986.

See also: al-Khwārizmī – al-Uqlīdisī – ʿUmar al-Khayyām – al-Karajī – Sharaf al-Dīn al-Ṭūsī – Thābit ibn Qurra – Ibn al-Haytham – Naṣīr al-Dīn al-Ṭūsī – al-Sijzī – Ibn Sahl – al-Qūhī – Ibn Muʿādh – al-Kāshī – *Qibla* – Astronomy in Islam – Geometry in Islam

MATHEMATICS IN JAPAN At various stages of Japanese history, mathematics developed as a direct consequence of contacts with foreign cultures, both Chinese and Western. Five successive waves of cultural influx may be delineated: (1) Chinese wave I, from the seventh to the end of the ninth century; (2) Chinese wave II, from the end of the sixteenth to the mid-nineteenth century; (3) Western wave I, 1543–1639 (4) Western wave II, 1720–1854, and (5) Western wave III, from 1854 onwards.

As far as may be surmised, the two Chinese waves developed independently and were separated by an interim period of semi-seclusion from continental influences during which Japanese mathematical activity subsided.

During Chinese wave I, Japanese mathematics did not depart significantly from that developed in China during the same period. This was also the case in Sui (589–618) and Tang (618–907) China. Japanese imperial authorities are said to have founded an elite school for training future accountants, fiscal officers, surveyors, calendar makers, and other such practitioners. The teaching of mathematics was based on the Chinese *Suanjing shishu* (The Ten Computational Canons) and not on Japanese autochthonous manuals. During that period and the following centuries, mathematics never developed beyond the rudiments. Fragmentary records such as those of a priest from the Kenninji temple show that ca. 1311, the Chinese *Jiuzhang suanshu* (Computational Prescriptions in Nine Chapters) from the Han dynasty was studied in Kyoto; certain texts known as *oraimono* (didactic exchanges of letters) contain, none the less a number of mathematical recreations such as the problem of *mamakodate* (lit. the standing stepchildren), a problem which supposes that a mother has fifteen true children and fifteen stepchildren; she must eliminate the latter by placing them in a circle in such a way that starting from some child and counting clockwise she eliminates every tenth child until at last only the true children are still standing. This problem is reminiscent of the Josephus puzzle, where children are replaced by Turks — to be eliminated — and Christians.

During Chinese wave II, a number of Chinese mathematical texts were imported into Japan on the occasion of the Japanese military expeditions in Korea of the 1590s.

One of them, the *Suanfa tongzong* (Comprehensive Treatise on Arithmetic), was a compendium of commercial arithmetic. First published in China in 1592, this manual was mostly intended for abacus calculations and was very often reprinted in China and Japan until the twentieth century. However, although strongly influenced by its Chinese model, Japanese arithmetic began to develop on its own. In particular, arithmeticians devised a new type of abacus (*soroban*). It was composed of five balls for the representation of the decimal digits of each order of units, instead of seven in the Chinese case. Thre were four balls valued at one unit each and one ball valued at five units, instead of five balls valued at one unit and two balls valued at five units. For the first time, the Japanese also wrote arithmetical books in their own language. One of them, the *Jingōki* (Treatise on Eternal Mathematical Truths) of Yoshida Mitsuyoshi (1578–1672) first published in 1627, became so popular that hundreds of plagiarized versions were published. In its 1641 version, the author included twelve problems left unsolved and bequeathed to posterity (*idai*). Subsequently, this kind of challenge was often issued and played an important role in the development of Japanese mathematics. At first, the custom was limited to more or less trivial arithmetical problems, but gradually amateur mathematicians imagined complex problems relating, for example, to the volume of intricate solids (such as those defined by the intersection of two other solids), or to chains of circles or spheres tangent to other circles, spheres, ellipses or ellipsoids, respectively. When solutions were found, the solvers published them and in their turn propounded new enigmas.

A development in mathematics more sophisticated than arithmetic, however, was triggered by the introduction into Japan of the *Suanxue qimeng* (Introduction to Computational Science), a Chinese algebraic textbook by Zhu Shijie published in 1299 in Yuan China. Subsequently forgotten in China for five centuries, but universally considered by present historians of Chinese mathematics as representative of the golden age of Chinese mathematics, this manual was first reprinted in Korea during the fifteenth century and in Japan in 1658. At first, the part of this text devoted to *tianyuan* (celestial origin) algebra was not well understood. This Chinese medieval technique was presented using a series of artificial problems, with some intermediary computations and a few cryptic and terse sentences apparently having more to do with magic than mathematics. Moreover, known and unknown quantities were not distinguished from each other by special symbols but only by the relative position of numbers on the counting board, so that both kinds of quantities were equally represented by mere numbers in all cases. Yet, twenty years later, certain scholars began to crack the code and realized that Zhu Shijie's algebra was in fact a powerful weapon capable of mechanically solving all sorts of intricate problems and particularly those of the *idai* new tradition. What is more, these efforts at last enabled Seki Takakazu, also called Seki Kowa (ca. 1742–1708), the son of a samurai who became a chief Palace accountant, to develop algebra well beyond the state he had found it in Zhu Shijie's manual.

Seki was so successful as a mathematician that he was retrospectively considered the father of a national tradition called *wasan*. That is literally Japanese (*wa*) mathematics (*san*) in contrast with *yosan*, Western (*yo*) mathematics; these two terms were coined ca. 1870. Although strongly influenced by Zhu Shijie, and more generally by Chinese models, Seki's work marks a certain rupture with mathematics conceived as collections of isolated problems. Concerning the elimination of unknowns between polynomial equations, he imagined a general method which resulted in an invention of determinants independently of Leibnitz. He also systematized the Chinese Hornerian methods for the numerical evaluation of the roots of polynomials, computed the decimals of π, and discovered the Bernoulli numbers. Last, but not least, he created a kind of notational algebra whose symbols consisted of dissections of Chinese characters which enabled mathematicians to perform literal (and not merely numerical) computations in writing rather than by using counting-rods as the Chinese did. In fact, Seki's algebra was not so different from that of Viète.

All this was improved by later mathematicians. In addition, questions relating to the precise evaluation of the circumference, arcs, and chord lengths became so important that they were conceived as a single domain called *enri* (circular principles). Later, this domain was generalized to the computation of all sorts of infinite series and to various curves in a way which more or less evokes Western calculus.

However, Chinese mathematics were not the sole source out of which Japanese mathematicians developed their own mathematics. During Western wave I a very limited amount of mathematical knowledge of Western origin (consisting perhaps only of some mathematical recreations) was introduced into Japan as a consequence of the diffusion of Catholicism by Jesuit missionaries. But edicts against Christianity were issued as early as 1612, and in 1630 a ban on the importation of Western books was decreed. During Western wave II, the ban was progressively removed and Western mathematical knowledge gradually permeated into Japan. In the period from 1720 to 1730, elementary geometry, plane and spherical trigonometry, and even logarithms became accessible through the medium of Chinese adaptations of European works such as the *Chongzhen lishu* (Chongzhen Period Compendium of Mathematical Astronomy), an encyclopedic work which marked the start of the reform of Chinese astronomy in 1644 on the basis of European knowledge. Later, the Japanese also came into contact with Western works written not in Chinese but in Dutch, the language of the Dutch merchants, the sole Westerners who had relations with Japan. This initiated the development of a new branch of knowledge called *rangaku* (Dutch learning). A consequence for mathematics of this new learning was the importation of the Dutch version of John Keill's (1671–1721) commentary on Newton's *Principia*.

Consequently, elements of European mathematics were integrated into *wasan* and new results were discovered. For example, Takebe Katahiro (1664–1734) developed an infinite series of the square of the arcsine. Takebe was a disciple of Seki and a Shogunal advisor also responsible for new developments in geography and astronomy. Ajima Naonobu, another distinguished *wasan* scholar, independently solved celebrated geometrical problems such as those of Steiner's chains of circles (circles mutually tangent inserted between two non-concentric circles situated one inside the other) or the so-called Malfatti's problem about three circles mutually tangent and inscribed in a triangle. Less anecdotally, Ajima and later Wada Yasushi (1787–1840) also elaborated methods for calculating areas and volumes as limits of infinitesimal rectangles or parallelepipeds in a way which is reminiscent of the construction of definite integrals. The achievements of Japanese autochthonous mathematics was thus quite high. A fundamental difference between Western and Japanese mathematics stems from the fact that the latter was essentially algebraic and developed independently of axiomatico-deductive reasoning.

While the the search for general and systematic methods concerned more and more mathematicians, an opposite trend towards solving highly artificial problems conceived as isolated puzzles became still more prominent. In fact, mathematics were mainly practiced by "hobbyists", members of rival schools or even mathematical sects. Works of wasanists were thus not much diffused. Original manuscripts were often copied by the disciples of some master (like Seki) and kept secret. Consequently, no consensus towards a uniformization of mathematical notations and concepts ever emerged, even though towards the end of the eighteenth century some mathematical works were eventually printed. More importantly, *wasan* studies were often condemned by Confucian elites who judged them restricted to "mental acrobatics" and thus futile and socially useless. No official schools devoted to the study of *wasan* were ever created. During the whole Edo period (1698–1868), *wasan* was often practiced by *rōnin* (lordless samurai) who traveled all over Japan and earned their living by teaching *wasan* in private academies or becoming private tutors at the service of rich merchants. Unexpectedly, owing to a surprising alliance between mathematics and religion, *wasan* became more and more visible even in the Japanese countryside. From the end of the seventeenth century, *wasan* adepts advertised the achievements of the groups they belonged to by means of votive tablets (*sangaku*). Hung in public view in Buddhist temples and Shinto shrines, these wooden tablets generally displayed problems and beautifully engraved geometric figures (often consisting of mutually tangent figures, especially circles and triangles) with numerical solutions but no intermediary calculations. Hundreds of tablets have survived, and it is still possible to admire them everywhere in Japan. Moreover, printed anthologies of such tablets have been published. Such a practice is unattested anywhere else in the world outside Japan.

Until 1853, despite the importation of a certain amount of knowledge from China and Europe, Japan essentially remained isolated. But the partisans of isolation did not succeed in imposing their policy after Commodore Matthew C. Perry of the United States had forced open Japan's ports. As early as 1855, an Office for Occidental Learning (*Yōgakusho*) was created; one year later, this office was renamed Office for Investigating Barbarian Documents (*Bansho shirabesho*). There, government appointed students learned to apply mathematics to navigation, ship building, armaments, and other military matters. But radical changes had to wait the Meiji restoration of 1868. With the advent of this utterly new era, Western learning was no longer confined to military affairs and other utilitarian goals. *Wasan* was completely abandoned to Western mathematics. In 1877, the University of Tokyo was founded. At first, four out of five professors were foreigners (one American and three French). The Japanese professor was Kikuchi Dairoku (1885–1917) who later became Minister of Education. Kikuchi was an outstanding student who had been trained at the *Bansho shirabesho* and sent to Cambridge University; he gave his mathematical lectures in English. None the less, a few years later, the best students of mathematics were appointed professors in their turn. In 1877, the Tokyo Mathematical Society was founded. Japanese mathematicians rapidly attained the same level as that of Western researchers. Many mathematicians, such as Takagi Teigi (1875–1960) who contributed to class-field theory and gained world-fame, were trained during this period.

JEAN-CLAUDE MARTZLOFF

REFERENCES

Fuji Tanki Daigaku (Fuji Junior College). *Sūgaku shi kenkyū* (Researches on the History of Mathematics), 1959 to the present.

Hirayama, Akira and Matsuoka, Motohisa, eds. *Naonobu Ajima's Complete Works*. Tokyo: Fuji Junior College, 1966 (in Japanese with English explanations).

Hirayama, Akira, Shimodaira, Kazuo and Hirose, Hideo, eds. *Takakazu Seki's Collected Works*. Osaka: Kyoiku Tosho, 1974 (in Japanese with English explanations).

Hirayama, Akira. *Wasan no tanjō* (The Birth of Wasan). Tokyo: Kōseisha, 1993.

Honda, Kin-ya. *Teiji Takagi: a Biography - On the 100th Anniversary of his Birth*. Tokyo: Seorsum Imperessum ex tom. XXIV, fasc. II Commentariorum Mathematicorum Universitatis Sancti Pauli, 1975, pp. 141–167.

Honda, Kin-ya. "A Survey of Japanese Mathematics During the Last Century." *Japanese Studies in the History of Science* 16: 1–16, 1977.

Martzloff, Jean-Claude. "A Survey of Japanese Publications on the History of Japanese Traditional Mathematics (Wasan) from the Last 30 Years." *Historia Mathematica* 17(4): 366–373, 1990.

Nihon Gakushiin (Japanese Academy). *Meiji zen Nihon sūgaku shi* (History of Pre-Meiji Japanese Mathematics). 5 vols. Tokyo: Iwanami Shoten, 1954–1960.

Nihon no sūgaku 100 nen shi henshū i-inkai (Committee for the Compilation of the History of Japanese Mathematics During the Last 100 Years). *Nihon no sūgaku 100 nen shi* (Japanese Mathematics During the Last 100 Years). 2 vols. Tokyo: Iwanami Shoten, 1983.

Ōya, Shinichi. *Wasan izen* (Japanese Mathematics Prior to Wasan), Tokyo: Chūō kōron sha, 1980.

Smith, David Eugene and Mikami, Yoshio. *A History of Japanese Mathematics*. Chicago: Open Court Publishing Company, 1914.

Sugimoto, Masayoshi and Swain, David L. *Science and Culture in Traditional Japan* AD *600–1854*. Cambridge, Massachusetts: The MIT Press, 1978.

See also: Takebe Katahiro – Ajima Naonobu – Magic and Science – Seki Kowa

MATHEMATICS IN KOREA During the Shilla dynasty (59 BC–AD 935), in AD 682, Korea established its official mathematical system under the influence of the mathematics of China (Tang dynasty) whose primary structure was algebra (theory of equations) and whose philosophical background was *yin-yang* (negative–positive) theory. It was an educational program designed to train professional mathematicians, and it prescribed the number of mathematicians to be trained and the length of study and curricula, and it remained the official mathematical system to the end of the Chosun dynasty in 1910.

The system soon died out in China. It was revived in the Song dynasty, but failed to become an official system again. Meanwhile, Japan set up its mathematics system under the influence of the Paekche and Shilla dynasties in Korea. The Taihorei of 710 clearly defined its contents, but this too soon became extinct. In China and Japan official mathematics ceased to exist, but civilian mathematics prospered. Korea, on the contrary, was fundamentally different from these two countries (see Table 1).

The arrangement of the Shilla curriculum here is hypothetical; originally only the names of the four subjects were given. The length of study for Korea, nine years, is for classical stuides; whether this applied to mathematics and other technical fields is yet to be explored.

In China during the Tang, there were different qualifications for the *guozixue* (National Academy) and mathematics and other fields. In Korea and Japan, qualifications also differed by field.

The Koryo dynasty (918–1392) continued using the Shilla mathematical system in a slightly changed version. Whereas the Koryo continued to use Nine Chapter Arithmetic, the Continuation Technique, and Three Opening Arithmetic, they discontinued Six Chapter Arithmetic and began to use Saga.

The Chosun dynasty (1392–1907) strengthened the system of its predecessors and the standard text books were completely changed. The texts include Sanmyon arithmetic, Yonghui arithmetic, Sanhak Kaemong arithmetic for Bureaucrats, and arithmetic for surveying.

In the Chosun dynasty, the study of mathematics was encouraged to fill administrative needs and was incorporated into the official system. During the reign of King Sejong (AD 1419–50), a Bureau of Mathematics and an Agency for Calendars were created, and mathematics was revived to match the level of the Koryo dynasty (936–1392). The positions of *Sanhak paksa* (Doctor of Mathematics), *Sanhak Kyosu* (Professor of Mathematics) and *Sansa* (mathematician) were created.

The Chosun dynasty attached great importance to mathematics from the beginning, and the bureaucrats in charge of the technical civil service examinations in ten fields began to play a greater role, ultimately forming the new social class of *chungin* (middle men). The *chungin* class is considered unique in the history of the world. (The term began to be used officially during the reign of King Sukchong (1675–1720).) As technocrats, they were recruited through comparatively low-level civil service examinations. Most of them came from the *chungin* class. It does not mean that a

Table 1: Mathematics education systems in Korea, China and Japan around the seventh century AD

Country	Enrollment age	Admission qualifications	Subjects	Length of study
Korea (Shilla) AD 682	15–30	Taesa (grade 12 in a hierarchy of 17 grades) and those who had no official positions	Six-chapter arithmetic Nine-chapter arithmetic Three opening arithmetic Continuation technique	9 years or longer
China (Tang) AD 624	14–19	Children of grade 8, and lower-grade public officials	Nine-chapter arithmetic Calculation of distance to a far-off island Sun Zi's arithmetic Arithmetic for five government bureaus Zhang Guijian's arithmetic Calculation and Gnomon Continuation technique Topics in number Three standard mathematics	7 years
Japan AD 710	13–15	Children of grade 8, and higher-grade public and local officials	Nine-chapter arithmetic Calculation of distance to a far-off island Sun Zi's arithmetic Three opening arithmetic Continuation technique Six-chapter arithmetic	7 years

son inherited his father's position; rather it seems that intermarriage among the *chungin* contributed to the preservation of the tradition. Among them, mathematicians showed an especially strong tendency toward preserving this hereditary tradition. The roster of successful candidates in the tests for mathematicians and the position of *Sanhak Sonsaeng* (mathematics teachers) lists a total of 1627 during 300 years from the fifteenth to the seventeenth centuries. Their fathers' occupations were as follows: 124 herbalists, 75 translators and 6 astronomers; the rest were mathematicians.

The mathematicians of the *chungin* class lived in a very closed society. As many of the books they wrote have been lost, and only fragmentary information is available, it is difficult to evaluate their achievements, either as a group or as individuals. A typical mathematician of the *chungin* is Hong Chong-ha, a professor of mathematics, who wrote an eight-volume (plus a supplement) *Kuilchip* (Nine Chapters on Arithmetic in One) which is still extant today. He was born in 1684 into a typical *chungin* family of mathematicians; his father, grandfather, great-grandfather, and his wife's father were all mathematicians.

From his book, we find the following. First, mathematicians of the day were quite unfamiliar with the course of events in China, whereas the literati of the *yangban* class maintained in direct contact with Chinese culture and with European culture through China. The mathematicians were bound by the old system and continued using only the traditionally handed-down manuals; they had no access to Chinese translations of Western mathematics books. Second *tianyuan shu* (Horner's approximation theorem for an equation with real coefficients) and the calculation rod thrived in Korea throughout the Chosun dynasty, long after they had ceased to be used and had been replaced by the abacus in China after the establishment of the Ming dynasty. Japanese mathematics in the Edo period (1603–1867) is called *wasan*. The origin of *wasan* is Chinese mathematics, but the Japanese replaced the calculation rod with hand writing and eventually developed it into symbolic algebra. Korean mathematicians were then isolated from the outside world and from the European mathematics which had already found its way into China, but they preserved traditional mathematics using the calculation rod.

One of the other mathematical trends in the Chosun dynasty was *yangbans* (nobles). Ch'oe Sok-chong was a mathematician of noble birth who was a great admirer of classical Chinese philosophy. As the author of *Kusuryak* (Concise Nine Chapter Arithmetic), similar in its style of description to that of early European monastic mathematics books, he was a "Boethius (480–525) of the Orient". Boethius's mathematics was theological, metaphysical and number-theory centered. Both gave a touch of mysticism to numbers, and Ch'oe studied magic squares of various types: circle, square, hexagon, etc. The hexagon denotes 'water' in Chinese traditional philosophy, and Korean mathematicians attempted to indicate some philosophical meanings by means of mathematics.

Culture and Mathematics

King Sejong wanted all the branches of Oriental learning — Confucianism, linguistics, music, astronomy, herbal medicine, and agriculture — to be in the service of his country. Unlike the Greek system of learning that branched off into mathematics, natural sciences, and metaphysics, the Oriental system tended to integrate all fields of learning into a whole. This was typified by classical Chinese studies (compare this with the Western trivium and quadrivium). For example, in the reform of music and the creation of the Korean alphabet, King Sejong remained true to the orthodox Oriental view of learning and was content with being a true inheritor of Eastern culture. His policy for the promotion of mathematics did not seek any new paradigm.

Sirak

Sirak is the Korean version of the neo-Confucian concept of "practical learning", similar in nature to *jitsugaku* in Japan and *shixue* in China. It was active for about three hundred years, from the mid-sixteenth to the mid-nineteenth centuries. Descriptions of some of the prominent mathematicians and their works follow. Yi Sugwang (1563–1628) wrote *Chibong yusol* (Chibong's Miscellany) which treats astronomy, geography, bureaucracy, belles-lettres, human behavior, technology, and even birds, animals, and insects.

Typical of this encyclopedic coverage is *Ojuyon munjangjon san'go* (An Oju's Multitude of Articles and Essays) by Yi Kyu-gyong (b. 1788) in sixty volumes, which contained 1400 entries relating to the problems of all ages and countries. He had a very practical outlook, and in a commentary on the original text of a geometry book, he regarded surveying as the primary purpose of its study.

Ch'oe Han-gi (1803–79) is said to have broken with the traditional position of neo-Confucianism and become an activist philosopher, adhering to thorough-going empiricism. But he too remained orthodox in his attitude to arithmetic, as is evident in his comment, "By the degree of knowledge one has acquired in arithmetic, we can judge one's insight; we can see whether one's attitude is reasonable or not by judging if one's reasoning is arithmetical."

In 1765 Hong Tae-yong (1731–83) visited a Catholic church in Beijing, China, and acquired first-hand knowledge of Western culture. He conversed with Hallerstein, the Chief Astronomer, and his deputy Gogeisl at a Chinese astronomical observatory, thereby broadening his knowledge of astronomy. His work *Tamhonso* (Tamhon's Writings) treats mathematics and astronomy in Volumes 4 to 6 of Book II. Hong consciously discussed the infinite; he was the first Korean to discuss an infinite Universe, and he also mentioned infinite decimals in discussing the value of π.

At the end of the Chosun dynasty a new movement in mathematics arose. Nam Pyong-gil (1820–59) and Yi Sang-hyok (1810–?), perhaps the two greatest arithmeticians at the time, did not belong to the *Sirak* school. None of the *sirak* scholars specialized in arithmetic. But Nam was born into a *yangban* family, and Yi was a professional arithmetician of the *chungin* lineage. They joined hands in the study of arithmetic.

Nam's works include many arithmetic books, but he made them look new by adding illustrated explanations. There is no trace of metaphysical view, a dominant characteristic of the works of other *yangban* scholars.

Yi Sang-hyok was a typical *chungin* arithmetician. After passing the national test, he was assigned to an astronomical observatory as a budget officer. Among his works are some astronomical books and the following books on arithmetic: *Iksan: Ch'agunbop monggu* (Winged Mathematics: Hypothetical Method for the Theory of Roots) and *Sanhak Kwan-gyon* (A Brief Survey of Arithmetic). The first title probably implies "mathematics of two wings', one wing referring to traditional arithmetic, and the other to modern arithmetic. *Ch'agunbop monggu* is an explanatory book on European algebraic equations, while *Sanhak Kwan-gyon* presents Yi's creative study of mathematics. Yi's single minded devotion to higher mathematics, ignoring the traditional patterns of thought when classic arithmetic was reviving, leads us to assume that this type of mathematical research may have prevailed among *chungin* scholars at that time.

Classical official Korean mathematics maintained its continuity without any fundamental changes from the seventh century (Shilla dynasty) to the beginning of the twentieth (the end of the Chosun dynasty).

The Chosun dynasty had three groups of mathematicians: the *chungin*, the *yangban* and the *sirak*. The *sirak* school searched for new mathematics as seen in collaboration between *yangban* and *chungin* mathematicians. But their study of European mathematics was limited to algebraic equations and geometry at best, and they never went much beyond the traditional Chosun dynasty mathematics even when they accepted European mathematics. The works of Hong Tae-yong, allegedly the most progressive of all *Sirak* mathematicians, differ from other classical manuals only in that they deal with practical applications of old principles.

At the end of the Chosun dynasty, some of the *chungin* mathematicians attempted an original study in mathematics, breaking with the old traditions. But they too were limited by the prevailing intellectual climate, which hindered understanding and assimilation of modern mathematics.

Korean mathematicians made little effort to change fundamentally the mathematics which originated in China, they did not yield to foreign influence, and they maintained the classical mathematics of the East to the last (to the end of the Chosun dynasty) even when Chinese and Japanese mathematics underwent drastic changes.

KIM YONG-WOON

REFERENCES

Primary sources

Korean:

Kyonggjuk taejon (National Code)

Munhon bigo (Record of Official Documents)

Samguk saki (A History of Three Kingdoms)

Sanhak ipkyok (A Roster of the Successful Examinees in the State Examinations for Mathematicians)

Sanhakja Palsebo (Genealogy of Mathematicians for Eight Generations)

Yijo Sirok (Authentic History of Yi Dynasty)

Japanese:

Meiji zen no nihon sugaku (Japanese Mathematics before Meiji). Tokyo: Iwanami, 1967.

Nihon shoki (The Chronicles of Japan).

Chinese:

Taitang luidian

Zhongguo kexuejishu-diangi tonghui (Complete Works on Chinese Science and Technology) Mathematics, vols. 1–5. Ed. Guo Shuchun. Beijing: Henan Publishing House, 1993.

Secondary sources

Kim, Y.W. "Introduction to Korean Mathematics History." *Korea Journal* 13(7): 16–23; (8): 26–32; (9): 35–9, 1973.

Kim, Y.W. *Han'guk suhak-sa* (A History of Korean Mathematics). Seoul: Yolhwa-dang, 1982.

Kim, Y.W. "Pan Paradigm and Korean Mathematics in Chosun Dynasty." *Korea Journal* 26(3): 24–46, 1986.

Kim, Y.W. and Kim, Y.G. *Kankoku sugakusi* (A History of Korean Mathematics). Tokyo: Maki Shoten, 1978.

See also: *Yinyang* – Sun Zi – Zhang Qiujian – Computation: Chinese Counting Rods – Magic Squares – Surveying – *Pi*

MATHEMATICS OF THE MAYA The common numerical notation of the Maya employed combinations of 'bars', having value 5, and 'dots', having value 1, to represent the vigesimal digits from 1 to 19. For example, the kneeling scribe in Figure 1 appears on a Classic Maya vase dating from around AD 750. From his armpit emanates a vegetative scroll containing bar and dot numerals. Beginning at the armpit, the sequence of numerals runs 13, 1, 2, 3, 4, 5, 6, 7, 8, and 9. Often, non-numerical crescents or other fillers were used to create a more aesthetic balance for the numeral.

In some records, these simple numerals are used to count objects, in particular to specify the quantity of offerings of a particular type. However, the overwhelming usage occurs in calendrical and chronological contexts. In the former, the numerals appear as coefficients in the Maya calendars of 260 and 365 days. The former consisted of the cycle of numbers from 1 to 13 paired with a cycle of 20 day names. As the days went by, both the number and the day name would simultaneously advance in their respective cycles, thus generating a sequence of 260 distinct dates. The second calendar was a civil year composed of eighteen 'months' of 20 days each and a residual 'month' of 5 days. The days within a given month were numbered consecutively from

Figure 1: A kneeling Maya scribe. (Drawing by M.P. Closs after Clarkson, 1978).

the first to the penultimate by prefixing the proper numeral to the appropriate month name. The last day was sometimes indicated by prefixing a special sign having the sense of 'end' to the month name. However, the more common practice was to prefix a sign signifying 'seating' to the following month and thereby identify the last day of one month as the seating day of the following month. Frequently, the Maya would describe a day by giving its dates in both the 260 and 365 day calendars. This generated a cycle of 18980 paired dates known as the 'calendar round'.

A characteristic of Maya writing is that it often uses many distinct signs to represent the same linguistic value. This is the case with numerical terms where, in addition to bar and dot numerals, the Maya represented numbers by head forms and even full body figures. Some examples of bar and dot numerals in calendrical contexts, with alternative representations of the same dates using head variant numerals, are illustrated in Figure 2. All of these examples come from the same ancient Maya city of Palenque and demonstrate the scribal variation to be found in Maya writing even at the same location. For a more extensive description and additional examples of Maya numeral forms, see Closs (1986).

The Maya frequently recorded the chronological interval separating two calendar dates. For this purpose they employed a vigesimal count of *tuns* (360 day periods) and separate counts of *winals* (20 day periods) and *k'ins* (days). A given interval would be expressed as a count of the respective time periods into which it could be minimally decomposed. Each of the time periods was represented by a characteristic sign and the appropriate count was indicated by a numerical prefix. In recording such chronological counts, it is necessary to have a symbol for zero so as to indicate an empty count of a particular period if this occurs. The Maya did have a zero symbol for that purpose. In fact, as with other signs employed in Maya writing, there are several variants of the zero symbol.

In many cases, and typically in the codices, Maya scribes would represent chronological counts using a system of positional notation. In this system, counts were written vertically with the lowest position being occupied by the count of *k'ins*, the next higher position by the count of *winals*, the next higher position by the count of *tuns*, and successively higher positions by the corresponding count of successively larger vigesimal multiples of the *tun*. Any chronological count could be recorded in this system by using the nineteen vigesimal bar and dot digits and a symbol for zero.

In the monumental inscriptions, it was customary to indicate the interval between two successive dates of the text by recording the chronological interval separating them. Such chronological counts are referred to as 'distance numbers' and might be added or subtracted to a given date to reach

Figure 2: Maya numerals in a calendrical context. Dates in the 260-day calendar: (a) 5 Lamat, Palenque, Palace Table, R4; (b) 5 Lamat, Palenque, Tablet of the 96 Glyphs, D4; (c) 9 Manik, Palenque, Dumbarton Oaks Panel 2, J; (d) 9 Manik, Palenque, Tablet of the 96 Glyphs, H1. Dates in the 365-day calendar: (e) 6 Xul, Palenque, Palace Tablet, N15; (f) 6 Xul, Palenque, Tablet of the 96 Glyphs, C5; (g) 15 Uo, Palenque, Tablet of the Slaves, H5a; (h) 15 Uo, Palenque, Tablet of the 96 glyphs, G2. [Drawing by M.P. Closs].

the following date. The initial date in an inscription is often anchored in an absolute chronology by recording the chronological count separating it from a common base date far in the past. These chronological records are referred to as Initial Series and have a characteristic format. An example of an Initial Series on Stela 1 at Pestac, employing positional notation and incorporating a zero, is illustrated in Figure 3. It shows a count of 9 baktuns (= 3600 tuns), 11 k'atuns (= 220 tuns), 12 tuns, 9 winals, and 0 k'ins (a total of 1,379,700 days or approximately 3777 years) and leads to a date in AD 665. This is the oldest securely dated Maya

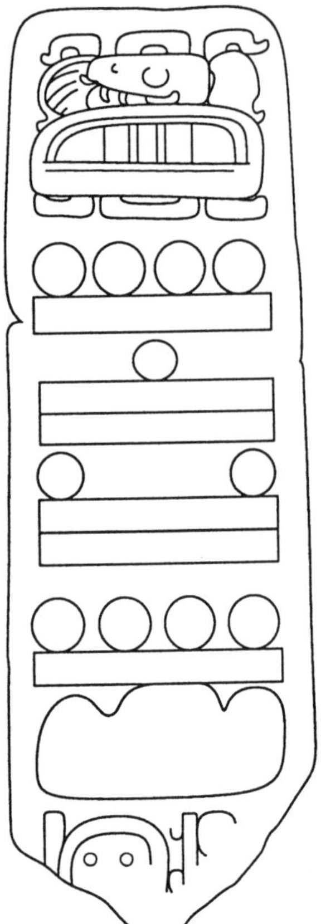

Figure 3: The front of Pestac, Stela 1, with an Initial Series of 9.11.12.9.0 employing positional notation and a zero sign. (Drawing by M.P. Closs).

Figure 4: A Maya ring number of 1.7.11 from the Dresden Codex, p. 58. [Drawing by M.P. Closs].

text incorporating the system of positional notation with a zero sign.

In non-positional contexts, the zero sign is found in Initial Series at a much earlier time. For example, the zero sign occurs on Stela 18 and 19 at Uaxactun, dating from AD 357. The most ancient non-positional chronological count of this type, but without a zero, is found on Stela 29 at Tikal, dedicated in AD 292. It is interesting to note that the system of positional notation is very ancient in Mesoamerica and is found on monuments that predate those of undisputed Maya origin. However, none of them exhibit a zero sign. The oldest of these monuments goes back to 36 BC.

In several instances, the Maya worked with negative chronological counts. These are used when counting backwards from the zero point (base date) of the absolute chronology. They are distinguished from positive counts by a peculiar notation that has engendered the term 'ring number'. An example of a ring number from the Dresden Codex, one of the few surviving Maya hieroglyphic books, is illustrated in Figure 4. The ring number of 1.7.11 is a negative chronological count with reference to the zero point of Maya chronology and leads to a mythological date.

Among the mathematical problems faced by the Maya were those of determining the calendar round date when a chronological interval is added to or subtracted from a given date and of determining the chronological interval between two given calendar round dates. The Maya also used other calendrical cycles of 4 days, 9 days, and 819 days. This led to other problems such as determining the station in the 819 day cycle closest to a given calendar round date. It is clear from artifactual evidence that the Maya performed their calculations using residue arithmetic.

The most sophisticated accomplishments in Maya mathematics are found in their arithmetically based astronomy. Here I will touch on three areas.

(1) From the earliest days, the Maya scribes kept track of the moon through a lunar calendar. We have a large number of moon age records giving the current age of the moon within a cycle of 29 or 30 days. If a moon age was greater than 19 days, a special symbol for 20 was employed along with a regular bar and dot numeral to make up the balance of the count. Occasionally, a similar notation was used in the chronological counts discussed above. Since a lunation is around 29.5 days, the Maya used varying proportions of 29-day and 30-day months to keep their moon age records in agreement with astronomical reality. This gives an idea as to how the Maya avoided the necessity of fractions in their integer-based arithmetic. We have a few examples of moon age calculations from contemporaneous dates into mythological time, thousands of years earlier, from which it has been possible to recover the moon-age formulas that were used. These are formulas of the type "x moons = y days" where x and y are integers. Dividing through by x, one can

Figure 5: An ancient Maya mathematics lecture. [Drawing by M.P. Closs from Kerr, 1989].

where x and y are integers. Dividing through by x, one can say that a particular Maya formula is equivalent to estimating an average lunation as y/x days. This can then be compared with modern estimates and the precision of the Maya results can be recognized. Nevertheless, it can be misleading because the Maya did not look at things in this way.

(2) The Dresden Codex contains a Venus table that permits the prediction of first appearance of the planet as morning star and as evening star over a period of about 104 years. The introductory page to the table has several features of mathematical interest. It has a mythological base date marked by a ring number. To this is added a companion number leading to a canonical base date of the table in the historic era. The companion number spans a period of 1,366,560 days (more than 3741 years). It has the prime factorization $2^5 \cdot 3^2 \cdot 5 \cdot 13 \cdot 73$ and so is highly divisible into a product of relatively small prime numbers. This is fairly typical of other companion numbers associated with ring numbers in the Dresden Codex. It is a contrived number constructed using a knowledge of residue classes. The page also includes a table of multiples of 2920 days. The importance of this number stems from the formula "*5 Venus years of 584 days = 8 solar years of 365 days*" commensurating the Venus and solar cycles. The existence of tables of multiples is habitual in the Dresden Codex and indicates that the Maya worked from such tables rather than depending on algorithms for multiplication. The table of multiples and other calculation factors provide a mechanism for advancing from the canonical base date of the Venus table into the Venus table proper at a contemporaneous date. Still other calculating factors on the same page permit the Venus table to be recycled over time so that it maintains its astronomical integrity over several hundreds of years, much longer than the actual span of the table.

(3) The Dresden Codex also contains an eclipse table that identifies dates of potential solar and lunar eclipses. It records a commensuration of the eclipse cycle with the sacred calendar of 260 days. This table also has mechanisms for recycling so that it can remain useful over many runs of the table. It is one of the great mathematical and astronomical achievements of the ancient Maya.

We also have some indications of geometrical knowledge among the Maya but nothing derived from written texts. Information that has been obtained comes from the study of architecture and site plans. For example, there are reasons to believe that an arrangement of three major temples at Tikal — at the vertices of an isosceles right triangle — is not coincidental. Many other alignments suggesting intentional geometrical concerns have been proposed at various Maya sites, but no adequate synthesis of the geometry involved has yet been achieved.

The idea of mathematics was sufficiently concrete in Maya thought that it has a presence in Maya art and iconography. In a number of almanacs in the Madrid Codex, deities are shown holding a vessel of black paint in one hand and a brush for painting or writing in the other. This type of scene is now known to depict the deity in the act of writing or painting. In at least two instances, an extra feature has been added to the scene. This consists of a scroll containing numbers issuing from the mouth of the deity. The added element indicates that the writing which is being performed is of a mathematical nature. By using this device, the Maya artist testifies to the visibility of mathematics as a discipline in Maya society.

Figure 6: Maya gods of mathematics and writing. [Drawing by M.P. Closs from Coe, 1978].

The notion that mathematics was also considered a distinct discipline within the scribal curriculum is demonstrated on a Classic Maya vase (ca. AD 750). The scene on one side of the vase, illustrated in Figure 5, shows a classroom scene in which Pauahtun, an aged god with a characteristic netted headdress, instructs two disciples in the mathematical art. Pauahtun, who is known to be a patron of scribes, is seated with a codex in front of him and a brush pen in his left hand. From his mouth issues a speech scroll containing the bar and dot numbers 11, 13, 12, 9, 8, and 7. The hieroglyphic caption behind the head of the first student is a name glyph. On the other side of this vase, there is a near identical classroom scene in which hieroglyphs are recorded in the speech scroll. In that case, the spoken text begins with a verb indicating 'to teach'.

A pair of deities in a detail from another Classic Maya vase is shown in Figure 6. The god on the right has the facial features of a monkey and carries a codex in his right hand. An effigy head rests on top of the opened codex. The god on the left rests one hand on the back of the previous figure and holds a conch shell ink-pot in the other. Of special interest is a vegetative scroll, containing bar and dot numerals, which emanates from his armpit. There is also a curl, with single digits, running down from his cheek. Coe has suggested that these deities are patrons of mathematics and writing. The pairing of the two deities in this manner distinguishes between mathematics and writing in Maya thought. It reinforces the idea that mathematics was recognized as a separate discipline.

It is apparent that a scribe who was a mathematical specialist must have mastered calendrical and chronological calculations. The most detailed information on Maya civilization at the time of the Spanish conquest is found in Bishop Landa's *Relacion de las Cosas de Yucatan*. This work was written shortly after the conquest and draws heavily on data provided by educated native informants, members of the former scribal class. Landa refers to the mathematical techniques of the Maya scribes as the computation of the katuns. He writes that this "was the science to which they gave the most credit, and that which they valued most and not all the priests knew how to describe it" (Tozzer, 1941). This informs us that not all scribes were mathematical specialists and that those who were had greater prestige. It supports the idea that mathematical specialists were recognized as a specialized subgroup of the scribal class. In this regard, the situation at the time of the conquest seems to have been no different than it was during the Classic period.

MICHAEL P. CLOSS

REFERENCES

Clarkson, Persis B. "Classic Maya Pictorial Ceramics: a Survey of Content and Theme." In *Papers on the Economy and Architecture of the Ancient Maya*. Ed. Raymond Sidrys. Institute of Archaeology, Monograph 7, Los Angeles: University of California, 1978, pp. 86–141.

Closs, Michael P. "The Mathematical Notation of the Ancient Maya." In *Native American Mathematics*. Ed. Michael P. Closs. Austin: University of Texas Press, 1986: pp. 291–369.

Coe, Michael D. *Lords of the Underworld: Masterpieces of Classic Maya Ceramics*. Princeton: Princeton University Press, 1978.

Kerr, Justin. *The Maya Vase Book: A Corpus of Rollout Photographs of Maya Vases*, vol. 1. New York: Kerr Associates, 1989.

Tozzer, Alfred M. *Landa's Relacion de las Cosas de Yucatan*. A translation, edited with notes. Cambridge, Massachussetts: Papers of the Peabody Museum, vol.18, Harvard University, 1941.

See also: Calendars – Astronomy – Eclipses

MATHEMATICS IN NATIVE NORTH AMERICA The mathematical development of the Native American peoples is highly variable among different cultural groups. In this article, I will treat two areas of interest: the wonderful diversity in the formation of number words and the use of tally systems to record information. Other articles in this volume give more explicit details on the mathematics of two particular indigenous groups, the Aztec and the Maya.

Number Words

Often numbers have a digital origin, indicating that counting began, or at least was remembered, by finger counting or by counting on the hands and feet. Here, I look at only a few examples. For a more extensive treatment, see Closs (1986).

The relationship between the method of finger counting used and the creation of number words may be very explicit. The Bacairi of Mato Grosso in Brazil have the number sequence shown below.

1	tokale	
2	ahage	
3	ahage tokale; ahewao	2 + 1; 3
4	ahage ahage	2 + 2
5	ahage ahage tokale	2 + 2 + 1
6	ahage ahage ahage	2 + 2 + 2

The second of the number words for 3 is not used more often than the form made up of 2 and 1, nor is it used in the formation of any of the higher number words. The word for 1 comes from the word for bow. It has been suggested that since each man has only one bow but many arrows, the bow came to exemplify 'oneness'. The word for 2 and the word for 'many' derive from the same source. Thus, the two basic terms in the number vocabulary have a non-digital origin. The number sequence is an example of an additive 2-system, that is, a system in which the terms are formed by using addition (implicitly) and groups of 2s.

In finger counting, a Bacairi starts with the little finger of the left hand and says *tokale*, grasps the adjacent finger and joins it with the little finger and says *ahage*, goes to the middle finger and, holding it separately beside the little finger and the ring finger, says *ahage tokale*, goes to the index finger and joining it to the middle finger says *ahage ahage*, grasps the thumb and says *ahage ahage tokale*, places the little finger of the right hand along side it and says *ahage ahage ahage*. He then goes to the remaining fingers of the right hand and touches each finger in turn while saying *mera* 'this one'. He continues by touching the toes of the left and right foot and each time says *mera*. If he is still not finished he grasps his hair and pulls it apart in all directions. The number 6 marks the end of the use of 2-groups in finger counting. Nevertheless, the finger counting still continues on to 20 by a straight one-to-one correspondence without the use of number words.

Number words are not always derived from words associated with counting on the fingers. Frequently, the meaning of even relatively small number words is transparent. In some languages the word for 1 is related to the first person pronoun. Often the word for 2 comes from roots denoting separation or pairs. Moreover, not only large numbers but also small ones may be formed by using arithmetical principles. In the simplest case, this takes the form of doubling but there are instances in which addition, subtraction, multiplication, and division are used. Number words may also be created in more exotic ways (see Table 1).

Larger number words may be formed by using superlatives or other expressions which are indefinite (see Table 2).

There are many number systems developed in the New World that follow unusual principles of grouping. The number sequence of the Coahuiltecan of Texas shown in Table 3 exhibits an extensive use of additive and multiplicative principles early in its development. It is also very unusual in the heavy reliance it places on 3-groups.

The word for 3 is an additive composite of the words for 2 and 1. That pattern, seen in the earlier 2-system, is not continued and new words are introduced for 4 and 5. The number 6 is expressed in two ways, either as a composite of 3 and 2 or by a new word. The next new term introduced is for 20. The number sequence rises from 6 to 20 by a regular use of 3-groups and the use of the arithmetic operations of addition and multiplication. Below 12 the development of the number system is not consistent. The Coahuiltecan sequence might be classified as a 2–3–20-system.

The number systems employed within closely related linguistic groups may also exhibit considerable variation. For example, the four Yukian languages of California belong to the same family but have very different number systems. In three of the four, the numbers up to 3 are related. However, from 4 on all the languages employ composite number terms whose meanings are completely different. Moreover, the methods of forming the numerals also differ since one

Table 1: Creation of number words

Apache	2	naki	from *ki-e* 'feet'
Micmac	2	tabu	'equal'
Omaha	2	nomba	'hands'
Micmac	3	tchicht	cognate with Delaware *tchitch* 'still more'
Abipones	4	geyenknute	'the ostrich's toes'
Yana	4	daumi	from *dau* 'to count'
Pawnee	5	sihuks	from *ishu* 'hand' and *huks* 'half'
Kutchin	6	neckh-kiethei	from *nackhai* '2' and *kiethei* '3'
(NW) Maidu	6	sai-tsoko	from *sapu* '3' and *tsoko* 'double'
(E) Pomo	7	kula-xotc	from *kula* and *xotc* '2'
(NW) Maidu	7	matsan-pene	from *ma-tsani* '5' and *pene* '2'
Crow	8	nupa-pik	from *upa* '2' and *pirake* '10'
Kansas	8	kiya-tuba	from *kiya* 'again' and *tuba* '4'
(NW) Maidu	8	tsoye-tsoko	from *tsoye* '4' and *tsoko* 'double'
(E) Pomo	8	koka-dol	from *ko* '2' and *dol* '4'
(NW) Maidu	9	tsoye-ni-masoko	'4 with 10' (that is, 4 towards 10)
(E) Pomo	9	hadagal-com	from *hadagal* '10' and *com* 'less'
Gabrieleño	10	wehes-mahar	from *wehe* '2' and *mahar* '5'
(NW) Maidu	10	ma-tsoko	from *ma* 'hand' and *tsoko* 'double'
(E) Pomo	10	hadagal-com	from *hadagal* '10' and *com* 'full'
Unalit	10	kolin	'upper half of the body'
(E) Pomo	11	hadagal-na-kali	from *hadagal* '10' and *kali* '1'
Cehiga	12	cape-nanba	from *cape* '6' and *nanba* '2'
(E) Pomo	12	hadagal-na-xotc	from *hadagal* '10' and *xotc* '2'
(NW) Maidu	13	sapwi-ni-hiwali	'3 with 15'

Table 2: Larger number words

Biloxi	1000	tsipitcya	'old man hundred'
Choctaw	1000	tahlepa siponki	'old hundred'
Delaware	1000	ngutti kittapachki	'great hundred'
Wiyot	1000	kucerawagaatoril piswak	'the counting runs out entirely once'
Kwakiutl	1000000	tlinhi	'number which cannot be counted'
Ojibway	1000000	ke-che me-das-wac	'great thousand'

of the four is an 8-system, two others are 5–10-systems and the fourth is a 5–20 system.

The number sequence of the Round Valley, or Yuki proper, dialect is shown in Table 4, together with an analysis of the numerals.

The Yuki number sequence is inextricably linked to their method of finger counting. Rather than counting the fingers themselves, the Yuki count the spaces between the fingers, in each of which, when the manipulation is possible, two twigs are laid. Except for the words for 1 and 2, common to all the Yukian languages, and the word for 3, common to all but one of the Yukian languages, the number words are descriptive of this method of counting. The number words have no relation to those used in the other related languages. From 9 to 15 the number words are formed by addition to a base of 8. Those from 10 to 15 include the term *sul* 'body', suggesting that 'body' represents the full count of 8 spaces between the fingers.

Many of the terms given for the larger numbers are residue representations as can be seen below.

16 = (8) + 8		24 = (16) + 8	
17 = (8) + 9		26 = (16) + 10	
18 = (8) + 10		35 = (16) + 19	
19 = (8) + 11		51 = (2×16) + 19	
20 = (8) + 12			

The above terms suggest that 16, as well as 8, may be used as a base for constructing the higher numbers. Indeed, the term for 64 is literally '4-pile-at' and 64 would be 4 piles of 16. It has been reported that 64 is also used as a higher unit in

Table 3: The number sequence of the Coahuiltecan of Texas

1	pil	
2	ajte	
3	ajti c pil	2 + 1
4	puguantzan	
5	juyopamáuj	
6	ajti c pil ajte; chicuas	3×2
7	puguantzan co ajti c pil	4 + 3
8	puguantzan ajte	4×2
9	puguantzan co juyopamauj	4 + 5
10	juyopamauj ajte	5×2
11	juyopamauj ajte co pil	(5×2) + 1
12	puguantzan ajti c pil	4×3
13	puguantzan ajti c pil co pil	4×3 + 1
14	puguantzan ajti c pil co ajte	4×3 + 2
15	juyopamauj ajti c pil	5×3
16	juyopamauj ajti c pil co pil	5×3 + 1
17	juyopamauj ajti c pil co ajte	5×3 + 2
18	chicuas ajti c pil	6×3
19	chicuas ajti c pil co pil	6×3 + 1
20	taiguaco	
30	taiguaco co juyopamauj ajte	20 + 10
40	taiguaco ajte	20×2
50	taiguaco ajte co juyopamauj ajte	20×2 + 10

Table 4: The number sequence of the Yuki (Round Valley dialect)

1	pa-wi	
2	op-i	
3	molm-i	
4	o-mahat, op-mahat	'two forks'
5	hui-ko	'middle-in'
6	mikas-tcil-ki	'even-*tcilki*'
7	mikas-ko	'even-in'
8	paum-pat	'one-flat'
9	hutcam-pawi-pan	'beyond-one-hang'
10	hutcam-opi-sul	'beyond-two-body'
11	molmi-sul	'three-body'
12	omahat-sul	'four-body'
13	huiko-sul	'five-body'
14	mikastcilki-sul	'six-body'
15	mikasko-sul	'seven-body'
16	hui-co(t), '8'	'middle-none', '8'
17	pawi-hui-luk, '9'	'one-middle-project', '9'
18	opi-hui-luk, '10'	'two-middle-project', '10'
19	molmi-hui-poi, '11'	'three-middle-project', '11'
20	omahat-hui-poi, '12'	'four-middle-project', '12'
24	'8'	'8'
26	'10'	'10'
35	'19'	'19'
51	'19'	'19'
64	omahat-tc-am-op	'four-pile-at'

the Yuki count. This has been taken as evidence that the Yuki had evolved a pure 8-system. Nevertheless, because of the term for 64, it seems better to classify it as an 8–16-system.

The Yuki number sequence illustrates that one may employ the fingers and hands in counting and not end up with a decimal or vigesimal system. This lesson could also have been drawn from the finger counting of the Bacairi who evolved a 2-system. It can also be seen that the Yuki had precise concepts of number and counting which went far beyond their formal number sequence. The existence of variant terms for the same number and the use of residue expressions for larger numbers attest to this.

The facility with which different indigenous peoples dealt with large numbers also shows considerable variability across the Americas. The Pomo of California have a deserved reputation as great counters. Large counts were commonly performed by them at the times of deaths and peace treaties. An example of such a count is related in a Pomo tale about the first bear shaman who gave 40,000 beads in pretended sympathy for the victim whose death he had caused. One investigator reports that his informant has observed counting in excess of 20,000.

Although the Pomo were able to express numbers reaching into the tens of thousands, the published lists of Dixon and Kroeber contain numbers which do not exceed 200. From these lists, I present the number sequence of the Eastern Pomo which is shown in Table 5. The second last column gives a simple analysis of the number words while the last column gives a second order arithmetical analysis so as to exhibit the structure of the numerals more simply.

It can be seen from the above number list that the Eastern Pomo employed a 5–(10)–20-system. The term '1-stick' is used to represent 20. The phenomenon of overcounting (that is, referring a count upwards to the next higher level) is also employed. For example, 50 is expressed as 10 towards 60 and 70 is expressed as 10 towards 80. Fortunately, the absence of higher numbers in the list has been remedied by Edwin Loeb. He provides the sequence for large counts among the Eastern Pomo in Table 6.

The Pomo number words are closely connected with the method of counting. For example, in counting small amounts, the word for 20 is *xai-di-lema-tek* 'full stick' and when that number is reached a stick is laid out for this primary unit. When 20 such sticks were accumulated they formed a larger unit of 400 that also was represented by a stick.

There is evidence that 1-stick represents counts of 20, 40, 80 or 100 in different Pomo number sequences. This simply indicates that variant grouping practices prevailed in different Pomo areas at different times. Using different sizes of groups to represent 1-stick would yield different number sequences even though the same techniques of stick counting were applied. There is evidence for distinct Pomo systems which may be classified as 5–20–400, 5–40–400, (5)–80–400 and (10)–100–400.

Tally Records

A tally is a simple method of representing the cardinal number of a group of objects by making a one-to-one correspondence between the objects being counted and special marks made by a counter. Today, many tallies are made by using automated counted devices. However, throughout history, tallies were made by more personalized methods such as marking vertical strokes on a flat surface, cutting notches on a stick, tying knots on a string, or placing pebbles in a bowl. It is the most ancient form of record keeping used by humans.

Biographical and chronological records formed of strings with simple knots tied in them existed among the interior Salish and neighboring tribes of southern British Columbia and the region about Yakima, Washington. More sophisticated knotted string records, *quipus*, were used by the Inca.

For this discussion, I consider three types of tally records employed by the Ojibway.

(1) The pictograph in Figure 1 was transcribed from the sides of a blazed pine tree found on the banks of a tributary of the Upper Mississippi in 1831. On the upper right is the totem of an Ojibway hunter who had encamped at that spot. It represents a fabulous animal called the copper-tailed bear. The two parallel lines beneath it, curved at each end, represent the hunter's canoe. The next sign, on the same side, below, is the totem of his companion, the cat-fish, and below that a representation of his canoe. The upper figure on the left represents the common black bear, the six figures below it denote six fish of the cat-fish species. The interpretation is

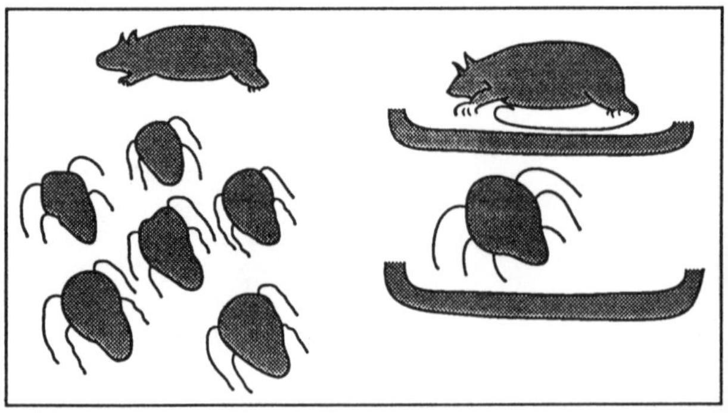

Figure 1: An Ojibway pictographic record [Drawing by M.P. Closs from Schoolcraft, 1851].

Table 5: The number sequence of the Eastern Pomo

1	kali		
2	xotc		
3	xomka		
4	dol		
5	lema		
6	tsadi	1-di	(5) + 1
7	kula-xotc	kula-2	(5) + 2
8	koka-dol	2-ka-4	2×4
9	hadagal-com	10-less	$10-(1)$
10	hadagal-tek	10-full	10
11	hadagal-na-kali	10 + 1	10 + 1
12	hadagal-na-xotc	10 + 2	10 + 2
13	hadagal-na-xomka	10 + 3	10 + 3
14	xomka-mar-com	3-mar-less	$(3\times5)-(1)$
15	xomka-mar-tek	3-mar-full	(3×5)
16	xomka-mar-na-kali	3-mar + 1	$(3\times5) + 1$
17	xomka-mar-na-xotc	3-mar + 2	$(3\times5) + 2$
18	xomka-mar-na-xomka	3-mar + 3	$(3\times5) + 3$
19	xai-di-lema-com	stick-di-5-less	$20-(1)$
20	xai-di-lema-tek	stick-di-5-full	20
21	xai-di-lema-na-kali	stick-di-5 + 1	20 + 1
30	na-hadagal	na-10	(20) + 10
40	xotsa-xai	2 sticks	2×20
50	hadagal-e-xomka-xai	10-e-3-sticks	10 towards (3×20)
60	xomka-xai	3 sticks	3×20
70	hadagal-ai-dola-xai	10-ai-4-sticks	10 towards (4×20)
80	dol-a-xai	4 sticks	4×20
90	hadagal-ai-lema-xai	10-ai-5-sticks	10 towards (5×20)
100	lema-xai	5 sticks	5×20
200	hadagal-a-xai	10 sticks	10×20

Table 6: The number sequence of the Eastern Pomo for large counts

80	dol-a-xai	4 sticks	4×20
100	lema-xai	5 sticks	5×20
200	hadagal-a-xai	10 sticks	10×20
300	xomka-mar-a-xai	15 sticks	15×20
400	kali-xai	1 (big) stick	400
500	kali-xai-wina-lema-xai	1 (big) stick + 5 sticks	$400 + (5\times20)$
800	xote-guma-wal	2 (big sticks)	2×400
2400	tsadi	6 (big sticks)	6×400
3600	hadagal-com	9 (big sticks)	9×400
4000	hadagal	10 (big sticks)	10×400

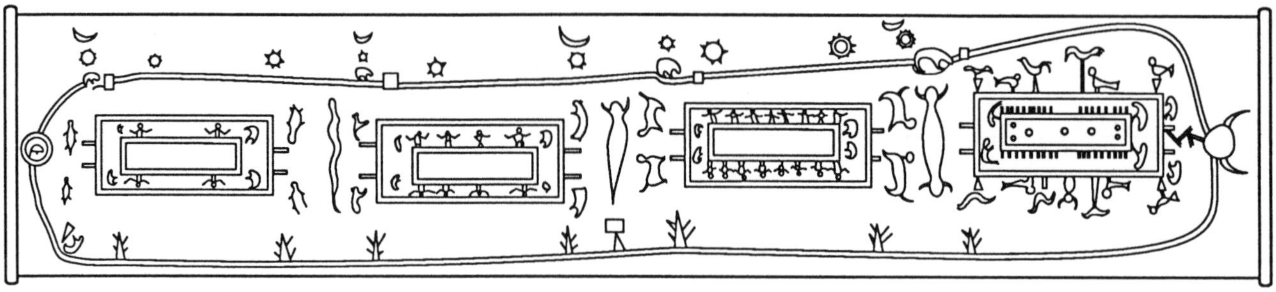

Figure 3: The Midé master scroll kp-1. [Drawing by M.P. Closs from Dewdney, 1975].

this: the two hunters, whose totems were cat-fish and copper-tailed bear, while encamped at the spot, killed a bear, and captured six cat-fish in the river. The record was designed to convey this piece of information to any of their people who should pass the locality.

(2) Tally records were also carved on Ojibway grave posts. Babesakundiba, or Man With Curled Hair, was the ruling chief of the Sandy Lake band of the Ojibway on the upper Mississippi. He died in the late 1840s, after a long life of usefulness and honor, and was buried on a prominent elevation, on the east bank of the river, where his grave and an ensign which waved over it were visible to all who

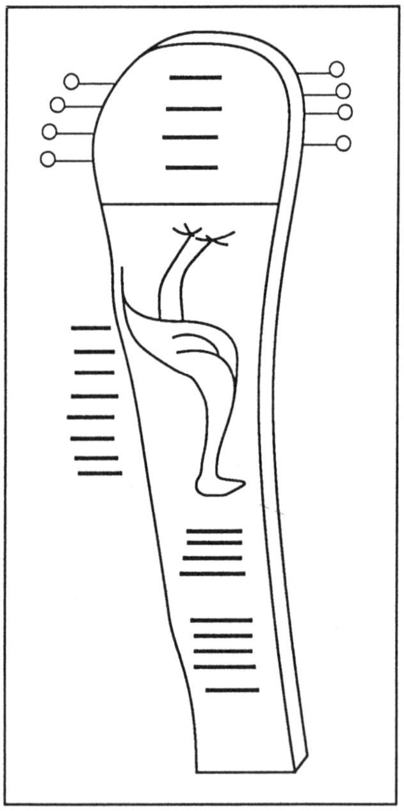

Figure 2: The grave post of Babesakundiba [Drawing by M.P. Closs from Schoolcraft, 1851].

navigated the stream. A drawing of his inscribed grave-post is shown in Figure 2.

The upside down bird denotes his family name, or clan, the crane. Four transverse lines above it signify that he had killed four of his enemies in battle. This fact was declared by the funeral orator at the time of the deceased's interment. At the same time, the orator dedicated the ghosts of the four men, whom the departed had killed in battle, and presented them to the dead chief, to accompany him to the land of the spirits. The four lines to the right and also the four corresponding lines to the left of the central marks denote eight eagle feathers and are commemorative of his bravery. Eight marks made across the edge of the inscription board signify that he had been a member of eight war parties. The nine other transverse marks below the sign of the crane signify that the orator who officiated at the funeral and drew the inscription had himself participated in nine war parties.

(3) The Midéwewin was a set of ceremonials conducted by an organized group of men and women among the Ojibway people who had occult knowledge of "killing" and "curing" by the use of herbs, missiles, medicine bundles, and other objects which possessed medicinal properties. The records and teaching of the Midéwewin were inscribed on birchbark scrolls. These form a body of pictographic material in which one can find various tallies and graphic notations exhibiting a ritual use of number among the Ojibway.

The scroll depicted in Figure 3 was used for instruction in the lore and rites of the Midéwewin. It is characterized by four rectangular floor plans corresponding to four lodges (and degrees of initiation) which a candidate had to pass through before achieving the status of a Midé master.

The inner rectangle in each lodge represents the path that is followed when processions are made around the interior of the lodge. There are four bear *manitos* (archetypal spirits) in each lodge, two guarding each entrance. In addition, there are 4 officials shown in the first lodge, 8 in the second, 16 in the third and 36 in the fourth.

Lurking between the lodges are evil manitos that block

the entrances. The first three lodges are each blocked by four manitos, two near each entrance. The fourth lodge is uniquely surrounded by twelve bird-like figures, possibly sky manitos, with an additional horned figure and two bear figures. These seem to be beneficent entities.

The network of lodges is bordered by the Ojibway universe. It begins with a small circle containing Bear and terminates with the horned symbol of Everlasting Life. Along the upper border, close to the upper left hand corner of each lodge, is a figure of Bear seated before his sacred drum. Outside of this border are lunar and solar symbols. Along the lower border is a sequence of trees which represents a forest.

The scroll emphasizes the number 4, sacred to the Ojibway, in many ways. It is the number of lodges, the number of bears in each lodge, the number of evil manitos adjacent to the entrance of the first three lodges, the number of bear and drum figures on the boundary, and the number of lunar signs. Also, the 12 bird manitos about the fourth lodge and the series of officials in the sequence 4, 8, 16, 36 are based on multiples of 4.

MICHAEL P. CLOSS

REFERENCES

Ascher, Marcia and Robert Ascher. *Code of the Quipu*. Ann Arbor: University of Michigan Press, 1981.

Closs, Michael P. "Native American Number Systems." In *Native American Mathematics*. Ed. Michael P. Closs. Austin: University of Texas Press, 1986, pp. 3–43.

Closs, Michael P. "Tallies and the Ritual Use of Number in Ojibway Pictography." In *Native American Mathematics*. Ed. Michael P. Closs. Austin: University of Texas Press, 1986, pp. 181–211.

Dewdney, Selwyn. *The Sacred Scrolls of the Souther Ojibway*. Toronto: University of Toronto Press, 1975.

Dixon, Roland B. and A. L. Kroeber. "Numeral Systems of the Languages of California." *American Anthropologist* 9: 663–690, 1907.

Harrington, M. R. "Some String Records of the Yakima." (Supplement to Leechman, 1921; see below.) *Indian Notes and Monographs*. New York: Museum of the American Indian, Heye Foundation, 1921.

Leechman, J. D. "String Records of the Northwest." *Indian Notes and Monographs*. New York: Museum of the American Indian, Heye Foundation, 1921.

Loeb, Edwin M. *Pomo Folkways*. Berkeley: University of California Publications in American Archaeology and Ethnology no. 19, 1926.

Schoolcraft, Henry R. *Historical and Statistical Information Respecting the History, Condition and Prospects of the Indian Tribes of the United States*. 6 volumes. Philadelphia: Lippincott, Grambo, 1851–1857.

See also: Quipu

MATHEMATICS IN THE PACIFIC The history of mathematics among the peoples living in New Guinea and the smaller islands of the Pacific is a rich and fascinating field of study. There is a vast diversity of mathematical systems among peoples who share many geographical and cultural elements. Additionally, many investigators have sought to understand how the various reckoning systems developed without a significant historical record in which to place the sometimes confusing tally methods. Complex and poorly understood cultural and linguistic traditions have slowed research efforts. In short, much remains to be learned.

Two aspects of this topic will be explored here: the mathematical reckoning of objects and the measurement of time. Counting objects has been studied with respect to a variety of body-count systems where many different bases were used depending on the objects counted and the natives doing the counting. Economic and cultural factors played a significant role in these systems. Local methods of tracking time closely followed natural cues, yet they often involved a certain mathematical aptitude possessed by only a handful of elders. The practice of passing this knowledge on would be an interesting study in itself.

Several topics typically associated with the history of western mathematics will be absent from this discussion. Complex formulas, abstract theories, and biographies of learned scholars have little relevance here. Nor will a long historical sequence of numerical developments be found. Rather, the traditional mathematics of several groups of Pacific Islanders, before European influence became widespread, will be introduced.

Based upon several surveys of numerous Pacific people groups, the following generalizations about their counting systems may be made. The words used to designate numbers essentially correspond to specific parts of the human body. This implies that, for the most part, the number of items counted rarely exceeded a few dozen, though efficient systems of counting several hundred items were occasionally developed. Therefore, most of these systems of enumeration were very practical and lacked such abstract concepts as infinity, symbolic representation of numbers, and the sometimes complex relationships found between items as represented in both ancient and modern mathematical formulas. This is not to say that the Pacific Islanders did not form abstract concepts. Rather, their mathematical reckoning systems simply met their basic daily needs for such tasks as building houses, paying debts, and planning important social events.

There are over one hundred languages in New Guinea alone and not all of them are well understood. However, each of the major linguistic groups has been surveyed. Table 1

Table 1: Comparative examples of mathematical terms used in Pacific body count systems (Laycock,1975; Bowers and Lepi, 1975).

	Body part				
No.	English	Kakoli	Melepa	Kewa	Crotty
1	little finger (thumb)	telu	dende	pameda	mend (ái)
2	ring (index) finger	talu	rakl	lapo	rab (um)
3	middle finger	yepoko	rakltika	repo	teb
4	index (ring) finger	kise	tembokaka	ki	kirúmend
5	thumb (little finger)	te(lu)pakara	pombi	kode	juúnjk
6	heel of palm	talupakara	pombinjkutl dende	kode lapo	t(r)oganjk
7	palm	yepokopakara	pombinjkutl rakl	kode repo	karánjk
8	wrist	enjgaki	enjkaka	ki lapo	tugurláb
9	forearm	rureponjga telu	pombi ti njkutl	ki lapona kode	tuguríb
10	radius	rureponjga talu	pombinraknjutl	ki lapona kode lapo	tuguréponje-méndai
11	ulna	rureponjga yepolo	pombinrakltikanjkutl	ki lapona kode repo	tuguréponje-ráb
12	elbow	rurepo	tembokakapoket	ki repo	tuguréponje-téb
13	lower upper arm	malapunjga telu	tembokakapoke pombin ti njuktl	ki repona kode	tuguréponje-garo
14	upper upper arm	malaounjga talu	tembokakapoket pombin rakl njkutl	ki repona kode lapo	mabonjeménd(ai)
15	edge of shoulder	malapunjga yepoko	tembokakapoket	ki repona kode repo	mabonjeráb

Table 2: The Mae Enga calendar

Local designation	Translation	Approximate month	Sample 'garden' event
wambu-mupa	(first-born)	January	Plant sugarcane
wambu-nenai	(first-lastborn)	February	Pandanus harvest
iki	(single)	March	Beans finished
ni-mupa	(gleaning-firstborn)	April	Sweet potato scarce
ni-nenai	(gleaning-lastborn)	May	Food scarce
pindi-mupa	(working-firstborn)	June	Burn off gardens
pindi-nenai	(working-lastborn)	July	Dig and fence garden
jambai-mupa	(thatching-firstborn)	August	Start planting
jambai-nenai	(thatching-lastborn)	September	Hasten planting corn
liu-mupa	(plucking-firstborn)	October	Trading trips
liu-nenai	(plucking-lastborn)	November	Plant bananas
kumba-mupa	(blotted outfirstborn)	December	Fights start

shows a sample, though small, which illustrates common body count systems.

Normally, body count systems begin with either the small finger or thumb, then proceed up one side of the body to a recognized central feature, such as the nose. From there, the count continues down the other side of the body until all the items in question have been tallied. If necessary, one may repeat the process.

The history of mathematics is better understood when considered in the cultural context of the community where it occurs. For instance, the Loboda people from Milne Bay use a counting system whereby they may count into the hundreds easily enough, but the enumeration of objects is relatively unimportant in this society, thus it rarely occurs.

A great deal of research into the mathematical systems used in the Pacific has involved discussion of the many numerical bases employed in this region. Not only do these bases vary between distinct people groups, but they also vary with the objects being counted. This was the same, for example, with the British system of pounds, shillings, and pence, which used multiple bases. For instance, with linear measurement alone, one encounters numerous bases. Base sixteen is typically used for inches, while base twelve is used for feet. Base three is employed for yards, and base 5280 is for miles. Liquid measures and weights involve other bases, all just as interesting.

In the Pacific, the Mailu of Papua's Central District normally use a base ten system. But when it comes to counting certain foods, such as taro, sweet potatoes, fish, and coconuts for a feast, the base used is four. People on the Duke of York Islands near Rabaul have a special set of terms for counting *diwara* (shell money), using sixty as the base. The Enga of the Western Highlands commonly use a base ten, but during the Te ceremonial exchange, they use a binary system to count pigs.

Other variations found in the Western Highlands include fractions, such as when counts of pandanus fruits (*Pandanus jiulianettii* Martelli) are made. The basic unit of measure consists of the pole used to carry a trimmed fruiting head bound to each end, *pupute*. The following numerical sequence serves as a tally of whole pandanus fruits: *pupu te mongo* (3 pandanus fruits), *pupu talu* (4 pandanus fruits), *pupu talu mongote* (5 pandanus fruits), *pupu yepoko* (6 pandanus fruits). If necessary, as in the case when there is not enough to go around, the distribution of pandanus fruits may be divided as follows: *pangare* (half), *ilisi* (slice), or *pikelele* (piece).

Numerical figuring in relation to time has not traditionally played a significant role in the development of counting systems found in the Pacific. Reckoning time rarely extends beyond a few days except in the planning of significant social events which may be months or even years in the future. Units of time typically follow lunar, solar, and seasonal cycles. There is rarely a need for nor a concern with precision. Time keeping systems are generally known by a few elders who pass on this information to younger men who have displayed a good match of interest and aptitude.

The Solomon Island society known as the Siuai of Bouganville marks events in time with lunar phases, yet there is little agreement as to the number of days from one phase to the next. These people recognize a relationship between lunar phases and various natural processes, such as a woman's menstrual cycle. In fact, a menstruating female is described as "going to the moon". The Mae Enga of New Guinea have similarly taken advantage of the apparent correlation between lunar phases and reproductive cycles. The time of a sow's farrowing, for example, is taken to be just four months from the time that a boar was observed mating with the female. Various crops are also monitored by the phases of the moon. A special unit of time is used to track gardening events, *Kana* (garden month). The Mae Enga calendar, as produced by Meggitt, is shown in Table 2.

In addition to gardening activities, time is tracked for important social events such as ceremonial dances. The Kewa people monitor a cycle of dances with the aid of a body count system. For example, if a cycle of dances is to begin in eleven months, it is counted as *komame roba suma* (elbow). The occurrence of the second dance some three months later is measured as *pesame roba suma* (shoulder). The third gathering, six to seven months later, is numbered as *rigame robasuna* (between-the-eyes). This sequence of festivals continues until the final "pig killing" feast is held, marked by the division of the wrist and little finger. More than mere social gatherings, these feasts are characterized by intense mathematical manipulations. One's prestige is often at stake with the count of shells, pigs, or whatever currency was appropriate. Having more objects to count, hence trade, is proudly and loudly displayed, for material wealth is considered equivalent to health, well being, and social security.

Pacific Islanders use a number of terms to describe periods of time shorter than the twenty-four hour day. These terms are generally qualitative, referring to dusk, dawn, etc., and, as such, are not suitable for tallying purposes. Many Pacific people groups have little regard for, or need to keep track of, relatively small increments of time, since their societies operate satisfactorily without doing so. People of the Kaegal Valley have terms to designate a few days prior to or after today, but there is little interest in a more elaborate system for tracking days of a week or weeks of a month, as one finds in Western cultures.

Occasionally, systems for tracking much longer periods of time are incorporated into the local vocabulary, such as when it becomes necessary to replant an abandoned garden plot. This event usually follows a twenty-nine year cycle, yet certain ecological cues are observed to mark the event rather than any artificial system. The local "time piece" of preference includes the size of certain second-growth vegetation which occupies the site in question. One would be hard pressed to claim that the artificial system used in Western cultures offers any advantages over local methods of reckoning time in light of the cultural context.

Regardless of the language or the cultural context in which it is found, the mathematics of the Pacific is as diverse as the people who live there. Much remains to be learned about how so many methods of enumeration developed alongside one another. Trade routes, social customs of prestige, as well as complex environmental factors must be considered together, within an oral tradition, to synthesize these influences and derive a whole understanding of mathematics in the Pacific prior to the advent of the Europeans.

WILLIAM T. JOHNSON

REFERENCES

Bowers, Nancy. "Kapauku Numeration: Reckoning, Racism, Scholarship, and Melanesian Counting Systems." *Journal of the Polynesian Society* 86:105–116, 1977.

Bowers, Nancy and Pundia Lepi. "Kaugel Valley Systems of Reckoning." *Journal of the Polynesian Society* 84: 309–324, 1975.

Franklin, Karl and Joice Franklin. "The Kewa Counting Systems." *Journal of the Polynesian Society* 71: 188–191, 1962.

Kluge, Theodor. *Die Zahlbegriffe der Australier, Papua und Batunger, nebst einer Einleitung uber die Zahl; ein Beitrag zur Geistesgeschichte des Menschen* (microfilmed typscript) Berlin, 1938.

Kluge, Theodor. *Die Zahlbegriffe der Sprachen Central und Sudost Asiens, Indonesiens, Mikronesiens und Polynesiens*. Berlin, 1941.

Laycock, D.C. "Observations On Number Systems and Semantics." In *New Guinea Area Languages and Language Study,* vol. 1, *Papuan Languages and the New Guinea Linguistic Scene*. Ed. S.A. Wurm. Canberra: Department of Linguistics, Research School of Pacific Studies, The Australian National University. 1975.

Meggitt, M.J. "Mae Enga Time Reckoning and Calendar, New Guinea." *Man* 58: 74–77, 1958.

Oliver, Douglas. *A Solomon Island Society; Kinship and Leadership Among the Siuai of Bouganville.* Cambridge, Massachusetts: Harvard University Press, 1955.

Smith, Geoff P. *Traditional Mathematics in Morobe*. Papua New Guinea:The Indigenous Mathematics Project, 1981.

Wolfers, Edward P. "The Original Counting Systems of Papua and New Guinea." *The Arithmetic Teacher* 18(2): 77–83, 1971.

MATHEMATICS, PRACTICAL AND RECREATIONAL

The geographical distribution of so-called "recreational" mathematical problems does not respect ideas about distinct mathematical cultures. The familiar conclusion is that they reflect "age-old cultural relations between Eastern and Western civilizations" (Hermelink, 1978). This inference is indubitably true but does not exhaust the matter. The reasons that precisely these kinds of problems reflect relations which are less visible in other mathematical sources are informative, both regarding the conditions and nature of mathematical activity in different civilizations and about the sense (or non-sense) of the concept of distinct mathematical cultures.

"Recreational problems" are pure in the sense that they do not deal with real applications, however much they speak in the idiom of everyday (some examples will be cited below). Nonetheless, their social basis is in the world of know-how, not that of know-why (productive versus theoretical knowledge, in Aristotle's terminology). The distinction between these two orientations of knowledge is of general validity but has particular implications for mathematics.

Beyond this distinction between orientations, it is also useful to distinguish two main ways in which "productive knowledge" is transmitted. One is through master-apprentice networks, through on-the-job training; the type of knowledge whose results may be labeled "subscientific", as is explained below. The other is through institutionalized schools, where training is separated from actual practice and taken care of by teachers whose own connection to the practice for which they prepare is reduced; the outcome may be labeled "scholasticized knowledge".

It is important not to identify practitioners' knowledge (whether subscientific or scholasticized) with practical knowledge alone. The difference has to do precisely with the influence of the social systems which carry the knowledge in question.

The larger part of the practitioners' fund of knowledge is evidently applicable in practice, at least according to the convictions of the social totality within which they function (whether *we* would regard the knowledge of seventeenth-century physicians as medically useless is irrelevant to the seventeenth-century existence and prestige of their profession). As far as this part is concerned, *problems* — *viz.* the problems which the craft or profession is supposed to deal with — are fundamental, and appropriate techniques have been developed which allow it to deal with these problems. But the training of future practitioners, even when done on the job, will have to start from simpler tasks than those taken care of by the master, in part from tasks which have been prepared with the special purpose of training the techniques which the apprentice should learn but which have no direct practical relevance. Here, techniques and methods are thus primary, and problems are secondary, derived from the techniques which are to be trained. Anybody familiar with school books on arithmetic will recognize the situation, and scholasticized systems are indeed those where problems constructed for training purposes dominate. Apprenticeship-based systems, for their part, will tend to train as much as possible on real, albeit simple tasks.

Scholasticized systems often make some use of recreational problems as a means to create variation. However, the genuine basis for the invention and spread of these problems — problems that distort everyday settings so as to create a striking or even absurd situation — is what we can call the subscientific knowledge system. These are oral in character, while scholasticized systems have always been geared to writing; and recreational problems are riddles for specialists, sharing with other riddles that eristic character which distinguishes oral cultures in general. Often, when they enter written problem collections or manuals, they contain phrases like "if you are an accomplished calculator, tell me . . ." — or they are presented as a way to impress the non-specialists who do not understand. In the pre-modern world, this type of knowledge was neither "folk" nor "popular", but a possession of the few to a significantly higher degree than scientific knowledge today.

Such remarks indicate how we are to understand the function of the "recreational" problems, which was not primarily recreational (whence the quotes) — no more recreational, indeed, than the potentially lethal riddle of the sphinx. They served as a means to display virtuosity, and thus, on one hand, to demonstrate the status of the profession as a whole as consisting of expert specialist, and, on the other, to let the single members of the profession stand out, and discover themselves, as accomplished calculators, surveyors, etc.

This function puts some constraints on the problems, which explains their character. They must arouse immediate interest, which explains the "recreational surface": if a camel is to transport grain from one place to another, being able to carry only one third of the grain in one turn, and seems as a consequence to devour exactly everything in the process, then the expert solution allowing a net transfer is impressive. A less striking formulation might provoke the reaction "so what?". The problems, furthermore, must appear to belong to the domain of the profession — skill in singing does not enhance the *professional* prestige of an accountant, does not demonstrate professional valor; according to their *form*, the problems have to be practical. But they must also be more difficult than the tasks that any average bungler in the profession performs without difficulty. This, together with the quest for the striking or absurd, is the reason that the problems are *pure in substance*, i.e. separated from real practice — and more truly so than the simplified problems of school teaching.

Like school problems, however, recreational problems are determined from methods, namely from the characteristic methods of the profession. Often they come from a peculiar trick (in the case of the camel, an intermediate stop and return) that will be known within the subculture of the profession but not outside.

Such tricks will often not be generalizable; moreover, as several mathematicians from the Islamic Middle Ages tell, the practitioners using them would often be ignorant of why their tricks worked. The purpose of posing and solving the problems is not to provide insight but to show off. The pure level of knowledge is thus neither a direct nor an indirect underpinning of practice, nor is it a critical reflection on the principles underlying practice. In this respect it differs fundamentally from what Aristotle and al-Fārābī would speak of as theoretical knowledge and from what we would call scientific knowledge.

But there is another point in the term "sub-scientific", namely that all levels of the subscientific knowledge system, the pure or recreational no less than the applied, have served as inspiration in the development of scientific mathematics. This process can be followed in the Greco-Islamic area (and its European offspring), in India, and in China; Japanese *wasan* may even offer an instance of direct transformation of a subscientific into a scientific system.

In the Greco-Islamic-pre-modern-European (the most sensible delimitation of "Western" culture if the long historical run is considered), in contrast, the theoretical knowledge system may have been socially segregated from the subscientific systems to an extent which in other cultural areas was reserved for different domains, as courtly music in Japan, or poetry as bound up with the writing system in China. While the distinction between different orientations of knowledge systems and the effect of scholastization versus apprenticeship learning hold in general, the social segregation and the high social prestige falling to philosophical and scientific knowledge in this broadly defined "Western" culture have tended to make the rupture between scientific and practitioners' mathematics more visible than elsewhere and to hide the whole subscientific complex from view. Another reason for its invisibility in the contemporary European orbit is evidently that the originally autonomous subscientific traditions were displaced from the late Renaissance onwards by methods that were ultimately derived from scientific mathematics. Increasingly since the late nineteenth century, this process occurs globally.

Few sources survive that tell how practical computation was performed in former times, in particular by practitioners not trained from school books. Moreover, most practical computations and geometrical constructions have relied on such simple techniques that questions of diffusion versus independent invention cannot be decided, which is precisely the reason that they could not serve as a basis for professional self-esteem. Even "wrong" formulae can be so intuitively near at hand that independent invention demonstrably happens time and again. For instance, Isabel Soto Cornejos has recently observed that semi-literate peasants in Chile reinvent the surveyors' formula for approximately rectangular areas (average length times average width) spontaneously. Only problems that are complex enough to serve as display are by the very fact also so characteristic that they may function as index fossils. This is why the recreational problems are crucial for the understanding of the subscientific knowledge system and its relations to the scholasticized and scientific systems.

Often the recreational problems go together in clusters which, although some exchange takes place, are relatively closed. As an example we may look at a cluster — better documented than most others — which appears to be connected to the community of merchants interacting along the Silk Road trading network, but which certainly also had an impact on local communities of accountants and calculators (whatever their social status and organization) wherever they were in contact with the merchants' community.

One favorite problem from this cluster was the doubling of unity, repeated either 30 or 64 times and ending by a summation. The first occurrence is in a text from Mari (Northern Babylonia) from the eighteenth century BC. Then it turns up in a papyrus from Roman Egypt, and next in a Carolingian problem collection (*Propositiones ad acuendos iuvenes*) based on material which had circulated since late antiquity in the Gallic region — all of these instances deal with 30 doublings. In the early ninth century, al-Khwārizmī wrote a treatise on the other (chess-board) variant, which appears to have passed via India, even though actual Indian attestations are later. In the following centuries, both versions are found in India, in the Islamic world, and from around 1200 even in Latin Europe.

Another problem from the cluster that was highly popular in the Islamic and European Middle Ages was the "purchase of a horse": three men go to the market in order to buy a horse. The first says that if he may have half the money belonging to the others he will be able to buy the horse; the second needs only one third, and the third only one fourth (the number of purchasers and the fractions may vary, but invariably they exhibit a striking pattern). Sometimes the price of the horse is told; sometimes the problem is left indeterminate and thus is not only striking but outright absurd (in the present case, any multiple of the set [10,22,26] will thus do as answer). The earliest Western source to state the problem (without the equestrian dress) is Diophantos' *Arithmetic*; slightly earlier is an occurrence with a different dress in the Chinese *Shushu jiuzhang* (Nine Chapters on the Arithmetic Art, first century AD). A passage in Book I of Plato's *Republic*, however, reveals that the problem could be expected to be familiar for the Athenian public of the fourth (and probably the fifth) century BC ("buying a horse collectively" is a situation where one needs an expert).

A third problem from the cluster is "the hundred fowls": Somebody is to buy one hundred fowls for one hundred monetary units, given that (e.g.) a rooster costs 5 units, and a hen 3, while chicks are sold three for one unit. It occurs in a Chinese fifth-century source, as well as the Carolingian *Propositiones*, and it is described by Abū Kāmil as widespread in his environment among people who do not understand the mathematical principles involved but just give one answer without adequate reasoning. A variant has been located by Jean Christianides in a Greco-Egyptian papyrus.

It is thus true that the distribution of these (and other) problems from the cluster bear witness of "age-old cultural relations between ... Civilizations". It is no less true, however, that they highlight the absence of cultural relations at other levels. The *Nine Chapters* as well as Plato's remark and the reflections of the subscientific corpus in Diophantos' *Arithmetic* demonstrate that the literate cultures of China and of the Mediterranean world (to take these as the paradigm) knew about the anonymous tradition, and would take over appropriate material without credit. The wholly different methods applied by the Chinese mathematicians and by Diophantos to solve the same problems also show that what they took over was inspiration, in particular problems, and not full mathematical structures. The high cultures did not really communicate with the low oral culture (no more in mathematics than in other cultural domains); they drew on them, and exploited them. Nor did they communicate with each other through the low cultures. Moreover, what happened at the level of the oral culture may not be adequately described as cultural relations. The very possibility of distinguishing particular clusters shows that even the oral culture had its sharp boundaries, across which only limited communication took place. But these boundaries did not coincide with the geographical boundaries between high cultures, and only in part were they at all geographical in nature. It was, so it seems, a common subscientific merchants' culture which inspired the Chinese and the Alexandrian mathematicians, and which thereby creates the illusion of general cultural relations (similarly, jugglers might move between China and ancient Rome; neither Roman nor Chinese gentlemen ever did).

The Silk Road cluster belonged with a calculators' and caravan merchants' culture. A cluster dealt with elsewhere in this volume, the surveyors' algebra, was carried by practical geometers from the Syro-Irano-Iraqi region. Even in this case, just enough evidence from the Greco-Roman world has survived to show directly that the tradition was known (one Greek papyrus, a problem in one agrimensor treatise). Indirect evidence shows that Diophantos drew inspiration even from this source, and that Book II of Euclid's *Elements* is closely connected to the tradition. For the present argument, however, the relation between the surveyors' algebra and Babylonian algebra is more important, as an unusually articulate instance of the relation between subscientific and scholasticized mathematics.

According to what can be concluded from combination of the evidence presented by the various written traditions that it inspired, the stock of quasi-algebraic recreational problems which were current among the practical geometers of the region was quite restricted:

- to find the area of a square when the area, the diagonal, the sum of the area and the side, or the sum of the area and all four sides was known;
- to find the sides of a rectangle when the sum of its sides

was known together with the area or the area augmented by the difference between the sides;

- the difference between the sides together with the area or the area augmented by both or by all four sides; or
- the diagonal together with the sum of or difference between the sides.

Each problem type was presented by only one or at most two examples, which would permit even those who did not understand to learn them by heart, and everything concerns the entities really present in the geometrical configuration — *the* area, not some multiple; the side, or the sides; etc. From indirect evidence we may assume that the diagonal of the 10×10-square was taken to be 14.

The technique was taken over by the Old Babylonian scribe school and developed into its central discipline — a quasi-algebra based on geometric cut-and-paste procedures. It still appears to have served professional self-esteem on the part of the scribes. But becoming a discipline (etymologically: a subject which is to be learned) it was systematized, and we find texts which vary the coefficients of both the area and the sides systematically, others which replace the length of a rectangle with the product of the length and the ratio between length and width, and the width with the product of the width and the ratio between width and length (formally the outcome is a problem of the sixth degree), and still others where the geometrical entities represent prices in artificial commercial problems. Mathematical irregularities like the diagonal of 14 have been eliminated.

Beyond professional pride, this systematically drilled technique served the purpose of normal school mathematics: through the introduction of coefficients and other complications, the discipline could function as a training ground for the use of the sexagesimal place value system, which was fundamental for the daily engineering and accounting practice of scribes (while all second-degree problems were completely artificial and pure). On the same occasion, the problems lost most of their recreational character.

But the introduction of systematic drill and the elimination of mathematical irregularities were not the only changes. The sixth-degree problems, however much they constitute a trivial extension, and the use of the geometrical technique as a representation of non-geometrical entities, have to be understood as systematic attempts to test the carrying capacity of the professional tools, including the trick of the quadratic completion that was the basis even of the surveyors' algebra. The outcome may be claimed to be really *an algebra*, while the "surveyors' algebra" was not — an algebra being understood as the application of a functionally abstract standard representation (in terms of x and y, Greek *arithmós*, or Arabic *šay*ᵓ (thing) and *māl* (possession) in the analysis of complex relationships involving entities of any kind. If the scribe school had possessed the intellectual drive for that, a transformation into scientific mathematics might have occurred, as it appears to have occurred in the case of *wasan*, and as it occurred — but then in interaction with already-present scientific mathematics — in the Islamic Middle Ages.

Lest anybody believe that the process of scholastization, with all its tediously repetitive drill, should be a characteristic of non-Western civilizations (or of some of them), one may point out that it reached a high point in the Humanist schools of the Renaissance and Early Modern period — the very cradle of the Western ideology.

JENS HØYRUP

REFERENCES

Hermelink, Heinrich. "Arabic Recreational Mathematics as a Mirror of Age-Old Cultural Relations Between Eastern and Western Civilizations." In *Proceedings of the First International Symposium for the History of Arabic Science, April 5–12, 1976.* Vol. II, Papers in European Languages. Ed. A. Y. al-Hassan et al. Aleppo: Institute for the History of Arabic Science, Aleppo University, 1978, pp. 44–92.

Høyrup, Jens. "Sub-Scientific Mathematics. Observations on a Pre-Modern Phenomenon." *History of Science* 28: 63–86, 1990.

Høyrup, Jens. "Algebra in the Scribal School — Schools in Babylonian Algebra?" *NTM. Schriftenreihe für Geschichte der Naturwissenschaften, Technik und Medizin*, N. S. 4: 201–218, 1993.

Høyrup, Jens. "Sub-Scientific Mathematics: Undercurrents and Missing Links in the Mathematical Technology of the Hellenistic and Roman World." In *Aufstieg und Niedergang der römischen Welt*, 2nd Series, vol. 37(3), forthcoming.

Ong, Walter J. *Orality and Literacy. The Technologizing of the Word*. London and New York: Methuen, 1982.

Tropfke, Johannes. *Geschichte der Elementarmathematik.* 4. Auflage. Band 1: *Arithmetik und Algebra.* Vollständig neu bearbeitet von Kurt Vogel, Karin Reich, Helmuth Gericke. Berlin: W. de Gruyter, 1980.

See also: Algebra, Surveyor's – Sexagesimal System – Knowledge Systems

MEDICAL ETHICS The medical ethics of non-Western cultures are not as cleanly differentiated from the rest of their religious and cultural value systems as they are in the West. There the traditional medical ethic of organized professional medicine is summarized in the Hippocratic Oath and the tradition surrounding it. This ethic that guides health care professionals is often in conflict with religious and philosophical traditions that provide more general ethical frameworks. By contrast, in non-Western cultures, medical ethical questions are often addressed in the core ethics literature of the group, such as religious texts and philosophical writings.

The ancient cultures of Asia provide an example. Generally, they turn to classical texts for their medical ethical insights.

China

Although Chinese medicine has a history of at least two thousand years, the first explicit medical ethical writing is usually attributed to Sun Simiao (also called Sun Simo, ca. AD 581–682). His "On the Absolute Sincerity of Great Physicians", which is part of the massive *Qianjin Yao Fang* (Important Prescriptions Worth a Thousand Pieces of Gold, or the Thousand Golden Remedies), is sometimes referred to as the Chinese Hippocratic Oath. Sun Simo, who is primarily associated with Daoist thought, but who also reflects Buddhist influences, is credited with differentiating medicine from more general social practices by demarcating a core group of "Great Physicians". This core group is held out as the normative ideal. The emergence of this concept of the great physician is thought to be the beginning of the professionalization of medicine.

Great Physicians practise the virtues of compassion and humaneness. They are committed to preserving life, a traditional Daoist moral orientation. In a manner typical of ancient Chinese medical ethical writings, but normally omitted from the Greek Hippocratic tradition, Sun Simiao says that the Great Physician "should not pay attention to status, wealth, age, neither should he question whether the particular person is attractive or unattractive, whether he is an enemy or a friend, whether he is Chinese or a foreigner, or finally, whether his is uneducated or educated. He should meet everyone on equal ground" (Unschuld, 1979). Paul Unschuld argues that this professionalization of the practice of medicine begins the effort to control the practice of medicine and its material and nonmaterial rewards.

About one hundred and fifty years later, Lu Zhi (AD 754–805) provided a Confucian response in his *Luxuan Gonglun*. It also emphasizes the virtues of humaneness and compassion, but, according to Unschuld's interpretation, the classical Confucian perspective resisted the professionalization of medicine, holding that medical skills and knowledge should be distributed among all people, not just specialized professionals. The theme of the duty to treat all in medical need regardless of status and concern over elitist professionalization occurs throughout ancient Chinese medical ethics. In the seventeenth century Li Ting (fl. AD 1615) wrote his often cited *Ten Maxims for Physicians; Ten Maxims for Patients*, which repeats these themes, beginning with the necessity of mastering Confucian teachings. While the Daoist and Buddhist strands of Chinese thought have reflected strong prohibitions on killing, generally including condemning of abortion and insisting on prolongation of life, Confucian views have been more tolerant of such practices.

India

India, like China, incorporated its medical ethical teachings within its classical philosophical/religious literature. Among the Vedic texts, the *Āyurveda*, initially developed beginning in the first millennium BC, contains the most important medical writings. Three such texts, the *Carakasaṃhitā*, the *Suśrutasaṃhitā*, and the *Vāgbhata*, include medical ethical writings. The first two also include oaths of initiation taken by students of medicine when they began their training.

The oldest, the *Carakasaṃhitā*, dates from the first century AD, but contains older material. Like the Hippocratic Oath, it requires the student to pledge loyalty to his teacher, to "dedicate thyself to me and regard me as thy chief". Reflecting Hindu reverence for animal species, it requires praying for the welfare of all creatures beginning with cows and Brahmanas, a large domestic fowl. In an unusual provision, though one understandable given the Hindu doctrine of karma, in which a person's conduct during the successive phases of his existence determines his destiny, the student pledges not to treat those "who are hated by the King or who are haters of the King", or those who are extremely abnormal, wicked, and of miserable character and conduct". It also proscribes treating "those who are at the point of death". In spite of this reluctance to treat those who are dying, Hindu medical ethics stresses the importance of the proscription against killing.

The Hindu notion of *ahiṃsā*, the avoidance of suffering, is another central theme in Hindu medical ethics. It is sometimes interpreted as requiring caution in order to avoid injuring, a concern often reflected in Western medical ethics with its slogan *primum non nocere*, first of all do no harm.

Buddhist Culture

Some of these notions are also reflected in the Buddhist medical ethic of various Asian countries. The prohibition on killing, the concern about *ahiṃsā* and a commitment to veracity all contribute to the Buddhist tradition of medical ethics. These commitments are often in tension both with indigenous pre-Buddhist religious traditions (Confucianism and Daoism in China, Hinduism in India, and Shintoism in Japan) as well as with modern Western culture. Pinit Ratanakul, commenting on the penetration of traditional Thai Buddhism by Western individualism, traces conflicts over four central themes of Buddhist medical ethics: veracity, non-injury to life, justice, and compassion. Hippocratic Western medical ethics has never manifested commitments to any of these until recent efforts have begun reflecting them at least in a modest way.

In Japan, Buddhist thought has provided the foundations for medical ethical thought. For example, in the sixteenth century a school of medicine, commonly known as the *Ri-shu* school, manifested classical Buddhist commitments. Students in this school were bound by *The 17 Rules of Enjuin*. Killing of any creature was proscribed. Even hunting and fishing were not acceptable. In contrast to the Hindu *Carakasaṃhitā*, the duty to rescue was extended even to those whom the physician disliked or hated. But virtuous acts were to be performed secretly so that they did not become known to people. Doing good deeds secretly was considered a part of virtue. In a fashion similar to the Hippocratic Oath, students were sworn to secrecy, being prohibited from disclosing any medical knowledge to outsiders. Unless a successor trained in this school was found, those of the Ri-shu school were even required to return all medical books to the school when a disciple ceased to practice.

In Japan, Buddhist medical ethics has survived in an uneasy tension with both modern Western thought and indigenous beliefs known as *kami no michi* (the way of Kami), or Shinto, according to the Chinese. For example, while killing is clearly condemned according to Buddhism, suicide and mercy killing receive more sympathetic assessment both by some proponents of the patients' rights movement imported from the West and by those reflecting traditional Shinto openness to *jyoshi* (love suicide) and *shinjyu* (group suicide). It is these latter influences that undoubtedly account for what could be called mercy killing in Japan.

The traditional influence is also seen in the resistance of Japan to the Western brain-oriented definitions of death. Japan is the only state that has not adopted such a legal definition. It is argued that the traditional Japanese notion of a life force penetrating the entire body clashes with Western notions of life related to brain or mental function.

The Near East

Near Eastern cultures have perspectives on medical ethics that date back at least to the second millennium BC. Zoroastrianism of ancient Persia saw the physician as a force for good in the struggle between good and evil. It is suggested that this provides a foundation for a medical ethic that would oppose euthanasia. In both Assyria-Babylonia and in Egypt, suicide was proscribed. There are records of abortifacient remedies in ancient Egypt, but there is doubt concerning abortion's legality. It was, however, by the middle of the second century BC prohibited by Assyrian law as it was in Persia.

The Babylonian Code of Hammurabi (1727 BC?) provided stiff penalties for incompetent surgery. The surgeon's hand was to be severed if surgery was performed on a nobleman that resulted in death or loss of an eye. If death was caused by a medical procedure on a slave, the physician was obliged to replace the slave. This "sliding scale" punishment was matched with a similar sliding-scale fee structure in both Assyro-Babylon and Persia. The higher the status of the patient, the higher the fee.

In Islam, medical ethics is grounded in the *Qur'ān*. A ninth century work *Adab al-tabib* (Practical Ethics of the Physician) reflects the Islamic synthesis of Hippocratic, Greek, and Arabic medicine. In the thirteenth century an Arabic version of the Hippocratic Oath is found in *Lives of Physicians* written by Ibn Abī Uṣaybi'ah. Nevertheless, the core of Islamic medical ethics is explicitly grounded in the *Qur'ān* and its central teaching, "There is no god but Allah, and Muhammad is Allah's Apostle." This has sometimes given rise to a kind of fatalism in Islamic folk medical ethics. It is sometimes reported that Muslims oppose medical manipulations such as birth control for this reason. But a sophisticated ethical framework in Islamic medical ethics includes a rigorous commitment to the preservation of life, and opposition to all killing including killing for mercy and abortion. "Whoever killeth a human being for other than manslaughter or corruption in the earth, it shall be as if he had killed all mankind, and whoso saveth the life of one, it shall be as if he had saved the life of all mankind" (*Qur'ān* 5:22). Islamic medicine, in contrast to Western Hippocratic medicine, is thoroughly theocentric.

This is reflected even in contemporary Islamic medical ethics. In January of 1981 Muslim scholars from throughout the world gathered in Kuwait at the First International Conference on Islamic Medicine. They produced the "Kuwait Document", a now-definitive Islamic Code of Medical Ethics. It contains twelve chapters outlining in detail Islamic positions supported by reference to the *Qur'ān*, culminating in "The Oath of the Doctor", a summary of the code. Among the prominent characteristics are a pledge to "protect human life in all stages and under all circumstances, doing my utmost to rescue it from death, malady, pain and anxiety". It also includes an exceptionless pledge of confidentiality that stands in contrast to the Hippocratic provision, which implies that some information ought to be disclosed to others. It also embodies a commitment to provide medical care for all, "near and far, virtuous and sinner and friend and enemy". Other chapters include an explicit recognition of the duty of the Islamic physician to society (a contrast to the exclusive focus on the individual in traditional Hippocratic medical ethics), further expounding on the sanctity of human life (including prohibitions on all abortion and mercy killing), and a full endorsement of the legitimacy of Islamic involvement in new biotechnological advances including organ transplantation. This last provision contrasts with the folk ethic that sometimes reflects a kind of fatalistic yielding to the power of Allah.

African Societies

Many African cultures have taken positions on matters related to medical ethics. Unfortunately, with the exception of ancient Egyptian views, relatively little is known about their specific medical ethical stances. In some sub-Saharan African tribal groups understandings of moral conduct related to life and death are closely tied to the religious culture (*Wiredu*). Since the cultures and languages are so diverse, it would be a mistake to assume that all African societies hold the same medical ethical views. Nevertheless some common patterns are reported. Although euthanasia is not widely discussed, it was and is reportedly practiced by family relatives for incurable and distressful mental illness and severe congenital malformations in many African societies. Likewise, abortion, though viewed with moral skepticism, appears to be practiced as it is elsewhere in the world.

Pre-Columbian Western Hemisphere

Among the most poorly understood medical ethical systems are those of the Pre-Columbian Western hemisphere. There has never been any formal analysis of the medical ethics of the great Inca, Aztec, or Maya cultures or of the Native American tribal groups of North America. Most of what is known pertains to views about abortion, suicide, human sacrifice, and related practices.

The Incas viewed children as an asset, and, while abortion was known (by means of fetal massage, beatings, and special drugs), it was punished by execution of both the woman who aborted and those who aided her. Nevertheless among the Inca human sacrifice was practiced. Male children could be demanded for sacrifice, although the practice was apparently rare. Sacrifice of one's young children was reportedly a last resort to attempt to win the favor of the spirits. The yearly sacrifice of two infants was a part of the ritual of the most important temple. In contrast to Greek and modern infanticide, Inca sacrifices had to be of children without blemish.

Human sacrifice was also practiced by the Mayas, either by "heart-rending" or by throwing the one to be sacrificed into a cenote or large sink hole. The Mayas also offered sacrifices of their own blood by piercing their cheeks, tongues, or lower lips. The blood was considered to express "vital principles". Suicides were held by the Mayas to be sacred, deserving of their own special heaven. Human sacrifice was more common among the Aztec as it was in the society of the Caribs. By contrast in these pre-Columbian cultures, there are few reports of killings of either infants or adults for the purpose of euthanasia or the sparing of the afflicted. The close integration of religion, magic, and medicine in these cultures makes the differentiation of a uniquely medical ethic implausible.

Inca doctor-sorcerers, called the *camasca*, or *soncoyoc*, are said to have employed deception in their practice of medicine. They earned entrance into the profession by telling stories of vivid dreams or miraculous recoveries from fatal diseases. Once they had obtained the status of *camasca* or *soncoyoc*, they staged dramatic procedures. One example is surgeries in which the doctor would claim that he had removed worms and stones — the supposed causes of disease — from the patient's body, when in actuality, he had recovered said items from no other place than his own pocket, using sleight of hand to make it seem as though they were being extracted from the body. In addition, the Inca physicians would create mixtures of herbs that they claimed to be deadly, in order to generate fear — and revenue — from their clientele. Despite all of this activity, to kill with magic charms was considered a serious crime. Like healers in many cultures, the practitioners were very exclusive about the practice of medicine, claiming that only they, the chosen ones, could practice it properly. Another source says that in some tribes, as in ancient China, herb healers were family based, and all professional information as to the nature of different herbs was kept as a family secret.

Medical ethical stances are derived from more general systems of ethics and belief. It is not surprising then to find that there are as many medical ethical systems as there are systems of ethical thought. To say that behavior in the medical sphere is ethical is, after all, simply saying that the behavior is ethical — at least according to some standard of ethics. Even those practicing medicine within Western culture will inevitably encounter patients who come from other cultures whose medical ethical values and beliefs may be quite different. Only by knowing the range of medical ethical systems will one's own positions be brought into focus.

ROBERT M. VEATCH

REFERENCES

Amundsen, Darrel W. "Medical Ethics, History of: Ancient Near East." In *Encyclopedia of Bioethics*. Ed. Warren T. Reich. New York: Free Press, 1978, vol. 2, pp. 880–84.

Cobo, Bernabe. *Inca Religion and Customs*. Trans. and ed. Roland Hamilton. Austin: University of Texas Press, 1990.

Desai, Prakash. "Medical Ethics in India." *Journal of Medicine and Philosophy* 13:231–55, 1988.

Hathout, Hassan. "Islamic Basis for Biomedical Ethics." In *Transcultural Dimensions in Medical Ethics*. Ed. Edmund D. Pellegrino, Patricia Mazzarella, and Pietro Corsi. Frederick, Maryland: University Publishing Group, 1992.

International Organization of Islamic Medicine. *Islamic Code of Medical Ethics*. [Kuwait]: International Organization of Islamic Medicine, 1981.

Kendall, Ann. *Everyday Life of the Incas*. New York: Putnam, 1973.

Kimura, Rihito. "Japan's Dilemma with the Definition of Death." *Kennedy Institute of Ethics Journal* 1:123–31, 1991.

Menon, A. and H. F. Haberman. "Oath of Initiation" (From the *Carakasaṃhitā*). *Medical History* 14: 295–96, 1970.

Osuntokun, B. O. "Biomedical Ethics in the Developing World: Conflicts and Resolutions." In *Transcultural Dimensions in Medical Ethics*. Ed. Edmund D. Pellegrino, Patricia Mazzarella, and Pietro Corsi. Frederick, Maryland: University Publishing Group, 1992, pp. 105–43.

Qiu, Ren-Zong. "Medicine–The Art of Humaneness: On Ethics of Traditional Chinese Medicine." *Journal of Medicine and Philosophy* 13:277–300, 1988.

Ratanakul, Pinit. "Bioethics in Thailand: The Struggle for Buddhist Solutions." *Journal of Medicine and Philosophy* 13:301–12, 1988.

Unschuld, Paul U. *Medical Ethics in Imperial China: A Study in Historical Anthropology*. Berkeley: University of California Press, 1979.

Veatch, Robert M., ed. *Cross Cultural Perspectives in Medical Ethics: Readings*. Boston: Jones and Bartlett, 1989.

Veatch, Robert M., and Carol G. Mason. "Hippocratic vs. Judeo-Christian Medical Ethics: Principles in Conflict." *The Journal of Religious Ethics* 15:86–105, 1987.

von Hagen, Victor Wolfgang. *The Ancient Sun Kingdoms of the Americas*. Cleveland: World Publishing Company, 1961.

MEDICAL ETHICS IN CHINA For most pre-modern civilizations with a certain degree of cultural sophistication, medical ethics were strikingly similar: hard work, and concern for the poor and needy. A good doctor not only excelled in his medical skills, but also cared for the sick and poor. Material benefits should not be his major pursuit. In the West, such tenets naturally conformed to Christian morality; in China, Confucianism, Buddhism, and Daoism mostly contributed to the formulation of such ethics.

Despite the universality of these basic ethical requirements, there were specific features in the historical development of Chinese medical ethics which can be said to have gone through several stages of formation: the mythical period of Antiquity (ca. 771 BC–AD 265); the period of Buddhist and Daoist influences of the early medieval era (265–960); the period of medical professional maturity of the late medieval era (960–1368); and lastly the late imperial period of Confucian influence (1368–1911).

In the period of Antiquity some of the basic tenets of medical ethics were put forward. From some of the earliest records of Chinese oracle bone writings and other texts, we know that medicine was then mixed with divination and magic. The shaman/doctor had a relatively high social position and a respectable one should, according to opinion of the time, have the quality of perseverance or stability (*heng*).

With the gradual emergence of professional doctors around the sixth century BC, clearer notions of medical ethics appeared. In the *Huangdi Neijing* (Inner Classic of Internal Medicine) attributed to the mythical Yellow Emperor, the basic medical classic compiled throughout a period of no less than 400 years beginning in the Warring States period (475 BC–221 BC), qualities such as erudition, rich experience, wisdom, humility, and hard work were evoked as necessary for a good doctor. Renown doctors of this period were reputed to have most or all of such qualities.

Another major quality said of the early doctors was their concern for the poor and lowly. They did not only cure the wealthy and the powerful, and profit was never their major pursuit.

Prominent doctors of this period, such as Bian Que (fifth century BC), Chunyu Yi (b. 205 BC), Zhang Zhongjing (second–third centuries AD), and Hua Tuo (d. ca. 208), were later to be venerated as deities of the medical profession, and symbolized the very virtues of good doctors. The deification of these doctors, especially Hua Tuo, was based not only on their exceptional medical skills, but also on their outstanding and legendary morality.

In the Early Medieval period, one witnessed the marked influence of Buddhism and Daoism, not only on medicine itself, but also on medical ethics. For many modern historians of Chinese medicine, true medical ethics were not systematically composed in China until the seventh century, when one of China's greatest doctors Sun Simiao (also called Sun Simo, 581–682) wrote systematically on the duties of a physician in his book *Qianjin Bao Yao* (The Thousand Golden Remedies). In this work Sun not only reiterated the qualities already mentioned for model doctors of Antiquity, but also emphasized the importance of retribution as a safeguard for good virtues: "Lao Zi, the father of Daoism, said, 'Open acts of kindness will be rewarded by man while secret acts of evil will be punished by God'. Retribution is very definite. A physician should not utilize his profession as a means for lusting. What he does to relieve distress will be duly rewarded by Providence." The notion of retribution, though here thought to be essentially Daoist, was in fact much influenced by Buddhism in this period.

The medical ethics put forward by Sun Simiao reflected other aspects of Buddhist–Daoist influences: to attend to all patients equally, disregarding their social and economic position, their age, and their physical appearances. Enemies and friends, foreigners and Chinese, the stupid and the wise should all be treated alike. All these categories should be considered as one single, general class by the doctor. This "equality" before medical care was something unfamiliar to Confucian morality but which has remained significant in medical ethics since the seventh century.

In the Late Medieval Period, Chinese medicine reached maturity, both as an institution and a clinical art . The formal division of medicine into different branches or specialties (notably *fuke*, the branch of medicine specializing in women's diseases; *erke*, in children's diseases; and *waike*, in bone fractures and operations) implied a finer division of labor amongst physicians, and this was also reflected in a more elaborate and systematic training program for officially approved doctors in this period, and increased state intervention in the management of medical resources. Though some scholars think that in this period medicine was "a respectable field of study ... but as a career and mode of life was highly controversial", one can still see the emergence of a certain medical professionalism, and with it, new elements in medical ethics.

One such element was the frequent caution against *yongyi* (common practitioners). The term was frequently evoked in Song medical texts. A *yongyi* is the antithesis of a *liangyi* (fine doctor), comparable to a *liangxiang* (fine minister). This comparison was first put forward by the famous Song minister Fan Zhongyan (989–1052), who had once considered becoming a doctor himself. The medical profession in the Song period took greater pains than before to draw lines between good and bad doctors, reflecting the emerging professional consciousness of doctors of the period.

However, the reputation of a fine doctor was not entirely based on his curing skills. A fine doctor now combined the religious selflessness and charity of the early Medieval period, as well as neo-Confucian ethics. One of the greatest *fuke* specialists of the Song, Chen Ziming (1109–1270) wrote that he came from a family of three generations of doctors, with an excellent private library of medical books. But his appetite for prescriptions and medical texts was so large that whenever he traveled in the southeast, he collected large quantities of texts which he studied in his free time. Obviously, a good doctor was now likened to a good Confucian scholar, for whom family tradition and textual learning were considered essential for professional success.

The increasing influence of Confucian morality on medical ethics was yet to be seen in the late imperial period. This was a period of relative political continuity and economic prosperity, accompanied by a significant growth in medical knowledge. However, this period was not one of institutional renovation and regulation in medicine. Consequently medical professionalism, in the sense of the maturation of a completely independent professional category, might have suffered. As Paul Unschuld has said, "The Chinese physician as a definable entity did not exist". The respectability of a physician was, besides his curing technique, increasingly linked to Confucian morality.

One prominent sixteenth-century physician, Xu Chunfu (1526–1596), wrote in his major work *Gujin Yitong Daquan* (Medical Tradition of the Past and of the Present) that "Confucianism and medicine cannot be separated". His contemporary, also a famous doctor, Gong Tingxian (1522–1619), put forward ten requirements for physicians, the first two being to cherish kindness and to understand Confucian principles. Another doctor, Li Ting (d. 1619), in his *Rules for Medical Studies* stated from the outset that, "Since medicine comes from Confucianism, unless one studies and understands the [basic] principles, one remains mediocre, vulgar and stupid ..." At the same time, detailed instructions on treating female patients were increasingly provided in order to avoid transgressing the Confucian principle of the separation of the sexes.

However, it was also prominent "Confucian doctors" of this period that further defined medical ethics in more systematic ways. The above mentioned Xu Chunfu was among the first to list the main "vices" of *yongyi*, including taking on the appearances of good doctors with a few tricks, extorting money out of patients, and freeing themselves of responsibility by any means in case of misdiagnosis leading to death. Another famous doctor of the seventeenth century wrote the first Medical Code in 1658. Yu Jiayan (ca. 1585–1664) wrote detailed technical diagnostic guidelines for professional doctors. He made the doctor responsible for all avoidable diagnostic errors, and in so doing, attempted to draw a clearer line constructed on technical considerations between a *liangyi* and a *yongyi*. In other words, development, though limited, was still perceivable in the professional consciousness of doctors despite the overwhelming influence of Confucianism in this period.

However, the incomplete growth of an autonomous medical profession inevitably left areas of ambiguities in medical ethics. One such is the transmission of "secret prescriptions". For some historians of medicine, "the dispensing of secret prescriptions was never considered to be unethical in China. Some even deemed it an honor for a physician to know a secret formula". However, there are indications that not all approved of this attitude. Among others Xu Youzhen (1407–1472), a scholar-official, accused contemporary doctors of preserving secret prescriptions for private profit. For him, effective prescriptions should be published in order to save more lives, and to be passed onto posterity. An early nineteenth-century physician, Bao Xiang'ao, wrote in the preface of his compilation of "effective prescriptions" that those who did not make effective prescriptions public were despicable.

This ethical ambiguity of the necessity of keeping prescriptions and healing techniques secret obviously was a consequence of the basic structure of the medical institution in late Imperial China: the lack of centralized control in the production of medical knowledge, either by the state or by a self- regulatory medical corps, and the accompanying uncontrolled distribution of medical resources in society. Each family or school of medicine thus had to preserve its share of the resources in order to be competitive in the field, though morally such an attitude was obviously questionable. Such a characteristic only encountered effective challenge when western medicine was introduced into China in the late nineteenth century, together with its ethics and institutions.

ANGELA KI CHE LEUNG

REFERENCES

He Zhaoxiong. *Zhongguo Yide Shi* (A History of Chinese Medical Ethics). Shanghai: Shanghai Yike Daxue Chubanshe, 1988.

Hymes, Robert. "Not Quite Gentlemen? Doctors in Sung and Yuan." *Chinese Science* 8: 33, 1987.

Li T'ao. "Medical Ethics in Ancient China." *Bulletin of the History of Medicine* 13(3): 268–277, 1943.

Unschuld, Paul U. *Medical Ethics in Imperial China*. Berkeley: University of California Press, 1979.

Zhou Yimou, ed. *Lidai Mingyi Lun Yide* (Medical Ethics Discussed by Doctors Throughout the Ages). Changsha: Hunan kexue jishu chubanshe, 1983.

See also: Bian Que – Sun Simo – Zhang Zhongjing – *Huangdi Neijing*

MEDICAL ETHICS IN INDIA The origins of medicine in India stretch back to antiquity. The Harrapan city culture flourished in and around the Indus Valley ca. 2500 BC; it is known for its elaborate bathhouses and drains and sewers built under the streets leading to soak pits. In the second millennium BC, the northwestern parts of India were host to a series of Indo-European immigrants and invaders from Central Asia. With them began the classical culture of India. Vedas, the sacred lore of the Indo-Europeans, celebrate the *Bhesaj*, one knowledgeable in medicinal herbs. One of the four Vedas, the *Atharvaveda* contains many chants, mantras, and herbal preparations to ward off evil, enemies, and diseases. The priest-physicians prescribed preparations of plants and herbs, and prayers and fasts for their patients. The Indian medical tradition, *Āyurveda*, meaning the science of vitality and long life, is considered a limb of the *Atharvaveda*.

A more formal system of medicine evolved from around the time of the Buddha (ca. 500 BC). It became organized in textual form in the first century AD, and reposes in a vast body of literature redacted and updated from that time to the present. There are six principal texts of the Āyurveda. The older three are the two compendia, *Carakasaṃhitā* and *Suśrutasaṃhitā*, named after the two legendary physicians, Caraka and Suśruta, and the *Aṣṭāṅgahṛdaya*, the eight-fold essence attributed to an eighth century physician named Vāgbhaṭa. The younger three are the *Mādhavanidāna* (ninth century), *Śārṅgadharasaṃhitā* (thirteenth or early fourteenth century), and *Bhāvaprakāś* Bhāvamiśra (sixteenth century). The word *caraka* also means one who moves about, and may have referred to the itinerant Buddhist and Jain monks who played a pioneering role in the evolution of the Indian medical tradition. In the realm of King Aṣoka (273–232 BC), who embraced Buddhist ideals, Buddist monasteries served as institutions, like hospitals and hospices, for the care of the sick and the dying.

The earliest medical writings known as the Bower manuscripts, discovered in a Buddhist Stupa in Kashgar (modern China), and translated by Rudolph Hoernle, are considered to have been written by Buddhist authors around AD 450. These texts contain medical treatises which describe the virtues of garlic in curing diseases and extending the life span, elixirs for a long life, ways of preparing medical mixtures, eye lotions, oils, enemas, aphrodisiacs, and procedures for the care of children. Early Indian medicine was carried to Tibet along with Buddhism and was best preserved there, as well as in China. Travelers to and from China, Greece, Persia, and Arabia contributed to the spread of Indian medicine outside India.

The basic assumptions of Indian medicine are rooted in the religious and philosophical traditions of India. Early developments exhibited great diversity in opinion and formulation in keeping with the diversity in Indian thought, tied to Hindu, Buddhist, or Jaina philosophies in various measures. Similarly the system allowed for significant geographic variation as knowledge spread through the subcontinent over a long period of time.

The medical ethics which are closely linked to these religious and philosophical perspectives (*darśanas*) reveal variable, shifting, and accommodating attitudes.

Āyurvedic constructs of the body and the self, central to

the medical enterprise, grew in tandem with the faith traditions. The primary vehicles of ayurvedic pathophysiology are the *doṣas* (humors): *vāyu* or *vāta* (wind), *pitta* (bile), and *kapha* (phlegm), and the *dhātus* (body substances). The three humors represent movement, heat, and moisture respectively in the body. The primary body substance, *rasa*, organic sap, is derived from food, moves throughout the body, is stored in various reservoirs, and is finally excreted as waste products. In processes of sequential transformation, the *dhātus*, flesh, fat, bone, marrow, and semen, are derived, semen being the purest and most vital product of this process.

The Indian system of medicine views health as a state of balance of body substances, *dhātusamya*, and illness as a state of disequilibrium. The body responds to many kinds of inputs: physical, as in food and drink, psychological, as in emotions of anger or jealousy, and social, as in affection, praise, or scorn. Each input is a potential source of a disease or a cure.

The theory of *guṇas* (lit. strands or qualities) introduces the notion of ethics as a material basis in the ayurvedic pathophysiology. Inherent and substantial, *sattva* (goodness), *rajas* (vitality or activity), and *tamas* (inertia) are qualities or traits found in all substances in various combinations. The balance determines the overall dispositions of persons, foods, activities, bodily substances, and so forth. *Sattva,* which is cool and light, produces calmness, purity, or virtue; *rajas,* which is hot and active, produces passion, happiness, or sorrow; and *tamas,* which is dark, heavy, and dull, produces sloth, stupidity, and evil. Contemplation, meditation, silence, devotion, and fasting promote goodness; love, battle, attachment, pleasure seeking, and emotionality enhance vitality. Sleep and idleness increase inertia. In a hierarchy of values, the *sattva* categories reign supreme and become less material, closer to the idea of *sat* (truth or essence), and often the same as the mind or self. The object of the therapeutic is to transform a person from lower to higher strands or qualities, which is accomplished through the prescription of foods and activities which build goodness. Thus the therapeutic and the ethical become coterminus.

In the Indian view life is not the opposite of death; birth is the opposite of death. Life begins when an embryo is formed out of the union of male and female germinal substances. Defining when human life begins was neither easy nor uniform and straightforward. Some texts maintained that life began with the aforesaid union, and others at the moment of quickening or the descent of the fetus into the pelvis; the latter was more frequently understood as a point of viability. Abnormal pregnancies, congenital deformities, multiple pregnancies, and infertility were explained in terms of defective germinal substances, unnatural coitus, failure in nourishment, or disturbances in humors in the mother or the fetus.

Among the religious obligations, having male progeny was imperative in order to secure a passage to the land of forefathers through the performance of funerary rites. In situations in which a woman failed to have a son, the man was to take another wife, or otherwise adopt a son. If the problem appeared to be male impotence or infertility, the husband's younger brother or another suitable man was to impregnate the wife (a custom called *niyoga*). Early medical texts elaborate on the ways of enhancing conception, and later texts discuss problems of contraception. Mythology also testifies to *in vitro* fertilization and embryo transfer.

The *Suśrutasaṃhitā* describes various forms of arrested fetal development or obstructed deliveries and describes ways of inducing labor and/or destroying the fetus, especially in the case of danger to the mother's life. A seventeenth century text also describes ways of inducing labor for purposes of abortion in cases of women in poor health, widows, and women of liberal morals.

In contemporary problems of medical ethics, no problem has caused as much furor as has amniocentesis. Preference for a male child, with an easily available technology to determine gender prenatally, has resulted in inordinate and indiscriminate use of abortions. Some states in India have enacted laws to restrict the scope of indications and use of amniocentesis.

There are three categories for the etiology of diseases in *Āyurveda*. External or invasive diseases are caused by foreign bodies, injuries, infestations, and possession by evil spirits. Internal diseases are disturbances of humors, in part caused by lapses in discretion, as in faulty or unseasonable diets, overexertion, sloth, sexual indulgence, or mental disturbances. In either case, the final pathway for the pathology of a disease is an imbalance of humors. The third category contains the diseases which are the fruits of *karma*, the operative principle of Hindu ethics. A very simple explanation might be "every action has a reaction" or "as you sow, so shall you reap", but the logic extends beyond one life. In *karma* theory, when a person dies his self moves to the other world, enveloped in the part material and part ethereal covering which carries the traces of all actions performed, and comes to determine its condition in the next life. Thus some diseases are the fruits of actions from past lives. The unseen hand of *karma* is invoked to explain the not so easily explicable. Events like epidemics and disasters are a result of bad actions of a whole community or the actions of a king.

Mental illnesses also arise from these etiologies: possession states, disturbances in humors, and lapses in discretion. Some disease states are also seen as the workings of time, as in aging.

Physicians in ancient India did consider *karma* in etiology, but they agreed that the passivity that results from assumptions of predetermination made the whole medical enterprise meaningless. Human effort was always a factor in the workings of *karma*, and caring and healing must be actively pursued by the physician. There was also a recognition of incurable diseases, in the face of which human effort was futile. The physician was prudent if he avoided heroic efforts to prevent the inevitable, which not only led to loss of income but also loss of prestige. If the case was hopeless, the physician was to do no more than attend to the nutrition of the dying patient, and even that might be withdrawn at the request of the family.

A category of "willed death" was also recognized in the various religious traditions and was understood to be different from suicide. Suicide was regarded as an act of desperation and willed death an act of determination. It involved permission of the religious order and was resorted to only when the quality of remaining life was likely to be poor.

The ayurvedic physician, called a *vaidya*, was esteemed for his powers but also shunned because of his contact with impurities such as body products, suppurative lesions, and corpses, and his mingling with common people. Taboos around touching ultimately resulted in palpation falling into disuse.

The physician was enjoined to strive constantly to acquire new knowledge, advance through practical experience, and enter into learned dialogues with practitioners from other places. His education began as an apprentice, with the teacher and pupil choosing each other. A good teacher was free of conceit, greed, and envy, and a student had to be calm, friendly, and without physical defects. Later on the *vaidya* became a sub-caste or occupational division, and the profession passed from father to son.

The *Caraksaṃhitā* contains an extensive list of ethical directives in the form of an oath to be taken by one entering medical practice. Among these were injunctions never to abandon a patient even if that interfered with one's livelihood, to be modest in dress and conduct, gentle, worthy, and wholesome. A physician must not enter a patient's house without permission, and be mindful of the peculiar customs of a household. He was to avoid women who belonged to others and maintain confidentiality.

Quacks and charlatans were known by their pretense and arrogance, boastfulness and superficial knowledge. The fate of their patients was worse than death. The *Caraksaṃhitā* says that one can survive a thunderbolt but not the medicine prescribed by quacks.

Medical ethics was an integral part of ancient Indian medicine. The texts addressed ethical issues that arose at both ends of life, birth and death. Their approach was pragmatic and flexible, and the purpose of alleviating an illness was always considered in the context of geographic locale, time (the era and the stages of a patient's life), and the particularities of a person. The physician's conduct was also to be always above reproach both in his professional and personal conduct.

PRAKASH N. DESAI

REFERENCES

Primary sources

Brihadaranyaka Upaniṣad. Trans. F. Max Muller in the *Upaniṣads* part 2. New York: Dover, 1962.

Carakasaṃhitā. 2 vols. Trans. Priyavrat Sharma. Varanasi, India: Chaukhambha Orientalia, 1981–83.

Mahābhārata, vols 1–3. Trans. J.A.B. Van Buitenen. Chicago: University of Chicago Press, 1973.

Suśrutasaṃhitā. Trans. Vaidya Jadavji Trikamji Acharya and Narayana Ram Acharya. Varanasi, India: Chaukhambha Orientalia, 1980.

Secondary sources

Chandrashekar, Sripati. *Abortion in a Crowded World: The Problem of Abortion with Special Reference to India*. London: Allen and Unwin, 1974.

Chattopadhyaya, Debiprasad. *Science and Society in Ancient India*. Calcutta: Research India Publications, 1977.

Dasgupta, Surendranath. *A History of Indian Philosophy*. 5 vols. London and New York: Cambridge University Press, 1922–1955.

Desai, Prakash. "Medical Ethics in India." *Journal of Medicine and Philosophy* 13: 231–255, 1988.

Desai, Prakash. *Health and Medicine in the Hindu Tradition*. New York: Crossroads, 1989.

Desai, Prakash. "Hinduism and Bioethics." In *Bioethics Yearbook*, vol. I, *Theological Developments in Bioethics*, 1988–1990. Eds. Baruch A. Brody et al. Dordrecht: Kluwer Academic Publishers, 1991, pp. 41–60.

Fujii, Masao. "Buddhism and Bioethics in India: A Tradition in Transition." In *Bioethics Yearbook*, vol. I, *Theological Developments in Bioethics*, 1988–1990. Eds. Baruch A. Brody et al. Dordrecht: Kluwer Academic Publishers, 1991, pp. 61–68.

Jolly, Julius. *Indian Medicine*. Trans. Chintamani Ganesh Kashikar. New Delhi: Munshiram Manoharlal, 1977.

Purushottama, Bilimoria. "The Jaina Ethic of Voluntary Death." *Bioethics* 6(4): 331–355, 1992.

Young, Katherine K. "Euthanasia: Traditional Hindu Views, and the Contemporary Debate." *In Hindu Ethics: Purity, Abortion, and Euthanasia*. Ed. Harold G. Coward, Julius J. Lipner, and Katherine K. Young. Albany: State University of New York Press, 1989, pp. 71–130.

See also: Medicine in India – Caraka – Suśruta

MEDICAL ETHICS IN ISLAM The ethical assumptions underlying the practice of medicine in Islam are inspired by two foundational texts: the *Qur'ān*, the divine message revealed to the Prophet Muḥammad; and the *Sunnah*, the paradigmatic life of the Prophet Muḥammad which complemented and exemplified the Quranic message. A secondary source of influence resulted from the Muslim conquest and expansion of the seventh to the eleventh centuries, when the cultural and scientific heritage of Antiquity was translated into Arabic and came selectively to be appropriated, refined and developed by Muslims. The integration of this heritage into Muslim civilization led to a new synthesis, not only of the science of medicine but also of the moral values supporting it. Although some of the medical traditions and values of Antiquity were sustained, they were set in Islamic contexts which allied their meaning and purpose to different goals.

The early centuries of Islam also represented a tremendous flowering of intellectual sciences. Muslims were encouraged by their faith and by the *Qur'ān* to engage actively in the pursuit of knowledge. This impetus translated itself into significant advances in the study of the natural sciences, cosmology, geography, mathematics, history, law, languages, and medicine to name some of the major disciplines. The emergance of Muslim medical ethics has to be situated within this matrix of scientific interest and intellectual commitment.

Muslim views regarding medicine evolved from that of a predominantly Arab cultural setting to that of a diverse and cosmopolitan civilization. The reconstruction in Muslim thought of a prophetic medicine, emphasizing the spiritual dimension and the role of faith in healing, represented an attempt to associate with the Prophet those practices of human life and healing promoted by the rise of a tradition of scientific medicine. The perception of Muslim adherence to prophetic medicine has often been misconstrued as promoting a kind of fatalism, but it would be misleading to extend this interpretation to the whole spectrum of medical ethics and values among Muslims. Prophetic medicine developed alongside professional medicine, which cultivated its own frame of reference for integrating ideas.

Adab is the Arabic term that best describes the pattern of medical ethics and conduct which evolved as a result of the synthesis of the Islamic message and Hellenistic thought. The term represents a set of cultural and moral assumptions articulating the linking of knowledge to appropriate behavior. Ishāq ibn ʿAlī al-Ruhāwi, a ninth-century Christian physician, translated one of the earliest texts dealing with medical ethics into Arabic. His work already reflects the vocabulary of Islam and a heritage of Prophetic religion integrated within a moral framework attributed to Galen and the ancient philosophers. According to al-Ruhāwi, medicine is a divine art granted to humanity by the Creator, whose healing role the physician imitates. Prayer, as the first and last act of the day, is recommended as part of the personal conduct and moral beliefs of physicians.

Within the larger harmony of moral and religious principles is the ethical concept of moderation. Muslim medicine did not accept the body/soul distinction of the Greek tradition but conceived of human beings as entities formed in symmetry and harmony. One of the most illustrious physicians of early Muslim history, Abū Zakariya al-Rāzī (Rhazes) argued for a healthy, moral life based on moderation between excessive indulgence and abstinence and an adherence to a life of knowledge pursued to balance physical, moral, and intellectual needs.

The *Qur'ān*'s message of social justice, developed into the notion of public welfare and charitable works toward the poor, influenced medical values in ways quite different from that of the Hippocratic tradition. The hospital, founded through royal acts of patronage or endowments, is an Islamic institution traceable to the eighth century. This larger social role, engendered by Islamic values of concern and care for the indigent, represents perhaps one major difference with the practice of Antiquity, where the emphasis lay mostly on the physician's concern and relationship with an individual patient.

The rise of the physicians and the medical ethics that guided the emergent profession in Muslim society cannot be separated from the general tenor of the civilization and the polity that had emerged after the founding and spread of Islam. The moral discussions and intellectual forces that emerged stimulated a concern for how moral and religious perspectives could be reconciled with intellectual modes of inquiry. But professional medicine was only one of several therapeutic systems available. As Sufism, the mystical and spiritual dimension of Islam, grew in influence and became institutionalized through a system of orders, the heads of these organizations came to be regarded as having the power to intercede in moments of crisis, including disease. In time this led to an extensive use of amulets, granting of prayers and recitations from the *Qur'ān* as aids to healing. Such traditions and practices came to occupy a relatively prominent place beside that of professional medicine during the pre-modern period, and continue to be influential in many Muslim societies.

Whereas the eighteenth and nineteenth centuries marked a new impulse in medical and ethical thinking in Europe, the events leading to the encounter of modern Western culture and political power with the world of Islam created a different context for the assimilation and practices of modern medicine. Muslims were faced with new tools and methods of healing which called into question all of the normative assumptions that previously had been integrated into their *adab* of medicine. The institutionalization of Western medical practices also resulted in moral and ethical consequences. Muslim responses reflected all of the ambiguities in the face of the challenge the new authority and its civilization brought. For some, Islam would serve as a moral sanctuary from which to combat secular ways of staving off disease. For others, it led to an acceptance of modernity and provided the impulse to shape it in conformity with the sources and perceptions of their past heritage. The process of change as it affected medical practice and value thus cannot be separated from the larger issues entailed by the colonial encounter and the patterns of medical education and practice, as well as attitudes to disease and healing that came in the wake of Western contact and influence. The measures instituted through European doctors and medicine eventually received acceptance by a new generation of professional physicians and led to the assimilation and emulation of the new practitioners and their values. The duality that resulted caused a failure to ground the new training and knowledge within Muslim theological moral and cultural contexts and separated the "new" physicians from traditional counterparts and their moral world. In many ways, this sense of duality and the effort to recover the center constitute the story of ethical issues facing twentieth-century Islam. The most important challenge however remains the formulation of an *adab* that will guide the practice of medicine while remaining engaged with its past ethical underpinning and taking account of the dramatic changes brought about by population growth, poverty, and national policy.

In the field of medicine as in other scientific and technological areas, Muslims are faced with complex choices and newly emerging issues that raise moral dilemmas. There is a greater need to harness resources that advance skills in moral reasoning and enhance sensitivity, and to integrate these into perspectives of medical education, institutional development and the Islamic commitment to ameliorate poverty and disease.

AZIM A. NANJI

REFERENCES

Hourani, G. *Reason and Tradition in Islamic Ethics*. Cambridge: Cambridge University Press, 1985.

Hovannisian, R., ed. *Ethics in Islam*. Malibu, California: Undena Publications, 1985.

Nanji, Azim. "Islamic Ethics." In *A Companion to Ethics*. Ed. Peter Singer. Cambridge: Blackwells, 1991.

Nasr, S.H. *Ideals and Realities in Islam*. Cambridge, Massachusetts: Beacon Press, 1972.

Rahman, F. *Health and Medicine in the Islamic Tradition*. New York: Crossroad, 1987.

See also: al-Rāzī

MEDICAL TEXTS IN CHINA The texts of Chinese medicine are extremely voluminous. They have a continuous history of more than two thousand years, of which little is known in the West. Standard works up to AD 1900 number around 190, and if popular writings are included this rises to at least 1500. There has been great activity in the half-century since the 1949 establishment of the New China, and a modern dictionary of traditional Chinese medicine written in Shanghai in 1988 refers to over four thousand individuals and eight thousand works. And yet there are barely a dozen translations of these texts. The whole corpus also includes works on massage, diet therapy, and therapeutic exercise similar to the contemporary popular *Taiji Quan* (Supreme Ultimate Boxing) and *Qi Gong* (Breathing Therapy). These have been largely omitted from the following entry, which describes, in chronological order, the most important texts on medical theory, acupuncture and moxibustion, and herbal medicine.

The earliest writings, which formed the cornerstone for all medicine in China, Japan, Korea and the Far East, are collected in the large compilation made during the Han Dynasty (202 BC–AD 220), named the *Huangdi Neijing* (The Yellow Emperor's Canon of Medicine). Extracting itself from the notion of ancestral curses and demonic attacks as the cause for illness, the reasoned tone of this compilation (and it is only one of many Han classics) laid a foundation for the next two thousand years of medical writing, and its influence soon spread to Korea, Japan, and beyond.

The basic theory of Chinese medicine, as spoken of in the *Huangdi Neijing*, involves the interplay and interfusion of *yin* (quiescent) and *yang* (active) forces, producing both health and disease. This alternation occurs both within the human body, and between the body and the natural world. Chinese medicine pays particular attention to natural rhythms and the *wuxing*, five fundamental elemental associations of vegetation, fire, soil, minerals, and fluids (wood, fire, earth, metal, and water), both within the natural world and as symbolic agents of change within the body. Illness is seen as a disorder of the *qi* (vital energy), perhaps enhanced by some external factor but primarily involving a deficiency within the individual.

The *Shennong Bencao* (Shennong's Herbal), appeared soon after the *Huangdi Neijing*. It is the earliest surviving materia medica, believed to have been compiled during the first century BC. In this work 365 kinds of medicinal substances are listed and divided up into three classes: major remedies, medium remedies, and minor remedies.

An almost contemporaneous and equally important text is the *Shanghan Lun* (Discussion on Cold-induced Disorders) which was written by Zhang Zhongjing (ca. AD 150–219) at the close of the Han, and which has proven to be one of the most influential texts in Chinese medicine. This was the first to advocate the analysis of medical conditions in accordance with the six channels (*taiyang, yangming, shaoyang, taiyin, shaoyin,* and *jueyin*) and eight syndromes (*yin/yang*, outer/inner, hot/cold, and excess/deficient). Along with the *Huangdi Neijing*, it formed a straightforward basis for the development of traditional Chinese medicine. Its companion text is the *Jingui Yaolue Fang Lun* (Concise Prescriptions from the Golden Casket, ca. 220), also by Zhang Zhongjing.

The importance of these two texts, the *Huangdi Neijing* and the *Shanghan Lun*, is that they illustrate the building up of a well worked out and heuristic, classificatory framework, involving the yin and yang, the five elements, the six channel types and the eight principle syndromes of diagnosis, into which the results of any doctor's medical observations could be fitted.

The first textbook on acupuncture and moxibustion was the *Zhenjiu Jiayi Jing* (A–Z of Acupuncture and Moxibustion), written by Huangfu Mi some time around AD 280. Much of this text actually repeats verbatim the *Huangdi Neijing*, but it also gives the first systematic description of acupuncture points, each listed according to its position on the body, and also their function. In addition, the first book on pulse diagnosis, the *Mai Jing* (Pulse Classic, ca. 300) appeared during this period. The author, Wang Shuhe, perfected the art of pulse taking, describing in detail nearly thirty separate categories of pulse. This work, along with Li Shizhen's *Binhu Maixue* (Pulse Studies of Master Binhu), written later during the Ming, formed the core texts for this unique method, which is characteristically Chinese.

It is recorded in the Tang Dynasty histories that the Imperial medical colleges based their teachings upon the *Huangdi Neijing* and Huangfu Mi's *Zhenjiu Jiayi Jing*. Among the books also at their disposal would have been the famous, and charmingly titled *Zhouhou Beiji Fang* (Emergency Remedies to Keep Up One's Sleeve, ca. 340), by the ardent Daoist Ge Hong. Ge Hong was also the compiler of the early alchemical and dietetic work *Bao Pu Zi* (The Work of Master Puzi).

Sun Simiao (also called Sun Simo), in 652, published the *Qianjin Yao Fang* (Remedies Worth their Weight in Gold) and its sequel the *Qianjin Yi Fang* (More Remedies Worth their Weight in Gold). These two volumes were extremely diverse, and together they summarized the medical achievements of the Tang, including practices as diverse as incantation, prayer, love philtres, exorcism, dietary restraints, as well as acupuncture and herbal remedies. They also contain the first known charts of the channels and points of the human body, with front, side and back views. Sun Simiao's work also contains the early *Yinhai Jingwei* (A Detailed Study of the Silver Sea) which describes eighty-one eye conditions and their treatment, although this work has also been attributed to an author from the Yuan dynasty.

During this period, the earliest official pharmacopoeia ever to be published was sponsored by the Tang government, based upon the work of the earlier Shennong Herbal. The *Tang Xinxiu Bencao* (Tang Newly Compiled Materia Medica) was complied by Su Jing along with twenty two other scholars, and it lists some 844 medical substances.

In AD 1027, at the establishment of the Song dynasty, two bronze statues were cast by edict of the Emperor, fashioned by a doctor, Wang Weiyi, who incorporated into them all the known acupuncture points. These statues were accompanied by his *Tongren Zhenjiu Shuxue Tu Jing* (Illustrated Canon of Acupuncture Points based upon the Bronze Figures, 1026). This text is still popular and used for point location in acupuncture colleges throughout the world. Also around this time appeared the earliest work on bone-setting, the *Lishang Suduan Mifang* (Secrets of Treating Wounds and Bone-setting, ca. 946), put together by Lin Daoren. This outlined the use of traction, fixation, reduction and reunion for fractures and dislocations.

The fixed attitude to traditional knowledge began to subside gradually during this period. With greater political stability and urbanization, medicine became more intensive and specialized. The number of medical publications during the Song exceeded those of all previous ages put together and it is largely the Song view, and Song editions of texts which are now extant. The work begun by Lin Yi, with his editing and reprinting between 1068 and 1077, produced the classical editions most in use today. This was also the century which initiated the massive Daoist patrology, the *Dao Zang* (Storehouse of the Dao).

The monumental *Shengji Zonglu* (Imperial Encyclopedia of Medicine) complied by a board of physicians under the emperor around AD 1111, was composed of 200 volumes. Other specialized works include the small pamphlet the *Yanglao Fengjin Shu* (Looking After the Aged, early eleventh century) by Chen Zhi, on the care and feeding of old people, and the famous monograph on beri-beri, the *Jiaoqi Chifa Zongyao* (Every Essential on the Treatment of Beri-beri, 1078) by Tong Zhi.

During this time, another influential book was the *Sanyin Jiyi Fang Lun* (Discourse on the Three Causes Ultimately for Any One Disease, 1174) by Zhen Yan. In this the cause of disease was seen as belonging to one of three categories either internal (including emotional disturbances), external (climatic change), or neither (malnutrition, over-feeding, animal or insect bites and stings, wounds, hoarseness through shouting, drowning, etc.). There was also the gynecological *Furen Da Quan Liang Fang* (Collection of Excellent Prescriptions for Women, AD 1237) by Chen Ziming, consisting of 260 articles in eight divisions, and numerous books on children's illnesses and surgery. One of the earliest was the *Xiaoer Yaozheng Zhijue* (Treatise on Pediatric Pharmaceutics, 1119) by Qian Yi. This small book exerted a profound influence on pediatrics, giving valuable insights into measles, scarlet fever, chickenpox and smallpox, along with innovative methods of diagnosis and treatment. Also, in 1461, the doctor Gou Bing produced the *Quan You Six Jiao* (Directions for Those Working in Pediatrics), in which indigestion or wrong feeding was held to be the probable cause of most ailments. The first treatise on forensic medicine also was published at this time, the *Xiyuan Jilu* (Instructions to Coroners, 1247), written by Song Zi. This too proved to be a foundation text.

Around this time medical colleges became well organized, and this setting of academic boundaries resulted in differing schools of thought. One considered disease to be caused by excessive heat in the body and advocated cooling medicines; another emphasized the use of purgatives and emetics; another extolled the use of tonics. The most influential book surviving from these debates was probably the *Piwei Lun* (Treatise on the Stomach and Spleen, ca. 1230) by the famous Li Dongyuan, who put all disease down to disorders in life style and the digestive tract. Also notable is the *Rumen Shiqin* (A Literati's Dutiful Care of his Parents) by Zhang Zihe (1156–1228), which advocated the use of sweating, emesis and purgation.

Arriving at the Ming dynasty (1368–1644) we find scholarship bearing remarkable results. *The Bencao Gangmu* (A Materia Medica Compendium, 1590) by Li Shizhen (1518–1593), is one of the most important works ever produced in China. It was a gigantic text in fifty-two volumes, listing nearly two thousand medical substances, including plants, minerals, and animal products, with over ten thousand prescriptions. It detailed the appearance, properties, methods of collection, preparation and function of each substance. This was a truly encyclopedic work commenting on all branches of natural history, botany, zoology, mineralogy, and metallurgy. Mention should also be made of its forerunner, the *Puji Fang* (Prescriptions for Universal Relief, 1406), produced by Teng Hong, which contained an astonishing sixty one thousand prescriptions.

Great textbooks on acupuncture also appeared during the Ming. Again these were truly comprehensive, reproducing the best of the old along with reworkings and selections from the new. Most notable are the *Zhenjiu Daquan* (Acupuncture and Moxibustion in its Grand Entirety, 1439) by Xu Feng, the skillfully compiled *Zhenjiu Juying* (Gatherings from Outstanding Acupuncturists, 1537) by Gao Wu, and the last in this tradition, the monmumental *Zhenjiu Dacheng* (Acupuncture and Moxibustion, the Grand Compendium), written by Yang Jizhou in 1601 — a book "still of the highest usefulness today" (Needham, 1980).

Mention should also be made of the reordering of the *Huangdi Neijing*, the *Lei Jing* (Classified Classic) by Zhang Jiebin, which appeared in 1624; with this work, and its illustrated appendices, the summit of Neijing scholarship had been reached.

During the later centuries little work of influence appeared. An exception is the *Yizong Jinjian* (Golden Mirror of Medicine, 1749) by Wu Qian. It is mostly made up of extracts, revisions and corrections of earlier writings. A government decree in 1822 actually eliminated acupuncture from the medical curriculum. And, although China's indigenous medicine survived, it met with disfavor from the established government until, as late as 1929, the Guomindang banned traditional Chinese medicine altogether. However, that same year, Mao Zedong wrote that both Chinese and Western medicines should be used to serve the population, and this has been the prevailing view ever since.

Since the founding of the People's Republic (1949) several medical colleges have been established, combining modern Western and traditional Chinese medicine, and an increasing number of texts have been re-edited and published. Many new areas of research have been opened up: acupuncture anesthesia, ear-needling, the discovery of new points, and the translation of texts has begun. Finally, to appreciate the resilience and accuracy of Chinese medical texts we can compare two extracts. One is from the first page of the first acupuncture book ever produced for the West (Beijing, 1980):

> The theory of Yin and Yang holds that every object or phenomenon in the universe consists of two opposite aspects, namely, Yin and Yang, which are at once in conflict and in interdependence; further, that this relation between Yin and Yang is the universal law of the material world, the principle and source of the existence of myriads of things, and the root cause for the flourishing and perishing of things.

The second is from the *Huangdi Neijing* (ca. 100 BC):

> Yin and Yang are the grand method of heaven and earth, the rule and pattern of the ten-thousand things, the father and mother of change and transformation, the fundamental origin of living and killing... (Ch.5)

The continuously resilient nature of Chinese medical texts could not be more clearly shown.

RICHARD BERTSCHINGER

REFERENCES

Bensoussan. A. *The Vital Meridian.* Edinburgh: Churchill Livingstone, 1992.

Concise Chinese-English Dictionary of Medicine. Beijing: People's Medical Publishing House, 1982.

Essentials of Chinese Acupuncture. Beijing: Foreign Languages Press, 1980.

Fu, Weikang. *The Story of Chinese Acupuncture and Moxibustion.* Beijing: Foreign Languages Press, 1975.

Fu, Weikang. *Traditional Chinese Medicine and Pharmacology.* Beijing: Foreign Languages Press, 1985.

Kaptchuk, Ted, J. *Chinese Medicine, The Web That Has No Weaver.* London: Rider, 1983.

Lu, Gweidjen and Needham, Joseph. *Celestial Lancets.* Cambridge: Cambridge University Press, 1980.

O'Conner, John and Dan Bensky, trans. *Acupuncture: A Comprehensive Text.* Chicago: Eastland Press, 1981.

Qiu, Mao-liang. *Chinese Acupuncture and Moxibustion.* Edinburgh: Churchill Livingstone, 1993.

Unschuld, Paul U. *Medicine in China.* California: University of California Press, 1985.

Ware, James R., trans. *Alchemy, Medicine and Religion in the China of AD 320 (The Bao Pu Zi).* New York: Dover, 1981.

Zhongyi Renwu Cidian, (A Dictionary of Personages in Traditional Chinese Medicine). Shanghai: Shanghai Book Publishing Company, 1988.

See also: *Huangdi Neijing* – Sun Simo – Zhang Zhongjing – Li Shizhen

MEDICINAL FOOD PLANTS The connection between diet and health is recognized by all human societies, and for many, healthful eating is a pivotal construct of their medical ideologies. How central a role food plays in disease prevention and therapeutics ranges along a continuum, one pole of which represents such indefinite notions as "healthy foods". This is a feature of all cuisines, although which particular foods are so regarded varies with both place and time. Toward the other end of the continuum, more specific explanations of the healthful nature of foods are based both on physical characteristics or effects and on abstract qualities.

Among the abstractions that people apply in their assessment of medicinal foods are paradigms of binary opposition such as heating–cooling and wet–dry — representations of intangible qualities that are not related to actual thermal or hydrous states. These are expressions of health conceptualized as a balance between certain key qualities that can be mediated by diet. In China, for example, cooling foods such as carrot and seaweed restore equilibrium in the case of rash, constipation, and other hot disorders. This general kind of food therapy also characterizes the medical traditions of India, Southeast Asia, Latin America, the Near East, parts of Africa, and much of Europe — although the attributes ascribed to particular foods and diseases vary across, and even within, cultures. In China these binary oppositions intercalate with older traditions that symbolize the more cosmic *yinyang* philosophy, take into account the theory of Five Phases (*wuxing*: earth, metal, fire, wood, water), and address the concept of *pu* (strengthening, patching up). These statements oversimplify, but they make the point that the healthful qualities of foods are manifold and are integrated in complex ways into different food cultures.

Regarding the physical characteristics of medicinal foods, the cultural dicta by which people interpret the appearance, taste/smell, or physical effects of foods are as varied as the religious beliefs, languages, marriage customs etc. that distinguish societies from one another. A common link between food plants and diseases is the taste that defines their curing properties. Consider this complex example. The medical tradition of a Hausa population in northern Nigeria advises that the treatment of measles be aimed at several stages, beginning with efforts to expel disease substance from the body's interior. As soon as the measles spots appear, bitter and astringent foods and medicines are sought to chase out internal sores so that the rash matures. Foods and medicines that occur later in the therapeutic progression are cold and aromatic because the illness likes those qualities. In the event that nausea and fever accompany the rash acid/sour tastes are indicated — to calm and cool. When there is evidence that all sores have been externalized (that is, when other internal signs such as fever and lymphatic inflammation have subsided), the rash is treated with astringent and emollient medicines, no longer with medicinal foods. In this example, taste features prominently in the selection of medicinal foods

Among another physical attributes by which people identify foods for medicinal use, color, location, texture, and shape may be evoked as part of a "doctrine of signatures". This fundamental tenet in plant selection maintains that these attributes are signs of a plant's intended use. Some of these associations conform as well to theories of sympathetic magic, which is based on the idea that like affects like. Thus, for example, in U.S. Colonial and Chinese medical traditions, the shape of walnut kernels suggested their effectiveness as a "brain tonic"; in Taiwan red foods such as tomato, red crab, and carrot are considered nutritious because red symbolizes prosperity and good fortune; in several

Southeast Asian cultures, where red protects against spirits, red eggs are used to ensure fertility; and in Ecuador the red flowers of the amaranth plant are made into a beverage to restore menstrual regularity and purify the blood. That other medical traditions identify plants with red flowers or leaves to fortify blood finds a parallel in the U.S., some decades ago, when "good red meat" (beef) defined the core of an "all American", healthy cuisine. Similarly, yellow plant foods are designated for such liver disorders as jaundice in which the skin and whites of the eye take on a yellowish tone, and plants whose leaves are shaped by several lobes are used in the treatment of lung disorders. These associations are not as simple as outlined here, because although appearance and other physical attributes are presented as the primary distinguishing feature of certain medicinal foods, they are understood to protect or interfere with disease processes in variously complicated ways that have to do with other characteristics that different medical cultures describe through such metaphors as "strength", "neutralization", and the like.

By the same token, foods also have negative health connotations, many of which have been systematically codified as food taboos, the symbolism of which is especially rich. Food prohibitions play a pivotal role in the traditional medicine of Thailand where five general taboo units subsume: meats — cow, water buffalo, pig, and chicken; frogs and scaleless fish; seafood; pickled food; and certain vegetables — cucumber, gourd, jackfruit. Further, specific foods and beverages are prohibited in particular conditions, and specific tastes such as sweet, fat, and sour are additionally proscribed in certain instances — for example, sour is discouraged because it is understood to result in the accumulation of pus, which should be avoided when one is treated for wounds. Similarly, for the Hausa in Nigeria, food plants (e.g. okra, cat's whiskers) which are otherwise appreciated for the gelatinous quality they impart to soup, are avoided when one has congestion from flu, and in the case of certain backaches that are understood to be caused by an accumulation of phlegm within the body. For altogether different associations, the Hausa consider that peanuts aggravate leprosy and offset the effect of medicines for anemia.

The cultural and social aspects of medicinal foods are especially apparent in circumstances in which their consumption extends beyond the sick person to include others. Among the Nekematigi of New Guinea, for instance, ceremonial therapeutic meals are shared by the entire community and consist of "all purpose" medicinal foods — ginger, for example — that are intrinsically healthful regardless of the cause of illness, and specific "antidotes" against the acts of enemy sorcery that accounts for most sickness in this society and that always originates from outside the village. If the illness persists, the meal is repeated, again involving the community and thus reaffirming solidarity in the face of malevolence introduced from outside the social body. Conversely, for Nigerian Hausa, whereas healthy foods such as fish and peanuts are shared among family members who eat from the same pot, specific medicinal foods are consumed only by the sick person. In other medical traditions the healer and patient consume the medicinal foods, usually at the same time, both to cement the therapeutic relationship and to empower the healer's empathy toward the patient. Since people everywhere recognize the relationship between a mother's health and that of her nursing infant, one is not surprised to find, as among Hausa, that when a baby is ill, medicinal foods are also consumed by the mother, and exclusively so if the infant has not yet begun to eat solid foods. And on the Western Pacific island Espiritu Santo (Vanuatu, formerly New Hebrides), food taboos for children who have become ill as a result of malevolent magic are extended to the parents. The general point made here is that the consumption of medicinal foods occurs in a variety of social contexts that range cross-culturally along a continuum of inclusiveness with respect to patients, their families, and their community.

The interpretation of these cultural expressions by social and health scientists has largely replicated a central idiom in Western thought — i.e. that food and medicine are separable categories. This is apparent, for example, in studies that inventory medicinal or food plants, overlooking the possibility that plants are not simply one or the other. In this way, whether or not a food is considered healthful depends on its intrinsic nutritional value. For the first half of the twentieth century nutrition scientists concentrated on identifying the nutrients that help the body's immune system fight infectious disease and whose deficiencies result in such discrete disorders as scurvy (vitamin C), rickets and osteomalacia (calcium, vitamin D), and beriberi (thiamin). (This can be related to the Chinese example above through observation that insufficiency of the vitamins found in carrots and other cooling vegetables typically eventuate in skin disorders.)

During the latter 1900s nutritional scientists focus on foods linked to chronic disorders, some revealing negative associations: foods high in salt, sugar, and fat have been implicated in hypertension, diabetes, and cardiovascular disorders respectively. Conversely, other nutrients have healthful effects on chronic diseases: dietary fiber — such as provided by whole-grain cereals, fruit skins, vegetables, legumes, and nuts — has been linked to diminished risk of colorectal and stomach cancers, high blood pressure, diabetes, high cholesterol, atherosclerosis, and coronary heart disease. Vitamin A — present in yellow and orange fruits and vegetables (carrots, bell pepper) and dark green leafy vegetables (spinach, kale) — has been reported to protect against lung and prostate cancers and to decrease the severity of measles. Vitamin E (found mostly in green leafy vegetables) and C (high in fruit, especially citrus) both have been linked to lower cancer risks as well.

In sum, to date the frame of reference has been the nutritional composition of medicinal foods; and apart from the conventional nutrients — vitamins, minerals, protein — foods have been regarded as chemically inert, thus of no salience to specific disease processes.

But foods are not chemically mundane. Most people are aware of toxins in potentially nutritious foods, so why not drugs? The line between toxin and medicine is in any case an uncertain one — toxic in what dose, and for whom? Many of the toxic "secondary" metabolites (allelochemicals) produced by plants are a primary defense against fungi and bacteria — microorganisms not unlike the pathogens with which human societies have to contend. Accordingly, people have taken advantage of these characteristics of plants, in some cases modifying that chemistry through selective breeding and special planting of food plants to exaggerate certain features and to reduce or eliminate others.

In view of this, a more recent development in medicinal food study is to focus on the drug-like qualities of plants. Attention has been paid especially to spices, perhaps because their role in the early history of Western medicine has made this category of plants the more likely subject of pharmacologic inquiry. In a biomedicine so strongly informed by the "germ theory of disease", the antimicrobial action of many spices merits attention; and because spices — like pharmaceuticals — are "small" (in quantity) and "powerful" (in smell, taste), they fit an allopathic model of healing. Evidence for pharmacologic activity in spices includes: capsaicin in chili pepper diminishes risk of cluster headaches and stomach cancer; sulfides in garlic and onion inhibit blood clotting and promote cardiovascular integrity; West African black pepper has anticonvulsant activity; ginger is effective in the treatment of motion sickness and nausea; cricetine in saffron has anti-atherosclerotic activity; vanillin and catechin in vanilla have, respectively, liver protective and anticariogenic actions; and caffeine, theophylline, and theobromine in chocolate elevate mood.

More recently attention has been directed as well to more ordinary foods: sulforaphane in broccoli protects against certain cancers; gamma-amino butyric acid in tomatoes is hypotensive; rhubarb has antibacterial activity; epicatechin gallate from tea leaves lowers serum cholesterol; and pigeon pea is helpful in diminishing the symptoms of sickle cell anemia.

These catalogs offer only an incomplete story of the medicinal potential of food; since much of what is reported is based only on laboratory studies of purified substances and/or on animal studies using healthy subjects. Thus, although some medicine-like action may be confirmed, one cannot be certain that this will indeed be the outcome when the plants are ingested by sick humans. One gains more insight by considering food use in its broadest physiological and cultural contexts, in order to take into account preparation and incidence and quantity of consumption.

In many traditional societies the same plants are used as both food and medicine, their function and preparation depending on how they are identified for a specified context of use. Over the last two decades, anthropologists have been especially adroit in detailing how foods and medicines overlap. This is significant on at least two levels. First we recognize that the use of plants in more than one context extends the range across which people are exposed to active constituents. Second, we begin to appreciate how very complex human–environment interactions are. For the latter it signifies that foods tend to be prepared in ways that diminish the risk of toxicity, especially cooking, thereby allowing the consumption of otherwise potentially dangerous foods in relatively large volume and at regular intervals. The same plants intended for drug use may be prepared differently — neutralizing pharmacologically active constituents only partially or not at all — and are consumed in small quantities, allowing the action of toxin against pathogen or symptom without overwhelming the human host.

Hausa peoples of northern Nigeria are a case in point that food plants overlap conspicuously with medicines. Ninety-six per cent of the 119 plants that one Hausa village identify as foods number as well among the 374 plants that make up the local pharmacopoeia. Hausa clearly distinguish foods from medicines by parts used, preparation, and, especially, intended outcome (nutritive or preventive/therapeutic). Further, Hausa have much more to say regarding the healthful qualities of a medicine compared to a food. While medicines address specific symptoms, dislodge phlegm, expel spirits, and so on, foods simply strengthen, fortify the blood, or promote growth. (Extending overlapping use beyond foods and medicines reveals that all of the twenty plants used cosmetically have medicinal uses, five are foods as well; of the sixteen plants used in personal hygiene, all are used as medicines, six are foods, and three are cosmetics.) The point again, is that attention to the different contexts of plant use accounts most fully for the range of people's exposure to the drug-like actions of botanicals.

An interesting question is raised in the rank ordering of use: medicine first or food first? These are not, of course mutually exclusive positions; pharmacopoeias and cuisines are created in increments, not all at once, so that one pattern or another may apply for a given plant. A sizeable number of plants used today as food were first appreciated for their medicinal qualities: throughout the Andes and

Mesoamerica the ritual and medicinal uses of red amaranth pigments were once more common than the use of these plants as a source of food grain. In the Mediterranean and Near East licorice was domesticated for medicinal use (it has estrogenic and antimicrobial activities), as were a variety of other plants that are used today as flavorings. Soybeans and *Mo-er* (black tree fungus) of Chinese cuisines were first cultivated as medicines. The latter demonstrates the kind of antiplatelet activity that might provide protection against atherosclerosis. In light of the especially uncertain distinction between medicine and food in Asian cultures, other plants also are likely to have been cultivated first as medicine. Drawing attention again to Hausa in Nigeria, a significant number of their wild plant foods have been "discovered" first as medicines. This is revealed by the inventory of Hausa medicinal plants, which is considerably more inclusive of local flora than lists of Hausa foods and other useful plants; the list of medicines includes all plants from those other categories, whereas the reverse is not the case. Hausa oral tradition further reinforces the primacy of medicinal use over food use for some plants. This perspective cannot be applied wholesale to all societies, since the cultural and environmental circumstances under which pharmacopoeias and cuisines are created involve different patterns of need and knowledge about particular plants. But this does challenge a conventional, unidirectional view that people learn about medicines only secondary to their search for food.

A comprehensive perspective on plant use better characterizes the interconnected histories of human food and medicine, and accommodates both "nonfood first" and "food first" models to explain the development of botanical knowledge. Thus, whereas it is customary to talk about the "traditional" medicines and foods of a people, region, or religion, in fact therapeutics and diet are dynamic. Transformations in the content and intent of medicines and diet are shaped in part by shifting concerns with health.

Finally, a timely issue relates the topic of medicinal foods to concern with environmental protection. Contemporary debates on the preservation of global biodiversity have focused on plants that development planners and national policies identify to be at risk. This cross cuts other international efforts directed at alleviating or preventing the effects of famines and lesser food shortages. Collectively these programs highlight staple foods in their efforts to assure that "important" and varied foods are available to the world's peoples, or that other "interesting" plants are preserved for their potential contribution to the pharmaceutical industry of the west. Thus, Western scientific paradigms and various economic and political agendas have been especially instrumental in shaping the direction of environmental protection and restoration efforts. These programs should be credited with the results achieved to date, as well as with their more recent inclusion of local populations in the planning of sustained efforts. In light of the present discussion of medicinal foods, one might wish to add certain refinements that assure that local cognitive categories and the specific contexts of plant use are given attention, including an examination of what the existing or restored diversity affords in human cultural terms. More specifically, to the extent that the same plants serve dietary and medicinal objectives, their significance to local populations is greater than what development planners might have considered; additional contexts of use elevate the local value of those plants even further.

NINA L. ETKIN

REFERENCES

Anderson, E.N. *The Food of China*. New Haven: Yale University Press, 1988.

Brunn, Viggo, and Trond Schumacher. *Traditional Herbal Medicine in Northern Thailand*. Berkeley: University of California Press, 1987.

Etkin, Nina L., ed. *Plants in Indigenous Medicine and Diet: Biobehavioral Approaches*. New York: Gordon and Breach Science Publishers (Redgrave), 1986.

Etkin, Nina L., ed. *Eating on the Wild Side: The Pharmacologic, Ecologic, and Social Implications of Using Noncultigens*. Tucson: University of Arizona Press, 1994.

Etkin, Nina L., and Paul J. Ross. "Should We Set a Place for Diet in Ethnopharmacology?" *Journal of Ethnopharmacology* 32:25–36.

Johns, Timothy. *With Bitter Herbs They Shall Eat It: Chemical Ecology and the Origins of Human Diet and Medicine*. Tucson: University of Arizona Press, 1990.

National Research Council. *Diet and Health: Implications for Reducing Chronic Disease Risk*. Washington, DC: National Academy Press, 1989.

Simoons, Frederick J. *Food in China: A Cultural and Historical Inquiry*. Boca Raton, Florida: CRC Press, 1991.

See also: Medicine – *Yinyang* – Five Phases – Ethnobotany

MEDICINE IN AFRICA Traditional medicine has been an integral part of African culture since time immemorial. African traditional medicine represents the sum of the peoples' medicinal knowledge as well as beliefs, skills, and practices used in diagnosing, preventing, or eliminating a physical, mental, or social disequilibrium. It is based on practical experiences and observations which have been handed down from one generation to another. Like other aspects of African culture, traditional medicine is dynamic and has, over the years, absorbed new concepts and treatment techniques and adjusted to a number of external challenges.

The traditional healing methods and health practices existing in Africa today are, therefore, the product of various cultural heritages which have been profoundly influenced by the interplay of a complex of social, economic, political, and biomedical factors.

Perhaps no aspect of African culture has received so much criticism from Western-oriented minds as the practice of traditional medicine. The fact that traditional medicine has nurtured and sustained Africans for years was blatantly ignored by Europeans when they arrived on the African continent. Prior to European settlement and contact, traditional medicine exerted powerful influence on the conduct of public and private affairs of various communities in Africa. But early European missionaries and colonial administrations suppressed traditional medicine, condemning it as satanic, primitive, and unscientific. Such an attitude, reinforced by the policy of Westernization of the medical care of the people, threats, and other forms of intimidation forced most traditional medical practitioners into clandestine practices and caused many to abandon their practices. The neglect of African traditional medicine as an important health asset has considerably contributed to the underdevelopment of indigenous pharmacopoeias and the loss of healers whose medical knowledge could have been tapped, improved, and developed. Currently, the greatest challenge to traditional medicine in Africa comes from Western-trained physicians, many of whom continue to treat traditional medical systems with disapproval and disdain. Many educated Africans also shy away from any contact with native healers in the open, though they clandestinely seek the services of healers for chronic and serious diseased states.

Today, African traditional medicine co-exists with the Western-model biomedical system. The majority of Africans still depend totally or partially on traditional healing methods. It is estimated that between 70 and 80% of the African population, especially those in rural areas, depend on the more readily available and relatively cheaper facilities of traditional medicine (Iwu, 1993). Even in urban communities, traditional medical practices are thriving, and most Africans still consult traditional healers, particularly for the more serious and/or chronic illnesses that defy Western hospital treatment.

Types of Traditional Healers

African traditional medicine is not a unified system. Several types of therapeutic systems are available, each with a distinctive approach to diagnosis and therapy. A fuller understanding requires, therefore, a recognition of several types of healing practices that exist, though in many aspects their therapeutic modalities may overlap. African traditional medical practices can be broadly categorized into those which are essentially "secular", employing skills passed on from generation to generation, and those which are "sacred", i.e. those which involve spiritual models of healing. Under these broad (but sometimes overlapping) categories a variety of healers exists. They include: priest-healers, faith-healers, cultists, herbalists, bone-setters, and traditional midwives. Charles Good has identified eleven distinct categories of medical and medico-religious skills in the Kilungu Hills in Kenya, while seven primary groups of healers who provide health care have been identified in the Kano State of Nigeria. They range from *mallam* (Quranic scholar) to *sarguwa* (midwife) (Stock,1981).

One of the dominant groups of healers in Africa is priest-healers (i.e. priests and priestesses). They become 'possessed' and employ techniques such as divination and ritual manipulation. They serve local deities or gods with whom they communicate while in a state of possession or by means of divining instruments which enable them to "see" or "hear". They can cure both organic and spiritually-caused diseases. The priests and priestesses are normally "called" or are coerced into the healing profession by an 'ancestor-sent illness'. Submission to the calling is regarded as the only way to survive or to find relief from the illness. Priest-healers undergo long apprenticeships because they have to learn how to harness a spiritual force as well as use herbal medicines for all kinds of somatic and spiritual health problems.

In addition to priest-healers, there is another group of religious-medical healers called faith-healers who claim that their powers come from God and support their claim and activities with the life, teaching, and works of Jesus Christ. A faith-healer is usually a head of an "Independent African Church". Some faith-healers, as they claim, become spiritualists through the divine inspirations received through dreams and "calls" from the Holy Spirit. Some study under established Spiritual Church leaders before founding their own Churches. The counterpart of faith-healers is another group of religious-medical practitioners generally known as *mallami* who attribute their healing prowess to Allah and dispense both herbal and prophetic medicines. Cultists are another example of sacred healers whose paraphernalia for healing range from simple Quranic charms to complex conglomerates of materia mystica. Some of the cultists exist for specific purposes: for example, to prevent theft, guard against cutlass wounds, or cure all kinds of illnesses. Among the Akan in Ghana, for example, the Tigare cult is well known for its apprehension of witches. Some cult-healers become possessed, though the element of possession is not a universal characteristic of all the healers who employ cults in their practice.

Among secular healers, the herbalist, sometimes referred

to as a traditional pharmacist, is the most visible in Africa. The practice of herbalism may be inherited from parents; it may be motivated by the desire for herbal knowledge and treatment. Some obtain the know-how through dreams or visions; others are motivated by a desire to make a living from the sale of herbal preparations. Herbalists are, therefore, a mixed group of traditional health practitioners without any standardized training. While some do receive long periods of apprenticeship, through direct instruction, observation, and collection of herbs and other ingredients, many of them have little or no formal training.

Other secular healers include traditional midwives who diagnose pregancy and provide a wide variety of care and advice relating to the period before and after birth. Traditional midwifery is usually not a full-time occupation. Another highly developed aspect of African traditional medicine is orthopedic surgery or bone-setting. Over the years, the art of repairing fractures and other orthopedic injuries has attained a level of success comparable to Western methods of orthopedic treatment. Traditional bone-setters are often capable of arresting the deterioration of gangrenous limbs that would normally require amputation in a modern hospital. In addition to bone-setting, there is the existence of such forms of surgery as tribal marks, clitoridectomy (a practice that is increasingly facing disapproval), removal of whitlow, male circumcision, and teeth extraction. A common type of a secular healer is also a plant drug peddler who moves from place to place selling diverse herbal preparations, most of which are prepared by renowned professional healers.

In spite of the heterogeneity of healers, several basic concepts, principles, and methods are common to the practice of traditional medicine in Africa. At the same time, personal, ethnic, and regional differences create unique place-to-place expressions of the art of healing. Traditional healers, especially the religious-medical specialists, tend to be highly individualistic in their use of cultural symbols and supporting paraphernalia, in their style of diagnosis, in recommendations regarding therapy, and in their dosage of remedies. Some healers focus upon one or a few illnesses, while others are generalists. Some are full-time practitioners, while others practice medicine as a secondary occupation. The level of popularity of healers also varies immensely. While some are frowned upon, others are very influential and nationally reputed and command much respect in their localities.

African traditional healers lost status and power during the colonial era. Even though not much of the past glory has been recouped since independence, the traditional healer still occupies a unique position in African society. The functions of a sacred healer, for example, are not limited to the diagnosis of diseases and the prescription of drugs; it is the sacred healer who provides the needed answers to the adversities imposed on a community by outside forces (such as curses, evil spirits, aggrieved ancestors, or witches) that are beyond the comprehension of ordinary people. It is the sacred healer who provides charms and prescribes the rituals to neutralize the effects of the enemy's charms, wards off evil spirits, and intercedes between the community to appease the gods and spirits of deceased ancestors. The sacred healer is consulted for advice on misfortunes and the intractable problems encountered in ordinary life. In addition to treating sick people, requests made to priest healers by patients usually include: personal protection, assistance in new enterprises, discovery of missing or stolen properties, increased financing, assistance in employment, and aid in romantic matters. Traditional medicine survives today because it is a culturally understood approach to managing a wide variety of physical and social ills. It is an integrated system of social control and curative procedure.

Disease Etiology, Diagnosis, and Treatment

One of the factors underlying the different diagnostic and treatment procedures within African traditional medicine is the different perceptions people have about the etiology of an illness. There is a false assumption that African medicine relies exclusively on magic, witchcraft, and necromancy. Illness, however, is considered by many Africans to be caused by either natural or spiritual forces or both. Several sacred healers, of course, question strongly any form of treatment that focuses on the organic diseased state and ignores the spiritual side of the disease. Most of them attribute sickness to supernatural forces including witchcraft, sorcery, ghosts, spirit disturbance, and the consequences of broken taboos and religious obligations. But this strong belief in spiritual causation of diseases and death somewhat blunts the empirical and objective diagnosis of organic ill health inherent in their practices. Herbalists, on the other hand, usually have a relatively rational empirical or naturalistic orientation to disease causation, and they are therefore not assisted in their diagnosis and treatment by a deity, spirit, or cult, and their healing procedure involves no form of divination or ritual manipulation.

The diagnosis of disease and illness is generally made on the basis of information derived through one or several procedures: observation, questioning regarding case history, and clinical examination, including pulsing and body temperature. A common diagnostic technique employed by sacred healers is divination, e.g. the casting of lots (using cowries, kola nuts, eggs, etc.), sand reading, and water- or mirror-gazing. Employment of one or a combination of these methods is dependent upon the etiological orientation and the characteristics of the different types of sacred healers.

The healers use medicines from local plants, minerals, or animal substances, prescribe exercises, incantations, dances, or music, and use sacrifices and purifying rituals of a religious nature — all to reflect the beliefs and attitudes of their community about what causes illness and how to prevent it. Faith-healers, like priest-healers, share a similar theory of ill health and disease causation. They heal through prayers, the laying on of hands, church services, immersion, fasting, incantations, and herbal medicines, as well as substantial elements of Western and Far East occultism. Several churches have "healing sanctuaries".

Inherent in all the different prognostic, diagnostic, and therapeutic techniques in traditional medicine employed is the concept of "holism", disease being viewed as a disequilibrium in the body as well as in the society. A multidimensional approach to treatment is therefore employed, based on the notion that all parts of the human body are interrelated and mutually affect each other and that the body is in constant interaction with, and also influenced by, the socio-cultural and physical environment. The relationship between healer and patient is usually close and intense.

African Pharmacopoeia

African medicinal agents consist of two groups: (1) those used for rituals, sacrifices, and other religious acts; and (2) those used as part of the process of treatment of diseases and the plant extracts, herbs, seeds, roots, leaves, juices, liquids, powders, bones, minerals, and other substances that are supposed to have organic effects directly on the patient. Wild animals and their by-products (hooves, skins, bones, feathers, tusks, etc.) form important ingredients used by sacred healers in the preparation of curative, preventive, and protective medicine. They use them to perform rituals and to invoke and appease gods and witches. In addition, a variety of wild animals forms an integral part of cultural and religious festivals and ceremonies, some of which seek to promote the good health of local people and communities. Faith-healers who conventionally have relied on the power of prayers, fasting, florida water, and spring water blessed as "Holy Water" for the treatment of patients, now do prescribe herbal medicines as well.

Herbalists also have a large armamentaria of herbal medicines administered in different forms. They depend primarily on the therapeutic qualities or the potency of medicine prepared from selected leaves, roots, or other parts of plants and animals, and minerals. The bark and roots are the most common plant parts employed; leaves, flowers, and fruits are also used in the preparation of remedies. Methods for the preparation of plant medicines include roasting, burning, soaking, pounding, and chewing the plant parts. Plant preparations may be taken orally, in liquid or in powder form, inhaled as a steam or smoke, snuffed, smeared on the body, and bathed in, or "injected" through incisions of superficial razor cuts. The varied methods used for the preparation of the remedies make standardization and reproducibility of results extremely difficult. The type and use of plant preparations are dictated by the nature of the illness and the plant part used.

African societies continue to experience tremendous changes as their ecological systems, economy, and socio-cultural life respond to the exigencies of development and population growth, as well as to the introduction of new forms of administrative systems, production techniques, and commercial activities. Several changes have occurred within the traditional medical system of Africa. There is widespread observation that traditional medicine is less and less attractive as a career and that traditional healers are not being and will not be turned out in sufficient numbers to make them viable major additions to health care services in Africa. There is a decline in the number of traditional midwives, herbalists, and cultists, because fewer new recruits are entering the healing occupation. The role of the traditional healer is being threatened by formal education, which is incompatible with traditional medicine as an occupational choice. But while these types of healers are declining in number, the same cannot be said about priest-healers and faith-healers.

Some studies have also shown that urban places in Africa are vigorous areas of traditional medical practices. In Kenya, Swaziland, Ghana, and other countries, the urban setting represents an area of growth for traditional medicine. The relative increase in the number of herbalists in urban areas is attributed in part to the high rate of demand for their services. In the city, it is often easier in terms of time, energy, and cost to resort to herbal treatment for slight organic malfunctions than to spend hours queueing for bio-medical treatment only to be prescribed with medicine which is not available at the hospital pharmacy. The mushrooming in recent years of many Prophets and Prophetesses of Sectarian or Independent African Churches has also increased the number of healers in African cities. If the present large following of these Churches, and the claim of successes in satisfying their members and clientele in both organic and pscho-social problems are good indicators, then there is every likelihood that the number of faith-healers will increase.

In recent years, scientific research into the pharmacodynamic properties and therapeutic potentials of many plant medicines has been intensified in Africa. Although such activities may result in discoveries of great medical and social value, the fact that such research is separated from the behav-oral and religious context of African culture and therapeutics may do little to advance understanding of how African cultures interpret the functions of or use plant medicines.

Another significant change within the traditional medical system is that there is a marked increase in the sale of herbal remedies, which has precipitated large scale harvesting of medicinal plants, factory-like production of herbal drugs, and animal poaching. The harvesting of herbal medicine in some areas is on a scale that is causing concern among conservation organizations and some rural herbalists. A vigorous trade now exists not only among some African countries but also between African countries and such countries as Canada, the United States, Germany, and Hong Kong. Commoditization of plant medicine (unknown until quite recently) is contributing to environmental degradation.

It must be noted, however, that the practice of traditional medicine contributes, to some extent, to the preservation of nature. For example, the activities of some cult organizations, priest-healers, and other medico-religious practitioners ritually protect certain wilderness areas in their localities through the creation of shrines, grooves, and other sacred spaces. Also, the religious aspects and belief systems related to African traditional medical practices tend to instill some fear in local people about certain natural landscapes. For example, in Uganda, among some lowland communities, mountains are feared because they are believed to harbor supernatural spirits and also diseases brought by the spirits of the mountains. In such areas, mountain resources are consequently often left unexploited.

Integrating Traditional Medicine and Biomedicine

For years, the relationship between African traditional medicine and biomedicine has been one of mutual antagonism, distrust, and contempt, even though there are many areas of common ground that could be integrated to provide much needed health care. The role of traditional medical practitioners in the deployment of public health services has been kept ambiguous in Africa, though Health ministries in many countries have been stressing the need to rehabilitate and harness the practices of healers into the national health care delivery system.

Despite their fundamentally distinct paradigms regarding disease, illness, and health and the different systems of etiology, diagnosis, and treatment, traditionally based therapies and biomedicine are used concurrently or simultaneously by many people in Africa. Establishing cooperation between traditional medical and biomedical systems will be very beneficial to the people. Cooperation has as yet not materialized due to such factors as paradigm conflict, restrictive attitudes held by some physicians toward traditional healers, doubts about the effectiveness of traditional therapies, and the illliteracy of many healers. There is support, however, among international agencies for the integration of the two systems of health care. The World Health Organization in particular has been at the forefront of such efforts. There is also a small group of biomedical practitioners who have developed interest in traditional medicine and would like to work with tradititonal healers. A substantial number of healers also have expressed the desire to increase their skills and knowledge through cooperation and exchange with biomedical workers.

The future of traditional medical practice is likely to be influenced by whether healers will be able to join forces and fight for a professional recognition from the modern medical community and various African governments. For many years, traditional medicine has played a tremendous role in the health care and general well-being of Africans. It will continue to do so, as it would be unwise to suggest that the role of healers will be replaced by the improvement in the availability and accessibility of biomedical services in Africa. The nature and characteristics of African traditional medicine will, however, likely change.

CHARLES ANYINAM

REFERENCES

Ampofo, O. and F.D. Johnson-Romauld. *Traditional Medicine and Its Role in the Development of Health Services in Africa*. Brazzaville: Regional Office for Africa, World Health Organization (WHO), 1978.

Anyinam, C.A. *Persistence With Change: A Rural-Urban Study of Ethnomedical Practices in Contemporary Ghana*. Ph.D. Thesis, Queen's University, Kingston, Ontario, Canada, 1987.

Good, C.M. *Ethnomedical Systems in Africa: Patterns of Traditional Medicine in Rural and Urban Kenya*. New York: Guilford Press, 1987.

Iwu, M.M. *Handbook of African Medicinal Plants*. London: CRC Press, 1993.

Last, M. and C.L. Chavunduka. *The Professionalisation of African Medicine*. Manchester: Manchester University Press, 1986.

MEDICINE IN CHINA The phrase Chinese medicine has a dual implication. It refers both to the complete medical system prevalent in contemporary China and to traditional or indigenous medicine, or TCM for short. The former includes three medical systems: the traditional, the biomedical or Western medicine as the Chinese call it, and integrated medicine. This article is concerned with the latter one.

Traditional Chinese medicine combines fighting against disease, keeping fit, and seeking longevity. It was created by all nationalities of the Chinese people and is the synthesis of all their medical systems. For historical reasons, TCM has been applied exclusively to the indigenous medical system created by the Han nationality. Logically, TCM should also include Tibetan, Mongolian, Korean, and Uyghur medicine, among others. Unfortunately, this interpretation would go against the common understanding of TCM. Hence, in this article we also use TCM in its narrow sense, referring to the Han medical system only.

TCM has a history of at least three to four millennia. Archaeological findings reveal that the application of fire in the palaeolithic age not only brought warmth and cooked food, which was beneficial to health, but also resulted in the invention of moxibustion therapy. The Neolithic age also saw ancient Chinese people applying the stone knife and "needles" for treating some external diseases. These passed through a long process of evolution from stone to needles made of bamboo, wood, porcelain, bronze, and ultimately metal. Shining gold and silver needles over two thousand years old have been unearthed. The last kind of needle, in fact, is the basis for the invention of the unique Channel system, though other factors may also have contributed. In terms of materia medica, legend has it that Shen Nong tasted and tried all kinds of plant herbs and other remedies from natural sources, beginning at the period of agriculture, about 7000–5000 BC. People of other nationalities also discovered some effective remedies, such as wine from highland barley for stopping hemorrhages, which was used by Tibetan people, and Cistanche Salsa, Koumiss for nourishing the body, which the Mongolian people use.

Early in the Xia and Shang Dynasty (twenty-first to eleventh centuries BC) some three thousand years ago, some characters related to medicine were inscribed, on bone and tortoise shells, including *yi* (medicine), *bing* (disease), and up to several hundred archaic characters relevant to the healing art. As early as the Zhou Dynasty (eleventh century BC–475 BC), medicine in the imperial court was divided into four departments: internal medicine, ulcerative (external) medicine, dietetic therapy, and veterinary medicine. TCM had already applied the four diagnostic methods — inspection, auscultation and olfaction, interrogation, and palpation (looking, listening and smelling, asking, and feeling) — by this period. Of these, palpation is the most worthy of mention. Chinese ancient physicians may have been the earliest to apply the art of pulse taking for medical purposes. In the *Shi Ji* (Historical Record), compiled by Sima Qian (b. 145 BC), the Herodotus of China, in the Western Han Dynasty (206 BC–AD 24), it is recorded that the physician Bian Que of the Warring States Period (475 BC–221 BC) was the first one to apply pulse taking.

The most famous medical classics were compiled in the Pre-Qin period (before 221 BC) and completed around the Han Dynasty (206 BC–AD 220). Among them, the most important and extant ones are *Huangdi Neijing* (Yellow Emperor's Inner Canon), *Shennong Bencao Jing* (Divine Husbandry's Classic of Herbology), *Nan Jing* (Classic of Questioning), *Shanghan Lun* (On Diseases Due to Cold Evil) and *Jingui Yaolüe* (Synopsis of Golden Chamber). These lay the foundation for clinical science with definite treatment principles and diagnostics. A case record was first formulated with a fixed pattern by Chunyu Yi of the Western Han Dynasty.

Chinese pharmacology reveals some outstanding achievements at this period. Shennong's Classic of Herbology presents many specific effective remedies. It sets up the theoretical basis of drug use, as well as collection, preservation, compounding, and method of administration. Therapeutic effects of specific drugs, such as *rhei* for catharsis, coptis root for asthma, seaweed for goiter, mercury for scabies, and many other are mentioned. All these have since been proved by modern techniques as correct and scientific. The famous surgeon of the Later Han Dynasty (AD 25–220), Hua Tuo, first applied *mafei* powder as an anesthetic for some major operations, including abdominal surgery. As early as the third century AD, the *Mai Jing* (Classic of Sphygmology), written by Wan Shuhe, recorded 24 kinds of pulse, touching the issues concerning heart rate, rhythmicity, condition of blood flow, texture of the artery, and the nature of blood itself such as viscosity and hemorrheology. Later, this Classic spread to the Arabic countries, and it is not surprising to note that in the *Canon of Medicine* by Ibn Sīnā (980–1037), Arabic pulsology has a lot of content in common with Wang Shuhe's. The Classic of Sphygmology has been translated into several foreign languages.

From the second century AD, medical disciplines were professionalized. The following are worth mentioning: *Zhenjiu Jiayi Jing* (A B Classic of Acupuncture and Moxibustion) by Huangfu Mi, *Leigong Paozhi lun* (Master Lei on Drug Processing) by Lei Xiao, and *Liujuanzi Guiyi Fang* (Liu Juanzi's Recipes Bequeathed from a Ghost) by Gong Qingxuan, a textbook of surgery in the fifth century. Clinical medicine developed greatly during the period of the third to tenth centuries. Ge Hong (265–341) was an expert in clinical medicine. His work, *Zouhou Beiji Fang* (Handbook of Prescription for Emergency) contained discoveries on tsutsugamushi and smallpox, inventions for the treatment of hydrophobia by applying the brain tissue of the mad dog, and treatment of malaria by squeezing the juice of artemisia, resulting in the extraction of a new effective anti-malaria remedy, artemisinin, in modern biomedicine. Moreover, new medical techniques such as abdominal paracentesis, catheterization, first aid therapy for foreign bodies in the esophagus, and chiropractic were introduced.

The Sui-Tang Dynasties (618–907) also saw several major medical issues. The Tai Yi Shu (Imperial Academy of Medicine) and the *Xin Xiu Ben Cao* (Newly Revised Herbology), 659) are respectively recognized as the earliest medical university and pharmacopoeia. Several important medical works were compiled in this period, representing the upgrading of medical science. Some are Cao Yuanfang's *Zhubing Yuan Hou Lun* (On Pathegenesis and Manifestations of All Diseases), which is the first elaboration on etiology, pathology, pathogenesis, and semiology in China. Sun Simiao's *Beiji Qianjin Yaofang* (Essential Recipes Worth a Thousand Gold) contained a great thesaurus of valuable recipes for many diseases, some still in use today. Wang Tao's *Waitai Miyao* (Clandestine Essentials from an Imperial Library Curator) recorded many effective recipes. During the Song Dynasty in the tenth to thirteenth centuries, a Jiaozheng Yishu Ju (Bureau for Reviewing Medical Publications) and Huimin Heji Ju (Bureau of Compounding Remedies for Benevolence) were set up by the Imperial Court. What should be mentioned here is the casting of two life-sized bronze human models for acupuncture and moxibustion in the year 1026, on which the acupoints and channels were cast on their surfaces. This is not only a valuable and sophisticated work for appreciation, but also a skilled teaching model which greatly enhanced the development of the art of acupuncture and moxibustion.

During the Jin-Yuan Dynasties (1115–1360), four main academic medical schools were established, namely the Schools of Cold-favoring, Spleen-Stomach Benefiting, Drastic Attack, and *Yin*-nourishing. Each had its own emphasis, both theoretical and practical. This greatly advanced the academic development of Chinese medicine.

The academic standard of Chinese medicine was further upgraded in the Ming-Qing Dynasty (1368–1910). A new school, the Wenbing Xuepai (Seasonal Febrile Disease School) evolved from the traditional Shanghan Xuepai (School of Disease Due to Cold Evil). This new school was devoted to acute infections. It successfully tackled many infectious diseases such as B-encephalitis, acute viral hepatitis, influenza, and other viral diseases. Another outstanding contribution of Chinese medicine in this period was the invention of a humanpox inoculation from which Edward Jenner's smallpox vaccination drew its inspiration. Inoculation should share the merit with vaccination in the global campaign of fighting to eradicate smallpox. The world-famous naturalist Li Shizhen (1518–1593) contributed his pharmacological knowledge to opening a new era in the history of Chinese herbology. His rich knowledge on natural science aroused the interests of the evolutionist Charles Darwin, who indirectly cited many biological examples as evidence supporting his theory of evolution.

Beginning with the middle of the nineteenth century, Western medicine, so-called by the Chinese people since it was introduced by Western medical missionaries, came to China, resulting in the formation of three different academic factions: the biomedical, the traditional or indigenous, and the convergent schools.

Through the ages, Chinese medicine applied and created a series of unique theoretical systems and practical techniques. The following is a brief introduction.

The philosophical concept of *yinyang* is based on the observation of contradiction in nature. This was coordinated into a *yinyang* theory for explaining the law of changes. In the medical circle, people made use of this idea to interpret the complex relationship between upper and lower, inner and outer, the body and nature and society. The equilibrium and harmony between these two aspects within the body is essential to and the base of the body's normal activities and functions. Conversely, once the harmony is broken, disorders of the body will develop, thus affecting normal physiological activities. Physiologically, *yin* refers to those tangible structures and *yang* to invisible functions. Thus blood itself belongs to *yin* while its circulation function falls under the category of *yang*. These, in TCM terms, fall under the categories of the so-called blood and *qi*. *Yin* and *yang* are mutually dependent. Without *qi*, blood will be stagnant and become a pathological entity, while *qi* attaches itself to the blood as its place to stay in. Without blood, *qi* will be "homeless". These ideas have direct bearing on the theory of treatment. Hence, *yin* and *yang* are mutually rooted, interdependent, and inversive.

Ancient philosophy states that the whole universe is constructed with five basic kinds of materials: wood, fire, earth, metal, and water. Each element has its own characteristics. As a microcosm, the human body, which is comparable to the universe, is also made up of these five elements. Like the universe, all the organs, tissues, functions, and systems can be compared with and assigned to one of the five categories. For instance, under the wood category, we have liver (*yin* viscera), gallbladder (*yang* viscera), eyes (sensory organ), sinews or tendons (tissue), sour (taste), wind (climatic factor), spring (season), anger (emotional state), and green (color). In the meantime, we have heart, small intestine, tongue, vessel, bitter, hot, summer, joy, and red under the fire category; spleen, stomach, mouth, flesh, sweet, damp, long-summer (the last month of summer), ratiocination, and yellow under the earth category; lung, large intestine, nose, hair, pungent, dry, autumn, sorrow, and white under the metal category; and kidney, urinary bladder, ear, bone, salty, cold, winter, apprehension, and black under the water category. The five categories have a dynamic rather than a static relationship, the order being wood, fire, earth, metal, water, and wood. Each category can also be conquered or restricted by another category, the order being wood (restricted by) metal, fire, water, earth, wood. Hence we have the title Five Phases or *wuxing*.

Both the *yinyang* Principle and the Five Phases theory are applied clinically for directing and interpreting physiology, pathology, diagnostics, treatment, or even prognosis, and they were proposed and completed some two thousand years ago. The *yunqi* (activity of *qi*) or the *wu yun liu qi* (Five Activities and Six Climatic Factors) theory or hypothesis investigates the influence of astronomical, atmospheric, and climatic factors on the human body and the occurrence of diseases. By five activities, it refers to the cyclic activities or movements of the five phases, wood, fire, earth, metal, and water, within the four seasons, while the six climatic factors refer to wind, cold, damp, dry, hot, and fire. This theory estimates the law of disease occurrence and yearly changes of weather with astronomy and the calendar as its parameters. In general, the yearly weather changes are wind in spring, hot in summer, damp in longsummer, autumn dry, and winter cold. Thus liver diseases are apt to occur in spring, heart diseases in summer, spleen diseases in late-summer, lung diseases in autumn, and kidney diseases in winter. The theory stressed the relations between weather and disease which, though a bit mechanical and controversial, has something to do with chrono-medicine. This theory reached its zenith a thousand years ago.

The theory of visceral manifestations deals with the physiology and pathophysiology of the five *yin* viscera: the heart, kidney, spleen, lung, and liver; the six *yang* viscera: the small intestinc, large intestine, stomach, bladder, gall-bladder, and triple burner (pancreas?); as well as the extraordinary viscera: the brain, marrow, bone, vessels, and uterus. The *yin* viscera function for the storage of essence and spirit of the body, while the *yang* viscera are responsible for the digestion, transformation, and transportation of residual materials. It is claimed that there exist mutually dependent inhibition relations among the *yin-yin*, *yin-yang*, and *yang-yang* viscera. Visceral manifestations also involve other body substances, including blood, saliva, mucus, sputum, body fluid, as well as body functions such as *qi*, spiritual forces, and genetic functions. The totality of the above-mentioned contents forms the visceral manifestations.

Visceral manifestations are closely tied to the theory of channels. Channels, the passages and tracts for the circulation of blood and *qi*, connect the outer with the inner part of the body and branch repeatedly to form a network spread over the whole body. Thus, through its connection, the whole body forms an organic whole. There are altogether twelve main or orthodox channels, each with its own underlying viscera, and eight extraordinary channels and collaterals and capillary networks. When channels are affected, their functions change accordingly, manifesting signs and symptoms by which a correct diagnosis can be made. Regulation of their functions through various stimulations, such as acupuncture, moxibustion, massage, electricity, and percussing, yields therapeutic results.

TCM stresses the importance of recognition of etiological factors, which have a direct bearing on the treatment and prognosis of disease. Harmonious relationships among the *yinyang* viscera themselves, and between the body and its environment, are crucial to the health of the body. All diseases occur on the basis of the disturbance or breaking down of this harmony. Diseases used to occur when the orthodox *qi* (body resistance) was defeated by the heteropathy (pathogenic evil). Orthodox *qi* is responsible for the body's resistance, preventing it from contracting a disease. Pathogenic evils include exogenous (climatic), endogenous (emotional), and others (trauma, accidental injuries, dietetic, behavior, etc.). The relative force between body resistance and pathogenic factors determines the result of their struggle, either keeping healthy or falling ill. The aim of treatment is to support the body's resistance and remove pathogenic evils so as to keep one in good health.

There are four diagnostic methods. Inspection includes the spirit, complexion, form, and status of the patient. A tongue picture is also essential, including the texture of the tongue and its coating, the sense organs, and the condition of excreta. These provide much information about the condition of the internal viscera. Auscultation refers to hearing the patient's voice, including speaking, respiration, and smelling the odor from the patient's body and excreta. Interrogation refers to questioning the patient and other respondents about his present illness, past history, and family history. Palpation includes feeling the pulse and other parts of the body. It is sometimes said that TCM puts its main stress on pulse taking, or even relies solely on pulse taking. This is not true. TCM emphasizes that an overall diagnosis should have all four diagnostic methods interpreted comprehensively with equal importance, instead of relying on any single one method.

The determination of syndrome manifestation (*bian zheng*) is the kernel of TCM clinical science. It is the process of analyzing the information and materials obtained from the four diagnostic methods, differentiating the causes, nature, location, stage of disease, and the reciprocal condition between the body and the pathogens. The result of this process is the identification of the type of syndrome manifestation which is crucial to therapy. Long experience enables Chinese physicians to form an ensemble of methods, or determination of syndrome manifestations. There are many methods. The important ones are the *ba gang* (Eight Rubrics), applied for all kinds of diseases; the Triple Burner Method for seasona febrile infectious diseases, and the Six Channel method for diseases due to cold evil.

Among them, the Eight Rubrics method is the most important and universally applied one. Eight Rubrics denotes outer and inner (location of illness), cold and hot (nature of illness), depletion and repletion (reciprocal condition between the body resistance and pathogens) and *yin* and *yang*. Within the Eight Rubrics, the *yinyang* is the key couple Rubric that dominates the other three. Furthermore, a timely, correct determination of syndrome manifestations, especially the Eight Rubrics, is the key to a reasonable and satisfactory therapy. The conception of Eight Rubrics has taken shape since the Han Dynasty.

The basic principle for treatment is the exploration of the root of disease, on which various therapeutic methods are based. The uniqueness of this is its flexibility, which varies with the analysis of the condition. Different treating measures may be given to the same disease because of different conditions manifested. Conversely, the same therapeutic measures may be administered for different diseases because of their common manifestations. The major principles include the regulation of *yinyang*, supporting the body's resistance, and removing pathogenic factors. During the treatment, differentiation of the true nature and superficial manifestation, emergency and steadiness, severeness and mildness are of paramount importance.

Moreover, therapy and medication in TCM should be adjusted to the season, the place, and the different individual. First, changes of climate in different seasons exert definite effects on the body. During summer, the pores in the skin and neighboring tissues are open; in winter, they are contracted and closed. Hence, while treating cold disease due to wind and cold evils, therapy in the summer should not apply too many drugs of a pungent and warm nature, in order to avoid profuse sweating, or outer depletion would result. In winter, for the same cold disorders, pungent and warm drugs can be applied in substantial amount, in order to expel the pathogens through perspiration without the risk of outer depletion. Since cold is very common in the north, for cases of external, pathogenic disorders, one can use a heavy dose of pungent and warm drugs for dispersion of pathogens through perspiration. Applying the same principle to the southern part where weather is generally hot, only a light dose of pungent and warm drugs can be used, to avoid profuse perspiration. Thirdly, all patients vary in sex, age, and body constitution. Moreover, women have the added complications of childbearing, menstruation, and vaginal discharge, while children have tender and delicate visceral systems. These conditions should be taken into account and carefully considered when prescribing. For instance, with a patient who is sensitive to cold with a constitution of cold tendency, cool or cold drugs should be used with caution, and vice versa for those of hot tendency.

As to concrete treating measures, basically there are eight therapeutic methods: diaphoresis, emesis, catharsis, mediation, warming, clearing, removing and benefiting. All these methods are applied not only for drug therapy but also for non-drug therapies such as acupuncture, moxibustion, massage, and others.

In terms of Chinese materia medica, all the drugs applied are natural products, including those from the plant and animal kingdoms as well as minerals. Most of the pharmacological knowledge is derived from practice. Pharmacological theory is summarized, again on the basis of experience. The theory is also unique. It claims that the potential of the drug comes from its "nature", composed of four *qi* and five flavors. The four *qi* are cool, cold, warm, and hot, while the five flavors refer to sour, salty, sweet, bitter, and pungent. The theory also includes channel tropisms, the functions of ascending, descending, floating, and sinking, as well as the toxicity of drugs. The nature of drugs is relevant to the condition of the disease as determined by diagnosis through differentiation of syndrome manifestations within the Eight Rubrics. Antagonistic therapy such as cool or cold drugs for heat disease, and warm or heat drugs for cold disease, is commonly used. Drugs with ascending nature are applied for heat disease of "collapsed" nature, such as gastroptosis and the like, whereas drugs with descending nature are applied for disease of uprushing or adverse ascending flow of normal *qi*, such as hiccough, belching, vertigo, and dizziness. The channel tropism of drugs is directly related to the channel attribution of disease. For instance, the primordial stage of influenza falls under the category of Taiyang Channel disorders, hence, drugs of Taiyang Channel tropism are to be applied. Compatibility of drug compounding and toxicity of drugs are all highly stressed in Chinese pharmacy. Precautions when administering drugs are also unique. The breaking of necessary precautions would lead to failure of even a correct treatment. All the drugs are prepared in various forms, including decoction, powder, paste, pills, bolus, ointment, patent drug, and also modern drug forms like injection and aerosol. TCM also pays attention to the time of taking drugs, claiming that this has a direct bearing on the chrono-physiology of the body and on its therapeutic efficacy.

The ingredients for compounds of Chinese medical recipes are differentiated into the "king", "ministers", "assistants", and "servants". The aim of this compounding is to focus on the mutual synergic and detoxifying action among the ingredients. As a result, the effect is much more satisfactory than single drug administration, and the toxicity is much ameliorated or even eliminated.

Acupuncture and moxibustion are special treating techniques as well as health care measures in TCM. Acupuncture refers to the needling of specific loci, the acupoints for stimulation, to regulate the disharmonious state and arouse the resistance potential of the body. Moxibustion refers to the application of a moxa roll or cone on the point or affected site instead of using a needle. As soon as the body is stimulated by these means, the afflicted *qi* and blood inside the channels and viscera are improved, activated, and regulated. Pathogens are expelled or eliminated, and normal physiological function is restored. The basic idea of acupuncture-moxibustion is also established on the same principles as TCM, i.e., treatment based on the differentiation of syndrome manifestations, though it also has its own special demand such as manipulation techniques.

Massage and *qigong* exercise are also integral parts of TCM. The former is performed by specific manipulation techniques on acupoints or specific locations, while the latter is a self-care method in which the patient consciously controls his/her own mind, body, and the circulation of *qi* through controlling one's breathing movement. It is said that a proficient exerciser may even direct the flowing of *qi* in his/her own body at will.

TCM pays attention to disease prevention, or so-called "pre-treating illness" in the Chinese term. Various measures are proposed for this purpose, among them self-care massage, Daoyin, Taiji boxing, hygienic measures, and breathing exercises.

Since the basic conception of traditional Chinese medicine took its shape several thousand years ago, it applied concepts which are rather abstract and vague, not tangible or perceptible, let alone quantitatively estimated. To meet the needs of modern investigation and understanding, over the decades TCM workers have been encouraged to integrate and interpret their knowledge by modern scientific concepts, means, and techniques. A new school, the School of Integrated Traditional and Western Medicine, has thus emerged. This has become one of the modern trends in the development of China's medical science.

CAI JINGFENG

REFERENCES

Cai, Jingfeng. *World Records in Chinese Medical History.* Changsha, China: Hunan People's Publishing House, 1984.

Chen, Bangxian. *A History of Chinese Medicine.* Beijing: Commercial Press, 1957.

Divine Husbandry's Classic of Herbology (Reprint). Beijing: People's Health Publishing House, 1956.

Editorial Committee of Chinese Great Encyclopedia. "Traditional Chinese Medicine" Volume of the *Chinese Great Encyclopedia.* Beijing: Publishing House of Great Encyclopedia, 1992.

Guangdong College of Chinese Medicine, et al. *Revised Outline of Chinese Medicine.* Beijing: People's Health Publishing House, 1972.

Huangfu, Mi. *A–B Classic of Acupuncture-Moxibustion* (Reprint). Beijing: People's Health Publishing House, 1956.

Needham, Joseph. *Science and Civilisation in China*, vols. I and II. Cambridge: Cambridge University Press, 1956.

Variorum of Yellow Emperor's Inner Canon: Plain Questions and Miraculous Pivot (Reprint). Shanghai: Shanghai Health Publishing House, 1957.

Wang, Jiusi. *Variorum of Classic of Questioning* (Reprint). Beijing: People's Health Publishing House, 1956.

Wang, Shuhe. *Classic of Sphygmology* (Reprint). Beijing: Commercial Press, 1955.

See also: Moxibustion – Bian Que – *Huangdi Neijing* – Medical Texts in China – *Yinyang* – Acupuncture – Ge Hong – *Shanghan lun* – *Jingui Yaolüe* – Wang Shuhe – Ibn Sīnā – Li Shizhen – Five Phases

MEDICINE IN CHINA: FORENSIC MEDICINE Forensic medicine refers to that part of medical science pertaining to legal and political affairs. In order to provide materials and evidence for trying and investigating cases, the discipline of forensic medicine deals with the issues of reconnoitering the scene, surveying the cadaver of the body, and studying the material evidence, poisonous substances, and other relevant items.

Chinese medico-jurisprudence makes its appearance in the Warring States period (475 BC–221 BC). *Li Ji* (Record of Rituals) and *Lüshi Chun Qiu* (Master Lü's Spring and Autumn Annals) both mention "investigating the wounds and trauma, inspecting and analyzing cases, reconnoitering and judging, and making decisions on lawsuits with justice." In 1975, some bamboo slips, later entitled "Qin Slips from Yunmeng", were unearthed from Yunmeng, Hubei Province in China. Within the slips, legal articles of the Qin Dynasty (221 BC–227 BC) and criminal cases were recorded. Seven of the twenty-two cases involved forensic medical jurisprudence, in which killing, hanging to death, chopped-off heads, abortion due to trauma, and leprosy are included with records from the scene. Foot, hand, and knee prints, as well as blood and stools are mentioned. Circumstantial descriptions are given on the differentiation between homicide and suicide by hanging. All these demonstrate that achievements in legal medicine appeared as early as two to three thousand years ago.

The earliest extant Chinese feudal code, *Tang Lü* (Law of the Tang Dynasty) was promulgated in 653. This stipulates that when investigating fake-illness, feigned death and injuries, false or incorrect reports will result in punishment in a grade next to that given to the swindler. For victims of illness, death, or injuries, a fake report would be given punishment equal to that of the sufferers themselves. Issues pertaining to autopsies of legal medical cases, including the severity of injury, fake illness, self-mutilation, administering abortion, disability, age, and critical diseases are also mentioned.

Medical jurisprudence developed a step forward in the Song Dynasty (960–1279). First, the officials responsible for the investigation of cases is stipulated in writing, saying that "for examining cadavers *Canjun* at the provincial level and *Xian Wei* at the county level are responsible. In case *Xianwei* is absent, officials of *Bu, Cheng* and *Jian* will be responsible in that order. The county governor himself should be responsible when all these staff members are absent." Regulations for responsibility and dereliction of duty, such as when those officials in charge are inaccessible, when there is a delayed presence at the scene, or when the wrong decision for the cause of death is made are also stipulated. *Gemu*, a compulsory regulation for the examiner, is mandated to avoid malpractice; this is claimed to be one of the important achievements in this field. This rigorous system for examination offers a firm basis for the advent of the prominent monograph on legal medicine, *Xi Yuan Ji Lu* (Collected Records of Washing Away the Wrong Cases) and other similar works. Prior to these monographs, other works on legal medicine made their appearance, including *Neishu Lu* (Record of Forgiveness) and *Jianyan Fa* (Method for Examination), which were unfortunately all lost. *Xi Yuan Ji Lu* is an epitome of the achievement in the discipline of Chinese medical jurisprudence. Its merits include records on

- the occurrence and distribution of cadaver speckle;
- conditions influencing the advent of putrefaction;
- the relationship between the cadaver manifestations and the duration after its death;
- types of rope used for hanging;
- the features of strangulation and how to differentiate those from self-hanging for suicide;
- the difference between drowning and suffocation by compressing someone's nose and mouth;
- the difference between fractures before and after death;
- the determination of fatal wounds; and
- various methods for determining different causes of death.

During the Yuan Dynasty (1271–1368), a formal pattern for examination of Confucian officials was announced. This is entitled *Jie An Shi* (Pattern for Winding Up a Case). It deals with issues relating to medical jurisprudence, including examination of the cadaver, biopsies for wounds and illnesses, and material evidence. This monograph is the first to combine the three integral portions of medical jurisprudence into a whole.

The Chinese made a significant contribution to legal medicine. Important Chinese monographs on legal medicine including *Xi Yuan Ji Lu Ping Yuan Lu* (Reassuring the Wrong Cases) and *Wu Yuan Lu* (Free of Wrong Cases) were translated into many foreign languages, including Korean, Japanese, French, German, Dutch, and English.

After 1911, performing autopsies became legal. When procurator and police officers are unable to ascertain the causes of death with a cadaver, an autopsy is performed by a medical practitioner rather than by the old-style *Wu Zuo* (cadaver examiner). This procedure is rigorously regulated.

The first Department of Forensic Medicine was established in the Medical College of Peiping University in 1930, while the First Institute of Forensic Medicine was established in Shanghai in 1932, and a Medical Jurisprudence Monthly was first published in China in 1934. A wealth of forensic medicine professionals have been cultivated, in order to increase public security and justice.

CAI JINGFENG

REFERENCES

Jia, Jingtao. "The Examination System in Ancient China." *Studies on Science of Law* (6): 59–63, 1980.

Jia, Jingtao. *The History of Forensic Medicine in Ancient China*. Beijing: Mass Press, 1984.

Song, Ci. *Washing Over of Wrong Cases* (Reprint). Shanghai: The Commercial Press, 1936.

MEDICINE IN EGYPT Humanity was concerned with sickness and death long before the Egyptians appeared along the banks of the Nile. But only from the beginning of pharaonic civilization, about 2900 BC, do we have evidence of how sickness and trauma were treated in the ancient world. It comes primarily from several monographs, or collations of sections of earlier monographs, written on papyrus. The major medical papyri are, in their probable chronological order:

- *Veterinary Papyrus of Kahun*, ca. 1900 BC, on the treatment of animals;
- *Gynecological Papyrus of Kahun*, ca. 1900 BC, a fragment of a monograph on the diseases of women;
- *Papyrus Edwin Smith*, ca. 1550 BC, part of a monograph on wounds that also includes a fragment of a work on the

heart and vessels called the *Secret Book of the Physician* that was probably composed ca. 3000 BC;

- *Papyrus Ebers*, ca. 1550 BC, a collection of remedies for several kinds of ailments that contains a longer version of the *Secret Book*;
- *Papyrus Hearst*, ca. 1550 BC, a less systematically organized collection of remedies that duplicates many of those in the Ebers Papyrus;
- *Papyrus Berlin 3038*, ca. 1350–1200 BC, another collection of drug recipes, with another version of the *Secret Book*;
- *Papyrus Chester Beatty*, ca. 1250–1150 BC, a fragment of an earlier monograph on diseases of the anus; and
- *Magical Papyrus of London and Leiden*, ca. third century AD, containing many examples of spells used in healing rites.

From the beginning, magic dominated Egyptian concepts of illness and its treatment. Empiric observations eventually entered therapeutic thinking, but magic was seldom far removed from it. The Egyptians attributed healing powers to a number of their local gods, while Thoth, scribe of the major gods and inventor of the arts and sciences, became a part-time god of medicine. Not until he was deified sometime after 525 BC did Imhotep, who had been the chief minister of the pharaoh Zoser (reigned 2630–2611 BC), and the inventor of the pyramid, assume the role of chief god of Egyptian medicine.

Throughout most of the three millennia of pharaonic history, three kinds of healers treated the sick. Two of them, magicians and priests, relied chiefly on magical or religious rituals. The third group, the lay physicians called *swnw* (pronounced something like "sounou"), relied chiefly on surgery and drugs, techniques that were also used by healing priests. How they learned their professional skills is not known, but most probably they learned from their fathers, and all were literate. Interestingly, not all were men: an inscription from the Old Kingdom period (2575–2134 BC) describes a woman named Peseshet as the Lady Director of the Lady Physicians.

Swnws usually accompanied armies into the field, and were employed at public works sites such as temples, pyramids, and quarries. The oldest known doctor's bill lists payments to a *swnw* who worked at the village of the workmen who built the tombs and temples of Thebes, in about 1165 BC. Although physicians were paid only about a third as much as construction workers, *swnws* sometimes achieved very high rank at court, and were rewarded accordingly.

The average age at marriage was twelve to thirteen for women, and fifteen to twenty for men. The fertility rate, at least among the few queens for whom we have data, was lower than might have been expected, about 3.5 births per woman, in part because children were not weaned until they were three years old. The average life expectancy at birth was probably about thirty years in the early Old Kingdom, and might have increased to as much as thirty-six years over the next 2500 years. Evidence of very old age is scarce among mummies simply because most Egyptians died in their thirties, before the diseases that are typical of old age could have developed. A few kings and priests lived into their nineties, but fewer than ten percent of Egyptians lived longer than forty years. By contrast, about 94 percent of modern Americans die after age forty, and their life expectancy is about 76. As in most pre-industrial cultures, both the population and the average life expectancy rose (and the death rate fell) when the food supply increased, as farmers increased their ability to exploit the Nile with improved irrigation systems.

Egyptians well understood the biological relationships among the testes, penis, semen, and pregnancy. However, because much of their anatomical knowledge came from cattle, they thought that semen was produced in the bone marrow, even if they also knew that removing the testes — castration — prevented any possibility of fatherhood. On the other hand, they knew nothing about the ovaries, and thought that women's role in reproduction was simply to nourish the fully formed seed that the father planted in the fertile uterus. Although the *swnw* did not attend births, he did treat women's medical problems. He attributed menstrual abnormalities to malpositions of the uterus, which meant that many "gynecological" treatments were designed to restore it to its proper position.

It has been difficult to learn much about the actual causes of death in ancient Egypt, despite the hundreds of mummies in the world's museums. Several have been studied non-destructively, using X-rays and CAT scans. Such non-invasive techniques can easily reveal diagnoses such as fractures, dislocations, calcified arteries, and gall stones, but such conditions need not be fatal *per se*. X-rays of 26 royal mummies from the New Kingdom (1550–1070 BC) period have revealed more about their age at death, their teeth, and about mummification techniques, than about their illnesses, and even less about the causes of their deaths. The few mummies that have been dissected have provided surprisingly little specific pathological information, partly because most Egyptians probably died of infections that left few anatomical traces, and partly because mummification usually destroyed potentially diagnostic tissues. On the other hand, Egyptians may have been so thinly scattered along the river that many infectious diseases could not easily have propagated themselves. Nevertheless, in some places at least, the population may have finally become sufficiently dense over the centuries to facilitate the spread of such diseases.

Not all evidence of illness in ancient Egypt has come

from human remains. A few statuettes show spinal distortions characteristic of tuberculosis of the spine, as do several mummies. Similarly, both a funeral monument and a mummified leg show the dropped foot deformity typical of poliomyelitis (although it could also have been a club foot). Identifiable illnesses are not found in the detailed pictures of everyday life that adorn tombs, since sickness had no useful role in the next life.

In addition to tuberculosis, Egyptians had many of the diseases we have, although in different proportions of the population, plus several protozoal and worm infestations that are still found along the banks of the Nile, such as trachoma and schistosomiasis. Most of the non-traumatic illnesses that occured in ancient Egypt have not been identified in modern terms, but the Ebers Papyrus shows that the *swnw* classified them by their anatomic location, such as the skin, hair, abdomen, limbs, genitalia, and so on.

The Edwin Smith Surgical Papyrus contains several "firsts" in the history of medicine, including the first written descriptions of any surgical procedures, and the earliest examples of inductive scientific reasoning. The 48 cases described in the monograph are organized anatomically, from the top of the head down through the neck to the upper arm and chest; the scribe stopped transcribing the text when he reached the section on the upper spine. Because many injuries described in the Smith Papyrus were probably battle wounds, its prognostic predictions can also be taken as early examples of systematic battlefield triage.

The original author of this surgical text classified injuries by their extent and severity, and by their localization in bone or flesh, and established a systematic procedure for dealing with each kind of problem. Each chapter is devoted to a different injury, beginning with clinically important phenomena that could be seen or palpated, followed by a diagnosis based on facts elicited during the examination, and concludes with a prognosis. The prognoses are in three standard forms: injuries that are treatable, those of uncertain outcome that the *swnw* will try to treat anyway, and those that are unlikely to respond to any treatment, such as depressed skull fractures and compound fractures. The latter prognosis usually leads to the recommendation that nature be allowed to take its course.

Ancient Egyptian surgery provided simple, practical solutions to a number of self-evident problems. Today, most of it would be called "minor surgery", such as incising and draining abscesses, removing superficial wens, tumors, and so on. Circumcision was practiced on adolescent boys, but by priests, not physicians. The *swnw* did not perform major operations such as amputations, although he did try to reduce simple fractures and dislocations. Penetrating wounds were drained and cleaned. The *swnw* used adhesive plasters, not sutures, to hold wound edges together.

He differentiated between "diseased", or infected, wounds, and "non-diseased" wounds, which we would call "clean". The Smith Papyrus recommends daily inspections so that dressings can be changed when necessary. Many wound ointments were made with honey or the green copper ore malachite. Recent experiments have shown that both substances could have been effective against the kinds of bacteria most often found in contaminated wounds, permitting them to begin healing of their own accord.

When it came to what we now call internal medicine, magicians and healing priests relied largely on spells and incantations to cure their patients, and on amulets, written spells, and repulsive materials like animal feces to drive away or prevent illness. Sometimes they relied on healing statues standing in sacred ponds, so that patients who bathed in the same water would be cured by the god to whom the statue was dedicated. Late in pharaonic history, a patient might be instructed to sleep in a temple, expecting that its god would send him a dream that would reveal his cure, a procedure called "incubation".

The *swnw* might not have understood the underlying pathology of non-surgical illness, but he had a logical, even if speculative, theory of disease. According to the medical historian Henry Sigerist, "Physiology began when man tried to correlate the action of food, air, and blood". The Egyptians saw the heart as the focal point of that correlation. It was the body's most important organ, and the seat of intelligence and emotion.

The *swnw*, who knew that air is vital to life, thought that it passed through the trachea to both the heart and the lungs. From the heart it travelled, in blood and along with other fluids, to other organs in primary afferent ducts called *metu*. From those organs a series of secondary *metu*, or efferent ducts, led to the surface of the body. The body's secretions and excretions, including phlegm, tears, semen, urine, and even a little blood, escaped through that second set of *metu*. This concept permitted the *swnw* to exploit a fairly plausible theory of disease. He could accept the *metu* as fact chiefly because Egyptians could not differentiate arteries, veins, nerves, and tendons anatomically. They thought that all of them were hollow *metu* that transported disease, in the form of a foul substance called *ukhedu*, to various organs, depending on how many *metu*, and which ones, were involved.

Because decay and foul odor characterize both normal feces and death, the potentially fatal *ukhedu* was thought to originate in the feces as the residue of incompletely digested food. Indeed, post-mortem putrefaction is most noticeable in the intestines. Thus, if any *ukhedu* were allowed to accumulate to overflowing within the intestines, the excess would overflow up into the *metu* that normally carry blood from the heart to the intestines, so that the excess travelled backward to the heart. From there it could then enter other primary or secondary *metu* and be carried to other organs. Once it had entered a given *metu*, the *ukhedu* in that vessel could destroy the blood in it, producing pus. When it reached other organs, the pus would settle in and produce disease in them. Thus, the appearance of pus on the surface of a wound was a favorable prognostic sign, inasmuch as it signified that it was escaping and not accumulating within the body. Many treatments were designed to help *ukhedu* escape from the body. For instance, because boils were obviously filled with pus, they were opened so that the dangerous *ukhedu* could escape. Similarly, the standard wound ointments made with malachite or honey would counteract *ukhedu* that surfaced in a wound.

Egyptian drugs included many animal parts. Almost no clinical selectivity was associated with any of them, although ostrich eggs were used somewhat selectively in diarrhea remedies. Most of the animal products used in drugs, such as fat and grease, were used principally as emollients in wound ointments. In addition, the blood of several species was thought to be modestly selective for hair problems and eye trauma. But most animal parts used in drugs were included chiefly because of their magical associations. For instance, since ravens are black, their blood was used to treat the Egyptians' black hair; stallion semen was used to restore sexual drive; and fish skulls appear in headache remedies.

By contrast, the majority of drugs recommended in the Ebers Papyrus was aimed at disorders of the gastrointestinal tract, followed by remedies for the eyes, limbs, and skin. The *swnw* classified accumulations of *ukhedu* in the *metu*-ducts as gastrointestinal disorders, because the *ukhedu* originated in the alimentary canal. Thus, cathartics were usually prescribed to help flush *ukhedu* out of the rectum before it could accumulate to dangerous levels. Although some remedies for disturbances within the *metu* or the intestines do promote bowel movements, many laxative or cathartic drugs were used to treat non-intestinal symptoms. For instance, the mild laxative aloe was sometimes prescribed for eye disease, and the strong laxative colocynth for respiratory ailments. Thus, it seems likely that the *swnw* thought that all cathartics were at least somewhat selective for removing *ukhedu* from both the rectum and the *metu* which distributed it in the body.

Constipation implied that accumulated *ukhedu* could not escape by its usual route through the anus. Indeed, an Egyptian might take strong laxatives three days a month just to prevent *ukhedu* from filling up, and overflowing from, his rectum into his *metu*, whence it might reach his heart. Although diarrhea was a frequent complaint, sometimes it was not treated at all, because plentiful stools implied that the bowels were being adequately emptied of the dangerous *ukhedu*.

Honey, the most popular of all drug ingredients mentioned in the papyri, was used not only in wound ointments but also in both laxatives and antidiarrheals, and in many other remedies. Since it was not used for any one kind of clinical problem more often than another, we can infer that honey was not used very selectively for symptoms associated with any particular organ system. Honey was probably regarded as selective for the *ukhedu*, especially when it appeared in wounds as pus.

The next most frequently used drug was called *djaret*. The word has not yet been convincingly translated, but it was clearly a plant product. Because it appears in about half of all prescriptions for diarrhea, *djaret* was probably thought to have some selectivity for that problem. In addition, it was included in a third of the prescriptions for eye diseases, and may have been thought to be at least modestly selective for them. However, whatever it was, *djaret* was also prescribed for many other disorders, so it must have been regarded as a multipurpose drug, even if as not as a panacea. Frankincense, too, was often prescribed for many illnesses, but it was aimed most selectively at pains of the head and limbs, and less often at intestinal problems. Like the mysterious *djaret*, an antimony ore was applied to eye problems.

The next most frequently used ingredients were applied fairly non-selectively, although some were often used in both laxative and antidiarrheal mixtures. The pulp of the colocynth gourd is a very powerful cathartic that was also used for the latter purpose. Figs were used for abdominal pains and urinary disorders. Malachite was aimed at eye problems; it could not be taken internally because copper salts cause vomiting. It probably first entered medical usage as a topical ointment after long usage as an eye cosmetic. The author of the Ebers Papyrus knew about the laxative property of castor oil, but he included it in only a very few recipes for cathartics; he thought it was better suited for making women's hair grow. Like many other Egyptian remedies, it, too, survived in Western medical usage until well into this century, but as a cathartic, not as a hair restorer. However, most of the 328 different drug ingredients mentioned in the Ebers Papyrus were used for almost any illness.

Although Egyptian physicians prescribed a wide variety of remedies, they appear merely to have dispensed predetermined remedies to patients with similar ailments; the *swnw* seems not to have treated each patient as an individual. That concept that would not be introduced until Greek medicine began to flourish in the fifth century BC. There is no evidence that any of the *swnw*'s remedies, save for malachite and honey, when applied topically, had any truly beneficial effect on the outcome of ancient patients' illnesses, nor is there any modern reason to think so. Although they may not have known it, ancient physicians were able to rely on their drugs, and even on their magical spells, because of the body's impressive ability to use mechanisms such as the immune and inflammatory responses to heal itself of most ordinary ailments. By contrast, Egyptian surgery probably did provide reasonably effective treatments for many traumatic injuries.

J. WORTH ESTES

REFERENCES

Breasted, James Henry. *The Edwin Smith Surgical Papyrus*, 2 vols. Chicago: University of Chicago Press, 1930.

Ebbell, B[endix], trans. *The Papyrus Ebers*. London: Oxford University Press, 1937.

Estes, J. Worth. *The Medical Skills of Ancient Egypt*. Rev. ed. Canton, Massachusetts: Science History Publications, 1993.

Harris, James E., and Edward F. Wente, eds. *An X-Ray Atlas of the Royal Mummies*. Chicago: University of Chicago Press, 1980.

Leake, Chauncey D. *The Old Egyptian Medical Papyri*. Lawrence: University of Kansas Press, 1952.

Leca, Ange-Pierre. *La Médecine Egyptienne au Temps des Pharaons*. Paris: Éditions Roger Dacosta, 1971.

Majno, Guido. *The Healing Hand: Man and Wound in the Ancient World*. Cambridge, Massachusetts: Harvard University Press. 2nd ed., 1992, pp. 69–140.

Sigerist, Henry E. *A History of Medicine*, vol. 1, *Primitive and Archaic Medicine*. New York: Oxford University Press, 1991, pp. 217–373.

MEDICINE IN INDIA: *ĀYURVEDA* The science and art of Āyurveda are integral parts of the cultural heritage which is preserved, fostered, and promoted in India and its neighboring countries for the preservation and promotion of positive health and the prevention and cure of disease. As a science, it is based on sound and rational principles of physiology, pathology, pharmacology, diagnostics, and therapeutics which have been critiqued, systematized, and generalized on the rigid principles of logic. Even today it is followed as a healing art, not only by rural and poor people but also by well-placed persons of learning in all walks of life, including affluent intellectuals who could otherwise easily afford to obtain the services of top-ranking physicians of modern Western medicine.

Literally meaning the "science of life", Āyurveda is often used in a narrow sense as a "system of medicine", which considerably dilutes and distorts its real scope and objective. Health according to Āyurveda is not only freedom from disease. According to Suśruta, one of the great early practitioners, it is a state of the individual where, in addition to harmony among the functional units (*doṣas*), digestive and metabolic mechanisms (*agnis*), structural elements (*dhātus*), and waste products (*malas*), a person should also be in an excellent state (*prasanna*) of the spirit (*ātman*), senses (*indriyas*), and mind (*manas*).

The history of Āyurveda is as old as that of the Vedas, as the former is considered to be one of the limbs (*aṅgas*) of the latter. It has eight specialized branches:

(1) *Kāya-cikitsā* (internal medicine),
(2) *Śalya-tantra* (surgery),
(3) *Śālākya-tantra* (treatment of diseases of the eyes, ears, nose, and throat),
(4) *Agada-tantra* (toxicology),
(5) *Bhūta-vidyā* (treatment of seizures by evil spirits and other mental disorders),
(6) *Bāla-tantra* (pediatrics),
(7) *Rasāyana-tantra* (geriatrics, including rejuvenation therapy), and
(8) *Vājikaraṇa-tantra* (sexology and the use of aphrodisiacs).

Several classics were composed for each of these branches, and each was widely practiced.

Āyurveda is a holistic science of life. Each part of the physique is interrelated with other parts for their proper functioning. The functions of the body are closely related to the mind and soul (spirit) of the individual. To maintain health, the body should be free from disease, the mind should be happy, and the person should be spiritually elevated. To cure diseases, the physician examines and treats the whole body and the mind as well as the spirit. Therefore, for the preservation of positive health and prevention as well as cure of diseases, several codes of conduct and religious rituals are prescribed, along with medicines and food. Thus, Āyurveda advocates the psychosomatic interrelation of health and disease.

Several categories of germs and organisms are said in Āyurveda to cause diseases like *kuṣṭha* (obstinate skin diseases, including leprosy). But these organisms are considered causative factors of a secondary nature, the primary cause being the disturbance in the equilibrium of the *doṣas* (three factors regulating the functioning of the body). As seeds germinate only when sown over fertile land, and decay if the land is barren, similarly germs, however virulent they may be, will not be able to multiply and cause a disease if the constituents of the body are in a state of equilibrium.

Āyurvedic therapies are prescribed not to destroy the invading germs (seeds), but to bring about equilibrium of the constituents by which these germs, finding the atmosphere hostile (barren field) for their survival, get destroyed.

In Āyurveda, drugs of vegetable origin, animal products, and metals, minerals, gems, and semi-precious stones are used for therapeutic purposes. These are used in their natural form and processed in order to obtain their whole extract or to make them non-toxic, palatable, and therapeutically more potent. Different parts of these drugs, like alkaloids, glucosides, and other active principles are not extracted for therapeutic use. According to Āyurveda, every drug has therapeutically useful parts that may produce toxic effects if used alone. The same drug, however, contains other parts that counteract these adverse effects. Therefore, the use of the whole drug is emphasized, and no isolated section or synthetic chemicals are used. Some ingredients used in recipes are no doubt toxic in their raw form. But these are processed and detoxified according to prescribed methods before being added to recipes. These recipes have been in constant and regular use for thousands of years, which testifies to the absence of any acute, subacute, or chronic toxicity and teratogenic effects. These recipes are equally useful for patients and healthy persons. In the former, while curing the disease, they produce immunity against future attacks, and therefore, instead of giving toxic side effects, they produce side benefits. In healthy individuals, they revitalize the body cells and stimulate the immune system as well as fortify the body's resistance against disease.

Considerable emphasis is laid upon proper diet in Āyurveda for the prevention and cure of disease. If a person consumes a proper diet, medicines are superfluous for him because he will not fall victim to diseases, and if he does, the ailment will get corrected easily. If, however, he does not follow a proper diet, he will not get cured in spite of the best medicines. Therefore, Āyurvedic works describe in detail the properties of various ingredients of diet and beverages.

According to Āyurveda, the individual (microcosm) is a replica of the universe (macrocosm). Virtually every phenomenon of the universe or *brahmāṇḍa* can be found to take place in the individual or *piṇḍa*, albeit in a subtle form. Therefore, every action of the individual in its turn has an impact on the environment as well as the universe. This intracosmic relationship which is the very foundation of Āyurveda is further elaborated in Tantra and Yoga.

The scope and utility of Āyurveda are not confined to any particular community or any political, geographical, economic, or sociocultural group. The principles of Āyurveda are equally useful for and applicable to all the people in different parts of the world. According to Caraka, an early physician (before 600 BC), the "science of life" has no limitations. Therefore, humility and relentless industry should characterize every endeavor of the physician to acquire knowledge. The entire world consists of teachers for the wise, and only a fool finds enemies in it. Therefore, according to him, the knowledge conducive to health, longevity, fame, and excellence coming even from an unfamiliar source (lit. enemy) should be received, assimilated, and utilized with earnestness. This approach of Āyurveda is evident from the descriptions in the classics. Several plants, animal products, and minerals which are not available in India are described in detail with reference to their therapeutic properties. The properties of several ingredients of food and beverages which are normally not used in India, such as meat, especially beef, are also described. The fundamental principles of Āyurveda are also described. These principles are applicable in any part of the world, and its practice can be modified according to regional requirements.

Pañcamahābhūta Theory

The body and the universe, according to Āyurveda, are composed of five *mahābhūtas* (basic elements), namely, *pṛthvī, ap, tejas vāyu*, and *ākāśa*. These terms are often mistranslated as earth, water, fire, and ether, which are misleading even for well-meaning critics of Āyurveda. These five categories of matter correspond to five senses, namely smell, taste, vision, touch, and hearing. The objective matters of the universe are subjectively comprehended by means of these five senses. These five states of matter (subjective series) and the five senses (objective series) are both the evolutive products of *prakṛti* (primordial-matter-stuff). These five *mahābhūtas* do not stand for the elements of the present day physical sciences as the mistranslation of these terms would have them, but the five classes of objects of our material universe correlated to the five senses by means of which one subjectively contacts the objective universe.

The body, as well as food and drugs, are all composed of these five *mahābhūtas*. It is disturbances in the equilibrium of these which cause disease. To correct the ailments so caused, appropriate *mahābhūtas* in the form of drugs and food must be used. Though the characteristic features of these *mahābhūtas* are described in detail in Āyurvedic texts, in practice it is extremely difficult to ascertain the precise requirements. Hence, their presence in the body is described in terms of *doṣas* (functional units of which there are three types), *dhātus* (tissue elements of which there are seven categories), and *malas* (waste products). Similarly, the mahābhautic composition of drugs, diet, and beverages is described in terms of *rasa* (taste), *guṇa* (attribute), *vīrya* (potency), *vipāka* (aftertaste), and *prabhāva* (specific action).

The *doṣas*, *vāyu*, *pitta*, and *kapha* are the three elementary functional units or principles on which the building and sustenance of the body depends. The individual remains healthy as long as these three elements remain in a state of equilibrium. But if this is disturbed beyond certain limits, the individual succumbs to disease and decay. The *doṣas* remain in two different forms in the body, namely, *sthúla* or gross, and *sūkṣma* or subtle. In their subtle state, they are *atīndriya* or beyond the normal cognition of the senses. Their normal and abnormal states are ascertained by the manifestation of their respective actions. These three *doṣas* control all the physical and psychic functions of an individual. Each one of these is further subdivided into five categories on the basis of their actions on different parts of the body.

The physical structure of the body, according to Āyurveda, is composed of seven categories of *dhātu*s or tissue elements:

- *rasa* (plasma, including chyle),
- *rakta* (blood, particularly the hemoglobin fraction of it),
- *māṃsa* (muscle tissue),
- *medas* (fat),
- *asthi* (bone tissue),
- *majjā* (bone marrow), and
- *śukra* (generative fluid, including sperm and ovum).

There are two categories of tissue elements, *poṣaka* (nutrients of tissues) and *poṣya* (stable tissues). The former moves through different channels to provide nourishment to the latter. The circulation of these nutrient tissue elements is controlled by *vāyu*, their metabolism including assimilation is controlled by *pitta*, and *kapha* helps in the maintenance of their cohesion and compactness. During the metabolic transformation of these tissue elements, a substance called *ojas* is produced as their essence. This *ojas* provides immunity or the power of resistance to attacks of disease. It is said that the diminution of this *ojas* (*ojaskṣaya*) is responsible for the disease called AIDS.

During the process of digestion and metabolism, several waste products are formed. Some of these serve a useful purpose, but ultimately they are eliminated from the body through feces, urine, sweat, and so on. Their regular elimination from the body is essential for the maintenance of good health. In Āyurvedic therapeutics, attention is always paid to the elimination of these wastes.

Āyurveda classifies individuals broadly into seven categories called *prakṛtis*. These are defined by the predominance of *vāyu*, *pitta*, and *kapha*, or their combinations, or by the state of equilibrium of them all. This trait of the individual is developed at the time of birth and continues throughout one's life. Individuals are also classified into sixteen categories, on the basis of their psychic disposition. The former seven categories are called *deha prakṛti* or physical constitution, and the latter sixteen categories are called *mānasa prakṛti* or psychic temperament. Knowledge of these *prakṛtis* is essential for selecting the appropriate food, drink, and other regimens for different categories of healthy individuals, and likewise for the treatment of their ailments.

As has been explained earlier, the ingredients of drugs, diet, and drinks are composed of five *mahābhūtas*. In order to facilitate examining the exact nature of these *mahābhūtas*, easy practical methods are described in Āyurveda in terms of *rasas* (tastes), of which there are six, twenty *guṇas* (attributes), *vīryas* (potencies) of which there are broadly two, and three *vipākas* (aftertastes). Āyurvedic texts also describe the specific and general actions of drugs.

There are several other unique concepts of Āyurveda, understanding of which is essential before appreciating the merits of this health science. Āyurveda is a repository of therapeutically useful and time-tested recipes for curing several obstinate and otherwise incurable ailments. But its prescriptions for the preservation and promotion of positive health and prevention of disease is what is especially useful to relieve human suffering.

BHAGWAN DASH

REFERENCES

Agniveśa. *Carakasaṃhitā*. Bombay: Nirnayasager Press, 1941.

Bhela. *Bhelasaṃhitā*. Delhi: Central Council of Research in Indian Medicine, 1977.

Dash, Bhagwan. *Fundamentals of Āyurvedic Medicine*. Delhi: Konark Publishers, 1992.

Dash, Bagwan, and Lalitesh Kashyap. *Materia Medica of Āyurveda*. Delhi: Concept Publishing Co., 1980.

Dash, Bagwan, and Lalitesh Kashyap. *Basic Principles of Āyurveda*. Delhi: Concept Publishing Co., 1980.

Śarmā, Śiva. *Āyurvedic Medicine: Past and Present*. Calcutta: Dabur, 1975.

Suśruta. *Suśrutasaṃhitā*. Varanasi: Chaukhamba Oriyantalia, 1980.

Vāgbhaṭa. *Aṣṭāṅgasaṅgraha*. Pune: Athvale, 1980.

Vāgbhaṭa. *Aṣṭāṅgahṛdaya*. Varanasi: Kṛṣṇadāsa Academy, 1982.

Vṛddhajīvaka. *Kāśyapasaṃhitā*. Varanasi: Chaukhamaba Sanskrit Series Office, 1953.

See also: Suśruta – Caraka

MEDICINE IN ISLAM An understanding of the conceptual aspects of Islamic medicine and ethnobotany requires an understanding of Islam itself. Islam as *Dīn al-Fiṭrah* — the natural way of life — has its own paradigms of knowledge. The Islamic view of reality provides a matrix in which central problems of knowledge are illuminated. Islam has its own ideas about the maintenance of health and the alleviation of disease which involve ideas regarding the nature of creation, the position of humankind, and the path of well-being. The health of an individual or a society can only be located within a context of nature, society and man, as medicine is a facet of an integral view of reality. *Tawḥīd* (Monotheism) is the first principle; it is declared five times a day in the call to prayers. Before Islam, numerous dieties were said to have influence over disease. Some were for relief in pregnancy and others for easy delivery or protection from epidemics. Monotheism abolished these ideas. The first verses of the *Qur'ān* ordered the Messenger to read, which demonstrates the importance of knowledge.

The Holy Book of Islam, the *Qur'ān*, contains several verses in which medical questions of a very general order are discussed, as well as many hygienic regulations and sanitary orders. The welfare of mothers and children is well regulated. Care of the fetus, nursing, wet-nursing, and maternal welfare are insisted upon. Orphans and the poor are well treated. Every person is responsible for the care of his body in general and of his senses in particular. Exposed parts of the body (face, forearms, feet, nose, eyes) are washed five times a day. The urethral orifice and anus are also washed after use. Alcoholic drinks are prohibited. Bathing is obligatory after a sexual act. Cleanliness is essential.

There are also many sayings of the Prophet Muḥammad (571–632 AD), called *ḥadīth*, on hygiene and sanitation. This body of sayings was systemized and became known as *al-Ṭibb al-Nabawī* (Medicine of the Prophet). The prophet says "There is a remedy for every malady and when the remedy is applied to the disease it is cured . . .". Islamic medicine realized the infectious nature of diseases and established a rule to control epidemics such as plague. The prophet says, "When you hear about it do not enter there and when it has broken out in a land and you were there, then don't run away from it". When Muslims became a community in Medina, after the immigration of the Prophet from Mecca, they began to develop a tradition of health and well-being. Medina provided the conditions for the unfolding of *Sharī'ah*, the Islamic way of life, of which medicine was an integral part. The prophet provided the foundation for a medical tradition that considered a human being in his totality; the spiritual, the psychological, and the physical were considered within the context of a social milieu.

The period that directly followed the time of the prophet is referred to as *al-Khilāfah al-Rāshida*, the right-guided rule. It was during this period that the medical center of Jundishapur became part of the Muslim lands and continued to flourish. New cities were built according to healthy principles.

In the forty-first year of the Islamic calendar (AD 661) Mu'awiya took over. The period that followed is generally referred to as the Umayyad Era. The Umayyad rule lasted until AH 132 (AD 750) in the east and AH 872 (AD 1492) in the west. It was during this period that translations of ancient medical works began. The Umayyad prince, Khālid ibn Yazīd, had a passion for medicine. He instructed a group of Greek scholars in Egypt to translate Greco-Egyptian medical literature into Arabic. The earliest hospital in Islam was that built in AD 707 by the Umayyad caliph al-Wālid ibn 'Abdal-Malik in Damascus. Wālid began a broad health care program. He had homes built for the blind and lepers. He isolated the lepers from other patients and provided medical facilities. The Hispanic Muslim areas of Cordoba and Granada became centers of learning. The rich and diverse flora of Spain was a contributing factor to the development of medical botany; in most cases we cannot separate medicine and botany as the majority of physicians were herbalists and vice versa. Physicians like Ibn al-Bayṭār, born in Malaga in AD 1197, spent his early life identifying and working on different plants. Other well-known physicians were Ibn Bājjah, Ibn Samghūn, and Abū'l-Ḥassan al-Andalusi, who also wrote extensively on plant remedies. The development of medicine in its varied forms was a particularly important contribution of Umayyad-Muslim Spain.

For five centuries starting from AH 132 (AD 750) the Abbasids dominated the sociopolitical life of the greater part of the Muslim world. There were major developments in medicine. The caliphate was moved to Baghdad, and as it developed and the Abbasid court grew wealthier, the Caliphs became renowned for their enlightened approach to diverse ideas. The ten Caliphs of the period were generous in their promotion of knowledge, especially al-Manṣūr (AD 754–775), Hārūn al-Rashīd (AD 786–802) and al-Ma'mūn (AD 813–833). A hospital, Bimaristan, was built and became the cradle of the Baghdad school of medicine. Here many physicians of Baghdad from the time of Bukhtīshū to the master of Arabic medicine, al-Rāzī, lectured and practiced. Countless manuscripts, particularly those written in Greek, were collected and stored in the Royal Library, *Bayt al-ḥikmah*, where scholars worked to translate them into Arabic.

As can be imagined, the effect on medical knowledge was far-reaching. Any idea consistent with Islam could be utilized by physicians, and they had in the library most of the knowledge of the time. Within a century Muslim physicians and scientists were writing original contributions to medical and botanical knowledge. The next three centuries saw the synthesis and creation of new drugs and therapies. One of the most illustrious figures of this period was Ibn Sīnā (Avicenna) known as Prince of Physicians. It was during the Abbasid period that the examination and licensing of physicians and pharmacists were formally organized. However , the central role of Baghdad became less important after it was devastated by Hulagu in AD 1240.

Before the invasion of Baghdad centers of medical learning had already been founded in other parts of the Muslim world. The core concepts and practices of Islamic medicine continued to be common to various areas, although there were characteristics unique to each locality. Persia and Samarqand continued to be sources of medical inspiration. In Egypt, under the Fāṭmids, Cairo became a center of learning and attracted competent physicians such as Maimonides from Cordoba, Spain, and Abd al-Raḥīm al-Dakhwār. Ibn al-Nafīs, who explained the minor circulation of the blood, was al-Dakhwār's student. Ottoman Turkey was also an important center of Islamic medical knowledge. The Sultans built both civil and military hospitals. One of the most distinguished physicians was Hakim Hae Pasa. A Turkish physician first used a smallpox vaccination in AD 1679. India and Pakistan produced some eminent physicians, such as Akbar Arazani who was one of the last physicians of the Mogul period, and Hakīm Hāfiḍ Muhammad Ajmal Khān after whom the alkaloid ajmaline was named.

There are many eminent medical figures from the Islamic world. Very brief descriptions of a few will be given here. See the more detailed entries in this *Encyclopaedia*.

Ḥunayn ibn Isḥaq (AH 194–263, AD 810–877). He was one of the Christian scholars who made an important contribution to the rise of Islamic sciences as a translator. Caliph al-Maʾmūn (AD 813–833) appointed him as chief in the official translating institute or the House of Wisdom (*Bayt al-ḥikmah*), and he later became court physician to Caliph al-Mutawakkil (AD 847–861).

Al-Rāzī (AH 250–318, AD 864–932). Abū Bakr Muḥammad Ibn Zakariyyā al-Rāzī, known to the west as Rhazes, was born in Persia. He was one of the greatest physicians of the Islamic world. He studied in Baghdad and was acquainted with Greek, Persian, and Indian medicines. The most celebrated of all the works of al-Rāzī is that on smallpox and measles. It was translated into Latin and later into other languages including English, being printed some forty times between 1498 and 1866. Rāzī became the director of the hospital in his native city of Rai, and later the director of the main hospital in Baghdad. He thus gained much practical experience, which played no small part in making him one of the greatest clinicians of the medieval period. His skill in prognosis, and his analysis of the symptoms of a disease, its manner of treatment and cure, made his case studies celebrated among later physicians.

Al-Rāzī was an author of more than 140 books. His greatest medical work was *Kitāb al-Ḥāwī fi 'l-Ṭibb* (Continens), an encyclopedia of medicine and surgery, a summary of all branches of the subject in twenty-four volumes. By 1542 five editions of this vast and costly work had appeared. *Al-Ḥāwī* was one of only nine books used in the medical faculty of the University of Paris.

Al-Zahrāwī (AD 936–1013). Abu'l-Qāsim Khalaf ibn ʿAbbas al-Zahrāwī or Albucasis, was one of the Muslim scientists who laid the foundations of modern surgery. He was the special physician of the eighth Umayyad Caliph (Abdul-Rahmān III). He was born in Zahra, near Cordoba. As soon as he completed his studies he took a job in the hospital. After years of work in the hospital he developed his own concepts and wrote three books. These books remained standard textbooks for nearly a thousand years. The most famous one of these was *al-Taṣrīf* (Manual for Medical Practitioners). Its most important aspect is that it showed 278 pictures of the equipment used for surgery.

Ibn-Sīnā (AH 371–428, AD 980–1037). Abū ʿAlī al-Ḥusayn ibn ʿAbdallāh ibn Sīnā, known to the west as Avicenna, was one of the greatest scholars of the Islamic world. His influence on European medicine has been great. Born near Bokhāra in a family devoted to learning, he received an excellent education. At the age of ten he had already mastered grammar and literature and knew the *Qurʾān* by heart. At the age of eighteen he had mastered all the sciences of his day. He was a man of enormous energy; he wrote 250 works of different lengths. The best known is *Al-Qānūn fi 'l-Ṭibb* (The Law of Medicine) which is the epitome of Islamic medicine. It was translated into Latin and taught for centuries in Western universities, being in fact one of the most frequently printed scientific texts in the Renaissance. Ibn Sīnā in his *Canon* concentrated Greek and Islamic medical knowledge. This medical encyclopedia embraced medicine, anatomy, physiology, pathology and pharmacopoeia. During the late fifteenth century AD it formed half the medical curriculum of European universities and at some medical faculties was a textbook until 1650.

Ibn Zuhr (AD 1113–1162). ʿAbd al-Malik ibn Zuhr, known as Avenzoar in the West, was a native of Seville and achieved widespread fame as a physician in Spain and North Africa. His chief work is *Al-Taysīr fi 'l Mudāwat wa 'l-Tadbīr* (Facilitation of Treatment). His interesting preparation of diets and medicines influenced the work of later European alchemists. From this work it is clear that in Ibn Zuhr's time physicians, surgeons, and apothecaries practiced separate professions.

Ibn Maymūn (AD 1135–1208). Abū ʿImrān Mūsā ibn Maymūn (Moses Maimonides) was born in Spain and spent most of his life in Cairo. His professional skill brought him fame and led to his appointment as leading physician to the Ayyubid court. The extensive range of Ibn Maymūn's work included criticisms of Aristotle, writings on the theory of medicine and books about the more practical matters of diet, hygiene, reptile poisons and their antidotes, asthma and hemorrhoids. His best medical work is *Aphorisms*. He had a substantial following throughout the Middle East and Mediterranean and his ideas were much valued by later generations outside the Arab world.

Ibn al-Bayṭār (AH 593–646, AD 1197–1249). Abū Muḥammad ʿAbdallāh Ibn al-Bayṭār was known as the "father of botany". He was born in Malaga and died in Damascus. He is considered to be the greatest of the Muslim botanists and pharmacologists. Several works have survived, including *Al-Mughnī fī-l-adwiya al-mufrada* (The Complete Book of Simple Drugs) in which he described more than 1400 medicinal drugs, a work of extraordinary erudition. His second book, *al-Jāmiʾ li-mufradāt al-adwiya wa'l-aghdhiya* (Compendium of Simple Drugs and Food), in which all that was known to the pharmacologists, as well as 300 drugs not previously described, was recorded alphabetically and discussed in detail.

In addition to those described above, three other physicians can be mentioned, who do not have separate entries in the *Encyclopaedia*.

Abū Ḥanīfah al-Daynūr (d. AH 281, AD 894). Aḥmad ibn Dawūd al-Daynūr al-Hanaf was born in Daynūr (Hamadān) where he spent most of his life. He was known in Arabic as al-Aʿshāb (the herbalist) due to his knowledge about herbs and their effects. His most famous book was *al-Nabāt* (The Plants), in six volumes. He mentioned 1120 plants, giving their different meanings in Arabic while keeping others with their Persian or Latin names. Al-Daynūr was also an astronomer. He wrote more than twenty books on different disciplines.

Al-Anṭākī (d. AH 1008, AD 1599). Dāʾūd ibn Omar al-Anṭākī died in Mecca during the pilgrimage. He was born in Antioch and lived in Dāhireyya in Egypt, where he wrote more than twenty-five books in pharmacy and medicine. Before residing in Egypt he traveled to many countries collecting information about plants and their medical uses. He wrote *Tadhkirah ūlu'l al-Albāb* (Reminder for the Wise), known as *Tadkarat Dawūd* (Treatise of Dawud), abridged in a small volume entitled *Tashīḍ al-Aḍhān* (Sharpening of Minds). His treatise contains a preface, four chapters and an epilogue. Subjects covered include the value of health, the necessity of treatment and diet, simple and compound medicines, anatomy and physiology, prognosis, dentistry, ophthalmology, and toxicology.

Those eminent figures are by no means the only ones to have noteworthy achievements in the fields of medicine and botany. The chain is very long; it contains among many others Ibn Juljul, al-Baghdādī, Ibn Rushd (Averroes), al-Qazwīnī, al-Andalusī and Ibn Bājjah.

MANSOUR SOLYMAN AL-SAID

REFERENCES

Chaudhry, Noor Hussein. "Islamic Tib Excellences and How to Reorganize and Acquire Them." *Bulletin of Islamic Medicine*, Vol. 1, 2nd ed. Kuwait: Ministry of Public Health and National Council for Culture, Arts and Letters, 1981, pp. 110–118.

Kamal, Hassan. *Encyclopaedia of Islamic Medicine*. General Egyptian Book Organization, 1975.

Khan, Muhammad Salim. *Islamic Medicine*. London: Routledge and Kegan Paul, 1986.

Nasr, Seyyed Hossein. *Science and Civilization in Islam*. Lahore, Pakistan: Suhail Academy, 1987.

Pearsall, Deborah, M. *Paleoethnobotany: A Handbook of Procedures*. San Diego: Academic Press, 1989.

Polunin, Miriam and C. Robbins. *The Natural Pharmacy*. London: Dorling Kindersley Book. 1992.

Sinclair, Michael J. *A History of Islamic Medicine*. London: Allied Medical Group, 1978.

See also: Magic and Science – Ottoman Science – Moses Maimonides – al-Rāzī – Ḥunayn ibn Isḥaq – Ibn Zuhr – Ibn Sīnā – al-Zahrāwī – Ibn al-Bayṭār – Ibn Rushd

MEDICINE IN JAPAN When considering the Japanese history of medicine, it is important not to project our ideas about medicine today onto the past. Whereas modern medicine worldwide is based on the language and methods of natural science, the medicine of pre-modern Japan consisted of many different languages and practices. Moreover, the historical development of medicine in pre-modern Japan must not be thought of as following a linear development that inevitably ended in the adoption of Western medicine. It is, rather, the story of theories and practices that unfolded according to their own historical logic, often in competition with each other and without theoretical or practical consistency. The fact that the Japanese did adopt Western medical ideas and practices during the premodern period simply reflects the ways in which some Japanese medical ideas and practices developed.

Throughout the pre-modern period, which in the case of Japan spans all of history until the mid-nineteenth century, numerous medical theories and practices, many of which were mutually contradictory, coexisted. Until modern times, most people depended primarily on folk remedies to treat their ailments. These consisted mostly of shamanistic rituals whose remnants can be found today in a small number of Shinto shrines and Buddhist temples. Physicians with a textually based theoretical training were rare in rural areas, where most Japanese people lived as peasants or fishers. However because few documentary sources exist on which a history of folk medicine could be based, historians have made the textual tradition the focus of their research. Hence it is also necessary to focus on this textual tradition here, but this is, of necessity, only a partial history.

Within these limits, the history of Japanese medicine can be divided into four periods: ancient, medieval, early modern, and modern. The ancient period spans from prehistory to the late twelfth century; the medieval period reaches from the beginning of the thirteenth century to the late sixteenth; and the early modern period encompasses the years from the late sixteenth until the mid nineteenth century. Because in many respects the history of Japanese medicine during the modern period parallels that of the Western world, this article deals primarily with the premodern period.

Prehistoric and Ancient Medicine

Based on skeletal remains, it is clear that the prehistoric inhabitants of the Japanese islands developed rudimentary medical practices. Some bone fractures, at least, were treated so that they healed cleanly. The earliest written records, which date from the early eighth century, recount the use of herbal remedies for some illnesses. However the same records also reflect the belief that most diseases were divine retribution for offending a deity or spirit; they were treated through exorcism, ritual ablution, and purification rituals. Some diseases, such as leprosy and tuberculosis, were also associated with ritual pollution, a belief whcih has found currency into modern times. Because this pollution was thought to be hereditary, these diseases often made it difficult for the persons who had them and members of their families to find marriage partners. When a disease reached epidemic proportion, it was thought to be caused by the more powerful deities that controlled forces of nature throughout the land, deities that the emperor or empress attempted to assuage with national purification rituals.

It is unclear when Chinese writing first reached Japan, but by the late fourth or early fifth century scholars from the Korean peninsula were tutoring members of the imperial family in the Chinese classics. The real rise of literacy came with the Japanese adoption of Buddhism from the sixth century; it can be surmised that the advent of textually-based medicine in Japan also dates from this time. By the seventh century, Buddhist monks from the continent were both practicing medicine and training Japanese monks to become practitioners themselves. Until the seventeenth century most trained medical practitioners retained the trappings, if not always the formal status, of Buddhist priests.

From the late seventh century the Japanese state adopted the legal codes of Tang China. These included provisions for government posts that specified the employment of various medical specialists, including internists, surgeons, acupuncturists, masseuses, exorcists, obstetricians, dentists, and pharmacologists. Although in China these posts were filled with scholars who passed required examinations, in Japan they soon became hereditary and remained so until the nineteenth century. This did not, however, preclude the adoption of new developments in medicine from the continent. Constant interaction with both Korea and China, where numerous Japanese Buddhist priests went to study, kept them informed of changing theories and practices, although the process of change was far slower than during more recent times.

Medicine during the Nara (710–794) and Heian (794–1185) periods remained the domain of court physicians and Buddhist priests. As in premodern Europe, internal medicine and surgery remained distinct practices. Internists had a high level of education and relatively high social status; surgeons, who mostly treated skin lesions, wounds, and fractures, had comparatively little theoretical training and low social status. These remained distinct until the late eighteenth century.

Internal medicine was based entirely on Chinese texts; the first Japanese medical text did not appear until the year 984. This was the *Ishimpō* (literally, Methods at the Heart of Medicine), which was written by the court physician Tanba Yasuyori (912–995) and remained an important text among some schools of medicine until the nineteenth century. This thirty-volume work contained information on internal medicine, pharmaceuticals, preventive practices, and other topics, and was based on over eighty Chinese texts written during the Sui (581–617) and Tang (618–907) dynasties.

Until the eighteenth century, anatomy envisioned the human body as containing five organs and six viscera, and physiology focused on their relationships with each other and the meridians found on the surface of the body. Visual observation of the internal organs played little role in the classical Chinese model of the human body, and organs including the brain, pancreas, thyroid, and adrenal glands played no role in the treatment of disease. Indeed, there was no word for the latter three organs, and visual depictions of human anatomy did not appear in the Japanese medical literature until the fourteenth century. Rather this model was conceptual, based on the Chinese notion of the five elements of fire, metal, water, wood, and earth. These were respectively associated with the heart, lungs, kidneys, liver, and spleen, and each organ was in turn associated with five colors, tastes, and seasons, the spleen being associated with the time of changes between seasons. Physicians interpreted diseases as resulting from imbalances within this system and treated them with herbal medicines, acupuncture, moxibustion, massage, and restrictions on diet and behavior. Yet this model did not completely replace ideas of disease as being caused by spirit possession, and the literature of the Heian period abounds in examples of Buddhist priests attempting to cure maladies through exorcism, a common practice until modern times.

Because the diagnostic criteria and disease categories of premodern Japan were so different from those used today it is difficult to establish with any accuracy in modern terms the diseases from which people suffered at the time. Yet it is certain that the most common epidemic diseases of premodern Japan, beginning with the Nara and Heian periods and continuing through the nineteenth century, were smallpox, measles, influenza, and enteric infections. Tuberculosis, malaria, and parasitic infections remained endemic during the premodern period. Despite a myth of cleanliness attributed to the Japanese, bathing and the regular washing of clothes were not common practices, making skin diseases common until modern times. In addition, kitchen areas frequently were far less than sanitary, and latrines and even graves were sometimes located close to water sources, contaminating them.

Although most medical folk beliefs are poorly documented, much is known concerning some popular beliefs related to what probably was tuberculosis. During the twelfth century, Chinese Daoist priests thought that a disease they called *zhuan shi* (*denshi* in Japanese), literally "transmission of the consumption bug", was caused by minute worms. According to Daoist texts, this disease passed through a cycle of six phases with turning points on certain calendar days, when the worms metamorphosed from one phase to the next; in the sixth phase they were thought to be highly contagious. A popular belief in these worms took root in Japan from the Heian period as part of the Kōshin folk religion. They were considered a deity's agents that reported a person's sins on the days of *kōshin*, which occurred once every sixty days in the calendar cycle. If a person abstained from sleep on those nights, the worms would remain dormant; otherwise they would divulge their host's sins to the deity, who punished people by causing consumption (*denshi*). At first, only the Heian-period aristocracy abstained from sleep on those nights, but the ritual spread throughout the country and was practiced in rural villages until the twentieth century.

Medieval Japanese Medicine

During the Kamakura period (1185–1333) warrior culture eclipsed the culture of the imperial court, but changes in medical theory and practice remained slow. The hereditary posts of court physicians remained in place, and the physicians who filled them were sometimes dispatched to treat the leaders of the military government in Kamakura. However these physicians had little opportunity to keep abreast of the changes in medicine that occurred in China, and in most of Japan Buddhist priests dominated medical practice.

The most important medical text of the Kamakura period was the *Don'isho* (which is untranslatable), written by Fujiwara Shōzen (1266–1337) in 1302. This work was significant for its visual representation of human anatomy, the first in Japan. Like previous Japanese medical texts, the *Don'isho* was also a compilation of Chinese sources, except for the section on leprosy, which was based entirely on Buddhist thought. This reflected current Japanese ideas concerning this disease. From the Heian period, leprosy was called a karmic disease (*gōbyō*), the result of sins committed in past lives. As in medieval Europe, leprosy was common during the middle ages in Japan and had much the same stigma; persons who developed leprosy became outcasts, shunned by their families and communities alike, finding succor only in the care of Buddhist monasteries. The stigma attached to leprosy did not change even into modern times. (The plague, another representative disease of medieval Europe, did not reach Japan until the late nineteenth century and was never a significant cause of mortality.)

The Muromachi (1333–1468) and Warring States (1468–1600) periods witnessed considerable changes in the theory and practice of medicine. During the Muromachi period Buddhist monks became increasingly knowledgeable in Confucian thought, and were influenced by the rise of Neo-Confucianism during the Song dynasty (960–1279) in China. This new current in Confucian thought emphasized the role of *qi* (vital force; *ki* in Japanese) and *li* (principle; *ri* in Japanese) in the order of nature, states, and individuals alike. The ideas of Neo-Confucianism entered the mainstream of Chinese medicine in the theories of Li Dongyuan and Zhu Danxi, who were active during the Yuan dynasty (1279–1368). Li and Zhu understood disease according to interrelationships between an individual's vital energy, environment, and behavior. In therapeutics they placed a new emphasis on emetics, purgatives, and medicines that caused a person to sweat. The ideas of Li and Zhu became influential in Japan from the early sixteenth century, primarily through the works of Tashiro Sanki (1465–1537), who studied medicine for twelve years in China. They then became established in the mainstream of Japanese medicine through the works of Manase Dōsan (1507–1574), who had studied under Sanki. Dōsan's descendants and followers remained highly influential, treating the warlords and hegemons of the Warring States period. Following the establishment of the Tokugawa *bakufu* in 1603 Dōsan's medical theories and practices became government-sanctioned medical orthodoxy (*hondō*). As such they remained at the core of some schools of medicine until the end of the premodern period, and had virtually no competitors until the beginning of the eighteenth century.

Dōsan and his followers had a powerful influence on diagnostics, which they standardized. Later and competing schools of medical thought only supplemented the diagnostic procedures delineated by Dōsan, which remained standard until the advent of Western diagnostics during the nineteenth century. He based diagnosis on a four-step method. Visual observation focused on the patient's skin color, weight, strength, condition of the hair, and inspections of sputum, feces, and urine. Aural observation included listening for responses indicating pain when the patient was examined, for the type of cough, and for sounds in the chest. The physician questioned the patient concerning appetite, waste elimination, emotional disposition, and the circumstances preceding the illness. Finally there came pulse diagnosis, a technique without parallel in Western medicine, which analyzed the strength, speed, location, and other aspects of the pulse.

Endemic warfare during the Warring States period stimulated new approaches to surgery, with a widespread need for specialists who could treat battle wounds. The arrival of the Westerners in East Asia during the sixteenth century was soon followed by the spread of firearms, whose wounds called for new forms of treatment. This led to the adoption of Portuguese surgical practices during the second half of the century in what was called the *Nanbanryū* (Southern Barbarian School). At this time, however, the Japanese adoption of Western medicine remained limited to practical measures for treating wounds, skin lesions, bone fractures, and dislocated joints, and contemporary European medical theory made little headway into Japanese medicine.

Early Modern Japanese Medicine

The defeat of the last major opponents to the hegemony of the warlord Tokugawa Ieyasu in 1600 marked the end of the Warring States period and the beginning of the Edo period (1600–1868), a time of peace and gradually increasing prosperity. The seventeenth century was a period of political, economic, and cultural stabilization and consolidation during which medical theories and practice changed little, dominated by the orthodoxy of Manase Dōsan and other established schools. A century of peace, however, ushered in a period of intellectual and cultural ferment that began during the last decade of the seventeenth century and continued into the nineteenth.

During the last half of the seventeenth century, Confucian scholars started to reject Neo-Confucian interpretations of the Chinese classics and focused instead on direct textual analysis. This trend appeared at the same time in medicine, with a number of physicians rejecting Song and Yuan dynasty interpretations of medical texts, emphasizing instead the direct reading of ancient Chinese medical works. Most of the leaders of this movement, which came to be called the School of Ancient Medicine (*Koihō*), lived in Kyoto, and included Gotō Konzan (1659–1733), Kagawa Shūan (1683–1755), Yoshimasu Tōdō (1702–1773), and Yamawaki Tōyō (1705–1762). Practitioners of the School of Ancient Medicine were by no means unified in their interpretations of either texts or phenomena, but most did emphasize practical methods to establish the validity of their ideas and methods. This was of momentous importance to the changes in Japanese medicine that followed during the rest of the premodern period.

Yamawaki Tōyō, a physician at the imperial court, was central to those changes. Tōyō questioned traditional Chinese interpretations of human anatomy and attempted to replace them with a view based on a passage in the *Zhou li* (Rites of Zhou), one of the early Chinese classics, which described the body as containing nine organs. To verify his view, Tōyō conducted the first public dissection of a human body in Japan in 1754, and published the results in the *Zōshi* (Anatomical Record) in 1759. Although Tōyō's nine-organ theory did not gain currency, his method of examining the body through dissections did. Thereafter, physicians in various parts of the country performed dissections and advanced other anatomical theories, none of which, however, replaced the Chinese theory of five organs and six viscera.

In 1771 three physicians, Sugita Genpaku (1733–1817), Maeno Ryōtaku (1723–1803), and Nakagawa Jun'an (1739–1786), witnessed the dissection of an executed criminal's corpse in Edo (Tokyo). Discussing the results afterward, they concurred that they could not consider themselves medically qualified without a true understanding of the structure of human anatomy, and that such an understanding could be gained only by translating the Dutch anatomy text which Genpaku and Ryōtaku had brought to this dissection. In 1774 they published the *Kaitai shinsho* [New Book of Anatomy], their translation of the *Ontleedkundige Tafelen*, or "Illustrated Anatomy", by the German physician Johann Adam Kulmus (1689–1745). This both started the study of Dutch medicine (*Ranpō*) in Japan and opened the door to European ideas by making the Dutch language widely accessible; after finishing this translation Ryōtaku then compiled a Dutch–Japanese dictionary.

During the first half of the nineteenth century, growing numbers of Japanese physicians studied Dutch medicine under Dutch and German physicians who offered instruction in Nagasaki. Until after the Meiji Restoration in 1868, when Western medicine became mandated as the sole basis for medical practice, various schools of Chinese medicine co-existed with Western medicine in Japan, but the door to Western medicine and science had been opened because of indigenous Japanese developments that had taken place during the eighteenth century, and in this respect the history of premodern medicine in Japan is unique in the world.

The spread of vaccination from the mid-nineteenth century the marks the beginning of modern medical practice in Japan. By the 1850s, vaccination had become common in many parts of the country. However it was not until after the Meiji Restoration in 1868 that modern medical education became instituted. Tokyo University was established in 1877, with its Medical School one of the initial departments. Thereafter, numerous Japanese students traveled to Europe, and particularly to Germany, to study medicine. Some continued to work in Europe as researchers and several, including Kitasato Shibasaburō (1852–1931), who co-discovered the tetanus antitoxin and plague bacillus, became internationally renowned; he was nominated for the first Nobel Prize in medicine. By the late nineteenth century modern medical schools had been established throughout the country and medical practitioners were required to hold state licenses.

During the twentieth century Japanese physicians have remained on the cutting edge of medical research in many fields. Since World War II, national health insurance has made medical care available to the entire population, helping to make the average life span the longest in the world.

WILLIAM D. JOHNSTON

REFERENCES

Bartholomew, James. "Science, Bureaucracy, and Freedom in Meiji and Taishō Japan." In *Conflict in Modern Japanese History*. Ed. Tetsuo Najita and J. Victor Koschmann. Princeton, New Jersey: Princeton University Press, 1982.

Bowers, John Z. *Western Medical Pioneers in Feudal Japan*. Baltimore: Johns Hopkins University Press, 1970.

Bowers, John Z. *When the Twain Meet: The Rise of Western Medicine in Japan*. Baltimore: Johns Hopkins University Press, 1980.

Farris, W. Wayne. "Diseases of the Premodern Period in Japan." In *The Cambridge World History of Human Disease*. Ed. Kenneth Kiple. Cambridge: Cambridge University Press, 1993, pp. 376–384.

Farris, W. Wayne. *Population, Disease, and Land in Early Japan: 645–900*. Cambridge: East Asian Monograph Series, Harvard University Press, 1985.

Jannetta, Ann Bowman. "Diseases of the Early Modern Period in Japan." In *The Cambridge World History of Human Disease*. Ed. Kenneth Kiple. Cambridge: Cambridge University Press, 1993, pp. 385–388.

Jannetta, Ann Bowman. *Epidemics and Mortality in Early Modern Japan*. Princeton: Princeton University Press, 1987.

Johnston, William D. *The Modern Epidemic: A History of Tuberculosis in Japan*. Cambridge: East Asian Monograph Series, Harvard University Press, 1995.

Kuriyama, Shigehisa. "Concepts of Disease in East Asia." In *The Cambridge World History of Human Disease*. Ed. Kenneth Kiple. Cambridge: Cambridge University Press, 1993, pp. 52–58.

Ohnuki-Tierney, Emiko. *Illness and Culture in Contemporary Japan*. Cambridge: Cambridge University Press, 1984.

Tatsukawa, Shōji. "Diseases of Antiquity in Japan." In *The Cambridge World History of Human Disease*. Ed. Kenneth Kiple. Cambridge: Cambridge University Press, 1993, pp. 373–375.

MEDICINE IN MESO AND SOUTH AMERICA Ancient America provides a unique case study for examining the independent development of medicinal practices and technologies in non-Western societal contexts. Initial European contacts with New World cultures of the early sixteenth century made clear that aboriginal medical systems and technologies embodied principles of a holistic — mental, somatic, spiritual, and supernatural — approach to healing. While the great centers of New World civilization provide our most complete record of medical practices and technologies, many localized native populations contributed to the extensive body of technical knowledge and expertise associated with herbal, chemical, surgical, extrasomatic, or ritual approaches to healing and public hygiene. From South American gold and other metal-based dental fillings, cranial trephination, post-cranial surgery, and coca-based anesthetics, to Mesoamerican intramedullar nails, medicinal enemas surgical sutures and cauterization, caesarean sections, topical anesthetics, poultices, and birth control, the list of ancient American medical practices and technologies is as impressive as it is extensive.

Because of the breadth and diversity of these practices in the Americas, we can only examine a narrow sampling specific to ancient Mesoamerica and Peru. The following discussion will move from a consideration of basic Native American concepts pertaining to the causes of disease to the examination of specific case studies concerning the development and sophistication of Native American practices The perspectives in question are drawn from contact era sixteenth-century accounts of the Aztec and Inca civilizations. One should bear in mind that the New World Inquisition inhibited and condemned the exercise of Native American medical practices. Through the entire duration of the colonial era (ca. AD 1521–1824) these practices were thought to be the work of sorcerers and other native practitioners in league with the Devil. European colonials actively sought to destroy ancient medical works, along with pagan practices and practitioners, throughout the contact era. While early chroniclers attempted to document such practices, they openly disparaged them, and every effort was made to minimize their significance by comparison with European practices of the time by way of blatantly ethnocentric and racist assumptions about the mental life and intellectual contributions and potential of America's aboriginal inhabitants. In those few instances where a concerted scientific effort was made to collect information on the medical practices of such groups as the Aztec (Sahagun, 1932), the distribution or publication of such works was prohibited for centuries.

Much of our knowledge of contact era medical practices is derived from the detailed chronicles compiled to document the cultural history of the Aztec and Inca civilizations. For example, medical anthropologist Bernard Ortiz de Montellano (1990) has subdivided Aztec concepts pertaining to the causes and treatment of disease into three categories: supernatural or religious, magical, and natural or physical. He indicates that the Aztec held a holistic world view pertaining to the causes and cures of disease, and refers us to the work of Mexican ethnohistorian, Lopez Austin: "the origin of illness is complex, including and often intertwining two types of causes: those that we would call natural — excesses, accidents, deficiencies, exposure to sudden temperature changes, contagions and the like — and those caused by the intervention of non-human beings or of human beings with more than normal powers. For example, a native could think that his rheumatic problems came from the supreme will of Titlacahuan, from the punishment sent by the tlaloque for not having performed a certain rite, from direct attack by a being who inhabited a certain spring, and from prolonged chilling in cold water; the native would not consider it all as a confluence of diverse causes but as a complex" (Lopez Austin, 1974; 216–217). This complex view required that the physician reconcile a variety of conceptual, spiritual, and physical dimensions in the course of diagnosis and treatment. Aztec doctors were required to balance herbal and other chemical treatments with interpretive models of causation ranging from the supernatural and magical to the natural, or a complex mix of both. The supernatural and magical, encompassing astrological interpretations such as those prevalent in sixteenth-century Europe, were of the greatest interest to early contact-era European chroniclers.

Botanical Knowledge

While the botanical repertoire of New World peoples is discussed elsewhere within this encyclopaedia, the relative significance of botanical specimens and knowledge to New World medical traditions necessitates brief consideration. It should be noted that recent research in this area makes clear the great contributions made by Native Americans (Schultes, 1994). The surviving Aztec herbal known as the *Codex Badianus* provides one of the most extensive listings of botanical specimens identified with the medication and treatment of a variety of ailments. However, in his efforts to make the herbal, authored by the Aztec doctor Martin de la Cruz, palatable to a European audience, Juan Badiano, the chronicler who prepared the document for submission to King Charles, modified it to incorporate European medical beliefs regarding the role played by temperature in illness, diagnosis, and treatment.

The large body of medicinal knowledge identified with the Americas, and subsequently adopted by European-based medical systems, included such ancient Native American medicinal and hallucinogenic substances as coca (*Erythroxylon coca*), mescaline (*Lophophora williamsii*), nicotine (*Nicotiana tabacum*), quinine (*Quina cinchona*), psilocyben (*Psilocybe mexicana*), dopamine (*Carnegiea gigantea*), anodyne analgesics (*Solandra guerrerensis*), the ergot alkoloid d-lysergic acid (*Ipomoea violacea*), and genipen-based antibacterial agents (*Chlorophora tinctoria*). To this list may be added medications and related chemicals and supplements ranging from N-dimethylhistamine to atropine, seratonin, tryptamine, kaempferol, prosopine, pectin, and camphor — to name but a few. These served Aztec physicians in a variety of capacities. Ortiz de Montellano has documented the medicinal properties of many of the herbal and chemical treatments administered by Aztec physicians. Included in that listing are diuretics, laxatives, sedatives, soporifics, purgatives, astringents, hemostats, hallucinogens, anesthetics, emetics, oxytocics, diaphoretics, and anthelmintics. Furthermore, there was a variety of antibiotic or antiseptic treatments for treating wounds, medicating infections and fractures, and performing surgery. These included the herbal vasoconstrictor *comelina pallida*, maguey or agave sap, for its hemolytic, osmotic, and detergent effects, hot urine in lieu of other available sources of sterile water, and mixtures of salt and honey which have been determined to provide enhanced antiseptic functions.

Recent studies of the "hidden chemical wealth of plants" used by the Native American tribes of the Amazon rain forest provide but one more point of departure for gauging the range and extent of pre-Columbian medical traditions (Schultes, 1994). In his recent summary of the pharmacology of the Kofan and Witoto tribes of the Amazon Basin, ethnobotanist Richard Evans Schultes has observed that "the forest peoples' acquaintance with plants is subtle as well as extensive. The Indians often distinguish "kinds" of a plant that appear indistinguishable, even to the experienced taxonomic botanist". This taxonomic acuteness extends to the level of being able to distinguish chemovars, or the basic chemical constitution of a specific subvariety, by visual inspection alone. Despite an estimated 80,000 species of higher plants in the Amazon, fewer than 10% have been "subjected to even superficial chemical analysis". Such a store of indigenous botanical knowledge recently prompted Schultes to ask "why not regard the Indians of the Amazon Basin as a kind of phytochemical rapid-assessment team already on the ground?"

Medical Specialists and Personnel

While it is clear from all accounts that herbal specialists existed in all regions of pre-Columbian America, the existence of a broader corps of trained medical specialists and personnel is less evenly documented. So intent were early contact-era European chroniclers on disparaging and discouraging pagan forms of medicine that much was done to reduce the role of medical specialists from Native American communities. In most instances, medical specialists ranging from herbalists to physicians were simply referred to as sorcerers or charlatans. Given the relatively impoverished state of European medicinal practices of the early sixteenth century, it is no wonder that most European chroniclers of Native American medical traditions expressed outright contempt for Native American physicians and their medical practices and traditions. Despite deliberate errors of omission and commission, surviving documents provide indications of the broad sophistication in Native American medicinal practices.

As for the documented existence of a scientific tradition with trained practitioners specialized in specific forms of surgical treatment, we are informed by archaeologist C.A. Burland that, in a region above Lima, Peru, there existed an ethnic enclave known as the Yauyos with whom the Inca collaborated in the training of specialists in the art of cranial trephination or skull surgery. The patient was drugged, and pain was alleviated by way of the application of direct pressure to nerve endings in the affected area. In such instances, trephination was used only as a last resort, whereby the diseased or smashed bone was cut away from the skull. Other forms of surgery included the "removal of a torn spleen, cleaning out of ulcers, and the cleaning and after-care of wounds" (Burland, 1967). Professional alliances between the Inca and Yauyo allowed for the exchange of technical knowledge and technicians. Accounts of the work of Yauyo physicians indicate that they were highly trained and had developed a formal discipline based on a corp of specialists. Apparently, herbal specialists of highland Peru were organized into "local confraternities", while medical specialists and physicians — *camasca* or *soncoyoc* — were sponsored by the Inca state and "highly trained within their own kind of collegiate discipline". This latter point is of paramount significance in establishing the existence of a specialized corps of medical personnel, and as such, the makings of a formally constituted and state-sponsored scientific tradition specialized in the treatment and trephination of the human cranium.

According to medical historian Gordon Schendel (1968), technical distinctions were made between "old" and "new" school physicians. At European contact, old school physicians were thought of as more traditional and more adept at conveying medical and spiritual beliefs pertaining to the art of healing, whereas the physicians of the new school engaged specialized methods and medical procedures. Among the Aztec, medical specialists included the *tlana-tepati-ticitl* or healer who "cured with medicines which were digested or applied on the skin" (Guzman Peredo, 1985), the *texoxotla-ticitl* or surgeon whose skills included bloodletting, and the *papiani-papamacani* or herbalist. Other terms utilized to identify Aztec medical specialists included *texoxtl*, or surgeons, and the *tlamatepatli* or medical interns of the *texoxtl* surgeons. The *tecoani* were the bloodletters; the *temixiuitiani* were the midwives. The *papiani* were the pharmacologists or herbal pharmacists. The *panamacani* are identified with pharmacognosists, or those individuals specialized in the identification, collection, and dispensing of herbal remedies, a specialty not unlike that of the plant pharmacologists of the Amazon Basin. Schendel has also documented the existence of a variety of specialists and areas of medical specialization, including internists, psychiatrists or psychotherapists, anesthesiologists, dermatologists, dentists, obstetricians, gynecologists, orthopediatricians, ophthamologists, urinogenital surgeons, and other practitioners specialized in the administration of tonsillectomy and embryotomy. In all areas of medical endeavor, specialists were held accountable for their actions and practices by their peers, as well as by the community at large.

Other specialists included those who used chiropractic methods, whereby "these doctors, in the case of falls, usually strip the patient and rub his flesh; they make him lie face downwards and step on his back. I have seen this myself, and I have heard patients say that they felt better . . . the pity of it is that there are even Spanish men and women who believe them [Aztec physicians] and are manipulated to serve their needs and evil" (Cervantes de Salazar 1936, as cited in Guzman Peredo, 1985). Spanish Friar Bernardino de Sahagun commended the technical expertise of Aztec physicians and noted that they "had great knowledge of vegetables; moreover, they knew how to perform bloodletting and to reduce dislocated bones and fractures. They made incisions. They healed sores and the gout. They cut the fleshy excrescence in cases of ophthalmia (inflamation of the eyes)".

Surgical Practices

As our discussion of medical specializations makes clear, pre-Columbian medical practices, particularly those pertaining to surgical methods, were comprehensive and sophisticated in scope. According to conquest period chronicles, "Aztec battle surgeons tended their wounded skillfully and healed them faster than did the Spanish surgeons ... [and] ... one area of clear Aztec superiority over the Spanish was the treatment of wounds. European wound treatment at that time consisted of cauterization with boiling oil and reciting of prayers while waiting for infection to develop the 'laudable pus' that was seen as a good sign" (as cited in Ortiz de Montellano, 1990). Medical historian Miguel Guzman Peredo cites Fray Bernardino de Sahagun: "Cuts and wounds on the nose after an accident had to be treated by suturing with hair from the head and by applying to the stitches and the wound white honey and salt. After this, if the nose fell off or if the treatment was a failure, an artificial nose took the place of the real one. Wounds on the lips had to be sutured with hair from the head, and afterwards melted juice from the maguey plant, called meulli, was poured on the wound; if, however, after the cure, an ugly blemish remained, an incision had to be made and the wound had to be burned and sutured again with hair and treated with melted meulli". The citation makes clear the availability of prosthetic or cosmetic devices, the use of sutures, and the application of maguey or agave sap as an antibiotic ointment.

The Aztec also maintained a complex typology for mapping human anatomy and physiology. They identified specific body parts, organs, and their respective biological functions, and employed anatomical terms for the articular surfaces and attachments of limbs, as for instance in the use of the terms *acolli*, *moliztli*, *maquechtli*, and *tlanquaitl*, for the articulation of the shoulder, elbow, wrist, and knee, respectively. According to Guzman Peredo, "those physicians had more than elementary concepts of the different organic functions. They knew, for example, of the circulation of the blood. They even became aware of the throbbing at the tip of the heart; this they called tetecuicaliztli. The radial pulse was called tlahuatl". Armed with such knowledge, the Aztec "used traction and countertraction to reduce fractures and sprains and splints to immobilize fractures" (Ortiz de Montellano, 1990). Perhaps one of the most significant medical innovations concerns the use of the intramedullar nail. Bernardino de Sahagun (1932) noted that in instances where bone fractures failed to heal, "the bone is exposed; a very resinous stick is cut; it is inserted within the bone, bound within the incision, covered over with the medicine mentioned". The intramedullar nail was not rediscovered by Western medicine until well into the twentieth century.

The most outstanding example of the empirical reliability and effectiveness of a pre-Columbian medical tradition centering on surgical applications was the use of cranial trephination or skull surgery. Examples of the practice have been documented from throughout South, Middle, and North America. An extensive review of this practice can be found elsewhere in the encyclopaedia.

Our review of pre-Columbian medicinal practices raises many more questions than can be addressed in this essay. Clearly, the European predilection for accomodating only that which suited prevailing eurocentric modes of thought contributed to the uneven documentation of significant Native American medical innovations and practices. A selected recounting of significant innovations should take into account practices centered on (a) holistic concepts of health, (b) state-sponsored public health programs, (c) an extensive body of anatomical terminology, (d) the existence of confraternal medical associations and state-sponsored medical corps, and specific practices centered on medical innovations such as those pertaining to (e) cranial trephination, (f) prosthetic and cosmetic devices, (g) antibiotic and antiseptic ointments and medications, (h) intramedullar nails, (i) formal procedures for the maintenance of dental health and hygiene, (j) psycho- and logo-therapeutic, or image-based, psychological approaches, and not surprisingly, (k) the largest pharmacological repertoire of effective and affective herbal and chemical remedies ever documented in the ancient world. Ultimately, any assessment of pre-Columbian medical traditions and innovations will need to contend with the fact that scholars have only just begun to scratch the surface of this New World of lost science and tradition.

RUBEN G. MENDOZA
JAY R. WOLTER

REFERENCES

Burland, C. A. *Peru Under the Incas*. New York: Putnam, 1967.

Classen, Constance. *Inca Cosmology and the Human Body*. Salt Lake City: University of Utah Press, 1993.

Cobo, Bernabe. *Inca Religion and Customs*. Trans. and Ed. Roland Hamilton. Austin: University of Texas Press, 1990.

Guerra, Francisco. *The Pre-Columbian Mind*. London: Seminar Press, 1971.

Guzman Peredo, Miguel. *Medicinal Practices in Ancient America*. Mexico: Ediciones Euroamericanas, 1985.

Lopez Austin. "Sahaguns's Work on the Medicine of the Ancient Nahuas: Possibilities for Study." In *Sixteenth Century Mexico*

— *The Work of Sahagun*, ed. M.S. Edmondson, pp. 205–224. Albuquerque: University of New Mexico, 1974.

Majno, Guido. *The Healing Hand: Man and Wound in the Ancient World*. Cambridge: Harvard University Press, 1975, reprinted 1991.

Ortiz de Montellano, Bernard R. *Aztec Medicine, Health, and Nutrition*. New Brunswick, New Jersey: Rutgers University Press, 1990.

Sahagun, Fray Bernardino de Sahagun. *A History of Ancient Mexico*. Trans. Fanny R. Bandelier. Nashville: Fisk University Press, 1932.

Schendel, Gordon. *Medicine in Mexico: From Aztec Herbs to Betatrons*. Written in collaboration with Jose Alvarez Amezquita and Miguel E. Bustamante. Austin: University of Texas Press, 1968.

Schultes, Richard Evans. "Burning the Library of Amazonia." *The Sciences* 34(2): 24–31. 1994.

See also: Trephination – Ethnobotany – Colonialism and Medicine

MEDICINE IN NATIVE NORTH AND SOUTH AMERICA

"Medicine" is an ambiguous word in American Indian connections. The first European missionaries and settlers learned that in the aboriginal languages the word corresponding to medicine could also be translated as supernatural power. A "medicine man" could be a man who healed a human being, but also a kind of a miracle man who through his connections with the supernatural world of gods and spirits was able to prophesy the future, locate lost articles or persons, bring on rain, attract game animals, call on the spirits, escort the newly dead to the other world, and many other things. In short, he was, and still is, a mediator between humankind and the spiritual world. It is through his equipment of supernatural power that he can cure the sick. In this article medicine will be understood as the way of curing and as a medicament, except in the compound medicine man, which will retain its old composite meaning. The role of the medicine man in curing will be further elucidated below.

All medical measures depend upon the ideas of the character of the disease. Slight injuries and mild diseases are not interpreted so much in terms of supernatural agencies at work, but dangerous diseases — or diseases considered to be dangerous — are mostly referred to supernatural intervention. Ghosts, unknown spirits, disease demons, witches, or taboo infringements are supposed to be responsible for cases of illness. The supernatural causation is a consequence of the fact that the basic harmony between humankind (or parts of humankind) and the sacred Universe has become upset. Prayers and propitiatory rituals may enclose all kinds of healing procedures but are more common where more grave diseases are concerned.

Whereas the supernatural element in curing is clearly present in the serious cases of disease and damage, natural knowledge plays a dominant role in the curing of ordinary wounds and diseases. Medicine men and women could relieve the pain and remove the diseases of this latter kind, but it was also common for old men and women to take care of them, at least in North America. These old people who had learned their medical arts during a long life and furthermore had particularly observed the traditional healing systems of the tribe, to a large extent used herbal medicines. They could therefore be called herbalists. It is not too much to say that herbs, for external or internal usage, were the most common medical cures in aboriginal America.

There were many different kinds of treatments. In Algonhian New England the colonists found that people who had been scalded had their sore skin washed with a strong decoction of tobacco, and thereafter a powder made from dried tobacco was sprinkled on to the wound. More to the south, in North Carolina, seeds of *Datura* were used for the same purpose. Bleeding was arrested with spiderwebs in large parts of North America. James Adair, who was a well-known trader among the Southeastern North American Indians in the latter part of the eighteenth century, reports that every Cherokee carried a variety of herbs and roots such as snake-root and wild plantain in his shot pouch as a remedy for the bites of poisonous snakes. The Indian chewed the root, swallowed a part of it, and applied another part to the wound. After some pain and contortions the man was relieved of the poison. Here we see how everybody was his/her own doctor.

From the same area the well-informed anthropologist John R. Swanton reports that the Creek drank a cold decoction of "Devil's shoe-string", or catgut (*Tephrosia virginiana*), to relieve themselves of bladder trouble. They also boiled the roots of sassafras (*Sassafras officinalis*) into a hot drink to get rid of bowel and stomach ache. Sassafras seems to have been a health medicament wherever it occurred. Venereal diseases were also cured with herbal decoctions.

Herbalism was a subject in Aztec schools, and some herbalists were even examined in the priest schools, the *Calmecac*. They were known for their empirical approach to health. Bernard Ortiz de Montellano writes that "the efficacy of Aztec wound treatments has been validated, as well as their accurate knowledge of the physiological activities of plants. Their extensive ethnobotanical knowledge and accurate taxonomy indicates that herbals may have existed and were taught in school, although no genuine pre-Columbian herbal has survived". As the same author points out, it would however be wrong to postulate a distinction between empirical and supernatural knowledge among the Aztecs, for "good doctors also included those who used mixed therapies and psycho-religious techniques". In this respect the old Mexican doctors remind us of North American medicine men.

In South America herbalism is mainly thought of as a medical subfield of the medicine man. Herbalist specialists among the Araucanians might know up to 250 medicinal plants. Erwin Ackerknecht, an authority on South American Indian medicine, has next to nothing to say about herbalism there. He points out, however, that such measures as massages, drugs, baths, bloodletting, diet and enemas do occur, and are "objectively effective". At the same time, the causation of such diseases, indeed, all diseases, is supernatural, he claims. Another expert, Alfred Métraux, insists that "light and common ailments and the sicknesses introduced by the Whites often were regarded as natural and were treated with drugs rather than by shamanistic means". He admits however that most diseases were attributed to supernatural causes. It is obvious that these questions have not been satisfactorily investigated. Several works indicate that herbalism was very widespread in South America, but whether there were — and are — herbalists who could be classed as inspirationally initiated medicine men is not clear.

A herbalist is, as seen in this article, a person who primarily deals with herbal medicine for internal or external usage. Secondarily he or she also handles stimulant drugs, emetics, and, in some cases, surgical operations. In North America the *peyote* (*Lophophora williamsii*), a cactus growing in the vicinity of the Rio Grande, is supposed to cure all kinds of diseases when eaten, drunk or smoked. Peyote is hallucinogenic but not narcotic; it is not habit-forming. A particular Peyote religion was formed at the end of the nineteenth century and spread over large areas of North America. The taking of peyote against disease occurs frequently in individual cases. In the cultic connections peyote is consumed because it gives supernatural blessings, including a medical cure. Peyote is powerful for many purposes, not just for medicine, although the medical reasons have been strongly supportive in the diffusion of the Peyote religion. It is very possible that other "herbs" have also had such a general effect because of their supernatural qualities. In South America narcotic beverages such as the *ayahuasca* prepared from the plant genus *Banisteriopsis* and the decoctions of *Datura arborea* are used in or after shamanic curing séances. Also here the drugs are active because of their spiritual force, and not because they have a specific medicinal content.

Some drugs have been favored since they have emetic qualities. In both North and South America poisonous or impure substances in the body are expelled from the mouth via emetics (and from the rectum via cathartics). The most well-known emetic is prepared from a particular holly, *Ilex vomitoria*, that grows in the southeastern part of North America. This drink, called "black drink" by the Whites because of its color, but "white drink" because of its supposed purifying qualities among the Indians of the Southeast, contains caffeine. Drunk in large quantities it provokes violent vomiting. This emetic was taken as a brew to produce purity before social and religious ceremonies — as ritual sweat baths in other areas — and in connection with diseases. The disease belonged to the impurities that were removed through the black drink.

There are some reputed cases of surgery in the old days; however, amputation seems to have been scarce. Scarification occurred in many places in North America. Thus, according to Frank Speck, the northeastern Algonkians tried to relieve pain by creating an exit for it. Skull surgery, or trepanation, is mentioned from both Americas. Some scholars have tried to find rational reasons behind such operations — head injuries, unconsciousness, and so on. Since they were performed in pre-Columbian times it is impossible to receive a definite answer. However, it seems more in conformity with American Indian thinking to pre-suppose an animistic model of explanation: the surgeons wanted to relieve the sick person from the spirit that plagued him.

The most common method was however to put some herb or bark on the wound or over the aching area.

If there is much uncertainty concerning the ideas of etiology and the nature of the healers in herbalism, there is more certainty of the medicine men and their disease ideology. As stated above, the medicine men and women, or those medicine men and women who are doctors, function as such when through their own or other healers' inspiration it can be stated that the disease is of supernatural origin. The medicine person was chosen by his (her) guardian spirit to conquer the malign spiritual influence behind the disease. His or her healing is dependent upon the power with which he/she was entrusted in the course of his/her calling.

The supernatural aspect of religious causation can mean many things, for instance, that witches on account of their supernatural powers, or sorcerers because of their magical manipulations, upset the normal health of individuals; or it can mean that transgression of tabooed places or actions, displeasing of the powers, or imbalance in the cosmic harmony cause the same result. Sometimes spirits and divinities introduce disease and wounds without any apparent reason. The immediate outcome of this supernatural line of action may be intrusion into a person's body of objects or spirits (which could be the same thing since objects may be inanimate manifestations of spirits, or instruments of spirits), and the loss of the soul which may be wandering around, or has been stolen by some witch or some spirit(s), usually the spirits of the dead. In the latter case the lost soul may be taken to the realm of the dead from which it is difficult for the medicine man to retrieve it and take it home. The soul entity that has been lost is usually a separable soul, mostly the free-soul, or the soul that sometimes distances itself from the body in dreams and trances. As long as the soul of vitality or, where it exists, the ego-soul remains with the body, the individual is alive; where all the souls (usually two souls, the free-soul and the body-soul) remain with the body the person is also alive. This is the general program, but many exceptions from this rule have been found among North and South American tribes.

There is a particular tendency among Native Americans to distinguish two definite systems of diseases behind the two causation theories. When the "intrusion" diagnosis is resorted to attention is directed to the body and its diseases. We can say that the patient's physical pain conducts the doctor. In "soul loss" however it is the sick person's mental state that stands in focus. His intellectual power fades away, fever or absent-mindedness rule, he languishes away, loses his consciousness, and so on. Scholars have until fairly recently tried to show that intrusion was more common in America, and therefore is an older diagnosis, whereas soul loss has had a more spotlike distribution and is therefore a younger diagnosis. As will soon be seen, the latter diagnosis corresponds to more difficult healing procedures. However, the more field research has proceeded the more cases of soul loss have been discovered. The present distribution of diagnoses confirm the generalization made here. Of course, in some regions we find intrusion as the dominating complex, in others, soul loss. But the presence of both diagnoses — for different types of disease — seems to be the original pattern. In our days soul loss is missing in many places, but the memory of its application by capable medicine men some decades ago is living.

In some places, for instance among the Navajo, the decision of the nature of the disease is left to particular diagnosticians who, in an inspirational state, are capable of finding out the roots of the disease. A serious disease to them is always referred to as being a break with the invisible world and its spirits; the diagnosis is never concentrated on the biological state of the individual, but on the acting spiritual forces. Among the Navajo the disease means that the afflicted person has fallen out with the balance and harmony of the Universe, so he has to be reintroduced by being identified with the supernatural powers in nocturnal rituals of up to nine nights. The officiant is a sacerdotal singer, not the diagnostician. The former could be termed a priest, the latter a shaman.

Shaman is a Tungusian word denoting those medicine men who perform their services in a trance or ecstasy. The word has since the eighteenth century become a technical term for medicine men who communicate with supernatural beings in states of trance. The Natives in America rarely have a term for differentiating shamans from ordinary medicine men; usually a shaman is said to be a stronger healer than others. Shamans occur, or have occurred, over most areas in America. The large majority of them, but far from all of them, handle diseases. When the recovery demands contacts with or intervention from supernatural powers the shaman appears on the scene in order to meet these powers in a trance situation.

All medicine men, shamans included, receive their doctor's powers through inheritance, spiritual calling, or vision quest. In the question of inheritance it is often possible to see who is going to become a doctor; he or she has a nervous psychic constitution or is reminiscent of a deceased medicine man. In South America it is not uncommon for a doctor to transfer his profession to his own sons or nephews. The calling of the spirits is more or less attached to the inheritance idea since the guardian spirits of a shaman try to find his successor in the same family. In the American Great Basin, California, and Gran Chaco are areas where shamans are called by the spirits. Sometimes this calling follows very aggressive lines, for instance, among the Mapuche of Southern Chile. Vision quests are used in the areas in eastern North America where ordinary individuals also seek guardian spirits in order to procure their powers and their protection. Shamanic spirits are stronger and more specialized in healing than other visionary spirits. The acquisition of spirits is connected with fasting and several days' staying out in the wilderness with the dangers of climate and wild animals. In South America the candidate smokes strong cigars and consumes *ayahuasca* wine in order to attain shamanic ecstasy. The novice also joins a medicine man school, or an experienced older shaman, in order to learn shamanic procedures and tricks.

The medicine man heals the patient partly through herbalism and partly through suction, blowing, massage, and wafting with feathers, to remove the disease object or disease spirit. In South America the medicine man may attain some degree of ecstasy by drinking *ayahuasca*, whereupon he blows thick clouds of tobacco into the patient's mouth. The disease object — an arrow, a dart — is sucked out by the medicine man and then regurgitated by him. In both Americas the medicine man often produces a little thing that he claims to have sucked out of the patient and that is supposed to have caused the disease.

In cases of soul loss the healing procedure is often more difficult and more dramatic. Let us however first of all state that the usual background of soul loss, the serious change of consciousness, in some places can be healed without too much difficulty on the part of the acting medicine man. Thus sometimes the soul may be called back by the medicine man, or sought by him and his associates in the neighbourhood of the camp or village. According to Rafael Karsten, the Jivaro Indians at the sources of the Amazon River know no remedy for fever diseases but destroy their own villages and leave them, apparently in the belief that not soul loss, but attacks of disease demons have hit the patients. In North America, for instance among the Yuma, it happens that soul loss is cured through blowing and suction, a proof that the intrusion model of disease diagnosis has formed a general curing pattern here.

The most common cure of soul loss is that the medicine man — who in this case operates as shaman — sinks into a trance to transgress the boundary between this and the other world. In North America this entrance into ecstasy is mostly brought about through auto-suggestion, and drumming, and in South America narcotics such as tobacco and ayahuasca effect the same state. The shaman's soul, and in some cultures his guardian spirit (it also happens that he is transformed into his guardian spirit) then departs on the long road to the land of the dead. Sometimes he manages to catch up with the fleeing soul and can then return it to its owner. If however the soul of the sick patient has reached the land of the dead the shaman has to risk his own life by seeking the fugitive soul in that realm and trying to persuade it to come back with him. In such cases the shaman has to fight violently with the mass of the departed who want to retain their newly arrived visitor. Usually the shaman wins the struggle and brings the patient's soul back home. He presses the soul against the patient's head or some other place, and after a short while the patient wakes up again. The shaman, who in most cases has been lying down as if he were dead during the soul journey, now also comes back to life.

Sometimes many shamans cooperate in a ritual drama in a so-called imitative shamanistic séance. An example of the latter is the voyage of shamans in a symbolical canoe to the land of the dead among the Puget Sound Coast Salish. The medicine men are equipped with boards which represent a canoe and paddles and act out the events of the voyage: the hardships on the journey, the battle with the dead, the release of the imprisoned soul, and its transporting home. Here it is not the question of a deep trance but a dramatic performance in the inspired state of an actor.

Collective medical contributions are also given in some agricultural societies, such as the Pueblo societies in New Mexico and Arizona. They have curing organizations constituted of people who have once been ill and healed by the same organizations. Animal spirits, so-called Beast gods, are the patrons of the medical societies among the Zuni Pueblo Indians. The societies are specialized in curing particular diseases, and use the methods adopted by medicine men healing patients suffering from "intrusion" diseases.

ÅKE HULTKRANTZ

REFERENCES

Ackerknecht, Erwin H. "Medical Practices." In *Handbook of South American Indians.* Ed. J.H. Steward. Washington: Bureau of American Ethnology, Bulletin 143, 1949, vol. 5, pp. 621–643.

Adair, James. *The History of the American Indians, Particularly Those Nations Adjoining to the Mississipi.* London: E. and C. Dill, 1775.

Hultkrantz, Åke. "Health, Religion, and Medicine in Native North American Traditions." In *Healing and Restoring: Health and Medicine in the World's Religious Traditions.* Ed. L.E. Sullivan. New York: Macmillan, 1989, pp. 327–358.

Hultkrantz, Åke. *Shamanic Healing and Ritual Drama: Health and Medicine in Native North American Religious Traditions.* New York: Crossroad, 1992.

Karsten, Rafael. "Zur Psychologie des indianischen Medizinmannes." *Zeitschrift für Ethnologie* 80(2): 170–177, 1955.

Métraux, Alfred. "Religion and Shamanism." In *Handbook of South American Indians.* Ed. J.H. Steward. Washington: Bureau of American Ethnology, Bulletin 143, 1949, vol.5. pp. 559–599.

Ortiz de Montellano, Bernard R. "Mesoamerican Religious Tradition and Medicine." In *Healing and Restoring: Health and Medicine in the World's Religious Traditions.* Ed. L.E. Sullivan. New York: Macmillan, 1989, pp. 359–394.

Sullivan, Lawrence E. *Icanchu's Drum: An Orientation to Meaning in South American Religions.* New York: Macmillan, 1988.

Sullivan, Lawrence E. "Religious Foundations of Health and Medical Power in South America." In *Healing and Restoring: Health and Medicine in the World's Religious Traditions.* Ed. L.E. Sullivan. New York: Macmillan, 1989, pp. 395–448.

Swanton, John R. "Religious Beliefs and Medical Practices of the Creek Indians." *42nd Annual Report of the Bureau of American Ethnology,* pp. 472–672. Washington, D.C., 1928.

Vogel, Virgil J. *American Indian Medicine.* Norman: University of Oklahoma Press, 1970.

See also: Religion and Science – Ethnobotany

MEDICINE IN THE PACIFIC ISLANDS In traditional Pacific societies medicine, the science and art dealing with the maintenance of health and the treatment of disease are closely interwoven with their indigenous religious beliefs and world views. These in turn are dynamic combinations of concepts and practices that have evolved in the context of the physical, cultural, and social environment in which the people have existed. The range and variety of traditional medical systems reflect the range and variety of their cultures. However for all of them, religion pervaded all aspects of life including the theory and practice of medicine.

The Pacific island traditions discussed in this account exclude those of the aboriginal people of Australia, and the Maoris of New Zealand. Most authorities estimate a total of about ten thousand islands in the Pacific, although the actual numbers quoted differ depending on the definition of island used. These islands cover less than one million square kilometers of land area scattered over more than thirty million square kilometers of ocean surface. The majority of the island groups lie within the tropics of Capricorn and Cancer. They are separated by vast expanses of ocean from the Mariana Islands to the northwest and the Hawaii Islands to the northeast, from Easter Islands to the southeast to Papua New Guinea to the west. The biota of Pacific islands display a high degree of endemism and a wide range of variability, from species rich continental high islands to the west to species-poor oceanic islands and atolls to the east and the north. Through extreme isolation both physically and in time, the people of these scattered island groups have evolved a wide range of cultures that reflect their diverse historical and ecological experiences. Over nine hundred different languages are spoken by the estimated 5.8 million inhabitants of the Pacific islands.

Indigenous Pacific islanders were classified by the early European explorers into three major cultural groups: The Melanesians (Papua New Guinea, Solomon Islands, Vanuatu, New Caledonia, and Fiji); the Polynesians (Tonga, Samoa, Cook Islands, French Polynesia, Tokelau, Tuvalu, Niue, Wallis & Futuna, Rotuma, Easter Islands, and Hawaii); and Micronesians (Kiribati, Nauru, Marshall Islands, Federated States of Micronesia, Belau, Guam, and Commonwealth of the Marianas). These terms are derived from the Greek words; *melanos* (black pigment), *polys* (many), *mikros* (small), and *nesos* (island). The three major cultural groupings are not mutually exclusive as there are many instances of intermingling of races and cultures. Archaeological evidence indicates that human occupation of the Pacific islands began at least 20,000 years ago from Southeast Asia. While the settlement of Melanesia continued from 20,000 to 10,000 years ago, migrations to Micronesia and Polynesia began only some 3000 to 4000 years ago. The often held theory that Pacific Islanders came in separate waves of "pure" races is refuted by evidence from latest archaeological and blood-type genetic studies. These indicate a more complex process of mixing and differentiation which is also reflected in the systems of indigenous medicine.

Twentieth century Pacific islands are, on the whole, quite cosmopolitan, receiving significant influences from cultures with which they have come into contact. British, French, German, Spanish, and American colonialists facilitated the migration into the Pacific of Asians, some Africans, and other ethnic groups. The widespread adoption of Christianity, and the introduction of modern scientific medicine, both first introduced by Europeans, have been significant influences in the changes to the nature of traditional medicine and the relative decrease in the customary respect associated with its use. On the other hand the inflow of a wide range of cultures with new settlers has added new ideas and expanded the repertoire of treatments applied by current practitioners of traditional medicine in many Pacific societies. What is today known as traditional medicine in the Pacific is a continually evolving product of the interaction of all of the above-mentioned factors. It continues to be practiced despite official discouragement in some areas and official disinterest in others.

This account can be no more than an overview of a widely variable practice of traditional medicine belonging to a great diversity of peoples who have developed systems of healing in a highly diverse physical setting and just as diverse cultural milieux. Many of the general observations made here would have exceptions or minor modifications from country to country and sometimes, even from island to island within one country.

In all traditional Pacific societies health and illness are considered to be the outward manifestation of the status of balance of spiritual power (sometimes known as *mana*) within a community. The maintenance of life and health is achieved through maintaining right relationships with other people in the community, with society as a whole, with the physical environment, and with the spirits identified by the community. The right relationship is associated with keeping an accepted balance of spiritual power between all these components. A person maintains good health when she/he and her/his close family members continue to fulfil obligations to family, community, and spirits. These obligations cover the social, economic, and political life of the community. In some communities these obligations include moral aspects of behavior, such as avoiding anger, adultery, jealousy, and theft. Obligations to spirits usually comprise a set of taboos (*tabus* or *kapus*). Obligations are mainly defined by a person's role in society and her/his position within the family. In some societies the role is determined by birth – particularly for males. For example the chiefs of the more hierarchical Polynesian societies are determined by birth; in a Melanesian society, a bigman (the equivalent of a small chief) often has to earn his position. In Micronesian societies too the roles are generally inherited.

Health is maintained when one carries out one's role faithfully, whether it be as fisherman, farmer, wife, priest or spokesperson for the community. The relationship with the physical environment is often governed by a system of taboos which cover sacred sites, sacred trees, and sacred animals both of the land and of the sea. The ultimate guardians of the society and its environment are the spirit gods worshipped as ancestors or respected as guardians of different aspects of the physical world. Violation of any of the rules that regulate relationships could incur the wrath of the spirits and therefore bring about some disturbance to health. In some cases illness may occur when good relationships between ordinary humans are disturbed without incurring the wrath of the spirit gods. The latter, however, may still be called upon to help put right the relationships and heal the sick. This happens for example in the ceremony of *ho'oponopono* in Hawaii. The extended family of the sick person, through confessions and prayers that may take several days, put right any relationships gone awry before physical treatments are sought.

Investigators of traditional medicine have usually focused on the treatment rather than the prevention of illness in traditional Pacific societies. There is therefore not as much information about the ancient beliefs and practices for the prevention of sickness, apart from keeping the taboos.

Traditional preventive medicine still widely practiced today is largely associated with women and children. Healers who deal with the health of women and children and who also include traditional midwives, much more than other healers, have a repertoire of herbal concoctions both for preventive as well as curative purposes. More than other healers also, they often prescribe special rules of diet for prevention of illness — particularly for women during pregnancy, childbirth, and lactation, as well as for babies and children.

Where belief in spirit illness is still prevalent, the use of special plant parts or herbal formulations for protection against such is also widespread. Often families or tribes have their own special formulae or plants.

In most Pacific societies it is recognized that illness occurs as a result of several factors: the transgression of rules for keeping harmonious relationships; the result of accidents or recognized physical causes; sorcery and spirit causes; or the exposure to sicknesses coming from outside of the community. Generally, illnesses are recognized in terms of their socially — or spirit — based causes rather than their biological nature. This raises difficulties with attempts to equate traditional medicine concepts and diagnoses with the concepts and diagnoses of modern medicine.

Relationships between a person and her/his social, physical, and spiritual environment are governed by systems of taboos which regulate social behavior, human relationships, food handling, and rights to natural areas and resources. In keeping these relationships one is accountable to different spirits. An offended spirit inflicts his/her power (*mana*) on the offender or a close relative resulting in sickness. Different societies recognize different classes and types of spirits that can cause sickness or help in healing. Generally however, there appear to be four to six major categories. Four common categories include:

- major spirit gods recognized by the whole community, such as the four great gods of Hawaii, or the great *vus* of Fiji, or the culture heroes of Madang in Papua New Guinea;
- minor spirits that belong to individual family lines;
- deceased relatives; and
- guardians of specific areas of sea or land.

Some societies recognize one powerful spirit above all others such as the god creator of the people of Madang, while a lot of communities also recognize the existence of both harmful and benevolent spirits that may be found anywhere.

Certain individuals who seek the assistance of specific spirits for purposes of magic or sorcery have to obey special sets of rules to maintain their relationships which when transgressed usually cause death.

The misfortune resulting from a transgression may befall either the person who transgresses or a close relative of hers/his. In both cases the more important part of the treatment is the righting of the relationship that had been transgressed. An illness resulting from a transgression may take any form of biological manifestation, or lead to psychological disorders. In some parts of the Pacific, particularly in East Polynesia (Hawaii, Tahiti, Cook Islands), the atonement for transgressions is performed first before any application of physical cures that address the biological nature of the illness. In some cases this atonement may comprise the total cure. In other societies, particularly in Melanesia, the treatment also includes application of herbal cures, massages, surgical operations or other biologically based treatments. In all cases, the atonement for transgressions equates with the balancing of power within the community by strengthening the patient and weakening the antagonistic forces. Illnesses associated with transgression of relationships or breaking of taboo may be manifested as physical or psychiatric disorders and may appear either suddenly or as a gradual process.

All Pacific island countries now have modern health systems that offer alternative health care. Pacific island peoples, now overwhelmingly Christian, largely use this system to treat their illnesses. Even in such cases, as particularly prevalent in Papua New Guinea for example, the steps to seek atonement are still taken as a necessary component of the cure.

Serious illnesses are usually identified as the result of sorcery when treatment both by modern medicine and traditional herbal concoctions fails to elicit a positive response. It is usually accepted that illness from sorcery will result in death unless treated through invoking a powerful spirit's help.

Accidental spirit illnesses cover a wide range of symptoms from minor skin ailments to madness. These are usually attributed to whims of evil spirits that may act on anyone without being solicited.

Illnesses that result from accidents or physical causes include cuts and sprains from careless activities, headaches from overexposure to a hot sun, diarrhea from eating stale food or drinking dirty water, skin disease from bathing in river water, fish poisoning, fracture from a fall, or cold from working long hours in wet clothes. Any of these could also be a manifestation of the transgression of right relationships, or of sorcery. If it is determined to be so, then the necessary ceremony or treatment for atonement or for sorcery is applied. Otherwise these are treated by purely physical means that vary from one community to another and from one healer to another.

Isolation kept Pacific islands relatively free of many diseases which have been introduced since contact with the outside world. Where the diseases are of fairly recent introduction they may be described as boat sickness, new sickness, or foreigners' sickness, such as in the case of bouts of influenza brought into the isolated island community of Tikopia in the Solomon Islands by visiting tourist boats. In other cases they may be described as whiteman's disease, or real disease. These latter descriptions are used particularly where the introduced diseases have been with the society for several generations. Illnesses in this category include measles, whooping cough, mumps, influenza, dengue fever, cholera, typhoid, and some venereal diseases.

Diagnosis in traditional medicine systems of Pacific peoples is often an iterative process with treatment. It is a process of therapeutic trial. Accounts of the nature of diagnosis by different authorities describe a lack of interest in the detailed physiological or biological nature of the illness. Rather the interest is in the treatment itself which then reveals the basic cause. Healers, usually specialists, often wait for the result of the treatment they give before confirming the diagnosis. If within the first three to five days the patient does not show the expected response, the healer may try another treatment or declare that the sickness is not one that s/he is a specialist of. In the latter case the patient sees some other healer. The process is: initial diagnosis and treatment; if this does not work, another diagnosis and treatment; should that not work, the process continues until the basic cause of the illness is finally identified as determined either by the treatment's working convincingly, or everyone concerned's agreeing with the diagnosis. Nowadays doctors trained in modern medicine become incorporated into this process of repeated diagnosis and treatment. In urban centers, Pacific islanders may seek hospital treatment first. If that fails, a traditional healer is often consulted and the iterative process of diagnosis and treatment continues. In rural areas where hospitals are difficult to reach, traditional healers may be consulted first, while the hospital doctor is only sought if the traditional system has not produced a positive result. Whether the hospital or private doctor is consulted before, in between, or after the traditional healers are consulted, is dependent entirely on the whims of the sick and her/his family. It is not uncommon for patients to take traditional herbal medicine in conjunction with medicine prescribed by modern doctors. This is particularly so when the diagnosis indicates a social or spirit based cause.

The identification and classification of illnesses in traditional medicine in the Pacific cannot be equated with the categories of illness recognized by the modern medicine system. Illnesses are generally classified according to what is believed to be the basic cause. Thus most societies recognize two major categories: those caused by purely physical means, and those caused by spirit powers. The different categories of these have already been explained. In many instances no systematic relationships appear to exist between the identified cause and the illness. Sometimes a range of seemingly unconnected ailments are all called by the same name; such as for example the *ira* in the Cook Islands and Tahiti, which is the *ila* in the rest of Polynesia; or the *macake* and the *kasi* in Fiji.

In some cases these categories group together ailments that are conceptualized to result from a similar biological defect. For example, the term *kasi* in Fiji is the name for ailments that are said to result from the cold penetrating to the muscles or bones, or to the internal organs and causing sharp pains. These include arthritic and rheumatic ailments. In Tonga the term *fasi* covers ailments that constitute an injury or break — sprains, fractures, or dislocations, and even breaks to internal organs. In other cases these inclusive terms indicate ailments that share the same basic cause. For example, accidental exposure to evil spirit forces may cause a wide range of sicknesses all called *baubau* in Fiji or sickness of talking spirits in Santo, Vanuatu. Otherwise the biological basis for a classification of traditionally recognized diseases is often unclear. The *ira* or *ila* of Polynesia is a collection of ailments that usually affect children but may also afflict adults. The manifestation of *ira* in the Cook Islands includes skin complaints, convulsions, lethargy, and pains in certain parts of the body. *Ila* in Tonga includes bad smelling stool, anal rash, birth marks, and skin discoloration.

There has been no written account of how healers classify illness on biological or physiological bases. Generally however some basic principles for classification, based on symptoms, may be drawn. These include regions of the body and sequence of stages as found for example in Rarotonga in the Cook Islands; or location in the body divided into internal organs, body muscles, skeleton, and skin, as found in Fiji.

The range of treatments offered by practitioners of traditional medicine in the Pacific include herbal medicine, surgery, bone setting, massage, spiritual healing, and magic.

Writings of early European settlers and missionaries to the Pacific indicate that the use of herbal medicine was much more widespread in Melanesia than in Polynesia and Micronesia. Pre-European contact mediated some transfer of knowledge of herbal cures as occurred for example between the Melanesian Fijians and the Polynesian Tongans. Throughout Polynesia, reports of early European settlers indicate that healers were priests who mediated between the living community and the spirit world. Accounts in the literature indicate a greater use of herbal medicine in traditional Hawaii than in other parts of Polynesia. Indeed some accounts claim that herbs may have initially been applied only externally in other Polynesian countries such as Tonga, Samoa, and Cook Islands. Since contact with the rest of the world, traditional herbalists have expanded their repertoire of treatment to incorporate much that they have learnt from outside sources. The association of herbal medicine skills with spirit powers is still strong in some societies, such as in parts of the Federated States of Micronesia. Here the knowledge of medicinal plants is a strictly guarded secret and plant treatments are usually accompanied with whispered incantations or chants. While herbalists in other parts of the Pacific may also whisper prayers, the practice is not as common.

Recent surveys of herbal cures used in several Pacific nations showed widespread use of herbs. A recent authority described over 150 plant species used in five countries of Polynesia. A total of over 340 species have been identified as medicinal plants used in Fiji while preliminary surveys in Solomon Islands and Kiribati recorded 143 and 42 species respectively. In most societies adults generally have some knowledge of herbal treatments. Most know of a basic stock of herbal treatments that may be equated with a first aid kit. However herbalists know much more than the rest of the community, often keeping the components of their specialist cures secret.

Herbal concoctions may be applied externally or taken orally. Leaves, stem, bark, and roots are all possible components. Concoctions may contain one or more species of plants. Melanesian herbalists tend to use more species in combination than do their Polynesian and Micronesian counterparts. This may be a reflection of the greater availability of species of plants to choose from in Melanesia which has a much richer flora than either Polynesia or Micronesia.

A significant proportion of the plants used for herbal treatments were introduced after European contact, often being wayside and garden weeds. Recent surveys for example show that in Kiribati, a Micronesian country, 13% of medicinal plants are post contact. In Tonga, a Polynesian country, the proportion is 27%, while in Melanesian Solomon Islands it is 13%. Some authorities argue that the widespread use of introduced plants, particularly those recently introduced, imply that many of the herbal treatments have been learnt from recent migrants into Pacific societies. Written accounts of Polynesian societies during first contact with Europeans indicate that Cook Islanders may have learnt much of their knowledge of herbal medicine from the Tahitians and from early European settlers, while the Samoans probably gained additional knowledge from Tongans and Fijians. Tongans are recorded to have learnt about herbal cures both from Fiji and from early Europeans. Tuvalu herbalists gained much from Fiji. Today's herbalists in Fiji enrich their indigenous repertoire with knowledge of herbs brought in by migrants from Asia, Europe, and other islands of the Pacific.

For ointments, dressings, salves, and droplets leaves are usually used with occasional application of bark latex or juice of fruit. Leaves may be crushed and the juice squeezed onto the affected area, or they may be heated and rubbed gently between the palms to soften them before being applied as dressing.

For baths, leaves are usually crushed in a large container and the patient bathed in the infusion. With fevers and extensive skin diseases, patients are often subjected to steam baths with crushed leaves in steamy-hot water. Some treatments expose the patient to smoke from burnt leaves.

For oral medicine, leaves, bark, roots, or whole plants may be used either singly or in combinations of two or more different species. The leaves are crushed in water. Roots and bark are pounded and mixed with a little water. Invariably water is used as the medium for the medicine. Sometimes it may be replaced by coconut water and occasionally by sea water. Some healers regularly advise bathing in sea water for certain ailments.

While in the majority of cases herbal medicines contain only plant parts, Micronesian healers in Pohnpei sometimes use pebbles in combination with herbs, while Melanesian healers in Fiji sometimes add fish bones to herbal concoctions.

Traditional surgeons, aptly called *matai ni yago* or carpenters of the body in Fiji, were much more prominent during times of tribal warfare before the introduction of Christianity. Nowadays surgical operations are left to the modern health system. Bloodletting and incisions are however still practiced in some societies for the treatment of body pains and minor swellings.

Records of the use of hot stones, smoke, and heated plant parts or hot cloth for the treatment of diseases are not as common as records of herbal treatments. Various forms of heat application are still used today.

Massage is a very common means of treatment for a wide range of complaints. These cover headaches, body aches, stomach aches, coughs and colds, diarrhea, sprains and fractures, or arthritic and rheumatic pains, and possession by spirits. Massage takes several forms: gentle stroking with the open palms; pressure application with the thumbs; kneading with the knuckles; and application of pressure through the feet and heels. Most adults can apply massage for the relief of tired muscles or minor body pains. Like herbalists, however, masseurs are usually specialists.

Traditional bone setters are found in most Pacific societies. These are still widely used in some societies today to treat fractures. For a serious compound fracture a bone setter may perform a simple operation to open up the wound, set the bones in place, treat it with herbs and cover it with more herbs to heal. For a simple fracture, some bone setters will manipulate the bone into place with herbal massage while others forbid any form of massage. These may use a rough splint where necessary to prevent misalignment of the bone. Otherwise they rely totally on oral administration of herbs to help the bones sit back in place and heal. These herbal concoctions sometimes contain bits of bones.

Spirit healing is still widely practiced today. The diviners and witch doctors who relieve illness through invoking the powers of the ancestral spirit gods are held in respect. Sorcerers who cause diseases by invoking the powers of the spirits are considered to be anti-Christian. The power of the spirit healers who invoke the assistance of the spirits to do good is often considered a gift from the Christian God.

Diviners use a variety of means both for diagnosis and for treatment. Cards, patterns made by leaves crushed between the palms, dreams, and even possession of the patient or the healer with spirits, may be used for diagnosis and for determining treatment. Diviners and witch doctors often also use herbs to treat the physical manifestation of the illness and to drive the offending spirits away. Herbs used for the latter in all societies are always those with strong odors.

Some societies claim to have certain people gifted with healing powers that are hereditary and do not involve the use of herbs or the soliciting of spirit assistance. For example in Fiji, descendants of the firewalkers are claimed to be able to heal serious burns by touch. Members of a certain family of the northern tribes of Fiji are supposed to be able to relieve fish poisoning merely by washing their hands in a basin of water and giving the water to the patient to drink.

SULIANA SIWATIBAU

REFERENCES

Christian Medical Commission, World Council of Churches. *Pacific Regional Consultation on the Christian Understanding of Health, Healing and Wholeness.* Geneva: World Council of Churches, 1981.

Eade, Elizabeth. *Preliminary Bibliography on Traditional Science and Technology in the Pacific Islands.* Suva, Fiji: Pacific Information Centre, University of the South Pacific, 1992.

Johnston, Emilie G. "A Review of Literature on Native Medicine in Micronesia with Emphasis on Guam and the Mariana Islands." *Guam Recorder* 5 (2): 60–65, 1975.

Parsons, Claire D., ed. *Healing Practices in the South Pacific.* Honolulu: University of Hawaii Press, 1985.

Sorcery, Healing and Magic in Melanesia. Melbourne, Australia: La Trobe University, 1982.

Tamson,R. *Bibliography on Medicinal Plants and Related Subjects.* South Pacific Commission, Technical Paper No.171. Noumea, New Caledonia: South Pacific Commission, 1974.

Whistler, Arthur W. *Polynesian Herbal Medicine.* Lawai, Hawaii: National Tropical Garden, 1992.

See also: Ethnobotany in the Pacific

MEDICINE: TALMUDIC MEDICINE Although whole books have been devoted to the study of talmudic medicine and quite a number of studies to specific related topics, only a small number of scholars in the field of medical history are aware of this rich corpus of knowledge. The core of the Talmud (the authoritative body of Jewish law and tradition), called *Mishnah* (divided into six tractates), was compiled between the second century BC and the second century AD. Two extensive commentaries and glosses were added to the basic text. One is the so-called Jerusalem Talmud, which was completed in the fifth century AD. The Babylonian Talmud which was much larger, was sealed in the sixth century. No medical texts from the ancient Hebrew-Jewish period have reached us, so the wealth of medical knowledge that is interspersed in the Talmud is the sole source of documentation in these matters. It should be made clear that these medical data are recorded by talmudic scholars in the midst and for the sake of legalistic discussions. In most cases few details are provided and only those that are relevant to the specific point under consideration.

The spectrum of medical knowledge covered in the Talmud is very wide. The field of anatomy (mainly of animals) is impressively represented, particularly in the tractate *Ḥullin* which deals with dietary laws. The inspection of slaughtered animals is one of the remarkable institutions of Jewish law, as it related to public health. In order to decide whether an animal was acceptable for consumption (*kasher*) or not (*taref*), anatomy and pathology had to be mastered to a considerable degree. Two examples will be given here. The number of bones is considered in the context of uncleanliness: it is ruled that everything contained in a tent (or room) in which there is a number of bones amounting to more than half of a corpse becomes unclean. The body is comprised of 248 bones (it says 'members', *evarim*), corresponding to the number of days of the lunar year, and 365 'sinews' (*gidim*) corresponding to the solar year [bab. Makkot 23b]. The Talmud mentions that once research was done on the corpse of a young female prostitute that had been executed by the (Roman?) authorities, and it was found that there were 251 bones. The Sages opined that the discrepancy was due to the fact that this was a young woman [bab. Bekhorot 45a]. Another opinion was that the number of bones may vary from 200 to 280 [Tosefta, Oholot 1:7]. Osteology was not very exact, even in 'academic' medicine. Galen speaks of "more than 200 bones" [De Form. Foet., Bk. 6]. Interestingly enough, Ibn Sīnā (Avicenna) and al-Zahrāwī (Abulcasis) both accepted the number 248, most probably in accordance with the lunar analogy and the macrocosmos/microcosmos similarity.

The second example is related to neuroanatomy. The case history featured a lamb which dragged its hind legs along, and the question was, "What kind of lesion does it have?" One sage said, "This is a case of sciatica" (Hebr. *shigrona*). Another replied, "Possibly this is a lesion of the spinal cord" (Hebr. *ḥut ha-shedrah*). The text then says, "They examined it (i.e. they performed an autopsy), and the second diagnosis was authenticated". This shows a remarkable experimental approach. The sages decided that if an animal dragged its legs it would be said to have sciatica, and such an animal would be permitted for consumption. [bab. Ḥullin 51a]. This is a very typical example of talmudic casuistry. The sages not only allow, they even advocate thorough examination of the case, but the decision is based on the opinion of the majority and on the most frequent occurrences.

The Talmud, in both versions, is full of observations of medical and historical interest pertaining to internal medicine, gynecology and obstetrics, dermatology, neuropsychiatry, surgery, traumatology, and most other specialties (including otology, ophthalmology, and dentistry). Other fields such as dietetics, preventive medicine, forensic medicine, public health, and materia medica are also widely represented. Instead of listing a catalog of items, we shall give a number of examples that stand out for their detailed description and/or for their originality.

Commenting on the scriptural verse [Deut. 7:15] which reads, "And the Lord will take away from thee all sickness . . .", the Talmud asks the question, "What is sickness?" A number of answers are provided by the Sages in both versions of the Talmud [bab. Baba Mezia 107b; jer. Shabbat 14:3]. In both versions we find three cardinal causes of diseases. One pertains to scientific medicine: the bile; one is related to popular medicine: the cold (or cold-and-warm); and one belongs to magic lore: the evil eye. Other agents that appear only in one version include: air (or wind); fever; abnormal or superfluous secretions; climatic factors (the *sharav* wind); obsession (one who is persuaded that he is sick), and carelessness (which is particularly stressed, thus enhancing the preventive aspects of medicine). Two other factors are mentioned elsewhere in the Talmud. One is blood (i.e. plethora): "At the head of all (causes of) diseases am I, the blood" [bab. Baba Bathra 58b]. The other is changes in one's habitual way of life [ibid. 126a]. Another version mentions changes in one's usual diet [bab. Kethubot 110b]. This particularly developed topic is characteristic of talmudic lore, as it includes popular beliefs, empirical notions, and scientific aspects. Humors (blood, bile), specific symptoms (fever), environmental causes (winds, cold), magical aspects (evil eye), psychic factors (obsession), and a heedless way of life (carelessness): these are part of a broad spectrum of the agents of sickness.

Gynecology and obstetrics are well represented in the talmudic corpus, particularly in the tractate *Niddah*. Menstruation (during, and seven days after which a woman cannot be approached by her husband), vaginal bleeding, recognition and duration of pregnancy, as well as a wealth of details related to embryology, sterility, and abortion, are only examples of the topics considered by the Sages. We shall again select two items: cesarean section, and embryotomy. Regarding cesarean section, it is stated [bab. Arakhin 7a], "If a woman dies during labor, the operation is performed even on the Sabbath day and the child is ripped out of the womb". A decree of this kind was already extant in the early Roman *Lex Regia* of Numa Pompilius. More challenging is the question of whether cesarian section was performed on a *living* mother in talmudic times, a procedure that has not been documented in adjoining contemporaneous cultures. We read [Mishnah, Bekhorot 8:3]: "A child born through the (abdominal) wall (Hebr. *yoze dofen*), and the one who comes (i.e. is born) after him, none of them are considered (legally) as being first-born". It seems that it was acceptable in the times of the *Mishnah* (second century AD) that a woman give birth by cesarean section, recover, and give birth later to another child in the normal way. Such a possibility seemed quite strange to Maimonides [see his commentary ad loc.], who tried painstakingly to devise a case that would make sense. There has been a lively and prolonged discussion among historians of medicine and talmudic scholars regarding the definition of *yoze dofen*. Some think it could mean extra-uterine pregnancy; others advocate abnormal birth through a perineal tear, or even through the anus. I am among those who think that this was a theoretical case based maybe on animal pathology [see ibid. Bekhorot 2:8], but the question is still open to discussion.

Embryotomy is mentioned in the *Mishnah* as well: "In case a woman experiences difficulty in labor (her life being in danger), her fetus should be cut to pieces inside her womb and extracted limb by limb" [Mishnah, Oholot 7:6]. It is clearly stated that the life of the mother has preference over that of the fetus. However, once his head, or the majority of his body, is out of his mother's body, the fetus may no longer be harmed, for one life cannot be put aside for the sake of another.

Circumcision was, according to the biblical narrative, first performed by Abraham, and was henceforward to be performed on every male child at the age of eight days [Gen. 17: 10–14]. The technical details of the surgical procedure are nowhere mentioned in the Bible; they are however discussed in the Talmud. What interests us here are some of the complications of the operation. It is stated that if two children of the same mother have died (from hemorrhage) as a result of circumcision, the third child should not be circumcised. It says further that if two children of one mother or one child each of two sisters dies, the third should not be circumcised.

The sages remark that there are families in which the blood is 'loose', whereas in others it is 'tied up'. This is most probably the first historical description of hemophilia and of its genetic transmission [bab. Shabbat 134a and Yebamot 64 b].

The operation that should be performed on a newborn baby which presents an imperforated anus is described in detail [Shabbat, ibid.]. It says that the membrane should be opened crosswise. This is noteworthy because a circular incision would indeed have been ineffective.

In accordance with Hippocratic medicine, the Talmudic sages considered that an eight-month baby could not be viable [bab. Shabbat 135 a]. A baby was usually breast fed for two years. Therefore if a nursing mother lost her husband, she was not allowed to marry again until two years (at least 18 months) had elapsed, for fear that she might become pregnant again and her milk production could be stopped [bab. Yebamot 42ab]. In order to prevent a nursing mother from becoming pregnant (which could stop lactation), some sages advocated the use of a pessary (*mokh*) during intercourse. Others promoted withholding intercourse altogether during lactation, which was repeatedly urged in ancient medical lore. Wet nurses were hired in case the mother could not nurse her child. The wet nurse was supposed to tend this nursling exclusively; she could not even nurse her own child [bab. Kethubot 60b]. The use of non-Jewish nurses was permitted. Some sources granted this permission without any condition [Tosefta, Niddah 2:5]; others asked that the nurse be under the parents' close supervision [Mishnah, Avodah Zarah 2:1]. Moreover the use of milk from animals, even from unclean beasts, was permitted, as the child was considered in deadly danger if he got no milk whatsoever.

Even a brief abstract of talmudic medicine cannot disregard the topic of public and personal hygiene. Dietetics are not exclusively centered on dietary laws. Several axiomatic statements, based on empiric and popular lore, are recorded. One should eat simply [bab. Shabbat 140b], slowly (carefully chewing food) [bab. Berakhot 54b], moderately [Shabbat 33a] and regularly. "Any change in one's usual diet is the beginning of bowel disease" [bab. Sanhedrin 101a]. Wine and meat are characteristic of a festive meal. [bab. Pesaḥim 109a]. Excess of meat is considered harmful; the priests in the Temple suffered from bowel diseases, as they consumed too much meat (from the offerings). There was even a 'specialist' for these diseases; his name was Ben Aḥijah [Mishnah Shekalim 5:1–2].

Fasts were instituted by the sages to commemorate great calamities such as the destruction of the Temple. The only fast that is of biblical foundation is the Day of Atonement (*Yom Kippur)*. Fasts were also initiated in case of oncoming epidemics, wars, floods, or drought. Such fasts did not usually exceed 24 hours [Mishnah, Ta'anit 3:3–5].

Personal hygiene included cleanliness as reflected in the laws of purity. Women took a ritual bath after menstruation and observed an additional seven days purification period. General hygiene is featured in a number of regulations on water supply, lavatories, bathing, care for the dying and burial, and the Sabbath rest.

Even after all this explanation, we can still ask the question: Is there actually a specifically talmudic medicine? I prefer the view that there is definitely a fascinating topic labeled "Medicine in the Talmud". These rich and multi-faceted data should take their legitimate rank in the history of both medicine and culture.

SAMUEL S. KOTTEK

REFERENCES

The Babylonian Talmud. Trans. I. Epstein and colleagues. London: Soncino Press, 1935–1952.

Codell Carter, K. "Causes of Disease and Death in the Babylonian Talmud." *Medizinhistorisches Journal* 26(1–2): 94–104, 1991.

Ebstein, Wilhelm. *Die Medizin im Neuen Testament und im Talmud*. Stuttgart: Enke, 1903.

Kottek, Samuel S. "Breast-feeding in Ancient Jewish Sources". *Assia* 7(3–4): 45–56, 1980 (in Hebrew).

Kottek, Samuel S. "Concepts of Disease in the Talmud." *Koroth* 9(1–2): 7–33, 1985.

Leibowitz, Joshua O. "Jews in Medicine." *Maḥanaim* 102: 18–35, 1969 [in Hebrew].

Preuss, Julius. *Biblical and Talmudic Medicine*. New York: Sanhedrin Press, 1978 [First German ed. 1911].

Rosner, Fred. *Medicine in the Bible and the Talmud*. New York: Ktav, 1977.

TRADITIONAL MEDICINE IN THAILAND In villages all over Thailand one finds people referred to as traditional doctors (*moo phaen booraan*), who use mainly herbs (*samunphraj*) to cure diseases. The sum of knowledge and practices of all these experts is called traditional Thai medicine. There is no doctor who knows the whole body of knowledge, and there exist many local and individual variations within Thailand both in the way diseases are diagnosed and in the way plants and other materia medica are used in the curing process. Traditional Thai medicine resembles that of its neighboring countries Burma, Laos, and Cambodia.

Traditional Thai medicine contains knowledge about the identification of plants (and minerals and animal components) and their curing properties, about the diagnosis, cause, and development of diseases, about prescriptions, and about the relevant incantations and ceremonies. The prescriptions, which contain the concrete relationship between specific diseases and the plant world, are considered to be the vital and most valuable part of the tradition. Normally the prescriptions contain from five to ten ingredients, but some contain as many as twenty or even fifty different ingredients. The prescriptions specify for each ingredient which part of the plant should be used — root, leaf, bark, fruit, etc. — and in what quantity. While much of the medical knowledge is transmitted orally, the prescriptions are written down. There exists a vast number of medical manuscripts in Thailand, both in temple libraries and in care of the herbalists, and almost all of these manuscripts are collections of prescriptions.

The medical tradition in Thailand is supposedly a chain of teacher–pupil relationships which goes back to the tradition's original teacher, Jiwaka Komarabhacca, the legendary physician of the Lord Buddha. Medical knowledge is restricted to those who have been ritually initiated into the tradition. Thus traditional medicine is, like other branches of traditional knowledge, surrounded with a certain amount of secrecy. This is particularly true for the incantations and the prescriptions which are regarded as the central pieces and most valuable parts of the medical tradition.

The traditional doctor is normally a specialist. He treats a certain disease or a group of related diseases. His career quite often starts with being sick himself, seeking out a traditional doctor, getting cured, and thereafter asking to be accepted as a student of that doctor and learn the prescription that cured him. Then he will start treating people with this particular disease. If he is successful, he may seek out more knowledge from other teachers, and thus gradually expand his curing range. Few traditional doctors, however, become true generalists and few become full-timers. To most of them curing remains a supplementary income. Still, a successful traditional doctor does receive respect and additional social status.

There are altogether probably several thousand plants (and ingredients from the mineral and animal kingdoms) in traditional Thai medical prescriptions which have a curing property of one kind or another. A traditional doctor does not need to be able to recognize any of these plants in nature, because there are a number of herbal stores where he can buy the ingredients he needs. Still it is convenient if the traditional doctor has the plants he uses most often ready at hand. Thus many herbalists do grow a number of medicinal plants in their gardens, and do have some botanical knowledge. A number of monks are traditional pharmacists or doctors, and in many Buddhist temples there is extensive gardening of medicinal plants.

There are also a number of traditional pharmacists who specialize in producing traditional medicine for sale. Thus people have easy access to ready-made traditional medicines. These drugs are cheaper than the Western drugs available, and although it is admitted that they may take a longer time to produce results than cosmopolitan medicine because they are weaker, it is also claimed that they have fewer side-effects than modern medicine.

Before any treatment can take place the disease has to be identified, that is, given a name, and the cause of the disease has to be established.

Many of the causes used by traditional Thai doctors are also recognized as causative factors by modern medicine. Still, traditional medicine operates with other types of causes, which modern science summarily refers to as superstition, namely spirits, black magic, and karma.

Certain epidemic or very contagious diseases, like plague, cholera, dysentery, tuberculosis, and polio, are said to be caused by unspecified spirits. There exist, as far as we know, no traditional medicines which claim to cure these diseases.

If the diagnosis is that a patient in some way has offended and angered a spirit, who in retaliation has caused the offender to fall sick, then the curing process may consist of ritual offerings only.

There are cases characterized by sudden violent and abusive behavior towards one's nearest relatives (typically daughter-in-law towards mother-in-law living in the same house). These cases may be attributed to possession by an evil spirit. A specialist is then called upon who will talk to the spirit and ask it who it is, where it lives and what it wants. If the spirit does not want to leave the body after offerings have been given, it is mercilessly beaten until it leaves. The point is that the patient is not held responsible for what she did or said during possession.

A similar type of asocial behavior, which is more introvert and unpredictable, may, after all other treatments have proved fruitless, be attributed to a good spirit, who willfully has inflicted this disease on the patient to force her to accept to become its medium. When the person accepts mediumship the spirit withdraws the disease, which immediately disappears.

Piercing or lancing pains in the joints, abdomen, or chest may, if unresponsive to treatment, be attributed to black magic. This means that an object — like a nail or a lump of skin — has been inserted into the patient's body by a doctor using special incantations, and acting on the instructions of an enemy of the patient. The only solution is to find a doctor who has even stronger incantations to counteract the original ones, and thus remove the object, or even send it back to attack its owner.

The Buddhist concept of karma may also be used as an ultimate causative factor, but only if all other explanations have failed. Using karma as the cause implies that the disease is a deserved and inescapable result of an action one has committed previously. People do not favor this explanation, because it means that cure is impossible and that one simply has to endure one's fate.

We have noted that food may be regarded as a cause of a disease. Many of the most common medicinal plants are in fact also used as herbs to season the food. The close relationship between food and disease is furthermore seen in the many food prohibitions which are part of the treatment of many diseases. These rules may contain prohibitions against meat, frogs, fish with skin, seafood, pickled food, certain vegetables, eggs, liquor, or foods with certain tastes — all depending on the disease in question. There are also specific dietary rules for the post-natal period.

The recitation of holy words plays an important role in many contexts in Thailand, and there is a strong belief that recitation of holy stanzas, if performed by the right persons, produces power and that this power can be transferred. Incantations are an integral part of traditional medicine and are in many cases recited over or blown onto the medicines and the patients by the traditional doctor. There are also treatments, for example against rabies, which consist of incantations only.

Traditional Thai doctors do not perform surgery or dissection, except for lancing abscesses. Thus their knowledge of the interior organs and how they function is superficial.

Massage and midwifery are separate branches of the local medical tradition.

Traditional saunas are also found, where damp from water containing herbs is led into a small, closed room. These steam baths are regarded as healthy in general, but can also be used medicinally to cure certain skin or respiratory diseases.

Thai culture has been influenced by India in many areas since ancient times, and this influence has been strongest among the elite at the court. This is also the case for traditional Thai medicine, which has developed a court tradition which claims Indian ancestry. Standardization of this court medical tradition was attempted in the 1890s, and has now become the officially recognized traditional medicine in Thailand, with schools in certain temples, printed textbooks, and official examinations. In order to practice traditional medicine legally one has to pass examinations from these schools, where courses in traditional pharmacy, midwifery, and general medical practice are offered. Women are allowed to participate on an equal footing with men. The Indian influence on the court medical tradition is most clearly seen on a linguistic and theoretical level: many of the concepts and the disease names are borrowed from Pali or Sanskrit, and central concepts, like the five elements (earth, water, wind, fire, and ether) and the three humors (bile, wind, and mucus) are of Indian origin. The fact is, though, that the whole theoretical structure lacks internal coherency and consistency, and that its relationship to actual medical practice is weak. Thus the theoretical structure functions as an overall frame of reference, but is not very useful in the treatment of concrete cases. Thus the royal tradition has an explicit theoretical framework which the village tradition does not have, but the actual medical practices of the two branches are still very similar.

For a casual visitor to Thailand, Chinese medicine is quite visible, especially in the towns, where one easily notices drugstores selling Chinese herbs and places where acupuncture treatment is offered. The Chinese custom of using the horns, the bones, or the bile of certain animals as aphrodisiacs contributes to the endangering of mammals like the rhinoceros, tiger, wild boar, bear, and certain snakes.

Western medicine has been present in Thailand since the 1830s, and the official medical system has, since the beginning of this century, been entirely based on cosmopolitan biomedicine. The official policy towards traditional medicine has generally been restrictive, although during the latter decades, certain attempts have been made to study and integrate the two systems. Pharmaceutical firms show particular interest in extracting medically active chemical components from traditional medical herbs, hoping to discover new drugs. Western medicine is also adopting traditional disease terms for modern disease concepts, thus eroding the traditional concepts and causing some confusion. Some herbalists have invented herbal medicines which they claim can cure new disease concepts, such as diabetes, cancer, and AIDS.

If we look at the patients' choice of curing methods in the countryside, we find, according to some statistics, that the large majority — some 95% — prefer Western medicine, while only a small minority — some 5% — choose traditional medicine. Furthermore, traditional medicine is in many cases resorted to only after Western medicine has failed. This means that patients who no longer can afford modern treatment or people with terminal, incurable, or long-term diseases will take refuge in traditional medicine. But there are also certain illnesses where traditional medicine has a reputation for being more effective than modern medicine, like hemorrhoids, calculus, and particularly diseases with 'wind' as a prominent symptom.

We should also note that people who are sick often go to several doctors and several pharmacies, both Western and traditional, and thus follow parallel treatments. This has the unfortunate consequence that they stuff themselves with too many drugs. It also makes it impossible to decide which medicine actually cured the disease.

Even if cosmopolitan medicine is dominating the medical scene in Thailand, it is likely that there always will be a corner for traditional medicine. And the growing awareness of and pride in one's national heritage, as well as the current popularity of 'nature' and everything 'natural' also contribute to strengthening the hand of traditional medicine.

VIGGO BRUN

REFERENCES

Brun, Viggo and Trond Schumacher. *Traditional Herbal Medicine in Northern Thailand*. Berkeley: University of California Press, 1987.

Golomb, Louis. *An Anthropology of Curing in Multiethnic Thailand*. Urbana and Chicago: University of Illinois Press, 1985.

Mulholland, Jean. *Medicine, Magic and Evil Spirits. Study of a Text on Thai Traditional Paediatrics*. Canberra: Australian National University, 1987.

MEDICINE WHEELS The Indians of the North American Plains, because of their nomadic ways, were not dependent upon permanent settlements and structures. Aside from rings of stone left after *tipis* were moved there is little evidence of construction left by these tribes. The noted exceptions to this generalization are the formations known as "Medicine Wheels". More than one hundred of these formations have been found in North America ranging through Saskatchewan, Alberta, Montana, North Dakota, Wyoming, and Northern Colorado. These sites are all located north to south, within a few hundred miles of the eastern boundary of the Rocky Mountains, suggesting the migratory patterns of the Native American culture of this area. They are distinguished from *tipi* rings and camp sites by their size and by the spokes or arms which radiate from a central cairn. These formations are found in isolated locations, inappropriate to community dwelling. Always of native stone, these "wheels" usually consist of a central hub of stone, and/or an inner circle, from which other stone lines radiate out toward the horizon. They are not meticulously constructed but rather seem like a design that a few workers could put together in a matter of days. However, the dimensions are sometimes substantial. The Bighorn Wheel in Wyoming, for example, has a central hub about twelve feet across; with radiating spokes ninety feet in length.

The existence of many of these stone formations has been known for at least a century. The Bighorn Medicine Wheel was reported in the 1880s by explorers and prospectors; by radio carbon analysis it has been dated AD 1600–1700. The first archaeologists to investigate the Bighorn site asked local Native Americans about its use or origin but initially none of the tribes claimed responsibility for its existence or knowledge of its use. Gradually stories emerged, but there is always the question of whether the informants were creating the sort of responses that they thought anthropologists wanted to hear. In point of fact, none of the indigenous groups, such as Sioux, Cheyenne, and Arapaho, have folk legends and traditions which either explain or validate the role of Medicine Wheels in their traditional cultures.

Until the 1970s, the generic explanation for Medicine Wheels was that they must have been of "ceremonial" use, though no one was forthcoming with any detailed account of just what ceremony. The Sioux and Cheyenne are known for their Sundance ceremonies, for which special circular medicine lodges were built. Some observers have been tempted to point out the spatial similarities between the circular medicine lodges built around a living tree used as the center pole with roof and walls radiating from that center and the medicine wheels with central cairn and radiating spokes. The symbolic similarity may be clear, but real intent and purpose remain unknown. So until the 1970s, Medicine Wheels were at most a curiosity of Plains Indian culture neither well known nor well explained. Then an integrative approach called "archaeoastronomy", linking the previously separate disciplines of archaeology and astronomy, emerged as a new way of examining ancient phenomena. Basically its unifying theme is to examine megalithic and prehistoric formations for astronomical significance. The best known example of this is Gerald Hawkins' work documenting the astronomical alignments of Stonehenge in Salisbury, England. Soon practitioners of both astronomy and archeology were looking with renewed interest at ancient megaliths around the world, including Medicine Wheels.

The astronomer John Eddy is most closely associated with an archeo-astronomical interpretation of North American medicine wheels. He, in collaboration with an archeological team, studied several medicine wheels, including the Bighorn Wheel in Wyoming and the Moose Mountain Wheel (radio carbon dating, AD 100–500) in Saskatchewan. His findings suggested that many such formations could have been used as landmarks in the rolling plains of an otherwise undistinguished terrain. The simplest of these seem only to point to other stone formations, in the same way guideposts mark long trails across isolated lands.

However, for both the Bighorn and Moose Mountain wheels, Eddy notes solar and stellar alignments that could have been used for solstice calculation. Normally solstice observation is associated with agricultural societies, but nomadic groups such as Plains Indians also needed techniques for anticipating the severe weather changes of the northern plains. While such an interpretation is not conclusive, various studies are now suggesting that the Plains Indians were very careful observers of the evening sky as well as the land. The Medicine Wheels may be indicators of their ability to use markings on the earth to note celestial changes in the sky and the corresponding journey of the seasons; all useful information for high plains travelers. In any case Medicine Wheels could have been of both ceremonial and astronomical utility since the world view of the plains tribes would have made no significant distinction between spiritual and stellar forms.

LAWRENCE TYLER

REFERENCES

Eddy, John A. "Astronomical Alignment of the Bighorn Medicine Wheel." *Science* 184: 1035–1043, 1974.

Eddy, John A. "Medicine Wheels and Plains Indian Astronomy." *Technology Review* 80 (2): 18–31, 1977.

Eddy, John A. "Mystery of the Medicine Wheels." *National Geographic Magazine* 151(1): 140–146, 1977.

Krupp, Edwin C., ed. *In Search of Ancient Astronomies*. New York: Doubleday, 1977.

Nikiforuk, Andrew. "Sacred Circles." *Canadian Geographic* 112: 50–60, 1992.

Tyler, Lawrence. "Megaliths, Medicine Wheels and Mandalas." *The Midwest Quarterly* 21(3): 290–305, 1980.

METALLURGY IN AFRICA A condensed discussion of African metallurgy is difficult because of the large size of the continent and the three thousand years over which it developed south of the Sahara desert. Furthermore, several metals were produced and used in Africa, and metal production involved many technological steps which were not necessarily used for each metal type (i.e. iron, copper, gold, tin). Iron production, for example, involved mining iron ore and smelting it to a bloom, a non-molten mass of metal intermixed with a waste product called slag. The bloom was then forged into objects by hammering, welding, and other processes. Some pre-industrial societies made cast iron, a molten form of iron, but there is little evidence for this technology in Africa. Copper and copper alloys, on the other hand, were often made by reducing ore into molten metal and pouring it into molds, or by hammer forging solid copper. Gold was hammered out from its original nugget form, or melted and cast. Pellets of tin were removed from the slag output of tin smelting and were then melted into ingots. Moreover, not all societies used the same metals or mastered the same manufacturing steps for any given metal. Some cultures specialized in iron smelting, while others forged iron blooms into objects. In copper producing areas, some societies had craftsmen who cast molten copper into molds, others hammered out unrefined copper, and still others specialized in drawing out copper wire.

Much of the current knowledge about African metallurgy concerns iron production. There are several reasons for this. Iron has the clearest presence in the archaeological record, iron ore is virtually ubiquitous across Africa, and, for centuries, it was a subject of interest to Greek and Arab travelers and, later, European explorers, missionaries, and scholars. Copper smelting and casting, as well as the production and use of bronze (an alloy of copper and tin), brass (an alloy of copper and zinc), and gold, have also received considerable attention. The more isolated occurrences of these technologies across Africa in the archaeological and ethnographic record, however, have resulted in less detailed reconstructions of their diverse histories and their significance to African cultures. Some tin production occurred in southern and western Africa, but this is the least studied indigenous metal and may have a relatively short history as compared to iron, copper, and gold.

The prehistory and history of African metallurgy come from several sources. Archaeological excavations often yield metal objects, and/or the physical remains of production centers (i.e. smelting furnaces containing slag, tuyères or blow pipes, and charcoal fuel; forging pits; casting crucibles; molds). Increasingly, archaeologists submit these remains to laboratory analysis to determine how an object was made, or the chemical and physical dynamics of an ancient metallurgical process, its environmental context, and its age by radiocarbon dating. Some scholars conduct collaborative research with village elders who still remember how to smelt and forge iron, or cast copper. Several projects have resulted in important films which underscore the complexity of many metallurgical operations, provide critical insights into poorly understood ancient practices in Africa and worldwide, and highlight the non-technical aspects of production. Some researchers also perform experimental reconstructions of metallurgical processes to understand further the thermodynamics involved. A final source of information is the archival record of numerous visitors to Africa over the centuries. The reports from the early twentieth century first featured the non-technical characteristics of African metallurgy, including esoteric knowledge, decorated furnaces, ritual, music, and taboos.

The following discussion begins with a brief overview of the prehistory of African metallurgy, then examines the technical diversity of metallurgical practices across Africa, and concludes with a look at its social and ideological components.

Although this review principally concerns Africa south of the Sahara desert, the earliest evidence of metal production and use was in Egypt. Copper was first used there around 5000–4000 BC and was being smelted by 3000 BC. Iron objects were rare through the Middle Bronze Age, but became more frequent during the New Kingdom after about 1570 BC. Iron smelting was practiced by the eighth century BC.

Ancient Egyptian metallurgy had no influence on the rest of the continent except to the immediate south in Nubia and the later kingdom of Kush, both in modern Sudan. An Egyptian outpost was established in Nubia to smelt local copper ores in 2600 BC, and Egyptians exploited Nubian gold from an early time. By the next millennium, Nubian craftsmen worked copper, bronze, silver, and gold. At Meroë (ca. 500 BC to AD 300), the capital of the Kushite state, there were craftsmen of copper, bronze, gold, and iron. The earliest iron slag from the site dates to the fifth century BC; domed, brick smelting furnaces, possibly of Roman influence, were used after about 200 BC.

There is controversy over two interrelated aspects of the origins of metallurgy in sub-Saharan Africa: (1) whether it was invented indigenously or introduced from elsewhere; and (2) which metal — copper or iron — was smelted first. Researchers looking for a 'natural' progression of pyrotechnological knowledge (i.e. simpler copper smelting to more complex iron smelting) to prove indigenous origins have been thwarted to date. Copper smelting only seems to have preceded iron smelting in Nubia and, during the early to mid first millennium BC, along the southern Sahara in Niger and Mauritania. Current evidence shows that iron appeared first or at the same time as copper in the rest of sub-Saharan Africa. Other than at Meroë, the earliest indications of iron smelting are in Nigeria (ca. 900–800 BC), Niger (ca. 500 BC), Rwanda/Burundi (ca. 700–500 BC), and Tanzania (ca. 300 BC).

Many archaeologists believe that knowledge of iron smelting was brought from abroad, but they cannot agree on the route of introduction. Several routes have been proposed: (1) Egypt to Meroë, then west and south; (2) from the Phoenician or Roman coast of North Africa across the Sahara desert; and, (3) via the Indian Ocean. The first has been ruled out because of the early evidence of iron smelting in West Africa. Little else can be resolved until issues over radiocarbon dating are settled, excavations are conducted in poorly explored regions, including the North African coast and Ethiopia, extensive historic linguistics are done, and the possibility that more than one route existed is carefully investigated. It is now recognized, however, that iron and copper were introduced to different regions of Africa at different times.

Iron working spread from the regions of early introduction in West Africa, Sudan, and East Africa to Southern Africa in 500–700 years. This rapid expansion was once thought to be linked to the movement of Bantu-speaking agriculturalists as they traveled south and east from their homeland in present-day Cameroon, over three thousand years ago. Recent archaeological and historic linguistic evidence discredits this theory. Traces of early iron smelting have been found in modern Gabon and Congo, but not in northeastern Zaïre. Historic linguistics suggest that iron working in East Africa probably had northeastern, not western, origins. The earliest evidence of copper smelting to the south of Cameroon, on the other hand, occurs by the fourth century AD in copper-rich areas on the Congo coast and in northern Zambia/southeastern Zaïre at about the same time as iron working.

Intensified use of iron, copper, bronze, gold, and brass during the first and second millennia AD was connected to, but not the cause of, the rise of African states, urbanization, and the development of long distance trade routes in some regions. For example, the city of Jenne-Jeno in present-day Mali was well-developed by the third century AD. It did not begin to receive gold from across the Sahara desert until around the eighth–ninth centuries AD or brass (a known import based on the non-indigenous zinc present in the alloy) until the ninth–tenth centuries AD. Jenne-Jeno then became a major trade center in metals. In southern Africa, a long distance trade in gold developed between people living in modern Zimbabwe and those in coastal towns on the Indian Ocean by the tenth century, on the back of an earlier trade in ivory and skins. A powerful state with its capital at Great Zimbabwe (AD 1275–1550) prospered, in large part, by taxing the gold mined and traded from this region. Bronze, a golden alloy of copper and tin, also was developed in the area around this time, but little is presently known about the technology and its local significance.

In many other parts of Africa at this time, relatively small quantities of iron and/or copper were produced, used, and traded locally. The diversity of mining, smelting, and forging techniques that developed over the last two millennia is extensive.

Metal ores, such as specular hematite, malachite, and galena, may have been mined many millennia ago in sub-Saharan Africa, but they were ground to powders for cosmetics. Once ores were mined for smelting, several techniques were developed depending on the type of ore exploited and the physical and technical constraints encountered.

A similar array of mining techniques was developed for copper, iron, gold, and tin across Africa. Some copper and gold ores differed from iron and tin ores in one critical way, however: they could be found in their native metallic state. Mining for native metals involved picking up workable nuggets, panning for and concentrating metal flakes in watery contexts, or digging shafts along seams of metal. This was often labor intensive, such as in ancient Zimbabwe where gold mining involved considerable digging, crushing the matrix rock, and then amassing the gold by panning.

African miners exploited the oxide minerals of copper, iron, and tin at or near the earth's surface for smelting. They dug narrow shafts until the ore was exhausted, the shaft was unsafe, or they reached water. Large open ditches were also excavated. Both techniques required metal tools and, in the latter case, fire-setting was also used to break up the rock. Often, the matrix rock had to be crushed to concentrate the ore. Cases also existed of panning to concentrate rich iron ores dispersed in sandy matrices. While many aspects of African metallurgy were performed and controlled by men, the labor of carrying the ore, panning, and working in the narrow shafts was often done by women and, probably, children.

A facet of African metallurgy that has long intrigued scholars is the bewildering diversity of iron smelting furnaces over time and across space. At a very general level, three types of furnaces were built: pit or bowl furnaces lacking walls and operated by bellows and tuyères; shaft or walled furnaces with a pit beneath, also operated by bellows and tuyères; and tall furnaces (ca. 2.5–4 meters), often without a pit, that used natural draft to stimulate combustion. Within each general type, there was enormous variation based on whether or not slag was tapped from the furnace during smelting, numbers of tuyères, bellows type, pit depth and diameter, height and shape of the furnace walls, building materials, decoration on the furnace walls, and presence/absence of interior furnace features.

Some scholars have theorized a chronological sequence for these general furnace types, from "primitive" pit furnaces in which wrought iron was made to shaft furnaces in which low grade steel was produced to natural draft furnaces. Regional and local variations would have then developed over time. Recent findings reveal that the earliest furnaces all had walled shafts, and that a heterogenous bloom, varying from soft wrought iron to high grade steels, could be produced in each furnace type. It is now generally thought that furnace variation was a response to certain local constraints, such as ore types, but primarily to the different sociocultural contexts into which iron smelting was brought or developed.

Less is known about African copper smelting because it was practiced less widely and has been the subject of less research and ethnographic observation. Most copper smelting furnaces were the shaft type, but considerable variation existed based on whether the molten copper was tapped directly from the furnace into molds or was melted in crucibles within a furnace. Pit size, building materials, wall height and shape, number of tuyères, bellows type, mold sizes and shapes, and wall decoration also varied.

Gold was not smelted, but melted in crucibles over an open fire. There is little evidence of gold extraction technologies in Africa except for occasional pieces of gold-encrusted crucibles from Zimbabwe and South Africa. Tin ore, like iron ore, had to be smelted or reduced in a furnace, but little is known about the processes used or when they were developed. An ethnographic account from Nigeria indicates that tin and iron ores were co-smelted and tin was removed from the slag. In South Africa, tin ore was smelted in tiny furnaces, pellets of tin were extracted from crushed slag, melted in crucibles, and cast in molds.

The fabrication techniques used to transform raw metal into objects were also diverse and varied by type of metal and region. Iron and steel, for instance, were always hammered into shape, but many types of hammers and anvils were used over the continent. Skill at welding together pieces of iron and/or steel and other joining techniques varied widely. Evidence that iron was drawn out into wire comes from East Africa and to the west in modern Angola, but the history of this technique is vague. Interestingly, no evidence of precolonial heat treatment of steel (i.e. quenching it in water) to harden and strengthen an object has been found.

Copper was manipulated in several ways: hammer forging solid metal into various forms; drawing out wire; casting molten copper into molds; and alloying copper with another metal, like tin to make bronze, and then casting the alloy (the ability to cast copper is improved by adding tin). Evidence to date suggests that a sophisticated lost-wax casting technology (intricate forms are made of wax and encased in clay; molten metal is poured into the mold which replaces the wax) was developed in West Africa by the ninth–tenth century AD. The site of Igbo Ukwu in Nigeria has yielded remarkable castings that were followed by later traditions at nearby Ife and Benin.

Igbo Ukwu also provides the earliest evidence of bronze which was used to make many lost-wax castings. Investigations into whether bronze alloying was an indigenous discovery in West Africa is ongoing. The only other region where bronze alloying developed was around present-day Zimbabwe, although little is known about the stimulus and dating of this technology. It probably arose after the beginning of the maritime trade along the Indian Ocean, perhaps in the eleventh–twelfth centuries AD, and was used to make wire, beads, and simple two-sided mold castings. Brass, the alloy of copper and zinc, on the other hand, was used extensively for lost-wax casting in West Africa beginning around AD 1000, although the metal itself was imported into Africa from across the Sahara Desert and along the West and East African coasts.

There is no evidence of lost-wax casting outside West Africa, although casting traditions did exist. Open-faced or one-sided molds seem to have been used to cast copper into bar shapes as early as the seventh century AD in present-day Zambia. In the Zaïre–Zambia copperbelt region, cross-shaped molds were made by the ninth–eleventh centuries AD in northern Zambia, but not until the fourteenth century in southeastern Zaïre. These ingots were then traded. Objects were made as the ingots or raw metal were hammer forged into sheet, rods, ribbon, and other shapes. Drawing copper into wire, a fabrication technique found in Central, Eastern, and Southern Africa, was used by the mid-second millennium. The history of its development and spread is unknown, although iron/steel draw plates for wire were found at Ingombe Ilede, Zambia, dating to the fourteenth–fifteenth centuries AD.

Gold objects tended to be produced by the same techniques as copper in a given region. Lost-wax casting of gold in West Africa, such as among the Asante of present-day Ghana, is particularly exquisite. Wire drawing of gold was also practiced in West Africa. At ancient Zimbabwe, gold was usually hammered to shape beads, wire, and sheet. Finally, there is no evidence that any objects were made of pure tin in Africa. Occasional ingots cast of tin are found in South Africa, presumably for used trade.

A critical part of African metallurgy that has not been discussed is the non-technical — the social organization of labor, the use of space, and the esoteric knowledge, rituals, taboos, special clothing, and music involved. The innumerable processes of African iron and copper smelting, in particular, were complicated mixtures of technique, special knowledge, and ritual controlled and designed by men to ensure success and circumvent danger and malevolence. This involved not only pleasing the ancestors, but countering the acts of sorcery that were perceived to be a threat to the process.

Central to this examination is the recognition that all social activity, including technology, must be explained and done so within a framework meaningful to the people involved. Questions concerning the sources of ore, how rocks become shiny metal, why some smelts are unsuccessful, and what slag is are answered in modern, Western societies by scientific principles of geology and engineering. In pre-colonial Africa, these questions were resolved through principles based largely on human physiology and social structure.

A compelling framework was offered recently to consider how many pre-colonial African societies explained metallurgical activities. This is based on two fundamental aspects of human experience — gender and age. Gender concerns the interaction of males and females through a life cycle, but focuses on one critical stage of life that is not shared. Women are capable of transformation and creation through their ability to give birth, a process that is interrupted by monthly periods of sterility and ends at menopause. The links between women, production, and reproduction are poignant. Male metal workers could only generate similar creative forces with which to transform rock into metal and then into objects by controlling and appropriating womens' natural abilities through symbol, metaphor, and ritual.

The axis of age encompasses the relationships between youth and elders, as well as between the living and the dead. In many African cultures, the human life cycle involves the accumulation of wisdom and power through adulthood. Greatest power is acquired as an ancestor. Thus, elders have the expertise and knowledge to demand and exploit the labor of youth in most activities, including metallurgy; the ancestors have ultimate power over all significant activities of the living, particularly its reproduction. Integral to mining, smelting, and fabricating metal objects, therefore, was gaining and maintaining ancestral approval.

The cultural influences of gender and age were most obvious during iron and copper smelting — technologies of transformation and creation with many opportunities for failure. These influences were manifested in highly diverse ways. Often, smelting involved rituals and song that simulated significant times in the life of a productive woman, such as marriage, pregnancy, and birth. The Fipa of Tanzania, for example, adorned and treated a newly built iron smelting furnace as a bride who would have many children. The furnace was perceived as a "wife" to the iron smelters in many African cultures, such as the Phoka of Malawi. Furthermore, various parts of furnaces were often given the same names as female body parts, particularly those related to sexuality and birth. The Shona of Zimbabwe, the Chokwe of Angola, and others were more explicit and built their furnaces as women. They decorated the walls with breasts and scarification, denoting fertility, and the bloom sometimes came out between leg-like projections. Rituals also were used to consecrate new iron forges or tools which drew analogies between the anvil/hammer (the most important tool of a smith) and a second wife, such as among the Nyoro of Uganda, or a child, such as among the Ondulu of Angola.

Both age and gender strongly affected the roles played and choices made during metal working, including the significant influence of ancestral spirits, the technical and ritual expertise of elders, the work load of the youth, and the exclusion of women. Although women were often miners in cultures with labor shortages, all women, particularly pregnant or menstruating ones, were excluded from smelting operations. Pre-pubescent girls and post-menopausal women, however, sometimes participated in pre-smelting rituals or cooked and transported food to the smelters. Furthermore, strong taboos existed to prevent men from having sexual relations prior to a smelt. Such behavior represented infidelity to the furnace, and adultery was often thought to cause miscarriages in pregnant women. Furnaces were usually placed far away from villages to minimize this potential threat.

These rules of participation were designed to please the ancestors by preventing the presence of forces — uncontrolled fertility and temporary sterility — that might jeopardize a productive metallurgical operation. Young girls and post-menopausal women were not threatening because of their lack of fertility and active sexuality. The rules also served to separate metalworkers from the general public as people with special knowledge and capabilities. Many accomplished metal workers became wealthy members of their societies, as well as important political figures.

Other forces influenced how a smelt proceeded and how failure was explained, including visible problems with the materials or technical steps used. In less obvious circumstances, however, a common explanation for failure was sorcery by jealous villagers or by competing metal-workers. Since iron, copper, bronze, brass, and gold workers were often relatively rich and powerful men, they sometimes became foci of envy. Actions to avert evil spells, therefore, involved meticulous attention to the preparation and placement of medicines in and around a furnace and, sometimes, a forge. Ethnographic and archaeological evidence reveal that the placement of medicines inside furnace pits has been practiced for two millennia. As a result of this integral part of the metallurgical process, metalworkers were often believed to be sorcerers and/or people with special powers. Particularly skilled metalworkers were regularly in demand by the general public for their abilities to heal and divine.

Two significant lessons — actually two sides of the same coin — may be learned from the study of African metallurgy: (1) a technology can exhibit tremendous variation through time and across a continent, and (2) a technology is a system that is at the same time technical, economic, social, ideological, and political. A technological system affects and is affected by the culture and society in which it operates such that a great diversity of associated behavior and knowledge may result over time and space. Unfortunately, all the complexity and variation of African metallurgy will never be fully appreciated and known, particularly as the elderly experts die with much of their precious knowledge untapped.

S. TERRY CHILDS

REFERENCES

Childs, S. Terry. "Style, Technology and Iron-smelting Furnaces in Bantu-speaking Africa." *Journal of Anthropological Archaeology* 10:332–59, 1991.

Childs, S. Terry and David J. Killick. "Indigenous African Metallurgy: Nature and Culture." *Annual Review of Anthropology* 22:317–37, 1993.

Cline, Walter. *Mining and Metallurgy in Negro Africa*. Menasha, Wisconsin: Banta, 1937.

David, Nicholas and Yves LeBléis. *Dokwaza, Last of the African Iron Masters*. (Film). University of Calgary: Department of Communications Media, 1988.

de Maret, Pierre. "The Smith's Myth and the Origin of Leadership in Central Africa." In *African Iron Working*. Ed. R. Haaland and R. Shinnie. Oslo: Norwegian University Press, 1985, pp. 73–87.

Herbert, Eugenia. *Red Gold of Africa*. Madison: University of Wisconsin Press, 1984.

Herbert, Eugenia. *Iron, Gender and Power: Rituals of Transformation in African Societies*. Bloomington: Indiana University Press, 1993.

McIntosh, Susan and Roderick McIntosh. "From Stone to Metal: New Perspectives on the Later Prehistory of West Africa." *Journal of World Prehistory* 2:89–133, 1988.

Miller, Duncan and Nikolaas van der Merwe. "Early Metal Working in Sub-Saharan Africa: A Review of Recent Research." *Journal of African History* 34: 1–36, 1994.

O'Neill, Peter, F. Mulhy, and Winifred Lambrecht. *Tree of Iron*. (Film.) Gainesville: Foundation for African Prehistory and Archaeology.

See also: Gender and Technology – Technology and Culture

METALLURGY IN CHINA The earliest metal relics ever found in China are two brass artifacts from the Yangshao Culture Ruins in Jiangzai Village, Lingtong County, Shaanxi Province. They date as far back as six thousand years ago, providing strong evidence that metallurgy germinated in China during that period. Considering that studies of possibly earlier metal relics are not done thoroughly enough, the problem of the origin of metallurgy in China has yet to be further explored.

By the latter part of the third millennium BC, metallurgy had come into being in a number of regions, and many kinds of metal materials — red copper, primitive brass, tin bronze, and lead-tin bronze — were already used for small implements and ornaments.

The Xia Dynasty (twenty-first century BC–sixteenth century BC) had evolved into the Bronze Age. In Erlitou Cultural Ruins, Yanshi County, Henan Providence, remains of foundry workshops have been discovered where bronze sacrificial vessels, weapons, implements, and casting moulds have been excavated. Apparently, the making of articles was done with casting as the main means.

The earliest copper mining and smelting ruins known in China are located in Tongling (meaning copper ridge), Ruichang County, Jiangxi Providence. Its mining can be traced back to the fourteenth century BC. The shafts and drifts were supported by timber frames. Also excavated were ore-dressing troughs and wood winches for hoisting. Further study reveals that during Shang-Zhou Periods (Shang Dynasty: sixteenth century BC–eleventh century BC; Western Zhou: eleventh century BC–771 BC; Eastern Zhou: 771 BC–221 BC) the mining of copper minerals had progressed to a rather large scale in the Liao River Valley and Yellow River Valley, especially in the middle and lower reaches of the Yangtze River. The smelting of copper was done in semi-continuous operation in shaft furnaces. By the late Western Zhou at the latest, copper sulfide minerals had been used to smelt copper.

The smelting technologies of lead and tin were mastered from the Shang Dynasty on. The lead vessels unearthed in Anyang have a purity of over 95%. Tin ore deposits were scattered in the Northeast, Northwest, and South China regions, especially in Jiangxi, Guangxi, and Yunnan, where tin reserves were very abundant. The early tin material must have come from these regions.

Alloying techniques improved greatly during the Shang Dynasty. The sacrificial vessels of imperial courts unearthed in Yin Ruins have a tin content of 18% or more. The book *Kao Gong Ji* of the Warring States (475 BC–221 BC) recorded something about the alloy proportioning of *liu qi* (six kinds of bronzes): *zhong* (bell), *ding* (cauldron), *fu* (hatchet), *ji* (halberd), *jian* (sword), *zu* (arrowhead), and *jing* (mirror). This indicates the craftsmen's clear understanding at that time that the mechanical performance of tin bronze varies with different contents of tin.

Shang-Zhou bronze culture was characterized by sacrificial vessels, complex in shape and delicate in design, which were mass-produced by casting. The key in achieving this characteristc without applying techniques such as the lost-wax process lay in the skillful use of composite pottery molds and various cast-joint technologies.

Beginning with the Spring and Autumn periods (771 BC–475 BC), it became in vogue to apply synthetically various shaping processes and decorating techniques: cast-joint, soldering, lost-wax process, forging, gilding, gold-plating, red-copper inlay, gold and silver inlay, engraving, etc. This significant change brought the manufacture of bronzes to a still higher level; the implements produced looked brighter and more colorful. As early as the late Neolithic Age, small-sized gold had already appeared . By the Warring States periods, there were more and more articles made from gold, and coins were first minted of silver. These may have been obtained by cupellation, a refining process in which metals are oxidized at high temperatures, and base metals are separated by absorption into the walls of a cupal, or porous cup made of bone ash. The production of mercury also reached a certain scale. It is recorded that a great amount of mercury used to preserve bodies from decay was found in the imperial grave of Emperor Qin.

It was far back in late Shang Dynasty that iron meteorites were used to be forged into blades, and cast with bronze into weapons. This process was passed down all along to the end of Western Zhou, perhaps providing impetus to the origin of certain iron-smelting technologies.

Archaeological excavations have shown that artificial iron-making may have begun in the late Western Zhou in China. It is worth noting that the smelting and casting of pig iron began only about two hundred years later, i.e. the late Spring and Autumn periods. Only another one hundred years had passed before pig iron was in wide use for casting production tools, especially for farming implements, thus marking the beginning of the Iron Ages in China.

Correspondingly, a series of outstanding inventions related to pig iron came springing up one after another. Two of the most important were the iron mold casting and cast-iron toughening techniques. In these, pig iron castings were changed into white-heart malleable cast iron or black-heart malleable iron through decarbonizing heat treatment or graphitizing heat treatment respectively. Others were the later ones of puddling iron (Western Han 206 BC–AD 8) and Guan-steel (made by smelting pig iron and puddling iron together, about the end of Eastern Han AD 25–AD 220). These revealed a technological course of development, which, though markedly different from that in ancient Europe, led to the same goal. This unique instance in the history of technology is interesting and thought provoking.

In China, many of the most important inventions in ancient metallurgy were made before the sixth century BC. After the fruits in the preceding times were digested and imbibed in the Tang Dynasty (AD 618–AD 907) and Song Dynasty (AD 960–AD 1279), metallurgical technologies in China took shape, leading to further prosperity and greater achievements. For example, in the southern regions there was full scale mining and smelting of iron minerals and others such as copper, tin, lead, gold, silver, and mercury. There were also large and extra-large castings and bronze or iron structures, etc.

It must be mentioned that the making of white copper (copper nickel alloy) was alreadly mastered as early as the fifth century, and that during the Han and Tang Dynasties and later, metallurgy in China had exchanges with and mutual impact on the surrounding regions. Some examples are the westward spreading of iron-casting skills and the influences of Persian artistry on gold or silver wares in the Tang Dynasty.

In the Song Dynasty, wet metallurgy was put into large-scale practice to extract pure copper, annually yielding about five hundred tons, which amounted to about one third of the total copper produced in the whole country.

About 1620 in the Ming Dynasty, it was possible to smelt zinc, which was used in great amount for minting coins. Antimony was smelted in the Ming Dynasty (AD 1368–AD 1644), but it was not recognized at the time as a new kind of metal and thus was mistaken for tin.

Recently, some unexpected archaeological discoveries— such as the bronze culture of the ancient Kingdom Shu and the bronze groups of the late Shang in Xinggan, Jiangxi Province — indicate that there are still many mysteries in the metallurgy of ancient China which remain to be disclosed. Those disclosures would contribute greatly to academic studies.

HUA JUEMING

REFERENCES

Barbard, N., and Sato Tamotsu. *Metallurigical Remains of Ancient China*. Tokyo: Nippon International Press, 1975.

Hua, Jueming. *Metallurgical Technologies in Ancient China*. Part 2 of *The History of the Development in World Metallurgy* Ed. Hua Jueming, et.al. Beijing: Science and Technology Literature Press, 1986.

Hua, Jueming. *An Anthology of Theses on the History of Smelting and Foundry in China*. Beijing: Wenwu Press, 1986.

Li, Cong. "Discussions on the Development of Iron & Steel Smelting Techniques during Pre-feudal Society of China." *Kaogu Xuebao* 2: 12–24, 1975.

Tylecote, R.F. *A History of Metallurgy*. London: The Metals Society, 1976. Chinese Edition, trans. Zhou Zengxiong and Hua Jueming. Beijing: Science and Technology Literature Press, 1985.

Zhou, Weijian, Lu Benshan, and Hua Jueming. "The Dating of Ancient Smelting Ruins in Copper Ridge, Ruichang City and Its Scientific Values." *Jiangxi Wenwu* 3: 1–12, 1991.

METALLURGY IN EGYPT Gold, silver, lead, and copper were among the metals exploited by Egyptians since the pre-dynastic period, prior to ca. 3100 BC. The main sources of these metals were the deposits in the ancient rocks of the Eastern Egyptian desert between the Nile and the Red Sea and in the Sinai. Iron implements, although present in Egypt from the twenty sixth dynasty (ca. 665–525 BC), did not become common until the Ptolemaic period, for ancient Egypt had no access to major sources of iron ore.

We lack written descriptions of the mining operations or expeditions in search of metals or ores, although we know that both methods were widely practiced as a way to procure metallic resources. Of course, some nuggets of relatively pure metals can be found, but these were certainly used up very soon and the ancient Egyptians must very quickly have learned to exploit deposits of metal ores. Mines might be either open shallow pits or underground tunnels following promising veins of ore. The remains of ancient mine shafts and miners' stone tools, as well as heaps of waste, still indicate many of these sites. There is some evidence that initial smelting (removing the metal from the ore matrix) was carried out at the mining site, but refining the metal was probably done in the workshops of the metalsmiths. By the New Kingdom period (ca. 1550–1085 BC), Egypt seems to have relied increasingly on imports of copper as well as of iron to meet its need for metallic implements.

Extraction of the metal from its ore involved such processes as roasting (heating the ore with charcoal in the open air), smelting in closed furnaces using the addition of coke and silicate particles to absorb impurities and aid in reducing the oxides to metals, or with a forced air blast through the molten slag, again to oxidize impurities and so free the metal. To achieve these effects, a high temperature was necessary. It was only achieved though the use of blowpipes or bellows to feed sufficient oxygen to the fire.

Pure copper is soft and so nearly useless for many purposes. But when hammered, it rapidly becomes much harder. This hardness can be removed by annealing (heating the metal above 500°C and allowing it to cool again). Many early edged implements were shaped by repeated hammering and annealing, ending with a final hammering to harden the cutting edge.

Other types of implements, such as an adze, are more easily formed by casting (pouring molten metal into a mold of an appropriate shape and allowing it to harden). For objects not requiring solidity, the "lost wax" method of casting used less of the precious metallic resources. In this process, the object was first modeled in clay or other heat-resistant medium. It was then coated with a layer of wax, which was in turn surrounded by yet another layer of clay or plaster, leaving only a small opening into the wax-filled layer. This mold was then heated so that the wax would melt and escape through the small drain. Through this opening, molten metal was then poured in.

Pure copper does not cast very easily because it tends to absorb gasses, creating air pockets in the finished product and so weakening the finished product. If, however, a bit of tin is added to the copper, the results are significantly improved. Alloying copper with tin to form bronze (about ten percent tin is ideal) also creates a metal considerably harder and more durable than pure copper itself. Bronze, therefore, was widely used for tools and weapons from the time of the Middle Kingdom (ca. 2150–1780 BC) until replaced by iron late in the New Kingdom period. Since tin is not found in commercial quanitities in Egypt, it is not certain whether bronze was imported as a semi-finished product or was internally produced with imported tin.

Metal tools such as chisels, knives, axe heads, and adzes are common in funerary collections and are often portrayed in Egyptian art. Another major use of metal was in the production of military weapons: daggers, swords, spears, and battle axes. Defensive armor, such as mail coats (made by riveting small bronze plates onto leather jerkins) first became common during the New Kingdom period. Protective helmets also appear to be a relatively late innovation, although some tomb paintings from the Ramesside period (ca. 1290–1225 BC) show foreign mercenaries equipped with protective head gear. A third important use of copper (and later, bronze) was in the production of domestic articles such as cauldrons, ewers, basins, and ladles. Mirrors of polished metal were common throughout the dynastic period. Pins, tweezers and razors have also been found by excavators.

The gold deposits of the Eastern Desert and Nubia were the largest in the ancient world, so perhaps the techniques for extracting both gold and its alloys, as well as the methods for working them, were discoveries of the ancient Egyptians. We cannot be certain, however, because the knowledge seems to have been handed down only verbally from father to son and from master to apprentice.

Sheets of metal, whether copper, bronze, gold, or silver, could be worked in a variety of techniques to produce decorative effects. Repoussé is worked from the back of the metal sheet so that the design stands out on the front in raised relief. Chasing is relief worked from the front of the sheet by hammering down the background while leaving the desired raised figures in the foreground. Engraving, also worked from the front of the piece, means cutting a groove into the metal from which the metal is removed. Tracing, which leaves a similar effect, does not remove any metal from the sheet.

Sometimes it is desirable to join together pieces of worked metal to form some object such as a piece of jewelry. The ancient Egyptians had learned how to use soldering techniques effectively. In soldering, a metal or alloy with a melting point lower than the pieces that one desires to join is allowed to flow along the seam and, when cooled, binds them together. Since the ancients rarely worked with pure gold or other metals, the choice of a suitable solder was very important. For example, gold and copper melt at nearly the same temperature (1083 and 1063°C respectively). If ten parts by weight of copper is added to ninety parts of gold, however, the mixture melts at 940°C. Thus, if one worked with relatively pure gold, this alloy could be used effectively as a solder. A modification of this technique, called sweating, involved coating the surfaces to be joined with a solder, placing the pieces in contact, and heating until the jointing compound melts, without adding additional solder. This often makes a neater join than soldering does.

There has been considerable debate over whether the ancient Egyptians knew how to draw wires. There exist, both physically and in tomb illustrations, numerous examples of what seem to be beads strung on wires. It is not clear, however, whether these are truly wires or merely thin strips cut from a sheet of metal using a chisel, for example, that have been rounded by friction with the beads which are being strung. Since the beads seem to have been manufactured by drilling a round hole through the stone or metal, the constant rubbing of this rounded object upon the metal strip may have produced its more rounded appearance.

Jewelry was an important part of the adornment of both gods and humans (men, women, and children). In ancient Egypt, people wore jewelry for a variety of purposes. Perhaps the most important was as amulets to protect the wearers from evil forces that seemed to surround them on every side. Gold itself was considered magical, identified by some with the flesh of the gods since gold, among the metals known to the ancient Egyptians, was least susceptible to tarnishing and change. Precious stones were also used, along with natural objects such as the claws of a ferocious creature (who may have been considered to embody the powers of the god). Jewels could also be used, perhaps as an extension of their magic power, to enhance the sexual attractiveness of the wearer. They could also, as today, serve as symbols of status or wealth. This was especially true for the king and members of the royal court. We have reports of pharaohs giving gold collars or other ornaments to their favorites as a mark of distinction. Jewelry, probably in its amulet form, was also essential to the burial customs of the ancient Egyptians. This is, indeed, one of our major source of exemplars of Egyptian jewelry. Finally, jewelry could indicate the power to carry out royal prerogatives delegated by the king. The man who held the king's signet ring or royal seal, for example, carried enormous responsibilities, both politically and religiously.

Jewelry making seems to have been an important activity in relatively few centers of culture and political power. Since gold was mainly the possession of royalty, it seems probable that the jewelers who supplied the royal household with golden ornaments enjoyed considerable social and economic status. Three overseers of goldsmiths had tombs at Thebes, as did two goldsmiths, indicating a fairly high social position. Of course, the workers and laborers who actually carried out the designs of the jewelers must have had a lower social status, but they were not of the lowest social class. Skilled workers in metal would always have been in demand, and such demand would translate into at least a modest level of social prestige and affluence.

GREGG DE YOUNG

REFERENCES

Aldred, C. *Jewels of the Pharaohs*. London: Thames and Hudson, 1978.

Andrews, C. *Ancient Egyptian Jewellery*. London: British Museum, 1990.

Forbes, R. J. *Metallurgy in Antiquity*. Leiden: Brill, 1950.

METALLURGY IN INDIA The ability to make iron from its ore marked a milestone in the history of civilization. On the one hand, it required a degree of sophistication in technology, because it involved working at the limits of temperatures that could be achieved with simple implements. On the other hand, it provided a material with wide ranging properties which offered a variety of possibilities for tool making. Until recent times therefore, the progress of a nation was often judged by the per capita consumption of iron and its alloy, steel.

Iron and steel are of great importance for the following reasons.

• Iron is the fourth most abundant element. Among metals only aluminum is more abundant. Five percent of the earth's crust is made of iron. Iron ores are also widely distributed. In India large and small deposits are distributed throughout the country, outside the Gangetic plains.

• Temperatures needed for iron smelting are easily attainable in a vigorously worked charcoal fire. They are, however, higher than those required for smelting some of the non-ferrous materials like copper, zinc, and tin, and therefore iron smelting needs great expertise.

• Iron, when mixed with the inexpensive alloy element, carbon, becomes steel. This alloying converts it into a material with a large possible range of properties from low strength, highly ductile low carbon steels to high strength, but less ductile, high carbon steel. At higher concentrations of carbon, iron forms an alloy with excellent casting properties which melts at a low temperature. This alloy is called cast iron.

• Iron can be easily welded by forging. Before fusion welding came into existence, this property was important.

The origin of the knowledge of iron and steel technology in India is still a matter of debate. Recent archaeological evidence, however, is making it increasingly apparent that iron making in India is of an independent origin and may even predate the iron age in many other civilizations. Chakrabarti and to some extent Hegde have presented extensive evidence to support this view.

The iron age probably started in India not later than the second half of the second millennium BC. Archaeological evidence pertaining to this period has come up in various places across the country, indicating that the origin might have been earlier. Though literary evidence also supports this hypothesis (e.g. the reference to a metal called *ayas* in the *Ṛgveda*), it is not conclusive.

It is undisputed that well before the beginning of the Christian era, artisans here had achieved a level of excellence, so that Indian steel was famous for its quality throughout the old world. There was extensive export trade even at that time. In the first millennium AD, the industry probably grew to large proportions. The Delhi pillar near the Kutub Minar, dated to the fourth to fifth centuries AD, is the classic example cited as evidence of the level of the iron industry in India.

The pillar, which is about twenty-five feet high and weighs several tons, has not rusted. Large iron objects of this age are found throughout the country — for example,

Figure 1: Iron pillar at Mehrauli.

there is the broken pillar at Dhar, Madhya Pradesh in central India (thirteenth century AD) which originally measured about fifteen meters high, and the ten meter pillar at Kodachadri Hills in Karnataka State in South India. Dharampal estimates that by the eighteenth century the iron and steel industry in India was very extensive indeed. There are several reports of British observers in Dharampal's compilation wherein the Indian product has been rated "better than the steel produced anywhere else in the World".

In the traditional iron making process, iron ore is reduced in a low shaft furnace built of clay. Iron ore and charcoal form the top, and air is blown by bellows from the bottom. Charcoal provides the fuel as well as the reducing agent for smelting. Temperatures between 1000°C and 1475°C are generated in the reaction zone. The reduction takes place as a gas–solid reaction. The reduced iron collects with some iron oxide to form liquid slag. The slag is tapped periodically through slag holes provided near the bottom of the furnace. At the end of a blowing cycle of three to six hours, the sponge iron bloom weighing 3–10 kg is extracted from the

Figure 2: Fired clay crucibles.

bottom of the furnace. It is then hammered to squeeze out any entrapped slag and to consolidate the iron.

A process of making special steels of exceptionally good quality by melting them into crucibles is probably unique to India and its vicinity. The special steels were traded extensively with other civilizations, especially for making swords and daggers and other implements where exceptional hardness and toughness were necessary. These swords became known as "Damascus swords" in Europe since they were obtained from the Middle East, and Damascus was the major trading center. Only later was it known that the Arabs bought the steel from India and fashioned it into various implements. The *wootz* steel from which these blades were made thus became famous. A very large industry of *wootz* making is known to have existed in the southern parts of the country until the beginning of the nineteenth century. This industry slowly died down during the nineteenth century. The process is now extinct.

British and European interest in Indian iron and steel has a long history. Several travelers' accounts such as one by Francis Buchanan provided detailed information on the extensive iron and steel industry around the seventeenth and eighteenth centuries. Crucible steel from India was also noted and described. In fact, technical papers and studies on various aspects of Indian iron and steel making continued until the turn of this century.

Starting from the turn of the eighteenth century, several efforts began to be made in England and later on in France, Italy, and much later in Russia to try to reproduce the Indian process of making steel. For a number of years various British scientists such as Pearson, as well as Stoddart and his student Faraday, tried to make *wootz* by the crucible process. Their experiments continued in England and France well into the 1850s and 1860s. They declined in importance in the 1860s, perhaps because of the general decline of interest in swords with the emergence of the gun. Also, by this time, the modern metallurgical processes for large-scale production of steel, using the Bessemer converter etc., had started to develop. In Russia these experiments continued well up to the turn of the twentieth century.

Around the turn of the century, the indigenous process had ceased to have any larger interest and the few observations and studies regarding it were by anthropologists and others concerned with exotic 'arts'. Ananda Coomara Swamy recorded the steel making process in Sri Lanka around 1908. In 1963 the National Metallurgical Laboratory at Jamshedpur hosted a symposium on the Delhi Iron Pillar. An interesting fact that then emerged was that the iron made by tribals in the early 1960s appeared to be comparable to the iron of the ancient Delhi Pillar from the point of view of corrosion resistance.

Beginning from the late 1950s a series of studies has been undertaken in England to characterize the various features of the bloomery process for producing wrought iron. The models taken up for reconstruction were from Africa or the earlier pre-medieval European designs. From the mid-1970s onwards there has also been a fresh spurt of interest in studying the traditional Indian process of making steel. Reacting to the discovery of a type of ultra-high carbon steel that was showing super plastic behavior, the historian Cyril Stanley Smith felt that in these new alloys one might have only "rediscovered" some properties of traditional Indian steel. The crucible process of reduction is perhaps much more complex than was initially thought and needs to be understood and characterized afresh.

A.V. BALASUBBRAMANIAN

REFERENCES

Bronson, Bennet. "The Making and Selling of Wootz: A Crucible Steel of India." *Archeomaterials* 1(1): 13–51, 1986.

Buchanan, Francis. *A Journey From Madras Through the Countries of Mysore, Canara, and Malabar.* 3 vols. Madras: Higgin Botham and Company, 1870.

Chakrabarti, D.K. *The Early Use of Iron in India.* New Delhi: Oxford University Press, 1992.

Dharampal. *Indian Science and Technology in the Eighteenth Century.* Hyderabad: Academy of Gandhian Studies, 1971, reprinted 1983.

Hadfield, Robert. "On Sinhalese Iron and Steel of Ancient Origin." *Proceedings of the Royal Society (London)* Series A, 86: 94–100, 1912.

Hegde, K.T.M. *An Introduction to Ancient Indian Metallurgy.* Bangalore: Geological Survey of India, 1991.

Holland, Thomas H. "On the Iron Ores and Iron-Industries of the Salem District." *Records of the Geological Survey of India* 25 (3): 135–159, 1892.

Lahiri, A.K., T. Banerjee, and B.R. Nijhawan. "Some Observations on Corrosion-Resistance of Ancient Delhi Iron Pillar and Present-Time Adivasi Iron Made By Primitive Methods." *National Metallurgical Laboratory Technical Journal* 5 (1): 46–54, 1963.

Lahiri, A.K., T. Banjeree, and B.R. Nijihawan. "Some Observations on Ancient Iron." *National Metallurgical Laboratory Technical Journal* 8: 32–33, 1967.

Pearson, George. "Experiments and Observations to Investigate the Nature of a Kind of Steel, Manufactured at Bombay and There called Wootz." *Philosophical Transactions* 85: 580–593, 1795.

Sherbey, Oleg D. " Damascus Steel Rediscovered?" *Transactions of the Iron and Steel Institute of Japan* 19: 381–390, 1979.

Yater, Wallace M. " The Legendary Steel of Damascus Part III." *The Anvil's Ring*, 1983–1984, pp. 1–17.

METALLURGY IN MESO AND NORTH AMERICA

Native American metallurgical technologies held great attraction for sixteenth century European empires seeking to stake claims in the American New World. While the technical feats and mastery of Native American metalworks have only recently generated systematic scientific inquiries, the attraction of Mesoamerican and Andean precious metals has resulted in the plundering of massive quantities of these metals — often in the form of jewelry and other materials — from ancient centers where they served both ceremonial and funerary functions. Precious metals craftsmanship was an ongoing enterprise at the time of the Spanish conquest of the Inca state in 1532; heirloom items, ritual caches, and funerary offerings fed the raging smelters of European outposts in the New World for the duration of the colonial period. The crafted booty delivered to Fernando Pizarro by the Inca nobility, as ransom for the Inca Emperor Atahualpa, occupied nine forges for four months in 1533; this, in order to smelt the over 26,000 pounds of silver and 13,420 pounds of gold jewelry and other reliquary collected as tribute at that time. For the Spaniards, gold and silver were the fuel that propelled the expansion of the Holy Roman Empire; while for the Inca, gold and silver symbolized the sweat of the sun and the tears of the moon, and thereby, the cosmologically-ordained male and female principles, respectively.

Much of the precious metal processed by early European colonists in the New World was smelted from intricate works of art crafted into jewelry sewn into elaborate blankets, capes, shawls, and other clothing, as well as worked into ritual and funerary art and furnishings, weapons of war, and medical and agricultural tools. Ancient and large-scale metallurgical workshops and industrial centers are known from the *patios de Indios* (native workshops) of Colombia, and from such ancient centers as Atzcapotzalco, Mexico, where craft guilds were a significant aspect of the economic and social landscape.

The growing international black market in antiquities has both hastened the looting of ancient sites containing pre-Columbian metal craft, and has, by extension, spurred the collection and preservation of these works as priceless relics of a bygone age. When one takes into account the fact that individual tombs from sites such as Batan Grande, Peru, contained as many as two hundred gold objects, much of which consisted of tall — 30 centimeter — gold beakers, it is no wonder that much of this legacy has already been destroyed. Other funerary chambers, like that of Tomb 107 at Monte Alban, Oaxaca, Mexico, contained a veritable treasure trove of over five hundred precious metal and stone objects recovered by Mexican archaeologists in the 1930s. Other archaeologically documented precious metals caches have been recovered from Zaachila, Oaxaca and Coclé, Panama. Where bronze craftsmanship is concerned, individual tombs have produced upwards of five hundred kilos of such metals worked from alloys of tin and arsenic-bronze. Given the notoriety of such discoveries among grave robbers and the general public, it should be of no surprise that looters persist in destroying ancient sites in search of treasures.

The recent discovery and subsequent excavation of the tomb of the warrior-priest of the ancient ceremonial center of Moche, Peru — which contained a collection of dynastic relics and funerary items of gold, silver, and alloys of copper and tin — was initially brought to the attention of the scientific community as a result of looting. Despite such destruction, archaeologists and other investigators continue to document the technical achievements of Native American metallurgists of Peru, Mesoamerica, Central, and North America.

In 1921, Swedish ethnographer, Erland Nordenskiold began research into technical secrets made apparent through metallurgical analysis of ancient metal objects from South America. Nordenskiold ascertained that pre-Columbian metal works were far more technically sophisticated than imagined at that time. He made a a number of observations concerning metallurgical problems, including those pertaining to the metallurgical composition of *tumbaga* (a complex alloy of copper and gold, or copper, gold, and silver), soldering with silver, and the welding of copper. The technical analyses completed by Nordenskiold were a precursor to the sophisticated analyses now undertaken by archaeologists, and metallurgists, using a variety of techniques ranging from replicative experiments — where the objects themselves are reproduced with known ancient methods and technologies under controlled conditions — to physical and chemical tests to ascertain the composition and construction of the alloys employed in the creation of pre-Columbian metal objects.

The body of ancient technical secrets relevant to the manufacture of pre-Columbian metal objects has grown considerably since the first European observations of Native American metallurgical techniques in the sixteenth century.

The Old Copper culture, which flourished from 3000 BC to 1000 BC along the shores of Lake Superior in North America, is credited with having introduced the earliest known metalworking tradition in the Americas. This tradition was the basis for later developments in the use of copper and related metals by Hopewellian and Mississippian peoples of the first-millennium AD. Old Copper culture sites have produced evidence of significant early metallurgical techniques, including cold hammering, hammer welding, annealing, and the production of socketed metal tools, conical points, knives, axes, chisels, awls, harpoon heads, and a variety of projectile point types derived from prototypes of stone, horn, shell, and bone. To this list we can add sheet metal, intricate sheet metal cut-outs, repoussé decoration, crimping, riveting, the gold sheathing of copper, gold and copper beads, and the hammer-welding of silver and copper, or "copper and meteoric iron to produce bimetallic objects" (Easby, Jr, 1966). This demonstrates an early, independent, and regional tradition in the art and science of metal craft.

The lost-wax, or *cire perdue*, casting process was first employed in the region of Colombia by 100 BC, but quickly spread into Ecuador, and lower Central America (Panama and Costa Rica), and was subsequently adopted in Peru and Mesoamerica by AD 800. Ultimately, according to historian Dudley T. Easby, Jr (1966), lost-wax casting achieved "its highest development in the Oaxacan area of Mexico, where during the 15th and 16th centuries AD Mixtec master craftsmen produced little hollow castings that are unrivaled for delicacy, realism and precision".

Where Mesoamerica is concerned, the North American tradition of cold hammering and annealing, and that of South America, consisting of cold hammering, annealing, and casting, inspired the initial development of three distinct Mesoamerican metallurgical traditions. These traditions include those that emerged in the areas of upper Central America or southern Mesoamerica (including southern Mexico, Guatemala, Honduras, and El Salvador); the Pacific coastal lowlands including the Tarascan culture area, and the Mexican Gulf Coast lowlands and Yucatan Peninsula which encompassed the ancient Huastec, Totonac, and Maya cultures. Archaeologist Dorothy Hosler (1986) argues that the relatively late adoption of metallurgy in Mesoamerica — after AD 700 — serves to explain the largely elite character of the Mesoamerican metallurgical tradition. While both South and North American metalcraft evolved from a utilitarian foundation centered on the manufacture of agricultural implements and other tools, trade and exchange in precious metals ultimately inspired the Mesoamerican metallurgical tradition. Hence, the wholesale and widespread adoption and exchange of metallic axe monies, tokens, and precious-metal objects.

While recent studies have yet to establish definitively the earliest dates identified with the origins of bronze metallurgy, Heather Lechtman (1986) argues that arsenic bronze was in use in northern Peru by the fourth-century AD. Tin-bronze originated in highland Bolivia by AD 700, and, by the Inca era (ca. AD 1450) spread throughout the areas identified with the modern states of Peru, Bolivia, Chile, and Argentina. Finally, metallic money — in the form of copper axe blades and tokens — appeared in Ecuador by AD 1000 and quickly spread throughout South, Central, and Middle America.

The holistic nature, independent development, and antiquity of Native American metals craftsmanship are only now beginning to be clarified. Metallurgical technologies that were developed by pre-Columbian craftsmen are far too numerous to discuss in any detail in this essay. However, a partial listing should provide some idea of the significance of this legacy. Those identified to date include, (a) the *cire perdue* or lost wax casting process, (b) surface metallurgy, depletion gilding, acid pickling, and tumbaga, (c) the application of organic reagents and binding emulsions, (d) arsenic, copper–arsenic, and tin–bronze casting, (e) copper–arsenic, tin, and bizmuth alloys, (f) silver chloride coatings, (g) gilt copper sheeting, gold and silver sheathing, sheet metal processing and fabrication, mechanical crimping, gold-leaf treatments, hammer-welding, and the raising of sheet metal vessels, (h) silver/silver–copper/spot solder and soldering, (i) copper soldering, brazing, and spot welding, (j) electrochemical replacement plating, (k) complex annealing, cold-hammer and anvil, binary and ternary alloy processing, and ground and hammered meteoric iron implements, (l) charcoal-fed ore reduction and air-blast smelting/refining furnaces, (m) iron ore or hematite flux, and the reduction of sulfide ores, (n) open cast, multi-component, and vented casting molds, and powdered carbon casting emulsions such as that of the Aztec *teculatl* (charcoal water), (o) mechanical and metallurgical joins including metal nails, rivets, staples, ribbon clips, strip clips, lacing, long sockets, short tabs, tab-and-slot and other metal fasteners, (p) repoussé and other embossed sheet metal applications, including cinnabar cloisonné, (q) solid-state diffusion bonding, sweat welding, fusion gilding or Sheffield plating, and cladding, (r) slush casting, (s) multi-component sheet metal miniatures, (t) color surface and powder metallurgy, (u) technologies for the manufacture of thin cast rods, wire coils, strip wire, wire-work surfacing and filigree, metal sequins, quad metal mosaics, architectural cramps, agricultural blades and implements, socketed chisels and related tools; and copper and bronze axe blades, metallic monies and tokens, (v) arsenic and tin–bronze implements including fish hooks, eyed needles, pins, depilatory tweezers, and surgical instruments such as *tumi* knives and blades, (w) the standardization of metal ingots and tools, (x) platinum processing and the sintering of refractory metals such as platinum, and finally, (y) a variety of prospecting methods, including shallow shaft mining, the strip mining of exposed deposits, and the placer mining of alluvial gold and platinum. According to Heather Lechtman, the tumbaga alloys alone "constitute the most significant contribution of the New World to the repertoire of alloy systems developed among ancient societies". It should be noted that the processing of platinum (which has a melting temperature of 3000 degrees Fahrenheit), was a feat accomplished by ancient Ecuadorian metallurgists by way of the "mixing of grains of platinum with gold dust" through a powder process identified with the "sintering of refractory metals" (Easby, Jr, 1966).

Electrochemical replacement plating and depletion gilding, developed by the Moche of Peru nearly two thousand years ago, allowed ancient Native Americans to plate precious metals on to semi-precious metals to a thickness of less than one micron. This electrochemical replacement process was not rediscovered until the twentieth century AD. Recent studies indicate that electrochemical replacement plating and depletion gilding or silvering "both involve sophisticated chemistry, and Precolumbian surface metallurgy is surely as much chemistry as it is metallurgy" (Lechtman, 1986). As Easby says, "the tale persists that the egotistical Benavento Cellini spent months trying to ascertain how an ancient Mexican craftsmen had fashioned a silver fish with gold scales and finally conceded that he was baffled". Unfortunately, as is the case with so many other aboriginal New World technological innovations, the very presence of such metallurgical traditions as that of the lost-wax casting process and tin and arsenic-copper bronze alloys was once taken to indicate that such technologies were introduced or diffused from the Old World.

Recent archaeological investigations underscore the paucity of information pertaining to metallurgy currently at our disposal, and the abundance of ancient archaeological materials that have yet to be studied in any systematic fashion. Unfortunately, the relatively recent and highly specialized nature of publications pertaining to pre-Columbian metallurgy have led some scholars to suppose that even the most ingenious ancient Native American technologies were little more than isolated or accidental instances of technical insight and ingenuity. Such scholarly perspectives are clearly artifacts borne of the relative scarcity of and limited access to information. As the body of studies grows, it is becoming clear that innovations in metallurgy, such as depletion gilding and electrochemical replacement plating, were far more ancient and widespread than once thought.

While the wholesale destruction of the pre-Columbian world has closed an important window on the cosmology and beliefs of its metallurgists, we can nevertheless advance interpretations as to the social and ritual significance of metals based on contact-period and ethnographic accounts.

Both Inca and Aztec craftsmen, and Native Americans more generally, identified precious metals — gold and silver — with the male and female principles. The alloying of copper and gold, or, copper, gold, and silver, which produced a red, pink, or golden metal known as *tumbaga*, was in turn identified with the menstrual flow and ambered moon. Among the contemporary metals craftsmen of west Africa, smelters are designed to symbolize the female sexual organs, while the metals themselves are thought symbolic of the male principle embodied in semen and other bodily fluids. The cosmological message inherent in the metal itself, when combined with the supernatural and religious icons and images, must surely have served to enhance the power and prestige of the bearer, while at the same time providing clear indications of that individual's identification with supernatural and cosmic forces. The use of metals in personal adornment and ritual attire served to convey the associations of the bearer with universal principles, within which gender ultimately served as a distinguishing characteristic of the individual and cosmos.

RUBEN G. MENDOZA

REFERENCES

Benson, Elizabeth P. *Pre-Columbian Metallurgy of South America*. Ed. E. P. Benson. Washington, D.C.: Dumbarton Oaks, 1979.

Bird, Junius B. "The 'Copper Man': A Prehistoric Miner and His Tools from Northern Chile." In *Pre-Columbian Metallurgy of South America*. Ed. E. P. Benson. Washington, D.C.: Dumbarton Oaks, 1979, pp. 105–132.

Easby, Jr, Dudley T. "Early Metallurgy in the New World." *Scientific American* 214 (4): 72–78, 1966. Reprinted in *New World Archaeology: Theoretical and Cultural Transformations*. Ed. Ezra B.W. Zubrow. San Francisco: W.H. Freeman, 1974.

Hosler, Dorothy. "The Metallurgy of Ancient West Mexico." In *The Beginning of the Use of Metals and Alloys*. Ed. Robert Maddin. Papers from the Second International Conference on the Beginning of the Use of Metals and Alloys, Zhengzhou, China, October 21–26, 1986, pp. 328–343.

Lechtman, Heather. "Issues in Andean Metallurgy." In *Pre-Columbian Metallurgy of South America*. Ed. E. P. Benson. Washington, D.C.: Dumbarton Oaks, 1979, pp. 1–40.

Lechtman, Heather, Antonieta Erlij, and Edward J. Barry, Jr. "New Perspectives on Moche Metallurgy: Techniques of Gilding Copper at Loma Negra, Northern Peru." *American Antiquity* 47(1): 3–30, 1982.

Lechtman, Heather. "Pre-Columbian Surface Metallurgy." *Scientific American* 250(6): 56–63, 1984.

Lechtman, Heather. "Traditions and Styles in Central Andean Metalworking." In *The Beginning of the Use of Metals and Alloys*. Ed. Robert Maddin. Papers from the Second International Conference on the Beginning of the Use of Metals and Alloys, Zhengzhou, China, October 21–26, 1986, pp. 344–378.

Lechtman, Heather. "The Production of Copper-arsenic Alloys in the Central Andes: Highland Ores and Coastal Smelter?" *Journal of Field Archaeology* 18(1): 43–76, 1991.

Linné, S. "Technical Secrets of the American Indians: The Huxley Memorial Lecture, 1957." *Journal of the Royal Anthropological Institute of Great Britain and Ireland* 37(2): 149–164, 1957.

Reader's Digest Association. "Clues to a Forgotten Past." In *Mysteries of the Ancient Americas: The New World Before Columbus*. Pleasantville, New York, 1986, pp. 166–187.

Shimada, Izumi. "Sican Metallurgy: Bronze Age." Lecture for the Department of Anthropology, University of Arizona, Tucson, April 11, 1988.

Shimada, Izumi, Stephen M. Epstein, and Alan K. Craig. "Batan Grande: A Prehistoric Metallurgical Center in Peru." *Science* 216: 952–959, 1982.

Shimada, Izumi, Stephen M. Epstein, and Alan K. Craig. "The Metallurgical Process in Ancient North Peru." *Archaeology* 36(5): 38–45, 1983.

METALLURGY IN SOUTH AMERICA Numerous pre-Hispanic artifacts of gold and silver craftsmanship testify to the exquisite skill of ancient Andean peoples in their production of metallic art. Prior to European arrival, ancient metallurgists prevailed in working gold, in winning silver and copper metal from a variety of rich ores, and in creating various sophisticated alloys. Ancient artisans triumphed in working these materials in ingenious ways to improve their performance and appearance, and in joining them to form complex composite pieces. A tremendous wealth of exquisitely crafted metal ornaments and metallic art was created through native talent. Today, the skill of pre-Hispanic Andean peoples in winning and working metals is revealed by early historical sources, archaeological research, and the remaining portion of metal objects that avoided the Conquistador's torch. Some of the finest examples come from the Moche and Chimu regions of northern Peru; they include funerary masks, breastplates, diadems, and crowns, some of which are inlaid with decorative stones of turquoise and chrysocolla.

Early historical sources are clear in their portrayal of native Andean peoples as skilled metallurgists. Among the first chroniclers, Cieza de Leon described the successful native process of smelting silver, using wind-blown furnaces (*huayras*). Historical records for Potosi are also clear in stating that it was through the work of native metallurgists that the silver wealth was first tapped. For nearly three decades, from 1545 to 1572, all silver production was the result of skilled Andean natives, who used thousands of wind-blown *huayra* furnaces to smelt the rich silver-lead ores. Such furnaces were still employed by Andean natives in the seventeenth century, at which time they were recorded by people familiar with Old World metallurgy, and similar devices have been employed well into the twentieth century.

Archaeological research today includes the discovery and study of metallurgical sites. Work at the site of Batan Grande in northern Peru reveals a centuries-long sequence of copper-bronze production, ending with Inca-period efforts just prior to the Spanish Conquest. Detailed research by such scholars as Izumi Shimada and John Merkel reveals the sophisticated nature of the smelting procedure, and reconstructs the various steps used in the metallurgical process. In addition, recent related efforts by Heather Lechtman focus on the source of the ores used in the production of the copper–arsenic alloys, arguing for a highland–coastal exchange. With regard to detailed archaeological investigations of sites where metal was crafted, rather than smelted, the research on metal craftsmanship at Chan Chan is an important contribution, which examines the activities carried out by metal smiths within a particular district of that pre-Incaic city. Also of note are investigations into the pre-Hispanic smelting facilities and processes in the South Andes. To date, the earliest metallurgical (copper) slags in this latter region come from the Wankarani site in highland Bolivia; they date between 250 and 1200 BC. Of particular note, research at the Ramaditas site in northern Chile reveals a highly skilled, natural draft technology operating in the Atacama region by 100 to 50 BC, where pre-Hispanic metallurgists were capable of achieving a good separation of copper metal from slag during production. These studies lend weight to the idea of a highly effective metal smelting technology in place in the South Andes during the first millennium BC.

In general, it is thought that gold working preceded copper smelting, gold being found in a natural metallic state in association with certain minerals. The earliest known gold-working kit and beaten gold work in the Andes dates to approximately 1500 BC from Waywaka, Peru. The first working of native copper would also be theoretically of a similarly early date.

The most extensive body of research on ancient Andean metallurgy deals with analyses of the metal artifacts themselves. Included here are laboratory studies on gilding, joining of metals, alloying, and the other techniques of artistry and design. When the Spanish began melting metal objects shortly after the Conquest, they noted with dismay that not all "golden" objects were in fact pure gold. Many specimens appeared to be gold on the surface, but were actually copper alloy in the core, which meant that they had little of their anticipated content of precious metal. Modern laboratory studies have succeeded in replicating the ancient techniques employed in gilding, as well as the casting, welding, and forming of exceptionally well-crafted objects (Lechtman et al., 1982, Tushingham et al., 1979). Andean craftsmen were adept in altering the appearance of an alloy, sometimes within the same piece, as in the case of Vicus nose ornaments. In some cases, their illusions seemed to accomplish the impossible, where different metals appeared to be welded together. These ancient metalsmiths exercised superb control over their artistic medium, in which the color of the metal was as important a factor as the iconography that the piece intended to convey.

As one shifts from the North Andes to the South Andes, there is a shift in the metallic art from complex and fully modelled pieces to artifacts of decorated sheet metal. In Bolivia, northern Chile, and northwestern Argentina, metallic art most often takes the form of objects made from flat metal. Discs, diadems, bracelets, rings, and pendants are common forms, all executed from hammered sheet. Casting is also known, but seems to be used primarily in the manufacture of axes and mace heads, i.e. non-decorative objects that required more substantial weight. The tradition of working sheet metal has a long antiquity, extending back three millennia.

Today, Andean natives remain active in working metal and crafting pieces of native art. Like their counterparts in the American Southwest, they melt coins for metal, rather than smelt ores as formerly done. The end product is most often geared towards tourist consumption, which means that it is generally of a form that is readily marketable. Still, there is a folk practice that persists among the native Andean peoples, one which has an extremely long history.

GRAY GRAFFAM

REFERENCES

Bakewell, Peter. *Miners of the Red Mountain: Indian Labor a Potosi, 1545–1650*. Albuquerque: University of New Mexico Press, 1984.

Barba, Alvaro Alonzo. *El Arte de los Metales*. Trans. Ross E Douglass and E. P. Mathewson. New York: Wiley, [1640] 1923

Cole, Jeffrey A. *The Potosi Mita, 1573–1700*. Stanford: Stanford University Press, 1985.

Gonzalez, Alberto Rex. *Las Placas Metalicas de los Andes del Sur* Mainz am Rhein: Verlag Philipp von Zabern, 1992.

Graffam, Gray, Mario Rivera, and Alvaro Carevic. "Copper Smelting in the Atacama: Ancient Metallurgy at the Ramaditas Site Northern Chile." In *In Quest of Mineral Wealth: Aboriginal and Colonial Mining and Metallurgy in Spanish America*. Ed. Alan

Craig and Robert West, Baton Rouge, LA: Geoscience Publications, 1994.

Grossman, Joel W. "An Ancient Gold Worker's Tool Kit." *Archaeology* 25: 270–275, 1972.

Lechtman, Heather. "The Production of Copper-Arsenic Alloys in the Central Andes: Highland Ores and Coastal Smelters?" *Journal of Field Archaeology* 18: 43–76, 1991.

Lechtman, Heather, Antonieta Erlij, and Edward J. Barry, Jr. "New Perspectives on Moche Metallurgy: Techniques of Gilding Copper at Loma Negra, Northern Peru." *American Antiquity* 47:3–30, 1982.

Niemeyer F., Hans, Miguel Cervellino, and Eduardo Munoz. "Vina del Cerro: Expresion Metalurgica Inka en el Valle de Copiapo." *Creces* 4(4): 32–35, 1983.

Ponce S., Carlos. *Las Culturas Wankarani y Chiripa y su Relacion con Tiwanaku*. La Paz: Editorial "Los Amigos del Libro", 1970.

Rodriguez O., Luis. "Pre-Columbian Metallurgy of the Southern Andes: a Regional Synthesis." In *Pre- Columbian Metallurgy*. Bogota: 45th International Congress of Americanists, Banco de la Republica, 1986, pp. 402–417.

Shimada, Izumi, Stephen M. Epstein, and Alan K. Craig. "The Metallurgical Process in Ancient North Peru." *Archaeology* 36(5): 38–45.

Shimada, Izumi, and John F. Merkel. "Copper-Alloy Metallurgy in Ancient Peru." *Scientific American* 265(1): 80–86, 1991.

Topic, John. "Craft Production in the Kingdom of Chimor." In *The Northern Dynasties: Kingship and Statecraft in Chimor*. Ed. Michael E. Moseley and Alana Cordy-Collins. Washington, D.C.: Dumbarton Oaks Research Library and Collection, 1990, pp. 145–176.

Tushingham, A. D., Ursula M. Franklin, and Christopher Toogood. *Studies in Ancient Peruvian Metalworking*. Toronto: Royal Ontario Museum of History, Technology, and Art, Monograph 3, 1979.

METEOROLOGY IN CHINA Chinese meteorology, here referring to the traditional meteorology which was used in China, has many unique characteristics. Although China began to adopt Western meteorological knowledge as it was introduced in the seventeenth century, Chinese traditional meteorology lasted two hundred and fifty years. Chinese meteorology can be described from four aspects.

Knowledge about Meteorological Phenomena

The Chinese recognized some meteorological phenomena three thousand years ago. In the inscriptions on borns or tortoise shells of the Shang Dynasty (ca. sixteenth–eleventh century BC), there were some words meaning rain, frost, snow, thunder, lightning, rainbow, etc. The Chinese identified the relationship between the rain and rainbow; they knew if there was a rainbow in the western sky in the morning, it would rain soon. The Book of Songs declared, "White dew is frost". In this case, frost must have been frozen dew. Two thousand years ago, the Chinese also recognized the six segments of a snowflake.

Wang Chong (AD 27–97) was one of the first to record meteorological phenomena. He said that the rain came from the ground, not from the sky, meaning that the rain came from the vapor rising from the ground. He also said that clouds and fog are omens of rain, there is dew in the summer but frost in the winter, and rain when it is hot but snow when cold. He understood that rain, dew, frost, and snow were formed of vapor from the ground at different temperatures.

The ancient Chinese had exact knowledge about rainbows, too. Kong Yingda (AD 574–648) who lived in the early time of the Tang Dynasty, pointed out that the rainbow was created by the sun's shining on the waterdrops. Sun Wanxian and Shen Kuo (AD 1030–1094) who lived in the Northern Song Dynasty, studied the rainbow too, and agreed with Kong Yingda. Sun Yanxian said that the rainbow was the reflection of the sun in the rain, created when the sun shines on the rain.

They also had a good deal of knowledge about the wind and the clouds. In the fourth century, it was recognized that the trade wind had twenty-four fans. Li Chunfeng (seventh century) recorded the wind as having ten grades according to its strength and twenty-four types according to its direction. Shen Kuo once recorded a land tornado in his book. "There is a tornado coming from the south-east in Enzhou and Wucheng. It looked like a huge sheep horn, and carried all the big trees. Quickly it disappeared in the sky". Sima Qian (b. 145 BC) divided clouds into three types according to their height from the ground. Eighteen pictures of clouds were drawn in the fourteenth century according to weather conditions. Later the number reached thirty-two.

Weather Forecasting

Many methods of weather forecasting were used in ancient China. The first is forecasting according to the air humidity. In the second century, the Chinese recognized the relationship between the sound of a musical instrument and the weather. Later, Wang Chong pointed out that it would rain as the strings of a *zheng* (an instrument in some ways similar to the zither) became slack. No later than the eighth century the Chinese recognized that many waterdrops appearing on a solid body with good heat-conductivity, or high temperature and great humidity, were all omens of the rain.

The second method is forecasting the weather according to optical phenomena such as rainbows, rosy clouds, and halos. Some records written in the nineteenth century proved that the natives of Fuzhou could predict a heavy rain and a great wind, even a typhoon on the sea, according to the position and height of the rainbow. In the eleventh century,

Kong Pingzhong pointed out that the morning glow is the omen of a rain, while the evening glow is the omen of a sunny day. Lou Yuanli, who lived in the Yuan Dynasty, stated that the solar halo was the omen of a rain, and the lunar halo of a wind. He also pointed out that the direction of a wind was the direction of the gap of the lunar halo.

The third method is forecasting according to the movement and patterns of clouds and fog. Clouds and fog are the bases of some weather phenomena such as rain and snow, so their height, patterns, and direction can be used to forecast the weather. Since the Tang Dynasty, there have been many such forecasts. Huang Zifa, who lived in the Tang Dynasty, once forecasted, "if there is some cloud moving against the wind, it will rain".

The fourth method is the forecast according to sounds and lightning. There were many weather related proverbs in ancient China. "If there is lightning in the southern sky, it will rain; if in the northern sky, it will not." (Lou Yuanli). "If the lightning is irregular, it will rain hard." "No rain, but thunder, go by boat, come by feet", meaning it would not rain for some days.

The fifth method is according to the activities of animals. Some animals are sensitive to weather changes. The ancient Chinese could forecast the weather according to their activities. Wang Chong said, "If it is going to rain, ants migrate, earthworms come from their holes". There was a proverb in the Tang Dynasty that said that if ants blocked up their holes, it would rain. Also ancient Chinese recognized that if birds' wings moved hard as they fly, it would rain.

Meteorological Survey and Instruments

The ancient Chinese conducted many meteorological surveys and invented many surveying instruments.

Wind and Surveying Instruments. Two thousand years ago, the Chinese used a surveying flag and Xiangfeng bird to judge the direction of the wind. A Xiangfeng bird was made of copper slices fixed on the top of a high pole. It could be revolved by the wind, and its head was always along the direction of the wind. At first, Xiangfang birds were used in meteorological observatories; later they were used in the government and private houses. Even now, some Xiangfeng birds can be found on the tops of some towers. Li Chunfeng recorded the method to measure the direction and speed of the wind using a chicken feather, i.e. to measure according to the moving direction and dip angle of the feather. In 1716, the Qing Government set up a meteorological network to survey the direction of the wind using surveying flags, which was the primary form of the modern meteorological network. However, the Chinese also began to use western wind surveying instruments at that time.

Precipitation and the Chinese Precipitation Gauge. Because precipitation was very important to agriculture and people's lives, in the Eastern Han Dynasty (AD 25–220) the court ordered that every noble government should report precipitation in the period from the beginning of the spring to the beginning of the autumn. Qin Jiushao recorded a kind of precipitation gauge — Tianchi Basin, which was widely used in 1247. The Western precipitation gauge and distiller were recorded in Chinese books of the eighteenth century.

Humidity and the Surveying Method. In the Western Han Dynasty, the Chinese invented a method to measure the air humidity by hanging a lump of earth and a bar of charcoal (or a feather). When the air was dry, the bar of charcoal (or the feather) was light; when the air was humid, it was heavy, but the earth had little change in its weight. By hanging a lump of earth and a bar of charcoal (or a feather) on the two ends of a staff separately and fixing a lifting string on the middle point, making the staff horizontal in the dry air, a humidometer was made. When the air became humid, the end which had the charcoal fell down. Huang Lüzhuang (AD 1626–?) invented a humidometer to indicate the air humidity by a moving needle. Later, Ferdinand Verbist (AD 1623–1688), a Belgian missionary, invented another one.

Atmospheric Temperature and Pressure. The Chinese paid much attention to atmospheric temperature. Wang Chong once pointed out that the atmospheric temperature changed during a day and affected rain and snow directly. Until recent centuries, the Chinese had not invented a thermograph. In the nineteenth century, Zou Boqi (AD1819–1869) recorded a barometer for the first time.

Achievements in Phenology

Phenology is the process of discovering meteorological laws by the regular activities of some animals and the regular changes of some plants. The Chinese began their study in phenology a long time ago; some records of phenological phenomena written three thousand years ago have been found. In Xiaxiaozheng (Lesser Annuary of the Xia Dynasty), there were some records of the turn of the months. The knowledge of the twenty-four divisions of the solar year in the traditional Chinese calendar was gained in the Spring and Autumn Period (770–476 BC) and the Warring States Period (475–221 BC). Every day marking the twenty-four divisions has relationships with agricultural activities. The Chinese usually arranged their agricultural work and other activities according to the twenty-four divisions of the solar year.

There are many records of phenological phenomena written during the period of the Western Han Dynasty to the Song and Yuan Dynasties (ca. second century BC–fourteenth cen

:ury AD). Lu Zuqian (AD 1137–1181) who lived in Northern Song Dynasty observed the phenological phenomena in Jinhua, Zhejiang province for nineteen months and made many records, including the blossoming of twenty-four kinds of flowers such as winter sweet, peach, plum, lotus, and chrysanthemum, and the first appearance of the spring warblers and the autumn insects.

No later than the Spring and Autumn Period the Chinese recognized that migrants' activities changed with the seasons. Shen Kuo recorded in his book that the natives of Hebei called the frost "information frost", because they knew there would be a frost when the wild white goose came.

There are many Chinese records of meteorological phenomena, and they are still used today.

LI DI

REFERENCES

Li Di. "Color Dispersion as Understood in Ancient China." *Wuli* (Physics) 5(3): 161–164, 1976 (in Chinese).

Li Di. "The Investigations of Meteorological Phenomena by Shen Kuo." *Scientia Atmospherica Sinica* 2: 159–161, 1977 (in Chinese).

Li Di. "On the Invention of Meteorological Instruments in Ancient China." *Scientia Atmospherica Sinica* 2(1): 85–88, 1978 (in Chinese).

Li Di. *New Source Materials for Meteorological Instruments, Science and Technology in Chinese Civilization*. Singapore: World Scientific Publishing Co., 1987, pp. 199–210.

Wang Jinguang and Hong Zhenhuan. "The Method to Measure Air Humidity in Ancient China." *The Collection of Papers on the History of Science* 9: 20–23, 1966 (in Chinese).

See also: Sima Qian – Shen Guo – Li Chunfeng – Qin Jiushao – Wang Chong – Surveying

METEOROLOGY IN INDIA The very first stage of civilization was marked by peoples' efforts to understand their surroundings and make use of their beneficial aspects. Their first action in this direction was to produce food, making use of the available water in the rivers and rainfall in the region. Though initially extreme phenomena like heavy rains, winds, cold and hot spells, droughts, and floods, appeared incomprehensible and hostile, early humans gradually sorted out their seasonal character and planned their agricultural operations accordingly. Thus began in a crude way the development of weather science all over the world.

In India, the development of this science commenced in the early Ṛgvedic period. That the heat of the sun lifts the water to the atmosphere which after some time comes down as rain was recognized by the Vedic seers at a very early stage. In order to explain the occurrence of rain during a restricted period of about two months in their region in extreme northwest India, they imagined that water was absorbed by the sun's rays in the vast ocean areas in the south during the winter season and the humid air carried northwards by the sun's rays. When the sun attains its extreme northward position and starts retracing its path, the humid air gets deflected near the foot of the Himalayas and brings rain from the east to their region. These moist easterlies replace the westerlies that were present in the region before the arrival of the monsoon rains. Whenever there was drought, they performed rituals to invoke the rain god. They believed that in Nature there is a feed of a substance called *soma* from above into the atmosphere, which aids the occurrence of rainfall. Therefore they fed into the ritual fire some substances like wild dates and some special types of grass which produced smoke and were believed to be effective in aiding rainfall.

The post-Vedic scholars developed the subject further, mainly working on the pregnancy concept of rainfall. They looked for symptoms in the winter season for the commencement of pregnancy and identified the characteristics of winter disturbances in their region as indicating the same. Working along these lines, they were able to observe weather very carefully during the pre-monsoon months and were able to define the course of events which go towards the nourishment of rain embryos and the delivery of good summer rainfall at the right time after 195 days. Any departure from the defined meteorological conditions during the growth period, such as too much rainfall, or snowfall, unfavorable winds, and temperature, was said to affect the quantity of rainfall delivered during the rainfall period. They also believed that hail would occur if the rain fetuses overstayed in the atmosphere. The moon's position with respect to the sun and the stars was believed to influence the formation of rain embryos. The moon was conceived as a replica of soma in the heavens, and soma was capable of fertilizing the atmosphere.

Based on such concepts and extensive observations, the post-Vedic scholars developed several rules of long range rainfall forecasting. If they were at all successful, it was certainly due to their capacity to observe day-to-day weather and individual weather elements, like clouds, temperature conditions, wind, rain, lightning, and thunder. They were extremely clever in mentally working out correlations based on observed data. For short- and medium-range forecasting they framed many rules of thumb based on winds, clouds, temperature, lightning, thunder, moisture in the atmosphere, behavior of people, animals, birds, snakes, worms, insects, trees, and plants, as well as visual impressions of the sun, moon, stars, and sky. They were so thorough with local weather that their capacity to forecast in the short- and medium-range was as high as that of any modern forecaster who does the same with sophisticated equipment and maps.

Measurement of rainfall in India dates back to the fourth century BC. A standard rain gauge was constructed around the third century BC, and this system of measurement was prevalent in North India for a very long time (third century BC to AD sixth century).

Well before the birth of Christ, the Arab dhows sailed across the Indian Ocean for trade purposes. Hippalus, a Greek pilot of the first century, sailed across the Arabian Sea for the first time. A handbook for merchants called *Periplus* was written by a Greek around AD 50. Subsequently, Arab geographers wrote many books giving details of Indian ocean voyages. Sidi Alis' *Mohit*, written around AD 1554, not only gives a map of the Indian Ocean area but also mentions the occurrence of monsoons at fifty distinct places.

With the arrival of more voyagers from the west in the Indian Ocean, a steady effort for systematically observing the wind, weather, and weather systems of the Indian Ocean commenced. In his first voyage from Melinda to Calicut in 1499, Vasco da Gama made use of the monsoon winds and reached his destination in just three weeks. William Dampier published many observations of Indian ocean weather and weather systems in his travel accounts. He was a sixteenth century buccaneer who lived and worked with some of the rowdiest pirates in history. But he was also an astute observer of nature in general and weather in particular. In his *Discourse on Winds and Breezes, Storms and Currents*, he deals with general wind systems throughout the world and their seasonal changes, which include the Southeast trades of the South Indian Ocean and Northeast and Southwest monsoons of the North Indian Ocean. During the seventeenth and eighteenth centuries, the military and trade activities of the European powers in the Indian Ocean waters increased.

Matthew Maury in his *Physical Geography of the Seas* (1874) explained the formation of monsoon winds as resulting from the heat of the plains and deserts of the Asian region. The following ideas about the mechanism of the southwest monsoon and its rainfall were generally agreed upon by the meteorologists of the nineteenth century.

The plains of North India get very hot during the summer, and the air over that region ascends and becomes light. As a result, air over the sea areas where the pressure is high both in the neighborhood of the Equator and south of it, moves towards the region of low pressure of the land. The southeast tradewinds, while moving northwards and crossing the equator, become southwest winds, owing to the rotation of the earth. Again, these southwest winds do not blow directly into the region of low pressure, but go around it in an anticlockwise direction. If one stands with one's back to the wind, the pressure to the left is lower than to the right in the northern hemisphere. In the southern hemisphere, the realtion is reversed. The copious precipitation of the west coast is due to the high mountains which run along the coast. The higher the mountains, the heavier the precipitation. The monsoon is sustained by the latent heat released during the precipitation, which adds more heat to the atmosphere, and therefore further rarefaction takes place. Strong winds blow into the region of heavy rainfall, since air from the neighboring regions rushes to occupy the space created by ascending air.

Meanwhile, more knowledge was added to the science of cyclones in the Indian Ocean. Henry Piddington made a monumental contribution to the science of storms. He was the first to coin the term "cyclone", which gained world usage later. In a series of papers he gave detailed accounts of many Indian Ocean cyclones. His best-seller at that time was the *Horn Book of Storms for the India and China Seas*, which was followed by another book called *Horn Book for the Law of Storms*, in which he explained the use of transparent horn cards provided in his book for finding out the center of cyclones.

Many Indian meteorologists led by Desai, Rao Koteswaram and Majumdar, worked on various aspects of the formation of cyclones. They investigated the role of the upper tropospheric flow patterns in the intensification, movement, and dissipation of tropical disturbances in the Indian Ocean. The availability of aircraft winds and satellite pictures enabled the meteorologists, such as Raman and Srinivasan, to study the low-level convergence and associated winds around the calm eye region of the cyclone, upper-level divergence, and the relation of the direction of movement to the upper-level winds. They also studied the influence of sea surface temperature on the formation of the cyclone.

As regards the southwest monsoon, the upper air observations of wind and temperature and also the newly formulated dynamical concepts enabled the meteorologists to understand many synoptic aspects of the monsoon. Many meteorologists studied the role of the easterly jet stream and the Tibetan high, the northward shift of the westerly jet stream, the advance of the intertropical convergence zone to northern India, and the extension of equatorial westerlies. Koteswaram and Flohn (1960) made important contributions in this field.

Today meteorology in India is a highly developed subject both from the research and service point of view. The countr has produced many skilled meteorologists whose expertise is on par with that of meteorologists of developed countries.

A.S. RAMANATHAN

REFERENCES

Blanford, H.F. *Climates and Weather of India, Ceylon and Burma*. New York: Macmillan Company, 1889.

Capper, J. *Observations on the Winds and Monsoons*. London: Debrett, 1801.

Dampier, William. *Voyages and Descriptions*. London: J. Knapton, 1699.

Das, P.K. *The Monsoons*. New Delhi: National Book Trust India, 1988.

Maury, Matthew. *Physical Geography of the Seas*. New York: Harper, 1861.

Piddington, H. *Horn Book of Storms for the India and China Seas*. Calcutta: Bishop's College Press, 1844.

Piddington, H. *Horn Book for the Law of Storms*. New York: Wiley, 1848.

Ramage, L.S. *Monsoon Meteorology*. New York: Academy Press, 1971.

Ramanathan, A.S. "Weather Science in Ancient India, I-VIII." *Indian Journal of History of Science* 21(1): 7–21, 1986; 22(1): 1–14, 1987; 22(3): 175–197 and 198–204, 1987; 22(4): 277–285, 1987.

Simpson, G.C. "The South West Monsoon." *Quarterly Journal of the Royal Meteorological Society of London*, XI–XII:199, 151–172, 1921.

Walker, G.T. "On the Meteorological Evidence for Supposed Changes of Climate in India." *Memoires of the Indian Meteorological Department*. 21: 1–22, 1910.

Walker, G.T. "Correlation in Seasonal Variation of Weather." *Memoires of the Indian Meteorological Department*. 24: 275–332, 1924.

See also: Navigation

METEOROLOGY IN THE ISLAMIC WORLD Medieval Islamic conceptions of nature and physical phenomena were partially based upon a translated accumulation of Greek thought. The translation efforts of Middle Eastern cultures contributed greatly to the preservation of Greek knowledge. Among the Greek philosophers who had conjectured upon the phenomena of the atmosphere, the most famous was Aristotle (384–322 BC), whose geoscience treatise in four books called *Meteorologica* dealt not only with atmospheric phenomena but also with the general terrestrial aspect (including geological, hydrological, and oceanographical ideas) of his systematic cosmology.

Some of the questions pondered were meteorological: whether the Milky Way and comets were of terrestrial or celestial origin, hail forming theories, the origin of wind, the relation of thunder and lightning, and optical theories f the rainbow and the halo.

With the ninth century came a stabilization of the long political turmoil after the Islamic conquests. Also, with the rise of the Abbasid Caliphate at Persian Baghdad, that civilization and India significantly influenced the seminal culture of the new Islamic empire. The legacy of Greek speculative philosophy also served as a stimulus to the expansion of the Islamic scholarly agenda. With the founding of a translation center within Caliph al-Maʾmūn's (813–833) *Bayt al-Ḥikma* (House of Wisdom) at Baghdad, the next three quarters of a century would transfer much of the Greek corpus into Arabic.

The *Meteorologica* would have been included among the Aristotelian translations. Evidently the first translation of the work into Arabic from the original Greek or Syriac of about 820 was that by the Jewish Arabic scholar Yaḥyā ibn-al-Biṭrīq (fl. ca. 820). Abū Naṣr al-Fārābī (ca. 870–950) had, following the Persian mathematician al-Kindī (ca. 801–ca. 866), adopted the Aristotelian classification of knowledge with study of nature under physics. In addition, the general scheme of Islamic knowledge defined philosophical science with seven subdisciplines of natural sciences, with meteorology as one of those. Though virtually lost, al-Fārābī's many large commentaries on Plato and Aristotle included the *Meteorologica* in his *Kitāb Iḥṣāʾ al-ʿulūm*. But he only discussed the traditional conception of the four elements of matter — earth, air, fire, and water — in noting the contents of the work.

Four particularly outstanding Islamic thinkers accentuate the tenth through the twelfth centuries. The great Afghani polymathic scholar Abū Rayḥān al-Bīrūnī (973–1048) wrote copiously, though again many works are known only by name. A rare linguist, who knew not only Arabic and Persian but also Turkish, Sanskrit, Hebrew, and Syriac, he steered clear of formal Aristotelian commentary, showing an observationally rich interest in the geosciences. His *Taḥdīd Nihāyāt al-amākin li-taṣḥīḥ masāfāt al-masākin* (Determination of the Coordinates of Cities) discussed fossils, physical geology, geography, and the ancient geodetic problem of finding the circumference of the earth. His *Kitāb al-tafhīm li-awāʾil ṣinā ʿat al-tanjīm* (The Book of Instruction in the Art of Astrology) also contained much on the sublunar world, including weather and climate over the known globe. Of the more conventional meteorological fare, he accepted basic ancient atmospheric ideas, including a variant view of the Milky Way as atmospheric smoky vapors screening the stars. The three most recognizable Islamic contributors to meteorology were: the Alexandrian mathematican/astronomer Ibn al-Haytham (Alhazen, 965–1039), the Arab-speaking Persian physician Ibn Sīnā (Avicenna, 980–1037), and the Spanish Moorish physician/jurist Ibn Rushd (Averröes, 1126–1198).

The commentaries of these great philosophers reflected the high end of Arabic evolution toward the dictates of observation based on the logic of both critical deductive and inductive reasoning. Ibn al-Haytham, particularly noted for his seminal experimentally based inductive reasoning, was the first outstanding medieval Arabic theorist of physical optics with important applications to meteorological phenomena. His prolific output included some twenty science treatises, including his great optical treatise *Kitāb al-manāẓir* (Book of Optics).

Among the optical discussions, the treatise contained his extensive experiments and findings on reflection and refraction, and his experiments on the rainbow mechanism, a phenomenon all the vogue as a physical problem in the Middle Ages. His rainbow findings were also reported in his *Qaws quzaḥ wa'l-ḥāla* (On the Rainbow and the Halo), a work not in the *Optics* nor available other than as a manuscript. Ibn al-Haytham's innovative experimental method entailed a laboratory to study the phenomenon of the earth, such as chemical compounds, as well as his optical studies. Aristotle had considered the rainbow a reflection phenomenon from clouds of uniform drops acting as a continuum surface like a convex mirror. Ibn al-Haytham, researching reflection of light from plane and curved mirrors, reasoned that the phenomenon was a case of reflection similar to a spherical concave mirror. He simulated the rainbow colors by transmitting sunlight through glass spheres of water, spherical concave mirrors representative of clouds, with the cloud still acting as a continuum. Unfortunately, he also decided that refraction had nothing to do with the phenomenon, considering the same mechanism for the lunar halo and solar corona. He also employed his ideas of reflection in dealing with a terrestrial Milky Way in one of his treatises, *al-Majarra* (On the Milky Way).

Ibn al-Haytham has been considered the first thinker to realize the refractive, i.e. light bending properties, of the atmosphere. The phenomenon had been continually hinted at since ancient times in the discrepancies found in observing celestial objects because of near horizon distortion of position and size. He wrote a short treatise, *Mas'ala fī Ikhtilāf al-naẓar* (A Question Relating to Parallax). And, in his *Fī Maʿrifat irtifāʿ al-ashkhāṣ al-qāʾima wa-aʿ midat al-jibāl wa irtifāʿ al-ghuyūm* (Determination of the Height of Erect Objects and the Altitudes of Mountains and of the Height of Clouds), he was evidently the first medieval thinker to use knowledge of refraction in theorizing by a convoluted geometry that the atmosphere was much lower than the ancients had estimated. His near contemporary Cordoban Ibn Muʿādh (ca. 989–1079) should also be mentioned in regard to atmospheric height for his singular hypothesis in his treatise — mistakenly attributed to Ibn al-Haytham — "On the Dawn" (evidently known only by the Latin translation by Gerard of Cremona *Liber de crepusculis*), also called "On Twilight and the Rising of Clouds". Sunlight before sunrise and after sunset are also phenomena of refraction, and Ibn Muʿādh estimated the angle of depression of the sun at dawn and evening twilights, arriving at the fairly accurate value of eighteen degrees by which the height of atmospheric moisture (believed responsible for twilights) and thus atmospheric height could be determined.

Ibn Sīnā and Ibn Rushd represented concerted Arabic commentary as it moved from the eleventh to the twelfth century. Though more noted for his varied contributions to medicine, Ibn Sīnā contributed to physical science in twenty volumes of general thought, *Kitāb al-ḥāṣil*. He also wrote on the seasons and climate in the *Kitāb al-Anwāʾ* (Book of Meteorological Qualities). Ibn Sīnā's meteorological significance centers on the rainbow mechanism and the medieval fascination with the origin of the Milky Way. Departing from Aristotle's cloud continuum and Ibn al-Haytham's spherical mirror analogy, he reasoned that the rainbow was the result of reflection from the total amalgamation of water drops — this being the key discovered later — supposedly released by clouds as they dissolved into rain. His observational prowess is seen in his explanation. The idea came to him by watching the diffraction of sunlight by water drops created by the watering of a garden in a bathhouse. He thought the Milky Way celestial in origin, voicing yet one more assent to a physical concept important to both meteorology and astronomy as an eventual point of redefinition of the ancient boundaries of celestial and terrestrial phenomena. Yet, as with thinkers to follow and into the late eighteenth century, the fact that the Milky Way was an expanse of stars and not a by-product of those stars, escaped him as well. Ibn Rushd also held a celestial opinion of the Milky Way, one more both analytical and worthy of further discussion.

Contrary to a later conception of Ibn Rushd as slavishly Aristotelian, we can say that his jurist's logic followed the exact Aristotelian order of nature as a model of systematic formulation. Ibn Rushd wrote both short introductions and extended larger commentaries, such as *Al-aṭār al-ʿAlwiyya*, which included textual discussion, arguments about the opinions of other commentators, and his own analysis.

The Milky Way was a more bothersome challenge. Ibn Rushd decided that Aristotle's theory was untenable since it depended on the reality of hot and dry exhalations which had not been proved. He also reasoned by the phenomenon of parallax (the apparent change in position of relatively close objects with change in position or view of an observer) that if the Milky Way were terrestrial (below the sphere of the moon) and thus relatively close, it would have a backdrop of different stars depending on where it was observed. In proving this, Ibn Rushd spent time observing and recording the positions of the Milky Way with respect to the stars in the constellation of Aquila from different locations. He found no change. He also noted that because the Milky Way was a constant phenomenon, whereas the exhalation was an ever changing one, the Milky Way appeared to be in the celestial sphere. But ultimately he kept the phenomenon terrestrial, calling it an atmospheric refracted image of light from a conglomerate of small stars seen from the perspective of earth.

Among later commentators was Abū'l-Faraj (Aboulfarag, d. 1286), the Nestorian bishop, who wrote a theory in 1279 of the Milky Way phenomenon in relation to fixed stars and constellations. He leaned toward considering the Milky Way as wholly consisting of stars and having nothing to do with terrestrial nature.

Islam's commentary also turned to its first generation of philosophers. Optical interests seemed to die with Ibn al-Haytham, until Quṭb al-Dīn al-Shīrāzī (1236–1311) and his student Kamāl al-Dīn (d. 1320) pursued a more critical look at his optics in *Tanqīḥ al-manāẓir* (Revision of Optics), which also delved into rainbow and halo theory. In analyzing Ibn al-Haytham, Kamāl al-Dīn initially looked to Ibn Sīnā's rainbow as a water drop reflection phenomenon, leading him to consider the water drops as analogous to transparent spheres of water. This was the breakthrough conclusion, allowing him to reason that two refractions took place, one on and one in a cloud drop in the rainbow optics. Kamāl al-Dīn used a better conceptual physics and geometry to explain the rainbow than Ibn al-Haytham had used.

The cultural devastation of the Mongol invasion of the thirteenth century punctuated the end of the Islamic golden age. Nonetheless, it left its intellectual legacy in North Africa and passed to Spain, where cosmopolitan Toledo served as a clearinghouse of translation for both Christian and Islamic scholars.

WILLIAM J. McPEAK

REFERENCES

Boyer, Carl B. *The Rainbow from Myth to Mathematics*. New York: T. Yoseloff, 1959.

Nasr, Seyyed Hossein. *An Introduction to Islamic Cosmological Doctrine*. Revised ed. New York: State University of New York Press, 1993.

Petraitis, Casimir. *The Arabic version of Aristotle's Meteorologica: A Critical Edition with an Introduction and Greek-Arabic Glossaries*. Beirut: Dar El-Machreq Editeurs, 1967.

Sabra, A. I. *The Optics of Ibn Al-Haytham. (Books I–III)*. London: The Warburg Institute, University of London, 1989.

Sarbra, A.I. "Ibn Haytham." In *Dictionary of Scientific Biography*. Ed. C.C. Gillispie. New York: Charles Scribner's Sons, 1970–1990, vol. 6, pp 189–210.

Smith, A. Mark. "The Latin Version of Ibn Muʿādh's Treatise 'On Twilight and the Rising of Clouds'." *Arabic Sciences and Philosophy* (2): 83–132, 1992.

Thorndike, Lynn. *History of Magic and Experimental Science*. 8 vols. 2nd ed. New York: Columbia University Press, 1964–66.

Urvoy, Dominique. *Ibn Rushd (Averroes)*. Trans. Olivia Stewart. London: Routledge, 1991.

Wickens, G. M. *Ibn Sina: Scientist and Philosopher*. Bristol: Burleigh Press, 1952.

See also: Ibn Sīnā – Ibn Rushd – al-Maʿmūn – al-Kindī – al-Bīrūnī – Ibn al-Haytham – Astronomy in Islam – Optics in Islam – al-Shirāzī – Kamāl al-Dīn

MILITARY TECHNOLOGY The term military technology is broad, and, as a subject restricted to non-Western cultures, potentially laden with analytical complexity. In fact, the contraints of a survey make it necessary to view the more technical innovations of the larger cultures rather than the myriad of variations on pointed weapons fashioned by essentially all peoples. Stimulated by environment and nature, the gamut of world cultures have used artistic and functional inventiveness in weaponry. Non-Western ancient military technology provided significant origins for Western military technology as well.

The first most significant line in the military technology progression was metallurgy of copper in the transitional period between the Neolithic and true Bronze Ages, approximately between 4500 BC (perhaps 5000 BC) and 3500 BC in the Near East arc from Mesopotamia to Egypt. Copper's cold malleability enabled the earliest metalworkers to beat, rather than fire it from the ore as with harder metals. In doing so, they could fashion a metal version of basic wood, bone, and stone pointed weapons: arrow tips, spears, and particularly, swords. This was followed by smelting (melting metal to separate out impurities) and founding (melting the purer metal for casting and molding). By about 3000 BC the general use of copper and experiment with its alloys (bronze with tin and brass with zinc) ushered in the Bronze Age to southwestern Asia over five centuries before general use. This was an essentially Near Eastern phenomenon, probably disseminated to India, Anatolia, and surrounding areas after this.

As far as we know today, the first great civilization of humanity was that of Sumer in southern Lower Mesopotamia after 4000 BC. Among so many accomplishments handed down to subsequent Mesopotamian civilizations and the west, one especially important one was worked copper alloys and probably bronze swords. Mesopotamian cast copper mace heads, the first technical use of metal, date from 2500 BC. About the same time Sumerian smiths were casting socketed axe heads. In the north, Semitic peoples to be called Akkadians, from which the Assyrian and Babylonian cultures developed, assimilated Sumerian technology. Before 2000 BC non-Semitic, Indo-European invaders from central Asia began various waves of infiltration from Asia through Asia Minor into Mesopotamia and on to India. All would leave their military mark. Among these were the Hittites from the northwest and later the Hurrians from the northeast and the Caucasuses–the one moving into Lower, the latter into Upper Mesopotamia.

The Hittites overran most of Asia Minor (Anatolia) after 2000 BC and about 1500 BC invaded Babylonia long enough to raze Babylon. Anatolia, a high plateau fringed by mountain ranges, was rich in mineral resources, among these gold and silver, but most importantly iron. The Hittites probably ushered in the early Iron Age by their use of this much harder metal in their weapons. In the general extent of southwestern Asia the Iron Age did not arrive until about 1000 BC, although a few Mesopotamian objects of perhaps smelted iron have been dated before 2200 BC, and some Egyptian work has been conjectured as even older. Iron was much superior to bronze in edged and projectile weapons and required higher temperature metallurgical processes of smelting iron ore and founding the crude metal. Although it was thought in some quarters of the last century that Egypt was the cradle of iron work, development of its metallurgy may have been contemporary with that of Asia Minor, considering abundant Egyptian iron resources. The use of iron also brought more effective defensive hardware, i.e. in armor and in horse trappings and the chariot.

The horse and the two-wheeled chariot were Hittite innovations to western Asian warfare. The horse brought mobility to tactical maneuvering on the battlefield for the specialized soldiers called cavalry. The chariot was introduced during the eighteenth century BC and likely by the Indo-European Aryans (Indo-Iranian, also metalworkers — perhaps early iron weapon users) who invaded Iran from the northeast at that time and influenced the Hittites and evidently held sway over the Hurrians. The chariot provided a further tactical edge, allowing a soldier or two soldiers to act in concert in inflicting multiple casualties at one time. As specialized warriors, the charioteers introduced military class rule to Near Eastern civilizations. With the added innovations of scythe-like blades on its wheels, the chariot also added the mass fear psychological factor to warfare. The Hittites took northern Syria in their clash with Egypt about 1400 BC, the latter having adopted the horse and chariot after their temporary defeat by the chariot tactics of the Hyksos, Amorite peoples who invaded Palestine about 1700 BC. These latter also contributed large fortification technology to the general mud wall military architecture pool of the Near East which started with the high curtain walls of ancient Jericho (8000 BC), the first example of specialist military architecture.

By the middle of the fourteenth century BC the Assyrians were able to take the military ascendancy in Upper Mesopotamia and eventually all of Mesopotamia by the late eleventh century BC, to become a great empire. By the eighth century BC the Assyrian army had reached an apex of coherency, a blueprint for the Persian army. Made up of both professional and militia soldiers, the Assyrian army equipped all troops with finely tempered iron weapons. They employed cavalry and chariots, archers (using the composite recurved or reinforced bow, found in the Middle East to 3000 BC), and slingers, who used the simplest, oldest missile weapon. Adding to their siege tactics, sappers (essentially meaning diggers at that time) were used in approaches to mud-walled defenses, as were battering rams and wheeled-platforms, equipped with shielding defenses against arrows, for rolling against such walls. By the sixth century BC the Persians had become heir to the Assyrian Empire and to the diverse military technology of the Near East. It remained dominant for two hundred years until the informal transition of east to west finally came face to face with the challenge of Greece under Alexander the Great in the middle of the fourth century BC.

In the Far East, Chinese civilization as far back at 2000 BC was characterized by a value placed on functional technology. The integration of the wall into Chinese cultural architecture was given a profound military expression in the Great Wall, which was started in 214 BC by the first emperor Shi Huangdi as a linking of earlier rampart walls. It was meant to keep out the north/northwestern invaders who would plague China for centuries. The crenellated, brick-faced wall still stands, stretching some 4000 miles and 30 feet (9 meters) high, with regular spaced square watchtowers 40 feet (12 meters) high with a 9–12 foot (3–4 meter) passageway through them. Along with their own cultural variations on basic weapons, the Chinese designed light hunting crossbows by the fifth century BC and were using them in combat by the second century BC.

The use of iron metallurgy continued to be the prime advance in military technology. Iron ore is plentiful all over the world. Variations of alloying iron with carbon in smelting processes, which included the introduction of air blasting to fan the fire (the forge) to high temperatures, meant that steel (iron with a small proportion of carbon) and its hardening were probably fairly contemporaneous with iron working (from 1000 BC). Although dating is indeterminate, the great deposits of iron in Central Africa and the proximity of the Egyptian influence point to limited iron and steel forging. Indian weapons of iron were prevalent by 500 BC. In fact, tempered steel was produced fairly early in India. Bars, rods, and plates of raw steel were exported throughout the Near and Middle East. Indian steel was used in the founding of blades of "watered steel" (the process of folding malleable steel over and over then beating it out). These light, high tensile strength curved (Damascus) blades enabled the effective long sweeping offensive draw cut, used by both the infantry and cavalry of western and central Asia down through the last century.

The development of the relatively simple smelting methods of steel and steel weaponry was disseminated eastward to southeast Asia via Indian colonization. Iron weaponry and working began independently in China about 500 BC and smelting of crude steel was fairly contemporaneous (about 400 BC). By the Middle Ages the effectiveness of the Mongolian steel sabre, influenced by Middle Eastern contacts, was supplanting the straight Chinese sword. Japanese iron weaponry, with Chinese influence, began about 200 BC, although the earliest relics date between the second and eighth centuries AD. The best of the distinctive long, slightly curved samurai steel blades date from the twelfth century, and progress to the fine temper-lined watered blades of later centuries. All these areas applied iron and steel technology to military accouterments and armour. The work of the Near and Middle East, China, and Japan was particularly artistic as the Middle Ages progressed.

Although the steel sword would remain the principal weapon of the great non-Western cultures, the destructive potential of gunpowder technology into the High Middle Ages was to affect the larger non-Western cultures as it did the West. The use of incendiaries was already ancient, most noticeably in China. The so-called "Greek fire", the generic term for a variety of mixtures based on naphtha (a petroleum distillate) added to sulphur, pitch, turpentine, tars, and oils (in modern interpretation, probably a suspension of metallic sodium, lithium, or potassium in a petroleum base), was perhaps in crude use by the fifth century BC. It is noted as being used by the Boeotian Greeks at the siege of Delium in 424 BC during the Peloponnesian War.

The historiographic origin of gunpowder, that is black powder, is still controversial. The gunpowder recipe itself is of uncertain origin, but its basic constituents are now generally first attributed to ninth century Chinese alchemists. It might also be the independent product of Islamic lands, most likely Moorish Spain by the mid-twelfth century. From there it perhaps moved to India where there may have been independent knowledge and use of the chemical ingredients from the late eleventh century. It was known in northern Europe by the early thirteenth century. The argument for an intermediary disseminator, the Eurasian Steppe lands, the European/Asain crossroad, to Islam and Europe, particularly by the thirteenth century Mongols is also plausible.

Explosive application of gunpowder in a weapon has also been controversial. Some theorists of Chinese primacy (Chinese toy rocket experiments for fireworks evolved early) date bamboo-tube hand guns or cannons and rockets for arrows and spears from AD 900–950. Various types of incendiary arrows, slings, and javelins, as well as incendiary and exploding bombs, grenades, and fire-balls are also attributed to the Chinese by the eleventh century. The historical point of military effectiveness of such devices remains uncertain. Widespread military use in China did not appear until the Song-Jin dynastic wars of the twelfth and thirteenth centuries. By the thirteenth century bomb technology with iron casings and large size was used by Chinese and Mongolian antagonists in land siege warfare. There is also evidence of time delay fusing using flintstone abraded against steel wheels to set off multiple mines in fourteenth century China. Rockets were introduced to Europeans during the Mongol western invasions at the Battle of Legnica in 1241.

Gunpowder weapons applications appeared about the middle of the thirteenth century in Muslim North Africa and Moorish Spain as crude iron and iron-reinforced wooden bucket mortars for flinging stones in fortifications warfare. Also, Moors were using effective rockets on Spanish soil by 1249. Thereafter some evidence shows that the evolution of mortars, cannons, and finally handheld firearms progressed with most tactical efficiency in Europe, although some historians date Chinese cannons of significant size and metal composition from as early as the tenth century. Non-Western applications were innovative in their own right. By the middle of the fourteenth century, cannons mounted on walls or on mobile carriages and cradles had replaced most of the traditional engines of war in both Europe and the Near East. And eastern projectiles ranged from stone balls to huge arrows with sheet-metal fins.

The growing threat to Eastern Europe, the Adriatic, and the Aegean by the ascendancy of the Ottoman Turks through the fourteenth century was furthered by their pursuing the use of artillery to challenge the weakening Byzantium Empire. A parallel was the thirteenth century Mongol challenge and conquest of China, with cavalry, siege tactics, and gunpowder technology. By the fifteenth century the Turks were casting — sometimes with the guidance of European renegades — huge bronze mortars and cannons, such as those used in the final siege and fall of Constantinople in 1453.

The Turks also turned to the Western matchlock arquebus,

the first gunpowder longarm, which was the single most important transitional pivot from medieval to modern warfare. The Ottoman domination over the Arab world influenced firearm dissemination to Arabia and North Africa where, unlike the more angular stock of Turkish and Persian guns, styles reflected Arab and Kabyle preferences.

The influence of the West on the Asian Pacific was initially felt in trade and subsequently in acquaintance with western gunpowder technology. Perhaps the most interesting case involved the Japanese, who quickly adapted the matchlock arquebus which the ubiquitous Portuguese traders, already established in China, brought in 1542. The Japanese matchlock was an austere but highly stylish weapon, smaller in size and caliber than western matchlocks with a spring design firing mechanism, which soon joined the traditional feudal weapon array and went on to change the tactical maneuvering of the civil warfare of the sixteenth century.

The Korean civilization provides an interesting development in Asian and world naval warfare at this point in military technological history. Located on a strategic peninsula, the Korean people endured centuries of piratical incursions from the Japanese islands on one hand and politically complex dynastic invasions from the Chinese mainland on the other. A sophisticated native culture, including science and technology (particularly, shipbuilding), was able to grow from the tenth century. Until 1592 peace and cultural advances continued. Then the Japanese general Toyotomi HideyoshiHideyoshi, Toyotomi unified Japan, calling for the invasion of China through Korea, which refused his passage. Although they had cannon, the Koreans did not have the matchlock longarms of the 200,000 Japanese invaders. The ensuing incursion was successful until 1593 when the Korean admiral Yisunsin invented what must be the first ironclad ship, evidently thin iron plating over a high, flattened oval-shaped ship of sixteen oars with circumference cannon ports. Burn and board-proof, a fleet of these "tortoise boats" was sent against and defeated a Japanese armada in Chinhai Bay. This triumph provided the impetus to drive the Japanese out.

The Chinese perpetuated their own hand cannon, large wall artillery technology, and shipboard cannon well into the nineteenth century. They adapted to the Portuguese style of longarm lock but designed their own pistol grip-like stock. Both features influenced the far away Malaysian peninsula gun style which itself influenced the intermediate region of the Gulf of Tonkin. On the under side of Asia, Indian matchlocks showed significant regional variations from both the Portuguese and Arabic initial introduction. Three basic subcontinent Indian matchlocks were joined by very stylized weapons from Ceylon (Sri Lanka). By the early seventeenth century the Ceylonese exceeded the Portuguese in the manufacture of musket size matchlocks, one type with a unique bifurcated scroll butt. The Burmese side of the Malaysian peninsula essentially used Indian matchlocks with local decorations.

Although more isolated non-Western peoples continued to use the matchlock (indeed, the Japanese did until the early nineteenth century), most succumbed to trade and import and adapted to the progression of firearms manufacturing and, just as significantly, ordinance technology in keeping with the single-minded exigencies of superiority in warfare. These latter factors inevitably and irrevocably set the new course of non-Western military technology as a dependent reflection of the West, a reflection all the more thought provoking in the modern shadows of nuclear and chemical weaponry.

WILLIAM J. McPEAK

REFERENCES

Bhakari, S.K. *Indian Warfare*. New Delhi: Munshiram Manoharlad 1981.

Bottomley, I. *Arms and Armour of the Samurai: the History of Weaponry in Ancient Japan*. New York: Crescent Books, 1988.

Creswell, K.A.C. *A Bibliography of Arms and Armour in Islam*. London: Royal Asiatic Society, 1956.

Held, Robert. *The Age of Firearms*. 2nd ed. Northfield, Illinois: Gun Digest Co., 1970.

Needham, Joseph. *Science and Civilisation in China*. vol. 5: *Chemistry and Chemical Technology*, pt. 7: *Military Technology: The Gunpowder Epic*. Cambridge: Cambridge University Press, 1986.

Oman, C.W.C. *The Wars of the Sixteenth Century*. Reprint of 193' ed. New York: E.P. Dutton, 1979.

Robinson, Charles A. Jr. *Ancient History From Prehistoric Times to the Death of Justinian*. New York: Macmillan, 1951.

Stone, George C. *A Glossary of the Construction, Decoration and Use of Arms and Armor in all Countries and in all Times*. Reprint of 1934 ed. New York: Brussel, 1966.

See also: Metallurgy – Gunpowder – Navigation

MOSES MAIMONIDES Moses Maimonides (1135–1204) is without a doubt the single most luminous figure in Jewish intellectual history since Talmudic times. He possessed professional expertise in most of the sciences of his day, most notably astronomy, mathematics, and medicine. Maimonides' early education in Spain, the country of his birth, seems to have stressed the exact sciences in particular. He refers in his writings to his studies with some students of Ibn Bājja. Furthermore, he edited and taught scientific text written by two Andalusians, Jābir ibn Aflaḥ and al-Muʾtamir ibn Hūd.

In matters astronomical Maimonides' chief contributions concern problems of cosmology and the first visibility of the lunar crescent. As to the former, Maimonides devotes an entire chapter (II, 24) of his philosophical *chef d'oeuvre*, *The Guide of the Perplexed* (Arabic *Dalālat al-Ḥāʾirīn*; Hebrew *Moreh ha-Nevukhim*), to a discussion of the various ways in which the then-accepted models for planets violate certain basic principles of Aristotelian natural philosophy, namely that all heavenly motions be uniform, circular, and about a stationary center. (This problem, by the way, seems to have vexed Andalusian thinkers in particular.) Maimonides surveys the proposed solutions of Ibn Bājja and Thābit ibn Qurra, but he finds no way out of the quandary. It remains a matter of debate among scholars whether Maimonides considered the problem insoluble, since the true workings of the heavens are a matter for metaphysics and hence beyond full understanding, or whether he felt the problem had a solution, indeed, one which would yield a system not very unlike the Ptolemaic models which he criticizes.

In the closing chapters of the section of his law code (*Mishneh Torah*) devoted to the sanctification of the new moon, Maimonides develops a full, sophisticated method for computing whether or not the crescent will be visible on the eve of the thirtieth day of the lunar month. One calculates the "arc of vision", which is the sum of the difference in right ascension between the true positions of the two luminaries, and two-thirds the latitude of the moon. If this arc is greater than 14°, or the sum of the arc and the elongation (the difference in ecliptic longitude between the two luminaries) is greater than or equal to 22°, the moon will be visible. As Maimonides himself avers, the method draws upon written sources, but some of the procedures have been simplified without doing damage to their accuracy.

Maimonides forcefully repudiated astrology. Like nearly all of his contemporaries, he acknowledged a gross physical effect which the motions and luminescence of the heavenly bodies exercise upon terrestrial processes. However, he rejected the notion, central to the astrology of his day, that the stars emanate any non-corporeal force, and he passionately urged that neither individuals nor nations allow themselves to be guided by astrological forecasts.

Maimonides was both a practicing physician and a medical author. According to his account, he traveled daily to treat the sick at court, and upon his return he found his waiting-room full of patients. His medical writings include condensations of the important works of Galen, and a number of original books and monographs. The final section of his own *Aphorisms* (*Fuṣūl Mūsā, Pirqei Moshe*) consists of a scathing critique of Galen's views on medicine and philosophy. Maimonides' medical writings display erudition, clear and concise formulations, and insight; however, his place in the history of medicine, particularly against the background of his contemporaries, remains to be determined.

Maimonides held definite opinions concerning the history and philosophy of science. Scientific teachings must be founded upon solid logical demonstrations. True, observations are vital, but purely empirical claims — those whose authenticity rests solely upon repeated observations, but cannot be placed within any logical framework — are not scientific. This point is made forcefully in his treatise on asthma, and it is one of the underpinnings of his rejection of astrology. Moreover, Maimonides held the view that science progresses in a cumulative fashion, through the refinement of existing data and the absorption of new information; there are no revolutionary leaps. Thus he was able to have it both ways with regard to unsolved issues, e.g. the question of the structure of the heavens. He took tactical advantage of the problem, using the cosmological quandary to attack the doctrine of the eternity of the universe (which rested on astronomical arguments), yet at the same time he felt confident enough in his basic understanding of the workings of the heavens to make use of that knowledge as a steppingstone in the path to knowledge of the Creator.

The most lasting influence of Maimonides, at least as far as his Jewish readership is concerned, was not in the specific scientific knowledge that he disseminated. Rather, his momentous contribution was to elevate the study of the sciences within the context of the religious life. According to Maimonides, the ritual performances and ethical demands of the Jewish tradition have as their goal the preparation of the individual for knowledge of God (to the extent that this is humanly possible), and mastery of the sciences is an indispensable step in this process of religious fulfilment. The observant Jew who follows the lead of Maimonides will regard the study of the sciences as a primary religious obligation.

Y. TZVI LANGERMANN

REFERENCES

Kraemer, Joel L. "Maimonides on Aristotle and the Scientific Method." In *Moses Maimonides and his Time*. Ed. E.L. Ormsby. Washington: Catholic University of America Press, 1989, pp. 53–88.

Langermann, Y.T. "The Mathematical Writings of Maimonides." *Jewish Quarterly Review* 75: 57–65, 1984.

Langermann, Y.T. "The 'True Perplexity': *The Guide of the Perplexed*, Part II, Chapter 24." In *Perpsectives on Maimonides. Philosophical and Historical Studies*. Ed. Joel L. Kraemer. Oxford: Oxford University Press, 1991, pp. 159–174.

Langermann, Y.T. "Maimonides' Repudiation of Astrology." *Maimonidean Studies* 2: 123–158, 1991.

Langermann, Y.T. "Maimonides on the Synochous Fever." *Israel Oriental Studies* 13: 175–198, 1993.

See also: Ibn Hūd – Jābir ibn Aflaḥ – Thābit ibn Qurra

MOUND CULTURES The subject of mound cultures is a vast topic involving numerous groups throughout the Eastern United States. Earthen mounds, one of the visible traits of these cultures, are located from the Gulf of Mexico to the Great Lakes with concentrations in the midwest along the Ohio and Mississippi River drainages. Numerous mound-building cultures were present across this area and through time, and the mounds served a variety of functions.

In some places conical mounds were built to inter the dead while flat-topped pyramidal mounds served as the foundations for important buildings, such as temples or chiefs' residences. Some of the better known mound sites are Cahokia, near St. Louis, Missouri (cultural phases and occupation between AD 800–1200) of the Mississippian tradition, Moundville, Alabama (a dominant center from AD 1250–1500), and those associated with the Hopewell Culture (ca. 200 BC–AD 400), centered in the Ohio Valley.

One of the most acrimonious debates of nineteenth century American archaeology concerned the origins of the mound builders of North America. The Europeans first noted the mounds in the late eighteenth century and quickly began arguments as to whether or not the Indians had constructed the structures. These continued until Cyrus Thomas' *Report of the Mound Explorations of the Bureau of Ethnology* (1894) demonstrated that Native Americans had built the mounds.

Styles and raw materials used by individuals varied between cultures based on location and time, but all of the people expressed their creativity and ingenuity through manufactured material artifacts. Indian technology included the actual mound construction, tool manufacture, pottery, and archaeoastronomy. A brief discussion of each technology follows.

Mound Construction

The types of soil and amount of material necessary for mound construction vary with each site whether they were built in prehistoric or historic time periods. The tons of material moved from the point of origin to a mound attest to the division of labor and orderliness of each culture. Individual basket loads denoting the means of transporting the raw material are often visible at excavations. The number of mounds at a single site vary greatly from one or two to ove
100 as evidenced at Cahokia.

One example of mound construction is the serpent mounc
from the Hopewellian culture which is nearly one-quarte
of a mile long. The people outlined the structure with smal
stones and lumps of clay and then dug up tons of yellow cla
and then buried their markers. This mound was not a plac
of burial, but a deliberate religious effigy with the result o
a "flawlessly modeled serpent, wriggling northward, mout
agape, trying to swallow a massive egg" (Ballantine, 1993)

Tool Manufacture

People of the mound cultures made projectile points for hunt
ing through the process known today as flintknapping. Th
knapper of a point used an antler hammer or stream cobbl
to remove flakes from the larger stone core. Smaller flake
were carefully removed as the work progressed on a singl
point flake. The final forms evident from some mound site
illustrate the meticulous work of highly skilled flintknap
pers across time and cultures. Changes in projectile poin
size probably reflect environmental changes that result
in variations in prey species and probably also in huntin
techniques.

For thousands of years prehistoric hunters used a spea
or javelin with a point attached to kill their prey. Forms o
points aid archaeologists in dating sites where some design
are similar across time, while others differentiate particula
cultures. Eventually the *atlatl* or spear thrower was intro
duced, which increased the casting distance and power o
the throws. The hunter held the *atlatl* which was shaped lik
a large crochet needle. The hooked end was inserted int
a shallow socket in the end of the spear opposite the poir
and hurled with a smooth gliding motion. The *atlatl* wa
made from available wood while the *atlatl* point was bon
or stone.

The technology of preparing meat after the kill also re
quired specialized tools including knives, scrapers, and cu
ters. Although similarities in design exist across time, ir
dividual types are indicative of different cultures. The pa
ent material tells much about a mound site and whether c
not the people were involved in trading. Some prehistori
mound sites were cultural trading centers such as Povert
Point (1500 BC) in northeastern Louisiana where numerou
raw materials were used by the people for crafting materi
artifacts. Trade routes were known to have spanned multip
state areas as the material artifacts were made from coppe
quartz, jasper, chert, and flint which were imported into th
area.

In addition to the technology for meat preparation, th
people of the mound cultures prepared the animal ski

for clothing and other utilitarian purposes. Construction of these items required technology for removing, preparing, and sewing the skins. A list of the tools used by people from the different mound cultures includes hammerstones, polishers, and whetstones. These tools required little if any modification of natural materials by the individual. However, once used, the alterations in shape and signs of wear indicate their uses. Other tools such as axes, drills, gouges, celts, and adzes had to be carefully shaped . Drills and gouges were used to make perforations in the skins, while celts and adzes were used for cutting.

Prehistoric technology also included the use of bone to manufacture different tools for the work around a community. The bone tools were made by breaking and grooving animal bones and then grinding the bone to shape the needed object. Fleshers, used to scrape the inside of fresh animal hides, were usually the lower leg bone of a large animal. Awls were used for perforating and sewing and along with small hammers and fish hooks were made from antlers or bone.

Pottery

The need to transport water and store food necessitated the use of containers for these purposes. Pottery making was an integral part of many of the later mound cultures. The earliest designs were simple and fewer vessels were made, but as time passed the designs and technology for pottery making resulted in works of great beauty and complexity. As with projectile points and other tools, the designs, shapes, and materials of the pottery crafted reveal specific information indicative of particular mound cultures.

In addition to everyday use, pottery was also part of ceremonial practices including burial. Whether a vessel or only a potsherd, the pottery yields valuable information about the technology of the cultures. Various tempering materials, including shell, bone, and sand, were mixed with the raw materials to strengthen it. In most mound cultures both decorated and undecorated pieces were crafted and the use of coloring slip was also part of the technology for some.

Decoration on the pottery, whether bowls, jars, or effigy pieces, was usually applied before the vessel was dried. A variety of methods were used which are significant because the individual expressions reflect change over time and culture. Decorations were often made by using the fingertip or a pointed stick or bone. Potters also used the cordwrapped technique which required wrapping a paddle or stick with a cord or woven material and then pressing it into the wet surface to create a pattern. Check-stamped decorating made use of carved bone or wood which was pressed or stamped into the object. As with the technologies of all the mound cultures, some individuals crafted pieces which are exquisite works of art illustrating the creativity and ingenuity of intelligent people from different mound cultures through centuries of time.

Archaeoastronomy

Native Americans are known for their close association with nature and heavens. The lives of the people from the mound cultures were also intertwined with the cycle of celestial bodies as they observed eclipses and the solstices, devised calendars for ceremonies, and established planting times. They left messages of their science and wisdom in their artifacts and the earthen works which are the visible legacy of their makers.

Two research questions related to Mississippian cultures have been studied by archaeologists. They concern measurement and units of measurements used by the mound builders/community planners, and archaeoastronomy. Studies have been conducted at Cahokia near St. Louis and in Arkansas at the Toltec Mound State Park to investigate the utilization of both orientation to celestial bodies and preconstruction engineering by the mound builders.

Results from a preliminary study by Sherrod and Rollingson show that within the Arkansas community there is a predetermined spacing of mounds. The unit is termed the Toltec Module and is measured in increments divisible by 47.5 meters. Alignments of the solstice, equinox, and stellar positions are evidence that the mound builders placed importance on the observation and knowledge of celestial phenomena and the mounds were positioned to mark these alignments permanently. Reconstruction planning of the community features including mound construction are evidenced by both standardized distance spacing and celestial alignment. "Interaction among many communities of the Mississippi River Valley may well have been widespread with the use of the Toltec Module a reflection of this interaction".

Questions go unanswered and debates continue concerning the mound cultures. What do the mounds mean and why were they built? What were the mechanisms which powered the large exchange and trading systems within cultures such as Poverty Point? Also debated is the issue of size of actual populations at the large ceremonial sites. The study of archaeoastronomy and standardized measurements are still in early research stages and thus the extent of Indian technology and meaning are still open for further investigation.

Recent research by Saunders, Allen, and Saucier of four mound complexes in Louisiana which predate Poverty Point raises the question as to whether mound construction technology diffused from a single area or independently developed within several cultures. In relation to actual construction of the mounds anywhere, did their makers build one at a time or were multiple mounds under construction at the same time?

A growing area of research which includes the mound cultures is the archaeology of gender (Gero and Conkey, 1991; Walde and Willows, 1991). Interpretation of archaeological sites has been dominated by views in which women and children were underrepresented if present at all in much research. As the mound cultures are studied from more equitable views, questions emerge on women's roles as tool makers and hunters. One theme which unifies the research is the theoretical outlook which views gender relationships as the fundamental structural component to social organization.

CONNIE H. NOBLES

REFERENCES

Ballantine, Betty, and Ian Ballantine, ed. *The Native Americans: An Illustrated History.* Atlanta: Turner Publishing, Inc., 1993.

Emerson, Thomas, and R. Barry Lewis, ed. *Cahokia and the Hinterlands.* Chicago: University of Illinois Press, 1991.

Gero, Joan, and Margaret Conkey, ed. *Engendering Archaeology: Women and Prehistory.* Oxford: Basil Blackwell, 1991.

Jennings, Jesse. *Prehistory of North America.* 3rd ed. Mountain View, California: Mayfield Publishing Company, 1989.

Saunders, Joe, Thurman Allen, and Roger Saucier. "Four Archaic? Mound Complexes in Northeast Louisiana." *Southeastern Archaeology* 13:134–153, 1994.

Sherrod, P. Clay, and Martha Rollingson. *Surveyors of the Ancient Mississippi Valley.* Arkansas: Arkansas Archeological Survey Research Series No. 28, 1987.

Thomas, Cyrus. "Report on the Mound Explorations of the Bureau of Ethnology." *Annual Report of the Bureau of American Ethnology,* 1894, pp. 17–743. Reprinted Washington, D.C.: Smithsonian Institution Press, 1985.

Walde, Dale, and Noreen D. Willows, ed. *The Archaeology of Gender: Proceedings of the 22nd Annual Chacmool Conference.* Calgary: The Archaeological Association of the University of Calgary, 1991.

Willey, Gordon. *An Introduction to American Archaeology.* Vol. 1. Englewood Cliffs, New Jersey: Prentice Hall, 1966.

See also: Technology and Gender

MOXIBUSTION Moxibustion, also spelled moxabustion, is a traditional East Asian therapeutic technique involving the burning of tinder made from the artemisia plant. The technique has three major variants. In one, small cones of the artemisia are burned directly on the skin; in a second, some intermediary substance — commonly a thin slice of garlic or ginger, or a layer of soybean paste — separates the tinder cone from the skin; and in a third, smoldering sticks of artemisia, about a half inch in diameter, are held about an inch to three inches away from the skin. The last two methods fall under the rubric of "warming" or "traceless" moxibustion: both heat the treated sites, but unlike the first technique, they leave no scars. There is, in addition, a hybrid combination of acupuncture and moxibustion in which a clump of artemisia is burned at the protruding end of an implanted needle.

The word *moxa* comes from the Japanese term for the artemisia tinder, *mogusa*. Though the term may have made its way into Portuguese as early as the sixteenth century, printed Western language accounts of the technique began to appear only in the 1670s. For a brief while, it enjoyed a minor vogue in Europe, particularly as a treatment for the gout, but ultimately it did not take root. Still, occasional theses on moxibustion continued to be presented at European medical faculties into the nineteenth century.

The details of moxibustion's origins in China are uncertain. In the most ambitious review of the subject to date, Keiji Yamada shows that the therapy originally had magical implications: the aim of cauterization was to drive out alien, noxious spirits. He argues that by the early Han dynasty (206 BC–AD 220), however, moxibustion had begun to assume a new, quite different identity. Gradually shedding its ties to demonic expulsion, it came to be conceived, instead, as a form of stimulus therapy. The purpose of burning now was to clear blockages in the flow of the body's own essences and rectify imbalances in the distribution of blood and vital breath (*qi*). Yamada's analysis of this change builds upon three theses: (1) that the key to the transformation of moxibustion was the "discovery", in the third century BC, of a series of *mo* — vessels thought to carry blood and vital breath throughout the body; (2) that the theory of the *mo* first arose in the context of moxibustion; and (3) that the practice of acupuncture followed, and was by made possible by, the discovery of the *mo* in moxibustion.

Yamada's account may not cover the full story; some evidence suggests that the experience of bloodletting also contributed to the rise of acupuncture. But two points are indisputable. The first is that the earliest descriptions of the *mo* — those in the Mawangdui manuscripts (third century BC) — concentrate exclusively on moxibustion, and don't mention needling. The second is that from the Han dynasty onward the histories of moxibustion and acupuncture were intimately intertwined. More often than not, acupuncture treatises were, at the same time, treatises on moxibustion. The titles of major traditional texts — from Huangfu Mi's *Zhenjiu jiayijing* (AD 282), through Wang Zhizhong's *Zhenjiu zisheng jing* (1220), to Gao Wu's *Zhenjiu jieyao* (1536) and Yang Jizhou's *Zhenjiu dacheng* (1601)–all evoke needling (*zhen*) and moxibustion (*jiu*) together in the same breath, as a compound, *zhenjiu*. The reason is clear: healing with moxa entailed burning artemisia along the same *mo*, and indeed on the same sites, needled in acupuncture.

Physicians did distinguish between the two therapies. For Yang Jizhou, in fact, the ability to discriminate between when and where to burn and not needle, and conversely, when and where to needle but not burn, marked the superior physician. Gao Wu's *Zhenjiu juying* (1537), for instance, names forty-five "forbidden points" for moxibustion (sites where treatment was to be avoided or at least pursued with special caution), but for acupuncture names only twenty-two. The two lists, moreover, have no points in common.

Between the seventh and thirteenth centuries, physicians also composed several treatises devoted to moxibustion alone. Cui Zhiti's *Guzhengbingjiufang* (640) discussed how to treat tubercular diseases with moxa, and Wenren Qinian's *Beiji jiufa* (1226) explained how to deploy moxibustion in emergencies; Zhuang Zhuo's *Gaohuang jiufa* (1128) detailed the special benefits of burning moxa on the so-called *gaohuang* points, whereas the *Mingtang jiujing* (seventh century) reviewed treatment sites more generally, identifying for each site the various ailments treatable by moxibustion.

By slight modifications in technique, moxibustion could be used either to tonify deficiencies in vital energy, or to disperse pathological excess. For example, to tonify, one simply allowed the moxa cone to burn down naturally; to disperse, the therapist blew gently upon the burning cone to make the heat more intense. Traditionally, however, the tendency was to deploy moxibustion primarily as a tonifying technique, and to favor it for treating chronic disorders.

People also turned to moxibustion to prevent illness. In his *Qianjin yaofang* (seventh century), Sun Simiao (Sun Simo) notes that officials going to the regions of Wu and Shu made sure always to keep several unhealed moxa spots on their bodies. This, they believed, protected them from a variety of epidemic diseases. More generally, it became proverbial wisdom that burning moxa regularly on special sites like the *sanli* points of the legs warded off sickness of all kinds, and promoted longevity.

Moxibustion's popularity as a prophylactic measure and as a treatment for chronic complaints drew theoretical support from the belief in its tonifying influence. But it also reflected a more basic fact: unlike needling, burning moxa didn't require sophisticated technical skill. Once patients learned where to burn — whether guided by illustrated books, tradition, or doctors, then they could treat themselves, or be treated by family members. Thus, while acupuncture and moxibustion shared common historical origins and a common understanding of the body, they diverged subtly in the sociology of their practice: whereas acupuncture remained largely the preserve of specialists, moxibustion tended to become part of popular self-treatment. Professional acupuncturists in East Asia today still make use of moxibustion, but so do many patients who have never been needled.

SHIGEHISA KURIYAMA

REFERENCES

Lu, Gwei-djen and Joseph Needham. *Celestial Lancets: a History and Rationale of Acupuncture and Moxa*. Cambridge: Cambridge University Press, 1980.

Yamada, Keiji. "Shinkyu no kigen." In *Shin hatsugen chugoku kagakushi shiryo no kenkyu. Ronko hen*. Ed. Keiji Yamada. Kyoto: Kyoto daigaku jimbun kagaku kenkyusho, 1985, pp. 3–78. (In Japanese).

See also: Huangfu Mi – Sun Simo – Acupuncture – Medicine in China – Medicine in Japan – Medical Texts in China

MUMMIES IN EGYPT Mummification was practiced in ancient Egypt to ensure the continued existence of the deceased. At death, several spirits were believed to be released, the most important of which were the Ka, the Ba, and the Akh. A person's fate in the afterlife, in the form of these three spirits, was believed to be directly tied to the continued existence of the physical body.

The Ka, appearing at birth, resembled the human physical body in all aspects. After death, it remained in the tomb with the mummified body, acting as a protective spirit, and fed on the daily offerings presented at the tomb.

The Ba embodied the personality and individual characteristics of the person. It also appeared at birth, but after death was believed to fly off to heaven, returning regularly to visit the Ka and the body. It sometimes seems to serve as "spiritual link" between the two.

The Akh, after a silent or dormant existence during the person's life, separated from the body at death and embarked on a journey through the land of the dead, never to return. All three spiritual elements were essential for the continued existence of the individual and their continued survival depended on the existence of the human body.

Egyptians in the pre-dynastic era, prior to 3100 BC, buried their dead in shallow graves under the hot desert sands. This left the body prey to wild animals and desert thieves. From the first dynasty (ca. 3100–2900 BC) on, the Egyptians built mud brick burial chambers, but these, too, proved unsatisfactory because the bodies gradually deteriorated under the action of moisture in these chambers.

By the end of the Old Kingdom (ca. 2600–2180 BC), the Egyptians had begun to embalm their dead through desiccation by means of dry natron (a mixture of sodium carbonate and bicarbonate). The technique was later termed "mummification" from the Persian word *mummiya*, meaning bitumen or pitch. Corpses embalmed by the Egyptians took on a blackened color, and this effect was mistakenly attributed to bitumen.

By the Middle Kingdom (ca. 2150–1780 BC), the process was perfected. After death, the body was taken to the "place of purification", where it was stripped and washed in a dilute natron solution. It was then moved to a special "embalming house" where it was placed on a large wooden board. The brain was broken into small pieces by a hooked utensil which was introduced through the nostrils and penetrated the cranial cavity by breaking through the ethmoid bone. The brain was then removed by a long delicate spoon and disposed of. The empty skull was filled with sawdust, resin, or resin-soaked linen.

An incision was then made on the left side of the abdomen and the liver, stomach and intestines removed. A puncture in the diaphragm also allowed for the removal of the lungs. The heart was left *in situ* as the center in which the good and evil deeds of the individual accumulated. (A light-weighted heart during the day of reckoning would ensure resurrection and an afterlife for the individual.)

The removed organs were washed with a natron solution, dried, and sealed in canopic jars (often with a solution of natron). These jars were eventually placed in the tomb with the deceased. The empty cavities were washed with palm wine and packed with sand, straw, or sawdust mixed with resin or bags of natron. The body was then placed on a slanting board and covered with dry natron for forty days to ensure total desiccation by osmosis (the skin acting as a semi-permeable membrane).

Following the dehydration of the body came a ceremonial washing and an anointing with oil. The cranial and body cavities were repacked with linen soaked in resin, the abdominal incision was closed, linen balls were placed in the eye-sockets. and the cheeks were padded with linen. The body was then coated with molten resin and subsequently wrapped, beginning from the toes, in strips of linen arranged in intricate patterns. In some cases, the body would be adorned with jewelry, and amulets conferring special protection on the mummy would be enclosed in the linen wrappings.

The entire embalming process took seventy days, after which the ceremony of the "Opening of the Mouth" took place. A priest would symbolically open the mouth of the deceased. This would be followed by an elaborate succession of actions and prayers.

The practice of mummification, with some variations in the process of preparation and wrapping, continued in Egypt until the fourth century AD, when Christianity had become the principal religion. The practice then steadily declined.

JEHANE RAGAI
GREGG DE YOUNG

REFERENCES

Andrews, Carol. *Mummies*. London: British Museum, 1984.

Cockburn, A. & E., eds. *Mummies, Disease, and Ancient Cultures*. Cambridge: Cambridge University Press, 1980.

Harris, J. E. and K. Weeks. *X-Raying the Pharaohs*. London: Macdonald, 1973.

Leda, A.-P. *The Cult of the Immortal*. London: Paladin, 1980.

Spenser. A. J. *Death in Ancient Egypt*. London: Penguin, 1982.

MUMMIES IN SOUTH AMERICA The Atacama desert, along the west coast of Chile and Peru, is an area of extreme aridity which has provided for unique preservation of human remains and cultural materials from thousands of years ago. The incredible preservation, especially of human mummies, has furnished us with a glimpse at mortuary traditions and rites associated with the first settlers beginning at least 9000 years ago, through to the arrival of the Spanish Conquistadors.

The Chinchorros fishermen, who lived from 7000 to about 1500 BC, are of special interest because from 5000 to 1700 BC, they practiced artificial mummification of their dead. The Chinchorro people lived in the Atacama from Ilo in Southern Peru to Antofagasta in Northern Chile, and occupied about nine hundred kilometers of the Pacific coast of South America. The area surrounding the modern city of Arica, in northern Chile, was where Chinchorro artificial mortuary practices originated. From this area artificial mummification customs (or intentional interventionary preservation) spread north and south. The bodies were prepared for the journey to the afterlife in remarkable ways. For example, some bodies were completely disarticulated and then wholly reassembled and sculpted. Various styles were practiced through time, such as Black, Red and Mud-Coated styles. All these styles have two things in common: human intervention in the preservation of the cadaver, and an extended body position for interment.

The Black Mummies were the oldest and most complex beginning about 5000 BC and lasting for about two millennia. To make the Black Mummies, the morticians completely cleaned and separated the deceased's bones and soft tissue. Subsequently, the skeletons were reconstructed, and the bodies rebuilt into statue-like rigid forms, using long sticks for internal reinforcement along the extremities and spine. Reed cords bound the bones and sticks together, and a light gray ash paste was applied to stuff and model the individual. The skin was often replaced, and sometimes pieces of sea lion skin were added when the person's own skin did not suffice after drying. Facial features and sexual organs were insinuated. A short wig of human hair was added to the head and secured with cords. Then the morticians painted the entire body with a black manganese paste, which was polished to a high sheen, hence the name Black Mummies.

In contrast, the Red Mummies often were made without disarticulation of the body. Instead incisions were made to remove organs. Long sharpened sticks were pushed under the skin of arms, legs, and the spine to add rigidity, and body cavities were stuffed. After suturing the incisions, the body was painted with red ocher, but the facial mask was often painted black. In a few cases the skin was replaced bandage style. A wig made of long black human hair was added to the head and secured with what looks like a red clay motorcycle helmet. This Red style appeared about 2000 BC and lasted about 500 years.

After the Red style, artificial mummification techniques were simplified. The bodies were not eviscerated; they were simply encased in a thick cement-like coating that prevented decomposition. This Mud-coated style lasted only a couple of centuries. After this period, ca. 1700 BC, the Chinchorro bodies were still buried in an extended position, but were preserved only through the desiccating forces of the environment; they were no longer artificially mummified.

Often the Chinchorro mummies were enshrouded in twined reed mats and buried in shallow pit graves in groups of about six bodies of various ages and both sexes. The cemeteries were located in the sandy coastal dunes beyond the reach of the tides. The few grave goods accompanying the dead were fishing lines, shellfish and cactus fishing hooks, harpoons, bone and stone tools, stone mortars, and gill nets. No individual had a substantially larger number of grave goods that would set him apart as socially above the others. The Chinchorros also received similar mummification and burial treatments regardless of age and sex. Even fetuses were mummified. Apparently everyone was treated equally, with the same mortuary treatment, as would be expected in an early egalitarian society.

Although Chinchorros were simple fisherfolk without knowledge of ceramics, agriculture, or loom weaving, their spiritual and religious life must have been highly sophisticated, as the mummies have demonstrated. The sophistication of the mortuary treatment, the repair of the mummies, and the millennial duration of their practices, all indicate that mummification was central to the social lives of Chinchorro people. It appeared they venerated the mummified

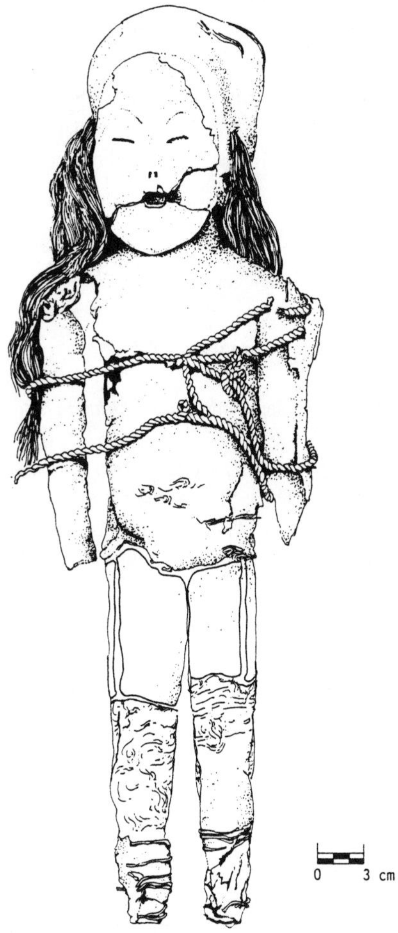

Figure 1: Chinchorro Child in the Red Mummy Style. Drawing by Raul Rocha. Used with permission of the author.

bodies of their ancestors by placing them on display for an extended period before burial. Perhaps they petitioned the mummies for blessings during their daily lives. Later South American cultures like the Incas revered their desiccated ancestor mummies (natural mummification). For the Inca the dried bodies of their ancestors were considered *Huacas* or deities that had the power to provide fertility, good crops, and happiness. The Inca brought food and drink to the dead and included the mummies in their religious celebrations. For the Incas and the Chinchorros, the mummies linked the real world with the spiritual world.

The Chinchorros did not vanish after their artificial mummification disappeared about 1700 BC. On the contrary, their descendants continued to thrive along the Pacific coast, with increased social and political complexity. However, for most areas, after 1500 BC, the dead were buried in a flexed or seated position, and became natural mummies by the desiccating action of the desert. Post-Chinchorro cultures developed and took advantage of new technologies such as agriculture, weaving, ceramics, and metallurgy, and now the dead were furnished with paraphernalia related to these achievements. Thus, from the numerous grave goods accompanying the dead, it can be seen that post-Chinchorro people also had powerful spiritual concerns about death and the afterlife.

BERNARDO ARRIAZA
VICKI CASSMAN

REFERENCES

Allison, M. Chile's Ancient Mummies. *Natural History* 10:75–81, 1985.

Arriaza, B. *Beyond Death: The Chinchorro Mummies of Ancient Chile.* Washington, D.C.: Smithsonian Institution Press, 1995.

Aufderheide, A., Muñoz, I. and Arriaza, B. "Seven Chinchorro Mummies and the Prehistory of Northern Chile." *American Journal of Physical Anthropology* 91:189–201, 1993.

Bittmann, B. "Revision del problema Chinchorro." *Chungara* 9: 46–79, 1982.

MUNĪŚVARA Munīśvara (b. 1603), son of Raṅganātha, was born into a family of reputed astronomers of several generations, who had migrated from their original home on the banks of river Godāvarī in the south to Varanasi in the north of India. Munīśvara's paternal uncle, Kṛṣṇa Daivajña, was patronized by the Mughal emperor Jehangir, who ruled from Delhi (1605–28). Elevating references by Munīśvara to Shahjehan, who succeeded Jehangir as emperor in 1628, and casting the horoscope of the time of Shahjehan's coronation are pointers to the continued royal patronage enjoyed by Munīśvara's family. In his commentary on the *Līlāvatī*, Munīśvara states that another name of his was Viśveśvara.

Munīśvara was a prolific writer, on both mathematics and astronomy, and wrote both original works and commentaries. The *Siddhāntasārvabhauma*, written in 1646, is his major work on astronomy. In twelve chapters, of which nine chapters constituted Part I, the work dealt with the subjects of a normal textbook. In Part II, the work dealt with the armillary sphere, astronomical instruments, and astronomical queries. He also composed a commentary on the work called *Āśayaprakāśinī*, which is dated 1650. On mathematics, Munīśvara has two works: *Pāṭīsāra* and *Gaṇitaprakāśa*. He was an admirer of Bhāskara II. His commentaries on Bhāskara's *Siddhāntaśiromaṇi*, entitled *Marīcī*, and on *Līlāvatī*, entitled *Nisṛṣṭārthadūtī*, are justly famous for their exhaustiveness, lucidity, and citations from earlier authors. He also commented on the *Pratodayantra* or *Cābukayantra*, a short work on an astronomical instrument used for the ascertainment of the time of the day, by Gaṇeśa Daivajña.

Munīśvara had professional detractors whose views differed from his. One was Raṅganātha, author of the manual *Siddhāntacūḍāmaṅi* (AD 1640), who, in a short work called *Bhaṅgīvibhaṅgī*, criticized Munīśvara's *Bhaṅgī* (Winding) method of computing true planets. This work was refuted by Munīśvara in his *Bhaṅgīvibhaṅgī-khaṇḍana*. Another was Ekanātha, an astronomer of Maharashtra origin, settled in Varanasi, who seems to have passed strictures on Munīśvara's exposition of three verses on declension (*krānti*) in Bhāskara's *Siddhāntaśiromaṇi*. Munīśvara refuted Ekanātha's criticism and established his views in a short work entitled *Ekanātha-mukhabhañjana* (A Slap in the Face of Ekanātha).

Though Munīśvara accepted Islamic trigonometry as an aid to studies in astronomy, he severely contradicted the theory of precession advocated by Kamalākara, against which Raṅganātha wrote a work entitled *Loha-gola-khaṇḍana*, which Munīśvara's cousin Gadādhara refuted in his *Lohagolasamarthana* (Refutation of the Loha-gola).

Characteristics that cannot be missed in Munīśvara's writings are the lucidity, chaste language, and the elegant style in which they are couched.

K.V. SARMA

REFERENCES

Primary sources

Siddhāntasārvabhauma of Munīśvara. Ed. Mitha Lal Ojha. Varanasi: Sampurnanand Sanskrit University, 1978.

Siddhāntaśiromaṇi of Bhāskara with the Commentary Marīcī of Munīśvara. Ed. Muralidhara Jha. Benares: E.J. Lazarus and Co., 1917.

Siddhāntaśiromaṇi of Bhāskara with the Commentary Marīcī of Munīśvara. Ed. Dattatreya Apte. Poona: Anandasrama Sanskrit Series, 1943.

Siddhāntaśiromaṇi of Bhāskara with the Commentary Marīcī of Munīśvara. Ed. Kedardatta Joshi. Varanasi: Banaras Hindu University, 1964.

Secondary sources

Dikshit, S.B. *Bhāratīya Jyotish Śāstra (History of Indian Astronomy)* Trans. R.V. Vaidya. Pt. II: *History of Astronomy During the Siddhantic and Modern Periods.* Calcutta: Positional Astronomy Centre, India Meterological Department, 1981.

Dvivedi, Sudhakara. *Gaṇaka Taraṅgiṇī or Lives of Hindu Astronomers.* Ed. Padmakara Dvivedi. Benares: Jyotish Prakash Press, 1933.

Pingree, David. *Census of the Exact Sciences in Sanskrit, Series A, vol. 4*. Philadelphia: American Philosophical Society, 1981.

See also: Mathematics in India – Astronomy in India – Precession of the Equinoxes – Kamalākara – Bhāskara II

AL-MUQADDASĪ Born at Al-Bayt al-Muqaddas (Jerusalem) in AD 946, al-Muqaddasī (Shams ad-Dīn Abū ʿAbdallāh Muḥammad ibn Aḥmad ibn Abī Bakr al-Bannāʾ), also called al-Maqdisī, made important contributions in regional geography. In his early days he studied Muslim history, especially its political and cultural aspects, civilization, religion, and jurisprudence. To seek knowledge he visited distinguished scholars, met the men of science, and studied in royal libraries at a fairly young age.

Among other Arab geographers of the time, al-Muqaddasī's definition of a region was probably the most original and produced one of the most valuable treatises in Arabic literature, *Kitāb aḥsan-al taqāsim fī maʿrifat al-aqālim* (The Best Divisions for the Classification of Regions). Though al-Muqaddasī belonged to the Balkhī school, he was critical of it and felt that they disregarded real geography. He argued that scientific geography must be based on observation. He critically examined the information presented by Al-Jayhānī, al-Balkhī, Ibn al-Faqīh, al-Jahiz and others and questioned their methods of acquiring information, their objectives, and their misrepresentations and selectivity of information. Then al-Muqaddasī stated, "I have endeavored not to repeat anything other writers have written, nor narrated any particulars they narrated, except where it was necessary, in order neither to deny their right nor myself to be guilty of plagiarism; for in any case those alone will be able to appreciate my book who examine the works of those authors or who have travelled through the country, and are men of education and intelligence".

Al-Muqaddasī claimed that no one who had treated geography before him adopted his method or provided the information he did. In order to achieve his objective, he traveled through the Muslim world, with the exception of Spain and Sindh, conversed with scholars, and waited on princes. He discussed matters with judges, studied under doctors of law, frequented the society of men of letters, and associated with people of all classes until he attained what he wanted.

In his book, *Aḥsan al-taqāsim*, al-Muqaddasī divided the Muslim empire into fourteen divisions, and treated the Arab world separately from the non-Arab world. Then he described the districts in each division, identifying their capitals and principal cities and giving their towns and villages in due order. Information which did not fit into either the Arab or non-Arab context was treated separately. While treating the regions in their entirety, Al-Muqadassī provided a regional framework, and can rightly be called the father of regional geography.

MUSHTAQUR RAHMAN

REFERENCES

Beasley, G. Raymond. *The Dawn of Modern Geography*, Vol. 1. London: John Murray, 1987.

LeStrange, G. *The Lands of the Eastern Caliphate*. Cambridge: Cambridge University Press, 1905.

Ranking, G.S.A. and R.F. Arzoo, trans. *Aḥsan-al taqāsim fī maʿrifat al-aqālim*. Calcutta: Bibliotheca Indica, 1897–1910.

Siddiqi, Akhtar Hussain. "Al-Muqaddasī's Treatment of Regional Geography." *International Journal of Islamic and Arabic Studies*. 4(2): 1–13, 1987.

See also: Geography in Islam – The Balkhī School of Arab Geographers

AL-MUʾTAMAN IBN HŪD Yūsuf Al-Muʾtaman ibn Hūd was the king of the kingdom of Zaragoza in Northern Spain from 1081 until his death in 1085. He lived in the Aljafería palace in Zaragoza, which was built by his father Aḥmad al-Muqtadir ibn Hūd (r. 1041–1081), and which is now the site of the Parliament of Aragon. Al-Muʾtaman was interested in mathematics, optics, and philosophy. He wrote a very long mathematical work, the *Kitāb al-Istikmāl* (Book of Perfection), of which large parts have recently been identified in four anonymous Arabic manuscripts in Copenhagen, Leiden, Cairo, and Damascus. A revised version of the whole *Istikmāl* has recently been discovered.

In the *Book of Perfection*, al-Muʾtaman divides most of pure mathematics according to a philosophical classification in five "species" (*anwāʿ*). Species 1 deals with arithmetic and the theory of numbers. Al-Muʾtaman summarizes the arithmetical books of Euclid's *Elements*, and he proves Thābit ibn Qurra's rule for amicable numbers. The remaining species 2–5 deal with geometry. Al-Muʾtaman summarizes the works of Greek authors such as the *Elements* and *Data* of Euclid (300 BC), *On the Sphere and Cylinder* of Archimedes (250 BC), and the *Conics* of Apollonius (200 BC). He does the same for Arabic authors as well, such as the *Quadrature of the Parabola* (miṣāḥat al-qaṭʿ al-mukāfī) by Ibrāhīm ibn Sinān (909–946) and the *Optics* (Kitāb al-Manāẓir), *On Analysis and Synthesis* (fīʾl-taḥlīl waʾl-tarkīb), and *On Given Things* (fīʾl-maʿlūmāt) by Ibn al-Haytham (965–ca. 1041). The summaries in the *Book of Perfection* show that al-Muʾtaman really understood these works. Often he was able to shorten and generalize their contents quite drastically.

Al-Muʾtaman does not mention his sources, and he does not tell us what his own contributions are. The *Book of Perfection* probably includes some original contributions, such as a construction of two mean proportionals between two given lines by means of a circle and a parabola. A few geometrical theorems occur in the *Book of Perfection* for the first time in history, such as the theorem of Ceva (hitherto named after Giovanni Ceva, who independently discovered it in 1678), and the general proof of the invariance of cross-ratios under a perspectivity (a special case was proved by Pappus of Alexandria in late antiquity). It seems that al-Muʾtaman intended to add to the *Book of Perfection* a second part on astronomy and optics, but he probably did not have the time to do so.

JAN P. HOGENDIJK

REFERENCES

Djebbar, A. "Deux mathématiciens peu connus de l'Espagne du XI^{e}siècle: al-Muʾtaman et Ibn Sayyid." In *Vestigia Mathematica. Studies in Medieval and Early Modern Mathematics in Honour of H.L.L. Busard*. Ed. M. Folkerts and J.P.Hogendijk. Amsterdam: Rodopi, 1993, pp. 79–91.

Djebbar, A. "La redaction de l'Istikmal d'al-Muʾtaman (XIe siècle) par Ibn Sartāq, un mathématicien des XIIIe–XIVe siècles." To appear in *Historia Mathematica*.

Hogendijk, J.P. "The Geometrical Parts of the *Istikmāl* of al-Muʾtaman ibn Hūd, an Analytical Table of Contents." *Archives Internationales d'Histoire des Sciences* 41: 207–281, 1991.

N

NAMORATUNGA Archaeoastronomy is the study of archaeological evidence documenting that ancient peoples made systematic observations of astronomical phenomena such as solstices and equinoxes (Robbins, 1990). Most typically, archaeoastronomical evidence is reflected in the positioning and alignments of buildings, earthworks and megaliths, such as at Stonehenge in England. At the latter site, the sun rises over the heel stone on the summer solstice, or first day of summer. In most cases, archaeoastronomical sites are thought to reflect religious beliefs in which the seasons of the year were being monitored for ceremonial purposes as well as for agricultural planning.

In Africa south of the Sahara, much less is known about the archaeological record relative to many other areas because of the comparative lack of intensive research. So the announcement of the discovery of the first archaeoastronmical site in 1978 was of considerable interest (Lynch and Robbins, 1978). The site of Namoratunga II, overlooking the barren western shore of Lake Turkana in Kenya has been compared to a miniature Stonehenge. Namoratunga II consists of a series of nineteen basalt pillars that have been placed in the ground at angles. The basalt used at Namoratunga II is locally available in the Losidok range. The alignment of pillars is surrounded by a stone circle and there is at least one nearby grave marked by upright slabs. This grave is similar to graves at the Namoratunga I site (though Namoratunga I lacks stone pillars) located far to the south adjacent to the Kerio river valley. Faint rock engravings which are thought to represent brand symbols used on domesticated animals occur on some of the pillars at Namoratunga I. Interestingly, the name, Namoratunga means "people of stone" in the local Turkana language, referring to a legend that the pillars (as well as upright grave slabs) were once people who were turned to stone while they were dancing because they mocked a spirit that had appeared to them. The Namoratunga sites are believed to date to approximately 300 BC.

Mark Lynch, who was working on his Ph.D. in Anthropology at Michigan State University, initially proposed that Namoratunga II was an archaeoastronomical site. Lynch believed that the site was most likely reflective of a calendar system similar to the traditional calendar used by eastern Cushitic speaking peoples of southern Ethiopia. Peoples such as the Borana have a 12-month 354-day calendar that is based on the rising of Triangulum, Pleiades, Bellatrix, Aldebaran, Central Orion, Saiph, and Sirius. In the Borana calendar, the rising of the above stars and constellations is related to phases of the moon. Lynch found that there were positive correlations between the rising of the seven stars or constellations used in the Cushitic calendar and the alignments of the stone pillars at Namoratunga II.

Figure 1: View of Namoratunga II site (Photo by L.H. Robbins).

Like many other archaeoastronomical sites including Stonehenge, the interpretation of Namoratunga II has been controversial. It has been argued that the original compass measurements were in error because of magnetite in the stone pillars (Soper,1982). For this reason, the view of the site as a calendar similar to the one used by the Borana people has been questioned. On the other hand, Laurence R. Doyle, of the NASA Ames Research Center, has re-measured the site and concluded that "the pillars were used for the specific purpose of aligning with the 300 BC positions of the Borana calendar stars". (Doyle and Wilcox, 1986)

LAWRENCE H. ROBBINS

REFERENCES

Doyle, L.R. and T.J. Wilcox "Statistical Analysis of Namoratunga: An Archaeoastronomical Site in Sub-Saharan Africa?" *Azania* 21: 125–129, 1986.

Lynch, B.M. and L.H. Robbins. "Namoratunga: The First Archaeoastronomical Evidence in Sub-Saharan Africa." *Science* 220: 766–768, 1978.

Robbins, L.H. *Stones, Bones and Ancient Cities.* New York: St Martin's Press, 1990.

Soper, R. "Archaeo-astronomical Cushites: Some Comments" (with comments by B.M. Lynch). *Azania* 17: 145–162, 1982.

See also: Astronomy in Africa

NANJING The *Nanjing* (Canon on Medical Difficulties) appeared probably in the first or second century AD. However, the Chinese believe it was written in the first or second century BC. Its authorship is unknown, although tradition has ascribed it to Qin Yueren, the usual sobriquet for Bian Qiao, the eminent Zhou Dynasty physician who is said to have lived in the sixth century BC. Attribution to an ancient legendary doctor is common in Chinese medical works. The most valued commentary is the *Nanjing Benyi* (The Genuine Significance of the Nanjing), compiled by Hua Shou in 1366. The book further develops the ideas of the *Huangdi Neijing* in question and answer format, revealing an increasing tendency for medical books to concentrate on acupuncture.

The following are outlined in the *Nanjing*:

- the criteria for pathology and health;
- the existence of *xu* (weak, deficient) or *shi* (full, excessive, toxic) conditions;
- the passage of disease through the various organs;
- pathology, especially that concerned with dissipation and cold-induced fevers;
- the attributes of a skilled or "divine" physician;
- the structure of the organs, including their weights and measurements; and
- needling technique.

Methods of tonification and sedation are particularly focused upon. Acupuncture points appear for the first time in various groupings, with a concentration on the *shu* (passage) points on the lower parts of the four limbs. One of its most influential ideas is the use of the points for particular categories of disease. Pathologies of the channels, and pulse diagnosis, are also touched upon.

The *Nanjing* contains 81 chapters. The *jingluo* (channels) appear in Chapters 23–29, the *shu*-points (passage points) on the limbs in Chapters 62–68, whilst Chapters 69–81 concentrate on needling. These sections show the development of extreme precision in needling, and a focused rationale during practice.

RICHARD BERTSCHINGER

REFERENCES

Lu, Gweidjen and Joseph Needham. *Celestial Lancets*. Cambridge: Cambridge University Press, 1980.

Unschuld, Paul U., trans. *The Nanjing*. California: University of California Press, 1990.

NĀRĀYAṆA PAṆḌITA Nārāyaṇa, the son of Nṛsiṃha (or Narasiṃha), was one of the major authorites on Indian mathematics after Bhāskara II. We do not know when or where he was born. He wrote two Sanskrit mathematical texts: the *Gaṇitakaumudī* on *pāṭī* (arithmetic) in 1356 (which is confirmed by the final verses of the book) and the *Bījagaṇitāvataṃsa* on *bīja* (algebra). Nārāyaṇa Paṇḍita was confused with another Nārāyaṇa, a commentator on the *Līlāvatī*.

The two books consist of rules (*sūtras*), examples (*udāharaṇas*) and commentary (*vāsanā*) thereon. It is in the *vāsanā* on the *Gaṇitakaumudī* but not in the *mūla* that a reference to the *Bījagaṇitāvataṃsa* is found. The *Gaṇitakaumudī* was published by P. Dvivedi in two volumes based on a single manuscript which had belonged to his late father. The numberings are not accurate. It consists of *paribhāsā* (metrology units), *parikarma* (basic operations) and fourteen *vyavahāras*: the traditional eight *vyavahāras*, *kuṭṭaka*, *vargaprakṛti* (indeterminate equations), calculations for fractions, rule for fractionizing, the net of numbers (combinatorics), and magic squares. A critical edition of the last two *vyavahāras* with an English translation and his own commentary was published by T. Kusuba in 1993. Nārāyaṇa's method of finding factors of a number given in the eleventh *vyavahāra* is equivalent to the one by the French mathematician Pierre de Fermat. The twelfth *vyavahāra* includes rules to express the number one as the sum of a number of unit fractions, which are similar to those given by Mahāvīra. The rules in the thirteenth chapter are modeled on those of the *Līlāvatī*, but are further advanced and can be compared to the rules for combinatorics in metrics and music. The *Gaṇitakaumudī* is the first Sanskrit mathematical text so far available that deals with magic squares.

Only the first portion of the *Bījagaṇitāvataṃsa*, based on a single and incomplete manuscript at Benares, has been published. The rules for *kuṭṭaka* and *vargaprakṛti* in the extant portion are similar to those in the *Ganitakaumudī* as well as those of Bhāskara II.

TAKANORI KUSUBA

REFERENCES

Primary sources

Gaṇitakaumudī. Ed. P. Dvivedi Benares, 1936–1942.

Bījagaṇitāvataṃsa. Ed. K.S. Shukla. Supplement to *Rtam* 1, pt. 2, 1969–1970.

Secondary sources

Cammann, Schuyler. "Islamic and Indian Magic Squares." *History of Religions* 7: 181–209 and 271–299, 1968 and 1969.

Datta, Bibhutibhusan. "Nārāyaṇa's Method for Finding Approximate Value of a Surd." *Bulletin of the Calucutta Mathematical Society* 23: 187–194, 1931.

Datta, Bibhutibhusan. "The Algebra of Nārāyaṇa." *Isis* 19: 472–485, 1933.

Datta, Bibhutibhusan, and A.N. Singh. *History of Hindu Mathematics,* 2 vols. Bombay: Asia Publishing House, 1935–38.

Kusuba, Takanori. *Combinatorics and Magic Squares in India.* Ph.D. Dissertation, Brown University, 1993.

Singh, Paramanand. "The So-called Fibonacci Numbers in Ancient and Medieval India." *Historia Mathematica* 12: 229–244, 1985.

NAṢĪR AL-DĪN AL-ṬŪSĪ Abū Ja‘far Muḥammad ibn Muḥammad ibn al-Ḥasan Naṣīr al-Dīn al-Ṭūsī was born on 17 February AD 1201 in Ṭūs in the northeastern Persian province of Khurāsān and died in Baghdad on 25 June 1274. He was a preeminent figure in medieval Islamic history, being a major participant in both political and intellectual life during a century that witnessed monumental changes in the Islamic world. Politically, he was active at the courts in Iran of both the Ismā‘īlīs, a Shī‘ite sect, and the Mongols who brought to an end both the formidable political power of the Ismā‘īlīs as well as the 500-year old ‘Abbāsid Caliphate. Intellectually, he played an even more important role, forging an intellectual synthesis that can be compared to that of Zhu Xi (d. AD 1200) in China and Thomas Aquinas (d. AD 1274) in the Latin West. Ṭūsī, though, was not simply a synthesizer and rejuvenator of the Hellenistic tradition in Islam; he also made innovative and significant contributions in science and mathematics.

Ṭūsī was born into a scholarly family of Imāmī (Twelver) Shī‘ites, one of the two major sectarian divisions of Islam. Although he received religious education at home, Naṣīr al-Dīn tells us in his autobiography, written when he was in his forties, that his father, "a worldly man", encouraged him to explore different sciences and listen to the masters of various sects and opinions. This he did and began studying the diverse branches of ancient science and philosophy, especially mathematics. Not content to remain in his hometown, he traveled while still a teenager to Nīsābūr, a major city in Khurāsān located to the west of Ṭūs, and became the student of a noted physician and a philosopher. It was at this point in his life that he began to study the works of his famous Persian predecessor Ibn Sīnā (Avicenna), whose works were influential in both the Middle East and Europe. He would later travel to Iraq and study both religious and secular subjects with several noted scholars. Though Ṭūsī's education was far from typical, one can gain some valuable lessons for trends that were occurring at the time. The fact that Ṭūsī was a Shī‘ite did not prevent him from studying with persons of different sectarian affiliations. Furthermore he was motivated to travel fairly widely to receive an education. Finally, he studied both religious subjects (*al-‘ulūm al-shar‘iyya*; "the Islamic sciences") and the rational sciences that had been appropriated into Islam (*al-‘ulūm al-awā’il*; "the ancient sciences"). The latter is important because Ṭūsī, especially in his later writings, felt that the ancient sciences — both philosophical and mathematical — could provide the means to transcend the religious disputes that had plagued Islam.

His formal education completed, Naṣīr al-Dīn would spend some 25 years (from the early 1230s until 1256) at the Persian courts of the Ismā‘īlīs, a powerful Shī‘ite group that had for a time vied for ascendancy in the Islamic world. (These were the Assassins of Crusader lore.) Whether Ṭūsī himself willingly converted to Ismā‘īlism or simply pretended to do so is a matter of dispute; whatever the true state of affairs, he did find a refuge from the ravages of the Mongol invasions and did produce some of his most important work during this period of his life.

After the fall of the last Ismā‘īlī stronghold at Alamūt in 1256, Naṣīr al-Dīn, by then a famous scholar, was enlisted by the Mongol conquerors of Iran into their entourage, becoming court astrologer as well as minister of religious endowments. He was also charged with building an astronomical observatory in the town of Marāgha in Ādharbayjān in northwest Iran. This was one of the most ambitious scientific institutions that had been established up to that time; because of the resources placed at his disposal (which included the religious endowments), Ṭūsī was able to oversee construction of the observatory and its instruments as well as a large library and school. A substantial number of scientists and students were attracted to Marāgha; indeed there are reports of Chinese astronomers on the staff. In 1274, Naṣīr al-Dīn left Marāgha with a group of his students for Baghdad and died there in that same year.

Works

Ṭūsī was one of the most prolific authors of the Islamic medieval period, and his works, written in both Persian and Arabic, number something over 150. (This does not include his poetry.) He wrote on both religious topics and on nonreligious or secular subjects that had been inherited in the main from ancient Greece. Here we have an example of an important trend that was beginning to take root in the thirteenth century, namely the breaking down of the earlier division between those who wrote exclusively in the "Islamic" or "Arabic" tradition and those who devoted themselves to the Greek tradition. Among Ṭūsī's religious writings are seminal works on Shī‘ite law (*fiqh*), dialectical theology (*kalām*), and Sufism. Works in the tradition of Greek philosophy included the very influential *Nāṣirian Ethics* and his commentary on

Ibn Sīnā's *al-Ishārāt wa'l-tanbīhāt* (Book of Directions and Remarks), which helped engender a renewed interest in his great Persian predecessor. Ṭūsī very much hoped that philosophy could provide a means toward rising above the many religious disputes that so racked Islam during his lifetime. In writing his commentary to Fakhr al-Dīn al-Rāzī's work on *kalām*, he wished to widen the opening Rāzī had made in allowing Greek philosophical concepts to enter the intellectual space of *kalām* while at the same time defending those concepts (as expounded by Ibn Sīnā) against Rāzī's attacks.

Naṣīr al-Dīn was especially attracted to the exact sciences, which, as Ptolemy had argued, provided a more sure means to truth than the murkier disciplines of physics and metaphysics. He thus devoted a considerable amount of his intellectual efforts to mathematics and astronomy. His recensions of Greek and early Islamic scientific works represent one of his most important contributions to the Islamic scientific tradition, an attempt to give vitality and renewed meaning to the translation movement of ninth century Baghdad. Because of the lack of an ongoing institutional structure for the teaching and perpetuation of science, Ṭūsī's reeditions, which often included insightful and original commentary, provided the means by which generations of students of late medieval Islam could assimilate the Greek scientific tradition, either with or without a teacher. These included Euclid's *Elements*, Ptolemy's *Almagest*, and the "Middle Books" of mathematics and astronomy with treatises by Euclid, Theodosius, Hypsicles, Autolycus, Aristarchus, Archimedes, Menelaus, Thābit ibn Qurra, and the Banū Mūsā.

In addition to this monumental role as textbook writer, Ṭūsī is also know for his original work, some of which occurs within these recensions. In mathematics, this included a highly sophisticated attempt to prove Euclid's parallels postulate, part of a long tradition that would eventually culminate in the nineteenth century with the realization that such "proofs" were not possible and that consistent non-Euclidian geometries using alternative postulates were constructible. In spherical trigonometry, Ṭūsī produced an important synthesis of earlier results of Islamic mathematicians; this marked a significant step in treating trigonometry as a discipline independent of astronomy, comparable in many respects with what was done later in Europe by Regiomontanus (1436–1476).

Ṭūsī, though, was most famous for his work in astronomy. In addition to a number of elementary treatises on practical astronomy, instruments, astrology, and cosmography (*hay'a*), many of which were meant for students, Ṭūsī also composed his *Zīj-i Īlkhānī*, a major astronomical handbook, for his Mongol patrons in Marāgha. Though not very original, and apparently compiled in haste without incorporating the Marāgha observations, it was destined to be widely used.

Ṭūsī's most original achievement was in theoretical astronomy. From a rather early date, Islamic astronomers had been disturbed by a number of inconsistencies in the Ptolemaic system, for example that some models violated the fundamental principle that all motions in the heavens should conform to the dictate of uniform circular motion. Ṭūsī responded to this by devising an astronomical model consisting of two rotating spheres the smaller of which was internally tangent to another twice as large. By having the smaller rotate twice as fast and in the opposite direction as the larger, Naṣīr al-Dīn was able to produce the linear oscillation of a given point, a property he was able to use in lunar and planetary models that could reproduce Ptolemaic accuracy while preserving uniform circular motion. Ṭūsī's new models had a decisive influence during at least another three or four centuries of late medieval Islamic astronomy, where they provided the starting point for numerous attempts to reform the Ptolemaic system. His device, which has been dubbed "the Ṭūsī couple", found its way into Sanskrit and Greek texts, and was influential in the work of several Renaissance astronomers, including Copernicus.

But Ṭūsī's most enduring influence, in fields as diverse as ethics, natural philosophy, mathematics, Sufism, astronomy, *kalām*, *fiqh*, music, mineralogy, and logic, was in the Eastern Islamic world, and in particular Persia, where his works continued to be studied and commented upon into the modern period.

F. JAMIL RAGEP

REFERENCES

Al-Daffa, Ali A. and John J. Stroyls. "Naṣīr al-Dīn al-Ṭūsī's Attempt to Prove the Parallel Postulate of Euclid." In *Studies in the Exact Sciences in Medieval Islam*. Ed. Ali A. Al-Daffa and John J. Stroyls. Chichester: John Wiley & Sons, 1984, pp. 31–59.

Ragep, F. J. *Naṣīr al-Dīn al-Ṭūsī's Memoir on Astronomy (al-tadhkira fī 'ilm al-hay'a)*. 2 vols. New York: Springer-Verlag, 1993.

Riḍawī, Muḥammad Mudarris. *Aḥwāl wa-āthār...Naṣīr al-Dīn*. Tehran: Farhang Iran, 1976 [in Persian].

Sayılı, Aydın. *The Observatory in Islam*. Ankara: The Turkish Historical Society, 1960.

Siddiqi, B. H. "Naṣīr al-Dīn al-Ṭūsī." In vol. I of *A History of Muslim Philosophy*. Ed. M. M. Sharif. Wiesbaden: Otto Harrassowitz, 1963, pp. 564–580.

See also: Marāgha – Ibn Sīnā – Astronomy in China – al-Rāzī – *Almagest* – *Hay'a* – Astronomy in the Islamic World – Marāgha

NĀṢIR-I KHUSRAW Abū Muʿīn Nāṣir ibn Khusraw al-Kabādhiyānī was a Persian philosopher, poet, and traveler. Born at al-Kabādhiyān in Transoxania (now in Tajikistan) in 1004, he was one of the founders of Ismāʿīlī theosophy and lived at Balkh and Ghazna (now in Afghanistan) at the court of the Ghaznewid sultans Maḥmūd and Masʿūd. After the Seljuqid conquest, he lived at Marw at the court of Seljuqid Chaghri Beg. He traveled from the Maghreb to India. In Egypt he became Ismaʿīlī and was made the Ismaʿīlī missionary in Persia and Transoxania. In this he was persecuted and was forced to take shelter in Yomghan in the Pamir mountains (now in Afghanistan). He wrote in both Persian and Arabic.

His main work is the *Safar-nāmeh* (Book on Travels), the diary of his journeys, containing an account of life in Egypt under the Fatimid Caliph al-Mustanṣir (1035–1094), and various geographic, ethnographic, and archaeological information.

In his philosophical treatise, *Kitāb zād al-musāfirīn* (Book Supply of of Travelers), which is a survey of Ismaʿīlī theosophy, there is information on the history of science and fragments of the philosophical treatises of Abū Bakr al-Rāzī (854–935) on space, time, and matter.

His philosophical *Kitāb jāmiʿ al-ḥikmatayn* (Book Joining Two Wisdoms) is devoted to the harmony between Greek philosophy and Ismaʿīlī theosophy. He also wrote a philosophical *Rowshanāʾ ī-nāmeh* (Book of Light) in six chapters, *Saʿādat-nāmeh* (Book of Happiness), *Kitāb gushāyish wa rahāyish* (Book of Unfettering and Liberation), *Wajh-i dīn* (Face of the Faith), and *Khān al-ikhwān* (Meal of the Brethren).

He also was the author of a non-extant mathematical treatise, *Gharāʾib al-ḥisāb wa ʿajāʾib al-ḥussāb* (Marvels of Arithmetic and Wonders of Calculators), containing two hundred problems with solutions and demonstrations to solve them correctly. In this treatise Nāṣir-i Khusraw complains of the absence of good mathematicians in Khurasan in his time. He died in ca. 1088.

BORIS ROSENFELD

REFERENCES

Primary sources

Naçir ed-Dîn ben Khosroû. "Le livre de la felicité." Ed. and trans. E. Faghan. *Zeitschrift der Deutschen Morgenlandischen Gesellschaft* 34: 643–764, 1880.

Nāṣir-i Khusraw. *Diary of a Journey Through Syria and Palestine.* Trans. G. Le Strange. London: Palestine Pilgrims' Text Society, 1893.

Nāṣir-i Khusraw. *Kitāb Wajh-i dīn.* Ed. T. Erani. Berlin: Dāviaānī, 1924.

Nāṣir-i Khusraw. *Khān al-ikhwān.* Cairo: Maṭbuʿah al-Maʿhad al-ʿIlmī al-Faransī hī al-Athār al-Shraqīyah, 1940.

Nāṣir-i Khusraw. *Six Chapters of Shish Fasl: also called Ravshanaʾi-nama.* Ed. and trans. W. Ivanov. Leiden: Brill, 1949.

Nāṣir-i Khusraw. *Kitāb Gushāʾish wa rahāʾish.* (The Book of Unfettering and Liberation). Bombay: Maṭbaʿat Qādirī Parīs, 1950.

Nāṣir-i Khusraw. *Sefer Nameh, Relation du voyage de Nassiri Khosrau.* Trans. Charles Schefer. Amsterdam: Philo Press, 1970.

Nāṣir-i Khusraw. *Kitāb Jāmiʿ al-hikmatayn.* Tehran: Kitabkhanah-i Tahuri, 1984.

Secondary sources

Matvievskaya, Galina P., and Rosenfeld, Boris A. *Mathematicians and Astronomers of Medieval Islam and their Works (8–17th c.)*, vol. 2. Moscow: Nauka, 1983, pp. 305–307 (In Russian).

Sarton, George. *Introduction to the History of Science*, vol. 1. Baltimore: Williams & Wilkins, 1927, pp. 768–769.

NAVIGATION IN AFRICA The physical geography and environment of the African continent precluded extensive water transportation on the interior waters, although light boats have been in use at least since the Middle Ages for fishing and transit on lakes and stretches of rivers between rapids and where permitted by depth or the tides. Navigation properly speaking was determined by three distinct regional environmental systems in the north, west, and east. Available historical data are uneven over time and area. Participation of African peoples in naval activities originating in Africa is not always attested by the sources. In the Mediterranean Sea, ancient Egyptian rulers, the kingdom of Israel, and Carthage made use of the sailing and shipbuilding skills of Phoenicians. Pharaonic Egypt, Israel, and ancient Ethiopia (Axum) used the Red Sea as part of a commercial system linking the eastern Mediterranean with the Middle East, East African coast, and India.

Under the Old Kingdom expeditions had been sent to the land of Punt (southwest coast of the Red Sea or the Somali coast facing Arabia) for gold, ivory, ebony, and myrrh. Ships built at the head of the Gulf of Suez traveled south under the Fifth and Sixth Dynasties (ca. 2470–ca. 2280 BC). A Red Sea shipwreck under the Middle Kingdom (ca. 2000–1800 BC) is recorded in the Egyptian *Story of the Shipwrecked Sailor*. Under the New Kingdom, expeditions were resumed by Queen Hatshepsut (ca. 1495 BC). According to Herodotus, the pharaoh Necho II (ca. 610–595 BC) was the first to attempt linking the Nile with the Red Sea by a canal. After calling off the construction he sent a fleet manned by a Phoenician crew to circumnavigate Africa clockwise. They took over three years to complete the journey, discovering in the process that Africa was surrounded by water. Some time before 480 BC, a counter-clockwise circumnavigation was attempted by Carthage. This voyage, described in the Greek document known as the *Periplus of Hanno*, may have discovered both the Canary and Cape Verde islands, subsequently forgotten and rediscovered again centuries later. In the north, Carthage reached out into the Atlantic as far as the British Isles and possibly the Baltic. Access to these routes was strictly controlled and outsiders were attacked to prevent competition. In the Mediterranean, Punic trade interests inevitably ran into conflict first with Greece and then with Rome. With the decline of Carthage, navigation on the Mediterranean effectively became part of the Western scientific and technological tradition.

Oceanic navigation in the African Atlantic is poorly documented. There is no evidence of seaworthy vessels being constructed in sub-Saharan West Africa in pre-modern times. However, several reports indicate that sailing in the Atlantic was occasionally attempted, usually resulting in failure. Juba II, king of Mauretania and a Roman ally, sent a fleet to the Canaries in about 25 BC and later had the Atlantic coast of Morocco explored. In the Islamic period, the route to the Canaries seems to have been forgotten, although the Arab geographers inherited knowledge of the islands from Ptolemy. The legend of a statue at Cadiz forbidding travel beyond the Pillars of Hercules was transferred to the Islands of the Blessed, or Eternal Isles, as the Arabs called them after Ptolemy. A report by the geographer al-Idrīsī (ca. 1154) tells of an attempted expedition to the Canaries under the Almoravid ruler Yūsuf ibn Tashfīn (1061–1106) who controlled parts of western Africa from the Senegal River to Morocco and Algiers. The admiral in charge having died before departure, the voyage never took place. Al-Idrīsī also describes an intriguing adventure originating in Lisbon. Some time before 1147, eighty explorers, *mugharrirūn*, built a large ship, loaded it with enough food and water for several months, and after sailing west and south for a total of 35 days came to an island where they were forced to land by light-skinned men in boats. The island's king, speaking through an Arabic interpreter, told of an expedition ordered into the sea by his father. After sailing "across" the sea for a month, that group had returned having found nothing. The explorers were then taken to the mainland reached after sailing with the west wind for "three days and nights". Their final rescue was effected by the Berbers of the Atlantic coast of Morocco.

Arabic accounts of the thirteenth and fourteenth centuries, especially those by Ibn Fātima (related by Ibn Saʿīd) and Ibn Fadl Allāh al-ʿUmarī suggest continuing but accidental maritime contacts between Morocco and Spain on the one hand, and black West Africans on the other. The availability of Arabic or Berber interpreters among peoples encountered by Muslims in these voyages undermines the speculation that these expeditions constitute an African discovery of the Americas. Of special interest is one report by al-ʿUmarī which describes an attempt at exploration of the Atlantic by a ruler of Mali, Abū Bakr II, who was the predecessor of the famous Mansā Mūsā (1307 or 1312–1327). An expedition of 200 light boats was sent down the Senegal River with provisions "to last many years". Only one vessel returned, with a story of the others disappearing in a violent current at sea. The persistent ruler equipped two thousand new boats (one thousand of them filled with stores of food and water) and departed on a second expedition, never to return. Some have suggested that Morocco was the intended destination but the text gives no such indication, citing "the extremes of the ocean" in the king's order. No such further attempts are known.

Pre-modern naval communications in Eastern Africa were subject to different conditions in the three subregions: the Red Sea, the East African coast with the adjacent islands, and Madagsacar with the Comoros. Numerous port cities on the coasts of Eritrea and northern Somalia are mentioned in the *Periplus of the Erythrean Sea* and in Ptolemy's *Geography* for the first and second centuries AD respectively. Adulis (ancient Berenice, at Massawa) became the principal port of the Ethiopian empire (Axum). Both Pseudo-Callisthenes (third century) and Cosmas Indicopleustes (sixth century) attest to the presence of Ethiopian merchants in India and Ceylon (Sri Lanka). However, it is not always clear in these or later reports who the sailors and navigators were. *The Periplus of the Erythrean Sea* specifies Arab supremacy on oceanic routes leading from Yemen to Africa and India. Early Islamic history records attacks on Arabian ports and shipping by Ethiopians, repulsed by the Arabs who established their firm control on the Red Sea by the early eighth century. However, Africans from the coasts of Ethiopia, the Horn, and the Dahlak islands continued to be involved in trade and piracy on the Red Sea. They also served as sailors on boats sailing the Indian Ocean owned by Arabs, Persians, Swahili, Indians, and even Portuguese. Massawa boats were described in the eighteenth century as "very slight and unsafe", but capable of carrying a "great weight". They were, as in previous centuries, of the sewn type with mat sails, equipped with a crew of seven or eight men.

Zeila and Barbara were major African counterparts to Aden, receiving fleets from Africa heading north or Indian and Persian Gulf vessels destined for Jedda. From Zeila and Massawa trade routes led inland, providing important access for Ethiopia to major commercial networks even during the centuries of its isolation from the world by a ring of Muslim principalities bordering on the sea. The port of Suakin attracted shipping largely because of Abyssinian gold. So many Ethiopians were engaged in oceanic shipping that an important social class of Habshis (i.e. Abyssinians) evolved in India's mixed coastal communities. Their presence at Calicut and Ceylon is first reported by Ibn Baṭṭūṭa in the fourteenth century. Sometimes associated with the East African immigrants called Sidis (who spoke Swahili and often originated as slaves from Mozambique and Mombasa), Habshis acquired special influence in the seventeenth to eighteenth centuries as sailors and soldiers protecting shipping from Gujarat in the northwest to Surat and even Bengal. One Sidi was the admiral of the Bijapur fleet and had under his jurisdiction the coast north and south of Janjira, the port attacked by the Moghuls in 1659. In the Bombay region in 1648 one area was controlled by a Habshi reponsible for maintenance of a marina for the purposes of transporting pilgrims to the Red Sea and protecting commerce. The office carried the title of *wazir* and was attained through merit. Some Habshis were non-Muslims. Habshi crews serving on Portuguese and Arab-owned ships often carried their families aboard. African sailors were apparently also involved in piracy on the west Indian coast.

In the southwestern part of the Indian ocean, sailing between the African mainland and the island of Madagascar was made difficult by the strong southward currents in the Mozambique channel and the weak reach of monsoons. Swahili legends of origin would put Muslim settlers on the Comoros as early as the eighth century, but the first trustworthy reports speak of Malay (*Wāqwāq* or *Qumr*) people reaching African destinations at a time when Mozambique was barely known to the Arabs and no evidence exists for African voyages from the mainland.

The earliest report of a Malay attack against a northern Swahili town (*Qanbalū*, on Pemba Island) dates back to 945. The Arabic book *Marvels of India* tells of a thousand boats attacking the city from the sea and surrounding channel. The assailants had plundered parts of Mozambique and an island six days' distance away. Their home was at the distance of one year's journey. In the thirteenth century Ibn al-Mujāwir speaks of an earlier, vaguely ancient, invasion of Aden by the Malagasy people (*al-Qumr*) who arrived in great numbers in boats to displace local fishermen. They were later expelled by Kushites from the Somali peninsula (*Barābir*). These people disappeared, and their skill had been lost, but a boat from *al-Qumr* arrived at Aden in AH 626/AD 1228–29. Ibn al-Mujāwir describes it as an outrigger vessel; he admires the sailing skill of these people who managed to sail from Madagascar to Aden in "one monsoon", instead of three stages involving layovers at Kilwa and Mogadishu. Around 1500 Aḥmad ibn Mājid includes the Comoros in the scope of Arab navigation. After the arrival of the Portuguese, and especially after hostilities developed in the north, Swahili migrations to northwest Madagascar and the Comoros expanded while the Malagasy navigation declined. Comorian sailors became occasional pirates and were involved in slaving raids against the mainland coast until the nineteenth century. Malagasy in turn attacked the Comoros from the eighteenth century, adapting for slaving raids large canoes otherwise used for whaling and capable of carrying thirty men. Some raids even reached across the Channel. Outrigger canoes were called *parabou*. There were also two-bridge galleys *coracores*, one-mast ships *pajalas* and *palans*, similar to small galleons.

The coast of East Africa proper, subject to the monsoon regime, is distinguished by the great number of coastal sites used as trading ports and fishing settlements. African participation in trans-oceanic navigation has not been documented for pre-Islamic or even early Islamic times but local inter-island shipping must have evolved sufficiently early to allow off-shore island settlement by at least the ninth century. According to al-Idrīsī (mid-twelfth century) the Zanj (East African Negroes) did not have sea-worthy vessels and their goods were exported by ships from Oman and elsewhere. By that time a network of sailing routes connected harbors from southern Somalia to northern Mozambique. The Swahili-speaking coastal Africans dominated local shipbuilding and trade but other groups, including the Bajun, provided sailors as well. Navigation on the Swahili coast carried many general characteristics of Indian Ocean navigation, including the sidereal rose, sailing calendar, some boat types, and terminology rich in loan-words from Arabic, Persian, and Indian languages. It also had some distinctive features and deeply affected the social culture of the coast. The best known is the maritime Swahili culture of the island city of Lamu (Kenya).

The sailing calendar was naturally based on monsoons within the solar year. The marine year began in early August, when the gusty southern *kusi* wind subsided. The new year's celebration, called *siku ya mwaka*, marked the resumption of shipping activity after a pause during May–July. The Persian word *Niruzi*, also used for this festival, gave rise to the false impression that this calendar is Persian in origin. Different sub-periods within the monsoon were best suited for plying the routes between Malindi and Madagascar, Kilwa to Sofala and back, Sawāhil (northern Tanzania) to Mogadishu, Aden and thence to Hormuz or Gujarat and the Maldives. Travel north was more difficult due to both the southwest monsoon and the strong contrary currents in channels between the mainland and the islands. The "normal" boundaries of Swahili travel were Cape Delgado in the south and Mogadishu in the north. In the fourteenth century the city of Pate briefly attempted to dominate the coast, but it always had to compete with its neighbor Lamu and the more remote Mogadishu, Malindi, Mombasa, and the powerful Kilwa. Regional and local rivalries involved sailing to foreign lands (Arabia or India) or to the mainland, and attacks and blockades from the sea. In island settings, it was generally faster, easier and safer to sail from one destination to another than travel overland well into the nineteenth century. The Portuguese arrival in East Africa did not noticeably change local shipbuilding and sailing practices, although this was not so elsewhere; the Eastern Arabian and Persian Gulf influences both pre- and post-dated this European infusion. Local ships still carried traditional trade goods and even revived with the rise in the slave trade before its suppression by the British in the late nineteenth century. By that time Zanzibar and Kilwa were the two leading ports of the coast, followed by Mombasa and Lamu.

Navigation proper was not highly developed. Given limited distances, boats mostly coasted. The steady daily and seasonal pattern of the winds made observation of local features such as banks, current speeds, and tides more important than instruments. Actual stars rather than compass rhumbs were used for determining latitude. Travel was slow, averaging 1 knot an hour, if rest days are included. In the fourteenth century Ibn Baṭṭūṭa was told that the area of Sawāhil was two weeks' sail from Mombasa. In the early nineteenth century it still took two weeks to reach Madagascar from Kenya with a fair northerly wind *kaskazi*. Traffic between Lamu and Somalia resumed immediately in August, reached its peak in November and died down in January. Foreign craft from the north began to arrive in November.

The sailing calendar was paralleled by a ritual calendar. At Lamu, a local shipowner held the hereditary office of *mkuu wa pwani* (master of the strand). This dignitary led the dhow launching, invariably accompanied by both an African *ngoma* festival and a Muslim ceremony. Other elders *wazee* gave the blessing and burned incense. The shipwright *fundi* was given a ceremonial present and the captain *nakhoda* declared by the *mkuu*. Ceremonial rites were performed in order to bring favorable sailing conditions or take wind out of an opponent's sail, break his mast, etc.

Today's dhows (*daus*) are sea-going sailing ships with a very large lateen which still may be rowed or poled when necessary. They serve as fishing vessels and may be flat-bottomed or keeled. The formerly prevalent *mtepe* (*dau la mtepe*), no longer in existence, was a type of sewn boat of up to thirty tons carrying a square sail of matting, double-ended with an upright mast. The local "big ship" is *jahazi*, a fairly recent type of sea-going coaster, single-masted, transom with an upright stem, with a cargo capacity of 25–60 tons. A smaller version of the same type used for intercoastal shipping is called *mashua*. *Buti* was a large (20–60 tons) coastal vessel type, now obsolete, with one mast with a slightly curved upright stem and square stern. *Baghala* is the most beautiful and largest (100–300 tons) of the dhows, a two-masted vessel with square stern, originally from the Persian Gulf. *Bum* is a layman's name for lateen-rigged sea-going dhows. The small keeled *msumari* (*dau la msumari*) also carries a lateen sail. The double-outrigger canoe *ngalawa* rarely occurs north of Mombasa. Lack of timber limits shipbuilding; local wood may suffice for plank-built boats of small dimensions but dug-out canoes have to be imported.

MARINA TOLMACHEVA

REFERENCES

Beckerleg, Susan. "Le réseau d'un Swahili constructeur de bateaux." In *Les Swahili entre Afrique et Arabie*. Ed. Françoise Le Guennec-Coppens and Pat Caplan. Paris and Nairobi: Karthala, 1991, pp. 107–27.

Chittick, Neville. "Sewn Boats in the Western Indian Ocean and Survivals in Somalia." *International Journal of Nautical Archaeology* 9(1): 73–76, 1980.

Chittick, Neville and Robert I. Rotberg, eds. *East Africa and the Orient*. New York: Africana, 1975.

Devisse, J. in collaboration with S. Labib. "Africa in Intercontinental Relations." In *General History of Africa*, vol. 4 *Africa from the Twelfth to the Sixteenth Century*. Ed. D.T. Niane. Paris and London: Heinemann for UNESCO, 1984, pp. 635–672.

Grosset-Grange, Henri. "La Côte africaine dans les routiers nautiques arabes au moment des grandes découvertes." *Azania* 13: 1–35, 1978.

Hornell, James. "Indian Boat Designs" and "The Sea-Going Mtepe and Dau of the Lamu Archipelago." *Mariner's Mirror* 27: 54–68, 1941.

Mauny, Raymond. *Les Navigations Médiévales sur les Côtes Sahariennes Antérieures à la Découverte Portugaise (1434)*. Lisbon: Centro de Estudios Historicos Ultramarinos, 1960.

Pankhurst, Richard. "The History of Ethiopia's Relations with India Prior to the Nineteenth Century." In *Congresso Internazionale di Studi Etiopici, 4th*. Rome, 1972, pp. 205–311.

Prins, A. H. J. *Sailing from Lamu; a Study of Maritime Culture in Islamic East Africa*. Assen: Van Gorcum, 1965.

Shumovskii, T. A. *Tri Neizvestnye Lotsii Akhmada ibn Madzhida, Arabskogo Lotsmana Vasko da Gamy*. Moscow and Leningrad: Akademiia Nauk SSSR, 1957.

See also: al-Idrīsī – Ibn Baṭṭūṭa

NAVIGATION IN CHINA Did the Chinese visit North America before Columbus? Did Chinese navigators land in Australia before Dampier and Cook? These two puzzles often pass our minds when we turn to the topic of Chinese navigation. In 1761 C.L.J. de Guignes wrote that in the early part of the sixth century a Chinese monk named Huishen had sailed to the west coast of North America. The story did not gain much support, but some interest in the issue was revived in the last two decades by the discovery of two pre-Columbian stone anchors off the Californian coast. One of the anchors is in the shape of an equilateral prism, and the other looks like a large cylinder with a hole bored through the center. Both were suggested to be of Chinese origin. As for Australia, aborigines along the northern coast still remember contacts with their northern neighbors who visited them annually in the past for the purpose of trade. Among them were the Macasarese and the Buginese; the former were preceded by the Baijini. The Baijini were said to be technologically advanced, with much lighter skin color; among the things they bartered for was *trepang*, from which the typical Chinese delicacy sea-slug was prepared. Hence they could be identified with the Chinese, the only people who knew how to prepare sea-slug. While we cannot yet arrive at a definite answer to the two questions it is possible to show that at least the technology and skill were already there in the past for Chinese ships to reach their destinations.

The Chinese must have navigated their rivers and coastal waters since the dawn of history. The character *zhou* for boat written on tortoise shells and buffalo shoulder-blades used for the purpose of divination during the Shang period (traditional dates 1766–1122 BC) took the form of a boat. Documentary records on navigation in ancient China are, however, rather scanty. Confucius himself once made a remark on venturing upon the open sea in a raft. Then comes the story about the emperor Qin Shihuangdi sponsoring expeditions to the East China Sea to seek the elixir of immortality during the third century BC. Modern archaeology has provided evidence that Chinese sailors had taken to the high sea at least by the third century BC. In 1976 the ruins of a large shipbuilding site of that period was discovered at Guangzhou in South China. The site could hold ships weighing between fifty and sixty tonnes and measuring up to thirty meters in length and about eight meters in width.

In the year 118 the Han astronomer Zhang Heng (78–139) noted that there were stars not visible to people in China itself, but that were seen by seafaring people. There were already a number of books on this subject then, but unfortunately none of them has survived. Those seafaring people could have steered their boats by the sun and the stars. Star catalogs were already available in China, even before Zhang Heng made observations with the new astronomical instrument he constructed. By about the seventh century maritime trade between China and the Arab countries saw Chinese and Arab sailors traveling between the ports of east and southwest Asia. Arab merchants traveled in Chinese boats. The large number of Tang potsherds of about the mid-eighth century discovered by the Japanese Idemitsu Archaeological Mission in 1966 at Fustat near Cairo and in Aidhab on the Red Sea coast of Egypt bear testimony to this trade. It could be during the mid-ninth century that Chinese navigators first made use of the magnetic compass. This could have taken place later, but not after the mid-eleventh century when Shen Gua (1031–1095) described the mariner's compass. This early form of compass consisted of a magnetic needle suspended by a thread.

With the invention of the mariner compass and increase in maritime trade the shipbuilding industry became highly developed during Song China (960–1279). In 1960 the remains of a Song wooden craft measuring twenty-four meters long and about four meters wide for navigating in rivers and canals was discovered in Yangzhou, Jiangsu province. Then in 1974 the remains of a *Fuchuan* — Fujian-built ship — was recovered in Quanzhou, Fujian province. Measuring over thirty-four meters and weighing about 374 tonnes, it was composed of thirteen water-tight compartments, the bulkheads. The shipbuilding industry continued to flourish during the time of the Mongols (1271–1368). Marco Polo gave an estimate of over 200,000 sailing vessels in China.

The most famous maritime exploration ever made by the Chinese was that of Zheng He (1371–1434). His fleets included some very large vessels. According to the *Mingshi* (Official History of the Ming Dynasty), the largest ones measured 134 meters in length and up to fifty-five meters in width. Modern scholars gave different estimates for their displacement, varying from over three thousand to just over eight hundred tonnes. In 1405 Zheng He made his first expedition to Southeast Asia and the Indian Ocean, employing a fleet of sixty-three ocean going junks. During the next thirty years he made six other expeditions to the Indian Ocean. His explorations preceded the Portuguese by several decades. For example, he visited Calicut in India in 1405, some ninety years before the latter arrived. He brought home strange animals like zebras and giraffes, and rare drugs, minerals, and other curiosities during his visits to Africa. He sailed along the eastern coast of that continent. One unanswered question is whether he had navigated round the Cape of Good Hope and ventured into Atlantic waters.

The late Ming military compendium *Wubeizhi* (Treatise on Armament Technology) compiled by Mao Yuanyi in 1628, contains the original schematic sailing charts tracing the routes followed by Zheng He and navigational diagrams indicating the star positions to be maintained during the voyages. The charts show the courses of the ships across the seas, with legends giving detailed compass-bearings, with distances expressed in number of watches. They also describe more interesting coastal features, showing half-tide rocks and shoals as well as ports and havens. For example in the description of the voyage from Bengal to Malé in the

Maldive Islands through Sri Lanka, it gives the polar elevation for every stage of the journey, such as mentioning that a certain mountain in Sri Lanka would be sighted when the polar altitude had sunk by a certain number of degrees. Hence Chinese sailors in the early fifteenth century already knew the method of finding and running down the latitude. For measuring the altitude of a star, Zheng He's crew probably made used of the cross-staff, which was an instrument used by Chinese surveyors as early as 1086. The traditional Chinese clepsydra of course would not function properly at sea. Needham suggests that they had an alternative, by simply using the incense or joss-stick, which was a popular method used for timekeeping.

The Chinese constructed many types of sailing vessels to suit different needs and environments. Navigating rapid shallow streams, sailing on big rivers and along the coast, transporting food and merchandise along the Grand Canal, and venturing further afield to the Indian Ocean would require different types of ships. Chinese sailing vessels differed from Western sailing ships in their hull structure and in their sails, as well as in their methods of propulsion. The word "junk" was first given to Chinese ships in records of the travels of Odoric of Pordenone and that of Ibn Baṭṭūṭa during the fourteenth century, but the origin of this word is uncertain. Needham suggests that it could have come either from the Chinese word *chuan* or from the cognate Javanese and Malay words *jong* and *ajong*. The term *sampan* is used for smaller Chinese sailing vessels. Undoubtedly it originated from the Chinese term *sanban*, which is written in two forms. One form literally means "three boards", indicating its size. In the nineteenth century the term referred to river gunboats rowed by eight oars on each side.

The oldest form of Chinese ship had a carvel-built hull without the keel, the stempost and the sternpost of European and Arabian ships. Generally, the Chinese junk could be either flat or slightly rounded at the bottom. However, there were also junks with a V-shaped bottom as in the case of the Fujian-built ships. The planking of the Chinese junk did not meet at a point at the stern and the stem, but stopped abruptly and was joined transversely by straight planks.

Also unlike its European counterpart, the Chinese junk had neither frames nor ribs; instead it was composed of solid transverse bulkheads, which gave rise to several water-tight compartments. This valuable shipbuilding technique could have been inspired by the bamboo that the Chinese used for some smaller rafts and for so many other purposes. Many types of bamboo raft still ply the rivers in China today, Some of them are made with the giant bamboo *dendrocalamus giganteus* that grows to a height of some twenty-five meters and has a diameter of about thirty centimeters. To enable the raft to slide over river rocks, the bow is usually bent upwards by heating. If a piece of bamboo is bisected longitudinally the nodal septa would represent the transverse bulkheads and the stem and stern transoms.

The sails used in Chinese ships were also different from those of their European counterparts. Different cultures developed different types of sails to take advantage of winds coming from different directions. For example the ancient Egyptians had the symmetrically hoisted square sail. The Chinese on the other hand had the lug-sail, which consisted of an upper sail support slanting upwards away from the mast, called the canted yard, and a horizontal sail support at the bottom called the horizontal boom. The sail was often strengthened by battens of bamboo, the ends of which were fastened to bolt-ropes suspended from the yard to take the weight of the sail. The sail was often made of bamboo matting and was kept flat and taut by using multiple sheets of matting. Several masts were used for larger vessels, the number varying from two to five. These masts would be staggered thwartwise such that one would not becalm another. The rake of a system of masts often radiated like the spine of a fan. The Chinese mat-and-batten sail had several advantages over the canvas sails used in European ships. It could manage with a material which, although not as strong as canvas, was aerodynamically more efficient. It offered more protection against tearing, and could easily be furled, as it would readily fall into pleats thus dispensing with having to send some crew member aloft to take in reefs.

The Chinese propelled their boats by punting with long poles and by rowing using steering oars and stern sweep. However, they also employed an ingenious method of mounting the oar approximately in the line of the main axis of the boat, and moving the oar from side to side about a fixed fulcrum. This was the Chinese *yaolu* that had fascinated many Western observers in the past. Louis Lecomte remarked that they made use of the *yaolu* as the fish did its tail.

The rudder is standard equipment in a sailing craft to control the direction of movement, although the same function may be performed by other means, such as by the oar and the paddle, albeit less efficiently. In the Western tradition the sternpost rudder is hung on pintle and dudgeon. What the Chinese did was to develop an axial rudder that could move up and down in guides. In shallow water it could be raised to its highest position to avoid damages, while in a heavy monsoon in the open sea it could be set in the lowest position for protection from the breaking water and to improve the ship's windward sailing quality. Sometimes the rudders were riddled with holes. These were fenestrated rudders. The holes were supposed to minimize the drag on the ship caused by the turbulence of the water flowing past the rudder, thus improving the hydrodynamics of the rudder. A ship would sometimes carry rudders of different sizes for use in different conditions. Sometimes instead of the rudder, the steering oar was developed into a long stern-sweep in boats that navigated rapid rivers and land locked waters. The rudder is only effective where there is relative motion between the boat and its surrounding water. When a boat comes down with the same speed as a descending rapid, the rudder does not operate. A long sternsweep, however, depends on the reaction to water resistance and controls the movement of the boat.

Thus Chinese ships ruled the waves in East Asian waters for a long period of time until the sixteenth century when the Portuguese arrived in East Asia and found no equal to their *calivers* outside the Atlantic. In the meantime from the days of the arrival of the latter until the second half of the twentieth century, the progress of modern science and technology has also left Chinese shipbuilding far behind. Although modern technology has already taken over from the traditional shipbuilding industry in China, we can still see Chinese junks and sampans navigating between ports in coastal waters in the China Sea and traditional rafts of all kinds sailing the rivers and the lakes in China today.

HO PENG YOKE

REFERENCES

Chen Guohua and Wu Tai. "Guanyu Quanzhouwan chutu haichuan de jige wenti" (Some Questions on the Remains of Sea-going Ships Discovered at Quanzhou Bay). *Wenwu* 4: 81–85, 1978.

Ho Peng Yoke. "Able and Adventurous." *Hemisphere* 21(6): 2–9, 1977.

Needham, Joseph. *Science and Civilisation in China* vol. 4 pt. 3. Cambridge: Cambridge University Press, 1971.

See also: Bamboo – Stars – Zheng Heng – Shen Gua

NAVIGATION IN THE INDIAN OCEAN AND RED SEA

Navigators in the Indian Ocean used the monsoons since before our era. The Chinese knowledge of monsoons was documented first, but the Indians and Middle Easterners benefited from them as well. The Greeks learned about sailing with monsoons between the Red Sea and India no later than the expedition of Nearchus (326–325 BC). A Roman port was established at Adulis to trade with India under Ptolemy III Euergetes (247–221 BC). The Greek *Periplus of the Erythrean Sea* (first century AD) attests to the Arab domination of routes between Arabia, East Africa, and India. With the rise of the Persian Sassanid Empire, Yemen, a crossroads of sea trade, became subject to rival interests of Persians, Byzantines, and Ethiopians. Persians seemed to control the navigation in the western part of the ocean until shortly before the rise of Islam. The revival and expansion of oceanic trade under the Abbasid caliphate (750–1258) must have involved not only Arabs and Persians but also coastal populations converted to Islam later, but the sources are not specific on this point. Participation of Indian Muslims as well as non-Muslims in navigation and piracy is recorded by Ibn Baṭṭūṭa in the fourteenth century. Islamic navigation in the Indian Ocean and the Red Sea and the Persian Gulf is often referred to as Arab navigation largely because the known sailing instructions and literary works describing methods of navigation are in Arabic. Early statements by some European scholars to the effect that the Arabs did not like or know the sea ignore the fact that Islam arose among northern Arabs at the time when south Arabians had accumulated many centuries' worth of sailing experience.

The Red Sea was the scene of early contacts between Muslims and Africa, especially Egypt and Ethiopia. It continued to play a role of conduit between the Mediterranean and the Indian Ocean and eventually carried heavy annual pilgrim traffic to Jedda and al-Jār (the port of Medina during much of the Middle Ages). Mocha was the main port in the south, Aqaba in the northeast; Qulzum (a major naval base) and Qusayr were prominent on the Egyptian coast, and ʿAidhāb on the African coast opposite Jedda. Outside the Bab el-Mandeb the ships stopped at Aden on the Arabian side or Zeila on the African side. Pirates found refuge on Dahlak and the smaller islands; once out in the Gulf of Aden, Socotran piracy was a threat.

During the Crusades the Red Sea became a scene of European attacks on Muslim shipping. Rulers of Egypt always tried to gain control of both coasts of the sea as well as Yemen. The Turkish conquest of Egypt and Yemen in the early sixteenth century made the Ottomans masters of the Red Sea and allowed them access to the Indian Ocean, where they tried to take over shipping routes leading to India, the Persian Gulf, and Africa. However, they were forced to yield to superior Portuguese force and later suffered naval intrusions into the Red Sea by both European and Indian (Gujarati) vessels. A major concern for Turkish authorities at Mocha and on the Ethiopian side (*eyalet of Habasha*) was the security and provisioning of the pilgrims to Mecca. Bombay, Goa, and Surat served as major ports for the Red Sea India trade, and Massawa had a colony of Indian merchants (*Banyans*) from the late sixteenth century.

Sailing from the north was relatively easy, although passing the tip of the Sinai Peninsula was feared because the winds from the Gulf of Aqaba and Gulf of Suez met there. Ships carried from the ocean to the Red Sea by the monsoon had to sail against northerly winds once past the strait of

Bab el-Mandeb. Jedda was the terminus of oceanic routes; transit further north had to use smaller boats. The only extant sailing instructions for the Red Sea cover the distance from Jedda to Aden (by Aḥmad ibn Mājid, ca. 1500). Latitude measurements taken by the stars could use the Polaris *Jāh* because of the northerly location. "Triangular instruments", not described otherwise, and quadrants are mentioned. Finding one's location was not difficult in confined waters, but navigation was dangerous because of numerous coral reefs, contrary winds, and currents. The journey from the north to the south end took thirty days (sailing by day only and coasting), but the ships of Saladin's navy could reach the speed of 4–5 knots. The north wind of the Red Sea *shamāl* reached the southern part only from May to September, coinciding with the short period when the prevailing wind in the Gulf of Aden was westerly, thus propelling ships into the Indian Ocean. The southeasterly wind of the Red Sea *azyab* reached half way up.

Navigation in the Indian Ocean was dominated by the monsoon (from Arabic *mawsim*, 'season'), a wind system that reverses direction seasonally. Both halves of the Indian ocean are subject to monsoon regime. The southwest monsoon *kaws* begins in March on the East African coast, slowly spreading eastwards. It reaches its maximum strength in June and blows across the ocean until October, bringing the heaviest rains to India in June and July and causing heavy swells which made landing difficult and even closed the ports. The northeast monsoon *azyab* originates from the Indian mainland in early October, reaching Zanzibar by late November. It makes it easy to sail almost directly from Malacca to Jedda as the wind continues into the Gulf of Aden. Between the monsoon periods, voyages were made in other directions, using variable winds and breezes. March to May are changeover months in the northwest corner of the ocean. In the Gulf of Aden the predominant non-monsoon wind is easterly. Travel from India to Africa had to be begun by early February. From Aden and Yemen one needed to leave in mid-October, and from northeast Arabia by late January, but one could not sail to Socotra during the same season. From Socotra to southeast Arabia one sailed in March–April, while India could be reached also by departing in May and during the August–September season *dāmānī*. Travel down the African coast was recommended from mid-November to April. From Bengal one had to leave westward by January; leaving from Malacca, Java, and Sumatra in February or March one could still reach Ceylon. Travel from Gujarat to Bengal and Indonesia began in April or late summer. October brought cyclones to the Bay of Bengal. The eastbound roundtrip journey from the Persian Gulf across the ocean took eighteen months. China-bound ships started from the Gulf in September or October, reached Kalah Bar in January and passed through the Strait of Malacca in time to use the southern monsoon in the Sea of China. Return to Malacca took place with the northeast monsoon between October and December; then ships could cross the Bay of Bengal in January and reach Arabia in February or March.

Navigation between the Middle East and China is confirmed by reports of an Arab embassy to China in the seventh century, a Persian settlement in the island of Hainan in 748, and a mixed Arab–Persian colony at Canton in the eighth-ninth centuries. In the seventh century Persian ships took twenty to thirty days to reach Sumatra from China, and one month from Ceylon to Palenbang. In the ninth century "Chinese ships" or "China ships" (that is, ships sailing to China) are reported in the Persian Gulf. Smaller boats brought goods from Basra and other ports to Siraf where they were reloaded on the large China boats. Ceylon (*Sarandīb*) was also visited and described by Arabic authors. Sea travel from the Persian Gulf to East African islands is mentioned by Arabic sources in the ninth century and described as routine in the tenth by the historian and traveler al-Mas'ūdī, although the country of Sofala (southern Tanzania and northern Mozambique) was then still poorly known. The Persian Gulf ports of Siraf and Hormuz as well as Oman were dominant on that route at the time; Aden emerged to prominence somewhat later. Some of the tales of Sindbad the Sailor (of Basra) originated in stories of Indian Ocean sailor and merchant adventures collected in the book *The Marvels of India* by Buzurg ibn Shāhriyār (ca. 950). Much of Marco Polo's return journey in the late thirteenth century must have taken place in Muslim boats and followed routes mentioned in this book. In the fourteenth century Ibn Baṭṭūṭa traveled by ship across the Red Sea and later to Africa, visiting Mogadishu, Mombasa, and Kilwa. On other occasions he sailed along the west coast of India and possibly to China. In India he encountered merchants from Cairo and Northwest Africa and fell victim to Indian pirates. He reported the presence of Chinese junks at Ceylon and planned to travel on one himself. The famous Chinese voyages led by the Ming court official Zheng He constitute the last known attempt by China to break into the Indian Ocean network. Two of these, in 1417–19 and 1421–22, reached Africa, visiting Malindi and the Horn of Africa. Even the arrival of the Portuguese caused only a disruption and reorganization of shipping. Lodovico Varthema (ca. 1510) and the early Portuguese sources note the continuing international presence at western Indian ports: Egyptians and "Moorish" mechants and ships from Hormuz, Arabia, Abyssinia, Kilwa, Malindi, Mombasa, and Mogadishu at Cambay, on the Malabar coast and the islands (Maldives and Laccadives). Early naval battles between the Portuguese and combined Muslim navies (e.g. at Diu in 1512) resulted in capture and destruction of numerous Muslim vessels

(In the sixteenth century only the Chinese had ships able to withstand attacks of the Portuguese galleons). However, the Portuguese soon realized that they would be unable to stop native shipping, and turned their efforts to diverting trade to ports which they controlled, carriage in their own ships, and taxation of all others. By the time the Dutch and the English arrived, an accommodation had been reached. However, intra-European competition, added to Christian–Muslim rivalry, contributed to pre-existing pirate activity, especially in the Persian Gulf and on the Gujarat and Malabar coasts. The maritime Muslim trade revived somewhat in the seventeenth-eighteenth centuries and declined again in the nineteenth century, at least in part due to increased British control over the routes and ports of the western Indian Ocean. Another factor was the growing dominance of European companies in the long-distance East–West trade and their penetration of local trade.

The sources for traditional Arab navigation date mostly from the late fifteenth through the sixteenth centuries, while our knowledge of ships and shipbuilding in the region is modern or contemporary. There are vague, scattered references to earlier sailing guides *rāhnāmaj*, devices and ships. Naval law is best known from the Malacca code of Shāh Mahmūd (1488–1530). A sixteenth-century Persian source lists twelve categories of crew members with job descriptions. The best information on navigation proper and sailing routes comes from the works of Aḥmad ibn Mājid of Julfar in Oman (d. ca. 1504) whose recognized expertise made him into a patron saint of Muslim sailors. A learned practitioner, he composed navigation manuals and sailing instructions, largely in verse, to ease memorization. From him we learn the names of several earlier pilots, dating back to the tenth-twelfth centuries. To these Ibn Mājid added the names of his own father and grandfather; apparently, the profession of pilot was hereditary but not highly regarded. It has been asserted by some scholars that Ibn Mājid was the pilot who guided Vasco da Gama from Malindi to Calicut, but this has been contested by others. Although he spent his life on the Indian Ocean, it appears that he was aware of the different methods of navigation in the Mediterranean. Other extant works on Islamic navigation belong to Sulaymān al-Mahrī (ca. 1511), a native of Shihr who wrote several practical and theoretical treatises, and the Ottoman writer Sidi Ali Çelebi who compiled a Turkish summary of the former two authors' work while moored in Gujarat in 1554 after a Portuguese attack on the Turkish Indian Ocean fleet originally commanded by the portolan-maker Pirī Reis.

Contemporary Arabic names of ships: *baghala*, *ganja*, *sanbūq*, and *jihāzi*, apply to vessels with square, transom sterns showing European influence. The older type is represented by vessels now called *būm* and *zārūq* — double-edged, coming to a point both at bow and stern. The name *sanbūq* was formerly applied to the small craft of the Red Sea; *jalbah* was the sewn boat typical of the Indian Ocean region. First mentioned in the *Periplus of the Erythrean Sea*, sewn boats were carvel-built, with planks edge-to-edge, and stitched with ropes of palm fiber. The timber was teak or coconut wood. They were leaky and frail but had an advantage over clinker-built boats with overlapping planks when striking a coral reef. A common legend explained that nails were not used because of the dangerous power of a magnetic rock somewhere in the middle of the ocean which could attract the nails. However, iron as well as stone anchors were used. The generic names for "ship" were *markab* and *safīna*. Indian pirates had *bārijas*; smaller boats *zawraq*, *qārib* and *dūnij* are also mentioned in medieval texts, but no particular shape is indicated. *Dau* is a generic name for lateen-rigged vessels; it is not used by the Arabs. The basis of classification was the form of the hull. The ships were usually one-masted, often without deck, with a cargo capacity of up to 200 tons. Erecting the mast and the rigging, and even making the sail was the responsibility of the owner *nakhōda* and the crew rather than the shipwright. The lateen sail associated with Arab ships probably evolved from a square sail on the Indian ocean; its use in the Mediterranean is first noted in the ninth century. African and Indian vessels continued using square sails of coconut matting into the twentieth century. The lateen is a tall, triangular fore-and-aft sail with the fore angle cut off to form a luff. It allows sailing into the wind by going on the tack, although Arab mariners preferred not to sail closer to the wind than 90°. It is possible that a second sail (topsail) was sometimes used. No reefing was done in strong wind, but the yard could be lowered. Two side rudders were originally used for steering, although by the thirteenth century the stern rudder was known. Sailing speeds averaging 1–3 knots were normal but could reach six knots under favorable wind.

The most important person on board ship was the *muʿallim* who served both as captain *rubbān* and pilot. He was hired for the voyage and allowed to carry merchandise as part of his pay. The shipmaster *nakhōda* was a merchant; ships were often owned by shareholders. We know Persian terms for eleven ranks of crew members; of these, the most important were the bo'sun *tandīl*, the ship's mate *sarhang*, the steersman *sukkan-gir*, and the look-out *panjarī*. Sailors were called *khallāsī* or *khārwah*. The captain was responsible not only for navigation but for the safety of passengers and goods as well. Among the necessities he carried were a nautical directory *rahnāmaj*, a measuring instrument *qiyās*, a bussole *huqqa* or *dīra*, lodestone *hajar*, lot *buld*, and lantern *fānūs*. The pilot's principal science consisted of knowing the coasts, winds and seasons, and his ship. Before departure, the Muslim prayer *Fātiha* was recited and an invocation was made to Khidr, the mythical patron saint of mariners.

By the sixteenth century Arabic sailing manuals listed over thirty different routes. Navigational books *dafātīr* and charts *suwar* carried on board are mentioned in the tenth century; Ibn Mājid calls his "chart" *qunbās*, but no charts have come to light. G.R. Tibbets argues that the Arab pilot plotted his course in his head and did not need a chart; besides, proper charts could not be made because the Arabs had no way of correctly determining longitude at sea. The winds and geography of the Indian ocean allowed the pilot to be guided roughly by the latitude of his destination (determined by the Pole star altitude). Once that was reached, he sailed down the latitude toward his goal. Another way was to keep to a recommended bearing until land was in sight and then make corrections. Extant Arab maps do not allow practical application to navigation. A Chinese chart based on the Zheng He expedition shows the routes from China to Hormuz, the Red Sea and Africa, but no measurements can be taken from it. Charts from Muslim Indian (Gujarati) nautical manuals *roz nāmah* of the seventeenth–eighteenth centuries show some European influence and use stellar compass bearings and Arab units of time-distance. A possibility of Chinese influence has been suggested as well. Considering that Arab information is already found on early sixteenth-century Portuguese maps, it is clear that the sharing of information among mariners created a truly international maritime culture drawing on indigenous and regional traditions and innovations.

The Arab system of nautical orientation evolved on the Indian ocean in the intertropical region, but probably north of the Equator; it may have been representative of all Indian Ocean sailing. It is based on a 32-rhumb *khann* sidereal rose *dīra* divided into eastern and western halves separated by the Polaris *Jāh* in the north and the South Pole *al-Qutb* in the south. The east and west divisions approximate the rising and setting of certain bright stars and constellations (Ursa Minor, Ursa Major, Cassiopeia, Capella, Vega, Arcturus, Pleiades, Altair, Orion, Sirius, Scorpio, Antares, Centaur, Canopus, Achernar). This system may have been in place by the ninth century. The bearings *majrā* were set by the actual stars, visible in the clear skies, not by the mathematically correct rhumbs. The compass was not unknown but rarely used or even carried. Star altitude *qiyās* was measured in units called *isbaʿ* (finger), supposed to correspond to the arc covered by the little finger of an outstreched hand. Its degree value measured $\frac{1}{2}$ of the distance from the Polaris to the true pole, and thus varied with precession. In 1394 one *isbaʿ* equalled 1°56′ but in 1550, 1°33′. The full circle of 360° corresponded alternatively to 210 or 224 *isbaʿ*. *Isbaʿ* also measured $\frac{1}{24}$ of a cubit. For longitude estimates, one *isbaʿ* equalled eight *zām*, each *zām* corresponding to the distance covered in three hours of sailing. A variety of other measurements, including something approximating triangulation, were calculated in these units. The altitude of the Polaris was supposed to be taken at its inferior elevation.

The instruments used for measurements included *kamāl*, *lawh*, and *bilistī*. The *kamāl* was a rectangle of horn or wood with a string through the middle. It was held against the horizon in an outstretched hand, with the cord held in the teeth by the knot. Knots tied at certain intervals on the cord corresponded to the varying arcs covered by the rectangle. Variations of this instrument included knots tied at intervals corresponding to locations on a set route, a set of boards *lawh* corresponding to different arc values fixed on the cord, and the cord being held to the nose. The *bilistī* was a later version of the *kamāl*, with a rod replacing the cord, and four sliders of different sizes; most likely it postdates the Portuguese arrival because its function is essentially that of the *balhestilha*. By the nineteenth century the system and the instruments had been largely driven out of use or forgotten, although the name and expertise of Aḥmad ibn Mājid were still respectifully remembered.

M.A. TOLMACHEVA

REFERENCES

Arunachalam, B. "The Haven-Finding Art in Indian Navigational Traditions and Cartography." In *The Indian Ocean: Explorations in History, Commerce, and Politics*. Ed. Satish Chandra. New Delhi: Sage Publications, 1987, pp. 191–221.

Chandra, Satish, ed. *The Indian Ocean: Explorations in History, Commerce and Politics*. New Delhi and London: Sage Publications, 1987.

Chaudhuri, K.N. *Trade and Civilization in the Indian Ocean: An Economic History from the Rise of Islam to 1750*. Cambridge: Cambridge University Press, 1985.

Clark, Alfred. "Medieval Arab Navigation on the Indian Ocean: Latitude Determinations." *Journal of the American Oriental Society* 113 (3): 360–73, 1993.

Ferrand, Gabriel. *Instructions Nautiques et Routiers Arabes et Portugais*. 3 vols. Paris: P. Guetner, 1921–28.

Hall, Kenneth R. *Maritime Trade and State Development in Early Southeast Asia*. Honolulu: University of Hawaii Press, 1985.

Hourani, George Fadlo. *Arab Seafaring in the Indian Ocean in Ancient and Early Medieval Times*. Princeton, New Jersey: Princeton University Press, 1951.

Hsu, Mei-ling. "Chinese Marine Cartography: Sea Charts of Pre-Modern China." *Imago Mundi* 40: 96–112, 1988.

Schwartzberg, Joseph E. "Nautical Maps." In *The History of Cartography*, vol. 2, book 1 *Cartography in the Traditional Islamic and South Asian Societies*. Ed. J. B. Harley and David Woodward. Chicago: University of Chicago Press, 1992, pp. 494–503.

Serjeant, R. B. "Star-calendars and an Almanac from South-west Arabia." *Anthropos* 49: 478–502, 1954.

Severin, Timothy. *The Sindbad Voyage*. London: Hutchinson, 1982.

Sidi Çelebi. "Extracts from the Mohit, that is the Ocean, a Turkish Work on Navigation in the Indian Seas, translated by J. Hammer-Purgstall." *Journal of the Asiatic Society of Bengal*, 1834: 545–53, 1836: 441–68, 1839: 805–12, 1838: 767–80, 1839: 823–30.

Tibbets, Gerald R. *Arab Navigation in the Indian Ocean before the Coming of the Portuguese*. London: Luzac for the Royal Asiatic Society of Great Britain and Ireland, 1971.

Tibbets, Gerald R. "The Role of Charts in Islamic Navigation in the Indian Ocean." In *The History of Cartography*, Vol. 2, Book 1 *Cartography in the Traditional Islamic and South Asian Societies*. Eds. J. B. Harley and David Woodward. Chicago: University of Chicago Press, 1992, pp. 256–62.

Tolmacheva, Marina. "On the Arab System of Nautical Orientation." *Arabica* 27 (2): 180–192, 1980.

Varadarajan, Lotika. "Traditions of Indigenous Navigation in Gujarat." *South Asia: Journal of South Asian Studies*, n.s., 3 (1): 28–35, 1980.

Villiers, Alan J. *Monsoon Seas: the Story of the Indian Ocean*. London and New York: McGraw-Hill, 1952.

Villiers, Alan J. *Sons of Sindbad: an Account of Sailing with the Arabs*. London: Hodder and Stroughton, 1940.

See also: Ibn Mājid – al-Masʿūdī – Ibn Baṭṭūṭa – Pirī Reis – Precession of the Equinoxes – Geography in the Islamic World – Maps in the Islamic World

NAVIGATION IN THE PACIFIC Three thousand years ago, Polynesian voyagers had already reached the small Pacific island chains of Tonga and Samoa. A thousand years later, following, as some scholars believe, the refinement of navigational capabilities and canoe design, they reached the far distant Marquesas, Tuamotus, and Society Islands. Geoffrey Irwin maintains on the basis of archaeological evidence and empirical analysis that there probably was no pause in the prehistoric settlement of eastern Polynesia. The rate of voyaging may very well have increased in accordance with the larger ocean gaps between archipelagoes. Voyagers eventually reached the small and very isolated Easter Island, may very well have established voyaging routes between the Society Islands and Hawaii, and may have even reached the coast of South America as is suggested, among other things, by the transmission of the sweet potato.

The questions of when the Pacific was settled are not as variable and questionable as those about how it was settled. The challenge of this issue, at least in terms of distance and the diminishing target areas of land as one moves east, becomes increasingly great as one can see by looking at a map of the Pacific Ocean. Several scholars consider all of these movements to have been the consequence of population demands, warfare, and the basic desire to explore and keep seeking. But the notion of great navigators who controlled their destiny and the destinies of those who put their trust in them, and who could ride the sea at will, prompted many articles and treatises. It also provided the easy and attractive answer to the haunting questions concerning the ancient settlement of Oceania.

S. Percy Smith, who wrote mostly between 1890 and 1920, promoted a migration of deliberate and controlled dimensions which he believed was most evident on the basis of Polynesian traditions and the inferences that one could make from them. He acknowledged the importance of applying the fields of ethnology, historical linguistics, and especially oral traditions to his theories on the ancient Aryan origins of the Polynesians. These alleged achievements included a voyage from Borneo to Hawaii in about AD 450 (the fact that voyagers continued on to Hawaii even after reaching hospitable islands along the way indicated to Smith that they were following the directions of a previous navigator), the reaching of Antarctic waters by two deliberate voyagers described in traditional histories, and the frequent voyaging between Tahiti and Hawaii (using the Line Islands as resting points) which eventually ceased in the fourteenth century, because the "boldest navigators of the race" had departed to New Zealand in the fourteenth century Great Fleet of the Maori people.

Elsdon Best, writing during the same time, also endowed navigators and voyagers in the Pacific with a perceptiveness of the environment and an adherence to the messages of traditions which made well planned, deliberate voyagers of settlement and communication a reality among the far flung islands of Oceania. While numerous early European explorers in the Pacific expressed wonderment in their journals at the existence of complex and populous societies on both low and high islands, which were at one stage thought to be remnants of a sunken continents occupied by survivors, very little was ever recorded on the actual means by which indigenous navigators dealt with practical problems related to the art of oceanic path-finding.

Peter H. Buck, during the next thirty years, considered more practical questions on canoe capabilities and types, and the evidence that ethnology in Polynesia suggested in relation to early Polynesian migration. He also asserted the probable movements and routes used by Polynesians to reach islands throughout Polynesia and the eastern regions of Micronesia and Melanesia. He recognized the relevance of Polynesian traditions and related genealogies for tracing Polynesian origins and for understanding early voyaging and settlement in the Pacific, with an emphasis on voyages between the Society Islands and New Zealand.

But these speculations and the assignment of great skills

to navigators who then sailed the Pacific with complete control, denied or superficially addressed issues. They thus failed to consider the diverse problems of both natural and human origin which confront a navigator without modern instruments on an open ocean. Were there explicit strategies followed for the initial exploration of islands to the east of the easier 'voyaging corridor' from Southeast Asia to Melanesia? If they were deliberate voyages with the intention of settlement, were such initial voyagers and those that were to follow one-way or two-way voyages? If two-way voyages were accomplished, were they done so regularly or perhaps even for the purposes of inter island communication? What might have been the distance limit for such intentional voyages of return? How did changes in climate and stellar positions over a long period of time affect the success rate of such voyages? Why did voyaging in some areas apparently cease before it did in other areas? Then, there is the question which has generated a significant amount of literature on the overall issue of oceanic voyaging and settlement. This involves numerous practical problems of reaching land, including the use of land indicating signs such as the flight of birds, the swell of waves generated from an unseen island or islands, the 'loom' of land produced from an island's white sand and a still lagoon reflecting the sun's glare, the ability to determine set and drift en route, the ability to steer and maintain a course from a known sequence of stars, and the ability to maintain or to eventually recover such a course when stars are obscured or the winds change.

Was it possible that early voyagers in the Pacific were capable of using island 'blocks' and 'screens' that are created by the ability to perceive and interpret these wave patterns emanating from unseen islands, the flight of certain sea birds (or 'bird zones'), types of fish, seaweed, water color, and clouds gathering over a distant island? These elements may have offered a thirty to forty mile radius around each island which essentially expanded the size of an island target. Such perceived circles ultimately overlap throughout significant areas of the Pacific and could have enabled navigators to sail for an archipelago from even great distances before seeking a specific island within the archipelago.

Voyaging researcher David Lewis in his numerous voyages with present-day indigenous navigators of the Pacific has also recorded the glowing plaques of bioluminesence that can indicate the existence of land eighty to one hundred miles away. But the elements of darkness and overcast skies have often been used by critics to maintain that many components of these blocks were not consistent enough to be dependable.

The academic granting of voyaging prowess to indigenous voyagers of the Pacific by these early writers and the lack of precise information on the practicality and effective means of using these natural elements to find land (not to mention the formidable question of how a course could be maintained over long distances before arriving at these island blocks), was underscored by Andrew Sharp's books and articles between 1956 and 1969 which dismissed the ability of Polynesians to exercise any significant control over their voyages. Sharp maintained that these voyagers were incapable of compensating for set and drift during the course of a voyage, of maintaining a course in accordance with star paths or movements when the night sky was obscured, and of being unable to comprehend their position and the position of their destination once they were driven off course by a gale or storm. According to Sharp, Polynesian traditions with names of distant islands and directions for reaching them were actually derived from European contact. Polynesians knew of distant archipelagoes. The Raiatean navigator Tupa'ia expressed this knowledge to Captain James Cook in 1769 whose assistants in turn created a map (apparently misunderstanding Tupa'ia's directions for north and south) that stretched from Fiji to the Marquesas. This merely demonstrated that Polynesians had traveled these distances but not necessarily as a result of deliberate and return voyages. Sharp also stressed that the apparent random distribution of several important plants and animals is simply reflective of the haphazard nature of Pacific settlement. While the accidental settlement of Polynesia, according to Sharp, did occur primarily from west to east (a direction which is generally agreed upon by Pacific scholars), winds and currents drove later voyagers in all directions. Although Sharp did allow for the ability of Polynesians (particularly in Tonga and the Tuamotu Archipelago) to undertake voyages of no more than three hundred miles without intervening islands with the intention of marking a return voyage home, the severe limitations of a noninstrumental navigational system nevertheless made even these voyages open to chance. The use of this "primitive navigation" by Pacific voyagers made them unable to protect themselves against any lateral displacement and did not enable them to derive much consistent and valuable information from horizon stars or to use stars to establish either latitude or longitude.

Thor Heyerdahl meanwhile challenged the general acceptance of an oceanic settlement pattern from west to east maintaining that archaeological, botanical, ethnological, biological, geographical, and navigational evidence supported a Peruvian movement from South America into Polynesia using balsa rafts with strategically arranged guara hardwood centerboards. He maintained that Spanish expeditions into Polynesia actually followed sea routes established by the Incas and often referred to traditional accounts of the voyage Yupaanqui Inca Tupacor Tupac Inca. He was said to have departed from Peru in search of legendary Pacific islands using the traditionally known island of Sala-y-Gómez as the primary landmark for beginning a voyage to the west, and to return nine to twelve months later with 'dark' people, which suggests that he may have voyaged as far west as Melanesia. In an attempt to prove that such voyages were possible, Heyerdahl and a crew sailed the Kon Tiki balsa raft from Peru to an eventual landfall on the atoll of Raroia in the Tuamotus in 1947. Heyerdahl, however, has been criticized by various scholars for, among other things, selectively neglecting evidence which strongly suggests a Southeast Asian origin of the Polynesians, his contention that "primitive asiatic craft" could not have successfully sailed from Southeast Asia to Polynesia, and for undermining or ignoring the fact that seasonal shifts in the prevailing winds of the Pacific do occur, making well planned voyages from west to east possible.

Sharp's works instilled what researcher Ben Finney characterizes as a "healthy corrective to some extravagant claims." The inferring of the abilities or inabilities of indigenous voyagers to control their destinies at sea on the basis of poor references in the journals of European explorers, simply produced theories which had little practical, at-sea evidence to support them. Ben Finney's and David Lewis' investigations into the navigational abilities of indigenous voyagers in the Pacific have made significant contributions to scholarship in this area.

Finney was and has been involved for a number of years in the building and sailing of the Hawaiian double canoe Hōkūle'a. Beginning with its first, non-instrumental voyage from Hawaii to Tahiti in 1976, it has provided an important component to Hawaiian cultural revival in the immediacy of its connections with the voyaging heritage of the Hawaiian people. It has also offered numerous opportunities for navigators (first the Satawalese navigator Mau Piailug and more recently the Hawaiian navigator Nainoa Thompson) to test and develop their use of various non-instrumental, navigational strategies to deal effectively with the voyaging problems of maintaining an accurate course under various natural conditions. This has also given other scholars the opportunity to apply these voyaging experiences to important questions related to the early settlement of Oceania, particularly the question of the movement from west to east which is normally a movement against prevailing winds. The Hōkūle'a 1985–1987 "Voyage of Rediscovery" in the Pacific engaged in experiments in west to east movements (in this case, from Samoa to Rarotonga via Aitutaki and from Rarotonga to Tahiti) to attempt to understand how voyagers effectively exploited periodic westerlies to cross the crucial gap between the Western Polynesia and Eastern Polynesia cultural provinces via the Samoa–Tahiti expanse. These methods for exploiting changes in the trade winds, often interrupted by low pressure systems, if they were undertaken by skilled navigators, would necessarily have required strategies for perhaps specific legs between western and eastern Polynesian islands.

Some items out of the rather voluminous literature on indigenous non-instrumental navigation and voyaging in the Pacific address the impact that climatic changes between the warming period of the Little Climatic Optimum (up to the twelfth and thirteenth centuries) and the advent of the Little Ice Age (approximately the fifteenth century) could have had on the occurrence of storms and thus the ability of voyagers to engage in long distant trips of exploration and settlement. Some scholars theorize that although voyagers from Eastern Polynesia were able to reach and settle in New Zealand by the early fourteenth century, this climatic change may have ultimately prevented them from going further to reach Australia and Tasmania. Some scholars have also conjectured on the impact that changing star patterns over several millennia may have on present-day investigations into this great achievement of human migration. At least one scholar has investigated how this change may provide a historical link between Arab and current Carolinian navigational star systems (Halpern, 1986).

It should be emphasized, however, that all this scholarship on indigenous navigational abilities in relation to the chronology and patterns of human settlement on Pacific islands is incomplete without incorporating the findings and theories of other areas of scholarship, particularly archaeology. While not necessarily being directly concerned with navigational capabilities, they often provide solid physical evidence of the past, from which theories on such migration and voyaging achievements can be made and developed and/or challenged. The part that archaeology plays in attempting to answer many of these questions cannot be understated, although the extensive work that has been done in the field and remains to be done cannot be given adequate summary in this article.

The literature on indigenous voyaging in the Pacific has also covered the use and possible use of several indigenously devised navigational aids — the best known of which are the Marshall Islands stick charts which a German writer, A. Schück, attempted to describe in 1887, noting European references to the existence of Marshallese maps as early as 1817. Europeans in the late nineteenth century noted the use of these charts which are basically intended, through unique connections of various sticks of wood and shells, to document wave patterns and swells in relation to specific islands in the two archipelagoes of Ratak and Ralik in The Marshall Islands. Several articles refer to the secrecy that was once associated with them, while Hops noted that only the Marshall Islands may have the natural prerequisites which make the existence of stick charts unique to them. Another device that has been explored in the literature has been the sacred calabash filled with water whose holes near its top allegedly enabled a navigator to establish the latitude of Hawaii. This was accomplished by lining up Polaris at a nineteen degree angle during a voyage from the south, and thus allowed a navigator to turn confidently to the west where he would eventually find Hawaii. Although confusion over a large calabash in the Bishop Museum in Hawaii added to the controversy, the use of the sacred calabash has generally been dismissed as being impractical. Others include the existence of nine coral stones on the northwestern end of Arorae atoll in Kiribati, perhaps used at one time by voyagers to set their courses to neighboring islands in Kiribati, a stone canoe in Beru (*kiribati*) for instructional purposes, a coconut wind compass with four directional holes used by Tahitian voyagers, the use of dials with wood hanging from strings with knots to indicate the position of stars, a bow hole and cord used for star course steering, the seeking of divination before a voyage through the tying of knots in palm leaves, and a cane filled with water to determine latitude and perhaps used in the Caroline Islands.

Research on the cognitive systems of indigenous navigators who still undertake voyages in the Pacific, particularly from the atolls of Satawal and Polowat in the central Caroline Islands of Micronesia, is far more extensive than these occasional references (with the exception of Marshallese stick charts) to these unique devises. Gladwin's groundbreaking study on thought processes involved in ixPolowatese navigational knowledge which he related to the academic test performance of economically underprivileged individuals in the United States, has been followed by numerous works concerned with the canoes, navigators, and navigational systems of Polowat and Satawal. There are also ethnographic concerns about cultural and social elements which enable non-instrumental indigenous navigation on these islands to survive in a modern technological world. While Western material influence has contributed to the decline of indigenous voyaging in areas where it was pursued until relatively recently, as has the actual government banning of voyaging, traditional navigation on these two atolls forms an integral part of their identity and society.

One well documented navigation technique found in the literature which discusses these esoteric systems is the technique of *etak*. The system of etak defies Western conceptualizations of movement in that a reference (sometimes imaginary island) to the side is seen as moving toward and past the voyager's canoe while the canoe remains stationary. The navigator is subsequently able to determine his position and progress in relation to stars and the etak segments that they provide. The survival of these systems of course depends upon how successfully they are transmitted to the young. The maintenance of star courses through songs and chants allowed for these courses to be used in recent six hundred mile voyages between Satawal/Polowat and Saipan in the Northern Mariana Islands, despite the fact that such voyages had apparently not been undertaken since the early part of the twentieth century.

We now know of this rich voyaging heritage which many Pacific islanders value in the context of their cultures today. And we are aware of the questions that relate to what must be seen as an astonishing achievement in human movement throughout a vast Pacific Ocean at a time when Europeans were still depending (and would continue to depend for a long time) upon the sight of a coast. It is little wonder why the literature on indigenous navigation voyaging in the Pacific has developed as it has since the latter portion of the nineteenth century.

NICHOLAS J. GOETZFRIDT

REFERENCES

Best, Elsdon. "Maori Voyagers and Their Vessels: How the Maori Explored the Pacific Ocean, and Laid Down the Sea Road for All Time." *Transactions of the Royal Society of New Zealand.* 48: 447–63, 1915.

Buck, Peter H. "The Value of Traditions in Polynesian Research." *Journal of the Polynesian Society* 35: 181–203, 1926.

Buck, Peter H. *Vikings of the Sunrise*. Philadelphia: J.B. Lippincott, 1938.

Finney, Ben. "A Voyage into Hawaii's Past." *Polynesian Voyaging Society*: 1–13, 1975.

Finney, Ben. *Hokule'a: The Way to Tahiti*. New York: Dodd, Mead & Company, 1979.

Finney, Ben. "Voyaging Against the Direction of the Trades: A Report on an Experimental Canoe Voyage from Samoa to Tahiti." *American Anthropologist* 90: 401–405, 1988.

Gladwin, Thomas. *East is a Big Bird: Navigation and Logic on Puluwat Atoll*. Cambridge, Massachusetts: Harvard University Press, 1970.

Goetzfridt, Nicholas J. *Indigenous Navigation and Voyaging in the Pacific: A Reference Guide*. New York: Greenwood Press, 1992.

Haddon, A.C. and James Hornell. *Canoes of Oceania*. Honolulu: Bishop Museum Press, 1975.

Halpern, Michael. "Sidereal Compasses: A Case for Carolinian-Arab Links." *Journal of the Polynesian Society* 95: 441–460, 1986.

Heyerdahl, Thor. *American Indians in the Pacific: The Theory Behind the Kon-Tiki Expedition*. London: Allen & Unwin, 1952.

Hops, A. "Über die Einmaligkeit der Marshall Stabkarten im Stillen Ozean." *Zeitschrift für Anthropologie, Ethnologie und Urgeschichte* 81: 104–110, 1956.

Irwin, Geoffrey. *The Prehistoric Exploration and Colonisation of the Pacific*. Cambridge: Cambridge University Press, 1992.

Lewis, David. *We, the Navigators: The Ancient Art of Landfinding in the Pacific*. Honolulu: The University Press of Hawaii, 1972.

Schück, A. Die Entwickelung Unseres Bekanntwerdens mit den Astronomischen, Geographischen und Nautischen Kenntnissen der Karolineninsulaner nebst Erklärung der Medos oder Segelkarten der MarshallInsulaner, im Westlichen Grossen Nord-Ocean." *Tijdschrift van het Koninklijke Nederlandsch Aardrijkskundig Genootschap te Amsterdam* 1(2): 226–251, 1887.

Sharp, Andrew. *Ancient Voyagers in the Pacific*. Harmondsworth, U.K.: Penguin Books, 1957.

Sharp, Andrew. *Ancient Voyagers in Polynesia*. Berkeley: University of California Press, 1964.

Smith, S. Percy. *Hawaiki: The Original Home of the Maori*. London: Whitcombe and Tombs, 1910.

Smith, S. Percy. *The Lore of the Whare-wananga or Teachings of the Maori College on Their History and Migrations*, etc. Part II, Te Kauwae-raro or 'Things Terrestrial. New Plymouth, New Zealand: Printed for the Polynesian Society by Thomas Avery, 1915.

See also: Maps and Mapmaking: Marshall Islands Stick Charts

NAVIGATION IN POLYNESIA Polynesian dispersal through the Pacific islands is one of the great maritime accomplishments of human history. In wooden boats held together with cord fashioned from coconut husks, and following the stars, sun, waves, and wind, they populated the vast triangle bounded by Hawaii, Rapanui (Easter Island), and Aotearoa (New Zealand).

The Polynesians' ancestors migrated from Island Southeast Asia to the Bismarck Archipelago in northeastern Papua New Guinea by 2000 BC, where they left their distinctive Lapita pottery. Around 1500 BC, they moved out into Fiji, and then to Tonga and Samoa. During the next two millennia they spread throughout the area now known as Polynesia. Later, a series of back-migrations populated western Polynesian "outliers" scattered through the regions commonly identified as Melanesia and Micronesia.

Some of these voyages were likely purposeful exploratory forays. Others were forced migrations resulting from population pressure or political and military conflict. Some were accidental drift voyages by sailors who had lost their way, or by fishermen swept out to sea. At each stage, return voyages probably were made for purposes of trade, marriage, and exchange of information.

Thinking about Polynesian seafaring abilities has shifted several times since early European contact. The French explorer Bougainville was so impressed by the canoes he witnessed in Samoa in 1768 that he called the archipelago "The Navigators' Islands". Britain's Captain James Cook noted that Polynesian canoes were often as fast and maneuverable as his ships, and he carefully recorded Raiatean chief Tupaia's impressive geographical knowledge. Even as late as the middle twentieth century, Sir Peter Buck (Te Rangi Hiroa), director of Hawaii's Bishop Museum, took Polynesian traditions of voyaging, exploration, migration, and settlement quite literally.

The tone of the discussion altered radically with the publication in 1957 of Andrew Sharp's *Ancient Voyagers in the Pacific*. Sharp argued that, in the absence of instruments, human beings lacked the ability to navigate successfully for more than a few hundred miles. In particular, he argued, one could not detect, accurately estimate, and correct for current or leeway drift. Over distances of the many hundreds (sometimes thousands) of miles claimed by Polynesian oral traditions, he argued, errors would have been compounded, making the prospect of intentional landfall almost nil. Therefore, exploration and settlement were almost certainly the result of accidental drift voyages by sailors blown off course or forced, against their will, to put to sea.

This view again shifted with the pioneering ethnographic work of Thomas Gladwin (1970) on the Micronesian atoll of Puluwat and David Lewis' more wide-ranging survey in 1972. They and others have documented voyages of hundreds of miles with accuracy comparable to that which can be achieved with modern instruments, as well as the techniques through which this is accomplished. Most of the active voyaging in recent times has been conducted by Micronesians, but the techniques are similar to those employed by Polynesians in the past. This is confirmed by comments from Polynesians recorded by early European explorers, and those few Polynesian communities where traditional wayfinding techniques have been retained in recent times.

A second line of evidence casting doubt on Sharp's drift voyage argument is provided by the direction of settlement. Linguistic and archaeological data make it clear that migration was predominantly west to east, against the prevailing winds. Recorded drift voyages, by contrast, are almost always in the opposite direction. Such voyages have certainly occurred, and more than a few sailors have been lost at sea. This however, cannot account for exploration and settlement of the Pacific.

A third strand of evidence comes from computer simulations. Importantly, in 1973, Levinson, Ward, and Webb demonstrated that Oceanic exploration and settlement could not have been a result of accidental drift, but must have resulted primarily from purposeful voyaging by skilled seafarers and navigators. They did this by taking into account prevailing winds and currents, and comparing them with the geographical relationships of various islands and archipelagoes.

The final line of evidence supporting the proficiency of Oceanic navigators comes from experimental voyages with replicas of traditional voyaging canoes, using traditional navigational techniques. During the 1980s, a group of Taumako (Duff) Islanders from the eastern Solomon Islands built a traditional voyaging canoe (*te puke*), which they successfully sailed to Guadalcanal, over 400 miles away.

Still more impressive are the exploits of the *Hōkūle'a*, a replica Hawaiian double-hulled voyaging canoe, built with modern materials but to traditional design, which has been sailed successfully through most of Polynesia. Its first major voyage was to Tahiti, following traditionally prescribed star paths, and performed under the guidance of Mau Piailug, a master navigator from Satawal Island in Micronesia. Later on, Nainoa Thompson, a Hawaiian trained largely by Piailug, navigated the *Hōkūle'a* without instruments through the major Polynesian archipelagoes, as far away as New Zealand. The next step is to build a voyaging canoe of traditional materials and repeat the feat in such a vessel. That project is now well underway.

Canoe Design

To travel safely on the vast Pacific Ocean requires seaworthy vessels. These had to be sturdy, maneuverable, able to resist leeway drift, and (to some degree) sail into the wind. For lengthy journeys, they had to be large enough to carry a good quantity of food, drink, and sometimes crops and livestock. This was accomplished differently in different parts of Oceania.

Most common were single-outrigger canoes ranging from small vessels paddled by one or two men (or occasionally women) for coastal travel and inshore fishing to the great Fijian *camakau* which might approach 100 feet in length and rival European sailing ships for speed. Resembling the *camakau* were voyaging canoes of several Polynesian outliers: e.g. Takuu (*vaka fai laa*), Nukumanu (*vaka hai laa*); and the Polynesian islands of the Santa Cruz group in the eastern Solomons (*puke*). Outlier Polynesians on Anuta still make single-outrigger sailing canoes which they have been known to sail hundreds of miles — to Vanikoro, northern Vanuatu, and recently to Santa Ana Island off Makira. In the outlier atolls, such as Nukumanu and Takuu, islanders have adapted the Fijian and Micronesian design of interchangeable bow and stern, where the canoe is tacked by moving the sail to the opposite end of the vessel, thereby always keeping the outrigger upwind. On other islands, such as Tikopia and Anuta, the more common Polynesian pattern was adopted. Bow and stern are distinct; the outrigger is always kept to the same side (most often, to port); and the canoe is sometimes sailed with the outrigger to the lee side of the craft.

Most imposing of the Polynesian voyaging canoes undoubtedly were the great double-hulled vessels of the Polynesian heartland. Such canoes have been reported from Rotuma and Tuvalu, south to Tonga and Aotearoa, and north to Hawaii. More specialized developments included the New Zealand Maori's impressive single-hulled, outriggerless canoes, designed to carry scores of warriors primarily in coastal waters. For an inventory of Pacific Islanders' canoe construction and design, Haddon and Hornell's encyclopedic treatment remains unparalleled.

Polynesian voyaging canoes could cover well over 100 miles a day with a favorable wind. The outrigger or double-hull design provides resistance to leeway drift without the need for a deep keel, which is a hazard sailing through a sea studded with shallow reefs. It is difficult, however, to beat into a head wind; and even the best outrigger canoes cannot sail better than a close reach or approximately 75° off the wind.

Navigational Techniques

The most important tool for Oceanic navigation is the stars. Near the equator, stars rise from the east, travel more or less straight overhead, and set toward the west. The expert navigator knows the relationship between the bearings of various islands and the rising and setting points of various stars. Ideally, he identifies a star that he knows rises directly over the target island and points the bow of the canoe toward that star. When the star rises too high to be a reliable guide, he turns his attention to another star following approximately the same trajectory. A sequence of such stars may well continue through the night, and it forms a star path.

In reality, it is unlikely for a sequence of stars to rise or set directly over the target island, so the navigator must set off the bow at an appropriate angle from the stars' actual position. In addition, the navigator must be able to estimate the direction and speed of wind, current, and leeway drift in determining a proper heading.

Wind strength and direction are not overly difficult to estimate as long as one can see the stars, moon, or sun; while leeway drift depends on the contours of the canoe's hull and heading in relation to the wind. It often can be gauged by the angle between the canoe and its wake.

Current is most difficult to estimate, but not impossible. First, a navigator knows the prevailing currents in the region that he plans to sail. These tend to be fairly constant at a particular time of year. The current's velocity is confirmed when the canoe is still near the point of embarkation, or when sailing over shallow reefs. The greatest danger is if currents change direction while one is in deep water, far from land. Such shifts are difficult to detect, but there are sometimes useful clues. Anutans, for example, say that when the current and wind are running in opposite directions, seas tend to be unusually steep and white-capped. Conversely, when wind and current coincide, the seas are exceptionally flat in relation to the strength of the wind. Thus, by feeling wind strength and seeing or feeling the condition of the sea, one often discerns the current's direction even in deep water.

Fixing one's position on the ocean is another major problem. Latitude can be gauged with some accuracy by identifying zenith stars: the point that is directly above the observer. In addition, the configuration of stars at a given latitude at the same time on any given night is identical regardless of longitude. Therefore, one can estimate latitude by observing the altitude at which specific stars rise and set. Longitude, however, cannot be gauged from observation of the sky and must be estimated by dead reckoning. This means that the navigator fixes the position of his canoe in relation to the islands in his universe by estimating the vessel's speed, direction, and how long it has been at sea.

This is difficult, requiring the navigator to keep a running total of his vessel's progress at all times. One consequence is that he cannot sleep, other than occasional brief naps, even on a lengthy journey. Sharp argued that the element of human error inherent in dead reckoning is cumulative and, over a distance of many hundreds of miles, would be too great to permit accurate landfall. Empirically, it has been demonstrated, however, that the error is random and tends to cancel itself out. Therefore, the longer the voyage, the smaller the probable error.

During the day, when stars are unavailable, Polynesian navigators use the sun. This is practical in early morning and late afternoon, when the sun is low on the horizon. Toward mid-day, however, it loses value as a navigational aid. The sky also loses its navigational value when it is obscured by clouds, whether in the day or night. Under such conditions, auxiliary techniques are needed.

Navigators are aware of the prevailing wind patterns. If the sky is overcast, they assume that winds will remain constant until the stars and sun can once again be seen.

Still more stable and reliable are swells. A regular ground swell is produced by winds from far away, blowing over a vast expanse of ocean. Because of this characteristic, it stays constant despite shifts in local wind direction. An expert navigator can distinguish the regular ground swell from choppier seas produced by local winds. Indeed, some Micronesians have demonstrated the ability to distinguish several swells caused simultaneously by different far-off wind patterns. Polynesian master navigators almost certainly could do the same. By holding a steady course in relation to the swell pattern, then, the navigator maintains his bearings until celestial bodies become visible again.

In addition, sea marks such as deep reefs that can be detected from the color of the water may help indicate one's route. Use of such marks has been reported ethnographically for Micronesia, and at least some Polynesian navigators undoubtedly employed similar techniques.

Sailing strategy, particularly for exploratory voyagers, was clearly to sail counter to prevailing winds. This was accomplished by waiting for a westerly shift at a time when the prevailing winds were from the east. If landfall was not made in a reasonable amount of time, it was expected that the wind would shift again, making for an easy return home. This accounts for the generally west to east direction of Polynesian migration.

Likewise, in historic times, Anutans like to sail at the beginning and end of the trade wind season, when breezes are relatively light and unstable, but the prevailing wind is from the southeast. During this period, they can usually find a period of a few days when wind of moderate strength is blowing in the preferred direction. Should it shift before they reach their destination, they can almost always return to their point of departure.

Geoffrey Irwin further suggests, plausibly although without definitive evidence, that long-distance voyagers employed "latitude sailing". This involves sailing to the latitude of the target island but at a point well to windward, then making landfall by running downwind. Sailors from all parts of the Pacific also minimized their risk by island-hopping, thereby breaking lengthy journeys into shorter segments. And they aimed, wherever possible, for groups of islands rather than isolated targets.

Land-finding

As the navigator approaches land, he uses a variety of techniques for homing in on his objective. These include attempting to make landfall during daylight to minimize the chance of either being swept onto an exposed reef or overrunning the target. Then, before land is visible, it may be located via one of several indicators.

Among the most important of these is the flight patterns of birds that roost on land at night and fly to sea to feed during the day. Reflected waves are shaped differently from waves produced directly by the wind, and they can, at times, be felt at distances of more than twenty miles from land. Refraction patterns are more difficult to decipher but are used by a few peoples. Clouds tend to accumulate around the peaks of high islands, indicating their presence; and a greenish tint to clouds can sometimes indicate the presence of an atoll when it still cannot be seen from a canoe.

Polynesia in comparison with other culture areas

A few writers have made much of the distinction between Polynesia and other Oceanic culture areas in terms of seafaring acumen. Edward Dodd, for example, has suggested that the comparatively short distances traversed by Melanesians and even Micronesians means their vessels and wayfinding techniques could be less sophisticated than those of Polynesians, who sometimes had to sail thousands of miles, e.g. on journeys from Tahiti to Hawaii or Aotearoa.

Clearly, the logistics of such lengthy voyages posed problems not faced by Micronesians sailing from the Carolines to Saipan — a distance of about five hundred miles. On such journeys, for example, the importance of dead reckoning increases and a smaller proportion of the voyage is guided by land-finding techniques such as reflected waves or flight patterns of birds. However, modern ethnographic evidence from around the Pacific and beyond supports the view that navigational techniques were fundamentally similar throughout Oceania. The transferability of old skills to new locations, and their ability to meet new challenges, is demonstrated by the success of Mau Piailug and Nainoa Thompson, piloting the *Hōkūle'a* through major portions of the Polynesian Triangle.

RICHARD FEINBERG

REFERENCES

Alkire, William H. *Lamotrek Atoll and Inter-island Socio-economic Ties*. Urbana: University of Illinois Press, 1965.

Buck, Sir Peter (Te Rangi Hiroa). *Vikings of the Sunrise*. New York: J. B. Lippincott, 1938.

Dodd, Edward. *The Island World of Polynesia*. Putney, Vermont: Windmill Hill Press, 1990.

Feinberg, Richard. *Polynesian Seafaring and Navigation: Ocean Travel in Anutan Culture and Society*. Kent, Ohio: Kent State University Press, 1988.

Feinberg, Richard, ed. *Seafaring in the Contemporary Pacific Islands: Studies in Continuity and Change*. DeKalb, IL: Northern Illinois University Press, 1995.

Feinberg, Richard. "Continuity and Change in Nukumanu Maritime Technology and Practice." In *Continuity and Change in Modern Seafaring*. Ed. Richard Feinberg, n.d.

Finney, Ben R. *Hōkūle'a: The Way to Tahiti*. New York: Dodd, Mead, 1979.

Finney, Ben R. "Anomalous Westerlies, El Niño, and the Colonization of Polynesia." *American Anthropologist* 87: 9–26, 1985.

Finney, Ben R., P. Frost, Richard Rhodes, and Nainoa Thompson. "Wait for the West Wind." *Journal of the Polynesian Society* 98: 261–302, 1989.

Gladwin, Thomas. *East is a Big Bird*. Cambridge, Massachusetts: Harvard University Press, 1970.

Haddon, A. C. and James Hornell. *Canoes of Oceania*. Honolulu: Bishop Museum Press, 1975. (Originally published as three volumes in 1936, 1937, and 1938).

Irwin, Geoffrey. *The Prehistoric Exploration and Colonisation of the Pacific*. Cambridge: Cambridge University Press, 1992.

Kyselka, Will. *An Ocean in Mind*. Honolulu: University of Hawaii Press, 1987.

Levinson, M., R. G. Ward, and J. W. Webb. *The Settlement of Polynesia: A Computer Simulation*. Minneapolis: University of Minnesota Press, 1973.

Lewis, David. *We, the Navigators*. Honolulu: University of Hawaii Press, 1972.

Sharp, Andrew. *Ancient Voyagers in the Pacific*. Harmondsworth, Middlesex: Penguin Books, 1957.

Thomas, Stephen D. *The Last Navigator*. New York: Henry Holt and Company, 1987.

AL-NAYRĪZĪ Born in Persia, al-Nayrīzī spent most of his life in Baghdad. He was a court astronomer/astrologer of caliph al-Muʿtaḍid (r. 892–902), to whom he dedicated several treatises, among them *Al-Risāla fī aḥdāth al-jaww* (Treatise on Meteorological Phenomena) or *Al-Risāla fī maʿrifat ālāt yuʿlamu bihā abʿād al-ashyāʾ al-shākhiṣa fī l-hawāʾ wal-latī ʿalā basīṭ al-arḍ wa-aghwār al-audīya wa 'l-abār wa-ʿurūḍ al-anhār* (Treatise on the Knowledge of Instruments through which Distances between Distinct Things in the Air or Set up on the Ground and the Depth of Valleys and Wells, and the Widths of Rivers can be known). He wrote commentaries on Ptolemy's (fl. ca. 127–167) *Almagest* and his astrological work *Tetrabiblos* and an astronomical handbook (*Zīj*) a longer and a shorter version. And he commented on Euclid's (fl. ca. 300 BC) *Elements* and his astronomical work *Phainomena*, and wrote independent works on the determination of the *qibla*, the direction of Mecca, the spherical astrolabe, methods for solving particular astronomical problems, and astrological subjects.

The longer version of his *Zīj* is said to have been based upon Indian astronomical tradition (*Sindhind*) and used data from the *Zīj* prepared for al-Maʾmūn (r. 813–833). This was criticized by Ibn Yūnus (d. 1009), who pointed out further differences in opinion, especially with respect to the theory of Mercury, the eclipse of the moon, and the parallax. His commentary on the *Almagest*, lost as are his handbooks and his commentary on the *Tetrabiblos*, was quoted by later authors like al-Bīrūnī or Ibn al-Haytham and even occasionally called the best work of this type.

His extant works on the *qibla* and the spherical astrolabe built on the works of earlier scholars are some of the very best summaries on the subject still available. Others of his extant works on astronomy, astrology, geodesy, and meteorology have not yet been seriously studied.

Al-Nayrīzī's commentary on the *Elements* translated by Gerard of Cremona (d. 1187) into Latin includes extracts of the lost Greek commentaries by Heron (second century?) and Simplikios (sixth century) and of Arabic treatises like Thābit Ibn Qurra's alternative proof of the Pythagorean theorem. He omitted Simplikios' own proof of the parallels postulate in favor of that by Aghānīs/Aghānyūs (Agapios?, fl. ca. 511), which is also preserved in his independently transmitted *Al-Risāla al-muṣādara al-mashhūra* (Treatise on the Proof of the Well-Known Postulate).

In Book V, al-Nayrīzī follows a theory of proportion adopted before him by al-Māhānī (d. ca. 880) and, perhaps, Thābit ibn Qurra, a theory based on definitions of ratio and proportion which compared the expansion of magnitudes in continued fractions.

The text of the *Elements* contained in the Arabic manuscripts of al-Nayrīzī's commentary was viewed for a long time as the second version made by al-Ḥajjāj ibn Yūsuf ibn Maṭar. Although it is derived from a text of the Ḥajjāj tradition, al-Nayrīzī evidently edited it using a text of the Isḥāq/Thābit tradition. He changed its language, didactical features, and even letter symbols used for diagrams. He incorporated references to earlier Euclidean theorems, definitions, postulates, or axioms as well as to propositions stated by the Greek commentators. In a similar manner, he edited those texts he added as comments.

SONJA BRENTJES

REFERENCES

Primary sources

Codex Leidensis 399,1. Euclidis Elementa ex interpretatione Al-Hadschdschadschii cum commentariis al-Narizii. Arabice et latine ediderunt R. O. Besthorn et J. L. Heiberg. Copenhagen: In Librarian Gyldendaliana, 1893, vols. 1 and 2. G. Junge, J., Raeder, W. Thomson. Copenhagen: 1932, vol. 3, pt. 2.

Schoy, Carl. "Abhandlung von al-Faḍl b. Ḥâtim an-Nairîzî: Über die Richtung der Qibla." *Sitzungsberichte der Bayerischen Akademie der Wissenschaften zu München. Mathematisch-physikalische Klasse*, 1922, pp. 55–68.

Seemann, H. "Das kugelförmige Astrolab nach den Mitteilungen von Alfons IX. von Kastilien und den vorhandenen arabischen Quellen." *Abhandlungen zur Geschichte der Naturwissenschaften und der Medizin* 7: 32–30, 1925.

Secondary sources

Dold-Samplonius, Yvonne. "al-Māhānī." In *Dictionary of Scientific Biography* vol. IX. Ed. Charles C. Gillispie. New York: Charles Scribner's Sons, 1974, pp. 21–22.

Jaouiche, Khalil. *La théorie des parallèles en pays d'Islam: contribution à la préhistoire des géometries non-euclidiennes.* Paris: J. Vrin, 1986.

Kennedy, E. S. "A Survey of Islamic Astronomical Tables." *Transactions of the American Philosophical Society*, n.s., 46 (2): 123–177, 1956.

Rosenfeld, Boris Abramovich. *A History of Non-Euclidean Geometry. Evolution of the Concept of a Geometric Space.* New York: Springer, 1988.

Sabra, A.S. "Simplicius's Proof of Euclid's Parallels Postulate." *Journal of the Warburg and Courtauld Institutes* 32:1–24, 1969.

Sabra, A.S. "al-Nayrīzī." In *Dictionary of Scientific Biography* vol. IX. Ed. Charles C. Gillispie. New York: Charles Scribner's Sons, 1974, pp. 5–7.

Sezgin, Fuat. *Geschichte des Arabischen Schrifttums.* Leiden: E. J. Brill, 1973, vol. V; 1974, vol. VI.

Youschkevitch, Adolph A. P. *Die Mathematik im Mittelalter.* Leipzig: Teubner Verlag, 1964.

See also: *Zīj* – Thābit ibn Qurra – *Almagest* – Ibn al-Haytham – al-Māhānī – Ibn Yūnus – al-Bīrūnī – *Qibla*

NAZCA LINES On the desert plains and the fringes of the valleys that form the heartland of the ancient Nasca culture of the south-central coast of Peru is a vast array of ground markings that are known collectively as the Nazca Lines. The most extensive concentration of these constructions, termed geoglyphs by archaeologists, is found on the Pampa de Nazca between the Ingenio and Nazca valleys. The pampa is an elevated desert plain 200 kilometers square, between 400 and 600 meters above sea level. It is severely dissected by dry watercourses formed by periodic flash floods. About 3.6 million square meters of this area are covered by geoglyphs. Lesser concentrations of markings also occur on the slopes of and on the plateaux and pampas between the other valleys of the Nazca drainage, particularly in the Palpa area.

Geoglyphs are found in many forms but can be grouped into four main classes: biomorphs, enclosures, lines, and geometric figures.

Biomorphs are stylized representations of various creatures, the commonest being birds among which are hummingbirds, the condor, pelican, and frigate bird. There are also depictions of a monkey, fox, lizard, spider, killer whale, and fish, as well as a few drawings of plants or flowers and bizarre combinations of body parts. The forms are outlined by a continuous line usually about 0.6 meter wide which never crosses and ends near where it starts. The greatest concentration of biomorphs occurs in the northwest corner of the Pampa de Nazca above the south edge of the Ingenio valley and close to a major regional center of the Nasca culture. Some biomorphs appear to be isolated drawings, but others, for example the monkey, seem to have a direct asso-

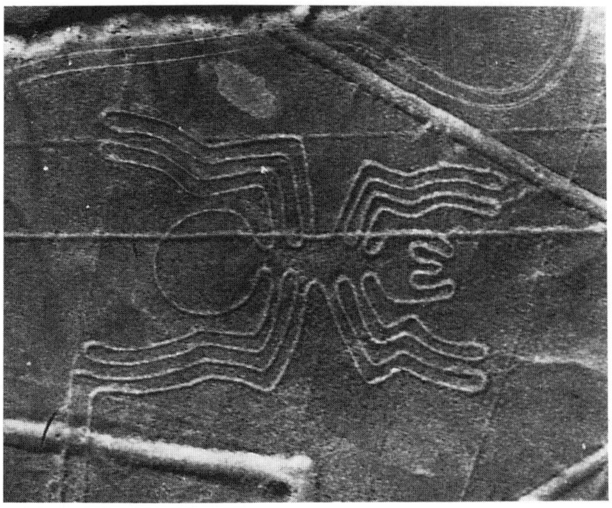

Figure 1: Spider biomorph with associated lines.

Figure 2: Trapezoid enclosure on intervalley ridge.

ciation with other geometric figures. Despite the impressive size of some drawings they are generally insignificant in scale when compared to many of the enclosures.

Enclosures are trapezoidal, triangular, quadrangular, and rectangular clearings defined by stone banks up to 1 meter high. They vary greatly in area and proportion, the largest occupying the space of several football fields. This class of geoglyph is the most frequently occurring form outside the Pampa de Nazca. Where they are built on the flat spurs between valleys they follow the long axis of the topography; on hillsides their narrow ends point towards the top of the hill; and where they are constructed in the bottom of a small, dry tributary valley their narrow ends point towards its head. About 300 are found across the Pampa itself, roughly two-thirds of which are oriented to the line of watercourses with their ends pointing upstream. Enclosures are commonly found in association with other forms of geoglyph such as line centers, spirals and zigzags.

Line centers consist of a series of straight lines converging on or radiating from usually a hillock or low mound on the periphery of the Pampa de Nazca. The lines vary in width and length, some running for several kilometers. Over ninety per cent of the geoglyphs found on the Pampa are associated with the sixty two line centers identified so far. However, only about a quarter of the centers are interconnected, suggesting that these features cannot be interpreted as a simple coherent system. Line center lines are frequently connected to enclosures.

Long lines, unconnected to line centers, running dead straight for several kilometers with no regard for topography are another feature of the Pampa and elsewhere, as are pairs of parallel straight lines.

The most prominent of the geometric forms found are spirals created by a single line along which it is possible to enter and exit in an unbroken progress. Other figures include zigzags and "fingers" consisting of a series of concentric semicircles formed by a single line continuously running parallel to itself in a set of arcs.

Contrary to some assertions, constructing the geoglyphs required no great engineering skill and no more complicated surveying equipment than simple poles and rope. The surface of the pampa is covered by rocks which are constantly subjected to daily extremes of temperature and humidity which leads through bacterial action to the formation of a characteristic dark-colored varnish on their upper parts. The underlying soil is lighter in color, and if the surface stones are removed a bright patch of ground is created, contrasting strongly with the darker, undisturbed desert floor. This contrast will persist for centuries, only gradually diminishing as the same weathering agencies act upon the newly-created surfaces. In most cases, the construction of a Nazca geoglyph involved nothing more than stripping the required width or area of its upper stones and regularly piling the rocks along the edge to form a boundary. Judging from some apparently incomplete examples, an intermediate stage in the construction of an enclosure was to collect the surface stones in a series of small piles on the surface of the intended clearing, probably in baskets rather than by sweeping, before removing them to the edges. It has been estimated that it would take ten people to clear 16,000 square meters in a week, and in a trial a spiral of ten meters diameter was built in less than two hours. The construction of the biomorphs probably required

Figure 3: General view of lines, Pampa de San José.

a preliminary model such as a sand drawing. The precise method whereby the final geoglyph depiction was scaled up from this model is unknown although several geometric systems have been suggested. Attempts to demonstrate the use of a standard unit of measurement in their formation have been unconvincing.

Several lines of evidence suggest that most geoglyphs were built by the peoples of the Nasca culture (ca. 200 BC–AD 600), although the practice may have been continued locally by later communities. Two radiocarbon dates from wooden posts supposedly associated with lines fall within the later Nasca period. Nasca pottery is so repetitively found on the surface of many enclosures as to indicate their construction and use during its currency. Radiocarbon dating of the organic material sealed under the varnish on rocks lifted by line builders indicates a similar date range. Additional support is given by the clear association of some geoglyphs with Nasca ceremonial architecture at the great pilgrimage center of Cahuachi and the lesser complex at Llipata.

Various explanations have been advanced for the Nazca Lines: that they were fields, the sites of textile workshops, a giant sports arena, and even markers and landing places for extraterrestrial spacecraft! The idea that they fulfilled primarily an astronomical function has been championed by the principal fieldworker on the Lines, Maria Reiche. In her view, the linear geoglyphs functioned to mark the solstices, the rising and setting of the moon, and the rise of the Pleiades. The multiplicity of some lines she explains as the attempts of ancient astronomers to compensate for the precession effect, the subtle shifts over time in star rising and setting positions. The biomorphs are regarded as representing constellations, for example the spider for Orion and the monkey for Ursa Major. The whole system was supposedly elaborated for calendric purposes, principally to predict the onset of the rains so essential to agriculture in the extremely arid Nazca environment.

Figure 4: Enclosures, lines, and geometric figures: Pampa de San José.

Reiche's claims for significant, repetitive astronomical alignments have not been supported by other studies. A more basic objection to her approach is that it projects on to the evidence a European, Northern Hemisphere perspective which ignores Andean peoples' astronomical perceptions. By using historical and ethnographic analogies it is possible to offer plausible explanations of the geoglyphs' forms and functions which make sense in the context of a general Andean religious tradition.

The location and form of the Lines leaves little doubt that they were built for ritual purposes. Many of the lines connected to line centers bear a general resemblance to Inca roads, and footpaths are still discernible within them, supporting the view that they were processional ways. The use of straight line paths has been shown to have been in widespread use up to the present day in Andean religious observances. The lines defining biomorphs and geometric figures are also laid out in a fashion which would allow uninterrupted procession through the figure. The core of the line centers and the enclosures were probably the foci of the processions where the principal rites and offerings were enacted. Depictions on Nasca pottery suggest that the participants in these rituals would have been specially attired, often wearing masks, and proceeded by rhythmic dancing to the accompaniment of drums and pipes. It is highly probable, given general Andean practice, that kinship groups (*ayllus*) in acts of communal labor would have been responsible for the construction and maintenance of particular lines and figures and for the organization of the relevant rites at a certain time of the year. The palimpsest of geoglyphs found can in part be explained by a practice of periodical renewal of the sacred site at a different location.

The basic concern of the Nasca peoples was the fertility of their valleys which was entirely dependent on the arrival of waters from the Andes to the east. It is not unlikely that peaks in these mountains were deified and perceived of as the sources of this life-force. The shape and orientation of many of the enclosures can be argued to be symbolic of the flow of water from the highlands. Similarly, certain geometric geoglyphs such as spirals and zigzags may have been symbols of a water cult.

The Nazca Lines were a product of intense, repetitive ritual activity over several centuries, concerned with guaranteeing the continuing fertility of the valley oases upon which human society is utterly dependent in southern Peru.

DAVID M. BROWNE

REFERENCES

Aveni, A. F., ed. *The Lines of Nazca*. Philadelphia: American Philosophical Society, 1990.

Hadingham, E. *Lines to the Mountain Gods: Nazca and the Mysteries of Peru*. New York: Random House, 1987.

Morrison, T. *Pathways to the Gods: the Mystery of the Andes Lines*. Lima: Andean Air Mail and Peruvian Times, 1977.

Reiche, M. I. *Mystery on the Desert*. Stuttgart: Heinrich Fink, 1968.

Silverman, H. and D. Browne. "New Evidence for the Date of the Nazca Lines." *Antiquity* 65: 208–220, 1991.

NĪLAKAṆṬHA SOMAYĀJI Nīlakaṇṭha Somayāji (AD 1444–1545) was one of the eminent astronomers of Kerala in South India. He was a pupil of Parameśvara, the promulgator of the Kerala Dṛgganita School of astronomy, and left after him a long line of astronomers. He was also a prolific writer on astronomy.

Nīlakaṇṭha's writings include *Golasāra* (Essence of Spherics), *Siddhāntadarpaṇa* (Mirror of the Laws of Astronomy) with his own commentary, *Candracchāyāgaṇita* (Computation of the Moon's Shadow), *Tantrasaṅgraha* (Resumé of Astronomy) in eight chapters, composed in AD 1501, Commentary on the *Āryabhaṭīya* of Āryabhaṭa, and *Jyotirmīmāṃsā* (Investigations on Astronomical Theories). Two of his works, *Grahaṇanirṇaya* and *Sundararājapraśnottara* are known from references but are yet to be retrieved.

While Nīlakaṇṭha's commentary on the *Āryabhaṭīya* is elaborate and exhaustive, and is infused with novel interpretations and new ideas, his *Jyotirmīmāṃsā* is of particular interest, in that it is wholly devoted to certain fundamental matters relating to Hindu astronomical theories and practices. It throws ample light on Nīlakaṇṭha's practical approach to astronomical studies. He says that there is no place, in a physical science like astronomy, for mythological explanations to astronomical phenomena. Enhanced by examples, the *Jyotirmīmāṃsā* emphasizes that astronomical computations should tally with actual observation. According to Nīlakaṇṭha, continued observation of planetary movements, experimentation, logical deduction, and enunciation of periodical corrections towards correlating computation with observation are the *sine qua non* of practical astronomy. Results obtained by the application of such corrections have to be verified at times of successive eclipses, which are visible celestial phenomena. Small errors that arise on account of observations being made from the surface of the earth, instead of from its center, are to to corrected by reducing the basic angular distances to the *dṛggola* (visible celestial sphere). Conversely, it is stated that in the case of the gnomonic shadow, it is the measure in the *dṛggola* that needs to be reduced to that of the *bhagola* (sphere of the zodiac). The *Jyotirmīmāṃsā* of Nīlakaṇṭha reveals him as an astronomer of extraordinary ingenuity and acumen.

K.V. SARMA

REFERENCES

Primary sources

Candracchāyāgaṇita (Computation of the Moon's Shadow). Ed. K.V. Sarma. Hoshiarpur: Vishveshvaranand Institute, 1976.

Commentary on the Āryabhaṭīya of Āryabhaṭa. Trivandrum: Trivandrum Sanskrit Series, 1930, 1931, 1957.

Golasāra (Essence of Spherics). Ed. K.V. Sarma. Hoshiarpur: Vishveshvaranand Institute, 1970.

Jyotirmīmāṃsā (Investigations on Astronomical Theories). Ed. K.V. Sarma. Hoshiarpur: Vishveshvaranand Institute, 1977.

Siddhāntadarpaṇa (Mirror of the Laws of Astronomy). Ed. K.V. Sarma. Hoshiarpur: Vishveshvaranand Institute, 1976.

Tantrasaṅgraha (Resumé of Astronomy). Ed. K.V. Sarma. Hoshiarpur: Vishveshvaranand Institute, 1977.

Secondary sources

Pingree, David. "Gārgyakerala Nīlakaṇṭha Somayājin." In *Census of the Exact Sciences in Sanskrit*. Ed. David Pingree. Philadelphia : American Philosophical Society, Series A, Vol. 3, 1976, pp.175–77; Vol. 4, p.142.

Sarma, K.V. "Gārgya-Kerala Nīlakaṇṭha Somayājin: The Bhāśyakāra of the Āryabhaṭīya." *Journal of Oriental Research* (Madras) 26 (1–4): 24–39, 1956–57.

NUMBER THEORY IN AFRICA Through the ages the peoples in Africa south of the Sahara invented hundreds of numeration systems, both spoken systems and symbolic ones that use body parts or objects to count or to represent numbers.

Verbal Numeration

The most common way to avoid the invention of completely new number words as one counts bigger quantities has been to compose new number words out of existing ones by relying on the arithmetical relationships between the involved numbers.

For example, in the Makhwa language spoken in northern Mozambique, one says *thanu na moza* (five plus one) to express six. Seven becomes *thanu na pili* (five plus two). To express twenty, one says *miloko mili*, tens two or 10×2. Thirty is *miloko miraru* (tens three). *Thanu* (5) and *nloko* (10) are the bases of the Makhwa system of numeration.

The most common bases in Africa are 10, 5, and 20. Some languages like Nyungwe (Mozambique) use only base ten. Others like Balante (Guinea Bissau) use 5 and 20 as bases. Verbal numeration in the Bété language of Côte d'Ivoire uses three bases: 5, 10, and 20. For instance, 56 is expressed as *golosso-ya-kogbo-gbeplo*, that is twenty times two plus ten (and) five (and) one:

	Number word	Structure
1	blo	
2	sô	
6	gbeplo	5+1
20	goloblo	20×1
40	golosso	20×2
50	golosso-ya-kogbo	20×2+10
56	golosso-ya-kogbo-gbeplo	20×2+10+5+1

The Bambara of Mali and Guinea have a ten–twenty system. The word for twenty, *mugan*, means one person; the word for forty, *debé,* means mat, referring to a mat on which husband and wife sleep together, and jointly they have forty digits.

The Bulanda of West Africa use six as a base: 7 is expressed as 6+1, 8 as 6+2, etc. The Adele (Togo) count *koro* (6), *koroke* (6+1=7), *nye* (8) and *nyeki* (8+1=9).

Among the Huku of Uganda the number words for 13, 14, and 15 may be formed by the addition of 1, 2, or 3 to twelve. For instance, 13 is expressed as *bakumba igimo*, meaning twelve plus one. The decimal alternatives 10+3, 10+4 and 10+5 were also known.

A particular case of the use of addition to compose number words occurs in the situation where both numbers are equal, or one of the two is equal to the other plus one. For instance, among the Mbai one counts from 6 to 9 in the following way: *mutu muta* (3+3), *sa do muta* (4+3), *soso* (4+4), and *sa dio mi* (4+5). Among the Sango (northern Zaire), 7 is expressed as *-na na-thatu* (4+3), 8 as *mnana* (4+4) and 9 as *-sano na-na* (5+4).

In several African languages, besides the additive and multiplicative principles, subtraction has also been used in forming number words. For example, in the Yoruba language of Nigeria, 16 is expressed as *eerin din logun* meaning four until one arrives at twenty:

	Number word	Structure
16	eerin din logun	4 until 20
17	eeta din logun	3 until 20
18	eegi din logun	2 until 20
19	ookan din logun	1 until 20
20	ogun	
21	ookan le logun	1+20
34	eerin le logban	4+30
35	aarun din logoji	5 until 20×2

Another example of the use of the subtractive principle may be found among the Luba-Hemba of Zaire. Seven is expressed as *habulwa mwanda* (lacking one until eight), and nine is *habulwa likumi*, lacking one until ten.

That spoken numeration systems may vary greatly in relatively small geographic regions is shown by the example of Guinea Bissau. The Bijagó have a pure decimal system; the Balante use a five-twenty system; the Manjaco have a decimal system with exceptional composite number words as 6+1 for 7 and 8+1 for 9; and the Felup use a ten–twenty system in which the duplicative principle is also employed in forms like 7 as 4+3 and 8 as 4+4.

Sometimes some number words are adjectives, while others are substantives. Where this happens, number word

Table 1: The *koti zigi* game

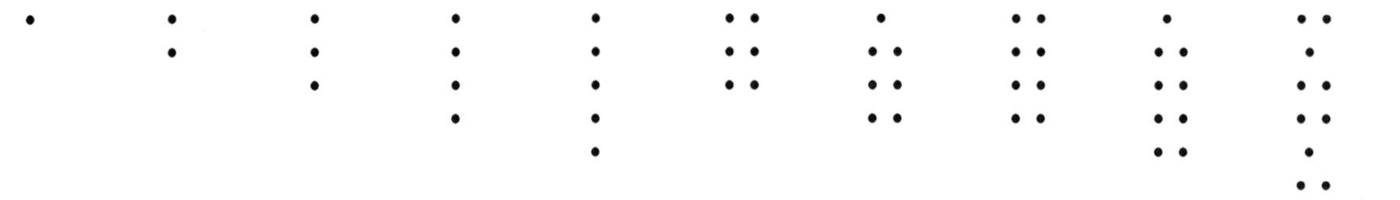

1	2	3	4	5	6	7	8	9	10

structures may appear that do not correspond directly to an addition, multiplication or subtraction. For instance, in the Tshwa language (central Mozambique) sixty is expressed as *thlanu wa makumi ni ginwe*, five times ten plus one more (ten).

In those contexts where it was necessary to have number words for relatively large numbers, there often appear completely new number words or ones that express a relationship with the base of the numeration system. For instance, among the Bangongo of Zaïre one says *kama* (100), *lobombo* (1000), *njuku* (10,000), *lukuli* (100,000), and *losenene* (1,000,000). Among the Ziba (Tanzania) one says *tsikumi* for 100, *lukumi* for 1000, and *kukumi* for ten thousand. All three terms are clearly related to *kumi* (10). Only the prefixes change.

Gesture Numeration

Gesture counting was common among many African peoples. The Yao (Malawi, Mozambique) represent 1, 2, 3, and 4 by pointing with the thumb of their right hand at 1, 2, 3 or 4 extended fingers of their left hand. Five is indicated by making a fist with the left hand. Six, seven, eight and nine are indicated by joining one, two, three or four extended fingers of the right hand to the left fist. Ten is represented by raising the fingers of both hands and joining the hands. On the contrary, the Makonde, who are also in the North of Mozambique, start counting on their right hand with the help of the index finger of the left hand. Five is indicated by making a fist with the right hand. For six to nine, the representation is symmetrical to that of one to four, that is, right and left hands change roles. Now the index finger of the right hand points at the fingers of the other hand. Ten is represented by joining two fists.

The method of gesture counting adopted by the Shambaa (Tanzania, Kenya) uses the duplicative principle. They indicate six by extending the three outer fingers of each hand, spread out; seven by showing four on the right hand and three on the left, and eight by showing four on each hand.

To express numbers greater than ten the Sotho (Lesotho) employ different men to indicate the hundreds, tens and units. For example, to represent 368, the first person raises three fingers of the left hand to represent three hundreds, the second one raises the thumb of the right hand to express six tens, and the third one raises three fingers of the right hand to express eight units. This is in fact a positional system, as it depends of the position of each man if he indicates units, tens, hundreds, thousands, etc.

Tally Devices

Many types of tally devices were used in Africa south of the Sahara. Two examples of widespread tallies are the following from Mozambique.

Among the Tswa, trees are used to record the age of children. After the birth of a child a cut is made on a trunk of a tree. Each year one adds a new cut until the person is old enough to count for him or herself. Tally sticks are used to control the number of animals in a herd. Each cut corresponds to one animal.

Among the Makonde, knotted strings were used. Suppose a man was going on an eleven day journey. He would tie eleven knots in a string and say to his wife, "This knot" (touching the first) "is today, when I am starting; tomorrow" (touching the second knot) "I shall be on the road, and I shall be walking the whole of second and third day, but here" (seizing the fifth knot) "I shall reach the end of the journey. I shall stay there the sixth day, and start for home on the seventh. Do not forget, wife, to undo a knot every day and on the tenth you will have to cook food for me; for see this is the eleventh day when I shall come back." Pregnant women used to tie a knot in a string at each full moon, to know when they were about to give birth. In order to register the age of a person, one uses two strings. A knot is tied in the first string at each full moon; once one has completed

1	i					
2	Ц	⌒	⟨			
3	Ш	⌃	Ɛ			
4	Ш	—	m			∂
5	ШШ	卐	○			6
6	ШШ	⌁⌁			⬡	6
7	ШШШ	▽	/			6c
8	ШШШ		×		⬡	6ε
9	ШШШШ	卐	\			6-
10	ШШШШ	卍	+			
11	ШШШШШ		ƆOƐ	□		
12		⌁⌁⌁				
13		⌁⌁⌁				-6-
14		▽▽				
15		卐卐卐	▣			
20		卍卍				

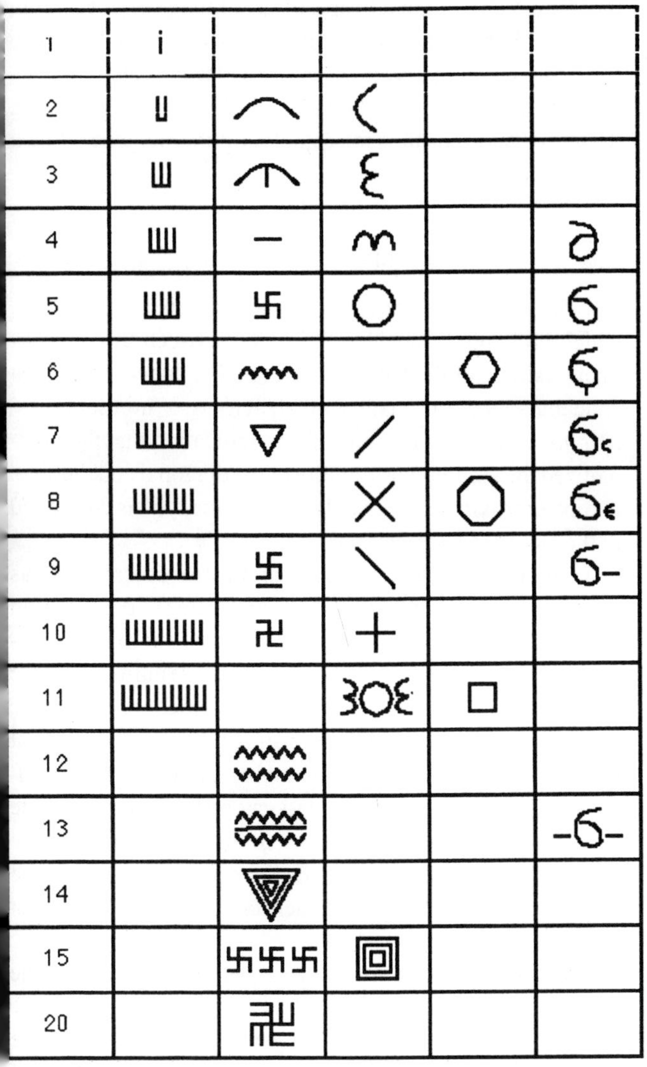

Figure 1: Examples of numerals on Akan gold weights (see Weights and Measures: Akan Gold Weights).

twelve knots, one ties a knot in a second string to mark the first year, etc.

Other Visual Numeration Systems

There exists a variety of numeration systems in Africa that are written in one way or another.

The *koti zigi* game is played by the Gbundi and Mende in Liberia and in the western parts of Côte d'Ivoire. The players form standardized patterns of stones on the ground to represent numbers. One observes that 6 is expressed 3+3, 7 as 3+1+3, 8 as 4+4, 9 as 4+1+4, and 10 as 5+5 or 4+2+4 (see Table 1).

The Fulani or Fulbe, a semi-nomadic pastoral people of Niger and northern Nigeria, place sticks in front of their

houses to indicate the number of cows or goats they possess. One hundred animals are represented by two short sticks placed on the ground in the form of a ∨. Two crossing sticks, ×, symbolize fifty animals. Four sticks in a vertical position, ||||, represent four; two sticks in a horizontal and three in a vertical position, — — |||, indicate twenty three animals. For example, the following was found in front of the house of a rich cattle owner:

∨ ∨ ∨ ∨ ∨ ∨ × ||

showing that he had 652 cows.

The Akan peoples (Côte d'Ivoire, Ghana, Togo) used money weights. That is to say, they used figurines in stone, metal or simply vegetable seeds as coins. The weight of a figurine was agreed to represent the monetary value that corresponded to a certain quantity of gold dust of the same weight. The figurines show animals, knots, stools, sandals, drums, etc. Figurines may also have diverse geometric forms such as step pyramids, stars, or cubes. Many display graphic signs representing numbers. Although in the languages spoken by the Akan peoples, like Anyi, Baoulé, Aboure, Attie and Ebrie, only base ten is used, base five is also found on the money weights:

$$5 = 6 \quad 6 = 5+1 = 6 \quad 7 = 5+2 = 6c \quad 8 = 5+3 = 6\epsilon \quad 9 = 5+4 = 6-$$

and

$$9 = 5+4 = 卐$$

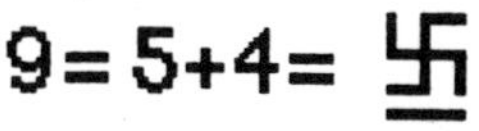

The symmetric structure of one of the symbols for 11 and of one for 13 may be observed:

$$11 = 3+5+3 = ƆOƐ \qquad 13 = 4+5+4 = \text{-}6\text{-}$$

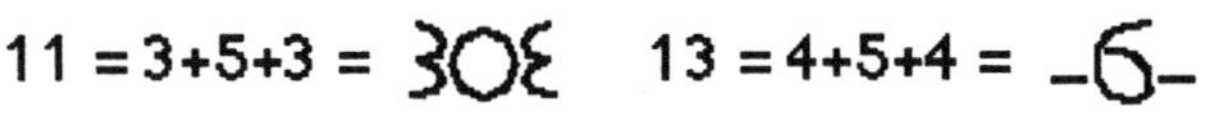

Duplication may be observed in the transition from

$$6 = ⌁⌁$$

to

$$12 = 6+6 = ⌁⌁⌁⌁$$

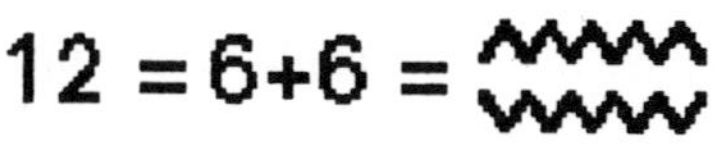

PAULUS GERDES

REFERENCES

Almeida, António. "Sobre a matemática dos indígenas da Guiné Portuguesa." *Boletim Cultural da Guiné Portuguesa* 6: 389–440, 1947.

Béart, Charles. *Jeux et jouets de l'ouest africain.* Dakar: IFAN, 1955.

Gerdes, Paulus. "On the History of Mathematics in Africa South of the Sahara." *AMUCHMA Newsletter* 9: 3–32, 1992.

Niangoran-Bouah, G. *The Akan World of Gold Weights.* Abidjan: Les Nouvelles Éditions Africaines, 1984.

Raum, Otto. *Arithmetic in Africa*. London: Evans Brothers, 1938.

Schmidl, Marianne: "Zahl und Zählen in Afrika." *Mitteillungen der Anthropologischen Gesellschaft* 35(3): 165–209, 1915.

Zaslavsky, Claudia. *Africa Counts: Number and Pattern in African Culture*. Brooklyn, New York: Lawrence Hill Books, 1979.

See also: Weights and Measures: Akan Gold Weights – Mathematics

NUMBER THEORY IN INDIA It is difficult to find 'number theory' in its proper sense in Indian mathematics. What I am going to describe below is how the Indians have treated kinds of numbers.

In the Vedas (ca. 1200–800 BC), the oldest Hindu literature, a number of numerical expressions occur. Their favorite numbers were three and seven as well as a hundred and a thousand. The largest number contained in their common list of names for powers of ten is 10^{12} (called *parārdha*). Later (by the fourth century AD), those names came to be employed for denoting decimal places, and became the nucleus of the Hindu list of decimal names (eighteen in number), while the Buddhists and the Jainas developed longer lists, which contained numbers as large as 10^{53} (*tallakṣaṇa*) or more.

The Jainas even speculated about different kinds of uncountable and infinite numbers (*Aṇuogaddārāiṃ*, between the third and the fifth centuries AD). They divided the whole set of "numbers (*saṃkhyā*) concerning counting (*gaṇanā*)" into three subsets: (1) countable (*saṃkhyeya*), (2) uncountable (*asaṃkhyeya*), and (3) infinite (*ananta*) numbers; and further divided the last two sets into three each (Table 1). The entire system of countable-uncountable-infinite depends upon the smallest number (a) of the "restrictively uncountable" set, which in turn is defined by means of white mustard seeds. The text is not very clear on this last point, but its intention was probably the same as what is meant by "the aleph zero" in modern mathematics, the smallest transfinite cardinal number.

Vedic stanzas also contain various series of numbers such as an integer series up to 200, an odd series (1,3,5, ...) up to 99 (accompanied by 100), an even series (2,4,6, ...) up to 100, and series made from multiples of four, five, ten, and twenty, up to 100, etc.

The Vedas tell us that only the gods Indra and Viṣṇu could divide a thousand equally into three, but it is a matter of argument how they did it. Natural fractions ($\frac{1}{2}$, $\frac{1}{4}$, $\frac{1}{8}$, and $\frac{1}{16}$) also occur in the Vedas.

The *Śulbasūtras* (ca. 600 BC and later), compendia of geometric knowledge related to the construction of various altars for the Vedic ritual, clearly state the so-called Pythagorean theorem: "The diagonal rope of an oblong produces both [areas] which its side and length produce separately." They explicitly mention several Pythagorean triples also: (3,4,5), (5,12,13), (8,15,17), (7,24,25), and (12,35,37). Moreover, they give an algorithm for calculating the diagonal d of a square whose side is a:

$$d = a + \frac{a}{3} + \frac{a}{3} \cdot \frac{1}{4} \cdot \frac{1}{34}$$

and call this value "one that has a difference" (*saviśeṣa*), which probably means the difference between this approximate value and the true one ($\sqrt{2}\,a$). The latter was called "one that makes [a square equivalent to] two [unit squares]" (*dvi-karaṇī*). There is, however, no indication that they recognized the incommensurability of the diagonal and the side. A root approximation formula of the same type is employed in the Bakhshālī Manuscript.

Later Indian mathematicians and astronomers, such as Varāhamihira (ca. AD 550) and Brahmagupta (AD 628), used the word *karaṇī* in the two contradictory (but mutually related) senses, the square root of a non-square number and the number whose square root should be obtained (or the square of any number), and easily performed the six arithmetical operations involving irrational numbers. Even the irrationality of *karaṇīs* may have been understood by Bhāskara of the seventh century, because he says, in his commentary on the *Āryabhaṭīya* (AD 629), "*Karaṇīs* have a size that cannot be stated exactly", although whether he proved it or not is not known.

Varāhamihira recognized zero as a number. In the *Bṛhatsaṃhitā* he added and subtracted zero in exactly the same way as he did other integers. Brahmagupta in the *Brāhmasphuṭasiddhānta* gave a complete set of rules for the six arithmetical operations involving zero as well as negative and irrational numbers. Thus by the seventh century AD the Indians acquired a very large domain of numbers including positive and negative integers (they accepted both the positive and the negative roots of a square number), fractions, irrational numbers, and zero, which enabled them to develop *bījagaṇita* ("seed mathematics") or algebra, the theory of equations.

In his *Āryabhaṭīya* (AD 499), Āryabhaṭa provided a solution (called *kuṭṭaka* or the pulverizer) to the linear indeterminate equation: $n = ax + r = by + s$, or $y = (ax + c)/b$. He "pulverized" (i.e. reduced) the coefficients a and b by means of their mutual divisions (the so-called Euclidean algorithm), and found a set of solutions to the reduced form by trial and error. Mahāvīra (ca. AD 850) removed the trial and error by carrying out the mutual divisions until the remainder became 1.

Brahmagupta treated indeterminate equations of the type

Table 1: Classification of numbers according to the Jainas

numbers for counting	countable		lowest:	2
			middle:	$3, 4, \cdots, a-2$
			highest:	$a-1$
	uncountable	restrictively uncountable	lowest:	$a\ (= \aleph_0?)$
			middle:	$a+1, a+2, \cdots, b-2$
			highest:	$b-1$
		properly uncountalbe	lowest:	$b = a^a$
			middle:	$b+1, b+2, \cdots, c-2$
			highest:	$c-1$
		uncountably uncountable	lowest:	$c = (b^2)^{b^2}$
			middle:	$c+1, c+2, \cdots, d-2$
			highest:	$d-1$
	infinite	restrictively infinite	lowest:	$d = c^c$
			middle:	$d+1, d+2, \cdots, e-2$
			highest:	$e-1$
		properly infinite	lowest:	$e = d^d$
			middle:	$e+1, e+2, \cdots, f-2$
			highest:	$f-1$
		infinitely infinite	lowest:	$f = (e^2)^{e^2}$
			middle:	$f+1, f+2, \cdots$
			highest:	does not exist

$Px^2 + t = y^2$. He showed, among other things, that this equation (called *vargaprakṛti* or the "square nature") can be solved for $t = 1$ if it is solved for $t = \pm 4, \pm 2$, or -1. Jayadeva (the eleventh century or before) gave a rule for arriving at a solution for $t = \pm 4, \pm 2$, or -1 from any solution for any t. Bhāskara of the twelfth century called it the "cyclic" (*cakravāla*) method.

Several rules given by Mahāvīra in his *Gaṇitasārasaṃgraha* indicate that he recognized two roots of a quadratic equation with one unknown, but all of his examples for those rules have two positive roots. (He, however, admits the negative root of a square number when he gives his rules for the six arithmetical operations). He was interested in the partition of numbers, and gave many rules for partitioning unity into the sum of unit fractions, a fraction into the sum of several fractions, etc. He also treated various mathematical progressions.

Śrīpati, perhaps for the first time in India, gave several rules for factorization in a chapter devoted to algebra in his astronomical work, *Siddhāntaśekhara* (ca. AD 1040). This topic, as well as the partition of numbers, mathematical progressions, combinatorics, and magic squares were highly developed by Nārāyaṇa in his *Gaṇitakaumudī* (AD 1356).

Bhāskara challenged various types of polynomial equations of the second and higher degrees (of special types) with the help of the "pulverizer" and the "square nature" in his work, *Bījagaṇita*, AD 1150.

Mādhava (fl. AD 1400) and his successors obtained, for the first time in the world, a number of power series for the circumference of a circle, sine, cosine, arctangent, etc. One of the most eminent scholars in his school was Nīlakaṇṭha. A great mathematician and reformer of Indian astronomy, he explicitly stated the incommensurability of the diameter and the circumference of a circle in his commentary on the *Āryabhaṭīya* (ca. AD 1540), although its proof is not found in his extant works.

TAKAO HAYASHI

REFERENCES

Datta, B. "The Jaina School of Mathematics." *Bulletin of the Calcutta Mathematical Society* 21: 115–145, 1929.

Datta, B. "Early Literary Evidences of the Use of the Zero in India." *American Mathematical Monthly* 33: 449–454, 1926, and 38: 566–572, 1931.

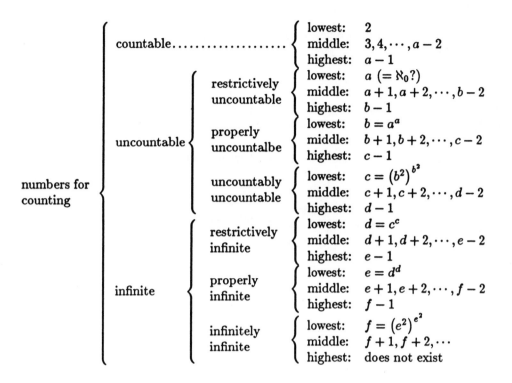

Datta, B. *The Science of the Śulba*. Calcutta: University of Calcutta, 1932.

Datta, B. "Vedic Mathematics." In *The Cultural Heritage of India*, vol. 3. Calcutta: Ramakrishna Centenary Committee, 1937, pp. 378–401.

Hayashi, T., Kusuba, T., and Yano, M. "The Correction of the Mādhava Series for the Circumference of a Circle." *Centaurus* 33: 149–174, 1990.

Michaels, A. *Beweisverfahren in der vedischen Sakralgeometrie*. Wiesbaden: Steiner, 1978.

Rajagopal, C.T. and M.S. Rangachari. "On an Untapped Source of Medieval Keralese Mathematics." *Archive for History of Exact Sciences* 18: 89–102, 1978.

Shukla, K.S. "Hindu Methods for Finding Factors or Divisors of a Number." *Gaṇita* 17: 109–117, 1966.

See also: Algebra in India – Arithmetic in India – *Śulbasūtras* – Geometry in India – Bakhshālī Manuscript – Varāhamihira – Brahmagupta – Nīlakaṇṭha – Mādhava – Magic Squares - Combinatorics in India – Nārāyaṇa – Śrīpati – Jayadeva – Mahāvīra – Bhāskara

NUMBER THEORY IN ISLAMIC MATHEMATICS Islamic number theory is characterized by two main developments, both of which stemmed from Greek knowledge. One of them is the relation between natural numbers and the sum of their proper divisors, the other the field of quadratic indeterminate equations.

Numbers and their Sums of Divisors

If N is a natural number, $\sigma(N)$ the sum of its divisors, and $s(N)$ the sum of the divisors without N itself, then $\sigma(N) = s(N) - N \geq 1$. The Greeks called N "abundant" if $s(N) > N$, "defective" if $s(N) < N$, and "perfect" if $s(N) = N$. Euclid had demonstrated that $N = 2^{m-1}(2^m - 1)$ is perfect if $2^m - 1$ is prime; we know today that it gives all even perfect numbers and possibly all perfect numbers. Thābit ibn Qurra (836–901) noticed that $2^{m-1} \cdot p$, m given and p prime, is defective or abundant depending on whether $p > 2^m - 1$ or $p < 2^m - 1$. In the same treatise, he provided for the first time a rule for finding a pair of "amicable" numbers, in which each number is equal to the sum of the proper divisors of the other: numbers N_1, N_2 such that $s(N_1) = N_2$ and $s(N_2) = N_1$. The Greeks knew of only one pair (220, 284). His rule, proved in true Euclidean manner, is: if $s = 3 \cdot 2^m - 1$, $t = 3 \cdot 2^{m-1} - 1$, $r = 9 \cdot 2^{2m-1} - 1$ $(m \neq 0, 1)$ are prime, then $2^m \cdot s \cdot t$ and $2^m \cdot r$ are amicable numbers. The severe restrictions on N_1 and N_2, however, limit the numerical application considerably.

Evidence of an apparently novel consideration appears in two incidental remarks made by al-Baghdādī (ca. 1000) concerning the solubility of $s(N) = k$, k given natural. It is asserted that $s(N) = k$ has no solution N for $k = 2$ and for $k = 5$. It is indeed true that, among all odd k, only 5 cannot be a sum of (proper) divisors, and a simple rule used by al-Baghdādī for finding numbers N_1, N_2 such that $s(N_1) = s(N_2)$ may be used to verify it (under assumption, though, of the still unproved Goldbach conjecture). If k is even, we know today that, besides 2, there is an infinite number of exceptions.

Quadratic Indeterminate Equations

Two Greek sources inspired Islamic mathematicians. The principal one is Diophantus's *Arithmetica*, of which seven of the original thirteen "books" (i.e. large chapters) were translated into Arabic. The other one is unknown to us today but was the origin of a set of indeterminate equations solved in Abū Kāmil's *Algebra* by Diophantine methods. Abū Kāmil says that such problems were the subject of discussion in his time. A good introduction to Diophantine methods occurs in al-Karajī's *Badīʿ*. The theoretical approach to the solution of finding a square which, increased or diminished by a given integer, produces a square in both cases, was treated in the tenth century by al-Khazīn; it later became the central subject of Leonardo Fibonacci's *Liber quadratorum*.

Other Indeterminate Equations

The above mentioned indeterminate equations always require positive rational solutions. Linear indeterminate equations with positive integral solutions were the subject of a separate treatise by Abū Kāmil, who searched for all such solutions for six typical pairs of linear equations with three to five unknowns. Of note also are attempts to prove the impossibility of $x^3 + y^3 = z^3$ in the tenth century.

JACQUES SESIANO

REFERENCES

Anbouba, Adel. "Un traité d'Abū Jaʿfar [al-Khazīn] sur les triangles rectangles numériques." *Journal for the History of Arabic Science* 3: 134–178, 1979.

Djafari Naini, Alireza. *Geschichte der Zahlentheorie in Orient*. Braunschweig: Klose, 1982.

Sesiano, Jacques. *Books IV to VII of Diophantus' Arithmetica*. New York: Springer, 1982.

Sesiano, Jacques. "Two Problems of Number Theory in Islamic Times." *Archive for History of Exact Sciences* 41 (3): 235–238, 1991.

Suter, Heinrich. "Das Buch der Seltenheiten der Rechenkunst von Abū Kāmil el-Miṣrī." *Bibliotheca Mathematica* Ser. 3, 11: 100–120, 1910/11.

Woepcke, Franz. "Notice sur une théorie ajoutée par Thâbit ben Korrah à l'arithmétique spéculative des Grecs." *Journal Asiatique* Ser. 4, 20: 420–429, 1852.

See also: Abū Kāmil – al-Karajī

O

OBSERVATORIES IN INDIA India has an ancient astronomical tradition. Information on its observatories is meager, however. It is certain that a number of prominent astronomers, patronized by kings, carried out their own observations, which are mentioned in *karaṇas*, or practical manuals. The places of such observations, if operated for a reasonable period of time, technically could be called observatories. A court astronomer, Śaṅkaranārāyaṇa (fl. 869), mentions such a place with instruments in the capital city of king Ravi Varmā of Kerala. Astronomers of the Islamic school of astronomy, such as ʿAbd al-Rashīd al-Yāqūtī (fifteenth century) report an observatory in the city of Jājilī in India. The Emperor Humayun (d. 1556) is said to have had a personal observatory at Kotah, near Delhi, where he himself took observations.

An ambitious program of building observatories was undertaken by Sawai Jai Singh (Savā'ī Jaya Siṃha), an astronomer–statesman of India. Between 1724 and 1735, Jai Singh built observatories at Delhi, Jaipur, Mathura, Varanasi, and Ujjain. His observatories, except for that of Mathura, still exist today in varying degrees of preservation. Sawai Jai Singh's purpose in building observatories was to update the existing planetary tables. Toward this purpose, he designed and built instruments of stone and masonry. These instruments may be classified into three main categories based on their precision which varies anywhere from $\pm 1'$ to a degree. Table 1 presents an inventory of his masonry instruments according to their precision, with the low precision instruments listed first. Table 2 lists instruments added after Sawai Jai Singh's death.

Jai Singh constructed fifteen different types of masonry instruments for his observatories. Of these, the *Samrāṭ yantra, Ṣaṣṭhāṃs'a, Dakṣinottara Bhitti, Jaya Prakāśa, Nāḍīvalaya*, and *Rāma yantras* are his most important instruments.

Samrāṭ Yantra

The *Samrāṭ yantra* or the "Supreme Instrument" is Jai Singh's most important creation. The instrument is basically an equinoctial sundial, which has been in use in one form or another for hundreds of years in different parts of the world.

The instrument consists of a meridian wall ABC, in the shape of a right triangle, with its hypotenuse or the gnomon

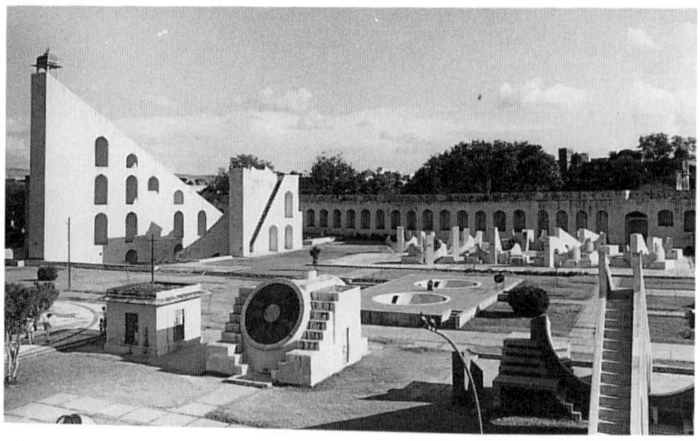

Jaipur Observatory of Sawai Jai Singh.

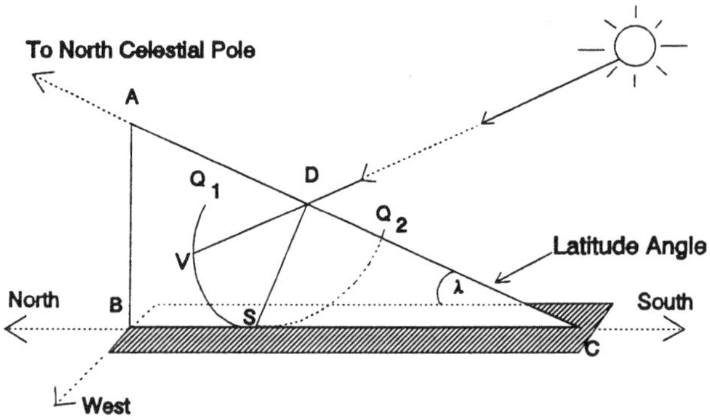

Figure 1: *Samrāṭ yantra*: principle and operation.

CA pointing toward the north celestial pole and its base BC horizontal along a north-south line. The angle ACB between the hypotenuse and the base equals the latitude λ of the place. Projecting upward from a point S near the base of the triangle are two quadrants SQ_1 and SQ_2 of radius DS. These quadrants are in a plane parallel to the equatorial plane. The center of the two "quadrant arcs" lies at point D on the hypotenuse. The length and radius of the quadrants are such that, if put together, they would form a semicircle in the plane of the equator.

The quadrants are graduated into equal-length divisions of time-measuring units, such as *ghaṭikās* and *palas*, according to the Hindu system, or hours, minutes and seconds, according to the Western system. The upper two ends Q_1 and Q_2 of the quadrants indicate either the 15-*ghaṭikā* marks for the Hindu system, or the 6 a.m. and the 6 p.m. marks according the Western system. The bottom-most point of both quadrants, on the other hand, indicates the zero *ghaṭikā* or

Table 1: Inventory of Jai Singh's masonry instruments

Instrument	No.	Location
Dhruvadarśaka Paṭṭikā (North Star Indicator)	1	Jaipur
Nāḍīvalaya (Equinoctial dial)	5	Jaipur (2), Varanasi, Ujjain, Mathura
Palabhā (Horizontal sundial)	2	Jaipur, Ujjain
Agrā (Amplitude instrument)	5	Delhi, Ujjain, Mathura
S'aṅku (Horizontal dial)	1	Mathura
Jaya Prakāśa (Hemispherical instrument)	2	Delhi, Jaipur
Rāma yantra (Cylindrical instrument)	2	Delhi, Jaipur
Rāśi valaya (Ecliptic dial)	12	Jaipur
Śara yantra (Celestial latitude dial)	1	Jaipur
Digaṃśa (Azimuth circle)	3	Jaipur, Varanasi, Ujjain
Kapāla (Hemispherical dial)	2	Jaipur
Samrāṭ (Equinoctial sundial)	6	Delhi, Jaipur (2), Varanasi (2), Ujjain
Ṣaṣṭhāṃśa (60 degree meridian chamber)	5	Delhi, Jaipur (4)
Dakṣinottara Bhitti (Meridian dial)	6	Delhi, Jaipur, Varanasi (2), Ujjain, Mathura

Table 2: Instruments added after Sawai Jai Singh's death

Instrument	No.	Location
1. Miśra yantra (Composite instrument)	1	Delhi
2. S'aṅku yantra (Vertical staff)	1	Ujjain
3. Horizontal Scale (known as the seat of Jai Singh)	1	Jaipur

12 noon. The hypotenuse or the gnomon edge AC is graduated to read the angle of declination. The declination scale is a tangential scale in which the division lengths gradually increase according to the tangent of the declination.

The primary object of a Samrāṭ is to indicate the apparent solar time or local time of a place. On a clear day, as the sun journeys from east to west, the shadow of the Samrāṭ gnomon sweeps the quadrant scales below from one end to the other. At a given moment, the time is indicated by the shadow's edge on a quadrant scale.

The time at night is measured by observing the hour angle of the star or its angular distance from the meridian. Because a Samrāṭ, like any other sundial, measures the local time or apparent solar time and not the "Standard Time" of a country, a correction has to be applied to its readings in order to obtain the standard time.

To measure the declination of the sun with a Samrāṭ, the observer moves a rod over the gnomon surface AC up or down until the rod's shadow falls on a quadrant scale below. The location of the rod on the gnomon scale then gives the declination of the sun. Declination measurement of a star or a planet requires the collaboration of two observers. One observer stays near the quadrants below and, sighting the star through a sighting device, guides the assistant, who moves a rod up or down along the gnomon scale. The assistant does this until the vantage point V on a quadrant edge below, the gnomon edge above where the rod is placed, and the star — all three — are in one line. The location of the rod on the gnomon scale then indicates the declination of the star.

Ṣaṣṭhāṁs'a

A *Ṣaṣṭhāṁs'a yantra* is a 60-degree arc built in the plane of meridian within a dark chamber. The instrument is used for measuring the declination, zenith distance, and the diameter of the sun. As the sun drifts across the meridian at noon, its pinhole image falling on the Ṣaṣṭhāṁs'a scale below enables the observer to measure the zenith distance, declination, and the diameter of the sun. The image formed by the pinhole on the scale below is usually quite sharp, such that at times even sunspots may be seen on it.

Dakṣinottara Bhitti Yantra

Dakṣinottara Bhitti yantra is a modified version of the meridian dial of the ancients. It consists of a graduated quadrant or a semicircle inscribed on a north–south wall. At the center of the arc is a horizontal rod. The instrument is used-for measuring the meridian altitude or the zenith distance of

Jaya Prakāśa at Jaipur. The *Nāḍīvalaya* is in the background.

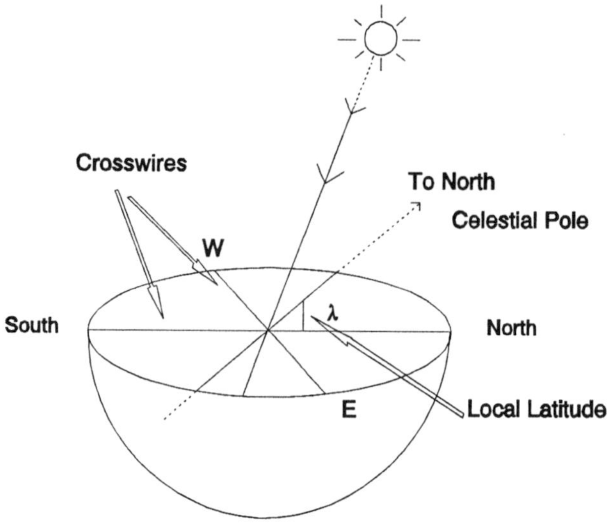

Figure 2: The principle of *Jaya Prakās'a* and *Kapāla yantras*.

an object such as the sun, the moon, or a planet. According to Jagannātha Samrāṭ, this was the instrument with which Jai Singh determined the obliquity of the ecliptic (the band of the zodiac through which the sun apparently moves in its yearly course), to be 23°28′ in 1729.

Jaya Prakāśa

The *Jaya Prakās'a* is a multipurpose instrument consisting of hemispherical surfaces of concave shape and inscribed with a number of arcs. These arcs indicate the local time, and also measure various astronomical parameters, such as the coordinates of a celestial body and ascendants, or a sign on the meridian. Jaya Prakās'a represents the inverted image of two coordinate systems, namely, the azimuth-altitude and the equatorial, drawn on a concave surface. For the azimuth-altitude system, the rim of the concave bowl indicates the horizon. Cardinal points are marked on the horizon, and cross wires are stretched between them. On a clear day, the shadow of the cross-wire falling on the concave surface below indicates the coordinates of the sun. Time is read by the shadow's angular distance from the meridian along a diurnal circle.

The instrument is built in two complementary halves, giving it the capacity for night observations. In the two halves the area between alternate hour circles is removed, and steps are provided in its place for the observer to move around freely for his readings. The space between identical hour circles of the two hemispheres is not removed, however. The sections left behind in the hemispheres complement each other. They do so in such a way that, if put together, they would form a complete hemispherical surface. For night observations the observer sights the object in the sky from the space between the sections. The observer obtains the object in the sky and the cross-wire in one line. The coordinates of the vantage points are then the coordinates of the object in the sky. Jai Singh built his Jaya Prakās'as only at Delhi and Jaipur. These instruments survive in varying degrees of preservation. The instrument at Delhi has a diameter of 8.33 m and that at Jaipur, 5.4 m.

Nāḍīvalaya

A Nāḍīvalaya consists of two circular plates fixed permanently on a masonry stand of convenient height above ground level. The plates are oriented parallel to the equatorial plane, and iron styles of appropriate length pointing toward the poles are fixed at their centers. The instrument Nāḍīvalaya is, in fact, an equinoctial sundial built in two halves, indicating the apparent solar time of the place.

The Nāḍīvalaya is an effective tool for demonstrating the passage of the sun across the celestial equator. On the vernal equinox and the autumnal equinox the rays of the sun fall parallel to the two opposing faces of the plates and illuminate them both. However, at any other time, only one or the other face remains in the sun. After the sun has crossed the equator around March 21, its rays illuminate the northern face for six months. After September 21, it is the southern face that receives the rays of the sun for the next six months. Jai Singh built Nāḍīvalayas at each of his observatory sites except Delhi.

Rāma yantras at Delhi.

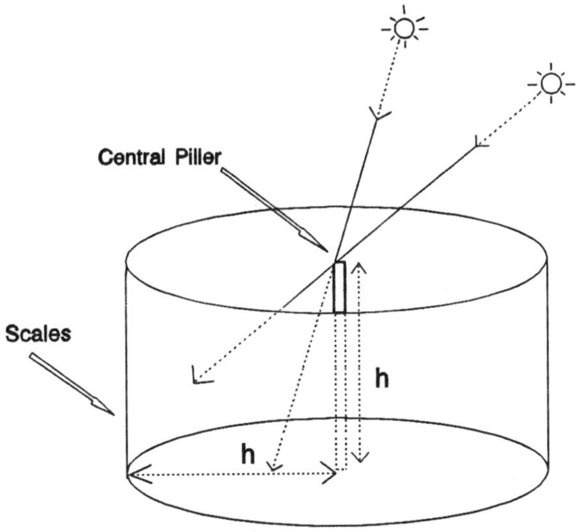

Figure 3: The principle of a *Rāma yantra*.

Rāma Yantra

The *Rāma yantra* is a cylindrical structure in two complementary halves that measure the azimuth and altitude of a celestial object, for example the sun. The cylindrical structure of Rāma yantra is open at the top, and its height equals its radius. The accompanying figure illustrates its principle and operation. To understand the principle, let us assume that the instrument is built as a single unit as illustrated.

The cylinder, as in the figure, is open at the top and has a vertical pole or pillar of the same height as the surrounding walls at the center. Both the interior walls and the floor of the structure are engraved with scales measuring the angles of azimuth and altitude. For measuring the azimuth, circular scales with their centers at the axis of the cylinder are drawn on the floor of the structure and on the inner surface of the cylindrical walls. The scales are divided into degrees and minutes. For measuring the altitude, a set of equally spaced radial lines is drawn on the floor.

These lines emanate from the central pillar and terminate at the base of the inner walls. Further, vertical lines are inscribed on the cylindrical wall, which begin at the wall's base and terminate at the top end. These lines may be viewed as the vertical extension of the radial lines drawn on the floor of the instrument.

In daytime the coordinates of the sun are determined by observing the shadow of the pillar's top end on the scales, as shown in the figure. The coordinates of the moon, when it is bright enough to cast a shadow, may also be read in a similar manner. However, if the moon is not bright enough, or if one wishes to measure the coordinates of a star or planet that does not cast a shadow, a different procedure is followed. To accomplish this, the instrument is built in two complementary units.

The two complementary units of a Rāma yantra may be viewed as if obtained by dividing an intact cylindrical structure into radial and vertical sectors. The units are such that if put together, they would form a complete cylinder with an open roof. The procedure for measuring the coordinates at night with a Rāma yantra is similar to the one employed for the Jaya Prakāśa. The observer works within the empty spaces between the radial sectors or between the walls of the instrument. Sighting from a vacant place, he obtains the object in the sky, the top edge of the pillar, and the vantage point in one line. The vantage point after appropriate interpolation gives the desired coordinates. If the vantage point lies within the empty spaces of the walls, well above the floor, the observer may have to sit on a plank inserted between the walls. The walls have slots built specifically for holding such planks.

VIRENDRA NATH SHARMA

REFERENCES

Garrett, A ff. and Guleri, Chandradhar, *The Jaipur Observatory and Its Builder*. Allahabad: Pioneer Press, 1902.

Kaye, G.R. *The Astronomical Observatories of Jai Singh*. New Delhi: Archaeological Survey of India, reprint, 1982.

Sharma, Virendra Nath. *Sawai Jai Singh and His Astronomy*. New Delhi: Motilal Banarasidass, 1995.

Sarma, Sreeramula Rajeswara. "Yantraprakāra of Sawai Jai Singh." Supplement to *Studies in History of Medicine and Science*, Vols. X and XI, New Delhi, 1986, 1987.

See also: Jai Singh – Gnomon in India – Time – Sundial – al-Yāqūtī

OBSERVATORIES IN THE ISLAMIC WORLD The observatory in Arabic/Islamic culture underwent considerable elaboration from its beginning as an imitation of earlier Greek observational posts. The first real observatories were founded under the patronage of the ʿAbbasid Caliph al-Ma'mūn (198–218 AH/AD 813–833), who also supported extensive translations from Greek into Arabic. The Shammāsīya and Qāsīyūn observatories had some permanent staff, but they are unidentified except for their names as reported by much later historians. The instruments used must have been versions of the Ptolemaic (Greek) equipment: the meridional armillary, for determining solstice points and the obliquity of the ecliptic; the plinth, used for the same purpose; the equinoctial armillary, used to determine the equinox points; the "parallactic instrument" (or "Ptolemy's Rulers"), to determine elevation on the zenith when heavenly objects, such as the moon, reached culmination; and the armillary or spherical astrolabe, for finding positions of heavenly bodies relative to a fixed or known celestial object.

In the centuries that followed, the physical equipment of observatories underwent a continuous evolution. The Arabic/Islamic observers sought precision in their work, and usually tried to obtain it by increasing the size of their instruments. Such large instruments had to be carefully fixed in place and required skilled craftsmen both for their construction and maintenance as well as a staff of observers and mathematicians skilled in reducing these observations to a mathematical model. In general, observation programs were initiated for correcting or improving the accuracy of earlier astronomical tables (*zīj*). The production of a *zīj* was useful not only for astrological/astronomical predictions but also for encouraging the collection of data that could support alternative models to Ptolemy's equant theory which was often opposed for its non-physical aspects.

Most observatories in Arabic/Islamic culture, because their operation revolved around large-scale expensive instruments, were supported by state funds, although there is evidence that some private institutions existed as well, such as the observatory of al-Battānī (d. 317 AH/AD 929) at Raqqa (274 or 275 AH–306 or 307 AH/AD 887–918). The majority, however, survived at the mercy of their royal patrons. Only a few observatories, such as the thirteenth century Marāgha Observatory, were assigned *waqf* income. Since the *waqf* was intended to be a grant of property in perpetuity for the sake of generating income for charitable institutions, this seems to imply that the Marāgha Observatory (as well as the Tabriz Observatory) was expected to exist for a long time and that it was seen as providing an essential public service. Perhaps this latter was related to the educational activities that accompanied the observational work.

The ʿAbbasid Caliph al-Ma'mūn (198–218 AH/AD 813–833) was the first ruler in Arabic/Islamic culture to devote state funds to the construction and maintenance of observatories. There are reports of observations made at Shammāsīya (in Baghdad) and at Mount Qāsīyūn (near Damascus) under his sponsorship. We know little about these early institutions. It seems that the Qāsīyūn Observatory was only intended to operate for one solar year, stretching over 216–217 AH/AD 831–832, perhaps as a continuation of the activities of the Shammāsīya installation. These sites, although they may have lacked buildings specifically constructed for observation, had both an observational staff and instrument makers. Their purpose seems to have been to make very accurate observations of the sun and moon only, although there are records of scattered observations of the other planets and fixed stars. The results of this work were presented in the *Mumtaḥan Zīj*.

Sharaf al-Dawla (372–378 AH/AD 982–989) built a major observatory in the garden of his Baghdad palace. It was housed in a building specifically constructed for that purpose and equipped with instruments, some of which seem to have been quite large. This institution marks an advance over the observatory building of al-Ma'mūn in several respects: (1) it had a specific physical structure, (2) it had a director/administrator, Abū Sahl al-Qūhī, the famous geometer and astronomer, in addition to a staff of astronomers and technicians, and (3) its research program seems to have extended to collecting data on the motions of all the planets. The research activity began with a great fanfare, but seems to have died out very quickly, perhaps due to the death of the royal sponsor.

The observatory at Marāgha (located south of Tabriz), was founded by the Mongol conqueror-prince, Hūlāgū in 657 AH/AD 1259. The construction, overseen by Mu'ayyad al-Din al-ʿUrḍi (from whose autobiographical account we learn many important details of the observatory and its operations), included several buildings and an extensive collection of instruments. There are also reports of a very large library. The observatory was staffed with a number of leading mathematicians and astronomers under the direction of Naṣīr al-Dīn al-Ṭūsī. (There are even reports of several Chinese astronomers at this observatory.) Their purpose was to collect accurate data to be used in constructing a new set of astronomical tables. The *Ilkhānī Zīj* was completed in 670 AH/AD 1271 after some twelve years of observation and calculation. After completion of the *zīj*, work at Marāgha seems to have slowed, although the institution continued to exist for quite some time.

Both in terms of the size and reputation of its astronomical staff and in terms of the amount and accuracy of its observational equipment, the Marāgha Observatory represents one of the highest developments in the history of the observatory within Islamic society. At the same time, it exhibited a number of unusual features. As noted previously, it was one of very few observatories to derive financial support from *waqf* funds. Not only does assignment of these *waqf* funds allow the institution greater longevity by making it less dependent upon the continued support of its founder, but it also may be taken to indicate a growing acceptance of observatories and their place within the culture. On the other hand, it is possible that *waqf* funds were assigned to the observatory in support of the advanced instruction in mathematical sciences which constituted another major feature of observatory activity.

Another major observatory was founded in 823 AH/AD 1420 in Samarqand by Ulugh Beg, governor of Khorasan, who was well-versed in mathematical and astronomical knowledge. Like the Marāgha Observatory, it had a staff of mathematicians, observational astronomers, instrument makers, and technicians, although in general it seems to have been a more compact operation. It contained numerous instruments, the best-known being a huge meridian arc enclosed within a large masonry structure. The institution enjoyed the active support and participation of Ulugh Beg until his assassination in 853 AH/AD 1449. Thus it functioned actively for nearly thirty years, in the course of which the *Zīj-i Jurjānī*, which was widely circulated in Arabic, Persian, and Turkish, was produced. Not only was this one of the longest-lived observational institutions, it was also the site of the most extensive collection of data on the fixed stars ever attempted in the Arabic/Islamic world.

In 983 AH/AD 1575, Taqī al-Dīn Muḥammed al-Rashīd ibn Maʿrūf, with the support of the Ottoman Grand Vizier, successfully petitioned the Sultan for permission to build an observatory in Istanbul in order to produce a new set of astronomical tables, since the older tables were outdated and needed revision to make them useful for the needs of his day. His request was successful, and in 985 AH/AD 1577, the Istanbul Observatory, with Taqī al-Dīn as its head, began operations. Like its predecessors, this institution had a staff of astronomers and supporting personnel, a library, and a permanent building housing instruments. Three years after its inception, the observatory was demolished. Contemporary reports differ on whether this action was undertaken with or without the support and encouragement of Taqī al-Dīn. Debate also continues about the possible influences of his work in observatory organization and instrument construction on Europe's greatest naked-eye observer, Tycho Brahe.

With the destruction of the Istanbul Observatory, the great pre-modern period of observatory development came to an end. The observatories of Jai Singh II, Maharaja of Jaipur (AD 1686–1740) at best represent only the dying embers of the tradition. These five observatories (located at Jaipur, Delhi, Benares, Ujjayin, and Mathura) contained enormous masonry instruments, many still extant today. Jai Singh's *Zīj-i Muḥāmmed Shāhī*, named for the emperor to whom it was dedicated, was largely patterned after the earlier *zīj* of Ulugh Beg. This has prompted some to conclude that these late observatories mark a period of decadence and derivative astronomy. This judgment is probably too harsh. These observatories, begun almost a century after the founding of the Paris Observatory (which marks the beginning of a new era in astronomical technique and organization in support of a new interpretation of celestial phenomena) stand today as monuments to a brilliant, but now abandoned, intellectual and social tradition within the cultural history of science.

GREGG DE YOUNG

REFERENCES

Kaye, G. R. *The Astronomical Observatories of Jai Singh*. Calcutta: Archaeological Survey of India, 1918.

Kennedy, E. S. *A Survey of Islamic Astronomical Tables*. Transactions of the American Philosophical Society, New Series. Volume 46, Part 2, 1956.

Mordtmann, J. H. "Das Observatorium des Taqi ed-din zu Pera." *Der Islam* 13: 82–96, 1923.

Sayılı, Aydın. *The Observatory in Islam and its Place in the General History of the Observatory*. 2nd ed. Ankara: Turk Tarih Kurumu Basimevi, 1988.

Sedillot, A. L. A. M. "Mémoire sur les Instruments Astronomiques des Arabes." *Memoires de l'Academie Royale des Inscriptions et Belles-Lettres de l'Institut de France*, Serie I. 1: 1–229, 1884.

Seemann, Hugo J. "Die Instrumente der Sternwarter zu Maragha." *Sitzungsberichte der physikalishmedizinischen Sozietät zu Erlangen* 60: 15–128, 1928.

See also: al-Maʿmūn – Armillary Spheres – Astrolabe – Astronomical instruments – *Zīj* – Astronomy – Astrology – al-Battānī – Marāgha – al-Qūhī – Ulugh Beg – Taqī al-Dīn – Jai Singh

OPTICS IN CHINESE SCIENCE Before AD 1911, optics in China went through four stages. The first stage was from remote antiquity to the Spring and Autumn Period (770 BC), the second ended in AD 220 (end of the Dong Han Dynasty), the third ended in AD 1380 (end of the Yuan Dynasty), and the fourth stage ended in 1911 (end of the Qing Dynasty).

In the first stage, the Chinese began to develop a philosophy of nature. The ideas of optics were in their infancy. In remote antiquity, the Chinese germinated basic knowledge of light sources, vision, shadow formation, and reflection. In addition, there were three inventions: artificial light sources, sundials called *Quibiao*, and reflectors. The artificial light sources or fire sources were obtained from striking stones, drilling wood, and focusing sunlight. The ancient sundial consisted of an elongated dial and one or two gnomons which were utilized to measure time and location. The reflectors included water mirrors (plane mirrors) and bronze mirrors (plane mirrors and convex mirrors). Even though all these achievements were superficial, they laid foundations for later studies on shadow formation and optical images.

The second stage took place within an important period for Chinese science and technology. During that time, optical technology developed very rapidly. As an example, techniques to make mirrors matured. Numerous studies led to deep understandings of light reflection and rectilinear propagation as recorded in the *Mo Jing* (Mohist Canon), which was written between 450 BC and 250 BC. There were eight sections in the book:

1. Processes of shadow formation and vanishing.
2. Umbras and penumbras.
3. Rectilinear propagation of light.
4. Sunlight reflection and formation of an inverted shadow.
5. Changes of shadow sizes (length and width).
6. Image formation and symmetries due to plane mirrors.
7. Two image variations from concave mirrors (erect image and inverted image).
8. One type of image formed by convex mirrors.

These eight propositions systematically described theories in geometrical optics. Technically, *Mo Jing* had achieved a level very similar to that of Euclid's *Optics*.

During the third stage, various optical phenomena were discovered by Chinese scholars. For example, aspects of atmospheric optics were studied in detail, including halo maps and rainbow formation. Since image formation had been a hot subject in Chinese optics, it was further advanced during this third stage. Most of the records can be found in a book called *Mengxi Bitan* (Meng Xi Essays) by the scientist Shen Guo (1032–1096), in the Song Dynasty (960–1279). Among hundreds of sections in the book, there were more than ten dealing with optics. Sections 44 and 330 are the most important. In Section 44, Shen discussed image formation by pinhole and concave mirrors in terms of a terminology called *ai*, or pinhole and focal point. He named such mathematical generalization *ge shu*. In Section 330, Shen discussed light penetrating mirrors, which were also called *tou-guang jian* or "magic mirrors". There were more than twenty characters inscribed on a magic mirror. When sunlight shone on the mirror, all the characters were clearly projected on to a wall. There were three magic mirrors in Shen's family. He also saw other mirrors in his friends' homes. However, some other extremely thin mirrors did not allow sunlight to pass through. The magic was in the inscription of faint lines on the back side of the mirror. (Based on Shen's descriptions, modern Chinese shops are able to reproduce such magic mirrors by several techniques.) Besides those two sections, Shen presented quantitative relationships between image sizes and curvatures of convex mirrors. He introduced a postmortem examination method in which a red light was utilized. He also attributed rainbows to the shadows of the sun during rain.

After Shen's work, another scientist, Zhao Youqin, carried out a famous optical experiment. Zhao lived in the thirteenth century, and recorded this experiment in his book, *Ge-Xiang Xin-Shu* (New Astronomy). In a chapter entitled "Pinhole Images", Zhao detailed a systematic study carried out in a two story house. There were two rooms on the first floor, one on the left and one on the right. To make two light sources in these two rooms, two boards were "planted" with thousands of candles. On the top of each light board, an additional covering board was placed. There was a hole in the center of the additional board. If the candles were lit, light could pass through the hole and then was projected on to a screen. The screen was either the fixed ceiling or an adjustable screen suspended from the ceiling so that the distance between the light source and the screen could be adjusted. The following observations were made in the experiment.

1. Shapes of the pinhole images are independent of the shapes and sizes of the small hole in the covering board.
2. The brightness of the image depends on the size of the hole. The larger the hole, the brighter the image.
3. When the source strength (number of candles) increases, the brightness of the image increases.
4. When the distance between the light source and the image screen increases, the image brightness decreases.

Zhao's experiment provided a good deal of information on pinhole images. From the experiment, he proposed this idea. On the image screen, there was a light spot corresponding to a single candle. If a thousand candles were lit on a source board, there would be a thousand images of the candles. These images would overlap each other. The final appearance of the image would depend on the distribution of lit candles.

In addition to basic optics, Zhao utilized his pinhole image theory to study eclipses of the sun and the moon and other astronomical phenomena.

The fourth stage marked the end of traditional Chinese optics. The new trend both continued the Chinese system and adapted the western system imported from Europe. During this period, a book called *Wuli Xiaozhi* (Small Encyclopedia of Physical Principles), was written by Fang Yizhi (1611–

1671). In the book, Fang pointed out that light travels in wave forms.

Several books and many articles on optics were translated from western languages to Chinese languages, such as *Yuan Jing Shuo* (Telescopium) by the German Missionary Johann Adam Schall Von Bell (1591–1666), and *On Optics* by Zhang Fuxi (d. 1862) and Englishman Joseph Edking (1823–1905). Another book entitled *Six Lectures on Light*, written by English physicist John Tyndall (1820–1893) was translated by Card T. Kreyer and Zhao Yuanyi in 1876.

Several monographs on optics were written by Chinese authors. In a book entitled *Jingjing Lingchi* (Treatise on Optics by an Untalented Scholar) by Zheng Fuguang (b. 1780), geometric optics was systematically introduced. In another monograph entitled *Geshu Bu* (Supplement to Geometric Optics) written by Zou Boqi (1819–1869), optical theorems and principles were discussed.

Several different kinds of optical instruments or devices were improved. For instance, an optical expert, Sun Yunqiu (seventeenth century), made spectacles, telescopes, microscopes, and distorting mirrors.

The Chinese achieved high levels of understanding in optics, as illustrated by *Mo Jing*, *Mengxi Bitan*, and *Gexiang Xinshu*. Their historical achievements were comparable to those of ancient Greece. The ancient Chinese paid attention to both theories and applications. A unique characteristic of Chinese optics is related to experimental approaches. All eight propositions in *Mo Jing* were based on experimental observations. Between the Qing and Han Dynasties, the Chinese mastered the concept of "focal distance" when they ignited fires from a spherical mirror. They created several novel devices such as the ice lens and open tube periscope. They observed similarities among rainbows, waterdrop dispersion, and crystal dispersion. Several Chinese scientists conducted high level experiments during their times, such as the "spherical mirror images" experiment by Shen Guo, and the "pinhole images" experiment by Zhao Youqin. The ancient Chinese optics system was based on empirical observations, which was short of theoretical abstraction and quantitative description. For example, no laws of reflection were proposed even after the phenomena of reflections had been observed for two thousand years. As Western optics was imported to China, the entire foundation of traditional Chinese optics was changed.

JINGUANG WANG
CAIWU WANG

REFERENCES

Graham, A.C. and Nathan Sivin. "A Systematic Approach to Mohist Optics." In *Chinese Science*. Ed. Shigera Nakayama and Nathan Sirin. Cambridge, Massachusetts: MIT Press, 1973, pp. 105–152.

Needham, Joseph. *Science and Civilisation in China*. Vol. 4, Part 1. Cambridge: Cambridge University Press, 1962, pp. 78–125.

Wang, Jinguang. "Zhao Youqing and His Research in Optics." *Journal of the History of Science and Technology* 12: 93–99, 1984 (in Chinese).

Wang, Jinguang. "Optics in China Based on Three Ancient Books." In *Science and Technology in Chinese Civilization*. Ed. Cheng-Yih Chen. New York: World Scientific Publishing, 1987, pp. 143–153.

Wang, Jinguang and Jun Wenren. "The Scientific Achievements of Shen Kuo." In *Research on Shen Kuo's Work*. Hangzhou: Zhejiang People's Publisher, 1985, pp. 64–123 (in Chinese).

Wang, Jinguang and Zhenhuan Hong. *History of Optics in China*. Changsha: Hunan Education Publisher, 1986 (in Chinese).

See also: Sundials – Shen Guo

OPTICS IN THE ISLAMIC WORLD The science of optics entered the Islamic world primarily through Greek sources, during the ninth-century transmission of ancient scientific and philosophical texts. But as distinct from many other fields, what emerged from a subdivision of geometry called *optika* was to transform its parent field quickly and irreversibly. Indeed, if one field were to be singled out in the Islamic world for having left the most influential mark on the development of a discipline, this might be *ʿilm al-manāẓir*, as Greek *optika* came to be called. During this particular phase of the discipline a decisive theory about the nature and manner of vision, which was the main subject of optics for some time, developed after a long period of debate. The domain of the field also expanded from a geometrical study of vision to one into which not only theories of light, mirrors, rainbows, and shadows, but also the psychology of visual perception, were integrated as common subjects of investigation. Some of these added inquiries produced impressive results, but as the field turned from a branch of mathematics into a discipline closer to physics, its methodology also left the purely geometrical world to enter an experimental realm.

Works which represent these developments include the *De aspectibus* of Yaʿqūb al-Kindī (d. AD 870), the *Kitāb al-manāẓir* of Ibn al-Haytham (d. AD 1039), and its rich commentary, *Tanqīḥ al-manāẓir* of Kamāl al-Dīn Fārisī (d. AD 1370).

A long list of optical works by these and other authors, composed in both Arabic and Persian, from the early ninth century all the way to about hundred and fifty years ago, are representative of the quantitative developments of the field (Kheiramdish, 1997).

Disciplinary Developments

One would expect the science of optics to have embraced a variety of phenomena from light and shadows, halos and rainbows, mirrors and burning instruments, some observable from the beginings of human history. But this was not always the case even in ancient and medieval times. For ancient Greek scholars, from mathematicians such as Euclid (ca. 300 BC) and Ptolemy (2nd century AD), to philosophers and physicians like Aristotle (ca. 400 BC) and Galen (2nd century AD), the focus of what was called *optika* and classified under geometry was vision, not light. Greek optics was primarily concerned with theories of vision, as the close association of the term *optika* with the eye indicates, and in the case of the geometrical tradition in optics — the only tradition treating the subject in independent works with that title — the proposed theories of vision were even expressed in terms of visual rays (*opseis*) extending from the eye to the object. Greek writings did go beyond the realm of direct vision, to include reflection from a polished surface (as in Euclid's *Optics,* or Hero of Alexandria's *Catoptrics)* or refraction through a different medium (as added in Ptolemy's *Optics*). But works titled *Optika* still dealt primarily with vision — leaving the domain of the field largely determined by the relevancy of its subjects to what was long considered the "most noble of the senses".

Early Islamic scholars inherited this particular orientation, which initially determined the focus of their own optical inquiries, even the classification and study of related subjects concurrently received from ancient sources, from reflection and refraction to shadows, rainbows, colors, sighting instruments, and burning mirrors. But there was soon a considerable change in optics' scope and profile. With the *Optics* of Ibn al-Haytham (also known as Alhazen, ca. 11th century AD) the extended disciplinary boundaries left an immediate trace on subsequent developments of a field, now called *Perspectiva*, which was largely carried over to the seventeenth century when optics acquired its current name and general character.

The change in the scope of *ʿilm al-manāẓir*, which occurred alongside other theoretical and methodological developments, was itself a gradual process. Major works on optics proper, often identifiable by the term *al-Manaẓir* in their titles (corresponding to the *Optika* in the titles of Euclid's and Ptolemy's texts), slowly came to expand the meaning of the study of vision. Works from the *De aspectibus (Ikhtilāf al-manāẓir?)* of Yaʿqūb al-Kindī (d. AD 870) and *Kitāb al-manāẓir* of Ibn al-Haytham, to *Taḥrīr al-manāẓir* of Naṣīr al-Dīn al-Ṭūsī (d. AD 1274) and *Tanqīḥ al-manāẓir* of Kamāl al-Dīn Fārisī (d. AD 1370), spanning a period of about five hundred years, well represent the changing disciplinary boundaries. Al-Kindī included in his *De aspectibus* a range of related discussions from shadows and mirrors to the clarity of perception, but left a number of others out. Ibn al-Haytham discussed in the seven books of his *Kitāb al-Manāẓir* a wide range of other subjects: the properties of light (and color) (Book I), visual perception and visual illusions (Books II and III), and reflection and refraction (Books IV–VII), but wrote separately on burning instruments, halos and rainbows, or *camera obscura*. Kamāl al-Dīn Fārisī's critical study of this same work, supplemented by a few of Ibn al-Haytham's independent optical writings (*On the Quality of Shadows, On the Form of the Eclipse,* and *On Light*), together with an examination of his own treatises (*On Burning Sphere* and *On the Rainbow and Halo*), expanded the topical range of his *Tanqīḥ al-manāẓir* to include all but the study of burning mirrors. The disciplinary developments are also reflected through a variety of related sources. Al-Fārābī's (d. AD 950) *Iḥṣāʾ al-ʿulūm* (Enumeration of the Sciences) describes *ʿilm al-manāẓir* as including not only the science of mirrors (*ʿilm al-marāyā*), but also such topics as the vision of distant objects and the application of vision in surveying. At the same time as the combination of the subjects of mirrors, or even burning mirrors with that of vision was not so uncommon, treatments of what are now considered optical problems often appeared in their own "proper place". Thus theories of light and perception were treated in philosophical works (following Aristotle), as in the related works of Ibn Sīnā (Avicenna, d. AD 1037), or Ibn Rushd (Averroes, d. AD 1198). Treatments related to the anatomy of the eye appeared in medical treatises (following Galen), as in the works of Ḥunayn ibn Isḥāq (d. AD 877). Such subjects as burning instruments were more commonly treated as part of the respective traditions of Diocles and Anthemius on *Burning Mirrors*, as in the case of Abū Saʿd Ibn Sahl (ca. AD 984). Shadows continued to be treated in books called *kutub al-aẓlāl* (Books on Shadows), as in Bīrūnī's (d. AD 1048) work devoted to this subject, and halos and rainbows often appeared in meteorological and astronomical literature, as in the case of Ibn Sīnā and Quṭb al-Dīn Shīrāẓī (d. A. D. 1311). It was not until the conscious effort of Shīrāẓī's student, Kamāl al-Dīn Fārisī, that most of these subjects were integrated into optics.

Theoretical Developments

At the very beginning of the text which set out to place the science of optics on a new foundation and was instantly recognized and utilized as the most complete work on the subject in Europe, Ibn al-Haytham wrote: "For two opposite doctrines, it is either the case that one of them is true and the other is false; or they are both false, . . . the truth being other than either of them; or they both lead to one thing which is the truth <in which case> each of the groups holding the two doctrines must have fallen short of completing their inquiry" (Sabra, 1989). Here, Ibn al-Haytham was reflecting on a long controversy in visual theory, which had taken shape between the mathematicians and physicists before him and for which he managed to find a permanent solution.

The Greek mathematicians had explained vision through the assumption of visual rays extended from the eye towards the visual object. The best known among these was Euclid, whose *Optics* represents a geometrical approach to the study of vision, and Ptolemy, who had added an experimental and psychological dimension in his own *Optics*. The Galenic version of the visual-ray hypothesis as contained in *De usu partium* included the anatomical aspects of vision and stressed the role of medium through its own physiological approach. Also, Aristotle's language of the reception of forms was adopted by the physicists, or natural philosophers, in matters regarding vision. Islamic theorists writing in the tradition of each of the major Greek philosophers on vision did not always fall strictly into one group or another. The visual-ray theorists (*aṣḥāb al-shuʿāʿ*) and the upholders of the theory of forms (*aṣḥāb al-inṭibāʿ*), to which Ibn al-Haytham repeatedly refers in his search for a systematic solution, often consisted of more than one group. The first included in addition to mathematicians (*taʿlimiyyūn*), the followers of Plato or Galen or even theologians (*mutakkalimūn*), while its immediate rival, natural phislophers (*ṭabīʿiyyūn*) covered both Aristotelians and atomists. The illuminationists (*ishrāqiyyūn*), who proposed an alternative form of perception altogether, had their own forms and variations. Ibn al-Haytham's solution to many complex problems raised by the first two groups was not only an attempt to determine such central questions as the direction of radiation in vision. The significance of his most celebrated work, *Kitāb al-Manāẓir*, goes much beyond treatments of the problems of vision. Having been translated into Latin in the late twelfth or early thirteenth century and then into Italian in the fourteenth century, the book had an impact on the epistemology of late medieval Europe, the linear perspective of Renaissance artists, and study of light all the way to the seventeenth century, with a list of figures including Roger Bacon, Witelo, Pecham, Ockham, Oresme, Ghiberti, Snellius, Kepler, Descartes, Barrow, and Huygens. Ibn al-Haytham's contribution to the study of light included the assumption that light requires a body (a medium) for its transmission, that its movement from the object to the eye is not of infinite speed, though too quick to be perceived by sense, and it is "easier" and "quicker" in rarer media. His non-traditional conception of a finite, imperceptible interval of time for the movement of light was defended and advanced by a few in the Latin West, and his appeal to mechanical analogies to explain reflection and refraction were adopted by many later thinkers (Sabra, 1962). The earlier independent researches of the tenth-century mathematician Abū Saʿd Ibn Sahl and his anticipation of Snell's law have been the subject of a recent study (Rashed, 1993). Earlier developments upon which Ibn al-Haytham's concepts of light and vision are clearly dependent include a principle now referred to as the "punctiform analysis of light radiation" (Lindberg, 1976). An ancient conception based on rectilinear radiation, it grew on the assumption of a one-to-one correspondence between points on a visible surface and the eye, laying the foundation for the geometry of sight of the Western perspectivists and of Kepler. The principle can already be traced not only in the works of earlier Arabic authors such al-Kindī, but even in the way Euclidean assumptions were translated into Arabic, thereby stretching the period and importance of theoretical developments to the early ninth century (Kheirandish, 1996). Immediate preoccupation with crucial questions about the exact nature of visual clarity and the structure of the visual cone (the role of central ray, the shape of its base), goes back to this same early period. Sometimes, being the direct result of adopted Arabic terminology, these questions were themselves the staring points for many later discussions in medieval Europe. Also, the limited understanding of refraction in some parts of the Islamic world, due to the poor reception of the *Optics* of both Ptolemy and Ibn al-Haytham, did not provide a major obstacle for important theoretical developments later, including the correct explanation of the theory of rainbow formation. Explanations of the primary and secondary bows in terms of the refraction of light in raindrops were offered by Quṭb al-Dīn Shirāzī and later refined through experimental means by his own student, Kamāl al-Dīn Fārisī in Islamic Persia, almost at the same time as Theodoric of Freiberg's (d. ca. AD 1310) theory of rainbows was being offered. The Islamic authors not only changed the face of theoretical problems related to optics, they also lay new methodological standards for optical inquiries.

Methodological Developments

The historical relationship between the sciences of optics and astronomy had a particularly important effect on the development of the methodological dimensions of the field. Euclid's *Optics* was among the Intermediate books (*mutawassiṭāt*), studied after his own *Elements* and before Ptolemy's *Almagest.* But optics was also to borrow from astronomy, though in a markedly different form and pace. The notion of *iʿtibār* (experiment) was adopted and transformed by Ibn al-Haytham from astronomical works to be used in optics, as a methodological measure to replace geometrical demonstrations related to light and vision with experimental ones. This, however, occurred only after the old Aristotelian superior–subordinate relationship of disciplines was replaced first by their cooperation (*ishtirāk*)(in the ninth century), and then by their combination (*tarkīb*)(in the eleventh century). So the full service of astronomy to optics came relatively late. The imported experimental dimension was a significant step for the methodological transformation of the science of optics (Sabra, 1989).

Before this transformation, the discipline's methodical guidelines relied heavily on the teachings of Aristotle. For Aristotle, optics was a branch of geometry and one of "the more physical of the mathematical sciences" (*Physics*, 194a 6–11). The method of applying geometrical demonstrations to propositions on vision was adopted by Euclid, and through him more directly by Arabic optical authors. But alternative methodological approaches had already begun with a few early Islamic scholars. Al-Kindī, who reminded the reader in the preface of his *De aspectibus* that "geometrical demonstrations would proceed in accordance with the requirement of physical things", supplied geometrical demonstrations with experimental ones throughout his text. About the same time, Qusṭā ibn Lūqā spoke in a text on optics and its application to mirrors about "the cooperation" of natural philosophy (from which we acquire sense perception) and geometry (for its geometrical demonstrations). Even in the early text of Aḥmad ibn ʿĪsā, specifically titled as being "in the tradition of Euclid", Euclidean examples were sometimes supplemented by "sensible examples " — those to be set up.

Ibn al-Haytham would continue to regard experimental optics as a mathematical inquiry. But in the course of his attempt to examine the study of vision and its foundations, the methodology of the mathematical science of optics was to be transformed radically. The synthesis (*tarkīb*) of physics (involving questions concerning the nature of light), and mathematics (dealing with manner of its propagation), following his explicit division between physical (*ṭabīʿiyya*) and mathematical (*taʿlīmiyya*) inquiries as two separate criteria, was at the heart of a new methodology which was to change the discipline. Ibn al-Haytham's contemporary, Ibn Sīnā, adopted a somewhat different methodological approach when he supplemented his critical remarks on previous treatments of rainbows with detailed observations of his own. Kamāl al-Dīn al-Fārisī, acknowledging the guidance of his predecessor, and inspired by his teacher, Quṭb al-Dīn al-Shīrāzī, added an unmistakably modern experimental touch to such observations. He reproduced an artificial object, such as a spherical globe filled with water to represent raindrops, as part of a conscious shift from traditional explanations of the phenomenon of the rainbow in terms of clouds acting as a concave mirror, to one in terms of the passage of light through a transparent sphere. By this later period, the field not only included subjects and theories more easily associated with optics today; it had also acquired the more modern methodological approach of "controlled" experimentation.

ELAHEH KHEIRANDISH

REFERENCES

Primary sources

Euclid. *Euclidis Opera Omnia*. Ed. J. Heiberg and H. Menge. Leipzig: Teubner, 1883–1916. English translation by H. E. Burton. "The Optics of Euclid." *Journal of the Optical Society of America* 35: 357–72, 1945.

Ibn al-Haytham. *Kitāb al-Manāẓir*. Ed. A.I. Sabra: *Al-Ḥasan Ibn al-Haytham, Kitāb al-Manāẓir: Books I–II–III <On Direct Vision>: Edited with Introduction, Arabic–Latin Glossaries and Concordance Tables by Abdelhamid I. Sabra*, Kuwait: The National Council for Arts and Letters, 1983; translation and commentary by A.I. Sabra, *The Optics of Ibn al-Haytham*, 2 vols. London: The Warburg Institute, 1989.

al-Fārisī, Kamāl al-Dīn. *Tanqīḥ al-Manāẓir li-dhawī alabṣār wa al-baṣā'ir*. 2 vols. Hyderabad: Maṭbaʿat Majlis Dāʾīrat al-Maʿārif,1347–48 (1928–30).

al-Kindī. *De aspectibus*. Ed. Björnbo, Axel Anthon, and Sebastian Vogl. "Al-Kindi, Tideus und Pseudo-Euklid: Drei optische Werke." Leipzig: Teubner, 1912.

Ptolemy. *L'Optique de Claude Ptolémée*. Ed. Albert Lejeune. Leiden: E. J. Brill,1989.

Secondary sources

Kheirandish, Elaheh. "The Arabic Version of Euclidean Optics: Transformations as Linguistic Problems in Transmission", *Tradition, Transmission, Transformation: Proceedings of Two Conferences on Pre-modern Science*. Held at the University of Oklahoma. Ed. F. Jamil Ragep and Sally P. Ragep with Stephen Livesy. Leiden, New York: E.J. Brill, Collection de travaux de l'Académie internationale d'histoire des sciences, 1996.

Kheirandish, Elaheh. *The Arabic Version of Euclid's Optics: Kitāb Uqlīdis fī Ikhtilāf al-manāẓir* (Arabic text and English translation with a historical commentary), 2 vols, New York: Springer-Verlag, forthcoming.

Lindberg, David, C. *Theories of Vision from Al-Kindi to Kepler*. Chicago: University of Chicago Press, 1976.

Meyerhof, Max. "Die Optik der Araber." *Zeitschrift für ophthalmologische Optik* 8:16–29, 42–54, 86–90, 1920.

Rashed, Roshdi. *Optique et mathématiques: Recherches sur l'histoire de la pensée scientifique en arabe*. London: Variorum, 1992.

Rashed, Roshdi. *Géométrie et dioptrique au Xe siècle: Ibn Sahl, al-Qūhī, et Ibn al-Haytham*. Paris: Les Belles Lettres, 1993.

Sabra, A. I. "Explanations of Optical Reflection and Refraction: Ibn al-Haytham, Descartes, Newton." *Actes du Dixième Congres*

Internationale d'Histoire des Sciences. Ithaca, 1962, pp. 551–554.

Sabra, A. I. "Manāẓir, or ‹Ilm al-manāẓir." In *Encyclopaedia of Islam*, vol. 6. Leiden: Brill, 1987, pp. 376–377.

Sabra, A. I. " Optics, Islamic." In *Dictionary of the Middle Ages*, vol. 9. Ed. Joseph R. Strayer. New York: Charles Scribner's Sons, 1987, pp. 240–247.

Sabra, A. I. *Optics, Astronomy and Logic: Studies in Arabic Sciences and Philosophy*. Aldershot, England: Variorum, 1994.

Schramm, Matthias. *Ibn al-Haythams Weg zur Physik*. Wiesbaden: F. Steiner, 1963.

See also: al-Kindī – Ibn al-Haytham – Naṣīr al-Dīn al-Ṭūsī – Qusṭā ibn Lūqā – Ibn Sinā – Ibn Sahl – al-Bīrūnī – al-Shīrāzī – Ḥunayn ibn Isḥāq – Ibn Rushd – Physics

OTTOMAN SCIENCE "Ottoman Science" is a term encompassing the scientific activities that occurred throughout the Ottoman epoch in the lands where the empire extended. The Ottoman Empire, which was established as a small principality at the turn of the fourteenth century, gradually expanded into the lands of the Byzantine Empire both in Anatolia and the Balkans. Its sovereignty reached the Arab world after 1517. It became the most powerful state of the Islamic world in a vast area extending from Central Europe to the Indian Ocean and persisted by keeping the balances of power with Europe. Following its defeat in World War I, the Ottoman Empire disintegrated in 1923.

Ottoman science emerged and developed on the basis of the scientific legacy and institutions of the pre-Ottoman Seljukid period in Anatolian cities, and benefited from the activities of scholars who came from Egypt, Syria, Iran, and Turkestan, which were the most important scientific and cultural centers of the time. The Ottomans brought a new dynamism to cultural and scientific life in the Islamic world and enriched it. Thus, the Islamic scientific tradition reached its climax in the sixteenth century. Besides the old centers of the Islamic civilization, new centers flourished, such as Bursa, Edirne, Istanbul, Skopje, and Sarajevo. The heritage which developed in this period constitutes the cultural identity and scientific legacy of present-day Turkey as well as several Middle Eastern, North African, and Balkan countries. This article aims to give an overview of the formation and development of Ottoman science in Anatolia and the scientific activities which expanded later from Istanbul, the capital of the empire, to Ottoman lands.

The Ottomans always sought solutions to the intellectual and practical problems they encountered in Islamic culture and science. But when the scientific and industrial revolutions occurred in Europe, a gap emerged between them and the Western world. Thus, Ottomans began to make some selective transfers from Western science, and gradually the scientific tradition began to change from "Islamic" to "Western". Ottoman science should therefore be studied under two headings; the classical Islamic tradition and the modern Western one. Although it is difficult to demarcate the two traditions in a clear cut way in the transition period, as the contacts became more frequent, the two periods were separated more clearly.

In the classical period, the *medrese* (*madrasa* in Arabic; college) was the source of science and education and the most important institution of learning in the Ottoman Empire. The Ottoman *medreses* continued their activities from the establishment of the state until approximately the turn of the twentieth century. The basic structure of the *medreses* remained the same within the framework of the Islamic tradition, but in terms of organization they underwent several changes in the Ottoman period. Starting with the first *medrese* established in 1331 in Iznik (Nicaea) by Orhan Bey, the second Ottoman sultan (1326–1362), all *medreses* had *waqfs* (public foundations) supporting their activities. The *waqfiyes* (the deed of endowment of a *waqf*) of important *medreses* stipulated that both the religious *(‹ulūm al-shar‹īa)* and rational sciences (*‹ulūm al-awā›il*) such as mathematics, astronomy, medicine, and physics be taught in these institutions. Besides the *ulema* who provided religious, scientific, and educational services, the *medreses* also trained the administrative and judicial personnel for bureaucratic posts. The *ulema*, members of the Muslim learned, cultural, and religious institution (*Ilmiye*) who played an important role in every aspect of social and official life, were recruited from the *medreses*. They had a twofold duty of interpreting and implementing Islamic law; the *mūftīs* fulfilled the first of these duties and the *qāḍīs* (magistrates) the second. The *ulema* were responsible for applying the *sharī‹a* (the sacred law of Islam) and *Qānūn* (Sultanic law) in the affairs of state. Starting from the reign of Mehmed II (Fātih, known as the Conqueror, 1451–1481), the number of *medreses* increased considerably; to facilitate differentiation among them, they were given ranks.

Shortly after Mehmed II conquered Istanbul, he built the Fātih *Külliye* (complex) which comprised a mosque located at the center, as well as colleges, a hospital, a *mektep* (elementary mosque school), a public kitchen, and other components located around the mosque. It set an example for similar edifices built by the sultans' successors and high-ranking members of the ruling class. The *Sahn-ı Semān Medreses* (Eight Court Colleges) of the Fātih Complex, comprising sixteen adjacent *medreses*, represented the first Ottoman *medreses* that had the structure of a university campus. Owing to the political stability and economic prosperity of the period of the Conqueror, distinguished scholars and artists of the Islamic world assembled in the capital of the empire. The Ottomans particularly protected the Muslim and Jewish scientists fleeing from the persecution that took place after the fall of Granada in 1492, providing them shelter in Ottoman lands. Moreover, as the *waqfs* which were the financial sources of *medreses* grew rich, scientific and educational life developed further.

The scholars who graduated from the *medreses* served as teachers, *qāḍīs* , kazaskers (military judges) and chief *müftīs*. Several physicians were trained and many patients were treated in the *darüşşifa* (hospital) of Fātih Complex which was active until the mid-nineteenth century. The Fātih Complex provided services for the society in various areas such as religion, education and science, health, and nourishment. From the second half of the nineteenth century, as the activities of the Fātih Complex became gradually ineffective, its various units, namely its hospital, its *tābhāne* (hospice), its *muvakkithāne* (timekeeper's office), caravanserai, and school fell out of service. Finally, when all *medreses* were closed in the Republican Period, its colleges, too, became inactive in 1924. The mosque of the Complex, however, has preserved its principal function to a considerable extent from its establishment until the present day.

The establishment of the Süleymāniye *Külliye* by Süleyman the Magnificent (1520–1566) in the sixteenth century marked the final stage in the development of the *medrese* system where, besides the conventional *medreses*, a specialized one named *Dārüttıb* (Medical College) was founded. Thus, for the first time in Ottoman history, in addition to the *şifahanes* (hospitals), an independent institution was established to provide medical education. The other specialized *medreses* established by the Ottomans were the *Dārülhadīs* and the *Dārülkurrā*. The *Dārülhadīs* had the highest rank in the *medrese* hierarchy.

In addition to the *medreses* which gave basic education, there were the institutions where medical sciences and astronomy were practiced and taught by the master-apprentice method. These were the *şifahanes*, the office of *müneccimbaşı* and the *muvakkithānes*.

The institutions which provided health services and medical education were called *darüşşifa*, *şifahane* or *bīmāristan*. The Seljukids had built *darüşşifas* in the cities of Konya, Sivas, and Kayseri. Similarly, the Ottomans built several *darüşşifas* in cities such as Bursa, Edirne, and Istanbul. Some Western sources mention that there were a great number in Istanbul in the sixteenth and seventeenth centuries. This indicates the importance that Ottomans attributed to *darüşşifas*. The Ottoman *darüşşifas* were not constructed as independent buildings, but as part of a *külliye*.

In the Ottoman palace administration, the person in charge of directing the astronomers was called *müneccimbaşı*, i.e. chief astronomer. The position of chief astronomer was established sometime between the late fifteenth and early sixteenth centuries. The chief astronomers were selected from among the *ulema* who were graduates of *medreses*. From the sixteenth century, they started to prepare calendars, fasting timetables, and horoscopes for the palace and prominent statesmen. Until 1800 the calendars were made according to the *Zīj* (astronomical handbook) of Uluğ Bey (Ulugh Beg); after that the *Zīj* of Jacques Cassini was used. The chief astronomer and sometimes a senior astronomer fixed the most propitious hour for important or trivial events such as imperial accessions, wars, imperial births, wedding ceremonies, the launching of ships, etc. Moreover, the chief astronomers followed extraordinary events related to astronomy such as the passage of comets, earthquakes, and fires as well as solar and lunar eclipses, and passed this information on to the palace with related interpretations. The administration of the *muvakkithānes* was also a duty of the chief astronomer. Besides these, the observatory founded in Istanbul was administered by the chief astronomer Taqī al-Dīn (d. 1585). Thirty-seven held the post of chief astronomer until the end of the empire in 1923. The office of *başmuvakkitlik* (chief of timekeepers) was established in 1927.

The timekeeper's offices (*muvakkithānes*) were public buildings located in the courtyards of mosques or *masjids* in almost every town. They were widely built by the Ottomans especially after the conquest of Istanbul. They were administered by the foundation (*waqf*) of the complex (*külliye*) and the persons who worked in the *muvakkithānes* were named *muvakkit*, meaning the person who kept the time, especially for the prayer hours. The major instruments used in the *muvakkithānes* were the following: quadrant, astrolabe, sextant, octant, hourglass, sundial, mechanical clock, and chronometer. Depending on the level of knowledge of the timekeepers, the *muvakkithānes* functioned as locations where astronomy was taught and also as simple observatories. Thus, some of the *muvakkithānes* were important for the education of chief astronomers. Indeed, quite a number of successful timekeepers rose to that rank. The timekeepers were appointed by the chief astronomer. The son of the deceased had priority, and if there were no son, a candidate would be selected by examination.

The Ottoman scientific literature in the classical period was produced mainly within the milieu of the *medrese*. Scholars compiled several original works and translations in the fields of religious sciences as well as mathematics, astronomy, and medicine, besides a great number of textbooks. These works were written in Arabic, Turkish, and Persian, the three languages called *elsine-i selâse* which Ottoman scholars knew. In the beginning, the literature was mostly written in Arabic, but from the fifteenth century onwards, Turkish was used more and more. From the eighteenth century, the majority of the scientific works were written in Turkish and upon the establishment of the first printing house in Istanbul in 1727, Ottoman Turkish became the most frequently used language in the transfer of modern sciences.

Bursalı Kadızāde-i Rūmī (d. 844 AH/AD 1440, known also as Qāḍī Zādeh al-Rūmī) made the first important contribution to the development of the Ottoman scientific tradition and literature. He flourished in Anatolia and settled in Samarkand after he compiled his first work. Qāḍī Zādeh wrote *Sharḥ Mulakhkhaṣ fī'l-Hay'a* (Commentary on the 'Compendium on Astronomy') and *Sharḥ Ashkāl al-Ta'sīs* (Commentary on 'The Fundamental Theorems') in Arabic in the fields of astronomy and mathematics and became the chief instructor at the Samarkand *medrese* and the director of the observatory founded by Ulugh Beg (d. 853/1449) in Samarkand. He was also the co-author of *Zīj-i Jurjānī* (The Astronomical Tables of Ulugh Beg) written in Persian. He simplified the calculation of the sine of a one degree arc in his work *Risāla fī Istikhrāj Jaybi Daraja Wāḥida* (Treatise on the Calculation of the Sine of a One Degree Arc). Qāḍī Zādeh's two students from Turkestan, Ali Kuşçu (d. 879/1474) and Fatḥullaḥ al-Shirwānī (d. 891/1486), influenced Ottoman science by disseminating mathematics and astronomy in the Ottoman Empire. In the preface of his work *Sharḥ Ashkāl al-Ta'sīs*, Qāḍī Zādeh indicated that the philosophers who ponder about the creation and the secrets of the universe, the jurists (*faqihs*) who give *fetvās* in religious matters, the officials who run the affairs of state, and the *qāḍīs* who deal with judicial matters should know geometry. Thus, he emphasized the necessity of science in philosophical, religious, and worldly matters. This understanding reflects a general characteristic of Ottoman science in the classical period. In the period of modernization, however, the Western concept of man's domination of nature through science and technology was foreign to Ottoman scholars.

Other astronomy books of this period included *Urjūza fī Manāzil al-Qamar wa Ṭulū'ihā* (Poem on the Mansions of the Moon and their Rising) and *Manẓūma fī Silk al-Nujūm* (Poem on the Orbits of the Stars) written by 'Abd al-Wahhāb ibn Jamāl al-Dīn ibn Yūsuf al-Maridānī in Arabic. The founder of the Marāgha school Naṣīr al-Dīn al-Ṭūsī's two books entitled *Risāla fi'l-Taqwīm* (Treatise on the Calendar) and *Sī Faṣl fi'l-Taqwīm* (Thirty Sections on the Calendar) were translated from Persian into Turkish. Ahmed-i Dâ'î (d. ca. 825/1421) is the translator of the second work.

During this period, Egypt was another source for Ottoman science. Hacı Paşa (Celaleddin Hıdır) (d. 1413 or 1417), a well-known physician of the time educated in Egypt, wrote two books in Arabic entitled *Shifā' al-Asqām wa Dawā' al-Ālām* (Treatment of Illnesses and the Remedy for Pains) and *Kitāb al-Ta'ālīm fi'l-Ṭibb* (Book on the Teaching of Medicine) which played an important part in the development of Ottoman medicine. He had many other works in Turkish and Arabic.

In medicine, the works of Sabuncuoğlu Şerefeddin (d. ca. 1468) are particularly important in the development of Ottoman medical literature and their influence on Safavid medicine. The first book on surgery that he wrote in Turkish entitled *Jarrāḥiyāt al-Khāniyya* (Treatise on Surgery of the Sultans) comprises the translation of Abu'l-Qāsım Zahrāwī's *al-Taṣrīf*, a self-contained handbook of the medical arts, and the three sections that he himself wrote. This work is much renowned in the history of Islamic medicine in that it illustrates surgical operations with miniatures for the first time. Besides the classical Islamic medical information, this work contains Turco-Mongolian and Far Eastern influences as well as the author's own experiences.

Ottoman science developed further owing to the personal interest of Mehmed II and the educational institutions which he established after the conquest of Istanbul. Consequently, some brilliant scholars emerged in the sixteenth century and made original contributions to science in this period. Mehmed the Conqueror patronized the Islamic scholars and at the same time ordered the Greek scholar from Trabzon, Georgios Amirutzes, and his son to translate Ptolemy's *Geography* into Arabic and to draw a world map. Mehmed II's interest in European culture had started while he was the crown prince settled in the Manisa Palace. In 1445, Italian humanist Ciriaco d'Ancona and other Italians who were in the palace taught him Roman and European history. While Patriarch Gennadious prepared his work on the Christian belief *I'tikādnāme* (The Book on Belief) for the sultan, Francesco Berlinghieri and Roberto Valtorio presented their works *Geographia* and *De re Militari*. Mehmed II also encouraged the scholars of his time to produce works in their special fields. For example, for the comparison of al-Ghazālī's criticisms of peripatetic philosophers regarding metaphysical matters, expressed in his work titled *Tahāfut al-Falāsifa* (The Incoherence of the Philosophers), and Ibn Rushd's answers to these criticisms in his work *Tahāfut al-Tahāfut* (The Incoherence of Inchoherence), he ordered two scholars, Hocazāde and 'Alā al-Dīn al-Ṭūsī, each to write a work on this subject.

No doubt the most notable scientist of the Conqueror's period is Ali Kuşçu, a representative of the Samarkand tradition. He wrote twelve works on mathematics and astronomy. One of them is his commentary on the *Zīj-i Uluğ Bey* in Persian. His two works in Persian, namely, *Risāla fi'l-*

Hay'a (Treatise on Astronomy) and *Risāla fi'l-Ḥisāb* (Treatise on Arithmetic) were taught in the Ottoman *medreses*. He rewrote these two works in Arabic with some additions under new titles, *al-Fatḥiyya* (Commemoration of Conquest) and *al-Muḥammadiyya* (The Book Dedicated to Sultan Muhammed), respectively. Another noteworthy scholar of the Bayezid II period (1481–1512) was Molla Lūtfi. He wrote a treatise about the classification of sciences titled *Mawḍūʿāt al-ʿUlūm* (Subjects of the Sciences) in Arabic and compiled a book on geometry titled *Taḍʿīf al-Madhbaḥ* (Duplication of the Cube) which was partly translated from Greek. Mīrīm Çelebi (d. 1525) who was a well-known astronomer and mathematician of this period and the grandson of Ali Kuşçu and Qāḍī Zādeh, contributed to the establishment of the scientific traditions of mathematics and astronomy and was renowned for the commentary he wrote on the *Zīj* of Ulugh Beg.

Some scholars who came from Andalusia also contributed to the Ottoman scientific literature. The Arabic medical and astronomical works of the Andalusian scholar ʿAbd al-Salām al-Muhtadī al-Muḥammadī (sixteenth century), who settled in Istanbul during the reign of Bayezid II and gave up his Jewish name Ilya ibn Abrām al-Yahūdī after embracing Islam, are examples of such contributions. In a treatise which he wrote in Hebrew and then translated into Arabic in 1503, he introduced the instrument called *al-Ḍābid*, which was his own invention, and stated that it was superior to the *Dhāt al-ḥalaq* (armillary sphere) invented by Ptolemy. This treatise illuminates an aspect of Ottoman scientific literature which is not much known.

Scientific literature developed considerably in the period of Sultan Süleyman the Magnificent. We find two major mathematical books in Turkish entitled *Jamāl al-Kuttāb wa Kamāl al-Ḥussāb* (Beauty of Scribes and Perfection of Accountants) and *ʿUmdat al-Ḥisāb* (Treatise on Arithmetic) by Naṣūḥ al-Silāḥī al-Matrāqī (d. 971/1564). His book in Turkish entitled *Beyān-ı Menāzil-i Sefer-i ʿIrakeyn* (Description of the Stopping Places on the Campaign to the Two Iraqs), related to geography, should also be mentioned. Mūsā ibn Hāmūn (d. 1554), one of the famous Jewish physicians of Andalusian descent, was appointed as Sultan Süleyman's physician and wrote the first Turkish and one of the earliest independent works on dentistry which is based on Greek, Islamic, and Uighur Turkish medical sources and in particular on Sabuncuoğlu Şerefeddin's works. In the sixteenth century, important works on astronomy were written by the representatives of the Egypt–Damascus tradition of astronomy-mathematics. The greatest astronomer of this period was Taqī al-Dīn al-Rāṣid (d. 1585) who combined the Egypt–Damascus and Samarkand traditions. He wrote more than thirty books in Arabic on the subjects of mathematics, astronomy, mechanics, and medicine.

Taqī al-Dīn came from Egypt to Istanbul in 1570. In 1571, he was appointed *müneccimbaşı* (chief astronomer) by Sultan Selīm II (1566–1574). Shortly after Sultan Murād III's (1574–1595) accession to the throne, he started the construction of the observatory of Istanbul. It is understood from his *Zīj* titled *Sidrat Muntahā'l-Afkār* (The Nabk Tree of the Extremity of Thoughts) that he made observations in the year 1573. It is generally agreed that the observatory was demolished on 4 Dhū'l-Ḥijja 987 corresponding to 22 January 1580. Therefore, it can be estimated that he carried out observations from 1573 until 1580.

In addition to the instruments of observation which were used until his time, Taqī al-Dīn invented new ones such as the *Mushabbaha bi'l-manāṭiq* (sextant) and *Dhāt al-awtār* in order to determine the equinoxes. Moreover, he also used mechanical clocks in his observations. When one compares the instruments of observation used by Tycho Brahe (1546–1601), a famous astronomer of this period, and those used by Taqī al-Dīn, one sees that they are very similar.

Taqī al-Dīn developed a different method of calculation to determine the latitudes and longitudes of stars by using Venus and the two stars near the ecliptic, i.e. Aldebaran and Spica Virginis. He determined that the magnitude of the annual movement of the Sun's apogee was 63″. Considering that the value known today is 61″, the method he used appears to be more precise than the methods of Copernicus (24″) and Tycho Brahe (45″).

Starting with Ptolemy in the second century AD and continuing until Copernicus in the sixteenth century, the Western world used chords for measuring angles. For this reason, the calculation of the value of the chord of 1° has been an important matter for astronomers. Thus, while Copernicus used the method based on the calculation of the chord of 2° that yielded an approximate value, Taqī al-Dīn used trigonometric functions such as the sine, cosine, tangent, and cotangent to measure the values of angles, in line with the tradition of Islamic astronomy. Inspired by Ulugh Beg, Taqī al-Dīn developed a different method to calculate the sine of 1°. Furthermore, he applied decimal fractions, which had been previously developed by Islamic mathematicians such as al-Uqlidīsī and al-Kashī, to astronomy and trigonometry, prepared sinus and tangent tables accordingly, and used them in his work titled *Jarīdat al-Durar wa Kharīdat al-Fikār*.

The first contact with Copernican astronomy in the Islamic world occurred around mid-seventeenth century when the Ottoman astronomer Tezkereci Köse Ibrāhim Efendi of Szigetvar translated a work by the French astronomer Noel Durret (d. ca. 1650). The introduction and spread of Copernicus' new heliocentric concept into the Ottoman world did not cause a conflict between religion and science, contrary to the case in Europe. This concept, which was first seen as a technical detail, was later preferred to Ptolemy's geocentric system and considered more suitable with respect to religion. However, the conflict between religion and science entered into Ottoman Turkish intellectual life around the end of the nineteenth century together with Western trends of thought such as positivism and biological materialism.

The Ottomans needed knowledge of geography in order to determine the borders of their continuously expanding territory and to establish control over the military and commercial activities in the Mediterranean, the Black Sea, the Red Sea, and the Indian Ocean. They made use both of the geographical works of classical Islam and of works of European origin. By adding their own observations, Ottoman geographers produced original works as well. The first source of the Ottoman knowledge about geography is the Samarkand tradition of astronomy and geography.

From the sixteenth century onwards, noteworthy geographical works were produced by Pirī Reis. In 1511, Pirī Reis drew his first map. This map is part of the world map prepared on a large scale. It was drawn on the basis of his own rich and detailed drafts and European maps including Columbus' map of America. This first Ottoman map which included preliminary information about the New World represents southwestern Europe, northwestern Africa, southeastern and Central America. It is a portolano, without latitude and longitude lines but with lines delineating coasts and islands. Pirī Reis drew his second map and presented it to Süleyman the Magnificent in 1528. Only the part which contains the North Atlantic Ocean and the then newly discovered areas of Northern and Central America is extant. Pirī Reis also wrote a book entitled *Kitāb-ı Bahriye* (Book of the Sea, 1521). In this work, Pirī Reis presents drawings and maps of the cities on the Mediterranean and Aegean coasts, and gives extensive information about navigation and nautical astronomy. Admiral Seydī Ali Reis (d. 1562), who wrote the work in Turkish titled *al-Muḥīṭ* (The Ocean), was a notable figure of the period in maritime geography. This work contains astronomical and geographical information necessary for long sea voyages and his own observations about the Indian Ocean.

Another work of the sixteenth century which contains information about the geographical discoveries and the New World is the book entitled *Tārih-i Hind-i Garbī* (History of Western India). This work, whose author is unknown, was presented to Sultan Murād III in 1583. It was based on Spanish and Italian geographical sources. It is important in showing that the geographical discoveries of the West were known to the Ottomans. The work has three parts; the third part, which is the most important and which comprises two thirds of the whole book, relates the adventures of Columbus, Balboa, Magellan, Cretes, and Pizarro during the sixty years from the voyage to America in 1492 until 1552. Apparently, cartography was organized as a profession in the Ottoman Empire; for example, in the seventeenth century, fifteen individuals were occupied with the art of surveying, in eight locations in Istanbul and nearby areas.

From the seventeenth century onwards, the new medical doctrines which were put forward by Paracelsus and his followers in the sixteenth century began to be observed in the Ottoman medical literature under the names of *Tıbb-ı cedīd* (new medicine) and *Tıbb-ı kimyāī* (chemical medicine), in the works of Ṣāliḥ ibn Naṣrullāh (d. 1669), ʿOmar ibn Sinan al-Iẓniki (eighteenth century), and ʿOmar Şifāī (d. 1742). Şemseddin Itākī's book on anatomy (1632) reflects the first influences of European anatomists. Ottoman medical literature carried both classical Islamic and European medical information side by side until the beginning of the nineteenth century when Şānīzāde Atāullah (d. 1826) wrote his work entitled *Hamse-i Şānīzāde* (Five Works of Şānīzāde) composed of four parts (physiology, pathology, surgery, and pharmacology) based totally on European sources without any reference to traditional medicine.

From the seventeenth century onwards, conditions were no longer conducive to the development of science because of the social and economic disruption resulting from the weakening of the central authority, dissolution of political stability, decreasing conquests, loss of land, influx of abundant American silver into Europe, and the diminishing revenues of the empire. The factors that had encouraged scholars to conduct scientific work disappeared and were replaced by the struggle to make a living. Disputes arose in the seventeenth century between the supporters of *salafī* Islam and mysticism among Ottoman intellectuals. The upholders of *salafīya*, who started the movement known as the Kadızādeli, had a negative attitude to philosophy and science which led to the regression of Ottoman science.

The famous Ottoman scholar and bibliographer Kâtip Çelebi (d. 1658), who is also known under the name of Hacı Halife, was one of the first Muslim intellectuals to notice the gap between the levels of scientific development of Europe and the Ottoman world. Kâtip Çelebi was able to approach analytically both classical Islamic culture and modern Western culture. He wrote in Arabic and Turkish on a variety of subjects. In history, he translated from Latin the *Chronik* of Johann Carion that he titled *Tārih-i Firengī Tercümesi* (Translation of European History) and compiled his *Ravnak al-Salṭana* (Splendor of the Sultanate) on the basis of works by authors such as Johannes Zouaras, Nicestas Acominate, Nicephorus Gregoras, and the Athenian Laonikas Chalcondyle. In the field of geography, he translated the *Atlas Minor* of Mercator and Hondius under the title *Lawamiʿ al-Nur fi Zulmat Atlas Minur* (Flashes of Light on the Darkness of Atlas Minor). Furthermore, in his work titled *Mīzān al-Haqq fī Ịkhtiyār al-Aḥaqq* (The Balance of Truth and the Choice of the Truest), Kâtip Çelebi criticized the intellectual life of his period.

The Ottoman world was the first environment with which Western science came into contact outside its own milieu, due to the close interaction and geographical proximity of the Ottomans with European countries. In the early periods when the Ottomans had contact with and transferred some Western technics (especially firearms, cartography, and mining) they also had some early contacts with Renaissance science (astronomy, medicine) through the emigrant Jewish scholars. Particularly in the early centuries, this interest of the Ottomans developed in a selective manner because of their feeling of superiority and their autarchic system. But functional transfers from European science developed gradually because of increasing needs, as the military, political, and economic balances turned against them. In these periods, the Ottomans required immediate transfers of science and technology to strengthen their military power. Thus, they established the imperial engineering schools at the end of the eighteenth century and the imperial medical school at the beginning of the nineteenth century. Major reforms known as the *Tanzīmāt* (1839) led to a shift in the process of selective transfer to include public ends and civilian objectives. In the second half of the nineteenth century individuals started to establish professional and learned associations similar to those in the West. These new corporate bodies with their legal statute and work procedures, which did not exist in the classical period, added a new dimension to Ottoman cultural and scientific life.

Ịshak Efendi (d. 1836), who was chief instructor in the Imperial School of Engineering, had a leading role in the transfer of modern science. Among his thirteen books, which he wrote using Western and particularly French sources, *Mecmūʿa-i ʿUlūm-i Riyāziye* (Compendium of Mathematical Sciences, four volumes) is of special importance, since it is the first attempt in any language of the Muslim world to present a comprehensive textbook on different sciences such as mathematics, physics, chemistry, astronomy, biology, botany, and mineralogy in one compendium. Ịshak Efendi's efforts to find the equivalents of the new scientific terminology and his influence on the transfer of modern science spread in other Islamic countries beyond Ottoman Turkey.

The Ottomans' interest was oriented towards practical ends and the application of scientific discoveries, while the three main aspects of modern Western science, namely theory, experiment, and research were not taken into consideration. This understanding was reflected in the educational and scientific policy of the Ottoman State before and during the *Tanzīmāt* period. The Ottomans made several attempts to establish an institution for higher education under the name of *Dārülfünūn* (House of sciences), apart from the *medrese*, in line with the model of the European university. However, they disregarded the importance of scientific research in the program of this institution and those of the previously established ones. For this reason, they were not as successful as their counterparts in Russia and Japan. The dimension of research was introduced to Ottoman scholarly circles upon the establishment of the Faculty of Sciences (1900) which started to function as a part of Istanbul University.

Ottoman contacts with European science and technology started with the purpose of fulfilling their needs, in a selective way. However, after a long process, they abandoned their own scientific traditions and began to think that development and progress could only be accomplished through Western science and technology.

EKMELEDDIN ỊHSANOĞLU

REFERENCES

Adıvar, Adnan. *Osmanlı Türklerinde Ịlim,* 5th ed. Ed. Aykut Kazancıgil and Sevim Tekeli. Istanbul: Remzi Kitabevi, 1983.

Aydüz, Salim. *Osmanlı Devleti'nde Müneccimbaşılık ve Müneccimbaşı.* M.A. thesis, Istanbul University, 1993.

Çavuşoğlu, Semiramis. *The Kadızadeli Movement: An Attempt of Şeri'at-Minded Reform in the Ottoman Empire.* Ph.D. dissertation, Princeton University, 1990.

Demir, Remzi. *Takıyüddin'in Ceridetü'd-Dürer ve Haridetü'l-Fiker Adlı Eseri ve Onun Ondalık Kesirleri Astronomi ve Trigonometriye Uygulaması.* Ph.D. dissertation, Ankara University, 1992.

Goodrich, Thomas C. *The Ottoman Turks and the New World, A Study of Tarih-i Hind-i Garbi and Sixteenth- Century Ottoman Americana.* Wiesbaden: Otto Harrassowitz, 1990.

Ịhsanoğlu, Ekmeleddin. *Başhoca Ịshak Efendi: Türkiye'de Modern Bilimin Öncüsü* (Chief Instructor Ishak Efendi: Pioneer of Modern Science in Turkey). Ankara: Kültür Bakanlığı Yayınları, 1989.

Ịhsanoğlu, Ekmeleddin. "Some Remarks on Ottoman Science and its Relation with European Science & Technology up to the End of the Eighteenth Century." Papers of the First Conference on the Transfer of Science and Technology between Europe and Asia since Vasco da Gama (1498–1998). *Journal of the Japan-Netherlands Institute* 3: 45–73, 1991.

Ịhsanoğlu, Ekmeleddin. "Introduction of Western Science to the Ottoman World: A Case Study of Modern Astronomy (1660–1860)." In *Transfer of Modern Science & Technology to the Muslim World.* Ed. Ekmeleddin Ịhsanoğlu. Istanbul: The Research Centre for Islamic History, Art and Culture, 1992, pp. 67–120.

İhsanoğlu, Ekmeleddin. "Ottoman Science in the Classical Period and Early Contacts with European Science and Technology." In *Transfer of Modern Science & Technology to the Muslim World.* Ed. Ekmeleddin İhsanoğlu. Istanbul: The Research Centre for Islamic History, Art and Culture, 1992, pp. 1–48.

İhsanoğlu, Ekmeleddin. "Ottomans and European Science." In *Science and Empires.* Ed. Patrick Petitjean, Catherine Jami and Anne Marie Moulin. Dordrecht: Kluwer Academic Publishers, 1992, pp. 37–48.

İhsanoğlu, Ekmeleddin. "Tanzimat Öncesi ve Tanzimat Dönemi Osmanlı Bilim ve Eğitim Anlayışı." *150. Yılında Tanzimat.* Ed. Hakkı Dursun Yıldız. Ankara: Türk Tarih Kurumu Basımevi, 1992, pp. 335–395.

İhsanoğlu, Ekmeleddin. "Başhoca Ishak Efendi, Pioneer of Modern Science in Turkey." In *Decision Making and Change In The Ottoman Empire.* Kirksville, Missouri: The Thomas Jefferson University Press at Northeast Missouri State University, 1993, pp. 157–168.

İnalcık, Halil. *The Ottoman Empire: The Classical Age, 1300–1600.* Trans. Norman Itzkowitz and Colin Imber. London: Weidenfeld and Nicolson, 1973.

Kâhya, Esin. *El-'Itaqi, The Treatise on Anatomy of Human Body.* Islamabad: National Hijra Council, 1990.

Miroğlu, İsmet. "İstanbul Rasathanesine Ait Belgeler." *Tarih Enstitüsü Dergisi* 3:75–82, 1973.

Russell, Gül. "The Owl and the Pussycat: The Process of Cultural Transmission in Anatomical Illustration." In *Transfer of Modern Science & Technology to the Muslim World.* Ed. Ekmeleddin İhsanoğlu. Istanbul: Research Centre for Islamic, History, Art and Culture, 1992, pp. 180–212.

Sarı, Nil & Zülfikar Bedizel. "The Paracelsusian Influence on Ottoman Medicine in the Seventeenth and Eighteenth Centuries." In *Transfer of Modern Science & Technology to the Muslim World.* Ed. Ekmeleddin İhsanoğlu. Istanbul: Research Centre for Islamic, History, Art and Culture, 1992, pp. 157–179.

Sayılı, Aydın. *The Observatory in Islam.* Ankara: Publications of the Turkish Historical Society, 1960.

Tekeli, Sevim. "Nasirüddin, Takiyüddin ve Tycho Brahe'nin Rasad Aletlerinin Mukayesesi." *Ankara Üniversitesi, Dil ve Tarih-Coğrafya Fakültesi Dergisi* 16(3–4): 301–393, 1958.

Tekeli, Sevim. "Onaltıncı Yüzyıl Trigonometri Çalışmaları Üzerine bir Araştırma, Copernicus ve Takiyüddin (Trigonometry in the Sixteenth Century, Copernicus and Taqi al Din)." *Erdem* 2(4): 219–272, 1986.

Terzioğlu, Arslan. *Moses Hamons Kompendium der Zahnheilkunde aus dem Anfang des 16. Jahrhunderts.* München, 1977.

Terzioğlu, Arslan. "Bîmâristan, İslâm dünyasında klasik hastahanelerin genel adı." *Türkiye Diyanet Vakfı İslâm Ansiklopedisi* 6: 163–198, 1992.

Unan, Fahri. *Kuruluştan Günümüze Fâtih Külliyesi.* Ph.D. dissertation, Hacettepe University, Ankara, 1993.

Ünver, A. Süheyl. "Osmanlı Türkleri İlim Tarihinde Muvakkithaneler." *Atatürk Konferansları 1971–72.* 5:217–257, 1975.

Uzel, İlter. *Dentistry on the Early Turkish Medical Manuscripts.* Ph.D. dissertation, University of Istanbul, 1979.

Uzel, İlter. *Şerefeddin Sabuncuoğlu. Cerrahiyyetü'l Haniyye.* Ankara: Publications of the Turkish Historical Society, 2 vols., 1992.

See also: Ulugh Beg – *Zīj* – Taqī al-Dīn – Pirī Reis – Maps and Mapmaking – Qāḍī Zādeh al-Rūmī – Rationality, Objectivity, and Method

P

PAKṢA *Pakṣa* (Half) refers, in Indian astronomy, to half the lunar month. The lunar month, which is the interval between two successive new moons (*amāvāsyā*) or full moons (*paurṇamāsyā*), is a visible natural phenomenon, which occurs at regular intervals and, as such, has been used for reckoning time in all ancient civilizations, including that of India. For the people of India of the Vedic times, the two *pakṣas*, one *śukla* (bright) and the other *kṛṣṇa* (dark), provided an easy and most convenient instrument for reckoning time. From actual observation, it was found that a *pakṣa* was completed in fifteen lunar days (*tithis*) and these were named after their serial numbers, as *prathamā* (first), *dvitīyā* (second), *tṛtīya* (third), *caturthī* (fourth) etc., up to *caturdaśī* (fourteenth), and the fifteenth was *amāvāsyā* or *paurṇamāsyā*, according to whether it was the dark fortnight or the bright fortnight.

The moon moves about thirteen degrees in its orbit, eastward, in a day of 24 hours, but since the sun also moves in the same direction, one degree, during the same period of 24 hours, the moon's resultant displacement, which constitutes the *tithi*, is about twelve degrees. A lunar month of thirty *tithis* would be around 29 and a half solar days, and in a solar year of 365 days, the moon would have completed its twelve months in about 354 days, and would have moved further by eleven *tithis*. In order to correlate and bring together the solar and lunar years, the astronomical practice is to expunge and leave uncounted one lunar month every three years, when the extra eleven *tithis* would have accumulated to one month. This expunged month is called *adhika-māsa* or 'added' or 'intercalary' month.

Time reckoning by *tithi* and *pakṣa* is very much in vogue, even today, among Hindus, for fixing auspicious times for ritual, religious, and social functions, and for horoscopic astrology.

K.V. SARMA

PAPER AND PAPERMAKING Paper is a felted sheet of plant fibers formed from a water suspension using a sieve-like screen. When water escapes and dries, the layer of interwoven fibers becomes a thin matted sheet and is called paper. It is one of the four great inventions of China, along with gunpowder, the compass, and printing. Traditionally, paper was thought to be invented by a eunuch, Cai Lun (ca. 60–121), in the year 105. But according to recent archaeological discoveries, paper made of hemp fibers from rags had already been made in the second century B.C. as a new type of writing material. After that papermaking was further improved and popularized in the second to the third centuries in China. During the third to the sixth centuries the raw materials for making paper were expanded. Apart from hemp, paper made from paper mulberry bark and rattan was produced. This marked an important stage in the history of papermaking. Meanwhile, paper converting was developed, such as sizing with starch, coating with white mineral powder and dyeing with various dyestuffs, especially cork tree bark (*Phellodendron amurens*) which makes paper yellow and moth-proof. Movable types of screens were generally used for making paper. The development of papermaking promoted the flourishing of education, culture, and science. Paper was widely used in daily life to make umbrellas, fans, kites, paper-cuts, toilet paper, and others.

The next important stage (sixth to tenth centuries) was the Tang dynasty (618–907), in which papermaking was greatly developed. In addition to the above mentioned kinds of paper, some new kinds appeared, such as bamboo paper and bark paper made of *Hibiscus mutabilis* and plants of the Thymelaeceae family, such as Daphne. Paper coated with colored wax or powder and wax, as well as that decorated with gold or silver dust, gelatin sized, water-marked, embossed, and marbled were also made in this period. Meanwhile, paper was used for clothing, furnishings, visiting and playing cards, lanterns, armor, window pasters, and in commercial activities as a medium of exchange. A large amount of paper was used for ceremonies and sacrificial purposes, as was the so-called 'fire paper'. The development of block printing also stimulated the manufacture of paper. Another achievement was the production of large-size paper more than 3 m long. The Tang dynasty truly entered the 'Age of Paper'.

After papermaking was perfected and popularized in China, it was spread in all directions throughout the world. It arrived first in Korea and Vietnam in the third to the fourth centuries, then in Japan and India in the seventh. Paper reached the Arabian world in the mid-eighth century and Africa in the tenth. The Arabs monopolized papermaking in the West for 500 years. Only in the twelfth century was it manufactured in Spain and France, then in Italy in the thirteenth. In the seventeenth century, paper was made in most European countries and even in America; in Australia it was produced in the nineteenth. Thus it took more than 1000 years for paper to spread from China to almost every part of the world. Paper and other Chinese inventions played a considerable role in the flourishing of medieval Arabic culture and in the development of European society after that

From the eighteenth century during the Industrial Revolution papermaking itself went through a number of technical reforms in Europe. Chemical pulp appeared in Europe in the mid-nineteenth century, although traditional handmade paper persisted by virtue of its artistic value and elegant appearance. It continues to be produced in China, Japan, Korea, and many Western countries including the United States. In fact, China remains the biggest supplier of handmade paper today. Well-known varieties include *xuan* paper from Anhui, mulberry bark paper from Zhejiang, hemp paper from Shan-xi, and bamboo paper from Sichuan.

PAN JIXING

REFERENCES

Needham, Joseph. *Science and Civilisation in China,* vol. 5, pt. 1, *Paper and Printing,* by Tsien Tsuen-Hsuin. Cambridge: Cambridge University Press, 1985.

Pan, Jixing. *Zhongguo Zaozhi Jishu Shigao* (History of Papermaking Technology in China). Beijing: Cultural Relics Press, 1979; English edition to be published by Foreign Languages Publishing House in Beijing.

Pan, Jixing. "On the Origin of Papermaking." *Bulletin of the International Association of Paper Historians* 2: 38–48, 1981.

PARAMEŚVARA Among Indian astronomers hailing from Kerala, Parameśvara (ca. AD 1360–1465), promulgator of the Kerala Dṛggaṇita School, and author of over a score of works on astronomy and astrology, holds a prominent place. He was a resident of Ālattūr, his house being situated on the northern bank of the river Bhāratappuzha, where he conducted his astronomical experiments and observations for over 55 years.

Parameśvara was a prolific writer. Some of his works still remain in manuscript form, while a few are yet to be found. His works on astronomy are:

1. *Dṛggaṇita*, (Computation True to Observation), AD 1431, his magnum opus, a practical manual, in two versions.
2. *Goladīpikā* I (Illumination on Spherics) , in 302 verses, on spherical astronomy.
3. *Goladīpikā* II (Illumination on Spherics), in four chapters, on the same subject as above, but different from it.
4. *Grahaṇamaṇḍana* (Ornament on Eclipses), in two versions, of 89 and 100 verses.
5. *Grahaṇanyāyadīpikā* (Illuminator on Rationale of Eclipses) in 85 verses.
6. *Grahaṇāṣṭaka* (Octad on Eclipses), on the computation of eclipses.
7. *Vākyakaraṇa*, on methods for the derivation of several astronomical tables.

Parameśvara also commented on several standard works on astronomy that were popular in Kerala. These include: *Āryabhaṭīya* of Āryabhaṭa, *Laghubhāskarīya* and *Mahābhāskarīya* of Bhāskara I, *Mahābhāskarīya-Bhāṣya* of Govindasvāmin, *Laghumānasa* of Muñjāla, *Sūryasiddhānta*, *Līlāvatī* of Bhāskara II, and his own *Goladīpikā* II. All these commentaries except that on *Līlāvatī* have been published:

Although he was primarily an author of astronomical works, Parameśvara also wrote on astrology:

1. *Ācārasaṅgraha*, a popular text.
2. *Jātakapaddhati*, in 44 verses.
3. Commentary on the *Muhūrtaratna* of Govinda.
4. Commentary on the *Jātakakarmapaddhati* of Śrīpati.
5. Commentary on the *Praśnaṣ-aṭapañcāśikā* of Pṛhuyaśas.
6. *Muhūrtāṣṭaka-dīpikā*, *Vākyadīpikā* and *Bhādīpikā*, mentioned by Parameśvara at the end of his commentary on the *Mahābhāskarīya.*

These works are not extant.

In his works Parameśvara evinces a refreshingly scientific outlook. He avers at the beginning of his *Grahaṇamaṇḍana* that he was setting out to compose the work after closely watching the movements and positions of the planets in the skies for a long time. At the beginning of his *Dṛggaṇita*, he says that, in real astronomy, computation should match observation. He enumerates a number of solar and lunar eclipses which he had observed between 1393 and 1432, and gives details about them in his *Grahaṇanyāyadīpikā*. He also points out the error between computed and observed readings and offers corrections and instructs that similar observations should be made at intervals and corrections enunciated for computation and observation to be identical.

K.V. SARMA

REFERENCES

Primary sources

Dṛggaṇita (Computation True to Observation). Ed. K.V. Sarma. Hoshiarpur: Vishveshvaranand Institute, 1963.

Goladīpikā I (Illumination on Spherics). Ed. T. Ganapati Sastri. Trivandrum: Trivandrum Sanskrit Series, No. 49, 1916.

Goladīpikā II (Illumination on Spherics). Ed. K.V. Sarma. Madras: Adyar Library and Research Centre, 1957.

Grahaṇāṣṭaka (Octad on Eclipses). Ed. K.V. Sarma. Madras: Kuppuswami Sastri Research Institute, 1959.

Grahaṇmaṇḍana (Ornament on Eclipses). Ed. K.V. Sarma. Hoshiarpur: Vishveshvaranand Institute, 1965.

Grahaṇanyāyadīpikā (Illuminator on Rationale of Eclipses). Ed. K.V. Sarma. Hoshiarpur: Vishveshvaranand Institute, 1966.

Jātakapaddhati. Ed. Kolatteri Sankara Menon. Trivandrum: Curator's Office, 1926.

Secondary sources

Pingree, David. "Eclipse Observations of Parameśvara between 1393 and 1432." *Journal of the American Oriental Society* 87:337–39, 1967.

Pingree, David. *Census of the Exact Sciences in Sanskrit*. Philadelphia: American Philosophical Society. Series A, Vol. 4, 1981.

Sarma, K.V. *A Bibliography of Kerala and Kerala-based Astronomy and Astrology*. Hoshiarpur: Vishveshvaranand Institute, 1972.

See also: Eclipses – Astronomy in India

PAULIŚA Pauliśa, of Greek origin, is the originator of the *Pauliśa Siddhānta*, one of the five systems of astronomy of the early centuries of the Christian era. These were selectively redacted in the *Pañcasiddhāntikā* of Varāhamihira, the prodigious Indian astronomer-astrologer of the sixth century AD According to a traditional verse attributed to the sage Kāśyapa, Pauliśa is one of the eighteen originators of Indian astronomical systems. Al-Bīrūnī, the Persian scholar, who sojourned in India from 1017 to 1030, stated that *Pauliśa Siddhānta* was written by Paulus-ul-Yunani, i.e. 'Paulus, the Greek'. Probably the Hindus prepared an Indianized Sanskrit *Pauliśa Siddhānta* on the basis of the Greek work. This original Sanskrit work is no longer available, and neither is the commentary on that work by Lāṭadeva which was referred to by Varāhamihira in *Pañcasiddhāntikā* I.3. However, the redaction of the *Siddhānta* in *Pañcasiddhāntikā* is fairly full and provides ample details about the nature and contents of the work. The Pauliśa system has certain things in common with the *Romaka Siddhānta* and with the *Vāsiṣṭha Siddhānta*, two of the other systems redacted in *Pañcasiddhāntikā*.

The epoch of the *Pauliśa Siddhānta*, which *Pañcasiddhāntikā* says is the same as that of *Romaka Siddhānta*, is the Hindu *caitra-śukla-pratipad* (first day of the bright fortnight of the month of Citrā) in the Śaka year 427 elapsed (*Pañcasiddhāntikā* I.8–10), which corresponds to mean sunset at Yavanapura (modern Alexandria in Egypt), or modern Sunday, March 20, 505. It is interesting that the same moment was adopted by Varāhamihira as the epoch of the *Saura Siddhānta* redacted in the *Pañcasiddhāntikā* (i.e. the old *Sūrya Siddhānta*). The Indian time then was 37–20 *nāḍīs* from mean sunrise in Ujjain, on Sunday, March 20, 505. (*Pañcasiddhāntikā* III.13). In *Pañcasiddhāntikā* III.13, it is stated that the *deśāntara-nāḍīs* (longitudinal difference in terms of time, expressed in *nāḍīs*) from Yavanapura to Ujjain is 7–20, while that to Varanasi is 9. The actual intervals according to modern calculations are 7–38 and 8–50, the difference being 18 *vināḍīs* and ten *vināḍīs*, respectively. One minute being equal to four *vināḍīs*, the difference works out only to 4.5 minutes and 2.5 minutes, which is remarkable.

At the commencement of the *Pañcasiddhāntikā* (I.4), Varāhamihira gives a pat to the *Pauliśa Siddhānta* for the accuracy of the *tithi* (lunar day) calculated according to it, though he adds that the *tithi* derived through the *Saura Siddhānta* is much more accurate. In fact, it is these two schools, the Pauliśa and the Saura, one representing the Greek and the other the Indian, that Varāhamihira depicts rather fully in *Pañcasiddhāntikā*, allotting them entire sections. Thus the entire third chapter of *Pañcasiddhāntikā* is devoted to the depiction of planetary computation and allied matters according to the *Pauliśa Siddhānta*, Chapter Five to the Moon's cusps, Chapter Six to lunar eclipses, Chapter Seven to solar eclipses, and much of the long Chapter Eighteen to the motion of the planets. Although the *Pauliśa Siddhānta* was based on a Greek original, the text was painstakingly Indianized, both in the matter of content and presentation.

K.V. SARMA

REFERENCES

Dikshit, S.B. *Bhāratīya Jyotish Sastra (History of Indian Astronomy)*. Trans. R.V. Vaidya. Calcutta: Positional Astronomy Centre, India Meteorological Department, 1981.

Dvivedi, Sudhakara. *Gaṇaka Taraṅgiṇī or Lives of Hindu Astronomers*. Ed. Padmakara Dvivedi. Benares: Jyotish Prakash Press, 1933.

Pañcasiddhāntikā of Varāhamihira Trans. T.S. Kuppanna Sastry. Ed. K.V. Sarma. Madras: P.P.S.T. Foundation, 1993.

The Pañcasiddhāntikā of Varāhamihira Ed. and trans. G. Thibaut and Sudhakara Dvivedi. Varanasi: Chowkhamba Sanskrit Series Office, 1968.

The Pañcasiddhāntikā of Varāhamihira. Ed. and trans. O. Neugebauer and David Pingree. Copenhagen: Munksgaard, 1970.

Pingree, David. "History of Mathematical Astronomy in India." In *Dictionary of Scientific Biography*, vol. 15. New York: Scribners, 1978, pp. 545–54.

See also: Varāhamihira – Astronomy in India – al-Bīrūnī – *Deśāntra*

PHYSICS Most histories of the exact sciences consider "physics" to have begun with the ancient Greeks, especially with the work of Aristotle (384–322 BC). Classical physics is usually said to have been developed during the European "Scientific Revolution" by figures like Galileo Galilei (1564–1642) and Isaac Newton (1642–1727), and modern physics is commonly introduced with accounts of the discovery of relativity and quantum mechanics around 1900. Few histories of physics consider developments outside Europe; and of those that do, non-Western physics is often presented as "derivative" of European work.

However, recent scholarship has made it clear that this European view is incomplete, and that numerous non-Western cultures, including those of China, India, the Islamic world, and others, developed sophisticated physical theories independently of Europe. In this expanded view, physics began in prehistoric times. From its origins in Africa, China, and India, physics developed to a surprisingly advanced level before its incorporation into the European scientific tradition.

Very little historical research into prehistoric African physics has so far been done, but the development of mathematics and astronomy has been dated back to very early times by archaeoastronomers such as Alexander Marshack. The bone tool handles of the Ishango culture of Lake Edward are inscribed with various collections of notches suggesting knowledge of abstract mathematical operations, and knowledge of astronomical events like new and full moons. These bones have been dated anywhere from 8500 to 25 000 BC. American physicist John Pappademos argues for a similar early start to African physics. He argues that the earliest dynamical experiments began with prehistoric African inventions like the spear and the bow and arrow, and that "the practical mastery of the principles of mechanics — the oldest branch of physics — grew as man learned to make flint weapons, tools, dwellings, boats, etc.". Similar arguments, of course, have been made concerning the prehistory of physics in other early cultures.

Pappademos further argues, based on archeological, linguistic, and historical evidence, that "ancient Egypt was essentially an African civilization", and upon this basis he ascribes priority to African cultures for the early development of astronomy, geometry, and measurement. "To the Egyptians", he concludes, "we owe the concepts of most of the fundamental physical quantities: distance, area, volume, weight, and time. Europe is indebted to Egypt for the invention of standards, units and methods for accurate measurements of all these quantities". If this view is correct, it sheds new light on the development of the sciences among the ancient Greeks, whose indebtedness to Mespotamian astronomy and Egyptian geometry has long been acknowledged.

China

A similar "prehistoric origin" may be claimed for physics in China, whose history has been dramatically advanced with the publication of Joseph Needham's multi-volume *Science and Civilisation in China*. Needham argues that traditional Chinese interest in physics was not particularly strong. Yet despite this, Chinese physics developed several interesting and original insights into various physical phenomena.

Chinese physicists, Needham says, were hampered by their lack of an atomic theory, used with great success by both early Greek and Indian, as well as later European scientists. However, their view of the physical universe as a "perfectly continuous whole" rather than as a collection of atoms meant that Chinese scholars were pioneers in understanding action-at-a-distance, and the connection of various forces. Chinese physics tended to emphasize waves rather than particles, and this dominance led to several early discoveries in magnetism, optics, dynamics, acoustics, and many other fields.

Among the earliest Chinese texts on physics are the writings of the Mohist schools of philosophy in the Warring States Period (480–221 BC). Followers of philosopher Mo Zi (ca. 479–381 BC), the Mohists made pioneering discoveries in statics and hydrostatics, dynamics, and especially optics, many of which are described in the *Mo Jing* (third–fourth centuries BC).

One of the areas of physics most successfully tackled by the Mohists was the field of optics. Now while Chinese optics never reached the theoretical level of medieval Arab sources, it is clear that optics in the East was at least as advanced as parallel developments in Ancient Greece.

The propositions of the *Mo Jing* tell us that Mohist physicists made considerable experimentation in their efforts to construct a coherent theory of light and optics. Mohist physicists clearly understood that light travels in straight lines, and by using fixed light-sources, screens with pinhole apertures, and possibly the camera obscura, were able to study the formation of inverted images and the idea of the focal point. The camera obscura is certainly known from the eighth century AD onwards, and is mentioned in the ninth century *Yuyang Za Zi* (*Miscellany of the Yu Yang Mountain Cave*), though the theoretical explanation of its operation is incorrect. By the *Meng Qi Bi Tan* (*Dream Pool Essays*) of Shen Gua in AD 1088 we see an active interest in experiments involving pinhole apertures in screens to pass cones of light into darkened rooms.

Back in the Warring States period, by examining combinations of plane mirrors, Mohists also studied the phenomenon of lateral inversion (left becoming right in an image). They measured refraction in different media by examining the apparent bending of a stick in a glass of water. In addition, they used concave and convex mirrors to determine the properties of real and inverted images, and to differentiate between what we now know as the center of curvature and the focal point.

Concave mirrors were also used by ancient Chinese physicists to start fires by sunlight. Bronze mirrors are described as far back as 672 BC; surviving dated mirrors occur from AD 6 to 10, and many exist from later Han times. Multiple reflections were studied in the second century BC, and Han accounts of concave "burning mirrors" in the *Huai Nan Zi* (Book of Huai-Nan) of 120 BC demonstrate they knew of the focal point of a concave mirror. Besides starting fires, concave mirrors were used with the moon in the later Han era to collect dew.

An interesting development of the fifth century AD was the use of "light-penetrating mirrors". These "magic mirrors" were inscribed on their backs with written characters; when held to the light, these otherwise-invisible letters would "shine through" the mirror and become legible. A 1932 investigation revealed these mirrors were made of more than one curvature, with the designs on the back imposing unequal strains on the mirror. The mirror would differ imperceptibly in thickness from place to place, with the convex reflecting surface on the front varying in curvature, and thicker areas having flatter curvatures than thinner areas. They were described in the eleventh century AD by Shen Gua.

Alongside mirrors, early Chinese opticians made considerable experimentation with lenses. Glass was known to the Zhou dynasty, which began several hundred years before the Warring States period. A glassmaking industry began around 500 BC, and by the AD 83 *Long Heng* (Discourses Weighed in the Balance) of Wang Chung, there are descriptions of glass lenses used to bring sunlight to a focus to start fires. In the third century AD *Bo Wu Zhi* (Notes on the Investigation of Things), burning lenses of rock crystal are mentioned. Finally, the 940 *Hua Shu* (*Book of Transformations*) of Tan Qiao describes the various real and inverted images produced by all four types of lenses, the planoconcave, planoconvex, biconcave and biconvex.

There also appears to be some evidence for an independent Chinese development of the telescope by Bo You and Sun Yun Qiu in the early seventeenth century. However, as for spectacles, though they are often considered a Chinese invention, they in fact originated in Europe, and were imported overland into China soon after their invention around AD 1286.

In mechanics, the Mohists performed many experiments to determine the properties of levers and balances. They defined forces and weights, knew the laws of levers, balances, and pulleys, and developed a simple version of what came to be known in Europe as the "Atwood's machine" (named after English physicist George Atwood in 1780). They also invented the steelyard, and used it in weights and measures. Mohists seem to have been familiar with the "parallelogram of forces", and may have been in possession of the full theory of equilibrium (ca. 287–212 BC).

Early Chinese physicists did not discuss theoretical centers of gravity, but they do appear to have done some practical research on centers of gravity in their "trick vessels"; that is, jugs which incline or sway depending on the amount of water they contain. Trick vessels were known to Confucius (551–479 BC), and were explicitly described in texts of the third century BC. Siphons were constructed since antiquity, and were used for irrigation around the end of the Eastern or Later Han dynasty (AD 25–220). They were later employed (ca. 450 AD) by Daoist scientists and inventors like Li Lan in their balanced clepsydra, or water-clocks. Clockmaking is an ancient art among the Chinese, and attained great precision in the eleventh century mechanical and astronomical clocks of Su Sung (AD 1090).

As for specific gravity, it was known to the ancient Chinese, and was discussed by the philosopher Mencius (374–ca. 291 BC), co-founder of Confucianism and one of the students of Confucius' grandson. "Archimedes' Principle" appears to have been known to Han technicians (AD 221–264), according to accounts in the *Zhou Li* (Record of the Institutions of the Zhou Dynasty). Han physicists and engineers also knew how to measure specific gravities in liquids, and were familiar with the use of the principle of bouyancy to weigh heavy objects.

Mohists approached the atomic theory in their discussion of the strengths of materials, but never articulated it clearly or developed its consequences. In the explanation of why a fiber breaks under tension, Master Lieh, in the *Lie Zu* book, argues the fiber is composed of unequally-strong or cohesive elements, so that a breaking-plane forms somewhere in the fiber. This view is in essence a particulate theory of matter, but Lie uses the example to argue for a continuous universe.

The *Mo Jing* contains some remarkable statements about the study of motion. Though Chinese physics did little theoretical dynamics, they did consider forces in some detail, and appear to have come remarkably close to the principle of inertia as stated by English physicist Isaac Newton (1642–1727) as the first law of motion in his *Principia Mathematica* of 1687. Newton stated that "every body continues in a state of rest, or uniform motion in a right [straight] line, unless it is compelled to change that state by forces impressed on it". According to Joseph Needham, the Mohists argued that

> motion is due to a kind of looseness [the absence of an opposing force]...the cessation of motion is due to the [opposing force] of a 'supporting pillar' [a force that changes the moving body's otherwise-permanent state of motion]...If there is no [opposing force] of a 'supporting pillar', the motion will never stop . . . If there is [some kind of] 'supporting pillar' [some other force interfering with the motion], and nevertheless the motion does not stop [it may still be called motion but it will not be a straight-line motion because there will have been a deflection].

Mohist Chinese physicists also investigated the relativity of motion, and motion along inclined planes or slopes; they also studied particular problems of moving spheres.

As for Chinese acoustics, its history extends back more than two millennia. Second century BC philosopher Dong Zhongshu studied sympathetic resonance in musical instruments; indeed the entire history of Chinese acoustics is intricately tied up with musical instruments, especially lutes and bells. Chinese physicists first classified sounds based on the four materials (stone, metal, bamboo, and skin) which made up the musical instruments producing them; later, they extended their classification to an eightfold scheme tied to the eight directions of the wind. Chinese musicians used resonance phenomena to tune their instruments, and were well aware of the fact that sound was caused by vibrations. Chinese military strategists even used resonance in the fourth century BC to construct primitive geophones to listen for the tunnelling of an enemy besieging a city.

Chinese investigation of magnetism and electricity is similarly ancient, but is better known in the West than acoustics. Almost all physics textbooks note Chinese priority in the magnetic compass. Lodestones and magnets are mentioned in the third century BC *Lo Shi Chun Qiu* (Master Lu's Spring and Autumn Annals), as well as in the later *Huai Nan Zi* (Book of (the Prince of) Huai-Nan). By the fifth century, Chinese physicists knew of the directive property of the lodestone, and had begun to measure magnetic forces; and by the eighth or ninth centuries, they had discovered both magnetic polarity and variation. Engineer-astronomer Shen Gua discussed both in his *Dream Pool Essays*; his descriptions indicate considerable experimentation on the magnet. As for the magnetic compass, it is clear that the Chinese were using it for geomancy and divination by the first century AD. Their earliest use of the compass in navigation is unclear; the first datable references occur in about the eleventh century, but it is certain that Chinese mariners used compasses many years earlier. As in Europe, Chinese physicists used both "wet" and "dry" compasses; the former were made by floating magnetized needles in bowls of water whose rims were inscribed with directions, and the latter by carefully mounting the needle on an upright pivot.

What about the physics of surface phenomena? As in the West, its study was long delayed in China. Of course, knowledge of friction was very old among both cultures. Ancient Chinese, for example, knew how to use friction to make fires; and the friction of wheels was discussed in the *Han Zou Li*. By the Tang dynasty (AD 618–906), prince Li Kao studied the fit of smooth surfaces, and in the thirteenth century, Zhou Mi studied monomolecular films. And, during the Sung dynasty (AD 960–1279) Zhang Shi-Nan's 1233 treatise, *Yu Huan Zhen Wen* (Things Seen and Heard on My Official Travels), discusses the ring-test to determine the quality of lacquer. The ring-test was a method of quality-control for centuries; it was not until 1878 that it was applied to measuring surface tension by finding the force to break the surface.

In China as in Europe, heat and combustion were a comparatively late interest. Making heat and sparks by rubbing wood and striking flint were techniques known to the ancient Chinese; and medieval texts give priority to the Chinese in the invention of the match (Northern Qi, AD 577). Candles were known to the Mohists in the Warring States period, and lamps since ancient times; but by the early eighth century AD. Chinese lampmakers had cleverly designed an oil-cooling water reservoir below the lamp to reduce evaporation. The use of natural gas for fuel goes back to the first few centuries BC, and Chinese inventors applied this knowledge on an industrial scale.

Some knowledge of the practical principles of steam power may go back as far as the second century BC, based on a passage in the undated Daoist alchemical and technical recipe book, the *Huai Nan Wan Bi Shu* (Ten Thousand Infallible Arts of the Prince of Huai-Nan). Interest in spontaneous combustion is also very old, and is discussed in accounts of the third century AD. Chinese physicists also looked at successive boiling phases of liquids (particularly water for tea).

On the theoretical side, Li Shizhen's sixteenth century *Bencao Gangmu* (Great Pharmacopoiea) of 1596 attempts to classify types of fire. Fire is of two varieties, *Yin* and *Yang*, with many subdivisions; in general, it has *qi* (energy) but not *chi* (matter). Shizhen could not classify luminescent "fires" like flourescence, phosphorescence, electro-luminescence, and piezo-luminescence.

Electro-luminescence was described by Chinese physicists of the third century AD. The *Bo Wuzhi* (Record of the Investigation of Things), of approximately AD 290, describes sparks created by contact with charged objects, such as drawing combs through the hair. As for other electrostatic phenomena, the ability of amber to attract objects when rubbed was known very early. Amber was first described by first century AD philosopher Wang Chong, and its properties were discussed in more detail around AD 500 by alchemist-physician Tao Hong Jing. But there was little progress in electrostatics until the eighteenth century. Meanwhile, piezo-luminescence, caused by rubbing certain types of crystals, was known in the Tang dynasty. Artificial phosphors may have been studied in the Sung period; the eleventh century *Xiangshan Yelu* (Rustic Notes from Xiang Shan) of monk Wen Rong explains the making of a phosphor from oyster shells. Some of these crystals and phosphors may have been imported overland from Arabic cultures, Needham argues; and if so, they may well have similarly influenced Indian physics on the way.

India

Like African science, there is still much historical research to be done on the growth and development of science in India. Hindu scientists have been justly celebrated for their early contributions to mathematics (including the Hindu-Arabic number system, the use of zero, trigonometry, series summation, algebraic operations, and so on) and astronomy (elaborate computational techniques to determine the positions of celestial objects), but until recently little has been written on their physics, which is both sophisticated and extensive.

Early Indian physics is concerned with the "doctrine of the five elements" (*pañcamahabhūtas*); that is, the interplay between the elements of earth, water, fire, air and a non-material substance (*pṛthvī*, *ap*, *tejas*, *vāyu* and *ākāśa*, respectively). Among the earliest treatises of the Ṛgveda period (2000–1500 BC) we see the emergence of primeval water as the "first cause" of all other elements, rather like the simple cosmologies of ancient Greek scientists like Thales (ca. 624–545 BC) and Anaximander (ca. 611–575 BC).

The five element theory was later developed by the Vaiśeṣika School of about 700–600 BC, which also began the study of atomism. The introduction of Buddhism and Jainism brought new inquiries into the problems of atomism, matter, and motion (ca. 600–200 BC); this was followed by the systematic formulation of atomic theories among Indian physicists between 200 BC and AD 400. Indian science made its greatest advances between the period AD 400 and 1200, especially among the Jaina and Buddhist schools, as well as the Nyāya-Vaiśeṣika commentators. These medieval Indian physicists refined atomic theory, and made detailed investigations into gravity, motion, and impetus.

Physicists first began to explain the natural world in terms of atomic theories after 700 BC. For the Vaiśeṣika School, atoms were eternal or indestructible, indivisible and infinitely small. These atoms (*pramāṇus*) had vibratory and rotational motions, along with inherent impulses to form binary molecules (*dvyaṇuka*). Various combinations of atoms made up all the diverse substances in our world. Interestingly enough, this theory explained chemical reactions by supposing that these substances were broken up into their original atoms (the *pramāṇu* state) before they could be recombined to make a new material.

Physical and chemical change, meanwhile, was closely linked to the theory of heat. The first century AD philosopher Vātsyāyana argued that chemical change occurs as a result of either internal or external heat; moreover, during combustion, heat trapped inside the fuel in a latent form is released, allowing chemical reactions to proceed. The *Kiranāvali* of Udayana ultimately traced all forms of heat back to their source in the Sun.

Ancient Hindus believed that both heat and light were composed of streams of infinitely small particles radiating in all directions from sources like the sun. These particles move at inconceivably high speeds, and, when they strike molecules, break them down into smaller atoms, promoting chemical change. This theory was described in detail in the eleventh century AD *Nyāyamanjari* of Jayanta.

Other factors governing matter and physical change included gravity (*gurutva*), fluidity (*dravatna*), viscosity (*sigdha*) and elasticity (*sthitishāpaka*). For the Vaiśeṣika, gravity was seen as a causal factor resulting in the fall of bodies; it is not a force in the modern sense. Moreover, only earth and water possess it. The later Nyāya-Vaiśeṣika school of physicists (ca. 400–1200) add that gravity could be cancelled out by the effects of conjunction (a object supported on a stand) or impetus (an arrow flying through the air). They argued that gravity is eternal for primeval atoms, but temporary for compounds.

Fluidity was of two types, natural and incidental, with the former reserved to water, the latter to water, earth, and fire. Viscosity governs cohesion and smoothness, and is a property of water, preventing particles in it from dispersing. Elasticity is a quality of earthy substances.

How did Indian physicists handle the problem of matter in motion? The Vaiśeṣika held that matter was naturally static or stationary. It could be moved by some physical quality, like gravity in a falling body, but this motion would be temporary; and perpetual motion was impossible. Interestingly enough, it seems that gravity was only required to set the falling object in motion. Once it had started moving, impetus is generated, which keeps it falling without any additional help from gravity.

The fifth century AD philosopher Praśastapāda made a detailed study of motion and impetus in his *Pradārtha dharma saṁgraha*, a revision of the much older *Vaiśeṣikasūtra*. He divided motion into five categories: *utkṣepaṇa*, *avakṣepaṇa*, *prasāraṇa*, *ākuñcana*, and *gamana*, or, respectively, lifting up, throwing down, expansion, contraction and action (gyrations, evacuations, quivering and the like). Motion can be caused, says Praśastapāda, by six things: gravity, fluidity, volitional effort, conjunction (such as impact or striking), impetus or unseen causes. His use of impetus (*vega*) is rather close to our modern momentum, and is illustrated by the examples of throwing a heavy shot or firing an arrow. When the shot is thrown, the first impelling push begins its motion; then the shot's internal impetus takes over, and maintains its motion. Once the impetus is exhausted, the shot falls back to the ground.

Praśastapāda also had a relatively modern conception of sound. He realized sound was produced by vibrations in air, and added that these vibrations travel "like a series of water-ripples . . . or a series of waves" to the ear.

Indian physicists had many contacts with Chinese developments in the East, but they were equally well connected to Arabic work in the West. The relations between Indian and Islamic science in the Medieval period has recently been the focus of much scholarly work (Needham, Chin, Filliozat, Sen).

Islam

Meanwhile, Islamic physicists, mathematicians and astronomers such as Ibn Sīnā, Ibn al-Haytham, al-Bīrūnī and al-Kindī have long been known to Western historians for their influences on scholastic science in the Medieval Latin West. Arabic scientists frequently criticized and extended the physics of Aristotle, Archimedes (ca. 287–212 BC), and other ancient physicists and often provided the original roots for many physical ideas, such as force and momentum.

In mechanics and dynamics, Muslim physicists developed several important concepts. Ibn Sīnā or Avicenna (980–1037) invoked the idea of *mayl* (Latin *inclinatio*, or inclination) to explain projectile motion; this theory was further developed by al-Baghdādī before its transmission to the Latin West. Ibn Sīnā and Ibn-al-Haytham developed the concept of momentum, and did pioneering research into the theory of gravity. Muslim philosophers knew that falling bodies accelerated independently of their mass, and that the attractive power between two objects increased as their masses got larger, but decreased with their separation.

Arabic physicists extensively studied the theory of weights, densities, bouyancy, and measures. They were familiar with the law of the lever and Archimedes' principle, and developed the balance as a scientific instrument to study topics such as specific weights. Thābit ibn Qurra (836–901) worked extensively on the law of the lever, and his ideas were later quite influential in the West. Meanwhile, al-Khāzinī's (fl. 1115–1130) celebrated *Kitāb mizān al-ḥikmah* (Book of the Balance of Wisdom) developed formulae for the specific and absolute weights of various alloys. Al-Khāzinī's work also discusses centers of gravity, hydrostatics, and other physical concepts.

Many Islamic scientists extended the theory of simple machines. Among these researchers were Baghdad's Banū Mūsā, and the many Arabic scientists involved with the development of various mechanical gadgets, automata, trick devices and other complex machines. This branch of physics, called *ʿilm al-ḥiyal*, was developed by the Banū Mūsā, the school of Ibn Sīnā, Ibn-al-Sāʿāti, and many more. It reached its climax in the famous *Kitāb fī maʿrifat al-ḥiyal al-handasiyyah* (Book of Knowledge of Ingenious Mechanical Devices) of al-Jazarī.

Among the most influential researches of Muslim physicists were their pioneering inquiries into light and optics, which virtually established this branch of physics as an organized discipline. Al-Kindī's (fl. 790–866) early treatises introduced Euclidean optics to the Medieval Latin West. His theories of vision influenced Islamic writers for several centuries, including al-Fārābi (d. 950), al-Ṭūsī (1201–1274), and many more.

Opticians such as Ibn Sīnā and al-Bīrūnī (973–ca. 1050) studied particular problems like the speed of light from ancient sources, usually declaring it to be finite. Meanwhile, doctors like al-Rāzī or Rhazes (ca. 854–935) and Ḥunayn ibn Isḥāq inquired into human eye physiology, making their own contributions to the theory of vision. However, the most significant figure in Muslim optics was the Egyptian physicist Ibn al-Haytham (ca. 965–1039), better known in the West as Alhazen.

Many historians consider Ibn al-Haytham to be the most important optician between Euclid and Johannes Kepler (1571–1630). Historian of science David C. Lindberg argues that "Alhazen was undoubtedly the most significant figure in the history of optics between antiquity and the seventeenth century". This is because Ibn al-Haytham successfully integrated not only competing mathematical and physical approaches to light in his new intromission theory of vision, but also harmonized his views on sight with the anatomical work of earlier Muslim physicians. A prolific writer, Ibn al-Haytham is credited with more than two hundred treatises on all branches of the sciences, including almost two dozen dealing with light and optics.

Ibn al-Haytham's most important optical work was his *Kitāb-al-manāẓir* (Book of Optics), which was translated into Latin as *De Aspectibus* or *Perspectiva*. This treatise had a profound effect on optics in the Latin West, influencing the work of Witelo (ca. 1230/5–1275), Roger Bacon (ca. 1220–1292), John Pecham (fl. 1250–1292), and many more. Traces of Ibn al-Haytham's theories are even found in the seventeenth century optical writings of Kepler and Newton.

Alongside his intromission theory, Ibn al-Haytham also studied various celestial and atmospheric phenomena such as twilight, refraction, atmospheric thickness, and optical illusions. He also contributed to the theory of catoptrics and dioptrics and extensively researched the laws of parabolic and spherical mirrors. "Alhazen's problem", which he solved geometrically, involved taking a spherical mirror and placing an object before it so that its image was reflected in the mirror; to find the point of reflection, one must solve the equivalent of a fourth degree equation. Ibn al-Haytham also demonstrated various reflection laws, including the significant finding that incident, normal and reflected rays all lie in the same plane. In his study of refraction, he discovered the equivalent of Pierre de Fermat's (1601–1665) "principle of least time": that a ray of light always takes the shortest and easiest path. Much of Ibn al-Haytham's work in this area was based on direct experimentation; glass cylinders immersed in water revealed refraction laws, and homemade lenses helped him determine the laws of magnification. Ibn al-Haytham even did early research into the camera obscura.

Islamic optics declined after Haytham's synthesis, though there was a revival of interest during the rise of the *ishrāq* (School of Illumination) in the thirteenth century. Al-Shīrāzī extended Ibn al-Haytham's account of the rainbow, arriving at essentially our modern understanding that its colors are due to a combination of light reflection and refraction through suspended water droplets. Al-Shīrāzī's explanation was confirmed experimentally by his student, al-Farīsī (d. ca. 1320). But by the fifteenth century, leadership in optics had passed to the West, and by the following century, European scientists were making discoveries in physics, astronomy, and mathematics that were to usher in the "Scientific Revolution".

JULIAN A. SMITH

REFERENCES

Bernal, J.D. *The Extension of Man: A History of Physics Before the Quantum.* Cambridge, Massachusetts: MIT Press, 1972.

Chin Keh-mu. "India and China: Scientific Exchange." In *Studies in the History of Science in India*, vol. 2. Ed. Debiprasad Chattopadhyaya. New Delhi: Barun Maitra, 1982 , p. 776–790.

Filliozat, J. "Ancient Relations between Indian and Foreign Astronomical Systems." In *Studies in the History of Science in India*, vol. 2. Ed. Debiprasad Chattopadhyaya. New Delhi: Barun Maitra, 1982 , p. 767–775.

Jaggi, O.P, ed. *History of Science and Technology in India*, vols. 1–15. Delhi: Atma Ram, 19691986.

Kuppuram, G, and Kumudamani, K., eds. *History of Science and Technology in India*, vols. 1–12. Delhi: Sundeep Prakashan, 1990.

Lindberg, David C. *Theories of Vision from Al-Kindi to Kepler.* Chicago: University of Chicago Press, 1976.

Marshack, Alexander. *The Roots of Civilization*. Mount Kisco, New York: Moyer Bell, 1991.

Nasr, Seyyed Hossein. *Science and Civilization in Islam*. Cambridge, Massachusetts: Harvard University Press, 1968.

Nasr, Seyyed Hossein. *Islamic Science: An Illustrated History.* Westerham, Kent: World of Islam Festival Publishing Company, 1976.

Needham, Joseph, *Science and Civilisation in China.* Cambridge: Cambridge University Press, 1954.

Pappademos, John. "An Outline of Africa's Role in the History of Physics." In *Blacks in Science: Ancient and Modern*. Ed. Ivan Van Sertima. London: Transaction Books, 1984, pp. 177–196.

Sen, S.N. "Interrelationship between Indian, Greek, Chinese and Arabic Astronomy," and other articles in *A Concise History of Science in India*. Ed. D.M. Bose. New Delhi: Indian National Science Academy, 1971.

Subbarayappa, B.V. "The Physical World: Views and Concepts." In *A Concise History of Science in India* Ed. D.M. Bose, S.N. Sen, and B.V. Subbarayappa. New Delhi: Indian National Science Academy, 1971, pp. 445–483.

Yisheng, Mao, ed. *Ancient China's Technology and Science.* Beijing: Institute of the History of Natural Sciences, China Academy of Sciences, 1983, pp. 124–175.

See also: Geomancy – Weights and Measures – Optics – Su Song – Clocks and Watches – Acoustics – Magnetism – Compasses – Gas – Atomism – Li Shizhen – *Yinyang* - Ibn Sīnā – Ibn al-Haytham – al-Kindī – al-Shīrāzī – al-Rāzī – Ḥunayn ibn Isḥāq – Banū Mūsā – Thābit ibn Qurra

PHYSICS IN CHINA An account of physics, as well as of any other science, in a pre-scientific revolution era or civilization does not present a meaningful picture without an understanding of the views of nature that are embedded in various natural philosophies that develop in each civilization and that are the most comprehensive account of natural phenomena before "physics" as a distinct discipline existed. In the West most of these natural philosophies had their origins in Greece, and included Pythagoreanism, Atomism, Platonic idealism and dualism, and Aristotelianism, each of which introduced distinct elements into Western science. However, they largely agreed that the final ground for understanding physical reality was substance, or "stuff". Additionally, in due course, rational proof, Euclidean geometry, reductionism, mechanistic materialism, and atomism became principal determinants of Western science.

In a similar way at about the middle of the first millennium BC various schools of natural philosophy arose in China. These included Daoism, Naturalism, Mohism, Legalism, and (much later) Sung Neo-Confucianism. The central concept of Daoism, systematized by Laozi (a contemporary of Aristotle) is the *Dao* (the way), the "way the universe works". It is the unifying principle of the universe, akin to the "One" of Parmenides. This natural, non-coercive, non-anthropomorphic principle encompasses all phenomena, and invites study of all aspects of nature. Human standards are irrelevant in nature; one is to subject oneself to the harmonies of nature (*wu wei*).

Naturalism, represented by its early proponent, Zou Yan (fl. 300 BC) introduced the doctrines of *Yin* and *Yang* and the *wuxing* (Five Elements or Phases). Yin and Yang are the two fundamental forces, the expression of the basic dualistic principle, two forces existing not in conflict, but in a balance or a cooperative tension of opposites. Yang represents light, heat, dryness, hardness, activity, heaven, sun, and south. Yin stands for dark, cold, wetness, softness, quiescence, earth, moon, and north. The Five Elements invite comparison with the four elements of Empedocles. In China the five are water, fire, wood, metal, and earth, representative less of substances than of processes of nature. The elements are further arranged into the Mutual Production Order: wood, fire, earth, metal, water, wood, etc., where each element produces the subsequent one. In the Mutual Conquest Order, each element dominates over the subsequent one.

The less prominent school of Mohism introduced more formal, abstract, and geometrical notions such as a dimensionless geometrical point in space and time. Somewhat in the same vein, the Legalists discussed formal logical propositions, paradoxes, and syllogisms. Curiously, modern Chinese writers during the Cultural Revolution were fond of heaping fulsome praise on the Legalists for their supposed kinship to the scientific materialism of Marxism.

Mohism and Legalism waned in importance, and after centuries of evolution indigenous Chinese natural philosophy reached its most mature form in the twelfth century AD in the Sung neo-Confucian synthesis, best represented by the natural philosopher Zhu Xi (AD 1131–1200). This school combined the attention to nature present in Daoism and Naturalism with the heretofore society-centered official Confucianism that governed human conduct and interaction. In this way the human world and the world of nature were unified and investigation progressed at each "level of integration" from inanimate to human. The synthesis introduced two other fundamental principles of *qi* and *li*, roughly similar to Aristotelean matter and form.

The picture or model of nature that emerges is quite different from the Western one. It is not mechanistic, legalistic, geometrical, linear, or even materialistic. The leading twentieth century student of Chinese science, Joseph Needham, described that picture of nature in China as being one of a harmoniously functioning organism, an interconnected, interdependent whole. Reductionism and analysis into parts will not yield an understanding of nature because each entity is defined by the role it plays in relation to the pattern it fits into, by the web of relationships it has. The principal element of explanation is not substance but relation and pattern. As in Chinese society, so also in nature there is "order without law". The legal metaphor is absent in statements about nature ("laws" of nature); the order of nature is not as that of citizens subject to universal law but as that of musicians in an ensemble or dancers in a pattern. For these reasons, traditional Chinese thought found no problem in retroactive causality, since temporal succession is unimportant compared to the interdependence of the system as a whole, as a pattern in space and time.

With a world picture so different from the Western one, it is difficult even to match traditional Chinese activity in some realm with "physics", itself a term that acquired a recognizable meaning only fairly recently. Nathan Sivin has proposed the following divisions for Chinese science: medicine, alchemy, astrology, geomancy, natural philosophy and physical studies, mathematics, harmonics and acoustics, and mathematical astronomy. Present-day "physics" corresponds to parts of "physical studies" and "harmonics". Also, the strong and weak branches of physics are reversed in China and in the West. There was no Chinese achievement in mechanics, dynamics or kinematics, nothing at all to rival Aristotle or the medieval precursors of Galileo. This is consistent with the absence in China of geometrical, corpuscular, and mechanistic ideas. Instead the preference was for continuum theories, pneumatic models, waves, vibrations, and resonances, all consistent with interrelation, pattern, and meaning only for a system as a totality. Thus, the highest achievements in indigenous Chinese physics were in the areas of optics, acoustics, and magnetism. Of the three, optics was the weakest, and ironically, its best practitioners were the Mohists, a school generally atypical of Chinese scientific philosophy. Although, lacking Greek geometry, Chinese optics never attained the level of medieval Islamic optics, it began at least as early as its Greek counterpart, with an account of light-sources, mirrors, and, later, lenses.

The fourth-century BC Mohist text *Mo Jing* describes shadow formation, the distinction between umbra and penumbra, and discusses the dependence of the size of the shadow on the position of the object and of the source of light. Properties of light passing through a pinhole are discussed, including the definition of a focal plane and the inversion of the image. Reflection from a plane mirror and from a combination of plane mirrors is treated. Refraction of light at the plane boundary between air and water is described, including the resulting phenomenon of the apparent reduction in depth of an object in water. Also, the formation of images by spherical concave and convex mirrors is discussed in some detail.

Of even greater antiquity in China was the practical use of burning-mirrors for igniting fires from the sun's rays. The earliest such mirrors, made of bronze, date from the Chinese Bronze Age, in the seventh century BC. A more difficult technological achievement was the production of lenses and burning-lenses. Rock-crystal, i.e. pure transparent quartz (SiO_2) was used; glass was produced as early as the sixth century BC, and lenses were in use from the time of the Han dynasty of the first century AD. By the tenth century AD the four fundamental types of lenses (plane-concave, biconcave, plano-convex, and biconvex) were produced and their properties analyzed. Eye-glasses and spectacles, as well as magnifying glasses were in use by the fifteenth century AD.

Progress in the fields of sound and acoustics followed naturally from the fundamental orientation of Chinese natural philosophy. The alternation between Yin and Yang implies a cyclic, recurring process between opposites, reminiscent of harmonic oscillation or wave motion. Also, sound was likened to waves in a pool, caused by friction, and to a resonance in the fundamental neo-Confucianist substance, *qi*.

In the description and classification of musical sound and in the construction of instruments, the Chinese displayed their characteristic tendency to emphasize the whole and the complex, instead of isolating the simple. To Western thought the primary attributes of sound are pitch (frequency) and loudness (amplitude). By contrast, Chinese acoustic theory concentrated from the beginning on timbre (the overtone structure) as the primary attribute or parameter describing sound. Sounds were classified according to timbre according to the "eight sources of sound": metal, stone, earth-clay, skin, silk-thread, wood, gourd, and bamboo. As can be expected, these timbres were correlated, like the five elements, with seasons, compass points, and, naturally, musical instruments.

The influence of the concept of pattern and organic unity was at work in the ready acceptance of the spontaneous resonances of two musical instruments, this being merely a corroboration of a resonance in the cosmic organism. In a similar vein, sounds were not classified by pitch; rather, the names of notes in the early five-note scale arose from the mutual arrangement of musical instruments in archaic ceremonies.

A distinctive later achievement of Chinese acoustic theory is less closely related to the organic view of nature. In the sixteenth century AD, Zhu Zaiyou (b. AD 1536) provided the earliest solution to the classic problems of rendering melodies in several keys on one musical instrument, without retuning the pitches. The natural scale is one where the frequencies of the sounds, and the associated lengths of strings in a musical instrument are in "just intonation", following Pythagorean ratios (e.g. 2:1 for octave to tonic, etc.). If the ratios are true for one scale they cannot be so for a scale that begins one note higher. Zhu's solution was to have frequencies of all adjacent notes, differing in pitch by half a tone in a twelve-note scale, differ by a ratio $2^{1/12}$ to 1. This system of equal temperament, in use on a modern piano, was introduced considerably later in Europe, becoming popular only in the eighteenth century, and is familiar to music-lovers from early examples of its use by composers, such as in Johann Sebastian Bach's *Well-Tempered Klavier*.

The most prominent achievement of traditional Chinese physics, however, was in magnetism. That phenomenon is totally harmonious with a model of the world as organism, with resonances and intangible influences on the parts from the pattern as a whole. Also, China, not knowing the Aristotelian concepts of natural place, natural and enforced motion, did not have to explain magnetic phenomena, as did the West, as deviations from this scheme. Even more to the point, action at a distance, or the concept of a field, was no problem in China, whereas Cartesian mechanicists saw these as throwbacks to scholasticism when used to account for gravitational, electrical, and magnetic phenomena.

Historically, the discovery of every significant magnetic phenomenon — attraction, directivity, polarity, thermoremanence, declination, as well as application to a working compass — considerably antedates comparable advances in the West. Chinese texts of the third century BC describe the simplest effect, the attraction of iron by natural magnetite (lodestone). The directional, north-south seeking property of the lodestone was observed about the first century AD and first demonstrated by a lodestone placed on a very smooth bronze surface. (Modern replicas of this arrangement have verified that friction can be made low enough to display the effect.) Subsequently, the lodestone was balanced on a pin, or floated on wood in water. In the early centuries of the first millennium AD the directive property of the lodestone was transferred to iron by rubbing. By the eleventh century AD the more sophisticated method of magnetization, thermoremanence, was in use, consisting of allowing a red-hot piece of iron to cool rapidly through the Curie point while it is oriented along the earth's magnetic field.

By about AD 1100, a useful magnetic compass was in use in navigation, although the traditional inward-looking agrarian economy and dependence on river, rather than ocean, transportation did not thereby lead to extensive global exploration. Beginning in the ninth century AD magnetic declination was observed. Subsequent recordings of declination chronicle a variation of declination over time, for example, the declination being to the east prior to about AD 1050, and to the west after AD 1050.

Philosophers of science differ on whether the fundamental concepts of Chinese natural philosophy, such as relation, pattern, and organism could have furnished an alternative, but adequate basis for a model of scientific explanation. On this question, in his extensive writings, Needham has probably overstated the case in favor of the Chinese potential for doing this.

In terms of transmissions from China to the West, the technological contributions are well-known; printing, gunpowder, and the magnetic compass are always at the head of the list. Many scholars also attribute to Chinese science a role in promoting the development, after Newton, of nonmechanistic concepts in physics, such as field, wave, fluids such as caloric and ether, to say nothing of the loss of distinction between matter and energy in relativity and between wave and particle in quantum mechanics.

R.A. URITAM

REFERENCES

Nakayama, Shigeru, and Nathan Sivin, eds. *Chinese Science, Explorations of an Ancient Tradition.* Cambridge, Massacusetts: MIT Press, 1973.

Needham, Joseph. *Science and Civilisation in China.* Volume 2: *History of Scientific Thought.* Cambridge: Cambridge University Press, 1956.

Needham, Joseph. *Science and Civilisation in China.* Volume 4: *Physics and Physical Technology.* Part I: *Physics.* Cambridge: Cambridge University Press, 1962.

Needham, Joseph. *Science in Traditional China: A Comparative Perspective.* Cambridge, Massachusetts: Harvard University Press, 1981.

Uritam, R.A. "Physics and the View of Nature in Traditional China." *American Journal of Physics* 43(2): 136–152, 1975.

See also: Environment and Nature in Chinese Thought – Atomism – *Yingyang* – Five Phases – Optics – Magnetism – East and West – Compasses – Acoustics – Zou Yan

PHYSICS IN INDIA Early Indian thinkers developed a number of theoretical systems which centered around two main themes: elements and atoms. Based upon a relatively broad review of the ancient philosophies, we will discover a rich scientific tradition in India.

Ancient India (2000 B.C–AD 800)

Early Indian explanations about the physical makeup of the universe were religious and philosophical in nature. The oldest literary record of this period, the *Ṛgveda*, presents several fundamental concepts related to physical science such as *ap* (primeval water), which was considered the basic element of matter. Gradually, the doctrine of five fundamental elements emerged, as seen in the Upaniṣadic literature of around 700 BC. The five elements were *pṛthvī* (earth), *ap* (water), *tejas* (fire), *vāyu* (air), and *ākāśa* (a non-material substance). The Sanskrit terms have a wider connotation than the English translation, so it is essential to present the original terms.

Much of this early literature was concerned with the attributes associated with each of these elements. They had both common as well as distinctive attributes, many coincident with the five senses, as follows: *pṛthvī* (earth) sound, touch, color, taste, and odor; *ap* (water) sound, touch, color, and taste; *tejas* (fire) sound, touch, and odor; *vāyu* (air) sound and touch; and *ākāśa* (non-material) sound. Different combinations of these five elements yielded certain products. The human body is a good example. The human embryo has energetic principles which are separated into form by *vāyu* (air). *Tejas* (fire) subsequently transforms the embryo. *Ap* (water) maintains moisture while *pṛthvī* (earth) gives it shape and size. Finally *ākāśa* expands the embryo and develops it.

Many of today's scholars question the notion that scientific inquiry only found fertile ground in the minds associated with the Greek tradition. In fact, a number of diverse systems of scientific inquiry were developing in India coincident with the Greek philosophies by 500 BC. Those which presented five fundamental elements were the *Sāṃkhya, Nyāya,* and *Vaiśeṣika* . The *Jaina, Bauddha,* and *Cārvāka* schools, like many of the Greek thinkers, presented four fundamental elements. It is difficult to say with certainty which school or culture developed these ideas first, therefore who influenced whom, or if they developed independently.

The *Nyāya-Vaiśeṣika* school, prominent among early Indian systems, was popularized by an individual who came to be known as Kaṇāda. He and his followers extended the concept of five elements and built a comprehensive theory of atoms. The first four elements of this system (*pṛthvī, ap, tejas,* and *vāyu*) were material and considered either eternal or temporal. The fifth element, *ākāśa*, was considered eternal only. The eternal form consisted of imperceptible atoms, while the temporal form arose when these atoms joined to create perceptible products. This view of five elements was part of a much larger conceptual picture, that of *dravya* (substance). There were nine types of substances. In addition to the five elements or substances already mentioned, there was *dik* (space), *kāla* (time), *ātman* (self), and *manas* (mind). Each of these substances was considered inseparable from its respective set of attributes. In other words, they were one and the same.

The word used for atom was *aṇu* or *pramāṇu*. The *Nyāya-Vaiśeṣika* school held that the atom was indestructible, indivisible, without magnitude, spherical, and in constant motion. Two atoms of the same substance could join to form a dyad. Atoms of different substances could not join together, yet they could play a supportive role in the combination of materially compatible atoms. The dyad was regarded as too small to be perceived. The smallest visible structure was a triad (three dyads). This structure was referred to as *trasareṇu* or *tryaṇuka* which was about the size of a speck floating through a sunbeam. The principle of causality was crucial to the *Nyāya-Vaiśeṣika* school. Atoms were the material cause for the dyad, the effect. The dyads were the cause for the production of a triad which was another effect. The individual atoms lost their causative property once the triad was formed. The reason for atoms joining in the first place was attributed to *aḍṛṣṭa*, an unseen force.

Jaina philosophers held that atoms were both cause and effect. They theorized that atoms joined to form aggregates in response to attractive and repulsive forces which were inherent characteristics of the atoms themselves. Jaina atoms were all of one class. There was no distinction, qualitatively or quantitatively, between types of atoms such as earth-atoms, water-atoms, etc. According to the Buddhists, the atom was indivisible, unable to be analyzed, invisible, inaudible, unable to be tested, and intangible. Neither were these atoms eternal. The Buddhists did not speak of atoms in terms of particles. They thought of them as a force of energy.

The *Nyāya-Vaiśeṣika* school was distinct from other Indian schools in that it placed much more emphasis on the attributes of matter. There were five general qualities possessed by all nine substances: number, dimension, distinctness, conjunction, and disjunction. Other qualities, more closely associated with modern physics included, *gurutva* (gravity), *dravatva* (fluidity), *sigdha* (viscosity), and *sthitishāpaka* (elasticity). Gravity, or the cause of falling, was not considered a force but a quality which resided in a whole object. No apparent correlation between gravity and the mass of a particular object was presented. Fluidity was a quality of only three substances — earth, water, and fire — and was of two kinds. Natural fluidity was a specific quality of water. Incidental fluidity was associated with fire in the case of some melted substances. Viscosity was specific to water, causing cohesion and smoothness. Elasticity was only a quality of earthy substances as in the branch of a tree which was caused to return to its original condition if displaced. Another fundamental concept considered by ancient Indian thinkers was that of motion.

The *Nyāya-Vaiśeṣika* concept of *karma* (motion) was represented by five actions: *utkṣepaṇa* (throwing upwards), *avakṣepaṇa* (throwing downwards), *prasāraṇa* (expansion), *ākuñcana* (contraction), and *gamana* (going). Only one kind of motion was considered possible at a time and a substance experienced motion only for a moment. This motion was subsequently destroyed or rendered ineffective once completed. The motion of free atoms made possible by *aḍṛṣta* (unseen force) caused the material world to be formed.

Among Indian physical concepts, that of *ākāśa* should not be overlooked. Its special quality was sound. Though considered a non-material substance, it was believed to play a role in the formation of material objects. A modern physics concept which some have equated with *ākāśa* is that of ether, which is conceived to be a vast expanse or continuum. Sound moved through *ākāśa* like ripples across the surface of a pond. Unlike the ripples, sound moved through *ākāśa* in a succession of points. The first sound caused the second, and once the second sound was created, the first was destroyed.

Heat and light were understood in relation to one of the five basic elements, *tejas* (fire). When an object was heated it went through a series of distinct changes, each one lasting a prescribed number of moments. A clay pot for example, if heated, would in the first moment experience the production of atomic motion. The second moment would be characterized by the destruction of the clay's original color. In the third and fourth moments one would find that the color red would be produced followed by the destruction of the atomic motion created earlier. A new type of atomic motion, creative in nature, occurred in the fifth moment. A series of two disjunctions between the atoms and *ākāśa* was followed by a conjunction in the sixth through eighth moments. The final two moments consisted of the formation of a red colored dyad followed by a triad. The ultimate source of all heat was thought to be the sun. Light, though often associated with heat in many modern physical systems, possessed many qualities, distinctive from heat, in the early Indian systems.

Light was thought to consist of rays which emanated from the eyes just as a candle casts its light throughout a room. If an obstruction prevented the rays from touching an object, then that object simply could not be perceived. Mirrors were thought to possess particular attributes of color which caused the light rays from one's eyes to return to one's face, upon striking the surface.

Medieval India (AD 800–1800)

It is a common misconception that there was a lack of scientific inquiry during the Middle Ages. While India experienced great change during this time, such as the introduction of Islam, there was a continued advance and assimilation of scientific information with an ever increasing number of collaborators. New cultural influences, changing technologies, and language barriers challenged as well as fostered scientific advances.

Physics was viewed as a distinct branch of study during the early part of the Middle Ages. Though experiments were still quite rare, use of this scientific technique began. A large number of texts were translated into different languages. A scholar named Ḥunayn ibn Isḥāq translated Aristotle's works into Syriac and his son Isḥāq ibn Ḥunayn followed suit, translating Euclid's *Elements*. Ideas were shared, expanded upon, and refined. A number of original inquiries were also initiated.

Among the more notable individuals describing physics research in India as well as Middle Asia at this time was Abū Rayḥān al-Bīrūnī. He authored many books devoted to the scientific achievements of the Indian people.

The latter part of this period experienced an influx of European influences. Educational institutions, scientific journals, and professional societies each played a role in the development of what was to become modern India. New political and economic structures, in the face of new language barriers between the educated few and the illiterate masses, threatened to disassemble the rich scientific heritage of the Indian people. Yet India assimilated the old and the new to emerge strong with an outstanding future of scientific inquiry before it. Much of that future rested in the hands of a few visionaries who understood the significance of science in India's future.

*Modern India (*AD *1800–Present)*

The outstanding efforts of many have contributed to India's rich past and promising future. The following few individuals are some of the physicists who have exemplified the level of achievement found throughout modern India.

Jagdish Chandra Bose (1858–1937) was a biophysicist who explored the response of plants and animals to electrical stimulation. Like many pioneering physicists active at the turn of the century, Bose used instruments of his own design in his work. He was elected a Fellow of the Royal Society of London in 1920. C.V. Raman (1888–1970) investigated optics, including diffraction, molecular scattering of light, and magneto-optics. He was awarded the Nobel Prize in 1930 for the discovery which bears his name, the Raman Effect. This pioneering work described the molecular scattering of light and explained, among other things, why the ocean appears blue. Finally, Jawaharlal Nehru, the first Prime Minister of India, worked hard to foster what he described as a "scientific temper". He believed that for science to succeed in modern India, as it had in its past, there had to be strong support for the scientific enterprise from all segments of society.

The history of physics in India is a vast subject, offering extensive opportunities for further study. It is hoped that this brief presentation has encouraged some to explore the ancient philosophies, medieval refinements, and modern achievements.

WILLIAM T. JOHNSON

REFERENCES

Divatia, A.S. "History of Accelerators in India." *Indian Journal of Physics* 62A (7):748–774, 1988.

Mitra, A.P. *Fifty Years of Radio Science in India*. New Delhi: Indian National Science Academy, 1984.

Panda, N.C. *Maya in Physics*. Delhi: Motilal Banarsidass, 1991.

Pingree, David. *Census of the Exact Sciences in Sanskrit*. Philadelphia: American Philosophical Society, 1970.

Rahman, A., M.A. Alvi, S.A. Khan Ghori, and K.V. Samba Murthy. *Science and Technology in Medieval India–A Bibliography of Source Materials in Sanskrit, Arabic and Persian*. New Delhi: Indian National Science Academy, 1982.

Rao, C.N.R. and H.Y. Mohan Ram, eds. *Science in India: 50 Years of the Academy*. New Delhi: Indian National Science Academy, 1985.

Romanovskaya, T.B. "The Interrelations Between India and Middle Asia in the Field of Physics in the Middle Ages." In *Indo-Soviet Seminar on Scientific and Technological Exchanges between India and Soviet Central Asia in Medieval Period, Proceedings, Bombay: November 7–12, 1981*. New Delhi: Indian National Science Academy, 1981.

Sachau, Edward C., ed. *Alberuni's India*. London: Kegan Paul, Trench, Trubner & Co. Ltd, 1914.

Subbarayappa, B.V. "The Physical World: Views and Concepts." In *A Concise History of Science in India*. Ed. D.M. Bose. New Delhi: Indian National Science Academy, 1971.

Venkataraman, G. *Journey into Light: Life and Science of C.V. Raman*. Bangalore: Indian Academy of Sciences, 1988.

See also: al-Bīrūnī – Atomisim – Ḥunayn ibn Isḥāq – Isḥāq ibn Ḥunayn

PHYSICS IN THE ISLAMIC WORLD With the rapid expansion of the Arabian tribes into western Asia and northern Africa in the early seventh century, the Islamic world fell heir to the documentary remains of Greek science and mathematics. Scientific research had ground to a halt in the Byzantine empire, and in a very real sense the intellectual center of gravity was transferred to the East, where, despite occasional upheavals, learning flourished under a series of enlightened patrons for over five hundred years.

In contemporary usage, the word "physics" denotes an autonomous and highly specialized branch of science which deals with the behavior of non-living systems, bringing them under the scope of the most general natural laws. There was no parallel to this discipline in medieval Islam; yet from the eighth century onwards there was a profusion of translations, commentaries, and learned writings on topics which would today fall under the broad heading of physics.

The contributions of Islamic thinkers to astronomy and mathematics were of the first importance. Optics, which would today be considered a branch of physics, was studied in close conjunction with both geometry and medicine, since various problems regarding vision remained unresolved in the Greek scientific literature. The contributions of Ibn al–Haytham to the problem of refraction are particularly noteworthy.

Astronomy and optics aside, Islamic scholars wrote profusely on the nature of matter, causality, the theory and practice of mechanics, and dynamics.

The Nature of Matter and Causal Laws

The dominant philosophical school in Islamic physics was the Aristotelians or *Falāsifa*, Prominent members of this school included al–Kindī (d. 873), al–Fārābī (d. ca. 950), Ibn Sīnā (Avicenna, d. 1037), and Ibn Bājja (d. 1138). While the *Falāsifa* held Aristotle in the highest respect, their work was somewhat syncretistic. The Greek philosophical corpus was often taken to be an integrated whole, and diverse elements of Plato and the Neoplatonists found their way into the synthesis of the Arabic Aristotelians. Al-Fārābī went so far as to write a book entitled *Jamʿ bayna raʾyay al-ḥakīmayn Aflāṭūn al-ilāhī wa-Arisṭūṭalīs* (The Harmony between Plato and Aristotle). Most Arabic scholars who sought to reconcile these strands of thought relied heavily on a work known as the *Theology* of Aristotle, in which Aristotle himself seemed to provide a bridge between his technical and logically rigorous work and the more literary and imaginative work of Plato. But modern scholarship has established that the *Theology* was wrongly attributed to Aristotle and is actually a compendium of the writings of the Neoplatonist Plotinus.

In physics, the doctrine held that every object could be analyzed as a composite of matter, which is the locus of its potency or possibilities, and form, which is the actualization of some of those potencies. Matter, like space itself, was believed to be infinitely divisible. Physical bodies behaved in qualitatively predictable ways because of their innate tendencies. Hence, a stone falls because it seeks its natural place at the center of the earth, and water seeks its lowest level in lakes and oceans because its attraction to the center of the earth is slightly less strong than that of the stone.

The infinite divisibility of space led to some perplexing dynamical difficulties, first noted by the Greek philosopher Zeno of Elea, which induced some Islamic physicists to question the coherence of the Aristotelian concept of space. Al-Bīrūnī (d. after 1050), a prominent critic of Aristotelian physics, posed this dilemma in a letter to Ibn Sīnā: If the sun is west of the moon in the sky, with a definite space between them, then even though the moon moves much faster than the sun, it should never be able to catch it. For the space between them can be conceived as divisible into an infinite number of parts; but how can a body moving with a finite speed cross an infinite number of spaces? In al-Bīrūnī's view, this paradoxical consequence reflected poorly on the Aristotelian concept of the infinite divisibility of space. After all, anyone can see that the moon does in fact overtake the sun.

A competing school of thought was inspired by Sunni theology, with its strict insistence on divine omnipotence. In this view, all bodies consist of a finite number of indivisible atoms, and the behavior of these atoms is not the consequence of any causal relations between them but is governed directly by the divine will. What appear to be causal connections are only regularities which we are in the habit of calling laws; there is no connection discoverable by reason between separate events, and Aristotelian explanations for the behavior of bodies in terms of natural tendencies are therefore hopelessly misguided.

The controversy between the theologians and the Aristotelians came to its climax in the *Tahāfut al-Falāsifa* (Self-Destruction of the Aristotelians) of al-Ghazālī (ca. 1058–1111), in which he maintained that philosophers could not know, for example, that fire causes cloth to burst into flame but only that the combustion of the cloth occurs at the moment that it makes contact with the flame. Ibn Rushd (ca. 1126–1198) responded with a *Tahāfut al-Tahāfut* (Self-destruction of the Self-destruction) in which he claimed that such skepticism regarding the efficacy of reason would undermine itself. In contemporary philosophy of science, a similar controversy is still raging over the nature of causality and our knowledge of causal connections, though the arguments arise not out of theological considerations but rather from Hume's sceptical doubts about induction and the verificationism of the logical positivists.

Mechanics

One of the simplest of mechanical devices, the lever, had been investigated mathematically by the Greek polymath Archimedes, who gave an elegant proof that the forces required to keep the unequal arms of a lever in balance were inversely proportionate to the length of the arms. In the course of his proof, Archimedes assumed that two equal weights located at different points on one arm of the lever produced the same effect as if they were both located at the midpoint between them — an assumption which is intuitively plausible but for which he gave no rigorous justification. Thābit ibn Qurra (d. 901) provided an intriguing argument for this assumption in his *Kitāb al-qarasṭūn* (Book of the Balance).

The balance was investigated very thoroughly in Islam. Al-Bīrūnī worked out a general method for determining the specific gravity of an irregular solid which he published in his *Maqala fi'l-ḥisab allatī bayn al-filizzāt wa'l-jawāhir fi'l-hajm* (Treatise on the Ratios Between the Volumes of Metals and Jewels). In later years, the best-known work on the subject was the *Kitāb mīzān al-ḥikma* (Book of the Balance of Wisdom) of al-Khāzinī (fl. ca. 1115–30), a careful though not very original treatise which covers the theory of balances (relying heavily on the work of al-Bīrūnī), the practical details of their construction, and their use for determining proportions of metals in alloys. In the course of his discussion, al-Khāzinī notes that heat decreases the density of metal objects. The *Kitāb* of al-Khāzinī is among other things a source of alchemical lore, and it is no accident that it became a standard text on the construction of equipment among alchemists in succeeding generations.

Islamic scientists were aware of more complex mechanical contraptions, and a few works described windmills, clocks, and other marvelous devices. The most lavishly illustrated volume on this subject, the *Kitāb fī ma'rifat al-ḥiyal al-handasiyya* (Book of Knowledge of Ingenious Mechanical Devices) of al-Jazarī, describes some fifty mechanical devices including several different types of fountains. But they made no systematic effort to apply the principles utilized in these devices on a broad scale. The knowledge of the principles of mechanics did not generate a technological revolution.

Dynamics

The Aristotelian theory of motion was a source of considerable difficulty for Islamic thinkers. Apart from the "natural motion" by which objects seek their natural place, the Aristotelians viewed motion as the result of the application of a force. In the absence of the force, motion ceases, as a cart grinds to a halt when the rope connecting it to the horse is broken. But this commonsense theory generates serious problems when applied to projectile motion. Why should a flung stone hurtle through the air when it leaves the hand? Why should an arrow continue to fly when it has left the bowstring? Aristotle's own suggestion was that somehow the air rushes around to the rear of the object in a way which prolongs its flight. But this was both vague and apparently irreconcilable with the facts; the flight of a javelin is not appreciately improved by design changes at its back end which would help it to "catch" the air.

In place of this unsatisfactory explanation, Ibn Sīnā and Abū'l-Barakāt al-Baghdādī (ca. 1080–ca. 1165) developed a theory of "violent inclination" which matches very closely the impetus theory first suggested by the Alexandrian Neoplatonist John Philoponus (d. ca. 570) and which was widely held until the seventeenth century. According to this theory, the agent of motion (the hand which flings a stone, for instance) imparts to the object a temporary inclination to continue moving, an inclination which partly counteracts the tendency of the object to seek its natural place and which is gradually dissipated through its flight.

Aristotle's notions regarding free fall also came under critical scrutiny by Islamic thinkers. According to Aristotle (*Physics*, Book 4, chapter 8), the motion of a falling object is directly proportional to its natural inclination (the tendency it has to seek its natural place) and inversely proportional to the resistance offered by the medium through which it moves, an account which strongly suggests the mathematical formulation

$$M \propto \frac{F}{R},$$

where M is motion, F is force and R is the resistance. Ibn Bājja (better known in the West as Avempace), though working within the broadly Aristotelian framework, criticized this view on the grounds that, if Aristotle's account were correct, unresisted motion (such as that of the heavenly bodies) would be instantaneous — which it clearly is not. In place of the Aristotelian law of motion, he suggests the formula

$$M \propto F - R,$$

which has the advantages that motion in the absence of resistance need not be instantaneous but rather proportional to the impressed force, and that when force and resistance are balanced, the result is not motion at all but cancellation of motion. Ibn Bājja's views here closely resemble some arguments advanced earlier by John Philoponus, but it was through Ibn Bājja's writings, preserved in a commentary on Aristotle's *Physics* by Ibn Rushd, that these views were transferred to the west and influenced the early work of Galileo.

Abū'l-Barakāt's reflections on motion in his *Kitāb al-Mu'tabar* (Book of What has been Established by Personal Reflection) are also remarkable as precursors of a radical departure from the Aristotelian view. In explaining the motion of a flung stone, he appeals not only to the dissipation of the violent inclination but also to a compounding of successive natural inclinations which increases the overall strength of the natural inclination during the fall. This idea of the compounding of the force through small intervals represents a serious advance over Aristotle's view that the distance travelled by an object in free fall per unit time is directly proportional to the weight of the object. It appears not to have occurred to Aristotle to consider the question of instantaneous velocity. Abū'l-Barakāt's explanation for free fall is thus one of the first hints that the constant action of a force might produce not merely motion but an acceleration.

TIMOTHY J. MCGREW

REFERENCES

Clagett, Marshall. *Greek Science in Antiquity*. New York: Barnes and Noble Books, 1994.

Fakhry, Majid. *Islamic Occasionalism*. London: George Allen and Unwin, 1958.

Lindberg, David, ed. *Science in the Middle Ages*. Chicago: University of Chicago Press, 1978.

Lindberg, David. *The Beginnings of Western Science*. Chicago: University of Chicago Press, 1992.

Moody, Ernest. *Studies in Medieval Philosophy, Science, and Logic*. Berkeley: University of California Press, 1975.

Nasr, Seyyed. *An Introduction to Islamic Cosmological Doctrines*. Cambridge, Massachusetts: Harvard University Press, 1964.

Nasr, Seyyed. *Islamic Science: An Illustrated Study*. London: World of Islam Festival Publishing Company, 1976.

See also: Mathematics in Islam – Astronomy in Islam – Optics – Ibn Sīnā – al-Jazarī – Atomism – al-Kindī – Ibn al-Haytham – Ibn Rushd – al-Bīrūnī – al-Khāzinī – Thābit ibn Qurra

***PI* IN CHINESE MATHEMATICS** In the earliest existing mathematical texts, *Zhoubi suanjing* (The Mathematical Classic of the Zhou Gnomon) and *Jiuzhang suanshu* (Nine Chapters on the Mathematical Art), the ratio of the circumference of the circle to its diameter, or π, was taken to be three. Liu Xin, who lived in the first century BC and was an astronomer and an expert on calendrical calculations, was mentioned in *Suishu* (Standard History of the Sui Dynasty) as being among the earliest to improve on this value. This arose from the construction of a standard measure called *jia lianghu* which he prepared for Emperor Wang Mang.

In his commentary on a problem concerning a sphere in *Jiuzhang suanshu*, Liu Hui of the third century referred to Zhang Heng's (AD 78–139) empirical deduction which implied a value of $\sqrt{10}$ for *pi*. The eighth century book, *Kaiyuan zhanjing* (Kai Yuan Treatise on Astrology), expressed Zhang Heng's estimation in the ratio of 736 to 232. A century later, Wang Fan improved this to the ratio of 142 to 45.

However, it was Wang Fan's contemporary, Liu Hui, who succeeded not only in attaining a better estimate for pi, but also in writing down the detailed method of how he arrived at the result.

His method commenced with a regular hexagon inscribed in a circle. This was extended to a twelve-sided polygon where the length of one side and its area were calculated from the hexagon based on the properties of the right-angled triangle. Next, a polygon of 24 sides was constructed and the length of its side and its area were calculated from the data of the dodecagon. Regular polygons inscribed in the circle were in this manner continuously constructed; the number of sides of one polygon was always double that of the previous polygon and the area of a polygon was formulated as one half the perimeter of the previous polygon multiplied by the radius of the circle. Liu Hui pointed out that if this process was repeated long enough, eventually there would be a polygon whose sides were so short that they would coincide with the circle. He concluded that this explained why the area of a circle was one half the circumference multiplied by the radius. He further pointed out that when the ratio of the circumference to the diameter was taken as 3 to 1, this value was in fact the ratio of the perimeter of the hexagon to the diameter.

Lui Hui reached a polygon of 192 sides when he calculated *pi* to lie between $3.14\frac{64}{625}$ and $3.14\frac{169}{625}$. From these figures he deduced an approximate value for *pi* of $\frac{157}{50}$ or 3.14. He gave another value of $\frac{3927}{1250}$ or 3.1416 which he said could be verified when calculations reached the polygon of 3072 sides.

Computations were performed through the rod numeral system which was not only a medium for whole numbers but also one for common fractions and decimal fractions. The knowledge of Lui Hui's method and the rod numeral system would enable the value of *pi* to be calculated to any desired degree of accuracy. It was therefore not surprising that Zu Chongzhi (AD 430–501), the most distinguished mathematician, astronomer and engineer of his time, found π to lie between 3.1415926 and 3.1415927 and also gave it a value of $\frac{355}{113}$. These values were not surpassed until a thousand years later when al-Kāshī evaluated *pi* correctly to sixteen decimal places. Zu's values were recorded in *Suishu* but no method was given, and Zu's own work had long been lost.

LAM LAY YONG

REFERENCES

He Shaogeng. "Method for Determining Segment Areas and Evaluation of π." In *Ancient China's Technology and Science.* Beijing: Foreign Languages Press, 1983, pp. 90–98.

Lam Lay Yong and Ang Tian Se. "Circle Measurements in Ancient China." *Historia Mathematica* 13: 325–340, 1986.

Li Yan and Du Shiran. *Chinese Mathematics. A Concise History.* Trans. J.N.Crossley and A.W.C. Lun. Oxford: Clarendon Press, 1987.

Needham, Joseph. *Science and Civilisation in China.* Vol.3: *Mathematics and the Sciences of the Heavens and the Earth.* Cambridge: Cambridge University Press, 1959.

Qian Baocong. "Zhong guo suan shu zhong zhi zhou lü yan jiu". (A Study of π in Chinese Mathematical Texts.) In *Qian Baocong ke xue shi lun wen xuan ji* (Selected Essays of Qian Baocong on the History of Chinese Science). Beijing: Kexue chubanshe, 1983, pp. 50–74.

Qian Baocong, ed. *Suanjing shishu* (Ten Mathematical Classics). Beijing: Zhonghua Shuju, 1963.

Suishu (Standard History of the Sui Dynasty). Compiled by Wei Zheng et al. AD 656. Beijing: Zhonghua Shuju, 1973.

See also: Liu Hui and the *Jiuzhang suanshu – Zhoubi suanjing* – Zhang Heng – Zu Chongzhi – al-Kāshī – Qian Baocong – Computation

PI IN INDIAN MATHEMATICS Undoubtedly pi or π is the most interesting number in mathematics, and its history will remain a never-ending story. It occurs in several formulas of mensuration and is variously involved in many branches of mathematics, including geometry, trigonometry, and analysis.

The earliest association of π is found in connection with the mensuration of a circle. The fact that the perimeter or circumference of any circle increases in proportion to its diameter was noted quite early. In other words, in every circle, perimeter/diameter = constant, or $p/d = \pi_1$, where π_1 is the same for all circles.

After knowing the perimeter, the area of the circle was often found by using the sophisticated relation

$$\text{area} = \left(\frac{p}{2}\right)\left(\frac{d}{2}\right) = \frac{pd}{4}.$$

But the earliest rules for determining area were of the form area = $(kd)^2$, where k is a constant prescribed variously. Both these methods imply that the area of a circle is proportional to the square of its diameter (or radius r), or area = $\pi_2 r^2$.

We know that π_2 is the same as π_1, but this was not always known. Similarly, π_3 may be defined from the volume of a sphere. In this article, the symbol π is used to denote all the above three values, as well as for their common value, which is now known to be not only an irrational but a transcendental number.

Since the Indus Valley script has not been deciphered successfully, we cannot say any final thing about the scientific knowledge of India of that time (about the third millennium BC). Some conjectures about the value of π used in the *Ṛgveda* (about the second millennium BC) have been made. Definite literary evidence is available from texts related to Vedāṅgas, especially the *Śulbasūtras* which contain much older traditional material. In the *Baudhāyana Śulbasūtra*, the oldest of them, the perimeter of a pit is mentioned to be three times its diameter, thereby implying $\pi_1 = 3$. This simplest approximation is found in almost all ancient cultures of the world. In India, it is found also in classical religious works such as *Mahābhārata* (Bhīṣmaparva, XII: 44), and certain *Purāṇas*, as well as in some Buddhists and Jaina canonical works.

Different approximations to π_2 are implied in the various Vedic rules for converting a square into a circle of equal area and vice-versa. If r is the radius of the circle equivalent to a square of side s, the usual *Śulba* rule is to take $r = s(2+\sqrt{2})/6$, which implies the approximation $\pi_2 = 18(3 - -2\sqrt{2}) = 3.088$. Recently, a new interpretation of the *Mānava Śulba Sūtra* has yielded the relation $r = 4s/5\sqrt{2}$, thereby implying $\pi_2 = \frac{25}{8} = 3.125$, which is the best *Śulba* approximation found so far.

The ancient Jaina School preferred the approximation $\pi = \sqrt{10}$, which they considered accurate and from which they derived the value $\pi = \sqrt{3^2+1} \approx 3 + \frac{1}{6} = \frac{19}{6}$. This value ($\sqrt{10}$) continued to be used in India not only by the Jainas but by others, such as Varāhamihira, Brahmagupta, and Śrīdhara, even when better values were known.

With Āryabhaṭa I (born AD 476), a new era of science began in India. In the *Āryabhaṭīya*, he gave a fine approximation of π_1, surpassing all older values. It contains the rule that the perimeter of a circle of diameter 20,000 is close to 62,832, so that $\pi_1 = \frac{62832}{20000} = 3.1416$ *nearly*, which is correct to four decimal places, and he still calls it close and not exact.

This value had a respectful place in Indian mathematics and exerted greater influence. How Āryabhaṭa obtained it is not known. It was known in China, but evidence of borrowing lacks documentary support. On the other hand, the two typically Indian values $\sqrt{10}$ and $\frac{62832}{20000}$ appear in many subsequent Arabic works.

In India, the Archimedian value $\frac{22}{7}$ for π first appeared in the lost part of Śrīdhara's *Pāṭī* (ca. AD 750). A Jaina writer, Vīrasena, quotes a peculiar rule in his commentary *Dhavalā* (AD 816). It is equivalent to $\rho = 3d + (16d + 16)/113$. If we leave out the redundant dimensionless number +16 in the brackets, this rule will imply a knowledge of the value $\pi_1 = \frac{355}{113}$ which was known in China to Zu Chongzhi (AD

429–500). In explicit form, this value is found in India much later, e.g. in the works of Narāyaṇa II, Nīlakaṇṭha, and others. The simplified or reduced form $\frac{3927}{1250}$ of Āryabhaṭa's value of π occurs in the works of Later Pauliśa, Lalla, Bhaṭṭotpala and the great Bhāskara II.

Most significant contributions on the computations of π were made by the mathematics of the late Āryabhaṭa School of South India. Mādhava of Saṅgamagrāma (ca. AD 1340 to 1425), the first great scholar and founder of the School, gave the value $\pi_1 = 2827, 4333, 8823, 3)/(9 \times 10^{1}1)$ as known to the "learned men". This value yields an approximation correct to eleven decimals. Mādhava also knew the series $\pi/4 = 1 - \frac{1}{3} + \frac{1}{5} - \frac{1}{7} + \ldots$ which was rediscovered in Europe in 1673 by Leibniz.

In discovering various series for π and in evolving techniques for improving their convergence, a great theoretical breakthrough was already attained in sixteenth-century India.

R.C. GUPTA

REFERENCES

Gupta, R.C. "Aryabhata I's Value of π." *Mathematics Education* 7(1): 17–20, 1973 (Sec. B).

Gupta, R.C. "Some Ancient Values of Pi and Their Use in India." *Mathematics Education* 9(1): 1–5, 1975 (Sec. B).

Gupta, R.C. "On the Values of π from the Bible." *Gaṇita Bhāratī* 10: 51–58, 1988.

Gupta, R.C. "New Indian Values of π from the Mānava Śulba Sūtra." *Centaurus* 31: 114–125, 1988.

Gupta, R.C. "The Value of π in the Mahābhārata." *Gaṇita Bhāratī* 12: 45–47, 1990.

Gupta, R.C. "Sundararāja's Improvements of Vedic Circle–Square Conversions." *Indian Journal of History and Science* 28(2): 81–101, 1993.

Hayashi, Takao, T. Kusuba, and M. Yano. "Indian Values for π Derived from Āryabhaṭa's Value." *Historia Scientiarum* 37: 1–16, 1989.

Marar, K. Mukunda, and C.T. Rajagopal. "On the Hindu Quadrature of the Circle." *Journal of the Bombay Branch of the Royal Asiatic Society* (New Series) 20: 65–82, 1944.

Rajagopal, C.T., and M.S. Rangachari. "On Medieval Kerala Mathematics." *Archive for History of Exact Sciences* 35(2): 91–99, 1986.

Smeur, A.J.E.M. "On the Value Equivalent to π in Ancient Mathematical Texts: A New Interpretation." *Archive for History of Exact Science* 6 (4): 249–270, 1970.

See also: *Śulbasūtras* – Varāhamihira – Brahmagupta – Śrīdhara – Āryabhaṭa – Nīlakaṇṭha – Pauliśa – Bhāskara – Mādhava

PIRĪ REIS Pirī Reis, Muhyī al-Dīn Pirī (1465–1554) was born in Gallipoli, a naval base along the Marmara coast. He was an important admiral, cartographer, and geographer of his time. His fame depends on two world maps and a book named *Kitāb-ı Bahriye*.

The First World Map (1513)

In 1929 a map was found in Topkapi Palace, on a piece of parchment 90×60 cm in size. It had been drawn in 1513, and later presented to Sultan Selim in 1517 in Egypt. This portion shows only the coast of South Western Europe, West Africa, the Middle East, and Central America. On the map, mountains were drawn in outlines, and the rivers marked with thick lines. Shallow places were indicated by red dots, and rocky places in the sea with crosses. Notes were added concerning different regions, and they were decorated with illustrations of special plants, and animals.

This map is a *portolano*, a navigation manual illustrated with charts, with no lines of longitude and latitude. Instead there are two wind roses, one in the North and the other in the South. Each of these roses is divided into 32 parts, and lines are extended beyond these roses; there are two scales indicating mileage. The lines that are extended from the wind roses and the scales are used in measuring the distances between the ports.

Pirī Reis made use of 34 maps in drawing his own. Of these, twenty had no dates. Eight of them were maps that were called *Jaferiyye* by Muslim geographers. Four were new maps drawn by the Portuguese, and one was the map of Christopher Columbus. Since this last is now lost, the only original document we have today is the map of Pirī Reis. When he was drawing the coasts of America, he remained faithful to Columbus' map, and copied it from several points. The Antilles and Cuba are shown as continents on his map, which was what Columbus believed. When Columbus was near the coast of Cuba in l494, he had the firm belief that Cuba was a continent. As a result, he had his conviction recorded by the notary public on board, Ferdinand Perez de Luna, and asked all the crew to sign it. Columbus had shown South America as a group of islands. Inspired by this, Pirī too drew lot of imaginary islands opposite Trinidad. This map is a very valuable historical map for two reasons. First of all, it was the most correct and scientific map of the time, and secondly it was the only map which was drawn using Columbus' map, of which neither the original nor the copy exists today.

The Second World Map (1528)

Fifteen years after his first map, Pirī Reis drew a second world map. Today we have only a small portion of it, 68×69 cm in size. On this portion, there are the northern parts of the Atlantic Ocean and the newly discovered regions of North and Central America.

The map starts with Greenland, in the north. Towards the south there are two pieces of land. The first is called Baccalo; the second one, further down, is called Terra Nova, and it is mentioned that these were discovered by the Portuguese. Further south there is the peninsula of Florida drawn quite correctly. The pieces of land at the side are the peninsulas of Honduras and Yucatan, discovered in 1517 and 1519, respectively.

Cuba and Haiti are drawn quite accurately. The errors on the previous map were corrected on this one.

There is a slight distortion on the map from the true positions as we know them today. This error was committed because of neglect in not taking into consideration the ten to thirteen degrees of difference in angle on contemporary compasses. This error is true for all Western-originated maps.

In this second map, the drawing of the coastlines shows greater improvement when compared with the inaccuracies of the first one. Only the parts of the world that were discovered were shown; the unexplored areas were left blank.

Kitāb-ı Bahriye

Pirī made a book of all his notes, revised and expanded it, and in 1525 presented it to Süleyman the Magnificient. He analyzed the geographic works of his time and, with his sharp faculties of observation, he set down everything he came across in his travels. In the preface of his book, he said that his purpose in writing such a book was to give information about ports, coasts, and islands by drawing them on maps known as portolanos in the west. However, he added that no matter how big the scales of maps were, it was impossible to show enough of the vital details on them. This lack of important information made him see the necessity of writing a book to compensate for this insufficiency. This then, is the novelty *Bahriye* brought to the science of navigation.

SEVIM TEKELI

REFERENCES

Adivar, Adnan. *Osmanli Türklerinde Ilim*. Istanbul: Remzi Kitabevi, Ankara Caddesi No. 93, 1982.

Afet, A. "Bir Türk Amirali, XVI. asrin büyük ceografi: Piri Reis." *Belleten* 1(2): 333–348, 1937.

Afetinan, A. *Life and Works of Pirī Reis*. Ankara: Türk Tarih Kurumu Basimevi, 1983.

Akçura, Yusuf. *Piri Reis Haritasi Hakkinda Izahname — Die Karte des Piri Reis — Piri Reis Map — Carte de Piri Reis*. Istanbul: Devlet, 1935.

Alpaqut, H. and F. Kurtoğlu. Mukaddime I-LV. *Pirī Reis: Kitāb-ı Bahriye*. Istanbul, 1935.

Kahle, P. *Piri Reis und seine Bahriye, Das türkishes segelhandbuch für das Mittelländische Meer vom Jahre 1521*. Berlin: de Gruyter, 1926.

Kahle, P. "Piri Reis und seine Bahriye." In *Beiträge zur historichen Geographie, Kulturgeographie, Ethnographie und Kartographie, vornehmlich des Orients*. Leipzig and Wien: Deuticke, 1929.

Kahle, P. *Die Verschollene Columbus–Karte von 1498 in einer Türkischen Weltkarte von 1513*. Berlin and Leipzig: de Gruyter, 1933.

Selen, S. "Piri Reis'in Şimali Amerika Haritasi: (Die Nord Amerika Karte des Piri Reis-1528)." *Belleten* 1(2): 515–528.

Senemoğlu, Y. *Kitāb-ı Bahriye* . 2 vols. Istanbul: 1001 Temel Eser, Tercuman Gazetesi, 1974.

Tekeli, Sevim, "The Oldest Map of Japan by a Türk, Mahmud of Kashgar: the Map of America by Pirī Reis." *Erdem* 1 (3): 665–683, 1985.

POTATO The Origins and Conservation of Potato Genetic Diversity When the conquistador Francisco Pizzarro and his men overran the Inca Empire in the 1530s they stumbled upon the potato, a vegetable with an annual crop value that today far exceeds that of all the gold the Spaniards ever removed from the New World. While the first potatoes that were taken to Europe in 1537 were regarded suspiciously or as just a curiosity, the potato is now the most important tuber crop globally, exceeded only in its contribution to human subsistence by the cereals wheat, rice, and maize.

What global humanity assumed from the Andeans was not just an important nutrient source, but also several millennia of accumulated empirical experimentation through which Amerindian farmers molded wild species of tuber-bearing members of the genus *Solanum* into the domesticated potato. This is a process determined by biological and environmental forces and by the intelligence and cultural traditions of these people in relation to plant biology and ecology. The Andeans made good use of the natural vegetative propagation method of potato plants, and altered it to their own advantage. They chose those tubers that were most favorable, and by planting the same preferred types year after year maintained their favorable characteristics.

Through time, human selection has resulted in specific changes in the potato plant. In comparing the potato of commerce, *Solanum tuberosum*, with its wild relatives, a botanist might first observe the changes in the length of the underground stems, or stolons, that bear the tubers. These are invariably short in modern cultigens. When wild and primitive plants are pulled up, the stolons are likely to break; tubers from plants with long stolons are more likely to be left in the ground, and less likely to be planted the next year. The result is an unconscious selection for shorter underground stems, and a plant that conveniently produces a mass of easily harvestable tubers close to its base. The typical wild potato is golf ball-sized or smaller. Selection for greater tuber size and plant yield improved the output compared to wild types.

In South America, several thousand different landraces and eight species of potatoes are presently cultivated. Farmers recognize them by differences in color, size, shape, and flavor as well as numerous agronomic traits. Most North American consumers are familiar with two or three colors of potatoes. In the Andes selection by humans has resulted in cultivars in which the skin and flesh color may be black, purple, red, yellow, or white, and spotted combinations of any of these colors. Many wild plants, in order to avoid being eaten by herbivores, produce unpleasant-tasting chemicals. Glycoalkaloids, steroidal nitrogen-containing compounds, are normally found in wild species of potatoes in concentrations high enough to be toxic or potentially fatal to humans or animals. These toxins have posed particular problems for Andean peoples and have been dealt with in various ways, including the selection of genes confering lower glycoalkaloid content.

Andeans make good use of the knowledge that different potato varieties grow best under different environmental conditions. Even if one type fails due to adverse weather or diseases caused by viral, bacterial, or fungal pathogens, others with specific resistances produce enough tubers to sustain the human population until the next season. Not surprisingly Andean potatoes are a source of invaluable germplasm from which plant breeders obtain sought after traits such as resistance to frost and disease.

Modern potato fields in the high Andes of Peru and Bolivia exist in the very locale from which wild plants were originally taken (the centre of diversity for potatoes) and wild plants, growing around fields, or even within fields, hybridize with cultivated plants. This creates a dynamic situation which produces, from seed, new genetic types. These plants will have valuable traits of the cultivated parent and perhaps valuable traits of the wild parent. Some of them will have a combination of characteristics superior to earlier types; they will often grow as weeds in fields and from them farmers continue to choose new varieties that constantly improve the nature of the cultivars.

Essential to this process of crop evolution is the intimate knowledge farmers have of their potatoes. Andean cultivators recognize clones of potatoes for their various qualities and may grow up to 35 different landraces together in one field. Aymara and Quechua Indians know tubers by name, e.g. *laram imilla* (blue girl), and have detailed taxonomies to describe the multitude of types.

The dynamic situation that created this diversity continues to the present day and is an essential resource both for ensuring human survival in the Andes and for maintaining potato genetic diversity. However, commercial interests which supplant this variation with new high-yielding varieties of potatoes, and introduce modern technologies, are bringing rapid change to the traditional agricultural system in the Andes; it is with alarm that scientists view the loss of potato landraces and the farming systems that guarantee the long-term survival of this crop upon which humankind is heavily dependent.

The conservation of biodiversity of both crop and wild species of plants is of global concern. However, the most obvious beneficiaries of the use of plant genes to produce new pharmaceuticals or improved crops are in the industrialized countries of the world. Fewer of the financial benefits derived from patented products are returned to the countries from which the germplasm was obtained. More importantly little recognition or compensation is provided to the communities who maintain the essential indigenous knowledge about plant properties. Indeed with economic development the marginalization of indigenous peoples in society is often increased rather than diminished; while indigenous peoples have traditions of conserving natural resources, changing economic conditions force them to adopt practices that are often destructive to their traditional systems.

In addition to the potato, the first peoples of the Americas were the domesticators of numerous crops including tomatoes, maize, and chili peppers. From their knowledge systems humanity learned the use of traditional medicines such as curare and ipecac which have provided valuable pharmaceuticals. Recognition of the true value of the intellectual property of indigenous people is the first step to settle the debt established in 1537; international governments and institutions need to seek in conjunction with these communities means to conserve *in situ* the valuable resources that are maintained within their cultural traditions.

TIMOTHY JOHNS

REFERENCES

Brush, Stephen B. "A Farmer-based Approach to Conserving Crop Germplasm." *Economic Botany* 45(2):153–165, 1991.

Hawkes, J.G. *The Potato: Evolution, Biodiversity, and Genetic Resources*. Blue Ridge Summit, Pennsylvania: Smithsonian Institution Press, 1990.

Johns, Timothy. *With Bitter Herbs They Shall Eat It: Chemical Ecology and the Origins of Human Diet and Medicine*. Tucson: The University of Arizona Press, 1990.

Salaman, Redcliffe N. *The History and Social Influence of the Potato*. Cambridge: Cambridge University Press, 1989.

Woolfe, Jennifer A. *The Potato in the Human Diet*. Cambridge: Cambridge University Press, 1987.

Zimmerer, Karl S., and David S. Douches. "Geographical Approaches to Crop Conservation: the Partitioning of Genetic Diversity in Andean Potatoes." *Economic Botany* 45(2):176–189, 1991.

See also: Agriculture – Environment and Nature in the Andes – Environment and Nature in Central America – Crops

PRECESSION OF THE EQUINOXES Precession of the equinoxes, which in Indian astronomy is called *ayana-calana* (shifting of the solstices), refers to the slow but continuous backward movement of the point of intersection of the ecliptic (which is a fixed circle) and the celestial equator (which keeps on moving backwards). This effectively means that the first point of Aries, which is the traditional point of the commencement of the Indian year, is really shifting backwards continuously. This shifting is, however, so slow, being a meagre 50.2 seconds per annum, that it comes to be noticed only when it accumulates over several years, and takes about 25,800 years to complete one circle.

Such a shift seems to have been noticed in India even during the Vedic times, as is evidenced by the Vedic priests' changing the beginning of their year backwards, from the constellation Mṛgaśiras (Delta Orionis) to the next previous constellation Rohiṇī (Alpha Tauri), and again backward, to Kṛittikā (Eta Tauri) in the course of time. But no measurement of it was made. Precession was not apparent during the time of astronomer Āryabhaṭa (AD 499), since at that time the First point of Aries coincided with the equinox. However, later astronomers noted it and also measured its rate. The first Indian astronomer who gave a rule for finding the value of precession seems to be Devācārya, who, in his *Karaṇaratna*, I.36 (AD 689), gives the rate of precession which works out to 47 seconds per annum. The magnitude of the precession, called *ayana-aṃśa* (degrees of precession) of a celestial body would be the angular distance between its computed longitude and the First point of Aries on the zodiac (which Hindu astronomy takes as fixed).

Hindu astronomers did not conceive the shifting of the equinoxes as a continuously regressing phenomenon, which it really is. They understood it as an oscillatory motion, with the beginning of the Kali era (3102 BC) as the 'zero-precession year', when the Sun and other planets were at the First point of Aries (Aśvinī), i.e. at the end of Zeta Piscium (Revatī). According to the *Sūryasiddhānta*, an oscillation of amplitude 27° to and fro would take 7200 years to complete. Thus, during the first 1800 years, the equinox moved forward by 27°, and during the next 1800 years it moved backward, coming back to the zero position at the end of the Kali year 3600 (which corresponds to AD 499, when Āryabhaṭa composed his work *Āryabhaṭīya*). It will continue to regress for 1800 years more, till AD 2299, when its forward motion for 1800 years would commence, again reaching the zero position at the end of 7200 years, covering, in all, 108°. Certain other texts, such as the *Karaṇaratna* and *Vākya-karaṇa* take the amplitude of the oscillation as 24°, the period for one complete oscillation being 7380 years. Among Indian astronomers, it was only Muñjāla (AD 932) who conceived precession as a continuous motion.

K.V. SARMA

REFERENCES

Dikshit, Sankar Balakrishna. "Ayana Calana or Displacement of Solsticial Points." In *Bhāratīya Jyotish Śāstra (History of Indian Astronomy)*. trans. R.V. Vaidya. Part II. New Delhi: Director General of Meteorology, 1981, pp. 205–21.

The Karaṇaratna of Devācārya. Ed. Kripa Shankar Shukla. Lucknow: Lucknow University, 1979.

Sastry, T.S. Kuppanna. "The Concept of Precession of the Equinoxes." In *Collected Papers on Jyotisha*. Tirupati: Kendriya Sanskrit Vidyapeetha, 1989, pp. 126—28.

PUTUMANA SOMAYĀJI Putumana Somayāji (ca. 1660–1740), author of the work *Karaṇapaddhati* (Methodology for Astronomical Manuals) was a *nampūtiri* (the appellation of the Brahmin community of Kerala in South India), who belonged to a family which bore the name Putu-mana (New House). He was of Ṛgvedic denomination, and secured the surname Soma-yāji by having performed the Vedic *Soma* sacrifice. His real name remains unknown. The date of Soma-yāji is surmised on the basis of the chronogram which gives the date of completion of his *Karaṇapaddhati* (as 17,65,653) of the Kali era, stated in the *kaṭapayādi* notation of expressing numerals, which is in the year A.D. 1732. Again, in the concluding verse, Somayāji states that he hailed from the village named Śivapura, which has been identified as Covvaram in Central Kerala, where his descendants still reside.

Somayāji was a prolific writer, mainly on astronomy and astrology, his only work in a different discipline being *Bahvṛcaprāyaścitta*, a treatise which prescribes expiations (*prayāścitta*) for lapses in the performance of rites and rituals by Bahvṛca (Ṛgvedic) Brahmins of Kerala. In addition to his major work, Somayāji is the author of several other works. In *Pañca-bodha* (Treatise on the Five), he briefly sets out computations at the times of *Vyatipāta* (an unsavory occasion), *Grahaṇa* (eclipse), *Chāyā*(measurements based on the gnomonic shadow), *Śṛṅgonnati* (elongation of the moon's horns), and *Mauḍhya* (retrograde motion of the planets), all of which are required for religious observances. His *Nyāyaratna* (Gems of Rationale), available in two slightly different versions, depicts the rationale of eight astronomical entities: true planet, declination, gnomic shadow, reverse shadow, eclipse, elongation of the moon's horns, retrograde motion of the planets, and *Vyatipāta*. Three short tracts on the computation of eclipses, including a *Grahaṇāṣṭaka* (Octad on Eclipses), are ascribed to Somayāji. He also composed a work called *Veṇvārohāṣṭaka* (Octad on the Ascent on the Bamboo), which prescribes methods for the computation of the accurate longitudes of the moon at very short intervals. A commentary in the Malayalam language on the *Laghumānasa* of Muñjāla is also ascribed to him. On horoscopy, Somayāji wrote a *Jātakādeśa-mārga* (Methods of Making Predictions on the Basis of Birth Charts), which is very popular in Kerala.

The *Karaṇa-paddhati*, in ten chapters, is his most important work. The work is not a manual prescribing computations; rather it enunciates the rationale behind such manuals. Towards the beginning of the work, the author states that he composed the book to teach how the several multipliers, divisors, and R sines pertaining to the different computations and the like are to be derived. Thus, the work is addressed not to the almanac maker but to the manual maker. All the topics necessary to make the daily almanac are not treated in *Karaṇapaddhati*, whereas several other items not pertaining to manuals are dealt with. The work takes as its basis the parameters and postulates of the Parahita system of astronomy promulgated by Haridatta, except for the section on eclipses, where the more accurate system from the *Dṛggaṇita* of Parameśvara is taken as the basis.

In chapter I, the *Karaṇapaddhati* sets out the planetary parameters, the computation of the mean planets, and the corrections to be applied thereto. Chapter II is devoted to the derivation, by means of pulverization (*kuṭṭaka*), of the multipliers and divisors necessary for planetary computations. Chapter III is concerned with various aspects of the computation of the moon, and Chapter IV deals with miscellaneous matters relating to the determination of suitable epochs for commencing computations. Chapter V mentions methods for correlating observed results with computed results, while chapter VI depicts such matters as the circle and the circumference, and the R sines pertaining to various entities for different purposes. Chapter VII is devoted to the rationale of the derivation of the mnemonics relating to the epicycles, their R sines, and allied matters. Chapter VIII describes varied derivations using the gnomonic shadow. Chapter IX is devoted to stellar declination and latitudes. The last chapter is concerned with certain derivations from the shadow of the gnomon.

As a work on astronomical rationale and exposition of the logic of several practices, the *Karaṇpaddhati* is highly important. A very significant statement made by the Putumana Somayāji, towards the close of Chapter V, is that the conception of aeons and eras, the measures therefore, and the various derivations based on them are not really true; they are only a means to compute the positions of celestial bodies and matters related to them. What is really important is the correlation of computation with observation, and to effect the same a practical astronomer is authorized to make changes as necessary. This is a highly significant statement, maybe even revolutionary, which was made in a forthright manner by an orthodox Hindu astronomer.

K.V. SARMA

REFERENCES

Karaṇa-paddhati of Putumana Samayāji. Ed. P.K. Koru. Cherp: Astro Printing and Publishing Co., 1953.

Karaṇa-paddhati of Putumana Samayāji. Ed. S.K. Nayar. Madras: Government Oriental Manuscripts Library, 1956.

Sarma, K.V. "Putumana Somayāji, an Astronomer of Kerala and his Hitherto Unknown Works." In *Proceedings of the 18th All India Oriental Conference*, Annamalainagar, 1955, pp. 562–64.

Walsh, C.M. "On the Quadrature of the Circle, and the Infinite Series of the Proportion of the Circumference to the Diameter Exhibited in the Four Sastras, the Tantra Sangraham, Yucti Bhasha, Carana Paddhati, and Sadratnamala." *Transactions of the Royal Asiatic Society of Great Britain and Ireland* 3(3): 509–23, 1835.

See also: Parameśvara – Muñjāla – Haridatta – Eclipses

PYRAMIDS The pyramids of ancient Egypt still stand beside the Nile bearing mute testimony to the civilization that constructed them nearly five thousand years ago. Considering the magnitude of this construction effort, it seems surprising that we have no information in extant records about either their planning or the construction process. We can only rely on study of the existing examples (a study limited to the external features only) and try to deduce from these external elements something of the methods used in their construction.

Pyramids are always situated west of the Nile, near the river yet high enough to be above flood stage, on solid bedrock that could be leveled and was free of defect so as to be able to support the enormous weight of the pyramid. The leveling process seems to have been carried out by enclosing the site with a mud wall and filling this enclosure with water. A network of trenches of uniform depth below the surface was then constructed. When the site was drained, the intervening material could easily be removed. Then, the site had to be laid out so that the faces of the pyramid would face the four cardinal directions. This must have involved some sort of astronomical technique, since the Egyptians seem not to have known the magnetic compass. One of the easiest techniques was to bisect the angle between the rising and setting point of one of the circumpolar stars, which would determine true north. Even this simple-sounding method demands the very accurate construction of an artificial horizon and an extraordinary precision of measurement in order to obtain the observed accuracy in pyramid orientations.

Large quantities of stone had to be quarried and transported to the construction site. Extracting large blocks of soft stone such as limestone posed no technical difficulty for the ancient Egyptians. Harder stone, such as granite, was considerably more difficult to work and posed immense transportation difficulties since most of it had to be brought from Upper Egypt. Whether transporting granite from Aswan or limestone from Tura and Giza, the preferred method was by water. The blocks of stone were floated to the point nearest the construction and then dragged to the construction site by teams of workers using sledges or rollers. There is still no conclusive evidence to indicate whether the stones were dressed at the quarry or at the pyramid site.

Once at the building site, the blocks were raised into position either by means of access ramps constructed for this purpose then dismantled and removed once the construction was complete, or by some sort of jacking mechanism. There is no evidence that the Egyptians ever used a block and tackle mechanism with pulleys to place large stones from the top, although this method may have been used with some smaller stones. How such ramps might have been arranged is largely a matter of conjecture. An early suggestion of helicoidal ramps spiraling upward around the outside of the pyramid has been superseded by the suggestion that a long paved supply ramp extended from near the river to the pyramid and was lengthened as the structure rose higher. Whether a pyramid was ever equipped with more than one such ramp is still a matter of debate. Whether the casing stones were put into position first or last in completing a particular course of stonework is also a matter of debate. Since the architects seem to have experimented with alternative techniques and to have adapted various practices over time, these and other positions seem to have some supporting evidence from one or another pyramid while others seem to follow alternative patterns. It is likely that there was no single method applied to all pyramid constructions.

The reason for such enormous constructions is also a matter of debate. It has generally been argued that they play a role in the cult of the Pharaoh. This cultic connection appears all the more likely because the pyramids were almost always physically connected to the mortuary temples of the pharaoh as part of a unified religious complex. There seems little evidence that the royal mummies were actually placed inside the pyramids, however. So, if they were not intended to be the last resting place of the Pharaonic remains, what were they for? An intriguing suggestion is that the pyramids represent large-scale public works intended to keep the largely agricultural labor force occupied during the annual inundation period. It has been suggested that this was more than an altruistic endeavor. It served to unify the state in a massive enterprise, thus cementing the recent political union between Upper and Lower Egypt.

GREGG DE YOUNG

REFERENCES

Arnold, D. *Building in Egypt*. Oxford: Oxford University Press, 1991.

Davidovits, J. and M. Morris. *The Pyramids: An Enigma Solved*. New York: Hippocrene Books, 1988.

Davis, A. R. *The Pyramid Builders of Ancient Egypt: A Modern Investigation of Pharaoh's Workforce*. London: Routledge & Kegan Paul, 1986.

Edwards, I. E. S. *The Pyramids of Egypt*. 2nd ed. London: Penguin, 1961.

Mendelsohn, K. *Riddle of the Pyramids*. New York: Praeger, 1974.

Q

QĀḌĪ ZĀDEH AL-RŪMĪ Qāḍī Zādeh was born in Bursa, Turkey about 765 AH/AD 1364. (Hence his name, al-Rūmī — from Byzantium.) His talent was early recognized and he was recommended to continue his studies of mathematics and mathematical astronomy among the scholars of Transoxiana, the center of such activities at that time. Consequently, about 814 AH/AD 1410, he went to the court of Ulugh Beg, governor of Samarqand and amateur astronomer, and joined his retinue.

In 824 AH/AD 1421, four years after establishing a major *madrasa* (school of advanced studies), Ulugh Beg decided to found an observatory in Samarqand in order to complete and improve the observational data gathered by Naṣīr al-Dīn al-Ṭūsī and his assistants in the famous *Zīj-i Ilkhānī*. Qāḍī Zādeh, along with the younger Ghiyāth al-Dīn Jamshīd al-Kāshī played an important role in both institutions of Ulugh Beg. When al-Kāshī died in 832 AH/AD 1429, Qāḍī Zādeh succeeded him as director of the observatory. At his death in 840 AH/AD 1436, the tables were still incomplete, and the work was completed under the direction of ʿAlī Kushjī and published as the *Zīj-i Jurjānī*. In addition to his role in producing these astronomical tables, Qāḍī Zādeh also authored *Sharḥ Mulakhkhaṣ fī'l-Hay'a* (Summary Commentary on Mathematical Astronomy), an influential commentary on ʿUmar al-Jaghmīnī's summary of astronomical knowledge.

In the field of mathematics, he is best known historically for his commentary on al-Samarqandī's *Ashkāl al-Ta'sīs* (Fundamental Theorems). His *Risālat al-Jayb* (Treatise on the Sine) displays a greater independence of thought, but even here he relies on the iterative techniques developed by his contemporary, al-Kāshī, to find the value of the sine of one degree.

GREGG DE YOUNG

REFERENCES

Dilgan, H. "Qadi Zada al-Rumi." In *Dictionary of Scientific Biography*. Ed. C. Gillispie. New York: Scribners, 1970/1981, vol. 10, pp. 227–229.

Souissi, M. *Ashkāl al-Ta'sīs lil-Samarqandī Sharḥ Qāḍī Zādeh al-Rūmī*. Tunis: al-Dar al-Tunisiyya lil-Nashr, 1984.

See also: Ulugh Beg – al-Kāshī – Naṣīr al-Dīn al-Ṭūsī

AL-QALAṢĀDĪ In the Islamic world of the fifteenth century there was a general halt in research and a lowering of the level of instruction. In mathematics this tendency resulted in a repetitive scientific production made up of poems or commentaries directed at an increasingly narrow public: teachers, *muwaqqits* (astronomers) charged with determining the hours of daily prayers, judicial functionaries, government bookkeepers, etc.

It is in light of this tendency that we can appreciate both the intellectual journey and the scientific contribution of al-Qalaṣādī, one of the last scholars of Andalusian origin who dedicated a large part of his life to disseminating — to the central Maghreb, to Ifriqiya, and to Egypt — what he knew of the Arab mathematical tradition of the preceding centuries.

ʿAlī ibn Muḥammad al-Qalaṣādī was born around 1412 in the Andalusian village of Basta. This small village, which is in the northeastern part of Grenada, was already prosperous in the time of the geographer al-Idrīsī (d. 1153) and continued to be so for two centuries.

We know nothing of al-Qalaṣādī's childhood except that he grew up in his birthplace where he learned the *Quʾrān* from his teacher Ibn ʿAzīz and where he received his first lessons in Arabic and probably also in mathematics. In the course of his adolescence he continued these studies and was taught grammar and mathematics from the arithmetic texts of Ibn al-Bannāʾ (d. 1321), as well as law, the science of inheritance, and calculation.

In 1436, at the age of 24, al-Qalaṣādī left his native village and began his first educational voyage, heading to Tlemcen, the first stop of a long journey to scientific centers in the Maghreb, al-Andalus, and Egypt. The educational voyage (*riḥla* in Arabic) was a medieval tradition, one of whose goals was to allow students to complete their education in various places, taking classes from famous professors.

Al-Qalaṣādī's stay in Tlemcan lasted eight years. His most serious study of mathematics was with az-Zaydūrī (d. 1441). He himself says that az-Zaydūrī taught him the contents of two important texts of Ibn al-Bannāʾ: the *Kitāb al-Uṣūl wa l-muqaddimāt fi l-jabr wa 'l-muqābala* (Book on the Foundations and Preliminaries in Algebra and Restauration) and the *Rafʿ al-ḥijāb ʿan wujūh aʿmāl al-hiṣab* (Lifting of Veil on the Science of Calculus Operations).

It also appears that the first scientific works of al-Qalaṣādī were written in Tlemcen, between 1436 and 1444. There were three commentaries on writings on the science of inheritance and a mathematics book entitled *al-Tabṣira* (The Book which Makes Things Intelligible).

In 1444, al-Qalaṣādī left Tlemcen for Tunis, which was at that time one of the most dynamic intellectual centers in the Maghreb. In the *madrasas* (colleges) where he studied, al-

Qalaṣādī took courses in Malikite law, the Arabic language, grammar, and the rational sciences. At the same time, he devoted time to his own writing. He published three books on calculation: the *Kashf al-jilbāb ʿan ʿilm al-ḥisāb* (Unfolding the Secrets of the Science of Calculation) in 1445, the *Qānūn fī al-ḥisāb* (Canon of Mathematics), and his commentary entitled *Inkishāf al-jilbāb ʿan qānūn al-ḥisāb* (Explaining the Canon of Mathematics). He also published two works on traditional sciences, *al-Kulliyyāt fī al farā ʾiḍ* (Collection on Successional Division), and a commentary on it. At the end of his stay in Tunis, in 1447, it appears that he wrote his epistle on irrational numbers entitled *Risāla fī dhawāt al-asmāʾ wa ʾl-munfaṣilāt* (Epistle on Binomials and Apothems).

On 30 May 1447, al-Qalaṣādī left Tunis and headed to Cairo and then to Mecca, where he wrote a commentary on the *Farā ʾiḍ* (Book on Successional Division) of Ibn al-Ḥājib on traditional science. Returning from the pilgrimage, he again stayed in Cairo, teaching and taking classes from several professors, notably Shams al-Dīn al-Samarqandī in rational sciences.

In April 1449, al-Qalaṣādī left Cairo and returned to al-Andalus. In 1451 he returned to his native village of Basta, and then went to Grenada, where he stayed for thirty years. There he studied with eminent professors like the astronomer Ibrāhīm Ibn Futūḥ, who specialized in the study of astrolabes.

A deterioration of the political climate of the interior of the kingdom of Grenada and increasingly worrisome menaces from outside forced al-Qalaṣādī to leave the Andalus once and for all in 1483. He settled finally in Béja in Ifriqiya where he died in 1486, just six years before the fall of Grenada.

Mathematical Works of al-Qalaṣādī

Like some of his contemporaries in al-Andalus and the Maghreb, al-Qalaṣādī wrote on many different subjects. In almost all disciplines he also, again like many of his contemporaries, published numerous commentaries on classical texts. But he is, to my knowledge, the only scholar of his era to have written so many works on mathematics.

The mathematical development of al-Qalaṣādī seems to have been centered basically on the science of calculation, on the processes of arithmetic and algebraic calculation and their application in the solving of abstract exercises and practical problems — e.g. problems of commercial transaction, monetary conversions, or the division of inheritances.

In algebra, he wrote chapters included in the arithmetical works already noted. In these chapters he treats algebraic operations and the resolution of the canonical equations of al-Khwārizmī. To that we must add his commentary on the algebraic poem of Ibn al-Yāsamīn.

In these texts one finds a peculiarity of the Maghrebian mathematical tradition, the utilization of a certain symbolism to express objects or algebraic concepts such as an unknown, different powers, or equality in equations. For a long time the invention of this symbolism was attributed to al-Qalaṣādī. But the results of recent research now permit us to affirm that this same algebraic symbolism was already in existence in this twelfth century, particularly in the work of Ibn al-Yāsamīn entitled *Talqīḥ al-afkār* (Fertilization of Thoughts), and that it was widely in use in the fourteenth century, in particular by Ibn Qunfudh (d. 1407). The work of al-Qalaṣādī bears witness to the persistence of these symbols and of their widespread use throughout the Maghreb.

In arithmetic, his writings deal essentially with the four arithmetic operations (applied to whole numbers, fractions, and irrational quadratics), with the extraction of the exact or approximate square root of a number, with the rule of three, with the method of false positives, and with other arithmetical procedures such as the breakdown of a number into prime factors and the calculation of the sums of series of natural numbers from their squares and cubes. Again, it is important to note that the techniques used by al-Qalaṣādī in these books are related to techniques already used in the works of al-Ḥaṣṣār (twelfth century) and of Ibn al-Bannāʾ.

The fact that these themes are present in the works of al-Qalaṣādī and his method of treating them leads to the conclusion that in the fifteenth century, the mathematical traditions of al-Andalus and of the Maghreb were unified by being based upon each other. Moreover, if one compares the work of al-Qalaṣādī with that of Ibn al-Bannāʾ, one notices a certain continuity both in form and content. Because of this, one can argue that al-Qalaṣādī is more Maghrebian than Andalusian. This stamp of the Maghreb on the education and work of a scholar from al-Andalus of the importance of al-Qalaṣādī offers proof, moreover, of the decline of scientific activity in al-Andalus in the fourteenth and fifteenth centuries.

On the other hand, when one compares his different works in mathematics, one does not see a noticeable evolution from one book to another, but rather different formulations of classic themes and techniques. Al-Qalaṣādī himself claimed that some of his books were only developments or summaries of previously published works. It is important to note that this process, which was not unique to al-Qalaṣādī and which was already in evidence at the end of the fourteenth century both in the Maghreb and in Egypt, only reflects the continuation in the fifteenth century of the slow decline of scientific activity in the Islamic city. In this difficult context the scientific aptitude of al-Qalaṣādī was not really able to flourish fully, and it is greatly to his credit that he contributed to maintaining the level of scientific activity of that of the fourteenth century in the Andalus and the Maghreb.

AHMED DJEBBAR

REFERENCES

Djebbar, A. *Recherches et Enseignement Mathématiques dans le Maghreb des XIIIe–XIVe siècles*. Paris: Université de Paris-Sud, Publications Mathématiques d'Orsay, 1980, pp. 41–54.

Djebbar, A. *Quelques aspects de l'algèbre dans la tradition mathématique arabe de L'Occident musulman*. In Actes du Premier Colloque maghrébin d'Alger sur l'Histoire des mathématiques arabes (Algeria, 1–3 Décembre 1986). Algeria: Maison du Livre, 1988, pp. 99–123.

Dozy, R. and M.J. De Geoje. *Edrisi, description de l'Afrique et de l'Espagne*. Leiden: Brill, 1968.

Woepcke, F. "Traduction du traité d'arithmétique d'Abou Haçan Ali Ben Mohammed Alkalaçadi." In *Recherches sur plusieurs ouvrages de Léonard de Pise*. Rome: Imprimerie des Sciences mathématiques et physiques, 1859.

Woepcke, F. "Recherches sur l'histoire des sciences mathématiques chez les orientaux: Notice sur des notations algébriques employées par les Arabes." *Journal Asiatique*, 5th series, 4: 348–384, 1854.

See also: al-Idrīsī – Ibn al-Bannāʾ

QANAT The qanat, invented in Iran, is a tunnel that allows a well in hilly country to supply water to users on lower ground by gravity flow.

Once the well has been sunk and has proved to be a good source of water, the route of the tunnel is carefully surveyed, and a series of further wells or guide shafts is dug, each one about 300 meters from the next. Excavation of the tunnel begins at the intended outlet and proceeds from the foot of one guide shaft to another. The depth of these shafts has previously been set, so that when the tunnel links them all together, it will have an even gradient of about 1:500. The head well may be anything from 15 to 100 meters deep, and tunnels are frequently 2000 meters or more in length.

As Andrew Watson points out, the history of the qanat can be clearly illustrated by a distribution map showing where water supplies of this kind are to be found. Before the time of Alexander the Great, i.e. before 330 BC, about the only qanats known to have existed were in Iran. A thousand years later, around AD 700, qanats had been constructed in the Arabian peninsula and Syria as well as in Persia. But then the Islamic people took up the qanat in a big way. Books were written on the subject in AD 840, and by al-Ḥasan al-Ḥāsib around AD 1000. Knowledge of the technique spread east to Afghanistan and west through North Africa to Morocco, Mauretania, and Spain.

Qanats continued to be developed in later centuries. By the sixteenth century, potteries were making earthenware rings for lining the tunnels where they passed through soft ground. In 1960, three-quarters of all water used in Iran came from qanats, with Tehran alone having thirty-six of them bringing water from hills several miles away.

The word "qanat" in Arabic is said to be the source of the English word "canal," which was initially used to mean a pipe or tube or tunnel carrying liquid; it retains that usage in anatomy — hence, alimentary canal.

ARNOLD PACEY

REFERENCES

Watson, Andrew M. *Agricultural Innovation in the Early Islamic World*. Cambridge: Cambridge University Press, 1983.

Wulff, Hans E. *The Traditional Crafts of Persia*. Cambridge, Massachusetts: MIT Press, 1966.

See also: Agriculture in Islam – Technology in Islam – Irrigation in Islam

QI In Chinese culture, *qi* is the common denominator that underlies all being and all behavior and that allows for their interaction and their interpenetration. *Qi* enables seeds, air, and soil to become plants, a mountain range to produce good fortune, or a human being to cultivate wisdom and human kindness. It is what makes possible the processes of chemistry and the researches of the alchemists. *Qi* is the fundamental Chinese articulation of what underlies the universe, what is the universe, and what is the reality of the universe.

Given the key position the concept of *qi* holds in explaining the process of change in Chinese scientific, medical, psychological, and spiritual thought, one is tempted to assume that it is one of the oldest and most central concepts in the Chinese world view. A look at the history of both the character and the concept shows otherwise.

The oldest inscriptions in Chinese, the bones used in the Shang dynasty oracle, are concerned with understanding the process of change in such uncertain areas as agriculture and illness. Nowhere in the more than 200,000 known extant oracle bones can one find the character *qi* or anything which might be its close antecedent or relative. Throughout the oracle bone inscriptions we see that, in Shang culture, change was brought about by the actions of spirits rather than by naturalistic forces.

The character *qi* is composed of an element which signifies vapors which rise to form clouds and another element which means grain. The character may mean vapors arising from cooked food or simply steam. At this level of the character's history *qi* is a simple notion. *Qi* is what makes the lid on a pot of boiling water rattle.

In the course of its development Chinese thought began to look for the origins of change in sources other than the spirits. For a time change, especially illness, was understood to come from the winds of the eight directions, each of which was responsible for causing particular diseases at particular seasons and each of which was personified as a deity. Although this model of change brought about by wind is a step toward a world of naturalistic causality and away from a world of spirit causality, it still has a strong component of the latter. It is probably, however, a major step in the development of the concept of *qi* which later comes to permeate all of Chinese thought.

By the time of Confucius (551–479 BC), the word *qi* appears in literature but it is rare and its meaning is confined to purely human qualities. The *Analects*, the record of Confucius' teachings, use the word four times, each of which is distinctly related to some aspect of human behavior, such as appetite, tone of voice, breath, and physical stamina. In none of these does it refer to anything beyond a specific physical quality of human life. At this stage in its history *qi* still shows its origin as a common, ordinary term.

The several centuries following Confucius were a critical time of change in Chinese culture. One of the key concepts to emerge at this time was *qi*. Mencius (371–289 BC), the follower of Confucius, refers to *qi* nineteen times and each time it is a concept of major philosophical import. *Zhangzi*, between 399 and 295 BC, a major Daoist text, refers even more frequently to *qi*. In this text we see one of the earliest and clearest uses of *qi* to refer to a natural force, something which, although an attribute of human beings, is also a power which makes things happen in the world. We see this conceptualization of *qi* (although still connected to wind) as a moving power in the universe in a famous passage from the *Zhangzi*.

> The great clod belches out breath (*qi*) and its name is wind. So long as it doesn't come forth nothing happens, but when it does, the ten thousand hollows begin crying wildly.

Here *qi* is a conceptual model for explaining and understanding the process of change in the observable world. The Shang world of a supernatural spirit causality which had developed into the world of semi-supernatural wind causality has now become a world of natural causality. The *qi* model explains the process of change and development in a naturalistic way which is conducive to the expansion of philosophic and scientific thought and inquiry.

By the late third century BC we see *qi* becoming a concept used throughout Chinese thought. This conceptualization/experiential model is used to explain the qualities of plants and animals, the uses of herbs, and the influences of climate, weather, and geographic location on people. From its use to explain the process of disease and illness comes psychological insights into character and personality types. Eventually, through philosophers such as Wang Yangming (1472–1529), *qi* becomes an essential quality which enables all change since, no matter how they seem to differ, all things share the same *qi*. It also becomes a moral spiritual quality allowing for the oneness of all being.

In recent times, possibly under the influence of western thought, *qi* has frequently been identified with the western concept of energy. Westerners are by no means the only ones to make this equation. To characterize *qi* as energy is to invoke a world view which the Chinese never had, a world view in which matter and energy are different things. While it is true that modern scientific theory no longer holds that matter and energy are separate entities the words have come, through several centuries of use, to connote, in common parlance, distinct and separate entities. To the Chinese matter was never inert; it always had dynamic and teleological properties.

As Western concepts of physics have caused some to think of *qi* as energy, nineteenth century western concepts of vitalism have caused it to be identified, especially in medicine, as some form of vital force, distinguishing living things. Historically, however, *qi* has been as much an attribute of a rock or river as of a person or animal. To call *qi* energy or life force is as erroneous as it is to call it matter. It is all and it is neither. *Qi* is the fundamental Chinese articulation of the interconectedness of the universe, of what the Chinese call the ten thousand things. *Qi* is the cause, process, and outcome of all activity in the cosmos.

STEVEN KLARER
TED J. KAPTCHUK

QIANJIN YAOFANG The *Qianjin Yaofang* (Remedies Worth their Weight in Gold), and its sequel the *Qianjin Yifang* (More Remedies Worth their Weight in Gold), were written by the physician and alchemist Sun Simiao (also called Sun Simo) in 652 and 682 respectively, during China's Tang dynasty.

These books are especially valuable for the breadth of their content. They described the medicine of the day, both the systematic acupuncture and herbal tradition, and the less organized folk tradition. Unfortunately neither has as yet been translated into a Western language.

Sun built on the work of Zhang Zhongjing's earlier *Shanghan Lun*, especially in advocating the careful selection, picking and preparation of herbs. In acupuncture he devised the system of *ahshi*! (ow! that's it!) points, or needling at the tender muscle nodules. These books also contained the author's newly verified and drawn charts of the channels and points on the body, with front, side, and back views — the earliest-known point charts to have been produced. The twelve channels were shown in five colors, while the eight extraneous channels were depicted in green.

Many of Sun's ideas were closer to Western notions of common sense. For instance he put disease down more to incorrect diet than to the influence of evil ghosts and spirits. He advocated general cleanliness to protect against the development of sores and ulcers, and he urged rinsing your mouth out several times after a meal to keep the teeth healthy and the breath fresh. During childbirth he said not more than two or three people should be allowed into the delivery room, to keep the mother quiet and the atmosphere calm, and he excluded anyone who had recently survived a fatal illness. He thus showed an awareness of complications arising in labor through the nervousness of the mother, and of infection spreading to the mother or newborn child. He also advocated the use of gentle exercise in imitation of animals in order to prevent disease and combat the process of aging. He introduced the treatment of night-blindness with goat or ox liver (which is rich in Vitamin A), the use of the thyroid glands of sheep or deer (equally rich in iodine) for a swollen thyroid, and abstaining from salt in cases of edema.

He innovated a clear ethical stance on medical practice, urging the doctor to be vigorous in investigating the disease and finding a suitable treatment. Many of his methods are in line with modern thinking. His work was equally valued in both Japan and Korea and works in these countries quote extensively from his tracts.

RICHARD BERTSCHINGER

REFERENCES

Fu, Weikang. *Traditional Chinese Medicine and Pharmacology*. Beijing: Foreign Languages Press, 1985.

See also: Sun Simo

QIBLA AND ISLAMIC PRAYER TIMES *Ṣalāt* (prayer) is one of five pillars of Islam. Islam requires all adult sane, healthy, and able-bodied Muslims to perform five daily prayers. The corresponding time intervals are commonly known as "Islamic prayer times".

Muslims are enjoined to face the *Kaʿba* in Mecca during their prayers. It is believed that *Kaʿba*, the cubical shaped edifice, was built by the Prophet Abraham. The direction to the *Kaʿba* from a place is known as its *qibla* direction. Mosques are built with a clear pointer in the qibla direction. Knowledge of the qibla direction is also important for certain other acts, including the call to prayer (*Adhān*), the ritual slaughter of animals for food, and burial of the dead. Muslim astronomers from early times dealt with the determination of the *qibla* and produced highly sophisticated trigonometric and geometric solutions.

The time intervals of the specified daily prayers are clearly specified. The intervals are exclusive of each other; the end of one time interval marks the beginning of the next. Besides regular prayers, there are supplementary prayers, and some of these follow specific time intervals. There are also certain periods of the day during which prayers are prohibited. All this reflects the need for a clear understanding of diurnal phenomena and the proper determination of the time of day. The way in which the various Islamic prayers are distributed throughout the day further emphasizes the treatment of day and night as a single time unit, in contrast to the practice of earlier civilizations in treating them separately. Also, the mutually exclusive time intervals for each regular prayer, which depend upon the specific (local) positions of the sun, introduced a new way of dividing the day.

The Islamic prayer times are so defined that they can be easily ascertained by observation on a clear day. However, under cloudy conditions, one has to resort to some sort of a time clock and mathematical computation. Since the performance of regular prayers within the specified time intervals is a most serious matter, advance determination of the intervals is of great importance. From the very beginning, Muslim scientists were engaged in mastering the techniques of positional and observational astronomy, including instrumentation, and went on to develop this and allied fields, dominating this field of science for some five hundred years. It is not surprising to find that almost all major astronomical works during that period incorporated a discussion relating to the Islamic prayer times, which came to be known as *ʿIlm al-mīqāt* or the times of 'fixed positions' of the sun. Closely associated with the Islamic prayer times is the matter of determining the direction of Mecca. Elaborate tables showing the direction of the *Kaʾba* from different parts of the world became an integral part of Islamic astronomical works.

The demarcation of various time intervals is based on the guidelines contained in the *Qurʿān* and elaborations given in the *Ḥadīth* (Traditions) of Prophet Muḥammad. Based on these, the time intervals were subsequently translated into more precise definitions by the theological jurists of the early centuries of Islam. Muslim astronomers helped translate these into equivalent astronomical definitions.

In Islam, the day begins at sunset and the time for the first prayer (*maghrib*) begins after the sun has set. The second (*ʿishāʾ*) and third (*fajr*) prayers then follow, the latter ending before the sun begins to rise. The fourth prayer (*ẓuhr*) is performed after midday and is followed by the fifth and last obligatory prayer (*ʾaṣr*) in the afternoon. The optional prayers performed after midnight (*tahajjud*) and a little after sunrise (*chasht*) have special importance. The time durations during which the sun rises, sets, and crosses the local meridian are prohibited times and no prayers are generally allowed during these periods. Thus, one has to determine the various portions of the day in a sequential and somewhat symmetrical way.

As we know, the 'true' solar day-length keeps changing through the seasons. Therefore, on a 24-hour civil clock, the solar phenomena, such as midday and midnight, keep moving backwards and forwards. However, as the Islamic time intervals are related directly to the true natural (solar) phenomena and not to the day length, it simply boils down to the daily determination of the times of various astronomical phenomena according to whatever day-time system is used, e.g. time-hours or time-degrees would be equally applicable. This situation obviously laid down strong foundations for mathematical-dynamical astronomy. Also we can recognize that the 'day-time' length depends upon the season and the latitude and varies from a winter minimum to a summer maximum. For Islamic fasting, which is done from dawn to sunset, some countries are favored (shortest duration) in one season and others in a different season. In Islam, this situation has been fully taken care of through the rotation of the Islamic (lunar) months through the seasons.

Muslim astronomers during the medieval period made considerable efforts to construct tables of Islamic prayer times according to the defined solar positions. The data on the solar position (azimuth) at different altitudes was also used for observationally determining the direction of the *qibla*. It was important to present the prayer times in the form of basic universal tables which would remain unchanged over a long time. These tables could then be used through secondary and smaller tables which could be recomputed if necessary.

These tables were universal, since the Muslims used a purely lunar calendar independent of solar calendar dates. For each new day/date, one simply needed to ascertain the solar longitude, by direct measurement or through the simultaneous measurement of positional parameters at meridian passage.

Muslim astronomers' interest in regulating daily time went beyond the determination of prayer times to include overall time measurement. Their tables contained information on various time parameters, such as length of daylight and duration of the fasting period (from morning twilight to sunset), auxiliary tables on various functions needed in computations, and special tables on parameters such as the solar azimuth and solar altitude in certain azimuths for use in *qibla* determination, besides supplementary information on such parameters as horizontal refraction. The prayer tables were generally produced with great computational skills and care and in a comprehensive way. A study of Cairo's tenth-century astronomer Ibn Yūnus' *Kitāb Ghayat al-intifa' fī ma'rifat al-da'ir wa-faḍlihi wa-l-samt min qibal al-irtifāʿ* (Very Useful Tables for Finding the Time since Sunrise, the Hour Angle, and the Solar Azimuth, from the Solar Altitude), one of the major *mīqāt* tables, shows that Ibn Yunūs' values were, in general, very carefully computed. The tabular works incorporating *ṣalāt* times and other auxiliary tables were prepared for various major localities of the Muslim world, such as Cairo, Mecca, Medinah, Jerusalem, Damascus, Tripoli, Aleppo, Tunis, Istanbul, Baghdad, and towns in West Africa, Andalusia and the Indian subcontinent. Later work by a Damascus astronomer, al-Khalīlī (fourteenth century), greatly expanded the dimension of *mīqāt* tables by latitude and longitude degree of the globe. David King's comparision of al-Khalīlī's data with modern calculations shows his work to be remarkably accurate. Al-Khalīlī's tables may thus be considered to present the first global data set on the *qibla*.

We noted how, through the ingenious use of solar longitude, the Tables were made time-invariant and independent of the solar calendar, and the computational system was further expanded to determine the solar altitude for a specific solar azimuth, giving the direction to Mecca. In this way, an accurate astronomical method was developed to meet an additional need. The remarkable achievements in mathematical astronomy mark the high points of Islamic astronomy.

In the period after the introduction of the mechanical clock, a 24-equal-hour civilian day took over from the 12-hour daytime and 12-hour nighttime day which had dominated civilian time-keeping systems well into the fourteenth century. Although Muslim scientists had started to utilize an Islamic equal hour day, the use of true solar time and sundials prevailed in non-European and Islamic countries for quite a while. The detailed tables of Islamic prayer times constructed for specific localities of the Muslim world in the early centuries have remained in effective use right into modern times.

We are aware that as we go towards the higher latitudes, the usual cycle of daylight, twilight, and night begins to change during certain parts of the year. First, the night gets shorter and shorter with the increasing latitude, until the evening twilight merges with the morning twilight. In other words, the clear distinction of the end of the evening twilight from the beginning (or end — depending how one looks at it) of morning twilight no longer exists, and the determination of times for the beginning of *ishā, fajr,* and fasting cannot be done through the conventional solar position basis. As one moves towards the higher latitudes, even the twilight period gets shorter, to the extent that the end of the day merges with the beginning of the day, so that the sun does not set (or rise) but remains continuously above the horizon. In a situation like this, with 24-hour daylight, how is one to determine the times of *fajr* and *maghrib* prayers and how is one to fast a 'non-ending daylight' day? This is one of the questions that needs to be tackled within the theological rather than the scientific arena.

MOHAMMAD ILYAS

REFERENCES

Hamidullah, M. *Introduction to Islam*. Riyadh: IIFSO, 1970.

Ilyas, M. *Astronomy of Islamic Prayer Times for the Twenty-first Century*. London: Mansell, 1989.

Ilyas, M. *A Modern Guide to Astronomical Calculations of Islamic Calendar, Times & Qibla*. Kuala Lumpur: Berita, 1984.

Kamal Eldin, H. *Praying Time Tables*. Riyadh: Imam Muḥammad Ibn Saud University, 1981.

King, D.A. "The Sacred Direction in Islam: A Study of the Interaction of Religion and Science." *Interdisciplinary Science Reviews* 10: 315–328, 1985.

King, D.A. "Science in the Service of Religion: The Case of Islam." *Impact of Science on Society* (159): 245–262, 1991.

Nasr, S.H. *Islamic Science: An Illustrated Study*. London: World of Islam Festival Trust, 1976.

Rehman, A. *Prayer: Its Significance and Benefits*. Singapore: Pustaka National, 1979.

See also: Ibn Yūnus – al-Khalīlī – Religion and Science in Islam I

QIN JIUSHAO Qin Jiushao was in all probability born in 1202 in Puzhou (Anyue). In his youth he was commander of the volunteers in his native place. In 1224–1225 his father was appointed in the capital of Southern Song, Hangzhou. In the preface to his *Shushu Jiuzhang* he says: "In my youth I was living in the capital, so that I was able to study in the Board of Astronomy; subsequently, I was instructed in mathematics by a recluse scholar". According to analysis of his mathematical work he must have studied the *Jiuzhang suanshu*. In 1226 he was staying with his father in Tongchuan (now Santai in Sizhuan). When the Mongols began their conquest of Sizhuan and destroyed its capital Chengdu in 1235, Qin was at the frontier as a military official.

According to his own preface it was during this period of distress that his mind went back to the mathematical studies started in his youth. He escaped calamities by going to the southeast. He was appointed subprefect in Qizhou (now Qichun in Hupei), where he behaved badly, causing a military revolt. Later he was appointed prefect in Hozhou (now Hoxian in Anhui), where he was responsible for the salt trade and sold salt illegally to the people, so that he left the southeast as a rich man, settling in Huzhou in Zhejiang. In 1244, he was appointed vice-administrator of the prefecture of Jian-kang fu (now Nanjing in Jiangsu). After three months he resigned because of his mother's death and returned to Huzhou. It was during this three-year mourning period that he wrote his mathematical work, which appeared in 1247. As a consequence he was recommended to the throne on account of his calendrical science, and he was allowed to take part in the examinations. We can only suppose that something went wrong, as we find him as advisor of the Directorate of Military Affairs in Jiankangfu from 1253 until 1259. However, after that he resigned and went back to his native place. He paid a visit to chancellor Jia Sidao, and he got an appointment as prefect of Qiongzhou (now Qiongshanxian) in far Hainan. But after a few months he had to leave Qiongzhou because he was impeached for corruption and exploitation of the people. He followed his friend Wu Qian, who was a naval officer in the district of Yin in Zhekiang, and in 1259 he was appointed assistant in the agricultural office there. Wu Qian subsequently became a minister, but when he was disgraced in 1260, chancellor Jia Sidao also collected data about Qin, and put him away in Meizhou (now Meixian in Guangtong), where he died in 1261 at the age of sixty.

ULRICH LIBBRECHT

REFERENCES

Primary sources

Chen, Zhen-sun *Zhi-zhai shu-lu jia-ti*. ca. 1250. Chapter 12, pp. 354f.

Qian, Daxin. *Shi-jia zahi-yang xin-lu*. Taipei: Taiwan shang wu yin shu guan, 1968 (Original ca. 1790).

Zhou, Mi *Gui-xin za-zhi, xu-ji*. Beijing: Zhong hua shuchu, 1988 (Original ca. 1290).

Secondary sources

Ho, Peng Yoke, "Ch'in Chiu-shao." In *Dictionary of Scientific Biography*. Ed. C.C. Gillispie. New York: Charles Scribner's Sons, 1970–1976, vol. III, pp. 249–256.

Libbrecht, Ulrich. "Biography of Ch'in Chiu-shao." In *Chinese Mathematics in the Thirteenth Century: The Shu-shu chiu-chang of Ch'in Chiu-shao*. Cambridge, Massachusetts: MIT Press, 1973, pp. 22–30.

See also: *Shushu jiuzhang* – Liu Hui and the *Jiuzhang suanshu* – Salt

QUADRANT The quadrant is an instrument which was used for measuring angles and for making astronomical calculations. Apart from those for the accurate observation of altitude, known to us almost entirely from written descriptions, the quadrants of the Arabic Middle Ages were portable. They fall into three categories: those carrying a projection, or two-dimensional representation, of the celestial sphere (the imaginary sphere containing the stars and celestial circles); those designed for astronomical calculations; and horary quadrants, i.e. those used specifically for telling the time by the altitude of the Sun. Most have a thread fixed at the vertex carrying a movable bead for transferring distance from the vertex from one point of the instrument to another. Of the first type the best known is the astrolabe quadrant, invented before 1100. This is essentially the plate and rete of a standard astrolabe overlaid on one another. The motion of a star (or the Sun) is imitated by the motion of the bead, which has been set by putting it over the representation of the star, about the vertex. In the *shakkāzīya* plate associated with the name of ʿAlī ibn Khalaf (eleventh-century Toledo) and taken over by al-Zarqāllu (Arzaquiel) celestial coordinates may be converted (e.g. ecliptical coordinates into equatorial) either by moving one projection over another or by using one twice and transferring one set of coordinates to the other by means of the thread and bead.

The second type of quadrant is closely related to graphical solutions of astronomical problems by ruler and compasses, the function of the compasses being taken over by the string and bead (or ruler) fixed at the vertex. Many astronomical calculations involve the use of sines of arcs in a standard circle (a sine is half the chord spanning double the arc) and these may be drawn as lines parallel to one side of the quadrant. Multiplication may be achieved by the help of scales and the thread-and-bead. The sine quadrant is well attested in the tenth century and was probably known in ninth-century Baghdad.

The horary quadrant was a quadrant on which "hour lines" were represented on the face of the instrument by circular arcs issuing from the vertex. The altitude of the Sun is taken by using sights on one side of the instrument and seeing where the thread, weighted with a plumb-bob, falls on the markings. When the position of the bead has been set for the day (e.g. by the noon altitude and the corresponding hour-line), the time can be read off from its position among the hour-lines. In some forms of the instrument a ruler or alidade takes over the functions of the thread and bead. The hours measured by the instrument are "seasonal" hours, for which 1 hour = $\frac{1}{12}$ time of daylight. The instrument, which is universal, is inaccurate in high latitudes. Underlying the instrument is an approximate formula, which probably came from India, used for calculating tables and also applied to a type of sine quadrant. Horary quadrants and related instruments were often provided with a cursor, or sliding plate carrying markings to facilitate the initial placing of the bead.

There were several further varieties of quadrant — e.g. horary quadrants for fixed latitudes — and also numerous mixed forms. Much of our knowledge of the rarer forms comes from the encyclopedic treatises of al-Marrākushī (thirteenth-century Cairo) and Ibn al-Sarrāj (fourteenth-century Aleppo). The "new" quadrant of Jacob ben Machir (thirteenth-century France), with its combination of horary quadrant and astrolabe projection, appears not to be of Arabic origin.

RICHARD P. LORCH

REFERENCES

Girke, D. and D. A. King. *An Approximate Trigonometrical Formula for Astronomical Timekeeping and Related Tables, Sundials and Quadrants from 800 to 1800*. Preprint Series No. 1, Institut für Geschiche der Naturwissenschaften, Univerisität Frankfurt, 1984.

Hahn, Nan L. *Medieval Mensuration*: "Quadrans Vetus and Geometrie Due Sunt Partes Principales ..." *Transactions of the American Philosophical Society* 72(8), 1982.

Lorch, Richard. "A Note on the Horary Quadrant." *Journal for the History of Arabic Science* 5: 115–120, 1981.

King, David A. "An Analog Computer for Solving Problems of Spherical Astronomy: the Shakkāzīya Quadrant of Jamāl al-Dīn al-Māridīnī." *Archives Internationales d'Histoire des Sciences* 24: 219–242, 1974.

Millás Vallicrosa, J. "La Introduccion del Cuadrante en Europa." *Isis* 17: 218–258, 1932.

Schmalzl, Peter. *Zur Geschichte des Quadranten bei den Arabern*. Munich: Druck der Salesianischen Offizin, 1929.

See also: Celestial Sphere – Astronomy – Time – Astrolabe – al-Zarqāllu

AL-QŪHĪ (OR AL-KŪHĪ) Abū Sahl Wayjan ibn Rustam Al-Qūhī (or al-Kūhī) probably originated from the village of Quh in the Iranian province of Tabaristan. He worked in

Baghdad under the Buwayhid Caliphs ʿAḍud al-Dawla and his son and successor Sharaf al-Dawla. In 969/970 al-Qūhī assisted at the observations of the solstices in Shiraz. These observations, ordered by ʿAḍud al-Dawla, were directed by Abū'l-Ḥusayn ʿAbd al-Raḥmān ibn ʿUmar al-Ṣūfī. In 988 al-Qūhī supervised astronomical observations in the garden of the palace of Sharaf al-Dawla in Baghdad in the company of several magistrates and respected scientists.

Some of al-Qūhī's contemporaries considered him to be the best geometer of his time; al-Khayyāmī held him in high esteem. In the geometrical writings known to us he mainly solved problems that would have led to equations of higher than the second degree. A note by al-Qūhī is added to Naṣīr al-Dīn al-Ṭūsī's redaction of Archimedes' *Sphere and Cylinder* in the Leiden manuscript, on how to construct a sphere segment equal in volume to a given sphere segment, and equal in surface area to a second sphere segment. This problem is similar to but more difficult than the constructions solved by Archimedes in *Sphere and Cylinder* II,6 and II,7. Al-Qūhī constructed the two unknowns, i.e. the radius of the sphere and the height of the segment, by intersecting an equilateral hyperbola with a parabola and rigorously discussed the conditions under which the problem is solvable.

Conic sections provided the tools for several problems, as in the classical problem of trisecting an angle. In the small treatise *Risāla fī qismat al-zāwiya* (On the Trisection of the Angle) al-Qūhī gave a purely Islamic solution by means of an orthogonal hyperbola. This solution was taken over by al-Sijzī. In *Risāla fī istikhrāj ḍil' al-musabba' al-mutasāwi 'l-aḍlāʾ* (On the Construction of the Regular Heptagon) the precise construction is more complete than the one attributed to Archimedes. Al-Qūhī's solution is based on finding a triangle with an angle ratio of 1 : 2 : 4. He constructed the ratio of the sides by intersecting a parabola and a hyperbola, with all parameters equal. Al-Sijzī, who claimed to follow the method of his contemporary Abū Sa'ied al-ʿAlā ibn Sahl, used the same principle. A second, different solution by al-Qūhī also exists. One of the most interesting examples of late tenth-century solutions is al-Qūhī's construction of an equilateral pentagon in a given square in *Risāla fī ʿamal mukhammas mutasāwi l-aḍlāʿ fī murabbaʿmaʿlūm* (On the Construction of an Equilateral Pentagon in a Known Square). This solution is based on Books I–VI of Euclid's *Elements*, Euclid's *Data*, and parts of Books I–III of Apollonius' *Conics*. The construction is remarkable, because it contains a proof of the focus-directrix property of a hyperbola with eccentricity $\varepsilon = 2$. It is reasonable to assume that al-Qūhī independently discovered and proved this property, thus going a step further than Apollonius.

In *Risāla fī istikhrāj misāḥat al-mujassam al-mukāfī* (On Measuring the Parabolic Body) al-Qūhī gave a somewha simpler and clearer solution than Archimedes had done. H said that he knew only Thābit ibn Qurra's treatise on this subject, and in three propositions showed a shorter and mor elegant method. Neither computed the paraboloids originating from the rotation of the parabola around an ordinate That was first done by Ibn al-Haytham, who was inspire by Thābit's and al-Qūhī's writings. Although he found al Qūhī's treatment incomplete, Ibn al-Haytham was neverthe less influenced by his thinking. Maybe the two met in Basra where al-Qūhī wrote his correspondence to Abū Isḥaq al Ṣābī and four books on centers of gravity. Analyzing the equation $x^3 + a = cx^2$, al-Qūhī concluded that it had a (positive) root if $a \leq 4c^3/27$. This result, already known to Archimedes, was not known to al-Khayyāmī, whose analysis is less accurate.

Al-Qūhī was the first to describe the so-called perfec compass, a compass with one leg of variable length fo drawing conic sections. In this clear and rather general work *Risāla fī'l birkar al-tāmm* (On the Perfect Compass), he firs described the method of constructing straight lines, circles and conic sections with this compass and then treated the theory. He concluded that one could now easily construc astrolabes, sundials, and similar instruments. In his *Kitā ṣan'at al-asṭurlāb* (On the Manufacture of the Astrolabe) al Qūhī used an original method for drawing azimuth circles based on an analemma, a procedure for reducing problems in three dimensions to two dimensions. A commentary on this work was written by Abū Sa'd al-ʿAlā ibn Sahl. Al-Qūhī' proofs for this construction were reproduced in an inferio form by Abū Naṣr Manṣūr ibn ʿIrāq, who highly esteemed al Qūhī, in *Risāla fī dawā'ir as-sumūt fī al-asṭurlāb* (Azimut Circles on the Astrolabe).

The correspondence between al-Qūhī and Abū Isḥaq al Ṣābī contains discussions of the possibility of curvilinea figures being equal to rectilinear figures, the meaning o "known ratio", and whether one can square a parabolic seg ment by exhausting it with triangles. The first letter espe cially gives impressive evidence for al-Qūhī's creativity i two theorems on centers of gravity of circular sectors an arcs. In the same correspondence he deducted a value of $\frac{2}{}$ for the ratio of the circumference of a circle to its diamete (π). This result was attacked by Abū Isḥaq and then, almos 150 years later, by Abū'l-Futūḥ Aḥmad ibn Muḥammad ib al-Sarī. The latter thought that al-Qūhī got swept away by er thusiasm about his result. Also the twenty seven propositior in *Hādhā mā wujida min ziyādat Abī Sahl ʿalā al-maqāla al-thāniiyah min kitāb Uqlīdis fī al-uṣūl* (Abū Sahl's Addi tions to Book II of Euclid's Elements) are rather weak an not very clearly stated. Probably, however, those addition if they were even written by al-Qūhī, were originally only marginal notes in his copy of Euclid's *Elements*.

YVONNE DOLD-SAMPLONIUS

REFERENCES

Abgrall, Philippe "Les Circles Tangents d'al-Qūhī." *Arabic Sciences and Philosophy* 5: 263–295, 1995.

Anbouba, Adel. "Construction de l'Heptagone Régulier par les Arabes au 4e Siècle H." *Journal of the History of Arabic Science* 2: 264–269, 1978.

Berggren, J. Lennart. "The Barycentric Theorems of Abū Sahl al-Qūhī." *Second International Symposium on the History of Arabic Science*, Aleppo, 1979.

Berggren, J. Lennart. "The Correspondence of Abū Sahl al-Qūhī and Abū Isḥaq al-Ṣābī: A Translation with Commentaries." *Journal of the History of Arabic Science* 7: 39–124, 1983.

Berggren, J. Lennart. "Medieval Islamic Methods for Drawing Azimuth Circles on the Astrolabe." *Centaurus* 34: 309–344, 1991.

Berggren, J. Lennart. "Abū Sahl al-Kūhi's Treatise on the Construction of the Astrolabe with Proof: Text, Translation and Commentary." *Physis* (New Series) 31: 141–252, 1994.

DeYoung, Gregg. "Abū Sahl's Additions to Book II of Euclid's Elements." *Zeitschrift für Geschichte der Arabisch-Islamischen Wissenschaften* 7: 73–135, 1991–92.

Dold-Samplonius, Yvonne. "al-Qūhī." In *Dictionary of Scientific Biography* vol. XI. Ed. Charles C. Gillispie. New York: Charles Scribner's Sons, 1970, pp. 239–241.

Hogendijk, Jan P. "Rearranging the Arabic Mathematical and Astronomical Manuscript Bankipore 2468." *Journal of the History of Arabic Science* 6: 133–159, 1982.

Hogendijk, Jan P. "Al-Kūhī's Construction of an Equilateral Pentagon in a Given Square." *Zeitschrift für Geschichte der Arabisch-Islamischen Wissenschaften* 1: 101–144, 1984.

Hogendijk, Jan P. "Greek and Arabic Constructions of the Regular Heptagon." *Archive for History of the Exact Sciences* 30:197–330, 1984.

Hogendijk, Jan P. *Ibn al-Haytham's "Completion of the Conics"*. Berlin and Heidelberg: Springer Verlag, 1984.

Knorr, Wilbur. *Textual Studies in Ancient and Medieval Geometry*. Boston, Basel, and Berlin: Birkhäuser, 1989.

Rashed, Roshdi. *Géométrie et dioptrique au Xe Siècle: Ibn Sahl, al-Qūhī et Ibn al-Haytham*. Paris: Les Belles Lettres, 1993.

Sesiano, Jacques. "Note sur trois théorèmes de Mécanique d'al-Qūhī et leur conséquence." *Centaurus* 22: 281–297, 1979.

See also: Abū'l Wafā᾽– al-Ṣāghānī – al-Maghribī – ʿUmar al-Khayyām – Naṣīr al-Dīn al-Ṭūsī – Ibn Sahl – Thābit ibn Qurra – Ibn al-Haytham – Conics – Astrolabe - al-Ṣūfī

QUIPU *Quipus* were the logical-numerical recording devices of the Incas, a civilization that dominated western South America from about 1400 to 1560. In almost every respect, the Incas are comparable to other early civilizations. They maintained a highly developed bureaucracy with strong centralized authority. They built an extensive road system, maintained a state religion, and supported a large mobile army. They constructed and regulated irrigation systems, built storage facilities, and organized distribution networks, and they forged communications linking cities and villages, up, down, and across a terrain that included mountains, tropics, desert and seacoast.

The Incas differed from other early civilizations in one crucial respect–they had no writing as we generally understand the term. In Inca civilization, quipus served the important ends that writing served elsewhere. Quipus are very general recording devices and were used to keep track of, for example, agricultural yields and population sizes. There is good reason to believe that they were used to follow astronomical events. Also, they may have been used for planning purposes, such as laying out structures to be built or scheduling amounts and types of goods to be moved from one location to another.

The Inca state had existed for about one hundred years when the Spanish arrived to conquer. A short time after that, Inca civilization collapsed. Knowledge of quipus was largely lost with the conquest. We have no sure way, for example, to assign a specific meaning or usage to a particular quipu. What we can know now comes from two sources: writings in Spanish, mostly by the conquerors; and quipus that were already in graves and then excavated at a much later date. Both sources pose problems of interpretation, but studies of the quipus themselves have proved most fruitful. Some 475–500 quipus are now located mostly in museums spread across three continents. About 400 of these have been studied and 200 are recorded in detail (Ascher and Ascher, 1978; 1988). The generalizations that follow are based largely on this quipu corpus.

A quipu is a spatial array of multicolored knotted cords. In general, a quipu has a main cord from which other cords are suspended. Most of the suspended cords fall in one direction (pendant cords); sometimes a few fall in the opposite direction (top cords). Often other cords (subsidiary cords) are suspended from some or all of the pendant or top cords. And, there can be subsidiaries of subsidiaries, and so on. (Notice in Figure 2 that the first pendant has two subsidiaries on the same level while the fourth pendant has two levels of subsidiaries.) Occasionally there is a single cord distinctively attached to the end of the main cord (dangle end cord). A quipu can be made up of from as few as three cords or as many as two thousand. On some, there are as many as ten levels of subsidiaries and, on others, as may as eighteen subsidiaries per level. All cord attachments are tight so that

Figure 1: A quipu in the collection of the Museo Nacional de Anthropología y Arquelogía, Lima, Peru. Photo by Marcia and Robert Ascher.

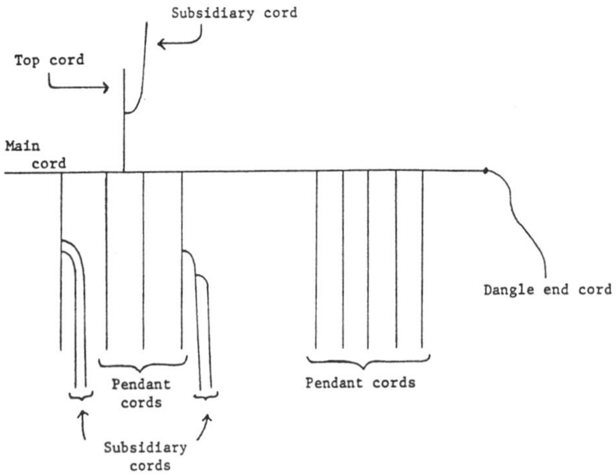

Figure 2: A schematic of a quipu.

the spacing between cords is fixed and serves to group or separate the cords. Color is another important means of associating or differentiating cords within a single quipu. For example, ten cords can be formed into two groups by having five yellow pendants followed by five red pendants, or by a five-color sequence repeated twice. In the latter case, each cord is not only associated with its group but also with the like-colored cord in the other group. In all, a complex logic of associations and distinctions is created by combining cord type, cord spacing, cord level, and cord color.

Within this logical array, numbers are represented by spaced clusters of knots on the cords. Only three types of knots appear: simple overhand knots, long knots of two or more turns, and figure-eight knots. Depending on the knot types and relative cluster positions, each cord can be interpreted as one number or multiple numbers. If it is one number, it is an integer in the base ten positional system. Each knot cluster represents a digit and each consecutive cluster, starting from the free end of the cord, is valued at one higher power of ten. The units position is always a long knot or a figure-eight knot; all other positions are clusters of simple knots. When, instead, the cord carries multiple numbers, they are still base ten integers, but each of the interspersed long knots or figure-eight knots signals the start of a new number.

Crucial to a base positional system is the concept of zero, namely that "nothing" is identified in some way. On the quipus, careful knot cluster alignment from cord to cord makes evident a position with no knots within a number. This is further supported by the identification of the units position by knot type. What is more, the color coding of the cords enables the distinction between a cord with value zero and an intentional omission or blank.

In Figure 3, the top cord carries the sum of the pendant cord numbers. When top cords appear, that is generally their role. This relationship between cord types and cord values was first observed by Leland Locke in 1923, corroborating the numerical interpretation of the knots. Consistent with the structural indicators of cord type, color, and placement many arithmetic relationships are found on quipus. There are, for example, groups that sum values in other groups which, in turn, are summed elsewhere, and sets of values with consistent ratios. There are even numerical relationships that involve several linked quipus. Some of the data structures are analogous to spreadsheets, matrices, and tree diagrams. But what makes the quipus a general recording device is that numbers are used as labels as well as magnitudes. This usage of numbers has recently become prevalent in our culture where, for example, 1-207-667-4854 identifies a type of phone call, a geographic region, a locale within that region, and a specific telephone within that locale.

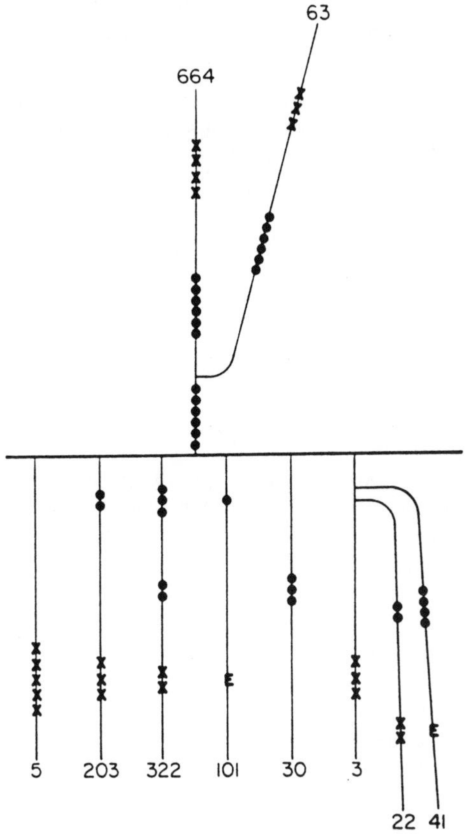

Figure 3: A schematic showing numbers represented by knots. (• = simple knot; × = one turn of a long knot; E = figure-eight knot.)

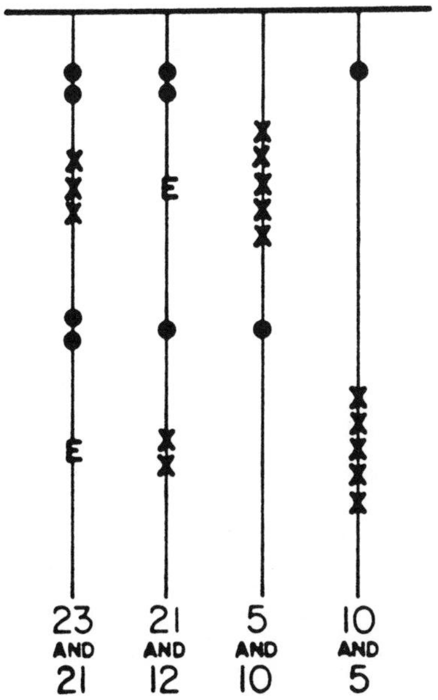

Figure 4: A schematic showing multiple numbers represented by knots. (• = simple knot; × = one turn of a long knot; E = figure-eight knot.)

The logical-numerical system of the quipus was sufficiently standardized to be read and interpreted by the community of trained quipu-makers. The quipus were not ad hoc personal mnemonic devices; they were the means of communication and record-keeping of a large bureaucratic state. In a very real sense, quipus are the quintessence of, or metaphor for, Inca civilization. Like the civilization itself, quipus emphasize regularity, spatial arrangement, and portability, and their construction is highly methodical and conservative.

MARCIA ASCHER
ROBERT ASCHER

REFERENCES

Ascher, Marcia and Robert Ascher. *Code of the Quipu: A Study in Media, Mathematics, and Culture*. Ann Arbor: University of Michigan Press, 1981.

Ascher, Marcia and Robert Ascher. *Code of the Quipu Databook*. Ann Arbor: University of Michigan Press, 1978 (Available only on microfiche from University Archivist, Cornell University, Ithaca, New York.)

Ascher, Marcia and Robert Ascher. *Code of the Quipu Databook II*. Ithaca: Ascher and Ascher, 1988 (Available only on microfiche from University Archivist, Cornell University, Ithaca, New York.)

Locke, Leland L. *The Ancient Quipu or Peruvian Knot Record*. New York: American Museum of Natural History, 1923.

Rowe, John. "The Inca Culture at the Time of the Spanish Conquest." In *Handbook of South American Indians, the Andean Indians*. Vol. 2. Ed. Julian H. Steward. Washington, D.C.: Smithsonian Institution, Bureau of American Ethnology, 1946, pp. 1–147.

QUSṬĀ IBN LŪQĀ Qusṭā ibn Lūqā is one of the important figures in the transmission of Greek scientific writings to the Muslim world and subsequently to the Latin west. The ancient Arab biographers say that he was a Christian of Greek origin from Baalbek (Heliopolis) in Lebanon; he visited the Byzantine empire and brought back Greek texts which he then translated into Syriac and Arabic. He worked for the Caliph al-Mustaʿīn (862–866) in Baghdad, where he may well have known al-Kindī and Thābit ibn Qurra; he died in Armenia (ca. AD 912/300 AH) where he had been an honored guest of the ruler Sanharib.

Qusṭā is important both for his translation and for his original works. The transition of Greek science and philosophy into the Arab world resulted in the development and preservation of traditions largely lost to the west. Most western medieval science owes a great debt to the Arab scholars who kept alive this vital link with ancient learning.

Qusṭā provided versions of Diophantos' *Arithmetica*, Hypsicles' *Liber ... de ascensionibus*, Autolycus' *De ortu et occasu*, Theodosius of Bithynia's *De sphaeris* and Hero of Alexandria's *Mechanics*; Qusṭā's Arabic version of this last work is the only text extant today. He is credited with other translations, notably some of the works of Aristotle with the commentaries of Alexander of Aphrodisias and John Philoponos, but they do not appear to have survived.

The biographers provide a list of over sixty original works by Qusṭā; the titles suggest that most are medical, but some are clearly works on mathematics, astronomy, logic, and natural science. Sezgin provides a list of the medical works of Qusṭā which are known to survive in manuscript. Very few of them have been printed, although interest in Qusṭā's treatises is increasing. His little monograph on infection (*Kitāb fī al-iʿdāʾ*) explains in terms of the medicine of the day how some diseases are contagious, and his *Viaticum*, explaining various principles of health and hygiene for travelers, are both now available. His work on sex has been published in Germany, but is not readily obtainable. A work on the efficacy of amulets is extant only in the Latin translation of Arnald of Villanova (called both *De physicis ligaturis* and *De incantatione*); the Arabic text of his work on the celestial globe still exists and there are in addition medieval translations of the treatise in Latin (*De sphaera solida*), Hebrew, Spanish, and Italian.

Probably the most famous and influential of all Qusṭā's works is the short treatise on the difference between the spirit and the soul (*Kitāb fi 'l-farq baina 'l-nafs wa 'l-rūh*), which, in the Latin translation of John of Seville, was used as an authority by almost all medieval Western physicians and by some of the philosophers. It distinguishes carefully between the immortal soul and the bodily 'spirit of life' within man. The idea that a material but gaseous substance existed inside the human frame and made possible the functions of life, sensation, and motion was an essential part of medical doctrine in the days before the discovery of electricity. Qusṭā's lucid explanation was received as a kind of gloss or commentary on Aristotle's *De anima*. It was frequently misattributed to Constantine the African.

E. RUTH HARVEY

REFERENCES

Edited texts

Daiber, Hans. "Qosta ibn Luqa (9. Jh.) über die Einteilung des Wissenschaften." *Zeitschrift fur Geschichte der arabisch-Islamischen Wissenschaften*. 6: 93–129, 1990.

Kitāb fī 'l-burhan ʿalā ʿamal-hisāb al-khataʾayn. In Heinrich Suter, "Die Abhandlung Qostā ben Lūqās und zwei andere anonyme uber die Rechnung mit zwei Fehlern and mit der angenommen Zahl." *Biblioteca Mathematica* 9(2): 111–122, 1908.

Qusṭā ibn Lūqā. *On Infection* (*Kitāb fī al-iʿdāʾ*). Edition and German translation. *Abhandlung über die Ansteckung,* hrsg. Stuttgart: Deutsche Morgenlandische Gesellschaft Abhandlungen für die Kunde des Morgenländes, 1987.

Qusṭā ibn Lūqā's Medical Regime for the Pilgrims to Mecca (Risāla fī tadbīr safar al-hajj). Ed. with English translation and commentary by G. Bos. Leiden: Brill, 1992.

Sbath, Paul. "Le Livre des Caractères de Qostā ibn Loûqā: Grand Savant et Célèbre Médecin au IXe Siècle." *Bulletin de l'Institute d'Egypte* 23:103–169, 1941.

Wilcox. J. "Qusṭā ibn Lūqā's Psycho-physiological Treatise On the Difference Between the Soul and the Spirit (Fi'l-farq baina 'l-nafs wa'l-rūh)." *Annals of Scholarship* 4: 57, 1987.

R

RAINWATER HARVESTING Throughout history people have lived in areas where there are few rivers and where the direct collection of rainwater from roofs, paved courtyards, hillsides, or rock surfaces is one of the best available methods for securing a water supply. By extending this principle to provide water for crops, early civilizations practiced agriculture much further into the semi-desert areas of Arabia, Sinai, North Africa, India, and Mexico than has been possible in modern times — and this is not explicable by changes in climate.

Agriculture in the Old World originated in climatically dry regions in the Middle East and may have depended to some degree on rainwater running off nearby slopes almost from the start. Evidence is lacking until a later period, however, when some of the most striking applications of rainwater harvesting were related to crop production in the Negev Desert between 200 BC and AD 700. One technique would be to dig a channel across a hillside to intercept water running downslope during storms. The water would be directed onto fields which, in the Negev, were carefully levelled and enclosed by bunds. Further west, steeper hillsides were used in Morocco, with cultivation on flat terraces formed behind stone retaining walls. In Tunisia, French travelers in the nineteenth century noted fruit trees being grown at the downslope end of small bunded rainwater catchment areas or microcatchments.

In India, one common technique is simply to build a bund across a gently sloping hillside, so that runoff flows originating from rainfall collect behind the bund, where water is left standing until the planting date for the crop approaches; then the land is drained, and the crop sown. This land behind the bund which is seasonally flooded and then later planted with a crop is known as an *ahar* in Bihar, or a *khadin* in Rajasthan. Although some *ahars* may be only one hectare in extent with a bund 100 meters long, others are very large and account for 800,000 hectares of cultivation in Bihar state. In desert areas of Rajasthan, there are many *khadins* of twenty hectares or more, some of them having been first constructed in the fifteenth century AD.

In North America, research on the modern potential of runoff farming methods has been stimulated by the realization that people living in what is now Mexico and the southwestern United States prior to European settlement had methods of directing rainwater from hillsides on to plots where crops were being raised, thereby making productive agriculture possible in an otherwise unpromising semi-arid environment. On lands occupied by the Hopi and Papago peoples in Arizona, fields were predominantly on alluvial valley soils below hillsides or gullies from which water could flow to the crops during rainstorms. Sites were chosen so that only minimal earthworks were needed to spread the water over the fields. These were short lengths of bund referred to as spreader dikes.

ARNOLD PACEY

REFERENCES

Bradfield, M. *The Changing Pattern of Hopi Agriculture*. London: Royal Anthropological Institute, 1971.

Evenari, M., L. Shanan, and N. Tadmor. *The Negev: The Challenge of a Desert*. 2nd ed. Cambridge, Massachusetts: Harvard University Press, 1982.

Kutsch, H. *Principal Features of a Form of Water-Concentrating Culture* (in Morocco). Trier: Geographisch Geselkschaft Trier, 1982.

Pacey, Arnold, and Adrian Cullis. *Rainwater Harvesting*. London: Intermediate Technology Publications, 1986.

See also: Agriculture – Irrigation in India

RAMANUJAN Srinivasa Ramanujan was born on 22 December 1887 in the home of his maternal grandmother in Erode, India, a small town located about 250 miles southwest of Madras. Soon thereafter, his mother returned with her son to her home in Kumbakonam, approximately 160 miles south-southwest of Madras. Ramanujan's father was a clerk in a cloth merchant's shop, and his mother took in local college students to augment the family's meager income.

Ramanujan's mathematical talent was recognized in grammar school, and he won prizes, usually books of English poetry, in recognition of his mathematical skills. At the age of fifteen, Ramanujan borrowed G. S. Carr's *Synopsis of Pure Mathematics* from the local Government College in Kumbakonam. This unusual book, written by a Cambridge tutor to teach students, contained approximately five thousand theorems, mostly without proofs, and was to serve as Ramanujan's primary source of mathematical knowledge.

With a scholarship, Ramanujan entered the Government College in Kumbakonam in 1904. However, by this time, he was completely absorbed with mathematics and would not study any other subject. Consequently, at the end of his first year, Ramanujan failed all of his exams, except mathematics. He lost his scholarship and therefore was unable to return to college.

For the next five years, working in isolation, Ramanujan devoted himself to mathematics. He worked on a slate, and because paper was expensive, recorded his mathematical discoveries without proofs in notebooks. During this time, he attempted once more to obtain a college education, at Pachaiyappa College in Madras, but his singular devotion to mathematics, and illness, deterred him again.

Having married S. Janaki in 1909, Ramanujan sought employment in 1910. For over a year, he was privately supported by R. Ramachandra Rao, as he gradually became known in the Madras area for his mathematical gifts. In 1912, Ramanujan became a clerk in the Madras Port Trust Office, and this was to be a watershed in his career, for the manager, S. Narayana Aiyar, and the Chairman, Sir Francis Spring, took a kindly interest in Ramanujan and encouraged him to write English mathematicians about his work.

On 16 January 1913, Ramanujan wrote to the famous English number theorist and analyst, G.H. Hardy. He and his colleague J.E. Littlewood examined the approximately sixty mathematical results communicated by Ramanujan and were astounded by his many beautiful and original claims. Hardy strongly encouraged Ramanujan to come to Cambridge, so that his mathematical talents could be fully developed. At first, Ramanujan was reluctant to accept the invitation, because of orthodox Brahmin beliefs that crossing the seas makes one unclean, but on 17 March 1914 Ramanujan sailed for England.

During the next three years Ramanujan achieved worldwide fame for his mathematical discoveries, some made in collaboration with Hardy. However, in the spring of 1917, Ramanujan became ill and was confined to nursing homes for the next two years. Tuberculosis, lead poisoning, and a vitamin deficiency were among the many diagnoses made, but a more recent examination of Ramanujan's symptoms points to hepatic amoebiasis. In 1919, he returned to India with the hope that a more favorable climate and more palatable food would restore his health. However, his condition worsened, and on 26 April 1920, Ramanujan passed away.

After Ramanujan's death, Hardy strongly urged that Ramanujan's notebooks be edited and published with his Collected Papers. Two English mathematicians, G.N. Watson and B.M. Wilson, devoted over ten years to proving the approximately three to four thousand theorems claimed by Ramanujan in his notebooks, but they never completed the task. It was not until 1957 that an unedited photostat edition of Ramanujan's notebooks was published. In 1977, B.C. Berndt, with the help of Watson and Wilson's notes, began to devote all of his research efforts toward editing the notebooks, and by early 1994 four volumes were published. In 1976, G. E. Andrews discovered a sheaf of 140 pages of Ramanujan's work, now called the "lost notebook", in the library at Trinity College, Cambridge. Andrews has now proved many of the 600 claims made in this "notebook", arising from the last year of Ramanujan's life.

Ramanujan made many beautiful discoveries in several areas of number theory and analysis, in particular, the theory of partitions, probabilistic number theory, highly composite numbers, arithmetical functions, elliptic functions, modular equations, modular forms, q-series, hypergeometric functions, asymptotic analysis, infinite series, integrals, continued fractions, and combinatorial analysis. His influence can be traced to many areas of contemporary mathematics; this is evident in the proceedings of major conferences commemorating Ramanujan on the one-hundredth anniversary of his birth. Although much of Ramanujan's work is quite deep, many of his original discoveries can be understood with a background of only high school mathematics. In particular, his several results on solving systems of equations, representing integers as sums of powers, and approximating π are elementary. In the past few decades, as more of Ramanujan's results have been unearthed, his already great reputation has soared even more.

Most biographical sketches of Ramanujan's life rely chiefly on the obituaries written by P.V. Seshu Aiyar and R. Ramachandra Rao, and the writings of Hardy. However, Robert Kanigel's biography is by far the most complete and detailed description of Ramanujan's life. Much can also be learned from Ramanujan's letters to Hardy, his family, and friends.

BRUCE C. BERNDT

REFERENCES

Andrews, George E., Richard A. Askey, Bruce C. Berndt, K. G. Ramanathan, and Robert A. Rankin. *Ramanujan Revisited*. Boston: Academic Press, 1988.

Berndt, Bruce C. *Ramanujan's Notebooks, Parts I-V*. New York: Springer Verlag, 1985, 1989, 1991, 1994, forthcoming.

Berndt, Bruce C., and Robert A. Rankin. *Ramanujan: Letters and Commentary*. Providence: American Mathematical Society, 1995.

Hardy, G. H. *Ramanujan*. Cambridge: Cambridge University Press, 1940 (reprinted New York: Chelsea, 1978).

Kanigel, Robert. *The Man Who Knew Infinity*. New York: Charles Scribner's Sons, 1991.

Ramanujan, Srinivasa. *Collected Papers*. Cambridge: Cambridge University Press, 1927 (reprinted New York: Chelsea, 1962).

Ramanujan, Srinivasa. *Notebooks*. Bombay: Tata Institute of Fundamental Research, 1957.

Ramanujan, Srinivasa. *The Lost Notebook and Other Unpublished Papers*. New Delhi: Narosa, 1988.

RATIONALE IN INDIAN MATHEMATICS Rationale in Hindu mathematics and astronomy is expressed by the terms *Yukti* and *Upapatti*, both meaning 'the logical principles implied'. It is characteristic of the Western scientific tradition, from the times of Euclid and Aristotle up to modern times, to enunciate and deduce using step by step reasoning. Such a practice is almost absent in the Indian tradition, even though the same background tasks of collecting and correlating data, identifying and analyzing methodologies, and arguing out possible answers, have to be gone through before arriving at results. However, in the final depiction, only the resultant formulae would be given, and that too in short, crisp aphoristic form, leaving out details of all the background work. Commentaries generally content themselves with explaining the text of the formulae and adding examples. This tendency towards selective depiction of results has resulted not only in blacking out the background but also in not understanding the mental working of the Indian scientist. It also throws into oblivion the methodologies that had evolved. For this reason, many Indian advances have been branded as unoriginal and borrowed.

The situation is, however, relieved to some extent by the presence of a few commentaries which took pains to explain elaborately the methodologies adopted by the original author and also set out the rationales of the formulae he enunciated. Among such commentaries might be mentioned:

Siddhāntadīpikā on Govindasvāmi's *Mahābhāskarīya-Bhāṣya* by Parameśvara (1360–1465),

Āryabhaṭīya-Bhāṣya by Nīlakaṇṭha Somayāji (b. 1444),

Yuktidīpikā on Nīlakaṇṭha Somayaji's *Tantrasaṅgraha*, by Śaṅkara Vāriyar (1500–1560),

Vāsanābhāṣya on Bhāskarācārya's *Siddhāntaśiromaṇi* by Nṛsiṃha Daivajña (1621),

Marīcī, again on Bhāskarācārya's *Siddhāntaśiromaṇi* by Munīśvara (1627),

A few texts devoted solely to the depiction of rationale are also known, such as the *Yuktibhāṣā* (Mathematical 'rationale in language' Malayalam) by Jyeṣṭhadeva (1500–1610), *Jyotirmīmāṃsā* (Investigations on Astronomical Theories) by Nīlakaṇṭha Somayāji (b. 1444), *Rāśigolasphuṭānīti* (True Longitude Computation on the Sphere of the Zodiac) by Acyuta Piṣāraṭi (1600), and *Karaṇapaddhati* (Methods of Astronomical Calculations) by Putumana Somayāji (1660–1740).

However, what is more significant is the occurrence of a number of short tracts giving mathematical and astronomical rationale which are available, some in print and several others in the form of manuscripts. These tracts take up individual topics of importance, analyze the technical principles involved therein, compare the procedures adopted in different texts, and often suggest revisions. To cite an example, the work *Gaṇitayuktayaḥ* (Rationales of Hindu Astronomy) contains a set of 27 tracts providing rationalistic exegeses on several topics including parallaxes of latitude and longitude, elevation of the moon's horns, constitution of epochs for new astronomical manuals, planetary deflections, and equation of the center. It is also noteworthy that some of these exegeses establish the originality of the methodologies and formulae depicted by Indian scientists.

K.V. SARMA

REFERENCES

Āryabhaṭīya of Āryabhaṭa with the Bhāṣya of Nīlakaṇṭha Somayāji. Trivandurm: Oriental Research Institute and Manuscripts Library. 1930, 1931, 1957.

Gaṇitayuktayaḥ: Rationales of Hindu Astronomy. Ed. K.V. Sarma. Hoshiarpur: Vishveshvaranand Institute, 1979.

Jyotirmīmāṃsā of Nīlakaṇṭha Somayāji. Ed. K.V. Sarma. Hoshiarpur: Vishveshvaranand Institute, 1977.

Mahābhāskarīya of Bhāskarācārya with the Bhāṣya of Govindasvāmin and the Super-commentary Siddhāntadīpikā of Parameśvara. Ed. T.S. Kuppanna Sastri. Madras: Government Oriental Manuscripts Library, 1957.

Rāśigolasphuṭānīti: True Longitude Computation on the Sphere of Zodiac. Ed. K.V. Sarma. Hoshiarpur: Vishveshvaranand Institute, 1977.

Sarma, K.V. *A History of the Kerala School of Hindu Astronomy.* Hoshiarpur: Vishveshvaranand Institute, 1972.

Srinivas, M.D. "The Methodology of Indian Mathematics and its Contemporary Relevance." *PPST Bulletin* (Madras) 12: 1–35, 1987.

RATIONALITY, OBJECTIVITY, AND METHOD Rationality, objectivity and method are three words that capture the essence of science and at the same time provide its mythic structure. Science has become the authoritative form of knowledge in the world today. The proof of its superiority lies in the ever expanding body of knowledge we have about reality. We can predict where, when, and how meteorites will collide with Jupiter, and we can build interplanetary spacecraft to send back signals recording the event. We can explore the atomic structure of chemicals and design new ones to suit whatever purpose we have in mind, ecstasy or health. We can explain the origins of everything in the universe down to the first few nanoseconds. The key, according to the myth, to this unparalleled success lies in science's embodiment of the highest form of rationality and objectivity in the scientific method. This mythical underpinning of science also provides the rationale for the celebration of modernism and the current domination of the West. This view is unself-consciously exemplified by the philosopher Ernest Gellner who claims, "If a doctrine conflicts with the acceptance of the superiority of scientific-industrial societies over others, then it really is out" (cited in Salmond, 1985).

Therein lies the first set of intrinsic problems and contradictions this myth conceals. Modernism is supposedly synonymous with development and social improvement, but it has become apparent in recent times that science and technology are no longer unalloyed agents of progress. They now seem to contribute significantly to the difficulties we are facing in environmental degradation, pollution, climate change, and waste disposal. The other equally difficult emergent problem for the mythological account of science is that it has been seen as quintessentially Western and as absent in the developing countries or an undeveloped possibility in the Islamic, Chinese, and Indian cultures. This can no longer be accepted as a simple fact but has now to be seen as an ideological marker in the creation of the "other". We have then a joint problem. On the one hand the future is not what it used to be, courtesy of the negative effects of science and technology, but we will inevitably rely on them nonetheless for their problem solving capacities. On the other hand it is becoming apparent that the grand project of modernism — a universal scientific culture — has failed and ought to be relinquished in favor of encouraging cultural diversity. Just as biological diversity has become recognized as an ecological necessity so too has our cultural survival come to be seen as dependent on a diversity of knowledge. Consequently "the central problem of social and political theory today is to decide the nature of communicatory reason between irreducibly different cultures" (Davidson, 1994). Given this double difficulty we need to ask ourselves if science can be reconstituted.

Rationality

It is in trying to explicate rationality, objectivity, the scientific method and the nature of science that the myth of science's transcendental supremacy really comes undone. Rationality is a deeply problematic concept. It is profoundly embedded in the hidden assumptions of late twentieth century occidentalism about what it is to be a knowing, moral, sane individual. Indeed it is so embedded that to be anything other than rational is to be ignorant, immoral, insane, or a member of an undifferentiated herd. Hence rationality cannot be treated as simply an epistemological concept about the conditions under which one can know something; it also carries ideological overtones, privileging certain ways of knowing over others. Rationality is a constituitive element in the moral economy.

Yet despite, or perhaps because of, this central role of rationality, there are no fully articulated rules or criteria for being rational in the acceptance of beliefs or in the pursuit of knowledge, nor is there a single type of rationality. This sense of incoherence in the concept reaches total intransigence in the recognition that ultimately there can be no rational justification for being rational. Nonetheless critical rationalism as advocated by Karl Popper has a primal persuasiveness in contemporary Western society. Mario Bunge has captured some of that self-evidentiality and variability in his seven desiderata for rationality.

1. Conceptual: minimizing fuzziness, vagueness, or imprecision.
2. Logical: striving for consistency, avoiding contradiction.
3. Methodological: questioning, doubting, criticizing, demanding proof or evidence.
4. Epistemological: caring for empirical support and compatibility with bulk of accepted knowledge.
5. Ontological: adopting a view consistent with science and technology.
6. Valuational: striving for goals which are worthy and attainable.
7. Practical: adopting means likely to attain the goals in view.

Indeed seeing them set out like this makes their denial seem irrational.

From a relativist's perspective there are no universal criteria of rationality, and even if the claim for rationality's governing role is weakened to talk of desiderata as Bunge does, Wittgenstein's point prevails: no body of rules can contain the rules for their application. Rationality consists in the application of locally agreed criteria in particular contexts. Hence it should be acknowledged that science, rather than exemplifying some transcendental rationality, has developed its own rationality that has in turn served to create a great divide between science and traditional beliefs. There is of course a middle ground, the so-called bridgehead argument which calls for the recognition that all humans must have practical rationality in common. Though this recognition has produced no universal truths and no universal criteria, it none the less provides the grounds for a commonality sufficient for communication and understanding across linguistic and cultural differences.

There is a tendency to talk of rationality as if it were a philosophical problem with no historical or sociological dimensions. This is to ignore the fact that the concept of the individual as a rational actor that is now so basic to Western ways of thought is not derived from first principles; it arose in conjunction with the development of modern science in the seventeenth century. This was a period which saw great debates over the appropriate forms of rationality between the Cartesian rationalists and the Baconian empiricists. Whether true knowledge was to be derived deductively from self-evident first principles or by observation and experiment, it had already been accepted that the acquisition of such knowledge was within the capacity of human individuals. The recognition that human reason and experience was not inherently limited and could be a source of knowledge re-emerged in the twelfth and thirteenth centuries in the West with the separation of the church from the state and with the development of secular law from the accompanying canon law. The development of this conception of rationality was not universal. For example it was not paralleled in Islamic society where men were denied rational agency; they were held to lack the capacity to change nature or to understand it. Knowledge was instead to be derived from traditional authority. This is not to deny that there has been any Islamic science or any Islamic discussion of rationality. On the contrary, there have been major achievements in Islamic science but in a radically different moral economy.

While the notion of the rational actor as unconstrained by circumstance or authority and moved only by logic and evidence has become embedded in our legal, economic, and scientific presuppositions, such an idealized conception is at variance with our lived reality both at the societal and the individual level. At the societal level, modern Western capitalism has become a bureaucratic system which, as Weber pointed out, relies on a calculative rationality. The administration and perpetuation of this system is crucially dependent on a system of rules from which legal and administrative calculations can be derived by professional objective experts. Hence modern science and capitalism are interdependent; they were co-produced on the basis of a calculability derived in part from rational structures of law and administration. In Weber's view it is the specific and peculiar rationalism of western culture that makes science unique to the west. Even if Weber was right, what needs further examination is how specific and peculiar that calculative rationalism is. That form of rationality has a number of interwoven components, for example the acceptance of written documents as evidence as opposed to oral testimony. This transition also occurred in the twelfth and thirteenth centuries but required the development of a "literate mentality" before it became self-evident that records and archives were more worthy of belief than the word of "twelve good men and true" (Street, 1984). Some, like Goody, have further argued that the accumulation of knowledge and the possibility of criticism and hence rationality are only possible in a literate culture. Similarly vision had to be rationalized to provide grammar or rules for the relationship between the representation of objects and their shapes as located in space. Yet another component of the form of rationality we equate with science was the acceptance of the validity of experimental evidence. Shapin and Schaffer have argued in *The Leviathan and the Air Pump* that Thomas Hobbes was able to dismiss Robert Boyle's experiments on the vacuum as private, local, and artifactual until Boyle was able to introduce a range of social, literary, and technical practices that enabled the knowledge produced in the isolation of the laboratory to be "virtually witnessed" and reproduced for other audiences and in other laboratories.

Rationality is not a particular human capacity. Rather there are forms and compounds of rationality which at the societal level, as Foucault has shown, are dependent on particular social and historical institutions, constituted through the interwoven practices, techniques, strategies and modes of calculation that traverse them. However, at the individual level we do not behave like the ends/means optimization calculators that economic rationalism would have us believe we are. We are at least as interested in meaning, significance, and personal values as we are in economic concerns. Nor are we quite the rational agents basing our knowledge on direct experience that the legal and philosophical theorists claim. A vast preponderance of our knowledge derives not from personal experience but from books, newspapers, journals, teachers, and experts. In other words our knowledge comes, directly or indirectly, from the testimony of others, in particular from those we trust. Thus our individual lived rationality is based in a range of social practices, traditions, and moralities that are suppressed and concealed in the portrayal of rationality as an ahistorical, universalistic form of reasoning exemplified by science.

Objectivity

Much the same can be said of objectivity. Objective knowledge is held to be the product of science that has established methods to ensure that individual, institutional, and cultural biases are eliminated. On closer examination objectivity is not characteristic of one special kind of knowledge — science. Rather it is the result of whatever institutionalized practices serve in a particular culture to create self-evident validity. Objective knowledge, in modern terms, is held to contrast with subjective knowledge. It is knowledge that is not local, that is not contingent on the circumstance, authority, or the perspective of the individual knower. However, the concept of objectivity, like that of rationality, is not immutable; it is an historic compound. In the seventeenth century objectivity meant "the thing insofar as it is known". The concept of aperspectival objectivity emerged in the moral and aesthetic philosophy of the late eighteenth century and spread to the natural sciences only in the mid-nineteenth century as result of the institutionalization of scientific life as a group rather than an individual activity.

This characterization of objectivity as the "view from nowhere" (Nagel, 1986) represents one of the essential contradictions of scientific knowledge production. Knowledge is necessarily a social product; it is the messy, contingent, and situated outcome of group activity. Yet in order to achieve credibility and authority in a culture that prefers the abstract over the concrete and that separates facts from values, knowledge has to be presented as unbiased and undistorted, as being without a place or a knower.

Objectivity, like democracy, is at best a worthy goal but one that is never capable of achievement. Since knowledge is the product of social processes it can never completely transcend the social. Objective knowledge cannot, for example, simply be knowledge which is unaffected by non-rational psychological forces, since scientists always have motivations even if they pursue knowledge for its own sake. Nor can objectivity be restricted to the avoidance of dogmatic commitment, because there have been scientists like Kepler whose obsession with regular solids and cosmic harmony led to his derivation of the laws of planetary motion. While it may be possible to avoid personal idiosyncrasy this can only be achieved through the establishment of communal or public knowledge. If knowledge is a communal product, then the question of how the community should be constituted arises. Should the scientific community be an essentially western institution? Consequently objective knowledge cannot simply be "value free" knowledge because what is counted as knowledge is itself a value. Similarly the criterion of practical effectiveness cannot determine objectivity since it too is based in community standards.

The only remaining possibilities for objective knowledge lie in the notions of correspondence with reality and experimental verification. There are well known difficulties here, since correspondent theories of truth and verification are dependent on empiricism and the scientific method, both of which have been subject to powerful criticism in the last half century or so. Essentially, what philosophers like Pierre Duhem, W.V.O. Quine and Ludwig Wittgenstein have argued is that our ways of knowing about the world are riddled with *indeterminacies*; which is to say that there is no set of procedures sufficiently powerful to determine which knowledge claims are absolutely true and certain, nor is there a certain way of grounding such claims. There are uncertainties inherent in all our ways of knowing that have to be bridged by a variety of practical and social strategies. Karl Popper and Thomas Kuhn have argued that neither deduction nor induction is capable of providing true and certain knowledge. Furthermore observations and theories are interrelated and hence neither can be an independent foundation for the other. Both Popper and Kuhn recognize that observations have point and meaning within the context or framework of a theory: we do not simply observe natural phenomena, we observe them in the light of some theory we already have in mind, or minimally with some set of expectations about what is interesting or what to look for. In this sense our observations of the world are "theory dependent" or "theory laden".

As Duhem points out, all theories are enmeshed in a web of other theories and assumptions. The apparent conflict between an experimental result and a particular hypothesis cannot conclusively lead to the rejection of that hypothesis, since the strongest conclusion that can be drawn is that the hypothesis under test and the web of theories and assumptions in which it is embedded cannot both be true. Since the experimental result by itself is insufficient to tell us where the flaw lies, we can always maintain a theory in the light of an apparently falsifying experiment if we are prepared to make sufficiently radical adjustments in the web of our assumptions. Conversely it is the case that for any given set of facts there is an indeterminably large number of possible theories that could explain them.

In addition to the problematic relationship of theory and observation there is a difficulty concerning the language in which our claims about the world are expressed. All propositions or observation statements contain descriptive predicates which imply a classification or categorization of the world based on postulated essences or natural kinds. We are stuck with some degree of circularity since we gain our knowledge about natural kinds from theories which are in turn based on observations. There is no neutral observation language; our only option is to recognize and acknowledge the conventional character of our linguistic classifications.

Scientific Method

Despite all these difficulties scientists are able to reach firm conclusions about the natural world. How is this possible? Many would claim that even though it may be logically true that there are an indefinite number of possible explanations for a given body of facts, in a given case there are typically a very restricted set of alternatives and there are adequate means of selecting the right theory, or at least the best possible theory in the circumstances, given the application of certain criteria. For example, we obviously desire theories which are internally coherent, consistent with other accepted theories, and simple rather than complex. However coherence, consistency, and simplicity as well as other criteria like plausibility have all proved notoriously difficult to express in a way which can be used to measure all theories in all circumstances. It is, none the less, very tempting from our twentieth century standpoint, imbued as we are with the scientific ethos, to suppose that there must be a particular set of rules, procedures, and criteria to which all scientists adhere. Taken together they should constitute the scientific method, and by diligent application of this method we should be able to arrive at all the scientific discoveries of our age. However, Paul Feyerabend in *Against Method*, for example, argues that no proposed set of rules and procedures has survived criticism or has been universally adopted by all scientists in all circumstances. Likewise he claims that in no case can it be shown that the success of science can be solely attributed to its adherence to the scientific method.

There can of course be endless debates about such claims, but so far no one has been able to identify the one scientific method which has been adopted in all the sciences. Compare for example theoretical physics (mainly mathematical) and biology (mainly observational). Nor is there a method which has been unilaterally accepted in a particular discipline. Compare again, Newton's espoused methodology ("I feign no hypotheses") with Einstein's (bold hypotheses and deductive tests). Further, particular instances of scientific practice under close examination reveal a pragmatic willingness to suspend or modify any particular version of the scientific method if necessary, as Feyerabend has shown in his analysis of Galileo and Copernicus. It seems then that there is no "single, invariant methodology of science". Instead there are series of complex interactions between method and practice. As science and technology develop, so the practitioners develop and negotiate the rules for doing them in a local and contingent fashion. New methodologies are propounded in order to provide support and credibility to a newly preferred scientific theory.

Scientific Practice

It could be argued that the indeterminacies of science and the lack of a specific scientific method are merely philosophical and theoretical problems, and that science is firmly grounded in experimental practice. The best way to know whether a particular knowledge claim about the world is true or false is to subject it to experimental test and then have somebody else repeat the experiment.

There are two kinds of related difficulties with this empirical approach. Experiments are inevitably performed on a simplified, artifactual, and isolated portion of reality. Hence the universal generalizations drawn from them are not indubitable. Cartwright (1983) goes so far as to claim that the Laws of Physics are, in effect, lies. Equally the effectiveness of experimental replication as the litmus test of truth is somewhat undermined by the role of skill or tacit knowledge in scientific practice. "The problem being that, since experimentation is a matter of skillful practice, it can never be clear whether a second experiment has been done sufficiently well to count as a check on the results of the first. Some further test is needed to test the quality of the experiment — and so forth" (Collins, 1985). This results in what Collins calls "the experimenter's regress". In the normal course this is resolved by social processes in which the judgment about whether to accept a particular result is based on the relevant community's evaluation of the skills of the experimenter in question. It becomes deeply problematic when the existence of the phenomenon itself is at issue, as in the case of gravity waves. An experiment showing the existence of such waves was followed by others seemingly denying their existence, or at least failing to detect them. Which was correct? The existence of gravity waves turns on the judgment of the community about competently performed experiments, and those judgments of competence are based on the accepted community knowledge about the nature of gravity waves. Replication then is not the test of their existence; rather it reflects the ability of the experimenter to achieve community standards of experimental practice.

Thus it would seem that conceptions of rationality, objectivity, and the scientific method cannot be derived from self-evident epistemological principles. They are instead embedded in the historically contingent processes of scientific practice, whereby the resistances and limitations of reality are encountered and accommodated. Science is a social activity that is essentially dependent on community and tradition, and the practice of science is governed by concrete, discrete, local traditions which resist rationalization.

The notion of a great divide between Western and so-called primitive knowledge systems has turned crucially on the question of rationality of science. If, as the arguments above suggest, science has a rationality of its own, but not one that is especially privileged, how do we both account for and deal with similarity and difference between cultures? How is it, on the one hand, that the peoples of the world are sufficiently alike to have universally developed complex languages, and yet those languages and their accompanying knowledge systems have produced profoundly different cultures? On the other hand, how are we to ensure communication and preserve cultural diversity?

In the entry on local knowledge it is claimed that the common element in all knowledge systems is their localness, and that their differences lie in the way that local knowledge is assembled through social strategies and technical devices for establishing equivalences and connections. It is no small reflexive irony that this entry on rationality and science and the encyclopaedia itself are dependent on unspoken assumptions concerning their credibility and authority which constitute a form of rationality. The encyclopaedia is based on the assumption that widely disparate knowledge can be meaningfully assembled into a volume without loss of coherence due to incommensurability. Furthermore it is assumed that the individual articles, dependent as they are on evidence, analysis, and argument, are capable of being read, understood, and utilized by readers from all cultures. In other words there is a strong resemblance between this encyclopaedia and the practice of science. Science is dependent on the assemblage of heterogeneous inputs, but that assemblage is not achieved by the application of logical and rational rules or conformity to a method or plan. Indeed it is not even dependent on a clearly articulated consensus. Rather the assemblage results from the work of negotiation and judgment that each of the participants puts in to create the equivalences and connections that produce order and meaning. Perhaps then it has to be acknowledged that there is a minimal rationality assumption, and that links between rationalities can be created by common human endeavor. So, given the lack of universal criteria of rationality the problem of working disparate knowledge systems together is one of creating a shared knowledge space in which equivalences and connections between differing rationalities can be constructed. Communication, understanding, equality, and diversity will not be achieved by others adopting Western information, knowledge, science, and rationality. It will only come from finding ways to work together in joint rationalities.

DAVID TURNBULL

REFERENCES

Albury, R. *The Politics of Objectivity*. Geelong: Deakin University Press, 1983.

Bunge, M. 1987. "Seven Desiderata for Rationality." In *Rationality: The Critical View*. Eds. J. Agassi and I. C. Jarvie. Dordrecht: Martinus Nijhoff, pp. 5–17.

Cartwright, N. *How The Laws of Physics Lie*. Oxford: Clarendon Press, 1983.

Collins, H. *Changing Order: Replication and Induction in Scientific Practice*. London: Sage, 1985.

Daston, L. "Objectivity and the Escape from Perspective." *Social Studies of Science* 22: 597–618, 1992.

Davidson, A. "Arbitrage." *Thesis Eleven* 38: 158–62, 1994.

Dean, M. *Critical and Effective History: Foucault's Methods and Historical Sociology*. London: Routledge, 1994.

Dear, P. "From Truth to Disinterestedness in the Seventeenth Century." *Social Studies of Science* 22: 619–31, 1992.

Duhem, P. *The Aim and Structure of Physical Theory*. Princeton: Princeton University Press, 1954.

Feyerabend, P. *Against Method: Outline of an Anarchist Theory of Knowledge*. London: Verso, 1978.

Finnegan, R. *Literacy and Orality: Studies in the Technology of Communication*. Oxford: Basil Blackwell, 1988.

Hesse, M. *Revolutions and Reconstructions in the Philosophy of Science*. Sussex: Harvester Press, 1980.

Hollis, M., and S. Lukes, eds. *Rationality and Relativism*. Oxford: Basil Blackwell, 1982.

Huff, T. *The Rise of Early Modern Science; Islam, China and the West*. Cambridge: Cambridge University Press, 1993.

Ivins, W. M. *Prints and Visual Communication*. Cambridge, Massachusetts: Harvard University Press, 1953.

King, M. D. "Reason, Tradition, and the Progressiveness of Science." In *Paradigms and Revolutions: Applications and Appraisals of Thomas Kuhn's Philosophy of Science*. Ed. G. Gutting. South Bend, Indiana: University of Notre Dame Press, 1980, pp. 97–116.

Kuhn, T. *The Structure of Scientific Revolutions*. 2nd ed. Chicago: University of Chicago Press, 1970.

Lave, J. *Cognition in Practice*. Cambridge: Cambridge University Press, 1988.

Nagel, T. *The View From Nowhere*. Oxford: Oxford University Press, 1986.

Pickering, A. "Objectivity and the Mangle of Practice." *Annals of Scholarship* 8: 409–425, 1991.

Popper, K. *Conjectures and Refutations: The Growth of Scientific Knowledge*. London: Routledge and Kegan Paul, 1963.

Quine, W. V. O. *From a Logical Point of View: Logico-Philosophical Essays*. New York: Harper Torchbooks, 1963.

Salmond, A.: Maori Epistemologies." In *Reason and Morality*. Ed. J. Overing. London: Tavistock, 1985, pp. 240–63.

Schuster, J. A., and R. Yeo, eds. *The Politics and Rhetoric of Scientific Method*. Dordrecht: Reidel, 1986.

Shapin, S. 1994. *A Social History of Truth: Civility and Science in 17th Century England*. Chicago: University of Chicago Press.

Shapin, S., and S. Schaffer. *Leviathan and the Air Pump: Hobbes, Boyle and the Experimental Life*. Princeton: Princeton University Press, 1985.

Street, B. V. *Literacy in Theory and Practice*. Cambridge: Cambridge University Press, 1984.

Weber, M. "Law, Rationalism and Capitalism." In *Law and Society*. Ed. C. M. Campbell and P. Wiles. Oxford: Martin Robertson, 1979, pp. 51–89.

Wittgenstein, L. *Philosophical Investigations*. Oxford: Blackwell, 1958.

See also: Knowledge Systems: Local Knowledge – Values – Magic and Science – Science as a Western Phenomenon

AL-RĀZĪ Abū Bakr Muḥammad Ibn Zakariyyā al Rāzī (Rhazes), the most original physician philosopher of his time, was born in al-Rayy (hence the name, al-Rāzī), now ruins near modern Teheran. Medieval historians wrote more on his copious bibliography than on his life.

Al-Bīrūnī (d. AD 1048 or 1050) writes in his *Risāla fī Fihrist Kutub Muḥammad Ibn Zakariyyā al-Rāzī* (Missive on the Index of Books of Muḥammad Ibn Zakariyyā al-Rāzī) that he was asked by a learned man to compile a bibliography of al-Rāzī. To accomplish this task, he examined many manuscripts, and recorded only those works in which he found al-Rāzī's name mentioned as author in the text. Al-Bīrūnī's method of research undoubtedly resulted in his failure to record some of al-Rāzī's genuine works.

Al-Rāzī advised doctors to practice in densely-populated cities, where they could benefit from the experience of many skillful physicians and have the chance to examine many patients. He himself moved from al-Rayy to Baghdad where, in his youth, he studied and practiced the art of healing at its hospital (*bīmāristān*). Later, he returned to al-Rayy, at the invitation of its ruler, Manṣūr ibn Isḥāq, to assume responsibility as Hosptial Director. To this ruler, al-Rāzī dedicated two books, *al-Almnṣūrī fi'l Ṭibb* (al-Manṣūrī on Medicine) and *al-Ṭibb al-Rūḥānī* (The Spiritual Physic). These were intended to unite the study of diseases of the body with those of the soul.

Having achieved fame in al-Rayy, al-Rāzī returned to Baghdad to become Head of its newly-founded hospital, named after its founder the Abbasid caliph al-Muʿtaḍid (d. AD 902). The last years of his life were spent in al-Rayy, where he suffered from glaucoma (*al-māʾ al-azraq*), and died in AD 932 or 925.

Al-Rāzī's book *Fi'l Shukūk ʿalā Jālīnūs* (Doubts about Galen), so far unpublished, is devoted to the criticism of twenty-eight of Galen's books, beginning with the *al-Burhān* (Demonstration) and ending with *Fi'l Nabḍ* (On the Pulse).

Before embarking on the criticism of Galen, he writes an apology in which he acknowledges his debt to his master, but says that "the art of physic is a philosophy which does not tolerate submission to any authority, nor does it accept any views, or yield to any dogmas without proper investigation". He says Galen himself supported this view. In the closing passage of his introduction, al-Rāzī affirms the validity of his book, saying that none of Galen's predecessors had escaped Galen's own scathing criticism.

To al-Rāzī, progress in scientific knowledge is inevitable. In his treatise *Fī Miḥnat al-Ṭabīb wa Taʿyīnih*, he says: "He who studies the works of the Ancients, gains the experience of their labour as if he himself had lived thousands of years spent on investigation."

Al-Rāzī also mentions *al-Jāmiʿ al-Kabīr* four times in his book *al-Murshid aw al-Fuṣūl* (The Guide or Aphorisms), which was written to serve as an introduction to medicine. *Al-Jāmiʿ al-Kabī* consists of twelve sections (*aqsām*), of which only two have been recently discovered in manuscripts: (*Ṣaydalat al Ṭibb*) (Pharmacology in Medicine) and *Fī Istinbāṭ al-Asmāʾ wa'l-Awzān wa'l-Makāyīl* (On Finding the Meaning of Unfamiliar Terms, Weights and Measures).

His medical prescriptions took into account the patients' social status. For the rich, princes, and rulers, the effective drugs had to be mixed with sweet vehicles, as explained in *al-Ṭibb al-Mulūkī* (The Royal Medicine). For the poor, he wrote a book of recipes entitled *Man lā Yaḥḍuruh al-Ṭabīb*), (Who has no Physician to Attend Him) also known by the title *Ṭibb al-Fuqarāʾ* (Medicine for the Poor).

Al-Rāzī argued that the medicinal properties attributed to various parts of animals, vegetables, and minerals should be recorded in books, which he did in his *Khawāṣṣ al-Ashyāʾ* (The Properties of Things). Such properties should neither be accepted nor discarded, unless experience (*al-tajriba*) proves them to be true or false. Physicians should not accept any property as authentic, unless it has been examined and tried.

In theory, al-Rāzī followed Galen, yet he found it necessary to correct him, and in practice, he revived the Hippocratic art of clinical observation. Having read the Hippocratic book *Abīdhīmyā* (Epidemics), he decided to write his own case histories, where he carefully recorded the name, age, sex, and profession of each patient. He also gave an early example of a clinical trial, when he divided his patients suffering from meningitis (*al-sirsām*), into two groups. He treated one group with bloodletting and intentionally, as a control, referred from applying venesection to the other group.

All these detailed case histories are extant in his private notes which became known after al-Rāzī's death as the *al-Ḥāwī fi'l Ṭibb* (Continens on Medicine). It should be considered a private library of a well read and highly educated physician-philosopher, not a book meant for publication. It is interesting to remark that illnesses which affected al-Rāzī himself are recorded in *al-Ḥāwī fi'l Ṭibb*. In a note, preceded with "*Lī* (mine)", he mentions how he cured an inflammation of his own uvula by gargling with strong vinegar; in another, he jots down the fact that he recovered from a swelling in the right testicle by taking emetics (*muqayyiʾāt*) for a long time.

Al-Rāzī's medical works had great influence on medical education in the Latin West. *Al-Ḥāwī fi'l Ṭibb* was rendered into Latin (*Liber Continens*) by a Sicilian Jew, Faraj Ibn Sālim (Farrajut) in A.D. 1279. It was printed five times between A.D. 1488 and 1542. His *Liber ad almansorem*, consisting of ten books, and his *Liber Regius* were very popular among medieval European practitioners. The seventh book of *Liber ad Almansorem* (On Surgery) and the ninth, entitled *Nonus Almansoris* (A General Book on Therapy) constituted a part of the medical curriculum in Western universities.

Al-Rāzī established an accurate differential diagnosis, based on clinical observations, between smallpox (*al-jadarī*) and measles (*al-ḥaṣba*). His book *Smallpox and Measles* was translated into Latin (*De variolis et morbillis*), and into other occidental languages and was printed about forty times between A.D. 1498 and 1866.

The subject matter in this book is quite original. First, al-Rāzī asserts that Galen knew of smallpox, yet failed to indicate its etiology and to prescribe any satisfactory therapy. Secondly, he lays down his own differential diagnosis by his vivid description of the pustules of smallpox and the rash of measles. In his prognosis of the course of smallpox, he recommends close attention to the heart, the pulse, respiration, and excreta. He outlines his own method of protecting the patient's eyes and elaborates on how to avoid deep facial scarring.

ALBERT Z. ISKANDAR

REFERENCES

Arberry, A.J. *The Spiritual Physick of Rhazes, translated from the Arabic*. London: John Murray, 1950.

Channing, Johannes, trans. *Rhazes de variolis et morbillis. Arabice et Latine*. London: G. Bowyer, 1766.

Greenhill, William Alexander, trans. *A Treatise on the Small-Pox and Measles*. London: 1848.

Iskandar, Albert Zaki. *A Study of al-Rāzī's Medical Writings, with Selected Texts and English Translations*. Ph.D. Thesis, University of Oxford.

Iskandar, Albert Zaki. "Rhazes' K. al-Murshid aw al-Fuṣūl (The Guide or Aphorisms), with Texts Selected from his Medical Writings." *Revue de l'Institut des Manuscrits Arabes* 7(1): 1–125 and 173–213, 1961.

Iskandar, Albert Zaki, "Taḥqīq fī Sinn al-Rāzī ʿind bidʾ Ishtighālih bi'l-Ṭibb." *al-Machriq* 54: 168–177, 1960.

Iskandar, Albert Zaki, "Kitāb Miḥnat al Ṭabīb li'l Rāzī." *al-Machriq* 54: 471–522, 1960.

Iskandar, Albert Zaki. "Al-Rāzī al-Ṭabīb al-Iklīnīkī." *al-Machriq* 56: 217–282, 1962.

Iskandar, Albert Zaki. "Al-Rāzī." In *The Cambridge History of Arabic Literature: Religion, Learning and Science in the ʿAbbasid Period*. Ed. M.J.L. Young. Cambridge: Cambridge University Press, 1990, pp. 370–377.

Kitābu'l-Ḥāwī fi't-Ṭibb (Continens of Rhazes, Encyclopaedia of Medicine by Abūl Bakr Muḥammad Ibn Zakariyyā ar-Rāzī (d. A.H. 313/A.D. 925), Hyderabad-Deccan: Dāʾiratu'l-Maʿārif-l-Osmania, 1955–1971, 23 vols.

Mayerhof, Max, "Thirty-Three Clinical Observations by Rhazes (*circa* 900 A.D.)" *Isis* 23: 321–356, 1955.

Müller, A. *Ibn Abī Uṣaybiʿa. ʿUyun al-Anbāʾ fī Ṭabaqāt al-Atibbāʾ*. Cairo and Königsberg: Būlāq, 1882–1884, 2 vols.

RELIGION AND SCIENCE IN CHINA "Religion and Science in China" is a controversial topic which requires some clarification before the discussion can begin. Because the first serious efforts to bring Chinese classical study to the Western academic world were initiated by Jesuit missionaries in the seventeenth century, it became a Western tradition to look at many Chinese cultural phenomena from a religious point of view and categorize them accordingly. For a long time in the West, the three main *Jiao* (systems of teachings and beliefs) of Chinese traditions, Confucianism, Daoism (Taoism), and Buddhism, were called three religions. However, since the reconstruction of Chinese classical study early in this century, Chinese scholars generally consider Confucianism and Classical Daoism philosophies, popular Daoism a religion, and Chinese Buddhism both a religion and a philosophy. Today in the West there is no consistent way of using these terms. All three are called either religions or philosophies or both, according to the idea or method or focus of the researcher. Furthermore, it is common practice to include other beliefs and activities in the study of religion, such as myths, rituals, customs, popular superstitions, and court divination rediscovered through mythological, linguistic, or archaeological findings. These new studies are in fact turning the topic of "religion and science" into a much broader topic of "culture and science".

On the other hand, the concept of "Chinese Science" is even more uncertain. There is a strong tendency both in the West and in present-day China to question the possibility of a different kind of science. The underlying implication is the belief that there is only one science for one universe — The Science, which is objective and universal. With this belief in mind, explicitly or implicitly, historians of science generally consider Western modern science as the best approximation of The Science, and use it as a measure to search for relevant materials in Chinese ancient relics and texts. By using the same measure, Chinese ancient ideas, theories, studies of nature, and activities dealing with nature are accordingly classified into "scientific subjects" and thereby evaluated. The same belief also provides criteria for distinguishing non-science from science. The danger of this viewpoint is that it will miss many of the real merits in cultures other than Western ones. One example is that the deep-set dichotomy of "body and soul (mind)" in the Western culture is typically a Greek tradition and may not be shared by other cultures. This difference will show itself especially in the different medical traditions.

For those who do not accept the idea of The Science, it becomes crucial to establish a reasonable or practical criterion for non-science and science. Research efforts in this regard have not been successful either. Another confusion comes from the fact that in ancient China achievements in technology and in science are not easily distinguishable. There are doubts about if it makes sense or even if it is possible to define these two concepts separately in the context of Chinese ancient culture.

With all these ambiguities in the main concepts and differences in the basic positions, the interaction between religion and science in China has not been formed into a regular study subject. In this article we will acknowledge that the concepts "Religion" and "Science" are both Western concepts used in a Chinese context. "Religion" means a system of beliefs and "Science" means knowledge, especially theoretical or systematic knowledge about nature, human beings, and life, but not about human beliefs, activities, and relationships. Thus we will be able to achieve some clarity in our discussion.

The question of why the scientific revolution did not take place in China has drawn much attention during recent decades. A generally accepted conclusion is still far away, and may not even be important. In this article we will explore the influence exerted by the main Chinese ideologies on the nature and development of Chinese science. Here "main Chinese ideologies" means Confucianism and Daoism. Also, "Chinese science" means and includes Chinese ideas, theories, and knowledge about nature (including humans as natural beings) which were accepted by Chinese scholars or professionals in corresponding historical periods.

In the West in ancient times, science was under the strong influence of religions, and in modern times, religions were under the influence and pressure of the development of science. In China, science went along a different road; it started under the decisive influence of early Confucianism and philosophical Daoism. These two traditions governed the development of Chinese science. There were no serious conflicts between these governing ideologies and Chinese science until China began to bring in Western science on a large scale in the late nineteenth century. The conflict between Confucianism and Western science at that time has been a much studied topic.

After Buddhism was introduced into China around the first century, it gradually became the primary organized religion. However, little can be found in the development of Chinese science that was affected by the development of Chinese Buddhism. One explanation of this phenomenon is that the characteristics of Chinese science, its subject matter and method, had been firmly established before the Qing dynasty (221–206 BC). The nature of that science is quite compatible with the philosophy and practice of certain kinds of Indian Buddhism. On the other hand, together with Confucianism and Daoism, Chinese science might have exerted a strong influence upon the development of Chinese Buddhism, especially the Chan sect (the origin of the Japanese "Zen"). Little research has been done on this topic.

Our main concern will be the influence of ancient Confucianism and Daoism on Chinese science. We will begin with ancient cosmological beliefs and the methodology associated with these beliefs. As Confucianism and Daoism are the two schools that have been most influential in the making of Chinese culture, have preserved a great quantity of texts and historical records for their teachings, and have been thoroughly studied, the common features of these two schools will provide a good starting point in searching the origin of Chinese science. After this, we will also discuss how the difference between these two schools influenced the development of Chinese science in their own way.

Chinese Ancient Cosmological Beliefs

It is important to examine the basic concepts that were used in both Confucian and Daoist texts. Those that concern Chinese science are the concepts of Heaven, Earth, Human, *Dao*, *Te*, and Fate. Though Chinese classics do not provide definitions for these concepts, we can study the way they were used in their textual contexts and interpret their meaning approximately by using Western concepts as follows: Heaven and Earth form the natural world that Human lives in. Heaven is the part that is farther away from Human but is more fundamental in the sense that it ultimately governs and decides everything that happens to Earth and Human. Earth is the part of the natural world that surrounds and interacts with Human directly. Human is the concept for every human being, who has a life that has a beginning, a growing process, and an end. The distinction between Heaven and Earth is of great importance in the interactive mechanism between Human and Nature. Human can act directly on Earth and change it and use it for his own purposes, but not on Heaven. Earth is governed by Heaven and not by Human; therefore, the way Human deals with Earth (e.g. agriculture and irrigation) can either be right or wrong according to whether it follows or violates Heaven's way. This sounds quite like the western idea of natural law and it seems compatible with modern scientific views, but there is a crucial difference. The Chinese "Heaven's way", which is called *Dao*, governs not only Earth, but also Human's life and society, and therefore it is a moral matter whether one follows or violates it. To follow the *Dao* is not only a matter of what to do to achieve a certain purpose, like what science tells us in the West, but also what purposes one should have in all his/her activities including governmental, social, familial, and individual activities. Thus, *Dao* as the most fundamental principle or cosmological law is objective and natural, and governs the whole world which in western categories splits into the natural world, society, and individual life.

To know *Dao* and therefore to act according to *Dao* is the goal of ancient Chinese study. The result of this study is to gain *Te*. *Te* is both knowledge in the sense of knowing what is right, and virtue in the sense of practicing what is right. Like knowledge and virtue in the West, *Te* can be both innate and acquired. That is, people are born with *Te*, to a greater or less degree, and can acquire and accumulate *Te* through study and practice in their lives. However, unlike the Western concepts of knowledge and virtue, *Te* cannot be defined by a general relationship, either functional or quantitative, nor be abstracted or theorized from its special context — the human conditions which define the reality of the individuals concerned.

The reality of an individual is his/her Fate, which is basically determined by Heaven. That means it is out of human reach to change or control it. To the contrary, Fate is the realization of *Dao* in each individual's life. It can be known either through study and practice, or by divination. Ancient Chinese divination, like a certain kind of Greek oracle, does not tell what will happen in the future but answers if actions are right or wrong according to Heaven's way. Therefore, the concept of "good fortune" is not defined by positive results to individual purposes, nor satisfaction of natural or whimsical human desires, but by the realization of the *Dao*. Therefore, ancient Chinese divination, typically represented by the *Yijing* (I Ching, Book of Changes), has a close relation with the ancient Chinese value system and moral principles as well as understandings of *Dao*, and thus exerts great influence on the development of Chinese science.

Now when we look back at these beliefs, the following conclusions are apparent:

(1) Because of the wide variety of its governed phenomena, and because of its moral nature, *Dao* is not a deterministic regularity which decides in every detail what happens in the world. Maybe it is, on the whole and in the long run, but certainly not in the form of an instant causal relationship in everyday phenomena which are available for measurement and observance. *Dao* is not a natural law which is impossible to violate.

(2) If *Dao* is ultimately deterministic, it can be seen in history. Therefore, to keep historical records is tremendously meaningful and important for the Chinese, and to study these records, what is called historiography in Western terminology, is especially so.

(3) If *Dao* is indeterministic in details, and can be violated by people for their wrong purposes, then it makes no sense to study *Dao* by mechanical experimenting, nor by searching underlying regularities in the observable movements of stars in heaven and things on earth. In other words, one can not study *Dao* through instant causal relationships, because in these relationships there are no observables which indicate whether what happened was on the *Dao* or against the *Dao*.

(4) Chinese divination has a nature of empirical knowledge which comes from generalization of phenomenological and historical observations, and a nature of theoretical knowledge which employs mathematical representation to calculate functional relations in the change of affairs. Not only does it exert great influence on the later development of Chinese science in its conceptual system and methodology, but it could be looked upon as the earliest achievement of Chinese science.

Confucianism

Confucianism is the English word for *Ru Jia* (the School of Scholars), which was founded by Confucius (551 BC–479 BC), the first private teacher in China. Confucius is the Latinized form of *Kong Fuzi*, a respectful way of addressing the master. *Kong* was his family name. He lived in a time when the empire was being broken up into numerous feudal states. It was a time of change, disorder, and degeneration of the old moralities. When he started as an officer and political reformer, Confucius' ideas of personal cultivation and moral government failed to attract the rulers of his time. So he turned to teaching and taught a large number of private students, preparing them to be good court ministers as well as good teachers. Thus, he managed to exert an influence by establishing a tradition of individual learning and private education. In the second century B. C. his ideas and teachings were made authoritative by the Han emperors, and the Classics Confucius edited and used in teaching became the only official textbooks in China until the beginning of the twentieth century.

Confucianism is *the* Chinese ideology. From the second century B. C. to the beginning of the twentieth century, it has been the orthodoxy. In a way, it is the Chinese religion, because it provides a system of beliefs (or disbeliefs) and values which calls for faith and acceptance from students, and also because it is more a way of life for students to follow than a body of knowledge for them to master. There are always scholars who spend their lives studying the text of Confucian Classics, but it is not through these studies that Confucianism exerts its influence on Chinese society and culture. In this regard, Confucianism is more comparable to western religions than philosophies.

However, Confucianism is not a religion in the western sense, because it has no doctrine nor reference to the supernatural. It gives no promise, no reward, and no consolation to its followers either in or beyond this life. It has no God, no church, and no organization. It is only a teaching. It teaches the follower how to be a noble man (a *Junzi*) living a noble life in his special social context.

The ultimate goal of Confucius' teaching, and therefore the goal of Confucian learning, is the benefit of the people. The essence of Confucianism lies in these two doctrines: (1) a moral government, i.e. governing by moral examples and education instead of by rules, law, and punishment, is the solution to social disorder and is to the highest benefit of the people; and (2) the cultivation of a group of *Junzi* (noble men) in a society is a practical road toward moral government.

A *Junzi* is an elite scholar. He is elite because he has higher moral standards and is therefore a better person than a common man (*Xiaoren*, the small men.) But he is not an elite in the Western sense, because his higher moral standards require that he has no social privileges but only duties and obligations. To be a *Junzi* is both to lead a noble life and to have a noble personality; it is a lifelong task of self-cultivation without personal gains. One will choose to be a *Junzi* because he knows it is his fate.

To be a *Junzi* is the immediate goal for a Confucian. Because Confucians believe that people's benefit will be served best by bettering the government, people are left alone to deal with their own material needs, their production, and their struggle with the natural surroundings. Confucians take it for granted that common people are naturally wise enough to deal with all that and take care of their own needs, and it is not *Junzi's* task to learn and develop knowledge and ability in this regard. Confucians believe that a moral society would naturally become prosperous.

The *Analects*, a basic Confucian Classic, recorded that Confucius once refused to teach his student agricultural knowledge, saying the elder farmers and gardeners were better than he in this regard. This shows that *Junzi's* learning should be limited to the scope of human relationships, and science or scientific knowledge has no value to *Junzi's* task of self-cultivation.

This Confucian attitude toward scientific knowledge has its roots in the belief that the world Human lives in is governed by the *Dao*, and the best thing Human can do is to follow the *Dao*. In Chinese cosmology it is a Human choice, therefore a moral matter, to follow or to go against the *Dao*. Therefore, the western task of changing the world according to man's ideas, or reforming the world to suit man's needs, would sound absurd and dangerous to the ancient Chinese, for it advocates going against the *Dao*. Human is a part of the universe, and not the governing part. Man is not the lord of the earth and cannot give orders to Nature. This is not to say that ancient Chinese did not make efforts to better their living conditions, but it does mean that they did not give these efforts high priority on the scale of social urgencies.

After Confucianism was made authoritative by Chinese emperors in the second century BC nearly all Chinese intellectuals became Confucians, who formed a elite class in Chinese society. Generally speaking, Confucians looked down on technical and scientific knowledge and achievements because these did not have any use in self-cultivation nor help in turning the government into a moral one. This attitude left the development and preservation of scientific knowledge in the hands of professionals, many of whom were illiterate and kept their skills and knowledge as a family tradition. With the exception of astronomy, mathematics, and medicine, Chinese ancient technological and scientific achievements were seldom recorded or their records preserved. When the British scientist and historian of science Joseph Needham started to search extensively through the extant texts for these records, and began to publish the results of his search in the multi-volumned *Science & Civilisation in China* (1954–), it shocked the world to see that Chinese themselves did not remember what they had achieved. Certainly the ancient Confucians would not value these "achievements" as much as modern Westerners do.

There is no doubt that this Confucian devaluation of scientific activity, knowledge, and achievements has been a significant hindrance to the development of Chinese science. However, Confucians did value the study of astronomy, mathematics, and medicine, though the former two were not necessarily as urgently needed by the people as the knowledge for agricultural production. The reason Confucians accepted them as a proper scholarly study is for their use in divination. Confucianism included divination in the basic teaching of *Li* (rites). Astronomy and mathematics were needed in making calendars, which were important for the agricultural economy. Medicine was an early tradition and developed into such a complicated system that it became an intellectual profession even before Confucianism gained its orthodox position.

Confucianism did allow for scientific and technological knowledge to grow among the professionals. The Confucian attitude toward science is to leave it alone. Science and technology are not Confucians' proper concerns, but they never go against them. Under the reign of Confucianism there were no known cases in which scientists and innovators were persecuted because of their ideas or inventions. Confucianism created unfavorable social attitudes and conditions for scientific and technological development in China, only because Chinese rulers made Confucianism orthodox and turned the intelligentsia into Confucians. But we can not oversimplify this situation and accuse Confucianism of smothering a scientific revolution which would otherwise have happened in China. When we study the three branches of science Confucianism valued, and compare them to western ones, it can be seen that they are metaphysically and methodologically different. What Confucianism did to Chinese science was external and only confined to social factors.

Daoism

The term *Daoism* in English has several meanings. In its basic meaning it is a translation of the Chinese *Dao Jia* (the school of *Dao*), which was represented by two Daoist classics: the *Lao Zi,* and the *Zhang Zi.* In this sense, Daoism (also called Classical Daoism or Daoist philosophy) is an ancient philosophical tradition. In its second meaning Daoism is a translation of the Chinese *Dao Jiao* (the religion of *Dao*, or Popular Daoism). It includes a variety of later organizational developments of the Daoist religion, mostly imitating the form of imported Buddhism but incorporating the ideas of the *Lao Zi* and *Zhang Zi.* It was often involved in political issues, sometimes was used to organize and mobilize peasants' uprisings, occasionally was used to seek favors from the court, and was thereby supported by Chinese emperors. It was a folk religion and Confucians looked down on it as low culture and superstition. In its third and very broad meaning, it is a vague concept that includes Classical Daoism, Popular Daoism, and some other schools that were similar to classical Daoism. These other schools include the School of *Yinyang* and the School of *Huang-Lao* (the school of the Yellow Emperor and *Laozi*.) It is in this last sense we are now using the term Daoism.

Generally speaking Daoists emphasize individual happiness rather than social welfare. They believe in *Dao*, in cutting down one's desires and wants, in living a simple, secluded, and natural life, and in cultivating spiritual as well as physical immortality. In the last aspect it is mysticism from a modern scientific point of view, but the method it uses in seeking the way of immortality, which according to Daoism should be realized in this life on earth, is very much empirical and congenial with the Western development of geography, health care, and elixir alchemy. Thus, the influence of Daoism on the development of Chinese science is substantial.

The central idea that exerts a strong impact on the development of Chinese science is the search for immortality in this life on earth. We do not know what the origin of this idea is and when its practice started. There are three approaches. The first is by cultivating one's *Qi*, the second by making special drugs such as the elixir of life, and the third by going deep into mountains or on the seas to find living immortals to learn from. The concept of *Qi* is one of the most ancient Chinese concepts about nature and life. It was used, in the sense of vitality, in the works of the warring-states philosophers, such as the *Mencius*, *Xun Zi, Zhuang Zi,* and *Lüshi Chunqiu* in the fourth and the third centuries BC. The earliest extant text that mentioned elixir drugs is *Hanfei Zi,* which was in the third century B. C. *Shi Ji* (Records of the Historian, compiled during the second and the first century BC) recorded that in the fourth and third century BC the rulers of *Chu* and *Yan* states sent envoys sailing to learn about the ways of immortality.

These three approaches led to three developments of Chinese science. The first is Chinese medicine, including health care techniques and *Qigong*. *Qigong* is an Eastern exercise which is based on the theory of interaction between body and spirit and the possibility of controlling one's mental state by manipulating one's body. The second is somewhat connected with the development of Chinese pharmacology, but more so with that of chemistry which led to the invention of gunpowder. The third helped to preserve ancient myths and legends, and initiated the need to exploit the outside world which Confucians neglected. The earliest Chinese geographical work, *Shan Hai Jing*, is basically the result of this effort.

The Daoist influence on Chinese science is more congenial with the criteria of modern science. Daoists believe in the power of true knowledge for their own purpose of immortality and that knowledge can be sought by experimenting or by learning from those who know. They accept that the truthfulness of that knowledge should be tested by the results of its applications. With their beliefs of *Yinyang* and *Wuxing* (Five Phases), which are not religious but ontological and methodological, Daoism is the Chinese system of belief that led to the flourishing of Chinese medical theory and health care practices, which might be considered the highest achievements of Chinese science.

WEIHANG CHEN

REFERENCES

Bodde, Derk. "The Attitude toward Science and Scientitic Method in Ancient China." *T'ien Hsia Monthly* 2:139–160, 1936.

Chan, Wing-tsit. "Neo-Confucianism and Chinese Scientific Thought." *Philosophy East and West* 6:309–332, 1957

Fung, Yulan. "Why China has no Science–An Interpretation of the History and Consequences of Chinese Philosophy." *The International Journal of Ethics* 32: 237–263, 1922.

Graham, A. C. "China, Europe, and the Origins of Modern Science: Needham's the Grand Titration." In *Chinese Science, Explorations of an Ancient Tradition.* Ed. Shigeru Nakayama and Nathan Sivin. Cambridge, Massachusetts: The MIT Press, 1973, pp. 45–69.

Huff, Toby E. *The Rise of Early Modern Science.* New York: Cambridge University Press, 1993, pp. 237–320.

Mungello, David E. "On the Significance of the Question 'Did China Have Science.'" *Philosophy East and West* 22: 467–478, 1972, and 23: 413–416, 1973.

Needham, Joseph. *Science and Civilisation in China.* Vol. V, Part IV. Cambridge: Cambridge University Press, 1980, pp. 210–323.

Nelson, Benjamin. "Sciences and Civilizations, 'East' and 'West', Joseph Needham and Max Weber." *Boston Studies in Philosophy of Science* 11: 445–493, 1974.

Sivin, Nathan. *Science and Technology in East Asia.* New York: Science History Publications, 1977, pp.11–21.

See also: Alchemy – Medicine in China – East and West: China in the Transmission of Knowledge East to West – Magic and Science – *Qi* – Five Phases (*Wuxing*) – *Yinyang* – Geomancy – Divination – Geography – Gunpowder – *Gaitian* – *Huntian*

RELIGION AND SCIENCE IN ISLAM I: TECHNICAL AND PRACTICAL ASPECTS In Islam, as in no other religion in human history, the performance of various aspects of religious ritual has been assisted by scientific procedures. The organization of the lunar calendar, the regulation of the astronomically defined times of prayer, and the determination of the sacred direction towards the Kaʿba in Mecca are topics of traditional Islamic science still of concern to Muslims today, and each has a history going back close to fourteen hundred years. But the techniques advocated by the scientists of medieval Islam on the one hand and by the scholars of religious law on the other were quite different, and our present knowledge of them is based mainly on research conducted during the past twenty years on one small fraction of the vast literary heritage of the Muslim peoples. To understand Muslim activity in this domain we must realize that there were two main traditions of astronomy in the Islamic Near East, folk astronomy and mathematical astronomy.

The Regulation of the Lunar Calendar

The Islamic calendar is strictly lunar. The beginnings and ends of the lunar months, in particular of the holy month of Ramadan, and various festivals throughout the twelve-month "year", are regulated by the first appearance of the lunar crescent. Since twelve lunar months add up to about 354 days, the twelve-month-cycles of the Islamic calendar begin some eleven days earlier each year, and the individual months move forward through the seasons.

For scholars of the sacred law, the month began with the first sighting of the crescent moon. This observation is a relatively simple affair, provided that one knows roughly where and when to look and the western sky is clear. Witnesses with exceptional eyesight were sent to locations that offered a clear view of the western horizon, and their sighting of the crescent determined the beginning of the month; otherwise they would repeat the process the next day. If the sky was cloudy, the calendar would be regulated by assuming a fixed number of days for the month just completed. Also, the crescent might be seen in one locality and not in another. Unfortunately the historical sources contain very little information on the actual practice of regulating the calendar.

Medieval astronomers, on the other hand, knew that the determination of the possibility of sighting on a given day was a complicated mathematical problem, involving knowledge of the positions of the sun and moon relative to each other and to the local horizon. The crescent will be seen after sunset on a given evening at the beginning of a lunar month if it is far enough away from the sun, and if it is high enough above the horizon not to be overpowered by the background sky glow. Conditions required to assure crescent visibility on most occasions can be determined by observations, but the formulation of a definitive set of conditions has defied even modern astronomers. The positions of the sun and moon must be investigated to see whether the assumed visibility conditions are satisfied, but, even if they are, the most ardent astronomer can be denied the excitement of sighting the crescent at the predicted time if clouds or haze on the western horizon restrict his view.

The earliest Muslim astronomers adopted a lunar visibility condition which they found in Indian sources. It was necessary to calculate the positions of the sun and moon from tables and then to calculate the difference in setting times over the local horizon. If the latter was 48 minutes or more, the crescent would be seen; if it was less, the crescent would not be seen. In the early ninth century the astronomer al-Khwārizmī compiled a table showing the minimum distances between the sun and moon (measured on the ecliptic) to ensure crescent visibility throughout the year, based on this condition and computed specifically for the latitude of Baghdad. During the following centuries Muslim astronomers not only derived far more complicated conditions for visibility determinations but also compiled highly sophisticated tables to facilitate their computations. Some of the leading Muslim astronomers proposed conditions involving three different quantities, such as the apparent angular separation of the sun and moon, the difference in their setting times over the local horizon, and the apparent lunar velocity. Annual ephemerides or almanacs gave information about the possibility of sighting at the beginning of each month.

The Regulation of the Five Daily Prayers

The times of the five daily prayers in Islam are defined in terms of astronomical phenomena dependent upon the position of the sun in the sky. More specifically, the times of daylight prayers are defined in terms of shadows, and those of night prayers in terms of twilight phenomena. They therefore vary with terrestrial latitude, and unless measured with respect to a local meridian, also with terrestrial longitude.

Because the months begin when the new moon is seen for the first time shortly after sunset, the Islamic day is considered to begin at sunset. Each of the five prayers may be performed during a specified interval of time, and the earlier during the interval the prayer is performed, the better. The day begins with the *maghrib* or sunset prayer. The second prayer is the *ishāʾ* or evening prayer, which begins at nightfall. The third is the *fajr* or dawn prayer, which begins at daybreak. The fourth is the *ẓuhr* or noon prayer, which begins shortly after astronomical midday when the sun has crossed the meridian. The fifth is the *ʿaṣr* or afternoon prayer, which begins when the shadow of any object has increased beyond its midday minimum by an amount equal to the length of the object casting the shadow. In some medieval circles, the *ẓuhr* prayer began when the shadow increase was one-quarter of the length of the object, and the *ʿaṣr* prayer continued until the shadow increase was twice the length of the object.

In the first few decades of Islam, the times of prayer were regulated by observation of shadow lengths by day and of twilight phenomena in the evening and early morning. Precisely how either the daylight or the nighttime prayers were regulated is unfortunately not clear from the available historical sources. Muezzins who performed the call to prayer from the minarets of mosques were chosen for their piety and the excellence of their voices, but their technical knowledge was limited.

On the other hand, the determination of the precise moments (expressed in hours and minutes, local time) when the prayers should begin, according to the standard definitions, required complicated mathematical procedures in spherical astronomy, that is, the study of problems associated with the apparent daily rotation of the celestial sphere. Accurate as well as approximate formulae for reckoning time of day or night from solar or stellar altitudes were available to Muslim scholars from Indian sources and these were improved and simplified by Muslim astronomers. Certain individual astronomers from the ninth century onwards applied themselves to the calculation of tables for facilitating the determination of the prayer times. The earliest known prayer-tables were prepared in the ninth century by al-Khwārizmī for the latitude of Baghdad. The first tables for finding the time of day from the solar altitude or the time of night from the altitudes of certain prominent fixed stars appeared in Baghdad in the ninth and tenth centuries. The extent to which these tables deriving from mathematical procedures were used before the thirteenth century is unknown. The earliest examples are contained in technical works which must have had fairly limited circulation; the muezzins certainly had no need of them. Only a professional astronomer could use the tables, together with some kind of observational instrument for measuring the sun's altitude and reckoning the passage of time.

It was not until the thirteenth century that the institution of the *muwaqqit* appeared in mosques and madrasas. These professional astronomers associated with a religious institution not only regulated the prayer times, but constructed instruments, wrote treatises on spherical astronomy, and gave instruction to students. In thirteenth-century Cairo, new tables were available, and these set the tone for astronomical timekeeping all over the Islamic world in the centuries that followed. In medieval Cairo there was a corpus of some 200 pages of tables available for timekeeping by the sun and for regulating the times of prayer; in numerous copies the tables are associated with Ibn Yūnus.

Impressive innovations in astronomical timekeeping were made in other medieval cities, especially Damascus, Tunis, and Taiz, although by the sixteenth century Istanbul had become the main center of this activity. Highly sophisticated tables of special trigonometric functions were compiled to solve problems of spherical astronomy for any latitude. Tables for finding the time of day from the solar altitude at any time of year were compiled for Cairo, as we have mentioned, and also for Damascus, Tunis, Taiz, Jerusalem, Maragha, Mecca, Edirne, and Istanbul. Medieval tables for regulating the times of prayer have been found for a series of localities between Fez in Morocco and Yarkand in China. Such tables have a history spanning the millennium from the ninth century to the nineteenth.

Astronomical tables for regulating the prayer times had to be used together with instruments; only in this way could one ascertain that the time advocated in the table had actually arrived. The most popular of these instruments were the astrolabe and the quadrant. Hundreds of Islamic astrolabes and several dozen quadrants are preserved in the museums of the world, only a small fraction of the instruments actually made by Muslim astronomers. An alternative means of regulating the daytime prayers was available to the Muslims in the form of the sundial. Many mosque sundials from the later period of Islamic astronomy survive to this day, though most are now non-functional.

The Determination of the Sacred Direction

The Kaʿba in Mecca was adopted as the focal point of the new religion since the *Qurʾān* advocates prayer towards it. For Muslims it is a physical pointer to the presence of God. Thus since the early seventh century Muslims have faced the Sacred Kaʿba in Mecca during their prayers. Mosques are built with the prayer-wall facing the Kaʿba, the direction being indicated by a *miḥrāb* or prayer-niche. In addition, certain ritual acts such as reciting the *Qurʾān*, announcing the call to prayer, and slaughtering animals for food, are to be performed facing the Kaʿba. Also Muslim graves and tombs were laid out so that the body would lie on its side and face the Kaʿba. (Modern burial practice is slightly different but still Mecca-oriented.) Thus the direction of the Kaʿba — called *qibla* in Arabic and all other languages of the Islamic commonwealth — is of prime importance in the life of every Muslim.

During the first two centuries of Islam, when mosques were being built from Andalusia to Central Asia, the Muslims had no truly scientific means of finding the qibla. Clearly they knew roughly the direction they had taken to reach wherever they were, and the direction of the road on which pilgrims left for Mecca could be, and, in some cases, actually was used as a qibla. But they also followed two basic procedures, observing tradition and developing a simple expedient. In the first case, some authorities observed that the Prophet Muḥammad when he was in Medina (north of Mecca) had prayed due south, and they advocated the general adoption of this direction for the qibla. This explains why many early mosques from Andalusia to Central Asia face south. Other authorities said that the Quranic verse quoted above meant standing precisely so that one faced the Kaʿba. Now the Muslims of Meccan origin knew that when they were standing in front of the walls or corners of the Kaʿba they were facing directions specifically associated with the risings and settings of the sun and certain fixed stars. They knew that the major axis of the rectangular base of the edifice points towards the rising point of Canopus, and the minor axis points towards summer sunrise and winter sunset. These assertions about the Kaʿba's astronomical alignments, found in newly-discovered medieval sources, have been confirmed by modern measurements.

In addition Arabic folklore associates the sides of the Kaʿba with the winds and rain. These features and associations cast new light on the origin of the edifice, and in a sense confirm the Muslim legend that the Kaʿba was built in the style of a celestial counterpart called *al-bayt al-maʿmūr*: indeed it seems to have been an architectural model of a pre-Islamic Arab cosmology in which astronomical and meteorological phenomena are represented. The religious association was achieved first by a number of statues of the gods of the pagan Arabs which were housed inside it. With the advent of Islam, these were removed, and the edifice has for close to 1400 years served for Muslims as a physical focus of their worship.

The corners of the Kaʿba were associated even in pre-Islamic times with the four main regions of the surrounding world: Syria, Iraq, the Yemen, and "the West". Some Muslim authorities said that to face the Kaʿba from Iraq, for example, one should stand in the same direction as if one were standing right in front of the north eastern wall of the Kaʿba. Thus the first Muslims in Iraq built their mosques with the prayer-walls towards winter sunset because they wanted the mosques to face the north eastern wall of the Kaʿba. Likewise the first mosques in Egypt were built with their prayer-walls facing winter sunrise so that the prayer-wall was "parallel" to the north western wall of the Kaʿba. Inevitably there were differences of opinion, and different directions were favored by particular groups. Indeed, in each major region of the Islamic world, there was a whole palette of directions used for the qibla. Only rarely do the orientations of medieval mosques correspond to the qiblas derived by computation. Recently some medieval texts have been identified which deal with the problem of the qibla in Andalusia, the Maghrib, Egypt, Iraq and Iran, and Central Asia. Their study has done much to clarify the orientation of mosques in these areas. In order that prayer in any reasonable direction be considered valid, some legal texts assert that while facing the *actual* direction of the Kaʿba (*ʿayn*) is optimal, facing the *general* direction of the Kaʿba (*jiha*) is also legally acceptable.

In various texts on folk astronomy, popular encyclopedias, and legal treatises, we find the notion of the world divided into sectors about the Kaʿba, with the qibla in each sector having an astronomically-defined direction. Some twenty different schemes have been discovered recently in the manuscript sources, attesting to a sophisticated tradition of sacred geography in Islam.

The earliest schemes of Islamic sacred geography date from the ninth century, but the main contributor to its development was a Yemeni legal scholar named Ibn Surāqa, who studied in Basra about the year 1000. Ibn Surāqa devised three different schemes of sacred geography, with the world arranged in 8, 11, and 12 sectors around the Kaʿba. Each sector of the world faces a particular section of the perimeter of the Kaʿba. Simpler versions of his 12-sector scheme occur in such popular geographical works as the *Taqwīm al-buldān* of Yāqūt al-Rūmī (ca. 1200) and the *Āthār al-bilād* of al-Qazwīnī (ca. 1250) as well as the encyclopedia *Ṣubḥ al-aʿshā* of al-Qalqashandī (ca. 1400). From the fifteenth century to the nineteenth, we find a proliferation of schemes with different numbers of divisions between eight and 72 divisions of the world around the Kaʿba.

Muslim astronomers from the eighth century onwards concerned themselves with the determination of the qibla as a problem of mathematical geography. This activity involved the measurement of geographical coordinates and the computation of the direction of one locality from another by procedures of geometry or trigonometry. The qibla at any locality was defined as the direction of Mecca along the great-circle on the terrestrial sphere.

Muslims inherited the Greek tradition of mathematical geography, together with Ptolemy's lists of localities and their latitudes and longitudes. By the early ninth century observations were conducted in order to measure the coordinates of Mecca and Baghdad as accurately as possible, with the express intention of computing the qibla at Baghdad. Indeed, the need to determine the qibla in different localities inspired much of the most sophisticated activity of the Muslim geographers (see below).

Once the geographical data are available, a mathematical procedure is necessary to determine the qibla. The earliest Muslim astronomers who considered this problem developed a series of approximate solutions, all adequate for most practical purposes, but in the early ninth century, if not before, an accurate solution by solid trigonometry was formulated. The accurate formulae derived by the Muslim astronomers from the ninth century onwards are impressive, and are mathematically equivalent to the modern formula. Muslim astronomers also compiled a series of tables displaying the qibla for each degree of latitude and longitude difference from Mecca, based on both approximate and exact formulae, the first of these being prepared in Baghdad in the ninth century.

Over the centuries, numerous Muslim scientists discussed the qibla problem, presenting solutions by spherical trigonometry, or reducing the three-dimensional situation to two dimensions and solving it by geometry or plane trigonometry. They also formulated solutions using calculating devices. But one of the finest medieval mathematical solutions to the qibla problem was reached in fourteenth-century Damascus: a table by al-Khalīlī displays the qibla for each degree of latitude from 10° to 56° and each degree of longitude from 1° to 60° east or west of Mecca, with entries correctly computed according to the accurate formula. This splendid table (rediscovered only in the early 1970s) was not widely known in later Muslim scientific circles. *Muwaqqit*s of later centuries wrote treatises about the determination of the qibla but did not mention this Syrian table. By the fourteenth century the correct values of the qibla of each major city had long been established (correct, that is, for the medieval coordinates used in the calculations). Simple qibla-indicators fitted with a magnetic compass and a gazetteer of localities and qiblas became common, and the modern variety represents a continuation of this tradition.

Some of the most important Muslim contributions to mathematical geography are to be found in a series of treatises by the early-eleventh-century scientist al-Bīrūnī. In one treatise he set out to determine for his patron the qibla at Ghazna (in what is now Afghanistan), first establishing the necessary geographical coordinates and then calculating the qibla by different procedures. In 1989 a Persian qibla-indicator, made in Isfahan about 1700, became available for study. It bears a cartographic grid so devised that one can read the direction and distance to Mecca directly. Mecca is at the center of the grid and one has only to lay the diametrical rule over any city marked on the map (between Spain and China, Europe and the Yemen) to read off the qibla on a circular scale around the grid and the distance on the diametral rule. The origin of this remarkable device is still under investigation; the first scholars to deal with mappings that preserve distance and direction were Ḥabash al-Ḥāsib (ca. 850) and al-Bīrūnī (ca. 1025). The tradition behind the Isfahan world map represents the the most sophisticated contribution to mathematical geography known between Antiquity and the Renaissance.

The alignment of medieval mosques reflects the fact that the astronomers were not always consulted on their orientation. But now that we know from textual sources which directions were used as a qibla in each major locality, we can not only better understand the mosque orientations but also recognize numerous cities in the Islamic world that can be said to be qibla-oriented. In some, such as Taza in Morocco and Khiva in Central Asia, the orientation of the main mosque dominates the orientation of the entire city. In the case of Cairo various parts of the city and its suburbs are oriented in three different qiblas. The new Fatimid city of al-Qāhira, founded in the tenth century, faces winter sunset, which was the qibla of the Companions of the Prophet who erected the first mosque in Egypt in nearby Fustat some three centuries previously. The later Mamluk "City of the Dead" faces the qibla of the astronomers. The predominant orientation of architecture in the suburb of al-Qarāfa is towards the south, another popular qibla. The splendid Mamluk mosques and madrasas built along the main thoroughfare of the old Fatimid city are aligned externally with the street plan, and internally with the qibla of the astronomers: one can observe the varying thickness of the walls when standing in front of the windows inside the mosque overlooking the street outside. This is an area of the history of urban development in the Islamic world which has only recently been studied for the first time, not least because, prior to the discovery of the textual evidence, it was by no means clear which directions were used as qiblas; even if a qibla at variance from the true qibla was clearly popular, it was not known why. The first accurate longitude values of localities in the Islamic world become available only with the systematic scientific cartographic surveys of the eighteenth and nineteenth centuries. Thus most of the accurately-computed qiblas of the medieval astronomers could be judged as being in error by a few degrees anyway.

Other Applications of Science to Daily Life

The Islamic laws of inheritance, based on prescriptions in the *Qur'ān*, are complicated, and their application involves some skill in arithmetic. Both legal scholars and certain mathematicians wrote on this subject, but only two or three simple works by legalists have been studied and until recently no research of consequence had been conducted on the large number of available sources. There is also a vast corpus of literature on weights, measures, and arithmetical techniques.

Muslims also developed geometric designs for the decoration of religious architecture and also secular artifacts. The acceptability of such ornamentation is discussed by various legal scholars, but their writings have yet to be properly studied. Only two Muslim mathematicians are known to have included remarks on geometric design in their writings, a fact which confirms the suspicion that this was an art passed down amongst the practitioners. Some years ago a manuscript of an artisan's manual with guidelines for generating numerous patterns was discovered but it has yet to be published.

The legal scholars of medieval Islam used methods for regulating the calendar and prayer times and for finding the sacred direction which were simple and adequate for practical purposes. Their ingenuity in coping with differences of opinion never lost sight of the basic purpose of Quranic and Prophetic injunctions. Some of the greatest of the Muslim scientists dealt with the calendar, prayer times, and the qibla, and in these areas, as in others, their mathematical creativity and their quest for greater accuracy was impressive. In later centuries (after the thirteenth), competent astronomers were appointed to the staffs of major mosques in order to advise on these specific subjects. But the solutions developed by Muslim scientists were invariably too complicated for widespread application in the medieval milieu. Although the scholars of the sacred law and the scientists proposed different solutions for the same individual problem, there are few records of serious discord between the two groups in the medieval sources. The legal scholars criticized mathematical astronomy mainly insofar as it was used by some as the handmaiden of astrology, which was an anathema to them. The scientists seldom spoke out against the simple procedures adopted by the legal scholars.

DAVID A. KING

REFERENCES

The Encyclopaedia of Islam. 2nd. ed., 8 vols. to date, Leiden: E. J. Brill, 1960 onwards.

Kennedy, Edward S. *A Commentary upon Biruni's Kitab Tahdid al-Amakin, An Eleventh Century Treatise on Mathematical Geography*. Beirut: American University of Beirut Press, 1973.

Kennedy, Edward S., et al. *Studies in the Islamic Exact Sciences*. Beirut: American University of Beirut Press, 1983.

King, David A. "The Orientation of Medieval Islamic Religious Architecture and Cities." *Journal for the History of Astronomy* 26: 253–274, 1995.

King, David A. *Islamic Mathematical Astronomy*. London: Variorum, 1986, 2nd rev. ed., Aldershot: Variorum, 1993.

King, David A. *Islamic Astronomical Instruments*. London: Variorum, 1987. Repr. Aldershot: Variorum, 1995

King, David A. *Astronomy in the Service of Islam*. Aldershot: Variorum, 1993.

King, David A. "A Survey of Medieval Islamic Shadow Schemes for Simple Time-Reckoning." *Oriens* 32: 191–249, 1990.

Rebstock, Ulrich. *Rechnen in islamischen Orient – Die literarischen Spuren der praktischen Rechenkunst*. Darmstadt, 1992.

See also: Astronomy – Calendars – *Qibla* and Islamic Prayer Times – Astrolabe – Quadrant – al-Khalīlī – Ibn Yūnus – al-Khwārizmī – Ottoman science – al-Bīrūnī – Maps and Mapmaking – Geometry – Weights and Measures

RELIGION AND SCIENCE IN ISLAM II: WHAT SCIENTISTS SAID ABOUT RELIGION AND WHAT ISLAM SAID ABOUT SCIENCE The interaction between religious life and scientific enterprise in the culture that evolved under the impact of Islam is one of the predominating themes in today's discussions and writings of Muslims all over the world. It has continued to be such ever since the beginning of the last century, when European forces, with their newly developed technological means, extended their rule over large areas with Muslim populations until then concerned only with the traditional values of their pre-industrial communities. Today's revivalist Muslim leaders, themselves often engineers or trained scientists and not theologians, fight for a new society based on a harmonious practice of Islam and science. Or governments build mosques with technological showpieces (e.g. minarets with laser beams as indicators of the *qibla*, the direction of Mecca, as recently in Morocco), and museums and universities dedicated to the "Islamic sciences". The scientific achievements of the West are considered to be the inheritance of great Muslim sages of the Middle Ages. Thus modern technology cannot be a dangerous evil because it belonged to the Muslims first; but it has been alienated by the West, truncated from its heart, which is the recognition of Allāh, the Creator of all beings. This was the view of the influential leader of the Muslim Brothers Sayyid Quṭb (1906–1966). Although he was executed under Nasser's government in 1966, his brief sermon on *This Faith, Islam* is still representative of some sections of opinion in the Muslim community.

Observers from the outside world, where the dispute between religion and science has largely turned into a relic of the past, if it does not lead into ethical discussions about the effects of scientific and technological progress on human values, find it difficult to do justice to this preoccupation of their Muslim neighbors, which they usually interpret in analogy to apparently similar ideologies in their own realm, such as fundamentalism in America's "Bible Belt". However, the religious as well as the scientific realities involved differ greatly on both sides. Instead, the insights of the history of science as well as that of religion should be drawn upon when the relationship between the two fields is to be treated objectively. As the history of science has brought to light a whole spectrum of changing methods and conceptualizations that over the centuries have been interpreted as "science", so also "religion" has been described by its historians as being practiced variously by just about every people, culture, time, or even individual. Generalizations about the religious factor in human activities, in the West always identified with dogmatic teachings, are just as false as those about an unhistorically monolithic science, but even more widespread. The surprisingly bitter reaction to the attack of some European powers and individuals will be more readily understood if the interaction of science and Islamic religion, from the outset centered on increasing knowledge, is rightly appreciated.

The case of scientific technology probably was chosen as a favored battlefield because the Western critics, proudly flying the flag of enlightenment, had previously used it as a weapon against the Islamic religion. Thus, cultural and technological backwardness became a religious issue for Muslim intellectuals. Curiously enough, it was the philosophical and scientific work of Ibn Rushd (Latin Averroes, 1126–1198), who as "the Commentator" (namely of Aristotle) had been accepted as master by the European scholars throughout the Middle Ages, which stood in the center of this modern attack on Islam. That he had not been able to influence his co-religionists in the Islamic world to the same extent as the Jewish and Christian schoolmen in Europe was the main reason that Islam in the last century was made responsible for the worst obstruction against scientific progress. The main exponent for this attack was Ernest Renan's *L'Islam et la science*, written in 1883; the Arabic writings of Faraḥ Anṭūn in the journal *Al-Jāmiʿa* (Cairo, 1902–1903), easily accessible for Egyptian intellectuals, made this attack on Islam even more effective. From then on, the relationship between science and religion, both seen in the crystallized and almost ahistorical notions of that time, was turned into the most crucial question for any global view of Islam in human history. The question was raised on more general grounds: "Why did the Muslims fall back, while the others made progress?" The more general and fundamental questions have stood in the foreground; the more particular ones, e.g. the Islamic position *vis-à-vis* alchemy, astrology, geocentric vs. heliocentric astronomy, or such consequential medical questions as anatomical dissections, preventive measures against epidemics, abortion, birth control, etc., have been strangely pushed into the background, although numerous treatises over the centuries have been written on them.

In defense, the beginnings of the spread of scientific activity in Islamic culture, and then its high period, were chiefly taken for reference, not the obvious stagnation in the later Middle Ages. The latter phenomenon has hardly been studied, and almost no convincing explanations have ever been proposed. But it is revealing that no responsibility for such an intrusion into scientific development has been assigned to the Islamic religious authorities, as has been the case of the Inquisition trials in Christian countries.

But, in confrontation with Medieval historians centered on Europe, it was not easy to reclaim the scientific heritage for the Muslims for this heritage was primarily seen as that of the ancient Greeks whose works had been translated into Arabic; and since these works, some centuries later, were again translated from Arabic into Latin, the Arabs apparently were assigned the humble role of transmitters only, like the merchants in an import and export business. For Renan, the cultural role of Islam was actually limited to the preservation of ancient culture for and transmission to Europe, where it would be revived. That such translations from pre-Islamic Hellenistic culture involved a heavy indebtedness was not denied by Arab historians. However, in their view this ancient heritage was integrated into an already established Islamic society. As it is today, with more sources to judge by, the roots of the scientific movement in Islam itself are more generally recognized as having exerted a stimulating influence on the introduction of the more complete scientific heritage of pre-Islamic cultures: the *Qurʾān* and its exegesis, jurisprudence, philology, etc.

The first line of defense was taken up by such rationalist reformers as Shaykh Muḥammad ʿAbduh (1849–1905). He argued that Quranic religion was in no way opposed to, but on the contrary totally in harmony with science. The guidelines of his argument were taken from the apologetic discussions about the necessity of miracles as testimonials for the true prophet: miracles supposedly revealed the divine nature of Jesus, the Son of God, while the Prophet Muḥammad had only referred to the factual evidence of the Arabic *Qur'ān* in support of his divine message. (Miracles like the splitting of the Moon by the Prophet, as they were described in popular literature, were apparently not considered in this argumentation of modernist theology.) That the fact of revelation, divine speech entering the limits of creation, or the *Qur'ān* as an inimitable revealed book, were proposed by the Prophet as the decisive miracle, or that creation and re-creation after death were described as miracles permeating the whole of human existence, all these truly 'miraculous' events apparently were not considered to interfere with the course of nature.

Unfortunately, ʿAbduh's apologetic argumentation was later exaggerated by his disciple Rashīd Riḍā (1865–1935) who claimed science for Islam. In his commentary of the revealed book, all knowledge was traced to the *Qur'ān*, even modern technological and medical inventions. For example, the Arabic *Jinn* were identified with the microbes of contemporary medicine. But progress soon let such hasty identifications appear outdated, and today this unhealthy spirit of "Islamization" is much less prevalent.

More common, and truer to the revealed Quranic texts, is the derivation of an original Muslim science from the so-called "sign-verses", i.e. those verses that prescribe the inquiry into such signs in creation as the sun, moon, stars, earth and sea, life in all its forms, in fact whatever enters the realm of human experience. Corresponding texts can be found in the works of Muslim scientists throughout Islamic history, in the *Taḥdīd nihāyāt al-amākin* (The Determination of the Coordinates of Positions for the Correction of Distances Between Cities) of the mathematician and astronomer al-Bīrūnī (973–1051), in the *Faṣl al-maqāl* (On the Harmony of Religion and Philosophy) of the philosopher Ibn Rushd (1126–1198), or the interviews of the living Nobel prize winner Abdus Salam. Such an inquiry, as should be noted against a frequent misunderstanding of Western scholars, was not to lead to higher developed proofs for the existence of God, for it would be contrary to the faith of Islam that such proofs should be needed. But Muslims of all times have interpreted the Quranic counsels of 'considering the wonders of creation' as the most effective promotion of scientific research. The results were not pre-determined; not even the existence of the Creator was imposed as the logical consequence of a binding argument. But the faithful scientist was to find himself placed in the more overwhelming presence of a personally approachable God Almighty. The inner spiritual motivations for an Islamic science, not the passing discoveries, were of the greatest significance.

How far along the way of an open and progressive science a Muslim scholar could be led by the sign-passages of the *Qur'ān* is best illustrated by Ibn Rushd's treatise *Faṣl al-maqāl*. Even though he was a high functionary of Islamic law at the court of the strict Almohad administration, Ibn Rushd wrote this work as a legal defense of objective and universalist scholarship. Beginning from the Quranic injunctions to respond with thought and inquiry to the wonders of creation, he gradually moved on to the defense of gradual progress beyond incomplete scientific results, relying in the process on the heritage of pre- and non-Islamic sages. Hence, the great lawyer could even accept preliminary results, teachings known to include errors. To tolerate such partially erroneous theories, while always proceeding towards deeper insights, was for him a basic consequence of the human condition. While not denying prophetic revelation as the source of the Islamic religious movement, Ibn Rushd with this open system may be said to have reached the highest peak of the Islamic scientific world view.

It cannot be surprising that natural research on the background of the Quranic finding of the Creator, whose inner nature is above all human speculation, in the signs he had placed for humanity into his creation, has sometimes been understood along Pantheistic lines. Knowledge did seem to acquire a higher character for salvation, or even a certain participation in the unitarian nature of the One God. Influential theologians like al-Ghazālī (1058–1111), but also leading scientists up to Abdus Salam in present times, insisted on Allāh's personal nature as Creator. Scientific disciplines, therefore, were to be used only for specific human purposes, and no one but God himself granted his highest blissful knowledge to the faithful who had clung to the fulfilment of his revealed will. Since God himself as Creator was also the author of all instruction, the growth of knowledge and the production of new inventions were not really possible for man restricted to his own created capacities.

But, as heir to a tradition of scholars who already had worked at filling a still Hellenistic science with Islamic spirit, al-Ghazālī also proposed the method of *Tafakkur*, a kind of meditation on the various phenomena of Creation, which could lead to greater insights and more extensive knowledge, under the constant guidance of the Almighty. Such a method may have been at the root of the widespread mysticism of nature in Islam. As S.H. Nasr has observed, some of the greatest scientists in Islam (e.g. Quṭb al-Dīn al-Shīrāzī, AD 1236–1311) are known to have practised Ṣūfism as well. But *Tafakkur* also presents a corrective for the unitarian view of Islamic science described above, Muslim thinkers asserted God's unity in all spheres of reality, thus reducing the multiplicity of created phenomena into a forced monism. The contrary can also be said, because by always keeping the spiritual goal of *taʿẓīm Allāh* (Magnification of God) in mind, the Muslim researcher, instead of closing it, would instead turn more and more leaves of Allāh's grand book of nature.

The linguistic particularity of the Arabic language, always concentrating on individuality (*shawādhdh*), naturally could enforce this tendency.

It has become a standard argument that Islam needed such sciences as astronomy, mathematics, and geography for ritual obligations like the pilgrimage to Mecca, the establishment of the *qibla* (direction to Mecca) of new mosques, or even any prayer according to the traditional prescriptions. But some scholars feel this point should not be overstressed: even a legal authority like Abū Ḥanīfa (d. 767) warned that spatial orientation was not essential for the submission of the worshipper to the Creator, and numerous religious leaders maintained a quite relaxed attitude when confronted with greater mathematical exactness in the setting of the *qibla*. On the other hand, quite simple devices were sufficient for these needs. The number of treatises on the mathematical problems of the *qibla* are not necessarily in proportion to its importance for the Muslim community.

Of greater significance, and easily demonstrable, is the promotion of mathematical works by the theological principles of Islamic law. As the mathematician and astronomer in the Biblical tradition was encouraged by the saying that God had created the whole world with numbers (Wisdom 11,20), so the Muslim was convinced that the divine lawgiver had left nothing to the arbitrary decisions of sinful mankind, not even the shares of inheritance to be distributed among the members of a family: "... the male shall receive the portion of two females ..." (Sūra 4, 175 ff.) The natural consequence was the discovery and perfecting of algebraic rules which future lawyers had to learn for the administration of bequests; in other words, some mathematical instruction even entered legal schools usually blamed for having marginalized the rational and natural sciences. There it achieved full recognition as the respectable "Science of shares" (*ʿilm al-farāʾiḍ*).

The most characteristic and generally applied argument in support of orthodoxy and orthopraxis in Islamic intellectual history has certainly been the principle of *Sunna*. With carefully examined traditional reports it had to be established that questionable teachings or practices were sanctioned by words or actions/omissions of the Prophet himself or his Companions, or, in the case of Shiite Islam, the Imams. As this had already been done for the concepts and arguments of the theologians and jurists, the procedure did not stop at science, where it would seem to have been singularly inappropriate. Thus one of the most copied treatises of the Middle Ages, an indicator of its popularity, was the *al-Hayʾa al-sanīya fī al-hayʾa al-sunnīya* by al-Suyūṭī (1445–1505), a cosmographical work totally made up of well-attested traditional fragments of early theories.

The apparently simplistic view that truth can only be what has been taught or practiced by respected members of the Muslim community, testified to by an interrupted chain of trustworthy witnessses, may have been grounded in the deeper conviction that all values are sanctioned by the harmony with the laws of a well-established community. Scientific progress cannot be judged good and sane if it does not fit into the proven structures of a justice-oriented society or a healthy environment. In Islam this social order and the harmony with a God-created nature are directly derived from religion; hence, their demands on the work of the scientists belong to the religious, not the ethical or secular, realm.

Jurisprudence, no doubt, has been the most important discipline in the intellectual formation of the Muslims throughout their history. Some historians, considering the numerical superiority of law teachers and their students at the colleges and universities in Muslim lands, have blamed the lawyers for having pushed the sciences to the margins of academic institutions. On the other hand, hardly any other religious community has been so deeply formed by the ideal of realizing all its vital functions according to the principles of justice. The work of the scientists has not remained unaffected by this ideal. Thus long before Simon Stevin in his *Weeghconst* (Theoretical Statics, 1586) prepared the 'democratization of science', which eventually gave it a special measure of vitality in Western nations, his Arab predecessor al-Khāzinī (fl. first half of the thirteenth century) had published in the *Kitāb mīzân al-ḥikma* a remarkable memorandum on science's being rooted in an all-encompassing justice. This is, however, an aspect of the interaction of science and faith in Islam which has hardly been studied, in spite of its significance for the appropriation of technology by widely spread popular organizations, guilds, and Sūfī brotherhoods, or the stagnation up to modern times.

ANTON M. HEINEN

REFERENCES

Abdus Salam, and Jacques Vauthier. *Abdus Salam un physicien. Prix Nobel de Physique 1979. Entretien avec Jacques Vauthier.* Paris: Beauchesne, 1990.

Bīrūnī: *Kitāb Taḥdīd nihāyāt al-amākin* (The Determination of the Coordinates of Positions for the Correction of Distances between Cities.) Trans. by Jamil Ali. Beirut: American University of Beirut, 1967.

Dhanani, Alnoor. *The Physical Theory of Kalām. Atoms, Space, and Void in Basrian Muʿtazilī Cosmology.* Leiden: E.J. Brill, 1994.

Frank, Richard. *Creation and the Coranic System. Al-Ghazālī and Avicenna.* Heidelberg: Heidelberger Akademie der Wissenschaften, 1992.

Ghazālī, Abū Ḥāmid. *Al-Munqidh min al-ḍalāl* (Freedom and Fulfillment). Trans. R.J. McCarthy. Boston: G.K. Hall & Co., 1980.

Heinen, Anton M. *Islamic Cosmology.* Beirut: Orient-Institut der Deutschen Morgenländischen Gesellschaft, 1982.

Ibn Rushd, Abū al-Walīd. *Faṣl al-maqāl fīmā bayn al-ḥikma wa al-sharīʾa min al-ittiṣāl* (Averroes on the Harmony of Religion and Philosophy). Trans. G.F. Hourani. London: Luzac & Co., 1961.

Khāzinī, ʿAbdarraḥmān. *Kitāb mīzān al-ḥikma* . Hyderabad: Dāʾirat al-Maʿārif al-ʿUthmānīya, 1359 H./1940. Trans. N. Khanikoff *Journal of the American Oriental Society* 6: 1–28, 1860.

King, David A. "*Makka.*" In *The Encyclopedia of Islam.* 2nd ed., vol. VI. Leiden: Brill, 1987, pp. 180–187.

Nasr, Seyyed H. *The Need for a Sacred Science.* Albany: State University of New York Press, 1993.

Renan, Ernest. "L'Islam et la Science." In *Oeuvres Complètes* , vol. 1. Paris: M. Lévy, 1883, pp. 946ff.

Rosenthal, Franz. *Das Fortleben der Antike im Islam.* Zurich: Artemis Verlag, 1965. English translation by Emile and Jenny Marmorstein: *The Classical Heritage in Islam.* London: Routledge & Kegan Paul, 1975.

See also: Religion and Science I – *Qibla* and Islamic Prayer Times – Ibn Rushd – *Hayʾa* – al-Suyūṭī – al-Khāzinī – Science as a Western Phenomenon

RELIGION AND SCIENCE IN THE NATIVE AMERICAS

If science may be said to be concerned with "observation, description, definition, classification, measurement, experimentation, generalization, explanation, prediction, evaluation, and control of the world", as it is in the *International Encyclopedia of the Social Sciences*, it is easy to conclude that all these practices have occurred among American Indians, mostly within a religious and mythological perspective. Some scholars presuppose that American Indians have lived in a natural–supernatural continuum where the distinction between this and the other world (however we care to define it) practically falls down. Others, including this author, are under the impression that the indigenous people postulated two experimental worlds, the natural world around them in everyday life, and the mysterious world of myth and religion which occasionally breaks into the natural world, for instance in dreams and rituals. In the long run this mysterious (supernatural) world seems to constitute the real, true world; at least some religious Native thinkers believe this is the case. It is not impossible that many Indian visionaries believed this without verbalizing their beliefs.

In any case, Indian observations and explanations have, in their own thinking, been linked to religious evaluations. Everyday reality was mostly experienced from a profane point of view where experiences followed each other in a foreseen way. Where this was not the case, or uncertain courses of events and unknown places are met, the other dimension is resorted to. Since the latter dimension is the most difficult to understand, and rules humankind's life, the interpreters of that world, medicine men, shamans and priests, and artisans that catch symbols and spiritual realities, are more important than experts on the everyday world. We can state that in Native America the range of spiritual experts has been much wider than that of technological experts.

The anthropologist Paul Radin has shown that there exists what he calls "two general types of temperament" among, in particular, American Indians: "the man of action and the thinker, the type which lives fairly exclusively on what might be called a motor level and the type that demands explanations and derives pleasure from some form of speculative thinking". It is characteristic that in order to describe the thinkers and their accomplishments he turns to the specialists on religion (and then a bit arbitrarily joins "historians", that is, mythological raconteurs, with such specialists of experimental religion as medicine men and shamans). American Indians are on the whole not skeptics, since dreams, hallucinations, and traditions of visits to the other world by eminent spiritualists give strength to their religious beliefs. Collective rituals give these beliefs a realistic stamp. Only when different religions collide and a relativism of values ensues does religious brooding, skepticism, and indifference take over. There is thus little space for other than religio-mythical incitements in penetrating the marginal recesses of the empirical world.

The dominance of religious thinking in what science today would call secular matters is obvious in the fields of cosmology and medicine. There are two great world pictures in America. One is certainly the older one, with Paleolithic-Mesolithic origins and connections with the hunting tribes in the Old World. It depicts a Universe where the world pillar or the world tree forms the axis which connects heaven and earth. It is often said that this pole or tree is situated in the middle of the world; its roots are fixed in the underworld and nourished by subterranean rivers or a well. Its top reaches the sky that rests on it. Sometimes the world column stretches through several skies until it reaches the highest sky. Symbols of the different species of animals living on different levels are supposed to dwell on the tree: lowest down snakes or fishes, on the earth level deer and buffalo, highest up birds like eagles and falcons. Some tribes relate that unborn souls inhabit the crown of the world tree. Other tribes believe that the world tree, that often has a forked top, is identical with the Milky Way which is "the backbone of the sky" and the road taken by the souls to the hereafter. The world has also formed the passageway downwards for the gods when they cared to visit the ground and upwards for the shamans when in their séances their souls went to the supernatural world to receive information of the gods (and, in some cases, to release a diseased person's captured soul from the realm of the dead). The world tree is ritually modeled in the middle pole of ceremonial structures, such as the Plains Indian Sun Dance hall, which is itself a ritual reproduction of the cosmos. The Sun Dance ceremony is a repetition of the cosmic creation.

The world tree expresses an inclination for a heavenly world. American high culture supplants the tree or pillar symbol with a temple mound, mostly a pyramid. On the top of this pyramid the gods have their habitats (read: statues), and many pyramids enclose the bodies of deceased princes. The sacred temple areas in Mexican pre-Columbian cities are modeled after the supposed heavenly geography.

The other world picture which has been distributed in agricultural areas has its parallels in the agrarian civilizations of the Old World. In this case the attention on the supernatural world is, at least partly, directed downwards, to the underworld. Here is the place from which man, like the plants, once in mythic time came up, often climbing on a reed. Myths even talk of four successive underworlds, one darker than the other, or qualified through different colors. Inundating waters or other difficulties started the evacuation. Myths tell us that valuable animals, like the buffalo in North America, also came up from the nether world. Men and animals appeared from a grotto which ever since has been said to be the center of the world. After death the living beings returned to the underworld, like the plant returns to it when it withers. The underworld realm of the dead is mostly the place to which shamans send out their soul to retrieve the lost soul of a sick patient. Some divinities, be it the lord-lady of the dead or the spirits of vegetation, may be found in the underworld.

These world pictures are supposed to be stable, as when the Lenape (Delaware Indian) Supreme Being with his hand holds the upper part of the world pole. However, in earthquake areas this is not the case. According to anthropologist Lowell J. Bean, the Cahuilla of Southern California believed "that all matter was subject to unpredictable change ... For example, dramatic changes in topography are vividly and frequently recalled ... This instability was true from the beginning. Creation of the earth and life itself was fraught with indecision, mistakes, and conflicts of power between the creator brothers". This is a neat example of how environmental changes are referred to mythical incidents.

Also in another topic, medicine and medical practice, religion outbids what we might consider a more "scientific" approach. Certainly, there is, even behind supernatural gestures, the beginning of a systematic and naturalist treatment of sick persons in a simple hunting milieu. The herbalists who supply herbs, cobwebs, sap, or leaves for wounds, or handle internal medicine for the patient, follow what they have learnt or what their own experiences have taught them. Observation and practice over the years thus guided their use of natural medicine, and today's science of laboratory medicine can sometimes corroborate the positive medical properties of these medicines. If anything the choice of useful medicaments points out the scientific endeavors of American Indians.

However, there are other doctors as well, and their curing is considered more important. Medicine men and shamans invoke supernatural powers in order to deliver the sick from a serious disease (or a disease which they comprehend as serious). Gods and spirits are then thought to be the active medical powers, and the disease itself may be a demoniac being. In other words, the more serious the disease, the more frequently supernatural powers have to be invoked. Of course, even simple herbalist methods may be accompanied by religious blessings.

Such a continuum between this world and the other world may, superficially seen, seem to make the distinction between them invalid, but this is not so. American Indians have had recourse to systems which would not satisfy Western logic; their logic is often created from the needs of the situation. In religion, completely reciprocally exclusive belief chains may be resorted to, depending on the situation. In the same manner, Indians may interpret phenomena as derived from this world or the other world, but also, in single cases, see a continuous line between them.

As a further example of such thinking the hierarchy of biological masters may be mentioned. In many places in both Americas animals and plants are supposed to be controlled by masters who rule over them and own them. These masters may be animals of some sort, or supernatural beings. The Mataco in Argentina and Bolivia believe that herbs and trees are subordinated to their "owners", certain animals and birds. Thus, the wild red pepper is owned by the red-eyed dove. The dove in its turn obeys a lord of the forest who is also the lord of honey. We learn that the latter is a supernatural being, whereas the dove is a natural bird. In this connection it should be pointed out that in many places, scholars have observed a folk classification of natural plants and animals. Indian tribes have constructed taxonomies with a series of hierarchical levels. In these cases it seems that only natural species have been segregated. It is another matter when Plains Indian medicine men arrange their guardian spirits, which appear in animal disguise, in an order according to their efficiency: they are ranked as spirits, not as animals, and may be included in a hierarchy. It is a remarkable fact that this hierarchical thinking occurs in societies which are otherwise characterized by their equality.

We have seen how the world picture, or the view of the Universe, is arranged according to religio-mythical models in original Native American thought. In addition to what was said above it can be noted that stars and planets are also included in these models. The sun and moon may be understood as divinities or symbols of divinities; the sun is often a manifestation of the Supreme Being, the moon a manifestation of Mother Earth, or the god (goddess) of vegetation. The planets and stars incarnate other supernatural beings, often the culture hero and hunting spirits. The Blackfoot and other tribes tell myths according to which the main actors are transformed into well-known stars when their adventures have finished. Among the Pawnee all dominant gods are stars, the chief one among them being the high god Tirawa who is identified as the North star. Thus, the stars are supposed to be living, spiritual beings, something which is also demonstrated in their movements over the sky. Such a portrait of the star charter is the presupposition of astrology. In the classical antiquity of the Old World some supreme power, or fate, ruled the star movements; in America, the prime movers could be the stars themselves, or some power behind the stars. In both worlds omens and policies were dictated by these powers, and thus what some call a pseudo-science, astrology, took form to find out the nature of these dictates.

Among the high civilizations, such as the Maya of southern Mexico and Guatemala, parts of the clergy made observations of the sky for astrological purposes. It has been said that in Mesoamerica and South America people tracked the positions of the stars in order to know how they themselves should behave, when they should plant, and when they should perform sacrifices, whereas in North America the signs from the heaven were an aid to good living, not frightening information. This distinction is probably too exaggerated. In North America there was great fear when, during eclipses of the sun or the moon, the celestial body was supposed to be swallowed up by some monster.

Astrology has, all over the world, paved the way for astronomy; this is also true in North and South America. The great annual festivals, the fertility ceremonies and other occasions for ritual celebrations could begin when star constellations were favorable. The Tapirape Indians of the Amazonian forest know that when the Pleiades disappear in the west the rainy season is over. The prehistoric northern Plains Indians constructed medicine wheels on mountain tops (or, if there were no mountains in the area, on the open plains) which archaeologists and astronomers interpret as calendar monuments, used for the determination of the summer solstice. For instance, the famous Medicine Wheel on the ridge of the Bighorn Mountains in Wyoming has the form of a wheel with a central cairn united with the peripheral stone ring through twenty-eight spokes of stone. Other cairns are situated close to the periphery. From one of the latter cairns one can perceive the sun over the central cairn on Midsummer's Day. There are reasons to assume that the structure also had cultic functions. Its ground plan mirrors the world picture with the world pillar in the middle of the world, as was described earlier. Among other things, the twenty-eight spokes correspond to the twenty-eight roof poles of the Sun Dance lodge. This might also correspond to the twenty-eight lunar mansions in Chinese, Islamic, and Indian astronomy.

In the matter of astronomical calculations the Maya were the masters. In order to arrive at exact computations they developed mathematics, used multiplication tables, invented the figure zero, and calculated the days of the year. With these means the Maya priest-astronomers managed to bring the Venus cycle into relation with the year, and they were also able to construct a table for predicting solar eclipses (although the latter were not always visible in the Maya area). Such eclipses were considered to be dangerous. The foreknowledge of the times they would appear made it possible for the priests to take action to help the threatened human beings.

Thus it appears that the beginning of science was motivated by its use for religious and ritual purposes. The Maya scholar, Eric Thompson, has this to say about the Maya intellectual achievements: "It is remarkable that the intellectual successes of the Maya were not (from our point of view) practical; they were the outcome of spiritual needs. The Maya astronomers strove for knowledge, not as an end in itself, but as a means of controlling fate, a kind of astrology. There was, he felt, an orderliness in the heavens to which the gods conformed; once that was learned, he could predict the future through exact knowledge of which gods held sway at any given time, and influence it by knowing when and whom to propitiate". It is also possible that the Maya writing with its multivalent hieroglyphs was first developed to serve sacerdotal ends. In any case, during the Classic Period, the Maya texts deal with astronomy, the connections between stars and gods, and associated ceremonies.

We could of course argue that there was in aboriginal America an appreciation of knowledge and learning that had no religious causation; but it can be scarcely understood as a scientific ambition. Skills in technology, and inventions (Inca road building, Maya corbelled vaults, Mexican wheel toys, etc.) rarely deserve this designation either. Inuit coast maps are a wonder of precision and may pass for art. Truly profane science was not part of the American Indian world.

ÅKE HULTKRANTZ

REFERENCES

Alvarsson, Jan-Åke. *The Mataco of the Gran Chaco.* Uppsala: Academiae Upsaliensis, 1988.

Aveni, Anthony F., ed. *Native American Astronomy.* Austin: University of Texas Press, 1977.

Bean, Lowell J. *Mukat's People: The Cahuilla Indians of Southern California.* Berkeley and Los Angeles: University of California Press, 1974.

Hultkrantz, Åke. *The Religions of the American Indians.* Berkeley and Los Angeles: University of California Press, 1979.

Nordenskiöld, Erland. "The American Indian as an Inventor." The Huxley Memorial Lecture for 1929. *Journal of the Royal Anthropological Institute* 59: 273–309, 1929.

Radin, Paul. *Primitive Man as Philosopher.* New York: Dover Publications, 1957.

Sturtevant, William C. "Studies in Ethnoscience." In *Transcultural Studies in Cognition.* Ed. A.K. Romney and R.G. D'Ándrade. American Anthropologist Special Publication, Menasha, Wisconsin, 1964, pp. 99–131.

Thompson, J. Eric S. *The Rise and Fall of Maya Civilization.* Norman: University of Oklahoma Press, 1954.

See also: Eclipses – Astronomy – Roads – Medicine – Medicine Wheels – Lunar Mansions – Mathematics – Magic and Science – Medicine in Native North America

ROAD NETWORKS IN ANCIENT NATIVE AMERICA

Pre-Columbian road networks in ancient America — specifically, the Southwestern United States, Mesoamerica, and Peru — provide a point of departure for exploring current issues in the study of non-Western technology. It should be noted that other major road networks have been documented for the Mississippian complex of eastern North America, Casas Grandes in Chihuahua, Mexico, and the Maya sacbe road networks of the Maya lowlands of Mesoamerica. These networks are currently the subject of ongoing investigations.

The three primary road networks of concern here are those of the North American Chaco Anasazi of west-central New Mexico, Mesoamerican road networks centered on La Quemada and Xochicalco, Mexico, and the Inca system of Peruvian South America. These networks vary in scale, with the La Quemada complex representing a highly integrated, valley-wide network, dating to the period between AD 700 and 900. The Chaco network covers a circuit of over 300 linear miles of ground-verified constructed road segments and features spanning the period from AD 920 to 1140. The Inca road network was expanded in a vigorous campaign to integrate all pre-existing road segments in the period between AD 1438 and 1532. This system incorporates over 2000 linear miles of roads into the single most massive archaeological feature in the Americas.

The American Southwest

The formalized road network of Chaco Canyon, New Mexico, connects the canyon core with peripheral pueblo communities lying at relatively great distance from the core area. Over five hundred miles of roads are indicated through the analysis of aerial photographs; however ground survey data are as yet incomplete, leaving unresolved the question of continuity in the network. The sociopolitical complexity of the Anasazi of the San Juan Basin is closely tied to the problem of road network continuity. If the road network was contiguous throughout, this would support the contention that Chaco served an administrative function as the center of a regional system extending from Southwestern Colorado to Southeastern Utah, into Western Arizona and Northwestern New Mexico. Lack of continuity in the road network indicates a less integrated system, and a more autonomous character for the outlying communities.

Functional aspects of the Chaco road system are as yet unclear. Functional considerations center on the nature and control of goods transported as well as with the degree of interaction between core and peripheral communities. Proposed interpretations include: (a) a defensive role for outlying pueblos, (b) protection of trade routes, (c) ceremonial functions, (d) political administration, and (e) the transport of subsistence-related goods. It has also been suggested that outlying pueblos served as rural sustaining communities directly affiliated with representative groups of the canyon core (Kincaid, 1983).

Chacoan road construction exhibits several characteristic features. Most notable is the absolute linearity of the roads, despite potential topographic obstacles. The great North Road deviates less than one degree from its northerly course in over 46 kilometers. Masonry road elements such as flanking walls, or curbs, ramps, and raised roadbeds also characterize the system. Several architectural features are associated with road segments. These range from simple cairns and shrines to residential room-blocks and Great House communities with Great Kiva ceremonial structures. Stylistically uniform, these features and structures employ Bonito-style construction of core-veneer masonry and massive walls.

There are no historical data available describing the Chacoan road system. To date, all interpretations have been based on archaeological research and remote sensing techniques. Gordon Vivian pioneered this research in 1948. From 1971 to 1977, the Remote Sensing Project of the Chaco Center — in conjunction with the National Park Service and the University of New Mexico — undertook investigations of the Chacoan road network. Their efforts focused on the use of ground verification studies for the extensive mapping of road alignments appearing in aerial photographs. From 1980 to 1983 the New Mexico Bureau of Land Management undertook a comprehensive effort to document the road network by means of ground verification.

Mesoamerica

Studies pertaining to Mesoamerican road networks are relatively recent, and most such studies have emphasized ground verification and mapping. The sites of Casas Grandes, La Quemada, Xochicalco, Teotihuacan, and the Maya lowlands have been the subject of most surveys to date, while the Tarascan and Sonoran regions have been the subject of the most recent surveys. The road networks of these regions provide a glimpse into the Protohistoric developments of the fourteenth through sixteenth centuries AD.

The area of La Quemada, Zacatecas, lies within a semi-arid region of Mesoamerica's Northern Frontier. This area is dominated by a series of hill-fort centers, each of which is characterized by the presence of civic-ceremonial precincts, agricultural terraces, and secondary centers with related defensive features. An extensive intra-regional network of roads connected the primary centers of La Quemada and Los Pilarillos with secondary centers, and in turn, isolated mound groups, isolated platforms, defensive positions or fortifications, and agricultural terraces. This system, by contrast with that of Chaco Canyon, served to integrate communities within a tightly constricted area. Road segments linking the La Quemada road network with regions beyond the valley have yet to be fully mapped or verified.

The specific road-related features identified with La Quemada include: (a) formal causeways or other elevated road segments; (b) road segments terminating at elevated platforms; (c) causeways of between five and seven meters to twelve and fourteen meters in width with road-bed elevations of between thirty and forty centimeters; (d) causeways that extend to wide, low platforms and stairways with precipitous descents down steep escarpments; (e) parallel road and canal or other irrigation segments; and finally (f) where the larger causeways are concerned, such roads are identified with defensive positions. At La Quemada, the aforementioned features are taken to represent a sociopolitical pattern geared to militarism or related defense activity.

Where the road networks of the Mesoamerican Epiclassic era (ca. AD 650–900) are concerned, the central Mexican city-state of Xochicalco, Morelos, provides a salient example. The site of Xochicalco, like its La Quemada contemporary, is situated in a strategically located acropolis-centered hill-fort locality. Most of the road segments identified with Xochicalco emanate from the hill-fort to other centers within the valley below, and as such, Xochicalco provides a characteristic example of roads as symbols of power, prestige, and levels of regional integration. In other words, this position holds that Mesoamerican road networks function primarily as mechanisms for social integration or as symbols of the power and prestige of social or religious elites (Hirth, 1991). The scale and magnitude of construction activity invested in the road systems of Mesoamerica have led some scholars to speculate that such roads were less functional and more ceremonial in character. Monumental construction and related public works are taken as indicators of political power, and thereby elite command of critical resources (such as slaves and construction personnel) in the valley during the pre-Columbian era.

Mesoamerican economic systems, by virtue of the nonexistence of draft animals, and thereby, wheeled vehicles, were limited to human burden bearers. Despite this limitation, such systems were every bit as formal as those of other Old and New World polities. Primitivist interpretations of New World economies have unnecessarily minimized the perceived scale of economic formations in Mesoamerica and the New World despite available ethnohistorical data. The extent and magnitude of investment in road construction, and scale of integration made evident by the contiguity in the system, is again a key indicator of the types of sociopolitical formations that once prevailed in Mesoamerica. The Aztec era provides only a hint at what prior civilizations engaged in the way of road network investment in construction and maintenance, and by any account, the investment was substantial and culturally significant. Robert Santley's recent projection of road networks for the Basin of Mexico presents us with a system that was highly integrated and equally complex.

The utility of a formalized — organized, linear, weatherproof — road network over which to maintain the movement of goods and services was critical to trade and exchange in Mesoamerica. Like the ancient informal paths that preceded them, formalized roads served to guide and expedite the movement of goods and services on both an intra- and inter-regional scale.

A typology of Aztec era roads was recorded by Fray Bernardino de Sahagun, the sixteenth century Franciscan cleric who documented such aspects of post-conquest central Mexican culture. Of the seven types of roads identified by Bernardino de Sahagun the ochpantli and oquetzalli road types have been isolated as representative of the Xochicalco road types. The first term pertains to the "main road" and the second to the "royal road" networks of the Aztec era. The well groomed roads of Xochicalco are interpreted as of the royal road type. This interpretation is based on the monumental nature of the associated architecture (primarily defensive in character), the cut-stone pavements, and the relatively large width of the roadbed itself. In addition, archaeologist Kenneth Hirth distinguishes between "road" and "thoroughfare". Roads are "transportation arteries at the regional level which connect two or more spatially separated sites"; thoroughfares are "streets and other communication corridors which organize space and/or direct traffic flow within the community" (Hirth, 1991). The fragmentary nature of the data precludes a region-wide economic and social interpretation of the Xochicalco network.

The Xochicalco road network exhibits a number of distinct architectural characteristics, including:

(a) stone-surfaced intra-site rough-cut mosaic pavements or roadbeds of 3–5 meter widths;
(b) intra-site masonry ramps;
(c) parallel platforms flanking key road segments or thoroughfares leading into the site core area;
(d) roads which bisect massive rampart walls with narrow gateways in order to access intra-site thoroughfares;
(e) intra-site pavements or roadbeds edged with cut-stone masonry blocks;
(f) stucco used along flanking masonry edging in the intra-site sectors;
(g) intra-site pavements made flush with surfaces traversed and modified with ramps – both smooth and stepped;
(h) lateral masonry walls flanking intra-site pavements;
(i) access causeways along circuits of defensive ditches or moats; and
(j) evidence for the use of a probable wooden bridge spanning a defensive moat adjacent to Cerro de la Bodega.

Transport architecture extending beyond the site perimeter consists of less formal construction techniques in addition to significantly less evidence of overall integration with the site core of Xochicalco.

Recent studies provide indications of a massive road network that connected the Classic period center of Teotihuacan with an ancient trade corridor linking that site with the Puebla Basin, and by extension, the Mexican Gulf lowlands. Unlike the more defensive character of the La Quemada and Xochicalco road networks, the Teotihuacan corridor linking Teotihuacan with a whole host of ancient mercantile centers in the Mesoamerican highlands and Gulf coastal plain served a key transport function for the movement of goods and services provisioned and sought by Teotihuacan.

The road-associated settlement hierarchy of the Teotihuacan Corridor, as well as the recent identification of specific features of transport network architecture along the Teotihuacan Corridor, is currently under study. Additional features of Mesoamerican transport network architecture include (a) causeways — such as the Street of the Dead — measuring some forty meters in width; (b) the presence of Momoztli structures, or elevated platforms or shrines (often laden with trade offerings) situated at the crux of important crossroads along the corridor; (c) an approximate inter-site road width of 24 meters, and 60 centimeter depth, for sites located on the Teotihuacan Corridor; (d) spur roads that connect road-associated settlements with the principal arteries of the Teotihuacan Corridor and with other key nodal points or market centers; (e) related architectural styles and construction along the course of the transport network; (f) roughly equidistant spacing of road-associated settlements at eight to nine kilometer intervals for specific Teotihuacan related centers; and (g) stepped ramps, road shrines, crossroads, and attendant roadbed linearity. The overall network is dendritic in structure, in that the main road connects with primary centers, which are in turn connected to a constellation of secondary centers sharing affinities with key terminals along the Teotihuacan Corridor.

Peru

The most intensively studied ancient road system of South America is that of the Inca Empire, extending over a vast area from Ecuador to Chile, from the coastal deserts to the Andean highlands, into the Bolivian forests. This network included some twenty thousand miles of roadways and encompassed forty thousand miles at the point of European contact. This road system constitutes "South America's largest contiguous archaeological remain", and provides valuable insights into the organization and technology that are hallmarks of the Inca state (Hyslop, 1984).

In contemplating the vastness of this road network, one is struck by the enormity of the Inca achievement. First, there is the effectiveness of the political infrastructure that organized and mobilized a labor force specialized in the building and maintenance of roadways. Second, the Inca were exceptionally efficient in their use of resources, both natural and human. Pre-existing roadways were frequently incorporated into the Inca network. Inca engineers did not seek to modify natural topography to any great extent unless required to do so in order to achieve safe passage. Where road modifications were concerned, locally available materials were utilized and work was performed by locally organized labor pools under the supervision of a road administrator. Third, the Inca brought to bear a vast body of technical knowledge and expertise, accumulated over the course of centuries among the diverse ethnic groups of the empire. Finally, the efficiency of the road network insured that Inca rulers could maintain communication with — and thereby the integration of — every corner of the empire.

Four primary state roads divided the empire into cosmologically-defined quadrants. The sacred geography associated with road networks reflected the Inca preoccupation with cosmology defined *vis-á-vis* extant social divisions. Cuzco, the political and administrative center of the Inca state, was linked via these sacred paths to the most ancient shrines and temples. Roadways not only defined the partitioning of sacred space, they also served as linkages in a massive transport and communication network.

Road characteristics varied considerably in terms of construction methods and the extent of labor investment. Construction methods varied with the highly fractured and variable topography of the region, and as such, a single road might include both road segments with very complex architectural features and evidence for intensive labor investment, and simple unpaved or earthen pathways. Ultimately, road construction varied with localized conditions, topographic obstacles, available materials and labor, and the importance of the specific road segment to ceremonial and ritual activities.

Coastal routes varied from simple "pole-roads" to broad thoroughfares with stone or adobe sidewalls. Pole-roads were paths marked by linear arrangements of wooden posts placed at intervals. High-altitude roads on non-agricultural lands were, by contrast, engineered for durability in order to withstand harsh climatic conditions. Such roads incorporated elaborate stone curbs, retaining walls, and paving. These later constructions are among the best-preserved roads in evidence for the Andes. Roads traversing agricultural fields are characterized by very high sidewalls — of one or two meters in height — designed to protect crops from travelers or livestock such as llamas and alpacas. Such walls were likewise built of stone or adobe. Agricultural roads were relatively narrow so as to minimize their impact on the availability and productivity of arable lands.

Where roads crossed wet plains or slopes, stone paving was used to preserve the roadbed. If flooding was unavoidable, the road was constructed on an elevated surface or causeway. Inca causeways were of rubble-core construction, with or without stone retaining walls. In those instances where existing slopes were less than ten degrees off horizontal, the Inca used a combination of stone or rock-cut ramps, stairways, or switchbacks in order to buffer ascent and descent. Depending on the steepness of the existing slope, structures ranged from a single retaining wall to a series of walls and benches built on the downhill side of the slope, and intended to serve as secondary retaining features.

Bridges are among the most spectacular structures engineered by the Inca. While sixteenth century European construction employed the principle of the arch in bridge construction, the Inca were able to span greater distances through a variety of engineering principles and techniques. Early Spanish explorers marveled at the sight of Inca suspension bridges which were unknown in Europe. Though made of woven and braided fiber coils, these suspended structures supported considerable pedestrian traffic, in addition to European horses and carts of the conquest period. Other types of spans built by the Inca included floating bridges supported by pontoons of reed bundles, wooden bridges manufactured from logs placed over stone abutments (with or without cantilevers), and a variety of stone bridges. Culverts were stone-lined troughs left open, or capped, if necessary for safe passage. Somewhat wider rivers were spanned by stone columns placed adjacent to one another, thereby forming "multi-cell" culverts which supported the overlying roadbed. Some stone bridges employed cantilevered abutments supporting very large stone pavers. Finally, natural rock formations were utilized as bridged areas where available.

An integral component of Inca era road networks was the *tambo* system. Tambos were multi-use structures or facilities maintained by the Inca state but administered locally. Hyslop estimates that between one and two thousand of these facilities were in use throughout the Inca empire. Tambos exhibit considerable variation in size, ranging from sites with a few isolated structures to large administrative centers representing a multitude of activity areas. Storage and lodging constitute the primary uses for which tambos were employed. Storage structures included silos, corrals, and adjoining rooms. Lodging in tambo facilities was temporary, and intended either for travelers or as permanent residence for local administrators. Tambos were also employed in the processing of raw materials, military and ceremonial activities, local administration, and craft production. Tambos were equidistantly spaced on any given route at intervals representing a day's travel time. According to recent Inca road surveys, tambos were spaced at intervals that average between 15 and 25 kilometers. The combined tambo system and vast complex of roadways formed a critical component of the state's infrastructure. This highly efficient transport and communication network supported the state's practical needs for efficient military mobilization, colonization, economic maintenance, and administration. In addition, the complex network of roads supported the ideological needs of the Inca hierarchy, providing a highly visible set of linkages between the people and the state.

RUBEN G. MENDOZA
GRETCHEN W. JORDAN

REFERENCES

Hirth, Kenneth. "Roads, Thoroughfares, and Avenues of Power at Xochicalco, Mexico." In *Ancient Road Networks and Settlement Hierarchies in the New World*. Ed. C.D. Trombold. Cambridge: Cambridge University Press, 1991, pp. 211–221.

Hyslop, John. *The Inka Road System*. New York: Academic Press, 1984.

Kincaid, Chris. *Chaco Roads Project, Phase I: A Reappraisal of Prehistoric Roads in the San Juan Basin*. Ed. C. Kincaid. Department of the Interior, Bureau of Land Management, New Mexico, 1983.

Sahagun, Fray Bernardino de. *Florentine Codex. General History of the Things of New Spain. Book 11, Earthly Things*. Trans. C. Dibble and A. Anderson. Santa Fe: School of American Research and the University of Utah, 1963.

Santley, Robert S. "The Structure of the Aztec Transport Network." In *Ancient Road Networks and Settlement Hierarchies in the New World*. Ed. C.D. Trombold. Cambridge: Cambridge University Press, 1991, pp. 198–210.

Trombold, Charles D., ed. *Ancient Road Networks and Settlement Hierarchies in the New World*. Cambridge: Cambridge University Press, 1991.

Trombold, Charles D. "Causeways in the Context of Strategic Planning in the La Quemada Region, Zacatecas, Mexico." In *Ancient Road Networks and Settlement Hierarchies in the New World*. Ed. C D. Trombold. Cambridge: Cambridge University Press, 1991, pp. 145–168.

Urton, Gary. *At the Crossroads of the Earth and the Sky: An Andean Cosmology*. Austin: University of Texas Press, 1981.

ROCKETS AND ROCKETRY Traditionally, a rocket is a flying device launched by direct-reaction using a solid black powder of high-nitrate composition. The appearance of rockets signified a revolution in the development of firearms. All modern rockets were gradually developed on the basis of traditional ones. In Ancient China the rocket was generally called *huojian* (fire-arrow) which sometimes meant incendiary arrow and sometimes rocket. Both of them were invented in China. It is necessary to differentiate one from the other. From 968, the classical fire-arrow was sent by low-nitrate gunpowder in paste form from a bow or crossbow. It was thus an incendiary arrow, although it was the earliest gunpowder weapon. Since the beginning of the twelfth century solid gunpowder of high-nitrate composition and fuses were made and used for making fireworks and firecrackers. Some of them were directional devices, such as the *di laoshu* (earth-rat), *qihuo* (flying fire), and others. This provided a necessary premise for making rockets. During 1127–1234 the invention and military use of rockets were perfected in China on the basis of fireworks technology. There were two types of early rockets having different names.

The so-called *pili pao* (thunder-bolt missile) used in the Battle of Caishi between the Song and Jin was actually the earliest rocket-propelled bomb in the world. It was an enlarged *ertijiao* (double-bang firecracker) and was made according to the principle of *qihuo* (flying fire) used by the Song people in an earlier time. The *fei huojiang* (flying fire-lance) used in the Battle of Kaifeng between the Jin and Mongols in 1232 was an ordinary rocket for setting on fire. Its rocket tube made of paper was fastened to the point of a lance or an arrow. During the thirteenth to fourteenth centuries the Song, Jin, and Mongol troops all used improved rockets to fight with each other. They even used the reaction weapon abroad. For instance, single rockets and multiple rocket-launchers with a common fuse were used by the Mongols in the Battle of Leignitz in Poland in 1241, known as the 'Chinese dragon belching fire' in the West. Winged rockets must have been made in about 1300, since it was described in the *Huolong Jing* (Fire-Dragon Manual, ca. 1350).

In the Ming dynasty (1368–1644) rocketry entered a new epoch. Two-stage rockets were designed and made at the beginning of the fourteenth century. There were various kinds of different rockets from single ordinary ones to multiple rocket-launchers and wheelbarrows, as well as winged rockets and rocket-propelled bombs. Their range reached 200–500 paces (330–825 meters). There was a special rocket troop in the Ming army and rocket weapons were used as standing weapons by land and naval forces and cavalry troops. It is interesting that a Chinese military officer, Wan Hu (fl. 1440–1495) built a flight device made of 47 big rockets for a flight experiment. He was the first in history to try to use rockets for flight as a means of transportation. All of these ancient achievements in the field of rocket technology can be found in the most important military work entitled *Wubei Zhi* (Treatise on Armament Technology) written by Mao Yuanyi (ca. 1570–1637) in 1621. After rocket technology was perfected in China it was spread to other parts of the world. It went first to the Arabian world in 1240–1260, then to Europe in 1260–1270. The term *rochetta* first appeared in Italian literature in 1330. In the eighteenth century rockets were popular in some major European countries. Rockets arrived in other Asian countries in the same time during the Mongol-Yuan period, first in Vietnam and Korea, then in India and Southeast Asia.

In 1805 the British military officer William Congreve (1772–1825) developed modern rockets made of iron tube with black powder, which had a range of 2300 meters. But after the mid-nineteenth century rocketry ceased to develop because modern cannons seemed to be more powerful and effective. Only since the beginning of the twentieth century have rockets been given more attention and studied carefully. As a result, various new types of rockets have been developed and used, especially after World War II. Joseph Needham says: "In this day and age, when man and vehicles have landed on the moon and when the exploration of outer space by means of rocket-propelled craft is opening before mankind, it is hardly necessary to expatiate upon what the Chinese started when they first made rockets fly."

PAN JIXING

REFERENCES

Needham, Joseph. *Science and Civilisation in China*, vol. 5, pt. 7, *Epic of Gunpowder*. Cambridge: Cambridge University Press, 1986, pp. 477–524.

Pan Jixing. *On Two Problems in the History of Science*. Kyoto: Doshisha University Press, 1986.

Pan Jixing. *Zhongguo Huojian Jishu Shigao* (The History of Rocket Technology in China). Beijing: Science Press, 1987. English edition will be published by Astronautic Publishing House in Beijing.

Pan Jixing. "On the Origin of Rockets." *T'oung Pao* 73: 2–15, 1987.

S

ṢADR AL-SHARĪʿAH Ṣadr al-Sharīʿah al-Thanī, ʿUbayd Allāh ibn Masʿūd (fl. Bukhārā, d. 747 AH/AD 1347) was a scholar and astronomer of the fourteenth century.

He belonged to a family of famous religious scholars; his great-great-grandfather, Ṣadr al-Sharīʿah al-Awwal, was referred to as the second Abū Ḥanīfa in recognition of his rank among the legal scholars of his time. Ṣadr al-Sharīʿah, moreover, was himself a traditional religious scholar: he wrote on the subjects of *Ḥanafī* positive law, principles of jurisprudence, Arabic grammar, rhetoric, theology, legal stipulations and contracts, and *Ḥadīth* (the spoken Traditions attributed to the prophet Muḥammed). In addition, he wrote a three-volume encyclopedic treatise, *Taʿdīl al-ʿUlūm* (The Adjustment of Sciences). The first of the three books is on logic, and the second is on theology (*kalām*). The questions discussed in these two books were not unfamiliar to theologians of the time, and it seems quite probable that a jurist would be informed in such areas of research. What is uncommon, however, is Ṣadr al-Sharīʿah's competence in astronomy, which he demonstrates in the third book of the above work.

This last section entitled *Kitāb Taʿdīl Hay'at al-Aflāk* (The Adjustment of the Configuration of the Celestial Spheres) was finished in the year 747 AH (AD 1347) shortly before the death of its author. The motive for writing it, as expressed by the author in the beginning of the book, was to resolve the problems of the longitudinal motion of the moon and the other planets, as well as the latitude motion. The resulting work is indeed a revision of the astronomical tradition from within its own existing framework. In this Ṣadr al-Sharīʿah was obviously driven by the momentum of the revisionist tradition of the thirteenth-century Maragha school, which aimed at reforming Ptolemaic astronomy.

Ṣadr al-Sharīʿah was especially influenced by two sources: the *Tadhkira* of Naṣīr al-Dīn al-Ṭusī (d. AD 1274), and the *Tuḥfa* of Quṭb al-Dīn al-Shīrāzī (d. AD 1311). Ṣadr al-Sharīʿah proposed to review critically the planetary models of his two predecessors, and in the course of this revision he proposed his own alternative configurations. The models which he purported to examine and replace were for: (1) the longitudinal motion of the moon; (2) the longitudinal motion of the superior planets; (3) the longitudinal motion of Mercury; and (4) the motions in latitude. The mathematical mechanisms utilized by Ṣadr al-Sharīʿah to solve the problems of these models were not new, but his ingenuity lay in bringing some of these different mechanisms together. In three of the above cases he succeeded in producing new models, and in at least two of these the resulting configurations seemed to resolve the theoretical problems of earlier models, while predicting the planetary positions in accordance with the Ptolemaic observations.

AHMAD DALLAL

REFERENCES

Brockelmann, C. *Geschichte der arabischen Litteratur*. Leiden: E.J. Brill, 1937. Vol. II, pp. 277–9, and Suppl. II, pp. 300–1.

Al-Dujaylī, ʿAbd al-Ṣāḥib 'Imrān. *A'lām al-ʿArab fi alʿUlūm wal Funūn*. 2 vols. Baghdad: Maṭba'at al-Nu'mān, 1966. Vol. II, pp. 162–3.

Ḥajī Khalīfa, Kātib Jelebī. *Kashf al-Ẓunūn ʿan Asāmi al-Kutub wal-Funūn*. Ed. Gustav Fluegel. 16 vols. London: Oriental Translation Fund, 1852. Vol. II, pp. 315, 417, 601, vol. III, p. 37, vol. IV, pp. 439, 440, vol. VI, pp. 373–6, 443, 458–66.

Ibn Baṭūṭa. *Voyages d'Ibn Batoutah*. Trans. C. Defrémery and B.R. Sanguinetti. 3 vols. Paris: Imprimerie nationale, 1977. Vol. III, p. 28.

Ibn Kuṭlūbughā, Zein-ad-dīn Kāsim. *Tāj al-Tarājim fi Tabaqāt al-Ḥanafiyya*. Ed. Gustav Flugel. Leipzig: F.A. Brockhaus, 1862, pp. 29, 30–115.

Ibn Quṭlūbughā. *Tāj al-Tarājim fī Ṭabaqāt al-Ḥanafiyyah*. Baghdād: Maktabat al-mathannā, 1962.

Kaḥḥālah, ʿUmar Riḍā. *Muʿjam al-Mu'allifīn*. 2 vols. Damascus: al-Maktaba al-ʿArabiyya, 1958, p. 246.

Al-Luknawi, ʿAbd al-Ḥayy. *Al-Fawā id al-Bahiyyah fī Tarājim al-Ḥanafiyya*. Banaras, India: Maktabat Nadwat al-Maʿārif, 1967, pp. 91–5.

Sarkīs, Yūsuf Ilyās. *Muʿjam al-Maṭbūʿāt al-ʿArabiyyah wal-Muʿarrabah*. Cairo: Maṭba ʿat Sarkīs, 1928, pp. 1199–200.

Suter, Heinrich. *Die Mathematiker und Astronomen der Araber und Ihre Werke*. Leipzig: B.G. Teubner, 1900, #404.

Tāshkoprūzāde, Aḥmad b. Muṣṭafā. *Miftāḥ al-Saʿāda wa Miṣbāḥ al-Siyāda*. 3 vols. Cairo: Dār al-Kutub al-Ḥadītha, 1968. Vol. I, pp. 60–1, vol. II, pp. 182, 191–2.

Ṭlās, Muḥammad Asʿad. *Al-Kashshāf ʿan Khazāʿin Kutub al-Awqāf*. Baghdad: 1953, pp. 61, 69, 70, 99, 100.

Zaydān, Jurjī. *Tārīkh Ādāb al-Lughah al-ʿArabiyyah*. 3 vols. Cairo: Maṭbaʿat al-Ḥilāl, 1913. Vol. III, p. 239.

Al-Zirkalī, Khayr al-Dīn. *Al-Aʿlām*. Beirut: Dār al-Kutub lil-Malāyīn, 1969, p. 354.

AL-ṢĀGHĀNĪ Abū Ḥāmid Aḥmad ibn Muḥammad al-Ṣāghānī al-Asṭurlābī was an Arabic astronomer and mathematician. He was born in Chaganian (Central Asia) and worked in Baghdad.

His main work was the *Kitāb fī'l-tasṭīḥ al-tāmm* (Book of the Perfect Projection on to a Plane) which is extant in two manuscripts. The treatise is devoted to a generalization of the stereographic projection of a sphere on to a plane, usually used in the making of astrolabes. This concerns the projection from a pole of the sphere on to the equatorial plane or a plane parallel to it. Under this projection, circles on the sphere are imaged on to the plane as circles or straight lines. The "perfect projection" invented by al-Ṣāghānī is the projection of the sphere from any point of its axis on to a plane orthogonal to the axis. Under this projection, circles on the sphere are imaged on the plane as conic sections (ellipses, hyperbolas, and parabolas) or straight lines. These descriptions of methods for conics constuction are important for geometry. In the treatise, al-Ṣāghānī considers the construction methods for images of different circles of the celestial sphere, such as the celestial equator and its parallels, the horizon and its almacantars, verticals, and one ecliptic (the band of the zodiac through which the sun apparently moves in its yearly course). The contents of the treatise are explained in detail by al-Bīrūnī (973–1048) in his *Astrolabes*.

Al-Ṣāghānī was also the author of two mathematical treatises, on the construction of a regular heptagon inscribed in a circle, and on the trisection of an angle. He also wrote three astronomical treatises. *Kitāb qawānīn ʿilm al-hay'a* (Book on Rules of the Science of Astronomy) is not extant, but al-Bīrūnī in his *Geodesy* mentioned measuring the value of the angle between the ecliptic and the celestial equator found by al-Ṣāghānī in Baghdad. The second, *Maqāla fī'l abʿād wa'l-ajrām* (Article on Distances and Volumes), dealt with the distances and volumes of planets and stars. The third work was *Fī'l sāʿāt al-maʿmūla ʿalā ṣafāʾiḥ al-asṭurlāb* (On Horary Lines Produced on the Tympanums of Astrolabes).

BORIS ROSENFELD

REFERENCES

Frank, Josef. "Zur Geschichte des Astrolabs." *Sitzungsberichte der Physikalisch-Medizinischen Sozietät zu Erlangen* 48–49: 275–305, 1918.

Matvievskaya, Galina P. and B.A. Rosenfeld. *Mathematicians and Astronomers of Medieval Islam and Their Works* (8th–17th C.) vol. 2. Moscow: Nauka, 1983. pp. 162–163 (in Russian).

Sarton, George. *Introduction to the History of Science*. Baltimore: Williams & Wilkins, 1927.

See also: Astrolabe

ṢĀʿID AL-ANDALUSĪ Abū al-Qāsim Ṣāʿid ibn Abū al-Walīd ibn ʿAbd al-Raḥmān ibn ʿUthmān al-Taghlibi, better known as Ṣāʿid al-Andalusī or Qāḍi Ṣāʿid, was born in Almeria in southern Spain in AD 1029. He was a philologist, natural philosopher, and historian as well as judge. As his name indicates, he was a member of the tribe of Taghlib, one of the largest tribes of Arabia. When the Arabs invaded Spain in AD 711, members of this tribe entered the country and prospered there.

Ṣāʿid was born into a well-to-do family, whose members spent much of their time and wealth in the quest of knowledge and education. His father occupied a highly respected position in the city of Cordoba, and his grandfather, ʿAbd al-Raḥmān, was a judge in Sidonia, Spain. After receiving his early education in Cordoba, Ṣāʿid toured Muslim Spain to further his education. For the same reason, at the age of seventeen, he moved to the city of Toledo.

Like most young Arab students, Ṣāʿid studied law, Islamic religion, Arabic language, and Arabic literature. Later in life, he specialized in the study of mathematics and observational astronomy. Among his most famous teachers were al-Fātiḥ Muḥammad ibn Yūsuf (d. AD 1059), Abū Walīd al-Waqshī (d. AD 1059), and Ibn Idris al-Tajibī, who taught him mathematics and astronomy.

Early in the eleventh century, Cordoba ceased to be the principal intellectual center of Spain, and the capital cities of the *Mulūk al-Ṭawā'if* (party-kings) tried to capture its former glory. One such city was Toledo, the capital of the princedom of Banū al-Nūn. When Ṣāʿid entered Toledo, it was governed by Yaḥyā ibn Dhi al-Nūn, who reigned from AD 1037 to 1074, and extended his kingdom to include Valencia and most of the eastern parts of Andalusia. During his reign, Toledo became an important literary and intellectual center. In addition to several authorities in Islamic sciences, Yaḥyā's court had several good mathematicians, well-known astronomers and astrologers, several men of medicine, poets, and geometers.

The number of eminent scholars, poets, and philosophers in a ruler's court was one of the status symbols of the time. Probably for that reason, Yaḥyā invited Ṣāʿid into his court and appointed him a judge (*qāḍi*).

Several of Ṣāʿid's students later became accomplished scholars. Among them was ʿAbd al-Bāqi ibn Baryal, who was in part responsible for the style and language of some of Ṣāʿid's writing. There was also the outstanding astronomer and mathematician Abū Isḥaq al-Zarqali (Arzaquiel) who, with the assistance and the guidance of his mentor, constructed the famed Toledan Tables. These astronomical tables, which were accurate calendars, were translated into Latin by Gerard of Cremona, and became the basis for the Marseilles Tables.

Ṣāʿid wrote several manuscripts on a variety of subjects. All of his works except for *Ṭabaqāt al-ʾUmam* (Book of the Categories of Nations) are lost. He mentioned in *Ṭabaqāt al-ʾUmam* that he wrote a document on observational astronomy, one on the history of nations, and another on their religions. He may have written a manuscript on the history of Andalusia and one on the history of Islam. *Ṭabaqāt al-ʾUmam* is an authoritative source and a precise reference that identifies the natural philosophers of Muslim Spain.

Ṣāʿid died in July, AD 1070. Ibn Yaḥyā al-Ḥadidī, the most illustrious dignitary in the court of Ibn Dhi al-Nūn, read his official obituary.

SEMA'AN I. SALEM

REFERENCES

Al-Andalusī, Ṣāʿid. *Ṭabaqāt al-ʾUmam*. Trans. Sema'an I. Salem and Alok Kumar. Austin: University of Texas Press, 1991.

Chejne, Anwar G. *Muslim Spain: Its History and Culture.* Minneapolis: University of Minnesota Press, 1974.

Glassé, Cyril. *The Concise Encyclopedia of Islam.* San Francisco: Harper and Row, 1989.

SALT Salt, its production, distribution and consumption, was an important element in the economies of all ancient societies and played a significant role in their societies, politics and culture. This article falls into two parts. First, the principle methods of producing salt before 1800 will be analyzed. Second, the particular methods in use in the four primary civilizations of Western Eurasia (excluding Europe itself), East Asia, sub-Saharan Africa and pre-Columbian America will be outlined, together with their characteristic means of distribution and levels of consumption.

Salt exists in nature in two forms. First, there is actual salt. This takes the form of rock deposits from evaporated seas which can be quarried or mined, spontaneous evaporation at the edge of the sea or saline lakes, and incrustation on the surface of saline soil, both of which may be collected. Second, there is potential salt. This takes the form principally of brine from the sea, lakes, springs or wells, but also of saline earth and the plants growing in it. From potential salt, actual salt can be made by a variety of techniques. In most non-Western cultures these techniques were more important as a source of salt than quarries, mines or collection. Evaporation of brine, whether from the sea, lakes, springs or wells, or artificially prepared from saline rock, earth or plants, accounted for 90% of all salt consumed before 1800. Brine could be made to yield salt in two ways. It could be boiled over an artifical fire and fueled in various ways: what the Chinese called *chien*. Or it could be evaporated naturally by the power of the sun and wind: what the Chinese called *shai*. Artificial boiling and natural evaporation produced different kinds of salt. Quicker, *chien* produced a small-crystal salt, or, if prolonged, cake salt. Slower, *shai* produced a large-crystal salt. Each method had sub-varieties whose differences varied the cost, color, consistency, purity and cleanliness of the salt produced.

Sub-varieties of artificial boiling may be classified according to the fuel used. Wood was commonest, but reeds, natural gas, coal and oil shale were also used. Fuel was the chief item of cost and its availability the principal constraint. The amount required depended on the salinity of the brine. To reduce it, brine salinity was frequently intensified either by preliminary exposure to sun and wind or by filtration through saline sand, earth or ash. Since artificial boiling was everywhere an older method than natural evaporation, it may have been this preliminary exposure which provided the idea of using sun and wind to effect the whole process.

Sub-varieties of natural evaporation may be classified according to the number and function of the basins used. First, there was single basin evaporation. Here, brine was run into a single basin, or battery of basins, and allowed to evaporate completely. Salt made in this way, however, was gritty, deliquescent, and possibly injurious to health, because the sodium chloride would be contaminated by the calcium and magnesium compounds contained in most brines and especially in sea water. Next, there was successive basin evaporation. Here, brine was run through a series of basins for catchment, condensation and crystallization. In this way, sodium chloride could be separated from the calcium and magnesium compounds through the different sedimentation rates of these chemicals. Thus the calcium compounds settled in the condensers, while salt was removed from the crystallizers before it could be contaminated by the magnesium compounds in the mother liquor which was then drawn off. By this triple process, a much purer sodium chloride was obtained. The process could be tuned to give different degrees of cleanliness and size of crystal by varying the frequency of harvest and the ratio of condensers to crystallizers. The more frequent the harvest, the more likely the adherence of the floor of the crystallizer to the salt, so the dirtier the product. The higher the ratio of condensers to crystallizers, the longer the process, so the bigger the crystals. The choice of frequency and ratio depended on cost, especially of land, convenience to the producer, climate, and consumer preference as to the kind of salt. Salt was a sophisticated industry.

Salt has been sought since the first agriculturalists found a vegetable diet bland and insipid compared to meat and fish. It was sought too for its supposed medicinal effects on the internal parasites with which early agriculuralists were endemically affected. Finally, as the "general of foods" it was sought to create flavor, cuisine and social distance. Salt was an adjunct of civilization. Although the body does require a minimum intake, the quantity is small and satisfied by normal diet, so that salt deficiency is rare in humans.

Most human consumption is in excess of physiological need. Culture, not nature, called for salt. It travelled by whatever form of transport was cheapest for bulk goods, which in premodernity meant by water, where available, rather than by land.

In the Ancient Near East and in its successor the Central Islamic lands, natural salt in the form of rock, spontaneous evaporation and incrustation was widely distributed. Salt never became big business and consumption was high. When Muslims moved beyond their homeland, however, they experienced a shortage of salt. The expansion of Islam to the Mediterranean, North Africa, India and Indonesia was therefore accompanied by the introduction of new techniques of salt production. In particular, Islam diffused the Chinese technique, invented around AD 500, of successive basin solar cum wind evaporation. That the technique was diffused rather than independently invented in a number of places is indicated by the presence everywhere initially of the original Chinese condensers to crystallizers ratio, itself functionally unnecessary and in some circumstances dysfunctional.

In China itself, before AD 500, salt had been acquired either, in the northwest, by the collection of natural deposit, or, in the southeast, by boiling of marine brine (*chien*). Initially *shai* was regarded as inferior to *chien*. Indeed, in the Mongol period it was temporarily abandoned in its original site. In the long run, however, its absence of fuel costs, plus a number of technical improvements, gave it complete victory over *chien*, save in Szechwan which enjoyed the advantage of natural gas and more saline brine. In India too, though the mines of the Salt Range continued to operate, successive basin solar evaporation became the chief form of production: principally at the Sambhar lake and on the west coast, but subsidiarily, with technical modifications, in the Tamil Nadu. Bengal long remained a center of *chien*, till its industry was overwhelmed by imports from Britain and the Red Sea, the latter being a *shai* center.

In pre-Columbian America and in sub-Saharan Africa, production remained limited and consumption low until the coming of the Europeans to the first, of Islam to the second. Some Matto Grosso Indians neither produced nor consumed salt (the same is true of the Maori in New Zealand). Elsewhere in America, supply was by collection, boiling, or single basin evaporation, though one archaeological site has raised the possibility of an independent Maya invention of the successive basin technique. In Africa, supply was similar, though the Saharan quarries and mines may have been developed before the coming of Islam. The greatest consumers of salt were the Islamic world and China. Despite the development of marine based salines at its extremities, Islam supplied itself mainly by land, using camels for transportation. China supplied itself mainly by water, by junk. There, because sources of supply, chiefly on the coast, were relatively few, salt did become big business. Indeed, until it was supplanted in the eighteenth century by textiles, it was the chief component of the premodern Chinese economy.

S.A.M. ADSHEAD

REFERENCES

Adshead, S.A.M. *Salt and Civilization*. London: Macmillan, 1992.

Aggarwal, S.C. *The Salt Industry in India*. New Delhi: Government of India Press, 1976.

Ewald, Ursula. *The Mexican Salt Industry 1560–1980*. Stuttgart: Gustav Fischer Verlag, 1985.

Hocquet, Jean-Claude. *Le Sel et le Pouvoir*. Paris: Albin Michel, 1985.

Lovejoy, Paul E. *Salt of the Desert Sun*. Cambridge: Cambridge University Press, 1986.

SALT IN CHINA In premodern China, hundreds of terms for different kinds of salt existed which were distinguished according to place of origin, type of deposit, method of production, shape, physical appearance and properties, taste, color, translucence, use, and function within the salt monopoly system. An important and authoritative scientific explanation for the existence of various salt deposits was provided by the *Shujing*. There, in the chapter "Hong fan", saltiness (*xian*) is related to the phase of Water, which by soaking and descending becomes salty. Another important explanation referred to cosmogonical thinking, stating that the clear and light (i.e. underground currents of sweet water and water of rivers and lakes) are to be found above, while the heavy and turbid (i.e. underground brine and sea water) below.

The economically and fiscally most important types of salt were, in descending order, sea salt (*haiyan*), lake salt (*chiyan*), well salt (*jingyan*), earth salt (*jianyan* or *tuyan*), and rock salt (*yayan*). In the late imperial period, sea salt accounted for more than eighty percent of the empire-wide production. The percentages for lake salt and well salt were about twelve and six percent respectively. The shares of earth and rock salt are quantitatively negligible.

The Chinese salt production sector reflects typical features of the history of production techniques in China. The far greatest amounts of salt, that is sea, lake, earth, and rock salt, were produced with the help of rather simple and old-established production techniques, which in many cases required a high input of human labor. Manpower also played an important role in well salt production, but, because of the use of deep-drilling methods and exploitation of natural gas resources, this sector contained a revolutionary technological potential. Finally, we can observe that, sometimes, sophisticated production techniques were abandoned, like the method of basin solar evaporation at the salt lake of Xiezhou in Hedong (Shanxi) during the thirteenth to the sixteenth centuries.

Several partly overlapping stages of development may be observed in the production of sea salt. Early sources mention that salt was produced by boiling seawater. It is, however, far from clear whether this statement refers to the rather uneconomic process of boiling seawater directly down or whether prior to boiling a concentrated brine had been prepared. Probably both methods co-existed.

The first written account of the production of concentrated brine comes from the late ninth century. A hollow was dug, covered with wood and bamboo twigs. Sands enriched in salt were heaped up over the hollow and were then leached out by the invading flood. The concentrated brine dripped into the hollow and was tested with grains of cooked rice which floated or descended according to the degree of salinity. Finally, the brine was boiled to salt in iron pans, or pans made out of woven bamboo strips and coated with a layer of clamshell lime. This became the basic method of sea salt production in China. Improvements were added in later ages. For instance, besides sand, ashes were used which were spread out on specially prepared fields, artificially sprinkled from time to time, and exposed to the sun, thus raising their content of salt particles. In this way, the amount of brine and its salinity could be increased. The sophisticated and labor-intensive process of later ages is fully described and illustrated in the *Aobo tu* of 1334 (see Figure 1).

Figure 1: Sea salt production from the *Aobo tu* of 1334, with the title "Carrying the ashes and pouring them into the leaching basin".

The production of sea salt by solar evaporation (*shai yan*) began probably in the late Song period. The term *shai yan* is, however, ambiguous. It may either denote the production of salt in successive solar evaporation basins, making use of the fact that common salt would crystallize *after* the precipitation in the condensers of the calcium salts contained in the brine and *before* the precipitation of the magnesium compounds. Or it may be connected with the traditional leaching process, in which the last step in the production process, the boiling of enriched brine, was simply replaced by solar evaporation. The latter technique may be considered a transitional step in the slow shift from the original method of boiling brine to the production of salt in successive solar evaporation basins. This transitional method had probably already begun in the late Song period. A special form developed in parts of eastern Zhejiang where from the late eighteenth century onwards the enriched brine was exposed to the sun in portable wooden trays.

Figure 2: The two methods of harvesting salt at the salt lake o Xiezhou, as shown in the *Zhenghe* pharmacopoeia of 1249.

While Adshead thinks that successive basin solar evaporation of sea salt first appeared in Changlu (Hebei) in 1522

Figure 3: A Sichuan salt well, as depicted in the Sichuan salt gazetteer of 1882.

at the latest, and then spread to Huaibei (1550), Shandung (1570), Fujian (1575) and other places, Liu Miao is of the opinion that it was not adopted at the sea coast before the middle or late sixteenth century. Moreover, there is clear evidence that Fujian was prior to Changlu which took over the method of successive basins from Fujian perhaps in the middle or late sixteenth century (Bai Guangmei, 1988; Lin Shuhan, 1992). From the seventeenth century onwards the method of solar evaporation in successive basins gradually replaced the traditional methods of preparing enriched brine for the final crystallization either with the help of fire or solar heat. Generally speaking, either method of solar evaporation resulted in economies of fuel and manpower.

The use of solar evaporation basin methods, however, did not originate at the sea coast, but at the salt lake of Xiezhou in southern Shanxi. In the beginning of Chinese culture, this lake must have been an important and contested source of salt supply. First, salters extracted naturally and spontaneously produced salt, a process which had always been considered an auspicious and divine phenomenon. In the late fifth or early sixth century at the latest artificial solar evaporation basins emerged. For both methods of harvesting (see Figure 2), the warm southern wind of the summer months was a decisive factor in bringing about crystallization. Salt production in basins was conceived basically as a kind of agricultural process which is suggested by the use of terms like "fields", "parterres", "ditches", "raised paths", and "sowing salt". From the late thirteenth to the late sixteenth century, however, solar evaporation basins were given up, and probably only naturally and spontaneously produced salt was extracted. Political reasons were partly responsible for this, because the method of solar evaporation basins was closely associated with the saline's corvée system. Solar evaporation basins were reintroduced in the late sixteenth century. A perennial problem was the protection of the salt lake's brine reservoir from intruding fresh water floods caused by inundating rivers. Dikes and spill-over areas were constructed for this purpose. One of these inundations silted up the lake's brine reservoir in 1757, so that thereafter wells and pits were constructed for getting brine.

Whether single basin or successive basin solar evaportation was carried out at Xiezhou is debatable. Successive basin solar evaporation was explicitly described for the first time in the early eighteenth century. Educated guesses about the actual date of invention or adoption of successive basin solar evaporation at Lake Xie, however, vary greatly. It is said to have been either the fifth or sixth century (Adshead, 1992), the eleventh century (Chai Jiguang, 1993) or the late sixteenth century (Vogel, Needham et al., forthcoming).

The production of well salt in Sichuan province can be divided roughly into two periods. From the third century BC to the eleventh century AD shaft wells were dug for exploiting the underground brine deposits. Recent research has shown that deep drilling was invented in the 1040s, thus inaugurating the era of deep drilled wells from the middle of the eleventh century up to the twentieth century. Salt produced by privately run deep drilled wells was cheaper and of better quality than that offered by the state managed shaft wells. The new type of well was drilled with a heavy iron bit which was fastened to a cable made out of bamboo strips notched together. The cable was attached to the drilling frame's lever beam. By jumping up and down the lever beam the drill was lifted and dropped, thus crushing the rock down in the well and creating a hole with a diameter of only about 13 cm. The upper part of the finished well was tubed with large bamboo stalks. From the late sixteenth century onwards it was also provided with a foundation of stone rings, and large tubes made out of wood were used. Slurry and brine were lifted with a bamboo tube which was equipped with a leather valve at its lower end. When entering the brine, the valve opened inwards, and it closed under the pressure of the brine when the tube was lifted. A great number of utensils were invented to remove obstructions that occurred during the process of drilling and brine hoisting. Deep drilling, hoisting, and repair operations developed in the course of time. This development was characterized by the use of heavier drill bits, more solid drilling frames, higher hoisting rigs, and larger hoisting tubes and hoisting drums, driven by up to four buffaloes. While in the mid-eleventh century a depth of about 200 m may have been sometimes reached, wells of the mid-eighteenth century may have been more than 500 m deep. And in 1835, the Xinghai well of Furong (*Ziliu jing*) measured more than 1000 m. The nineteenth-century Furong wells (see Figure 3) were much more successful than their predecessors in the tapping of subterranean brine and natural gas deposits.

Earth salt was produced in northern Shanxi and eastern Hebei in a way similar to producing sea salt, that is by leaching earth containing salt particles and by boiling down the resulting brine. The exploitation of rock salt was reported from Gansu and Shaanxi, where a kind of red-colored salt was collected in rocky caves.

HANS ULRICH VOGEL

REFERENCES

Adshead, S. A. M. "From Chien to Shai: Chronology of a Technological Revolution." *Journal of the Economic and Social History of the Orient* 33: 80–104, 1990.

Adshead, S. A. M. *Salt and Civilization*. Houndmills and London: Macmillan, 1992.

Bai Guangmei. "Zhongguo gudai haiyan shengchan kao" (An Investigation into the Production of Sea Salt in Ancient China). *Yan ye shi yan jiu* 1: 49–63, 1988.

Chai Ji-Guang. "Guanyu Yuncheng yanchi zhi yan xiliechi jiq shengchan gongyi jishu xingcheng shichi de tantao — jianda Deguo Haidebao daxue Fu Hansi jiaoshou" (On the Successive Solar Evaporation Basins for Salt Production at the Salt Lake o Yuncheng and the Period of Formation of this Production Technique — A reply to Professor Hans Ulrich Vogel of Heidelberg University in Germany). *Yan ye shi yan jiu* 1: 60–63, 1993.

Lin Shu-han. "Zhongguo haiyan shengchanshi shang sanc zhongda jishu gexin" (The Three Important Technological Innovations in the History of Sea Salt Production in China). *Zhongguo kejishi ziliao* 13 (2): 3–8, 1992.

Liu Miao. "Mingdai haiyan zhifa gao" (An Investigation into the Production Methods of Sea Salt During the Ming period). *Yan ye shi yan jiu* 4: 58–72, 1988.

Sun, E-Tu Zen and Shiou-Chuan Sun, trans. *T'ien-kung k'ai-wu Chinese Technology in the Seventeenth Century, by Sung Ying Hsing*. University Park and London: Pennsylvania State University Press, 1966.

Vogel, Hans Ulrich. *Naturwissenschaften und Technik im vormodernen China: Historie, Historiographie und Ideologie*, Kurseinheit 3: *Geschichte der Tiefbohrtechnik im vormodernen China Invention, Innovation, Diffusion, Transmission*. Hagen: Fern Universität Gesamthochschule, 1993.

Vogel, Hans Ulrich, Joseph Needham, et al. "The Salt Industry" Section 37 of *Science and Civilisation in China*. Cambridge Cambridge University Press, forthcoming.

Yoshida Tora. *Salt Production Techniques in Ancient China: Th Aobo Tu*. Trans. Hans Ulrich Vogel. Leiden: Brill, 1993.

See also: Natural Gas in China

SALT IN INDIA As India is one of the oldest civilizations, is no wonder that salt was produced in ancient India and tha one finds its mention in ancient scriptures. The salt industr flourished as a cottage industry for centuries. The word fo salt is *Lāvaṇa* in Sanskrit. A passage from *Arthaśāstra*, book dealing with the history of the Mauryan period (300 B.C), says that salt manufacture was even at that distant date supervised by a state official named Lavanadhyaksa and the business was carried out under a system of licences granted on the payment of fixed fees or part of the output. This tradition, handed down from Hindu kings of old, is even now followed with variations by the Government of India through its Salt Department. The history of many countries shows connections with salt; in India salt was used by Mahatma Gandhi as a tool to win independence, when he completed his famous 400-km march to the sea at Dandi on 6th April 1930. The development of the Indian salt industry will now be briefly examined in two periods, pre- and post-independence.

Salt was prepared in ancient times and until British rule from sources like seawater, subsoil and lake saline water, rock salt deposists, and water extracts of saline soils (particularly in Uttar Pradesh) mainly by solar evaporation. If required, artificial evaporation was resorted to. The salt was produced in the coastal regions of Bengal, Bombay, Madras and the Rann of Kutch and in the inland regions of Rajasthan, Uttar Pradesh, and Central India from saline brines of lakes and as rock salt in Punjab. The salt industry provided a source of revenue to the rulers in the respective regions, and it received protection and encouragement from them. However, the industry faced a sort of setback and discouragement from the British rulers who not only raised the taxes and levies as early as 1768 but also later started importing salt around 1835–36 from European countries, Aden, and other places. The salt industry in Bengal province first felt the impact of the British policy which then affected the salt industry not only in other parts of their empire but also in Goa, which was under Portugese rule. The quantity of salt imported, mainly in Bengal, in the pre-independence period ranged between 0.4 and 0.6 million tons annually.

However the British rulers reconsidered their drastic policies around 1930–31, which helped to revive the Indian salt industry, and salt production picked up. Changes and improvements in salt production methods, particularly from sea water, based on scientific principles were brought about by an Indian chemical engineer, Kapilram Vakil. Before establishing India's first large scale marine salt farm at Mithapur in 1927 with the support of the Maharaja of Baroda, he had earlier (1919–20) studied in detail the status of the salt industry in the eastern states of Bengal and Orissa. Thus production of salt in large salt works established on the basis of scientific knowledge began in India before independence.

After Independence, the majority of rock salt producing areas went to Pakistan. Salt is now mainly produced from sea water, subsoil brines, and inland lake brines. India exports salt; the imports are around 8000 to 15,000 tons per year in recent years and it is mostly as rock salt from Pakistan.

Efforts have to be made to improve the quality of salt produced from both sea water and subsoil brines in the field itself, and this is one of the challenges faced by the Indian salt industry. The industry is highly labor intensive and thus has a low output. It is necessary to upgrade the technologies, adopt mechanization, and improve transportation, loading, and port facilities. This is another challenge to be tackled by the Indian salt industry,an ancient industry that must adapt to changing times.

S.D. GOMKALE

REFERENCES

Aggarwal, S.C. *The Salt Industry in India*. Government of India, 1976.

Annual Reports of the Salt Department, Ministry of Industry, Government of India.

AL-SAMARQANDĪ Shams al-Dīn al-Samarqandī, as his name indicates, was from Samarqand, in Uzbekistan. We know few biographical details of his life with any certainty. He is believed to have been active during the last half of the thirteenth century, since he composed a star calendar for 675 AH/AD 1276–1277 to accompany his summary of astronomy entitled *Al-Tadhkira fī'l-Hay'a* (Synopsis of Mathematical Astronomy). It was during this period that Naṣīr al-Dīn al-Ṭūsī gathered many leading intellectuals together at his observatory in Marāgha, but al-Samarqandī was not among them.

Al-Samarqandī's earliest contributions were in the field of logic, but he is best known to historians of science for his brief tract, *Kitāb Ashkāl al-Ta'sīs* (Book of the Fundamental Theorems), which gives thirty-five key propositions from Euclid's *Elements* along with condensed demonstrations. Of special interest has been the attempt to prove Euclid's parallels postulate, the fifth postulate of Book I, included here. (Once thought to be the work of al-Samarqandī, this demonstration is now frequently ascribed to Athīr al-Dīn al-Abhārī.) The treatise is extremely concise, and has historically been better known in the commentary composed by Qāḍī Zādeh al-Rūmī.

GREGG DE YOUNG

REFERENCES

Dilgan, H. "Demonstration du Ve postulat d'Euclide par Shams-ed-din Samarkandi." *Revue d'Histoire des Sciences et leurs Applications* 13: 191–196, 1960.

Dilgan. H. "Al-Samarqandī, Shams al-Dīn Muḥammad ibn Ashraf al-Ḥusaynī." In *Dictionary of Scientific Biography*. Vol. 12. Ed. C. Gillispie. New York: Scribners, 1970–1981, p. 91.

Jaouiche, K. *La Théorie des Parallèles en Pays d'Islam*. Paris: Vrin, 1986.

Souissi, M. *Ashkāl al-Taʾsīs lil-Samarqandī Sharḥ Qāḍī Zādeh al-Rūmī*. Tunis: Dar al-Tunisiyya lil-Nashr, 1984.

SAMŪʿĪL IBN ʿABBĀS AL-MAGHRIBĪ Samūʿīl (or Samau'al) ibn ʿAbbās, also called Abū Nasr Samūʿīl ibn Yahyā ibn ʿAbbās, was a Jew whose father's family origins were in Fez in Morocco. Samūʿīl is often called al-Maghribī, "the westerner", but he spent his life in Iraq, Syria, Kurdistan, and Azerbaijan, and died in Maragha in about 1175 (570 A.H.). He was famous as a mathematician and physician, and left a number of works on both subjects. In 1163 he underwent a conversion to Islam and subsequently wrote a polemical treatise against the Jews; in an autobiographical appendage to this work he provided a very interesting account of his careful upbringing, extensive education, and the dreams which prompted his conversion. Samūʿīl practiced medicine at Baghdad (where he wrote some of his works) and Maragha, where he served a number of local princes and important personages.

The early Arab biographers ascribe many works to Samūʿīl, but not all of these are extant. Steinschneider and Brockelmann list seventeen treatises, some of which are works not mentioned by the biographers but which are ascribed to Samūʿīl in manuscript; some attributions are doubtful.

The only extant medical treatise attributed to Samūʿīl is an elaborate work on sex and gynecology entitled *Nuzhat al ashāb fī muʾāsharat al-ahbāb*: (The Entertainment of Friends and the Dealings of Lovers), often referred to in catalogs as *De Coitu*. Works of this kind are not infrequent in the Arabian medical tradition; Samūʿīl's treatise was written for the ruler of Hisn Kayfa, Muḥammad ibn Qara Arslan, and was famous enough to have survived in a number of widely scattered manuscripts; it has never been printed. An account of the work may be found in Leclerc. It contains two books: the first deals with sex, both homosexual and heterosexual, sexual desire, diet, and hygiene; it contains an account of the physical and moral characteristics of women from different countries. The second book deals with such topics as impotence, aphrodisiacs, conception, sterility, and female ailments. There are manuscripts in Paris, Berlin, the Escorial, Leipzig, and Istanbul. The Berlin catalog lists (in Arabic) all the chapter headings of the *Nuzhat.*

Some of the treatises on mathematical subjects ascribed to Samūʿīl by the early biographers are no longer extant, but copies of his *al-Tabsira fīʾ ilm al-ḥisāb* (Introduction to Arithmetic) survive. Both Leiden and Oxford have manuscript copies of a work entitled *Fī kashf ʾawar al munajjimin* (On the Errors of Astronomers), in which Samūʿīl defends the idea of scientific progress, citing an arithmetical work of his own, *Al-bahir,* as an example of an advance on ancient mathematics. Brockelmann says that a manuscript of a work entitled *Al-bahir fī ʾilm al-ḥisāb* attributed to Samūʿīl exists in Istanbul; perhaps this is the one referred to in *Errors of the Astronomers*. Another work on arithmetic, *Al-mujiz al-mardawi fī al-ḥisāb*, also exists in a single manuscript in Istanbul.

The Arabic text of Samūʿīl's polemic against Judaism, *Ifḥām al-Yahūd* (Silencing the Jews), together with his autobiographical addition, has been edited and translated into English by Perlmann, who provides in his introduction the best account of Samūʿīl and his works. Perlmann points out that the work *Al-Ajwiba alfakhira raddan ʾan il-milla al-kafira* ascribed to Samūʿīl by Brockelmann derives from Samūʿīl's work but is not by him. Another polemical work against the Jews, the *Contra Judeos* of Samuel Marochitanus which was translated from Arabic into Latin in 1339, has often been ascribed to Samūʿīl ibn Abbas; it contrasts Judaism unfavorably with Christianity, not Islam, and is clearly the work of another author.

E. RUTH HARVEY

REFERENCES

al-Maghribī, Samual ibn Yahyā. *Ifhām al-Yahūd* (Silencing the Jews). Ed. and trans. Mosche Perlmann. New York: American Academy for Jewish Research, 1964.

Fritsch, Erdmann. *Islam und Christentum in Mittelalter*. Breslau: Müller & Seifert, 1930.

Ibn Abī Uṣaybiʾah. *ʾUyūn al-Anbā fī ṭabaqāt al-aṭibbāʿ*. Beirut: Dār Maktabat al-Ḥayāh, 1965, pp. 471–2.

Leclerc, Lucien. *Histoire de la Médecine Arabe*. Paris: Leroux 1876; reprinted New York: Franklin, 1971.

Rosenthal, F. "Al-Asturlabi and as-Samaw'al on Scientific Progress." *Osiris* 9: 555–564, 1950.

Schreiner, Martin. "Samau'al b. Jahja al-Magribi und seine Schrift 'Ifham al-Jahud'." *Monatsschrift für Geschichte und Wissenschaft des Judenthums* 42, 1898.

Steinschneider, Moritz. *Die Arabische Literatur der Juden*. Frankfurt: Kauffmann, 1902, reprinted Hildesheim: Olms, 1964.

ŚAṄKARA VĀRIYAR Śaṅkara Vāriyar (ca. 1500–1560) was a brilliant expositor of astronomy and mathematics who hailed from Kerala in the south of India. He belonged to the *vāriyar* community which was professionally assigned to certain peripheral duties in temples. Śaṅkara was a direc disciple of the well-known Kerala astronomer Nīlakaṇṭha Somyāji (b. 1444). Among other teachers whom he mentions in his works are Dāmodara (ca. 1450–1550), Jyeṣṭhadeva (ca. 1500–1610) and Citrabhānu (ca. 1475–1550). Śaṅkara mentions that his commentary on the work *Pañcabodha* was completed on the day 16,92,972 of the Kali era, which occurs in AD 1534, from which his date is definitely ascertained.

Śaṅkara Vāriyar wrote an advanced astronomical manual entitled *Karaṇasāra*, to which he has added a gloss of his own. But he is better known for his elaborate commentaries on such standard works as the *Līlāvatī* of Bhāskarācārya, the *Tantra-saṅgraha* of Nīlakaṇṭha Somayāji, and the anonymous *Pañcabodha*. The two commentaries, the one on the *Līlāvatī*, called *Kriyākramakarī* (Sequential Evolution of Mathematical Procedures) and that on *Tantrasaṅgraha* called *Yuktidīpikā* (Light on Astronomical Rationale), are highly significant writings in Indian mathematical literature. After giving the meaning of each textual verse or group of verses from these two texts, Śaṅkara sets out the background and evolution of the enunciation commented on, the different aspects thereof, and the step by step derivation of the relevant formula or procedure. This exposition of rationale, couched in simple verses and often running to several pages, serves to give an exposure to Indian mathematical thinking and open up the normally unexpressed mental working of the Indian mathematician, which has led most modern historians of mathematics to presume that much of Indian mathematics and astronomy was borrowed and not original. The rationale elaborated here relates among other things, to the summations of series, the circle and the irrationality of π and pulverization in the field of mathematics; and the theory of epicycles, ascensional differences, rising of the signs, and problems based on the gnomic shadow, in astronomy. These commentaries are valuable also for the information on earlier authors like Mādhava of Saṅgamagrāma and their theories.

K.V. SARMA

REFERENCES

Līlāvatī of Bhāskarācārya with *Kriyākramakarī of* Śaṅkara and Nārāyaṇa. Ed. K.V. Sarma. Hoshiarpur: Vishveshvaranand Vedic Research Institute, 1975.

Tantrasaṅgraha of Nīlakaṇṭha Somayāji with *Yuktidīpikā* and *Laghuvivṛti* of Śaṅkara. Ed. K.V. Sarma. Hoshiarpur: Vishveshvaranand Vishva Bandhu Institute of Sanskrit and Indological Studies, Punjab University, 1977.

See also: Nīlakaṇṭha Somyaji – Rationale in Indian Mathematics

ŚATĀNANDA Śatānanda, son of Śaṅkara and author of the popular astronomical manual *Bhāsvatī*, was a resident of Puruṣottamapurī, the modern city of Puri, in Orissa. The *Bhāsvatī* was written in the Śaka year 1021, AD 1099. The epochal constants are also given for Puri, instead of for Ujjain as is normal in astronomical manuals. In 82 pithy verses distributed in eight chapters, the *Bhāsvatī* sets out the several computations required for preparing the daily almanac. There is a traditional statement that eclipses computed according to the *Bhāsvatī* would be exact. Towards the beginning of the work, the author avers that he was composing the work on the basis of the (Old) *Sūryasiddhānta* condensed by Varāhamihira in his *Pañcasiddhāntikā*. There is also a pun in the word *Bhāsvatī*, based on the above, since *bhāsvān* means *Sūrya* (Sun). There is a pun on the word *Śatānanda* as well; the literal meaning of the word is "one who revels in hundreds", and in this work the author used the centismal system for commencing the epochal position and specified several of the multipliers and divisors in computation in terms of hundreds. There are several recensions of the work and some manuscripts add an *Uttara-Bhāsvatī*.

Śatānanda introduced certain other innovations in his work. For the computation of Mean planets, he took not the *ahar-gaṇa* (number of elapsed days), but the *Varṣa-gaṇa* (number of elapsed years). As the commencement of the year, he did not take the Mean *Meṣādi* (Aries ingress), but the True *Meṣādi*, which is advantageous in certain aspects. Another speciality of Śatānanda's is that, as mentioned above, he adopted the centismal system for commencing epochal positions and for specifying multipliers and divisors. The positions of the sun and the moon are stated in terms of *nakṣatras* (constellations) and not in *rāśis*, which is normally the case in texts of this type. Then again, Śatānanda took AD 528 as the zero precession year and the rate of precession as one minute per annum.

The popularity of Śatānanda's work in North and Northeast India is attested to by the presence of a large number of manuscripts of *Bhāsvatī* in these regions and by the fact that most of his commentators hail from there. Some of the principal commentators are Aniruddha (*Śiśubodhinī*, 1495), Mādhava (*Mādhavī*, 1525), Acyuta (*Ratnamālā*, ca. 1530), Kuvera Miśra (*Ṭīkā*, 1685), Rāmakṛṣṇa (*Tattvaprakāśikā*, 1739), and Yogindra (1742).

K.V. SARMA

REFERENCES

Dikshit, S.B. *Bharatiya Jyotish Sastra (History of Indian Astronomy)* Trans. R.V. Vaidya. Pt. II: *History of Astronomy During the Siddhantic and Modern Periods*. Calcutta: Positional Astronomy Centre, Indian Meterological Department, 1981.

Dvivedi, Sudhakara. *Gaṇaka Taraṅgiṇī: Lives of Hindu Astronomers.* Ed. Padmakara Dvivedi. Benares: Jyotish Prakash Press, 1933, pp. 33–34.

Pingree, David. *Jyotiḥśāstra: Astral and Mathematical Literature.* vol. IV, Fasc. 4 of *A History of Indian Literature.* Ed. Jan Gonda. Wiesbaden: Otto Harrassowitz, 1981.

Śrīman-Śatānanda-viracita Bhāsvatī with Sanskrit and Hindi commentaries by Matr Prasad Pandeya. Ed. Ramajanma Mishra. Varanasi: Chaukhamba Sanskrit Samsthan, 1985.

See also: Varāhamihira – *Sūryasīddhānta* – Lunar Mansions – Precession of the Equinoxes – Astronomy in India

SCIENCE AS A WESTERN PHENOMENON Philosophers, historians, and sociologists of science all accept as a basic postulate that science is essentially Western. This postulate is still conditioning contemporary scientific ideologies. This article analyzes the characteristics, history, and validity of this doctrine by means of a confrontation with one of the non-Western scientific contributions: science written in Arabic.

Classical science is essentially European, and its origins are directly traceable to Greek philosophy and science; this tenet has survived intact through numerous conflicts of interpretation over the last two centuries. Almost without exception, the philosophers accepted it. Kant, as well as Comte, the neo-Kantians as well as the neopositivists, Hegel as well as Husserl, the Hegelians and the phenomenologists as well as the Marxists, all acknowledge this postulate as the basis of their interpretations of Classical Modernity. Even until our time, the names of Bacon, Descartes, and Galileo (sometimes omitting the first, and sometimes adding a number of others), are cited as so many markers on the road to a revolutionary return to Greek science and philosophy. This return was understood by all to be both the search for a model and the rediscovery of an ideal. One might impute this unanimity to the philosophers' zeal to pass beyond the immediate data of history, to their wish for radical insight, or to their effort to seize what Husserl describes as "the original phenomenon *(Urphänomen)* which characterizes Europe from the spiritual point of view". One would expect that the position taken by those who have stuck more closely with the facts of the history of science would be quite different, but such is not the case. This same postulate is adopted by the historians of science as a point of departure for their work, and especially for their interpretations. Whether they interpret the advent of classical science as the product of a break with the Middle Ages, whether they defend the thesis of continuity without breaking or cutting, or whether they adopt an eclectic position, the majority of historians agree in accepting this postulate more or less implicitly.

Today, in spite of the works of many scholars on the history of Arabic and Chinese science, in spite of the wide representation of non-Western scientists in *Dictionary of Scientific Biography*, the works of the historians rest on an identical fundamental concept: in its modernity as well as in its historical context, classical science is a work of European man alone. Furthermore, it is essentially the means by which this branch of humanity is defined. Occasionally the existence of a certain practical science in other cultures might be acknowledged; nevertheless, it rests outside history, or is integrated into it only to the extent of its contributions to the essentially European sciences. These are only technical supplements which do not modify the intellectual configuration or the spirit of the latter in any way. The image given of Arabic science constitutes an excellent illustration of this approach. Essentially it consists of a conservatory of the Greek patrimony, transmitted intact or enriched by technical innovation to the legitimate heirs of ancient science. In all cases, scientific activity outside Europe is badly integrated into the history of the sciences; rather, it is the object of an ethnography of science whose translation into university study is nothing more than Orientalism.

The effects of this doctrine are not limited to the domain of science, its history, and its philosophy. It is at the center of the debate between modernism and tradition. As was the case in eighteenth-century Europe, we find in certain Mediterranean and Asian countries of today, that science (which is qualified as European) is identified with modernism. Our purpose here is not to redress wrongs, nor to oppose to that science qualified as European an alleged Eastern science. It is simply a matter of understanding the significance of the European determination of the concept of classical science, grasping the reasons for it and measuring its importance.

We shall begin by sketching the history of this view of European science and then estimate its effects. We shall limit ourselves to posing the problem and advancing several hypotheses, and we also add these two restrictions: the only non-European science considered is one which was produced by various cultures, by scholars of different beliefs and religions, all of whom wrote their science principally, if not exclusively, in Arabic. As for the tenets of the history of the sciences, we shall most often cite those of the French historians.

The concept of a European science is already present in the works of the historians and philosophers of the eighteenth century. In the debate of the Ancients and the Moderns, scholars and philosophers referred to science to define modernity where one combines reason and experience. Historical induction intended to give its concrete determinations to this dogmatic debate, so as to render the superiority of the Moderns indisputable. But the West was already being identified as Europe, and "Oriental wisdom" was already counterpoised against the natural philosophy of the post-Newtonian West, such as we find in Montesquieu's *Persian Letters* (1721).

Classical science is European and Western only to the degree that it represents a stage in the continuous and regular development of humanity. The *Discours Préliminaire* of Abbé Bossut in Diderot's *Encyclopédie Méthodique* offers an illustration of this concept. Dividing the history of the progress of the exact sciences into three periods, this tableau allows conjecture, alleged facts, and facts to intermingle. Its initial postulate is that "... all of the eminent peoples of the ancient world liked and cultivated mathematics. The most distinguished among them are the Chaldeans, the Egyptians, the Chinese, the Indians, the Greeks, the Romans, the Arabs, etc. ... in modern times, the western nations of Europe". Classical science is European and Western because, writes Abbé Bossut, "... the progress made by the western nations of Europe in the sciences from the sixteenth century to our times utterly effaces those of other peoples."

The concept of Western science changed in nature and extent at the turn of the nineteenth century. With what Edgar Quinet called the "Oriental Renaissance", the conceptualization was completed in its anthropological dimension in the last century. This Oriental Renaissance ended by discrediting science in the East.

If it is true that the eighteenth-century concept still survived here and there, from the first years of the nineteenth century the materials and ideas of Oriental studies contributed the most to the makeup of the historical themes of the different philosophies. In Germany as well as in France, the philosophers adopted Oriental studies for diverse reasons in accordance with an identical representation: the East and the West do not oppose each other as geographical, but as historical positivities; this opposition is not limited to a period of history, but goes back to the essence of each term. In this regard, *Lessons on the History of Philosophy* and other works of Hegel can be invoked. Also at this time, as is shown by the French Restoration philosophers, the themes of the "call of the East" and the "return to the Orient" appear, which translate as a reaction against science, and more generally, against Rationalism. But it is with the advent and growth of the German philological school that the notion of science as a Western phenomenon was regarded as having been endowed with the scientific, and no longer purely philosophical, support which had been lacking until then.

This influence also extended into mythological and religious studies. For example, Friedrich von Schlegel distinguishes two classes of language: the flexional Indo-European languages, and others. The former are "noble", the latter less perfect. Sanskrit, and consequently, German, considered the closest to it, is "... a systematic language and perfect from its conception"; it is "... the language of a people not composed of brutes, but of clear intelligence". There is nothing surprising in this; with the advent of the German school we are already in the realm of classifying mentalities. From now on everything is in place for effecting the passage from the history of languages to history-through-languages.

The comparative study of religions and myths is developed around the middle of the century by A. Kuhn and Max Müller in particular. The classification of mentalities is perfected. It is from the basis of these tenets and dating from this period that one of the most important efforts to establish the notion of science as Western and European in an allegedly scientific manner is elaborated. This project achieves its full extent in France in the work of Ernest Renan.

For Renan, civilization is divided between Aryans and Semites; the historian only has to evaluate their contributions in a differential and comparative manner. The notion of race would constitute the foundation of historiography. By race, one meant the whole of the "... aptitudes and instincts which are recognizable solely through linguistics and the history of religions". In the last analysis, it is for reasons attributable to the Semitic languages rather than the Semites themselves that they did not and could not have either philosophy or science. "The Semitic race", writes Renan, "is distinguished almost exclusively by its negative traits: it has neither mythology, epic poetry, science, philosophy, fiction, plastic arts nor civil life". The Aryans, whatever their origin, define the West and Europe at one and the same time. Arabic science is, "... a reflection of Greece, combined with Persian and Indian influences".

The historians of science borrowed not only their representation of the Western essence of science from this tradition, but also some of their methods for describing and commenting on the evolution of science. Thus, they applied themselves to discovering the concepts and methods of science and to following their genesis and propagation through philological analyses of the terms and on the basis of the texts at their disposal. Like the historian of myths or of religion, the historian of the sciences must be a philologist as well. In France, the situation is such that philosophers borrow Renan's interpretation and, often, even his terminology. Even though this brand of anthropology has already been abandoned by historians, they nevertheless preserve and propagate a series of inferences engendered by it. These can be enumerated as follows:

(1) Just as science in the East did not leave any consequential traces in Greek science, Arabic science has not left any traces of consequence in classical science. In both cases, the discontinuity was such that the present could no longer recognize itself in its abandoned past.

(2) Science subsequent to that of the Greeks is strictly dependent upon it. According to Duhem, "... Arabic science only reproduced the teachings that it received from Greek science". In a general fashion, Tannery reminds us that the more one examines the Hindu and Arabic scholars, "... the more they appear dependent upon the Greeks ... (and) ... quite inferior to their predecessors in all respects".

(3) Whereas Western science addresses itself to theoretical fundaments, Oriental science, even in its Arabic period, is defined essentially by its practical aims.

(4) The distinctive mark of Western science is its conformity to rigorous standards; in contrast, Oriental science in general, and Arabic science in particular, lets itself be carried away by empirical rules and methods of calculation, neglecting to verify the soundness of each step on its path. The case of Diophantus illustrates this idea perfectly: as a mathematician, said Tannery, "... Diophantus is hardly Greek". But when he compares the *Arithmetics* of Diophantus to Arabic algebra, Tannery writes that the latter "... in no way rises above the level achieved by Diophantus."

(5) The introduction of experimental norms which, according to historians, totally distinguishes Hellenistic science from classical science, is solely the achievement of Western science.

Thus it is to Western science alone that we owe both the concept and experimentation. We are not going to oppose this ideology to another. We propose simply to confront some of these elements with the facts of the history of science, beginning with algebra and concluding with the crucial problem of the relationships between mathematics and experimentation.

Algebra

As with the other Arabic sciences, algebra had practical aims, a flair for calculation, and an absence of rigorous standards. It is precisely this that allowed Tannery to make his statement. Bourbaki took this as his authorization to exclude the Arabic period when he retraced the evolution of algebra. The historical writings of the modern mathematician Dieudonné are significant; between the Greek prehistory of algebraic geometry and Descartes, he finds only a void, which, far from being frightening, is ideologically reassuring. Some historians cite al-Khwārizmī, his definition of algebra and his solution of the quadratic equation, but it is generally to reduce Arabic algebra to its initiator. This restriction misconstrues the history of algebra, which in actuality does not show a simple extension of the work of al-Khwārizmī in the West, but an attempt at theoretical and technical overtaking of his achievements. Moreover, this overtaking is not the result of a number of individual works, but the outcome of genuine traditions. The first of these traditions had conceived the particular project of arithmetizing the algebra inherited from al-Khwārizmī and his immediate successors. The second one, in order to surmount the obstacle of the solution by radicals of third and fourth degree equations, formulated in its initial stage a geometric theory of equations, subsequently to change viewpoint and study known curves by means of their equations. In other words, this tradition engaged itself explicitly in the first research in algebraic geometry.

As we have said, the first tradition had proposed arithmetizing the inherited algebra. This theoretical program was inaugurated at the end of the tenth century by al-Karajī, and is thus summarized by one of his successors, al-Samaw'al (d. 1176): "to operate on unknowns as the arithmeticians work on known quantities".

The execution is organized into two complementary stages. The first is to apply the operations of elementary arithmetic to algebraic expressions systematically; the second is to consider algebraic expressions independently from that which they can represent so as to be able to apply them to operations which, up to that point, had been restricted to numbers. Nevertheless, a program is defined not only by its theoretical aims, but also by the technical difficulties which it must confront and resolve. One of the most important of these was the extension of abstract algebraic calculation. At this stage, the mathematicians of the eleventh and twelfth centuries obtained some results which unjustly are attributed to the mathematicians of the fifteenth and sixteenth centuries. Among these are the extension of the idea of an algebraic power to its inverse after defining the power of zero in a clear fashion, the rule of signs in all its general aspects, the formula of binomials and the tables of coefficients, the algebra of polynomials, and above all, the algorithm of division, and the approximation of whole fractions by elements of the algebra of polynomials.

In a second period, the algebraists intended to apply this same extension of algebraic calculation to irrational algebraic expressions. Al-Karajī questioned how to operate by means of multiplication, division, addition, subtraction, and extraction of roots on irrational quantities. To answer this question the mathematicians gave, for the first time, an algebraic interpretation of the theory contained in Book X of the *Elements*. This book was regarded by Pappus, as well as by Ibn al-Haytham much later, as a geometry book, because of the traditional fundamental separation between continuous and discontinuous magnitudes. With the school of al-Karajī, a better understanding of the structure of real algebraic numbers is achieved.

In addition, the works of this algebraic tradition opened the route to new research on the theory of numbers and numerical analysis. An examination of numerical analysis, for example, reveals that after renewing algebra through arithmetic, the mathematicians of the eleventh and twelfth centuries also effected a return movement to arithmetic to search for an applied extension of the new algebra. It is true that the arithmeticians who preceded the algebraists of the eleventh and twelfth centuries extracted square and cube roots, and had formulas of approximation for the same powers. But, lacking an abstract algebraic calculation, they could generalize neither their results, their methods, nor their algorithms. With the new algebra, the generalization of algebraic calculation became a constituent of numerical analysis which, until then, had only been a sum of procedures, if not prescriptions. It is in the course of this double movement which is established between algebra and arithmetic that the mathematicians of the eleventh and twelfth centuries achieved results which are still wrongly attributed to the mathematicians of the fifteenth and sixteenth centuries. This is the case with the method attributed to Viète for the resolution of numerical equations, the method ascribed to Ruffini-Horner, the general methods of approximation, in particular that which D. T. Whiteside designates by the name of al-Kāshī-Newton, and finally, the theory of decimal fractions. In addition to methods, which were to be reiterative and capable of leading in a recursive manner to approximations, the mathematicians of the eleventh and twelfth centuries also formulated new procedures of demonstration such as complete induction.

We have just seen that the concept of polynomials is among those elaborated by the algebraist arithmeticians from the end of the tenth century. This tradition of algebra as the "arithmetic of unknowns", to use the expression of the time, opened the road toward another algebraic tradition which was initiated by ʿUmar al-Khayyām (eleventh century), and renewed at the end of the twelfth century by Sharaf al-Dīn al-Ṭūsī. While the former formulated a geometric theory of equations for the first time, the latter left his mark on the beginnings of algebraic geometry.

The immediate predecessors to al-Khayyām, such as al-Bīrūnī, al-Māhānī, and Abū al-Jūd, had already been able, in contrast to the Alexandrian mathematicians and precisely because of the concept of the polynomial, to treat the problems of solids in terms of third degree equations. But al-Khayyām was the first to address these unpondered questions: can one reduce the problems of straight lines, planes, and solids to equations of a corresponding degree, on the one hand, and on the other, re-order the group of third degree equations to seek, in the absence of a solution by factoring, solutions which can be reached through means of the intersection of auxiliary curves? To answer these questions, al-Khayyām is led to formulate the geometric theory of equations of a third or lesser degree. His successor, al-Ṭūsī, did not delay in changing perspective; far from adhering to geometric figures, he thought in terms of functional relations and studied curves by means of equations. Even if al-Ṭūsī still solved equations by means of auxiliary curves, in each case the intersection of the curves is demonstrated algebraicly by means of their equations. This is important, since the systematic use of these proofs introduces into the practice instruments which were already available to the mathematical analysts of the tenth century: affine transformations, the study of the maxima of algebraic expressions, and with the aid of what will later be regarded as derivatives, the study of the upper bounds and lower bounds of roots. It is in the course of these studies and in applying these methods that al-Ṭūsī grasps the importance of the discriminant of the cubic equation and gives the so-called Cardan formula just as it is found in the *Ars Magna*. Finally, without enlarging any further on the results which were obtained, we can say that both on the level of results as well as that of style, we find al-Khayyām and al-Ṭūsī fully in the field allegedly pioneered by Descartes.

If we exclude these traditions and justify this exclusion by invoking the practical and computational aims of the Arab mathematicians and an absence of rigorous standards of proof in their work, we can say that the history of classical algebra is the work of the Renaissance.

Among the mathematical disciplines, algebra is not a unique case. To varying degrees, trigonometry, geometry, and infinitesimal determinations are likewise illustrative of the preceding analysis. In a more general sense, optics, statistics, mathematical geography, and astronomy are also no exception. Recent works in the history of astronomy render Tannery's understanding of the Arab astronomers and the interpretations which he gives of them manifestly outmoded, if not erroneous. But since we assigned ourselves the task of examining the doctrine of the Western nature of classical science, we shall restrict our discussion to an essential component of this doctrine, experimentation.

Experimentation

In fact, is not the cleavage between the two periods of Western science, the Greek period and the Renaissance, often marked by the introduction of experimental norms? Undoubtedly the general agreement of the philosophers, historians, and sociologists of science stops here; the divergences become apparent as soon as they attempt to define the meaning, the implications, and the origins of these experimental norms. The origins are linked in one case to the current of Augustinian-Platonism, in another to the Christian tradition, and particularly to the dogma of Incarnation, in a third case to the engineers of the Renaissance, in a fourth to the *Novum Organum* of Francis Bacon, and finally, in a fifth, to Gilbert, Harvey, Kepler, and Galileo. Some of these attitudes superimpose upon one another, become entangled or contradictory, but they all converge on one point: the occidental nature of the new norms.

Nevertheless, as early as the nineteenth century, historians and philosophers such as Alexander von Humboldt in Germany and Cournot in France diverge from this predominating position to attribute to the Arab period the origins of experimentation. It is difficult to analyze the origins or the beginnings of experimentation correctly, since no study has been made of the interrelations of the different traditions and the different themes to which the concept of experimentation has been applied. Perhaps it would be in writing such a history, especially a history of the term itself, that one could give an accounting of the multiplicity of uses and ambiguities of the concept. For this analysis two histories are needed: the history of the relationship between art and science and that of the links between mathematics and physics.

With the history of the relationship between science and art, we are in a position to understand when, why, and how it became accepted that knowledge can emanate from demonstrations and from the rules of practice at the same time, and that a body of knowledge possesses the stature of a science while, at the same time, it is conceived in its possibilities of practical realization with an external purpose. The traditional opposition between science and art seems likely to be the work of the intellectual currents of the Arabic period. Certainly one fact is striking: whether we are dealing with Muslim traditionalists, rationalist theologians, scholars of different fields, and even philosophers of the Hellenistic tradition such as al-Kindī or al-Fārābī, all contribute to the weakening of the traditional differentiation between science and art. In other respects, this general trait is at the origin of the opinion of some historians regarding the practical spirit and realistic imagination of the Arab scholars. Knowledge is accepted as scientific without its conforming either to the Aristotelian or the Euclidean scheme. This new concept of the stature of science promoted the dignity of scientific understanding of disciplines which traditionally were confined to the domain of art, such as alchemy, medicine, pharmacology, music, or lexicography. Whatever might be the importance of this concept, it could only lead to an extension of empirical research and to a diffuse notion of experimentation. One does witness the multiplication and systematic use of empirical procedures: the classifications of the botanists and the linguists, the control experiments of the doctors and alchemists, and the clinical observations and comparative diagnostics of the physicians. But it was not until new links were established between mathematics and physics that such a diffuse notion of experimentation acquired the dimension that determines it, a regular and systematic component of the proof. Primarily it is in Ibn al-Haytham's work in the field of optics where the emergence of this new dimension can be perceived.

With Ibn al-Haytham the break is established with optics as the geometry of vision or light. Experimentation had indeed become a category of the proof. The successors of Ibn al-Haytham, such as al-Fārisī, adopted experimental norms in their optical research, such as that performed on the rainbow. What did Ibn al-Haytham understand by experimentation? We will find in his work as many meanings of this word and as many functions served by experimentation as there are links between mathematics and physics. A thorough look at his texts indicates that the term and its derivatives belong to several superimposed systems, and are not likely to be discerned through simple philological analysis. But if attention is fixed on the content rather than the lexical form, one can distinguish several types of relationships between mathematics and physics which allow one to spot the corresponding functions of the idea of experimentation. In fact, the links between mathematics and physics are established in several ways; even if they are not specifically treated by Ibn al-Haytham, they underlie his work and are amenable to analysis.

As for the field of geometric optics, which was reformed by Ibn al-Haytham himself, the only link between mathematics and physics is a similarity of structures. Owing to his definition of a light ray, Ibn al-Haytham was able to formulate his theory on the phenomena of propagation, including the important phenomenon of diffusion, so that they relate perfectly to geometry. Then several experiments were devised to assure technical verification of the propositions. Experiments were designed to prove the laws and rules of geometrical optics. The work of Ibn al-Haytham attests to two important facts which are often insufficiently stressed: first of all, some of his experiments were not simply designed to verify qualitative assertions, but also to obtain quantitative results; in the second place, the apparatus devised by Ibn al-Haytham, which was quite varied and complex, is not limited to that of the astronomers.

In physical optics one encounters another type of relationship between mathematics and physics and therefore a second meaning of experimentation. Without opting for an atomistic theory, Ibn al-Haytham states that light, or as he writes, "the smallest of the lights", is a material thing, external to vision, which moves in time, changes its velocity according to its medium, follows the easiest path, and diminishes in its intensity depending on its distance from its source. Mathematics is introduced into physical optics at this stage by means of analogies established between the systems of movement of a heavy body and those of the reflection and refraction of light. This previous mathematical treatment of the concepts of physics permitted them to be transferred to an experimental plane. Although this situation on the experimental level might be somewhat approximate in nature, it nevertheless furnishes a level of existence to ideas which are syntactically structured, but semantically indeterminate, such as Ibn al-Haytham's scheme of the movements of a projectile.

A third type of experimentation, which was not practiced by Ibn al-Haytham himself but was made possible by his own reform and his discoveries in optics, appears at the beginning of the fourteenth century in the work of his successor al-Fārisī. The links established between mathematics and physics aim to construct a model and to reduce by geometric means the propagation of light in a natural object to its propagation in an artificial object. The problem is to define for propagation, between the natural and the artificial object, some analogical correspondences which were genuinely certain of mathematical status. For example, they built a model of a massive glass sphere filled with water to explain the rainbow. In this case, experimentation serves the function of simulating the physical conditions of a phenomenon that can be studied neither directly nor completely. The three types of experimentation studied all reveal themselves both as a means of verification and as furnishing a plane of material existence to ideas which are syntactically structured. In the three cases, the scientist must realize an object physically in order to handle it conceptually. Thus, in the most elementary example of rectilinear propagation, Ibn al-Haytham does not consider any arbitrarily chosen opening of a black box, but rather specific ones, in accordance with specific geometric relationships, in order to realize as precisely as possible his concept of a ray.

To recapitulate several points:

(1) The tenet of the Western nature of classical science which was launched in the eighteenth century owes to the orientalism of the nineteenth century the image that we now recognize.

(2) On the one hand, the opposition between East and the West underlies the critique of science and rationalism in general; on the other, it excludes the scientific production of the Orient from the history of science both *de facto* and *de jure*. An absence of rigor is invoked, as well as the computational attributes and the practical aims of science written in Arabic, to justify this effective debarment from the history of science.

(3) This tenet reveals a disdain for the data of history as well as a creative capacity for ideological interpretation, which are admitted as evidence for ideas that raise many more problems than they solve. Thus we have the notion of a Scientific Renaissance, when in several disciplines everything indicates that there was merely a reactivation. These pieces of pseudo-evidence quickly become conceptual bases for a philosophy or sociology of science, as well as the departure point for theoretical elaborations in the history of science.

We must ask ourselves if the moment has not arrived to abandon this characterization of classical science and its still lively traces in the writing of history, to restore to the profession of the historian of science the objectivity required of it, to ban the clandestine importation and diffusion of uncontrolled ideologies, to refrain from all reductionist tendencies which favor similarities at the expense of differences, and to be wary of miraculous events in history. The neutrality of the historian is not an a priori ethical value; it can only be the product of patient work which will not be duped by the myths which the East and West have engendered. Above all, it is necessary to cast out the periodization everywhere admitted in the history of science. The term used for classical algebra or classical optics, for example, will integrate the works which extended from the tenth to the seventeenth centuries. Consequently, it will re-align not only the idea of the classical sciences but also that of medieval science. The classical sciences will then reveal themselves as the product of the Mediterranean which was the hub of exchanges among all civilizations of the ancient world. Only then will the historian of science be able to enlighten the debate over modernism and tradition.

Science as a Western Phenomenon: Postscript

The sixteen years since the publication of the original French edition of the text above has been a very fertile period for the study of the history of Islamic science. Indeed, we have witnessed an unprecedented rebirth of this discipline. Texts have been written and translated, new collections have appeared, and specialized reviews and journals have been published. These have offered historians the possibility of developing and comparing their research findings with facts. The task remains huge, and we are only at the beginning, but at least this new growth of information puts to rest the argument of ignorance.

With all this new information, one would have expected historians and philosophers to rectify the impressions and ideas they had inherited from the nineteenth century. One would have thought that the doctrine of Western science which we have described and analyzed here would have disappeared along with the props on which that doctrine was based. Indeed we were beginning to see a growing tendency to break with this doctrine and its implications. Then, for reasons which are extraneous to science and its history, images of Islamic society — if not of Islam itself — arose, according to which it was seen as irrational and intolerant and thus a society foreign to science. The aging doctrine was naturally given new life because of these images. How is it possible, under these conditions, to reconcile such an image of Islamic society with scientific results obtained from the heart of that same society? It was enough to back up the preceding doctrine with another, the doctrine of double marginality: with regard to the society which saw the development of science, and with regard to the history of the sciences. Thus, one could still write in 1992, "We must remember that at an advanced level the foreign sciences had never found a stable institutional home in Islam", or "Greek learning never found a secure institutional home in Islam, as it was eventually to do in the universities of medieval Christendom" (Lindberg, 1992). As for the second marginality, we have already described how it works. Thus, we are back where we started and the doctrine of the Westernness of science is saved. Undoubtedly, these ideological views are beginning to give way, weakened by new research findings. And even if they are still capable of slowing down the acceptance of facts, it is not widespread, and it certainly will not last much longer.

ROSHDI RASHED

REFERENCES

al-Khayyām. *L'algèbre d'Omar Alkhayyāmī.* Ed. F. Woepcke. Paris: B. Duprat, 1851

al-Samaw'al. *al-Bāhir en algèbre.* Ed. R. Rashed. Damascus: University of Damascus, 1972.

Cournot, A. A. *Considérations sur la marche des idées et des évenements dans les temps modernes.* Paris: Vrin, 1973.

Dieudonné, J. *Cours de géométrie algébrique.* Paris: Presses universitaires de France, 1974.

Duhem, P. *Le système du monde.* 10 vols. Paris: A. Hermann, 1913–1959.

Encyclopédie méthodique. Paris: Pancoucke, 1784.

Hegel, G.W.F. *Lectures on the History of Philosophy.* New York: Humanities Press, 1968.

Kojève, A. "The Christian Origin of Modern Science." In *Mélanges Alexandre Koyré.* Paris: Hermann,1964, vol. II, pp. 295–306.

Lindberg, David. *The Beginning of Western Science.* Chicago: University of Chicago Press, 1992, p. 182.

Montesquieu. *Oeuvres complètes.* Paris: Gallimard, 1964.

Rashed, R. "L'extraction de la racine $n^{ième}$ et l'invention des fractions décimales." *Archive for History of Exact Sciences* 18(3): 191, 1978.

Rashed, R. "Résolution des équations numériques et algèbre: Sharaf al-Dī al-Tasi, Viète." *Archive for History of Exact Sciences* 12(3): 244, 1974.

Renan, E. *Histoire général et système comparé des langues sémitiques.* Paris: Imprimerie impériale, 1863.

Renan, E. *Nouvelles considérations sur le caractère générale des peuples sémitiques.* Paris: Imprimerie impériale, 1859.

Sarton, G. *The Incubation of Western Culture in the Middle East.* Washington: Library of Congress, 1951.

Schlegel, F. "Essai sur la langue et la philosophie des Indiens." In *The Philosophy of History.* London: Bell & Daldy, 1871.

Tannery, P. *La Géométrie grecque.* Paris: Gauthier-Villars, 1887.

See also: Western Dominance – East and West – Algebra – Ibn al-Haytham – Optics – ‘Umar al-Khayyām

SEKI KOWA Seki Kowa, also called Takakazu, was born about 1642, probably in Fujioka, Gumma, Japan. Seki is his surname: Kowa is the Chinese reading of the characters; the Japanese reading is Takakazu. He was born into a samurai warrior family, the second son of Uchiyama Nagaakira; he became a son-in-law of Seki Gorozaemon and became a member of the Seki family. In 1678, he became an auditor of the Shogun's family, then became a landlord of a 300 person village (300 *Koku*). The family records of landlords were normally kept by the Shogun government. But since Seki Kowa's son-in-law Shinshichiro lost his position, owing to his gambling activities, we cannot fix even his birth year exactly.

Seki studied mathematics under Takahara Yoshitane. Takahara Yoshitane was a disciple of Mori Shigeyoshi, who wrote the *Warizansho* (Books of Division, 1622) the second mathematical book in Japanese. But Takahara Yoshitane was little known for his mathematical works. Basically Seki taught himself mathematics, using famous texts of Chinese and Japanese mathematicians, such as the *Yang Hui Suan Fa* (Yang Ḥui's Method of Computation, 1275), and Sawaguchi Kazuyuki's *Sampo Ketsugi Sho* (Solving Mathematical Questions, 1659). He amended works of other mathematicians, and founded the *Seki-ryu* school. He is generally considered the founder of Japanese mathematics.

The *Tianyuanshu* (lit. Celestial Element Method, Chinese Algebra System) had already been introduced from China by Zhu Shijie's *Suan Xue Qi Meng* (Introduction to Mathematics Studies, 1299). This system was for the *suanzi* (also called *chou*: counting rods) in the Song and Yuan dynasties in China. In the Ming dynasty, however, new counting tools such as the *suanpan* (abacus) became popular, since no mathematicians understood the *Tianyuanshu* even in China. Japanese mathematicians taught the Tianyuanshu using a Korean edition of the *Suan Xue Qi Meng*, and Sawaguchi Kazuyuki was the first Japanese mathematician to master it.

Traditional Chinese mathematics used counting tools,

Table 1: Shosa-ho's table, Chapter 1 of the *Katsuyo Sampo* (Essential Mathematics, 1712)

	1	1											
x^2	1	2	(1)										
x^3	1	3	3	(1)									
x^4	1	4	6	4	(1)								
x^5	1	5	10	10	5	(1)							
x^6	1	6	15	20	15	6	(1)						
x^7	1	7	21	35	35	21	7	(1)					
x^8	1	8	28	56	70	56	28	8	(1)				
x^9	1	9	36	84	126	126	84	36	9	(1)			
x^{10}	1	10	45	120	210	252	210	120	45	10	(1)		
x^{11}	1	11	55	165	330	462	462	330	165	55	11	(1)	
x^{12}	1	12	66	220	495	792	924	792	495	220	66	12	(1)
power	1	2	3	4	5	6	7	8	9	10	11	12	13
(+1)	1	$\frac{1}{2}$	$\frac{1}{6}$	0	$-\frac{1}{30}$	0	$-\frac{1}{42}$	0	$-\frac{1}{30}$	0	$\frac{5}{66}$	0	

suanzi, or *suanpan*; they had no algebraic symbols. The position of rods (or beads of the *suanpan*) indicated numbers and the power of unknown numbers. Therefore this system could only operate with unknown numbers. In his book *Kaiho Sanshiki* (Formulae of Solving Higher Degree Equations), Seki Kowa created the *Tenzan-jutsu* (Japanese algebra system), which could operate with unknown formulae, making it possible to solve complicated problems.

Seki called his method *Endan-jutsu* or *Boshoho* (Methods of Writing by the Side). Matsunaga Yoshisuke (1693–1744), a mathematician of the *Seki-ryu* school, renamed it the *Tenzan-jutsu*. Seki indicated unknown numbers using Chinese characters, and created the symbols of calculations and powers.

His works cover the following categories:

(1) Japanese algebra system;
(2) solution of higher degree equations;
(3) properties (e.g. number of solutions) of higher degree equations;
(4) infinite series;
(5) approximate values of fractions;
(6) indeterminate equations;
(7) the method of interpolation;
(8) obtaining Bernoulli numbers by *Ruisai Shosa-ho*;
(9) computing area of polygons;
(10) principle of the circle;
(11) Newton's formula by *Kyusho* method;
(12) computing the area of rings;
(13) conic curved lines;
(14) magic squares and magic circles; and
(15) *Mamakodate* (Josephus question) and *Metsuke-ji* (the game of finding a Chinese character).

Many of his mathematical inventions were analogous to the mathematics of European mathematicians of his and later centuries. He discovered determinants ten years before Leibnitz, solved numerical algebraic equations by a method similar to the Ruffini-Horner method, discovered the conditions for the existence of positive and negative roots of polynomials, discovered a method for finding maxima and minima, invented continued fractions, solved some Diophantine fractions, and discovered the Bernoulli numbers. In 1712, one year before Bernoulli's *Ars Conjectandi* (1713), Seki obtained Bernoulli numbers by *Ruisai Shosa-ho*. The example in Table 1 is characteristic of his mathematics.

Using this table, Seki computed to higher powers and obtained the fraction numbers of the last line of the table:

$$\lambda_0 = 1, \quad \lambda_1 = \frac{1}{2}, \quad \lambda_2 = \frac{1}{6}, \quad \lambda_3 = 0, \ldots$$

These numbers fulfil the following formula:

$$\sum_{i=1}^{n} i^{\sigma-1} = \frac{1}{p}\sum_{i=0}^{p-1}\left(\lambda_1\binom{p}{i}n^{\sigma-1}\right).$$

These are Bernoulli's numbers.

In the *Katsuyo sampo*, he also calculated the number π and the volume of a sphere by an original integral method called *enri* (method of circles), found Newton's interpolation formula, discovered some properties of ellipses and of spirals, and discovered what we call the Pappus-Gulden theorem. He computed much larger numbers than other mathematicians had done before, and obtained general rules by the inductive method.

Seki died in Edo (now Tokyo), on December 5, 1708 [October 24, 1708, in the Japanese calendar].

SHIGERU JOCHI
BORIS ROSENFELD

REFERENCES

Primary sources

Collected Works. English translation by J. Sudo. Osaka: Kyoiku tosho, 1974.

Hatsubi Sampo (Mathematical Methods for Finding Details). Edo, 1674.

Katsuyo Sampo (Essential Mathematics). Edo, 1712.

Juji Hatsumei (Comments of the Works and Days Calendar) 1680.

Jujireki-kyo Rissei no Ho (Methods of Manual Tables of the Works and Days Calendar). 1681.

Taisei Sampo (Complete Works of Seki Kowa). 1710.

Secondary sources

Fujiwara, Matsusaburo. *Meiji-zen Nihon Sugakushi* (Mathematics in Japan before the Meiji Era). 5 vols. Tokyo: Iwanami Shoten, 1954.

Hirayama, Akira. *Seki Kowa*. Tokyo: Koseikaku Kosei-sha,1959.

Hirayama, Akira et al, eds. *Seki Kowa Zenshu*. Osaka: Osaka Kyoiku Tosho, 1974.

Mikami, Yoshio. *The Development of Mathematics in China and Japan*. Leipzig: Teubner, 1913, and New York: Chelsea, 1961.

Shimodaira, Kazuo. *Wasan no Rekishi* (History of Japanese Mathematics in the Edo Period). 2 vols. Tokyo: Fuji Junior College Press, 1965–70.

SEXAGESIMAL SYSTEM The sexagesimal system was an ancient system of counting, calculation, and numerical notation that used powers of sixty much as the decimal system uses powers of ten. Rudiments of the ancient system survive in vestigial form in our division of the hour into sixty minutes and the minute into sixty seconds. Origins of the system of counting are essentially irretrievable because they lie in the period before the invention of writing. They are associated with the ancient Sumerians, whose language incorporated the only known system of sexagesimal counting. The Sumerians were probably also the inventors of the system of calculation, which seems to have its origin in a system of counters (small clay objects also referred to as tokens or calculi) representing the units 1, 10, 60, 60×10, 60^2, $60^2 \times 10$, and 60^3. This counter-calculation system may go back as far as ca. 3500 BC. The method of numerical notation apparently originated from the system of counters by adapting it to a form that could be represented in writing when the first writing system was invented ca. 3000 BC.

By 2000 BC this had developed into a system of sexagesimal place notation that functioned like our decimal system but with several important distinctions. Whereas the decimal system uses ten unique symbols, plus a period or comma to separate fractions from integers, the fully developed sexagesimal system of notation used only two unique symbols (those for 1 and 10) and essentially ignored our distinction between fractions and integers. These features, as well as lack of a symbol for zero, reflect its origins in the system of calculating with counters: zero would not have been represented in a system of counters (obviously, one does not count something that does not exist), and unique symbols for the integers 1–60 would have been excluded for practical reasons. Lack of a distinction between fractions and integers goes back to the fact that fractions are essentially alien to prehistoric systems of counting. Names for small quantities that we would call fractions are based on subdivisions of the weight system (*mina* and *shekel*) and were clearly not thought of as fractions in our sense of the word. In keeping with this picture is the symbol for medial zero that appears in Babylonian mathematical texts of the late first millennium BC. This so-called zero is really nothing but a "spacer" symbol and reflects the practice (occasionally observable in surviving texts) of leaving a blank space where we would place a zero.

During the third millennium BC, this Sumerian sexagesimal system was adopted, along with other features of Sumerian culture, by semitic-speaking Akkadians, and the Akkadian language, which — like Sumerian — was written in cuneiform script on clay tablets, served as the vehicle for its diffusion and preservation. As an essential part of the education of scribes, it was still very much alive when Greeks conquered the Near East in the time of Alexander the Great. The two centuries of Greek presence in Mesopotamia (330–129 BC) facilitated the incorporation of Babylonian mathematical astronomy into the Western tradition, and with it also came knowledge of the Babylonian sexagesimal system. This, however, seems never to have been used systematically by Greeks outside of Babylonia. Division of the circle into 360 degrees is a vestige of the Babylonian system that has been transmitted to us by the Greeks.

Modern knowledge of the existence of an ancient sexagesimal system goes back as far as the recovery of Greek mathematics, beginning in the Renaissance. Little, however, was known about its true character until the middle of the nineteenth century when decipherment of cuneiform writing revealed the system in its developed form of the first millennium BC. Speculations about its origins have tended to postulate conscious creation, as opposed to gradual evolution. The 360-day year and choice of 60 as base (because 1, 2, 3, and 5 are all prime factors) are among these theories, all equally without evidence. Only in the twentieth century has the Sumerian system of counting become relatively well understood, and the role of clay counters as prototypes of number symbols has only become clear in the last decade.

MARVIN A. POWELL

REFERENCES

Powell, Marvin A. *Sumerian Numeration and Metrology*. Ph. D. Dissertation, University of Minnesota, 1971.

Powell, Marvin A. "The Origin of the Sexagesimal System: The Interaction of Language and Writing." *Visible Language* 6: 5–18, 1972.

Powell, Marvin A. "The Antecedents of Old Babylonian Place Notation and the Early History of Babylonian Mathematics." *Historia Mathematica* 3: 417–439, 1976.

Powell, Marvin A. "Metrology and Exact Sciences in Ancient Mesopotamia." In *Civilizations of the Ancient Near East.* Ed. J. Sasson et al. New York: Scribners, 1995.

SHANGHAN LUN The *Shanghan Lun* (Treatise on Cold-induced Fevers) by Zhang Zhongjing was written about AD 210. It is the first important practical work in Chinese medicine that has come down to us. The legend is that within a period of ten years nearly two-thirds of Zhang's village died of illness, mostly from infectious diseases. Zhang, seeing so much suffering, forced himself to find a way to save them, and looked up the ancient prescriptions and read the ancient books. At the end of the third century, after the political turmoil accompanying the breakup of the Han dynasty, the book was reorganized and edited. In the eleventh century, the Song scholar Lin Yi further recompiled and divided it into two sections: the *Shanghan Lun* as we know it today and the *Jingui Yaolue Fang Lun* (Treatise on the Golden Casket Collection of Prescriptions).

Most admirable in Zhang was his scientific approach to medicine. He berated doctors who used secret remedies to dupe people, saying "what looks magnificent on the surface may be rotten underneath". He attached great importance to close observation and early treatment of conditions, adjusting the remedies to suit the disease and changing the medicine as the cure progressed. He deplored the frequent use of purgatives. As a mild laxative he prescribed pig's bile mixed with vinegar poured into the rectum through a bamboo tube. He observed the fluctuation in temperature during fevers, and devised a model for the progression of disease based upon the separation of *Yin* and *Yang* into *Shaoyang, Taiyang, Yangming, Shaoyin, Taiyin, Jueyin*. This has been a most influential categorization in Chinese medical literature.

Earlier works had recorded the reaction of the pulse to pathological conditions, but Zhang extended his theory to included the pulse's reaction to the effects of drugs.

About two hundred simple herbal medicines were then known, of which the most common were emetics, purgatives, and anti-pyretics. For instance rhubarb, sodium sulphate, and croton plant were used as purgatives. For reducing fever there was scutellaria, cinnamon bark, ephedra, and bupleurum. As emetics the stalks of various gourds or melons, and parts of the gardenia were recommended. Such remedies are now in use throughout the world. Rhubarb, for instance, was introduced into Europe in the fifteenth century, and imported up to the nineteeth century. The action of ephedra has been mimicked by modern drugs and is invaluable in asthma.

As analgesics and sedatives, various parts of the aconite plant and almond were used. Almond extracts are still used to relieve coughing. Aconite was toxic, but with extreme care it could be useful as a cardiac tonic and nervine. Ginger, licorice, dates, onion roots, and orange peel were used as digestives.

Artificial respiration is mentioned and the use of it in suicides by hanging. In cases of poisoning, water was forced down the victim's throat to wash out the stomach — much as in modern hospitals. In diet, Zhang also forbade the use of food derived from animals which had died from disease, and stressed the importance of proper storage and hygiene.

His whole work is a clear example of early scientific method. His manner, most obviously suited to herbal medicine, has been largely adopted by the modern Chinese in their presentation of Chinese traditional medicine. With its differentiation of disease by signs and symptoms, it guides the student into a clear etiology, differentiation of condition, treatment plan, and prescription — whether treatment by acupuncture, herbal medicine, massage, diet or therapeutic exercise is intended.

RICHARD BERTSCHINGER

REFERENCES

Fu, Weikang. *Traditional Chinese Medicine and Pharmacology.* Beijing: Foreign Languages Press, 1985.

Lu, Gweidjen and Needham, Joseph. *Celestial Lancets.* Cambridge: Cambridge University Press, 1980.

Zhang, Zhongjing. *Treatise on Febrile Diseases Caused by Cold.* Trans. Luo Xiwen. Beijing: New World Press, 1986.

See also: Zhang Zhongjing

SHARAF AL-DĪN AL-ṬŪSĪ Little is known about the life of Sharaf al-Dīn al-Muẓaffar ibn Muḥammad ibn al-Muẓaffar al-Ṭūsī. He was born in Ṭūs (Iran), probably around 1135. He spent part of his life in Syria and Iraq, and died in Iran around 1213. He had a number of pupils, including the famous polymath Kamāl al-Dīn ibn Yūnus.

Sharaf al-Dīn al-Ṭūsī was one of the best mathematicians of the Arabic tradition. His most important work is a treatise on cubic equations, in which he develops the theory quite beyond the limits reached by al-Khayyām. The Arabic mathematicians only considered positive coefficients, and they therefore distinguished various types of cubic equations, such as $x^3 + bx = c$, $x^3 + c = bx$, $x^3 = bx + c$, $x^3 + ax^2 = bx + c$, $x^3 + bx = ax^2 + c$ and so on. For all these types, ʿUmar al- Khayyām had indicated a geometrical solution of a positive root x (negative and zero roots were not taken into account). For al-Khayyām, the coefficients a, b, c were not in the first place numbers, but a is a line segment, b is a rectangle, and c is a parallelepiped. Using these data, al-Khayyām constructed two conic sections, and he supposed that these conics intersected in a point P. He then constructed x by means of P. It is obvious that this procedure only works if point P exists. For some types of cubic equations (for example $x^3 + bx = c$ and $x^3 = bx + c$) it is easily proven that a point P always exists, so the equation always has a root x. For other types of cubic equations (such as $x^3 + c = bx$) P may or may not exist, depending on the coefficients of the equation. The only way in which al-Khayyām could find out whether point P existed was by actually drawing the conic sections, which could be done with limited accuracy (it is unclear whether al-Khayyām ever constructed a root x this way, because the subject seems to have been of theoretical interest only). Thus it was not precisely clear when a cubic equation like $x^3 + c = bx$ had a root.

For the particular case of $x^3+c = ax^2$, some predecessors of al-Khayyām had shown that a root x exists if and only if $c \leq (4/27)a^3$. Thus these mathematicians knew in advance whether a root existed, and they only needed the conic sections to construct x. Sharaf al-Dīn showed that similar (but much more complicated) conditions exist for all types of cubic equations which do not always have positive roots. He meticulously proved the correctness of these conditions by means of the methods of "geometrical algebra" used by Euclid in Book II of the *Elements*. He then gave a geometrical construction of the roots x, and (unlike al-Khayyām) a numerical procedure, which could be used to approximate the roots with any desired accuracy. In the theory of cubic equations, no further progress was made until the discovery of the algebraical solution in Italy after 1500.

In addition to his treatise on equations, Sharaf al-Dīn wrote two short works on elementary geometry. He also invented the "linear astrolabe", consisting of a staff with a plumb line and two cords. The staff contained the markings of the North–South line of an ordinary astrolabe. Although this instrument is easier to construct than an ordinary astrolabe, it does not look nice and it is not user-friendly. Thus it is not surprising that no examples have survived.

JAN P. HOGENDIJK

REFERENCES

Anbouba, A. "Al-Ṭūsī, Sharaf al-Dīn." In *Dictionary of Scientific Biography*, vol.XIII. Ed. C.G. Gillispie. New York: Scribner's, 1976, pp. 514–517.

Hogendijk, Jan P. "Sharaf al-Dīn al-Ṭūsī on the Number of Positive Roots of Cubic Equations." *Historia Mathematica* 16: 69–85, 1989.

al-Ṭūsī, Sharaf al-Dīn. *Oeuvres Mathématiques*. Ed. and trans. R. Rashed. Paris: Les belles lettres, 1985.

See also: ʿUmar al-Khayyām

SHEN GUA The Chinese scholar, statesman, and scientist Shen Gua was born in 1031 in present-day Hangzhou (Zhejiang). Born into a family of gentry, he entered civil service in 1054, holding minor posts in different provinces where he distinguished himself in a series of projects involving water control. After passing the national doctoral examinations in 1063 he moved to the capital, Kaifeng, and in 1072 was named Director of the Astronomy Bureau. In that position he reorganised the Bureau, planned a program of astronomical observations, and devised a calendar that was effective from 1075 to 1092. His work as astronomer alternated with missions to the provinces, where he was again able to display his skills in hydraulic engineering. The relief maps he made in 1074 (using wooden plates, sawdust impregnated with glue, and melted wax) are among the most celebrated specimens of Chinese cartography.

The whole of Shen's career was influenced by his partnership with Wang Anshi (1021–1086), the leader of the "New Policies" (*xinfa*) group which advocated a reform program aimed at strengthening the central government, regulating finances by law, and exploiting nature for the state's benefit. Shen's memorials to Emperor Shenzong (r. 1067–1084) on a variety of subjects — including military strategy and tax levies — won approbation in a dozen instances. In a related development, Shen was sent as an envoy to the Khitan court in 1075, and, in a feat of diplomacy, was able to put at least a temporary end to the borderland skirmishes and the repeated Khitan claims of Chinese land.

Following his diplomatic victory, Shen Gua was appointed Finance Commissioner in 1077, but that very year also marked the turning point of his political fortune. The opponents of the "New Policies" had succeeded in pressing Wang Anshi into retirement. Shen was in turn charged with duplicity in his fiscal policies, and demoted. Three years later, partially rehabilitated, he was sent to the Shenxi province as Commissioner for Prefectural and Military Affairs. In 1081 he obtained an important military victory against the Tanguts. One year later, however, he was held responsible for the disastrous defeat that the Chinese troops suffered in the same area, and put on probation.

Shen spent the following years completing a map of the territories under Chinese control, for which he was commissioned by the Emperor in 1076. When he submitted it to the throne in 1088, his rewards included the right of choosing his residence. He thus moved to the "Dream Brook", a garden estate that he had bought some years earlier near present-day Zhenjiang (Jiangsu), and devoted most of his time to writing, until his death in 1095.

Shen Gua's writings include treatises on administration, military strategy, astronomy, medicine, poetry, ritual, and several other subjects. His best known work, entitled after the name of his retreat, is the *Mengqi bitan* (Brush Talks from Dream Brook). Written in the final years of his life, and supplemented by two shorter works probably published posthumously, it consists of about five hundred jottings on disparate matters, usually short and organized into topics. The original edition was in 30 chapters, of which 26 are extant. Of special importance from the point of view of the history of science are the sections pertaining to mathematics, astronomy, meteorology, optics, physics, geology, and medicine.

To give only a few examples, chosen among the more important parts of his work but clearly inadequate to reflect the richness of its contents, in the field of astronomy Shen Gua described solar and lunar eclipses, studied the moon phases, and provided reasons for the apparent planetary motions. During his appointment as Director of the Astronomy Bureau, moreover, he modified the design of the sighting tube pointing to the Pole Star in the armillary sphere. In the field of geology he recognized the origin of fossils, and gave an account of the role of erosion. Other jottings are concerned with magnetism, the production of inverted images on concave mirrors, the formation of rainbows, the fall of a meteorite, and the process of printing. All this adds to his skills in cartography and hydraulic engineering.

As several scholars have remarked, Shen Gua never attempted to organize his observations into a general theory. His work is, in many ways, the product of an exceptionally bright mind, motivated by a sharp curiosity in observing, understanding, and describing whatever phenomenon aroused his interest. This was by no means limited to natural processes or to man-made techniques. One chapter of the *Mengqi bitan* deals with supernatural matters, which the author treats with the same sympathy he accords to the other fields that attracted his attention. For a proper assessment of Shen Gua's personality and production, the "scientific" features of his works should be evaluated without neglecting their own context or disregarding other features which we might label as "unscientific".

From another point of view, Shen Gua reveals the shortcomings inherent in the equation between Daoism and scientific progress on one hand, and Confucianism and scientific conservatism on the other, which has influenced some recent work in the history of Chinese science. While Daoists were not interested in nature *per se* (nature was before all an "image", *xiang*, of the absolute principle, the *Dao*), Shen Gua shows that interest in natural facts could be, and indeed was, actively pursued by a "Confucian" scholar and state administrator.

FABRIZIO PREGADIO

REFERENCES

Billeter, Jean François. "Florilège des Notes du Ruisseau (Mengqi bitan) de Shen Gua (1031–1095)." *Asiatische Studien / Études Asiatiques* 47: 389–451, 1993.

Brenier, Joël, et al. "Shen Gua (1031–1091) et les Sciences." *Revue d'Histoire des Sciences et de Leurs Applications* 42: 333–351, 1989.

Fu Daiwie. "A Contextual and Taxonomic Study of the 'Divine Marvels' and 'Strange Occurrences' in the *Mengxi bitan*." *Chinese Science* 11: 3–35, 1993–94.

Holzman, Donald. "Shen Kua and his *Meng-ch'i pi-t'an*." *T'oung Pao* 46: 260–292, 1958.

Mengqi bitan jiaozheng [A Critical Edition of the *Mengqi bitan*]. Ed. Hu Daojing. Shanghai: Guji Chubanshe, 1985.

Needham, Joseph, et al. *Science and Civilisation in China*. Cambridge: Cambridge University Press, 1954–.

Sivin, Nathan. "Shen Kua." In *Dictionary of Scientific Biography*. Ed. Charles C. Gillispie. New York: Charles Scribner's Sons, 1975, vol. XII, pp. 369–393.

See also: Calendars in China – Astronomy in China – Maps and Mapmaking in China – Armillary Spheres in China

SHIBUKAWA HARUMI Shibukawa Harumi (1639–1715) was a Japanese astronomer and calendar-reformer whose distinguished service in calendar reform led to his appointment in 1685 as official astronomer.

The Chinese luni-solar Xuan-ming calendar, adopted in Japan in 862, had not been reformed for more than eight hundred years. Over the centuries the discrepancy in the length of a solar year had increased so that by Harumi's time there was a two-day delay in the winter solstice.

With able diplomatic skill, Harumi urged that the Shogunate government undertake calendar reform. In 1669 he began conducting astronomical observations, probably the first systematic observations made in Japan. Following the procedures of traditional Chinese astronomy, he set up a gnomon and measured the lengths of shadows at various points before and after the winter solstice, in order to calculate the time of occurrence. He was especially interested in the Shoushi calendar of the Yuan dynasty (1279–1368), a crowning achievement of calendrical astronomy adopted in China in 1282, and his observations were based upon its methods.

In his reference to Western theories, Harumi based his information on scanty sources and came to regard Western astronomers as "barbarians who may have theories but cannot prove methods". It is regrettable that sufficient material for evaluating Western theories was not available to him.

During the eighteenth century Japanese astronomy altered its orientation from China to the West. Harumi belonged to the first generation of astronomers who, with only limited knowledge of Western astronomy, began evaluating the merits of both systems.

NAKAYAMA SHIGERU

REFERENCES

Nakayama, Shigeru. *A History of Japanese Astronomy: Chinese Background and Western Impact*. Cambridge, Massachusetts: Harvard University Press, 1969.

Tadashi, Nishiuchi. *Shibukawa Harumi no kenkyū* (A Study on Shibukawa Harumi). Tokyo: Kinsesisha, 1940.

See also: Calendars

AL-SHĪRĀZĪ Quṭb al-Dīn Maḥmūd ibn Masʿūd al-Shīrāzī (1236–1311) was a Persian scientist and philosopher. He was born into a medical family in Shiraz in central Iran. His father was a staff physician and ophthalmologist at the newly established Muẓaffarī hospital. As a child he was the apprentice of his father, who died when Shīrāzī was fourteen. Having assumed his father's position at the hospital, he continued his studies with several other local teachers of medicine and the rational and religious sciences. At this time he began the study of the first book, *al-Kullīyāt* (On General Principles) of Ibn Sīnā's *Qānūn fī 'l-Ṭibb* (Canon of Medicine), the leading medical textbook of the Middle Ages. His father had also initiated him into the Suhrawardī order of Sufis (Islamic mystics).

By the time he reached adulthood, he was in need of advanced medical instruction that could not be provided by the teachers of a provincial city like Shiraz. It would have been normal for him to have gone off earlier to one of the major Islamic cities for advanced studies, but in 1253 the Mongols had invaded the Islamic lands of Central and Southwestern Asia, sacking Baghdad in 1258, so Shīrāzī was confined to his home city until peace was reestablished.

In 1259 Hulegu, the Mongol ruler of Iran, gave a large grant to the famous scientist Naṣīr al-Dīn al-Ṭūsī (1201–74) to pay for the preparation of a new set of astronomical tables (*zīj*) for use in astrological calculations. Ṭūsī established an observatory in the town of Marāgha, at that time the Mongol capital, and brought together a brilliant team of scientists and scholars. Shīrāzī came to the observatory as a student soon after it was founded. While he was disappointed by Ṭūsī's lack of knowledge about medical theory, he learned a great deal of mathematics, astronomy, and philosophy from him and his faculty, and he soon became the most important student at the observatory. Ṭūsī took him along on his travels and introduced him at court. Shīrāzī also apparently spent some time studying in Qazvīn, Khorasan, and Baghdad with various teachers.

Around 1270 he left the observatory and went to Konya in Anatolia, where he met the famous Sufi poet Rūmī and studied *ḥadīth* (the sayings of Muḥammad) with Ṣadr al-Dīn Qūnawī, the leading disciple of the mystical philosopher Ibn ʿArabī. Soon after he was appointed *qāḍī* (religious judge) of the Anatolian cities of Sivas and Malatya. This was evidently a sinecure to allow him to pursue his scientific work, for his first major work, *Nihāyat al-Idrāk fī Dirāyat al-Aflāk* (The Highest Attainment in the Knowledge of the Spheres), a technical work on mathematical astronomy, was published in Sivas in 1281 and was dedicated to the vizier of the Mongol ruler. Other works on astronomy and mathematics soon followed.

In 1282 he was appointed to a diplomatic mission to Egypt. Though the mission failed in its political objectives, Shīrāzī found three complete commentaries on the first book of Ibn Sīnā's *Qānūn*, along with glosses and other sources. With this new material in hand, he was finally able to achieve his youthful goal of mastering the intricacies of this work. Shortly after his return to Anatolia he published a large commentary on the *Qānūn*. He published second and third editions in 1294 and in 1310, a few months before his death.

Of his personality we are told that he had a sharp wit, and indeed he was a stock character in a certain genre of joke for several centuries thereafter. He was expert in chess and prestidigitation and was a lively conversationalist and lecturer. He was an authority on musical theory, which in the Middle Ages was considered a branch of mathematics, and was a fine player of the *ribāb*, a forerunner of the violin. He seems to have grown more concerned with religious matters as he grew older, although he was a Sufi all his life.

He eventually settled — interrupted by several exiles — in Tabriz, at that time the capital of Mongol Iran. He was an intimate of the court. The funds that supported his scientific work came from a series of viziers and petty rulers. In addition to his works on astronomy and medicine, he wrote on mathematics, philosophy, and the Islamic religious sciences. In his last years he spent less time on the rational sciences and more on religious subjects. He died in relative poverty, having given almost everything he had to charity and to his students and not yet having received the large payment promised for the third edition of his commentary on Ibn Sīnā's *Qānūn*. He was given a lavish funeral by a wealthy student.

Shīrāzī was typical in most respects of Islamic scientists. He was a polymath, interested not only in the corpus of science inherited from the Greeks but also in the religious sciences of Islam. Of the rational sciences it was astronomy and medicine that found the most ready market—astronomy as the handmaid of astrology or for timekeeping in the mosques. The other rational sciences, particularly mathematics and philosophy, were supported by their more practical subordinate sciences. Sophisticated practitioners of the rational sciences generally drifted to the royal courts, the only reliable sources of funding for such subjects, and became involved in the life and politics of the court—and in its perils.

JOHN WALBRIDGE

REFERENCES

Nasr, Seyyed Hossein. "Quṭb al-Dīn Shīrāzī." In *Dictionary of Scientific Biography,* vol. IX. Ed. Charles Gillispie. New York: Charles Scribner's Sons, 1975, pp. 247–253.

Saliba, George. *A History of Arabic Astronomy: Planetary Theories during the Golden Age of Islam.* New York: New York University Press, 1994.

Shīrāzī, Quṭb al-Dīn. *Sharḥ Ḥikmat al-Ishrāq.* Tehran: 1895–97.

Shīrāzī, Quṭb al-Dīn. *Durrat al-Tāj li-Ghurrat al-Dubāj.* Tehran: Majlis, 1938–45.

Varjāvand, Parvīz. *Kāvush-i Raṣd-khāna-i Marāgha* [Excavation of the Marāgha Observatory]. Tehran: Amīr Kabīr, 1987.

Walbridge, John. *The Science of Mystic Lights: Quṭb al-Dīn Shīrāzī and the Illuminationist Tradition in Islamic Philosophy.* Cambridge, Massachusetts: Harvard Center for Middle East Studies, 1992.

See also: Marāgha – Naṣīr al-Dīn al-Ṭūsī – *Zīj* – Ibn Sīnā – Ibn al-ʿArabī – al-ʿUrḍī – Optics – Astronomy

SHIZUKI TADAO Shizuki Tadao was a Japanese natural philosopher, born in 1760. He was adopted by the Shizuki family, whose head was a government interpreter from the Dutch. In 1776 he became an assistant interpreter, succeeding his adoptive father, but he retired from that post the following year because of ill health. After leaving public service, he spend the rest of his life in the private study of Dutch books and in contemplation. He died in 1806.

He wrote books about the Dutch language and partially translated the Dutch translation of Engelbert Kaempher's *Geschichte und Beschreibung von Japan* (History and Description of Japan). But Tadao's major work was *Rekisho shinsho* (Introduction to True Physics), a compilation of his own theories on natural philosophy, inspired by his translation of the Dutch version of John Keill's *Introductio ad veram physicam* and the *Introductio ad veram astronomiam.*

In advocating Newtonianism, Keill's work had a polemical tone, and more importantly and quite unlike the readjusted interpretations of later authors, it dealt in abstractions and included a great many elements of natural philosophy. The book especially suited Tadao's inclination toward natural philosophy and Tadao thus became the first Newtonian in the East. He established unitary *qi* (energy) and its dual function (rarefaction and condensation) as the basis for his natural philosophy. He was not successful, however, in relating the natural philosophy that he derived from the theory of monistic *qi* to Newtonian dynamics.

Tadao also raised the question of why the planets rotate and revolve in the same direction in planes not greatly inclined to the ecliptic (the band of the zodiac through which the sun apparently moves in its yearly course). He proposed a hypothesis concerning the formation of the planetary system, which resembles the celebrated hypothesis of Kant and Laplace.

NAKAYAMA SHIGERU

REFERENCES

Nakayama, Shigeru. *A History of Japanese Astronomy: Chinese Background and Western Impact.* Cambridge, Massachusetts:

Tadao, Shizuki. *Rekisho Shinsho* (New Treasise on Calendrical Phenomena). Japan: Heibon-sha, 1956. Harvard University Press, 1969.

SHUSHU JIUZHANG The *Shushu Jiuzhang* (Nine Books on Mathematics) is a Chinese mathematical work, written by Qin Jiushao and published in 1247. This book is one of the highlights of Chinese mathematics of the Sung-Yüan-period, 960–1368.

The work is not organized according to mathematical methods, but to practical applications, with the exception of Chapter 1 (*ta-yen*). The nine chapters are:

(1) Indeterminate analysis;
(2) Heavenly phenomena, which dealt with chronological and meteorological questions;
(3) "Boundaries of fields", which covered surveying;
(4) "Telemetry" or measuring at a distance, a kind of proto-trigonometry;
(5) Taxes and levies of service;
(6) "Money and grain", i.e. taxes;
(7) Fortifications and buildings;
(8) Military affairs; and
(9) Commercial affairs.

An investigation of the socioeconomic information contained in the work proved that all the information was correct and related to the Southern Sung society, especially regarding money and currency (including paper money), credit systems, commercial life, harmonious purchase (*hoti*), transportation problems, construction of dikes, taxes and levies of service (*fuyi*), architecture, military affairs, and astronomy. This proves that the work was written for practical use, although it was, partly because of the Mongol invasion of China, never used as a handbook for practical training.

The basis for the common arithmetical procedures was the ancient mathematical classic, the *Jiuzhang Suanshu*. There was one method, the *ta-yen* rule, which Qin Jiushao learned when he worked in the Board of Astronomy. This method is preserved only in Qin's work.

As for notation and terminology, it is important to note that the zero was printed for the first time in this book, so that we have a complete decimal place-value system. The arithmetical algorithms are the traditional ones, as are the geometrical methods and trigonometrical procedures. The most important part of the work is the "algebra". This dealt with simultaneous linear equations, determinants, and series and progressions. Also covered were numerical equations of higher degree, an extension of the method for square and cube root extraction used in the *Jiuzhang Suanshu*. This method is the same as the Horner-Ruffini procedure. The most important part is undoubtedly the Chinese Remainder Theorem, solved by means of the *ta-yen* rule. It is the solution of the problem:

$$N = r_1(\text{mod}_1) = r_2(\text{mod}_2) = \cdots = r_n(\text{mod}m_n).$$

It is remarkable that Qin Jiushao was able to solve problems where the moduli are not relatively prime — a method not known to Euler and Gauss, and only solved definitively by Stieltjes in 1890.

ULRICH LIBBRECHT

REFERENCES

Libbrecht, U. *Chinese Mathematics in the Thirteenth Century, The Shushu chiu-chang of Ch'in Chiu-shao*. Cambridge, Massachusetts: MIT Press, 1973.

Wu Wen-jen, ed. *Qin Jiu-shao yu Shushu jiuzhang*. Beijing: Beijing shifan daxue chubanshe, 1987.

See also: Qin Jiushao – Liu Hui and the *Jiuzhang shanshu*

AL-SIJZĪ Al-Sijzī, Abū Sa'īd Aḥmad ibn Muḥammad ibn ‹Abd al-Jalīl, was born in Persia, ca. 945 and died ca. 1020. The name al-Sijzī indicates that he was a native of Sijistān in southeastern Iran and southwestern Afghanistan. This is confirmed, for example, when al-Sijzī refers in his *al-Mudkhal ilā ‹ilm al-handasa* (Introduction to Geometry) to a planetarium which he had constructed in Sijistān. Al-Sijzī was already active in 963, when he copied a manuscript of Pappus' *Introduction to Mechanics* (Book VIII of the Collection), and still active in 998, when he completed a treatise on the proof of the plane transversal theorem.

Al-Sijzī's father, Abū'l-Ḥusayn Muḥammad ibn ‹Abd al-Jalīl, was also interested in geometry and astronomy, and al-Sijzī addressed some of his works to him, e.g. *Risālat fī Khawāṣṣ al-qubba az-zā'ida wa-l-mukāfiya* (On Parabolic and Hyperbolic Cupolas), completed in 972. Around 969 al-Sijzī spent some time at the Buwayhid court in Shirāz and assisted in 969/970 at the observations of the solstices conducted by ‹Abd al-Raḥmān ibn ‹Umar al-Ṣūfī. There he met a number of important geometers and astronomers, including Abū'l Wafā›, Abū Sahl al-Qūhī, and Naẓif ibn Yumn (d. ca. 990).

Al-Sijzī was active in astrology and had a vast knowledge of the literature. He usually compiled and tabulated, adding his own critical commentary, as with three works by Abū Ma‹shar and the second of the five books ascribed to Zoroaster. He uses Sassanid material and sources from the time of Hārūn al-Rashīd and from the late Umayyad period for a book on general astrology and its history. In his work on horoscopes he gives tables based on Hermes, Ptolemy, Dorotheus, and "the Moderns". Al-Sijzī's tables, together with those of Ptolemy, are quoted by Iḥtiyāzu' l-Dīn Muḥammad in his *Judicial Astrology*. The treatise *Kitāb fī Qawānīn mizājāt al-asṭurlāb al-shimālī ma‹a l-janūbī* (On the Astrolabe) deals with the different kinds of astrolabe retes (a circular plate with many holes used on the astrolabe to indicate the positions of the principal fixed stars) with which al-Sijzī was familiar. This treatise is used by al-Bīrūnī in his *Istīʿāb al-wujūh al-mumkina fī ṣan'at al-asṭurlāb* (Book on the Possible Methods for Constructing the Astrolabe).

Al-Sijzī wrote at least 45 geometrical treatises, of which some 35 are extant, and about fourteen astronomical treatises. He was well-read and had many contacts with his contemporaries. A number of his works are therefore of unusual historical interest. In *Risālat fī Qismat az-zāwiya al-mustaqīmat al-khaṭṭain bi-ṭalāṭat aqsām mutasāwiya* (On the Division of the Angle into Three Equal Parts) al-Sijzī describes a number of problems, to which various other writers had reduced the problem of the trisection of the angle. The method of "the Ancients" by means of a neusis was not considered a legitimate construction by al-Sijzī. His own construction, by intersecting a circle and a hyperbola, is a variation of the solution by al-Qūhī. The treatise ends with five problems of al-Bīrūnī. Regarding al-Sijzī's construction of the heptagon, this is, according to Jan Hogendijk, to a great extent the history of a dispute between two young geometers, al-Sijzī and Abū'l-Jūd, who were both engaged in plagiarism. In the meantime al-Qūhī solved the problem in an elegant manner.

For the *Geometrical Annotations* (Kitāb fī l-masā'ili l-mukhṭārati llatī jarat baynahu wa-bayna muhandisī Shīrāz wa-Khurāsān wa-taʿlīqātihi, Book on the Selected Problems which were Currently being Discussed by him and the Geometers of Shirāz and Khorāsān, and his (own) Annotations) al-Sijzī had the example of Ibrāhīm ibn Sinān's *al-Masā'il al-mukhṭāra* (Selected Problems) in mind. A number of the problems and solutions are clearly influenced by, or adapted from the *Selected Problems*, but al-Sijzī's treatise is on the whole on a lower level.

The *Misāḥat al-ukar bi-l-ukar* (Book of the Measurement of Spheres by Spheres) is about a surrounding sphere which contains in its interior up to three mutually tangent spheres. Al-Sijzī determines the volume of the solid which results when one deletes from the surrounding sphere all points that belong to the spheres in its interior. He expresses this volume as the volume of a new sphere, and he determines the radius of this new sphere in terms of the radii of the spheres used in the definition of the solid. Al-Sijzī's proofs are trivial consequences of identities for line segments, proved geometrically. The treatise contains twelve propositions, of which proposition 11 and its proof are false for (three-dimensional) spheres. The proposition holds in dimension four. Perhaps al-Sijzī had four-dimensional spheres in mind, although he does not use them elsewhere in the treatise; perhaps he made a mistake.

His small treatise *Risālat fī Kaifīyat taṣauwur al-khaṭṭain alladhain yaqrubān wa-lā yaltaqiyān* (On the Asymptote, or How to Conceive Two Lines Which Approach Each Other but do not Meet, if one Extends Them All the Way to Infinity) is devoted to Apollonius II,14. Some cases, he says, can be solved as explained in his *Kitāb fī tashīl as-subul li-stikhrāj al-ashkāl al-handasīya* (On Facilitating the Roads to the Geometrical Propositions); for others a philosophical method is needed, as Proclus has shown in the definitions of his *Elements of Physics*. The treatise ends with the case where the two asymptotes are two hyperbolas. In *Risālat fī anna 'l-ashkāl kullahā min al-dā'ira* (On [the fact] that all Figures are Derived from the Circle), a treatise until recently attributed to Naṣr ibn ʿAbdallāh, al-Sijzī describes one of the few instruments that finds the *qibla* (the direction of Mecca) geometrically. He also wrote an original treatise on the construction of a conic compass.

In the introduction to *Risālat fī'l-shakl al-qaṭṭāʿ* (The Transversal Figure), written before 969, al-Sijzī says that he wrote the work having seen neither Thābit ibn Qurra's *Kitāb fī'l-shakl al-mulaqqab bi-l-qaṭṭāʿ* (Book of the Transversal) nor any other work on the topic, except Ptolemy's *Almagest*. The treatise begins with enunciations and proofs of two lemmas, which also appear, in different terms, in the *Almagest*. Following the two lemmas al-Sijzī enunciates and proves his twelve propositions. Aware of the astronomical applications of the theorem he evidently saw the need to provide a complete mathematical basis for these uses. The details of the proofs of all twelve theorems are carried out according to a uniform procedure. This makes the treatise a step towards recognizing the mathematical discipline of trigonometry.

YVONNE DOLD-SAMPLONIUS

REFERENCES

Anbouba, Adel. "Construction de l'Heptagone Régulier par les Arabes au 4e Siècle H." *Journal of History of Arabic Science* 2: 264–269, 1978.

Berggren, J. Len. "Al-Sijzī on the Transversal Figure." *Journal of the History of Arabic Science* 5: 23–36, 1981.

Crozet, Pascal. "L'Idée de Dimension Chez al-Sijzī." *Arabic Sciences and Philosophy* 3: 251–286, 1993.

Dold-Samplonius, Yvonne. "al-Sijzī." In *Dictionary of Scientific Biography*, vol. 12. Ed. Charles C. Gillispie. New York: Charles Scribner's Sons, 1975, pp. 431–432.

Hogendijk, Jan P. "How Trisections of the Angle were Transmitted from Greek to Islamic Geometry." *Historia Mathematica* 8: 417–438, 1981.

Hogendijk, Jan P. "Greek and Arabic Constructions of the Regular Heptagon." *Archive for History of Exact Sciences* 30: 197–330, 1984.

Hogendijk, Jan P. "New Findings of Greek Geometrical Fragments in the Arabic Tradition." In *Symposium Graeco-Arabicum II.*

Amsterdam: B.R. Grüner, 1989, pp. 22–23.

Knorr, Wilbur R. *Textual Studies in Ancient and Medieval Geometry*. Boston and Berlin: Birkhauser, 1989.

Rashed, Roshdi. "Al-Sijzī et Maomonide: Commentaire Mathématique et Philosophique de la Proposition II-14 des Coniques d'Apollonius." *Archives Internationales d'Histoire des Sciences* 37: 263–296, 1987.

Rosenfeld, Boris A., R. S. Safarov, and E. I. Slavutin. "The Geometric Algebra of as-Sijzī." *Istoriko-Matematicheskie Issledovaniia* 29: 321–325, 349, 1985. (in Russian)

Saidan, Aḥmad S. *Rāsa'il Ibn Sinān* (The Works of Ibrāhīm ibn Sinān, with two more tracts, 1. by Thābit ibn Qurra, 2. by Al-Sijzī [i.e. On Facilitating the Roads to the Geometrical Propositions]). Kuwait: Qism al-Turāth al-ʿArabī, 1983. (in Arabic)

See also: Astrolabe – Astrology – Abū Maʿshar – al-Qūhī – Ibrāhīm ibn Sinān - *Qibla – Almagest*

SILK AND THE LOOM In China the raw material silk dominated textile technology as a whole, and dictated the development of looms and their technology in particular. Thus the weaving methods and the patterns depended to some extent on the loom technology available.

The biological prerequisites for silk production on a commercially significant scale are domesticated silkworms *Bombyx mori* L. (sericaria) [*jiacan*] which belong to the family Bombycidae in the order of the Lepidoptera nocturno and various mulberry trees of the family of the Moraceae, such as the *Morus alba* Linn. (*baisang*), *Morus bombycis* Koidz. (*jingsang*), and *Morus alba* var. *multicaulis* (*lusang*), commonly called the domestic mulberry (*jiasang*). Silkworms are generally fed with the fresh leaves of the mulberry tree.

Not only were these two prerequisites present from an early stage in China, as can be seen from pictographs on bone inscriptions of the Shang era, but their biological connection was also recognized. Various archaeological discoveries, fragments of textile, and imprints show that in many areas of Neolithic China a form of silk technology already existed which clearly belonged to a long and practically, perhaps even experimentally, oriented agricultural tradition. The production and treatment of silk had begun many years before and by then had already reached a high technical level. So far the earliest find of silk in China is dated between 2850 and 2650 BC. The earliest cocoons of the domesticated spinner are from the Tang dynasty (618–907).

To understand the importance of silk as far as weaving and the development of weaving patterns are concerned, it is essential to look at the genuine properties of a single silk thread. Unlike other textile fibers silk is not spun but reeled. Silk-reeling (*sao si*) involves taking up the ends of fibers from several cocoons and combining them into a thread. It takes place just before the moth tries to break through the cocoon. This timing is crucial if endless silk threads are required. The length of the cocoon-filaments which can be obtained varies between 700 and 900 m, but without a silk-winding instrument, a reel, and a water basin in which to float the cocoons, silk-reeling cannot be performed. Through the centuries hand-driven silk-reeling was practiced as a seasonal business in small-scale family production. From the terminology used it is certain that prototypes of silk reels and silk-reeling frames (*saochechuang*) were known in the Zhou period. The earliest detailed description of the treadle-operated, highly mechanized silk-reeling frame is from ca. 1090.

Loom weaving, the production of even-textured fabric, requires that the warp threads be kept at a consistently high level of tension. The warp threads (*jing*) are those which, in parallel, run lengthwise across the loom; the weft threads (*wei*) are introduced laterally at right angles to them. A high weaving tension necessitates the use of comparatively elastic warp threads of a kind not naturally found in raw silk. This elasticity is obtained through winding or throwing or else doubling and twisting the thread. Various spinning techniques could spin the thread in the requisite manner. In China the spindle wheel (*fangche*) had been in use since the middle of the first millennium BC. The spindle wheel was not only very efficient in the spooling of threads, which means transferring a thread from a reel to a spindle, but also most useful in the processes of doubling and quilling, when an empty spindle is filled with a twisted thread which now combines two or three threads in one. Much has been published about the origin of the spindle wheel in the East and in India, but all the pictorial and textual evidence available points to China as the geographical area where the spindle wheel originated in the context of silk technology, especially the spooling and doubling process, as pictured on some Han stone reliefs. It took a long time before the spindle wheel made its appearance in Europe in the thirteenth century. All these inventions and innovations in the processing of silk threads have to be seen in the context of the treadle-operated loom (*jiaotaji*) which was equipped with two treadles in order to mechanize the opening (*kaikou*) of the shed. Thus a clear shed for the shuttle (*suokou*) was formed through which the weft thread could be passed. Such a loom was employed to weave simple tabby fabrics to standards on a large scale. From its structure it may be described as a combined horizontal-vertical or oblique warp sheet loom (*xiezhiji*). Depictions of the loom on stone reliefs from Han times which were found in Shandong, Jiangsu, and Sichuan provide ample evidence of this type of loom. It took many centuries, most probably until the medieval age, before a treadle-operated loom was set up in Europe. A more complicated loom with a treadle mechanism for the forming of the shed and a system of draw cords (*tihuaxian*) or pattern rods (*wen'gan*) which looped the warp threads required for the pattern already existed in Zhou times.

If maximum benefit was to be got from the exceptional length of the silk threads, then it was only logical to dress the loom with a silk warp. For this one needed a warp beam (*jingzhou*). The width of the fabric was determined by the length of the warp beam, and by the ideal working width for the weaver, who not only inserted the shuttle containing the weft but also operated the treadles with her feet so as to form the sheds. In addition, she would have to perform other operations in order to create the particular pattern. On account of these factors, most fabrics of the Han era as well as later generally had a width of about 50 cm (two *chi* and two *cun*). The length varied between 9 and 12 m. The qualities of silk as an ideal warp thread and hence too the dominance of the warp in the weave resulted in the weave pattern being created by accentuating the warp threads. From the Zhou era (1045–221 BC) until the early Tang era (618–907), therefore, Chinese silk possesses a warp pattern unlike that of other cultures, which, due to differences in the raw material, was produced on wholly different looms and which is weft patterned. Only in Tang times did the patterning technique change to the weft, which resulted in more colorful fabrics. Although it is known from a few silk fragments that the weft pattern was practiced in early China the change in weaving technique in Tang times was certainly due to influence from the region of Samarkand or Persia.

Several types of looms were already in use and operated as early as Han times (206 BC–AD 220). They are documented in historical works from Han and Jin times, and especially in the agricultural and encyclopedic works from the Song, Yuan, Ming and Qing dynasties (960–1911).

The loom which combined seventy or even more movable shafts (healds) with the function of many treadles (*duozong duonie*) was widespread. Although this type of loom was difficult to set up, it was operated until recently in Sichuan province. The big advantage of this type of loom was that one weaver could operate it easily and quickly. Furthermore its advanced and highly efficient patterning equipment qualified it for the production of so-called brocades and other complicated weaves.

Another type of loom which is not clearly distinguished in the Chinese publications from the above mentioned loom made use of figuring healds or pattern rods (*tihuazong, wen'gan*), which were manually lifted. Textual sources mention up to 120 patterning devices. This ingenious hand operated patterning device was already used on the backstrap loom (*yaoji*) in very early times. The simple backstrap loom originally had no wooden frame but consisted exclusively of functional parts. The weaver sat on the ground. In many old publications it is depicted with a wooden framework. It was used until recently. The earliest weaving implements which may be ascribed to such a loom were excavated from the site of Hemudu, datable to ca. 4300 BC. It may be justified to assume that the patterning technique with pattern rods which can be created in a sitting position on the ground is older than the use of looms with shafts and treadles which require a wooden frame. Although the simple type of backstrap loom was an excellent device for weaving on a comparatively small scale, it had its obvious limitations which were set by the length and width of the circular warp. In the view of Chinese textile historians this loom with pattern rods belongs to a loom category of either an early primitive type or of ethnic minorities. Only after the horizontal backstrap loom had been equipped with a real warp beam, supported by a wooden framework, which helped to facilitate its operation, can it be regarded as a loom which worked economically and which was suitable for silk weaving with patterning on a large scale. In all probability and in the opinion of a few eminent Western weaving specialists, such a loom with two shafts producing a tabby ground and a system of pattern rods of the heddle rod type that were lifted by the weaver as required by the pattern was the type of loom used and preferred to all others for the production of complex patterned weaves before Tang times in China.

The third type of loom before Han times was a forerunner of the Chinese drawloom. It employed a string heddle patterning device (*shuzong tihua zhuangzhi*) and a simple treadle operated shedding device. During Eastern Han times its equipment was improved in such a way that the whole "machine" (*ji*) gained the structural and functional features of a real drawloom (*hualou tihuaji*). This drawloom was equipped with a figure or harness tower and all the lifting devices for the warp threads required. Furthermore there were lifting and depressing shafts. It was operated by a weaver and a drawgirl. This type of loom, which was fashioned in a large variety, dominated the production of high quality polychrome patterned silk fabrics until the late nineteenth century when Western looms were introduced into China. Some Western historians of weaving technology have ruled out the use of a drawloom in Han times in China and suggested that it was invented somewhere in the region of Samarkand, Persia or even the Near East. To support their notion they practically demonstrated that even difficult polychrome weaves could have been produced on pattern rod looms. However the "Rhapsody on Women Weavers" bears witness to the fact that the drawloom with all its characteristic features was already in use in the big workshops of China as early as the first half of the second century AD. The earliest technical description and illustrations on the subject were published in the *Ziren yizhi* (Traditions of the Joiners' Craft) of 1264. It took roughly one thousand years before a simplified version of the Chinese drawloom was set up in medieval Europe.

DIETER KUHN

REFERENCES

Becker, John, and Donald Wagner. *Pattern and Loom. A Practical Study of the Development of Weaving Techniques in China, Western Asia and Europe*. Copenhagen: Rhodos, 1987.

Chen, Weiji, ed. *History of Textile Technology of Ancient China*. New York: Science Press, 1992.

Fangzhi. Beijing: Zhongguo dabaike quanshu chubanshe, 1984.

Flanagan, J.F. "Figured Fabrics." In *A History of Technology*. Ed. Charles Singer et al. Oxford: Clarendon Press, 1957, vol. 3, pp. 187–205.

Forbes, R.J. *Studies in Ancient Technology*. Leiden: Brill, 1956.

Geijer, Agnes. *A History of Textile Art*. London: Philip Wilson, 1979.

Kuhn, Dieter. *Die Webstühle des Tzu-jen i-chih aus der Yuan-Zeit*. Wiesbaden: Steiner, 1977.

Kuhn, Dieter. "Spinning and Reeling." In *Science and Civilisation in China*. Vol. 5, Part 9. Ed. Joseph Needham. Cambridge: Cambridge University Press, 1988.

Kuhn, Dieter. *Zur Entwicklung der Webstuhltechnologie im alten China*. Heidelberg: Edition Forum, 1990.

Riboud, Krishna. "A Closer View of Early Chinese Textiles." In *Studies in Textile History. In Memory of Harold B. Burnham*. Ed. Veronika Gervers. Toronto: Royal Ontario Museum, 1977.

Schaefer, Gustav. "The Hand-loom of the Middle Ages and the Following Centuries." *Ciba Review* II 16: 542–577, 1938.

See also: Textiles

SINĀN IBN THĀBIT Abū Saʿīd Sinān ibn Thābit ibn Qurra was born in Baghdad ca. 880. He was the son of the translator and scientist Thābit ibn Qurra (ca. 830–901), who dedicated his *al-Dhakhīra fī ʿilm al-ṭibb* (Treasury of Medicine) to him, and the father of the mathematician and astronomer Ibrāhīm ibn Sinān ibn Thābit (908–946). His family came from Ḥarrān, crossroads of caravan routes and cultures on the Upper Euphrates, and the religious center of the Sabian sect. One of the most famous physicians of his time, he directed the hospitals and medical administration of Baghdad under the reign of the Abbassid Caliph al-Muqtadir (908–932). He also served as physician to him and to his successors al-Qāhir (932–934) and al-Rāḍī (934–940). Under the former's persecution of the Sabians Sinān had to convert to Islam. He later fled to Khurāsān, and returned when al-Rāḍī came into power.

Sinān's work, as catalogued by Ibn al-Qifṭī, can be divided into three categories: historical–political, mathematical, and astronomical; no medical texts are mentioned. None of it is extant. In the first category he wrote a description of the life at the court of Caliph al-Muʿtaḍid (892–902), his father's protector, and a sketch for a government along the lines of Plato's *Republic*. Both treatises are mentioned by al-Maʿsūdī, who criticizes the latter, adding that Sinān should rather have occupied himself with topics within his competence.

Four mathematical treatises are listed, of which two cannot be by Sinān, since the dates given were in the second half of the tenth century. Of the remaining two, one is connected with Archimedes' *On Triangles*, whereas the other consists of an improvement, with additions, of the book by Aqāṭun (Yāqūt = Euclid), *On Elements of Geometry*.

As to the third category, only the content of the *Kitāb al-Anwāʾ*, dedicated to al-Muʿtaḍid, is somewhat known through excerpts given by al-Bīrūnī in the *al-Āthār al-bāqiya min al-Qurūn al khāliya* (Chronology of Ancient Nations). A *Kitāb al-Anwāʾ* is a kind of almanac describing the astro-meteorological properties (*anwāʾ*) of the individual days. Sinān's almanac is fundamentally based on an anonymous Arabic translation of Ptolemy's *Phaseis*, in which he provides the times of the rising and setting of prominent fixed stars. Other sources are Ibn Māsawayh (d. 857) and Hippocrates, as well as personal contributions by Sinān. According to al-Bīrūnī, Sinān also relates an Egyptian theory and one by Hipparchus, on where to fix the beginnings of the seasons. Another treatise, seemingly also of astro-meteorological content, is on the assignment of the planets to the days of the week, composed for the Sabian Abū Isḥāq Ibrāhīm ibn Hilāl (ca. 924–994). The seven planets were important in Sabian religion; each one had its own temple. Ibn al-Qifṭī lists several works on Sabian rites and religion. Sinān died, probably in Baghdad, in 943.

YVONNE DOLD-SAMPLONIUS

REFERENCES

Biesterfeld, H.H. "Ğālīnūs Quwā n-nafs." *Der Islam* 63: 126, 1986.

Dold-Samplonius, Yvonne. "Sinān ibn Thābit ibn Qurra." In *Dictionary of Scientific Biography* vol. XII. New York: Charles Scribner's Sons, 1970, pp. 447–448.

Dold-Samplonius, Yvonne. *Book of Assumptions by Aqāṭun*. Thesis, Municipal University of Amsterdam, 1977 (to be reprinted in the series *Algorismus*, München).

Samsó, Julio and Blas Rodríquez. "Las "Pháseis" de Ptolemeo el 'Kitāb al-Anwāʾ' de Sinān b. Ṯābit." *al-Andalus* 41: 15–48, 1976.

See also: Thābit ibn Qurra – Astronomy in Islam – al-Bīrūnī

SIYUAN YUJIAN This book (The Jade Mirror of the Four Unknowns, also translated as Precious Mirror of the Four Elements) by Zhu Shijie, dating from 1303, appears to represent the most advanced achievement of the mathematical tradition of polynomial algebra or the "procedure of the celestial unknown" which developed in north China in the second half of the thirteenth century. Jock Hoe describes the history of its reception among mathematicians and historians in both East and West. The *Siyuan yujian* shares the same basic features with the first book that has come down to us on this topic, Li Ye's (Li Zhi's) *Ceyuan haijing* (*Sea-Mirror of the Circle Measurements*, 1248). The mathematical knowledge it contains is presented in the form of problems, the first four of which lay the foundations for the whole book by gradually presenting, though extremely concisely, how to use polynomials with up to four indeterminates to solve given problems. Mirroring the intimate relationship that the right-angled triangle had with the topic of algebraic equations, and later of polynomials, from the very first appearance of these matters in China, these problems all concern the right-angled triangle, as do 101 of the 284 problems that follow in the *Siyuan yujian*. Once a problem is set, Zhu suggests the choice of one or several unknowns to form polynomials on the basis of the conditions given in the terms of the problem. In the case when one unknown suffices, like Li Ye, Zhu forms two *polynomial expressions*, representing the same quantity, and thus by elimination between them establishes an algebraic equation, a root of which is the unknown sought. Since Zhu does not explain the algorithms, similar to the so-called Ruffini-Horner ones, that can be used to determine this root, he seems to have assumed the reader's knowledge of this procedure. They are found in Qin Jiushao's *Shushu jiuzhang* (*Mathematics in Nine Chapters*, 1247) and seem to be common knowledge in the milieu developing the "procedure of the celestial unknown". Yet Zhu elaborates them further, especially in cases where the result is a rational number; he then systematically introduces techniques to expand the root sought.

These polynomials he uses are written in the form of an array of numbers, as in Figure 1, where the indeterminate and its successive positive or negative powers receive a positional expression. Note that the writing of a polynomial differs from the writing of an equation, as in Figure 2, in that the constant term of the polynomial is indicated, being accompanied by a character, thus allowing the polynomial to contain any power of the indeterminate.

When more than one unknown is needed, Zhu designs several equations involving them, which he represents by tables of numbers, where again position is used to express the various powers of the unknowns. For two unknowns, the vertical direction refers to one unknown whereas the horizontal one refers to the other; the surface of the plane suffices for any monom to be represented by its coefficient written in an appropriate place on the counting board, as in Figure 3. Yet for three or four unknowns, limitations occur in the representation of all possible monoms and new positions on the counting board are brought into play, but in a less systematic way, as in Figure 4.

In contrast to equations with one unknown, such equations need the constant term to be indicated by a character. Once these equations are determined, elimination is performed between them until an equation with only one unknown is obtained, which is then solved. Several techniques are used for this purpose, the most basic one resembling

太
0
729
-81
-9
1

Figure 1: $729x^2 - 81x^3 - 9x^4 + x^5 = 0$.

0
-3888
729
-81
-9
1

Figure 2: $x^5 - 9x^4 - 81x^3 + 729x^2 - 3888 = 0$.

2	0	太
-1	2	0
0	0	0
0	0	1

Figure 3: $x^2 + 2xy + 2y^2 - y^2x = 0$.

0	-1	太	-1
-1	0	-1	0

Figure 4: $xyz - xy^2 - x - y - z = 0$.

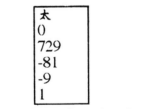

the elimination between linear equations put into play by the algorithm to solve systems of simultaneous linear equations presented in the Han book *Jiuzhang suanshu* (The Nine Chapters on Mathematical Procedures). The difference lies in the fact that now the terms involved in the procedure of elimination are polynomials, and no longer numbers. Stress must be placed on a conceptual gap. In contrast to the case for one unknown, the only one Li Ye dealt with, the elimination no longer bears on polynomials representing a given *quantity*, but rather on equations with several unknowns representing a *relationship* between the unknowns.

In the rest of the book, this apparatus is applied to all kinds of mathematical situations, such as geometry and summation of series, involving various applications to economic or social life.

KARINE CHEMLA

REFERENCES

Du Shiran. "Zhu Shijie yanjiu" (Research on Zhu Shijie). In *Song Yuan Shuxueshi Lunwenji* (*Collected Essays on the History of Mathematics during the Song and Yuan Dynasties)*. Ed. Qian Baocong. Beijing: Kexue chubanshe, 1966, pp. 166–209.

Ho Peng Yoke. "Chu Shih-chieh." In *Dictionary of Scientific Biography*, vol. 3. Ed. Charles Gillispie. New York: Charles Scribner's Sons, 1970–80, pp. 265–71.

Hoe, Jock. *Les Systèmes d'Équations Polynômes dans le* Siyuan yujian *(1303)*. Paris: Institut des Hautes Études Chinoises, 1977.

Juschkewitsch, Adolf Pavlovitch. *Geschichte der Mathematik im Mittelalter*. Leipzig: Teubner, 1964.

Qian Baocong. *Zhongguo Shuxue Shi (History of Mathematics in China*), 2nd ed. Beijing, 1981.

See also: Zhu Shijue – Mathematics in China – Li Zhi – Qin Jiushao – Liu Hui and the *Jiuzhang Suanshu*

SONG CHI Song Chi was born in Jianyang in 1186 of the Southern Song Dynasty and died in 1249 in Guangzhou. He was a forensic scientist, an expert in legal medicine.

A native of Jianyang (now Jianyang County, Fujian Province), Song Chi was born to a family of middle-class bureaucrats. His father was Song Huifu. He was well-educated, and passed the feudal examination at the highest (imperial) level. He held several posts, including being the confidential secretary of local government, county magistrate, and provincial officials until he was promoted to the provincial official responsible for law suits and jail affairs. At that time, most feudal officials did not take their responsibilities seriously, especially those involving the law, resulting in the pile-up of unsolved law suits. Many were legal offenders themselves. Song ordered deadlines to sort out all these cases. Eventually, he cleared a backlog of over two hundred cases.

During his tenure, he set up a whole series of rules for examiners of the wounded or the dead. He stressed that legal medical experts or officials must, under the accompaniment of an examiner, arrive at the spot as soon as possible, and that no fumigation or incense was to be used. The officer or expert should examine the case in person, instead of hearing the examiner's report. He also claimed that "the most important issue in criminality is the death sentence, which should be carefully understood at the very beginning of the case by serious investigation". He also pointed out that the malfeasance of prison officials always begins with a tiny error during the investigation. Lack of experience and cursory investigation are the direct causes of wrong cases and unjust verdicts. He criticized or even punished those fraudulent examiners whenever cases were in question, and he pondered carefully and conscientiously so that few wrong conclusions were drawn. Based on his long-term practical experience, he compiled, in 1247, the famous *Xiyuan Jilu* (Collected Records of Washing Away the Wrong Cases, also translated as The Washing Away of Wrongs) which is the earliest work in medico-jurisprudence both in China and abroad. This book has been translated into several Western languages.

HONG WULI

REFERENCES

Gao, Mingxuan. "The First Work on Forensic Medicine." In *Achievements of Ancient Chinese Science & Technology*. Beijing: Chinese Youth Press, 1978, pp. 474–479.

Jia, Jingtao. *A History of Ancient Chinese Forensic Medicine*. Beijing: Mass Publishing House, 1984.

Song, Ci. *The Washing Away of Wrongs* . Shanghai: Shanghai Science & Technology Press, 1981.

See also: Forensic Medicine in China

SONG YINGXING AND THE *TIANGONG KAIWU* Song Yingxing (alias Changgeng, 1587–ca. 1666) was a great scientist and thinker of the seventeenth century. He was born in Fengxin county in Jiangxi province in 1587. After graduating from the county school he became *jüren* or a successful candidate in the imperial examination at the provincial level in 1615. He took part in the highest imperial examinations in Beijing six times, but failed to become a *jinshi*. After that he decided to give up Confucianism and turned to the practical knowledge of natural science. From 1635–163 he taught at Fenyi county school; then he was appointed the prison officer of Tingzhou prefecture in Fujian province from 1638 to 1640. In 1643 he worked as the magistrate of Bozhou prefecture in Anhui province. After the Ming dynasty (1368–1644) was overthrown by Manchu-Qing rulers, Song lived at home as a hermit until his death in about 1666. He had extensive knowledge in both humanistic and natural sciences. He traveled from the south to the north to obtain practical knowledge of agricultural and industrial techniques as well as political and economical information on Chinese society. This made it possible for him to write more than ten books, most of which were completed at Fenyi from 1636 to 1637.

The *Tiangong Kaiwu* (Exploitation of Products from Nature by Means of a Combination of Artificial Skills and Natural Power, also translated as The Creations of the Nature and Man, 1637), is his most important work. It consists of 18 chapters on 33 departments of agricultural and industrial production including grain planting, sericulture, spinning and weaving, dyeing, salt and sugar making, vegetable oils, distiller's yeasts, ceramics and porcelain, metallurgy and casting, paper and ink making, coal, alum and copper producing, carriages and shipbuilding, cold weapons and firearms, and pearl and jade exploiting. Apart from detailed descriptions of the above mentioned techniques, there are 123 useful illustrations in this book. It recorded many advanced technical achievements obtained from the distant past until the Ming dynasty and made a comprehensive generalization of them. Some had been rarely talked about in previous works, such as improving paddy and silkworm varieties through artificial hybridization, smelting zinc from calamine, smelting wrought iron from pig iron, producing alloys of copper and zinc, etc. This work also developed many advanced technical ideas, especially the *Tiangong Kaiwu* thought which emphasized the coordination between mankind and nature as well as artificial skills and natural power in order to exploit various products from nature for people's use.

Not only was the *Tiangong Kaiwu* welcomed in China; it also spread to Japan and Korea from the seventeenth to the eighteenth centuries and became a popular reference work for Japanese and Korean scholars during the Edo period (1608–1868) and the Yi dynasty (1392–1910). It was reprinted in Osaka in 1771 and translated into Japanese in the 1950s. During 1830–1840 five chapters of it were translated into French, then into English and German. The part on sericulture was cited by Charles Darwin and highly valued. In 1869 many chapters on industrial techniques were collected in the *Industries anciennes et modernes de l'Empire Chinois* published in Paris. The first four chapters on agriculture were translated into German in 1964; two years later the whole book was published in English in the United States. It was regarded as an 'encyclopédie technique' by French sinologist Stanislas Julien of the nineteenth century, and Joseph Needham, the British scholar, called Song a "Chinese Agricola" or "Chinese Diderot". It has now become a world-famous classical scientific work.

Among Song's existing works there are also the *Ye Yi* (Proposals to the Court from Common People, 1636), *Si Lian Shi* (Poems on Praising the Good and Pitying the Foolish, 1637), *Lun Qi* (On the *Qi*, 1637) and *Tan Tian* (On Celestial Bodies, 1637). The *Ye Yi* reflected the author's political and economic thought. He put forward a plan of political reform for the Ming court for averting the political and economical crisis in the society. He considered that social wealth was created by the labor and that currency should not be the definition of wealth. He believed that increasing social wealth meant producing more consumable goods. Therefore, his ideas are similar to that of Adam Smith.

The *Lun Qi* is a work on natural philosophy. It explains the formation and composition of all things by means of a combination of the theory on the *qi* and the new theory of *yinyang* and five elements. He introduced a new conception of *xing* (form or shape) between the *qi* and five elements, and he explained the material unity in the composition between organic and inorganic substances. He also discussed the idea of conservation of matter in this work. It also dealt with acoustical problems, and pointed out that the formation of sound is the result of the vibration of substances and the medium of the spread of sound is the air. Song considered that the spread of sound in the air should be in the form of a water wave; he thus put forward a preliminary conception of a sound wave.

In his astronomical work, the *Tan Tian*, he criticized a traditional theory that the solar eclipse was connected with human affairs. He thought that all things including the sun are always changing. He said: "Therefore, the sun today is no longer the same one as yesterday". Most of Song's advanced philosophical thought was further developed by another great philosopher, Wang Fuzhi (alias Chuanshan, 1619–1692). Those in academic circles are familiar with his scientific achievements, but much less familiar with his political, economic, and philosophical thoughts, because those works were only found in the mid-twentieth century and have not been carefully studied yet.

PAN JIXING

REFERENCES

Pan, Jixing. *Mingdai Kexuejia Song Yingxing* (Biography of Song Yingxing, the Scientist of the Ming Dynasty). Beijing: Science Press, 1981.

Pan, Jixing. *Tiangong Kaiwu Jiaozhu yu Yanjiu* (Collations, Explanations and Researches of the *Tiangong Kaiwu*). Chengdu: Bashu (Sichuan) Publishing House, 1989.

Pan, Jixing. *Tiangong Kaiwu Yizhu* (Translation into modern Chinese and explanations of the *Tiangong Kaiwu*). Shanghai: Classical Works Publishing House, 1993.

Song Yingxing. *T'ienkung K'aiwu. Chinese Technology in the 17th Century.* Trans. E-Tu Zen Sun et al. College Park: Pennsylvania State University Press, 1966.

Yabuuti, K. *Tienko Kaibutsu no Kenkyu* (Translation and Researches of the *Tiangong Kaiwu*). Tokyo: Kosai Sha, 1953.

SPHUJIDHVAJA Sphujidhvaja was apparently of Greek descent, and flourished in Western India under the patronage of the Kṣatrapa ruler Rudradāman II. He wrote the extensive genethlialogical manual entitled *Yavanajātaka* (Horoscopy of the Greeks) in AD 270, which shows the position and influence of the stars at one's birth. Towards the close of his work, Sphujidhvaja says that, before him, in AD 150, the great Greek genethlialogist Yavaneśvara redacted into Sanskrit prose a Greek astrological work, so that it could be studied by those who did not know Greek, and that he, Sphujidhvaja is composing a versified redaction of the work of Yavaneśvara. The work reveals Sphujidhvaja as a competent scholar, a master of Sanskrit versification, and an expert genethlialogist.

Sphujidhvaja states that he composed the work in 4000 verses. But the only manuscript of the work available today contains only 2300 verses. In this imperfect manuscript, the first few sections are numbered, but not so the subsequent ones, which number, in all, 79. Possibly several sections are lost. In these 79 sections, the work covers a large number of aspects of horoscopy and natural astrology, including:

- Zodiacal signs and planets, their icons, nature, and relationships;
- Iconography of *horās* and decons;
- Astrology of conception, birth, and nature;
- Horoscopes;
- Planetary placements, and combinations affecting human beings;
- Prediction on the basis of questions;
- Reconstruction of lost horoscopes;
- Omens and dreams; and
- Military astrology.

The several aspects of each of these topics are looked at from different points of view, and intimations, indications and predictions based thereon are stated categorically.

It is interesting that, although based ultimately on a Greek text, there was substantial Indianization effected in the *Yavanajātaka.* This was accomplished by using Indian equivalents to Greek terms, adopting Hindu deities in place of Greek ones, incorporating the names of Hindu castes and professional orders, mentioning local manners and customs, and the like. *Yavanajātaka* was looked upon as authority by all later Indian genethlialogists and used as such even in the foremost of Indian texts like *Bṛhajjātaka* and *Bṛhatsaṃhitā* of Varāhamihira, *Sārāvalī* of Kalyāṇavarman, *Praśnavidyā* of Bādarāyaṇa, and many others.

The significance of *Yavanajātaka* stems from another point as well. The work refers to contemporary social orders, professions, religious classes and groups, items of ordinary use, manners and customs, dress, and a host of other things related to the life and society of the times. The information provided by the work on these subjects makes it a good source for the study of the culture and civilization of India during the early centuries of the Christian era.

K.V. SARMA

REFERENCES

Bṛhajjātaka of Varāhamihira with the commentary of Bhaoṭṭotpala. Bombay : Jnanadarpana Press, 1874.

Bṛhatsaṃhitā by Varāhamihirācārya, with the commentary of Bhaoṭṭotpala. 2 pts. Ed. Avadha Vihari Tripathi. Varanasi: Varanaseya Sanskrit University, 1968.

Pingree, David. "The Yavanajātaka of Sphujidhvaja." *Journal of Oriental Research* (Madras) 31(1–4): 26–32, 1961–62.

Sārāvalī of Kalyāṇavarman. Ed. V. Subrahmanya Sastri. Bombay: Nirṇayasāgar Press, 1928.

The Yavanajātaka of Sphujidhvaja. Ed. and trans. David Pingree. Cambridge, Massachusetts: Harvard University Press, 1978.

See also: Astrology in India – Yavaneśvara – Varāhamihira

ŚRĪDHARA The mathematical works of Śrīdhara were very popular and made him quite famous. In spite of his great popularity, some controversies have been raised about his life, work, and time, such as whether he was a Hindu or a Jaina, and whether he wrote before or after Mahāvīra (ninth century AD). Some uncertainties exist because Śrīdhara's works are not fully extant. Often he is confused with other authors of the same name. Here we shall give views which are now generally accepted.

Like so many ancient Indian authors, Śrīdhara did not provide any information about himself in his works. Other sources have not been helpful in finding any glimpse of his personal life. So we do not know his parents or teachers.

or even where he was born, educated, or worked. But some evidence shows that he was a Saivite Hindu. An example in his *Pāṭīgaṇita* is about the payment for the worship of the five-faced Hindu god Śiva. He starts his *Triśatikā* with a homage to the same god. However, in a manuscript which has a possible commentary and additional examples written by an apparently Jaina writer, the word "Jinam" is found in place of "Śivam". This led to the claim that Śrīdhara was a Jaina, but other evidence does not support this view.

More serious is the controversy about Śrīdhara's time. That he lived after Brahmagupta is evident from the fact that he literally quoted (and criticized) a rule from Brahmagupta's *Brāhmasphuṭasiddhānta* (AD 628). The most significant question in this connection is whether Śrīdhara wrote before or after Mahāvīra, whose date of ca. AD 850 is certain. There are similarities in the works of these two mathematicians. Most scholars believe that indications generally place Śrīdhara before Mahāvīra. A new fact has come to light recently; David Pingree found that a rule of Śrīdhara's had been quoted by Govindasvāmin (about AD 800–850) in his commentary on the *Horāśāstra* of Parāśara (under 14.97). This latest and direct evidence once again supports the generally accepted view of placing Śrīdhara in the eighth century or ca. AD 750.

Śrīdhara is known to be the author of the following three works on mathematics:

(1) *Pāṭīgaṇita* (on arithmetic and mensuration);
(2) *Pāṭīgaṇita-sāra* (an epitome of the above); and
(3) *Bījagaṇita* (on algebra).

The *Pāṭīgaṇita* is a standard Indian treatise on practical mathematics meant "for the use of the people". It is also called *Bṛhat-Pāṭi* (Bigger Pāṭi) and *Navaśatī* (Having Nine Hundred) because it is believed to have nine hundred stanzas. Unfortunately, it is not extant in full; the available text contains only two hundred fifty-one stanzas. In terms of topics, treatment of definitions, logistics, mixtures of things, and series is available in full, but that of the plane figures is incomplete. Many other treated topics are mentioned in the list of contents, but they are totally missing from the manuscript. These were excavations, piles of bricks, sawn pieces of timber, heaps of grain, shadows, and the mathematics of zero (whose loss is quite sad).

The *Pāṭīgaṇita-sāra* is also called *Triśatī* or *Triśatikā* because it was a "Collection of 300 verses". It was the author's own summary of the larger work on the subject.

Śrīdhara's *Bījagaṇita* (Algebra) seems to be lost completely. We know about it from a statement of Bhāskara II (AD 1150) who also quoted a rule from it. The rule gives a method for solving any quadratic equation and has also been quoted by others. Many other rules from different works of Śrīdhara have also been quoted by various writers. This shows the popularity of his works.

A recently discovered work called *Gaṇita-Pañcaviṁśī* is also stated to be from the pen of Śrīdhara, but it may not be his genuine work. Similarly, a number of astronomical, astrological, and other works authored by different persons of the same name have been ascribed wrongly. Confusions were created both by the similarity of names and also by the proximity of their dates.

In India, ten has been the base of counting since very early times. But the number and names of decuple terms used in the decimal numeration were at variation throughout the ancient period. Later on, the decuple terms were used to denote the notational places when the positional system was developed, and their number was standardized to eighteen (which was a traditionally sacred number). Śrīdhara gave a definite list of eighteen terms which became standard in Indian mathematics. It runs thus: *eka*, *daśa*, *śata*, *saharsa*, *ayuta*, *lakṣa*, *prayuta*, *Koṭi* ($= 10^7$), *arbuda*, *abja* (or *abda*), *kharva*, *nikharva*, *mahāsaroja*, *śaṅku* (or *śaṅkha)*, *saritāṁpati*, *antya*, *madhya*, and *parārdha* ($= 10^{17}$).

Among the arithmetical rules discussed in the *Pāṭīgaṇita, Sūtra* 49–50 gives a formula for finding the time in which a sum lent out at simple interest will be paid back by equal monthly installments. Under the *Sūtra* 63–64 he presents the famous problem of a Hundred Fowls, in which we have to solve the indeterminate equations in integers.

For mensurations related to a circle, Śrīdhara used the approximation $\pi = \sqrt{10}$, a very ancient Indian value. But there is some evidence to show that he used $\pi = \frac{22}{7}$ in the lost part of *Pāṭīgaṇita*.

Another of Śrīdhara's great achievements was his mensuration of the volume of a sphere.

R.C. GUPTA

REFERENCES

Asthana, Usha. *Ācārya Śrīdhara and his Triśatikā*. Ph.D. thesis, Lucknow University, 1960.

Datta, B. "On the Relation of Mahāvīra to Śrīdhara." *Isis* 17: 25–33, 1932.

Dvivedi, Sudhakara, ed. *Triśatikā of Śrīdhara*. Benares: Chowkhamba Sanskrit Book Depot, 1899.

Ganitanand. "On the Date of Śrīdhara." *Gaṇita Bhāratī* 9: 54–56, 1987.

Pingree, David. "The Gaṇitapañcaviṁśī of Śrīdhara" (with text). In *Ludwik Sternbach Felicitation Volume*. Ed. J.P. Sinha. Lucknow: Akhila Bharatiya Sanskrit Parishad, 1979, pp. 887–909.

Ramanujacharia, N. and G.R. Kaye. "The Triśatikā of Śrīdharācārya" (with translation). *Bibliotheca Mathematica Series 3*, 13: 203–217, 1912–1913.

Shukla, K.S., ed. and trans. *The Pāṭīgaṇita of Śrīdharācārya*. Lucknow: Lucknow University, 1959.

Singh, Sabal. "Time of Śrīdharācārya." *Annals of the Bhandarkar Oriental Research Institute* 21: 267–272, 1950.

See also: Mahāvīra – Algebra in India – Mathematics in India – Bhāskara

ŚRĪPATI Śrīpati was the most prominent Indian mathematician of the eleventh century. He was the son of Nāgadeva and lived in Mahārāṣṭra. He flourished during the period from AD 1039–1056.

General appreciation for Śrīpati's fame as a mathematician is based on his arithmetic, *Gaṇitatilaka*, and the two mathematical chapters of his astronomical work entitled *Siddhāntaśekhara*. The thirteenth chapter, the *Vyktagaṇitādhyāya*, contains arithmetical rules, series, mensuration, and shadow reckoning. The fourteenth chapter, the *Avyaktagaṇitādhyāya*, is one of the few extant Hindu works on algebra. The only edition (by Kapadia) of the *Gaṇitatilaka* contains a valuable commentary by Siṁhatilaka (ca. AD 1275).

Besides including in his works selected rules of his predecessors, Śrīpati enriched Indian mathematics by giving some improved and some original rules. The *Gaṇitatilaka* contains the earliest known treatment of simple addition and subtraction in any Indian work. Also included is the earliest version of our angular method of addition used to check the accuracy of a sum and simplify addition. The *Vyktagaṇitādhyāya* gives an improved rule for the volume of an excavation.

The *Avyaktagaṇitādhyāya* contains most of Śrīpati's original ideas and rules. In it, he presents the idea of an extensive system of symbolism in algebra, the only Hindu treatment of the cubing of signed numbers, and explicit recognition of the nature of imaginary quantities that is second only to Mahāvīra's. He also provides the earliest versions of our ordinary method of solving simple linear equations by using inverse operations and the earliest formal treatment of factorization. The rules not only give the ordinary method of factoring based on successive division, but also an additional method for factoring a non-square number by expressing it as the difference of two squares.

Śrīpati also displayed his mastery of indeterminate equations by giving an improved rule for solution of the factum and original rules for the solution of the pulverizer. He also described an original method to obtain rational solutions of the square-nature.

All Śrīpati's works are in verses. The *Siddhāntaśekhara* contains only rules, but the *Gaṇitatilaka* is written in an autocratic style of teaching strategy, in which a rule is followed by plentiful exercises. The book is quite secular in nature. According to the author's testimony, it was written for public use. It contains arithmetical and commercial rules, and some problems solvable by simple linear, quadratic, and radical equations. A garland problem and numerous other fanciful problems included in the book make it pleasurable and interesting. The practical and aesthetic values of the *Gaṇṇitatilaka* cannot be underestimated.

Śrīpati, however, also had some weaknesses. His mathematics of division by zero is all wrong. In the *Vyaktagaṇitādhyāya*, he uses the term *caturbūja* to mean a square, a quadrilateral in general, a cyclic quadrilateral, a quadrilateral with equal altitudes, and an isosceles trapezoid. Because of this inconsistency, his rules on mensuration of quadrilaterals are hard to interpret.

KRIPANATH SINHA

REFERENCES

Kapadia, H.R. *Gaṇitatilaka by Śrīpati*. Baroda: Gackwad's Oriental Series No. 78, 1937.

Misra, Babuaji. *Siddhāntaśekhara by Śrīpati, Pt. II*. Calcutta: Calcutta University Press, 1947.

Sinha, Kripanath. "Śrīpati: An Eleventh-Century Indian Mathematician." *Historia Mathematica* 12: 25–44, 1985.

See also: Algebra in India – Mathematics in India – Mahāvīra

STARS IN CHINESE SCIENCE In Chinese astronomy, stars were called *Hengxing* (fixed stars) or *Jingxing* (warp stars), while planets were called *Xingxing* (movable stars) or *Weixing* (web stars). The traditional Chinese sky was composed of 283 constellations which included 1464 individual stars. This unique system was first established by Chen Zhuo, an astronomer of the Three Kingdom period (AD 220–280), who collected knowledge about stars from the Han time and summarized it in a new catalog of stars. According to Chen Zhuo, these constellations belonged to three separate astronomical schools: 93 constellations belonged to Shi Shi (Master Shi), 118 to Gan Shi (Master Gan), and 44 to Wuxian Shi (Master of Wuxian). There were also twenty-eight lunar mansions (*Xiu*), which were very important in the Chinese sky. Stars of the three astronomical schools were preserved in the *Kaiyuan Zhanjing* (Treatise on Astrology) of the Kaiyuan period (AD 729) from the Tang Dynasty (AD 618–907). After Chen Zhuo little was changed of the Chinese system of constellations.

The observation of stars started as early as civilization began. Among the oracle bone inscriptions from the Shang Dynasty (ca. sixteenth–eleventh centuries BC) some stars were mentioned. In the *Shangshu* (Historical Classic), the paragraph concerning four cardinal asterisms has generally been agreed to be the record of the observation of stars from before the twenty-first century BC. But before the Han Dynasties (106 BC–AD 220), no description of the total sky existed; only thirty-eight star names were mentioned in pre-Han literature. These stars were either of the twenty-eight *Xiu* or were very popular stars such as *Beidou* (Ursa Major), *Niulang* (α Aql), and *Zhinü* (Vega, α Lyr). They appeared in folklore and poems. During the Warring States period (480–222 BC) there lived two famous astronomers, Shi Shen and Gan De, who were founders of two of the abovementioned astronomical schools. But what were called Shi Shi's and Gan Shi's constellations did not seem to exist in their own time; they were later developments of the nomenclature of stars during the Han. The earliest descriptions of the total sky was in the *Tianguan Shu* (Monograph on Heavenly Officers) by Sima Qian (145–86 BC) in which ninety-one constellations including about five hundred stars were mentioned. The earliest star catalog, giving coordinates of ninety-two leading stars of Shi Shi's constellations and of twenty-eight determinative stars of the twenty-eight *Xiu*, were observed a few decades after Sima Qian. That is, the evidence that shows that Shi Shi's constellations were scientifically established during the Former Han (206 BC–AD 25). Shi Shi's constellations consisted of the brightest stars in the sky. They served as the referential system to constellations of other schools. Most constellations of Gan Shi and Wuxian Shi were just fill-ins among Shi Shi's constellations. It is quite certain that the constellations of the three schools obtained their complete forms during the Han times. Chen Zhuo's work was to summarize them.

Constellations in the Chinese sky are much smaller than the Greek ones. A constellation usually includes several stars linked together with imaginary lines. Some constellations are even composed of single stars. Constellations were called inner ones or outer ones according to their location to the North or South of the equator.

The twenty-eight *Xiu* were the basis of the Chinese sky. They spread near the equator and served as one of the dimensions of the Chinese polar-equatorial coordinate system. One coordinate corresponding to the Right Ascension was given as the hour distance to the determinative star of a certain *Xiu*, which was almost always the most western star in the *Xiu*. With great circles linking the determinative stars and the North Pole, the celestial sphere was divided as segments of an orange. Constellations were described in the sequence of the *Xiu* as they were located.

Historical research shows that the twenty-eight *Xiu* are much earlier than the whole constellations of the three astronomical schools. Some star names seemed to be direct derivations from the mythological, legendary, or astrological meaning of the *Xiu*. A complete set of the names of the twenty-eight *Xiu* was discovered on a lacquered box cover excavated from a tomb dated from 433 BC. The *Xiu* were formed into four groups of seven which were called Four Images. The Four Images corresponded to the four cardinal points in the sky and the four seasons of the year.

Ancient India and the Islamic world also had similar lunar lodge systems. It is certain that these three systems were originally related. But when and where they were first constructed and how they diffused is still an unsolved problem.

Another feature of the Chinese sky is the existence of the three-wall system. There were three enclosed areas in the sky: *Ziwei Yuan*, *Taiwei Yuan*, and *Tianshi Yuan*. The meaning of *Yuan* is a wall, which was formed by stars surrounding the enclosed area. *Ziwei Yuan* indicated that polar area which included fifteen constellations, and the wall was formed with fifteen stars. *Tianshi Yuan* and *Taiwei Yuan* indicated two other groups of constellations between *Ziwei Yuan* and the ecliptic (the band of the zodiac through which the sun apparently moves in its yearly course). The former had thirteen constellations, and the wall was formed with twenty-two stars; the latter also had thirteen constellations, but the wall was formed with ten stars. The names of the three-wall system might have existed early, but their use to indicate groups of constellations was rather late, probably after Chen Zhuo.

Since the Tang Dynasty, the three *Yuan* and the twenty-eight *Xiu* became the main structure by which the stars were organized. The famous astronomer Li Chunfeng (AD 602–670) used this system to describe the sky in *Jinshu Tianwenzhi* (The Astronomical Chapter of the History of the Jin Dynasty, AD 265–420).

The names of the Chinese constellations provide a unique document to be studied. The way stars were named was completely different from the Greek. Stars were organized into a heavenly human society. In the sky you can see:

- emperor, queen, and princes;
- royal court and clan;
- imperial bureaucracy and administration;
- all kinds of buildings and facilities, even a toilet in the sky;
- military installation, armies and weapons;
- traffic and transportation;
- rituals, ceremonies, and pictures of social life;
- philosophical and religious concepts;
- mythological and legendary figures; and
- local states and geographical features.

It is a total cultural complexity projected in the sky. In the polar area, for example, the *Ziwei Yuan* was furnished with an emperor, queen, princes, royal high-ranking officials, and facilities in a court. There was one constellation called the North Pole, *Beiji*, the brightest star of which was called *Di* the Emperor. The *Ziwei Yuan* was just like a royal palace; it took the polar area and all the stars turned round it. It reflected the social order of the imperial society of ancient China, as Confucius analogized: "He who is a governor because of his Virtue may be compared to the North Pole Star, sticking to its place and all stars turning about it". In other parts of the sky, social life pictures were reflected. The *Tianshi Yuan* was a heavenly market, in which all kinds of shops were marked. The Autumn harvest of peasants, the imperial hunt ceremony, and cavalry troops could also be found in the Chinese sky.

Because the appearance of a certain part of the sky is related to seasons, and because it also changes with the precession of the sky, seasonal considerations may help to estimate the time when these earthly happenings were projected in the sky. By and large this practice happened during the Han and was carried on with the impetus given by the philosophy of *Tian Ren Gan Ying* (The Interaction between Heaven and Human Beings) which came to a climax during the Han and which urged a new type of astrology by means of division of the sky. The sky was constructed as a heavenly imperial society so that astrologers could interpret imperial society straightforwardly in correlation with the terrestrial society. Stars in the Chinese sky provided a basis for ancient Chinese astrology.

Throughout the history of China, there has been continuous observation of stars. Although the constellations have been kept basically unchanged, new measurements of the position of the stars were made in almost every dynasty. Star maps were made to study the sky. The earliest existent star map is from the *Dunhuang* collection of manuscripts of the Tang Dynasty. More accurate and scientific star maps are from the Song Dynasty (AD 960–1279). The Suzhou Astronomical Chart on a stone plate made in AD 1274 is well known. Almost all stars in the Chinese sky were carved quite precisely on the planisphere. The star map was based on the observations made during the Song. Even the supernova of the year of 1054 in Taurus was marked on the star map.

Since the late Ming Dynasty (AD 1368–1644), European astronomy began to influence China. Star catalogs and star maps were made in the way Westerners did. Traditional stars were identified with the sky and more stars were observed. But the framework of the traditional constellations was not abandoned. Newly measured stars were counted as supplementary stars in old constellations. New constellations were added only in the southern sky, where stars had not been registered by ancient Chinese astronomers. The pioneer work of this kind was done by Xu Guangqi (AD 1562–1633) and the German Jesuit Adam Schall von Bell (AD 1592–1666) in *Chongzhen Lishu* (Calendrical Treatise of Chongzhen Period, AD 1628) at the end of the Ming Dynasty. In this book, 23 new constellations including 126 stars were constructed in the southern hemisphere of the sky. Their work paved the way for the Chinese sky to be absorbed into world astronomy.

SUN XIAOCHUN

REFERENCES

Ho Peng Yoke. *Astronomical Chapters of Jinshu*. Paris and the Hague: Mouton & Co., 1966.

Pan Nai. *Zhongguo Hangxing Guanceshi* (A History of Star Observation in China). Shanghai: Scholar Press, 1989.

Qutan Xida (Gautama Siddharta). *Kaiyuan Zhanjing* (Treatise on Astrology of the Kaiyuan Period). AD 729.

Sun Xiaochun. *A Study of the Chinese Sky During the Han: Constellations of Shi Shen Shi, Gan De Shi and Wu Xian Shi*. Doctoral thesis, Amsterdam and Beijing, 1993.

See also: Lunar Mansions – Gan De – Sima Qian – Astrology in China – Astronomy in China

STARS IN ARABIC-ISLAMIC SCIENCE The Arabs in pre-Islamic times — mostly Bedouins, in the Arabian peninsula — already had a good knowledge of the stars. They used certain stars or asterisms for orientation in their desert travels, to fix seasons, and to predict weather, especially rain. For the last two purposes they developed a system of so-called *anwāʾ* (sing. *nawʾ*), in which a star or asterism was observed setting shortly before sunrise and another simultaneously rising, just opposite (the latter one was called *raqīb*, with respect to the first one). At an unknown time and through unknown channels they received — most probably from India — the system of the twenty-eight lunar mansions (*manāzil al-qamar*), stars or asterisms along the path of the moon, near the ecliptic (the band of the zodiac through which the sun apparently moves in its yearly course). The indigenous *anwāʾ* were then merged with the lunar mansions, and the twenty-eight mansions were given names from among the former *anwāʾ*. All the five planets visible to the naked eye seem to have been known, since the Arabic language has old indigenous names for them: Mercury – *ʿuṭarid*, Venus – *al-zuhara*, Mars – *al-mirrīkh*, Jupiter – *al-mushtarī*, and Saturn – *zuḥal*. Also, fragments of the Babylonian zodiac seem to have reached the Arabs in pre-Islamic times, though their location in the sky was different from that in Babylonian and Greek (and modern) astronomy; the figure corresponding to Gemini (*al-jawzāʾ*), for example, was located in the stars of Orion, or the figure corresponding to Aquarius (*al-dalw*) in the Pegasus square. Such star lore was also much used in classical Arabic poetry, in pre-Islamic times and afterwards. From the ninth century on, Arabic lexicographers and philologists collected the dispersed elements of old Arabic folk astronomy and put them together in special monographs (called *Kutub al-anwāʾ*, Books of the *anwāʾ*). From these, more than three hundred old Arabic names for stars and asterisms could be recovered.

A new development began after the Muslim conquest of the lands of the Middle East. Now the Arabs came into contact with Greek science as transmitted and practiced in learned circles, mostly by Christian monks and scholars, in the area. After a period of undocumented first contacts, from the eighth century on, Greek scientific texts were translated into Arabic in great numbers. Scientific astronomy in the Islamic world therefore was largely built on Greek knowledge, with additional elements received from India and Persia. Greek cosmology and Greek theories of the planets, the fixed stars, and other celestial phenomena formed the basis of Arabic-Islamic astronomy. Later Muslim astronomers continued to work with this material, developed it, and introduced many corrections and improvements. But the "Copernican revolution" did not reach the Islamic world; Arabic-Islamic astronomy continued to be Ptolemaic into the nineteenth century.

On the (fixed) stars, the main source for the Arabs was Ptolemy's *Almagest*, which contains in Books VII and VIII a catalog of 1025 stars arranged in 48 constellations and listed with their ecliptical coordinates, longitude and latitude, and magnitude. These 48 constellations (which also live on, in modified form, in modern astronomy) and the derived nomenclature of the 1025 known stars were used by Muslim astronomers through all times, in books as well as on instruments (astrolabes, celestial globes, quadrants). To make the catalog valid for all times, Ptolemy proposed to add to the longitudes of the stars a constant of 1 degree in 100 years, to correct for precession. The Muslim astronomers very soon (in the ninth century) found a better value, one degree in $66\frac{2}{3}$ years; later also one degree in 70 years and other values were used. The pictorial figures of the constellations were also basically taken from Greek models; only details of the physiognomy of human figures and clothing were more or less adapted to contemporary local styles.

The main Arabic work on the fixed stars and the constellations was the *Book on the Constellations* by the Persian astronomer Abu'l-Ḥusayn al-Ṣūfī (903–986); al-Ṣūfī anxiously followed the Arabic versions of the *Almagest*, added criticisms based on his own observations, but established his star catalog exactly according to the data given in the *Almagest* (adding a constant of 12°42′ to the longitudes, for precession, for his epoch AD 964). For each constellation al-Ṣūfī added two drawings, one showing the figure as seen in the sky, and the other showing it as seen on the celestial globe (where the figures appear as seen from outside the globe, with the right and left sides reversed against the sky view). Most Islamic globe makers, through all centuries, followed al-Ṣūfī's models in depicting the constellation figures.

Several Islamic astronomers established star catalogs in the manner of the *Almagest*: al-Battānī (epoch 880, precession constant added 11°10′, al-Ṣūfī, al-Bīrūnī (epoch 1031, precession constant 13°0′), and Ulugh Beg (epoch 1437, own observations, but some stars taken over from al-Ṣūfī). Besides these big catalogs, smaller star lists were drawn up by very many astronomers, for use on astrolabes or other instruments. Most of the catalogs and star lists, however, were not obtained through observation; they were adapted from the *Almagest* or derived catalogs by adding the precession constant to the longitudes. A small list of twenty-four stars obtained through independent observation was established by the astronomers of the caliph al-Maʾmūn for the epoch 214 Hijra = AD 829–830; also other star tables were afterwards derived from it by adding the precession constant.

Other celestial phenomena, such as comets or meteor showers, were not considered by the Arabic-Islamic astronomers. Following Greek cosmological theory, they supposed these to belong to the sublunar sphere, i.e. the space between the Earth's surface and the moon, and not to the sphere of the stars. The fixed stars in their opinion were all situated on the farthest, eighth sphere, beyond the seven spheres of the planets and the sun and moon. It was mostly in chronicles and in astrological works that such phenomena were registered, and they were usually regarded as bad omens indicating evil events.

Through Latin translations in late tenth and mainly in the twelfth century in Spain, many star names of Arabic origin entered the West and continue to be used in Western astronomy, mostly in heavily distorted spellings, today.

PAUL KUNITZSCH

REFERENCES

Almagest, star catalog, Arabic and Latin versions: Claudius Ptolemäus. *Der Sternkatalog des Almagest. Die arabisch-mittelalterliche Tradition*. Ed. P. Kunitzsch. i–iii. Wiesbaden: Harrassowitz, 1986, 1990, 1991.

al-Battānī. *Al-Battānī sive Albatenii Opus astronomicum*. Ed. and trans. C.A. Nallino, vols. 1–3. Milan: Reale Osservatorio di Brera, 1899–1907.

Kennedy, E.S. "A Survey of Islamic Astronomical Tables." *Transactions of the American Philosophical Society*, N.S. 46: 123–177, 1956.

Knobel, E.B. *Ulugh Beg's Catalogue of Stars*. Washington: The Carnegie Institution, 1915.

Kunitzsch, P. *Arabische Sternnamen in Europa*. Wiesbaden: Harrassowitz, 1959.

Kunitzsch, P. *The Arabs and the Stars*. Northampton: Variorum Reprints, 1986.

Kunitzsch, P. "al-Manāzil (Lunar Mansions)." In *Encyclopaedia of Islam*, 2nd ed., vol. vi. Leiden: Brill, 1991, pp. 374–376.

Kunitzsch, P. "Minṭaḳat al-Burūdj (The Zodiac)." In *Encyclopaedia of Islam*, 2nd ed., vol. vii. Leiden: Brill, 1991, pp. 81–87.

Kunitzsch, P. "al-Nudjūm (The Stars)." In *Encyclopaedia of Islam*, 2nd ed., vol. viii. Leiden: Brill, 1993, pp. 97–105.

Kunitzsch, Paul, and Tim Smart. *Short Guide to Modern Star Names and Their Derivations*. Wiesbaden: Harrassowitz, 1986.

Pellat, Ch. "al-Anwāʾ." In *Encyclopaedia of Islam*, 2nd ed., vol. i. Leiden: Brill, 1960, pp. 523f.

Savage-Smith, Emilie, and Andrea P.A. Bellioli. *Islamicate Celestial Globes; Their History, Construction, and Use*. Washington, D.C.: Smithsonian Institution Press, 1985.

See also: *Almagest* – Precession of the Equinoxes – al-Ṣūfī – Celestial Vault and Sphere – Maps and Mapmaking: Islamic Celestial Maps – Ulugh Beg – al-Battānī – Lunar Mansions

INCA STONEMASONRY

> And what one admires most is that, although these [stones] in the wall I am talking about are not cut straight but are very uneven in size and shape among themselves, they fit together with incredible precision without mortar (Acosta, lib. 6, cap. 14; 1962:297).

Like Father Acosta in the seventeenth century, the modern traveler stands in awe before the Inca walls of huge stones, some weighing over a hundred metric tons, tightly fitted together. They wonder about how a people who did not know the wheel and had no iron tools transported and dressed stones of that size, and mated them one to another in such a way that, as the saying goes, not even the blade of a knife could be inserted into the joints. Answers to these questions have been replete with speculations ranging from the intervention of extraterrestrial beings, to the application of an herb, the juice of which softened the stone, to the more prosaic investment of an immense labor force. Recent studies of Inca construction, however, demonstrate that the awe-inspiring, tightly-fit megalithic masonry was well within the technological reach of the Incas (Protzen 1985, 1986, 1993). It is to be hoped that the fantastic tales of extraterrestrial intervention will be put to rest once and for all.

Quarrying

The Incas did not practice quarrying in the proper sense. The stone is neither cut off a rock face nor is it detached from bedrock by undercutting. The Incas gained their building stone either by picking suitable blocks out of a rockfall or by prying it loose from fragmented rock faces. Large blocks of five or more tons were only minimally shaped in the quarries; all the fine work was done at the construction site. Smaller building blocks were dressed on most sides in the quarries, and only the finishing and the fitting to other stones was reserved for the site.

Cutting and Dressing

To cut and dress stones the Incas used simple river cobbles of various sizes as hammers. These tools and their fragments are found in abundance in the ancient quarries scattered among roughed out building blocks and in the quarrying waste. The hammerstones are easily distinguished from other stones by both their shape and their petrological characteristics; they are water-worn rounded stones of materials different from the quarried stone and the surrounding bedrock. The hammers come in different sizes; some are as small as an egg, others are the size of a football, and others still are two to three times the size of an American football. The largest of these hammerstones were used to break up and roughly shape the raw stones, the medium sized to dress the faces of the building blocks, and the smallest to draft and cut their edges and corners. The technique involved is exactly as Garcilaso de la Vega "El Inca" described it: "The quarryfolks, . . ., who had no other instruments to work the stones, but some black cobbles they called hihuana (sic. for hihuaya), with which they dress the stones by bruising rather than cutting." (lib II, cap. XXVIII; 1976:119, tomo I).

Indeed, when hitting the workpiece straight on with a hammer stone one crushes the rock producing little more than dust. However, if one increases the angle of impact to about 15 to 20°, little chips can be torn off. By further increasing the angle to some 45° by imparting a twist to the hammer just before impact larger chips can be removed thus accelerating the process considerably. The impact of the hammer leaves a small pitmark on the workpiece. Such pitmarks can be observed on every face of every building block in every Inca wall of cut stones, regardless of the building blocks' material. Smaller, finer pitmarks found along the edges of the building blocks indicate that smaller stones were used to cut the edges. The particular technique in cutting edges requires that the edge be shaped by hitting the workpiece with grazing blows directed away from the workpiece, which results in corners with dihedral angles larger than 90°. It is these obtuse angles that account for the characteristic beveled joints of Inca cut stonemasonry and that brings about the chiaroscuro effect.

Extensive experiments have demonstrated that the process is relatively easy, effective and precise, and not as time consuming as one might assume. Twenty quarry people working side by side could rough out a block 4.5 meters long, 3.2 meters wide, and 1.7 meters high — the dimensions of one of the largest building blocks in an ancient Inca quarry near Ollantaytambo — in less than fifteen days.

Fitting

The most intriguing question about Inca cut stonemasonry concerns the precision fitting of the blocks. It has been repeatedly argued that the Inca stonemasons ground the blocks into place using a mixture of sand and water. The evidence, however, does not support this hypothesis. Where walls have been dismantled or fallen apart one finds the exact imprint of the stones which have been removed or fallen off. Very often the shape of the imprints determines a unique position for the stone it once accommodated. To grind in a stone, however, requires that the stone can move freely along a path in at least one direction. Thus, the ground stone would fit in any, and not just one, position along that path Furthermore, if the stones had been ground into place the joints should show signs of abrasion, but they do not. Instead one finds the typical pitmarks which result from pounding.

The imprints seen on dismantled walls indicate that when laying a wall, the Inca stonemasons left the top face of every new course uncut until it was to receive a new course. The bottom face of the block of the new course was cut first, and a suitable bed was carved out of the course already in place. A similar approach was used for the rising joints; it is the side of the block already *in situ* that was carved to match the precut shape of its new neighbor. The technique used appears to have been one of trial and error. The masons started by outlining the shape of the new stone atop the course it was to be fitted to and proceeded to carve out a bed. By trying the fit time and again the masons obtained a perfect match. Granted, the trial and error technique is a tedious one, and perhaps not very convincing if one considers megalithic building blocks of several dozens of tons. However, it works and does not postulate the use of tools and machinery of which no traces have been found. The Incas had plenty of time and manpower at hand. Since they moved huge blocks over many kilometers, it is not inconceivable that they were capable of setting up a stone several times to achieve the desired fit. It is of course also conceivable that the Inca stonemasons knew of another technique to transfer the shape of one stone to another without actually trying it in successive steps. Such a technique has in fact been proposed by Vincent Lee. Inspired by log cabin builders, Lee suggests that the shape of one stone was scribed onto the other with an ingenious but simple device consisting of a stick and a plumb bob. Although very plausible, Lee's hypothesis has yet to be corroborated by field work.

Antecedents

John Hemming argued emphatically that the high skills of the Inca stonemasons "cannot have been developed in the century or less of Inca ascendance. Architecture anywhere in the world evolves from precursors". Here, the precedent is Tiahuanaco at the south end of Lake Titicaca. Tiahuanaco has some of the world's finest stonemasonry: carefully fitted rectangular ashlars, tightly joined polygonal masonry, and exquisitely worked monolithic gates and statues. It is said that Pachakuti was so impressed by this stone work when he first saw Tiahuanaco, that he advised his architects to take careful note of how this masonry was made because he wanted Cuzco to be built in the same manner. It is also said that he imported Qolla laborers from the Lake Titicaca area to work on construction projects (Sarmiento de Gamboa cap. 40; 1943:111–112). But because Tiahuanaco culture developed at least a millennium before the Incas, and may have reached its peak some five to six centuries before Pachakuti's rise to power, and nothing similar has been built in the interim, one may question why the Qolla laborer should remember any of the ancient techniques.

Transporting

The one problem that still defies everybody's imagination is how the Incas transported and heaved enormous building stones. Rough abrasion marks found on building blocks at Ollantaytambo suggest that the stones were simply dragged along the ground. From these marks it is even possible to determine the direction in which the blocks traveled. The dragging of big stones is consistent with chroniclers' reports: "These Indians used to move very large stones with muscle power, pulling them with many long ropes of lianas and leaf fibers, . . ., and they [the stones] are so big that fifteen yokes of oxen could not pull them." (Gutierrez de Santa Clara lib. 3, cap. 63; 1904–1929:550). And dragging big blocks involves large transportation crews: "Four thousand of them were breaking stones and extracting stones; six thousand were hauling them with big ropes of hide and leaf fibers; . . ." (Cieza de León cap. LI; 1967:170).

Six thousand people dragging a stone is not unreasonable; many Old World transportation problems were solved with very large work forces. Engelbach, for example, calculated that it would have taken six thousand men to haul the unfinished obelisk at Aswan from the quarries to the Nile. The questions raised by a transportation crew as numerous as this, which remain largely unanswered, are how the crew was harnessed to the stones and how it negotiated turns on the narrow roads on the steep slopes of the Andes.

JEAN-PIERRE PROTZEN

REFERENCES

Acosta, Joseph de. *Historia Natural y Moral de las Indias* [1590]. 3 vols. Ed. Edmundo O'Gorman. 2nd ed. Mexico: Fondo de Cultura Económica, 1962.

Cieza de León, Pedro de. *El Señorio de los Incas* (Second part of the Crónica del Perú, 1553). Lima: Instituto de Estudios Peruanos, 1967.

Engelbach, Reginald. *The Problem of the Obelisks, from a Study of the Unfinished Obelisk at Aswan.* London: T. Fisher Unwin, Ltd, 1923.

Garcilaso de la Vega, Inca. *Comentarios Reales de los Incas* [1604]. Ed. Aurelio Miro Quesada. Caracas: Bibliotheca Ayacucho, Vol. 5 and 6, 1976.

Gutierrez de Santa Clara, Pedro. *Noticias de las Guerras Civiles de Perú (1544–1548) y de Otros Ssucesos de las Indias* [1603]. Colección de Libros y Documentos referentes a la Historia de América, vols. II,III,IV,X,XX,XXI. Madrid: Libreria General de Victoriano Suarez, 1904–29.

Hemming, John. *Monuments of the Incas*. Albuquerque: University of New Mexico Press, 1980.

Lee, Vincent R. "The Building of Sacsayhuaman." *Ñawpa Pacha* 24: 49–60, 1986.

Protzen, Jean-Pierre. "Inca Quarrying and Stonecutting." *Journal of the Society of Architectural Historians* 44 (2):161–182, 1985.

Protzen, Jean-Pierre. "Inca Stonemasonry." *Scientific American* 254 (2): 94–105, 1986.

Protzen, Jean-Pierre. *Inca Architecture and Construction at Ollantaytambo*. New York: Oxford University Press, 1993.

Rowe, John Howland. "Inca Culture at the Time of the Spanish Conquest." In *Handbook of South American Indians*, vol. 2. Ed. Julian H. Steward. *Bureau of American Ethnology Bulletin* 143. Washington, D.C., 1946, pp. 183–330.

SUANXUE QIMENG The *Suanxue qimeng* (Introduction to Mathematical Studies), the first book composed by Zhu Shijie, a Chinese mathematician of the Yuan dynasty, was published in 1299. It differs from his second one, the *Siyuan yujian* (1303), in that its mathematical content presents a stronger continuity with the Chinese mathematical tradition. Most of the topics treated in it can be traced back to the Han classic which founded this tradition, the *Jiuzhang suanshu* (Nine Chapters on Mathematical Procedures, hereafter called The Nine Chapters), probably completed between the first century BC and the first century AD. Indeed, Zhu Shijie deals there with the rule of three, the distribution into unequal parts, and their composition or reiteration along the lines opened by the sixth of The Nine Chapters. He also describes how to perform the same computations with fractions and presents algorithms to obtain areas and volumes, even though he considers some forms not found in The Nine Chapters. The rule of excess and deficit, which was transmitted to the West where it became known as the rule of false double position, receives in his *Suanxue qimeng* a presentation as complete as in The Nine Chapters, even though it had been partially forgotten and sometimes transmitted in an imperfect state during the time elapsed since the composition of the latter. Moreover, Zhu Shijie devotes a chapter to an improved presentation of the algorithm for the solution of systems of simultaneous equations given by The Nine Chapters and comparable to the so-called Gaussian pivot method.

In addition to this, the *Suanxue qimeng* includes the treatment of topics that had emerged in the previous decades in Northern China, Zhu Shijie's homeland, such as the "procedure of the celestial unknown", a technique that puts into play polynomial algebra to obtain an equation to solve a given problem.

Yet, in contrast to The Nine Chapters, Zhu Shijie, in a first part of the book, makes clear which algorithms can be used to perform the basic arithmetical computations on the counting board and includes new procedures for such computations when particular numbers are involved. It is interesting that some of the procedures he described could be used later without change when the abacus replaced the counting board. In this respect as in others, such as the treatment of topics like the summation of all kinds of series, the inclusion of tables of unit conversion, a style of practicing mathematics in close contact with the problems of socio-economic life, or the use of certain new technical terms, the *Suanxue qimeng* repeatedly evokes the writings of Yang Hui, a mathematician who had lived slightly earlier in Southern China. Indeed, in contrast to the period before the unification of China under Mongol rule, when mathematics seems to have developed independently in the north and the south of China, the time of unification when Zhu Shijie worked might have allowed the partial merging of these two traditions; his book may be our first reflection of this.

KARINE CHEMLA

REFERENCES

Du Shiran. "Zhu Shijie yanjiu" (Research on Zhu Shijie). In *Song Yuan Shuxueshi Lunwenji* (*Collected Essays on the History of Mathematics during the Song and Yuan Dynasties*). Ed. Qian Baocong. Beijing: Kexue chubanshe, 1966, pp. 166–209.

Ho Peng Yoke. "Chu Shih-chieh." In *Dictionary of Scientific Biography*, vol. 3. Ed. Charles Gillispie. New York: Charles Scribner's Sons, 1970–80, pp. 265–71.

Lam Lay Yong. "Chu Shih-chieh's *Suan-hsüeh ch'i-meng* (Introduction to Mathematical Studies)." *Archive for History of Exact Sciences* 21: 1–31, 1979.

See also: Computation: Chinese Counting Rods – Liu Hui and the *Jiuzhang Suanshu*

AL-ṢŪFĪ Al-Ṣūfī, Abu'l-Ḥusayn ʿAbd al-Raḥmān ibn ʿUmar, was born in 903 in Rayy, near modern Tehran, and died in 986.

Al-Ṣūfī was an astronomer in the Arabic-Islamic area. He was of Persian origin, but wrote in Arabic, the language of all science in that time. He is best renowned, and became most influential, through his *Kitāb ṣuwar al-kawākib al-thābita* (Book on the Constellations), written around 964. Knowledge of the fixed stars in Greek-based Arabic-Islamic astronomy was mainly derived from Ptolemy's catalog of 1025 stars arranged in 48 constellations contained in his *Almagest* (ca. AD 150). Al-Ṣufī re-examined Ptolemy's values of the star coordinates and magnitudes. In his book, he described all the stars catalogued by Ptolemy, adding his criticism in each individual case. However, in the tables added to his book he nevertheless faithfully rendered Ptolemy's traditional values, adding a constant of 12°42′, for precession, to Ptolemy's longitudes. Only the magnitudes were given according to al-Ṣūfī's own observation. For each constellation he added two drawings, one showing the figure as seen in the sky, the other as seen on the celestial globe. His book and his drawings served as models for work on the fixed stars in the Arabic–Islamic world for many centuries, and became known even in medieval Europe, where his constellation drawings were imitated in a series of Latin astronomical manuscripts (though no veritable Latin translation of his book was made). Apart from this book, al-Ṣūfī left treatises on the use of the astrolabe and the celestial globe, an introduction to astrology, and a short geometrical treatise. His name lives on, Latinized as Azophi, as a name for a crater on the Moon.

PAUL KUNITZSCH

REFERENCES

Kunitzsch, Paul. *Peter Apian und Azophi: Arabische Sternbilder in Ingolstadt im frühen 16. Jahrhundert*. Munich: Beck, 1986.

Kunitzsch, Paul. "The Astronomer Abu'l-Ḥusayn al-Ṣūfī and his Book on the Constellations." In *The Arabs and the Stars*. Northampton: Variorum Reprints, 1989, item XI.

Kunitzsch, Paul. "Al-Ṣūfī." In *Dictionary of Scientific Biography*, vol. 13. Ed. C.C. Gillispie. New York: Scribners, 1976, pp. 149—150.

Kunitzsch, Paul. "Abd-al-Raḥmān b. ʿOmar Ṣūfī." In *Encyclopaedia Iranica*, I, Fasc. 2. London: Routledge & Kegan Paul, 1982, pp. 148–149.

Schjellerup, H.C.F.C., trans. *Description des étoiles fixes par Abd-al-Rahman al-Sufi*. St. Petersburg 1874; repr. Frankfurt am Main: Institut für Geschichte der Arabisch-Islamischen Wissenschaften, 1986.

SUGAR IN LATIN AMERICA The sucrose from sugar cane is identical to that in beet sugar and other sugars such as glucose, dextrose, fructose, levulose, and lactose. The chemical formula for sucrose is the oligosaccharide $C_{12}H_{22}O_{11}$. After refining, sugar is 99.9% pure.

Sugar cane was first cultivated and processed in India thousands of years ago. Gradually, medieval trade routes brought the product and its cultivation to the Mediterranean basin and Madeira where it languished on a small scale until its introduction to the Caribbean and Brazil. Columbus took it to Hispaniola on his second voyage in 1493. Cortés took it to New Spain, Balboa to Central America, and Pizarro to Peru. It was in Brazil, however, that the efforts of Duarte Coelho Pereira converted his captaincy of Pernambuco into a prosperous sugar-producing center fashioned after the *engenhos* (sugar works) of Madeira from whence the first cuttings and processing technology came. This became the model for an industry so vital to Portugal they courted Dutch capital to finance the African slave trade to supply the necessary labor. In exchange, the Lowlands received the crude sugar for refining and distribution. So profitable was this arrangement that the Portuguese eventually abandoned their enterprises in the Far East for the Brazilian milk cow, as Emperor João III fondly referred to the flourishing colonial sugar industry.

Aside from the introduction of large-scale cultivation and processing, the techniques of sugar production changed little since the early days of Venetian and Dutch refining and trading monopolies. Cultivation was usually from small cuttings or rattoons planted in evenly spaced furrows and allowed to mature for periods from six to eighteen months — depending on the type of cane, the time and amount of rainfall or irrigation, temperature, storms, quality of soil, disease, and pests.

Plantation owners and managers were constantly alert for crop dangers and began to employ technicians to advise them. They experimented with irrigation systems, tested new strains of cane from both seed and rattoon, and tried crop rotation. Harvest had to follow immediately upon maturation. Normally this was by hand-held machetes wielded by slave laborers. Sometimes they used slash-and-burn techniques.

Harvesting was simultaneously accompanied by manually loading the syrup-filled stalks into mule-drawn carts for transport to the sugar houses. The cane was then hand-fed through the vertical or horizontal cylindrical presses which were originally adapted from wooden olive presses. These were two- or three-roller, metal-clad, devices with gears powered by humans or animals (*trapiches*) or water (*engenhos*). The extracted syrup flowed through wooden troughs or, in some cases where capital and topography permitted, through Roman-style aqueducts to the boiling sheds. Large *engenhos* comprised thousands of acres with thousands of residents in self sufficient hacienda communities. More modest enterprises relied on hand-carrying the crude syrup to nearby pots for boiling.

Early production systems were ecologically complete, for the crushed stalks (or bagasse) were dried and used as fuel to fire the boiling vats. The ashes and extraneous leaves and stalks were tilled back into the soil as fertilizer and also served as animal fodder. Slaves found that unprocessed molasses from pieces of stalk that were not disposed of quickly underwent inversion and began fermentation; this primitive rum was used as currency to reward slaves and keep them cooperative but was not recognized as a commercial product for export.

Until the nineteenth century, the juice from the raw cane went through a series of open-pot boilings at extremely high temperatures. This process was watched very carefully to avoid fire and to avert burning the sugar. Slaves ladled off the scum of debris which flocculated with animal grease at the top; other impurities settled to the bottom in sedimentation. Lime was then added to help clarify the syrup. This purification, also known as defecation, removed the maximum number of impurities. The clear syrup was then placed in evaporating pans. The sludge of megasse sugar was next scooped into hanging clay cones with open bottoms to allow the heavy brown molasses to drip out. The resulting crude sugar was about 96% pure and fit for transport in casks or in loaves.

Shipped to European ports, crude sugar was sometimes sold directly, but most was transshipped and refined by a process of reheating in huge vats to which water and bone carbon filters were added. After a series of washings, the clear liquid was evaporated in copper pans and allowed to crystallize or granulate — the finer the quality, the whiter the crystal. As with the processing of raw sugar, refining would remain unmodified until the nineteenth century introduced steam vacuums and centrifuges to the finishing processes.

Throughout the colonial period, every finishing process was done by the mother country or its agents, usually in the Lowlands. Therefore, refining always occurred in Europe. The most efficient refineries were scattered around Antwerp which remained the capital for finishing sugar until war razed the production centers and blockaded trade. Britain, France, and Germany then built their own refineries. Long before Britain took possession of sugar-producing lands, England monopolized refining and distribution industries. This availability of mass production coupled with a growing popular taste for sweetened coffee and bread with treacle (molasses) provided an amenable consumer base for the confection in England and on the continent which demanded the colonial product.

Having pirated the rattoons and seeds of Brazilian and Spanish haciendas, European powers sought to monopolize refining and trade. In 1662 Catherine of Braganza married Charles II of England; her dowry included former East Indian territories such as Bombay and Tangiers and also a taste for tea which swept the court with exceeding popularity and would eventually displace both coffee and cocoa as England's hot, sweetened beverage of choice.

The Caribbean was an entrepôt of plantation islands. England, Spain, Portugal, France, and Holland all engaged in slave trade with Africa, and all quickly learned the commercial value not only of sugar but also of its intermediary byproducts, molasses and rum. After the molasses dripped off the crude muscovado sugar, the syrup was fermented and distilled into rum on the islands as early as 1530 and used as currency to purchase African slaves. Molasses was shipped directly to North American ports where New England distilleries enjoyed a burgeoning business. From the time of the capture of Jamaica in 1655, rum was regularly issued to British sailors, for it lasted longer than the traditional ale on high seas at certain temperatures. In 1754 it was watered down to grog and freely rationed until the 1960s. The Sugar Interest monopolized Parliament and led to colonial uprisings which reshaped the empire.

Britain and its competitors consolidated their territories and introduced technological advances. As early as 1794, steam was used to power a three-roller press in Jamaica. A similar steam-driven mill operated in Cuba in 1797 — ten years before the first successful steamboat. These machines were neither uniformly nor extensively adopted and operated imperfectly. Early fixed iron rollers often clogged or broke when the woody stalks of older cane were fed through the press. Shutdowns were costly, for the syrup in cane spoiled if not processed within three days of harvest. Mechanical repairs and waiting for replacement parts also impeded production.

The vacuum pan was invented by Edward Charles Howard in England in 1813, but it was not introduced to the colonies until 1835. This pan admitted steam through holes in the pan's bottom which was lined by a filtering cloth and capped by a dome connected to an air pump with a condenser. Interior heating coils were added by Daniel Wilson, and in 1828 Louisiana planter Valcour Aime implemented these improvements in his sugar houses and also introduced the use of the polarimeter to value raw sugars in cane. Despite patent disputes a system of multiple evaporation in six to ten successive pans was widely adopted and diversely adapted throughout the world. Penzoldt's centrifugal machine invented in Germany in 1837 for drying wool was modified for sugar by Sir Henry Bessemer. Cane processing technology was adapted for beet sugar. In London the Fairrie family experimented with decolorizing agents for their refinery and discovered animal charcoal.

With the mechanization of sugar in the latter nineteenth century came the professionalization of technical advisors and managers. Technical assistants had to know how to assemble equipment correctly, keep it in good repair, and operate it. Agricultural experts were consulted to identify the best strains of cane for a particular climate or soil. Hawaiian planters of the 1850s maintained an experimental station which offered soil analyses, pathology reports, and pest controls. Entomological research at the British Museum identified natural predators and prescribed counter parasites.

The first mechanized harvests were successful in Australia. In the 1960s, Cuba contracted with Soviet engineers to develop a combine harvester. The self-propelled model ruined the rattoons and the ground for several seasons. Eventually the Cubans developed a harvester but technical problems diverted its production to Australia in 1984.

The sucrochemical industry experimented with research and development. Cane sucrose sales were eroded by beet sugar, high fructose corn sweeteners (HFCS) in processed foods, and synthetic sweeteners. New applications included gasohol, furniture, paper, and plastics. Brazil was the only successful producer of commercial gasohol although Australia and the US successfully converted their sugar industries to gasohol production during World War II.

The original sugar-producing countries of Latin America and the Caribbean still produce their crops of precious cane, but unstable governments, lack of capital, natural disaster, and inefficient management hinder technological growth.

JUDITH VIDAL

REFERENCES

Abbott, George C. *Sugar*. London: Routledge, 1990.

Ellis, Ellen Deborah. *An Introduction to the History of Sugar as a Commodity*. Ph.D. dissertation. Philadelphia: Winston, 1905.

Ely, Roland T. *Cuando Reinaba Su Majestad El Azúcar.* Buenos Aires: Sudamericana, 1963.

Lowndes, A. G. *South Pacific Enterprise: The Colonial Sugar Refining Company*. Sydney: Angus, 1956.

Mintz, Sidney W. *Sweetness and Power: The Place of Sugar in Modern History.* New York: Penguin, 1985.

Ortiz, Fernando. *Contrapunteo Cubano del Tabaco y el Azucar.* Barcelona: Ariel, 1973.

Rolph, George M. *Something about Sugar: Its History, Growth, Manufacture and Distribution.* San Francisco: Newbegin, 1917.

See also: Crops

ŚULBASŪTRAS The *Śulbasūtras* are manuals which prescribe the construction of different types of fire altars. Every householder was instructed, by the Vedic religion, to maintain a sacred fire, primarily for daily worship and offerings. These fires, called *gārhapatya* (domestic), *dakṣiṇa* (southern), and *āhavanīya* (oblatory), were to be maintained, without their ever going out, in altars of different designs, such as circular, semi-circular, square, rectangular, and triangular. Seasonal and special worships required altars with elaborate designs like that of an eagle with outstretched wings. The size, shape, direction, position, and the number and measure of the bricks used, and also the increase in the measure, for extraneous reasons, were all specified in the *Śulbasūtras*. These traditional practices are referred to in the *Ṛgveda*, the earliest of the Vedas, and elaborated in the *Yajurveda*. Later they came to be written down as manuals, supplementing the regular texts depicting sacrifices, and also in independent texts called *Śulbasūtras*. Adherents of different Vedas and Vedic schools had different *śulba* texts, named after the authors of these texts and pertaining to the Veda which they advocated. This is how the *śulba* texts named after Āpastamba, Bodhāyana, Kātyāyana, Mānava, Maitrāyaṇa, Varāha, Hiraṇyakeśin, Satyāṣaḍha, Vādhūla, and Laugākṣi, pertaining to one or the other of the schools of the *Yajurveda*, and *Maśaka Śulbasūtra* pertaining to the *Sāmaveda* came into being.

The *Śulbasūtras*, meaning pithy aphoristic statements (*sūtra*) for work with string (*śulba*), prescribes, by means of addition, subtraction, multiplication, division, and squaring, simple rules not only for the construction of circles, squares, rectangles, triangles, wheel-shapes, trapezia, and rhombi, but also for extending these figures by specific proportions as required in different sacrifices. It also specifies methods to reduce a circle to a square of equal area, and vice versa. A fine approximation of the value of the root of two is contained in the rule given in the *Baudhāyana-Śulbasūtra*: "Increase the measure of the side (of a square) by its third part, and the third part by its fourth part. The fourth part is decreased by its own thirty-fourth part. (The approximate diagonal will result)." The rule gives the approximation:

$$\text{Root } 2 = 1 + \frac{1}{3} + \frac{1}{3 \times 4} - \frac{1}{3 \times 34},$$

i.e. $\frac{577}{408}$, i.e. 1.4142157, which is very approximate to the correct value of the root of 2.

The *Śulbasūtras* specify or give geometrical constructions for the following: (1) to draw a straight line at a right angle to a given line; (2) to construct a square on a given side; (3) to construct a rectangle of given sides; (4) to construct an isosceles trapezium of given base, top and altitude; (5) to construct a square equal to the sum of two squares; (6) to construct a square equal to the difference of two squares; (7) to construct a square equal to a given rectangle; (8) to construct a rectangle with a given side and equal in area to a given square; (9) to construct a square equal in area to a given isosceles triangle; (10) to construct a square equal in area to a rhombus; (11) to construct a square equal in area to a given circle; and (12) to construct a circle equal in area to a given square.

It would seem that the sacrificial hall of the Vedic Indians formed, as it were, the workshop and laboratory to formulate and develop their geometry.

K.V. SARMA

REFERENCES

Amma, T.A. Sarasvati. *Geometry in Ancient and Medieval India*. Delhi: Motilal Banarsidass, 1979.

Datta, B. and A.N. Singh. "Hindu Geometry." *Indian Journal of History of Science* 18: 121–88. 1980.

Gupta, R.C. "Baudhāyana's Value of Root 2." *Mathematics Education* 6B: 77–79, 1972.

Gupta, R.C. "Vedic Mathematics from the *Śulbasūtras*." *Indian Journal of Mathematics Education* 9 (2): 1–10, 1989.

Sen, S.N. and A.K. Bag, Ed. and Trans. *The Śulbasūtras*. New Delhi: Indian National Science Academy, 1983.

van der Waerden, B.L. *Geometry and Algebra in Ancient Civilizations*. Berlin: Springer-Verlag, 1983.

See also: Geometry in India – Baudhāyana

SUN SIMO The Chinese physician and medical author Sun Simo (alternative spelling: Sun Simiao) was a native of Huayan, in modern Shensi. His biography is so much a composite of fact and legend that it is even impossible either to substantiate or invalidate his traditional dates (581–682). From official, autobiographical, and hagiographic sources it emerges that he retired at an early age on Mount Taibai, not far from his birthplace. He repeatedly declined imperial summonses and official titles, but was almost certainly in Emperor Gaozong's retinue from 659 to 674, when he retired on account of illness. He seems to have spent an extended period in Sichuan, which may explain why many legends that concern him are located in that area. After his death he was venerated as *Yaowang* or "King of Medicine" (by which he is still known) in temples dedicated to him.

Sun Simo is the author of two of the most important Chinese medical compilations, the *Qianjin fang* (Prescriptions Worth a Thousand or, Remedies Worth their Weight in Gold), also known as *Beiji qianjing yaofang* (Important Prescriptions Worth a Thousand, for Urgent Need), and the *Qianjin yifang* (Revised Prescriptions Worth a Thousand). The former, in thirty chapters, was completed soon after the middle of the seventh century (apparently in 652). The latter, also in thirty chapters, dates from the late seventh century. Both works are preserved in editions derived from versions published in the eleventh century, when they were edited to be used as textbooks in the Imperial Academy of Medicine. In these texts, Sun provides an extended compendium of contemporary medical knowledge, arranged in sections dealing with such subjects as pharmacology, etiology, gynecology, pediatrics, dietetics, acupuncture, moxibustion, and specific diseases. Both texts include a wide selection of prescriptions (about 5300 in the *Yaofang*, about 2000 in the *Yifang*).

Among the many points of interest in these compilations, three deserve special mention. The first is the priority that Sun Simo accords to gynecology and pediatrics, the two branches of medicine with which he deals first in both works. The second is the importance given to medical ethics, reflected in this well known passage from the first chapter of the *Qianjin yaofang*: "When someone comes to look for help, a doctor should not question rank or wealth, age or beauty, nor should he have personal feelings towards that person, his race, or his mental capacities. He should treat all his patients as equal, as though they were his own closest relatives." The influence of Sun's medical ethics spread beyond China, reaching Korea and Japan through quotations of relevant passages in texts of these two countries.

A third aspect is Sun's relationship with Daoism (Taoism) and Buddhism. The nature of his involvement with the Way of the Celestial Masters (*Tianshi Dao*, one of the main traditions of liturgical Daoism) is debated. In the two chapters entitled *Jinjing* (Book of Interdictions) of the *Qianjin yifang*, Sun quotes formulas used in exorcist rituals by the

Celestial Masters. This raises the issue of how he gained access to them. His interest in Daoism is also reflected in the chapter on "Nourishing the Vital Principle" (*Yangxing*), and in another extant text on physiological disciplines which is attributed to him, the *Sheyang zhenzhong fang* (Pillowbook of Methods for Nourishing [the vital principle]).

Another source which points to Sun's relationship with Daoism is the *Taiqing danjing yaojue* (Essential Instructions from the Books on the Elixirs of the Great Purity). This text —available in an excellent English translation—consists of a collection of alchemical methods, probably derived from the Six Dynasties compilations centered around the now lost *Taiqing jing* or Book of the Great Purity, one of the main early alchemical canons. Although Sun's authorship cannot be definitively proved, we know from his own witness that he was involved in the compounding of elixirs around AD 610. Among the medical disorders which he experienced, of which he left a first-hand account in his medical works, is intoxication due to elixir ingestion.

In addition to Daoism, recent research has pointed out Sun Simo's close connection with Buddhism. For example, he refers to Indian massage techniques, and mentions methods for the treatment of beriberi from works edited by Buddhist monks. Perhaps under the influence of Tiantai disciplines, he also introduced meditation in his medical practice. Moreover, the above mentioned "Jinjing" section of the *Qianjing yifang* includes incantatory formulas in Sanskrit. The main factor behind these Buddhist elements may have been Sun's interest in the doctrines of the Huayan school. Some passages of his texts, especially those on medical ethics, acquire new meaning if read in this light.

FABRIZIO PREGADIO

REFERENCES

Despeux, Catherine. *Préscriptions d'Acuponcture Valant Milles Onces d'Or*. Paris: Guy Trédaniel, 1987. Translation of the chapter on acupuncture of the *Qianjing fang*.

Engelhart, Ute. "*Qi* for Life: Longevity in the Tang." In *Taoist Meditation and Longevity Techniques*. Ed. Livia Kohn, in collaboration with Yoshinobu Sakade. Ann Arbor: Center For Chinese Studies, University of Michigan, 1989, 263–296.

Sakade Yoshinobu. "Sun Simiao et le Bouddhisme." *Kansai Daigaku bungaku ronshû* 42.1: 81–98, 1992.

Sivin, Nathan. "A Seventh-Century Chinese Medical Case History." *Bulletin of the History of Medicine* 41: 267–273, 1967.

Sivin, Nathan. *Chinese Alchemy: Preliminary Studies*. Cambridge, Massachusetts: Harvard University Press, 1968.

Unschuld, Paul. *Medicine in China; A History of Ideas*. Berkeley: University of California Press, 1985.

See also: Chinese Medicine – *Qianjin Fang* – Acupuncture – Moxibustion – Medical Ethics – Alchemy in China

SUN ZI Of all the mathematicians in ancient China, we know the least about the life of Sun Zi. The fact that he was given the honorific designation of *Zi* (Master) after his surname led some people like Zhu Zunyi (AD 1629–1709) to identify him with Sun Wu, a celebrated tactician in the sixth century BC. Then Ruan Yuan (AD 1764–1849) assigned him to the late Zhou period of the third century BC. With regard to the text written by Sun Zi, Dai Zhen, the eighteenthth century scholar and mathematician, citing from internal evidence, argued that the text could not have been written earlier than the Han dynasty at the turn of the Christian era. Though this text by Sun Zi was listed in all the bibliographical chapters of the *Sui Shu* (Standard History of the Sui dynasty), the *Jiu Tang Shu* (Old Standard History of the Tang Dynasty), and the *Xin Tang Shu* (New Standard History of the Tang Dynasty), there was no mention of its author. This shows that as early as the middle of the seventh century AD, no one seemed to know who Sun Zi was.

Sun Zi had neither high political position nor influential social standing to merit a place in official history. He appeared to be merely a scholar with some Buddhist inclinations as is evidenced in a problem mentioning the length of a *sutra*. The mention of *wei qi* (encirclement chess), and the taxation in terms of silk floss in other problems indicates the text of Sun Zi could not have been written earlier than the third century AD and later than the fifth century. Thus we assume he lived around AD 400.

Sun Zi Suan Jing is the earliest existing text to provide a description of the rod numerals and their operations. It also gives the names of large and small numbers, tables of measures for length, weight, and capacity, as well as densities for metals. The most famous problem is the oldest example of the remainder problem which reads as follows:

> "Now there are an unknown number of things. If we count by threes, there is a remainder 2; if we count by fives, there is a remainder 3; if we count by sevens, there is a remainder 2. Find the number of things."

The problem has since evolved into what is now known as the Chinese Remainder Problem. Written in modern form, the problem can be expressed thus:

$$N \equiv 2 (\text{mod } 3) \equiv 3 (\text{mod } 5) \equiv 2 (\text{mod } 7).$$

It appeared that Sun Zi's solution of the remainder problem had been extensively employed by the ancient astronomers to perfom complicated computations of the calendar. By the thirteenth century, the problem was fully developed by Qin Jiushao with a sophisticated method of solution which also tackled the difficult case where the moduli were not relatively prime.

ANG TIAN SE

REFERENCES

Lam Lay Yong and Ang Tian Se. *Fleeting Footsteps: Tracing the Conception of Arithmetic and Algebra in Ancient China.* Singapore: World Scientific Publishing Co., 1992.

Mikami, Yoshio. *The Development of Mathematics in China and Japan.* New York: Chelsea Publishing Company, 2nd ed., 1974.

Qian, Baozong, ed. *Suanjing Shi Shu* (Ten Mathematical Classics). Beijing: Zhonghua Shuju, 1963.

Ruan, Yuan. *Chouren Zhuan* (Biographies of Mathematicians and Astronomers). Vol. 1. Shanghai: Shangwu Yinshuguan, 1955.

See also: Computation: Chinese Counting Rods – Qin Jiushao

SUNDIALS IN CHINA The measurement of time has an long history in China. Besides clepsydrae and water clocks, sundials and incense clocks were the only timekeepers employed in the epoch preceding the introduction of mechanical timepieces.

The earliest form of timekeeper used in China appears to be the sundial. Although the use of a simple vertical pole, employed to determine the epoch of solstices by measuring the length of the sun's shadow, appears to go back to Shang times (ca. 1500–1000 BC), it is more difficult to establish the epoch in which the shadow of a simple style was used in China to measure time. Perhaps the most ancient reference is the one made by Joseph Needham et al., contained in the *Shi Ji* (Historical Record), which goes back to the beginning of the fifth century BC. This reference, as well as those in subsequent epochs, makes no mention of the typology of these ancient sundials. Among later apparent references to the building of sundials of the equatorial type, there is the one contained in the *Sui Shu* (History of the Sui Dynasty), which refers to a "graduated sundial" the shadow of which reproduced the motion of the heavens. It was built by Ge Heng, a master of astronomical learning and instrument maker active in about the middle of the third century AD. One of the first descriptions of a Chinese equatorial sundial that has come down to us is contained in the *Duxing Zazhi* (Miscellaneous Records of the Lone Watcher) by Zeng Minxing (ca. AD 1176) in which, among other references, it is stated that Zen Nanzhong, master of astronomy and a highly skilled instrument maker, who was active in Chiangsi province, invented and built many water clocks and sundials: "he also made two wooden dials with diagrams on them. One was set upon a wooden support for reading (the hours by) the sun's shadow" (cited in Needham, 1986).

But the only extant ancient objects that appear to have been built and used as accurate timekeepers are the two so-called Han sundials, dated as former Han, perhaps going back to the second century BC. These two objects have been interpreted in different ways. The most probable interpretation is, however, the one proposed by White and Milman in 1938. According to them the base-plate of these sundials was inclined in the equatorial plane, and the central gnomon was not a style but a rectangular plate connected with a perimetral T-shaped gnomon by a bronze bridge. Time would thus be shown by the position of the shadow of the peripheral gnomon in line with the bridge according to the graduation engraved on the face of the sundial. The season would be shown by the height of the shadow of the T-shaped gnomon on the central rectangular plate. Upon these ancient, supposedly equatorial solar timekeepers the graduations appear on one face only.

The most remote mention of the existence of Chinese equatorial sundials with surfaces graduated and a gnomon extending both above and below the equatorial plane plate appears to go back to the twelfth century. The *Duxing Zazhi*, mentioned above, reports the following passage: "So at Yunchang (Chiangsi province) he (Zeng Nanzhong) made a *Gui Yingtu* (diagram of the sun shadow) and constructed a sundial (*gui*). The dial was supported on posts so that it was high towards the south and low towards the north (i.e. in the plane of the equator). The gnomon pierced the dial at the center, one end pointing to the north pole and the other to the south pole. After the spring equinox, one had to look for the shadow on the side facing the north pole, and after the autumn equinox one found it on the other (the under) side" (Needham, 1986). Numerous examples of late (Ming and Ching dynasties) Chinese equatorial sundials graduated on both sides have come down to us. The dials existing in the imperial palace at Beijing are of this kind.

Besides fixed equatorial sundials, two portable equatorial sundials made of lacquered wood, with inscriptions in red and black, or black only, were constructed in China, at least as far back as the Sung period (tenth–twelfth centuries). The largest production of these portable sundials appears to have been concentrated somewhere around the nineteenth century. Their construction is quite simple. They are generally composed of a rectangular piece of wood as a base plate with a compass needle having an azimuth graduation for meridian orientation set in it. A second, mobile lunette (the plate dial) is connected to the board at the middle.

The lunette is thus half the size of the base plate. The plate dial is graduated and may be raised or lowered to any desired angle so that the gnomon, fixed at right angles at the center of the graduation, may point at the pole whatever the latitude of the observer may be. The dial is then fixed in position by means of a ratchet prop. It is interesting to note that the latitude scale materialized by the position of the ratchet is given not in degrees or by the names of cities, but by the twenty-four fortnightly periods (*chi*) into which the ecliptic (the band of the zodiac through which the sun apparently moves in its yearly course) was subdivided in traditional Chinese astronomy. The hour lines of the equinoctial dial plate are also expressed in the so-called Chinese "double-hours" associated with the twelve ideograms of the duodenary cycle. In fact, despite their everyday, popular origin, these time-keepers seem to embody elements of the long tradition of Chinese geomancy or topomancy, divination by means of configurations of earth.

So we must believe that Chinese equinoctial sundials belong to the pure Chinese chronometric tradition, an interpretation that takes us back to the origin of the Chinese science known as *Feng shui* (geomancy, wind and water), based on the use of the geomantic compass, and of the ancient divinatory system based on the *Yijing* (I Ching, Book of Changes).

The first portable (and fixed?) horizontal solar time pieces were introduced into China starting from the end of the sixteenth century by Matteo Ricci and other Jesuit missionaries. Indeed, Father Ricci himself, in the preface to his memoirs, wrote: "As for their clocks, there are some which use water, and others the fire of certain perfumed fibers made all of the same size; besides this they make others with wheels which are moved by sand, but all of them are very imperfect. Of sundials they have only the equinoctial type, but do not know well how to adjust it for the position (i.e. the latitude) in which it is placed" (cited in Needham). The structure of these Chinese horizontal sundials which have come down to us is perfectly similar to that of the European "diptych" sundials, whose center of diffusion in Europe appears to have been Nuremberg in the late sixteenth century. Chinese horizontal sundials were not constructed before the end of the Ming dynasty (second half of the seventeenth century), and we have significant evidence showing that this kind of sundial was also employed more for geomantic purposes than as simple timekeepers.

Besides the two types of portable sundials so far mentioned, few specimens of a third kind of sundial, probably horizontal, consisting of a rectangular ivory plate, each one ruled for a different latitude, with hour and declination lines engraved, exist in China. The time of day and the period of the year were presumably indicated by the shadow of a style fitted into a central hole.

Unfortunately, given our still insufficient knowledge on the latter, as on other Chinese sundials previously discussed, it is often difficult to know exactly on what working principle they were based and the precise uses for which they were constructed and put to use. Even more uncertain is the epoch in which many portable Chinese sundials, still extant in museums and on exhibition, were built, where they were built, and by whom.

EDOARDO PROVERBIO

REFERENCES

Bedini, Silvio A. "The Scent of Time." *Transactions of the American Philosophical Society* 56(5): 3–51, 1963.

Gouk, Penelope. *The Ivory Sundials of Nuremberg*. Cambridge: Whipple Museum of the History of Science, 1988.

Iannacone, Isaia. "Il 'Feng Shui' e la Bussola Geomantica all'Istituto e Museo di Storia della Scienza di Firenze." *Nuncius* 2: 205–219, 1990.

Needham, Joseph. *Science and Civilisation in China* Cambridge: Cambridge University Press, Vol. III, 1962.

Needham, Joseph, Wang Ling, and Derek J. de Solla Price. *Heavenly Clockwork*. Cambridge: Cambridge University Press, 1986.

Proverbio, Edoardo and Giuliano Bertuccioli. "On a Singular Chinese Portable Sundial." *Nuncius* 1: 47–58, 1986.

Wgite, W.C. and P. M. Millman. "An Ancient Chinese Sundial." *Journal of the Royal Astronomical Society* 3: 417–431, 1938.

See also: Ge Heng – Geomancy

SUNDIALS IN ISLAM The first sundials were formed by nomads who made holes in the tops of their tents. The sunbeam entered the tent through the hole and reached the floor or walls at different places at different hours. The time could be determined by the positions of these spots. From these developed the mural sextant described by the Islamic scientist and traveler al-Bīrūnī (973–1048) in his *Ḥikāya al-āla al-musammā al-suds al-Fakhrī* (Information on the Instrument Called the Fakhri Sextant) and instruments in the observatories of Ulugh Beg (1394–1449) in Samarqand and of Jai Singh (1686–1743) in Jaipur, Delhi and, other cities of India. In these instruments the sunbeams entered the instrument through the hole in its top and reach a special scale in its lower part.

The ancient Greeks invented sundials with gnomons, so that the time could be determined by the position of the shadow of the end of the gnomon on the plane or curved surface. The theory of sundials with gnomons obtained the name *Gnomonic*. Analogous sundials were used by ancient Arabs, Indians, and Chinese. Al-Bīrūnī in *Kitāb fī ifrād al-maqāl fī amr al-ẓlāl* (The Exhaustive Treatise on Shadows) wrote about all of these. All three nations measured the lengths of the shadows of a gnomon in parts: the Greeks in sixtieth parts, Islamic scientists in twelfth parts called feet, Indians in seventh or 6.5th parts called fingers. The Greek division of the gnomon into sixty parts came from the tradition of Babylonian astronomers. The Arabic division is explained by their travels in deserts where they used their own bodies as gnomons and measured their lengths with their own feet. The Indians used the horizontal palms of their hands and erect fingers, and they measured the lengths of the shadows of their fingers on their palms with other fingers.

The most detailed descriptions of different kinds of sundials and of their theory appeared in the medieval Islamic countries. The first known to us, although non extant, was written in Baghdad by Ibrāhim al-Fāzārī (d. ca. 777). It was called *Kitāb al-miqyās li'l-zawāl* (Book of the Gnomon for the Noon). The first extant Arabic treatise on sundials was also written in Baghdad by Muḥammad al-Khwārizmī (ca. 780–ca. 950) with the title *ʿAmal al-sāʿāt fī basīṭ al-rukhāma* (Construction of Hour [lines] on the Plane of a Sundial). The sundial described in this treatise consisted of a horizontal marble board (the literal meaning of the word *rukhāma* is "marble board") and a gnomon (*miqyās*). When the sun accomplishes a visible diurnal circle on the celestial sphere, the shadow of the end of the gnomon describes an arc of a conic or (in the equinoctial days) a segment of a straight line on the plane of the sundial. On this plane the arcs of conics for the days of the entry of the sun in different zodiacal signs (beginnings of months of the zodiacal calendar) and a segment of the straight line for the equinoctial days are drawn. (For Baghdad these conics are hyperbolas). The hour lines are lines joining the points of these arcs and segments corresponding to the same hours of a day (in this treatise hour lines are straight lines). In the treatise the construction of this sundial is described and the tables for the functions necessary for this construction for different latitudes are given (the most detailed tables for 33° (Baghdad) and 34° (Samarra) are given). In al-Khwārizmī's treatises *ʿAmal al-miknasa* (Construction of al-miknasa) and *ʿAmal al-mukḥula li'l-sāʿāt* (Construction of al-mukḥula for [determination of] Hours) different kinds of sundials are described. These include a horizontal plane sundial with a vertical gnomon used in *kanīsas*–churches or heathen temples — and a sundial with a conical surface whose name literally means "vessel for storing antimony (*kuḥl*)" because of its likeness to this vessel.

In the treatise by Thābit ibn Qurra (836–901) *Maqāla fī ṣifa al-ashkāl allatī taḥduthu bi mamarr ṭaraf ẓill al-miqyās fī saṭḥ al-ʿufq fī kull yawm wa fī kull balad* (Article on the Description of Figures Obtained at the Passing of the End of the Shadow of a Gnomon on a Horizontal Plane on any Day and in any Town), the horizontal sundial can be used at any latitude when the trajectories of the end of the shadow of the gnomon are arcs not only of hyperbolas but also of parabolas, ellipses, and circles. In his *Kitāb fī ālāt al-sāʿāt allatī tusammā rukhāmāt* (Book on Horary Instruments Called Sundials) all kinds of plane sundials — horizontal, vertical, and oblique — are described.

Different kinds of sundials which secured the timely calls of Muslims for prayers are described in Arabic *zīj*es (astronomical handbooks with tables) and in special treatises written by *muwaqqits* (timekeepers). Many kinds of sundials are described in al-Bīrūnī's book, *The Exhaustive Treatise on Shadows*, mentioned above. This is also true in the book by al-Ḥasan al-Marrākishī (d. 1262) *Kitāb jāmiʿ al-mabādī wa'l-ghāyāt fī ʿilm al-mīqāt* (Book of the Collection of Beginnings and Results in the Science of the Determination of Time). The time on sundials was measured not in astronomical hours equal to 1/24th of the day and night but in temporal hours equal to 1/12th of the day.

In later times oblique (not horizontal or vertical) sundials called *al-munḥarifa* were often used. Of particular interest is the treatise on these sundials written by the *muwaqqit* of the mosque al-Azhar in Cairo Sibṭ al-Māridīnī (1423–ca.1495) *Risāla fī'l-munḥarifa wa'l-shākhis* (Treatise on an Oblique [sundial] and a Pole).

BORIS A. ROSENFELD

REFERENCES

al-Khwārizmī. "Construction of Hours on the Plane of Sundials." Trans. J. al-Dabbagh, commentary by J. al-Dabbagh and B.A.Rosenfeld. *Muhammad ibn Musa al-Khwarizmi. On the 1200th Anniversary of his Birthday*. Moscow: Nauka, 1983, pp. 221–234 (in Russian).

Schoy, Carl. *Gnomonik der Araber (Die Geschichte der Zetmessung und der Uhren)*. Berlin: Springer, 1923.

Schoy, Carl. "Sonnenuhren in der spätarabischen Astronomie." *Isis* 5: 332–360, 1924.

Thābit ibn Qurra. "Ein Werk über ebene Sonnenuhren." In *Quellen und Studien zur Geschichte der Mathematik, Astronomie und Physik*. Abt.A. Bd. 4. Berlin: Springer, 1936.

See also: Gnomon – al-Bīrūnī – al-Khwārizmī – Time – Qibla and Islamic Prayer Times – al-Māridīnī – Thābit ibn Qurra

SURVEYING Surveying is the mathematical science that incorporates the application of geometric principles with concepts of measurement in order to delineate the forms, position, and extent of terrestrial features or man-made structures. It is an ancient activity that has been used by all urbanized societies. The establishment and maintenance of land boundaries, the construction of walls, and the lay out of irrigation systems and aqueducts rely on the use of some kind of surveying. Extent structures such as Stonehenge in England, the Great Wall of China, Machu Picchu in the Andes of Peru, the Great Pyramids of Giza, or the temples of Angkor in Kampuchea testify to their builders' knowledge and use of surveying. However, all too often, in the ancient world, surveying activities were the closely guarded prerogative of an elite, members of a priestly or bureaucratic class. Knowledge of their mathematical discipline was usually transmitted orally from master to student. Such practices, combined with the ravishes of time, account for few existing documented records of old surveying practices and techniques. Therefore, much of what can be gleaned about surveying in early non-Western societies is speculative in nature and rests on extant archaeological and architectural evidence.

The most basic of surveying activities include the determination of straight horizontal and vertical lines and the establishment of a level plane upon which structures can be erected or reference slopes established. Instruments to achieve these goals are strikingly simple: straight horizontal lines can be obtained through the use of sighting poles or stakes and retained by the use of stretched cords or ropes; straight vertical lines are obtained by the use of a plumb bob, a weight suspended from a string, and a level plane can be constructed with the assistance of a leveling device such as a water-filled trench.

Rope Stretching

Herodotus (ca. 484–425 BC), the Greek historian, attributed the origins of geometry to the Nile Valley of Egypt, where priest-surveyors stretched ropes to mark out land boundaries. These "rope-stretchers", or as Herodotus called them *harpedonaptae*, are the first historically recognized surveyors. Rope stretching activities also took place in Babylonia where ancient clay tablets mention the act of "stretching a field", that is using a rope to determine the dimensions of an agricultural field and denoting a particular individual as "the dragger of the rope", a surveyor. It appears that these early Egyptian and Babylonian land measuring activities were prompted by the need for royal levies. However, in the early Indus Valley civilization, rope stretching served another need. The *Śulbasūtras* (Rules of the Cord) compiled between the fifth to eighth centuries BC supply geometric prescriptions for the construction of ritual altars. These prescriptions were based on rope-stretching and were carried out by the Vedic priests. Later Pali literature makes mention of "rope holders" (surveyors) in reference to land measurements. In all early urbanized societies, rope stretching provided the basis for surveying.

Egypt

Surveying activities in Egypt certainly preceded the fifth century BC observations of Herodotus. An inscription on the Palermo Stone dating from the Old Kingdom period of Egyptian history (ca. 3000 BC) notes the existence of land surveys. Tomb inscriptions of about the same period mention the existence of land registry offices. A wall scene in the tomb of Menna at Thebes depicts surveyors at work. The scene shows two men measuring a field of corn with a long cord on which knots are marked at intervals of about 4 or 5 cubits. A standard measuring cord or rope of this period was 100 cubits (52.5 meters) long. To obtain accurate vertical lines as well as to sight over long distances, Egyptian surveyors use the *merkhet.* This instrument also existed in Egypt from the earliest times; it consisted of a short plumb-line and plummet hanging from a holder that contained a sighting slit. Thus alignments could be made on distant objects. Merkhets were employed in the orientation of temples in a process called the "stretching of the measuring cord". Egyptian surveyors also employed two types of levels, the water level via a water filled trench which was suitable for large scale leveling, and the plumb-bob level erected with the aid of a wooden, right isosceles triangular frame work. Modern surveys have affirmed that the ancient Egyptians obtained very accurate results using their simple tools: the foundation for the Great Pyramid of Giza is almost perfectly level; boundary markers on the sides of the Nile River are aligned over long distances and a very accurate system of nilometers (flood gauges), were established along the Nile from its delta to the First Cataract, a marvelous feat of leveling.

Babylonia

Large scale construction projects existed in the Tigris-Euphrates region as early as 2300 BC. The accomplishments of Gudea of Lagash, an engineer or architect of this time, is commemorated by statues which depict him holding a tablet containing scaled plans for a structure. These plans are superimposed on a rectangular grid. Both the use of scaling and a rectangular reference grid indicate the existence of a high level of surveying skill at this time. Further, extensive systems of irrigation channels relied on the establishment of adequate gradients or slope determined by leveling techniques. Fragments of Babylonian clay astrolabes dating back to the second millennium BC have been found. With such instruments, observers of the time could determine angles of inclination; however, it is believed that these astrolabes were employed for astronomical sightings rather than terrestrial surveying.

Ancient India

Archaeological evidence from such sites as Mohenjo-Daro and Harappa, cities of the early Indus civilization (3500–2500 BC) indicate that city planning principles were followed. Buildings were uniform in appearance, and roads were laid out at right angles to each other. The existence of sewerage systems as well as flowing aqueducts testify to a knowledge and use of leveling principles. Builders of the cities of Mohenjo-Daro and Harappa knew surveying. Linear measuring scales found at excavation sites in the region indicate the early Indus peoples employed a decimal system of measurement. This system was based on a "Mohenjo-Daro inch" of 0.67 cm. Further, at Lethal, the remains of a sighting instrument were found. The instrument consists of a hollow shell with four slits cut into its sides. These slits are situated at right angles to each other and allow for perpendicular sighting as would be necessary in surveying a rectangular road system. During the later *Śulbasūtra* period (800 BC), bamboo poles, *saṇku*, were used for measuring and laying out circular regions, and a standardized chain or measuring rope, *rajju*, was employed. The *Śulbasūtra* texts describe an extensive mathematics supporting its rope stretching surveying techniques. Included in this mathematics was a theory of similar triangles.

While works of later Indian mathematical authors primarily concerned applications of mathematics to astronomy, some also included mathematical information for surveyors. Āryabhaṭa I (476–550) in his *Āryabhaṭiya* (ca. 499) discusses procedures for finding areas and volumes of plane figures and solids. Brahmagupta (ca. AD 628) in his *Brahmasphutasiddhānta* (Correct Astronomical System of Brahma) provides much information relevant to the needs of surveyors including specific computation procedures necessary for working with a shadow gnomon, a sighting staff employed for inclined sightings, thus incorporating a concept of angle into surveying activities.

China and the Far East

Early Chinese society was river based. Settling along the banks of the Yangtze and Hwang Ho rivers, the Chinese people harnessed and controlled the rivers by a system of dikes, canals and irrigation channels. These construction projects required a knowledge and use of land surveying. Surveying was openly recognized as an important societal activity; folk hero Fu Xi and legendary emperor-engineer Yu the Great were often depicted holding surveying instruments.

Discussions and illustrations in extant texts and reference works provide some knowledge of the instruments used by early Chinese surveyors. Calibrated sighting poles, *biao* were used in conjunction with sighting tubes, *wang tong*, or sighting boards, *ce shi pai*. A water level, *zhun*, was employed for leveling and a bamboo measure tape, *bu che*, devised for chaining land. A primary surveying instrument was the L-shaped set square or gnomon, *ju*. The earliest documented reference to surveying is found in the *Zhoubi Suanjing* (The Arithmetical Classic of the Gnomon and the Circular Paths of Heaven, ca. 100 BC–AD 100) where in a fanciful conversation between Zhou Gong, a duke of the Zhou dynasty (ca. 1030–221 BC) and the Grand Prefect Shang Gao, the duke advises the Prefect in the use of the set square. More substantial information on the mathematics of surveying is supplied in *Jiuzhang suanshu* (The Nine Chapters on the Mathematical Art, ca. 100 BC). The work contains nine chapters on specific applications of mathematics. Chapter 1, "Field Measurements"; Chapter 5, "Construction Consultations" and Chapter 9 "*Gougu*" (right-triangle) are directly concerned with surveying computations.

In AD 263, the scholar-official Liu Hui wrote a commentary on the *Jiuzhang* and revised much of its contents. He paid particular attention to the ninth chapter and extended its collection of problems to allow for more surveying situations which involved the obtaining of measurements to inaccessible points. Liu stressed a technique called *chong cha* (double difference) requiring two distinct sighting observations from separate locations. At the beginning of the Tang dynasty (AD 618–906), Liu's problems involving double differences were separated from the *Jiuzhang* and made into an independent mathematical work on surveying, the *Haidao suanjing* (Sea Island Mathematical Manual). In AD 656, the Royal Academy established an official curriculum to be used for the training of state officials. The *Haidao* was included among the ten mathematical works to be studied. Later, this curriculum was adopted in Japan and Korea where the instructions of the *Haidao* provided the basis for surveying. One of the most accomplished feats of early Chinese surveying was begun in the year 724 when the State Astronomical Bureau of the Tang dynasty initiated the first meridian survey in the ancient world. Under the supervision of the scholar Yixing, thirteen observation stations were established near the meridian 114°E and between latitudes 29°N to 52°N. Observations were taken over a period of several years. The expedition determined the angular attitude

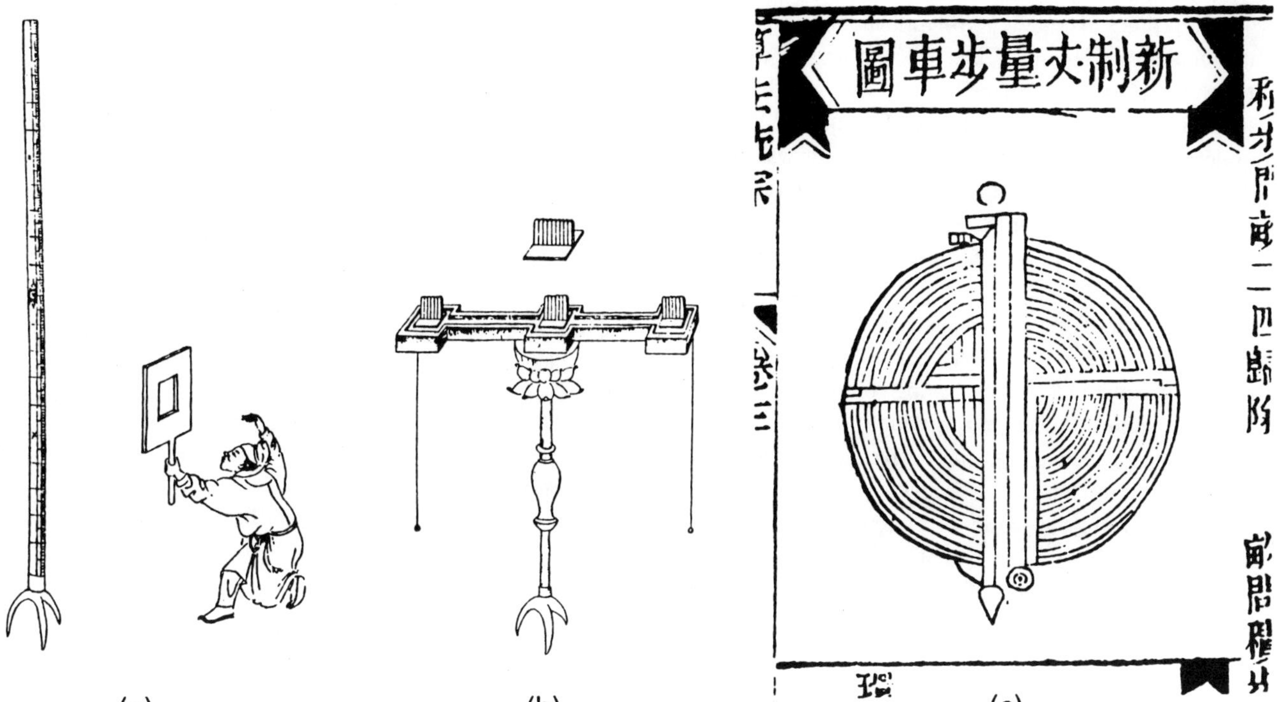

(a) (b) (c)

Figure 1: Classical Chinese surveying instruments. (a) *Biao*, calibrated sighting pole; (b) water level, *zhun*; (c) *bu che*, a bamboo measuring tape. From Frank Swetz, *The Sea Island Mathematical Manual: Surveying and Mathematics in Ancient China*. Penn State University Press, 1992. Used with permission of the author.

of the north celestial pole above the horizon and recorded the length of shadows at noon for the summer and winter solstices and equinoxes.

The *Haidao suanjing's* problems became the basis of later works which considered surveying computations. In 1247, the mathematician Qin Jiushao published *Shushu jiuzhang* (Mathematical Treatise in Nine Sections). Three of its nine chapters concerned survey applications: Chapter 4 was entitled "Surveying"; Chapter 7, "Architecture"; and Chapter 8, "Military Matters" which concerned the use of surveying techniques for observing an enemy from a distance. His contemporary, Yang Hui wrote *Tian mu bi lei cheng chu jie fa* (Practical Rules of Arithmetic for Surveying). Toward the end of the Ming dynasty, Western influence began to penetrate China. In 1582, the Italian Jesuit Matteo Ricci arrived in Macao and eventually made his way to Beijing where he used his mathematical and scientific knowledge to win favor with the court. Ricci collaborated with the scholar Xu Guangqi to publish *Celiang fayi* (Essentials of Surveying) (1607–1608). This book introduced contemporary European survey and land measurement methodology to China. Xu himself was a skilled surveyor and wrote an appendix to the *Essentials of Surveying*. It appeared in 1608 under the title *Celiang i tung* (Similarities and Differences [Between Chinese and European] Surveying Techniques). From this period onward, surveying in China combined both European and traditional theories and practices.

The Islamic World

Islam did not emerge as an intellectual and political force until about AD 726 with the establishment of the ʿAbbasid Caliphate. From their capital in Baghdad, the early ʿAbbasid caliphs patroned the collection of scientific works and used this acquired knowledge to consolidate their religious and political empire. Their era was marked by the building of new canals, bridges and aqueducts and by the reconstruction of old Babylonian irrigation systems. Surveying was used in this work. While Muslims became the heirs of Babylonian, Egyptian and Greek surveying theory, they soon became accomplished practitioners and innovators in their own right. Religious prescriptions required daily prayers toward Mecca; in turn, mosques had to be constructed facing Mecca. Thus determining the *qibla*, or direction of Mecca from a given location, became an important task for surveyors.

Muslim surveyors used several methods of leveling: the plumb-bob level was employed as well as leveling poles. Al-Khāzinī (ca. 840) and Ibn-al-ʿAwwām wrote on the use of leveling poles. The latter wrote a handbook for farmers that included the layout of fields. Right angles were laid out with the use of an L-shaped square, *kunija*. Abu'l Wafā (940–998) wrote about the use of such a square. The most complete early Muslim treatise on surveying was written by Muḥammad ibn al-Hassan al-Hasib al-Karajī (ca. 1000). It was entitled *Kitāb al ʿuqūd wa'l abniyah* (Of Vaults and Building). Al-Karajī discussed both the mathematical and practical aspects of surveying and provided specific instructions for the surveying of tunnels and underground aqueducts, *qanat*. Later writer Abū Saqr al-Qabisi introduced trigonometric methods into surveying computations.

The one area of surveying in which Muslim scientists and craftsman excelled was the design and utilization of measuring instruments. Their knowledge of the astrolabe was obtained from the works of Ptolemy. The first noted Muslim maker of astrolabes was al-Fazārī (d. ca. 777) who worked under the patronage of al-Mansur. By the end of the eighth century a number of scholars were producing works on the construction and use of the astrolabe. The most famous of these scholars was Māshā'allāh (762–ca. 815), a Jew who worked under Islamic patronage. His *Kitāb sana ʿat al-asturlāb wa l-ʿanal bihā* (Book of the Construction of an Astrolabe and Its Use) became the authoritative reference of its time. At a later date (ca. 1380), Chaucer used Māshā 'allāh's theories in his European introduction of the astrolabe as a scientific instrument. It was Muslim efforts that resulted in the astrolabe becoming a valued instrument in land surveying activities. Similarly, the Jacob's staff, a popular medieval instrument for determining planar angles, is believed to have reached Europe via Muslim sources and may have had its origins in navigational methods used by early Muslim traders. The oldest actual description of this instrument comes from a navigator's manual, the *Mohit*, written by Sidi al-Chelebri, captain of the Turkish fleet under Sultan Suleimann in 1554.

Pre-Columbian America

Although no written records exist to document the surveying knowledge of early native American civilizations, archaeological sites testify to these peoples' application and understanding of surveying techniques and principles. The city planning and construction carried out by the Olmec, Maya, Teotihuacan, Toltec, and Aztec peoples of South and Central America indicate that they undertook some surveying activities. For example, the Aztec capital of Tenochtitlan was laid out according to a grid system. It contained markets, palaces and streets and canals and held a population of approximately 200,000 people in the year 1521. Temples and ball courts were oriented to the four cardinal directions. Aqueducts brought fresh water into the city from many miles distant. Some of the Aztec surveying instruments are known by name; plumb line, *temetzlepilolli*; the water level, *atezcath*; set square, *tlanacazanimi*; measuring scale, *tlahuahuanoloni* and the construction compass, *tlayolloanaloni*.

FRANK J. SWETZ

REFERENCES

Amma, T.A. Sarasvati. *Geometry in Ancient and Medieval India*. Delhi: Motlal Banarsidass, 1979.

Berggren, J.L. *Episodes in the Mathematics of Medieval Islam*. New York: Springer-Verlag, 1986.

Guerra, Francisco. "Aztec Science and Technology." *History of Science* 8: 32–52, 1969.

Hedquist, Bruce. "On the History of Land Surveying in China." *Surveying and Mapping* 35: 251–254, 1975.

Joseph, George Gheverghese. *The Crest of the Peacock: Non-European Roots of Mathematics*. London: I.B. Tauris, 1991.

Kiely, Edmond. *Surveying Instruments: Their History and Classroom Use*. New York: Teacher's College Press, Columbia University, 1947.

Rayner, W. H. "Surveying in Ancient Times." *Civil Engineering* 9: 612–614, 1939.

Swetz, Frank J. *The Sea Island Mathematical Manual: Surveying and Mathematics in Ancient China*. University Park: The Pennsylvania State University Press, 1992.

See also: *Śulbasūtras* – Āryabhaṭa – Brahmagupta – *Zhoubi Suanjing* – Liu Hui and the *Jiuzhang Suanshu* – The *Gougu* Theorem – Liu Hui – Qin Jiushao – al-Khāzinī – Māshā'allāh – *Qibla* – City Planning – Irrigation – Abu'l Wafā – al-Karaji – *Qanat* – Astrolabe

SŪRYASIDDHĀNTA The *Sūryasiddhānta* is a complete work on Hindu astronomy and is more popular and widely studied in North India than in the South. In order to enhance its prestige and antiquity, it is stated in the text itself (I.29) that it had been communicated by a representative of the God Sun to Maya, several thousand years ago. However, both internal and external evidence show that the work was composed between AD 600 and 1000. In about five hundred verses, distributed through fourteen chapters, the work deals with all aspects of Hindu astronomy, and also cosmology, geography, astronomical instruments, and time-reckoning. It follows the midnight day-reckoning.

The contents of the work are comprehensive. Chapter I speaks about the circumstances that led to the composition of the work, the aeons and aeonary revolutions of the planets, the principles underlying the computation of the planets and the nodes, the position of the planets at the beginning of the current aeon, the Prime meridian and local time, and the inclination of the orbits of the planets. The time units given in the work are more in conformity with the *Purāṇas* than with other texts on Hindu astronomy. Chapter II deals with the computation of the true motion and the true longitudes of the planets. Chapter III is devoted to the determination of the directions, place and time, and also the calculation of the precession of the equinoxes. It is noteworthy that the *Sūryasiddhānta* is the earliest available Indian text which contains a discussion of the calculation of the precession of the equinoxes. Chapters IV and V deal in detail with lunar and solar eclipses and their computation. Chapters VII and VIII are concerned with the conjunction of one planet with another, and a planet with a star, including their observation. Chapter IX deals with the determination of the heliacal rising of the planets and Chapter X with the phases of the moon. Chapter XI is concerned with the phenomenon of *pāta*. Cosmology and geography of the worlds and of the earth occur in Chapter XII. The construction of the armillary sphere and its working, and a mention of the main astronomical instruments form the subject matter of Chapter XIII. Chapter XIV enumerates and defines nine types of reckoning time, such as lunar, solar, siderial, tropical, etc.

The *Sūryasiddhānta* is indebted to earlier astronomers like Āryabhaṭa (b. 476) and Brahmagupta (b. 598) in certain matters like the inclination of the planetary orbits, the tabular Sines, etc. It is also to be noted that the *Sūryasiddhānta* does not include chapters on arithmetic, algebra, or astronomical problems, which are generally included in texts of this type.

The popularity of the *Sūryasiddhānta* is clear from the very large number of manuscripts of the work found throughout the land. About 35 commentaries on the work, written by scholars from different regions of India, and about 20 works, including planetary tables and manuals, based in the *Sūryasiddhānta*, have been identified. The popularity of the work is also reflected by the number of almanacs prepared on the basis of the *Sūryasiddhānta*.

K.V. SARMA

REFERENCES

Dikshit, S.B. *Bharatiya Jyotish Sastra*. (*History of Indian Astronomy*). Pt. II. Trans. R.V. Vaidya. Calcutta: Positional Astronomy Centre, India Meteorological Department, 1981.

Pingree, David. *Jyotiḥśāstra: Astral and Mathematical Literature*. Wiesbaden: Otto Harrassowitz, 1981.

The Sūryasiddhānta: A Text-book of Hindu Astronomy. Trans. with notes by Ebenezer Burgess. Delhi: Motilal Banarsidass, 1989.

The Sūryasiddhānta with the Commentary of Parameśvara. Ed. Kripa Shankar Shukla. Lucknow: Department of Mathematics and Astronomy, Lucknow University, 1957.

See also: Astronomy in India – Precession of the Equinoxes – Armillary Spheres in India – Astronomical Instruments – Āryabhaṭa – Brahmagupta

SUŚRUTA Suśruta is the author of one of the three major Ayurvedic works, the *Suśrutasaṃhitā*. He was the son of Viśvāmitra and a student of Divodāsa Dhanvantarī, the King of Kāśi (now Varanasi) who flourished prior to 1000 BC.

It was the practice in ancient times to engage military physicians who accompanied the kings and army commanders to the battlefield to treat wounded soldiers. Therefore, there was a need to train surgeons. One of the eight specialized brances of Āyurveda is surgery (*śalya-tantra*), and Suśruta's work stands foremost in this field.

The *Suśrutasaṃhitā* consists of 186 chapters which are divided into six sections as follows:

1. forty-six chapters in the *Sūtra* section dealing with the principles and practices of surgery, and a description of food and drinks;
2. sixteen chapters in the *Nidāna* section dealing with the diagnosis of important surgical ailments;
3. ten chapters in the *Śārīra* section dealing with the creation of the universe, embryology, obstetrics and anatomy;
4. forty chapters in the *Cikistā* section dealing with the treatment of surgical ailments;
5. eight chapters in the *Kalpa* section dealing with toxicology;
6. sixty-six chapters in the *Uttara* section dealing with the diagnosis and treatment of the diseases of the eye, ear, nose, and throat, paediatrics, seizures by evil spirits, and internal diseases in general.

Though the primary object of this work is to deal with surgery, it can be seen from the above that all eight specialized branches of Āyurveda are covered in it. According to Suśruta, any one who wants to be highly skilled in surgery should be thoroughly versed in anatomy by dissecting a dead body. He describes the selection of dead bodies, their preservation, and different methods of examining various parts of the body. He describes different types of blunt and sharp instruments, sixty different steps to be followed for healing wounds, plastic surgery, and operative procedures for the removal of stones in the urinary tract. According to Suśruta, there can be no more virtuous act for a physician or surgeon than to alleviate human suffering.

BHAGWAN DASH

REFERENCES

Bhishagratna, Kunjālal. *An English Translation of the Suśrutasaṃhitā*. Varanasi: Chowkhamba Sanskrit Series Office, 1991.

Kutumbiah, P. *Ancient Indian Medicine*. Bombay: Orient Longmans, 1962.

Mukhopadhyaya, Girindranath. *History of Indian Medicine*. New Delhi: Oriental Books Reprint Corporation, 1974.

Sankaran, P.S., and P.J. Deshpande."Suśruta." In *Cultural Leaders of India: Scientists*. Ed. V. Raghavan. New Delhi: Publication Division, Ministry of Information and Broadcasting, Government of India, 1976, pp. 44–72.

Śarmā, Priyavrata. *Indian Medicine in Classical Age*. Varanasi: Chowkhamba Sanskrit Series Office, 1972.

Suśruta. *Suśrutasaṃhitā*. Varanasi: Chaukamba Orientalia, 1980.

See also: Medical Ethics in India – *Āyurveda*

AL-SUYŪṬĪ As-Suyūṭī (1445–1505) wrote on just about every discipline that had been recognized, in his time, as having its own method and subject matter. What is most noteworthy is that *hay'a* (cosmology, cosmography, but also astronomy) was among them. The author was in his own estimation an authority in the traditional Arabic and Islamic sciences (*al-ʿulūm al-ʿnaqliyyah*, transmitted sciences), especially in grammar, jurisprudence, and in tradition (*ḥadīth*, the Sayings of the Prophet), but in no way in those "Sciences of the Ancients" (*al-ʿulūm al-ʿaqliyyah*), which had entered the libraries of Islamic culture through numerous translations from Greek, Syriac, Middle Persian, etc. He even expressed his special hatred for philosophy and logic. The title of his treatise, *al-Hayʾa al-sanīya fī l-hayʾa al-sunnīya*, already reveals the challenge he had in mind: his was to be *the* Islamic cosmology, based on authentic Islamic traditions, the *Sunna*, which so conveniently rhymed with *sanīya* (brilliant, magnificent, glorious). The choice of the technical term *al-hay'a* in the title implies that the author is offering an alternative to those cosmologies based on the principles and methods of pre-Islamic astronomers. As such it is a parallel of his book *al-Ṭibb al-nabawī* (Prophetic Medicine). As the author says himself, it was his goal "that those with intelligence might rejoice, and those with eyes take heed". Actually, the great number of extant manuscripts in our libraries proves that as-Suyūṭī's *Hay'a* attracted more attention than most other contemporary books on the cosmos.

This work is a collection of fragmentary descriptions and explanations of such natural phenomena as the sun, moon, and stars in their celestial spheres, lands and seas, winds and clouds, etc. The distinctive feature is, however, that all these fragments are authenticated in the traditional manner with chains of trustworthy authorities which connect them with Quranic revelation or Prophetic wisdom. As a result, some of the earliest theories about cosmological entities and natural phenomena have been preserved that may elucidate the world views prevalent among the young Muslim community before they were developed under the impact of the translations from pre-Islamic cultures. It remains doubtful whether even in the Middle Ages a mythical theory about the winds, because of the traditional authorities, would have been accepted with the same truth claim as a modern one. But the fact that the authorities were already interested in the phenomena of nature and cosmos may have opened rather than closed the eyes of the student.

ANTON M. HEINEN

REFERENCES

Goldziher, Ignaz. "Zur Charakteristik Jelāl ud-dīn us-Sujūṭī's und seiner literarischen Thätigkeit." *Sitzungsberichte der Koeniglichen Akademie der Wissenschaften, Philosophisch-historische Klasse* 69:17, 1871.

Heinen, Anton M. *Islamic Cosmology. A Study of as-Suyūṭī's al-Hayʾa as-saniya fī l-Haʾya as-sunnīya with critical edition, translation, and commentary.* Beirut: In Kommission bei Franz Steiner Verlag Wiesbaden, 1982.

Sartain, E.M. *Jalāl al-Dīn al-Suyūṭī*. 2 vols. Cambridge: Cambridge University Press, 1975.

See also: *Hay'a*

SWIDDEN Swidden, also known as "slash and burn", "long fallow", and *roça* in Brazil, is a system of agriculture that involves clearing small areas within a forest, burning the slash, and planting for one to five years. The plot is then abandoned for 25 to 200 years — long enough for the forest to reclaim the cleared area. In contemporary Latin America this system may support as many as fifty million people.

In spite of the labor involved in clearing the forest, it is less than that required in clearing brush or turning under grass sod. This explains why fields are abandoned until the forest has regrown. In the tropics most nutrient and organic matter is locked up in the plants and trees, rather than in the soil. Thus, ash from the burned slash also frees nutrients for the crops.

A typical swidden planting mimics the diversity of the forest, with a variety of crops interspersed among each other, or planted in closely spaced zones within the field. In the Amazon-Orinoco drainages one may encounter ten to fifteen crops being grown together. Once crops are planted, they require only one or two weedings before harvest. In the tropics, harvest can extend over many months; just enough is taken from the field to provide for a few days or a week at a time. In a sense, then, the crop is stored in the field, rather than in a storehouse. The labor required in swidden is minimal, averaging two to four hours a day (though perhaps concentrated into five or six hours two or three times a week).

Swidden crops are often limited to starchy staples, vegetables and fruits, producing little protein. Thus subsistence is typically supplemented by fishing and/or hunting. Only in Mesoamerica, where beans are important, is the need for supplementary animal protein reduced somewhat.

Swidden agriculture is often characterized as unproductive and destructive of soil fertility. In fact, during the first one to three years it is generally more productive than most permanent fields. The ash provides ready nutrients and there is little competition with other plants. The return on labor investment is even more rewarding.

After a few years productivity typically declines. Traditionally this was blamed on exhaustion of the soil. Extensive research indicates that this is due not so much to declining soil fertility, as to invasion by weeds and grasses which compete with crops for space and nutrients. A further motivation for abandonment is that this tangled growth becomes a magnet for vermin, insects and disease. Thus the farmer logically seeks out a new spot in virgin forest to clear another plot free of problems.

Contrary to popular belief, this does not necessitate periodic village movement. Robert Carneiro has shown that five hundred people practicing swidden agriculture could subsist permanently on six thousand acres located within a three mile radius of their village. Other explanations based on social, political, or religious factors must be sought for the practice of moving villages.

There is much debate about the "backward" and "deleterious" effects of swidden agriculture. When practiced by a sparse, widely dispersed population it does not seem to harm the ecosystem, and may in fact prove beneficial by opening up the forest canopy and stimulating new growth. Some ecologists believe that the equilibrium state under aboriginal conditions was grounded on periodic swidden clearing. On the other hand, when population growth intensifies clearing to a point where the forest can no longer regenerate itself, the result will be widespread environmental degradation.

KARL H. SCHWERIN

REFERENCES

Beckerman, Stephen. "Does the Swidden Ape the Jungle?" *Human Ecology* 11(1):1–12, 1983.

Boserup, Ester. *The Conditions of Agricultural Growth. The Economics of Agrarian Change under Population Pressure.* Chicago: Aldine, 1965.

Carneiro, Robert L. "Slash-and-Burn Cultivation Among the Kuikuru and its Implications for Cultural Development in the Amazon Basin." In *The Evolution of Horticultural Systems in Native South America, Causes and Consequences. Antropológica*, Supplement no. 2. Caracas: Sociedad de Ciencias Naturales La Salle, 1961, pp. 47–67.

T

AL-ṬABARĪ Abū'l -Ḥasan ʿAlī ibn Sahl Rabbān al-Ṭabarī was born in the environs of the city of Marw, in the province of Khurasān in Persia (presently Mary, in Turkmenistan), about AD 783, before the reign of the Abbasid Caliph Hārūn al-Rashīd (786–809). His father Sahl was a prominent citzen of great learning and a highly placed state official. As a religious leader in the Syriac speaking community, he was reverently called *Rabbān* (from the Aramaic for teacher), and had far-reaching knowledge in theology, philosophy, and medicine.

Sahl took a special interest in his son's upbringing, providing him with the best available educational opportunities. ʿAli read the best Syriac books and excelled in learning. When he was 14, he turned to medicine. He concentrated there, because he realized he could help the sick and the needy.

From the Marw region, he moved to Tabaristān (Māzandarān, south of the Caspian Sea). Thus he became known as al-Ṭabarī. He was appointed counselor-secretary-scribe to the Prince-Sultan Māzyār ibn Qārin. When the latter rebelled against the Abbasid's authority in open revolt, Caliph al-Muʿtaṣim sent a powerful army that crushed the mutiny and killed Māzyār in AD 839.

Al-Ṭabarī spoke of it later on as "a tragic episode" which left deep scars that remained until late in his life. Meanwhile, he was summoned to the Caliph's court at the new capital, Sāmarrāʾ. Under the Caliph's influence, al-Ṭabarī renounced his Christian faith and embraced Islam. He continued as a physician during the remaining part of al-Muʿtaṣim's life, and remained there under his successor, his brother al-Wāthiq (843–47).

His good fortune came with the rise of Caliph al-Mutawakkil (AD 847–61). He was promoted to the position of physician-in-ordinary, and also became the Caliph's counselor and trusted companion (*nadīm*). In appreciation, al-Ṭabarī dedicated his best and largest medical compendium, *Firdaws al-Ḥikmah* (Paradise of Wisdom) to his patron in AD 850. This was the first and most comprehensive medical encyclopedia of its kind in Islam. It took him twenty years to complete. In the introduction, the author stated that the work was to be useful to his medical students as well as to practitioners. He listed five points concerning the importance of the art of medicine:

1. It brings relief and healing to the sick, and consolation to the weary.
2. It successfully diagnoses and skillfully treats ailments, even unseen diseases not easy to discover or observe.
3. It is needed by all, regardless of age, gender, or wealth.
4. It is among the noblest of all callings.
5. The words *ṭibb*, *ṭibābah*, *mu'āssāh*, and *usāt* all relate to medicine and its healing processes.

Al-Ṭabarī then mentioned four virtues that all physicans had to possess in order to be successful and esteemed: *al-rifq* (leniency and kindness), *raḥmah* (mercy and compassion), *qanāʿah* (contentedness and gratification), and *ʿafāf* (chastity with simplicity).

Firdaws was divided into seven sections in thirty treatises, composed of 360 chapters in all. They ranged from cosmogony, the nature of man, embryology, and anatomy, to materia medica, psychotherapy, pathology, and surgery. Other topics included theoretical and practical medicine in the Greek and Indian traditions, and rules of conduct with insistence on strict adherence to the highest ethical standards.

Another noteworthy literary contribution was his book *al-Dīn wā'l-Dawlah* (Religion and the State). It represents a defense as well as an exposition of the religion of Islam, the Holy Qurʾān, and the Holy Prophet Muḥammad. It seems temperate, rational, and objective in style and tone, and appears free from misgivings or barren argumentation. It also abounds with quotations from the Bible (in the Syriac version).

In these two works, al-Ṭabarī shed much light on the development and progress of the religious, philosopical, and medical advancements during the first two quarters of the ninth century AD. He died in Sāmarrā ca. 858.

The *Kitāb al-Dīn* was soon eclipsed during the Islamic middle ages, because of a lack of interest in such studies as comparative religion. Only in the twentieth century was the book edited more than once. *Firdaws*, however, has continued to enjoy a good reputation, with a wide circulation throughout the Islamic world. Both works can now be considered classical literary works.

SAMI K. HAMARNEH

REFERENCES

Al-Ṭabarī. *Firdaws al-Ḥikmah*. Ed. M.Z. Ṣiddīqī. Berlin: Sonne, 1928, and Karachi: Hamdard Press, 1981.

Bouyges, M. "ʿAlī al-Ṭabarī." *Mélanges de l'Université St. Joseph* 28: 69–114, 1949–50.

Brockelmann, Carl. "*Firdaws al-Ḥikmah* of ʿAlī al-Ṭabarī." *Zeitschrift für Semitistik und verwandte Gebiete* 8: 270–76, 1932.

Browne, Edward G. *Arabian Medicine*. Cambridge: Cambridge University Press, 1921.

Kitāb al-Din wā'l–Dawlah. Ed. A. Mingana. Manchester: University Press, 1923, and Cairo: al-Muqtaṭaf Press, 1923.

Leclerc, Lucien. *Histoire de la Médecine Arabe*. Paris: Leroux, 1876 and Rabat, Morocco: Ministry of Islamic Affairs, 1980.

Meyerhof, Max. "ʿAlī al-Ṭabarī's "Paradise of Wisdom": One of the Oldest Arabic Compendiums of Medicine." *Isis* 16: 6–54, 1931.

Schmucker, W. *Die Materia Medica im Firdaws al-Ḥikmah des Tabari*. Bonn: University of Bonn, 1969.

Siggel, Alfred. "Gynäkologie und Embryologie aus dem Parad der Weisheit des al-Ṭabarī." *Quellen und Studien zur Geschichte der Naturwissenschaften und der Medizin* 8(1/2): 216–72, 1941.

TAKEBE KATAHIRO Takebe Katahiro was born in 1664 at Edo (now Tokyo). His father, Takebe Naotsune, was a *Yuhitsu* (secretary) of the Shogun. In 1676, when he was thirteen years old, he and his elder brother Takebe Kataaki (1661–1716) became pupils of Seki Kowa (d. 1708) and studied mathematics. The Takebe brothers and Seki Kowa were colleagues in the Shogun's government, and their families were the same rank: 300 *koku* (landlord of a village of 300 persons).

Takebe's mathematical works are in three fields. One concerns completing the *tenzan-jutsu* or *endan-jutsu* (lit. addition and subtraction methods, Japanese algebra system), which was created by Seki Kowa. In the second work Takebe created the *tetsu-jutsu* (inductive methods). Using these methods he obtained the formula of $(\arcsin\theta)^2$. In the third, for computing the approximate value of fractions, he solved the Diophanine equations using the *reiyaku-jutsu* (continual division method). Takebe also worked in the fields of astronomy and geography.

In 1683, Takebe wrote his maiden work, *Kenki Sampo* (Studies for Mathematical Methods). The book contained counter-arguments for Saji Ippei's *Sampo Nyumon* (Introduction to Mathematical Methods, 1680). Saji criticized Seki Kowa's *tenzan-jutsu* system in the *Hatsubi Sampo* (Mathematical Methods for Finding Details, 1674) and solved Ikeda Masoki's problems in the *Sugaku Jojo Orai* (Textbook of Mathematical Multiplication and Division, 1672). Takebe made good use of the *tenzan-jutsu* method for solving Ikeda's remainder problems.

Next, Takebe commented on Seki Kowa's *Hatsubi Sampo* and published the *Sampo Endan Genkai* (Commentaries for Japanese Algebra System), in 1685. This is one of the best books for studying the *tenzan-jutsu* method.

Chinese mathematicians in the Song and Yuan dynasties used counting rods to solve higher degree equations of more than the fourth degree. There were two color symbols in the counting rods: red rods were plus and black were minus. They had no symbols for the power; the position of the rods on the counting board indicated the powers. Therefore, the system could not indicate complex expressions. For example, $1/(x-1)$, that is $(x-1)^{-1}$, was very difficult to indicate by that system. Seki Kowa abandoned the counting rods system, and created algebraic symbols, which used Chinese characters for calculation with figures. This must be the first evidence of the progress of Japanese mathematics from the Chinese.

Takebe commented on the most famous Chinese algebraic text in Japan at that time, *Suanxue Qimeng* (Introduction to Mathematical Studies, Zhu Shijie, 1299) and published the *Sangaku Keimo Genkai Taisei* (Complete Works of Commentaries on Suan Xue Qi Meng) in 1690.

Seki Kowa and the Takebe brothers started to compile an edition of the mathematical works of *Seki-ryu*, Seki Kowa's school. After Seki died, Takebe Katashiro continued this work, and published *Taisei Sankyo* (Complete Mathematical Manual) about 1710.

Takebe's main work is the *tetsu-jutsu* method, a sort of inductive method. He started to compute small natural numbers, then predicted infinite numbers. The computation was helped by the algebraic symbols of the *tenzan-jutsu* method. Using these methods, Takebe obtained the formula of $(\arcsin\theta)^2$. He computed the length of curve AB (hereafter s) using the diameter d and the length of straight line AB (h).

Takebe set up $d = 10$ and $h = 10^{-5}$. Then letting the half point of straight line AB be C_2, and the half point of curve AB be B_2, he computed the length of AB_2 (h_2). Then he computed the length of AB_4 as h_4, and continued to compute h_8, h_{16}, h_{32} and h_{64}. Takebe computed h_∞ using a sort of infinity series *zoyaku-jutsu* (extra division method), and obtained h_∞.

$$\left(\frac{s}{2}\right)^2 = 10^{-4} \times 1.000\,000\,333\,333\,411\,111\,225\,396\,906$$
$$666\,728\,234\,776\,947\,959\,587\ldots$$

$$h(\mathrm{AC_2B}) = 10^{-5},$$
$$d(\mathrm{B_2OD}) = 10^{1}.$$

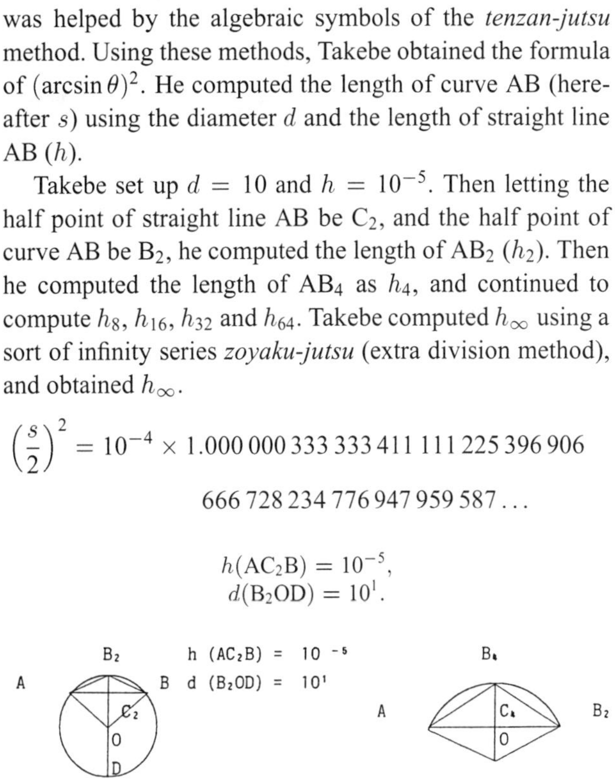

Second, he indicated this value using h and d. The power 10^{-4} is h by d, and the approximate value of the coefficient is 1, therefore,

$$\left(\frac{s}{2}\right)^2 = 1hd + 10^{-10} \times 0.333\,333\,511\,111\ldots.$$

Then he used the same method. The power 10^{-10} is h^2, and the approximate value of the coefficient is $\frac{1}{3}$ using the *reiyaku-jutsu* method,

$$\left(\frac{s}{2}\right)^2 = 1hd + \frac{1}{3h^2} + 10^{-16} \times 0.177\,777\,992\ldots.$$

Takebe continued to compute as above, and he set the series as

$$\left(\frac{s}{2}\right)^2 = A_0 + A_1 + A_2 + A_3 + A_4 + \cdots.$$

He obtained

$$A_0 = hd, \qquad \frac{A_1}{A_0} = a_1\left(\frac{h}{d}\right),$$

$$\frac{A_2}{A_1} = a_2\left(\frac{h}{d}\right), \qquad \frac{A_3}{A_2} = a_3\left(\frac{h}{d}\right), \qquad \ldots$$

$$a_1 = \frac{1}{3}, \quad a_2 = \frac{8}{15}, \quad a_3 = \frac{9}{14}, \quad a_4\frac{32}{45}, \quad \ldots$$

He predicted them to the general series of a_n, which was

$$A_n = \frac{2n^2}{(n+1)(n+2)}.$$

Therefore Takebe obtained the formula

$$\left(\frac{s}{d}\right)^2 = 2\sum_{n=0}^{\infty}\left(\frac{(n! \times 2^n)^2}{(2n+2)!} \times \frac{h^{n+1}}{d^{n-1}}\right).$$

Takebe had no notion of triangle functions. However, if we set

$$\theta = \sqrt{\frac{h}{d}},$$

we can obtain $\arcsin\theta = s/(2d)$. Therefore, his formula has the same value as the formula

$$(\arcsin\theta)^2 = 2\sum_{n=0}^{\infty}\left(\frac{(n! \times 2^n)^2}{(2n+2)!} \times \theta^{2n+2}.\right.$$

This work was described in the *Fukyu Tetsu-jutsu* (Inductive Methods) in 1722, and the special manuscript was sent to to Shogun in 1730.

The key method of *tetsu-jutsu* was computing the approximate value of decimal fractions. He named it the *reiyaku-jutsu*, which used the Euclidean algorithm to compute the value of continuing fractions. For example, to try to compute the approximative value of π using this method, we set;

$$\pi_n = 3.1415926, \text{ or } 3 + \frac{1,415,926}{10,000,000}.$$

First, divide 10,000,000 by 1,415,926; the quotient q_1 is 7 and the remainder r_1 is 88,518. Next, divide the former divisor 1,415,926 by the remainder r_1; the quotient of q_2 is 15 and the remainder r_2 is 88,156. Next, divide the former divisor r_1 by the newer remainder r_2, and the quotient q_3 is 1 and the remainder r_3 is 362. Continuing this algorithm, the quotients are:

$$\{q_1, q_2, q_3, \ldots, q_n\} = \{7, 15, 1, 243, 1, 1, 9, 1, 1, 4\},$$
$$p_1 = 3 + 1/7 - 22/7 > \pi_n,$$
$$p_2 = 3 + 1/(7 + 1/15) = 333/106 < \pi_n,$$
$$p_3 = 3 + 1/(7 + 1/(15 + 1/1)) = 355/113 > \pi_n.$$
$$p_{2k} = q_0 +$$
$$1/(q_1 + 1/(q_2 + 1/q_3 + 1/(q_4 + \cdots + 1/q_{2n}))) < \pi_n,$$
$$p_{2k+1} = q_0 +$$
$$1/(q_1 + 1/q_2 + 1/(q_3 + 1/(q_4 + \cdots + 1/q_{2n+1}))) > \pi_n.$$

The value of p_1 and p_2 had already been computed by the Chinese mathematician, Zu Chongzhi (429–500). Takebe concluded that Zu Chongzhi also invented the same method as his own *reiyaku-jutsu* and Zu Chongzhi's significant figures were seven decimal places, the same as the above computation. He admired Zu Chongzhi and named his book *Fukyu Tetsu-jutsu*; the title was connected with Zu Chongzhi's *Zhui Shu*.

In 1723, Takebe made a map of Japan under the order of Tokugawa Yoshimune, the eighth Shogun, this map is now lost. That year Takebe became a *Yoriai* (adviser) of the Shogun.

Shogun Yoshimune already permitted the import of foreign scientific books in 1720, if they were not related with missionaries. Thus the knowledge of Western astronomy was imported in some Chinese translations. Takebe and his student, Nakane Genkei (1662–1733) translated Mei Juecheng's *Li Suan Quan Shu* (Complete Works of Calendar and Mathematics, 1723) and sent the manuscript (Japanese name *Rekisan Zensho*) to Yoshimune in 1733. It was published and read by many Japanese scholars. Kepler's newest opinion, the elliptical orbit of planets, however, was hidden by the missionaries who advised Mei Juecheng. It became known to the Japanese after Asada Goryu (1734–1799) translated Kepler's (Part 2 of) *Li Suan Quan Shu*.

Takebe was held in honor by Yoshimune, and he became a *Hoi* (Knight), and held successively better positions. In 1733, he retired, and he received a life annuity of 300 *koku*. He died on July 20, 1739 [August 24, 1739 in the present calendar] at Edo.

SHIGERU JOCHI

REFERENCES

Fujiwara Matsusaburo. *Meiji-zen Nihon Sugakushi* (Mathematics in Japan before the Meiji Era). 5 vols. Tokyo: Iwanami Shoten, 1954.

Hirayama Akira et al., eds. *Seki Kowa Zenshi*. Osaka: Osaka Kyoiku Tosho, 1974.

Mikami Yoshio. *The Development of Mathematics in China and Japan*. 2nd ed. New York: Chelsea, 1913.

Shimodaira Kazuo. *Wasan no Rekishi* (History of Japanese Mathematics in the Edo Period). 2 vols. Tokyo: Fuji Junior College Press, 1965–70.

Takebe Kataaki. *Rokkaku Sasaki Yamanouchi-ryu Takebe-shi Denki* (Biographies of Takebe Families). Edo, 1715.

See also: Computation: The Chinese Rod Numeral System – Asada Goryu – Seki Kowa – Mathematics in Japan

TANG SHENWEI Tang Shenwei was a medical practitioner of the eleventh century. He was a native of Chongqing, Sichuan Province with a surname Shenyuan, and later moved to Chengdu of the same province. He was born to a family of many generations of physicians and was an expert in medical science. During the Yuanyou period (1086–1094) of the Song Dynasty, his tutor was Li Duanbo. Tang inherited his tutor's medical knowledge and became quite adept at treatment. He had a lofty medical virtue, so that he responded to any patient's call and never refused to see a patient, whether rich or poor. The only payment he asked was knowledge about a certain herb or an effective recipe.

He was especially conversant in the herbal art. He compiled a 32-volume *Herbology of Classified Syndromes* which was based on the combination of two other existing herbological works, the *Jiayou Bencao* (Jiayou Herbology) and *Tujing Bencao* (Illustrated Classic of Herbology). He added a large amount of new material extracted from the classics of philosophy, history, and other branches of the natural and social sciences, as well as the Buddhist canon, with a total number of 1746 herbal drugs. In the field of traditional pharmacology, in addition to absorbing the knowledge inherited from earlier practitioners, he was full of initiative and creativity. Most of his experience was derived from his own long-term practice. He enriched the traditional herbological work by adding processing methods and effective recipes for each herb. Meanwhile, he was also a proficient clinical physician. He created a new style of combining medical practice with herbal knowledge to form the principle of "verifying the drug by recipes", which was quite helpful to clinical practitioners, thus pushing medical science forward a step further. His work was treated as an officially promulgated herbological book and circulated for several hundred years. His working methodology and epistemology were praised and copied by later scholars in the same field. Tang exerted an especially profound influence on the work of the great naturalist of the Ming Dynasty, Li Shizhen.

HONG WULI

REFERENCES

Shang, Zhijun, et al. *Essentials of Chinese Materia Medica of Successive Ages*. Beijing: Publishing House of Science and Technology Literatures, 1989.

Tang, Shenwei. *Classified Herbology from Historical Classics in Daguan Period*. Wuchang: Kefengshi's Xylographic edition, 1904.

Tang, Shenwei. *Newly Revised Alternate Classified Herbology from Historical Classics in Zhenghe Period*. Beijing: People's Health Publishing House, 1957

See also: Medical Ethics in China – Li Shizhen

TAO HONGJING The Daoist (Taoist) master, alchemist, and pharmacologist Tao Hongjing was born in 456 near modern Nanjing. He served in various positions at the courts of the Liu Song and Qi dynasties until 492. In that year he retired to Mount Mao, the seat of Shangqing or Supreme Purity, a Daoist tradition based on meditation and visualization techniques. The retreat he built on the mountain was to remain the center of his activities until his death in 536.

After his initiation into Daoism around 485, Tao set out to recover the original manuscripts, dating from about one century before, that contained the revelations at the source of the Shangqing tradition. Tao authenticated and edited the manuscripts, and wrote extended commentaries on them. This undertaking resulted in two texts completed in ca. 500, the *Zhengao* (Declarations of the Perfected) and the *Dengzhen yinjue* (Concealed Instructions on the Ascent to Perfection, only partially preserved). These and other works make Tao Hongjing the first systematizer of Shangqing Daoism, of which he became the ninth patriarch.

Since the establishment of the Liang dynasty in 502, Tao enjoyed the favor of Emperor Wu (r. 502–549), on whom he exerted remarkable influence. Shortly after, he began to devote himself to alchemical practices under imperial patronage. His main biographical source, written in the Tang period, has left a vivid account of these endeavours. Along with scriptural sources they testify to the importance of alchemy within the Shangqing tradition, which represents the first known instance of close links between alchemy and an established Daoist movement.

A third text on which Tao Hongjing worked during his retirement on Mount Mao was the *Bencao jing jizhu*, a commentary on the earliest known Chinese pharmacopoeia, the

Shennong bencao. The original text contained notes on 365 drugs. To these Tao added 365 more, taken from a corpus of writings that he refers to as "Separate Records of Eminent Physicians". Tao's arrangement of the materia medica was also innovative. He divided drugs into six broad categories (minerals, plants, mammals, etc.), and retained the three traditional classes of the *Shennong bencao* only as subdivisions within each section. In a further group he classified the "drugs that have a name but are no longer used [in pharmacology]." Tao's commentary discusses the nomenclature, notes changes in the geographical distribution, and identifies varieties; it also includes references to the Daoist *Xianjing* (Books of the Immortals) and to alchemical practices. With the exception of a manuscript of the preface found at Dunhuang, the *Bencao jing jizhu* is lost as an independent text, but has been reconstructed based on quotations in later sources.

FABRIZIO PREGADIO

REFERENCES

Honzōkyō shūchū (*Bencao jing jizhu*; Collected Commentaries on the *Shennong bencao*). Reconstruction by Mori Risshi (ca. 1850), reedited and published by Okanishi Tameto. Osaka: Minami Osaka Insatsu Center, 1972.

Needham, Joseph, et al. *Science and Civilisation in China*, vol. VI: *Biology and Botanical Technology*, part 1: *Botany*. Cambridge: Cambridge University Press, 1986.

Robinet, Isabelle. *Taoist Meditation. The Mao-shan Tradition of Great Purity*. Albany: State University of New York Press, 1993.

Strickmann, Michel. "On the Alchemy of T'ao Hung-ching." In *Facets of Taoism. Essays in Chinese Religion*. Ed. Holmes Welch and Anna Seidel. New Haven and London: Yale University Press, 1981, pp. 123–192.

Unschuld, Paul. *Medicine in China; A History of Pharmaceutics*. Berkeley, Los Angeles, and London: University of California Press, 1986.

See also: Alchemy in China – *Shennong Bencao*

TAQĪ AL-DĪN Taqī al–Dīn, Muḥammad al–Rāṣid ibn Maʿrūf, was a mathematician and astronomer. He was born in Damascus in 1521 and died in about 1585. He wrote several books on mathematics, astronomy, optics and theology.

The most important of his books are: *Jabr wa 'l-Muqābala* (Algebra), *Bughyat al-Ṭullāb min ʿilm al-Ḥisāb* (The Desire of Students for Arithmetic), *Sidrat al-Muntahā fī al-Afkār* (The Nabk Tree of the Extremity of Thoughts), and *Ālāt-i Raṣdīya li Zīj-i Shāhinshāhīya* (Observational Instruments of the Emperor's Catalogue), *Al-Kawākib al-Durrīya fī Bengamāt al-Dawrīya* (The Brightest Stars for the Construction of Mechanical Clocks).

From the point of view of the history of Ottoman science, the most important event of the sixteenth century was the foundation of the Istanbul Observatory, which Taqī founded under the sponsorship of Murād III (1574–1595). This observatory was an elaborate building which contained dwelling places, a library, and offices for the astronomers. It was conceived as one of the largest of the observatories of Islam and was comparable to Tycho Brahe's (1546–1601) Uranienborg Observatory built in 1576, equipped with the best instruments of his time in Europe. There is a striking similarity between the instruments of Tycho Brahe and those of Taqī al-Dīn.

The instruments of the observatory included the following. First there were those originally constructed by Ptolemy: the armillary sphere, the paralactic ruler, and the astrolabe. Then there were those invented by Muslim astronomers, such as the azimuthal and mural quadrants. Taqī al-Dīn invented the *mushabbaha bi 'l manāṭiq* (sextant, an instrument with cords for the determination of the equinoxes), which was also an important invention of Tycho Brahe. In addition, he built a wooden quadrant for the measurement of azimuths and elevations, and clocks for the measurement of right ascensions of the stars. The latter was one of the most important discoveries in the field of practical astronomy in the sixteenth century, because in the beginning clocks were not accurate enough to be used for astronomical purposes.

In *The Astronomical Instruments of the Emperor's Catalogue* the author says, "The ninth instrument is an observational clock". The following statement is taken from Ptolemy: "I could have freedom of action if I were able to measure the time accurately. Now our master Taqī al-Dīn, with the help of God, upon the instructions of the Sultan, planned the observational clocks". In *The Nabk Tree of the Extremity of Thoughts* Taqī al-Dīn says, "We constructed a mechanical clock with three dials which show the hours, the minutes, and the seconds. We divided each minute into five seconds". On the basis of his observations, Taqī al-Dīn prepared astronomical catalogues and books.

Hipparchos (second century BC) used the intervals of seasons for the calculation of the solar parameters. But the variation of the declinations around the tropics in one day rendered difficult the correct determination of the beginning of the seasons. In spite of this difficulty, the method was used for a long time. After him, al-Bīrūnī (d. ca. 1048), Copernicus (1473–1543), and Tycho Brahe were interested in this subject, and used a new method called "three point obsvervation". Taqī al-Dīn, a contemporary of Tycho Brahe, says the following in *The Nabk Tree*: "The moderns follow the method of three points observation, two of them being in opposition in the ecliptic and the third in any desired place." This method was an important contribution to astronomy. By using this method, Copernicus, Tycho Brahe, and Taqī al-Dīn calculated the eccentricity of the orbit of the Sun, and yearly mean motion of the apogee. According to Copernicus the eccentricity is 1p 56′; according to Tycho Brahe it is 2p 9′, and according to Taqī al-Dīn it is 2p 0′ 34″ 6‴ 53⁗ 41‴‴ 8‴‴′. As compared to modern calculation, Taqī al-Dīn's is the most accurate value. According to Copernicus the annual motion of the apogee is 24″; to Tycho Brahe it is 45″, and to Taqī al-Dīn it is 63″. Its real value is 61″. As far as world astronomy is concerned, Taqī al-Dīn's results can be said to be the most precise in the calculation of solar parameters.

The next important contribution of Taqī al-Dīn concerns the use of decimal fractions, the system of numerals formed from initial letters, used in the Hellenic world. This system hindered the development of algebra.

Al-Khwārizmī (d. 801) presented the decimal system which was inspired by Indians to the Islamic world. The aplication of this to fractions started with Abū'l-Ḥasan Aḥmad ibn Ibrāhīm al-Uqlīdīsī and continued with al-Kāshī (d. 1437). But its application to astronomic and trigonometric tables was realized by Taqī al-Dīn. Thus the tables of his *zīj* named Kharīdat al-Durar (*unbored pearl*) and a *zīj* were prepared using the decimal system and decimal fractions.

SEVIM TEKELI

REFERENCES

Sayılı, Aydın. "Ala ad Dīn al Mansûr's Poems on the Istanbul Observatory." *Belleten* 20: 429–484, 1956.

Sayılı, Aydın. *The Observatory in Islam*. Ankara: Publications of the Turkish Historical Society, 1960.

Süheyl Ünver. *Istanbul Rasathanesi*. Ankara: Publications of the Turkish Historical Society, 1969.

Tekeli, Sevim. "Nasirüddin, Takiyüddin ve Tycho Brahe'nin Rasad Aletlerinin Mukayesesi. *Ankara Üniversitesi, Dil ve Tarih-Coğrafya Fakültesi Dergisi* 16(3–4): 301–393, 1958.

Tekeli, Sevim. "Al-Urdī'nin 'Risalet ün fi Keyfiyet il-Ersâd' adli Makalesi, "The Article on the Quality of the Observations" of Al-Urdī." *Arastirma* 8: 1–169, 1970.

Tekeli, Sevim. "Solar Parameters and Certain Observational Methods of Taqī al Dīn and Tycho Brahe." In *Ithaca-26 VIII-2, IX*. Paris: Hermann, vol. 2, 1962, pp. 623–626.

Tekeli, Sevim. *16. Asirda Osmanlilarda Saat ve Takiyüddin'in 'Mekanik Saat Konstrüksüyonuna Dair En Parlak Yildizlar' adli Eseri (The Clock in the Ottoman Empire in the Sixteenth Century and Taqī al Dīn's 'The Brightest Stars For the Construction of the Mechanical Clocks*'). Ankara: Ankara Üniversitesi Basimevi, 1966.

See also: Observatories – Clocks and Watches – Ottoman Science – Astronomical Instruments – Quadrant

TECHNOLOGY The development of technology has usually been socially determined, changing in direction as different social formations have emerged. For example, there are important distinctions to be made between, first, small communities of farmers or dispersed groups of hunters with technologies outstanding chiefly for their adaptation to local *environments*; second, the larger kingdoms and empires, whose technology tended to be *engineering-centered*; and third, trading communities and merchants or entrepreneurs with *production-centered* technologies.

An initial problem in discussing technologies of the first kind is nomenclature. J.S. Weiner, who has discussed the remarkable protective clothing developed by the Inuit for life in the Arctic, has described these people as "the great pioneers of micro-climatological bioengineering". This tribute to the control of body heat loss achieved by Inuit clothing is well deserved but places the skills of Arctic people within the wrong frame of reference. Their technology was environment-centered, not engineering-centered. On the one hand, they were adapting to a very demanding environment; on the other, they were making very efficient use of environmental resources. Thus, the animals hunted for meat were also a source of oil for lamps, bone for making needles and other tools, and the skins, intestines, and fibers from which clothing was made.

Archaeological evidence suggests that the invention of tailored skin clothing adequate to enable people to winter in the Arctic was accomplished in Siberia around 2000 BC. After that, it was several centuries before oil lamps, houses built of snow, and dog sleds were developed by the people of the Dorset Culture, based on Baffin Island. Later still, improved boats and harpoons were developed by the Alaskan Inuit.

All these inventions, quite clearly, were the work of small, dispersed groups without centralized organization, and all were responses to an exacting environment. Perhaps what we ought also to remember is that for the Inuit, as for most non-Western peoples, the environment was endowed with personal and spiritual meanings, and with magical qualities. It was not merely the object of detached, if concerned, analysis that it has become for most Westerners. Thus, their technology was based on an organic world view.

The same points can be made about peoples inhabiting other environments, such as rain forests, deserts, steppes, or coastal and island situations. The efficiency with which resources were used by such people is illustrated by the estimate that in AD 1500 the Amazon basin supported a somewhat larger population than it did in 1990, yet without the extent of destruction of forest and fishery resources now taking place. Fruits, nuts, leaves, seeds, and roots were gathered for use as food and medicine. Fish and animals were caught, often using poison tipped spears, but with less ecological disturbance.

In some rain forests, notably in Central America as well as in Asia and Africa, people developed a form of agriculture in which a great variety of crops was raised while extensive tree cover was retained. Cereals such as corn in the Americas or rice in Southeast Asia were grown in forest glades with shade tolerant vegetables under trees. Those trees that were left unfelled or were newly planted would be selected according to whether they produced fruit, nuts, timber, fodder, or *materia medica*. Thus, an artificial rain forest was formed in a system often referred to as forest farming, forest interculture, layered gardening or, more recently, agroforestry. The key point — whatever the name — is that there would always be tree cover, and usually other vegetation, to protect the soil from erosion and to retain plant nutrients. By contrast, where rain forest is cleared for conventional western style agriculture, soils become degraded very rapidly.

In Africa, rain forest crop production became widespread only after the banana and Asian yam were introduced on the east coast by Indonesian traders and colonists around AD 400–500. In quite a short space of time, these crops were adopted by many gardeners across the continent, enabling populations in forested areas to increase considerably. In Central America, archaeologists have now shown that the Maya culture used forest farming with high yielding nut trees and manioc (cassava) as major crops around AD 600–900.

Other environment-centered technologies were those of people living in hot, dry, semi-desert conditions who evolved sophisticated water conservation techniques. Some of these made use of small structures, such as check dams in gullies and spreader dikes on flatter ground — for example, in northern Mexico and Arizona — or else stone lines on contours (in the African Sahel). Others practiced the careful planning of fields in relation to rainwater catchment areas, and the construction of bunds to channel water and hold it on the land. Among the most elaborate rainwater harvesting systems of this kind were those of Nabataean farmers in the Negev Desert in Israel from about 200 BC, and comparable systems in Morocco and elsewhere in North Africa. Small dams, river diversions, and the Persian *qanat* (wells linked by a tunnel) were other means of providing irrigation water to crops on a small scale.

Consideration of environment-centered techniques such as these offers a distinct perspective on early technology quite unlike the conventional view based on the materials from which tools were made, namely, stone, bronze, and iron. That view seeks to discover a pattern of tool use and metal-working skill common to all human societies, and ignores the environmental particularism emphasized here. It parallels the modern assumption that the principles of technology are independent of cultural values and underestimates the different ways in which human societies explored the varying surroundings in which they lived. By contrast, a perspective based on environmental adaptation allows one to recognize the extraordinary sophistication of some peoples who nominally remained in the "Stone Age" even in the early twentieth century, including the Inuit and some rain forest communities.

Viewed from another angle, however, the Neolithic period in the later Stone Age does seem to have been a time of particularly important innovation, at least for western Asia. It is associated with the first domestication of crops and livestock, and hence the invention of agriculture, probably associated with the earliest used of some of the water conservation techniques previously discussed. Complementing these innovations were others, such as pottery (and soon after, the potter's wheel), grain milling, tool-making, and textiles (with the looms on which they were woven). Many of these innovations relate to the domestic sphere of life, and it is likely, therefore, that many of the inventors concerned were women.

It would be a mistake, however, to assume that the invention of agriculture, the domestication of plants, and the appearance of pottery or textiles were unique events, each occurring only once in human history, in one part of Asia and then diffusing to other regions. Independent inventions undoubtedly occurred, just as different crops were domesticated in various places and at different times. For example, about 1500 BC, corn (i.e. maize) and pottery were both known in Mexico, but not in Peru, where beans and peppers were cultivated. Thus, the two cultures may have developed agriculture independently of one another. Moreover, some plants were still being newly domesticated in the twentieth century, for one reaction to deforestation in Africa has been to cultivate a number of fruits and vegetables that people had collected from the wild until the loss of forest cover threatened their existence. Reports from Kenya show how modern gardeners, usually women, have selected from wild varieties and have produced strains adapted for garden conditions.

Engineering-centered technologies developed in a very different context. They were characteristic mainly of kingdoms or empires whose labor resources and administration made large scale construction works possible. One view of the origins of engineering is that once agriculture was established, irrigation became necessary in many of the warm, dry countries of western and southern Asia to produce sufficient food for growing populations. The dams and canal systems necessary for irrigation could not have been constructed, it is argued, without recruiting a large labor force, and that in turn could not have been done except in centrally organized kingdoms. In this view, the need for irrigation canals and other hydraulic works dictated the formation of centralized government administrations with the coercive power and management skills needed to organize large scale construction works.

This theory about how hydraulic civilizations evolved does not apply everywhere, however, since empirical evidence shows that in many areas where irrigation was practiced, farmers did the essential earth moving and engineering work themselves, or with their neighbors through local systems of cooperation, but without central organization. This seems to be true for rice culture in China, according to Francesca Bray. Some of the biggest centrally organized engineering schemes prior to AD 605, involving thousands of laborers, were not for irrigation but for construction of transport canals for carrying grain to the capital. However, flood control and irrigation in north China did depend on large scale works, so the evidence is not clear cut.

In Mesopotamia and Egypt, much early irrigation was also on a small scale, dependent on manually operated water raising devices such as the *shaduf*. The type of crops grown and the seasons of cultivation meant that water requirements in early agriculture were much less than after the agricultural revolution of the Islamic period (after AD 700), when crops that demanded more water (including rice) were more widely grown. In Mesopotamia, large scale works were needed at an early date, but in most places it is hard to argue that the requirements of hydraulic management were sufficiently exacting to determine the development of centrally organized states. On the contrary, the prior existence of centralized political power made possible a wide range of construction programs that included fortifications, monuments, and temples, as well as hydraulic works, and it is impossible to say which kind of engineering came first. One of the earliest sites to provide evidence is Jericho, where defensive walls, but also big water tanks, have been found dating from before 6000 BC. The pyramids of Egypt were among the largest construction projects every conceived when they were begun about 2600 BC. But a comparable amount of labor may have gone into embankments and canals built at about the same time for flood control and irrigation by the Sumerians in Mesopotamia. Thus, the evidence does not give a clear picture of hydraulic requirements, rather than other construction works, forcing the pace in either administration or engineering.

That brings us to a second hypothesis about how engineering-centered technologies evolved, suggested by Lewis Mumford and some like-minded commentators. This is the view that such technologies were related to the invention of institutionalized warfare. In early societies, conflicts were sometimes wantonly violent, but some groups had customs that enabled one side to signal a surrender before serious injuries were inflicted. Among a few isolated peoples in Africa and Oceania, conflicts between communities were still like this until recently, and observers note that, if somebody was killed, the combatants were so shocked that fighting immediately stopped. A ceremonial burying of spears and an exchange of cattle or captives would then settle outstanding issues.

Organized, lethal warfare had to be invented, perhaps during the Neolithic period, and possibly following the invention of agriculture, since agriculture made possible denser populations within which tensions could be greater. Also, agriculture produced the economic surplus needed to pay for arms and fortifications. The suggestion is that military institutions evolved from about this time and provided a model of how large bodies of men could be recruited, disciplined, and supervised while constructing fortifications, pyramids or hydraulic works. Many instances can be quoted of the administration of hydraulic installations based on military routines. In later times, at Merv in northern Iran, a water storage dam below the city was looked after by four hundred "guards" who minutely regulated and recorded all outflows of water. The head of the water office had more authority than the local police chief, and when repairs to the dam or canals were necessary or when new works were needed, he could call up ten thousand people to do the work.

The development of warfare provided much other stimulus to engineering-centered technology. Mechanized fighting may be said to have begun with the clumsy, four-wheeled Sumerian chariot of about 2400 BC. The Hittites, whose empire overlapped Asia Minor and Syria, invented some of the first effective siege engines around 1600 BC and introduced the first iron weapons soon after. Horses were initially used mainly for drawing chariots, their full potential for warfare only emerging much later with the invention of the stirrup and its associated harness. This freed the rider's arms to use weapons while he remained securely in the saddle.

The most famous of military inventions was, of course, gunpowder, known by AD 900, used in rockets by about 1100, and in guns before 1288. These inventions, developed within Chinese military institutions, were stimulated by earlier Indian discoveries about the chemical behavior of saltpeter, and by the use of incendiary weapons for naval warfare in Southeast Asia. The latter often depended on petroleum, obtained from surface oil seepages on the island of Sumatra. Although guns evolved more rapidly in Europe after 1300, China was the source of many ideas for gunpowder siege weapons used in the Arab world, and there were transfers of Chinese firearms technology to Korea and Thailand.

Considerable resources were needed for manufacture of the heavy cannons that had become widespread by 1500, and only the larger Asian empires could afford them in any number — but those that could were then able to expand their territory and consolidate their power. In India, the Mughal Empire's arsenals and gun-casting capability were important for its expansion between 1526 and 1700. In Persia also, the new technology had a centralizing, consolidating influence during the reign of Shah Abbas (1587–1627), though his empire depended less on guns than did the powerful Ottoman or Turkish Empire, an exporter of guns and know-how. Given that most cannon were cast in bronze, W.H. McNeill has called this period a "second bronze age", referring to "gunpowder empires" not only in connection with Turkish, Persian, and Mughal rule, but also in relation to China and the Russian Empire in Asia.

The ships used in naval warfare and trade in Southeast Asia were of sewn construction, with rattan fiber holding together carefully fitted timbers in a flexible hull suited to landing on sandy beaches. They usually had outriggers and tripod masts to support sails, and regularly crossed the Indian Ocean. The Chinese, meanwhile, were building larger ships whose wooden hulls were held together with iron nails and had watertight bulkheads. China was also developing inland water transport on a large scale. The canals had ramps and spillways for moving boats between one level and another, but also incorporated the first pound locks and lock gates, introduced in AD 983 (a precisely dated invention).

One Western bias in understanding technology is a tendency to focus on the wheel and machines using wheels. But wheels did not always mean progress. Wheeled vehicles were used in the north and west of Africa around 500 BC, regularly crossing the Sahara Desert. Yet they went out of use later when the camel was introduced, because camels provided much more efficient means of transport in a region with sandy deserts and no paved roads. For parallel reasons, wheeled vehicles were not much used in southern Asia. In Central America, rollers and wheeled toys have been found, but in a region where there were not animals capable of pulling carts, no purpose could be served by developing such devices.

The potter's wheel (and later lathes working on similar principles) were probably the first machines, around 3000 BC, to use the wheel. Much later, a basic water powered corn mill evolved, with a vertical axle and propeller like blades below the millstones. One view is that this was a Greek invention made about the time of Alexander the Great. Another view, however, is that it arose out of the Hittite tradition of engineering which had earlier pioneered so much military equipment. After the Hittite empire collapsed around 1200 BC, some aspects of its culture survived in Syria, and itinerant craftsmen perhaps reached Mesopotamia and Iran. It was probably in one of these countries that water mills originated. Since the same type of mill appeared in China not long afterwards, one might reasonably look for its origins close to trade routes with China. As for the introduction of the mill into Greece, this could be a by-product of Alexander's expedition through Iran to the Indus River in 330–323 BC.

Some thousand years later, Iran was certainly where an early type of windmill was invented. It worked in much the same way as the vertical axle water mill, though mounted in a tower with vents in the walls to catch the wind.

Mumford sees these inventions as part of the tradition of innovation stemming from small dispersed communities rather than from centrally organized states. However, at Baghdad, which had about one million inhabitants in AD 1000, corn milling was carried out by a series of floating mills on the Tigris River which operated continuously, night and day. The water wheels were of the later undershot type, driving millstones through wooden gears. Other uses of water power in Iraq and Iran by AD 1000 were in sugar cane crushing mills, fulling mills, and mills for preparing the pulp for papermaking. Paper manufacturing had been introduced at Baghdad in AD 794, probably with the help of Chinese workmen, and pulping was water powered from about AD 950.

In all these instances, it would seem, we are no longer dealing either with the environment-centered technology of dispersed communities, nor with the labor intensive engineering of powerful centralized states. Rather, we are seeing the emergence of production-centered technologies in expanding industries. An even more striking example can be cited from northern China, associated with the use of blast furnaces for iron production. Methods of achieving high temperatures in furnaces were highly developed in China's porcelain industry as well as in metal smelting. Chinese iron masters had pioneered blast furnaces capable of melting large batches of metal, and output rose steadily to a peak in AD 1078. Moreover, the furnaces were owned and run by independent entrepreneurs whose market oriented, proto-capitalist operations are seen by McNeill as marking a turning point in world history.

More evidence that technology in China was moving into a new phase is provided by the great flowering of mechanical devices that were introduced there in the three or four centuries prior to 1250. They included spinning wheels, silk winding machines, and most striking of all, water-powered mills for spinning and winding thread of a local type comparable to linen. Some improvements in textile technology were due to women innovators, and many businesses were run in an independent, entrepreneurial style.

By contrast, much manufacturing in the major Asian empires was directed toward meeting state requirements for armaments, or the needs of royal palaces for luxury textiles, porcelain, and furniture. The most characteristic unit of production was the royal factory or *karkhana* in Mughal India, or the specialist porcelain factory or arsenal in China. At one time, four thousand silk workers were employed in karkhanas in Delhi. Some may have been conscripted, and the factories were run in rather the same way as the labor intensive, state-run engineering works mentioned previously. Such factories were highly successful in meeting government requirements for arms or the court's demand for luxury goods, but could not respond readily to a varying market demand. Production for purposes of trade, and especially for export, flourished most markedly where merchants and other entrepreneurs could function independently. By 1700, India had become the world's greatest exporter of textiles, sending its cotton cloth to Europe, Africa, and many destinations in Asia. Both Indian and foreign merchants were involved, sometimes financed by the many Parsi and Gujarati bankers to be found in ports such as Surat.

It should not be thought, though, that proto-capitalist industrial organization was always favorable to the use of machines, water powered or otherwise. The high quality of Indian textiles was achieved by painstaking handwork and a very fine division of labor. British observers noted that cotton cloth would sometimes be worked on by four people for every one employed on the same tasks in Europe. Thus, the remarkably fine muslin produced in Bengal was the result of many detailed processes which would be simplified and reduced to a single operation in the West.

Indian cloth was also noted for the fastness and brilliance of the colors with which it was dyed, mostly using vegetable dyes, such as indigo and madder. However, some inorganic substances were also needed as mordants, and this gave rise to a significant chemical industry. For example, the alum used in madder dyeing was produced in Rajasthan by processing broken shale tipped as waste around copper mines. The shale was steeped in water. Aluminum compounds were separated from copper salts by differential crystallization. Then a reaction with saltpeter produced the alum. These chemical processes were operated on a substantial scale but using labor intensive methods and minimal equipment. Merchants and banks had ample funds to invest in the textile trade, but because wages were low, there was little advantage in financing better production equipment. Instead, most of the capital available was put into building ships to carry exported cloth. Thus, Indian shipbuilding developed to a very high level in the eighteenth century. Ships were built for Indian merchants and foreign traders, and from 1800 for the British navy. Meanwhile, cotton cloth, dyestuffs, and chemicals continued to be produced by laborious manual techniques.

Similarly in China, there were many flourishing industries, often organized on proto-capitalist lines, but the interest in mechanization evident before 1300 was not sustained, and indeed, some machines went out of use. Silks, porcelain, and cotton goods were produced in quantity, but mainly by labor intensive methods in an economy where wages were tending to fall. But, as in India, the production technologies involved are of considerable interest. Moreover, they are quite distinct from the environment-centered and engineering technologies discussed earlier. It is of particular relevance to observe that printing evolved as a production technology in China, and by 1600 had export markets in Korea and Japan. But the content of the books being printed was dominated by the literary interests of the bureaucrats who served the imperial government. Thus, relatively few books were of significance for the dissemination of technical information. It is worth recalling, though, that the first known printed book, the *Diamond Sutra*, an extract from the Buddhist scriptures, dates from AD 868, and that by the eleventh century a few books on agriculture were being printed. In AD 1044, Zeng Gongliang published the *Wujing zongyao* (Collection of Military Techniques) that quoted the formula for a weak gunpowder mixture, and in 1313 a remarkable work appeared — *Nong Shu* (Treatise on Agriculture) by Wang Zhen — which, among other things, described water powered spinning mills.

But whatever limited writings there were on technical subjects, either in Islamic manuscripts or Chinese printed books, technology depended far more on knowledge, skill, and technique passed from one craft worker to another through processes of observation, personal contact, and apprenticeship. Nonverbal habits of thought and of getting the "feel" for how a process should work were undoubtedly central for innovation and learning.

Nonverbal communication, often in the form of a visual or experimental dialogue, was also the means whereby techniques were passed from one community or culture to another. Confronted with an unfamiliar product or process, a craft worker would not always copy it directly, but would "question" it to try and work out how the same result might be achieved by more familiar means, or how it might be reinterpreted to suit his or her community's culture or resources. Moreover, this dialogue or questioning approach could at times prompt entirely new lines of thought or innovation. For example, between the sixth and tenth centuries AD, the Chinese picked up ideas from India and Persia about chemistry, dyestuffs, textile printing, windmills, suspension bridges, and other matters. These were developed in entirely new ways as a result of being questioned and reinterpreted rather than being copied, and so contributed to some of the best known of Chinese inventions, including gunpowder and printing.

Between 1100 and 1800, and especially after their voyages of discovery began around 1450, Europeans were continually picking up ideas from other cultures: about water clock mechanisms, chemistry, cooking, and glassmaking from Islamic sources; about gunpowder, firearms, printing, and porcelain manufacture from China; and about metallurgy, cotton textiles, and dyestuffs from India. Few of these techniques were directly copied in any detail, but they became part of a dialogue in which Westerners did most of the questioning and learning — and in which they reinterpreted many techniques in terms of the Western enthusiasm for machines and mechanization.

This source of stimulus contributed significantly to the technology of the European industrial revolution, but that in turn brought a change in attitudes and relationships. A sense of technological superiority coincided with a tendency to use non-Western countries as markets for factory made goods and as sources of raw materials. Local manufactures in many parts of the non-Western world therefore declined. One striking feature of the late twentieth century has been a new openness in the West for dialogue with the environment-centered technologies practiced in the rest of the world. But, again, Westerners are reinterpreting everything they learn in terms of their own outlook, discarding the organic and holistic world views of other peoples and fitting everything into an analytical, scientific frame of reference. Thus, Inuit technology has been discussed in terms of "ice alloys" and "bioengineering" whilst indigenous rainforest technologies are being reinvented as "ethnobotany" and "agroforestry".

ARNOLD PACEY

REFERENCES

Bray, Francesca. "Agriculture." In *Science and Civilisation in China*, vol. 6, part 2. Ed. Joseph Needham. Cambridge: Cambridge University Press, 1984.

Dharampal. *Indian Science and Technology in the Eighteenth Century*. Delhi: Impex India, 1971.

al-Hassan, A.Y., and D.R. Hill. *Islamic Technology: An Illustrated History*. Paris: UNESCO, and Cambridge: Cambridge University Press, 1986.

McNeill, W.H. *The Pursuit of Power: Technology, Armed Force and Society*. Chicago: Chicago University Press, 1982.

Mumford, Lewis. *The Myth of the Machine: Technics and Human Development*. New York: Harcourt, Brace, Jovanovich, 1966.

Needham, Joseph. *Science and Civilisation in China*. Vol. 4, part 2, *Mechanical Engineering* (co-author Wang Ling). Cambridge: Cambridge University Press, 1965. Vol. 4, part 3, *Civil Engineering and Nautics* (co-authors Wang Ling and Lu Gwei-Djen), 1971.

Pacey, Arnold. *Technology in World Civilization*. Cambridge, Massachusetts: MIT Press, 1990.

Richards, Paul. *Indigenous Agricultural Revolution: Ecology and Food Production in West Africa*. London: Hutchinson, 1985.

Watson, Andrew M. *Agricultural Innovation in the Early Islamic World*. Cambridge: Cambridge University Press, 1983.

Weiner, J.S. *Man's Natural History*. London: Weidenfeld and Nicolson, 1971.

See also: Agriculture – Textiles – Ethnobotany – Agroforestry – Military Technology – Gunpowder – Paper and Papermaking – Colonialism and Science – Pyramids – Crops – *Qanat* – Irrigation – East and West – Western Dominance

TECHNOLOGY AND CULTURE General accounts of the history of science and technology (or, more narrowly, of inventions) are scarce. The few that are available are also of fairly recent origin: obviously, the idea of a history of science (where science has been identified with Galilean science) and technology (identified with industrial technology) could not have appeared much earlier than this century. Not many people even know that the word "scientist" was first used by William Whewell in 1833.

Also, most available histories have remained the work of western scholars. This has not been an entirely happy circumstance. On the contrary, it has afflicted these histories with certain methodological and other infirmities which have had the effect of reducing them to mythological works. This is especially so when they are studied with regard to aspects of the history of science, technology, and medicine in the non-Western world.

One of the first is a history of technology and engineering written by Dutch historian R.J. Forbes. Forbes's work appeared in 1950 under the title *Man the Maker*. In it, he conceded that technology was the work of humankind as a whole, and that "no part of the world can claim to be more innately gifted than any other part". A few years thereafter, Forbes produced his rich and prodigiously detailed *Studies in Ancient Technology* which set out a remarkable description of the different technologies of Asia, Africa, pre-Colombian America, and Europe. However, it is in *The Conquest of Nature* that his Eurocentric assumptions came to the fore: in that work (as the title itself indicates), Forbes went on to subsume the technological experience of people from diverse cultures under a philosophical anthropology that was unmistakably Western, if not Biblical — the domination of nature myth originating in Genesis. And after a discussion about the grievous consequences of a seriously flawed modern technology, he ended his book promising redemption from the technological genie through the Christian event of Easter. How does one prescribe a text like this to Hindu, Chinese, or Arab readers?

Another influential work of about the same period is *A History of Western Technology* by a German scholar, Friedrich Klemm. In it, Klemm provided a picture of technological development in the West in which non-Western ideas and inventions had no hand at all. The English translation which appeared in 1959 barely mentions Joseph Needham's work on China in the bibliography. Klemm could not have substantiated his interpretation of Western technological development unless he consciously played down non-western technology. In fact, the only quote on Chinese technology in Klemm's book is from the *Guan yin zi*, the work of a Daoist mystic of the eighth century AD: Klemm used it to prove why the alleged religiously colored oriental rejection of the world in China could not have provided a stimulus for the emergence of science and technology in that country.

This distorting Eurocentric perspective continued to hold sway even over the more standard (five volume) *A History of Technology* edited by Charles Singer, E.J. Holmyard and A.R. Hall. The first volume of this work appeared in the same year as Joseph Needham's *Science and Civilisation in China*, and the editors themselves acknowledged that up to the period of the Middle Ages in Europe, China had the most sophisticated fund of technological expertise. Three of the Singer volumes dealt with pre-industrial technology, where logically China (and India) should have been given major space and Western technological development would have appeared in proper perspective in the nature of an appendix. However, Chinese, Indian, and other technologies were ignored and Western technology made the focus of the exercise.

In addition to manifesting such ignorance of non-Western science and technology, these histories suffered from another methodological limitation: they restricted themselves to a record of artifacts and machines disembodied from the latter's social and cultural contexts. The problem was eventually recognized by some Western scholars themselves.

These histories are evidence that the western scholars associated with them proved incapable of stepping out of their cultural cages, either knowingly or involuntarily. Either way, this eroded the credibility of their work as it exhibited both their lack of objectivity and their general incompetence when called upon to deal with societies other than theirs.

They show that our dominant descriptive and evaluatory ideas of technology and culture both in the Western and non-Western world have been formulated over the past couple of centuries with reference to the West's experience of these phenomena. Concepts and categories reflected from a limited area of human experience have been indiscriminately used to explain and assess the rest of the world.

New frameworks are therefore inevitable. We are in the post-traditional, post-colonial, post-modern age. But unless the outmoded intellectual environment that engendered this subjective and tunnel-visioned output is rigorously dissected, analyzed, and then jettisoned, the new frameworks needed for the alternative histories and encyclopaedias intending to take their place are in danger of turning into copies of the old.

There are two preliminary aspects of this intellectual mal-development that need elucidation. First, there is the perception of humankind as *homo faber*, a tool-making animal, which is basically a reflection of fairly recent Western experience with the machine. Fascinated by the bewildering profusion of tools and machines, Western historians began to look at the ability to produce these as a special field with its own history and set out to create a distinct species of man in the image of *homo faber*. This scholarly creation had its repercussions in encouraging the overestimation of the singularity or uniqueness of Western culture in comparison with others (although all cultures are unique and incommensurate). The elaborate, embarrassing exercise in culture-narcissism soon became routine since it was not to be challenged for nearly a century. (It is important to point out that the *homo faber* idea is quite recent to humankind: it is consistently absent in not just other cultures, but even within a large part of the West itself).

For instance, it was taken for granted that the system of production that got generated in the last century and a half in the West was the only one with any significance simply because in the light of the present — and to all appearances — it had apparently emerged as the dominant one. Therefore, its past was the only one worth considering. This notion was in turn bolstered by another: a self-generating model of technological development in which the historian attempted to trace the evolution of modern science and technology by working backwards to the experiences and ways of thinking characteristic of Mediterranean antiquity. Thus the roots of modern technology were shown to be exclusively founded on the work of Greek and Roman thinkers, mathematicians, engineers and observers of nature with no input from any other culture areas or people.

This brings us to the second aspect we have alluded to above, and this concerns the relationship between knowledge and power and the impact of this on interpretations of technology and culture. Throughout history, knowledge has generally remained closely linked with interests. Even when encyclopaedias, for instance, have traditionally sold themselves on the Francis Bacon principle that "knowledge is power", they too have continued to reflect an undeclared, equally influential, political principle — that "power is knowledge".

The intrusion of Europeans into non-European societies and the gradual establishing of political dominance and inequality between societies stimulated the inauguration of a new discourse about such societies. Political dominance came to be as routinely and unabashedly expressed in the form of knowledge as it was through the barrel of guns. Edward Said has already written on the invention of the discourse on "orientalism" and its direct political uses. But there are less controversial discourses that have had even larger repercussions only now being acknowledged. As a result, much academic knowledge in the Western world about the non-Western world, particularly the latter's technology traditions, remains not only distorted or contaminated by the ethnic concerns, goals, theories, obsessions, and peculiar assumptions of Western scholars and universities; it is still largely defined, legitimized, and decided by them irrespective of whether there is any concurrence from the non-Western world.

The combination of these two aspects proved deadly: the emerging conception of Western man alone as *homo faber*, once it took firm roots within the situation of political dominance, rendered any appreciation of technique elsewhere — technique not necessarily reflected only in tools or machines — difficult and often impossible. In fact, the combination helped inaugurate its very own dark age. For it generated among Western (and not a few non-Western) scholars several major assumptions concerning technology and culture. We shall discuss three of these.

The first emerged in relation to Western man's attitude towards the past, particularly with regard to pre-industrial technology. *Homo faber* exercised his new found power over the past by deriding it: this is reflected in the rewriting of history from today's perspective in which the past is seen as mere prelude to the present. Earlier technological innovations are considered primitive precursors of later developments. Here we have a good example of the parochialism of the modern/Western mind as it proceeds to take experiences of technology and culture exclusive not just to the late twentieth century but to extremely small segments of the world population and makes these the basis for investigating, analyzing, assessing, and judging the general activities of human societies over hundreds of years. This was the case even when such societies were not so technologically enamored, dependent, or controlled as some of them seem to be now.

The second assumption relates to humankind's so-called unique propensities for technology when compared with that of the animal world, an uncritical theory best summed up in a single word: speciesism. After deciding on the issue of the comparative technological competence of all living species in its own favor, the West came to the conclusion that the rest of creation, because inferior, was expendable if so required to further its own scheme of things.

But it is the third assumption that concerns us most seriously here: it is the idea that Western man can be equally distinguished from non-Western societies as well on the ground that the latter, like the animal and other "lower" species, also lacked technological development as it emerged in the West.

This idea was appropriately reflected in academia in the emergence of two new sciences: the discipline of sociology, which focused on so-called advanced societies and their flair for technology; and the subject of anthropology which occupied itself with non-Western cultures, limited to primitive or pre-industrial tools. Anthropology's political origins have been blandly asserted by Claude Levi-Strauss in his controversial Smithsonian lecture:

> Anthropology is not a dispassionate science like astronomy, which springs from the contemplation of things at a distance. It is the outcome of a historical process which has made the larger part of mankind subservient to the other, and during which millions of innocent human beings have had their resources plundered and their institutions and beliefs destroyed, whilst they themselves were ruthlessly killed, thrown into bondage, and contaminated by diseases they were unable to resist. Anthropology is daughter to this era of violence: its capacity to assess more objectively the facts pertaining to the human condition reflects, on the epistemological level, a state of affairs in which onc part of mankind treated the other as an object.

It is within such an imperialist context that the histories and technological experience of non-Western societies could be written off or ignored: the latter, after all, were conquered peoples. When technology is seen through an anthropological prism, the emerging picture is bound to be far removed in character from a scenario that emerges from a sociological perspective. What is more, it is bound to be even more far removed from reality itself.

Some impression of that reality is discernible in the period before political dominance began to corrupt the objectivity of knowledge. Before the so-called "voyages of discovery", though non-Europeans were conceived as fantastic, wild, opulent, even monstrous, they were rarely considered inferior or backward; and even the actual European encounter with the scientific, technologic and medical traditions of non-Western societies was different from what eventually became the stuff of politically directed myths. In fact, from the day that the Portuguese mariner Vasco da Gama landed in India until almost three centuries later, Asia had a larger and more powerful impact on Europe than is normally recognized. Donald Lach has appropriately titled the first volume in his *Asia in the Making of Europe* "The Century of Wonder". It was not without reason that an Englishman of the time addressed the Indian Emperor by describing himself as "the smallest particle of sand, John Russel, President of the East India Company with his forehead at command rubbed on the ground". Nor can we forget that the first presents offered by da Gama to the King of Calicut included some striped cloth, hats, strings of coral beads, wash basins and jars of oil and honey. The king's officers naturally found them laughable.

It would take a few more decades before the Europeans landing in the Indian subcontinent would notice anything beyond gold and spices. But by 1720 and for a period of up to a hundred years, a new category of observers came visiting, some from newly formed learned societies in England. Their detailed reports were a result of the European quest for useful knowledge in different fields.

In his pioneering volume, *Indian Science and Technology in the Eighteenth Century*, the Indian historian Dharampal includes several accounts from these observers which describe among others the Indian techniques of inoculation against smallpox and plastic surgery. (While the first was eventually banned by the English, the latter was learnt, adopted, and developed). The accounts also document Indian processes like the making of ice, mortar, and waterproofing for the bottoms of ships; water mills, agricultural implements like the drill plough, water harvesting and irrigation works, and the manufacture of iron and of a special steel called *wootz*.

More techniques (like those involved, for instance, in the manufacture of Indian textiles) are described in *DeColonizing History* (Alvares, 1991) and *Science and Technology in Indian Culture* (Rahman, 1984). But even this documentation, impressive as it is, is now recognized to be but the tip of the proverbial iceberg.

The Chinese, like the Indians at Calicut, had a similar experience with an embassy and its gifts from London. The edict of Qian Long to the embassy is worth quoting: "There is nothing we lack, as your principal envoy and others have themselves observed. We have never set much store on strange or ingenious objects, nor do we need any of your country's manufactures" (Fairbank, 1971).

Immediately after the encounter, the graph of European reaction rises with esteem and wonder; and then, as political conquest and overlordship increase, the graph alters course and begins to record increasing denigration. A remarkable transformation of image thus takes place as the political relationship between Europe and non-European societies changes to the advantage of the former, rendering the Europeanization of the world picture almost an act of divine will.

By 1850, political dominance over the non-Western world was clearly installing distorted ideas not only about that part of the world but rebounding to distort Western man's image of himself as well. Already by 1835, for instance, the British had acquired a flattering notion of their own civilization (Victorian England was seen to be at the top of the pyramid of civilization) and a thorough-going contempt for Asia.

This contempt finds expression in the famous Minute of Lord Babington Macaulay:

> I have never found one amongst them (the orientalists) who could deny that a single shelf of a good European library was worth the whole native literature of India and Arabia . . . It is, I believe, no exaggeration to say that all the historical information which has been collected from all the books written in the Sanskrit language is less valuable than what may be found in the most paltry abridgment used at preparatory schools in England. In every branch of physical or moral philosophy the relative position of the two nations is nearly the same.

Dharampal has produced an interesting record of these assessments of science and technology in India among Western observers as the relationship between India and Britain changed to Britain's advantage.

Regarding the question of Indian astronomy, he discusses the case of Prof. John Playfair, Professor of Mathematics at the University of Edinburgh and an academician of distinction. Playfair studied the accumulated European information then available on Indian astronomy and arrived at the conclusion that the Indian astronomical observations pertaining to the period 3102 years BC appeared to be correct in every text. This accuracy could only have been achieved either through complex astronomical calculations by the Indians or by direct observation in the year 3102 BC. Playfair chose the latter. Opting for the former would have meant admitting that "there had arisen a Newton among Brahmins to discover that universal principle which connects, not only the most distant regions of space, but the most remote periods of duration, and a De La Grange, to trace, through the immensity of both its most subtle and complicated operations."

Similar attitudes prevailed concerning the knowledge of how Indians produced *wootz*. J.M. Heath, founder of the Indian Iron and Steel Company and later prominently connected with the development of the steel industry in Sheffield, wrote ". . . iron is converted into case steel by the natives of India, in two hours and a half, with an application of heat that in this country, would be considered quite inadequate to produce such an effect; while at Sheffield it requires at least four hours to melt blistered steel in wind-furnaces of the best construction, although the crucibles in which the steel is melted, are at a white heat when the metal is put into them, and in the Indian process, the crucibles are put into the furnace quite cold".

However, Health would not admit that the Indian practice was based on knowledge "of the theory of operations", simply because "the theory of it can only be explained by the lights of modern chemistry".

By the beginning of this century, the Western mind had already convinced itself that Western science and philosophy were the only approach to metaphysical truth ever attained by the human species and that the Christian religion provided wisdom and insight incumbent on all people everywhere to believe.

The result is reflected in the output of academia: a "history of art" turned out to be nothing but a history of European art and a "history of ethics" a history of Western ethics. While European music was music, everything else remained mere anthropology. The contemporary evaluation of human activity in the West as compared with the non-Western world was unabashedly provided by the late Jacob Bronowski in the Ascent of Man in words almost echoing Macaulay in 1837:

> We have to understand that the world can only be grasped by action, not by contemplation. The hand is more important than the eye. We are not one of those resigned, contemplative civilizations of the Far East or the Middle Ages, that believed that the world has only to be seen and thought about and who practiced no science in the form that is characteristic for us. We are active; and indeed we know, as something more than a symbolic accident in the evolution of man, that it is the hand that drives the subsequent evolution of the brain. We find tools today made by man before he became man. Benjamin Franklin in 1778 called man a 'tool-making animal' and that is right.

Now, there were obviously perverse consequences of such a view: scholars in several non-western societies, schooled in an educational system imposed on their societies through the colonial establishment, readily incorporated similar ideas about their own histories. In an article in *Nature* 35 years ago, Joseph Needham had to chide a native scholar of Thailand for claiming that his own people had not made any contribution to science despite compelling evidence to the contrary. Nevertheless, the colonization project succeeded in convincing many non-European intellectuals and scholars that only the West was active. They facilely accepted the idea that activity *per se* was desirable compared to judicious or necessary activity; that only the West was capable of thinking in the abstract sense. If this opinion were carried to its logical conclusion, it would appear that if the rest of humankind had survived for hundreds of years, this must be due to some form of manna falling providentially from the heavens.

The damage done by these years of extremely ideological scholarship and a ruinous ethnocentrism to the history of technology was bad enough. Predictably, the impression of an empty technological wilderness invented by Western scholarship about non-Western societies had a parallel, simultaneous, destructive impact on the assessment of their cultures as well. So insidious was the nature of this outrageous assumption regarding Western and non-Western abilities, that even Joseph Needham, Mark Elwin, Abdur Rahman and a host of other scholars participated in pointless debates which often took it for granted. One major debate, for example, focused on why China (and India) did not produce either modern science or an industrial revolution on the European pattern, especially since Chinese technology had already reached a level of sophistication not yet attained in any other part of the world as late as the fifteenth century.

Attempted answers compared and contrasted the internal conditions within Chinese society with those within Europe; the argument eventually succeeded in establishing the conclusion that no scientific or industrial revolution occurred in China because the social conditions in China were not the same as those within Europe. Thereafter, a host of cultural and social factors were dragged out of context and labeled probable "obstacles" either to the development of technology or modern science.

A critique of the three assumptions we have surveyed above therefore becomes compelling and inevitable, if we are to eschew their myriad fallacies in future. We shall take each in turn.

The idea that the past was merely a prelude, and a primitive one at that, may come naturally to anyone who has begun to feel that the present era of technical change is inevitable. Yet future societies may assess their past (our today) basing themselves on values other than those celebrating mere technical change. Already mindless technical change and built-in technological obsolescence have been assaulted by several global thinkers on the ground of ecological unsustainability and resource scarcity. It would also be wrong to think that because man did not have technology as he now does, he was necessarily impoverished. If there is anything the past gives us it is this positive impression of survival in all kinds of environmental scenarios. There is also evidence of more widely dispersed creativity when man was not submerged by technology than there is today. In many areas of human experience, we are yet to match even the technological achievements of the past which were driven by values other than mere complexity for its own sake or profit.

A similar argument may be used against the assumption that humankind is the only tool-making species there is. Several naturalists and ethologists have documented the diversity of nature's schemes at fabrication; most notably, Felix Paturi in his *Nature, Mother of Invention* and Karl von Frisch in *Animal Architecture*. Scholars like Lewis Mumford have gone further in stating quite bluntly that in their expression of certain technical abilities other species have for long been more knowledgeable than man.

> Insects, birds and animals, for example, have made far more radical innovations in the fabrication of containers, with their intricate nests and bowers, their geometric beehives, their urbanoid anthills, and termitaries, their beaver lodges, than man's ancestors had achieved in the making of tools until the emergence of homo sapiens. In short, if technical proficiency alone were sufficient to identify and foster intelligence, man was for long a dullard, compared with many other species.

Niko Tinbergen, another ethologist, after years of close observation of other species, has come to the following conclusion: "It was said that 1. animals cannot learn; 2. animals cannot conceptualize; 3. cannot plan ahead; 4. cannot use, much less make tools; 5. it was said they have no language; 6. they cannot count; 7. they lack artistic sense; 8. they lack all ethical sense." All of these statements, says Tinbergen, are untrue.

It cannot be said therefore that, in contrast with other species, humankind alone is a tool-maker. Thus the attempt to distinguish man from other living species because of his tool-making capacity is now seen to be a result of limited knowledge and unwarranted assumption of qualitative discontinuities between human beings and other species. It will also be useful to recall here that the ability to fabricate and organize is not a singular human trait — it is an intrinsic feature of nature since nature can exist only in a given form, whether at its most primary constituents at the sub-atomic level or even at the level of crystalline structures or the multiple tiers of a primary forest.

However, it is the third assumption — concerning the West's genius for technology and the rest of the world's incompetence in the same department — that contains the greatest mythological component of them all.

As we shall presently see, such an assumption has not only no historical basis, it is in fact contrary to historical and even to contemporary evidence. As for the gift of Greek rationality, suffice it to say that for two thousand years it gave no technological advantage to those who had it over those who did not. On the contrary, major scientific concepts, technological artifacts, tools, and instruments emerged in cultures that had nothing to do with either Greece or Rome.

The other problem with this assumption is it cannot even cope with the long established view that the science and technology traditions of most societies, particularly so of the West, are in significant ways mixed traditions. Even the little that we know about it indicates that the cross-cultural borrowing of technics and technology is impressive. Thus a very large number of critical inventions from both India and China helped fill significant gaps in the technological development of the West. A simple example from Francis Bacon's work will suffice to illustrate this point. He wrote:

> It is well to observe the force and virtue and consequences of discoveries. These are to be seen nowhere more conspicuously than in those three which were unknown to the ancients, and of which the origin, though recent, is obscure and inglorious; namely, printing, gunpowder, and the magnet. For these three have changed the whole face and state of things throughout the world, the first in literature, the second in warfare, and the third in navigation; whence have followed innumerable changes; inasmuch that no empire, no sect, no star, seems to have exerted greater power and influence in human affairs than these mechanical discoveries.

Now all these three mechanical discoveries were Chinese. Yet here again Western scholars have found it hard to acknowledge their origin. Borrowing of techniques from India is easily documented as well.

The documentation of technology in other cultures is only beginning. For example, it was only in 1974 that Sang Woon Jeon's *Science and Technology in Korea* appeared. There is as yet no major record of technology in Africa or South America though there is now available a large volume of documented evidence that both areas were rich in tools and techniques, from metallurgy to textiles.

In India, the other large storehouse of useful and appropriate tools (some still in productive use), the most extensive documentation of technology has only recently commenced, sparked in part by Dharampal's *Indian Science and Technology in the Eighteenth Century* and the work of scholars like Abdur Rahman.

The immediate impact of these re-invigorated investigations, stimulated largely by political independence, is a fresh debate over the issue of technology and culture: the old assumption of one technology and one culture in which others are seen to make a few, presumably inconsequential, contributions, is in tatters. Whatever its own pretensions to be the only viable culture, the West is finally being seen by non-Western societies as only one among several: a certain balance between cultures gets restored even though economic inequality persists. In fact, in some cases the pendulum has swung to the other side with cultures unabashedly resuming their traditions. There has naturally been a reverberation in the climate of ideas.

Changes in perception of this kind have already come about in other academic disciplines. To cite just one example, world histories were once written as if Europe were the center of the planet, if not the cosmos. There has been progress since: Geoffrey Barraclough and Leon Stavrianos, for instance, have both succeeded in producing comprehensive histories which avoid the older Eurocentric perspectives.

But even assuming we are able to produce, culture by culture, a fairly objective and comprehensive record of science, technology, and medicine, we would still be uncomfortably close to the pet obsession and perception of the present epoch. If there is anything the recent past has shown us, it is that we can be all too zealous judges in our own cause. We continue to celebrate uncritically our technological feats even when we know that the principal criterion of success for any species (and the human species is no exception) is primarily its ability to survive.

Therefore, it may be best not to get trapped in the debates on what is basically a sub-history: the history of slave-machines or automation or the recent machine-propensity of some cultures. After all, the *homo faber* concept is itself a distorted reflection of the natural endowments of the human species, an example of reductionist thinking. We know now that reductionism readily distorts knowledge, often pauperizes it, but rarely enhances it.

What is required in the circumstances then is a paradigm shift. I would like to suggest this can be achieved by replacing the heavily loaded term "technology" (too close identified with externalized objects) with the more neutral word, "technique". Technique has a larger ambit than technology and does not necessarily express itself only in the form of tools or artifacts. For the moment, we may define it briefly as every culture's distinct means of achieving its purposes. The natural propensity of human beings is to rely on technique, not technology, for while it has been proven that we can survive without technology, we cannot survive without technique.

Thus there can be no technique without culture, no culture without technique. An investigation into a culture's techniques is bound to be considerably more difficult than the recording of a culture's artifacts. The important gain here would be that we would begin with a more democratic assumption: that there is no culture without a system of techniques. Such a postulate would inoculate us effectively against methodological, ethnocentric, and other fatal flaws the *homo faber* concept was both parent and heir to. It would nip in the bud any undesirable future forays into cultural narcissism or ethnocentric discourse.

If this is indeed so, the more logical assumption would be that every culture has relied on a corresponding system of techniques that has guaranteed survival. Understood in this way, it makes far better sense to talk of Western technique (even though today largely expressed in the form of technology), or African, Indian, Chinese, Maya, or Arabic techniques of survival (in which technology may not be given that importance for fairly valid reasons). But even a relatively low importance given to technology could never mean a poverty of technique. The idea that the human species is technique-natured could be empirically falsified if a human society could be found that lacked technique — and not just machines or artifacts.

Technique, then, is nothing but the permanent but dynamic expression of an individual culture. Cultures can only express themselves or survive through technique: the alternative is chaos. Non-human species may be guided in the exercise of technique by inflexible inborn patterns of behavior. But the human world is as rigorously bound by the controls imposed by the symbolic universe that emerged as a substitute for weakened instinctual patterns. Myth, for instance, is technique. Interaction with (or manipulation of) nature may take place either through myth-making, scientific construction, and myriad other ways. All are expressions of the symbolic universe human beings inherit because they are human beings.

Thus we share the necessity of functioning through technique — not just through tools — with other living species — from the mammoth geobiological processes of Gaia to the cross-pollination of the rice plant to species of bird and animal, some of which, like the bower bird, are more prone to technology than others. Thus the so-called potter-wasp is known for its technique in constructing what we human beings culturally recognize as pots: however, the small vessels are the conclusion of technique: without it, there would be no "pot", no propagation and, therefore, no survival.

This will also explain why so-called primitive societies are often more complex in their social-cultural arrangements — their rich fund of botanical knowledge, slash-and-burn techniques, elaborate myths are as much an expression of technique — than modern societies.

Our new paradigm — based on a thorough-going analysis of technique — will enable us to concentrate more effectively on those aspects of human experience in non-western societies where there may be appropriate development of technology (as in India and China) but a superabundance yet of technique. A large number of these, particularly in India or in the Islamic cultures, may be located squarely within the domain of the sacred. They would be unintelligible outside such a framework of understanding.

We shall also observe in such societies that even where there is sophisticated technology, it retains an unobtrusive (not invasive) character. This can be seen from the merged outlines of Arab architecture to the irrigation works of South India or Sri Lanka.

An encyclopaedia of non-Western science, technology, and medicine may restrict its scheme to a bare description of the evolution of machines or artifacts incompetently covered by earlier conventional Western works, but it must do so guided by the background of the larger canvas of technique. Here the scholar will eventually examine theories of language in the same detailed manner as he would the culture and preparation of food or the control of breath — all extremely detailed sciences in India and China; he would examine irrigation and animal husbandry techniques, the domestication of cultivars of crop plants, record the elaborate knowledge of plants and of the human body, and seek to understand theories of cosmic phenomena and of the behavior of annual events like the monsoons.

The aim of the historian is to describe the nature of this individual system and not place it within a hierarchical ordering of societies. His task is to document this immense richness, not endeavor to swamp and drive it into oblivion on the questionable assumption that Western technology is the only direction that human technique will take. The growing anxiety over Western technology is closely associated with the threat it is perceived to pose to the fate of the planet and to survival. We may have to examine its history clinically to diagnose why it has generated the kind of problems it poses for humankind. Here, only a proper study of technique and culture within non-Western societies (and not as Forbes hoped, the event of Easter) will bring some balance and provide urgent clues to the origins of what Jamal-ud-din described as the illness of occidentosis, the plague of the West.

CLAUDE ALVARES

REFERENCES

Alvares, Claude. *De-Colonizing History: Technology and Culture in India, China and the West: 1492 to the Present Day*. Goa: The Other India Press, 1991.

Barraclough, G. *An Introduction to Contemporary History*. London: Penguin, 1967–1974.

Dharampal. *Indian Science and Technology in the Eighteenth Century*. Delhi: Impex India, 1971.

Goody, J. *Technology, Tradition and the State in Africa*. Cambridge: Cambridge University Press, 1980.

Jeon Sang-woon. *Science and Technology in Korea*. Cambridge, Massachusetts: MIT Press, 1974.

Lach, Donald F and E. Van Kley. *Asia in the Making of Europe*. Chicago: University of Chicago Press, 1965.

Levi-Strauss, Claude. "The Scope of Anthropology." *Current Anthropology* 7(2): 112–123, 1966.

Mumford, Lewis. *The Myth of the Machine*. 2 vols. New York: Harcourt, Brace and World, 1967–1970.

Nasr, S.H. *Science and Civilization in Islam*. Cambridge, Massachusetts: Harvard University Press, 1968.

Nasr, S.H. *Islamic Science: an Illustrated Study*. London: World of Islam Festival Publishing Co., 1976.

Needham, J. et al. *Science and Civilisation in China*. Cambridge: Cambridge University Press, 1954.

Rahman, A., ed. *Science and Technology in Indian Culture–a Historical Perspective*. Delhi: NISTADS, 1984.

Said, Edward. *Orientalism*. London: Routledge & Kegan Paul, 1978.

Sardar, Z., ed. *The Touch of Midas: Science, Values and Environment in Islam and the West*. Manchester: Manchester University Press, 1983.

Stavrianos, L.S. *Global Rift*. New York: Morow, 1981.

Teng, Ssu-Yu. *China's Response to the West: A Documentary Survey, 1939–1923*. Cambridge, Massachusetts: Harvard University Press, 1954.

Tinbergen, N. "The Cultural Ape." *TLS: Times Literary Supplement*, 28 February 1975, p. 217.

See also: East and West – Colonialism and Science – Environment and Nature – Textiles – Metallurgy – Technology – Food Technology – Ethnobotany – Irrigation – Agriculture

TECHNOLOGY IN THE ISLAMIC WORLD Hydraulic engineering was the most important technology of medieval Islam. Irrigation schemes supported a thriving agriculture and such schemes involved the construction of canals and *qanats* (see below), dams and water-raising machines. At the other end of the chain, water power was needed to process the raw agricultural products. In this article, we will begin with a survey of hydraulic engineering which will give us an overview of many of the significant aspects of Islamic technology.

Following this, a brief summary of a few other technologies will help us to appreciate the richness and diversity of Islamic material culture. The brevity of this summary is because of limited space and should not mislead readers into thinking that these technologies were insignificant, either in the social life of Islam or in the history of technology.

In Egypt, basin irrigation was in general use. This consists of leveling large plots of land adjacent to a river or canal, and surrounding each plot by dikes. When the river reaches a certain level the dikes are breached, allowing the water to inundate the plots. It remains there until the fertile sediment has settled, whereupon the surplus is drained back into the watercourse.

The regime of the Nile, with the predictable arrival of the flood, made Egypt particularly suitable for basin irrigation. Elsewhere in the Islamic world, however, perennial irrigation was, and is, practiced extensively on all the major river systems. As the name implies, it consists of watering crops regularly throughout the growing season by leading the water into small channels which form a matrix over the field. Water from a main artery — a river or a major canal — is diverted into supply canals, then into smaller irrigation canals, and so on to the fields. In many cases the systems operate entirely by gravity flow, but water-raising machines were used to overcome obstacles such as high banks, whether they were natural or artificial. Perennial irrigation from wells, again using water-raising machines, is also extensively practiced in the Islamic world.

After the advent of Islam a number of existing systems were extended in order to cater for the needs of newly-founded cities. Completely new systems were also constructed. In central Iraq the Muslims inherited the network constructed by their predecessors, the Sasānid dynasty of Persia. The major expansion occurred after the foundation of Baghdad in AD 762. The great Nahrawān canal, which left the east bank of the Tigris a short distance below Takrīt, was extended southwards.

The rivers ʿUzaym and Diyāla, which discharge into the Nahrawān from the east, were dammed by the Muslim engineers to provide water for a huge irrigated area. Further south, to the west of the Shaṭṭ al-ʿArab, the city of Basra was founded in the seventh century. Originally it was simply an army encampment consisting of reed huts, but soon it grew into a great city which had no rival in Islam before the foundation of Baghdad. Gradually, in the last decades of the seventh and the first decades of the eighth centuries a vast network of canals was constructed to serve the demands of a thriving agriculture.

Other important systems were based upon the rivers in the province of Khurāsān, then much larger than the eastern Iranian province of the same name. Some of the most impressive networks were based upon the river Murghāb, upon which stands the city of Marv. In the tenth century the system was controlled by a specially appointed Amir who had 10,000 men under him, each with an appointed task. This Amir is said to have had more authority than the prefect of Marv. Other important schemes in the east were those centred on the cities of Bukhārāʾ and Samarqand and on the lower reaches of the Amu Darya (Oxus) river.

In Spain, the Muslims greatly extended the irrigation schemes that had been constructed by the Romans and Visigoths. Syrian irrigation technology — large contingents of the conquering armies were from Syria — was applied in Spain, where the climatic and hydraulic conditions of the rivers of the north were very similar to those in Syria. The irrigation systems, such as those along the Gudalquivir river and those in the province of Valencia, were the basis of the agricultural prosperity of Muslim Spain. Indeed, systems such as that in Valencia, basically unchanged since Islamic times, continue to serve the needs of the province to the present day.

A particular technique, originating in Armenia about the eighth century BC, and still in widespread use in modern Iran, was the *qanat*, an almost horizontal underground conduit that conducts water from an aquifer to the place where it is needed. For the preparatory work an experienced surveyor carefully examines the alluvial fans, the terrestrial equivalent of a river delta formation, in the general location of the proposed *qanat*, looking for traces of seepage on the surface and often for a barely noticeable change in vegetation, before deciding where the trial well is to be dug. When a successful trial well has been excavated, it is now known as the mother well. The surveyor levels from this point to the outlet of the qanat, marking the positions and levels of the proposed ventilation shafts at 30- to 50-meter intervals. These shafts also provide for excavation of the spoil. A skilled artisan (*muqannī*) then begins work with his assistants by driving the conduit into the alluvial fan, starting at the mouth. At first the conduit is an open channel, but it soon becomes a tunnel. Another team sinks ventilation shafts ahead of the tunnellers, and laborers haul up the spoil to the surface through these shafts by means of a windlass. The tunnel is about one meter wide by one and a half meters high; in soft soil hoops of baked clay, oval in shape, have to be used as reinforcement. As the work nears the mother well, great care has to be taken in case a muqanni misjudges the distance and strikes the full well, in which case he might be swept away by the sudden flow.

The construction of canals, qanats, and other public works required the assistance of skilled surveyors. The earlier methods of levelling were rather slow and tedious. This involved stretching a string between two staffs divided into graduations and held vertically. A level was suspended to the center of the string. One type, for example, was an inverted isosceles triangular frame suspended by hooks to the string. Fixed to the center of the horizontal leg was a plumb line. One end of the string was moved up or down the staff until the plumb line passed through the inverted apex of the triangle, at which point the string was level. The difference in level between the two staffs was noted. In the eleventh century a level was introduced akin to our modern instrument but without, of course, any telescopic or electronic aids. A thin copper tube rotating on a circular plate that was suspended to a "gallows" was aimed horizontally on to a level staff. The reading gave the difference in level between the two points. The rises and falls along the route of the survey, summed algebraically, gave the level difference required. Triangulation, by which the heights and distances of objects, whether accessible or not, could be determined was usually carried out by the astrolabe. Quantity surveying methods were very similar to those in use today. The amount of excavation for a canal, for instance, was first calculated. Then, by applying known unit rates, the engineer worked out the number of excavators, laborers, foremen and supervisors. From these results he prepared a Bill of Quantities which was used for measuring progress and paying the contractor.

The main purpose of dams was to divert water from rivers into irrigation systems. Our knowledge of Muslim dams comes from two sources: reports in the works of Muslim geographers, and the examination of Muslim dams which have survived to the present day. In the east, one of the most impressive dams, known as the Band-i-Amir, was built in 960 over the river Kūr in Iran, between Shirāz and Persepolis. As described by the tenth-century geographer al-Muqaddasī, it was used to irrigate 300 villages. At each side of the dam, downstream, were ten water-raising wheels and ten water mills, all given extra power by the head of water impounded behind the dam. The dam was constructed throughout of masonry blocks, and the joints were made of cement mortar strengthened with lead dowels. The dam still exists and it is not surprising, given the solidity of its construction, that it has had such a long and useful life.

Also in Iran, at Kebar, about fifteen miles south of Qum, is a dam which is very important in the history of dam construction. Built in the thirteenth century, it is the first known example of the true arch dam. It did not depend for its resistance to water pressure on gravity but was built as an arch, its convexity pointing upstream, and its sides were anchored securely into the rocky sides of the gorge in which it was built. The forces were transferred to the abutments, and it was considerably more slender than a gravity dam across a similar river.

A number of important Muslim dams were built in the Iberian peninsula. These included the dam at Cordoba with an overall length of 1400 feet. Downstream from the dam were three millhouses that each contained four mills, as the geographer al-Idrīsī reported when he saw them in the twelfth century. Also below the dam is a large waterwheel. There are a series of eight dams on the river Turia in the province of Valencia. They are very securely built in order to withstand the sudden dangerous floods to which the river is subjected. They are also provided with desilting sluices, without which the canal intakes would soon become hopelessly choked.

These dams and their associated canals continue to provide for the needs of the province at the present time. It has been shown in modern measurements that the eight dams and their associated canals have a total capacity slightly less than that of the river. This, of course, raises the question whether or not the Muslims were able to gauge a river and then design their dams and canals to match. There seems no reason to doubt that they indeed had that capability.

Given that dam building had been an established practice since the times of the Sumerians and ancient Egyptians, it is not easy to isolate those elements that were Muslim innovations. On present evidence, it seems certain that the introduction of desilting sluices, the arch dam and hydropower were all Muslim inventions. The Muslims also probably perfected the technique of gauging rivers.

All the main water-raising machines were in existence in the Middle East before the advent of Islam. These included the Archimedean screw, the well windlass, the *shādūf*, the *sāqiya* and the *noria*. Because of their survival right up to the present day, and/or their importance in the development of machine technology, the last three of these are the most significant. The *shādūf* is illustrated as early as 2500 BC in Akkadian reliefs, and it has remained in use throughout the world until now. Its success is due to its simplicity and its efficiency. It can be easily constructed by the village carpenter using local materials, and for low lifts it delivers substantial quantities of water. In consists of a long wooden pole suspended at a fulcrum to a wooden beam supported on columns. At the end of the short arm of the lever is a counterweight made of stone or clay. The bucket is suspended to the other end by a rope. The operator lowers the bucket and allows it to fill. It is then raised by the action of the counterweight and its contents are discharged into an irrigation ditch or a head tank.

The *sāqiya* was invented in Hellenistic Egypt, probably in the third century BC. It is operated by an animal, usually a donkey, walking in a circle. On its shoulders and neck the animal wears a collar harness that transmits the power through two traces to a double-tree fastened to a drawbar. The drawbar passes through a hole in an upright shaft. This shaft carries the lantern-pinion, which is a type of gear-wheel consisting of two wooden discs separated by pins, the spaces between the pins being entered by the cogs of a vertical gear. This vertical gear has the cogs on one side of its disc and these protrude from the other side to form the wheel that carries the chain-of-pots, or potgarland. The component is therefore known as the potgarland wheel. It is erected on a horizontal axle over a well or other water source. The pots fill with water at the bottom of their travel and discharge at the top, like the shaduf, into a head tank or irrigation ditch. The saqiya was transitted by the Muslims from the Middle East to Spain and eventually to the New World. In some areas it has retained its popularity to the present day.

The *noria* is perhaps the most significant, from a technical viewpoint, of the traditional water-raising machines. Being driven by water, it is self-acting and requires the presence of neither man nor animal for its operation. Essentially it is a large wheel constructed of timber. At intervals, paddles project outside the rim of the wheel, which is divided into compartments. The noria is provided with an iron axle and this is housed in bearings that are installed on columns over a running stream. As the wheel is rotated by the impact of the water on its paddles, the compartments fill with water at the bottom of their travel and discharge their contents at the top, usually into an aqueduct.

The point of origin of the noria is unknown, but it was certainly known in the Roman world and in China by the first century BC. There is a possibility, therefore, that it was invented somewhere in the highlands of southwest Asia. The large noria at Murcia in Spain is still in operation, as are norias in various parts of the world, where they are often able to compete successfully with modern pumps.

Water wheels, as used in mills, were of three types: vertical undershot, vertical overshot and horizontal. The first type was a paddle wheel mounted directly in a running stream In the overshot wheel the rim was divided into compartments into which the water was directed from above, usually from an artificial channel or leat. Both vertical wheels required a pair of gears in order to transmit the motion of the water wheel to a vertical axle that led up to the millstones. In the horizontal type a jet of water from a reservoir was directed on to the vanes of the wheel, on the top of whose axle the millstones were installed. Various methods were used to increase the power and hence the output of mills. The wheel could be located in the piers of bridges or on boats moored on rivers, in both cases to take advantage of the increased rate of flow in midstream. In the Basra area in the tenth century there were mills that were operated by the ebb tide — about a century before the first mention of tidal mills in Europe. Apart from the production of flour from grains, water mills were also used for industrial purposes such as papermaking, the fulling of cloth and the crushing of metallic ores.

Bridges of all types were important in the Islamic world not only for communications but also for protecting the banks of canals from damage due to fording people and animals. In the mountains of Central Asia, such as the Tien Shan and Hindu Kush ranges, both cantilever and suspension bridges were used for crossing ravines. The great rivers of Egypt, Iraq and Transoxiana were usually crossed by bridges of boats. In many cases, however, rivers and other obstacles were crossed by masonry arch bridges built of dressed stone or, sometimes, especially in Iran, of kiln-burnt bricks.

Two bridges in western Iran, built in the tenth century, are constructed with pointed arches. This type of arch, which was to be such an important element in European Gothic architecture from the twelfth century onwards, had appeared in Syria and Egypt during the seventh and eighth centuries, but only as a decorative feature. These Iranian bridges are the earliest known occurrence of the pointed arch as a load-bearing component.

Chemical technology was a highly developed profession in medieval Islam. The most important writer on the subject was Muḥammad ibn Zakariyyā᾽ al-Rāzī. Although he wrote a number of alchemical books, his major work, *Kitāb al-asrār* (The Book of Secrets), written early in the tenth century, leaves us with the impression of a powerful mind, much more interested in practical chemistry than in theoretical alchemy. The *Book of Secrets* contains a comprehensive list of pieces of equipment, many of which are still in use today. The chemical processes described or mentioned by al-Rāzī include distillation, calcination, solution, evaporation, crystallization, sublimation, filtration and amalgamation. Arabic works were undoubtedly an important influence upon the development of European chemistry. Evidence for this influence is the abundance of Arabic words in the chemical vocabularies of European languages. Examples in English are alkali, alchemy, alcohol, elixir, naphtha and many others.

Mining played an important part in the economic life of the Islamic world. Spread throughout this vast area were mines for all the important metals, precious stones and essential commodities such as salt, coal and petroleum. The oilfields at Baku were exploited by the Muslims on a commercial scale by the ninth century. In the thirteenth century Marco Polo reported that a hundred shiploads could be taken from Baku at one time.

Iron and steel were, of course, the most important metals for many industrial and military purposes. Cast iron was described about 1040 by the great scientist al-Bīrūnī — it was exported to many countries as a raw material. There was also an extensive steel industry which produced, among other grades, the famous Damascus steel. As far as we can ascertain at present, this came from a number of Islamic centers in the Middle East and Central Asia. But it was certainly produced in Damascus itself.

We have been able to mention only the more important areas in which Muslim engineers and technologists were pre-eminent for several centuries. In many cases Islamic ideas were transmitted to Europe, but it is important not to evaluate Islamic technologies only with regard to their contribution to their European counterparts. Technologists in the Islamic world were responding to the needs of society and were extremely successful in a number of fields.

DONALD R. HILL

REFERENCES

Goblot, Henri. *Les Qanats, Une Technique D'Acquisition de L'Eau.* Paris: Mouton Éditeur, 1979.

Hassan, Ahmad Y, and Donald R. Hill. *Islamic Technology.* Cambridge, Cambridge University Press, 1986.

Hill, Donald R. *Islamic Science and Technology*. Edinburgh: Edinburgh University Press, 1993.

al-Karajī, Abū Bakr Muḥammad. *Inbāṭ al-miyāh al-Khaffiyya* (Search for Hidden Waters). Hyderabad: Maṭba ᶜat Dā ᶜirat al-Ma ᶜārif al-ᶜUthmāniyah, 1945.

al-Muqaddasī, Shams al-Dīn. *Aḥsān al-tagāsīm fī macrifat al-aqālīm* (The Best System for the Knowledge of the Regions). Ed. M.J. de Goeje. Leiden: Brill, 1906

Nordon, M. *Histoire de l'Hydralique 2 : L'Eau Démontreé* Paris: Masson, 1992.

al-Rāzī. *Kitāb al-asrār* (The Book of Secrets). Trans. and annotated by Julius Ruska. Berlin: Springer, 1937.

Smith, Norman. *A History of Dams*. London: Peter Davies, 1971.

Schiøler, Thorkild. *Roman and Islamic Water-Lifting Wheels*. Odense: Odense University Press, 1973.

See also: *Qanat* – Irrigation – al-Muqaddasī – al-Karajī – Surveying

TECHNOLOGY IN THE NEW WORLD The indigenous peoples of the New World were excellent builders. Although their construction tools were limited they managed to produce pyramids, temples, and other public buildings in the Andean Highlands, the lowland jungles of Central America and the Yucatan, and the central highlands of Mexico. There were ceremonial ballcourts built throughout the Caribbean as well as in what is now Mexico, Guatemala, and Belize. Construction in North America included the mound sites of the Midwest and the Southeast and the cliff dwellings of the Southwest. While estimated populations at these sites varied from the hundreds to the thousands these were all urban societies and civilizations.

These master builders relied on their own concepts, designs, craftsmen, and techniques. In spite of claims of trans-Pacific, trans-African and other exogenous contacts, the evidence is sparse and less than convincing. The civilizations of the New World evolved on their own with only limited contacts with one another over space and time. Their achievements as well as their failures are entirely their own, including matters of technology.

The New World societies concentrated on identifying, applying, and improving instrumental technologies, whether for agriculture, irrigation, construction, crafts, or other utilitarian uses. The legacy that has survived is one of warrior-kings, royal courts, artisans cum farmers, priests and scribes, builders and artists.

The peopling of the New World began approximately 15,000 years ago when climatic changes made migration from Siberia and East Asia possible across the Bering Straits. It took another 8000 years for these small wandering human bands to reach Patagonia and spread throughout most of Middle and South America. Somehow they managed to keep fertility above infant mortality as they diffused on foot throughout the New World. Hunters and gatherers, they relied on stone age worked flint edges to kill, strip, and consume their prey. Their principal weapon, originating in Asia, was the *atlatl* or spear thrower improved with flexible sheets and stone weights. Bows and arrows were introduced later, also perhaps from Asia via the Arctic, and became a favorite weapon for hunting bison on the Great Plains. When the Tainos paddled island by island from Lake Maracaibo up the Caribbean Archipelago (AD 1000–1400) they were armed with an array of bows, arrows, stone, bone, and flint knives, *atlatls* and other weapons.

Those who undertook the Great Journey throughout the New World survived through environmental adaptation as they fanned out across two continents. Those who adapted to the Amazon Basin continued in a hunter-gatherer mode, learning the skills needed to survive in the tropical rain forest. The Arawaks, Tainos, and other voyagers to the Caribbean also relied primarily on hunting and gathering adapted to coastal fishing. Latecomers who remained in the Arctic or sub-Arctic turned to hunting, fishing, shelter, and other technologies to survive harsh winters.

Agriculture probably arose about 3000 BC simultaneously in the Andean highlands and Central Mexico, first as a supplement to hunting and gathering and as a response to growing populations. Gatherers observed which seeds yielded what plants and began to cultivate on a trial and error basis. Teosinte, a high-protein precursor of corn, was first cultivated in Mexico, while potatoes were an early ecologically suitable cultigen in the colder Andes. The list of cultivated edible plants soon expanded to include manioc, squash, peppers, pineapples, and other New World originals, especially several varieties of beans. The Tainos learned to grow tobacco and to cure it as snuff in their Caribbean islands, which became valuable trade items.

Agricultural technologies evolved to include the use of digging sticks, irrigation canals, terracing and ridging, intercropping, and seed beds. Lacking draught animals fires were set to clear new land. The most sophisticated technique was that of the *milpas* practiced by the Aztecs who built silted mounds of fertile run-off materials.

Food processing and storage technologies were also innovated and widely diffused. The stone *metate*, or mortar and pestle, became the basic Mesoamerican tool for crushing corn. Ethnobotanical experience produced a variety of herbs, spices, and for the Mexicans chocolate to add to largely vegetable diets. As towns and cities grew, especially in the Andes and Mesoamerica, hunting declined in favor of limited domestication of chickens, turkey, pond fish, and small animals. However, animal protein was hard to come by in all the predominantly agricultural societies which relied on trade and food storage to supplement local corn, beans, manioc, and other staples.

Transport was a severe obstacle for all the New World peoples. Dogs were used to pull sleds and traverse with limited loads in the Arctic and on the Great Plains. Alpacas and llamas were semi-domesticated and used for limited local transport in the Andes. However, everywhere terrain was rugged, surfaced roads were few and far between, and runners and head-loads were the primary mode of transport. Sailing vessels were unknown and the canoes of the Tainos and others had to hug shorelines for safety. Although the concept of the wheel was known, and toy wheels were used in the Andes, the lack of roads and the natural obstacles impeded the development of wheeled vehicles. The Spanish introduction of the horse and the cow revolutionized transport in the New World. Trade previously was confined to what could be carried on men's backs. Thus the Aztec long-distance traders, *pochotes*, specialized in high-value items such as jewelry, ceremonial feathers, and weapons.

Confined to limited home markets, New World societies tended to invest their agricultural surpluses in exquisite crafts. As royal courts developed around ceremonial centers so did specialized artisans capable of meeting quality demands, as well as items for daily use. The oldest discovered textiles date back 3600 years to the Peruvian coast and were made of cotton with intricate woven designs. Dyed cotton for personal clothing became widely diffused with pendants and ornaments often added. Excavation of Andean and other tombs has revealed the importance attached to clothing and jewelry and the high skills of the artisans. The Olmecs of the Mexican Southeast were the first to work in jade and turquoise as well as to create massive stone carvings. Metallurgy, especially in the form of masks, was highly accomplished in Mexico as well as among the peoples of what is now Costa Rica, Colombia, and Peru.

Ceramics and basket and gourd work were closely associated with the emergence of agriculture. Initial demands were for items for carrying and storage. However, in the Andes and Mexico, as techniques improved, ceramics became an important ceremonial art form. Funerary jars, urns, water carriers, ceremonial masks, and other items are regularly depicted in sculptures and paintings and found in Aztec, Maya, and other tombs. Pre-Incan and Incan Andean societies also took pride in their varied and elaborate ceramics, baskets, and gourds.

Ceremonials were held largely outdoors, while palaces were for storage purposes. Most common people, including craftsmen, lived, ate, and slept in simple mud and brick dwellings. Their lords and masters lived in larger and more solid enclosures but with similar creature comforts. Building skills were directed at large-scale public constructions to honor the deities and/or fortresses for defense purposes. While kings lived in palaces with a few of their retainers many members of the court had to accept more humble accommodations. The benign tropical climate of the Caribbean contributed to nearly everyone's relying on hammocks and huts made of straw.

The Aztecs alone tackled the sewage and other waste disposal problems of man made cities. They instituted barges along the canals of their capital at Mexico City, and also made an effort to keep water potable. At the Inca capital of Cuzco public hygiene was minimal as it had been in the Maya city-states. It may have been the scarcity of domesticated animals that reduced disease vectors in these highly crowded urban sites.

As builders the peoples of the New World concentrated their efforts on specific times and places. These included the Olmec (1000–600 BC), Tectihuacan (AD 0–650), Toltec (1000–1300 BC), and Aztec (AD 1400–1530) civilizations in Mexico, the Mayas (AD 0–900) in Guatemala and Mexico, and the Moche (AD 0–600), the Nazca (AD 0–600), the Chimu (AD 1300–1420) and the Inca (AD 1400–1530) in Peru. They were able to construct magnificent and lasting buildings in environments as different as lowland jungles and highland mountains. They moved massive amounts of stone and carved and fitted it exactly with simple tools, as well as working with wood. Their architects designed vast open spaces and compelling interiors. Their multi-story edifices reached up to the heavens, and served sometimes as astronomical observation sites, while commanding the world below. Their equivalents are neither the medieval cathedrals of Europe nor the Acropolis of Greece. Instead they are the most significant continuing testimony to the uniqueness of the these New World civilizations and peoples.

While we can directly experience and marvel at these sites it is much more difficult to penetrate the intellectual worlds of their builders. The Mayas, prior to their collapse in AD 800–900, had a sophisticated writing system which was lost, as were many of their ideas about astronomy and mathematics. Similarly, the Nazcas of coastal Northern Peru who vanished around AD 600 may have had more advanced ideas about astronomy than their predecessors. The unevenness of historical experience and its inherently non-linear nature is one of the sobering lessons of the history of science and technology in the New World. The Mayas alone invented a writing system which took nearly 450 years after the European Conquest to decipher. The Incas adapted a system of counting called the *quipu* for purposes of accounts and storage but never extended it into a mathematical base. The Aztecs excelled at urban planning, irrigation, and public health but showed little interest in writing.

The peoples of the New World were builders, agriculturalists, and craftsmen rather than scholars or theologians. The European conquerors came from an age of iron and steel with wooden ships, navigation, steel swords, guns, explosives, and literacy. They brought horses, cattle, and smallpox, measles, diphtheria, trachoma, whooping cough, chicken pox, bubonic plague, typhoid fever, scarlet fever, amoebic dysentery, and influenza.

AARON SEGAL

REFERENCES

Aveni, Anthony. *Skywatchers of Ancient Mexico*. Austin: University of Texas Press, 1980.

Bauer, Brian and David Osborne. *Astronomy and Empire in the Ancient Andes*, University of Peru Press, 1996.

Bernal, Ignacio. *The Olmec World*. Los Angeles: University of California Press, 1976.

Coe, Michael D. *Breaking the Maya Code*. New York: Thames and Hudson, 1992.

Coe, Michael, Dean Snow and Elizabeth Benson. *Atlas of Ancient America*. New York: Facts on File, 1986.

Fagan, Brian M. *The Great Journey: The Peopling of Ancient America*. London: Thames and Hudson, 1987.

Fiedel, Stuart J. *Prehistory of the Americas*. 2nd edition. New York: Cambridge University Press, 1992.

Hudson, Charles. *The Southeastern Indians*. Knoxville: University of Tennessee Press, 1992.

Rouse, Irving. *The Tainos: Rise and Decline of the People Who Greeted Columbus*. New Haven: Yale University Press, 1992.

Schele, Linda and Mary E. Miller. *The Blood of Kings*. New York: Braziller, 1986.

Weber, David J. *The Spanish Frontier in North America*. New Haven: Yale University Press, 1992.

See also: *Quipu* – Mathematics – Inca – City Planning – Mound Cultures – Potatoes – Crops – Agriculture – Food – Technology – Sugar – Animal Domestication – Metallurgy – Textiles – Nazca Lines – Roads – Stonemasonry

TELESCOPE The telescope is an optical instrument used to make distant objects appear nearer and larger; it consists of one or more tubes with a series of lenses, mirrors, or both (refracting, reflecting and catadioptric models respectively), through which light rays are collected, brought to a focus, and magnified. Despite their great variety, all telescopes have two basic parts: the objective, which intercepts and focuses the incoming light, and the mounting, which supports the objective. The telescope, which revolutionized astronomical research in the seventeenth century, is usually considered to be a European invention, though parallel discoveries occurred elsewhere, and some of its elements were developed earlier by other cultures.

The traditional Western account of the telescope's invention is quite complex. The use of hollow sighting tubes without lenses to observe stars had been known to the ancient Greeks, being recorded by Aristotle (384–322 BC) and the geographer Strabo (ca. 63 BC–ca. 19 AD). Lenses are mentioned by the playwright Aristophanes (ca. 450–388 BC), and mirrors were investigated by Heron of Alexandria (ca. 60 AD). The geometer Euclid (ca. 300 BC) and the engineer Archimedes (ca. 287–212 BC) studied both. The investigation of lenses, mirrors and sighting tubes continued throughout the Medieval period. The invention of spectacles, an important telescope precursor, occurred between 1285–1300 by an unknown Italian glassmaker, but Alexandro della Spina (d. 1313) and his friend Salvino d'Armati (d. 1317) are most often credited. Meanwhile, English Franciscan friar Roger Bacon (1211–1294) wrote extensively of spectacles and other optical devices, and influenced many later scientists with his research into the laws of reflection and refraction.

An enormous number of sixteenth century figures developed various combinations of perspective glasses and mirrors to enlarge distant objects, including the English father and son surveying team of Leonard (ca. 1520–ca. 1571) and Thomas Digges (1546?–1595), mathematician William Bourne (d.1583), Italian natural philosopher Giambattista della Porta (1534/5–1615), and Dutch mathematician Cornelius Drebbel (1572–1634). But the lack of documentary evidence has thus far left historians divided over whether these instruments were true telescopes or just magnifying glasses.

The first unambiguous accounts of the European telescope come from Holland around 1608. Most credit the Middleburg spectacle maker Hans Lippershey (ca. 1570–1619) with the first refracting telescope; but there are rival claims by Middleburg optician Zacharias Jansen (1580–ca. 1638), and Alkmaar inventor Jacob Metius/Adriaanzoon (1571–1635).

Early telescopes were first applied to military operations. But over the winter of 1609–1610, Italian physicist and astronomer Galileo Galilei (1564–1642) dramatically improved Lippershey's primitive device and turned it to the sky. His announcements of lunar valleys and mountains, sunspots, the phases of Venus, the moons of Jupiter, and much more (published in the March, 1610 *Sidereal Messenger*) thoroughly shook the ancient earth-centered cosmology of Alexandrian astronomer Claudius Ptolemy (ca. 150 AD). Similar observations were made between 1609–1612 by English polymath Thomas Harriot (1560–1621). The word "telescope", meanwhile, was coined in 1611, and was first printed in Julius Caesar Lagalla's (1576–1624) *Lunar Phenomena* of 1612.

Early Galilean telescopes, with one bi-convex and one bi-concave lens, suffered from both chromatic and spherical aberration, which left blurry images. The German astronomer Johannes Kepler (1571–1630) designed the improved Keplerian telescope of two bi-convex lenses in 1611, but problems remained. It was not until 1636, when the Minorite friar Marin Mersenne (1588–1648) suggested replacing lenses with mirrors, that real progress could be made. Reflecting telescopes were designed by Scottish mathematician James Gregory (1638–1675) in 1663, and built a few years later by both Isaac Newton (1642–1727) and French physicist Guillame/Nicholas Cassegrain (fl. 1672); these models (Newtonian and Cassegrain) are still widely used today.

During the period 1610–1650, the telescope was carried by European explorers to all regions of the earth; frequently Jesuit missionaries exported it to non-western cultures. However, there is evidence that other societies developed both the telescope and its various parts independently of Europe. Glass originated in Egypt about 3500 BC and was first produced on a large scale by the Phoenicians, and lenses dating back to pre-Greek times (2000 BC) have been discovered in Crete and Asia Minor/Mesopotamia.

Meanwhile, Arabic astronomers played a significant role in paving the way for the telescope. The mathematical laws of optical reflection and refraction in glass and other media were thoroughly investigated and extended by the physicist-astronomer Ibn al-Haytham (Alhazen, 965–ca. 1040). Ibn al-Haytham knew how to use spherical glass segments to magnify objects, and was familiar with spherical aberration and techniques to calculate the focal lengths of lenses and mirrors. Much of this knowledge was subsequently transmitted to Europe through Latin editions of his work, and especially through the texts of his Polish disciple, Witelo of Silesia (ca. 1230–ca. 1275). Bacon also learned much from Ibn al-Haytham.

Sighting tubes were also early employed by Arabic astronomers; al-Battāni (858–929) used them in his Raqqa observatory. Naṣīr al-Dīn al-Ṭūsī attached one to a sextant at his Maragha observatory in 1259 to study the sun.

Meanwhile, Joseph Needham argues that Chinese opticians may have discovered the elements of the telescope either before, or parallel to, the Europeans. Though glass was developed fairly late (ca. 500 BC), quartz rock-crystal had been in use since ancient times; and lenses of both materials were employed to focus solar rays to start fires. Needham also claims the Chinese used smoky rock-crystal to observe sunspots and eclipses. An important text of Tan Qiao (the *Hua Shu* or Book of Transformations of about AD 940) records the use of plano-concave, plano-convex, bi-concave and bi-convex lenses to make objects larger, smaller, upright and inverted. Spectacles have often been considered a Chinese invention, but this is based on a garbled text; they were brought overland from Europe by 1300. As for catoptrics, Mohist opticians of the fourth century BC studied both concave and convex mirrors, and were familiar with real and inverted images, focal points and refraction in different media. The use of concave mirrors to ignite tinder by focusing sunlight goes back to the Chinese Bronze Age; one of the earliest citations is dated 672 BC.

Chinese sighting tubes made of bamboo are similarly old; the most important early reference is the (approximate) sixth century BC *Shu Jing* (*Historical Classic*). By the *Huia Nan Zi* (*Book of Huai-Nan*) of 120 BC, they were widely used in land surveying and triangulation; and by the twelfth century, they were standard navigational equipment for taking astronomical observations aboard ships.

Though many of these developments were contemporaneous with European discoveries, there are two parts of the telescope in which Chinese scientists can claim undisputed priority: the invention of equatorial mounting, and the development of clock drives, both of which are standard equipment on today's telescopes.

Astronomer Guo Shoujing (fl. 1270) developed the equatorial mounting now used on all modern telescopes for his "simplified instrument", which was basically a dissected armillary sphere related to the torquetum, an Arabic invention which had been recently imported (ca. 1267) to China from Persia. With an equatorial mounting, a rotation about only one axis (one parallel to the earth's polar axis) was needed to follow the curved paths of the stars.

Chinese physicists and engineers anticipated the fourteenth century mechanical clocks and clock-driven astronomical instruments of Europe by over a thousand years. Mathematician, geographer and astronomer Zheng Heng (ca. 78–ca. 142) constructed what Needham calls the "grand ancestor of all clock drives", a rotating bronze armillary sphere powered by a system of gears attached to a waterwheel. This device was used to predict the positions of various stars and planets. The later clock-driven armillary sphere of Su Sung (AD 1090) added a sighting tube to permit direct observation of a star as it moved over prolonged periods, resulting in a true "clock drive" for astronomy.

But despite Chinese priority in clockwork and mounting, just as in Europe, it was not until the seventeenth century that the disparate elements of the telescope were joined together. Though Chinese texts describe European telescopic discoveries from 1615 onwards, and Jesuit Father Terrentius/Johannes Schreck brought a telescope to China in 1618, Needham maintains that enough documentary evidence exists to assert a parallel, independent discovery by Suchow opticians Po You and Sun Yun Qiu. Between about 1620–1650, this pair constructed not only early telescopes, but also compound microscopes, magnifying glasses, searchlights, magic lanterns and other instruments. Meanwhile, Sun Yun Qiu wrote a text on these and other optical devices, entitled *Jingshi* (*History of Optick Glasses*). But the first Chinese book on the telescope specifically was the *Yuan Jing Shuo* (*Far Seeing Optick Glass*) of Tang Ruo-Wang (Adam Schall von Bell) in 1626. *Yuanjing* became the standard term for the telescope, and by 1635 it was being widely used by Chinese artillery in battle.

Telescopes appear to have come to Indian astronomers fairly late. Indian craftsmen made glassware, lenses, and mirrors throughout the ancient and medieval periods. And *gola yantra* or armillary spheres, first mentioned in the *Āryabhaṭīya* of Āryabhaṭa I of Kusumpura (b. 476 AD), had clock-drives (powered by clepsydra) for stellar observation by the *Bhatadipika* of Parameśvara (ca. 1432).

But it was not until the seventeenth century that primitive telescopes were used in India. The celebrated astronomer and instrument-maker Sawai Jai Singh (1686/8–1743) employed telescopes to "show that Mercury and Venus get their light from the Sun as the Moon does" (*Zica-i Muhammad Shahi*); and though they revealed Saturn to be an irregularly-shaped "oval", these telescopes were apparently not good enough to resolve the rings, which had already been accomplished by Dutch physicist Christiaan Huygens (1629–1695) several decades previously, in 1659. Indian achievements in this area thus far appear to be rather derivative, but much research still needs to be done in this area.

JULIAN A. SMITH

REFERENCES

Abdi, W.H., et al. *Interaction between Indian and Central Asian Science and Technology in Medieval Times*. New Delhi: Indian National Science Academy, 1990.

Jaggi, O.P. *History of Science, Technology and Medicine in India*, esp. vol. 6–7. Delhi: Atma Ram, 1977–1986.

King, David A. *Islamic Astronomical Instruments*. London: Variorum, 1987.

King, Henry C. *Geared to the Stars: the Evolution of Planetariums, Orreries and Astronomical Clocks*. Toronto: University of Toronto Press, 1978.

Kuppuram, G., and K. Kumudamini. *History of Science and Technology in India*, 1–12. Delhi: Sundeep Prakashan, 1990

Needham, Joseph. *Science and Civilisation in China*. Cambridge: Cambridge University Press, 1954.

Needham, Joseph. *Heavenly Clockwork*. Cambridge: Cambridge University Press, 1960.

Needham, Joseph. *Clerks and Craftsmen in China and the West*. Cambridge: Cambridge University Press, 1970.

Sayılı, Aydın. *The Observatory in Islam*. Ankara: Turkish Historical Society, 1960.

See also: Jai Singh – Ibn al-Haytham – al-Battāni – Naṣīr al-Dīn al-Ṭūsī – Observatories – Maragha – Guo Shujing – Zhang Heng – Clocks and Watches – Parameśvara – Āryabhaṭa

TEXTILES IN AFRICA Historians of African textiles now have at their disposal a wide range of oral and written documented sources. These describe significant centers of textile production over time, the raw materials, implements, and techniques used by the various cloth producers, varieties of fabric and techniques of dyeing and coloration, symbolic expression as reflected in the finished products, and the many functions for which the latter were used. The process of technological transfer within various parts of the continent and the elaborate structure of guilds and schools of apprenticeship are also better known to us as a result of the systematic collection of oral history in some areas.

From the Northeast African Nile region to West Africa and elsewhere, travel reports, missionary reports, and even autobiographies have provided details about aspects of the development of cloth making techniques. Herodotus in his travels in Egypt as far as the first cataract obtained specific knowledge about the implements used by the Egyptians and the existing division of labor. Such information has complemented the archaeological evidence brought forward by teams of Egyptologists as they continue to come in contact with numerous reams of cloth in the mummified corpses and burial goods which the Egyptian nobility, like some of their counterparts in West Africa, sent along with the dead to the afterlife.

In the case of Ethiopia, travel reports by missionaries, explorers, and travelers such as Francisco Alvarez, Jerome Lobo, Charles Poncet, James Bruce and Henry Salt, provide direct and indirect information on the variety of textiles in the Axumite and post Axumite realm. For West Africa in the eighteenth and nineteenth centuries, Mungo Park, Barth, and Baikie complement the perspectives diffused in the missionary reports of Trotter, Allen, and Crowther, who emphasized that the people of Onitsha, Eastern Nigeria, like their counterparts elsewhere, "manufacture their own clothes generally plain or fanciful with cotton grown in their farms." (Crowther, 1968)

Archaeological reports have been no less useful than the eyewitness travel and missionary accounts cited. Reams of linen and cotton have been found in Egyptian tombs, and the Sudan, also one of the Nile Valley kingdoms, has yielded cloth and looms dating back to 500 BC. The Igbo-Ukwu finds of Eastern Nigeria included thousands of artifacts, among which was cloth dated to the ninth century. The Bandiagara cliffs of Mali have yielded textile products dated to the eleventh century. The evidence from the excavated pits complements the various collective recollections reflected in poetry, song, and narrative accounts no less than the honorific codes and titles and range of linguistic terms accorded textile specialists and their products. Some city states and towns in the African continent gave their names to a product line, as was the case of Akwete and Okenne in Eastern and Central Nigeria, West Africa.

Whether in renowned Nigerian textile centers such as Kano, Iseyin, Bida or Akwete and Okenne, or in the Wolof empire of Senegal, the Bambara kingdoms of Mali, the Mossi Empire of Burkina Faso, or the Baule polities of the Ivory Coast, we can identify some basic raw materials, production instruments and techniques, as well as common tendencies relevant to textile not only in West Africa, where the latter regions were located, but also in Central, East and Southern Africa. Raw materials were derived from vegetable or animal products and generally involved wool, camel hair, flax, cotton, the leaves of the raffia palm (*Raphia rufia* or *Raphia vinifera*), silk from cocoons, and bark from the baobab tree. In all the above cases, with the exception of bark, which was hammered into shape, there developed over time sophisticated spinning, weaving, and dyeing techniques which included the gradual invention and improvement over time of vertical and horizontal frames, lower and upper beams, beaters, shed sticks, hecklers, shuttles, and templars, all various components of the horizontal and vertical looms produced across the continent.

African textiles, by the nineteenth century, included a wide range of fabrics, each influenced by the base material from which the thread was made, the texture of the thread, the width of the strip woven cloth, the alignment of the thread, and the intensity of inlays and the dyeing procedure, whether starch resist or not. Indigo, guinea corn stalk, the bark of the locust tree, the leaves of the tombolo tree, combined with ash potash and any of a long list of colorants were used to produce blue, red, buff, rust, or brown and other colors. Dyes were derived from experimentation with vegetable and other products. The famous *Kente* cloth of strip woven silk (sometimes interlaced with threads of gold and very often confined to the Ashanti nobility), *Sanyan*, Western Nigeria's silk derived fabric, *Adrinkra*, the hand-printed cloth of the Ivory Coast, and *Sotiba* of Senegal, are some of the various types of textile products which have become household names in the continent.

A wide range of symbols was reflected in cloth through the representation of motifs of special shapes, figures, dimensions, and sizes. These were either impressionistically done by the use of geometrical shapes and symbols or were made to reflect naturalistic images derived from African cosmologies and indigenous belief systems. It was common for Dahomean quilts to portray images of the founding fathers of particular dynasties. A lion represented King Gelele and a buffalo King Gezo, whilst a representation of a ship was the symbol of King Agaza — all historic figures in the making of the Dahomean (Beninois) state. More recently, worldly events ranging from the American soap opera Dallas to statements about the prevailing economic reform programs have been coded into fabric. A study of African textile over the centuries yields information not only about changes in technical expertise and accretional gains made from the intra-regional and inter-regional exchange of ideas, but also the prevailing lifestyles, philosophies, and world views of Africans in various parts of the continent. Textiles in themselves provide a rich source of historical information. They were associated with many activities and had many uses. They were a medium of exchange in barter, or units of currency in a wide range of commercial transactions. Tax payment was collected in the form of textiles and so too tribute. Since fabric differed in cost and the degree of technical expertise associated with various types, it was easy for it to become a symbol of wealth, extravagance, and conspicuous consumption. Cloth very often was a symbol of class affiliation. Saddle cloth and the overall accoutrements of the horse included specialized fabric, as did the panoply associated with Ashanti royalty. Cloth had special burial functions. Specialized fabric was associated with shrines of the dead. Marital gifts, whether pre-nuptial or not, tended to include cloth as part of the expected dowry. Decorative and symbolic objectives were matched by functional and practical ones. Cloth was used for protective purposes against the elements, as sheets and covers, and also to make rice bags, purses, or tents.

The reproduction of knowledge systems was done through institutionalized apprentice systems of guilds, each of which adhered to strict codes of conduct and behavior. Fees in cash or kind were paid by the apprentice who was expected to "gain freedom" after periods agreed to in the context of elaborate ceremonies. These guilds themselves had a hierarchy of officials who in many cases were of significant political clout. The teaching and training of spinners, weavers, and dyers was therefore organized in the context of established custom and practice, even though there was a general tendency for specialization to be restricted to specific lineages within various regions.

There is evidence that Africans imported cloth in the context of the trans-Saharan trade as well as the trans-Atlantic, but these importations were complementary to a wide range of indigenous textile products. It was not unusual for various political elites to flaunt some of the imported textiles along with indigenous products, or for textile specialists to unravel an epic of imported fabric with the objective of isolating particular reams of thread. The British industrial revolution had specific implications for African textiles, given the fact that factory produced cloth was cheaper with the mechanization of spinning and weaving. The successful transfer of textile technology from India to Britain by the nineteenth century meant that the British factory system was able to churn out some relatively attractive textile products for the African consumer. This, however, did not lead to the destruction of the indigenous textile sector which continued to be a dominant component of the informal sector well into the twentieth century, despite some ill-intentioned legislation and policies during the period of colonial rule.

Contemporary Africa is home to a wide variety of indigenous textiles. Local factory-produced cloth has had to compete with Taiwanese, Indonesian, and Dutch imports, most of which are imitations of indigenous African fabric. IMF and World Bank conditionalities tend to call for the removal of duties on imports and therefore undermine those local textile centers, some of which lack the productive capability to compete on the open market. In spite of these trends, however, there is every indication that African textile producers will continue to have a large share in the market and continue to experiment, innovate and produce the high quality textiles historically associated with the continent.

GLORIA T. EMEAGWALI

REFERENCES

Adler, P., and B. Barnard. *African Majesty: The Textile Art of the Ashanti and Ewe*. London: Thames and Hudson, 1992.

Afigbo, A., and C. Okeke. *Weaving Tradition in Igboland.* Lagos: Nigeria Magazine, 1985.

Andah, Bassey. *Nigeria's Indigenous Technology*. Nigeria: Ibadan University Press, 1992.

Crowther, Samuel. *Niger Expedition 1857*. London: Dawson, 1968.

Picton, J., and J. Mack. *African Textiles: Looms, Weaving, and Design*. London: British Museum, 1979.

Thomas-Emeagwali, G., ed. *African Systems of Science, Technology and Art*. London: Karnak, 1993.

"West African Textiles." Special Issue of *African Arts*, vol. 25, July 3, 1992.

TEXTILES IN CHINA A textile is here understood as a woven fabric made from a wide range of raw materials. Some of these were silk, cotton, wool, and the bast fibers ramie (*Boehmeria nivea L.*), hemp (*Cannabis sativa L.*), and the bean- or vine-creeper (*Pueraria thunbergiana*), which possessed the greatest economic and cultural importance in pre-Mongol (prior to AD 1279) China. The development of textiles is described here on the basis of archaeological evidence and from the view of its cultural importance.

The origin of weaving can be traced back to basketry and matting techniques from the Neolithic. Impressions on ceramic sherds and fragments from the archaeological sites of Banpocun near Xi'an in Shaanxi and from Hemudu in Yuyao county in Zhejiang prove that they existed in China as early as the fifth millennium BC. The earliest finds of textiles are from southeast China. There is a complicated fragment of fabric, made of *Pueraria thunbergiana*, found at Caoxieshan in Jiangsu province and dated to the fourth millennium BC. Its structure features a combination of hand twining and loom weaving techniques. A number of silk belts, a piece of silk, and a scrap of ramie cloth were discovered in a bamboo box excavated from the site of Qianshanyang in Wuxing in northern Zhejiang, dated to ca. 2750 BC.

Small stone-cut and jade figures were found in various Anyang sites from the the Shang period (sixteenth–eleventh centuries BC). They clearly show "textile ornaments" which give an impression of Shang clothing and fashion. Apart from furs and leather, especially suitable as winter clothing and used for ceremonial occasions, hemp and silk were tailored. Tailoring does not mean cutting pieces of fabric so as to make them fit but simply sewing pieces of fabric together, only allowing a few corrections. It was probably the Shang people who introduced right side fastening and a hair-dressing style which could be distinguished from the barbarians. Most garments including lined garments were buttoned on the right side under the right arm. Very often the overlap of a robe was unbuttoned. *Obi*-like waist-belts kept the coat and the undergarments in position. Women's dress included a skirt. Women had their hair styled in a cylindrical shape, and flat rather high round caps were common. Upperclass people wore gaiters and shoes that curled upwards to the toe. Geometrically patterned fabrics (*T-, hui* and *lozenge*-patterns) were exclusively used as borders of garments, at the openings of the sleeves, at the collar, and along the overlap, for girdles and waist-belts, caps and hats. Colorful painted and embroidered silk was available to the higher echelons of Shang society. Certain important weaving techniques were applied to produce silk fabrics as early as Shang times:

1. Tabby weaves with threads of almost identical diameter and a thread-count of warp and weft between 8:7 and 75:50 per centimeter;
2. Warp-faced tabby, also called rep or rib, where the number of warp-threads is roughly double the number of weft-threads per centimeter;

3. Monochrome tabby patterned with twill (3/1), very often named twill damask (*wenqi*). The ground weave is a tabby weave, and the pattern is woven in twill weave with the warp-threads forming the pattern;
4. Tabby crêpe (*zhou* or *hu*) with a thread-count of ca. 30:30 per centimeter. The warp-threads show a twist with 2500–3000 twists per meter. The weft-threads are twisted together from several threads in an *S*-twist showing 2100–2500 twists per meter. The strong twisting of the threads causes the crêpe effect;
5. crossed-warp weave technology was known and used.

The most important and spectacular textiles of the Zhou period (1045–221 BC) discovered so far are from central and southern China, especially from Jingzhou in Hubei and from the region of Changsha in Hunan. In Western Zhou times (1045–771 BC) the *jin* brocade appeared, an outstanding weave which on the one hand was produced on a rather complicated loom and on the other hand asked for expert craftsmanship. This so-called brocade was a new type of warp-faced compound weave with the warp divided into at least two series, normally of different colors. Even picks of weft interlace with the warp either in tabby or twill. Although this weaving method produced polychrome silk fabrics with a colorful shiny and mostly geometric pattern on the surface, its repertoire was still limited by the weaving technology at the time. Weavers exhausted the technical possibilities of their looms and composed scenes and figures which were evidently intended to be pictorial descriptions, but they are still symmetric with straight lines and cornered outlines. Embroidery helped to make the patterns of tigers, phoenix, dragons, birds, and blossoms appear more lively. The patterns were arranged in various ways adorning the silk robes of the fourth and third centuries BC.

In 1972 outstanding textile fabrics and garments totalling more than one hundred objects were unearthed at tomb no. 1 at Mawangdui in Changsha. The tomb belonged most probably to the Lady Dai (d. 168 BC). Among the well preserved fabrics and garments there were more than a dozen robes of various make, such as eleven floss-wadded robes, one lined robe, three unlined robes, several blouses and skirts, two pairs of socks, four pairs of shoes, three pairs of gloves, pillow covers, forty-six rolls of single-width silk fabrics, and many more items of daily life. The textiles exhibit an unrivaled excellence in weaving skills, the mastering of pattern design, and imagination in applying all sorts of patterning techniques. Favorite weaves were thin and loosely-structered gauzes (*sha*) and lozenge-patterned leno (*luo*), a fabric of open structure which is made by crossing warp yarns.

The most sophisticated weaving techniques and looms were used to produce brocade with small geometric patterns. A new technical dimension of weaving becomes obvious with the pile-loop brocade, a velvet-like fabric with geometric patterns of different sizes. Forty silk garments are embroidered with the colorful and curvilinear designs of *xin qi* (abiding faith), *changshou* (longevity), *cheng yun* (riding on the clouds), and various plants and cloud patterns. Embroidery as a patterning technique lost some of its importance when by Eastern Han times (AD 25–220) the variety of weaving patterns was finally extended to include mythological beasts, birds, fishes, flowers, all sorts of four-legged creatures, and Chinese characters.

The brocade manufactured from Han to early Tang times (seventh century) was woven with warp-faced patterns. The colors of the previously dyed warp threads mounted on the looms dominated the patterns. The patterns could be as wide as the width of the fabric (ca. 50 cm) but their length was rather limited. Probably during the eighth century of the Tang dynasty a dramatic innovation took place. The weft-faced patterning method as it had already been practiced in a few woollen textiles from Han times was now widely applied to silk weaving. The advantage of weft-patterning was that the colors of the pattern produced by the weft could easily be changed, which resulted in larger and more vivid pattern units. Furthermore the dressing of the loom was facilitated. After the Tang dynasty, the weft-faced patterning method gained predominance in brocades and in other weaves. At the same time the cultural influence of Central Asia became evident in textile patterns. Apart from several hundred fragments of silks unearthed in 1987 from the underground palace of Famen Temple near Xi'an, most textile finds from the Tang period were discovered in the dry desert region of Turfan (Xinjiang province) and in the cave temples of Dunhuang.

If satin weave (*duanwen*) is classified as an irregular twill (*xiewen*), then satin could have been created as early as Tang times. Whereas the interlacing points of warp and weft in twill weaves are arranged in a continuous oblique line, those points in satin do not form a line but are evenly distributed, thus allowing long floats of the threads which give the fabric a glittering and at the same time smooth appearance. The French name *satin* was derived from the word *zaituni* which was used by Persian merchants to name the city of Citong at the coast of Fujian, another name for the famous commercial center of Quanzhou in the Song period (AD 960–1279). Several well preserved complete sets of official robes, garments of various types, underwear, and other textile items were found in three Song tombs. The tomb of Huang Sheng (1226–1243) in Fuzhou in Fujian province, who was the daughter of an official and married to an imperial clansman, contained 354 textile items, of which 201 were articles of clothing. The textiles are of top quality. The weavers and textile printers made use of the most advanced techniques of their time in producing figured leno (*hualuo*), gauze (*sha*), crêpe (*zhou*), and figured twill silks (*ling*). Even a few satin weaves (*duan*) are described. More than thirty textile items were discovered in the tomb of the student of the Imperial College Zhou Yu (1222–1261) in Jiangsu province, and the recently excavated tomb of Mme. Zhou (1240–1274) in Jiangxi province yielded 329 items of textiles. Among the many regional brocades produced, the pure red brocade from Sichuan (*Shu jin*), with a formidable array of realistically depicted designs, was most famous.

In northern China, where the Liao dynasty reigned from AD 916 until 1125, and in the Song empire, the use of various types of tapestry (*kesi*) became highly fashionable. Among the textiles recovered from the Liao tomb of Yemaotai, dated between 959 and 986, there was a shroud made of silk tapestry in gold threads with a powerful design of dragons. The forerunners of this *kesi* tapestry technique can be traced back to the *zhicheng* technique of Han times, an intricate inlaid pattern produced by the weft yarn employing the swivel weaving method. During the Yuan and the Ming dynasties the use of various types of weaves with gold threads (*jinjin*) increased. The weavers of Ming times, especially the craftsmen of the cloud pattern brocade (*yunjin*) from Nanjing mastered a swivel weaving method (*zhuanghua*) making use of colored wefts to form a pattern on a fabric of various weaves. In many cases a glossy satin served as ground for the colorful swivel weave.

Three types of pile fabrics were produced on a large scale in Ming and Qing dynasties. The *Zhang* satin, originally from Zhangzhou in Fujian, was a figured warp pile fabric. From early Qing times until the end of the eighteenth century its main centers of production were Nanjing and Suzhou. The *Zhang rong* was a velvet where in order to produce a pattern the loops in the pattern area were cut. Thus the velvet pattern stood out on a ground of loops. The *Jian rong* was a cut pile fabric made of black silk threads produced on looms in Nanjing.

For the dragon robes (*longpao*) of the imperial Ming and Qing courts, the formal Manchu court robes (*chaofu*), and the semi-formal coats (*qifu*) all patterning techniques known at the time were employed. This applies especially to the robes manufactured during the reign of the Qianlong emperor after 1759. Many of the old weaving techniques were handed down from generation to generation and can still be found in China.

DIETER KUHN

REFERENCES

Cammann, Schuyler. *China's Dragon Robes*. New York: The Ronald Press, 1952.

Chen Weiji, ed. *Zhongguo fangzhi kexue jishu shi (gudai bufen)*. Beijing: Kexue chubanshe, 1984.

Chen Weiji, ed. *History of Textile Technology of Ancient China*. New York: Science Press, 1992.

Gao Hanyu, ed. *Zhongguo lidai zhiranxiu tulu* [Chinese Patterns through the Ages]. Hong Kong: Commercial Press, 1986.

Institute of the History of Natural Sciences, Chinese Academy of Science. *Ancient China's Technology and Science*. Beijing: Foreign Languages Press, 1983.

Kuhn, Dieter. "The Silk-Workshops of the Shang Dynasty (16th-11th Century BC)." In *Explorations in the History of Science and Technology in China*. Ed. Li Guohao et al. Shanghai: Guji chubanshe, 1982.

Kuhn, Dieter. *Die Song-Dynastie (960 bis 1279): eine neue Gesellschaft im Spiegel ihrer Kultur*. Weinheim: Acta humaniora VCH, 1987.

Kuhn, Dieter. *Spinning and Reeling*. vol.5, part 9 of *Science and Civilisation in China*.. Ed. Joseph Needham. Cambridge: Cambridge University Press, 1988.

Kuhn, Dieter. "Tombs and Textiles." *Needham Research Institute Newsletter* 5: 2–3, 1989.

Kuhn, Dieter. *Zur Entwicklung der Webstuhltechnologie im alten China*. Heidelberg: Edition Forum, 1990.

Shanghaishi fangzhi kexue yanjiuyuan, Shanghaishi sichou gongye gongsi, eds. *Changsha Mawangdui yihao Han mu chutu fangzhipin de yanjiu*. Beijing: Wenwu chubanshe, 1980.

Vollmer, John E. *In the Presence of the Dragon Throne. Qing Dynasty Costume in the Royal Ontario Museum*. Toronto: Royal Ontario Museum, 1976.

Zhongguo dabaike quanshu. Fangzhi. Beijing: Zhongguo dabaike quanshu chubanshe, 1984.

Zhongguo meishu quanji. Gongyi meishu bian 6 yinran zhixiu (shang). Beijing: Wenwu chubanshe, 1985.

Zhongguo meishu quanji. Gongyi meishu bian 7 yinran zhixiu (xia). Beijing: Wenwu chubanshe, 1986.

See also: Silk and the Loom

TEXTILES IN EGYPT Egyptian textiles during the Dynastic era, since 3100 BC, were almost exclusively of linen, although wool was not unknown. (Cotton and silk were introduced only later.) Linen was produced in three basic grades: royal linen, thin cloth, and smooth cloth. Production of royal linen, the highest grade, was a royal monopoly. Its manufacture took place both in the royal harem and in workshops associated with state temples. These workshops were supervised from the harem and were obligated to provide specific amounts of linen annually to the Pharaoh. This linen would then be used for sale or barter to support both the royal household and the temples.

The Egyptians did not ordinarily color their linen because most of the dyes known to them were not color fast. When one desired to dye a piece of cloth, it was generally necessary to treat the material first with a mordant (one of several substances, such as alum, that would both adhere to the cloth and accept dying) then with the dye. Thus, the natural color of the Egyptian linen cloth ranged from near white (from young, immature flax plants) to golden brown (from fully mature flax).

When the Egyptians harvested flax, they pulled it up by the roots, rather than cutting the stalks near the ground. The stalks were first drawn through a comb-like device in order to separate the stems into fibers. These were soaked, then beaten and sometimes scraped, so as to separate the woody parts from the long flexible fibers. A final combing prepared the fibers for being made into threads.

Once dried, the flax fibers were rolled together (usually between the palm and the left thigh) forming a loosely twisted strand which was wound into loose balls on a pottery reel and stored in clay or basketry containers until ready for spinning. The Egyptians used three different kinds of spindles in the spinning process. The supported spindle was rotated by being rolled on the thigh of the spinner. The grasped spindle was rotated between the palms of the hands. The most popular, however, was the suspended spindle. This spindle was set spinning, then allowed to drop and swing freely, its rotary motion being maintained by the weight of the whorl (the drum-shaped or domed attachment to the thin shaft of the spindle) which acted as a small fly-wheel. If a heavier, thicker thread were desired, two or more spun threads could be twined together.

When the thread was ready for weaving, a number of warp threads were attached between two beams fixed parallel to one another. In the simplest weaving, the weft thread is passed over and under alternate warp threads. On its return, the process is reversed so that the weft thread passes on the opposite side of each warp thread. A flat stick is used to beat down and compress the weft thread together. There are, of course, many variations on this basic pattern, some of them of considerable complexity, by which a variety of patterns were developed for decorative purposes.

The earliest looms were simple ground looms in which the two beams supporting the warp threads were placed on the ground and held in position by pegs. At about the beginning of the New Kingdom (ca. 1550–1085 BC) we also begin to find examples of the vertical-framed loom, in which the two beams are incorporated into a less portable rectangular wooden frame. (There is some speculation that this innovation may have been imported into Egypt at the time of the Hyksos, ca. 1700–1550 BC).

Woven cloth was cut either with a metal shears or a knife when the ancient Egyptians wished to fashion a piece of clothing. These pieces were sewn together using a fine linen thread and a needle of wood, bone, or metal (usually copper or brass, but occasionally gold or silver). The pieces were held together during sewing by pins of the same composition. Both needles and pins tended to have larger dimensions than modern examples. Pins were frequently capped by a loop head, which may have served a decorative purpose. Seams (and hems) were frequently rolled and secured by a rather crude whipping stitch. Seams which might be subjected to greater wear, as in articles of clothing that might be frequently laundered, might be joined instead by flat seams. Still other techniques could be employed for decorative effect.

Fine linen was apparently quite valuable. Considerable effort was often expended to repair damaged garments. Usually repairs were in the form of darning (re-weaving), rather than patching. The more ordinary grades of linen, however, were repaired with considerably less effort. A frayed edge might be bound carefully with whipping stitches, but more serious damage often led to the object being discarded and replaced.

Based on artistic evidence, it appears that men were most frequently involved in the cultivation and harvesting of flax and in the preparation of the fibers for use. Both men and women (and even children) participated in spinning the fibers into linen threads. But only women are shown working the ground looms, while only men are depicted working on the vertical looms. Whether this represents a true division of labor or merely an artistic convention is unclear.

GREGG DE YOUNG

REFERENCES

Forbes, R.J. *Studies in Ancient Technology*, Vol. IV. Leiden: Brill, 1964.

Hall, Rosalind. *Egyptian Textiles*. Aylesbury: Shire, 1986.

TEXTILES IN INDIA India may be described as one of the ancient centers of the cotton textile industry, since early evidence of cloth has been found in prehistoric archaeological sites. The spinning and weaving of cloth was very much a part of everyday life in ancient India. The loom is used as poetic imagery in several ancient texts. The *Atharvaveda* says that day and night spread light and darkness over the earth as the weavers throw a shuttle over the loom. The Hindu God Vishnu is called *tantuvardhan* or "weaver" because he is said to have woven the rays of the sun into a garment for himself.

It is interesting to note that in the third or second century BC, when the cotton industry in India was in a flourishing state, in Europe cotton was still virtually unknown. The Greek scholar Herodotus thought that cotton was a kind of animal hair like sheep's hair. At the beginning of the Christian era, Indian textiles figure prominently in the trade with Rome. Arrian, the Roman historian, testifies to the export of dyed cloth from Masulia (Masulipatnam on the Coromandel coast), Poduca (Pondicherry), Argaru (Uraiyūr in Tanjavūr district, Tamil Nāḍu) and other places in south India. Legend has it that Indian cloth was purchased in Rome for its weight in gold. The quality of Indian dyeing too was proverbial in the ancient world, and in St. Jerome's bible, Job says that wisdom is more enduring than the dyed colors of India. Indian textiles even passed into Roman vocabulary as is seen by the use as early as 200 BC of a Latin word for cotton, Carbasina, derived from the Sanskrit *kārpasa*.

The history of Indian textiles constitutes one of the most fascinating and at the same time tragic chapters in Indian history. In the sixteenth century the Portuguese first set foot on Indian shores and were followed in quick succession by the Dutch, English, and French. For the next hundred years "Indian cotton was king" and Europe was in the grip of what economic historians describe as "the calico craze". Indian textiles were used in the Middle East, Africa, and Europe not merely as dress material but also as coverlets, bed spreads, and wall hangings. The joint English sovereigns William and Mary are described as having landed in England in 1689, resplendent in Indian calico. Daniel Defoe, the author of *Robinson Crusoe*, commented that Indian calico, which at one time was thought fit to be used only as doormats, was now being used to adorn royalty.

However, there was a dramatic reversal of fortune in the eighteenth century. The cotton revolution in England rendered redundant the products of Indian handlooms. The first ban on Indian textiles was imposed by the British crown in 1700 and repeatedly after that. By the end of the century, instead of Indian cloth being exported abroad, the Indian market was flooded by the machine-made cloth of Manchester and Lancashire. Around the same period, India was hit by one of the worst famines beginning in the late seventeenth century and continuing through the eighteenth century. The words attributed to Lord Bentinck, the Governor of India in the 1830s, that "the bones of the weavers are bleaching the plains of India" are a dramatic but apt description of the fate of the Indian weavers.

The eclipse suffered by the Indian textile industry lasted until the early twentieth century until its grand revival under Gandhi, who initiated the *khādi* movement. The *charkhā*, or Indian spinning wheel, and *khādi*, or homespun cloth, became symbolic of the Indian struggle for independence. Foreign cloth was burnt in the public squares and the Indian spinning wheel became a part of the home of every Indian patriot.

Since Independence, a sea change has occurred in the traditional Indian textile industry. The changeover to power looms and jet looms and the introduction of computer designs is setting new traditions in Indian textiles. In the course of its historical vicissitudes, the Indian textile industry has gone through a process of change as well as cultural assimilation.

Indian Textile Technology

The first process in the weaving of a cloth is warping and sizing, and in India this is done in the open. Bamboo sticks, about one hundred and twenty in number, are fixed upright in the street or what is called the warping grove, at a distance of a cubit from one another. Rows of women walk up and down the line, each carrying a wooden spindle in the left hand and a bamboo wand in the right. As they walk, they intertwine the threads between the split bamboos. These threads are then stretched horizontally from tree to tree, evenly washed with rice starch and carefully brushed. The right amount of tension in the warp in required to prevent the yarn from breaking while on the loom.

In India spinning was and still is almost exclusively the occupation of women. More specifically, this was the sole occupation of destitute women and widows. It is interesting that this corresponds to the English notion of the 'spinster' as one who has to spin for her livelihood since she has no one to support her.

The earliest looms in use in India were either the pit loom or the vertical loom. The *Atharvaveda*, probably compiled in the early pre-Christian era, says, "A man weaves it, ties it up; a man hath borne it upon the firmament. These pegs propped up the sky; the chants, they made shuttles for weaving . . . (sic)." However, the most common type of loom in use was the horizontal pit loom in which the loom is placed inside an earthen pit and is operated with foot treadles. By depressing the pedal with one foot and raising the other, one set of threads get depressed and the weaving shed is formed through which the throw shuttle is shot across by hand. References to such looms are scattered throughout ancient and medieval inscriptions. Around the fifteenth century one begins to get reference to the draw loom. This would consist of several levers and so enable the weaving of complex patterns. The introduction of the fly shuttle in the 1930s towards the end of British rule in India resulted in the partial mechanization of loom technology and in another three decades this was followed by the introduction of the jacquard. Nowadays, partially mechanized looms, power looms, and jet looms are displacing the traditional Indian handloom.

Traditional Indian Costumes

Different types of cloth are worn and woven in the different parts of India, since this is a vast land with varying climates. Generally, men tend to wear a longish lower cloth of about one and a half yards in length, called *dhōti* or *lungi* in the north and *veshti* in south India, while the traditional upper cloth consists of a single piece of cloth called *aṅgavastra*. However, in hot weather, men generally go without the upper cloth. In many parts of northern India, men also wear a head gear against the dust and heat. This is especially true of desert regions like Rajasthān. The Indian women wear large skirts or loose trousers called *salwār* and longish or short jackets. Alternately, they wear a six-yard piece called a *sāri* and a blouse for the upper part. In the colder parts of India, such as the Himalayan mountain ranges and Kushmir, the garments are thicker and more elaborate, including warm woolen shawls and heavy jackets. It is noteworthy that in antiquity, stitched garments such as shirts, trousers and blouses were hardly ever worn in India. In the ancient sculptures and paintings such as the ones at Amarāvati or Brahadīśvaram, it is only the menials, palace attendants, common soldiers and dancing girls, all of them belonging to the lower echelons of society, who are depicted wearing stitched garments. Such garments are never depicted on the upper classes or royalty nor on the images of gods and goddesses. A plausible reason may be the association of impurity and pollution with stitched cloth.

Colors and Designs in Textiles

Traditional Indian textiles reflect the Indian ethos. There is an aura of religion and romance around Indian weaving. Everything is significant — the colors chosen, the motifs, and the wearing occasion. Crimson or shades of red are very auspicious and worn by women on the occasion of their marriage as well as by ceremonial priests in certain parts of India, such as the Madhvā Brāhmins of Karnātaka. White represents purity and ochre, renunciation, and these are the colors worn by Hindu widows as well as ascetics. Yellow and green denote fertility and prosperity and are worn in the spring. Black is considered inauspicious, although pregnant women in south India wear black, perhaps to ward off the evil eye. As late as the eighteenth century, coloring was done entirely through vegetable dyes such as madder and indigo, although now dyers have almost entirely switched over to chemical dyes except in the case of highly specialized textiles like the *kalamkāris*.

The earliest designs on Indian textiles seem to have been geometrical. A twelfth century Sanskrit text called the *Mānasōllāsa* described textiles designed with dots, circles, squares and triangles. The depiction of flora and fauna was related to religion and popular beliefs. The lotus, which has great spiritual significance in Hinduism, and the mango design are among the most popular Indian motifs. Swan, peacock, parrot and elephant are also commonly depicted. The tree of life, which symbolizes fertility and prosperity, is another auspicious motif. All these designs are patterned on the loom itself and it may take a handloom weaver working on an ordinary frame loom as long as thirty days to weave an elaborate six-yard *sāri* with designs and gold lace. As the weaver weaves, he also sings the special loom songs, a tradition which has now almost entirely died out except perhaps in some interior weaving villages in Uttar Pradesh or the remote south. These loom songs tell of the glory of particular weaving castes or they are full of esoteric religious metaphors describing god as the eternal weaver, weaving the web of life, and the human body as the cloth he has woven.

Textile Varieties

Traditional Indian textiles are unique and unparalleled for their beauty. The *jāmdāni* is an elaborate textile which is woven with multiple shuttles and resembles tapestry work. Floral motifs called *bootis* in gold or silver lace are scattered over the body with heavy gold lace on the borders. The most striking of these designs is the *pannā hazāra*, literally a thousand emeralds, in which the flowers shimmer and gleam all over the sari. The Benarsi sāris called *Kiṁkhābs* woven in Uttar Pradesh are legendary for their loveliness, although Benaras in the north and Kāñchipuram in the south were traditionally associated with pure cotton rather than silk. It was the British who introduced sericulture in Kanchipuram in the nineteenth century. Gadhwāl and Venkaṭagiri saris of Andhra and the Īrkal saris of Karnataka specialize in rich gold borders and heavy panel-like *pallūs* (that portion of the sāri which is draped over the shoulder). Another variety is the tie and dye (called variously *bandini*, *ikāt*, or *chungdi*) produced in Rajasthān, Orissa, Andhra Pradesh, and Madurai where the fiber is tie-dyed before weaving. A unique Andhra textile is the *telia*, which was soaked in oil before weaving and catered exclusively to the West Asian market because it was woven to suit desert conditions. This textile appeared in the sixteenth century with Muslim rule and died out with the collapse of the Islamic empires. Another textile which became popular in the mughal period was the *mashroo* (also the *himroo*) in which cotton was used in the warp and silk in the weft. Initially these were used as Islamic prayer mats by the Mughal nobility who were forbidden by

Islamic tenets to use any animal product. They therefore contrived the mashroo which enabled them to have their comfort without violating the religious tenet against the use of pure silk.

Textiles also form an important part of temple ceremonials such as the flag cloth hoisted in temples, the garments put on the deity, the cloth covering the chariots in which the deities are taken out on a procession and the ritual dance costumes. The *kalamkāri* cloth of Andhra, in which mythological stories are sketched minutely on cloth with a fine pen as well as the *Nādhadwāra pichwāis* of Rajasthan, are of this genre. In India it is also the practice among wandering groups of minstrels to render dramatic narrations of mythological stories, and the elaborately painted screens used on these occasions form an important aspect of traditional Indian textiles.

VIJAYA RAMASWAMY

REFERENCES

Desai, Clena. *Ikat Textiles of India*. Bombay: Marg Publications, 1965.

Irwin, John. "A Bibliography of the Indian Textile Industry." *Journal of Indian Textile History*, part II, 1956.

Irwin, John and B. Catherine Brett. *Origins of Chintz*. London: Her Majesty's Stationery Office, 1970.

Ramaswamy, Vijaya. *Textiles and Weavers in Medieval South India*. New Delhi: Oxford University Press, 1985.

See also: Colonialism and Science

TEXTILES IN MESOAMERICA Because textiles are rare among the artifacts of cultures known to us only archaeologically, any analysis of them, or of the technology of their production, must be pieced together from indirect sources. In Mesoamerica, while archaeological finds attest to the existence of various textile technologies at specific dates, much of our understanding of the subject comes from ancient Maya, Mixtec, and Aztec books, stone sculpture, painted pottery, murals, clay figurines, European documents from the time of the Conquest, and modern textile traditions.

Though the physical environment of Mesoamerica generally precludes the survival of perishable artifacts, textiles have none the less survived from certain areas. The majority, mostly small fragments, have been found in dry caves in the arid regions of Mexico. In the humid southern lowlands very little has survived, some 2500 carbonized fragments dredged from protecting mud at the bottom of the Sacred Well at Chichen Itza constituting the most important single find.

The oldest textiles to survive in Mesoamerica are cordage, netting, and basketry, worked in vegetal fibers other than cotton, with early examples dating to at least 7000 BC in Oaxaca and to 5000 BC in central and northern Mexico. Such textiles are fashioned from the leaves and stems of various plants worked without the benefit of a loom. In instances where fibers were extracted from their plant sources, they were probably spun by rolling them together between the hand and thigh, as no tools for spinning have been found. Spindle whorls and evidence of loom woven textiles appear much later.

The earliest evidence for loom weaving in Mesoamerica consists of fabric-impressed ceramics datable to 1500 BC, at which time spindle whorls also begin to appear. Woven cotton fragments follow soon after. These early woven textiles are worked in plain weave and almost certainly were created on a backstrap loom. From these simple beginnings, a sophisticated textile industry developed.

The weaving process begins with the selection and preparation of suitable materials. The predominant fiber of woven Mesoamerican textiles was cotton, *Gossypium* spp., with both annual and perennial varieties reported at the time of the Conquest. Cotton was cultivated in at least two colors, white and brown, and traded throughout ancient Mesoamerica. Other fibers, generically termed istle (from the Nahuatl *ixtli*), were drawn from the leaves and stems of many plants of more local distribution, including *Agave* spp. and *Apocynum* spp. Both cotton and istle required much preparation prior to spinning and weaving. Cotton was carefully picked over, its numerous small seeds and other vegetal debris removed by hand and the mass of fiber then fluffed or beaten to produce a uniform, smooth mass. Istle, on the other hand, was toasted, split, and scraped to remove the plant flesh from fibers which were then washed, dried, and combed. No animal fibers are known to have been used in ancient Mesoamerica. Textile headdresses incorporating human hair, however, have been found.

Both cotton and istle were spun on the spindle, a simple device consisting of a "whorl" wedged onto the end of a straight tapered shaft. A mass of fiber was attached to the shaft, which was then twirled, imparting twist. The whorl helped the device to spin easily and for an extended time, and provided weight against which the fiber could be drawn out and twisted into thread. Different fibers required different types of spindles and sizes of whorls. Cotton, because of its very short fibers, was spun on a small spindle, the lower end of which was supported in a bowl or on the ground. The longer istle fibers required a larger, heavier spindle. Yarns of any size could be created and some as fine as 0.005mm in diameter have been recovered. Yarns could be used singly, or two or more individually spun strands could be retwisted

Figure 1: An Aztec woman instructs her seven-year-old daughter in the art of spinning cotton on a supported spindle. The child's ration of one-and-a-half tortillas is shown between them. From *Mendoza Codex*, page 59. [From a photo of the codex.]

Figure 2: A Maya almanac illustrating the goddess *Sak Na* preparing her warp. From *Madrid Codex*, page 102. [Drawing after Villacorta.]

("plied") together. A mixture of cornmeal and water was likely applied to some yarn to smooth and strengthen it for weaving.

While there is little archaeological evidence for the use of dyes, Post-Classic pictorial codices and reports of the *conquistadores* both attest to the vivid colors of Mesoamerican textiles. Plant dyes probably included indigo (*Indigofera anil*) for blue, brazilwood (*Caesalpinia* spp.) for red, logwood (*Haematoxylon campechianum*) for black or blue, annatto (*Bixa orellana*) for orange, and many other leaves, seeds, roots, barks, and fruits. Two animal dyes were important: on the Pacific coast, purple was extracted from small molluscs of the genera *Thais*, *Murex*, and especially *Purpura*, while in arid regions, cochineal (*Coccus cacti*, a scale insect of the order Hemiptera) was cultivated on the prickly-pear cacti (*Opuntia* spp.). The beautiful red produced by cochineal was much admired by the Spanish, and the dyestuff became a major export to Europe following the Conquest. Mineral pigments, such as ochre, iron pyrite, cinnabar, carbon, copper sulfate, and "Maya blue" clay were known, and there is some evidence for the use of mordants which make dyes more permanent. Dyes and pigments were applied by painting, by stamping or rolling with figured clay implements, and by immersion dyeing. Patterns were produced during dyeing through the application of resists and through tying-off of sections to create undyed areas of yarn or fabric. Fabric incorporating tie-dyed warp, called *jaspe*, is still produced in Guatemala today.

Once suitable yarns were spun and dyed, the process of setting up the loom began. The *Florentine Codex*, a documentary of Aztec life and custom written by the Spanish friar Sahagún just after the Conquest, illustrates the weaver's equipment, including spindle, warpboard, loom sticks, backstrap, and batten. It also describes the training of weavers, from the presentation of the newborn girl with miniature weaving equipment, to the day of a woman's death when her loom and spindle were burned in the funeral pyre to await her in the afterworld. In the *Mendoza Codex*, created by Aztec scribes at the order of the first Viceroy of New Spain, there is a section devoted to a mother's training of her daughter in domestic chores. It portrays the child's instruction in spinning (Figure 1) and weaving, from ages 3 through 14, her food rations, and punishments for unacceptable work.

Information about weaving is also contained in Maya books, whose almanacs provide endlessly repeating prognostications for the timing of certain quotidian activities. One almanac in the *Madrid Codex* depicts the process of preparing the warp for the loom (Figure 2). The weaving goddess, named *Sak Na* in the accompanying hieroglyphic text, is shown seated cross-legged with her left hand against a horizontal frame supported by at least two vertical posts and with lines stretched between their projecting upper parts. Her raised right hand holds an inverted spindle from which thread reaches to the frame before her. While there have been varying interpretations of the activity portrayed in this illustration, the overhead texts clearly read: *Sinah u chuch Sak Na*, or "She strings her warpboard, *Sak Na*", describing in ancient Mayan the process depicted. This manner of preparing the warp yarn, by measuring and stretching it between vertical posts on a warping frame, was in use at the time of the Conquest and survives to this day. It is the initial phase of preparing a backstrap loom (Figure 3) for weaving.

As the warp yarns are measured, they are wound alternately to one side and the other of a pair of vertical posts, thus creating a lease which maintains their order and allows

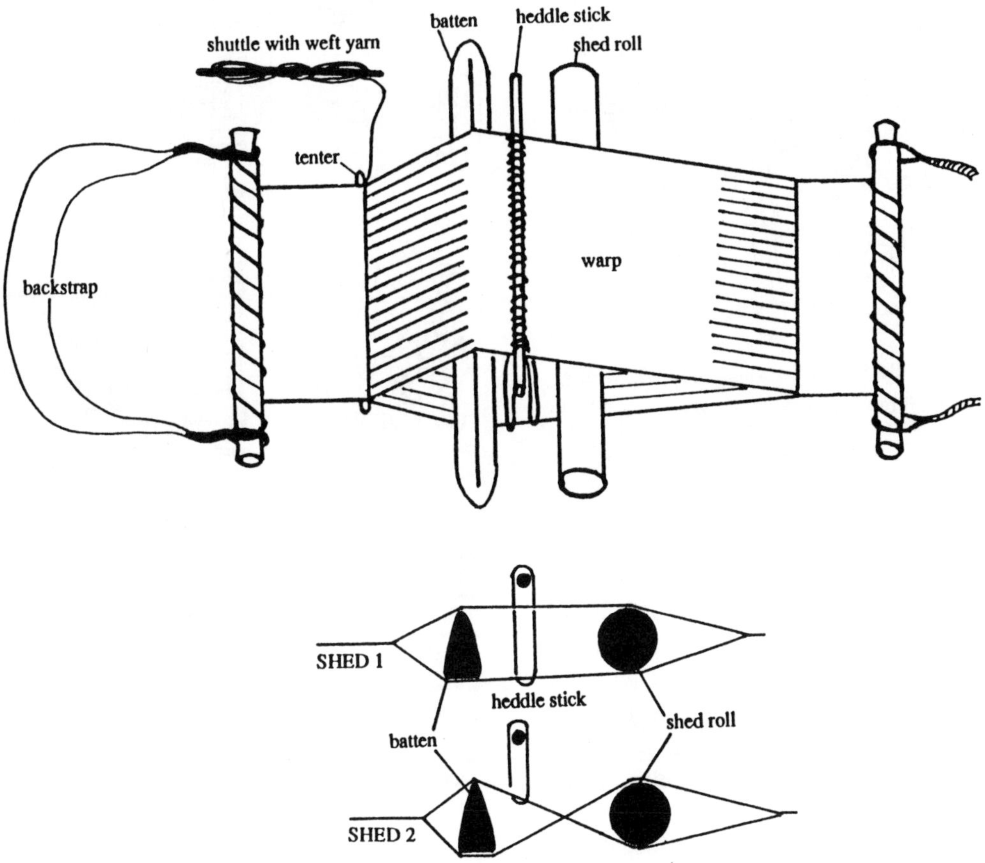

Figure 3: The structure of a backstrap loom, showing component parts and how they are used to create two sheds for weaving. [Drawing by Carolyn Jones.]

the easy selection of alternate threads. Once wound, the opposite ends of the warp are secured to bars which form the ends of the loom, and a large, smooth, rounded stick (the shed roll) is placed through one side of the lease. The shed roll allows for the lifting of those threads which travel over it, one half of the total warp threads in alternating order, creating the first opening or shed. The remaining threads, which travel under the shed roll, are individually secured to a second stick (the heddle). By pushing the shed roll back and lifting the heddle, the other half of the warp threads are raised, creating the second shed. This is the mechanism by which Mesoamerican weavers created plain woven fabric.

Before the loom is ready for weaving, additional implements are required. A long, heavy, straight-sided wooden stick (the batten) which can be turned on its side to enlarge the shed through which yarn (the weft, carried on a shuttle) will pass, and which is used to beat that weft into place, is employed continuously during weaving. A thin stick (the tenter) is usually attached to the fore-edge of the weaving to regulate the fabric width. Finally, the far end of the loom is tied to a tree or post, while the close end is attached to a strap which travels around the weaver's hips as she sits on the ground. By adjusting the position of her body, the weaver controls the tension of the warp of her loom.

This action is illustrated in the *Florentine, Mendoza, Matritense, Dresden*, and *Madrid Codices*, and in clay figurines from Maya burials on the island of Jaina. The *Madrid Codex* contains an almanac comprised of two episodes (Figure 4). The first is illustrated with a crude drawing of a skirted female on her knees, her left hand supporting a back-strap loom, her raised right hand inserting the batten. Over the scene is an explanatory text that reads: *Och-i ti te' Ch'ul Na Che'el*, or "Divine *Na Che'el* (another name for the Maya weaving goddess) weaves at the post."

Simple in construction, the backstrap loom was well suited to the Mesoamerican woman's environment. Spinning and weaving were but two of her many daily chores, and the easy portability of the backstrap loom allowed its be-

Figure 4: A Maya almanac illustrating the goddess *Na Che'el* weaving at her backstrap loom. From *Madrid Codex*, page 102. [Drawing after Villacorta.]

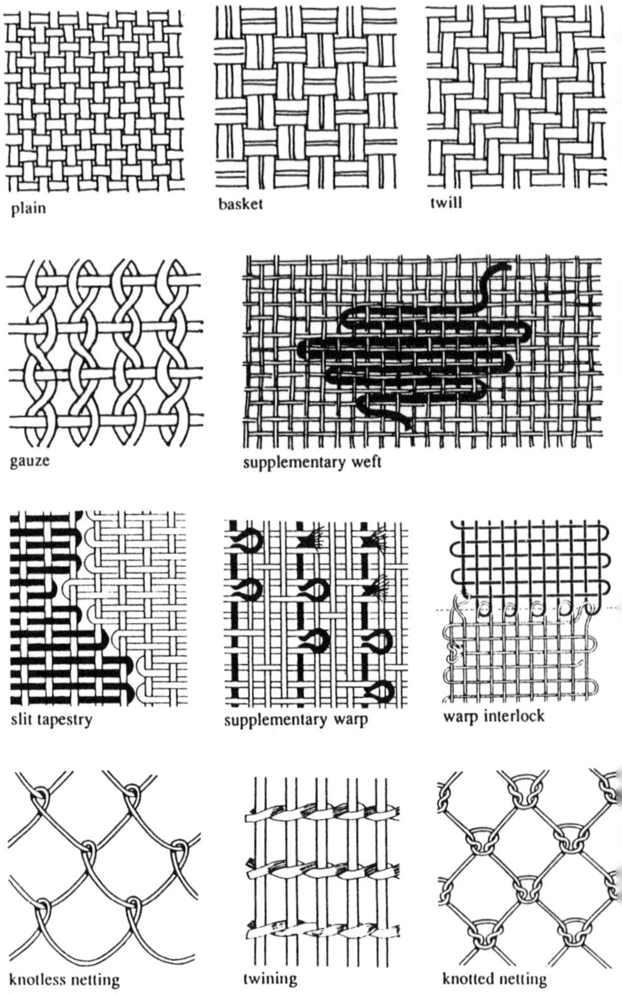

Figure 5: Some structures known from ancient Mesoamerican textiles. The bottom row provides non-woven techniques for comparison with the loom-woven structures shown above. [Drawing by Carolyn Jones.]

ing rolled up for safekeeping when not in use, or set up when and where convenient. The simplicity of the equipment, however, did not preclude complex weaving. Extant fabric fragments display a wide variety of techniques. Among the simple weaves are plain, semi-basket, basket, twill, and gauze. More complex weaves include supplementary weft brocade, tapestry, inlay, supplementary warp patterning and pile, warp interlock, and layer exchange double cloth (Figure 5).

The structure of the backstrap loom determined several qualities of the fabric to be woven. The size of the finished cloth was determined when the loom was set up, and was limited in both width (by the weaver's armspan and her need to position herself at the center of the loom for proper tensioning), and length (the longer the warp, the more cumbersome and difficult it was to weave). When a large fabric was required, it was woven in small rectangles and pieced together, usually without cutting. Mesoamerican fabrics could include four finished edges and tended to show more warp than weft, with warp counts as high as 78 threads to the inch recorded.

In addition to weaving, ancient Mesoamericans were familiar with twining, braiding, plaiting, knotted and knotless netting, sewing, and embroidery. Methods for making leather, felt, bark cloth, and paper were known. At the time of the conquest, new materials and techniques were introduced, including needle knitting and the use of sheep wool and silk. In some areas, the treadle loom came into use, and men entered the field of textile production. These different technologies survive side-by-side with backstrap weaving today.

The importance of textiles, and particularly of weaving, to the ancient cultures of Mesoamerica must not be underestimated. Weaving was not simply a means of producing necessities of daily life, but was an expression of the Mesoamerican world view. The act of weaving was seen as a basic creative force, analogous to the original creation of the world in Mesoamerican myth. The symbolic significance of the act of weaving lives on in the complex patterns of the modern Maya woman's *huipil* (or traditional poncho-like garment), each of which is a cosmogram that places the weaver at the center of the universe. Many designs used today are notably similar to symbols depicted in carved stone representations of textiles from the ninth century, thus displaying remarkable continuity with a lengthy and rich Mesoamerican textile tradition.

CAROLYN JONES
TOM JONES

REFERENCES

Johnson, Irmgard Weitlander. "Basketry and Textiles." In *The Handbook of Middle American Indians* Volume 10: *Archaeology of Northern Mesoamerica* Part 1. Ed. Gordon F. Ekholm and Ignacio Bernal. Austin: University of Texas Press, 1971, pp. 297–321.

King, Mary Elizabeth. "The Prehistoric Textile Industry of Mesoamerica." In *The Junius B. Bird Pre-Columbian Textile Conference*. Ed. Ann Pollard Rowe, Elizabeth P. Benson, and Anne-Louise Schaffer. Washington D.C.: The Textile Museum and Dumbarton Oaks, 1979, pp. 265–78.

Mahler, Joyce. "Garments and Textiles of the Maya Lowlands." In *The Handbook of Middle American Indians,* Volume 3: *Archaeology of Southern Mesoamerica* Part 2. Ed. Gordon R. Willey. Austin: University of Texas Press, 1965, pp. 581–593.

O'Neale, Lila M. *Textiles of Highland Guatemala*. Washington D.C.: Carnegie Institution (Publication 567). [Reprinted by Johnson Reprint in 1976.]

Randall, Jeanne L. and Edwin M. Shook. *Bibliography of Mayan Textiles*. Guatemala: Museo Ixchel, 1993.

Schevill, Margot Blum, J. Catherine Berlo, and E. B. Dwyer, eds. *Textile Traditions in Mesoamerica and the Andes*. New York: Garland Publishing, 1991.

TEXTILES IN SOUTH AMERICA Unlike most parts of the world where discoveries of ancient textiles are unusual, the Pacific desert coast and dry western Andean slopes have preserved enormous quantities of textiles, wood, feathers, plant material, and other usually perishable artifacts. Most of the best preserved textiles were originally part of burial furnishings left in ancient cemeteries stretching from central Peru to northern Chile. The far north Andean coast (modern Colombia, Ecuador, and northern Peru) encounters periodic torrential showers which have destroyed most ancient remains. Few textiles have survived from the highland wet and dry climate or the rainforest regions of the eastern Andean slopes stretching to the Atlantic ocean.

It is apparent that desert conditions and the careful preparation of tombs are the two elements most responsible for the preservation of ancient Andean textiles. Beginning around 3200 BC and continuing until the conquest of the Andes in AD 1532, it is possible to reconstruct textile technology in coastal regions and occasionally to witness highland development through textiles preserved in coastal sites.

Archaeologists have designated the Cotton Preceramic (3200–1800 BC) as a period when a variety of coastal cultures developed cotton twining, looping, and other non-loomed textiles before the use of the heddle loom or ceramics. Major discoveries at Huaca Prieta, La Galgada (a highland site with remarkable preservation), and in Asia have determined that not only were large quantities of cotton textiles used, but that some were elaborately patterned. Images on Preceramic cotton textiles include condors with outstretched wings, two-headed snakes, and humans or deities with splayed feet, all subjects with similar presentation to images which continue in the art of many subsequent Andean cultures. Designs are created through transposing twined elements, often combining a change in regular yarn movement with alternating colored yarns. Cotton dyes include indigo for blue and unidentified, red, yellow, and brown colors. Thicker plant fibers were twisted into sturdy twined mats, while cotton was used for looped caps and twined mantles.

Beginning in the Initial Period (1800–800 BC) and the following Early Horizon (800–0 BC), Peruvian textiles developed through the use of the heddle loom. Hundreds of painted Early Horizon cotton plain-weave textiles discovered near Karwa on the Peruvian south coast identified fabrics with designs very similar to stone carvings from the north highland pilgrimage site of Chavin de Huantar. This important and well-preserved painted ritual cloth provided evidence of the ways in which textiles were used to transport religious imagery throughout the Andes beginning in very early periods.

Camelid fibers, sometimes termed "wool" yarn, from the hair of the non-domesticated guanaco and vicuna or the domesticated llama and alpaca, were introduced into coastal weaving during the Early Horizon. Camelid hair is more easily dyed than cotton and its introduction into coastal technology is usually based in its application in brilliant colors for weft-patterned structures. The first all-wool tapestry textiles appear in coastal cemeteries during this period and it is likely that these identify an ancient highland wool technology rarely preserved in the highland regions of natural camelid habitat.

The famous Paracas textiles were woven and embroidered on the southern Peruvian coast during the Early Horizon. On the Paracas Peninsula, elite individuals were buried with hundreds of embroidered shirts, skirts, mantles, feather fans, and golden objects all wrapped inside enormous plain woven cotton winding cloth. The multi-colored wool embroideries depicted repeating human, deity, and animal images and were executed in stem stitch on plain-woven fabric. Paracas textiles also included a unique type of three-dimensional embroidery using the cross-knit loop stitch to embelish borders with polychrome images. The technique was created with a single cactus spine needle as a continuous form of crossed looping.

The Nazca culture, which followed the Paracas on the south coast in the following Early Intermediate Period (AD 0–500), continued one of the most elaborate weaving technologies ever known, with textiles in double and triple cloth, warp and weft patterning, oblique interlacing, and fine garments woven with both discontinuous warp and weft. Andean loomed textiles are characterized in their tradition of four finished selvedges. Individual finished cloth webs were sewn together to complete a garment and were rarely cut.

Very little is known of north coast weaving during the Early Intermediate Period, the time of Moche cultural development. The few surviving Moche textiles identify a technology principally based in the use of cotton and a narrow backstrap loom like that employed on the south coast. But structurally, Moche textiles are distinguished through the use of fine un-plied cotton yarns in twilled and weft-brocaded structures. Moche tapestry is woven with both cotton and dyed camelid fiber weft over a cotton warp using a non-interlocking weft which creates vertical slits between different color areas. Apparently slit-tapestry was common to coastal cultures both north and south.

Although known for only a handful of large weft-interlocked tapestries or tapestry fragments and triple-cloth narrow bands, the Early Intermediate Recuay culture of the Peruvian north highlands developed a distinctive textile tradition. Recuay tapestries were woven on a wide loom of more than seven feet with a short warp of no more than two feet. Brilliantly dyed red and yellow wool yarns are characteristic of Recuay textiles, and tapestries use the highland weft-interlocked structure which leaves no openings between areas of different color. In the south central Andes, weft-interlocked tapestries uncovered on the Chilean north coast have been attributed to the Alto Ramirez culture of the southern highlands. Alto Ramirez textiles exhibit a decided preference for the use of blue and red dyed yarns almost certainly identifying an ancient source and knowledge of indigo dyeing in the southern highlands.

The following Middle Horizon (AD 500–1000) marks a break with previous cultural development. The Peruvian Huari culture located near the modern city of Ayacucho appears to have controlled the central Andean highlands and coast, while the south central Andes was allied to the site of Tiwanaku. Tapestry was the distinctive Middle Horizon medium woven in slit techniques with cotton on the coast and with interlocking camelid fiber wefts in the highlands. Local, coastal tapestry continued to be woven on the narrow backstrap loom. Highland Tiwanaku and Huari tapestries preserved in coastal desert burials identify the use of the wide loom with narrow warp like that used in Recuay tapestry. Huari shirts were woven in two parts and seamed down the center leaving an opening for the head and neck. The few Tiwanaku shirts discovered in northern Chilean desert cemeteries were patterned with similar images but were woven in a single panel with a neck opening woven through discontinuous wefts. Headgear was always used as an important Andean badge of identity. Huari and Tiwanaku officials wore a knotted hat with four peaks or points on the top. Tiwanaku fourpointed or cornered hats created polychrome designs in lark's head knots while similar Huari hats were patterned with knots with tufts of wool pile in each knot.

By AD 1000 the highland centers of Huari and Tiwanaku had collapsed and individual cultures began clearly to establish local identities noted in regional textile styles. Warp-patterned structures such as complementary and supplementary-warp weaves were commonly woven in the following Late Intermediate Period (AD 1000–1450). Textiles are often characterized as having repetitive, small-scale imagery in gauze, painted cotton, weft-brocades, and double-woven fabrics. South coast weavers shaped bags and shirts through the selective addition of warp yarns during the construction process. North coast weavers often wove exotic bird feathers into the cloth, creating shirts and other garments with one face completely covered with feather patterns. During this period coastal weaving especially exploited the full potential of natural native colored cottons which were woven in contrasts of white and natural red-brown, beige, and grey.

The highland Aymara weaving tradition was consolidated in the south central Andes during the Late Intermediate Period after the fall of Tiwanaku. Aymara textiles are characterized by elegantly striped warp-faced shirts, woman's dresses, head cloths, and mantles in natural colored camelid fiber yarns often dyed blue, green, and red.

The Late Horizon (AD 1450–1534) again marks the period when local cultures were brought under highland control, this time that of the monolithic Inca state with its capital at Cuzco in the southern Andes. Inca textile patterns are strictly geometric and non-figurative and the most valued cloth was weft-interlocked tapestry woven on the wide tapestry loom with a short warp. Some of the best preserved Inca textiles have been discovered as miniature offerings covering gold and silver male and female figurines and left on the tops of Andean mountain peaks from Ecuador to Argentina. Male garments include a tapestry tunic, a mantle, a bag with a carrying strap, and a large feather headdress. Female garments include a large wrap-around dress and a narrow, highly patterned belt. Most of these garments are woven entirely in camelid fiber and are colored with red and yellow dyes.

Following the conquest of the Inca state by European conquerors, European methods of textile manufacture were introduced throughout the Andes. The spinning wheel was adopted for workshop production, and the treadle loom was constructed for the manufacture of yardage to be sewn into non-Andean style tailored clothing. Needle knitting was introduced and is now regularly used for sweaters and knitted caps in local communities. Felting was introduced for wide-brimmed hats which have now become part of indigenous community dress in many regions.

While foreign clothing styles and techniques replaced local garment manufacture in coastal regions, European methods never replaced highland traditions in many indigenous communities in Colombia, Ecuador, Peru, and Bolivia. Today, some areas have maintained camelid-fiber spinning practices using the drop spindle and the backstrap loom or the staked ground loom to produce native four-selvedge garments. Men are the principal weavers in indigenous communities in Colombia and Ecuador, and women weave in Peru and Bolivia. In the southern Andes men weave on the European treadle loom. All community members spin, but spinning is generally considered women's work. Many communities continue to express local identities through handwoven patterns and specific color combinations which are worn daily or for community rituals. Traditional four-selvedged handwoven garments are also worn with European-style vests, pants, sweaters, or skirts in many areas. Some communities have specialized in the decoration of textiles with sewing machine embroidery.

Outside of the Andean region, a few Brazilian tribes and groups living in the Amazonian areas of Ecuador, Peru, and Bolivia, have continued lowland traditions using barkcloth, elaborate feather headdresses, and oblique-interlaced bags and narrow bands, textile traditions which may reflect ancient lowland origins never preserved in these wet regions.

South America continues as the native home to herds of guanaco, vicuna, llamas, and alpacas with an export industry in camelid fiber and manufactured textiles. Peruvian cotton is valued for its luster and is enjoying a revival in interest in native natural-colored cotton yarns.

AMY OAKLAND RODMAN

REFERENCES

Cason, Marjorie and Adele Cahlander. *The Art of Bolivian Highland Weaving*. New York: Watson-Guptill, 1976.

d'Harcourt, Raoul. *Textiles of Ancient Peru and Their Techniques*. Seattle: University of Washington Press, 1962.

Emery, Irene. *The Primary Structures of Fabric*. Washington, D.C.: The Textile Museum, 1966.

Murra, John. "Cloth and its Function in the Inca State." *American Anthropologist* 64: 710–728, 1962.

Ravines, Rogger H. *Technologia Andina*. Lima: Instituto de Estudios Peruanos, 1978.

Rowe, Ann P. *Warp-patterned Weaves of the Andes*. Washington, D.C.: The Textile Museum, 1977.

Rowe, Ann P. *The Junius B. Bird Conference on Andean Textiles*. Washington, D.C.: The Textile Museum, 1986.

Rowe, Ann P., Elizabeth Benson, and Anne-Louise Schaffer. *The Junius B. Bird Pre-Columbian Textile Conference*. Washington, D.C.: The Textile Museum and Dumbarton Oaks, 1979.

THĀBIT IBN QURRA Abū'l-Ḥasan Thābit ibn Qurra ibn Marwān al-Ḥarrānī al-Ṣābiʾ (836–901), was a Syrian mathematician, astronomer, physicist, physician, geographer, philosopher, historian, and translator from Greek and Syriac into Arabic. His scientific treatises were written primarily in Arabic and partly in Syriac. He was born in Kafartūtha near Ḥarrān (now Altinbaşak in Southern Turkey) and was a student in Ḥarrān. Ḥarrānians, the descendants of the ancient nation Mitanni in the Hellenistic age, were hellenized, and their ancient religion of star-worship was deeply connected to Greek philosophy. In the Arab caliphate Ḥarrānians called themselves Ṣābians since the Ṣābian religion was one permitted by the *Qurʾān*. Ḥarrān University was founded in the fifth century in Alexandria as a school of philosophy and medicine. After the Arab conquest it was moved to Antiochia and later to Ḥarrān where, under the influence of Ḥarrānian traditions, astronomy and mathematics were taught, and it became a university.

At first Thābit ibn Qurra worked in Kafartūtha as a money-changer. Here the Baghdad mathematician Muḥammad ibn Mūsā ibn Shākir met him and invited him to Baghdad, where Muḥammad and his brothers Aḥmad and al-Ḥasan, the Banū Mūsā, became his teachers. Later he worked at the court of the caliphs in Baghdad and in Surra man ra'a (Samarra) as a physician and astronomer. His position as caliph's physician allowed him to keep his heathen religion. His son Sinān ibn Thābit and grandson Ibrāhīm ibn Sinān also were mathematicians, astronomers, and physicians in Baghdad.

Thābit ibn Qurra's contributions to science covered many different disciplines, from mathematics to philosophy. In mathematics, he was a translator or editor of translations of many works of Euclid, Archimedes, Apollonius, Theodosius, and Menelaus. Many of these are extant only in these translations. These translations, together with the geometric treatise of Thābit's teachers, the brothers Banū Mūsa, and his *Kitāb al-mafrūḍāt* (Book of Assumptions) constituted the so called "middle books" which were studied between Euclid's *Elements* and Ptolemy's *Almagest*.

Two of Thābit's treatises on parallel lines were first written in Syriac, the first under the title *Ktovo al-hay da-tren surte trishe kad mettapkin al bshir men tarten gonowoto dagʿin bahdode* (Book [in which is proved] that Two Lines Produced Under Angles Which are Less Than Two Right Angles Will Meet). The second is called "the second book on the same topic". Both these treatises are extant only in the Arabic translations made by Thābit himself. The ideas of these treatises were further developed by Ibn al-Haytham (965–ca. 1050), ʿUmar al-Khayyām (1048–1131) and Naṣīr al-Dīn al-Ṭūsī (1201–1274) and later led to the discovery of non-Euclidean geometry.

Thābit's *Kitāb fī taʾlīf al-nusub* (Book on Composition of Ratios) was devoted to the theory of compound ratios. This theory later led to the notion of real numbers and to the discovery of differential calculus.

Other work covered such subjects as a simple proof of the Menelaus theorem (the first theorem of spherical trigonometry), mensuration of plane and solid figures, and solutions of different problems of integral calculus. His books contained some proofs of the Pythagorean theorem and its generalization, and dealt with the subject of amicable numbers, in which each number is equal to the sum of the divisors of the other.

In the field of astronomy, Thābit was the editor of the translation of Ptolemy's *Almagest* and the author of many treatises on the movement of the sun and moon, sundials, visibility of the new moon, and celestial spheres. In his treatise "On the Motion of the Eighth Sphere", extant only in Latin translation (*De motu octave sphere*), he added the ninth to Ptolemy's eight spheres and proposed the theory of "trepidation" to explain the precession of equinoxes. The fragments of Thābit's Syriac *Ktovo d'pulog d'yumoto d'shoboʿo al koukbe shabʿe* (Book on the Subdivision of Seven Days of the Week according to Seven Planets) are extant in Bar Hebraeus' *Chronography*. In this book the planets are designated by their Babylonian and Greek names; this subdivision was known to Romans and Indians and is the source of the names of the days of the week in many European and Asian languages.

Thābit also wrote books on mechanics and physics. His *Kitāb al-qarasṭūn* (Book on Lever Balance) discusses the conditions for equilibrium of different kinds of levers. In his *Kitāb fī masāʾil al-mushawwiqa* (Book on Interesting Questions) Thābit tries to explain the phenomenon of the camera obscura. This attempt was erroneous, but it led Ibn al-Haytham in the *Book on Forms of Eclipses* and al-Bīrūnī (973–1048) in *The Exhaustive Treatise on Shadows* to the solution of this problem. He also studied the problems of acoustics in *Mas'ala fī'l-mūsīqā* (Question on Music) which is an extant fragment of his great *Book on Music*.

In geography and medicine, Thābit revised works of Ptolemy (*Kitāb ṣūra al-arḍ*, Book of the Picture of Earth) and Galen. He was the author of the fundamental *Kitāb al-dhakhīra fī ʿilm al-ṭibb* (Book of Treasure in the Science of Medicine).

In philosophy, Thābit emphasized that integer numbers were abstractions of objects of counting and criticized Aristotle who rejected actual infinity. In his commentaries to Aristotle's *Metaphysics* he considered the problem of the "first motor" and argued with Aristotle's opinion that the essence is immobile. Many of Thābit's treatises are devoted to problems of religion, in which he is critical of both Christianity and Islam.

BORIS ROSENFELD

REFERENCES

Carmody, Francis J. *The Astronomical Works of Thābit ibn Qurra*. Berkeley and Los Angeles: University of California Press, 1960.

Matvievskaya, Galina P. and Boris A. Rosenfeld. *Mathematicians and Astronomers of Medieval Islam and their Works (8–17th centuries)*, vol. 2. Moscow: Nauka, 1983, pp. 85–103 (in Russian).

Rosenfeld, Boris, and Ashot Grigorian. "Thābit ibn Qurra." In *Dictionary of Scientific Biography*, vol. 13. Ed. Charles Gillispie. New York: Scribners, 1976, pp. 288–295.

Thābit ibn Qurra. *Kitāb fī'l-aʿdād al-mutaḥābba*. Amman, 1977.

Thābit ibn Qurra. *Mathematical Treatises*. Ed. B.A. Rosenfeld. Moscow: Nauka, 1984 (in Russian).

Thābit ibn Qurra. *Oeuvres d'Astronomie*. Texte établi et traduit par R. Morelon. Paris: Les Belles Lettres, 1987.

See also: Banū Mūsā – Sinān ibn Thābit – Ibrāhīm ibn Sinān – *Elements* – *Almagest* – Ibn al-Haytham – al-Bīrūnī – ʿUmar al-Khayyām – Naṣīr al-Dīn al-Ṭūsī

TIME The dual classification of cultural systems of time as Western or non-Western is an oversimplification that limits our understanding of how time is created, represented, measured, and practiced in different cultures. The premises underlying dualistic thinking about time reveal its limitations.

First, dualistic thinking adopts the premise that Western time is linear (irreversible), abstract, quantitative, and homogeneous. In contrast, non-Western time is cyclical (reversible), concrete, qualitative, and heterogeneous. There is a tendency to draw too sharp distinctions based on characteristics presumed to be in opposition. Underlying these are further distinctions between *them* (oriental, primitive, oral, preindustrial) and *us* (occidental, modern, literate, industrial). Embedded in these implicit distinctions is the assumption that non-Western time is to be thought about within a conceptual frame that implicitly adopts western time as a standard for perception and evaluation of non-Western time. In a word, dualistic thinking about time has been Eurocentric.

Second, dualistic thinking rests on the premise that Western and non-Western cultures are different and unrelated totalities, and that time is unitary within each. The unitarian premise leads to generalizations such as believing that non-western cultures have a cyclical as opposed to a linear concept of time. The result is that diversity is underreported and undertheorized.

Third, dual systems of culture and time suffer from an ahistorical and overly formal foundation to knowledge. Western societies have 'history' and a consciousness of being in time, in contrast to non-Western societies which have 'myth' and a sense only of relational time that renders events mere epiphenomena of structure. Myth is not about the causation of before and after but about a formal means for making all the elements and characters ever-present.

Fourth, the ahistoricist premise also leads to the view that non-western societies have an awareness of themselves in time only to the extent that westerners bring such awareness to them, either in the form of historical consciousness or as a practical matter of the efficient and productive use of time as a resource.

Fifth, dualistic thinking rests on the premise that time is a thing that is either present or absent in a culture. A dichotomy between western *linear* time and non-Western *cyclical* time appears as real and coerces people to behave and think in a particular way. But the terms of this dualism are highly abstract and refer to no particular period, bounded space, social organization, or real people. The dichotomy bears little relation to the cultural construction of time in the everyday lives of people in a specific region, a particular period, or known conditions of migration, diffusion, and contact.

The situation is complex. While no cultural constructions of time have remained unaffected by European expansion, it is presumptuous to think that we know what aspects of European time (which varied in different periods and places) influenced a bewildering variety of religious communities, empires, and kin-ordered societies, each with its own historically constituted temporal rhythms. Indeed, examining the real complexities of culture and history raises important issues about temporal dynamics only to the extent that we abandon dualistic thinking.

An alternative perspective, one that anthropologists, historians, and archaeologists increasingly embrace, pays attention to multiple constructions of time in single social formations, the extent to which different times are articulated, and the temporal dynamics of development and change in the rhythms of everyday life.

While some languages lack a generic word for time, there is abundant evidence for the universality of concepts of duration and succession in the linguistic and cultural practices of every people. The French sociologist Emile Durkheim expressed the opinion that it is hardly possible to think about time "... without the processes by which we divide it, measure it or express it with objective signs". These processes, he concluded, are social in origin. The anthropologist Edmund Leach, following Durkheim's lead, noted that "We talk of measuring time, as if it were a concrete thing waiting to be measured; but in fact we create time by creating intervals in social life. Until we have done this there is no time to be measured." To these ideas about time we need to add those of another French sociologist, Georges Gurvitch, who pointed out that every social formation has a multiplicity of times. Not only are there different collective representations of time among different cultures, but different time systems co-exist within a single social formation.

These ideas, that time is culturally constructed, socially embedded, and multiple, have become guiding principles in cultural studies of time. The other articles on time in this encyclopaedia illustrate the complexity and diversity of time systems from the multiplicity of time perspective. Brief comment on a number of issues will clarify the present direction in cultural studies of time systems.

1. Acceptance of the multiplicity of time within single social formations has led to the rejection of attempts to classify whole societies, not to mention a whole class of societies such as 'Western' or 'non-Western' by any single subsystem of time. Comparison remains important, but comparativists now recognize that they must seek out similar subsystems in different cultures. Leopold Howe, for example, disputes Clifford Geertz's conclusion that the Balinese calendrical system does not represent a flow by which the passage of time can be measured, that the calendar represents a concept of time that is non-durational, and that Balinese experience of time is of "islands" or points that are unconnected (an extreme form of discontinuity, imprecision, and non-countability). Geertz asserts that the particulate nature of days implies a non-durational concept of time. Such a description of Balinese time carries an implicit comparison with our own, based on dualistic assumptions that go unexamined in both. Howe provides evidence that Balinese count days and are forever referring to past and future in terms of such counts. Geertz confused an attribute of members of a class with the class itself—each day is qualitatively different from others, but days fit into a succession that is endlessly repeated. Each name day partakes of being a day that *qua* day has the attribute of countability. The Maya had no such confusion. They viewed time as a flow that could be measured and also endowed days with qualities dependent upon the work of gods and men.

2. Recognition of the multiplicity of time shifts attention away from gross comparisons of presumed-to-be-different systems of time toward contextualized descriptions and similarities of subsystems in apparently different cultures. There

has been a concomitant shift from classification to process, raising questions about temporal dynamics. How *does* time get constructed? What *are* the processes of development and change that account for the multiplicity of time and the degree of articulation among subsystems? Cultural studies of time have taken several directions, none of which is mutually exclusive. Nancy Munn challenges, from a phenomenological perspective, Evans-Pritchard's conclusion that Nuer time amounts to a creation of a static and non-developmental experience of time. His structural models of Nuer time overlook Nuer accounts of their bodily activity of procreation, their own subjective experience of time in terms of layers of ash on the fires, their ability to plan for events in the future, and other evidence of developmental constructs of time in their projects and practices. Peter Rigby, who has researched the age-set system of the pastoral Ilparakuyo Maasai in Kenya, challenges, from a historical materialist perspective, Evans-Pritchard's failure to address the degree of articulation of Nuer multiplicity of time. Had Evans-Pritchard taken into account the historical consciousness of Nuer, he would have found that all the elements of time-reckoning can be placed in a single conceptual frame arising from the historical development of Nuer within a specific social formation. Rigby, drawing on his own work and that of others, attempts such a synthesis for the Ilaparakuyo Maasai.

3. Cultural studies of time that explore the underlying temporal dynamics of how a multiplicity of time is constructed have paid increasing attention to the politics of time. Cultural constructions of time are part of the political economy of development. The creation of time requires agents and agencies temporalizing their respective projects in struggles for power and legitimacy. Farriss' attempt to reconcile cyclical and linear concepts within the same calendrical system in Maya culture rests on an interpretation of Mayan political transformation from city-states to rule by dynastic lineages. With the demise of Maya elites, subsystems of time wither and die.

4. The attention paid to the politics of time has raised anew issues about the reflexive study of time in different cultures. Johannes Fabian has criticized anthropologists for failing to recognize that theories which distance 'other' cultures by placing them in an-other time unwittingly use time as an instrument of political domination. Apparently disinterested studies of difference are revealed to be interested studies of 'the other', a form of cultural politics. The solution is to recognize that people of different cultures are contemporaries and use that as a point of departure for all conceptual frames concerning the politics of time. A reflexive approach to cultural construction of time would overcome Rigby's concern that most descriptions of time in 'different' cultures are really implicit comparisons using hegemonic bourgeois time as a standard. Most of us reckon time in our daily lives by systems other than the formal-mathematical or chronological ones deployed by an army of scientists, historians, and anthropologists. Such reflexivity, or self-awareness, of our own multiple concepts of time would be salutary for advancing our understanding of the cultural construction of time in different cultures.

HENRY J. RUTZ

REFERENCES

Aveni, Anthony. *Empires of Time: Calendars, Clocks, and Cultures*. New York: Basic Books, 1989

Evans-Pritchard, E.E. *The Nuer*. New York: Oxford University Press, 1940, reprinted 1970.

Fabian, Johannes. *Time and the Other*. New York: Columbia University Press, 1983.

Gell, Alfred. *The Anthropology of Time: Cultural Constructions of Temporal Maps and Images*. Providence, Rhode Island: Berg, 1992.

Howe, Leopold E.A. "The Social Determination of Knowledge: Maurice Bloch and Balinese Time." *Man* (N.S.) 16: 220–234, 1981.

Geertz, Clifford. "Person, Time and Conduct in Bali." In *The Interpretation of Cultures*. New York: Basic Books, 1973, pp. 360–411.

Munn, Nancy. "The Cultural Anthropology of Time: A Critical Essay." *Annual Review of Anthropology* 21: 93–123, 1992.

Rutz, Henry J., ed. *The Politics of Time*. Washington, D.C.: American Anthropological Association, 1992.

Rutz, Henry J. "Meaning of Time in Primitive Societies." In *Encyclopedia of Time*. Ed. Samuel L. Macey. New York: Garland, 1994, pp. 371–372.

Rutz, Henry J. "Primitive Time-Reckoning." In *Encyclopedia of Time*. Ed. Samuel L. Macey. New York: Garland, 1994, pp. 496–497.

See also: Calendars

TIME IN AFRICA In every society, one source by which time is divided, measured, and expressed by objective signs is the mutual obligations and principles that structure social relations between persons and groups. Among the most important are kinship. In those societies in which kin communities encompass the widest range of social relations, time is expressed as duration and succession of kin-ordered activity. Perhaps the best known case is that of the Nuer, a Nilotic pastoral people who live in villages of the upper reaches of the Nile in Sudan.

E.E. Evans-Pritchard has given us a rich account of ways in which village kin communities conceptualize and reckon time as an aspect of social structure. Among the multiplicity of times apprehended by Nuer, three of the most important subsystems are (1) a series of social activities articulated with a series of ecological events by which one is used to measure the duration and succession of the other; (2) a series of age-sets that measures intervals and succession in the life cycle; and (3) a hierarchy of segmented kin groups that measures lapsed time, i.e. a form of collective memory of the "past" that constitutes the conceptual frame of Nuer history.

Nuer livelihood is based primarily on cattle herding and the cultivation of millet, a dual adaptation to vast stretches of marsh and savannah in a climate that necessitates a transhumant existence between village settlements on knolls during the rainy season and dry season cattle camps near rivers. A change of season is the objective sign for a change in settlement pattern, accompanied by changes in ceremonial and social activities. Most collective rituals, such as marriage and male initiation, take place in wet season villages, but the dry season cattle camps, when youth disperse to small camps in search of pasturage near sources of water, are viewed as a time of freedom and courting. In a rare comment on the experience of seasonal rhythms, Evans-Pritchard states that the pace of time speeded up during the wet season and slowed down during the dry season, an effect of the density of social activities and the significance the Nuer attributed to them.

Although the Nuer have no word for a generic "year", they reckon cycles of wet and dry seasons, giving names to each, and divide the cycle into twelve named "moons" of unequal duration that mark the succession from wet to dry season and back again. The duration of months depends less upon lunar observance (though they are familiar with phases of the moon) than the agricultural and pastoral activities associated with ecological phases within seasons. Nuer adjust the series of ecological activities to fit social contingencies. If the Nuer were still in their dry camps, waiting for rain, then they would say that they were still in the named "moon" prior to the onset of rains. If they were still engaged in rituals associated with the wet season villages, then they would reckon the name of the moon by the social activities appropriate to it. The result is an annual calendar embedded in a cordinated parallel series of ecological and social rhythms, the durations of which vary within and between cycles. Working one series against the other, Nuer rename months, skip months altogether, or even omit "years" in their recounting of socially significant events over a number of seasonal cycles.

Nuer can count seasonal cycles, months, and days (there is no "week"), but they tend not to. Chronology or a sense of lapsed time with respect to livelihood is culturally suppressed, giving the illusion of endless recurrence. Days, months, and seasons remain socially embedded and unarticulated as an abstract and numerical system independent of the concrete activities that lend the system meaning. The Nuer have no generic word for time.

Biological facts of individual maturation and aging, like ecological and physical facts, are given social significance in every society. But their division, measurement, and expression in objective signs are contingent and vary from one society to another.

Among Nuer, the reproduction of cattle, or cattle genealogies, is the basis of a concept of male personhood. Individual Nuer men are fiercely egalitarian and quick to defend themselves against perceived affronts to their manhood. The natural reproduction of cattle, combined with a culturally constituted series in the life cycle of male individuals — for example the transition from boyhood to manhood marked by the gift of a first cow, the transition to husband marked by a gift of cattle at marriage (a transfer from the groom's to the wife's kin), and a further transfer of cattle marking the transition to 'fatherhood' with the birth of each successive child — regulate the rhythms of interpersonal kinship, articulating the pedigree of specific cows with the genealogy and life history of individuals.

Cattle genealogies mark the passage of time in the lives of individual male Nuer, a process that in its very nature is directional and irreversible. But Nuer also have organized individual male maturation as a collective problem of social significance. Village boys of comparable age (between 14–16 years) are initiated into a named age-set during an open period. A "Man of the Cattle" is responsible for closing a period, "thereby dividing the sets". The initiation consists of a painful ritual whereby each boy is cut to the bone across the forehead six times, producing deep scars called *gar*. His reward is a gift of his first cow and a taboo on milking. The only point of articulation between the seasonal and age-set time systems is the timing of the initiation, which occurs at the end of the wet season when food is plentiful. Proximate villages tend to coordinate periods despite independent declarations in each. Regions tend to share at least some age-set names (but vary the sequence), which differ from one part of Nuerland to another.

In pastoral Kenya and parts of Bantu East Africa, the age-set/generation-set system consists of age-sets organized into three distinct generation-sets of boys, warriors, and elders, with a pronounced taboo on warriors marrying. It is hard to avoid the conclusion that Nuer age-sets have something to do with the collective control of elders over young men. The military and administrative functions of Nuer age-sets, however, are attenuated and appear to have more to do with regulating the domestic obligations of manhood.

Nuer conceive of the age-set system as fixed at six living sets. Using a kinship idiom, they break the sets down into "brothers" who are equals, "fathers" who are elders in authority, and "sons" over whom they have authority. But age-set names are not repeated and Nuer do not perceive the age-set system as cyclical (some Bantu groups do). Age-sets whose members have died are not remembered or recounted. Even when a few old men of the most senior age-sets are still alive, junior age-sets are merging them with members of adjacent age-sets. The system is designed for the living, intended to mark lapsed time by the intervals between open and closed periods that define the age of one group of males relative to another in a weak form of gerontocracy. Like individual biography inscribed in cattle gifts and their genealogies, a sense of an irreversible flow of time is given in the fact that a named age-set "changes its position in relation to the whole system, passing through points of relative juniority and seniority".

All human communities develop collective representations of a past that embody a historical consciousness of the community *qua* community. Nuer have embedded their sense of historical consciousness in a model of a lineage system that creates the illusion of a fixed temporal horizon to events such as migration, tribal war, and territorial expansion that might otherwise be perceived as a succession of unique events in an irreversible "before and after" construction of past and future. The correct view of Nuer history is that it is ever-present and performative in the reciprocal rituals of lineage segments.

The segmentary principle of membership in a Nuer lineage is common descent through males from a recognized ancestor. The way Nuer think about tribal history is structured by how they think about the hierarchical relations between lineages, which also form the basis for claims to a division of territory. Reckoning political relations by lineages is relative because it depends on the particular person who is selected as the point of departure in tracing descent. The largest lineages, those which incorporate the widest latitude of members connected by descent from a common ancestor, are termed clans, and they in turn are comprised of lineages with several lines, each of which is a smaller lineage tracing descent through less distant ancestors to incorporate a narrower range of members.

The lineage system is employed by living Nuer to reckon their common past. Although every man is a potential founding ancestor, the social fiction is that the living are organized into segments in such a way that there are only ten to twelve ancestors at the maximum social distance, and three to four at the minimum. This fixed system of reckoning is in contrast to the real events of Nuer history, which is one of human propagation inscribed in Nuer genealogical reckoning, expansion and contraction of territory, and the actual bifurcation and amalgamation of lineage segments. But the Nuer fit such events into a structure of lineages with a time depth of ten to twelve generations when there is no reason to presume that they have existed for so short a time. The Nuer express their political history as a fixed structure of relations between lineage segments. They prefer a stylization of history to a record of the endless march of unique events in an irreversible flow of time. The absence of writing and the strength of oral culture no doubt have something to do with it. The content of relations changes and is irreversible, for example who the ancestors are or where a particular named unit fits into the hierarchy, but the structure remains constant. Eternal returns, with their prophecies of foretelling and fulfilment, or dynastic genealogies that record the unfolding succession of heroic deeds from some fixed starting point into a receding time horizon, are not the stuff of Nuer social history.

The Nuer are a tribal state without centralized authority. Their kin-ordering and genealogical reckoning are the key to their construction of time, which contrasts greatly with that of empires. In those states where imperial elites (military and religious) ruled over vast areas and innumerable village communities from imperial centers, there developed an awareness of time itself as an objective instrument of power and legitimation.

HENRY J. RUTZ

REFERENCES

Evans-Pritchard, E.E. *The Nuer*. New York: Oxford University Press, 1940, reprinted 1970.

Rigby, Peter. "Time and Historical Consciousness: The Case of Ilparakuyo Maasai." *Comparative Studies in Society and History* 25 (3): 428–456, 1983.

TIME IN CHINA Chinese methods of calculating time are of great antiquity. According to the *Shi Ji* (Book of Records), as early as 2254 BC. Emperor Yao employed astronomers to calculate solstices and equinoxes and predict seasonal change so that farmers would know when to plant crops. Oracle bones dating to ca. 1200–1181 BC attest to the fact that Shang Dynasty Chinese calculated time using a sixty-day divinatory calendar that still is in widespread use. The early development of methods of measuring time was not entirely endogenous to China; cultures throughout the ancient world exchanged astronomical ideas and data.

Striking similarities exist between calendrical systems in widely-separated regions, including parallels between the form and names of the Chinese and Maya divinatory calendars. At least as early as the first century BC the Chinese used a luni-solar calendar resembling the standardized Babylonian calendar developed in the fourth century BC. The similarities suggest borrowing, and it is possible that the Babylonians were the original inspiration for the Chinese soli-lunar calendar. In the Sui and Tang dynasties (589–960), Indian astronomers resided in China, and during the Yuan dynasty (1280–1368) Chinese collaborated with Persian and Arab astronomers. The technology of observation and measurement included an observatory built in Beijing during the Yuan dynasty. Contrasting with the astronomical system developed in Greek and medieval European astronomy, Chinese astronomy was polar and equatorial rather than ecliptic.

Although at times the Chinese borrowed astronomical ideas and technologies, these became incorporated into a system of time reckoning rooted in Chinese society. Sociologists have observed that time is a symbolic structure that represents a society's collective rhythms. In the Chinese calendar, some elements are based on astronomical cycles, while others have been shaped by the temporal rhythms of social life.

Chinese notions of the seasonal cycle are integrated into a system of classification based on the dualism of *yang* and *yin*. While *yang* and *yin* are often defined in terms of the complementary dualism of male and female, in Chinese they originally referred to sun and shadow, the very elements used to measure the changing seasons. For Chinese metaphysicians, *yang* and *yin* came to denote primal cosmic forces that interacted to generate a cycle of five phases (*wuxing*) which were identified with five primary elements (wood, fire, metal, water, and earth). These phases and their associated elements were in turn identified with the seasons. Thus Chinese classified spring with wood, summer with fire, autumn with metal, and winter with water, while earth was associated with the midpoint of the year.

This model of cyclical process associated the five phases with five colors (green, red, yellow, white, and black), directions (east, south, the center, west, and north), organs of the body, tastes, planets, virtues, passions, etc. Chinese elaborated rituals designed to control this cyclical process. Throughout the yearly cycle, the Emperor performed rites to inaugurate the seasons, and his performance gave concrete expression to the association of season, direction, color, and ritual. Chinese thought that by means of these ritual performances, the emperor could ensure harmony in the universe, and interpreted natural calamities as evidence of his loss of the "mandate of heaven" (Granet, 1934). The cosmological framework of *yin* and *yang* and the five phases also informed Daoist alchemy, and still is fundamental to the Daoist rituals of popular religious culture.

Chinese philosophers and scientists appreciated the importance of natural cycles, and also linked the cycles of time with the cycles of human life. Not only did they link cyclic ideas of time to the liturgical calendar that regulated rites performed in the imperial court and its temples, but they also linked it to the microcosm of the body. In contrast with the taxonomic thinking that informs much Western science, Chinese explored patterns of function *through* time to make sense of various experiences, including ones related to health and illness. The associations of the five phases applied not only to the yearly cycle, but also to the cycle of the month and the day. For example, Chinese medical practitioners thought that the "seasons of the day" could predict crisis periods for physical disorders affecting different parts of the body.

While cyclical time had key significance, Chinese also had a well-developed concept of continuous time. This concept found expression in "continuity history-writing", in which historians sought to chronicle causal sequences of events in history. While official historians recorded the objective facts of history, they also critiqued the past in a form of "praise and blame" historiography that sought to discern the logic of a *Dao* (path or way) that had its origins in Heaven. Thus the rulers of China employed history, like divination, as a means to gain insight into the logic of events, and as a guide to appropriate action.

In an agricultural society, charting seasonal change was crucial, and one responsibility of the emperor was promulgation of the luni-solar calendar. Since promulgation of a calendar was a political responsibility and privilege, astronomy was an orthodox Confucian science. Every year a Board of Mathematicians led by an Imperial Astronomer (who was a Minister of State) prepared the calendar. After the Emperor approved the new almanac, high-ranking officials received it with great ceremony. The Emperor, who had exclusive rights to promulgate the calendar, also bestowed the almanac upon China's vassal states as a mark of favor. During the Qing Dynasty (1644–1911) over two million almanacs were officially printed each year, and excess copies were available for sale. Despite this imperial monopoly, pirated editions circulated widely. Rebellious feudal lords sometimes expressed their withdrawal of allegiance by issuing a new calendar.

The Chinese used the calendar as an instrument to predict not only the best time for planting, but also the best time to initiate action. As a consequence, astronomy was interwoven with astrology and divination. Details of the zodiac are still published in the almanac, and the Chinese may consult this source to predict dates that are propitious or unpropitious for human undertakings such as weddings.

The most ancient method of time-keeping is the sixty day divinatory calendar, evidence for which appears on oracle-bones in the late Shang dynasty (ca. 2300–1181 BC). The cycle is composed of two interlocking sets of cyclical characters that are combined to form sixty unique two-syllable names. The "Ten Heavenly Stems" (*tian gan*) revolve in concert with the "Twelve Earthly Branches" (*di zhi*), and their combination generated sixty cyclical names. These sixty combinations were arranged in six 10-day groupings called *xun*. Since the first century BC, years have been named after the consecutive days of the divinatory calendar, producing a repeating 60 year cycle. In the late Shang dynasty, the *xun* of the dyadic divinatory cycle was the basis for ordering sacrifices to royal ancestors. A full ritual cycle was completed in 360 days, increased to 370 by the addition of an intercalary *xun* every second cycle. This arrangement created a calendar that approximated the length of a solar year.

The Twelve Earthly Branches correspond to the twelve signs of the Chinese Solar Zodiac, which also name twelve constellations used to calculate the position of the sun every month. The zodiac animals (six wild or mythical and six domestic) are the Rat, Ox, Tiger, Rabbit, Dragon, Snake, Horse, Sheep, Monkey, Cock, Dog, and Pig. They are divided between passive *yin* and active *yang* symbols, and name the years in a repeating twelve year cycle. Chinese popular wisdom has it that persons who are born in a particular year have personalities shaped by the associated zodiac animal. Individuals should avoid marriage with persons born in the zodiac years of antagonistic animals, and also seek to avoid having children during such years.

The task facing early astronomers who sought to invent a calendar for use in predicting the seasons involved reconciling lunar and solar cycles. The regular cycle of the moon provided a readily observed celestial clock, but the lunar year does not correspond with the seasons, since twelve lunar months produces a cycle of only 354 days. The yearly cycle of the sun produces a cycle of $365\frac{1}{4}$ days, but it does not predict the full moon. The 365 day Gregorian calendar follows the solar year alone, with an extra day added every leap year to adjust for the fraction. The Chinese, however, sought to reconcile solar and lunar calendars.

By the eighth century BC, Chinese could calculate and record eclipses, and they knew of the Metonic Cycle. This was a nineteen year cycle at whose beginning and end the sun and moon are in the same relative position to each other. Discovery of this cycle led to a system of coordination of lunar and solar cycles that resembles the Babylonian standardized calendar, and some authors have suggested that this system was perhaps borrowed rather than invented (for details of debates see Needham 1959:171–177).

During nineteen solar years, there were 235 lunar months, and Chinese astronomers divided these into twelve years that were twelve lunar months long, and seven years that were thirteen lunar months long. The thirteen month years were produced by duplicating one month. The inventor of this calendar introduced that intercalary month in such a way as to ensure that the winter solstice always occurred in the eleventh lunar month, the summer solstice in the fifth, the spring equinox in the second, and the autumn equinox in the eighth. In addition, the introduction of an intercalary month ensured that the Chinese New Year always fell on the second new moon after the winter solstice, between January 20th and February 19th.

Before the adoption of the luni-solar calendar, Chinese dynasties inaugurated their reign by adopting a different month as the beginning of the civil year. The Manchus are said to have followed this tradition when Emperor Kang Xi in 1669 adopted the revised calendar proposed to him by the Jesuit Father Verbiest. The Republican Government expressed its break with imperial rule by adopting the Gregorian calendar, which was favored because of its universality. Its adoption also eliminated a powerful source of imperial authority, since monopolistic control of the calendar contributed to a mystification of Manchu power. None the less, the traditional luni-solar almanac still regulates the cycle of community festivals and the rituals of ancestor worship.

JEAN DEBERNARDI

REFERENCES

Bredon, Juliet and Igor Mitrophanow. *The Moon Year*. Kelly and Walsh, 1927. Reprinted Hong Kong: Oxford University Press, 1982.

DeBernardi, Jean. "Space and Time in Chinese Religious Culture." *History of Religions* 31(3): 247–268, 1992.

Farquhar, Judith. "Time and Text: Approaching Chinese Medical Practice through Analysis of a Published Case." In *Paths to Asian Medical Knowledge*. Ed. C.A. Leslie and A. Young. Berkeley: University of California Press, 1992.

Granet, Marcel. *La Pensée Chinoise*. Paris: La Renaissance du Livre, 1934. Reprinted Paris: Éditions Albin Michel, 1968.

Needham, Joseph. "Time and Eastern Man, The Henry Myers Lecture, 1964". In *The Great Titration: Science and Society in East and West*. Ed. Joseph Needham. London: Allen and Unwin, 1969.

Needham, Joseph with the collaboration of Wang Ling. *Science and Civilisation in China, Vol. 2. History of Scientific Thought.* Cambridge: Cambridge University Press, 1956.

Needham, Joseph with the collaboration of Wang Ling. *Science and Civilisation in China,* Vol. 3: *Mathematics and the Sciences of the Heavens and the Earth.* Cambridge: Cambridge University Press, 1959.

Schipper, Kristopher. *The Daoist Body.* Trans. K. C. Duval. Berkeley: University of California Press, 1993.

Sivin, Nathan. "The Theoretical Background of Elixir Alchemy." In *Science and Civilisation in China,* vol. 5, pt. 4. Ed. Joseph Needham et al. Cambridge: Cambridge University Press, 1980, pp. 210–305.

Sivin, Nathan. "On the Limits of Empirical Knowledge in the Traditional Chinese Sciences." In *Time, Science, and Society in China and the West: The Study of Time V*. Ed. J. T. Fraser, N. Lawrence, and F. C. Haber. Amherst: The University of Massachusetts Press, 1986, pp. 151–169.

Sivin, Nathan. "Chinese Conceptions of Time." *Earlham Review* 1:82–92.

Smith, Richard J. *Fortune Tellers and Philosophers: Divination in Traditional Chinese Society*. Boulder: Westview Press, 1991.

Whittaker, Gordon. *Calendar and Script in Protohistorical China and Mesoamerica, A Comparative Study of Day Names and Their Signs*. Bonn: Holos Verlag, 1990.

See also: *Yinyang* – Five Phases – Geomancy – Astrology – Lunar Mansions – Calendars – Stars in Chinese Science – Zodiac – Divination in China

TIME IN INDIA The ancient Indians acknowledged their gratitude to Nature with elaborate rituals centered around Fire (*Agni*). The positions of fiery objects in the sky which were the prime agents of Nature decided when the Gods had to be propitiated. Even today, the temporal coordinates of important rituals like the sacred thread ceremony, marriages, and anniversaries are scheduled and announced to the gods in astronomical terms and not by the Gregorian calendar. The householder (*grihasta*) had to establish the sacrificial fire (*Agnyadhāna*) on the first day of the waning moon, or perhaps on the full moon. But dissenting opinions always existed. The *Satapatha Brāhmaṇa* (Commentary of the Hundred Ways), for example, urged the householder to install the sacrificial fire when he so desired.

The *Satapatha Brāhmaṇa* is the oldest available prose record of these ancient religious practices. It is next only to the *Ṛigveda* and its accompanying commentary (Brāhmaṇa), the *Aitareya*. These texts record the development of five centuries of fire rituals. The *Ṛigveda* has been variously dated from 4500 to 2500 BC. According to the tradition of the age, the teacher (*guru*) conveyed knowledge orally to the student (*shishya*) who learned by rote. All astronomical works evolved over the centuries without the limitations of a script and adapting to contemporary conditions. Terminology also changed so much that the original Vedic texts and commentaries became obscure.

The *Ṛigveda* refers to a *Yuga* of four years (*samvatsara*) as did the Aztec and Egyptian calendars. Intercalary months were inserted in the middle and at the end of the *Yuga* to catch up with the seasons. Dirghatma, the son of Mamata, the first known astronomer of the Vedic era, recorded the movements of the sun, moon and planets over forty years and found that the lunar year (*Chandravarsha*) of twelve months (*Chandramāsa*) with thirty days in each leaves a gap of 21 days in the four year *Yuga*. He therefore fixed the *Yuga* to be a unit of four times 365 days with the insertion of intercalary days.

Linking seasonal periodicity with the lunar cycle caused trouble for all ancient astronomers. The Vedic sages laid down four kinds of years: the lunar year of 354 days, the tropical year of 365.25 days, the civil (*sāvaṇa*) year of 360 days, and the sidereal year of 366 days. The lunar year was allowed to retrograde through the seasons to come into coincidence at the end of thirty sidereal years; however some schools preferred to insert intercalary days for this purpose. The Egyptians also divided their year into 36 decans each of ten days, and it is now believed that these systems derived from an early Aryan civilization.

Around 1200 BC, the *Yuga* cycle of five sidereal years of 366 days seems to have become generally accepted. The new calendar was documented around 900 BC by Lagadha who lived near Srinagar in Kashmir. His *Vedāṅga Jyotiṡa* is a compendium of astronomical rules that have been extant since perhaps 1200 BC. As was the practice then, the work is written in Sanskrit verse to facilitate commitment to memory. The *Vedānga Jyotiṡa* names the five years of the *Yuga* as *Samvatsara*, *Parivatsara*, *Idavatsara*, *Anuvatsara*, and *Vatsara*. The *Yuga* began at the white half of the month of January (*Magha*) and ended with the dark half of December (*Pausa*). The year was bisected according to the northward (*uttarāyaṇa*) and southward progessions (*dakshiṇāyaṇa*) of the sun. Each *āyana* then comprised 183 civil days (*sāvana divasa*) measured from sunrise to sunrise. Each day was found to increase by one *prastha* during the *uttarāyaṇa* while the night shortened by the same amount. This trend was reversed during the *dakshiṇāyaṇa*. Since it is also stated that the increase or decrease of daylight during one *āyana* is equal to six muhurtas, the *prastha* is 1.5738 of our present day minutes. This peculiar unit of time appears in the *Vishṇu Purāṇa* which predated Hellenic science, as well as in the later Jain work on astronomy, the *Sūryaprajnāpathi*. The *Sūryasiddāntha* also divides the 24 hour day into 21600 *prāna*, perhaps because a heavenly body transits one minute of arc in this period. Being related to the seasonal variation of the length of the day the *prastha* is 23.61 *prāna*.

By the end of the fifth century AD, at least five great texts (*Siddānthas*) and thirteen minor texts (*Siddānthikas*) on astronomy had been inscribed on dried palm leaves or on birch bark. The texts themselves were frustratingly obscure but commentaries and explanations were provided by the astrologers of the period. The great astronomer, Varāhamihira (AD 500), who summarized the systems of astronomy in his Five Texts (*Pancha Siddānthika*), admitted that their common source was the Text of the Forefathers (*Pitāmaha Siddāntha*). Distinct similarities observed between the verses of the *Vedāṅga Jyotiṣa* and of the *Pitāmaha Siddāntha* indicate that the *Pitāmaha Siddāntha* was a development from much earlier times. It is therefore incorrect to state that Indian astronomy began only in AD 500 with imported knowledge from Mesopotamia, Greece, or Persia.

After the missionary Padmasambhava took Buddhism over the Himalayan mountains to Tibet and China, the pilgrim Yuan-Zhuang traveled in India from AD 629 to 645 and collected 657 Buddhist texts written in various Indian languages. In his *Abhidharma Shunzheng Lilun*, he describes units of time then prevalent in India. The day was divided into the forenoon, noon and afternoon while the night had three watches (*trimāya*). The shortest interval of time was the *kshaṇa*, the time taken by a woman to spin one *hsun* of thread; it also referred to an 'instant' or the blink of an eye. This term is still in colloquial use.

Instruments to measure time documented by Varāhamihira and others around AD 500 included the sundial, water clocks, and sand clocks. The sundial (*chhāya yantra*) was elementary in construction as compared to later European models where the time could be read directly off the dial. A vertical staff (*yashti* or *shanku*) 12 digits (*aṅgulas*) long was fixed at the center of a horizontal circle of radius equal to the length of the staff. This celestial circle represents the intersection of the celestial sphere with the horizontal plane. The shadow of the staff would cross the celestial circle at two points on an east–west line. A square (*chaturasra*) was then drawn circumscribing the circle, touching it at the four cardinal points. The eastern and western sides of the square were then graduated in digits. The projection of the shadow of the staff when the sun was upon the equator (*vishuvadbha*) determined the latitude of the place. Further measurements on the shadow of the staff enabled the chronographers to compute the time of day. The main instrument (*nara yantra*) was a larger and more precise version of the sundial.

The *Siddāntha Śiromani* also describes simpler variants of the sundial such as the *yashti yantra*. Here the celestial circle is drawn as for the *chhāya yantra*. From tables existing at that time or from observations conducted over the previous years, the radius of the diurnal circle of the day (*dyujya*) was calculated from the declination of the sun on that day. A second and smaller circle of this radius was then drawn on the floor concentric with the earlier celestial circle. The circumference of this diurnal circle was then divided into sixty equal parts representing the division of the daily solar time of revolution into sixty *nādis*. The staff was set but not rigidly fixed at the center, and to ascertain the time of the day, it was pointed towards the sun in such a manner that it cast no shadow. The extremity of the staff then represented the position of the sun on the celestial sphere. The distance from the tip of the staff to the point of sunrise or sunset on the celestial circle was measured. The arc subtended on the diurnal circle by this chord length gave, in *nādis*, the time since sunrise or until sunset.

The wheel (*chakra*) was a hand-held instrument to obtain the altitude of the sun and its zenith distance and so compute the time since sunrise. Even smaller instruments such as the arc (*dhanus*) were in use at that time.

Brahminaic rituals require not only the absolute time as determined by stellar and planetary position but also a stop watch to regulate the course of various sacrifices. The water clock (*jala yantra*) served this purpose. A leaking water vessel, the equivalent of the Greek clepsydra, was the first version. This, however, proved to be very inaccurate as the vessel takes infinite time to empty completely. As an improvement, water leaking at a constant rate from a reservoir was collected in a pot of volume exactly one *prastha*. In due course the *prastha* became a unit of time.

The sinking water clock (*ghaṭi*) proved to be more accurate and was frequently in use in the early eleventh century and up to the arrival of the mechanical clocks from Europe. A hemispherical copper vessel (*kapala*) six digits high and with a capacity of about 3.13 liters (60 *palas*) with a hole at the bottom to admit a gold pin four digits long and weighing approximately 0.183 gram (3.33 *māsha*) was placed on water. The vessel sank in one *ghaṭi* or *ghaṭika* sixty times in a day. This clock is used even now in some temples and for orthodox rituals.

As the measurement of time in ancient India developed, astronomers and astrologers became powerful and influential. Varāhamihira observed that a king who did not honor an astrologer was destined to destruction; that neither a thousand elephants nor four thousand horses could accomplish as much as a single astrologer. Time keeping, so essential to astronomy, then evolved into an esoteric science.

Division of the day according to various sources.

The Puranas (? BC)
15 nimesha = 1 kastha
30 kastha = 1 kala
30 kala = 1 muhurta
30 muhurta = 1 day or nycthemeron

The Sūrya Siddāntha (*<900* BC)
10 gurvakshara = 1 prana
1 prana = 1 vinadi
60 vinadi = 1 nādi, nādika or ghaṭika
60 nādi = 1 day

The Vedāṅga Jyotiṣa (AD *<400*)
5 aksharas = 1 kastha
124 kasthas = 1 kala = 305/201 prasthas
10.05 kalas = 1 nadika
2 nadikas = 1 muhurta = 30.5 prasthas

30 muhurtas = 1 day

The traveler Yuan Zhuang (AD *629*)
120 kshaṇas = 1 tatkshana
60 tatkshaṇas = 1 lava
30 lavas = 1 muhurta
5 muhurtas = 1 'time'
6 'times' = 1 nycthemeron

The Jain saint Mahāviracharya (*ca.* AD *850*)
7 Uchchhavas = 1 stoka
7 stokas= 1 lav
38.5 lav = 1 ghaṭi
2 ghatis= 1 muhurta
30 muhurtas = 1 day
15 days = 1 paksha
2 pakshas = 1 masa
2 masa = 1 season
3 seasons = 1 āyana
2 āyanas = 1 year.

The traveler al-Bīrūnī (AD *1030*)
6 prana = 1 cashaka
15 cashaka = 1 kshaṇa
4 kshaṇa = 1 ghaṭi
60 ghaṭi = 1 nycthemeron

NATARAJA SARMA

REFERENCES

Sarma, Nataraja. "Measures of Time in Ancient India." *Endeavour* 16(4): 185–188, 1991.

Satya Prakash. *Founders of Sciences in Ancient India*. Delhi: Research Institute of Ancient Scientific Studies, 1965.

Shastri, A.M. *India as Seen in the Bṛihat Saṁhita of Varāhamihira*. Delhi: Motilal Banarsidass, 1969.

Sūrya Siddāntha with a commentary by Ranganātha. Ed. P. Gangooly; Trans. E. Burgess. Calcutta: University of Calcutta, 1935.

Thibaut, G. "Contributions to the Explanation of the Vedānga Jyothiśa." *Journal of the Asiatic Society of Bengal* 46: 411, 1877.

Watters. T. *On Yuang-Chwang's Travels in India*. Delhi: M. Manoharlal, reprinted 1973.

See also: Astronomy – *Sūrya Siddhānta* – Varāhamihira – Gnomon in India – Clocks and Watches

TIME IN THE ISLAMIC WORLD Islam is now a major global religion. It has its roots, however, in the Arabian peninsula of the early seventh century. Its core elements were implanted in the early years and all later developments relate, one way or another, to that core.

In pre-Islamic Arabia some elements seem to have related to time (*dahr*) as fate, an underlying force directing human and natural destiny. It was conceived as a power existing eternally and responsible for the happiness or agony of humanity. Thus the *Qur'ān* refers to those who "say ... 'nothing but time can destroy us.' " [45:24/23]

While time was generally seen as an impersonal force, some people sought to identify God with time or, worse still by Islamic standards, to use this concept to deny God's existence. The Apostolic Tradition (*ḥadīth*) records Muḥammad's saying that Allah commanded men not to blame *dahr* "for I [God] am *dahr*". Later there were various groups of radical thinkers, often only vaguely defined and known primarily from polemic references, who asserted "the eternity of the course of time". Accordingly, because they denied both Creation and Judgment, the beginning and end of time, they were considered atheists. The term *dahrīya*, therefore, came to encompass a broad spectrum of dissenters advocating some form of atheism and/or hedonistic materialism.

Islam universally views time as created. It is God's handiwork which, like all Creation, is under God's command. God acts in and through history but, being eternal, time does not encompass him. Nature is also his creation but, while it manifests his greatness, it in no way encompasses Allah.

According to al-Bayḍāwī (d. 1282) in the *Anwār at-tanzīl wa-asrār at-ta'wīl* (The Lights of the Revelation and the Secrets of the Interpretation), time consists of *mudda*, the period of revolution of the sphere from beginning to end (i.e. the totality of time); *az-Zamān*, a gross subdivision of *mudda* into long periods of time (e.g. specific historical eras such as dynastic reigns); and *al-Waqt*, a fine subdivision of *zamān* into definite points of time or short intervals (e.g. precise times of the five obligatory prayers).

There are some within the Islamic tradition who assert that time, begun at Creation, has no necessary end. Al-Ghazālī (1058–1111) speaking in the *Tahāfut al-falāsifah* (Inconsistency of the Philosophers) for the mainstream tradition, teaches that religious dogma, for example the Day of Judgment, points to a finite end to time.

Measurement of Time

The two most significant historical eras, in the consciousness of Muslims, are the Age of Ignorance (*al-Jāhilīya*), the dark age in Arabia before the revelation of the *Qur'ān*, and the Islamic age stretching from the formation of the Islamic community in AD 622 until the end of history at Judgment.

There is evidence which suggests that most ancient Arabs followed a pure lunar calendar. About two centuries before the Prophet, apparently under the influence of Jewish civilization, many in Arabia adopted a luni-solar calendar. After Muḥammad's transfer to Medina (the *hijra*), an absolute lunar calculation was mandated.

The *hijrī* year consists, in theory, of twelve months of twenty-nine days, twelve hours, forty-four minutes and three seconds. In fact, for various practical reasons, the lunar months are computed variously as having either twenty-nine or thirty days. The *hijrī* year is about eleven days shorter than the solar year. There is no intercalation to the solar. Any particular *hijrī* date, over a period of approximately thirty three years, will move through all four solar seasons.

A number of Islamic societies have introduced a solar or luni-solar calendar (e.g. pre-revolutionary Iran, Kemalist Turkey). None of these has achieved any universal acceptance among the world's Muslims. It is accepted, further, that such a calendar has no support in the *Qur'ān* or the prophetic *Sunnah*.

Specific points in the yearly cycle are sanctified and celebrated in various ways. There are many local holidays and celebrations, and certain times are acknowledged as primary Muslim holy times. The first is the month of Ramaḍān during which the Muḥammad's prophetic career was begun. For the entire month Muslims abstain from food, drink, smoking, and sexual contact. The fast begins each day at the moment that there is sufficient light to distinguish white from black threads. It ends each night after the sun has completely set. Islam's "lesser feast", ʿId al-Fiṭr (breakfast), brings release from the month-long abstention.

The "greater feast" is the ʿId al-Aḍḥā (Feast of Sacrifice) on the tenth of Dhū 'l-Hijjah. It celebrates the end of the Holy Pilgrimage (*Hajj*) and reenacts Abraham's sacrifice of a ram in place of his son (*Qur'ān* 37:102). The primary sacrifice is made by the pilgrims in the valley of Minā, near Mecca. Muslims all over the world also offer such a sacrifice.

In Muḥammad's time the Arabs counted a seven day week. There is evidence that this is a relatively late practice imported from Babylonia (or, perhaps, by way of resident Jewish communities who observed the weekly Sabbath). Earlier time flow was divided according to weather and similar seasonal changes. Under Islam the week was given significance by assigning a special status to Friday. It is not a day of rest but rather a day of communal prayer.

Time as a Moral Dimension

By [the token of] Time [through the Ages].
Verily Man is in loss.
Except such as have faith and do righteous deeds and [join together]
in the mutual teaching of truth and of patience and constancy.

Qur'ān 103:1–3.

There are numerous Quranic references to man's time and God's time. "He rules [all] affairs from the heavens to the earth: in the end will [all affairs] go up to Him, on a Day, the space whereof will be [as] a thousand years of your reckoning [32:5]." Creation, as well, is said to have been effected in six days [32:4]. Exegetes often emphasize that the days of Creation, before the placement of the sun, refer to God's days reckoned as a thousand man years (and, in 70:4, to 50,000 years). Muslim scholars reiterate that Allah is in no way bound by time; he is totally transcendent.

The point is that throughout the course of human history God maintains full control over his creation. In his mercy Allah provided man with a window to his will and commanded him in the straight path. When time ends, at Judgement, man will face his time compacted to God's scale. "... In the immense future all affairs will go up to Him, for He will be the Judge, and His restoration of all values will be as a day or an hour or the twinkling of an eye; and yet to our idea it will be a thousand years." (Yūsuf Alī)

There is an implicit ambivalence as regards the pursuit of scientific and technological knowledge. In so far as the Muslim centers his consciousness on God, his greatness and omnipotence, he can easily succumb to a fatalistic acceptance of what comes to him. He then asserts God's absolute will and denies causality in nature. Without cause and effect there is no necessary natural order. In so far as man sees time as his sphere of action, he can hear the Quranic command to perceive God in his creation. Exploration of the world becomes an act of moral compliance.

Time's purpose is to serve as an arena of moral action. It is the place which God has established for the exercise of will. Judgement and consequent assignment to paradise or hellfire will mean the end of time, its sole purpose being the setting for man's surrender.

M.S. STERN

REFERENCES

The Encyclopaedia of Islam, 1st ed. Leiden; Brill, 1913–1936. See entries "Miqāt", "Taʿrīkh", and "Zamān".

The Holy Qur'ān. Text, translation, and commentary by Abdullāh Yūsuf Alī. 2 vols. New York: Hafner, 1938.

Lazarus-Yafeh, Hava. *Some Religious Aspects of Islam*. Leiden: Brill, 1981.

Mernissi, Fatima. "The Muslim and Time." In *The Veil and the Male Elite*. Reading, Massachusetts: Addison-Wesley, 1991, pp. 15–24.

von Grunebaum, Gustave E. *Muhammedan Festivals*. New York: Schuman, 1951.

TIME IN MAYA CULTURE Classic Maya culture developed from the second to the tenth centuries AD and expanded to encompass all of the Yucatan peninsula including the modern countries of Guatemala, Belize, and portions of El Salvador and Honduras. Unlike the Aztec and Inca states, which developed highly centralized imperial cities, the Maya developed city-states. Elites intermarried, formed alliances, and shared a complex calendar.

While all empires have created multiple times to connect village community to imperial center, and earthly city to heavenly city, the historian Nancy Farriss has noted that "From their earliest recorded history, the Maya displayed an intense interest, bordering on obsession, in measuring and recording the passage of time". Anthony Aveni observes that "The first bits of information in any Maya inscription are about time. What sets the Maya apart is not the number of time units they devised, or even their complexity; rather, it is their preoccupation with 'commensurateness' — perfecting the way time cycles interlock or fit together". Over a millennium since the climax of classic Maya culture, and five centuries after the Spanish conquest, Maya continue to practice their own time, albeit in modified form.

The Maya had a military state with a literate elite. Our information about their multiplicity of time comes from their own writings. Although Maya hieroglyphic writing died out, Maya elites were able to maintain an unbroken continuity from their system of writing to that of Spanish script. The language survived in many colonial documents from the sixteenth century on. Today two million people still speak more than a dozen dialects of the Mayan language.

The Maya calendar is the centerpiece of Maya concepts of time and its measurement. It consists of the conjunction of cycles upon cycles of time of long duration. Like the Gregorian calendar we use today, the Maya calendrical system was the culmination of many different cultural influences over millennia. The conception of time underlying the calendar is one of movement. Maya referred to time as a flow and perceived a need for both human and divine intervention in maintaining it. In Maya culture, people actively construct their sense of time, a point to which I return below with reference to daykeepers and their role in the regulation of everyday life.

The flow of time is incorporated into an all encompassing cyclical pattern. Conjunction of complex forces mark cycles of different duration. The Body Count of twenty is the basic unit of Mesoamerican counting (ten fingers plus ten toes). The classic Maya week consists of twenty named days, each propitious or unpropitious for particular activities. The twenty day week was combined with a separate series of thirteen numbered days to create the Sacred Round of two hundred and sixty days. Later, a Solar Year was calculated consisting of eighteen twenty day months plus five days. The solar year was articulated with the sacred round to create the Calendar Round of fifty-two years. The Maya also developed a Katun Round consisting of thirteen twenty year periods for a duration of two hundred and sixty years. Despite an obsession with commensurability that led Maya priests into astronomy and cosmology, not all the cycles were articulated with each other. Late in the political development of pre-Conquest Maya culture a Long Count of five millennia appeared. Today, Maya daykeepers reckon the rhythms of everyday life by interpreting the solar and divinatory calendars.

Although Maya viewed time as a flow, it is also time with a content. Each day was named after a god and was qualitatively different from every other. Furthermore, the content of time was regulated by a conjunction of forces. Those days that were at the conjunction of different cycles had a greater force than other days. The larger the cycle, the greater the number of forces intermeshing from different sub-cycles. The Maya depicted time on stone carvings or drawings as weighty numbers carried by gods.

Intermeshing cycles produce a succession of days whose events are linear, unique, and irreversible. But this apparent linearity is an illusion because "no matter how unique a pattern of events may seem, no matter how long the sequence, it will eventually be repeated, when the governing forces of all the cycles of different dimensions coincide in one huge cycle" (Farriss, 1987).

Why these cycles? Maya observers were not disinterested observers of the heavens. Cosmic time links common people to rulers through the latter's' association with the cosmos. The construction of cycles appears to be associated with the problems of legitimacy and political order. Returning to the same point, at least logically, is the ultimate form of reassurance against the reality of present conflict, chaos, and events that seem to be getting out of hand. As Farris says, "Accounts of the famine in the books of Chilam Balam show less concern with starvation than with the fact that people flee into the forest to eat roots and other wild food ... In other words, people cease to live in their customary fashion ... Political conflicts are condemned because they disrupt the established hierarchy, and upstarts and invaders are vilified for breaking the rules". Encompassing cycles provide a cosmic charter of legitimacy of a magnitude rarely conceived in the history of cultures.

The predictable order of cosmic time-reckoning, unfortunately, depended upon human agency for execution. "It is unlikely that the Maya priests claimed to comprehend all the possible combinations of divine forces governing the subordinate movements of so immense a projection." Past, present, and future are written in a single bound book, but one that admits of many interpretations.

How did the dual conception of cyclical and linear time work out in practice? Both serve as charters for sociopolitical arrangements, but each in its own way. It appears that classic Maya elites confined linear concepts to relatively short term duration out of choice (they possessed the literary or numeracy skills to create historical memory). Among the most important were the katun round and the chronicles written by members of Maya royal lineages during the colonial period.

The katun round, consisting of a recurring wheel (time is often depicted as a wheel in Maya stone carvings) of thirteen twenty year counts for a total of two hundred and sixty years, recorded the important events of royal lineages. The interesting aspect of the katun round is its duality of both history and prophecy. Each year has its own name together with characteristic and appropriate events. The future is contained in the past, and therefore Maya could look to a record of the past as a guide to the future. The same pattern of events for a given year recurs every two hundred and sixty years. In Chilam Balam texts, logically associated events always occur in the same katun regardless of their inner chronology. It is as if all invasions were taken out of their chronological sequence and placed in their cosmological reckoning. Even references to the most recent "descent" — the Spanish conquest — are woven into the pre-Hispanic mix.

In addition to general reference to recurrent patterns of events there is also remarkable specificity to katun accounts, which name specific people and places, details of real famines, invasions, migrations, exiles, and rivalries. The specific content is about unique and irreversible events. The chronicles appeal to secular principles as the basis for claims to legitimacy, for example descent from previous rulers. The twenty year katun cycle may have served as a calendar for rotation and succession of ruling families in post-classic Maya. In contrast, books of Chilam Balam that record cycles stress quality of time — joy, misery, abundance, and the place of cosmic and divine forces in their production. The sacred personages of the calendar rounds merge with their worldly actors.

There is some evidence that the Maya calendar evolved toward a dominance of linear over cyclical concepts of time during the post classic period as a consequence of the emergence of dynastic rule and its dominance over the more autonomous city-states of the classic period. Unlike the calendrical system of cycles and their conjunction, which have no particular starting date, the Long Count system measured elapsed time from a starting point that would extend to a time horizon of five thousand two hundred years before its repetition. This new calendar represented a radical shift from the dominance of cyclical time over linear time because its duration was long enough to encompass the ancient Calendar Round of fifty-two years and more recent katun round of two hundred and sixty years. Specific details of human affairs accompany the Long Count and a new emphasis on written genealogies supports the hypothesis of a shift to dynastic rule.

If it is unlikely that ancient Maya priests could comprehend all the divine forces converging on particular days or years, how much less likely is it that ordinary people could use the Maya calendar to take control of their own lives? A partial answer appears in the guise of Maya daykeepers. Barbara Tedlock has given us an account of time among Quiche Maya in the town of Momostenango, highland Guatemala. There daykeepers use an integrated solar year and the sacred round to divine whether a day is good or bad for particular activities. The appropriate determination of qualities to a day are specific to a particular client and require active participation on the part of both daykeeper and client.

Daykeepers use a set of practices referred to as "the speaking of the day" and "the speaking of the blood". The qualities of the twenty name days are based on stipulated but vague mnemonic phrases that require interpretation by the daykeeper. Other factors specific to a client enter the picture, including the client's character, part of which is determined by the particular day on which a child is born.

The daykeeper achieves understanding by sortilege, or divination by lots. She or he mixes, grabs, and arranges piles of seeds and crystals into a pattern, all the while counting and interpreting the two hundred and sixty days of the divinatory calendar. The blood is considered to be an active substance that sends signals or "speaks" to the daykeeper. The rapport between daykeeper and client are also thought to affect the divining of a day as good or bad for a particular activity. Clients come to a chosen daykeeper with a specific question about such probable occurrences as illness, accident, land disputes, building houses, inheritance, business transactions, travel, marriage, adultery, quarrels, dreams, births, deaths, and interpretation of omens. The divination is performed for the specific question posed by a particular client.

HENRY J. RUTZ

REFERENCES

Aveni, Anthony. *Empires of Time: Calendars, Clocks, and Cultures*. New York: Basic Books, 1989

Farriss, Nancy. "Remembering the Future, Anticipating the Past: History, Time, and Cosmology Among the Maya of Yucatan." *Comparative Studies in Society and History* 29 (3): 566–593, 1987.

Tedlock, Barbara. *Time and the Highland Maya*. Albuquerque: University of New Mexico Press, 1982.

See also: Calendars – Long Count – Writing – Mathematics

TIME AND NATIVE AMERICANS: TIME IN THE PUEBLO WORLD The Pueblo people of the United States Southwest believe in a cosmos in which nature functions with the active cooperation of humankind. The proper ceremonies must be carried out at the proper time so that the cosmic order is sustained. Traditional doctrines held that there were correct times for planting, harvesting, hunting, ceremonies, and many other activities — all embedded in a sense of sacred time. The right times for these crucial undertakings are established by astronomical observations to regulate the ritual calendar. The cycles of the sun and moon set the rhythm of Pueblo time.

Sacred time is ordered with different levels of periodicities. The longest appears to be the seasonal year. We have very little evidence — almost all ambiguous — that the Pueblos kept long counts or tallies greater than a year. Until the twentieth century, with the intrusion of European concepts of time, no indigenous interest appears in tracking cycles over many years. The yearly cycle is all important; after its end, a new, equally unique one begins. Within the seasonal span, ceremonies occur in a fixed sequence, in which the completion of one sets the stage for the next. The observation of the phases of the moon marked the subdivision of the year. Within a night, the start and end of some ceremonies are flagged by observations of the positions of certain stars, especially the Pleiades as seen from the smokehole of a *kiva*, a structure used as a ceremonial room.

The shortest unit of time reckoned by the Pueblos is the day, which begins with sunrise. The day has vague subdivisions into loose "hours", most of which are noted near sunrise and sunset by the color of the sky. About the only time of day noticed with any care is that of noon, which is typically observed by the length and directions of shadows cast by the edges of walls, buildings, trees, or sticks embedded in the ground.

A religious official has the responsibility and the authority to set the ceremonial dates within the ritual cycle. The crucial task entrusted to this official is the *forecasting* of the correct date by making *anticipatory* astronomical observations. The dates are announced ahead of time so that the people of the pueblo can enter into proper preparations (abstinence, practicing songs and dances, preparing costumes and special foods) so that the ceremony may be carried out with the "good heart" and be effective.

Typically, one official — the Sun Priest — performs the sun-watching for the seasonal cycle, which includes the planting schedule. He does so from a sun-watching station that is usually located within or close to the pueblo. From this fixed spot, the Sun Priest keeps a horizon calendar (usually at sunrise using an observation of the first gleam) against the horizon profile. He knows from experience that when the sun rises (or sets) against a certain horizon feature, so many days will elapse until, say, the winter solstice. Then he uses tally markers (such as a knotted cord) to count down to that day, with an announcement made to the pueblo a few days ahead of the celebratory date. This anticipatory technique allows the Sun Priests to forecast the dates of the solstices to within one day of the astronomically correct date when the sun appears to "stand still" at its sunrise point. The observations made about two weeks in advance catch the sun when the angular speed of its sunrise position is large enough to be discerned by the naked eye on a day-to-day basis. Hence, this practice neatly solves both a cultural and astronomical problem.

Moon watching regulates timings between the seasonal and daily ones. The official responsible for tracking the phases of the moon generally is *not* the one who watches the sun. A Pueblo month begins with the observation of the first visible crescent and ends with the last. The days of invisibility are not typically counted. Calendar sticks were used to tally the observed months.

The months of the year are counted starting with the first before the winter solstice ceremony. The first five (or six) are named (usually after seasonal characteristics); then these names are repeated for the next five or six. The lunar calendar needs to be synchronized with the solar one, which is accomplished by adding an intercalary month (which can be as short as four days!) at the end of the regular count or after the winter solstice.

Although Pueblo people feared eclipses of the sun and the moon, they had no way to forecast them. In part, this lack was related to their disinterest in keeping counts longer than a year. We also find no evidence for knowledge of the 18.6-year lunar standstill cycle, even though the moon certainly was observed rising and setting. The historic Pueblo culture did not seem to attach any importance to this astronomical cycle; so it was ignored.

The ancestors of the Pueblo people were the Anasazi, who inhabited what is now the greater Four Corners area of the U.S. Southwest as a settled agricultural people. During the height of the classic Anasazi culture some thousand years ago, two major sites developed: Mesa Verde and Chaco Canyon. A regional center developed in the San Juan basin around Chaco by AD 1110; as many as 70 communities dispersed among 50,000 square kilometers were integrated into a socioeconomic and ritual network connected by a road system.

How did the Anasazi regulate and track time? By using ethnographic analogy, we can generate hypotheses about Anasazi astronomy that can be checked against the archaeological record. The most important cycle of time for the Pueblos is the seasonal year, which is typically tracked by horizon calendars. Anticipatory observations are made about two weeks prior to the solstices. Of the Anasazi sites field-tested so far, about one-third have reasonable horizon calendars.

A secondary strategy employed in the historic Pueblos used a special opening or window to cast sunlight on an opposing wall. The best Anasazi analogs to this type of interior observational technique occur in Hovenweep National Monument at Hovenweep Castle, Unit Type House, and the Cajon Group ruins with small portals. Hovenweep Castle and Cajon ruins work at sunset, Unit Type House just after sunrise. In all three locations, the horizon profiles lack the relief for tracking horizon calendars. For all, the throw from the shadow-casting edges is large enough so that the linear motion on the receiving wall is typically a few centimeters per day from the sun's angular positional change at sunrise or sunset. The evidence so far hints that interior light and shadow casting played an important role in Anasazi seasonal time keeping.

Exterior light and shadow casting was not an important time keeping technique in historic times. It may have been used more extensively by the Anasazi, though perhaps more for ritual than calendric purposes. The three-slab site atop Fajada Butte in Chaco Canyon has been much investigated; so has the Holly site at Hovenweep. In both these cases, light and shadow cast by rock edges fall onto panels of rock art at important seasonal times of the year. These exterior sites generally suffer from a lack of resolving power for reliable calendric forecasting, if we apply the standard achieved by the historic Pueblo Sun Priests — within a day of the astronomical date of the solstices. Hence, these sites may have served as sun shrines for offerings to the sun with a light and shadow display that served a commemorative rather than a calendric function.

We have no firm evidence so far of prehistoric monthly lunar calendars, though such a count of days would have a unique sequence tied to the phases of the moon. No rock art in the Anasazi region has been demonstrated to display the measure of a lunar tally.

Occasional claims have been made that the Anasazi noted the 18.6-year standstill cycle of the moon. The hypothesis of attention to this interval has been put forth for the Fajada three-slab site and for Chimney Rock Archaeological Area in Colorado (among a few others). The most convincing case is made for Chimney Rock, where the natural rock pillars (after which the site is named) act as a natural foresight when viewed from the site of Chimney Rock Pueblo, a Chacoan outlier, which was occupied from about 1075 to 1175.

The orientation allows a spectacular anticipation and forecasting of the major lunar standstills. For about two and a half years prior to the standstill, the moon rises between the pillars for one or two days per month. Starting after the summer solstice, the moon appears as a waxing crescent. Finally, near the winter solstice, the full moon stands between the pillars at around sunset. Hence, the priest in charge of the moon watching could forecast the date of the standstill and also the full moon nearest the winter solstice — an important conjunction among the historic Pueblos.

MICHAEL ZEILIK

REFERENCES

Williamson, Ray A. *Living the Sky: The Cosmos of the American Indian.* Boston: Houghton Mifflin, 1984.

Zeilik, Michael. "Keeping the Sacred and Planting Calendar: Archaeoastronomy in the Pueblo Southwest." In *World Archaeoastronomy*. Ed. A. F. Aveni. Cambridge: Cambridge University Press, 1989, pp. 143–166.

See also: Calendar – Long Count – Astronomy – Roads

TONGREN ZHENJIU SHUXUE TUJING The *Tongren Zhenjiu Shuxue Tujing* (Illustrated Canon of Acupuncture Points based upon the Bronze Figure) was written by Wang Weiji, and published in AD 1026. It has been a baseline in the identification of acupuncture points ever since. It is used not only in Chinese medicine, but also in acupressure as developed recently in the West, in *shiatsu* (pressure point massage) in Japan during this century, and in all acupuncture-related disciplines.

By the Song Dynasty (AD 960–1126), after much copying and recopying, the locations of the points and channels had become confused. Because of this and the greater stability brought about by the Song reunification of the empire, the medical scholar Wang Weiji reorganized and collated all the then available material in order to locate and define the points precisely.

He produced a book which very quickly became the authoritative text throughout the country. It gave very detailed and accurate information concerning the points, the channels, the depths and effects of needling. The total number of points named in the *Huangdi Neijing* (Yellow Emperor's Canon of Medicine) had been 160; in the *Zhenjiu Jiayi Jing* (A–Z of Acupuncture and Moxibustion) it was 349. But now it was raised to 354. Not long afterwards, the full text of this book was cut onto two stone tablets, each some two metres high and seven metres in length, which were erected in the Song capital, Kaifeng, where they could be read by everyone, or where ink-impressions could be made from them.

At the same time Wang Weiji directed the casting of two life-size bronze figures — hence the title of his book-which were completed in 1027.

These bronze figures are the very earliest of their kind. They were hollow and had the exact locations and names of the acupuncture points marked on their surface. It is noted in the Song histories that the purpose of the figures was for them to be covered in beeswax and then filled with water. When a student palpated and punctured a point correctedly, a stream of water would flow out, reminiscent of the energy-stream (or *qi*) tapped by that particular point.

The later history of these figures and tablets is also revealing. After two hundred years, when the Yuan Dynasty moved the capital to what is now Beijing, the bronze figures and tablets were also moved, to take pride of place in the temple at the new Imperial Medical College. However, much use had been made of the figures, and the tablets were so worn that they were indecipherable. Reproductions were therefore made again. During excavations from 1965 to 1971 in Beijing, fragments of the original Song tablets were discovered, and their texts shows that the tradition has been faithfully and accurately preserved to the present day. This book is used as the foundation text for the listing of points in the *Essentials of Chinese Acupuncture* (Beijing, 1980) and other texts which have all been recently produced in China and which are aimed at the interested acupuncture audience in the West.

RICHARD BERTSCHINGER

REFERENCES

Essentials of Chinese Acupuncture. Beijing: Foreign Languages Press, 1980.

Fu, Weikang. *The Story of Chinese Acupuncture and Moxibustion*. Beijing: Foreign Languages Press, 1975.

Lu, Gweidjen and Needham, Joseph. *Celestial Lancets*. Cambridge: Cambridge University Press, 1980.

TREPHINATION Perhaps one of the least understood surgical practices performed in ancient America concerns cranial trephination (also spelled "trepanation"), or the surgical modification of the skull. While much of the prevailing literature is largely dated, an older generation of theories tended to ascribe the practice to primitive wonderment and notions of spirit release and possession. This is despite a substantial body of technical evidence available from trephined specimens and a large and specialized body of surgical instruments and procedures documented in archaeological and contact-period historical contexts. According to medical historian Guido Majno, trephination was used as a cure for a number of ailments including skull fractures and related trauma, epileptic seizures, and insanity. In fact, many of those skulls examined to date bear evidence of blunt trauma or pathology that may provide a more direct indicator of why the practice was carried out in the first place.

While ancient Peru provides the largest body of trephined specimens available for study, and the clearest evidence of an established medical tradition in skull surgery, other important specimens have been reported from throughout the Americas. Documented examples, with evidence of osteitis, have been reported from the Mexican states of Chihuahua and Oaxaca, and archaeological sites in Lamy, New Mexico, and Accokeek, Maryland, in the United States, where trephined surgical openings are often no more than two centimeters in diameter. Reported trephined specimens from Columbia County, Georgia, archaeological zones along the Skeena River, and in the sites of Eburne and Lytton, British Columbia and Kodiak Island, Alaska, range between 3.8 and 6.0 centimeters in diameter. If all the putative cases of trephination reported in these examples are a reliable indication of areas in which the practice was known, then cranial trephination was quite widespread.

Majno has identified five primary techniques of trephination in specimens from both Peru and Mesoamerica. These include the most ancient method on record — the scraping-away of that bone just underlying the scalp — as well as other techniques such as grooving, boring, cutting, sawing, and rectangular intersection incision, the most predominant method employed, in which cut, sawed, or abraded grooves were used to perforate the skull in a rectangular pattern. Once the intersection incisions were connected to each other, a rectangular or octagonally-shaped plate of bone remaining within the intersection was removed with a spatula-like device. Schendel (1968) notes that "the trepanning technique used in Mexico was to punch a series of small holes in the skull outlining the fracture, or the area to be removed, then to cut between those holes and lift off the depressed section of cranial bone". After this initial procedure, the exposed brain was protected with a thin plate of hardwood and cotton pads. When trephination was not deemed suitable because of the extent of skull trauma, the damaged portion of skull was encased in a protective plaster cast. There were two types of plaster cast. The first was a mixture of feathers, egg whites, and resin, and the second was prepared from a mixture of animal ash, blood, and egg whites.

Specialized tools and instruments employed by Inca

physicians in the trephination of the skull included *tumi* knives crafted from alloyed and annealed, or metalurgically-hardened, copper and bronze metals. The *tumi* knife consisted of a crescent-shaped blade, at the midpoint of which was attached a cylindrical metal handle or other appendage. *Tumi* knives used in trephination included razor sharp, serrated, or other sawtooth-edged crescent-shaped scalpels. Additional instruments included bronze perforators, drills, and chisels. The basic surgical instrumentation utilized in cranial trephination long predated the rise of the Inca state, whose government came to sponsor and control this specialized field of medical endeavor (Burland, 1967). Cranial trephination, and both *tumi* knives and bronze perforators, are frequently depicted on the carved and painted surfaces of vessels of the Mochica civilization of the north coast of Peru (ca. AD 500).

In one collection of 273 skulls from Peru, 47 bore evidence of having been trephined between one and five times. According to Majno, the skulls studied thus far suggest that the survival rate was near one hundred percent. Only a few cases of osteomyelitis, or bone deterioration, were documented in the cranial collection he examined. Those cases appear to have been associated primarily with the size of the original injury, and whether or not the trauma was located at the base of the skull. Apparently, patients suffering blunt trauma over the cerebral cortex or brain stem were least likely to benefit from treatment by trephination. In other recent studies, projected survival rates, based on the examination of the presence of osteitis, range from 62.5% in one study that found healing or evidence of osteitis in 250 of 400 skulls examined, to between 23.4% and 55.3% in other studies (Froeschner, 1992). These latter figures are questionable in that the studies cited by Froeschner do not distinguish post-mortem trephination from that performed on severely traumatized patients. Froeschner's review also fails to acknowledge the existence of even the most basic metal instruments and surgical kits documented for the practice of trephination in ancient Peru, claiming instead that "primitive tools" were the mainstay of the art of Inca trephining.

It is clear that specialized techniques and treatments contributed to the diversity and variation in surgical localities known from trephined specimens alone. In one analysis of a trephined skull collection consisting of 112 specimens (the Tello collection housed at the Peabody Museum of Harvard), 53.6% of the skulls, or 60 specimens, bore trephined areas in the frontal area of the skull (26 of which were situated in the frontal bone, 12 in the region of the bregma, and another 12 crossing over the left coronal suture); 33%, or 37 specimens, were trephined in the parietal area (18 of which bore surgical openings over the left parietal, while 15 crossed the sagital suture), and 13.4%, or 15 specimens, were trephined over the occipital area (with 7 in the region of the lambda, and 4 in the occipital bone). Most trephined specimens from the Harvard collection bear surgical openings within the frontal areas of the cranium, and again, trephined openings are most highly correlated with blunt trauma and other pathologies. The incidence of epileptic seizures among Inca era peoples was documented by both European-contact chroniclers and contact-era Inca scribes. This fact may provide directions for future inquiry into the origins of specific types of trephination among Inca and related peoples.

Based on a worldwide survey of ancient skull surgery, Majno has concluded that the Inca were the "Masters of the Art of Trepanning". He also noted that the survival rate for patients in Inca times was better than for such rates in modern times. He says that modern survival rates are the inverse of what they had been in earlier Inca times, and concludes that modern humans complicated the task of skull surgery by introducing new sources of infection and disease. In turn, modern physicians were trained to perpetuate otherwise antiquated traditions of skull surgery based on flawed nineteenth century assumptions and beliefs. European medical belief systems regarding infection, surgical procedure, anesthetics, and basic hygiene necessitated a constantly expanding and contracting repertoire of experimental procedures which further complicated survival rates. As a result, the fledgling early twentieth century version of modern skull surgery suffered many casualties.

RUBEN G. MENDOZA

REFERENCES

Burland, C.A. *Peru Under the Incas.* New York: Putnam, 1967.

Froeschner, Elsie H. "Historical Vignette: Two Examples of Ancient Skull Surgery." *Journal of Neurosurgery* 76(3): 550–552, 1992.

Guzman Peredo, Miguel. *Medicinal Practices in Ancient America.* Mexico: Ediciones Euroamericanas, 1985.

Majno, Guido. *The Healing Hand: Man and Wound in the Ancient World.* Cambridge: Harvard University Press, 1975, reprinted 1991.

Schendel, Gordon. *Medicine in Mexico: From Aztec Herbs to Betatrons.* Written in collaboration with Jose ALvarez Amezquita and Miguel E. Bustamante. Austin: University of Texas Press, 1968.

Stewart, T. Dale. "Stone Age Skull Surgery: A General Review with Emphasis on the New World." *Annual Reports of the Smithsonian Institution, 1957*, pp. 469–591, 1958.

See also: Medicine in Mesoamerica

TRIBOLOGY Tribology, the study of the phenomena and mechanisms of friction, lubrication, and wear of surfaces in relative motion, can be traced back to remote antiquity in China, where it developed significantly.

Sliding bearings were discovered in ancient mechanisms, especially in carts which operated at high speeds and carried great loads. In order to reduce friction and damage, Chinese mechanics lubricated bearings on their carts. This was recorded in the earliest poetry collection entitled *Bei Fen Quan Shui* (The Bei Wind and Spring Water) as follows:

The grease used is sufficient to lubricate the axle shaft.
On the shaft end, we carefully check the pin bolt.
Quickly, send me home driving the cart.
Quickly, send me back to my Wei country town.

It proves that before the book *Shi Jing* (Poems and Odes) appeared — before 1100–600 BC — lubricant was widely applied in China. Owing to their rhythmic quality, the verses were repeatedly quoted and talked about, and often appeared in later books as well.

By the time of the Chun Qui (Spring–Autumn) period (500 BC) in the Lu and Qi states, there was a group of special officers called *Jinche*, whose duty was to check the pin bolts on the shaft ends. They were probably the earliest tribology supervisors. This demonstrates that the ancient Chinese already knew well that the better the lubrication, the faster and easier ran the cart.

Early lubricant was called *zhi* or *gao*. According to research on ancient characters, *zhi* was the fat of horned animals, and that of hornless animals was called *gao*. The lubricant used might be sheep oil, which was easier to get. Since zhi or gao is in a solid state at normal temperature, it had to be heated and melted before it was used. In 300 BC petroleum began to be used.

Metal bushes appeared in China quite early. There were relevant records about their use in *Wuzi Yebing* (Wuzi casting), which shows that they appeared no later than 400 BC. The metal bush was made of two layers: outside and inside. The outer one was called *gong*, which turned together with the wheel, and the inside one was *jian*, which turned together with the shaft. Between them a kind of lubricant was provided, and the cart ran quickly and easily. In addition to carts, the metal bushes were applied in other mechanisms of antiquity, and they appeared in a variety of forms. In the Han Dynasty, they were already used quite widely. This is illustrated by the discovery of hundreds of mold sets of metal bushes in county Wen in Henan province. Most of them were still undamaged. They reflect the level of manufacturing as well as the scale of their production and application in the Han Dynasty.

The ancients took other measures to reduce friction and damage. Most mechanisms were made of wood. In the *Kao Gongji* (The Artificer's Record), three principles of wood selection were listed. (1) The wood must be smooth, without joints; (2) it must be tough and wear-resistant, and (3) it must be thick enough and easy to revolve. In the sixteenth century a book entitled *Tian Gong Kaiwu* (Exploitation of the Works of Nature) was written, in which a further summary on the basis of experience was made. It analyzed the advantages and shortcomings of various woods, and provided an example: if wood works for too long, it gives out heat. In the field of structural design, the *Kao Gongji* and other books analyzed the relationship between the axis path and the length of the shaft, and pointed out that when the cart and load were different, the interior axis of the wheel must keep a fixed proportion to the length of the shaft.

There was one other achievement of note in this field. In the Yuan Dynasty (thirteenth century), a ball bearing which was circle-formed and consisted of four balls was successfully applied to an astronomical instrument made by Guo Shoujing.

LU JING-YAN

TRIGONOMETRY IN INDIA Trigonometry offers one of the most remarkable examples of transmission of the exact sciences in antiquity and the Middle Ages. Originating in Greece, it was transmitted to India and, with several modifications, passed into the Islamic world. After further development it found its way to medieval Europe.

The very term "sine" illustrates the process of transmission. The Greek word for "chord" (εὐθεῖα, literally "a straight line [subtending an arc]") was translated into Sanskrit as *jīva* or *jyā* ("string of a bow") from the similarity of its appearance. The former word was phonetically translated into Arabic as *jyb*, which was vocalized as *jayb* (meaning "fold" in Arabic), and this was again translated into Latin as *sinus*, an equivalent to the English sine.

It was by tracking back along this stream of transmission that the first chord table ascribed to Hipparchus (fl. 150 BC) was successfully recovered by G. J. Toomer in 1973 from an Indian sine table (compare Tables 1 and 2 below). Toomer showed that some numerical values ascribed to Hipparchus in the *Almagest* of Ptolemy (fl. AD 150) could only be explained by hypothesizing the use of this reconstructed table.

According to this reconstruction, Hipparchus used 6875 "minutes" as the length of the diameter (1) of the base circle, in other words, as the greatest chord subtending the half circle ($= R\text{crd}180° = 6875$). This number is the result of rounding after dividing 21,600 minutes (360°) by the value of $\pi = 3;8,30$ (in this article we follow the convention:

integer and fraction are separated by a semicolon, the former is in decimal form and the latter is in sexagesimal form with commas to separate the places.) In India 3438 "minutes", namely, the rounded half of D, was used as the length of the radius (R), which is the largest "half chord" (*jyārdha* or *ardhajyā*).

Thus the relation between the Greek chord and the Indian sine can be expressed as:

$$AB = 2AH \qquad R\text{crd}2\alpha = 2R\sin\alpha. \tag{1}$$

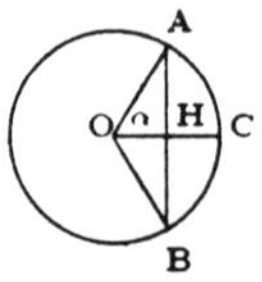

Plane trigonometry was the essential tool for mathematical astronomy in India. All the astronomical texts in Sanskrit either give a kind of sine table or presuppose one. On the other hand, trigonometry was studied only as a part of astronomy and it was never an independent subject of mathematics. Furthermore since they were not aware of spherical trigonometry, Indian astronomers developed the method called *chedyaka* in which the sphere was projected on to a plane.

Table 1: Hipparchus' chord table reconstructed

Number	α	$R\text{crd}\alpha$
1	7;30	450
2	15	897
3	22;30	1341
4	30	1779
5	37;30	2210
6	45	2631
7	52;30	3041
8	60	3438
9	67;30	3820
10	75	4185
11	82;30	4533
12	90	4862
13	97;30	5169
14	105	5455
15	112;30	5717
16	120	5954
17	127;30	6166
18	135	6352
19	142;30	6510
20	150	6641
21	157;30	6743
22	165	6817
23	172;30	6861
24	180	6875

Sine Table with $R = 3438$

The earliest Indian sine table with $R = 3438$ is found in Āryabhaṭa's book on astronomy, *Āryabhaṭīya* (AD 499). It should be remembered that "table" here does not mean that the numbers are actually arranged in a tabular form, i.e. in lines and columns. As is usually the case with Sanskrit scientific texts, all the numbers are expressed verbally in verse. For brevity's sake, Āryabhaṭa gives only the tabular differences (Δ, the fourth column of Table 2). This is the standard sine table in ancient India. Exactly the same table is found in the *Jiuzhili*, a Chinese text on the Indian calendar written in AD 718 by an astronomer of Indian descent, but it did not have any influence on Chinese mathematics.

The values of sines were geometrically derived from $R\sin 90°$ (= R) and $R\sin 30°$ (= R) by two formulas:

$$R\sin(90° - \alpha) = \sqrt{R^2 - (R\sin\alpha)^2}, \tag{2}$$

$$R\sin\frac{\alpha}{2} = \sqrt{\left\{\frac{R - R\sin(90° - \alpha)}{2}\right\}^2 + \left(\frac{R\sin\alpha}{2}\right)^2}. \tag{3}$$

It is worth noting here that the first tabular sine was equated with the arc (3;45° = 225′) which it subtends. This means that, when α' is small enough, the approximation $\sin\alpha' \approx \alpha'$ can be applied.

What is more remarkable in Āryabhaṭa is that he gives an alternative method for computing tabular differences by means of the formula:

$$\Delta_{n+1} = \Delta_n - (\Delta_1 - \Delta_2)\frac{J_n}{J_1},$$

where $J_n = R\sin n(3;45°)$. The formula, after several centuries of misunderstanding, was correctly interpreted by a south Indian astronomer Nīlakaṇṭha (born in 1444), one of the most distinguished scholars belonging to the Mādhava school (see below). When suitable values of $(\Delta_1 - \Delta_2)$ and J_1 are used, this formula produces very good values for the rest of R sins. It seems that Āryabhaṭa's sine values were comuted by this second method rather than by the geometrical method.

Sine Table with $R = 120$

There is another kind of Indian sine table which uses $R = 120$. The table is found in the *Pañcasiddhāntikā* of Varāhamihira, a younger contemporary of Āryabhaṭa. This table is closely related to the Greek chord table with $R = 60$ which is offered in Ptolemy's *Almagest*. Because of the relation (1) given above, all the numerical values in the chord

Table 3: Comparison of the first four and last four values in the *Pañcasiddhāntikā* with those of Ptolemy

		Varāhamihira with $R = 120$	Ptolemy with $R = 60$
Number	α (°)	$R \sin\alpha$	$R\text{crd}2\alpha$
1	3;45	7;51	7;50,54
2	7;30	15;40	15;39,47
3	11;15	23;25	23;24,39
4	15	31;4	31;3,30
...	...	...	...
21	78;45	117;42	117;41,40
22	82;30	118;59	118;58,25
23	86;15	119;44	119;44,36
24	90	120	120

Table 2: Indian sine table with $R = 3438$

Number	α	$R \sin\alpha$	Δ
1	3;45	225	225
2	7;30	449	224
3	11;15	671	222
4	15	890	219
5	18;45	1105	215
6	22;30	1315	210
7	26;15	1520	205
8	30	1719	199
9	33;45	1910	191
10	37;30	2093	183
11	41;15	2267	174
12	45	2431	164
13	48;45	2585	154
14	52;30	2728	143
15	56;15	2859	131
16	60	2978	119
17	63;45	3084	106
18	67;30	3177	93
19	71;15	3256	79
20	75	3321	65
21	78;45	3372	51
22	82;30	3409	37
23	86;15	3431	22
24	90	3438	7

Table 4: Mādhavaś sine table

Number	α (°)	$R \sin\alpha$
1	3;45	0224;50,22
2	7;30	0448;42,58
3	11;15	0670;40,16
4	15	0889;45,15
5	18;45	1105;01,39
6	22;30	1315;34,07
7	26;15	1520;28,35
8	30	1718;52,24
9	33;45	1909;54,35
10	37;30	2092;46,03
11	41;15	2266;39,50
12	45	2430;51,15
13	48;45	2548;38,06
14	52;30	2727;20,52
15	56;15	2858;22,55
16	60	2977;10,34
17	63;45	3038;13,17
18	67;30	3176;03,50
19	71;15	3255;18,22
20	75	3320;36,30
21	78;45	3321;41,29
22	82;30	3408;20,11
23	86;15	3430;23,11
24	90	3437;44,48

table with $R = 60$ can be transferred directly to the sine table with $R = 120$. Table 3 compares the first four and the last four values in the *Pañcasiddhāntikā* with Ptolemy's corresponding ones. The fractional parts after the semicolon in both tables are expressed sexagesimally. It seems that Varāhamihira's values were the results of rounding the numbers in the second fractional place of a chord table similar to that of Ptolemy.

In these earlier Indian sine tables, only the twenty-four values in the first quadrant are given, with the interval of 3°45′. Although Indians knew and used cosines (*koṭijyā*), they had no need of tabulating them because they knew that cosines could be derived from sines by the relation (2) above. On the other hand they were interested in the versed sine (*śara* in Sanskrit, meaning "arrow" or *utkramajyā*, "sine of the reversed order", CH in Fig. 1) which is defined as:

$$R\text{vers}\alpha = R - R\sin(90^\circ - \alpha).$$

Using this relation, Brahmagupta (seventh century) simplified some formulas. For instance, formula (3) was rewritten as:

$$R\sin\frac{\alpha}{2} = \sqrt{\frac{D \times R\text{vers}\alpha}{4}}. \qquad (3')$$

A table of versed sines can be easily obtained by adding Δs in Table 2 successively from the bottom upward (namely, in the "reversed order").

Brahmagupta computed anew 24 sines with $R = 3270$ in the *Brāhmasphuṭasiddhānta*. Elsewhere in this book and in the *Khaṇḍakhādyaka*, he offers a sine table with $R = 150$ and with the interval of 15°. This small table generates remarkably correct sine values when his ingenious method of second order interpolation is applied.

Bhāskara II

An improved version of the traditional sine table was prepared by Bhāskara II (b. 1114). In the chapter "Derivation of Sines" (*Jyotpatti*) of his *Siddhāntaśiromaṇi* he introduces two new values:

$$R\sin 36^\circ = \sqrt{\frac{5-\sqrt{5}}{8}}\,R,$$

$$R\sin 18^\circ = \sqrt{\frac{\sqrt{5}-1}{4}}\,R.$$

With these two values and formulas (2) and (3) above, he obtains $R\sin 3n^\circ$ (where $n = 1, 2, 3, \ldots, 30$). Further he combines them with the approximate value

$$R\sin 1^\circ \approx 60'$$

using the new formula:

$$R\sin(\alpha \pm \beta) = \frac{R\sin\alpha R\cos\beta \pm R\cos\alpha R\sin\beta}{R}, \qquad (4)$$

which is equivalent to the modern formula:

$$\sin(\alpha \pm \beta) = \sin\alpha\cos\beta \pm \cos\alpha\sin\beta.$$

Thus he could obtain sines for all the integer degrees of a quadrant. Formula (4) was unknown to Indians before Bhāskara II, while a chord version of the same formula was known to Ptolemy.

Trigonometry underwent a remarkable development in the early fifteenth century on the western coast of south India (the modern state of Kerala). The person who initiated this development was Mādhava (fl. ca. 1380/1420) of Saṅgamagrāma (near modern Cochin). His important works on astronomy and mathematics are now lost, but we know his achievements from the books of his successors. A sine table ascribed to him is quoted in Nīlakaṇṭha's commentary on the *Āryabhaṭīya* (Table 4).

A couple of verses, which are often quoted by the students of the Mādhava school and which are ascribed to Mādhava himself by Nīlakaṇṭha, give the method of computing sines. The method can be expressed as:

$$R\sin\theta = \theta - \frac{\theta^3}{3!R^2} + \frac{\theta^5}{5!R^4} - \frac{\theta^7}{7!R^6} + \frac{\theta^9}{9!R^8} - \cdots$$

With $R = 1$ this is equivalent to Newton's

$$\sin\theta = \theta - \frac{\theta^3}{3!} + \frac{\theta^5}{5!} - \frac{\theta^7}{7!} + \frac{\theta^9}{9!} - \cdots.$$

Similar power series for cosine and versed sine are ascribed to Mādhava.

MICHIO YANO

REFERENCES

Gupta, R.C. "South Indian Achievements in Medieval Mathematics." *Gaṇita-Bhāratī* 9(1–4): 15–40, 1987.

Pingree, David and David Gold. "A Hitherto Unknown Sanskrit Work Concerning Mādhava's Derivation of the Power Series for Sine and Cosine." *Historia Scientiarum* 42: 49–65, 1991.

Toomer, Gerald J. "The Chord Table of Hipparchus and the Early History of Greek Trigonometry." *Centaurus* 18: 6–28, 1973.

Toomer, Gerald J. *Ptolemy's Almagest.* New York: Springer Verlag, 1984.

See also: Almagest – Decimal System – Sexagesimal System – Āryabhaṭa – Mādhava – Varāhamihira – Brahmagupta –Śrīpati – Nīlakaṇṭha

TRIGONOMETRY IN ISLAMIC MATHEMATICS

Trigonometry is the connecting link between mathematics and astronomy, between the way calendars are calculated, the gnomon, and the sundial. In the Islamic world, the calculation of spherical triangles was necessary to carry out ritual customs. The *qibla*, the direction to Mecca, was indicated next to the hour lines on all public sundials.

The first trigonometric problems appeared in the field of spherical astronomy. Around the year 773 one of the Indian *siddhāntas* (astronomy books) was made known in Baghdad. The Indian astronomers Varāhamihira (fifth century) and Brahmagupta (sixth century) solved different problems in spherical astronomy by means of rules equivalent to a general sine theorem for a spherical triangle ABC with sides a, b, c and angles A, B, C (where angle A is opposite to side a, etc.), namely $(\sin A / \sin a) = (\sin B / \sin b) = (\sin C / \sin c)$ and to the cosine theorem for the same triangle $\cos a = \cos b \cos c + \sin b \sin c \cos A$.

In the ninth century Ptolemy's *Almagest* and Menelaus' *Spherics* were also translated, and commentaries were written to these works. Many trigonometrical problems were solved in Ptolemy's *Almagest*, in which Menelaus' theorem on the spherical complete quadrilateral was used. The cases of this theorem used by Ptolemy are equivalent to the sine and tangent theorems for a right-angled spherical triangle. The *Almagest*, the *Spherics* and the Indian *siddhāntas* formed the basis on which Arab mathematicians built their trigonometry.

The ancient Greek astronomers only used one trigonometric function, the chord of an arc. The "theorem of Ptolemy", which is equivalent to the formula for the sine of the sum of the angles, forms, together with the formula for the chord of the half arc, the basis for the chord table in the *Almagest*. The Indian people replaced the chord with the sine, introduced the cosine and the versed sine, and compiled a small table of sine values. The Arabic mathematicians progressively made trigonometry into a science independent of its (astronomical) context.

Applications of trigonometry analogous to those in the Indian *siddhāntas* are found in the astronomical works of al-Khwārizmī. An analogous geometric construction for finding the azimuth according to the rule formulated in al-Khwārizmī's third treatise was provided by al-Māhānī (ca. 825–888) in his *Treatise on the Determination of the Azimuth at Any Time and in Any Place*. The rules equivalent to the spherical sine and cosine theorems were also used by Thābit ibn Qurra in his *Book on Horary Instruments Called Sundials*. With Ḥabash the applications of the tangent and cotangent functions went beyond the usual applications in the theory of sundials. The introduction of the tangent and cotangent and their application in astronomy was a novelty. The names *zill* (shadow) and *zill maʿqus* (reversed shadow) apparently are translations from Sanskrit. In the case of a vertical gnomon, al-Ḥabas expressed the cosecant as the "diameter of the shadow" for a given height of the sun, i.e. as a hypotenuse. He computed a table for the cosecant with steps of one degree.

For a long time, the chord was used along with the sine. A theory of these magnitudes is found in the work of al-Battānī (ca. 858–929). In his astronomical work *Islaḥ al-Majisṭī* (The Perfection of the Almagest), he systematically employed the trigonometric functions sine and versed sine with arguments between 0° and 180°. Since the cosine is defined as the sine of the complement of the angle, and since no negative numbers are used, the versed sine is defined in the second quadrant as a sum of two quantities. The elements of trigonometry are set forth in an even more systematic way in the *Kitāb al-Kāmil* (Perfect Book) of Abū'l-Wafā᾿ (940–997/8). He defined several trigonometric functions in the circle with radius 1. The trigonometrical tangent function is defined as a line on a tangent to the circle.

The proof of the general spherical sine theorem was given by Abū'l-Wafā᾿ in his *al-Majisṭī* (Almagest), by his pupil Abū Naṣr ibn ʿIrāq (d. 1036) in the *Risāla fī maʿrifa al-qisī al-falakiyya* (Treatise on the Determination of Celestial Arcs), and by al-Khujandī (d. ca. 1000) in the *Kitāb fī al-sāʿāt al-māḍiyya fī al-layl* (Book on Past Hours in the Night). The history of the discovery of this theorem was described by al-Bīrūnī, the pupil of Ibn ʿIrāq and al-Khujandī, in the *Kitāb maqālīd ʿilm al-hay'a* (Book on the Keys of Astronomy).

The use of trigonometry was expanded through al-Bīrūnī (973–1048). He is the author of the *Masʿūdic Canon*, which is a summary of the results from the works of many predecessors and of personal observations and calculations. It comprises eleven books. Book III is dedicated to trigonometry. It has calculations equivalent to the formulas for the sine of the sum of two angles, the sine of the differences between two angles, and the sine of the double angle. It also includes the solution of cubic equations and the division of angles into three parts, and the sine rule of plane trigonometry: $(\sin A/a) = (\sin B/b) = (\sin C/c)$. (The plane cosine theorem $a^2 = b^2 + c^2 - 2bc \cos A$ is equivalent to two of Euclid's theorems).

Another important scholar in the area of trigonometry was Naṣīr al-Dīn al-Ṭūsī (1201–1274). His principal work was *Kitāb al-shakl al-qaṭṭāʿ* (Book on the Secant Figure, also known as Treatise on the Complete Quadrilateral). It was written in Persian and translated by the author into Arabic in 1260, possibly for the needs of the observatory of Maragha. In five books, it contains a full system of trigonometrical formulas for plane and spherical triangles. If any three elements of such a triangle are given, the other three elements can be found by the theory explained in this work, which also contains the notion of the polar triangle A′B′C′ of a spherical triangle ABC ($A' = 180° - a$, $B' = 180° - b$, $C' = 180° - c$). This work played an important role in the development of mathematics in Europe.

SAMI CHALHOUB
BORIS A. ROSENFELD

REFERENCES

Bruins, Evert M. "Ptolemaic and Islamic Trigonometry: The Problem of the Qibla." *Journal for the History of Arabic Science* 9(2): 45–68, 1991.

Debarnot, Marie-Thérese. *Kitāb Maqālīd ʿIlm al-Hay'a: la Trigonométrie Sphérique chez les Arabes de l'Est à la Fin du Xe Siècle*. Damascus: Institut français de Damas, 1985.

Juschkewitsch, A.P. *Geschichte der Mathematik im Mittelalter*. Teubner: Leipzig, 1964.

Kennedy, E.S. *A Survey of Islamic Astronomical Tables*. Philadelphia, Pennsylvania: American Philosophical Society, 1956.

von Braunmühl. Anton. *Vorlesungen über Geschichte der Trigonometrie*. Leipzig, 1900; Repr. Wiesbaden: Sändig, 1971.

See also: *Qibla* – Varāhamihira – Brahmagupta – al-Khwārizmī – al-Māhānī – Thābit ibn Qurra – al-Battānī – Abū'l-Wafā'– al-Khujandī – al-Bīrūnī – Naṣīr al-Dīn al-Ṭūsī

U

ULUGH BĒG Ulugh Bēg, Mīrzā Muḥammad ibn Shārhrukh ibn Tīmūr Ulugh Bēg Guragān, 1394–1449, was the ruler of Samarqand (now in Uzbekistan) and an astronomer, mathematician, and poet. His nickname, Ulugh Bēg, means "great prince". He was the grandson of the great conqueror Tīmūr. In 1417, as a pupil of Qāḍī Zādeh al-Rūmī of Bursa (Turkey), he opened in Samarqand the *madrasa* (school) where al-Rūmī was the teacher. In 1425 he founded an astronomical observatory and invited Jamshīd al-Kāshī to be its director. After al-Kāshī's death, the head of the observatory became ʿAlī al-Qūshjī. Ulugh Bēg was killed by enemies of enlightenment. After his death his observatory was destroyed, and al-Qūshjī fled to Turkey.

The main work of Ulugh Bēg was a book of astronomical tables known as the *Zīj-i Ulugh Bēg* or *Zīj-i jadīd-i Guragānī* (New Zij of Guragan) written together with al-Kāshī, al-Rūmī, and al-Qūshjī. The tables are written in Persian and are extant in many manuscripts in Persian and Arabic translations. The work consists of four books of trigonometrical, astronomical, geographical, and astrological tables.

Book I covered the subject of calendars, including the Muslim lunar calendar, the pre-Islamic Persian solar calendar, and the Chinese-Uyghur calendar. Book II dealt with spherical astronomy, covering sine and versed sine, shadows (tangents and cotangents), spherical coordinates on the celestial sphere (equatorial, ecliptical, and horizontal coordinates), geographical coordinates, as well as spherical distances between stars and the direction of *qibla* (to Mecca). In Book III Ulugh Bēg considered planetary and stellar astronomy, such as the motion of the sun, moon, and planets, the distances of the sun and the moon from the center of the world, and the equalization of astrological houses. Finally, Book IV dealt with astronomical calculations.

Ulugh Bēg also was the author of the mathematical *Risāla fī istikhrāj jayb daraja wāḥida* (Treatise on the Determination of Sine of One Degree). Originally the treatise was wrongly ascribed to al-Rūmī; Ulugh Bēg's authorship was established on the basis of the information in al-Birjandī's commentaries to the *Zīj* of Ulugh Bēg.

BORIS ROSENFELD

REFERENCES

Kennedy, E.S. "A Letter of Jamshīd al-Kāshī to his Father." In *Studies in the Islamic Exact Sciences*. Beirut: American University, 1983, pp. 722–744.

Knobel, E.B., ed. *Ulugh Beg's Catalogue of Stars*. Washington, D.C.: Carnegie Institution, 1917.

Matvievskaya, Galina P. and B.A. Rosenfeld. *Mathematicians and Astronomers of Medieval Islam and Their Works (8th–17th c.)*. Moscow: Nauka, 1983, pp. 492–495 (in Russian).

Sayılı, Aydın. *The Observatory in Islam*. Istanbul: Turk Tarih Kurumu, 1960.

Sédillot, L.A. *Prolegomènes des Tables Astronomiques d'Oloug Beg*. Paris: Didot frères, 1847 and 1853.

Ulugh Bēg. "Treatise on Determination of Sine of One Degree (Fragment)." Trans. B.A. Rosenfeld. In *Reader in the History of Mathematics, Arithmetic, Algebra, Number Theory, and Geometry*. Ed. I.G. Bashmakova et al. Moscow: Prosveshchenie, 1975, pp. 79–82 (in Russian).

See also: al-Kāshī – Observatories in Islam – *Zīj* – al-Rūmī – *Qibla*

AL-UQLĪDISĪ Al-Uqlīdisī, Abū'l-Ḥasan Aḥmad ibn Ibrāhīm, wrote an Arithmetic (*Kitāb al-Fuṣūl fī'l-ḥisāb al-hindī*) in Damascus in 952–53. This is a sizable compendium and remarkable as the earliest arithmetic extant in Arabic.

The first part explains the place-value system, the four arithmetical operations, and the extraction of square roots for integers and fractions, both common and sexagesimal. Numerous examples are given. This part is supposed to be accessible to a large audience. The second part develops the earlier topics and adds curiosities or different methods. The third part would seem to be the result of the author's experience in teaching; it consists of explanations and questions with their answers concerning some difficulties the reader might have met in the first two parts. The fourth part contains some digressions about the changes Indian arithmetic undergoes when one uses ink and paper (since Indian computations were made on the dust abacus). In this part al-Uqlīdisī also explains (according to him, better than his predecessors) how to extract cube roots.

Al-Uqlīdisī was concerned with the applicability of arithmetic. How original his work was we do not know. He often claims originality or at least superiority of his teaching, but so do his contemporaries. He does not claim originality, however, for the most important feature of his *Arithmetic*, the first occurrence of decimal fractions (besides the usual common and sexagesimal ones). He uses a mark placed over the last integral unit in order to indicate the separation from the subsequent, decimal part.

JACQUES SESIANO

REFERENCES

Saidan, A.S. *Al-fuṣūl fī'l-ḥisāb al-hindī*. Amman: al-Lajnah al-Urdunnīyah lil-Taʿrīb wa'l-Nashr wa'l-Tarjaman, 1973. [Arabic text].

Saidan, A.S. *The Arithmetic of al-Uqlīdisī*. Dordrecht: Reidel, 1978. [English translation].

Saidan, Ahmad. "The Earliest Extant Arabic Arithmetic." *Isis* 57: 475–490, 1966.

AL-ʿURḌĪ Muʾayyad al-Dīn ibn Barmak al-ʿUrḍī al-Dimashqī (thirteenth century) was an astronomer, architect, and engineer. He as born in Damascus and first worked in Syria. He did some hydraulic engineering in Damascus and constructed an astronomical instrument for the ruler of Hims (Emessa), al-Manṣūr Ibrāhīm (1239–1245). He also taught *handasa* (architecture or geometry).

In or soon after 1259 he was in Maragha, the capital of the Mongol conqueror Hulagu Khan, the grandson of Genghis Khan. He was one of four astronomers who worked with Naṣīr al-Dīn al-Ṭūsi, the founder of the Maragha observatory. He participated in the organization, building, and construction of the instruments of this observatory, and in the building of a mosque and a palace in Maragha.

The best known al-ʿUrḍī's works is *Risāla fī kayfiyya al-arṣād wa mā yukhtāj ilā ʿilmihi wa ʿamalihi min al-ṭuruq al-muwaddiya ilā maʿrifa ʿawdāt al-kawākib* (Modes of Astronomical Observations and the Theoretical and Practical Knowledge Needed to Make Them, and the Methods Leading to Understanding the Regularities of the Stars). Manuscripts of this treatise are extant in Istanbul, Paris, and Teheran. The treatise contains the description of eleven of the most important astronomical instruments of the Maragha observatory which he himself mainly constructed: (1) mural quadrant, (2) armillary sphere, (3) solstitial armilla, (4) equinoctial armilla, (5) Hipparchus' diopter (alidade), (6) instrument with two quadrants, (7) instrument with two limbs, (8) instruments to determine sines and azimuths, (9) instruments to determine sines and versed sines, (10) "the perfect instrument" built by him in Syria, and (11) parallactic ruler (after Ptolemy).

Al-ʿUrḍī was also the author of three astronomical treatises: *Kitāb al-hay'a* (Book on Astronomy) on the motion of planets; *Risāla fī ʿamal al-kura al-kāmila* (Treatise on Construction of the Perfect Sphere); and *Risāla fī 'l-taʿrīf al-buʿd bayna markaz al-shams wa 'l-awj* (Treatise on the Determination of the Distance between the Center of the Sun and the Apogee).

Al-ʿUrḍī's sons Shams al-Dīn and Muḥammad also worked in the Maragha observatory. Muḥammad (ca. 1280) constructed a celestial globe 150 mm in diameter, which is extant in the Mathematical Salon in Dresden.

BORIS A. ROSENFELD

REFERENCES

Jourdain, Amable. *Mémoire sur l'Observatoire de Méragha et les Instruments Employées pour y Observer*. Paris: Joubert, 1810.

Matvievskaya, Galina P. and B.A. Rosenfeld. *Mathematicians and Astronomers of Medieval Islam and Their Works (8th-17th c.)*, vol. 2. Moscow: Nauka, 1983, pp. 414–415. (in Russian)

Sarton, George. *Introduction to the History of Science*, vol. 2. Baltimore: Williams & Wilkins, 1931, pp. 1013–1014.

Sayılı, Aydın. *The Observatory in Islam*. Istanbul: Turk Tarih Kurumu, 1960.

See also: Maragha – Observatories – Armillary Sphere – Naṣīr al-Dīn al-Ṭūsi

V

VĀKYAKARAṆA *Vākyakaraṇa*, apocryphally ascribed to Vararuci, is an astronomical manual produced in about AD 1300 which was very popular in South India, especially in Tamil Nadu and the adjoining regions. The work is so called because it was a *karaṇa* (astronomical manual) which used computational tables where the numbers are expressed in mnemonic sentences (*vākyas*), phrases, and words. Its is also called *Vākyapañcādhyāyī*, because it contains *pañca adhyāyas* (five chapters). Until recently, when the *Nautical Almanac* came to be used in South India for the computation of the Hindu almanac with its 'five limbs' (*pañca-aṅga*): the lunar day, weekday, asterism, *yoga* and *karaṇa*, the *Vākyakaraṇa* was widely used for that purpose.

The author of the *Vākyakaraṇa* hailed from Kariśaila or Kānchī in Tamil Nadu, as he himself states in his work. He also states that he based his work on the writings of Bhāskara I (AD 629), the exponent of the school of astronomy promulgated by Āryabhaṭa (b. AD 476), and the works of Haridatta (AD 683), author of *Grahacāranibandhana* and *Mahāmārganibandhana*. The *Vākyakaraṇa* was elaborately expounded by Śundararāja (AD 1500) who had contacts with the Kerala astronomer Nīlakaṇṭha Śomayāji. The date of *Vākyakaraṇa* is determined as ca. AD 1300, on the basis of epochs which the author gives for the computation of the planets.

In five chapters, the *Vākyakaraṇa* deals with all aspects of astronomy required for the preparation of the Hindu almanac. Chapter I is concerned with the computation of the sun, the moon, and the moon's nodes, and Chapter II with that of the planets. Chapter III is devoted to problems involving time, position, and direction, and other preliminaries like the precession of the equinoxes. The computation of the lunar and solar eclipses is the concern of Chapter IV. Chapter V is devoted to the computation of the conjunction of the planets, and of planets and stars.

For the computation of the moon, the *Vākyakaraṇa* employs the 248 moon sentences of the ancient astronomer Vararuci. But for the five planets, Mars, Mercury, Jupiter, Venus, and Saturn, the author himself computed 82 tables devoted to the different planetary cycles, containing, in all, 2075 mnemonic sentences (*Kujādi-pañca-graha-mahā-vākyas*).

The results obtained through the computations enunciated in the *Vākyakaraṇa* are not very accurate, going by modern standards, but they were accurate enough for the determination of auspicious times and other matters required in the routine life of orthodox Hindus. Besides, there was the saving of much time and labor in working with the simple methods advocated in the work.

K.V. SARMA

REFERENCES

Candravākyāni (Moon Sentences of Vararuci). Ed. as Appendix II in *Vākyakaraṇa*. Ed. T.S. Kuppanna Sastri and K.V. Sarma. Madras: Kuppuswami Sastri Research Institute, 1962, pp. 125–34.

Kujādi-pañcagraha-mahāvākyāni (Long Sentences of the Planets Kuja etc.). Ed. as Appendix III in *Vākyakaraṇa* Ed. T.S. Kuppanna Sastri and K.V. Sarma. Madras: Kuppusawmi Sastri Research Institute, 1962, pp. 135–249.

Vākyakaraṇa with the commentary *Laghuprakāśikā* by Sundararāja. Ed. T.S. Kuppanna Sastri and K.V. Sarma. Madras: Kuppuswami Sastri Research Institute, 1962.

See also: Bhāskara I – Haridatta – Āryabhaṭa – Nīlakaṇṭha Śomayāji – Precession of the Equinoxes – Lunar Mansions

VALUES AND SCIENCE Is science value free? If "science", as Lord Rutherford is reported to have said, "is what scientists do" then scientists would have to be superhuman to keep their values out of what they do in their laboratories. We cannot clinically isolate our cultural and ethical (and through them our historical) baggage from our human activities. Values play an important part in what we select to do or not do and how we actually do it. In as far as science is a human activity, it is subject to the strengths and weaknesses of all human activities.

However, in Western tradition, up to quite recently, scientists were seen as quasi-religious supermen, heroically battling against all odds to discover the truth. Also, the truths they wrestled out of nature were said to be absolute, objective, value-free, and universal. The idea of scientists as dedicated hermit-like lone researchers is now obsolete. Nowadays, science is an organized, institutionalized, and industrialized venture. The days when individual scientists, working on their own, and often in their garden sheds, made original discoveries are really history. Virtually all science today is big science requiring huge funding, large, sophisticated, and expensive equipment, and hundreds of scientists working on minute problems. As such, science has become a unified system of research and application, with funding at one end and the end-product of science, often technology, at the other.

Values enter this system in a number of ways. The first point of entry is the selection of the problem to be investigated. The choice of the problem, who makes the choice and on what grounds, is the principle point of influence of society, political realities of power, prejudice, and value systems on even the "purest" science. Often, it is the source of funding that defines what problem is to be investigated. If the funding is coming from government sources then it will reflect the priorities of the government — whether space exploration is more important than health problems of the inner city poor, or nuclear power or solar energy should be developed further. Private sector funding, mainly from multinationals, is naturally geared towards research that would eventually bring dividends in terms of hard cash. Some eighty percent of research in the United States is funded by what is called the "military–industrial complex" and is geared towards producing both military and industrial applications.

Subjectivity thus enters science in terms of what is selected for research which itself depends on where the funding is coming from. But values also play an important part in what is actually seen as a problem, what questions are asked and how they are answered. For example, cancer rather than diabetes may be seen as a problem even though they may both claim the same number of victims. Here both political and ideological concerns, as well as public pressure, can make one problem invisible while focusing attention on another. Moreover, if, for example, the problem of cancer is defined as finding a cure then the benefits of the scientific research accrue to certain groups, particularly the pharmaceutical companies. But if the function of scientific research is seen as eliminating the problems of cancer from society, then another group benefits from the efforts of research: the emphasis here shifts to investigating diet, smoking, polluting industries, and the like. Similarly, if the problems of the developing countries are seen in terms of population, then research is focused on reproductive systems of Third World women, methods of sterilization, and new methods of contraceptives. However, if poverty is identified as the main cause of the population explosion then research would take a totally different direction: the emphasis would have to shift to investigating ways and means of eliminating poverty, developing low cost housing, basic and cheap health delivery systems, and producing employment generating (rather than profit producing) technologies. The benefits of scientific research would go to the Third World poor rather than Western institutions working on developing new methods of contraceptives and companies selling these contraceptives to developing countries. Thus both the selection of problems and also their framing in a particular way are based on value criteria.

It can be legitimately argued that these factors are external to science, that within science, the scientific method ensures neutrality and objectivity by following a strict logic — observation, experimentation, deduction, and value-free conclusion. But scientists do not make observations in isolation. All observations take place within a well defined theory. The observations, and the data collection that goes with them, are designed either to refute a theory or provide support for it, and theories themselves are not plucked out of the air. Theories exist within paradigms — that is a set of beliefs and dogmas. The paradigms provide a grand framework within which theories are developed and make sense, and observations themselves have validity only within specific theories. Thus, all observations are theory laden, and theories themselves are based on paradigms which in turn are burdened with cultural baggage. All of which raises the question: can there ever be such things as value-neutral "objective facts"? Studies of scientists working in laboratories have shown that the scientific method, by and large, is a myth. Researchers seldom follow it in the linear fashion that exists in the text books. Neither do they ascertain new "facts" suddenly out of the blue. And the same holds true of the laws of nature they are supposed to be discovering. It appears that scientists do not actually "discover" laws of nature; they manufacture them. Scientific knowledge advances by a process of manufacturing which involves thousands and thousands of workers assembling "facts" which through peer review and other procedures end as description or laws of nature.

Value judgments are also at the very heart of a common element of scientific technique: statistical inference. When it comes to measuring risks, scientists can never give a firm answer. Statistical inferences cannot be stated in terms of "true" or "false" statements. When statisticians test a scientific hypothesis they have to go for a level of "confidence". Different problems are conventionally investigated to different confidence-limits. Whether the limit is 95 or 99.9 per cent depends on the values defining the investigations, the costs, and weight placed on social, environmental, or cultural consequences. In most cases, the importance given to social and environmental factors determines the limits of confidence and the risks involved in a hazardous scientific endeavor. For example, when a chemical plant is placed in an area with an aware and politically active citizenry the risks are worked out to a high level of confidence. However, when one is located in an area where the citizens themselves are ignorant of the dangers and do not command political power, the confidence levels are much more relaxed. The people of Bhopal and Chernobyl know this to their cost.

But it is not just in its institutions and method that science is value laden. The very assumptions of science about nature, universe, time, and logic are ethnocentric. In modern science, nature is seen as hostile, something to be dominated. The Western "disenchantment of nature" was a crucial element in the shift from the medieval to the modern mentality, from feudalism to capitalism, from Ptolemaic to Galilean astronomy, and from Aristotelian to Newtonian physics. In this picture, "Men" stand apart from nature, on a higher level, ready to subjugate and "torture" her, as Francis Bacon declared, in order to wrestle out her secrets. This view of nature contrasts sharply with how nature is seen in other cultures and civilizations. In Chinese culture, for example, nature is seen as an autonomous self-organizing entity which includes humanity as an integral part. In Islam, nature is a trust, something to be respected and cultivated and people and environment are a continuum — an integrated whole. The conception of laws of nature in modern science drew on both Judeo-Christian religious beliefs and the increasing familiarity in early modern Europe with centralized royal authority, with royal absolutism. The idea that the universe is a great empire, ruled by a divine logos, is, for example, quite incomprehensible both to the Chinese and the Hindus. In these traditions the universe is a cosmos to which humans relate directly and which echoes their concerns. Similarly, while modern science sees time as linear, other cultures view it as cyclic as in Hinduism or as a tapestry weaving the present with eternal time in the Hereafter as in Islam. While modern science operates on the basis of either/or Aristotelian logic (X is either A or non-A), in Hinduism logic can be fourfold or even sevenfold. The fourfold Hindu logic (X is neither A, nor non-A, nor both A and non-A, nor neither A nor non-A) is both symbolic as well as a logic of cognition and can achieve a precise and unambiguous formulation of universal statements without quantification. Thus the metaphysical assumptions of modern science make it specifically Western in its main characteristics.

The metaphysical assumptions of modern science are also reflected in its contents. For example, certain laws of science, as Indian physicists have began to demonstrate, are formulated in an ethnocentric and racist way. The Second Law of Thermodynamics, so central to classical physics, is a case in point: due to its industrial origins the Second Law presents a definition of efficiency that favors high temperatures and the allocation of resources to big industry. Work done at ordinary temperature is by definition inefficient. Both nature and the non-Western world become losers in this new definition. For example, the monsoon, transporting millions of tons of water across a subcontinent is "inefficient" since it does its work at ordinary temperatures. Similarly, traditional crafts and technologies are designated as inefficient and marginalized. In biology, social Darwinism is a direct product of the laws of evolutionary theories. Genetic research appears to be obsessed with how variations in genes account for differences among people. Although we share between 99.7 (unrelated people) and 100 per cent (monozygotic twins) of our genes, genetic research has been targeted towards the minute percentage of genes that are different in order to discover correlations between genes and skin color, sex or "troublesome" behavior. Enlightened societal pressures often push the racist elements of science to the sidelines. But the inherent metaphysics of science ensures that they reappear in new disguise. Witness how eugenics keeps reappearing with persistent regularity. The rise of IQ tests, behavioral conditioning, fetal research, and sociobiology are all indications of the racial bias inherent in modern science.

Given the Eurocentric assumptions of modern science, it is not surprising that the way in which its benefits are distributed and its consequences are accounted for are themselves ethnocentric. The benefits are distributed disproportionally to already over-advantaged groups in the West and their allies elsewhere, and the costs disproportional to everyone else. When scientific research improves the military, agriculture, manufacturing, health, or even the environment, the benefits and expanded opportunities science makes possible are distributed predominantly to already privileged people of European descent, while the costs are dumped on the poor, racial and ethnic minorities, women, and people located at the periphery of global economic and political networks. Science in developing countries has persistently reflected the priorities of the West, emphasizing the needs and requirements of middle class western society, rather than the wants and conditions of their own society. In over five decades of science development, most of the Third World countries have nothing to show for it. The benefits of science just refuse to trickle down to the poor.

But modern science is not only culturally biased towards the West: it represents the values of a particular class and gender in Western societies. As feminist scholars have shown, science in the West has systematically marginalized women. Women, on the whole, are not interested in research geared towards military ends, or torturing animals in the name of progress, or working on machines that put one's sisters out of work. But more than that, even the least likely fields and aspects of science bear the fingerprints of androcentric projects. Physics and logic, the prioritizing of mathematics and abstract thought, the so-called standards of objectivity, good method and rationality — feminist critique has revealed androcentric fingerprints in all. This is the case, for example, in the mechanistic model of early modern astronomy and physics, in modern particle physics, and in the coding of reason as part of ideal masculinity. The focus on quantitative measurements, variable analysis, impersonal and excessively abstract conceptual schemes is both a distinctively masculine tendency and one that serves to hide its own gendered character. Science has tried to hide its own masculine nature in other ways, by, for example, making women themselves objects in scientific investigation. It was not entirely accidental that sexology became a major science at the same time as women in the West were fighting for the vote and equal rights in education and employment. A number of studies have shown that scientific work done by women is invisible to men even when it is objectively indistinguishable from men's work. Thus, it appears that neither social status within science nor the results of research are actually meant to be neutral or socially impartial. Instead, the discourse of value-neutrality, objectivity, and social impartiality appears to serve projects of domination and control.

The history of science bears this out. The evolution of Western science can be traced back to the period when Europe began its imperial adventure. Science and empire developed and grew together, each enhancing and sustaining the other. In India, for example, European science served as a handmaiden to colonialism. The British needed better navigation so they built observatories and kept systematic records of their voyages. The first sciences to be established in India were, not surprisingly, geography and botany. Western science progressed primarily because of the military, economic, and political power of Europe, focusing on describing and explaining those aspects of nature that promoted European power, particularly the power of the upper classes. The disinterested commitment of European scientists to the pursuit of truths had little to do with the development of science. The subordination of the blacks in the ideology of the black "child/savage" and the confinement of the white women in the cult of "true womanhood" emerged in this period and are both by-products of the Empire. While the blacks were assigned animal and brutish qualities the white women were elevated and praised for their morality. While the blacks were segregated and enslaved, the women were placed in narrow circles of domestic life and in conditions of dependency. Racist and androcentric evolutionary theories were developed to explain human behavior and canonized in the history of human evolution. The origins of Western, middle class social life, where men go out to do what men have to do, and women tend the babies and look after the kitchen, are to be found in the bonding of "man-the-hunter"; in the early phases of evolution women were the gatherers and men went out to bring in the beef. Now this theory is based on little more than the discovery of chipped stones that are said to provide evidence for the male invention of tools for use in the hunting and preparation of animals. However, if one looks at the same stones with different cultural perceptions, say one where women are seen as the main providers of the group — and we know that such cultures exist even today — you can argue that these stones were used by women to kill animals, cut corpses, dig up roots, break down seed pods, or hammer and soften tough roots to prepare them for consumption. A totally different hypothesis emerges and the course of the whole evolutionary theory changes.

Thus the cultural, racial, and gender bias of modern science can be easily distinguished when it is seen from the perspective of non-Western cultures, marginalized minorities, and women. The kinds of questions science asks when seeking to explain nature's regularities and underlying causal tendencies, the kinds of data it generates and appeals to as evidence for different types of questions, the hypotheses that it offers as answers to these questions, the distance between evidence and the hypothesis in each category, and how these distances are traversed — all have the values of white middle class men embedded in them. Put simply, this implies a relativism in science as in any other sphere of human knowledge. However, most scientists do not look kindly towards criticism, or sociological, philosophical, historical, and anthropological studies which highlight science's value laden nature. Relativism is anathema to scientists: many believe that they are engaged in revealing nature's absolute truths. Science, they argue, is special and different from any other body of knowledge: it is counter-intuitive and rarely a matter of common sense. Some propagandists for science have even suggested that the entire discipline of the sociology of knowledge is a conspiracy of the academic left against science.

The idolization and mystification of science, the insistence on its value neutrality and objectivity, is an attempt not only to direct our attention away from its subjective nature but also from the social and hierarchical structure of science. Whenever we think of "the scientists" we imagine white men in white coats: the sort of chaps we see in advertisements for washing power and skin care preparations, standing in a busy laboratory behind a Bunsen burner and distillation equipment telling us how the application of science has led to a new and improved soap or cold cream. This view of scientists is not far from reality. True power in science belongs to white, middle-aged men of upper classes. Everyone else working in science — women, minorities, black men, white men of lower classes, and third world researchers — are actually basically rank and file laboratory workers. The social hierarchy within science by and large preserves absolute social status, the social status scientific workers hold in the larger society. The people who make decisions in science, who decide what research is to be done, what questions are going to be asked, and how the research is going to be done are a highly selective, tiny minority. These people have the right background, the contacts to get the necessary appointments, and then further contacts to secure funding for their research projects. The actual execution of scientific research, the grinding and repetitive laboratory work, is rarely done by the same person who conceptualizes that research; even the knowledge of how to conduct research is rarely possessed by those who actually do it. This is why the dominant (Western) social policy agendas and the conception of what is significant among scientific problems are so similar. This is why the values and agendas important to white, middle class men pass through the scientific process to emerge intact in the results of research as implicit and explicit policy recommendations. This is why modern science has become an instrument of control and manipulation of non-Western cultures, marginalized minorities, and women.

Even if we were to ignore all other arguments and evidence, the very claim of modern science to be value-free and neutral would itself mark it as an ethnocentric and a distinctively Western enterprise. Both claiming and maximizing cultural neutrality is itself a specific western cultural value: non-Western cultures do not value neutrality for its own sake but emphasize and encourage the connection between knowledge and values. By deliberately trying to hide its values under the carpet, by pretending to be neutral, by attempting to monopolize the notion of absolute truth, Western science has transformed itself into a dominant and dominating ideology.

ZIAUDDIN SARDAR

REFERENCES

Alvares, Claude. *Science, Development and Violence*. Delhi: Oxford University Press, 1992.

Aronowitz, Stanley. *Science As Power*. London: Macmillan, 1988.

Collins, Harry and Trevor Pinch. *The Golem: What Everyone Should Know About Science*. Cambridge: Cambridge University Press, 1993.

Feyerabend, Paul. *Against Method*. London: NLB, 1975.

Feyerabend, Paul. *Science in a Free Society*. London: Verso, 1978.

Funtowicz, Silvio and J.R. Ravetz. *Uncertainty and Quality in Science for Policy*. Dordrecht: Kluwer, 1990.

Goonatilake, Susantha. *Aborted Discovery: Science and Creativity in the Third World*. London: Zed, 1984.

Gross, Paul R. and Norman Levitt. *Higher Superstition: The Academic Left and Its Quarrels with Science*. Baltimore: Johns Hopkins University Press, 1994.

Harding, Sandra. *The Science Question in Feminism*. Milton Keynes: Open University Press, 1986.

Harding, Sandra, ed. *The "Racial" Economy of Science*. Bloomington: Indiana University Press, 1993.

Harding, Sandra. "Is Science Multicultural? Challenges, Resources, Opportunities, Uncertainties." In *Multiculturalism: A Critical Reader*. Ed. David Theo Goldberg. Oxford: Blackwell, 1994, pp. 344–370.

Jacob, Margaret, ed. *The Politics of Western Science*. Atlantic Highlands, New Jersey: Humanities Press, 1994.

Knorr-Cetina, Karin. *The Manufacture of Knowledge*. Oxford: Pergamon, 1981.

Kuhn, T. S. *The Structure of Scientific Revolutions*. Chicago: University of Chicago Press, 1962; 2nd ed., 1972.

Lakatos, Imre and Alan Musgrove. *Criticism and the Growth of Knowledge*. Cambridge: Cambridge University Press, 1970.

McNeil, Maureen, ed. *Gender and Expertise*. London: Free Association Books, 1987.

Mitroff, Ian. *The Subjective Side of Science*. Amsterdam: Elsevier, 1974.

Moraze, Charles. *Science and the Factors of Inequality*. Paris: UNESCO, 1979.

Nandy, Ashis, ed. *Science and Violence*. Delhi: Oxford University Press, 1988.

Ravetz, J. R. *Scientific Knowledge and Its Social Problems*. Oxford: Oxford University Press, 1971.

Ravetz, J. R. "Science and Values." In *The Touch of Midas: Science, Values and the Environment in Islam and the West*. Ed. Ziauddin Sardar. Manchester: Manchester University Press, 1982, pp. 43–53.

Ravetz, J. R. *The Merger of Knowledge With Power*. London: Mansell, 1990.

Rouse, Joseph. *Knowledge and Power*. Ithaca: Cornell University Press, 1987.

Sardar, Ziauddin, ed. *The Touch of Midas: Science, Values and the Environment in Islam and the West*. Manchester: Manchester University Press, 1982.

Sardar, Ziauddin. "Conquests, Chaos, Complexity: The Other in Modern and Postmodern Science." *Futures* 26(6): 665–682, 1994.

Seshadri, C.V. *Equity is Good Science*. Madras: Murugappa Chettier Research Centre, 1993.

Walpert, Lewis. *The Unnatural Nature of Science*. London: Faber and Faber, 1992.

See also: Colonialism and Science – Technology and Culture – Western Dominance – Environment and Nature

VARĀHAMIHIRA Varāhamihira, who flourished in Ujjain, in Central India, during the sixth century, was perhaps the greatest exponent of the twin disciplines of astronomy and astrology in India. A master of all three branches of the disciplines astronomy, natural astrology, and horoscopic astrology, he was a prolific writer whose works number more than a dozen, some of which are extensive.

Varāhamihira was born in Kāpitthaka, present-day Kapitha, in Uttar Pradesh, known also as Saṅkāśya and mentioned as a great center of learning by the Chinese pilgrim Yuan Chwang as Kah-pi-t'a. He was the son of Ādityadāsa, and a Śakadvīpī brāhmaṇa of the Maga sect who were sun worshippers. His renown has caused several legends, both Hindu and Jain, being woven round his birth, growth, and predictive propensity. A legend has it that he was one of the nine luminaries of the court of King Chandragupta II Vikramāditya, but the definitively known date of Varāhamihira goes against this identification. Varāhamihira's patron has now been identified as King Mahārājādhirāja Dravyavardhana who ruled over Ujjain during the middle of the sixth century. Varāhamihira also possessed high poetic talents, so that some of the later rhetoricians have extracted verses from his writings to illustrate poetic qualities. Indeed, the fame of Varāhamihira has even induced several places in India to be named after him.

It is a characteristic of Varāhamihira that he produced larger and shorter versions of most of his works. His only work on astronomy is *Pañcasiddhāntikā* (The Five Basic Texts), being a redaction of select topics from the basic texts of five earlier astronomical schools: *Vāsiṣṭha*, *Paitāmaha*, *Romaka*, *Pauliśa* and *Saura*, in 443 verses, in the form of a work manual.

Varāhamihira wrote several astrological works. In his *Bṛhajjātaka* (Large Horoscopy), called also *Horāśāstra* (Science of Hours), in 25 chapters, containing about 400 verses, he treated all conceivable topics on the subject. This work has been a model for later works and is highly popular even today. In this work, Varāhamihira exhibits his understanding of Greek astrology and employs the Sanskritized forms of a number of Greek terms. His *Laghujātaka* (Shorter Horoscopy) is an abridged form of the previous work. On marriage and the prediction of auspicious times to marry, Varāhamihira composed two works, the *Bṛhadvivāha-paṭala* (Larger Treatise on Marriage), and its abridgement, the *Svalpa-vivāha-paṭala*. Prognostication on military marches and domestic journeys were treated in three works: *Bṛhad-Yoga-Yātrā* (Larger Course on Expedition), in 34 chapters, *Yoga-yātrā* (Course on Expedition), and *Svalpa-yātrā* (Shorter Course on Expedition).

On natural astrology, Varāhamihira's major work is the *Bṛhatsaṃhitā* (Large Compendium), the *Svalpa-saṃhitā* (Shorter Compendium) and *Vaṭakaṇikā* (Short Text), the last two known only through profuse quotations in later works. The *Bṛhatsaṃhitā*, in 106 chapters, is an encyclopedic compendium on numerous subjects relating to life and nature, such as physical astronomy, geography, calendar, meteorology, flora, portents, agriculture, economics, politics, physiognomy, engineering, botany, industry, zoology, erotica, gemology, hygiene, omens, and prognostication on the basis of asterisms, lunar days, etc. A fund of material on applied science, physical observation, and deduction on the basis of statistics and experimentation went into the production of this work which is perhaps unparalleled in the early literature of the world. Varāhamihira's works, set in precise terminology and in graceful language, have been models for writers of later times, not only in astronomy and astrology, but in other disciplines as well.

K.V. SARMA

REFERENCES

Bṛhad-yātrā of Varāhamihira. Ed. David Pingree. Madras: Government Oriental Manuscripts Library, 1972.

Bṛhajjātakam of Varāhamihira. Ed. and trans. V. Subrahmanya Sastri, 2nd ed. Bangalore: K.R. Krishnamurthy, 1971.

Bṛhat-saṃhitā of Varāhamihira. Ed. and trans. N.Ramakrishna Bhat. Delhi: Motilal Banarsidass. 2 Pts. 1982.

Fleet, J.F. *The Topographical List of Bṛhat-saṃhitā* . Ed. Kalyan Kumar Dasgupta. Calcutta: Semushi, 1973.

Pañcasiddhāntikā of Varāhamihira. Ed. and trans. T.S.K. Sastry and K.V. Sarma. Madras: PPST Foundation, 1993.

The Pañcasiddhāntikā of Varāhamihira. Ed. and trans. Otto Neugebauer and David Pingree. Copenhagen: Munksgaard. 2 Pts., 1970, 1971.

Shastri, Ajay Mitra. *India as Seen in the Bṛhat-saṃhitā of Varāhamihira*. Delhi: Motilal Banarsidass, 1969.

Shastri, Ajay Mitra. *Varāhamihira and his Times*. Jodhpur: Kusumanjali Prakashan, 1991.

See also: Astronomy in India – Astrology in India

VAṬEŚVARA Vaṭeśvara (b. AD 880), son of Mahadatta, hailed from Ānandapura or Vaḍanagar, in Gujarat in Western India, a great center of learning of the time. The *Vaṭeśvara-siddhānta*, composed by Vaṭeśvara in AD 904, is one of the largest and most comprehensive works on astronomy, a work which throws much light on the theories, methodologies, and processes of Indian astronomers until the tenth century AD. It was one of the standard works for the study of the discipline, and several of the rules enunciated by Vaṭeśvara were adopted by later astronomers like Śrīpati (eleventh century) and Bhāskara II (b. 1114). It is also noteworthy that the Persian scholar and polymath al-Bīrūnī, who came to India early in the eleventh century, referred to Vaṭeśvara and cited some of his views in his own writings. Vaṭeśvara also wrote a *Karaṇasāra*, known only through references, and a *Gola* (Treatise on Spherics), which is available only in part, the existing portion dealing with a graphical demonstration of planetary motion, construction of the armillary sphere, spherical rationale, and the nature of the terrestrial globe.

The *Vaṭeśvara-siddhānta*, in 1326 verses set out in eight chapters, deals exhaustively with all aspects of Indian astronomy, besides bringing in a number of methodologies, short-cuts, and interpretations. Chapter I, on Mean motion, depicts the astronomical parameters, time-measures, calculation of the aeonary days, and computation of the mean planet in three ways: using parameters, using a cut-off date, and by the orbital method. The longitude corrections to be applied to mean longitude are also dealt with here. Chapter II, on true motion, deals with the corrections to be applied to the mean planet by the epicyclic theory, by the eccentric theory, and by the use of the R sine table. Different items relevant to the almanac are also set out here. Chapter III treats in detail the three problems in diurnal motion. Chapters IV and V deal with the computation of lunar and solar eclipses. Chapter VI treats of heliacal rising, chapter VII with elevation of lunar horns, and chapter VIII with the conjunction of celestial bodies.

Again, a unique feature of the work, found in no other work, consists in Vaṭeśvara's dividing each chapter into small sections earmarked for different topics. Unlike other works, rules are formulated for all possible alternatives of a theory or practice. It is interesting that sets of problems are posted at the ends of chapters for the student to solve, as in textbooks of modern days. The work is also characterized by the depiction of novel interpretations and methodologies practised by early Indian astronomers. In fact, the invaluable service done by Vaṭeśvara through his *Siddhānta* lies in his presenting the achievements of the Indian astronomers from the sixth century to the tenth century and methodically documenting the astronomical knowledge of the Hindus during those centuries.

K.V. SARMA

REFERENCES

Sastry, T.S. Kuppanna. "The System of the Vaṭeśvara Siddhānta." *Indian Journal of History of Science* 4 (1–2): 135–43, 1969. Reprinted in *Collected Papers on Jyotisha*. Ed. T.S. Kuppanna Sastry. Tirupati: Kendriya Sanskrit Vidyapeetha, 1989, pp. 76–88.

Shukla, K.S. "Hindu Astronomer Vaṭeśvara and his works." *Ganita* 23(2): 65–74, 1972.

Vaṭeśvara-siddhānta and Gola of Vaṭeśvara, 2 vols. Ed. K. S. Shukla. New Delhi: Indian National Science Academy, 1986.

See also: Śrīpati – Bhāskara – Astronomy in India – al-Bīrūnī

W

WANG CHONG The Chinese philosopher Wang Chong was born in Shangyu (modern Zhejiang) in AD 27. According to his biographies, he came from a poor family and devoted most of his life to teaching, but also held minor posts in the state administration. Later he retired to compose the work by which we know him today, the *Lunheng* (Balanced Discussions). This extensive treatise in eighty-five chapters (one of which is lost) was completed in AD 82 or 83, about fifteen years before the author's death.

Wang Chong lived about one century after Confucianism had emerged as imperial ideology. In this process, the "rationalist" and socially-minded philosophy taught by Confucius had been integrated with cosmological doctrines extraneous to the letter of his teaching. The school of thought to which Wang Chong belonged propounded, on the contrary, a reading of the classic texts devoid of esoteric interpretations.

In his work, Wang Chong analyzes, with a strongly sceptical and even iconoclastic spirit, ideas expounded by earlier thinkers and beliefs shared by the people of his time (e.g. the recourse to divination, the belief in ghosts, the search for physical immortality, and the idea of an individual spirit that persists after death). Wang's typical procedure is to bring out contradictions in the anecdotes and accounts that he first quotes in full. Through an exemplary logical method, he often does not hesitate to take as true one detail that he has previously refuted, if this may serve to invalidate a different detail.

Philosophically, Wang Chong maintained that all phenomena arise spontaneously, and are not expressions of Heaven's will. Related to this was his opposition to the belief in prophecies and portents, through which Heaven was deemed to legitimate or censure rulers, and assent to or dissent from their policies. While Wang Chong rejected the blend of these forms of esoterism with Confucianism, he fully accepted the metaphysical and cosmogonical doctrines traditionally placed under the egis of Daoism. In this way, he anticipated some of the new developments in post-Han Confucianism.

FABRIZIO PREGADIO

REFERENCES

Chan Wing-tsit. *A Source Book in Chinese Philosophy*. Princeton: Princeton University Press, 1963.

Forke, Alfred. *Lun-heng. Philosophical Essays of Wang Ch'ung*. Vol. I, London: Luzac and Co., 1907; Vol. II, Berlin: Georg Reimer, 1911.

Fung Yu-lan, *A History of Chinese Philosophy*, vol. II: *The Period of Classical Learning*. Princeton: Princeton University Press, 1953.

Lunheng jiaoshi (A Critical and Annotated Edition of the *Lunheng*). Ed. Huang Hui. Changsha: Shangwu Yinshuguan, 1938.

Needham, Joseph. *Science and Civilisation in China*, vol. II: *History of Scientific Thought*. Cambridge: Cambridge University Press, 1956.

WANG XIAOTONG There is no record of Wang Xiaotong's early life nor his year of death. We estimate that he flourished from the second half of the sixth century to the first half of the seventh century. The little we know of him is from a memorial he presented to Emperor Gaozu of the early Tang dynasty (AD 618–906) on the occasion of the submission of his mathematical text to the throne. The mathematical text he submitted was known as *Jigu Suanjing* (Continuation of Ancient Mathematics) and was subsequently selected as a prescribed text for imperial examinations in AD 656. In his memorial, which is now attached to his mathematical text, he mentioned that he had studied mathematics from a very young age. He studied the *Jiuzhang Suanshu* (Nine Chapters in Mathematical Art) thoroughly and had great admiration for Liu Hui's in-depth commentary on the text. On account of his mathematical acumen, Wang Xiaotong was appointed as an instructor in the Department of Mathematics, and later as a deputy director in the Astronomical Bureau. In AD 623, together with Zu Xiaosun, an official of the board of Civil Office, he was appointed to re-examine the adequacy of the current calendar.

Composed by Fu Renjun and promulgated for use since AD 619, the calendar had on several occasions been found to be losing accuracy in the predictions of solar and lunar eclipses. Based on the structure of the Kaihuang calendar composed by Zhang Bin of the previous Sui dynasty (AD 581–618), Wang Xiaotong criticized the adoption of the *ding shuo* method and the precession of equinoxes in the current calendar. The critique sparked a three-year debate between Fu Renjun and Wang Xiaotong, culminating in a submission of a proposal for rectifications consisting of more than thirty errors to the Astronomer-Royal. This does not necessarily reflect Wang Xiaotong's achievement in calendrical science. On the contrary, his view of adhering to the traditional model without taking into consideration the uneven apparent motion of the sun and the precession of equinoxes in calendrical calculations was a retrogressive one. His contribution lies in mathematics.

From the late Han dynasty in the first century onwards,

Chinese mathematicians were familiar with quadratic equations and their solutions. But it was not until the appearance of the *Jigu Suanjing* that equations of the third degree were presented. They arose because of the special needs of engineers, architects, and surveyors, of the Sui dynasty. There are altogether twenty practical problems in the *Jigu Suanjing* consisting of a problem (No.1) on calendrical calculation, six problems (Nos. 2–6 and 8) on engineering constructions, seven problems (Nos. 7, 9–14) on the volume of granaries, and six problems (Nos. 15–20) on right-angled triangles. The solutions of most of the problems involved equations of the third degree. For example, problem No. 15 says: "There is a right-angled triangle, the product of two sides of which is $706\frac{1}{50}$ and whose hypotenuse is greater than the first side by $36\frac{9}{10)}$. Find the lengths of the three sides." Wang's solution amounts to the formation of the cubic equation as follows:

$$x^3 + \frac{S}{2}x^2 = \frac{P^2}{2S},$$

where P is the product and S the surplus. As a matter of fact, most of the problems in the last three categories involved the use of the equation

$$x^3 + Ax^2 + Bx = C,$$

where A, B, and C are positive numbers. Wang Xiaotong provided the rules for the arrangement of the equations in all these problems but did not explain the procedure for arriving at such equations. He also did not discuss the equations of higher degrees. Numerical equations of degrees higher than the third degree occurred first in the work of Qin Jiushao around AD 1245.

ANG TIAN SE

REFERENCES

Guo, Shuchun. "Wang Xiaotong." *Zhongguo Gudai Kexuejia Zhuanji* (Biographies of Ancient Chinese Scientists). Ed. Du Shiran. Beijing: Kexue Chubanshe, 1992, pp. 317–319.

Mikami, Yoshio. *The Development of Mathematics in China and Japan*, 2nd ed. New York: Chelsea Publishing Company, 1974.

Qian, Baozong, ed. *Suanjing Shi Shu* (Ten Mathematical Classics). Beijing: Zhonghua Shuju, 1963.

Ruan, Yuan (AD 1799). *Chouren Zhuan* (Biographies of Mathematicians and Astronomers), vol. 1. Shanghai: Shangwu Yinshuguan, 1955.

See also: Liu Hui and the *Jiuzhang Suanshu* – Calendars – Qin Jiushao

WANG XICHAN Wang Xichan (July 23, 1628–October 18, 1682), sometimes known by his literary name, Xiao'an, was from Wujiang, Jiangsu province, China. When he was sixteen, the Ming Dynasty (1368–1644) collapsed and the door to social advancement through the imperial examination system was suddenly closed. After several unsuccessful suicide attempts, Wang abandoned his hopes for an official career and became one of the most distinguished Ming loyalists in his area. He seems to have survived by teaching literature, although his main interest was in astronomy, which he had studied on his own since his youth. Despite both poverty and illness, Wang made unremitting efforts to observe the heavens, calculate planetary positions, and write about astronomy.

Beginning in the late Ming Dynasty, Western astronomy had been disseminated intermittently into China, and subsequently it was adopted by the Qing (1644–1911) rulers. In 1646, a number of astronomical treatises, earlier written or translated by Western missionaries serving the Ming court, were published together by order of the Qing emperor under the general title *Xi Yang Xin Fa Lin Shu* (Astronomical Treatises of the New Methods of the West, 1646). Unfortunately, the "New Methods" introduced by the missionaries did not reflect the advanced achievements of modern astronomy in seventeenth-century Europe. Even worse, there were some defects and internal contradictions, especially in the parts dealing with the cosmological theory of planetary motions.

In his *Xiao An Xin Fa* (New Method of Xiao'an, completed in 1633), Wang argued that all of the Western techniques could be reconciled with classical Chinese schemes and therefore could be used to revive the lost traditional astronomy of ancient China. By means of trigonometry, which was not used in traditional Chinese astronomy and mathematics, Wang created a series of methods to calculate ecliptic positions and predict planetary occultations. Forty years later, in his *Wu Xing Xing Du Jie* (On the Angular Motion of the Five Planets, completed in 1673), Wang criticized the contradictions in the *Xi Yang Xin Fa Li Shu* and proposed instead his own model of the planetary motions which differed from both the Aristotelian-Ptolemaic and the Tychonic geostatic models. In addition, he attributed the planetary motions to an attractive force radiating from the outermost moving sphere (i.e. the *primum mobile*). Wang also suggested that there were planets inside the orbit of Mercury which might account for the appearance of sunspots. All of these thoughts, along with some of his methods, were heuristic and had considerable influence on his successors.

In general, Wang was one of a few pioneers who responded to Western science by conscientiously studying it; he was open to critical acceptance of its major ideas when suitably reinterpreted for use in seventeenth-century China. Nevertheless, some of Wang's arguments were clearly exaggerated. For instance, he assumed that Western astronomy had in fact originated in ancient China, but such claims were due to his radical nationalism and traditional reluctance to recognize any innovations from the West.

LIU DUN

REFERENCES

Jiang, Xiaoyuan. "Wang Xichan." In *Zhong Guo Gu Dai Ke Xue Jia Zhuan Ji* (Biography of Ancient Chinese Scientists), vol. 2. Ed. Shiran Du. Beijing: Science Press, 1993, pp. 1005–1015.

Sivin, Nathan. "Wang Hsi-Shan" (Wang Xichan). In *Dictionary of Scientific Biography*, vol. 14. Ed. Charles C. Gillispie. New York: Charles Scribner's Sons, 1976, pp. 159–168.

Wang, Xichan. *Xiao An Xin Fa* (New Methods of Xiao'an, completed in 1663). Shanghai: Commercial Press, 1936.

Wang, Xichan. *Wu Xing Xing Du Jie* (On the Angular Motions of the Five Planets, completed in 1673). Shanghai: Commerical Press, 1939.

Xi, Zezong. "Shi Lun Wang Xi Chan De Tian Wen Gong Zuo" (An Essay on the Astronomical Work of Wang Xichan). *Ke Xue Shi Ji Kan* 6:53–65, 1963.

WEI BOYANG Although many Chinese works relate the name of Wei Boyang to the origins of alchemy, nothing is known about him from a historical point of view, and his figure may be entirely legendary. Some sources, which place him in the second century, relate that he came from Shangyu (in modern Zhejiang) and transmitted his teaching to Xu Congshi, who in turn handed it down to Chunyu (alternative spelling: Shunyu) Shutong. The text deemed to embody the gist of this transmission is the *Zhouyi cantong qi* (Agreement of the Three According to the Book of Changes), a scripture in verses that is attributed to Wei Boyang. The commentaries usually interpret the "Three" as Heaven, Earth, and Man, or as Daoism, cosmology, and alchemy. The *Book of Changes* is the *Zhouyi* (or *I Ching*, also spelled *Yijing*), the renowned divination manual, which also includes a classical exposition of cosmology in its section commonly known in the West as "Great Treatise".

The genesis of the *Cantong qi* is as obscure as its putative author. Current research tends to consider that an originally Han text on cosmology may have been elaborated into an exposition of alchemical doctrines, perhaps at different dates until the seventh century. The very few references to a *Cantong qi* before that time seem to refer to the original cosmological treatise, whose circulation may conceivably have been restricted due to its association with an unofficial and proscribed body of writings propounding esoteric interpretations of the Confucian classics.

Since the late Tang, however, the *Cantong qi* began almost abruptly to exert a far greater influence in the history of Chinese alchemy than that of any other text. Hidden in the highly allusive language, and the thick layers of symbols and images that characterise this scripture, is an exposition of the doctrine that inspired a variety of commentaries (about thirty of which are extant) and other texts, written between the eighth and the nineteenth centuries, in both Daoist and Neo-Confucian traditions. One of the most well known exegeses was written by the Neo-Confucian philosopher Zhu Xi, who commended the high literary quality of the text.

The *Cantong qi* would hardly be intelligible without recourse to the related literature. Although it was occasionally read as a treatise on cosmology, most commentaries agree that it articulates the theories found in the tradition of the *Book of Changes*, applying them to the alchemical work. The *gua* (trigrams and hexagrams) of the *Book of Changes* are used both to construct a cosmological model and, at the same time, to represent phases of the process performed by the alchemist. The various possible levels of reading of the text possibly mean that it was used within both the main traditions of alchemy in China, i.e. *waidan* or "external alchemy" and *neidan* or "internal alchemy".

FABRIZIO PREGADIO

REFERENCES

Fukui Kōjun. "A Study of Chou-i Ts'an-t'ung-ch'i." *Acta Asiatica* 27: 19–32, 1974.

Ho Peng Yoke. "The System of the *Book of Changes* and Chinese Science." *Japanese Studies in the History of Science* 11: 23–39, 1972.

Needham, Joseph, Ho Peng-Yoke, and Lu Gwei-Djen. *Science and Civilisation in China*, vol. V: *Chemistry and Chemical Technology*, part 3: *Spagyrical Discovery and Invention: Historical Survey, from Cinnabar Elixirs to Synthetic Insulin*. Cambridge: Cambridge University Press, 1976, pp. 50–75.

Partington, J.R. "Ancient Chinese Treatise on Alchemy." *Nature* 136: 287–288, 1935.

Pregadio, Fabrizio. "Alchemical Doctrines and the Textual History of the *Zhouyi cantong qi*." *Cahiers d'Extrème-Asie* 8, 1995.

"*Shūeki sandōkei bunshō tsū shingi kōhon* (Collated text of the *Zhouyi cantong qi fenzhang tong zhenyi*)." Ed. Imai Usaburō. *Tōkyō Kyōiku Daigaku Bungakubu kiyō* 57 (= *Kokubungaku kanbungaku ronsō* vol. 11, 1966, pp. 1–89.

"The Representation of Time in the *Zhouyi cantong qi*. Cahiers d'Extreme-Asie 8: 155–173, 1995.

Wu Lu-ch'iang, and Tenney L. Davis. "An Ancient Chinese Treatise on Alchemy Entitled Ts'an T'ung Ch'i." *Isis* 18: 210–289, 1932.

See also: Alchemy in China – *Yijing* – Zhu Xi – Divination

WEIGHTS AND MEASURES IN AFRICA: AKAN GOLD WEIGHTS The subject of gold weights is complex and multidimensional. It can be understood only when placed in the context of its original cultural environment, which was linked intimately to gold for its physical substance and to the package (*dja*) in which it came for its socio-cultural identity.

The Akan country, on the Gulf of Guinea in West Africa, was and is an area of gold deposits. This metal is both feared and worshipped. Well before the first contacts with Europeans in the fifteenth century, the Akan people used gold dust as a medium of exchange. However, the concept of "gold weights" came from Western traders (Dutch, Portuguese, English, and French).

The first question to ask is: are the figures really weights? In describing these figures, the Baolé use several different terms: *Dja-yôbwê, sika-yôbwê, shin-dra-yôbwê, nsangan-yôbwê, ngwa-yôbwê*. Let us look at the meaning of each of them.

- *Dja-yôbwê* (dja stone). Here the term refers to the contents; it designates what is contained in the package or *dja*. These original elements would be in stone and would be concerned with different realms of human knowledge deemed worthy of interest.
- *Sika-yôbwê* (stone of gold) money. In this second case, the term describes a price or a monetary total. These elements would deal with economics, calculation, and mathematics.
- *Nsangan-yôbwê* (stone of a fine). This term describes the price of a fine, a tax, or a tribute, and is concerned with elements of economics and finance.
- *Ahindr-yôbwê* (proverb stone). This term evokes the notion of speech, thought, and discourse, the fields of literature and philosophy.
- *Ngwa-yôbwê* (game stone). This term refers to the figures comprised of graphic signs. The exercise of reading or deciphering the signs was considered a kind of game, involving elements of imitation and intellectual training.

As we see, the term "weights" appears nowhere in these descriptions.

The *dja* or *sanaa* was originally a package made of animal skin or thick cloth in which the Akan placed their figurines and certain accessories. It seems to have been a kind of encyclopedia written in miniature figures. The package and its contents symbolized the economic power of the living king and the spirits of the dead sovereigns. The act of taking possession of the *dja* signified for a new ruler that he was assuming the power to raise taxes, impose fines, and take measures to increase the state treasury. For his subjects, knowing that the ruler was in possession of the *dja* of the state meant that they judged him capable of administering the financial and economic heritage of the country.

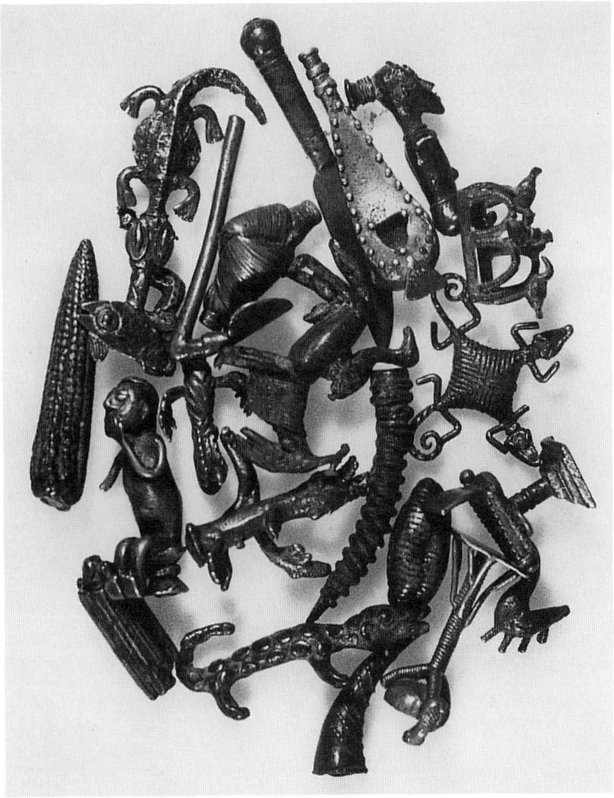

Weights with Figurative Elements.

Many different objects can be found in a *dja*, but we will mention here only the essential, original ones.

The Equipment is a group of accessories used to manipulate the gold power. They consist of: (1) balance scales, used to determine the weight value of different quantities of gold dust used as a medium of exchange; (2) spoons to put the gold dust on the plates of the scales; (3) boxes of many different kinds, their covers often decorated with graphic designs. They are used to hold gold dust already weighed or about to be weighed; (4) winnowing baskets used to separate and rid the gold powder of any impurities; and (5) sieves to separate the different grains of gold dust.

The weights provide knowledge of the weight and monetary value of the quantity of gold powder placed in the plate of the scale. There are three kinds of weights — figurative weights, weights with graphic designs, and geometric weights. Gold weights were (usually) made of an original alloy whose composition was similar to that of bronze and brass. However, there are also weights made of silver, copper, and solid gold.

The weights were made by the Tounfouê, an artisans'

Weights with Graphic Designs. Photographs by the author. Used with his permission.

Weights with Geometric Elements.

group, different from blacksmiths and jewelers. These artisans used the lost wax method to produce the weights.

The Akan used a system of computing weight consisting of eleven units. It began at *dama* and ended with *bèna*. It was possible to multiply *bèna* by infinity and the values went from single to double or were multiplied by two. There were three series of weights — small, medium, and large. They could be added and multiplied.

The small weights series consisted of ten monetary units and was used for all sorts of small transactions:

ba = unit = 0.148 grams
ba (*gnon*) = ba × 2
ba (*nsan*) = ba × 3
ba (*nan*) = ba × 4
ba (*nou*) = ba × 5
ba (*nzien*) = ba × 6
ba (*nzo*) = ba × 7
ba (*motchué*) = ba × 8
ba (*brou*) = ba × 10

The medium weights series consisted of seven units. The computation is done from simple to double, and each unit has multiples and submultiples.

Assan = 4 m.v.
Gbangbandia = 4 m.v.
Tya = 5 m.v.
Anui = 5 m.v.
Gua = 5 m.v.
Anan = 5 m.v.
Tyasue = 5 m.v.
Total = 33 m.v.

These seven units comprise 33 monetary values. The smallest value is *méteba* which equals 12 *ba* or 1.77 grams. The largest value is the *ta*, which equals 348 *ba* or 51 grams of gold.

In practice, the system worked as follows. For example, the *gua*, the fifth unit, comprised the following five monetary values:

Météba = 12 ba = 1.77 grams of gold
Adjratchui = 24 ba = 3.55 grams of gold
Tra = 48 ba = 7.54 grams of gold
These are all sub-multiples of *gua*.
Gua = 96 ba = 14.20 grams of gold (Unit of this series)
Guagnan = 192 ba = 28.40 grams of gold (Multiple of *gua*)

The large weight and monetary values series had only three units. They were:

Banda = 384 ba = 56.80 grams of gold
Banna = 432 ba = 67.44 grams of gold
Pereguan = 478 ba = 71.92 grams of gold.

In considering the weights and numeric representation, we will consider here only those weights with graphic signs which correspond with calculation and mathematics.

Concerning the signs and marks, the anthropologist François H. Abel has written: "A. Amélékia, a well known man named Diénélou confirmed for me that the Ancients knew how to read from the weights ... In the village of Lomo-north, in the region of Toumodi, the village chief knew that the signs on the weights had meaning."

C. Savary, the Director of the Department of Black Africa in the Museum of Ethnography in Geneva has written: "Each weight is the product of two signs written on it ... Reading it is sometimes simple, but often difficult. This is because some Black Africans had a different concept for numeric figuration and for the representation of the product of two numbers. [Also] zero did not exist ...". In the system, figures and numbers are represented by vertical and horizontal lines, such as marks and arrows similar to those still seen in charcoal in the houses of African villages.

Commenting in 1605 on the Akan system of accounting and calculating, the Dutch explorer and historian Pieter de Marées made this remark: "The Negroes have weights of copper and tin which they have cast themselves, and, although they do not divide in the same way we do, it comes out the same, and the accounting is always correct."

For the people of the Akan civilization, the gold weights contained in the *dja* constitute a fundamental cultural text, comparable to the Christian *Bible*, the Islamic *Qu'rān*, or the Hindu *Veda*. It is within this sacred package that they had consigned, in idiogrammatic letters and signs, their knowledge and values to be passed on to posterity. In this way, their descendants would not have to reinvent that which had been known by their ancestors and which constituted the foundation of their civilization.

GEORGES NIANGORAN-BOUAH

REFERENCES

Abel, H. "Les Poids à Peser l'Or en Côte d'Ivoire."*Bulletin de l'Institut Français d'Afrique Noire* 16B: 23–95, 1952.

de Marées, P. *Description et Récit Historial du Riche Royaume d'Or de Guinée*. Amsterdam: C. Claeffon, 1605.

Garrard, T.F. *Akan Weights and Gold Trade*. London and New York: Longmans Group, 1980.

Gluck, J. *Die Goldgewichte von Oberguinea*. Heidelberg: Carl Winter's Universitat's Buchhandlung, 1937.

Niangoran-Bouah, G. *L'Univers Akan des Poids à Peser l'Or*. 3 vols. Abidian: NEA–MLB, 1984, 1985, and 1987.

Savary, C. "Poids à Peser l'Or du Musée d'Ethnographie de Genève." *Bulletin Annual du Musée d'Ethnographie de Genève*, no. II, 1969.

Zeller, R. *Die Gold Gewichte von Asante (Westaftica)* Leipzig and Berlin: Baessler Archiv (Beiheft III), 1912.

WEIGHTS AND MEASURES IN BURMA: THE ROYAL ANIMAL-SHAPED WEIGHTS OF THE BURMESE EMPIRES The royal animal-shaped weights of the Burmese Empires appear to be unique not only as a weight system, but also as one of the most important artifact series, especially in a country with but few durable artifacts. Their shape motifs are mythically leonine (lion-like), elephantine, anserine (goose-like), and gallinaceous (poultry-like). They symbolize Buddhist and pre-Buddhist beliefs, and the Burmese institution of monarchy and its continuity, dynastic, and imperial changes.

For the purpose of this account the Burmese empires comprise those regions of the Southeast Asian peninsula now known as Myanmar (Burma) and northern Thailand (Siam).

Buddhist art in India gave rise to the Burmese choices of animal motifs to use on the weights. The anserine weight, in a tenth century Indian style, apparently was chosen first during a period of Theravāda revival. However, the motifs were ancient even then, originating before the Burmese entered Burma. Shamanistic and animistic influences, both of the post-1000 BC steppes and China, are also evident. Even more obvious is the use of Chinese models and execution for most of the style groups. Moreover, the decision to use Buddhistic symbolic animals may also have been influenced by the enduring association of Indian and Chinese Mahāyāna Buddhists with commerce.

The mass units and scales reached Burma from India before the twelfth century AD, but India itself obtained them from Achaemenid Persia, which in turn, obtained them from Assyria and Babylonia. However, with the growth of Chinese trade in Southeast Asia from the thirteenth century onwards and the adoption in Burma of Chinese mass standards, the Indian mass system was displaced. (A mass unit is defined here as that which was used in multiples to form weights mainly used for heavy trade goods and in fractions mainly for bullion or small amounts of other valuable items, the multiples and fractions together forming the mass scale).

A Burmese term often given to the animal-shaped weights is *shway arlay* (gold weights) which may have been derived from the expression *sri arlay* (king's weights). However, in the East the word gold has customarily been associated with royalty. Also, until the late eighteenth century the weights of Burma were mainly used on behalf of the king, who enjoyed monopolies in trade and was the chief import/export broker.

Of the three main weight shapes, the leonine and anserine were used throughout the empires, while those of elephant shapes were used only in north Siam. The first two shapes were still in use in 1970 along the more remote routes that pass from northeast Burma into Yunnan, north Siam, north Laos and the Thai cantons of Vietnam.

Most commonly, the diameters and heights of the weights do not exceed about 70×120 mm. In shape they occur in three leonine- and seven bird-weight style groups which can be placed in stylistic sequence. A sequence in the elephant-shaped weight styles has not been identified with certainty.

The leonine shape, called *to* (taw = "royal" in Burmese) is a stylized combination of the parts of four animals. The lion of west and south Asia supplied the model for the open-mouth, bearded head and probably for the crouching torso. The horns were either bull-like or were adapted from the antlered muntjac deer of eastern and southern Asia. The tail of two of the styled groups is usually lion-like, but it is occasionally horse-like. With one exception, all the tails are characterized by the artificially-raised tail base of the ancient Yunnanese horse. The legs and feet were modeled upon those of the elephant. All these animal representations are present on the abacus of the Asokan pillar at Sarnath.

The standing bird shapes of six of the bird style groups are in the form of a stylized Chinese mandarin drake (*aix galericulata*) having a knobbed crest and known as a *hintha* (Pali). The model for the squatting/brooding, gallinaceous style group has not been identified.

The elephant shapes are naturalistic. In Siam, the weights of elephant shape, known as *chang* in Thai, are difficult to separate from those figurines of similar shape intended for other purposes. The animal representations stand on usually pyramid-shaped bases which may be rectangular, octagonal, circular, or hexagonal.

Frequently, impressed signs 4–19 mm in size are found, usually on the front or right side of the base. In shape they may be script-like circular, square, bird- or feline-like or in patterns normally of 4, 5, 6, 8, and 9 rays, diverging from a central point. These also can be placed in sequence.

The weights are made of a metal alloy, the essential components being copper (50–80%), lead (3–30%) and tin (0–20%). In some of the nineteenth century weights zinc may amount to as much as 35%. The alloy, formally known as *ganza* (Tamil *kamsa*), was imported either in the form of broken vessels from Canton or as the coins known as *cash* (Sanskrit, *karsha*) from Yunnan.

The *kyat* was the mass unit of the *to* and *hintha* weights and varied with time from about 14 grams in the fifteenth

Figure 1: Weights of different shapes. Photograph by the author.

century to over 16 grams in the eighteenth century. The 20 *kyat* (320 grams) weights constitute about 50% of all weights, while weights with a mass of about 250 *kyats* (3,750 grams) amount to about 2%. About one-half of the elephant weights weigh less than 45 grams and 2% have a mass greater than 130 grams. The average mass unit of the upper part of the gallinaceous mass scale is about 11.2 grams and that of the lower part is about 13.9 grams. The average mass unit of the elephant-shaped weights is about 12.7 grams.

The essentially decimal mass scale of the *to* and *hintha* weights, based on the mass unit of one *kyat* is $\frac{1}{8}$ (or $\frac{1}{10}$); $\frac{1}{4}$ (or $\frac{1}{5}$); $\frac{1}{2}$; 1; 2; 5; 10; 20; 50; 100; 250. The elephant-shaped weights occur on the following scale: $\frac{1}{8}$; $\frac{1}{4}$; $\frac{1}{2}$; 1; 2; 5; 10; 20; 40; ?; 60. The gallinaceous scale, at about 37 grams, separates into an upper binary scale (1; 2; 4; 8; (?); 32 and what may be a lower trinary scale, $\frac{1}{9}$; $\frac{1}{3}$; 1. Bengalese and Chinese mass scales of the period are similar in part.

99% of the *to* and *hintha* weights fall within $\pm 12\%$ of the mean mass of a particular style group while about 51% of them fall within $\pm 2\%$. These variations lie in the making of the weights not in additions, subtractions, oxidation, nor wear. The accuracy of the elephant weights (and figurines) is less.

During the nineteenth century the weights were made at a village near the capital under the supervision of the Chief Minister. These standard weights were stored in the treasury and issued to officials in other towns. It was a criminal offense to use weights other than those made "at the palace". Special sets of weights were kept for the purposes of comparison and the settling of disputes. Though only two nation-wide "standardizations" have been recorded, the earliest in the eleventh century, nevertheless it was the required duty of each new monarch, upon ascending the throne, to verify the weights.

Copies of animal-shaped weights, usually somewhat

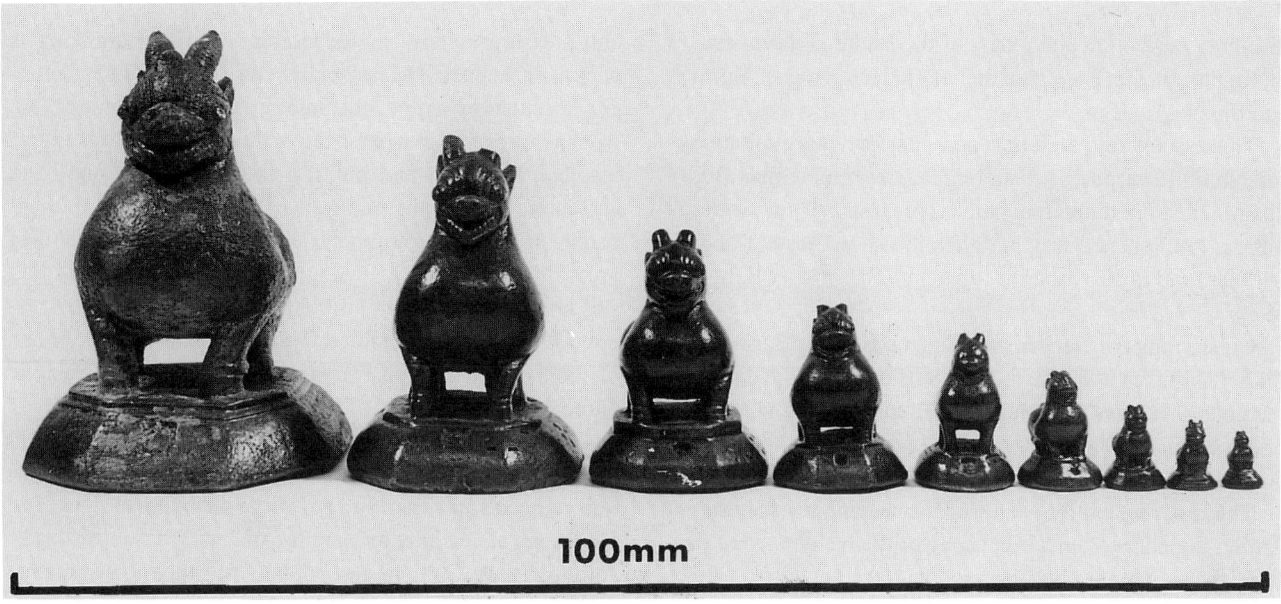

Figure 2: Weights of different sizes. Photograph by the author.

crude, are still made today in eastern Burma (Shan States) using techniques similar to those of thirteenth century AD and second century BC Burma and Yunnan. The weights were made by the lost-wax technique, using clay to surround the wax model and metal molds to shape the wax model. The high-melting point wax used was produced in Yunnan by an insect.

Among the goods imported from India which may have required weighing were opium, indigo, and mercury. From Yunnan came gold leaf, silver ignots, copper-alloy, cash, salt, white insect wax, cinnabar, and tea. Burma exported gems, including pearls and coral, costly medicines, musk camphor, gums, resins, waxes, ivory, rhinoceros horn, and rare woods. Considerable trade would have been done with other Southeast Asian states, e.g. tin from Malaya and Laos, arsenic, lead, and silver from the Shan States, zinc from Laos.

Seeds and a Chinese type of equal-arm balance were used for the weighing of small highly valuable articles and the steelyard or datchin for large masses. For moderately valuable articles, e.g. foreign silver coins and ingots, the animal-shaped weights would have been used.

Until after the middle of the eighteenth century ordinary village commerce was conducted by barter, counting, purchase with cowries, and by measurement of volume or length. Weights were rarely necessary. However, from the last quarter of the eighteenth century onwards, it was the ordinary Burman, rather than the king's officers, who made use of the weights, mainly to weigh currency ingots. This change was caused by the Burmese wars of the time. These resulted in the cessation of the supplies of copper alloy and cowries, led to the Burmese capture of the large lead and silver mines of the Shan States and so to the copying of the Chinese practice of using chopped lead as a low value currency. As a result of the increased need for weights to weigh the abundance of lead and silver ignots, there was a flood of copies from the formerly Burmese north Siam region, now occupied by south Siam (Ayuthya). This new domination also led to the replacement or adjustment of the animal-shaped weights to agree with those of south Siam.

Each of the weight shapes together with its base, and each of the parts of the weights, was intended to convey a particular meaning. The meanings had to be understood by illiterate speakers of many different languages. Though they were intended to be understood primarily by animists and Buddhists, they would also have been understood by Confucianists, Daoists, and others. Characteristic of animist/Buddhist Southeast Asia (though originating much further away in space and time) was the belief in the earthly, semi-divine king. This is what the *to* weight was intended to symbolize. It combined the physical characteristics of a potential Buddha (*bodhisattva*) and a universal monarch (*cakravartin*). These, in turn conveyed the idea of secular power both to animist and Buddhist, being associated with earth, fertility, and healing, with legitimacy of rule and righteousness of conquest (a prerogative of a universal monarch). New styles of "*to*" weights were issued only at the times of the "righteous conquests" of a Burmese king, usually at the beginning of a dynasty with the accompanying empire-building.

The *hintha* weight symbolizes, among other things, the

heavenly perfection and purity of the Buddhist faith, especially that of the Theravāda belief characteristic of Burma and Siam.

The association of feline and anserine representations for symbolic purposes has an origin more remote than Buddhism. The lion–duck association is present on some Asokan pillars. Lion-shaped weights were in use in Assyria about 1500 BC and duck-shaped weights in Sumeria and Babylonia.

Concerning the elephant weights the kings of Burma and Siam (Ayuthya) were the incarnations of Indra, the chief of the gods, who was ancient before Hinduism, and who rode on an elephant. Thus, one reason (among several) for the choice of the elephant was to symbolize these kings.

The most generally useful of the techniques for establishing the chronological sequence of the weights were the style and sign sequences; the relation of the *to* weight styles, through their symbolism, to the dates of the dynastic changes and territorial expansions; the sequence of increase in the average unit mass of each style group; the dates of the unit masses obtained from the weighings recorded by many European traders from AD 1515, and their relation to the average unit masses of the style groups.

DONALD GEAR
JOAN GEAR

REFERENCES

Annandale, N. "Weighing Apparatus from the Southern Shan States." *Memoirs of the Asiatic Society of Bengal* 5: 195–205, 1917.

Braun, Rolfe and Ilse Braun. *Opium Weights*. London: Braun, 1983.

Decourdemanche, J. *Traite des Monnaies, Mesures et Poids Anciens et Modernes de l'Inde et de la Chine*. Paris: Institute Ethnographique International de Paris, 1913.

Fraser-Lu, Sylvia. "Burmese Opium Weights." *Arts of Asia* 1: 73–81, 1982.

Gear, Donald and Joan Gear. *Earth to Heaven: The Royal Animal-shaped Weights of the Burmese Empires*. London: Twinstar, 1992.

Mollat, Helmut. "Die Standardformen der Tiergewichte Birmas." *Baessler Archiv* N.F. 32: 405–440, 1984.

Temple, Richard. "Currency and Coinage among the Burmese." *Indian Antiquary*. 1897, 4 articles; 1898, 9 articles; 1919, 15 articles; 1927; 1 article; 1928, 5 articles.

WEIGHTS AND MEASURES IN CHINA Metrology means the "science of weights and measures". But when referring to premodern periods the term "knowledge of weights and measures" is more appropriate. In a more concrete sense, metrology is the art of calculation with number, weight, and measure units in the economic and fiscal domains as well as in science. "Metrosophy" may be defined as "number speculation within cosmological philosophemes", but from a more inclusive perspective the relationships between magical, religious, and political thought on the one hand and metrology on the other also have to be taken into account. In both metrology and metrosophy, numbers are of great importance. In metrology, they are used for defining and counting measure and weight units, while in metrosophy they serve as basic stuff for the creation of systems of number symbolisms, magical numbers, and correlative numerologies.

For the study of metrosophy and metrology, two different types of sources are basically available. First, we may mention sources of the tradition type, like the metrosophical and metrological sections in chapters of the dynastic histories. Although these sections also contain information of a metrological nature, it is clear that they are principally metrosophical in character. The second important types of sources are concrete remains, especially the relatively great number of real ancient weights and measures which either have been excavated at archaeological sites or which survived by having been handed down from generation to generation. While the sources of the tradition type are closely related with metrosophical thought and number systems, concrete remains inform us almost exclusively of metrological aspects and realities. The tension between these two clusters of information and data is of particular importance and is rewarding to be investigated.

Chapter 21A in the *Hanshu*, written in the first century AD, contains one of the first surviving texts that gives evidence of a system integrating metrology with such diverse fields as numerology, musical pitch, astronomical phenomena, astronomy, cosmology, historiography, and political, ethical, and moral thought. The textual arrangement of this chapter makes clear that metrology belonged to the traditional Chinese science of mathematical harmonics (*lü* or *lülü*).

Chapter 21A of the *Hanshu* states that harmonics consist of five categories: number, musical pitch, length measures, capacity measures, and weights and balances. All five had their origin in the *huangzhong* pitch-pipe, which had been created and standardized by the legendary Yellow Emperor. Obviously, the antiquity of these conceptual and technical devices was an important criterion for their purported eternal validity. Within the five categories a certain hierarchy is expressed, ranking numbers above, followed, in descending order, by pitch, length, capacity, and weight measures. Numbers are obviously of primary importance, because they provided the basic counting units for the other four categories. Metrology was grounded on a numerological system combining the numbers of the Five Phases (*Wuxing*) and other

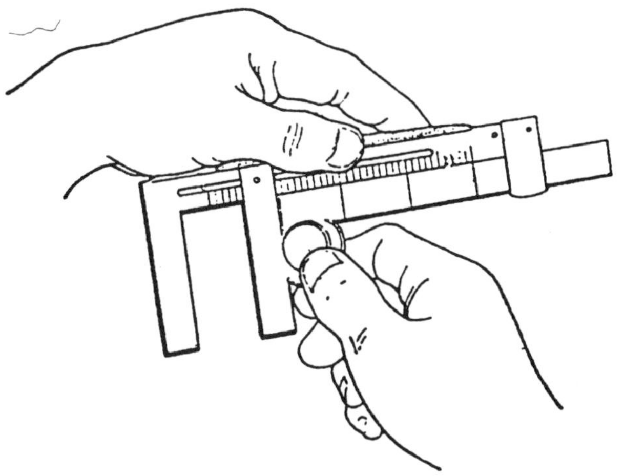

Figure 1: Handling the Wang Mang proto sliding callipers. From Liu Dong-rui, "The Earliest Slide Ruler in the world — the Bronze Slideruler of the Xin Dynasty (AD 923)." *Bulletin of the Museum of Chinese History* 1:96, 1979 (in Chinese).

numerological concepts with the numbers derived from the changes of *yin* and *yang* as they are described in the *Yijing* (I Ching Book of Changes) and its commentaries. Moreover, weights and measures were conceived as an integrated system. With grains of medium-sized black millet, the length and capacity of the *huangzhong* pitch-pipe were measured and thus the basic length and capacity measure units derived. The weight of the grain filling of the pitch-pipe also served as the basis for fixing the weight units.

Concrete remains of ancient Chinese weights and measures are of great importance for supplementing the information contained in the largely normative written sources. A spectacular example are the proto sliding callipers of the Wang Mang period (AD 9–23), of which apparently three pairs survived.

The intaglio inscription on the movable part says: "Manufactured at the first day of the first month of the first year of the Shijian guo reign-period." This corresponds to March 15, AD 9, the first day of Wang Mang's interregnum. The front side of the fixed part of the sliding callipers is divided into four inches (*cun*), that of the movable part into five inches. In addition, the four inches of the fixed part are further subdivided into ten parts (*fen*) each. With the Wang Mang proto sliding callipers outer diameters of less than four inches could be measured. The most conspicuous difference between modern sliding callipers and the Wang Mang piece is that the latter does not have the differential scale of the vernier.

No less spectacular is the great number of other Chinese metrological remains. For instance, more than 240 ancient length measures are known, which can serve as a basis for ascertaining the "true" length of a length measure unit of a given period. This is no easy task as may be shown in the case of the length measure unit *chi* of the Eastern Han period. Modern measuring has shown that the actual lengths of 85 concrete *chi* measures of this period vary from 20.49 cm to 24.73 cm. Obviously, state efforts at unification and standardization of weights and measures were of limited success and collided with diverging interests having their origin in different regional, local, political, economic, fiscal, and social conditions. Small wonder that in the *Suishu*, the dynastic history of the Sui dynasty written in the early seventh century, fifteen different *chi* measures are listed.

The development of the types of weighing apparatuses can be divided into three stages. From the fifth century BC to the second century AD balances with equal arms prevailed. In the period from the third to the ninth centuries the steelyard with unequal arms appeared and was used together with the ancient balances. Finally, the third stage from the tenth to the early twentieth centuries was characterized by the emergence of the highly accurate *deng* steelyard and the introduction of a decimal system of weighing units. An important promoter in the introduction and improvement of the *deng* steelyard was the eunuch Liu Chenggui (950–1013). Liu was in charge of the court treasury and fiscal affairs. The purpose of the introduction of the *deng* steelyard was to reduce disputes in the weighing of precious metals, which arose between different officials involved in fiscal affairs.

HANS ULRICH VOGEL

REFERENCES

Ferguson, John C. "Chinese Foot Measure." *Monumenta Serica* 6: 357–382, 1941.

Guo Zhengzhong. "The Deng Steelyards of the Song Dynasty (960–1279): In Commemoration of the One Thousandth Anniversary of their Manufacture by Liu Chenggui." In *Une activité universelle: peser et mesurer à travers les âges*. Ed. Jean-Claude Hocquet. Caen: Editions du Lys, 1994, pp. 297–306.

Iwata Shigeo. "The Changes in Linear Measures in China and Japan." In *Acta Metrologiae Historicae: Travaux du IIIe Congrès International de la Métrologie Historique*. Linz, 7–9 Oct. 1983. Linz: Trauner Verlag, 1985, pp. 117–137.

Jun Wenren and James M. Hargett. "The Measures Li and Mou during the Song, Liao, and Jin Dynasties." *Bulletin of Sung-Yuan Studies* 21: 8–30, 1989.

Loewe, Michael. "The Measurement of Grain during the Han Dynasty." *Toung Bao* 49: 64–95, 1961.

Qiu Guangming. *Zhonguo lidai duliangheng kao* (Investigations into the Length, Capacity, and Weight Measures of China Through the Ages). Beijing: Kexue chubanshe, 1992.

Vogel, Hans Ulrich. "Metrology and Metrosophy in Premodern China: A Brief Outline of the State of the Field." In *Une activité universelle: Peser et mesurer à travers les âges*. Ed. Jean-Claude Hocquet. Caen: Editions du Lys, 1994, pp. 315–332.

Vogel, Hans Ulrich. "Aspects of Metrosophy and Metrology during the Han Period." *Extrême-Orient, Extrême-Occident*, 16: 135–152, 1994.

Vogel, Hans Ulrich. "Zur Frage der Genauigkeit antiker Längenmaße und deren interkulturelle Zusammenhänge im Lichte chinesischer metrologischer Sachüberreste." In Rainer S. Elkar, u.a.: *Vom rechten Maß der Dinge: Beiträge zur Wirtschafts- und Sozialgeschichte. Festschrift für Harald Witthöft zum 65. Geburtstag*. Scripta Mercaturae Verlag, St. Katharinen, 1996.

Vogel, Hans Ulrich, Guo Zhengzhong, and Qiu Guangming. *Bibliography of Works on Historical Metrology Relating to China* [in preparation].

See also: Acoustics – Five Phases – *Yinyang*

WEIGHTS AND MEASURES IN EGYPT The gradual change from a nomadic to a settled existence, which began about 8000 BC, marked humankind's determination to shape the environment to their requirements. Needing units of measure for both building and agriculture, people chose the most readily available and useable references — simple parts of the human body in various positions. The need for uniform measurements was recognized, and as early as 3000 BC official reference standards of length, volume, and weight were being maintained in temples and royal palaces in Egypt, Mesopotamia, and around the eastern seaboard of the Mediterranean.

The standards developed in these ancient cultures moved westward, mostly as a result of trade, to the Greek and Roman empires, thence to Gaul and Britain via the Roman conquest. Egypt, by virtue of its geography a great trading nation and of its climate a preserver of archaeological treasures, has provided an incomparable record of this metrological heritage.

Measures of Length

Throughout the ancient Middle East the basic unit of linear measure was the cubit, which was defined as the length of the forearm from point of elbow to extended fingertips. In Mesopotamia the cubit was generally subdivided into 2 feet each of 3 or 4 palms, a palm having 4 digits. In Egypt, however, the most ubiquitous standard was the royal cubit. It was subdivided into 7 palms (28 digits), these divisions having significant mystical relationship to the four 7-day phases of the 28-day lunar month by which time was reckoned. The basic sub-unit was the digit (fingerwidth): 4 digits = a palm; 5 digits = a hand; 12 digits (outstretched forefinger tip to little finger tip) = a small span ($\frac{1}{3}$ cubit); 14 digits (outstretched thumb tip to little finger tip) = 1 large span ($\frac{1}{2}$ cubit). Measurements smaller than a cubit were expressed in digits (up to 10), in palms and digits, or, rarely, as $\frac{1}{2}$ or $\frac{1}{3}$ cubit.

Surveys of the Great Pyramid confirm its construction according to the royal cubit of 20.62 in (52.4 cm). The skill of the Egyptian workmen is shown by the fact that the sides of its base vary no more than 0.05% from the mean length of 440 cubits (9069.45 in or 230,364 m). The Egyptians also employed a short cubit of 6 palms (24 digits) = 17.68 in (44.9 cm) for general purposes including calibration of the Nilometers (masonry stairwells leading down to the river at points where the priests recorded the rise of the annual floods).

Egyptian scales are square rods with one face half-beveled to form a reading surface. The fractions of the digit are shown by dividing the first into halves, the second into thirds, and so on down to sixteenths. The oldest extant cubit rods date from the Eighteenth Dynasty, 1550–1307 BC; none are longer than the double cubit, 41.2 in. Most of those used by the workmen are made of wood with deeply cut notches of division. The stone cubit rods are generally ceremonial, belonging to the temples and crowded with inscriptions referring to the religious aspects of the digits as connected with the gods and with the signs of the nomes. On these the calibrations are often carelessly marked.

Land was measured by a linear unit called khet or khet u nu (reel of cord) 100 royal cubits in length. Measuring cords were made of palm-fiber or flax-fiber and knotted at 5-cubit intervals. For long distances the foot (12 digits) and the pace (double step) were used; 2 paces measured heel to toe = the extended arms (4 cubits), and 100 paces = a stade. This distance, later used by the Greeks in the Olympic foot races, became known as a stadium.

indexmeasures of area and length indexmeasures of area and length, cubit, Egyptian indexmeasures of area and length, digit indexmeasures of area and length, foot indexmeasures of area and length, khet, khet u nu indexmeasures of area and length, palm indexmeasures of area and length, remen indexmeasures of area and length, setat indexmeasures of area and length, span

Measures of Area

The basic unit for reckoning areas of fields and nomes was the setat, the square of the khet containing, therefore, 10,000 square cubits. Such an area was called a cubit of land, but was visualized as 100 parallel strips of land each 1×100 cubits. Ten setats were called a thousand of land. In devising ways to re-establish field boundaries after the annual flooding of the

Nile, the Egyptians discovered the rudiments of geometry. A new unit formed from the diagonal of a square with sides equal to the royal cubit was found to be divisible by 40 digits of 0.73 in, twice the length of the remen (forearm from elbow to clenched knuckles) and consequently named the double remen. Half of the area of 100×100 cubits was also called remen in land measure. By having two standards, one the diagonal on the square of the other, it was possible to denote areas in squares equal to one-half or double the area of others.

Measures of Time

Through studies based primarily on the movements of the heavenly bodies in relation to the regular flooding of the Nile the Egyptians established two calendars. The priests followed an exact year of $365\frac{1}{4}$ days, while the civil year was subdivided into 3 seasons of 4 months each having 30 days — inundation, cultivation, and harvest — plus 5 days-upon-the-year.

Measures of Volume

In establishing standard units of capacity for barter, taxation, and medication convenient reference was made to human factors such as a mouthful (tablespoonful), handful, cupful (two handfuls) and sack- (hide-) ful for easy carriage. The principal Egyptian standards from small to large were the *ro, hin, hekat*, and *khar*.

Medications were prescribed in terms of the ro (mouthful, tablespoonful) = $\frac{1}{2}$ fl oz (14.5 cc). By the usual Egyptian method of doubling this standard produces 2 ro = 1 handful (fl oz, jigger); 4 ro = 2 handfuls (jack, jackpot); 8 ro = 4 handfuls (gill); 16 ro = 2 hands cuppedful (cup); 32 ro = jugful (pint, hin.); 64 ro = pitcherful (quart) and, by extension, the old English measures including the gallon, peck, bushel, barrel, hogshead, and tun.

The hin for liquids was subdivided dimidially down to $\frac{1}{32}$ = 1 ro. There was also $\frac{1}{3}$ hin known as the khay, thirds being a peculiarity of Egyptian metrology occurring also in the subdivision of the cubit and qedet standards. There were two principal standards for larger volumes. Multiples of the hin were decimal: 10 hin = 1 *hekat* (4.7 liters). The hekat was the official corn measure: hekat = 10 hin; 30 hekats = 1 cu cubit. The khar, orginally a hide or sackful, was a special standard = 20 hekats (94 liters). Subdivisions of the hekat in the dimidial series $\frac{1}{2}$ to $\frac{1}{64}$ (= 5 ro, an amount frequently prescribed for internal medications) were always written in symbols now known as the Horus-eye notation. The Rhind Mathematical Papyrus, attributed to the Twelfth Dynasty (1991–1783 BC), provides the following conversion table: 9600 ro = 300 hin = 30 hekat = $\frac{3}{2}$ khar = 1 cubic cubit (37 U.S. gallons).

The early use of additional capacity standards is documented by a tomb painting of the Third Dynasty (2649–2565 BC). Two series of cylindrical capacity measures, one of copper for measuring wine and oil and one of hooped wood staves for grain, have been shown to conform to a mixed series in binary progression based on the standards of both the Egyptian hin and the Syro-Phoenician kotyle (21.4 cu in or 350 cc). Other standards of volume used in Egypt include the Syro-Babylonian log (33.1 cu in or 542 cc), the Attic kotyle (17.2 cu in or 282 cc), and the Persian kapetis (74.9 cu in or 1227 cc).

Measures of Weight

The practice of weighing began with the discovery of metal-working. The earliest weights are of a small order, suggesting that at first only small quantities had to be weighed. The Egyptians developed two indigenous standards and as an international trading center employed at least six more. The Egyptian national standard was based on the qedet, which varied over time from 138 to 152 gr giving a median value of 144 gr (9.33 gm). It was multiplied decimally: 10 qedet = 1 deben; 10 deben = 1 sep, as first recorded in a Third Dynasty tomb painting. The unit was sometimes subdivided into thirds, as evidenced by a series of weights from graves of the First Dynasty (2920–2770 BC). The qedet is the basis of nearly all statements of weight from the Eighteenth Dynasty onward. The qedet and deben were also known as the kite and uten when applied to pieces of copper used as a medium of exchange.

Less well known but more influential in world affairs was the beqa, which was associated with the weighing of gold and silver from predynastic times and used in Egypt for nearly 4000 years. Although the standard is seldom mentioned in the Egyptian inscriptions, numerous beqa weights have survived. These are sometimes marked with the hieroglyph for gold (nub), perhaps recalling its origin in the Nubian gold-fields. As in most middle eastern standards, the basic unit was called a shekel; 50 beqa shekels = 1 mina. The beqa was multiplied decimally to 2000 beqa shekels and subdivided dimidially down to $\frac{1}{16}$. Over time the several distinct values merged into a single standard of 192 gr (12.4 gm) which was adopted by the Greeks around 700 B.C as the standard of Aegina. From it evolved two of the Roman pounds (the silver denarius and the gold aureaus coinage). In the seventh century AD it was adopted by the Arabic Empire for bulk gold, and it ultimately became the basis for the English troy weight standard.

Other standards in use in Egypt included the Palestinian

peyem, 120 gr (7.78 gm); Mesopotamian daric, 129 gr (8.36 gm); Attic stater, 135 gr (8.75 gm); Syrian necef, 160 gr (10.37 gm); Persian khoirine, 178 gr (11.53 gm); and Phoenician sela, 218 gr (14.13 gm).

All the balances used in Egypt before the Roman occupation in 30 BC were equilateral, doubtless inspired by the yokes with which men balanced heavy loads on their shoulders. Because they were made of perishable wood until about 1550 BC, the only records of their design and use come from reliefs, paintings, and inscriptions. From the Third through the Sixth dynasties these show a simple beam, rectangular in section, pierced vertically for support at the fulcrum and for the hangers, which were single cords hooked at the end to hold baskets or bags for gold dust. During this period the major innovations included specialized tongs to hold metal vases, counterweights of predetermined mass, and tongues and plummets to help in sighting against the vertical. Twelfth Dynasty tomb paintings show hangers with multiple cords, making possible the shallow concave pans used still. Balances were employed only for weighing gold, silver, copper, and lapis lazuli, the minerals used by the craftsmen or presented as offerings or tribute. The weights were made of various hard stones in a succession of geometric forms readily identifiable by Dynasty.

Improved technology for bronzeworking produced startling changes in Egyptian metrological tools during the brilliant Eighteenth Dynasty. Weights were cast of bronze in the shapes of animals — chiefly the gazelle, ox, and hippopotamus. Massive ceremonial balances, often with the head of a god as a finial, were featured in reliefs, paintings, and inscriptions concerning actual events as well as in the allegorical Books of the Dead, in which the heart of the deceased was weighed against the feather of truth as a prerequisite for admission to paradise. The beams, which could be of wood or bronze, were circular in section, with the end suspensions emerging from the beam ends, making a more accurate instrument that could be adjusted for wear. The cumbersome wood tongues evolved into lightweight bronze pointers. For the first time public weighers and shopkeepers weighed commodities against rings of gold, silver, and copper in both international and domestic trade. This improved Egyptian balance became the standard weighing instrument of the Near East until Roman times, but beginning in the Nineteenth Dynasty the weights once more assumed simple geometric shapes, first of bronze and later of stone.

Glassmaking, another technological innovation of the Eighteenth Dynasty, was applied to metrology with the production of a few weights during the Roman and Byzantine periods. The Arabs, after their conquest of the country in AD 640, continued the tradition, issuing a fine series of reference weights for their coinage. These are circular discs of translucent glass in various colors, molded with a thick rim surrounding a central depression that carries religious and civil inscriptions in Arabic. The Arabic standards of metrology continued to be dominant through the nineteenth century with standards varying by city and by commodity. The metric system was made permissible in Egypt in 1873.

RUTH HENDRICKS WILLARD

REFERENCES

Bell, Barbara. "The Oldest Records of the Nile Floods." *Geographical Journal* 136:569–573, 1970.

Berriman, A. E. *Historical Metrology*. New York: E. P. Dutton, 1953. Reprinted New York: Greenwood Press, 1969.

Edwards, I. E. S. *The Pyramids of Egypt*. Hammondsworth, Middlesex: Penguin Books, 1947, reprinted 1977.

Glanville, S. R. K. "Weights and Balances in Ancient Egypt." *Proceedings of the Royal Institution of Great Britain* 29:10–40,1937.

Griffith, F. L. "Notes on Egyptian Weights and Measures." *Proceedings of the Society of Biblical Archaeology* 14: 403–40, 1892; 15: 301–15, 1893.

Leake, Chauncey D. *The Old Egyptian Medical Papyri*. Lawrence: University of Kansas Press, 1952.

Lexikon der Agyptologie. 6 vols. Wiesbaden, 1972–1984. "Masse und Gewichte," by Wolfgang H. Helck; "Masse und Gewichte," by Sven V. Vleming; "Waage," by Erva Martin-Pardey.

Lucas, A. *Ancient Egyptian Materials and Industries*. 4th ed. London: Edward Arnold, 1926, reprinted 1962.

Petrie, W. M. F. *Glass Stamps and Weights; Ancient Weights and Measures*. 2 vols. Egypt: The British School of Archaeology, 1926. Reprinted Warminster, England: Aris & Phillips, 1974.

Skinner, F. G. *Weights and Measures*. London: Her Majesty's Stationery Office, 1967.

See also: Calendars in Egypt – Pyramids

WEIGHTS AND MEASURES OF THE HEBREWS

Weights

The earliest Biblical reference to a unit of weight is in Genesis 23:16: "Abraham came to an agreement with him and weighed out . . . four hundred shekels (of silver) of the standard recognized by merchants". Abraham was a new immigrant in Canaan. The *shekel* units which he used to weigh out the silver for Ephron the Hittite may have been those of his native Mesopotamia, with which he would have been most familiar.

The Mesopotamian *shekel* weighed about 8.3–8.5 g. Sixty *shekels* = one *maneh* (Greek: Mna, mana. English translations of the Bible often render *maneh* as "pound") of 497–

508 g. Sixty *maneh* = one *kikar* (Greek: Talanton; English: talent) of 29.8–30.5 kg. This system, which was originally Sumerian, in the fourth or third millennium BC, was used with little change for thousands of years by the peoples of Mesopotamia; the terms SH-Q-L, M-N-H, and K-K-R are Semitic in origin and were first used by the Akkadians. Systems similar in structure, though varying in detail and in the absolute mass of the units, were used by most of the peoples of the Eastern Mediterranean, well into Roman Imperial times.

The term *shekel* is, in short, a generic one, and in this context could apply to any one of a number of units (including the Mesopotamian) in use in Canaan in the Bronze Age. Since the price of 400 *shekels* for the Cave of Machpelah was quoted by Ephron himself (Genesis 23:14), these may very well have been *shekels* of the Syrian standard (also called Ugaritic), weighing about 9.4 g. Weights on this standard were used throughout Canaan and northern Syria, and are found at Bronze Age sites in the Aegean sea, as far north as Troy. The standard also reached Egypt, where its unit became the *qedet*. Three thousand Syrian *shekels* made one talent.

Genesis 24:22 says: "... the man took a gold nose-ring weighing half a *shekel*, and two bracelets for her wrists weighing ten *shekels* ..." The Hebrew word rendered in English as "half a *shekel*" is *beka'*. The word means "half", and appears on certain Iron Age weights described below.

The next mention of units of weight in the Bible is in the Book of Exodus. Several hundred years have passed; the Children of Israel, now a numerous people, have received the Law at Mt. Sinai. This includes instructions for the payment of taxes and the performance of religious ritual, which call for clearly defined units of measurement. The first specifically Hebrew unit of weight is *shekel ha-kodesh*, which the King James version translates as "shekel of the sanctuary" and the New English Bible as "*shekel* by the sacred standard", "twenty *gerahs* to the *shekel*" (Exodus 30:13). This unit is referred to repeatedly (Exodus 30:24, 38:24, 25, 26, Leviticus 5:15, 27:25, etc.). Reference is now also made to *talents* of silver and gold: the talent weighed 3000 *shekels* (Exodus 38: 25, 26). There are no surviving artifacts of this period from which the absolute mass of these units might be inferred.

It should be stressed that the *shekel ha-kodesh* and its fractions were used by the Hebrews for religious purposes, but not necessarily in commerce. In the ancient world, as in medieval Europe, there were many weight-systems in simultaneous use in international trade. Weights representing all the major systems of the ancient Near East are found at all important archaeological sites in Israel, and were obviously in use there. These systems are often linked by networks of interrelationships, with multiples of different units having a common mass and being used interchangeably.

The oldest extant weights bearing Hebrew inscriptions date from the period of the Kingdom of Judah (tenth–sixth centuries BC). They are normally of limestone, dome-shaped with a flat bottom; a few bronze specimens exist. There are several types, all rare. The total number of known specimens of all groups is a few dozen. They are:

- *NECEF*. Average mass about 9.8 g. The *necef* is not mentioned in the Bible. The name means "half": it may have originally been half of some more ancient unit. Scholars today identify the *necef* with the Canaanite/Syrian/Aegean unit already described. It is also equivalent to one Egyptian *qedet* and to four-fifths of the royal *shekel* of Judah (see below).
- *BEKA'*. This word also means "half" (Genesis 24: 22 – see above). The mean mass of this group is about 6 g. The *beka'* is one-half of the "shekel of the king" first mentioned in II Samuel 14:26.
- *PYM* (Petrie: *PEYEM*). Mean mass about 7.8 g. The word appears once in the Bible (I Samuel 13:21) : "The charge was two-thirds of a *shekel* (Hebrew: *pym*) for (sharpening) ploughshares and mattocks, and one-third (of a *shekel*) for sharpening the axes and setting the goads". The *pym* is two-thirds of the royal *shekel*.
- *SHEKEL*. Weights of this group are inscribed, not with a word, but a sign γ. Its meaning has been variously interpreted as being an official adjuster's mark, a symbol of the bag (*tsror*) in which lump silver was often carried, and a stylized representation of the royal scarab emblem of the kings of Judah. Sir Flinders Petrie read it as a monogram composed of the Greek letters *chi* χ and *omicron* o, and on this basis, erroneously attributed these weights to a (supposed) system which he called Khoirine. The mean mass of the *shekel*, calculated from all known marked specimens, is about 11.4 g.

The *shekel* sign is accompanied by a mark indicating the denomination: 1, 2, 4, 8, 16, 24. Since numbers of *shekels* are given in the Bible in multiples of fifty, this binary multiplication seems illogical; it is explained by the fact that this *shekel* is equal in mass to one and a quarter Egyptian *qedets*. Hence, four *shekels* = five *qedet*, eight *shekels* = ten *qedet*, etc. The denomination marks on the *shekel* weights are, in fact (apart from simple strokes to indicate 1 and 2), Egyptian hieratic decimal numerals, the sign for 4 *shekels* meaning in Egyptian 5, that for 8 meaning 10, and so on. These weights thus could be conveniently used interchangeably to weigh either in *shekels* or *qedets*. In fact, it can be seen that all the units here described are closely related. The Canaanite/Syrian/Aegean *maneh* of about 940 g can be divided into either 80 royal *shekels*, 100 necef (or *qedet*), 120 *pym*, or 160 *beka'*, providing simple arithmetical relationships between the systems prevalent in different regions.

The name "royal *shekel*" is confirmed by the existence of a two-*shekel* weight inscribed in Hebrew "L-M-L-KH" (of the king). A series of fractional weights also exists, both for the *shekel* and for the *necef*, showing that both were divided into twenty units (*gerah*). The denominations of these are also indicated by Egyptian hieratic numerals.

Scholars attribute most of these weights to some time near the end of the seventh century BC, leading some of them to suggest that the religious reforms of Josiah (II Kings 22, 23; II Chronicles 34, 35) were accompanied by a reform of weights and measures. There is no direct documentary evidence of this. In fact, biblical references are less than enlightening. Ezekiel, foreseeing the rebuilding of the Temple, orders that the *maneh* shall consist of "twenty, and twenty-five, and fifteen (i.e. 60) *shekels*". This is so inconsistent with other data that the translators of the New English Bible assume a misspelling in the Hebrew text, and render *essrim* as "ten" (Ezekiel 45:10), giving fifty *shekels* to the *maneh* as before. After the Temple was rebuilt, the people of Israel pledged to pay an annual tax of one-third *shekel*, not one-half as in Moses' time (Nehemiah 10:32).

Nothing more is heard of the *shekel ha-kodesh*; it is last mentioned in Numbers 18:16. Yet the correct weight of the Temple tax was a subject of deep concern to the Jewish religious authorities; it is discussed at length in the Mishnah, the text book of Jewish laws. By the first century BC it had been decided that all religious taxes should be paid in "*shekels* of the sacred standard, of the *maneh* of Tyre". The Tyrian *shekel* at that time weighed a little over 14 g. and was famous for its purity of metal and uniformity of weight. No other coin would do. This is the reason why Jesus found so many money-changers in the Temple precinct (Matthew 21:12): Jewish pilgrims arriving from foreign parts and wishing to pay their dues needed Tyrian *shekels* in order to do so.

From AD 66 to 70, when Jerusalem was besieged by the Romans, and Tyrian coin was no longer obtainable, the *shekels* struck throughout the siege by the embattled Jews (probably for religious use) were of the Tyrian standard.

The use of Hebrew weights and measures for secular commerce almost certainly ended with the fall of the Kingdom of Judah and the destruction of the Temple by Nebuchadnezzar in 586 BC. There is no evidence from the successive periods of Persian, Hellenistic, and Roman domination of the Land of Israel (or from times of relative independence, under the Hasmonean and Herodian dynasties) of the use of any specially Jewish units. Rabbinical discussions of weights and measures usually center on attempts to define Biblical units in terms of contemporary non-Jewish systems, in order to ensure the proper performance of the Mitzvoth (commandments).

Measures of Length and Area

The Hebrew unit of linear measure was the *ammah* or cubit — the length of a man's forearm, from the elbow to the tip of the middle finger. Deuteronomy 3:11 actually uses the expression *le-ammat ish* — "by the forearm of an adult male". The measurement of length by units based on parts of the human body is universal among ancient cultures.

The Bible mentions no unit greater than a cubit. *Kaneh ha-midah*, the surveyor's measuring rod (or reed), was six cubits in length (Ezekiel 40:5). The measurements of the inner temple are given by Ezekiel as 500 by 500 *kaneh* (Ezekiel 42:16–20): the Authorized Version translates these as "reeds", the New English Bible as "cubits". Surveyor's measurements of up to 25,000 cubits are on record (Ezekiel 45:1). Long distances are expressed in terms of a day's journey, *derekh yom* (Genesis 30:35; Numbers 11:31; II Kings 3:9).

The Hebrews used two cubits, one a hand's breadth (*tefah* or *tofah*) longer than the other (Ezekiel 40:5, 43:13). The longer cubit was used in the rebuilding of the Second Temple (Ezekiel 40–48), and also, presumably, in building the Temple of Solomon, since the dimensions quoted for the Holy of Holies and the Sanctuary are the same in both cases (20×20 cubits and 20×40 cubits respectively: I Kings 6:14–28, Ezekiel 41:1–4).

The use of two cubits suggests an association with Egyptian linear measure. The greater ("Royal") Egyptian cubit was of seven hands, each hand of four fingers (also four spans of seven fingers each). Its length of about 52.5 cm is attested by many surviving measuring rods and by records of the dimensions of surviving monuments. The lesser Egyptian cubit was of six hands.

The Biblical fractional units of length are the span (*zeret* – Exodus 28:16, 39:9; I Samuel 17:4), the hand (*tefah* or *tofah* – Exodus 25:25, 37:12; I Kings 7:26, II Chronicles 4:5), and the finger (*etzba* – Jeremiah 52:21). The only direct clue to the absolute value of the Hebrew cubit is the Siloam inscription (ca. 700 BC) from which C.R. Conder estimates a value of 17.1 inches (43.4 cm); he warns, however, that this estimate may be in error by 10% or more.

The Mishnah (second–third century AD) states that the greater cubit was of six hands, the lesser of five; the hand is of four fingers. Six hands of Egyptian dimensions would measure 45 cm. Most rabbinical estimates, however, are in the range 58–66 cm.

There are no specially named units of area in the Bible. Ezekiel, discussing the layout of rebuilt Jerusalem, speaks of "twenty-five thousand cubits square (*revi'it*)" (Ezekiel 48:20).

Measures of Cubic Capacity

"... your bushel (Heb. *'ephah*) and your gallon (Heb. *bath*) shall be honest. There shall be one standard for each, taking each as the tenth of a *homer*, and the *homer* shall have its fixed standard" (Ezekiel 45:11). Solid and liquid measure are thus given a common basis. The volumetric measures mentioned in the Bible are: *Ephah* (Leviticus 19:36; Deuteronomy 25:14 ("measure"); Judges 6:19; Ezekiel 45:24); *S'ah, seah* (Genesis 18:6; I Samuel 25:18; I Kings 18:32 (always "measure"); *Kab, cab* (II Kings 6:25); *Omer* (Exodus 17:36 – "An *omer* is one tenth of an *ephah*"); these are always associated with solid measure. *Letekh* (Hosea 3:2 – AV: "a half *homer* of barley"; NEB: "a measure of wine") is mentioned only once. *Homer* – (Ezekiel 45:11, see above) and *Kor, cor* (I Kings 4:22, 5:11) are associated with both solid and liquid measure. *Bath* (I Kings 7:26, 38; I Chronicles 2:10) and *Log* (Leviticus 14:10, 12, 15, 21, 24) are associated with liquid measure. The relations between any of these measures are only mentioned once or twice. The Mishnah gives a table relating some of them: 1 *ephah* = 3 *se'ah*; 1 *se'ah* = 6 *kab*; 1 *kab* = 4 *log*; 1 *log* = 6 eggs. The water displaced by one egg is about 55 ml (there are differing rabbinical views on the precise quantity), which would give the *log* a volume of 330 ml and the *ephah* and *bath* each about 24 l.

Attempts to establish mathematical relationships between ancient units of length, mass, and volume were popular a century ago, but today are not taken seriously. Reliance on post-Biblical authors such as Josephus Flavius is also risky: Josephus lived more than a thousand years after King Solomon, and is more likely to have been concerned with giving his (mainly non-Jewish) reading public a notion of orders of magnitude than with achieving a maximum of mensural precision. Hebrew weights and measures will always remain a subject fraught with uncertainty, which will be dispelled only when archaeologists are able to recover the actual instruments of measurement.

LIONEL HOLLAND

REFERENCES

Barkay, G. "A Group of Iron-Age Scale Weights." *Israel Exploration Journal* 28(4): 209–217, pp. 33–34, 1978.

Conder, C.R. "Hebrew Weights and Measures." In *The Palestine Exploration Fund Quarterly Statement*. London, 1902, pp.175–195.

De Sgani, Leah. "Shitot ha-Mishkal be-Eretz Israel" (Weight Systems in the Land of Israel). In *Perakim be-Toldot Ha-Miskhar Be-Eretz Israel* (Chapters in the History of Commerce in the Land of Israel). Ed. B. Keidar, T. Dothan, and S. Safrai. Jerusalem: Ben-Zvi Memorial, 1990 (in Hebrew).

Kletter, Raz. "The Inscribed Weights of the Kingdom of Judah". *Tel Aviv Journal of the Institute of Archaeology of Tel Aviv University* 18: 121–163, 1991.

Petrie, W.M.F. *Ancient Weights and Measures*. Dept. of Egyptology, University College, London, 1926.

Scott, R.B.Y. "The Shekel Sign on Stone Weights." *BASOR* 153: 32–35, 1959.

Scott, R.B.Y. "Shekel-Fraction Marking on Hebrew Weights." *BASOR* 173:53–64, 1964.

Scott, R.B.Y. "The Scale-Weights from Ophel, 1963–64." *Palestine Exploration Quarterly* 96: 128–139, pp. xxiii–xxiv, 1965.

Scott, R.B.Y. "The N-S-F Weights from Judah." *Bulletin of the American Society for Oriental Research* 200: 62–66, 1970.

WEIGHTS AND MEASURES IN THE INDUS VALLEY The golden era of the Indus civilization in ancient India extended from 2300–1750 BC. This vast civilization had significant uniformity and standardization in its material culture, as reflected in its town planning, building construction, pottery, metallurgy, and system of weights and measures.

The prosperity of the Indus cities depended to a large extent on trade. Many raw materials were brought by land and sea routes from within and outside the Indus valley. Inland trade must have extended beyond the Baluchistan to Afghanistan and the Iranian highlands on the one hand and to the Punjab and Aravalli hills on the other. Overseas trade covered the Makran and Persian Gulf ports on the west and the Gujarat and Konkan ports, if not those of the Malabar coast in the south. The writing system has not yet been deciphered, and the names of their measuring units are still unknown.

Length

Fragments of linear measures excavated from Mohenjo-daro and Lothal are made of shell and ivory. The average length of a unit is estimated to have been 67.6 cm. The linear measures were graduated using the decimal system. Some of the precise linear measures were graduated even to the one-hundredth and one-four hundredth of the unit. A fragment of bronze linear measure, which graduated to one-half (9.34 mm) of the later unit, *angula* (digit) was also discovered in Harappa.

Area and Volume

The use of area measures should have been quite common. There is, however, no piece of evidence which could be connected with an area measure. In the excavations carried out at

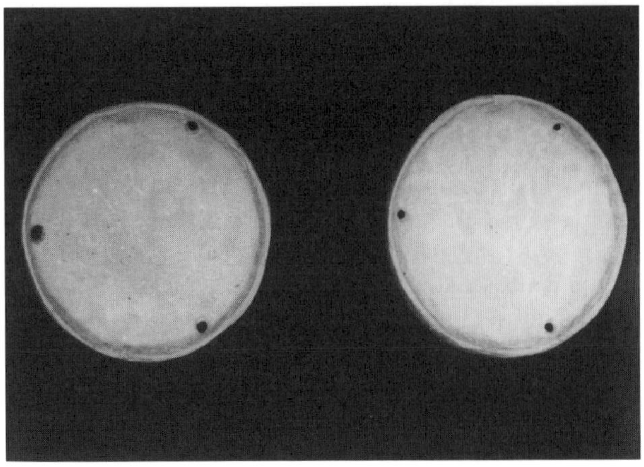

Figure 1: Balance pans. Photograph by the Mainichi Newspapers, 1961. Used with permission.

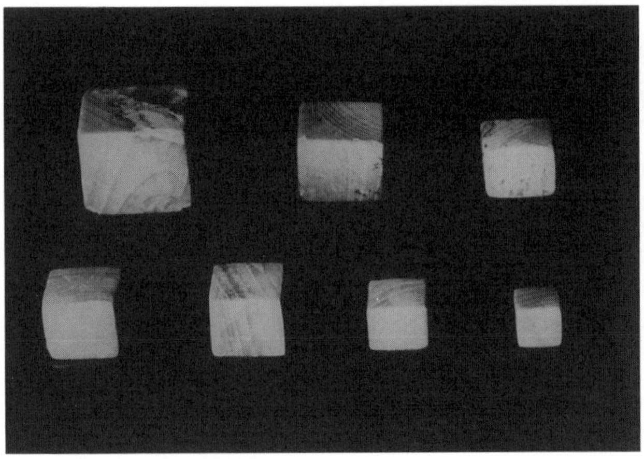

Figure 2: Weights. Photograph by the Mainichi Newspapers, 1961. Used with permission.

various sites of the Indus civilization a variety of pots made of clay, and sometimes of metal, has been discovered. No systematic determinations of the volumes of pottery seem, however, to have been made.

Mass

The oldest known weight in the Indus measuring system was excavated from Dashli Tepe, south Turkmenia in Russia. This era dates back to the fifth millennium BC. The other three weights belong to the fourth millennium BC and were discovered in northern Iran. These weights belong to pre-Indus civilization. In the third millennium BC the Indus measuring system was further developed in the ancient regions of Iran and Afghanistan.

A total of 558 weights were excavated from Mohenjo-daro, Harappa, and Chanhu-daro, not including defective weights. They did not find statistically significant differences between weights that were excavated from five different layers each about 1.5 m in depth. This was evidence that strong control existed for at least a five-hundred-year period. The 13.7 g weight seems to be one of the units used in the Indus valley. The notation was based on the decimal system. Eighty-three percent of the weights which were excavated from the above three cities were cubic, and sixty-eight percent were made of chert.

Balance pans were made of copper, bronze, and ceramics. A bronze beam was found with the two pans in Mohenjo-daro. The fulcrum was the cord-pivot type (Figure 1).

The measuring system used in the Indus valley was different from the Mesopotamian and Egyptian measuring systems, but the sensitivity of precision balances used in these regions is assumed to have been comparable. The weights excavated from Taxila (sixth century BC–seventh century AD) descend from the system of weights used in the Indus civilization (Figure 2).

SHIGEO IWATA

REFERENCES

Bhardwaj, H.C. *Aspects of Ancient Indian Technology*. Delhi: Motilal Banarsidass, 1979.

Hori, Akira. "A Consideration of the Ancient Near Eastern Systems of Weights." *Orient* 22: 16–36, 1986.

Iwata, Shigeo. "On the Standard Deviation of the Weights of Indus Civilization." *Bulletin of the Society for Near Eastern Studies in Japan* 27 (2): 13–26, 1974.

Iwata, Shigeo. "Development of Sensitivity of the Precision Balances." *Travaux Du 1er Congres International De La Metrologie Historique*. Zagreb: Jugoslavenska academija znanosti i umjetnosti, Historijski Zavod, pp. 1–25 + Fig. 1, 1975.

Mainkar, V.B. "Metrology in the Indus Civilization." In *Frontiers of the Indus Civilization*. Ed. B.B.Lal and S.P. Gupta. New Delhi: Books & Books, 1984, pp. 141–151.

Rao, S.R. *Lothal and the Indus Civilization*. London: Asia Publishing House, 1973.

WEIGHTS AND MEASURES IN ISLAM In the sphere of Islamic influence the Quranic injunction "to give full measure and to weigh with the right scales" (*Qur'ān* 17:35), led to a systematic way of measuring. With the determination of the fundamental relation of the Islamic weight system — the *dirham* has a ratio to the *mithqāl* of 10:7 — Islamic law introduced one single standard. Everyday use polished this ratio into the handier one of 3:2.

Through the Quranic revelation, the measure of time assumed a dual character. The natural solar year gave way to the ritual lunar year (29–30 days of the month) except in the fields of public administration and astronomical science.

Thus the development of systems of measure was influenced by the interplay between cultural tradition and the order of authorities. Until the tenth century, the spread of Islam brought about an intercontinental economic and cultural sphere which amalgamated measuring standards of Egyptian, Arabic, Greek, Roman-Byzantine, Mesopotamian, and Persian origin. This resulted in a multiplicity of regional and functional systems of measurement, which were constantly modified by power politics.

Following is a summary of some of the standards developed:

Length Standards

1 *ashl* (rope) = 1 *silsila* (chain) = 10 *bāb* = 10 *qaṣaba* (pipe) = 60 *dhirā'* (ell) = 360 *qabaḍa* (fist) = 1440 *iṣba'* (fingers) = 3600 fals = 8640 *sha'īr* (kernels of barley).

Volume Standards

1 *azāla* = 1 *kurr* = 100 *dhirā'* = 6000 *qafīz* = 172,800 *qabaḍ* = 11,059,200 *iṣba'*.

Weight Standards

(a) Coin Weights: 1 *dirham* = 6 *dāniq* = 20/24 *qirāṭ* = 60/72 *ḥabba* = 240 *aruzza* = 600 *fals/dirham* = 6/12 *dāniq* = 10 *'ashīr* = 24 *ṭassūj* = 60 *ḥabba/'ashīr* = 96 *fals*.
(b) Weights by Merchandise: 1 *kurr* = 60 *qafīz*= 480 *makkūk* = 600 *'ushr* = 1400 *kīlaja* = 6570 *rub'* (fourth) = 7200 *raṭl* = 11,520 *ṭumn* (eighth).

The basis of all length measures and their derivatives is the 'black ell' taken from the elbow of an Abbasid nilometer (or from the elbow of a black slave) from the ninth century, which measured 54.04 cm. There are almost thirty variants of this ell, some varying 30-fold from the original.

The basis for the weight standards is the *mithqāl*, whose coin weight can be defined by eighth century glass weights of 4233 grams. For coins and goods, *mithqāl* weights are recorded between 3 and 6 g. From these developed the multifarious *raṭl*-value, the most used weight for wares between North Africa and central Asia. There exists a specialized literature by region on the art of administrating the systematization of the units of measure. The most important instrument is the proportionality factor through which access to other regional specific or ware specific measures was regulated.

Since the tenth century mathematicians became aware of the confusing relationships and tried to systematize the usage of measuring units in popular treatises, with the aim of standardizing and facilitating the conversions used by customs, market, and tax offices. This resulted for the first time in varying physical sizes, measuring units, and measuring numbers. Thus, stimulated by theoretical manuals, methods of measuring became popularized. This had some retroactive effects on the practice of Islamic laws.

ULRICH REBSTOCK

REFERENCES

Cahen, C. "Quelques Problèmes Économiques et Fiscaux de l'Iraq Buyide d'après un Traité Mathématique." *Annales de l'Institut d'Études Orientales* 10: 326–363, 1952.

Ehrenkreutz, A.S. "The Kurr System in Medieval Iraq." *Journal of the Economic and Social History of the Orient* 5: 305–314, 1962.

Hinz, W. *Islamische Masse und Gewichte. Umgerechnet ins metrische System*. Leiden: E.J. Brill, 1955.

Miles, G.C. *Early Arabic Glass Weights and Stamps*. New York: American Numismatic Society, 1948.

Rebstock, U. *Rechnen im islamischen Orient*. Darmstadt: Wissenschaftliche Buchgesellschaft, 1992.

WEIGHTS AND MEASURES IN JAPAN Several units of measure were used in ancient Japan, including *ki* (digit), *tsuka* (palm), *ata* (span), and *hiro* (fathom). It is not yet clear which measuring system was used most frequently. It seems that Chinese metrology had begun gradually penetrating into Japan by the seventh century. Japanese metrology, which developed from Chinese metrology, was called the *shakukan* system.

Length

The basic Chinese linear measure was almost constant at 23 cm from the fifth century BC to the first century AD. Various civil disturbances in the fifth and sixth centuries had the effect of lengthening the linear measure substantially to 29 cm. Under the Tang dynasty, a law was enacted mandating the use of two methods based on large and small linear measures. The large scale was 1.2 times as long as the small scale, which we refer to as the ancient linear measure. The small scale was used for music, astronomy, and ceremonial items.

Japan introduced this Chinese measuring system in 701. The large scale later became known as *kanejaku*, which refers to an L-shaped ruler used by architects. The small scale gradually dropped out of use. Linear measurement tools were mainly made of wood, though ivory was also used in ancient times. The largest unit of length in ancient times was the *ri*, which was five to seventy *cho* (a *cho* was 109.1 m). The *ri* was set at 36 *cho* in the seventeenth century. One *skahu* was one-tenth of a *jo*. Metal linear measurement tools began appearing in the medieval era.

The average value of one *skahu* was 29.6 cm in the seventh century, 29.8 cm in the eighth century, 29.9 cm in the ninth century, 30.0 cm in the tenth century, 30.1 cm in the twelfth century, 30.2 cm in the fourteenth century, and 30.3 cm in the eighteenth century. The rate of elongation was slower than in China.

Several linear measures for sewing appeared in the medieval era. The length of one *skahu* ranged from 1.15 to 1.27 times that of a *kanejaku*. These linear measures for sewing were neatly arranged into two kinds of scales. One scale, 1.20 times as long as a *kanejaku*, was called *gofukujaku*, while the other was 1.25 times as long as the *kanejaku* and called a *kujirajaku*. The *gofukujaku* gradually dropped out of use in the first half of the eighteenth century and was finally abolished in 1875.

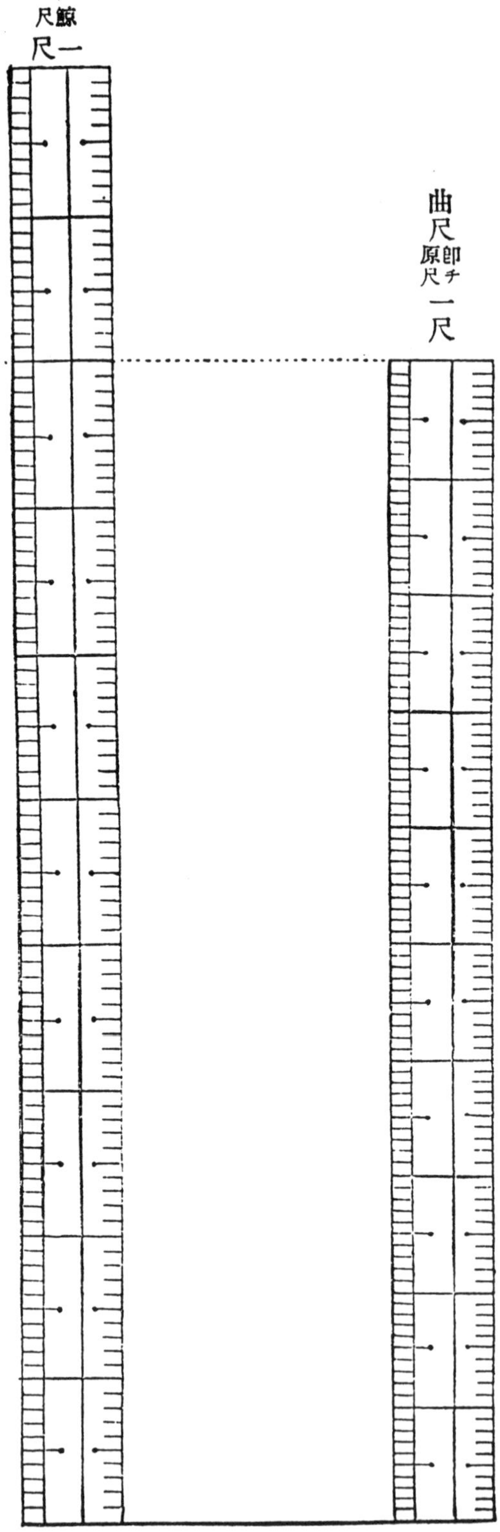

Kujirajaku Kanejaku

Figure 1: Linear Measure (Drawing by The Ministry of Finance. Doryoko Shurai Hyo (The Classification Table of Weights and Measures) Genbei Kinokuniya, 1875, chos 1, 4–6, 9–11.

Area

The earliest area unit we know of was the *shiro*, which was in use before the sixth century. One shiro was thirty *skahu* square. In 646, new units were defined: the *bu*, which was six *skahu* square; the *tan*, equal to 360 *bu*; and the *cho*, equal to ten *tan*. One *tan* has been equal to 300 *bu* since the end of the sixteenth century. For land surveying purposes, one *bu* was defined as six *skahu* five *sun* square, rather than six *skahu* square. At the end of the sixteenth century, a *bu* was defined as six *skahu* three *sun* square. Despite this, the *bu* was commonly considered to be six *skahu* square, and was finally defined as such in 1891. A *se* was ten times as large as a *bu*, which was also referred to as a *tsubo*. These names appeared in the medieval era. Land surveyers made their measurements using flax string measures and wood measuring poles.

Figure 2: Land Surveying (From *Tokugawa Bakufu Kenchi Yoryaku* [Essential Summary of the Government of Prefectures, Tokugawa Shogunate]. Tokyo: Kashiwa Shobo, 1915).

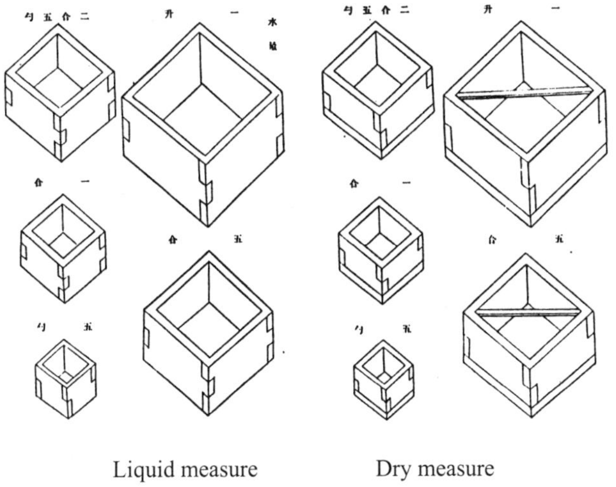

Figure 3: Measures (Drawing by The Ministry of Finance. Doryoko Shurai Hyo (The Classification Table of Weights and Measures) Genbei Kinokuniya, 1875, chos 1, 4–6, 9–11.

Volume

The names for most units of volume also came from China. The basic unit was the *sho*. One *sho* averaged 720 cm^3 in the eighth century, 800 cm^3 in the ninth century, 970 cm^3 in the eleventh century, 1200 cm^3 in the fifteenth century, and 1650 cm^3 in the sixteenth century. Since the seventeenth century, the average value has remained constant at 1804 cm^3. During the medieval era, different measures of volume were used for goods, taxes, and financial transactions.

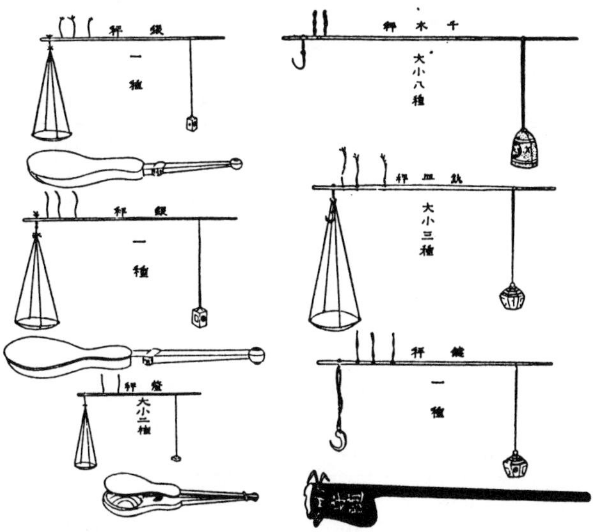

Figure 4: Steelyard (Drawing by The Ministry of Finance. Doryoko Shurai Hyo (The Classification Table of Weights and Measures) Genbei Kinokuniya, 1875, chos 1, 4–6, 9–11.

The Edo-Shogunate (1603–1867) established the East and West Measure Guilds in 1655, and exerted strong control over the new standard. Use of *sho* continued until 1958, when the *skahu-kan* system was abolished.

Rectangular-shaped measures were made of wood. The brims of many dry measures were protected by bamboo or iron planks. Most of the large dry measures also featured a narrow iron bar positioned diagonally across the top. Round wooden sticks were used to level off the tops of measures. Liquid measures had lacquered surfaces to prevent the contents from leaking.

Standard sizes for measures have been one *to*, seven *sho*, five *sho*, one *sho*, five *go*, two *go* five *skahu*, and one *go* since 1655, and one *to*, five *sho*, one *sho*, five *go*, two *go* five *skahu*, one *go*, and five *skahu* since 1875.

Mass

Between the eighth century BC and the fifth century AD, the Chinese expressed mass in terms of the *liang*, which was equal to 14 g. In this measuring system, one *jin* was equal to sixteen *liang*, and one *liang* to twenty-four *zhu*. The civil disturbances of the fifth and sixth centuries also had their effect on the *liang*, which trippled in mass. Under the Tang dynasty, a law stipulating two measuring scales was enacted, which fixed the mass of the new large *liang* at three times that of the mass of the older *liang*, which became used only for medicine.

Japan adopted this Chinese measuring system as well,

Table 1: Conversion tables of Japanese weights and measures (1891–1958)

Length

ri	*cho*	*ken*	*skahu*	*sun*	*bu*	*rin*	*mo*	Metric system
1	36	2160	12,960	129,600	1,296,000	12,960,000	129,600,000	3.927 km
	1	60	360	3600	36,000	360,000	3,600,000	109.1 m
		1	6	60	600	6000	60,000	1.818 m
			1	10	100	1000	10,000	30.30 cm
				1	10	100	1000	3.030 cm
					1	10	100	0.303 cm
						1	10	0.0303 cm
							1	0.00303 cm

Area

cho	*tan*	*se*	*bu (tsubo)*	*go*	*skahu*	Metric system
1	10	100	3000	30,000	300,000	9917 m^2
	1	10	300	3000	30,000	991.7 m^2
		1	30	300	3000	99.17 m^2
			1	10	100	3.306 m^2
				1	10	0.331 m^2
					1	0.0331 m^2

Volume

koku	*to*	*sho*	*go*	*skahu*	Metric system
1	10	100	1000	10,000	180,400 cm^3
	1	10	100	1000	18,040 cm^3
		1	10	100	1804 cm^3
			1	10	180.4 cm^3
				1	18.04 cm^3

Mass

kan	*ryo*	*monme*	*fun*	*rin*	*mo*	Metric system
1	100	1000	10,000	100,000	1,000,000	3.750 kg
	1	10	100	1000	10,000	37.50 g
		1	10	100	1000	3.750 g
			1	10	100	0.375 g
				1	10	0.0375 g
					1	0.00375 g

changing the names of *jin*, *liang*, and *zhu* to the *kin*, *ryo*, and *shu*, but retaining the same characters. The absolute mass of the *ryo* in the large scale decreased year by year, while that in the small scale became rather unclear. The average mass of one *ryo* was 42 g in the eighth century, 41 g in the tenth century, 40 g in the twelfth century, and 37.47 g at the end of the sixteenth century. Under the strong control of the Edo-Shogunate, the average mass of the *ryo* remained almost unchanged until the end of the nineteenth century. The *kan* and *monme* appeared in the medieval era. One *kan* was equal to one hundred *ryo*, and one *monme* equal to one-tenth of a *ryo*. These units gradually became standard.

Chinese use of the steelyard dates back to at least several centuries BC. The steelyard was also used in Japan, where the Edo-Shogunate established the East and West Steelyard Guilds in 1613. Steelyard beams were made of white oak, persimmon wood, ivory, horn, and bone. The beam had from one fulcrum to eight fulcra, which were made using string, cord, or rope. A copper pan or bronze hook was installed on the end of the beam, and the bob-weight was made of

Figure 5: Balance (Drawing by Shigeo Iwata (1981)).

Transition to the Metric System

In 1886, Japan signed the Treaty of the Meter, and promulgated the Law of Weights and Measures. As a result, the *skahu-kan* system and metric system began to be used jointly in 1891. The conversion coefficients between these two measuring systems were also fixed at the same time (Table 1). The law went into formal effect in 1893. In 1903, the Central Inspection Institute of Weights and Measures was established in Tokyo. Japan also adopted the units of the foot-pound system as legal in 1909.

A Measurement Law was passed by the Diet in 1951, and went into effect on March 1, 1952. It mandated that measuring units be based on the metric system. The *skahu-kan* system and foot-pound system were abolished in 1958.

A new Measurement Law was promulgated on May 20, 1993, and went into effect on November 1, 1993. The law made a sweeping revision of the regulations by adopting the Système Internationale d'Unités. The units of the other measuring systems on commercial transactions and certifications will be abolished by the end of the twentieth century.

SHIGEO IWATA

bronze or iron. Steelyards were used for most commercial transactions.

Balance makers did not belong to a guild, so standards for making balances were never formulated. The most famous balance makers were members of the Nakabori family, who lived mainly in the western counties of Japan. The balance beam was suspended by its pointer from the center of a box-mounted rectangular wooden stand. A central pin was used for the main fulcrum, and during the final stages of weighing, bearing friction was overcome by tapping on the pin with a small wooden hammer. Balances were mainly used in the manufacture of gold, silver, and copper coins, and in money-changing transactions.

Beginning in the seventeenth century, users of the balance had to be supplied with weights made by the Goto family. Two sets of weights were available: one consisted of weights of 20, 10, 5, 4, 3, 2, 1 *ryo*, 5, 4, 3, 2, 1 *monme*, and 5, 4, 3, 2, 1 *fun*, and the other was identical to the first but for the addition of fifty and thirty *ryo* weights. Weights were made of bronze, and engraved on the surface with the inspector's mark and maker's crest, leaves and flowers of the Japanese paulownia. The Goto family belonged to the Guild of Goldsmiths, and were inspectors authorized by the feudal government. They examined user's weights periodically by comparing them with standard weights.

REFERENCES

Iwata, Shigeo. "Changes in Mass Standards in Modern Japan." *Bulletin of the Society of Historical Metrology, Japan.* 1(1): 5–9, 1979.

Iwata, Shigeo. "Japaneses Scales and Weights." *Equilibrium* 1981, pp. 319–326.

Iwata, Shigeo. "Changes in Linear Measures in China and Japan." In *Acta Metrologiae Historicae I*. Ed. Gustav Otruba. Linz: Rudolf Trauner Verlag, 1985, pp. 117–137, 488–489.

Iwata, Shigeo. "Changes in the Chinese Standard Mass Unit." In *Acta Metrologiae Historicae II*. Ed. Harald Witthöft and Cornelius Neutsch. Linz: Rudolf Trauner Verlag, 1989, pp. 117–129.

Iwata, Shigeo. "Changes in the Japanese Weights and Measures Standards under the Influence of China." *The 15th Symposium on Historical Metrology*. Japan, November 6, 1992, pp. 1–17.

WEIGHTS AND MEASURES IN MEXICO Mexico is said to have used an incredible variety of weights and measures changing according to the region and the period: The different uses of the weights and measures increased the complexity of the situation. The task of the historian consists of restoring the ancient systems and understanding their use by posing the questions: Who was using the weights and measures and for what purpose? The more powerful of the two parties in the exchange would impose, upon the other, modes of measuring and counting to his own advantage.

Indian Measures

In 1520, Hernan Cortes had the opportunity to see the market of Tlalecolco and the transactions which went on there. He wrote, "Everything is sold by count and measure, and up till now I haven't seen anything sold by weight." Antonio de Herrera y Tordessilas reported that the Indians "have measures for everything" and that the judges controlled the weights and measures and punished the cheaters; but his information has not been confirmed. However, the level of development attained by the Mesoamerican people before the Conquest makes it very probable that they had an ensemble of measures and standards used in the apportionment of land, commercial transactions, levying of taxes, construction, pharmacy, etc., and that these measures constituted a unified and coherent system.

After the Conquest, the Spanish imposed Andalusian and Castilian measures such as *brazas*, *codos*, and *palmos*, that the Indians adopted easily, as, like their own, they were anthropometric. The result of this adoption was that Mesoamerican symbols of measure were transcribed in European terms on the pictograms of indigenous manuscripts written in the Nahuatl language. Bars represented digits, little black circles were for twenties, and other signs represented hands or feet. Thus, for example, the Indians gave the name *tlamamalli* to the load (*carga*) that a man (porter, *tameme*) could carry on his shoulders during a day's work. In the *novohispano* system, this load had the name *tameme*, by a simple movement of language and meaning. It was equivalent to a Spanish half-*fanega* or to two *arrobas*. This indigenous load varied in weight, in the function of the weighed merchandise, and with the strength of the carrier: for cocoa it was equivalent to three *xiquipilli* of 8000 grains each.

Among the measures of capacity, of great diversity, the smallest was the *centlachipinilli*, which literally means "a small something" (*una gota de algo*) and which was itself a multiple of *centlachipiniltontli* (*una gotilla de algo*). There were also small measures used in pharmacy, such as spoonsful (*cemixcolli*, *cenxumatli*, etc.). *Testal* designated the quantity of cornmeal necessary to make a tortilla for one person.

To measure dry materials, in particular for grains, they were often able to use measures equivalent to those imported from Spain: the *acalli*, for example, of which we do not know the weight, was part of an arithmetic system with the *cencaauhacalli* equal to a half-*fanega* and with the *cuauhacaltontili* equal to a *celemin*. Such equivalent approaches make it possible to conclude that the conquerors tried to impose their own system while at the same time adopting the native terms.

Measures of length were used for the preparation of clothing, especially for ponchos, made of bands of different widths. Most measures of length were land measures, based on proportions of the human body: *cemizteltl*, *cemmapilli*, and *cemmatl* literally a finger, a palm, and a span. This "span" designated the height of a man, from foot to raised hand, but it also measured the diagonal, from the left foot to the raised right hand, and it stood for three Spanish *varas* (yards or cubits). Such a dimension proves that the span was a geometric measure and not an anthropometric one; it was obtained by attaching an anthropometric measure such as a digit, measured by its length, to a coefficient. European measures of length did not have the same way of varying, in which a "foot" could vary between 28 and 60 cm. However, for land measures, certain spans were smaller, and the *vara* was therefore equal to two spans. In Tula, there was a division of land in order to reform the tax law, and "each Indian was given one hundred *varas* in length and twenty in width. Each one of these *varas* made two spans; this is the measure which the Indians use". It is possible that this text makes reference to an Indian *vara* and a Spanish *braza*. Standards of measure and length existed, and, to measure land, poles or cords were used, just as European surveyors did.

Road measures also existed, which were used by travelers, soldiers, and merchants. These were the *cennecehuilli* and the *cennetlalolli*, which Spanish authors tended to translate as "leagues".

Weights and Measures of New Spain

In the sources of the eighteenth century the range of different names involved — *carga*, *fanega*, and *arroba* — hides the numerical relations which link multiples and submultiples. The *carga* is made up of two *fanegas*, composed in turn of six to eight *arrobas*. In general the historian goes from the largest to the smallest, whereas in practice, the peasant, the muleteer, or the trader proceeds in the reverse order. In the same way that the metric *quintal* is made up of grams and kilograms between which the factor ten governs all relations, in the eighteenth century, two and six (equals two multiplied by three) governed the relations of the *carga* with its submultiples. The decimal metric system is a scientific system, mathematic in character, invariant, exterior to humans. In contrast, previous systems were based on people or on observations concerning them, or on their work and capacity to complete certain tasks with the aid of animals. That is, if a mule must climb steep paths with its load, the muleteer compensates for this by lowering the weight of the load by one or two *arrobas*. At times, the sources mention *cargas pequeñas* in contrast to *cargas regulares de mula*. "As long as freighting relied chiefly on pack trains and carts, the weight of the loads was determined by distances to the customer, the availability of pasture *en route*, as well as by the nature of roads and tracks. In the case of steep mountain paths, broken country or long distances, the *carga*, the traditional load of a mule, could be lowered in weight. So a mule load could comprise one and a half or two *fanegas*. In 1832 the Governor of Oaxaca stated that because of the poor roads the *carga* could never consist of more than eight *arrobas*" (Ewald, 1985). Generally the custom of mentioning the number of *arrobas* making up a *fanega* or a *carga* does not eliminate the uncertainties which surround all quantitative data. In 1751, the *carga* of salt in Tehuantepec consisted of 14 *arrobas*. Another eighteenth century source stated that moist Tehuantepec salt was 16 *arrobas*, while drier salt was computed at 12–14 *arrobas*.

The dampness or dryness of goods being transported markedly affects their weight. The presence of water increases the weight of light goods such as grains, but it lessens the weight of heavier ones, in particular salt. In Tehuantepec, the variation in weight often reached a quarter. Other variations were also noticeable between the different regions: "In eighteenth century Yucatan, one *fanega* of old salt weighed approximately 9 *arrobas*, one *fanega* of new salt 10–11 *arrobas*" (Ewald). The measure known as *carga* in the valley of Mexico was often mistakenly referred to as *fanega* in Yucatan.

But there is another trap: Ralph Roys, for example, noted that a *fanega* "is variously defined as 1.6 bushels, and a load of 100 pounds of grain", which is an example of recent efforts to align traditional measurements with those of the national market governed by the decimal metric system, whilst at the same time conserving the ancient names. But "this statement has led a number of scholars to assume that a *fanega* weighs 100 pounds, which is not necessarily the case. The *fanega*, like the bushel, is a measure of capacity, not weight. Hence, the weight of a *fanega* will vary enormously depending on the goods being measured by this standard. A *fanega* of corn may weigh close to 100 pounds, but salt is much heavier. An extensive survey of the literature on weights and measures used in the Maya area — from Yucatan to El Salvador suggests that a *fanega* of salt commonly weighed about 115 kg" (Andrews, 1983). It is important not to confuse measures of capacity, which are volumes, and units of weight which measure mass: weighing the contents of a volumetric measure can only lead to considerable distortions, since each body has a specific weight by which it distinguishes itself from all others.

The foreigners who traveled or worked in nineteenth-century Mexico reported that a *fanega* of salt weighed approximately 70 kg and a *carga* 140 kg. J. Buschmann stated that the *carga* consisted of 138 kg. In 1912–1913, however, the business records of Salinas, Mexico calculated the *carga* of Colima salt at 161 kg. In 1916–17, the *carga* was again lowered to the more customary 140 kg. The nineteenth century sources gave 300 pounds as the equivalent of a *carga*. In summary, here are some of the results found in the historical literature on the subject.

In Mexico, on the central plateau (Tehuantepec salt):

carga	1		
fanega	2	1	
arroba, dry salt	14	7	1
arroba, moist salt	16	8	1
pounds, dry salt	300	150	21,4
kg	138	69	

In Yucatan, the substitution of the *fanega* for the *carga* does not modify the system.

fanega	1		
arroba, old salt	9	1	
arroba, new salt	11		
pounds	300	33.33	1
kg	138		0.460

The historian will recognize the mule load of three hundred pounds and the sixteen ounce pound, such familiar territory that he would think he was in Castile in the time of the Catholic Kings.

JEAN-CLAUDE HOCQUET

REFERENCES

Alonso, M. M.. *Medidas Indígenas de Longitud*. Mexico: CIESAS, 1984.

Andrews, Anthony P. *Maya Salt Production and Trade*. Tucson: University of Arizona Press, 1983.

Castillo Farreras, V. "Unidades Nahuas de Medida." In *Estudios de Cultura Nahuatl*, vol. 10. Mexico: Instituto de Invesitgaciones Historicas, Universidad Nacional Autonoma de Mexico, 1972.

Ewald, Ursula. *The Mexican Salt Industry, 1560–1980. A Study in Change*. Stuttgart: G. Fischer, 1985.

Garcia Acosta, Virginia. *Los Precios del Trigo en la Historia Colonial de Mexico*. Mexico DF: Centro de Investigaciones y Estudios Superiores en Antropologia Social, 1988.

Gual Camarena, Miguel. *El Primer Manual Hispanico de Mercaderia (Siglo XIV)*. Barcelona: Consejo Superior de Investigaciones Cientificas, 1981.

Hocquet, Jean Claude. *Anciens Systèmes de Poids et Mesures en Occident*. London: Variorum Reprints, 1992.

Roys, Ralph L. *The Political Geography of the Yucatan Maya*. Washington D.C.: Carnegie Institution of Washington, 1957.

Villena, Leonardo. "Weights and Measures in Islamic Spain." *Acta Metrologiae Historicae* 1985, pp. 298–303.

WEIGHTS AND MEASURES IN PERU Ancient Peru covered most of the Andes, where the Peruvian culture lasted for more than ten thousand years. Because most of the region was in the highlands, more than three thousand meters high, their metrology was slightly different from other civilizations'. Also, they had no form of writing until 1532. For calculation and for recording metrological units decimally, they used *quipus*, spatial arrays of multicolored knotted cords. The names of the units and the values differed significantly depending on the period, the people, and the region.

Length

Peruvians used linear measurements which were related to human measurement patterns. Examples are listed in Table 1.

Peruvians chewed lime and *lipta* with coca leaves as a stimulant. They invented a measure called *cocada*, which measures how far one can walk while under the drug's influence. This distance is three kilometers on level ground or two kilometers on an uphill slope when carrying a 45 kg load for 35–40 minutes in Pataz county.

A twelfth century pre-Inca site in Puruchuco, located about six kilometers from Lima, was recently restored. The linear measurements of each part of the structure showed that the average stature of men from head to toe was 1.701 m, while the height from the middle fingertip to one's toe was 2.106 m. They used a geometrical design of proportions based on the golden section (1:1.618) between the sides of the rectangles.

Area

The Aymara called the area where they could raise one or two heads of cattle the *callapa*. *Tupu* was the Incan name for an area of cultivated land or pasture where a married couple could live. The Inca called the area, which was equal to 625 m^2 *huiri*. The was the area a man could cultivate in one day. The Cuzco called the area equal to 282 m^2 of sweet potatoes or corn *papacancha*.

Volume

The Peruvians used a hand held bowl to measure small portions of grain. They also used gourds, dried pumpkins, pots, and jars to measure things. The names of the units and the values are shown in Table 2.

The Quechua called a broad crate filled with coca or red pepper a *runcu*. One half of a *runcu* was called *checta runcu*, one fourth was called *cutmu*, and one eighth was called *sillcu*.

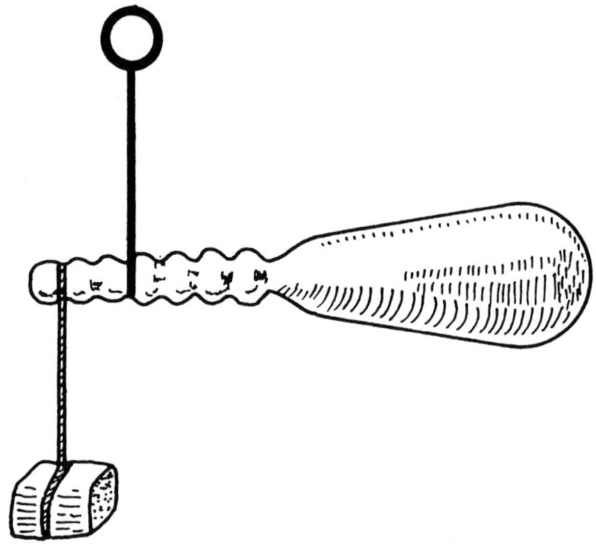

Figure 1: Bismar. Adapted by the author from M. Uhle, "La balance romaine au Pérou." *Journal de la Société des américanistes de Paris* 17:335, 1925.

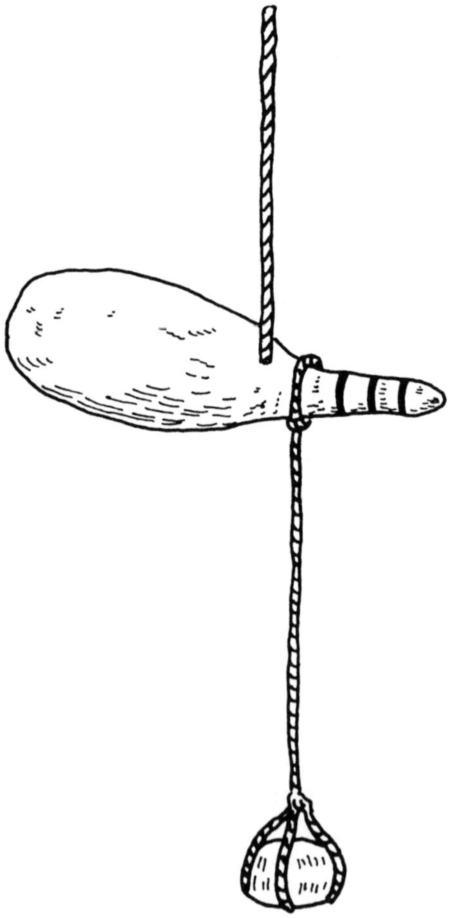

Figure 2: Movable Load Scale. Adapted by the author from P. Rivet, "La balance romaine au Pérou." *L'Anthropologie* 33:535, 1923.

Table 1: Examples of Peruvian linear measurements

	Quechua	Aymara	Inca	Ratio	Metric system
Handbreadth		*ttkhlli*		0.051	8.5 cm
Link	*yaku*	*vicu*		0.107	18 cm
Span	*capa*	*chia*		0.125	21 cm
Cubit	*cuchuch tupu*			0.25	42 cm
Yard	*sikya*			0.50	84 cm
Pace		*chillque*		0.89	150 cm
Fathom	*rikra*			1.00	168 cm
		loca		10^2	168 m
		ecca		10^3	1.68 km
	cocoda			$1.19–1.79\times10^3$	2–3 km
League	*tupu*		*tupu*	3.33×10^3	5.6 km
		yapu		5.00×10^3	8.4 km
		chuta (sayhua)		10^4	16.8 km
			guamanin	10^5	168 km

Table 2: Hand-held bowl measure of the Peruvians

	Quechua	Aymara	Ratio	Metric system
Both hands holding a bowl	*poktoy*	*iuu*	0.043	300 cm^3
		laqui	1.00	7000 cm^3
	pokcha	*hullu*	2.00	14,000 cm^3

Mass

Because fewer weights have been discovered in the Peruvian region than in other civilizations, it has been difficult to estimate the mass standard accurately. We assume that the average mass of unit was 23.1 g. The notation was based on the decimal system. The maximum unit was assumed to be 23.1 kg, which is about eighty percent of the maximum units from other civilizations, 27–30 kg in Egypt, Mesopotamia, Indus and China. This is probably because of the low concentration of oxygen in the atmosphere in the Andean highlands.

Thirty-nine percent of the weights were made of stone, thirty two percent of iron, twenty four percent of lead, and five percent were made of other nonferrous metals. One-third of the weights were globe shaped; the others were conical, cylindrical, or in spindle and other irregular forms.

All types of scales, namely balance, steelyard, bismar, and movable load, were discovered in the northwest region of South America. The fulcrum was the cord-pivot type. Balance beams were made of wood, horn, and bone with humans, monkeys, birds, and geometrical patterns engraved on them. The two items supporting the load and weights were mostly nets, and partly pans of metals. The balance might have been invented by the ninth century.

Miquel de Estete, who attended Francisco Pizarro, found a steelyard on the coast of Ecuador in 1531. The steelyard was graduated from the middle of the beam to the end, and had a bob-weight suspended from the arm. The steelyard was used to weigh gold and silver. The bismar was discovered at the market in Tarma, 170 kilometers northeast of Lima. The total length was 27.8 cm, and the capacity was assumed to be 924 g. The movable load scale, called *wipe* in the actual place, was discovered in Huarochiri, ninety kilometers to the east of Lima. The scale was made of wood, and was used to weigh coca leaves.

SHIGEO IWATA

REFERENCES

Iwata, Shigeo. "Ancient Peruvian Mass Standard and Scales." *Bulletin of the Society of Historical Metrology, Japan* 7(1): 23–33, 1985.

Nordenskiöld, Erland. "Emploi de la Balance Romaine en Amérique du Sud avant la Conquête." *Journal de la Société des Américanistes de Paris nouvelle série* 13:169–171, 1921

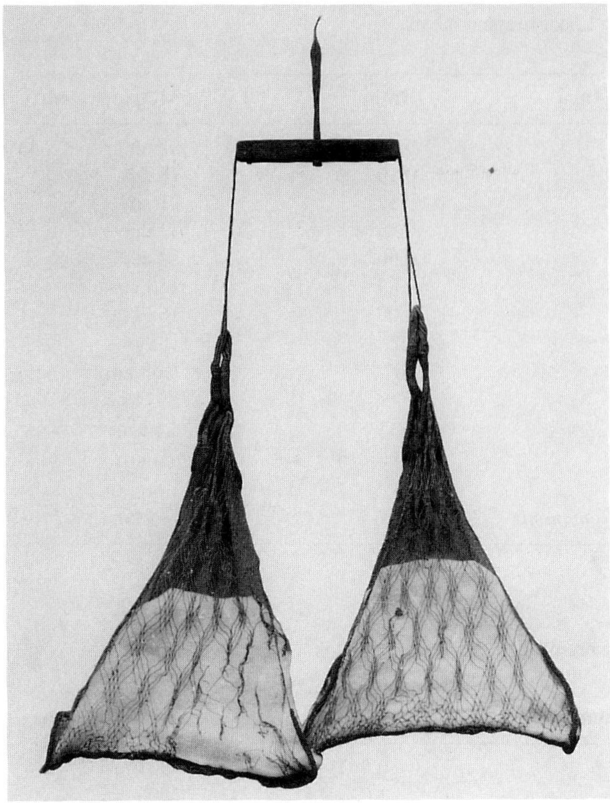

Figure 3: Balance. From the Museo Amano.

Rivet, P. "La Balance Romaine au Pérou." *l'Anthropologie* 33: 535–538, 1923.

Rostworowski de Diez Canseco, Maria. *Pesos y Medidas en el Perú Pre-Hispánico*. Lima: Imprenta Minerva, 1960, pp.1–27.

Rostworowski de Diez Canseco, Maria. "Mediciones y Computos en el Antiguo Peru." *Cuadernos Prehispanicos* 6:21–48, 1978.

Uhle, Max. "La Balance Romaine au Pérou." *Journal de la Société des Américanistes de Paris nouvelle série* 17: 335–336, 1925.

Wakeham Dasso, Roberto S. *Puruchuco, Investigacion Arquitectonica*. Lima: Universidad Nacional De Ingenieria, 1–25+6, 1976.

See also: *Quipus*

WESTERN DOMINANCE: WESTERN SCIENCE AND TECHNOLOGY IN THE CONSTRUCTION OF IDEOLOGIES OF COLONIAL DOMINANCE Scientific curiosity was a major motive behind Western overseas expansion from the fifteenth century onward and technological innovations, particularly in shipbuilding, navigational instruments and firearms, made that expansion possible. But early European explorers and conquistadors did not rely heavily on evidence of scientific or technological achievements as gauges of the worth of the peoples they encountered or as explanations for their growing dominance in the Americas and maritime Africa and Asia. In encounters with the great centers of civilization in Africa and Asia, European superiority in these endeavors was highly selective, marginal, or in many areas non-existent. In fact, travelers to China and the Indian subcontinent in the early centuries of expansion were as likely to dwell on the technological deficiencies of the West, when compared to these great civilizations, as to boast of European advantages. As in India, China, and Japan, the Europeans were able to make little headway into the heartlands of the Islamic world in this era. That standoff and the fact that they had borrowed so heavily from the scientific learning and technology of Muslim cultures in the centuries of Europe's emergence as a global force, rendered it unlikely that material standards would supplant the long contested religious differences that the Europeans had employed to set themselves off from and above the followers of Islam.

In Africa, disease and geographical barriers and the power of coastal kingdoms prevented the Europeans from translating their technological edge into significant conquests. Failure to move into the African interior also meant that the Europeans had only the vaguest notions about African epistemologies or understandings of the natural world, which were usually dismissed as superstition or fetishism. In sharp contrast to their experience in Africa, European invaders encountered few disease barriers in the Americas. In fact, diseases from smallpox to the measles became powerful allies of the Spanish conquistadors in their assaults on the heavily populated and highly advanced civilizations of Mesoamerica and the Andean highlands. The long isolation of the Amerindian peoples from the Afroeurasian people and cultures left them highly vulnerable to both the microbes borne and the iron-age technology wielded by the European invaders. None the less, the Spanish tended to attribute their startling successes in battle against seemingly overwhelming numbers of Aztec or Inca adversaries and the rapid conquests that followed to supernatural forces and to the superiority of their militant brand of Christianity over the "heathen" faiths of the indigenous inhabitants of the "New" World.

Lacking in-depth knowledge of the epistemologies and scientific learning of most of the peoples they encountered in the early centuries of overseas expansion and often enjoying only very selective (but at times critical) technological advantages over them, European explorers, missionaries, and Crown or Company officials were unlikely to rely on material standards to judge the level of development attained by other cultures or to compare overseas civilizations to Europe itself. Until at least the end of the seventeenth century, religious beliefs, or the Europeans' certitude that they possessed vastly superior understandings of the transcendent world, predominated as the gauge by which other cultures and peoples were assessed and ranked. Additional cultural variables, such as the position and treatment of women, were frequently cited as evidence of advancement or savagery; and physical features, especially skin color, were sometimes emphasized in attempts to distinguish and rank the peoples encountered overseas.

Scientific and technological gauges of past attainments and present abilities remained peripheral to most evaluations of the peoples and cultures that the Europeans encountered as they expanded across the globe. None the less, signs of material advancement — the existence of large cities, sophisticated techniques of fortress construction, or evidence of complex scientific instruments — were often noted and even cited to support arguments regarding the level of development achieved by different peoples. In two areas in particular, in the perception and measurement of time and in perspectives on space associated with artistic and mathematical advances of the Italian Renaissance, European overseas observers began to see a clear divide between the West and all other civilizations and cultures. In addition, as early as the sixteenth century, European commentators began to rank African cultures beneath those of Asia and the Americas, not so much because of skin color or other physical features, as has often been argued, but due to what was perceived as a markedly lower level of material culture in Africa than that found by European travelers and traders in India, China, or central Mexico.

Although ideologies justifying overseas expansion and the domination of non-European peoples from the fourteenth to the early eighteenth century were rooted in religious belief and were generally culture- rather than racially-oriented, material accomplishment, including the assumed capacity for invention and scientific thinking, was increasingly associated with racist defenses of the enslavement of Africans. Defenders of the slave trade sought to counter the abolitionists' objections with often lurid descriptions of the alleged savagery of African life and the debased level of African material culture. Along with skin color and other physical differences, racist writers, such as Samuel Estwick and Dominique Lamiral, emphasized material backwardness and ignorance of the workings of the natural world as proof of the subhuman nature of Africans that justified their subjugation as slaves.

By the last decades of the eighteenth century, this rather broad association between racial ideology, material culture, and the defense of slavery, was refined and enhanced by the rise of racist theories allegedly grounded in scientific experimentation and reasoning. Physicians and ethnologists devised a variety of measurements — from skull size and shape to genital configurations — in efforts to prove that there were innate physical, mental and moral differences between human racial groups. The fact that the measurements reflected a priori assumptions and were based on small and suspect samples, and that even the racial categories themselves were hotly contested, did not prevent "scientific" racism from winning widespread support from European scientists, social commentators, and politicians throughout the nineteenth century. The tenets of scientific racism were popularized among the middle and working classes by practitioners of phrenology, whose booths could be found at county fairs and on the promenades of seaside resort towns, and by the pulp press, where allegedly scientific proofs of European racial superiority were linked to social evolutionist arguments for imperialist expansion.

By the first decades of the nineteenth century, scientific and technological gauges of human achievement and worth were clearly in the ascendant. Earlier measures of the level of development achieved by different cultures continued to be cited. Religious belief, for example, remained of paramount importance to missionaries active in overseas lands. But even missionaries increasingly linked Europe's advances in the sciences and invention to the rhetoric of Christian proselytization. Ignoring past and contemporary tensions between science and religion in Europe itself, prominent missionaries, such as the Abbé Boilat and David Livingstone, argued that Christian culture had been particularly receptive to scientific investigation and technological innovation, and that conversion to Christianity would promote the scientific and material development of colonized peoples. The growing numbers of ethnologists and professionally trained anthropologists, who found in the colonies relatively safe and fertile environments for their research, also privileged non-scientific or technological standards, such as modes of political organization or gender relations, in assessing the level of development attained by African, Asian, Amerindian, and Pacific Island peoples. But evidence of material culture was increasingly linked to societal advance, and indigenous "superstitions" or at best magical beliefs contrasted with the scientific mindset that was seen as typical of the educated West.

A number of factors account for the dominance of material standards, particularly those linked to science and technological innovation, in nineteenth-century ideologies of European global hegemony. Most critically, the transformations wrought by industrialization from the middle of the eighteenth century in England and somewhat later in Belgium, Germany, France, and Italy made the gap in scientific and technological capacity and material development between western Europe and non-Western societies increasingly apparent to European and non-European observers alike. Maxim guns, steamboats, and railway lines carried elements of Europe's industrialization to colonized areas, and champions of imperialist expansion reasoned that these wonders could not help but impress subjugated peoples with the unprecedented degree to which European societies had advanced over their own. Not only was European superiority in science and technology obvious, it could be empirically tested in ways that claims of higher religious understanding or moral probity could not. Europeans had vastly more firepower, could produce incomparably greater quantities of goods much more rapidly, and could move both these products and themselves about the globe with much greater speed and comfort than any other people, including the once highly touted Chinese. In an age when what were held to be scientific proofs were authoritative, attainments that could be measured statistically were viewed as the most reliable gauges of human ability and social development. Mechanical principles and mathematical propositions could be tested; cast iron or steel bridge spans could be compared for size and strength with the stone or wooden structures of non-Western societies; and human skulls could be quantified in seemingly infinite ways to assess the highly variable mental capacity of the "races of man".

The preeminence gained in the nineteenth century by scientific and technological standards of human worth and ability not only bolstered proponents of theories of European racial supremacy, but it also proved vital to various formulations of the civilizing mission ideology that both inspired and rationalized European imperialist expansion from the early 1800s to 1914. Chauvinistic politicians in the metropoles and imperial proconsuls in the colonies increasingly stressed the importance of the diffusion of Western science and technology to what they viewed as the benighted peoples and backward lands that had come under European control. Proponents of the civilizing mission confidently predicted that the world would be remade in the image of industrializing Europe. Given Europe's material advancement, it was seen as appropriate that Europe and North America serve as the sources of capital, both machine and financial; of entrepreneurial, scientific, and managerial expertise; and of manufactured goods for the rest of the globe. In this view, the non-Western world, including both areas that had been formally colonized and those that had come under the informal sway of the Great Powers, were best suited to provide abundant and cheap land, labor, and raw materials that were required to fuel the industrial economies of Europe and North America.

According to the "improvers" or non-racist advocates of the civilizing mission, the spread of Western education among colonized peoples — emphasizing the inculcation of at least rudimentary Western scientific learning and technological skills — would provide the critical means by which the material level of non-Western societies would gradually be raised. Though they approved of the diffusion of essential Western technology to overseas areas under the paternalist supervision of European colonizers, racist apologists for imperialism had little faith in the ability of subjugated peoples to master the sciences or engineering of the West. Thus, they envisioned the period of European "tutelage" extending for centuries, if not indefinitely, into the future.

The non-Western peoples who were the targets of the European colonial enterprise were very often awed and overwhelmed by their initial encounters with the science and technology of the industrializing West. Whether they were indigenous leaders resisting the growing encroachments of European forces or scribes and merchants who allied themselves with the invaders, the colonized could not help but be impressed with the clear and increasing advantages in power that the Europeans gained through their superior capacity to tap the resources of the natural world, to produce material goods, and to devise more deadly weapons. As surveys taken as late as the post-World War II era, such as those which form the basis for G. Jahoda's *White Man*, the science and technology of the colonizers gave them an aura of magical power among the colonized masses in many areas. Though Western-educated Africans, Asians, or Polynesians were likely to scoff at such expressions of popular admiration, most came to accept that Western science and technology were not only on the whole superior to their own but essential for the future "development" of societies they hoped someday to rule. Therefore, nationalist critiques of imperial domination often deplored the fact that colonialism had severely constricted the flow of science and technology from the West to dominated areas, and demanded that technical education and scientific facilities for indigenous peoples be expanded and improved.

By the last decades of the nineteenth century, however, a number of influential African, Asian, and Caribbean thinkers were mounting cogent challenges both to notions of European racial superiority based on evidence of scientific and technological achievement and to the advisability of the wholesale transformation of non-Western cultures and societies along Western, industrial lines. Much of this resistance to the hegemonic ideologies of the Western colonizers focused on efforts to reassert and revitalize indigenous epistemologies, modes of social organization, and approaches to the natural world. Thinkers such as Vivekananda and Aurobindo Ghosh contrasted an Indian spiritualism with the deadening abstractions of Western materialism. African writers, such as the Caribbean-born Edward Blyden, deplored the devastating impact of the Atlantic slave trade on African cultures and celebrated the Africans' strong sense of community, reverence for and care of the elderly, and sophisticated artistic creations.

Ironically, these defenses of colonized cultures were buttressed by contemporary European anthropological studies, usually carried out under the auspices of colonial administrations; by the intense interest in "Oriental" religions fashionable among European intellectuals in the decades before World War I; and by the "discovery" of the abstract power of African masks and other forms of "primitive" artistic expression by Picasso, Derain, Matisse, and other avant-garde artists in the early 1900s. Inadvertently, however, the works of these first generations of Indian and African critics of European hegemonic ideologies often validated the very materialistic standards they sought to contest. For example, Indian thinkers, particularly Vivekananda, repeatedly stressed the scientific accomplishments of India's ancient civilizations, while African and West Indian writers, most notably Anetor Firmin, claimed Egypt, with its impressive engineering and architectural feats, as a civilization that black Africans had done much to build.

With the coming of the First World War, non-Western critics of what had been characterized as the excessively rationalistic, impersonal, and materialistic West found numerous, and highly vocal, European intellectual allies. The horrific trench slaughter on the Western Front and the multitude of ways that Western scientific knowledge and experimentation were harnessed to the war effort as a whole raised profound doubts for noted thinkers, such as Paul Valery, Sigmund Freud, and Georges Duhamel, about the long-assumed progressive nature of Western science and technology.

After such a savage and suicidal war, the tenets of the civilizing mission rang hollow, and "scientific" racist thought came under increasing assault in both western Europe and the United States. Collaboration with indigenous elites was increasingly stressed in the governance of the colonies, and a rhetoric of science and technology as agents of development through cooperation with indigenous peoples permeated colonial policy making.

European doubts about the directions taken by the industrial West and its global hegemony gave new impetus to African, Asian, and Caribbean critiques of European global hegemony. René Maran's *Prix Goncourt*-winning novel, *Batouala*, mocked the pretensions of racial superiority held by European colonizers, and idealized village life in French West Africa. The poets of the *négritude* movement, most powerfully L.S. Senghor and Aimé Césaire, mourned the suffering and destruction wrought by European science and technology in Africa and the lands of the slave diaspora, and inverted racist epithets by exulting blackness, intuition, affinity for the natural world, and indifference to inventiveness. Aurobindo Ghosh and Rabindranath Tagore viewed the war as the fulfilment of earlier Indian prophecies of a coming cataclysm in the aggressive and materialist West, and proof of importance of India's spiritual mission in the modern age. Mohandas Gandhi also cited the senseless violence and colossal destructiveness of the war in support of his sweeping assaults on industrial society. In the decades after the war, he sought to formulate for India (and implicitly for other colonized areas) a community-centered, low tech, and conservationist alternative to industrialized society as it had developed in the West. In this same period, Gandhi also worked out a strategy for confrontational but non-violent protest that repeatedly proved an effective antidote to the advanced technologies of repression employed by Western overlords in the decades of decolonization from the 1920s onward.

Though battered and under assault, the scientific and technological underpinnings of ideologies of Western dominance survived the crisis of two global wars and the powerful critiques of Gandhi and the *négritude* writers largely due to the emergence of the United States as *the* global power from the 1920s onwards. Entering the First World War late and just when new technologies had restored a war of motion and decision, Americans continued their long-standing infatuation with science and technology after the conflict. Faith in the essentially progressive nature and beneficence of science and technology informed the development (later called modernization) theory that came to dominate both American thinking on colonial issues and that of the European and Japanese imperialist rivals of the United States. After World War II, America's chief rival, the Soviet Union, also championed a rhetoric of development that privileged science and large-scale industrialization. With modernization theory (in a number of capitalist and socialist versions) in the ascendant, non-Western alternatives to social development and economic well-being, such as that formulated by Mohandas Gandhi, were marginalized or openly spurned by the Western-educated elites that governed the new states that emerged from the collapsing colonial empires. Though alternative approaches have gained significant support in some of these new nations, most notably India and Tanzania, international agencies and Western and non-Western planners continue to rely upon modernization schemes, based overwhelmingly on Western precedents, to solve the problems of poverty and growing wealth differentials and the demographic and environmental dilemmas that have been building on a global basis for centuries.

MICHAEL ADAS

REFERENCES

Adas, Michael. *Machines as the Measure of Men: Science, Technology and Ideologies of Western Dominance*. Ithaca, New York: Cornell University Press, 1989.

Cohen, William B. *The French Encounter with Africans: White Response to Blacks, 1530–1880*. Bloomington: University of Indiana Press, 1980.

Curtin, Philip. *The Image of Africa: British Ideas and Action, 1780–1850*. Madison: University of Wisconsin Press, 1964.

Gould, Stephen J. *The Mismeasure of Man*. New York: Norton, 1981.

Irele, Abiola. "Négritude or Black Cultural Nationalism." *The Journal of Modern African Studies* 3(3): 321–48, 1965.

Irele, Abiola. "Négritude–Literature and Ideology." *The Journal of Modern African Studies* 3(4): 499–526, 1965.

Jahoda, G. *Whiteman: A Study of the Attitudes of the Africans to Europeans in Ghana Before Independence*. Oxford: Oxford University Press, 1961.

Jordan, Winthrop D. *White Over Black: American Attitudes Toward the Negro, 1550–1812*. Chapel Hill: University of North Carolina Press, 1968.

Leclerc, Gérard. *Anthropologie et Colonialisme: Essai sur l'Histoire de l'Africanisme*. Paris: Fayard, 1972.

Nandy, Ashis. *Science, Hegemony & Violence: A Requiem for Modernity*. Delhi: Oxford University Press, 1988.

Stephan, Nancy. *The Idea of Race in Science: Great Britain 1800–1960*. Hamden, Connecticut: Greenwood, 1982.

See also: Magic and Science – Colonialism and Science – Science as a Western Phenomenon – East and West – Technology and Culture

WINDPOWER There is little doubt that the first practical use of the wind as an energy source other than as the motive power for sailing ships occurred in the East. Those of Persia were probably the first but precisely when is uncertain. According to a story of al-Ṭabarī writing around AD 850, and later writers, the second orthodox Caliph, ʿUmar ibn al-Khaṭṭāb, was murdered in AD 644 by a captured Persian technician, Abū Lu'lu'a, who claimed to be able to construct mills driven by the power of the wind and was bitter about the taxes he had to pay. This early date cannot be confirmed but Arabic geographers of the tenth century all confirm the existence of windmills in the region of Seistan in north eastern Iran. For example, al-Masʿūdī around AD 950 wrote of the wind driving mills and raising water from streams.

Nothing has survived to show how such mills pumped water but later drawings of one type of windmill suggest that it could have been derived from horizontal corn-grinding watermills. In the lower part of such watermills, the water fell down a chute and struck blades placed radially around a vertical shaft. The top of the shaft passed into a higher room and carried the upper millstone and so rotated it above the bedstone which was set on a floor in the middle of the mill. When developed into a windmill, the wind rotor was situated in the bottom of the mill. The building was constructed with four "loop holes" to direct the wind on to the blades from whichever quarter the wind might be coming. There might be eight or ten blades, and the wind was directed by the loop holes onto one side of the rotor. In this way the sails were pushed round on that side away from the direction of the wind while those advancing into the wind on the other side were shielded by the walls around the loop holes. The grinding stones were placed in an upper room with the upper stone turned by an extension of the rotor shaft in the 'underdrift' manner. None of these mills has survived.

The type which may still be in use today had the grinding stones situated in a room below the rotor with the sails above. This necessitated a change to the 'overdrift' method of driving the stones where the spindle passed down through the bedstone to a bearing underneath which had to support the weight of both the upper stone and the sails. It was possible to disconnect the driving shaft at the place where it was connected to the upper stone at the 'rynd' to allow for the stones being separated for dressing. The layout for grinding was similar to other corn mills with a hopper mounted on the wall from which the grain was fed into the central hole of the upper stone through a chute or shoe while the flour was passed out around the circumference of the stones.

The advantage of this second layout was that the rotors with the sails could be greatly enlarged. Mills near Seistan and on the borders of Iran and Afghanistan might have rotors approximately 5 m (16.4 ft) high by 3 m (9.8 ft) in diameter. There could be six or eight sails which had wooden framing that was interlaced with straw or covered with wooden boards. The upper parts of these mills were built with one wall that directed the wind on to one side of the rotor while another wall shielded the other half. Sometimes matting screens were erected to help channel the winds to the sails. There were no brakes, but more screens might be placed across the slots between the walls to regulate the wind reaching the sails; to secure the mill when out of use, the rotor and upper millstone were lowered to rest on the bedstone. A wind speed of 22.40 m/sec (50 m.p.h.) was needed to drive these mills at only 30 r.p.m. Such a mill, working with intermittent wind for about four months of the year, would grind enough flour for about fifteen families. Seistan was known as the land of the winds and between mid-June to mid-October the wind regularly blew from the north for a period of 120 days. The mills were built in line to face these winds and the famous example at Neh had a long row of 75 mills.

Because these horizontal mills needed such strong and regular winds to power them, they did not spread in this form much beyond the borders of Iran and Afghanistan. They were invented earlier than the Western vertical type but it is doubtful whether there is any connection between the two. Those in the West, which have been described as the 'full admission, axial flow type', first appeared around AD 1150 possibly either in the southeast of England, the northwest of France, or Flanders. However these Persian mills were probably the source of the later Chinese mills and possibly the Tibetan wind-powered prayer wheels.

The wind-powered Tibetan prayer wheels were not surrounded by elaborate buildings to guide the wind onto the rotors. The most common form had a vertical axle with horizontal spokes at its top, on the ends of which were fixed sails shaped like cups so that they caught the wind on one side of the rotor, but were smoothed or curved to present less resistance on the other. The wind would be caught in the concave part of the cup, or a curved sheet, during half the circle to turn the rotor, while in the other half, the convex or streamlined side would be advancing into the wind and so present less resistance. Power output was minimal but was sufficient to turn the prayer wheels which was the only use to which they were applied.

Prayer cylinders designed for automatic repetition of the Buddhist mantra are unlikely to have been produced before the reign of K'ri-srong-Ide-brtsan in AD 755 to 797 when Buddhism conquered Tibet. It is unlikely that such cylinders were turned by the wind at this period, although no doubt there were prayer flags fluttering in the wind from around this time. Early in the twelfth century, a new fashion for mechanical piety swept China, but again it seems doubtful whether this included wind-powered prayer wheels, which must therefore be placed later, and certainly therefore after the Persian mills.

More is known about the adoption of the windmill in China. Once again early dates for this have been suggested, but these are doubtful. In about AD 1230, Yelu Chu Zai was captured by Jinghi Khan and became his minister. He was an extremely good scholar, administrator and mathematician. An accurate description of the Persian windmill has been discovered in his memoirs, with a comment on how good it would be if the Chinese used it. A Chinese book of the seventeenth century, the *Zhu Qi Tu Shu*, describes the windmill as if it were a European invention, which could be a mistake for Persia. In China, these mills were used for raising water and evolved into an entirely different form from the ones in Persia or Tibet.

Chinese horizontal windmills are still used today along the eastern sea coast north of the Yangtze and in the region of Thangku and Taku near Tientsin to operate chain pumps through gearing for raising salt water for salt pans or fresh water for irrigation. Once again they have no elaborate structures for directing the wind onto the rotors but are closely linked to the way the sails of junks operate. Canvas-covered sails are mounted at the ends of radial arms, each with its own mast, in such a way that they can be spread to catch the wind on the side of the rotor turning away from the wind and be feathered to present the least resistance on the side turning against the wind. The sails can pivot on these masts, and, as on a junk, the sail extends to the front of the mast. The longer, or driving, side of the sail is tied to the rotor framework by a piece of rope of such a length that, when the sail is turning with the wind, it is held into the wind, but can rotate out of the wind when advancing against the wind. The angle of the sail can be set by the length of rope to a position in which the wind can do useful work on it for more than 180 degrees.

These mills must have been the most efficient of the horizontal types developed before the twentieth century, but all horizontal windmills suffer from the same problem, that only a small part of the wind rotor can be used to its maximum efficiency at any time. The theoretical maximum power coefficient for a simple horizontal windmill is only one third, but in practice it will be much less. These are the reasons why the horizontal windmill has not been developed further and few examples remain at work today.

In the eastern Mediterranean, at some period during the Middle Ages, vertical windmills appeared with a different type of sails from those normally used in the West. The horizontal wind shaft was extended in front of the mill like a ship's bowsprit so that ropes from it could help to stay six or eight radiating spars. The sails were triangular pieces of canvas like the sails of a modern yacht and worked in the same way. The leading edge of each sail was attached to the spar, round which it could be wrapped to reef it in strong winds. The free corner was secured by a rope in much the same way as a ship's boom. These sails were much lighter than those of a conventional western windmill and have found a new application in water pumping mills in some developing countries today.

RICHARD L. HILLS

REFERENCES

Harverson, Michael. *Persian Windmills*. The Hague: International Molinological Society, 1991.

Hills, Richard Leslie. *Power from Wind, A History of Windmill Technology*. Cambridge: Cambridge University Press, 1994.

Needham, Joseph. *Science and Civilisation in China.* Vol. 1, *Introductory Orientations*, and Vol. IV, *Physics and Physical Technology,* Part II, *Mechanical Engineering.* Cambridge: Cambridge University Press, 1954 and 1965.

White, Lynn. *Medieval Technology and Social Change.* Oxford: Clarendon Press, 1962.

WRITING OF THE MAYAS Some dozen or so writing systems have been identified as having been in use at one time or another in ancient Mesoamerica, all sharing a number of general traits that include: (1) pictographic signs, clearly derived from things seen, used in combination with wholly abstract signs, the derivations of which remain obscure; (2) a reading order that generally proceeds from top to bottom and from left to right; (3) a vigesimal number system (base 20) with a dot representing the unit, generally in conjunction with individualized signs for numbers of higher orders of magnitude, though in certain areas, in conjunction with bars for '5' and other signs for '0' to express higher numbers; and (4) a shared, compound, cyclical calendar of 260×365 (18,980) days' duration, expressed with readily identifiable calendrical elements such as numbers, day names, year signs and periods of various magnitude.

The most sophisticated writing system — and certainly the most intriguing — of Mesoamerica, and that which will serve as the basis for this discussion, is the 'hieroglyphic' writing of the Maya region. Examples of Maya hieroglyphic writing are known from many sources: from carved monumental inscriptions in stone and wood, painted stucco-coated, screen-fold, paper books, or 'codices', painted and carved ceramic pots, cups, bowls, and dishes, and from other portable artifacts of stone, shell, ceramic, and bone. The visual style, syntax, and content of the writing was shaped by both the nature of the medium upon which it appeared and the message to be conveyed.

Though open to doubt for many years, there is no question today of the Maya writing system not being true writing. It follows linguistically logical rules of syntax, phonetic construction, and semantic expression, although with the same willful departure from those rules as that of creative writers of our own culture — even, at times, to the point of whimsical defiance. The Maya hieroglyphs were capable of accomodating the full range of both the sounds and the syntactical structures of spoken Mayan. Indeed, the relatively narrow range of subject matter of extant Maya writing is surely more of a reflection of its differential survival on imperishable artifacts than of any limitations in the writing system itself.

The typical Maya monument inscription is composed of a number of tightly-formed glyph-blocks arranged in vertical columns and horizontal rows, with the glyph-blocks formed of glyphic elements of two general sorts: relatively large ones — known as 'main signs' — generally positioned at the center of glyph-blocks and occupying most of the area (Figure 1), and smaller ones — known as 'affixes' — often positioned before, atop, after, or beneath the main signs, and sometimes (in the absence of a main sign) forming clusters of their own (Figure 2). In his 1962 *Catalog of Maya Hieroglyphs*, Eric Thompson attempted a classification of all the known glyphs of his time, assigning to 'affixes' the numbers 1–370, and to 'main signs' the numbers 501–856, with what he called 'portraits' receiving the numbers 1000a–1087. The study and discussion of Maya writing has benefited enormously by the preferred use of these T-numbers (as they are called) over the arbitrary nick-names and labored descriptions of individual glyphs that once necessarily dominated their study.

Nevertheless, epigraphic study since the publication of Thompson's *Catalog* has shown that this main sign/affix distinction is an artificial one. For while it is true that certain glyph designs seem only to have been employed as main signs and that certain others were used exclusively as affixes, a considerable number of glyphs are known to function in both categories with no apparent change in their semantic or phonetic value. This is easily seen in the variety of ways in which the well-known *bakab* title is recorded in the inscriptions (Figure 3). Size, then, is not necessarily correlated with significance.

While not entirely true, it is generally the case that both the reading order of the glyph-blocks that comprise a text (typically laid out in a grid-like arrangement of columns and rows) (Figure 4) and the internal elements of the glyph-blocks themselves proceed from upper-left to lower-right (Figure 5).

Texts composed of more than one column and one row usually read from the upper left corner, two glyphs from each row, descending two columns at a time to the bottom. The next two columns are read in the same manner, and so forth, to the end of the text. The reading order of uneven or exceptionally disposed texts necessarily depart from the ideal form just described (Figure 4). Departure from the general principle of upper-left to lower-right is probably more common in the reading order of glyphic elements within the glyph-blocks than in the organization of the latter in the texts. This is, no doubt, due in some part to what were apparently culturally prescribed arrangements of certain hackneyed terms such as the **AHPO** superfix on emblem-glyph titles that identify lords of specific Maya cities (Figure 5g) or the **-NAL** superfix on certain Maya place-names (Figure 5h), in greater part to the flexibility of the glyphic elements

Figure 1: Main signs with T-numbers.

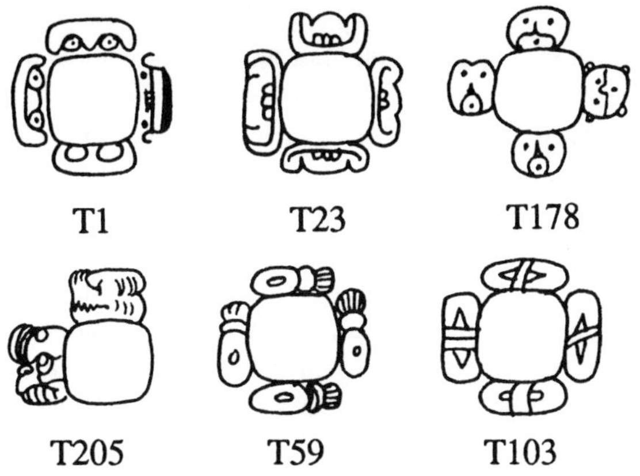

Figure 2: Affixes with T-numbers.

Figure 3: Three spellings of the title *bakab*. All are constructed *ba-ka-ba*. Note the variation in size of the elements, substitutions and the use of head variants.

in both size and shape, and in large part to the creativity of individual scribes.

Though an argument over whether the hieroglyphic signs were of ideographic, morphemic, or phonetic value raged for decades in the world of Maya scholarship, today there is general agreement that there are glyphs of both morphemic and phonetic kinds, and that there is at least one glyph that always functions as a semantic determinitive. About 180 distinct glyphic elements have so far been demonstrated to carry the phonetic values of some 77 consonant-vowel (cv)

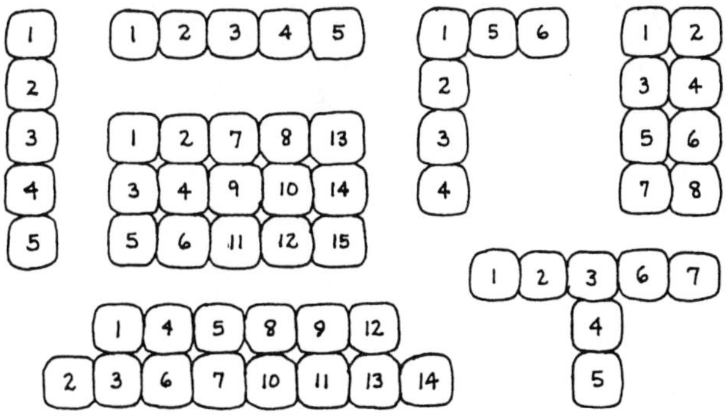

Figure 4: Reading order for texts.

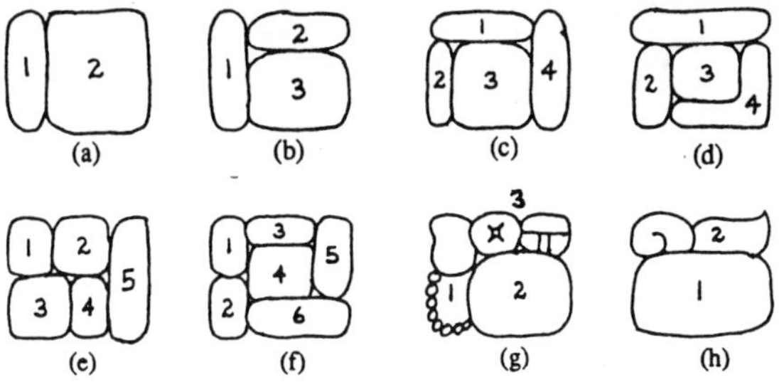

Figure 5: Reading order for the elements within a glyph-block.

syllables, leaving undetected but 23 of the 100 probable sounds employed in the ancient script. At the same time, in addition to the signs for the twenty day names, about ninety glyphic elements have been demonstrated to represent consonant–vowel–consonant (c–v–c), or even more complex, morphemes. Thus, words can be recorded by using a single morphemic sign (should there exist one that signifies the desired word), by assembling a phonetic construction from cv-signs that sound it out, in which case the final vowel sound is dropped, or by combining the two, perhaps using a cv-sign as a (silent) phonetic complement to reinforce the pronunciation of the intended word.

In addition to the aforementioned options available to the Maya scribe, there were four further features of the writing system. (1) Allographs are multiple signs that were used interchangeably to express the same phoneme. The sound *u*, for example, could be expressed by any one of at least a dozen different allographs (Figure 6a).

(2) Puns are single signs used to express homophonous words of multiple semantic value. As an example, the Mayan word *chan* means 'snake', 'sky', and 'four'. The glyphs for these three meanings could be used interchangeably to convey the sound *chan*, with the intended semantic value, which was sometimes multiple, to be inferred from the context (Figure 6b).

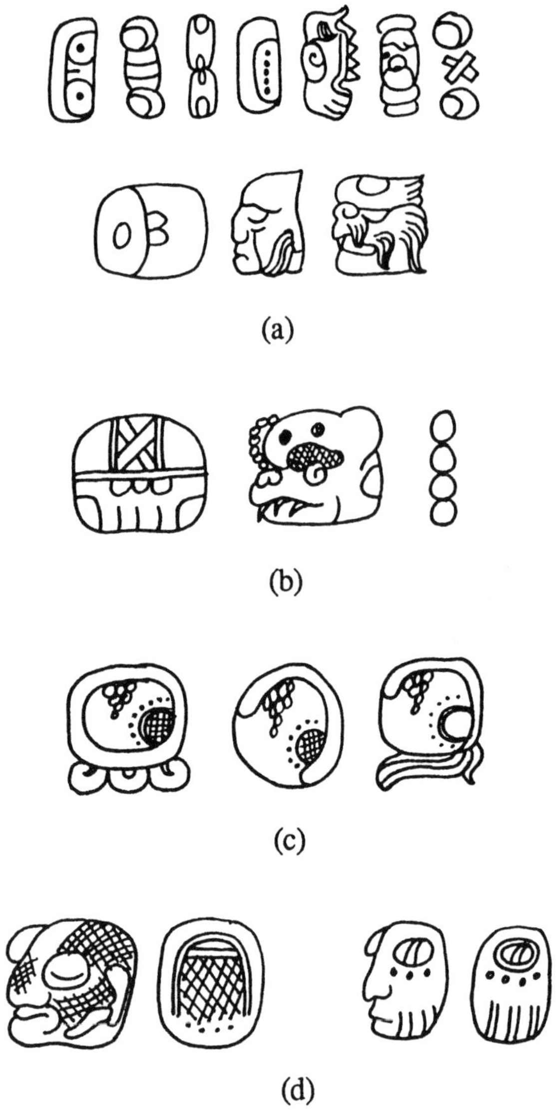

Figure 6: (a) Ten allographs for the phoneme *u*. (b) Three signs for **chan**: 'sky,' 'snake' and '4'. (c) Polyvalency of T528, as the day name *Kawak*, the phoneme *ku*, and with phonetic complement **ni** to spell **TUN**-(ni). (d) Head variants for the phonemes **pa** and **ba**.

(3) Polyvalent signs are single signs of multiple phonetic and/or semantic value which, again, are best understood from their context. Thus when the sign T528 appears framed by the day-name cartouche (serving as a semantic determinant), it is understood to represent the name of the day *Kawak*; when it appears suffixed with T116, the sign for **ni** (serving as a phonetic complement), it is understood to represent the term **tun** ('stone' or 'year'); and when it appears with no affixes, it *usually* represents the syllable *ku* (Figure 6c).

(4) Head-variants are main signs modified from their generally rounded forms into the profile heads of persons or monsters that retain their diagnostic markings (Figure 6d).

The range of possible sizes of the various glyphic elements, their substitutibility and variety of appearance (anthropomorphic, zoomorphic, and abstract), their use as cv syllables, cvc morphemes and phonetic complements, all go toward accounting for the rich variety of visual expression that characterizes the Maya hieroglyphic writing system. Though many words could be written with a single glyph, those same words could also be expressed in other ways. As an example, the word for *balam* (jaguar) was variously expressed by using the sign that represents both the word and the concept (Figure 7a), by using the signs for the syllables *ba*, *la*, and *ma* (Figure 7b), by using the sign for **BALAM** with *ma* as a phonetic complement (Figure 7c), or by using the **BALAM** sign with both **ba** and **ma** complements (Figure 7d).

Similarly, *pakal* (shield) was expressed by using a sign that portrays a shield (Figure 8a), by the signs for the three syllables *pa*, *ka*, and *la* (Figure 8b–c), or by the **PAKAL** sign with a *la* phonetic complement (Figure 8d). It will be noted that, in spite of their difference in visual effect, both the second and third of these examples are phonetic constructions of the syllables *pa-ka-la* — the second example substituting a head-variant of the more usual *pa*-glyph.

Despite these general principles, there were words for which there was no single sign and the organization of which often defied the general upper-left-to-lower-right reading order. The word **yichnal** (together with) is a particularly interesting example of such a construction, in part because it always includes the previously mentioned -**NAL** ending superfixed to its main sign (Figure 9). The latter might be either the standard sign for phonetic *chi* or the 'torso-and-left-arm' sign that represents the sound **YICH**. The several ways of recording **yichnal**, with the elements read in the usual order, were **yi**-*NAL-chi-la* (Figure 9a), **yi-NAL**-*chi* (Figure 9b), **yi-NAL-YICH** (Figure 9c), *NAL-yi-chi* (Figure 9d), **NAL-YICH-la** (Figure 9e), **NAL-YICH** (Figure 9f), and *NAL-yi-chi-la* (Figure 9g).

Among the tasks to which a Maya scribe might be called upon to apply his skills was the execution of commissions to inscribe in stone the dynastic history and military achievements of his city's royal family. Commissions of this sort would typically involve the recording of a variety of genealogical relationships such as *Ah Balam yal Na Chan* (Sir Jaguar is the child of Lady Sky), where the two names are separated by a glyph block that defines their relationship, with the subordinate person named first (Figure 10a). A military encounter might be expressed with two phrases such as *Chukah Pakal. Pakal u bak Ah Balam* (Pakal was captured. Pakal was the captive of Sir Balam), the first being a typical

BALAM
balam
balam
'jaguar'
(a)

ba-la-ma
balam(a)
balam
'jaguar'
(b)

BALAM-ma
balam(na)
balam
'jaguar'
(c)

ba-BALAM-ma
(ba)***balam***(ma)
balam
'jaguar'
(d)

Figure 7: Four ways of spelling **balam** (jaguar).

PAKAL
pakal
pakal
'shield'
(a)

pa-ka-la
pakal(a)
pakal
'shield'
(b)

pa-ka-la
pakal(a)
pakal
'shield'
(c)

PAKAL-la
pakal(la)
pakal
'shield'
(d)

Figure 8: Four ways of spelling **pakal** (shield).

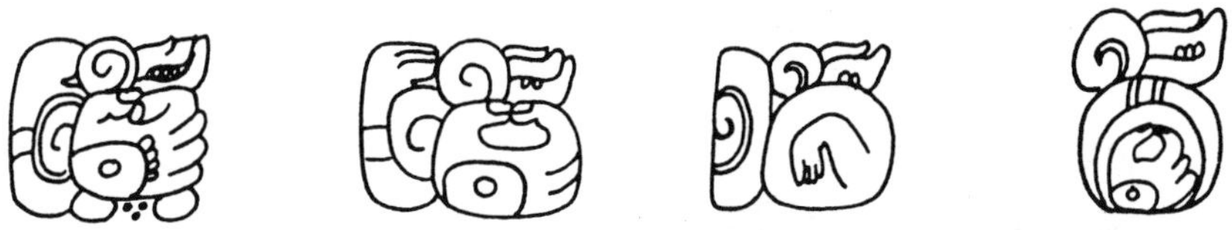

yi-NAL-chi-la
yichnal(la)
yichnal
'together with'
(a)

yi-NAL-chi
yichnal
yichnal
'together with'
(b)

yi-NAL-YICH
(yi)**yichnal**
yichnal
'together with'
(c)

Nal-yi-chi
(yi)**yichnal**
yichnal
'together with'
(d)

NAL-YICH-la
yichnal(la)
yichnal
'together with'
(e)

NAL-YICH
yichnal
yichnal
'together with'
(f)

NAL-yi-chi-la
(yi)**yichnal**(la)
yichnal
'together with'
(g)

Figure 9: Seven ways of spelling **yichnal** (together with).

Figure 10: The construction of Maya sentences.

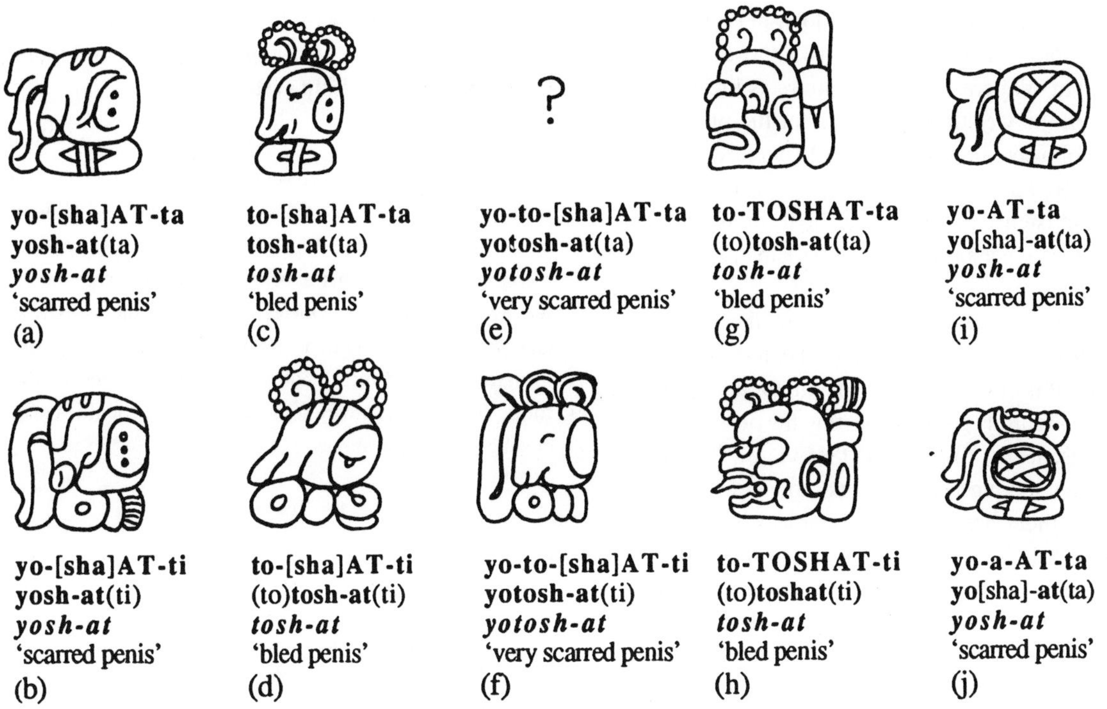

Figure 11: Semantic substitutions in the 'penis title'.

verb-initial Mayan sentence followed by the subject, the second defining another relationship with the identical syntax as the earlier example (Figure 10b). Accession to political office might be recorded as *Chumwan ti ahawel Ah Balam* (Sir Jaguar was seated in the Lordship), in another typical verb-initial Mayan sentence where a positional verb is followed by a prepositional phrase with the subject named last (Figure 10c). Were our scribe to turn from his work on public monuments to the designing of a personal book for and about Lady Sky, he might record such sentences as **Sinah u chuch Na Chan** (Lady Sky strung her warpboard) or **Och-i ti te' Na Chan** (Lady Chan wove at the post), the first a verb-initial sentence with a transitive verb followed by a possessed object and the subject (Figure 10d), and the second a verb-initial sentence with the verb followed by a prepositional phrase and the subject (Figure 10e). It should be clear from the preceding examples of inscriptions on public monuments that Maya hieroglyphic writing is indeed a true writing system, fully capable of recording the language it represents.

However, for all of the marvelous achievements of recent Maya epigraphy, much work remains. Many glyphs are undeciphered and problems still linger amongst glyphs for which solutions have been offered. An understanding of the kinds of difficulties encountered by epigraphers in deciphering individual glyphs can be gained from a close look at a fairly common lordly title, the five basic forms of which preclude ready demonstration of their homophony, and suggest instead the presence of both phonetic and semantic substitutions. Known to scholars as the 'penis' title, the three most widely-distributed forms consist of a main sign that has the appearance of a profile representation of male genitalia infixed with the sign for **sha** and suffixed alternatively with **-ti** or **-ta** affixes (Figure 11a–f). Desirable as it might be to have a homophonous reading for the three forms — and their interchangeability in the texts is beyond doubt — the conflicting sounds of their initiating prefixes would appear to preclude any such solution. One form is prefixed with the sign for **yo-** (Figure 11a–b), a second with **to-** (Figure 11c–d), and a third with both **yo-** and **to-** (Figure 11e–f). Because these substituting sounds are prefixed, the words that they signify must begin with the sounds **yo**, **to**, and **yoto**, respectively. The likelihood that the male genitalia-main sign is intended to convey the sound **AT** (the pan-Mayan word for penis), is confirmed by the interchangeable **-ti** and **-ta** suffixes that invariably accompany it, apparently serving as phonetic complements. If, then, the **sha** infix is read before the presumed sound of the main sign, the sound sequence should be **sha-AT**(ti/ta), or the not too promising **shat**. However, if the previously mentioned **yo-** prefix is added to this, the result is **yoshat**, or **yosh-at** (scarred penis); if **to-** is added, the result is **toshat**, or **tosh-at** (bled penis); and if **yo-to** is added, the result is **yotoshat**, or **yotosh-at** (very scarred penis). Though any such title might strike those of us with a background in Western European culture as bizarre, nevertheless given the wealth of evidence for the presence of penis-perforation ceremonies in which Maya lords engaged and the occasional portrayal of a scarred penis on figures in Maya art, the existence of a hieroglyphic form of a title that testified to a lord's personal sacrifice perhaps ought not to surprise anyone.

Understanding of the remaining two of the five forms which this title assumes is less certain. The first of these is a head-variant that is always superfixed with **to-** and postfixed with **-ti** or **-ta** (Figure 11g–h), suggesting the possibility that the head represents the full term **TOSHAT**, or **TOSH-AT**, with both affixes serving as phonetic complements. The second (of which there are but two examples) is a phonetic construction that is in clear contradiction to the previously discussed **yosh-at** renderings: one reads **yo-ta-ta**, the other **yo-a-ta-ta** (Figure 11i–11j). Not only is the **ta** sound doubled in each, but there is also nothing present that could be construed to represent **sha**, the necessary middle consonant of **yosh-at**. Nor is there an easy resolution of this difficulty; either the previously proposed **yosh-at**, **tosh-at**, and **yotosh-at** readings are in error and the arguments in support of them faulty, or the Maya scribe that recorded the two purely phonetic constructions (they are both at Copan) failed to include all the sounds of the word he intended to record, or — and this is a solution that preserves both the earlier **yosh-at**, **tosh-at**, and **yotosh-at** readings and the literary reputation of the Copan scribe — the scribe deliberately omitted the middle (**sh**) consonant, as is thought to have been done with other two-syllable words. One can only hope that further research and further field discoveries will someday throw stronger light on this and similar difficulties. Nor has there ever been a time when Maya scholars were more optimistic about the possibilities of their ultimate decipherment of this poetically conceived, strikingly beautiful, and wonderfully complex ancient writing system.

TOM JONES

POSTSCRIPT: Since the preceding was written and submitted to the publisher, an alternative to the reading of the penis-title presented above has been proposed by Linda Schele based upon a recently recognized example that appears to include a sign for the sound **ho**. Schele proposes a **yoh-at** reading for the **yo**-prefixed form and phonetic construction, a **toh-at** reading for the **to**-prefixed form (and its head-form), but is silent regarding the **yo-to**-prefixed form. This should convey some idea of the difficulties raised in the closing paragraphs above.

REFERENCES

Coe, Michael D. *Breaking the Maya Code.* New York: Thames and Hudson, 1992.

Harris, John F., and Stephen K. Stearns. *Understanding Maya Inscriptions: A Hieroglyph Handbook.* Philadelphia: University of Pennsylvania Museum, 1992.

Houston, Stephen D. *Reading the Past: Maya Glyphs.* Berkeley: University of California Press, 1989.

Jones, Tom and Carolyn. *Notebook for Maya Hieroglyphic Writing Workshops.* Arcata, California: U Mut Maya, 1994.

Kelley, David H. *Deciphering the Maya Script.* Austin: University of Texas Press, 1976.

Marcus, Joyce. *Mesoamerican Writing Systems: Propaganda, Myth, and History in Four Ancient Civilizations.* Princeton, New Jersey: Princeton University Press, 1992.

Thompson, J. Eric S. *Maya Hieroglyphic Writing: An Introduction.* Norman: University of Oklahoma Press, 1960.

See also: Cuneiform

X

XIAHOU YANG Much of the life story of Xiahou Yang remains unknown, and so is the date of *Xiahou Yang Suanjing* (Xiahou Yang's Mathematical Classic), a text purported to have been written by him. When Zhang Qiujian wrote his mathematical text some time between AD 468 and 486, he criticized the inaccuracy of the solution of a 'rectangular granary' problem from Xiahou Yang's work. This shows that Xiahou Yang was either Zhang's contemporary or his predecessor. The text also mentioned a change of volume standard which took place in AD 425, thus affirming that both Xiahou Yang and Zhang Qiujian were mathematicians of the Northern Wei dynasty (AD 386–535) and that the text of Xiahou Yang was written between AD 425 and 468.

The *Xiahou Yang Suanjing* appeared to be an important text as it was selected by Yu Zhining in AD 656 for use in the National Academy during the Tang dynasty (AD 618–906). In A.D 662, the text was annotated by Li Chunfeng along with nine others for use in the imperial examinations. However, when the text was reprinted and included as one of the *Ten Mathematical Classics* in AD 1084 during the succeeding Song dynasty, it appeared to have been rewritten, probably by Han Yan (AD 780–804) in the Tang dynasty. This is evident in the text itself where a number of statutes governing the Tang systems of land, taxation, and official hierarchy were mentioned.

The present version of the text has three chapters consisting of 82 problems. It deals with percentages and roots, as well as the ordinary logistic operations. Xiahou Yang seemed to have understood both the positive and negative properties of powers of 10 such as $10{,}000 = 10^4$, and $1/10{,}000 = 10^{-4}$. He also seemed to have understood the developed conception of decimals. In one problem, he suggested the conversion of the last four metrological units to simply writing down the numbers in four successive places as they would be expressed in modern decimal notation. Though Xiahou Yang provided some modified forms for multiplication and division in some problems, he also repeated some inaccurate solutions found in the *Wu Cao Suanjing* (Mathematical Classics of the Five Government Departments) of the fourth century, the most glaring of which was the rule for determining the area of a figure consisting of two congruent trapezoids and a quadrilateral.

ANG TIAN SE

REFERENCES

Needham, Joseph. *Science and Civilisation in China*, vol.3. Cambridge: Cambridge University Press, 1959.

Qian, Baozong. *Qian Baozong Kexueshi Lunwenji* (Collection of Qian Baozong's Essays on the History of Science). Beijing: Kexue Chubanshe, 1983.

Ruan, Yuan (AD 1799). *Chouren Zhuan* (Biographies of Mathematicians and Astronomers), vol. 1. Shanghai: Shangwu Yinshuguan, 1955.

Van Hee, L. "The Arithmetical Classic of Hsiahou Yang." *American Mathematical Monthly* 31: 235, 1924.

XU YUE Xu Yue, whose literary name is Gonghe, was a native of Donglai (in present Ye District, Shandong Province). Very little is known of his background except that he was a follower of Liu Hong, an eminent mathematician and calendar expert of the second century. He studied mathematics and astronomy under Liu Hong and had frequent discussions with him and the Astronomer-Royal of the Astronomical Bureau on matters pertaining to calendrical astronomy. He was said to have written a commentary on the *Jiuzhang Suanshu* (Nine Chapters on Mathematical Art) and a mathematical text known as *Shu Shu Ji Yi* (Memoir on Some Traditions of Mathematical Art) around AD 190. The commentary had long been lost while the *Shu Shu Ji Yi* was handed down with a commentary written by Zhen Luan (fl. AD 570). There are a number of Buddhist references in Zhen Luan's commentary.

The *Shu Shu Ji Yi* was written in a terse and obscure style, tinged with Daoism and divination. Xu Yue mentioned that the mathematics he learnt from Liu Hong was that transmitted directly by a Daoist adept called Tian-Mu Xiansheng (Mr Eye-of-Heaven). The text mentions fourteen old methods of calculation, many of them bearing the names taken from the *Yijing* (*I Ching*, Book of Changes). Of special interest is one called "ball-arithmetic", a trough-and-ball instrument similar to an abacus. Three other methods also involved the use of "balls". The first involved the use of one ball per column moving up and down a board or a trough. The second used two balls of two different colors, one relating to a y-axis on the left and the other to a y-axis on the right. Both these methods seem to have utilized a map-like grid of lines, similar to that of Cartesian coordinates. The third utilized balls of three different colors for use on three horizontal positions only. Joseph Needham has commented that this system of calculation "shows an interesting appreciation of coordinate relationships".

Another interesting aspect of *Shu Shu Ji Yi* is its reference to large numbers and to their representation in three forms of arithmetical series. Xu Yue called them the "upper, middle,

Table 1: The three classes of large numbers following Xu Yue

	wan	*yi*	*zhao*	*jing*	*gai*	*zi*	*rang*	*gou*	*jian*	*zheng*	*zai*
Upper	10^4	10^8	10^{16}	10^{32}	–	–	–	–	–	–	–
Middle	10^4	10^8	10^{12}	10^{16}	10^{20}	10^{24}	10^{28}	10^{32}	10^{36}	10^{40}	10^{44}
Lower	10^4	10^5	10^6	10^7	10^8	10^9	–	–	–	–	–

and lower" classes of numbers, all beginning with *wan* (10^4). The three classes of large numbers may be summarized in Table 1.

The *Shu Shu Ji Yi* also gives a description of the "calculations of the nine balls", which according to the commentary provided by Zhen Luan was a simple magic square of order three.

Though brief, obscure, and couched in religious nuances, Xu Yue's *Shu Shu Ji Yi* was selected as a prescribed mathematical text for imperial examinations during the Tang dynasty (AD 618–906). It was a difficult text, and candidates had to learn it by heart in order to pass the examinations.

ANG TIAN SE

REFERENCES

Li, Yan. *Zhongguo Gudai Shuxue Shiliao* (Materials for the Study of the History of Ancient Chinese Mathematics). Shanghai: Shanghai Kexue Jishu Chubanshe, 1963.

Needham, Joseph. *Science and Civilisation in China*, vol.3. Cambridge: Cambridge University Press, 1959.

Qian, Baozong, ed. *Suanjing Shi Shu* (Ten Mathematical Classics). Beijing: Zhonghua Shuju, 1963.

Ruan, Yuan (AD 1799). *Chouren Zhuan* (Biographies of Mathematicians and Astronomers), vol. 1. Shanghai: Shangwu Yinshuguan, 1955.

See also: Liu Hong – *Jiuzhang Suanshu* – *Yijing* – Magic Squares

Y

YAḤYĀ IBN ABĪ MANṢŪR Yaḥyā ibn Abī Manṣūr was an astronomer from an important family of Persian scientists (his father was an astrologer). Yaḥyā spent his life casting horsocopes and trying to determine the positions of the stars with precision. He began working for al-Faḍl ibn Sahl, the vizier of the Caliph al-Maʾmūn and, after his death, for the Caliph himself. He taught the Banū Mūsā and died near Aleppo, Syria in AD 832.

Yaḥyā was the director of a group of astronomers working under the Caliph al-Maʾmūn, in the observatories of Shamāsiyya in Baghdad and Dayr Murrān in Damascus. The scholars included in the group were al-Marwarrūdhī, al-Khwārizmī, Sanad ibn ʿAlī, the Banū Mūsā and al-Jawharī. They measured one degree of meridian by using two different procedures. The results of the observations were recorded in various *zīj*es (astronomical handbooks with tables), one of which is the *Zīj al-mumtaḥan* (The Tested Tables), that was written by Yaḥyā himself. The observations ended when Yaḥyā and the Caliph died almost simultaneously.

Yaḥyā's *zīj* is preserved in one manuscript (Escorial 927) which is badly bound and comprises many folios that are not from Yaḥyā's work. It contains an explanation of calendars and chronological eras. The elements to calculate planetary longitudes are the result of the combination of Hindu-Iranian and Ptolemaic methods.

Yaḥyā's tables exerted a great influence on later astronomers: Thābit ibn Qurra wrote an introduction to them, Ibn Yūnus adapted them to Egypt, and Ibn al-Zarqāllu derived from them certain values such as the inclination of the ecliptic.

EMILIA CALVO

REFERENCES

Kennedy, E.S. "A Survey of Islamic Astronomical Tables." *Transactions of the American Philosophical Society* 46(2): nos. 15, 51, 1956.

Sayılı, A. *The Observatory in Islam*. Ankara: Türk Tarih Kurumu Basmevi, 1960.

Vernet, Juan. "Yaḥyā ibn Abī Manṣūr." In *Dictionary of Scientific Biography*, vol. XIV. Ed. Charles C. Gillispie. New York: Charles Scribner's Sons, 1976, pp. 537–538.

Yaḥyā ibn Abī Manṣūr. *The Verified Astronomical Tables for the Caliph al-Maʾmūn* (in Arabic). Ed. Fuat Sezgin. Frankfurt: Publications of the Institute for the History of Arabic-Islamic Sciences, 1986.

YANG HUI Yang Hui (fl. 1261–1275), a native of Qiantang (in modern Hangzhou), ranks with Qin Jiushao (ca. 1202–ca.1261), Li Zhi (1192–1279), and Zhu Shijie (fl. 1280–1303) as one of the four great mathematicians during the golden age of mathematics in thirteenth-century China. Of the four we know the least about Yang Hui's life, yet he has left behind a much larger number of writings than the others, showing that his mathematical interest also covered a wider field than any of his contemporaries known to us.

In the year 1261 he wrote the *Xiangjie jiuzhang suanfa* (Detailed Analysis in the Mathematical Rules in the Nine Chapters and Their Reclassifications). The original work comprised twelve chapters and an Appendix, which included an arrangement of numbers which he attributed to the eleventh-century mathematician Jia Xian and which in 1654 came to be known as Pascal's Triangle. Although the original work is no longer extant in its entirety, sections of it are quoted in the early fifteenth-century Imperial Compendium, the *Yongle dadian*.

Yang Hui also wrote the *Xiangjie suanfa* (Mathematical Methods with Detailed Explanations) on the processes of multiplication and division. Then in 1262 he produced an elementary text for beginners called the *Riyong suanfa* (Everyday Mathematics). In 1274 he wrote the *Chengqu tongbian suanbao* (Precious Reckoner for Variations in Multiplications and Divisions). In 1275 he wrote the *Tianmou bilei chengqu jiefa* (Practical Rules of Arithmetic for Surveying). The same year he assembled some old and forgotten materials and published the *Xugu zheqi suanfa* (Continuation of Ancient Mathematical Methods for Elucidating the Strange Properties of Numbers). This is the earliest Chinese monograph that we have on magic squares. The last three works mentioned above were published together in 1378 under a single title *Yang hui suanfa* (Yang Hui's Mathematical Methods). A full study of this book was undertaken by Lam Lay Yong in 1977.

Like his contemporaries Yang Hui used numerical equations of higher degree. His *sanchengfang* (quartic root) method is very similar to that rediscovered by Horner and Ruffini in the early nineteenth century. He employed freely the concepts of "dummies" and substitution in his algebra. At the same time his attention was on the practical side, always trying to provide the quickest working method. It seems that Chinese mathematicians were quite active during the eleventh and twelfth centuries, but none of their writings remains. At least we have come to know something about them through Yang Hui's writings.

HO PENG YOKE

REFERENCES

Ho Peng Yoke. "Yang Hui, Thirteenth-Century Chinese Mathematician." In *Dictionary of Scientific Biography*, vol. 4. Ed. Chalres C. Gillispie. New York: Charles Scribner's Sons, 1976, pp. 538–546.

Lam Lay Yong. *A Critical Study of the Yang Hui Suan Fa*. Singapore: University of Singapore Press, 1977.

Needham, Joseph. *Science and Civilisation in China*, vol 3. Cambridge: Cambridge University Press, 1969.

See also: Magic Squares – Qin Jinshao – Li Zhi – Zhu Shijie

YA‘QŪB IBN ṬĀRIQ Ya‘qūb ibn Ṭāriq was one of the earliest astronomers in the Islamic tradition. Very little is known about his life. He worked in the last quarter of the eighth century, when Arabic astronomy was not yet influenced by Ptolemy's *Almagest*. The works of Ya‘qūb are lost, but some fragments have come down to us in later astronomical treatises and handbooks. The following information is based on such fragments. Ya‘qūb authored three works:

1. A *Zīj* (astronomical handbook with tables), which was influenced by Indian and pre-Islamic Persian astronomy. For example, he took the zero meridian and the mean motion of the planets from India, and he seems to have adopted the planetary equations from a Persian source. In the lunar visibility theory, Ya‘qūb used an Indian method to measure the width of the lunar crescent in "digits". His lunar visibility table is extant.

2. The *Tarkīb al-Aflāk* (Composition of the Heavenly Spheres) is summarized in some detail in Al-Bīrūnī's *India*. In this work Ya‘qūb discussed the size of the earth and the distances of the planets.

3. In the *Kitāb al-‘Ilal* (Book of Reasons), Ya‘qūb explained trigonometrical rules, which he probably used in his *zīj* . A few citations of this work are given by Al-Bīrūnī. Ya‘qūb based his trigonometrical calculations on a sine table with base 3438.

JAN P. HOGENDIJK

REFERENCES

Hogendijk, Jan P. "New Light on the Lunar Visibility Theory of Ya‘qūb ibn Ṭāriq." *Journal of Near Eastern Studies* 47: 95–104, 1988.

Kennedy, E.S. "The Lunar Visibility Theory of Ya‘qūb ibn Ṭāriq." *Journal of Near Eastern Studies* 27: 126–132, 1968.

Pingree, D. "The Fragments of the Works of Ya‘qūb ibn Ṭāriq." *Journal of Near Eastern Studies* 27: 97–125, 1968.

Sezgin, F. *Geschichte des arabischen Schrifttums*, vol. 6. Leiden: Brill, 1978, pp. 124–127.

YAVANEŚVARA The name Yavaneśvara, meaning Lord of Greeks, referring to one of Greek descent, is said to be the author of a number of Indian astrological works. He is mentioned for the first time by Sphujidhvaja who wrote his *Yavanajātaka* in AD 240. Towards the close of his book, Sphujidhvaja states that his work was based on that of Yavaneśvara. Yavaneśvara, who had been blessed by the Sun-god, is stated to have rendered a Greek work on genethlialogy into Sanskrit in AD 150, at the instance of the ruler of the land. This work shows the position and influence of the stars at one's birth. The date mentioned by Sphujidhvaja is a little after the Kṣatrapa dynasty of Greek descent had established itself in the region of Saurashtra and Gujarat and ruled with its capital at Ujjain. The patron of Yavaneśvara has been identified, on the basis of coins and inscriptions, as Rudradāman I. Sphujidhvaja states that he (Rudradāman I) was quite conversant with the said Greek work.

Though the original work of Yavaneśvara is apparently lost, it is possible to get an idea of its extent and contents from *Yavanajātaka* which is its redaction. The *Yavanajātaka*, as it is available now, is an extensive work in 79 chapters, and takes under its purview a large variety of topics on horoscopy and natural astrology, including the delineation of the planets, their lords, their characteristics, the major and minor influences they exert on human beings at different periods, predictions relating to professions, experience of happiness and sorrow on account of planetary combinations, and predictions of the future on the basis of questions, omens, and military astrology. Most of these would have been depicted by Yavaneśvara as well. This is borne out also by several later texts, which are ascribed to Yavaneśvara. Among such works might be mentioned: The *Candrābharaṇahorā*, an extensive work in 101 chapters, prevalent in South India, and a shorter work with the same title prevalent in North India; two texts with the title *Yavanasaṃhitā*; a work called *Bhāvadīpikā* or *Bhāvādhyāya*, making predictions on the basis of the placement of the planets in the horoscope; *Nakṣatracūḍāmaṇi*, being predictions on the basis of the twenty-seven constellations; and, a *Yavanapārijāta* depicting the results of good and evil deeds in life. It is clear from this that a regular school of the Yavana tradition of astrology had developed in India.

What is significant is that in all these texts which ostensibly go under the authorship of 'Yavana', Hellenistic practices have been Indianized, both in the matter of content and presentation. Hindu caste distinctions and social orders are duly taken note of in making predictions; Hindu deities and their descriptions are duly invoked, and suitable modifications are made to their Greek counterparts. In effect, these texts seem to be wholly indigenous but for the ascription of their authorship to Yavana (Greek).

K.V. SARMA

REFERENCES

Kane, P.V. "Yavaneśvara and Utpala." *Journal of the Asiatic Society* 30(1): 1–8, 1955.

Majumdar, M.R. *Historical and Cultural Chronology of Gujarat.* Baroda: Maharaja Sayajirao University, 1960.

Pingree, David. "The Yavanajātaka of Sphujidhvaja." *Journal of Oriental Research* (Madras) 31(1–4): 16–31, 1961–62.

The Yavanajātaka of Sphujidhvaja. Ed. David Pingree. Cambridge, Massachusetts: Harvard University Press, 1976.

See also: Astrology in India

YINYANG *Yinyang* is a Chinese term composed of two words. *Yin* means the shade and *yang*, the sunshine. It is the most important concept in the traditional Chinese way of thinking, the basic concept for classification and explanation in Chinese science, medicine, and philosophy.

The word *yinyang* designates a special method for looking at things as well as for describing the phenomena of the world, including the human body. *Yin* and *yang* make a pair which exist together. In their co-existence, these two oppose each other in their different functions and at the same time help each other. They both contribute and are necessary to the other's existence. All phenomena can be looked at and described using the *yinyang* paradigm. This is the typical methodology of Chinese science.

The concept of *yinyang* can be compared with the Western dialectic method. However, in Chinese science this way of thinking has developed into a very complicated system of knowledge, in which *yin* and *yang* are represented in every real object. Therefore, *yinyang* is more than a way of thinking; it is also a theory that provides guidelines for human action.

The use of the concept of *yinyang* can be best illustrated by examples. The earliest example which can be found in extant Chinese ancient texts is from the Confucian classic *Yijing* (*I Ching*, The Book of Changes). It is not to be found in the *Yijing* proper, but in the commentaries (so-called *Yizhuan*, *I Chuan*, or the Ten Wings.) It is generally accepted that *Yizhuan* was complied during the Warring States period (the fifth to the third century BC), so that is the earliest date we can attribute to the origin of the concept of *yinyang*. *Yizhuan* uses *yinyang* to explain the *Yijing* proper in such a way that the binary notation for every possible movement (change) in the real world is explained as representing the actual acting of the two opposing factors (or forces in the Western sense of the word) that substantiate the situation under consideration. Thus in the relation of men and women, *yang* represents man and *yin* represents woman. Likewise in the relation of heaven and earth, of sun and moon, of day and night, etc., *yang* represents the former and *yin* the latter. Furthermore, *yang* and *yin* are used to represent opposite qualities including natural attributes like hot and cold, hard and soft, bright and dark, odd and even, as well as human dispositions like aggressiveness and submissiveness, or activeness and inactiveness.

Two points need to be emphasized here. The first is that the designation of *yang* or *yin* to an object is only effective in certain relationships. Relative to the woman, the man is *yang*. But there are other traditionally important relations that man comes into, such as that of heaven and man, of man and gods, of man and his fate, or of individual man and his family or society. In any of these, "man is *yang*" does not apply. Therefore, the traditional designation of *yang* and *yin* forms the hard core of a knowledge about the world. The second is that the use of the concept of *yinyang* is a pragmatic one. To say some thing or some quality is *yang* and the opposite is *yin* will be meaningful only if this designation classifies these two objects or qualities into two existing categories into which many other things or qualities are already classified. A wrong designation will make a real difference in human action when using this knowledge of classification to make a decision according to a certain theory.

According to Chinese historical records, especially the authoritative *Shi Ji* (Records of Historians) by Si Maqian (b. 145 or 135 B. C.), during the Warring States period one of the six most important schools of ideas was called the *yinyang* school. It is generally accepted that, beside its political and philosophical relevance, this school brought about two important developments, which somewhat related to each other throughout the two thousand years afterwards. One is the Daoist practice of alchemy and physical immortality; the other is Chinese traditional medicine, or more accurately, Chinese body science.

From both the Chinese traditional viewpoint and from the modern Western point of view, the most important use of the *yinyang* concept can be found in the theory and practice of Chinese traditional medicine. The earliest extant medical literature was unearthed in 1973 from the Han Tomb No. 3 at Mawangtui, Changsha, Hunan Province, China. There are records which show that these texts were buried in the tomb with the dead Marquis in the year 168 BC. In these texts the *yinyang* concept is used as the most basic classifying concept. It is the same with the *Huangdi Neijing* (Yellow Emperor's Classic of Internal Medicine), which contains medical essays edited in the first centuries AD. From these texts we know that the *yinyang* concept was used in Chinese medical theory much earlier than the second century BC.

In Chinese medical theory, *yin* and *yang* are two different kinds of *qi* (literally "breath" but meaning "vital force"), which function within the human body contradictorily and complementarily, and thus make the body alive. Human health is attained by keeping these two *qi* in a good, natural balance. All health problems are caused by the lost of this balance. All medical theories, diagnoses, and therapies can be understood as practical knowledge of how to classify the imbalance into a certain system of categories and how to use different means, which are classified into corresponding categories, to retrieve the natural balance. For a simplified example, two common health problems are diagnosed as *yang*-deficiency – *yin*-excess and *yin*-deficiency – *yang*-excess, which correspond to cold and heat syndromes. In diagnosis these two syndromes are related to two categories of symptoms which include every possible change in the pulse, the complexion, the tongue, the breath, the excrement and urine, etc., of the patient. In therapy, herbal medicine, or the method and location of acupuncture points are correspondingly classified.

There are two other concepts which are closely connected with the *yinyang* concept. They are the *wuxing* (five phases), which is extremely important in Chinese science, and the *taiji* (the ultimate beginning), which is more important in Chinese philosophy. The *yinyang* concept is also the basic concept used in Chinese fortune telling such as astrology (*suanming*, literally "calculating the fate") and geomancy (*fengshui*, literally "wind and waters").

WEIHANG CHEN

REFERENCES

Graham, A. C. *Yin-Yang and the Nature of Correlative Thinking.* Singapore: National University of Singapore, 1986.

Needham, Joseph. *Science and Civilisation in China*, vol.2. Cambridge: Cambridge University Press, 1956.

Porkert, Manfred. *The Theoretical Foundations of Chinese Medicine: Systems of Correspondence.* Cambridge, Massachusetts: MIT Press, 1974.

Xia, Songlin. *Images of Heaven and Man: An Introduction to the History of Doctrines of Yinyang and Wuxing.* (in Chinese) Jinan, China: Shandong Literary Press, 1989.

Yosida, Mitukuni "The Chinese Concept of Nature." In *Chinese Science: Explorations of an Ancient Tradition.* Ed. Shigeru Nakayama and Nathan Sivin. Cambridge, Massachusetts: MIT Press, 1973, pp. 71–89.

See also: Five Phases – Astrology – Geomancy – Medicine in China – Divination in China – *Huangdi Neijing*

YOGA Yoga is one of the six principal systems of Indian thinking known as *darśanas*. The word *darśana* is derived from the Sanskrit root *drs*, meaning "to see". Fundamentally, *darśana* means "view" or "a particular way of viewing". Yoga, as one of the six *darśanas* has its source in the Vedas. In the traditional Indian view these are called *vaidika*, or Vedic *darśanas*. These are: *nyāya*, *vaiśeṣika*, *saṅkhyā*, *yoga*, *mīmāṁsā*, and *vedānta* (there are other *Darśanas* that do *not* accept the supremacy of the Vedas, such as Buddhism and Jainism). While the source of Yoga was the Vedas, Yoga was formalized by Patañjali, one of the great Indian sages. His classic text is *Yoga Sūtra* (Aphorisms on Yoga). Although there are many other major treatises on Yoga that postdate Patañjali's, his work is the most authoritative.

All the *darśanas* proclaim that it is their aim to help human beings achieve clarity and balance of perception and action. Yoga is unique in as much as it offers practical suggestions and guidelines to achieve this end. According to the tenets of Yoga, human beings are under the influence of *avidyā*, which is what prevents correct perceptive analysis. Sage Patañjali suggests practical ways to reduce and remove *avidyā*. In his *Yoga Sūtra*, three things are suggested to help us explore the meaning of Yoga and therefore feel *avidyā*. These are *tāpas*, *svādhyāya*, and *īśvara pranidhāna*. *Tāpas* is a means by which we keep ourselves fit and clean. Often *tāpas* is defined as penance, mortification, and dietary austerity, but what is meant is the practice of *āsana* (postures), *prāṇāyama* (control of the breath), and other disciplines. These practices aid in the removal of impurities from our systems. In so doing we gain control of our whole system. It is the same principle as heating gold to purify it.

The next part of Yoga is *svādhyāya*, the study of the self. Where are we? What are we? What is our relationship to the world? It is not enough to keep ourselves fit; we should know who we are and how we relate to others. This is not easy because we do not have an actual mirror for our minds as we do for our bodies. We must use reading, study, discussion, and reflection as a mirror to the mind.

The third means of exploration is *īśvara-pranidhāna*. It is usually defined as "love of God" but it also means "quality of action". We must carry out our jobs, and all our actions must be done with quality. Since we can never be certain of the fruits of our labors, it is better to remain slightly detached from them and pay more attention to the actions themselves.

Together, these three cover the whole of human action: fitness, inquiry, and quality of action. Taken together, these practices are known as *Kriyā Yoga*, the Yoga of action. Yoga is not passive. We must be involved in life, and preparation is necessary for this involvement.

Patañjali's Yoga is sometimes called *Aṣṭāṅga* Yoga, which literally means Eightfold Yoga. These eight are *yama*, *niyama*, *āsana*, *prāṇāyāma*, *pratyāhāra*, *dhārāṇa*, *dhyāna*, and *samādhi*.

Patañjali considers five different attitudes (*yamas*) or relationships between an individual and "the outside". The first is *ahimsā*. While the word *himsā* means injury or cruelty, *ahimsā* means more than merely the absence of *himsā*. It means kindness, consideration, or thoughtful consideration of people or things. The next *yama* is called *satya*, "to speak the truth". The third *yama* is *asteya*. *Steya* means "to steal"; *asteya,* the opposite, means if we are in a situation where people trust us, we will not take advantage of them. The next yama is *brahmacarya.* The word is composed of the root *car* (to move) and *brahma* (the truth). If we move towards the understanding of truth, and sensual pleasures get in the way, we must keep our direction and not become lost. The last *yama* is *aparigraha*, "hands off". *Parigraha* is the opposite of the word *dana*, which means "to give". *Aparigraha* means " to receive exactly what is appropriate".

Niyamas, like *yamas*, are attitudes and are not to be taken as actions or practices. The five *niyamas* are more intimate in the sense that they are the attitudes we have towards ourselves. The first *niyama* is *śauca*, or cleanliness. There are two parts to this, external and internal. External *śauca* has to do with simply keeping ourselves clean. Internal *śauca* has to do with cleanliness of the internal organs and mind. The practice of *āsanas* or *prāṇāyāma* could be an internal *śauca.* The second *niyama* is *santosa*, a feeling of contentment. The next is *tāpas,* a word we have already discussed. With *tāpas* the idea is to bring out *asuddhi,* "dirt" inside the body. *Svādhyāya* is the fourth *niyama.* As we defined it earlier, *sva* means self; *adhyaya* means study or inquiry. Actually, *adhyaya* means to go near. *Svādayāya* means to go near yourself, that is, to study yourself. Any study, reflection, or contact that helps us understand more about ourselves is *svādhyāya.* The last *niyama* has also been mentioned before. *Īśvarapraṇidhāna* means " to leave all our actions at the feet of the Lord". Since our actions often come from *avidyā* it is possible that they might go wrong. That is why contentment is so important. This attitude suggests that we have done our best, and can leave the fruits of our actions in the hands of something higher than ourselves.

The third *aṅga* is *āsana*. In the theory of *āsana* practice, there are two aspects, *sukha* and *sthira.* We must be comfortable and at ease (*sukta*) and we must be steady and alert (*sthira*). We must be involved and at the same time attentive. Yoga suggests ways to achieve these qualities in *āsana.* The fourth *aṅga* is *prāṇāyama* which is conscious regulated breathing. *Pratyāhāra* , the fifth *aṅga,* involves the senses. The word *āhāra* means "food". *Pratyāhāra* means "withdrawing from that on which we are feeding". This refers to the senses: when the senses refrain from "feeding" on their objects, that is *pratyāhāra.*

Dhāraṇā comes from the root *dhr*, "to hold". *Dhāraṇā* occurs when we create a condition so that the mind, normally going in a hundred different directions, is directed towards one point. *Dhāraṇā* is a step leading towards *dhyāna.* In *dhāraṇā* the mind is moving in one direction; nothing else has happened. In *dhyāna*, when we become involved with a particular thing and we begin to investigate it, there is a link between ourselves and this object; that is, there is a perceptual and continuous communication between the object and our mind. This communication is called *dhyāna.* Further, when we become so involved with an object that our mind completely merges with it, that is called *samādhi.* In *samādhi* we are almost absent; we become one with that object.

There are many varieties of Yoga. Some people say that *dhyāna* is the means to *Jñāna Yoga.* In this context, this means inquiry about the truth, the real understanding that we attain in a state of *samādhi.* Inquiry in which we hear, then reflect, and then gradually see the truth, is *Jñāna Yoga*. In the *Yoga Sūtra* it is said that in the state of mind where there is no *avidyā*, automatically there is *Jñāna.*

Bhakti Yoga comes from the root *bhak* which means "to serve that which is higher than ourselves". This means an attitude of devotion. In *Mantra Yoga*, a teacher who knows us very well might give us a *mantra* which has a particular connotation because of the way it has been arranged. If that *mantra* is repeated in a certain way, if we are aware of its meaning, and perhaps if we want to use a particular image, *Mantra Yoga* brings about the same effect as *Jñāna* or *Bhakti Yoga.* In *Rāja Yoga,* the word *Rāja* means "the king who is always in a state of bliss, who is always smiling". Any process through which we achieve greater understanding of the mysterious and the obscure is *Rāja Yoga.* In the Vedas there are many references to the word *Rāja* in relation to *īśvara*.

It is best to explain *Laya Yoga* in a context of *samādhi,* When the meditator completely merges with the object of meditation, that is *Laya*. We merge with the object and nothing else exists.

In recent times, much has been written about Hinduism, and a lot of it pertains to and derives from the viewpoint of

Vedānta. It is important to see that the viewpoint of Yoga differs in some crucial respects from the viewpoint of Vedānta. Brahman is considered the "Pāramārthika Satya" or ultimate truth, and the world we live in and experience through our senses is granted the status of truth at an operational level, i.e. "Vyāvahārika Satya". In a sense, this carries the implication that the world is false and illusory. According to Yoga, everything we see, experience, and feel is not an illusion but is true and real. This concept is called *satvāda.*

Everything, including *avidyā,* dreams, and even fancy and imagination, is real. However, all these are constantly in a state of flux. This concept of change is called *pariṇāmavāda.* In Yoga, although everything we see and experience is true and real, changes do occur either in character or in content.

A.V. BALASUBRAMANIAN

REFERENCES

Aranya, Swami Hariharananda. *Yoga Philosophy of Patañjali.* Trans. P.N. Mukherji. Calcutta: University of Calcutta Press, 1977.

The *Bhagavadgītā.* Trans. S. Radhakrishnan. New York: Harper and Row, 1948.

The *Gheraṇḍa Saṃhita.* Trans. Sris Chandra Vasu. New York: AMS Press, 1974.

Haṭhayogapradīpikā of Svātmārāma. Trans. Srinivasa Iyangar. Madras: Adyar Library and Research Centre, 1972.

Yoga Yājñavalkya. Trans. Sri Prahlad Divanji. Bombay: Royal Asiatic Society, 1954.

YUKTIBHĀṢĀ OF JYEṢṬHADEVA Jyeṣṭhadeva (fl. 1500–1610) was a Nambūthiri Brahmin from the Ālattūr village, an important Brahmin settlement near Cochin. He was probably a student of Dāmodara, the son of Parameśvara, who also taught Nīlakaṇṭha Somayāji. His fame rests on the authorship of one of the most important texts of the Kerala school of mathematics and astronomy, the *Yuktibhāṣā* (An Exposition of the Rationale [of mathematics and astronomy]) also called *Gaṇita-nyāya-saṅgraha* (Compendium of Mathematical Rationale). It is a unique work on the rationale of Hindu mathematics and astronomy as it was understood in medieval India. It is unique in the sense that it is neither a textbook nor a commentary, but a work which is wholly devoted to a systematic exposition of mathematical rationale, written in Malayalam, the local language of Kerala.

Born about 1500, Jyeṣṭhadeva probably composed the *Yuktibhāṣā* about 1530, since it is known that a little after 1534, Śaṅkara Vāriyar, another contemporary astronomer, used it in his commentaries. Another work of Jyeṣṭhadeva, the *Dṛkkaraṇam*, also in Malayalam, was composed in AD 1608.

At the outset of the work, the author states that he is attempting "to set out in full the rationale useful for understanding the planetary motion according to the *Tantrasaṅgraha* of Nīlakaṇṭha Somayāji" (b. 1444). But he actually goes much beyond that and subjects to rationalistic analysis the entire gamut of mathematics and astronomy. In fact, he takes up the treatment from the very fundamentals, the concept of numeration and the theory of numbers. The work is made up of two divisions, each one divisible into several sequential chapters.

The first deals with the following subjects: (1) the eight fundamental operations, from simple addition to the roots of sums and differences of squares, wherein several methods, including diagrammatic solutions are offered; (2) algebraic problems; and (3) operations on fractions. The other chapters deal with: (4) the general nature of the Rule of Three (direct proportion) and (5) application of the Rule of Three in the computation of mean planets; (6) elaborate rationalizations of the circumference of the circle; and the last chapter: (7) the rationales of the derivation of the R sines, R versed sines, and their addition, properties of cyclic quadrilaterals, and the surface area and volume of a sphere. Many of the rationales are demonstrated both algebraically and geometrically.

Part Two is devoted to the exposition of rationales in astronomy, including (1) the computation of mean and true planets by means of two types of epicycles, supplemented by corrections; (2) the celestial sphere, the related great circles and secondaries, the precession of the equinoxes, and the armillary sphere; (3) declination, right ascension, and related matters; (4) problems related to spherical triangles; (5) problems connected with direction and shadow; (6) computation of the rising and setting points of the ecliptic, the ecliptic having constant variation. A direct method, enunciated in Indian astronomy for the first time, is used; (7) eclipses and the attendant parallax corrections; (8) the *Vyatipāta*, which is the moment when the sun and the moon have equal declinations, but in different quadrants; and (9) reduction of computed results to observation and with the phases of the moon.

Some points are of special interest in the *Yuktibhāṣā.* One is the rationale for three or four steps for true planets, although the result can also be obtained in two steps; the derivation of inverse declination and inverse right ascension; novel solutions for some of the problems on spherical triangles and on shadows; refinements for parallax corrections; and an alternate method with a novel correction for the computation of the moment of *Vyatipāta.* A noteworthy characteristic is its elucidating rationale from the fundamentals, first setting out the axioms and postulates involved, then developing the arguments and methodologies step by step.

It might also be noted here that Śaṅkara Vāriyar provided a valuable service by incorporating rationales from the *Yuktibhāṣā* into Sanskrit, the language of scholars. This he did in two elaborate commentaries called *Kriyākramakarī* on the *Līlāvatī* by Bhāskarācārya, and *Yuktidīpikā* on the *Tantrasaṅgraha*, a work on astronomy by Nīlakaṇṭha Somayāji. There is also a highly corrupt rendering of *Yuktibhāṣā* into Sanskrit.

K.V. SARMA

REFERENCES

Balagangadharan, K. "Mathematical Analysis in Medieval Kerala." In *Scientific Heritage of India: Mathematics.* Ed. K.G. Poulos. Tripunithutua: Government Sanskrit College, 1991, pp. 29–42.

Līlāvatī of Bhāskarācārya with Kriyākramakarī of Śaṅkara and Nārāyaṇa. Ed. K.V. Sarma. Hoshiarpur: Vishveshvaranand Vedic Research Institute, 1975.

Rajagopal, C.T., and M.S. Rangachari. "On an Untapped Source of Medieval Keralese Mathematics." *Archive for History of Exact Sciences* 18: 89–101, 1978.

Rajagopal, C.T., and M.S. Rangachari. "On Medieval Kerala Mathematics." *Archive for History of Exact Sciences*, 18: 89–101, 1978.

Sarma, K.V. and S. Hariharan. "Yuktibhāṣā: A Book of Rationales in Indian Mathematics and Astronomy, An Analytical Appraisal." *Indian Journal of History and Science* 26(2): 185–207, 1991.

Tantrasaṅgraha of Nīlakaṇṭha Somayāji with Yuktidīpikā and Laghuvivṛti of Śaṅkara (An Elaborate Exposition of the Rationale of Hindu Astronomy). Ed. K.V. Sarma. Hoshiarpur: Vishveshvaranand Vishva Bandhu Institute of Sanskrit and Indological Studies, Punjab University, 1977.

Yuktibhāṣā, Pt. I. *Mathematics.* Ed. Ramavarma (Maru) Thampuran and A.R. Akhileswarayyar. Trissivaperur: Mangalodayam Limited, 1948.

See also: Nīlakaṇṭha Somayāji – Parameśvara – Rationale in Indian Mathematics – Śaṅkara Vāriyar

Z

ZACUT, ABRAHAM Abraham bar Samuel bar Abraham Zacut was the most prominent astronomer of the late Middle Ages in the Iberian Peninsula. He was born, probably in 1452, in Salamanca (Spain), a Castilian town whose university had a renowned chair of astrology.

Although he is mainly known for his astronomical activity, Zacut wrote on other subjects such as lexicography (*Hosefot leséfer ha'Aruk*) and history (*Séfer Yuḥasin*). Other treatises have been ascribed to him, but there is no adequate evidence to support this claim.

Zacut's most outstanding work is *Ha-ḥibbur ha-gadol* (The Great Compilation), written at the request of Gonzalo de Vivero, bishop of Salamanca. The *Ha-ḥibbur ha-gadol* is composed of some fifty astronomical tables and the canons explaining their use. The tables have the year 1473 as radix, and they are arranged for the Christian calendar and the meridian of Salamanca. The tables give the positions of the sun, moon, and the five planets presented in the form of an almanac. Of special interest are the tables listing the day-by-day positions of the sun (for 4 years from 1473), and the moon (for 31 years from 1473). They are calculated according to the Alfonsine Tables. Another table lists all true syzygies for 31 years; it is structured on the cycle of 767 syzygies discovered by the Catalan astronomer Jacob ben David Bonjorn, author of astronomical tables for Perpignan (1361), and a follower of Levi ben Gerson (1288–1344). Both these astronomers are mentioned, among others, in the canons of the *Ha-ḥibbur*, together with Ptolemy, and the Castilian king Alfonso X (1252–1284).

Zacut finished *Ha-ḥibbur* around 1478; three years later it was translated from Hebrew to Castilian, with the help of Zacut himself, by Juan de Salaya, who held the chair of astrology at the University of Salamanca.

Following the request of his new protector, Juan de Zúñiga, Master of the Order of Alcántara, in 1486 Zacut wrote a work on medical astrology: *Tratado de las influencias del cielo* (Treatise on the Influence of the Heavens), followed by a short text on eclipses: *Juicio de los eclipses* (Judgment on Eclipses). Two other works on astronomy have been attributed to Zacut: *'Oṣar ḥayyim* (Treasure of Life), and *Mishpaṭé ha'isteganin* (Judgments of Astrology).

Because of the expulsion of the Jews promulgated by the rulers of Spain (1492), Zacut moved to Portugal and entered the service of King João II as an astronomer and chronicler.

A version of Zacut's *Ha-ḥibbur* was published in Leiria, Portugal under the title *Tabulae tabularum coelestium motuum sive Almanach Perpetuum* (1496). The Portuguese, José Vizinho, was responsible for this version, which contains a summary of the canons of *Ha-ḥibbur*, and most of its tables. Actually two editions came out of the printing house that year: one with the canons in Castilian, the translation of which was made by Vizinho, the other in Latin. The *Almanach Perpetuum* played a significant role in the navigation projects of the kingdom of Portugal, and especially in the preparation of Vasco da Gama's expedition, for whom Zacut drew up tables for the solar declination for four years.

After the practice of Judaism was declared illegal in Portugal in 1496, Zacut traveled to Northern Africa, and settled in Tunis. He adapted his tables for the year 1501, and he compiled another set of tables, beginning with year 1513, arranged for the Jewish calendar and the meridian of Jerusalem, where he then lived. Zacut probably died in Damascus in 1515.

JOSÉ CHABÁS

REFERENCES

Cantera Burgos, Francisco. "El Judío Salmantino Abraham Zacut." In *Revista de la Academia de Ciencias Exactas, Físicas y Naturales*. Madrid: Bermejo, 1931, pp. 63–98.

Cantera Burgos, Francisco. *Abraham Zacut*. Madrid: M. Aguilar, 1935.

Carvalho, Joaquim de. "Dois inéditos de Abraham Zacuto." *Revista de Estudios Hebráicos* (Lisboa) I: 1–54, 1927.

Goldstein, Bernard R. "The Hebrew Astronomical Tradition: New Sources." *Isis* 72: 237–251, 1981.

See also: Levi Ben Gerson – Alfonso X

ZERO Mathematics today owes its existence in part to the discovery of zero. For the purpose of calculation it needed a short symbol, which is at present denoted by a small circle in nearly every part of the world. In India in the early period the form of short symbol which represented zero (*śūnya*) was both a dot and a small circle.

In the Vedic literature, in the *Amarakosa*, zero (*śūnyam bindau*) was represented by a dot. The form was also suggested by the word *kṣudra* (very small) in the *Atharvaveda*. The word *randhra*, used in the Hindu *Gaṇita Sastraka*, indicated a small hole. There are many examples of zero being represented as a dot in the Kashmirian *Atharvaveda*, both in the marginal notes and in the text itself.

In the marginal notes, the numbers are as follows:

Symbol	Number	Folio	Page
◇ • •	100	100b	192
◇ • ◇	101	101b	194
◇2•	170	170b	310

The symbols in the marginal notes represented the Folio numbers. In most of the pages of the book there are examples of the above type. In the text itself the following numbers are found:

1	◇
2	
3	
4	
5	
6	
7	
8	
9	
10	◇•

Small circles in pairs have been used to denote blank spaces in the text. In Folio No. 178a, page 323 in the last line, pairs of small circles are found as follows:

○○ ○○ ○○ ○○

The numbers in the marginal notes may be taken from a later period, but the symbols in the text itself must be from the time of the *Atharvaveda* (500 BC).

In the Bakhshālī Manuscript, the oldest extant manuscript in Indian mathematics (AD 200), the symbol  represents the number 4000, while the symbol  represents 500,000,000. The fourth zero in the manuscript is a small circle; the others are dots. In the Shahpur Stone image Inscription of Aditya Sena Bihar, India (AD 672), the symbol 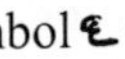 stands for 60, and zero is represented as a dot. In the Malay Inscription at Katakapur (AD 686), symbol 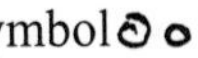 represents the number 608 (Śāka), and zero is represented as a circle with a deep circumference.

In the above examples it is found that zero has been exhibited either as a dot or as a small circle to fit in numbers in the place value scale, or to represent absence. In the inscriptions at Sambor, Palembang, and Kotakapur, which were Indian colonies of the Far East, the numbers have been written to represent the Sāka era. Hence undoubtedly they also represent an Indian origin.

Now the question arises as to why zero has been represented as dot or a small circle and not as a square or rectangle or anything else. Two reasons can be assigned; one is spiritual or metaphysical; the other is physical or atomic.

In the spiritual sphere, *śūnya,* or zero or nothing, the symbol of absence of everything, identifies itself with *Nirguṇa* Brahma, the absence of all qualities. The no-quality (*Nirguṇa*) in Brahma represents the fact that He is not guided by any of the qualities or constituents of nature. But at the same time Brahma is the source of all qualities, energy, power, and strength. Similarly *śūnya* or zero itself signifies absence when placed independently, but it represents fullness when it is placed in the decimal system of numeration. (Placing zeros on the right side of a number increases the value to an infinite step).

Since the conception of *śūnya* identifies with the conception of Brahma both in absence and fullness the symbol of *śūnya* (zero) must be guided by the symbol of Brahma. The conception of Brahma lies in meditation on a particular point or small circle in the space between the two eyebrows.

Swami Sivananda has given concentration the name 'one-pointedness'. Concentration can also start by fixing one's gaze on a black dot on the wall and later on a bright light first of the size of a pinpoint and later of the size of a sun coming out from the space in between the two eyebrows. Hence the symbol of Brahma can taken to be a dot or a small circle, and therefore the symbol of mathematical zero is either a point or a small circle.

There are also physical reasons for the symbol for zero. Planets seen from the earth look just like dots. From Vedic times Indians excelled in astronomical observations as we can see in the *Vedāṅga Jyotiṣa* (1200 BC). So the physical reason for zero's being represented as a dot or a small circle lies in the fact that the sun, moon, and planets were seen as dots or small circles to an observer on the earth.

Now why should a planet be chosen to represent zero? It is because of the other interpretation of mathematical zero, the absence of atoms. The idea of the atom was mentioned in ancient times in the Buddhist work *Lalita Vistara* (500 BC), where the diameter of a *parmāṇu* (molecule) was given as $1.32\times 7^{-}$ inches, whereas at present the diameter of an atom is 2×10^{-8} cm. The nucleus of the atom has a radius of $1.37\times 10^{-}$ cm. The quantity is so small that it is a *kṣudra* (minute), a synonym for zero in the *Athravaveda*. Thus the absence in mathematical zero is guided by the absence inherent in the smallness of an atom or its nucleus. The concept of the fullness of mathematical zero is also present in the infinite motion of the planets and of the electrons around the nucleus of an atom.

The electron has no mass in the material sense, and the mass which it has developed from electrical energy is negligible, having a radius of $1.875\times 10^{-}$ cm. This guides the concept of mathematical zero. The movement of electrons around the nucleus with a very high velocity and in an infinite motion is identical with the concept of fullness of mathematical zero which guides the numbers to move in an infinite journey like 10, 100, 100, 10,000 to 10^n.

Thus the double interpretations, absence and fullness, of mathematical zero are identical with the double interpretations in a planet, and the symbol of zero is guided by the symbol of a planet which is observed either as a dot or a small circle and whose path is almost circular.

Zero has a double meaning in Vedic literature. Etymologically the word *śūnya* comes from the word *śūna* (*śūna* + *yat*). The word *śūna* has two meanings. One is the killing of animals or the slaughter house, which represents absence; the other is increase, which leads to the conception of fullness.

The synonyms of zero, *randhra*, *tuccha*, *kṣudra*, and *ritka* project a concept of nothingness, while the synonyms *vyoma*, *diba*, *akasa*, *antariksa*, and *jaladhra patha* mean infinite expanse of sky. *Purna* means full, and *ananta* means infinite. *Drabinam* and *balam* mean strength, vigor, and force.

Zero, though it signifies nothing, has a full voltage battery charge when used in the decimal place value system.

The German mathematician B.L. Van der Waerden opines that the symbol of zero as a small circle came from the first letter 'o' of the Greek word *ouden* meaning nothing. This claim can be compared with a parallel claim (apparently convincing but far from the truth) that 4 in Brahmi Numerals written as च comes from the first letter च of the word च (our four in English). Similarly in India one may say that the symbol ५ (5) in Deonagri script comes from the first letter of the word पाँच (five), ६ (6) in Deonagri comes from the first letter of the word छः (six). But these are not so. Every number symbol has a heritage and a path through which it has come down to its present form. That 'o' is the first letter of the word is just a coincidence. Rather the symbol of a small circle (○) was used for the numbers ten, seventy, and a hundred in Greece.

R.N. MUKHERJEE

REFERENCES

Atharvaveda. Ed. Maruice Bloomfield. Strasbourg: K.J. Treubner, 1899, reprinted 1971.

Bag, A.K. "Symbol for Zero in Mathematical Notation in India." *Boletin de la Academia Nacional de Ciencias* 48: 251, 1970.

Bose, D.M., ed. *A Concise History of Science in India.* New Delhi: Indian National Science Academy, 1971.

Datta, B. and A.N. Singh. *History of Hindu Mathematics*, vol. 1. Lahore: Motilal Banarsi Das, 1935.

Kaye, G.R. *Bakhshālī Manuscript. A Study in Mediaeval Mathematics.* Calcutta: Government of India, 1927–1933.

Van der Waerden B.L. *Science Awakening.* Oxford: Oxford University Press, 1961.

See also: Bakhshālī Manuscript

ZHANG HENG Zhang Heng was a Chinese astronomer, writer, and machinist (AD 78–139). As a man of noble descent, he assumed many official positions, and from AD 115, twice was the head of the imperial astronomical organization for a total of fifteen years. He was a versatile man who contributed greatly to Chinese astronomy.

The title of one of his works is *Ling Xian* (Mystical Laws); it is very possibly an outline of astronomy–astrology, but only part of the beginning has been handed down. He made an astronomical instrument like a planetarium to demonstrate the motion of the sun, moon, and planets with stars in the background, and to give the correct time. According to historical records he provided this instrument with an automatic mechanical installation, but the detailed information has not been preserved.

Zhang Heng made a wonderful invention for earthquake measurement. In AD 132 he invented a seismoscope called *Hou Feng Di Dong Yi* (Instrument for Earthquake), and at Luoyang (the capital of the empire) he successfully determined that an earthquake had occurred at a place in the northwest of China more than a thousand kilometers from the capital. We have the following information about this seismograph.

This instrument was cast in copper. Its main body is a standing egg-shaped shell with an erect pendulum in the center. It links eight dragons each of whom, with some dexterous mechanical installations, keeps a small copper ball in its mouth. When an earthquake happens somewhere, the dragon in that direction spits out its ball, so that people know where the earthquake has occurred.

This famous seismoscope was not preserved, so scholars have tried to reproduce it since the nineteenth century. Many reconstructive schemes on the basis of modern seismological principles have been published. Up to now, the most successful was by Wang Zhenduo (1963) of which a model has been made.

Zhang Heng is also an important writer. His *Si Chou Shi* (Four Chapters of Distressed Poems) and *Gui Tian Fu* (To Live in Seclusion) are literary masterpieces and his *Tong Shen Ge* (Song of Love) is one of the most important early documents for the history of Chinese sexual culture.

JIANG XINOYUAN

REFERENCES

Fan Ye. *How Han Shu* (The History of the Eastern Han Dynasty), vol. 10. Beijing: Zhonghua Press, 1965.

Wang Zhenduo. "The Reconstruction of the Hou Feng Di Dong Yi (Seismograph). Invented by Zhang Heng of the Eastern Hang Dynasty." In *Papers in Technical Archaeology* (Wang's collected papers). Beijing: Cultural Relics Publishing House, 1989, pp. 287–344.

Zhang Heng. *Zhang Heng Ji* (Collected Works of Zhang Heng). Shanghai: Shanghai Classics Publishing House, 1986.

ZHANG QIUJIAN SUANJING *Zhang Qiujian suanjing* (The Mathematical Classic of Zhang Qiujian) is the only known work of the fifth century mathematician, Zhang Qiujian. It is one of ten mathematical books known collectively as *Suanjing shishu* (Ten Mathematical Classics). In AD 656 when mathematics was included in the official examinations, these ten outstanding works, which covered a period of over a thousand years, were specially selected as textbooks.

Jiuzhang suanshu (Nine Chapters on the Mathematical Art) and *Sun Zi suanjing* (The Mathematical Classic of Sun Zi) are two of these texts that precede *Zhang Qiujian suanjing*. All three works share a large number of common topics. Though *Sun Zi suanjing* was intended as a primer, it now provides significant documentation of the rod numerals and their use in multiplication, division and other mathematical methods. *Jiuzhang suanshu* is undoubtedly the most important and influential of the early mathematical texts. Both works occupy prominent places in the history of Chinese mathematics with one showing the initial learning stage and the other manifesting an evolutional culmination. *Zhang Qiujian suanjing* demonstrates the continuation of the development of mathematics from here, and provides an important bridge of knowledge on the evolution of traditional mathematics to its apogee in the thirteenth century. It is all the more important as such surviving works are few and far between.

Zhang Qiujian regarded fractions of primary importance; he mentioned the difficulty of the subject in his preface and provided more complicated problems than those in *Jiuzhang suanshu*. There are only very brief general descriptions of finding the square root and the cube root of a number in *Jiuzhang suanshu* and these are written with the use of technical phrases. It is a relief to scholars to find that examples are employed in *Zhang Qiujian suanjing* to illustrate the step by step procedures of both methods and also to show how the solution of a non-integral approximation is obtained.

Zhang Qiujian continued with the tradition of the development of arithmetic by supplying problems involved with what is known in the west as the Rule of Three and extending to problems concerned with proportions, compound proportions, and proportional parts. Other problems are involved with relative speed, different shaped frustums and granaries, the Rule of False Position, the arithmetical progression, the computation of interests and taxes, similar right angled triangles, and the quadratic equation. There are a number of problems concerned with the well known *fang cheng* method, which is the Chinese way of solving a set of simultaneous linear equations. The last problem of the book is the famous problem of a hundred fowls. In this problem, somebody is to buy one hundred fowls for one hundred monetary units, given that a rooster costs 5 units, a hen 3, and chicks are sold three for one unit. Even though the answers given in the book are correct, it is not clear how the answers were obtained.

Zhang Qiujian's lucid style has been a great help to the scholar's understanding of how the ancient mathematicians manipulated the rod numerals to arrive at the answer. The existing edition of the book has three chapters with ninety two problems; the last portion of the second chapter and the beginning of the third chapter are incomplete. The book has been translated into English by Ang Tian Se.

Zhang Qiujian suanjing has an important place in the world history of mathematics: it is one of those rare books before AD 500 that manifests the upward development of mathematics due fundamentally to the notations of the numeral system and the common fraction. The numeral system has a place value notation with ten as base, and the concise notation of the common fraction is the one we still use today.

LAM LAY YONG

REFERENCES

Ang Tian Se. "A Study of the Mathematical Manual of Chang Ch'iu-Chien." Unpublished M.A. dissertation, University of Malaya, 1969.

Li Yan and Du Shiran. *Chinese Mathematics. A Concise History.* Trans. J. N. Crossley and A. W. C. Lun. Oxford: Clarendon Press, 1987.

Needham, Joseph. *Science and Civilisation in China.* Vol.3: *Mathematics and the Sciences of the Heavens and the Earth.* Cambridge: Cambridge University Press, 1959.

Qian Baocong, ed. *Suanjing shishu* (Ten Mathematical Classics). Beijing: Zhong hua shu ju, 1963.

See also: Computation: Chinese Counting Rods – Liu Hui and the *Jiuzhang Suanshu*

ZHANG ZHONGJING The Chinese physician and medical author Zhang Ji (ca. 150–220), also known as Zhang Zhongjing, was born in the Nanyang commandery (in modern Henan). As stated in the preface to his main work, he was prompted to study medicine and collect prescriptions by the spread of epidemic diseases which caused the death of several members of his own clan. The preface adds that he was Governor (*taishou*) of Changsha (modern Hunan), but this detail is not independently confirmed by other sources.

Zhang Ji's main work, completed in the first or second decade of the second century, was originally entitled *Shanghan zabing lun* (Treatise on Cold Damage and Miscellaneous Disorders) and included sixteen chapters. The rather intricate bibliographic history of the text cast doubts on its authenticity. About one century after its compilation, Wang Shuhe (ca. 265–317) produced a revised, expanded version. The final six chapters, lost by the Song period, were replaced on the basis of an abridged version, and published as a separate work. The two main parts in which the text is extant today are entitled *Shanghan lun* (Treatise on Cold Damage) and *Jingui yaolüe* (Essentials from the Golden Casket), respectively. The latter is based on the "Miscellaneous Disorders" section of the original text.

The *Shanghan lun* is the first Chinese medical text devoted to a specific etiology. It is concerned with a class of diseases symptomatized by acute fevers, and named "Cold Damage" after one of their factors. Their diagnosis is based on the identification of symptoms (*zheng* or "evidence") that indicate the stage reached by the disease in its progression towards the vital centers of the body. These stages of development are defined according the Six Warps (*liujing*) system, one of the frameworks used to the describe the cyclical flow of energy within the body (Others are *qi* and *wuxing*, Five Agents or Phases). The text emphasizes the practical rather than the theoretical aspects of healing. Therapy is based on the administration of medicines (the whole work includes more than 300 prescriptions for specific diseases), and to a lesser extent on acupuncture, moxibustion, and other methods such as baths and massages.

Zhang Ji's work gave rise to the second most important tradition within Chinese classical medicine after the one represented by the texts of the "Inner Canon of the Yellow Emperor" corpus (*Huangdi neijing*). Although Zhang refers to this corpus in the preface, even the most basic notions of the Inner Canon — e.g. the Five Agents or Phases, the visceral systems (*zangfu*), and the circulation tracts (*jing*) of acupuncture — are absent from his work. As has been pointed out, this is not necessarily a sign of alternative or competing traditions. An extended commentatorial tradition, in fact, has tried to bring together the respective basic notions. Only recent research has shown the ways and extent to which the general framework of the two works differs.

FABRIZIO PREGADIO

REFERENCES

Ågren, Hans. "Chinese Traditional Medicine: Temporal Order and Synchronous Events." In *Time, Science, and Society in China and the West*. Ed. J.T. Fraser, N. Lawrence, and F.C. Haber. Amherst: The University of Massachusetts Press, 1986, pp. 211–218.

Despeux, Catherine. *Shanghan lun. Traité des "Coups de Froid"*. Paris: Éditions de la Tisserande, 1985. This translation is preferable to the English version by Luo Xiwen, *Treatise on Febrile Diseases Caused by Cold (Shanghan lun) by Zhang Zhongjing*. Beijing: New World Press, 1986.

Sivin, Nathan. *Traditional Medicine in Contemporary China*. Ann Arbor: Center for Chinese Studies, The University of Michigan, 1987.

Unschuld, Paul. *Medicine in China; A History of Ideas*. Berkeley: University of California Press, 1985.

See also: *Shanghan Lun* – Five Phases – *Qi* – Acupuncture – Moxibustion – Medicine in China – *Huangdi Neijing*

ZHENJIU DACHENG The *Zhenjiu Dacheng* (Great Compendium of Acupuncture and Moxibustion) is a digest of acupuncture and moxibustion writings from the Han to the Ming dynasties. As such, it forms the greatest comprehensive survey of acupuncture in China ever to be produced during classical times. Much of its greatness is due to its selection of works from the Song and Yuan. It was edited by Yang Jizhou in 1601, as the culmination of his life's work. He was then close on eighty years old and obviously still active.

An early title for the work had been "Secrets of the Dark Inner Mechanism of Acupuncture and Moxibustion for the Preservation of Health". In this text Yang Jizhou included mainly the traditions of his own family, but later he enlarged the book to include earlier literature, and so changed the title. Generally he selected the best of the old, and reedited it, adding a certain amount of his own material where it suited.

The Grand Compendium comprises some ten volumes, covering all aspects of acupuncture and moxibustion. It also contains an appendix on massage techniques particularly suited for children, entitled *Biaoying Shenshu* (The Divine Art of Protecting Infants), taken from an earlier book which is now lost. The work includes listings of channels and points, many point-formulae for medical conditions, particular systems for choosing points, refined needling techniques such as those developed during the Song, Yuan, and Ming, detailed formulae for calculating the precise time and exact point for treatment; glossaries of point names; and selections from the classics, especially the *Huangdi Neijing* and its slighter companion the *Nan Jing.* In all this constitutes a great wealth of historical material.

Particularly striking are the exquisite verse poems selected for the second volume. These include the "Ode to the Golden Needle", which focuses on needling techniques, the *Liuzhu Zhiwei Fu* (Ode to Intricacies in the Circulating Flow), which outlines the Law of Midday–Midnight in which the choice of points is determined by the particular hour and day of treatment, and the *Biaoyou Fu* (Ode to the Streamer out of the Dark), the longest and most comprehensive of these poems, which contains an extensive commentary by Yang Jizhou.

A gradual increase in prosperity was spawned by better social conditions during the Ming. These in turn created an interest in the old ideas. Undoubtedly influential on Yang's work were the *Zhenjiu Sunan Yaozhi* (Essentials of Acupuncture and Moxibustion, 1536) and the *Zhenjiu Juying* (Gatherings of Outstanding Acupuncturists, 1546), both issued by Gao Wu during the earlier century. Gao Wu's work was particularly commendable as he made a stand against any superstitions in medicine. Where he was uncertain he would simply state it, without comment. From Gao Wu's work Yang took both an uncluttered approach and a grand overview of the acupuncture tradition. In many places Yang Jizhou pays debt to Gao Wu by copying his selection of classical extracts exactly.

Another great influence was the *Shenying Jing* (Classic of Divine Resonance, 1425) by Chen Hui. From this work Yang Jizhou borrowed his simple and practical approach, cutting great swathes through confusing systems of point selection by adhering strictly to the Divine Resonance's printed lists. This compendium is an invaluable treasure for all acupuncturists and has become well-known abroad as well as in China.

RICHARD BERTSCHINGER

REFERENCES

Bertschinger, Richard. *The Golden Needle, and other Odes of Traditional Acupuncture.* Edinburgh: Churchill Livingstone, 1991. (A translation of Book Two of the *Zhenjiu Dacheng.*)

Lu, Gweidjen and Joseph Needham. *Celestial Lancets.* Cambridge: Cambridge University Press, 1980.

Ta'o, Lee. "Chinese Medicine During the Ming Dynasty." *Chinese Medical Journal* 76: 285–304, 1958.

See also: Acupuncture – Moxibustion – Medicine in China

ZHENJIU JIAYIJING The *Zhenjiu Jiayijing* (The A–Z Canon on Acupuncture and Moxibustion), was written by Huangfu Mi (AD 215–282), during the short-lived Jin Dynasty. Needham dates the book as being finished some time between AD 256 and AD 282.

This is the oldest surviving text devoted exclusively to acupuncture and moxibustion. It was produced after the fall of the Han nation, during a time of violent turmoil and disruption in Chinese society. Huangfu Mi summarized the knowledge of his day concerning acupuncture and moxibustion, and ordered the many texts then current. He drew from both sections of the *Huangdi Neijing* as well as the *Mingtang Kongxue Zhenjiu Zhiyao* (Therapeutic Essentials of the Acupuncture Points Revealed in the Clear-lit Hall), of unknown authorship, a book now lost. Huangfu Mi took up medicine in middle age because of his own family's illness and his own arthritis.

This book's most impressive character lies in its systematic outline of the subject (as revealed in the title). The many points are grouped under the various channels to which they belong, with number, name, point location and tips for finding each. Additionally there are listings of the illnesses, or syndromes, for which each can be used, detailing how deep the needle should be inserted and the number of breaths for which it should be retained in position. Tonification or reducing techniques are also described, as well as angling the needle shaft along or against the direction of energy flow down the channel. In addition, Huangfu mentions the use of moxa, the number of cones for each point being duly recorded. All this had never been so accurately or clearly laid out before.

This book, quite justifiably, has had an enormous influence throughout China and abroad. It became *de rigueur* reading for the Japanese imperial colleges from the seventh century, and found its way to Korea even earlier. Its texts have been the foundation of acupuncture teaching and practice for some seventy generations in the East. As yet, there has been no translation into English.

The book's 128 chapters comprise: medical theory, the examination and diagnosis; characteristics and pathology of the channels; a complete catalogue of the points; pulse diagnosis; needling technique; outline of pathology and transmission of disease; types of fevers; pains, swelling, coughing, rheumatism, wasting diseases and dumbness; eye, nose and throat problems; women's and children's illness; and appendices.

Huangfu Mi's approach, in particular, stressed the need for a proper adjustment of the patient's emotions. He knew that physical and environmental factors played a part in illness, that treatment should proceed with great care, and that signs and symptoms should be precisely differentiated. All shows a clear and intelligent approach to practice.

Very little has changed in the location and identification

of acupuncture points since the publication of this text some 1800 years ago. For instance, it is mentioned in the preface to the *Essentials of Chinese Acupuncture*, the first acupuncture book produced by the Chinese for a Western audience.

RICHARD BERTSCHINGER

REFERENCES

Essentials of Chinese Acupuncture. Beijing: Foreign Languages Press, 1980.

Fu, Weikang. *The Story of Chinese Acupuncture and Moxibustion*. Beijing: Foreign Languages Press, 1975.

Lu, Gweidjen and Joseph Needham. *Celestial Lancets*. Cambridge: Cambridge University Press, 1980.

See also: Huangfu Mi

ZHOUBI SUANJING The *Zhoubi suanjing* (frequently also romanized as *Zhoubei suanjing*), is an anonymous Chinese work on astronomy and mathematics. It is a composite work which probably reached its final form in the first century AD under the Western Han dynasty.

The contents deal mainly with calendrical astronomy, and the shape and size of heaven and earth according to the *gaitian* (umbrella [-like] heaven) cosmography. It opens with a brief dialogue between the Duke of Zhou (fl. ca. 1000 BC) and Shang Gao, an otherwise unknown figure, who explains the use of the try-square *ju* and refers briefly to a round heaven lying above a square earth. In connection with the try-square Shang Gao refers cryptically to the relationship between the sides of a 3–4–5 right triangle. The text then continues with a long and fairly coherent section which may be the original core of the book.

In this section an otherwise unknown Chen Zi discusses the use of a gnomon (a vertical pole) for observing the shadow cast by the noon sun. It is claimed that the shadow of a pole 80 *cun* (inches) long changes by one *cun* for every 1,000 *li* [about 400 miles] moved north or south in relation to the point on the earth where the sun is directly overhead. The use of this rule, which does not correspond to actual observation, enables the dimensions of the heavens to be calculated. Most of the rest of the book contains calculations relating to the operation of luni-solar calendrical systems of the *si fen* (quarter [remainder]) type. The text bears an important commentary by Zhao Shuang, probably written ca. AD 270–280, and has further annotations by Li Chunfeng, who prepared an edition for use in the State College in AD 656. The book was first printed under the Song dynasty in AD 1084.

CHRISTOPHER CULLEN

REFERENCES

Cullen, Christopher. *Astronomy and Mathematics in Ancient China*. Cambridge: Cambridge University Press, 1996.

See also: *Gaitian*

ZHU SHIJIE Virtually nothing is known of the life of this Yuan mathematician, except the little we can learn from the prefaces to his two books, *Suanxue qimeng* (*Introduction to Mathematical Studies*, 1299) and *Siyuan yujian* (*The Jade Mirror of the Four Unknowns*, 1303). The first preface was written by Zhao Cheng and the second one by Mo Ruo and Zu Yi. From those we can surmise that Zhu came from Yanshan, near modern Beijing, that he traveled around China for more than twenty years in the last decades of the thirteenth century, after China had been unified under the Mongol rule, and that numerous students came to study with him.

His books attest to the fact that he inherited — from the northern milieu of the Hebei and Shanxi provinces which developed the "procedure of the celestial unknown" — how to use polynomial algebra to build an equation with which to solve a problem. This achievement seems to have been unknown in south China and benefited from the potentialites of the counting instrument on which it is based: the counting surface on which numbers were represented with counting-rods which could be moved or changed in their values throughout the computations. The earliest book that has come down to us attest this procedure, though it is probably not the first to have dealt with it, dates from 1248. It is Li Ye's (Li Zhi's) *Ceyuan haijing* (Sea Mirror of the Circle Measurements). Yet Zhu Shijie does not mention this author or any other one. On the basis of extensions to polynomials with two and three indeterminates "that the procedure of the celestial unknown had received previously to him", according to Zu Yi's preface, Zhu Shijie developed it to what was to remain an unsurpassed peak in Yuan and Ming China: the use of polynomial algebra with four indeterminates to establish an equation to solve a given problem. As Jock Hoe has stressed, the language and notations with which Zhu Shijie developed this algebra can, in their conciseness and precision, stand comparison with aspects of modern algebraic symbolism. Still, he notices some ambiguous modes of expression that do not make it entirely independent from the context. This very language might help us to make the historical connections between Zhu Shijie and such previous authors as Li Ye more precise, since as far as topics such as right-angled triangles are concerned, both use the same terminology. On the other hand, as regards the expression of formulas, they present differences even though their languages share the same basic features. For the expression of formulas or algorithms, Li Ye uses a language whose syntax is restricted in such a way that the understanding of a formula is unambiguous, which makes it context-free, in contrast to Zhu Shijie's.

Moreover, Zhu's writings demonstrate that he shared bits of knowledge (for instance, finite difference procedures for the summation of series, a topic to whose development he made crucial contributions) with the authors of the calendar *Shoushili* commissioned by the Mongol Court, the foremost being Guo Shoujing.

In addition to this — is it a reflection of his travels in south China — his books reveal his interest in algorithms for quick, daily-life computations, or for mercantile and bureaucratic mathematics. In the surviving writings from this period, such topics had mainly been treated by mathematicians from south China, such as Yang Hui. At the same time other topics developed in south China, such as research connected with the Chinese remainder theorem, seem not to have found their way into his writings. Still, Du Shiran called attention to the fact that Zhu seems to represent a kind of synthesis of traditions that, for at least decades, had been developing in isolation from each other in North and in South China, though they shared a common basis in the same fundamental characteristics and knowledge, including algorithms to find "the root" of any algebraic equation with positive, negative, or null coefficients.

Zhu Shijie's books do not seem to have had any impact on the development of mathematics in China; not only do we not find any mention of his name or achievements after their publication, but the books themselves would eventually be lost, to be found again in China and reprinted only at the beginning of the nineteenth century. This merely reflects the progressive loss of mathematical knowledge from the Song-Yuan period. In contrast to other writings of the same period, his *Introduction to Mathematical Studies* was to influence the development of mathematics in Korea and in Japan. It was probably printed in 1433 in Korea, where it became a textbook for studying mathematics, and was recovered thanks to a Korean edition of 1660. Moreover, its introduction to Japan in the second half of the seventeenth century, where it was reprinted and commented upon many times, allowed Japanese mathematicians to reconstruct the procedure of the celestial unknown, by that time forgotten in China, and to improve it until it became a tool in some ways comparable to the algebra developed in Europe during the same period. A. Horiuchi studies this aspect of the mathematics developed in Japan with great care and compares it with elements of the contemporary European algebra. Indeed, Japanese mathematicians' use of the counting board for their computations was probably instrumental in that respect, whereas this instrument had fallen into oblivion and was completely replaced by the abacus in China.

KARINE CHEMLA

REFERENCES

Du Shiran. "Zhu Shijie yanjiu" (Research on Zhu Shijie). In *Song Yuan Shuxueshi Lunwenji* (*Collected Essays on the History of Mathematics during the Song and Yuan Dynasties*). Ed. Qian Baocong. Beijing: Kexue chubanshe, 1966, pp. 166–209.

Ho Peng Yoke. "Chu Shih-chieh." In *Dictionary of Scientific Biography*, vol. 3. Ed. Charles Gillispie. New York: Charles Scribner's Sons, 1970–80, pp. 265–71.

Hoe, Jock. *Les Systèmes d'Équations Polynômes dans le Siyuan yujian (1303)*. Paris: Institut des Hautes Études Chinoises, 1977.

Horiuchi, Annick. *Les Mathématiques Japonaises à l'Époque d'Edo*. Paris: Vrin, 1994.

Juschkewitsch, Adolf Pavlovitch. *Geschichte der Mathematik im Mittelalter*. Leipzig: Teubner, 1964.

Qian Baocong. *Zhongguo Shuxue Shi (History of Mathematics in China)*, 2nd ed. Beijing, 1981.

See also: Computation: Chinese Counting Rods – Liu Hui and the *Jiuzhang Suanshu* – Guo Shoujing – Li Zhi – Mathematics in Korea – Mathematics in Japan – Yang Hui

ZĪJ The term *zīj* is used everywhere in the study of Islamic culture to signify an astronomical handbook. These handbooks consist of a collection of astronomical tables together with such textual material as the reader would need in using the tables. The material is often divided into the following sections.

1. Calendrical conversion;
2. Mean motions of the sun, moon, and planets;
3. Equations of the sun, moon, and planets;
4. Positions of fixed stars;
5. Trigonometrical tables (sine, tangent);
6. Spherical astronomy;
7. Parallax;
8. Eclipses of the sun and moon;
9. Geographical coordinates;
10. Astrological quantities.

The underlying theoretical model of planetary equations was almost invariably Ptolemy's, and to a large extent *zījes* represent a continuation of the *Handy Tables* of Ptolemy (ca. AD 140). However the earliest Arabic *zīj* is that of al-Khwārizmī (ca. AD 830), which was based on procedures derived from the Indian treatise *Brāhmasphuṭasiddhānta* of Brahmagupta (AD 628). A large number of such works survive intact in Arabic and Persian manuscripts, and many others are known by name only. The most notable modern edition (Nallino, 1907) is that of the *Zīj* al-Ṣābiʾ of al-Battānī (fl. AD 880). For the long Islamic period the *zīj*es are a vital repository of data, primarily through the constant improvements in the parameters of mean motions reflecting new observations, and also of improvements in mathematical methods.

In most *zīj*es the textual material was generally restricted to instructions in the use of the tables. Larger astronomical treatises, such as the *Qānūn al-Masʿūdī* of al-Bīrūnī (ca. AD 1030), included all the material which would be found in a *zīj*, but went further in its detailed treatment of the whole subject.

A number of *zīj*es were translated into Latin and Greek, and initially at least, the term *zīj* was transcribed as *ezich*, *ezeig*, etc. (= *al-zīj*), and ζηζι respectively. In Latin the term was soon replaced by *tabulae*, while in Greek one finds *συταξιφ*, reminiscent of Ptolemy's *Almagest*.

The term *zīj* is originally Middle Persian, where it means 'stretched cord'. The sense 'astronomical handbook' goes back to the early sixth century when Sanskrit works were introduced into Iran. The name *zīj* may have arisen as a literal translation of the Sanskrit term *tantra* (from *tan* 'to stretch') literally 'warp, loom', but which is used in the sense of 'system' or 'text book'. The word is singled out by Varāhamihira (ca. 580) as the name of the branch of astronomy which is concerned with planetary calculations. In a Middle Persian tract of the ninth century, the *Epistles of Manuščihr*, there are references to the *zīg ī hindūg* (Indian Astronomy), and also to a *zīg ī šahriyārān* (Royal Astronomy), a Sasanid compilation probably of the sixth century, referred to later by Arabic authors as the *Zīj al-Shātroyārān*. The former may be one of the works of Āryabhaṭa (ca. AD 520), which were referred to as Tantras by an early commentator.

RAYMOND MERCIER

REFERENCES

Kennedy, E.S. "A Survey of Islamic Astronomical Tables." *Transactions of the American Philosophical Society* 46: 123–176, 1956.

Kennedy, E.S. "The Sasanian Astronomical Handbook Zīj-i Shāh and the Astrological Doctrine of 'Transit' (mamarr)." *Journal of the American Oriental Society* 78: 246–262, 1958.

Mercier, R.P. "Astronomical Tables in the Twelfth Century." In *Adelard of Bath, an English Scientist and Arabist of the Early Twelfth Century.* Ed. Charles Burnett. London: The Warburg Institute, 1987, pp. 87–118.

Nallino, C.A. *Al-Battānī sive Albatenii Opus Astronomicum.* 3 vols, Milan: Reale Observatorio, 1899–1907.

See also: Astronomy in Islam

ZODIAC IN INDIA The signs of the zodiac originated in Mesopotamia. In the first stage of development, twelve constellations along the ecliptic (i.e. the apparent course of the sun in the sky) were roughly marked out and each was named after the animal whose shape it resembled. Later, with the need for a rigid coordinate system for planetary positions, the zodiacal sign assumed a new meaning: the length of 30 degrees along the ecliptic, so that twelve equal signs comprised a complete circuit (360 degrees) of the ecliptic. The change from the older irregular constellations to the signs of regular spacing took place somewhere around 500 BC.

In the cuneiform texts the ecliptic coordinates were sidereally fixed, and the vernal equinox was several degrees off "the first point of Aries": at the tenth degree of Aries in System A and at the eighth in System B. The Mesopotamian idea of twelve zodiacal signs of equal length was transmitted to Greece about 300 BC, where the iconography of the signs was modified by their mythology.

With the discovery of the precession of equinoxes by Hipparchus in about 150 BC, the significance of the zodiacal signs changed drastically. The first point of Aries was equated with the vernal equinox. Since this is in constant retrograde motion relative to the fixed stars (the shift being about 51 minutes of arc per year), the original relation between the constellations and signs was completely severed and the zodiacal signs became a purely mathematical reference system.

With the development of astrology, which preserved the old association of zoomorphic shape with zodiacal signs (except Libra), the zodiacal signs assumed new meanings. They were classified in various ways: by sex, the ownership of the house of planets, seasons, tastes, four humors, four elements, the governorship over the parts of body, plants, animals, geographical regions, etc.

All these ideas were transmitted to India in the second century of the Christian era. The very Sanskrit names of the zodiacal signs show that they were translated from Greek. In some texts even phonetic translations of Greek words are found. The earliest Sanskrit text which contains a list

Table 1: The signs of the zodiac

Degrees	English	Sanskrit
0	Aries	meṣa
30	Taurus	vṛṣan
60	Gemini	mithuna
90	Cancer	karkaṭa
120	Leo	siṃha
150	Virgo	kanyā
180	Libra	tulā
210	Scorpio	vṛścika
240	Sagittarius	dhanus or dhanvin
270	Capricorn	makara or mṛga
300	Aquarius	kumbha
330	Pisces	mīna

of these names is the *Yavanajātaka* (ca. AD 269), a Sanskrit version of a Greek book on horoscopic astrology.

Three iconographic modifications in the process of transmission are worth mentioning. (1) While Gemini are the twin boys in Western iconography, *mithuna* in Sanskrit is a couple consisting of a male and a female, and this was interpreted as "husband and wife" in Chinese texts on Buddhist astrology. (2) The Sanskrit word *makara* stands for a kind of sea monster, and *mṛga* for a "forest animal" such as a deer. Thus Capricorn was divided into two separate animals. (3) The word *dhanvin* (one who has a bow) is a better translation of Sagittarius (archer), but the simpler *dhanus* (bow) without a human figure is more frequently used in Sanskrit texts.

In spite of the similarity of the names, the astronomical meaning of Indian zodiacal signs is different from that of the Western ones, because the precession of the equinoxes was not taken into account in India, and the first point (*meṣādi*) of the ecliptic coordinates was sidereally fixed some time in the third or fourth century AD.The difference (*ayanāṃśa*) between the vernal equinox and the *meṣādi*, which has accumulated in the present day (1994), is about 23°40′; thus the sun's entry into *meṣa* now falls on the 14th of April, and the *makarasaṃkrānti*, originally a winter solstice festival, on the 15th of January.

This seemingly conservative attitude is closely related to the Indian system of naming the lunar month. A year is divided into twelve solar months by the sun's entry (*saṃkrānti*) into a new zodiacal sign. The lunar month is named after the *saṃkrānti* which falls during that month. For example, the lunar month Caitra is defined as the month during which the sun's entry into *meṣa* occurs. The full moon of that month has to be located near the diametrically opposite point on the ecliptic, that is, at the lunar mansion *citrā*. Thus, in order to keep the relation of the month name and the constellation name, they had to stick to the sidereal (*nirayana*) system even at the sacrifice of the correspondence between the seasons and month names.

The Western system has ignored the original association between constellations and zodiacal signs, and the word Aries, for example, has two meanings, one as an actual constellation and the other as the first thirty degrees in the ecliptic longitude. The former is used in astronomy and the latter in astrology. This is not the case in the traditional Indian system.

In south India and Nepal the solar month is still used in the civil calendar. Since a solar month is the time during which the true sun stays in one zodiacal sign, the length of a month varies from 28 to 32 days.

As mentioned above, Indian zodiacal signs were transmitted to China in the eighth century by Buddhist astrology and ultimately to Japan in the ninth century. The iconography of the Indian zodiacal signs was preserved in the star *maṇḍalas* (especially in the temples belonging to the Shingon sect) which were used in the ritual of worshipping the planetary deities.

YANO MICHIO

REFERENCES

Neugebauer, Otto. *A History of Ancient Mathematical Astronomy*, 3 vols. New York: Springer, 1975.

Pingree, David. *Yavanajātaka of Sphujidhvaja*, 2 vols. Cambridge, Massachusetts: Harvard University Press, 1976–1978.

See also: Precession of the Equinoxes – Lunar Mansions – Astrology in India

ZODIAC IN ISLAMIC ASTRONOMY The zodiac, i.e. the band or zone around the sky through which the sun, the moon and the planets travel in their apparent revolutions, was first established by the Babylonians. They formed the constellations along this zone and instituted its division into twelve equal portions, the "signs". This knowledge was then handed on, in a northern branch of transmission, to the Greeks and through them further on into modern astronomy. In a southern branch of transmission, some of the zodiacal constellations reached the Arabs in the Arabian peninsula. In their folk astronomy they knew some of the zodiacal constellations — though not the complete system of twelve — which, in their astronomical lore, were sometimes located differently. For example the constellation *al-jawzā'* (standing for Gemini) is located in the Greek (and modern) Orion, *al-dalw* (Aquarius) in Pegasus, and *al-ḥūt* (Pisces) in Andromeda.

With the reception of Greek astronomy, through the translation of the most important Greek writings in astronomy and astrology, the system of the twelve zodiacal constellations and the twelve signs was also received and continued to be used by Arabic-Islamic astronomers and astrologers. For some constellations the translators introduced new names (derived from Greek) besides the names inherited from the old Arabs. The names for the twelve constellations of the zodiac in Arabic are:

1. Aries: *al-ḥamal*;
2. Taurus: *al-thawr*;
3. Gemini: *al-jawzāʾ* (old), *al-tawʾamān* (transl.);
4. Cancer: *al-saraṭān*;
5. Leo: *al-asad* (both old and transl.);
6. Virgo: *al-sunbula* (old), *al-ʿadhrāʾ* (transl.);
7. Libra: *al-mīzān*;
8. Scorpio: *al-ʿaqrab* (both old and transl.);
9. Sagittarius: *al-qaws* (old), *al-rāmī* (transl.);
10. Capricornus: *al-jady*;
11. Aquarius: *al-dalw* (old), *sākib al-māʾ* (transl.);
12. Pisces: *al-ḥūt* (old), *al-samakatān* (transl.).

The zodiac was treated by the Arabic-Islamic astronomers and astrologers according to the teachings of the Greeks.

When Arabic astronomical and astrological works were translated into Latin in Europe in the Middle Ages, the Arabic names of the twelve zodiacal constellations were also sometimes borrowed. But they did not gain the same popularity as the Arabic star names and the names of the lunar mansions; most Latin authors used the well-known Latin names for them.

PAUL KUNITZSCH

REFERENCES

Kunitzsch, P. and W. Hartner. "Minṭaḳat al-Burūdj." In *Encyclopaedia of Islam*, new edition. Leiden: Brill, vol. 7, 1993, pp. 81–87.

ZOU YAN Our only source on Zou's life that has any chance of being reliable is the entry in Sima Qian's universal history *Shiji*, completed ca. 90 BC. According to this, Zou (fl. 250 BC) was a successful member of the group of wandering scholars who moved from one feudal court to another in China's Warring States period (fifth to third centuries BC), offering expertise in statecraft and useful arts. He came from the northeastern seaboard state of Qi, where he became a member of the "academy" set up by the ruler of that state to house his more favored scholarly retainers. Although he is said to have written several voluminous books none of them has survived.

According to the *Shiji*, Zou "looked into the rise and fall of the Yin and Yang" and wrote on their "strange transformations". His method of reasoning "started by checking some small thing, and then extrapolated to a large scale". On this basis he is said to have reasoned his way back to the state of affairs "before the origin of heaven and earth" on the basis of his knowledge of more recent times. Similarly, he claimed that China only occupied one-ninth of a large continent, which was itself only one out of nine giant land masses surrounded by a vast ocean.

One source of the seventh century AD (Li Shan's commentary on the Wen Xuan anthology) quotes a work said to be by Zou Yan, in which it is stated that he taught that the successive dynasties ruling in China had each come to power by virtue of the cyclical dominance in the cosmos of one of the *wude* (Five Powers) of earth, wood, metal, fire, and water. (These are more commonly called the *wuxing*, Five Phases.) While such a view is not inconsistent with Zou Yan's thought as known from the *Shiji*, it is by no means certain that he originated it. Whatever its origins, this theory of the cosmological determination of political dominance was adopted by the first emperor of the Qin dynasty when he came to the throne in 221 BC, and remained important for several centuries thereafter.

CHRISTOPHER CULLEN

REFERENCES

Needham, Joseph. *Science and Civilisation in China*, vol. 2. Cambridge: Cambridge University Press, 1954, pp 232–244.

ZU CHONGZHI Zu Chongzhi, whose literary name was Wenyuan, was born in Jiankang (present Nanjing of Jiangsu Province) in AD 429 into a family of bureaucrats. His great grandfather, a native of Hebei Province, was an official during the Eastern Jin dynasty (AD 317–420), and his grandfather and father were officials of the Northern and Southern dynasties (AD 479–581). Interestingly enough, the Zu family had for successive generations been involved in studies of astronomy and calendrical science. Subsequently, they moved south and settled in Jiankang, the political and economic center of the fifth and sixth centuries. From a very young age, Zu Chongzhi received instruction in a variety of subjects. He was particularly interested in mathematical astronomy. While examining many past and existing astronomical systems, he found many errors and discrepancies. This prompted him to compile a calendrical system known as the Daming Calendar (Calendar of Great Brightness). However, when he presented this calendar to the throne for promulgation in AD 462, he met with vehement opposition from the emperor's minister Dai Faxin. Zu Chongzhi's innovation and improvement in the astronomical system was branded as "distorting the truth about heaven and violating the teaching of the classics". The Daming calendar was shelved for about half a century before it was finally promulgated for official use in AD 510 when Zu Chongzhi had been dead for ten years.

The Daming calendar was the epitome of Zu Chongzhi's mathematical and astronomical skills. He was the first person to take into account the fact of precession of the equinoxes discovered by Yu Xi, an astronomer in the fourth century AD. By applying the precession of equinoxes, Zu Chongzhi differentiated the tropical year from the sidereal year (the time at which the sun's center, departing eastward from the ecliptic meridian of a given star, returns to that meridian). As a matter of fact, he determined several astronomical constants with remarkable accuracy. For example, he gave a value of 365.24281481 days for the tropical year, only about 50 seconds off the modern value. He also gave for the first time a value of 27.21233 days for the length of a nodal month, the modern value being 27.21222 days. For planetary motion, he found that the planet Jupiter completes seven and one-twelfth circuits of the heavens in every seven cycles of 12 years. This gives Jupiter a sidereal period of 11.859 years, which differs from the modern value by only 1 part in 4000.

In order to make the lunar year tally with the solar year, ancient Chinese calendar-makers inserted seven intercalary months every 19 years. This rule of intercalation had been in use for more than a thousand years until AD 412 when Zhao Fei changed it to 221 intercalary months in every 600 years. Zu Chongzhi did not find the rule satisfactory and improved it by introducing 144 intercalations in every 391 years.

The most significant contribution by Zu Chongzhi is in the evaluation of π. He gave two approximating ratios for π: a coarse one of $\frac{22}{7}$ and a fine one of $\frac{355}{113}$. The true ratio, according to him, lay between 3.1415926 and 3.1415927.

Zu Chongzhi also wrote a mathematical text known by the title *Zhui Shu*. This text, prescribed as a textbook for advanced students of mathematics in the official examinations of the Tang dynasty, unfortunately was lost in the wars of the early twelfth century. Only fragments of the text still exist in the Calendrical and Astronomical Chapters of the Sui and Tang dynasties. Zu Chongzhi was said to have worked out an evaluation of the volume of a sphere. He considered a sphere as a pile of circles of varying sizes and arrived at a formula, which in modern mathematical language, could be expressed as $V = \frac{4}{3}\pi r^3$.

Zu Chongzhi was a versatile person. Apart from being a celebrated astronomer and mathematician, he was well-versed in engineering. He produced a "south-pointing vehicle" worked by a set of five differential cog-wheels, a "thousand-mile boat" propelled by paddle wheels, a hydraulic mortar-mill, worked by a combination of separate water-driven mortars and mills, and a brass ruler for measuring the pitch of musical instruments.

Zu Chongzhi's remarkable achievements are acknowledged internationally. A crater on the reverse side of the moon, just south of the Sea of Moscow, is named after him.

ANG TIAN SE

REFERENCES

Du, Shiran. "Zhu Chongzhi." *Zhongguo Gudai Kexuejia Zhuanji* (Biographies of Ancient Chinese Scientists). Ed. Du Shiran. Beijing: Kexue chubanshe, 1992, pp. 221–234.

Li, Di. *Da Kexuejia Zhu Chongzhi* (Zhu Chongzhi, the Great Scientist). Shanghai: Renmin chubanshe, 1962.

Ruan, Yuan (AD 1799). *Chouren Zhuan* (Biographies of Mathematicians and Astronomers), vol. 1. Shanghai: Shangwu Yinshuguan, 1955.

Shen, Yo (AD 500). *Song Shu* (Standard History of the (Liu) Song Dynasty), Ch.13. Beijing: Zhonghua Shuju, 1972.

See also: Precession of the Equinoxes – *Pi* in Chinese Mathematics

LIST OF AUTHORS

George Abraham Annanagar, Madras, India
Gnomon in India

Bala Achi National Museum, Abuja, Nigeria
Construction Techniques in Africa

Michael Adas Department of History, Rutgers University, New Brunswick, New Jersey, U.S.A.
Colonialism and Science
Western Dominance: Western Science and Technology in the Construction of Ideologies of Colonial Dominance

Samuel A.M. Adshead Department of History, University of Canterbury, Christchurch, New Zealand
Salt

Bilal Ahmad Department of Geography, University of Iowa, Iowa City, Iowa, U.S.A.
Ibn Baṭṭūṭa

S.M. Ahmad Al Al-Bayt University, Jubayha, Amman, Jordan
Ibn Khurdādhbih
Ibn Mājid
al-Idrīsī

Mansour Solyman al-Said Department of Pharmacognosy, College of Pharmacy, King Saud University, Riyadh, Saudi Arabia
Medicine in Islam

Claude Alvares The Other India Bookstore, Goa, India
Irrigation in India and Sri Lanka
Technology and Culture

Mehdi Aminrazavi Department of Philosophy, Mary Washington College, Fredericksburg, Virginia, U.S.A.
Ibn Sīnā (Avicenna)

Gene Ammarell Department of Sociology and Anthropology, Ohio University, Athens, Ohio, U.S.A.
Astronomy in the Indo-Malay Archipelago

Munawar A. Anees Periodica Islamica, Kuala Lumpur, Malaysia
Ibn Ẓuhr

Tian Se Ang Department of Language Studies, Faculty of Arts, Edith Cowan University, Mount Lawley, Western Australia
Acupuncture
Five Phases (*Wuxing*)
Sun Zi
Wang Xiaotong
Xiahou Yang
Xu Yue
Zu Chongzhi

Charles Anyinam Department of Geography, University of Toronto, Toronto, Ontario, Canada
Medicine in Africa

Roger Arnaldez Paris, France
Ibn Buṭlān

R.K. Arora International Board for Plant Genetic Resources, New Delhi, India
Agriculture in India

Bernardo Arriaza Department of Anthropology and Ethnic Studies, University of Nevada, Las Vegas, Nevada, U.S.A.
Mummies in South America

Marcia Ascher Ithaca College, Ithaca, New York, U.S.A.
Ethnomathematics
Quipu

Robert Ascher Cornell University, Ithaca, New York, U.S.A.
Quipu

Kamel Arifin Mohd Atan Department of Mathematics, University of Agriculture Malaysia, Selangor, Malaysia
Algebra in Islamic Mathematics

George Atiyeh Library of Congress, Near East Section, Washington, DC, U.S.A.
al-Kindī

A.V. Balasubramanian Centre for Indian Knowledge Systems, Madras, India
Knowledge Systems in India
Metallurgy in India
Yoga

Thomas J. Bassett Department of Geography, University of Illinois at Urbana-Champaign, Champaign, Illinois, U.S.A.
Maps and Mapmaking in Africa

Frances Berdan Department of Anthropology, California State University, San Bernardino, California, U.S.A.
Aztec Science

Kenneth J.E. Berger San Diego, California, U.S.A.
Environment and Nature: China

Bruce Berndt Department of Mathematics, University of Illinois, Urbana, Illinois, U.S.A.
Ramanujan

Richard Bertschinger Montacute, Somerset, England
Huangdi Neijing
Medical Texts in China
Nanjing
Qianjin Yaofang
Shanghan Lun
Tongren Zhenjiu Shuxue Tujing
Zhenjiu Dacheng
Zhenjiu jiayijing

Lucie Bolens Faculty of Letters, Department of General History, University of Geneva, Switzerland
Agriculture in the Islamic World
Irrigation in the Islamic World

Annie L. Booth Faculty of Natural Resources and Environmental Studies, University of Northern British Columbia, Prince George, British Columbia, Canada
Environment and Nature: Native North America

Ahmed Bouzid Center for the Study of Science in Society, Virginia Polytechnic Institute, Blacksburg, Virginia, U.S.A.
al-Farghānī

Francesca Bray Manchester University, Manchester, England
Agriculture in China

Sonja Brentjes Karl-Sudhoff-Institut, Universität Leipzig, Leipzig, Germany; and Institute for Advanced Study, Princeton, New Jersey, U.S.A.
Elements–Reception of Euclid's *Elements* in the Islamic World
al-Jawharī
al-Nayrīzī

David Browman Department of Anthropology, Washington University, St. Louis, Missouri, U.S.A.
Environment and Nature: the Andes

David M. Browne Royal Commission on Ancient and Historical Monuments in Wales, Aberystwyth, Wales
Nazca Lines

Viggo Brun East Asian Institute, University of Copenhagen, Denmark
Medicine in Thailand

Amriah Buang Universiti Kebangsaan Malaysia, Selangor, Malaysia
Geography in the Islamic World

Anthony R. Butler Department of Chemistry, University of St. Andrews, Fife, Scotland
Chemistry in China

Charles Butterworth Department of Government and Politics, University of Maryland, College Park, Maryland, U.S.A.
Ibn Khaldūn

Cai Jingfeng China Institute for the History of Medicine and Medical Literature, China Academy of Traditional Chinese Medicine, Beijing, China
Chinese Medicine
Medicine in China: Forensic Medicine

Emilia Calvo Department of Arabic Philology, University of Barcelona, Spain
Abū'l-Fidā᾽
al-Damīrī
Ibn al-Bannā᾽

Ibn al-Bayṭār
Ibn al-Zarqāllu
Ibn Ḥawqal
Ibn Juljul
Ibn Tibbon
Ibn Wāfid
al-Majrīṭī
Yaḥyā ibn Abī Manṣūr

Vicki Cassman Department of Anthropology and Ethnic Studies, University of Nevada, Las Vegas, Nevada, U.S.A.
Mummies in South America

José Chabás Brussels, Belgium
Zacut, Abraham

A.K. Chakravarty Department of Mathematics, Mahishadal Raj College, Midnapore, West Bengal, India
Calendars in India

Sami Chalhoub Institute for the History of Arabic Science, University of Aleppo, Syria
Trigonometry in Islamic Mathematics

D.P. Chattopadhyaya Project of History of Indian Science, Philosophy and Culture, Calcutta, India
Environment and Nature: Indian

Karine Chemla Centre National de Recherches Scientifiques (CNRS), Histoire des Sciences, Paris, France; and Wissenschaftskolleg, Berlin, Germany
Siyuan Yujian
Suanxue Qimeng
Zhu Shijie

Joseph Chen Department of Physics, University of California, San Diego, La Jolla, California, U.S.A.
Acoustics in Chinese Culture
Magnetism in Chinese Culture

Weihang Chen Hampshire College, Amherst, Massachusetts, U.S.A.
Yinyang
Religion and Science in China

S. Terry Childs Conservation Analytical Laboratory, Smithsonian Institution, Washington, DC, U.S.A.
Metallurgy in Africa

William C. Chittick Department of Comparative Studies, Program in Religious Studies, State University of New York at Stony Brook, Stony Brook, New York, U.S.A.
Ibn al-ʿArabī

W.C. Clarke Eumundi, Queensland, Australia
Agriculture in the Pacific
Agroforestry in the Pacific

Michael P. Closs University of Ottawa, Department of Mathematics, Ottawa, Ontario, Canada
Mathematics in Native North America
Mathematics of the Aztecs
Mathematics of the Maya

Christopher Cullen Needham Research Institute, Cambridge, England; and Department of History, School of Oriental and African Studes, London, England
Clocks: Astronomical Clocks in China
Gaitian
Huntian
Zhoubi Suanjing
Zou Yan

Roger B. Culver Colorado State University, Fort Collins, Colorado, U.S.A.
Astronomy

Ahmad Dallal
Ṣadr al-Sharīʿah

Ubiratan D'Ambrosio University of Campinas, Cidade Universitaria 'Zeferina Vaz', Campinas, Brazil
Colonialism and Science in the Americas
Americas: Native American Science

Bhagwan Dash Āyurveda Rasashala Clinic, Delhi, India
Alchemy in India
Ātreya
Caraka
Medicine in India: *Āyurveda*
Suśruta

Joseph W. Dauben City University of New York, Herbert H. Lehman College, Bronx, New York, U.S.A.
The *Gougu* Theorem

William Davenport University of Pennsylvania Museum, Philadelphia, Pennsylvania, U.S.A.
Maps and Mapmaking: Marshall Island Stick Charts

Jean Debernardi Department of Anthropology, University of Alberta, Edmonton, Alberta, Canada
Time in Chinese Thought

Catherine Delano Smith Institute of Historical Research, University of London, London, England
Maps and Mapmaking in Asia (Prehistoric)

P.N. Desai University of Illinois College of Medicine, Chicago, Illinois, U.S.A.
Medical Ethics in India

Gregg DeYoung Science Department, American University in Cairo, Egypt
Astronomy in Egypt
Calendars in Egypt
al-Ḥajjāj
Ikhwān al-Ṣafā'
Isḥāq ibn Ḥunayn
Maragha
Metallurgy in Egypt
Mummies in Egypt
Observatories in the Islamic World
Pyramids
Qāḍī Zādeh al-Rūmī
al-Samarqandī
Textiles in Egypt

Alnoor Dhanani Lexington, Massachusetts, U.S.A.
Atomism in Islamic Thought

Ahmed Djebbar Garches, France
Combinatorics in Islamic Mathematics
Ibn al-Yāsamīn
Mathematics of Africa: Maghreb
al-Qalaṣādī

Yvonne Dold-Samplonius Institute of Mathematics, University of Heidelberg, Germany
Abū'l-Wafā'
al-Mahānī
al-Qūhī
al-Sijzī
Sinān ibn Thābit

Salimata Doumbia Institut de Recherches Mathématiques, Abidjan, Côte d'Ivoire
Mathematics in Africa: West African Games

Laurence R. Doyle SETI Institute, NASA Ames Research Center, Moffett Field, California, U.S.A.
Astronomy in Africa

Ashok Dutt Department of Geography, University of Akron, Akron, Ohio, U.S.A.
Geography in India

Gloria T. Emeagwali History Department, Central Connecticut State University, New Britain, Connecticut, U.S.A.
Colonialism and Science in Africa
Textiles in Africa

J. Worth Estes Department of Pharmacology and Experimental Therapeutics, Boston University School of Medicine, Boston, Massachusetts, U.S.A.
Medicine in Egypt

Nina Etkin Anthropology Department, University of Hawaii, Honolulu, Hawaii, U.S.A.
Medicinal Food Plants

Gillian R. Evans History Department, Cambridge University, Cambridge, England
Abacus

Harmut Fähndrich Eidgenössische Technische Hochschule, Swiss Federal Institute of Technology, Bern, Switzerland
Ibn Jumayʾ

Richard Feinberg Kent State University, Kent, Ohio, U.S.A.
Navigation in Polynesia

Feng Li-sheng Institute for the History of Science, Inner Mongolia Normal University, Huhehot, Inner Mongolia Autonomous Region, China
Chinese Minorities

Edward W. Frank Santa Monica, California, U.S.A.
Astronomy in Africa

Donald and Joan Gear Cato Ridge, South Africa
Weights and Measure in Burma: The Royal Animal-shaped Weights of the Burmese Empires

Paulus Gerdes Universidade Pedagogica, Maputo, Mozambique
Geometry in Africa: Sona Geometry
Mathematics in Africa South of the Sahara
Number Theory in Africa

Thomas F. Glick Department of History, Boston University, Boston, Massachusetts, U.S.A.
Leo the African

Christopher Glidewell Department of Chemistry, University of St. Andrews, Fife, Scotland
Chemistry in China

Nicholas J. Goetzfridt Madison, Wisonsin, U.S.A.
Navigation in the Pacific Islands

S.D. Gomkale Central Salt and Marine Chemicals Research Institute, Bhavnagar, Gujarat, India
Salt in India

Susantha Goonatilake
East and West: India in the Transmission of Knowledge East to West

Gray Graffam Department of Anthropology, Trent University, Peterborough, Ontario, Canada; and Department of Anthropology, University of Toronto, Toronto, Ontario, Canada;
Irrigation in South America
Metallurgy in the Andes

Paul Gregory Centre for Urban and Community Studies, University of Toronto, Toronto, Ontario, Canada
City Planning in India

Y. Guergour Département de Mathématiques, École Normale Supérieure, Kouba, Algeria
Ibn Qunfudh

Guo Shirong Institute for the History of Science, Inner Mongolia Normal University, Huhehot, Inner Mongolia Autonomous Region, China
Chinese Minorities

R.C. Gupta Ganita Bharati Academy, Ras Bahar Colony, Jhansi, India
Āryabhaṭa
Baudhāyana
Bhāskara II
Brahmagupta
Mādhava
Mahāvīra
Pi in Indian mathematics
Śrīdhara

Sami Khalaf Hamarneh ISTAC, P.O. Box 11961, 50672 Kuala Lumpur, Malaysia
Ibn al-Majūsī
Ibn al-Quff (al-Karakī)
al-Ṭabarī

S. Nomanul Haq Department of Religious Studies, Brown University, Providence, Rhode Island, U.S.A.
Jābir ibn Ḥayyān

Hairuddin Harun Faculty of Science, University of Malaya, Kuala Lumpur, Malaysia
Colonialism and Medicine in Malaysia
Colonialism and Science in the Malay World

E. Ruth Harvey Centre for Medieval Studies, University of Toronto, Toronto, Ontario, Canada
Ibn Hubal
Ibn Sarabi (Serapion)
Ibn Ṭufayl
al-Jurjānī
Qusṭā ibn Lūqā
Samū'īl ibn 'Abbas al-Maghribī

Takao Hayashi Science and Engineering Research Institute, Doshisha University, Kyoto, Japan
Algebra in India: *Bījagaṇita*
Arithmetic in India: *Pāṭīgaṇita*
Bakhshālī Manuscript
Combinatorics in Indian Mathematics
Number Theory in India

Roslynn D. Haynes English Department, University of New South Wales, Kensington, New South Wales, Australia
Astronomy of the Australian Aboriginal People

Anton M. Heinen München, Germany
Religion and Science in Islam II: What Scientists Said About Religion And What Islam Said About Science
al-Suyūṭī

Martha L. Henderson The Evergreen State University, Olympia, Washington, U.S.A.
Geography of Native North Americans

Donald R. Hill Deceased
Banū Mūsa
Clocks and Watches
al-Jazarī
Technology in the Islamic World

Richard L. Hills Institute of Science and Technology, University of Manchester, Manchester, England
Windpower

Ho Peng Yoke Needham Research Institute, Cambridge, England
Astrology in China
Astronomy in China
China
Ge Hong
Gunpowder
Guo Shoujing
Li Zhi
Liu Hui and the *Jiuzhang Suanshu*
Magic Squares in China
Navigation in China
Yang Hui

J.C. Hocquet Villeneuve d'Ascq, France
Weights and Measures in Mexico

Jan P. Hogendijk Institute of Mathematics, State University of Utrecht, Utrecht, The Netherlands
Abū Ja'far
Conics
Mathematics in Islam
al-Mu'taman ibn Hūd
Sharaf al-Dīn al-Ṭūsī
Ya'qūb ibn Ṭāriq

Lionel Holland Hadera, Israel
Weights and Measures of the Hebrews

Hong Wuli China Institute for the History of Medicine and Medical Literature, China Academy of Traditional Chinese Medicine, Beijing, China
Chen Yan
Huangfu Mi
Li Gao
Song Ci
Tang Shenwei

Jens Høyrup Roskilde University Centre, Roskilde, Denmark

Algebra: Surveyors' Algebra
Geometry in the Near and Middle East
Mathematics, Practical And Recreational

Mei-ling Hsu Department of Geography, University of Minnesota, Minneapolis, Minnesota, U.S.A.
Geography in China

Hua Jueming Institute for the History of Natural Science, Chinese Academy of Science, Beijing, China
Metallurgy in China

Huang, H. T. Needham Research Institute, Cambridge, England
Food Technology in China

Huang Yi-long Institute of History, National Tsinghua University, Hsinchu, Taiwan
Gan De

Nancy Hudson-Rodd Department of Community Health Nursing, Edith Cowan University, Churchlands, Western Australia
Geographical Knowledge

Åke Hultkrantz Department of Comparative Religion, University of Stockholm, Lidingö, Sweden
Medicine in Native North America
Religion and Science in Native North America

Ekmeleddin İhsanoğlu Research Centre for Islamic History, Art and Culture, Istanbul, Turkey
Ottoman Science

Mohammad Ilyas Astronomy and Atmospheric Research Unit, University of Science Malaysia, Penang, Malaysia
Calendars in Islam
Qibla and Islamic Prayer Times

Albert Z. Iskandar Wellcome Institute, London, England
Ḥunayn Ibn Isḥāq
Ibn al-Nafīs
Ibn Riḍwān
Ibn Rushd (Averroes)
al-Rāzī

Mat Rofa bin Ismail Department of Mathematics, University of Agriculture Malaysia, Selangor, Malaysia
Algebra in Islamic Mathematics

Shigeo Iwata Tokyo, Japan
Weights and Measures in the Indus Valley
Weights and Measures in Japan
Weights and Measures in Peru

Danielle Jacquart Sciences Historiques et Philologiques, École Pratique des Hautes Études, Paris, France
Ibn Māsawayh Yūḥannā

S.K. Jain National Botanical Research Institute, Lucknow, Uttar Pradesh, India
Ethnobotany in India

Jennifer Jay Department of History and Classics, University of Alberta, Edmonton, Alberta, Canada
Li Chunfeng

Jeon Sang-woon Sungshin Women's University, Seoul, South Korea
Korean Science

Jiang Xiaoyuan Shanghai Observatory, Chinese Academy of Sciences, Shanghai, China
Armillary Spheres in China
Liu Hong
Luoxia Hong
Zhang Heng

Shigeru Jochi Japanese Center, Ming Chuan College, Taipei, Taiwan
Aida Yasuaki
Ajima Naonobu
Hazama Shigetomi
Seki Kowa
Takebe Katahiro

Timothy Johns Centre for Nutrition and the Environment of Indigenous Peoples, Macdonald Campus of McGill University, Ste. Anne de Bellevue, Quebec, Canada
Potato

William T. Johnson University Libraries, Texas Tech University, Lubbock, Texas, U.S.A.
Mathematics in the Pacific
Physics in India

William D. Johnston Department of History, Wesleyan University, Middletown, Connecticut, U.S.A.
Medicine in Japan

Karen Louise Jolly Department of History, University of Hawaii at Mãnoa, Honolulu, Hawaii, U.S.A.
Magic and Science

Carolyn Jones U Mut Maya, Bayside, California, U.S.A.
Textiles in Mesoamerica

Tom Jones College of Arts and Humanities, Humboldt State University, Arcata, California, U.S.A.
Calendars in Mesoamerica
Textiles in Mesoamerica
Writing of the Maya

Gretchen W. Jordan Department of Anthropology, University of Colorado at Denver, Denver, Colorado, U.S.A.
Roads in Native North America

George Gheverghese Joseph Faculty of Econometrics and Social Statistics, University of Manchester, Manchester, England
Mathematics
Mathematics in India
Geometry in India

Carol F. Justus Linguistics Research Center, University of Texas, Austin, Texas, U.S.A.
Cuneiform

Ted Kaptchuk Boston, Massachusetts, U.S.A.
Qi

Ahmet T. Karamustafa Department of Asian and Near Eastern Languages and Literature, Washington University, St. Louis, Missouri, U.S.A.
Maps and Mapmaking: Islamic Terrestrial Maps

Margarita Kay Tucson, Arizona, U.S.A.
Childbirth

Jeanne Kay Faculty of Environmental Studies, University of Waterloo, Waterloo, Ontario, Canada
Environment and Nature: the Hebrew People

M.S. Khan Calcutta, India
India: Medieval Science and Technology

Elaheh Kheirandish Department of History of Science, Harvard University, Cambridge, Massachusetts, U.S.A.
Optics in Islamic Science

Kim Yong Woon Department of Mathematics, Hanyang University, Seoul, Korea
Mathematics in Korea

David A. King Institute for History of Science, Johann Wolfgang Goethe University, Frankfurt am Main, Germany
Astronomy in the Islamic World
Astronomical Instruments in the Islamic World
Ibn Yūnus
al-Khalīlī
Maps and Mapmaking: Islamic, Mecca-centered Maps
al-Māridīnī, Jamāl al-Dīn and Badr al-Dīn
Religion and Science in Islam I: Technical and Practical Aspects

Steven Klarer Boston, Massachusetts, U.S.A.
Qi

Christoph Koerbs Berlin, Germany
East and West: China in the Transmission of Knowledge East to West

Samuel S. Kottek Faculty of Medicine, Hebrew University of Jerusalem, Jerusalem, Israel
Medicine in the Talmud
Abraham Ibn Ezra

E.C. Krupp Griffith Observatory, Los Angeles, California, U.S.A.
Astronomy in Native North America

Dieter Kuhn Institute of Sinology, University of Würzberg, Germany
Silk and the Loom
Textiles in China

Deepak Kumar National Institute of Science, Technology and Development Studies, New Delhi, India
Colonialism and Science in India

Paul Kunitzsch Institut für Semitistik, Universität München, München, Germany
Almagest: Its Reception and Transmission in the Islamic World
Ibn Qutayba
Lunar Mansions in Islamic Astronomy
al-Maʾmūn
Stars in Arabic-Islamic Science
al-Ṣūfī
Zodiac in Islam

Shigehisa Kuriyama International Research Center for Japanese Studies, Kyoto, Japan
Moxibustion

Takanori Kusuba Osaka University of Economics, Osaka City, Japan
Nārāyaṇa Paṇḍita

Chi-Wan Lai Neurology Department, University of Kansas Medical Center, Kansas City, Kansas, U.S.A.
Epilepsy in Chinese Medicine

Yen-Huei C. Lai Neurology Department, University of Kansas Medical Center, Kansas City, Kansas, U.S.A.
Epilepsy in Chinese Medicine

Lam Lay Yong Department of Mathematics, National University of Singapore, Kent Ridge, Singapore
Algebra in China
Computation: Chinese Counting Rods
Pi in Chinese Mathematics
Zhang Qiujian Suanjing

Y. Tzvi Langermann Institute of Microfilmed Hebrew Manuscripts, Jewish National and University Library, Jerusalem, Israel
Abū'l-Barakāt
Astronomy of the Hebrew People
Moses Maimonides

Gari Ledyard Department of East Asian Languages and Cultures, Columbia University, New York, New York, U.S.A.
Maps and Mapmaking in Korea

Richard Lemay New York, New York, U.S.A.
Abū Maʿshar
Astrology in Islam
Māʾshāʾallāh

Angela Ki Che Leung ISSP Sun Yat-Sen Institute for Social Sciences and Philosophy, Academia Sinica, Nankang, Taipei, Taiwan
Medical Ethics in China

Tony Lévy CNRS (Centre National de la Recherche Scientifique), Centre d'Histoire des Sciences et des Philosophies Arabes et Mediévales, Vanves, France
Bar Ḥiyya
Ḥasdai Crescas
Mathematics of the Hebrews

G. Malcolm Lewis Sheffield, S. Yorkshire, England
Maps and Mapmaking of the Native North Americans

Ch'iao-p'ing Li Department of Chemistry, University of St. Andrews, Fife, Scotland
Chemistry in China

Li Di Institute for the History of Science, Inner Mongolia Normal University, Huhehot, Inner Mongolia Autonomous Region, China
Meteorology in China

Liao Yuqun Institute for the History of Natural Science, Chinese Academy of Science, Beijing, China
Ben Cao Gang Mu
Bian Que

Ulrich Libbrecht Kluisbergen, Belgium
Mathematics in China
Qin Jiushao
Shushu Jiuzhang

Liu Dun Institute for the History of Natural Science, Chinese Academy of Sciences, Beijing, China
Wang Xichan

Richard P. Lorch München, Germany

Jābir ibn Aflaḥ
Quadrant

Lu Jing-Yan Associate Professor of Engineering, Tongji University, Shanghai, China
Tribology in China

Murdo J. Macleod Department of History, University of Florida, Gainesville, Florida, U.S.A.
Crops in Pre-Columbian America
Dyes

Martha J. Macri Department of Native American Studies and Anthropology, University of California Davis, Davis, California, U.S.A.
Astronomy in Mesoamerica

Tim McGrew Department of Philosophy, Western Michigan University, Kalamazoo, Michigan, U.S.A.
Physics in the Islamic World

William J. McPeak Lake Forest, California, U.S.A.
Meteorology in Islam
Military Technology

Vincent H. Malmström Department of Geography, Dartmouth College, Hanover, New Hampshire, U.S.A.
Geography in Mesoamerica
Long Count
Magnetism in Mesoamerica

J.L. Mancha Facultad de Filosofía, Universidad de Sevilla, Sevilla, Spain
Levi Ben Gerson

Bala V. Manyam Department of Neurology, Southern Illinois University School of Medicine, Springfield, Illinois, U.S.A.
Epilepsy in Indian Medicine

S. Parvez Manzoor Department of the Study of Religion and Worldviews, University of Uppsala, Uppsala Sweden
Environment and Nature: Islam

Jean-Claude Martzloff Centre National de la Recherche Scientifique (CNRS), Saint-Denis, France
Approximation Formulae in Chinese Mathematics
Geometry in China
Li Shanlan
Mathematics in Japan

Zeina Matar Eutin, Germany
Ibn ʿAbbād
Ibn Ṭāwūs

Ruben Mendoza Department of Anthropology, California State University at Monterey Bay, Monterey, California, U.S.A.
City Planning: Maya City Planning
Medicine in Meso and South America
Metallurgy in Meso and North America
Road Networks in Ancient Native North America
Trephination

Raymond P. Mercier Needingworth, St. Ives, Cambridgeshire, England
Geodesy
Ibn al-Aʿlam
Zīj

Georges Métailié Centre National de la Recherche Scientifique (CNRS) and Musée National d'Histoire Naturel (MNHN), Paris, France
Ethnobotany in China

Yoshimasa Michiwaki Maebashi City College of Technology, Maebashi, Japan
Geometry in Japan
Magic Squares in Japanese Mathematics

Daniel E. Moerman Behavioral Sciences Department, University of Michigan-Dearborn, Dearborn, Michigan, U.S.A.
Ethnobotany in Native North America

R.N. Mukherjee Bihar, India
Zero

Barbara Mundy Hoboken, New Jersey, U.S.A.
Maps and Mapmaking in Mesoamerica

Nakayama Shigeru School of International Business Administration, Kanagawa University, Kanagawa, Japan
Asada Goryu
Calendars in East Asia

Ino Tadataka
Japanese Science
Shibukawa Harumi
Shizuki Tadao

Azim A. Nanji Department of Religion, University of Florida, Gainesville, Florida, U.S.A.
Medical Ethics in Islam

Georges Niangoran-Bouah Faculté des Lettres et Sciences Humaines, Université d'Abidjan, Côte d'Ivoire
Weights and Measures in Africa: Akan Gold Weights

Connie Nobles Southeastern Louisiana University, Hammond, Louisiana, U.S.A.
Mound Cultures

Yukio Ōhashi Tokyo, Japan
Astronomical Instruments in India
Astronomy in Tibet

Richard Okagbue National University of Science and Technology, Bulawayo, Zimbabwe
Food Technology in Africa

Harold Olofson Graduate School and Area Research and Training Center, University of San Carlos, Cebu City, Philippines
Agroforestry

Arnold Pacey Addingham, Ilkley, West Yorkshire, England
Agroforestry in Africa
Engineering
Gender and Technology
Qanat
Rainwater Harvesting
Technology

Pan Jixing Institute for the History of Natural Science, Chinese Academy of Science, Beijing, China
Paper and Papermaking
Rockets and Rocketry
Song Yingxing and *Tiangong Kaiwu*

Deepa Pande Economics Department, Kumaon University, Naini Tal, India
Forestry in India

Cornelis Plug Department of Psychology, University of South Africa, Pretoria, South Africa
Celestial Vault and Sphere

Malcolm Potts School of Public Health, University of California, Berkeley, California, U.S.A.
Abortion

Marvin A. Powell Department of History, Northern Illinois University, DeKalb, Illinois, U.S.A.
Sexagesimal System

Fabrizio Pregadio Venice, Italy
Alchemy in China
Chao Yuanfang
Huangdi Jiuding Shendan Jing
Li Shizhen
Shen Gua
Sun Simo
Tao Hongjing
Wang Chong
Wei Boyang
Zhang Zhongjing

Jean-Pierre Protzen Department of Architecture, College of Environmental Design, University of California at Berkeley, Berkeley, California, U.S.A.
City Planning: Inca City Planning
Inca Stonemasonry

Edoardo Proverbio Osservatorio Astronomico, Milano, Italia
Gnomon in China

Jehane Ragai Science Department, American University in Cairo, Egypt
Calendars in Egypt
Mummies in Egypt

F. Jamil Ragep Department of the History of Science, University of Oklahoma, Norman, Oklahoma, U.S.A.
al-Hāshimī
Hay'a
Naṣīr al-Dīn al-Ṭūsī

Mushtaqur Rahman Department of Geography, College of Liberal Arts and Sciences, Iowa State University, Ames, Iowa, U.S.A.
al-Masʿūdī
al-Muqaddasī

A.S. Ramanathan Kuppuswami Sastri Research Institute, Madras, India
Meteorology in India

Vijaya Ramaswamy Indian Institute of Advanced Study, Simla, India
Textiles in India

Roshdi Rashed Centre d'Histoire des Sciences et des Philosophies Arabes et Mediévales, C.N.R.S.–E.P.H.E., Paris, France; and University of Tokyo, Tokyo, Japan
Ibn al-Haytham (Alhazen)
Ibn Sahl
Ibrāhīm ibn Sinān
Science as a Western Phenomenon

Ulrich Rebstock Albert-Ludwigs-Universität Freiburg, Orientalisches Seminar, Freiburg, Germany
Weights and Measures in Islam

Paul Richards Group for Technology and Agrarian Development, Agricultural University, Wageningen, The Netherlands
Agriculture in Africa

James Ritter Département des Mathématiques, Université de Paris, Saint Denis, France
Mathematics in Egypt

Lawrence H. Robbins Department of Anthropology, Michigan State University, East Lansing, Michigan, U.S.A.
Namoratunga

Amy Oakland Rodman California State University, Hayward, California, U.S.A.
Textiles in South America

Christine M. Rodrigue Department of Geography and Planning, California State University, Chico, California, U.S.A.
Animal Domestication

Boris A. Rosenfeld Mathematics Department, Eberly College of Science, Pennsylvania State University, University Park, Pennsylvania, U.S.A.
al-Fazārī
Geometry in the Islamic World
Gnomon in Islam
al-Kāshī
al-Kharaqī
ʿUmar al-Khayyām
Nāṣir-i Khusraw
al-Ṣāghānī
Thābit ibn Qurra
Trigonometry in Islamic Mathematics
Ulugh Bēg
al-ʿUrḍī

Henry Rutz Department of Anthropology, Hamilton College, Clinton, New York, U.S.A.
Time
Time and the Maya
Time in Africa

Seiji Sakiura Sapporo, Hokkaido, Japan
Agriculture in Japan

Sema'am I. Salem Physics and Astronomy Department, California State University at Long Beach, Long Beach, California, U.S.A.
Alphabet
Muslim Spain
Ṣāʿid al-Andalusī

Abdul Latif Samian Universiti Kebangsaan Malaysia, Selangor de Malaysia
al-Bīrūnī

Julio Samsó Departamento de Arabe, Universidad de Barcelona, Barcelona, Spain

Abū'l-Ṣalt
Alfonso X
al-Battānī
al-Biṭrūjī
Ibn al-Hā'im
Ibn al-Kammād
Ibn al-Raqqām
Ibn Isḥāq al-Tūnisī
al-Khāzinī

Ziauddin Sardar London, England
Environment and Nature: Islam
Islamic Science: the Contemporary Debate
Values and Science

K.V. Sarma Adyar Library and Research Centre, Adyar, Madras, India
Acyuta Piṣāraṭi
Armillary Spheres in India
Astronomy in India
Calculus
Candraśekhara Sāmanta
Decimal Notation
Deśāntara
Devācārya
Haridatta
Jagannātha Samrāṭ
Jayadeva
Kamalākara
Lalla
Lunar Mansions in Indian Astronomy
Mahādeva
Mahendra Sūri
Makaranda
Munīśvara
Nīlakaṇṭha Somayāji
Pakṣa
Parameśvara
Pauliśa
Precession of the Equinoxes
Putumana Somayāji
Rationale in Indian Mathematics
Śaṅkara Vāriyar
Śatānanda
Sphujidhvaja
Śulbasūtras
Sūryasiddhānta
Vākyakaraṇa
Varāhamihira
Vaṭeśvara
Yavaneśvara
Yuktibhāṣā of Jyeṣṭhadeva

Nataraja Sarma Vigyan Society, New Bombay, India
Time in India

Emilie Savage-Smith Wellcome Unit for the History of Medicine, University of Oxford, Oxford, England
Geomancy in the Islamic World
Globes
Maps and Mapmaking: Celestial Islamic Maps

Richard Evans Schultes Melrose, Massachusetts, U.S.A.
Ethnobotany
Ethnobotany in Mesoamerica

Joseph E. Schwartzberg Department of Geography, University of Minnesota, Minneapolis, Minnesota, U.S.A.

Maps and Mapmaking in India
Maps and Mapmaking in Southeast Asia
Maps and Mapmaking in Tibet

Karl H. Schwerin Department of Anthropology, University of New Mexico, Albuquerque, New Mexico, U.S.A.
Agriculture in South and Central America
Swidden

Aaron Segal El Paso, Texas, U.S.A.
Technology in the New World

Jacques Sesiano Départment des Mathématiques, École Polytechnique Fédérale, Lausanne, Switzerland
Abū Kāmil
al-Karajī
al-Khwārizmī
Magic Squares in Islamic Mathematics
Number Theory in Islamic Mathematics
al-Uqlīdisī

Virendra Nath Sharma Department of Physics and Astronomy, University of Wisconsin Fox Valley, Menasha, Wisconsin, U.S.A.
Jai Singh
Observatories in India

Akhtar H. Siddiqi Department of Geography, Indiana State University, Terre Haute, Indiana, U.S.A.
al-Bīrūnī: Geographical Contributions

Yadhu N. Singh College of Pharmacy, South Dakota State University, Brookings, South Dakota, U.S.A.
Ethnobotany in the Pacific

Kripanath Sinha Department of Education, University of Kalyani, West Bengal, India
Śrīpati

Suliana Siwatibau Port Vila, Vanuatu
Medicine of the Pacific Islands

A. Mark Smith Department of History, University of Missouri, Columbia, Missouri, U.S.A.
Ibn Muʿādh

Julian A. Smith Institute for the History and Philosophy of Science and Technology, University of Toronto, Toronto, Ontario, Canada
Architecture in the Islamic World
Arithmetic in Islamic Mathematics
Astrolabe
Compass
Physics
Telescope

Michael E. Smith Department of Anthropology, State University of New York, Albany, New York, U.S.A.
City Planning: Aztec City Planning

Richard J. Smith Department of History, Rice University, Houston, Texas, U.S.A.
Divination in China

Lawrence Souder Department of Rhetoric and Communication, Temple University, Philadelphia, Pennsylvania, U.S.A.
Ḥabash al-Ḥāsib

Leslie E. Sponsel Department of Anthropology, University of Hawaii, Honolulu, Hawaii, U.S.A.
Environment and Nature: Amazon
Environment and Nature in Buddhist Thought

F. Richard Stephenson Department of Physics, University of Durham, Durham, England
Eclipses
Lunar Mansions in Chinese Astronomy
Maps and Mapmaking: Celestial East Asian Maps

M.S. Stern Department of Religion,University of Manitoba, Winnipeg, Canada
Time in Islam

Sun Xiaochun Institute for the History of Science, Chinese Academy of Sciences, Beijing, China
Stars in Chinese Science

Sun Xiaoli The Fourth University of Military Medicine, Xi-an, China
Isa Tarjaman
Li Bing

Frank J. Swetz The Pennsylvania State University Harrisburg, Middletown, Pennsylvania, U.S.A.

Geometry
Surveying

Richard C. Taylor Department of Philosophy, Marquette University, Milwaukee, Wisconsin, U.S.A.
East and West: Islam in the Transmission of Knowledge East to West

Sevim Tekeli Ankara, Turkey
al-Khujandī
Pirī Reis
Taqī al–Dīn

Gerald R. Tibbets Eynsham, Oxford, England
The Balkhī School of Arab Geographers

Christopher Toll University of Copenhagen, Denmark
al-Hamdānī

Marina A. Tolmacheva Department of History, Washington State University, Pullman, Washington, U.S.A.
East and West: Africa in the Transmission of Knowledge East to West
Environment and Nature: Africa
Navigation in Africa
Navigation in the Indian Ocean and the Red Sea

Vijaya Narayan Tripathi Varanasi, India
Astrology in India

John Allen Tucker Department of History and Philosophy, University of North Florida, Jacksonville, Florida
Environment and Nature: Japan

Dorothy Tunbridge Australian Institute for Aboriginal Studies, Canberra, Australia
Environment and Nature: the Australian Aboriginal People

David Turnbull School of Social Inquiry, Arts Faculty, Deakin University, Geelong, Victoria, Australia
Bamboo
Knowledge Systems: Local Knowledge
Maps and Mapmaking of the Australian Aboriginal People
Rationality, Objectivity and Method

B.L. Turner II Graduate School of Geography and George Perkins Marsh Institute, Clark University, Worcester, Massachusetts, U.S.A.
Agriculture of the Maya

Christine Tuschinsky Institut für Ethnologie, Universität Hamburg, Hamburg, Germany
Jamu

Lawrence Tyler Office of International Affairs, Western Michigan University, Kalamazoo, Michigan, U.S.A.
Medicine Wheels

Kazutaka Unno Mie, Japan
Maps and Mapmaking in Japan

Ryan A. Uritam Department of Physics, Boston College, Chestnut Hill, Massachusetts, U.S.A.
Physics in China

Edwin J. Van Kley Calvin College, Grand Rapids, Michigan, U.S.A.
East and West

Robert M. Veatch Kennedy Institute of Ethics, Georgetown University, Washington, DC, U.S.A.
Medical Ethics

Judith Vidal University of Delaware, Newark, Delaware, U.S.A.
Food Technology in Latin America
Sugar in Latin America

Hans Ulrich Vogel Seminar für Sinologie und Koreanistik der Universität Tübingen, Tübingen, Germany
Bitumen in Premodern China
Natural Gas in China
Salt in China
Weights and Measures in China

John Walbridge Indiana University, Bloomington, Indiana, U.S.A.
al-Shīrāzī

Jinguang Wang Hangzhou University, Hangzhou, Zhejiang Province, China
Optics in China

Caiwu Wang Tyndall Air Force Base, Florida, U.S.A.
Optics in China

Helen Watson-Verran Department of History and Philosophy of Science, University of Melbourne, Victoria, Australia
Knowledge Systems of the Australian Aboriginal People
Mathematics of the Australian Aboriginal People

John Whitmore University of Michigan, Ann Arbor, Michigan, U.S.A.
Maps and Mapmaking in Vietnam

Ruth Hendricks Willard San Francisco, California, U.S.A.
Weights and Measures in Egypt

Denis Wood School of Design, North Carolina State University, Raleigh, North Carolina, U.S.A.
Maps and Mapmaking

Yang DiSheng Tsing Hua University, Institute of Humanities, Beijing, China
Knowledge Systems in China

Yano Michio International Institute for Linguistic Sciences, Kyoto Sangyo University, Kyoto, Japan
Kūshyār ibn Labbān
Trigonometry in Indian Mathematics
Zodiac in India

Cordell D.K. Yee St. Johns College, Annapolis, Maryland, U.S.A.
Maps and Mapmaking in China

H.K. Yoon Department of Geography, University of Auckland, Auckland, New Zealand
Geomancy in China
Maps and Mapmaking: Chinese Geomantic Maps

Michael Zeilik Department of Physics and Astronomy, University of New Mexico, Albuquerque, New Mexico, U.S.A.
Time in Native North America

R. Tom Zuidema Department of Anthropology, University of Illinois at Urbana-Champaign, Urbana, Illinois, U.S.A.
Knowledge Systems of the Incas
Time and the Incas

INDEX

Entries printed capitals refer to the articles in the *Encyclopaedia*.

In the index all Arabic names starting with "al-" have been placed under the letter A.

To increase the readability of the index the diacritical marks have not been included. For the correct spelling, the reader is refered to the main text.

A

B

C

D

E

F

G

H

I

J

K

L

M

N

O

P

Q

R

S

T

U

V

W

X

Y

Z